FUNDAMENTAL CONSTANTS

Constant	Symbol	Value		
			Power of 10	Units
Speed of light	c	2.997 924 58*	10^{8}	m s^{-1}
Elementary charge	e	1.602 176 565	10^{-19}	C
Planck's constant	h	6.626 069 57	10^{-34}	J s
	$\hbar = h/2\pi$	1.054 571 726	10^{-34}	J s
Boltzmann's constant	k	1.380 6488	10^{-23}	J K^{-1}
Avogadro's constant	N_A	6.022 141 29	10^{23}	mol^{-1}
Gas constant	$R = N_A k$	8.314 4621		J K^{-1} mol^{-1}
Faraday's constant	$F = N_A e$	9.648 533 65	10^{4}	C mol^{-1}
Mass				
Electron	m_e	9.109 382 91	10^{-31}	kg
Proton	m_p	1.672 621 777	10^{-27}	kg
Neutron	m_n	1.674 927 351	10^{-27}	kg
Atomic mass constant	m_u	1.660 538 921	10^{-27}	kg
Vacuum permeability	μ_0	4π*	10^{-7}	J s^{2} C^{-2} m^{-1}
Vacuum permittivity	$\varepsilon_0 = 1/\mu_0 c^2$	8.854 187 817	10^{-12}	J^{-1} C^{2} m^{-1}
	$4\pi\varepsilon_0$	1.112 650 056	10^{-10}	J^{-1} C^{2} m^{-1}
Bohr magneton	$\mu_B = e\hbar/2m_e$	9.274 009 68	10^{-24}	J T^{-1}
Nuclear magneton	$\mu_N = e\hbar/2m_p$	5.050 783 53	10^{-27}	J T^{-1}
Proton magnetic moment	μ_p	1.410 606 743	10^{-26}	J T^{-1}
g-Value of electron	g_e	2.002 319 304		
Magnetogyric ratio				
Electron	$\gamma_e = -g_e e/2m_e$	−1.760 859 630	10^{11}	C kg^{-1}
Proton	$\gamma_p = 2\mu_p/\hbar$	2.675 222 004	10^{8}	C kg^{-1}
Bohr radius	$a_0 = 4\pi\varepsilon_0\hbar^2/e^2 m_e$	5.291 772 109	10^{-11}	m
Rydberg constant	$\tilde{R}_\infty = m_e e^4/8h^3 c\varepsilon_0^2$	1.097 373 157	10^{5}	cm^{-1}
	$hc\tilde{R}_\infty/e$	13.605 692 53		eV
Fine-structure constant	$\alpha = \mu_0 e^2 c/2h$	7.297 352 5698	10^{-3}	
	α^{-1}	1.370 359 990 74	10^{2}	
Stefan–Boltzmann constant	$\sigma = 2\pi^5 k^4/15h^3c^2$	5.670 373	10^{-8}	W m^{-2} K^{-4}
Standard acceleration of free fall	g	9.806 65*		m s^{-2}
Gravitational constant	G	6.673 84	10^{-11}	N m^{2} kg^{-2}

* Exact value. For current values of the constants, see the National Institute of Standards and Technology (NIST) website.

Atkins'
PHYSICAL CHEMISTRY

Eleventh edition

Peter Atkins
Fellow of Lincoln College,
University of Oxford,
Oxford, UK

Julio de Paula
Professor of Chemistry,
Lewis & Clark College,
Portland, Oregon, USA

James Keeler
Senior Lecturer in Chemistry and
Fellow of Selwyn College,
University of Cambridge,
Cambridge, UK

OXFORD
UNIVERSITY PRESS

OXFORD
UNIVERSITY PRESS

Great Clarendon Street, Oxford, OX2 6DP,
United Kingdom

Oxford University Press is a department of the University of Oxford.
It furthers the University's objective of excellence in research, scholarship,
and education by publishing worldwide. Oxford is a registered trade mark of
Oxford University Press in the UK and in certain other countries

Eighth edition 2006
Ninth edition 2009
Tenth edition 2014

Impression: 5

Published in the United States of America by Oxford University Press
198 Madison Avenue, New York, NY 10016, United States of America

British Library Cataloguing in Publication Data
Data available

Library of Congress Control Number: 2017950918

ISBN 978–0–19–876986–6

Printed in Great Britain by Bell and Bain Ltd, Glasgow

The cover image symbolizes the structure of the text, as a collection of Topics that merge into a unified whole. It also symbolizes the fact that physical chemistry provides a basis for understanding chemical and physical change.

PREFACE

Our *Physical Chemistry* is continuously evolving in response to users' comments and our own imagination. The principal change in this edition is the addition of a new co-author to the team, and we are very pleased to welcome James Keeler of the University of Cambridge. He is already an experienced author and we are very happy to have him on board.

As always, we strive to make the text helpful to students and usable by instructors. We developed the popular 'Topic' arrangement in the preceding edition, but have taken the concept further in this edition and have replaced chapters by Focuses. Although that is principally no more than a change of name, it does signal that groups of Topics treat related groups of concepts which might demand more than a single chapter in a conventional arrangement. We know that many instructors welcome the flexibility that the Topic concept provides, because it makes the material easy to rearrange or trim.

We also know that students welcome the Topic arrangement as it makes processing of the material they cover less daunting and more focused. With them in mind we have developed additional help with the manipulation of equations in the form of annotations, and *The chemist's toolkits* provide further background at the point of use. As these Toolkits are often relevant to more than one Topic, they also appear in consolidated and enhanced form on the website. Some of the material previously carried in the 'Mathematical backgrounds' has been used in this enhancement. The web also provides a number of sections called *A deeper look*. As their name suggests, these sections take the material in the text further than we consider appropriate for the printed version but are there for students and instructors who wish to extend their knowledge and see the details of more advanced calculations.

Another major change is the replacement of the 'Justifications' that show how an equation is derived. Our intention has been to maintain the separation of the equation and its derivation so that review is made simple, but at the same time to acknowledge that mathematics is an integral feature of learning. Thus, the text now sets up a question and the *How is that done?* section that immediately follows develops the relevant equation, which then flows into the following text.

The worked *Examples* are a crucially important part of the learning experience. We have enhanced their presentation by replacing the 'Method' by the more encouraging *Collect your thoughts*, where with this small change we acknowledge that different approaches are possible but that students welcome guidance. The *Brief illustrations* remain: they are intended simply to show how an equation is implemented and give a sense of the order of magnitude of a property.

It is inevitable that in an evolving subject, and with evolving interests and approaches to teaching, some subjects wither and die and are replaced by new growth. We listen carefully to trends of this kind, and adjust our treatment accordingly. The topical approach enables us to be more accommodating of fading fashions because a Topic can so easily be omitted by an instructor, but we have had to remove some subjects simply to keep the bulk of the text manageable and have used the web to maintain the comprehensive character of the text without overburdening the presentation.

This book is a living, evolving text. As such, it depends very much on input from users throughout the world, and we welcome your advice and comments.

PWA
JdeP
JK

USING THE BOOK

TO THE STUDENT

For this eleventh edition we have developed the range of learning aids to suit your needs more closely than ever before. In addition to the variety of features already present, we now derive key equations in a helpful new way, through the *How is that done?* sections, to emphasize how mathematics is an interesting, essential, and integral feature of understanding physical chemistry.

Innovative structure

Short Topics are grouped into Focus sections, making the subject more accessible. Each Topic opens with a comment on why it is important, a statement of its key idea, and a brief summary of the background that you need to know.

TOPIC 2A Internal energy

> ➤ Why do you need to know this material?
> The First Law of thermodynamics is the foundation of the discussion of the role of energy in chemistry. Wherever the generation or use of energy in physical transformations or chemical reactions is of interest, lying in the background are the concepts introduced by the First Law.
>
> ➤ What is the key idea?
> The total energy of an isolated system is constant.
>
> ➤ What do you need to know already?
> This Topic makes use of the discussion of the properties of gases (Topic 1A), particularly the perfect gas law. It builds on the definition of work given in *The chemist's toolkit* 6.

For the purposes of thermodynamics, the universe is divided into two parts, the **system** and its surroundings. The system is the part of the world of interest. It may be a reaction vessel, an engine, an electrochemical cell, a biological cell, and so on. The **surroundings** comprise the region outside the system and are where measurements are made. The type of system depends on the characteristics of the boundary that divides it from the

For example, a closed system can expand and thereby raise a weight in the surroundings; a closed system may also transfer energy to the surroundings if they are at a lower temperature. An **isolated system** is a closed system that has neither mechanical nor thermal contact with its surroundings.

2A.1 Work, heat, and energy

Although thermodynamics deals with observations on bulk systems, it is immeasurably enriched by understanding the molecular origins of these observations.

(a) Operational definitions

The fundamental physical property in thermodynamics is work: **work** is done to achieve motion against an opposing force (*The chemist's toolkit* 6). A simple example is the process of raising a weight against the pull of gravity. A process does work if in principle it can be harnessed to raise a weight somewhere in the surroundings. An example of doing work is the expansion of a gas that pushes out a piston: the motion of the piston can in principle be used to raise a weight. Another example is a chemical reaction in a cell, which leads to an electric

Notes on good practice

Our 'Notes on good practice' will help you avoid making common mistakes. Among other things, they encourage conformity to the international language of science by setting out the conventions and procedures adopted by the International Union of Pure and Applied Chemistry (IUPAC).

A note on good practice An *allotrope* is a particular molecular form of an element (such as O_2 and O_3) and may be solid, liquid, or gas. A *polymorph* is one of a number of solid phases of an element or compound.

The number of phases in a system is denoted P. A gas, or a gaseous mixture, is a single phase ($P = 1$), a crystal of a sub-

Resource section

The *Resource section* at the end of the book includes a table of useful integrals, extensive tables of physical and chemical data, and character tables. Short extracts of most of these tables appear in the Topics themselves: they are there to give you an idea of the typical values of the physical quantities mentioned in the text.

Contents

Checklist of concepts

A checklist of key concepts is provided at the end of each Topic, so that you can tick off the ones you have mastered.

Checklist of concepts

☐ 1. The **physical state** of a sample of a substance, its physical condition, is defined by its physical properties.

☐ 2. **Mechanical equilibrium** is the condition of equality of pressure on either side of a shared movable wall.

PRESENTING THE MATHEMATICS

How is that done?

You need to understand how an equation is derived from reasonable assumptions and the details of the mathematical steps involved. This is accomplished in the text through the new *'How is that done?'* sections, which replace the *Justifications* of earlier editions. Each one leads from an issue that arises in the text, develops the necessary mathematics, and arrives at the equation or conclusion that resolves the issue. These sections maintain the separation of the equation and its derivation so that you can find them easily for review, but at the same time emphasize that mathematics is an essential feature of physical chemistry.

How is that done? 4A.1 Deducing the phase rule

The argument that leads to the phase rule is most easily appreciated by first thinking about the simpler case when only one component is present and then generalizing the result to an arbitrary number of components.

Step 1 *Consider the case where only one component is present*

When only one phase is present ($P = 1$), both p and T can be varied independently, so $F = 2$. Now consider the case where two phases α and β are in equilibrium ($P = 2$). If the phases are in equilibrium at a given pressure and temperature, their chemical potentials must be equal:

The chemist's toolkits

The chemist's toolkits, which are much more numerous in this edition, are reminders of the key mathematical, physical, and chemical concepts that you need to understand in order to follow the text. They appear where they are first needed. Many of these Toolkits are relevant to more than one Topic, and a compilation of them, with enhancements in the form of more information and brief illustrations, appears on the web site. **www.oup.com/uk/pchem11e/**

The chemist's toolkit 2 Properties of bulk matter

The state of a bulk sample of matter is defined by specifying the values of various properties. Among them are:

The **mass**, m, a measure of the quantity of matter present (unit: kilogram, kg).

The **volume**, V, a measure of the quantity of space the sample occupies (unit: cubic metre, m^3).

The **amount of substance**, n, a measure of the number of specified entities (atoms, molecules, or formula units) present (unit: mole, mol).

Annotated equations and equation labels

We have annotated many equations to help you follow how they are developed. An annotation can take you across the equals sign: it is a reminder of the substitution used, an approximation made, the terms that have been assumed constant, an integral used, and so on. An annotation can also be a reminder of the significance of an individual term in an expression. We sometimes colour a collection of numbers or symbols to show how they carry from one line to the next. Many of the equations are labelled to highlight their significance.

$U_m(T) = U_m(0) + N_A\langle\varepsilon^V\rangle$

$d(1/f)/dx = -(1/f^2)df/dx$ used twice

$$C_{V,m}^V = \frac{dN_A\langle\varepsilon^V\rangle}{dT} = R\theta^V \frac{d}{dT}\frac{1}{e^{\theta^V/T}-1} = R\left(\frac{\theta^V}{T}\right)^2 \frac{e^{\theta^V/T}}{(e^{\theta^V/T}-1)^2}$$

By noting that $e^{\theta^V/T} = (e^{\theta^V/2T})^2$, this expression can be rearranged into

$$C_{V,m}^V = Rf(T) \quad f(T) = \left(\frac{\theta^V}{T}\right)^2\left(\frac{e^{-\theta^V/2T}}{1-e^{-\theta^V/T}}\right)^2$$

Vibrational contribution to $C_{V,m}$ (13E.3)

Checklists of equations

A handy checklist at the end of each topic summarizes the most important equations and the conditions under which they apply. Don't think, however, that you have to memorize every equation in these checklists.

Checklist of equations

Property	Equation
Gibbs energy of mixing	$\Delta_{mix}G = nRT(x_A \ln x_A + x_B \ln x_B)$
Entropy of mixing	$\Delta_{mix}S = -nR(x_A \ln x_A + x_B \ln x_B)$

SETTING UP AND SOLVING PROBLEMS

Brief illustrations

A *Brief illustration* shows you how to use an equation or concept that has just been introduced in the text. It shows you how to use data and manipulate units correctly. It also helps you to become familiar with the magnitudes of quantities.

Brief illustration 3B.1

When the volume of any perfect gas is doubled at constant temperature, $V_f/V_i = 2$, and hence the change in molar entropy of the system is

$$\Delta S_m = (8.3145\,\mathrm{J\,K^{-1}\,mol^{-1}}) \times \ln 2 = +5.76\,\mathrm{J\,K^{-1}\,mol^{-1}}$$

Examples

Worked *Examples* are more detailed illustrations of the application of the material, and typically require you to assemble and deploy the relevant concepts and equations.

We suggest how you should collect your thoughts (that is a new feature) and then proceed to a solution. All the worked *Examples* are accompanied by Self-tests to enable you to test your grasp of the material after working through our solution as set out in the *Example*.

Example 1A.1 **Using the perfect gas law**

In an industrial process, nitrogen gas is introduced into a vessel of constant volume at a pressure of 100 atm and a temperature of 300 K. The gas is then heated to 500 K. What pressure would the gas then exert, assuming that it behaved as a perfect gas?

Collect your thoughts The pressure is expected to be greater on account of the increase in temperature. The perfect gas

Discussion questions

Discussion questions appear at the end of every Focus, and are organised by Topic. These questions are designed to encourage you to reflect on the material you have just read, to review the key concepts, and sometimes to think about its implications and limitations.

FOCUS 3 The Second and Third Laws

Assume that all gases are perfect and that data refer to 298.15 K unless otherwise stated.

TOPIC 3A Entropy

Discussion questions

D3A.1 The evolution of life requires the organization of a very large number of molecules into biological cells. Does the formation of living organisms violate the Second Law of thermodynamics? State your conclusion clearly and present detailed arguments to support it.

D3A.2 Discuss the significance of the terms 'dispersal' and 'disorder' in the context of the Second Law.

D3A.3 Discuss the relationships between the various formulations of the Second Law of thermodynamics.

Exercises

E3A.1(a) Consider a process in which the entropy of a system increases by $125\,\mathrm{J\,K^{-1}}$ and the entropy of the surroundings decreases by $125\,\mathrm{J\,K^{-1}}$. Is the process spontaneous?
E3A.1(b) Consider a process in which the entropy of a system increases by $105\,\mathrm{J\,K^{-1}}$ and the entropy of the surroundings decreases by $95\,\mathrm{J\,K^{-1}}$. Is the process spontaneous?

E3A.2(a) Consider a process in which 100 kJ of energy is transferred reversibly and isothermally as heat to a large block of copper. Calculate the change in entropy of the block if the process takes place at (a) 0 °C, (b) 50 °C.
E3A.2(b) Consider a process in which 250 kJ of energy is transferred reversibly and isothermally as heat to a large block of lead. Calculate the change in entropy of the block if the process takes place at (a) 20 °C, (b) 100 °C.

E3A.3(a) Calculate the change in entropy of the gas when 15 g of carbon dioxide gas are allowed to expand isothermally from $1.0\,\mathrm{dm^3}$ to $3.0\,\mathrm{dm^3}$ at 300 K.
E3A.3(b) Calculate the change in entropy of the gas when 4.00 g of nitrogen is allowed to expand isothermally from $500\,\mathrm{cm^3}$ to $750\,\mathrm{cm^3}$ at 300 K.

E3A.4(a) Calculate the change in the entropies of the system and the surroundings, and the total change in entropy, when a sample of nitrogen gas of mass 14 g at 298 K doubles its volume in (a) an isothermal reversible expansion, (b) an isothermal irreversible expansion against $p_{ex} = 0$, and (c) an adiabatic reversible expansion.
E3A.4(b) Calculate the change in the entropies of the system and the surroundings, and the total change in entropy, when the volume of a sample of argon gas of mass 2.9 g at 298 K increases from $1.20\,\mathrm{dm^3}$ to $4.60\,\mathrm{dm^3}$ in (a) an isothermal reversible expansion, (b) an isothermal irreversible expansion against $p_{ex} = 0$, and (c) an adiabatic reversible expansion.

E3A.5(a) In a certain ideal heat engine, 10.00 kJ of heat is withdrawn from the hot source at 273 K and 3.00 kJ of work is generated. What is the temperature of cold sink?
E3A.5(b) In an ideal heat engine the cold sink is at 0 °C. If 10.00 kJ of heat is withdrawn from the hot source and 3.00 kJ of work is generated, at what temperature is the hot source?

E3A.6(a) What is the efficiency of an ideal heat engine in which the hot source is at 100 °C and the cold sink is at 10 °C?
E3A.6(b) An ideal heat engine has a hot source at 40 °C. At what temperature must the cold sink be if the efficiency is to be 10 per cent?

Problems

P3A.1 A sample consisting of 1.00 mol of perfect gas molecules at 27 °C is expanded isothermally from an initial pressure of 3.00 atm to a final pressure of 1.00 atm in two ways: (a) reversibly, and (b) against a constant external pressure of 1.00 atm. Evaluate q, w, ΔU, ΔH, ΔS, ΔS_{surr}, and ΔS_{tot} in each case.

P3A.2 A sample consisting of 0.10 mol of perfect gas molecules is held by a piston inside a cylinder such that the volume is $1.25\,\mathrm{dm^3}$; the external pressure is constant at 1.00 bar and the temperature is maintained at 300 K by a thermostat. The piston is released so that the gas can expand. Calculate (a) the volume of the gas when the expansion is complete; (b) the work done when the gas expands; (c) the heat absorbed by the system. Hence calculate ΔS_{tot}.

P3A.3 Consider a Carnot cycle in which the working substance is 0.10 mol of perfect gas molecules, the temperature of the hot source is 373 K, and that of the cold sink is 273 K; the initial volume of gas is $1.00\,\mathrm{dm^3}$, which doubles over the course of the first isothermal stage. For the reversible adiabatic stages it may be assumed that $VT^{3/2} = \text{constant}$. (a) Calculate the volume of the gas after Stage 1 and after Stage 2 (Fig. 3A.8). (b) Calculate the volume of gas after Stage 3 by considering the reversible adiabatic compression from the starting point. (c) Hence, for each of the four stages of the cycle, calculate the heat transferred to or from the gas. (d) Explain why the work done is equal to the difference between the heat extracted from the hot source and that deposited in the cold sink. (e) Calculate the work done over the cycle and hence the efficiency η. (f) Confirm that your answer agrees with the efficiency given by eqn 3A.9 and that your values for the heat involved in the isothermal stages are in accord with eqn 3A.6.

P3A.4 The Carnot cycle is usually represented on a pressure–volume diagram (Fig. 3A.8), but the four stages can equally well be represented on temperature–entropy diagram, in which the horizontal axis is entropy and the vertical axis is temperature; draw such a diagram. Assume that the temperature of the hot source is T_h and that of the cold sink is T_c, and that the volume of the working substance (the gas) expands from V_A to V_B in the first isothermal stage. (a) By considering the entropy change of each stage, derive an expression for the area enclosed by the cycle in the temperature–entropy diagram. (b) Derive an expression for the work done over the cycle. (*Hint:* The work done is the difference between the heat extracted from the hot source and that deposited in the cold sink; or use eqns 3A.7 and 3A.9) (c) Comment on the relation between your answers to (a) and (b).

Exercises and problems

Exercises and Problems are also provided at the end of every Focus and organised by Topic. Exercises are designed as relatively straightforward numerical tests; the Problems are more challenging and typically involve constructing a more detailed answer. The Exercises come in related pairs, with final numerical answers available online for the 'a' questions. Final numerical answers to the odd-numbered Problems are also available online.

Integrated activities

At the end of every Focus you will find questions that span several Topics. They are designed to help you use your knowledge creatively in a variety of ways.

THERE IS A LOT OF ADDITIONAL MATERIAL ON THE WEB

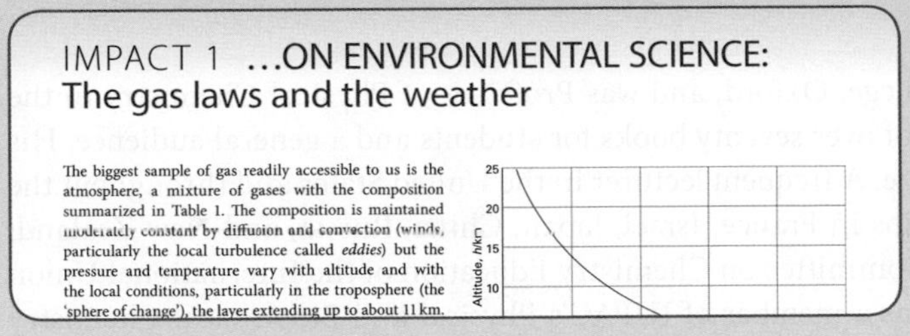
IMPACT 1 …ON ENVIRONMENTAL SCIENCE: The gas laws and the weather

The biggest sample of gas readily accessible to us is the atmosphere, a mixture of gases with the composition summarized in Table 1. The composition is maintained moderately constant by diffusion and convection (winds, particularly the local turbulence called *eddies*) but the pressure and temperature vary with altitude and with the local conditions, particularly in the troposphere (the 'sphere of change'), the layer extending up to about 11 km.

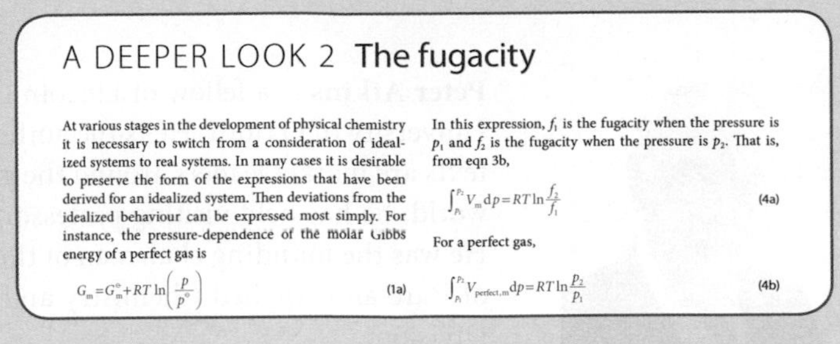
A DEEPER LOOK 2 The fugacity

At various stages in the development of physical chemistry it is necessary to switch from a consideration of idealized systems to real systems. In many cases it is desirable to preserve the form of the expressions that have been derived for an idealized system. Then deviations from the idealized behaviour can be expressed most simply. For instance, the pressure-dependence of the molar Gibbs energy of a perfect gas is

$$G_m = G_m^{\ominus} + RT\ln\left(\frac{p}{p^{\ominus}}\right) \quad (1a)$$

In this expression, f_1 is the fugacity when the pressure is p_1 and f_2 is the fugacity when the pressure is p_2. That is, from eqn 3b,

$$\int_{p_1}^{p_2} V_m \mathrm{d}p = RT\ln\frac{f_2}{f_1} \quad (4a)$$

For a perfect gas,

$$\int_{p_1}^{p_2} V_{\text{perfect},m} \mathrm{d}p = RT\ln\frac{p_2}{p_1} \quad (4b)$$

'Impact' sections

'Impact' sections show how physical chemistry is applied in a variety of modern contexts. They showcase physical chemistry as an evolving subject. **www.oup.com/uk/pchem11e/**

A deeper look

These online sections take some of the material in the text further and are there if you want to extend your knowledge and see the details of some of the more advanced derivations **www.oup.com/uk/pchem11e/**

Group theory tables

Comprehensive group theory tables are available to download.

Molecular modelling problems

Files containing molecular modelling problems can be downloaded, designed for use with the Spartan Student™ software. However they can also be completed using any modelling software that allows Hartree–Fock, density functional, and MP2 calculations. The site can be accessed at **www.oup.com/uk/pchem11e/**.

TO THE INSTRUCTOR

We have designed the text to give you maximum flexibility in the selection and sequence of Topics, while the grouping of Topics into Focuses helps to maintain the unity of the subject. Additional resources are:

Figures and tables from the book

Lecturers can find the artwork and tables from the book in ready-to-download format. These may be used for lectures without charge (but not for commercial purposes without specific permission).

Key equations

Supplied in Word format so you can download and edit them.

Lecturer resources are available only to registered adopters of the textbook. To register, simply visit **www.oup.com/uk/pchem11e/** and follow the appropriate links.

SOLUTIONS MANUALS

Two solutions manuals have been written by Peter Bolgar, Haydn Lloyd, Aimee North, Vladimiras Oleinikovas, Stephanie Smith, and James Keeler.

The *Student's Solutions Manual* (ISBN 9780198807773) provides full solutions to the 'a' Exercises and to the odd-numbered Problems.

The *Instructor's Solutions Manual* provides full solutions to the 'b' Exercises and to the even-numbered Problems (available to download online for registered adopters of the book only).

ABOUT THE AUTHORS

Photograph by Natasha Ellis-Knight.

Peter Atkins is a fellow of Lincoln College, Oxford, and was Professor of Physical Chemistry in the University of Oxford. He is the author of over seventy books for students and a general audience. His texts are market leaders around the globe. A frequent lecturer in the United States and throughout the world, he has held visiting professorships in France, Israel, Japan, China, Russia, and New Zealand. He was the founding chairman of the Committee on Chemistry Education of the International Union of Pure and Applied Chemistry and was a member of IUPAC's Physical and Biophysical Chemistry Division.

Julio de Paula is Professor of Chemistry at Lewis & Clark College. A native of Brazil, he received a B.A. degree in chemistry from Rutgers, The State University of New Jersey, and a Ph.D. in biophysical chemistry from Yale University. His research activities encompass the areas of molecular spectroscopy, photochemistry, and nanoscience. He has taught courses in general chemistry, physical chemistry, biophysical chemistry, inorganic chemistry, instrumental analysis, environmental chemistry, and writing. Among his professional honours are a Christian and Mary Lindback Award for Distinguished Teaching, a Henry Dreyfus Teacher-Scholar Award, and a Cottrell Scholar Award from the Research Corporation for Science Advancement.

Photograph by Nathan Pitt, ©University of Cambridge.

James Keeler is a Senior Lecturer in Chemistry at the University of Cambridge, and Walters Fellow in Chemistry at Selwyn College, Cambridge. He took his first degree at the University of Oxford and continued there for doctoral research in nuclear magnetic resonance spectroscopy. Dr Keeler is Director of Teaching for undergraduate chemistry, and teaches courses covering a range of topics in physical and theoretical chemistry.

ACKNOWLEDGEMENTS

A book as extensive as this could not have been written without significant input from many individuals. We would like to reiterate our thanks to the hundreds of people who contributed to the first ten editions. Many people gave their advice based on the tenth edition, and others, including students, reviewed the draft chapters for the eleventh edition as they emerged. We wish to express our gratitude to the following colleagues:

Andrew J. Alexander, *University of Edinburgh*
Stephen H. Ashworth, *University of East Anglia*
Mark Berg, *University of South Carolina*
Eric Bittner, *University of Houston*
Melanie Britton, *University of Birmingham*
Eleanor Campbell, *University of Edinburgh*
Andrew P. Doherty, *Queen's University of Belfast*
Rob Evans, *Aston University*
J.G.E. Gardeniers, *University of Twente*
Ricardo Grau-Crespo, *University of Reading*
Alex Grushow, *Rider University*
Leonid Gurevich, *Aalborg University*
Ronald Haines, *University of New South Wales*
Patrick M. Hare, *Northern Kentucky University*
John Henry, *University of Wolverhampton*
Karl Jackson, *Virginia Union University*
Carey Johnson, *University of Kansas*
George Kaminski, *Worcester Polytechnic Institute*
Scott Kirkby, *East Tennessee State University*
Kathleen Knierim, *University of Louisiana at Lafayette*
Jeffry Madura, *University of Pittsburgh*
David H. Magers, *Mississippi College*
Kristy Mardis, *Chicago State University*
Paul Marshall, *University of North Texas*
Laura R. McCunn, *Marshall University*
Allan McKinley, *University of Western Australia*
Joshua Melko, *University of North Florida*
Yirong Mo, *Western Michigan University*
Gareth Morris, *University of Manchester*
Han J. Park, *University of Tennessee at Chattanooga*
Rajeev Prabhakar, *University of Miami*
Gavin Reid, *University of Leeds*
Chad Risko, *University of Kentucky*
Nessima Salhi, *Uppsala University*
Daniel Savin, *University of Florida*
Richard W. Schwenz, *University of Northern Colorado*
Douglas Strout, *Alabama State University*
Steven Tait, *Indiana University*
Jim Terner, *Virginia Commonwealth University*
Timothy Vaden, *Rowan University*
Alfredo Vargas, *University of Sussex*
Darren Walsh, *University of Nottingham*
Collin Wick, *Louisiana Tech University*
Shoujun Xu, *University of Houston*
Renwu Zhang , *California State University*
Wuzong Zhou, *St Andrews University*

We would also like to thank Michael Clugston for proofreading the entire book, and Peter Bolgar, Haydn Lloyd, Aimee North, Vladimiras Oleinikovas, Stephanie Smith, and James Keeler for writing a brand new set of solutions. Last, but by no means least, we acknowledge our two commissioning editors, Jonathan Crowe of Oxford University Press and Jason Noe of OUP USA, and their teams for their assistance, advice, encouragement, and patience.

BRIEF CONTENTS

FULL CONTENTS

CONVENTIONS

To avoid intermediate rounding errors, but to keep track of values in order to be aware of values and to spot numerical errors, we display intermediate results as *n.nnn*… and round the calculation only at the final step.

Blue terms are used when we want to identify a term in an equation. An entire quotient, numerator/denominator, is coloured blue if the annotation refers to the entire term, not just to the numerator or denominator separately.

LIST OF TABLES

LIST OF *THE CHEMIST'S TOOLKITS*

LIST OF MATERIAL PROVIDED AS *A DEEPER LOOK*

LIST OF *IMPACTS*

PROLOGUE Energy, temperature, and chemistry

Energy is a concept used throughout chemistry to discuss molecular structures, reactions, and many other processes. What follows is an informal first look at the important features of energy. Its precise definition and role will emerge throughout the course of this text.

The transformation of energy from one form to another is described by the laws of **thermodynamics**. They are applicable to bulk matter, which consists of very large numbers of atoms and molecules. The 'First Law' of thermodynamics is a statement about the quantity of energy involved in a transformation; the 'Second Law' is a statement about the dispersal of that energy (in a sense that will be explained).

To discuss the energy of individual atoms and molecules that make up samples of bulk matter it is necessary to use **quantum mechanics**. According to this theory, the energy associated with the motion of a particle is 'quantized', meaning that the energy is restricted to certain values, rather than being able to take on any value. Three different kinds of motion can occur: *translation* (motion through space), *rotation* (change of orientation), and *vibration* (the periodic stretching and bending of bonds). Figure 1 depicts the relative sizes and spacing of the energy states associated with these different kinds of motion of typical molecules and compares them with the typical energies of electrons in atoms and molecules. The allowed energies associated with translation are so close together in normal-sized containers that they form a continuum. In contrast, the separation between the allowed electronic energy states of atoms and molecules is very large.

The link between the energies of individual molecules and the energy of bulk matter is provided by one of the most important concepts in chemistry, the **Boltzmann distribution**. Bulk matter consists of large numbers of molecules, each of which is in one of its available energy states. The total number of molecules with a particular energy due to translation, rotation, vibration, and its electronic state is called the 'population' of that state. Most molecules are found in the lowest energy state, and higher energy states are occupied by progressively fewer molecules. The Boltzmann distribution gives the population, N_i, of any energy state in terms of the energy of the state, ε_i, and the absolute temperature, T:

$$N_i \propto e^{-\varepsilon_i/kT}$$

In this expression, k is *Boltzmann's constant* (its value is listed inside the front cover), a universal constant (in the sense of having the same value for all forms of matter). Figure 2 shows the Boltzmann distribution for two temperatures: as the temperature increases higher energy states are populated at the expense of states lower in energy. According to the Boltzmann distribution, the temperature is the single parameter that governs the spread of populations over the available energy states, whatever their nature.

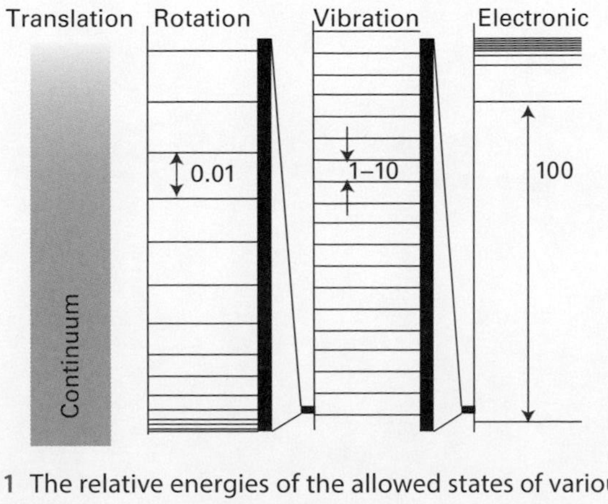

Figure 1 The relative energies of the allowed states of various kinds of atomic and molecular motion.

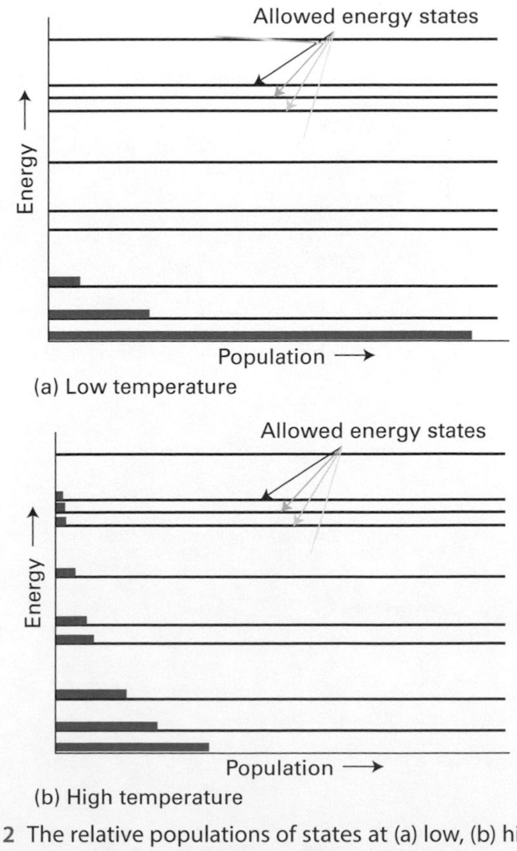

Figure 2 The relative populations of states at (a) low, (b) high temperature according to the Boltzmann distribution.

The Boltzmann distribution, as well as providing insight into the significance of temperature, is central to understanding much of chemistry. That most molecules occupy states of low energy when the temperature is low accounts for the existence of compounds and the persistence of liquids and solids. That highly excited energy levels become accessible at high temperatures accounts for the possibility of reaction as one substance acquires the ability to change into another. Both features are explored in detail throughout the text.

You should keep in mind the Boltzmann distribution (which is treated in greater depth later in the text) whenever considering the interpretation of the properties of bulk matter and the role of temperature. An understanding of the flow of energy and how it is distributed according to the Boltzmann distribution is the key to understanding thermodynamics, structure, and change throughout chemistry.

FOCUS 1

The properties of gases

A gas is a form of matter that fills whatever container it occupies. This Focus establishes the properties of gases that are used throughout the text.

1A The perfect gas

This Topic is an account of an idealized version of a gas, a 'perfect gas', and shows how its equation of state may be assembled from the experimental observations summarized by Boyle's law, Charles's law, and Avogadro's principle.

1B The kinetic model

A central feature of physical chemistry is its role in building models of molecular behaviour that seek to explain observed phenomena. A prime example of this procedure is the development of a molecular model of a perfect gas in terms of a collection of molecules (or atoms) in ceaseless, essentially random motion. As well as accounting for the gas laws, this model can be used to predict the average speed at which molecules move in a gas, and its dependence on temperature. In combination with the Boltzmann distribution (see the text's *Prologue*), the model can also be used to predict the spread of molecular speeds and its dependence on molecular mass and temperature.

1C Real gases

The perfect gas is a starting point for the discussion of properties of all gases, and its properties are invoked throughout thermodynamics. However, actual gases, 'real gases', have properties that differ from those of perfect gases, and it is necessary to be able to interpret these deviations and build the effects of molecular attractions and repulsions into the model. The discussion of real gases is another example of how initially primitive models in physical chemistry are elaborated to take into account more detailed observations.

Web resources What is an application of this material?

The perfect gas law and the kinetic theory can be applied to the study of phenomena confined to a reaction vessel or encompassing an entire planet or star. In *Impact* 1 the gas laws are used in the discussion of meteorological phenomena—the weather. *Impact* 2 examines how the kinetic model of gases has a surprising application: to the discussion of dense stellar media, such as the interior of the Sun.

TOPIC 1A The perfect gas

> ➤ **Why do you need to know this material?**
>
> Equations related to perfect gases provide the basis for the development of many relations in thermodynamics. The perfect gas law is also a good first approximation for accounting for the properties of real gases.
>
> ➤ **What is the key idea?**
>
> The perfect gas law, which is based on a series of empirical observations, is a limiting law that is obeyed increasingly well as the pressure of a gas tends to zero.
>
> ➤ **What do you need to know already?**
>
> You need to know how to handle quantities and units in calculations, as reviewed in *The chemist's toolkit* 1. You also need to be aware of the concepts of pressure, volume, amount of substance, and temperature, all reviewed in *The chemist's toolkit* 2.

The properties of gases were among the first to be established quantitatively (largely during the seventeenth and eighteenth centuries) when the technological requirements of travel in balloons stimulated their investigation. These properties set the stage for the development of the kinetic model of gases, as discussed in Topic 1B.

1A.1 Variables of state

The **physical state** of a sample of a substance, its physical condition, is defined by its physical properties. Two samples of the same substance that have the same physical properties are in the same state. The variables needed to specify the state of a system are the amount of substance it contains, n, the volume it occupies, V, the pressure, p, and the temperature, T.

(a) Pressure

The origin of the force exerted by a gas is the incessant battering of the molecules on the walls of its container. The collisions are so numerous that they exert an effectively steady force, which is experienced as a steady pressure. The SI unit of pressure, the *pascal* (Pa, 1 Pa = 1 N m^{-2}), is introduced in *The chemist's toolkit* 1. Several other units are still widely used (Table 1A.1). A pressure of 1 bar is the **standard pressure** for reporting data; it is denoted $p^{\ominus}$.

Table 1A.1 Pressure units*

Name	Symbol	Value
pascal	Pa	1 Pa = 1 N m^{-2}, 1 kg m^{-1} s^{-2}
bar	bar	**1 bar = 10^5 Pa**
atmosphere	atm	**1 atm = 101.325 kPa**
torr	Torr	**1 Torr = (101 325/760) Pa** = 133.32… Pa
millimetres of mercury	mmHg	1 mmHg = 133.322… Pa
pounds per square inch	psi	1 psi = 6.894 757… kPa

* Values in bold are exact.

If two gases are in separate containers that share a common movable wall (Fig. 1A.1), the gas that has the higher pressure will tend to compress (reduce the volume of) the gas that has lower pressure. The pressure of the high-pressure gas will fall as it expands and that of the low-pressure gas will rise as it is compressed. There will come a stage when the two pressures are equal and the wall has no further tendency to move. This condition of equality of pressure on either side of a movable wall is a state of **mechanical equilibrium** between the two gases. The pressure of a gas is therefore an indication of whether a container that contains the gas will be in mechanical equilibrium with another gas with which it shares a movable wall.

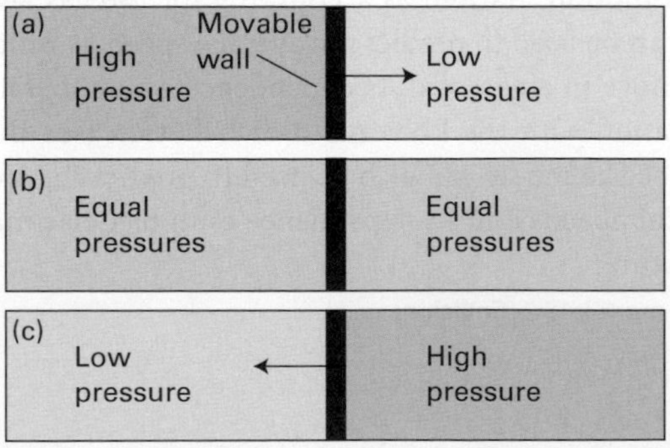

Figure 1A.1 When a region of high pressure is separated from a region of low pressure by a movable wall, the wall will be pushed into one region or the other, as in (a) and (c). However, if the two pressures are identical, the wall will not move (b). The latter condition is one of mechanical equilibrium between the two regions.

The chemist's toolkit 1 Quantities and units

The result of a measurement is a **physical quantity** that is reported as a numerical multiple of a unit:

$$\text{physical quantity} = \text{numerical value} \times \text{unit}$$

It follows that units may be treated like algebraic quantities and may be multiplied, divided, and cancelled. Thus, the expression (physical quantity)/unit is the numerical value (a dimensionless quantity) of the measurement in the specified units. For instance, the mass m of an object could be reported as $m = 2.5\,\text{kg}$ or $m/\text{kg} = 2.5$. In this instance the unit of mass is 1 kg, but it is common to refer to the unit simply as kg (and likewise for other units). See Table A.1 in the *Resource section* for a list of units.

Although it is good practice to use only SI units, there will be occasions where accepted practice is so deeply rooted that physical quantities are expressed using other, non-SI units. By international convention, all physical quantities are represented by oblique (sloping) letters (for instance, m for mass); units are given in roman (upright) letters (for instance m for metre).

Units may be modified by a prefix that denotes a factor of a power of 10. Among the most common SI prefixes are those listed in Table A.2 in the *Resource section*. Examples of the use of these prefixes are:

$$1\,\text{nm} = 10^{-9}\,\text{m} \qquad 1\,\text{ps} = 10^{-12}\,\text{s} \qquad 1\,\mu\text{mol} = 10^{-6}\,\text{mol}$$

Powers of units apply to the prefix as well as the unit they modify. For example, $1\,\text{cm}^3 = 1\,(\text{cm})^3$, and $(10^{-2}\,\text{m})^3 = 10^{-6}\,\text{m}^3$. Note that $1\,\text{cm}^3$ does not mean $1\,\text{c}(\text{m}^3)$. When carrying out numerical calculations, it is usually safest to write out the numerical value of an observable in scientific notation (as $n.nnn \times 10^n$).

There are seven SI base units, which are listed in Table A.3 in the *Resource section*. All other physical quantities may be expressed as combinations of these base units. *Molar concentration* (more formally, but very rarely, *amount of substance concentration*) for example, which is an amount of substance divided by the volume it occupies, can be expressed using the derived units of $\text{mol}\,\text{dm}^{-3}$ as a combination of the base units for amount of substance and length. A number of these derived combinations of units have special names and symbols. For example, force is reported in the derived unit newton, $1\,\text{N} = 1\,\text{kg}\,\text{m}\,\text{s}^{-2}$ (see Table A.4 in the *Resource section*).

The pressure exerted by the atmosphere is measured with a *barometer*. The original version of a barometer (which was invented by Torricelli, a student of Galileo) was an inverted tube of mercury sealed at the upper end. When the column of mercury is in mechanical equilibrium with the atmosphere, the pressure at its base is equal to that exerted by the atmosphere. It follows that the height of the mercury column is proportional to the external pressure.

The pressure of a sample of gas inside a container is measured by using a pressure gauge, which is a device with properties that respond to the pressure. For instance, a *Bayard–Alpert pressure gauge* is based on the ionization of the molecules present in the gas and the resulting current of ions is interpreted in terms of the pressure. In a *capacitance manometer*, the deflection of a diaphragm relative to a fixed electrode is monitored through its effect on the capacitance of the arrangement. Certain semiconductors also respond to pressure and are used as transducers in solid-state pressure gauges.

(b) Temperature

The concept of temperature is introduced in *The chemist's toolkit* 2. In the early days of thermometry (and still in laboratory practice today), temperatures were related to the length of a column of liquid, and the difference in lengths shown when the thermometer was first in contact with melting ice and then with boiling water was divided into 100 steps called 'degrees', the lower point being labelled 0. This procedure led to the **Celsius scale** of temperature. In this text, temperatures on the Celsius scale are denoted θ (theta) and expressed in *degrees Celsius* (°C). However, because different liquids expand to different extents, and do not always expand uniformly over a given range, thermometers constructed from different materials showed different numerical values of the temperature between their fixed points. The pressure of a gas, however, can be used to construct a **perfect-gas temperature scale** that is independent of the identity of the gas. The perfect-gas scale turns out to be identical to the **thermodynamic temperature scale** (Topic 3A), so the latter term is used from now on to avoid a proliferation of names.

On the thermodynamic temperature scale, temperatures are denoted T and are normally reported in *kelvins* (K; not °K). Thermodynamic and Celsius temperatures are related by the exact expression

$$T/\text{K} = \theta/{}^\circ\text{C} + 273.15 \qquad \text{Celsius scale [definition]} \qquad (1\text{A}.1)$$

This relation is the current definition of the Celsius scale in terms of the more fundamental Kelvin scale. It implies that a difference in temperature of 1 °C is equivalent to a difference of 1 K.

Brief illustration 1A.1

To express 25.00 °C as a temperature in kelvins, eqn 1A.1 is used to write

$$T/\text{K} = (25.00\,^\circ\text{C})/^\circ\text{C} + 273.15 = 25.00 + 273.15 = 298.15$$

The chemist's toolkit 2 Properties of bulk matter

The state of a bulk sample of matter is defined by specifying the values of various properties. Among them are:

The **mass**, m, a measure of the quantity of matter present (unit: kilogram, kg).

The **volume**, V, a measure of the quantity of space the sample occupies (unit: cubic metre, m^3).

The **amount of substance**, n, a measure of the number of specified entities (atoms, molecules, or formula units) present (unit: mole, mol).

The amount of substance, n (colloquially, 'the number of moles'), is a measure of the number of specified entities present in the sample. 'Amount of substance' is the official name of the quantity; it is commonly simplified to 'chemical amount' or simply 'amount'. A mole is currently defined as the number of carbon atoms in exactly 12 g of carbon-12. (In 2011 the decision was taken to replace this definition, but the change has not yet, in 2018, been implemented.) The number of entities per mole is called **Avogadro's constant**, N_A; the currently accepted value is $6.022 \times 10^{23}\,mol^{-1}$ (note that N_A is a constant with units, not a pure number).

The **molar mass of a substance**, M (units: formally $kg\,mol^{-1}$ but commonly $g\,mol^{-1}$) is the mass per mole of its atoms, its molecules, or its formula units. The amount of substance of specified entities in a sample can readily be calculated from its mass, by noting that

$$n = \frac{m}{M} \quad \text{Amount of substance}$$

A note on good practice Be careful to distinguish atomic or molecular mass (the mass of a single atom or molecule; unit: kg) from molar mass (the mass per mole of atoms or molecules; units: $kg\,mol^{-1}$). *Relative* molecular masses of atoms and molecules, $M_r = m/m_u$, where m is the mass of the atom or molecule and m_u is the atomic mass constant (see inside front cover), are still widely called 'atomic weights' and 'molecular weights' even though they are dimensionless quantities and not weights ('weight' is the gravitational force exerted on an object).

A sample of matter may be subjected to a **pressure**, p (unit: pascal, Pa; $1\,Pa = 1\,kg\,m^{-1}\,s^{-2}$), which is defined as the force, F, it is subjected to, divided by the area, A, to which that force is applied. Although the pascal is the SI unit of pressure, it is also common to express pressure in bar ($1\,bar = 10^5\,Pa$) or atmospheres ($1\,atm = 101\,325\,Pa$ exactly), both of which correspond to typical atmospheric pressure. Because many physical properties depend on the pressure acting on a sample, it is appropriate to select a certain value of the pressure to report their values. The **standard pressure** for reporting physical quantities is currently defined as $p^{\ominus} = 1\,bar$ exactly.

To specify the state of a sample fully it is also necessary to give its **temperature**, T. The temperature is formally a property that determines in which direction energy will flow as heat when two samples are placed in contact through thermally conducting walls: energy flows from the sample with the higher temperature to the sample with the lower temperature. The symbol T is used to denote the **thermodynamic temperature** which is an absolute scale with $T = 0$ as the lowest point. Temperatures above $T = 0$ are then most commonly expressed by using the **Kelvin scale**, in which the gradations of temperature are expressed in kelvins (K). The Kelvin scale is currently defined by setting the triple point of water (the temperature at which ice, liquid water, and water vapour are in mutual equilibrium) at exactly 273.16 K (as for certain other units, a decision has been taken to revise this definition, but it has not yet, in 2018, been implemented). The freezing point of water (the melting point of ice) at 1 atm is then found experimentally to lie 0.01 K below the triple point, so the freezing point of water is 273.15 K.

Suppose a sample is divided into smaller samples. If a property of the original sample has a value that is equal to the sum of its values in all the smaller samples (as mass would), then it is said to be **extensive**. Mass and volume are extensive properties. If a property retains the same value as in the original sample for all the smaller samples (as temperature would), then it is said to be **intensive**. Temperature and pressure are intensive properties. Mass density, $\rho = m/V$, is also intensive because it would have the same value for all the smaller samples and the original sample. All molar properties, $X_m = X/n$, are intensive, whereas X and n are both extensive.

Note how the units (in this case, °C) are cancelled like numbers. This is the procedure called 'quantity calculus' in which a physical quantity (such as the temperature) is the product of a numerical value (25.00) and a unit (1 °C); see *The chemist's toolkit* 1. Multiplication of both sides by K then gives $T = 298.15\,K$.

A note on good practice The zero temperature on the thermodynamic temperature scale is written $T = 0$, not $T = 0\,K$. This scale is absolute, and the lowest temperature is 0 regardless of the size of the divisions on the scale (just as zero pressure is denoted $p = 0$, regardless of the size of the units, such as bar or pascal). However, it is appropriate to write 0 °C because the Celsius scale is not absolute.

1A.2 Equations of state

Although in principle the state of a pure substance is specified by giving the values of n, V, p, and T, it has been established experimentally that it is sufficient to specify only three of these variables since doing so fixes the value of the fourth variable.

That is, it is an experimental fact that each substance is described by an **equation of state**, an equation that interrelates these four variables.

The general form of an equation of state is

$$p = f(T,V,n) \qquad \text{General form of an equation of state} \qquad (1A.2)$$

This equation states that if the values of n, T, and V are known for a particular substance, then the pressure has a fixed value. Each substance is described by its own equation of state, but the explicit form of the equation is known in only a few special cases. One very important example is the equation of state of a 'perfect gas', which has the form $p = nRT/V$, where R is a constant independent of the identity of the gas.

The equation of state of a perfect gas was established by combining a series of empirical laws.

(a) The empirical basis

The following individual gas laws should be familiar:

Boyle's law: $pV = \text{constant, at constant } n, T$ (1A.3a)

Charles's law: $V = \text{constant} \times T, \text{ at constant } n, p$ (1A.3b)

$p = \text{constant} \times T, \text{ at constant } n, V$ (1A.3c)

Avogadro's principle:

$V = \text{constant} \times n \text{ at constant } p, T$ (1A.3d)

Boyle's and Charles's laws are examples of a **limiting law**, a law that is strictly true only in a certain limit, in this case $p \to 0$. For example, if it is found empirically that the volume of a substance fits an expression $V = aT + bp + cp^2$, then in the limit of $p \to 0$, $V = aT$. Many relations that are strictly true only at $p = 0$ are nevertheless reasonably reliable at normal pressures ($p \approx 1$ bar) and are used throughout chemistry.

Figure 1A.2 depicts the variation of the pressure of a sample of gas as the volume is changed. Each of the curves in the graph corresponds to a single temperature and hence is called an **isotherm**. According to Boyle's law, the isotherms of gases are hyperbolas (a curve obtained by plotting y against x with xy = constant, or y = constant/x). An alternative depiction, a plot of pressure against 1/volume, is shown in Fig. 1A.3. The linear variation of volume with temperature summarized by Charles's law is illustrated in Fig. 1A.4. The lines in this illustration are examples of **isobars**, or lines showing the variation of properties at constant pressure. Figure 1A.5 illustrates the linear variation of pressure with temperature. The lines in this diagram are **isochores**, or lines showing the variation of properties at constant volume.

A note on good practice To test the validity of a relation between two quantities, it is best to plot them in such a way that they should give a straight line, because deviations from a straight line are much easier to detect than deviations from a curve. The development of expressions that, when plotted, give a straight line is a very important and common procedure in physical chemistry.

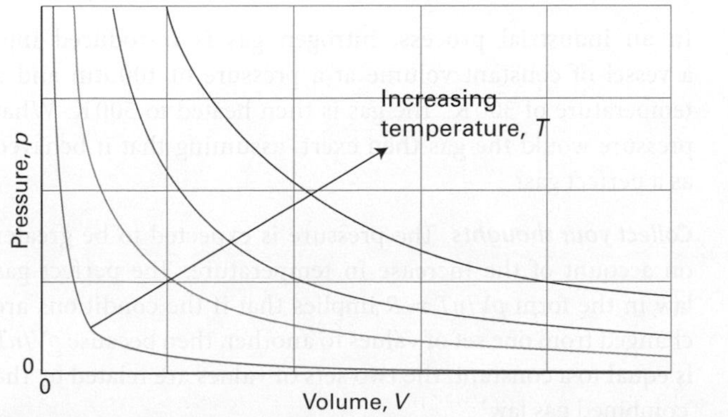

Figure 1A.2 The pressure–volume dependence of a fixed amount of perfect gas at different temperatures. Each curve is a hyperbola (pV = constant) and is called an isotherm.

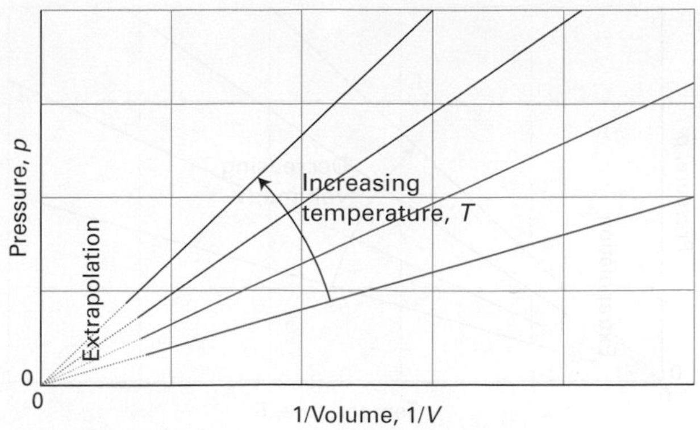

Figure 1A.3 Straight lines are obtained when the pressure of a perfect gas is plotted against $1/V$ at constant temperature. These lines extrapolate to zero pressure at $1/V = 0$.

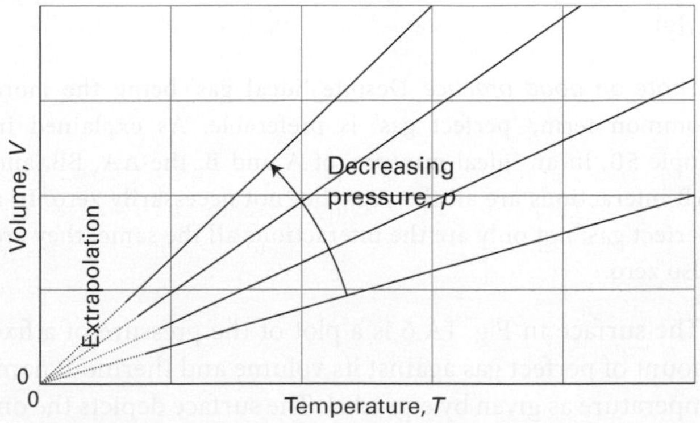

Figure 1A.4 The variation of the volume of a fixed amount of a perfect gas with the temperature at constant pressure. Note that in each case the isobars extrapolate to zero volume at $T = 0$, corresponding to $\theta = -273.15\,°\text{C}$.

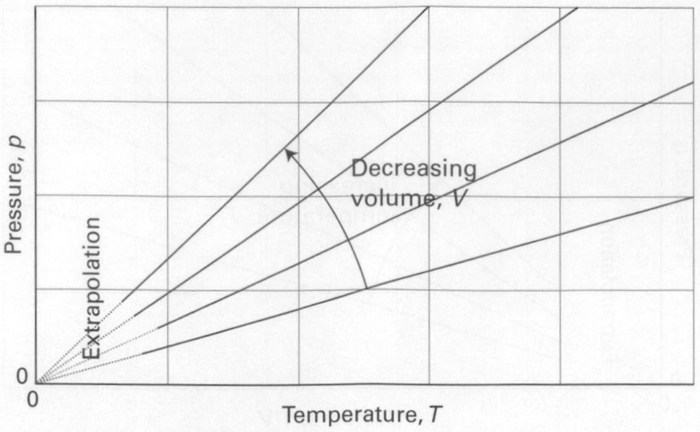

Figure 1A.5 The pressure of a perfect gas also varies linearly with the temperature at constant volume, and extrapolates to zero at $T = 0$ (−273.15 °C).

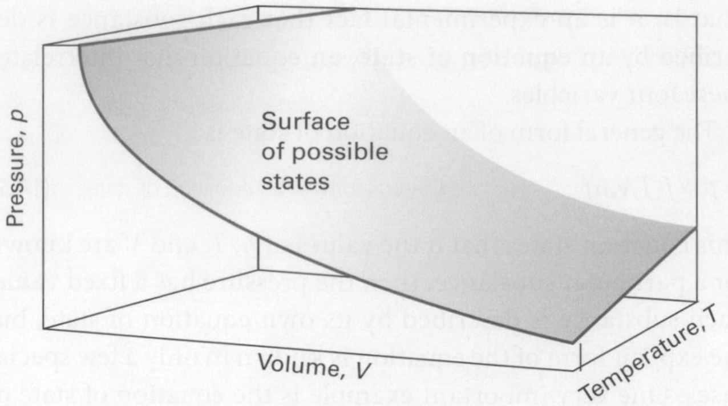

Figure 1A.6 A region of the p,V,T surface of a fixed amount of perfect gas. The points forming the surface represent the only states of the gas that can exist.

The empirical observations summarized by eqn 1A.3 can be combined into a single expression:

$$pV = \text{constant} \times nT$$

This expression is consistent with Boyle's law (pV = constant) when n and T are constant, with both forms of Charles's law ($p \propto T$, $V \propto T$) when n and either V or p are held constant, and with Avogadro's principle ($V \propto n$) when p and T are constant. The constant of proportionality, which is found experimentally to be the same for all gases, is denoted R and called the (molar) **gas constant**. The resulting expression

$$pV = nRT \qquad \text{Perfect gas law} \qquad (1A.4)$$

is the **perfect gas law** (or *perfect gas equation of state*). It is the approximate equation of state of any gas, and becomes increasingly exact as the pressure of the gas approaches zero. A gas that obeys eqn 1A.4 exactly under all conditions is called a **perfect gas** (or *ideal gas*). A **real gas**, an actual gas, behaves more like a perfect gas the lower the pressure, and is described exactly by eqn 1A.4 in the limit of $p \to 0$. The gas constant R can be determined by evaluating $R = pV/nT$ for a gas in the limit of zero pressure (to guarantee that it is behaving perfectly).

A note on good practice Despite 'ideal gas' being the more common term, 'perfect gas' is preferable. As explained in Topic 5B, in an 'ideal mixture' of A and B, the AA, BB, and AB interactions are all the same but not necessarily zero. In a perfect gas, not only are the interactions all the same, they are also zero.

The surface in Fig. 1A.6 is a plot of the pressure of a fixed amount of perfect gas against its volume and thermodynamic temperature as given by eqn 1A.4. The surface depicts the only possible states of a perfect gas: the gas cannot exist in states that do not correspond to points on the surface. The graphs in Figs. 1A.2 and 1A.4 correspond to the sections through the surface (Fig. 1A.7).

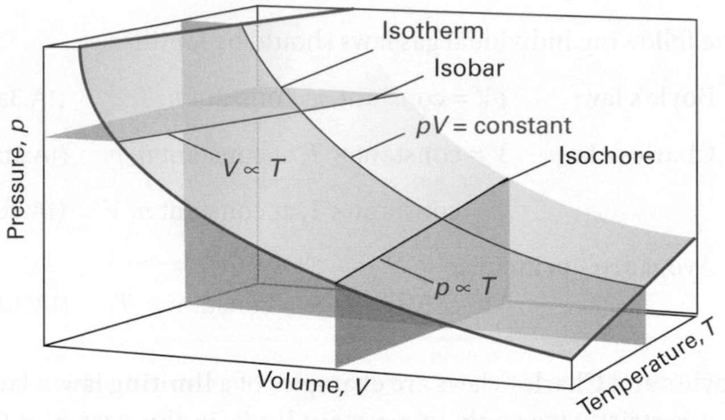

Figure 1A.7 Sections through the surface shown in Fig. 1A.6 at constant temperature give the isotherms shown in Fig. 1A.2. Sections at constant pressure give the isobars shown in Fig. 1A.4. Sections at constant volume give the isochores shown in Fig. 1A.5.

Example 1A.1 **Using the perfect gas law**

In an industrial process, nitrogen gas is introduced into a vessel of constant volume at a pressure of 100 atm and a temperature of 300 K. The gas is then heated to 500 K. What pressure would the gas then exert, assuming that it behaved as a perfect gas?

Collect your thoughts The pressure is expected to be greater on account of the increase in temperature. The perfect gas law in the form $pV/nT = R$ implies that if the conditions are changed from one set of values to another, then because pV/nT is equal to a constant, the two sets of values are related by the 'combined gas law'

$$\frac{p_1V_1}{n_1T_1} = \frac{p_2V_2}{n_2T_2} \qquad \text{Combined gas law} \qquad (1A.5)$$

This expression is easily rearranged to give the unknown quantity (in this case p_2) in terms of the known. The known and unknown data are summarized as follows:

	n	p	V	T
Initial	Same	100 atm	Same	300 K
Final	Same	?	Same	500 K

The solution Cancellation of the volumes (because $V_1 = V_2$) and amounts (because $n_1 = n_2$) on each side of the combined gas law results in

$$\frac{p_1}{T_1} = \frac{p_2}{T_2}$$

which can be rearranged into

$$p_2 = \frac{T_2}{T_1} \times p_1$$

Substitution of the data then gives

$$p_2 = \frac{500\,\text{K}}{300\,\text{K}} \times (100\,\text{atm}) = 167\,\text{atm}$$

Self-test 1A.1 What temperature would result in the same sample exerting a pressure of 300 atm?

Answer: 900 K

The perfect gas law is of the greatest importance in physical chemistry because it is used to derive a wide range of relations that are used throughout thermodynamics. However, it is also of considerable practical utility for calculating the properties of a gas under a variety of conditions. For instance, the molar volume, $V_m = V/n$, of a perfect gas under the conditions called **standard ambient temperature and pressure** (SATP), which means 298.15 K and 1 bar (i.e. exactly 10^5 Pa), is easily calculated from $V_m = RT/p$ to be 24.789 $\text{dm}^3\,\text{mol}^{-1}$. An earlier definition, **standard temperature and pressure** (STP), was 0 °C and 1 atm; at STP, the molar volume of a perfect gas is 22.414 $\text{dm}^3\,\text{mol}^{-1}$.

The molecular explanation of Boyle's law is that if a sample of gas is compressed to half its volume, then twice as many molecules strike the walls in a given period of time than before it was compressed. As a result, the average force exerted on the walls is doubled. Hence, when the volume is halved the pressure of the gas is doubled, and pV is a constant. Boyle's law applies to all gases regardless of their chemical identity (provided the pressure is low) because at low pressures the average separation of molecules is so great that they exert no influence on one another and hence travel independently. The molecular explanation of Charles's law lies in the fact that raising the temperature of a gas increases the average speed of its molecules. The molecules collide with the walls more frequently and with greater impact. Therefore they exert a greater pressure on the walls of the container. For a quantitative account of these relations, see Topic 1B.

(b) Mixtures of gases

When dealing with gaseous mixtures, it is often necessary to know the contribution that each component makes to the total pressure of the sample. The **partial pressure**, p_J, of a gas J in a mixture (any gas, not just a perfect gas), is defined as

$$p_J = x_J p \qquad \text{Partial pressure [definition]} \qquad (1A.6)$$

where x_J is the **mole fraction** of the component J, the amount of J expressed as a fraction of the total amount of molecules, n, in the sample:

$$x_J = \frac{n_J}{n} \qquad n = n_A + n_B + \cdots \qquad \text{Mole fraction [definition]} \qquad (1A.7)$$

When no J molecules are present, $x_J = 0$; when only J molecules are present, $x_J = 1$. It follows from the definition of x_J that, whatever the composition of the mixture, $x_A + x_B + \cdots = 1$ and therefore that the sum of the partial pressures is equal to the total pressure:

$$p_A + p_B + \cdots = (x_A + x_B + \cdots)p = p \qquad (1A.8)$$

This relation is true for both real and perfect gases.

When all the gases are perfect, the partial pressure as defined in eqn 1A.6 is also the pressure that each gas would exert if it occupied the same container alone at the same temperature. The latter is the original meaning of 'partial pressure'. That identification was the basis of the original formulation of **Dalton's law**:

> The pressure exerted by a mixture of gases is the sum of the pressures that each one would exert if it occupied the container alone. — Dalton's law

This law is valid only for mixtures of perfect gases, so it is not used to define partial pressure. Partial pressure is defined by eqn 1A.6, which is valid for all gases.

Example 1A.2 Calculating partial pressures

The mass percentage composition of dry air at sea level is approximately N_2: 75.5; O_2: 23.2; Ar: 1.3. What is the partial pressure of each component when the total pressure is 1.20 atm?

Collect your thoughts Partial pressures are defined by eqn 1A.6. To use the equation, first calculate the mole fractions of the components, by using eqn 1A.7 and the fact that the amount of atoms or molecules J of molar mass M_J in a sample of mass m_J is $n_J = m_J/M_J$. The mole fractions are independent of the total mass of the sample, so choose the latter to be exactly 100 g (which makes the conversion from mass percentages very easy). Thus, the mass of N_2 present is 75.5 per cent of 100 g, which is 75.5 g.

The solution The amounts of each type of atom or molecule present in 100 g of air, in which the masses of N_2, O_2, and Ar are 75.5 g, 23.2 g, and 1.3 g, respectively, are

$$n(N_2) = \frac{75.5\,g}{28.02\,g\,mol^{-1}} = \frac{75.5}{28.02}\,mol = 2.69\,mol$$

$$n(O_2) = \frac{23.2\,g}{32.00\,g\,mol^{-1}} = \frac{23.2}{32.00}\,mol = 0.725\,mol$$

$$n(Ar) = \frac{1.3\,g}{39.95\,g\,mol^{-1}} = \frac{1.3}{39.95}\,mol = 0.033\,mol$$

The total is 3.45 mol. The mole fractions are obtained by dividing each of the above amounts by 3.45 mol and the partial pressures are then obtained by multiplying the mole fraction by the total pressure (1.20 atm):

	N_2	O_2	Ar
Mole fraction:	0.780	0.210	0.0096
Partial pressure/atm:	0.936	0.252	0.012

Self-test 1A.2 When carbon dioxide is taken into account, the mass percentages are 75.52 (N_2), 23.15 (O_2), 1.28 (Ar), and 0.046 (CO_2). What are the partial pressures when the total pressure is 0.900 atm?

Answer: 0.703, 0.189, 0.0084, and 0.00027 atm

Checklist of concepts

☐ 1. The **physical state** of a sample of a substance, its physical condition, is defined by its physical properties.

☐ 2. **Mechanical equilibrium** is the condition of equality of pressure on either side of a shared movable wall.

☐ 3. An **equation of state** is an equation that interrelates the variables that define the state of a substance.

☐ 4. Boyle's and Charles's laws are examples of a **limiting law**, a law that is strictly true only in a certain limit, in this case $p \to 0$.

☐ 5. An **isotherm** is a line in a graph that corresponds to a single temperature.

☐ 6. An **isobar** is a line in a graph that corresponds to a single pressure.

☐ 7. An **isochore** is a line in a graph that corresponds to a single volume.

☐ 8. A **perfect gas** is a gas that obeys the perfect gas law under all conditions.

☐ 9. **Dalton's law** states that the pressure exerted by a mixture of (perfect) gases is the sum of the pressures that each one would exert if it occupied the container alone.

Checklist of equations

Property	Equation	Comment	Equation number
Relation between temperature scales	$T/K = \theta/°C + 273.15$	273.15 is exact	1A.1
Perfect gas law	$pV = nRT$	Valid for real gases in the limit $p \to 0$	1A.4
Partial pressure	$p_J = x_J p$	Valid for all gases	1A.6
Mole fraction	$x_J = n_J/n$ $n = n_A + n_B + \cdots$	Definition	1A.7

TOPIC 1B The kinetic model

➤ **Why do you need to know this material?**

This material illustrates an important skill in science: the ability to extract quantitative information from a qualitative model. Moreover, the model is used in the discussion of the transport properties of gases (Topic 16A), reaction rates in gases (Topic 18A), and catalysis (Topic 19C).

➤ **What is the key idea?**

A gas consists of molecules of negligible size in ceaseless random motion and obeying the laws of classical mechanics in their collisions.

➤ **What do you need to know already?**

You need to be aware of Newton's second law of motion, that the acceleration of a body is proportional to the force acting on it, and the conservation of linear momentum (*The chemist's toolkit* 3).

In the **kinetic theory** of gases (which is sometimes called the *kinetic-molecular theory*, KMT) it is assumed that the only contribution to the energy of the gas is from the kinetic energies of the molecules. The kinetic model is one of the most remarkable—and arguably most beautiful—models in physical chemistry, for from a set of very slender assumptions, powerful quantitative conclusions can be reached.

1B.1 The model

The kinetic model is based on three assumptions:

1. The gas consists of molecules of mass m in ceaseless random motion obeying the laws of classical mechanics.
2. The size of the molecules is negligible, in the sense that their diameters are much smaller than the average distance travelled between collisions; they are 'point-like'.
3. The molecules interact only through brief elastic collisions.

The chemist's toolkit 3 Momentum and force

The **speed**, v, of a body is defined as the rate of change of position. The **velocity**, $\boldsymbol{v}$, defines the direction of travel as well as the rate of motion, and particles travelling at the same speed but in different directions have different velocities. As shown in Sketch 1, the velocity can be depicted as an arrow in the direction of travel, its length being the speed v and its components v_x, v_y, and v_z along three perpendicular axes. These components have a sign: $v_x = +5\,\mathrm{m\,s^{-1}}$, for instance, indicates that a body is moving in the positive x-direction, whereas $v_x = -5\,\mathrm{m\,s^{-1}}$ indicates that it is moving in the opposite direction. The length of the arrow (the speed) is related to the components by Pythagoras' theorem: $v^2 = v_x^2 + v_y^2 + v_z^2$.

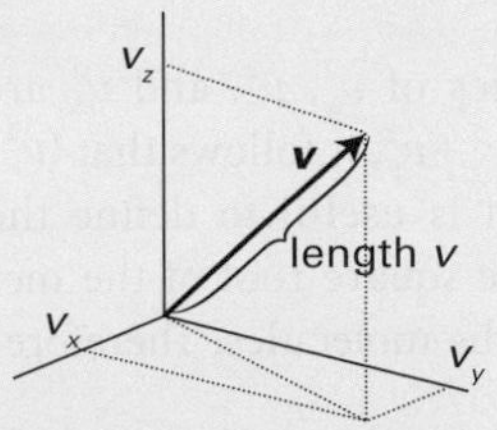

Sketch 1

The concepts of classical mechanics are commonly expressed in terms of the **linear momentum**, $\boldsymbol{p}$, which is defined as

$$\boldsymbol{p} = m\boldsymbol{v} \qquad \text{Linear momentum [definition]}$$

Momentum also mirrors velocity in having a sense of direction; bodies of the same mass and moving at the same speed but in different directions have different linear momenta.

Acceleration, $\boldsymbol{a}$, is the rate of change of velocity. A body accelerates if its speed changes. A body also accelerates if its speed remains unchanged but its direction of motion changes. According to Newton's **second law of motion**, the acceleration of a body of mass m is proportional to the force, $\boldsymbol{F}$, acting on it:

$$\boldsymbol{F} = m\boldsymbol{a} \qquad \text{Force}$$

Because $m\boldsymbol{v}$ is the linear momentum and $\boldsymbol{a}$ is the rate of change of velocity, $m\boldsymbol{a}$ is the rate of change of momentum. Therefore, an alternative statement of Newton's second law is that the force is equal to the rate of change of momentum. Newton's law indicates that the acceleration occurs in the same direction as the force acts. If, for an isolated system, no external force acts, then there is no acceleration. This statement is the **law of conservation of momentum**: that the momentum of a body is constant in the absence of a force acting on the body.

An **elastic collision** is a collision in which the total translational kinetic energy of the molecules is conserved.

(a) Pressure and molecular speeds

From the very economical assumptions of the kinetic model, it is possible to derive an expression that relates the pressure and volume of a gas.

How is that done? 1B.1 Using the kinetic model to derive an expression for the pressure of a gas

Consider the arrangement in Fig. 1B.1, and then follow these steps.

Step 1 *Set up the calculation of the change in momentum*

When a particle of mass m that is travelling with a component of velocity v_x parallel to the x-axis collides with the wall on the right and is reflected, its linear momentum changes from mv_x before the collision to $-mv_x$ after the collision (when it is travelling in the opposite direction). The x-component of momentum therefore changes by $2mv_x$ on each collision (the y- and z-components are unchanged). Many molecules collide with the wall in an interval Δt, and the total change of momentum is the product of the change in momentum of each molecule multiplied by the number of molecules that reach the wall during the interval.

Step 2 *Calculate the change in momentum*

Because a molecule with velocity component v_x travels a distance $v_x\Delta t$ along the x-axis in an interval Δt, all the molecules within a distance $v_x\Delta t$ of the wall strike it if they are travelling towards it (Fig. 1B.2). It follows that if the wall has area A, then all the particles in a volume $A \times v_x\Delta t$ reach the wall (if they are travelling towards it). The number density of particles is nN_A/V, where n is the total amount of molecules in the container of volume V and N_A is Avogadro's constant. It follows that the number of molecules in the volume $Av_x\Delta t$ is $(nN_A/V) \times Av_x\Delta t$.

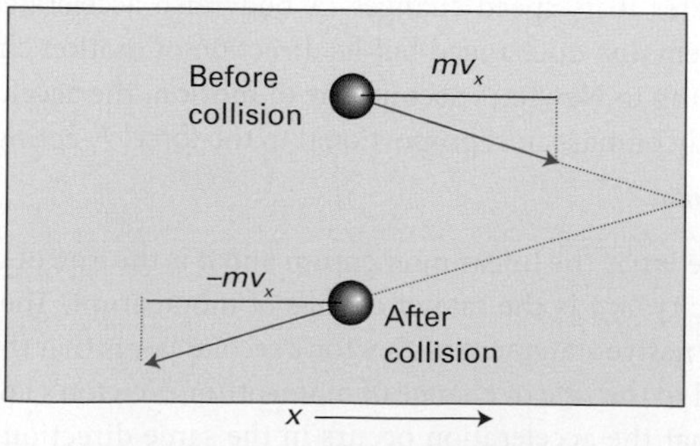

Figure 1B.1 The pressure of a gas arises from the impact of its molecules on the walls. In an elastic collision of a molecule with a wall perpendicular to the x-axis, the x-component of velocity is reversed but the y- and z-components are unchanged.

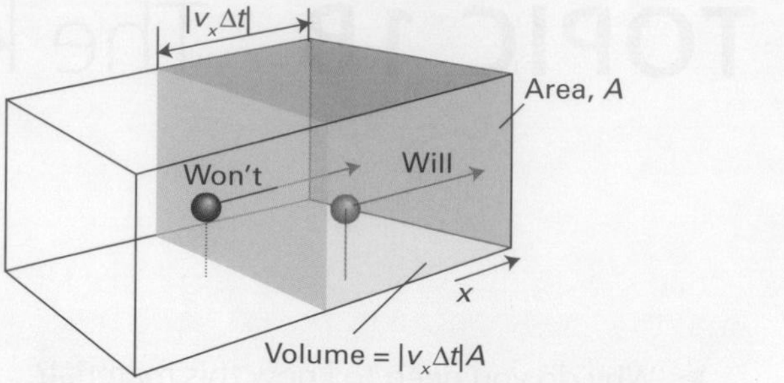

Figure 1B.2 A molecule will reach the wall on the right within an interval of time Δt if it is within a distance $v_x\Delta t$ of the wall and travelling to the right.

At any instant, half the particles are moving to the right and half are moving to the left. Therefore, the average number of collisions with the wall during the interval Δt is $\frac{1}{2}nN_AAv_x\Delta t/V$. The total momentum change in that interval is the product of this number and the change $2mv_x$:

$$\text{Momentum change} = \frac{nN_AAv_x\Delta t}{2V} \times 2mv_x$$

$$= \frac{n\overbrace{mN_A}^{M}Av_x^2\Delta t}{V} = \frac{nMAv_x^2\Delta t}{V}$$

Step 3 *Calculate the force*

The rate of change of momentum, the change of momentum divided by the interval Δt during which it occurs, is

$$\text{Rate of change of momentum} = \frac{nMAv_x^2}{V}$$

According to Newton's second law of motion this rate of change of momentum is equal to the force.

Step 4 *Calculate the pressure*

The pressure is this force $(nMAv_x^2/V)$ divided by the area (A) on which the impacts occur. The areas cancel, leaving

$$\text{Pressure} = \frac{nMv_x^2}{V}$$

Not all the molecules travel with the same velocity, so the detected pressure, p, is the average (denoted $\langle ... \rangle$) of the quantity just calculated:

$$p = \frac{nM\langle v_x^2 \rangle}{V}$$

The average values of v_x^2, v_y^2, and v_z^2 are all the same, and because $v^2 = v_x^2 + v_y^2 + v_z^2$, it follows that $\langle v_x^2 \rangle = \frac{1}{3}\langle v^2 \rangle$.

At this stage it is useful to define the **root-mean-square speed**, v_{rms}, as the square root of the mean of the squares of the speeds, v, of the molecules. Therefore

$$v_{rms} = \langle v^2 \rangle^{1/2} \quad \text{Root-mean-square speed [definition]} \quad (1B.1)$$

The mean square speed in the expression for the pressure can therefore be written $\langle v_x^2\rangle=\frac{1}{3}\langle v^2\rangle=\frac{1}{3}v_{\rm rms}^2$ to give

$$pV=\tfrac{1}{3}nMv_{\rm rms}^2 \qquad (1B.2)$$

Relation between pressure and volume [KMT]

This equation is one of the key results of the kinetic model. If the root-mean-square speed of the molecules depends only on the temperature, then at constant temperature

$$pV=\text{constant}$$

which is the content of Boyle's law. The task now is to show that the right-hand side of eqn 1B.2 is equal to nRT.

(b) The Maxwell–Boltzmann distribution of speeds

In a gas the speeds of individual molecules span a wide range, and the collisions in the gas ensure that their speeds are ceaselessly changing. Before a collision, a molecule may be travelling rapidly, but after a collision it may be accelerated to a higher speed, only to be slowed again by the next collision. To evaluate the root-mean-square speed it is necessary to know the fraction of molecules that have a given speed at any instant. The fraction of molecules that have speeds in the range v to $v + dv$ is proportional to the width of the range, and is written $f(v)dv$, where $f(v)$ is called the **distribution of speeds**. An expression for this distribution can be found by recognizing that the energy of the molecules is entirely kinetic, and then using the Boltzmann distribution to describe how this energy is distributed over the molecules.

How is that done? 1B.2 Deriving the distribution of speeds

The starting point for this derivation is the Boltzmann distribution (see the text's *Prologue*).

Step 1 *Write an expression for the distribution of the kinetic energy*

The Boltzmann distribution implies that the fraction of molecules with velocity components v_x, v_y, and v_z is proportional to an exponential function of their kinetic energy: $f(v) = Ke^{-\varepsilon/kT}$, where K is a constant of proportionality. The kinetic energy is

$$\varepsilon=\tfrac{1}{2}mv_x^2+\tfrac{1}{2}mv_y^2+\tfrac{1}{2}mv_z^2$$

Therefore, use the relation $a^{x+y+z}=a^xa^ya^z$ to write

$$f(v)=Ke^{-(mv_x^2+mv_y^2+mv_z^2)/2kT}=Ke^{-mv_x^2/2kT}e^{-mv_y^2/2kT}e^{-mv_z^2/2kT}$$

The distribution factorizes into three terms as $f(v)=f(v_x)f(v_y)f(v_z)$ and $K=K_xK_yK_z$, with

$$f(v_x)=K_xe^{-mv_x^2/2kT}$$

and likewise for the other two coordinates.

Step 2 *Determine the constants K_x, K_y, and K_z*

To determine the constant K_x, note that a molecule must have a velocity component somewhere in the range $-\infty < v_x < \infty$, so integration over the full range of possible values of v_x must give a total probability of 1:

$$\int_{-\infty}^{\infty} f(v_x)dv_x=1$$

(See *The chemist's toolkit* 4 for the principles of integration.) Substitution of the expression for $f(v_x)$ then gives

$$1=K_x\overbrace{\int_{-\infty}^{\infty}e^{-mv_x^2/2kT}dv_x}^{\text{Integral G.1}}=K_x\left(\frac{2\pi kT}{m}\right)^{1/2}$$

Therefore, $K_x=(m/2\pi kT)^{1/2}$ and

$$f(v_x)=\left(\frac{m}{2\pi kT}\right)^{1/2}e^{-mv_x^2/2kT} \qquad (1B.3)$$

The expressions for $f(v_y)$ and $f(v_z)$ are analogous.

Step 3 *Write a preliminary expression for $f(v_x)f(v_y)f(v_z)dv_xdv_ydv_z$*

The probability that a molecule has a velocity in the range v_x to $v_x + dv_x$, v_y to $v_y + dv_y$, v_z to $v_z + dv_z$, is

$$f(v_x)f(v_y)f(v_z)dv_xdv_ydv_z=\left(\frac{m}{2\pi kT}\right)^{3/2}\overbrace{e^{-mv_x^2/2kT}e^{-mv_y^2/2kT}e^{-mv_z^2/2kT}}^{e^{-m(v_x^2+v_y^2+v_z^2)/2kT}}\times dv_xdv_ydv_z$$

$$=\left(\frac{m}{2\pi kT}\right)^{3/2}e^{-mv^2/2kT}dv_xdv_ydv_z$$

where $v^2=v_x^2+v_y^2+v_z^2$.

Step 3 *Calculate the probability that a molecule has a speed in the range v to $v + dv$*

To evaluate the probability that a molecule has a speed in the range v to $v + dv$ regardless of direction, think of the three velocity components as defining three coordinates in 'velocity space', with the same properties as ordinary space except that the axes are labelled (v_x, v_y, v_z) instead of (x, y, z). Just as the volume element in ordinary space is $dxdydz$, so the volume element in velocity space is $dv_xdv_ydv_z$. The sum of all the volume elements in ordinary space that lie at a distance r from the centre is the volume of a spherical shell of radius r and thickness dr. That volume is the product of the surface area of the shell,

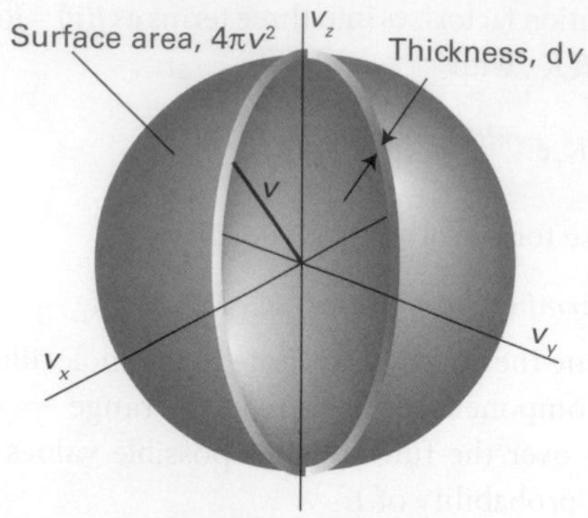

Figure 1B.3 To evaluate the probability that a molecule has a speed in the range v to $v + dv$, evaluate the total probability that the molecule will have a speed that is anywhere in a thin shell of radius $v = (v_x^2 + v_y^2 + v_z^2)^{1/2}$ and thickness dv.

$4\pi r^2$, and its thickness dr, and is therefore $4\pi r^2 dr$. Similarly, the analogous volume in velocity space is the volume of a shell of radius v and thickness dv, namely $4\pi v^2 dv$ (Fig. 1B.3). Now, because $f(v_x)f(v_y)f(v_z)$, the term in blue in the last equation, depends only on v^2, and has the same value everywhere in a shell of radius v, the total probability of the molecules possessing a speed in the range v to $v + dv$ is the product of the term in blue and the volume of the shell of radius v and thickness dv. If this probability is written $f(v)dv$, it follows that

$$f(v)dv = 4\pi v^2 dv \left(\frac{m}{2\pi kT}\right)^{3/2} e^{-mv^2/2kT}$$

and $f(v)$ itself, after minor rearrangement, is

$$f(v) = 4\pi \left(\frac{m}{2\pi kT}\right)^{3/2} v^2 e^{-mv^2/2kT}$$

Because $R = N_A k$ (Table 1B.1), $m/k = mN_A/R = M/R$, it follows that

$$f(v) = 4\pi \left(\frac{M}{2\pi RT}\right)^{3/2} v^2 e^{-Mv^2/2RT} \quad (1B.4)$$

Maxwell–Boltzmann distribution [KMT]

The function $f(v)$ is called the **Maxwell–Boltzmann distribution of speeds**. Note that, in common with other distribution functions, $f(v)$ acquires physical significance only after it is multiplied by the range of speeds of interest.

Table 1B.1 The (molar) gas constant*

R	
8.314 47	J K^{-1} mol^{-1}
8.205 74 × 10^{-2}	dm^3 atm K^{-1} mol^{-1}
8.314 47 × 10^{-2}	dm^3 bar K^{-1} mol^{-1}
8.314 47	Pa m^3 K^{-1} mol^{-1}
62.364	dm^3 Torr K^{-1} mol^{-1}
1.987 21	cal K^{-1} mol^{-1}

* The gas constant is now defined as $R = N_A k$, where N_A is Avogadro's constant and k is Boltzmann's constant.

The chemist's toolkit 4 Integration

Integration is concerned with the areas under curves. The **integral** of a function $f(x)$, which is denoted $\int f(x)dx$ (the symbol $\int$ is an elongated S denoting a sum), between the two values $x = a$ and $x = b$ is defined by imagining the x-axis as divided into strips of width δx and evaluating the following sum:

$$\int_a^b f(x)dx = \lim_{\delta x \to 0} \sum_i f(x_i)\delta x$$

Integration [definition]

As can be appreciated from Sketch 1, the integral is the area under the curve between the limits a and b. The function to be integrated is called the **integrand**. It is an astonishing mathematical fact that the integral of a function is the inverse of the differential of that function. In other words, if differentiation of f is followed by integration of the resulting function, the result is the original function f (to within a constant).

The integral in the preceding equation with the limits specified is called a **definite integral**. If it is written without the limits specified, it is called an **indefinite integral**. If the result of carrying out an indefinite integration is $g(x) + C$, where C is a constant, the following procedure is used to evaluate the corresponding definite integral:

$$I = \int_a^b f(x)dx = \{g(x)+C\}\Big|_a^b = \{g(b)+C\} - \{g(a)+C\}$$
$$= g(b) - g(a)$$

Definite integral

Note that the constant of integration disappears. The definite and indefinite integrals encountered in this text are listed in the *Resource section*. They may also be calculated by using mathematical software.

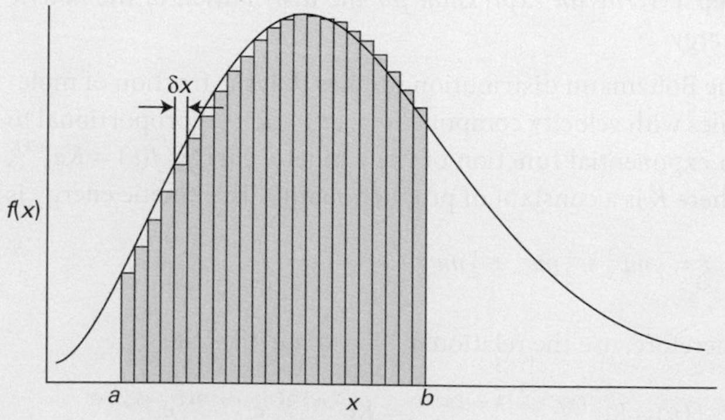

Sketch 1

The important features of the Maxwell–Boltzmann distribution are as follows (and are shown pictorially in Fig. 1B.4):

- Equation 1B.4 includes a decaying exponential function (more specifically, a Gaussian function). Its presence implies that the fraction of molecules with very high speeds is very small because e^{-x^2} becomes very small when x is large.
- The factor $M/2RT$ multiplying v^2 in the exponent is large when the molar mass, M, is large, so the exponential factor goes most rapidly towards zero when M is large. That is, heavy molecules are unlikely to be found with very high speeds.
- The opposite is true when the temperature, T, is high: then the factor $M/2RT$ in the exponent is small, so the exponential factor falls towards zero relatively slowly as v increases. In other words, a greater fraction of the molecules can be expected to have high speeds at high temperatures than at low temperatures.
- A factor v^2 (the term before the e) multiplies the exponential. This factor goes to zero as v goes to zero, so the fraction of molecules with very low speeds will also be very small whatever their mass.
- The remaining factors (the term in parentheses in eqn 1B.4 and the 4π) simply ensure that, when the fractions are summed over the entire range of speeds from zero to infinity, the result is 1.

Physical interpretation

(c) Mean values

With the Maxwell–Boltzmann distribution in hand, it is possible to calculate the mean value of any power of the speed by evaluating the appropriate integral. For instance, to evaluate the fraction, F, of molecules with speeds in the range v_1 to v_2 evaluate the integral

$$F(v_1,v_2)=\int_{v_1}^{v_2} f(v)\mathrm{d}v \quad (1B.5)$$

This integral is the area under the graph of f as a function of v and, except in special cases, has to be evaluated numerically by using mathematical software (Fig. 1B.5). The average value of v^n is calculated as

$$\langle v^n\rangle=\int_0^{\infty} v^n f(v)\mathrm{d}v \quad (1B.6)$$

In particular, integration with $n=2$ results in the mean square speed, $\langle v^2\rangle$, of the molecules at a temperature T:

$$\langle v^2\rangle=\frac{3RT}{M} \quad \text{Mean square speed [KMT]} \quad (1B.7)$$

It follows that the root-mean-square speed of the molecules of the gas is

$$v_{\mathrm{rms}}=\langle v^2\rangle^{1/2}=\left(\frac{3RT}{M}\right)^{1/2} \quad \text{Root-mean-square speed [KMT]} \quad (1B.8)$$

which is proportional to the square root of the temperature and inversely proportional to the square root of the molar mass. That is, the higher the temperature, the higher the root-mean-square speed of the molecules, and, at a given temperature, heavy molecules travel more slowly than light molecules.

The important conclusion, however, is that when eqn 1B.8 is substituted into eqn 1B.2, the result is $pV = nRT$, which is the equation of state of a perfect gas. This conclusion confirms that the kinetic model can be regarded as a model of a perfect gas.

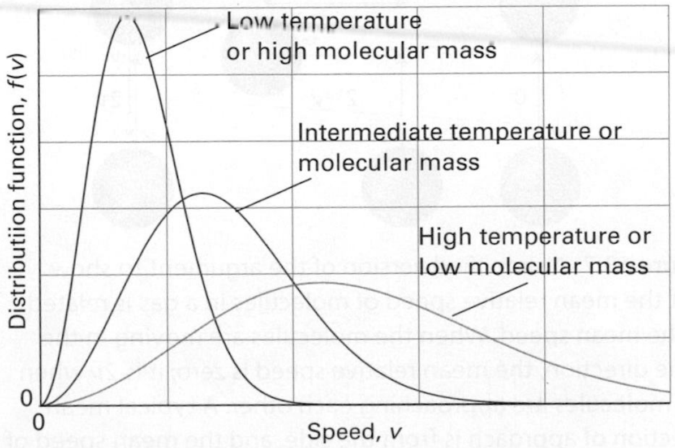

Figure 1B.4 The distribution of molecular speeds with temperature and molar mass. Note that the most probable speed (corresponding to the peak of the distribution) increases with temperature and with decreasing molar mass, and simultaneously the distribution becomes broader.

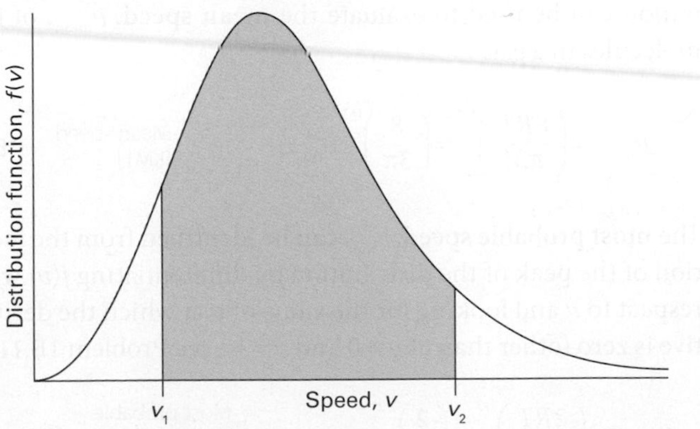

Figure 1B.5 To calculate the probability that a molecule will have a speed in the range v_1 to v_2, integrate the distribution between those two limits; the integral is equal to the area under the curve between the limits, as shown shaded here.

Example 1B.1 Calculating the mean speed of molecules in a gas

Calculate v_{rms} and the mean speed, v_{mean}, of N_2 molecules at 25 °C.

Collect your thoughts The root-mean-square speed is calculated from eqn 1B.8, with M = 28.02 g mol^{-1} (that is, 0.028 02 kg mol^{-1}) and T = 298 K. The mean speed is obtained by evaluating the integral

$$v_{mean} = \int_0^\infty v f(v)\,dv$$

with $f(v)$ given in eqn 1B.3. Use either mathematical software or the integrals listed in the *Resource section* and note that 1 J = 1 kg m^2 s^{-2}.

The solution The root-mean-square speed is

$$v_{rms} = \left\{\frac{3\times(8.3145\,\mathrm{J\,K^{-1}\,mol^{-1}})\times(298\,\mathrm{K})}{0.028\,02\,\mathrm{kg\,mol^{-1}}}\right\}^{1/2} = 515\,\mathrm{m\,s^{-1}}$$

The integral required for the calculation of v_{mean} is

$$v_{mean} = 4\pi\left(\frac{M}{2\pi RT}\right)^{3/2}\overbrace{\int_0^\infty v^3 e^{-Mv^2/2RT}\,dv}^{\text{Integral G.4}}$$
$$= 4\pi\left(\frac{M}{2\pi RT}\right)^{3/2}\times\tfrac{1}{2}\left(\frac{2RT}{M}\right)^2 = \left(\frac{8RT}{\pi M}\right)^{1/2}$$

Substitution of the data then gives

$$v_{mean} = \left(\frac{8\times(8.3145\,\mathrm{J\,K^{-1}\,mol^{-1}})\times(298\,\mathrm{K})}{\pi\times(0.028\,02\,\mathrm{kg\,mol^{-1}})}\right)^{1/2} = 475\,\mathrm{m\,s^{-1}}$$

Self-test 1B.1 Confirm that eqn 1B.7 follows from eqn 1B.6.

As shown in *Example* 1B.1, the Maxwell–Boltzmann distribution can be used to evaluate the **mean speed,** v_{mean}, of the molecules in a gas:

$$v_{mean} = \left(\frac{8RT}{\pi M}\right)^{1/2} = \left(\frac{8}{3\pi}\right)^{1/2} v_{rms} \qquad \text{Mean speed [KMT]} \qquad (1B.9)$$

The **most probable speed,** v_{mp}, can be identified from the location of the peak of the distribution by differentiating $f(v)$ with respect to v and looking for the value of v at which the derivative is zero (other than at $v = 0$ and $v = \infty$; see Problem 1B.11):

$$v_{mp} = \left(\frac{2RT}{M}\right)^{1/2} = \left(\frac{2}{3}\right)^{1/2} v_{rms} \qquad \text{Most probable speed [KMT]} \qquad (1B.10)$$

Figure 1B.6 summarizes these results.

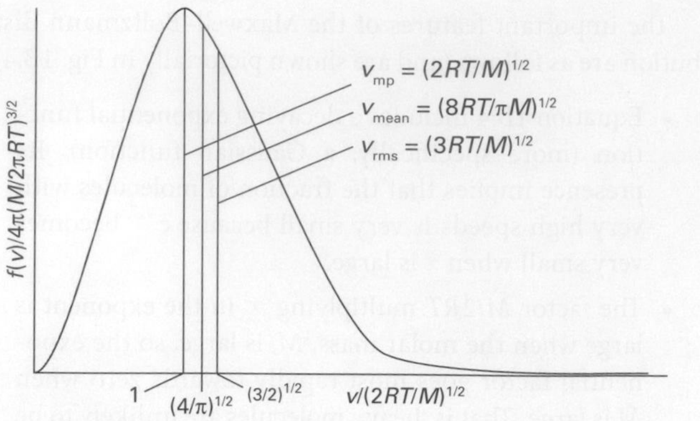

Figure 1B.6 A summary of the conclusions that can be deduced from the Maxwell distribution for molecules of molar mass M at a temperature T: v_{mp} is the most probable speed, v_{mean} is the mean speed, and v_{rms} is the root-mean-square speed.

The **mean relative speed,** v_{rel}, the mean speed with which one molecule approaches another of the same kind, can also be calculated from the distribution:

$$v_{rel} = 2^{1/2} v_{mean} \qquad \text{Mean relative speed [KMT, identical molecules]} \qquad (1B.11a)$$

This result is much harder to derive, but the diagram in Fig. 1B.7 should help to show that it is plausible. For the relative mean speed of two dissimilar molecules of masses m_A and m_B:

$$v_{rel} = \left(\frac{8kT}{\pi\mu}\right)^{1/2} \qquad \mu = \frac{m_A m_B}{m_A + m_B} \qquad \text{Mean relative speed [perfect gas]} \qquad (1B.11b)$$

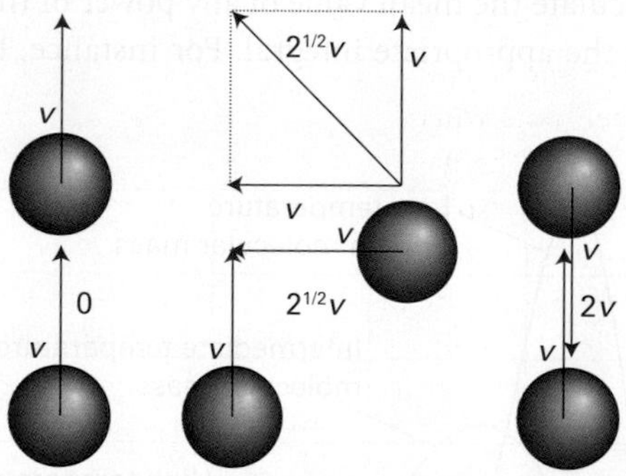

Figure 1B.7 A simplified version of the argument to show that the mean relative speed of molecules in a gas is related to the mean speed. When the molecules are moving in the same direction, the mean relative speed is zero; it is $2v$ when the molecules are approaching each other. A typical mean direction of approach is from the side, and the mean speed of approach is then $2^{1/2}v$. The last direction of approach is the most characteristic, so the mean speed of approach can be expected to be about $2^{1/2}v$. This value is confirmed by more detailed calculation.

Brief illustration 1B.1

As already seen (in *Example* 1B.1), the mean speed of N_2 molecules at 25 °C is 475 m s^{-1}. It follows from eqn 1B.11a that their relative mean speed is

$$v_{rel} = 2^{1/2} \times (475\,\mathrm{m\,s^{-1}}) = 671\,\mathrm{m\,s^{-1}}$$

1B.2 Collisions

The kinetic model can be used to develop the qualitative picture of a perfect gas, as a collection of ceaselessly moving, colliding molecules, into a quantitative, testable expression. In particular, it provides a way to calculate the average frequency with which molecular collisions occur and the average distance a molecule travels between collisions.

(a) The collision frequency

Although the kinetic model assumes that the molecules are point-like, a 'hit' can be counted as occurring whenever the centres of two molecules come within a distance d of each other, where d, the **collision diameter**, is of the order of the actual diameters of the molecules (for impenetrable hard spheres d is the diameter). The kinetic model can be used to deduce the **collision frequency**, z, the number of collisions made by one molecule divided by the time interval during which the collisions are counted.

How is that done? 1B.3 Using the kinetic model to derive an expression for the collision frequency

Consider the positions of all the molecules except one to be frozen. Then note what happens as this one mobile molecule travels through the gas with a mean relative speed v_{rel} for a time Δt. In doing so it sweeps out a 'collision tube' of cross-sectional area $\sigma = \pi d^2$, length $v_{rel}\Delta t$ and therefore of volume $\sigma v_{rel}\Delta t$ (Fig. 1B.8). The number of stationary molecules with centres inside the collision tube is given by the volume V of the tube multiplied by the number density $\mathcal{N} = N/V$, where N is the total number of molecules in the sample, and is $\mathcal{N}\sigma v_{rel}\Delta t$. The collision frequency z is this number divided by Δt. It follows that

$$z = \sigma v_{rel} \mathcal{N} \quad \text{Collision frequency [KMT]} \qquad (1B.12a)$$

The parameter σ is called the **collision cross-section** of the molecules. Some typical values are given in Table 1B.2.

An expression in terms of the pressure of the gas is obtained by using the perfect gas equation and $R = N_A k$ to write the number density in terms of the pressure:

$$\mathcal{N} = \frac{N}{V} = \frac{nN_A}{V} = \frac{nN_A}{nRT/p} = \frac{pN_A}{RT} = \frac{p}{kT}$$

Then

$$z = \frac{\sigma v_{rel} p}{kT} \quad \text{Collision frequency [KMT]} \qquad (1B.12b)$$

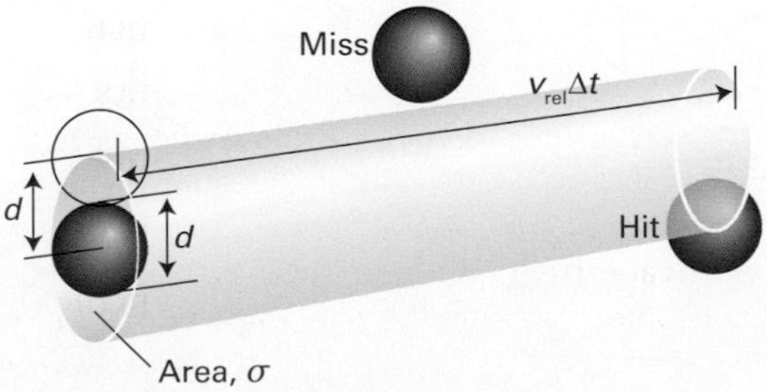

Figure 1B.8 The basis of the calculation of the collision frequency in the kinetic theory of gases.

Table 1B.2 Collision cross-sections*

	σ/nm^2
C_6H_6	0.88
CO_2	0.52
He	0.21
N_2	0.45

* More values are given in the *Resource section*.

Equation 1B.12a shows that, at constant volume, the collision frequency increases with increasing temperature, because most molecules are moving faster. Equation 1B.12b shows that, at constant temperature, the collision frequency is proportional to the pressure. The greater the pressure, the greater the number density of molecules in the sample, and the rate at which they encounter one another is greater even though their average speed remains the same.

Brief illustration 1B.2

For an N_2 molecule in a sample at 1.00 atm (101 kPa) and 25 °C, from *Brief illustration* 1B.1 v_{rel} = 671 m s^{-1}. Therefore, from eqn 1B.12b, and taking σ = 0.45 nm^2 (corresponding to 0.45×10^{-18} m^2) from Table 1B.2,

$$z = \frac{(0.45\times10^{-18}\,\mathrm{m^2})\times(671\,\mathrm{m\,s^{-1}})\times(1.01\times10^5\,\mathrm{Pa})}{(1.381\times10^{-23}\,\mathrm{J\,K^{-1}})\times(298\,\mathrm{K})} = 7.4\times10^9\,\mathrm{s^{-1}}$$

so a given molecule collides about 7×10^9 times each second. The timescale of events in gases is becoming clear.

(b) The mean free path

The **mean free path**, λ (lambda), is the average distance a molecule travels between collisions. If a molecule collides with a frequency z, it spends a time $1/z$ in free flight between collisions, and therefore travels a distance $(1/z)v_{rel}$. It follows that the mean free path is

$$\lambda = \frac{v_{rel}}{z} \quad \text{Mean free path [KMT]} \quad (1B.13)$$

Substitution of the expression for z from eqn 1B.12b gives

$$\lambda = \frac{kT}{\sigma p} \quad \text{Mean free path [perfect gas]} \quad (1B.14)$$

Doubling the pressure shortens the mean free path by a factor of 2.

Brief illustration 1B.3

From *Brief illustration* 1B.1 $v_{rel} = 671\,\text{m s}^{-1}$ for N_2 molecules at 25 °C, and from *Brief illustration* 1B.2 $z = 7.4\times10^9\,\text{s}^{-1}$ when the pressure is 1.00 atm. Under these circumstances, the mean free path of N_2 molecules is

$$\lambda = \frac{671\,\text{m s}^{-1}}{7.4\times10^9\,\text{s}^{-1}} = 9.1\times10^{-8}\,\text{m}$$

or 91 nm, about 10^3 molecular diameters.

Although the temperature appears in eqn 1B.14, in a sample of constant volume, the pressure is proportional to T, so T/p remains constant when the temperature is increased. Therefore, the mean free path is independent of the temperature in a sample of gas provided the volume is constant. In a container of fixed volume the distance between collisions is determined by the number of molecules present in the given volume, not by the speed at which they travel.

In summary, a typical gas (N_2 or O_2) at 1 atm and 25 °C can be thought of as a collection of molecules travelling with a mean speed of about $500\,\text{m s}^{-1}$. Each molecule makes a collision within about 1 ns, and between collisions it travels about 10^3 molecular diameters.

Checklist of concepts

☐ 1. The **kinetic model** of a gas considers only the contribution to the energy from the kinetic energies of the molecules.

☐ 2. Important results from the model include expressions for the pressure and the **root-mean-square speed**.

☐ 3. The **Maxwell–Boltzmann distribution of speeds** gives the fraction of molecules that have speeds in a specified range.

☐ 4. The **collision frequency** is the average number of collisions made by a molecule in an interval divided by the length of the interval.

☐ 5. The **mean free path** is the average distance a molecule travels between collisions.

Checklist of equations

Property	Equation	Comment	Equation number
Pressure of a perfect gas from the kinetic model	$pV = \frac{1}{3}nMv_{rms}^2$	Kinetic model of a perfect gas	1B.2
Maxwell–Boltzmann distribution of speeds	$f(v) = 4\pi(M/2\pi RT)^{3/2}v^2e^{-Mv^2/2RT}$		1B.4
Root-mean-square speed	$v_{rms} = (3RT/M)^{1/2}$		1B.8
Mean speed	$v_{mean} = (8RT/\pi M)^{1/2}$		1B.9
Most probable speed	$v_{mp} = (2RT/M)^{1/2}$		1B.10
Mean relative speed	$v_{rel} = (8kT/\pi\mu)^{1/2}$ $\mu = m_Am_B/(m_A+m_B)$		1B.11b
The collision frequency	$z = \sigma v_{rel}p/kT, \sigma = \pi d^2$		1B.12b
Mean free path	$\lambda = v_{rel}/z$		1B.13

TOPIC 1C Real gases

➤ **Why do you need to know this material?**

The properties of actual gases, so-called 'real gases', are different from those of a perfect gas. Moreover, the deviations from perfect behaviour give insight into the nature of the interactions between molecules.

➤ **What is the key idea?**

Attractions and repulsions between gas molecules account for modifications to the isotherms of a gas and account for critical behaviour.

➤ **What do you need to know already?**

This Topic builds on and extends the discussion of perfect gases in Topic 1A. The principal mathematical technique employed is the use of differentiation to identify a point of inflexion of a curve (*The chemist's toolkit* 5).

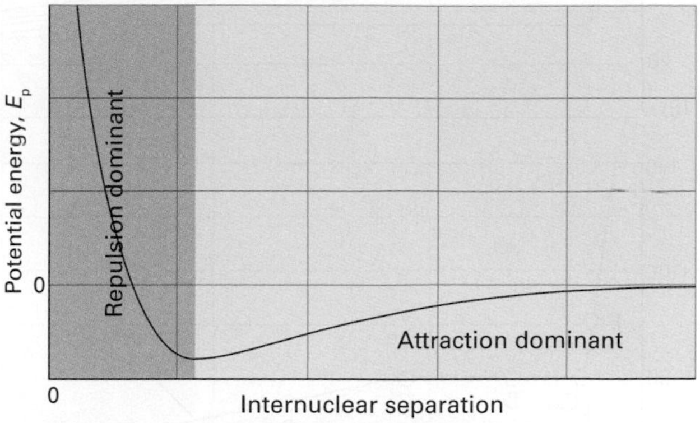

Figure 1C.1 The dependence of the potential energy of two molecules on their internuclear separation. High positive potential energy (at very small separations) indicates that the interactions between them are strongly repulsive at these distances. At intermediate separations, attractive interactions dominate. At large separations (far to the right) the potential energy is zero and there is no interaction between the molecules.

Real gases do not obey the perfect gas law exactly except in the limit of $p \to 0$. Deviations from the law are particularly important at high pressures and low temperatures, especially when a gas is on the point of condensing to liquid.

1C.1 Deviations from perfect behaviour

Real gases show deviations from the perfect gas law because molecules interact with one another. A point to keep in mind is that repulsive forces between molecules assist expansion and attractive forces assist compression.

Repulsive forces are significant only when molecules are almost in contact: they are short-range interactions, even on a scale measured in molecular diameters (Fig. 1C.1). Because they are short-range interactions, repulsions can be expected to be important only when the average separation of the molecules is small. This is the case at high pressure, when many molecules occupy a small volume. On the other hand, attractive intermolecular forces have a relatively long range and are effective over several molecular diameters. They are important when the molecules are fairly close together but not necessarily touching (at the intermediate separations in Fig. 1C.1). Attractive forces are ineffective when the molecules are far apart (well to the right in Fig. 1C.1). Intermolecular forces are also important when the temperature is so low that the molecules travel with such low mean speeds that they can be captured by one another.

The consequences of these interactions are shown by shapes of experimental isotherms (Fig. 1C.2). At low pressures, when the sample occupies a large volume, the molecules are so far apart for most of the time that the intermolecular forces play no significant role, and the gas behaves virtually perfectly. At moderate pressures, when the average separation of the molecules is only a few molecular diameters, the attractive forces dominate the repulsive forces. In this case, the gas can be expected to be more compressible than a perfect gas because the forces help to draw the molecules together. At high pressures, when the average separation of the molecules is small, the repulsive forces dominate and the gas can be expected to be less compressible because now the forces help to drive the molecules apart.

Consider what happens when a sample of gas initially in the state marked A in Fig. 1C.2b is compressed (its volume is reduced) at constant temperature by pushing in a piston. Near A, the pressure of the gas rises in approximate agreement with Boyle's law. Serious deviations from that law begin to appear when the volume has been reduced to B.

At C (which corresponds to about 60 atm for carbon dioxide), all similarity to perfect behaviour is lost, for suddenly the

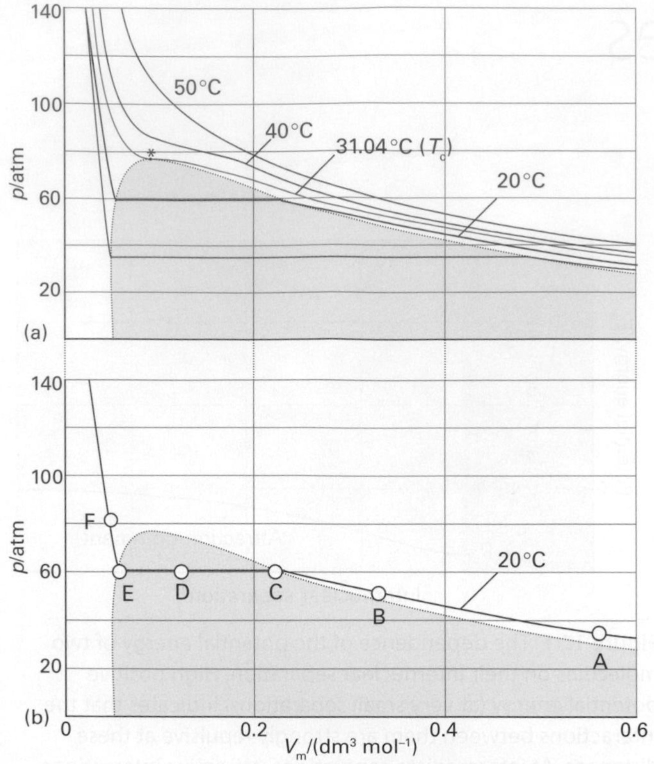

Figure 1C.2 (a) Experimental isotherms of carbon dioxide at several temperatures. The 'critical isotherm', the isotherm at the critical temperature, is at 31.04 °C (in blue). The critical point is marked with a star. (b) As explained in the text, the gas can condense only at and below the critical temperature as it is compressed along a horizontal line (such as CDE). The dotted black curve consists of points like C and E for all isotherms below the critical temperature.

piston slides in without any further rise in pressure: this stage is represented by the horizontal line CDE. Examination of the contents of the vessel shows that just to the left of C a liquid appears, and there are two phases separated by a sharply defined surface. As the volume is decreased from C through D to E, the amount of liquid increases. There is no additional resistance to the piston because the gas can respond by condensing. The pressure corresponding to the line CDE, when both liquid and vapour are present in equilibrium, is called the **vapour pressure** of the liquid at the temperature of the experiment.

At E, the sample is entirely liquid and the piston rests on its surface. Any further reduction of volume requires the exertion of considerable pressure, as is indicated by the sharply rising line to the left of E. Even a small reduction of volume from E to F requires a great increase in pressure.

(a) The compression factor

As a first step in understanding these observations it is useful to introduce the **compression factor**, Z, the ratio of the measured molar volume of a gas, $V_m = V/n$, to the molar volume of a perfect gas, V_m°, at the same pressure and temperature:

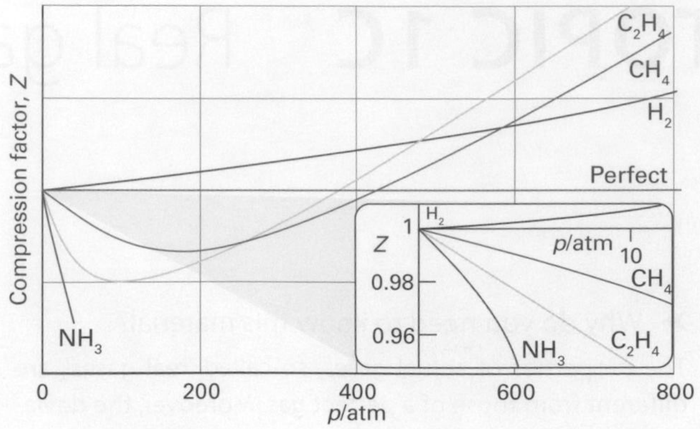

Figure 1C.3 The variation of the compression factor, Z, with pressure for several gases at 0 °C. A perfect gas has $Z = 1$ at all pressures. Notice that, although the curves approach 1 as $p \to 0$, they do so with different slopes.

$$Z = \frac{V_m}{V_m^\circ} \qquad \text{Compression factor [definition]} \qquad (1C.1)$$

Because the molar volume of a perfect gas is equal to RT/p, an equivalent expression is $Z = pV_m/RT$, which can be written as

$$pV_m = RTZ \qquad (1C.2)$$

Because for a perfect gas $Z = 1$ under all conditions, deviation of Z from 1 is a measure of departure from perfect behaviour.

Some experimental values of Z are plotted in Fig. 1C.3. At very low pressures, all the gases shown have $Z \approx 1$ and behave nearly perfectly. At high pressures, all the gases have $Z > 1$, signifying that they have a larger molar volume than a perfect gas. Repulsive forces are now dominant. At intermediate pressures, most gases have $Z < 1$, indicating that the attractive forces are reducing the molar volume relative to that of a perfect gas.

Brief illustration 1C.1

The molar volume of a perfect gas at 500 K and 100 bar is $V_m^\circ = 0.416\,dm^3\,mol^{-1}$. The molar volume of carbon dioxide under the same conditions is $V_m = 0.366\,dm^3\,mol^{-1}$. It follows that at 500 K

$$Z = \frac{0.366\,dm^3\,mol^{-1}}{0.416\,dm^3\,mol^{-1}} = 0.880$$

The fact that $Z < 1$ indicates that attractive forces dominate repulsive forces under these conditions.

(b) Virial coefficients

At large molar volumes and high temperatures the real-gas isotherms do not differ greatly from perfect-gas isotherms.

Table 1C.1 Second virial coefficients, $B/(\mathrm{cm^3\,mol^{-1}})$*

	Temperature	
	273 K	600 K
Ar	−21.7	11.9
CO_2	−149.7	−12.4
N_2	−10.5	21.7
Xe	−153.7	−19.6

* More values are given in the *Resource section*.

The small differences suggest that the perfect gas law $pV_\mathrm{m} = RT$ is in fact the first term in an expression of the form

$$pV_\mathrm{m} = RT(1 + B'p + C'p^2 + \cdots) \quad (1C.3a)$$

This expression is an example of a common procedure in physical chemistry, in which a simple law that is known to be a good first approximation (in this case $pV_\mathrm{m} = RT$) is treated as the first term in a series in powers of a variable (in this case p). A more convenient expansion for many applications is

$$pV_\mathrm{m} = RT\left(1 + \frac{B}{V_\mathrm{m}} + \frac{C}{V_\mathrm{m}^2} + \cdots\right) \quad \text{Virial equation of state} \quad (1C.3b)$$

These two expressions are two versions of the **virial equation of state**.[1] By comparing the expression with eqn 1C.2 it is seen that the term in parentheses in eqn 1C.3b is just the compression factor, Z.

The coefficients B, C, …, which depend on the temperature, are the second, third,… **virial coefficients** (Table 1C.1); the first virial coefficient is 1. The third virial coefficient, C, is usually less important than the second coefficient, B, in the sense that at typical molar volumes $C/V_\mathrm{m}^2 \ll B/V_\mathrm{m}$. The values of the virial coefficients of a gas are determined from measurements of its compression factor.

Brief illustration 1C.2

To use eqn 1C.3b (up to the B term) to calculate the pressure exerted at 100 K by 0.104 mol O_2(g) in a vessel of volume 0.225 dm^3, begin by calculating the molar volume:

$$V_\mathrm{m} = \frac{V}{n_{\mathrm{O_2}}} = \frac{0.225\,\mathrm{dm^3}}{0.104\,\mathrm{mol}} = 2.16\,\mathrm{dm^3\,mol^{-1}} = 2.16\times10^{-3}\,\mathrm{m^3\,mol^{-1}}$$

Then, by using the value of B found in Table 1C.1 of the *Resource section*,

$$p = \frac{RT}{V_\mathrm{m}}\left(1 + \frac{B}{V_\mathrm{m}}\right)$$

$$= \frac{(8.3145\,\mathrm{J\,mol^{-1}\,K^{-1}})\times(100\,\mathrm{K})}{2.16\times10^{-3}\,\mathrm{m^3\,mol^{-1}}}\left(1 - \frac{1.975\times10^{-4}\,\mathrm{m^3\,mol^{-1}}}{2.16\times10^{-3}\,\mathrm{m^3\,mol^{-1}}}\right)$$

$$= 3.50\times10^5\ \mathrm{Pa, or\ 350\,kPa}$$

where 1 Pa = 1 J m^{-3}. The perfect gas equation of state would give the calculated pressure as 385 kPa, or 10 per cent higher than the value calculated by using the virial equation of state. The difference is significant because under these conditions $B/V_\mathrm{m} \approx 0.1$ which is not negligible relative to 1.

An important point is that although the equation of state of a real gas may coincide with the perfect gas law as $p \to 0$, not all its properties necessarily coincide with those of a perfect gas in that limit. Consider, for example, the value of $\mathrm{d}Z/\mathrm{d}p$, the slope of the graph of compression factor against pressure (see *The chemist's toolkit* 5 for a review of derivatives and differentiation). For a perfect gas $\mathrm{d}Z/\mathrm{d}p = 0$ (because $Z = 1$ at all pressures), but for a real gas from eqn 1C.3a

$$\frac{\mathrm{d}Z}{\mathrm{d}p} = B' + 2pC' + \cdots \to B' \text{ as } p \to 0 \quad (1C.4a)$$

However, B' is not necessarily zero, so the slope of Z with respect to p does not necessarily approach 0 (the perfect gas value), as can be seen in Fig. 1C.4. By a similar argument (see *The chemist's toolkit* 5 for evaluating derivatives of this kind),

$$\frac{\mathrm{d}Z}{\mathrm{d}(1/V_\mathrm{m})} \to B \text{ as } V_\mathrm{m} \to \infty \quad (1C.4b)$$

Because the virial coefficients depend on the temperature, there may be a temperature at which $Z \to 1$ with zero slope at low pressure or high molar volume (as in Fig. 1C.4). At this temperature, which is called the **Boyle temperature**, T_B, the properties of the real gas do coincide with those of a per-

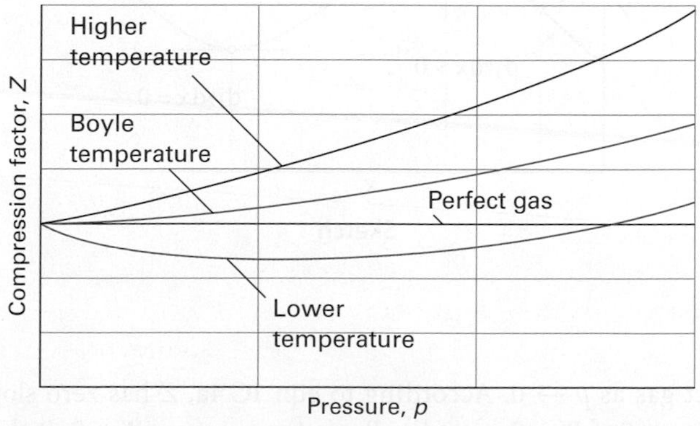

Figure 1C.4 The compression factor, Z, approaches 1 at low pressures, but does so with different slopes. For a perfect gas, the slope is zero, but real gases may have either positive or negative slopes, and the slope may vary with temperature. At the Boyle temperature, the slope is zero at $p = 0$ and the gas behaves perfectly over a wider range of conditions than at other temperatures.

[1] The name comes from the Latin word for force. The coefficients are sometimes denoted B_2, B_3, ….

The chemist's toolkit 5 Differentiation

Differentiation is concerned with the slopes of functions, such as the rate of change of a variable with time. The formal definition of the **derivative**, df/dx, of a function $f(x)$ is

$$\frac{df}{dx}=\lim_{\delta x\to 0}\frac{f(x+\delta x)-f(x)}{\delta x} \quad \text{First derivative [definition]}$$

As shown in Sketch 1, the derivative can be interpreted as the slope of the tangent to the graph of $f(x)$ at a given value of x. A positive first derivative indicates that the function slopes upwards (as x increases), and a negative first derivative indicates the opposite. It is sometimes convenient to denote the first derivative as $f'(x)$. The **second derivative**, d^2f/dx^2, of a function is the derivative of the first derivative (here denoted f'):

$$\frac{d^2f}{dx^2}=\lim_{\delta x\to 0}\frac{f'(x+\delta x)-f'(x)}{\delta x} \quad \text{Second derivative [definition]}$$

It is sometimes convenient to denote the second derivative f''. As shown in Sketch 2, the second derivative of a function can be interpreted as an indication of the sharpness of the curvature of the function. A positive second derivative indicates that the function is ∪ shaped, and a negative second derivative indicates that it is ∩ shaped. The second derivative is zero at a **point of inflection**, where the first derivative changes sign.

The derivatives of some common functions are as follows:

$$\frac{d}{dx}x^n=nx^{n-1}$$

$$\frac{d}{dx}e^{ax}=ae^{ax}$$

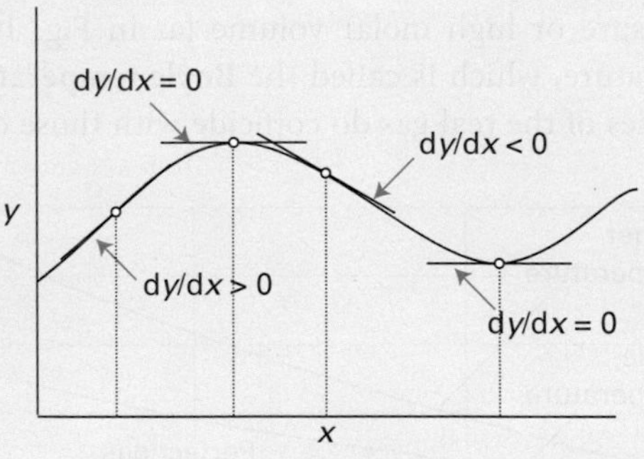

Sketch 1

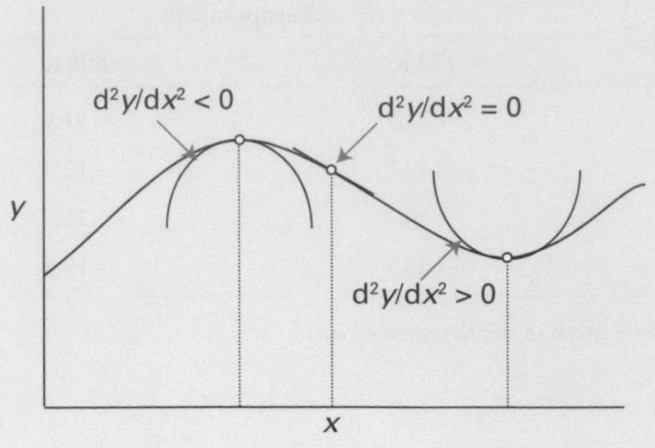

Sketch 2

$$\frac{d}{dx}\sin ax=a\cos ax \qquad \frac{d}{dx}\cos ax=-a\sin ax$$

$$\frac{d}{dx}\ln ax=\frac{1}{x}$$

It follows from the definition of the derivative that a variety of combinations of functions can be differentiated by using the following rules:

$$\frac{d}{dx}(u+v)=\frac{du}{dx}+\frac{dv}{dx}$$

$$\frac{d}{dx}uv=u\frac{dv}{dx}+v\frac{du}{dx}$$

$$\frac{d}{dx}\frac{u}{v}=\frac{1}{v}\frac{du}{dx}-\frac{u}{v^2}\frac{dv}{dx}$$

It is sometimes convenient to differentiate with respect to a function of x, rather than x itself. For instance, suppose that

$$f(x)=a+\frac{b}{x}+\frac{c}{x^2}$$

where a, b, and c are constants and you need to evaluate $df/d(1/x)$, rather than df/dx. To begin, let $y = 1/x$. Then $f(y) = a + by + cy^2$ and

$$\frac{df}{dy}=b+2cy$$

Because $y = 1/x$, it follows that

$$\frac{df}{d(1/x)}=b+\frac{2c}{x}$$

fect gas as $p \to 0$. According to eqn 1C.4a, Z has zero slope as $p \to 0$ if $B' = 0$, so at the Boyle temperature $B' = 0$. It then follows from eqn 1C.3a that $pV_m \approx RT_B$ over a more extended range of pressures than at other temperatures because the first term after 1 (i.e. $B'p$) in the virial equation is zero and $C'p^2$ and higher terms are negligibly small. For helium $T_B = 22.64$ K; for air $T_B = 346.8$ K; more values are given in Table 1C.2.

(c) Critical constants

There is a temperature, called the **critical temperature**, T_c, which separates two regions of behaviour and plays a special role in the theory of the states of matter. An isotherm slightly below T_c behaves as already described: at a certain pressure, a liquid condenses from the gas and is distinguishable from it by

Table 1C.2 Critical constants of gases*

	p_c/atm	V_c/(cm^3 mol^{-1})	T_c/K	Z_c	T_B/K
Ar	48.0	75.3	150.7	0.292	411.5
CO_2	72.9	94.0	304.2	0.274	714.8
He	2.26	57.8	5.2	0.305	22.64
O_2	50.14	78.0	154.8	0.308	405.9

* More values are given in the *Resource section.*

the presence of a visible surface. If, however, the compression takes place at T_c itself, then a surface separating two phases does not appear and the volumes at each end of the horizontal part of the isotherm have merged to a single point, the **critical point** of the gas. The pressure and molar volume at the critical point are called the **critical pressure**, p_c, and **critical molar volume**, V_c, of the substance. Collectively, p_c, V_c, and T_c are the **critical constants** of a substance (Table 1C.2).

At and above T_c, the sample has a single phase which occupies the entire volume of the container. Such a phase is, by definition, a gas. Hence, the liquid phase of a substance does not form above the critical temperature. The single phase that fills the entire volume when $T > T_c$ may be much denser than considered typical for gases, and the name **supercritical fluid** is preferred.

Brief illustration 1C.3

The critical temperature of oxygen, 155 K, signifies that it is impossible to produce liquid oxygen by compression alone if its temperature is greater than 155 K. To liquefy oxygen the temperature must first be lowered to below 155 K, and then the gas compressed isothermally.

1C.2 The van der Waals equation

Conclusions may be drawn from the virial equations of state only by inserting specific values of the coefficients. It is often useful to have a broader, if less precise, view of all gases, such as that provided by an approximate equation of state.

(a) Formulation of the equation

The equation introduced by J.D. van der Waals in 1873 is an excellent example of an expression that can be obtained by thinking scientifically about a mathematically complicated but physically simple problem; that is, it is a good example of 'model building'.

How is that done? 1C.1 Deriving the van der Waals equation of state

The repulsive interaction between molecules is taken into account by supposing that it causes the molecules to behave as small but impenetrable spheres, so instead of moving in a volume V they are restricted to a smaller volume $V - nb$, where nb is approximately the total volume taken up by the molecules themselves. This argument suggests that the perfect gas law $p = nRT/V$ should be replaced by

$$p=\frac{nRT}{V-nb}$$

when repulsions are significant. To calculate the excluded volume, note that the closest distance of approach of two hard-sphere molecules of radius r (and volume $V_{\text{molecule}} = \frac{4}{3}\pi r^3$) is $2r$, so the volume excluded is $\frac{4}{3}\pi(2r)^3$, or $8V_{\text{molecule}}$. The volume excluded per molecule is one-half this volume, or $4V_{\text{molecule}}$, so $b \approx 4V_{\text{molecule}}N_A$.

The pressure depends on both the frequency of collisions with the walls and the force of each collision. Both the frequency of the collisions and their force are reduced by the attractive interaction, which acts with a strength proportional to the number of interacting molecules and therefore to the molar concentration, n/V, of molecules in the sample. Because both the frequency and the force of the collisions are reduced by the attractive interactions, the pressure is reduced in proportion to the square of this concentration. If the reduction of pressure is written as $a(n/V)^2$, where a is a positive constant characteristic of each gas, the combined effect of the repulsive and attractive forces is the **van der Waals equation**:

$$p=\frac{nRT}{V-nb}-a\frac{n^2}{V^2} \quad \text{(1C.5a)}$$

van der Waals equation of state

The constants a and b are called the **van der Waals coefficients**, with a representing the strength of attractive interactions and b that of the repulsive interactions between the molecules. They are characteristic of each gas and taken to be independent of the temperature (Table 1C.3). Although a and b are not precisely defined molecular properties, they correlate with physical properties that reflect the strength of intermolecular interactions, such as critical temperature, vapour pressure, and enthalpy of vaporization.

Table 1C.3 van der Waals coefficients*

	a/(atm dm^6 mol^{-2})	b/(10^{-2} dm^3 mol^{-1})
Ar	1.337	3.20
CO_2	3.610	4.29
He	0.0341	2.38
Xe	4.137	5.16

* More values are given in the *Resource section.*

Brief illustration 1C.4

For benzene $a = 18.57\,\text{atm}\,\text{dm}^6\,\text{mol}^{-2}$ ($1.882\,\text{Pa}\,\text{m}^6\,\text{mol}^{-2}$) and $b = 0.1193\,\text{dm}^3\,\text{mol}^{-1}$ ($1.193 \times 10^{-4}\,\text{m}^3\,\text{mol}^{-1}$); its normal boiling point is 353 K. Treated as a perfect gas at $T = 400\,\text{K}$ and $p = 1.0\,\text{atm}$, benzene vapour has a molar volume of $V_\text{m} = RT/p = 33\,\text{dm}^3\,\text{mol}^{-1}$, so the criterion $V_\text{m} \gg b$ for perfect gas behaviour is satisfied. It follows that $a/V_\text{m}^2 \approx 0.017\,\text{atm}$, which is 1.7 per cent of 1.0 atm. Therefore, benzene vapour is expected to deviate only slightly from perfect gas behaviour at this temperature and pressure.

Equation 1C.5a is often written in terms of the molar volume $V_\text{m} = V/n$ as

$$p = \frac{RT}{V_\text{m} - b} - \frac{a}{V_\text{m}^2} \qquad \text{(1C.5b)}$$

Example 1C.1 Using the van der Waals equation to estimate a molar volume

Estimate the molar volume of CO_2 at 500 K and 100 atm by treating it as a van der Waals gas.

Collect your thoughts You need to find an expression for the molar volume by solving the van der Waals equation, eqn 1C.5b. To rearrange the equation into a suitable form, multiply both sides by $(V_\text{m} - b)V_\text{m}^2$, to obtain

$$(V_\text{m} - b)V_\text{m}^2 p = RTV_\text{m}^2 - (V_\text{m} - b)a$$

Then, after division by p, collect powers of V_m to obtain

$$V_\text{m}^3 - \left(b + \frac{RT}{p}\right)V_\text{m}^2 + \left(\frac{a}{p}\right)V_\text{m} - \frac{ab}{p} = 0$$

Although closed expressions for the roots of a cubic equation can be given, they are very complicated. Unless analytical solutions are essential, it is usually best to solve such equations with mathematical software; graphing calculators can also be used to help identify the acceptable root.

The solution According to Table 1C.3, $a = 3.610\,\text{dm}^6\,\text{atm}\,\text{mol}^{-2}$ and $b = 4.29 \times 10^{-2}\,\text{dm}^3\,\text{mol}^{-1}$. Under the stated conditions, $RT/p = 0.410\,\text{dm}^3\,\text{mol}^{-1}$. The coefficients in the equation for V_m are therefore

$$b + RT/p = 0.453\,\text{dm}^3\,\text{mol}^{-1}$$
$$a/p = 3.61 \times 10^{-2}\,(\text{dm}^3\,\text{mol}^{-1})^2$$
$$ab/p = 1.55 \times 10^{-3}\,(\text{dm}^3\,\text{mol}^{-1})^3$$

Therefore, on writing $x = V_\text{m}/(\text{dm}^3\,\text{mol}^{-1})$, the equation to solve is

$$x^3 - 0.453x^2 + (3.61 \times 10^{-2})x - (1.55 \times 10^{-3}) = 0$$

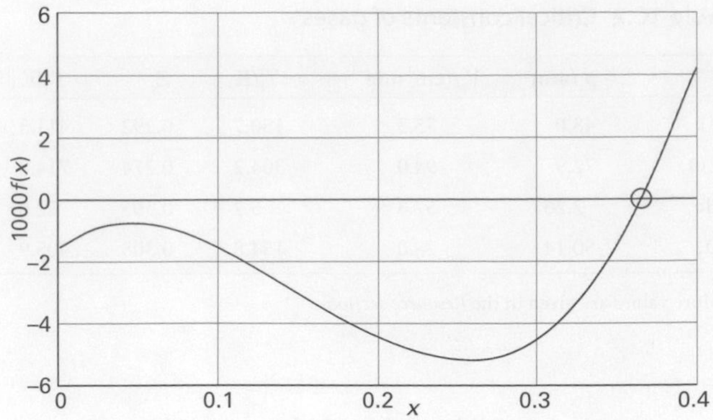

Figure 1C.5 The graphical solution of the cubic equation for V in *Example* 1C.1.

The acceptable root is $x = 0.366$ (Fig. 1C.5), which implies that $V_\text{m} = 0.366\,\text{dm}^3\,\text{mol}^{-1}$. The molar volume of a perfect gas under these conditions is $0.410\,\text{dm}^3\,\text{mol}^{-1}$.

Self-test 1C.1 Calculate the molar volume of argon at 100 °C and 100 atm on the assumption that it is a van der Waals gas.

Answer: $0.298\,\text{dm}^3\,\text{mol}^{-1}$

(b) The features of the equation

To what extent does the van der Waals equation predict the behaviour of real gases? It is too optimistic to expect a single, simple expression to be the true equation of state of all substances, and accurate work on gases must resort to the virial equation, use tabulated values of the coefficients at various temperatures, and analyse the system numerically. The advantage of the van der Waals equation, however, is that it is analytical (that is, expressed symbolically) and allows some general conclusions about real gases to be drawn. When the equation fails another equation of state must be used (some are listed in Table 1C.4), yet another must be invented, or the virial equation is used.

The reliability of the equation can be judged by comparing the isotherms it predicts with the experimental isotherms in Fig. 1C.2. Some calculated isotherms are shown in Figs. 1C.6 and 1C.7. Apart from the oscillations below the critical temperature, they do resemble experimental isotherms quite well. The oscillations, the **van der Waals loops**, are unrealistic because they suggest that under some conditions an increase of pressure results in an increase of volume. Therefore they are replaced by horizontal lines drawn so the loops define equal areas above and below the lines: this procedure is called the **Maxwell construction** (1). The van der Waals coefficients, such as those in Table 1C.3, are found by fitting the calculated curves to the experimental curves.

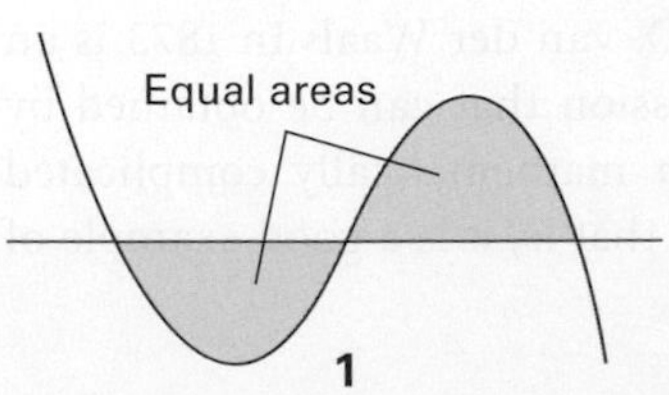

1

Table 1C.4 Selected equations of state

	Equation	Reduced form*	Critical constants p_c	V_c	T_c
Perfect gas	$p=\frac{nRT}{V}$				
van der Waals	$p=\frac{nRT}{V-nb}-\frac{n^2a}{V^2}$	$p_r=\frac{8T_r}{3V_r-1}-\frac{3}{V_r^2}$	$\frac{a}{27b^2}$	$3b$	$\frac{8a}{27bR}$
Berthelot	$p=\frac{nRT}{V-nb}-\frac{n^2a}{TV^2}$	$p_r=\frac{8T_r}{3V_r-1}-\frac{3}{T_rV_r^2}$	$\frac{1}{12}\left(\frac{2aR}{3b^3}\right)^{1/2}$	$3b$	$\frac{2}{3}\left(\frac{2a}{3bR}\right)^{1/2}$
Dieterici	$p=\frac{nRTe^{-na/RTV}}{V-nb}$	$p_r=\frac{T_re^{2(1-1/T_rV_r)}}{2V_r-1}$	$\frac{a}{4e^2b^2}$	$2b$	$\frac{a}{4bR}$
Virial	$p=\frac{nRT}{V}\left\{1+\frac{nB(T)}{V}+\frac{n^2C(T)}{V^2}+\cdots\right\}$				

* Reduced variables are defined as $X_r = X/X_c$ with $X = p$, V_m, and T. Equations of state are sometimes expressed in terms of the molar volume, $V_m = V/n$.

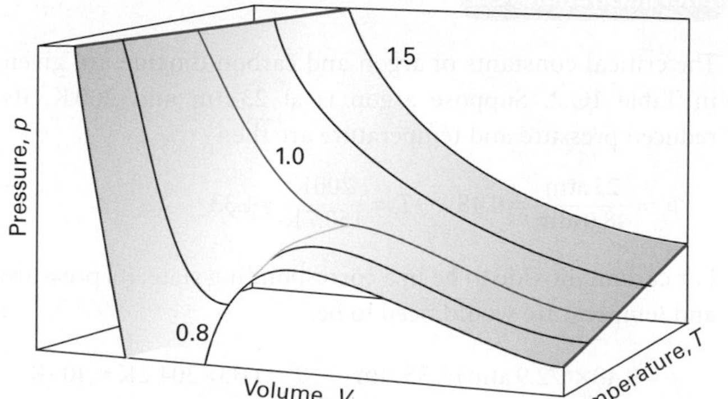

Figure 1C.6 The surface of possible states allowed by the van der Waals equation. The curves drawn on the surface are isotherms, labelled with the value of T/T_c, and correspond to the isotherms in Fig. 1C.7.

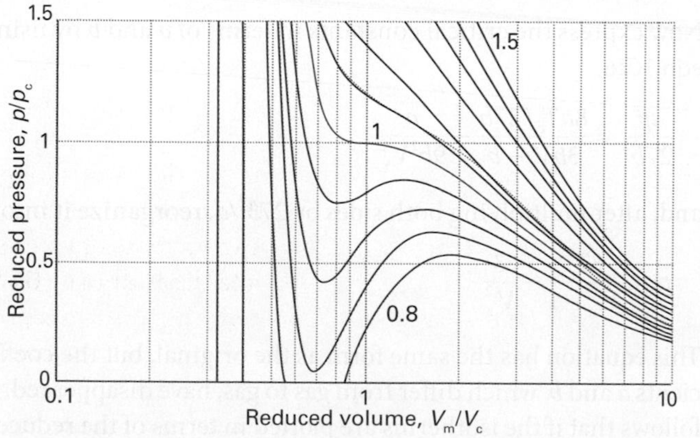

Figure 1C.7 Van der Waals isotherms at several values of T/T_c. The van der Waals loops are normally replaced by horizontal straight lines. The critical isotherm is the isotherm for $T/T_c = 1$, and is shown in blue.

The principal features of the van der Waals equation can be summarized as follows.

1. Perfect gas isotherms are obtained at high temperatures and large molar volumes.

When the temperature is high, RT may be so large that the first term in eqn 1C.5b greatly exceeds the second. Furthermore, if the molar volume is large in the sense $V_m >> b$, then the denominator $V_m - b \approx V_m$. Under these conditions, the equation reduces to $p = RT/V_m$, the perfect gas equation.

2. Liquids and gases coexist when the attractive and repulsive effects are in balance.

The van der Waals loops occur when both terms in eqn 1C.5b have similar magnitudes. The first term arises from the kinetic energy of the molecules and their repulsive interactions; the second represents the effect of the attractive interactions.

3. The critical constants are related to the van der Waals coefficients.

For $T < T_c$, the calculated isotherms oscillate, and each one passes through a minimum followed by a maximum. These extrema converge as $T \rightarrow T_c$ and coincide at $T = T_c$; at the critical point the curve has a flat inflexion (2). From the properties of curves, an inflexion of this type occurs when both the first and second derivatives are zero. Hence, the critical constants can be found by calculating these derivatives and setting them equal to zero at the critical point:

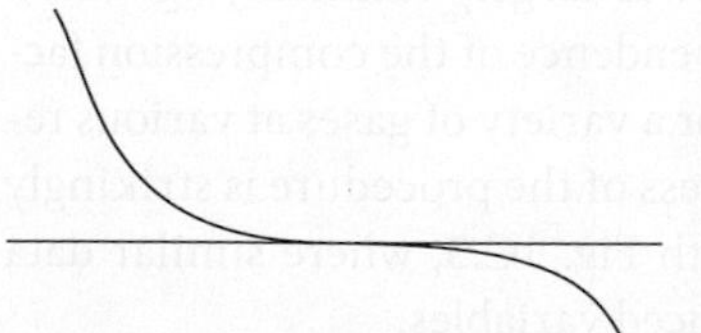

2

$$\frac{dp}{dV_m}=-\frac{RT}{(V_m-b)^2}+\frac{2a}{V_m^3}=0$$

$$\frac{d^2p}{dV_m^2} = \frac{2RT}{(V_m - b)^3} - \frac{6a}{V_m^4} = 0$$

The solutions of these two equations (and using eqn 1C.5b to calculate p_c from V_c and T_c; see Problem 1C.12) are

$$V_c = 3b \qquad p_c = \frac{a}{27b^2} \qquad T_c = \frac{8a}{27bR} \tag{1C.6}$$

These relations provide an alternative route to the determination of a and b from the values of the critical constants. They can be tested by noting that the **critical compression factor**, Z_c, is predicted to be

$$Z_c = \frac{p_c V_c}{RT_c} = \frac{3}{8} \tag{1C.7}$$

for all gases that are described by the van der Waals equation near the critical point. Table 1C.2 shows that although $Z_c < \frac{3}{8} = 0.375$, it is approximately constant (at 0.3) and the discrepancy is reasonably small.

(c) The principle of corresponding states

An important general technique in science for comparing the properties of objects is to choose a related fundamental property of the same kind and to set up a relative scale on that basis. The critical constants are characteristic properties of gases, so it may be that a scale can be set up by using them as yardsticks and to introduce the dimensionless **reduced variables** of a gas by dividing the actual variable by the corresponding critical constant:

$$V_r = \frac{V_m}{V_c} \qquad p_r = \frac{p}{p_c} \qquad T_r = \frac{T}{T_c} \qquad \text{Reduced variables [definition]} \tag{1C.8}$$

If the reduced pressure of a gas is given, its actual pressure is calculated by using $p = p_r p_c$, and likewise for the volume and temperature. Van der Waals, who first tried this procedure, hoped that gases confined to the same reduced volume, V_r, at the same reduced temperature, T_r, would exert the same reduced pressure, p_r. The hope was largely fulfilled (Fig. 1C.8). The illustration shows the dependence of the compression factor on the reduced pressure for a variety of gases at various reduced temperatures. The success of the procedure is strikingly clear: compare this graph with Fig. 1C.3, where similar data are plotted without using reduced variables.

The observation that real gases at the same reduced volume and reduced temperature exert the same reduced pressure is called the **principle of corresponding states**. The principle is only an approximation. It works best for gases composed of spherical molecules; it fails, sometimes badly, when the molecules are non-spherical or polar.

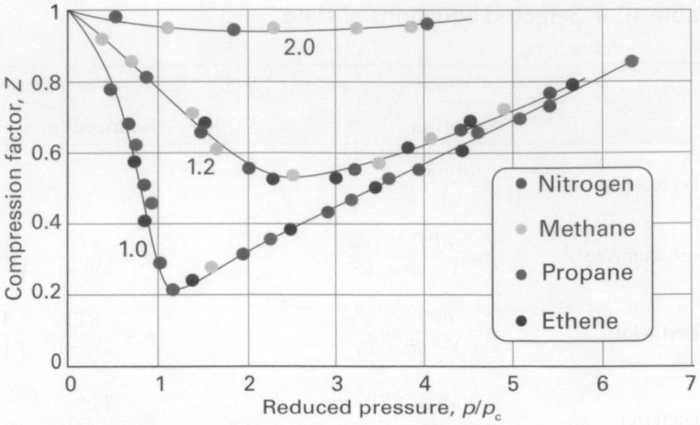

Figure 1C.8 The compression factors of four gases plotted using reduced variables. The curves are labelled with the reduced temperature $T_r = T/T_c$. The use of reduced variables organizes the data on to single curves.

Brief illustration 1C.5

The critical constants of argon and carbon dioxide are given in Table 1C.2. Suppose argon is at 23 atm and 200 K, its reduced pressure and temperature are then

$$p_r = \frac{23\,\text{atm}}{48.0\,\text{atm}} = 0.48 \qquad T_r = \frac{200\,\text{K}}{150.7\,\text{K}} = 1.33$$

For carbon dioxide to be in a corresponding state, its pressure and temperature would need to be

$$p = 0.48 \times (72.9\,\text{atm}) = 35\,\text{atm} \qquad T = 1.33 \times 304.2\,\text{K} = 405\,\text{K}$$

The van der Waals equation sheds some light on the principle. When eqn 1C.5b is expressed in terms of the reduced variables it becomes

$$p_r p_c = \frac{RT_r T_c}{V_r V_c - b} - \frac{a}{V_r^2 V_c^2}$$

Now express the critical constants in terms of a and b by using eqn 1C.6:

$$\frac{ap_r}{27b^2} = \frac{8aT_r/27b}{3bV_r - b} - \frac{a}{9b^2V_r^2}$$

and, after multiplying both sides by $27b^2/a$, reorganize it into

$$p_r = \frac{8T_r}{3V_r - 1} - \frac{3}{V_r^2} \tag{1C.9}$$

This equation has the same form as the original, but the coefficients a and b, which differ from gas to gas, have disappeared. It follows that if the isotherms are plotted in terms of the reduced variables (as done in fact in Fig. 1C.7 without drawing attention to the fact), then the same curves are obtained whatever the gas. This is precisely the content of the principle of corresponding states, so the van der Waals equation is compatible with it.

Looking for too much significance in this apparent triumph is mistaken, because other equations of state also accommodate the principle. In fact, any equation of state (such as those in Table 1C.4) with two parameters playing the roles of *a* and *b* can be manipulated into a reduced form. The observation that real gases obey the principle approximately amounts to saying that the effects of the attractive and repulsive interactions can each be approximated in terms of a single parameter. The importance of the principle is then not so much its theoretical interpretation but the way that it enables the properties of a range of gases to be coordinated on to a single diagram (e.g. Fig. 1C.8 instead of Fig. 1C.3).

Checklist of concepts

☐ 1. The extent of deviations from perfect behaviour is summarized by introducing the **compression factor**.

☐ 2. The **virial equation** is an empirical extension of the perfect gas equation that summarizes the behaviour of real gases over a range of conditions.

☐ 3. The isotherms of a real gas introduce the concept of **critical behaviour**.

☐ 4. A gas can be liquefied by pressure alone only if its temperature is at or below its **critical temperature**.

☐ 5. The **van der Waals equation** is a model equation of state for a real gas expressed in terms of two parameters, one (*a*) representing molecular attractions and the other (*b*) representing molecular repulsions.

☐ 6. The van der Waals equation captures the general features of the behaviour of real gases, including their critical behaviour.

☐ 7. The properties of real gases are coordinated by expressing their equations of state in terms of **reduced variables**.

Checklist of equations

Property	Equation	Comment	Equation number
Compression factor	$Z = V_m/V_m^\circ$	Definition	1C.1
Virial equation of state	$pV_m = RT(1 + B/V_m + C/V_m^2 + \cdots)$	B, C depend on temperature	1C.3b
van der Waals equation of state	$p = nRT/(V - nb) - a(n/V)^2$	a parameterizes attractions, b parameterizes repulsions	1C.5a
Reduced variables	$X_r = X/X_c$	$X = p$, V_m, or T	1C.8

FOCUS 1 The properties of gases

TOPIC 1A The perfect gas

Discussion questions

D1A.1 Explain how the perfect gas equation of state arises by combination of Boyle's law, Charles's law, and Avogadro's principle.

D1A.2 Explain the term 'partial pressure' and explain why Dalton's law is a limiting law.

Exercises

E1A.1(a) Express (i) 108 kPa in torr and (ii) 0.975 bar in atmospheres.
E1A.1(b) Express (i) 22.5 kPa in atmospheres and (ii) 770 Torr in pascals.

E1A.2(a) Could 131 g of xenon gas in a vessel of volume $1.0\,dm^3$ exert a pressure of 20 atm at 25 °C if it behaved as a perfect gas? If not, what pressure would it exert?
E1A.2(b) Could 25 g of argon gas in a vessel of volume $1.5\,dm^3$ exert a pressure of 2.0 bar at 30 °C if it behaved as a perfect gas? If not, what pressure would it exert?

E1A.3(a) A perfect gas undergoes isothermal compression, which reduces its volume by $2.20\,dm^3$. The final pressure and volume of the gas are 5.04 bar and $4.65\,dm^3$, respectively. Calculate the original pressure of the gas in (i) bar, (ii) atm.
E1A.3(b) A perfect gas undergoes isothermal compression, which reduces its volume by $1.80\,dm^3$. The final pressure and volume of the gas are 1.97 bar and $2.14\,dm^3$, respectively. Calculate the original pressure of the gas in (i) bar, (ii) torr.

E1A.4(a) A car tyre (an automobile tire) was inflated to a pressure of $24\,lb\,in^{-2}$ ($1.00\,atm = 14.7\,lb\,in^{-2}$) on a winter's day when the temperature was −5 °C. What pressure will be found, assuming no leaks have occurred and that the volume is constant, on a subsequent summer's day when the temperature is 35 °C? What complications should be taken into account in practice?
E1A.4(b) A sample of hydrogen gas was found to have a pressure of 125 kPa when the temperature was 23 °C. What can its pressure be expected to be when the temperature is 11 °C?

E1A.5(a) A sample of 255 mg of neon occupies $3.00\,dm^3$ at 122 K. Use the perfect gas law to calculate the pressure of the gas.
E1A.5(b) A homeowner uses $4.00 \times 10^3\,m^3$ of natural gas in a year to heat a home. Assume that natural gas is all methane, CH_4, and that methane is a perfect gas for the conditions of this problem, which are 1.00 atm and 20 °C. What is the mass of gas used?

E1A.6(a) At 500 °C and 93.2 kPa, the mass density of sulfur vapour is $3.710\,kg\,m^{-3}$. What is the molecular formula of sulfur under these conditions?
E1A.6(b) At 100 °C and 16.0 kPa, the mass density of phosphorus vapour is $0.6388\,kg\,m^{-3}$. What is the molecular formula of phosphorus under these conditions?

E1A.7(a) Calculate the mass of water vapour present in a room of volume $400\,m^3$ that contains air at 27 °C on a day when the relative humidity is 60 per cent. *Hint:* Relative humidity is the prevailing partial pressure of water vapour expressed as a percentage of the vapour pressure of water vapour at the same temperature (in this case, 35.6 mbar).
E1A.7(b) Calculate the mass of water vapour present in a room of volume $250\,m^3$ that contains air at 23 °C on a day when the relative humidity is 53 per cent (in this case, 28.1 mbar).

E1A.8(a) Given that the mass density of air at 0.987 bar and 27 °C is $1.146\,kg\,m^{-3}$, calculate the mole fraction and partial pressure of nitrogen and oxygen assuming that (i) air consists only of these two gases, (ii) air also contains 1.0 mole per cent Ar.
E1A.8(b) A gas mixture consists of 320 mg of methane, 175 mg of argon, and 225 mg of neon. The partial pressure of neon at 300 K is 8.87 kPa. Calculate (i) the volume and (ii) the total pressure of the mixture.

E1A.9(a) The mass density of a gaseous compound was found to be $1.23\,kg\,m^{-3}$ at 330 K and 20 kPa. What is the molar mass of the compound?
E1A.9(b) In an experiment to measure the molar mass of a gas, $250\,cm^3$ of the gas was confined in a glass vessel. The pressure was 152 Torr at 298 K, and after correcting for buoyancy effects, the mass of the gas was 33.5 mg. What is the molar mass of the gas?

E1A.10(a) The densities of air at −85 °C, 0 °C, and 100 °C are $1.877\,g\,dm^{-3}$, $1.294\,g\,dm^{-3}$, and $0.946\,g\,dm^{-3}$, respectively. From these data, and assuming that air obeys Charles' law, determine a value for the absolute zero of temperature in degrees Celsius.
E1A.10(b) A certain sample of a gas has a volume of $20.00\,dm^3$ at 0 °C and 1.000 atm. A plot of the experimental data of its volume against the Celsius temperature, θ, at constant p, gives a straight line of slope $0.0741\,dm^3\,°C^{-1}$. From these data alone (without making use of the perfect gas law), determine the absolute zero of temperature in degrees Celsius.

E1A.11(a) A vessel of volume $22.4\,dm^3$ contains 2.0 mol $H_2(g)$ and 1.0 mol $N_2(g)$ at 273.15 K. Calculate (i) the mole fractions of each component, (ii) their partial pressures, and (iii) their total pressure.
E1A.11(b) A vessel of volume $22.4\,dm^3$ contains 1.5 mol $H_2(g)$ and 2.5 mol $N_2(g)$ at 273.15 K. Calculate (i) the mole fractions of each component, (ii) their partial pressures, and (iii) their total pressure.

Problems

P1A.1 A manometer consists of a U-shaped tube containing a liquid. One side is connected to the apparatus and the other is open to the atmosphere. The pressure p inside the apparatus is given $p = p_{ex} + \rho gh$, where p_{ex} is the external pressure, ρ is the mass density of the liquid in the tube, $g = 9.806\,m\,s^{-2}$ is the acceleration of free fall, and h is the difference in heights of the liquid in the two sides of the tube. (The quantity ρgh is the *hydrostatic pressure* exerted by a column of liquid.) (i) Suppose the liquid in a manometer is mercury, the external pressure is 760 Torr, and the open side is 10.0 cm higher than the side connected to the apparatus. What is the pressure in the apparatus? The mass density of mercury at 25 °C is $13.55\,g\,cm^{-3}$. (ii) In an attempt to determine an accurate value of the gas constant, R, a student heated a container of volume $20.000\,dm^3$ filled with 1.485 g of helium gas to 500 °C and measured the pressure as 183.2 cm in a manometer filled with water at 25 °C. Calculate the value of R from these data. The mass density of water at 25 °C is $0.997\,07\,g\,cm^{-3}$.

P1A.2 Recent communication with the inhabitants of Neptune have revealed that they have a Celsius-type temperature scale, but based on the melting point (0 °N) and boiling point (100 °N) of their most common substance, hydrogen. Further communications have revealed that the Neptunians know about perfect gas behaviour and they find that in the limit of zero pressure, the value of pV is 28 dm^3 atm at 0 °N and 40 dm^3 atm at 100 °N. What is the value of the absolute zero of temperature on their temperature scale?

P1A.3 The following data have been obtained for oxygen gas at 273.15K. From the data, calculate the best value of the gas constant R.

p/atm	0.750 000	0.500 000	0.250 000
V_m/(dm^3 mol^{-1})	29.8649	44.8090	89.6384

P1A.4 Charles's law is sometimes expressed in the form $V = V_0(1 + \alpha\theta)$, where θ is the Celsius temperature, α is a constant, and V_0 is the volume of the sample at 0 °C. The following values for have been reported for nitrogen at 0 °C:

p/Torr	749.7	599.6	333.1	98.6
$10^3\alpha$/°C^{-1}	3.6717	3.6697	3.6665	3.6643

For these data estimate the absolute zero of temperature on the Celsius scale.

P1A.5 Deduce the relation between the pressure and mass density, ρ, of a perfect gas of molar mass M. Confirm graphically, using the following data on methoxymethane (dimethyl ether) at 25 °C, that perfect behaviour is reached at low pressures and find the molar mass of the gas.

p/kPa	12.223	25.20	36.97	60.37	85.23	101.3
ρ/(kg m^{-3})	0.225	0.456	0.664	1.062	1.468	1.734

P1A.6 The molar mass of a newly synthesized fluorocarbon was measured in a gas microbalance. This device consists of a glass bulb forming one end of a beam, the whole surrounded by a closed container. The beam is pivoted, and the balance point is attained by raising the pressure of gas in the container, so increasing the buoyancy of the enclosed bulb. In one experiment, the balance point was reached when the fluorocarbon pressure was 327.10 Torr; for the same setting of the pivot, a balance was reached when CHF_3 (M = 70.014 g mol^{-1}) was introduced at 423.22 Torr. A repeat of the experiment with a different setting of the pivot required a pressure of 293.22 Torr of the fluorocarbon and 427.22 Torr of the CHF_3. What is the molar mass of the fluorocarbon? Suggest a molecular formula.

P1A.7 A constant-volume perfect gas thermometer indicates a pressure of 6.69 kPa at the triple point temperature of water (273.16 K). (a) What change of pressure indicates a change of 1.00 K at this temperature? (b) What pressure indicates a temperature of 100.00 °C? (c) What change of pressure indicates a change of 1.00 K at the latter temperature?

P1A.8 A vessel of volume 22.4 dm^3 contains 2.0 mol H_2(g) and 1.0 mol N_2(g) at 273.15 K initially. All the H_2 then reacts with sufficient N_2 to form NH_3. Calculate the partial pressures of the gases in the final mixture and the total pressure.

P1A.9 Atmospheric pollution is a problem that has received much attention. Not all pollution, however, is from industrial sources. Volcanic eruptions can be a significant source of air pollution. The Kilauea volcano in Hawaii emits 200–300 t (1 t = 10^3 kg) of SO_2 each day. If this gas is emitted at 800 °C and 1.0 atm, what volume of gas is emitted?

P1A.10 Ozone is a trace atmospheric gas which plays an important role in screening the Earth from harmful ultraviolet radiation, and the abundance of ozone is commonly reported in *Dobson units*. Imagine a column passing up through the atmosphere. The total amount of O_3 in the column divided by its cross-sectional area is reported in Dobson units with 1 Du = 0.4462 mmol m^{-2}. What amount of O_3 (in moles) is found in a column of atmosphere with a cross-sectional area of 1.00 dm^2 if the abundance is 250 Dobson units (a typical midlatitude value)? In the seasonal Antarctic ozone hole, the column abundance drops below 100 Dobson units; how many moles of O_3 are found in such a column of air above a 1.00 dm^2 area? Most atmospheric ozone is found between 10 and 50 km above the surface of the Earth. If that ozone is spread uniformly through this portion of the atmosphere, what is the average molar concentration corresponding to (a) 250 Dobson units, (b) 100 Dobson units?

P1A.11‡ In a commonly used model of the atmosphere, the atmospheric pressure varies with altitude, h, according to the *barometric formula:*

$$p = p_0 e^{-h/H}$$

where p_0 is the pressure at sea level and H is a constant approximately equal to 8 km. More specifically, $H = RT/Mg$, where M is the average molar mass of air and T is the temperature at the altitude h. This formula represents the outcome of the competition between the potential energy of the molecules in the gravitational field of the Earth and the stirring effects of thermal motion. Derive this relation by showing that the change in pressure dp for an infinitesimal change in altitude dh where the mass density is ρ is $dp = -\rho g dh$. Remember that ρ depends on the pressure. Evaluate (a) the pressure difference between the top and bottom of a laboratory vessel of height 15 cm, and (b) the external atmospheric pressure at a typical cruising altitude of an aircraft (11 km) when the pressure at ground level is 1.0 atm.

P1A.12‡ Balloons are still used to deploy sensors that monitor meteorological phenomena and the chemistry of the atmosphere. It is possible to investigate some of the technicalities of ballooning by using the perfect gas law. Suppose your balloon has a radius of 3.0 m and that it is spherical. (a) What amount of H_2 (in moles) is needed to inflate it to 1.0 atm in an ambient temperature of 25 °C at sea level? (b) What mass can the balloon lift (the payload) at sea level, where the mass density of air is 1.22 kg m^{-3}? (c) What would be the payload if He were used instead of H_2?

P1A.13‡ Chlorofluorocarbons such as CCl_3F and CCl_2F_2 have been linked to ozone depletion in Antarctica. In 1994, these gases were found in quantities of 261 and 509 parts per trillion by volume (World Resources Institute, *World resources* 1996–97). Compute the molar concentration of these gases under conditions typical of (a) the mid-latitude troposphere (10 °C and 1.0 atm) and (b) the Antarctic stratosphere (200 K and 0.050 atm). *Hint:* The composition of a mixture of gases can be described by imagining that the gases are separated from one another in such a way that each exerts the same pressure. If one gas is present at very low levels it is common to express its concentration as, for example, 'x parts per trillion by volume'. Then the volume of the separated gas at a certain pressure is $x \times 10^{-12}$ of the original volume of the gas mixture at the same pressure. For a mixture of perfect gases, the volume of each separated gas is proportional to its partial pressure in the mixture and hence to the amount in moles of the gas molecules present in the mixture.

P1A.14 At sea level the composition of the atmosphere is approximately 80 per cent nitrogen and 20 per cent oxygen by mass. At what height above the surface of the Earth would the atmosphere become 90 per cent nitrogen and 10 per cent oxygen by mass? Assume that the temperature of the atmosphere is constant at 25 °C. What is the pressure of the atmosphere at that height? *Hint:* Use a barometric formula, see Problem P1A.11, for each partial pressure.

‡ These problems were supplied by Charles Trapp and Carmen Giunta.

TOPIC 1B The kinetic model

Discussion questions

D1B.1 Specify and analyse critically the assumptions that underlie the kinetic model of gases.

D1B.2 Provide molecular interpretations for the dependencies of the mean free path on the temperature, pressure, and size of gas molecules.

D1B.3 Use the kinetic model of gases to explain why light gases, such as He, are rare in the Earth's atmosphere but heavier gases, such as O_2, CO_2, and N_2, once formed remain abundant.

Exercises

E1B.1(a) Determine the ratios of (i) the mean speeds, (ii) the mean translational kinetic energies of H_2 molecules and Hg atoms at 20 °C.
E1B.1(b) Determine the ratios of (i) the mean speeds, (ii) the mean translational kinetic energies of He atoms and Hg atoms at 25 °C.

E1B.2(a) Calculate the root-mean-square speeds of H_2 and O_2 molecules at 20 °C.
E1B.2(b) Calculate the root-mean-square speeds of CO_2 molecules and He atoms at 20 °C.

E1B.3(a) Use the Maxwell–Boltzmann distribution of speeds to estimate the fraction of N_2 molecules at 400 K that have speeds in the range 200–210 m s^{-1}. *Hint:* The fraction of molecules with speeds in the range v to $v + dv$ is equal to $f(v)dv$, where $f(v)$ is given by eqn 1B.4.
E1B.3(b) Use the Maxwell–Boltzmann distribution of speeds to estimate the fraction of CO_2 molecules at 400 K that have speeds in the range 400–405 m s^{-1}. See the hint in Exercise E1B.3(a).

E1B.4(a) What is the relative mean speed of N_2 and H_2 molecules in a gas at 25 °C?
E1B.4(b) What is the relative mean speed of O_2 and N_2 molecules in a gas at 25 °C?

E1B.5(a) Calculate the most probable speed, the mean speed, and the mean relative speed of CO_2 molecules at 20 °C.
E1B.5(b) Calculate the most probable speed, the mean speed, and the mean relative speed of H_2 molecules at 20 °C.

E1B.6(a) Evaluate the collision frequency of H_2 molecules in a gas at 1.00 atm and 25 °C.
E1B.6(b) Evaluate the collision frequency of O_2 molecules in a gas at 1.00 atm and 25 °C.

E1B.7(a) Assume that air consists of N_2 molecules with a collision diameter of 395 pm. Calculate (i) the mean speed of the molecules, (ii) the mean free path, (iii) the collision frequency in air at 1.0 atm and 25 °C.
E1B.7(b) The best laboratory vacuum pump can generate a vacuum of about 1 nTorr. At 25 °C and assuming that air consists of N_2 molecules with a collision diameter of 395 pm, calculate at this pressure (i) the mean speed of the molecules, (ii) the mean free path, (iii) the collision frequency in the gas.

E1B.8(a) At what pressure does the mean free path of argon at 20 °C become comparable to the diameter of a 100 cm^3 vessel that contains it? Take $\sigma = 0.36\ nm^2$.
E1B.8(b) At what pressure does the mean free path of argon at 20 °C become comparable to 10 times the diameters of the atoms themselves? Take $\sigma = 0.36\ nm^2$.

E1B.9(a) At an altitude of 20 km the temperature is 217 K and the pressure is 0.050 atm. What is the mean free path of N_2 molecules? ($\sigma = 0.43\ nm^2$).
E1B.9(b) At an altitude of 15 km the temperature is 217 K and the pressure is 12.1 kPa. What is the mean free path of N_2 molecules? ($\sigma = 0.43\ nm^2$).

Problems

P1B.1 A rotating slotted-disc apparatus consists of five coaxial 5.0 cm diameter discs separated by 1.0 cm, the radial slots being displaced by 2.0° between neighbours. The relative intensities, I, of the detected beam of Kr atoms for two different temperatures and at a series of rotation rates were as follows:

v/Hz	20	40	80	100	120
I (40 K)	0.846	0.513	0.069	0.015	0.002
I (100 K)	0.592	0.485	0.217	0.119	0.057

Find the distributions of molecular velocities, $f(v_x)$, at these temperatures, and check that they conform to the theoretical prediction for a one-dimensional system for this low-pressure, collision-free system.

P1B.2 Consider molecules that are confined to move in a plane (a two-dimensional gas). Calculate the distribution of speeds and determine the mean speed of the molecules at a temperature T.

P1B.3 A specially constructed velocity-selector accepts a beam of molecules from an oven at a temperature T but blocks the passage of molecules with a speed greater than the mean. What is the mean speed of the emerging beam, relative to the initial value? Treat the system as one-dimensional.

P1B.4 What, according to the Maxwell–Boltzmann distribution, is the proportion of gas molecules having (i) more than, (ii) less than the root mean square speed? (iii) What are the proportions having speeds greater and smaller than the mean speed? *Hint:* Use mathematical software to evaluate the integrals.

P1B.5 Calculate the fractions of molecules in a gas that have a speed in a range Δv at the speed nv_{mp} relative to those in the same range at v_{mp} itself. This calculation can be used to estimate the fraction of very energetic molecules (which is important for reactions). Evaluate the ratio for $n = 3$ and $n = 4$.

P1B.6 Derive an expression for $\langle v^n \rangle^{1/n}$ from the Maxwell–Boltzmann distribution of speeds. *Hint:* You will need the integrals given in the *Resource section*, or use mathematical software.

P1B.7 Calculate the escape velocity (the minimum initial velocity that will take an object to infinity) from the surface of a planet of radius R. What is the value for (i) the Earth, $R = 6.37 \times 10^6$ m, $g = 9.81$ m s^{-2}, (ii) Mars, $R = 3.38 \times 10^6$ m, $m_{Mars}/m_{Earth} = 0.108$. At what temperatures do H_2, He, and O_2 molecules have mean speeds equal to their escape speeds? What proportion of the molecules have enough speed to escape when the temperature is (i) 240 K, (ii) 1500 K? Calculations of this kind are very important in considering the composition of planetary atmospheres.

P1B.8 Plot different Maxwell–Boltzmann speed distributions by keeping the molar mass constant at 100 g mol^{-1} and varying the temperature of the sample between 200 K and 2000 K.

P1B.9 Evaluate numerically the fraction of O_2 molecules with speeds in the range 100 m s^{-1} to 200 m s^{-1} in a gas at 300 K and 1000 K.

P1B.10 The maximum in the Maxwell–Boltzmann distribution occurs when $df(v)/dv = 0$. Find, by differentiation, an expression for the most probable speed of molecules of molar mass M at a temperature T.

P1B.11 A methane, CH_4, molecule may be considered as spherical, with a radius of 0.38 nm. How many collisions does a single methane molecule make if 0.10 mol CH_4(g) is held at 25 °C in a vessel of volume 1.0 dm^3?

TOPIC 1C Real gases

Discussion questions

D1C.1 Explain how the compression factor varies with pressure and temperature and describe how it reveals information about intermolecular interactions in real gases.

D1C.2 What is the significance of the critical constants?

D1C.3 Describe the formulation of the van der Waals equation and suggest a rationale for one other equation of state in Table 1C.4.

D1C.4 Explain how the van der Waals equation accounts for critical behaviour.

Exercises

E1C.1(a) Calculate the pressure exerted by 1.0 mol C_2H_6 behaving as a van der Waals gas when it is confined under the following conditions: (i) at 273.15 K in 22.414 dm^3, (ii) at 1000 K in 100 cm^3. Use the data in Table 1C.3 of the *Resource section*.

E1C.1(b) Calculate the pressure exerted by 1.0 mol H_2S behaving as a van der Waals gas when it is confined under the following conditions: (i) at 273.15 K in 22.414 dm^3, (ii) at 500 K in 150 cm^3. Use the data in Table 1C.3 of the *Resource section*.

E1C.2(a) Express the van der Waals parameters $a = 0.751$ atm dm^6 mol^{-2} and $b = 0.0226$ dm^3 mol^{-1} in SI base units (kg, m, s, and mol).

E1C.2(b) Express the van der Waals parameters $a = 1.32$ atm dm^6 mol^{-2} and $b = 0.0436$ dm^3 mol^{-1} in SI base units (kg, m, s, and mol).

E1C.3(a) A gas at 250 K and 15 atm has a molar volume 12 per cent smaller than that calculated from the perfect gas law. Calculate (i) the compression factor under these conditions and (ii) the molar volume of the gas. Which are dominating in the sample, the attractive or the repulsive forces?

E1C.3(b) A gas at 350 K and 12 atm has a molar volume 12 per cent larger than that calculated from the perfect gas law. Calculate (i) the compression factor under these conditions and (ii) the molar volume of the gas. Which are dominating in the sample, the attractive or the repulsive forces?

E1C.4(a) In an industrial process, nitrogen is heated to 500 K at a constant volume of 1.000 m^3. The mass of the gas is 92.4 kg. Use the van der Waals equation to determine the approximate pressure of the gas at its working temperature of 500 K. For nitrogen, $a = 1.352$ dm^6 atm mol^{-2}, $b = 0.0387$ dm^3 mol^{-1}.

E1C.4(b) Cylinders of compressed gas are typically filled to a pressure of 200 bar. For oxygen, what would be the molar volume at this pressure and 25 °C based on (i) the perfect gas equation, (ii) the van der Waals equation? For oxygen, $a = 1.364$ dm^6 atm mol^{-2}, $b = 3.19 \times 10^{-2}$ dm^3 mol^{-1}.

E1C.5(a) Suppose that 10.0 mol C_2H_6(g) is confined to 4.860 dm^3 at 27 °C. Predict the pressure exerted by the ethane from (i) the perfect gas and (ii) the van der Waals equations of state. Calculate the compression factor based on these calculations. For ethane, $a = 5.507$ dm^6 atm mol^{-2}, $b = 0.0651$ dm^3 mol^{-1}.

E1C.5(b) At 300 K and 20 atm, the compression factor of a gas is 0.86. Calculate (i) the volume occupied by 8.2 mmol of the gas molecules under these conditions and (ii) an approximate value of the second virial coefficient B at 300 K.

E1C.6(a) The critical constants of methane are $p_c = 45.6$ atm, $V_c = 98.7$ cm^3 mol^{-1}, and $T_c = 190.6$ K. Calculate the van der Waals parameters of the gas and estimate the radius of the molecules.

E1C.6(b) The critical constants of ethane are $p_c = 48.20$ atm, $V_c = 148$ cm^3 mol^{-1}, and $T_c = 305.4$ K. Calculate the van der Waals parameters of the gas and estimate the radius of the molecules.

E1C.7(a) Use the van der Waals parameters for chlorine in Table 1C.3 of the *Resource section* to calculate approximate values of (i) the Boyle temperature of chlorine from $T_B = a/Rb$ and (ii) the radius of a Cl_2 molecule regarded as a sphere.

E1C.7(b) Use the van der Waals parameters for hydrogen sulfide in Table 1C.3 of the *Resource section* to calculate approximate values of (i) the Boyle temperature of the gas from $T_B = a/Rb$ and (ii) the radius of an H_2S molecule regarded as a sphere.

E1C.8(a) Suggest the pressure and temperature at which 1.0 mol of (i) NH_3, (ii) Xe, (iii) He will be in states that correspond to 1.0 mol H_2 at 1.0 atm and 25 °C.

E1C.8(b) Suggest the pressure and temperature at which 1.0 mol of (i) H_2O (ii) CO_2, (iii) Ar will be in states that correspond to 1.0 mol N_2 at 1.0 atm and 25 °C.

E1C.9(a) A certain gas obeys the van der Waals equation with $a = 0.50$ m^6 Pa mol^{-2}. Its molar volume is found to be 5.00×10^{-4} m^3 mol^{-1} at 273 K and 3.0 MPa. From this information calculate the van der Waals constant b. What is the compression factor for this gas at the prevailing temperature and pressure?

E1C.9(b) A certain gas obeys the van der Waals equation with $a = 0.76$ m^6 Pa mol^{-2}. Its molar volume is found to be 4.00×10^{-4} m^3 mol^{-1} at 288 K and 4.0 MPa. From this information calculate the van der Waals constant b. What is the compression factor for this gas at the prevailing temperature and pressure?

Problems

P1C.1 What pressure would 4.56 g of nitrogen gas in a vessel of volume 2.25 dm^3 exert at 273 K if it obeyed the virial equation of state up to and including the first two terms?

P1C.2 Calculate the molar volume of chlorine gas at 350 K and 2.30 atm using (a) the perfect gas law and (b) the van der Waals equation. Use the answer to (a) to calculate a first approximation to the correction term for attraction and then use successive approximations to obtain a numerical answer for part (b).

P1C.3 At 273 K measurements on argon gave $B = -21.7$ cm^3 mol^{-1} and $C = 1200$ cm^6 mol^{-2}, where B and C are the second and third virial coefficients in the expansion of Z in powers of $1/V_m$. Assuming that the perfect gas law holds sufficiently well for the estimation of the molar volume, calculate the compression factor of argon at 100 atm and 273 K. From your result, estimate the molar volume of argon under these conditions.

P1C.4 Calculate the volume occupied by 1.00 mol N_2 using the van der Waals equation expanded into the form of a virial expansion at (a) its critical temperature, (b) its Boyle temperature. Assume that the pressure is 10 atm throughout. At what temperature is the behaviour of the gas closest to that of a perfect gas? Use the following data: $T_c = 126.3$ K, $T_B = 327.2$ K, $a = 1.390$ dm^6 atm mol^{-2}, $b = 0.0391$ dm^3 mol^{-1}.

P1C.5‡ The second virial coefficient of methane can be approximated by the empirical equation $B(T) = a + \mathrm{e}^{-c/T^2}$, where $a = -0.1993\,\text{bar}^{-1}$, $b = 0.2002\,\text{bar}^{-1}$, and $c = 1131\,\text{K}^2$ with $300\,\text{K} < T < 600\,\text{K}$. What is the Boyle temperature of methane?

P1C.6 How well does argon gas at 400 K and 3 atm approximate a perfect gas? Assess the approximation by reporting the difference between the molar volumes as a percentage of the perfect gas molar volume.

P1C.7 The mass density of water vapour at 327.6 atm and 776.4 K is $133.2\,\text{kg m}^{-3}$. Given that for water $a = 5.464\,\text{dm}^6\,\text{atm mol}^{-2}$, $b = 0.03049\,\text{dm}^3\,\text{mol}^{-1}$, and $M = 18.02\,\text{g mol}^{-1}$, calculate (a) the molar volume. Then calculate the compression factor (b) from the data, and (c) from the virial expansion of the van der Waals equation.

P1C.8 The critical volume and critical pressure of a certain gas are $160\,\text{cm}^3\,\text{mol}^{-1}$ and 40 atm, respectively. Estimate the critical temperature by assuming that the gas obeys the Berthelot equation of state. Estimate the radii of the gas molecules on the assumption that they are spheres.

P1C.9 Estimate the coefficients a and b in the Dieterici equation of state from the critical constants of xenon. Calculate the pressure exerted by 1.0 mol Xe when it is confined to $1.0\,\text{dm}^3$ at 25 °C.

P1C.10 For a van der Waals gas with given values of a and b, identify the conditions for which $Z < 1$ and $Z > 1$.

P1C.11 Express the van der Waals equation of state as a virial expansion in powers of $1/V_m$ and obtain expressions for B and C in terms of the parameters a and b. The expansion you will need is $(1-x)^{-1} = 1 + x + x^2 + \cdots$. Measurements on argon gave $B = -21.7\,\text{cm}^3\,\text{mol}^{-1}$ and $C = 1200\,\text{cm}^6\,\text{mol}^{-2}$ for the virial coefficients at 273 K. What are the values of a and b in the corresponding van der Waals equation of state?

P1C.12 The critical constants of a van der Waals gas can be found by setting the following derivatives equal to zero at the critical point:

$$\frac{\mathrm{d}p}{\mathrm{d}V_m} = -\frac{RT}{(V_m - b)^2} + \frac{2a}{V_m^3} = 0$$

$$\frac{\mathrm{d}^2 p}{\mathrm{d}V_m^2} = \frac{2RT}{(V_m - b)^3} - \frac{6a}{V_m^4} = 0$$

Solve this system of equations and then use eqn 1C.5b to show that p_c, V_c and T_c are given by eqn 1C.6.

P1C.13 A scientist proposed the following equation of state:

$$p = \frac{RT}{V_m} - \frac{B}{V_m^2} + \frac{C}{V_m^3}$$

Show that the equation leads to critical behaviour. Find the critical constants of the gas in terms of B and C and an expression for the critical compression factor.

P1C.14 Equations 1C.3a and 1C.3b are expansions in p and $1/V_m$, respectively. Find the relation between B, C and B', C'.

P1C.15 The second virial coefficient B' can be obtained from measurements of the mass density ρ of a gas at a series of pressures. Show that the graph of p/ρ against p should be a straight line with slope proportional to B'. Use the data on methoxymethane in Problem P1A.5 to find the values of B' and B at 25 °C.

P1C.16 The equation of state of a certain gas is given by $p = RT/V_m + (a + bT)/V_m^2$, where a and b are constants. Find $(\partial V_m/\partial T)_p$.

P1C.17 Under what conditions can liquid nitrogen be formed by the application of pressure alone?

P1C.18 The following equations of state are occasionally used for approximate calculations on gases: (gas A) $pV_m = RT(1 + b/V_m)$, (gas B) $p(V_m - b) = RT$. Assuming that there were gases that actually obeyed these equations of state, would it be possible to liquefy either gas A or B? Would they have a critical temperature? Explain your answer.

P1C.19 Derive an expression for the compression factor of a gas that obeys the equation of state $p(V - nb) = nRT$, where b and R are constants. If the pressure and temperature are such that $V_m = 10b$, what is the numerical value of the compression factor?

P1C.20 What would be the corresponding state of ammonia, for the conditions described for argon in Brief illustration 1C.5?

P1C.21‡ Stewart and Jacobsen have published a review of thermodynamic properties of argon (R.B. Stewart and R.T. Jacobsen, *J. Phys. Chem. Ref. Data* **18**, 639 (1989)) which included the following 300 K isotherm.

p/MPa	0.4000	0.5000	0.6000	0.8000	1.000
V_m/(dm^3 mol^{-1})	6.2208	4.9736	4.1423	3.1031	2.4795
p/MPa	1.500	2.000	2.500	3.000	4.000
V_m/(dm^3 mol^{-1})	1.6483	1.2328	0.98357	0.81746	0.60998

(a) Compute the second virial coefficient, B, at this temperature. (b) Use non-linear curve-fitting software to compute the third virial coefficient, C, at this temperature.

P1C.22 Use the van der Waals equation of state and mathematical software or a spreadsheet to plot the pressure of 1.5 mol CO_2(g) against volume as it is compressed from $30\,\text{dm}^3$ to $15\,\text{dm}^3$ at (a) 273 K, (b) 373 K. (c) Redraw the graphs as plots of p against $1/V$.

P1C.23 Calculate the molar volume of chlorine on the basis of the van der Waals equation of state at 250 K and 150 kPa and calculate the percentage difference from the value predicted by the perfect gas equation.

P1C.24 Is there a set of conditions at which the compression factor of a van der Waals gas passes through a minimum? If so, how does the location and value of the minimum value of Z depend on the coefficients a and b?

FOCUS 1 The properties of gases

Integrated activities

I1.1 Start from the Maxwell–Boltzmann distribution and derive an expression for the most probable speed of a gas of molecules at a temperature T. Go on to demonstrate the validity of the equipartition conclusion that the average translational kinetic energy of molecules free to move in three dimensions is $\frac{3}{2}kT$.

I1.2 The principal components of the atmosphere of the Earth are diatomic molecules, which can rotate as well as translate. Given that the translational kinetic energy density of the atmosphere is $0.15\,\text{J cm}^{-3}$, what is the total kinetic energy density, including rotation?

I1.3 Methane molecules, CH_4, may be considered as spherical, with a collision cross-section of $\sigma = 0.46\,\text{nm}^2$. Estimate the value of the van der Waals parameter b by calculating the molar volume excluded by methane molecules.

FOCUS 2

The First Law

The release of energy can be used to provide heat when a fuel burns in a furnace, to produce mechanical work when a fuel burns in an engine, and to generate electrical work when a chemical reaction pumps electrons through a circuit. Chemical reactions can be harnessed to provide heat and work, liberate energy that is unused but which gives desired products, and drive the processes of life. Thermodynamics, the study of the transformations of energy, enables the discussion of all these matters quantitatively, allowing for useful predictions.

2A Internal energy

This Topic examines the ways in which a system can exchange energy with its surroundings in terms of the work it may do or have done on it, or the heat that it may produce or absorb. These considerations lead to the definition of the 'internal energy', the total energy of a system, and the formulation of the 'First Law' of thermodynamics, which states that the internal energy of an isolated system is constant.

2A.1 Work, heat, and energy; 2A.2 The definition of internal energy; 2A.3 Expansion work; 2A.4 Heat transactions

2B Enthalpy

The second major concept of the Focus is 'enthalpy', which is a very useful book-keeping property for keeping track of the heat output (or requirements) of physical processes and chemical reactions that take place at constant pressure. Experimentally, changes in internal energy or enthalpy may be measured by techniques known collectively as 'calorimetry'.

2B.1 The definition of enthalpy; 2B.2 The variation of enthalpy with temperature

2C Thermochemistry

'Thermochemistry' is the study of heat transactions during chemical reactions. This Topic describes methods for the determination of enthalpy changes associated with both physical and chemical changes.

2C.1 Standard enthalpy changes; 2C.2 Standard enthalpies of formation; 2C.3 The temperature dependence of reaction enthalpies; 2C.4 Experimental techniques

2D State functions and exact differentials

The power of thermodynamics becomes apparent by establishing relations between different properties of a system. One very useful aspect of thermodynamics is that a property can be measured indirectly by measuring others and then combining their values. The relations derived in this Topic also apply to the discussion of the liquefaction of gases and to the relation between the heat capacities of a substance under different conditions.

2D.1 Exact and inexact differentials; 2D.2 Changes in internal energy; 2D.3 Changes in enthalpy; 2D.4 The Joule–Thomson effect

2E Adiabatic changes

'Adiabatic' processes occur without transfer of energy as heat. This Topic describes reversible adiabatic changes involving perfect gases because they figure prominently in the presentation of thermodynamics.

2E.1 The change in temperature; 2E.2 The change in pressure

Web resource What is an application of this material?

A major application of thermodynamics is to the assessment of fuels and their equivalent for organisms, food. Some thermochemical aspects of fuels and foods are described in *Impact* 3 on the website of this text.

TOPIC 2A Internal energy

➤ Why do you need to know this material?

The First Law of thermodynamics is the foundation of the discussion of the role of energy in chemistry. Wherever the generation or use of energy in physical transformations or chemical reactions is of interest, lying in the background are the concepts introduced by the First Law.

➤ What is the key idea?

The total energy of an isolated system is constant.

➤ What do you need to know already?

This Topic makes use of the discussion of the properties of gases (Topic 1A), particularly the perfect gas law. It builds on the definition of work given in *The chemist's toolkit* 6.

For the purposes of thermodynamics, the universe is divided into two parts, the system and its surroundings. The **system** is the part of the world of interest. It may be a reaction vessel, an engine, an electrochemical cell, a biological cell, and so on. The **surroundings** comprise the region outside the system and are where measurements are made. The type of system depends on the characteristics of the boundary that divides it from the surroundings (Fig. 2A.1). If matter can be transferred through the boundary between the system and its surroundings the system is classified as **open**. If matter cannot pass through the boundary the system is classified as **closed**. Both open and closed systems can exchange energy with their surroundings.

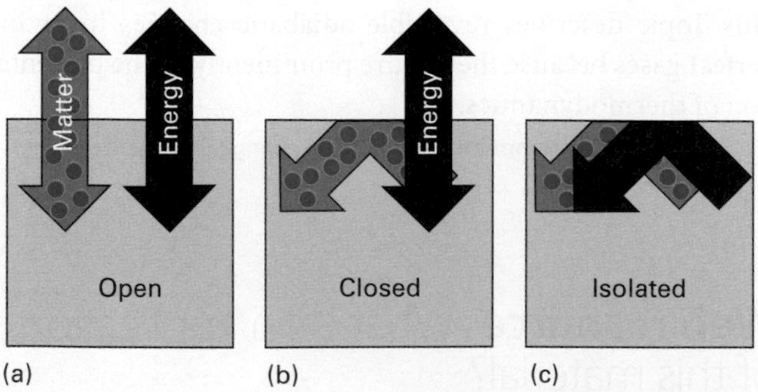

Figure 2A.1 (a) An open system can exchange matter and energy with its surroundings. (b) A closed system can exchange energy with its surroundings, but it cannot exchange matter. (c) An isolated system can exchange neither energy nor matter with its surroundings.

For example, a closed system can expand and thereby raise a weight in the surroundings; a closed system may also transfer energy to the surroundings if they are at a lower temperature. An **isolated system** is a closed system that has neither mechanical nor thermal contact with its surroundings.

2A.1 Work, heat, and energy

Although thermodynamics deals with observations on bulk systems, it is immeasurably enriched by understanding the molecular origins of these observations.

(a) Operational definitions

The fundamental physical property in thermodynamics is work: **work** is done to achieve motion against an opposing force (*The chemist's toolkit* 6). A simple example is the process of raising a weight against the pull of gravity. A process does work if in principle it can be harnessed to raise a weight somewhere in the surroundings. An example of doing work is the expansion of a gas that pushes out a piston: the motion of the piston can in principle be used to raise a weight. Another example is a chemical reaction in a cell, which leads to an electric current that can drive a motor and be used to raise a weight.

The **energy** of a system is its capacity to do work (see *The chemist's toolkit* 6 for more detail). When work is done on an otherwise isolated system (for instance, by compressing a gas or winding a spring), the capacity of the system to do work is increased; in other words, the energy of the system is increased. When the system does work (when the piston moves out or the spring unwinds), the energy of the system is reduced and it can do less work than before.

Experiments have shown that the energy of a system may be changed by means other than work itself. When the energy of a system changes as a result of a temperature difference between the system and its surroundings the energy is said to be transferred as **heat**. When a heater is immersed in a beaker of water (the system), the capacity of the system to do work increases because hot water can be used to do more work than the same amount of cold water. Not all boundaries permit the transfer of energy even though there is a temperature difference between the system and its surroundings. Boundaries that do permit the transfer of energy as heat are called **diathermic**; those that do not are called **adiabatic**.

The chemist's toolkit 6 Work and energy

Work, w, is done when a body is moved against an opposing force. For an infinitesimal displacement through ds (a vector), the work done *on the body* is

$$dw_{body} = -F\cdot ds \quad \text{Work done on body [definition]}$$

where $F\cdot ds$ is the 'scalar product' of the vectors F and ds:

$$F\cdot ds = F_x dx + F_y dy + F_z dz \quad \text{Scalar product [definition]}$$

The energy lost as work by the system, dw, is the negative of the work done on the body, so

$$dw = F\cdot ds \quad \text{Work done on system [definition]}$$

For motion in one dimension, $dw = F_x dx$, with $F_x < 0$ (so $F_x = -|F_x|$) if it opposed the motion. The total work done along a path is the integral of this expression, allowing for the possibility that F changes in direction and magnitude at each point of the path. With force in newtons (N) and distance in metres, the units of work are joules (J), with

$$1\,J = 1\,N\,m = 1\,kg\,m^2\,s^{-2}$$

Energy is the capacity to do work. The SI unit of energy is the same as that of work, namely the joule. The rate of supply of energy is called the **power** (P), and is expressed in watts (W):

$$1\,W = 1\,J\,s^{-1}$$

A particle may possess two kinds of energy, kinetic energy and potential energy. The **kinetic energy**, E_k, of a body is the energy the body possesses as a result of its motion. For a body of mass m travelling at a speed v,

$$E_k = \tfrac{1}{2}mv^2 \quad \text{Kinetic energy [definition]}$$

Because $p = mv$ (*The chemist's toolkit* 3 of Topic 1B), where p is the magnitude of the linear momentum, it follows that

$$E_k = \frac{p^2}{2m} \quad \text{Kinetic energy [definition]}$$

The **potential energy**, E_p, (and commonly V, but do not confuse that with the volume!) of a body is the energy it possesses as a result of its position. In the absence of losses, the potential energy of a stationary particle is equal to the work that had to be done on the body to bring it to its current location. Because $dw_{body} = -F_x dx$, it follows that $dE_p = -F_x dx$ and therefore

$$F_x = -\frac{dE_p}{dx} \quad \text{Potential energy [relation to force]}$$

If E_p increases as x increases, then F_x is negative (directed towards negative x, Sketch 1). Thus, the steeper the gradient (the more strongly the potential energy depends on position), the greater is the force.

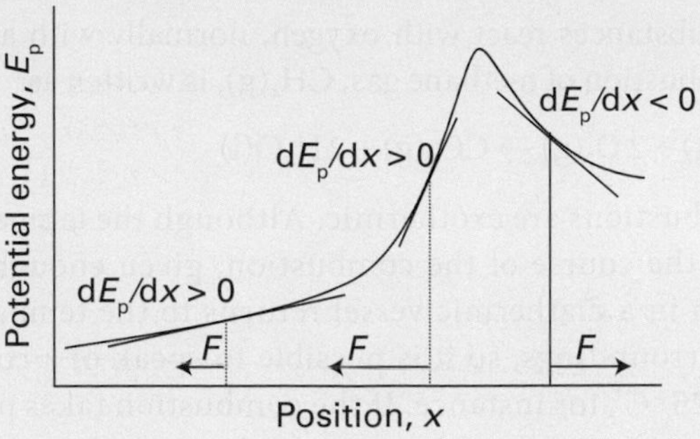

Sketch 1

No universal expression for the potential energy can be given because it depends on the type of force the body experiences. For a particle of mass m at an altitude h close to the surface of the Earth, the gravitational potential energy is

$$E_p(h) = E_p(0) + mgh \quad \text{Gravitational potential energy}$$

where g is the **acceleration of free fall** (g depends on location, but its 'standard value' is close to $9.81\,m\,s^{-2}$). The zero of potential energy is arbitrary. For a particle close to the surface of the Earth, it is common to set $E_p(0) = 0$.

The **Coulomb potential energy** of two electric charges, Q_1 and Q_2, separated by a distance r is

$$E_p = \frac{Q_1 Q_2}{4\pi\varepsilon r} \quad \text{Coulomb potential energy}$$

The quantity ε (epsilon) is the **permittivity**; its value depends upon the nature of the medium between the charges. If the charges are separated by a vacuum, then the constant is known as the **vacuum permittivity**, ε_0 (epsilon zero), or the **electric constant**, which has the value $8.854 \times 10^{-12}\,J^{-1}\,C^2\,m^{-1}$. The permittivity is greater for other media, such as air, water, or oil. It is commonly expressed as a multiple of the vacuum permittivity:

$$\varepsilon = \varepsilon_r \varepsilon_0 \quad \text{Permittivity [definition]}$$

with ε_r the dimensionless **relative permittivity** (formerly, the *dielectric constant*).

The **total energy** of a particle is the sum of its kinetic and potential energies:

$$E = E_k + E_p \quad \text{Total energy [definition]}$$

Provided no external forces are acting on the body, its total energy is constant. This central statement of physics is known as the **law of the conservation of energy**. Potential and kinetic energy may be freely interchanged, but their sum remains constant in the absence of external influences.

An **exothermic process** is a process that releases energy as heat. For example, combustions are chemical reactions in which substances react with oxygen, normally with a flame. The combustion of methane gas, $CH_4(g)$, is written as:

$$CH_4(g) + 2\,O_2(g) \rightarrow CO_2(g) + 2\,H_2O(l)$$

All combustions are exothermic. Although the temperature rises in the course of the combustion, given enough time, a system in a diathermic vessel returns to the temperature of its surroundings, so it is possible to speak of a combustion 'at 25 °C', for instance. If the combustion takes place in an adiabatic container, the energy released as heat remains inside the container and results in a permanent rise in temperature.

An **endothermic process** is a process in which energy is acquired as heat. An example of an endothermic process is the vaporization of water. To avoid a lot of awkward language, it is common to say that in an exothermic process energy is transferred 'as heat' to the surroundings and in an endothermic process energy is transferred 'as heat' from the surroundings into the system. However, it must never be forgotten that heat is a process (the transfer of energy as a result of a temperature difference), not an entity. An endothermic process in a diathermic container results in energy flowing into the system as heat to restore the temperature to that of the surroundings. An exothermic process in a similar diathermic container results in a release of energy as heat into the surroundings. When an endothermic process takes place in an adiabatic container, it results in a lowering of temperature of the system; an exothermic process results in a rise of temperature. These features are summarized in Fig. 2A.2.

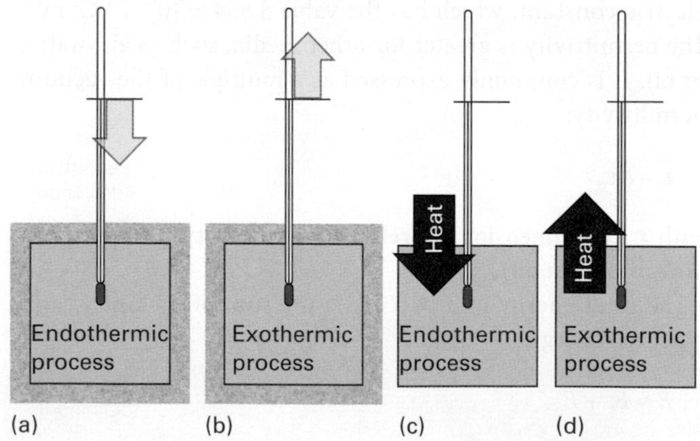

Figure 2A.2 (a) When an endothermic process occurs in an adiabatic system, the temperature falls; (b) if the process is exothermic, then the temperature rises. (c) When an endothermic process occurs in a diathermic container, energy enters as heat from the surroundings (which remain at the same temperature), and the system remains at the same temperature. (d) If the process is exothermic, then energy leaves as heat, and the process is isothermal.

(b) The molecular interpretation of heat and work

In molecular terms, heating is the transfer of energy that makes use of disorderly, apparently random, molecular motion in the surroundings. The disorderly motion of molecules is called **thermal motion**. The thermal motion of the molecules in the hot surroundings stimulates the molecules in the cooler system to move more vigorously and, as a result, the energy of the cooler system is increased. When a system heats its surroundings, molecules of the system stimulate the thermal motion of the molecules in the surroundings (Fig. 2A.3).

In contrast, work is the transfer of energy that makes use of organized motion in the surroundings (Fig. 2A.4). When a weight is raised or lowered, its atoms move in an organized way (up or down). The atoms in a spring move in an orderly way when it is wound; the electrons in an electric current

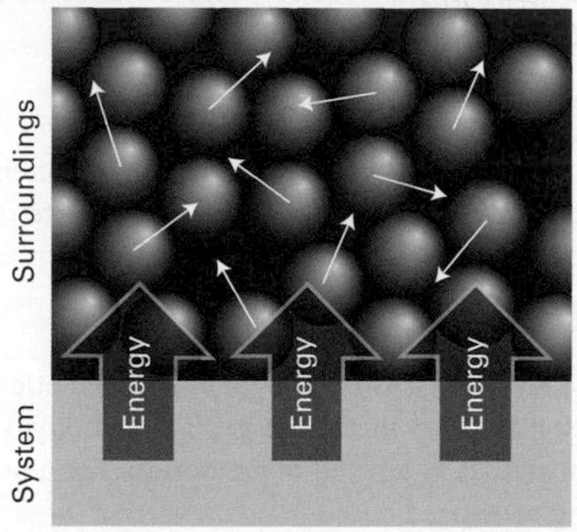

Figure 2A.3 When energy is transferred to the surroundings as heat, the transfer stimulates random motion of the atoms in the surroundings. Transfer of energy from the surroundings to the system makes use of random motion (thermal motion) in the surroundings.

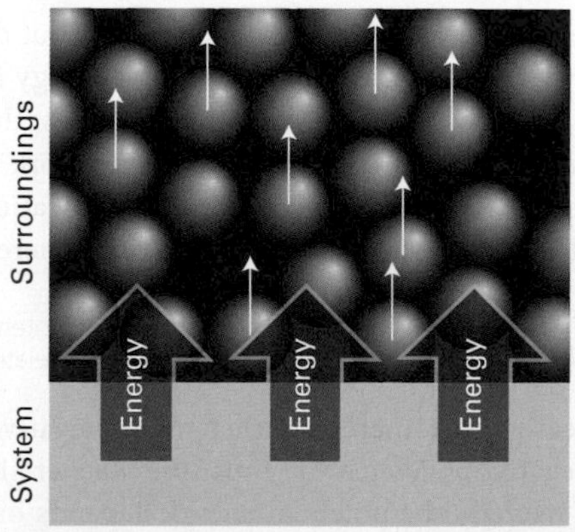

Figure 2A.4 When a system does work, it stimulates orderly motion in the surroundings. For instance, the atoms shown here may be part of a weight that is being raised. The ordered motion of the atoms in a falling weight does work on the system.

move in the same direction. When a system does work it causes atoms or electrons in its surroundings to move in an organized way. Likewise, when work is done on a system, molecules in the surroundings are used to transfer energy to it in an organized way, as the atoms in a weight are lowered or a current of electrons is passed.

The distinction between work and heat is made in the surroundings. The fact that a falling weight may stimulate thermal motion in the system is irrelevant to the distinction between heat and work: work is identified as energy transfer making use of the organized motion of atoms in the surroundings, and heat is identified as energy transfer making use of thermal motion in the surroundings. In the compression of a gas in an adiabatic enclosure, for instance, work is done on the system as the atoms of the compressing weight descend in an orderly way, but the effect of the incoming piston is to accelerate the gas molecules to higher average speeds. Because collisions between molecules quickly randomize their directions, the orderly motion of the atoms of the weight is in effect stimulating thermal motion in the gas. The weight is observed to fall, leading to the orderly descent of its atoms, and work is done even though it is stimulating thermal motion.

2A.2 The definition of internal energy

In thermodynamics, the total energy of a system is called its **internal energy**, *U*. The internal energy is the total kinetic and potential energy of the constituents (the atoms, ions, or molecules) of the system. It does not include the kinetic energy arising from the motion of the system as a whole, such as its kinetic energy as it accompanies the Earth on its orbit round the Sun. That is, the internal energy is the energy 'internal' to the system. The change in internal energy is denoted by ΔU when a system changes from an initial state i with internal energy U_i to a final state f of internal energy U_f:

$$\Delta U = U_f - U_i \qquad (2A.1)$$

A convention used throughout thermodynamics is that $\Delta X = X_f - X_i$, where X is a property (a 'state function') of the system.

The internal energy is a **state function**, a property with a value that depends only on the current state of the system and is independent of how that state has been prepared. In other words, internal energy is a function of the variables that determine the current state of the system. Changing any one of the state variables, such as the pressure, may result in a change in internal energy. That the internal energy is a state function has consequences of the greatest importance (Topic 2D).

The internal energy is an extensive property of a system (a property that depends on the amount of substance present; see *The chemist's toolkit* 2 in Topic 1A) and is measured in joules ($1\,J = 1\,kg\,m^2\,s^{-2}$). The molar internal energy, U_m, is the internal energy divided by the amount of substance in a system, $U_m = U/n$; it is an intensive property (a property independent of the amount of substance) and is commonly reported in kilojoules per mole ($kJ\,mol^{-1}$).

(a) Molecular interpretation of internal energy

A molecule has a certain number of motional degrees of freedom, such as the ability to move through space (this motion is called 'translation'), rotate, or vibrate. Many physical and chemical properties depend on the energy associated with each of these modes of motion. For example, a chemical bond might break if a lot of energy becomes concentrated in it, for instance as vigorous vibration. The internal energy of a sample increases as the temperature is raised and states of higher energy become more highly populated.

The 'equipartition theorem' of classical mechanics, introduced in *The chemist's toolkit* 7, can be used to predict the contributions of each mode of motion of a molecule to the total energy of a collection of non-interacting molecules (that is, of a perfect gas, and providing quantum effects can be ignored).

The chemist's toolkit 7 The equipartition theorem

The Boltzmann distribution (see the *Prologue*) can be used to calculate the average energy associated with each mode of motion of an atom or molecule in a sample at a given temperature. However, when the temperature is so high that many energy levels are occupied, there is a much simpler way to find the average energy, through the **equipartition theorem**:

> For a sample at thermal equilibrium the average value of each quadratic contribution to the energy is $\frac{1}{2}kT$.

A 'quadratic contribution' is a term that is proportional to the square of the momentum (as in the expression for the kinetic energy, $E_k = p^2/2m$; *The chemist's toolkit* 6) or the displacement from an equilibrium position (as for the potential energy of a harmonic oscillator, $E_p = \frac{1}{2}k_f x^2$). The theorem is a conclusion from classical mechanics and for quantized systems is applicable only when the separation between the energy levels is so small compared to kT that many states are populated. Under normal conditions the equipartition theorem gives good estimates for the average energies associated with translation and rotation. However, the separation between vibrational and electronic states is typically much greater than for rotation or translation, and for these types of motion the equipartition theorem is unlikely to apply.

Brief illustration 2A.1

An atom in a gas can move in three dimensions, so its translational kinetic energy is the sum of three quadratic contributions:

$$E_{\text{trans}} = \tfrac{1}{2}mv_x^2 + \tfrac{1}{2}mv_y^2 + \tfrac{1}{2}mv_z^2$$

The equipartition theorem predicts that the average energy for each of these quadratic contributions is $\tfrac{1}{2}kT$. Thus, the average kinetic energy is $E_{\text{trans}} = 3 \times \tfrac{1}{2}kT = \tfrac{3}{2}kT$. The molar translational energy is therefore $E_{\text{trans,m}} = \tfrac{3}{2}kT \times N_A = \tfrac{3}{2}RT$. At 25°C, $RT = 2.48\,\text{kJ mol}^{-1}$, so the contribution of translation to the molar internal energy of a perfect gas is $3.72\,\text{kJ mol}^{-1}$.

The contribution to the internal energy of a collection of perfect gas molecules is independent of the volume occupied by the molecules: there are no intermolecular interactions in a perfect gas, so the distance between the molecules has no effect on the energy. That is,

> The internal energy of a perfect gas is independent of the volume it occupies.

The internal energy of interacting molecules in condensed phases also has a contribution from the potential energy of their interaction, but no simple expressions can be written down in general. Nevertheless, it remains true that as the temperature of a system is raised, the internal energy increases as the various modes of motion become more highly excited.

(b) The formulation of the First Law

It has been found experimentally that the internal energy of a system may be changed either by doing work on the system or by heating it. Whereas it might be known how the energy transfer has occurred (if a weight has been raised or lowered in the surroundings, indicating transfer of energy by doing work, or if ice has melted in the surroundings, indicating transfer of energy as heat), the system is blind to the mode employed. That is,

> Heat and work are equivalent ways of changing the internal energy of a system.

A system is like a bank: it accepts deposits in either currency (work or heat), but stores its reserves as internal energy. It is also found experimentally that if a system is isolated from its surroundings, meaning that it can exchange neither matter nor energy with its surroundings, then no change in internal energy takes place. This summary of observations is now known as the **First Law of thermodynamics** and is expressed as follows:

> The internal energy of an isolated system is constant.
>
> First Law of thermodynamics

It is not possible to use a system to do work, leave it isolated, and then come back expecting to find it restored to its original state with the same capacity for doing work. The experimental evidence for this observation is that no 'perpetual motion machine', a machine that does work without consuming fuel or using some other source of energy, has ever been built.

These remarks may be expressed symbolically as follows. If w is the work done on a system, q is the energy transferred as heat to a system, and ΔU is the resulting change in internal energy, then

$$\Delta U = q + w \qquad \text{Mathematical statement of the First Law} \qquad (2A.2)$$

Equation 2A.2 summarizes the equivalence of heat and work for bringing about changes in the internal energy and the fact that the internal energy is constant in an isolated system (for which $q = 0$ and $w = 0$). It states that the change in internal energy of a closed system is equal to the energy that passes through its boundary as heat or work. Equation 2A.2 employs the 'acquisitive convention', in which w and q are positive if energy is transferred to the system as work or heat and are negative if energy is lost from the system.[1] In other words, the flow of energy as work or heat is viewed from the system's perspective.

Brief illustration 2A.2

If an electric motor produces 15 kJ of energy each second as mechanical work and loses 2 kJ as heat to the surroundings, then the change in the internal energy of the motor each second is $\Delta U = -2\,\text{kJ} - 15\,\text{kJ} = -17\,\text{kJ}$. Suppose that, when a spring is wound, 100 J of work is done on it but 15 J escapes to the surroundings as heat. The change in internal energy of the spring is $\Delta U = 100\,\text{J} - 15\,\text{J} = +85\,\text{J}$.

A note on good practice Always include the sign of ΔU (and of ΔX in general), even if it is positive.

2A.3 Expansion work

The way is opened to powerful methods of calculation by switching attention to infinitesimal changes in the variables that describe the state of the system (such as infinitesimal change in temperature) and infinitesimal changes in the internal energy dU. Then, if the work done on a system is dw and the energy supplied to it as heat is dq, in place of eqn 2A.2, it follows that

$$dU = dq + dw \qquad (2A.3)$$

[1] Many engineering texts adopt a different convention for work: $w > 0$ if energy is used to do work in the surroundings.

The ability to use this expression depends on being able to relate dq and dw to events taking place in the surroundings.

A good starting point is a discussion of **expansion work**, the work arising from a change in volume. This type of work includes the work done by a gas as it expands and drives back the atmosphere. Many chemical reactions result in the generation of gases (for instance, the thermal decomposition of calcium carbonate or the combustion of hydrocarbons), and the thermodynamic characteristics of the reaction depend on the work that must be done to make room for the gas it has produced. The term 'expansion work' also includes work associated with negative changes of volume, that is, compression.

(a) The general expression for work

The calculation of expansion work starts from the definition in *The chemist's toolkit* 6 with the sign of the opposing force written explicitly:

$$\mathrm{d}w = -|F|\mathrm{d}z \quad \text{Work done [definition]} \quad (2A.4)$$

The negative sign implies that the internal energy of the system doing the work decreases when the system moves an object against an opposing force of magnitude $|F|$, and there are no other changes. That is, if $\mathrm{d}z$ is positive (motion to positive z), $\mathrm{d}w$ is negative, and the internal energy decreases ($\mathrm{d}U$ in eqn 2A.3 is negative provided that $\mathrm{d}q = 0$).

Now consider the arrangement shown in Fig. 2A.5, in which one wall of a system is a massless, frictionless, rigid, perfectly fitting piston of area A. If the external pressure is p_{ex}, the magnitude of the force acting on the outer face of the piston is $|F| = p_{ex}A$. The work done when the system expands through a distance $\mathrm{d}z$ against an external pressure p_{ex}, is $\mathrm{d}w = -p_{ex}A\mathrm{d}z$. The quantity $A\mathrm{d}z$ is the change in volume, $\mathrm{d}V$, in the course of the expansion. Therefore, the work done when the system expands by $\mathrm{d}V$ against a pressure p_{ex} is

$$\mathrm{d}w = -p_{ex}\mathrm{d}V \quad \text{Expansion work} \quad (2A.5a)$$

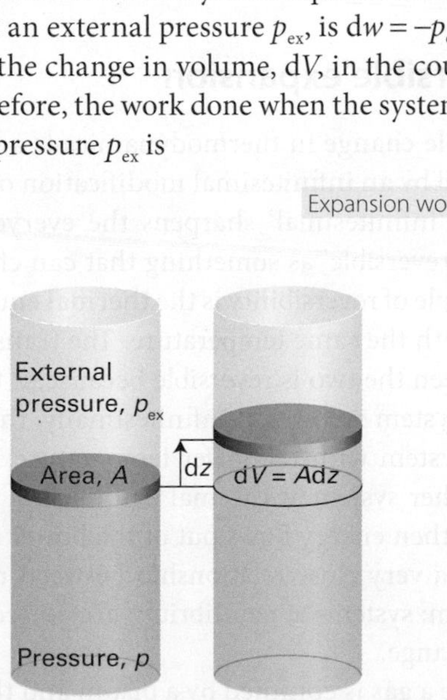

Figure 2A.5 When a piston of area A moves out through a distance $\mathrm{d}z$, it sweeps out a volume $\mathrm{d}V = A\mathrm{d}z$. The external pressure p_{ex} is equivalent to a weight pressing on the piston, and the magnitude of the force opposing expansion is $p_{ex}A$.

Table 2A.1 Varieties of work*

Type of work	dw	Comments	Units†
Expansion	$-p_{ex}\mathrm{d}V$	p_{ex} is the external pressure	Pa
		$\mathrm{d}V$ is the change in volume	m^3
Surface expansion	$\gamma\mathrm{d}\sigma$	γ is the surface tension	$N\ m^{-1}$
		$\mathrm{d}\sigma$ is the change in area	m^2
Extension	$f\mathrm{d}l$	f is the tension	N
		$\mathrm{d}l$ is the change in length	m
Electrical	$\phi\mathrm{d}Q$	ϕ is the electric potential	V
		$\mathrm{d}Q$ is the change in charge	C
	$Q\mathrm{d}\phi$	$\mathrm{d}\phi$ is the potential difference	V
		Q is the charge transferred	C

* In general, the work done on a system can be expressed in the form $\mathrm{d}w = -|F|\mathrm{d}z$, where $|F|$ is the magnitude of a 'generalized force' and $\mathrm{d}z$ is a 'generalized displacement'.

† For work in joules (J). Note that 1 N m = 1 J and 1 V C = 1 J.

To obtain the total work done when the volume changes from an initial value V_i to a final value V_f it is necessary to integrate this expression between the initial and final volumes:

$$w = -\int_{V_i}^{V_f} p_{ex}\mathrm{d}V \quad (2A.5b)$$

The force acting on the piston, $p_{ex}A$, is equivalent to the force arising from a weight that is raised as the system expands. If the system is compressed instead, then the same weight is lowered in the surroundings and eqn 2A.5b can still be used, but now $V_f < V_i$. It is important to note that it is still the external pressure that determines the magnitude of the work. This somewhat perplexing conclusion seems to be inconsistent with the fact that the gas *inside* the container is opposing the compression. However, when a gas is compressed, the ability of the *surroundings* to do work is diminished to an extent determined by the weight that is lowered, and it is this energy that is transferred into the system.

Other types of work (e.g. electrical work), which are called either **non-expansion work** or **additional work**, have analogous expressions, with each one the product of an intensive factor (the pressure, for instance) and an extensive factor (such as a change in volume). Some are collected in Table 2A.1. The present discussion focuses on how the work associated with changing the volume, the expansion work, can be extracted from eqn 2A.5b.

(b) Expansion against constant pressure

Suppose that the external pressure is constant throughout the expansion. For example, the piston might be pressed on by the atmosphere, which exerts the same pressure throughout the expansion. A chemical example of this condition is the expansion of a gas formed in a chemical reaction in a container

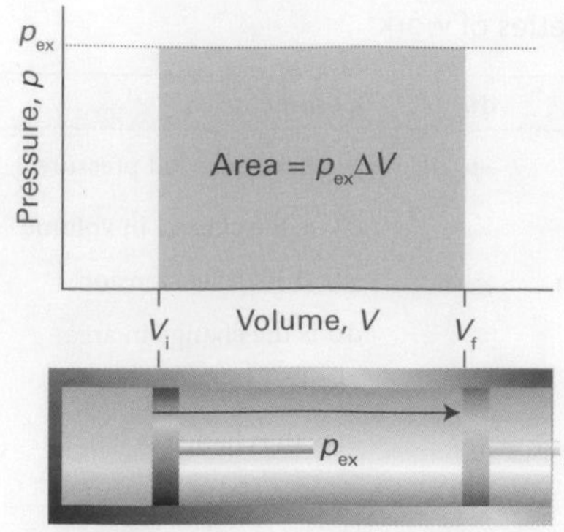

Figure 2A.6 The work done by a gas when it expands against a constant external pressure, p_{ex}, is equal to the shaded area in this example of an indicator diagram.

that can expand. Equation 2A.5b is then evaluated by taking the constant p_{ex} outside the integral:

$$w = -p_{ex}\int_{V_i}^{V_f} dV = -p_{ex}(V_f - V_i)$$

Therefore, if the change in volume is written as $\Delta V = V_f - V_i$,

$$w = -p_{ex}\Delta V \qquad \text{Expansion work [constant external pressure]} \qquad (2A.6)$$

This result is illustrated graphically in Fig. 2A.6, which makes use of the fact that the magnitude of an integral can be interpreted as an area. The magnitude of w, denoted $|w|$, is equal to the area beneath the horizontal line at $p = p_{ex}$ lying between the initial and final volumes. A p,V-graph used to illustrate expansion work is called an **indicator diagram**; James Watt first used one to indicate aspects of the operation of his steam engine.

Free expansion is expansion against zero opposing force. It occurs when $p_{ex} = 0$. According to eqn 2A.6, in this case

$$w = 0 \qquad \text{Work of free expansion} \qquad (2A.7)$$

That is, no work is done when a system expands freely. Expansion of this kind occurs when a gas expands into a vacuum.

Example 2A.1 Calculating the work of gas production

Calculate the work done when 50 g of iron reacts with hydrochloric acid to produce $FeCl_2(aq)$ and hydrogen in (a) a closed vessel of fixed volume, (b) an open beaker at 25 °C.

Collect your thoughts You need to judge the magnitude of the volume change and then to decide how the process occurs. If there is no change in volume, there is no expansion work however the process takes place. If the system expands against a constant external pressure, the work can be calculated from eqn 2A.6. A general feature of processes in which a condensed phase changes into a gas is that you can usually neglect the volume of a condensed phase relative to the volume of the gas it forms.

The solution In (a) the volume cannot change, so no expansion work is done and $w = 0$. In (b) the gas drives back the atmosphere and therefore $w = -p_{ex}\Delta V$. The initial volume can be neglected because the final volume (after the production of gas) is so much larger and $\Delta V = V_f - V_i \approx V_f = nRT/p_{ex}$, where n is the amount of H_2 produced. Therefore,

$$w = -p_{ex}\Delta V \approx -p_{ex} \times \frac{nRT}{p_{ex}} = -nRT$$

Because the reaction is $Fe(s) + 2\,HCl(aq) \rightarrow FeCl_2(aq) + H_2(g)$, 1 mol H_2 is generated when 1 mol Fe is consumed, and n can be taken as the amount of Fe atoms that react. Because the molar mass of Fe is 55.85 g mol^{-1}, it follows that

$$w = -\frac{50\,\text{g}}{55.85\,\text{g}\,\text{mol}^{-1}} \times (8.3145\,\text{J}\,\text{K}^{-1}\,\text{mol}^{-1}) \times (298\,\text{K})$$

$$\approx -2.2\,\text{kJ}$$

The system (the reaction mixture) does 2.2 kJ of work driving back the atmosphere.

Comment. The magnitude of the external pressure does not affect the final result: the lower the pressure, the larger is the volume occupied by the gas, so the effects cancel.

Self-test 2A.1 Calculate the expansion work done when 50 g of water is electrolysed under constant pressure at 25 °C.

Answer: −10 kJ

(c) Reversible expansion

A **reversible change** in thermodynamics is a change that can be reversed by an infinitesimal modification of a variable. The key word 'infinitesimal' sharpens the everyday meaning of the word 'reversible' as something that can change direction. One example of reversibility is the **thermal equilibrium** of two systems with the same temperature. The transfer of energy as heat between the two is reversible because, if the temperature of either system is lowered infinitesimally, then energy flows into the system with the lower temperature. If the temperature of either system at thermal equilibrium is raised infinitesimally, then energy flows out of the hotter system. There is obviously a very close relationship between reversibility and equilibrium: systems at equilibrium are poised to undergo reversible change.

Suppose a gas is confined by a piston and that the external pressure, p_{ex}, is set equal to the pressure, p, of the confined gas. Such a system is in **mechanical equilibrium** with its surroundings because an infinitesimal change in the external pressure in either direction causes changes in volume in opposite directions. If the external pressure is reduced infinitesimally, the gas expands slightly. If the external pressure is increased infinitesimally, the gas contracts slightly. In either case the change is reversible in the thermodynamic sense. If, on the other hand, the external pressure is measurably greater than the internal pressure, then decreasing p_{ex} infinitesimally will not decrease it below the pressure of the gas, so will not change the direction of the process. Such a system is not in mechanical equilibrium with its surroundings and the compression is thermodynamically irreversible.

To achieve reversible expansion p_{ex} is set equal to p at each stage of the expansion. In practice, this equalization could be achieved by gradually removing weights from the piston so that the downward force due to the weights always matches the changing upward force due to the pressure of the gas or by gradually adjusting the external pressure to match the pressure of the expanding gas. When $p_{ex} = p$, eqn 2A.5a becomes

$$dw = -p_{ex}dV = -pdV \qquad \text{Reversible expansion work} \qquad (2A.8a)$$

Although the pressure inside the system appears in this expression for the work, it does so only because p_{ex} has been arranged to be equal to p to ensure reversibility. The total work of reversible expansion from an initial volume V_i to a final volume V_f is therefore

$$w = -\int_{V_i}^{V_f} p dV \qquad (2A.8b)$$

The integral can be evaluated once it is known how the pressure of the confined gas depends on its volume. Equation 2A.8b is the link with the material covered in Focus 1 because, if the equation of state of the gas is known, p can be expressed in terms of V and the integral can be evaluated.

(d) Isothermal reversible expansion of a perfect gas

Consider the isothermal reversible expansion of a perfect gas. The expansion is made isothermal by keeping the system in thermal contact with its unchanging surroundings (which may be a constant-temperature bath). Because the equation of state is $pV = nRT$, at each stage $p = nRT/V$, with V the volume at that stage of the expansion. The temperature T is constant in an isothermal expansion, so (together with n and R) it may be taken outside the integral. It follows that the work of isothermal reversible expansion of a perfect gas from V_i to V_f at a temperature T is

$$w = -nRT \overbrace{\int_{V_i}^{V_f} \frac{dV}{V}}^{\text{Integral A.2}} = -nRT \ln\frac{V_f}{V_i} \qquad \text{Work of isothermal reversible expansion [perfect gas]} \qquad (2A.9)$$

Brief Illustration 2A.3

When a sample of 1.00 mol Ar, regarded here as a perfect gas, undergoes an isothermal reversible expansion at 20.0 °C from $10.0\,dm^3$ to $30.0\,dm^3$ the work done is

$$w = -(1.00\,mol)\times(8.3145\,J\,K^{-1}\,mol^{-1})\times(293.2\,K)\ln\frac{30.0\,dm^3}{10.0\,dm^3}$$
$$= -2.68\,kJ$$

When the final volume is greater than the initial volume, as in an expansion, the logarithm in eqn 2A.9 is positive and hence $w < 0$. In this case, the system has done work on the surroundings and there is a corresponding negative contribution to its internal energy. (Note the cautious language: as seen later, there is a compensating influx of energy as heat, so overall the internal energy is constant for the isothermal expansion of a perfect gas.) The equations also show that more work is done for a given change of volume when the temperature is increased: at a higher temperature the greater pressure of the confined gas needs a higher opposing pressure to ensure reversibility and the work done is correspondingly greater.

The result of the calculation can be illustrated by an indicator diagram in which the magnitude of the work done is equal to the area under the isotherm $p = nRT/V$ (Fig. 2A.7). Superimposed on the diagram is the rectangular area obtained for irreversible expansion against constant external pressure fixed at the same final value as that reached in the reversible expansion. More work is obtained when the expansion is reversible (the area is greater) because matching the external pressure to the internal pressure at each stage of the process ensures that none of the pushing power of the system is wasted. It is not possible to obtain more work than that for

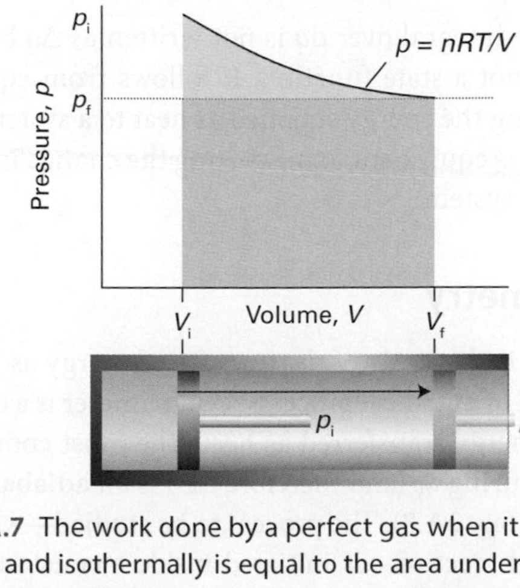

Figure 2A.7 The work done by a perfect gas when it expands reversibly and isothermally is equal to the area under the isotherm $p = nRT/V$. The work done during the irreversible expansion against the same final pressure is equal to the rectangular area shown slightly darker. Note that the reversible work done is greater than the irreversible work done.

the reversible process because increasing the external pressure even infinitesimally at any stage results in compression. It can be inferred from this discussion that, because some pushing power is wasted when $p > p_{ex}$, the maximum work available from a system operating between specified initial and final states is obtained when the change takes place reversibly.

2A.4 Heat transactions

In general, the change in internal energy of a system is

$$dU = dq + dw_{exp} + dw_{add} \quad (2A.10)$$

where dw_{add} is work in addition ('add' for additional) to the expansion work, dw_{exp}. For instance, dw_{add} might be the electrical work of driving a current of electrons through a circuit. A system kept at constant volume can do no expansion work, so in that case $dw_{exp} = 0$. If the system is also incapable of doing any other kind of work (if it is not, for instance, an electrochemical cell connected to an electric motor), then $dw_{add} = 0$ too. Under these circumstances:

$$dU = dq \quad \text{Heat transferred at constant volume} \quad (2A.11a)$$

This relation can also be expressed as $dU = dq_V$, where the subscript implies the constraint of constant volume. For a measurable change between states i and f along a path at constant volume,

$$\overbrace{\int_i^f dU}^{U_f - U_i} = \overbrace{\int_i^f dq_V}^{q_V}$$

which is summarized as

$$\Delta U = q_V \quad (2A.11b)$$

Note that the integral over dq is not written as Δq because q, unlike U, is not a state function. It follows from eqn 2A.11b that measuring the energy supplied as heat to a system at constant volume is equivalent to measuring the change in internal energy of the system.

(a) Calorimetry

Calorimetry is the study of the transfer of energy as heat during a physical or chemical process. A **calorimeter** is a device for measuring energy transferred as heat. The most common device for measuring q_V (and therefore ΔU) is an **adiabatic bomb calorimeter** (Fig. 2A.8). The process to be studied—which may be a chemical reaction—is initiated inside a constant-volume container, the 'bomb'. The bomb is immersed in a stirred water bath, and the whole device is the calorimeter. The calorimeter is also immersed in an outer water bath. The water in the calorimeter and of the outer bath are both monitored and adjusted to the same temperature. This arrangement ensures that there is no net loss of heat from the calorimeter to the surroundings (the bath) and hence that the calorimeter is adiabatic.

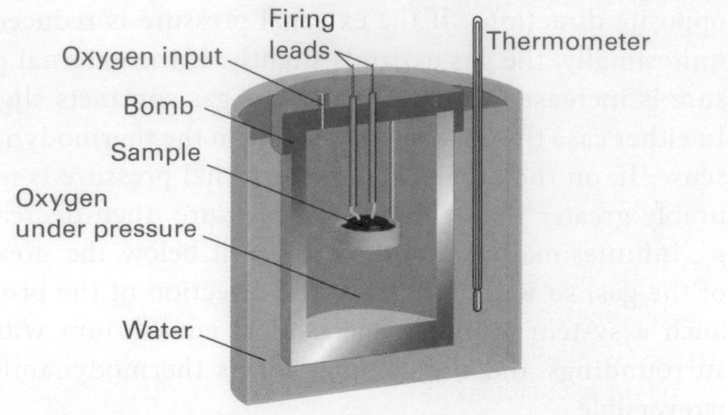

Figure 2A.8 A constant-volume bomb calorimeter. The 'bomb' is the central vessel, which is strong enough to withstand high pressures. The calorimeter is the entire assembly shown here. To ensure adiabaticity, the calorimeter is immersed in a water bath with a temperature continuously readjusted to that of the calorimeter at each stage of the combustion.

The change in temperature, ΔT, of the calorimeter is proportional to the energy that the reaction releases or absorbs as heat. Therefore, q_V and hence ΔU can be determined by measuring ΔT. The conversion of ΔT to q_V is best achieved by calibrating the calorimeter using a process of known output and determining the **calorimeter constant**, the constant C in the relation

$$q = C\Delta T \quad (2A.12)$$

The calorimeter constant may be measured electrically by passing a constant current, I, from a source of known potential difference, $\Delta\phi$, through a heater for a known period of time, t, for then (*The chemist's toolkit* 8)

$$q = It\Delta\phi \quad (2A.13)$$

Brief illustration 2A.4

If a current of 10.0 A from a 12 V supply is passed for 300 s, then from eqn 2A.13 the energy supplied as heat is

$$q = (10.0\,\text{A}) \times (300\,\text{s}) \times (12\,\text{V}) = 3.6 \times 10^4\,\text{A V s} = 36\,\text{kJ}$$

The result in joules is obtained by using $1\,\text{A V s} = 1\,(\text{C s}^{-1})\,\text{V s} = 1\,\text{C V} = 1\,\text{J}$. If the observed rise in temperature is 5.5 K, then the calorimeter constant is $C = (36\,\text{kJ})/(5.5\,\text{K}) = 6.5\,\text{kJ K}^{-1}$.

Alternatively, C may be determined by burning a known mass of substance (benzoic acid is often used) that has a known heat output. With C known, it is simple to interpret an observed temperature rise as a release of energy as heat.

The chemist's toolkit 8 Electrical charge, current, power, and energy

Electrical charge, Q, is measured in *coulombs*, C. The fundamental charge, e, the magnitude of charge carried by a single electron or proton, is approximately 1.6×10^{-19} C. The motion of charge gives rise to an **electric current**, I, measured in coulombs per second, or *amperes*, A, where $1\,A = 1\,C\,s^{-1}$. If the electric charge is that of electrons (as it is for the current in a metal), then a current of 1 A represents the flow of 6×10^{18} electrons (10 μmol e^-) per second.

When a current I flows through a potential difference $\Delta\phi$ (measured in volts, V, with $1\,V = 1\,J\,C^{-1} = 1\,W\,A^{-1}$), the power, P, is

$$P = I\Delta\phi$$

It follows that if a constant current flows for a period t the energy supplied is

$$E = Pt = It\Delta\phi$$

Because $1\,A\,V\,s = 1\,(C\,s^{-1})\,V\,s = 1\,C\,V = 1\,J$, the energy is obtained in joules with the current in amperes, the potential difference in volts, and the time in seconds. That energy may be supplied as either work (to drive a motor) or as heat (through a 'heater'). In the latter case

$$q = It\Delta\phi$$

(b) Heat capacity

The internal energy of a system increases when its temperature is raised. This increase depends on the conditions under which the heating takes place. Suppose the system has a constant volume. If the internal energy is plotted against temperature, then a curve like that in Fig. 2A.9 may be obtained. The slope of the tangent to the curve at any temperature is called the **heat capacity** of the system at that temperature. The **heat capacity at constant volume** is denoted C_V and is defined formally as

$$C_V = \left(\frac{\partial U}{\partial T}\right)_V \qquad \text{Heat capacity at constant volume [definition]} \qquad (2A.14)$$

(Partial derivatives and the notation used here are reviewed in *The chemist's toolkit* 9.) The internal energy varies with the temperature and the volume of the sample, but here only its variation with the temperature is important, because the volume is held constant (Fig. 2A.10), as signified by the subscript V.

Brief illustration 2A.5

In *Brief illustration* 2A.1 it is shown that the translational contribution to the molar internal energy of a perfect monatomic gas is $\frac{3}{2}RT$. Because this is the only contribution to the internal energy, $U_m(T) = \frac{3}{2}RT$. It follows from eqn 2A.14 that

$$C_{V,m} = \frac{\partial}{\partial T}\left\{\tfrac{3}{2}RT\right\} = \tfrac{3}{2}R$$

The numerical value is 12.47 J K^{-1} mol^{-1}.

Heat capacities are extensive properties: 100 g of water, for instance, has 100 times the heat capacity of 1 g of water (and therefore requires 100 times the energy as heat to bring about the same rise in temperature). The **molar heat capacity at constant volume**, $C_{V,m} = C_V/n$, is the heat capacity per mole of substance, and is an intensive property (all molar quantities are intensive). For certain applications it is useful to know the

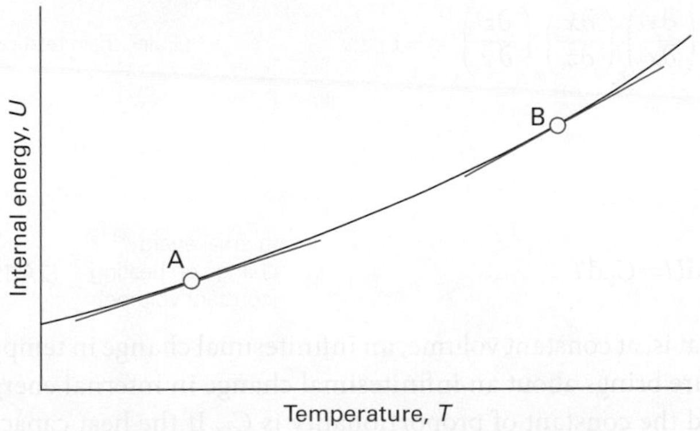

Figure 2A.9 The internal energy of a system increases as the temperature is raised; this graph shows its variation as the system is heated at constant volume. The slope of the tangent to the curve at any temperature is the heat capacity at constant volume at that temperature. Note that, for the system illustrated, the heat capacity is greater at B than at A.

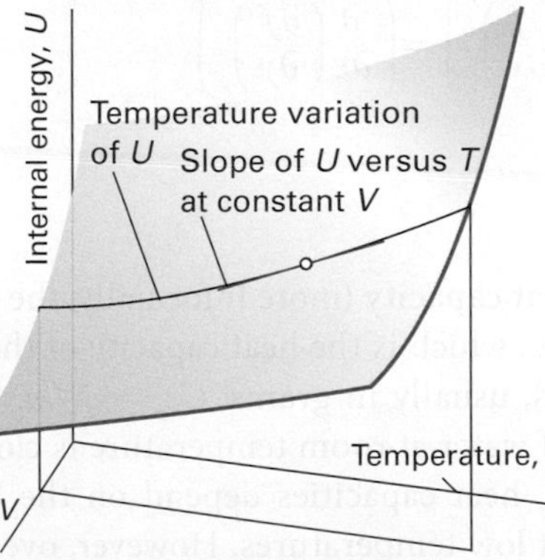

Figure 2A.10 The internal energy of a system varies with volume and temperature, perhaps as shown here by the surface. The variation of the internal energy with temperature at one particular constant volume is illustrated by the curve drawn parallel to the temperature axis. The slope of this curve at any point is the partial derivative $(\partial U/\partial T)_V$.

The chemist's toolkit 9 Partial derivatives

A **partial derivative** of a function of more than one variable, such as $f(x,y)$, is the slope of the function with respect to one of the variables, all the other variables being held constant (Sketch 1). Although a partial derivative shows how a function changes when one variable changes, it may be used to determine how the function changes when more than one variable changes by an infinitesimal amount. Thus, if f is a function of x and y, then when x and y change by dx and dy, respectively, f changes by

$$df = \left(\frac{\partial f}{\partial x}\right)_y dx + \left(\frac{\partial f}{\partial y}\right)_x dy$$

where the symbol ∂ ('curly d') is used (instead of d) to denote a partial derivative and the subscript on the parentheses indicates which variable is being held constant.

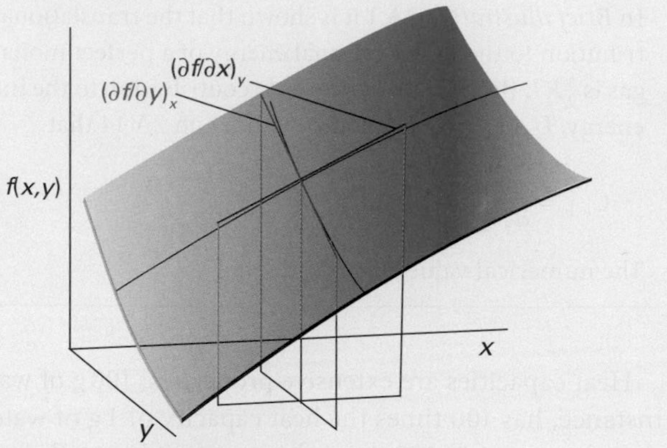

Sketch 1

The quantity df is also called the **differential** of f. Successive partial derivatives may be taken in any order:

$$\left(\frac{\partial}{\partial y}\left(\frac{\partial f}{\partial x}\right)_y\right)_x = \left(\frac{\partial}{\partial x}\left(\frac{\partial f}{\partial y}\right)_x\right)_y$$

For example, suppose that $f(x,y) = ax^3y + by^2$ (the function plotted in Sketch 1) then

$$\left(\frac{\partial f}{\partial x}\right)_y = 3ax^2y \qquad \left(\frac{\partial f}{\partial y}\right)_x = ax^3 + 2by$$

Then, when x and y undergo infinitesimal changes, f changes by

$$df = 3ax^2y\,dx + (ax^3 + 2by)\,dy$$

To verify that the order of taking the second partial derivative is irrelevant, form

$$\left(\frac{\partial}{\partial y}\left(\frac{\partial f}{\partial x}\right)_y\right)_x = \left(\frac{\partial(3ax^2y)}{\partial y}\right)_x = 3ax^2$$

$$\left(\frac{\partial}{\partial x}\left(\frac{\partial f}{\partial y}\right)_x\right)_y = \left(\frac{\partial(ax^3+2by)}{\partial x}\right)_y = 3ax^2$$

Now suppose that z is a variable on which x and y depend (for example, x, y, and z might correspond to p, V, and T). The following relations then apply:

Relation 1. When x is changed at constant z:

$$\left(\frac{\partial f}{\partial x}\right)_z = \left(\frac{\partial f}{\partial x}\right)_y + \left(\frac{\partial f}{\partial y}\right)_x\left(\frac{\partial y}{\partial x}\right)_z$$

Relation 2

$$\left(\frac{\partial y}{\partial x}\right)_z = \frac{1}{(\partial x/\partial y)_z}$$

Relation 3

$$\left(\frac{\partial x}{\partial y}\right)_z = -\left(\frac{\partial x}{\partial z}\right)_y\left(\frac{\partial z}{\partial y}\right)_x$$

Combining Relations 2 and 3 results in the **Euler chain relation**:

$$\left(\frac{\partial y}{\partial x}\right)_z\left(\frac{\partial x}{\partial z}\right)_y\left(\frac{\partial z}{\partial y}\right)_x = -1 \qquad \text{Euler chain relation}$$

specific heat capacity (more informally, the 'specific heat') of a substance, which is the heat capacity of the sample divided by its mass, usually in grams: $C_{V,s} = C_V/m$. The specific heat capacity of water at room temperature is close to $4.2\,\mathrm{J\,K^{-1}\,g^{-1}}$. In general, heat capacities depend on the temperature and decrease at low temperatures. However, over small ranges of temperature at and above room temperature, the variation is quite small and for approximate calculations heat capacities can be treated as almost independent of temperature.

The heat capacity is used to relate a change in internal energy to a change in temperature of a constant-volume system. It follows from eqn 2A.14 that

$$dU = C_V dT \qquad \text{Internal energy change on heating [constant volume]} \qquad (2A.15a)$$

That is, at constant volume, an infinitesimal change in temperature brings about an infinitesimal change in internal energy, and the constant of proportionality is C_V. If the heat capacity is independent of temperature over the range of temperatures of interest, then

$$\Delta U = \int_{T_1}^{T_2} C_V\,dT = C_V\int_{T_1}^{T_2} dT = C_V\overbrace{(T_2 - T_1)}^{\Delta T}$$

A measurable change of temperature, ΔT, brings about a measurable change in internal energy, ΔU, with

$$\Delta U = C_V \Delta T$$ Internal energy change on heating [constant volume] (2A.15b)

Because a change in internal energy can be identified with the heat supplied at constant volume (eqn 2A.11b), the last equation can also be written as

$$q_V = C_V \Delta T \quad (2A.16)$$

This relation provides a simple way of measuring the heat capacity of a sample: a measured quantity of energy is transferred as heat to the sample (by electrical heating, for example) under constant volume conditions and the resulting increase in temperature is monitored. The ratio of the energy transferred as heat to the temperature rise it causes ($q_V/\Delta T$) is the constant-volume heat capacity of the sample. A large heat capacity implies that, for a given quantity of energy transferred as heat, there will be only a small increase in temperature (the sample has a large capacity for heat).

Brief illustration 2A.6

Suppose a 55 W electric heater immersed in a gas in a constant-volume adiabatic container was on for 120 s and it was found that the temperature of the gas rose by 5.0 °C (an increase equivalent to 5.0 K). The heat supplied is (55 W) × (120 s) = 6.6 kJ (with 1 J = 1 W s), so the heat capacity of the sample is

$$C_V = \frac{6.6\,\text{kJ}}{5.0\,\text{K}} = 1.3\,\text{kJ}\,\text{K}^{-1}$$

Checklist of concepts

☐ 1. **Work** is the process of achieving motion against an opposing force.

☐ 2. **Energy** is the capacity to do work.

☐ 3. An **exothermic process** is a process that releases energy as heat.

☐ 4. An **endothermic process** is a process in which energy is acquired as heat.

☐ 5. **Heat** is the process of transferring energy as a result of a temperature difference.

☐ 6. In molecular terms, work is the transfer of energy that makes use of organized motion of atoms in the surroundings and heat is the transfer of energy that makes use of their disorderly motion.

☐ 7. **Internal energy**, the total energy of a system, is a state function.

☐ 8. The internal energy increases as the temperature is raised.

☐ 9. The **equipartition theorem** can be used to estimate the contribution to the internal energy of each classically behaving mode of motion.

☐ 10. The **First Law** states that the internal energy of an isolated system is constant.

☐ 11. Free expansion (expansion against zero pressure) does no work.

☐ 12. A **reversible change** is a change that can be reversed by an infinitesimal change in a variable.

☐ 13. To achieve **reversible expansion**, the external pressure is matched at every stage to the pressure of the system.

☐ 14. The energy transferred as heat at constant volume is equal to the change in internal energy of the system.

☐ 15. **Calorimetry** is the measurement of heat transactions.

Checklist of equations

Property	Equation	Comment	Equation number
First Law of thermodynamics	$\Delta U = q + w$	Convention	2A.2
Work of expansion	$dw = -p_{ex}dV$		2A.5a
Work of expansion against a constant external pressure	$w = -p_{ex}\Delta V$	$p_{ex} = 0$ for free expansion	2A.6
Reversible work of expansion of a gas	$w = -nRT\ln(V_f/V_i)$	Isothermal, perfect gas	2A.9
Internal energy change	$\Delta U = q_V$	Constant volume, no other forms of work	2A.11b
Electrical heating	$q = It\Delta\phi$		2A.13
Heat capacity at constant volume	$C_V = (\partial U/\partial T)_V$	Definition	2A.14

TOPIC 2B Enthalpy

➤ **Why do you need to know this material?**

The concept of enthalpy is central to many thermodynamic discussions about processes, such as physical transformations and chemical reactions taking place under conditions of constant pressure.

➤ **What is the key idea?**

A change in enthalpy is equal to the energy transferred as heat at constant pressure.

➤ **What do you need to know already?**

This Topic makes use of the discussion of internal energy (Topic 2A) and draws on some aspects of perfect gases (Topic 1A).

The change in internal energy is not equal to the energy transferred as heat when the system is free to change its volume, such as when it is able to expand or contract under conditions of constant pressure. Under these circumstances some of the energy supplied as heat to the system is returned to the surroundings as expansion work (Fig. 2B.1), so $\mathrm{d}U$ is less than $\mathrm{d}q$. In this case the energy supplied as heat at constant pressure is equal to the change in another thermodynamic property of the system, the 'enthalpy'.

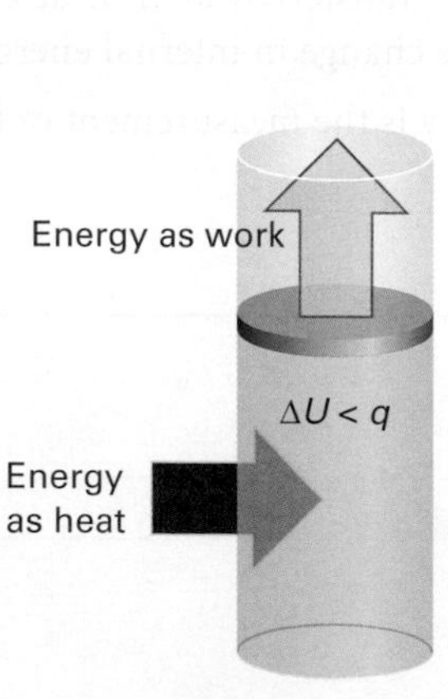

Figure 2B.1 When a system is subjected to constant pressure and is free to change its volume, some of the energy supplied as heat may escape back into the surroundings as work. In such a case, the change in internal energy is smaller than the energy supplied as heat.

2B.1 The definition of enthalpy

The **enthalpy**, H, is defined as

$$H = U + pV \quad \text{Enthalpy [definition]} \quad (2B.1)$$

where p is the pressure of the system and V is its volume. Because U, p, and V are all state functions, the enthalpy is a state function too. As is true of any state function, the change in enthalpy, ΔH, between any pair of initial and final states is independent of the path between them.

(a) Enthalpy change and heat transfer

An important consequence of the definition of enthalpy in eqn 2B.1 is that it can be shown that the change in enthalpy is equal to the energy supplied as heat under conditions of constant pressure.

How is that done? 2B.1 Deriving the relation between enthalpy change and heat transfer at constant pressure

In a typical thermodynamic derivation, as here, a common way to proceed is to introduce successive definitions of the quantities of interest and then apply the appropriate constraints.

Step 1 *Write an expression for $H + \mathrm{d}H$ in terms of the definition of H*

For a general infinitesimal change in the state of the system, U changes to $U + \mathrm{d}U$, p changes to $p + \mathrm{d}p$, and V changes to $V + \mathrm{d}V$, so from the definition in eqn 2B.1, H changes by $\mathrm{d}H$ to

$$H + \mathrm{d}H = (U + \mathrm{d}U) + (p + \mathrm{d}p)(V + \mathrm{d}V)$$
$$= U + \mathrm{d}U + pV + p\mathrm{d}V + V\mathrm{d}p + \mathrm{d}p\mathrm{d}V$$

The last term is the product of two infinitesimally small quantities and can be neglected. Now recognize that $U + pV = H$ on the right (in blue), so

$$H + \mathrm{d}H = H + \mathrm{d}U + p\mathrm{d}V + V\mathrm{d}p$$

and hence

$$\mathrm{d}H = \mathrm{d}U + p\mathrm{d}V + V\mathrm{d}p$$

Step 2 *Introduce the definition of $\mathrm{d}U$*

Because $\mathrm{d}U = \mathrm{d}q + \mathrm{d}w$ this expression becomes

$$\mathrm{d}H = \mathrm{d}q + \mathrm{d}w + p\mathrm{d}V + V\mathrm{d}p$$

Step 3 *Apply the appropriate constraints*

If the system is in mechanical equilibrium with its surroundings at a pressure p and does only expansion work, then $dw = -pdV$, which cancels the other pdV term, leaving

$$dH = dq + Vdp$$

At constant pressure, $dp = 0$, so

$$dH = dq \text{ (at constant pressure, no additional work)}$$

The constraint of constant pressure is denoted by a p, so this equation can be written

$$dH = dq_p \quad \text{Heat transferred at constant pressure [infinitesimal change]} \quad (2B.2a)$$

This equation states that, provided there is no additional (non-expansion) work done, *the change in enthalpy is equal to the energy supplied as heat at constant pressure*.

Step 4 *Evaluate ΔH by integration*

For a measurable change between states i and f along a path at constant pressure, the preceding expression is integrated as follows

$$\overbrace{\int_i^f dH}^{H_f - H_i} = \overbrace{\int_i^f dq}^{q_p}$$

Note that the integral over dq is not written as Δq because q, unlike H, is not a state function and $q_f - q_i$ is meaningless. The final result is

$$\Delta H = q_p \quad \text{Heat transferred at constant pressure [measurable change]} \quad (2B.2b)$$

Brief illustration 2B.1

Water is heated to boiling under a pressure of 1.0 atm. When an electric current of 0.50 A from a 12 V supply is passed for 300 s through a resistance in thermal contact with the water, it is found that 0.798 g of water is vaporized. The enthalpy change is

$$\Delta H = q_p = It\Delta\phi = (0.50\,\text{A}) \times (300\,\text{s}) \times (12\,\text{V})$$
$$= 0.50 \times 300\,\text{J} \times 12$$

where 1 A V s = 1 J. Because 0.798 g of water is (0.798 g)/(18.02 g mol^{-1}) = (0.798/18.02) mol H_2O, the enthalpy of vaporization per mole of H_2O is

$$\Delta H_m = \frac{0.50\times12\times300\,\text{J}}{(0.798/18.02)\,\text{mol}} = +41\,\text{kJ}\,\text{mol}^{-1}$$

(b) Calorimetry

An enthalpy change can be measured calorimetrically by monitoring the temperature change that accompanies a physical or chemical change at constant pressure. A calorimeter for

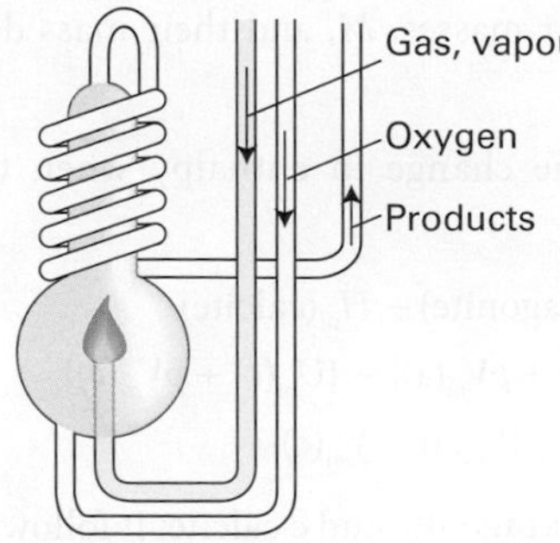

Figure 2B.2 A constant-pressure flame calorimeter consists of this component immersed in a stirred water bath. Combustion occurs as a known amount of reactant is passed through to fuel the flame, and the rise of temperature is monitored.

studying processes at constant pressure is called an **isobaric calorimeter**. A simple example is a thermally insulated vessel open to the atmosphere: the energy released as heat in the reaction is monitored by measuring the change in temperature of the contents. For a combustion reaction an **adiabatic flame calorimeter** may be used to measure ΔT when a given amount of substance burns in a supply of oxygen (Fig. 2B.2). The most sophisticated way to measure enthalpy changes, however, is to use a *differential scanning calorimeter* (DSC), as explained in Topic 2C. Changes in enthalpy and internal energy may also be measured by non-calorimetric methods (Topic 6C).

One route to ΔH is to measure the internal energy change by using a bomb calorimeter (Topic 2A), and then to convert ΔU to ΔH. Because solids and liquids have small molar volumes, for them pV_m is so small that the molar enthalpy and molar internal energy are almost identical ($H_m = U_m + pV_m \approx U_m$). Consequently, if a process involves only solids or liquids, the values of ΔH and ΔU are almost identical. Physically, such processes are accompanied by a very small change in volume; the system does negligible work on the surroundings when the process occurs, so the energy supplied as heat stays entirely within the system.

Example 2B.1 Relating ΔH and ΔU

The change in molar internal energy when $CaCO_3$(s) as calcite converts to its polymorph aragonite is +0.21 kJ mol^{-1}. Calculate the difference between the molar enthalpy and internal energy changes when the pressure is 1.0 bar. The mass densities of the polymorphs are 2.71 g cm^{-3} (calcite) and 2.93 g cm^{-3} (aragonite).

Collect your thoughts The starting point for the calculation is the relation between the enthalpy of a substance and its internal energy (eqn 2B.1). You need to express the difference between the two quantities in terms of the pressure and the difference of their molar volumes. The latter can be calculated from their molar masses, M, and their mass densities, ρ, by using $\rho = M/V_m$.

The solution The change in enthalpy when the transition occurs is

$$\begin{aligned}\Delta H_m &= H_m(\text{aragonite}) - H_m(\text{calcite}) \\ &= \{U_m(a) + pV_m(a)\} - \{U_m(c) + pV_m(c)\} \\ &= \Delta U_m + p\{V_m(a) - V_m(c)\}\end{aligned}$$

where a denotes aragonite and c calcite. It follows by substituting $V_m = M/\rho$ that

$$\Delta H_m - \Delta U_m = pM\left(\frac{1}{\rho(a)} - \frac{1}{\rho(c)}\right)$$

Substitution of the data, using $M = 100.09\,\text{g mol}^{-1}$, gives

$$\begin{aligned}\Delta H_m - \Delta U_m &= (1.0\times10^5\,\text{Pa})\times(100.09\,\text{g mol}^{-1}) \\ &\quad\times\left(\frac{1}{2.93\,\text{g cm}^{-3}} - \frac{1}{2.71\,\text{g cm}^{-3}}\right) \\ &= -2.8\times10^5\,\text{Pa cm}^3\,\text{mol}^{-1} = -0.28\,\text{Pa m}^3\,\text{mol}^{-1}\end{aligned}$$

Hence (because $1\,\text{Pa m}^3 = 1\,\text{J}$), $\Delta H_m - \Delta U_m = -0.28\,\text{J mol}^{-1}$, which is only 0.1 per cent of the value of ΔU_m.

Comment. It is usually justifiable to ignore the difference between the molar enthalpy and internal energy of condensed phases except at very high pressures when $p\Delta V_m$ is no longer negligible.

Self-test 2B.1 Calculate the difference between ΔH and ΔU when 1.0 mol Sn(s, grey) of density $5.75\,\text{g cm}^{-3}$ changes to Sn(s, white) of density $7.31\,\text{g cm}^{-3}$ at 10.0 bar.

Answer: $\Delta H - \Delta U = -4.4\,\text{J}$

In contrast to processes involving condensed phases, the values of the changes in internal energy and enthalpy might differ significantly for processes involving gases. The enthalpy of a perfect gas is related to its internal energy by using $pV = nRT$ in the definition of H:

$$H = U + pV = U + nRT \qquad (2B.3)$$

This relation implies that the change of enthalpy in a reaction that produces or consumes gas under isothermal conditions is

$$\Delta H = \Delta U + \Delta n_g RT \qquad \text{Relation between } \Delta H \text{ and } \Delta U \text{ [isothermal process, perfect gas]} \qquad (2B.4)$$

where Δn_g is the change in the amount of gas molecules in the reaction. For molar quantities, replace Δn_g by $\Delta \nu_g$.

Brief illustration 2B.2

In the reaction $2\,H_2(g) + O_2(g) \rightarrow 2\,H_2O(l)$, 3 mol of gas-phase molecules are replaced by 2 mol of liquid-phase molecules, so $\Delta n_g = -3\,\text{mol}$ and $\Delta \nu_g = -3$. Therefore, at 298 K, when $RT = 2.5\,\text{kJ mol}^{-1}$, the enthalpy and internal energy changes taking place in the system are related by

$$\Delta H_m - \Delta U_m = (-3)\times RT \approx -7.5\,\text{kJ mol}^{-1}$$

Note that the difference is expressed in kilojoules, not joules as in Example 2B.1. The enthalpy change is smaller than the change in internal energy because, although energy escapes from the system as heat when the reaction occurs, the system contracts as the liquid is formed, so energy is restored to it as work from the surroundings.

2B.2 The variation of enthalpy with temperature

The enthalpy of a substance increases as its temperature is raised. The reason is the same as for the internal energy: molecules are excited to states of higher energy so their total energy increases. The relation between the increase in enthalpy and the increase in temperature depends on the conditions (e.g. whether the pressure or the volume is constant).

(a) Heat capacity at constant pressure

The most frequently encountered condition in chemistry is constant pressure. The slope of the tangent to a plot of enthalpy against temperature at constant pressure is called the **heat capacity at constant pressure** (or *isobaric heat capacity*), C_p, at a given temperature (Fig. 2B.3). More formally:

$$C_p = \left(\frac{\partial H}{\partial T}\right)_p \qquad \text{Heat capacity at constant pressure [definition]} \qquad (2B.5)$$

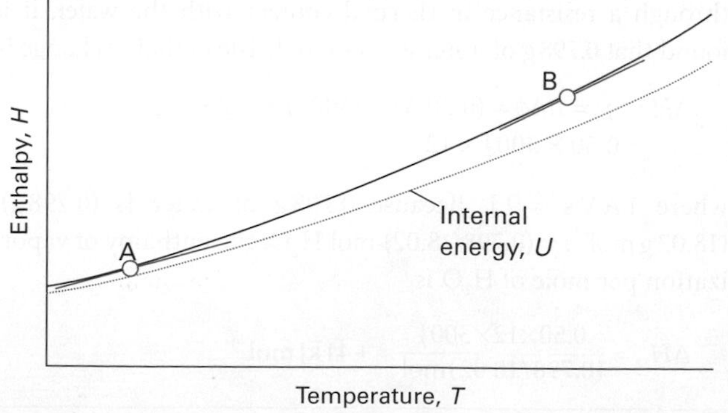

Figure 2B.3 The constant-pressure heat capacity at a particular temperature is the slope of the tangent to a curve of the enthalpy of a system plotted against temperature (at constant pressure). For gases, at a given temperature the slope of enthalpy versus temperature is steeper than that of internal energy versus temperature, and $C_{p,m}$ is larger than $C_{V,m}$.

The heat capacity at constant pressure is the analogue of the heat capacity at constant volume (Topic 2A) and is an extensive property. The **molar heat capacity at constant pressure**, $C_{p,m}$, is the heat capacity per mole of substance; it is an intensive property.

The heat capacity at constant pressure relates the change in enthalpy to a change in temperature. For infinitesimal changes of temperature, eqn 2B.5 implies that

$$dH = C_p dT \text{ (at constant pressure)} \quad (2B.6a)$$

If the heat capacity is constant over the range of temperatures of interest, then for a measurable increase in temperature

$$\Delta H = \int_{T_1}^{T_2} C_p \mathrm{d}T = C_p \int_{T_1}^{T_2} \mathrm{d}T = C_p \overbrace{(T_2 - T_1)}^{\Delta T}$$

which can be summarized as

$$\Delta H = C_p \Delta T \text{ (at constant pressure)} \quad (2B.6b)$$

Because a change in enthalpy can be equated to the energy supplied as heat at constant pressure, the practical form of this equation is

$$q_p = C_p \Delta T \quad (2B.7)$$

This expression shows how to measure the constant-pressure heat capacity of a sample: a measured quantity of energy is supplied as heat under conditions of constant pressure (as in a sample exposed to the atmosphere and free to expand), and the temperature rise is monitored.

The variation of heat capacity with temperature can sometimes be ignored if the temperature range is small; this is an excellent approximation for a monatomic perfect gas (for instance, one of the noble gases at low pressure). However, when it is necessary to take the variation into account for other substances, a convenient approximate empirical expression is

$$C_{p,m} = a + bT + \frac{c}{T^2} \quad (2B.8)$$

The empirical parameters a, b, and c are independent of temperature (Table 2B.1) and are found by fitting this expression to experimental data.

Table 2B.1 Temperature variation of molar heat capacities, $C_{p,m}/(\mathrm{J\,K^{-1}\,mol^{-1}}) = a + bT + c/T^2$*

	a	$b/(10^{-3}\,\mathrm{K^{-1}})$	$c/(10^5\,\mathrm{K^2})$
C(s, graphite)	16.86	4.77	−8.54
CO_2(g)	44.22	8.79	−8.62
H_2O(l)	75.29	0	0
N_2(g)	28.58	3.77	−0.50

* More values are given in the *Resource section*.

Example 2B.2 Evaluating an increase in enthalpy with temperature

What is the change in molar enthalpy of N_2 when it is heated from 25 °C to 100 °C? Use the heat capacity information in Table 2B.1.

Collect your thoughts The heat capacity of N_2 changes with temperature significantly in this range, so you cannot use eqn 2B.6b (which assumes that the heat capacity of the substance is constant). Therefore, use eqn 2B.6a, substitute eqn 2B.8 for the temperature dependence of the heat capacity, and integrate the resulting expression from 25 °C (298 K) to 100 °C (373 K).

The solution For convenience, denote the two temperatures T_1 (298 K) and T_2 (373 K). The required relation is

$$\int_{H_m(T_1)}^{H_m(T_2)} \mathrm{d}H_m = \int_{T_1}^{T_2} \left(a + bT + \frac{c}{T^2} \right) \mathrm{d}T$$

By using Integral A.1 in the *Resource section* for each term, it follows that

$$H_m(T_2) - H_m(T_1) = a(T_2 - T_1) + \tfrac{1}{2}b(T_2^2 - T_1^2) - c\left(\frac{1}{T_2} - \frac{1}{T_1}\right)$$

Substitution of the numerical data results in

$$H_m(373\,\mathrm{K}) = H_m(298\,\mathrm{K}) + 2.20\,\mathrm{kJ\,mol^{-1}}$$

Comment. If a constant heat capacity of 29.14 J K^{-1} mol^{-1} (the value given by eqn 2B.8 for T = 298 K) had been assumed, then the difference between the two enthalpies would have been calculated as 2.19 kJ mol^{-1}, only slightly different from the more accurate value.

Self-test 2B.2 At very low temperatures the heat capacity of a solid is proportional to T^3, and $C_{p,m} = aT^3$. What is the change in enthalpy of such a substance when it is heated from 0 to a temperature T (with T close to 0)?

Answer: $\Delta H_m = \frac{1}{4}aT^4$

(b) The relation between heat capacities

Most systems expand when heated at constant pressure. Such systems do work on the surroundings and therefore some of the energy supplied to them as heat escapes back to the surroundings as work. As a result, the temperature of the system rises less than when the heating occurs at constant volume. A smaller increase in temperature implies a larger heat capacity, so in most cases the heat capacity at constant pressure of a system is larger than its heat capacity at constant volume. As shown in Topic 2D, there is a simple relation between the two heat capacities of a perfect gas:

$$C_p - C_V = nR \quad \text{Relation between heat capacities [perfect gas]} \quad (2B.9)$$

It follows that the molar heat capacity of a perfect gas is about $8\,\mathrm{J\,K^{-1}\,mol^{-1}}$ larger at constant pressure than at constant volume. Because the molar constant-volume heat capacity of a monatomic gas is about $\frac{3}{2}R = 12\,\mathrm{J\,K^{-1}\,mol^{-1}}$ (Topic 2A), the difference is highly significant and must be taken into account. The two heat capacities are typically very similar for condensed phases, and for them the difference can normally be ignored.

Checklist of concepts

☐ 1. Energy transferred as heat at constant pressure is equal to the change in **enthalpy** of a system.

☐ 2. Enthalpy changes can be measured in a constant-pressure calorimeter.

☐ 3. The **heat capacity at constant pressure** is equal to the slope of enthalpy with temperature.

Checklist of equations

Property	Equation	Comment	Equation number
Enthalpy	$H = U + pV$	Definition	2B.1
Heat transfer at constant pressure	$\mathrm{d}H = \mathrm{d}q_p$, $\Delta H = q_p$	No additional work	2B.2
Relation between ΔH and ΔU at a temperature T	$\Delta H = \Delta U + \Delta n_g RT$	Molar volumes of the participating condensed phases are negligible	2B.4
Heat capacity at constant pressure	$C_p = (\partial H/\partial T)_p$	Definition	2B.5
Relation between heat capacities	$C_p - C_V = nR$	Perfect gas	2B.9

TOPIC 2C Thermochemistry

> ➤ **Why do you need to know this material?**
>
> Thermochemistry is one of the principal applications of thermodynamics in chemistry. Thermochemical data provide a way of assessing the heat output of chemical reactions, including those involved with the combustion of fuels and the consumption of foods. The data are also used widely in other chemical applications of thermodynamics.
>
> ➤ **What is the key idea?**
>
> Reaction enthalpies can be combined to provide data on other reactions of interest.
>
> ➤ **What do you need to know already?**
>
> You need to be aware of the definition of enthalpy and its status as a state function (Topic 2B). The material on temperature dependence of reaction enthalpies makes use of information about heat capacities (Topic 2B).

The study of the energy transferred as heat during the course of chemical reactions is called **thermochemistry**. Thermochemistry is a branch of thermodynamics because a reaction vessel and its contents form a system, and chemical reactions result in the exchange of energy between the system and the surroundings. Thus calorimetry can be used to measure the energy supplied or discarded as heat by a reaction, with q identified with a change in internal energy if the reaction occurs at constant volume (Topic 2A) or with a change in enthalpy if the reaction occurs at constant pressure (Topic 2B). Conversely, if ΔU or ΔH for a reaction is known, it is possible to predict the heat the reaction can produce.

As pointed out in Topic 2A, a process that releases energy as heat is classified as exothermic, and one that absorbs energy as heat is classified as endothermic. Because the release of heat into the surroundings at constant pressure signifies a decrease in the enthalpy of a system, it follows that an exothermic process is one for which $\Delta H < 0$; such a process is **exenthalpic**. Conversely, because the absorption of heat from the surroundings results in an increase in enthalpy, an endothermic process has $\Delta H > 0$; such a process is **endenthalpic**:

exothermic (exenthalpic) process: $\Delta H < 0$
endothermic (endenthalpic) process: $\Delta H > 0$

2C.1 Standard enthalpy changes

Changes in enthalpy are normally reported for processes taking place under a set of standard conditions. The **standard enthalpy change**, $\Delta H^{\ominus}$, is the change in enthalpy for a process in which the initial and final substances are in their standard states:

> The **standard state** of a substance at a specified temperature is its pure form at 1 bar. *Specification of standard state*

For example, the standard state of liquid ethanol at 298 K is pure liquid ethanol at 298 K and 1 bar; the standard state of solid iron at 500 K is pure iron at 500 K and 1 bar. The definition of standard state is more sophisticated for solutions (Topic 5E). The standard enthalpy change for a reaction or a physical process is the difference in enthalpy between the products in their standard states and the reactants in their standard states, all at the same specified temperature.

An example of a standard enthalpy change is the *standard enthalpy of vaporization*, $\Delta_{vap}H^{\ominus}$, which is the enthalpy change per mole of molecules when a pure liquid at 1 bar vaporizes to a gas at 1 bar, as in

$$H_2O(l) \rightarrow H_2O(g) \qquad \Delta_{vap}H^{\ominus}(373\,K) = +40.66\,kJ\,mol^{-1}$$

As implied by the examples, standard enthalpies may be reported for any temperature. However, the conventional temperature for reporting thermodynamic data is 298.15 K. Unless otherwise mentioned or indicated by attaching the temperature to $\Delta H^{\ominus}$, all thermodynamic data in this text are for this conventional temperature.

A note on good practice The attachment of the name of the transition to the symbol Δ, as in $\Delta_{vap}H$, is the current convention. However, the older convention, ΔH_{vap}, is still widely used. The current convention is more logical because the subscript identifies the type of change, not the physical observable related to the change.

(a) Enthalpies of physical change

The standard molar enthalpy change that accompanies a change of physical state is called the **standard enthalpy of transition** and is denoted $\Delta_{trs}H^{\ominus}$ (Table 2C.1). The **standard enthalpy of vaporization**, $\Delta_{vap}H^{\ominus}$, is one example. Another is

Table 2C.1 Standard enthalpies of fusion and vaporization at the transition temperature* $\Delta_{trs}H^{\ominus}/(kJ\,mol^{-1})$

	T_f/K	Fusion	T_b/K	Vaporization
Ar	83.81	1.188	87.29	6.506
C_6H_6	278.61	10.59	353.2	30.8
H_2O	273.15	6.008	373.15	40.656 (44.016 at 298 K)
He	3.5	0.021	4.22	0.084

* More values are given in the *Resource section.*

the **standard enthalpy of fusion,** $\Delta_{fus}H^{\ominus}$, the standard molar enthalpy change accompanying the conversion of a solid to a liquid, as in

$$H_2O(s) \rightarrow H_2O(l) \qquad \Delta_{fus}H^{\ominus}(273\,K) = +6.01\,kJ\,mol^{-1}$$

As in this case, it is sometimes convenient to know the standard molar enthalpy change at the transition temperature as well as at the conventional temperature of 298 K. The different types of enthalpy changes encountered in thermochemistry are summarized in Table 2C.2.

Because enthalpy is a state function, a change in enthalpy is independent of the path between the two states. This feature is of great importance in thermochemistry, because it implies that the same value of $\Delta H^{\ominus}$ will be obtained however the change is brought about between specified initial and final states. For example, the conversion of a solid to a vapour can be pictured either as occurring by sublimation (the direct conversion from solid to vapour),

$$H_2O(s) \rightarrow H_2O(g) \qquad \Delta_{sub}H^{\ominus}$$

Table 2C.2 Enthalpies of reaction and transition

Transition	Process	Symbol*
Transition	Phase $\alpha \rightarrow$ phase β	$\Delta_{trs}H$
Fusion	$s \rightarrow l$	$\Delta_{fus}H$
Vaporization	$l \rightarrow g$	$\Delta_{vap}H$
Sublimation	$s \rightarrow g$	$\Delta_{sub}H$
Mixing	Pure $\rightarrow$ mixture	$\Delta_{mix}H$
Solution	Solute $\rightarrow$ solution	$\Delta_{sol}H$
Hydration	$X^{\pm}(g) \rightarrow X^{\pm}(aq)$	$\Delta_{hyd}H$
Atomization	Species(s, l, g) $\rightarrow$ atoms(g)	$\Delta_{at}H$
Ionization	$X(g) \rightarrow X^{+}(g) + e^{-}(g)$	$\Delta_{ion}H$
Electron gain	$X(g) + e^{-}(g) \rightarrow X^{-}(g)$	$\Delta_{eg}H$
Reaction	Reactants $\rightarrow$ products	$\Delta_{r}H$
Combustion	Compound(s, l, g) + $O_2(g) \rightarrow CO_2(g)$ + H_2O(l, g)	$\Delta_{c}H$
Formation	Elements $\rightarrow$ compound	$\Delta_{f}H$
Activation	Reactants $\rightarrow$ activated complex	$\Delta^{\ddagger}H$

* IUPAC recommendations. In common usage, the process subscript is often attached to ΔH, as in ΔH_{trs} and ΔH_f. All are molar quantities.

or as occurring in two steps, first fusion (melting) and then vaporization of the resulting liquid:

$$\begin{array}{lll} & H_2O(s) \rightarrow H_2O(l) & \Delta_{fus}H^{\ominus} \\ & H_2O(l) \rightarrow H_2O(g) & \Delta_{vap}H^{\ominus} \\ \text{Overall:} & H_2O(s) \rightarrow H_2O(g) & \Delta_{fus}H^{\ominus} + \Delta_{vap}H^{\ominus} \end{array}$$

Because the overall result of the indirect path is the same as that of the direct path, the overall enthalpy change is the same in each case (1), and (for processes occurring at the same temperature)

$$\Delta_{sub}H^{\ominus} = \Delta_{fus}H^{\ominus} + \Delta_{vap}H^{\ominus} \qquad (2C.1)$$

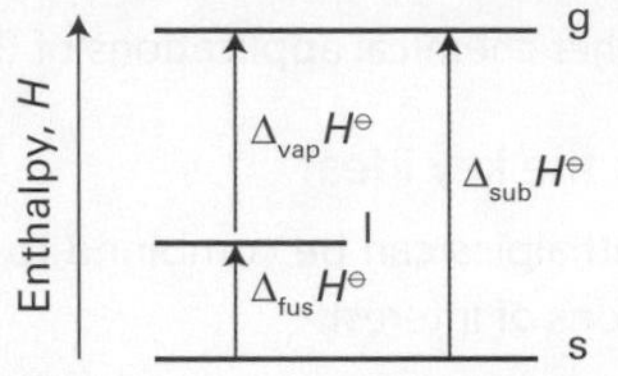

1

It follows that, because all enthalpies of fusion are positive, the enthalpy of sublimation of a substance is greater than its enthalpy of vaporization (at a given temperature).

Another consequence of H being a state function is that the standard enthalpy change of a forward process is the negative of its reverse (2):

$$\Delta H^{\ominus}(A \rightarrow B) = -\Delta H^{\ominus}(A \leftarrow B) \qquad (2C.2)$$

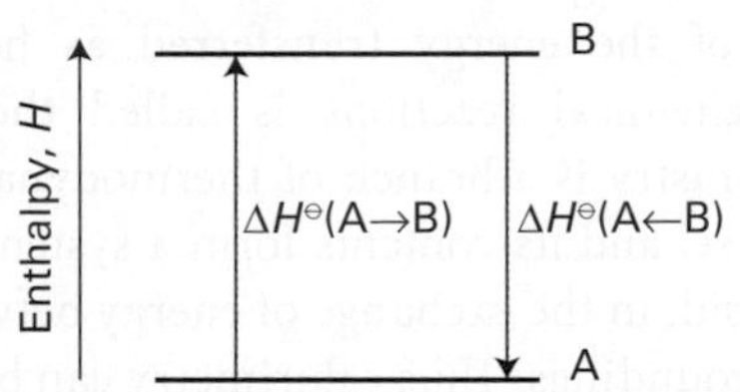

2

For instance, because the enthalpy of vaporization of water is $+44\,kJ\,mol^{-1}$ at 298 K, the enthalpy of condensation of water vapour at that temperature is $-44\,kJ\,mol^{-1}$.

(b) Enthalpies of chemical change

There are two ways of reporting the change in enthalpy that accompanies a chemical reaction. One is to write the **thermochemical equation**, a combination of a chemical equation and the corresponding change in standard enthalpy:

$$CH_4(g) + 2\,O_2(g) \rightarrow CO_2(g) + 2\,H_2O(l) \quad \Delta H^{\ominus} = -890\,kJ$$

$\Delta H^{\ominus}$ is the change in enthalpy when reactants in their standard states change to products in their standard states:

Pure, separate reactants in their standard states
$\rightarrow$ pure, separate products in their standard states

Except in the case of ionic reactions in solution, the enthalpy changes accompanying mixing and separation are insignificant in comparison with the contribution from the reaction itself. For the combustion of methane, the standard value refers to the reaction in which 1 mol CH_4 in the form of pure methane gas at 1 bar reacts completely with 2 mol O_2 in the form of pure oxygen gas to produce 1 mol CO_2 as pure carbon dioxide at 1 bar and 2 mol H_2O as pure liquid water at 1 bar; the numerical value quoted is for the reaction at 298.15 K.

Alternatively, the chemical equation is written and the **standard reaction enthalpy**, $\Delta_r H^\ominus$ (or 'standard enthalpy of reaction') reported. Thus, for the combustion of methane at 298 K, write

$$CH_4(g) + 2O_2(g) \rightarrow CO_2(g) + 2H_2O(l) \quad \Delta_r H^\ominus = -890\,\text{kJ mol}^{-1}$$

For a reaction of the form 2 A + B → 3 C + D the standard reaction enthalpy would be

$$\Delta_r H^\ominus = \{3H_m^\ominus(C) + H_m^\ominus(D)\} - \{2H_m^\ominus(A) + H_m^\ominus(B)\}$$

where $H_m^\ominus(J)$ is the standard molar enthalpy of species J at the temperature of interest. Note how the 'per mole' of $\Delta_r H^\ominus$ comes directly from the fact that molar enthalpies appear in this expression. The 'per mole' is interpreted by noting the stoichiometric coefficients in the chemical equation. In this case, 'per mole' in $\Delta_r H^\ominus$ means 'per 2 mol A', 'per mol B', 'per 3 mol C', or 'per mol D'. In general,

$$\Delta_r H^\ominus = \sum_{\text{Products}} \nu H_m^\ominus - \sum_{\text{Reactants}} \nu H_m^\ominus \qquad \text{Standard reaction enthalpy [definition]} \qquad (2C.3)$$

where in each case the molar enthalpies of the species are multiplied by their (dimensionless and positive) stoichiometric coefficients, ν. This formal definition is of little practical value, however, because the absolute values of the standard molar enthalpies are unknown; this problem is overcome by following the techniques of Section 2C.2a.

Some standard reaction enthalpies have special names and significance. For instance, the **standard enthalpy of combustion**, $\Delta_c H^\ominus$, is the standard reaction enthalpy for the complete oxidation of an organic compound to CO_2 gas and liquid H_2O if the compound contains C, H, and O, and to N_2 gas if N is also present.

Brief illustration 2C.1

The combustion of glucose is

$$C_6H_{12}O_6(s) + 6O_2(g) \rightarrow 6CO_2(g) + 6H_2O(l)$$
$$\Delta_c H^\ominus = -2808\,\text{kJ mol}^{-1}$$

The value quoted shows that 2808 kJ of heat is released when 1 mol $C_6H_{12}O_6$ burns under standard conditions (at 298 K). More values are given in Table 2C.3.

Table 2C.3 Standard enthalpies of formation and combustion of organic compounds at 298 K*

	$\Delta_f H^\ominus/(\text{kJ mol}^{-1})$	$\Delta_c H^\ominus/(\text{kJ mol}^{-1})$
Benzene, $C_6H_6(l)$	+49.0	−3268
Ethane, $C_2H_6(g)$	−84.7	−1560
Glucose, $C_6H_{12}O_6(s)$	−1274	−2808
Methane, $CH_4(g)$	−74.8	−890
Methanol, $CH_3OH(l)$	−238.7	−721

* More values are given in the *Resource section*.

(c) Hess's law

Standard reaction enthalpies can be combined to obtain the value for another reaction. This application of the First Law is called **Hess's law**:

> The standard reaction enthalpy is the sum of the values for the individual reactions into which the overall reaction may be divided. *Hess's law*

The individual steps need not be realizable in practice: they may be 'hypothetical' reactions, the only requirement being that their chemical equations should balance. The thermodynamic basis of the law is the path-independence of the value of $\Delta_r H^\ominus$. The importance of Hess's law is that information about a reaction of interest, which may be difficult to determine directly, can be assembled from information on other reactions.

Example 2C.1 Using Hess's law

The standard reaction enthalpy for the hydrogenation of propene,

$$CH_2{=}CHCH_3(g) + H_2(g) \rightarrow CH_3CH_2CH_3(g)$$

is −124 kJ mol⁻¹. The standard reaction enthalpy for the combustion of propane,

$$CH_3CH_2CH_3(g) + 5O_2(g) \rightarrow 3CO_2(g) + 4H_2O(l)$$

is −2220 kJ mol^{-1}. The standard reaction enthalpy for the formation of water,

$$H_2(g) + \tfrac{1}{2}O_2(g) \rightarrow H_2O(l)$$

is −286 kJ mol^{-1}. Calculate the standard enthalpy of combustion of propene.

Collect your thoughts The skill you need to develop is the ability to assemble a given thermochemical equation from others. Add or subtract the reactions given, together with any others needed, so as to reproduce the reaction required. Then add or subtract the reaction enthalpies in the same way.

The solution The combustion reaction is

$$C_3H_6(g) + \tfrac{9}{2}O_2(g) \rightarrow 3CO_2(g) + 3H_2O(l)$$

This reaction can be recreated from the following sum:

	$\Delta_r H^{\ominus}/(\text{kJ mol}^{-1})$
$C_3H_6(g) + H_2(g) \rightarrow C_3H_8(g)$	−124
$C_3H_8(g) + 5\,O_2(g) \rightarrow 3\,CO_2(g) + 4\,H_2O(l)$	−2220
$H_2O(l) \rightarrow H_2(g) + \frac{1}{2}O_2(g)$	+286
$C_3H_6(g) + \frac{9}{2}O_2(g) \rightarrow 3\,CO_2(g) + 3\,H_2O(l)$	−2058

Self-test 2C.1 Calculate the standard enthalpy of hydrogenation of liquid benzene from its standard enthalpy of combustion (−3268 kJ mol^{-1}) and the standard enthalpy of combustion of liquid cyclohexane (−3920 kJ mol^{-1}).

Answer: −206 kJ mol^{-1}

2C.2 Standard enthalpies of formation

The **standard enthalpy of formation**, $\Delta_f H^{\ominus}$, of a substance is the standard reaction enthalpy for the formation of the compound from its elements in their reference states:

The **reference state** of an element is its most stable state at the specified temperature and 1 bar.

Specification of reference state

For example, at 298 K the reference state of nitrogen is a gas of N_2 molecules, that of mercury is liquid mercury, that of carbon is graphite, and that of tin is the white (metallic) form. There is one exception to this general prescription of reference states: the reference state of phosphorus is taken to be white phosphorus despite this allotrope not being the most stable form but simply the most reproducible form of the element. Standard enthalpies of formation are expressed as enthalpies per mole of molecules or (for ionic substances) formula units of the compound. The standard enthalpy of formation of liquid benzene at 298 K, for example, refers to the reaction

$$6\,C(s,\text{graphite}) + 3\,H_2(g) \rightarrow C_6H_6(l)$$

and is +49.0 kJ mol^{-1}. The standard enthalpies of formation of elements in their reference states are zero at all temperatures because they are the enthalpies of such 'null' reactions as $N_2(g) \rightarrow N_2(g)$. Some enthalpies of formation are listed in Tables 2C.4 and 2C.5 and a much longer list will be found in the *Resource section*.

The standard enthalpy of formation of ions in solution poses a special problem because it is not possible to prepare a solution of either cations or anions alone. This problem is overcome by defining one ion, conventionally the hydrogen ion, to have zero standard enthalpy of formation at all temperatures:

$$\Delta_f H^{\ominus}(H^+,aq) = 0 \qquad \text{Ions in solution [convention]} \qquad (2C.4)$$

Brief illustration 2C.2

If the enthalpy of formation of HBr(aq) is found to be −122 kJ mol^{-1}, then the whole of that value is ascribed to the formation of $Br^-(aq)$, and $\Delta_f H^{\ominus}(Br^-,aq) = -122\,\text{kJ mol}^{-1}$. That value may then be combined with, for instance, the enthalpy of formation of AgBr(aq) to determine the value of $\Delta_f H^{\ominus}(Ag^+,aq)$, and so on. In essence, this definition adjusts the actual values of the enthalpies of formation of ions by a fixed value, which is chosen so that the standard value for one of them, $H^+(aq)$, is zero.

Conceptually, a reaction can be regarded as proceeding by decomposing the reactants into their elements in their reference states and then forming those elements into the products. The value of $\Delta_r H^{\ominus}$ for the overall reaction is the sum of these 'unforming' and forming enthalpies. Because 'unforming' is the reverse of forming, the enthalpy of an unforming step is

Table 2C.4 Standard enthalpies of formation of inorganic compounds at 298 K*

	$\Delta_f H^{\ominus}/(\text{kJ mol}^{-1})$
$H_2O(l)$	−285.83
$H_2O(g)$	−241.82
$NH_3(g)$	−46.11
$N_2H_4(l)$	+50.63
$NO_2(g)$	+33.18
$N_2O_4(g)$	+9.16
NaCl(s)	−411.15
KCl(s)	−436.75

* More values are given in the *Resource section*.

Table 2C.5 Standard enthalpies of formation of organic compounds at 298 K*

	$\Delta_f H^{\ominus}/(\text{kJ mol}^{-1})$
$CH_4(g)$	−74.81
$C_6H_6(l)$	+49.0
$C_6H_{12}(l)$	−156
$CH_3OH(l)$	−238.66
$CH_3CH_2OH(l)$	−277.69

* More values are given in the *Resource section*.

the negative of the enthalpy of formation (3). Hence, in the enthalpies of formation of substances, there is enough information to calculate the enthalpy of any reaction by using

$$\Delta_r H^\ominus = \sum_{\text{Products}} \nu \Delta_f H^\ominus - \sum_{\text{Reactants}} \nu \Delta_f H^\ominus \quad \text{Standard reaction enthalpy [practical implementation]} \quad (2C.5a)$$

3

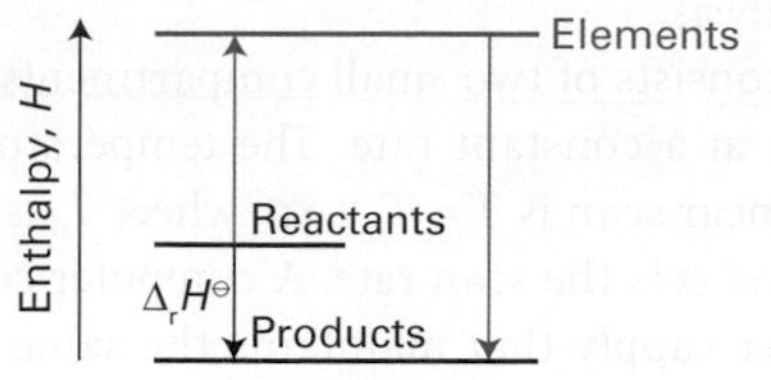

where in each case the enthalpies of formation of the species that occur are multiplied by their stoichiometric coefficients. This procedure is the practical implementation of the formal definition in eqn 2C.3. A more sophisticated way of expressing the same result is to introduce the **stoichiometric numbers** ν_J (as distinct from the stoichiometric coefficients) which are positive for products and negative for reactants. Then

$$\Delta_r H^\ominus = \sum_J \nu_J \Delta_f H^\ominus(J) \quad (2C.5b)$$

Stoichiometric *numbers*, which have a sign, are denoted ν_J or $\nu(J)$. Stoichiometric *coefficients*, which are all positive, are denoted simply ν (with no subscript).

Brief illustration 2C.3

According to eqn 2C.5a, the standard enthalpy of the reaction $2\,HN_3(l) + 2\,NO(g) \rightarrow H_2O_2(l) + 4\,N_2(g)$ is calculated as follows:

$$\begin{aligned}\Delta_r H^\ominus &= \{\Delta_f H^\ominus(H_2O_2,l) + 4\Delta_f H^\ominus(N_2,g)\} \\ &\quad - \{2\Delta_f H^\ominus(HN_3,l) + 2\Delta_f H^\ominus(NO,g)\} \\ &= \{-187.78 + 4(0)\}\,\text{kJ mol}^{-1} \\ &\quad - \{2(264.0) + 2(90.25)\}\,\text{kJ mol}^{-1} \\ &= -896.3\,\text{kJ mol}^{-1}\end{aligned}$$

To use eqn 2C.5b, identify $\nu(HN_3) = -2$, $\nu(NO) = -2$, $\nu(H_2O_2) = +1$, and $\nu(N_2) = +4$, and then write

$$\Delta_r H^\ominus = \Delta_f H^\ominus(H_2O_2,l) + 4\Delta_f H^\ominus(N_2,g) - 2\Delta_f H^\ominus(HN_3,l) - 2\Delta_f H^\ominus(NO,g)$$

which gives the same result.

2C.3 The temperature dependence of reaction enthalpies

Many standard reaction enthalpies have been measured at different temperatures. However, in the absence of this information, standard reaction enthalpies at different temperatures can be calculated from heat capacities and the reaction enthalpy at some other temperature (Fig. 2C.1). In many cases heat capacity data are more accurate than reaction enthalpies. Therefore, providing the information is available, the procedure about to be described is more accurate than the direct measurement of a reaction enthalpy at an elevated temperature.

It follows from eqn 2B.6a ($dH = C_p dT$) that, when a substance is heated from T_1 to T_2, its enthalpy changes from $H(T_1)$ to

$$H(T_2) = H(T_1) + \int_{T_1}^{T_2} C_p\,dT \quad (2C.6)$$

(It has been assumed that no phase transition takes place in the temperature range of interest.) Because this equation applies to each substance in the reaction, the standard reaction enthalpy changes from $\Delta_r H^\ominus(T_1)$ to

$$\Delta_r H^\ominus(T_2) = \Delta_r H^\ominus(T_1) + \int_{T_1}^{T_2} \Delta_r C_p^\ominus\,dT \quad \text{Kirchhoff's law} \quad (2C.7a)$$

where $\Delta_r C_p^\ominus$ is the difference of the molar heat capacities of products and reactants under standard conditions weighted by the stoichiometric coefficients that appear in the chemical equation:

$$\Delta_r C_p^\ominus = \sum_{\text{Products}} \nu C_{p,m}^\ominus - \sum_{\text{Reactants}} \nu C_{p,m}^\ominus \quad (2C.7b)$$

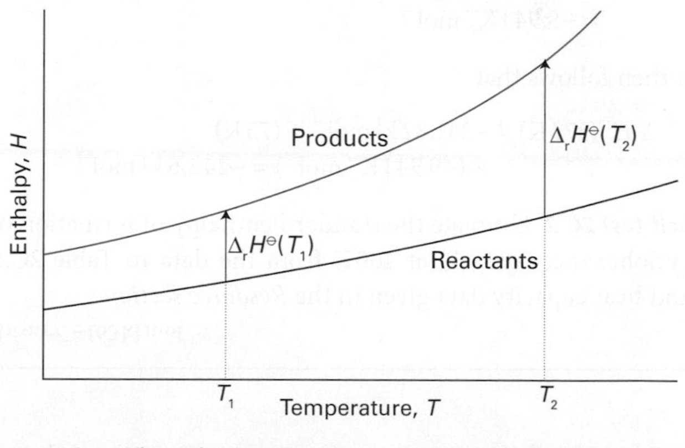

Figure 2C.1 When the temperature is increased, the enthalpy of the products and the reactants both increase, but may do so to different extents. In each case, the change in enthalpy depends on the heat capacities of the substances. The change in reaction enthalpy reflects the difference in the changes of the enthalpies of the products and reactants.

or, in the notation of eqn 2C.5b,

$$\Delta_r C_p^{\ominus} = \sum_J \nu_J C_{p,m}^{\ominus}(J) \qquad (2C.7c)$$

Equation 2C.7a is known as **Kirchhoff's law**. It is normally a good approximation to assume that $\Delta_r C_p^{\ominus}$ is independent of the temperature, at least over reasonably limited ranges. Although the individual heat capacities might vary, their difference varies less significantly. In some cases the temperature dependence of heat capacities is taken into account by using eqn 2C.7a. If $\Delta_r C_p^{\ominus}$ is largely independent of temperature in the range T_1 to T_2, the integral in eqn 2C.7a evaluates to $(T_2 - T_1)\Delta_r C_p^{\ominus}$ and that equation becomes

$$\Delta_r H^{\ominus}(T_2) = \Delta_r H^{\ominus}(T_1) + \Delta_r C_p^{\ominus}(T_2 - T_1) \qquad (2C.7d)$$

Integrated form of Kirchhoff's law

Example 2C.2 Using Kirchhoff's law

The standard enthalpy of formation of $H_2O(g)$ at 298 K is −241.82 kJ mol^{-1}. Estimate its value at 100 °C given the following values of the molar heat capacities at constant pressure: $H_2O(g)$: 33.58 J K^{-1} mol^{-1}; $H_2(g)$: 28.84 J K^{-1} mol^{-1}; $O_2(g)$: 29.37 J K^{-1} mol^{-1}. Assume that the heat capacities are independent of temperature.

Collect your thoughts When $\Delta_r C_p^{\ominus}$ is independent of temperature in the range T_1 to T_2, you can use the integrated form of the Kirchhoff equation, eqn 2C.7d. To proceed, write the chemical equation, identify the stoichiometric coefficients, and calculate $\Delta_r C_p^{\ominus}$ from the data.

The solution The reaction is $H_2(g) + \frac{1}{2}O_2(g) \rightarrow H_2O(g)$, so

$$\Delta_r C_p^{\ominus} = C_{p,m}^{\ominus}(H_2O,g) - \{C_{p,m}^{\ominus}(H_2,g) + \tfrac{1}{2}C_{p,m}^{\ominus}(O_2,g)\}$$
$$= -9.94\ \text{J K}^{-1}\,\text{mol}^{-1}$$

It then follows that

$$\Delta_r H^{\ominus}(373\ \text{K}) = -241.82\ \text{kJ mol}^{-1} + (75\ \text{K}) \times (-9.94\ \text{J K}^{-1}\,\text{mol}^{-1}) = -242.6\ \text{kJ mol}^{-1}$$

Self-test 2C.2 Estimate the standard enthalpy of formation of cyclohexane, $C_6H_{12}(l)$, at 400 K from the data in Table 2C.5 and heat capacity data given in the *Resource section*.

Answer: −163 kJ mol^{-1}

2C.4 Experimental techniques

The classic tool of thermochemistry is the calorimeter (Topics 2A and 2B). However, technological advances have been made that allow measurements to be made on samples with mass as little as a few milligrams.

(a) Differential scanning calorimetry

A **differential scanning calorimeter** (DSC) measures the energy transferred as heat to or from a sample at constant pressure during a physical or chemical change. The term 'differential' refers to the fact that measurements on a sample are compared to those on a reference material that does not undergo a physical or chemical change during the analysis. The term 'scanning' refers to the fact that the temperatures of the sample and reference material are increased, or scanned, during the analysis.

A DSC consists of two small compartments that are heated electrically at a constant rate. The temperature, T, at time t during a linear scan is $T = T_0 + \alpha t$, where T_0 is the initial temperature and α is the scan rate. A computer controls the electrical power supply that maintains the same temperature in the sample and reference compartments throughout the analysis (Fig. 2C.2).

If no physical or chemical change occurs in the sample at temperature T, the heat transferred to the sample is written as $q_p = C_p\Delta T$, where $\Delta T = T - T_0$ and C_p is assumed to be independent of temperature. Because $T = T_0 + \alpha t$, it follows that $\Delta T = \alpha t$. If a chemical or physical process takes place, the energy required to be transferred as heat to attain the same change in temperature of the sample as the control is $q_p + q_{p,\text{ex}}$.

The quantity $q_{p,\text{ex}}$ is interpreted in terms of an apparent change in the heat capacity at constant pressure, from C_p to $C_p + C_{p,\text{ex}}$ of the sample during the temperature scan:

$$C_{p,\text{ex}} = \frac{q_{p,\text{ex}}}{\Delta T} = \frac{q_{p,\text{ex}}}{\alpha t} = \frac{P_{\text{ex}}}{\alpha} \qquad (2C.8)$$

where $P_{\text{ex}} = q_{p,\text{ex}}/t$ is the excess electrical power necessary to equalize the temperature of the sample and reference compartments. A DSC trace, also called a **thermogram**, consists of

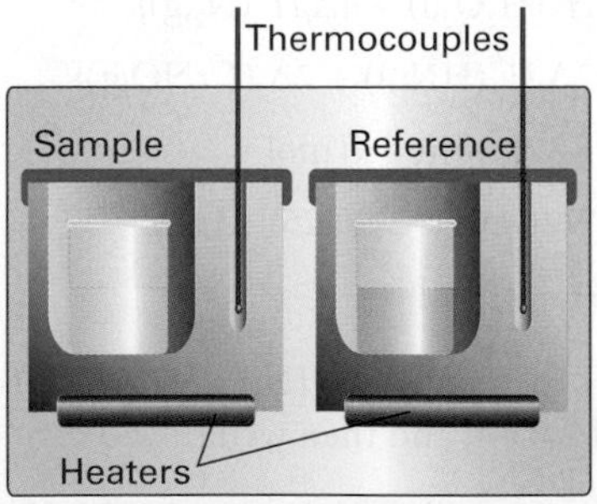

Figure 2C.2 A differential scanning calorimeter. The sample and a reference material are heated in separate but identical metal heat sinks. The output is the difference in power needed to maintain the heat sinks at equal temperatures as the temperature rises.

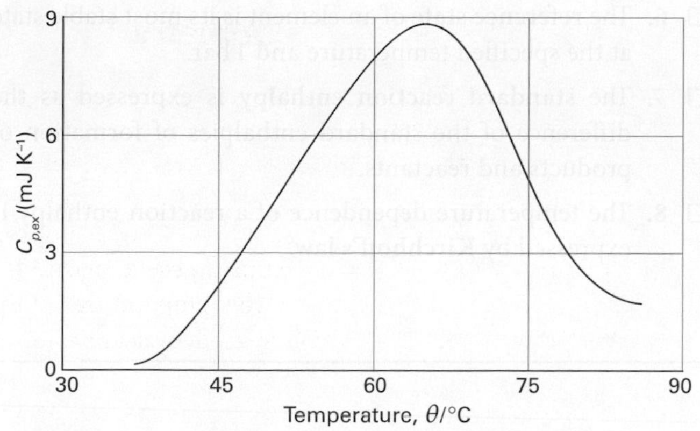

Figure 2C.3 A thermogram for the protein ubiquitin at pH = 2.45. The protein retains its native structure up to about 45 °C and then undergoes an endothermic conformational change. (Adapted from B. Chowdhry and S. LeHarne, *J. Chem. Educ.* 74, 236 (1997).)

a plot of $C_{p,ex}$ against T (Fig. 2C.3). The enthalpy change associated with the process is

$$\Delta H = \int_{T_1}^{T_2} C_{p,ex}\, dT \tag{2C.9}$$

where T_1 and T_2 are, respectively, the temperatures at which the process begins and ends. This relation shows that the enthalpy change is equal to the area under the plot of $C_{p,ex}$ against T.

(b) Isothermal titration calorimetry

Isothermal titration calorimetry (ITC) is also a 'differential' technique in which the thermal behaviour of a sample is compared with that of a reference. The apparatus is shown in Fig. 2C.4. One of the thermally conducting vessels, which have a volume of a few cubic centimetres, contains the reference (water for instance) and a heater rated at a few milliwatts. The second vessel contains one of the reagents, such as a solution of a macromolecule with binding sites; it also contains a heater. At the start of the experiment, both heaters are activated, and then precisely determined amounts (of volume of about a cubic millimetre) of the second reagent are added to the reaction cell. The power required to maintain the same temperature differential with the reference cell is monitored.

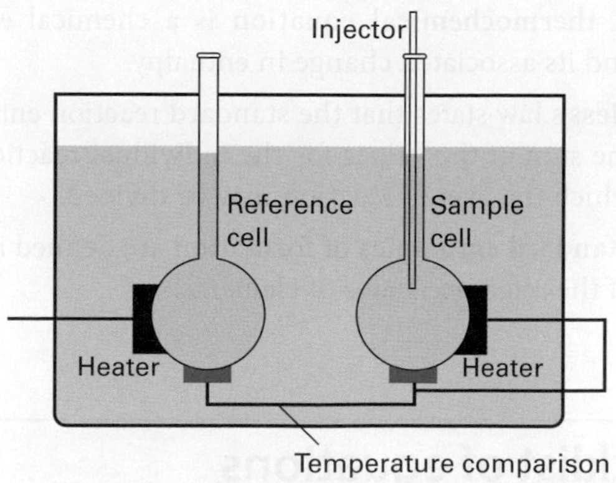

Figure 2C.4 A schematic diagram of the apparatus used for isothermal titration calorimetry.

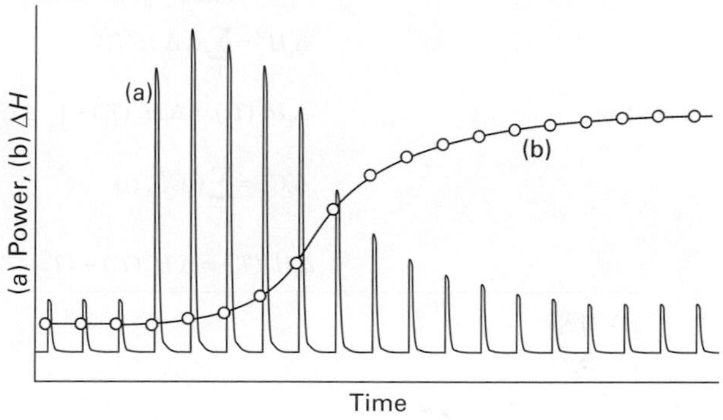

Figure 2C.5 (a) The record of the power applied as each injection is made, and (b) the sum of successive enthalpy changes in the course of the titration.

If the reaction is exothermic, less power is needed; if it is endothermic, then more power must be supplied.

A typical result is shown in Fig. 2C.5, which shows the power needed to maintain the temperature differential: from the power and the length of time, Δt, for which it is supplied, the heat supplied, q_i, for the injection i can be calculated from $q_i = P_i\Delta t$. If the volume of solution is V and the molar concentration of unreacted reagent A is c_i at the time of the ith injection, then the change in its concentration at that injection is Δc_i and the heat generated (or absorbed) by the reaction is $V\Delta_r H\Delta c_i = q_i$. The sum of all such quantities, given that the sum of Δc_i is the known initial concentration of the reactant, can then be interpreted as the value of $\Delta_r H$ for the reaction.

Checklist of concepts

☐ 1. The **standard enthalpy of transition** is equal to the energy transferred as heat at constant pressure in the transition under standard conditions.

☐ 2. The **standard state** of a substance at a specified temperature is its pure form at 1 bar.

☐ 3. A **thermochemical equation** is a chemical equation and its associated change in enthalpy.

☐ 4. **Hess's law** states that the standard reaction enthalpy is the sum of the values for the individual reactions into which the overall reaction may be divided.

☐ 5. **Standard enthalpies of formation** are defined in terms of the reference states of elements.

☐ 6. The **reference state** of an element is its most stable state at the specified temperature and 1 bar.

☐ 7. The **standard reaction enthalpy** is expressed as the difference of the standard enthalpies of formation of products and reactants.

☐ 8. The temperature dependence of a reaction enthalpy is expressed by **Kirchhoff's law.**

Checklist of equations

Property	Equation	Comment	Equation number
The standard reaction enthalpy	$\Delta_r H^\ominus = \sum_{\text{Products}} \nu\Delta_f H^\ominus - \sum_{\text{Reactants}} \nu\Delta_f H^\ominus$	ν: stoichiometric coefficients; ν_J: (signed) stoichiometric numbers	2C.5
	$\Delta_r H^\ominus = \sum_J \nu_J \Delta_f H^\ominus(J)$		
Kirchhoff's law	$\Delta_r H^\ominus(T_2) = \Delta_r H^\ominus(T_1) + \int_{T_1}^{T_2} \Delta_r C_p^\ominus \mathrm{d}T$		2C.7a
	$\Delta_r C_p^\ominus = \sum_J \nu_J C_{p,\mathrm{m}}^\ominus(J)$		2C.7c
	$\Delta_r H^\ominus(T_2) = \Delta_r H^\ominus(T_1) + (T_2 - T_1)\Delta_r C_p^\ominus$	If $\Delta_r C_p^\ominus$ independent of temperature	2C.7d

TOPIC 2D State functions and exact differentials

➤ **What do you need to know this material?**

Thermodynamics has the power to provide relations between a variety of properties. This Topic introduces its key procedure, the manipulation of equations involving state functions.

➤ **What is the key idea?**

The fact that internal energy and enthalpy are state functions leads to relations between thermodynamic properties.

➤ **What do you need to know already?**

You need to be aware that the internal energy and enthalpy are state functions (Topics 2B and 2C) and be familiar with the concept of heat capacity. You need to be able to make use of several simple relations involving partial derivatives (*The chemist's toolkit* 9 in Topic 2A).

A **state function** is a property that depends only on the current state of a system and is independent of its history. The internal energy and enthalpy are two examples. Physical quantities with values that do depend on the path between two states are called **path functions**. Examples of path functions are the work and the heating that are done when preparing a state. It is not appropriate to speak of a system in a particular state as possessing work or heat. In each case, the energy transferred as work or heat relates to the path being taken between states, not the current state itself.

A part of the richness of thermodynamics is that it uses the mathematical properties of state functions to draw far-reaching conclusions about the relations between physical properties and thereby establish connections that may be completely unexpected. The practical importance of this ability is the possibility of combining measurements of different properties to obtain the value of a desired property.

2D.1 Exact and inexact differentials

Consider a system undergoing the changes depicted in Fig. 2D.1. The initial state of the system is i and in this state the internal energy is U_i. Work is done by the system as it expands

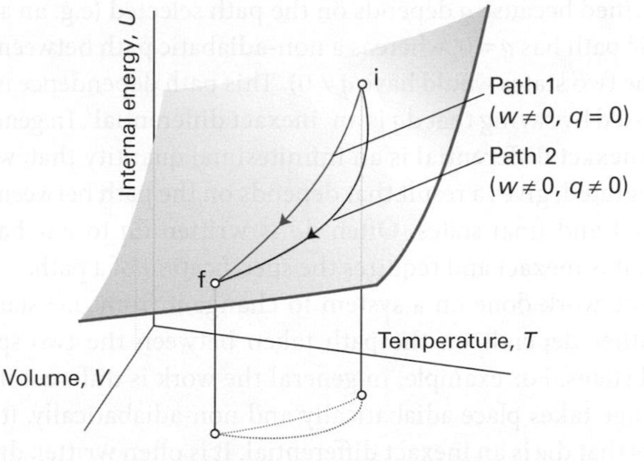

Figure 2D.1 As the volume and temperature of a system are changed, the internal energy changes. An adiabatic and a non-adiabatic path are shown as Path 1 and Path 2, respectively: they correspond to different values of q and w but to the same value of ΔU.

adiabatically to a state f. In this state the system has an internal energy U_f and the work done on the system as it changes along Path 1 from i to f is w. Notice the use of language: U is a property of the state; w is a property of the path. Now consider another process, Path 2, in which the initial and final states are the same as those in Path 1 but in which the expansion is not adiabatic. The internal energy of both the initial and the final states are the same as before (because U is a state function). However, in the second path an energy q' enters the system as heat and the work w' is not the same as w. The work and the heat are path functions.

If a system is taken along a path (e.g. by heating it), U changes from U_i to U_f, and the overall change is the sum (integral) of all the infinitesimal changes along the path:

$$\Delta U = \int_i^f \mathrm{d}U \qquad (2D.1)$$

The value of ΔU depends on the initial and final states of the system but is independent of the path between them. This path-independence of the integral is expressed by saying that $\mathrm{d}U$ is an 'exact differential'. In general, an **exact differential** is an infinitesimal quantity that, when integrated, gives a result that is independent of the path between the initial and final states.

When a system is heated, the total energy transferred as heat is the sum of all individual contributions at each point of the path:

$$q=\int_{i,path}^{f} dq \qquad (2D.2)$$

Notice the differences between this equation and eqn 2D.1. First, the result of integration is q and not Δq, because q is not a state function and the energy supplied as heat cannot be expressed as $q_f - q_i$. Secondly, the path of integration must be specified because q depends on the path selected (e.g. an adiabatic path has $q = 0$, whereas a non-adiabatic path between the same two states would have $q \neq 0$). This path dependence is expressed by saying that dq is an 'inexact differential'. In general, an **inexact differential** is an infinitesimal quantity that, when integrated, gives a result that depends on the path between the initial and final states. Often dq is written đq to emphasize that it is inexact and requires the specification of a path.

The work done on a system to change it from one state to another depends on the path taken between the two specified states. For example, in general the work is different if the change takes place adiabatically and non-adiabatically. It follows that dw is an inexact differential. It is often written đw.

Example 2D.1 Calculating work, heat, and change in internal energy

Consider a perfect gas inside a cylinder fitted with a piston. Let the initial state be T,V_i and the final state be T,V_f. The change of state can be brought about in many ways, of which the two simplest are the following:

- Path 1, in which there is free expansion against zero external pressure;
- Path 2, in which there is reversible, isothermal expansion.

Calculate w, q, and ΔU for each process.

Collect your thoughts To find a starting point for a calculation in thermodynamics, it is often a good idea to go back to first principles and to look for a way of expressing the quantity to be calculated in terms of other quantities that are easier to calculate. It is argued in Topic 2B that the internal energy of a perfect gas depends only on the temperature and is independent of the volume those molecules occupy, so for any isothermal change, $\Delta U = 0$. Also, $\Delta U = q + w$ in general. To solve the problem you need to combine the two expressions, selecting the appropriate expression for the work done from the discussion in Topic 2A.

The solution Because $\Delta U = 0$ for both paths and $\Delta U = q + w$, in each case $q = -w$. The work of free expansion is zero (eqn 2A.7 of Topic 2A, $w = 0$); so in Path 1, $w = 0$ and therefore $q = 0$ too. For Path 2, the work is given by eqn 2A.9 of Topic 2A ($w = -nRT \ln(V_f/V_i)$) and consequently $q = nRT \ln(V_f/V_i)$.

Self-test 2D.1 Calculate the values of q, w, and ΔU for an irreversible isothermal expansion of a perfect gas against a constant non-zero external pressure.

Answer: $q = p_{ex}\Delta V$, $w = -p_{ex}\Delta V$, $\Delta U = 0$

2D.2 Changes in internal energy

Consider a closed system of constant composition (the only type of system considered in the rest of this Topic). The internal energy U can be regarded as a function of V, T, and p, but, because there is an equation of state that relates these quantities (Topic 1A), choosing the values of two of the variables fixes the value of the third. Therefore, it is possible to write U in terms of just two independent variables: V and T, p and T, or p and V. Expressing U as a function of volume and temperature turns out to result in the simplest expressions.

(a) General considerations

Because the internal energy is a function of the volume and the temperature, when these two quantities change, the internal energy changes by

$$dU = \left(\frac{\partial U}{\partial V}\right)_T dV + \left(\frac{\partial U}{\partial T}\right)_V dT \qquad \text{General expression for a change in } U \text{ with } T \text{ and } V \qquad (2D.3)$$

The interpretation of this equation is that, in a closed system of constant composition, any infinitesimal change in the internal energy is proportional to the infinitesimal changes of volume and temperature, the coefficients of proportionality being the two partial derivatives (Fig. 2D.2).

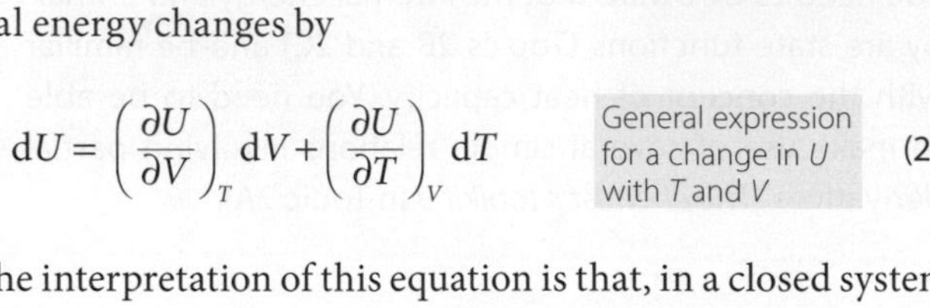

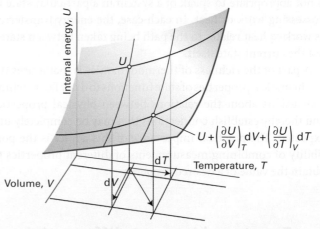

Figure 2D.2 An overall change in U, which is denoted dU, arises when both V and T are allowed to change. If second-order infinitesimals are ignored, the overall change is the sum of changes for each variable separately.

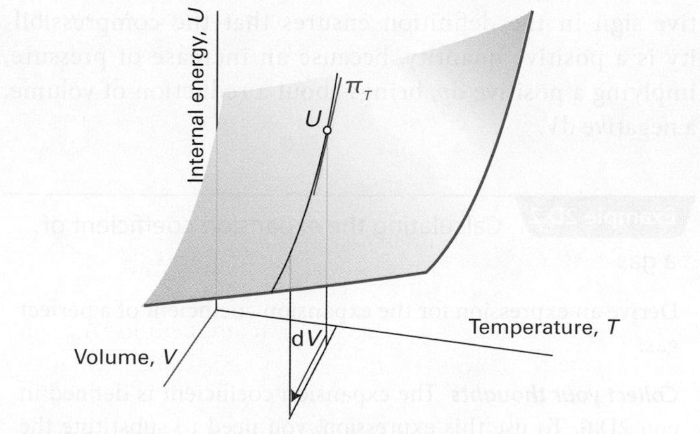

Figure 2D.3 The internal pressure, π_T, is the slope of U with respect to V with the temperature T held constant.

In many cases partial derivatives have a straightforward physical interpretation, and thermodynamics gets shapeless and difficult only when that interpretation is not kept in sight. The term $(\partial U/\partial T)_V$ occurs in Topic 2A, as the constant-volume heat capacity, C_V. The other coefficient, $(\partial U/\partial V)_T$, denoted π_T, plays a major role in thermodynamics because it is a measure of the variation of the internal energy of a substance as its volume is changed at constant temperature (Fig. 2D.3). Because π_T has the same dimensions as pressure but arises from the interactions between the molecules within the sample, it is called the **internal pressure**:

$$\pi_T = \left(\frac{\partial U}{\partial V}\right)_T \quad \text{Internal pressure [definition]} \quad (2D.4)$$

In terms of the notation C_V and π_T, eqn 2D.3 can now be written

$$dU = \pi_T dV + C_V dT \quad (2D.5)$$

It is shown in Topic 3E that the statement $\pi_T = 0$ (i.e. the internal energy is independent of the volume occupied by the sample) can be taken to be the definition of a perfect gas, because it implies the equation of state $pV \propto T$. In molecular terms, when there are no interactions between the molecules, the internal energy is independent of their separation and hence independent of the volume of the sample and $\pi_T = 0$. If the gas is described by the van der Waals equation with a, the parameter corresponding to attractive interactions, dominant, then an increase in volume increases the average separation of the molecules and therefore raises the internal energy. In this case, it is expected that $\pi_T > 0$ (Fig. 2D.4). This expectation is confirmed in Topic 3E, where it is shown that $\pi_T = na/V^2$.

James Joule thought that he could measure π_T by observing the change in temperature of a gas when it is allowed to expand into a vacuum. He used two metal vessels immersed in a water bath (Fig. 2D.5). One was filled with air at about

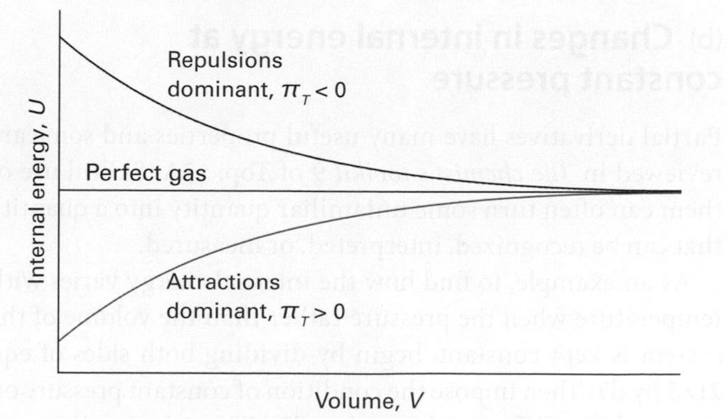

Figure 2D.4 For a perfect gas, the internal energy is independent of the volume (at constant temperature). If attractions are dominant in a real gas, the internal energy increases with volume because the molecules become farther apart on average. If repulsions are dominant, the internal energy decreases as the gas expands.

22 atm and the other was evacuated. He then tried to measure the change in temperature of the water of the bath when a stopcock was opened and the air expanded into a vacuum. He observed no change in temperature.

The thermodynamic implications of the experiment are as follows. No work was done in the expansion into a vacuum, so $w = 0$. No energy entered or left the system (the gas) as heat because the temperature of the bath did not change, so $q = 0$. Consequently, within the accuracy of the experiment, $\Delta U = 0$. Joule concluded that U does not change when a gas expands isothermally and therefore that $\pi_T = 0$. His experiment, however, was crude. The heat capacity of the apparatus was so large that the temperature change, which would in fact occur for a real gas, is simply too small to measure. Joule had extracted an essential limiting property of a gas, a property of a perfect gas, without detecting the small deviations characteristic of real gases.

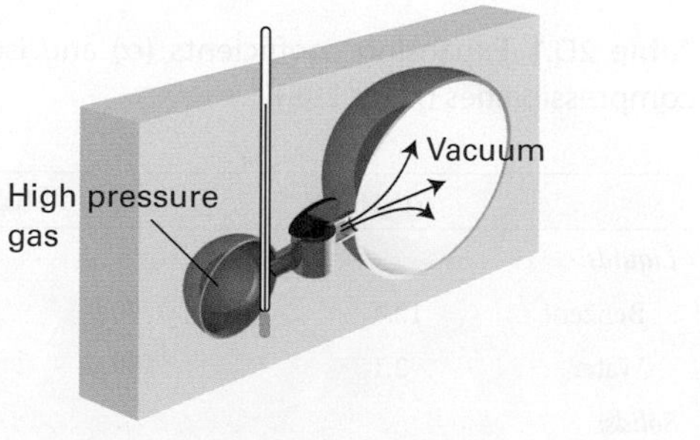

Figure 2D.5 A schematic diagram of the apparatus used by Joule in an attempt to measure the change in internal energy when a gas expands isothermally. The heat absorbed by the gas is proportional to the change in temperature of the bath.

(b) Changes in internal energy at constant pressure

Partial derivatives have many useful properties and some are reviewed in *The chemist's toolkit* 9 of Topic 2A. Skilful use of them can often turn some unfamiliar quantity into a quantity that can be recognized, interpreted, or measured.

As an example, to find how the internal energy varies with temperature when the pressure rather than the volume of the system is kept constant, begin by dividing both sides of eqn 2D.5 by dT. Then impose the condition of constant pressure on the resulting differentials, so that dU/dT on the left becomes $(\partial U/\partial T)_p$. At this stage the equation becomes

$$\left(\frac{\partial U}{\partial T}\right)_p = \pi_T\left(\frac{\partial V}{\partial T}\right)_p + C_V$$

As already emphasized, it is usually sensible in thermodynamics to inspect the output of a manipulation to see if it contains any recognizable physical quantity. The partial derivative on the right in this expression is the slope of the plot of volume against temperature (at constant pressure). This property is normally tabulated as the **expansion coefficient**, α, of a substance, which is defined as

$$\alpha = \frac{1}{V}\left(\frac{\partial V}{\partial T}\right)_p \qquad \text{Expansion coefficient [definition]} \qquad (2D.6)$$

and physically is the fractional change in volume that accompanies a rise in temperature. A large value of α means that the volume of the sample responds strongly to changes in temperature. Table 2D.1 lists some experimental values of α. For future reference, it also lists the **isothermal compressibility**, κ_T (kappa), which is defined as

$$\kappa_T = -\frac{1}{V}\left(\frac{\partial V}{\partial p}\right)_T \qquad \text{Isothermal compressibility [definition]} \qquad (2D.7)$$

The isothermal compressibility is a measure of the fractional change in volume when the pressure is increased; the negative sign in the definition ensures that the compressibility is a positive quantity, because an increase of pressure, implying a positive dp, brings about a reduction of volume, a negative dV.

Table 2D.1 Expansion coefficients (α) and isothermal compressibilities (κ_T) at 298 K*

	$\alpha/(10^{-4}\,K^{-1})$	$\kappa_T/(10^{-6}\,bar^{-1})$
Liquids:		
Benzene	12.4	90.9
Water	2.1	49.0
Solids:		
Diamond	0.030	0.185
Lead	0.861	2.18

* More values are given in the *Resource section*.

Example 2D.2 Calculating the expansion coefficient of a gas

Derive an expression for the expansion coefficient of a perfect gas.

Collect your thoughts The expansion coefficient is defined in eqn 2D.6. To use this expression, you need to substitute the expression for V in terms of T obtained from the equation of state for the gas. As implied by the subscript in eqn 2D.6, the pressure, p, is treated as a constant.

The solution Because $pV = nRT$, write

$$\alpha = \frac{1}{V}\left(\frac{\partial V}{\partial T}\right)_p \overset{pV=nRT}{=} \frac{1}{V}\left(\frac{\partial (nRT/p)}{\partial T}\right)_p = \frac{1}{V}\times\frac{nR}{p} = \frac{nR}{pV} \overset{pV=nRT}{=} \frac{nR}{nRT} = \frac{1}{T}$$

The physical interpretation of this result is that the higher the temperature, the less responsive is the volume of a perfect gas to a change in temperature.

Self-test 2D.2 Derive an expression for the isothermal compressibility of a perfect gas.

Answer: $\kappa_T = 1/p$

Introduction of the definition of α into the equation for $(\partial U/\partial T)_p$ gives

$$\left(\frac{\partial U}{\partial T}\right)_p = \alpha\pi_T V + C_V \qquad (2D.8)$$

This equation is entirely general (provided the system is closed and its composition is constant). It expresses the dependence of the internal energy on the temperature at constant pressure in terms of C_V, which can be measured in one experiment, in terms of α, which can be measured in another, and in terms of the internal pressure π_T. For a perfect gas, $\pi_T = 0$, so then

$$\left(\frac{\partial U}{\partial T}\right)_p = C_V \qquad (2D.9)$$

That is, although the constant-volume heat capacity of a perfect gas is defined as the slope of a plot of internal energy against temperature at constant volume, for a perfect gas C_V is also the slope of a plot of internal energy against temperature at constant pressure.

Equation 2D.9 provides an easy way to derive the relation between C_p and C_V for a perfect gas (they differ, as explained in Topic 2B, because some of the energy supplied as heat escapes back into the surroundings as work of expansion when the volume is not constant). First, write

$$C_p - C_V = \overbrace{\left(\frac{\partial H}{\partial T}\right)_p}^{\text{Definition of } C_p} - \overbrace{\left(\frac{\partial U}{\partial T}\right)_p}^{\text{eqn 2D.9}}$$

Then introduce $H = U + pV = U + nRT$ into the first term and obtain

$$C_p - C_V = \left(\frac{\partial (U+nRT)}{\partial T}\right)_p - \left(\frac{\partial U}{\partial T}\right)_p = nR \quad \text{(2D.10)}$$

The general result for any substance (the proof makes use of the Second Law, which is introduced in Focus 3) is

$$C_p - C_V = \frac{\alpha^2 TV}{\kappa_T} \quad \text{(2D.11)}$$

This relation reduces to eqn 2D.10 for a perfect gas when $\alpha = 1/T$ and $\kappa_T = 1/p$. Because expansion coefficients α of liquids and solids are small, it is tempting to deduce from eqn 2D.11 that for them $C_p \approx C_V$. But this is not always so, because the compressibility κ_T might also be small, so α^2/κ_T might be large. That is, although only a little work need be done to push back the atmosphere, a great deal of work may have to be done to pull atoms apart from one another as the solid expands.

Brief illustration 2D.1

The expansion coefficient and isothermal compressibility of water at 25 °C are given in Table 2D.1 as $2.1 \times 10^{-4}\,\text{K}^{-1}$ and $49.0 \times 10^{-6}\,\text{bar}^{-1}$ ($4.90 \times 10^{-10}\,\text{Pa}^{-1}$), respectively. The molar volume of water at that temperature, $V_m = M/\rho$ (where ρ is the mass density), is $18.1\,\text{cm}^3\,\text{mol}^{-1}$ ($1.81 \times 10^{-5}\,\text{m}^3\,\text{mol}^{-1}$). Therefore, from eqn 2D.11, the difference in molar heat capacities (which is given by using V_m in place of V) is

$$C_{p,\text{m}} - C_{V,\text{m}} = \frac{(2.1\times10^{-4}\,\text{K}^{-1})^2 \times (298\,\text{K}) \times (1.81\times10^{-5}\,\text{m}^3\,\text{mol}^{-1})}{4.90\times10^{-10}\,\text{Pa}^{-1}}$$
$$= 0.485\,\text{Pa}\,\text{m}^3\,\text{K}^{-1}\,\text{mol}^{-1} = 0.485\,\text{J}\,\text{K}^{-1}\,\text{mol}^{-1}$$

For water, $C_{p,\text{m}} = 75.3\,\text{J}\,\text{K}^{-1}\,\text{mol}^{-1}$, so $C_{V,\text{m}} = 74.8\,\text{J}\,\text{K}^{-1}\,\text{mol}^{-1}$. In some cases, the two heat capacities differ by as much as 30 per cent.

2D.3 Changes in enthalpy

A similar set of operations can be carried out on the enthalpy, $H = U + pV$. The quantities U, p, and V are all state functions; therefore H is also a state function and $\text{d}H$ is an exact differential. It turns out that H is a useful thermodynamic function when the pressure can be controlled: a sign of that is the relation $\Delta H = q_p$ (eqn 2B.2b). Therefore, H can be regarded as a function of p and T, and the argument in Section 2D.2 for the variation of U can be adapted to find an expression for the variation of H with temperature at constant volume.

How is that done? 2D.1 Deriving an expression for the variation of enthalpy with pressure and temperature

Consider a closed system of constant composition. Because H is a function of p and T, when these two quantities change by an infinitesimal amount, the enthalpy changes by

$$\text{d}H = \left(\frac{\partial H}{\partial p}\right)_T \text{d}p + \left(\frac{\partial H}{\partial T}\right)_p \text{d}T$$

The second partial derivative is C_p. The task at hand is to express $(\partial H/\partial p)_T$ in terms of recognizable quantities. If the enthalpy is constant, then $\text{d}H = 0$ and

$$\left(\frac{\partial H}{\partial p}\right)_T \text{d}p = -C_p \text{d}T \quad \text{at constant } H$$

Division of both sides by $\text{d}p$ then gives

$$\left(\frac{\partial H}{\partial p}\right)_T = -C_p\left(\frac{\partial T}{\partial p}\right)_H = -C_p\mu$$

where the **Joule–Thomson coefficient**, μ (mu), is defined as

$$\mu = \left(\frac{\partial T}{\partial p}\right)_H \quad \text{Joule–Thomson coefficient [definition]} \quad \text{(2D.12)}$$

It follows that

$$\text{d}H = -\mu C_p \text{d}p + C_p \text{d}T \quad \text{The variation of enthalpy with temperature and pressure} \quad \text{(2D.13)}$$

Brief illustration 2D.2

The Joule–Thomson coefficient for nitrogen at 298 K and 1 atm (Table 2D.2) is $+0.27\,\text{K}\,\text{atm}^{-1}$. (Note that μ is an intensive property.) It follows that the change in temperature the gas undergoes when its pressure changes by −10 atm under isenthalpic conditions is

$$\Delta T \approx \mu\Delta p = +(0.27\,\text{K}\,\text{atm}^{-1}) \times (-10\,\text{atm}) = -2.7\,\text{K}$$

Table 2D.2 Inversion temperatures (T_I), normal freezing (T_f) and boiling (T_b) points, and Joule–Thomson coefficients (μ) at 1 atm and 298 K*

	T_I/K	T_f/K	T_b/K	μ/(K atm^{-1})
Ar	723	83.8	87.3	
CO_2	1500	194.7	+1.10	+1.11 at 300 K
He	40	4.2	4.22	−0.062
N_2	621	63.3	77.4	+0.27

* More values are given in the *Resource section*.

2D.4 The Joule–Thomson effect

The analysis of the Joule–Thomson coefficient is central to the technological problems associated with the liquefaction of gases. To determine the coefficient, it is necessary to measure the ratio of the temperature change to the change of pressure, $\Delta T/\Delta p$, in a process at constant enthalpy. The cunning required to impose the constraint of constant enthalpy, so that the expansion is **isenthalpic**, was supplied by James Joule and William Thomson (later Lord Kelvin). They let a gas expand through a porous barrier from one constant pressure to another and monitored the difference of temperature that arose from the expansion (Fig. 2D.6). The change of temperature that they observed as a result of isenthalpic expansion is called the **Joule–Thomson effect**.

The 'Linde refrigerator' makes use of the Joule–Thomson effect to liquefy gases (Fig. 2D.7). The gas at high pressure is allowed to expand through a throttle; it cools and is circulated past the incoming gas. That gas is cooled, and its subsequent expansion cools it still further. There comes a stage when the circulating gas becomes so cold that it condenses to a liquid.

(a) The observation of the Joule–Thomson effect

The apparatus Joule and Thomson used was insulated so that the process was adiabatic. By considering the work done at each stage it is possible to show that the expansion is isenthalpic.

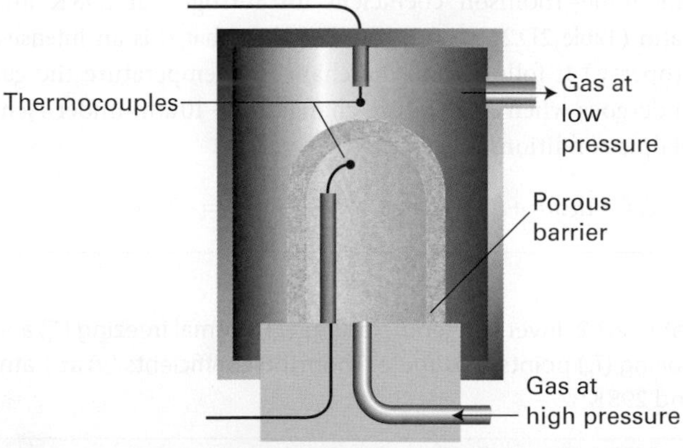

Figure 2D.6 The apparatus used for measuring the Joule–Thomson effect. The gas expands through the porous barrier, which acts as a throttle, and the whole apparatus is thermally insulated. As explained in the text, this arrangement corresponds to an isenthalpic expansion (expansion at constant enthalpy). Whether the expansion results in a heating or a cooling of the gas depends on the conditions.

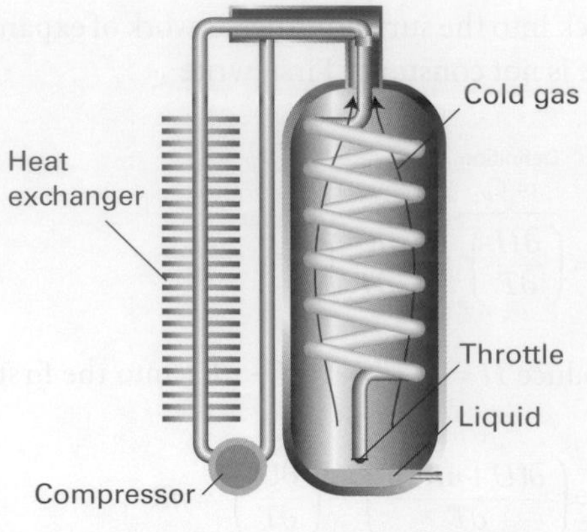

Figure 2D.7 The principle of the Linde refrigerator is shown in this diagram. The gas is recirculated, and so long as it is beneath its inversion temperature it cools on expansion through the throttle. The cooled gas cools the high-pressure gas, which cools still further as it expands. Eventually liquefied gas drips from the throttle.

How is that done? 2D.2 Establishing that the expansion is isenthalpic

Because all changes to the gas occur adiabatically, $q = 0$ and, consequently, $\Delta U = w$.

Step 1 *Calculate the total work*

Consider the work done as the gas passes through the barrier by focusing on the passage of a fixed amount of gas from the high pressure side, where the pressure is p_i, the temperature T_i, and the gas occupies a volume V_i (Fig. 2D.8). The gas

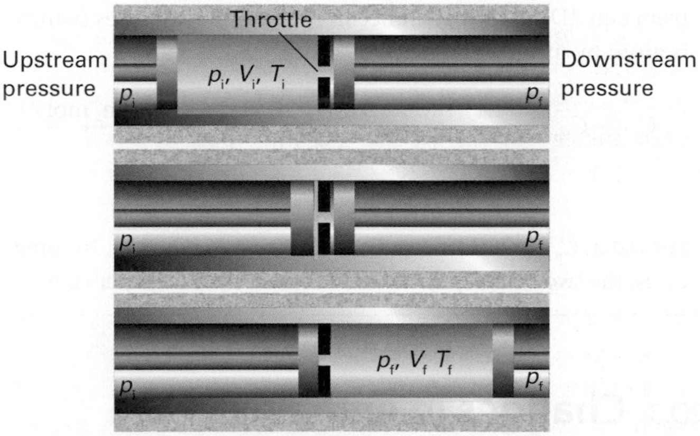

Figure 2D.8 The thermodynamic basis of Joule–Thomson expansion. The pistons represent the upstream and downstream gases, which maintain constant pressures either side of the throttle. The transition from the top diagram to the bottom diagram, which represents the passage of a given amount of gas through the throttle, occurs without change of enthalpy.

emerges on the low pressure side, where the same amount of gas has a pressure p_f, a temperature T_f, and occupies a volume V_f. The gas on the left is compressed isothermally by the upstream gas acting as a piston. The relevant pressure is p_i and the volume changes from V_i to 0; therefore, the work done on the gas is

$$w_1 = -p_i(0 - V_i) = p_iV_i$$

The gas expands isothermally on the right of the barrier (but possibly at a different constant temperature) against the pressure p_f provided by the downstream gas acting as a piston to be driven out. The volume changes from 0 to V_f, so the work done on the gas in this stage is

$$w_2 = -p_f(V_f - 0) = -p_fV_f$$

The total work done on the gas is the sum of these two quantities, or

$$w = w_1 + w_2 = p_iV_i - p_fV_f$$

Step 2 *Calculate the change in internal energy*

It follows that the change of internal energy of the gas as it moves adiabatically from one side of the barrier to the other is

$$U_f - U_i = w = p_iV_i - p_fV_f$$

Step 3 *Calculate the initial and final enthalpies*

Reorganization of the preceding expression, and noting that $H = U + pV$, gives

$$U_f + p_fV_f = U_i + p_iV_i \text{ or } H_f = H_i$$

Therefore, the expansion occurs without change of enthalpy.

For a perfect gas, $\mu = 0$; hence, the temperature of a perfect gas is unchanged by Joule–Thomson expansion. This characteristic points clearly to the involvement of intermolecular forces in determining the size of the effect.

Real gases have non-zero Joule–Thomson coefficients. Depending on the identity of the gas, the pressure, the relative magnitudes of the attractive and repulsive intermolecular forces, and the temperature, the sign of the coefficient may be either positive or negative (Fig. 2D.9). A positive sign implies that dT is negative when dp is negative, in which case the gas cools on expansion. However, the Joule–Thomson coefficient of a real gas does not necessarily approach zero as the pressure is reduced even though the equation of state of the gas approaches that of a perfect gas. The coefficient behaves like the properties discussed in Topic 1C in the sense that it depends on derivatives and not on p, V, and T themselves.

Gases that show a heating effect ($\mu < 0$) at one temperature show a cooling effect ($\mu > 0$) when the temperature is below their upper **inversion temperature**, T_I (Table 2D.2, Fig. 2D.10). As indicated in Fig. 2D.10, a gas typically has two inversion temperatures.

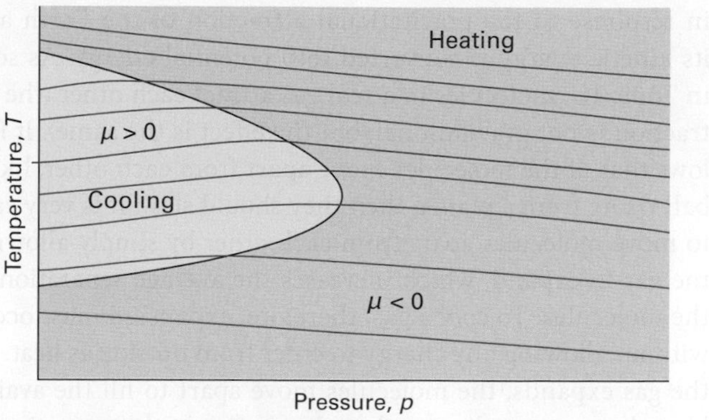

Figure 2D.9 The sign of the Joule–Thomson coefficient, μ, depends on the conditions. Inside the boundary, the blue area, it is positive and outside it is negative. The temperature corresponding to the boundary at a given pressure is the 'inversion temperature' of the gas at that pressure. Reduction of pressure under adiabatic conditions moves the system along one of the isenthalps, or curves of constant enthalpy (the blue lines). The inversion temperature curve runs through the points of the isenthalps where their slope changes from negative to positive.

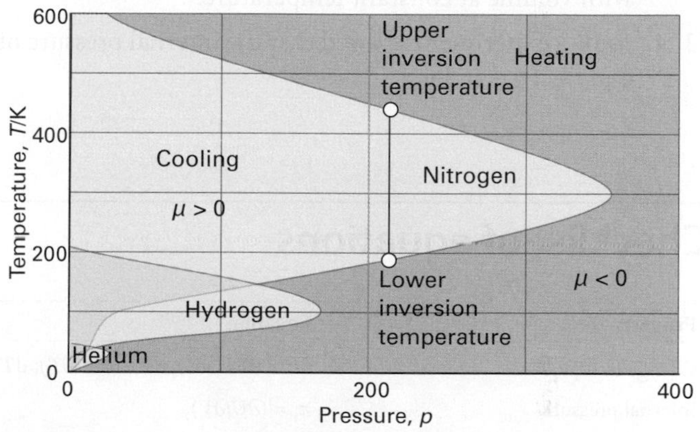

Figure 2D.10 The inversion temperatures for three real gases, nitrogen, hydrogen, and helium.

(b) The molecular interpretation of the Joule–Thomson effect

The kinetic model of gases (Topic 1B) and the equipartition theorem (*The chemist's toolkit* 7 of Topic 2A) jointly imply that the mean kinetic energy of molecules in a gas is proportional to the temperature. It follows that reducing the average speed of the molecules is equivalent to cooling the gas. If the speed of the molecules can be reduced to the point that neighbours can capture each other by their intermolecular attractions, then the cooled gas will condense to a liquid.

Slowing gas molecules makes use of an effect similar to that seen when a ball is thrown up into the air: as it rises it slows in response to the gravitational attraction of the Earth and its kinetic energy is converted into potential energy. As seen in Topic 1C, molecules in a real gas attract each other (the attraction is not gravitational, but the effect is the same). It follows that, if the molecules move apart from each other, like a ball rising from a planet, then they should slow. It is very easy to move molecules apart from each other by simply allowing the gas to expand, which increases the average separation of the molecules. To cool a gas, therefore, expansion must occur without allowing any energy to enter from outside as heat. As the gas expands, the molecules move apart to fill the available volume, struggling as they do so against the attraction of their neighbours. Because some kinetic energy must be converted into potential energy to reach greater separations, the molecules travel more slowly as their separation increases, and the temperature drops. The cooling effect, which corresponds to $\mu > 0$, is observed in real gases under conditions when attractive interactions are dominant ($Z < 1$, where Z is the compression factor defined in eqn 1C.1, $Z = V_m/V_m^\circ$), because the molecules have to climb apart against the attractive force in order for them to travel more slowly. For molecules under conditions when repulsions are dominant ($Z > 1$), the Joule–Thomson effect results in the gas becoming warmer, or $\mu < 0$.

Checklist of concepts

- ☐ 1. The quantity $\mathrm{d}U$ is an exact differential, $\mathrm{d}w$ and $\mathrm{d}q$ are not.
- ☐ 2. The change in internal energy may be expressed in terms of changes in temperature and volume.
- ☐ 3. The **internal pressure** is the variation of internal energy with volume at constant temperature.
- ☐ 4. **Joule's experiment** showed that the internal pressure of a perfect gas is zero.
- ☐ 5. The change in internal energy with pressure and temperature is expressed in terms of the internal pressure and the heat capacity and leads to a general expression for the relation between heat capacities.
- ☐ 6. The **Joule–Thomson effect** is the change in temperature of a gas when it undergoes isenthalpic expansion.

Checklist of equations

Property	Equation	Comment	Equation number
Change in $U(V,T)$	$\mathrm{d}U = (\partial U/\partial V)_T\mathrm{d}V + (\partial U/\partial T)_V\mathrm{d}T$	Constant composition	2D.3
Internal pressure	$\pi_T = (\partial U/\partial V)_T$	Definition; for a perfect gas, $\pi_T = 0$	2D.4
Change in $U(V,T)$	$\mathrm{d}U = \pi_T\mathrm{d}V + C_V\mathrm{d}T$	Constant composition	2D.5
Expansion coefficient	$\alpha = (1/V)(\partial V/\partial T)_p$	Definition	2D.6
Isothermal compressibility	$\kappa_T = -(1/V)(\partial V/\partial p)_T$	Definition	2D.7
Relation between heat capacities	$C_p - C_V = nR$	Perfect gas	2D.10
	$C_p - C_V = \alpha^2 TV/\kappa_T$		2D.11
Joule–Thomson coefficient	$\mu = (\partial T/\partial p)_H$	For a perfect gas, $\mu = 0$	2D.12
Change in $H(p,T)$	$\mathrm{d}H = -\mu C_p\mathrm{d}p + C_p\mathrm{d}T$	Constant composition	2D.13

TOPIC 2E Adiabatic changes

> ➤ **Why do you need to know this material?**
>
> Adiabatic processes complement isothermal processes, and are used in the discussion of the Second Law of thermodynamics.
>
> ➤ **What is the key idea?**
>
> The temperature of a perfect gas falls when it does work in an adiabatic expansion.
>
> ➤ **What do you need to know already?**
>
> This Topic makes use of the discussion of the properties of gases (Topic 1A), particularly the perfect gas law. It also uses the definition of heat capacity at constant volume (Topic 2A) and constant pressure (Topic 2B) and the relation between them (Topic 2D).

The temperature falls when a gas expands adiabatically (in a thermally insulated container). Work is done, but as no heat enters the system, the internal energy falls, and therefore the temperature of the working gas also falls. In molecular terms, the kinetic energy of the molecules falls as work is done, so their average speed decreases, and hence the temperature falls too.

2E.1 The change in temperature

The change in internal energy of a perfect gas when the temperature is changed from T_i to T_f and the volume is changed from V_i to V_f can be expressed as the sum of two steps (Fig. 2E.1). In the first step, only the volume changes and the temperature is held constant at its initial value. However, because the internal energy of a perfect gas is independent of the volume it occupies (Topic 2A), the overall change in internal energy arises solely from the second step, the change in temperature at constant volume. Provided the heat capacity is independent of temperature, the change in the internal energy is

$$\Delta U = (T_f - T_i)C_V = C_V\Delta T$$

Because the expansion is adiabatic, $q = 0$; then because $\Delta U = q + w$, it follows that $\Delta U = w_{ad}$. The subscript 'ad' denotes an

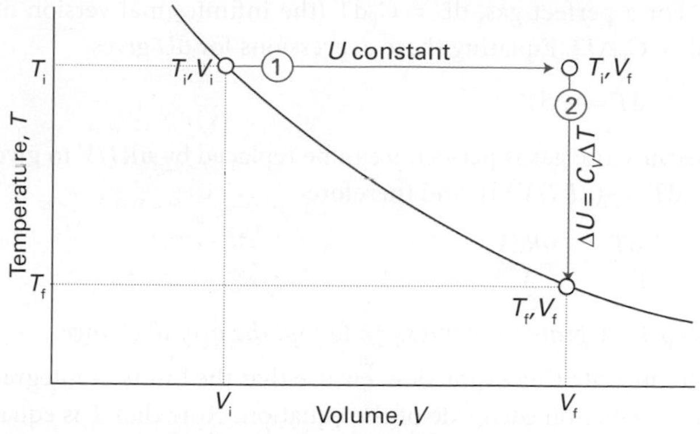

Figure 2E.1 To achieve a change of state from one temperature and volume to another temperature and volume, treat the overall change as composed of two steps. In the first step, the system expands at constant temperature; there is no change in internal energy if the system consists of a perfect gas. In the second step, the temperature of the system is reduced at constant volume. The overall change in internal energy is the sum of the changes for the two steps.

adiabatic process. Therefore, by equating the two expressions for ΔU,

$$w_{ad} = C_V\Delta T \qquad \text{Work of adiabatic change [perfect gas]} \qquad (2E.1)$$

That is, the work done during an adiabatic expansion of a perfect gas is proportional to the temperature difference between the initial and final states. That is exactly what is expected on molecular grounds, because the mean kinetic energy is proportional to T, so a change in internal energy arising from temperature alone is also expected to be proportional to ΔT. From these considerations it is possible to calculate the temperature change of a perfect gas that undergoes reversible adiabatic expansion (reversible expansion in a thermally insulated container).

How is that done? 2E.1 Deriving an expression for the temperature change in a reversible adiabatic expansion

Consider a stage in a reversible adiabatic expansion of a perfect gas when the pressure inside and out is p. When considering reversible processes, it is usually appropriate to consider infinitesimal changes in the conditions, because pressures and temperatures typically change during the process. Then follow these steps.

Step 1 *Write an expression relating temperature and volume changes*

The work done when the gas expands reversibly by dV is $dw = -pdV$. This expression applies to any reversible change, including an adiabatic change, so specifically $dw_{ad} = -pdV$. Therefore, because $dq = 0$ for an adiabatic change, $dU = dw_{ad}$ (the infinitesimal version of $\Delta U = w_{ad}$).

For a perfect gas, $dU = C_V dT$ (the infinitesimal version of $\Delta U = C_V \Delta T$). Equating these expressions for dU gives

$$C_V dT = -pdV$$

Because the gas is perfect, p can be replaced by nRT/V to give $C_V dT = -(nRT/V)dV$ and therefore

$$\frac{C_V dT}{T} = -\frac{nRdV}{V}$$

Step 2 *Integrate the expression to find the overall change*

To integrate this expression, ensure that the limits of integration match on each side of the equation. Note that T is equal to T_i when V is equal to V_i, and is equal to T_f when V is equal to V_f at the end of the expansion. Therefore,

$$C_V \int_{T_i}^{T_f} \frac{dT}{T} = -nR \int_{V_i}^{V_f} \frac{dV}{V}$$

where C_V is taken to be independent of temperature. Use Integral A.2 in each case, and obtain

$$C_V \ln \frac{T_f}{T_i} = -nR \ln \frac{V_f}{V_i}$$

Step 3 *Simplify the expression*

Because $\ln(x/y) = -\ln(y/x)$, the preceding expression rearranges to

$$\frac{C_V}{nR} \ln \frac{T_f}{T_i} = \ln \frac{V_i}{V_f}$$

Next, note that $C_V/nR = C_{V,m}/R = c$ and use $\ln x^a = a \ln x$ to obtain

$$\ln\left(\frac{T_f}{T_i}\right)^c = \ln \frac{V_i}{V_f}$$

This relation implies that $(T_f/T_i)^c = (V_i/V_f)$ and, upon rearrangement,

$$T_f = T_i\left(\frac{V_i}{V_f}\right)^{1/c} \quad c = C_{V,m}/R \qquad \text{(2E.2a)}$$

Temperature change [reversible adiabatic expansion, perfect gas]

By raising each side of this expression to the power c and reorganizing it slightly, an equivalent expression is

$$V_i T_i^c = V_f T_f^c \quad c = C_{V,m}/R \qquad \text{(2E.2b)}$$

Temperature change [reversible adiabatic expansion, perfect gas]

This result is often summarized in the form VT^c = constant.

Brief illustration 2E.1

Consider the adiabatic, reversible expansion of 0.020 mol Ar, initially at 25 °C, from 0.50 dm^3 to 1.00 dm^3. The molar heat capacity of argon at constant volume is 12.47 J K^{-1} mol^{-1}, so $c = 1.501$. Therefore, from eqn 2E.2a,

$$T_f = (298\,\text{K}) \times \left(\frac{0.50\,\text{dm}^3}{1.00\,\text{dm}^3}\right)^{1/1.501} = 188\,\text{K}$$

It follows that $\Delta T = -110$ K, and therefore, from eqn 2E.1, that

$$w_{ad} = \{(0.020\,\text{mol}) \times (12.47\,\text{J K}^{-1}\,\text{mol}^{-1})\} \times (-110\,\text{K}) = -27\,\text{J}$$

Note that temperature change is independent of the amount of gas but the work is not.

2E.2 The change in pressure

Equation 2E.2a may be used to calculate the pressure of a perfect gas that undergoes reversible adiabatic expansion.

How is that done? 2E.2 Deriving the relation between pressure and volume for a reversible adiabatic expansion

The initial and final states of a perfect gas satisfy the perfect gas law regardless of how the change of state takes place, so $pV = nRT$ can be used to write

$$\frac{p_i V_i}{p_f V_f} = \frac{T_i}{T_f}$$

However, $T_i/T_f = (V_f/V_i)^{1/c}$ (eqn 2E.2a). Therefore,

$$\frac{p_i V_i}{p_f V_f} = \left(\frac{V_f}{V_i}\right)^{1/c}, \text{ so } \frac{p_i}{p_f}\left(\frac{V_i}{V_f}\right)^{\frac{1}{c}+1} = 1$$

For a perfect gas $C_{p,m} - C_{V,m} = R$ (Topic 2B). It follows that

$$\frac{1}{c} + 1 = \frac{1+c}{c} = \frac{R + C_{V,m}}{C_{V,m}} = \frac{C_{p,m}}{C_{V,m}} = \gamma$$

and therefore that

$$\frac{p_i}{p_f}\left(\frac{V_i}{V_f}\right)^{\gamma} = 1$$

which rearranges to

$$p_f V_f^{\gamma} = p_i V_i^{\gamma} \qquad \text{(2E.3)}$$

Pressure change [reversible adiabatic expansion, perfect gas]

This result is commonly summarized in the form pV^{γ} = constant.

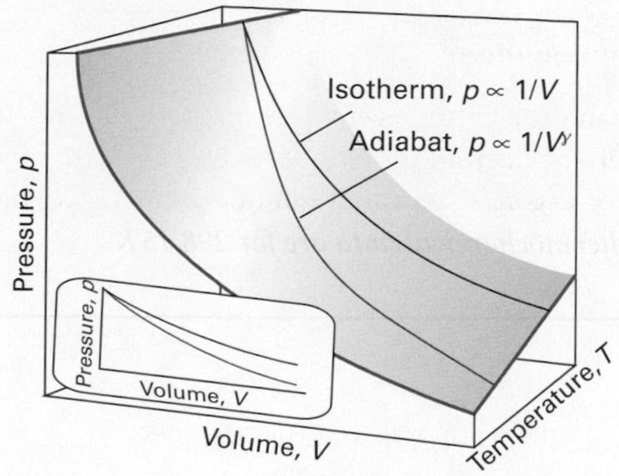

Figure 2E.2 An adiabat depicts the variation of pressure with volume when a gas expands adiabatically and, in this case, reversibly. Note that the pressure declines more steeply for an adiabat than it does for an isotherm because in an adiabatic change the temperature falls.

For a monatomic perfect gas, $C_{V,m} = \frac{3}{2}R$ (Topic 2A), and $C_{p,m} = \frac{5}{2}R$ (from $C_{p,m} - C_{V,m} = R$), so $\gamma = \frac{5}{3}$. For a gas of nonlinear polyatomic molecules (which can rotate as well as translate; vibrations make little contribution at normal temperatures), $C_{V,m} = 3R$ and $C_{p,m} = 4R$, so $\gamma = \frac{4}{3}$. The curves of pressure versus volume for adiabatic change are known as **adiabats**, and one for a reversible path is illustrated in Fig. 2E.2. Because $\gamma > 1$, an adiabat falls more steeply ($p \propto 1/V^\gamma$) than the corresponding isotherm ($p \propto 1/V$). The physical reason for the difference is that, in an isothermal expansion, energy flows into the system as heat and maintains the temperature; as a result, the pressure does not fall as much as in an adiabatic expansion.

Brief illustration 2E.2

When a sample of argon (for which $\gamma = \frac{5}{3}$) at 100 kPa expands reversibly and adiabatically to twice its initial volume the final pressure will be

$$p_f = \left(\frac{V_i}{V_f}\right)^\gamma p_i = \left(\frac{1}{2}\right)^{5/3} \times (100\,\text{kPa}) = 31\,\text{kPa}$$

For an isothermal expansion in which the volume doubles the final pressure would be 50 kPa.

Checklist of concepts

☐ 1. The temperature of a gas falls when it undergoes an adiabatic expansion in which work is done.

☐ 2. An **adiabat** is a curve showing how pressure varies with volume in an adiabatic process.

Checklist of equations

Property	Equation	Comment	Equation number
Work of adiabatic expansion	$w_{ad} = C_V\Delta T$	Perfect gas	2E.1
Final temperature	$T_f = T_i(V_i/V_f)^{1/c}$	Perfect gas, reversible adiabatic expansion	2E.2a
	$c = C_{V,m}/R$		
	$V_iT_i^c = V_fT_f^c$		2E.2b
Adiabats	$p_fV_f^\gamma = p_iV_i^\gamma$		2E.3
	$\gamma = C_{p,m}/C_{V,m}$		

FOCUS 2 The First Law

Assume all gases are perfect unless stated otherwise. Unless otherwise stated, thermochemical data are for 298.15 K.

TOPIC 2A Internal energy

Discussion questions

D2A.1 Describe and distinguish the various uses of the words 'system' and 'state' in physical chemistry.

D2A.2 Describe the distinction between heat and work in thermodynamic terms and, by referring to populations and energy levels, in molecular terms.

D2A.3 Identify varieties of additional work.

D2A.4 Distinguish between reversible and irreversible expansion.

D2A.5 How may the isothermal expansion of a gas be achieved?

Exercises

E2A.1(a) Use the equipartition theorem to estimate the molar internal energy of (i) I_2, (ii) CH_4, (iii) C_6H_6 in the gas phase at 25 °C.

E2A.1(b) Use the equipartition theorem to estimate the molar internal energy of (i) O_3, (ii) C_2H_6, (iii) SO_2 in the gas phase at 25 °C.

E2A.2(a) Which of (i) pressure, (ii) temperature, (iii) work, (iv) enthalpy are state functions?

E2A.2(b) Which of (i) volume, (ii) heat, (iii) internal energy, (iv) density are state functions?

E2A.3(a) A chemical reaction takes place in a container fitted with a piston of cross-sectional area 50 cm^2. As a result of the reaction, the piston is pushed out through 15 cm against an external pressure of 1.0 atm. Calculate the work done by the system.

E2A.3(b) A chemical reaction takes place in a container fitted with a piston of cross-sectional area 75.0 cm^2. As a result of the reaction, the piston is pushed out through 25.0 cm against an external pressure of 150 kPa. Calculate the work done by the system.

E2A.4(a) A sample consisting of 1.00 mol Ar is expanded isothermally at 20 °C from 10.0 dm^3 to 30.0 dm^3 (i) reversibly, (ii) against a constant external pressure equal to the final pressure of the gas, and (iii) freely (against zero external pressure). For the three processes calculate q, w, and ΔU.

E2A.4(b) A sample consisting of 2.00 mol He is expanded isothermally at 0 °C from 5.0 dm^3 to 20.0 dm^3 (i) reversibly, (ii) against a constant external pressure equal to the final pressure of the gas, and (iii) freely (against zero external pressure). For the three processes calculate q, w, and ΔU.

E2A.5(a) A sample consisting of 1.00 mol of perfect gas atoms, for which $C_{V,\mathrm{m}} = \frac{3}{2}R$, initially at $p_1 = 1.00$ atm and $T_1 = 300$ K, is heated reversibly to 400 K at constant volume. Calculate the final pressure, ΔU, q, and w.

E2A.5(b) A sample consisting of 2.00 mol of perfect gas molecules, for which $C_{V,\mathrm{m}} = \frac{5}{2}R$, initially at $p_1 = 111$ kPa and $T_1 = 277$ K, is heated reversibly to 356 K at constant volume. Calculate the final pressure, ΔU, q, and w.

E2A.6(a) A sample of 4.50 g of methane occupies 12.7 dm^3 at 310 K. (i) Calculate the work done when the gas expands isothermally against a constant external pressure of 200 Torr until its volume has increased by 3.3 dm^3. (ii) Calculate the work that would be done if the same expansion occurred reversibly.

E2A.6(b) A sample of argon of mass 6.56 g occupies 18.5 dm^3 at 305 K. (i) Calculate the work done when the gas expands isothermally against a constant external pressure of 7.7 kPa until its volume has increased by 2.5 dm^3. (ii) Calculate the work that would be done if the same expansion occurred reversibly.

Problems

P2A.1 Calculate the molar internal energy of carbon dioxide at 25 °C, taking into account its translational and rotational degrees of freedom.

P2A.2 A generator does work on an electric heater by forcing an electric current through it. Suppose 1 kJ of work is done on the heater and in turn 1 kJ of energy as heat is transferred to its surroundings. What is the change in internal energy of the heater?

P2A.3 An elastomer is a polymer that can stretch and contract. In a perfect elastomer the force opposing extension is proportional to the displacement x from the resting state of the elastomer, so $|F| = k_f x$, where k_f is a constant. But suppose that the restoring force weakens as the elastomer is stretched, and $k_f(x) = a - bx^{1/2}$. Evaluate the work done on extending the polymer from $x = 0$ to a final displacement $x = l$.

P2A.4 An approximate model of a DNA molecule is the 'one-dimensional freely jointed chain', in which a rigid unit of length l can make an angle of only 0° or 180° with an adjacent unit. In this case, the restoring force of a chain extended by $x = nl$ is given by

$$F = \frac{kT}{2l}\ln\left(\frac{1+\nu}{1-\nu}\right) \qquad \nu = \frac{n}{N}$$

where k is Boltzmann's constant, N is the total number of units, and $l = 45$ nm for DNA. (a) What is the magnitude of the force that must be applied to extend a DNA molecule with $N = 200$ by 90 nm? (b) Plot the restoring force against ν, noting that ν can be either positive or negative. How is the variation of the restoring force with end-to-end distance different from that predicted by Hooke's law? (c) Keeping in mind that the difference in end-to-end distance from an equilibrium value is $x = nl$ and, consequently, $dx = l dn = Nl d\nu$, write an expression for the work of extending a DNA molecule. *Hint:* You must integrate the expression for w. The task can be accomplished best with mathematical software.

P2A.5 As a continuation of Problem P2A.4, (a) show that for small extensions of the chain, when $\nu \ll 1$, the restoring force is given by

$$F \approx \frac{\nu kT}{l} = \frac{nkT}{Nl}$$

(b) Is the variation of the restoring force with extension of the chain given in part (a) different from that predicted by Hooke's law? Explain your answer.

P2A.6 Suppose that attractions are the dominant interactions between gas molecules, and the equation of state is $p = nRT/V - n^2a/V^2$. Derive an expression for the work of reversible, isothermal expansion of such a gas. Compared with a perfect gas, is more or less work done *on the surroundings* when it expands?

P2A.7 Calculate the work done during the isothermal reversible expansion of a van der Waals gas (Topic 1C). Plot on the same graph the indicator diagrams (graphs of pressure against volume) for the isothermal reversible expansion of (a) a perfect gas, (b) a van der Waals gas in which $a = 0$ and $b = 5.11 \times 10^{-2}\,\mathrm{dm^3\,mol^{-1}}$, and (c) $a = 4.2\,\mathrm{dm^6\,atm\,mol^{-2}}$ and $b = 0$. The values selected exaggerate the imperfections but give rise to significant effects on the indicator diagrams. Take $V_i = 1.0\,\mathrm{dm^3}$, $V_f = 2.0\,\mathrm{dm^3}$, $n = 1.0\,\mathrm{mol}$, and $T = 298\,\mathrm{K}$.

P2A.8 A sample consisting of 1.0 mol $CaCO_3(s)$ was heated to 800 °C, at which temperature the solid decomposed to CaO and CO_2. The heating was carried out in a container fitted with a piston that was initially resting on the solid. Calculate the work done during complete decomposition at 1.0 atm. What work would be done if instead of having a piston the container was open to the atmosphere?

P2A.9 Calculate the work done during the isothermal reversible expansion of a gas that satisfies the virial equation of state (eqn 1C.3b) written with the first three terms. Evaluate (a) the work for 1.0 mol Ar at 273 K (for data, see Table 1C.3) and (b) the same amount of a perfect gas. Let the expansion be from 500 cm^3 to 1000 cm^3 in each case.

P2A.10 Express the work of an isothermal reversible expansion of a van der Waals gas in reduced variables (Topic 1C) and find a definition of reduced work that makes the overall expression independent of the identity of the gas. Calculate the work of isothermal reversible expansion along the critical isotherm from V_c to xV_c.

TOPIC 2B Enthalpy

Discussion questions

D2B.1 Explain the difference between the change in internal energy and the change in enthalpy accompanying a process.

D2B.2 Why is the heat capacity at constant pressure of a substance normally greater than its heat capacity at constant volume?

Exercises

E2B.1(a) When 229 J of energy is supplied as heat at constant pressure to 3.0 mol Ar(g) the temperature of the sample increases by 2.55 K. Calculate the molar heat capacities at constant volume and constant pressure of the gas.

E2B.1(b) When 178 J of energy is supplied as heat at constant pressure to 1.9 mol of gas molecules, the temperature of the sample increases by 1.78 K. Calculate the molar heat capacities at constant volume and constant pressure of the gas.

E2B.2(a) Calculate the value of $\Delta H_m - \Delta U_m$ for the reaction $N_2(g) + 3\,H_2(g) \rightarrow 2\,NH_3(g)$ at 298 K.

E2B.2(b) Calculate the value of $\Delta H_m - \Delta U_m$ for the reaction $C_6H_{12}O_6(s) + 6\,O_2(g) \rightarrow 6\,CO_2(g) + 6\,H_2O(l)$ at 298 K.

E2B.3(a) The constant-pressure heat capacity of a sample of a perfect gas was found to vary with temperature according to the expression $C_p/(\mathrm{J\,K^{-1}}) = 20.17 + 0.3665(T/\mathrm{K})$. Calculate q, w, ΔU, and ΔH when the temperature is raised from 25 °C to 100 °C (i) at constant pressure, (ii) at constant volume.

E2B.3(b) The constant-pressure heat capacity of a sample of a perfect gas was found to vary with temperature according to the expression $C_p/(\mathrm{J\,K^{-1}}) = 20.17 + 0.4001(T/\mathrm{K})$. Calculate q, w, ΔU, and ΔH when the temperature is raised from 25 °C to 100 °C (i) at constant pressure, (ii) at constant volume.

E2B.4(a) When 3.0 mol O_2 is heated at a constant pressure of 3.25 atm, its temperature increases from 260 K to 285 K. Given that the molar heat capacity of O_2 at constant pressure is 29.4 J K^{-1} mol^{-1}, calculate q, ΔH, and ΔU.

E2B.4(b) When 2.0 mol CO_2 is heated at a constant pressure of 1.25 atm, its temperature increases from 250 K to 277 K. Given that the molar heat capacity of CO_2 at constant pressure is 37.11 J K^{-1} mol^{-1}, calculate q, ΔH, and ΔU.

Problems

P2B.1 Benzene is heated to boiling under a pressure of 1.0 atm with a 12 V source operating at an electric current of 0.50 A. For how long would a current need to be supplied in order to vaporize 10 g of benzene? The molar enthalpy of vaporization of benzene at its boiling point (353.25 K) is 30.8 kJ mol^{-1}.

P2B.2 The heat capacity of air is much smaller than that of liquid water, and relatively modest amounts of heat are therefore needed to change the temperature of air. This is one of the reasons why desert regions, though very hot during the day, are bitterly cold at night. The molar heat capacity of air at 298 K and 1.00 atm is approximately 21 J K^{-1} mol^{-1}. Estimate how much energy is required to raise the temperature of the air in a room of dimensions 5.5 m × 6.5 m × 3.0 m by 10 °C. If losses are neglected, how long will it take a heater rated at 1.5 kW to achieve that increase, given that 1 W = 1 J s^{-1}?

P2B.3 The following data show how the standard molar constant-pressure heat capacity of sulfur dioxide varies with temperature:

T/K	300	500	700	900	1100	1300	1500
$C^{\ominus}_{p,m}/(\mathrm{J\,K^{-1}\,mol^{-1}})$	39.909	46.490	50.829	53.407	54.993	56.033	56.759

By how much does the standard molar enthalpy of $SO_2(g)$ increase when the temperature is raised from 298.15 K to 1500 K? *Hint:* Fit the data to an expression of the form of $C^{\ominus}_{p,m}(T) = a + bT + c/T^2$, note the values of the coefficients, then use the approach in *Example* 2B.2 to calculate the change in standard molar enthalpy.

P2B.4 The following data show how the standard molar constant-pressure heat capacity of ammonia depends on the temperature. Use mathematical software to fit an expression of the form of eqn 2B.8 to the data and determine the values of *a*, *b*, and *c*. Explore whether it would be better to express the data as $C_{p,m} = \alpha + \beta T + \gamma T^2$, and determine the values of these coefficients.

T/K	300	400	500	600	700	800	900	1000
$C^{\ominus}_{p,m}/(\mathrm{J\,K^{-1}\,mol^{-1}})$	35.678	38.674	41.994	45.229	48.269	51.112	53.769	56.244

P2B.5 A sample consisting of 2.0 mol CO_2 occupies a fixed volume of 15.0 dm^3 at 300 K. When it is supplied with 2.35 kJ of energy as heat its temperature increases to 341 K. Assuming that CO_2 is described by the van der Waals equation of state (Topic 1C), calculate w, ΔU, and ΔH.

TOPIC 2C Thermochemistry

Discussion questions

D2C.1 A simple air-conditioning unit for use in places where electrical power is not available can be made by hanging up strips of fabric soaked in water. Explain why this strategy is effective.

D2C.2 Describe two calorimetric methods for the determination of enthalpy changes that accompany chemical processes.

D2C.3 Distinguish between 'standard state' and 'reference state', and indicate their applications.

D2C.4 The expressions 'heat of combustion' and 'heat of vaporization' are used commonly, especially in the earlier literature. Why are the expressions 'enthalpy of combustion' and 'enthalpy of vaporization' more appropriate?

Exercises

E2C.1(a) For tetrachloromethane, $\Delta_{vap}H^{\ominus} = 30.0\,\mathrm{kJ\,mol^{-1}}$. Calculate q, w, ΔH, and ΔU when 0.75 mol $CCl_4(l)$ is vaporized at 250 K and 1 bar.

E2C.1(b) For ethanol, $\Delta_{vap}H^{\ominus} = 43.5\,\mathrm{kJ\,mol^{-1}}$. Calculate q, w, ΔH, and ΔU when 1.75 mol $C_2H_5OH(l)$ is vaporized at 260 K and 1 bar.

E2C.2(a) The standard enthalpy of formation of ethylbenzene is $-12.5\,\mathrm{kJ\,mol^{-1}}$. Calculate its standard enthalpy of combustion.

E2C.2(b) The standard enthalpy of formation of phenol is $-165.0\,\mathrm{kJ\,mol^{-1}}$. Calculate its standard enthalpy of combustion.

E2C.3(a) Given that the standard enthalpy of formation of HCl(aq) is $-167\,\mathrm{kJ\,mol^{-1}}$, what is the value of $\Delta_f H^{\ominus}(\mathrm{Cl^-, aq})$?

E2C.3(b) Given that the standard enthalpy of formation of HI(aq) is $-55\,\mathrm{kJ\,mol^{-1}}$, what is the value of $\Delta_f H^{\ominus}(\mathrm{I^-, aq})$?

E2C.4(a) When 120 mg of naphthalene, $C_{10}H_8(s)$, was burned in a bomb calorimeter the temperature rose by 3.05 K. Calculate the calorimeter constant. By how much will the temperature rise when 150 mg of phenol, $C_6H_5OH(s)$, is burned in the calorimeter under the same conditions? ($\Delta_c H^{\ominus}(C_{10}H_8,s) = -5157\,\mathrm{kJ\,mol^{-1}}$.)

E2C.4(b) When 2.25 mg of anthracene, $C_{14}H_{10}(s)$, was burned in a bomb calorimeter the temperature rose by 1.75 K. Calculate the calorimeter constant. By how much will the temperature rise when 125 mg of phenol, $C_6H_5OH(s)$, is burned in the calorimeter under the same conditions? ($\Delta_c H^{\ominus}(C_{14}H_{10},s) = -7061\,\mathrm{kJ\,mol^{-1}}$.)

E2C.5(a) Given the reactions (1) and (2) below, determine (i) $\Delta_r H^{\ominus}$ and $\Delta_r U^{\ominus}$ for reaction (3), (ii) $\Delta_f H^{\ominus}$ for both HCl(g) and H_2O(g), all at 298 K.

(1) $H_2(g) + Cl_2(g) \rightarrow 2\,HCl(g)$ $\quad \Delta_r H^{\ominus} = -184.62\,\mathrm{kJ\,mol^{-1}}$

(2) $2\,H_2(g) + O_2(g) \rightarrow 2\,H_2O(g)$ $\quad \Delta_r H^{\ominus} = -483.64\,\mathrm{kJ\,mol^{-1}}$

(3) $4\,HCl(g) + O_2(g) \rightarrow 2\,Cl_2(g) + 2\,H_2O(g)$

E2C.5(b) Given the reactions (1) and (2) below, determine (i) $\Delta_r H^{\ominus}$ and $\Delta_r U^{\ominus}$ for reaction (3), (ii) $\Delta_f H^{\ominus}$ for both HI(g) and H_2O(g), all at 298 K.

(1) $H_2(g) + I_2(s) \rightarrow 2\,HI(g)$ $\quad \Delta_r H^{\ominus} = +52.96\,\mathrm{kJ\,mol^{-1}}$

(2) $2\,H_2(g) + O_2(g) \rightarrow 2\,H_2O(g)$ $\quad \Delta_r H^{\ominus} = -483.64\,\mathrm{kJ\,mol^{-1}}$

(3) $4\,HI(g) + O_2(g) \rightarrow 2\,I_2(s) + 2\,H_2O(g)$

E2C.6(a) For the reaction $C_2H_5OH(l) + 3\,O_2(g) \rightarrow 2\,CO_2(g) + 3\,H_2O(g)$, $\Delta_r U^{\ominus} = -1373\,\mathrm{kJ\,mol^{-1}}$ at 298 K. Calculate $\Delta_r H^{\ominus}$.

E2C.6(b) For the reaction $2\,C_6H_5COOH(s) + 15\,O_2(g) \rightarrow 14\,CO_2(g) + 6\,H_2O(g)$, $\Delta_r U^{\ominus} = -772.7\,\mathrm{kJ\,mol^{-1}}$ at 298 K. Calculate $\Delta_r H^{\ominus}$.

E2C.7(a) From the data in Table 2C.4 of the *Resource section*, calculate $\Delta_r H^{\ominus}$ and $\Delta_r U^{\ominus}$ at (i) 298 K, (ii) 478 K for the reaction C(graphite) + $H_2O(g) \rightarrow CO(g) + H_2(g)$. Assume all heat capacities to be constant over the temperature range of interest.

E2C.7(b) Calculate $\Delta_r H^{\ominus}$ and $\Delta_r U^{\ominus}$ at 298 K and $\Delta_r H^{\ominus}$ at 427 K for the hydrogenation of ethyne (acetylene) to ethene (ethylene) from the enthalpy of combustion and heat capacity data in Tables 2C.3 and 2C.4 of the *Resource section*. Assume the heat capacities to be constant over the temperature range involved.

E2C.8(a) Estimate $\Delta_r H^{\ominus}(500\,\mathrm{K})$ for the reaction C(graphite) + $O_2(g) \rightarrow CO_2(g)$ from the listed value of the standard enthalpy of formation of CO_2(g) at 298 K in conjunction with the data on the temperature-dependence of heat capacities given in Table 2B.1.

E2C.8(b) Estimate $\Delta_r H^{\ominus}(750\,\mathrm{K})$ for the reaction $N_2(g) + H_2(g) \rightarrow NH_3(g)$ from the listed value of the standard enthalpy of formation of NH_3(g) at 298 K in conjunction with the data on the temperature-dependence of heat capacities given in Table 2B.1.

Problems

P2C.1 An average human produces about 10 MJ of heat each day through metabolic activity. If a human body were an isolated system of mass 65 kg with the heat capacity of water, what temperature rise would the body experience? Human bodies are actually open systems, and the main mechanism of heat loss is through the evaporation of water. What mass of water should be evaporated each day to maintain constant temperature?

P2C.2 Predict the output of energy as heat from the combustion of 1.0 dm^3 of octane at 298 K and 1 bar. Its mass density is 0.703 g cm^{-3}.

P2C.3 The standard enthalpy of combustion of cyclopropane is $-2091\,\mathrm{kJ\,mol^{-1}}$ at 25 °C. (a) From this information and enthalpy of formation data for CO_2(g) and H_2O(l), calculate the enthalpy of formation of cyclopropane.

(b) The enthalpy of formation of propene is +20.42 kJ mol^{-1}. Calculate the enthalpy of isomerization of cyclopropane to propene.

P2C.4 From the following data, determine $\Delta_f H^\ominus$ for diborane, B_2H_6(g), at 298 K:

$$(1)\quad B_2H_6(g) + 3O_2(g) \rightarrow B_2O_3(s) + 3H_2O(g) \qquad \Delta_r H^\ominus = -1941\ \mathrm{kJ\,mol^{-1}}$$

$$(2)\quad 2B(s) + \tfrac{3}{2}O_2(g) \rightarrow B_2O_3(s) \qquad \Delta_r H^\ominus = -2368\ \mathrm{kJ\,mol^{-1}}$$

$$(3)\quad H_2(g) + \tfrac{1}{2}O_2(g) \rightarrow H_2O(g) \qquad \Delta_r H^\ominus = -241.8\ \mathrm{kJ\,mol^{-1}}$$

P2C.5 A sample of the sugar D-ribose ($C_5H_{10}O_5$) of mass 0.727 g was placed in a calorimeter and then ignited in the presence of excess oxygen. The temperature rose by 0.910 K. In a separate experiment in the same calorimeter, the combustion of 0.825 g of benzoic acid, for which the internal energy of combustion is −3251 kJ mol^{-1}, gave a temperature rise of 1.940 K. Calculate the enthalpy of formation of D-ribose.

P2C.6 For the reaction $Cr(C_6H_6)_2(s) \rightarrow Cr(s) + 2C_6H_6(g)$, $\Delta_r U^\ominus(583\ \mathrm{K}) =$ +8.0 kJ mol^{-1}. Find the corresponding reaction enthalpy and estimate the standard enthalpy of formation of $Cr(C_6H_6)_2$(s) at 583 K.

P2C.7‡ Kolesov et al. reported the standard enthalpy of combustion and of formation of crystalline C_{60} based on calorimetric measurements (V.P. Kolesov et al., *J. Chem. Thermodynamics* **28**, 1121 (1996)). In one of their runs, they found the standard specific internal energy of combustion to be −36.0334 kJ g^{-1} at 298.15 K. Compute $\Delta_c H^\ominus$ and $\Delta_f H^\ominus$ of C_{60}.

P2C.8‡ Silylene (SiH_2) is a key intermediate in the thermal decomposition of silicon hydrides such as silane (SiH_4) and disilane (Si_2H_6). H.K. Moffat et al. (*J. Phys. Chem.* **95**, 145 (1991)) report $\Delta_f H^\ominus(SiH_2) = +274\ \mathrm{kJ\,mol^{-1}}$. Given that $\Delta_f H^\ominus(SiH_4) = +34.3\ \mathrm{kJ\,mol^{-1}}$ and $\Delta_f H^\ominus(Si_2H_6) = +80.3\ \mathrm{kJ\,mol^{-1}}$, calculate the standard enthalpy changes of the following reactions:

(a) $SiH_4(g) \rightarrow SiH_2(g) + H_2(g)$

(b) $Si_2H_6(g) \rightarrow SiH_2(g) + SiH_4(g)$

P2C.9 As remarked in Problem P2B.4, it is sometimes appropriate to express the temperature dependence of the heat capacity by the empirical expression $C_{p,m} = \alpha + \beta T + \gamma T^2$. Use this expression to estimate the standard enthalpy of combustion of methane to carbon dioxide and water vapour at 500 K. Use the following data:

	$\alpha/(\mathrm{J\,K^{-1}\,mol^{-1}})$	$\beta/(\mathrm{mJ\,K^{-2}\,mol^{-1}})$	$\gamma/(\mathrm{\mu J\,K^{-3}\,mol^{-1}})$
CH_4(g)	14.16	75.5	−17.99
CO_2(g)	26.86	6.97	−0.82
O_2(g)	25.72	12.98	−3.862
H_2O(g)	30.36	9.61	1.184

P2C.10 Figure 2.1 shows the experimental DSC scan of hen white lysozyme (G. Privalov et al., *Anal. Biochem.* **79**, 232 (1995)) converted to joules (from calories). Determine the enthalpy of unfolding of this protein by integration of the curve and the change in heat capacity accompanying the transition.

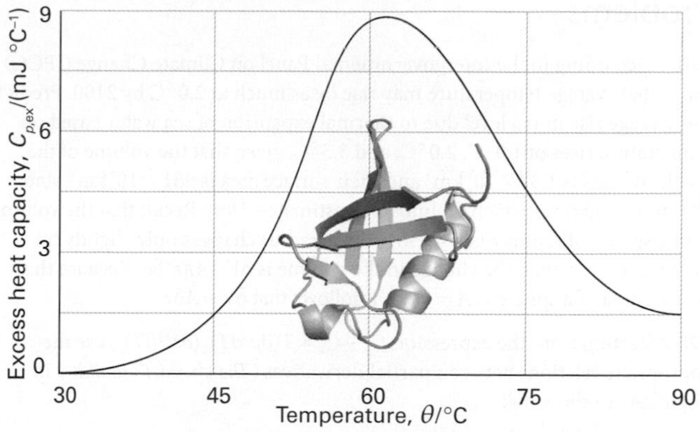

Figure 2.1 The experimental DSC scan of hen white lysozyme.

P2C.11 In biological cells that have a plentiful supply of oxygen, glucose is oxidized completely to CO_2 and H_2O by a process called *aerobic oxidation*. Muscle cells may be deprived of O_2 during vigorous exercise and, in that case, one molecule of glucose is converted to two molecules of lactic acid ($CH_3CH(OH)COOH$) by a process called *anaerobic glycolysis*. (a) When 0.3212 g of glucose was burned at 298 K in a bomb calorimeter of calorimeter constant 641 J K^{-1} the temperature rose by 7.793 K. Calculate (i) the standard molar enthalpy of combustion, (ii) the standard internal energy of combustion, and (iii) the standard enthalpy of formation of glucose. (b) What is the biological advantage (in kilojoules per mole of energy released as heat) of complete aerobic oxidation compared with anaerobic glycolysis to lactic acid?

TOPIC 2D State functions and exact differentials

Discussion questions

D2D.1 Suggest (with explanation) how the internal energy of a van der Waals gas should vary with volume at constant temperature.

D2D.2 Explain why a perfect gas does not have an inversion temperature.

Exercises

E2D.1(a) Estimate the internal pressure of water vapour at 1.00 bar and 400 K, treating it as a van der Waals gas, when $\pi_T = a/V_m^2$. You may simplify the problem by assuming that the molar volume can be predicted from the perfect gas equation.

E2D.1(b) Estimate the internal pressure of sulfur dioxide at 1.00 bar and 298 K, treating it as a van der Waals gas, when $\pi_T = a/V_m^2$. You may simplify the problem by assuming that the molar volume can be predicted from the perfect gas equation.

E2D.2(a) For a van der Waals gas, $\pi_T = a/V_m^2$. Assuming that this relation applies, calculate ΔU_m for the isothermal expansion of nitrogen gas from an initial volume of 1.00 dm^3 to 20.00 dm^3 at 298 K. What are the values of q and w?

‡ These problems were supplied by Charles Trapp and Carmen Giunta.

E2D.2(b) Repeat Exercise E2D.2(a) for argon, from an initial volume of $1.00\,dm^3$ to $30.00\,dm^3$ at 298 K.

E2D.3(a) The volume of a certain liquid varies with temperature as

$$V = V'\{0.75 + 3.9 \times 10^{-4}(T/K) + 1.48 \times 10^{-6}(T/K)^2\}$$

where V' is its volume at 300 K. Calculate its expansion coefficient, α, at 320 K.

E2D.3(b) The volume of a certain liquid varies with temperature as

$$V = V'\{0.77 + 3.7 \times 10^{-4}(T/K) + 1.52 \times 10^{-6}(T/K)^2\}$$

where V' is its volume at 298 K. Calculate its expansion coefficient, α, at 310 K.

E2D.4(a) The isothermal compressibility, κ_T, of water at 293 K is $4.96 \times 10^{-5}\,atm^{-1}$. Calculate the pressure that must be applied in order to increase its density by 0.10 per cent.

E2D.4(b) The isothermal compressibility, κ_T, of lead at 293 K is $2.21 \times 10^{-6}\,atm^{-1}$. Calculate the pressure that must be applied in order to increase its density by 0.10 per cent.

E2D.5(a) Use data from the *Resource section* to evaluate the difference $C_{p,m} - C_{V,m}$ in molar heat capacities for liquid benzene at 298 K.

E2D.5(b) Use data from the *Resource section* to evaluate the difference $C_{p,m} - C_{V,m}$ in molar heat capacities for liquid ethanol at 298 K.

Problems

P2D.1‡ According to the Intergovernmental Panel on Climate Change (IPCC) the global average temperature may rise by as much as 2.0 °C by 2100. Predict the average rise in sea level due to thermal expansion of sea water based on temperature rises of 1.0 °C, 2.0 °C, and 3.5 °C, given that the volume of the Earth's oceans is $1.37 \times 10^9\,km^3$ and their surface area is $361 \times 10^6\,km^2$; state the approximations which go into your estimates. *Hint:* Recall that the volume V of a sphere of radius r is $V = \frac{4}{3}\pi r^3$. If the radius changes only slightly by δr, with $\delta r << r$, then the change in the volume is $\delta V \approx 4\pi r^2 \delta r$. Because the surface area of a sphere is $A = 4\pi r^2$, it follows that $\delta V \approx A\delta r$.

P2D.2 Starting from the expression $C_p - C_V = T(\partial p/\partial T)_V(\partial V/\partial T)_p$, use the appropriate relations between partial derivatives (*The chemist's toolkit* 9 in Topic 2A) to show that

$$C_p - C_V = \frac{T(\partial V/\partial T)_p^2}{(\partial V/\partial p)_T}$$

Use this expression to evaluate $C_p - C_V$ for a perfect gas.

P2D.3 (a) Write expressions for dV and dp given that V is a function of p and T and p is a function of V and T. (b) Deduce expressions for $d\ln V$ and $d\ln p$ in terms of the expansion coefficient and the isothermal compressibility.

P2D.4 Rearrange the van der Waals equation of state, $p = nRT/(V - nb) - n^2a/V^2$ (Topic 1C) to give an expression for T as a function of p and V (with n constant). Calculate $(\partial T/\partial p)_V$ and confirm that $(\partial T/\partial p)_V = 1/(\partial p/\partial T)_V$.

P2D.5 Calculate the isothermal compressibility and the expansion coefficient of a van der Waals gas (see Problem P2D.4). Show, using Euler's chain relation (*The chemist's toolkit* 9 in Topic 2A), that $\kappa_T R = \alpha(V_m - b)$.

P2D.6 The speed of sound, c_s, in a perfect gas of molar mass M is related to the ratio of heat capacities γ by $c_s = (\gamma RT/M)^{1/2}$. Show that $c_s = (\gamma p/\rho)^{1/2}$, where ρ is the mass density of the gas. Calculate the speed of sound in argon at 25 °C.

P2D.7‡ A gas obeying the equation of state $p(V - nb) = nRT$ is subjected to a Joule–Thomson expansion. Will the temperature increase, decrease, or remain the same?

P2D.8 Use the fact that $(\partial U/\partial V)_T = a/V_m^2$ for a van der Waals gas (Topic 1C) to show that $\mu C_{p,m} \approx (2a/RT) - b$ by using the definition of μ and appropriate relations between partial derivatives. *Hint:* Use the approximation $pV_m \approx RT$ when it is justifiable to do so.

P2D.9‡ Concerns over the harmful effects of chlorofluorocarbons on stratospheric ozone have motivated a search for new refrigerants. One such alternative is 1,1,1,2-tetrafluoroethane (refrigerant HFC-134a). A compendium of thermophysical properties of this substance has been published (R. Tillner-Roth and H.D. Baehr, *J. Phys. Chem. Ref. Data* **23**, 657 (1994)) from which properties such as the Joule–Thomson coefficient μ can be computed. (a) Compute μ at 0.100 MPa and 300 K from the following data (all referring to 300 K):

p/MPa	0.080	0.100	0.12
Specific enthalpy/(kJ kg^{-1})	426.48	426.12	425.76

(The specific constant-pressure heat capacity is $0.7649\,kJ\,K^{-1}\,kg^{-1}$.) (b) Compute μ at 1.00 MPa and 350 K from the following data (all referring to 350 K):

p/MPa	0.80	1.00	1.2
Specific enthalpy/(kJ kg^{-1})	461.93	459.12	456.15

(The specific constant-pressure heat capacity is $1.0392\,kJ\,K^{-1}\,kg^{-1}$.)

TOPIC 2E Adiabatic changes

Discussion questions

D2E.1 On a p against V plot, why are adiabats steeper than isotherms?

D2E.2 Why do heat capacities play a role in the expressions for adiabatic expansion?

Exercises

E2E.1(a) Use the equipartition principle to estimate the values of $\gamma = C_p/C_V$ for gaseous ammonia and methane. Do this calculation with and without the vibrational contribution to the energy. Which is closer to the experimental value at 25 °C?

E2E.1(b) Use the equipartition principle to estimate the value of $\gamma = C_p/C_V$ for carbon dioxide. Do this calculation with and without the vibrational contribution to the energy. Which is closer to the experimental value at 25 °C?

E2E.2(a) Calculate the final temperature of a sample of argon of mass 12.0 g that is expanded reversibly and adiabatically from $1.0\,dm^3$ at 273.15 K to $3.0\,dm^3$.

E2E.2(b) Calculate the final temperature of a sample of carbon dioxide of mass 16.0 g that is expanded reversibly and adiabatically from $500\,cm^3$ at 298.15 K to $2.00\,dm^3$.

E2E.3(a) A sample consisting of 1.0 mol of perfect gas molecules with $C_V = 20.8\,\mathrm{J\,K^{-1}}$ is initially at 4.25 atm and 300 K. It undergoes reversible adiabatic expansion until its pressure reaches 2.50 atm. Calculate the final volume and temperature, and the work done.

E2E.3(b) A sample consisting of 2.5 mol of perfect gas molecules with $C_{p,\mathrm{m}} = 20.8\,\mathrm{J\,K^{-1}\,mol^{-1}}$ is initially at 240 kPa and 325 K. It undergoes reversible adiabatic expansion until its pressure reaches 150 kPa. Calculate the final volume and temperature, and the work done.

E2E.4(a) A sample of carbon dioxide of mass 2.45 g at 27.0 °C is allowed to expand reversibly and adiabatically from 500 cm^3 to 3.00 dm^3. What is the work done by the gas?

E2E.4(b) A sample of nitrogen of mass 3.12 g at 23.0 °C is allowed to expand reversibly and adiabatically from 400 cm^3 to 2.00 dm^3. What is the work done by the gas?

E2E.5(a) Calculate the final pressure of a sample of carbon dioxide that expands reversibly and adiabatically from 67.4 kPa and 0.50 dm^3 to a final volume of 2.00 dm^3. Take $\gamma = 1.4$.

E2E.5(b) Calculate the final pressure of a sample of water vapour that expands reversibly and adiabatically from 97.3 Torr and 400 cm^3 to a final volume of 5.0 dm^3. Take $\gamma = 1.3$.

Problems

P2E.1 Calculate the final temperature, the work done, and the change of internal energy when 1.00 mol NH_3(g) at 298 K is used in a reversible adiabatic expansion from 0.50 dm^3 to 2.00 dm^3.

P2E.2 The constant-volume heat capacity of a gas can be measured by observing the decrease in temperature when it expands adiabatically and reversibly. The value of $\gamma = C_p/C_V$ can be inferred if the decrease in pressure is also measured and the constant-pressure heat capacity deduced by combining the two values. A fluorocarbon gas was allowed to expand reversibly and adiabatically to twice its volume; as a result, the temperature fell from 298.15 K to 248.44 K and its pressure fell from 202.94 kPa to 81.840 kPa. Evaluate $C_{p,\mathrm{m}}$.

FOCUS 2 The First Law

Integrated activities

I2.1 Give examples of state functions and discuss why they play a critical role in thermodynamics.

I2.2 The thermochemical properties of hydrocarbons are commonly investigated by using molecular modelling methods. (a) Use software to predict $\Delta_c H^\ominus$ values for the alkanes methane through pentane. To calculate $\Delta_c H^\ominus$ values, estimate the standard enthalpy of formation of C_nH_{2n+2}(g) by performing semi-empirical calculations (e.g. AM1 or PM3 methods) and use experimental standard enthalpy of formation values for CO_2(g) and H_2O(l). (b) Compare your estimated values with the experimental values of $\Delta_c H^\ominus$ (Table 2C.3 of the *Resource section*) and comment on the reliability of the molecular modelling method. (c) Test the extent to which the relation $\Delta_c H^\ominus = \text{constant} \times \{M/(\mathrm{g\,mol^{-1}})\}^n$ holds and determine the numerical values of the constant and n.

I2.3 It is often useful to be able to anticipate, without doing a detailed calculation, whether an increase in temperature will result in a raising or a lowering of a reaction enthalpy. The constant-pressure molar heat capacity of a gas of linear molecules is approximately $\frac{7}{2}R$ whereas that of a gas of nonlinear molecules is approximately $4R$. Decide whether the standard enthalpies of the following reactions will increase or decrease with increasing temperature:

(a) $2\,H_2(g) + O_2(g) \rightarrow 2\,H_2O(g)$

(b) $CH_4(g) + 2\,O_2(g) \rightarrow CO_2(g) + 2\,H_2O(g)$

(c) $N_2(g) + 3\,H_2(g) \rightarrow 2\,NH_3(g)$

I2.4 The molar heat capacity of liquid water is approximately $9R$. Decide whether the standard enthalpy of the first two reactions in the preceding exercise will increase or decrease with a rise in temperature if the water is produced as a liquid.

I2.5 As shown in *The chemist's toolkit* 9 in Topic 2A, it is a property of partial derivatives that

$$\left(\frac{\partial}{\partial y}\left(\frac{\partial f}{\partial x}\right)_y\right)_x = \left(\frac{\partial}{\partial x}\left(\frac{\partial f}{\partial y}\right)_x\right)_y$$

Use this property and eqn 2A.14 to write an expression for $(\partial C_V/\partial V)_T$ as a second derivative of U and find its relation to $(\partial U/\partial V)_T$. Then show that $(\partial C_V/\partial V)_T = 0$ for a perfect gas.

I2.6 The heat capacity ratio of a gas determines the speed of sound in it through the formula $c_s = (\gamma RT/M)^{1/2}$, where $\gamma = C_p/C_V$ and M is the molar mass of the gas. Deduce an expression for the speed of sound in a perfect gas of (a) diatomic, (b) linear triatomic, (c) nonlinear triatomic molecules at high temperatures (with translation and rotation active). Estimate the speed of sound in air at 25 °C.

I2.7 Use mathematical software or a spreadsheet (a) to calculate the work of isothermal reversible expansion of 1.0 mol CO_2(g) at 298 K from 1.0 dm^3 to 3.0 dm^3 on the basis that it obeys the van der Waals equation of state; (b) explore how the parameter γ affects the dependence of the pressure on the volume when the expansion is reversible and adiabatic and the gas is perfect. Does the pressure–volume dependence become stronger or weaker with increasing volume?

FOCUS 3

The Second and Third Laws

Some things happen naturally, some things don't. Some aspect of the world determines the **spontaneous** direction of change, the direction of change that does not require work to bring it about. An important point, though, is that throughout this text 'spontaneous' must be interpreted as a natural *tendency* which might or might not be realized in practice. Thermodynamics is silent on the rate at which a spontaneous change in fact occurs, and some spontaneous processes (such as the conversion of diamond to graphite) may be so slow that the tendency is never realized in practice whereas others (such as the expansion of a gas into a vacuum) are almost instantaneous.

3A Entropy

The direction of change is related to the *distribution of energy and matter*, and spontaneous changes are always accompanied by a dispersal of energy or matter. To quantify this concept we introduce the property called 'entropy', which is central to the formulation of the 'Second Law of thermodynamics'. That law governs all spontaneous change.

3B Entropy changes accompanying specific processes

This Topic shows how to use the definition of entropy change to calculate its value for a number of common physical processes, such as the expansion of a gas, a phase transition, and heating a substance.

3C The measurement of entropy

To make the Second Law quantitative, it is necessary to measure the entropy of a substance. The measurement of heat capacities, and the energy transferred as heat during physical processes, makes it possible to determine the entropies of substances. The discussion in this Topic also leads to the 'Third Law of thermodynamics', which relates to the properties of matter at very low temperatures and is used to set up an absolute measure of the entropy of a substance.

3D Concentrating on the system

One problem with dealing with the entropy is that it requires separate calculations of the changes taking place in the system and the surroundings. Providing certain restrictions on the system can be accepted, that problem can be overcome by introducing the 'Gibbs energy'. Indeed, most thermodynamic calculations in chemistry focus on the change in Gibbs energy rather than the entropy change itself.

3E Combining the First and Second Laws

In this Topic the First and Second Laws are combined, which leads to a very powerful way of applying thermodynamics to the properties of matter.

Web resources What are the applications of this material?

The Second Law is at the heart of the operation of engines of all types, including devices resembling engines that are used to cool objects. See *Impact* 4 on the website of this book for an application to the technology of refrigeration. Entropy considerations are also important in modern electronic materials for they permit a quantitative discussion of the concentration of impurities. See *Impact* 5 for a note about how measurement of the entropy at low temperatures gives insight into the purity of materials used as superconductors.

TOPIC 3A Entropy

➤ Why do you need to know this material?

Entropy is the concept on which almost all applications of thermodynamics in chemistry are based: it explains why some physical transformations and chemical reactions are spontaneous and others are not.

➤ What is the key idea?

The change in entropy of a system can be calculated from the heat transferred to it reversibly; a spontaneous process in an isolated system is accompanied by an increase in entropy.

➤ What do you need to know already?

You need to be familiar with the First-Law concepts of work, heat, and internal energy (Topic 2A). The Topic draws on the expression for work of expansion of a perfect gas (Topic 2A) and on the changes in volume and temperature that accompany the reversible adiabatic expansion of a perfect gas (Topic 2E).

What determines the direction of spontaneous change? It is not a tendency to achieve a lower energy, because the First Law asserts that the total energy of the universe does not change in any process. It turns out that the direction is determined by the manner in which energy and matter are distributed. This concept is made precise by the Second Law of thermodynamics and made quantitative by introducing the property known as 'entropy'.

3A.1 The Second Law

The role of the distribution of energy and matter can be appreciated by thinking about a ball bouncing on a floor. The ball does not rise as high after each bounce because some of the energy associated with its motion spreads out—is dispersed—into the thermal motion of the particles in the ball and the floor. The direction of spontaneous change is towards a state in which the ball is at rest with all its energy dispersed into disorderly thermal motion of the particles in the surroundings (Fig. 3A.1).

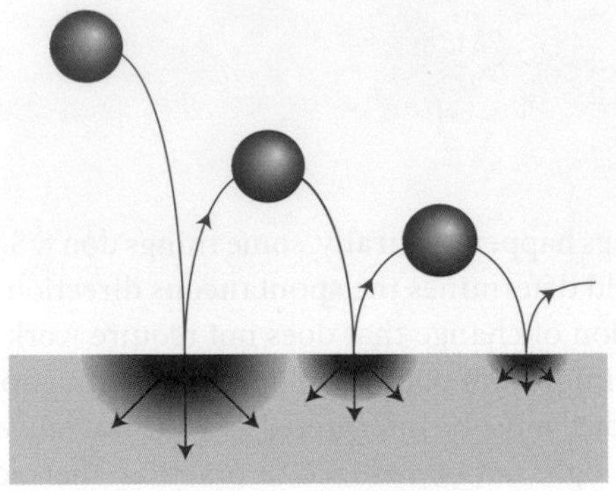

Figure 3A.1 The direction of spontaneous change for a ball bouncing on a floor. On each bounce some of its energy is degraded into the thermal motion of the atoms of the floor, and that energy then disperses. The reverse process, a ball rising from the floor as a result of acquiring energy from the thermal motion of the atoms in the floor, has never been observed to take place.

A ball resting on a warm floor has never been observed to start bouncing as a result of energy transferred to the ball from the floor. For bouncing to begin, something rather special would need to happen. In the first place, some of the thermal motion of the atoms in the floor (the surroundings) would have to accumulate in a single, small object, the ball (the system). This accumulation requires a spontaneous localization of energy from the myriad vibrations of the atoms of the floor into the much smaller number of atoms that constitute the ball (Fig. 3A.2). Furthermore, whereas the thermal motion is random, for the ball to move upwards its atoms must all move in the same direction. The localization of random, disorderly motion as directed, orderly motion is so unlikely that it can be dismissed as virtually impossible.[1]

The signpost of spontaneous change has been identified: *look for the direction of change that leads to the dispersal of energy.* This principle accounts for the direction of change of the bouncing ball, because its energy is spread out as thermal motion of the atoms of the floor. The reverse process is not spontaneous because it is highly improbable that energy will become localized, leading to uniform motion of the ball's atoms.

[1] Orderly motion, but on a much smaller scale and continued only very briefly, is observed as *Brownian motion*, the jittering motion of small particles suspended in a liquid or gas.

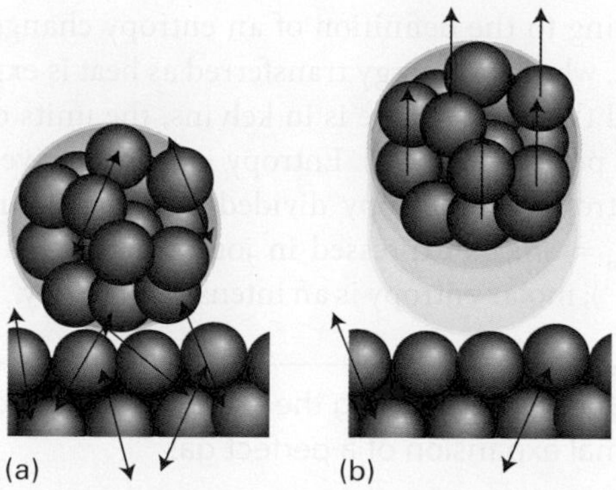

Figure 3A.2 (a) A ball resting on a warm surface; the atoms are undergoing thermal motion (vibration, in this instance), as indicated by the arrows. (b) For the ball to fly upwards, some of the random vibrational motion would have to change into coordinated, directed motion. Such a conversion is highly improbable.

Matter also has a tendency to disperse. A gas does not contract spontaneously because to do so the random motion of its molecules would have to take them all into the same region of the container. The opposite change, spontaneous expansion, is a natural consequence of matter becoming more dispersed as the gas molecules are free to occupy a larger volume.

The **Second Law of thermodynamics** expresses these conclusions more precisely and without referring to the behaviour of the molecules that are responsible for the properties of bulk matter. One statement was formulated by Kelvin:

> No process is possible in which the sole result is the absorption of heat from a reservoir and its complete conversion into work.

Statements like this are commonly explored by thinking about an idealized device called a *heat engine* (Fig. 3A.3(a)). A heat engine consists of two reservoirs, one hot (the 'hot source') and one cold (the 'cold sink'), connected in such a way that some of the energy flowing as heat between the two reservoirs can be converted into work. The Kelvin statement implies that it is not possible to construct a heat engine in which all the heat drawn from the hot source is completely converted into work (Fig. 3A.3(b)): all working heat engines must have a cold sink. The Kelvin statement is a generalization of the everyday observation that a ball at rest on a surface has never been observed to leap spontaneously upwards. An upward leap of the ball would be equivalent to the spontaneous conversion of heat from the surface into the work of raising the ball.

Another statement of the Second Law is due to Rudolf Clausius (Fig. 3A.4):

> Heat does not flow spontaneously from a cool body to a hotter body.

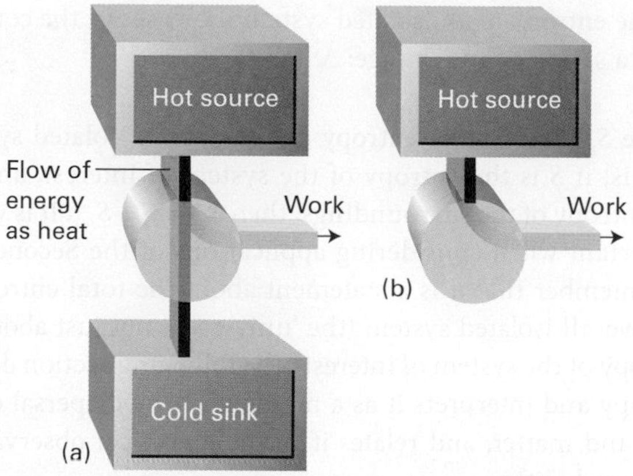

Figure 3A.3 (a) A heat engine is a device in which energy is extracted from a hot reservoir (the hot source) as heat and then some of that energy is converted into work and the rest discarded into a cold reservoir (the cold sink) as heat. (b) The Kelvin statement of the Second Law denies the possibility of the process illustrated here, in which heat is changed completely into work, there being no other change.

To achieve the transfer of heat to a hotter body, it is necessary to do work on the system, as in a refrigerator. Although they appear somewhat different, it can be shown that the Clausius statement is logically equivalent to the Kelvin statement. One way to do so is to show that the two observations can be summarized by a single statement.

First, the system and its surroundings are regarded as a single (and possibly huge) isolated system sometimes referred to as 'the universe'. Energy can be transferred within this isolated system between the actual system and its surroundings, but none can enter or leave it. Then the Second Law is expressed in terms of a new state function, the **entropy**, S:

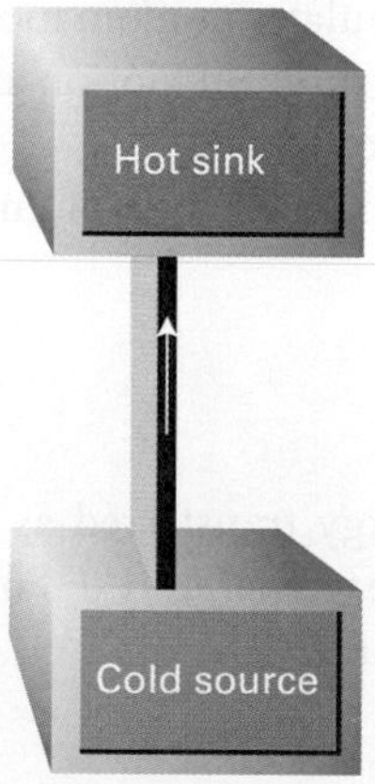

Figure 3A.4 According to the Clausius statement of the Second Law, the process shown here, in which energy as heat migrates from a cool source to a hot sink, does not take place spontaneously. The process is not in conflict with the First Law because energy is conserved.

The entropy of an isolated system increases in the course of a spontaneous change: $\Delta S_{\text{tot}} > 0$

where S_{tot} is the total entropy of the overall isolated system. That is, if S is the entropy of the system of interest, and S_{sur} the entropy of the surroundings, then $S_{\text{tot}} = S + S_{\text{sur}}$. It is vitally important when considering applications of the Second Law to remember that it is a statement about the total entropy of the overall isolated system (the 'universe'), not just about the entropy of the system of interest. The following section defines entropy and interprets it as a measure of the dispersal of energy and matter, and relates it to the empirical observations discussed so far.

In summary, the First Law uses the internal energy to identify *permissible* changes; the Second Law uses the entropy to identify which of these permissible changes are *spontaneous.*

3A.2 The definition of entropy

To make progress, and to turn the Second Law into a quantitatively useful expression, the entropy change accompanying various processes needs to be defined and calculated. There are two approaches, one classical and one molecular. They turn out to be equivalent, but each one enriches the other.

(a) The thermodynamic definition of entropy

The thermodynamic definition of entropy concentrates on the change in entropy, dS, that occurs as a result of a physical or chemical change (in general, as a result of a 'process'). The definition is motivated by the idea that a change in the extent to which energy is dispersed in a disorderly way depends on how much energy is transferred as heat, not as work. As explained in Topic 2A, heat stimulates random motion of atoms whereas work stimulates their uniform motion and so does not change the extent of their disorder.

The thermodynamic definition of entropy is based on the expression

$$dS = \frac{dq_{\text{rev}}}{T} \qquad \text{Entropy change [definition]} \qquad (3A.1a)$$

where q_{rev} is the energy transferred as heat reversibly to the system at the absolute temperature T. For a measurable change between two states i and f,

$$\Delta S = \int_{\text{i}}^{\text{f}} \frac{dq_{\text{rev}}}{T} \qquad (3A.1b)$$

That is, to calculate the difference in entropy between any two states of a system, find a *reversible* path between them, and integrate the energy supplied as heat at each stage of the path divided by the temperature at which that heat is transferred.

According to the definition of an entropy change given in eqn 3A.1a, when the energy transferred as heat is expressed in joules and the temperature is in kelvins, the units of entropy are joules per kelvin (J K^{-1}). Entropy is an extensive property. Molar entropy, the entropy divided by the amount of substance, $S_{\text{m}} = S/n$, is expressed in joules per kelvin per mole (J K^{-1} mol^{-1}); molar entropy is an intensive property.

Example 3A.1 Calculating the entropy change for the isothermal expansion of a perfect gas

Calculate the entropy change of a sample of perfect gas when it expands isothermally from a volume V_{i} to a volume V_{f}.

Collect your thoughts The definition of entropy change in eqn 3A.1b instructs you to find the energy supplied as heat for a reversible path between the stated initial and final states regardless of the actual manner in which the process takes place. The process is isothermal, so T can be treated as a constant and taken outside the integral in eqn 3A.1b. Moreover, because the internal energy of a perfect gas is independent of its volume (Topic 2A), $\Delta U = 0$ for the expansion. Then, because $\Delta U = q + w$, it follows that $q = -w$, and therefore that $q_{\text{rev}} = -w_{\text{rev}}$. The work of reversible isothermal expansion is calculated in Topic 2A. Finally, calculate the change in molar entropy from $\Delta S_{\text{m}} = \Delta S/n$.

The solution The temperature is constant, so eqn 3A.1b becomes

$$\Delta S = \frac{1}{T}\int_{\text{i}}^{\text{f}} dq_{\text{rev}} = \frac{q_{\text{rev}}}{T}$$

From Topic 2A the reversible work in an isothermal expansion is $w_{\text{rev}} = -nRT\ln(V_{\text{f}}/V_{\text{i}})$, hence $q_{\text{rev}} = nRT\ln(V_{\text{f}}/V_{\text{i}})$. It follows, after dividing q_{rev} by T, that

$$\Delta S = nR\ln\frac{V_{\text{f}}}{V_{\text{i}}} \quad \text{and} \quad \Delta S_{\text{m}} = R\ln\frac{V_{\text{f}}}{V_{\text{i}}}$$

Self-test 3A.1 Calculate the change in entropy when the pressure of a fixed amount of perfect gas is changed isothermally from p_{i} to p_{f}. What is the origin of this change?

Answer: $\Delta S = nR\ln(p_{\text{i}}/p_{\text{f}})$; the change in volume when the gas is compressed or expands.

To see how the definition in eqn 3A.1a is used to formulate an expression for the change in entropy of the surroundings, ΔS_{sur}, consider an infinitesimal transfer of heat dq_{sur} from the system to the surroundings. The surroundings consist of a reservoir of constant volume, so the energy supplied to them by heating can be identified with the change in the internal energy of the surroundings, dU_{sur}.[2] The internal energy is a state function, and dU_{sur} is an exact differential. These properties

[2] Alternatively, the surroundings can be regarded as being at constant pressure, in which case $dq_{\text{sur}} = dH_{\text{sur}}$.

imply that dU_{sur} is independent of how the change is brought about and in particular it is independent of whether the process is reversible or irreversible. The same remarks therefore apply to dq_{sur}, to which dU_{sur} is equal. Therefore, the definition in eqn 3A.1a can be adapted simply by deleting the constraint 'reversible' and writing

$$dS_{sur} = \frac{dq_{sur}}{T_{sur}} \quad \text{Entropy change of the surroundings} \quad (3A.2a)$$

Furthermore, because the temperature of the surroundings is constant whatever the change, for a measurable change

$$\Delta S_{sur} = \frac{q_{sur}}{T_{sur}} \quad (3A.2b)$$

That is, regardless of how the change is brought about in the system, reversibly or irreversibly, the change of entropy of the surroundings is calculated simply by dividing the heat transferred by the temperature at which the transfer takes place.

Equation 3A.2b makes it very simple to calculate the changes in entropy of the surroundings that accompany any process. For instance, for any adiabatic change, $q_{sur} = 0$, so

$$\Delta S_{sur} = 0 \quad \text{Adiabatic change} \quad (3A.3)$$

This expression is true however the change takes place, reversibly or irreversibly, provided no local hot spots are formed in the surroundings. That is, it is true (as always assumed) provided the surroundings remain in internal equilibrium. If hot spots do form, then the localized energy may subsequently disperse spontaneously and hence generate more entropy.

Brief illustration 3A.1

To calculate the entropy change in the surroundings when 1.00 mol $H_2O(l)$ is formed from its elements under standard conditions at 298 K, use $\Delta_f H^\ominus = -286\,kJ\,mol^{-1}$ from Table 2C.4. The energy released as heat from the system is supplied to the surroundings, so $q_{sur} = +286\,kJ$. Therefore,

$$\Delta S_{sur} = \frac{2.86\times10^5\,J}{298\,K} = +960\,J\,K^{-1}$$

This strongly exothermic reaction results in an increase in the entropy of the surroundings as energy is released as heat into them.

You are now in a position to see how the definition of entropy is consistent with Kelvin's and Clausius's statements of the Second Law and unifies them. In Fig. 3A.3(b) the entropy of the hot source is reduced as energy leaves it as heat. The transfer of energy as work does not result in the production of entropy, so the overall result is that the entropy of the (overall isolated) system decreases. The Second Law asserts that such a process is not spontaneous, so the arrangement shown in Fig. 3A.3(b) does not produce work. In the Clausius version, the entropy of the cold source in Fig 3A.4 decreases when energy leaves it as heat, but when that heat enters the hot sink the rise in entropy is not as great (because the temperature is higher). Overall there is a decrease in entropy and so the transfer of heat from a cold source to a hot sink is not spontaneous.

(b) The statistical definition of entropy

The molecular interpretation of the Second Law and the 'statistical' definition of entropy start from the idea, introduced in the *Prologue*, that atoms and molecules are distributed over the energy states available to them in accord with the Boltzmann distribution. Then it is possible to predict that as the temperature is increased the molecules populate higher energy states. Boltzmann proposed that there is a link between the spread of molecules over the available energy states and the entropy, which he expressed as[3]

$$S = k \ln \mathcal{W} \quad \text{Boltzmann formula for the entropy} \quad (3A.4)$$

where k is Boltzmann's constant ($k = 1.381\times10^{-23}\,J\,K^{-1}$) and $\mathcal{W}$ is the number of **microstates**, the number of ways in which the molecules of a system can be distributed over the energy states for a specified total energy. When the properties of a system are measured, the outcome is an average taken over the many microstates the system can occupy under the prevailing conditions. The concept of the number of microstates makes quantitative the ill-defined qualitative concepts of 'disorder' and 'the dispersal of matter and energy' used to introduce the concept of entropy: a more disorderly distribution of matter and a greater dispersal of energy corresponds to a greater number of microstates associated with the same total energy. This point is discussed in much greater detail in Topic 13E.

Equation 3A.4 is known as the **Boltzmann formula** and the entropy calculated from it is called the **statistical entropy**. If all the molecules are in one energy state there is only one way of achieving this distribution, so $\mathcal{W} = 1$ and, because $\ln 1 = 0$, it follows that $S = 0$. As the molecules spread out over the available energy states, $\mathcal{W}$ increases and therefore so too does the entropy. The value of $\mathcal{W}$ also increases if the separation of energy states decreases, because more states become accessible. An example is a gas confined to a container, because its translational energy levels get closer together as the container expands (Fig. 3A.5; this is a conclusion from quantum theory which is verified in Topic 7D). The value of $\mathcal{W}$, and hence the entropy, is expected to increase as the gas expands, which is in accord with the conclusion drawn from the thermodynamic definition of entropy (*Example* 3A.1).

[3] He actually wrote S = k log W, and it is carved on his tombstone in Vienna.

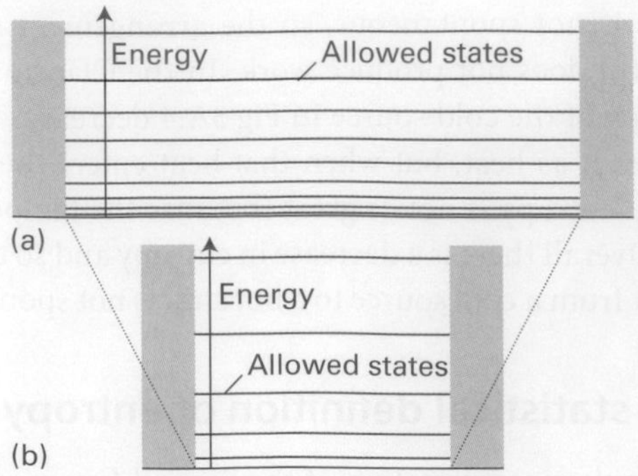

Figure 3A.5 When a container expands from (b) to (a), the translational energy levels of gas molecules in it come closer together and, for the same temperature, more become accessible to the molecules. As a result the number of ways of achieving the same energy (the value of $\mathcal{W}$) increases, and so therefore does the entropy.

The molecular interpretation of entropy helps to explain why, in the thermodynamic definition given by eqn 3A.1, the entropy change depends inversely on the temperature. In a system at high temperature the molecules are spread out over a large number of energy states. Increasing the energy of the system by the transfer of heat makes more states accessible, but given that very many states are already occupied the proportionate change in $\mathcal{W}$ is small (Fig. 3A.6). In contrast, for a system at a low temperature fewer states are occupied, and so the transfer of the same energy results in a proportionately larger increase in the number of accessible states, and hence a larger increase in $\mathcal{W}$. This argument suggests that the change in entropy for a given transfer of energy as heat should be greater at low temperatures than at high, as in eqn 3A.1a.

There are several final points. One is that the Boltzmann definition of entropy makes it possible to calculate the absolute value of the entropy of a system, whereas the thermodynamic definition leads only to values for a change in entropy. This point is developed in Focus 13 where it is shown how to relate values of S to the structural properties of atoms and molecules. The second point is that the Boltzmann formula cannot readily be applied to the surroundings, which are typically far too complex for $\mathcal{W}$ to be a meaningful quantity.

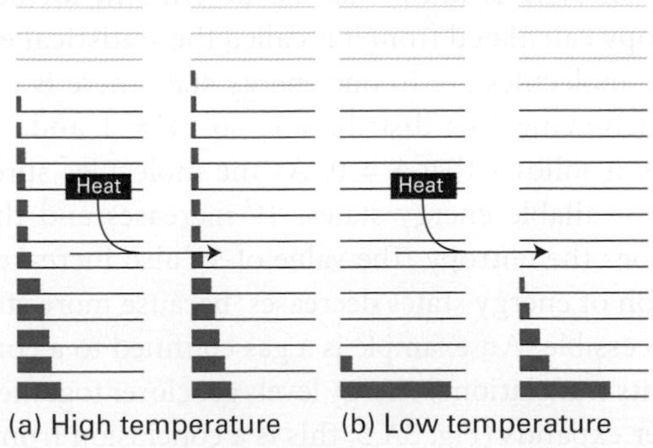

Figure 3A.6 The supply of energy as heat to the system results in the molecules moving to higher energy states, so increasing the number of microstates and hence the entropy. The increase in the entropy is smaller for (a) a system at a high temperature than (b) one at a low temperature because initially the number of occupied states is greater.

3A.3 The entropy as a state function

Entropy is a state function. To prove this assertion, it is necessary to show that the integral of dS between any two states is independent of the path between them. To do so, it is sufficient to prove that the integral of eqn 3A.1a round an arbitrary cycle is zero, for that guarantees that the entropy is the same at the initial and final states of the system regardless of the path taken between them (Fig. 3A.7). That is, it is necessary to show that

$$\oint \mathrm{d}S = \oint \frac{\mathrm{d}q_{\mathrm{rev}}}{T} = 0 \qquad (3A.5)$$

where the symbol $\oint$ denotes integration around a closed path. There are three steps in the argument:

1. First, to show that eqn 3A.5 is true for a special cycle (a 'Carnot cycle') involving a perfect gas.
2. Then to show that the result is true whatever the working substance.
3. Finally, to show that the result is true for any cycle.

(a) The Carnot cycle

A **Carnot cycle**, which is named after the French engineer Sadi Carnot, consists of four reversible stages in which a gas (the working substance) is either expanded or compressed in

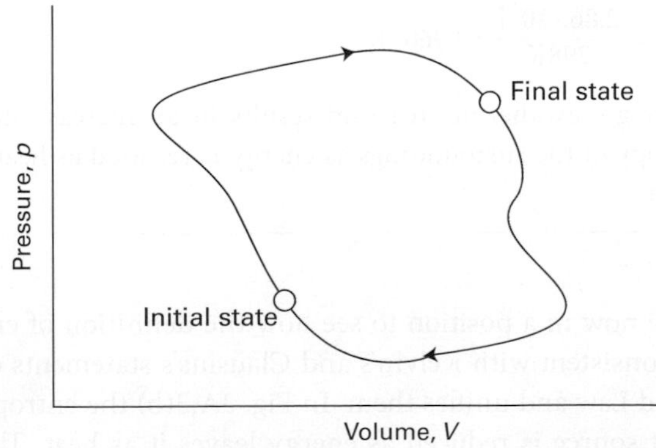

Figure 3A.7 In a thermodynamic cycle, the overall change in a state function (from the initial state to the final state and then back to the initial state again) is zero.

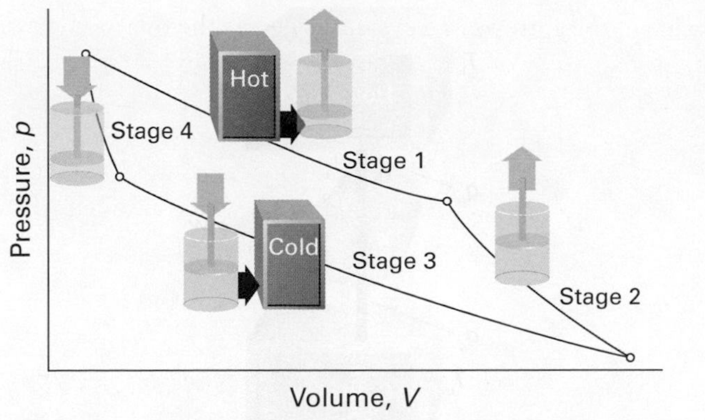

Figure 3A.8 The four stages which make up the Carnot cycle. In stage 1 the gas (the working substance) is in thermal contact with the hot reservoir, and in stage 3 contact is with the cold reservoir; both stages are isothermal. Stages 2 and 4 are adiabatic, with the gas isolated from both reservoirs.

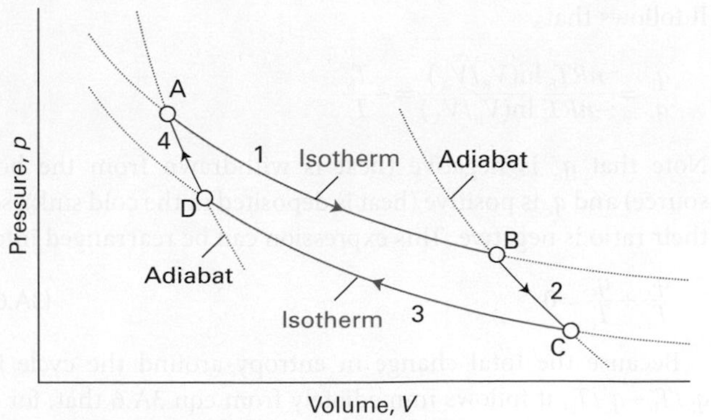

Figure 3A.9 The basic structure of a Carnot cycle. Stage 1 is the isothermal reversible expansion at the temperature T_h. Stage 2 is a reversible adiabatic expansion in which the temperature falls from T_h to T_c. Stage 3 is an isothermal reversible compression at T_c. Stage 4 is an adiabatic reversible compression, which restores the system to its initial state.

various ways; in two of the stages energy as heat is transferred to or from a hot source or a cold sink (Fig. 3A.8).

Figure 3A.9 shows how the pressure and volume change in each stage:

1. The gas is placed in thermal contact with the hot source (which is at temperature T_h) and undergoes reversible isothermal expansion from A to B; the entropy change is q_h/T_h, where q_h is the energy supplied to the system as heat from the hot source.
2. Contact with the hot source is broken and the gas then undergoes reversible adiabatic expansion from B to C. No energy leaves the system as heat, so the change in entropy is zero. The expansion is carried on until the temperature of the gas falls from T_h to T_c, the temperature of the cold sink.
3. The gas is placed in contact with the cold sink and then undergoes a reversible isothermal compression from C to D at T_c. Energy is released as heat to the cold sink; the change in entropy of the system is q_c/T_c; in this expression q_c is negative.
4. Finally, contact with the cold sink is broken and the gas then undergoes reversible adiabatic compression from D to A such that the final temperature is T_h. No energy enters the system as heat, so the change in entropy is zero.

The total change in entropy around the cycle is the sum of the changes in each of these four steps:

$$\oint dS = \frac{q_h}{T_h} + \frac{q_c}{T_c}$$

The next task is to show that the sum of the two terms on the right of this expression is zero for a perfect gas and so confirming, for that substance at least, that entropy is a state function.

How is that done? 3A.1 Showing that the entropy is a state function for a perfect gas

First, you need to note that a reversible adiabatic expansion (stage 2 in Fig. 3A.9) takes the system from T_h to T_c. You can then use the properties of such an expansion, specifically VT^c = constant (Topic 2E), to relate the two volumes at the start and end of the expansion. You also need to note that energy as heat is transferred by reversible isothermal processes (stages 1 and 3) and, as derived in *Example* 3A.1, for a perfect gas

$$\overbrace{q_h = nRT_h \ln\frac{V_B}{V_A}}^{\text{stage 1}} \qquad \overbrace{q_c = nRT_c \ln\frac{V_D}{V_C}}^{\text{stage 3}}$$

Step 1 *Relate the volumes in the adiabatic expansions*

For a reversible adiabatic process the temperature and volume are related by VT^c = constant (Topic 2E). Therefore

for the path D to A (stage 4): $V_A T_h^c = V_D T_c^c$

for the path B to C (stage 2): $V_C T_c^c = V_B T_h^c$

Multiplication of the first of these expressions by the second gives

$$V_A V_C T_h^c T_c^c = V_D V_B T_h^c T_c^c$$

which, on cancellation of the temperatures, simplifies to

$$\frac{V_D}{V_C} = \frac{V_A}{V_B}$$

Step 2 *Establish the relation between the two heat transfers*

You can now use this relation to write an expression for energy discarded as heat to the cold sink in terms of V_A and V_B

$$q_c = nRT_c \ln\frac{V_D}{V_C} = nRT_c \ln\frac{V_A}{V_B} = -nRT_c \ln\frac{V_B}{V_A}$$

It follows that

$$\frac{q_h}{q_c} = \frac{nRT_h \ln(V_B/V_A)}{-nRT_c \ln(V_B/V_A)} = -\frac{T_h}{T_c}$$

Note that q_h is negative (heat is withdrawn from the hot source) and q_c is positive (heat is deposited in the cold sink), so their ratio is negative. This expression can be rearranged into

$$\frac{q_h}{T_h} + \frac{q_c}{T_c} = 0 \qquad (3A.6)$$

Because the total change in entropy around the cycle is $q_h/T_h + q_c/T_c$, it follows immediately from eqn 3A.6 that, for a perfect gas, this entropy change is zero.

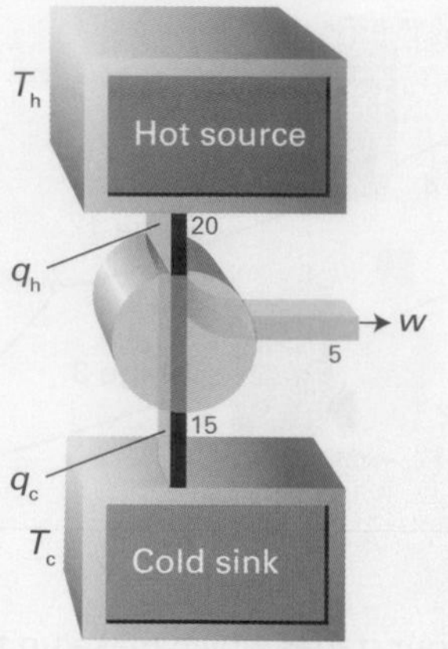

Figure 3A.10 In a heat engine, an energy q_h (for example, $|q_h| = 20\,kJ$) is extracted as heat from the hot source and q_c is discarded into the cold sink (for example, $|q_c| = 15\,kJ$). The work done by the engine is equal to $|q_h| - |q_c|$ (e.g. 20 kJ − 15 kJ = 5 kJ).

Brief illustration 3A.2

The Carnot cycle can be regarded as a representation of the changes taking place in a heat engine in which part of the energy extracted as heat from the hot reservoir is converted into work. Consider an engine running in accord with the Carnot cycle, and in which 100 J of energy is withdrawn from the hot source ($q_h = -100\,J$) at 500 K. Some of this energy is used to do work and the remainder is deposited in the cold sink at 300 K. According to eqn 3A.6, the heat deposited is

$$q_c = -q_h \times \frac{T_c}{T_h} = -(-100\,J) \times \frac{300\,K}{500\,K} = +60\,J$$

This value implies that 40 J was used to do work.

It is now necessary to show that eqn 3A.5 applies to any material, not just a perfect gas. To do so, it is helpful to introduce the **efficiency**, η (eta), of a heat engine:

$$\eta = \frac{\text{work performed}}{\text{heat absorbed from hot source}} = \frac{|w|}{|q_h|} \qquad \text{Efficiency [definition]} \qquad (3A.7)$$

Modulus signs (|...|) have been used to avoid complications with signs: all efficiencies are positive numbers. The definition implies that the greater the work output for a given supply of heat from the hot source, the greater is the efficiency of the engine. The definition can be expressed in terms of the heat transactions alone, because (as shown in Fig. 3A.10) the energy supplied as work by the engine is the difference between the energy supplied as heat by the hot source and that returned to the cold sink:

$$\eta = \frac{|q_h| - |q_c|}{|q_h|} = 1 - \frac{|q_c|}{|q_h|} \qquad (3A.8)$$

It then follows from eqn 3A.6, written as $|q_c|/|q_h| = T_c/T_h$ that

$$\eta = 1 - \frac{T_c}{T_h} \qquad \text{Carnot efficiency} \qquad (3A.9)$$

Brief illustration 3A.3

A certain power station operates with superheated steam at 300 °C ($T_h = 573\,K$) and discharges the waste heat into the environment at 20 °C ($T_c = 293\,K$). The theoretical efficiency is therefore

$$\eta = 1 - \frac{293\,K}{573\,K} = 0.489$$

or 48.9 per cent. In practice, there are other losses due to mechanical friction and the fact that the turbines do not operate reversibly.

Now this conclusion can be generalized. The Second Law of thermodynamics implies that *all reversible engines have the same efficiency regardless of their construction.* To see the truth of this statement, suppose two reversible engines are coupled together and run between the same hot source and cold sink (Fig. 3A.11). The working substances and details of construction of the two engines are entirely arbitrary. Initially, suppose that engine A is more efficient than engine B, and that a setting of the controls has been chosen that causes engine B to acquire energy as heat q_c from the cold sink and to release a certain quantity of energy as heat into the hot source. However, because engine A is more efficient than engine B, not all the work that A produces is needed for this process and the difference can be used to do work. The net result is that the cold reservoir is unchanged, work has been done, and the hot reservoir has lost a certain amount of energy. This outcome is contrary to the Kelvin statement of the Second Law, because some heat has been converted directly into work. Because the conclusion is contrary to experience, the initial assumption that engines A and B can have different efficiencies must be false. It follows

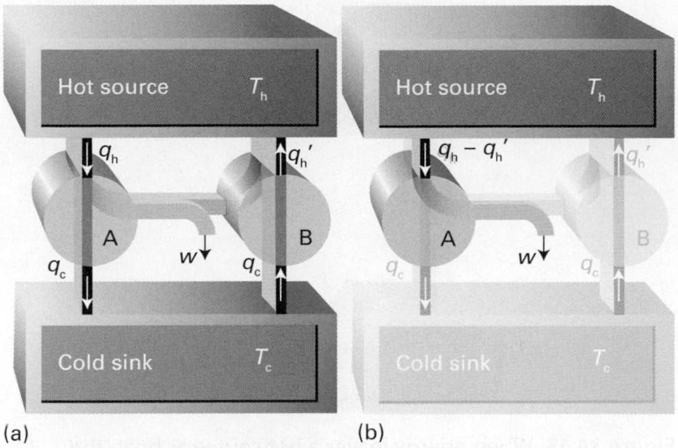

Figure 3A.11 (a) The demonstration of the equivalence of the efficiencies of all reversible engines working between the same thermal reservoirs is based on the flow of energy represented in this diagram. (b) The net effect of the processes is the conversion of heat into work without there being a need for a cold sink. This is contrary to the Kelvin statement of the Second Law.

that the relation between the heat transfers and the temperatures must also be independent of the working material, and therefore that eqn 3A.9 is true for any substance involved in a Carnot cycle.

For the final step of the argument note that any reversible cycle can be approximated as a collection of Carnot cycles. This approximation is illustrated in Fig. 3A.12, which shows three Carnot cycles A, B, and C fitted together in such a way that their perimeter approximates the cycle indicated by the

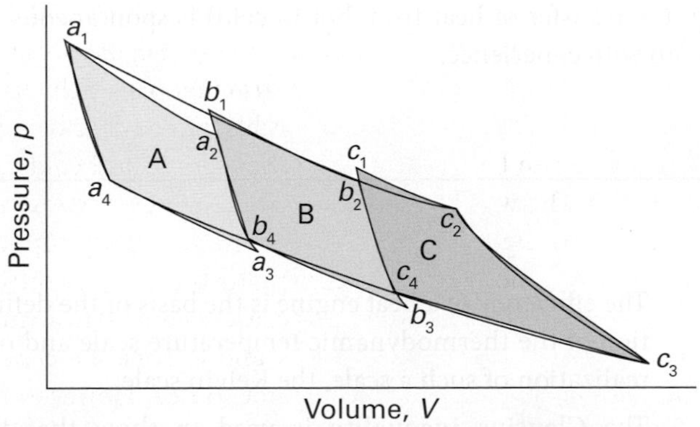

Figure 3A.12 The path indicated by the purple line can be approximated by traversing the *overall perimeter* of the area created by the three Carnot cycles A, B, and C; for each individual cycle the overall entropy change is zero. The entropy changes along the adiabatic segments (such as a_1–a_4 and c_2–c_3) are zero, so it follows that the entropy changes along the isothermal segments of any one cycle (such as a_1–a_2 and a_3–a_4) cancel. The entropy change resulting from traversing the overall perimeter of the three cycles is therefore zero.

purple line. The entropy change around each individual cycle is zero (as already demonstrated), so the sum of entropy changes for all the cycles is zero. However, in the sum, the entropy change along any individual path is cancelled by the entropy change along the path it shares with the neighbouring cycle (because neighbouring paths are traversed in opposite directions). Therefore, all the entropy changes cancel except for those along the perimeter of the overall cycle and therefore the sum q_{rev}/T around the perimeter is zero.

The path shown by the purple line can be approximated more closely by using more Carnot cycles, each of which is much smaller, and in the limit that they are infinitesimally small their perimeter matches the purple path exactly. Equation 3A.5 (that the integral of dq_{rev}/T round a general cycle is zero) then follows immediately. This result implies that dS is an exact differential and therefore that S is a state function.

(b) The thermodynamic temperature

Suppose an engine works reversibly between a hot source at a temperature T_h and a cold sink at a temperature T, then it follows from eqn 3A.9 that

$$T = (1 - \eta)T_h \qquad (3A.10)$$

This expression enabled Kelvin to define the **thermodynamic temperature scale** in terms of the efficiency of a heat engine: construct an engine in which the hot source is at a known temperature and the cold sink is the object of interest. The temperature of the latter can then be inferred from the measured efficiency of the engine. The **Kelvin scale** (which is a special case of the thermodynamic temperature scale) is currently defined by using water at its triple point as the notional hot source and defining that temperature as 273.16 K exactly.[4]

(c) The Clausius inequality

To show that the definition of entropy is consistent with the Second Law, note that more work is done when a change is reversible than when it is irreversible. That is, $|dw_{rev}| \geq |dw|$. Because dw and dw_{rev} are negative when energy leaves the system as work, this expression is the same as $-dw_{rev} \geq -dw$, and hence $dw - dw_{rev} \geq 0$. The internal energy is a state function, so its change is the same for irreversible and reversible paths between the same two states, and therefore

$$dU = dq + dw = dq_{rev} + dw_{rev}$$

and hence $dq_{rev} - dq = dw - dw_{rev}$. Then, because $dw - dw_{rev} \geq 0$, it follows that $dq_{rev} - dq \geq 0$ and therefore $dq_{rev} \geq dq$. Division

[4] The international community has agreed to replace this definition by another that is independent of the specification of a particular substance, but the new definition has not yet (in 2018) been implemented.

by T then results in $\mathrm{d}q_{\mathrm{rev}}/T \ge \mathrm{d}q/T$. From the thermodynamic definition of the entropy ($\mathrm{d}S = \mathrm{d}q_{\mathrm{rev}}/T$) it then follows that

$$\mathrm{d}S \ge \frac{\mathrm{d}q}{T} \quad \text{Clausius inequality} \quad (3A.11)$$

This expression is the **Clausius inequality**. It proves to be of great importance for the discussion of the spontaneity of chemical reactions (Topic 3D).

Suppose a system is isolated from its surroundings, so that $\mathrm{d}q = 0$. The Clausius inequality implies that

$$\mathrm{d}S \ge 0 \quad (3A.12)$$

That is, *in an isolated system the entropy cannot decrease when a spontaneous change occurs*. This statement captures the content of the Second Law.

The Clausius inequality also implies that spontaneous processes are also necessarily irreversible processes. To confirm this conclusion, the inequality is introduced into the expression for the total entropy change that accompanies a process:

$$\mathrm{d}S_{\mathrm{tot}} = \overbrace{\mathrm{d}S}^{\ge \mathrm{d}q/T} + \overbrace{\mathrm{d}S_{\mathrm{sur}}}^{-\mathrm{d}q/T} \ge 0$$

where the inequality corresponds to an irreversible process and the equality to a reversible process. That is, a spontaneous process ($\mathrm{d}S_{\mathrm{tot}} > 0$) is an irreversible process. A reversible process, for which $\mathrm{d}S_{\mathrm{tot}} = 0$, is spontaneous in neither direction: it is at equilibrium.

Apart from its fundamental importance in linking the definition of entropy to the Second Law, the Clausius inequality can also be used to show that a familiar process, the cooling of an object to the temperature of its surroundings, is indeed spontaneous. Consider the transfer of energy as heat from one system—the hot source—at a temperature T_{h} to another system—the cold sink—at a temperature T_{c} (Fig. 3A.13). When

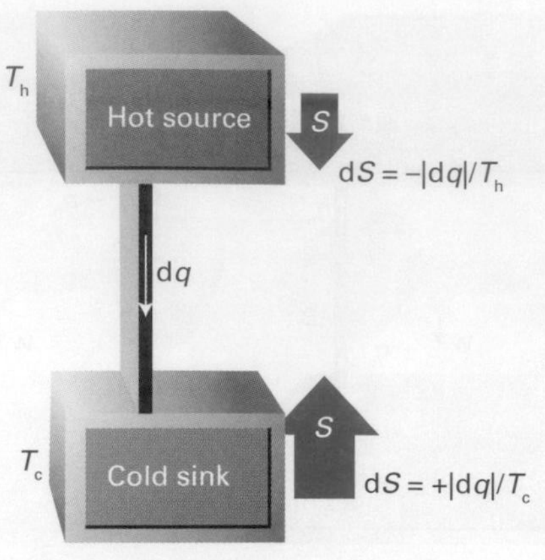

Figure 3A.13 When energy leaves a hot source as heat, the entropy of the source decreases. When the same quantity of energy enters a cooler sink, the increase in entropy is greater. Hence, overall there is an increase in entropy and the process is spontaneous. Relative changes in entropy are indicated by the sizes of the arrows.

$|\mathrm{d}q|$ leaves the hot source (so $\mathrm{d}q_{\mathrm{h}} < 0$), the Clausius inequality implies that $\mathrm{d}S \ge \mathrm{d}q_{\mathrm{h}}/T_{\mathrm{h}}$. When $|\mathrm{d}q|$ enters the cold sink the Clausius inequality implies that $\mathrm{d}S \ge \mathrm{d}q_{\mathrm{c}}/T_{\mathrm{c}}$ (with $\mathrm{d}q_{\mathrm{c}} > 0$). Overall, therefore,

$$\mathrm{d}S \ge \frac{\mathrm{d}q_{\mathrm{h}}}{T_{\mathrm{h}}} + \frac{\mathrm{d}q_{\mathrm{c}}}{T_{\mathrm{c}}}$$

However, $\mathrm{d}q_{\mathrm{h}} = -\mathrm{d}q_{\mathrm{c}}$, so

$$\mathrm{d}S \ge -\frac{\mathrm{d}q_{\mathrm{c}}}{T_{\mathrm{h}}} + \frac{\mathrm{d}q_{\mathrm{c}}}{T_{\mathrm{c}}} = \left(\frac{1}{T_{\mathrm{c}}} - \frac{1}{T_{\mathrm{h}}}\right)\mathrm{d}q_{\mathrm{c}}$$

which is positive (because $\mathrm{d}q_{\mathrm{c}} > 0$ and $T_{\mathrm{h}} \ge T_{\mathrm{c}}$). Hence, cooling (the transfer of heat from hot to cold) is spontaneous, in accord with experience.

Checklist of concepts

☐ 1. The **entropy** is a signpost of spontaneous change: the entropy of the universe increases in a spontaneous process.

☐ 2. A change in entropy is defined in terms of reversible heat transactions.

☐ 3. The **Boltzmann formula** defines entropy in terms of the number of ways that the molecules can be arranged amongst the energy states, subject to the arrangements having the same overall energy.

☐ 4. The **Carnot cycle** is used to prove that entropy is a state function.

☐ 5. The **efficiency** of a heat engine is the basis of the definition of the thermodynamic temperature scale and one realization of such a scale, the Kelvin scale.

☐ 6. The **Clausius inequality** is used to show that the entropy of an isolated system increases in a spontaneous change and therefore that the Clausius definition is consistent with the Second Law.

☐ 7. Spontaneous processes are irreversible processes; processes accompanied by no change in entropy are at equilibrium.

Checklist of equations

Property	Equation	Comment	Equation number
Thermodynamic entropy	$dS = dq_{rev}/T$	Definition	3A.1a
Entropy change of surroundings	$\Delta S_{sur} = q_{sur}/T_{sur}$		3A.2b
Boltzmann formula	$S = k \ln \mathcal{W}$	Definition	3A.4
Carnot efficiency	$\eta = 1 - T_c/T_h$	Reversible processes	3A.9
Thermodynamic temperature	$T = (1 - \eta)T_h$		3A.10
Clausius inequality	$dS \geq dq/T$		3A.11

TOPIC 3B Entropy changes accompanying specific processes

> **➤ Why do you need to know this material?**
>
> The changes in entropy accompanying a variety of basic physical processes occur throughout the application of the Second Law to chemistry.
>
> **➤ What is the key idea?**
>
> The change in entropy accompanying a process is calculated by identifying a reversible path between the initial and final states.
>
> **➤ What do you need to know already?**
>
> You need to be familiar with the thermodynamic definition of entropy (Topic 3A), the First-Law concepts of work, heat, and internal energy (Topic 2A), and heat capacity (Topic 2B). The Topic makes use of the expressions for the work and heat transactions during the reversible, isothermal expansion of a perfect gas (Topic 2A).

The thermodynamic definition of entropy change given in eqn 3A.1,

$$dS = \frac{dq_{rev}}{T} \qquad \Delta S = \int_i^f \frac{dq_{rev}}{T} \qquad \text{Entropy change [definition]} \qquad (3B.1a)$$

where q_{rev} is the energy supplied reversibly as heat to the system at a temperature T, is the basis of all calculations relating to entropy in thermodynamics. When applied to the surroundings, this definition implies eqn 3A.2b, which is repeated here as

$$\Delta S_{sur} = \frac{q_{sur}}{T_{sur}} \qquad \text{Entropy change of surroundings} \qquad (3B.1b)$$

where q_{sur} is the energy supplied as heat to the surroundings and T_{sur} is their temperature; note that the entropy change of the surroundings is the same whether or not the process is reversible or irreversible for the system. The total change in entropy of an (overall) isolated system (the 'universe') is

$$\Delta S_{tot} = \Delta S + \Delta S_{sur} \qquad \text{Total entropy change} \qquad (3B.1c)$$

The entropy changes accompanying some physical changes are of particular importance and are treated here. As explained in Topic 3A, a spontaneous process is also irreversible (in the thermodynamic sense) and a process for which $\Delta S_{tot} = 0$ is at equilibrium.

3B.1 Expansion

In Topic 3A (specifically *Example* 3A.1) it is established that the change in entropy of a perfect gas when it expands isothermally from V_i to V_f is

$$\Delta S = nR\ln\frac{V_f}{V_i} \qquad \text{Entropy change for the isothermal expansion of a perfect gas} \qquad (3B.2)$$

Because S is a state function, the value of ΔS *of the system* is independent of the path between the initial and final states, so this expression applies whether the change of state occurs reversibly or irreversibly. The logarithmic dependence of entropy on volume is illustrated in Fig. 3B.1.

The *total* change in entropy, however, does depend on how the expansion takes place. For any process the energy lost as heat from the system is acquired by the surroundings, so $dq_{sur} = -dq$. For the reversible isothermal expansion of a perfect gas $q_{rev} = nRT\ln(V_f/V_i)$, so $q_{sur} = -nRT\ln(V_f/V_i)$, and consequently

$$\Delta S_{sur} = -\frac{q_{rev}}{T} = -nR\ln\frac{V_f}{V_i} \qquad (3B.3a)$$

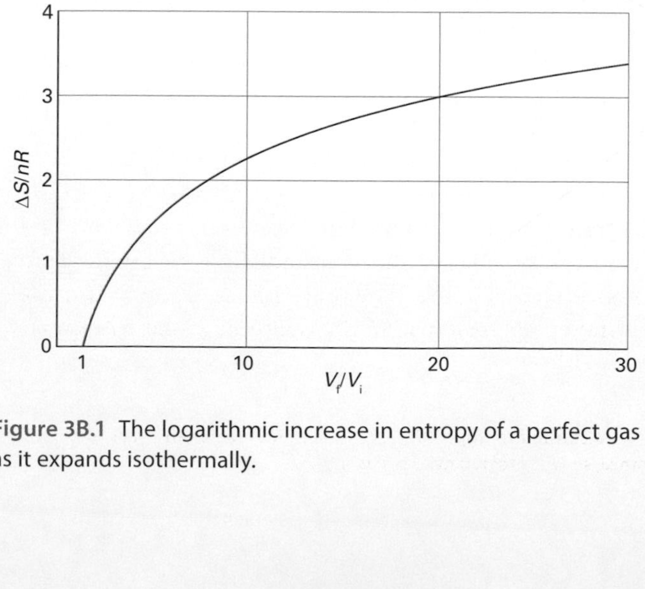

Figure 3B.1 The logarithmic increase in entropy of a perfect gas as it expands isothermally.

This change is the negative of the change in the system, so $\Delta S_{tot} = 0$, as expected for a reversible process. If, on the other hand, the isothermal expansion occurs freely (if the expansion is into a vacuum) no work is done ($w = 0$). Because the expansion is isothermal, $\Delta U = 0$, and it follows from the First Law, $\Delta U = q + w$, that $q = 0$. As a result, $q_{sur} = 0$ and hence $\Delta S_{sur} = 0$. For this expansion doing no work the total entropy change is therefore given by eqn 3B.2 itself:

$$\Delta S_{tot} = nR\ln\frac{V_f}{V_i} \quad (3B.3b)$$

In this case, $\Delta S_{tot} > 0$, as expected for an irreversible process.

Brief illustration 3B.1

When the volume of any perfect gas is doubled at constant temperature, $V_f/V_i = 2$, and hence the change in molar entropy of the system is

$$\Delta S_m = (8.3145\,J\,K^{-1}\,mol^{-1}) \times \ln 2 = +5.76\,J\,K^{-1}\,mol^{-1}$$

If the change is carried out reversibly, the change in entropy of the surroundings is $-5.76\,J\,K^{-1}\,mol^{-1}$ (the 'per mole' meaning per mole of gas molecules in the sample). The total change in entropy is 0. If the expansion is free, the change in molar entropy of the gas is still $+5.76\,J\,K^{-1}\,mol^{-1}$, but that of the surroundings is 0, and the total change is $+5.76\,J\,K^{-1}\,mol^{-1}$.

3B.2 Phase transitions

When a substance freezes or boils the degree of dispersal of matter and the associated energy changes reflect the order with which the molecules pack together and the extent to which the energy is localized. Therefore, a transition is expected to be accompanied by a change in entropy. For example, when a substance vaporizes, a compact condensed phase changes into a widely dispersed gas, and the entropy of the substance can be expected to increase considerably. The entropy of a solid also increases when it melts to a liquid.

Consider a system and its surroundings at the **normal transition temperature**, T_{trs}, the temperature at which two phases are in equilibrium at 1 atm. This temperature is 0 °C (273 K) for ice in equilibrium with liquid water at 1 atm, and 100 °C (373 K) for water in equilibrium with its vapour at 1 atm. At the transition temperature, any transfer of energy as heat between the system and its surroundings is reversible because the two phases in the system are in equilibrium. Because at constant pressure $q = \Delta_{trs}H$, the change in molar entropy *of the system* is[1]

$$\Delta_{trs}S = \frac{\Delta_{trs}H}{T_{trs}} \quad \text{Entropy of phase transition [at } T_{trs}\text{]} \quad (3B.4)$$

If the phase transition is exothermic ($\Delta_{trs}H < 0$, as in freezing or condensing), then the entropy change of the system is negative. This decrease in entropy is consistent with the increased order of a solid compared with a liquid, and with the increased order of a liquid compared with a gas. The change in entropy of the surroundings, however, is positive because energy is released as heat into them. At the transition temperature the total change in entropy is zero because the two phases are in equilibrium. If the transition is endothermic ($\Delta_{trs}H > 0$, as in melting and vaporization), then the entropy change of the system is positive, which is consistent with dispersal of matter in the system. The entropy of the surroundings decreases by the same amount, and overall the total change in entropy is zero.

Table 3B.1 lists some experimental entropies of phase transitions. Table 3B.2 lists in more detail the standard entropies of vaporization of several liquids at their normal boiling points. An interesting feature of the data is that a wide range of liquids give approximately the same standard entropy of vaporization (about $85\,J\,K^{-1}\,mol^{-1}$): this empirical observation is called **Trouton's rule**. The explanation of Trouton's rule is that a similar change in volume occurs when any liquid evaporates and becomes a gas. Hence, all liquids can be expected to have similar standard entropies of vaporization.

Liquids that show significant deviations from Trouton's rule do so on account of strong molecular interactions that result

Table 3B.1 Standard entropies of phase transitions, $\Delta_{trs}S^\ominus/(J\,K^{-1}\,mol^{-1})$, at the corresponding normal transition temperatures*

	Fusion (at T_f)	Vaporization (at T_b)
Argon, Ar	14.17 (at 83.8 K)	74.53 (at 87.3 K)
Benzene, C_6H_6	38.00 (at 279 K)	87.19 (at 353 K)
Water, H_2O	22.00 (at 273.15 K)	109.1 (at 373.15 K)
Helium, He	4.8 (at 1.8 K and 30 bar)	19.9 (at 4.22 K)

* More values are given in the *Resource section*.

Table 3B.2 The standard enthalpies and entropies of vaporization of liquids at their boiling temperatures*

	$\Delta_{vap}H^\ominus/(kJ\,mol^{-1})$	$\theta_b/°C$	$\Delta_{vap}S^\ominus/(J\,K^{-1}\,mol^{-1})$
Benzene	30.8	80.1	87.2
Carbon tetrachloride	30	76.7	85.8
Cyclohexane	30.1	80.7	85.1
Hydrogen sulfide	18.7	−60.4	87.9
Methane	8.18	−161.5	73.2
Water	40.7	100.0	109.1

* More values are given in the *Resource section*.

[1] According to Topic 2C, $\Delta_{trs}H$ is an enthalpy change per mole of substance, so $\Delta_{trs}S$ is also a molar quantity.

in a partial ordering of their molecules. As a result, there is a greater change in disorder when the liquid turns into a vapour than for when a fully disordered liquid vaporizes. An example is water, where the large entropy of vaporization reflects the presence of structure arising from hydrogen bonding in the liquid. Hydrogen bonds tend to organize the molecules in the liquid so that they are less random than, for example, the molecules in liquid hydrogen sulfide (in which there is no hydrogen bonding). Methane has an unusually low entropy of vaporization. A part of the reason is that the entropy of the gas itself is slightly low (186 J K^{-1} mol^{-1} at 298 K; the entropy of N_2 under the same conditions is 192 J K^{-1} mol^{-1}). As explained in Topic 13B, fewer translational and rotational states are accessible at room temperature for molecules with low mass and moments of inertia (like CH_4) than for molecules with relatively high mass and moments of inertia (like N_2), so their molar entropy is slightly lower.

Brief illustration 3B.2

There is no hydrogen bonding in liquid bromine and Br_2 is a heavy molecule which is unlikely to display unusual behaviour in the gas phase, so it is safe to use Trouton's rule. To predict the standard molar enthalpy of vaporization of bromine given that it boils at 59.2 °C, use Trouton's rule in the form

$$\Delta_{vap}H^{\ominus} = T_b \times (85\,\mathrm{J\,K^{-1}\,mol^{-1}})$$

Substitution of the data then gives

$$\Delta_{vap}H^{\ominus} = (332.4\,\mathrm{K}) \times (85\,\mathrm{J\,K^{-1}\,mol^{-1}}) = +2.8 \times 10^4\,\mathrm{J\,mol^{-1}} = +28\,\mathrm{kJ\,mol^{-1}}$$

The experimental value is +29.45 kJ mol^{-1}.

3B.3 Heating

The thermodynamic definition of entropy change in eqn 3B.1a is used to calculate the entropy of a system at a temperature T_f from a knowledge of its entropy at another temperature T_i and the heat supplied to change its temperature from one value to the other:

$$S(T_f) = S(T_i) + \int_{T_i}^{T_f} \frac{dq_{rev}}{T} \quad (3B.5)$$

The most common version of this expression is for a system subjected to constant pressure (such as from the atmosphere) during the heating, so then $dq_{rev} = dH$. From the definition of constant-pressure heat capacity (eqn 2B.5, $C_p = (\partial H/\partial T)_p$) it follows that $dH = C_p dT$, and hence $dq_{rev} = C_p dT$. Substitution into eqn 3B.5 gives

$$S(T_f) = S(T_i) + \int_{T_i}^{T_f} \frac{C_p dT}{T} \quad (3B.6)$$

Entropy variation with temperature [constant p]

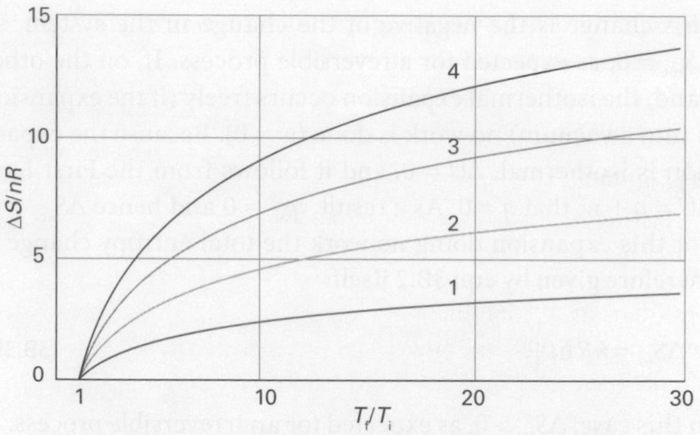

Figure 3B.2 The logarithmic increase in entropy of a substance as it is heated at either constant volume or constant pressure. Different curves are labelled with the corresponding value of C_m/R, taken to be constant over the temperature range. For constant volume conditions $C_m = C_{V,m}$, and at constant pressure $C_m = C_{p,m}$.

The same expression applies at constant volume, but with C_p replaced by C_V. When C_p is independent of temperature over the temperature range of interest, it can be taken outside the integral to give

$$S(T_f) = S(T_i) + C_p \int_{T_i}^{T_f} \frac{dT}{T} = S(T_i) + C_p \ln\frac{T_f}{T_i} \quad (3B.7)$$

with a similar expression for heating at constant volume. The logarithmic dependence of entropy on temperature is illustrated in Fig. 3B.2.

Brief illustration 3B.3

The molar constant-volume heat capacity of water at 298 K is 75.3 J K^{-1} mol^{-1}. The change in molar entropy when it is heated from 20 °C (293 K) to 50 °C (323 K), supposing the heat capacity to be constant in that range, is therefore

$$\Delta S_m = S_m(323\,\mathrm{K}) - S_m(293\,\mathrm{K}) = (75.3\,\mathrm{J\,K^{-1}\,mol^{-1}}) \times \ln\frac{323\,\mathrm{K}}{293\,\mathrm{K}} = +7.34\,\mathrm{J\,K^{-1}\,mol^{-1}}$$

3B.4 Composite processes

In many processes, more than one parameter changes. For instance, it might be the case that both the volume and the temperature of a gas are different in the initial and final states. Because S is a state function, the change in its value can be calculated by considering any reversible path between the initial and final states. For example, it might be convenient to split the path into two steps: an isothermal expansion to the final volume, followed by heating at constant volume to the final temperature. Then the total entropy change when both variables change is the sum of the two contributions.

Example 3B.1 Calculating the entropy change for a composite process

Calculate the entropy change when argon at 25 °C and 1.00 bar in a container of volume 0.500 dm^3 is allowed to expand to 1.000 dm^3 and is simultaneously heated to 100 °C. (Take the molar heat capacity at constant volume to be $\frac{3}{2}R$.)

Collect your thoughts As remarked in the text, you can break the overall process down into two steps: isothermal expansion to the final volume, followed by heating at constant volume to the final temperature. The entropy change in the first step is given by eqn 3B.2 and that of the second step, provided C_V is independent of temperature, by eqn 3B.7 (with C_V in place of C_p). In each case you need to know n, the amount of gas molecules, which can be calculated from the perfect gas equation and the data for the initial state by using $n = p_iV_i/RT_i$.

The solution The amount of gas molecules is

$$n = \frac{(1.00\times10^5\,\text{Pa})\times(0.500\times10^{-3}\,\text{m}^3)}{(8.3145\,\text{J K}^{-1}\,\text{mol}^{-1})\times298\,\text{K}} = 0.0201\ldots\,\text{mol}$$

From eqn 3B.2 the entropy change in the isothermal expansion from V_i to V_f is

$$\Delta S(\text{Step 1}) = nR\ln\frac{V_f}{V_i}$$
$$= 0.0201\ldots\,\text{mol}\times(8.3145\,\text{J K}^{-1}\,\text{mol}^{-1})\ln\frac{1.000\,\text{dm}^3}{0.500\,\text{dm}^3}$$
$$= +0.116\ldots\,\text{J K}^{-1}$$

From eqn 3B.6, the entropy change in the second step, heating from T_i to T_f at constant volume, is

$$\Delta S(\text{Step 2}) = nC_{V,\text{m}}\ln\frac{T_f}{T_i} = \tfrac{3}{2}nR\ln\frac{T_f}{T_i}$$
$$= \tfrac{3}{2}\times(0.0201\ldots\,\text{mol})\times(8.3145\,\text{J K}^{-1}\,\text{mol}^{-1})\ln\frac{373\,\text{K}}{298\,\text{K}}$$
$$= +0.0564\ldots\,\text{J K}^{-1}$$

The overall entropy change of the system, the sum of these two changes, is

$$\Delta S = 0.116\ldots\,\text{J K}^{-1} + 0.0564\ldots\,\text{J K}^{-1} = +0.173\,\text{J K}^{-1}$$

Self-test 3B.1 Calculate the entropy change when the same initial sample is compressed to 0.0500 dm^3 and cooled to −25 °C.

Answer: $-0.43\,\text{J K}^{-1}$

Checklist of concepts

☐ 1. The entropy of a perfect gas increases when it expands isothermally.

☐ 2. The change in entropy of a substance accompanying a change of state at its transition temperature is calculated from its enthalpy of transition.

☐ 3. The increase in entropy when a substance is heated is calculated from its heat capacity.

Checklist of equations

Property	Equation	Comment	Equation number
Entropy of isothermal expansion	$\Delta S = nR\ln(V_f/V_i)$	Perfect gas	3B.2
Entropy of transition	$\Delta_{trs}S = \Delta_{trs}H/T_{trs}$	At the transition temperature	3B.4
Variation of entropy with temperature	$S(T_f) = S(T_i) + C\ln(T_f/T_i)$	The heat capacity, C, is independent of temperature and no phase transitions occur; $C = C_p$ for constant pressure and C_V for constant volume.	3B.7

TOPIC 3C The measurement of entropy

> **➤ Why do you need to know this material?**
>
> For entropy to be a quantitatively useful concept it is important to be able to measure it: the calorimetric procedure is described here. The Third Law of thermodynamics is used to report the measured values.
>
> **➤ What is the key idea?**
>
> The entropy of a perfectly crystalline solid is zero at $T = 0$.
>
> **➤ What do you need to know already?**
>
> You need to be familiar with the expression for the temperature dependence of entropy and how entropies of phase changes are calculated (Topic 3B). The discussion of residual entropy draws on the Boltzmann formula for the entropy (Topic 3A).

The entropy of a substance can be determined in two ways. One, which is the subject of this Topic, is to make calorimetric measurements of the heat required to raise the temperature of a sample from $T = 0$ to the temperature of interest. There are then two equations to use. One is the dependence of entropy on temperature, which is eqn 3B.6 reproduced here as

$$S(T_2) = S(T_1) + \int_{T_1}^{T_2} \frac{C_p(T)}{T}\mathrm{d}T \qquad \text{Entropy and temperature} \qquad (3C.1a)$$

The second is the contribution of a phase change to the entropy, which according to eqn 3B.4 is

$$\Delta S(T_{\mathrm{trs}}) = \frac{\Delta_{\mathrm{trs}}H(T_{\mathrm{trs}})}{T_{\mathrm{trs}}} \qquad \text{Entropy of phase transition} \qquad (3C.1b)$$

where $\Delta_{\mathrm{trs}}H(T_{\mathrm{trs}})$ is the enthalpy of transition at the transition temperature T_{trs}. The other method, which is described in Topic 13E, is to use calculated parameters or spectroscopic data to calculate the entropy by using Boltzmann's statistical definition.

3C.1 The calorimetric measurement of entropy

According to eqn 3C.1a, the entropy of a system at a temperature T is related to its entropy at $T = 0$ by measuring its heat capacity C_p at different temperatures and evaluating the integral. The entropy of transition for each phase transition that occurs between $T = 0$ and the temperature of interest must then be included in the overall sum. For example, if a substance melts at T_f and boils at T_b, then its molar entropy at a particular temperature T above its boiling temperature is given by

$$S_\mathrm{m}(T) = S_\mathrm{m}(0) + \overbrace{\int_0^{T_\mathrm{f}} \frac{C_{p,\mathrm{m}}(\mathrm{s},T')}{T'}\mathrm{d}T'}^{\text{Heat solid to its melting point}} + \overbrace{\frac{\Delta_{\mathrm{fus}}H}{T_\mathrm{f}}}^{\text{Entropy of fusion}}$$
$$+ \overbrace{\int_{T_\mathrm{f}}^{T_\mathrm{b}} \frac{C_{p,\mathrm{m}}(\mathrm{l},T')}{T'}\mathrm{d}T'}^{\text{Heat liquid to its boiling point}} + \overbrace{\frac{\Delta_{\mathrm{vap}}H}{T_\mathrm{b}}}^{\text{Entropy of vaporization}} \qquad (3C.2)$$
$$+ \overbrace{\int_{T_\mathrm{b}}^{T} \frac{C_{p,\mathrm{m}}(\mathrm{g},T')}{T'}\mathrm{d}T'}^{\text{Heat vapour to the final temperature}}$$

The variable of integration has been changed to T' to avoid confusion with the temperature of interest, T. All the properties required, except $S_\mathrm{m}(0)$, can be measured calorimetrically, and the integrals can be evaluated either graphically or, as is now more usual, by fitting a polynomial to the data and integrating the polynomial analytically. The former procedure is illustrated in Fig. 3C.1: the area under the curve of $C_{p,\mathrm{m}}(T)/T$ against T is the integral required. Provided all measurements are made at 1 bar on a pure material, the final value is the **standard entropy**, $S^\ominus(T)$; division by the amount of substance, n, gives the **standard molar entropy**, $S^\ominus_\mathrm{m}(T) = S^\ominus(T)/n$. Because $\mathrm{d}T/T = \mathrm{d}\ln T$, an alternative procedure is to evaluate the area under a plot of $C_{p,\mathrm{m}}(T)$ against $\ln T$.

Brief illustration 3C.1

The standard molar entropy of nitrogen gas at 25 °C has been calculated from the following data:

	Contribution to $S_m^{\ominus}/(\mathrm{J\,K^{-1}\,mol^{-1}})$
Debye extrapolation	1.92
Integration, from 10 K to 35.61 K	25.25
Phase transition at 35.61 K	6.43
Integration, from 35.61 K to 63.14 K	23.38
Fusion at 63.14 K	11.42
Integration, from 63.14 K to 77.32 K	11.41
Vaporization at 77.32 K	72.13
Integration, from 77.32 K to 298.15 K	39.20
Correction for gas imperfection	0.92
Total	192.06

Therefore, $S_m^{\ominus}(298.15\,\mathrm{K}) = S_m(0) + 192.1\,\mathrm{J\,K^{-1}\,mol^{-1}}$. The Debye extrapolation is explained in the next paragraph.

One problem with the determination of entropy is the difficulty of measuring heat capacities near $T = 0$. There are good theoretical grounds for assuming that the heat capacity of a non-metallic solid is proportional to T^3 when T is low (see Topic 7A), and this dependence is the basis of the **Debye extrapolation** (or the *Debye T^3 law*). In this method, C_p is measured down to as low a temperature as possible and a curve of the form aT^3 is fitted to the data. The fit determines the value of a, and the expression $C_{p,m}(T) = aT^3$ is then assumed to be valid down to $T = 0$.

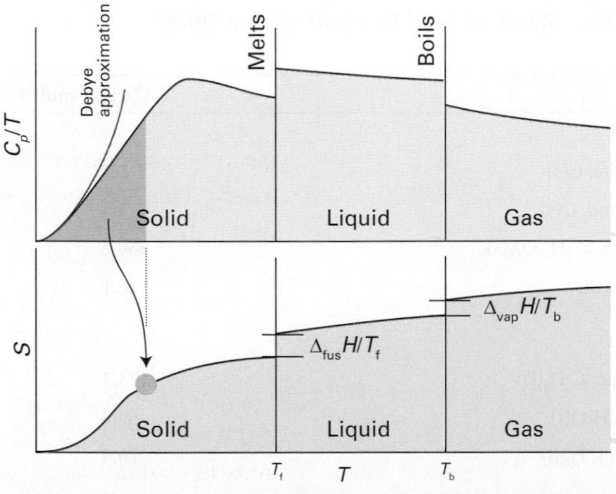

Figure 3C.1 The variation of C_p/T with the temperature for a sample is used to evaluate the entropy, which is equal to the area beneath the upper curve up to the corresponding temperature, plus the entropy of each phase transition encountered between $T = 0$ and the temperature of interest. For instance, the entropy denoted by the yellow dot on the lower curve is given by the dark shaded area in the upper graph.

Example 3C.1 Calculating the entropy at low temperatures

The molar constant-pressure heat capacity of a certain non-metallic solid at 4.2 K is $0.43\,\mathrm{J\,K^{-1}\,mol^{-1}}$. What is its molar entropy at that temperature?

Collect your thoughts Because the temperature is so low, you can assume that the heat capacity varies with temperature according to $C_{p,m}(T) = aT^3$, in which case you can use eqn 3C.1a to calculate the entropy at a temperature T in terms of the entropy at $T = 0$ and the constant a. When the integration is carried out, it turns out that the result can be expressed in terms of the heat capacity at the temperature T, so the data can be used directly to calculate the entropy.

The solution The integration required is

$$S_m(T) = S_m(0) + \int_0^T \frac{aT'^3}{T'}\mathrm{d}T' = S_m(0) + a\overbrace{\int_0^T T'^2\mathrm{d}T'}^{\text{Integral A.1}}$$
$$= S_m(0) + \tfrac{1}{3}aT^3 = S_m(0) + \tfrac{1}{3}C_{p,m}(T)$$

from which it follows that

$$S_m(4.2\,\mathrm{K}) = S_m(0) + 0.14\,\mathrm{J\,K^{-1}\,mol^{-1}}$$

Self-test 3C.1 For metals, there is also a contribution to the heat capacity from the electrons which is linearly proportional to T when the temperature is low; that is, $C_{p,m}(T) = bT$. Evaluate its contribution to the entropy at low temperatures.

Answer: $S_m(T) = S_m(0) + C_{p,m}(T)$

3C.2 The Third Law

At $T = 0$, all energy of thermal motion has been quenched, and in a perfect crystal all the atoms or ions are in a regular, uniform array. The localization of matter and the absence of thermal motion suggest that such materials also have zero entropy. This conclusion is consistent with the molecular interpretation of entropy (Topic 3A) because there is only one way of arranging the molecules when they are all in the ground state, which is the case at $T = 0$. Thus, at $T = 0$, $\mathcal{W} = 1$ and from $S = k\ln\mathcal{W}$ it follows that $S = 0$.

(a) The Nernst heat theorem

The **Nernst heat theorem** summarizes a series of experimental observations that turn out to be consistent with the view that the entropy of a regular array of molecules is zero at $T = 0$:

> The entropy change accompanying any physical or chemical transformation approaches zero as the temperature approaches zero: $\Delta S \rightarrow 0$ as $T \rightarrow 0$ provided all the substances involved are perfectly ordered.
>
> *Nernst heat theorem*

Brief illustration 3C.2

The entropy of the transition between orthorhombic sulfur, α, and monoclinic sulfur, β, can be calculated from the transition enthalpy (402 J mol⁻¹) at the transition temperature (369 K):

$$\Delta_{\rm trs}S(369\,{\rm K}) = S_{\rm m}(\beta,369\,{\rm K}) - S_{\rm m}(\alpha,369\,{\rm K})$$
$$= \frac{\Delta_{\rm trs}H}{T_{\rm trs}} = \frac{402\,{\rm J\,mol^{-1}}}{369\,{\rm K}} = 1.09\,{\rm J\,K^{-1}\,mol^{-1}}$$

The entropies of the α and β allotropes can also be determined by measuring their heat capacities from $T = 0$ up to $T = 369$ K. It is found that $S_{\rm m}(\alpha,369\,{\rm K}) = S_{\rm m}(\alpha,0) + 37\,{\rm J\,K^{-1}\,mol^{-1}}$ and $S_{\rm m}(\beta,369\,{\rm K}) = S_{\rm m}(\beta,0) + 38\,{\rm J\,K^{-1}\,mol^{-1}}$. These two values imply that at the transition temperature

$$\Delta_{\rm trs}S(369\,{\rm K}) = \{S_{\rm m}(\beta,0) + 38\,{\rm J\,K^{-1}\,mol^{-1}}\} - \{S_{\rm m}(\alpha,0) + 37\,{\rm J\,K^{-1}\,mol^{-1}}\}$$
$$= S_{\rm m}(\beta,0) - S_{\rm m}(\alpha,0) + 1\,{\rm J\,K^{-1}\,mol^{-1}}$$

On comparing this value with the one above, it follows that $S_{\rm m}(\beta,0) - S_{\rm m}(\alpha,0) \approx 0$, in accord with the theorem.

It follows from the Nernst theorem that, if the value zero is ascribed to the entropies of elements in their perfect crystalline form at $T = 0$, then all perfect crystalline compounds also have zero entropy at $T = 0$ (because the change in entropy that accompanies the formation of the compounds, like the entropy of all transformations at that temperature, is zero). This conclusion is summarized by the **Third Law of thermodynamics**:

> The entropy of all perfect crystalline substances is zero at $T = 0$.
>
> *Third Law of thermodynamics*

As far as thermodynamics is concerned, choosing this common value as zero is a matter of convenience. As noted above, the molecular interpretation of entropy justifies the value $S = 0$ at $T = 0$ because at this temperature $\mathcal{W} = 1$.

In certain cases $\mathcal{W} > 1$ at $T = 0$ and therefore $S(0) > 0$. This is the case if there is no energy advantage in adopting a particular orientation even at absolute zero. For instance, for a diatomic molecule AB there may be almost no energy difference between the arrangements …AB AB AB… and …BA AB BA… in a solid, so $\mathcal{W} > 1$ even at $T = 0$. If $S(0) > 0$ the substance is said to have a **residual entropy**. Ice has a residual entropy of 3.4 J K⁻¹ mol⁻¹. It stems from the arrangement of the hydrogen bonds between neighbouring water molecules: a given O atom has two short O–H bonds and two long O···H bonds to its neighbours, but there is a degree of randomness in which two bonds are short and which two are long.

(b) Third-Law entropies

Entropies reported on the basis that $S(0) = 0$ are called **Third-Law entropies** (and commonly just 'entropies'). When the substance is in its standard state at the temperature T, the **standard (Third-Law) entropy** is denoted $S^\ominus(T)$. A list of values at 298 K is given in Table 3C.1.

The **standard reaction entropy**, $\Delta_{\rm r}S^\ominus$, is defined, like the standard reaction enthalpy in Topic 2C, as the difference between the molar entropies of the pure, separated products and the pure, separated reactants, all substances being in their standard states at the specified temperature:

$$\Delta_{\rm r}S^\ominus = \sum_{\rm Products} \nu S_{\rm m}^\ominus - \sum_{\rm Reactants} \nu S_{\rm m}^\ominus \qquad \text{Standard reaction entropy [definition]} \qquad (3C.3a)$$

In this expression, each term is weighted by the appropriate stoichiometric coefficient. A more sophisticated approach is to adopt the notation introduced in Topic 2C and to write

$$\Delta_{\rm r}S^\ominus = \sum_{\rm J} \nu_{\rm J} S_{\rm m}^\ominus({\rm J}) \qquad (3C.3b)$$

where the $\nu_{\rm J}$ are signed (+ for products, − for reactants) stoichiometric numbers. Standard reaction entropies are likely to be positive if there is a net formation of gas in a reaction, and are likely to be negative if there is a net consumption of gas.

Table 3C.1 Standard Third-Law entropies at 298 K*

	$S_{\rm m}^\ominus/({\rm J\,K^{-1}\,mol^{-1}})$
Solids	
Graphite, C(s)	5.7
Diamond, C(s)	2.4
Sucrose, $C_{12}H_{22}O_{11}$(s)	360.2
Iodine, I_2(s)	116.1
Liquids	
Benzene, C_6H_6(l)	173.3
Water, H_2O(l)	69.9
Mercury, Hg(l)	76.0
Gases	
Methane, CH_4(g)	186.3
Carbon dioxide, CO_2(g)	213.7
Hydrogen, H_2(g)	130.7
Helium, He(g)	126.2
Ammonia, NH_3(g)	192.4

* More values are given in the *Resource section*.

Brief illustration 3C.3

To calculate the standard reaction entropy of $H_2(g) + \frac{1}{2}O_2(g) \rightarrow H_2O(l)$ at 298 K, use the data in Table 2C.4 of the *Resource section* to write

$$\Delta_r S^\ominus = S_m^\ominus(H_2O,l) - \{S_m^\ominus(H_2,g) + \tfrac{1}{2}S_m^\ominus(O_2,g)\}$$
$$= 69.9\,J\,K^{-1}\,mol^{-1} - \{130.7 + \tfrac{1}{2}(205.1)\}\,J\,K^{-1}\,mol^{-1}$$
$$= -163.4\,J\,K^{-1}\,mol^{-1}$$

The negative value is consistent with the conversion of two gases to a compact liquid.

A note on good practice Do not make the mistake of setting the standard molar entropies of elements equal to zero: they have non-zero values (provided $T > 0$).

Just as in the discussion of enthalpies in Topic 2C, where it is acknowledged that solutions of cations cannot be prepared in the absence of anions, the standard molar entropies of ions in solution are reported on a scale in which by convention the standard entropy of the H^+ ions in water is taken as zero at all temperatures:

$$S^\ominus(H^+,aq) = 0 \qquad \text{Ions in solution [convention]} \qquad (3C.4)$$

Table 2C.4 in the *Resource section* lists some values of standard entropies of ions in solution using this convention.[1] Because the entropies of ions in water are values relative to the hydrogen ion in water, they may be either positive or negative. A positive entropy means that an ion has a higher molar entropy than H^+ in water and a negative entropy means that the ion has a lower molar entropy than H^+ in water. Ion entropies vary as expected on the basis that they are related to the degree to which the ions order the water molecules around them in the solution. Small, highly charged ions induce local structure in the surrounding water, and the disorder of the solution is decreased more than in the case of large, singly charged ions. The absolute, Third-Law standard molar entropy of the proton in water can be estimated by proposing a model of the structure it induces, and there is some agreement on the value $-21\,J\,K^{-1}\,mol^{-1}$. The negative value indicates that the proton induces order in the solvent.

Brief illustration 3C.4

The standard molar entropy of $Cl^-(aq)$ is $+57\,J\,K^{-1}\,mol^{-1}$ and that of $Mg^{2+}(aq)$ is $-128\,J\,K^{-1}\,mol^{-1}$. That is, the molar entropy of $Cl^-(aq)$ is $57\,J\,K^{-1}\,mol^{-1}$ higher than that of the proton in water (presumably because it induces less local structure in the surrounding water), whereas that of $Mg^{2+}(aq)$ is $128\,J\,K^{-1}\,mol^{-1}$ lower (presumably because its higher charge induces more local structure in the surrounding water).

[1] In terms of the language introduced in Topic 5A, the entropies of ions in solution are actually *partial molar entropies*, for their values include the consequences of their presence on the organization of the solvent molecules around them.

(c) The temperature dependence of reaction entropy

The temperature dependence of entropy is given by eqn 3C.1a, which for the molar entropy becomes

$$S_m(T_2) = S_m(T_1) + \int_{T_1}^{T_2} \frac{C_{p,m}(T)}{T}\,dT$$

This equation applies to each substance in the reaction, so from eqn 3C.3 the temperature dependence of the standard reaction entropy, $\Delta_r S^\ominus$, is

$$\Delta_r S^\ominus(T_2) = \Delta_r S^\ominus(T_1) + \int_{T_1}^{T_2} \frac{\Delta_r C_p^\ominus}{T}\,dT \qquad (3C.5a)$$

where $\Delta_r C_p^\ominus$ is the difference of the molar heat capacities of products and reactants under standard conditions weighted by the stoichiometric numbers that appear in the chemical equation:

$$\Delta_r C_p^\ominus = \sum_J \nu_J C_{p,m}^\ominus(J) \qquad (3C.5b)$$

Equation 3C.5a is analogous to Kirchhoff's law for the temperature dependence of $\Delta_r H^\ominus$ (eqn 2C.7a in Topic 2C). If $\Delta_r C_p^\ominus$ is independent of temperature in the range T_1 to T_2, the integral in eqn 3C.5a evaluates to $\Delta_r C_p^\ominus \ln(T_2/T_1)$ and

$$\Delta_r S^\ominus(T_2) = \Delta_r S^\ominus(T_1) + \Delta_r C_p^\ominus \ln\frac{T_2}{T_1} \qquad (3C.5c)$$

Brief illustration 3C.5

The standard reaction entropy for $H_2(g) + \frac{1}{2}O_2(g) \rightarrow H_2O(g)$ at 298 K is $-44.42\,J\,K^{-1}\,mol^{-1}$, and the molar heat capacities at constant pressure of the molecules are $H_2O(g)$: $33.58\,J\,K^{-1}\,mol^{-1}$; $H_2(g)$: $28.84\,J\,K^{-1}\,mol^{-1}$; $O_2(g)$: $29.37\,J\,K^{-1}\,mol^{-1}$. It follows that

$$\Delta_r C_p^\ominus = C_{p,m}^\ominus(H_2O,g) - C_{p,m}^\ominus(H_2,g) - \tfrac{1}{2}C_{p,m}^\ominus(O_2,g)$$
$$= -9.94\,J\,K^{-1}\,mol^{-1}$$

This value of $\Delta_r C_p^\ominus$ is used in eqn 3C.5c to find $\Delta_r S^\ominus$ at another temperature, for example at 373 K

$$\Delta_r S^\ominus(373\,K) = -44.42\,J\,K^{-1}\,mol^{-1} + (-9.94\,J\,K^{-1}\,mol^{-1}) \times \ln\frac{373\,K}{298\,K}$$
$$= -46.65\,J\,K^{-1}\,mol^{-1}$$

Checklist of concepts

☐ 1. Entropies are determined calorimetrically by measuring the heat capacity of a substance from low temperatures up to the temperature of interest and taking into account any phase transitions in that range.

☐ 2. The **Debye extrapolation** (or the *Debye T^3-law*) is used to estimate heat capacities of non-metallic solids close to $T = 0$.

☐ 3. The **Nernst heat theorem** states that the entropy change accompanying any physical or chemical transformation approaches zero as the temperature approaches zero: $\Delta S \rightarrow 0$ as $T \rightarrow 0$ provided all the substances involved are perfectly ordered.

☐ 4. The **Third Law of thermodynamics** states that the entropy of all perfect crystalline substances is zero at $T = 0$.

☐ 5. The **residual entropy** of a solid is the entropy arising from disorder that persists at $T = 0$.

☐ 6. **Third-law entropies** are entropies based on $S(0) = 0$.

☐ 7. The **standard entropies of ions in solution** are based on setting $S^\ominus(H^+,aq) = 0$ at all temperatures.

☐ 8. The **standard reaction entropy**, $\Delta_r S^\ominus$, is the difference between the molar entropies of the pure, separated products and the pure, separated reactants, all substances being in their standard states.

Checklist of equations

Property	Equation	Comment	Equation number
Standard molar entropy from calorimetry	See eqn 3C.2	Sum of contributions from $T = 0$ to temperature of interest	3C.2
Standard reaction entropy	$\Delta_r S^\ominus = \sum_{\text{Products}} \nu S_m^\ominus - \sum_{\text{Reactants}} \nu S_m^\ominus$	ν: (positive) stoichiometric coefficients;	3C.3
	$\Delta_r S^\ominus = \sum_J \nu_J S_m^\ominus(J)$	ν_J: (signed) stoichiometric numbers	
Temperature dependence of the standard reaction entropy	$\Delta_r S^\ominus(T_2) = \Delta_r S^\ominus(T_1) + \int_{T_1}^{T_2} (\Delta_r C_p^\ominus / T)\mathrm{d}T$		3C.5a
	$\Delta_r S^\ominus(T_2) = \Delta_r S^\ominus(T_1) + \Delta_r C_p^\ominus \ln(T_2/T_1)$	$\Delta_r C_p^\ominus$ independent of temperature	3C.5c

TOPIC 3D Concentrating on the system

> ➤ Why do you need to know this material?
>
> Most processes of interest in chemistry occur at constant temperature and pressure. Under these conditions, thermodynamic processes are discussed in terms of the Gibbs energy, which is introduced in this Topic. The Gibbs energy is the foundation of the discussion of phase equilibria, chemical equilibrium, and bioenergetics.
>
> ➤ What is the key idea?
>
> The Gibbs energy is a signpost of spontaneous change at constant temperature and pressure, and is equal to the maximum non-expansion work that a system can do.
>
> ➤ What do you need to know already?
>
> This Topic develops the Clausius inequality (Topic 3A) and draws on information about standard states and reaction enthalpy introduced in Topic 2C. The derivation of the Born equation makes use of the Coulomb potential energy between two electric charges (*The chemist's toolkit* 6 in Topic 2A).

Entropy is the basic concept for discussing the direction of natural change, but to use it the changes in both the system and its surroundings must be analysed. In Topic 3A it is shown that it is always very simple to calculate the entropy change in the surroundings (from $\Delta S_{\text{sur}} = q_{\text{sur}}/T_{\text{sur}}$) and this Topic shows that it is possible to devise a simple method for taking this contribution into account automatically. This approach focuses attention on the system and simplifies discussions. Moreover, it is the foundation of all the applications of chemical thermodynamics that follow.

3D.1 The Helmholtz and Gibbs energies

Consider a system in thermal equilibrium with its surroundings at a temperature T. When a change in the system occurs and there is a transfer of energy as heat between the system and the surroundings, the Clausius inequality (eqn 3A.11, $\text{d}S \geq \text{d}q/T$) reads

$$\text{d}S - \frac{\text{d}q}{T} \geq 0 \qquad (3\text{D}.1)$$

This inequality can be developed in two ways according to the conditions (of constant volume or constant pressure) under which the process occurs.

(a) Criteria of spontaneity

First, consider heating at constant volume. Under these conditions and in the absence of additional (non-expansion) work $\text{d}q_V = \text{d}U$; consequently

$$\text{d}S - \frac{\text{d}U}{T} \geq 0$$

The importance of the inequality in this form is that it expresses the criterion for spontaneous change solely in terms of the state functions of the system. The inequality is easily rearranged into

$$T\text{d}S \geq \text{d}U \qquad \text{(constant } V\text{, no additional work)} \qquad (3\text{D}.2)$$

If the internal energy is constant, meaning that $\text{d}U = 0$, then it follows that $T\text{d}S \geq 0$, but as $T > 0$, this relation can be written $\text{d}S_{U,V} \geq 0$, where the subscripts indicate the constant conditions. This expression is a criterion for spontaneous change in terms of properties relating to the system. It states that in a system at constant volume and constant internal energy (such as an isolated system), the entropy increases in a spontaneous change. That statement is essentially the content of the Second Law.

When energy is transferred as heat at constant pressure and there is no work other than expansion work, $\text{d}q_p = \text{d}H$. Then eqn 3D.1 becomes

$$T\text{d}S \geq \text{d}H \qquad \text{(constant } p\text{, no additional work)} \qquad (3\text{D}.3)$$

If the enthalpy is constant as well as the pressure, this relation becomes $T\text{d}S \geq 0$ and therefore $\text{d}S \geq 0$, which may be written $\text{d}S_{H,p} \geq 0$. That is, in a spontaneous process the entropy of the system at constant pressure must increase if its enthalpy remains constant (under these circumstances there can then be no change in entropy of the surroundings).

The criteria of spontaneity at constant volume and pressure can be expressed more simply by introducing two more thermodynamic quantities. One is the **Helmholtz energy**, A, which is defined as

$$A = U - TS \qquad \text{Helmholtz energy [definition]} \qquad (3\text{D}.4\text{a})$$

The other is the **Gibbs energy**, G:

$$G = H - TS \qquad \text{Gibbs energy [definition]} \qquad (3D.4b)$$

All the symbols in these two definitions refer to the system.

When the state of the system changes at constant temperature, the two properties change as follows:

$$\text{(a) } dA = dU - TdS \qquad \text{(b) } dG = dH - TdS \qquad (3D.5)$$

At constant volume, $TdS \geq dU$ (eqn 3D.2) which, by using (a), implies $dA \leq 0$. At constant pressure, $TdS \geq dH$ (eqn 3D.3) which, by using (b), implies $dG \leq 0$. Using the subscript notation to indicate which variables are held constant, the criteria of spontaneous change in terms of dA and dG are

$$\text{(a) } dA_{T,V} \leq 0 \qquad \text{(b) } dG_{T,p} \leq 0 \qquad \text{Criteria of spontaneous change} \qquad (3D.6)$$

These criteria, especially the second, are central to chemical thermodynamics. For instance, in an endothermic reaction H increases, $dH > 0$, but if such a reaction is to be spontaneous at constant temperature and pressure, G must decrease. Because $dG = dH - TdS$, it is possible for dG to be negative provided that the entropy of the system increases so much that TdS outweighs dH. Endothermic reactions are therefore driven by the increase of entropy of the system, which overcomes the reduction of entropy brought about in the surroundings by the inflow of heat into the system in an endothermic process ($dS_{sur} = -dH/T$ at constant pressure). Exothermic reactions are commonly spontaneous because $dH < 0$ and then $dG < 0$ provided TdS is not so negative that it outweighs the decrease in enthalpy.

(b) Some remarks on the Helmholtz energy

At constant temperature and volume, a change is spontaneous if it corresponds to a decrease in the Helmholtz energy: $dA_{T,V} \leq 0$. Such systems move spontaneously towards states of lower A if a path is available. The criterion of *equilibrium*, when neither the forward nor reverse process has a tendency to occur, is $dA_{T,V} = 0$.

The expressions $dA = dU - TdS$ and $dA_{T,V} \leq 0$ are sometimes interpreted as follows. A negative value of dA is favoured by a negative value of dU and a positive value of TdS. This observation suggests that the tendency of a system to move to lower A is due to its tendency to move towards states of lower internal energy and higher entropy. However, this interpretation is false because the tendency to lower A is solely a tendency towards states of greater overall entropy. *Systems change spontaneously if in doing so the total entropy of the system and its surroundings increases, not because they tend to lower internal energy.* The form of dA may give the impression that systems favour lower energy, but that is misleading: dS is the entropy change of the system, $-dU/T$ is the entropy change of the surroundings (when the volume of the system is constant), and their total tends to a maximum.

(c) Maximum work

As well as being the signpost of spontaneous change, a short argument can be used to show that the *change in the Helmholtz energy is equal to the maximum work obtainable from a system at constant temperature.*

How is that done? 3D.1 Relating the change in the Helmholtz energy to the maximum work

To demonstrate that maximum work can be expressed in terms of the change in Helmholtz energy, you need to combine the Clausius inequality $dS \geq dq/T$ in the form $TdS \geq dq$ with the First Law, $dU = dq + dw$, and obtain

$$dU \leq TdS + dw$$

The term dU is smaller than the sum of the two terms on the right because dq has been replaced by TdS, which in general is larger than dq. This expression rearranges to

$$dw \geq dU - TdS$$

It follows that the most negative value of dw is obtained when the equality applies, which is for a reversible process. Thus a reversible process gives the maximum amount of energy as work, and this maximum work is given by

$$dw_{max} = dU - TdS$$

Because at constant temperature $dA = dU - TdS$ (eqn 3D.5), it follows that

$$dw_{max} = dA \qquad \text{Maximum work [constant } T\text{]} \qquad (3D.7)$$

In recognition of this relation, A is sometimes called the 'maximum work function', or the 'work function'.[1]

When a measurable isothermal change takes place in the system, eqn 3D.7 becomes $w_{max} = \Delta A$ with $\Delta A = \Delta U - T\Delta S$. These relations show that, depending on the sign of $T\Delta S$, not all the change in internal energy may be available for doing work. If the change occurs with a decrease in entropy (of the system), in which case $T\Delta S < 0$, then $\Delta U - T\Delta S$ is not as negative as ΔU itself, and consequently the maximum work is less than ΔU. For the change to be spontaneous, some of the energy must escape as heat in order to generate enough entropy in the surroundings to overcome the reduction in entropy in the system (Fig. 3D.1). In this case, Nature is demanding a tax

[1] *Arbeit* is the German word for work; hence the symbol A.

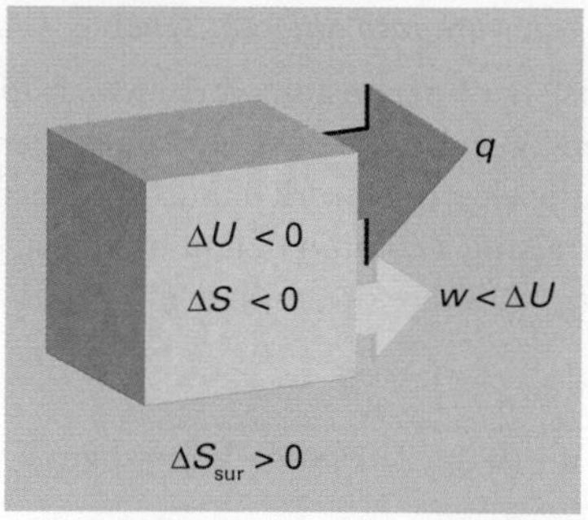

Figure 3D.1 In a system not isolated from its surroundings, the work done may be different from the change in internal energy. In the process depicted here, the entropy of the system decreases, so for the process to be spontaneous the entropy of the surroundings must increase, so energy must pass from the system to the surroundings as heat. Therefore, less work than ΔU can be obtained.

on the internal energy as it is converted into work. This interpretation is the origin of the alternative name 'Helmholtz free energy' for A, because ΔA is that part of the change in internal energy free to do work.

Further insight into the relation between the work that a system can do and the Helmholtz energy is to recall that work is energy transferred to the surroundings as the uniform motion of atoms. The expression $A = U - TS$ can be interpreted as showing that A is the total internal energy of the system, U, less a contribution that is stored as energy of thermal motion (the quantity TS). Because energy stored in random thermal motion cannot be used to achieve uniform motion in the surroundings, only the part of U that is not stored in that way, the quantity $U - TS$, is available for conversion into work.

If the change occurs with an increase of entropy of the system (in which case $T\Delta S > 0$), $\Delta U - T\Delta S$ is more negative than ΔU. In this case, the maximum work that can be obtained from the system is greater than ΔU. The explanation of this apparent paradox is that the system is not isolated and energy may flow in as heat as work is done. Because the entropy of the system increases, a reduction of the entropy of the surroundings can be afforded yet still have, overall, a spontaneous process. Therefore, some energy (no more than the value of $T\Delta S$) may leave the surroundings as heat and contribute to the work the change is generating (Fig. 3D.2). Nature is now providing a tax refund.

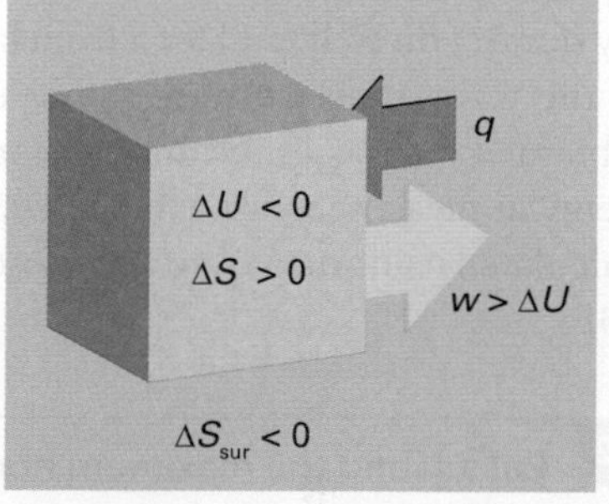

Figure 3D.2 In this process, the entropy of the system increases; hence some reduction in the entropy of the surroundings can be tolerated. That is, some of their energy may be lost as heat to the system. This energy can be returned to them as work, and hence the work done can exceed ΔU.

Example 3D.1 Calculating the maximum available work

When 1.000 mol $C_6H_{12}O_6$ (glucose) is oxidized completely to carbon dioxide and water at 25 °C according to the equation $C_6H_{12}O_6(s) + 6\,O_2(g) \rightarrow 6\,CO_2(g) + 6\,H_2O(l)$, calorimetric measurements give $\Delta_r U = -2808\,\text{kJ mol}^{-1}$ and $\Delta_r S = +182.4\,\text{J K}^{-1}\,\text{mol}^{-1}$ at 25 °C and 1 bar. How much of this change in internal energy can be extracted as (a) heat at constant pressure, (b) work?

Collect your thoughts You know that the heat released at constant pressure is equal to the value of ΔH, so you need to relate $\Delta_r H$ to the given value of $\Delta_r U$. To do so, suppose that all the gases involved are perfect, and use eqn 2B.4 ($\Delta H = \Delta U + \Delta n_g RT$) in the form $\Delta_r H = \Delta_r U + \Delta\nu_g RT$. For the maximum work available from the process use $w_{max} = \Delta A$ in the form $w_{max} = \Delta_r A$.

The solution (a) Because $\Delta\nu_g = 0$, $\Delta_r H = \Delta_r U = -2808\,\text{kJ mol}^{-1}$. Therefore, at constant pressure, the energy available as heat is $2808\,\text{kJ mol}^{-1}$. (b) Because $T = 298\,\text{K}$, the value of $\Delta_r A$ is

$$\Delta_r A = \Delta_r U - T\Delta_r S = -2862\,\text{kJ mol}^{-1}$$

Therefore, the complete oxidation of 1.000 mol $C_6H_{12}O_6$ at constant temperature can be used to produce up to 2862 kJ of work.

Comment. The maximum work available is greater than the change in internal energy on account of the positive entropy of reaction (which is partly due to there being a significant increase in the number of molecules as the reaction proceeds). The system can therefore draw in energy from the surroundings (so reducing their entropy) and make it available for doing work.

Self-test 3D.1 Repeat the calculation for the combustion of 1.000 mol $CH_4(g)$ under the same conditions, using data from Table 2C.3 and that $\Delta_r S$ for the reaction is $-243\,\text{J K}^{-1}\,\text{mol}^{-1}$ at 298 K.

Answer: $|q_p| = 890\,\text{kJ}$, $|w_{max}| = 813\,\text{kJ}$

(d) Some remarks on the Gibbs energy

The Gibbs energy (the 'free energy') is more common in chemistry than the Helmholtz energy because, at least in laboratory chemistry, changes occurring at constant pressure are more common than at constant volume. The criterion $dG_{T,p} \leq 0$ carries over into chemistry as the observation that, *at constant temperature and pressure, chemical reactions are spontaneous in the direction of decreasing Gibbs energy*. Therefore, to decide whether a reaction is spontaneous, the pressure and temperature being constant, it is necessary to assess the change in the Gibbs energy. If G decreases as the reaction proceeds, then the reaction has a spontaneous tendency to convert the reactants into products. If G increases, the reverse reaction is spontaneous. The criterion for equilibrium, when neither the forward nor reverse process is spontaneous, under conditions of constant temperature and pressure, is $dG_{T,p} = 0$.

The existence of spontaneous endothermic reactions provides an illustration of the role of G. In such reactions, H increases, the system rises spontaneously to states of higher enthalpy, and $dH > 0$. Because the reaction is spontaneous, $dG < 0$ despite $dH > 0$; it follows that the entropy of the system increases so much that TdS outweighs dH in $dG = dH - TdS$. Endothermic reactions are therefore driven by the increase of entropy of the system, and this entropy change overcomes the reduction of entropy brought about in the surroundings by the inflow of heat into the system ($dS_{sur} = -dH/T$ at constant pressure). Exothermic reactions are commonly spontaneous because $dH < 0$ and then $dG < 0$ provided TdS is not so negative that it outweighs the decrease in enthalpy.

(e) Maximum non-expansion work

The analogue of the maximum work interpretation of ΔA, and the origin of the name 'free energy', can be found for ΔG. By an argument like that relating the Helmholtz energy to maximum work, it can be shown that, at constant temperature and pressure, the change in Gibbs energy is equal to the maximum additional (non-expansion) work.

How is that done? 3D.2 Relating the change in Gibbs energy to maximum non-expansion work

Because $H = U + pV$ and $dU = dq + dw$, the change in enthalpy for a general change in conditions is

$$dH = dq + dw + d(pV)$$

The corresponding change in Gibbs energy ($G = H - TS$) is

$$dG = dH - TdS - SdT = dq + dw + d(pV) - TdS - SdT$$

Step 1 *Confine the discussion to constant temperature*

When the change is isothermal $dT = 0$; then

$$dG = dq + dw + d(pV) - TdS$$

Step 2 *Confine the change to a reversible process*

When the change is reversible, $dw = dw_{rev}$ and $dq = dq_{rev} = TdS$, so for a reversible, isothermal process

$$dG = \overbrace{TdS}^{dq_{rev}} + dw_{rev} + d(pV) - TdS = dw_{rev} + d(pV)$$

Step 3 *Divide the work into different types*

The work consists of expansion work, which for a reversible change is given by $-pdV$, and possibly some other kind of work (for instance, the electrical work of pushing electrons through a circuit or of raising a column of liquid); this additional work is denoted $dw_{add,rev}$. Therefore, with $d(pV) = pdV + Vdp$,

$$dG = \overbrace{(-pdV + dw_{add,rev})}^{dw_{rev}} + \overbrace{pdV + Vdp}^{d(pV)} = dw_{add,rev} + Vdp$$

Step 4 *Confine the process to constant pressure*

If the change occurs at constant pressure (as well as constant temperature), $dp = 0$ and hence $dG = dw_{add,rev}$. Therefore, at constant temperature and pressure, $dw_{add,rev} = dG$. However, because the process is reversible, the work done must now have its maximum value, so it follows that

$$dw_{add,max} = dG \quad \text{Maximum non-expansion work [constant } T, p] \qquad (3D.8)$$

For a measurable change, the corresponding expression is $w_{add,max} = \Delta G$. This is particularly useful for assessing the maximum electrical work that can be produced by fuel cells and electrochemical cells (Topic 6C).

3D.2 Standard molar Gibbs energies

Standard entropies and enthalpies of reaction (which are introduced in Topics 2C and 3C) can be combined to obtain the **standard Gibbs energy of reaction** (or 'standard reaction Gibbs energy'), $\Delta_r G^{\ominus}$:

$$\Delta_r G^{\ominus} = \Delta_r H^{\ominus} - T\Delta_r S^{\ominus} \quad \text{Standard Gibbs energy of reaction [definition]} \qquad (3D.9)$$

The standard Gibbs energy of reaction is the difference in standard molar Gibbs energies of the products and reactants in their standard states for the reaction as written and at the temperature specified.

Calorimetry (for ΔH directly, and for S from heat capacities) is only one of the ways of determining Gibbs energies. They may also be obtained from equilibrium constants (Topic 6A) and electrochemical measurements (Topic 6D), and for gases they may be calculated using data from spectroscopic observations (Topic 13E).

Example 3D.2 Calculating the maximum non-expansion work of a reaction

How much energy is available for sustaining muscular and nervous activity from the oxidation of 1.00 mol of glucose molecules under standard conditions at 37 °C (blood temperature)? The standard entropy of reaction is $+182.4\,J\,K^{-1}\,mol^{-1}$.

Collect your thoughts The non-expansion work available from the reaction at constant temperature and pressure is equal to the change in standard Gibbs energy for the reaction, $\Delta_r G^\ominus$. To calculate this quantity, you can (at least approximately) ignore the temperature dependence of the reaction enthalpy, and obtain $\Delta_r H^\ominus$ from Table 2C.4 (where the data are for 25 °C, not 37 °C), and substitute the data into $\Delta_r G^\ominus = \Delta_r H^\ominus - T\Delta_r S^\ominus$.

The solution Because the standard reaction enthalpy is $-2808\,\mathrm{kJ\,mol^{-1}}$, it follows that the standard reaction Gibbs energy is

$$\Delta_r G^\ominus = -2808\,\mathrm{kJ\,mol^{-1}} - (310\,\mathrm{K}) \times (182.4\,\mathrm{J\,K^{-1}\,mol^{-1}}) = -2865\,\mathrm{kJ\,mol^{-1}}$$

Therefore, $w_{\mathrm{add,max}} = -2865\,\mathrm{kJ}$ for the oxidation of 1 mol glucose molecules, and the reaction can be used to do up to 2865 kJ of non-expansion work.

Comment. To place this result in perspective, consider that a person of mass 70 kg needs to do 2.1 kJ of work to climb vertically through 3.0 m; therefore, at least 0.13 g of glucose is needed to complete the task (and in practice significantly more).

Self-test 3D.2 How much non-expansion work can be obtained from the combustion of 1.00 mol CH_4(g) under standard conditions at 298 K? Use $\Delta_r S^\ominus = -243\,\mathrm{J\,K^{-1}\,mol^{-1}}$.

Answer: 818 kJ

(a) Gibbs energies of formation

As in the case of standard reaction enthalpies (Topic 2C), it is convenient to define the **standard Gibbs energies of formation**, $\Delta_f G^\ominus$, the standard reaction Gibbs energy for the formation of a compound from its elements in their reference states, as specified in Topic 2C. Standard Gibbs energies of formation of the elements in their reference states are zero, because their formation is a 'null' reaction. A selection of values for compounds is given in Table 3D.1. The standard Gibbs energy of a reaction is then found by taking the appropriate combination:

Table 3D.1 Standard Gibbs energies of formation at 298 K*

	$\Delta_f G^\ominus/(\mathrm{kJ\,mol^{-1}})$
Diamond, C(s)	+2.9
Benzene, C_6H_6(l)	+124.3
Methane, CH_4(g)	−50.7
Carbon dioxide, CO_2(g)	−394.4
Water, H_2O(l)	−237.1
Ammonia, NH_3(g)	−16.5
Sodium chloride, NaCl(s)	−384.1

* More values are given in the *Resource section*.

$$\Delta_r G^\ominus = \sum_{\text{Products}} \nu \Delta_f G^\ominus - \sum_{\text{Reactants}} \nu \Delta_f G^\ominus \quad \text{Standard Gibbs energy of reaction [practical implementation]} \quad (3D.10a)$$

In the notation introduced in Topic 2C,

$$\Delta_r G^\ominus = \sum_J \nu_J \Delta_f G^\ominus(J) \quad (3D.10b)$$

where the ν_J are the (signed) stoichiometric numbers in the chemical equation.

Brief illustration 3D.1

To calculate the standard Gibbs energy of the reaction $CO(g) + \frac{1}{2}O_2(g) \rightarrow CO_2(g)$ at 25 °C, write

$$\begin{aligned}\Delta_r G^\ominus &= \Delta_f G^\ominus(CO_2,g) - \{\Delta_f G^\ominus(CO,g) + \tfrac{1}{2}\Delta_f G^\ominus(O_2,g)\}\\ &= -394.4\,\mathrm{kJ\,mol^{-1}} - \{(-137.2) + \tfrac{1}{2}(0)\}\,\mathrm{kJ\,mol^{-1}}\\ &= -257.2\,\mathrm{kJ\,mol^{-1}}\end{aligned}$$

As explained in Topic 2C the standard enthalpy of formation of H^+ in water is by convention taken to be zero; in Topic 3C, the absolute entropy of H^+(aq) is also by convention set equal to zero (at all temperatures in both cases). These conventions are needed because it is not possible to prepare cations without their accompanying anions. For the same reason, the standard Gibbs energy of formation of H^+(aq) is set equal to zero at all temperatures:

$$\Delta_f G^\ominus(H^+,aq) = 0 \quad \text{Ions in solution [convention]} \quad (3D.11)$$

This definition effectively adjusts the actual values of the Gibbs energies of formation of ions by a fixed amount, which is chosen so that the standard value for one of them, H^+(aq), has the value zero.

Brief illustration 3D.2

For the reaction

$$\tfrac{1}{2}H_2(g) + \tfrac{1}{2}Cl_2(g) \rightarrow H^+(aq) + Cl^-(aq) \quad \Delta_r G^\ominus = -131.23\,\mathrm{kJ\,mol^{-1}}$$

the value of $\Delta_r G^\ominus$ can be written in terms of standard Gibbs energies of formation as

$$\Delta_r G^\ominus = \Delta_f G^\ominus(H^+,aq) + \Delta_f G^\ominus(Cl^-,aq)$$

where the $\Delta_f G^\ominus$ of the elements on the left of the chemical equation are zero. Because by convention $\Delta_f G^\ominus(H^+,aq) = 0$, it follows that $\Delta_r G^\ominus = \Delta_f G^\ominus(Cl^-,aq)$ and therefore that $\Delta_f G^\ominus(Cl^-,aq) = -131.23\,\mathrm{kJ\,mol^{-1}}$.

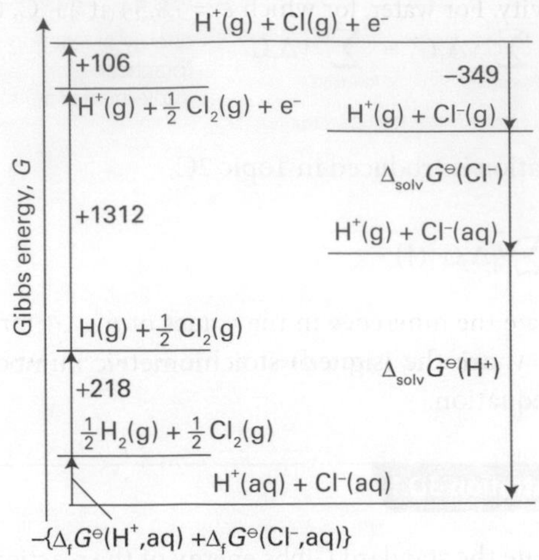

Figure 3D.3 A thermodynamic cycle for discussion of the Gibbs energies of hydration and formation of chloride ions in aqueous solution. The changes in Gibbs energies around the cycle sum to zero because G is a state function.

The factors responsible for the Gibbs energy of formation of an ion in solution can be identified by analysing its formation in terms of a thermodynamic cycle. As an illustration, consider the standard Gibbs energy of formation of Cl^- in water. The formation reaction $\frac{1}{2}H_2(g) + \frac{1}{2}Cl_2(g) \rightarrow H^+(aq) + Cl^-(aq)$ is treated as the outcome of the sequence of steps shown in Fig. 3D.3 (with values taken from the *Resource section*). The sum of the Gibbs energies for all the steps around a closed cycle is zero, so

$$\Delta_f G^\ominus(Cl^-,aq) = 1287\,kJ\,mol^{-1} + \Delta_{solv}G^\ominus(H^+) + \Delta_{solv}G^\ominus(Cl^-)$$

The standard Gibbs energies of formation of the gas-phase ions are unknown and have been replaced by energies and electron affinities and the assumption that any differences from the Gibbs energies arising from conversion to enthalpy and the inclusion of entropies to obtain Gibbs energies in the formation of H^+ are cancelled by the corresponding terms in the electron gain of Cl. The conclusions from the cycles are therefore only approximate. An important point to note is that the value of $\Delta_f G^\ominus$ of Cl^- is not determined by the properties of Cl alone but includes contributions from the dissociation, ionization, and hydration of hydrogen.

(b) The Born equation

Gibbs energies of solvation of individual ions may be estimated on the basis of a model in which solvation is expressed as an electrostatic property.

How is that done? 3D.3 Developing an electrostatic model for solvation

The model treats the interaction between the ion and the solvent using elementary electrostatics: the ion is regarded as a charged sphere and the solvent is treated as a continuous medium (a continuous dielectric). The key step is to use the result from Section 3D.1(e) to identify the Gibbs energy of solvation with the work of transferring an ion from a vacuum into the solvent. That work is calculated by taking the difference of the work of charging an ion when it is in the solution and the work of charging the same ion when it is in a vacuum.

The derivation uses concepts developed in *The chemist's toolkit* 6 in Topic 2A, where it is seen that the Coulomb potential energy of two point electric charges Q_1 and Q_2 separated by a distance r in a medium with permittivity ε is

$$V(r) = \frac{Q_1 Q_2}{4\pi\varepsilon r}$$

The energy of this interaction may also be expressed in terms of the **Coulomb potential** ϕ that the point charge Q_2 experiences at a distance r from the point charge Q_1. Then $V(r) = Q_2\phi(r)$, with

$$\phi(r) = \frac{Q_1}{4\pi\varepsilon r}$$

With the distance r in metres and the charge Q_1 in coulombs (C), the potential is obtained in $J\,C^{-1}$. By definition, $1\,J\,C^{-1} = 1\,V$ (volt), so ϕ can also be expressed in volts.

Step 1 *Obtain an expression for charging a spherical ion to its final value in a medium*

The Coulomb potential, ϕ, at the surface of a sphere (representing the ion) of radius r_i and charge Q is the same as the potential due to a point charge at its centre, so

$$\phi(r_i) = \frac{Q}{4\pi\varepsilon r_i}$$

The work of bringing up a charge dQ to the sphere is $\phi(r_i)\mathrm{d}Q$. If the charge number of the ion is z_i, the total work of charging the sphere from 0 to $z_i e$ is

$$w = \int_0^{z_i e}\phi(r_i)\mathrm{d}Q = \frac{1}{4\pi\varepsilon r_i}\int_0^{z_i e} Q\mathrm{d}Q = \frac{z_i^2 e^2}{8\pi\varepsilon r_i}$$

This electrical work of charging, when multiplied by Avogadro's constant, N_A, is the molar Gibbs energy for charging the ions.

Step 2 *Apply the result to solution and a vacuum*

The work of charging an ion in a vacuum is obtained by setting $\varepsilon = \varepsilon_0$, the vacuum permittivity. The corresponding value for charging the ion in a medium is obtained by setting $\varepsilon = \varepsilon_r\varepsilon_0$, where ε_r is the relative permittivity of the medium.

$$w(\text{vacuum}) = \frac{z_i^2 e^2}{8\pi\varepsilon_0 r_i} \qquad w(\text{medium}) = \frac{z_i^2 e^2}{8\pi\varepsilon_r\varepsilon_0 r_i}$$

Step 3 *Identify the Gibbs energy of solvation as the work needed to move the ion from a vacuum into the medium*

It follows that the change in molar Gibbs energy that accompanies the transfer of ions from a vacuum to a solvent is the difference of these two expressions for the work of charging:

$$\Delta_{\text{solv}}G^{\ominus} = \frac{z_i^2 e^2 N_A}{8\pi\varepsilon r_i} - \frac{z_i^2 e^2 N_A}{8\pi\varepsilon_0 r_i} = \frac{z_i^2 e^2 N_A}{8\pi\varepsilon_r\varepsilon_0 r_i} - \frac{z_i^2 e^2 N_A}{8\pi\varepsilon_0 r_i}$$

A minor rearrangement of the right-hand side gives the **Born equation**:

$$\Delta_{\text{solv}}G^{\ominus} = -\frac{z_i^2 e^2 N_A}{8\pi\varepsilon_0 r_i}\left(1 - \frac{1}{\varepsilon_r}\right) \qquad (3\text{D}.12\text{a})$$

Born equation

Note that $\Delta_{\text{solv}}G^{\ominus} < 0$, and that $\Delta_{\text{solv}}G^{\ominus}$ is strongly negative for small, highly charged ions in media of high relative permittivity. For water, for which $\varepsilon_r = 78.54$ at 25 °C, the Born equation becomes

$$\Delta_{\text{solv}}G^{\ominus} = -\frac{z_i^2}{r_i/\text{pm}} \times 6.86 \times 10^4 \text{ kJ mol}^{-1} \qquad (3\text{D}.12\text{b})$$

Brief illustration 3D.3

To estimate the difference in the values of $\Delta_{\text{solv}}G^{\ominus}$ for Cl^- and I^- in water at 25 °C, given their radii as 181 pm and 220 pm, respectively, write

$$\Delta_{\text{solv}}G^{\ominus}(Cl^-) - \Delta_{\text{solv}}G^{\ominus}(I^-) = -\left(\frac{1}{181} - \frac{1}{220}\right) \times 6.86 \times 10^4 \text{ kJ mol}^{-1}$$
$$= -67 \text{ kJ mol}^{-1}$$

Checklist of concepts

- ☐ 1. The Clausius inequality implies a number of **criteria for spontaneous change** under a variety of conditions which may be expressed in terms of the properties of the system alone; they are summarized by introducing the Helmholtz and Gibbs energies.
- ☐ 2. A **spontaneous process** at constant temperature and volume is accompanied by a decrease in the Helmholtz energy.
- ☐ 3. The change in the Helmholtz energy is equal to the **maximum work** obtainable from a system at constant temperature.
- ☐ 4. A spontaneous process at constant temperature and pressure is accompanied by a decrease in the Gibbs energy.
- ☐ 5. The change in the Gibbs energy is equal to the **maximum non-expansion work** obtainable from a system at constant temperature and pressure.
- ☐ 6. **Standard Gibbs energies of formation** are used to calculate the standard Gibbs energies of reactions.
- ☐ 7. The standard Gibbs energies of formation of ions may be estimated from a thermodynamic cycle and the **Born equation**.

Checklist of equations

Property	Equation	Comment	Equation number
Criteria of spontaneity	$dS_{U,V} \geq 0$	Subscripts show which variables are held constant, here and below	
	$dS_{H,p} \geq 0$		
Helmholtz energy	$A = U - TS$	Definition	3D.4a
Gibbs energy	$G = H - TS$	Definition	3D.4b
Criteria of spontaneous change	(a) $dA_{T,V} \leq 0$ (b) $dG_{T,p} \leq 0$	Equality refers to equilibrium	3D.6
Maximum work	$dw_{\max} = dA$, $w_{\max} = \Delta A$	Constant temperature	3D.7
Maximum non-expansion work	$dw_{\text{add,max}} = dG$, $w_{\text{add,max}} = \Delta G$	Constant temperature and pressure	3D.8
Standard Gibbs energy of reaction	$\Delta_r G^{\ominus} = \Delta_r H^{\ominus} - T\Delta_r S^{\ominus}$	Definition	3D.9
	$\Delta_r G^{\ominus} = \sum_J \nu_J \Delta_f G^{\ominus}(J)$	Practical implementation	3D.10b
Ions in solution	$\Delta_f G^{\ominus}(H^+, aq) = 0$	Convention	3D.11
Born equation	$\Delta_{\text{solv}}G^{\ominus} = -(z_i^2 e^2 N_A/8\pi\varepsilon_0 r_i)(1 - 1/\varepsilon_r)$	Solvent treated as a continuum and the ion as a sphere	3D.12a

TOPIC 3E Combining the First and Second Laws

> ➤ Why do you need to know this material?
>
> The First and Second Laws of thermodynamics are both relevant to the behaviour of bulk matter, and the whole force of thermodynamics can be brought to bear on a problem by setting up a formulation that combines them.
>
> ➤ What is the key idea?
>
> The fact that infinitesimal changes in thermodynamic functions are exact differentials leads to relations between a variety of properties.
>
> ➤ What do you need to know already?
>
> You need to be aware of the definitions of the state functions *U* (Topic 2A), *H* (Topic 2B), *S* (Topic 3A), and *A* and *G* (Topic 3D). The mathematical derivations in this Topic draw frequently on the properties of partial derivatives, which are described in *The chemist's toolkit* 9 in Topic 2A.

The First Law of thermodynamics may be written $dU = dq + dw$. For a reversible change in a closed system of constant composition, and in the absence of any additional (non-expansion) work, $dw_{rev} = -pdV$ and (from the definition of entropy) $dq_{rev} = TdS$, where p is the pressure of the system and T its temperature. Therefore, for a reversible change in a closed system,

$$dU = TdS - pdV \quad \text{The fundamental equation} \quad (3E.1)$$

However, because dU is an exact differential, its value is independent of path. Therefore, the same value of dU is obtained whether the change is brought about irreversibly or reversibly. Consequently, this equation *applies to any change—reversible or irreversible—of a closed system that does no additional (non-expansion) work*. This combination of the First and Second Laws is called the **fundamental equation**.

The fact that the fundamental equation applies to both reversible and irreversible changes may be puzzling at first sight. The reason is that only in the case of a reversible change may TdS be identified with dq and $-pdV$ with dw. When the change is irreversible, $TdS > dq$ (the Clausius inequality) and $-pdV > dw$. The sum of dw and dq remains equal to the sum of TdS and $-pdV$, provided the composition is constant.

3E.1 Properties of the internal energy

Equation 3E.1 shows that the internal energy of a closed system changes in a simple way when either S or V is changed ($dU \propto dS$ and $dU \propto dV$). These simple proportionalities suggest that U is best regarded as a function of S and V. It could be regarded as a function of other variables, such as S and p or T and V, because they are all interrelated; but the simplicity of the fundamental equation suggests that $U(S,V)$ is the best choice.

The *mathematical* consequence of U being a function of S and V is that an infinitesimal change dU can be expressed in terms of changes dS and dV by

$$dU = \left(\frac{\partial U}{\partial S}\right)_V dS + \left(\frac{\partial U}{\partial V}\right)_S dV \quad (3E.2)$$

The two partial derivatives (see *The chemist's toolkit* 9 in Topic 2A) are the slopes of the plots of U against S at constant V, and U against V at constant S. When this expression is compared term-by-term to the *thermodynamic* relation, eqn 3E.1, it follows that for systems of constant composition,

$$\left(\frac{\partial U}{\partial S}\right)_V = T \quad \left(\frac{\partial U}{\partial V}\right)_S = -p \quad (3E.3)$$

The first of these two equations is a purely thermodynamic definition of temperature as the ratio of the changes in the internal energy (a First-Law concept) and entropy (a Second-Law concept) of a constant-volume, closed, constant-composition system. Relations between the properties of a system are starting to emerge.

(a) The Maxwell relations

An infinitesimal change in a function $f(x,y)$ can be written $df = gdx + hdy$ where g and h may be functions of x and y. The mathematical criterion for df being an exact differential (in the sense that its integral is independent of path) is that

$$\left(\frac{\partial g}{\partial y}\right)_x = \left(\frac{\partial h}{\partial x}\right)_y \quad (3E.4)$$

This criterion is derived in *The chemist's toolkit* 10. Because the fundamental equation, eqn 3E.1, is an expression for

The chemist's toolkit 10 Exact differentials

Suppose that df can be expressed in the following way:

$$df = g(x,y)dx + h(x,y)dy$$

Is df is an exact differential? If it is exact, then it can be expressed in the form

$$df = \left(\frac{\partial f}{\partial x}\right)_y dx + \left(\frac{\partial f}{\partial y}\right)_x dy$$

Comparing these two expressions gives

$$\left(\frac{\partial f}{\partial x}\right)_y = g(x,y) \quad \left(\frac{\partial f}{\partial y}\right)_x = h(x,y)$$

It is a property of partial derivatives that successive derivatives may be taken in any order:

$$\left(\frac{\partial}{\partial y}\left(\frac{\partial f}{\partial x}\right)_y\right)_x = \left(\frac{\partial}{\partial x}\left(\frac{\partial f}{\partial y}\right)_x\right)_y$$

Taking the partial derivative with respect to x of the first equation, and with respect to y of the second gives

$$\left(\frac{\partial}{\partial y}\left(\frac{\partial f}{\partial x}\right)_y\right)_x = \left(\frac{\partial g(x,y)}{\partial y}\right)_x \quad \left(\frac{\partial}{\partial x}\left(\frac{\partial f}{\partial y}\right)_x\right)_y = \left(\frac{\partial h(x,y)}{\partial x}\right)_y$$

By the property of partial derivatives these two successive derivatives of f with respect to x and y must be the same, hence

$$\left(\frac{\partial g(x,y)}{\partial y}\right)_x = \left(\frac{\partial h(x,y)}{\partial x}\right)_y$$

If this equality is satisfied, then $df = g(x,y)dx + h(x,y)dy$ is an exact differential. Conversely, if it is known from other arguments that df is exact, then this relation between the partial derivatives follows.

an exact differential, the functions multiplying dS and dV (namely T and $-p$) must pass this test. Therefore, it must be the case that

$$\left(\frac{\partial T}{\partial V}\right)_S = -\left(\frac{\partial p}{\partial S}\right)_V \qquad \text{A Maxwell relation} \quad (3E.5)$$

A relation has been generated between quantities which, at first sight, would not seem to be related.

Equation 3E.5 is an example of a **Maxwell relation**. However, apart from being unexpected, it does not look particularly interesting. Nevertheless, it does suggest that there might be other similar relations that are more useful. Indeed, the fact that H, G, and A are all state functions can be used to derive three more Maxwell relations. The argument to obtain them runs in the same way in each case: because H, G, and A are state functions, the expressions for dH, dG, and dA satisfy relations like eqn 3E.4. All four relations are listed in Table 3E.1.

Table 3E.1 The Maxwell relations

State function	Exact differential	Maxwell relation
U	$dU = TdS - pdV$	$\left(\frac{\partial T}{\partial V}\right)_S = -\left(\frac{\partial p}{\partial S}\right)_V$
H	$dH = TdS + Vdp$	$\left(\frac{\partial T}{\partial p}\right)_S = \left(\frac{\partial V}{\partial S}\right)_p$
A	$dA = -pdV - SdT$	$\left(\frac{\partial p}{\partial T}\right)_V = \left(\frac{\partial S}{\partial V}\right)_T$
G	$dG = Vdp - SdT$	$\left(\frac{\partial V}{\partial T}\right)_p = -\left(\frac{\partial S}{\partial p}\right)_T$

Example 3E.1 Using the Maxwell relations

Use the Maxwell relations in Table 3E.1 to show that the entropy of a perfect gas is linearly dependent on $\ln V$, that is, $S = a + b\ln V$.

Collect your thoughts The natural place to start, given that you are invited to use the Maxwell relations, is to consider the relation for $(\partial S/\partial V)_T$, as that differential coefficient shows how the entropy varies with volume at constant temperature. Be alert for an opportunity to use the perfect gas equation of state.

The solution From Table 3E.1,

$$\left(\frac{\partial S}{\partial V}\right)_T = \left(\frac{\partial p}{\partial T}\right)_V$$

Now use the perfect gas equation of state, $pV = nRT$, to write $p = nRT/V$:

$$\left(\frac{\partial p}{\partial T}\right)_V = \left(\frac{\partial(nRT/V)}{\partial T}\right)_V = \frac{nR}{V}$$

At this point, write

$$\left(\frac{\partial S}{\partial V}\right)_T = \frac{nR}{V}$$

and therefore, at constant temperature,

$$\int dS = nR\int \frac{dV}{V} = nR\ln V + \text{constant}$$

The integral on the left is S + constant, which completes the demonstration.

Self-test 3E.1 How does the entropy depend on the volume of a van der Waals gas? Suggest a reason.

Answer: S varies as $nR\ln(V - nb)$; molecules in a smaller available volume

(b) The variation of internal energy with volume

The internal pressure, π_T (introduced in Topic 2D), is defined as $\pi_T = (\partial U/\partial V)_T$ and represents how the internal energy changes as the volume of a system is changed isothermally; it plays a central role in the manipulation of the First Law. By using a Maxwell relation, π_T can be expressed as a function of pressure and temperature.

How is that done? 3E.1 Deriving a thermodynamic equation of state

To construct the partial differential $(\partial U/\partial V)_T$ you need to start from eqn 3E.2, divide both sides by dV, and impose the constraint of constant temperature:

$$\overbrace{\left(\frac{\partial U}{\partial V}\right)_T}^{\pi_T} = \overbrace{\left(\frac{\partial U}{\partial S}\right)_V}^{T}\left(\frac{\partial S}{\partial V}\right)_T + \overbrace{\left(\frac{\partial U}{\partial V}\right)_S}^{-p}$$

Next, introduce the two relations in eqn 3E.3 (as indicated by the annotations) and the definition of π_T to obtain

$$\pi_T = T\left(\frac{\partial S}{\partial V}\right)_T - p$$

The third Maxwell relation in Table 3E.1 turns $(\partial S/\partial V)_T$ into $(\partial p/\partial T)_V$, to give

$$\pi_T = T\left(\frac{\partial p}{\partial T}\right)_V - p \qquad \text{(3E.6a)}$$

A thermodynamic equation of state

Equation 3E.6a is called a **thermodynamic equation of state** because, when written in the form

$$p = T\left(\frac{\partial p}{\partial T}\right)_V - \pi_T \qquad \text{(3E.6b)}$$

it is an expression for pressure in terms of a variety of thermodynamic properties of the system.

Example 3E.2 Deriving a thermodynamic relation

Show thermodynamically that $\pi_T = 0$ for a perfect gas, and compute its value for a van der Waals gas.

Collect your thoughts Proving a result 'thermodynamically' means basing it entirely on general thermodynamic relations and equations of state, without drawing on molecular arguments (such as the existence of intermolecular forces). You know that for a perfect gas, $p = nRT/V$, so this relation should be used in eqn 3E.6. Similarly, the van der Waals equation is given in Table 1C.4, and for the second part of the question it should be used in eqn 3E.6.

The solution For a perfect gas write

$$\left(\frac{\partial p}{\partial T}\right)_V = \left(\frac{\partial nRT/V}{\partial T}\right)_V = \frac{nR}{V}$$

Then, eqn 3E.6 becomes

$$\pi_T = T\overbrace{\left(\frac{\partial p}{\partial T}\right)_V}^{(\partial p/\partial T)_V = nR/V} - p = \overbrace{\left(\frac{nRT}{V}\right)}^{p} - p = 0$$

The equation of state of a van der Waals gas is

$$p = \frac{nRT}{V-nb} - a\frac{n^2}{V^2}$$

Because a and b are independent of temperature,

$$\left(\frac{\partial p}{\partial T}\right)_V = \left(\frac{\partial nRT/(V-nb)}{\partial T}\right)_V = \frac{nR}{V-nb}$$

Therefore, from eqn 3E.6,

$$\pi_T = \frac{nRT}{V-nb} - p = \frac{nRT}{V-nb} - \overbrace{\left(\frac{nRT}{V-nb} - a\frac{n^2}{V^2}\right)}^{p} = a\frac{n^2}{V^2}$$

Comment. This result for π_T implies that the internal energy of a van der Waals gas increases when it expands isothermally, that is, $(\partial U/\partial V)_T > 0$, and that the increase is related to the parameter a, which models the attractive interactions between the particles. A larger molar volume, corresponding to a greater average separation between molecules, implies weaker mean intermolecular attractions, so the total energy is greater.

Self-test 3E.2 Calculate π_T for a gas that obeys the virial equation of state (Table 1C.4), retaining only the term in B.

Answer: $\pi_T = RT^2(\partial B/\partial T)_V/V_m^2$

3E.2 Properties of the Gibbs energy

The same arguments that were used for U can also be used for the Gibbs energy, $G = H - TS$. They lead to expressions showing how G varies with pressure and temperature and which are important for discussing phase transitions and chemical reactions.

(a) General considerations

When the system undergoes a change of state, G may change because H, T, and S all change:

$$\mathrm{d}G = \mathrm{d}H - \mathrm{d}(TS) = \mathrm{d}H - T\mathrm{d}S - S\mathrm{d}T$$

Because $H = U + pV$,

$$dH = dU + d(pV) = dU + pdV + Vdp$$

and therefore

$$dG = dU + pdV + Vdp - TdS - SdT$$

For a closed system doing no non-expansion work, dU can be replaced by the fundamental equation $dU = TdS - pdV$ to give

$$dG = TdS - pdV + pdV + Vdp - TdS - SdT$$

Four terms now cancel on the right, and so for a closed system in the absence of non-expansion work and at constant composition

$$dG = Vdp - SdT \quad \text{The fundamental equation of chemical thermodynamics} \quad (3E.7)$$

This expression, which shows that a change in G is proportional to a change in p or T, suggests that G may be best regarded as a function of p and T. It may be regarded as the **fundamental equation of chemical thermodynamics** as it is so central to the application of thermodynamics to chemistry. It also suggests that G is an important quantity in chemistry because the pressure and temperature are usually the variables that can be controlled. In other words, G carries around the combined consequences of the First and Second Laws in a way that makes it particularly suitable for chemical applications.

The same argument that led to eqn 3E.3, when applied to the exact differential $dG = Vdp - SdT$, now gives

$$\left(\frac{\partial G}{\partial T}\right)_p = -S \qquad \left(\frac{\partial G}{\partial p}\right)_T = V \quad \text{The variation of } G \text{ with } T \text{ and } p \quad (3E.8)$$

These relations show how the Gibbs energy varies with temperature and pressure (Fig. 3E.1).

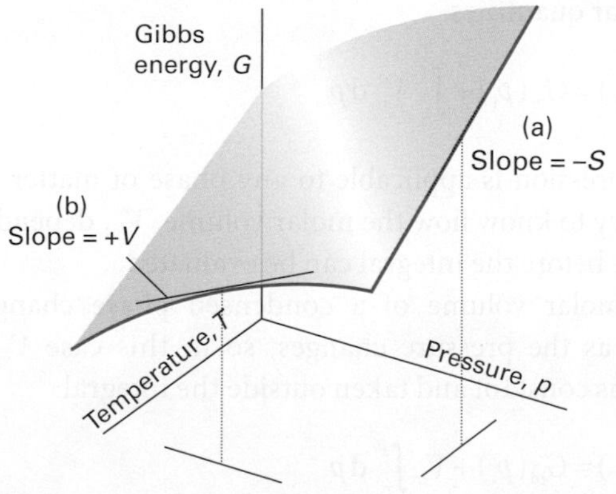

Figure 3E.1 The variation of the Gibbs energy of a system with (a) temperature at constant pressure and (b) pressure at constant temperature. The slope of the former is equal to the negative of the entropy of the system and that of the latter is equal to the volume.

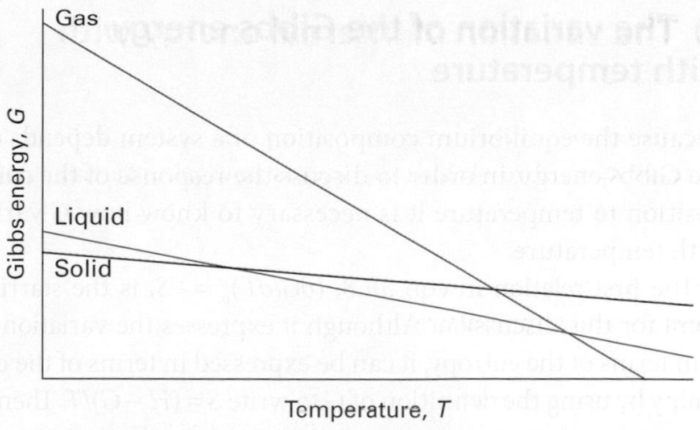

Figure 3E.2 The variation of the Gibbs energy with the temperature is determined by the entropy. Because the entropy of the gaseous phase of a substance is greater than that of the liquid phase, and the entropy of the solid phase is smallest, the Gibbs energy changes most steeply for the gas phase, followed by the liquid phase, and then the solid phase of the substance.

The first implies that:

- Because $S > 0$ for all substances, G always *decreases* when the temperature is raised (at constant pressure and composition).
- Because $(\partial G/\partial T)_p$ becomes more negative as S increases, G decreases most sharply with increasing temperature when the entropy of the system is large.

Physical interpretation

Therefore, the Gibbs energy of the gaseous phase of a substance, which has a high molar entropy, is more sensitive to temperature than its liquid and solid phases (Fig. 3E.2).

Similarly, the second relation implies that:

- Because $V > 0$ for all substances, G always *increases* when the pressure of the system is increased (at constant temperature and composition).
- Because $(\partial G/\partial p)_T$ increases with V, G is more sensitive to pressure when the volume of the system is large.

Physical interpretation

Because the molar volume of the gaseous phase of a substance is greater than that of its condensed phases, the molar Gibbs energy of a gas is more sensitive to pressure than its liquid and solid phases (Fig. 3E.3).

Brief illustration 3E.1

The mass density of liquid water is $0.9970\,\mathrm{g\,cm^{-3}}$ at 298 K. It follows that when the pressure is increased by 0.1 bar (at constant temperature), the molar Gibbs energy changes by

$$\Delta G_m \approx \left(\frac{\partial G_m}{\partial p}\right)_T \Delta p = V_m \Delta p = \overbrace{\frac{18.0\,\mathrm{g\,mol^{-1}}}{0.9970\times10^6\,\mathrm{g\,m^{-3}}}}^{V_m} \times (0.1\times10^5\,\mathrm{N\,m^{-2}})$$

$$= +0.18\,\mathrm{J\,mol^{-1}}$$

(b) The variation of the Gibbs energy with temperature

Because the equilibrium composition of a system depends on the Gibbs energy, in order to discuss the response of the composition to temperature it is necessary to know how G varies with temperature.

The first relation in eqn 3E.8, $(\partial G/\partial T)_p = -S$, is the starting point for this discussion. Although it expresses the variation of G in terms of the entropy, it can be expressed in terms of the enthalpy by using the definition of G to write $S = (H - G)/T$. Then

$$\left(\frac{\partial G}{\partial T}\right)_p = \frac{G-H}{T} \quad (3E.9)$$

In Topic 6A it is shown that the equilibrium constant of a reaction is related to G/T rather than to G itself. With this application in mind, eqn 3E.9 can be developed to show how G/T varies with temperature.

How is that done? 3E.2 Deriving an expression for the temperature variation of G/T

First, note that

$d(fg)/dx = f(dg/dx) + g(df/dx)$

$$\left(\frac{\partial G/T}{\partial T}\right)_p = \frac{1}{T}\left(\frac{\partial G}{\partial T}\right)_p + G\underbrace{\frac{d(1/T)}{dT}}_{-1/T^2} = \frac{1}{T}\left(\frac{\partial G}{\partial T}\right)_p - \frac{G}{T^2}$$

$$= \frac{1}{T}\left\{\left(\frac{\partial G}{\partial T}\right)_p - \frac{G}{T}\right\}$$

Now replace the term $(\partial G/\partial T)_p$ on the right by eqn 3E.9

$(\partial G/\partial T)_p = (G - H)/T$

$$\left(\frac{\partial G/T}{\partial T}\right)_p = \frac{1}{T}\left\{\left(\frac{\partial G}{\partial T}\right)_p - \frac{G}{T}\right\} = \frac{1}{T}\left\{\frac{G-H}{T} - \frac{G}{T}\right\} = \frac{1}{T}\left\{\frac{-H}{T}\right\}$$

from which follows the **Gibbs–Helmholtz equation**

$$\left(\frac{\partial G/T}{\partial T}\right)_p = -\frac{H}{T^2} \quad (3E.10)$$

Gibbs–Helmholtz equation

The Gibbs–Helmholtz equation is most useful when it is applied to changes, including changes of physical state, and chemical reactions at constant pressure. Then, because $\Delta G = G_f - G_i$ for the change of Gibbs energy between the final and initial states, and because the equation applies to both G_f and G_i,

$$\left(\frac{\partial \Delta G/T}{\partial T}\right)_p = -\frac{\Delta H}{T^2} \quad (3E.11)$$

This equation shows that if the change in enthalpy of a system that is undergoing some kind of transformation (such as vaporization or reaction) is known, then how the corresponding change in Gibbs energy varies with temperature is also known. This turns out to be a crucial piece of information in chemistry.

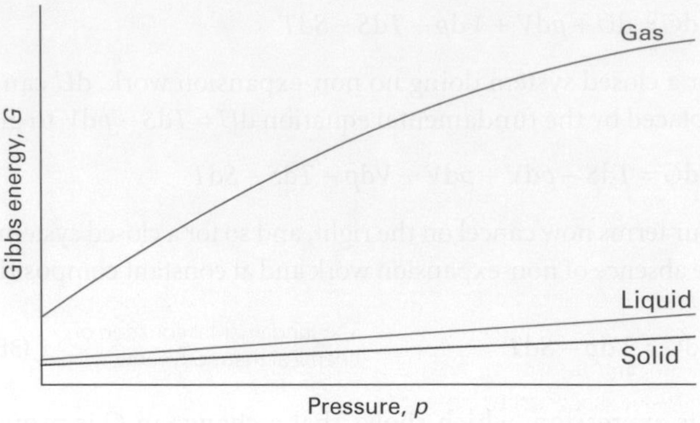

Figure 3E.3 The variation of the Gibbs energy with the pressure is determined by the volume of the sample. Because the volume of the gaseous phase of a substance is greater than that of the same amount of liquid phase, and the volume of the solid phase is smallest (for most substances), the Gibbs energy changes most steeply for the gas phase, followed by the liquid phase, and then the solid phase of the substance. Because the molar volumes of the solid and liquid phases of a substance are similar, their molar Gibbs energies vary by similar amounts as the pressure is changed.

(c) The variation of the Gibbs energy with pressure

To find the Gibbs energy at one pressure in terms of its value at another pressure, the temperature being constant, set $dT = 0$ in eqn 3E.7, which gives $dG = Vdp$, and integrate:

$$G(p_f) = G(p_i) + \int_{p_i}^{p_f} V\,dp \quad (3E.12a)$$

For molar quantities,

$$G_m(p_f) = G_m(p_i) + \int_{p_i}^{p_f} V_m\,dp \quad (3E.12b)$$

This expression is applicable to any phase of matter, but it is necessary to know how the molar volume, V_m, depends on the pressure before the integral can be evaluated.

The molar volume of a condensed phase changes only slightly as the pressure changes, so in this case V_m can be treated as constant and taken outside the integral:

$$G_m(p_f) = G_m(p_i) + V_m\int_{p_i}^{p_f} dp$$

That is,

$$G_m(p_f) = G_m(p_i) + (p_f - p_i)V_m \quad (3E.13)$$

Molar Gibbs energy [incompressible substance]

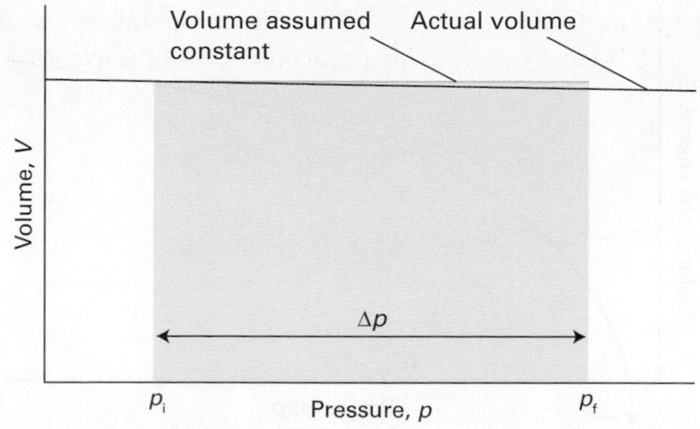

Figure 3E.4 At constant temperature, the difference in Gibbs energy of a solid or liquid between two pressures is equal to the rectangular area shown. The variation of volume with pressure has been assumed to be negligible.

The origin of the term $(p_f - p_i)V_m$ is illustrated graphically in Fig. 3E.4. Under normal laboratory conditions $(p_f - p_i)V_m$ is very small and may be neglected. Hence, the Gibbs energies of solids and liquids are largely independent of pressure. However, in geophysical problems, because pressures in the Earth's interior are huge, their effect on the Gibbs energy cannot be ignored. If the pressures are so great that there are substantial volume changes over the range of integration, then the complete expression, eqn 3E.12, must be used.

Example 3E.3 Evaluating the pressure dependence of a Gibbs energy of transition

Suppose that for a certain phase transition of a solid $\Delta_{trs}V = +1.0\,cm^3\,mol^{-1}$ independent of pressure. By how much does that Gibbs energy of transition change when the pressure is increased from 1.0 bar (1.0×10^5 Pa) to 3.0 Mbar (3.0×10^{11} Pa)?

Collect your thoughts You need to start with eqn 3E.12b to obtain expressions for the Gibbs energy of each of the phases 1 and 2 of the solid

$$G_{m,1}(p_f) = G_{m,1}(p_i) + \int_{p_i}^{p_f} V_{m,1}\,dp$$

$$G_{m,2}(p_f) = G_{m,2}(p_i) + \int_{p_i}^{p_f} V_{m,2}\,dp$$

Then, to obtain $\Delta_{trs}G = G_{m,2} - G_{m,1}$ subtract the second expression from the first, noting that $V_{m,2} - V_{m,1} = \Delta_{trs}V$:

$$\Delta_{trs}G(p_f) = \Delta_{trs}G(p_i) + \int_{p_i}^{p_f} \Delta_{trs}V\,dp$$

Use the data to complete the calculation.

The solution Because $\Delta_{trs}V_m$ is independent of pressure,

$$\Delta_{trs}G(p_f) = \Delta_{trs}G(p_i) + \overbrace{\Delta_{trs}V}^{\text{constant}} \int_{p_i}^{p_f} dp = \Delta_{trs}G(p_i) + \Delta_{trs}V(p_f - p_i)$$

Inserting the data and using $1\,Pa\,m^3 = 1\,J$ gives

$$\begin{aligned}\Delta_{trs}G(3\,\text{Mbar}) &= \Delta_{trs}G(1\,\text{bar}) + (1.0 \times 10^{-6}\,m^3\,mol^{-1}) \\ &\quad\times (3.0 \times 10^{11}\,Pa - 1.0 \times 10^5\,Pa) \\ &= \Delta_{trs}G(1\,\text{bar}) + 3.0 \times 10^2\,kJ\,mol^{-1}\end{aligned}$$

Self-test 3E.3 Calculate the change in G_m for ice at $-10\,°C$, with density $917\,kg\,m^{-3}$, when the pressure is increased from 1.0 bar to 2.0 bar.

Answer: $+2.0\,J\,mol^{-1}$

The molar volumes of gases are large, so the Gibbs energy of a gas depends strongly on the pressure. Furthermore, because the volume also varies markedly with the pressure, the volume cannot be treated as a constant in the integral in eqn 3E.12b (Fig. 3E.5).

For a perfect gas, substitute $V_m = RT/p$ into the integral, note that T is constant, and find

$$G_m(p_f) = G_m(p_i) + RT \overbrace{\int_{p_i}^{p_f} \frac{1}{p}\,dp}^{\text{Integral A.2}} = G_m(p_i) + RT \ln\frac{p_f}{p_i} \quad (3E.14)$$

This expression shows that when the pressure is increased tenfold at room temperature, the molar Gibbs energy increases by $RT\ln 10 \approx 6\,kJ\,mol^{-1}$. It also follows from this equation that if $p_i = p^{\ominus}$ (the standard pressure of 1 bar), then the molar Gibbs energy of a perfect gas at a pressure p (set $p_f = p$) is related to its standard value by

$$G_m(p) = G_m^{\ominus} + RT\ln\frac{p}{p^{\ominus}} \quad \text{Molar Gibbs energy [perfect gas, constant } T] \quad (3E.15)$$

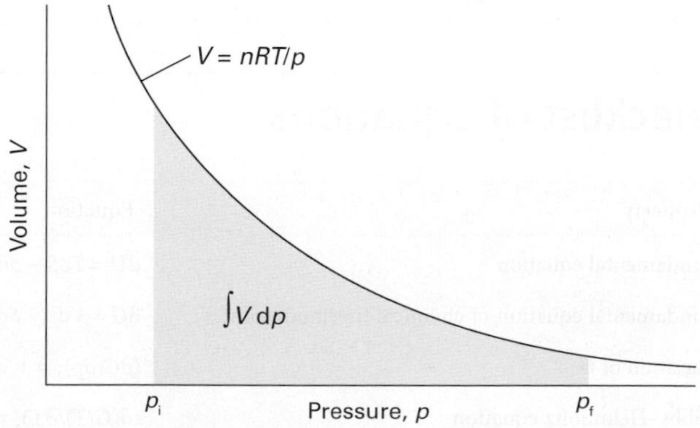

Figure 3E.5 At constant temperature, the change in Gibbs energy for a perfect gas between two pressures is equal to the area shown below the perfect-gas isotherm.

Brief illustration 3E.2

When the pressure is increased isothermally on water vapour (treated as a perfect gas) from 1.0 bar to 2.0 bar at 298 K, then according to eqn 3E.15

$$G_m(2.0\,\text{bar}) = G_m^\ominus(1.0\,\text{bar}) + (8.3145\,\text{J}\,\text{K}^{-1}\,\text{mol}^{-1})$$
$$\times(298\,\text{K})\times\ln\left(\frac{2.0\,\text{bar}}{1.0\,\text{bar}}\right)$$
$$= G_m^\ominus(1.0\,\text{bar}) + 1.7\,\text{kJ}\,\text{mol}^{-1}$$

Note that whereas the change in molar Gibbs energy for a condensed phase is a few joules per mole, for a gas the change is of the order of kilojoules per mole.

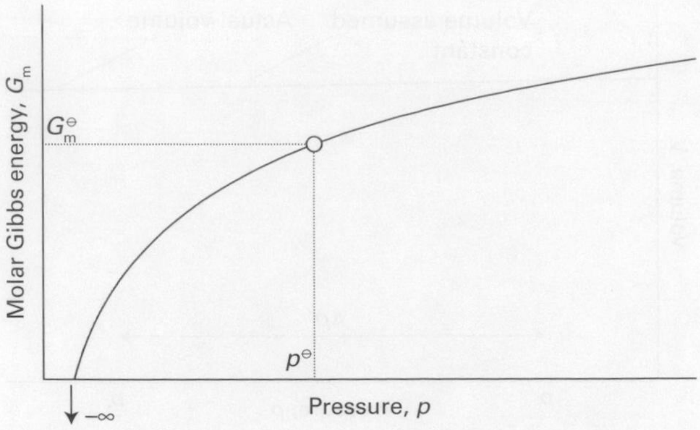

Figure 3E.6 At constant temperature, the molar Gibbs energy of a perfect gas varies as $\ln p$, and the standard state is reached at $p^\ominus$. Note that, as $p \to 0$, the molar Gibbs energy becomes negatively infinite.

The logarithmic dependence of the molar Gibbs energy on the pressure predicted by eqn 3E.15 is illustrated in Fig. 3E.6. This very important expression applies to perfect gases (which is usually a good approximation).

Checklist of concepts

☐ 1. The **fundamental equation**, a combination of the First and Second Laws, is an expression for the change in internal energy that accompanies changes in the volume and entropy of a system.

☐ 2. Relations between thermodynamic properties are generated by combining thermodynamic and mathematical expressions for changes in their values.

☐ 3. The **Maxwell relations** are a series of relations between partial derivatives of thermodynamic properties based on criteria for changes in the properties being exact differentials.

☐ 4. The Maxwell relations are used to derive the **thermodynamic equation of state** and to determine how the internal energy of a substance varies with volume.

☐ 5. The variation of the Gibbs energy of a system suggests that it is best regarded as a function of pressure and temperature.

☐ 6. The Gibbs energy of a substance decreases with temperature and increases with pressure.

☐ 7. The variation of Gibbs energy with temperature is related to the enthalpy by the **Gibbs–Helmholtz equation**.

☐ 8. The Gibbs energies of solids and liquids are almost independent of pressure; those of gases vary linearly with the logarithm of the pressure.

Checklist of equations

Property	Equation	Comment	Equation number
Fundamental equation	$dU = TdS - pdV$	No additional work	3E.1
Fundamental equation of chemical thermodynamics	$dG = Vdp - SdT$	No additional work	3E.7
Variation of G	$(\partial G/\partial p)_T = V$ and $(\partial G/\partial T)_p = -S$	Composition constant	3E.8
Gibbs–Helmholtz equation	$(\partial(G/T)/\partial T)_p = -H/T^2$	Composition constant	3E.10
Pressure dependence of G_m	$G_m(p_f) = G_m(p_i) + V_m(p_f - p_i)$	Incompressible substance	3E.13
	$G_m(p_f) = G_m(p_i) + RT\ln(p_f/p_i)$	Perfect gas, isothermal	3E.14
	$G_m(p) = G_m^\ominus + RT\ln(p/p^\ominus)$	Perfect gas, isothermal	3E.15

FOCUS 3 The Second and Third Laws

Assume that all gases are perfect and that data refer to 298.15 K unless otherwise stated.

TOPIC 3A Entropy

Discussion questions

D3A.1 The evolution of life requires the organization of a very large number of molecules into biological cells. Does the formation of living organisms violate the Second Law of thermodynamics? State your conclusion clearly and present detailed arguments to support it.

D3A.2 Discuss the significance of the terms 'dispersal' and 'disorder' in the context of the Second Law.

D3A.3 Discuss the relationships between the various formulations of the Second Law of thermodynamics.

Exercises

E3A.1(a) Consider a process in which the entropy of a system increases by $125\,\mathrm{J\,K^{-1}}$ and the entropy of the surroundings decreases by $125\,\mathrm{J\,K^{-1}}$. Is the process spontaneous?

E3A.1(b) Consider a process in which the entropy of a system increases by $105\,\mathrm{J\,K^{-1}}$ and the entropy of the surroundings decreases by $95\,\mathrm{J\,K^{-1}}$. Is the process spontaneous?

E3A.2(a) Consider a process in which 100 kJ of energy is transferred reversibly and isothermally as heat to a large block of copper. Calculate the change in entropy of the block if the process takes place at (i) 0 °C, (ii) 50 °C.

E3A.2(b) Consider a process in which 250 kJ of energy is transferred reversibly and isothermally as heat to a large block of lead. Calculate the change in entropy of the block if the process takes place at (i) 20 °C, (ii) 100 °C.

E3A.3(a) Calculate the change in entropy of the gas when 15 g of carbon dioxide gas are allowed to expand isothermally from $1.0\,\mathrm{dm^3}$ to $3.0\,\mathrm{dm^3}$ at 300 K.

E3A.3(b) Calculate the change in entropy of the gas when 4.00 g of nitrogen is allowed to expand isothermally from $500\,\mathrm{cm^3}$ to $750\,\mathrm{cm^3}$ at 300 K.

E3A.4(a) Calculate the change in the entropies of the system and the surroundings, and the total change in entropy, when a sample of nitrogen gas of mass 14 g at 298 K doubles its volume in (i) an isothermal reversible expansion, (ii) an isothermal irreversible expansion against $p_{ex} = 0$, and (iii) an adiabatic reversible expansion.

E3A.4(b) Calculate the change in the entropies of the system and the surroundings, and the total change in entropy, when the volume of a sample of argon gas of mass 2.9 g at 298 K increases from $1.20\,\mathrm{dm^3}$ to $4.60\,\mathrm{dm^3}$ in (i) an isothermal reversible expansion, (ii) an isothermal irreversible expansion against $p_{ex} = 0$, and (iii) an adiabatic reversible expansion.

E3A.5(a) In a certain ideal heat engine, 10.00 kJ of heat is withdrawn from the hot source at 273 K and 3.00 kJ of work is generated. What is the temperature of the cold sink?

E3A.5(b) In an ideal heat engine the cold sink is at 0 °C. If 10.00 kJ of heat is withdrawn from the hot source and 3.00 kJ of work is generated, at what temperature is the hot source?

E3A.6(a) What is the efficiency of an ideal heat engine in which the hot source is at 100 °C and the cold sink is at 10 °C?

E3A.6(b) An ideal heat engine has a hot source at 40 °C. At what temperature must the cold sink be if the efficiency is to be 10 per cent?

Problems

P3A.1 A sample consisting of 1.00 mol of perfect gas molecules at 27 °C is expanded isothermally from an initial pressure of 3.00 atm to a final pressure of 1.00 atm in two ways: (a) reversibly, and (b) against a constant external pressure of 1.00 atm. Evaluate q, w, ΔU, ΔH, ΔS, ΔS_{surr}, and ΔS_{tot} in each case.

P3A.2 A sample consisting of 0.10 mol of perfect gas molecules is held by a piston inside a cylinder such that the volume is $1.25\,\mathrm{dm^3}$; the external pressure is constant at 1.00 bar and the temperature is maintained at 300 K by a thermostat. The piston is released so that the gas can expand. Calculate (a) the volume of the gas when the expansion is complete; (b) the work done when the gas expands; (c) the heat absorbed by the system. Hence calculate ΔS_{tot}.

P3A.3 Consider a Carnot cycle in which the working substance is 0.10 mol of perfect gas molecules, the temperature of the hot source is 373 K, and that of the cold sink is 273 K; the initial volume of gas is $1.00\,\mathrm{dm^3}$, which doubles over the course of the first isothermal stage. For the reversible adiabatic stages it may be assumed that $VT^{3/2} = \text{constant}$. (a) Calculate the volume of the gas after Stage 1 and after Stage 2 (Fig. 3A.8). (b) Calculate the volume of gas after Stage 3 by considering the reversible adiabatic compression from the starting point. (c) Hence, for each of the four stages of the cycle, calculate the heat transferred to or from the gas. (d) Explain why the work done is equal to the difference between the heat extracted from the hot source and that deposited in the cold sink. (e) Calculate the work done over the cycle and hence the efficiency η. (f) Confirm that your answer agrees with the efficiency given by eqn 3A.9 and that your values for the heat involved in the isothermal stages are in accord with eqn 3A.6.

P3A.4 The Carnot cycle is usually represented on a pressure–volume diagram (Fig. 3A.8), but the four stages can equally well be represented on a temperature–entropy diagram, in which the horizontal axis is entropy and the vertical axis is temperature; draw such a diagram. Assume that the temperature of the hot source is T_h and that of the cold sink is T_c, and that the volume of the working substance (the gas) expands from V_A to V_B in the first isothermal stage. (a) By considering the entropy change of each stage, derive an expression for the area enclosed by the cycle in the temperature–entropy diagram. (b) Derive an expression for the work done over the cycle. (*Hint:* The work done is the difference between the heat extracted from the hot source and that deposited in the cold sink; or use eqns 3A.7 and 3A.9.) (c) Comment on the relation between your answers to (a) and (b).

P3A.5 A heat engine does work as a result of extracting energy as heat from the hot source and discarding some of it into the cold sink. Such an engine can also be used as a *heat pump* in which heat is extracted from a cold source; some work is done *on* the engine and thereby converted to heat which is added to that from the cold source before being discarded into the hot sink. (a) Assuming that the engine is perfect and that the heat transfers are reversible, use the Second Law to explain why it is not possible for heat to be extracted from the cold source and discarded into the hot sink without some work being done on the engine. (b) Assume that the hot sink is at temperature T_h and the cold source at T_c, and that heat of magnitude $|q|$ is extracted from the cold source. Use the Second Law to find the magnitude of the work $|w|$ needed to make it possible for heat of magnitude $|q| + |w|$ to be discarded into the hot sink.

P3A.6 Heat pumps can be used as a practical way of heating buildings. The ground itself can be used as the cold source because at a depth of a few metres the temperature is independent of the air temperature; in temperate latitudes the ground temperature is around 13 °C at a depth of 10 m. On a cold day it is found that to keep a certain room at 18 °C a heater rated at 5 kW is required. Assuming that an ideal heat pump is used, and that all heat transfers are reversible, calculate the power needed to maintain the room temperature. Recall that $1\,W = 1\,J\,s^{-1}$. *Hint:* See the results from Problem P3A.5.

P3A.7 Prove that two reversible adiabatic paths can never cross. Assume that the energy of the system under consideration is a function of temperature only. *Hint:* Suppose that two such paths can intersect, and complete a cycle with the two paths plus one isothermal path. Consider the changes accompanying each stage of the cycle and show that they conflict with the Kelvin statement of the Second Law.

TOPIC 3B Entropy changes accompanying specific processes

Discussion question

D3B.1 Account for deviations from Trouton's rule for liquids such as water, mercury, and ethanol. Is their entropy of vaporization larger or smaller than $85\,J\,K^{-1}\,mol^{-1}$? Why?

Exercises

E3B.1(a) Use Trouton's rule to predict the enthalpy of vaporization of benzene from its normal boiling point, 80.1 °C.
E3B.1(b) Use Trouton's rule to predict the enthalpy of vaporization of cyclohexane from its normal boiling point, 80.7 °C.

E3B.2(a) The enthalpy of vaporization of trichloromethane (chloroform, $CHCl_3$) is $29.4\,kJ\,mol^{-1}$ at its normal boiling point of 334.88 K. Calculate (i) the entropy of vaporization of trichloromethane at this temperature and (ii) the entropy change of the surroundings.
E3B.2(b) The enthalpy of vaporization of methanol is $35.27\,kJ\,mol^{-1}$ at its normal boiling point of 64.1 °C. Calculate (i) the entropy of vaporization of methanol at this temperature and (ii) the entropy change of the surroundings.

E3B.3(a) Estimate the increase in the molar entropy of $O_2(g)$ when the temperature is increased at constant pressure from 298 K to 348 K, given that the molar constant-pressure heat capacity of O_2 is $29.355\,J\,K^{-1}\,mol^{-1}$ at 298 K.
E3B.3(b) Estimate the change in the molar entropy of $N_2(g)$ when the temperature is lowered from 298 K to 273 K, given that $C_{p,m}(N_2) = 29.125\,J\,K^{-1}\,mol^{-1}$ at 298 K.

E3B.4(a) The molar entropy of a sample of neon at 298 K is $146.22\,J\,K^{-1}\,mol^{-1}$. The sample is heated at constant volume to 500 K; assuming that the molar constant-volume heat capacity of neon is $\frac{3}{2}R$, calculate the molar entropy of the sample at 500 K.
E3B.4(b) Calculate the molar entropy of a constant-volume sample of argon at 250 K given that it is $154.84\,J\,K^{-1}\,mol^{-1}$ at 298 K; the molar constant-volume heat capacity of argon is $\frac{3}{2}R$.

E3B.5(a) Two copper blocks, each of mass 1.00 kg, one at 50 °C and the other at 0 °C, are placed in contact in an isolated container (so no heat can escape) and allowed to come to equilibrium. Calculate the final temperature of the two blocks, the entropy change of each, and ΔS_{tot}. The specific heat capacity of copper is $0.385\,J\,K^{-1}\,g^{-1}$ and may be assumed constant over the temperature range involved. Comment on the sign of ΔS_{tot}.
E3B.5(b) Calculate ΔS_{tot} when two iron blocks, each of mass 10.0 kg, one at 100 °C and the other at 25 °C, are placed in contact in an isolated container and allowed to come to equilibrium. The specific heat capacity of iron is $0.449\,J\,K^{-1}\,g^{-1}$ and may be assumed constant over the temperature range involved. Comment on the sign of ΔS_{tot}.

E3B.6(a) Calculate ΔS (for the system) when the state of 3.00 mol of gas molecules, for which $C_{p,m} = \frac{5}{2}R$, is changed from 25 °C and 1.00 atm to 125 °C and 5.00 atm.
E3B.6(b) Calculate ΔS (for the system) when the state of 2.00 mol of gas molecules, for which $C_{p,m} = \frac{7}{2}R$, is changed from 25 °C and 1.50 atm to 135 °C and 7.00 atm.

E3B.7(a) Calculate the change in entropy of the system when 10.0 g of ice at −10.0 °C is converted into water vapour at 115.0 °C and at a constant pressure of 1 bar. The molar constant-pressure heat capacities are: $C_{p,m}(H_2O(s)) = 37.6\,J\,K^{-1}\,mol^{-1}$; $C_{p,m}(H_2O(l)) = 75.3\,J\,K^{-1}\,mol^{-1}$; and $C_{p,m}(H_2O(g)) = 33.6\,J\,K^{-1}\,mol^{-1}$. The standard enthalpy of vaporization of $H_2O(l)$ is $40.7\,kJ\,mol^{-1}$, and the standard enthalpy of fusion of $H_2O(l)$ is $6.01\,kJ\,mol^{-1}$, both at the relevant transition temperatures.
E3B.7(b) Calculate the change in entropy of the system when 15.0 g of ice at −12.0 °C is converted to water vapour at 105.0 °C at a constant pressure of 1 bar. For data, see the preceding exercise.

Problems

P3B.1 Consider a process in which 1.00 mol $H_2O(l)$ at −5.0 °C solidifies to ice at the same temperature. Calculate the change in the entropy of the sample, of the surroundings and the total change in the entropy. Is the process spontaneous? Repeat the calculation for a process in which 1.00 mol $H_2O(l)$ vaporizes at 95.0 °C and 1.00 atm. The data required are given in Exercise E3B.7(a).

P3B.2 Show that a process in which liquid water at 5.0 °C solidifies to ice at the same temperature is not spontaneous (*Hint:* calculate the total change in the entropy). The data required are given in Exercise E3B.7(a).

P3B.3 The molar heat capacity of trichloromethane (chloroform, $CHCl_3$) in the range 240 K to 330 K is given by $C_{p,m}/(J\,K^{-1}\,mol^{-1}) = 91.47 + 7.5 \times 10^{-2}(T/K)$.

Calculate the change in molar entropy when $CHCl_3$ is heated from 273 K to 300 K.

P3B.4 The molar heat capacity of $N_2(g)$ in the range 200 K to 400 K is given by $C_{p,m}/(J\,K^{-1}\,mol^{-1}) = 28.58 + 3.77 \times 10^{-3}(T/K)$. Given that the standard molar entropy of $N_2(g)$ at 298 K is 191.6 J K^{-1} mol^{-1}, calculate the value at 373 K. Repeat the calculation but this time assuming that $C_{p,m}$ is independent of temperature and takes the value 29.13 J K^{-1} mol^{-1}. Comment on the difference between the results of the two calculations.

P3B.5 Find an expression for the change in entropy when two blocks of the same substance and of equal mass, one at the temperature T_h and the other at T_c, are brought into thermal contact and allowed to reach equilibrium. Evaluate the change in entropy for two blocks of copper, each of mass 500 g, with $C_{p,m} = 24.4$ J K^{-1} mol^{-1}, taking $T_h = 500$ K and $T_c = 250$ K.

P3B.6 According to Newton's law of cooling, the rate of change of temperature is proportional to the temperature difference between the system and its surroundings:

$$\frac{dT}{dt} = -\alpha(T - T_{sur})$$

where T_{sur} is the temperature of the surroundings and α is a constant.
(a) Integrate this equation with the initial condition that $T = T_i$ at $t = 0$.
(b) Given that the entropy varies with temperature according to $S(T) - S(T_i) = C\ln(T/T_i)$, where T_i is the initial temperature and C the heat capacity, deduce an expression entropy of the system at time t.

P3B.7 A block of copper of mass 500 g and initially at 293 K is in thermal contact with an electric heater of resistance 1.00 kΩ and negligible mass. A current of 1.00 A is passed for 15.0 s. Calculate the change in entropy of the copper, taking $C_{p,m} = 24.4$ J K^{-1} mol^{-1}. The experiment is then repeated with the copper immersed in a stream of water that maintains the temperature of the copper block at 293 K. Calculate the change in entropy of the copper and the water in this case.

P3B.8 A block of copper ($C_{p,m} = 24.44$ J K^{-1} mol^{-1}) of mass 2.00 kg and at 0 °C is introduced into an insulated container in which there is 1.00 mol $H_2O(g)$ at 100 °C and 1.00 atm. Assuming that all the vapour is condensed to liquid water, determine: (a) the final temperature of the system; (b) the heat transferred to the copper block; and (c) the entropy change of the water, the copper block, and the total system. The data needed are given in Exercise E3B.7a.

P3B.9 The protein lysozyme unfolds at a transition temperature of 75.5 °C and the standard enthalpy of transition is 509 kJ mol^{-1}. Calculate the entropy of unfolding of lysozyme at 25.0 °C, given that the difference in the molar constant-pressure heat capacities upon unfolding is 6.28 kJ K^{-1} mol^{-1} and can be assumed to be independent of temperature. (*Hint:* Imagine that the transition at 25.0 °C occurs in three steps: (i) heating of the folded protein from 25.0 °C to the transition temperature, (ii) unfolding at the transition temperature, and (iii) cooling of the unfolded protein to 25.0 °C. Because the entropy is a state function, the entropy change at 25.0 °C is equal to the sum of the entropy changes of the steps.)

P3B.10 The cycle involved in the operation of an internal combustion engine is called the *Otto cycle* (Fig. 3.1). The cycle consists of the following steps: (1) Reversible adiabatic compression from A to B, (2) reversible constant-volume pressure increase from B to C due to the combustion of a small amount of fuel, (3) reversible adiabatic expansion from C to D, and (4) reversible constant-volume pressure decrease back to state A. Assume that the pressure, temperature, and volume at point A are p_A, T_A, and V_A, and likewise for B–D; further assume that the working substance is 1 mol of perfect gas diatomic molecules with $C_{V,m} = \frac{5}{2}R$. Recall that for a reversible adiabatic expansion (such as step 1) $V_A T_A^c = V_B T_B^c$, where $c = C_{V,m}/R$, and that for a perfect gas the internal energy is only a function of the temperature.

(a) Evaluate the work and the heat involved in each of the four steps, expressing your results in terms of $C_{V,m}$ and the temperatures T_A–T_D.

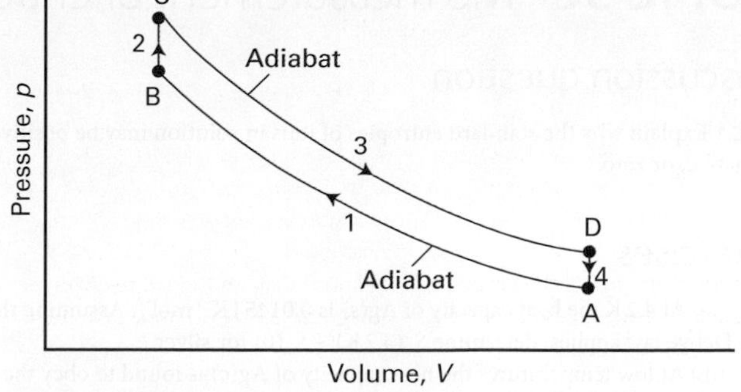

Figure 3.1 The Otto cycle.

(b) The efficiency η is defined as the modulus of the work over the whole cycle divided by the modulus of the heat supplied in step 2. Derive an expression for η in terms of the temperatures T_A–T_D.

(c) Use the relation between V and T for the reversible adiabatic processes to show that your expression for the efficiency can be written $\eta = 1 - (V_B/V_A)^{1/c}$ (*Hint:* recall that $V_C = V_B$ and $V_D = V_A$.)

(d) Derive expressions, in terms of $C_{V,m}$ and the temperatures, for the change in entropy (of the system and of the surroundings) for each step of the cycle.

(e) Assuming that $V_A = 4.00$ dm^3, $p_A = 1.00$ atm, $T_A = 300$ K, and that $V_A = 10V_B$ and $p_C/p_B = 5$, evaluate the efficiency of the cycle and the entropy changes for each step. (*Hint*: for the last part you will need to find T_B and T_D, which can be done by using the relation between V and T for the reversible adiabatic process; you will also need to find T_C which can be done by considering the temperature rise in the constant volume process.)

P3B.11 When a heat engine is used as a refrigerator to lower the temperature of an object, the colder the object the more work that is needed to cool it further to the same extent.

(a) Suppose that the refrigerator is an ideal heat engine and that it extracts a quantity of heat $|dq|$ from the cold source (the object being cooled) at temperature T_c. The work done on the engine is $|dw|$ and as a result heat $(|dq| + |dw|)$ is discarded into the hot sink at temperature T_h. Explain how the Second law requires that, for the process to be allowed, the following relation must apply:

$$\frac{|dq|}{T_c} = \frac{|dq| + |dw|}{T_h}$$

(b) Suppose that the heat capacity of the object being cooled is C (which can be assumed to be independent of temperature) so that the heat transfer for a change in temperature dT_c is $dq = CdT_c$. Substitute this relation into the expression derived in (a) and then integrate between $T_c = T_i$ and $T_c = T_f$ to give the following expression for the work needed to cool the object from T_i to T_f as

$$w = CT_h\left|\ln\frac{T_f}{T_i}\right| - |C(T_f - T_i)|$$

(c) Use this result to calculate the work needed to lower the temperature of 250 g of water from 293 K to 273 K, assuming that the hot reservoir is at 293 K ($C_{p,m}(H_2O(l)) = 75.3$ J K^{-1} mol^{-1}). (d) When the temperature of liquid water reaches 273 K it will freeze to ice, an exothermic process. Calculate the work needed to transfer the associated heat to the hot sink, assuming that the water remains at 273 K (the standard enthalpy of fusion of H_2O is 6.01 kJ mol^{-1} at the normal freezing point). (e) Hence calculate the total work needed to freeze the 250 g of liquid water to ice at 273 K. How long will this take if the refrigerator operates at 100 W?

P3B.12 The standard molar entropy of $NH_3(g)$ is 192.45 J K^{-1} mol^{-1} at 298 K, and its heat capacity is given by eqn 2B.8 with the coefficients given in Table 2B.1. Calculate the standard molar entropy at (a) 100 °C and (b) 500 °C.

TOPIC 3C The measurement of entropy

Discussion question

D3C.1 Explain why the standard entropies of ions in solution may be positive, negative, or zero.

Exercises

E3C.1(a) At 4.2 K the heat capacity of Ag(s) is 0.0145 J K^{-1} mol^{-1}. Assuming that the Debye law applies, determine $S_m(4.2\,K) - S_m(0)$ for silver.

E3C.1(b) At low temperatures the heat capacity of Ag(s) is found to obey the Debye law $C_{p,m} = aT^3$, with $a = 1.956 \times 10^{-4}$ J K^{-4} mol^{-1}. Determine $S_m(10\,K) - S_m(0)$ for silver.

E3C.2(a) Use data from Tables 2C.3 and 2C.4 to calculate the standard reaction entropy at 298 K of

(i) $2\,CH_3CHO(g) + O_2(g) \rightarrow 2\,CH_3COOH(l)$
(ii) $2\,AgCl(s) + Br_2(l) \rightarrow 2\,AgBr(s) + Cl_2(g)$
(iii) $Hg(l) + Cl_2(g) \rightarrow HgCl_2(s)$

E3C.2(b) Use data from Tables 2C.3 and 2C.4 to calculate the standard reaction entropy at 298 K of

(i) $Zn(s) + Cu^{2+}(aq) \rightarrow Zn^{2+}(aq) + Cu(s)$
(ii) sucrose $[C_{12}H_{22}O_{11}(s)] + 12\,O_2(g) \rightarrow 12\,CO_2(g) + 11\,H_2O(l)$

E3C.3(a) Calculate the standard reaction entropy at 298 K when 1 mol $NH_3(g)$ is formed from its elements in their reference states.

E3C.3(b) Calculate the standard reaction entropy at 298 K when 1 mol $N_2O(g)$ is formed from its elements in their reference states.

Problems

P3C.1 At 10 K $C_{p,m}(Hg(s)) = 4.64$ J K^{-1} mol^{-1}. Between 10 K and the melting point of Hg(s), 234.3 K, heat capacity measurements indicate that the entropy increases by 57.74 J K^{-1} mol^{-1}. The standard enthalpy of fusion of Hg(s) is 2322 J mol^{-1} at 234.3 K. Between the melting point and 298.0 K, heat capacity measurements indicate that the entropy increases by 6.85 J K^{-1} mol^{-1}. Determine the Third-Law standard molar entropy of Hg(l) at 298 K.

P3C.2 The measurements described in Problem P3C.1 were extended to 343.9 K, the normal boiling point of Hg(l). Between the melting point and the boiling point, heat capacity measurements indicate that the entropy increases by 10.83 J K^{-1} mol^{-1}. The standard enthalpy of vaporization of Hg(l) is 60.50 kJ mol^{-1} at 343.9 K. Determine the Third-Law standard molar entropy of Hg(g) at 343.9 K (you will need some of the data from Problem P3C.1).

P3C.3 The molar heat capacity of lead varies with temperature as follows:

T/K	10	15	20	25	30	50
$C_{p,m}$/(J K^{-1} mol^{-1})	2.8	7.0	10.8	14.1	16.5	21.4
T/K	70	100	150	200	250	298
$C_{p,m}$/(J K^{-1} mol^{-1})	23.3	24.5	25.3	25.8	26.2	26.6

(a) Use the Debye T^3-law and the value of the heat capacity at 10 K to determine the change in entropy between 0 and 10 K. (b) To determine the change in entropy between 10 K and 298 K you will need to measure the area under a plot of $C_{p,m}/T$ against T. This measurement can either be done by counting squares or by using mathematical software to fit the data to a simple function (for example, a polynomial) and then integrating that function over the range 10 K to 298 K. Use either of these methods to determine the change in entropy between 10 K and 298 K. (c) Hence determine the standard Third-Law entropy of lead at 298 K, and also at 273 K.

P3C.4 The molar heat capacity of anhydrous potassium hexacyanoferrate(II) varies with temperature as follows:

T/K	10	20	30	40	50	60
$C_{p,m}$/(J K^{-1} mol^{-1})	2.09	14.43	36.44	62.55	87.03	111.0
T/K	70	80	90	100	110	150
$C_{p,m}$/(J K^{-1} mol^{-1})	131.4	149.4	165.3	179.6	192.8	237.6
T/K	160	170	180	190	200	
$C_{p,m}$/(J K^{-1} mol^{-1})	247.3	256.5	265.1	273.0	280.3	

Determine the Third-Law molar entropy at 200 K and at 100 K.

P3C.5 Use values of standard enthalpies of formation, standard entropies, and standard heat capacities available from tables in the *Resource section* to calculate the standard enthalpy and entropy changes at 298 K and 398 K for the reaction $CO_2(g) + H_2(g) \rightarrow CO(g) + H_2O(g)$. Assume that the heat capacities are constant over the temperature range involved.

P3C.6 Use values of enthalpies of formation, standard entropies, and standard heat capacities available from tables in the *Resource section* to calculate the standard enthalpy and entropy of reaction at 298 K and 500 K for $\frac{1}{2}N_2(g) + \frac{3}{2}H_2(g) \rightarrow NH_3(g)$. Assume that the heat capacities are constant over the temperature range involved.

P3C.7 The compound 1,3,5-trichloro-2,4,6-trifluorobenzene is an intermediate in the conversion of hexachlorobenzene to hexafluorobenzene, and its thermodynamic properties have been examined by measuring its heat capacity over a wide temperature range (R.L. Andon and J.F. Martin, *J. Chem. Soc. Faraday Trans. I* 871 (1973)). Some of the data are as follows:

T/K	14.14	16.33	20.03	31.15	44.08	64.81
$C_{p,m}$/(J K^{-1} mol^{-1})	9.492	12.70	18.18	32.54	46.86	66.36
T/K	100.90	140.86	183.59	225.10	262.99	298.06
$C_{p,m}$/(J K^{-1} mol^{-1})	95.05	121.3	144.4	163.7	180.2	196.4

Determine the Third-Law molar entropy of the compound at 100 K, 200 K, and 300 K.

P3C.8‡ Given that $S_m^\ominus = 29.79$ J K^{-1} mol^{-1} for bismuth at 100 K and the following tabulated heat capacity data (D.G. Archer, *J. Chem. Eng. Data* **40**, 1015 (1995)), determine the standard molar entropy of bismuth at 200 K.

T/K	100	120	140	150	160	180	200
$C_{p,m}$/(J K^{-1} mol^{-1})	23.00	23.74	24.25	24.44	24.61	24.89	25.11

Compare the value to the value that would be obtained by taking the heat capacity to be constant at 24.44 J K^{-1} mol^{-1} over this range.

P3C.9 At low temperatures there are two contributions to the heat capacity of a metal, one associated with lattice vibrations, which is well-approximated by

‡ These problems were provided by Charles Trapp and Carmen Giunta.

the Debye T^3-law, and one due to the valence electrons. The latter is linear in the temperature. Overall, the heat capacity can be written

$$C_{p,\mathrm{m}}(T) = \overbrace{aT^3}^{\text{Debye}} + \overbrace{bT}^{\text{electronic}}$$

The molar heat capacity of potassium metal has been measured at very low temperatures to give the following data

T/K	0.20	0.25	0.30	0.35	0.40	0.45	0.50	0.55
$C_{p,\mathrm{m}}$/($\mathrm{J\,K^{-1}\,mol^{-1}}$)	0.437	0.560	0.693	0.838	0.996	1.170	1.361	1.572

(a) Assuming that the expression given above for the heat capacity applies, explain why a plot of $C_{p,\mathrm{m}}(T)/T$ against T^2 is expected to be a straight line with slope a and intercept b. (b) Use such a plot to determine the values of the constants a and b. (c) Derive an expression for the molar entropy at temperature T. (*Hint:* you will need to integrate $C_{p,\mathrm{m}}(T)/T$.) (d) Hence determine the molar entropy of potassium at 2.0 K.

P3C.10 At low temperatures the heat capacity of a metal is the sum of a contribution due to lattice vibrations (the Debye term) and a term due to the valence electrons, as given in the preceding problem. For sodium metal $a = 0.507 \times 10^{-3}\,\mathrm{J\,K^{-4}\,mol^{-1}}$ and $b = 1.38 \times 10^{-3}\,\mathrm{J\,K^{-2}\,mol^{-1}}$. Determine the temperature at which the Debye contribution and the electronic contribution to the entropy of sodium are equal. At higher temperatures, which contribution becomes dominant?

TOPIC 3D Concentrating on the system

Discussion questions

D3D.1 The following expressions establish criteria for spontaneous change: $\mathrm{d}A_{T,V} < 0$ and $\mathrm{d}G_{T,p} < 0$. Discuss the origin, significance, and applicability of each criterion.

D3D.2 Under what circumstances, and why, can the spontaneity of a process be discussed in terms of the properties of the system alone?

Exercises

E3D.1(a) Calculate values for the standard reaction enthalpies at 298 K for the reactions in Exercise E3C.2(a) by using values of the standard enthalpies of formation from the tables in the *Resource section*. Combine your results with the standard reaction entropies already calculated in that Exercise to determine the standard reaction Gibbs energy at 298 K for each.

E3D.1(b) Calculate values for the standard reaction enthalpies at 298 K for the reactions in Exercise E3C.2(b) by using values of the standard enthalpies of formation from the tables in the *Resource section*. Combine your results with the standard reaction entropies already calculated in that Exercise to determine the standard reaction Gibbs energy at 298 K for each.

E3D.2(a) Calculate the standard Gibbs energy of reaction for $4\,HI(g) + O_2(g) \rightarrow 2\,I_2(s) + 2\,H_2O(l)$ at 298 K, using the values of standard entropies and enthalpies of formation given in the *Resource section*.

E3D.2(b) Calculate the standard Gibbs energy of the reaction $CO(g) + CH_3CH_2OH(l) \rightarrow CH_3CH_2COOH(l)$ at 298 K, using the values of standard entropies and enthalpies of formation given in the *Resource section*. The data for $CH_3CH_2COOH(l)$ are $\Delta_f H^\ominus = -510\,\mathrm{kJ\,mol^{-1}}$, $S_\mathrm{m}^\ominus = 191\,\mathrm{J\,K^{-1}\,mol^{-1}}$ at 298 K.

E3D.3(a) Calculate the maximum non-expansion work per mole of CH_4 that may be obtained from a fuel cell in which the chemical reaction is the combustion of methane under standard conditions at 298 K.

E3D.3(b) Calculate the maximum non-expansion work per mole of C_3H_8 that may be obtained from a fuel cell in which the chemical reaction is the combustion of propane under standard conditions at 298 K.

E3D.4(a) Use values of the relevant standard Gibbs energies of formation from the *Resource section* to calculate the standard Gibbs energies of reaction at 298 K of

(i) $2\,CH_3CHO(g) + O_2(g) \rightarrow 2\,CH_3COOH(l)$
(ii) $2\,AgCl(s) + Br_2(l) \rightarrow 2\,AgBr(s) + Cl_2(g)$
(iii) $Hg(l) + Cl_2(g) \rightarrow HgCl_2(s)$

E3D.4(b) Use values of the relevant standard Gibbs energies of formation from the *Resource section* to calculate the standard Gibbs energies of reaction at 298 K of

(i) $Zn(s) + Cu^{2+}(aq) \rightarrow Zn^{2+}(aq) + Cu(s)$
(ii) sucrose $[C_{12}H_{22}O_{11}(s)] + 12\,O_2(g) \rightarrow 12\,CO_2(g) + 11\,H_2O(l)$

E3D.5(a) The standard enthalpy of combustion of liquid ethyl ethanoate (ethyl acetate, $CH_3COOC_2H_5$) is $-2231\,\mathrm{kJ\,mol^{-1}}$ at 298 K and its standard molar entropy is $259.4\,\mathrm{J\,K^{-1}\,mol^{-1}}$. Calculate the standard Gibbs energy of formation of the compound at 298 K.

E3D.5(b) The standard enthalpy of combustion of the solid glycine (the amino acid, NH_2CH_2COOH) is $-969\,\mathrm{kJ\,mol^{-1}}$ at 298 K and its standard molar entropy is $103.5\,\mathrm{J\,K^{-1}\,mol^{-1}}$. Calculate the standard Gibbs energy of formation of glycine at 298 K. Note that the nitrogen-containing species produced on combustion is taken to be $N_2(g)$.

Problems

P3D.1 A perfect gas is contained in a cylinder of fixed volume and which is separated into two sections A and B by a frictionless piston; no heat can pass through the piston. Section B is maintained at a constant temperature of 300 K; that is, all changes in section B are isothermal. There are 2.00 mol of gas molecules in each section and the constant-volume heat capacity of the gas is $C_{V,\mathrm{m}} = 20\,\mathrm{J\,K^{-1}\,mol^{-1}}$, which can be assumed to be constant. Initially $T_\mathrm{A} = T_\mathrm{B} = 300\,\mathrm{K}$, $V_\mathrm{A} = V_\mathrm{B} = 2.00\,\mathrm{dm^3}$. Energy is then supplied as heat to Section A so that the gas in A expands, pushing the piston out and thereby compressing the gas in section B. The expansion takes place *reversibly* and the final volume in section B is $1.00\,\mathrm{dm^3}$. Because the piston is free to move, the pressures in sections A and B are always equal; recall, too, that for a perfect gas the internal energy is a function of only the temperature.

(a) Calculate the final pressure of the gas and hence the temperature of the gas in section A. (b) Calculate the change in entropy of the gas in section A (*Hint:* you can think of the process as occurring in a constant volume step and then a constant temperature step).

(c) Calculate the entropy change of the gas in section B. (d) Calculate the change in internal energy for each section. (e) Use the values of ΔS and ΔU that you have already calculated to calculate ΔA for section B; explain why it is not possible to do the same for section A. (f) Given that the process is reversible, what does this imply about the total ΔA for the process (the sum of ΔA for section A and B)?

P3D.2 In biological cells, the energy released by the oxidation of foods is stored in adenosine triphosphate (ATP or ATP^{4-}).The essence of ATP's action is its ability to lose its terminal phosphate group by hydrolysis and to form adenosine diphosphate (ADP or ADP^{3-}):

$$ATP^{4-}(aq) + H_2O(l) \rightarrow ADP^{3-}(aq) + HPO_4^{2-}(aq) + H_3O^+(aq)$$

At pH = 7.0 and 37 °C (310 K, blood temperature) the enthalpy and Gibbs energy of hydrolysis are $\Delta_r H = -20\,\text{kJ mol}^{-1}$ and $\Delta_r G = -31\,\text{kJ mol}^{-1}$, respectively. Under these conditions, the hydrolysis of 1 mol $ATP^{4-}(aq)$ results in the extraction of up to 31 kJ of energy that can be used to do non-expansion work, such as the synthesis of proteins from amino acids, muscular contraction, and the activation of neuronal circuits in our brains. (a) Calculate and account for the sign of the entropy of hydrolysis of ATP at pH = 7.0 and 310 K. (b) Suppose that the radius of a typical biological cell is 10 μm and that inside it 1×10^6 ATP molecules are hydrolysed each second. What is the power density of the cell in watts per cubic metre ($1\,\text{W} = 1\,\text{J s}^{-1}$)? A computer battery delivers about 15 W and has a volume of 100 cm³. Which has the greater power density, the cell or the battery? (c) The formation of glutamine from glutamate and ammonium ions requires $14.2\,\text{kJ mol}^{-1}$ of energy input. It is driven by the hydrolysis of ATP to ADP mediated by the enzyme glutamine synthetase. How many moles of ATP must be hydrolysed to form 1 mol glutamine?

P3D.3 Construct a cycle similar to that in Fig. 3D.3 to analyse the reaction $\frac{1}{2}H_2(g) + \frac{1}{2}I_2(s) \rightarrow H^+(aq) + I^-(aq)$ and use it to find the value of the standard Gibbs energy of formation of $I^-(aq)$. You should refer to the tables in the *Resource section* for relevant values of the Gibbs energies of formation. As in the text, the standard Gibbs energy for the process $H(g) \rightarrow H^+(g) + e^-(g)$ should be approximated by the ionization energy, and that for $I(g) + e^-(g) \rightarrow I^-(g)$ by the electron affinity. The standard Gibbs energy of solvation of H^+ can be taken as $-1090\,\text{kJ mol}^{-1}$ and of I^- as $-247\,\text{kJ mol}^{-1}$.

P3D.4 The solubility of an ionic solid such as NaCl can be explored by calculating the standard Gibbs energy change for the process $NaCl(s) \rightarrow Na^+(aq) + Cl^-(aq)$. Consider this process in two steps: (1) $NaCl(s) \rightarrow Na^+(g) + Cl^-(g)$ and then (2) $Na^+(g) + Cl^-(g) \rightarrow Na^+(aq) + Cl^-(aq)$. Estimate $\Delta_r G^\ominus$ for the first step given that $\Delta_r H^\ominus = 787\,\text{kJ mol}^{-1}$ and the following values of the absolute entropy: $S_m^\ominus(Na^+(g)) = 148\,\text{J K}^{-1}\text{mol}^{-1}$, $S_m^\ominus(Cl^-(g)) = 154\,\text{J K}^{-1}\text{mol}^{-1}$, $S_m^\ominus(NaCl(s)) = 72.1\,\text{J K}^{-1}\text{mol}^{-1}$ (all data at 298 K). The value of $\Delta_r G^\ominus$ for the second step can be found by using the Born equation to estimate the standard Gibbs energies of solvation. For these estimates, use $r(Na^+) = 170\,\text{pm}$ and $r(Cl^-) = 211\,\text{pm}$. Hence find $\Delta_r G^\ominus$ for the overall process and comment on the value you find.

P3D.5 Repeat the calculation in Problem P3D.4 for LiF, for which $\Delta_r H^\ominus = 1037\,\text{kJ mol}^{-1}$ in step 1 and with the following values of the absolute entropy: $S_m^\ominus(Li^+) = 133\,\text{J K}^{-1}\text{mol}^{-1}$, $S_m^\ominus(F^-) = 145\,\text{J K}^{-1}\text{mol}^{-1}$, $S_m^\ominus(LiF(s)) = 35.6\,\text{J K}^{-1}\text{mol}^{-1}$ (all data at 298 K). Use $r(Li^+) = 127\,\text{pm}$ and $r(F^-) = 163\,\text{pm}$.

P3D.6 From the Born equation derive an expression for $\Delta_{solv}S^\ominus$ and $\Delta_{solv}H^\ominus$ (*Hint:* $(\partial G/\partial T)_p = -S$). Comment on your answer in the light of the assumptions made in the Born model.

TOPIC 3E Combining the First and Second Laws

Discussion questions

D3E.1 Suggest a physical interpretation of the dependence of the Gibbs energy on the temperature.

D3E.2 Suggest a physical interpretation of the dependence of the Gibbs energy on the pressure.

Exercises

E3E.1(a) Suppose that 2.5 mmol of perfect gas molecules initially occupies 42 cm³ at 300 K and then expands isothermally to 600 cm³. Calculate ΔG for the process.

E3E.1(b) Suppose that 6.0 mmol of perfect gas molecules initially occupies 52 cm³ at 298 K and then expands isothermally to 122 cm³. Calculate ΔG for the process.

E3E.2(a) The change in the Gibbs energy of a certain constant-pressure process is found to fit the expression $\Delta G/\text{J} = -85.40 + 36.5(T/\text{K})$. Calculate the value of ΔS for the process.

E3E.2(b) The change in the Gibbs energy of a certain constant-pressure process is found to fit the expression $\Delta G/\text{J} = -73.1 + 42.8(T/\text{K})$. Calculate the value of ΔS for the process.

E3E.3(a) The change in the Gibbs energy of a certain constant-pressure process is found to fit the expression $\Delta G/\text{J} = -85.40 + 36.5(T/\text{K})$. Use the Gibbs–Helmholtz equation to calculate the value of ΔH for the process.

E3E.3(b) The change in the Gibbs energy of a certain constant-pressure process is found to fit the expression $\Delta G/\text{J} = -73.1 + 42.8(T/\text{K})$. Use the Gibbs–Helmholtz equation to calculate the value of ΔH for the process.

E3E.4(a) Estimate the change in the Gibbs energy of 1.0 dm³ of liquid octane when the pressure acting on it is increased from 1.0 atm to 100 atm. Given that the mass density of octane is $0.703\,\text{g cm}^{-3}$, determine the change in the molar Gibbs energy.

E3E.4(b) Estimate the change in the Gibbs energy of 100 cm³ of water when the pressure acting on it is increased from 100 kPa to 500 kPa. Given that the mass density of water is $0.997\,\text{g cm}^{-3}$, determine the change in the molar Gibbs energy.

E3E.5(a) The change in the molar volume accompanying fusion of solid CO_2 is $-1.6\,\text{cm}^3\,\text{mol}^{-1}$. Determine the change in the molar Gibbs energy of fusion when the pressure is increased from 1 bar to 1000 bar.

E3E.5(b) The change in the molar volume accompanying fusion of solid benzene is $0.5\,\text{cm}^3\,\text{mol}^{-1}$. Determine the change in Gibbs energy of fusion when the pressure is increased from 1 bar to 5000 bar.

E3E.6(a) Calculate the change in the molar Gibbs energy of a perfect gas when its pressure is increased isothermally from 1.0 atm to 100.0 atm at 298 K.

E3E.6(b) Calculate the change in the molar Gibbs energy of a perfect gas when its pressure is increased isothermally from 50.0 kPa to 100.0 kPa at 500 K.

Problems

P3E.1 (a) By integrating the Gibbs–Helmholtz equation between temperature T_1 and T_2, and with the assumption that ΔH is independent of temperature, show that

$$\frac{\Delta G(T_2)}{T_2}=\frac{\Delta G(T_1)}{T_1}+\Delta H\left(\frac{1}{T_2}-\frac{1}{T_1}\right)$$

where $\Delta G(T)$ is the change in Gibbs energy at temperature T. (b) Using values of the standard Gibbs energies and enthalpies of formation from the *Resource section*, determine $\Delta_r G^\ominus$ and $\Delta_r H^\ominus$ at 298 K for the reaction $2\,CO(g) + O_2(g) \rightarrow 2\,CO_2(g)$. (c) Hence estimate $\Delta_r G^\ominus$ at 375 K.

P3E.2 Calculate $\Delta_r G^\ominus$ and $\Delta_r H^\ominus$ at 298 K for $N_2(g) + 3\,H_2(g) \rightarrow 2\,NH_3(g)$. Then, using the result from Problem P3E.1 (a), estimate $\Delta_r G^\ominus$ at 500 K and at 1000 K.

P3E.3 At 298 K the standard enthalpy of combustion of sucrose is $-5797\,\text{kJ mol}^{-1}$ and the standard Gibbs energy of the reaction is $-6333\,\text{kJ mol}^{-1}$. Estimate the additional non-expansion work that may be obtained by raising the temperature to blood temperature, 37 °C. (*Hint:* use the result from Problem P3E.1 to determine $\Delta_r G^\ominus$ at the higher temperature.)

P3E.4 Consider gases described by the following three equations of state:

(a) perfect: $p=\dfrac{RT}{V_m}$

(b) van der Waals: $p=\dfrac{RT}{V_m-b}-\dfrac{a}{V_m^2}$

(c) Dieterici: $p=\dfrac{RTe^{-a/RTV_m}}{V_m-b}$

Use the Maxwell relation $(\partial S/\partial V)_T = (\partial p/\partial T)_V$ to derive an expression for $(\partial S/\partial V)_T$ for each equation of state. For an isothermal expansion, compare the change in entropy expected for a perfect gas and for a gas obeying the van der Waals equation of state: which has the greatest change in entropy and how can this conclusion be rationalized?

P3E.5 Only one of the four Maxwell relations is derived in the text. Derive the remaining three to give the complete set listed in Table 3E.1. Start with the definition of H ($H = U + pV$), form the exact differential ($dH = dU + pdV + Vdp$), and then substitute $dU = TdS - pdV$. The resulting expression gives rise to a Maxwell relation in a way analogous to how eqn 3E.5 arises from eqn 3E.1. Repeat the process, starting with the definitions of A and then G, to give the remaining two Maxwell relations.

P3E.6 Suppose that S is regarded as a function of p and T so that

$$dS=\left(\frac{\partial S}{\partial p}\right)_T dp+\left(\frac{\partial S}{\partial T}\right)_p dT$$

Use $(\partial S/\partial T)_p = C_p/T$ and an appropriate Maxwell relation to show that $TdS = C_p dT - \alpha TVdp$, where the expansion coefficient, α, is defined as $\alpha = (1/V)(\partial V/\partial T)_p$. Hence, show that the energy transferred as heat, q, when the pressure on an incompressible liquid or solid is increased by Δp in a reversible isothermal process is given by $q = -\alpha TV\Delta p$. Evaluate q when the pressure acting on 100 cm^3 of mercury at 0 °C is increased by 1.0 kbar. ($\alpha = 1.82 \times 10^{-4}\,\text{K}^{-1}$.)

P3E.7 The pressure dependence of the molar Gibbs energy is given by $(\partial G_m/\partial p)_T = V_m$. This problem involves exploring this dependence for a gas described by the van der Waals equation of state

$$p=\frac{RT}{V_m-b}-\frac{a}{V_m^2}$$

(a) Consider first the case where only the repulsive term is significant; that it, $a = 0$, $b \neq 0$. Rearrange the equation of state into an expression for V_m, substitute it into $(\partial G_m/\partial p)_T = V_m$, and then integrate so as to obtain an expression for the pressure dependence of G_m. Compare your result with that for a perfect gas. (b) Now consider the case where only the attractive terms are included; that is, $b = 0$, $a \neq 0$. The equation of state then becomes a quadratic equation in V_m. Find this equation and solve it for V_m. Approximate the solution by assuming that $pa/R^2T^2 \ll 1$ and using the expansion $(1-x)^{1/2} \approx 1-\frac{1}{2}x$, which is valid for $x \ll 1$. Hence find an expression for the pressure dependence of G_m, and interpret the result. (c) For CO_2, $a = 3.610$ atm dm^6 mol^{-2}, $b = 4.29 \times 10^{-2}$ dm^3 mol^{-1}. Use mathematical software to plot G_m as a function of pressure at 298 K for a perfect gas and the two cases analysed above. (Use $R = 8.2057 \times 10^{-2}$ dm^3 atm K^{-1} mol^{-1}.)

P3E.8‡ Nitric acid hydrates have received much attention as possible catalysts for heterogeneous reactions that bring about the Antarctic ozone hole. Worsnop et al. (*Science* **259**, 71 (1993)) investigated the thermodynamic stability of these hydrates under conditions typical of the polar winter stratosphere. They report thermodynamic data for the sublimation of mono-, di-, and trihydrates to nitric acid and water vapours, $HNO_3{\cdot}nH_2O(s) \rightarrow HNO_3(g) + nH_2O(g)$, for $n = 1$, 2, and 3. Given $\Delta_r G^\ominus$ and $\Delta_r H^\ominus$ for these reactions at 220 K, use the Gibbs–Helmholtz equation to compute $\Delta_r G^\ominus$ for each at 190 K.

n	1	2	3
$\Delta_r G^\ominus/(\text{kJ mol}^{-1})$	46.2	69.4	93.2
$\Delta_r H^\ominus/(\text{kJ mol}^{-1})$	127	188	237

FOCUS 3 The Second and Third Laws

Integrated activities

I3.1 A sample consisting of 1.00 mol gas molecules is described by the equation of state $pV_m = RT(1 + Bp)$. Initially at 373 K, it undergoes Joule–Thomson expansion (Topic 2D) from 100 atm to 1.00 atm. Given that $C_{p,m} = \frac{5}{2}R$, $\mu = 0.21\,\text{K atm}^{-1}$, $B = -0.525(\text{K}/T)\,\text{atm}^{-1}$, and that these are constant over the temperature range involved, calculate ΔT and ΔS for the gas.

I3.2 Discuss the relation between the thermodynamic and statistical definitions of entropy.

I3.3 Use mathematical software or an electronic spreadsheet to:

(a) Evaluate the change in entropy of 1.00 mol CO_2(g) on expansion from 0.001 m^3 to 0.010 m^3 at 298 K, treated as a van der Waals gas.

(b) Plot the change in entropy of a perfect gas of (i) atoms, (ii) linear rotors, (iii) nonlinear rotors as the sample is heated over the same range under conditions of constant volume and then constant pressure.

(c) Allow for the temperature dependence of the heat capacity by writing $C = a + bT + c/T^2$, and plot the change in entropy for different values of the three coefficients (including negative values of c).

(d) Show how the first derivative of G, $(\partial G/\partial p)_T$, varies with pressure, and plot the resulting expression over a pressure range. What is the physical significance of $(\partial G/\partial p)_T$?

(e) Evaluate the fugacity coefficient (see *A deeper look* 2 on the website for this book) as a function of the reduced volume of a van der Waals gas and plot the outcome for a selection of reduced temperatures over the range $0.8 \le V_r \le 3$.

FOCUS 4

Physical transformations of pure substances

Vaporization, melting (fusion), and the conversion of graphite to diamond are all examples of changes of phase without change of chemical composition. The discussion of the phase transitions of pure substances is among the simplest applications of thermodynamics to chemistry, and is guided by the principle that, at constant temperature and pressure, the tendency of systems is to minimize their Gibbs energy.

4A Phase diagrams of pure substances

One type of phase diagram is a map of the pressures and temperatures at which each phase of a substance is the most stable. The thermodynamic criterion for phase stability leads to a very general result, the 'phase rule', which summarizes the constraints on the equilibria between phases. In preparation for later chapters, this rule is expressed in a general way that can be applied to systems of more than one component. This Topic also introduces the 'chemical potential', a property that is at the centre of discussions of mixtures and chemical reactions. The Topic then describes the interpretation of the phase diagrams of a representative selection of substances.

4B Thermodynamic aspects of phase transitions

This Topic considers the factors that determine the positions and shapes of the phase boundaries. The expressions derived show how the vapour pressure of a substance varies with temperature and how the melting point varies with pressure.

Web resource What is an application of this material?

The properties of carbon dioxide in its supercritical fluid phase can form the basis for novel and useful chemical separation methods, and have considerable promise for the synthetic procedures adopted in 'green' chemistry. Its properties and applications are discussed in *Impact* 6 on the website of this book.

TOPIC 4A Phase diagrams of pure substances

➤ Why do you need to know this material?

Phase diagrams summarize the behaviour of substances under different conditions, and identify which phase or phases are the most stable at a particular temperature and pressure. Such diagrams are important tools for understanding the behaviour of both pure substances and mixtures.

➤ What is the key idea?

A pure substance tends to adopt the phase with the lowest chemical potential.

➤ What do you need to know already?

This Topic builds on the fact that the Gibbs energy is a signpost of spontaneous change under conditions of constant temperature and pressure (Topic 3D).

One of the most succinct ways of presenting the physical changes of state that a substance can undergo is in terms of its 'phase diagram'. This material is also the basis of the discussion of mixtures in Focus 5.

4A.1 The stabilities of phases

Thermodynamics provides a powerful framework for describing and understanding the stabilities and transformations of phases, but the terminology must be used carefully. In particular, it is necessary to understand the terms 'phase', 'component', and 'degree of freedom'.

(a) The number of phases

A **phase** is a form of matter that is uniform throughout in chemical composition and physical state. Thus, there are the solid, liquid, and gas phases of a substance, as well as various solid phases, such as the white and black allotropes of phosphorus, or the aragonite and calcite polymorphs of calcium carbonate.

A note on good practice An *allotrope* is a particular molecular form of an element (such as O_2 and O_3) and may be solid, liquid, or gas. A *polymorph* is one of a number of solid phases of an element or compound.

The number of phases in a system is denoted P. A gas, or a gaseous mixture, is a single phase ($P = 1$), a crystal of a substance is a single phase, and two fully mixed liquids form a single phase.

Brief illustration 4A.1

A solution of sodium chloride in water is a single phase ($P = 1$). Ice is a single phase even though it might be chipped into small fragments. A slurry of ice and water is a two-phase system ($P = 2$) even though it is difficult to map the physical boundaries between the phases. A system in which calcium carbonate undergoes the thermal decomposition $CaCO_3(s) \rightarrow CaO(s) + CO_2(g)$ consists of two solid phases (one consisting of calcium carbonate and the other of calcium oxide) and one gaseous phase (consisting of carbon dioxide), so $P = 3$.

Two metals form a two-phase system ($P = 2$) if they are immiscible, but a single-phase system ($P = 1$), an alloy, if they are miscible (and actually mixed). A solution of solid B in solid A—a homogeneous mixture of the two miscible substances—is uniform on a molecular scale. In a solution, atoms of A are surrounded by atoms of A and B, and any sample cut from the sample, even microscopically small, is representative of the composition of the whole. It is therefore a single phase.

A dispersion is uniform on a macroscopic scale but not on a microscopic scale, because it consists of grains or droplets of one substance in a matrix of the other (Fig. 4A.1). A small sample could come entirely from one of the minute grains of pure A and would not be representative of the whole. A dispersion therefore consists of two phases.

(b) Phase transitions

A **phase transition**, the spontaneous conversion of one phase into another phase, occurs at a characteristic **transition temperature**, T_{trs}, for a given pressure. At the transition temperature

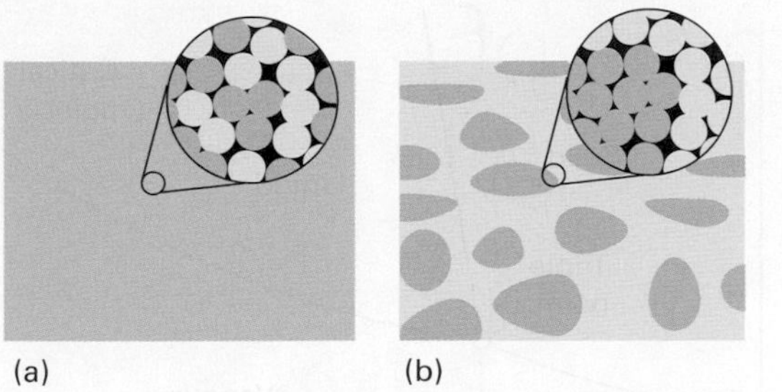

Figure 4A.1 The difference between (a) a single-phase solution, in which the composition is uniform on a molecular scale, and (b) a dispersion, in which microscopic regions of one component are embedded in a matrix of a second component.

the two phases are in equilibrium and the Gibbs energy of the system is a minimum at the prevailing pressure.

Brief illustration 4A.2

At 1 atm, ice is the stable phase of water below 0 °C, but above 0 °C liquid water is more stable. This difference indicates that below 0 °C the Gibbs energy decreases as liquid water changes into ice, but that above 0 °C the Gibbs energy decreases as ice changes into liquid water. The numerical values of the Gibbs energies are considered in the next *Brief illustration*.

The detection of a phase transition is not always straightforward as there may be nothing to see, especially if the two phases are both solids. **Thermal analysis**, which takes advantage of the heat that is evolved or absorbed during a transition, can be used. Thus, if the phase transition is exothermic and the temperature of a sample is monitored as it cools, the presence of the transition can be recognized by a pause in the otherwise steady fall of the temperature (Fig. 4A.2). Similarly, if a sample is heated steadily and the transition is endothermic, there will

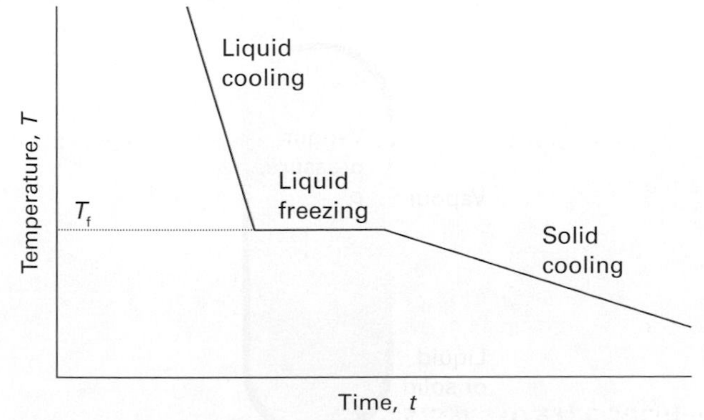

Figure 4A.2 A cooling curve at constant pressure. The flat section corresponds to the pause in the fall of temperature while an exothermic transition (freezing) occurs. This pause enables T_f to be located even if the transition cannot be observed visually.

be a pause in the temperature rise at the transition temperature. Differential scanning calorimetry (Topic 2C) is also used to detect phase transitions, and X-ray diffraction (Topic 15B) is useful for detecting phase transitions in a solid, because the two phases will have different structures.

As always, it is important to distinguish between the thermodynamic description of a process and the rate at which the process occurs. A phase transition that is predicted by thermodynamics to be spontaneous might occur too slowly to be significant in practice. For instance, at normal temperatures and pressures the molar Gibbs energy of graphite is lower than that of diamond, so there is a thermodynamic tendency for diamond to change into graphite. However, for this transition to take place, the C atoms must change their locations, which, except at high temperatures, is an immeasurably slow process in a solid. The discussion of the rate of attainment of equilibrium is a kinetic problem and is outside the range of thermodynamics. In gases and liquids the mobilities of the molecules allow phase transitions to occur rapidly, but in solids thermodynamic instability may be frozen in. Thermodynamically unstable phases that persist because the transition is kinetically hindered are called **metastable phases**. Diamond is a metastable but persistent phase of carbon under normal conditions.

(c) Thermodynamic criteria of phase stability

All the following considerations are based on the Gibbs energy of a substance, and in particular on its molar Gibbs energy, G_m. In fact, this quantity plays such an important role in this Focus and elsewhere in the text that it is given a special name and symbol, the **chemical potential**, μ (mu). For a system that consists of a single substance, the 'molar Gibbs energy' and the 'chemical potential' are exactly the same: $\mu = G_m$. In Topic 5A the chemical potential is given a broader significance and a more general definition. The name 'chemical potential' is also instructive: as the concept is developed it will become clear that μ is a measure of the potential that a substance has for undergoing change. In this Focus, and in Focus 5, it reflects the potential of a substance to undergo physical change. In Focus 6, μ is the potential of a substance to undergo chemical change.

The discussion in this Topic is based on the following consequence of the Second Law (Fig. 4A.3):

> At equilibrium, the chemical potential of a substance is the same in and throughout every phase present in the system.

Criterion for phase equilibrium

To see the validity of this remark, consider a system in which the chemical potential of a substance is μ_1 at one location and μ_2 at another location. The locations may be in the same or in different phases. When an infinitesimal amount dn of the substance is transferred from one location to the other, the Gibbs energy of the system changes by $-\mu_1$dn (i.e. dG = $-G_{m,1}$dn) when material is removed from location 1. It changes

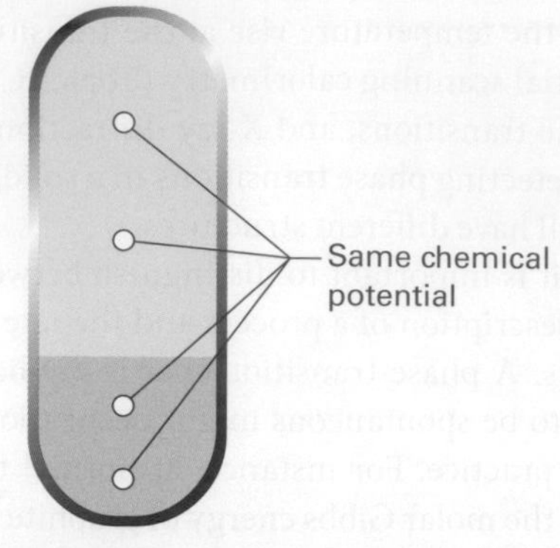

Figure 4A.3 When two or more phases are in equilibrium, the chemical potential of a substance (and, in a mixture, a component) is the same in each phase, and is the same at all points in each phase.

by $+\mu_2 dn$ (i.e. $dG = G_{m,2}dn$) when that material is added to location 2. The overall change is therefore $dG = (\mu_2 - \mu_1)dn$. If the chemical potential at location 1 is higher than that at location 2, the transfer is accompanied by a decrease in G, and so has a spontaneous tendency to occur. Only if $\mu_1 = \mu_2$ is there no change in G, and only then is the system at equilibrium.

Brief illustration 4A.3

The standard molar Gibbs energy of formation of water vapour at 298 K (25 °C) is −229 kJ mol⁻¹, and that of liquid water at the same temperature is −237 kJ mol⁻¹. It follows that there is a decrease in Gibbs energy when water vapour condenses to the liquid at 298 K, so condensation is spontaneous at that temperature (and 1 bar).

4A.2 Phase boundaries

The **phase diagram** of a pure substance shows the regions of pressure and temperature at which its various phases are thermodynamically stable (Fig. 4A.4). In fact, any two intensive variables may be used (such as temperature and magnetic field; in Topic 5A mole fraction is another variable), but this Topic focuses on pressure and temperature. The lines separating the regions, which are called **phase boundaries** (or *coexistence curves*), show the values of p and T at which two phases coexist in equilibrium and their chemical potentials are equal. A single phase is represented by an *area* on a phase diagram.

(a) Characteristic properties related to phase transitions

Consider a liquid sample of a pure substance in a closed vessel. The pressure of a vapour in equilibrium with the liquid

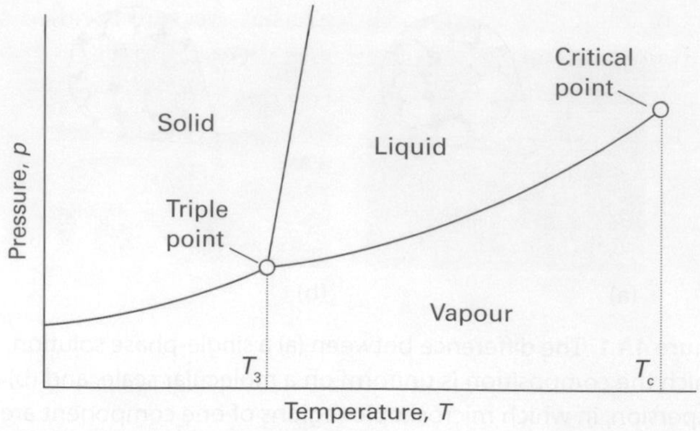

Figure 4A.4 The general regions of pressure and temperature where solid, liquid, or gas is stable (that is, has minimum molar Gibbs energy) are shown on this phase diagram. For example, the solid phase is the most stable phase at low temperatures and high pressures.

is its vapour pressure (the property introduced in Topic 1C; Fig. 4A.5). Therefore, the liquid–vapour phase boundary in a phase diagram shows how the vapour pressure of the liquid varies with temperature. Similarly, the solid–vapour phase boundary shows the temperature variation of the **sublimation vapour pressure**, the vapour pressure of the solid phase. The vapour pressure of a substance increases with temperature because at higher temperatures more molecules have sufficient energy to escape from their neighbours.

When a liquid is in an open vessel and subject to an external pressure, it is possible for the liquid to vaporize from its surface. However, only when the temperature is such that the vapour pressure is equal to the external pressure will it be possible for vaporization to occur throughout the bulk of the liquid and for the vapour to expand freely into the surroundings. This condition of free vaporization throughout the liquid is called **boiling**. The temperature at which the vapour pres-

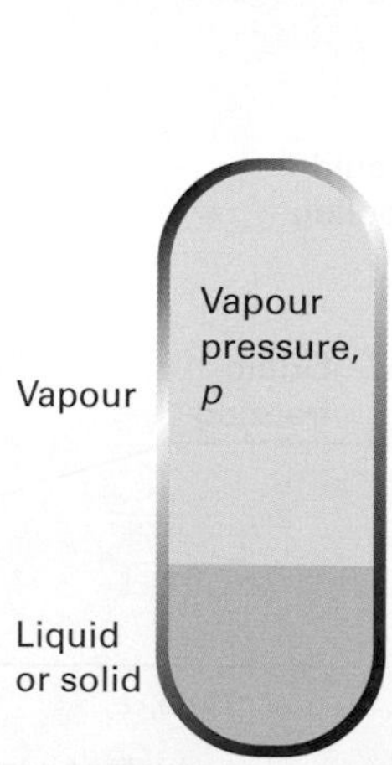

Figure 4A.5 The vapour pressure of a liquid or solid is the pressure exerted by the vapour in equilibrium with the condensed phase.

sure of a liquid is equal to the external pressure is called the **boiling temperature** at that pressure. For the special case of an external pressure of 1 atm, the boiling temperature is called the **normal boiling point**, T_b. With the replacement of 1 atm by 1 bar as standard pressure, there is some advantage in using the **standard boiling point** instead: this is the temperature at which the vapour pressure reaches 1 bar. Because 1 bar is slightly less than 1 atm (1.00 bar = 0.987 atm), the standard boiling point of a liquid is slightly lower than its normal boiling point. For example, the normal boiling point of water is 100.0 °C, but its standard boiling point is 99.6 °C.

Boiling does not occur when a liquid is heated in a rigid, closed vessel. Instead, the vapour pressure, and hence the density of the vapour, rises as the temperature is raised (Fig. 4A.6). At the same time, the density of the liquid decreases slightly as a result of its expansion. There comes a stage when the density of the vapour is equal to that of the remaining liquid and the surface between the two phases disappears. The temperature at which the surface disappears is the **critical temperature**, T_c, of the substance. The vapour pressure at the critical temperature is called the **critical pressure**, p_c. At and above the critical temperature, a single uniform phase called a **supercritical fluid** fills the container and an interface no longer exists. That is, above the critical temperature, the liquid phase of the substance does not exist.

The temperature at which, under a specified pressure, the liquid and solid phases of a substance coexist in equilibrium is called the **melting temperature**. Because a substance melts at exactly the same temperature as it freezes, the melting temperature of a substance is the same as its **freezing temperature**. The freezing temperature when the pressure is 1 atm is called the **normal freezing point**, T_f, and its freezing point when the pressure is 1 bar is called the **standard freezing point**. The normal and standard freezing points are negligibly different for most purposes. The normal freezing point is also called the **normal melting point**.

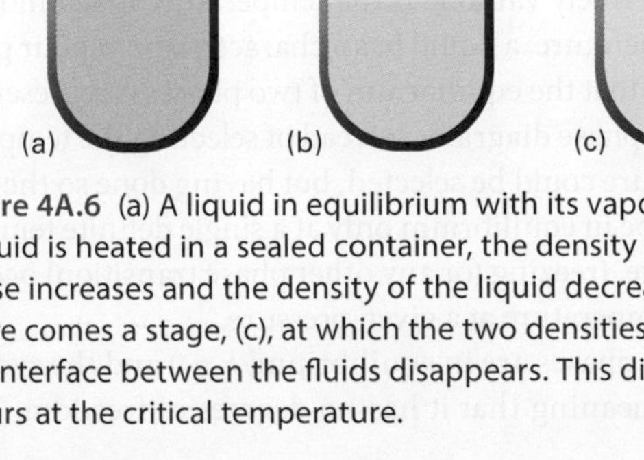

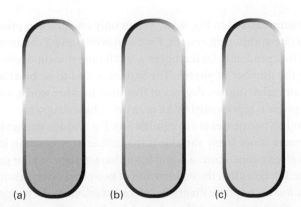

Figure 4A.6 (a) A liquid in equilibrium with its vapour. (b) When a liquid is heated in a sealed container, the density of the vapour phase increases and the density of the liquid decreases slightly. There comes a stage, (c), at which the two densities are equal and the interface between the fluids disappears. This disappearance occurs at the critical temperature.

There is a set of conditions under which three different phases of a substance (typically solid, liquid, and vapour) all simultaneously coexist in equilibrium. These conditions are represented by the **triple point**, a point at which the three phase boundaries meet. The temperature at the triple point is denoted T_3. The triple point of a pure substance cannot be changed: it occurs at a single definite pressure and temperature characteristic of the substance.

As can be seen from Fig. 4A.4, the triple point marks the lowest pressure at which a liquid phase of a substance can exist. If (as is common) the slope of the solid–liquid phase boundary is as shown in the diagram, then the triple point also marks the lowest temperature at which the liquid can exist.

Brief illustration 4A.4

The triple point of water lies at 273.16 K and 611 Pa (6.11 mbar, 4.58 Torr), and the three phases of water (ice, liquid water, and water vapour) coexist in equilibrium at no other combination of pressure and temperature. This invariance of the triple point was the basis of its use in the now superseded definition of the Kelvin scale of temperature (Topic 3A).

(b) The phase rule

In one of the most elegant arguments in the whole of chemical thermodynamics, J.W. Gibbs deduced the **phase rule**, which gives the number of parameters that can be varied independently (at least to a small extent) while the number of phases in equilibrium is preserved. The phase rule is a general relation between the variance, F, the number of components, C, and the number of phases at equilibrium, P, for a system of any composition. Each of these quantities has a precisely defined meaning:

- The **variance** (or *number of degrees of freedom*), F, of a system is the number of intensive variables that can be changed independently without disturbing the number of phases in equilibrium.
- A **constituent** of a system is any chemical species that is present.
- A **component** is a *chemically independent* constituent of a system.
- The number of components, C, in a system is the minimum number of types of independent species (ions or molecules) necessary to define the composition of all the phases present in the system.

Brief illustration 4A.5

A mixture of ethanol and water has two constituents. A solution of sodium chloride has three constituents: water, Na^+ ions, and Cl^- ions, but only two components because the numbers of Na^+ and Cl^- ions are constrained to be equal by the requirement of charge neutrality.

The relation between these quantities, which is called the **phase rule**, is established by considering the conditions for equilibrium to exist between the phases in terms of the chemical potentials of all the constituents.

How is that done? 4A.1 Deducing the phase rule

The argument that leads to the phase rule is most easily appreciated by first thinking about the simpler case when only one component is present and then generalizing the result to an arbitrary number of components.

Step 1 *Consider the case where only one component is present*

When only one phase is present ($P = 1$), both p and T can be varied independently, so $F = 2$. Now consider the case where two phases α and β are in equilibrium ($P = 2$). If the phases are in equilibrium at a given pressure and temperature, their chemical potentials must be equal:

$$\mu(\alpha; p,T) = \mu(\beta; p,T)$$

This equation relates p and T: when the pressure changes, the changes in the chemical potentials are different in general, so in order to keep them equal, the temperature must change too. To keep the two phases in equilibrium only one variable can be changed arbitrarily, so $F = 1$.

If three phases of a one-component system are in mutual equilibrium, the chemical potentials of all three phases (α, β, and γ) must be equal:

$$\mu(\alpha; p,T) = \mu(\beta; p,T) = \mu(\gamma; p,T)$$

This relation is actually *two* equations $\mu(\alpha; p,T) = \mu(\beta; p,T)$ and $\mu(\beta; p,T) = \mu(\gamma; p,T)$, in which there are *two* variables: pressure and temperature. With two equations for two unknowns, there is a single solution (just as the pair of algebraic equations $x + y = xy$ and $3x - y = xy$ have the single, fixed solutions $x = 2$ and $y = 2$). There is therefore only one single, unchangeable value of the pressure and temperature as a solution. The conclusion is that there is no freedom to choose these variables, so $F = 0$.

Four phases cannot be in mutual equilibrium in a one-component system because the three equalities

$$\mu(\alpha; p,T) = \mu(\beta; p,T),\ \mu(\beta; p,T) = \mu(\gamma; p,T),\ \text{and}\ \mu(\gamma; p,T) = \mu(\delta; p,T)$$

are three equations with only two unknowns (p and T), which are not consistent because no values of p and T satisfy all three equations (just as the three equations $x + y = xy$, $3x - y = xy$, and $4x - y = 2xy^2$ have no solution).

In summary, for a one-component system ($C = 1$) it has been shown that: $F = 2$ when $P = 1$; $F = 1$ when $P = 2$; and $F = 0$ when $P = 3$. The general result is that for $C = 1$, $F = 3 - P$.

Step 2 *Consider the general case of any number of components, C*

Begin by counting the total number of intensive variables. The pressure, p, and temperature, T, count as 2. The composition of a phase is specified by giving the mole fractions of the C components, but as the sum of the mole fractions must be 1, only $C - 1$ mole fractions are independent. Because there are P phases, the total number of composition variables is $P(C - 1)$. At this stage, the total number of intensive variables is $P(C - 1) + 2$.

At equilibrium, the chemical potential of a component J is the same in every phase:

$$\mu_J(\alpha; p,T) = \mu_J(\beta; p,T) = \cdots \text{ for } P \text{ phases}$$

There are $P - 1$ equations of this kind to be satisfied for each component J. As there are C components, the total number of equations is $C(P - 1)$. Each equation reduces the freedom to vary one of the $P(C - 1) + 2$ intensive variables. It follows that the total number of degrees of freedom is

$$F = P(C - 1) + 2 - C(P - 1)$$

The right-hand side simplifies to give the phase rule in the form derived by Gibbs:

$$F = C - P + 2 \qquad \text{The phase rule} \qquad (4A.1)$$

The implications of the phase rule for a one-component system, when

$$F = 3 - P \qquad \text{The phase rule } [C = 1] \qquad (4A.2)$$

are summarized in Fig. 4A.7. When only one phase is present in a one-component system, $F = 2$ and both p and T can be varied independently (at least over a small range) without changing the number of phases. The system is said to be **bivariant**, meaning having two degrees of freedom. In other words, a single phase is represented by an *area* on a phase diagram.

When two phases are in equilibrium $F = 1$, which implies that pressure is not freely variable if the temperature is set; indeed, at a given temperature, a liquid has a characteristic vapour pressure. It follows that the equilibrium of two phases is represented by a *line* in the phase diagram. Instead of selecting the temperature, the pressure could be selected, but having done so the two phases would be in equilibrium only at a single definite temperature. Therefore, freezing (or any other phase transition) occurs at a definite temperature at a given pressure.

When three phases are in equilibrium, $F = 0$ and the system is **invariant**, meaning that it has no degrees of freedom. This

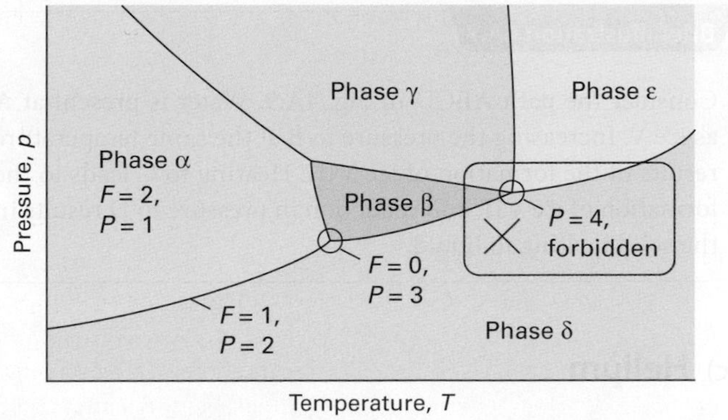

Figure 4A.7 The typical regions of a one-component phase diagram. The lines represent conditions under which the two adjoining phases are in equilibrium. A point represents the unique set of conditions under which three phases coexist in equilibrium. Four phases cannot mutually coexist in equilibrium when only one component is present.

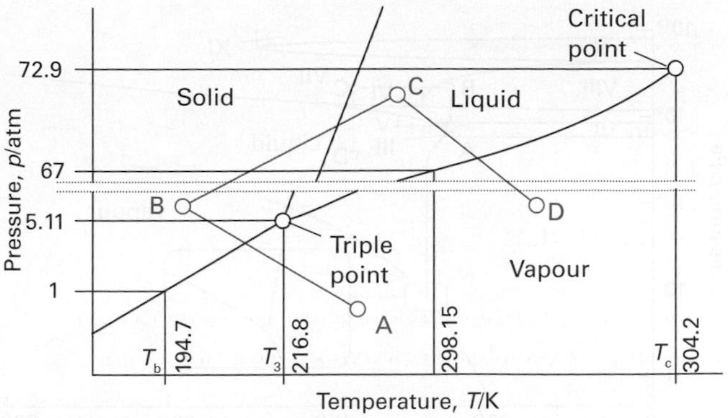

Figure 4A.8 The experimental phase diagram for carbon dioxide; note the break in the vertical scale. As the triple point lies at pressures well above atmospheric, liquid carbon dioxide does not exist under normal conditions; a pressure of at least 5.11 atm must be applied for liquid to be formed. The path ABCD is discussed in *Brief illustration* 4A.6.

special condition can be established only at a definite temperature and pressure that is characteristic of the substance and cannot be changed. The equilibrium of three phases is therefore represented by a *point*, the triple point, on a phase diagram. Four phases cannot be in equilibrium in a one-component system because F cannot be negative.

4A.3 Three representative phase diagrams

Carbon dioxide, water, and helium illustrate the significance of the various features of a phase diagram.

(a) Carbon dioxide

Figure 4A.8 shows the phase diagram for carbon dioxide. The features to notice include the positive slope (up from left to right) of the solid–liquid phase boundary; the direction of this line is characteristic of most substances. This slope indicates that the melting temperature of solid carbon dioxide rises as the pressure is increased. Notice also that, as the triple point lies above 1 atm, the liquid cannot exist at normal atmospheric pressures whatever the temperature. As a result, the solid sublimes when left in the open (hence the name 'dry ice'). To obtain the liquid, it is necessary to exert a pressure of at least 5.11 atm. Cylinders of carbon dioxide generally contain the liquid or compressed gas; at 25 °C that implies a vapour pressure of 67 atm if both gas and liquid are present in equilibrium. When the gas is released through a tap (which acts as a throttle) the gas cools by the Joule–Thomson effect, so when it emerges into a region where the pressure is only 1 atm, it condenses into a finely divided snow-like solid. That carbon dioxide gas cannot be liquefied except by applying high pressure reflects the weakness of the intermolecular forces between the nonpolar carbon dioxide molecules (Topic 14B).

Brief illustration 4A.6

Consider the path ABCD in Fig. 4A.8. At A the carbon dioxide is a gas. When the temperature and pressure are adjusted to B, the vapour condenses directly to a solid. Increasing the pressure and temperature to C results in the formation of the liquid phase, which evaporates to the vapour when the conditions are changed to D.

(b) Water

Figure 4A.9 shows the phase diagram for water. The liquid–vapour boundary in the phase diagram summarizes how the vapour pressure of liquid water varies with temperature. It also summarizes how the boiling temperature varies with pressure: simply read off the temperature at which the vapour pressure is equal to the prevailing atmospheric pressure. The solid (ice I)–liquid boundary shows how the melting temperature varies with the pressure. Its very steep slope indicates that enormous pressures are needed to bring about significant changes. Notice that the line has a negative slope (down from left to right) up to 2 kbar, which means that the melting temperature falls as the pressure is raised.

The reason for this almost unique behaviour can be traced to the decrease in volume that occurs on melting: it is more favourable for the solid to transform into the liquid as the pressure is raised. The decrease in volume is a result of the very open structure of ice: as shown in Fig. 4A.10, the water molecules are held apart, as well as together, by the hydrogen bonds

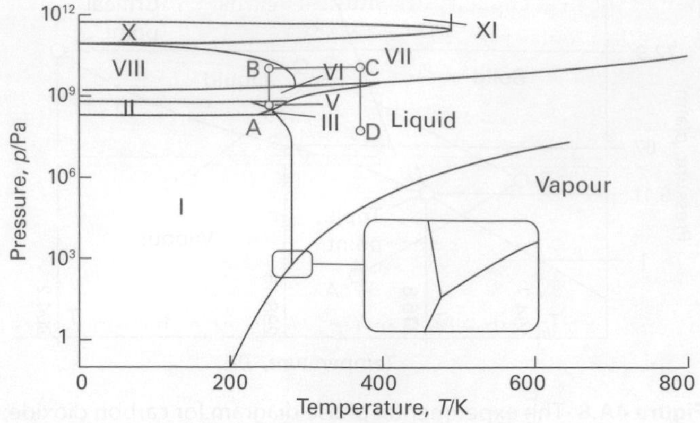

Figure 4A.9 The phase diagram for water showing the different solid phases, which are indicated with Roman numerals I, II, …; solid phase I (ice I) is ordinary ice. The path ABCD is discussed in *Brief illustration* 4A.7.

between them, but the hydrogen-bonded structure partially collapses on melting and the liquid is denser than the solid. Other consequences of its extensive hydrogen bonding are the anomalously high boiling point of water for a molecule of its molar mass and its high critical temperature and pressure.

The diagram shows that water has one liquid phase but many different solid phases other than ordinary ice ('ice I'). Some of these phases melt at high temperatures. Ice VII, for instance, melts at 100 °C but exists only above 25 kbar. Two further phases, Ice XIII and XIV, were identified in 2006 at −160 °C but have not yet been allocated regions in the phase diagram. Note that several more triple points occur in the diagram other than the one where vapour, liquid, and ice I coexist. Each one occurs at a definite pressure and temperature that cannot be changed. The solid phases of ice differ in the arrangement of the water molecules: under the influence of very high pressures, hydrogen bonds buckle and the H_2O molecules adopt different arrangements. These polymorphs of ice may contribute to the advance of glaciers, for ice at the bottom of glaciers experiences very high pressures where it rests on jagged rocks.

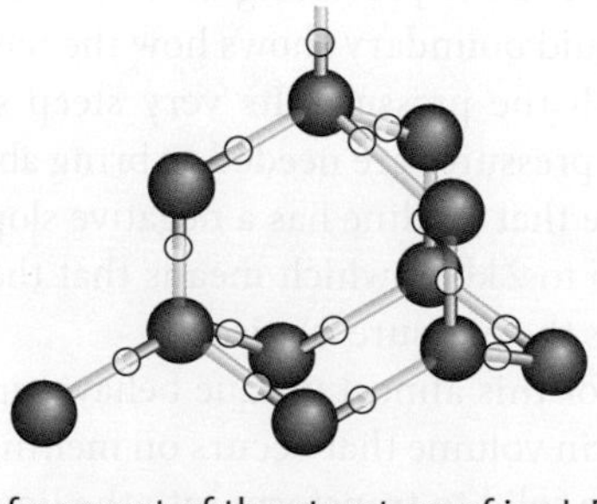

Figure 4A.10 A fragment of the structure of ice I. Each O atom is linked by two covalent bonds to H atoms and by two hydrogen bonds to a neighbouring O atom, in a tetrahedral array.

Brief illustration 4A.7

Consider the path ABCD in Fig. 4A.9. Water is present at A as ice V. Increasing the pressure to B at the same temperature results in the formation of ice VIII. Heating to C leads to the formation of ice VII, and reduction in pressure to D results in the solid melting to liquid.

(c) Helium

The two isotopes of helium, ^{3}He and ^{4}He, behave differently at low temperatures because ^{4}He is a boson whereas ^{3}He is a fermion, and are treated differently by the Pauli principle (Topic 8B). Figure 4A.11 shows the phase diagram of helium-4. Helium behaves unusually at low temperatures because the mass of its atoms is so low and there are only very weak interactions between neighbours. At 1 atm, the solid and gas phases of helium are never in equilibrium however low the temperature: the atoms are so light that they vibrate with a large-amplitude motion even at very low temperatures and the solid simply shakes itself apart. Solid helium can be obtained, but only by holding the atoms together by applying pressure.

Pure helium-4 has two liquid phases. The phase marked He-I in the diagram behaves like a normal liquid; the other phase, He-II, is a **superfluid.** It is so called because it flows without viscosity.[1] The liquid–liquid phase boundary is called the **λ-line** (lambda line) for reasons related to the shape of a

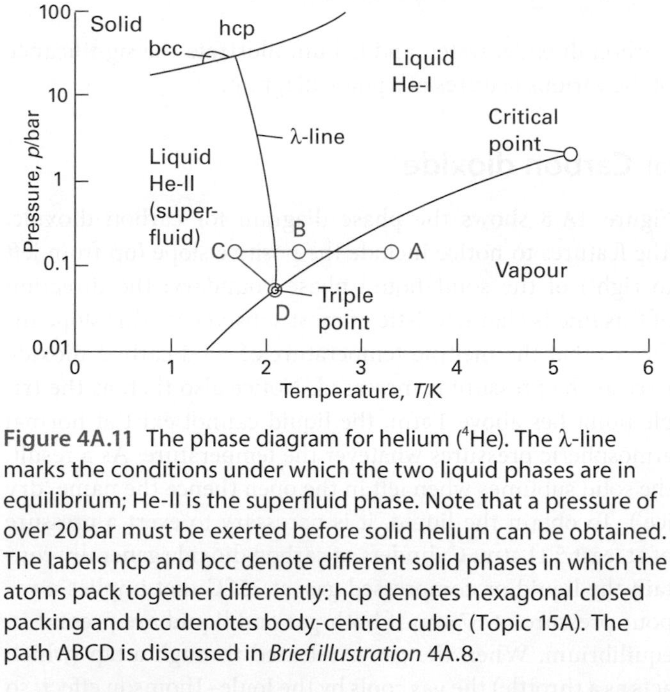

Figure 4A.11 The phase diagram for helium (^{4}He). The λ-line marks the conditions under which the two liquid phases are in equilibrium; He-II is the superfluid phase. Note that a pressure of over 20 bar must be exerted before solid helium can be obtained. The labels hcp and bcc denote different solid phases in which the atoms pack together differently: hcp denotes hexagonal closed packing and bcc denotes body-centred cubic (Topic 15A). The path ABCD is discussed in *Brief illustration* 4A.8.

[1] Water might also have a superfluid liquid phase.

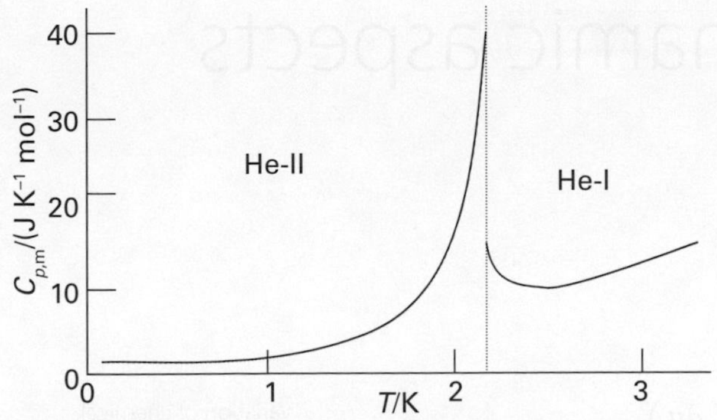

Figure 4A.12 The heat capacity of superfluid He-II increases with temperature and rises steeply as the transition temperature to He-I is approached. The appearance of the plot has led the transition to be described as a λ-transition and the line on the phase diagram to be called a λ-line.

plot of the heat capacity of helium-4 against temperature at the transition temperature (Fig. 4A.12).

Helium-3 also has a superfluid phase. Helium-3 is unusual in that melting is exothermic ($\Delta_{fus}H < 0$) and therefore (from $\Delta_{fus}S = \Delta_{fus}H/T_f$) at the melting point the entropy of the liquid is lower than that of the solid.

Brief illustration 4A.8

Consider the path ABCD in Fig. 4A.11. At A, helium is present as a vapour. On cooling to B it condenses to helium-I, and further cooling to C results in the formation of helium-II. Adjustment of the pressure and temperature to D results in a system in which three phases, helium-I, helium-II, and vapour are in mutual equilibrium.

Checklist of concepts

☐ 1. A **phase** is a form of matter that is uniform throughout in chemical composition and physical state.

☐ 2. A **phase transition** is the spontaneous conversion of one phase into another.

☐ 3. The thermodynamic analysis of phases is based on the fact that at equilibrium, the chemical potential of a substance is the same throughout a sample.

☐ 4. A **phase diagram** indicates the values of the pressure and temperature at which a particular phase is most stable, or is in equilibrium with other phases.

☐ 5. The **phase rule** relates the number of variables that may be changed while the phases of a system remain in mutual equilibrium.

Checklist of equations

Property	Equation	Comment	Equation number
Chemical potential	$\mu = G_m$	For a single substance	
Phase rule	$F = C - P + 2$	F is the variance, C the number of components, and P the number of phases	4A.1

TOPIC 4B Thermodynamic aspects of phase transitions

> ➤ Why do you need to know this material?
>
> Thermodynamic arguments explain the appearance of phase diagrams and can be used to make predictions about the effect of pressure on phase transitions. They provide insight into the properties that account for the behaviour of matter under different conditions.
>
> ➤ What is the key idea?
>
> The effect of temperature and pressure on the chemical potential of a substance in each phase depends on its molar entropy and molar volume, respectively.
>
> ➤ What do you need to know already?
>
> You need to be aware that phases are in equilibrium when their chemical potentials are equal (Topic 4A) and that the variation of the molar Gibbs energy of a substance depends on its molar volume and entropy (Topic 3E). The Topic makes use of expressions for the entropy of transition (Topic 3B) and of the perfect gas law (Topic 1A).

As explained in Topic 4A, the thermodynamic criterion for phase equilibrium is the equality of the chemical potentials of each substance in each phase. For a one-component system, the chemical potential is the same as the molar Gibbs energy ($\mu = G_m$). In Topic 3E it is explained how the Gibbs energy varies with temperature and pressure:

$$dG = -SdT \text{ at constant pressure;}$$
$$dG = Vdp \text{ at constant temperature}$$

These expressions also apply to the molar Gibbs energy, and therefore to the chemical potential. By using the notation of partial derivatives (*The chemist's toolkit* 9 in Topic 2A) they can be expressed as

$$\left(\frac{\partial \mu}{\partial T}\right)_p = -S_m \qquad \text{Variation of chemical potential with } T \text{ [constant } p\text{]} \qquad (4B.1a)$$

$$\left(\frac{\partial \mu}{\partial p}\right)_T = V_m \qquad \text{Variation of chemical potential with } p \text{ [constant } T\text{]} \qquad (4B.1b)$$

By combining the equality of chemical potentials of a substance in each phase with these expressions for the variation of μ with temperature and pressure it is possible to deduce how phase equilibria respond to changes in the conditions.

4B.1 The dependence of stability on the conditions

At sufficiently low temperatures the solid phase of a substance commonly has the lowest chemical potential and is therefore the most stable phase. However, the chemical potentials of different phases depend on temperature to different extents (because the molar entropy of each phase is different), and above a certain temperature the chemical potential of another phase (perhaps another solid phase, a liquid, or a gas) might turn out to be lower. Then a transition to the second phase becomes spontaneous and occurs if it is kinetically feasible.

(a) The temperature dependence of phase stability

Because $S_m > 0$ for all substances above $T = 0$, eqn 4B.1a shows that the chemical potential of a pure substance decreases as the temperature is raised. That is, a plot of chemical potential against temperature slopes down from left to right. It also implies that because $S_m(g) > S_m(l)$, the slope is steeper for gases than for liquids. Because it is almost always the case that $S_m(l) > S_m(s)$, the slope is also steeper for a liquid than the corresponding solid. These features are illustrated in Fig. 4B.1. The steeper slope of $\mu(l)$ compared with that of $\mu(s)$ results in $\mu(l)$ falling below $\mu(s)$ when the temperature is high enough; then the liquid becomes the stable phase, and melting is spontaneous. The chemical potential of the gas phase plunges steeply downwards as the temperature is raised (because the molar entropy of the vapour is so high), and there comes a temperature at which it lies below that of the liquid. Then the gas is the stable phase and vaporization is spontaneous.

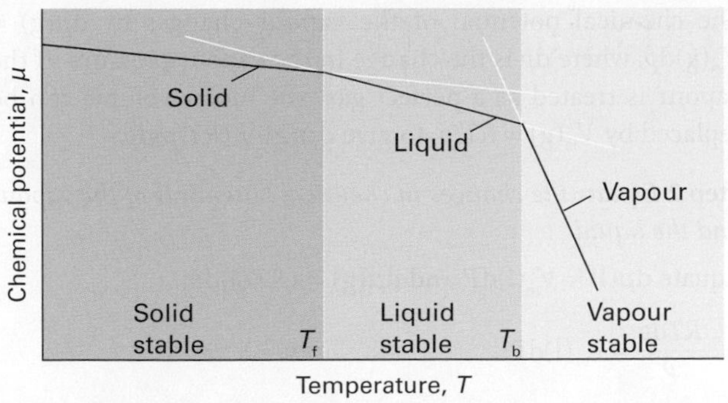

Figure 4B.1 The schematic temperature dependence of the chemical potential of the solid, liquid, and gas phases of a substance (in practice, the lines are curved). The phase with the lowest chemical potential at a specified temperature is the most stable one at that temperature. The transition temperatures, the freezing (melting) and boiling temperatures (T_f and T_b, respectively), are the temperatures at which the chemical potentials of the two phases are equal.

Brief illustration 4B.1

The standard molar entropy of liquid water at 100 °C is 86.8 J K^{-1} mol^{-1} and that of water vapour at the same temperature is 195.98 J K^{-1} mol^{-1}. It follows that when the temperature is raised by 1.0 K the changes in chemical potential are

$$\Delta\mu(\mathrm{l}) \approx -S_\mathrm{m}(\mathrm{l})\Delta T = -87\,\mathrm{J\,mol^{-1}}$$
$$\Delta\mu(\mathrm{g}) \approx -S_\mathrm{m}(\mathrm{g})\Delta T = -196\,\mathrm{J\,mol^{-1}}$$

At 100 °C the two phases are in equilibrium with equal chemical potentials. At 101 °C the chemical potential of both vapour and liquid are lower than at 100 °C, but the chemical potential of the vapour has decreased by a greater amount. It follows that the vapour is the stable phase at the higher temperature, so vaporization will be spontaneous.

(b) The response of melting to applied pressure

Equation 4B.1b shows that because $V_\mathrm{m} > 0$, an increase in pressure raises the chemical potential of any pure substance. In most cases, $V_\mathrm{m}(\mathrm{l}) > V_\mathrm{m}(\mathrm{s})$, so an increase in pressure increases the chemical potential of the liquid phase of a substance more than that of its solid phase. As shown in Fig. 4B.2(a), the effect of pressure in such a case is to raise the freezing temperature slightly. For water, however, $V_\mathrm{m}(\mathrm{l}) < V_\mathrm{m}(\mathrm{s})$, and an increase in pressure increases the chemical potential of the solid more than that of the liquid. In this case, the freezing temperature is lowered slightly (Fig. 4B.2(b)).

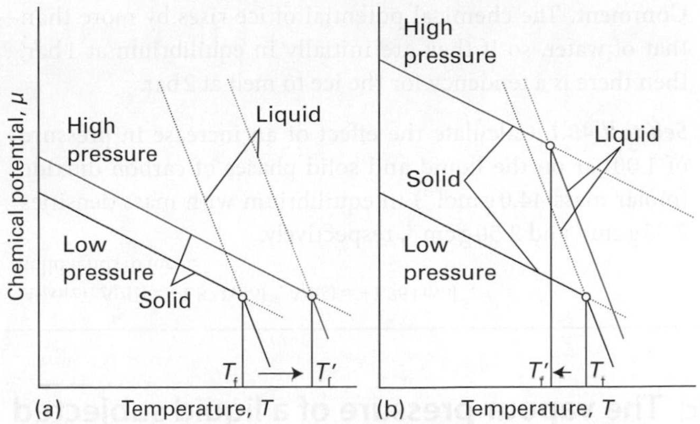

Figure 4B.2 The pressure dependence of the chemical potential of a substance depends on the molar volume of the phase. The lines show schematically the effect of increasing pressure on the chemical potential of the solid and liquid phases (in practice, the lines are curved), and the corresponding effects on the freezing temperatures. (a) In this case the molar volume of the solid is smaller than that of the liquid and μ(s) increases less than μ(l). As a result, the freezing temperature rises. (b) Here the molar volume is greater for the solid than the liquid (as for water), μ(s) increases more strongly than μ(l), and the freezing temperature is lowered.

Example 4B.1 Assessing the effect of pressure on the chemical potential

Calculate the effect on the chemical potentials of ice and water of increasing the pressure from 1.00 bar to 2.00 bar at 0 °C. The mass density of ice is 0.917 g cm^{-3} and that of liquid water is 0.999 g cm^{-3} under these conditions.

Collect your thoughts From $\mathrm{d}\mu = V_\mathrm{m}\mathrm{d}p$, you can infer that the change in chemical potential of an incompressible substance when the pressure is changed by Δp is $\Delta\mu = V_\mathrm{m}\Delta p$. Therefore, you need to know the molar volumes of the two phases of water. These values are obtained from the mass density, ρ, and the molar mass, M, by using $V_\mathrm{m} = M/\rho$. Then $\Delta\mu = M\Delta p/\rho$. To keep the units straight, you will need to express the mass densities in kilograms per cubic metre (kg m^{-3}) and the molar mass in kilograms per mole (kg mol^{-1}), and use 1 Pa m^3 = 1 J.

The solution The molar mass of water is 18.02 g mol^{-1} (i.e. 1.802×10^{-2} kg mol^{-1}); therefore, when the pressure is increased by 1.00 bar (1.00×10^5 Pa)

$$\Delta\mu(\text{ice}) = \frac{(1.802\times10^{-2}\,\mathrm{kg\,mol^{-1}})\times(1.00\times10^{5}\,\mathrm{Pa})}{917\,\mathrm{kg\,m^{-3}}} = +1.97\,\mathrm{J\,mol^{-1}}$$

$$\Delta\mu(\text{water}) = \frac{(1.802\times10^{-2}\,\mathrm{kg\,mol^{-1}})\times(1.00\times10^{5}\,\mathrm{Pa})}{999\,\mathrm{kg\,m^{-3}}}$$
$$= +1.80\,\mathrm{J\,mol^{-1}}$$

Comment. The chemical potential of ice rises by more than that of water, so if they are initially in equilibrium at 1 bar, then there is a tendency for the ice to melt at 2 bar.

Self-test 4B.1 Calculate the effect of an increase in pressure of 1.00 bar on the liquid and solid phases of carbon dioxide (molar mass 44.0 g mol^{-1}) in equilibrium with mass densities 2.35 g cm^{-3} and 2.50 g cm^{-3}, respectively.

Answer: $\Delta\mu(\mathrm{l}) = +1.87\ \mathrm{J\,mol^{-1}}$, $\Delta\mu(\mathrm{s}) = +1.76\ \mathrm{J\,mol^{-1}}$; solid tends to form.

(c) The vapour pressure of a liquid subjected to pressure

Pressure can be exerted on the condensed phase mechanically or by subjecting it to the applied pressure of an inert gas (Fig. 4B.3). In the latter case, the **partial vapour pressure** is the partial pressure of the vapour in equilibrium with the condensed phase. When pressure is applied to a condensed phase, its vapour pressure rises: in effect, molecules are squeezed out of the phase and escape as a gas. The effect can be explored thermodynamically and a relation established between the applied pressure P and the vapour pressure p.

How is that done? 4B.1 Deriving an expression for the vapour pressure of a pressurized liquid

At equilibrium the chemical potentials of the liquid and its vapour are equal: $\mu(\mathrm{l}) = \mu(\mathrm{g})$. It follows that, for any change that preserves equilibrium, the resulting change in $\mu(\mathrm{l})$ must be equal to the change in $\mu(\mathrm{g})$; therefore, $\mathrm{d}\mu(\mathrm{g}) = \mathrm{d}\mu(\mathrm{l})$.

Step 1 *Express changes in the chemical potentials that arise from changes in pressure*

When the pressure P on the liquid is increased by $\mathrm{d}P$, the chemical potential of the liquid changes by $\mathrm{d}\mu(\mathrm{l}) = V_\mathrm{m}(\mathrm{l})\mathrm{d}P$.

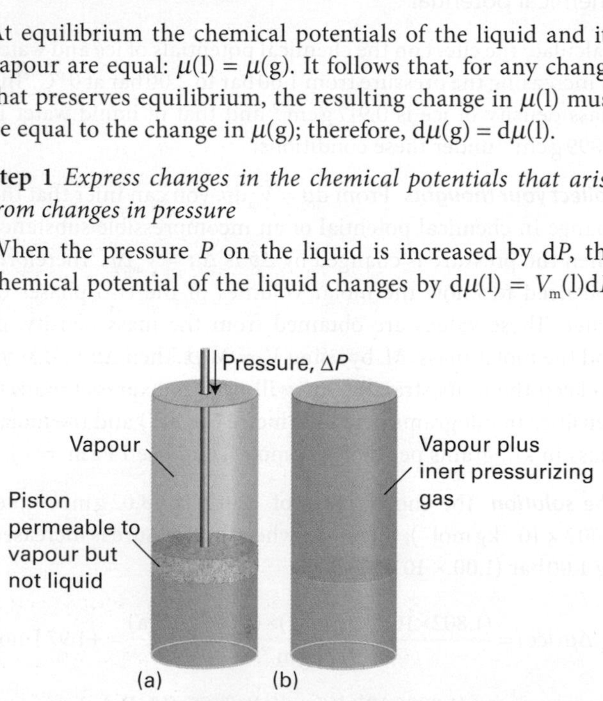

Figure 4B.3 Pressure may be applied to a condensed phase either (a) by compressing it or (b) by subjecting it to an inert pressurizing gas. When pressure is applied, the vapour pressure of the condensed phase increases.

The chemical potential of the vapour changes by $\mathrm{d}\mu(\mathrm{g}) = V_\mathrm{m}(\mathrm{g})\mathrm{d}p$, where $\mathrm{d}p$ is the change in the vapour pressure. If the vapour is treated as a perfect gas, the molar volume can be replaced by $V_\mathrm{m}(\mathrm{g}) = RT/p$, to give $\mathrm{d}\mu(\mathrm{g}) = (RT/p)\mathrm{d}p$.

Step 2 *Equate the changes in chemical potentials of the vapour and the liquid*

Equate $\mathrm{d}\mu(\mathrm{l}) = V_\mathrm{m}(\mathrm{l})\mathrm{d}P$ and $\mathrm{d}\mu(\mathrm{g}) = (RT/p)\mathrm{d}p$:

$$\frac{RT\mathrm{d}p}{p} = V_\mathrm{m}(\mathrm{l})\mathrm{d}P$$

Be careful to distinguish between P, the total pressure, and p, the partial vapour pressure.

Step 3 *Set up the integration of this expression by identifying the appropriate limits*

When there is no additional pressure acting on the liquid, P (the pressure experienced by the liquid) is equal to the normal vapour pressure p^*, so when $P = p^*$, $p = p^*$ too. When there is an additional pressure ΔP on the liquid, so $P = p + \Delta P$, the vapour pressure is p (the value required). Provided the effect of pressure on the vapour pressure is small (as will turn out to be the case) a good approximation is to replace the p in $p + \Delta P$ by p^* itself, and to set the upper limit of the integral to $p^* + \Delta P$. The integrations required are therefore as follows:

$$RT\int_{p^*}^{p}\frac{\mathrm{d}p'}{p'} = \int_{p^*}^{p^*+\Delta P} V_\mathrm{m}(\mathrm{l})\mathrm{d}P$$

(In the first integral, the variable of integration has been changed from p to p' to avoid confusion with the p at the upper limit.)

Step 4 *Carry out the integrations*

Divide both sides by RT and assume that the molar volume of the liquid is the same throughout the small range of pressures involved:

$$\overbrace{\int_{p^*}^{p}\frac{\mathrm{d}p'}{p'}}^{\text{Integral A.2}} = \frac{1}{RT}\int_{p^*}^{p^*+\Delta P} V_\mathrm{m}(\mathrm{l})\mathrm{d}P = \frac{V_\mathrm{m}(\mathrm{l})}{RT}\overbrace{\int_{p^*}^{p^*+\Delta P}\mathrm{d}P}^{\text{Integral A.1}}$$

Both integrations are straightforward, and lead to

$$\ln\frac{p}{p^*} = \frac{V_\mathrm{m}(\mathrm{l})}{RT}\Delta P$$

which (by using $\mathrm{e}^{\ln x} = x$) rearranges to

$$p = p^*\mathrm{e}^{V_\mathrm{m}(\mathrm{l})\Delta P/RT} \quad (4\mathrm{B}.2)$$

Effect of applied pressure ΔP on partial vapour pressure p

One complication that has been ignored is that, if the condensed phase is a liquid, then the pressurizing gas might dissolve and change its properties. Another complication is that the gas-phase molecules might attract molecules out of the liquid by the process of **gas solvation**, the attachment of molecules to gas-phase species.

Brief illustration 4B.2

For water, which has mass density 0.997 g cm^{-3} at 25 °C and therefore molar volume 18.1 cm^3 mol^{-1}, when the applied pressure is increased by 10 bar (i.e. $\Delta P = 1.0 \times 10^6$ Pa)

$$\ln\frac{p}{p^*} = \frac{V_m(l)\Delta P}{RT} = \frac{(1.81\times10^{-5}\,\text{m}^3\,\text{mol}^{-1})\times(1.0\times10^6\,\text{Pa})}{(8.3145\,\text{J}\,\text{K}^{-1}\,\text{mol}^{-1})\times(298\,\text{K})}$$
$$= 0.0073\ldots$$

where 1 J = 1 Pa m^3. It follows that $p = 1.0073p^*$, an increase of only 0.73 per cent.

4B.2 The location of phase boundaries

The precise locations of the phase boundaries—the pressures and temperatures at which two phases can coexist—can be found by making use once again of the fact that, when two phases are in equilibrium, their chemical potentials must be equal. Therefore, when the phases α and β are in equilibrium,

$$\mu(\alpha; p,T) = \mu(\beta; p,T) \qquad (4B.3)$$

Solution of this equation for p in terms of T gives an equation for the phase boundary (the coexistence curve).

(a) The slopes of the phase boundaries

Imagine that at some particular pressure and temperature the two phases are in equilibrium: their chemical potentials are then equal. Now p and T are changed infinitesimally, but in such a way that the phases remain in equilibrium: after these changes, the chemical potentials of the two phases change but remain equal (Fig. 4B.4). It follows that the change in the chemical potential of phase α must be the same as the change in chemical potential of phase β, so $d\mu(\alpha) = d\mu(\beta)$.

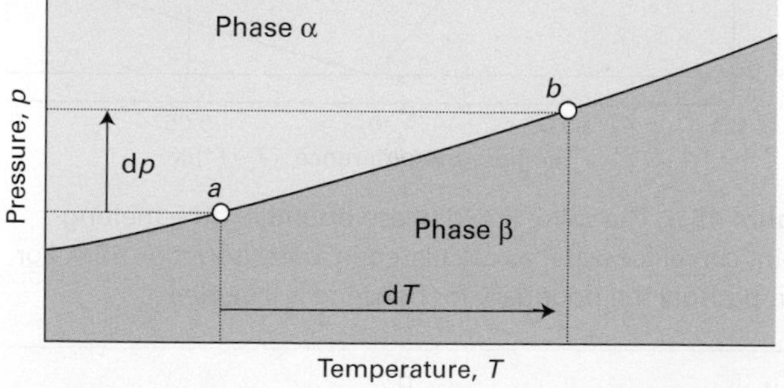

Figure 4B.4 When pressure is applied to a system in which two phases are in equilibrium (at a), the equilibrium is disturbed. It can be restored by changing the temperature, so moving the state of the system to b. It follows that there is a relation between dp and dT that ensures that the system remains in equilibrium as either variable is changed.

Equation 3E.7 ($dG = Vdp - SdT$) gives the variation of G with p and T, so with $\mu = G_m$, it follows that $d\mu = V_m dp - S_m dT$ for each phase. Therefore the relation $d\mu(\alpha) = d\mu(\beta)$ can be written

$$V_m(\alpha)dp - S_m(\alpha)dT = V_m(\beta)dp - S_m(\beta)dT$$

where $S_m(\alpha)$ and $S_m(\beta)$ are the molar entropies of the two phases, and $V_m(\alpha)$ and $V_m(\beta)$ are their molar volumes. Hence

$$\{S_m(\beta) - S_m(\alpha)\}dT = \{V_m(\beta) - V_m(\alpha)\}dp$$

The change in (molar) entropy accompanying the phase transition, $\Delta_{trs}S$, is the difference in the molar entropies $\Delta_{trs}S = S_m(\beta) - S_m(\alpha)$, and likewise for the change in (molar) volume, $\Delta_{trs}V = V_m(\beta) - V_m(\alpha)$. Therefore,

$$\Delta_{trs}S dT = \Delta_{trs}V dp$$

This relation turns into the **Clapeyron equation:**

$$\frac{dp}{dT} = \frac{\Delta_{trs}S}{\Delta_{trs}V} \qquad \text{Clapeyron equation} \qquad (4B.4a)$$

The Clapeyron equation is an exact expression for the slope of the tangent to the phase boundary at any point and applies to any phase equilibrium of any pure substance. It implies that thermodynamic data can be used to predict the appearance of phase diagrams and to understand their form. A more practical application is to the prediction of the response of freezing and boiling points to the application of pressure, when it can be used in the form obtained by inverting both sides:

$$\frac{dT}{dp} = \frac{\Delta_{trs}V}{\Delta_{trs}S} \qquad (4B.4b)$$

Brief illustration 4B.3

For water at 0 °C, the standard volume of transition of ice to liquid is −1.6 cm^3 mol^{-1}, and the corresponding standard entropy of transition is +22 J K^{-1} mol^{-1}. The slope of the solid–liquid phase boundary at that temperature is therefore

$$\frac{dT}{dp} = \frac{-1.6\times10^{-6}\,\text{m}^3\,\text{mol}^{-1}}{22\,\text{J}\,\text{K}^{-1}\,\text{mol}^{-1}} = -7.3\times10^{-8}\,\frac{\text{K}}{\text{J}\,\text{m}^{-3}}$$
$$= -7.3\times10^{-8}\,\text{K}\,\text{Pa}^{-1}$$

which corresponds to −7.3 mK bar^{-1}. An increase of 100 bar therefore results in a lowering of the freezing point of water by 0.73 K.

(b) The solid–liquid boundary

Melting (fusion) is accompanied by a molar enthalpy change $\Delta_{fus}H$, and if it occurs at a temperature T the molar entropy of melting is $\Delta_{fus}H/T$ (Topic 3B); all points on the phase boundary correspond to equilibrium, so T is in fact a transition temperature, T_{trs}. The Clapeyron equation for this phase transition then becomes

$$\frac{dp}{dT}=\frac{\Delta_{fus}H}{T\Delta_{fus}V} \quad \text{Slope of solid–liquid boundary} \quad (4B.5)$$

where $\Delta_{fus}V$ is the change in molar volume that accompanies melting. The enthalpy of melting is positive (the only exception is helium-3); the change in molar volume is usually positive and always small. Consequently, the slope dp/dT is steep and usually positive (Fig. 4B.5).

The equation for the phase boundary is found by integrating dp/dT and assuming that $\Delta_{fus}H$ and $\Delta_{fus}V$ change so little with temperature and pressure that they can be treated as constant. If the melting temperature is T^* when the pressure is p^*, and T when the pressure is p, the integration required is

$$\int_{p^*}^{p}dp=\frac{\Delta_{fus}H}{\Delta_{fus}V}\overbrace{\int_{T^*}^{T}\frac{dT}{T}}^{\text{Integral A.2}}$$

Therefore, the approximate equation of the solid–liquid boundary is

$$p=p^*+\frac{\Delta_{fus}H}{\Delta_{fus}V}\ln\frac{T}{T^*} \quad (4B.6)$$

This equation was originally obtained by yet another Thomson—James, the brother of William, Lord Kelvin.

When T is close to T^*, the logarithm can be approximated by using the expansion $\ln(1+x)=x-\frac{1}{2}x^2+\cdots$ (see *The chemist's toolkit* 12 in Topic 5B) and neglecting all but the leading term:

$$\ln\frac{T}{T^*}=\ln\left(1+\frac{T-T^*}{T^*}\right)\approx\frac{T-T^*}{T^*}$$

Therefore

$$p\approx p^*+\frac{\Delta_{fus}H}{T^*\Delta_{fus}V}(T-T^*) \quad (4B.7)$$

This expression is the equation of a steep straight line when p is plotted against T (as in Fig. 4B.5).

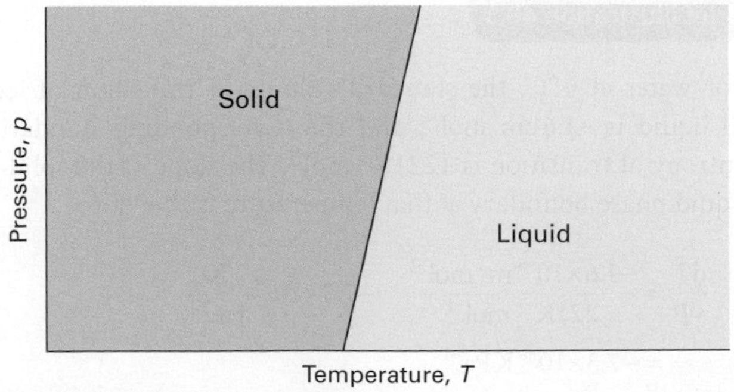

Figure 4B.5 A typical solid–liquid phase boundary slopes steeply upwards. This slope implies that, as the pressure is raised, the melting temperature rises. Most substances behave in this way, water being the notable exception.

Brief illustration 4B.4

The enthalpy of fusion of ice at 0 °C (273 K) and 1 bar is 6.008 kJ mol^{-1} and the volume of fusion is $-1.6\,\text{cm}^3\,\text{mol}^{-1}$. It follows that the solid–liquid phase boundary is given by the equation

$$p\approx1.0\times10^5\,\text{Pa}+\frac{6.008\times10^3\,\text{J mol}^{-1}}{(273\,\text{K})\times(-1.6\times10^{-6}\,\text{m}^3\,\text{mol}^{-1})}(T-T^*)$$
$$\approx1.0\times10^5\,\text{Pa}-1.4\times10^7\,\text{Pa K}^{-1}(T-T^*)$$

That is,

$$p/\text{bar}=1-140(T-T^*)/\text{K}$$

with T^* = 273 K. This expression is plotted in Fig. 4B.6.

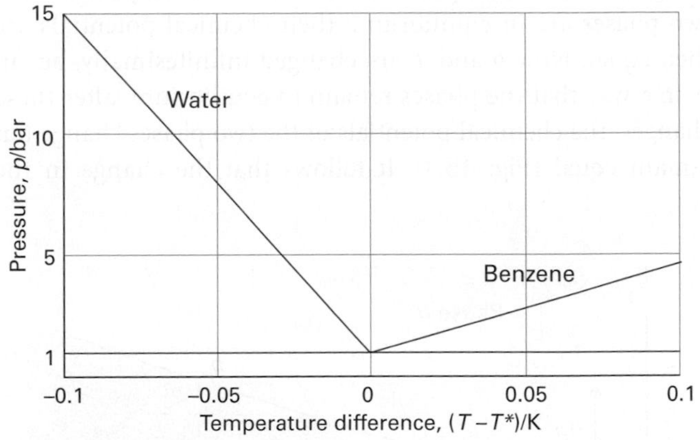

Figure 4B.6 The solid–liquid phase boundary (the melting point curve) for water as calculated in *Brief illustration* 4B.4. For comparison, the boundary for benzene is included.

(c) The liquid–vapour boundary

The entropy of vaporization at a temperature T is equal to $\Delta_{vap}H/T$ (as before, all points on the phase boundary correspond to equilibrium, so T is a transition temperature, T_{trs}), so the Clapeyron equation for the liquid–vapour boundary can therefore be written

$$\frac{dp}{dT}=\frac{\Delta_{vap}H}{T\Delta_{vap}V} \quad \text{Slope of liquid–vapour boundary} \quad (4B.8)$$

The enthalpy of vaporization is positive and $\Delta_{vap}V$ is large and positive, so dp/dT is positive, but much smaller than for the solid–liquid boundary. Consequently dT/dp is large, and the boiling temperature is more responsive to pressure than the freezing temperature.

Example 4B.2 Estimating the effect of pressure on the boiling temperature

Estimate the typical size of the effect of increasing pressure on the boiling point of a liquid.

Collect your thoughts To use eqn 4B.8 you need to estimate the right-hand side. At the boiling point, the term $\Delta_{vap}H/T$ is Trouton's constant (Topic 3B). Because the molar volume of a gas is so much greater than the molar volume of a liquid, you can write $\Delta_{vap}V=V_m(g)-V_m(l)\approx V_m(g)$ and take for $V_m(g)$ the molar volume of a perfect gas (at low pressures, at least). You will need to use $1\,J = 1\,Pa\,m^3$.

The solution Trouton's constant has the value $85\,J\,K^{-1}\,mol^{-1}$. The molar volume of a perfect gas is about $25\,dm^3\,mol^{-1}$ at 1 atm and near but above room temperature. Therefore,

$$\frac{dp}{dT}\approx\frac{85\,J\,K^{-1}\,mol^{-1}}{2.5\times10^{-2}\,m^3\,mol^{-1}}=3.4\times10^3\,Pa\,K^{-1}$$

This value corresponds to $0.034\,atm\,K^{-1}$ and hence to $dT/dp = 29\,K\,atm^{-1}$. Therefore, a change of pressure of +0.1 atm can be expected to change a boiling temperature by about +3 K.

Self-test 4B.2 Estimate dT/dp for water at its normal boiling point using the information in Table 3B.2 and $V_m(g) = RT/p$.

Answer: $28\,K\,atm^{-1}$

Because the molar volume of a gas is so much greater than the molar volume of a liquid, $\Delta_{vap}V \approx V_m(g)$ (as in Example 4B.2). Moreover, if the gas behaves perfectly, $V_m(g) = RT/p$. These two approximations turn the exact Clapeyron equation into

$$\frac{dp}{dT}=\frac{\Delta_{vap}H}{T(RT/p)}=\frac{p\Delta_{vap}H}{RT^2}$$

By using $dx/x = d\ln x$, this expression can be rearranged into the **Clausius–Clapeyron equation** for the variation of vapour pressure with temperature:

$$\frac{d\ln p}{dT}=\frac{\Delta_{vap}H}{RT^2} \quad \text{Clausius–Clapeyron equation} \quad (4B.9)$$

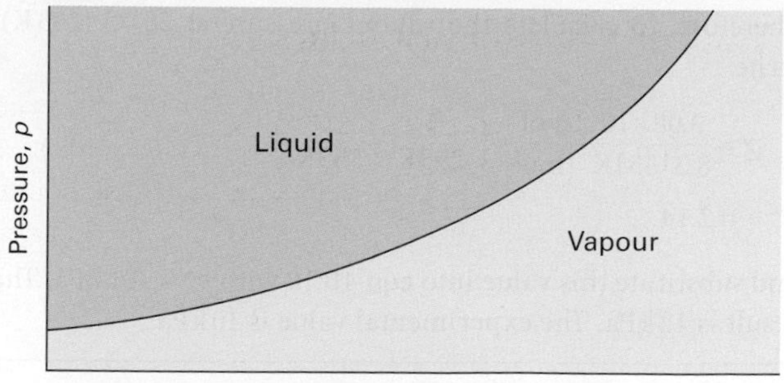

Figure 4B.7 A typical liquid–vapour phase boundary. The boundary can be interpreted as a plot of the vapour pressure against the temperature. This phase boundary terminates at the critical point (not shown).

Like the Clapeyron equation (which is exact), the Clausius–Clapeyron equation (which is an approximation) is important for understanding the appearance of phase diagrams, particularly the location and shape of the liquid–vapour and solid–vapour phase boundaries. It can be used to predict how the vapour pressure varies with temperature and how the boiling temperature varies with pressure. For instance, if it is also assumed that the enthalpy of vaporization is independent of temperature, eqn 4B.9 can be integrated as follows:

$$\overbrace{\int_{\ln p^*}^{\ln p} d\ln p}^{\text{Integral A.1, with } x=\ln p}=\frac{\Delta_{vap}H}{R}\overbrace{\int_{T^*}^{T}\frac{dT}{T^2}}^{\text{Integral A.1}}$$

hence

$$\ln\frac{p}{p^*}=-\frac{\Delta_{vap}H}{R}\left(\frac{1}{T}-\frac{1}{T^*}\right)$$

where p^* is the vapour pressure when the temperature is T^*, and p the vapour pressure when the temperature is T. It follows that

$$p=p^*e^{-\chi} \quad \chi=\frac{\Delta_{vap}H}{R}\left(\frac{1}{T}-\frac{1}{T^*}\right) \quad (4B.10)$$

Equation 4B.10 is plotted as the liquid–vapour boundary in Fig. 4B.7. The line does not extend beyond the critical temperature, T_c, because above this temperature the liquid does not exist.

Brief illustration 4B.5

Equation 4B.10 can be used to estimate the vapour pressure of a liquid at any temperature from knowledge of its normal boiling point, the temperature at which the vapour pressure is 1.00 atm (101 kPa). The normal boiling point of benzene is 80 °C (353 K) and (from Table 3B.2) $\Delta_{vap}H^{\ominus} = 30.8\,kJ\,mol^{-1}$.

Therefore, to calculate the vapour pressure at 20 °C (293 K), write

$$\chi = \frac{3.08\times10^4\,\mathrm{J\,mol^{-1}}}{8.3145\,\mathrm{J\,K^{-1}\,mol^{-1}}}\left(\frac{1}{293\,\mathrm{K}}-\frac{1}{353\,\mathrm{K}}\right)$$
$$=2.14\ldots$$

and substitute this value into eqn 4B.10 with $p^* = 101$ kPa. The result is 12 kPa. The experimental value is 10 kPa.

A note on good practice Because exponential functions are so sensitive, it is good practice to carry out numerical calculations like this without evaluating the intermediate steps and using rounded values.

(d) The solid–vapour boundary

The only difference between the solid–vapour and the liquid–vapour boundary is the replacement of the enthalpy of vaporization by the enthalpy of sublimation, $\Delta_{\mathrm{sub}}H$. Because the enthalpy of sublimation is greater than the enthalpy of vaporization (recall that $\Delta_{\mathrm{sub}}H = \Delta_{\mathrm{fus}}H + \Delta_{\mathrm{vap}}H$), at similar temperatures the equation predicts a steeper slope for the sublimation curve than for the vaporization curve. These two boundaries meet at the triple point (Fig. 4B.8).

Brief illustration 4B.6

The enthalpy of fusion of ice at the triple point of water (6.1 mbar, 273 K) is negligibly different from its standard enthalpy of fusion at its freezing point, which is 6.008 kJ mol^{-1}. The enthalpy of vaporization at that temperature is 45.0 kJ mol^{-1} (once again, ignoring differences due to the pressure not being 1 bar). The enthalpy of sublimation is therefore 51.0 kJ mol^{-1}. Therefore, the equations for the slopes of (a) the liquid–vapour and (b) the solid–vapour phase boundaries at the triple point are

$$\text{(a)}\ \frac{\mathrm{d}\ln p}{\mathrm{d}T} = \frac{45.0\times10^3\,\mathrm{J\,mol^{-1}}}{(8.3145\,\mathrm{J\,K^{-1}\,mol^{-1}})\times(273\,\mathrm{K})^2} = 0.0726\,\mathrm{K^{-1}}$$

$$\text{(b)}\ \frac{\mathrm{d}\ln p}{\mathrm{d}T} = \frac{51.0\times10^3\,\mathrm{J\,mol^{-1}}}{(8.3145\,\mathrm{J\,K^{-1}\,mol^{-1}})\times(273\,\mathrm{K})^2} = 0.0823\,\mathrm{K^{-1}}$$

The slope of ln p plotted against T is greater for the solid–vapour boundary than for the liquid–vapour boundary at the triple point.

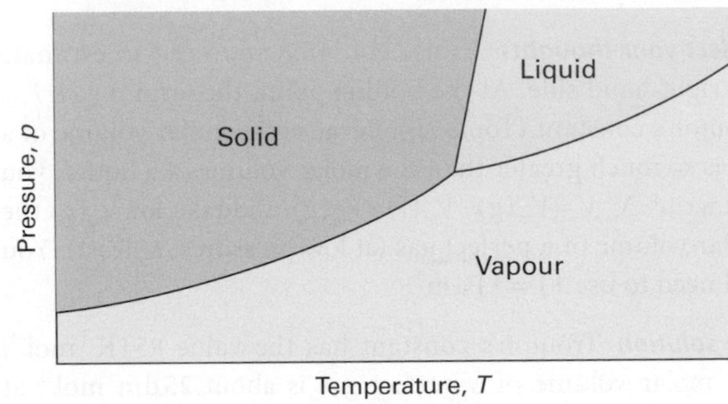

Figure 4B.8 At temperatures close to the triple point the solid–vapour boundary is steeper than the liquid–vapour boundary because the enthalpy of sublimation is greater than the enthalpy of vaporization.

Checklist of concepts

☐ 1. The chemical potential of a substance decreases with increasing temperature in proportion to its molar entropy.

☐ 2. The chemical potential of a substance increases with increasing pressure in proportion to its molar volume.

☐ 3. The vapour pressure of a condensed phase increases when pressure is applied.

☐ 4. The **Clapeyron equation** is an exact expression for the slope of a phase boundary.

☐ 5. The **Clausius–Clapeyron equation** is an approximate expression for the boundary between a condensed phase and its vapour.

Checklist of equations

Property	Equation	Comment	Equation number
Variation of μ with temperature	$(\partial\mu/\partial T)_p = -S_{\mathrm{m}}$	$\mu = G_{\mathrm{m}}$	4B.1a
Variation of μ with pressure	$(\partial\mu/\partial p)_T = V_{\mathrm{m}}$		4B.1b
Vapour pressure in the presence of applied pressure	$p = p^* e^{V_{\mathrm{m}}(\mathrm{l})\Delta P/RT}$	$\Delta P = P - p^*$	4B.2
Clapeyron equation	$\mathrm{d}p/\mathrm{d}T = \Delta_{\mathrm{trs}}S/\Delta_{\mathrm{trs}}V$		4B.4a
Clausius–Clapeyron equation	$\mathrm{d}\ln p/\mathrm{d}T = \Delta_{\mathrm{vap}}H/RT^2$	Assumes $V_{\mathrm{m}}(\mathrm{g}) >> V_{\mathrm{m}}(\mathrm{l})$ or $V_{\mathrm{m}}(\mathrm{s})$, and vapour is a perfect gas	4B.9

FOCUS 4 Physical transformations of pure substances

TOPIC 4A Phase diagrams of pure substances

Discussion questions

D4A.1 Describe how the concept of chemical potential unifies the discussion of phase equilibria.

D4A.2 Why does the chemical potential change with pressure even if the system is incompressible (i.e. remains at the same volume when pressure is applied)?

D4A.3 Explain why four phases cannot be in equilibrium in a one-component system.

D4A.4 Discuss what would be observed as a sample of water is taken along a path that encircles and is close to its critical point.

Exercises

E4A.1(a) How many phases are present at each of the points *a–d* indicated in Fig. 4.1a?

E4A.1(b) How many phases are present at each of the points *a–d* indicated in Fig. 4.1b?

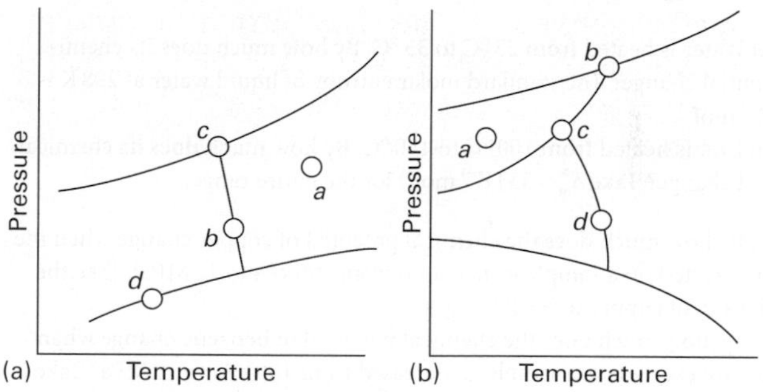

Figure 4.1 The phase diagrams referred to in (a) Exercise 4A.1(a) and (b) Exercise 4A.1(b).

E4A.2(a) The difference in chemical potential of a particular substance between two regions of a system is $+7.1\,\mathrm{kJ\,mol^{-1}}$. By how much does the Gibbs energy change when 0.10 mmol of that substance is transferred from one region to the other?

E4A.2(b) The difference in chemical potential of a particular substance between two regions of a system is $-8.3\,\mathrm{kJ\,mol^{-1}}$. By how much does the Gibbs energy change when 0.15 mmol of that substance is transferred from one region to the other?

E4A.3(a) What is the maximum number of phases that can be in mutual equilibrium in a two-component system?

E4A.3(b) What is the maximum number of phases that can be in mutual equilibrium in a four-component system?

E4A.4(a) In a one-component system, is the condition $P = 1$ represented on a phase diagram by an area, a line or a point? How do you interpret this value of P?

E4A.4(b) In a one-component system, is the condition $P = 2$ represented on a phase diagram by an area, a line or a point? How do you interpret this value of P?

E4A.5(a) Refer to Fig. 4A.8. Which phase or phases would you expect to be present for a sample of CO_2 at: (i) 200 K and 2.5 atm; (ii) 300 K and 4 atm; (iii) 310 K and 50 atm?

E4A.5(b) Refer to Fig. 4A.9. Which phase or phases would you expect to be present for a sample of H_2O at: (i) 100 K and 1 atm; (ii) 300 K and 10 atm; (iii) 273.16 K and 611 Pa?

Problems

P4A.1 Refer to Fig. 4A.8. Describe the phase or phases present as a sample of CO_2 is heated steadily from 100 K: (a) at a constant pressure of 1 atm; (b) at a constant pressure of 70 atm.

P4A.2 Refer to Fig. 4A.8. Describe the phase or phases present as the pressure on a sample of CO_2 is steadily increased from 0.1 atm: (a) at a constant temperature of 200 K; (b) at a constant temperature of 310 K; (c) at a constant temperature of 216.8 K.

P4A.3 For a one-component system draw a schematic labelled phase diagram given that at low T and low p, only phase γ is present; at low T and high p, only phase β is present; at high T and low p, only phase α is present; at high T and high p, only phase δ is present; phases γ and δ are never in equilibrium. Comment on any special features of your diagram.

P4A.4 For a one-component system draw a schematic labelled phase diagram given that at low T and low p, phases α and β are in equilibrium; as the temperature and pressure rise there comes a point at which phases α, β, and γ are all in equilibrium; at high T and high p, only phase γ is present; at low T and high p, only phase α is present. Comment on any special features of your diagram.

TOPIC 4B Thermodynamic aspects of phase transitions

Discussion questions

D4B.1 What is the physical reason for the decrease of the chemical potential of a pure substance as the temperatures is raised?

D4B.2 What is the physical reason for the increase of the chemical potential of a pure substance as the pressure is raised?

D4B.3 How may differential scanning calorimetry (DSC) be used to identify phase transitions?

Exercises

E4B.1(a) The standard molar entropy of liquid water at 273.15 K is 65 J K^{-1} mol^{-1}, and that of ice at the same temperature is 43 J K^{-1} mol^{-1}. Calculate the change in chemical potential of liquid water and of ice when the temperature is increased by 1 K from the normal melting point. Giving your reasons, explain which phase is thermodynamically the more stable at the new temperature.

E4B.1(b) Repeat the calculation in Exercise E4B.1(a) but for a decrease in temperature by 1.5 K. Giving your reasons, explain which phase is thermodynamically the more stable at the new temperature.

E4B.2(a) Water is heated from 25 °C to 35 °C. By how much does its chemical potential change? The standard molar entropy of liquid water at 298 K is 69.9 J K^{-1} mol^{-1}.

E4B.2(b) Iron is heated from 100 °C to 150 °C. By how much does its chemical potential change? Take $S_m^{\ominus}$ = 53 J K^{-1} mol^{-1} for the entire range.

E4B.3(a) By how much does the chemical potential of copper change when the pressure exerted on a sample is increased from 100 kPa to 10 MPa? Take the mass density of copper to be 8960 kg m^{-3}.

E4B.3(b) By how much does the chemical potential of benzene change when the pressure exerted on a sample is increased from 100 kPa to 10 MPa? Take the mass density of benzene to be 0.8765 g cm^{-3}.

E4B.4(a) Pressure was exerted with a piston on water at 20 °C. The vapour pressure of water when the applied pressure is 1.0 bar is 2.34 kPa. What is its vapour pressure when the pressure on the liquid is 20 MPa? The molar volume of water is 18.1 cm^3 mol^{-1} at 20 °C.

E4B.4(b) Pressure was exerted with a piston on molten naphthalene at 95 °C. The vapour pressure of naphthalene when the applied pressure is 1.0 bar is 2.0 kPa. What is its vapour pressure when the pressure on the liquid is 15 MPa? The mass density of naphthalene at this temperature is 1.16 g cm^{-3}.

E4B.5(a) The molar volume of a certain solid is 161.0 cm^3 mol^{-1} at 1.00 atm and 350.75 K, its melting temperature. The molar volume of the liquid at this temperature and pressure is 163.3 cm^3 mol^{-1}. At 100 atm the melting temperature changes to 351.26 K. Calculate the enthalpy and entropy of fusion of the solid.

E4B.5(b) The molar volume of a certain solid is 142.0 cm^3 mol^{-1} at 1.00 atm and 427.15 K, its melting temperature. The molar volume of the liquid at this temperature and pressure is 152.6 cm^3 mol^{-1}. At 1.2 MPa the melting temperature changes to 429.26 K. Calculate the enthalpy and entropy of fusion of the solid.

E4B.6(a) The vapour pressure of dichloromethane at 24.1 °C is 53.3 kPa and its enthalpy of vaporization is 28.7 kJ mol^{-1}. Estimate the temperature at which its vapour pressure is 70.0 kPa.

E4B.6(b) The vapour pressure of a substance at 20.0 °C is 58.0 kPa and its enthalpy of vaporization is 32.7 kJ mol^{-1}. Estimate the temperature at which its vapour pressure is 66.0 kPa.

E4B.7(a) The vapour pressure of a liquid in the temperature range 200–260 K was found to fit the expression $\ln(p/\text{Torr}) = 16.255 - (2501.8\,\text{K})/T$. What is the enthalpy of vaporization of the liquid?

E4B.7(b) The vapour pressure of a liquid in the temperature range 200–260 K was found to fit the expression $\ln(p/\text{Torr}) = 18.361 - (3036.8\,\text{K})/T$. What is the enthalpy of vaporization of the liquid?

E4B.8(a) The vapour pressure of benzene between 10 °C and 30 °C fits the expression $\log(p/\text{Torr}) = 7.960 - (1780\,\text{K})/T$. Calculate (i) the enthalpy of vaporization and (ii) the normal boiling point of benzene.

E4B.8(b) The vapour pressure of a liquid between 15 °C and 35 °C fits the expression $\log(p/\text{Torr}) = 8.750 - (1625\,\text{K})/T$. Calculate (i) the enthalpy of vaporization and (ii) the normal boiling point of the liquid.

E4B.9(a) When benzene freezes at 1 atm and at 5.5 °C its mass density changes from 0.879 g cm^{-3} to 0.891 g cm^{-3}. The enthalpy of fusion is 10.59 kJ mol^{-1}. Estimate the freezing point of benzene at 1000 atm.

E4B.9(b) When a certain liquid (with M = 46.1 g mol^{-1}) freezes at 1 bar and at −3.65 °C its mass density changes from 0.789 g cm^{-3} to 0.801 g cm^{-3}. Its enthalpy of fusion is 8.68 kJ mol^{-1}. Estimate the freezing point of the liquid at 100 MPa.

E4B.10(a) Estimate the difference between the normal and standard melting points of ice. At the normal melting point, the enthalpy of fusion of water is 6.008 kJ mol^{-1}, and the change in molar volume on fusion is −1.6 cm^3 mol^{-1}.

E4B.10(b) Estimate the difference between the normal and standard boiling points of water. At the normal boiling point the enthalpy of vaporization of water is 40.7 kJ mol^{-1}.

E4B.11(a) In July in Los Angeles, the incident sunlight at ground level has a power density of 1.2 kW m^{-2} at noon. A swimming pool of area 50 m^2 is directly exposed to the Sun. What is the maximum rate of loss of water? Assume that all the radiant energy is absorbed; take the enthalpy of vaporization of water to be 44 kJ mol^{-1}.

E4B.11(b) Suppose the incident sunlight at ground level has a power density of 0.87 kW m^{-2} at noon. What is the maximum rate of loss of water from a lake of area 1.0 ha? (1 ha = 10^4 m^2.) Assume that all the radiant energy is absorbed; take the enthalpy of vaporization of water to be 44 kJ mol^{-1}.

E4B.12(a) An open vessel containing water stands in a laboratory measuring 5.0 m × 5.0 m × 3.0 m at 25 °C; the vapour pressure of water at this temperature is 3.2 kPa. When the system has come to equilibrium, what mass of water will be found in the air if there is no ventilation? Repeat the calculation for open vessels containing benzene (vapour pressure 13.1 kPa) and mercury (vapour pressure 0.23 Pa).

E4B.12(b) On a cold, dry morning after a frost, the temperature was −5 °C and the partial pressure of water in the atmosphere fell to 0.30 kPa. Will the frost sublime? The enthalpy of sublimation of water is 51 kJ mol^{-1}. (*Hint:* Use eqn 4B.10 to calculate the vapour pressure expected for ice at this temperature; for p^* and T^* use the values for the triple point of 611 Pa and 273.16 K.)

E4B.13(a) Naphthalene, $C_{10}H_8$, melts at 80.2 °C. If the vapour pressure of the liquid is 1.3 kPa at 85.8 °C and 5.3 kPa at 119.3 °C, use the Clausius–Clapeyron equation to calculate (i) the enthalpy of vaporization, (ii) the normal boiling point, and (iii) the entropy of vaporization at the boiling point.

E4B.13(b) The normal boiling point of hexane is 69.0 °C. Estimate (i) its enthalpy of vaporization and (ii) its vapour pressure at 25 °C and at 60 °C. (*Hint:* You will need to use Trouton's rule.)

E4B.14(a) Estimate the melting point of ice under a pressure of 50 bar. Assume that the mass density of ice under these conditions is approximately $0.92\,g\,cm^{-3}$ and that of liquid water is $1.00\,g\,cm^{-3}$. The enthalpy of fusion of water is $6.008\,kJ\,mol^{-1}$ at the normal melting point.

E4B.14(b) Estimate the melting point of ice under a pressure of 10 MPa. Assume that the mass density of ice under these conditions is approximately $0.915\,g\,cm^{-3}$ and that of liquid water is $0.998\,g\,cm^{-3}$. The enthalpy of fusion of water is $6.008\,kJ\,mol^{-1}$ at the normal melting point.

Problems

P4B.1 Imagine the vaporization of 1 mol $H_2O(l)$ at the normal boiling point and against 1 atm external pressure. Calculate the work done by the water vapour and hence what fraction of the enthalpy of vaporization is spent on expanding the vapour. The enthalpy of vaporization of water is $40.7\,kJ\,mol^{-1}$ at the normal boiling point.

P4B.2 The temperature dependence of the vapour pressure of solid sulfur dioxide can be approximately represented by the relation $\log(p/Torr) = 10.5916 - (1871.2\,K)/T$ and that of liquid sulfur dioxide by $\log(p/Torr) = 8.3186 - (1425.7\,K)/T$. Estimate the temperature and pressure of the triple point of sulfur dioxide.

P4B.3 Prior to the discovery that freon-12 (CF_2Cl_2) is harmful to the Earth's ozone layer it was frequently used as the dispersing agent in spray cans for hair spray etc. Estimate the pressure that a can of hair spray using freon-12 has to withstand at 40 °C, the temperature of a can that has been standing in sunlight. The enthalpy of vaporization of freon-12 at its normal boiling point of −29.2 °C is $20.25\,kJ\,mol^{-1}$; assume that this value remains constant over the temperature range of interest.

P4B.4 The enthalpy of vaporization of a certain liquid is found to be $14.4\,kJ\,mol^{-1}$ at 180 K, its normal boiling point. The molar volumes of the liquid and the vapour at the boiling point are $115\,cm^3\,mol^{-1}$ and $14.5\,dm^3\,mol^{-1}$, respectively. (a) Use the Clapeyron equation to estimate dp/dT at the normal boiling point. (b) If the Clausius–Clapeyron equation is used instead to estimate dp/dT, what is the percentage error in the resulting value of dp/dT?

P4B.5 Calculate the difference in slope of the chemical potential against temperature on either side of (a) the normal freezing point of water and (b) the normal boiling point of water. The molar entropy change accompanying fusion is $22.0\,J\,K^{-1}\,mol^{-1}$ and that accompanying evaporation is $109.9\,J\,K^{-1}\,mol^{-1}$. (c) By how much does the chemical potential of water supercooled to −5.0 °C exceed that of ice at that temperature?

P4B.6 Calculate the difference in slope of the chemical potential against pressure on either side of (a) the normal freezing point of water and (b) the normal boiling point of water. The mass densities of ice and water at 0 °C are $0.917\,g\,cm^{-3}$ and $1.000\,g\,cm^{-3}$, and those of water and water vapour at 100 °C are $0.958\,g\,cm^{-3}$ and $0.598\,g\,dm^{-3}$, respectively. (c) By how much does the chemical potential of water vapour exceed that of liquid water at 1.2 atm and 100 °C?

P4B.7 The enthalpy of fusion of mercury is $2.292\,kJ\,mol^{-1}$ at its normal freezing point of 234.3 K; the change in molar volume on melting is $+0.517\,cm^3\,mol^{-1}$. At what temperature will the bottom of a column of mercury (mass density $13.6\,g\,cm^{-3}$) of height 10.0 m be expected to freeze? The pressure at a depth d in a fluid with mass density ρ is ρgd, where g is the acceleration of free fall, $9.81\,m\,s^{-2}$.

P4B.8 Suppose $50.0\,dm^3$ of dry air at 25 °C was slowly bubbled through a thermally insulated beaker containing 250 g of water initially at 25 °C. Calculate the final temperature of the liquid. The vapour pressure of water is approximately constant at 3.17 kPa throughout, and the heat capacity of the liquid is $75.5\,J\,K^{-1}\,mol^{-1}$. Assume that the exit gas remains at 25 °C and that water vapour is a perfect gas. The standard enthalpy of vaporization of water at 25 °C is $44.0\,kJ\,mol^{-1}$. (*Hint:* Start by calculating the amount in moles of H_2O in the $50.0\,dm^3$ of air after it has bubbled through the liquid.)

P4B.9 The vapour pressure, p, of nitric acid varies with temperature as follows:

θ/°C	0	20	40	50	70	80	90	100
p/kPa	1.92	6.38	17.7	27.7	62.3	89.3	124.9	170.9

Determine (a) the normal boiling point and (b) the enthalpy of vaporization of nitric acid.

P4B.10 The vapour pressure of carvone ($M = 150.2\,g\,mol^{-1}$), a component of oil of spearmint, is as follows:

θ/°C	57.4	100.4	133.0	157.3	203.5	227.5
p/Torr	1.00	10.0	40.0	100	400	760

Determine (a) the normal boiling point and (b) the enthalpy of vaporization of carvone.

P4B.11‡ (a) Starting from the Clapeyron equation, derive an expression, analogous to the Clausius–Clapeyron equation, for the temperature variation of the vapour pressure of a solid. Assume that the vapour is a perfect gas and that the molar volume of the solid is negligible in comparison to that of the gas. (b) In a study of the vapour pressure of chloromethane, A. Bah and N. Dupont-Pavlovsky (*J. Chem. Eng. Data* **40**, 869 (1995)) presented data for the vapour pressure over solid chloromethane at low temperatures. Some of that data is as follows:

T/K	145.94	147.96	149.93	151.94	153.97	154.94
p/Pa	13.07	18.49	25.99	36.76	50.86	59.56

Estimate the standard enthalpy of sublimation of chloromethane at 150 K.

P4B.12 The change in enthalpy dH resulting from a change in pressure dp and temperature dT is given by $dH = C_p dT + V dp$. The Clapeyron equation relates dp and dT at equilibrium, and so in combination the two equations can be used to find how the enthalpy changes along a phase boundary as the temperature changes and the two phases remain in equilibrium. (a) Show that along such a boundary $d\Delta_{trs}H = \Delta_{trs}C_p dT + (\Delta_{trs}H/T)dT$, where $\Delta_{trs}H$ is the enthalpy of transition and $\Delta_{trs}C_p$ the difference of molar heat capacity accompanying the transition. (b) Show that this expression can also be written $d(\Delta_{trs}H/T) = \Delta_{trs}C_p d\ln T$. (*Hint:* The last part is most easily approached by starting with the second expression and showing that it can be rewritten as the first.)

P4B.13 In the 'gas saturation method' for the measurement of vapour pressure, a volume V of gas at temperature T and pressure P, is bubbled slowly through the liquid that is maintained at the same temperature T. The mass m lost from the liquid is measured and this can be related to the vapour pressure in the following way. (a) If the molar mass of the liquid is M, derive an expression for the mole fraction of the liquid vapour. (*Hint:* If it is assumed to be a perfect gas, the amount in moles of the input gas can be found from its pressure, temperature and volume.) (b) Hence derive an expression for the partial pressure of the liquid vapour, assuming that the gas remains at the total pressure P after it has passed through the liquid. (c) Then show that the

‡These problems were supplied by Charles Trapp and Carmen Giunta.

vapour pressure p is given by $p = AmP/(1 + Am)$, where $A = RT/MPV$. (d) The gas saturation method was used to measure the vapour pressure of geraniol ($M = 154.2\,\mathrm{g\,mol^{-1}}$) at 110 °C. It was found that, when 5.00 dm^3 of nitrogen at 760 Torr was passed slowly through the heated liquid, the loss of mass was 0.32 g. Calculate the vapour pressure of geraniol.

P4B.14 The vapour pressure of a liquid in a gravitational field varies with the depth below the surface on account of the hydrostatic pressure exerted by the overlying liquid. The pressure at a depth d in a fluid with mass density ρ is ρgd, where g is the acceleration of free fall (9.81 m s^{-2}). Use this relation to adapt eqn 4B.2 to predict how the vapour pressure of a liquid of molar mass M varies with depth. Estimate the effect on the vapour pressure of water at 25 °C in a column 10 m high.

P4B.15 The 'barometric formula', $p = p_0 e^{-a/H}$, where $H = 8$ km, gives the dependence of the pressure p on the altitude, a; p_0 is the pressure at sea level, assumed to be 1 atm. Use this expression together with the Clausius–Clapeyron equation to derive an expression for how the boiling temperature of a liquid depends on the altitude (*Hint:* The boiling point is when the vapour pressure is equal to the external pressure.) Use your result to predict the boiling temperature of water at 3000 m. The normal boiling point of water is 373.15 K and you may take that the standard enthalpy of vaporization as 40.7 kJ mol^{-1}.

P4B.16 Figure 4B.1 gives a schematic representation of how the chemical potentials of the solid, liquid, and gaseous phases of a substance vary with temperature. All have a negative slope, but it is unlikely that they are straight lines as indicated in the illustration. Derive an expression for the curvatures, that is, the second derivative of the chemical potential with respect to temperature, of these lines. Is there any restriction on the value this curvature can take? For water, compare the curvature of the liquid line with that for the gas in the region of the normal boiling point. The molar heat capacities at constant pressure of the liquid and gas are 75.3 J K^{-1} mol^{-1} and 33.6 J K^{-1} mol^{-1}, respectively.

FOCUS 4 Physical transformations of pure substances

Integrated activities

I4.1 Construct the phase diagram for benzene near its triple point at 36 Torr and 5.50 °C from the following data: $\Delta_{\mathrm{fus}}H = 10.6\,\mathrm{kJ\,mol^{-1}}$, $\Delta_{\mathrm{vap}}H = 30.8\,\mathrm{kJ\,mol^{-1}}$, $\rho(\mathrm{s}) = 0.891\,\mathrm{g\,cm^{-3}}$, $\rho(\mathrm{l}) = 0.879\,\mathrm{g\,cm^{-3}}$.

I4.2‡ In an investigation of thermophysical properties of methylbenzene R.D. Goodwin (*J. Phys. Chem. Ref. Data* **18**, 1565 (1989)) presented expressions for two phase boundaries. The solid–liquid boundary is given by

$$p/\mathrm{bar} = p_3/\mathrm{bar} + 1000(5.60 + 11.727x)x$$

where $x = T/T_3 - 1$ and the triple point pressure and temperature are $p_3 = 0.4362\,\mu\mathrm{bar}$ and $T_3 = 178.15$ K. The liquid–vapour curve is given by

$$\ln(p/\mathrm{bar}) = -10.418/y + 21.157 - 15.996y + 14.015y^2 - 5.0120y^3 + 4.7334(1-y)^{1.70}$$

where $y = T/T_c = T/(593.95\,\mathrm{K})$. (a) Plot the solid–liquid and liquid–vapour phase boundaries. (b) Estimate the standard melting point of methylbenzene. (c) Estimate the standard boiling point of methylbenzene. (The equation you will need to solve to find this quantity cannot be solved by hand, so you should use a numerical approach, e.g. by using mathematical software.) (d) Calculate the standard enthalpy of vaporization of methylbenzene at the standard boiling point, given that the molar volumes of the liquid and vapour at the standard boiling point are 0.12 dm^3 mol^{-1} and 30.3 dm^3 mol^{-1}, respectively.

I4.3 Proteins are polymers of amino acids that can exist in ordered structures stabilized by a variety of molecular interactions. However, when certain conditions are changed, the compact structure of a polypeptide chain may collapse into a random coil. This structural change may be regarded as a phase transition occurring at a characteristic transition temperature, the *melting temperature*, T_m, which increases with the strength and number of intermolecular interactions in the chain. A thermodynamic treatment allows predictions to be made of the temperature T_m for the unfolding of a helical polypeptide held together by hydrogen bonds into a random coil. If a polypeptide has N amino acid residues, $N - 4$ hydrogen bonds are formed to form an α-helix, the most common type of helix in naturally occurring proteins (see Topic 14D). Because the first and last residues in the chain are free to move, $N - 2$ residues form the compact helix and have restricted motion. Based on these ideas, the molar Gibbs energy of unfolding of a polypeptide with $N \geq 5$ may be written as

$$\Delta_{\mathrm{unfold}}G = (N-4)\Delta_{\mathrm{hb}}H - (N-2)T\Delta_{\mathrm{hb}}S$$

where $\Delta_{\mathrm{hb}}H$ and $\Delta_{\mathrm{hb}}S$ are, respectively, the molar enthalpy and entropy of dissociation of hydrogen bonds in the polypeptide. (a) Justify the form of the equation for the Gibbs energy of unfolding. That is, why are the enthalpy and entropy terms written as $(N-4)\Delta_{\mathrm{hb}}H$ and $(N-2)\Delta_{\mathrm{hb}}S$, respectively? (b) Show that T_m may be written as

$$T_m = \frac{(N-4)\Delta_{\mathrm{hb}}H}{(N-2)\Delta_{\mathrm{hb}}S}$$

(c) Plot $T_m/(\Delta_{\mathrm{hb}}H_m/\Delta_{\mathrm{hb}}S_m)$ for $5 \leq N \leq 20$. At what value of N does T_m change by less than 1 per cent when N increases by 1?

I4.4‡ A substance as well-known as methane still receives research attention because it is an important component of natural gas, a commonly used fossil fuel. Friend et al. have published a review of thermophysical properties of methane (D.G. Friend, J.F. Ely, and H. Ingham, *J. Phys. Chem. Ref. Data* **18**, 583 (1989)), which included the following vapour pressure data describing the liquid–vapour phase boundary.

T/K	100	108	110	112	114	120	130	140	150	160	170	190
p/MPa	0.034	0.074	0.088	0.104	0.122	0.192	0.368	0.642	1.041	1.593	2.329	4.521

(a) Plot the liquid–vapour phase boundary. (b) Estimate the standard boiling point of methane. (c) Compute the standard enthalpy of vaporization of methane (at the standard boiling point), given that the molar volumes of the liquid and vapour at the standard boiling point are 3.80×10^{-2} dm^3 mol^{-1} and 8.89 dm^3 mol^{-1}, respectively.

I4.5‡ Diamond is the hardest substance and the best conductor of heat yet characterized. For these reasons, it is used widely in industrial applications that require a strong abrasive. Unfortunately, it is difficult to synthesize diamond from the more readily available allotropes of carbon, such as graphite. To illustrate this point, the following approach can be used to estimate the pressure required to convert graphite into diamond at 25 °C (i.e. the pressure at which the conversion becomes spontaneous). The aim is to find an expression for $\Delta_r G$ for the process graphite → diamond as a function of the applied pressure, and then to determine the pressure at which the Gibbs energy change becomes negative. (a) Derive the following expression for the pressure variation of $\Delta_r G$ at constant temperature

$$\left(\frac{\partial \Delta_r G}{\partial p}\right)_T = V_{\mathrm{m,d}} - V_{\mathrm{m,gr}}$$

where $V_{\mathrm{m,gr}}$ is the molar volume of graphite and $V_{\mathrm{m,d}}$ that of diamond. (b) The difficulty with dealing with the previous expression is that the V_m depend on the pressure. This dependence is handled as follows. Consider $\Delta_r G$ to be a function of pressure and form a Taylor expansion about $p = p^\ominus$:

$$\Delta_r G(p) = \Delta_r G(p^\ominus) + \overbrace{\left(\frac{\partial \Delta_r G}{\partial p}\right)_{p=p^\ominus}}^{\text{A}}(p-p^\ominus) + \frac{1}{2}\overbrace{\left(\frac{\partial^2 \Delta_r G}{\partial p^2}\right)_{p=p^\ominus}}^{\text{B}}(p-p^\ominus)^2$$

where the derivatives are evaluated at $p = p^\ominus$ and the series is truncated after the second-order term. Term A can be found from the expression in part (a) by using the molar volumes at $p^\ominus$. Term B can be found by using a knowledge of the isothermal compressibility of the solids, $\kappa_T = -(1/V)(\partial V/\partial p)_T$. Use this definition to show that at constant temperature

$$\frac{\partial^2 \Delta_r G}{\partial p^2} = \frac{\partial}{\partial p}(V_{\mathrm{m,d}} - V_{\mathrm{m,gr}}) = -\kappa_{T,\mathrm{d}} V_{\mathrm{m,d}} + \kappa_{T,\mathrm{gr}} V_{\mathrm{m,gr}}$$

where $\kappa_{T,\mathrm{d}}$ and $\kappa_{T,\mathrm{gr}}$ are the isothermal compressibilities of diamond and graphite, respectively. (c) Substitute the results from (a) and (b) into the expression for $\Delta_r G(p)$ in (b) and hence obtain an expression for $\Delta_r G(p)$ in terms of the isothermal compressibilities and molar volumes under standard conditions. (d) At 1 bar and 298 K the value of $\Delta_r G$ for the transition graphite → diamond is +2.8678 kJ mol^{-1}. Use the following data to estimate the pressure at which this transformation becomes spontaneous. Assume that κ_T is independent of pressure.

	Graphite	Diamond
$V_s/(\mathrm{cm^3\,g^{-1}})$ at 1 bar	0.444	0.284
$\kappa_T/\mathrm{kPa^{-1}}$	3.04×10^{-8}	0.187×10^{-8}

FOCUS 5

Simple mixtures

Mixtures are an essential part of chemistry, either in their own right or as starting materials for chemical reactions. This group of Topics deals with the rich physical properties of mixtures and shows how to express them in terms of thermodynamic quantities.

5A The thermodynamic description of mixtures

The first Topic in this Focus develops the concept of chemical potential as an example of a partial molar quantity and explores how to use the chemical potential of a substance to describe the physical properties of mixtures. The underlying principle to keep in mind is that at equilibrium the chemical potential of a species is the same in every phase. By making use of the experimental observations known as Raoult's and Henry's laws, it is possible to express the chemical potential of a substance in terms of its mole fraction in a mixture.

5A.1 **Partial molar quantities;** 5A.2 **The thermodynamics of mixing;** 5A.3 **The chemical potentials of liquids**

5B The properties of solutions

In this Topic, the concept of chemical potential is applied to the discussion of the effect of a solute on certain thermodynamic properties of a solution. These properties include the lowering of vapour pressure of the solvent, the elevation of its boiling point, the depression of its freezing point, and the origin of osmotic pressure. It is possible to construct a model of a certain class of real solutions called 'regular solutions', which have properties that diverge from those of ideal solutions.

5B.1 **Liquid mixtures;** 5B.2 **Colligative properties**

5C Phase diagrams of binary systems: liquids

One widely employed device used to summarize the equilibrium properties of mixtures is the phase diagram. The Topic describes phase diagrams of systems of liquids with gradually increasing complexity. In each case the phase diagram for the system summarizes empirical observations on the conditions under which the liquid and vapour phases of the system are stable.

5C.1 **Vapour pressure diagrams;** 5C.2 **Temperature–composition diagrams;** 5C.3 **Distillation;** 5C.4 **Liquid–liquid phase diagrams**

5D Phase diagrams of binary systems: solids

In this Topic it is seen how the phase diagrams of solid mixtures summarize experimental results on the conditions under which the liquid and solid phases of the system are stable.

5D.1 **Eutectics;** 5D.2 **Reacting systems;** 5D.3 **Incongruent melting**

5E Phase diagrams of ternary systems

Many modern materials (and ancient ones too) have more than two components. This Topic shows how phase diagrams are extended to the description of systems of three components and how to interpret triangular phase diagrams.

5E.1 **Triangular phase diagrams;** 5E.2 **Ternary systems**

5F Activities

The extension of the concept of chemical potential to real solutions involves introducing an effective concentration called an 'activity'. In certain cases, the activity may be interpreted in terms of intermolecular interactions. An important example is an electrolyte solution. Such solutions often deviate considerably from ideal behaviour on account of the strong, long-range interactions between ions. This Topic shows how a model can be used to estimate the deviations from ideal behaviour when the solution is very dilute, and how to extend the resulting expressions to more concentrated solutions.

5F.1 **The solvent activity**; 5F.2 **The solute activity**; 5F.3 **The activities of regular solutions**; 5F.4 **The activities of ions**

Web resources What is an application of this material?

Two applications of this material are discussed, one from biology and the other from materials science, from among the huge number that could be chosen for this centrally important field. *Impact* 7 shows how the phenomenon of osmosis contributes to the ability of biological cells to maintain their shapes. In *Impact* 8, phase diagrams of the technologically important liquid crystals are discussed.

TOPIC 5A The thermodynamic description of mixtures

➤ Why do you need to know this material?

Chemistry deals with a wide variety of mixtures, including mixtures of substances that can react together. Therefore, it is important to generalize the concepts introduced in Focus 4 to deal with substances that are mingled together.

➤ What is the key idea?

The chemical potential of a substance in a mixture is a logarithmic function of its concentration.

➤ What do you need to know already?

This Topic extends the concept of chemical potential to substances in mixtures by building on the concept introduced in the context of pure substances (Topic 4A). It makes use of the relation between the temperature dependence of the Gibbs energy and entropy (Topic 3E), and the concept of partial pressure (Topic 1A). Throughout this and related Topics various measures of concentration of a solute in a solution are used: they are summarized in *The chemist's toolkit* 11.

The consideration of mixtures of substances that do not react together is a first step towards dealing with chemical reactions (which are treated in Topic 6A). At this stage the discussion centres on **binary mixtures**, which are mixtures of two components, A and B. In Topic 1A it is shown how the partial pressure, which is the contribution of one component to the total pressure, is used to discuss the properties of mixtures of gases. For a more general description of the thermodynamics of mixtures other analogous 'partial' properties need to be introduced.

5A.1 Partial molar quantities

The easiest partial molar property to visualize is the 'partial molar volume', the contribution that a component of a mixture makes to the total volume of a sample.

(a) Partial molar volume

Imagine a huge volume of pure water at 25 °C. When a further 1 mol H_2O is added, the volume increases by 18 cm^3 and it follows that the molar volume of pure water is 18 $cm^3\,mol^{-1}$. However, upon adding 1 mol H_2O to a huge volume of pure ethanol, the volume is found to increase by only 14 cm^3. The reason for the different increase in volume is that the volume occupied by a given number of water molecules depends on the identity of the molecules that surround them. In the latter case there is so much ethanol present that each H_2O molecule is surrounded by ethanol molecules. The network of hydrogen bonds that normally hold H_2O molecules at certain distances from each other in pure water does not form; as a result the H_2O molecules are packed more tightly and so increase the volume by only 14 cm^3. The quantity 14 $cm^3\,mol^{-1}$ is the 'partial molar volume' of water in pure ethanol. In general, the **partial molar volume** of a substance A in a mixture is the change in volume per mole of A added to a large volume of the mixture.

The partial molar volumes of the components of a mixture vary with composition because the environment of each type of molecule changes as the composition changes from pure A to pure B. This changing molecular environment, and the consequential modification of the forces acting between molecules, results in the variation of the thermodynamic properties of a mixture as its composition is changed. The partial molar volumes of water and ethanol across the full composition range at 25 °C are shown in Fig. 5A.1.

The partial molar volume, V_J, of a substance J at some general composition is defined formally as follows:

$$V_J = \left(\frac{\partial V}{\partial n_J}\right)_{p,T,n'} \qquad \text{Partial molar volume [definition]} \qquad (5A.1)$$

where the subscript n' signifies that the amounts of all other substances present are constant. The partial molar volume is the slope of the plot of the total volume as the amount of J is changed, the pressure, temperature, and amount of the other components being constant (Fig. 5A.2). Its value depends on the composition, as seen for water and ethanol.

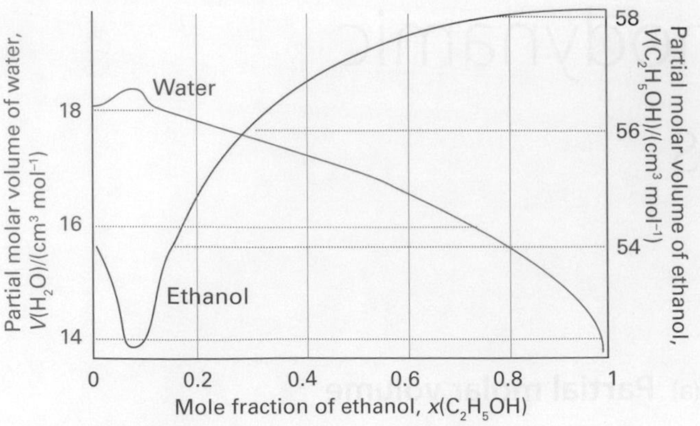

Figure 5A.1 The partial molar volumes of water and ethanol at 25 °C. Note the different scales (water on the left, ethanol on the right).

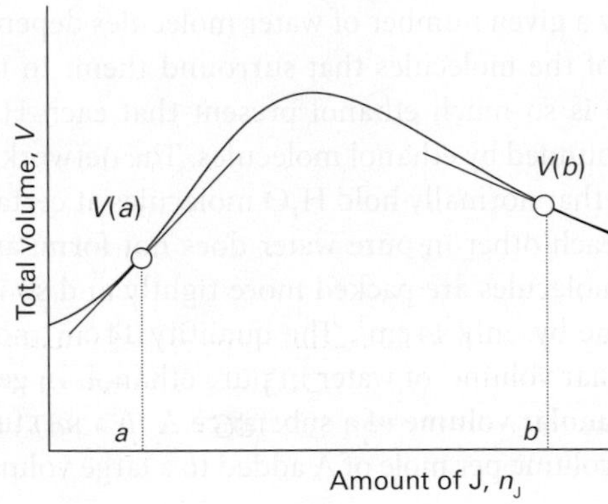

Figure 5A.2 The partial molar volume of a substance is the slope of the variation of the total volume of the sample plotted against the amount of that substance. In general, partial molar quantities vary with the composition, as shown by the different slopes at *a* and *b*. Note that the partial molar volume at *b* is negative: the overall volume of the sample decreases as A is added.

A note on good practice The IUPAC recommendation is to denote a partial molar quantity by $\bar{X}$, but only when there is the possibility of confusion with the quantity X. For instance, to avoid confusion, the partial molar volume of NaCl in water could be written $\bar{V}$(NaCl,aq) to distinguish it from the total volume of the solution, V.

The definition in eqn 5A.1 implies that when the composition of a binary mixture is changed by the addition of dn_A of A and dn_B of B, then the total volume of the mixture changes by

$$dV = \left(\frac{\partial V}{\partial n_A}\right)_{p,T,n_B} dn_A + \left(\frac{\partial V}{\partial n_B}\right)_{p,T,n_A} dn_B \qquad (5A.2)$$
$$= V_A dn_A + V_B dn_B$$

This equation can be integrated with respect to n_A and n_B provided that the amounts of A and B are both increased in such a way as to keep their ratio constant. This linkage ensures that the partial molar volumes V_A and V_B are constant and so can be taken outside the integrals:

$$V = \int_0^{n_A} V_A\, dn_A + \int_0^{n_B} V_B\, dn_B = V_A \int_0^{n_A} dn_A + V_B \int_0^{n_B} dn_B \qquad (5A.3)$$
$$= V_A n_A + V_B n_B$$

Although the two integrations are linked (in order to preserve constant relative composition), because V is a state function the final result in eqn 5A.3 is valid however the solution is in fact prepared.

Partial molar volumes can be measured in several ways. One method is to measure the dependence of the volume on the composition and to fit the observed volume to a function of the amount of the substance. Once the function has been found, its slope can be determined at any composition of interest by differentiation.

Example 5A.1 Determining a partial molar volume

A polynomial fit to measurements of the total volume of a water/ethanol mixture at 25 °C that contains 1.000 kg of water is

$$v = 1002.93 + 54.6664z - 0.363\,94z^2 + 0.028\,256z^3$$

where $v = V/\text{cm}^3$, $z = n_E/\text{mol}$, and n_E is the amount of CH_3CH_2OH present. Determine the partial molar volume of ethanol.

Collect your thoughts Apply the definition in eqn 5A.1, taking care to convert the derivative with respect to n to a derivative with respect to z and keeping the units intact.

The solution The partial molar volume of ethanol, V_E, is

$$V_E = \left(\frac{\partial V}{\partial n_E}\right)_{p,T,n_W} = \left(\frac{\partial (V/\text{cm}^3)}{\partial (n_E/\text{mol})}\right)_{p,T,n_W} \frac{\text{cm}^3}{\text{mol}}$$
$$= \left(\frac{\partial v}{\partial z}\right)_{p,T,n_W} \text{cm}^3\,\text{mol}^{-1}$$

Then, because

$$\frac{dv}{dz} = 54.6664 - 2(0.363\,94)z + 3(0.028\,256)z^2$$

it follows that

$$V_E/(\text{cm}^3\,\text{mol}^{-1}) = 54.6664 - 0.727\,88z + 0.084\,768z^2$$

Figure 5A.3 shows a graph of this function.

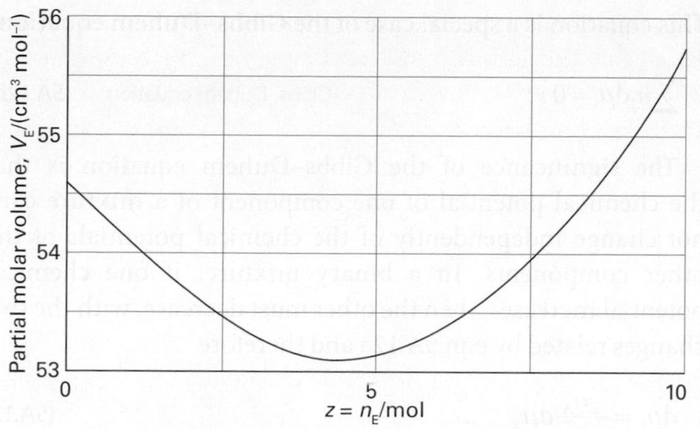

Figure 5A.3 The partial molar volume of ethanol as expressed by the polynomial in *Example* 5A.1.

Self-test 5A.1 At 25 °C, the mass density of a 50 per cent by mass ethanol/water solution is 0.914 g cm^{-3}. Given that the partial molar volume of water in the solution is 17.4 cm^3 mol^{-1}, what is the partial molar volume of the ethanol?

Answer: 56.4 cm^3 mol^{-1} by using eqn 5A.3; 54.6 cm^3 mol^{-1} by the formula above

Molar volumes are always positive, but partial molar quantities need not be. For example, the limiting partial molar volume of $MgSO_4$ in water (its partial molar volume in the limit of zero concentration) is −1.4 cm^3 mol^{-1}, which means that the addition of 1 mol $MgSO_4$ to a large volume of water results in a decrease in volume of 1.4 cm^3. The mixture contracts because the salt breaks up the open structure of water as the Mg^{2+} and SO_4^{2-} ions become hydrated, so the structure collapses slightly.

(b) Partial molar Gibbs energies

The concept of a partial molar quantity can be broadened to any extensive state function. For a substance in a mixture, the chemical potential is *defined* as the partial molar Gibbs energy:

$$\mu_J = \left(\frac{\partial G}{\partial n_J}\right)_{p,T,n'} \quad \text{Chemical potential [definition]} \quad (5A.4)$$

where n' is used to denote that the amounts of all other components of the mixture are held constant. That is, the chemical potential is the slope of a plot of Gibbs energy against the amount of the component J, with the pressure, temperature, and the amounts of the other substances held constant (Fig. 5A.4). For a pure substance $G = n_J G_{J,m}$, and from eqn 5A.4 it follows that $\mu_J = G_{J,m}$: in this case, the chemical potential is simply the molar Gibbs energy of the substance, as is used in Topic 4A.

By the same argument that led to eqn 5A.3, it follows that the total Gibbs energy of a binary mixture is

$$G = n_A\mu_A + n_B\mu_B \quad (5A.5)$$

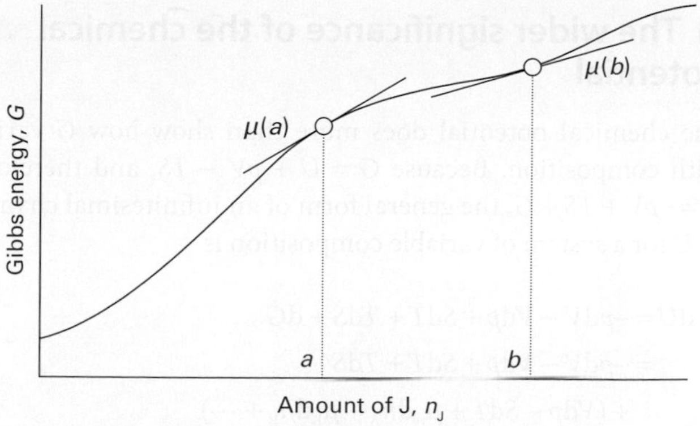

Figure 5A.4 The chemical potential of a substance is the slope of the total Gibbs energy of a mixture with respect to the amount of substance of interest. In general, the chemical potential varies with composition, as shown for the two values at *a* and *b*. In this case, both chemical potentials are positive.

where μ_A and μ_B are the chemical potentials at the composition of the mixture. That is, the chemical potential of a substance, multiplied by the amount of that substance present in the mixture, is its contribution to the total Gibbs energy of the mixture. Because the chemical potentials depend on composition (and the pressure and temperature), the Gibbs energy of a mixture may change when these variables change, and for a system of components A, B, …, eqn 3E.7 ($dG = Vdp - SdT$) for a general change in G becomes

$$dG = Vdp - SdT + \mu_A dn_A + \mu_B dn_B + \cdots \quad \text{Fundamental equation of chemical thermodynamics} \quad (5A.6)$$

This expression is the **fundamental equation of chemical thermodynamics**. Its implications and consequences are explored and developed in this and the next Focus.

At constant pressure and temperature, eqn 5A.6 simplifies to

$$dG = \mu_A dn_A + \mu_B dn_B + \cdots \quad (5A.7)$$

As established in Topic 3D, under the same conditions $dG = dw_{add,max}$. Therefore, at constant temperature and pressure,

$$dw_{add,max} = \mu_A dn_A + \mu_B dn_B + \cdots \quad (5A.8)$$

That is, additional (non-expansion) work can arise from the changing composition of a system. For instance, in an electrochemical cell the chemical reaction is arranged to take place in two distinct sites (at the two electrodes) and the electrical work the cell performs can be traced to its changing composition as products are formed from reactants.

(c) The wider significance of the chemical potential

The chemical potential does more than show how G varies with composition. Because $G = U + pV - TS$, and therefore $U = -pV + TS + G$, the general form of an infinitesimal change in U for a system of variable composition is

$$\begin{aligned} dU &= -pdV - Vdp + SdT + TdS + dG \\ &= -pdV - Vdp + SdT + TdS \\ &\quad + (Vdp - SdT + \mu_A dn_A + \mu_B dn_B + \cdots) \\ &= -pdV + TdS + \mu_A dn_A + \mu_B dn_B + \cdots \end{aligned}$$

This expression is the generalization of eqn 3E.1 (that $dU = TdS - pdV$) to systems in which the composition may change. It follows that at constant volume and entropy,

$$dU = \mu_A dn_A + \mu_B dn_B + \cdots \tag{5A.9}$$

and hence that

$$\mu_J = \left(\frac{\partial U}{\partial n_J}\right)_{S,V,n'} \tag{5A.10}$$

Therefore, not only does the chemical potential show how G changes when the composition changes, it also shows how the internal energy changes too (but under a different set of conditions). In the same way it is possible to deduce that

$$\text{(a)}\ \mu_J = \left(\frac{\partial H}{\partial n_J}\right)_{S,p,n'} \qquad \text{(b)}\ \mu_J = \left(\frac{\partial A}{\partial n_J}\right)_{T,V,n'} \tag{5A.11}$$

Thus, μ_J shows how all the extensive thermodynamic properties U, H, A, and G depend on the composition. This is why the chemical potential is so central to chemistry.

(d) The Gibbs–Duhem equation

Because the total Gibbs energy of a binary mixture is given by eqn 5A.5 ($G = n_A\mu_A + n_B\mu_B$), and the chemical potentials depend on the composition, when the compositions are changed infinitesimally the Gibbs energy of a binary system is expected to change by

$$dG = \mu_A dn_A + \mu_B dn_B + n_A d\mu_A + n_B d\mu_B$$

However, at constant pressure and temperature the change in Gibbs energy is given by eqn 5A.7. Because G is a state function, these two expressions for dG must be equal, which implies that at constant temperature and pressure

$$n_A d\mu_A + n_B d\mu_B = 0 \tag{5A.12a}$$

This equation is a special case of the **Gibbs–Duhem equation**:

$$\sum_J n_J d\mu_J = 0 \qquad \text{Gibbs–Duhem equation} \tag{5A.12b}$$

The significance of the Gibbs–Duhem equation is that the chemical potential of one component of a mixture cannot change independently of the chemical potentials of the other components. In a binary mixture, if one chemical potential increases, then the other must decrease, with the two changes related by eqn 5A.12a and therefore

$$d\mu_B = -\frac{n_A}{n_B} d\mu_A \tag{5A.13}$$

Brief illustration 5A.1

If the composition of a mixture is such that $n_A = 2n_B$, and a small change in composition results in μ_A changing by $\Delta\mu_A = +1\,\text{J mol}^{-1}$, μ_B will change by

$$\Delta\mu_B = -2 \times (1\,\text{J mol}^{-1}) = -2\,\text{J mol}^{-1}$$

The same line of reasoning applies to all partial molar quantities. For instance, changes in the partial molar volumes of the species in a mixture are related by

$$\sum_J n_J dV_J = 0 \tag{5A.14a}$$

For a binary mixture,

$$dV_B = -\frac{n_A}{n_B} dV_A \tag{5A.14b}$$

As seen in Fig. 5A.1, where the partial molar volume of water increases, the partial molar volume of ethanol decreases. Moreover, as eqn 5A.14b implies, and as seen from Fig. 5A.1, a small change in the partial molar volume of A corresponds to a large change in the partial molar volume of B if n_A/n_B is large, but the opposite is true when this ratio is small. In practice, the Gibbs–Duhem equation is used to determine the partial molar volume of one component of a binary mixture from measurements of the partial molar volume of the second component.

Example 5A.2 Using the Gibbs–Duhem equation

The experimental values of the partial molar volume of K_2SO_4(aq) at 298 K are found to fit the expression

$$v_B = 32.280 + 18.216z^{1/2}$$

where $v_B = V_{K_2SO_4}/(\text{cm}^3\,\text{mol}^{-1})$ and z is the numerical value of the molality of K_2SO_4 ($z = b/b^{\ominus}$; see *The chemist's toolkit* 11). Use the Gibbs–Duhem equation to derive an equation for the molar volume of water in the solution. The molar volume of pure water at 298 K is 18.079 cm^3 mol^{-1}.

Collect your thoughts Let A denote H_2O, the solvent, and B denote K_2SO_4, the solute. Because the Gibbs–Duhem equation for the partial molar volumes of two components implies that $dv_A = -(n_B/n_A)dv_B$, v_A can be found by integration:

$$v_A = v_A^\star - \int_0^{v_B} \frac{n_B}{n_A} dv_B$$

where $v_A^\star = V_A^\star/(\text{cm}^3\,\text{mol}^{-1})$ is the numerical value of the molar volume of pure A. The first step is to change the variable of integration from v_B to $z = b/b^\ominus$; then integrate the right-hand side between $z = 0$ (pure A) and the molality of interest.

The solution It follows from the information in the question that, with $B = K_2SO_4$, $dv_B/dz = 9.108z^{-1/2}$. Therefore, the integration required is

$$v_A = v_A^\star - 9.108\int_0^{b/b^\ominus} \frac{n_B}{n_A} z^{-1/2} dz$$

The amount of A (H_2O) is $n_A = (1\,\text{kg})/M_A$, where M_A is the molar mass of water, and $n_B/(1\,\text{kg})$, which then occurs in the ratio n_B/n_A, will be recognized as the molality b of B:

$n_A = (1\,\text{kg})/M_A$ $n_B/(1\,\text{kg}) = b$

$$\frac{n_B}{n_A} = \frac{n_B}{(1\,\text{kg})/M_A} = \frac{n_B M_A}{1\,\text{kg}} = bM_A = zb^\ominus M_A$$

Hence

$$v_A = v_A^\star - 9.108 M_A b^\ominus \int_0^{b/b^\ominus} z^{1/2} dz$$

$$= v_A^\star - \tfrac{2}{3}(9.108 M_A b^\ominus)(b/b^\ominus)^{3/2}$$

It then follows, by substituting the data (including $M_A = 1.802 \times 10^{-2}\,\text{kg}\,\text{mol}^{-1}$, the molar mass of water), that

$$V_A/(\text{cm}^3\,\text{mol}^{-1}) = 18.079 - 0.1094(b/b^\ominus)^{3/2}$$

The partial molar volumes are plotted in Fig. 5A.5.

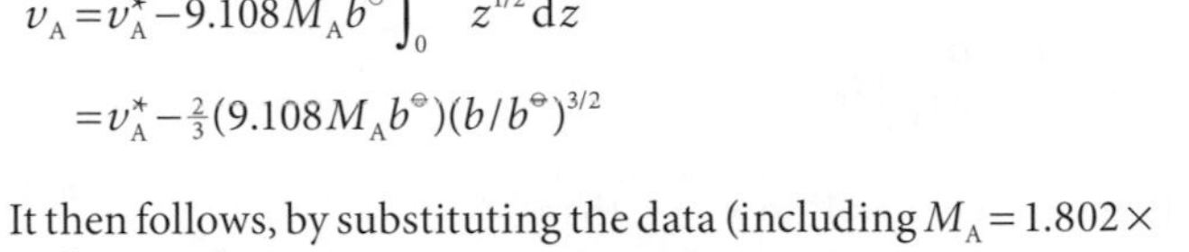

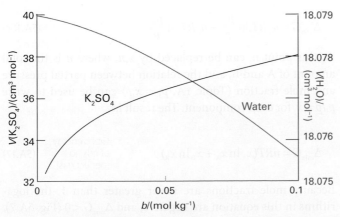

Figure 5A.5 The partial molar volumes of the components of an aqueous solution of potassium sulfate.

Self-test 5A.2 Repeat the calculation for a salt B for which $V_B/(\text{cm}^3\,\text{mol}^{-1}) = 6.218 + 5.146z - 7.147z^2$ with $z = b/b^\ominus$.

Answer: $V_A/(\text{cm}^3\,\text{mol}^{-1}) = 18.079 - 0.0464z^2 + 0.0859z^3$

5A.2 The thermodynamics of mixing

The dependence of the Gibbs energy of a mixture on its composition is given by eqn 5A.5, and, as established in Topic 3E, at constant temperature and pressure systems tend towards lower Gibbs energy. This is the link needed in order to apply thermodynamics to the discussion of spontaneous changes of composition, as in the mixing of two substances. One simple example of a spontaneous mixing process is that of two gases introduced into the same container. The mixing is spontaneous, so it must correspond to a decrease in G.

(a) The Gibbs energy of mixing of perfect gases

Let the amounts of two perfect gases in the two containers before mixing be n_A and n_B; both are at a temperature T and a pressure p (Fig. 5A.6). At this stage, the chemical potentials of the two gases have their 'pure' values, which are obtained by applying the definition $\mu = G_m$ to eqn 3E.15 ($G_m(p) = G_m^\ominus + RT\ln(p/p^\ominus)$):

$$\mu = \mu^\ominus + RT\ln\frac{p}{p^\ominus} \quad \text{Variation of chemical potential with pressure [perfect gas]} \quad (5A.15a)$$

where $\mu^\ominus$ is the **standard chemical potential**, the chemical potential of the pure gas at 1 bar.

The notation is simplified by replacing $p/p^\ominus$ by p itself, for eqn 5A.15a then becomes

$$\mu = \mu^\ominus + RT\ln p \quad (5A.15b)$$

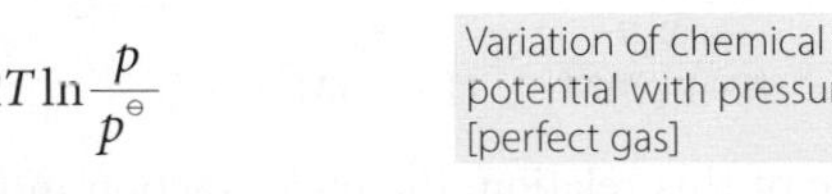

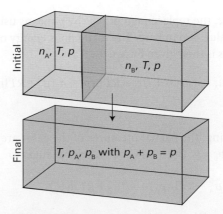

Figure 5A.6 The arrangement for calculating the thermodynamic functions of mixing of two perfect gases.

The chemist's toolkit 11 Measures of concentration

Let A be the solvent and B the solute. The **molar concentration** (informally: 'molarity'), c_B or [B], is the amount of solute molecules (in moles) divided by the volume, V, of the solution:

$$c_B = \frac{n_B}{V}$$

It is commonly reported in moles per cubic decimetre ($\mathrm{mol\,dm^{-3}}$) or, equivalently, in moles per litre ($\mathrm{mol\,L^{-1}}$). It is convenient to define its 'standard' value as $c^{\ominus} = 1\,\mathrm{mol\,dm^{-3}}$.

The **molality**, b_B, of a solute is the amount of solute species (in moles) in a solution divided by the total mass of the solvent (in kilograms), m_A:

$$b_B = \frac{n_B}{m_A}$$

Both the molality and mole fraction are independent of temperature; in contrast, the molar concentration is not. It is convenient to define the 'standard' value of the molality as $b^{\ominus} = 1\,\mathrm{mol\,kg^{-1}}$.

1. The relation between molality and mole fraction

Consider a solution with one solute and having a total amount n of molecules. If the mole fraction of the solute is x_B, the amount of solute molecules is $n_B = x_B n$. The mole fraction of solvent molecules is $x_A = 1 - x_B$, so the amount of solvent molecules is $n_A = x_A n = (1 - x_B)n$. The mass of solvent, of molar mass M_A, present is $m_A = n_A M_A = (1-x_B)nM_A$. The molality of the solute is therefore

$$b_B = \frac{n_B}{m_A} = \frac{x_B n}{(1-x_B)nM_A} = \frac{x_B}{(1-x_B)M_A}$$

The inverse of this relation, the mole fraction in terms of the molality, is

$$x_B = \frac{b_B M_A}{1+b_B M_A}$$

2. The relation between molality and molar concentration

The total mass of a volume V of *solution* (not solvent) of mass density ρ is $m = \rho V$. The amount of solute molecules in this volume is $n_B = c_B V$. The mass of solute present is $m_B = n_B M_B = c_B V M_B$. The mass of solvent present is therefore $m_A = m - m_B = \rho V - c_B V M_B = (\rho - c_B M_B)V$. The molality is therefore

$$b_B = \frac{n_B}{m_A} = \frac{c_B V}{(\rho - c_B M_B)V} = \frac{c_B}{\rho - c_B M_B}$$

The inverse of this relation, the molar concentration in terms of the molality, is

$$c_B = \frac{b_B \rho}{1+b_B M_B}$$

3. The relation between molar concentration and mole fraction

By inserting the expression for b_B in terms of x_B into the expression for c_B, the molar concentration of B in terms of its mole fraction is

$$c_B = \frac{x_B \rho}{x_A M_A + x_B M_B}$$

with $x_A = 1 - x_B$. For a dilute solution in the sense that $x_B M_B \ll x_A M_A$,

$$c_B \approx \left(\frac{\rho}{x_A M_A}\right) x_B$$

If, moreover, $x_B \ll 1$, so $x_A \approx 1$, then

$$c_B \approx \left(\frac{\rho}{M_A}\right) x_B$$

In practice, the replacement of $p/p^{\ominus}$ by p means using the numerical value of p in bars. The total Gibbs energy of the separated gases is then given by eqn 5A.5 as

$$G_i = n_A\mu_A + n_B\mu_B = n_A(\mu_A^{\ominus} + RT\ln p) + n_B(\mu_B^{\ominus} + RT\ln p) \quad (5A.16a)$$

After mixing, the partial pressures of the gases are p_A and p_B, with $p_A + p_B = p$. The total Gibbs energy changes to

$$G_f = n_A(\mu_A^{\ominus} + RT\ln p_A) + n_B(\mu_B^{\ominus} + RT\ln p_B) \quad (5A.16b)$$

The difference $G_f - G_i$, the **Gibbs energy of mixing**, $\Delta_{mix}G$, is therefore

$$\Delta_{mix}G = n_A RT\ln\frac{p_A}{p} + n_B RT\ln\frac{p_B}{p} \quad (5A.16c)$$

At this point n_J can be replaced by $x_J n$, where n is the total amount of A and B, and the relation between partial pressure and mole fraction (Topic 1A, $p_J = x_J p$) can be used to write $p_J/p = x_J$ for each component. The result is

$$\Delta_{mix}G = nRT(x_A\ln x_A + x_B\ln x_B) \quad \text{Gibbs energy of mixing [perfect gas]} \quad (5A.17)$$

Because mole fractions are never greater than 1, the logarithms in this equation are negative, and $\Delta_{mix}G < 0$ (Fig. 5A.7). The conclusion that $\Delta_{mix}G$ is negative for all compositions confirms that perfect gases mix spontaneously in all proportions.

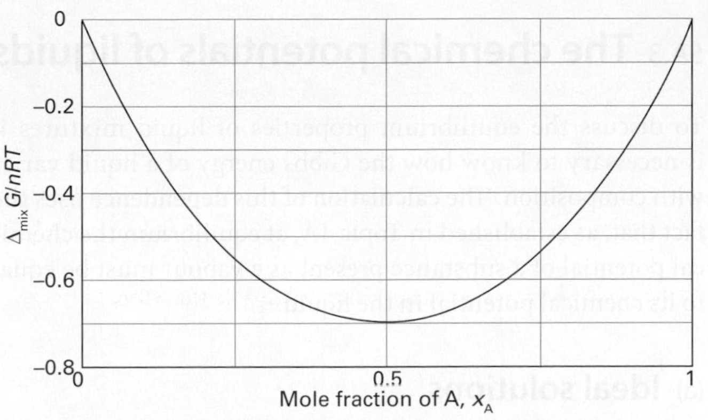

Figure 5A.7 The Gibbs energy of mixing of two perfect gases at constant temperature and pressure, and (as discussed later) of two liquids that form an ideal solution. The Gibbs energy of mixing is negative for all compositions, so perfect gases mix spontaneously in all proportions.

Example 5A.3 Calculating a Gibbs energy of mixing

A container is divided into two equal compartments (Fig. 5A.8). One contains 3.0 mol H_2(g) at 25 °C; the other contains 1.0 mol N_2(g) at 25 °C. Calculate the Gibbs energy of mixing when the partition is removed. Assume that the gases are perfect.

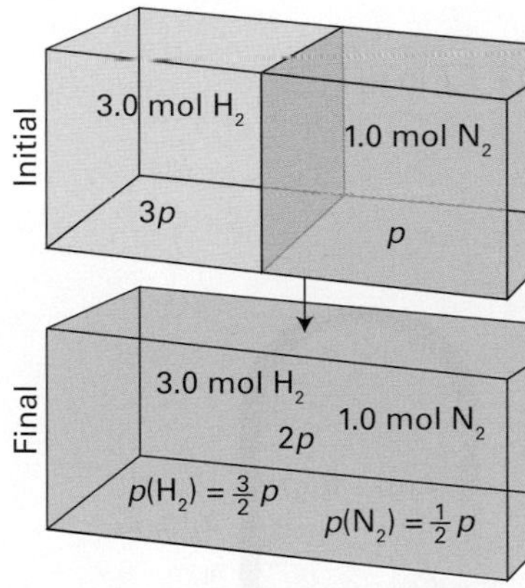

Figure 5A.8 The initial and final states considered in the calculation of the Gibbs energy of mixing of gases at different initial pressures.

Collect your thoughts Equation 5A.17 cannot be used directly because the two gases are initially at different pressures, so proceed by calculating the initial Gibbs energy from the chemical potentials. To do so, calculate the pressure of each gas: write the pressure of nitrogen as p, then the pressure of hydrogen as a multiple of p can be found from the gas laws. Next, calculate the Gibbs energy for the system when the partition is removed. The volume occupied by each gas doubles, so its final partial pressure is half its initial pressure.

The solution Given that the pressure of nitrogen is p, the pressure of hydrogen is $3p$. Therefore, the initial Gibbs energy is

$$\begin{aligned}G_i &= (3.0\,\text{mol})\{\mu^{\ominus}(H_2) + RT\ln 3p\}\\ &\quad + (1.0\,\text{mol})\{\mu^{\ominus}(N_2) + RT\ln p\}\end{aligned}$$

When the partition is removed and each gas occupies twice the original volume, the final total pressure is $2p$. The partial pressure of nitrogen falls to $\frac{1}{2}p$ and that of hydrogen falls to $\frac{3}{2}p$. Therefore, the Gibbs energy changes to

$$\begin{aligned}G_f &= (3.0\,\text{mol})\{\mu^{\ominus}(H_2) + RT\ln \tfrac{3}{2}p\}\\ &\quad + (1.0\,\text{mol})\{\mu^{\ominus}(N_2) + RT\ln \tfrac{1}{2}p\}\end{aligned}$$

The Gibbs energy of mixing is the difference of these two quantities:

$$\begin{aligned}\Delta_{mix}G &= (3.0\,\text{mol})RT\ln\frac{\frac{3}{2}p}{3p} + (1.0\,\text{mol})RT\ln\frac{\frac{1}{2}p}{p}\\ &= -(3.0\,\text{mol})RT\ln 2 - (1.0\,\text{mol})RT\ln 2\\ &= -(4.0\,\text{mol})RT\ln 2 = -6.9\,\text{kJ}\end{aligned}$$

Comment. In this example, the value of $\Delta_{mix}G$ is the sum of two contributions: the mixing itself, and the changes in pressure of the two gases to their final total pressure, $2p$. Do not be misled into interpreting this negative change in Gibbs energy as a sign of spontaneity: in this case, the pressure changes, and $\Delta G < 0$ is a signpost of spontaneous change only at constant temperature and pressure. When 3.0 mol H_2 mixes with 1.0 mol N_2 at the same pressure, with the volumes of the vessels adjusted accordingly, the change of Gibbs energy is −5.6 kJ. Because this value is for a change at constant pressure and temperature, the fact that it is negative does imply spontaneity.

Self-test 5A.3 Suppose that 2.0 mol H_2 at 2.0 atm and 25 °C and 4.0 mol N_2 at 3.0 atm and 25 °C are mixed by removing the partition between them. Calculate $\Delta_{mix}G$.

Answer: −9.7 kJ

(b) Other thermodynamic mixing functions

In Topic 3E it is shown that $(\partial G/\partial T)_p = -S$. It follows immediately from eqn 5A.17 that, for a mixture of perfect gases initially at the same pressure, the entropy of mixing, $\Delta_{mix}S$, is

$$\Delta_{mix}S = -\left(\frac{\partial \Delta_{mix}G}{\partial T}\right)_p = -nR(x_A\ln x_A + x_B\ln x_B)$$

Entropy of mixing [perfect gases, constant T and p] (5A.18)

Because $\ln x < 0$, it follows that $\Delta_{mix}S > 0$ for all compositions (Fig. 5A.9).

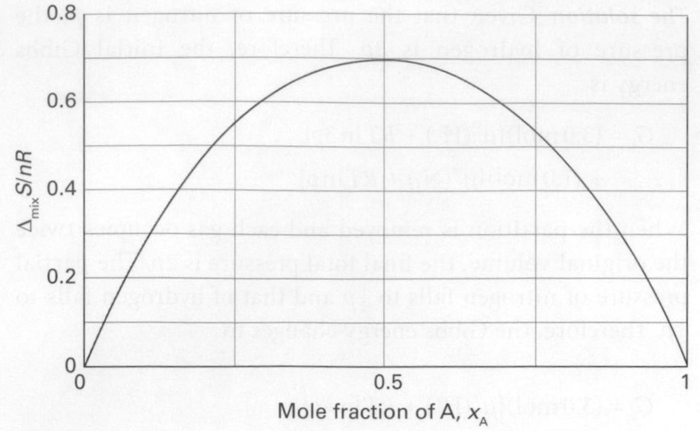

Figure 5A.9 The entropy of mixing of two perfect gases at constant temperature and pressure, and (as discussed later) of two liquids that form an ideal solution. The entropy increases for all compositions, and because there is no transfer of heat to the surroundings when perfect gases mix, the entropy of the surroundings is unchanged. Hence, the graph also shows the total entropy of the system plus the surroundings; because the total entropy of mixing is positive at all compositions, perfect gases mix spontaneously in all proportions.

Brief illustration 5A.2

For equal amounts of perfect gas molecules that are mixed at the same pressure, set $x_A = x_B = \frac{1}{2}$ and obtain

$$\Delta_{mix}S = -nR\{\tfrac{1}{2}\ln\tfrac{1}{2}+\tfrac{1}{2}\ln\tfrac{1}{2}\} = nR\ln 2$$

with n the total amount of gas molecules. For 1 mol of each species, so $n = 2\,\text{mol}$,

$$\Delta_{mix}S = (2\,\text{mol}) \times R\ln 2 = +11.5\,\text{J K}^{-1}$$

An increase in entropy is expected when one gas disperses into the other and the disorder increases.

Under conditions of constant pressure and temperature, the **enthalpy of mixing**, $\Delta_{mix}H$, the enthalpy change accompanying mixing, of two perfect gases can be calculated from $\Delta G = \Delta H - T\Delta S$. It follows from eqns 5A.17 and 5A.18 that

$$\Delta_{mix}H = 0 \quad \text{Enthalpy of mixing [perfect gases, constant } T \text{ and } p] \qquad (5A.19)$$

The enthalpy of mixing is zero, as expected for a system in which there are no interactions between the molecules forming the gaseous mixture. It follows that, because the entropy of the surroundings is unchanged, the whole of the driving force for mixing comes from the increase in entropy of the system.

5A.3 The chemical potentials of liquids

To discuss the equilibrium properties of liquid mixtures it is necessary to know how the Gibbs energy of a liquid varies with composition. The calculation of this dependence uses the fact that, as established in Topic 4A, at equilibrium the chemical potential of a substance present as a vapour must be equal to its chemical potential in the liquid.

(a) Ideal solutions

Quantities relating to pure substances are denoted by a superscript *, so the chemical potential of pure A is written μ_A^* and as $\mu_A^*(\text{l})$ when it is necessary to emphasize that A is a liquid. Because the vapour pressure of the pure liquid is p_A^* it follows from eqn 5A.15b that the chemical potential of A in the vapour (treated as a perfect gas) is $\mu_A^\ominus + RT\ln p_A$ (with p_A to be interpreted as $p_A/p^\ominus$). These two chemical potentials are equal at equilibrium (Fig. 5A.10), so

$$\underbrace{\mu_A^*(\text{l})}_{\text{liquid}} = \underbrace{\mu_A^\ominus(\text{g}) + RT\ln p_A^*}_{\text{vapour}} \qquad (5A.20a)$$

If another substance, a solute, is also present in the liquid, the chemical potential of A in the liquid is changed to μ_A and its vapour pressure is changed to p_A. The vapour and solvent are still in equilibrium, so

$$\mu_A(\text{l}) = \mu_A^\ominus(\text{g}) + RT\ln p_A \qquad (5A.20b)$$

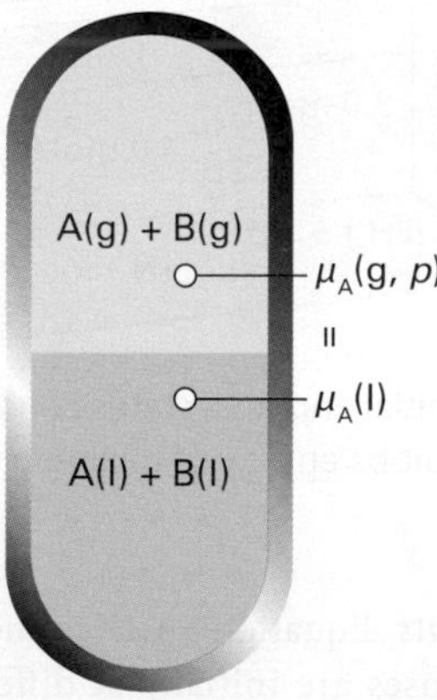

Figure 5A.10 At equilibrium, the chemical potential of the gaseous form of a substance A is equal to the chemical potential of its condensed phase. The equality is preserved if a solute is also present. Because the chemical potential of A in the vapour depends on its partial vapour pressure, it follows that the chemical potential of liquid A can be related to its partial vapour pressure.

The next step is the combination of these two equations to eliminate the standard chemical potential of the gas, $\mu_A^\ominus(g)$. To do so, write eqn 5A.20a as $\mu_A^\ominus(g) = \mu_A^*(l) - RT\ln p_A^*$ and substitute this expression into eqn 5A.20b to obtain

$$\mu_A(l) = \overbrace{\mu_A^*(l) - RT\ln p_A^*}^{\mu_A^\ominus(g)} + RT\ln p_A = \mu_A^*(l) + RT\ln\frac{p_A}{p_A^*} \quad (5A.21)$$

The final step draws on additional experimental information about the relation between the ratio of vapour pressures and the composition of the liquid. In a series of experiments on mixtures of closely related liquids (such as benzene and methylbenzene), François Raoult found that the ratio of the partial vapour pressure of each component to its vapour pressure when present as the pure liquid, p_A/p_A^*, is approximately equal to the mole fraction of A in the liquid mixture. That is, he established what is now called **Raoult's law**:

$$p_A = x_A p_A^* \quad \text{Raoult's law [ideal solution]} \quad (5A.22)$$

This law is illustrated in Fig. 5A.11. Some mixtures obey Raoult's law very well, especially when the components are structurally similar (Fig. 5A.12). Mixtures that obey the law throughout the composition range from pure A to pure B are called **ideal solutions**.

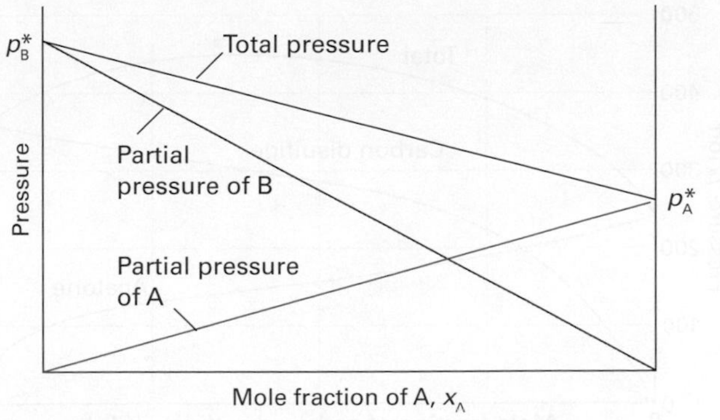

Figure 5A.11 The partial vapour pressures of the two components of an ideal binary mixture are proportional to the mole fractions of the components, in accord with Raoult's law. The total pressure is also linear in the mole fraction of either component.

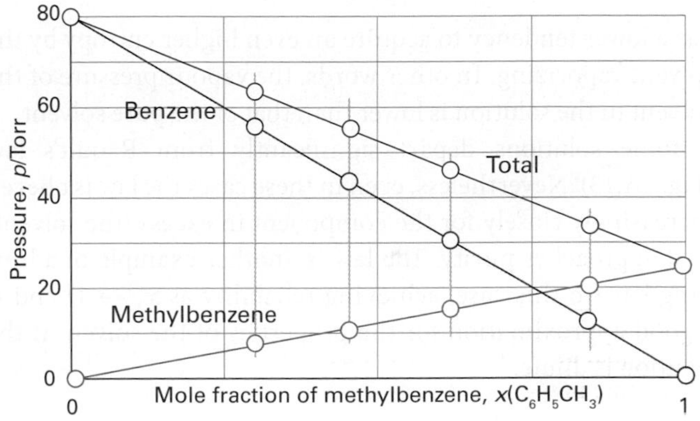

Figure 5A.12 Two similar liquids, in this case benzene and methylbenzene (toluene), behave almost ideally, and the variation of their vapour pressures with composition resembles that for an ideal solution.

Brief illustration 5A.3

The vapour pressure of pure benzene at 25 °C is 95 Torr and that of pure methylbenzene is 28 Torr at the same temperature. In an equimolar mixture $x_{benzene} = x_{methylbenzene} = \frac{1}{2}$ so the partial vapour pressure of each one in the mixture is

$$p_{benzene} = \tfrac{1}{2} \times 95\,\text{Torr} = 47.5\,\text{Torr}$$

$$p_{methylbenzene} = \tfrac{1}{2} \times 28\,\text{Torr} = 14\,\text{Torr}$$

The total vapour pressure of the mixture is 61.5 Torr. Given the two partial vapour pressures, it follows from the definition of partial pressure (Topic 1A) that the mole fractions in the vapour are

$$x_{vap,benzene} = (47.5\,\text{Torr})/(61.5\,\text{Torr}) = 0.77$$

and

$$x_{vap,methylbenzene} = (14\,\text{Torr})/(61.5\,\text{Torr}) = 0.23$$

The vapour is richer in the more volatile component (benzene).

For an ideal solution, it follows from eqns 5A.21 and 5A.22 that

$$\mu_A(l) = \mu_A^*(l) + RT\ln x_A \quad \text{Chemical potential [ideal solution]} \quad (5A.23)$$

This important equation can be used as the *definition* of an ideal solution (so that it implies Raoult's law rather than stemming from it). It is in fact a better definition than eqn 5A.22 because it does not assume that the vapour is a perfect gas.

The molecular origin of Raoult's law is the effect of the solute on the entropy of the solution. In the pure solvent, the molecules have a certain disorder and a corresponding entropy; the vapour pressure then represents the tendency of the system and its surroundings to reach a higher entropy. When a solute is present, the solution has a greater disorder than the pure solvent because a molecule chosen at random might or might not be a solvent molecule. Because the entropy of the solution is higher than that of the pure solvent, the solution

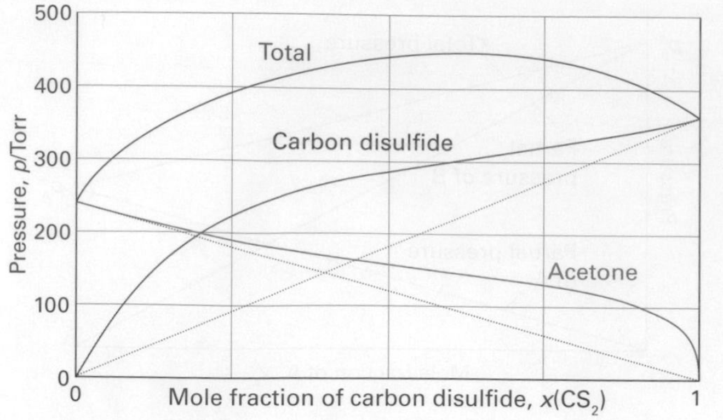

Figure 5A.13 Strong deviations from ideality are shown by dissimilar liquids (in this case carbon disulfide and acetone (propanone)). The dotted lines show the values expected from Raoult's law.

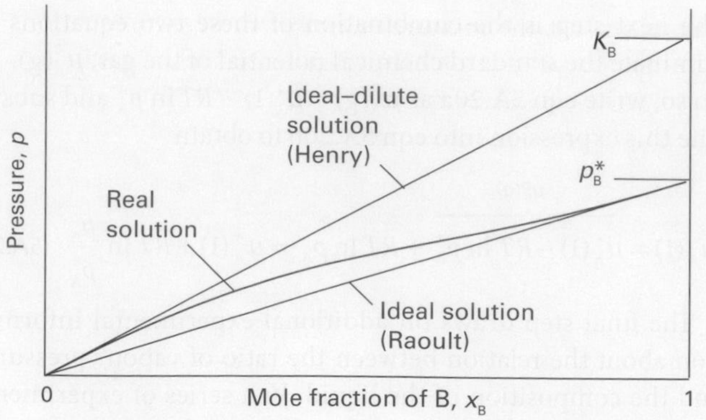

Figure 5A.14 When a component (the solvent) is nearly pure, it has a vapour pressure that is proportional to the mole fraction with a slope p_B^* (Raoult's law). When it is the minor component (the solute) its vapour pressure is still proportional to the mole fraction, but the constant of proportionality is now K_B (Henry's law).

has a lower tendency to acquire an even higher entropy by the solvent vaporizing. In other words, the vapour pressure of the solvent in the solution is lower than that of the pure solvent.

Some solutions depart significantly from Raoult's law (Fig. 5A.13). Nevertheless, even in these cases the law is obeyed increasingly closely for the component in excess (the solvent) as it approaches purity. The law is another example of a limiting law (in this case, achieving reliability as $x_A \to 1$) and is a good approximation for the properties of the solvent if the solution is dilute.

(b) Ideal-dilute solutions

In ideal solutions the solute, as well as the solvent, obeys Raoult's law. However, William Henry found experimentally that, for real solutions at low concentrations, although the vapour pressure of the solute is proportional to its mole fraction, the constant of proportionality is not the vapour pressure of the pure substance (Fig. 5A.14). **Henry's law** is:

$$p_B = x_B K_B \quad \text{Henry's law [ideal-dilute solution]} \quad (5A.24)$$

In this expression x_B is the mole fraction of the solute and K_B is an empirical constant (with the dimensions of pressure) chosen so that the plot of the vapour pressure of B against its mole fraction is tangent to the experimental curve at $x_B = 0$. Henry's law is therefore also a limiting law, achieving reliability as $x_B \to 0$.

Mixtures for which the solute B obeys Henry's law and the solvent A obeys Raoult's law are called **ideal-dilute solutions**. The difference in behaviour of the solute and solvent at low concentrations (as expressed by Henry's and Raoult's laws, respectively) arises from the fact that in a dilute solution the solvent molecules are in an environment very much like the one they have in the pure liquid (Fig. 5A.15). In contrast, the solute molecules are surrounded by solvent molecules,

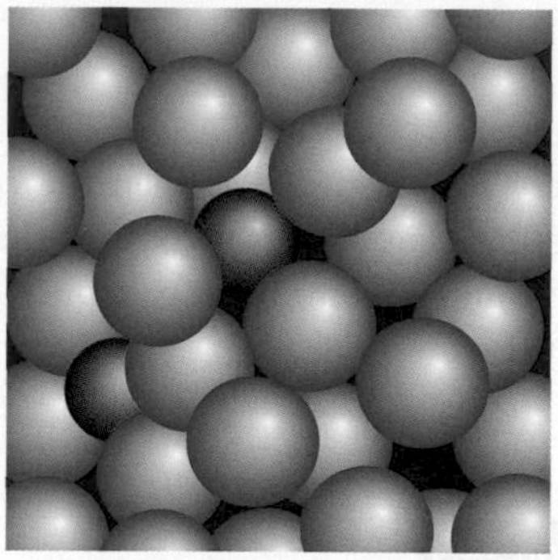

Figure 5A.15 In a dilute solution, the solvent molecules (the blue spheres) are in an environment that differs only slightly from that of the pure solvent. The solute particles (the red spheres), however, are in an environment totally unlike that of the pure solute.

which is entirely different from their environment when it is in its pure form. Thus, the solvent behaves like a slightly modified pure liquid, but the solute behaves entirely differently from its pure state unless the solvent and solute molecules happen to be very similar. In the latter case, the solute also obeys Raoult's law.

Example 5A.4 Investigating the validity of Raoult's and Henry's laws

The vapour pressures of each component in a mixture of propanone (acetone, A) and trichloromethane (chloroform, C) were measured at 35 °C with the following results:

x_C	0	0.20	0.40	0.60	0.80	1
p_C/kPa	0	4.7	11	18.9	26.7	36.4
p_A/kPa	46.3	33.3	23.3	12.3	4.9	0

Confirm that the mixture conforms to Raoult's law for the component in large excess and to Henry's law for the minor component. Find the Henry's law constants.

Collect your thoughts Both Raoult's and Henry's laws are statements about the form of the graph of partial vapour pressure against mole fraction. Therefore, plot the partial vapour pressures against mole fraction. Raoult's law is tested by comparing the data with the straight line $p_J = x_J p_J^*$ for each component in the region in which it is in excess (and acting as the solvent). Henry's law is tested by finding a straight line $p_J = x_J K_J$ that is tangent to each partial vapour pressure curve at low x, where the component can be treated as the solute.

The solution The data are plotted in Fig. 5A.16 together with the Raoult's law lines. Henry's law requires $K_A = 24.5$ kPa for acetone and $K_C = 23.5$ kPa for chloroform.

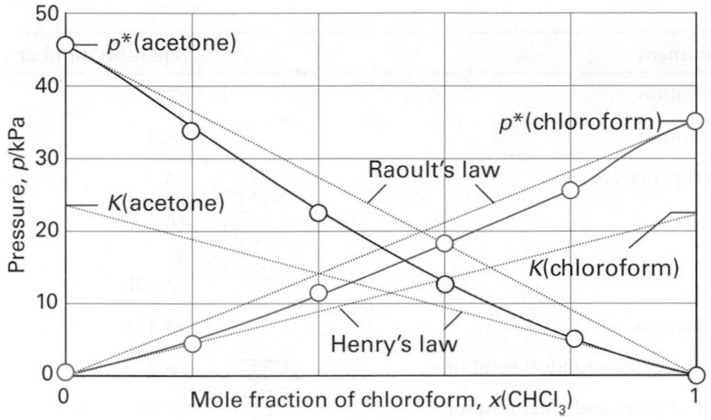

Figure 5A.16 The experimental partial vapour pressures of a mixture of chloroform (trichloromethane) and acetone (propanone) based on the data in *Example* 5A.4. The values of K are obtained by extrapolating the dilute solution vapour pressures, as explained in the *Example*.

Comment. Notice how the system deviates from both Raoult's and Henry's laws even for quite small departures from $x = 1$ and $x = 0$, respectively. These deviations are discussed in Topic 5E.

Self-test 5A.4 The vapour pressure of chloromethane at various mole fractions in a mixture at 25 °C was found to be as follows:

x	0.005	0.009	0.019	0.024
p/kPa	27.3	48.4	101	126

Estimate the Henry's law constant for chloromethane.

Answer: 5 MPa

For practical applications, Henry's law is expressed in terms of the molality, b, of the solute, $p_B = b_B K_B$. Some Henry's law data for this convention are listed in Table 5A.1. As well as providing a link between the mole fraction of the solute and its partial pressure, the data in the table may also be used to calculate gas solubilities. Knowledge of Henry's law constants for gases in blood and fats is important for the discussion of respiration, especially when the partial pressure of oxygen is abnormal, as in diving and mountaineering, and for the discussion of the action of gaseous anaesthetics.

Table 5A.1 Henry's law constants for gases in water at 298 K*

	K/(kPa kg mol^{-1})
CO_2	3.01×10^3
H_2	1.28×10^5
N_2	1.56×10^5
O_2	7.92×10^4

* More values are given in the *Resource section*.

Brief illustration 5A.4

To estimate the molar solubility of oxygen in water at 25 °C and a partial pressure of 21 kPa, its partial pressure in the atmosphere at sea level, write

$$b_{O_2} = \frac{p_{O_2}}{K_{O_2}} = \frac{21\,\text{kPa}}{7.9\times10^4\,\text{kPa kg mol}^{-1}} = 2.7\times10^{-4}\,\text{mol kg}^{-1}$$

The molality of the saturated solution is therefore 0.27 mmol kg^{-1}. To convert this quantity to a molar concentration, assume that the mass density of this dilute solution is essentially that of pure water at 25 °C, or $\rho = 0.997$ kg dm^{-3}. It follows that the molar concentration of oxygen is

$$[O_2] = b_{O_2}\rho = (2.7\times10^{-4}\,\text{mol kg}^{-1})\times(0.997\,\text{kg dm}^{-3}) = 0.27\,\text{mmol dm}^{-3}$$

Checklist of concepts

☐ 1. The **partial molar volume** of a substance is the contribution to the volume that a substance makes when it is part of a mixture.

☐ 2. The **chemical potential** is the partial molar Gibbs energy and is the contribution to the total Gibbs energy that a substance makes when it is part of a mixture.

☐ 3. The chemical potential also expresses how, under a variety of different conditions, the thermodynamic functions vary with composition.

☐ 4. The **Gibbs–Duhem equation** shows how the changes in chemical potentials (and, by extension, of other partial molar quantities) of the components of a mixture are related.

☐ 5. The **Gibbs energy of mixing** is negative for perfect gases at the same pressure and temperature.

☐ 6. The **entropy of mixing** of perfect gases initially at the same pressure is positive and the **enthalpy of mixing** is zero.

☐ 7. **Raoult's law** provides a relation between the vapour pressure of a substance and its mole fraction in a mixture.

☐ 8. An **ideal solution** is a solution that obeys Raoult's law over its entire range of compositions; for real solutions it is a limiting law valid as the mole fraction of the species approaches 1.

☐ 9. **Henry's law** provides a relation between the vapour pressure of a solute and its mole fraction in a mixture; it is the basis of the definition of an ideal–dilute solution.

☐ 10. An **ideal–dilute solution** is a solution that obeys Henry's law at low concentrations of the solute, and for which the solvent obeys Raoult's law.

Checklist of equations

Property	Equation	Comment	Equation number
Partial molar volume	$V_J = (\partial V/\partial n_J)_{p,T,n'}$	Definition	5A.1
Chemical potential	$\mu_J = (\partial G/\partial n_J)_{p,T,n'}$	Definition	5A.4
Total Gibbs energy	$G = n_A\mu_A + n_B\mu_B$	Binary mixture	5A.5
Fundamental equation of chemical thermodynamics	$dG = Vdp - SdT + \mu_A dn_A + \mu_B dn_B + \cdots$		5A.6
Gibbs–Duhem equation	$\sum_J n_J d\mu_J = 0$		5A.12b
Chemical potential of a gas	$\mu = \mu^{\ominus} + RT\ln(p/p^{\ominus})$	Perfect gas	5A.15a
Gibbs energy of mixing	$\Delta_{mix}G = nRT(x_A\ln x_A + x_B\ln x_B)$	Perfect gases and ideal solutions	5A.17
Entropy of mixing	$\Delta_{mix}S = -nR(x_A\ln x_A + x_B\ln x_B)$	Perfect gases and ideal solutions	5A.18
Enthalpy of mixing	$\Delta_{mix}H = 0$	Perfect gases and ideal solutions	5A.19
Raoult's law	$p_A = x_A p_A^{*}$	True for ideal solutions; limiting law as $x_A \to 1$	5A.22
Chemical potential of component	$\mu_A(l) = \mu_A^{*}(l) + RT\ln x_A$	Ideal solution	5A.23
Henry's law	$p_B = x_B K_B$	True for ideal–dilute solutions; limiting law as $x_B \to 0$	5A.24

TOPIC 5B The properties of solutions

➤ **Why do you need to know this material?**

Mixtures and solutions play a central role in chemistry, and so it is important to understand how their compositions affect their thermodynamic properties, such as their boiling and freezing points. One very important physical property of a solution is its osmotic pressure, which is used, for example, to determine the molar masses of macromolecules.

➤ **What is the key idea?**

The chemical potential of a substance in a mixture is the same in every phase in which it occurs.

➤ **What do you need to know already?**

This Topic is based on the expression derived from Raoult's law (Topic 5A) in which chemical potential is related to mole fraction. The derivations make use of the Gibbs–Helmholtz equation (Topic 3E) and the effect of pressure on chemical potential (Topic 5A). Some of the derivations are the same as those used in the discussion of the mixing of perfect gases (Topic 5A).

Thermodynamics can provide insight into the properties of liquid mixtures, and a few simple ideas can unify the whole field of study.

5B.1 Liquid mixtures

The development here is based on the relation derived in Topic 5A between the chemical potential of a component (which here is called J, with J = A or B in a binary mixture) in an ideal mixture or solution, μ_J, its value when pure, μ_J^*, and its mole fraction in the mixture, x_J:

$$\mu_J = \mu_J^* + RT\ln x_J \qquad \text{Chemical potential [ideal solution]} \qquad (5B.1)$$

(a) Ideal solutions

The Gibbs energy of mixing of two liquids to form an ideal solution is calculated in exactly the same way as for two gases (Topic 5A). The total Gibbs energy before the liquids are mixed is

$$G_i = n_A\mu_A^* + n_B\mu_B^* \qquad (5B.2a)$$

where the * denotes the pure liquid. When they are mixed, the individual chemical potentials are given by eqn 5B.1 and the total Gibbs energy is

$$G_f = n_A(\mu_A^* + RT\ln x_A) + n_B(\mu_B^* + RT\ln x_B) \qquad (5B.2b)$$

Consequently, the Gibbs energy of mixing, the difference of these two quantities, is

$$\Delta_{mix}G = nRT(x_A\ln x_A + x_B\ln x_B) \qquad \text{Gibbs energy of mixing [ideal solution]} \qquad (5B.3)$$

where $n = n_A + n_B$. As for gases, it follows that the ideal entropy of mixing of two liquids is

$$\Delta_{mix}S = -nR(x_A\ln x_A + x_B\ln x_B) \qquad \text{Entropy of mixing [ideal solution]} \qquad (5B.4)$$

Then from $\Delta_{mix}G = \Delta_{mix}H - T\Delta_{mix}S$ it follows that the ideal enthalpy of mixing is zero, $\Delta_{mix}H = 0$. The ideal volume of mixing, the change in volume on mixing, is also zero. To see why, consider that, because $(\partial G/\partial p)_T = V$ (eqn 3E.8), $\Delta_{mix}V = (\partial\Delta_{mix}G/\partial p)_T$. But $\Delta_{mix}G$ in eqn 5B.3 is independent of pressure, so the derivative with respect to pressure is zero, and therefore $\Delta_{mix}V = 0$.

Equations 5B.3 and 5B.4 are the same as those for the mixing of two perfect gases and all the conclusions drawn there are valid here: because the enthalpy of mixing is zero there is no change in the entropy of the surroundings so the driving force for mixing is the increasing entropy of the system as the molecules mingle. It should be noted, however, that solution ideality means something different from gas perfection. In a perfect gas there are no interactions between the molecules. In ideal solutions there are interactions, but the average energy of A–B interactions in the mixture is the same as the average energy of A–A and B–B interactions in the pure liquids. The variation of the Gibbs energy and entropy of mixing with composition is the same as that for gases (Figs. 5A.7 and 5A.9); both graphs are repeated here (as Figs. 5B.1 and 5B.2).

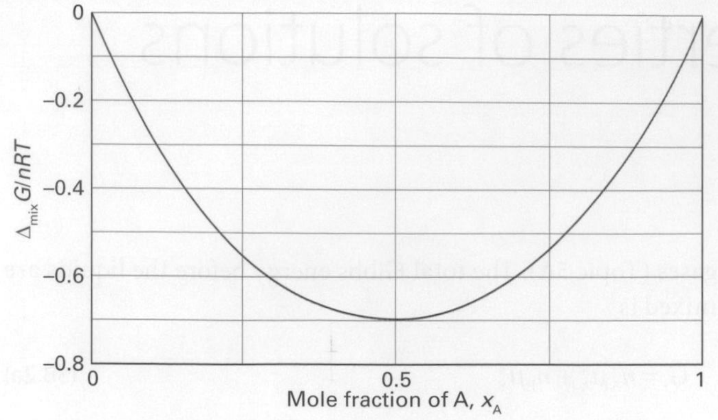

Figure 5B.1 The Gibbs energy of mixing of two liquids that form an ideal solution.

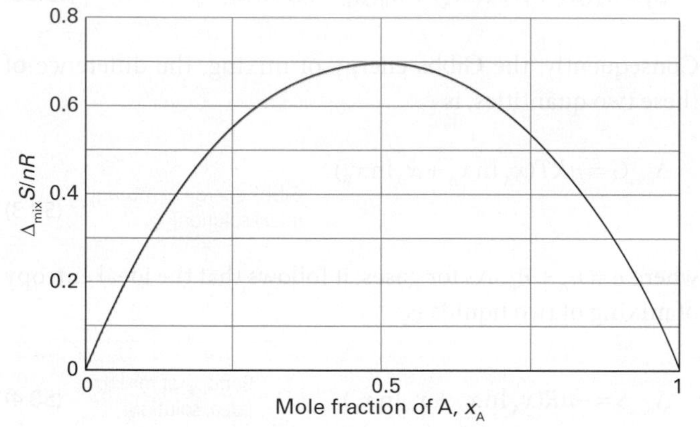

Figure 5B.2 The entropy of mixing of two liquids that form an ideal solution.

A note on good practice It is on the basis of this distinction that the term 'perfect gas' is preferable to the more common 'ideal gas'. In an ideal solution there are interactions, but they are effectively the same between the various species. In a perfect gas, not only are the interactions the same, but they are also zero. Few people, however, trouble to make this valuable distinction.

Brief illustration 5B.1

Consider a mixture of benzene and methylbenzene, which form an approximately ideal solution, and suppose 1.0 mol C_6H_6(l) is mixed with 2.0 mol $C_6H_5CH_3$(l). For the mixture, $x_{benzene} = 0.33$ and $x_{methylbenzene} = 0.67$. The Gibbs energy and entropy of mixing at 25 °C, when $RT = 2.48\,\mathrm{kJ\,mol^{-1}}$, are

$$\Delta_{mix}G/n = (2.48\,\mathrm{kJ\,mol^{-1}}) \times (0.33\ln 0.33 + 0.67\ln 0.67)$$
$$= -1.6\,\mathrm{kJ\,mol^{-1}}$$

$$\Delta_{mix}S/n = -(8.3145\,\mathrm{J\,K^{-1}\,mol^{-1}}) \times (0.33\ln 0.33 + 0.67\ln 0.67)$$
$$= +5.3\,\mathrm{J\,K^{-1}\,mol^{-1}}$$

The enthalpy of mixing is zero (presuming that the solution is ideal).

Real solutions are composed of molecules for which the A–A, A–B, and B–B interactions are all different. Not only may there be enthalpy and volume changes when such liquids mix, but there may also be an additional contribution to the entropy arising from the way in which the molecules of one type might cluster together instead of mingling freely with the others. If the enthalpy change is large and positive, or if the entropy change is negative (because of a reorganization of the molecules that results in an orderly mixture), the Gibbs energy of mixing might be positive. In that case, separation is spontaneous and the liquids are immiscible. Alternatively, the liquids might be **partially miscible**, which means that they are miscible only over a certain range of compositions.

(b) Excess functions and regular solutions

The thermodynamic properties of real solutions are expressed in terms of the **excess functions**, X^E, the difference between the observed thermodynamic function of mixing and the function for an ideal solution:

$$X^E = \Delta_{mix}X - \Delta_{mix}X^{ideal} \qquad \text{Excess function [definition]} \qquad (5B.5)$$

The **excess entropy**, S^E, for example, is calculated by using the value of $\Delta_{mix}S^{ideal}$ given by eqn 5B.4. The excess enthalpy and volume are both equal to the observed enthalpy and volume of mixing, because the ideal values are zero in each case.

Figure 5B.3 shows two examples of the composition dependence of excess functions. Figure 5B.3(a) shows data for

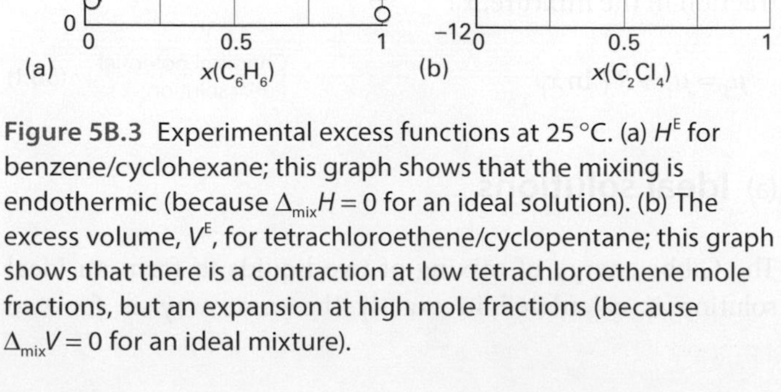

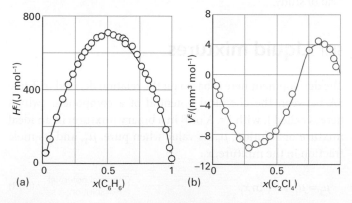

Figure 5B.3 Experimental excess functions at 25 °C. (a) H^E for benzene/cyclohexane; this graph shows that the mixing is endothermic (because $\Delta_{mix}H = 0$ for an ideal solution). (b) The excess volume, V^E, for tetrachloroethene/cyclopentane; this graph shows that there is a contraction at low tetrachloroethene mole fractions, but an expansion at high mole fractions (because $\Delta_{mix}V = 0$ for an ideal mixture).

a benzene/cyclohexane mixture: the positive values of H^E, which implies that $\Delta_{mix}H > 0$, indicate that the A–B interactions in the mixture are less attractive than the A–A and B–B interactions in the pure liquids. The symmetrical shape of the curve reflects the similar strengths of the A–A and B–B interactions. Figure 5B.3(b) shows the composition dependence of the excess volume, V^E, of a mixture of tetrachloroethene and cyclopentane. At high mole fractions of cyclopentane, the solution contracts as tetrachloroethene is added because the ring structure of cyclopentane results in inefficient packing of the molecules, but as tetrachloroethene is added, the molecules in the mixture pack together more tightly. Similarly, at high mole fractions of tetrachloroethene, the solution expands as cyclopentane is added because tetrachloroethene molecules are nearly flat and pack efficiently in the pure liquid, but become disrupted as the bulky ring cyclopentane is added.

Deviations of the excess enthalpy from zero indicate the extent to which the solutions are non-ideal. In this connection a useful model system is the **regular solution**, a solution for which $H^E \neq 0$ but $S^E = 0$. A regular solution can be thought of as one in which the two kinds of molecules are distributed randomly (as in an ideal solution) but have different energies of interaction with each other. To express this concept more quantitatively, suppose that the excess enthalpy depends on composition as

$$H^E = n\xi RTx_Ax_B \tag{5B.6}$$

where ξ (xi) is a dimensionless parameter that is a measure of the energy of A–B interactions relative to that of the A–A and B–B interactions. (For H^E expressed as a molar quantity, discard the n.) The function given by eqn 5B.6 is plotted in Fig. 5B.4; it resembles the experimental curve in Fig. 5B.3a. If $\xi < 0$, then mixing is exothermic and the A–B interactions are more favourable than the A–A and B–B interactions. If $\xi > 0$,

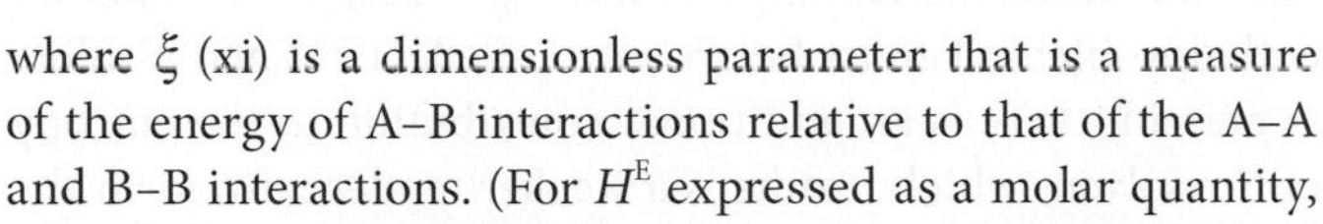

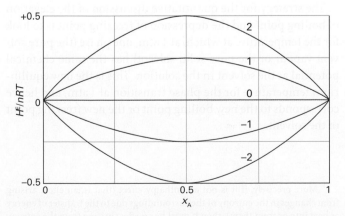

Figure 5B.4 The excess enthalpy according to a model in which it is proportional to $\xi x_A x_B$, for different values of the parameter ξ.

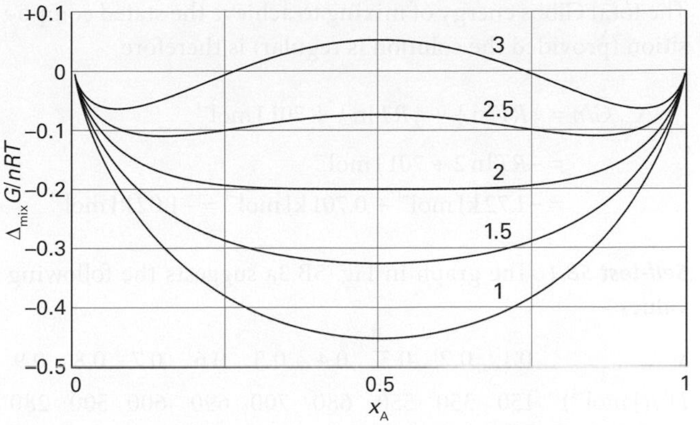

Figure 5B.5 The Gibbs energy of mixing for different values of the parameter ξ.

then the mixing is endothermic. Because the entropy of mixing has its ideal value for a regular solution, the Gibbs energy of mixing is

$$\begin{aligned}\Delta_{mix}G &= \overbrace{n\xi RTx_Ax_B}^{\Delta_{mix}H} - T\overbrace{[-nR(x_A\ln x_A + x_B\ln x_B)]}^{\Delta_{mix}S} \\ &= nRT(x_A\ln x_A + x_B\ln x_B + \xi x_Ax_B)\end{aligned} \tag{5B.7}$$

Figure 5B.5 shows how $\Delta_{mix}G$ varies with composition for different values of ξ. The important feature is that for $\xi > 2$ the graph shows two minima separated by a maximum. The implication of this observation is that, provided $\xi > 2$, the system will separate spontaneously into two phases with compositions corresponding to the two minima, because such a separation corresponds to a reduction in Gibbs energy. This point is developed in Topic 5C.

Example 5B.1 Identifying the parameter for a regular solution

Identify the value of the parameter ξ that would be appropriate to model a mixture of benzene and cyclohexane at 25 °C, and estimate the Gibbs energy of mixing for an equimolar mixture.

Collect your thoughts Refer to Fig. 5B.3a and identify the value of the maximum in the curve; then relate it to eqn 5B.6 written as a molar quantity ($H^E = \xi RTx_Ax_B$). For the second part, assume that the solution is regular and that the Gibbs energy of mixing is given by eqn 5B.7.

The solution In the experimental data the maximum occurs close to $x_A = x_B = \frac{1}{2}$ and its value is close to 701 J mol^{-1}. It follows that

$$\xi = \frac{H^E}{RTx_Ax_B} = \frac{701\,\text{J mol}^{-1}}{(8.3145\,\text{J K}^{-1}\,\text{mol}^{-1})\times(298\,\text{K})\times\frac{1}{2}\times\frac{1}{2}} = 1.13$$

The total Gibbs energy of mixing to achieve the stated composition (provided the solution is regular) is therefore

$$\Delta_{\text{mix}}G/n = \tfrac{1}{2}RT\ln\tfrac{1}{2} + \tfrac{1}{2}RT\ln\tfrac{1}{2} + 701\,\text{J mol}^{-1}$$
$$= -RT\ln 2 + 701\,\text{J mol}^{-1}$$
$$= -1.72\,\text{kJ mol}^{-1} + 0.701\,\text{kJ mol}^{-1} = -1.02\,\text{kJ mol}^{-1}$$

Self-test 5B.1 The graph in Fig. 5B.3a suggests the following values:

x	0.1	0.2	0.3	0.4	0.5	0.6	0.7	0.8	0.9
$H^{\text{E}}/(\text{J mol}^{-1})$	150	350	550	680	700	690	600	500	280

Use a curve-fitting procedure to fit these data to an expression of the form in eqn 5B.6 written as $H^{\text{E}}/n = Ax(1-x)$.

Answer: The best fit is with $A = 690\,\text{J mol}^{-1}$

5B.2 Colligative properties

A **colligative property** is a physical property that depends on the relative number of solute particles present but not their chemical identity ('colligative' denotes 'depending on the collection'). They include the lowering of vapour pressure, the elevation of boiling point, the depression of freezing point, and the osmotic pressure arising from the presence of a solute. In dilute solutions these properties depend only on the number of solute particles present, not their identity.

In this development, the solvent is denoted by A and the solute by B. There are two assumptions. First, the solute is not volatile, so it does not contribute to the vapour. Second, the solute does not dissolve in the solid solvent: that is, the pure solid solvent separates when the solution is frozen. The latter assumption is quite drastic, although it is true of many mixtures; it can be avoided at the expense of more algebra, but that introduces no new principles.

(a) The common features of colligative properties

All the colligative properties stem from the reduction of the chemical potential of the liquid solvent as a result of the presence of solute. For an ideal solution (one that obeys Raoult's law, Topic 5A; $p_{\text{A}} = x_{\text{A}}p_{\text{A}}^{*}$), the reduction is from μ_{A}^{*} for the pure solvent to $\mu_{\text{A}} = \mu_{\text{A}}^{*} + RT\ln x_{\text{A}}$ when a solute is present ($\ln x_{\text{A}}$ is negative because $x_{\text{A}} < 1$). There is no direct influence of the solute on the chemical potential of the solvent vapour and the solid solvent because the solute appears in neither the vapour nor the solid. As can be seen from Fig. 5B.6, the reduction in chemical potential of the solvent implies that the liquid–vapour equilibrium occurs at a higher temperature (the boiling point is raised) and the solid–liquid equilibrium occurs at a lower temperature (the freezing point is lowered).

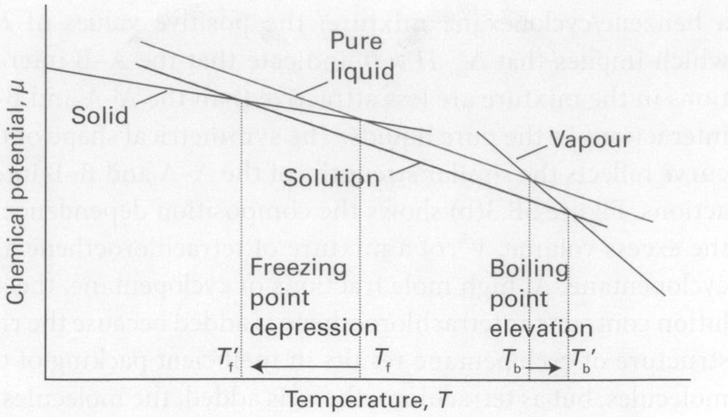

Figure 5B.6 The chemical potential of the liquid solvent in a solution is lower than that of the pure liquid. As a result, the temperature at which the chemical potential of the solvent is equal to that of the solid solvent (the freezing point) is lowered, and the temperature at which it is equal to that of the vapour (the boiling point) is raised. The lowering of the liquid's chemical potential has a greater effect on the freezing point than on the boiling point because of the angles at which the lines intersect.

The molecular origin of the lowering of the chemical potential is not the energy of interaction of the solute and solvent particles, because the lowering occurs even in an ideal solution (for which the enthalpy of mixing is zero). If it is not an enthalpy effect, it must be an entropy effect.[1] When a solute is present, there is an additional contribution to the entropy of the solvent which results is a weaker tendency to form the vapour (Fig. 5B.7). This weakening of the tendency to form a vapour lowers the vapour pressure and hence raises the boiling point. Similarly, the enhanced molecular randomness of the solution opposes the tendency to freeze. Consequently, a lower temperature must be reached before equilibrium between solid and solution is achieved. Hence, the freezing point is lowered.

The strategy for the quantitative discussion of the elevation of boiling point and the depression of freezing point is to look for the temperature at which, at 1 atm, one phase (the pure solvent vapour or the pure solid solvent) has the same chemical potential as the solvent in the solution. This is the new equilibrium temperature for the phase transition at 1 atm, and hence corresponds to the new boiling point or the new freezing point of the solvent.

[1] More precisely, if it is not an enthalpy effect (that is, an effect arising from changes in the entropy of the surroundings due to the transfer of energy as heat into or from them), then it must be an effect arising from the entropy of the system.

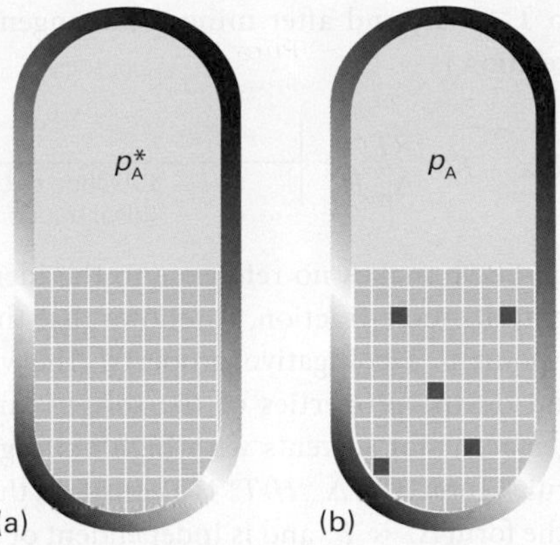

Figure 5B.7 The vapour pressure of a pure liquid represents a balance between the increase in disorder arising from vaporization and the decrease in disorder of the surroundings. (a) Here the structure of the liquid is represented highly schematically by the grid of squares. (b) When solute (the dark green squares) is present, the disorder of the condensed phase is higher than that of the pure liquid, and there is a decreased tendency to acquire the disorder characteristic of the vapour.

(b) The elevation of boiling point

The equilibrium of interest when considering boiling is between the solvent vapour and the solvent in solution at 1 atm (Fig. 5B.8). The equilibrium is established at a temperature for which

$$\mu_A^*(g) = \mu_A^*(l) + RT\ln x_A \qquad (5B.8)$$

where $\mu_A^*(g)$ is the chemical potential of the pure vapour; the pressure of 1 atm is the same throughout, and will not be written explicitly. It can be shown that a consequence of this relation is that the normal boiling point of the solvent is raised and that in a dilute solution the increase is proportional to the mole fraction of solute.

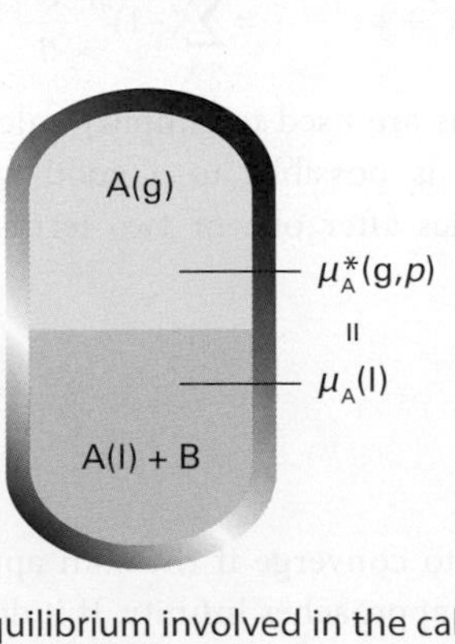

Figure 5B.8 The equilibrium involved in the calculation of the elevation of boiling point is between A present as pure vapour and A in the mixture, A being the solvent and B a non-volatile solute.

How is that done? 5B.1 Deriving an expression for the elevation of the boiling point

The starting point for the calculation is the equality of the chemical potentials of the solvent in the liquid and vapour phases, eqn 5B.8. The strategy then involves examining how the temperature must be changed to maintain that equality when solute is added. You need to follow these steps.

Step 1 *Relate* $\ln x_A$ *to the Gibbs energy of vaporization*

Equation 5B.8 can be rearranged into

$$\ln x_A = \frac{\mu_A^*(g) - \mu_A^*(l)}{RT} = \frac{\Delta_{vap}G}{RT}$$

where $\Delta_{vap}G$ is the (molar) Gibbs energy of vaporization of the pure solvent (A).

Step 2 *Write an expression for the variation of* $\ln x_A$ *with temperature*

Differentiating both sides of the expression from Step 1 with respect to temperature and using the Gibbs–Helmholtz equation (Topic 3E, $(\partial(G/T)/\partial T)_p = -H/T^2$) to rewrite the term on the right gives

$$\frac{d\ln x_A}{dT} = \frac{1}{R}\frac{d(\Delta_{vap}G/T)}{dT} = -\frac{\Delta_{vap}H}{RT^2}$$

The change in temperature dT needed to maintain equilibrium when solute is added and the change in $\ln x_A$ by $d\ln x_A$ are therefore related by

$$d\ln x_A = -\frac{\Delta_{vap}H}{RT^2}dT$$

Step 3 *Find the relation between the measurable changes in* $\ln x_A$ *and* T *by integration*

To integrate the preceding expression, integrate from $x_A = 1$, corresponding to $\ln x_A = 0$ (and when $T = T^*$, the boiling point of pure A) to x_A (when the boiling point is T). As usual, to avoid confusing the variables of integration with the final value they reach, replace $\ln x_A$ by $\ln x_A'$ and T by T':

$$\int_0^{\ln x_A} d\ln x_A' = -\frac{1}{R}\int_{T^*}^{T}\frac{\Delta_{vap}H}{T'^2}dT'$$

The left-hand side integrates to $\ln x_A$, which is equal to $\ln(1 - x_B)$. The right-hand side can be integrated if the enthalpy of vaporization is assumed to be constant over the small range of temperatures involved, so can be taken outside the integral:

$$\ln(1-x_B) = -\frac{\Delta_{vap}H}{R}\overbrace{\int_{T^*}^{T}\frac{1}{T'^2}dT'}^{\text{Integral A.1 with } n=-2}$$

Therefore

$$\ln(1-x_B)=\frac{\Delta_{vap}H}{R}\left(\frac{1}{T}-\frac{1}{T^*}\right)$$

Step 4 *Approximate the expression for dilute solutions*

Suppose that the amount of solute present is so small that $x_B \ll 1$; the approximation $\ln(1-x) \approx -x$ (*The chemist's toolkit* 12) can then be used. It follows that

$$x_B=\frac{\Delta_{vap}H}{R}\left(\frac{1}{T^*}-\frac{1}{T}\right)$$

Finally, because the increase in the boiling point is small, $T \approx T^*$, it also follows that

$$\frac{1}{T^*}-\frac{1}{T}=\frac{T-T^*}{TT^*}\approx\frac{T-T^*}{T^{*2}}=\frac{\Delta T_b}{T^{*2}}$$

with $\Delta T_b = T - T^*$. The previous equation then becomes

$$x_B=\frac{\Delta_{vap}H}{R}\times\frac{\Delta T_b}{T^{*2}} \quad \text{(5B.9a)}$$

which confirms that the elevation of boiling point and the mole fraction of solute are proportional to each other.

Step 5 *Rearrange the expression*

The calculation has shown that the presence of a solute at a mole fraction x_B causes an increase in normal boiling point from T^* to $T^* + \Delta T$, and after minor rearrangement of eqn 5B.9a the relation is

$$\Delta T_b = K x_B \quad K=\frac{RT^{*2}}{\Delta_{vap}H} \quad \text{(5B.9b)}$$

Elevation of boiling point [ideal solution]

Because eqn 5B.9b makes no reference to the identity of the solute, only to its mole fraction, it follows that the elevation of boiling point is a colligative property. The value of ΔT does depend on the properties of the solvent, and the biggest changes occur for solvents with high boiling points. By Trouton's rule (Topic 3B), $\Delta_{vap}H/T^*$ is a constant; therefore eqn 5B.9b has the form $\Delta T \propto T^*$ and is independent of $\Delta_{vap}H$ itself. If $x_B \ll 1$ it follows that the mole fraction of B is proportional to its molality, b (see *The chemist's toolkit* 11 in Topic 5A). Equation 5B.9b can therefore be written as

$$\Delta T_b = K_b b \quad \text{(5B.9c)}$$

Boiling point elevation [empirical relation]

where K_b is the empirical **boiling-point constant** of the solvent (Table 5B.1).

Table 5B.1 Freezing-point (K_f) and boiling-point (K_b) constants*

	K_f/(K kg mol^{-1})	K_b/(K kg mol^{-1})
Benzene	5.12	2.53
Camphor	40	
Phenol	7.27	3.04
Water	1.86	0.51

* More values are given in the *Resource section*.

The chemist's toolkit 12 Series expansions

A function $f(x)$ can be expressed in terms of its value in the vicinity of $x = a$ by using the **Taylor series**

$$f(x)=f(a)+\left(\frac{df}{dx}\right)_a(x-a)+\frac{1}{2!}\left(\frac{d^2f}{dx^2}\right)_a(x-a)^2+\cdots$$

$$=\sum_{n=0}^{\infty}\frac{1}{n!}\left(\frac{d^nf}{dx^n}\right)_a(x-a)^n \quad \text{Taylor series}$$

where the notation $(\ldots)_a$ means that the derivative is evaluated at $x = a$ and $n!$ denotes a **factorial** defined as

$$n! = n(n-1)(n-2)\ldots1, \quad 0! \equiv 1 \quad \text{Factorial}$$

The **Maclaurin series** for a function is a special case of the Taylor series in which $a = 0$. The following Maclaurin series are used at various stages in the text:

$$(1+x)^{-1}=1-x+x^2-\cdots=\sum_{n=0}^{\infty}(-1)^n x^n$$

$$e^x=1+x+\tfrac{1}{2}x^2+\cdots=\sum_{n=0}^{\infty}\frac{x^n}{n!}$$

$$\ln(1+x)=x-\tfrac{1}{2}x^2+\tfrac{1}{3}x^3-\cdots=\sum_{n=1}^{\infty}(-1)^{n+1}\frac{x^n}{n}$$

Series expansions are used to simplify calculations, because when $|x| \ll 1$ it is possible, to a good approximation, to terminate the series after one or two terms. Thus, provided $|x| \ll 1$,

$$(1+x)^{-1} \approx 1-x$$

$$e^x \approx 1+x$$

$$\ln(1+x) \approx x$$

A series is said to **converge** if the sum approaches a finite, definite value as n approaches infinity. If it does not, the series is said to **diverge**. Thus, the series expansion of $(1+x)^{-1}$ converges for $|x| < 1$ and diverges for $|x| \geq 1$. Tests for convergence are explained in mathematical texts.

Brief illustration 5B.2

The boiling-point constant of water is $0.51\,\mathrm{K\,kg\,mol^{-1}}$, so a solute present at a molality of $0.10\,\mathrm{mol\,kg^{-1}}$ would result in an elevation of boiling point of only 0.051 K. The boiling-point constant of benzene is significantly larger, at $2.53\,\mathrm{K\,kg\,mol^{-1}}$, so the elevation would be 0.25 K.

(c) The depression of freezing point

The equilibrium now of interest is between pure solid solvent A and the solution with solute present at a mole fraction x_B (Fig. 5B.9). At the freezing point, the chemical potentials of A in the two phases are equal:

$$\mu_A^*(s)=\mu_A^*(l)+RT\ln x_A \qquad (5B.10)$$

where $\mu_A^*(s)$ is the chemical potential of pure solid A. The only difference between this calculation and the last is the appearance of the chemical potential of the solid in place of that of the vapour. Therefore the result can be written directly from eqn 5B.9b:

$$\Delta T_f=K'x_B \quad K'=\frac{RT^{*2}}{\Delta_{fus}H} \qquad \text{Freezing point depression} \qquad (5B.11)$$

where T^* is the freezing point of the pure liquid, ΔT_f is the freezing point depression, T^*-T, and $\Delta_{fus}H$ is the enthalpy of fusion of the solvent. Larger depressions are observed in solvents with low enthalpies of fusion and high melting points. When the solution is dilute, the mole fraction is proportional to the molality of the solute, b, and it is common to write the last equation as

$$\Delta T_f=K_f b \qquad \text{Freezing point depression [empirical relation]} \qquad (5B.12)$$

where K_f is the empirical **freezing-point constant** (Table 5B.1).

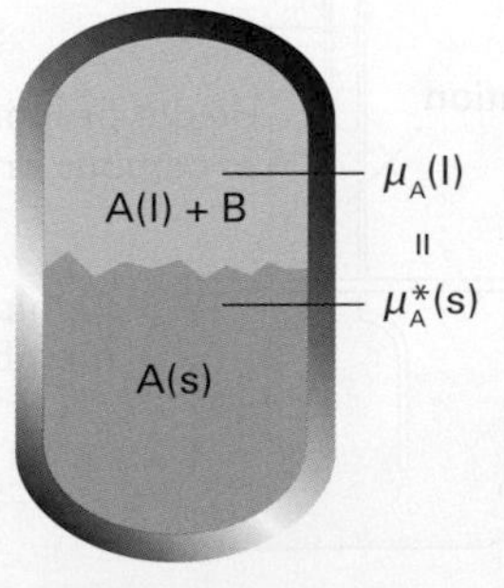

Figure 5B.9 The equilibrium involved in the calculation of the lowering of freezing point is between A present as pure solid and A in the mixture, A being the solvent and B a solute that is insoluble in solid A.

Brief illustration 5B.3

The freezing-point constant of water is $1.86\,\mathrm{K\,kg\,mol^{-1}}$, so a solute present at a molality of $0.10\,\mathrm{mol\,kg^{-1}}$ would result in a depression of freezing point of only 0.19 K. The freezing-point constant of camphor is significantly larger, at $40\,\mathrm{K\,kg\,mol^{-1}}$, so the depression would be 4.0 K.

(d) Solubility

Although solubility is not a colligative property (because solubility varies with the identity of the solute), it may be estimated in a similar way. When a solid solute is left in contact with a solvent, it dissolves until the solution is saturated. **Saturation** is a state of equilibrium, with the undissolved solute in equilibrium with the dissolved solute. Therefore, in a saturated solution the chemical potential of the pure solid solute, $\mu_B^*(s)$, and the chemical potential of B in solution, μ_B, are equal (Fig. 5B.10). Because the latter is related to the mole fraction in the solution by $\mu_B=\mu_B^*(l)+RT\ln x_B$, it follows that

$$\mu_B^*(s)=\mu_B^*(l)+RT\ln x_B \qquad (5B.13)$$

This expression is the same as the starting equation of the last section, except that the quantities refer to the solute B, not the solvent A. It can be used in a similar way to derive the relation between the solubility and the temperature.

How is that done? 5B.2 Deriving a relation between the solubility and the temperature

In the present case, the goal is to find the mole fraction of B in solution at equilibrium when the temperature is T. Therefore, start by rearranging eqn 5B.13 into

$$\ln x_B=\frac{\mu_B^*(s)-\mu_B^*(l)}{RT}=-\frac{\Delta_{fus}G}{RT}$$

As in the derivation of eqn 5B.9, differentiate both side of this equation with respect to T to relate the change in composition

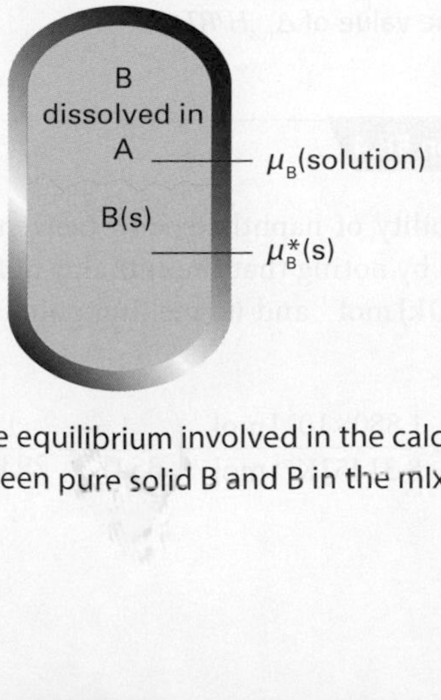

Figure 5B.10 The equilibrium involved in the calculation of the solubility is between pure solid B and B in the mixture.

to the change in temperature, and use the Gibbs–Helmholtz equation. Then integrate the resulting expression from the melting temperature of B (when $x_B = 1$ and $\ln x_B = 0$) to the temperature of interest (when x_B has a value between 0 and 1):

$$\int_0^{\ln x_B} \mathrm{d}\ln x_B' = \frac{1}{R}\int_{T_f}^{T} \frac{\Delta_{fus}H}{T'^2}\mathrm{d}T'$$

where $\Delta_{fus}H$ is the enthalpy of fusion of the solute and T_f is its melting point.

In the final step, suppose that the enthalpy of fusion of B is constant over the range of temperatures of interest, and take it outside the integral. The result of the calculation is then

$$\ln x_B = \frac{\Delta_{fus}H}{R}\left(\frac{1}{T_f} - \frac{1}{T}\right) \quad (5B.14)$$

Ideal solubility

This equation is plotted in Fig. 5B.11. It shows that the solubility of B decreases as the temperature is lowered from its melting point. The illustration also shows that solutes with high melting points and large enthalpies of melting have low solubilities at normal temperatures. However, the detailed content of eqn 5B.14 should not be treated too seriously because it is based on highly questionable approximations, such as the ideality of the solution. One aspect of its approximate character is that it fails to predict that solutes will have different solubilities in different solvents, for no solvent properties appear in the expression.

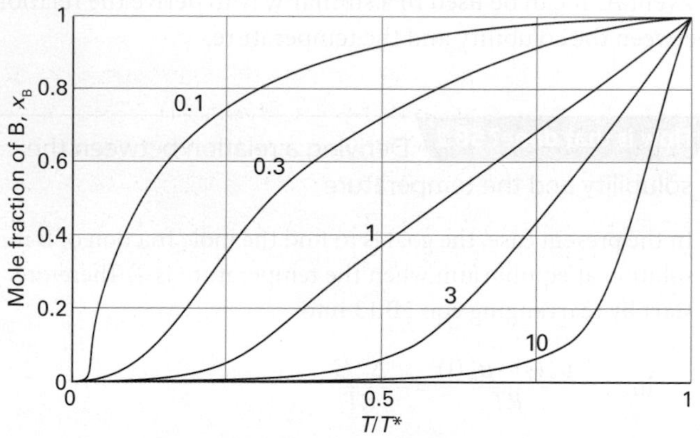

Figure 5B.11 The variation of solubility, the mole fraction of solute in a saturated solution, with temperature; T^* is the freezing temperature of the solute. Individual curves are labelled with the value of $\Delta_{fus}H/RT^*$.

Brief illustration 5B.4

The ideal solubility of naphthalene in benzene is calculated from eqn 5B.14 by noting that the enthalpy of fusion of naphthalene is 18.80 kJ mol^{-1} and its melting point is 354 K. Then, at 20 °C,

$$\ln x_{naphthalene} = \frac{1.880\times10^4\,\mathrm{J\,mol^{-1}}}{8.3145\,\mathrm{J\,K^{-1}\,mol^{-1}}}\left(\frac{1}{354\,\mathrm{K}} - \frac{1}{293\,\mathrm{K}}\right) = -1.32\ldots$$

and therefore $x_{naphthalene} = 0.26$. This mole fraction corresponds to a molality of 4.5 mol kg^{-1} (580 g of naphthalene in 1 kg of benzene).

(e) Osmosis

The phenomenon of **osmosis** (from the Greek word for 'push') is the spontaneous passage of a pure solvent into a solution separated from it by a **semipermeable membrane**, a membrane permeable to the solvent but not to the solute (Fig. 5B.12). The **osmotic pressure**, Π (uppercase pi), is the pressure that must be applied to the solution to stop the influx of solvent. Important examples of osmosis include transport of fluids through cell membranes, dialysis, and **osmometry**, the determination of molar mass by the measurement of osmotic pressure. Osmometry is widely used to determine the molar masses of macromolecules.

In the simple arrangement shown in Fig. 5B.13, the opposing pressure arises from the column of solution that the osmosis itself produces. Equilibrium is reached when the pressure

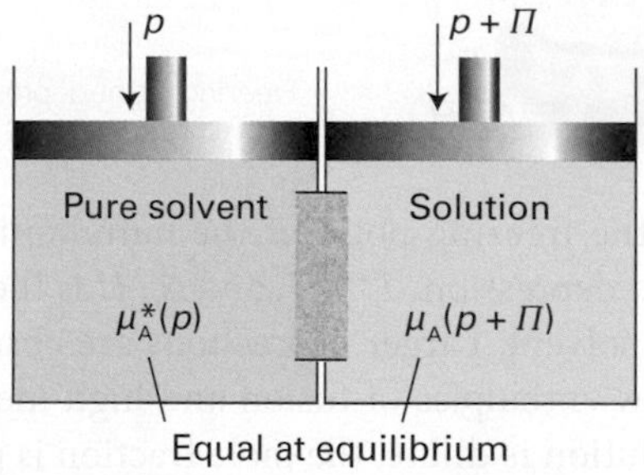

Figure 5B.12 The equilibrium involved in the calculation of osmotic pressure, Π, is between pure solvent A at a pressure p on one side of the semipermeable membrane and A as a component of the mixture on the other side of the membrane, where the pressure is $p + \Pi$.

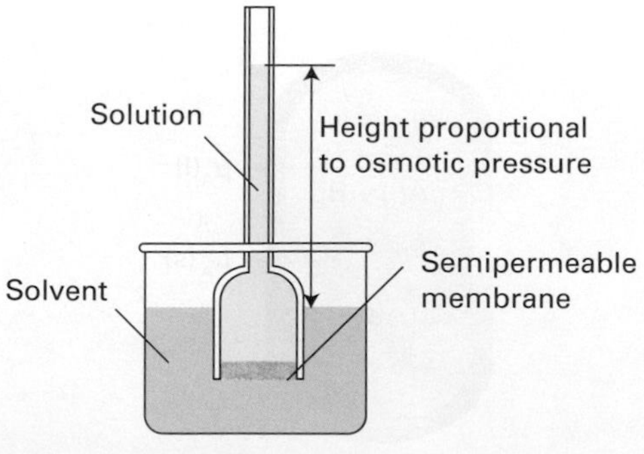

Figure 5B.13 In a simple version of the osmotic pressure experiment, A is at equilibrium on each side of the membrane when enough has passed into the solution to cause a hydrostatic pressure difference.

due to that column matches the osmotic pressure. The complicating feature of this arrangement is that the entry of solvent into the solution results in its dilution, and so it is more difficult to treat than the arrangement in Fig. 5B.12, in which there is no flow and the concentrations remain unchanged.

The thermodynamic treatment of osmosis depends on noting that, at equilibrium, the chemical potential of the solvent must be the same on each side of the membrane. The chemical potential of the solvent is lowered by the solute, but is restored to its 'pure' value by the application of pressure. The challenge in this instance is to show that, provided the solution is dilute, the extra pressure to be exerted is proportional to the molar concentration of the solute in the solution.

How is that done? 5B.3 Deriving a relation between the osmotic pressure and the molar concentration of solute

On the pure solvent side the chemical potential of the solvent, which is at a pressure p, is $\mu_A^\star(p)$. On the solution side, the chemical potential is lowered by the presence of the solute, which reduces the mole fraction of the solvent from 1 to x_A. However, the chemical potential of A is raised on account of the greater pressure, $p + \Pi$, that the solution experiences. Now follow these steps, and be prepared to make a number of approximations by supposing that the solution is dilute ($x_B << 1$).

Step 1 *Write an expression for the chemical potential of the solvent in the solution*

At equilibrium the chemical potential of A is the same in both compartments:

$$\mu_A^\star(p) = \mu_A(x_A, p + \Pi)$$

The presence of solute is taken into account in the normal way by using eqn 5B.1:

$$\mu_A(x_A, p + \Pi) = \mu_A^\star(p + \Pi) + RT\ln x_A$$

By combining these two expressions it follows that

$$\mu_A^\star(p) = \mu_A^\star(p + \Pi) + RT\ln x_A$$

and therefore

$$\mu_A^\star(p + \Pi) = \mu_A^\star(p) - RT\ln x_A$$

Step 2 *Evaluate the effect of pressure on the chemical potential of the solvent*

The effect of pressure is taken into account by using eqn 3E.12b,

$$G_m(p_f) = G_m(p_i) + \int_{p_i}^{p_f} V_m \mathrm{d}p$$

written as

$$\mu_A^\star(p + \Pi) = \mu_A^\star(p) + \int_p^{p+\Pi} V_m \mathrm{d}p$$

where V_m is the molar volume of the pure solvent A. On substituting $\mu_A^\star(p + \Pi) = \mu_A^\star(p) - RT\ln x_A$ into this expression and cancelling the $\mu_A^\star(p)$, it follows that

$$-RT\ln x_A = \int_p^{p+\Pi} V_m \mathrm{d}p \qquad (5B.15)$$

Step 3 *Evaluate the integral*

Suppose that the pressure range in the integration is so small that the molar volume of the solvent is a constant. Then the right-hand side of eqn 5B.15 simplifies to

$$\int_p^{p+\Pi} V_m \mathrm{d}p = V_m \int_p^{p+\Pi} \mathrm{d}p = V_m \Pi$$

which implies that

$$-RT\ln x_A = V_m \Pi$$

On the left-hand side of this expression, $\ln x_A$ may be replaced by $\ln(1 - x_B)$, and if it is assumed that the solution is dilute $\ln(1 - x_B) \approx -x_B$ (*The chemist's toolkit* 12), then

$$-RT\ln x_A = -RT\ln(1 - x_B) \approx RTx_B$$

The equation then becomes

$$RTx_B = \Pi V_m$$

Step 4 *Simplify the expression for the osmotic pressure for dilute solutions*

When the solution is dilute, $x_B \approx n_B/n_A$, and therefore $RTn_B \approx n_A \Pi V_m$. Moreover, $n_A V_m = V$, the total volume of the solvent, so $RTn_B \approx \Pi V$. At this stage n_B/V can be recognized as the molar concentration [B] of the solute B. It follows that for dilute solutions the osmotic pressure is given by

$$\Pi = [\mathrm{B}]RT \qquad (5B.16)$$

van 't Hoff equation

This relation, which is called the **van 't Hoff equation**, is valid only for ideal solutions. However, one of the most common applications of osmometry is to the measurement of molar masses of macromolecules, such as proteins and synthetic polymers. As these huge molecules dissolve to produce solutions that are far from ideal, it is assumed that the van 't Hoff equation is only the first term of a virial-like expansion, much like the extension of the perfect gas equation to real gases (in Topic 1C) to take into account molecular interactions:

$$\Pi = [\mathrm{J}]RT\{1 + B[\mathrm{J}] + \cdots\} \qquad \text{Osmotic virial expansion} \qquad (5B.17)$$

(The solute is denoted as J to avoid too many different Bs in this expression.) The additional terms take the non-ideality into account; the empirical constant B is called the **osmotic virial coefficient**. When it is possible to ignore corrections beyond the term depending on B, the osmotic pressure is written as

$$\Pi = [\mathrm{J}]RT\{1 + B[\mathrm{J}]\} \quad \text{or} \quad \Pi/[\mathrm{J}] = RT + BRT[\mathrm{J}] \qquad (5B.18)$$

It follows that the osmotic virial coefficient may be calculated from the slope, BRT, of a plot of $\Pi/[\mathrm{J}]$ against [J], as shown in Fig. 5B.14a.

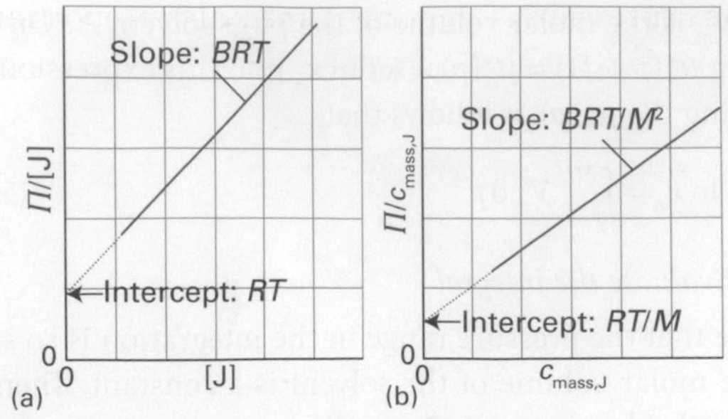

Figure 5B.14 The plot and extrapolation made to analyse the results of an osmometry experiment using (a) the molar concentration and (b) the mass concentration.

Example 5B.2 Using osmometry to determine the molar mass of a macromolecule

The osmotic pressures of solutions of a polymer, denoted J, in water at 298 K are given below. Determine the molar mass of the polymer.

$c_{\text{mass,J}}/(\text{g dm}^{-3})$	1.00	2.00	4.00	7.00	9.00
Π/Pa	27	70	197	500	785

Collect your thoughts This example is an application of eqn 5B.18, but as the data are in terms of the mass concentration, that equation must first be converted. To do so, note that the molar concentration [J] and the mass concentration $c_{\text{mass,J}}$ are related by $[\text{J}] = c_{\text{mass,J}}/M$, where M is the molar mass of J. Then identify the appropriate plot and the quantity (it will turn out to be the intercept on the vertical axis at $c_{\text{mass,J}} = 0$) that gives you the value of M.

The solution To express eqn 5B.18 in terms of the mass concentration, substitute $[\text{J}] = c_{\text{mass,J}}/M$ and obtain

$$\overbrace{\frac{\Pi M}{c_{\text{mass,J}}}}^{\Pi/[\text{J}]} = RT + \overbrace{\frac{BRTc_{\text{mass,J}}}{M}}^{BRT[\text{J}]} + \cdots$$

Division through by M gives

$$\overbrace{\frac{\Pi}{c_{\text{mass,J}}}}^{y} = \overbrace{\frac{RT}{M}}^{=\text{intercept}} + \overbrace{\left(\frac{BRT}{M^2}\right)}^{+\text{slope}} \overbrace{c_{\text{mass,J}}}^{x} + \cdots$$

It follows that, by plotting $\Pi/c_{\text{mass,J}}$ against $c_{\text{mass,J}}$, the results should fall on a straight line with intercept RT/M on the vertical axis at $c_{\text{mass,J}} = 0$. The following values of $\Pi/c_{\text{mass,J}}$ can be calculated from the data:

$c_{\text{mass,J}}/(\text{g dm}^{-3})$	1.00	2.00	4.00	7.00	9.00
$(\Pi/\text{Pa})/(c_{\text{mass,J}}/\text{g dm}^{-3})$	27	35	49.2	71.4	87.2

The intercept with the vertical axis at $c_{\text{mass,J}} = 0$ (which is best found by using linear regression and mathematical software) is at

$$\frac{\Pi/\text{Pa}}{c_{\text{mass,J}}/(\text{g dm}^{-3})} = 19.8$$

which rearranges into

$$\Pi/c_{\text{mass,J}} = 19.8\ \text{Pa g}^{-1}\,\text{dm}^3$$

Therefore, because this intercept is equal to RT/M,

$$M = \frac{RT}{19.8\ \text{Pa g}^{-1}\,\text{dm}^3} = \frac{RT}{1.98\times10^{-2}\ \text{Pa g}^{-1}\,\text{m}^3}$$

It follows that

$$M = \frac{(8.3145\ \text{J K}^{-1}\,\text{mol}^{-1})\times(298\ \text{K})}{1.98\times10^{-2}\ \text{Pa g}^{-1}\,\text{m}^3} = 1.25\times10^5\ \text{g mol}^{-1}$$

(1 J = 1 Pa m³)

The molar mass of the polymer is therefore 125 kg mol⁻¹.

Comment. Note that once M is known, the coefficient B can be determined from the slope of the graph, which is equal to BRT/M^2, as shown in Fig. 5B.14b.

Self-test 5B.2 The osmotic pressures of solutions of poly(vinyl chloride), PVC, in dioxane at 25 °C were as follows:

$c_{\text{mass,J}}/(\text{g dm}^{-3})$	0.50	1.00	1.50	2.00	2.50
Π/Pa	33.6	35.2	36.8	38.4	40.0

Determine the molar mass of the polymer.

Answer: 77 kg mol^{-1}

Checklist of concepts

☐ 1. The **Gibbs energy of mixing** of two liquids to form an ideal solution is calculated in the same way as for two perfect gases.

☐ 2. The **enthalpy of mixing** for an ideal solution is zero and the Gibbs energy is due entirely to the entropy of mixing.

☐ 3. A **regular solution** is one in which the entropy of mixing is the same as for an ideal solution but the enthalpy of mixing is non-zero.

☐ 4. A **colligative property** depends only on the number of solute particles present, not their identity.

☐ 5. All the colligative properties stem from the reduction of the chemical potential of the liquid solvent as a result of the presence of solute.

☐ 6. The **elevation of boiling point** is proportional to the molality of the solute.

☐ 7. The **depression of freezing point** is also proportional to the molality of the solute.

☐ 8. The **osmotic pressure** is the pressure that when applied to a solution prevents the influx of solvent through a semipermeable membrane.

☐ 9. The relation of the osmotic pressure to the molar concentration of the solute is given by the **van 't Hoff equation** and is a sensitive way of determining molar mass.

Checklist of equations

Property	Equation	Comment	Equation number
Gibbs energy of mixing	$\Delta_{mix}G = nRT(x_A \ln x_A + x_B \ln x_B)$	Ideal solutions	5B.3
Entropy of mixing	$\Delta_{mix}S = -nR(x_A \ln x_A + x_B \ln x_B)$	Ideal solutions	5B.4
Enthalpy of mixing	$\Delta_{mix}H = 0$	Ideal solutions	
Excess function	$X^E = \Delta_{mix}X - \Delta_{mix}X^{ideal}$	Definition	5B.5
Regular solution	$H^E = n\xi RTx_Ax_B$	Model; $S^E = 0$	5B.6
Elevation of boiling point	$\Delta T_b = K_b b$	Empirical, non-volatile solute	5B.9c
Depression of freezing point	$\Delta T_f = K_f b$	Empirical, solute insoluble in solid solvent	5B.12
Ideal solubility	$\ln x_B = (\Delta_{fus}H/R)(1/T_f - 1/T)$	Ideal solution	5B.14
van 't Hoff equation	$\Pi = [B]RT$	Valid as $[B] \to 0$	5B.16
Osmotic virial expansion	$\Pi = [J]RT\{1 + B[J] + \cdots\}$	Empirical	5B.17

TOPIC 5C Phase diagrams of binary systems: liquids

➤ Why do you need to know this material?

The separation of complex mixtures is a common task in the chemical industry. The information needed to formulate efficient separation methods is contained in phase diagrams, so it is important to be able to interpret them.

➤ What is the key idea?

The phase diagram of a liquid mixture can be understood in terms of the variation with temperature and pressure of the composition of the liquid and vapour in mutual equilibrium.

➤ What do you need to know already?

It would be helpful to review the interpretation of one-component phase diagrams and the phase rule (Topic 4A). This Topic also draws on Raoult's law (Topic 5A) and the concept of partial pressure (Topic 1A).

One-component phase diagrams are described in Topic 4A. The phase equilibria of binary systems are more complex because composition is an additional variable. However, they provide very useful summaries of phase equilibria for both ideal and empirically established real systems. This Topic focuses on binary mixtures of liquids. The phase diagrams of liquid–solid mixtures are discussed in Topic 5D.

5C.1 Vapour pressure diagrams

The partial vapour pressures of the components of an ideal solution of two volatile liquids are related to the composition of the liquid mixture by Raoult's law (Topic 5A):

$$p_A = x_A p_A^* \qquad p_B = x_B p_B^* \tag{5C.1}$$

where p_J^*, with J = A, B, is the vapour pressure of pure J and x_J is the mole fraction of J in the liquid. The total vapour pressure p of the mixture is therefore

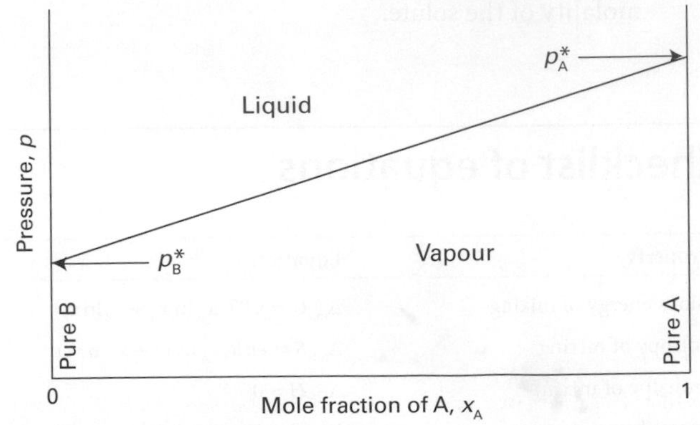

Figure 5C.1 The variation of the total vapour pressure of a binary mixture with the mole fraction of A in the liquid when Raoult's law is obeyed.

$x_B = 1 - x_A$

$$p = p_A + p_B = x_A p_A^* + x_B p_B^* = p_B^* + (p_A^* - p_B^*)x_A$$

Total vapour pressure (5C.2)

This expression shows that the total vapour pressure (at some fixed temperature) changes linearly with the composition from p_B^* to p_A^* as x_A changes from 0 to 1 (Fig. 5C.1).

The compositions of the liquid and vapour that are in mutual equilibrium are not necessarily the same. Common sense suggests that the vapour should be richer in the more volatile component. This expectation can be confirmed as follows. If the mole fractions of the components in the vapour are y_J with J = A and B, then their partial pressures are $p_J = y_J p$, with p the total pressure. Therefore

$$y_A = \frac{p_A}{p} \qquad y_B = \frac{p_B}{p} \tag{5C.3}$$

Provided the mixture is ideal, the partial pressures and the total pressure may be expressed in terms of the mole fractions in the liquid by using eqn 5C.1 for p_J and eqn 5C.2 for the total vapour pressure p. The result of combining these relations is

$$y_A = \frac{x_A p_A^*}{p_B^* + (p_A^* - p_B^*)x_A} \qquad y_B = 1 - y_A$$

Composition of vapour (5C.4)

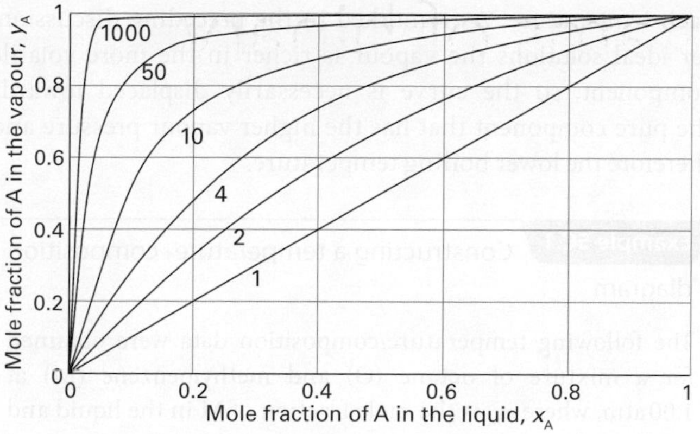

Figure 5C.2 The mole fraction of A in the vapour of a binary ideal solution expressed in terms of its mole fraction in the liquid, calculated using eqn 5C.4 for various values of p_A^*/p_B^*. For A more volatile than B ($p_A^*/p_B^* > 1$), the vapour is richer in A compared with the liquid..

Figure 5C.2 shows the composition of the vapour plotted against the composition of the liquid for various values of $p_A^*/p_B^* \geq 1$. Provided that $p_A^*/p_B^* > 1$, then $y_A > x_A$: the vapour is richer than the liquid in the more volatile component. Note that if B is not volatile, so $p_B^* = 0$ at the temperature of interest, then it makes no contribution to the vapour ($y_B = 0$).

Brief illustration 5C.1

The vapour pressures of pure benzene and methylbenzene at 20 °C are 75 Torr and 21 Torr, respectively. The composition of the vapour in equilibrium with an equimolar liquid mixture ($x_{\text{benzene}} = x_{\text{methylbenzene}} = \frac{1}{2}$) is

$$y_{\text{benzene}} = \frac{\frac{1}{2}\times(75\text{ Torr})}{21\text{ Torr} + (75\text{ Torr} - 21\text{ Torr})\times\frac{1}{2}} = 0.78$$

$$y_{\text{methylbenzene}} = 1 - 0.78 = 0.22$$

The partial vapour pressure of each component is

$$p_{\text{benzene}} = \tfrac{1}{2}\times(75\text{ Torr}) = 37.5\text{ Torr}$$

$$p_{\text{methylbenzene}} = \tfrac{1}{2}\times(21\text{ Torr}) = 10.5\text{ Torr}$$

and the total vapour pressure is the sum of these two values, 48 Torr.

Equations 5C.2 and 5C.4 can be combined to express the total vapour pressure in terms of the composition of the vapour.

How is that done? 5C.1 Deriving an expression for the total vapour pressure of a binary mixture in terms of the composition of the vapour

Equation 5C.4 can be rearranged as follows to express x_A in terms of y_A. First, multiply both sides by $p_B^* + (p_A^* - p_B^*)x_A$ to obtain

$$p_B^* y_A + (p_A^* - p_B^*)x_A y_A = x_A p_A^*$$

Then collect terms in x_A:

$$p_B^* y_A = \{p_A^* + (p_B^* - p_A^*)y_A\}x_A$$

which rearranges to

$$x_A = \frac{p_B^* y_A}{p_A^* + (p_B^* - p_A^*)y_A}$$

From eqn 5C.2 and the expression for x_A,

$$p = p_B^* + (p_A^* - p_B^*)x_A = p_B^* + \frac{(p_A^* - p_B^*)p_B^* y_A}{p_A^* + (p_B^* - p_A^*)y_A}$$

Finally, after some algebra,

$$p = \frac{p_A^* p_B^* + (p_B^* - p_A^*)p_B^* y_A + (p_A^* - p_B^*)p_B^* y_A}{p_A^* + (p_B^* - p_A^*)y_A}$$

which simplifies to

$$p = \frac{p_A^* p_B^*}{p_A^* + (p_B^* - p_A^*)y_A} \qquad (5C.5)$$

Total vapour pressure

This expression is plotted in Fig. 5C.3.

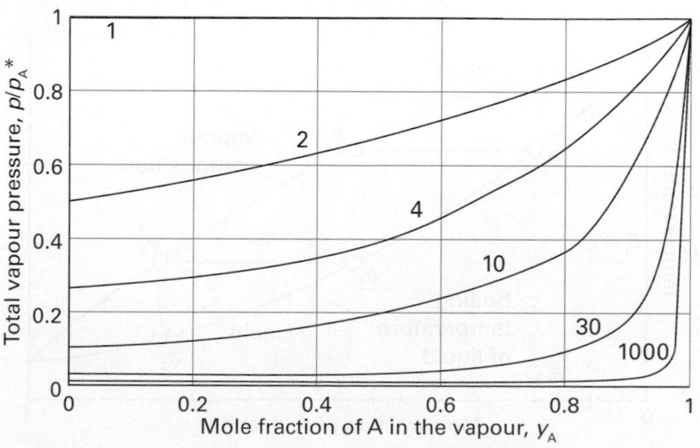

Figure 5C.3 The dependence of the vapour pressure of the same system as in Fig. 5C.2, but expressed in terms of the mole fraction of A in the vapour by using eqn 5C.5. Individual curves are labelled with the value of p_A^*/p_B^*.

5C.2 Temperature–composition diagrams

A **temperature–composition diagram** is a phase diagram in which the boundaries show the composition of the phases that are in equilibrium at various temperatures (and a given pressure, typically 1 atm). An example is shown in Fig. 5C.4. Note that the liquid phase lies in the lower part of the diagram. Temperature–composition diagrams are central to the discussion of distillation. In the following discussion, it will be best to keep in mind a system consisting of a liquid and its vapour confined inside a cylinder fitted with a movable piston that exerts a constant pressure, which in most cases is 1 atm. In this arrangement, the liquid and its vapour are in equilibrium at the normal boiling point of the mixture.

(a) The construction of the diagrams

Although in principle a temperature–composition diagram could be constructed from vapour-pressure diagrams by examining the temperature dependence of the vapour pressures of the components and identifying the temperature at which the total vapour pressure becomes equal to 1 atm (or whatever ambient pressure is of interest), they are normally constructed from empirical data on the composition of the phases in equilibrium at each temperature.

Provided the ambient pressure is 1 atm, the points representing liquid/vapour equilibrium for each of the pure liquid components are their normal boiling points. The line labelled 'Liquid' displays the boiling temperature (the temperature at which the total vapour pressure is 1 atm) of the mixture across the range of compositions. The line labelled 'Vapour' is the composition of the vapour in equilibrium with the liquid at each temperature. As remarked in the preceding discussion, for ideal solutions the vapour is richer in the more volatile component, so the curve is necessarily displaced towards the pure component that has the higher vapour pressure and therefore the lower boiling temperature.

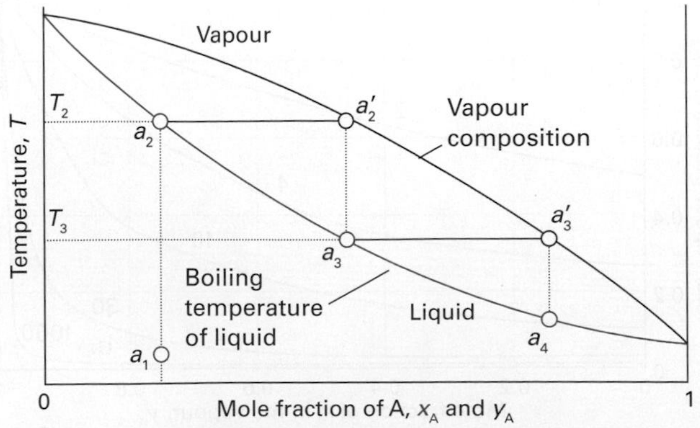

Figure 5C.4 The temperature–composition diagram corresponding to an ideal mixture with the component A more volatile than component B. As described in Section 5C.2, successive boilings and condensations of a liquid originally of composition a_1 lead to a condensate that is pure A.

Example 5C.1 Constructing a temperature–composition diagram

The following temperature/composition data were obtained for a mixture of octane (O) and methylbenzene (M) at 1.00 atm, where x_M is the mole fraction of M in the liquid and y_M the mole fraction in the vapour at equilibrium.

θ/°C	110.9	112.0	114.0	115.8	117.3	119.0	121.1	123.0
x_M	0.908	0.795	0.615	0.527	0.408	0.300	0.203	0.097
y_M	0.923	0.836	0.698	0.624	0.527	0.410	0.297	0.164

The boiling points are 110.6 °C and 125.6 °C for M and O, respectively. Plot the temperature/composition diagram for the mixture.

Collect your thoughts Plot the composition of each phase (on the horizontal axis) against the temperature (on the vertical axis). The two boiling points give two further points corresponding to $x_M = 1$ and $x_M = 0$, respectively. Use a spreadsheet or mathematical software to draw the phase boundaries.

The solution The points are plotted in Fig. 5C.5. The two sets of points are fitted to the polynomials $a + bz + cz^2 + dz^3$ with $z = x_M$ for the liquid line and $z = y_M$ for the vapour line.

$$\text{For the liquid line: } \theta/°\text{C} = 125.422 - 22.9494x_M + 6.64602x_M^2 + 1.32623x_M^3$$

$$\text{For the vapour line: } \theta/°\text{C} = 125.485 - 11.9387y_M - 12.5626y_M^2 + 9.36542y_M^3$$

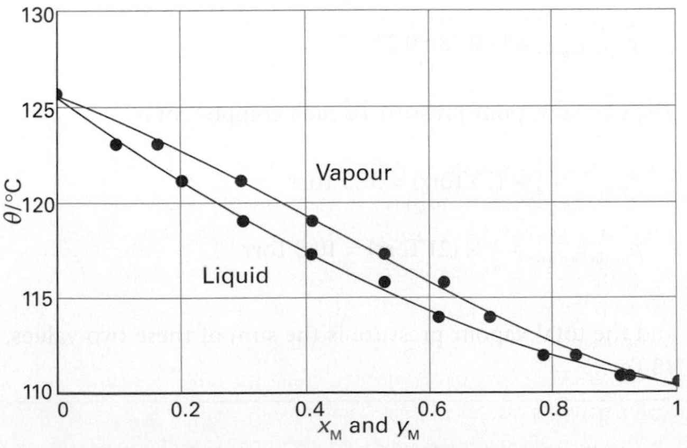

Figure 5C.5 The plot of data and the fitted curves for a mixture of octane (O) and methylbenzene (M) in *Example* 5C.1.

Self-test 5C.1 Repeat the analysis for the following data on hexane and heptane:

$\theta/°C$	65	66	70	77	85	100
x_{hexane}	0	0.20	0.40	0.60	0.80	1
y_{hexane}	0	0.02	0.08	0.20	0.48	1

Answer: Fig. 5C.6

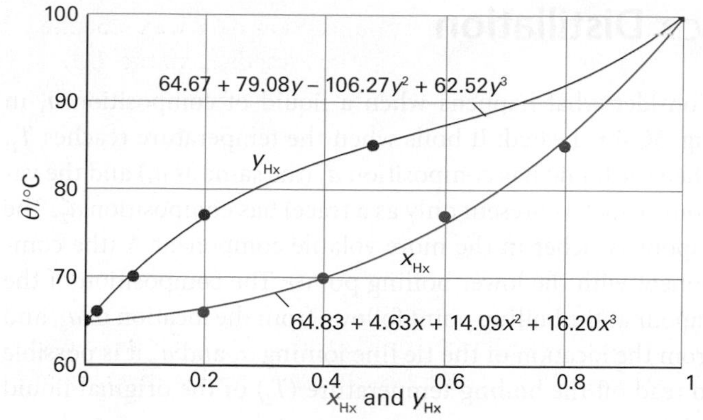

Figure 5C.6 The plot of data and the fitted curves for a mixture of hexane (Hx) and heptane in *Self-test* 5C.1.

(b) The interpretation of the diagrams

The horizontal axis of the diagram denotes the value of the mole fraction x_A when interpreting the 'Liquid' line and the mole fraction y_A when interpreting the 'Vapour' line, as illustrated in *Example* 5C.1. That is, a vertical line at x_A intersects the 'Liquid' line at the boiling point of the mixture as it was prepared. The horizontal line at that temperature, which is called a **tie line**, intersects the 'Vapour' line at a composition that represents the mole fraction y_A of A in the vapour phase in equilibrium with the boiling liquid. When appropriate, the horizontal axis will be labelled z_A and interpreted as x_A or y_A according to which line, 'Liquid' or 'Vapour' respectively, is of interest.

A point in the diagram below the 'Liquid' line at a given temperature corresponds to the mixture being at a temperature below its boiling point. If the ambient pressure is 1 atm, which is greater than the vapour pressure at that temperature, the entire sample is liquid and x_A is its composition. Similarly, if a point is above the 'Vapour' line at a given temperature, then that temperature is above the boiling point of the mixture, its vapour pressure is greater than 1 atm, and the entire sample is a vapour with a composition that is the same as that of the original mixture (because it has become entirely vapour). If the temperature is such that the point lies on the 'Liquid' curve, then the liquid and its vapour are in equilibrium and the composition of the vapour is represented by noting where the tie line meets the 'Vapour' curve. Note that the phase boundary (the 'coexistence curve') representing the frontier between the

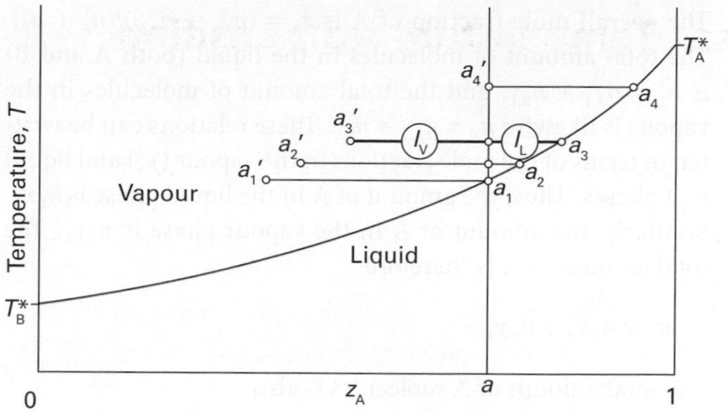

Figure 5C.7 The points of the temperature–composition diagram discussed in the text. The vertical line through a is an isopleth, a line of constant composition of the entire system.

regions where either the liquid or the vapour is the more stable phase is the 'Liquid' line: the 'Vapour' line simply provides additional information.

Points that lie between the two lines do provide additional information if the horizontal axis denotes the *overall* composition of the mixture in equilibrium at a given temperature rather than the liquid or vapour composition separately. Thus, consider what happens when a mixture in which the mole fraction of A is z_A is heated. The overall composition does not change regardless of how much liquid vaporizes, so the system moves up the vertical line at a in Fig. 5C.7. Such a vertical line is called an **isopleth** (from the Greek words for 'equal abundance').

At a_1 the liquid boils and initially is in equilibrium with its vapour of composition a_1', as given by the tie line. This vapour is richer in the more volatile component (B), so the liquid is depleted in B. Being richer in A, the boiling point of the remaining liquid moves to a_2 and the composition of the vapour in equilibrium with that liquid changes to a_2'. Further heating migrates the composition of the liquid further towards pure A, the boiling point rises and the composition of the vapour changes accordingly to a_3'. At a_4' the composition of the vapour is the same as the overall composition of the mixture, which implies that all the liquid has vaporized. Above that temperature, only vapour is present and has the initial overall composition.

It is also possible to predict the abundances of liquid and vapour at any stage of heating when the temperature and overall composition correspond to a point between the 'Liquid' and 'Vapour' lines, where a liquid of one composition is in equilibrium with a vapour of another composition.

How is that done? 5C.2 Establishing the lever rule

If the amount of A molecules in the vapour is $n_{A,V}$ and the amount in the liquid phase is $n_{A,L}$, the total amount of A molecules is $n_A = n_{A,L} + n_{A,V}$ and likewise for B molecules.

The overall mole fraction of A is $z_A = (n_{A,L} + n_{A,V})/(n_A + n_B)$. The total amount of molecules in the liquid (both A and B) is $n_L = n_{A,L} + n_{B,L}$, and the total amount of molecules in the vapour is likewise $n_V = n_{A,V} + n_{B,V}$. These relations can be written in terms of the mole fractions in the vapour (y_A) and liquid (x_A) phases. Thus, the amount of A in the liquid phase is $n_L x_A$. Similarly, the amount of A in the vapour phase is $n_V y_A$. The total amount of A is therefore

$$n_A = n_L x_A + n_V y_A$$

The total amount of A molecules is also

$$n_A = n z_A = n_L z_A + n_V z_A$$

By equating these two expressions it follows that $n_L x_A + n_V y_A = n_L z_A + n_V z_A$, and therefore

$$n_L \overbrace{(z_A - x_A)}^{l_L} = n_V \overbrace{(y_A - z_A)}^{l_V}$$

As shown in Fig. 5C.8, with $z_A - x_A$ defined as the 'length' l_L, and $y_A - z_A$ defined as the 'length' l_V, this relation can be expressed as the **lever rule**:

$$n_L l_L = n_V l_V \qquad (5C.6)$$

Lever rule

The lever rule applies to any phase diagram, not only to liquid–vapour equilibria.

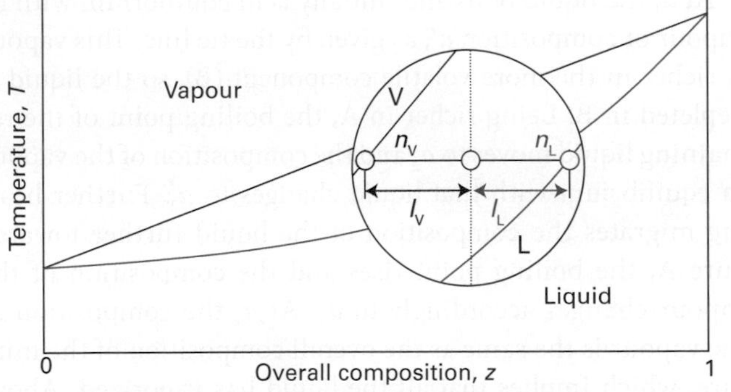

Figure 5C.8 The lever rule. The distances l_V and l_L are used to find the proportions of the amounts of the vapour and liquid present at equilibrium. The lever rule is so called because a similar rule relates the masses at two ends of a lever to their distances from a pivot (in that case $m_V l_V = m_L l_L$ for balance).

Brief illustration 5C.2

In the case illustrated in Fig. 5C.7, because $l_V \approx 2l_L$ at the tie line at a_3, the amount of molecules in the liquid phase is about twice the amount of molecules in the vapour phase. At a_1 in Fig. 5C.7, the ratio l_V/l_L is almost infinite for this tie line, so n_L/n_V is also almost infinite, and there is only a trace of vapour present. When the temperature is raised to a_2, the value of l_V/l_L is about 6.9, so $n_L/n_V \approx 0.15$ and the amount of molecules present in the liquid is about 0.15 times the amount in the vapour. When the temperature has increased to a_4 and $l_V/l_L \approx 0$ there is only a trace of liquid present.

5C.3 Distillation

Consider what happens when a liquid of composition a_1 in Fig. 5C.4 is heated. It boils when the temperature reaches T_2. Then the liquid has composition a_2 (the same as a_1) and the vapour (which is present only as a trace) has composition a_2'. The vapour is richer in the more volatile component A (the component with the lower boiling point). The composition of the vapour at the boiling point follows from the location of a_2, and from the location of the tie line joining a_2 and a_2' it is possible to read off the boiling temperature (T_2) of the original liquid mixture.

(a) Simple and fractional distillation

In a **simple distillation**, the vapour is withdrawn and condensed. This technique is used to separate a volatile liquid from a non-volatile solute or solid. In **fractional distillation**, the boiling and condensation cycle is repeated successively. This technique is used to separate volatile liquids.

Consider what happens if the vapour at a_2' in Fig. 5C.4 is condensed, and then this condensate (of composition a_3) is reheated. The phase diagram shows that this mixture boils at T_3 and yields a vapour of composition a_3', which is even richer in the more volatile component. That vapour is drawn off, and the first drop condenses to a liquid of composition a_4. The cycle can then be repeated until in due course almost pure A is obtained in the vapour and pure B remains in the liquid.

The efficiency of a fractionating column is expressed in terms of the number of **theoretical plates**, the number of effective vaporization and condensation steps that are required to achieve a condensate of given composition from a given distillate.

Brief illustration 5C.3

To achieve the degree of separation shown in Fig. 5C.9a, the fractionating column must correspond to three theoretical plates. To achieve the same separation for the system shown in Fig. 5C.9b, in which the components have more similar normal boiling points, the fractionating column must be designed to correspond to four theoretical plates.

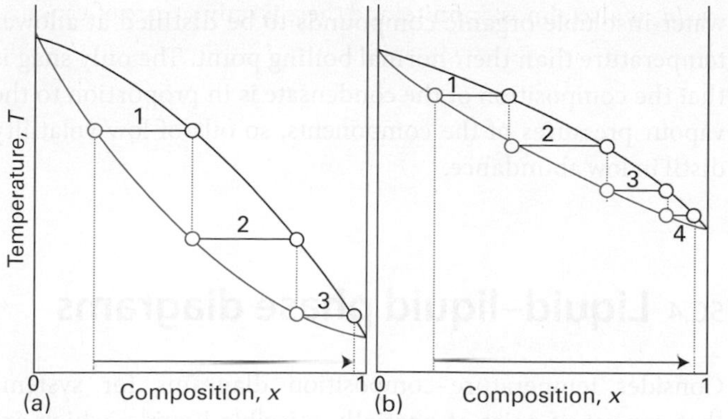

Figure 5C.9 The number of theoretical plates is the number of steps needed to bring about a specified degree of separation of two components in a mixture. The two systems shown correspond to (a) 3, (b) 4 theoretical plates.

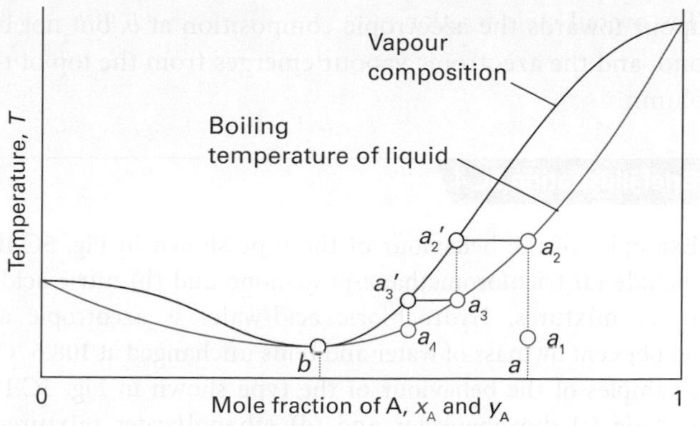

Figure 5C.11 A low-boiling azeotrope. When the mixture at *a* is fractionally distilled, the vapour in equilibrium in the fractionating column moves towards *b* and then remains unchanged.

(b) **Azeotropes**

Although many liquids have temperature–composition phase diagrams resembling the ideal version shown in Fig. 5C.4, in a number of important cases there are marked deviations. A maximum in the phase diagram (Fig. 5C.10) may occur when the favourable interactions between A and B molecules reduce the vapour pressure of the mixture below the ideal value and so raise its boiling temperature: in effect, the A–B interactions stabilize the liquid. In such cases the excess Gibbs energy, G^E (Topic 5B), is negative (more favourable to mixing than ideal). Phase diagrams showing a minimum (Fig. 5C.11) indicate that the mixture is destabilized relative to the ideal solution, the A–B interactions then being unfavourable; in this case, the boiling temperature is lowered. For such mixtures G^E is positive (less favourable to mixing than ideal), and there may be contributions from both enthalpy and entropy effects.

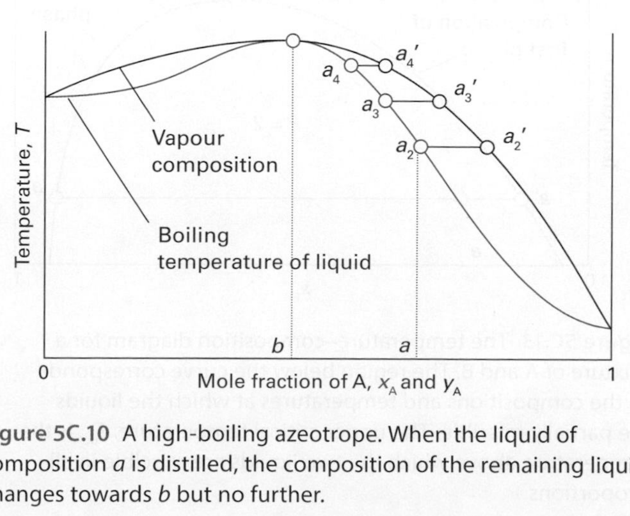

Figure 5C.10 A high-boiling azeotrope. When the liquid of composition *a* is distilled, the composition of the remaining liquid changes towards *b* but no further.

Deviations from ideality are not always so strong as to lead to a maximum or minimum in the phase diagram, but when they do there are important consequences for distillation. Consider a liquid of composition *a* on the right of the maximum in Fig. 5C.10. The vapour (at a_2') of the boiling mixture (at a_2) is richer in A. If that vapour is removed (and condensed elsewhere), then the remaining liquid will move to a composition that is richer in B, such as that represented by a_3, and the vapour in equilibrium with this mixture will have composition a_3'. As that vapour is removed, the composition of the boiling liquid shifts to a point such as a_4, and the composition of the vapour shifts to a_4'. Hence, as evaporation proceeds, the composition of the remaining liquid shifts towards B as A is drawn off. The boiling point of the liquid rises, and the vapour becomes richer in B. When so much A has been evaporated that the liquid has reached the composition *b*, the vapour has the same composition as the liquid. Evaporation then occurs without change of composition. The mixture is said to form an **azeotrope**.[1] When the azeotropic composition has been reached, distillation cannot separate the two liquids because the condensate has the same composition as the azeotropic liquid.

The system shown in Fig. 5C.11 is also azeotropic, but shows its azeotropy in a different way. Suppose we start with a mixture of composition a_1, and follow the changes in the composition of the vapour that rises through a fractionating column (essentially a vertical glass tube packed with glass rings to give a large surface area). The mixture boils at a_2 to give a vapour of composition a_2'. This vapour condenses in the column to a liquid of the same composition (now marked a_3). That liquid reaches equilibrium with its vapour at a_3', which condenses higher up the tube to give a liquid of the same composition, which we now call a_4. The fractionation therefore shifts the

[1] The name comes from the Greek words for 'boiling without changing'.

vapour towards the azeotropic composition at *b*, but not beyond, and the azeotropic vapour emerges from the top of the column.

Brief illustration 5C.4

Examples of the behaviour of the type shown in Fig. 5C.10 include (a) trichloromethane/propanone and (b) nitric acid/water mixtures. Hydrochloric acid/water is azeotropic at 80 per cent by mass of water and boils unchanged at 108.6 °C. Examples of the behaviour of the type shown in Fig. 5C.11 include (c) dioxane/water and (d) ethanol/water mixtures. Ethanol/water boils unchanged when the water content is 4 per cent by mass and the temperature is 78 °C.

(c) Immiscible liquids

Consider the distillation of two immiscible liquids, such as octane and water. At equilibrium, there is a tiny amount of A dissolved in B, and similarly a tiny amount of B dissolved in A: both liquids are saturated with the other component (Fig. 5C.12(a)). As a result, the total vapour pressure of the mixture is close to $p = p_A^* + p_B^*$. If the temperature is raised to the value at which this total vapour pressure is equal to the atmospheric pressure, boiling commences and the dissolved substances are purged from their solution. However, this boiling results in a vigorous agitation of the mixture, so each component is kept saturated in the other component, and the purging continues as the very dilute solutions are replenished. This intimate contact is essential: two immiscible liquids heated in a container like that shown in Fig. 5C.12(b) would not boil at the same temperature. The presence of the saturated solutions means that the 'mixture' boils at a lower temperature than either component would alone because boiling begins when the total vapour pressure reaches 1 atm, not when either vapour pressure reaches 1 atm. This distinction is the basis of **steam distillation**, which enables some heat-sensitive, water-insoluble organic compounds to be distilled at a lower temperature than their normal boiling point. The only snag is that the composition of the condensate is in proportion to the vapour pressures of the components, so oils of low volatility distil in low abundance.

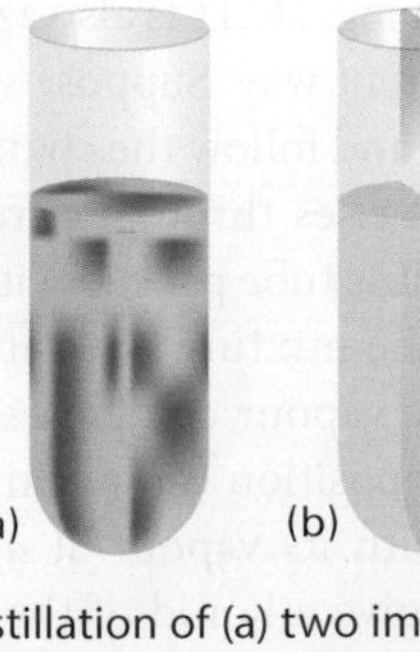

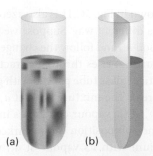

Figure 5C.12 The distillation of (a) two immiscible liquids is quite different from (b) the joint distillation of the separated components, because in the former, boiling occurs when the sum of the partial pressures equals the external pressure.

5C.4 Liquid–liquid phase diagrams

Consider temperature–composition diagrams for systems that consist of pairs of **partially miscible** liquids, which are liquids that do not mix in all proportions at all temperatures. An example is hexane and nitrobenzene. The same principles of interpretation apply as to liquid–vapour diagrams.

(a) Phase separation

Suppose a small amount of a liquid B is added to a sample of another liquid A at a temperature T'. Liquid B dissolves completely, and the binary system remains a single phase. As more B is added, a stage comes at which no more dissolves. The sample now consists of two phases in equilibrium with each other, the most abundant one consisting of A saturated with B, the minor one a trace of B saturated with A. In the temperature–composition diagram drawn in Fig. 5C.13, the composition of the former is represented by the point a' and that of the latter by the point a''. The relative abundances of the two phases are given by the lever rule. When more B is added the composition *a* moves to the right on the diagram, A dissolves in the added B slightly, and the compositions of the two phases in equilibrium remain a' and a''. As yet more B is added, composition *a* moves further to the right and eventually crosses the phase

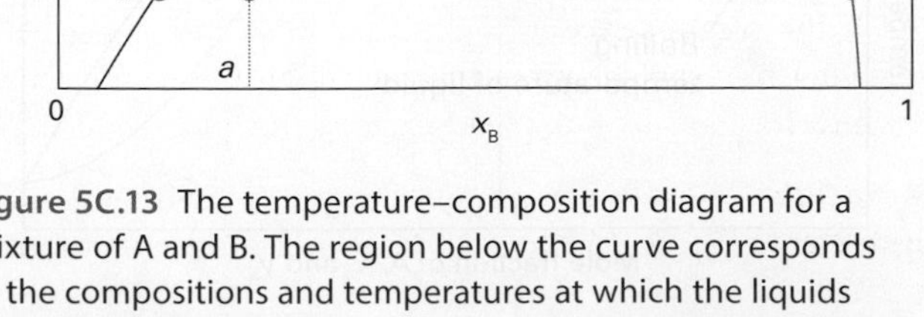

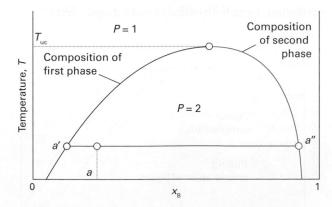

Figure 5C.13 The temperature–composition diagram for a mixture of A and B. The region below the curve corresponds to the compositions and temperatures at which the liquids are partially miscible. The upper critical temperature, T_{uc}, is the temperature above which the two liquids are miscible in all proportions.

boundary into the one-phase region. So much B is now present that it can dissolve all the A and the system reverts to a single phase. The addition of more B now simply dilutes the solution, and from then on a single phase remains.

The composition of the two phases at equilibrium varies with the temperature. For the system shown in Fig. 5C.13, raising the temperature increases the miscibility of A and B. The two-phase region therefore becomes narrower because each phase in equilibrium is richer in its minor component: the A-rich phase is richer in B and the B-rich phase is richer in A. The entire phase diagram can be constructed by repeating the observations at different temperatures and drawing the envelope of the two-phase region.

Example 5C.2 Interpreting a liquid–liquid phase diagram

The phase diagram for the system nitrobenzene/hexane at 1 atm is shown in Fig. 5C.14. A mixture of 50 g of hexane (0.59 mol C_6H_{14}) and 50 g of nitrobenzene (0.41 mol $C_6H_5NO_2$) was prepared at 290 K. What are the compositions of the phases, and in what proportions do they occur? To what temperature must the sample be heated in order to obtain a single phase?

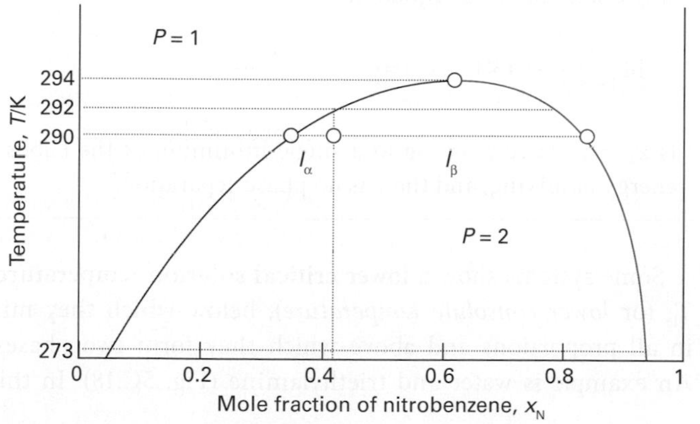

Figure 5C.14 The temperature–composition diagram for hexane and nitrobenzene at 1 atm, with the points and lengths discussed in the text.

Collect your thoughts The compositions of phases in equilibrium are given by the points where the tie line at the relevant temperature intersects the phase boundary. Their proportions are given by the lever rule (eqn 5C.6). The temperature at which the components are completely miscible is found by following the isopleth upwards and noting the temperature at which it enters the one-phase region of the phase diagram.

The solution Denote hexane by H and nitrobenzene by N, then refer to Fig. 5C.14. The mole fraction of N in the mixture is 0.41/(0.41 + 0.59) = 0.41. The point $x_N = 0.41$, $T = 290\,\text{K}$ occurs in the two-phase region of the phase diagram. The horizontal tie line cuts the phase boundary at $x_N = 0.35$ and $x_N = 0.83$, so those are the compositions of the two phases.

According to the lever rule, the ratio of amounts of each phase, which are now denoted α and β, is equal to the ratio of the distances l_α and l_β:

$$\frac{n_\alpha}{n_\beta}=\frac{l_\beta}{l_\alpha}=\frac{0.83-0.41}{0.41-0.35}=\frac{0.42}{0.06}=7$$

That is, there is about 7 times more hexane-rich phase than nitrobenzene-rich phase. Heating the sample to 292 K takes it into the single-phase region. Because the phase diagram has been constructed experimentally, these conclusions are not based on any assumptions about ideality. They would be modified if the system were subjected to a different pressure.

Self-test 5C.2 Repeat the problem for 50 g of hexane and 100 g of nitrobenzene at 273 K.

Answer: $x_N = 0.09$ and 0.95 in ratio 1:1.3; 294 K

(b) Critical solution temperatures

The **upper critical solution temperature**, T_{uc} (or *upper consolute temperature*), is the highest temperature at which phase separation occurs. Above the upper critical temperature the two components are fully miscible. This temperature exists because the greater thermal motion overcomes any potential energy advantage in molecules of one type being close together. An example is the nitrobenzene/hexane system shown in Fig. 5C.14.

The thermodynamic interpretation of the upper critical solution temperature focuses on the Gibbs energy of mixing and its variation with temperature. The simple model of a real solution (specifically, of a regular solution) discussed in Topic 5B results in a Gibbs energy of mixing that behaves as shown in Fig. 5C.15.

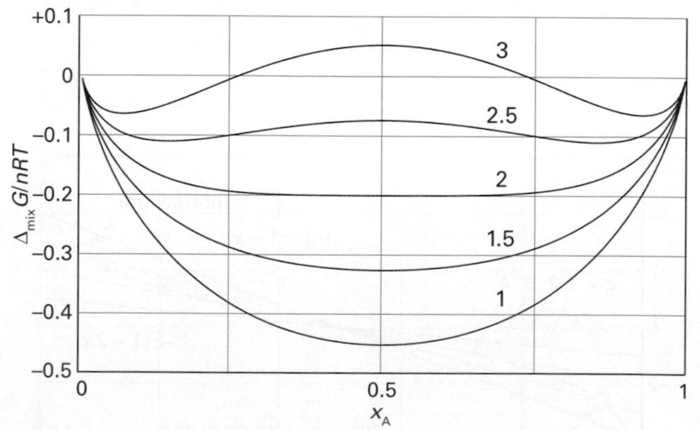

Figure 5C.15 The temperature variation of the Gibbs energy of mixing of a system composed of two components that are partially miscible at low temperatures. When two minima are present in one of these curves, the system separates into two phases with compositions corresponding to the position of the local minima. This illustration is a duplicate of Fig. 5B.5.

Provided the parameter ξ introduced in eqn 5B.6 ($H^E = n\xi RTx_Ax_B$) is greater than 2, the Gibbs energy of mixing has a double minimum. As a result, for $\xi > 2$ phase separation is expected to occur. The compositions corresponding to the minima are obtained by looking for the conditions at which $\partial\Delta_{mix}G/\partial x_A = 0$. A simple manipulation of eqn 5B.7 ($\Delta_{mix}G = nRT(x_A \ln x_A + x_B \ln x_B + \xi x_A x_B)$, with $x_B = 1 - x_A$) shows that

$$\left(\frac{\partial\Delta_{mix}G}{\partial x_A}\right)_{T,p} = nRT\left(\frac{\partial\{x_A\ln x_A + (1-x_A)\ln(1-x_A) + \xi x_A(1-x_A)\}}{\partial x_A}\right)_{T,p}$$

$$= nRT\{\ln x_A + 1 - \ln(1-x_A) - 1 + \xi(1-2x_A)\}$$

$$= nRT\left\{\ln\frac{x_A}{1-x_A} + \xi(1-2x_A)\right\}$$

The Gibbs-energy minima therefore occur where

$$\ln\frac{x_A}{1-x_A} = -\xi(1-2x_A) \qquad (5C.7)$$

This equation is an example of a 'transcendental equation', an equation that does not have a solution that can be expressed in a closed form. The solutions (the values of x_A that satisfy the equation) can be found numerically by using mathematical software or by plotting the terms on the left and right against x_A for a choice of values of ξ and identifying the values of x_A where the plots intersect, which is where the two expressions are equal (Fig. 5C.16). The solutions found in this way are plotted in Fig. 5C.17. As ξ decreases, the two minima move together and merge when $\xi = 2$.

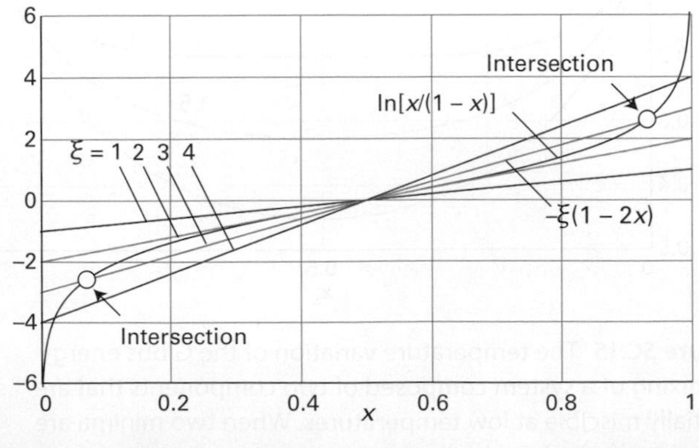

Figure 5C.16 The graphical procedure for solving eqn 5C.7. When $\xi < 2$, the only intersection occurs at $x = 0$. When $\xi \geq 2$, there are two solutions (those for $\xi = 3$ are marked).

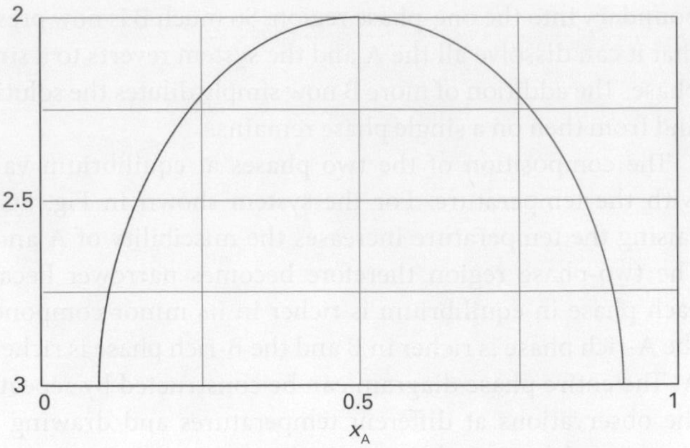

Figure 5C.17 The location of the phase boundary as computed on the basis of the ξ-parameter model introduced in Topic 5B.

Brief illustration 5C.5

In the system composed of benzene and cyclohexane treated in *Example* 5B.1 it is established that $\xi = 1.13$, so a two-phase system is not expected. That is, the two components are completely miscible at the temperature of the experiment. The single solution of the equation

$$\ln\frac{x_A}{1-x_A} + 1.13(1-2x_A) = 0$$

is $x_A = \frac{1}{2}$, corresponding to a single minimum of the Gibbs energy of mixing, and there is no phase separation.

Some systems show a **lower critical solution temperature**, T_{lc} (or *lower consolute temperature*), below which they mix in all proportions and above which they form two phases. An example is water and triethylamine (Fig. 5C.18). In this

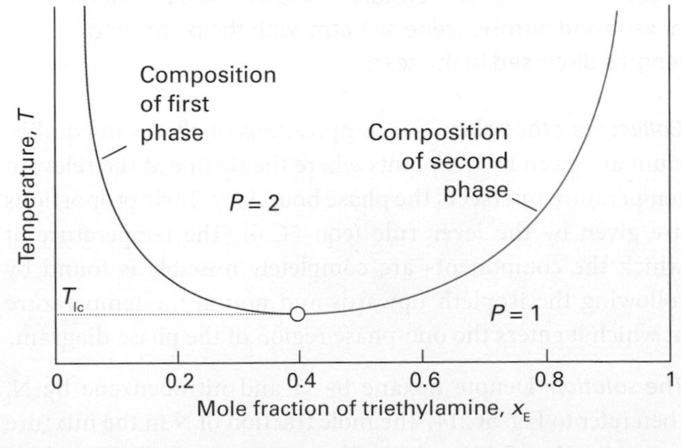

Figure 5C.18 The temperature–composition diagram for water and triethylamine. This system shows a lower critical solution temperature at 292 K. The labels indicate the interpretation of the boundaries.

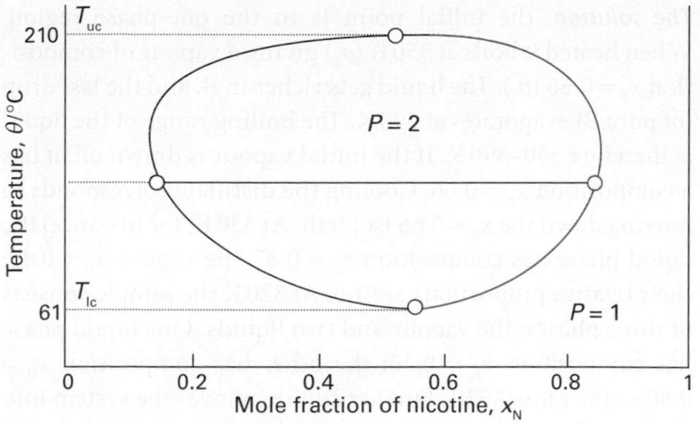

Figure 5C.19 The temperature–composition diagram for water and nicotine, which has both upper and lower critical temperatures. Note the high temperatures for the liquid (especially the water): the diagram corresponds to a sample under pressure.

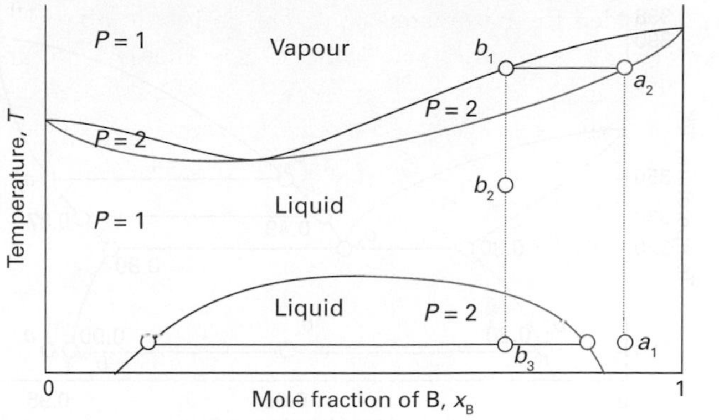

Figure 5C.20 The temperature–composition diagram for a binary system in which the upper critical solution temperature is less than the boiling point at all compositions. The mixture forms a low-boiling azeotrope.

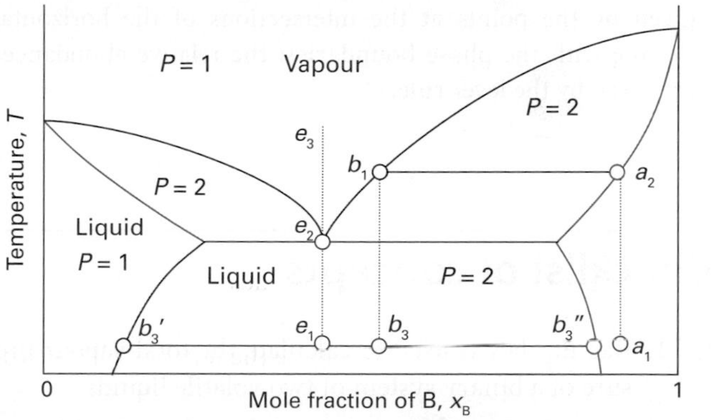

Figure 5C.21 The temperature–composition diagram for a binary system in which boiling occurs before the two liquids are fully miscible.

case, at low temperatures the two components are more miscible because they form a weak complex; at higher temperatures the complexes break up and the two components are less miscible.

Some systems have both upper and lower critical solution temperatures. They occur because, after the weak complexes have been disrupted, leading to partial miscibility, the thermal motion at higher temperatures homogenizes the mixture again, just as in the case of ordinary partially miscible liquids. The most famous example is nicotine and water, which are partially miscible between 61 °C and 210 °C (Fig. 5C.19).

(c) The distillation of partially miscible liquids

Consider a pair of liquids that are partially miscible and form a low-boiling azeotrope. This combination is quite common because both properties reflect the tendency of the two kinds of molecule to avoid each other. There are two possibilities: one in which the liquids become fully miscible before they boil; the other in which boiling occurs before mixing is complete.

Figure 5C.20 shows the phase diagram for two components that become fully miscible before they boil. Distillation of a mixture of composition a_1 leads to a vapour of composition b_1, which condenses to the completely miscible single-phase solution at b_2. Phase separation occurs only when this distillate is cooled to a point in the two-phase liquid region, such as b_3. This description applies only to the first drop of distillate. If distillation continues, the composition of the remaining liquid changes. In the end, when the whole sample has evaporated and condensed, the composition is back to a_1.

Figure 5C.21 shows the second possibility, in which there is no upper critical solution temperature. The distillate obtained from a liquid initially of composition a_1 has composition b_3 and is a two-phase mixture. One phase has composition b_3' and the other has composition b_3''.

The behaviour of a system of composition represented by the isopleth e in Fig. 5C.21 is interesting. A system at e_1 forms two phases, which persist (but with changing proportions) up to the boiling point at e_2. The vapour of this mixture has the same composition as the liquid (the liquid is an azeotrope). Similarly, condensing a vapour of composition e_3 gives a two-phase liquid of the same overall composition. At a fixed temperature, the mixture vaporizes and condenses like a single substance.

Example 5C.3 Interpreting a phase diagram

State the changes that occur when a mixture of composition $x_B = 0.95$ (a_1) in Fig. 5C.22 is boiled and the vapour condensed.

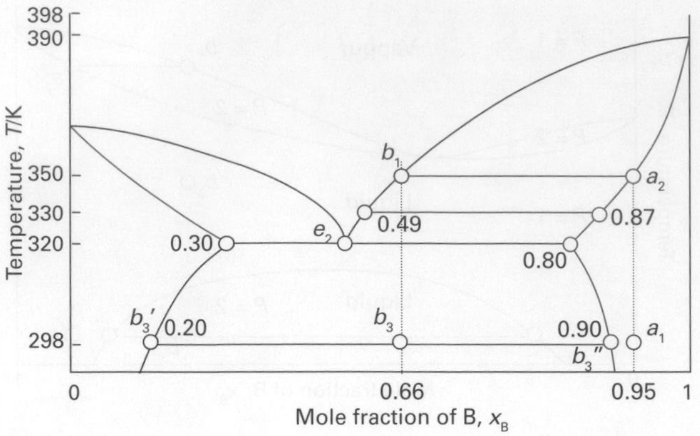

Figure 5C.22 The points of the phase diagram in Fig. 5C.20 that are discussed in *Example* 5C.3.

Collect your thoughts The area in which the point lies gives the number of phases; the compositions of the phases are given by the points at the intersections of the horizontal tie line with the phase boundaries; the relative abundances are given by the lever rule.

The solution The initial point is in the one-phase region. When heated it boils at 350 K (a_2) giving a vapour of composition $x_B = 0.66$ (b_1). The liquid gets richer in B, and the last drop (of pure B) evaporates at 390 K. The boiling range of the liquid is therefore 350–390 K. If the initial vapour is drawn off, it has a composition $x_B = 0.66$. Cooling the distillate corresponds to moving down the $x_B = 0.66$ isopleth. At 330 K, for instance, the liquid phase has composition $x_B = 0.87$, the vapour $x_B = 0.49$; their relative proportions are 1:6. At 320 K the sample consists of three phases: the vapour and two liquids. One liquid phase has composition $x_B = 0.30$; the other has composition $x_B = 0.80$ in the ratio 0.52:1. Further cooling moves the system into the two-phase region, and at 298 K the compositions are 0.20 and 0.90 in the ratio 0.82:1. As further distillate boils over, the overall composition of the distillate becomes richer in B. When the last drop has been condensed the phase composition is the same as at the beginning.

Self-test 5C.3 Repeat the discussion, beginning at the point $x_B = 0.4$, $T = 298$ K.

Checklist of concepts

☐ 1. Raoult's law is used to calculate the total vapour pressure of a binary system of two volatile liquids.

☐ 2. A **temperature–composition diagram** is a phase diagram in which the boundaries show the composition of the phases that are in equilibrium at various temperatures.

☐ 3. The composition of the vapour and the liquid phase in equilibrium are located at each end of a **tie line**.

☐ 4. The **lever rule** is used to deduce the relative abundances of each phase in equilibrium.

☐ 5. Separation of a liquid mixture by **fractional distillation** involves repeated cycles of boiling and condensation.

☐ 6. An **azeotrope** is a liquid mixture that evaporates without change of composition.

☐ 7. Phase separation of partially miscible liquids may occur when the temperature is below the **upper critical solution temperature** or above the **lower critical solution temperature**; the process may be discussed in terms of the model of a regular solution.

Checklist of equations

Property	Equation	Comment	Equation number
Composition of vapour	$y_A = x_A p_A^*/\{p_B^* + (p_A^* - p_B^*)x_A\}$ $y_B = 1 - y_A$	Ideal solution	5C.4
Total vapour pressure	$p = p_A^* p_B^*/\{p_A^* + (p_B^* - p_A^*)y_A\}$	Ideal solution	5C.5
Lever rule	$n_L l_L = n_V l_V$ (liquid and vapour phase at equilibrium)	In general, $n_\alpha l_\alpha = n_\beta l_\beta$ for phases α and β	5C.6

TOPIC 5D Phase diagrams of binary systems: solids

> ➤ **Why do you need to know this material?**
> Phase diagrams of solid mixtures are used widely in materials science, metallurgy, geology, and the chemical industry to summarize the composition of the various phases of mixtures, and it is important to be able to interpret them.
>
> ➤ **What is the key idea?**
> A phase diagram is a map showing the conditions under which each phase of a system is the most stable.
>
> ➤ **What do you need to know already?**
> It would be helpful to review the interpretation of liquid–liquid phase diagrams and the significance of the lever rule (Topic 5C).

This Topic considers systems where solid and liquid phases might both be present at temperatures below the boiling point.

5D.1 Eutectics

Consider the two-component liquid of composition a_1 in Fig. 5D.1. The changes that occur as the system is cooled may be expressed as follows:

- $a_1 \rightarrow a_2$. The system enters the two-phase region labelled 'Liquid + B'. Pure solid B begins to come out of solution and the remaining liquid becomes richer in A.
- $a_2 \rightarrow a_3$. More of the solid B forms and the relative amounts of the solid and liquid (which are in equilibrium) are given by the lever rule (Topic 5C). At this stage there are roughly equal amounts of each. The liquid phase is richer in A than before (its composition is given by b_3) because some B has been deposited.
- $a_3 \rightarrow a_4$. At the end of this step, there is less liquid than at a_3, and its composition is given by e_2. This liquid now freezes to give a two-phase system of pure B and pure A.

Physical interpretation

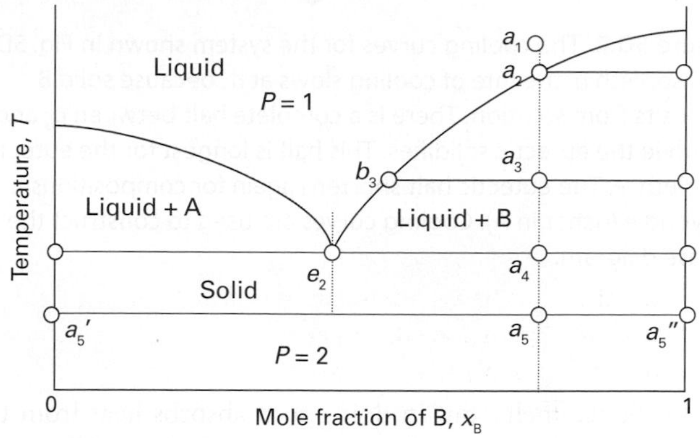

Figure 5D.1 The temperature–composition phase diagram for two almost immiscible solids and their completely miscible liquids. Note the similarity to Fig. 5C.21. The isopleth through e_2 corresponds to the eutectic composition, the mixture with lowest melting point.

The isopleth (constant-composition line) at e_2 in Fig. 5D.1 corresponds to the **eutectic** composition, the mixture with the lowest melting point.[1] A liquid with the eutectic composition freezes at a single temperature, without previously depositing solid A or B. A solid with the eutectic composition melts, without change of composition, at the lowest temperature of any mixture. Solutions of composition to the right of e_2 deposit B as they cool, and solutions to the left deposit A: only the eutectic mixture (apart from pure A or pure B) solidifies at a single definite temperature without gradually unloading one or other of the components from the liquid.

One eutectic that was technologically important until replaced by modern materials is a formulation of solder in which the mass composition is about 67 per cent tin and 33 per cent lead and melts at 183 °C. The eutectic formed by 23 per cent NaCl and 77 per cent H_2O by mass melts at −21.1 °C. When salt is added to ice under isothermal conditions (for example, when spread on an icy road) the mixture melts if the temperature is above −21.1 °C (and the eutectic composition has been achieved). When salt is added to ice under adiabatic conditions (for example, when added to ice in a vacuum

[1] The name comes from the Greek words for 'easily melted'.

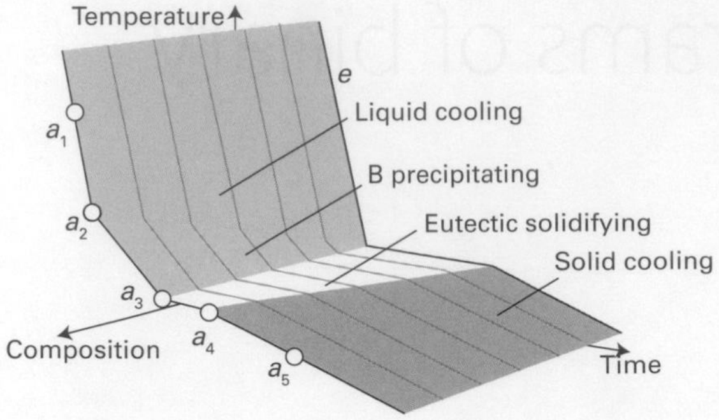

Figure 5D.2 The cooling curves for the system shown in Fig. 5D.1. For isopleth *a*, the rate of cooling slows at a_2 because solid B deposits from solution. There is a complete halt between a_3 and a_4 while the eutectic solidifies. This halt is longest for the eutectic isopleth, *e*. The eutectic halt shortens again for compositions beyond *e* (richer in A). Cooling curves are used to construct the phase diagram.

flask) the ice melts, but in doing so it absorbs heat from the rest of the mixture. The temperature of the system falls and, if enough salt is added, cooling continues down to the eutectic temperature. Eutectic formation occurs in the great majority of binary alloy systems, and is of great importance for the microstructure of solid materials. Although a eutectic solid is a two-phase system, it crystallizes out in a nearly homogeneous mixture of microcrystals. The two microcrystalline phases can be distinguished by microscopy and structural techniques such as X-ray diffraction (Topic 15B).

Thermal analysis is a very useful practical way of detecting eutectics. How it is used can be understood by considering the rate of cooling down the isopleth through a_1 in Fig. 5D.1. The liquid cools steadily until it reaches a_2, when B begins to be deposited (Fig. 5D.2). Cooling is now slower because the solidification of B is exothermic and retards the cooling. When the remaining liquid reaches the eutectic composition, the temperature remains constant until the whole sample has solidified: this region of constant temperature is the **eutectic halt**. If the liquid has the eutectic composition *e* initially, the liquid cools steadily down to the freezing temperature of the eutectic, when there is a long eutectic halt as the entire sample solidifies (like the freezing of a pure liquid).

Brief illustration 5D.1

Figure 5D.3 shows the phase diagram for the binary system silver/tin. The regions have been labelled to show which each one represents. When a liquid of composition *a* is cooled, solid silver with dissolved tin begins to precipitate at a_1 and the sample solidifies completely at a_2.

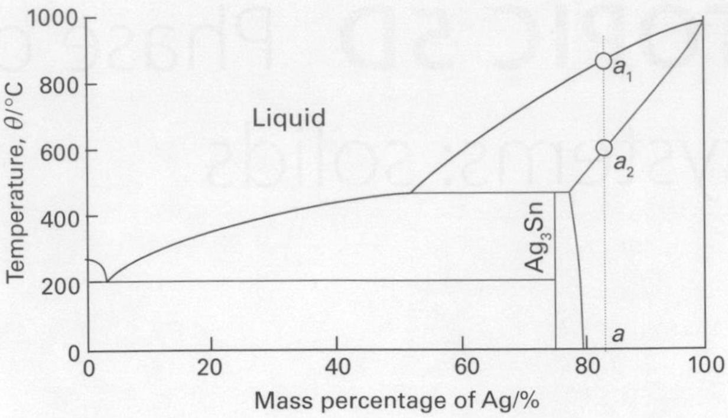

Figure 5D.3 The phase diagram for silver/tin discussed in *Brief illustration* 5D.1.

Monitoring the cooling curves at different overall compositions gives a clear indication of the structure of the phase diagram. The solid–liquid boundary is given by the points at which the rate of cooling changes. The longest eutectic halt gives the location of the eutectic composition and its melting temperature.

5D.2 Reacting systems

Many binary mixtures react to produce compounds, and technologically important examples of this behaviour include the Group 13/15 (III/V) semiconductors, such as the gallium arsenide system, which forms the compound GaAs. Although three constituents are present, there are only two components because GaAs is formed from the reaction Ga + As → GaAs. To illustrate some of the principles involved, consider a system that forms a compound C that also forms eutectic mixtures with the species A and B (Fig. 5D.4).

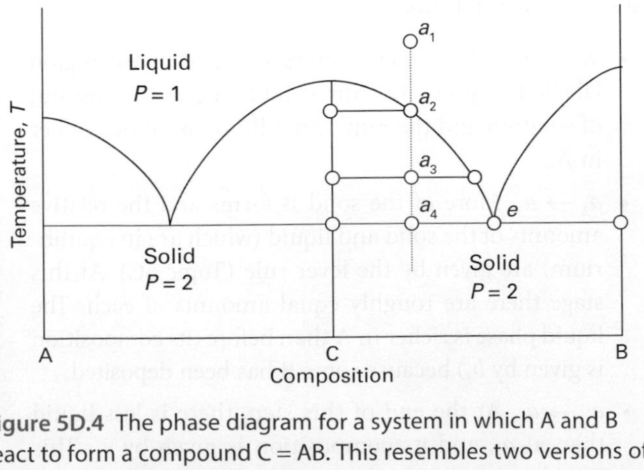

Figure 5D.4 The phase diagram for a system in which A and B react to form a compound C = AB. This resembles two versions of Fig. 5D.1 in each half of the diagram. The constituent C is a true compound, not just an equimolar mixture.

A system prepared by mixing an excess of B with A consists of C and unreacted B. This is a binary B,C system, which in this illustration is supposed to form a eutectic. The principal change from the eutectic phase diagram in Fig. 5D.1 is that the whole of the phase diagram is squeezed into the range of compositions lying between equal amounts of A and B ($x_B = 0.5$, marked C in Fig. 5D.4) and pure B. The interpretation of the information in the diagram is obtained in the same way as for Fig. 5D.1. The solid deposited on cooling along the isopleth a is the compound C. At temperatures below a_4 there are two solid phases, one consisting of C and the other of B. The pure compound C melts **congruently**, that is, the composition of the liquid it forms is the same as that of the solid compound.

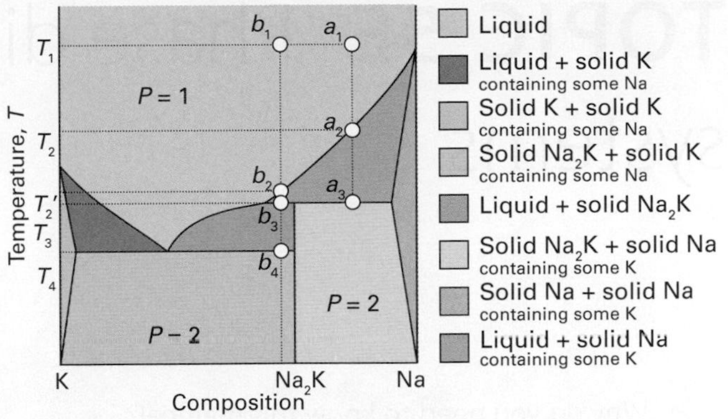

Figure 5D.5 The phase diagram for an actual system (sodium and potassium) like that shown in Fig. 5D.4, but with two differences. One is that the compound is Na_2K, corresponding to A_2B and not AB as in that illustration. The second is that the compound exists only as the solid, not as the liquid. The transformation of the compound at its melting point is an example of incongruent melting.

5D.3 Incongruent melting

In some cases the compound C is not stable as a liquid. An example is the alloy Na_2K, which survives only as a solid (Fig. 5D.5). Consider what happens as a liquid at a_1 is cooled:

Physical interpretation

- $a_2 \to a_3$. A solid solution rich in Na is deposited, and the remaining liquid is richer in K.
- Below a_3. The sample is now entirely solid and consists of a solid solution rich in Na and solid Na_2K.

Now consider the isopleth through b_1:

- $b_1 \to b_2$. No obvious change occurs until the phase boundary is reached at b_2 when a solid solution rich in Na begins to deposit.
- $b_2 \to b_3$. A solid solution rich in Na deposits, but at b_3 a reaction occurs to form Na_2K: this compound is formed by the K atoms diffusing into the solid Na.
- At b_3, three phases are in mutual equilibrium: the liquid, the compound Na_2K, and a solid solution rich in Na. The horizontal line representing this three-phase equilibrium is called a **peritectic line**. At this stage the liquid Na/K mixture is in equilibrium with a little solid Na_2K, but there is still no liquid compound.
- $b_3 \to b_4$. As cooling continues, the amount of solid compound increases until at b_4 the liquid reaches its eutectic composition. It then solidifies to give a two-phase solid consisting of a solid solution rich in K and solid Na_2K.

Physical interpretation

If the solid is reheated, the sequence of events is reversed. No liquid Na_2K forms at any stage because it is too unstable to exist as a liquid. This behaviour is an example of **incongruent melting**, in which a compound melts into its components and does not itself form a liquid phase.

Checklist of concepts

- ☐ 1. At the **eutectic composition** the liquid phase solidifies without change of composition.
- ☐ 2. A **peritectic line** in a phase diagram represents an equilibrium between three phases.
- ☐ 3. In **congruent melting** the composition of the liquid a compound forms is the same as that of the solid compound.
- ☐ 4. During **incongruent melting**, a compound melts into its components and does not itself form a liquid phase.

TOPIC 5E Phase diagrams of ternary systems

➤ **Why do you need to know this material?**

Ternary phase diagrams have become important in materials science as more complex materials are investigated, such as the ceramics found to have superconducting properties.

➤ **What is the key idea?**

A phase diagram is a map showing the conditions under which each phase of a system is the most stable.

➤ **What do you need to know already?**

It would be helpful to review the interpretation of two-component phase diagrams (Topics 5C and 5D) and the phase rule (Topic 4A). The interpretation of the phase diagrams presented here uses the lever rule (Topic 5C).

Consider the phases of a **ternary system**, a system with three components so $C = 3$. In terms of the phase rule (Topic 4A), $F = 5 - P$. If the system is restricted to constant temperature and pressure, two degrees of freedom are discarded and $F'' = 3 - P$. If two phases are present ($P = 2$), then $F'' = 1$ and the system has one degree of freedom: changing the amount of one component results in changes in the amounts of the other two components. This condition is represented by an area in the phase diagram. If three phases are present ($P = 3$), then $F'' = 0$, and the system is represented by a single point on the phase diagram.

Lines in ternary phase diagrams represent conditions under which two phases may coexist. Two phases are in equilibrium when they are connected by tie lines, as in binary phase diagrams.

5E.1 Triangular phase diagrams

The mole fractions of the three components of a ternary system satisfy $x_A + x_B + x_C = 1$. A phase diagram drawn as an equilateral triangle ensures that this property is satisfied automatically because the sum of the distances to a point inside an equilateral triangle of side 1 and measured parallel to the edges is equal to 1 (Fig. 5E.1).

Figure 5E.1 shows how this approach works in practice. The edge AB corresponds to $x_C = 0$, and likewise for the other two edges. Hence, each of the three edges corresponds to one of

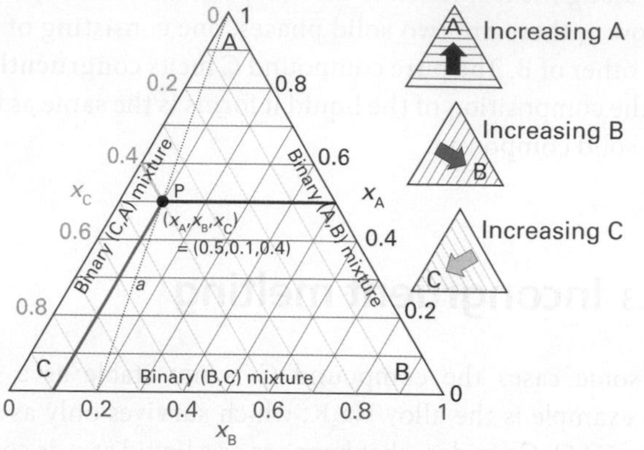

Figure 5E.1 The triangular coordinates used for the discussion of three-component systems. Each edge corresponds to a binary system. All points along the dotted line *a* correspond to mole fractions of C and B in the same ratio.

the three binary systems (A,B), (B,C), and (C,A). An interior point corresponds to a system in which all three components are present. The point P, for instance, represents $x_A = 0.50$, $x_B = 0.10$, $x_C = 0.40$.

Any point on a straight line joining the A apex to a point on the opposite edge (the dotted line *a* in Fig. 5E.1) represents a composition that is progressively richer in A the closer the point is to the A apex, but for which the concentration ratio B:C remains constant. Therefore, to represent the changing composition of a system as A is added, draw a line from the A apex to the point on BC representing the initial binary system. Any ternary system formed by adding A then lies at some point on this line.

Brief illustration 5E.1

The following points are represented on Fig. 5E.2.

Point	x_A	x_B	x_C
a	0.20	0.80	0
b	0.42	0.26	0.32
c	0.80	0.10	0.10
d	0.10	0.20	0.70
e	0.20	0.40	0.40
f	0.30	0.60	0.10

Note that the points *d*, *e*, and *f* have $x_A/x_B = 0.50$ and lie on a straight line.

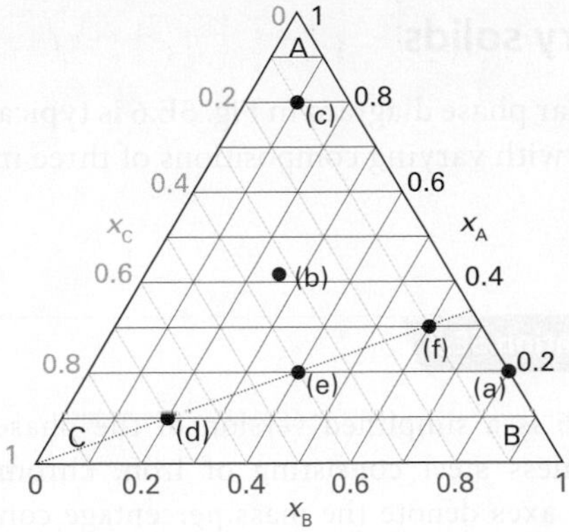

Figure 5E.2 The points referred to in *Brief illustration* 5E.1.

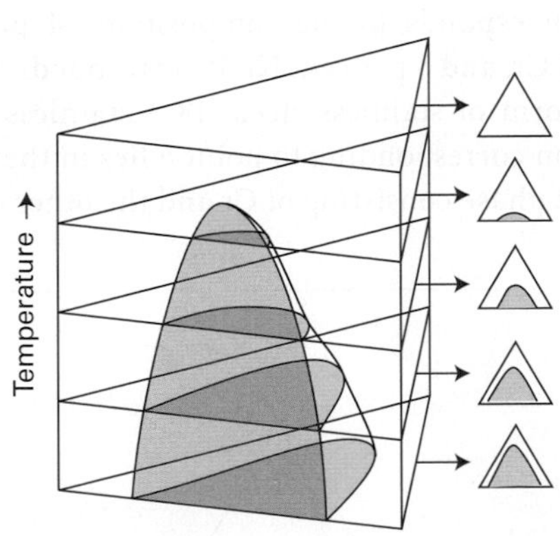

Figure 5E.3 When temperature is included as a variable, the phase diagram becomes a triangular prism. Horizontal sections through the prism correspond to the triangular phase diagrams being discussed and illustrated in Fig. 5E.1.

A single triangle represents the equilibria when one of the discarded degrees of freedom (the temperature, for instance) has a certain value. Different temperatures give rise to different equilibrium behaviour and therefore different triangular phase diagrams. Each one may therefore be regarded as a horizontal slice through a three-dimensional triangular prism, such as that shown in Fig. 5E.3.

5E.2 Ternary systems

Ternary phase diagrams are widely used in metallurgy and materials science. Although they can become quite complex, they can be interpreted in much the same way as binary diagrams.

(a) Partially miscible liquids

The phase diagram for a ternary system in which W (in due course: water) and E (in due course: ethanoic acid (acetic acid)) are fully miscible, E and T (in due course: trichloromethane (chloroform)) are fully miscible, but W and T are only partially miscible is shown in Fig. 5E.4. This illustration is for the system water/ethanoic acid/trichloromethane at room temperature, which behaves in this way:

- The two fully miscible pairs, (E,W) and (E,T), form single-phase regions.
- The (W,T) system (along the base of the triangle) has a two-phase region.

The base of the triangle corresponds to one of the horizontal lines in a two-component phase diagram. The tie lines in the two-phase regions are constructed experimentally by determining the compositions of the two phases that are in equilibrium, marking them on the diagram, and then joining them with a straight line.

A single-phase system is formed when enough E is added to the binary (W,T) mixture. This effect is illustrated by following the line *a* in Fig. 5E.4:

- a_1. The system consists of two phases and the relative amounts of the two phases can be read off by using the lever rule.
- $a_1 \rightarrow a_2$. The addition of E takes the system along the line joining a_1 to the E apex. At a_2 the solution still has two phases, but there is slightly more W in the largely T phase (represented by the point a_2'') and more T in the largely W phase (a_2') because the presence of E helps both to dissolve. The phase diagram shows that there is more E in the W rich phase than in the T-rich phase (a_2' is closer than a_2'' to the E apex).
- $a_2 \rightarrow a_3$. At a_3 two phases are present, but the T-rich layer is present only as a trace (lever rule).
- $a_3 \rightarrow a_4$. Further addition of E takes the system towards and beyond a_4, and only a single phase is present.

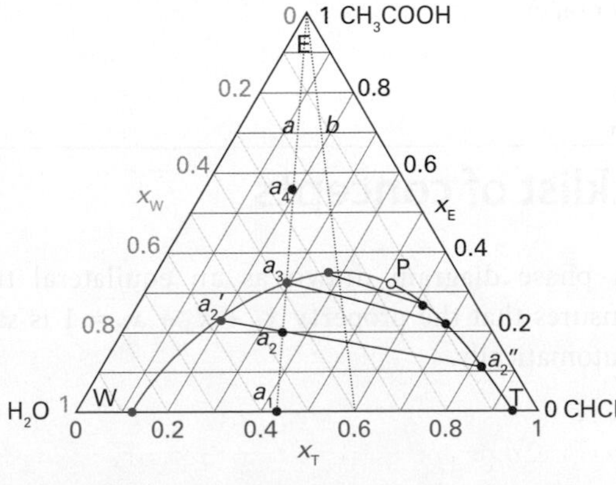

Figure 5E.4 The phase diagram, at fixed temperature and pressure, of the three-component system ethanoic acid (E), trichloromethane (T), and water (W). Only some of the tie lines have been drawn in the two-phase region. All points along the line *a* correspond to trichloromethane and water present in the same ratio.

Brief illustration 5E.2

Consider a mixture of water (W in Fig. 5E.4) and trichloromethane (T) with $x_W = 0.40$ and $x_T = 0.60$, and ethanoic acid (E) is added to it. The relative proportions of W and T remain constant, so the point representing the overall composition moves along the straight line *b* from $x_T = 0.60$ on the base to the ethanoic acid apex. The initial composition is in a two-phase region: one phase has the composition $(x_W, x_T, x_E) = (0.05, 0.95, 0)$ and the other has composition $(x_W, x_T, x_E) = (0.88, 0.12, 0)$. When sufficient ethanoic acid has been added to raise its mole fraction to 0.18 the system consists of two phases of composition (0.07, 0.82, 0.11) and (0.57, 0.20, 0.23) in the ratio 1:3.

The point marked P in Fig. 5E.4 is called the **plait point**: at this point the compositions of the two phases in equilibrium become identical. It is yet another example of a critical point. For convenience, the general interpretation of a triangular phase diagram is summarized in Fig. 5E.5.

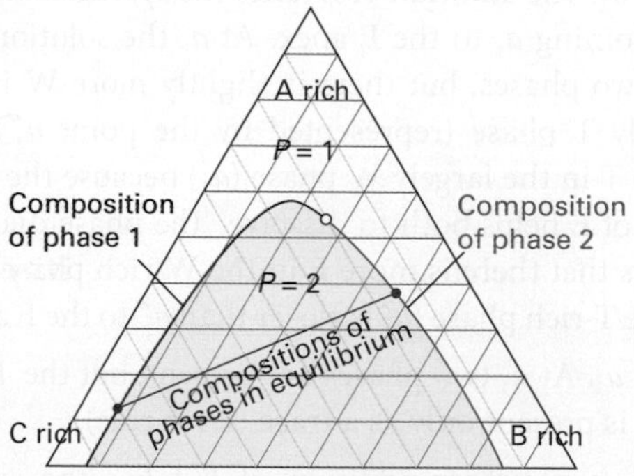

Figure 5E.5 The interpretation of a triangular phase diagram. The region inside the curved line consists of two phases, and the compositions of the two phases in equilibrium are given by the points at the ends of the tie lines (the tie lines are determined experimentally).

(b) Ternary solids

The triangular phase diagram in Fig. 5E.6 is typical of that for a solid alloy with varying compositions of three metals, A, B, and C.

Brief illustration 5E.3

Figure 5E.6 is a simplified version of the phase diagram for a stainless steel consisting of iron, chromium, and nickel. The axes denote the mass percentage compositions instead of the mole fractions, but as the three percentages add up to 100 per cent, the interpretation of points in the triangle is essentially the same as for mole fractions. The point *a* corresponds to the composition 74 per cent Fe, 18 per cent Cr, and 8 per cent Ni. It corresponds to the most common form of stainless steel, '18-8 stainless steel'. The composition corresponding to point *b* lies in the two-phase region, one phase consisting of Cr and the other of the alloy γ-FeNi.

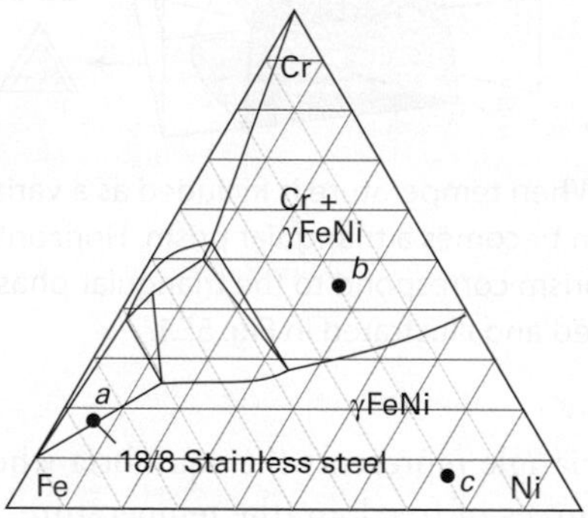

Figure 5E.6 A simplified triangular phase diagram of the ternary system represented by a stainless steel composed of iron, chromium, and nickel.

Checklist of concepts

☐ 1. A phase diagram drawn as an equilateral triangle ensures that the property $x_A + x_B + x_C = 1$ is satisfied automatically.

☐ 2. At the **plait point**, the compositions of the two phases in equilibrium are identical.

TOPIC 5F Activities

> **➤ Why do you need to know this material?**
>
> The concept of an ideal solution is a good starting point for the discussion of mixtures, but to understand real solutions it is important to be able to describe deviations from ideal behaviour and to express them in terms of molecular interactions.
>
> **➤ What is the key idea?**
>
> The activity of a species, its effective concentration, helps to preserve the form of the expressions derived on the basis of ideal behaviour but extends their reach to real mixtures.
>
> **➤ What do you need to know already?**
>
> This Topic is based on the expression for chemical potential of a species derived from Raoult's and Henry's laws (Topic 5A). It also uses the formulation of a model of a regular solution introduced in Topic 5B.

This Topic shows how to adjust the expressions developed in Topics 5A and 5B to take into account deviations from ideal behaviour. As in other Topics collected in this Focus, the solvent is denoted by A, the solute by B, and a general component by J.

5F.1 The solvent activity

The general form of the chemical potential of a real or ideal solvent is given by a straightforward modification of eqn 5A.21 ($\mu_A = \mu_A^* + RT\ln(p_A/p_A^*)$), where p_A^* is the vapour pressure of pure A and p_A is the vapour pressure of A when it is a component of a solution. The solvent in an ideal solution obeys Raoult's law (Topic 5A, $p_A = x_A p_A^*$) at all concentrations and the chemical potential is expressed as eqn 5A.23 ($\mu_A = \mu_A^* + RT\ln x_A$). The form of this relation can be preserved when the solution does not obey Raoult's law by writing

$$\mu_A = \mu_A^* + RT\ln a_A \qquad \text{Activity of solvent [definition]} \qquad (5F.1)$$

The quantity a_A is the **activity** of A, a kind of 'effective' mole fraction.

Because eqn 5F.1 is true for both real and ideal solutions, comparing it with $\mu_A = \mu_A^* + RT\ln(p_A/p_A^*)$ gives

$$a_A = \frac{p_A}{p_A^*} \qquad \text{Activity of solvent [measurement]} \qquad (5F.2)$$

There is nothing mysterious about the activity of a solvent: it can be determined experimentally simply by measuring the vapour pressure and then using this relation.

Brief illustration 5F.1

The vapour pressure of 0.500 mol dm^{-3} KNO_3(aq) at 100 °C is 99.95 kPa, and the vapour pressure of pure water at this temperature is 1.00 atm (101 kPa). It follows that the activity of water in this solution at this temperature is

$$a_A = \frac{99.95\,\text{kPa}}{101\,\text{kPa}} = 0.990$$

Because all solvents obey Raoult's law more closely as the concentration of solute approaches zero, the activity of the solvent approaches the mole fraction as $x_A \to 1$:

$$a_A \to x_A \text{ as } x_A \to 1 \qquad (5F.3)$$

A convenient way of expressing this convergence is to introduce the **activity coefficient**, γ (gamma), by the definition

$$a_A = \gamma_A x_A \quad \gamma_A \to 1 \text{ as } x_A \to 1 \qquad \text{Activity coefficient of solvent [definition]} \qquad (5F.4)$$

at all temperatures and pressures. The chemical potential of the solvent is then

$$\mu_A = \mu_A^* + RT\ln x_A + RT\ln \gamma_A \qquad \text{Chemical potential of solvent} \qquad (5F.5)$$

The standard state of the solvent is established when $x_A = 1$ (the pure solvent) and the pressure is 1 bar.

5F.2 The solute activity

The problem with defining activity coefficients and standard states for solutes is that they approach ideal–dilute (Henry's law) behaviour as $x_B \to 0$, not as $x_B \to 1$ (corresponding to pure solute).

(a) Ideal–dilute solutions

A solute B that satisfies Henry's law (Topic 5A) has a vapour pressure given by $p_B = K_B x_B$, where K_B is an empirical constant. In this case, the chemical potential of B is

$$\mu_B = \mu_B^* + RT\ln\frac{p_B}{p_B^*} \overset{p_B = K_B x_B}{=} \mu_B^* + RT\ln\frac{K_B x_B}{p_B^*} = \overbrace{\mu_B^* + RT\ln\frac{K_B}{p_B^*}}^{\mu_B^\ominus} + RT\ln x_B \qquad (5F.6)$$

Both K_B and p_B^* are characteristics of the solute, so the two blue terms may be combined to give a new standard chemical potential, $\mu_B^\ominus$

$$\mu_B^\ominus = \mu_B^* + RT\ln\frac{K_B}{p_B^*} \qquad (5F.7)$$

It then follows that the chemical potential of a solute in an ideal–dilute solution is related to its mole fraction by

$$\mu_B = \mu_B^\ominus + RT\ln x_B \qquad (5F.8)$$

If the solution is ideal, $K_B = p_B^*$ (Raoult's law) and eqn 5F.7 reduces to $\mu_B^\ominus = \mu_B^*$, as expected.

Brief illustration 5F.2

In *Example* 5A.4 it is established that in a mixture of propanone and trichloromethane at 298 K $K_{\text{propanone}} = 24.5\,\text{kPa}$, whereas $p^*_{\text{propanone}} = 46.3\,\text{kPa}$. It follows from eqn 5F.7 that

$$\mu^\ominus_{\text{propanone}} = \mu^*_{\text{propanone}} + RT\ln\frac{24.5\text{ kPa}}{46.3\text{ kPa}}$$

$$= \mu^*_{\text{propanone}} + (8.3145\,\text{J K}^{-1}\,\text{mol}^{-1})\times(298\,\text{K})\times\ln\frac{24.5}{46.3}$$

$$= \mu^*_{\text{propanone}} - 1.58\,\text{kJ mol}^{-1}$$

and the standard value differs from the value for the pure liquid by $-1.58\,\text{kJ mol}^{-1}$.

(b) Real solutes

Real solutions deviate from ideal–dilute, Henry's law behaviour. For the solute, the introduction of a_B in place of x_B in eqn 5F.8 gives

$$\mu_B = \mu_B^\ominus + RT\ln a_B \qquad \text{Chemical potential of solute [definition]} \qquad (5F.9)$$

The standard state remains unchanged in this last stage, and all the deviations from ideality are captured in the activity a_B.

It remains true that $\mu_B = \mu_B^* + RT\ln(p_B/p_B^*)$, but now, from eqn 5F.7 written as $\mu_B^* = \mu_B^\ominus - RT\ln(K_B/p_B^*)$ it follows that

$$\mu_B = \overbrace{\mu_B^\ominus - RT\ln\frac{K_B}{p_B^*}}^{\mu_B^*} + RT\ln\frac{p_B}{p_B^*} = \mu_B^\ominus + RT\ln\frac{p_B}{K_B}$$

Comparison of this expression with eqn 5F.9 identifies the activity a_B as

$$a_B = \frac{p_B}{K_B} \qquad \text{Activity of solute [measurement]} \qquad (5F.10)$$

As for the solvent, it is sensible to introduce an activity coefficient through

$$a_B = \gamma_B x_B \qquad \text{Activity coefficient of solute [definition]} \qquad (5F.11)$$

Now all the deviations from ideality are captured in the activity coefficient γ_B. Because the solute obeys Henry's law ($p_B = K_B x_B$) as its concentration goes to zero. It follows that

$$a_B \to x_B \text{ and } \gamma_B \to 1 \text{ as } x_B \to 0 \qquad (5F.12)$$

at all temperatures and pressures. Deviations of the solute from ideality disappear as its concentration approaches zero.

Example 5F.1 Measuring activity

Use the following information to calculate the activity and activity coefficient of trichloromethane (chloroform, C) in propanone (acetone, A) at 25 °C, treating it first as a solvent and then as a solute.

x_C	0	0.20	0.40	0.60	0.80	1
p_C/kPa	0	4.7	11	18.9	26.7	36.4
p_A/kPa	46.3	33.3	23.3	12.3	4.9	0

Collect your thoughts For the activity of chloroform as a solvent (the Raoult's law activity), write $a_C = p_C/p_C^*$ and $\gamma_C = a_C/x_C$. For its activity as a solute (the Henry's law activity), write $a_C = p_C/K_C$ and $\gamma_C = a_C/x_C$ with the new activity.

The solution Because $p_C^* = 36.4\,\text{kPa}$ and $K_C = 23.5\,\text{kPa}$ (from *Example* 5A.4), construct the following tables. For instance, at $x_C = 0.20$, in the Raoult's law case $a_C = (4.7\,\text{kPa})/(36.4\,\text{kPa}) = 0.13$ and $\gamma_C = 0.13/0.20 = 0.65$; likewise, in the Henry's law case, $a_C = (4.7\,\text{kPa})/(23.5\,\text{kPa}) = 0.20$ and $\gamma_C = 0.20/0.20 = 1.0$.

From Raoult's law (chloroform regarded as the solvent):

x_C	0	0.20	0.40	0.60	0.80	1
a_C	0	0.13	0.30	0.52	0.73	1.00
γ_C		0.65	0.75	0.87	0.92	1.00

From Henry's law (chloroform regarded as the solute):

x_C	0	0.20	0.40	0.60	0.80	1
a_C	0	0.20	0.47	0.80	1.14	1.55
γ_C	1	1.00	1.17	1.34	1.42	1.55

These values are plotted in Fig. 5F.1. Notice that $\gamma_C \to 1$ as $x_C \to 1$ in the Raoult's law case, but that $\gamma_C \to 1$ as $x_C \to 0$ in the Henry's law case.

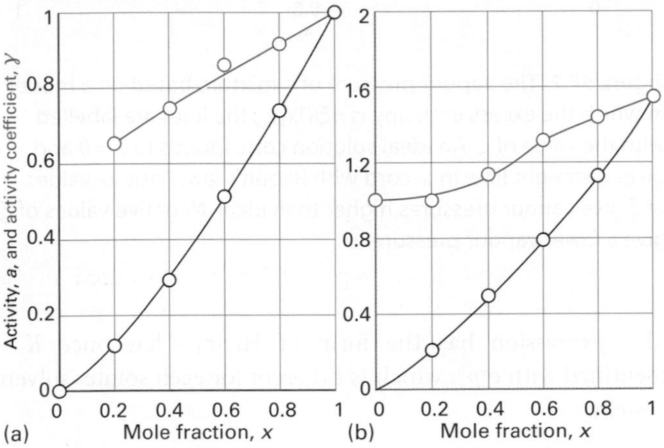

Figure 5F.1 The variation of activity and activity coefficient for a trichloromethane/propanone (chloroform/acetone) mixture with composition according to (a) Raoult's law, (b) Henry's law.

Self-test 5F.1 Calculate the activities and activity coefficients for acetone according to the two conventions (use $p_A^* = 46.3\,\text{kPa}$ and $K_A = 24.5\,\text{kPa}$)

Answer: At $x_A = 0.60$, for instance $a_R = 0.50$; $\gamma_R = 0.84$; $a_H = 0.95$, $\gamma_H = 1.59$

(c) Activities in terms of molalities

The selection of a standard state is entirely arbitrary and can be chosen in a way that suits the description of the composition of the system best. Because compositions are often expressed as molalities, b, in place of mole fractions (see *The chemist's toolkit* 11 in Topic 5A) it is then convenient to write

$$\mu_B = \mu^{\ominus}{}_B + RT\ln\frac{b_B}{b^{\ominus}} \qquad (5F.13)$$

where $\mu_B^{\ominus}$ has a different value from the standard value introduced earlier. According to this definition, the chemical potential of the solute has its standard value $\mu_B^{\ominus}$ when the molality of B is $b^{\ominus}$ (i.e. at 1 mol kg^{-1}). Note that as $b_B \to 0$, $\mu_B \to -\infty$; that is, as the solution becomes diluted, so the solute becomes increasingly thermodynamically stable. The practical consequence of this result is that it is very difficult to remove the last traces of a solute from a solution.

As before, deviations from ideality are incorporated by introducing a dimensionless activity a_B and a dimensionless activity coefficient γ_B, and writing

$$a_B = \gamma_B\frac{b_B}{b^{\ominus}}, \text{ where } \gamma_B \to 1 \text{ as } b_B \to 0 \qquad (5F.14)$$

at all temperatures and pressures. The standard state remains unchanged in this last stage and, as before, all the deviations from ideality are captured in the activity coefficient γ_B. The final expression for the chemical potential of a real solute at any molality is then

$$\mu_B = \mu^{\ominus}{}_B + RT\ln a_B \qquad (5F.15)$$

5F.3 The activities of regular solutions

The concept of regular solutions (Topic 5B) gives further insight into the origin of deviations from Raoult's law and its relation to activity coefficients. The starting point is the model expression for the excess enthalpy (eqn 5B.6, $H^E = n\xi RTx_Ax_B$) and its implication for the Gibbs energy of mixing for a regular solution (eqn 5B.7, $\Delta_{mix}G = nRT\{x_A\ln x_A + x_B\ln x_B + \xi x_Ax_B\}$). On the basis of this model it is possible to develop expressions for the activity coefficients in terms of the parameter ξ.

How is that done? 5F.1 Developing expressions for the activity coefficients of a regular solution

The Gibbs energy of mixing to form an ideal solution is given in eqn. 5B.3:

$$\Delta_{mix}G = nRT\{x_A\ln x_A + x_B\ln x_B\}$$

The corresponding expression for a non-ideal solution is

$$\Delta_{mix}G = nRT\{x_A\ln a_A + x_B\ln a_B\}$$

This relation follows in the same way as for an ideal mixture but with activities in place of mole fractions. However, in Topic 5B.7 it is established (in eqn 5B.7) that for a regular solution

$$\Delta_{mix}G = nRT\{x_A\ln x_A + x_B\ln x_B + \xi x_Ax_B\}$$

The last two equations can be made consistent as follows. First replace each activity by γ_Jx_J:

$$\begin{aligned}\Delta_{mix}G &= nRT\{x_A\ln x_A\gamma_A + x_B\ln x_B\gamma_B\}\\ &= nRT\{x_A\ln x_A + x_B\ln x_B + x_A\ln\gamma_A + x_B\ln\gamma_B\}\end{aligned}$$

For consistency, the sum of the two terms in blue must be equal to $\xi x_A x_B$, which can be achieved by writing $\ln\gamma_A = \xi x_B^2$ and $\ln\gamma_B = \xi x_A^2$, because then

$$x_A \overbrace{\ln\gamma_A}^{\xi x_B^2} + x_B \overbrace{\ln\gamma_B}^{\xi x_A^2} = \xi x_A x_B^2 + \xi x_B x_A^2 = \xi\overbrace{(x_A + x_B)}^{1} x_A x_B = \xi x_A x_B$$

It follows that the activity coefficients of a regular solution are given by what are known as the **Margules equations**:

$$\ln\gamma_A = \xi x_B^2 \qquad \ln\gamma_B = \xi x_A^2 \qquad (5F.16)$$

Margules equations

Note that the activity coefficients behave correctly for dilute solutions: $\gamma_A \to 1$ as $x_B \to 0$ and $\gamma_B \to 1$ as $x_A \to 0$. Also note that A and B are treated here as equal components of a mixture, not as solvent and solute.

At this point the Margules equations can be used to write the activity of A as

$$a_A = \gamma_A x_A = x_A e^{\xi x_B^2} = x_A e^{\xi(1-x_A)^2} \qquad (5F.17)$$

($\gamma_A = e^{\xi x_B^2}$; $x_B = 1 - x_A$)

with a similar expression for a_B. The activity of A, though, is just the ratio of the vapour pressure of A in the solution to the vapour pressure of pure A (eqn 5F.2, $a_A = p_A/p_A^*$), so

$$p_A = p_A^* x_A e^{\xi(1-x_A)^2} \qquad (5F.18)$$

This function is plotted in Fig. 5F.2, and interpreted as follows:

Physical interpretation

- When $\xi = 0$, corresponding to an ideal solution, $p_A = p_A^* x_A$, in accord with Raoult's law.
- Positive values of ξ (endothermic mixing, unfavourable solute–solvent interactions) give vapour pressures higher than for an ideal solution.
- Negative values of ξ (exothermic mixing, favourable solute–solvent interactions) give a vapour pressure lower than for an ideal solution.

All the plots of eqn 5F.18 approach linearity and coincide with the Raoult's law line as $x_A \to 1$ and the exponential function in eqn 5F.18 approaches 1. When $x_A \ll 1$, eqn 5F.18 approaches

$$p_A = p_A^* x_A e^{\xi} \qquad (5F.19)$$

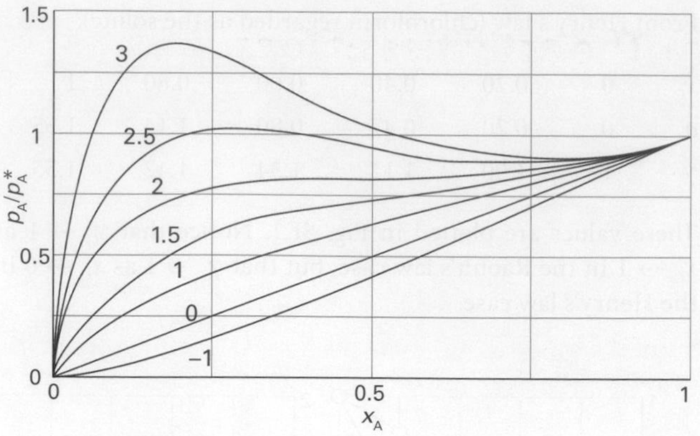

Figure 5F.2 The vapour pressure of a mixture based on a model in which the excess enthalpy is $n\xi RTx_A x_B$; the lines are labelled with the value of ξ. An ideal solution corresponds to $\xi = 0$ and gives a straight line, in accord with Raoult's law. Positive values of ξ give vapour pressures higher than ideal. Negative values of ξ give a lower vapour pressure.

This expression has the form of Henry's law once K is identified with $e^{\xi}p_A^*$, which is different for each solute–solvent system.

Brief illustration 5F.3

In *Example* 5B.1 of Topic 5B it is established that $\xi = 1.13$ for a mixture of benzene and cyclohexane at 25 °C. Because $\xi > 0$ the vapour pressure of the mixture is expected to be greater than its ideal value. The total vapour pressure of the mixture is therefore

$$p = p_{\text{benzene}}^* x_{\text{benzene}} e^{1.13(1-x_{\text{benzene}})^2} + p_{\text{cyclohexane}}^* x_{\text{cyclohexane}} e^{1.13(1-x_{\text{cyclohexane}})^2}$$

This expression is plotted in Fig. 5F.3, using $p_{\text{benzene}}^* = 10.0\,\text{kPa}$ and $p_{\text{cyclohexane}}^* = 10.4\,\text{kPa}$.

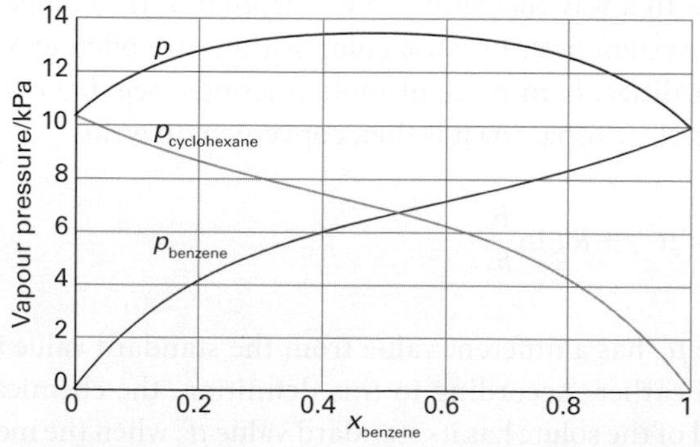

Figure 5F.3 The computed vapour pressure curves for a mixture of benzene and cyclohexane at 25 °C as derived in *Brief illustration* 5F.3.

5F.4 The activities of ions

Interactions between ions are so strong that the approximation of replacing activities by molalities is valid only in very dilute solutions (less than 1 mmol kg^{-1} in total ion concentration), and in precise work activities themselves must be used.

If the chemical potential of the cation M^+ is denoted μ_+ and that of the anion X^- is denoted μ_-, the molar Gibbs energy of the ions in the electrically neutral solution is the sum of these partial molar quantities. The molar Gibbs energy of an *ideal* solution of such ions is

$$G_m^{ideal} = \mu_+^{ideal} + \mu_-^{ideal} \quad (5F.20)$$

with $\mu_J^{ideal} = \mu_J^{\ominus} + RT\ln x_J$. However, for a *real* solution of M^+ and X^- of the same molality it is necessary to write $\mu_J = \mu_J^{\ominus} + RT\ln a_J$ with $a_J = \gamma_J x_J$, which implies that $\mu_J = \mu_J^{ideal} + RT\ln \gamma_J$. It then follows that

$$G_m = \mu_+ + \mu_- = \mu_+^{ideal} + \mu_-^{ideal} + RT\ln\gamma_+ + RT\ln\gamma_-$$
$$= G_m^{ideal} + RT\ln\gamma_+\gamma_- \quad (5F.21)$$

All the deviations from ideality are contained in the last term.

(a) Mean activity coefficients

There is no experimental way of separating the product $\gamma_+\gamma_-$ into contributions from the cations and the anions. The best that can be done experimentally is to assign responsibility for the non-ideality equally to both kinds of ion. Therefore, the 'mean activity coefficient' is introduced as the geometric mean of the individual coefficients, where the geometric mean of x^p and y^q is $(x^p y^q)^{1/(p+q)}$. For a 1,1-electrolyte $p = 1$, $q = 1$ and the required geometric mean is

$$\gamma_\pm = (\gamma_+\gamma_-)^{1/2} \quad (5F.22)$$

The individual chemical potentials of the ions are then written

$$\mu_+ = \mu_+^{ideal} + RT\ln\gamma_\pm \qquad \mu_- = \mu_-^{ideal} + RT\ln\gamma_\pm \quad (5F.23)$$

The sum of these two chemical potentials is the same as before, eqn 5F.21, but now the non-ideality is shared equally.

To generalize this approach to the case of a compound M_pX_q that dissolves to give a solution of p cations and q anions from each formula unit, the molar Gibbs energy of the ions is written as the sum of their partial molar Gibbs energies (i.e. their chemical potentials):

$$G_m = p\mu_+ + q\mu_- = G_m^{ideal} + pRT\ln\gamma_+ + qRT\ln\gamma_- \quad (5F.24)$$

The **mean activity coefficient** can now be defined in a more general way as

$$\gamma_\pm = (\gamma_+^p\gamma_-^q)^{1/s} \qquad s = p + q \qquad \text{Mean activity coefficient [definition]} \quad (5F.25)$$

and the chemical potential of each ion written as

$$\mu_i = \mu_i^{ideal} + RT\ln\gamma_\pm \quad (5F.26)$$

(b) The Debye–Hückel limiting law

The long range and strength of the Coulombic interaction between ions means that it is likely to be primarily responsible for the departures from ideality in ionic solutions and to dominate all the other contributions to non-ideality. This domination is the basis of the **Debye–Hückel theory** of ionic solutions, which was devised by Peter Debye and Erich Hückel in 1923. The following is a qualitative account of the theory and its principal conclusions. For a quantitative treatment, see *A deeper look* 1 on the website for this text.

Oppositely charged ions attract one another. As a result, anions are more likely to be found near cations in solution, and vice versa (Fig. 5F.4). Overall, the solution is electrically neutral, but near any given ion there is an excess of counter ions (ions of opposite charge). Averaged over time, counter ions are more likely to be found near any given ion. This time-averaged, spherical haze around the central ion, in which counter ions outnumber ions of the same charge as the central ion, has a net charge equal in magnitude but opposite in sign to that on the central ion, and is called its **ionic atmosphere**. The energy, and therefore the chemical potential, of any given central ion are lowered as a result of its electrostatic interaction with its ionic atmosphere. This lowering of energy appears as the difference

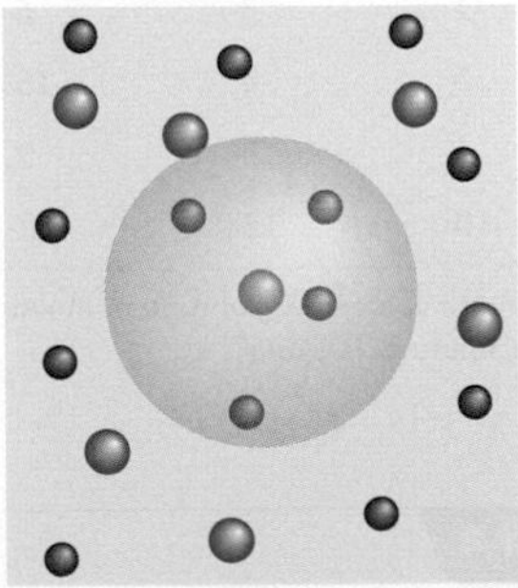

Figure 5F.4 The model underlying the Debye–Hückel theory is of a tendency for anions to be found around cations, and of cations to be found around anions (one such local clustering region is shown by the shaded sphere). The ions are in ceaseless motion, and the diagram represents a snapshot of their motion. The solutions to which the theory applies are far less concentrated than shown here.

between the molar Gibbs energy G_m and the ideal value of the molar Gibbs energy G_m^{ideal} of the solute, and hence can be identified with the term $RT\ln\gamma_{\pm}$ in eqn 5F.21. The stabilization of ions by their interaction with their ionic atmospheres is part of the explanation why chemists commonly use dilute solutions, in which the stabilization is minimized, to achieve precipitation of ions from electrolyte solutions.

The model leads to the result that at very low concentrations the activity coefficient can be calculated from the **Debye–Hückel limiting law**

$$\log\gamma_{\pm} = -A|z_+z_-|I^{1/2} \quad \text{Debye–Hückel limiting law} \quad (5F.27)$$

where $A = 0.509$ for an aqueous solution at 25 °C and I is the dimensionless **ionic strength** of the solution:

$$I = \tfrac{1}{2}\sum_i z_i^2(b_i/b^{\ominus}) \quad \text{Ionic strength [definition]} \quad (5F.28)$$

In this expression z_i is the charge number of an ion i (positive for cations and negative for anions) and b_i is its molality. The ionic strength occurs widely, and often as its square root (as in eqn 5F.27) wherever ionic solutions are discussed. The sum extends over all the ions present in the solution. For solutions consisting of two types of ion at molalities b_+ and b_-,

$$I = \tfrac{1}{2}(b_+z_+^2 + b_-z_-^2)/b^{\ominus} \quad (5F.29)$$

The ionic strength emphasizes the charges of the ions because the charge numbers occur as their squares. Table 5F.1 summarizes the relation of ionic strength and molality in an easily usable form.

Table 5F.1 Ionic strength and molality, $I = kb/b^{\ominus}$

k	X^-	X^{2-}	X^{3-}	X^{4-}
M^+	1	3	6	10
M^{2+}	3	4	15	12
M^{3+}	6	15	9	42
M^{4+}	10	12	42	16

For example, the ionic strength of an M_2X_3 solution of molality b, which is understood to give M^{3+} and X^{2-} ions in solution, is $15b/b^{\ominus}$.

Brief illustration 5F.4

The mean activity coefficient of 5.0 mmol kg^{-1} KCl(aq) at 25 °C is calculated by writing $I = \tfrac{1}{2}(b_+ + b_-)/b^{\ominus} = b/b^{\ominus}$, where b is the molality of the solution (and $b_+ = b_- = b$). Then, from eqn 5F.27,

$$\log\gamma_{\pm} = -0.509 \times (5.0 \times 10^{-3})^{1/2} = -0.03\ldots$$

Hence, $\gamma_{\pm} = 0.92$. The experimental value is 0.927.

Table 5F.2 Mean activity coefficients in water at 298 K*

$b/b^{\ominus}$	KCl	$CaCl_2$
0.001	0.966	0.888
0.01	0.902	0.732
0.1	0.770	0.524
1.0	0.607	0.725

* More values are given in the *Resource section*.

The name 'limiting law' is applied to eqn 5F.27 because ionic solutions of moderate molalities may have activity coefficients that differ from the values given by this expression, but all solutions are expected to conform as $b \to 0$. Table 5F.2 lists some experimental values of activity coefficients for salts of various valence types. Figure 5F.5 shows some of these values plotted against $I^{1/2}$, and compares them with the theoretical straight lines calculated from eqn 5F.27. The agreement at very low molalities (less than about 1 mmol kg^{-1}, depending on charge type) is impressive and convincing evidence in support of the model. Nevertheless, the departures from the theoretical curves above these molalities are large, and show that the approximations are valid only at very low concentrations.

(c) Extensions of the limiting law

When the ionic strength of the solution is too high for the limiting law to be valid, the activity coefficient may be estimated

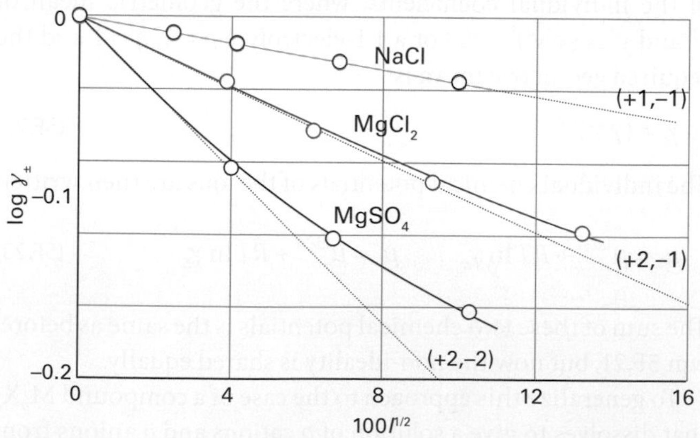

Figure 5F.5 An experimental test of the Debye–Hückel limiting law. Although there are marked deviations for moderate ionic strengths, the limiting slopes (shown as dotted lines) as $I \to 0$ are in good agreement with the theory, so the law can be used for extrapolating data to very low molalities. The numbers in parentheses are the charge numbers of the ions.

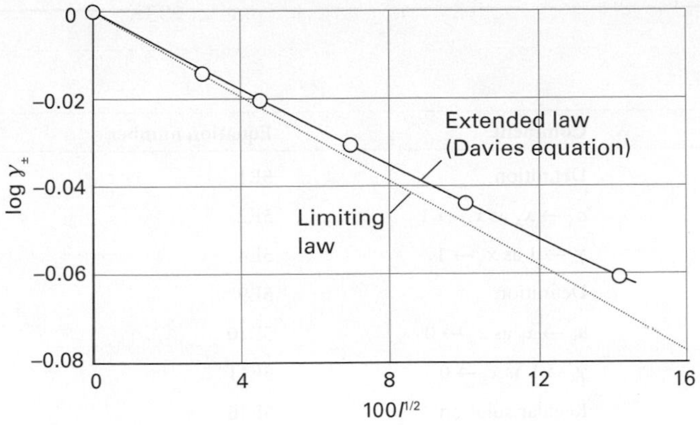

Figure 5F.6 The Davies equation gives agreement with experiment over a wider range of molalities than the limiting law (shown as a dotted line), but it fails at higher molalities. The data are for a 1,1-electrolyte.

from the **extended Debye–Hückel law** (sometimes called the *Truesdell–Jones equation*):

$$\log\gamma_\pm = -\frac{A|z_+z_-|I^{1/2}}{1+BI^{1/2}} \qquad \text{Extended Debye–Hückel law} \qquad (5F.30a)$$

where B is a dimensionless constant. A more flexible extension is the **Davies equation** proposed by C.W. Davies in 1938:

$$\log\gamma_\pm = -\frac{A|z_+z_-|I^{1/2}}{1+BI^{1/2}} + CI \qquad \text{Davies equation} \qquad (5F.30b)$$

where C is another dimensionless constant. Although B can be interpreted as a measure of the closest approach of the ions, it (like C) is best regarded as an adjustable empirical parameter. A graph drawn on the basis of the Davies equation is shown in Fig. 5F.6. It is clear that eqn 5F.30b accounts for some activity coefficients over a moderate range of dilute solutions (up to about 0.1 mol kg^{-1}); nevertheless it remains very poor near 1 mol kg^{-1}.

Current theories of activity coefficients for ionic solutes take an indirect route. They set up a theory for the dependence of the activity coefficient of the solvent on the concentration of the solute, and then use the Gibbs–Duhem equation (eqn 5A.12a, $n_A d\mu_A + n_B d\mu_B = 0$) to estimate the activity coefficient of the solute. The results are reasonably reliable for solutions with molalities greater than about 0.1 mol kg^{-1} and are valuable for the discussion of mixed salt solutions, such as sea water.

Table 5F.3 Activities and standard states: a summary*

Component	Basis	Standard state	Activity	Limits
Solid or liquid		Pure, 1 bar	$a = 1$	
Solvent	Raoult	Pure solvent, 1 bar	$a = p/p^*$, $a = \gamma x$	$\gamma \to 1$ as $x \to 1$ (pure solvent)
Solute	Henry	(1) A hypothetical state of the pure solute	$a = p/K$, $a = \gamma x$	$\gamma \to 1$ as $x \to 0$
		(2) A hypothetical state of the solute at molality $b^\ominus$	$a = \gamma b/b^\ominus$	$\gamma \to 1$ as $b \to 0$
Gas	Fugacity†	Pure, a hypothetical state of 1 bar and behaving as a perfect gas	$f = \gamma p$	$\gamma \to 1$ as $p \to 0$

* In each case, $\mu = \mu^\ominus + RT\ln a$.
† Fugacity is discussed in *A deeper look* 2 on the website for this text.

Checklist of concepts

☐ 1. The **activity** is an effective concentration that preserves the form of the expression for the chemical potential. See Table 5F.3.

☐ 2. The chemical potential of a solute in an ideal–dilute solution is defined on the basis of Henry's law.

☐ 3. The activity of a solute takes into account departures from Henry's law behaviour.

☐ 4. The **Margules equations** relate the activities of the components of a model regular solution to its composition. They lead to expressions for the vapour pressures of the components of a regular solution,

☐ 5. **Mean activity coefficients** apportion deviations from ideality equally to the cations and anions in an ionic solution.

☐ 6. An **ionic atmosphere** is the time average accumulation of counter ions that exists around an ion in solution.

☐ 7. The **Debye–Hückel theory** ascribes deviations from ideality to the Coulombic interaction of an ion with the ionic atmosphere around it.

☐ 8. The **Debye–Hückel limiting law** is extended by including two further empirical constants.

Checklist of equations

Property	Equation	Comment	Equation number
Chemical potential of solvent	$\mu_A = \mu_A^* + RT\ln a_A$	Definition	5F.1
Activity of solvent	$a_A = p_A/p_A^*$	$a_A \to x_A$ as $x_A \to 1$	5F.2
Activity coefficient of solvent	$a_A = \gamma_A x_A$	$\gamma_A \to 1$ as $x_A \to 1$	5F.4
Chemical potential of solute	$\mu_B = \mu_B^\ominus + RT\ln a_B$	Definition	5F.9
Activity of solute	$a_B = p_B/K_B$	$a_B \to x_B$ as $x_B \to 0$	5F.10
Activity coefficient of solute	$a_B = \gamma_B x_B$	$\gamma_B \to 1$ as $x_B \to 0$	5F.11
Margules equations	$\ln\gamma_A = \xi x_B^2$, $\ln\gamma_B = \xi x_A^2$	Regular solution	5F.16
Vapour pressure	$p_A = p_A^* x_A e^{\xi(1-x_A)^2}$	Regular solution	5F.18
Mean activity coefficient	$\gamma_\pm = (\gamma_+^p \gamma_-^q)^{1/s}$, $s = p + q$	Definition	5F.25
Debye–Hückel limiting law	$\log\gamma_\pm = -A\lvert z_+z_-\rvert I^{1/2}$	Valid as $I \to 0$	5F.27
Ionic strength	$I = \frac{1}{2}\sum_i z_i^2(b_i/b^\ominus)$	Definition	5F.28
Davies equation	$\log\gamma_\pm = -A\lvert z_+z_-\rvert I^{1/2}/(1+BI^{1/2})+CI$	A, B, C empirical constants	5F.30b

FOCUS 5 Simple mixtures

TOPIC 5A The thermodynamic description of mixtures

Discussion questions

D5A.1 Explain the concept of partial molar quantity, and justify the remark that the partial molar properties of a solute depend on the properties of the solvent too.

D5A.2 Explain how thermodynamics relates non-expansion work to a change in composition of a system.

D5A.3 Are there any circumstances under which two (real) gases will not mix spontaneously?

D5A.4 Explain how Raoult's law and Henry's law are used to specify the chemical potential of a component of a mixture.

D5A.5 Explain the molecular origin of Raoult's law and Henry's law.

Exercises

E5A.1(a) A polynomial fit to measurements of the total volume of a binary mixture of A and B is

$$v = 987.93 + 35.6774x - 0.459\,23x^2 + 0.017\,325x^3$$

where $v = V/\text{cm}^3$, $x = n_B/\text{mol}$, and n_B is the amount of B present. Derive an expression for the partial molar volume of B.
E5A.1(b) A polynomial fit to measurements of the total volume of a binary mixture of A and B is

$$v = 778.55 - 22.5749x + 0.568\,92x^2 + 0.010\,23x^3 + 0.002\,34x^4$$

where $v = V/\text{cm}^3$, $x = n_B/\text{mol}$, and n_B is the amount of B present. Derive an expression for the partial molar volume of B.

E5A.2(a) The volume of an aqueous solution of NaCl at 25 °C was measured at a series of molalities b, and it was found to fit the expression $v = 1003 + 16.62x + 1.77x^{3/2} + 0.12x^2$ where $v = V/\text{cm}^3$, V is the volume of a solution formed from 1.000 kg of water, and $x = b/b^{\ominus}$. Calculate the partial molar volume of the components in a solution of molality 0.100 mol kg^{-1}.
E5A.2(b) At 18 °C the total volume V of a solution formed from $MgSO_4$ and 1.000 kg of water fits the expression $v = 1001.21 + 34.69(x - 0.070)^2$, where $v = V/\text{cm}^3$ and $x = b/b^{\ominus}$. Calculate the partial molar volumes of the salt and the solvent in a solution of molality 0.050 mol kg^{-1}.

E5A.3(a) Suppose that $n_A = 0.10n_B$ and a small change in composition results in μ_A changing by $\delta\mu_A = +12\,\text{J mol}^{-1}$, by how much will μ_B change?
E5A.3(b) Suppose that $n_A = 0.22n_B$ and a small change in composition results in μ_A changing by $\delta\mu_A = -15\,\text{J mol}^{-1}$, by how much will μ_B change?

E5A.4(a) Consider a container of volume 5.0 dm^3 that is divided into two compartments of equal size. In the left compartment there is nitrogen at 1.0 atm and 25 °C; in the right compartment there is hydrogen at the same temperature and pressure. Calculate the entropy and Gibbs energy of mixing when the partition is removed. Assume that the gases are perfect.
E5A.4(b) Consider a container of volume 250 cm^3 that is divided into two compartments of equal size. In the left compartment there is argon at 100 kPa and 0 °C; in the right compartment there is neon at the same temperature and pressure. Calculate the entropy and Gibbs energy of mixing when the partition is removed. Assume that the gases are perfect.

E5A.5(a) The vapour pressure of benzene at 20 °C is 10 kPa and that of methylbenzene is 2.8 kPa at the same temperature. What is the vapour pressure of a mixture of equal masses of each component?
E5A.5(b) At 90 °C the vapour pressure of 1,2-dimethylbenzene is 20 kPa and that of 1,3-dimethylbenzene is 18 kPa. What is the composition of the vapour of an equimolar mixture of the two components?

E5A.6(a) The partial molar volumes of propanone (acetone) and trichloromethane (chloroform) in a mixture in which the mole fraction of $CHCl_3$ is 0.4693 are 74.166 cm^3 mol^{-1} and 80.235 cm^3 mol^{-1}, respectively. What is the volume of a solution of mass 1.000 kg?
E5A.6(b) The partial molar volumes of two liquids A and B in a mixture in which the mole fraction of A is 0.3713 are 188.2 cm^3 mol^{-1} and 176.14 cm^3 mol^{-1}, respectively. The molar masses of the A and B are 241.1 g mol^{-1} and 198.2 g mol^{-1}. What is the volume of a solution of mass 1.000 kg?

E5A.7(a) At 25 °C, the mass density of a 50 per cent by mass ethanol–water solution is 0.914 g cm^{-3}. Given that the partial molar volume of water in the solution is 17.4 cm^3 mol^{-1}, calculate the partial molar volume of the ethanol.
E5A.7(b) At 20 °C, the mass density of a 20 per cent by mass ethanol–water solution is 968.7 kg m^{-3}. Given that the partial molar volume of ethanol in the solution is 52.2 cm^3 mol^{-1}, calculate the partial molar volume of the water.

E5A.8(a) At 300 K, the partial vapour pressures of HCl (i.e. the partial pressures of the HCl vapour) in liquid $GeCl_4$ are as follows:

x_{HCl}	0.005	0.012	0.019
p_{HCl}/kPa	32.0	76.9	121.8

Show that the solution obeys Henry's law in this range of mole fractions, and calculate Henry's law constant at 300 K.
E5A.8(b) At 310 K, the partial vapour pressures of a substance B dissolved in a liquid A are as follows:

x_B	0.010	0.015	0.020
p_B/kPa	82.0	122.0	166.1

Show that the solution obeys Henry's law in this range of mole fractions, and calculate Henry's law constant at 310 K.

E5A.9(a) Calculate the molar solubility of nitrogen in benzene exposed to air at 25 °C; the partial pressure of nitrogen in air is calculated in *Example* 1A.2 of Topic 1A.
E5A.9(b) Calculate the molar solubility of methane at 1.0 bar in benzene at 25 °C.

E5A.10(a) Use Henry's law and the data in Table 5A.1 to calculate the solubility (as a molality) of CO_2 in water at 25 °C when its partial pressure is (i) 0.10 atm, (ii) 1.00 atm.
E5A.10(b) The mole fractions of N_2 and O_2 in air at sea level are approximately 0.78 and 0.21. Calculate the molalities of the solution formed in an open flask of water at 25 °C.

E5A.11(a) A water carbonating plant is available for use in the home and operates by providing carbon dioxide at 5.0 atm. Estimate the molar concentration of CO_2 in the carbonated water it produces.
E5A.11(b) After some weeks of use, the pressure in the water carbonating plant mentioned in the previous exercise has fallen to 2.0 atm. Estimate the molar concentration of CO_2 in the carbonated water it produces at this stage.

Problems

P5A.1 The experimental values of the partial molar volume of a salt in water are found to fit the expression $v_B = 5.117 + 19.121x^{1/2}$, where $v_B = V_B/(\text{cm}^3\,\text{mol}^{-1})$ and x is the numerical value of the molality of B ($x = b/b^{\ominus}$). Use the Gibbs–Duhem equation to derive an equation for the molar volume of water in the solution. The molar volume of pure water at the same temperature is 18.079 cm³ mol⁻¹.

P5A.2 Use the Gibbs–Duhem equation to show that the partial molar volume (or any partial molar property) of a component B can be obtained if the partial molar volume (or other property) of A is known for all compositions up to the one of interest. Do this by proving that

$$V_B = V_B^* - \int_{V_A^*}^{V_A} \frac{x_A}{1-x_A} dV_A$$

where the x_A are functions of the V_A. Use the following data (which are for 298 K) to evaluate the integral graphically to find the partial molar volume of propanone dissolved in trichloromethane at $x = 0.500$.

$x(CHCl_3)$	0	0.194	0.385	0.559	0.788	0.889	1.000
$V_m/(\text{cm}^3\,\text{mol}^{-1})$	73.99	75.29	76.50	77.55	79.08	79.82	80.67

P5A.3 Consider a gaseous mixture with mass percentage composition 75.5 (N_2), 23.2 (O_2), and 1.3 (Ar). (a) Calculate the entropy of mixing when the mixture is prepared from the pure (and perfect) gases. (b) Air may be taken as a mixture with mass percentage composition 75.52 (N_2), 23.15 (O_2), 1.28 (Ar), and 0.046 (CO_2). What is the change in entropy from the value calculated in part (a)?

P5A.4 For a mixture of methylbenzene (A) and butanone in equilibrium at 303.15 K, the following table gives the mole fraction of A in the liquid phase, x_A, and in the gas phase, y_A, as well as the total pressure p. Take the vapour to be perfect and calculate the partial pressures of the two components. Plot them against their respective mole fractions in the liquid mixture and find the Henry's law constants for the two components.

x_A	0	0.0898	0.2476	0.3577	0.5194	0.6036
y_A	0	0.0410	0.1154	0.1762	0.2772	0.3393
p/kPa	36.066	34.121	30.900	28.626	25.239	23.402

x_A	0.7188	0.8019	0.9105	1
y_A	0.4450	0.5435	0.7284	1
p/kPa	20.6984	18.592	15.496	12.295

P5A.5 The mass densities of aqueous solutions of copper(II) sulfate at 20 °C were measured as set out below. Determine and plot the partial molar volume of $CuSO_4$ in the range of the measurements.

$m(CuSO_4)$/g	5	10	15	20
$\rho/(\text{g cm}^{-3})$	1.051	1.107	1.167	1.230

where $m(CuSO_4)$ is the mass of $CuSO_4$ dissolved in 100 g of solution.

P5A.6 Haemoglobin, the red blood protein responsible for oxygen transport, binds about 1.34 cm³ of oxygen per gram. Normal blood has a haemoglobin concentration of 150 g dm⁻³. Haemoglobin in the lungs is about 97 per cent saturated with oxygen, but in the capillary is only about 75 per cent saturated. What volume of oxygen is given up by 100 cm³ of blood flowing from the lungs in the capillary?

P5A.7 Use the data from *Example* 5A.1 to determine the value of b at which V_E has a minimum value.

TOPIC 5B The properties of solutions

Discussion questions

D5B.1 Explain what is meant by a regular solution; what additional features distinguish a real solution from a regular solution?

D5B.2 Would you expect the excess volume of mixing of oranges and melons to be positive or negative?

D5B.3 Explain the physical origin of colligative properties.

D5B.4 Identify the feature that accounts for the difference in boiling-point constants of water and benzene.

D5B.5 Why are freezing-point constants typically larger than the corresponding boiling-point constants of a solvent?

D5B.6 Explain the origin of osmosis in terms of the thermodynamic and molecular properties of a mixture.

D5B.7 Colligative properties are independent of the identity of the solute. Why, then, can osmometry be used to determine the molar mass of a solute?

Exercises

E5B.1(a) Predict the partial vapour pressure of HCl above its solution in liquid germanium tetrachloride of molality 0.10 mol kg⁻¹. For data, see Exercise E5A.8(a).

E5B.1(b) Predict the partial vapour pressure of the component B above its solution in A in Exercise E5A.8(b) when the molality of B is 0.25 mol kg⁻¹. The molar mass of A is 74.1 g mol⁻¹.

E5B.2(a) The vapour pressure of benzene is 53.3 kPa at 60.6 °C, but it fell to 51.5 kPa when 19.0 g of a non-volatile organic compound was dissolved in 500 g of benzene. Calculate the molar mass of the compound.

E5B.2(b) The vapour pressure of 2-propanol is 50.00 kPa at 338.8 °C, but it fell to 49.62 kPa when 8.69 g of a non-volatile organic compound was dissolved in 250 g of 2-propanol. Calculate the molar mass of the compound.

E5B.3(a) The addition of 100 g of a compound to 750 g of CCl_4 lowered the freezing point of the solvent by 10.5 K. Calculate the molar mass of the compound.

E5B.3(b) The addition of 5.00 g of a compound to 250 g of naphthalene lowered the freezing point of the solvent by 0.780 K. Calculate the molar mass of the compound.

E5B.4(a) Estimate the freezing point of 200 cm³ of water sweetened by the addition of 2.5 g of sucrose. Treat the solution as ideal.

E5B.4(b) Estimate the freezing point of 200 cm³ of water to which 2.5 g of sodium chloride has been added. Treat the solution as ideal.

E5B.5(a) The osmotic pressure of an aqueous solution at 300 K is 120 kPa. Estimate the freezing point of the solution.

E5B.5(b) The osmotic pressure of an aqueous solution at 288 K is 99.0 kPa. Estimate the freezing point of the solution.

E5B.6(a) Calculate the Gibbs energy, entropy, and enthalpy of mixing when 0.50 mol C_6H_{14} (hexane) is mixed with 2.00 mol C_7H_{16} (heptane) at 298 K. Treat the solution as ideal.

E5B.6(b) Calculate the Gibbs energy, entropy, and enthalpy of mixing when 1.00 mol C_6H_{14} (hexane) is mixed with 1.00 mol C_7H_{16} (heptane) at 298 K. Treat the solution as ideal.

E5B.7(a) What proportions of hexane and heptane should be mixed (i) by mole fraction, (ii) by mass in order to achieve the greatest entropy of mixing?

E5B.7(b) What proportions of benzene and ethylbenzene should be mixed (i) by mole fraction, (ii) by mass in order to achieve the greatest entropy of mixing?

E5B.8(a) The enthalpy of fusion of anthracene is 28.8 kJ mol^{-1} and its melting point is 217 °C. Calculate its ideal solubility in benzene at 25 °C.

E5B.8(b) Predict the ideal solubility of lead in bismuth at 280 °C given that its melting point is 327 °C and its enthalpy of fusion is 5.2 kJ mol^{-1}.

E5B.9(a) A dilute solution of bromine in carbon tetrachloride behaves as an ideal dilute solution. The vapour pressure of pure CCl_4 is 33.85 Torr at 298 K. The Henry's law constant when the concentration of Br_2 is expressed as a mole fraction is 122.36 Torr. Calculate the vapour pressure of each component, the total pressure, and the composition of the vapour phase when the mole fraction of Br_2 is 0.050, on the assumption that the conditions of the ideal dilute solution are satisfied at this concentration.

E5B.9(b) The vapour pressure of a pure liquid A is 23 kPa at 20 °C and the Henry's law constant of B in liquid A is 73 kPa. Calculate the vapour pressure of each component, the total pressure, and the composition of the vapour phase when the mole fraction of B is 0.066 on the assumption that the conditions of the ideal–dilute solution are satisfied at this concentration.

E5B.10(a) At 90 °C, the vapour pressure of methylbenzene is 53.3 kPa and that of 1,2-dimethylbenzene is 20.0 kPa. What is the composition of a liquid mixture that boils at 90 °C when the pressure is 0.50 atm? What is the composition of the vapour produced?

E5B.10(b) At 90 °C, the vapour pressure of 1,2-dimethylbenzene is 20 kPa and that of 1,3-dimethylbenzene is 18 kPa What is the composition of a liquid mixture that boils at 90 °C when the pressure is 19 kPa? What is the composition of the vapour produced?

E5B.11(a) The vapour pressure of pure liquid A at 300 K is 76.7 kPa and that of pure liquid B is 52.0 kPa. These two compounds form ideal liquid and gaseous mixtures. Consider the equilibrium composition of a mixture in which the mole fraction of A in the vapour is 0.350. Calculate the total pressure of the vapour and the composition of the liquid mixture.

E5B.11(b) The vapour pressure of pure liquid A at 293 K is 68.8 kPa and that of pure liquid B is 82.1 kPa. These two compounds form ideal liquid and gaseous mixtures. Consider the equilibrium composition of a mixture in which the mole fraction of A in the vapour is 0.612. Calculate the total pressure of the vapour and the composition of the liquid mixture.

E5B.12(a) It is found that the boiling point of a binary solution of A and B with $x_A = 0.6589$ is 88 °C. At this temperature the vapour pressures of pure A and B are 127.6 kPa and 50.60 kPa, respectively. (i) Is this solution ideal? (ii) What is the initial composition of the vapour above the solution?

E5B.12(b) It is found that the boiling point of a binary solution of A and B with $x_A = 0.4217$ is 96 °C. At this temperature the vapour pressures of pure A and B are 110.1 kPa and 76.5 kPa, respectively. (i) Is this solution ideal? (ii) What is the initial composition of the vapour above the solution?

Problems

P5B.1 Potassium fluoride is very soluble in glacial acetic acid (ethanoic acid) and the solutions have a number of unusual properties. In an attempt to understand them, freezing-point depression data were obtained by taking a solution of known molality and then diluting it several times (J. Emsley, *J. Chem. Soc. A*, 2702 (1971)). The following data were obtained:

b/(mol kg^{-1})	0.015	0.037	0.077	0.295	0.602
ΔT/K	0.115	0.295	0.470	1.381	2.67

Calculate the apparent molar mass of the solute and suggest an interpretation. Use $\Delta_{fus}H = 11.4$ kJ mol^{-1} and $T_f^* = 290$ K for glacial acetic acid.

P5B.2 In a study of the properties of an aqueous solution of $Th(NO_3)_4$ by A. Apelblat, D. Azoulay, and A. Sahar (*J. Chem. Soc. Faraday Trans., I*, 1618, (1973)), a freezing-point depression of 0.0703 K was observed for an aqueous solution of molality 9.6 mmol kg^{-1}. What is the apparent number of ions per formula unit?

P5B.3[1] Comelli and Francesconi examined mixtures of propionic acid with various other organic liquids at 313.15 K (F. Comelli and R. Francesconi, *J. Chem. Eng. Data* **41**, 101 (1996)). They report the excess volume of mixing propionic acid with tetrahydropyran (THP, oxane) as $V^E = x_1x_2\{a_0 + a_1(x_1 - x_2)\}$, where x_1 is the mole fraction of propionic acid, x_2 that of THP, $a_0 = -2.4697$ cm^3 mol^{-1}, and $a_1 = 0.0608$ cm^3 mol^{-1}. The density of propionic acid at this temperature is 0.971 74 g cm^{-3}; that of THP is 0.863 98 g cm^{-3}. (a) Derive an expression for the partial molar volume of each component at this temperature. (b) Compute the partial molar volume for each component in an equimolar mixture.

[1] These problems were provided by Charles Trapp and Carmen Giunta.

P5B.4‡ Equation 5B.14 indicates, after it has been converted into an expression for x_B, that solubility is an exponential function of temperature. The data in the table below gives the solubility, S, of calcium ethanoate in water as a function of temperature.

θ/°C	0	20	40	60	80
S/(g/100 g solvent)	36.4	34.9	33.7	32.7	31.7

Determine the extent to which the data fit the exponential $S = S_0e^{\tau/T}$ and obtain values for S_0 and τ. Express these constants in terms of properties of the solute.

P5B.5 The excess Gibbs energy of solutions of methylcyclohexane (MCH) and tetrahydrofuran (THF) at 303.15 K were found to fit the expression

$$G^E = RTx(1-x)\{0.4857 - 0.1077(2x-1) + 0.0191(2x-1)^2\}$$

where x is the mole fraction of MCH. Calculate the Gibbs energy of mixing when a mixture of 1.00 mol MCH and 3.00 mol THF is prepared.

P5B.6 The excess Gibbs energy of a certain binary mixture is equal to $gRTx(1-x)$ where g is a constant and x is the mole fraction of a solute B. Find an expression for the chemical potential of B in the mixture and sketch its dependence on the composition.

P5B.7 The molar mass of a protein was determined by dissolving it in water, and measuring the height, h, of the resulting solution drawn up a capillary tube at 20 °C. The following data were obtained.

c/(mg cm^{-3})	3.221	4.618	5.112	6.722
h/cm	5.746	8.238	9.119	11.990

The osmotic pressure may be calculated from the height of the column as $\Pi = h\rho g$, taking the mass density of the solution as $\rho = 1.000\,\text{g cm}^{-3}$ and the acceleration of free fall as $g = 9.81\,\text{m s}^{-2}$. Determine the molar mass of the protein.

P5B.8‡ Polymer scientists often report their data in a variety of units. For example, in the determination of molar masses of polymers in solution by osmometry, osmotic pressures are often reported in grams per square centimetre (g cm^{-2}) and concentrations in grams per cubic centimetre (g cm^{-3}). (a) With these choices of units, what would be the units of R in the van 't Hoff equation? (b) The data in the table below on the concentration dependence of the osmotic pressure of polyisobutene in chlorobenzene at 25 °C have been adapted from J. Leonard and H. Daoust (*J. Polymer Sci.* 57, 53 (1962)). From these data, determine the molar mass of polyisobutene by plotting Π/c against c. (c) 'Theta solvents' are solvents for which the second osmotic coefficient is zero; for 'poor' solvents the plot is linear and for good solvents the plot is nonlinear. From your plot, how would you classify chlorobenzene as a solvent for polyisobutene? Rationalize the result in terms of the molecular structure of the polymer and solvent. (d) Determine the second and third osmotic virial coefficients by fitting the curve to the virial form of the osmotic pressure equation. (e) Experimentally, it is often found that the virial expansion can be represented as

$$\Pi/c = RT/M\,(1 + B'c + gB'^2c^2 + \dots)$$

and in good solvents, the parameter g is often about 0.25. With terms beyond the second power ignored, obtain an equation for $(\Pi/c)^{1/2}$ and plot this quantity against c. Determine the second and third virial coefficients from the plot and compare to the values from the first plot. Does this plot confirm the assumed value of g?

$10^{-2}(\Pi/c)/(\text{g cm}^{-2}/\text{g cm}^{-3})$	2.6	2.9	3.6	4.3	6.0	12.0
$c/(\text{g cm}^{-3})$	0.0050	0.010	0.020	0.033	0.057	0.10

$10^{-2}(\Pi/c)/(\text{g cm}^{-2}/\text{g cm}^{-3})$	19.0	31.0	38.0	52	63
$c/(\text{g cm}^{-3})$	0.145	0.195	0.245	0.27	0.29

P5B.9‡ K. Sato, F.R. Eirich, and J.E. Mark (*J. Polymer Sci., Polym. Phys.* 14, 619 (1976)) have reported the data in the table below for the osmotic pressures of polychloroprene ($\rho = 1.25\,\text{g cm}^{-3}$) in toluene ($\rho = 0.858\,\text{g cm}^{-3}$) at 30 °C. Determine the molar mass of polychloroprene and its second osmotic virial coefficient.

$c/(\text{mg cm}^{-3})$	1.33	2.10	4.52	7.18	9.87
$\Pi/(\text{N m}^{-2})$	30	51	132	246	390

P5B.10 Use mathematical software or an electronic spreadsheet, draw graphs of $\Delta_{\text{mix}}G$ against x_A at different temperatures in the range 298–500 K. For what value of x_A does $\Delta_{\text{mix}}G$ depend on temperature most strongly?

P5B.11 Use mathematical software or an electronic spreadsheet to reproduce Fig. 5B.4. Then fix ξ and vary the temperature. For what value of x_A does the excess enthalpy depend on temperature most strongly?

P5B.12 Derive an expression for the temperature coefficient of the solubility, dx_B/dT, and plot it as a function of temperature for several values of the enthalpy of fusion.

P5B.13 Calculate the osmotic virial coefficient B from the data in *Example* 5B.2.

TOPIC 5C Phase diagrams of binary systems: liquids

Discussion questions

D5C.1 Draw a two-component, temperature–composition, liquid–vapour diagram featuring the formation of an azeotrope at $x_B = 0.333$ and complete miscibility. Label the regions of the diagrams, stating what materials are present, and whether they are liquid or gas.

D5C.2 What molecular features determine whether a mixture of two liquids will show high- and low-boiling azeotropic behaviour?

D5C.3 What factors determine the number of theoretical plates required to achieve a desired degree of separation in fractional distillation?

Exercises

E5C.1(a) The following temperature–composition data were obtained for a mixture of octane (O) and methylbenzene (M) at 1.00 atm, where x is the mole fraction in the liquid and y the mole fraction in the vapour at equilibrium.

θ/°C	110.9	112.0	114.0	115.8	117.3	119.0	121.1	123.0
x_M	0.908	0.795	0.615	0.527	0.408	0.300	0.203	0.097
y_M	0.923	0.836	0.698	0.624	0.527	0.410	0.297	0.164

The boiling points are 110.6 °C and 125.6 °C for M and O, respectively. Plot the temperature–composition diagram for the mixture. What is the composition of the vapour in equilibrium with the liquid of composition (i) $x_M = 0.250$ and (ii) $x_O = 0.250$?

E5C.1(b) The following temperature/composition data were obtained for a mixture of two liquids A and B at 1.00 atm, where x is the mole fraction in the liquid and y the mole fraction in the vapour at equilibrium.

θ/°C	125	130	135	140	145	150
x_A	0.91	0.65	0.45	0.30	0.18	0.098
y_A	0.99	0.91	0.77	0.61	0.45	0.25

The boiling points are 124 °C for A and 155 °C for B. Plot the temperature/composition diagram for the mixture. What is the composition of the vapour in equilibrium with the liquid of composition (i) $x_A = 0.50$ and (ii) $x_B = 0.33$?

E5C.2(a) Figure 5.1 shows the phase diagram for two partially miscible liquids, which can be taken to be that for water (A) and 2-methylpropan-1-ol (B). Describe what will be observed when a mixture of composition $x_B = 0.8$ is heated, at each stage giving the number, composition, and relative amounts of the phases present.

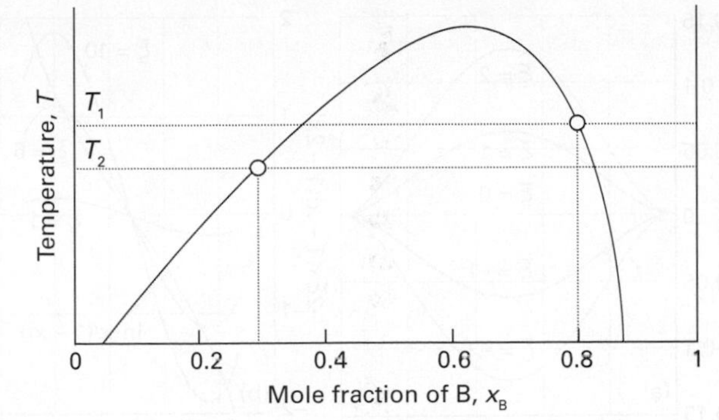

Figure 5.1 The phase diagram for two partially miscible liquids.

E5C.2(b) Refer to Fig. 5.1 again. Describe what will be observed when a mixture of composition $x_B = 0.3$ is heated, at each stage giving the number, composition, and relative amounts of the phases present.

E5C.3(a) Phenol and water form non-ideal liquid mixtures. When 7.32 g of phenol and 7.95 g of water are mixed together at 60 °C they form two immiscible liquid phases with mole fractions of phenol of 0.042 and 0.161. (i) Calculate the overall mole fraction of phenol in the mixture. (ii) Use the lever rule to determine the relative amounts of the two phases.

E5C.3(b) Aniline, $C_6H_5NH_2$, and hexane, C_6H_{14}, form partially miscible liquid–liquid mixtures at temperatures below 69.1 °C. When 42.8 g of aniline and 75.2 g of hexane are mixed together at a temperature of 67.5 °C, two separate liquid phases are formed, with mole fractions of aniline of 0.308 and 0.618. (i) Determine the overall mole fraction of aniline in the mixture. (ii) Use the lever rule to determine the relative amounts of the two phases.

E5C.4(a) Hexane and perfluorohexane show partial miscibility below 22.70 °C. The critical concentration at the upper critical temperature is $x = 0.355$, where x is the mole fraction of C_6F_{14}. At 22.0 °C the two solutions in equilibrium have $x = 0.24$ and $x = 0.48$, respectively, and at 21.5 °C the mole fractions are 0.22 and 0.51. Sketch the phase diagram. Describe the phase changes that occur when perfluorohexane is added to a fixed amount of hexane at (i) 23 °C, (ii) 22 °C.

E5C.4(b) Two liquids, A and B, show partial miscibility below 52.4 °C. The critical concentration at the upper critical temperature is $x = 0.459$, where x is the mole fraction of A. At 40.0 °C the two solutions in equilibrium have $x = 0.22$ and $x = 0.60$, respectively, and at 42.5 °C the mole fractions are 0.24 and 0.48. Sketch the phase diagram. Describe the phase changes that occur when B is added to a fixed amount of A at (i) 48 °C, (ii) 52.4 °C.

Problems

P5C.1 The vapour pressures of benzene and methylbenzene at 20 °C are 75 Torr and 21 Torr, respectively. What is the composition of the vapour in equilibrium with a mixture in which the mole fraction of benzene is 0.75?

P5C.2 Dibromoethene (DE, $p^*_{DE} = 22.9$ kPa at 358 K) and dibromopropene (DP, $p^*_{DP} = 17.1$ kPa at 358 K) form a nearly ideal solution. If $z_{DE} = 0.60$, what is (a) p_{total} when the system is all liquid, (b) the composition of the vapour when the system is still almost all liquid.

P5C.3 Benzene and methylbenzene (toluene) form nearly ideal solutions. Consider an equimolar solution of benzene and methylbenzene. At 20 °C the vapour pressures of pure benzene and methylbenzene are 9.9 kPa and 2.9 kPa, respectively. The solution is boiled by reducing the external pressure below the vapour pressure. Calculate (a) the pressure when boiling begins, (b) the composition of each component in the vapour, and (c) the vapour pressure when only a few drops of liquid remain. Assume that the rate of vaporization is low enough for the temperature to remain constant at 20 °C.

P5C.4‡ 1-Butanol and chlorobenzene form a minimum boiling azeotropic system. The mole fraction of 1-butanol in the liquid (x) and vapour (y) phases at 1.000 atm is given below for a variety of boiling temperatures (H. Artigas et al., *J. Chem. Eng. Data* **42**, 132 (1997)).

T/K	396.57	393.94	391.60	390.15	389.03	388.66	388.57
x	0.1065	0.1700	0.2646	0.3687	0.5017	0.6091	0.7171
y	0.2859	0.3691	0.4505	0.5138	0.5840	0.6409	0.7070

Pure chlorobenzene boils at 404.86 K. (a) Construct the chlorobenzene-rich portion of the phase diagram from the data. (b) Estimate the temperature at which a solution whose mole fraction of 1-butanol is 0.300 begins to boil. (c) State the compositions and relative proportions of the two phases present after a solution initially 0.300 1-butanol is heated to 393.94 K.

P5C.5 Figure 5.2 shows the experimentally determined phase diagrams for the nearly ideal solution of hexane and heptane. (a) Indicate which phases are present in each region of the diagram. (b) For a solution containing 1 mol each of hexane and heptane molecules, estimate the vapour pressure at 70 °C when vaporization on reduction of the external pressure just begins. (c) What is the vapour pressure of the solution at 70 °C when just one drop of liquid remains? (d) Estimate from the figures the mole fraction of hexane in the liquid and vapour phases for the conditions of part b. (e) What are the mole fractions for the conditions of part c? (f) At 85 °C and 760 Torr, what are the amounts of substance in the liquid and vapour phases when $z_{heptane} = 0.40$?

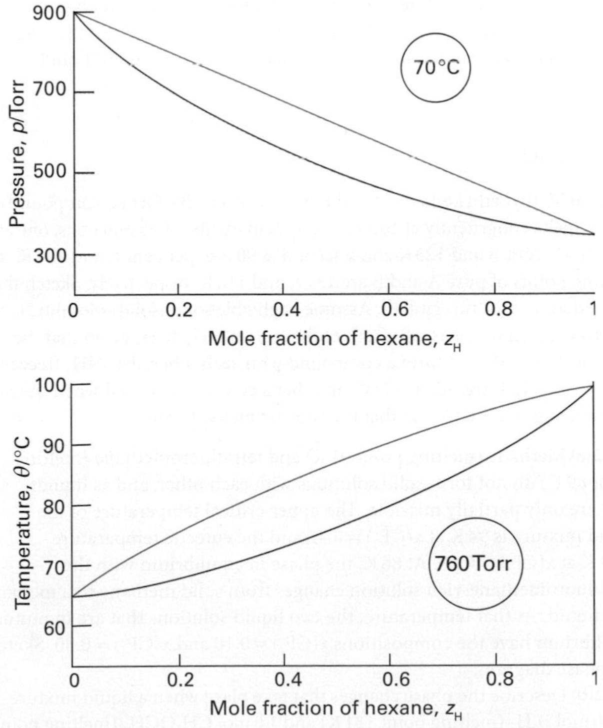

Figure 5.2 Phase diagrams of the solutions discussed in Problem P5C.5.

P5C.6 Suppose that in a phase diagram, when the sample was prepared with the mole fraction of component A equal to 0.40 it was found that the compositions of the two phases in equilibrium corresponded to the mole fractions $x_{A,\alpha} = 0.60$ and $x_{A,\beta} = 0.20$. What is the ratio of amounts of the two phases?

P5C.7 To reproduce the results of Fig. 5C.2, first rearrange eqn 5C.4 so that y_A is expressed as a function of x_A and the ratio p_A^*/p_B^*. Then plot y_A against x_A for several values of ratio $p_A^*/p_B^* > 1$.

P5C.8 To reproduce the results of Fig. 5C.3, first rearrange eqn 5C.5 so that the ratio p/p_A^* is expressed as a function of y_A and the ratio p_A^*/p_B^*. Then plot p_A/p_A^* against y_A for several values of $p_A^*/p_B^* > 1$.

P5C.9 In the system composed of benzene and cyclohexane treated in *Example* 5B.1 it is established that $\xi = 1.13$, so the two components are completely miscible at the temperature of the experiment. Would phase separation be expected if the excess enthalpy were modelled by the expression $H^E = \xi RTx_A^2x_B^2$ (Fig. 5.3a)? *Hint:* The solutions of the resulting equation for the minima of the Gibbs energy of mixing are shown in Fig. 5.3b.

P5C.10 Generate the plot of ξ at which $\Delta_{mix}G$ is a minimum against x_A by one of two methods: (a) solve the transcendental equation $\ln\{x/(1-x)\} + \xi(1-2x) = 0$ numerically, or (b) plot the first term of the transcendental equation against the second and identify the points of intersection as ξ is changed.

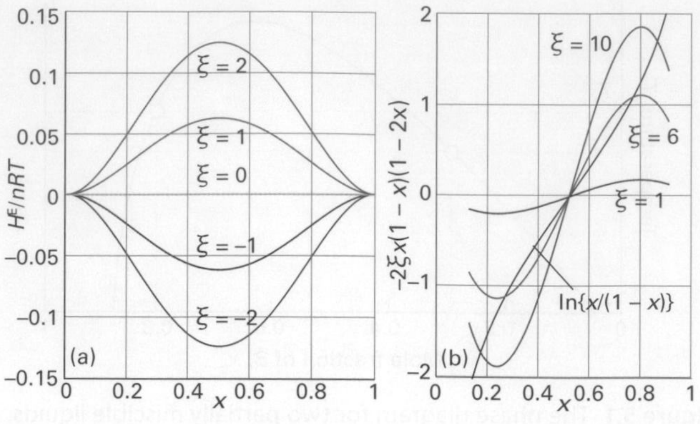

Figure 5.3 Data for the benzene–cyclohexane system discussed in Problem P5C.9.

TOPIC 5D Phase diagrams of binary systems: solids

Discussion questions

D5D.1 Draw a two-component, temperature–composition, solid–liquid diagram for a system where a compound AB forms and melts congruently, and there is negligible solid–solid solubility. Label the regions of the diagrams, stating what materials are present and whether they are solid or liquid.

D5D.2 Draw a two-component, temperature–composition, solid–liquid diagram for a system where a compound of formula AB_2 forms that melts incongruently, and there is negligible solid–solid solubility.

Exercises

E5D.1(a) Methyl ethyl ether (A) and diborane, B_2H_6 (B), form a compound which melts congruently at 133 K. The system exhibits two eutectics, one at 25 mol per cent B and 123 K and a second at 90 mol per cent B and 104 K. The melting points of pure A and B are 131 K and 110 K, respectively. Sketch the phase diagram for this system. Assume negligible solid–solid solubility.

E5D.1(b) Sketch the phase diagram of the system NH_3/N_2H_4 given that the two substances do not form a compound with each other, that NH_3 freezes at −78 °C and N_2H_4 freezes at +2 °C, and that a eutectic is formed when the mole fraction of N_2H_4 is 0.07 and that the eutectic melts at −80 °C.

E5D.2(a) Methane (melting point 91 K) and tetrafluoromethane (melting point 89 K) do not form solid solutions with each other, and as liquids they are only partially miscible. The upper critical temperature of the liquid mixture is 94 K at $x(CF_4) = 0.43$ and the eutectic temperature is 84 K at $x(CF_4) = 0.88$. At 86 K, the phase in equilibrium with the tetrafluoromethane-rich solution changes from solid methane to a methane-rich liquid. At that temperature, the two liquid solutions that are in mutual equilibrium have the compositions $x(CF_4) = 0.10$ and $x(CF_4) = 0.80$. Sketch the phase diagram.

E5D.2(b) Describe the phase changes that take place when a liquid mixture of 4.0 mol B_2H_6 (melting point 131 K) and 1.0 mol CH_3OCH_3 (melting point 135 K) is cooled from 140 K to 90 K. These substances form a compound $(CH_3)_2OB_2H_6$ that melts congruently at 133 K. The system exhibits one eutectic at $x(B_2H_6) = 0.25$ and 123 K and another at $x(B_2H_6) = 0.90$ and 104 K.

E5D.3(a) Refer to the information in Exercise E5D.2(a) and sketch the cooling curves for liquid mixtures in which $x(CF_4)$ is (i) 0.10, (ii) 0.30, (iii) 0.50, (iv) 0.80, and (v) 0.95.

E5D.2(b) Refer to the information in Exercise E5D.2(b) and sketch the cooling curves for liquid mixtures in which $x(B_2H_6)$ is (i) 0.10, (ii) 0.30, (iii) 0.50, (iv) 0.80, and (v) 0.95.

E5D.4(a) Indicate on the phase diagram in Fig. 5.4 the feature that denotes incongruent melting. What is the composition of the eutectic mixture and at what temperature does it melt?

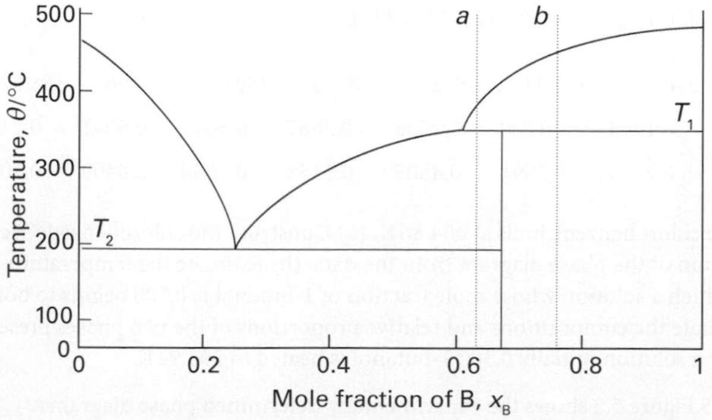

Figure 5.4 The temperature-composition diagram discussed in Exercises E5D.4(a), E5D.5(a), and E5D.6(b).

E5D.4(b) Indicate on the phase diagram in Fig. 5.5 the feature that denotes incongruent melting. What is the composition of the eutectic mixture and at what temperature does it melt?

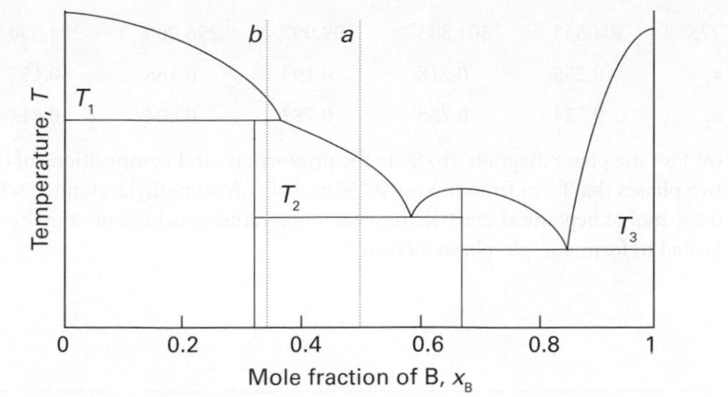

Figure 5.5 The temperature–composition diagram discussed in Exercises E5D.4(b) and E5D.5(b).

E5D.5(a) Sketch the cooling curves for the isopleths *a* and *b* in Fig. 5.4.

E5D.5(b) Sketch the cooling curves for the isopleths *a* and *b* in Fig. 5.5.

E5D.6(a) Use the phase diagram in Fig. 5D.3 to state (i) the solubility of Ag in Sn at 800 °C and (ii) the solubility of Ag_3Sn in Ag at 460 °C, (iii) the solubility of Ag_3Sn in Ag at 300 °C.

E5D.6(b) Use the phase diagram in Fig. 5.4 to state (i) the solubility of B in A at 500 °C and (ii) the solubility of AB_2 in A at 390 °C, (iii) the solubility of AB_2 in B at 300 °C.

Problems

P5D.1 Uranium tetrafluoride and zirconium tetrafluoride melt at 1035 °C and 912 °C respectively. They form a continuous series of solid solutions with a minimum melting temperature of 765 °C and composition $x(ZrF_4) = 0.77$. At 900 °C, the liquid solution of composition $x(ZrF_4) = 0.28$ is in equilibrium with a solid solution of composition $x(ZrF_4) = 0.14$. At 850 °C the two compositions are 0.87 and 0.90, respectively. Sketch the phase diagram for this system and state what is observed when a liquid of composition $x(ZrF_4) = 0.40$ is cooled slowly from 900 °C to 500 °C.

P5D.2 Phosphorus and sulfur form a series of binary compounds. The best characterized are P_4S_3, P_4S_7, and P_4S_{10}, all of which melt congruently. Assuming that only these three binary compounds of the two elements exist, (a) draw schematically only the P/S phase diagram plotted against x_S. Label each region of the diagram with the substance that exists in that region and indicate its phase. Label the horizontal axis as x_S and give the numerical values of x_S that correspond to the compounds. The melting point of pure phosphorus is 44 °C and that of pure sulfur is 119 °C. (b) Draw, schematically, the cooling curve for a mixture of composition $x_S = 0.28$. Assume that a eutectic occurs at $x_S = 0.2$ and negligible solid–solid solubility.

P5D.3 Consider the phase diagram in Fig. 5.6, which represents a solid–liquid equilibrium. Label all regions of the diagram according to the chemical species exist in that region and their phases. Indicate the number of species and phases present at the points labelled *b, d, e, f, g,* and *k*. Sketch cooling curves for compositions $x_B = 0.16$, 0.23, 0.57, 0.67, and 0.84.

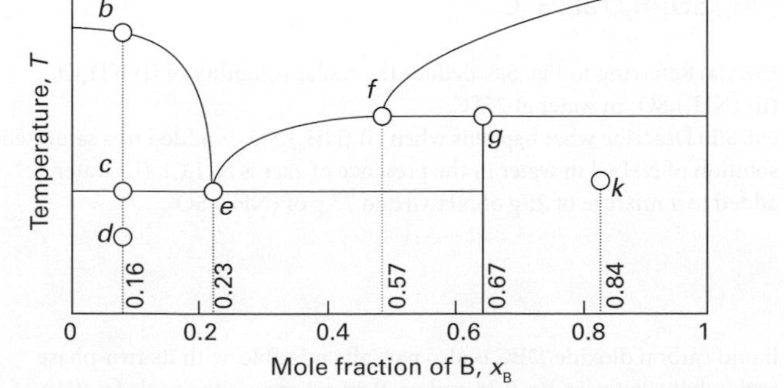

Figure 5.6 The temperature–composition diagram discussed in Problem P5D.3.

P5D.4 Sketch the phase diagram for the Mg/Cu system using the following information: $\theta_f(Mg) = 648$ °C, $\theta_f(Cu) = 1085$ °C; two intermetallic compounds are formed with $\theta_f(MgCu_2) = 800$ °C and $\theta_f(Mg_2Cu) = 580$ °C; eutectics of mass percentage Mg composition and melting points 10 per cent (690 °C), 33 per cent (560 °C), and 65 per cent (380 °C). A sample of Mg/Cu alloy containing 25 per cent Mg by mass was prepared in a crucible heated to 800 °C in an inert atmosphere. Describe what will be observed if the melt is cooled slowly to room temperature. Specify the composition and relative abundances of the phases and sketch the cooling curve.

P5D.5‡ The temperature/composition diagram for the Ca/Si binary system is shown in Fig. 5.7. (a) Identify eutectics, congruent melting compounds, and incongruent melting compounds. (b) A melt with composition $x_{Si} = 0.20$ at 1500 °C is cooled to 1000 °C, what phases (and phase composition) would be at equilibrium? Estimate the relative amounts of each phase. (c) Describe the equilibrium phases observed when a melt with $x_{Si} = 0.80$ is cooled to 1030 °C. What phases, and relative amounts, would be at equilibrium at a temperature (i) slightly higher than 1030 °C, (ii) slightly lower than 1030 °C?

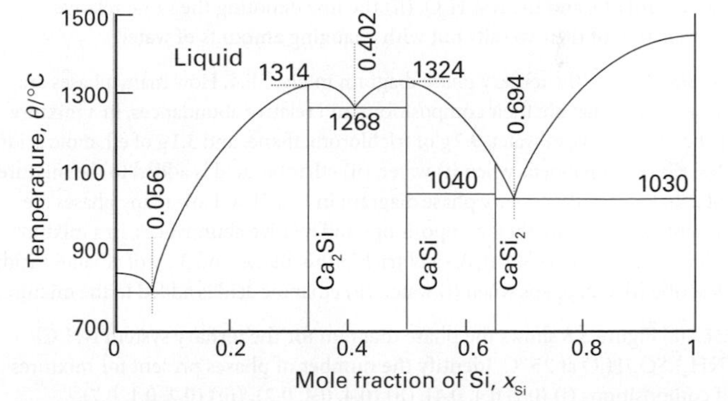

Figure 5.7 The temperature–composition diagram for the Ca/Si binary system.

P5D.6 Iron(II) chloride (melting point 677 °C) and potassium chloride (melting point 776 °C) form the compounds $KFeCl_3$ and K_2FeCl_4 at elevated temperatures. $KFeCl_3$ melts congruently at 399 °C and K_2FeCl_4 melts incongruently at 380 °C. Eutectics are formed with compositions $x = 0.38$ (melting point 351 °C) and $x = 0.54$ (melting point 393 °C), where x is the mole fraction of $FeCl_2$. The KCl solubility curve intersects the A curve at $x = 0.34$. Sketch the phase diagram. State the phases that are in equilibrium when a mixture of composition $x = 0.36$ is cooled from 400 °C to 300 °C.

P5D.7‡ An, Zhao, Jiang, and Shen investigated the liquid–liquid coexistence curve of *N,N*-dimethylacetamide and heptane (X. An et al., *J. Chem.*

Thermodynamics 28, 1221 (1996)). Mole fractions of *N,N*-dimethylacetamide in the upper (x_1) and lower (x_2) phases of a two-phase region are given below as a function of temperature:

T/K	309.820	309.422	309.031	308.006	306.686	304.553	301.803	299.097	296.000	294.534
x_1	0.473	0.400	0.371	0.326	0.293	0.255	0.218	0.193	0.168	0.157
x_2	0.529	0.601	0.625	0.657	0.690	0.724	0.758	0.783	0.804	0.814

(a) Plot the phase diagram. (b) State the proportions and compositions of the two phases that form from mixing 0.750 mol of *N,N*-dimethylacetamide with 0.250 mol of heptane at 296.0 K. To what temperature must the mixture be heated to form a single-phase mixture?

TOPIC 5E Phase diagrams of ternary systems

Discussion questions

D5E.1 What is the maximum number of phases that can be in equilibrium in a ternary system?

D5E.2 Does the lever rule apply to a ternary system?

D5E.3 Could a regular tetrahedron be used to depict the properties of a four-component system?

D5E.4 Consider the phase diagram for a stainless steel shown in Fig. 5E.6. Identify the composition represented by point *c*.

Exercises

E5E.1(a) Mark the following features on triangular coordinates: (i) the point (0.2, 0.2, 0.6), (ii) the point (0, 0.2, 0.8), (iii) the point at which all three mole fractions are the same.

E5E.1(b) Mark the following features on triangular coordinates: (i) the point (0.6, 0.2, 0.2), (ii) the point (0.8, 0.2, 0), (iii) the point (0.25, 0.25, 0.50).

E5E.2(a) Mark the following points on a ternary phase diagram for the system $NaCl/Na_2SO_4\cdot10H_2O/H_2O$: (i) 25 per cent by mass NaCl, 25 per cent $Na_2SO_4\cdot10H_2O$, and the rest H_2O, (ii) the line denoting the same relative composition of the two salts but with changing amounts of water.

E5E.2(b) Mark the following points on a ternary phase diagram for the system $NaCl/Na_2SO_4\cdot10H_2O/H_2O$: (i) 33 per cent by mass NaCl, 33 per cent $Na_2SO_4\cdot10H_2O$, and the rest H_2O, (ii) the line denoting the same relative composition of the two salts but with changing amounts of water.

E5E.3(a) Refer to the ternary phase diagram in Fig. 5E.4. How many phases are present, and what are their compositions and relative abundances, in a mixture that contains 2.3 g of water, 9.2 g of trichloromethane, and 3.1 g of ethanoic acid? Describe what happens when (i) water, (ii) ethanoic acid is added to the mixture.

E5E.3(b) Refer to the ternary phase diagram in Fig. 5E.4. How many phases are present, and what are their compositions and relative abundances, in a mixture that contains 55.0 g of water, 8.8 g of trichloromethane, and 3.7 g of ethanoic acid? Describe what happens when (i) water, (ii) ethanoic acid is added to the mixture.

E5E.4(a) Figure 5.8 shows the phase diagram for the ternary system $NH_4Cl/(NH_4)_2SO_4/H_2O$ at 25 °C. Identify the number of phases present for mixtures of compositions (i) (0.2, 0.4, 0.4), (ii) (0.4, 0.4, 0.2), (iii) (0.2, 0.1, 0.7), (iv) (0.4, 0.16, 0.44). The numbers are mole fractions of the three components in the order (NH_4Cl, $(NH_4)_2SO_4$, H_2O).

E5E.4(b) Refer to Fig. 5.8 and identify the number of phases present for mixtures of compositions (i) (0.4, 0.1, 0.5), (ii) (0.8, 0.1, 0.1), (iii) (0, 0.3,0.7), (iv) (0.33, 0.33, 0.34). The numbers are mole fractions of the three components in the order (NH_4Cl, $(NH_4)_2SO_4$, H_2O).

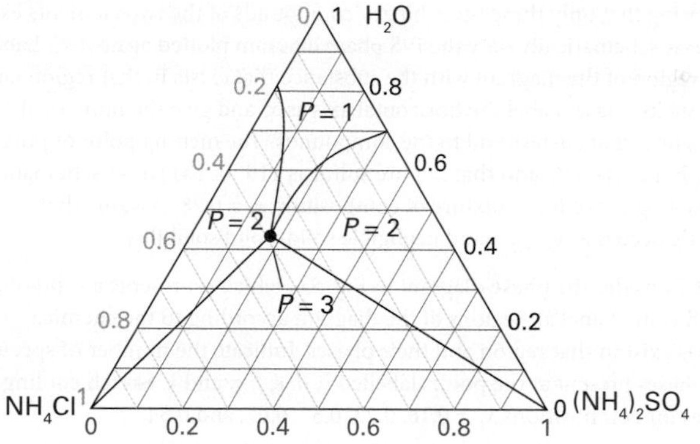

Figure 5.8 The phase diagram for the ternary system $NH_4Cl/(NH_4)_2SO_4/H_2O$ at 25 °C.

E5E.5(a) Referring to Fig. 5.8, deduce the molar solubility of (i) NH_4Cl, (ii) $(NH_4)_2SO_4$ in water at 25 °C.

E5E.5(b) Describe what happens when (i) $(NH_4)_2SO_4$ is added to a saturated solution of NH_4Cl in water in the presence of excess NH_4Cl, (ii) water is added to a mixture of 25 g of NH_4Cl and 75 g of $(NH_4)_2SO_4$.

Problems

P5E.1 At a certain temperature, the solubility of I_2 in liquid CO_2 is $x(I_2) = 0.03$. At the same temperature its solubility in nitrobenzene is 0.04. Liquid carbon dioxide and nitrobenzene are miscible in all proportions, and the solubility of I_2 in the mixture varies linearly with the proportion of nitrobenzene. Sketch a phase diagram for the ternary system.

P5E.2 The binary system nitroethane/decahydronaphthalene (DEC) shows partial miscibility, with the two-phase region lying between $x = 0.08$ and $x = 0.84$, where x is the mole fraction of nitroethane. The binary system liquid carbon dioxide/DEC is also partially miscible, with its two-phase region lying between $y = 0.36$ and $y = 0.80$, where y is the mole fraction of DEC. Nitroethane and liquid carbon dioxide are miscible in all proportions. The addition of liquid carbon dioxide to mixtures of nitroethane and DEC increases the range of miscibility, and the plait point is reached when the mole fraction of CO_2 is 0.18 and $x = 0.53$. The addition of nitroethane to mixtures of carbon dioxide and DEC also results in another plait point at $x = 0.08$ and $y = 0.52$. (a) Sketch the phase diagram for the ternary system. (b) For some binary mixtures of nitroethane and liquid carbon dioxide the addition of arbitrary amounts of DEC will not cause phase separation. Find the range of concentration for such binary mixtures.

P5E.3 Prove that a straight line from the apex A of a ternary phase diagram to the opposite edge BC represents mixtures of constant ratio of B and C, however much A is present.

TOPIC 5F Activities

Discussion questions

D5F.1 What are the contributions that account for the difference between activity and concentration?

D5F.2 How is Raoult's law modified so as to describe the vapour pressure of real solutions?

D5F.3 Summarize the ways in which activities may be measured.

D5F.4 Why do the activity coefficients of ions in solution differ from 1? Why are they less than 1 in dilute solutions?

D5F.5 Describe the general features of the Debye–Hückel theory of electrolyte solutions.

D5F.6 Suggest an interpretation of the additional terms in extended versions of the Debye–Hückel limiting law.

Exercises

E5F.1(a) The vapour pressure of water in a saturated solution of calcium nitrate at 20 °C is 1.381 kPa. The vapour pressure of pure water at that temperature is 2.3393 kPa. What is the activity of water in this solution?

E5F.1(b) The vapour pressure of a salt solution at 100 °C and 1.00 atm is 90.00 kPa. What is the activity of water in the solution at this temperature?

E5F.2(a) Substances A and B are both volatile liquids with $p_A^* = 300$ Torr, $p_B^* = 250$ Torr, and $K_B = 200$ Torr (concentration expressed in mole fraction). When $x_A = 0.900$, $p_A = 250$ Torr, and $p_B = 25$ Torr. Calculate the activities of A and B. Use the mole fraction, Raoult's law basis system for A and the Henry's law basis system for B. Go on to calculate the activity coefficient of A.

E5F.2(b) Given that $p^*(H_2O) = 0.023\,08$ atm and $p(H_2O) = 0.022\,39$ atm in a solution in which 0.122 kg of a non-volatile solute ($M = 241$ g mol^{-1}) is dissolved in 0.920 kg water at 293 K, calculate the activity and activity coefficient of water in the solution.

E5F.3(a) By measuring the equilibrium between liquid and vapour phases of a propanone(P)/methanol(M) solution at 57.2 °C at 1.00 atm, it was found that $x_P = 0.400$ when $y_P = 0.516$. Calculate the activities and activity coefficients of both components in this solution on the Raoult's law basis. The vapour pressures of the pure components at this temperature are: $p_P^* = 105$ kPa and $p_M^* = 73.5$ kPa. (x_P is the mole fraction in the liquid and y_P the mole fraction in the vapour.)

E5F.3(b) By measuring the equilibrium between liquid and vapour phases of a solution at 30 °C at 1.00 atm, it was found that $x_A = 0.220$ when $y_A = 0.314$. Calculate the activities and activity coefficients of both components in this solution on the Raoult's law basis. The vapour pressures of the pure components at this temperature are: $p_A^* = 73.0$ kPa and $p_B^* = 92.1$ kPa. (x_A is the mole fraction in the liquid and y_A the mole fraction in the vapour.)

E5F.4(a) Suppose it is found that for a hypothetical regular solution that $\xi = 1.40$, $p_A^* = 15.0$ kPa and $p_B^* = 11.6$ kPa. Draw plots similar to Fig. 5F.3.

E5F.4(b) Suppose it is found that for a hypothetical regular solution that $\xi = -1.40$, $p_A^* = 15.0$ kPa and $p_B^* = 11.6$ kPa. Draw plots similar to Fig. 5F.3.

E5F.5(a) Calculate the ionic strength of a solution that is 0.10 mol kg^{-1} in KCl(aq) and 0.20 mol kg^{-1} in $CuSO_4$(aq).

E5F.5(b) Calculate the ionic strength of a solution that is 0.040 mol kg^{-1} in $K_3[Fe(CN)_6]$(aq), 0.030 mol kg^{-1} in KCl(aq), and 0.050 mol kg^{-1} in NaBr(aq).

E5F.6(a) Calculate the masses of (i) $Ca(NO_3)_2$ and, separately, (ii) NaCl to add to a 0.150 mol kg^{-1} solution of KNO_3(aq) containing 500 g of solvent to raise its ionic strength to 0.250.

E5F.6(b) Calculate the masses of (i) KNO_3 and, separately, (ii) $Ba(NO_3)_2$ to add to a 0.110 mol kg^{-1} solution of KNO_3(aq) containing 500 g of solvent to raise its ionic strength to 1.00.

E5F.7(a) Estimate the mean ionic activity coefficient of $CaCl_2$ in a solution that is 0.010 mol kg^{-1} $CaCl_2$(aq) and 0.030 mol kg^{-1} NaF(aq) at 25 °C.

E5F.7(b) Estimate the mean ionic activity coefficient of NaCl in a solution that is 0.020 mol kg^{-1} NaCl(aq) and 0.035 mol kg^{-1} $Ca(NO_3)_2$(aq) at 25 °C.

E5F.8(a) The mean activity coefficients of HBr in three dilute aqueous solutions at 25 °C are 0.930 (at 5.00 mmol kg^{-1}), 0.907 (at 10.0 mmol kg^{-1}), and 0.879 (at 20.0 mmol kg^{-1}). Estimate the value of B in the Davies equation.

E5F.8(b) The mean activity coefficients of KCl in three dilute aqueous solutions at 25 °C are 0.927 (at 5.00 mmol kg^{-1}), 0.902 (at 10.0 mmol kg^{-1}), and 0.816 (at 50.0 mmol kg^{-1}). Estimate the value of B in the Davies equation.

Problems

P5F.1‡ Francesconi, Lunelli, and Comelli studied the liquid–vapour equilibria of trichloromethane and 1,2-epoxybutane at several temperatures (*J. Chem. Eng. Data* **41**, 310 (1996)). Among their data are the following measurements of the mole fractions of trichloromethane in the liquid phase (x_T) and the vapour phase (y_T) at 298.15 K as a function of total pressure.

p/kPa	23.40	21.75	20.25	18.75	18.15	20.25	22.50	26.30
x_T	0	0.129	0.228	0.353	0.511	0.700	0.810	1
y_T	0	0.065	0.145	0.285	0.535	0.805	0.915	1

Compute the activity coefficients of both components on the basis of Raoult's law.

P5F.2 Use mathematical software or a spreadsheet to plot p_A/p_A^* against x_A with $\xi = 2.5$ by using eqn 5F.18 and then eqn 5F.19. Above what value of x_A do the values of p_A/p_A^* given by these equations differ by more than 10 per cent?

P5F.3 The mean activity coefficients for aqueous solutions of NaCl at 25 °C are given below. Confirm that they support the Debye–Hückel limiting law and that an improved fit is obtained with the Davies equation.

b/(mmol kg^{-1})	1.0	2.0	5.0	10.0	20.0
$\gamma_\pm$	0.9649	0.9519	0.9275	0.9024	0.8712

P5F.4 Consider the plot of log $\gamma_\pm$ against $I^{1/2}$ with $B = 1.50$ and $C = 0$ in the Davies equation as a representation of experimental data for a certain MX electrolyte. Over what range of ionic strengths does the application of the Debye–Hückel limiting law lead to an error in the value of the activity coefficient of less than 10 per cent of the value predicted by the extended law?

FOCUS 5 Simple mixtures

Integrated activities

I5.1 The table below lists the vapour pressures of mixtures of iodoethane (I) and ethyl ethanoate (E) at 50 °C. Find the activity coefficients of both components on (a) the Raoult's law basis, (b) the Henry's law basis with iodoethane as solute.

x_I	0	0.0579	0.1095	0.1918	0.2353	0.3718
p_I/kPa	0	3.73	7.03	11.7	14.05	20.72
p_E/kPa	37.38	35.48	33.64	30.85	29.44	25.05

x_I	0.5478	0.6349	0.8253	0.9093	1.0000
p_I/kPa	28.44	31.88	39.58	43.00	47.12
p_E/kPa	19.23	16.39	8.88	5.09	0

I5.2 Plot the vapour pressure data for a mixture of benzene (B) and ethanoic acid (E) given below and plot the vapour pressure/composition curve for the mixture at 50 °C. Then confirm that Raoult's and Henry's laws are obeyed in the appropriate regions. Deduce the activities and activity coefficients of the components on the Raoult's law basis and then, taking B as the solute, its activity and activity coefficient on a Henry's law basis. Finally, evaluate the excess Gibbs energy of the mixture over the composition range spanned by the data.

x_E	0.0160	0.0439	0.0835	0.1138	0.1714
p_E/kPa	0.484	0.967	1.535	1.89	2.45
p_B/kPa	35.05	34.29	33.28	32.64	30.90

x_E	0.2973	0.3696	0.5834	0.6604	0.8437	0.9931
p_E/kPa	3.31	3.83	4.84	5.36	6.76	7.29
p_B/kPa	28.16	26.08	20.42	18.01	10.0	0.47

I5.3‡ Chen and Lee studied the liquid–vapour equilibria of cyclohexanol with several gases at elevated pressures (J.-T. Chen and M.-J. Lee, *J. Chem. Eng. Data* **41**, 339 (1996)). Among their data are the following measurements of the mole fractions of cyclohexanol in the vapour phase (y) and the liquid phase (x) at 393.15 K as a function of pressure.

p/bar	10.0	20.0	30.0	40.0	60.0	80.0
y_{cyc}	0.0267	0.0149	0.0112	0.00947	0.00835	0.00921
x_{cyc}	0.9741	0.9464	0.9204	0.892	0.836	0.773

Determine the Henry's law constant of CO_2 in cyclohexanol, and compute the activity coefficient of CO_2.

I5.4‡ The following data have been obtained for the liquid–vapour equilibrium compositions of mixtures of nitrogen and oxygen at 100 kPa.

T/K	77.3	78	80	82	84	86	88	90.2
$100x(O_2)$	0	10	34	54	70	82	92	100
$100y(O_2)$	0	2	11	22	35	52	73	100
$p^*(O_2)$/Torr	154	171	225	294	377	479	601	760

Plot the data on a temperature–composition diagram and determine the extent to which it fits the predictions for an ideal solution by calculating the activity coefficients of O_2 at each composition.

I5.5 For the calculation of the solubility c of a gas in a solvent, it is often convenient to use the expression $c = Kp$, where K is the Henry's law constant. Breathing air at high pressures, such as in scuba diving, results in an increased concentration of dissolved nitrogen. The Henry's law constant for the solubility of nitrogen is 0.18 μg/(g H_2O atm). What mass of nitrogen is dissolved in 100 g of water saturated with air at 4.0 atm and 20 °C? Compare your answer to that for 100 g of water saturated with air at 1.0 atm. (Air is 78.08 mol per cent N_2.) If nitrogen is four times as soluble in fatty tissues as in water, what is the increase in nitrogen concentration in fatty tissue in going from 1 atm to 4 atm?

I5.6 Dialysis may be used to study the binding of small molecules to macromolecules, such as an inhibitor to an enzyme, an antibiotic to DNA, and any other instance of cooperation or inhibition by small molecules attaching to large ones. To see how this is possible, suppose inside the dialysis bag the molar concentration of the macromolecule M is [M] and the total concentration of small molecule A is $[A]_{in}$. This total concentration is the sum of the concentrations of free A and bound A, which we write $[A]_{free}$ and $[A]_{bound}$, respectively. At equilibrium, $\mu_{A,free} = \mu_{A,out}$, which implies that $[A]_{free} = [A]_{out}$, provided the activity coefficient of A is the same in both solutions. Therefore, by measuring the concentration of A in the solution outside the bag, the concentration of unbound A in the macromolecule solution can be found and, from the difference $[A]_{in} - [A]_{free} = [A]_{in} - [A]_{out}$, the concentration of bound A. Now explore the quantitative consequences of the experimental arrangement just described. (a) The average number of A molecules bound to M molecules, ν, is

$$\nu = \frac{[A]_{bound}}{[M]} = \frac{[A]_{in} - [A]_{out}}{[M]}$$

The bound and unbound A molecules are in equilibrium, $M + A \rightleftharpoons MA$. Recall from introductory chemistry that the equilibrium constant for binding, K, may be written as

$$K = \frac{[MA]c^{\ominus}}{[M]_{free}[A]_{free}}$$

Now show that

$$K = \frac{\nu c^{\ominus}}{(1-\nu)[A]_{out}}$$

(b) If there are N *identical* and *independent* binding sites on each macromolecule, each macromolecule behaves like N separate smaller macromolecules, with the same value of K for each site. It follows that the average number of A molecules per site is ν/N. Show that, in this case, the *Scatchard equation*

$$\frac{\nu c^{\ominus}}{[A]_{out}} = KN - K\nu$$

is obtained. (c) To apply the Scatchard equation, consider the binding of ethidium bromide (E^-) to a short piece of DNA by a process called *intercalation*, in which the aromatic ethidium cation fits between two adjacent DNA base pairs. An equilibrium dialysis experiment was used to study the binding of ethidium bromide (EB) to a short piece of DNA. A 1.00 μmol dm^{-3} aqueous solution of the DNA sample was dialyzed against an excess of EB. The following data were obtained for the total concentration of EB, [EB]/(μmol dm^{-3}):

Side without DNA	0.042	0.092	0.204	0.526	1.150
Side with DNA	0.292	0.590	1.204	2.531	4.150

From these data, make a Scatchard plot and evaluate the intrinsic equilibrium constant, *K*, and total number of sites per DNA molecule. Is the identical and independent sites model for binding applicable?

I5.7 The form of the Scatchard equation given in Integrated activity I5.6 applies only when the macromolecule has identical and independent binding sites. For non-identical independent binding sites, *i*, the Scatchard equation is

$$\frac{\nu c^{\ominus}}{[\mathrm{A}]_{\mathrm{out}}}=\sum_i \frac{N_i K_i}{1+K_i[\mathrm{A}]_{\mathrm{out}}/c^{\ominus}}$$

Plot $\nu/[\mathrm{A}]$ for the following cases. (a) There are four independent sites on an enzyme molecule and the intrinsic binding constant is $K = 1.0\times10^7$. (b) There are a total of six sites per polymer. Four of the sites are identical and have an intrinsic binding constant of 1×10^5. The binding constants for the other two sites are 2×10^6.

I5.8 The addition of a small amount of a salt, such as $(NH_4)_2SO_4$, to a solution containing a charged protein increases the solubility of the protein in water. This observation is called the *salting-in effect*. However, the addition of large amounts of salt can decrease the solubility of the protein to such an extent that the protein precipitates from solution. This observation is called the *salting-out effect* and is used widely by biochemists to isolate and purify proteins. Consider the equilibrium $\mathrm{PX}_\nu(\mathrm{s}) \rightleftharpoons \mathrm{P}^{\nu+}(\mathrm{aq}) + \nu\mathrm{X}^-(\mathrm{aq})$, where $P^{\nu+}$ is a polycationic protein of charge ν+ and X^- is its counterion. Use Le Chatelier's principle and the physical principles behind the Debye–Hückel theory to provide a molecular interpretation for the salting-in and salting-out effects.

I5.9 The *osmotic coefficient* ϕ is defined as $\phi = -(x_A/x_B)\ln a_A$. By writing $r = x_B/x_A$, and using the Gibbs–Duhem equation, show that the activity of B can be calculated from the activities of A over a composition range by using the formula

$$\ln\frac{a_B}{r}=\phi-\phi(0)+\int_0^r \frac{\phi-1}{r}\mathrm{d}r$$

I5.10 Show that the osmotic pressure of a real solution is given by $\Pi V = -RT\ln a_A$. Go on to show that, provided the concentration of the solution is low, this expression takes the form $\Pi V = \phi RT[\mathrm{B}]$ and hence that the osmotic coefficient ϕ (which is defined in Problem I5.9) may be determined from osmometry.

I5.11 Show that the freezing-point depression of a real solution in which the solvent of molar mass M has activity a_A obeys

$$\frac{\mathrm{d}\ln a_A}{\mathrm{d}(\Delta T)}=-\frac{M}{K_\mathrm{f}}$$

and use the Gibbs–Duhem equation to show that

$$\frac{\mathrm{d}\ln a_\mathrm{B}}{\mathrm{d}(\Delta T)}=-\frac{1}{b_\mathrm{B}K_\mathrm{f}}$$

where a_B is the solute activity and b_B is its molality. Use the Debye–Hückel limiting law to show that the osmotic coefficient (ϕ, Problem I5.9) is given by $\phi = 1-\frac{1}{3}A'I$ with $A' = 2.303A$ and $I = b/b^{\ominus}$.

binding of [illegible] (EB) to a short piece of DNA. A 1.00 mmol dm^{-3} aqueous solution of the DNA sample was dialysed against an excess of EB. The following data were obtained for the total concentration of EB, [EB]/(μmol dm^{-3}):

Side without DNA	0.042	0.092	0.204	0.526	1.150
Side with DNA	0.292	0.590	1.204	2.531	4.150

From these data, make a Scatchard plot and evaluate the intrinsic equilibrium constant, K, and total number of sites per DNA molecule. Is the identical and independent sites model for binding applicable?

15.7 The form of the Scatchard equation given in [illegible] is applicable only when the macromolecule has identical and independent binding sites. For non-identical independent binding sites, the Scatchard equation is

$$\frac{\nu}{[A]_{out}} = \sum_i \frac{N_i K_i}{1+K_i[A]}$$

Plot $\nu/[A]$ for the following cases. (a) There are four independent sites on an enzyme molecule and the intrinsic binding constant is $K = 1.0 \times 10^{7}$. (b) There are a total of six sites per polymer. Four of the sites are identical and have an intrinsic binding constant of 1×10^{5}. The binding constants for the other two sites are 2×10^{6}.

15.8 The addition of a small amount of a salt, such as $(NH_4)_2SO_4$, to a solution containing a charged protein increases the solubility of the protein in water. This observation is called the *salting-in effect*. However, the addition of large amounts of salt can decrease the solubility of the protein to such an extent that the protein precipitates from solution. This observation is called the *salting-out effect* and is used widely by biochemists to isolate and purify proteins. Consider the equilibrium $PX_\nu(s) \rightleftharpoons P^{\nu+}(aq) + \nu X^-(aq)$, where $P^{\nu+}$ is a polycationic protein of charge $\nu+$ and X^- is its counterion. Use Le Chatelier's principle and the physical principles behind the Debye–Hückel theory to provide a molecular interpretation for the salting-in and salting-out effects.

15.9 The osmotic coefficient ϕ is defined as $\phi = -(x_A/x_B)\ln a_A$. By writing $r = x_B/x_A$, and using the Gibbs–Duhem equation, show that the activity of B can be calculated from the activities of A over a composition range by using the formula

$$\ln \frac{a_B}{r} = \phi - \phi(0) + \int_0^r \frac{\phi - 1}{r}\,dr$$

15.10 Show that the osmotic pressure of a real solution is given by $\Pi V = -RT \ln a_A$. Go on to show that, provided the concentration of the solution is low, this expression takes the form $\Pi V = \phi RT[B]$ and hence that the osmotic coefficient ϕ (which is defined in Problem 15.9) may be determined from osmometry.

15.11 Show that the freezing-point depression of a real solution in which the solvent of molar mass M has activity a_A obeys

$$\frac{d \ln a_A}{d(\Delta T)} = -\frac{M}{K_f}$$

and use the Gibbs–Duhem equation to show that

$$\frac{d \ln a_B}{d(\Delta T)} = -\frac{1}{b_B K_f}$$

where a_B is the solute activity and b_B is its molality. Use the Debye–Hückel limiting law to show that the osmotic coefficient (ϕ, Problem 15.9) is given by $\phi = 1 - \frac{1}{3}A'I^{1/2}$ with $A' = 2.303A$ and $I = b/b^{\ominus}$.

FOCUS 6

Chemical equilibrium

Chemical reactions tend to move towards a dynamic equilibrium in which both reactants and products are present but have no further tendency to undergo net change. In some cases, the concentration of products in the equilibrium mixture is so much greater than that of the unchanged reactants that for all practical purposes the reaction is 'complete'. However, in many important cases the equilibrium mixture has significant concentrations of both reactants and products.

6A The equilibrium constant

This Topic develops the concept of chemical potential and shows how it is used to account for the equilibrium composition of chemical reactions. The equilibrium composition corresponds to a minimum in the Gibbs energy plotted against the extent of reaction. By locating this minimum it is possible to establish the relation between the equilibrium constant and the standard Gibbs energy of reaction.

6A.1 **The Gibbs energy minimum**; 6A.2 **The description of equilibrium**

6B The response of equilibria to the conditions

The thermodynamic formulation of equilibrium establishes the quantitative effects of changes in the conditions. One very important aspect of equilibrium is the control that can be exercised by varying the conditions, such as the pressure or temperature.

6B.1 **The response to pressure**; 6B.2 **The response to temperature**

6C Electrochemical cells

Because many reactions involve the transfer of electrons, they can be studied (and utilized) by allowing them to take place in a cell equipped with electrodes, with the spontaneous reaction forcing electrons through an external circuit. The electric potential of the cell is related to the reaction Gibbs energy, so its measurement provides an electrical procedure for the determination of thermodynamic quantities.

6C.1 **Half-reactions and electrodes**; 6C.2 **Varieties of cells**; 6C.3 **The cell potential**; 6C.4 **The determination of thermodynamic functions**

6D Electrode potentials

Electrochemistry is in part a major application of thermodynamic concepts to chemical equilibria as well as being of great technological importance. As elsewhere in thermodynamics, electrochemical data can be reported in a compact form and applied to problems of chemical significance, especially to the prediction of the spontaneous direction of reactions and the calculation of equilibrium constants.

6D.1 **Standard potentials**; 6D.2 **Applications of standard potentials**

Web resources What is an application of this material?

The thermodynamic description of spontaneous reactions has numerous practical and theoretical applications. One is to the discussion of biochemical processes, where one reaction drives another (*Impact* 9). Ultimately that is why we have to eat, for the reaction that takes place when one substance is oxidized can drive non-spontaneous reactions, such as protein synthesis, forward. Another makes use of the great sensitivity of electrochemical processes to the concentration of electroactive materials, and leads to the design of electrodes used in chemical analysis (*Impact* 10).

TOPIC 6A The equilibrium constant

> ➤ **Why do you need to know this material?**
>
> Equilibrium constants lie at the heart of chemistry and are a key point of contact between thermodynamics and laboratory chemistry. To understand the behaviour of reactions you need to see how equilibrium constants arise and understand how thermodynamic properties account for their values.
>
> ➤ **What is the key idea?**
>
> At constant temperature and pressure, the composition of a reaction mixture tends to change until the Gibbs energy is a minimum.
>
> ➤ **What do you need to know already?**
>
> Underlying the whole discussion is the expression of the direction of spontaneous change in terms of the Gibbs energy of a system (Topic 3D). This material draws on the concept of chemical potential and its dependence on the concentration or pressure of the substance (Topic 5A). You need to know how to express the total Gibbs energy of a mixture in terms of the chemical potentials of its components (Topic 5A).

As explained in Topic 3D, the direction of spontaneous change at constant temperature and pressure is towards lower values of the Gibbs energy, G. The idea is entirely general, and in this Topic it is applied to the discussion of chemical reactions. At constant temperature and pressure, a mixture of reactants has a tendency to undergo reaction until the Gibbs energy of the mixture has reached a minimum: that condition corresponds to a state of chemical equilibrium. The equilibrium is dynamic in the sense that the forward and reverse reactions continue, but at matching rates. As always in the application of thermodynamics, spontaneity is a *tendency*: there might be kinetic reasons why that tendency is not realized.

6A.1 The Gibbs energy minimum

The equilibrium composition of a reaction mixture is located by calculating the Gibbs energy of the reaction mixture and then identifying the composition that corresponds to minimum G.

(a) The reaction Gibbs energy

Consider the equilibrium $A \rightleftharpoons B$. Even though this reaction looks trivial, there are many examples of it, such as the isomerization of pentane to 2-methylbutane and the conversion of L-alanine to D-alanine.

If an infinitesimal amount $d\xi$ of A turns into B, the change in the amount of A present is $dn_A = -d\xi$ and the change in the amount of B present is $dn_B = +d\xi$. The quantity ξ (xi) is called the **extent of reaction**; it has the dimensions of amount of substance and is reported in moles. When the extent of reaction changes by a measurable amount $\Delta\xi$, the amount of A present changes from $n_{A,0}$ to $n_{A,0} - \Delta\xi$ and the amount of B changes from $n_{B,0}$ to $n_{B,0} + \Delta\xi$. In general, the amount of a component J changes by $\nu_J\Delta\xi$, where ν_J is the stoichiometric number of the species J (positive for products, negative for reactants). For example, if initially 2.0 mol A is present and after a period of time $\Delta\xi = +1.5$ mol, then the amount of A remaining is 0.5 mol. The amount of B formed is 1.5 mol.

The **reaction Gibbs energy**, $\Delta_r G$, is defined as the slope of the graph of the Gibbs energy plotted against the extent of reaction:

$$\Delta_r G = \left(\frac{\partial G}{\partial \xi}\right)_{p,T} \quad \text{Reaction Gibbs energy [definition]} \qquad (6A.1)$$

Although Δ normally signifies a *difference* in values, here it signifies a *derivative*, the slope of G with respect to ξ. However, to see that there is a close relationship with the normal usage, suppose the reaction advances by $d\xi$. The corresponding change in Gibbs energy is

$$dG = \mu_A dn_A + \mu_B dn_B = -\mu_A d\xi + \mu_B d\xi = (\mu_B - \mu_A)d\xi$$

This equation can be reorganized into

$$\left(\frac{\partial G}{\partial \xi}\right)_{p,T} = \mu_B - \mu_A$$

That is,

$$\Delta_r G = \mu_B - \mu_A \qquad (6A.2)$$

and $\Delta_r G$ can also be interpreted as the difference between the chemical potentials (the partial molar Gibbs energies) of the reactants and products *at the current composition of the reaction mixture*.

Because chemical potentials vary with composition, the slope of the plot of Gibbs energy against extent of reaction, and

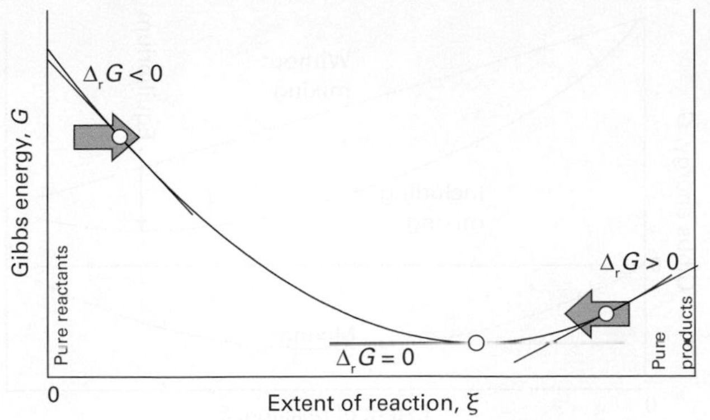

Fig. 6A.1 As the reaction advances, represented by the extent of reaction ξ increasing, the slope of a plot of total Gibbs energy of the reaction mixture against ξ changes. Equilibrium corresponds to the minimum in the Gibbs energy, which is where the slope is zero.

therefore the reaction Gibbs energy, changes as the reaction proceeds. The spontaneous direction of reaction lies in the direction of decreasing G (i.e. down the slope of G plotted against ξ). Thus, the reaction A → B is spontaneous when $\mu_A > \mu_B$, whereas the reverse reaction is spontaneous when $\mu_B > \mu_A$. The slope is zero, and the reaction is at equilibrium and spontaneous in neither direction, when

$$\Delta_r G = 0 \qquad \text{Condition of equilibrium} \qquad (6A.3)$$

This condition occurs when $\mu_B = \mu_A$ (Fig. 6A.1). It follows that if the composition of the reaction mixture that ensures $\mu_B = \mu_A$ can be found, then that will be the composition of the reaction mixture at equilibrium. Note that the chemical potential is now fulfilling the role its name suggests: it represents the potential for chemical change, and equilibrium is attained when these potentials are in balance.

(b) Exergonic and endergonic reactions

The spontaneity of a reaction at constant temperature and pressure can be expressed in terms of the reaction Gibbs energy:

If $\Delta_r G < 0$, the forward reaction is spontaneous.
If $\Delta_r G > 0$, the reverse reaction is spontaneous.
If $\Delta_r G = 0$, the reaction is at equilibrium.

A reaction for which $\Delta_r G < 0$ is called **exergonic** (from the Greek words for 'work producing'). The name signifies that, because the process is spontaneous, it can be used to drive another process, such as another reaction, or used to do non-expansion work. A simple mechanical analogy is a pair of weights joined by a string (Fig. 6A.2): the lighter of the pair of weights will be pulled up as the heavier weight falls down. Although the lighter weight has a natural tendency to move down, its coupling to the heavier weight results in it being raised. In biological cells, the oxidation of carbohydrates acts as the heavy weight that drives other reactions forward and results in the formation of proteins from amino acids, muscle contraction, and brain activity. A reaction for which $\Delta_r G > 0$ is called **endergonic** (signifying 'work consuming'); such a reaction can be made to occur only by doing work on it.

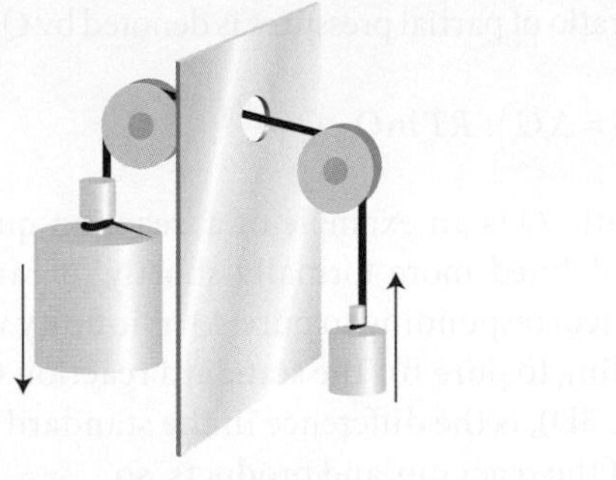

Fig. 6A.2 If two weights are coupled as shown here, then the heavier weight will move the lighter weight in its non-spontaneous direction: overall, the process is still spontaneous. The weights are the analogues of two chemical reactions: a reaction with a large negative ΔG can force another reaction with a smaller ΔG to run in its non-spontaneous direction.

Brief illustration 6A.1

The reaction Gibbs energy of a certain reaction is $-200\,\text{kJ mol}^{-1}$, so the reaction is exergonic, and in a suitable device (a fuel cell, for instance) operating at constant temperature and pressure, it could produce 200 kJ of electrical work for each mole of reaction events. The reverse reaction, for which $\Delta_r G = +200\,\text{kJ mol}^{-1}$ is endergonic and at least 200 kJ of work must be done to achieve it, perhaps through electrolysis.

6A.2 The description of equilibrium

With the background established, it is now possible to apply thermodynamics to the description of chemical equilibrium.

(a) Perfect gas equilibria

When A and B are perfect gases, eqn 5A.15a ($\mu = \mu^{\ominus} + RT\ln(p/p^{\ominus})$) can be used to write

$$\Delta_r G = \mu_B - \mu_A = (\mu_B^{\ominus} + RT\ln\frac{p_B}{p^{\ominus}}) - (\mu_A^{\ominus} + RT\ln\frac{p_A}{p^{\ominus}})$$
$$= \Delta_r G^{\ominus} + RT\ln\frac{p_B}{p_A} \qquad (6A.4)$$

If the ratio of partial pressures is denoted by Q, then it follows that

$$\Delta_r G = \Delta_r G^\ominus + RT\ln Q \quad Q = \frac{p_B}{p_A} \tag{6A.5}$$

The ratio Q is an example of a 'reaction quotient', a quantity to be defined more formally shortly. It ranges from 0 when $p_B = 0$ (corresponding to pure A) to infinity when $p_A = 0$ (corresponding to pure B). The standard reaction Gibbs energy, $\Delta_r G^\ominus$ (Topic 3D), is the difference in the standard molar Gibbs energies of the reactants and products, so

$$\Delta_r G^\ominus = G_m^\ominus(B) - G_m^\ominus(A) = \mu_B^\ominus - \mu_A^\ominus \tag{6A.6}$$

Note that in the definition of $\Delta_r G^\ominus$, the Δ_r has its normal meaning as the difference 'products − reactants'. As seen in Topic 3D, the difference in standard molar Gibbs energies of the products and reactants is equal to the difference in their standard Gibbs energies of formation, so in practice $\Delta_r G^\ominus$ is calculated from

$$\Delta_r G^\ominus = \Delta_f G^\ominus(B) - \Delta_f G^\ominus(A) \tag{6A.7}$$

At equilibrium $\Delta_r G = 0$. The ratio of partial pressures, the reaction quotient Q, at equilibrium has a certain value K, and eqn 6A.5 becomes

$$0 = \Delta_r G^\ominus + RT\ln K$$

which rearranges to

$$RT\ln K = -\Delta_r G^\ominus \quad K = \left(\frac{p_B}{p_A}\right)_{\text{equilibrium}} \tag{6A.8}$$

This relation is a special case of one of the most important equations in chemical thermodynamics: it is the link between tables of thermodynamic data, such as those in the *Resource section* and the chemically important 'equilibrium constant', K (again, a quantity that will be defined formally shortly).

Brief illustration 6A.2

The standard Gibbs energy for the isomerization of pentane to 2-methylbutane at 298 K, the reaction $CH_3(CH_2)_3CH_3(g) \rightarrow (CH_3)_2CHCH_2CH_3(g)$, is close to $-6.7\,kJ\,mol^{-1}$ (this is an estimate based on enthalpies of formation; its actual value is not listed). Therefore, the equilibrium constant for the reaction is

$$K = e^{-(-6.7\times10^3\,J\,mol^{-1})/(8.3145\,J\,K^{-1}\,mol^{-1})\times(298\,K)} = e^{2.7\ldots} = 15$$

In molecular terms, the minimum in the Gibbs energy, which corresponds to $\Delta_r G = 0$, stems from the Gibbs energy of mixing of the two gases. To see the role of mixing, consider the reaction A → B. If only the enthalpy were important, then H, and therefore G, would change linearly from its value for pure reactants to its value for pure products. The slope of this straight line is a constant and equal to $\Delta_r G^\ominus$ at all stages of the reaction and there is no intermediate minimum in the graph (Fig. 6A.3). However, when the entropy is taken into account, there is an additional contribution to the Gibbs energy that is given by eqn 5A.17 ($\Delta_{mix}G = nRT(x_A\ln x_A + x_B\ln x_B)$). This expression makes a U-shaped contribution to the total Gibbs energy. As can be seen from Fig. 6A.3, when it is included there is an intermediate minimum in the total Gibbs energy, and its position corresponds to the equilibrium composition of the reaction mixture.

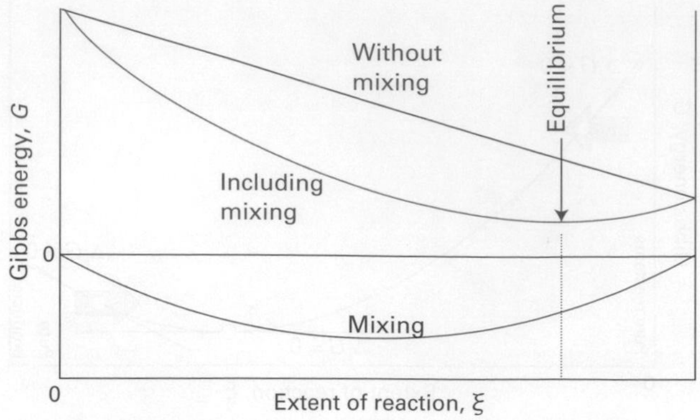

Fig. 6A.3 If the mixing of reactants and products is ignored, the Gibbs energy changes linearly from its initial value (pure reactants) to its final value (pure products) and the slope of the line is $\Delta_r G^\ominus$. However, as products are produced, there is a further contribution to the Gibbs energy arising from their mixing (lowest curve). The sum of the two contributions has a minimum, which corresponds to the equilibrium composition of the system.

It follows from eqn 6A.8 that, when $\Delta_r G^\ominus > 0$, $K < 1$. Therefore, at equilibrium the partial pressure of A exceeds that of B, which means that the reactant A is favoured in the equilibrium. When $\Delta_r G^\ominus < 0$, $K > 1$, so at equilibrium the partial pressure of B exceeds that of A. Now the product B is favoured in the equilibrium.

A note on good practice A common remark is that 'a reaction is spontaneous if $\Delta_r G^\ominus < 0$'. However, whether or not a reaction is spontaneous at a particular composition depends on the value of $\Delta_r G$ at that composition, not $\Delta_r G^\ominus$. The forward reaction is spontaneous ($\Delta_r G < 0$) when $Q < K$ and the reverse reaction is spontaneous when $Q > K$. It is far better to interpret the sign of $\Delta_r G^\ominus$ as indicating whether K is greater or smaller than 1.

(b) The general case of a reaction

To extend the argument that led to eqn 6A.8 to a general reaction, first note that a chemical reaction may be expressed symbolically in terms of stoichiometric numbers as

$$0 = \sum_J \nu_J J \quad \text{Chemical equation [symbolic form]} \tag{6A.9}$$

where J denotes the substances and the ν_J are the corresponding stoichiometric numbers in the chemical equation, which are positive for products and negative for reactants. In the reaction $2A+B \to 3C+D$, for instance, $\nu_A=-2$, $\nu_B=-1$, $\nu_C=+3$, and $\nu_D=+1$.

With these points in mind, it is possible to write an expression for the reaction Gibbs energy, $\Delta_r G$, at any stage during the reaction.

How is that done? 6A.1 Deriving an expression for the dependence of the reaction Gibbs energy on the reaction quotient

Consider a reaction with stoichiometric numbers ν_J. When the reaction advances by $d\xi$, the amounts of reactants and products change by $dn_J = \nu_J d\xi$. The resulting infinitesimal change in the Gibbs energy at constant temperature and pressure is

$$dG=\sum_J \mu_J dn_J=\sum_J \mu_J \nu_J d\xi=\left(\sum_J \mu_J \nu_J\right) d\xi$$

It follows that

$$\Delta_r G=\left(\frac{\partial G}{\partial \xi}\right)_{p,T}=\sum_J \nu_J \mu_J$$

Step 1 *Write the chemical potential in terms of the activity*

To make progress, note that the chemical potential of a species J is related to its activity by eqn 5F.9 ($\mu_J = \mu_J^{\ominus} + RT\ln a_J$). When this relation is substituted into the expression for $\Delta_r G$ the result is

$$\Delta_r G= \overbrace{\sum_J \nu_J \mu_J^{\ominus}}^{\Delta_r G^{\ominus}} +RT\sum_J \nu_J \ln a_J$$

$$= \Delta_r G^{\ominus}+RT\sum_J \nu_J \ln a_J = \Delta_r G^{\ominus}+RT\sum_J \ln a_J^{\nu_J}$$

Because $\ln x + \ln y + \cdots = \ln xy \ldots$, it follows that

$$\sum_i \ln x_i=\ln\left(\prod_i x_i\right)$$

The symbol Π denotes the product of what follows it (just as Σ denotes the sum). The expression for the Gibbs energy change then simplifies to

$$\Delta_r G= \Delta_r G^{\ominus}+RT\ln\prod_J a_J^{\nu_J}$$

Step 2 *Introduce the reaction quotient*

Now define the reaction quotient as

$$Q=\prod_J a_J^{\nu_J} \qquad \text{Reaction quotient [definition]} \qquad (6A.10)$$

Because reactants have negative stoichiometric numbers, they automatically appear as the denominator when the product is written out explicitly and this expression has the form

$$Q=\frac{\text{activities of products}}{\text{activities of reactants}} \qquad \text{Reaction quotient [general form]} \qquad (6A.11)$$

with the activity of each species raised to the power given by its stoichiometric coefficient.

It follows that the expression for the reaction Gibbs energy simplifies to

$$\Delta_r G=\Delta_r G^{\ominus}+RT\ln Q \qquad \text{Reaction Gibbs energy at an arbitrary stage} \qquad (6A.12)$$

Brief illustration 6A.3

Consider the reaction $2A+3B \to C+2D$, in which case $\nu_A = -2$, $\nu_B=-3$, $\nu_C=+1$, and $\nu_D=+2$. The reaction quotient is then

$$Q=a_A^{-2}a_B^{-3}a_C a_D^2=\frac{a_C a_D^2}{a_A^2 a_B^3}$$

As in Topic 3D, the standard reaction Gibbs energy is calculated from

$$\Delta_r G^{\ominus}= \sum_{\text{Products}} \nu\Delta_f G^{\ominus}- \sum_{\text{Reactants}} \nu\Delta_f G^{\ominus} \qquad \text{Reaction Gibbs energy [practical implementation]} \qquad (6A.13a)$$

where the ν are the (positive) stoichiometric coefficients. More formally,

$$\Delta_r G^{\ominus}= \sum_J \nu_J \Delta_f G^{\ominus}(J) \qquad \text{Reaction Gibbs energy [formal expression]} \qquad (6A.13b)$$

where the ν_J are the (signed) stoichiometric numbers.

At equilibrium, the slope of G is zero: $\Delta_r G=0$. The activities then have their equilibrium values and

$$K=\left(\prod_J a_J^{\nu_J}\right)_{\text{equilibrium}} \qquad \text{Equilibrium constant [definition]} \qquad (6A.14)$$

This expression has the same form as Q but is evaluated using equilibrium activities. From now on, the 'equilibrium' subscript will not be written explicitly, but it will be clear from the context that Q is defined in terms of the activities at an arbitrary stage of the reaction and K is the value of Q at equilibrium. An equilibrium constant K expressed in terms of activities is called a **thermodynamic equilibrium constant**. Note that, because activities are dimensionless, the thermodynamic equilibrium constant is also dimensionless. In elementary applications, the activities that occur in eqn 6A.14 are often replaced as follows:

State	Measure	Approximation for a_J	Definition
Solute	molality	$b_J/b_J^{\ominus}$	$b^{\ominus} = 1\,\text{mol kg}^{-1}$
	molar concentration	$[J]/c^{\ominus}$	$c^{\ominus} = 1\,\text{mol dm}^{-3}$
Gas phase	partial pressure	$p_J/p^{\ominus}$	$p^{\ominus} = 1\,\text{bar}$
Pure solid, liquid		1 (exact)	

Note that the activity is 1 for pure solids and liquids, so such substances make no contribution to Q even though they might appear in the chemical equation. When the approximations are made, the resulting expressions for Q and K are only approximations. The approximation is particularly severe for electrolyte solutions, for in them activity coefficients differ from 1 even in very dilute solutions (Topic 5F).

Brief illustration 6A.4

The equilibrium constant for the heterogeneous equilibrium $CaCO_3(s) \rightleftharpoons CaO(s) + CO_2(g)$ is

$$K = a_{CaCO_3(s)}^{-1} a_{CaO(s)} a_{CO_2(g)} = \frac{\overbrace{a_{CaO(s)}}^{1} a_{CO_2(g)}}{\underbrace{a_{CaCO_3(s)}}_{1}} = a_{CO_2(g)}$$

Provided the carbon dioxide can be treated as a perfect gas, go on to write

$$K = p_{CO_2}/p^{\ominus}$$

and conclude that in this case the equilibrium constant is the numerical value of the equilibrium pressure of CO_2 above the solid sample.

At equilibrium $\Delta_r G = 0$ in eqn 6A.12 and Q is replaced by K. The result is

$$\Delta_r G^{\ominus} = -RT \ln K \quad \text{Thermodynamic equilibrium constant} \quad (6A.15)$$

This is an exact and highly important thermodynamic relation, for it allows the calculation of the equilibrium constant of any reaction from tables of thermodynamic data, and hence the prediction of the equilibrium composition of the reaction mixture.

Example 6A.1 Calculating an equilibrium constant

Calculate the equilibrium constant for the ammonia synthesis reaction, $N_2(g) + 3\,H_2(g) \rightleftharpoons 2\,NH_3(g)$, at 298 K, and show how K is related to the partial pressures of the species at equilibrium when the overall pressure is low enough for the gases to be treated as perfect.

Collect your thoughts Calculate the standard reaction Gibbs energy from eqn 6A.13 and use its value in eqn 6A.15 to evaluate the equilibrium constant. The expression for the equilibrium constant is obtained from eqn 6A.14, and because the gases are taken to be perfect, replace each activity by the ratio $p_J/p^{\ominus}$, where p_J is the partial pressure of species J.

The solution The standard Gibbs energy of the reaction is

$$\Delta_r G^{\ominus} = 2\Delta_f G^{\ominus}(NH_3,g) - \{\Delta_f G^{\ominus}(N_2,g) + 3\Delta_f G^{\ominus}(H_2,g)\}$$
$$= 2\Delta_f G^{\ominus}(NH_3,g) = 2 \times (-16.45\,\text{kJ mol}^{-1})$$

Then,

$$\ln K = -\frac{2\times(-1.645\times10^4\,\text{J mol}^{-1})}{(8.3145\,\text{J K}^{-1}\text{mol}^{-1})\times(298\,\text{K})} = \frac{2\times1.645\times10^4}{8.3145\times298} = 13.2\ldots$$

Hence, $K = 5.8 \times 10^5$. This result is thermodynamically exact. The thermodynamic equilibrium constant for the reaction is

$$K = \frac{a_{NH_3}^2}{a_{N_2} a_{H_2}^3}$$

and has the value just calculated. At low overall pressures, the activities can be replaced by the ratios $p_J/p^{\ominus}$ and an approximate form of the equilibrium constant is

$$K = \frac{(p_{NH_3}/p^{\ominus})^2}{(p_{N_2}/p^{\ominus})(p_{H_2}/p^{\ominus})^3} = \frac{p_{NH_3}^2 p^{\ominus 2}}{p_{N_2} p_{H_2}^3}$$

Self-test 6A.1 Evaluate the equilibrium constant for $N_2O_4(g) \rightleftharpoons 2\,NO_2(g)$ at 298 K.

Answer: $K = 0.15$

Example 6A.2 Estimating the degree of dissociation at equilibrium

The *degree of dissociation* (or *extent of dissociation*, α) is defined as the fraction of reactant that has decomposed; if the initial amount of reactant is n and the amount at equilibrium is n_{eq}, then $\alpha = (n - n_{eq})/n$. The standard reaction Gibbs energy for the decomposition $H_2O(g) \rightarrow H_2(g) + \frac{1}{2}O_2(g)$ is $+118.08\,\text{kJ mol}^{-1}$ at 2300 K. What is the degree of dissociation of H_2O at 2300 K when the reaction is allowed to come to equilibrium at a total pressure of 1.00 bar?

Collect your thoughts The equilibrium constant is obtained from the standard Gibbs energy of reaction by using eqn 6A.15, so your task is to relate the degree of dissociation, α, to K and then to find its numerical value. Proceed by expressing the equilibrium compositions in terms of α, and solve for α in terms of K. Because the standard reaction Gibbs energy is large and positive, you can anticipate that K will be small, and hence that $\alpha \ll 1$, which opens the way to making approximations to obtain its numerical value.

The solution The equilibrium constant is obtained from eqn 6A.15 in the form

$$\ln K = -\frac{\Delta_r G^\ominus}{RT} = -\frac{1.1808\times10^5\,\mathrm{J\,mol^{-1}}}{(8.3145\,\mathrm{J\,K^{-1}\,mol^{-1}})\times(2300\,\mathrm{K})}$$

$$= -\frac{1.1808\times10^5}{8.3145\times2300} = -6.17\ldots$$

It follows that $K = 2.08 \times 10^{-3}$. The equilibrium composition is expressed in terms of α by drawing up the following table:

	$H_2O \rightarrow$	H_2 +	$\frac{1}{2}O_2$	
Initial amount	n	0	0	
Change to reach equilibrium	$-\alpha n$	$+\alpha n$	$+\frac{1}{2}\alpha n$	
Amount at equilibrium	$(1-\alpha)n$	αn	$\frac{1}{2}\alpha n$	Total: $(1+\frac{1}{2}\alpha)n$
Mole fraction, x_J	$\frac{1-\alpha}{1+\frac{1}{2}\alpha}$	$\frac{\alpha}{1+\frac{1}{2}\alpha}$	$\frac{\frac{1}{2}\alpha}{1+\frac{1}{2}\alpha}$	
Partial pressure, p_J	$\frac{(1-\alpha)p}{1+\frac{1}{2}\alpha}$	$\frac{\alpha p}{1+\frac{1}{2}\alpha}$	$\frac{\frac{1}{2}\alpha p}{1+\frac{1}{2}\alpha}$	

where, for the entries in the last row, $p_J = x_J p$ (eqn 1A.6) has been used. The equilibrium constant is therefore

$$K = \frac{p_{H_2} p_{O_2}^{1/2}}{p_{H_2O}} = \frac{\alpha^{3/2} p^{1/2}}{(1-\alpha)(2+\alpha)^{1/2}}$$

In this expression, p has been used in place of $p/p^\ominus$, to simplify its appearance. Now make the approximation that $\alpha \ll 1$, so $1-\alpha \approx 1$ and $2+\alpha \approx 2$, and hence obtain

$$K \approx \frac{\alpha^{3/2} p^{1/2}}{2^{1/2}}$$

Under the stated conditions, $p = 1.00\,\mathrm{bar}$ (that is, $p/p^\ominus = 1.00$), so $\alpha \approx (2^{1/2}K)^{2/3} = 0.0205$. That is, about 2 per cent of the water has decomposed.

A note on good practice Always check that the approximation is consistent with the final answer. In this case, $\alpha \ll 1$ in accord with the original assumption.

Self-test 6A.2 For the same reaction, the standard Gibbs energy of reaction at 2000 K is $+135.2\,\mathrm{kJ\,mol^{-1}}$. Suppose that steam at 200 kPa is passed through a furnace tube at that temperature. Calculate the mole fraction of O_2 present in the output gas stream.

Answer: 0.00221

(c) The relation between equilibrium constants

Equilibrium constants in terms of activities are exact, but it is often necessary to relate them to concentrations. Formally, it is necessary to know the activity coefficients γ_J (Topic 5F), and then to use $a_J = \gamma_J x_J$, $a_J = \gamma_J b_J/b^\ominus$, or $a_J = \gamma_J[J]/c^\ominus$, where x_J is a mole fraction, b_J is a molality, and [J] is a molar concentration. For example, if the composition is expressed in terms of molality for an equilibrium of the form $A + B \rightleftharpoons C + D$, where all four species are solutes, then

$$K = \frac{a_C a_D}{a_A a_B} = \frac{\gamma_C \gamma_D}{\gamma_A \gamma_B} \times \frac{b_C b_D}{b_A b_B} = K_\gamma K_b \qquad (6A.16)$$

The activity coefficients must be evaluated at the equilibrium composition of the mixture (for instance, by using one of the Debye–Hückel expressions, Topic 5F), which may involve a complicated calculation, because the activity coefficients are known only if the equilibrium composition is already known. In elementary applications, and to begin the iterative calculation of the concentrations in a real example, the assumption is often made that the activity coefficients are all so close to unity that $K_\gamma = 1$. Given these difficulties, it is common in elementary chemistry to assume that $K \approx K_b$, which allows equilibria to be discussed in terms of the molalities (or molar concentrations) themselves.

A special case arises when the equilibrium constant of a gas-phase reaction is to be expressed in terms of molar concentrations instead of the partial pressures that appear in the thermodynamic equilibrium constant. Provided the gases are perfect, the p_J that appear in K can be replaced by $[J]RT$, and

$$K = \prod_J a_J^{\nu_J} = \prod_J \left(\frac{p_J}{p^\ominus}\right)^{\nu_J} = \prod_J [J]^{\nu_J} \left(\frac{RT}{p^\ominus}\right)^{\nu_J}$$

$$= \prod_J [J]^{\nu_J} \times \prod_J \left(\frac{RT}{p^\ominus}\right)^{\nu_J}$$

(Products can always be factorized in this way: *abcdef* is the same as *abc* × *def*.) The (dimensionless) equilibrium constant K_c is defined as

$$K_c = \prod_J \left(\frac{[J]}{c^\ominus}\right)^{\nu_J} \qquad \text{K_c for gas-phase reactions [definition]} \qquad (6A.17)$$

It follows that

$$K = K_c \times \prod_J \left(\frac{c^\ominus RT}{p^\ominus}\right)^{\nu_J} \qquad (6A.18a)$$

With $\Delta\nu = \sum_J \nu_J$, which is easier to think of as ν(products) − ν(reactants), the relation between K and K_c for a gas-phase reaction is

$$K = K_c \times \left(\frac{c^\ominus RT}{p^\ominus}\right)^{\Delta\nu} \qquad \text{Relation between K and K_c for gas-phase reactions} \qquad (6A.18b)$$

For numerical calculations, note that $p^\ominus/c^\ominus R$ evaluates to 12.03 K.

Brief illustration 6A.5

For the reaction $N_2(g)+3H_2(g)\rightarrow 2NH_3(g)$, $\Delta\nu=2-3-1=-2$, so

$$K=K_c\times\left(\frac{T}{12.03\,K}\right)^{-2}=K_c\times\left(\frac{12.03\,K}{T}\right)^{2}$$

At 298.15 K the relation is

$$K=K_c\times\left(\frac{12.03\,K}{298.15\,K}\right)^{2}=\frac{K_c}{614.2}$$

so $K_c = 614.2K$. Note that both K and K_c are dimensionless.

(d) Molecular interpretation of the equilibrium constant

Deeper insight into the origin and significance of the equilibrium constant can be obtained by considering the Boltzmann distribution of molecules over the available states of a system composed of reactants and products (see the *Prologue* to this text). When atoms can exchange partners, as in a reaction, the species present include atoms bonded together as molecules of both reactants and products. These molecules have their characteristic sets of energy levels, but the Boltzmann distribution does not distinguish between their identities, only their energies. The available atoms distribute themselves over both sets of energy levels in accord with the Boltzmann distribution (Fig. 6A.4). At a given temperature, there will be a specific distribution of populations, and hence a specific composition of the reaction mixture.

It can be appreciated from the illustration that, if the reactants and products both have similar arrays of molecular energy levels, then the dominant species in a reaction mixture at equilibrium is the species with the lower set of energy levels (Fig. 6A.4(a)). However, the fact that the Gibbs energy occurs in the expression for the equilibrium constant is a signal that entropy plays a role as well as energy. Its role can be appreciated by referring to Fig. 6A.4. Figure 6A.4(b) shows that, although the B energy levels lie higher than the A energy levels, in this instance they are much more closely spaced. As a result, their total population may be considerable and B could even dominate in the reaction mixture at equilibrium. Closely spaced energy levels correlate with a high entropy (Topic 13E), so in this case entropy effects dominate adverse energy effects. This competition is mirrored in eqn 6A.15, as can be seen most clearly by using $\Delta_r G^\ominus = \Delta_r H^\ominus - T\Delta_r S^\ominus$ and writing it in the form

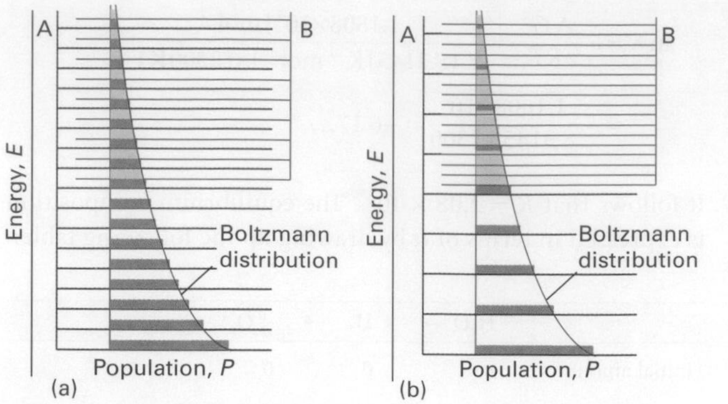

Fig. 6A.4 The Boltzmann distribution of populations over the energy levels of two species A and B. The reaction A → B is endothermic in this example. In (a) the two species have similar densities of energy levels: the bulk of the population is associated with the species A, so that species is dominant at equilibrium. In (b) the density of energy levels in B is much greater than that in A, and as a result, even though the reaction A → B is endothermic, the population associated with B is greater than that associated with A, so B is dominant at equilibrium.

$$K=e^{-\Delta_r H^\ominus/RT}\,e^{\Delta_r S^\ominus/R} \tag{6A.19}$$

Note that a positive reaction enthalpy results in a lowering of the equilibrium constant (that is, an endothermic reaction can be expected to have an equilibrium composition that favours the reactants). However, if there is positive reaction entropy, then the equilibrium composition may favour products, despite the endothermic character of the reaction.

Brief illustration 6A.6

From data provided in the *Resource section* it is found that for the reaction $N_2(g) + 3H_2(g) \rightleftharpoons 2NH_3(g)$ at 298 K, $\Delta_r G^\ominus = -32.9\,\mathrm{kJ\,mol^{-1}}$, $\Delta_r H^\ominus = -92.2\,\mathrm{kJ\,mol^{-1}}$, and $\Delta_r S^\ominus = -198.8\,\mathrm{J\,K^{-1}\,mol^{-1}}$. The contributions to K are therefore

$$K=e^{-(-9.22\times10^4\,\mathrm{J\,mol^{-1}})/(8.3145\,\mathrm{J\,K^{-1}\,mol^{-1}})\times(298\,\mathrm{K})}$$
$$\times e^{(-198.8\,\mathrm{J\,K^{-1}\,mol^{-1}})/(8.3145\,\mathrm{J\,K^{-1}\,mol^{-1}})}$$
$$=e^{37.2\ldots}\times e^{-23.9\ldots}$$

Note that the exothermic character of the reaction encourages the formation of products (it results in a large increase in entropy of the surroundings) but the decrease in entropy of the system as H atoms are pinned to N atoms opposes their formation.

Checklist of concepts

☐ 1. The **reaction Gibbs energy** $\Delta_r G$ is the slope of the plot of Gibbs energy against extent of reaction.

☐ 2. Reactions that have $\Delta_r G < 0$ are classified as **exergonic**, and those with $\Delta_r G > 0$ are classified as **endergonic**.

☐ 3. The **reaction quotient** is a combination of activities used to express the current value of the reaction Gibbs energy.

☐ 4. The **equilibrium constant** is the value of the reaction quotient at equilibrium.

Checklist of equations

Property	Equation	Comment	Equation number
Reaction Gibbs energy	$\Delta_r G = (\partial G/\partial \xi)_{p,T}$	Definition	6A.1
Reaction Gibbs energy	$\Delta_r G = \Delta_r G^{\ominus} + RT\ln Q,\ Q = \prod_J a_J^{\nu_J}$	Evaluated at arbitrary stage of reaction	6A.12
Standard reaction Gibbs energy	$\Delta_r G^{\ominus} = \sum_{\text{Products}} \nu\Delta_f G^{\ominus} - \sum_{\text{Reactants}} \nu\Delta_f G^{\ominus} = \sum_J \nu_J \Delta_f G^{\ominus}(J)$	ν are positive; ν_J are signed	6A.13
Equilibrium constant	$K = \left(\prod_J a_J^{\nu_J}\right)_{\text{equilibrium}}$	Definition	6A.14
Thermodynamic equilibrium constant	$\Delta_r G^{\ominus} = -RT\ln K$		6A.15
Relation between K and K_c	$K = K_c(c^{\ominus}RT/p^{\ominus})^{\Delta\nu}$	Gas-phase reactions; perfect gases	6A.18b

TOPIC 6B The response of equilibria to the conditions

> ➤ **Why do you need to know this material?**
>
> Chemists, and chemical engineers designing a chemical plant, need to know how the position of equilibrium will respond to changes in the conditions, such as a change in pressure or temperature. The variation with temperature also provides a way to determine the standard enthalpy and entropy of a reaction.
>
> ➤ **What is the key idea?**
>
> A system at equilibrium, when subjected to a disturbance, tends to respond in a way that minimizes the effect of the disturbance.
>
> ➤ **What do you need to know already?**
>
> This Topic builds on the relation between the equilibrium constant and the standard Gibbs energy of reaction (Topic 6A). To express the temperature dependence of *K* it draws on the Gibbs–Helmholtz equation (Topic 3E).

The equilibrium constant for a reaction is not affected by the presence of a catalyst. As explained in detail in Topics 17F and 19C, catalysts increase the rate at which equilibrium is attained but do not affect its position. However, it is important to note that in industry reactions rarely reach equilibrium, partly on account of the rates at which reactants mix and products are extracted. The equilibrium constant is also independent of pressure, but as will be seen, that does not necessarily mean that the composition at equilibrium is independent of pressure. The equilibrium constant does depend on the temperature in a manner that can be predicted from the standard reaction enthalpy.

6B.1 The response to pressure

The equilibrium constant depends on the value of $\Delta_r G^\ominus$, which is defined at a single, standard pressure. The value of $\Delta_r G^\ominus$, and hence of *K*, is therefore independent of the pressure at which the equilibrium is actually established. In other words, at a given temperature, *K* is a constant.

The effect of pressure depends on how the pressure is applied. The pressure within a reaction vessel can be increased by injecting an inert gas into it. However, so long as the gases are perfect, this addition of gas leaves all the partial pressures of the reacting gases unchanged: the partial pressure of a perfect gas is the pressure it would exert if it were alone in the container, so the presence of another gas has no effect on its value. It follows that pressurization by the addition of an inert gas has no effect on the equilibrium composition of the system (provided the gases are perfect).

Alternatively, the pressure of the system may be increased by confining the gases to a smaller volume (that is, by compression). Now the individual partial pressures are changed but their ratio (raised to the various powers that appear in the equilibrium constant) remains the same. Consider, for instance, the perfect gas equilibrium $A(g) \rightleftharpoons 2B(g)$, for which the equilibrium constant is

$$K = \frac{p_B^2}{p_A p^\ominus}$$

The right-hand side of this expression remains constant when the mixture is compressed only if an increase in p_A cancels an increase in the *square* of p_B. This relatively steep increase of p_A compared to p_B will occur if the equilibrium composition shifts in favour of A at the expense of B. Then the number of A molecules will increase as the volume of the container is decreased and the partial pressure of A will rise more rapidly than can be ascribed to a simple change in volume alone (Fig. 6B.1).

The increase in the number of A molecules and the corresponding decrease in the number of B molecules in the equilibrium $A(g) \rightleftharpoons 2B(g)$ is a special case of a principle proposed by the French chemist Henri Le Chatelier, which states that:

> A system at equilibrium, when subjected to a disturbance, tends to respond in a way that minimizes the effect of the disturbance.

Le Chatelier's principle

The principle implies that, if a system at equilibrium is compressed, then the reaction will tend to adjust so as to mini-

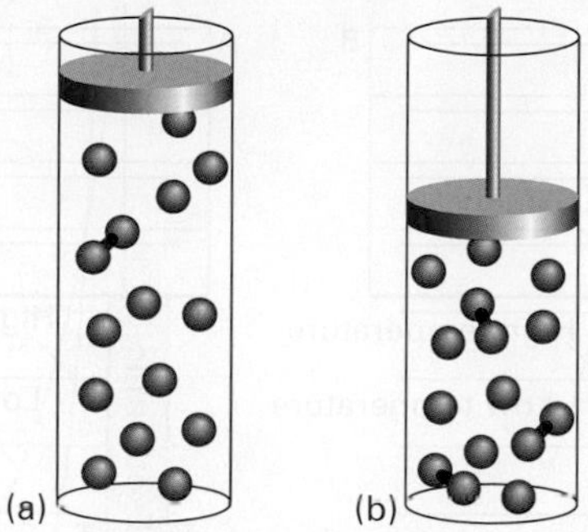

Figure 6B.1 When a reaction at equilibrium is compressed (from *a* to *b*), the reaction responds by reducing the number of molecules in the gas phase (in this case by producing the dimers represented by the linked spheres).

mize the increase in pressure. This it can do by reducing the number of particles in the gas phase, which implies a shift $A(g) \leftarrow 2B(g)$.

To treat the effect of compression quantitatively, suppose that there is an amount n of A present initially (and no B). At equilibrium the amount of A is $(1-\alpha)n$ and the amount of B is $2\alpha n$, where α is the **degree of dissociation** of A into 2B. It follows that the mole fractions present at equilibrium are

$$x_A = \frac{n_A}{n_{tot}} = \frac{(1-\alpha)n}{(1-\alpha)n+2\alpha n} = \frac{1-\alpha}{1+\alpha} \qquad x_B = \frac{2\alpha}{1+\alpha}$$

The equilibrium constant for the reaction is

$$K = \frac{p_B^2}{p_A p^\ominus} = \frac{x_B^2 p^2}{x_A p p^\ominus} = \frac{4\alpha^2(p/p^\ominus)}{1-\alpha^2}$$

where p is the total pressure. This expression rearranges to

$$\alpha = \left(\frac{1}{1+4p/Kp^\ominus}\right)^{1/2} \qquad (6B.1)$$

This formula shows that, even though K is independent of pressure, the amounts of A and B do depend on pressure (Fig. 6B.2). It also shows that as p is increased, α decreases, in accord with Le Chatelier's principle.

Brief illustration 6B.1

To predict the effect of an increase in pressure on the composition of the ammonia synthesis at equilibrium, $N_2(g) + 3H_2(g) \rightleftharpoons 2NH_3(g)$, note that the number of gas molecules decreases (from 4 to 2). Le Chatelier's principle predicts that an increase in pressure favours the product. The equilibrium constant is

$$K = \frac{p_{NH_3}^2 p^{\ominus 2}}{p_{N_2} p_{H_2}^3} = \frac{x_{NH_3}^2 p^2 p^{\ominus 2}}{x_{N_2} x_{H_2}^3 p^4} = \frac{x_{NH_3}^2 p^{\ominus 2}}{x_{N_2} x_{H_2}^3 p^2} = K_x \times \frac{p^{\ominus 2}}{p^2}$$

where K_x is the part of the equilibrium constant expression that contains the equilibrium mole fractions of reactants and products (note that, unlike K itself, K_x is not an equilibrium constant). Therefore, doubling the pressure must increase K_x by a factor of 4 to preserve the value of K.

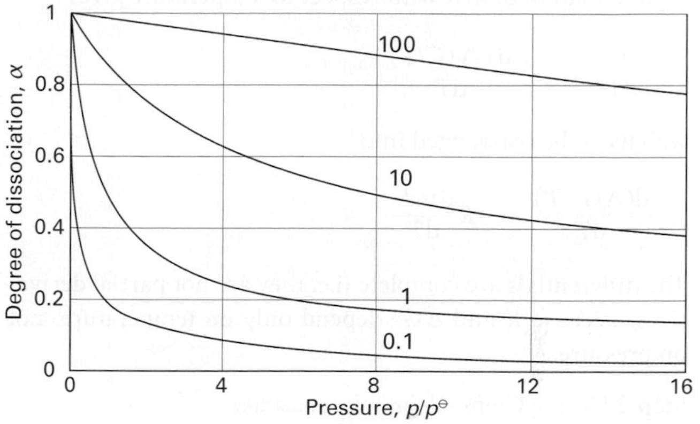

Figure 6B.2 The pressure dependence of the degree of dissociation, α, at equilibrium for an $A(g) \rightleftharpoons 2B(g)$ reaction for different values of the equilibrium constant K (the line labels). The value $\alpha = 0$ corresponds to pure A; $\alpha = 1$ corresponds to pure B.

6B.2 The response to temperature

Le Chatelier's principle predicts that a system at equilibrium tends to shift in the endothermic direction if the temperature is raised, for then energy is absorbed as heat and the rise in temperature is opposed. Conversely, an equilibrium can be expected to tend to shift in the exothermic direction if the temperature is lowered, for then energy is released and the reduction in temperature is opposed. These conclusions can be summarized as follows:

Exothermic reactions: increased temperature favours the reactants.

Endothermic reactions: increased temperature favours the products.

(a) The van 't Hoff equation

The response to temperature can be explored quantitatively by deriving an expression for the slope of a plot of the equilibrium constant (specifically, of ln K) as a function of temperature.

How is that done? 6B.1 Deriving an expression for the variation of ln *K* with temperature

The starting point for this derivation is eqn 6A.15 ($\Delta_r G^\ominus = -RT\ln K$), in the form

$$\ln K = -\frac{\Delta_r G^\ominus}{RT}$$

Now follow these steps.

Step 1 *Differentiate the expression for* ln *K*

Differentiation of ln *K* with respect to temperature gives

$$\frac{d\ln K}{dT} = -\frac{1}{R}\frac{d(\Delta_r G^\ominus/T)}{dT}$$

which can be rearranged into

$$\frac{d(\Delta_r G^\ominus/T)}{dT} = -R\frac{d\ln K}{dT}$$

The differentials are complete (i.e. they are not partial derivatives) because *K* and $\Delta_r G^\ominus$ depend only on temperature, not on pressure.

Step 2 *Use the Gibbs–Helmholtz equation*

To develop the preceding equation, use the Gibbs–Helmholtz equation (eqn 3E.10, $d(G/T)/dT = -H/T^2$) in the form

$$\frac{d(\Delta_r G^\ominus/T)}{dT} = -\frac{\Delta_r H^\ominus}{T^2}$$

where $\Delta_r H^\ominus$ is the standard reaction enthalpy at the temperature *T*. Combining this equation with the expression from Step 1 gives

$$R\frac{d\ln K}{dT} = \frac{\Delta_r H^\ominus}{T^2}$$

which rearranges into

$$\frac{d\ln K}{dT} = \frac{\Delta_r H^\ominus}{RT^2} \qquad (6B.2)$$

van 't Hoff equation

Equation 6B.2 is known as the **van 't Hoff equation**. For a reaction that is exothermic under standard conditions ($\Delta_r H^\ominus < 0$), it implies that $d\ln K/dT < 0$ (and therefore that $dK/dT < 0$). A negative slope means that ln *K*, and therefore *K* itself, decreases as the temperature rises. Therefore, in line with Le Chatelier's principle, in the case of an exothermic reaction the equilibrium shifts away from products. The opposite occurs in the case of endothermic reactions.

Insight into the thermodynamic basis of this behaviour comes from the expression $\Delta_r G^\ominus = \Delta_r H^\ominus - T\Delta_r S^\ominus$ written in the form $-\Delta_r G^\ominus/T = -\Delta_r H^\ominus/T + \Delta_r S^\ominus$. When the reaction is exothermic, $-\Delta_r H^\ominus/T$ corresponds to a positive change of entropy of the surroundings and favours the formation of products. When the temperature is raised, $-\Delta_r H^\ominus/T$ decreases and the increasing entropy of the surroundings has a less important role. As a result, the equilibrium lies less to the right. When the reaction is endothermic, the contribution of the unfavourable change of entropy of the surroundings is reduced if the temperature is raised (because then $\Delta_r H^\ominus/T$ is smaller), and the reaction then shifts towards products.

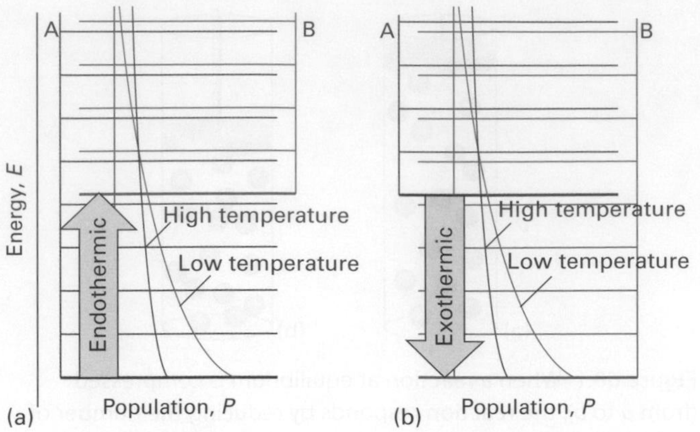

Figure 6B.3 The effect of temperature on a chemical equilibrium can be interpreted in terms of the change in the Boltzmann distribution with temperature and the effect of that change in the population of the species. (a) In an endothermic reaction, the population of B increases at the expense of A as the temperature is raised. (b) In an exothermic reaction, the opposite happens.

These remarks have a molecular basis that stems from the Boltzmann distribution of molecules over the available energy levels (see the *Prologue* to this text). The typical arrangement of energy levels for an endothermic reaction is shown in Fig. 6B.3a. When the temperature is increased, the Boltzmann distribution adjusts and the populations change as shown. The change corresponds to an increased population of the higher energy states at the expense of the population of the lower energy states. The states that arise from the B molecules become more populated at the expense of the A molecules. Therefore, the total population of B states increases, and B becomes more abundant in the equilibrium mixture. Conversely, if the reaction is exothermic (Fig. 6B.3b), then an increase in temperature increases the population of the A states (which start at higher energy) at the expense of the B states, so the reactants become more abundant.

Example 6B.1 Measuring a standard reaction enthalpy

The data below show the temperature variation of the equilibrium constant of the reaction $Ag_2CO_3(s) \rightleftharpoons Ag_2O(s) + CO_2(g)$. Calculate the standard reaction enthalpy of the decomposition.

T/K	350	400	450	500
K	3.98×10^{-4}	1.41×10^{-2}	1.86×10^{-1}	1.48

Collect your thoughts You need to adapt the van 't Hoff equation into a form that corresponds to a straight line. So note that $d(1/T)/dT = -1/T^2$, which implies that $dT = -T^2 d(1/T)$. Then, after cancelling the T^2, eqn 6B.2 becomes

$$-\frac{d\ln K}{d(1/T)} = \frac{\Delta_r H^\ominus}{R}$$

Therefore, provided the standard reaction enthalpy can be assumed to be independent of temperature, a plot of $-\ln K$ against $1/T$ should be a straight line of slope $\Delta_r H^\ominus/R$. The actual dimensionless plot is of $-\ln K$ against $1/(T/\mathrm{K})$, so equate $\Delta_r H^\ominus/R$ to *slope* × K.

The solution Draw up the following table:

T/K	350	400	450	500
$(10^3\,\mathrm{K})/T$	2.86	2.50	2.22	2.00
$-\ln K$	7.83	4.26	1.68	−0.392

These points are plotted in Fig. 6B.4. The slope of the graph is $+9.6\times10^3$, and it follows from *slope* × K = $\Delta_r H^\ominus/R$ that

$$\Delta_r H^\ominus = (+9.6\times10^3\,\mathrm{K})\times R = +80\,\mathrm{kJ\,mol^{-1}}$$

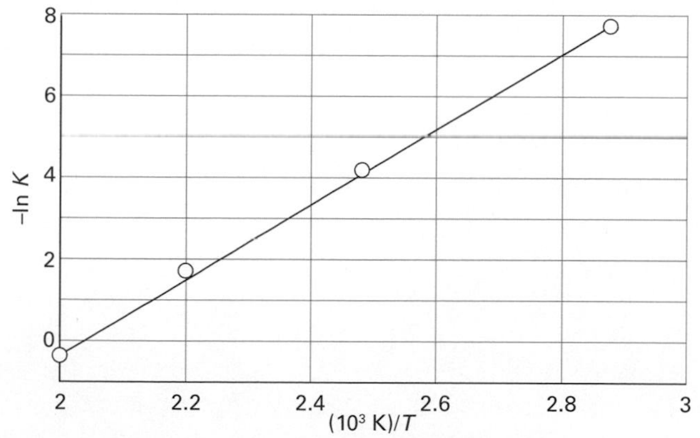

Figure 6B.4 When $-\ln K$ is plotted against $1/T$, a straight line is expected with slope equal to $\Delta_r H^\ominus/R$ if the standard reaction enthalpy does not vary appreciably with temperature. This is a non-calorimetric method for the measurement of standard reaction enthalpies. The data plotted are from *Example* 6B.1.

Self-test 6B.1 The equilibrium constant of the reaction $2\,SO_2(g) + O_2(g) \rightleftharpoons 2\,SO_3(g)$ is 4.0×10^{24} at 300 K, 2.5×10^{10} at 500 K, and 3.0×10^4 at 700 K. Estimate the standard reaction enthalpy at 500 K.

Answer: $-200\,\mathrm{kJ\,mol^{-1}}$

The temperature dependence of the equilibrium constant provides a non-calorimetric method of determining $\Delta_r H^\ominus$. A drawback is that the standard reaction enthalpy is actually temperature-dependent, so the plot is not expected to be perfectly linear. However, the temperature dependence is weak in many cases, so the plot is reasonably straight. In practice, the method is not very accurate, but it is often the only one available.

(b) The value of K at different temperatures

To find the value of the equilibrium constant at a temperature T_2 in terms of its value K_1 at another temperature T_1, integrate eqn 6B.2 between these two temperatures:

$$\ln K_2 - \ln K_1 = \frac{1}{R}\int_{T_1}^{T_2}\frac{\Delta_r H^\ominus}{T^2}\,dT \qquad (6B.3)$$

If $\Delta_r H^\ominus$ is supposed to vary only slightly with temperature over the temperature range of interest, it may be taken outside the integral. It follows that

$$\ln K_2 - \ln K_1 = \frac{\Delta_r H^\ominus}{R}\overbrace{\int_{T_1}^{T_2}\frac{1}{T^2}\,dT}^{\text{Integral A.1, } n=-2}$$

and therefore that

$$\ln K_2 - \ln K_1 = -\frac{\Delta_r H^\ominus}{R}\left(\frac{1}{T_2}-\frac{1}{T_1}\right) \qquad \text{Temperature dependence of } K \qquad (6B.4)$$

Brief illustration 6B.2

To estimate the equilibrium constant for the synthesis of ammonia at 500 K from its value at 298 K (5.8×10^5 for the reaction written as $N_2(g) + 3\,H_2(g) \rightleftharpoons 2\,NH_3(g)$), use the standard reaction enthalpy, which can be obtained from Table 2C.4 in the *Resource section* by using $\Delta_r H^\ominus = 2\Delta_f H^\ominus(NH_3,g)$, and assume that its value is constant over the range of temperatures. Then, with $\Delta_r H^\ominus = -92.2\,\mathrm{kJ\,mol^{-1}}$, from eqn 6B.4 it follows that

$$\ln K_2 = \ln(5.8\times10^5) - \left(\frac{-9.22\times10^4\,\mathrm{J\,mol^{-1}}}{8.3145\,\mathrm{J\,K^{-1}\,mol^{-1}}}\right)\times\left(\frac{1}{500\,\mathrm{K}}-\frac{1}{298\,\mathrm{K}}\right)$$
$$= -1.7\ldots$$

That is, $K_2 = 0.17$, a lower value than at 298 K, as expected for this exothermic reaction.

Checklist of concepts

☐ **1.** The thermodynamic equilibrium constant is independent of the presence of a catalyst and independent of pressure.

☐ **2.** The response of composition to changes in the conditions is summarized by **Le Chatelier's principle**.

☐ **3.** The dependence of the equilibrium constant on the temperature is expressed by the **van 't Hoff equation** and can be explained in terms of the distribution of molecules over the available states.

Checklist of equations

Property	Equation	Comment	Equation number
van 't Hoff equation	$\mathrm{d}\ln K/\mathrm{d}T = \Delta_r H^\ominus/RT^2$		6B.2
	$\mathrm{d}\ln K/\mathrm{d}(1/T) = -\Delta_r H^\ominus/R$	Alternative version	
Temperature dependence of equilibrium constant	$\ln K_2 - \ln K_1 = -(\Delta_r H^\ominus/R)(1/T_2 - 1/T_1)$	$\Delta_r H^\ominus$ assumed constant	6B.4

TOPIC 6C Electrochemical cells

➤ **Why do you need to know this material?**

One very special case of the material treated in Topic 6B, with enormous fundamental, technological, and economic significance, concerns reactions that take place in electrochemical cells. Moreover, the ability to make very precise measurements of potential differences ('voltages') means that electrochemical methods can be used to determine thermodynamic properties of reactions that may be inaccessible by other methods.

➤ **What is the key idea?**

The electrical work that a reaction can perform at constant pressure and temperature is equal to the reaction Gibbs energy.

➤ **What do you need to know already?**

This Topic develops the relation between the Gibbs energy and non-expansion work (Topic 3D). You need to be aware of how to calculate the work of moving a charge through an electrical potential difference (Topic 2A). The equations make use of the definition of the reaction quotient *Q* and the equilibrium constant *K* (Topic 6A).

An **electrochemical cell** consists of two **electrodes**, or metallic conductors, in contact with an **electrolyte**, an ionic conductor (which may be a solution, a liquid, or a solid). An electrode and its electrolyte comprise an **electrode compartment**; the two electrodes may share the same compartment. The various kinds of electrode are summarized in Table 6C.1. Any 'inert metal' shown as part of the specification is present to act as a source or sink of electrons, but takes no other part in the reaction other than perhaps acting as a catalyst for it. If the electrolytes are different, the two compartments may be joined by a **salt bridge**, which is a tube containing a concentrated electrolyte solution (for instance, potassium chloride in agar jelly) that completes the electrical circuit and enables the cell to function. A **galvanic cell** is an electrochemical cell that produces electricity as a result of the spontaneous reaction occurring inside it. An **electrolytic cell** is an electrochemical cell in which a non-spontaneous reaction is driven by an external source of current.

Table 6C.1 Varieties of electrode

Electrode type	Designation	Redox couple	Half-reaction
Metal/metal ion	$M(s)|M^+(aq)$	M^+/M	$M^+(aq) + e^- \rightarrow M(s)$
Gas	$Pt(s)|X_2(g)|X^+(aq)$	X^+/X_2	$X^+(aq) + e^- \rightarrow \frac{1}{2}X_2(g)$
	$Pt(s)|X_2(g)|X^-(aq)$	X_2/X^-	$\frac{1}{2}X_2(g) + e^- \rightarrow X^-(aq)$
Metal/insoluble salt	$M(s)|MX(s)|X^-(aq)$	$MX/M,X^-$	$MX(s) + e^- \rightarrow M(s) + X^-(aq)$
Redox	$Pt(s)|M^+(aq),M^{2+}(aq)$	M^{2+}/M^+	$M^{2+}(aq) + e^- \rightarrow M^+(aq)$

6C.1 Half-reactions and electrodes

It will be familiar from introductory chemistry courses that **oxidation** is the removal of electrons from a species, **reduction** is the addition of electrons to a species, and a **redox reaction** is a reaction in which there is a transfer of electrons from one species to another. The electron transfer may be accompanied by other events, such as atom or ion transfer, but the net effect is electron transfer and hence a change in oxidation number of an element. The **reducing agent** (or *reductant*) is the electron donor; the **oxidizing agent** (or *oxidant*) is the electron acceptor. It should also be familiar that any redox reaction may be expressed as the difference of two reduction **half-reactions**, which are conceptual reactions showing the gain of electrons. Even reactions that are not redox reactions may often be expressed as the difference of two reduction half-reactions. The reduced and oxidized species in a half-reaction form a **redox couple**. A couple is denoted Ox/Red and the corresponding reduction half-reaction is written

$$\mathrm{Ox} + \nu e^- \rightarrow \mathrm{Red} \qquad (6C.1)$$

Brief illustration 6C.1

The dissolution of silver chloride in water $AgCl(s) \rightarrow Ag^+(aq) + Cl^-(aq)$, which is not a redox reaction, can be expressed as the difference of the following two reduction half-reactions:

$$AgCl(s) + e^- \rightarrow Ag(s) + Cl^-(aq)$$
$$Ag^+(aq) + e^- \rightarrow Ag(s)$$

The redox couples are $AgCl/Ag,Cl^-$ and Ag^+/Ag, respectively.

It is often useful to express the composition of an electrode compartment in terms of the reaction quotient, Q, for the half-reaction. This quotient is defined like the reaction quotient for the overall reaction (Topic 6A, $Q=\prod_J a_J^{\nu_J}$), but the electrons are ignored because they are stateless.

Brief illustration 6C.2

The reaction quotient for the reduction of O_2 to H_2O in acid solution, $O_2(g) + 4H^+(aq) + 4e^- \rightarrow 2H_2O(l)$, is

$$Q=\frac{a_{H_2O}^2}{a_{H^+}^4 a_{O_2}} \approx \frac{p^{\ominus}}{a_{H^+}^4 p_{O_2}}$$

The approximations used in the second step are that the activity of water is 1 (because the solution is dilute) and the oxygen behaves as a perfect gas, so $a_{O_2} \approx p_{O_2}/p^{\ominus}$.

The reduction and oxidation processes responsible for the overall reaction in a cell are separated in space: oxidation takes place at one electrode and reduction takes place at the other. As the reaction proceeds, the electrons released in the oxidation $\text{Red}_1 \rightarrow \text{Ox}_1 + \nu e^-$ at one electrode travel through the external circuit and re-enter the cell through the other electrode. There they bring about reduction $\text{Ox}_2 + \nu e^- \rightarrow \text{Red}_2$. The electrode at which oxidation occurs is called the **anode**; the electrode at which reduction occurs is called the **cathode**. In a galvanic cell, the cathode has a higher potential than the anode: the species undergoing reduction, Ox_2, withdraws electrons from its electrode (the cathode, Fig. 6C.1), so leaving a relative positive charge on it (corresponding to a high potential). At the anode, oxidation results in the transfer of electrons to the electrode, so giving it a relative negative charge (corresponding to a low potential).

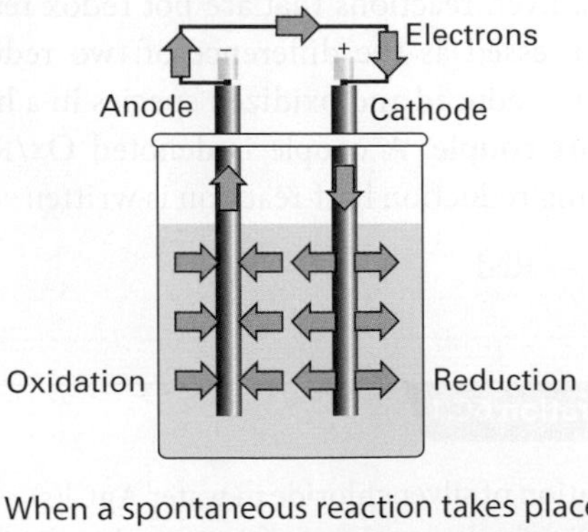

Figure 6C.1 When a spontaneous reaction takes place in a galvanic cell, electrons are deposited in one electrode (the site of oxidation, the anode) and collected from another (the site of reduction, the cathode), and so there is a net flow of current which can be used to do work. Note that the + sign of the cathode can be interpreted as indicating the electrode at which electrons enter the cell, and the – sign of the anode is where the electrons leave the cell.

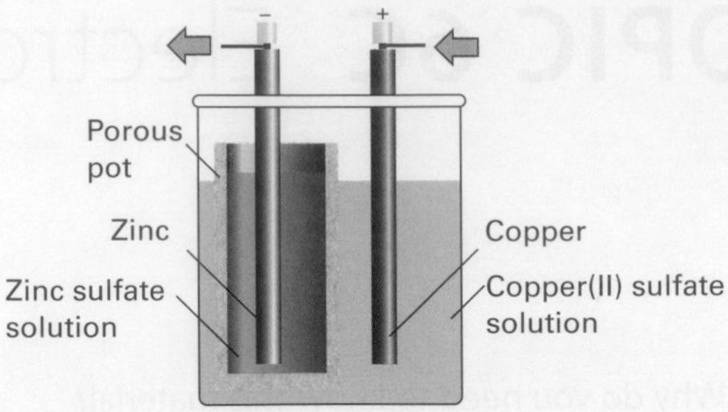

Figure 6C.2 One version of the Daniell cell. The copper electrode is the cathode and the zinc electrode is the anode. Electrons leave the cell from the zinc electrode and enter it again through the copper electrode.

6C.2 Varieties of cells

The simplest type of cell has a single electrolyte common to both electrodes (as in Fig. 6C.1). In some cases it is necessary to immerse the electrodes in different electrolytes, as in the 'Daniell cell' in which the redox couple at one electrode is Cu^{2+}/Cu and at the other is Zn^{2+}/Zn (Fig. 6C.2). In an **electrolyte concentration cell**, the electrode compartments are identical except for the concentrations of the electrolytes. In an **electrode concentration cell** the electrodes themselves have different concentrations, either because they are gas electrodes operating at different pressures or because they are amalgams (solutions in mercury) or analogous materials with different concentrations.

(a) Liquid junction potentials

In a cell with two different electrolyte solutions in contact, as in the Daniell cell, there is an additional source of potential difference across the interface of the two electrolytes. This contribution is called the **liquid junction potential**, E_{lj}. Another example of a junction potential is that at the interface between different concentrations of hydrochloric acid. At the junction, the mobile H^+ ions diffuse into the more dilute solution. The bulkier Cl^- ions follow, but initially do so more slowly, which results in a potential difference at the junction. The potential then settles down to a value such that, after that brief initial period, the ions diffuse at the same rates. Electrolyte concentration cells always have a liquid junction; electrode concentration cells do not.

The contribution of the liquid junction to the potential difference can be reduced (to about 1–2 mV) by joining the electrolyte compartments through a salt bridge (Fig. 6C.3). The reason for the success of the salt bridge is that, provided the ions dissolved in the jelly have similar mobilities, then the liquid junction potentials at either end are largely independent of the concentrations of the two dilute solutions, and so nearly cancel.

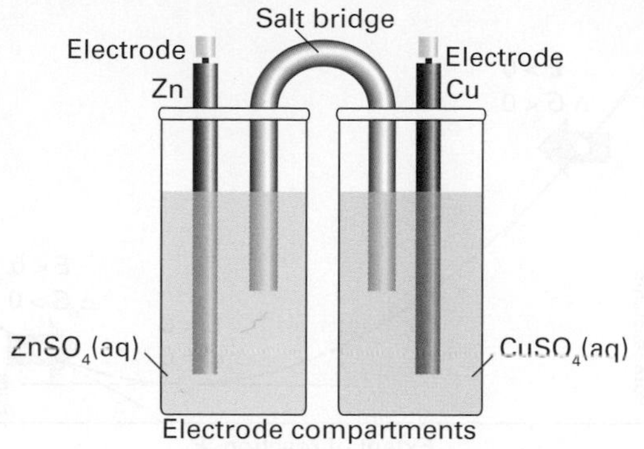

Figure 6C.3 The salt bridge, essentially an inverted U-tube full of concentrated salt solution in a jelly, has two opposing liquid junction potentials that almost cancel.

(b) Notation

The following notation is used for electrochemical cells:

|	An interface between components or phases
⋮	A liquid junction
‖	An interface for which it is assumed that the junction potential has been eliminated

Brief illustration 6C.3

A cell in which two electrodes share the same electrolyte is

$$\mathrm{Pt(s)|H_2(g)|HCl(aq)|AgCl(s)|Ag(s)}$$

The cell in Fig. 6C.2 is denoted

$$\mathrm{Zn(s)|ZnSO_4(aq)⋮CuSO_4(aq)|Cu(s)}$$

The cell in Fig. 6C.3 is denoted

$$\mathrm{Zn(s)|ZnSO_4(aq)||CuSO_4(aq)|Cu(s)}$$

An example of an electrolyte concentration cell in which the liquid junction potential is assumed to be eliminated is

$$\mathrm{Pt(s)|H_2(g)|HCl(aq,}b_1\mathrm{)||HCl(aq,}b_2\mathrm{)|H_2(g)|Pt(s)}$$

6C.3 The cell potential

The current produced by a galvanic cell arises from the spontaneous chemical reaction taking place inside it. The **cell reaction** is the reaction in the cell written on the assumption that the right-hand electrode is the cathode, and hence the assumption that the spontaneous reaction is one in which reduction is taking place in the right-hand compartment. If the right-hand electrode is in fact the cathode, then the cell reaction is spontaneous as written. If the left-hand electrode turns out to be the cathode, then the reverse of the corresponding cell reaction is spontaneous.

To write the cell reaction corresponding to a cell diagram, first write the right-hand half-reaction as a reduction. Then subtract from it the left-hand reduction half-reaction (because, by implication, that electrode is the site of oxidation). If necessary, adjust the number of electrons in the two half-reactions to be the same.

Brief illustration 6C.4

For the cell $\mathrm{Zn(s)|ZnSO_4(aq)||CuSO_4(aq)|Cu(s)}$ the two electrodes and their reduction half-reactions are

Right-hand electrode: $\mathrm{Cu^{2+}(aq) + 2\,e^- \rightarrow Cu(s)}$

Left-hand electrode: $\mathrm{Zn^{2+}(aq) + 2\,e^- \rightarrow Zn(s)}$

The same number of electrons is involved in each half-reaction. The overall cell reaction is the difference Right − Left:

$$\mathrm{Cu^{2+}(aq) + 2\,e^- - Zn^{2+}(aq) - 2\,e^- \rightarrow Cu(s) - Zn(s)}$$

which, after cancellation of the $2\,e^-$, rearranges to

$$\mathrm{Cu^{2+}(aq) + Zn(s) \rightarrow Cu(s) + Zn^{2+}(aq)}$$

(a) The Nernst equation

A cell in which the overall cell reaction has not reached chemical equilibrium can do electrical work as the reaction drives electrons through an external circuit. The work that a given transfer of electrons can accomplish depends on the potential difference between the two electrodes. When the potential difference is large, a given number of electrons travelling between the electrodes can do a lot of electrical work. When the potential difference is small, the same number of electrons can do only a little work. A cell in which the overall reaction is at equilibrium can do no work, and then the potential difference is zero.

According to the discussion in Topic 3D, the maximum non-expansion work a system can do at constant temperature and pressure is given by eqn 3D.8 ($w_{\mathrm{add,max}} = \Delta G$). In electrochemistry, the additional (non-expansion) work is identified with electrical work, w_e: the system is the cell, and ΔG is the Gibbs energy of the cell reaction, $\Delta_r G$. Because maximum work is produced when a change occurs reversibly, it follows that, to draw thermodynamic conclusions from measurements of the work that a cell can do, it is necessary to ensure that the cell is operating reversibly. Moreover, it is established in Topic 6A that the reaction Gibbs energy is actually a property relating, through the term $RT\ln Q$, to a specified composition of the reaction mixture. Therefore, the cell must be operating reversibly at a specific, constant composition. Both these conditions are achieved by measuring the potential difference generated by the cell when it is balanced by an exactly opposing source of potential difference so that the cell reaction occurs reversibly, the composition is constant, and no current flows: in effect, the cell reaction is poised for change, but not actually changing. The resulting potential difference is called the **cell potential**, E_{cell}, of the cell.

A note on good practice The cell potential was formerly, and is still widely, called the *electromotive force* (emf) of the cell. IUPAC prefers the term 'cell potential' because a potential difference is not a force.

As this introduction has indicated, there is a close relation between the cell potential and the reaction Gibbs energy. It can be established by considering the electrical work that a cell can do.

How is that done? 6C.1 Establishing the relation between the cell potential and the reaction Gibbs energy

Consider the change in G when the cell reaction advances by an infinitesimal amount $d\xi$ at some composition. From Topic 6A, specifically the equation $\Delta_r G = (\partial G/\partial \xi)_{T,p}$, it follows that (at constant temperature and pressure)

$$dG = \Delta_r G d\xi$$

The maximum non-expansion (electrical) work, w_e, that the reaction can do as it advances by $d\xi$ at constant temperature and pressure is therefore

$$dw_e = \Delta_r G d\xi$$

This work is infinitesimal, and the composition of the system is virtually constant when it occurs.

Suppose that the reaction advances by $d\xi$, then $\nu d\xi$ electrons must travel from the anode to the cathode, where ν is the stoichiometric coefficient of the electrons in the half-reactions into which the cell reaction can be divided. The total charge transported between the electrodes when this change occurs is $-\nu e N_A d\xi$ (because $\nu d\xi$ is the amount of electrons in moles and the charge per mole of electrons is $-eN_A$). Hence, the total charge transported is $-\nu F d\xi$ because $eN_A = F$, Faraday's constant. The work done when an infinitesimal charge $-\nu F d\xi$ travels from the anode to the cathode is equal to the product of the charge and the potential difference, E_{cell} (see Table 2A.1, the entry $dw = \phi dQ$):

$$dw_e = -\nu F E_{cell} d\xi$$

When this relation is equated to the one above ($dw_e = \Delta_r G d\xi$), the advancement $d\xi$ cancels, and the resulting expression is

$$-\nu F E_{cell} = \Delta_r G \qquad (6C.2)$$

The cell potential

This equation is the key connection between electrical measurements on the one hand and thermodynamic properties on the other. It is the basis of all that follows.

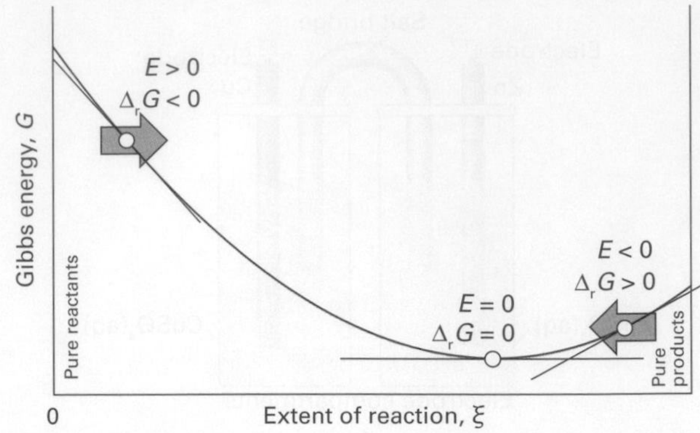

Figure 6C.4 A spontaneous reaction occurs in the direction of decreasing Gibbs energy. When expressed in terms of a cell potential, the spontaneous direction of change can be expressed in terms of the cell potential, E_{cell}. The cell reaction is spontaneous as written when $E_{cell} > 0$. The reverse reaction is spontaneous when $E_{cell} < 0$. When the cell reaction is at equilibrium, the cell potential is zero.

It follows from eqn 6C.2 that, by knowing the reaction Gibbs energy at a specified composition, the cell potential is known at that composition. Note that a negative reaction Gibbs energy, signifying a spontaneous cell reaction, corresponds to a positive cell potential, one in which a voltmeter connected to the cell shows that the right-hand electrode (as in the specification of the cell, not necessarily how the cell is arranged on the bench) is the positive electrode. Another way of looking at the content of eqn 6C.2 is that it shows that the driving power of a cell (that is, its potential difference) is proportional to the slope of the Gibbs energy with respect to the extent of reaction (the significance of $\Delta_r G$). It is plausible that a reaction that is far from equilibrium (when the slope is steep) has a strong tendency to drive electrons through an external circuit (Fig. 6C.4). When the slope is close to zero (when the cell reaction is close to equilibrium), the cell potential is small.

Brief illustration 6C.5

Equation 6C.2 provides an electrical method for measuring a reaction Gibbs energy at any composition of the reaction mixture: simply measure the cell potential and convert it to $\Delta_r G$. Conversely, if the value of $\Delta_r G$ is known at a particular composition, then it is possible to predict the cell potential. For example, if $\Delta_r G = -1.0 \times 10^2\,\text{kJ mol}^{-1}$ and $\nu = 1$, then (using $1\,\text{J} = 1\,\text{C V}$):

$$E_{cell} = -\frac{\Delta_r G}{\nu F} = -\frac{(-1.0\times10^5\,\text{J mol}^{-1})}{1\times(9.6485\times10^4\,\text{C mol}^{-1})} = 1.0\,\text{V}$$

The reaction Gibbs energy is related to the composition of the reaction mixture by eqn 6A.12 ($\Delta_r G = \Delta_r G^\ominus + RT\ln Q$). It follows, on division of both sides by $-\nu F$ and recognizing that $\Delta_r G/(-\nu F) = E_{cell}$, that

$$E_{cell} = -\frac{\Delta_r G^\ominus}{\nu F} - \frac{RT}{\nu F}\ln Q$$

The first term on the right is written

$$E^\ominus_{cell} = -\frac{\Delta_r G^\ominus}{\nu F} \quad \text{Standard cell potential [definition]} \quad (6C.3)$$

and called the **standard cell potential**. That is, the standard cell potential is the standard reaction Gibbs energy expressed as a potential difference (in volts). It follows that

$$E_{cell} = E^\ominus_{cell} - \frac{RT}{\nu F}\ln Q \quad \text{Nernst equation} \quad (6C.4)$$

This equation for the cell potential in terms of the composition is called the **Nernst equation**; the dependence that it predicts is summarized in Fig. 6C.5.

Through eqn 6C.4, the standard cell potential can be interpreted as the cell potential when all the reactants and products in the cell reaction are in their standard states, for then all activities are 1, so $Q = 1$ and $\ln Q = 0$. However, the fact that the standard cell potential is merely a disguised form of the standard reaction Gibbs energy (eqn 6C.3) should always be kept in mind and underlies all its applications.

Brief illustration 6C.6

Because RT/F = 25.7 mV at 25 °C, a practical form of the Nernst equation at this temperature is

$$E_{cell} = E^\ominus_{cell} - \frac{25.7\,\text{mV}}{\nu}\ln Q$$

It then follows that, for a reaction in which $\nu = 1$, if Q is increased by a factor of 10, then the cell potential decreases by 59.2 mV.

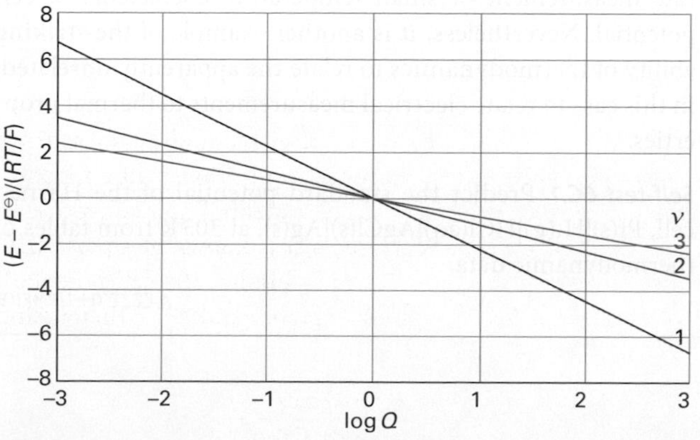

Figure 6C.5 The variation of cell potential with the value of the reaction quotient for the cell reaction for different values of ν (the number of electrons transferred). At 298 K, RT/F = 25.69 mV, so the vertical scale refers to multiples of this value.

An important feature of a standard cell potential is that it is unchanged if the chemical equation for the cell reaction is multiplied by a numerical factor. A numerical factor increases the value of the standard Gibbs energy for the reaction. However, it also increases the number of electrons transferred by the same factor, and by eqn 6C.3 the value of $E^\ominus_{cell}$ remains unchanged. A practical consequence is that a cell potential is independent of the physical size of the cell. In other words, the cell potential is an intensive property.

(b) Cells at equilibrium

A special case of the Nernst equation has great importance in electrochemistry and provides a link to the discussion of equilibrium in Topic 6A. Suppose the reaction has reached equilibrium; then $Q = K$, where K is the equilibrium constant of the cell reaction. However, a chemical reaction at equilibrium cannot do work, and hence it generates zero potential difference between the electrodes of a galvanic cell. Therefore, setting $E_{cell} = 0$ and $Q = K$ in the Nernst equation gives

$$E^\ominus_{cell} = \frac{RT}{\nu F}\ln K \quad \text{Equilibrium constant and standard cell potential} \quad (6C.5)$$

This very important equation (which could also have been obtained more directly by substituting eqn 6A.15, $\Delta_r G^\ominus = -RT\ln K$, into eqn 6C.3) can be used to predict equilibrium constants from measured standard cell potentials.

Brief illustration 6C.7

Because the standard potential of the Daniell cell is +1.10 V, the equilibrium constant for the cell reaction $Cu^{2+}(aq) + Zn(s) \rightarrow Cu(s) + Zn^{2+}(aq)$, for which $\nu = 2$, is $K = 1.5 \times 10^{37}$ at 298 K. That is, the displacement of copper by zinc goes virtually to completion. Note that a cell potential of about 1 V is easily measurable but corresponds to an equilibrium constant that would be impossible to measure by direct chemical analysis.

6C.4 The determination of thermodynamic functions

The standard potential of a cell is related to the standard reaction Gibbs energy through eqn 6C.3 (written as $-\nu FE^\ominus_{cell} = \Delta_r G^\ominus$). Therefore, this important thermodynamic quantity can be obtained by measuring $E^\ominus_{cell}$. Its value can then be used to calculate the Gibbs energy of formation of ions by using the convention explained in Topic 3D, that $\Delta_f G^\ominus(H^+,aq) = 0$.

Brief illustration 6C.8

The reaction taking place in the cell

$$\mathrm{Pt(s)|H_2(g)|H^+(aq)||Ag^+(aq)|Ag(s)} \qquad E^{\ominus}_{\mathrm{cell}} = +0.7996\,\mathrm{V}$$

is

$$\mathrm{Ag^+(aq)} + \tfrac{1}{2}\mathrm{H_2(g)} \rightarrow \mathrm{H^+(aq)} + \mathrm{Ag(s)} \quad \Delta_r G^{\ominus} = -\Delta_f G^{\ominus}(\mathrm{Ag^+,aq})$$

Therefore, with $\nu = 1$,

$$\begin{aligned}\Delta_f G^{\ominus}(\mathrm{Ag^+,aq}) &= -(-FE^{\ominus}) \\ &= (9.6485\times 10^4\,\mathrm{C\,mol^{-1}})\times(0.7996\,\mathrm{V}) \\ &= +77.15\,\mathrm{kJ\,mol^{-1}}\end{aligned}$$

which is in close agreement with the value in Table 2C.4 of the *Resource section*.

The temperature coefficient of the standard cell potential, $\mathrm{d}E^{\ominus}_{\mathrm{cell}}/\mathrm{d}T$, gives the standard entropy of the cell reaction. This conclusion follows from the thermodynamic relation $(\partial G/\partial T)_p = -S$ derived in Topic 3E and eqn 6C.3, which combine to give

$$\frac{\mathrm{d}E^{\ominus}_{\mathrm{cell}}}{\mathrm{d}T} = \frac{\Delta_r S^{\ominus}}{\nu F} \qquad \text{Temperature coefficient of standard cell potential} \qquad (6C.6)$$

The derivative is complete (not partial) because $E^{\ominus}_{\mathrm{cell}}$, like $\Delta_r G^{\ominus}$, is independent of the pressure. This is an electrochemical technique for obtaining standard reaction entropies and through them the entropies of ions in solution.

Finally, the combination of the results obtained so far leads to an expression for the standard reaction enthalpy:

$$\Delta_r H^{\ominus} = \Delta_r G^{\ominus} + T\Delta_r S^{\ominus} = -\nu F\left(E^{\ominus}_{\mathrm{cell}} - T\frac{\mathrm{d}E^{\ominus}_{\mathrm{cell}}}{\mathrm{d}T}\right) \qquad (6C.7)$$

This expression provides a non-calorimetric method for measuring $\Delta_r H^{\ominus}$ and, through the convention $\Delta_f H^{\ominus}(\mathrm{H^+,aq}) = 0$, the standard enthalpies of formation of ions in solution (Topic 2C).

Example 6C.1 Using the temperature coefficient of the standard cell potential

The standard potential of the cell $\mathrm{Pt(s)|H_2(g)|HBr(aq)|AgBr(s)|Ag(s)}$ was measured over a range of temperatures, and the data were found to fit the following polynomial:

$$E^{\ominus}_{\mathrm{cell}}/\mathrm{V} = 0.071\,31 - 4.99\times 10^{-4}(T/\mathrm{K} - 298) - 3.45\times 10^{-6}(T/\mathrm{K} - 298)^2$$

The cell reaction is $\mathrm{AgBr(s)} + \tfrac{1}{2}\mathrm{H_2(g)} \rightarrow \mathrm{Ag(s)} + \mathrm{HBr(aq)}$, and has $\nu = 1$. Evaluate the standard reaction Gibbs energy, enthalpy, and entropy at 298 K.

Collect your thoughts The standard Gibbs energy of reaction is obtained by using eqn 6C.3 after evaluating $E^{\ominus}_{\mathrm{cell}}$ at 298 K and by using $1\,\mathrm{V\,C} = 1\,\mathrm{J}$. The standard reaction entropy is obtained by using eqn 6C.6, which involves differentiating the polynomial with respect to T and then setting $T = 298\,\mathrm{K}$. The standard reaction enthalpy is obtained by combining the values of the standard Gibbs energy and entropy.

The solution At $T = 298\,\mathrm{K}$, $E^{\ominus}_{\mathrm{cell}} = 0.071\,31\,\mathrm{V}$, so

$$\begin{aligned}\Delta_r G^{\ominus} &= -\nu F E^{\ominus}_{\mathrm{cell}} = -(1)\times(9.6485\times 10^4\,\mathrm{C\,mol^{-1}})\times(0.071\,31\,\mathrm{V}) \\ &= -6.880\times 10^3\,\mathrm{C\,V\,mol^{-1}} = -6.880\,\mathrm{kJ\,mol^{-1}}\end{aligned}$$

The temperature coefficient of the standard cell potential is

$$\frac{\mathrm{d}E^{\ominus}_{\mathrm{cell}}}{\mathrm{d}T} = -4.99\times 10^{-4}\,\mathrm{V\,K^{-1}} - 2(3.45\times 10^{-6})(T/\mathrm{K} - 298)\,\mathrm{V\,K^{-1}}$$

At $T = 298\,\mathrm{K}$ this expression evaluates to

$$\frac{\mathrm{d}E^{\ominus}_{\mathrm{cell}}}{\mathrm{d}T} = -4.99\times 10^{-4}\,\mathrm{V\,K^{-1}}$$

So, from eqn 6C.6 the standard reaction entropy is

$$\begin{aligned}\Delta_r S^{\ominus} = \nu F\frac{\mathrm{d}E^{\ominus}_{\mathrm{cell}}}{\mathrm{d}T} &= (1)\times(9.6485\times 10^4\,\mathrm{C\,mol^{-1}})\times(-4.99\times 10^{-4}\,\mathrm{V\,K^{-1}}) \\ &= -48.1\,\mathrm{J\,K^{-1}\,mol^{-1}}\end{aligned}$$

The negative value stems in part from the elimination of gas in the cell reaction. It then follows that

$$\begin{aligned}\Delta_r H^{\ominus} = \Delta_r G^{\ominus} + T\Delta_r S^{\ominus} &= -6.880\,\mathrm{kJ\,mol^{-1}} \\ &\quad + (298\,\mathrm{K})\times(-0.0481\,\mathrm{kJ\,K^{-1}\,mol^{-1}}) \\ &= -21.2\,\mathrm{kJ\,mol^{-1}}\end{aligned}$$

Comment. One difficulty with this procedure lies in the accurate measurement of small temperature coefficients of cell potential. Nevertheless, it is another example of the striking ability of thermodynamics to relate the apparently unrelated, in this case to relate electrical measurements to thermal properties.

Self-test 6C.1 Predict the standard potential of the Harned cell, $\mathrm{Pt(s)|H_2(g)|HCl(aq)|AgCl(s)|Ag(s)}$, at 303 K from tables of thermodynamic data.

Answer: +0.2222 V

Checklist of concepts

☐ 1. A **cell reaction** is expressed as the difference of two reduction **half-reactions**; each one defines a redox couple.

☐ 2. **Galvanic cells** can have different electrodes or electrodes that differ in either the electrolyte or electrode concentration.

☐ 3. A **liquid junction potential** arises at the junction of two electrolyte solutions.

☐ 4. The **cell potential** is the potential difference measured under reversible conditions. The cell potential is positive if a voltmeter shows that the right-hand electrode (in the specification of the cell) is the positive electrode.

☐ 5. The **Nernst equation** relates the cell potential to the composition of the reaction mixture.

☐ 6. The **standard cell potential** may be used to calculate the standard Gibbs energy of the cell reaction and hence its equilibrium constant.

☐ 7. The temperature coefficient of the standard cell potential is used to measure the standard entropy and standard enthalpy of the cell reaction.

Checklist of equations

Property	Equation	Comment	Equation number
Cell potential and reaction Gibbs energy	$-\nu FE_{\text{cell}} = \Delta_r G$	Constant temperature and pressure	6C.2
Standard cell potential	$E^{\ominus}_{\text{cell}} = -\Delta_r G^{\ominus}/\nu F$	Definition	6C.3
Nernst equation	$E_{\text{cell}} = E^{\ominus}_{\text{cell}} - (RT/\nu F)\ln Q$		6C.4
Equilibrium constant of cell reaction	$E^{\ominus}_{\text{cell}} = (RT/\nu F)\ln K$		6C.5
Temperature coefficient of cell potential	$\mathrm{d}E^{\ominus}_{\text{cell}}/\mathrm{d}T = \Delta_r S^{\ominus}/\nu F$		6C.6

TOPIC 6D Electrode potentials

> ➤ **Why do you need to know this material?**
>
> A very powerful, compact, and widely used way to report standard cell potentials is to ascribe a potential to each electrode. Electrode potentials are used in chemistry to assess the oxidizing and reducing power of redox couples and to infer thermodynamic properties, including equilibrium constants.
>
> ➤ **What is the key idea?**
>
> Each electrode of a cell can be supposed to make a characteristic contribution to the cell potential; redox couples with low electrode potentials tend to reduce those with higher potentials.
>
> ➤ **What do you need to know already?**
>
> This Topic develops the concepts in Topic 6C, so you need to understand the concept of cell potential and standard cell potential; it also makes use of the Nernst equation. The measurement of standard potentials makes use of the Debye–Hückel limiting law (Topic 5F).

As explained in Topic 6C, a galvanic cell is a combination of two electrodes. Each electrode can be considered to make a characteristic contribution to the overall cell potential. Although it is not possible to measure the contribution of a single electrode, the potential of one of the electrodes can be defined as zero, so values can be assigned to others on that basis.

6D.1 Standard potentials

The specially selected electrode is the **standard hydrogen electrode** (SHE):

$$\mathrm{Pt(s)|H_2(g)|H^+(aq)} \quad E^\ominus = 0 \text{ at all temperatures}$$

Standard potentials [convention] (6D.1)

To achieve standard conditions, the activity of the hydrogen ions must be 1 (i.e. pH = 0) and the pressure of the hydrogen gas must be 1 bar.[1] The **standard potential**, $E^\ominus(\mathrm{X})$, of another redox couple X is then equal to the cell potential in which it forms the right-hand electrode and the standard hydrogen electrode is the left-hand electrode:

$$\mathrm{Pt(s)|H_2(g)|H^+(aq)||X} \quad E^\ominus(\mathrm{X}) = E^\ominus_{\mathrm{cell}}$$

Standard potentials [convention] (6D.2)

The standard potential of a cell of the form L||R, where L is the left-hand electrode of the cell as written (not as arranged on the bench) and R is the right-hand electrode, is then given by the difference of the two standard (electrode) potentials:

$$\mathrm{L||R} \quad E^\ominus_{\mathrm{cell}} = E^\ominus(\mathrm{R}) - E^\ominus(\mathrm{L})$$

Standard cell potential (6D.3)

A list of standard potentials at 298 K is given in Table 6D.1, and longer lists in numerical and alphabetical order are in the *Resource section*.

Table 6D.1 Standard potentials at 298 K*

Couple	$E^\ominus$/V
$\mathrm{Ce^{4+}(aq) + e^- \rightarrow Ce^{3+}(aq)}$	+1.61
$\mathrm{Cu^{2+}(aq) + 2\,e^- \rightarrow Cu(s)}$	+0.34
$\mathrm{AgCl(s) + e^- \rightarrow Ag(s) + Cl^-(aq)}$	+0.22
$\mathrm{H^+(aq) + e^- \rightarrow \frac{1}{2}H_2(g)}$	0
$\mathrm{Zn^{2+}(aq) + 2\,e^- \rightarrow Zn(s)}$	−0.76
$\mathrm{Na^+(aq) + e^- \rightarrow Na(s)}$	−2.71

* More values are given in the *Resource section*.

Brief illustration 6D.1

The cell $\mathrm{Ag(s)|AgCl(s)|HCl(aq)|O_2(g)|Pt(s)}$ can be regarded as formed from the following two electrodes, with their standard potentials taken from the *Resource section*:

Electrode	Half-reaction	Standard potential
R: $\mathrm{Pt(s)|O_2(g)|H^+(aq)}$	$\mathrm{O_2(g) + 4H^+(aq) + 4e^- \rightarrow 2H_2O(l)}$	+1.23 V
L: $\mathrm{Ag(s)|AgCl(s)|Cl^-(aq)}$	$\mathrm{AgCl(s) + e^- \rightarrow Ag(s) + Cl^-(aq)}$	+0.22 V
	$E^\ominus_{\mathrm{cell}} =$	+1.01 V

[1] Strictly speaking the *fugacity*, which is the equivalent of activity for a gas (see *A deeper look* 2 on the website for this text), should be 1. This complication is ignored here, which is equivalent to assuming perfect gas behaviour.

(a) The measurement procedure

The procedure for measuring a standard potential can be illustrated by considering a specific case, the silver/silver chloride electrode. The measurement is made on the 'Harned cell':

$$\mathrm{Pt(s)|H_2(g)|HCl(aq,}b\mathrm{)|AgCl(s)|Ag(s)}$$

$$\tfrac{1}{2}\,\mathrm{H_2(g) + AgCl(s) \rightarrow HCl(aq) + Ag(s)}$$

$$E^{\ominus}_{\text{cell}} = E^{\ominus}(\mathrm{AgCl/Ag,Cl^-}) - E^{\ominus}(\mathrm{SHE}) = E^{\ominus}(\mathrm{AgCl/Ag,Cl^-}),\ \nu = 1$$

for which the Nernst equation is

$$E_{\text{cell}} = E^{\ominus}(\mathrm{AgCl/Ag,Cl^-}) - \frac{RT}{F}\ln\frac{a_{\mathrm{H^+}}a_{\mathrm{Cl^-}}}{a_{\mathrm{H_2}}^{1/2}}$$

If the hydrogen gas is at the standard pressure of 1 bar, then $a_{\mathrm{H_2}} = 1$. For simplicity, writing the standard potential of the AgCl/Ag,Cl⁻ electrode as $E^{\ominus}$, turns this equation into

$$E_{\text{cell}} = E^{\ominus} - \frac{RT}{F}\ln a_{\mathrm{H^+}}a_{\mathrm{Cl^-}}$$

The activities in this expression can be written in terms of the molality b of HCl(aq) through $a_{\mathrm{H^+}} = \gamma_{\pm}b/b^{\ominus}$ and $a_{\mathrm{Cl^-}} = \gamma_{\pm}b/b^{\ominus}$, as established in Topic 5F:

$$E_{\text{cell}} = E^{\ominus} - \frac{RT}{F}\ln\frac{\gamma_{\pm}^2 b^2}{b^{\ominus 2}}$$

$$= E^{\ominus} - \frac{2RT}{F}\ln\frac{b}{b^{\ominus}} - \frac{2RT}{F}\ln\gamma_{\pm}$$

and therefore

$$E_{\text{cell}} + \frac{2RT}{F}\ln\frac{b}{b^{\ominus}} = E^{\ominus} - \frac{2RT}{F}\ln\gamma_{\pm}$$

From the Debye–Hückel limiting law for a 1,1-electrolyte (eqn 5F.27, $\log\gamma_{\pm} = -A|z_+z_-|I^{1/2}$), it follows that as $b \rightarrow 0$

$$\log\gamma_{\pm} = -A|z_+z_-|I^{1/2} = -A(b/b^{\ominus})^{1/2}$$

Therefore, because $\ln x = \ln 10 \log x$,

$$\ln\gamma_{\pm} = \ln 10 \log\gamma_{\pm} = -(A\ln 10)\,(b/b^{\ominus})^{1/2}$$

The equation for E_{cell} then becomes

$$E_{\text{cell}} + \frac{2RT}{F}\ln\frac{b}{b^{\ominus}} = E^{\ominus} + \frac{2ART\ln 10}{F}\left(\frac{b}{b^{\ominus}}\right)^{1/2} \quad \text{as } b \rightarrow 0$$

With the term in blue denoted C, this equation becomes.

$$\overbrace{E_{\text{cell}} + \frac{2RT}{F}\ln\frac{b}{b^{\ominus}}}^{y} = \overbrace{E^{\ominus}}^{\text{intercept}} + \overbrace{C \times \left(\frac{b}{b^{\ominus}}\right)^{1/2}}^{\text{slope}\times x} \qquad (6D.4)$$

where C is a constant. To use this equation, which has the form $y = \text{intercept} + \text{slope} \times x$ with $x = (b/b^{\ominus})^{1/2}$, the expression on the left is evaluated at a range of molalities, plotted against $(b/b^{\ominus})^{1/2}$, and extrapolated to $b = 0$. The intercept at $b^{1/2} = 0$ is the value of $E^{\ominus}$ for the silver/silver-chloride electrode. In precise work, the $(b/b^{\ominus})^{1/2}$ term is brought to the left, and a higher-order correction term from extended versions of the Debye–Hückel law (Topic 5F) is used on the right.

Example 6D.1 Evaluating a standard potential

The potential of the Harned cell at 25 °C has the following values:

$b/(10^{-3}b^{\ominus})$	3.215	5.619	9.138	25.63
E_{cell}/V	0.520 53	0.492 57	0.468 60	0.418 24

Determine the standard potential of the silver/silver chloride electrode.

Collect your thoughts As explained in the text, evaluate $y = E_{\text{cell}} + (2RT/F)\ln(b/b^{\ominus})$ and plot it against $(b/b^{\ominus})^{1/2}$; then extrapolate to $b = 0$.

The solution To determine the standard potential of the cell, draw up the following table, using $2RT/F = 0.051\,39$ V:

$b/(10^{-3}b^{\ominus})$	3.215	5.619	9.138	25.63
$\{b/(10^{-3}b^{\ominus})\}^{1/2}$	1.793	2.370	3.023	5.063
E_{cell}/V	0.520 53	0.492 57	0.468 60	0.418 24
y/V	0.2256	0.2263	0.2273	0.2299

The data are plotted in Fig. 6D.1; as can be seen, they extrapolate to $E^{\ominus} = +0.2232$ V (the value obtained, to preserve the precision of the data, by linear regression).

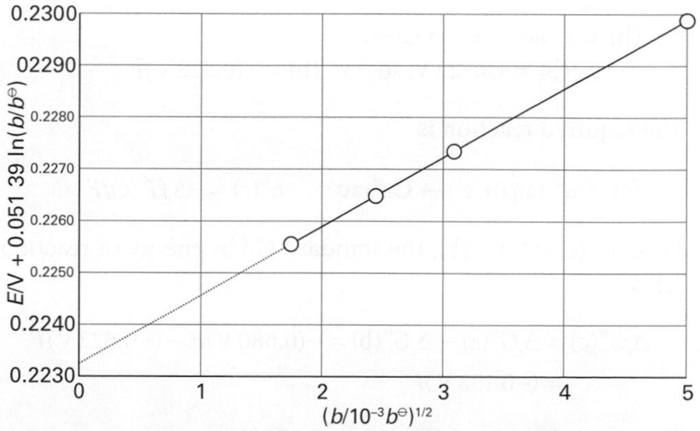

Figure 6D.1 The plot and the extrapolation used for the experimental measurement of a standard cell potential. The intercept at $b^{1/2} = 0$ is $E^{\ominus}_{\text{cell}}$.

Self-test 6D.1 The following data are for the cell $\mathrm{Pt(s)|H_2(g)|HBr(aq,}b\mathrm{)|AgBr(s)|Ag(s)}$ at 25 °C and with the hydrogen gas at 1 bar. Determine the standard cell potential.

$b/(10^{-4}b^{\ominus})$	4.042	8.444	37.19
E_{cell}/V	0.469 42	0.436 36	0.361 73

Answer: +0.071 V

(b) Combining measured values

The standard potentials in Table 6D.1 may be combined to give values for couples that are not listed there. However, to do so, it is necessary to take into account the fact that different couples might correspond to the transfer of different numbers of electrons. The procedure is illustrated in the following *Example*.

Example 6D.2 Evaluating a standard potential from two others

Given that the standard potentials of the Cu^{2+}/Cu and Cu^{+}/Cu couples are +0.340 V and +0.522 V, respectively, evaluate $E^{\ominus}(Cu^{2+},Cu^{+})$.

Collect your thoughts First, note that reaction Gibbs energies may be added (as in a Hess's law analysis of reaction enthalpies). Therefore, you should convert the $E^{\ominus}$ values to $\Delta_r G^{\ominus}$ values by using eqn 6C.3 ($-\nu FE^{\ominus} = \Delta_r G^{\ominus}$), add them appropriately, and then convert the overall $\Delta_r G^{\ominus}$ to the required $E^{\ominus}$ by using eqn 6C.3 again. This roundabout procedure is necessary because, as seen below, although the factor F cancels (and should be kept in place until it cancels), the factor ν in general does not cancel.

The solution The electrode half-reactions are as follows:

(a) $Cu^{2+}(aq) + 2e^- \rightarrow Cu(s)$
$E^{\ominus}(a) = +0.340\,V$, so $\Delta_r G^{\ominus}(a) = -2(0.340\,V)F$

(b) $Cu^{+}(aq) + e^- \rightarrow Cu(s)$
$E^{\ominus}(b) = +0.522\,V$, so $\Delta_r G^{\ominus}(b) = -(0.522\,V)F$

The required reaction is

(c) $Cu^{2+}(aq) + e^- \rightarrow Cu^{+}(aq) \qquad E^{\ominus}(c) = -\Delta_r G^{\ominus}(c)/F$

Because (c) = (a) − (b), the standard Gibbs energy of reaction (c) is

$$\Delta_r G^{\ominus}(c) = \Delta_r G^{\ominus}(a) - \Delta_r G^{\ominus}(b) = -(0.680\,V)F - (-0.522\,V)F = (-0.158\,V)F$$

Therefore, $E^{\ominus}(c) = -\Delta_r G^{\ominus}(c)/F = +0.158\,V$.

Self-test 6D.2 Evaluate $E^{\ominus}(Fe^{3+},Fe^{2+})$ from $E^{\ominus}(Fe^{3+},Fe)$ and $E^{\ominus}(Fe^{2+},Fe)$.

Answer: +0.76 V

The generalization of the calculation in the *Example* is

$$\nu_c E^{\ominus}(c) = \nu_a E^{\ominus}(a) - \nu_b E^{\ominus}(b) \qquad \text{Combination of standard potentials} \qquad (6D.5)$$

with the ν_r the stoichiometric coefficients of the electrons in each half-reaction.

6D.2 Applications of standard potentials

Cell potentials are a convenient source of data on equilibrium constants and the Gibbs energies, enthalpies, and entropies of reactions. In practice the standard values of these quantities are the ones normally determined.

(a) The electrochemical series

For two redox couples, Ox_L/Red_L and Ox_R/Red_R, and the cell

$$L||R = Ox_L/Red_L||Ox_R/Red_R$$

$$Ox_R + \nu e^- \rightarrow Red_R \quad Ox_L + \nu e^- \rightarrow Red_L \qquad \text{Cell convention} \qquad (6D.6a)$$

$$E^{\ominus}_{cell} = E^{\ominus}(R) - E^{\ominus}(L)$$

the cell reaction

$$R - L: Red_L + Ox_R \rightarrow Ox_L + Red_R \qquad (6D.6b)$$

has $K > 1$ if $E^{\ominus}_{cell} > 0$, and therefore if $E^{\ominus}(L) < E^{\ominus}(R)$. Because in the cell reaction Red_L reduces Ox_R, it follows that

> Red_L has a thermodynamic tendency (in the sense $K > 1$) to reduce Ox_R if $E^{\ominus}(L) < E^{\ominus}(R)$.

More briefly: low reduces high.

Table 6D.2 shows a part of the **electrochemical series**, the metallic elements (and hydrogen) arranged in the order of their reducing power as measured by their standard potentials in aqueous solution. A metal low in the series (with a lower standard potential) can reduce the ions of metals with higher standard potentials. This conclusion is qualitative. The quantitative value of K is obtained by doing the calculations described previously and reviewed below.

Brief illustration 6D.2

Zinc lies above magnesium in the electrochemical series, so zinc cannot reduce magnesium ions in aqueous solution. Zinc can reduce hydrogen ions, because hydrogen lies higher in the series. However, even for reactions that are thermodynamically favourable, there may be kinetic factors that result in very slow rates of reaction.

(b) The determination of activity coefficients

Once the standard potential of an electrode in a cell is known, it can be used to determine mean activity coefficients by measuring the cell potential with the ions at the concentration of

Table 6D.2 The electrochemical series*

Least strongly reducing
Gold (Au^{3+}/Au)
Platinum (Pt^{2+}/Pt)
Silver (Ag^{+}/Ag)
Mercury (Hg^{2+}/Hg)
Copper (Cu^{2+}/Cu)
Hydrogen (H^{+}/H_2)
Tin (Sn^{2+}/Sn)
Nickel (Ni^{2+}/Ni)
Iron (Fe^{2+}/Fe)
Zinc (Zn^{2+}/Zn)
Chromium (Cr^{3+}/Cr)
Aluminium (Al^{3+}/Al)
Magnesium (Mg^{2+}/Mg)
Sodium (Na^{+}/Na)
Calcium (Ca^{2+}/Ca)
Potassium (K^{+}/K)
Most strongly reducing

* The complete series can be inferred from Table 6D.1 in the *Resource section*.

interest. For example, in the Harned cell analysed in Section 6D.1, the mean activity coefficient of the ions in hydrochloric acid of molality b is obtained from the relation

$$E_{\text{cell}} + \frac{2RT}{F}\ln\frac{b}{b^{\ominus}} = E^{\ominus} - \frac{2RT}{F}\ln\gamma_{\pm}$$

which can be rearranged into

$$\ln\gamma_{\pm} = \frac{E^{\ominus} - E_{\text{cell}}}{2RT/F} - \ln\frac{b}{b^{\ominus}} \qquad (6D.7)$$

Brief illustration 6D.3

The data in *Example* 6D.1 include the fact that $E_{\text{cell}} = 0.468\,60\ \text{V}$ when $b = 9.138 \times 10^{-3} b^{\ominus}$. Because $2RT/F = 0.051\,39\ \text{V}$, and in the *Example* it is established that $E^{\ominus}_{\text{cell}} = 0.2232\ \text{V}$, the mean activity coefficient at this molality is

$$\ln\gamma_{\pm} = \frac{0.2232\,\text{V} - 0.46860\,\text{V}}{0.05139\,\text{V}} - \ln(9.138\times10^{-3}) = -0.0799\ldots$$

Therefore, $\gamma_{\pm} = 0.9232$.

(c) The determination of equilibrium constants

The principal use for standard potentials is to calculate the standard potential of a cell formed from any two electrodes and then to use that value to evaluate the equilibrium constant of the cell reaction. To do so, construct $E^{\ominus}_{\text{cell}} = E^{\ominus}(\text{R}) - E^{\ominus}(\text{L})$ and then use eqn 6C.5 of Topic 6C ($E^{\ominus}_{\text{cell}} = (RT/\nu F)\ln K$, arranged into $\ln K = \nu FE^{\ominus}_{\text{cell}}/RT$).

Brief illustration 6D.4

A *disproportionation reaction* is a reaction in which a species is both oxidized and reduced. To study the disproportionation $2\,Cu^{+}(aq) \rightarrow Cu(s) + Cu^{2+}(aq)$ at 298 K, combine the following electrodes:

R: $Cu(s)\vert Cu^{+}(aq)$	$Cu^{+}(aq) + e^{-} \rightarrow Cu(s)$	$E^{\ominus}(\text{R}) = +0.52\ \text{V}$
L: $Pt(s)\vert Cu^{2+}(aq),Cu^{+}(aq)$	$Cu^{2+}(aq) + e^{-} \rightarrow Cu^{+}(aq)$	$E^{\ominus}(\text{L}) = +0.16\ \text{V}$

The cell reaction is therefore $2\,Cu^{+}(aq) \rightarrow Cu(s) + Cu^{2+}(aq)$, and the standard cell potential is

$$E^{\ominus}_{\text{cell}} = 0.52\ \text{V} - 0.16\ \text{V} = +0.36\ \text{V}$$

Now calculate the equilibrium constant of the cell reaction. Because $\nu = 1$, from eqn 6C.5 with $RT/F = 0.025\,693\ \text{V}$,

$$\ln K = \frac{0.36\,\text{V}}{0.025\,693\,\text{V}} = 14.0\ldots$$

Hence, $K = 1.2 \times 10^{6}$.

Checklist of concepts

☐ 1. The **standard potential** of a couple is the potential of a cell in which the couple forms the right-hand electrode and the left-hand electrode is a standard hydrogen electrode, all species being present at unit activity.

☐ 2. The **electrochemical series** lists the metallic elements in the order of their reducing power as measured by their standard potentials in aqueous solution: low reduces high.

☐ 3. The difference of the cell potential from its standard value is used to measure the activity coefficient of ions in solution.

☐ 4. Standard potentials are used to calculate the standard cell potential and then to calculate the equilibrium constant of the cell reaction.

Checklist of equations

Property	Equation	Comment	Equation number
Standard cell potential from standard potentials	$E^{\ominus}_{\text{cell}} = E^{\ominus}(\text{R}) - E^{\ominus}(\text{L})$	Cell: L‖R	6D.3
Combined standard potentials	$\nu_c E^{\ominus}(\text{c}) = \nu_a E^{\ominus}(\text{a}) - \nu_b E^{\ominus}(\text{b})$		6D.5

FOCUS 6 Chemical equilibrium

TOPIC 6A The equilibrium constant

Discussion questions

D6A.1 Explain how the mixing of reactants and products affects the position of chemical equilibrium.

D6A.2 What is the physical justification for not including a pure liquid or solid in the expression for an equilibrium constant?

Exercises

E6A.1(a) Consider the reaction $A \rightarrow 2B$. Initially 1.50 mol A is present and no B. What are the amounts of A and B when the extent of reaction is 0.60 mol?

E6A.1(b) Consider the reaction $2A \rightarrow B$. Initially 1.75 mol A and 0.12 mol B are present. What are the amounts of A and B when the extent of reaction is 0.30 mol?

E6A.2(a) When the reaction $A \rightarrow 2B$ advances by 0.10 mol (i.e. $\Delta\xi = +0.10$ mol) the molar Gibbs energy of the system changes by $-6.4\,kJ\,mol^{-1}$. What is the Gibbs energy of reaction at this stage of the reaction?

E6A.2(b) When the reaction $2A \rightarrow B$ advances by 0.051 mol (i.e. $\Delta\xi = +0.051$ mol) the molar Gibbs energy of the system changes by $-2.41\,kJ\,mol^{-1}$. What is the Gibbs energy of reaction at this stage of the reaction?

E6A.3(a) Classify the formation of methane from its elements in their reference states as exergonic or endergonic under standard conditions at 298 K.

E6A.3(b) Classify the formation of liquid benzene from its elements in their reference states as exergonic or endergonic under standard conditions at 298 K.

E6A.4(a) Write the reaction quotient for $A + 2B \rightarrow 3C$.

E6A.4(b) Write the reaction quotient for $2A + B \rightarrow 2C + D$.

E6A.5(a) Write the equilibrium constant for the reaction $P_4(s) + 6H_2(g) \rightleftharpoons 4PH_3(g)$, with the gases treated as perfect.

E6A.5(b) Write the equilibrium constant for the reaction $CH_4(g) + 3Cl_2(g) \rightleftharpoons CHCl_3(l) + 3HCl(g)$, with the gases treated as perfect.

E6A.6(a) Use data found in the *Resource section* to decide which of the following reactions have $K > 1$ at 298 K: (i) $2CH_3CHO(g) + O_2(g) \rightleftharpoons 2CH_3COOH(l)$, (ii) $2AgCl(s) + Br_2(l) \rightleftharpoons 2AgBr(s) + Cl_2(g)$

E6A.6(b) Use data found in the *Resource section* to decide which of the following reactions have $K < 1$ at 298 K: (i) $Hg(l) + Cl_2(g) \rightleftharpoons HgCl_2(s)$, (ii) $Zn(s) + Cu^{2+}(aq) \rightleftharpoons Zn^{2+}(aq) + Cu(s)$

E6A.7(a) One reaction has a standard Gibbs energy of $-320\,kJ\,mol^{-1}$ and a second reaction has a standard Gibbs energy of $-55\,kJ\,mol^{-1}$, both at 300 K. What is the ratio of their equilibrium constants at 300 K?

E6A.7(b) One reaction has a standard Gibbs energy of $-200\,kJ\,mol^{-1}$ and a second reaction has a standard Gibbs energy of $+30\,kJ\,mol^{-1}$, both at 300 K. What is the ratio of their equilibrium constants at 300 K?

E6A.8(a) The standard Gibbs energy of the reaction $N_2(g) + 3H_2(g) \rightarrow 2NH_3(g)$ is $-32.9\,kJ\,mol^{-1}$ at 298 K. What is the value of $\Delta_r G$ when Q = (i) 0.010, (ii) 1.0, (iii) 10.0, (iv) 100 000, (v) 1 000 000? Estimate (by interpolation) the value of K from the values you calculate. What is the actual value of K?

E6A.8(b) The standard Gibbs energy of the reaction $2NO_2(g) \rightarrow N_2O_4(g)$ is $-4.73\,kJ\,mol^{-1}$ at 298 K. What is the value of $\Delta_r G$ when Q = (i) 0.10, (ii) 1.0, (iii) 10, (iv) 100? Estimate (by interpolation) the value of K from the values you calculate. What is the actual value of K?

E6A.9(a) At 2257 K and 1.00 bar total pressure, water is 1.77 per cent dissociated at equilibrium by way of the reaction $2H_2O(g) \rightleftharpoons 2H_2(g) + O_2(g)$. Calculate K.

E6A.9(b) For the equilibrium, $N_2O_4(g) \rightleftharpoons 2NO_2(g)$, the degree of dissociation, α, at 298 K is 0.201 at 1.00 bar total pressure. Calculate K.

E6A.10(a) Establish the relation between K and K_c for the reaction $H_2CO(g) \rightleftharpoons CO(g) + H_2(g)$.

E6A.10(b) Establish the relation between K and K_c for the reaction $3N_2(g) + H_2(g) \rightleftharpoons 2HN_3(g)$.

E6A.11(a) In the gas-phase reaction $2A + B \rightleftharpoons 3C + 2D$, it was found that, when 1.00 mol A, 2.00 mol B, and 1.00 mol D were mixed and allowed to come to equilibrium at 25 °C, the resulting mixture contained 0.90 mol C at a total pressure of 1.00 bar. Calculate (i) the mole fractions of each species at equilibrium, (ii) K, and (iii) $\Delta_r G^\ominus$.

E6A.11(b) In the gas-phase reaction $A + B \rightleftharpoons C + 2D$, it was found that, when 2.00 mol A, 1.00 mol B, and 3.00 mol D were mixed and allowed to come to equilibrium at 25 °C, the resulting mixture contained 0.79 mol C at a total pressure of 1.00 bar. Calculate (i) the mole fractions of each species at equilibrium, (ii) K, and (iii) $\Delta_r G^\ominus$.

E6A.12(a) The standard reaction Gibbs energy of the isomerization of borneol ($C_{10}H_{17}OH$) to isoborneol in the gas phase at 503 K is $+9.4\,kJ\,mol^{-1}$. Calculate the reaction Gibbs energy in a mixture consisting of 0.15 mol of borneol and 0.30 mol of isoborneol when the total pressure is 600 Torr.

E6A.12(b) The equilibrium pressure of H_2 over solid uranium and uranium hydride, UH_3, at 500 K is 139 Pa. Calculate the standard Gibbs energy of formation of $UH_3(s)$ at 500 K.

E6A.13(a) The standard Gibbs energy of formation of $NH_3(g)$ is $-16.5\,kJ\,mol^{-1}$ at 298 K. What is the corresponding reaction Gibbs energy when the partial pressures of the N_2, H_2, and NH_3 (treated as perfect gases) are 3.0 bar, 1.0 bar, and 4.0 bar, respectively? What is the spontaneous direction of the reaction in this case?

E6A.13(b) The standard Gibbs energy of formation of $PH_3(g)$ is $+13.4\,kJ\,mol^{-1}$ at 298 K. What is the corresponding reaction Gibbs energy when the partial pressures of the H_2 and PH_3 (treated as perfect gases) are 1.0 bar and 0.60 bar, respectively? What is the spontaneous direction of the reaction in this case?

E6A.14(a) For $CaF_2(s) \rightleftharpoons Ca^{2+}(aq) + 2F^-(aq)$, $K = 3.9 \times 10^{-11}$ at 25 °C and the standard Gibbs energy of formation of $CaF_2(s)$ is $-1167\,kJ\,mol^{-1}$. Calculate the standard Gibbs energy of formation of $CaF_2(aq)$.

E6A.14(b) For $PbI_2(s) \rightleftharpoons Pb^{2+}(aq) + 2I^-(aq)$, $K = 1.4 \times 10^{-8}$ at 25 °C and the standard Gibbs energy of formation of $PbI_2(s)$ is $-173.64\,kJ\,mol^{-1}$. Calculate the standard Gibbs energy of formation of $PbI_2(aq)$.

Problems

P6A.1 The equilibrium constant for the reaction, $I_2(s) + Br_2(g) \rightleftharpoons 2\,IBr(g)$ is 0.164 at 25 °C. (a) Calculate $\Delta_r G^\ominus$ for this reaction. (b) Bromine gas is introduced into a container with excess solid iodine. The pressure and temperature are held at 0.164 atm and 25 °C, respectively. Find the partial pressure of IBr(g) at equilibrium. Assume that all the bromine is in the gaseous form and that the vapour pressure of iodine is negligible. (c) In fact, solid iodine has a measurable vapour pressure at 25 °C. In this case, how would the calculation have to be modified?

P6A.2 Calculate the equilibrium constant of the reaction $CO(g) + H_2(g) \rightleftharpoons H_2CO(g)$ given that, for the production of liquid methanal (formaldehyde), $\Delta_r G^\ominus = +28.95\,\text{kJ mol}^{-1}$ at 298 K and that the vapour pressure of methanal is 1500 Torr at that temperature.

P6A.3 A sealed container was filled with 0.300 mol $H_2(g)$, 0.400 mol $I_2(g)$, and 0.200 mol HI(g) at 870 K and total pressure 1.00 bar. Calculate the amounts of the components in the mixture at equilibrium given that $K = 870$ for the reaction $H_2(g) + I_2(g) \rightleftharpoons 2\,HI(g)$.

P6A.4‡ Nitric acid hydrates have received much attention as possible catalysts for heterogeneous reactions that bring about the Antarctic ozone hole. Standard reaction Gibbs energies at 190 K are as follows:

(i) $H_2O(g) \rightarrow H_2O(s)$ $\qquad \Delta_r G^\ominus = -23.6\,\text{kJ mol}^{-1}$

(ii) $H_2O(g) + HNO_3(g) \rightarrow HNO_3 \cdot H_2O(s)$ $\qquad \Delta_r G^\ominus = -57.2\,\text{kJ mol}^{-1}$

(iii) $2\,H_2O(g) + HNO_3(g) \rightarrow HNO_3 \cdot 2H_2O(s)$ $\qquad \Delta_r G^\ominus = -85.6\,\text{kJ mol}^{-1}$

(iv) $3\,H_2O(g) + HNO_3(g) \rightarrow HNO_3 \cdot 3H_2O(s)$ $\qquad \Delta_r G^\ominus = -112.8\,\text{kJ mol}^{-1}$

Which solid is thermodynamically most stable at 190 K if $p_{H_2O} = 0.13\,\mu\text{bar}$ and $p_{HNO_3} = 0.41\,\text{nbar}$? *Hint:* Try computing $\Delta_r G$ for each reaction under the prevailing conditions. If more than one solid form spontaneously, then examine $\Delta_r G$ for the conversion of one solid to another.

P6A.5 Express the equilibrium constant of a gas-phase reaction $A + 3\,B \rightleftharpoons 2\,C$ in terms of the equilibrium value of the extent of reaction, ξ, given that initially A and B were present in stoichiometric proportions. Find an expression for ξ as a function of the total pressure, p, of the reaction mixture and sketch a graph of the expression obtained.

P6A.6 Consider the equilibrium $N_2O_4(g) \rightleftharpoons 2\,NO_2(g)$. From the tables of data in the *Resource section*, assess the contributions of $\Delta_r H^\ominus$ and $\Delta_r S^\ominus$ to the value of K at 298 K.

TOPIC 6B The response to equilibria to the conditions

Discussion questions

D6B.1 Suggest how the thermodynamic equilibrium constant may respond differently to changes in pressure and temperature from the equilibrium constant expressed in terms of partial pressures.

D6B.2 Account for Le Chatelier's principle in terms of thermodynamic quantities. Could there be exceptions to Le Chatelier's principle?

D6B.3 Explain the molecular basis of the van 't Hoff equation for the temperature dependence of K.

Exercises

E6B.1(a) Dinitrogen tetroxide is 18.46 per cent dissociated at 25 °C and 1.00 bar in the equilibrium $N_2O_4(g) \rightleftharpoons 2\,NO_2(g)$. Calculate K at (i) 25 °C, (ii) 100 °C given that $\Delta_r H^\ominus = +56.2\,\text{kJ mol}^{-1}$ over the temperature range.

E6B.1(b) Molecular bromine is 24 per cent dissociated at 1600 K and 1.00 bar in the equilibrium $Br_2(g) \rightleftharpoons 2\,Br(g)$. Calculate K at (i) 1600 K, (ii) 2000 K given that $\Delta_r H^\ominus = +112\,\text{kJ mol}^{-1}$ over the temperature range.

E6B.2(a) From information in the *Resource section*, calculate the standard Gibbs energy and the equilibrium constant at (i) 298 K and (ii) 400 K for the reaction $PbO(s,\text{red}) + CO(g) \rightleftharpoons Pb(s) + CO_2(g)$. Assume that the standard reaction enthalpy is independent of temperature.

E6B.2(b) From information in the *Resource section*, calculate the standard Gibbs energy and the equilibrium constant at (i) 25 °C and (ii) 50 °C for the reaction $CH_4(g) + 3\,Cl_2(g) \rightleftharpoons CHCl_3(l) + 3\,HCl(g)$. Assume that the standard reaction enthalpy is independent of temperature. At 298.15 K $\Delta_f G^\ominus(CHCl_3(l)) = -73.7\,\text{kJ mol}^{-1}$ and $\Delta_f H^\ominus(CHCl_3(l)) = -134.1\,\text{kJ mol}^{-1}$.

E6B.3(a) The standard reaction enthalpy of $Zn(s) + H_2O(g) \rightarrow ZnO(s) + H_2(g)$ is approximately constant at $+224\,\text{kJ mol}^{-1}$ from 920 K up to 1280 K. The standard reaction Gibbs energy is $+33\,\text{kJ mol}^{-1}$ at 1280 K. Estimate the temperature at which the equilibrium constant becomes greater than 1.

E6B.3(b) The standard enthalpy of a certain reaction is approximately constant at $+125\,\text{kJ mol}^{-1}$ from 800 K up to 1500 K. The standard reaction Gibbs energy is $+22\,\text{kJ mol}^{-1}$ at 1120 K. Estimate the temperature at which the equilibrium constant becomes greater than 1.

E6B.4(a) The equilibrium constant of the reaction $2\,C_3H_6(g) \rightleftharpoons C_2H_4(g) + C_4H_8(g)$ is found to fit the expression $\ln K = A + B/T + C/T^2$ between 300 K and 600 K, with $A = -1.04$, $B = -1088\,\text{K}$, and $C = 1.51 \times 10^5\,\text{K}^2$. Calculate the standard reaction enthalpy and standard reaction entropy at 400 K.

E6B.4(b) The equilibrium constant of a reaction is found to fit the expression $\ln K = A + B/T + C/T^3$ between 400 K and 500 K with $A = -2.04$, $B = -1176\,\text{K}$, and $C = 2.1 \times 10^7\,\text{K}^3$. Calculate the standard reaction enthalpy and standard reaction entropy at 450 K.

E6B.5(a) Calculate the percentage change in K_x for the reaction $H_2CO(g) \rightleftharpoons CO(g) + H_2(g)$ when the total pressure is increased from 1.0 bar to 2.0 bar at constant temperature.

E6B.5(b) Calculate the percentage change in K_x for the reaction $CH_3OH(g) + NOCl(g) \rightleftharpoons HCl(g) + CH_3NO_2(g)$ when the total pressure is increased from 1.0 bar to 2.0 bar at constant temperature.

E6B.6(a) The equilibrium constant for the gas-phase isomerization of borneol ($C_{10}H_{17}OH$) to its isomer isoborneol at 503 K is 0.106. A mixture consisting of 7.50 g of borneol and 14.0 g of isoborneol in a container of volume 5.0 dm^3 is heated to 503 K and allowed to come to equilibrium. Calculate the mole fractions of the two substances at equilibrium.

E6B.6(b) The equilibrium constant for the reaction $N_2(g) + O_2(g) \rightleftharpoons 2\,NO(g)$ is 1.69×10^{-3} at 2300 K. A mixture consisting of 5.0 g of nitrogen and 2.0 g of oxygen in a container of volume 1.0 dm^3 is heated to 2300 K and allowed to come to equilibrium. Calculate the mole fraction of NO at equilibrium.

E6B.7(a) What is the standard enthalpy of a reaction for which the equilibrium constant is (i) doubled, (ii) halved when the temperature is increased by 10 K at 298 K?

‡ These problems were supplied by Charles Trapp and Carmen Giunta.

E6B.7(b) What is the standard enthalpy of a reaction for which the equilibrium constant is (i) doubled, (ii) halved when the temperature is increased by 15 K at 310 K?

E6B.8(a) Estimate the temperature at which the equilibrium constant for the decomposition of $CaCO_3$(s, calcite) to CO_2(g) and CaO(s) becomes 1; assume $p_{CO_2} = 1$ bar.

E6B.8(b) Estimate the temperature at which the equilibrium constant for $CuSO_4 \cdot 5H_2O(s) \rightarrow CuSO_4(s) + 5H_2O(g)$ becomes 1; assume $p_{H_2O} = 1$ bar.

E6B.9(a) The dissociation vapour pressure of a salt $A_2B(s) \rightleftharpoons A_2(g) + B(g)$ at 367 °C is 208 kPa but at 477 °C it has risen to 547 kPa. For the dissociation reaction of A_2B(s), calculate (i) the equilibrium constant, (ii) the standard reaction Gibbs energy, (iii) the standard enthalpy, and (iv) the standard entropy of dissociation, all at 422 °C. Assume that the vapour behaves as a perfect gas and that $\Delta_r H^\ominus$ and $\Delta_r S^\ominus$ are independent of temperature in the range given.

E6B.9(b) Solid ammonium chloride dissociates according to $NH_4Cl(s) \rightarrow NH_3(g) + HCl(g)$. The dissociation vapour pressure of NH_4Cl at 427 °C is 608 kPa but at 459 °C it has risen to 1115 kPa. Calculate (i) the equilibrium constant, (ii) the standard reaction Gibbs energy, (iii) the standard enthalpy, (iv) the standard entropy of dissociation, all at 427 °C. Assume that the vapour behaves as a perfect gas and that $\Delta_r H^\ominus$ and $\Delta_r S^\ominus$ are independent of temperature in the range given.

Problems

P6B.1 The equilibrium constant for the reaction $N_2(g) + 3H_2(g) \rightleftharpoons 2NH_3(g)$ is 2.13×10^6 at 288 K and 1.75×10^5 at 308 K. Calculate the standard reaction enthalpy, assuming it to be constant over this temperature range.

P6B.2 Consider the dissociation of methane, CH_4(g), into the elements H_2(g) and C(s, graphite). (a) Given that $\Delta_f H^\ominus(CH_4, g) = -74.85$ kJ mol^{-1} and that $\Delta_f S^\ominus = -80.67$ J K^{-1} mol^{-1} at 298 K, calculate the value of the equilibrium constant at 298 K. (b) Assuming that $\Delta_r H^\ominus$ is independent of temperature, calculate K at 50 °C. (c) Calculate the degree of dissociation, α, of methane at 298 K and a total pressure of 0.010 bar. (d) Without doing any numerical calculations, explain how the degree of dissociation for this reaction will change as the pressure and temperature are varied.

P6B.3 The equilibrium pressure of H_2 over U(s) and UH_3(s) between 450 K and 715 K fits the expression $\ln(p/\text{Pa}) = A + B/T + C\ln(T/\text{K})$, with $A = 69.32$, $B = -1.464 \times 10^4$ K, and $C = -5.65$. Find an expression for the standard enthalpy of formation of UH_3(s) and from it calculate $\Delta_f C_p^\ominus$.

P6B.4 Use the following data on the reaction $H_2(g) + Cl_2(g) \rightleftharpoons 2HCl(g)$ to determine the standard reaction enthalpy:

T/K	300	500	1000
K	4.0×10^{31}	4.0×10^{18}	5.1×10^{8}

P6B.5 The degree of dissociation, α, of CO_2(g) into CO(g) and O_2(g) at high temperatures and 1 bar total pressure was found to vary with temperature as follows:

T/K	1395	1443	1498
$\alpha/10^{-4}$	1.44	2.50	4.71

Assume $\Delta_r H^\ominus$ to be constant over this temperature range, and calculate K, $\Delta_r G^\ominus$, $\Delta_r H^\ominus$, and $\Delta_r S^\ominus$ at 1443 K. Make any justifiable approximations.

P6B.6 The standard reaction enthalpy for the decomposition of $CaCl_2 \cdot NH_3$(s) into $CaCl_2$(s) and NH_3(g) is nearly constant at +78 kJ mol^{-1} between 350 K and 470 K. The equilibrium pressure of NH_3 in the presence of $CaCl_2 \cdot NH_3$ is 1.71 kPa at 400 K. Find an expression for the temperature dependence of $\Delta_r G^\ominus$ in the same range.

P6B.7 Ethanoic acid (acetic acid) was evaporated in container of volume 21.45 cm^3 at 437 K and at an external pressure of 101.9 kPa, and the container was then sealed. The mass of acid present in the sealed container was 0.0519 g. The experiment was repeated with the same container but at 471 K, and it was found that 0.0380 g of the acid was present. Calculate the equilibrium constant for the dimerization of the acid in the vapour, and the standard enthalpy of the dimerization reaction.

P6B.8 The dissociation of I_2(g) can be monitored by measuring the total pressure, and three sets of results are as follows:

T/K	973	1073	1173
$100p$/atm	6.244	6.500	9.181
$10^4 n_{I_2}$	2.4709	2.4555	2.4366

where n_{I_2} is the amount of I_2 molecules introduced into a container of volume 342.68 cm^3. Calculate the equilibrium constants of the dissociation and the standard enthalpy of dissociation assuming it to be constant over the range of temperatures.

P6B.9‡ The 1980s saw reports of $\Delta_f H^\ominus(SiH_2)$ ranging from 243 to 289 kJ mol^{-1}. If the standard enthalpy of formation is uncertain by this amount, by what factor is the equilibrium constant for the formation of SiH_2 from its elements uncertain at (a) 298 K, (b) 700 K?

P6B.10 Fuel cells show promise as power sources for automobiles. Hydrogen and carbon monoxide have been investigated for use in fuel cells, so their solubilities, s, in molten salts are of interest. Their solubilities in a molten $NaNO_3/KNO_3$ mixture were found to fit the following expressions:

$$\log(s_{H_2}/\text{mol cm}^{-3}\,\text{bar}^{-1}) = -5.39 - \frac{768}{T/\text{K}}$$

$$\log(s_{CO}/\text{mol cm}^{-3}\,\text{bar}^{-1}) = -5.98 - \frac{980}{T/\text{K}}$$

Calculate the standard molar enthalpies of solution of the two gases at 570 K.

P6B.11 Find an expression for the standard reaction Gibbs energy at a temperature T' in terms of its value at another temperature T and the coefficients a, b, and c in the expression for the molar heat capacity listed in Table 2B.1 ($C_{p,m} = a + bT + c/T^2$). Evaluate the standard Gibbs energy of formation of H_2O(l) at 372 K from its value at 298 K.

P6B.12 Derive an expression for the temperature dependence of K_c for a general gas-phase reaction.

TOPIC 6C Electrochemical cells

Discussion questions

D6C.1 Explain why reactions that are not redox reactions may be used to generate an electric current.

D6C.2 Distinguish between galvanic and electrolytic cells.

D6C.3 Explain the role of a salt bridge.

D6C.4 Why is it necessary to measure the cell potential under zero-current conditions?

D6C.5 Identify contributions to the cell potential when a current is being drawn from the cell.

Exercises

You will need to draw on information from Topic 6D to complete the answers.

E6C.1(a) Write the cell reaction and electrode half-reactions and calculate the standard potential of each of the following cells:

(i) $Zn(s)|ZnSO_4(aq)||AgNO_3(aq)|Ag(s)$
(ii) $Cd(s)|CdCl_2(aq)||HNO_3(aq)|H_2(g)|Pt(s)$
(iii) $Pt(s)|K_3[Fe(CN)_6](aq),K_4[Fe(CN)_6](aq)||CrCl_3(aq)|Cr(s)$

E6C.1(b) Write the cell reaction and electrode half-reactions and calculate the standard potential of each the following cells:

(i) $Pt(s)|Cl_2(g)|\ HCl(aq)||K_2CrO_4(aq)|Ag_2CrO_4(s)|Ag(s)$
(ii) $Pt(s)|Fe^{3+}(aq),Fe^{2+}(aq)||Sn^{4+}(aq),Sn^{2+}(aq)|Pt(s)$
(iii) $Cu(s)|Cu^{2+}(aq)||Mn^{2+}(aq),H^+(aq)|MnO_2(s)|Pt(s)$

E6C.2(a) Devise cells in which the following are the reactions and calculate the standard cell potential in each case:

(i) $Zn(s) + CuSO_4(aq) \rightarrow ZnSO_4(aq) + Cu(s)$
(ii) $2\,AgCl(s) + H_2(g) \rightarrow 2\,HCl(aq) + 2\,Ag(s)$
(iii) $2\,H_2(g) + O_2(g) \rightarrow 2\,H_2O(l)$

E6C.2(b) Devise cells in which the following are the reactions and calculate the standard cell potential in each case:

(i) $2\,Na(s) + 2\,H_2O(l) \rightarrow 2\,NaOH(aq) + H_2(g)$
(ii) $H_2(g) + I_2(g) \rightarrow 2\,HI(aq)$
(iii) $H_3O^+(aq) + OH^-(aq) \rightarrow 2\,H_2O(l)$

E6C.3(a) Use the Debye–Hückel limiting law and the Nernst equation to estimate the potential of the cell $Ag(s)|AgBr(s)|KBr(aq, 0.050\,mol\,kg^{-1})||Cd(NO_3)_2(aq, 0.010\,mol\,kg^{-1})|Cd(s)$ at 25 °C.

E6C.3(b) Consider the cell $Pt(s)|H_2(g,p^\ominus)|HCl(aq)|AgCl(s)|Ag(s)$, for which the cell reaction is $2\,AgCl(s) + H_2(g) \rightarrow 2\,Ag(s) + 2\,HCl(aq)$. At 25 °C and a molality of HCl of $0.010\,mol\,kg^{-1}$, $E_{cell} = +0.4658\,V$. (i) Write the Nernst equation for the cell reaction. (ii) Calculate $\Delta_r G$ for the cell reaction. (iii) Assuming that the Debye–Hückel limiting law holds at this concentration, calculate $E^\ominus(AgCl/Ag,Cl^-)$.

E6C.4(a) The standard potential of a Daniell cell, with cell reaction $Zn(s) + Cu^{2+}(aq) \rightarrow Zn^{2+}(aq) + Cu(s)$, is 1.10 V at 25 °C. Calculate the corresponding standard reaction Gibbs energy.

E6C.4(b) The cell reaction for the 'Bunsen cell' is $Zn(s) + 2\,NO_3^-(aq) + 4\,H^+(aq) \rightarrow Zn^{2+}(aq) + 2\,H_2O(l) + 2\,NO_2(g)$. The standard cell potential at 25 °C is −0.040 V. Calculate the electrical work that can be done by the cell.

E6C.5(a) By how much does the cell potential change when Q is decreased by a factor of 10 for a reaction in which $\nu = 2$ at 298 K?

E6C.5(b) By how much does the cell potential change when Q is increased by a factor of 5 for a reaction in which $\nu = 3$ at 298 K?

Problems

You will need to draw on information from Topic 6D to complete the answers.

P6C.1 A fuel cell develops an electric potential difference from the chemical reaction between reagents supplied from an outside source. What is the standard potential of a cell fuelled by (a) hydrogen and oxygen, (b) the combustion of butane at 1.0 bar and 298 K?

P6C.2 Calculate the value of $\Delta_f G^\ominus(H_2O,l)$ at 298 K from the standard potential of the cell $Pt(s)|H_2(g)|HCl(aq)|O_2(g)|Pt(s)$, $E^\ominus_{cell} = +1.23\,V$.

P6C.3 Although the hydrogen electrode may be conceptually the simplest electrode and is the basis for the choice of reference potential in electrochemical systems, it is cumbersome to use. Therefore, several substitutes for it have been devised. One of these alternatives is the quinhydrone electrode (quinhydrone, $Q\cdot QH_2$, is a complex of quinone, $C_6H_4O_2 = Q$, and hydroquinone, $C_6H_4O_2H_2 = QH_2$), where the concentrations of $Q\cdot QH_2$ and QH_2 are equal to each other. The electrode half-reaction is $Q(aq) + 2\,H^+(aq) + 2\,e^- \rightarrow QH_2(aq)$, $E^\ominus = +0.6994\,V$. If the cell $Hg(s)|Hg_2Cl_2(s)|HCl(aq)|Q\cdot QH_2|Au(s)$ is prepared, and the measured cell potential is +0.190 V, what is the pH of the HCl solution?

P6C.4 State what is expected to happen to the cell potential when the specified changes are made to the following cells. Confirm your prediction by using the Nernst equation in each case.

(a) The molar concentration of silver nitrate in the left-hand compartment is increased in the cell $Ag(s)|AgNO_3(aq,m_L)||AgNO_3(aq,m_R)|Ag(s)$.
(b) The pressure of hydrogen in the left-hand compartment is increased in the $Pt(s)|H_2(g,p_L)|HCl(aq)|H_2(g,p_L)|Pt(s)$.
(c) The pH of the right-hand compartment is decreased in the cell $Pt(s)|K_3[Fe(CN)_6](aq),K_4[Fe(CN)_6](aq)||Mn^{2+}(aq),H^+(aq)|MnO_2(s)|Pt(s)$.
(d) The concentration of HCl is increased in the cell $Pt(s)|Cl_2(g)|HCl(aq)||HBr(aq)|Br_2(l)|Pt(s)$.
(e) Some iron(III) chloride is added to both compartments of the cell $Pt(s)|Fe^{3+}(aq),Fe^{2+}(aq)||Sn^{4+}(aq),Sn^{2+}(aq)|Pt(s)$
(f) Acid is added to both compartments of the cell $Fe(s)|Fe^{2+}(aq)||Mn^{2+}(aq),H^+(aq)|MnO_2(s)|Pt(s)$.

TOPIC 6D Electrode potentials

Discussion questions

D6D.1 Describe a method for the determination of the standard potential of a redox couple.

D6D.2 Suggest reasons why a glass electrode can be used for the determination of the pH of an aqueous solution.

Exercises

E6D.1(a) Calculate the equilibrium constants of the following reactions at 25 °C from standard potential data:

(i) $Sn(s) + Sn^{4+}(aq) \rightleftharpoons 2\,Sn^{2+}(aq)$
(ii) $Sn(s) + 2\,AgCl(s) \rightleftharpoons SnCl_2(aq) + 2\,Ag(s)$

E6D.1(b) Calculate the equilibrium constants of the following reactions at 25 °C from standard potential data:

(i) $Sn(s) + CuSO_4(aq) \rightleftharpoons Cu(s) + SnSO_4(aq)$
(ii) $Cu^{2+}(aq) + Cu(s) \rightleftharpoons 2\,Cu^+(aq)$

E6D.2(a) The standard potential of the cell Ag(s)|AgI(s)|AgI(aq)|Ag(s) is +0.9509 V at 25 °C. Calculate the equilibrium constant for the dissolution of AgI(s).

E6D.2(b) The standard potential of the cell $Bi(s)|Bi_2S_3(s)|Bi_2S_3(aq)|Bi(s)$ is +0.96 V at 25 °C. Calculate the equilibrium constant for the dissolution of $Bi_2S_3(s)$.

E6D.3(a) (i) Use the information in the *Resource section* to calculate the standard potential of the cell $Ag(s)|AgNO_3(aq)||Cu(NO_3)_2(aq)|Cu(s)$ and the standard Gibbs energy and enthalpy of the cell reaction at 25 °C. (ii) Estimate the value of $\Delta_r G^\ominus$ at 35 °C.

E6D.3(b) Calculate the standard potential of the cell Pt(s)|cystine(aq), cysteine(aq)|| $H^+(aq)|O_2(g)|Pt(s)$ and the standard Gibbs energy of the cell reaction at 25 °C. Use $E^\ominus = -0.34$ V for cystine(aq) + $2\,H^+(aq) + 2\,e^- \rightarrow$ 2 cysteine(aq).

E6D.4(a) Can mercury produce zinc metal from aqueous zinc sulfate under standard conditions?

E6D.4(b) Can chlorine gas oxidize water to oxygen gas under standard conditions in basic solution?

Problems

6D.1 Tabulated thermodynamic data can be used to predict the standard potential of a cell even if it cannot be measured directly. The standard Gibbs energy of the reaction $K_2CrO_4(aq) + 2\,Ag(s) + 2\,FeCl_3(aq) \rightarrow Ag_2CrO_4(s) + 2\,FeCl_2(aq) + 2\,KCl(aq)$ is -62.5 kJ mol^{-1} at 298 K. (a) Calculate the standard potential of the corresponding galvanic cell and (b) the standard potential of the $Ag_2CrO_4/Ag,CrO_4^{2-}$ couple.

6D.2 A fuel cell is constructed in which both electrodes make use of the oxidation of methane. The left-hand electrode makes use of the complete oxidation of methane to carbon dioxide and liquid water; the right-hand electrode makes use of the partial oxidation of methane to carbon monoxide and liquid water. (a) Which electrode is the cathode? (b) What is the cell potential at 25 °C when all gases are at 1 bar?

6D.3 One ecologically important equilibrium is that between carbonate and hydrogencarbonate (bicarbonate) ions in natural water. (a) The standard Gibbs energies of formation of $CO_3^{2-}(aq)$ and $HCO_3^-(aq)$ are -527.81 kJ mol^{-1} and -586.77 kJ mol^{-1}, respectively. What is the standard potential of the $HCO_3^-/CO_3^{2-},H_2$ couple? (b) Calculate the standard potential of a cell in which the cell reaction is $Na_2CO_3(aq) + H_2O(l) \rightarrow NaHCO_3(aq) + NaOH(aq)$. (c) Write the Nernst equation for the cell, and (d) predict and calculate the change in cell potential when the pH is changed to 7.0 at 298 K.

6D.4 The potential of the cell $Pt(s)|H_2(g,p^\ominus)|HCl(aq,b)|Hg_2Cl_2(s)|Hg(l)$ has been measured with high precision with the following results at 25 °C:

b/(mmol kg^{-1})	1.6077	3.0769	5.0403	7.6938	10.9474
E/V	0.600 80	0.568 25	0.543 66	0.522 67	0.505 32

Determine the standard cell potential and the mean activity coefficient of HCl at these molalities. (Make a least-squares fit of the data to the best straight line.)

6D.5 For a hydrogen/oxygen fuel cell, with an overall four-electron cell reaction $2\,H_2(g) + O_2(g) \rightarrow 2\,H_2O(l)$, the standard cell potential is +1.2335 V at 293 K and +1.2251 V at 303 K. Calculate the standard reaction enthalpy and entropy within this temperature range.

6D.6 The standard potential of the AgCl/Ag,Cl$^-$ couple fits the expression

$$E^\ominus/\mathrm{V} = 0.236\,59 - 4.8564 \times 10^{-4}(\theta/°\mathrm{C}) - 3.4205 \times 10^{-6}(\theta/°\mathrm{C})^2 + 5.869 \times 10^{-9}(\theta/°\mathrm{C})^3$$

Calculate the standard Gibbs energy and enthalpy of formation of Cl$^-$(aq) and its standard entropy at 298 K.

FOCUS 6 Chemical equilibrium

Integrated activities

I6.1‡ Thorn et al. (*J. Phys. Chem.* **100**, 14178 (1996)) carried out a study of $Cl_2O(g)$ by photoelectron ionization. From their measurements, they report $\Delta_f H^\ominus(Cl_2O) = +77.2$ kJ mol^{-1}. They combined this measurement with literature data on the reaction $Cl_2O\,(g) + H_2O(g) \rightarrow 2\,HOCl(g)$, for which $K = 8.2 \times 10^{-2}$ and $\Delta_r S^\ominus = +16.38$ J K^{-1} mol^{-1}, and with readily available thermodynamic data on water vapour to report a value for $\Delta_f H^\ominus$(HOCl). Calculate that value. All quantities refer to 298 K.

I6.2 Given that $\Delta_r G^\ominus = -212.7$ kJ mol^{-1} for the reaction $Zn(s) + Cu^{2+}(aq) \rightarrow Zn^{2+}(aq) + Cu(s)$ in the Daniell cell at 25 °C, and $b(CuSO_4) = 1.00 \times 10^{-3}$ mol kg^{-1} and $b(ZnSO_4) = 3.00 \times 10^{-3}$ mol kg^{-1}, calculate (a) the ionic strengths of the solutions, (b) the mean ionic activity coefficients in the compartments, (c) the reaction quotient, (d) the standard cell potential, and (e) the cell potential. (Take $\gamma_+ = \gamma_- = \gamma_\pm$ in the respective compartments. Use the Debye–Hückel limiting law.)

I6.3 Consider the cell, $Zn(s)|ZnCl_2(0.0050\,\mathrm{mol\,kg^{-1}})|Hg_2Cl_2(s)|Hg(l)$, for which the cell reaction is $Hg_2Cl_2(s) + Zn(s) \rightarrow 2\,Hg(l) + 2\,Cl^-(aq) + Zn^{2+}(aq)$. The cell potential is +1.2272 V, $E^\ominus(Zn^{2+},Zn) = -0.7628$ V, and $E^\ominus(Hg_2Cl_2,Hg) = +0.2676$ V. (a) Write the Nernst equation for the cell. Determine (b) the standard cell potential, (c) $\Delta_r G$, $\Delta_r G^\ominus$, and K for the cell reaction, (d) the mean ionic activity and activity coefficient of $ZnCl_2$ from the measured cell potential, and (e) the mean ionic activity coefficient of $ZnCl_2$ from the Debye–Hückel limiting law. (f) Given that $(\partial E_{cell}/\partial T)_p = -4.52 \times 10^{-4}$ V K^{-1}, Calculate $\Delta_r S$ and $\Delta_r H$.

I6.4 Careful measurements of the potential of the cell $Pt|H_2(g,p^\ominus)|$ NaOH(aq,0.0100 mol kg^{-1}),NaCl(aq, 0.011 25 mol kg^{-1})|AgCl(s)|Ag(s) have been reported. Among the data is the following information:

θ/°C	20.0	25.0	30.0
E_{cell}/V	1.04774	1.04864	1.04942

Calculate pK_w at these temperatures and the standard enthalpy and entropy of the autoprotolysis of water at 25.0 °C. Recall that K_w is the equilibrium constant for the autoprotolysis of liquid water.

I6.5 Measurements of the potential of cells of the type $Ag(s)|AgX(s)|MX(b_1)|M_xHg|MX(b_2)|AgX(s)|Ag(s)$, where M_xHg denotes an amalgam and the electrolyte is LiCl in ethylene glycol, are given below for M = Li and X = Cl. Estimate the activity coefficient at the concentration marked * and then use this value to calculate activity coefficients from the measured cell potential at the other concentrations. Base your answer on the Davies equation (eqn 5F.30b) with $A = 1.461$, $B = 1.70$, $C = 0.20$, and $I = b/b^\ominus$. For b_2 = 0.09141 mol kg^{-1}:

$b_1/(\text{mol kg}^{-1})$	0.0555	0.09141	0.1652	0.2171	1.040	1.350*
E/V	−0.0220	0.0000	0.0263	0.0379	0.1156	0.1336

I6.6‡ The table below summarizes the potential of the cell $\text{Pd(s)}|\text{H}_2\text{(g, 1 bar)}|\text{BH(aq}, b\text{), B(aq}, b\text{)}|\text{AgCl(s)}|\text{Ag(s)}$. Each measurement is made at equimolar concentrations of 2-aminopyridinium chloride (BH) and 2-aminopyridine (B). The data are for 25 °C and it is found that $E^{\ominus} = 0.222\,51$ V. Use the data to determine $\text{p}K_a$ for the acid at 25 °C and the mean activity coefficient ($\gamma_\pm$) of BH as a function of molality (b) and ionic strength (I). Use the Davies equation (eqn 5F.30b) with $A = 0.5091$ and B and C are parameters that depend upon the ions.

$b/(\text{mol kg}^{-1})$	0.01	0.02	0.03	0.04	0.05
$E_{\text{cell}}(25\,°\text{C})/\text{V}$	0.74452	0.72853	0.71928	0.71314	0.70809
$b/(\text{mol kg}^{-1})$	0.06	0.07	0.08	0.09	0.10
$E_{\text{cell}}(25\,°\text{C})/\text{V}$	0.70380	0.70059	0.69790	0.69571	0.69338

Hint: Use mathematical software or a spreadsheet.

I6.7 Read *Impact* 9 on the website of this text before attempting this problem. Here you will investigate the molecular basis for the observation that the hydrolysis of adenosine triphosphate (ATP) to adenosine diphosphate (ADP) is exergonic at pH = 7.0 and 310 K. (a) It is thought that the exergonicity of ATP hydrolysis is due in part to the fact that the standard entropies of hydrolysis of polyphosphates are positive. Why would an increase in entropy accompany the hydrolysis of a triphosphate group into a diphosphate and a phosphate group? (b) Under identical conditions, the Gibbs energies of hydrolysis of H_4ATP and $MgATP^{2-}$, a complex between the Mg^{2+} ion and ATP^{4-}, are less negative than the Gibbs energy of hydrolysis of ATP^{4-}. This observation has been used to support the hypothesis that electrostatic repulsion between adjacent phosphate groups is a factor that controls the exergonicity of ATP hydrolysis. Provide a rationale for the hypothesis and discuss how the experimental evidence supports it. Do these electrostatic effects contribute to the $\Delta_r H$ or $\Delta_r S$ terms that determine the exergonicity of the reaction? *Hint:* In the $MgATP^{2-}$ complex, the Mg^{2+} ion and ATP^{4-} anion form two bonds: one that involves a negatively charged oxygen belonging to the terminal phosphate group of ATP^{4-} and another that involves a negatively charged oxygen belonging to the phosphate group adjacent to the terminal phosphate group of ATP^{4-}.

I6.8 Read *Impact* 9 on the website of this text before attempting this problem. To get a sense of the effect of cellular conditions on the ability of adenosine triphosphate (ATP) to drive biochemical processes, compare the standard Gibbs energy of hydrolysis of ATP to ADP (adenosine diphosphate) with the reaction Gibbs energy in an environment at 37 °C in which pH = 7.0 and the ATP, ADP, and P_i^- concentrations are all 1.0 mmol dm^{-3}.

I6.9 Read *Impact* 9 on the website of this text before attempting this problem. Under biochemical standard conditions, aerobic respiration produces approximately 38 molecules of ATP per molecule of glucose that is completely oxidized. (a) What is the percentage efficiency of aerobic respiration under biochemical standard conditions? (b) The following conditions are more likely to be observed in a living cell: $p_{CO_2} = 5.3\times10^{-2}$ atm, $p_{O_2} = 0.132$ atm, [glucose] = 5.6 pmol dm^{-3}, [ATP] = [ADP] = [P_i] = 0.10 mmol dm^{-3}, pH = 7.4, T = 310 K. Assuming that activities can be replaced by the numerical values of molar concentrations, calculate the efficiency of aerobic respiration under these physiological conditions. (c) A typical diesel engine operates between T_c = 873 K and T_h = 1923 K with an efficiency that is approximately 75 per cent of the theoretical limit of $1 - T_c/T_h$ (see Topic 3A). Compare the efficiency of a typical diesel engine with that of aerobic respiration under typical physiological conditions (see part b). Why is biological energy conversion more or less efficient than energy conversion in a diesel engine?

I6.10 In anaerobic bacteria, the source of carbon may be a molecule other than glucose and the final electron acceptor is some molecule other than O_2. Could a bacterium evolve to use the ethanol/nitrate pair instead of the glucose/O_2 pair as a source of metabolic energy?

I6.11 The standard potentials of proteins are not commonly measured by the methods described in this chapter because proteins often lose their native structure and function when they react on the surfaces of electrodes. In an alternative method, the oxidized protein is allowed to react with an appropriate electron donor in solution. The standard potential of the protein is then determined from the Nernst equation, the equilibrium concentrations of all species in solution, and the known standard potential of the electron donor. This method can be illustrated with the protein cytochrome *c*. The one-electron reaction between cytochrome *c*, cyt, and 2,6-dichloroindophenol, D, can be followed spectrophotometrically because each of the four species in solution has a distinct absorption spectrum. Write the reaction as $\text{cyt}_{\text{ox}} + \text{D}_{\text{red}} \rightleftharpoons \text{cyt}_{\text{red}} + \text{D}_{\text{ox}}$, where the subscripts 'ox' and 'red' refer to oxidized and reduced states, respectively. (a) Consider $E^{\ominus}_{\text{cyt}}$ and $E^{\ominus}_{\text{D}}$ to be the standard potentials of cytochrome *c* and D, respectively. Show that, at equilibrium, a plot of $\ln([\text{D}_{\text{ox}}]_{\text{eq}}/[\text{D}_{\text{red}}]_{\text{eq}})$ versus $\ln([\text{cyt}_{\text{ox}}]_{\text{eq}}/[\text{cyt}_{\text{red}}]_{\text{eq}})$ is linear with slope of 1 and y-intercept $F(E^{\ominus}_{\text{cyt}} - E^{\ominus}_{\text{D}})/RT$, where equilibrium activities are replaced by the numerical values of equilibrium molar concentrations. (b) The following data were obtained for the reaction between oxidized cytochrome *c* and reduced D in a pH 6.5 buffer at 298 K. The ratios $[\text{D}_{\text{ox}}]_{\text{eq}}/[\text{D}_{\text{red}}]_{\text{eq}}$ and $[\text{cyt}_{\text{ox}}]_{\text{eq}}/[\text{cyt}_{\text{red}}]_{\text{eq}}$ were adjusted by titrating a solution containing oxidized cytochrome *c* and reduced D with a solution of sodium ascorbate, which is a strong reductant. From the data and the standard potential of D of 0.237 V, determine the standard potential cytochrome *c* at pH 6.5 and 298 K.

$[\text{D}_{\text{ox}}]_{\text{eq}}/[\text{D}_{\text{red}}]_{\text{eq}}$	0.00279	0.00843	0.0257	0.0497	0.0748	0.238	0.534
$[\text{cyt}_{\text{ox}}]_{\text{eq}}/[\text{cyt}_{\text{red}}]_{\text{eq}}$	0.0106	0.0230	0.0894	0.197	0.335	0.809	1.39

I6.12‡ The dimerization of ClO in the Antarctic winter stratosphere is believed to play an important part in that region's severe seasonal depletion of ozone. The following equilibrium constants are based on measurements on the reaction $2\,\text{ClO(g)} \rightarrow (\text{ClO})_2\text{(g)}$.

T/K	233	248	258	268	273	280
K	4.13×10^{8}	5.00×10^{7}	1.45×10^{7}	5.37×10^{6}	3.20×10^{6}	9.62×10^{5}
T/K	288	295	303			
K	4.28×10^{5}	1.67×10^{5}	6.02×10^{4}			

(a) Derive the values of $\Delta_r H^{\ominus}$ and $\Delta_r S^{\ominus}$ for this reaction. (b) Compute the standard enthalpy of formation and the standard molar entropy of $(\text{ClO})_2$, given $\Delta_f H^{\ominus}(\text{ClO,g}) = +101.8$ kJ mol^{-1} and $S^{\ominus}_{\text{m}}(\text{ClO,g}) = 226.6$ J K^{-1} mol^{-1}.

FOCUS 7

Quantum theory

It was once thought that the motion of atoms and subatomic particles could be expressed using 'classical mechanics', the laws of motion introduced in the seventeenth century by Isaac Newton, for these laws were very successful at explaining the motion of everyday objects and planets. However, a proper description of electrons, atoms, and molecules requires a different kind of mechanics, 'quantum mechanics', which is introduced in this Focus and applied widely throughout the text.

7A The origins of quantum mechanics

Experimental evidence accumulated towards the end of the nineteenth century showed that classical mechanics failed when it was applied to particles as small as electrons. More specifically, careful measurements led to the conclusion that particles may not have an arbitrary energy and that the classical concepts of a particle and wave blend together. This Topic shows how these observations set the stage for the development of the concepts and equations of quantum mechanics in the early twentieth century.

7B Wavefunctions

In quantum mechanics, all the properties of a system are expressed in terms of a wavefunction which is obtained by solving the equation proposed by Erwin Schrödinger. This Topic focuses on the interpretation of the wavefunction, and specifically what it reveals about the location of a particle.

7C Operators and observables

A central feature of quantum theory is its representation of observables by 'operators', which act on the wavefunction and extract the information it contains. This Topic shows how operators are constructed and used. One consequence of their use is the 'uncertainty principle', one of the most profound departures of quantum mechanics from classical mechanics.

7D Translational motion

Translational motion, motion through space, is one of the fundamental types of motion treated by quantum mechanics. According to quantum theory, a particle constrained to move in a finite region of space is described by only certain wavefunctions and can possess only certain energies. That is, quantization emerges as a natural consequence of solving the Schrödinger equation and the conditions imposed on it. The solutions also expose a number of non-classical features of particles, especially their ability to tunnel into and through regions where classical physics would forbid them to be found.

7E Vibrational motion

This Topic introduces the 'harmonic oscillator', a simple but very important model for the description of vibrations. It shows that the energies of an oscillator are quantized and that an oscillator may be found at displacements that are forbidden by classical physics.

7F Rotational motion

The constraints on the wavefunctions of a body rotating in two and three dimensions result in the quantization of its energy.

In addition, because the energy is related to the angular momentum, it follows that angular momentum is also restricted to certain values. The quantization of angular momentum is a very important aspect of the quantum theory of electrons in atoms and of rotating molecules.

7F.1 **Rotation in two dimensions;** 7F.2 **Rotation in three dimensions**

Web resources What is an application of this material?

Impact 11 highlights an application of quantum mechanics which still requires much research before it becomes a useful technology. It is based on the expectation that a 'quantum computer' can carry out calculations on many states of a system simultaneously, leading to a new generation of very fast computers. 'Nanoscience' is the study of atomic and molecular assemblies with dimensions ranging from 1 nm to about 100 nm, and 'nanotechnology' is concerned with the incorporation of such assemblies into devices. *Impact* 12 explores quantum mechanical effects that show how the properties of a nanometre-sized assembly depend on its size.

TOPIC 7A The origins of quantum mechanics

➤ **Why do you need to know this material?**

Quantum theory is central to almost every explanation in chemistry. It is used to understand atomic and molecular structure, chemical bonds, and most of the properties of matter.

➤ **What is the key idea?**

Experimental evidence led to the conclusion that energy can be transferred only in discrete amounts, and that the classical concepts of a 'particle' and a 'wave' blend together.

➤ **What do you need to know already?**

You should be familiar with the basic principles of classical mechanics, especially momentum, force, and energy set out in *The chemist's toolkits* 3 (in Topic 1B) and 6 (in Topic 2A). The discussion of heat capacities of solids makes light use of material in Topic 2A.

The classical mechanics developed by Newton in the seventeenth century is an extraordinarily successful theory for describing the motion of everyday objects and planets. However, late in the nineteenth century scientists started to make observations that could not be explained by classical mechanics. They were forced to revise their entire conception of the nature of matter and replace classical mechanics by a theory that became known as **quantum mechanics**.

7A.1 Energy quantization

Three experiments carried out near the end of the nineteenth century drove scientists to the view that energy can be transferred only in discrete amounts.

(a) Black-body radiation

The key features of electromagnetic radiation according to classical physics are described in *The chemist's toolkit* 13. It is observed that all objects emit electromagnetic radiation over a range of frequencies with an intensity that depends on the temperature of the object. A familiar example is a heated metal bar that first glows red and then becomes 'white hot' upon further heating. As the temperature is raised, the colour shifts from red towards blue and results in the white glow.

The chemist's toolkit 13 Electromagnetic radiation

Electromagnetic radiation consists of oscillating electric and magnetic disturbances that propagate as waves. The two components of an electromagnetic wave are mutually perpendicular and are also perpendicular to the direction of propagation (Sketch 1). Electromagnetic waves travel through a vacuum at a constant speed called the **speed of light**, c, which has the defined value of exactly $2.997\,924\,58 \times 10^8\,\mathrm{m\,s^{-1}}$.

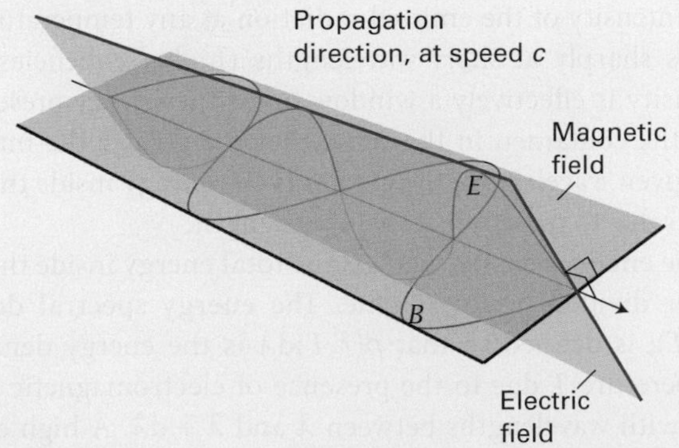

Sketch 1

A wave is characterized by its **wavelength**, λ (lambda), the distance between consecutive peaks of the wave (Sketch 2). The classification of electromagnetic radiation according to its wavelength is shown in Sketch 3. Light, which is electromagnetic radiation that is visible to the human eye, has a wavelength in the range 420 nm (violet light) to 700 nm (red light). The properties of a wave may also be expressed in terms of its **frequency**, ν (nu), the number of oscillations in a time interval divided by the duration of the interval. Frequency is reported in hertz, Hz, with $1\,\mathrm{Hz} = 1\,\mathrm{s^{-1}}$ (i.e. 1 cycle per second). Light spans the frequency range from 710 THz (violet light) to 430 THz (red light).

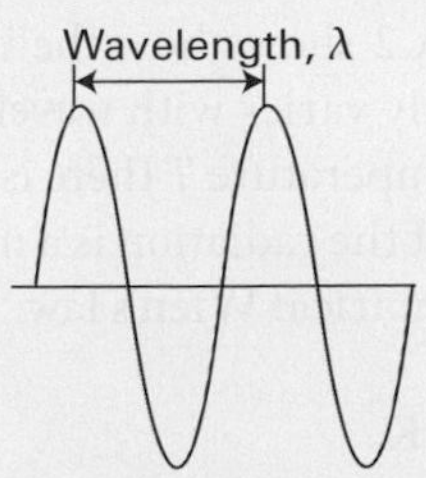

Sketch 2

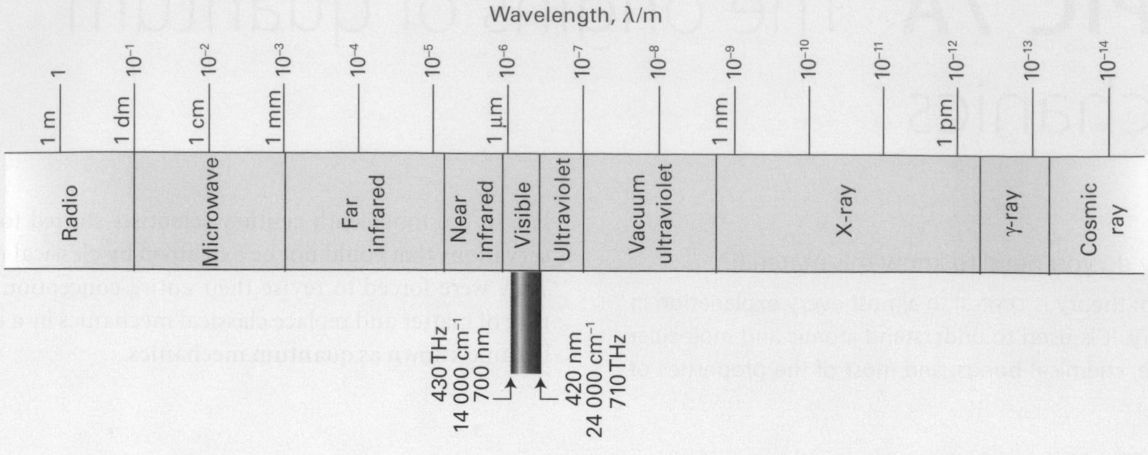

Sketch 3

The wavelength and frequency of an electromagnetic wave are related by:

$$c = \lambda\nu$$ The relation between wavelength and frequency in a vacuum

It is also common to describe a wave in terms of its **wavenumber**, $\tilde{\nu}$ (nu tilde), which is defined as

$$\tilde{\nu} = \frac{1}{\lambda} \text{ or equivalently } \tilde{\nu} = \frac{\nu}{c}$$ Wavenumber [definition]

Thus, wavenumber is the reciprocal of the wavelength and can be interpreted as the number of wavelengths in a given distance. In spectroscopy, for historical reasons, wavenumber is usually reported in units of reciprocal centimetres (cm^{-1}). Visible light therefore corresponds to electromagnetic radiation with a wavenumber of 14 000 cm^{-1} (red light) to 24 000 cm^{-1} (violet light).

Electromagnetic radiation that consists of a single frequency (and therefore single wavelength) is **monochromatic**, because it corresponds to a single colour. *White light* consists of electromagnetic waves with a continuous, but not uniform, spread of frequencies throughout the visible region of the spectrum.

A characteristic property of waves is that they interfere with one another, which means that they result in a greater amplitude where their displacements add and a smaller amplitude where their displacements subtract (Sketch 4). The former is called 'constructive interference' and the latter 'destructive interference'. The regions of constructive and destructive interference show up as regions of enhanced and diminished intensity. The phenomenon of **diffraction** is the interference caused by an object in the path of waves and occurs when the dimensions of the object are comparable to the wavelength of the radiation. Light waves, with wavelengths of the order of 500 nm, are diffracted by narrow slits.

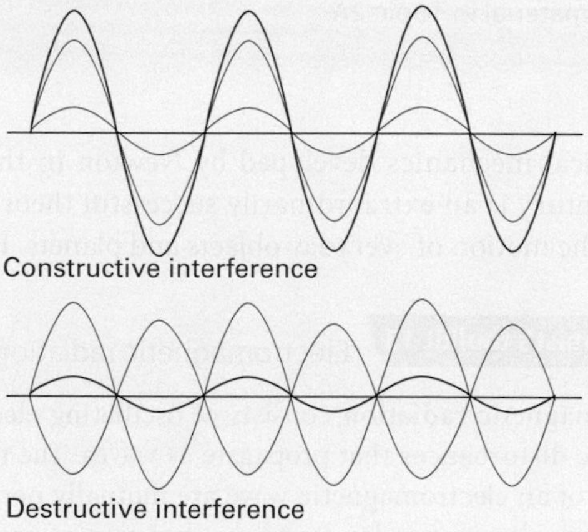

Sketch 4

The radiation emitted by hot objects is discussed in terms of a **black body**, a body that emits and absorbs electromagnetic radiation without favouring any wavelengths. A good approximation to a black body is a small hole in an empty container (Fig. 7A.1). Figure 7A.2 shows how the intensity of the radiation from a black body varies with wavelength at several temperatures. At each temperature T there is a wavelength, λ_{max}, at which the intensity of the radiation is a maximum, with T and λ_{max} related by the empirical **Wien's law**:

$$\lambda_{max}T = 2.9\times10^{-3}\,\text{mK} \qquad \text{Wien's law} \quad (7A.1)$$

The intensity of the emitted radiation at any temperature declines sharply at short wavelengths (high frequencies). The intensity is effectively a window on to the energy present inside the container, in the sense that the greater the intensity at a given wavelength, the greater is the energy inside the container due to radiation at that wavelength.

The **energy density**, $\mathcal{E}(T)$, is the total energy inside the container divided by its volume. The **energy spectral density**, $\rho(\lambda,T)$, is defined so that $\rho(\lambda,T)\mathrm{d}\lambda$ is the energy density at temperature T due to the presence of electromagnetic radiation with wavelengths between λ and $\lambda + \mathrm{d}\lambda$. A high energy

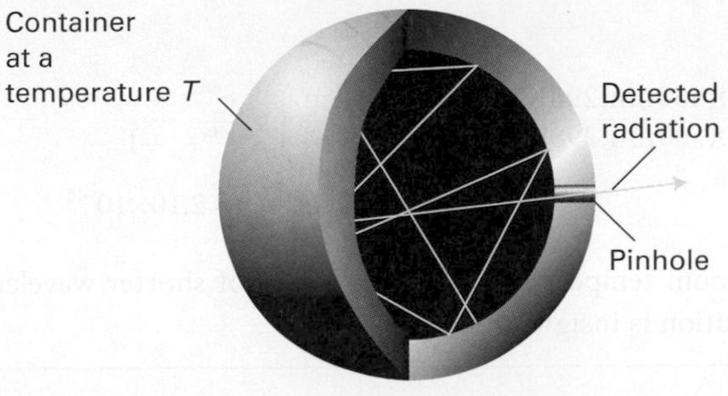

Figure 7A.1 Black-body radiation can be detected by allowing it to leave an otherwise closed container through a pinhole. The radiation is reflected many times within the container and comes to thermal equilibrium with the wall. Radiation leaking out through the pinhole is characteristic of the radiation inside the container.

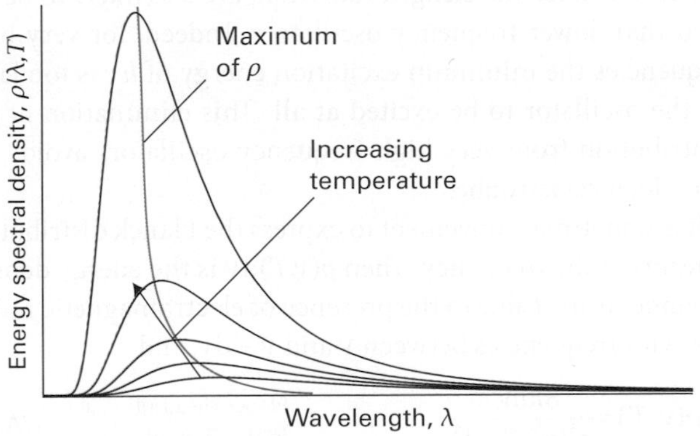

Figure 7A.2 The energy spectral density of radiation from a black body at several temperatures. Note that as the temperature increases, the maximum in the energy spectral density moves to shorter wavelengths and increases in intensity overall.

spectral density at the wavelength λ and temperature T simply means that there is a lot of energy associated with wavelengths lying between λ and $\lambda + d\lambda$ at that temperature. The energy density is obtained by summing (integrating) the energy spectral density over all wavelengths:

$$\mathcal{E}(T)=\int_0^\infty \rho(\lambda,T)\mathrm{d}\lambda \qquad (7A.2)$$

The units of $\mathcal{E}(T)$ are joules per metre cubed ($\mathrm{J\,m^{-3}}$), so the units of $\rho(\lambda,T)$ are $\mathrm{J\,m^{-4}}$. Empirically, the energy density is found to vary as T^4, an observation expressed by the **Stefan–Boltzmann law**:

$$\mathcal{E}(T) = \text{constant} \times T^4 \qquad \text{Stefan–Boltzmann law} \qquad (7A.3)$$

with the constant equal to $7.567 \times 10^{-16}\,\mathrm{J\,m^{-3}\,K^{-4}}$.

The container in Fig. 7A.1 emits radiation that can be thought of as oscillations of the electromagnetic field stimulated by the oscillations of electrical charges in the material of the wall. According to classical physics, every oscillator is excited to some extent, and according to the equipartition principle (*The chemist's toolkit* 7 in Topic 2A) every oscillator, regardless of its frequency, has an average energy of kT. On this basis, the physicist Lord Rayleigh, with minor help from James Jeans, deduced what is now known as the **Rayleigh–Jeans law**:

$$\rho(\lambda,T)=\frac{8\pi kT}{\lambda^4} \qquad \text{Rayleigh–Jeans law} \qquad (7A.4)$$

where k is Boltzmann's constant ($k = 1.381 \times 10^{-23}\,\mathrm{J\,K^{-1}}$).

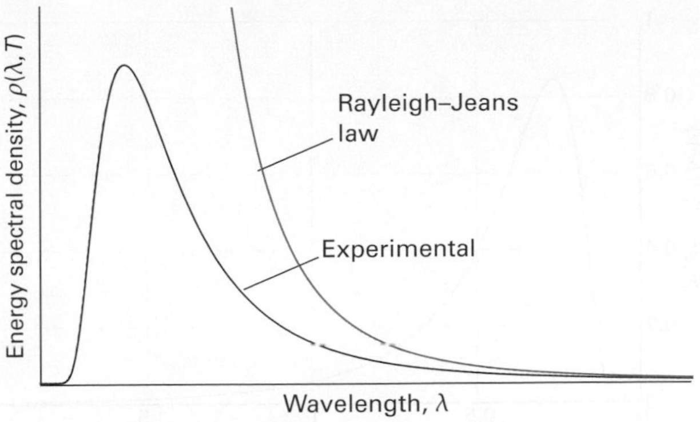

Figure 7A.3 Comparison of the experimental energy spectral density with the prediction of the Rayleigh–Jeans law (eqn 7A.4). The latter predicts an infinite energy spectral density at short wavelengths and infinite overall energy density.

The Rayleigh–Jeans law is not supported by the experimental measurements. As is shown in Fig. 7A.3, although there is agreement at long wavelengths, it predicts that the energy spectral density (and hence the intensity of the radiation emitted) increases without going through a maximum as the wavelength decreases. That is, the Rayleigh–Jeans law is inconsistent with Wien's law. Equation 7A.4 also implies that the radiation is intense at very short wavelengths and becomes infinitely intense as the wavelength tends to zero. The concentration of radiation at short wavelengths is called the **ultraviolet catastrophe**, and is an unavoidable consequence of classical physics.

In 1900, Max Planck found that the experimentally observed intensity distribution of black-body radiation could be explained by proposing that the energy of each oscillator is limited to discrete values. In particular, Planck assumed that for an electromagnetic oscillator of frequency ν, the permitted energies are integer multiples of $h\nu$:

$$E = nh\nu \qquad n = 0, 1, 2, \ldots \qquad (7A.5)$$

In this expression h is a fundamental constant now known as **Planck's constant**. The limitation of energies to discrete values is called **energy quantization**. On this basis Planck was able to derive an expression for the energy spectral density which is now called the **Planck distribution**:

$$\rho(\lambda,T)=\frac{8\pi hc}{\lambda^5(\mathrm{e}^{hc/\lambda kT}-1)} \qquad \text{Planck distribution} \qquad (7A.6a)$$

This expression is plotted in Fig. 7A.4 and fits the experimental data very well at all wavelengths. The value of h, which is an

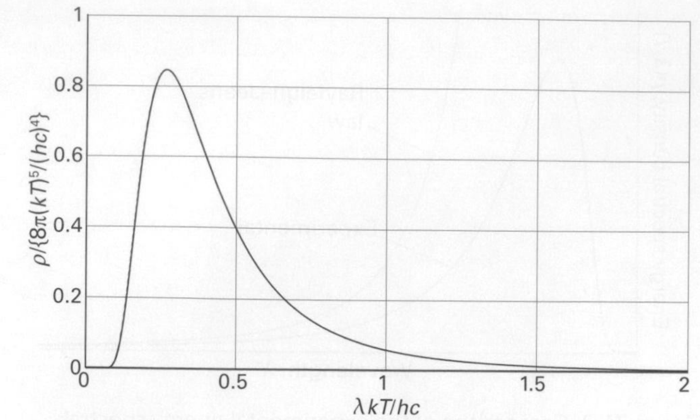

Figure 7A.4 The Planck distribution (eqn 7A.6a) accounts for the experimentally determined energy distribution of black-body radiation. It coincides with the Rayleigh–Jeans distribution at long wavelengths.

undetermined parameter in the theory, can be found by varying its value until the best fit is obtained between the eqn 7A.6a and experimental measurements. The currently accepted value is $h = 6.626 \times 10^{-34}$ J s.

For short wavelengths, $hc/\lambda kT >> 1$, and because $e^{hc/\lambda kT} \to \infty$ faster than $\lambda^5 \to 0$ it follows that $\rho \to 0$ as $\lambda \to 0$. Hence, the energy spectral density approaches zero at short wavelengths, and so the Planck distribution avoids the ultraviolet catastrophe. For long wavelengths in the sense $hc/\lambda kT << 1$, the denominator in the Planck distribution can be replaced by (see *The chemist's toolkit* 12 in Topic 5B)

$$e^{hc/\lambda kT} - 1 = \left(1 + \frac{hc}{\lambda kT} + \cdots\right) - 1 \approx \frac{hc}{\lambda kT}$$

When this approximation is substituted into eqn 7A.6a, the Planck distribution reduces to the Rayleigh–Jeans law, eqn 7A.4. The wavelength at the maximum can be found by differentiation, and is given by $\lambda_{max}T$ = constant, in accord with Wien's law; the value of the constant found in this way, $hc/5k$, agrees with the experimentally determined value. Finally, the total energy density is

$$\mathcal{E}(T) = \int_0^\infty \frac{8\pi hc}{\lambda^5(e^{hc/\lambda kT}-1)}\mathrm{d}\lambda = aT^4 \quad \text{with} \quad a = \frac{8\pi^5 k^4}{15(hc)^3} \qquad (7A.7)$$

which is finite and agrees with the Stefan–Boltzmann law (eqn 7A.3), including predicting the value of its constant correctly.

Brief illustration 7A.1

Consider eqn 7A.6a with $\lambda_1 = 450$ nm (blue light) and $\lambda_2 = 700$ nm (red light), and $T = 298$ K. It follows that

$$\frac{hc}{\lambda_1 kT} = \frac{(6.626\times10^{-34}\,\mathrm{J\,s})\times(2.998\times10^{8}\,\mathrm{m\,s^{-1}})}{(450\times10^{-9}\,\mathrm{m})\times(1.381\times10^{-23}\,\mathrm{J\,K^{-1}})\times(298\,\mathrm{K})} = 107.2\ldots$$

$$\frac{hc}{\lambda_2 kT} = \frac{(6.626\times10^{-34}\,\mathrm{J\,s})\times(2.998\times10^{8}\,\mathrm{m\,s^{-1}})}{(700\times10^{-9}\,\mathrm{m})\times(1.381\times10^{-23}\,\mathrm{J\,K^{-1}})\times(298\,\mathrm{K})} = 68.9\ldots$$

and

$$\frac{\rho(450\,\mathrm{nm},298\,\mathrm{K})}{\rho(700\,\mathrm{nm},298\,\mathrm{K})} = \left(\frac{700\times10^{-9}\,\mathrm{m}}{450\times10^{-9}\,\mathrm{m}}\right)^5 \times \frac{e^{68.9\ldots}-1}{e^{107.2\ldots}-1}$$

$$= 9.11\times(2.30\times10^{-17}) = 2.10\times10^{-16}$$

At room temperature, the proportion of shorter wavelength radiation is insignificant.

There is a single reason why Planck's approach is successful but Rayleigh's is not. Instead of allowing each oscillator to have the same average energy, regardless of its frequency, Planck used the Boltzmann distribution (see the *Prologue* to this text) to argue that higher frequency oscillators, which generate shorter wavelength radiation, are less likely to be excited than lower frequency oscillators. Indeed, for very high frequencies the minimum excitation energy of $h\nu$ is too large for the oscillator to be excited at all. This elimination of the contribution from very high frequency oscillators avoids the ultraviolet catastrophe.

It is sometimes convenient to express the Planck distribution in terms of the frequency. Then $\rho(\nu,T)\mathrm{d}\nu$ is the energy density at temperature T due to the presence of electromagnetic radiation with frequencies between ν and $\nu + \mathrm{d}\nu$, and

$$\rho(\nu,T) = \frac{8\pi h\nu^3}{c^3(e^{h\nu/kT}-1)} \qquad \text{Planck distribution in terms of frequency} \qquad (7A.6b)$$

(b) Heat capacity

When energy is supplied as heat to a substance its temperature rises; the heat capacity (Topic 2A) is the constant of proportionality between the energy supplied and the temperature rise ($C = \mathrm{d}q/\mathrm{d}T$ and, at constant volume, $C_{V,\mathrm{m}} = (\partial U_\mathrm{m}/\partial T)_V$). Experimental measurements made during the nineteenth century had shown that at room temperature the molar heat capacities of many monatomic solids are about $3R$, where R is the gas constant.[1] However, when measurements were made at much lower temperatures it was found that the heat capacity decreased, tending to zero as the temperature approached zero.

Classical physics was unable to explain this temperature dependence. The classical picture of a solid is of atoms oscillating about fixed positions, with the expectation that each oscillating atom will have the same average energy kT. This model predicts that a solid consisting of N atoms, each free to oscillate in three dimensions, will have energy $U = 3NkT$ and hence heat capacity $C_V = (\partial U/\partial T)_V = 3Nk$. The molar heat capacity is therefore predicted to be $3N_\mathrm{A}k$ which, recognizing that $N_\mathrm{A}k = R$, is equal to $3R$ at all temperatures. In 1905, Einstein suggested applying Planck's hypothesis and supposing that each oscillating atom

[1] The gas constant occurs in the context of solids because it is actually the more fundamental Boltzmann's constant in disguise: $R = N_\mathrm{A}k$.

could have an energy $nh\nu$, where n is an integer and ν is the frequency of the oscillation. Einstein went on to show by using the Boltzmann distribution that each oscillator is unlikely to be excited to high energies and at low temperatures few oscillators can be excited at all. As a consequence, because the oscillators cannot be excited, the heat capacity falls to zero. The quantitative result that Einstein obtained (as shown in Topic 13E) is

$$C_{V,\mathrm{m}}(T) = 3Rf_\mathrm{E}(T), \quad f_\mathrm{E}(T) = \left(\frac{\theta_\mathrm{E}}{T}\right)^2 \left(\frac{\mathrm{e}^{\theta_\mathrm{E}/2T}}{\mathrm{e}^{\theta_\mathrm{E}/T}-1}\right)^2$$

Einstein formula (7A.8a)

In this expression θ_E is the **Einstein temperature**, $\theta_\mathrm{E} = h\nu/k$.

At high temperatures (in the sense $T \gg \theta_\mathrm{E}$) the exponentials in f_E can be expanded as $\mathrm{e}^x = 1 + x + \cdots$ and higher terms ignored (*The chemist's toolkit* 12 in Topic 5B). The result is

$$f_\mathrm{E}(T) = \left(\frac{\theta_\mathrm{E}}{T}\right)^2 \left\{\frac{1+\theta_\mathrm{E}/2T+\cdots}{(1+\theta_\mathrm{E}/T+\cdots)-1}\right\}^2 \approx \left(\frac{\theta_\mathrm{E}}{T}\right)^2 \left\{\frac{1}{\theta_\mathrm{E}/T}\right\}^2 \approx 1 \qquad (7A.8b)$$

and the classical result ($C_{V,\mathrm{m}} = 3R$) is obtained. At low temperatures (in the sense $T \ll \theta_\mathrm{E}$), $\mathrm{e}^{\theta_\mathrm{E}/T} \gg 1$ and

$$f_\mathrm{E}(T) \approx \left(\frac{\theta_\mathrm{E}}{T}\right)^2 \left(\frac{\mathrm{e}^{\theta_\mathrm{E}/2T}}{\mathrm{e}^{\theta_\mathrm{E}/T}}\right)^2 = \left(\frac{\theta_\mathrm{E}}{T}\right)^2 \mathrm{e}^{-\theta_\mathrm{E}/T} \qquad (7A.8c)$$

The strongly decaying exponential function goes to zero more rapidly than $1/T^2$ goes to infinity; so $f_\mathrm{E} \to 0$ as $T \to 0$, and the heat capacity approaches zero, as found experimentally. The physical reason for this success is that as the temperature is lowered, less energy is available to excite the atomic oscillations. At high temperatures many oscillators are excited into high energy states leading to classical behaviour.

Figure 7A.5 shows the temperature dependence of the heat capacity predicted by the Einstein formula and some experimental data; the value of the Einstein temperature is adjusted to obtain the best fit to the data. The general shape of the curve is satisfactory, but the numerical agreement is in fact quite poor. This discrepancy arises from Einstein's assumption that all the atoms oscillate with the same frequency. A more sophisticated treatment, due to Peter Debye, allows the oscillators to have a range of frequencies from zero up to a maximum. This approach results in much better agreement with the experimental data and there can be little doubt that mechanical motion as well as electromagnetic radiation is quantized.

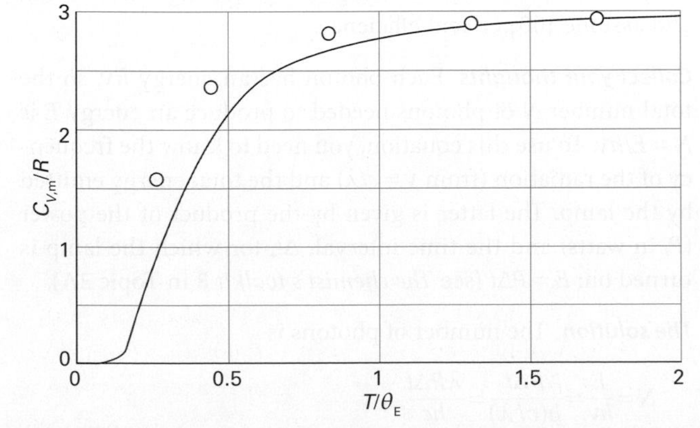

Figure 7A.5 Experimental low-temperature molar heat capacities (open circles) and the temperature dependence predicted on the basis of Einstein's theory (solid line). His equation (eqn 7A.8) accounts for the dependence fairly well, but is everywhere too low.

(c) Atomic and molecular spectra

The most compelling and direct evidence for the quantization of energy comes from **spectroscopy**, the detection and analysis of the electromagnetic radiation absorbed, emitted, or scattered by a substance. The record of the variation of the intensity of this radiation with frequency (ν), wavelength (λ), or wavenumber ($\tilde{\nu} = \nu/c$, see *The chemist's toolkit* 13) is called its **spectrum** (from the Latin word for appearance).

An atomic emission spectrum is shown in Fig. 7A.6, and a molecular absorption spectrum is shown in Fig. 7A.7. The obvious feature of both is that radiation is emitted or absorbed at a series of discrete frequencies. This observation can be understood if the energy of the atoms or molecules is also confined to discrete values, because then the energies that a molecule can discard or acquire are also confined to discrete values (Fig. 7A.8). If the energy of an atom or molecule decreases by ΔE, and this energy is carried away as radiation, the frequency of the radiation ν and the change in energy are related by the **Bohr frequency condition**:

$$\Delta E = h\nu \qquad \text{Bohr frequency condition} \quad (7A.9)$$

A molecule is said to undergo a **spectroscopic transition**, a change of state, and as a result an emission 'line', a sharply defined peak, appears in the spectrum at frequency ν.

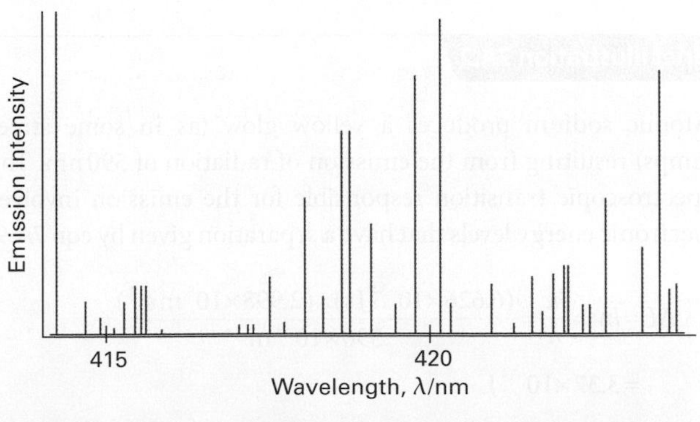

Figure 7A.6 A region of the spectrum of radiation emitted by excited iron atoms consists of radiation at a series of discrete wavelengths (or frequencies).

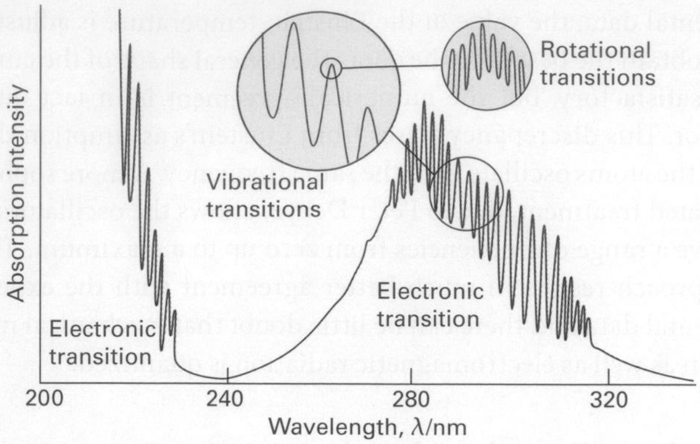

Figure 7A.7 A molecule can change its state by absorbing radiation at definite frequencies. This spectrum is due to the electronic, vibrational, and rotational excitation of sulfur dioxide (SO_2) molecules. The observation of discrete spectral lines suggests that molecules can possess only discrete energies, not an arbitrary energy.

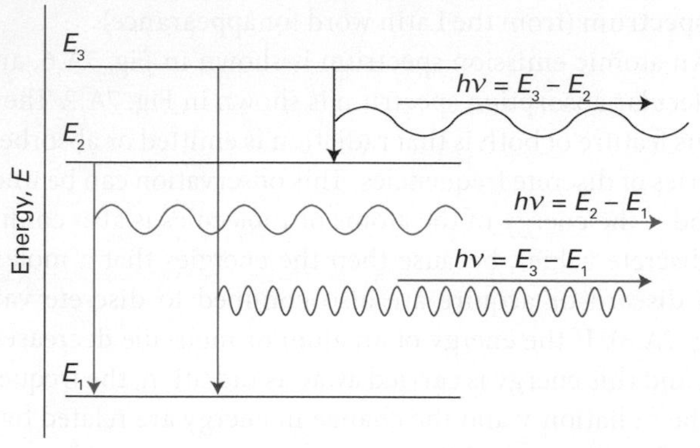

Figure 7A.8 Spectroscopic transitions, such as those shown in Fig. 7A.6, can be accounted for by supposing that an atom (or molecule) emits electromagnetic radiation as it changes from a discrete level of high energy to a discrete level of lower energy. High-frequency radiation is emitted when the energy change is large. Transitions like those shown in Fig. 7A.7 can be explained by supposing that a molecule (or atom) absorbs radiation as it changes from a low-energy level to a higher-energy level.

Brief illustration 7A.2

Atomic sodium produces a yellow glow (as in some street lamps) resulting from the emission of radiation of 590 nm. The spectroscopic transition responsible for the emission involves electronic energy levels that have a separation given by eqn 7A.9:

$$\Delta E = h\nu = \frac{hc}{\lambda} = \frac{(6.626\times10^{-34}\,\text{J s})\times(2.998\times10^{8}\,\text{m s}^{-1})}{590\times10^{-9}\,\text{m}}$$

$$=3.37\times10^{-19}\,\text{J}$$

This energy difference can be expressed in a variety of ways. For instance, multiplication by Avogadro's constant results in an energy separation per mole of atoms, of 203 kJ mol^{-1}, comparable to the energy of a weak chemical bond.

7A.2 Wave–particle duality

The experiments about to be described show that electromagnetic radiation—which classical physics treats as wave-like—actually also displays the characteristics of particles. Another experiment shows that electrons—which classical physics treats as particles—also display the characteristics of waves. This wave-particle **duality**, the blending together of the characteristics of waves and particles, lies at the heart of quantum mechanics.

(a) The particle character of electromagnetic radiation

The Planck treatment of black-body radiation introduced the idea that an oscillator of frequency ν can have only the energies 0, $h\nu$, $2h\nu$, ... This quantization leads to the suggestion (and at this stage it is only a suggestion) that the resulting electromagnetic radiation of that frequency can be thought of as consisting of 0, 1, 2, ... particles, each particle having an energy $h\nu$. These particles of electromagnetic radiation are now called **photons**. Thus, if an oscillator of frequency ν is excited to its first excited state, then one photon of that frequency is present, if it is excited to its second excited state, then two photons are present, and so on. The observation of discrete spectra from atoms and molecules can be pictured as the atom or molecule generating a photon of energy $h\nu$ when it discards an energy of magnitude ΔE, with $\Delta E = h\nu$.

Example 7A.1 Calculating the number of photons

Calculate the number of photons emitted by a 100 W yellow lamp in 1.0 s. Take the wavelength of yellow light as 560 nm, and assume 100 per cent efficiency.

Collect your thoughts Each photon has an energy $h\nu$, so the total number N of photons needed to produce an energy E is $N = E/h\nu$. To use this equation, you need to know the frequency of the radiation (from $\nu = c/\lambda$) and the total energy emitted by the lamp. The latter is given by the product of the power (P, in watts) and the time interval, Δt, for which the lamp is turned on: $E = P\Delta t$ (see *The chemist's toolkit* 8 in Topic 2A).

The solution The number of photons is

$$N = \frac{E}{h\nu} = \frac{P\Delta t}{h(c/\lambda)} = \frac{\lambda P \Delta t}{hc}$$

Substitution of the data gives

$$N = \frac{(5.60\times10^{-7}\,\text{m})\times(100\,\text{J s}^{-1})\times(1.0\,\text{s})}{(6.626\times10^{-34}\,\text{J s})\times(2.998\times10^{8}\,\text{m s}^{-1})} = 2.8\times10^{20}$$

A note on good practice To avoid rounding and other numerical errors, it is best to carry out algebraic calculations first, and to substitute numerical values into a single, final formula. Moreover, an analytical result may be used for other data without having to repeat the entire calculation.

Self-test 7A.1 How many photons does a monochromatic (single frequency) infrared rangefinder of power 1 mW and wavelength 1000 nm emit in 0.1 s?

Answer: 5×10^{14}

So far, the existence of photons is only a suggestion. Experimental evidence for their existence comes from the measurement of the energies of electrons produced in the **photoelectric effect**, the ejection of electrons from metals when they are exposed to ultraviolet radiation. The experimental characteristics of the photoelectric effect are as follows:

- No electrons are ejected, regardless of the intensity of the radiation, unless its frequency exceeds a threshold value characteristic of the metal.
- The kinetic energy of the ejected electrons increases linearly with the frequency of the incident radiation but is independent of the intensity of the radiation.
- Even at low radiation intensities, electrons are ejected immediately if the frequency is above the threshold value.

Figure 7A.9 illustrates the first and second characteristics.

These observations strongly suggest that in the photoelectric effect a particle-like projectile collides with the metal and, if the kinetic energy of the projectile is high enough, an electron is ejected. If the projectile is a photon of energy $h\nu$ (ν is the frequency of the radiation), the kinetic energy of the electron is E_k, and the energy needed to remove an electron from the metal, which is called its **work function**, is Φ (uppercase phi), then as illustrated in Fig. 7A.10, the conservation of energy implies that

$$h\nu = E_k + \Phi \quad \text{or} \quad E_k = h\nu - \Phi \qquad \text{Photoelectric effect} \qquad (7A.10)$$

This model explains the three experimental observations:

- Photoejection cannot occur if $h\nu < \Phi$ because the photon brings insufficient energy.
- The kinetic energy of an ejected electron increases linearly with the frequency of the photon.
- When a photon collides with an electron, it gives up all its energy, so electrons should appear as soon as the collisions begin, provided the photons have sufficient energy.

A practical application of eqn 7A.10 is that it provides a technique for the determination of Planck's constant, because the slopes of the lines in Fig. 7A.9 are all equal to h.

The energies of photoelectrons, the work function, and other quantities are often expressed in the alternative energy unit the **electronvolt** (eV): 1 eV is defined as the kinetic energy acquired when an electron (of charge $-e$) is accelerated from rest through a potential difference $\Delta\phi = 1$ V. That kinetic energy is $e\Delta\phi$, so

$$E_k = e\Delta\phi = (1.602\times10^{-19}\,\text{C})\times 1\,\text{V} = 1.602\times10^{-19}\,\text{C V} = 1\,\text{eV}$$

Because 1 C V = 1 J, it follows that the relation between electronvolts and joules is

$$1\,\text{eV} = 1.602\times10^{-19}\,\text{J}$$

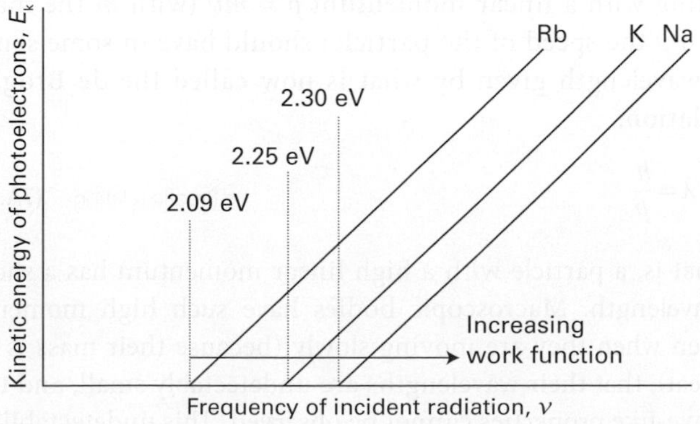

Figure 7A.9 In the photoelectric effect, it is found that no electrons are ejected when the incident radiation has a frequency below a certain value that is characteristic of the metal. Above that value, the kinetic energy of the photoelectrons varies linearly with the frequency of the incident radiation.

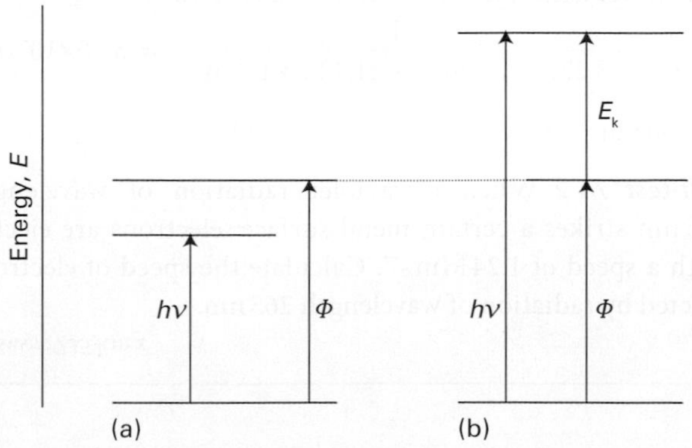

Figure 7A.10 The photoelectric effect can be explained if it is supposed that the incident radiation is composed of photons that have energy proportional to the frequency of the radiation. (a) The energy of the photon is insufficient to drive an electron out of the metal. (b) The energy of the photon is more than enough to eject an electron, and the excess energy is carried away as the kinetic energy of the photoelectron (the ejected electron).

Example 7A.2 Calculating the longest wavelength capable of photoejection

A photon of radiation of wavelength 305 nm ejects an electron with a kinetic energy of 1.77 eV from a metal. Calculate the longest wavelength of radiation capable of ejecting an electron from the metal.

Collect your thoughts You can use eqn 7A.10, rearranged into $\Phi = h\nu - E_k$, to compute the work function because you know the frequency of the photon from $\nu = c/\lambda$. The threshold for photoejection is the lowest frequency at which electron ejection occurs without there being any excess energy; that is, the kinetic energy of the ejected electron is zero. Setting $E_k = 0$ in $E_k = h\nu - \Phi$ gives the minimum photon frequency as $\nu_{min} = \Phi/h$. Use this value of the frequency to calculate the corresponding wavelength, λ_{max}.

The solution The minimum frequency for photoejection is

$$\nu_{min} = \frac{\Phi}{h} = \frac{h\nu - E_k}{h} \overset{\nu = c/\lambda}{=} \frac{c}{\lambda} - \frac{E_k}{h}$$

The longest wavelength that can cause photoejection is therefore

$$\lambda_{max} = \frac{c}{\nu_{min}} = \frac{c}{c/\lambda - E_k/h} = \frac{1}{1/\lambda - E_k/hc}$$

Now substitute the data. The kinetic energy of the electron is

$$E_k = 1.77\,\text{eV} \times (1.602 \times 10^{-19}\,\text{J eV}^{-1}) = 2.83\ldots \times 10^{-19}\,\text{J}$$

so

$$\frac{E_k}{hc} = \frac{2.83\ldots \times 10^{-19}\,\text{J}}{(6.626 \times 10^{-34}\,\text{J s}) \times (2.998 \times 10^{8}\,\text{m s}^{-1})} = 1.42\ldots \times 10^{6}\,\text{m}^{-1}$$

Therefore, with $1/\lambda = 1/305\,\text{nm} = 3.27\ldots \times 10^{6}\,\text{m}^{-1}$,

$$\lambda_{max} = \frac{1}{(3.27\ldots \times 10^{6}\,\text{m}^{-1}) - (1.42\ldots \times 10^{6}\,\text{m}^{-1})} = 5.40 \times 10^{-7}\,\text{m}$$

or 540 nm.

Self-test 7A.2 When ultraviolet radiation of wavelength 165 nm strikes a certain metal surface, electrons are ejected with a speed of $1.24\,\text{Mm s}^{-1}$. Calculate the speed of electrons ejected by radiation of wavelength 265 nm.

Answer: $735\,\text{km s}^{-1}$

(b) The wave character of particles

Although contrary to the long-established wave theory of radiation, the view that radiation consists of particles had been held before, but discarded. No significant scientist, however, had taken the view that matter is wave-like. Nevertheless, experiments carried out in 1927 forced people to consider that possibility. The crucial experiment was performed by Clinton Davisson and Lester Germer, who observed the diffraction of electrons by a crystal (Fig. 7A.11). As remarked in *The chemist's toolkit* 13, diffraction is the interference caused by an object in the path of waves. Davisson and Germer's success was a lucky accident, because a chance rise of temperature caused their polycrystalline sample to anneal, and the ordered planes of atoms then acted as a diffraction grating. The Davisson–Germer experiment, which has since been repeated with other particles (including α particles, molecular hydrogen, and neutrons), shows clearly that particles have wave-like properties. At almost the same time, G.P. Thomson showed that a beam of electrons was diffracted when passed through a thin gold foil.

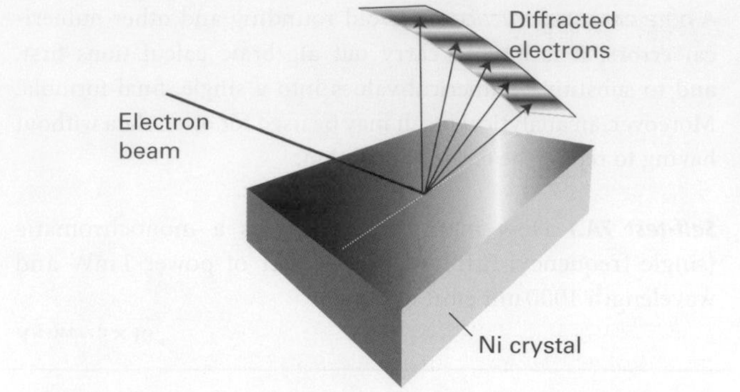

Figure 7A.11 The Davisson–Germer experiment. The scattering of an electron beam from a nickel crystal shows a variation in intensity characteristic of a diffraction experiment in which waves interfere constructively and destructively in different directions.

Some progress towards accounting for wave–particle duality had already been made by Louis de Broglie who, in 1924, suggested that any particle, not only photons, travelling with a linear momentum $p = mv$ (with m the mass and v the speed of the particle) should have in some sense a wavelength given by what is now called the **de Broglie relation**:

$$\lambda = \frac{h}{p} \qquad \text{de Broglie relation} \qquad (7A.11)$$

That is, a particle with a high linear momentum has a short wavelength. Macroscopic bodies have such high momenta even when they are moving slowly (because their mass is so great), that their wavelengths are undetectably small, and the wave-like properties cannot be observed. This undetectability is why classical mechanics can be used to explain the behaviour of macroscopic bodies. It is necessary to invoke quantum mechanics only for microscopic bodies, such as atoms and molecules, in which masses are small.

Example 7A.3 Estimating the de Broglie wavelength

Estimate the wavelength of electrons that have been accelerated from rest through a potential difference of 40 kV.

Collect your thoughts To use the de Broglie relation, you need to know the linear momentum, p, of the electrons. To calculate the linear momentum, note that the energy acquired by an electron accelerated through a potential difference $\Delta\phi$ is $e\Delta\phi$, where e is the magnitude of its charge. At the end of the period of acceleration, all the acquired energy is in the form of kinetic energy, $E_k = \frac{1}{2}m_e v^2 = p^2/2m_e$. You can therefore calculate p by setting $p^2/2m_e$ equal to $e\Delta\phi$. For the manipulation of units use $1\,\text{V C} = 1\,\text{J}$ and $1\,\text{J} = 1\,\text{kg m}^2\,\text{s}^{-2}$.

The solution The expression $p^2/2m_e = e\Delta\phi$ implies that $p = (2m_e e\Delta\phi)^{1/2}$ then, from the de Broglie relation $\lambda = h/p$,

$$\lambda = \frac{h}{(2m_e e\Delta\phi)^{1/2}}$$

Substitution of the data and the fundamental constants gives

$$\lambda = \frac{6.626\times10^{-34}\,\text{J s}}{\{2\times(9.109\times10^{-31}\,\text{kg})\times(1.602\times10^{-19}\,\text{C})\times(4.0\times10^{4}\,\text{V})\}^{1/2}}$$
$$= 6.1\times10^{-12}\,\text{m}$$

or 6.1 pm.

Comment. Electrons accelerated in this way are used in the technique of electron diffraction for imaging biological systems and for the determination of the structures of solid surfaces (Topic 19A).

Self-test 7A.3 Calculate the wavelength of (a) a neutron with a translational kinetic energy equal to kT at 300 K, (b) a tennis ball of mass 57 g travelling at 80 km h^{-1}.

Answer: (a) 178 pm, (b) 5.2×10^{-34} m

Checklist of concepts

☐ 1. A **black body** is an object capable of emitting and absorbing all wavelengths of radiation without favouring any wavelength.

☐ 2. An electromagnetic field of a given frequency can take up energy only in discrete amounts.

☐ 3. Atomic and molecular spectra show that atoms and molecules can take up energy only in discrete amounts.

☐ 4. The **photoelectric effect** establishes the view that electromagnetic radiation, regarded in classical physics as wave-like, consists of particles (photons).

☐ 5. The diffraction of electrons establishes the view that electrons, regarded in classical physics as particles, are wave-like with a wavelength given by the **de Broglie relation**.

☐ 6. **Wave–particle duality** is the recognition that the concepts of particle and wave blend together.

Checklist of equations

Property	Equation	Comment	Equation number
Wien's law	$\lambda_{max}T = 2.9\times10^{-3}\,\text{m K}$		7A.1
Stefan–Boltzmann law	$\mathcal{E}(T) = \text{constant}\times T^4$		7A.3
Planck distribution	$\rho(\lambda,T) = 8\pi hc/\{\lambda^5(e^{hc/\lambda kT}-1)\}$ $\rho(\nu,T) = 8\pi h\nu^3/\{c^3(e^{h\nu/kT}-1)\}$	Black-body radiation	7A.6
Einstein formula for heat capacity of a solid	$C_{V,m}(T) = 3Rf_E(T)$ $f_E(T) = (\theta_E/T)^2\{e^{\theta_E/2T}/(e^{\theta_E/T}-1)\}^2$	Einstein temperature: $\theta_E = h\nu/k$	7A.8
Bohr frequency condition	$\Delta E = h\nu$		7A.9
Photoelectric effect	$E_k = h\nu - \Phi$	Φ is the work function	7A.10
de Broglie relation	$\lambda = h/p$	λ is the wavelength of a particle of linear momentum p	7A.11

TOPIC 7B Wavefunctions

> ➤ **Why do you need to know this material?**
>
> Wavefunctions provide the essential foundation for understanding the properties of electrons in atoms and molecules, and are central to explanations in chemistry.
>
> ➤ **What is the key idea?**
>
> All the dynamical properties of a system are contained in its wavefunction, which is obtained by solving the Schrödinger equation.
>
> ➤ **What do you need to know already?**
>
> You need to be aware of the shortcomings of classical physics that drove the development of quantum theory (Topic 7A).

In classical mechanics an object travels along a definite path or trajectory. In quantum mechanics a particle in a particular state is described by a **wavefunction**, ψ (psi), which is spread out in space, rather than being localized. The wavefunction contains all the dynamical information about the object in that state, such as its position and momentum.

7B.1 The Schrödinger equation

In 1926 Erwin Schrödinger proposed an equation for finding the wavefunctions of any system. The **time-independent Schrödinger equation** for a particle of mass m moving in one dimension with energy E in a system that does not change with time (for instance, its volume remains constant) is

$$-\frac{\hbar^2}{2m}\frac{d^2\psi}{dx^2}+V(x)\psi=E\psi \qquad \text{Time-independent Schrödinger equation} \qquad (7B.1)$$

The constant $\hbar = h/2\pi$ (which is read h-cross or h-bar) is a convenient modification of Planck's constant used widely in quantum mechanics; $V(x)$ is the potential energy of the particle at x. Because the total energy E is the sum of potential and kinetic energies, the first term on the left must be related (in a manner explored later) to the kinetic energy of the particle. The Schrödinger equation can be regarded as a fundamental postulate of quantum mechanics, but its plausibility can be demonstrated by showing that, for the case of a free particle, it is consistent with the de Broglie relation (Topic 7A).

How is that done? 7B.1 Showing that the Schrödinger equation is consistent with the de Broglie relation

The potential energy of a freely moving particle is zero everywhere, $V(x) = 0$, so the Schrödinger equation (eqn 7B.1) becomes

$$\frac{d^2\psi}{dx^2}=-\frac{2mE}{\hbar^2}\psi$$

Step 1 *Find a solution of the Schrödinger equation for a free particle*

A solution of this equation is $\psi = \cos kx$, as you can confirm by noting that

$$\frac{d^2\psi}{dx^2}=\frac{d^2\cos kx}{dx^2}=-k^2\cos kx=-k^2\psi$$

It follows that $-k^2 = -2mE/\hbar^2$ and hence

$$k=\left(\frac{2mE}{\hbar^2}\right)^{1/2}$$

The energy, which is only kinetic in this instance, is related to the linear momentum of the particle by $E = p^2/2m$ (*The chemist's toolkit* 6 in Topic 2A), so it follows that

$$k=\left(\frac{2m(p^2/2m)}{\hbar^2}\right)^{1/2}=\frac{p}{\hbar}$$

The linear momentum is therefore related to k by $p = k\hbar$.

Step 2 *Interpret the wavefunction in terms of a wavelength*

Now recognize that a wave (more specifically, a 'harmonic wave') can be described mathematically by a sine or cosine function. It follows that $\cos kx$ can be regarded as a wave that goes through a complete cycle as kx increases by 2π. The wavelength is therefore given by $k\lambda=2\pi$, so $k=2\pi/\lambda$. Therefore, the linear momentum is related to the wavelength of the wavefunction by

$$p=k\hbar=\frac{2\pi}{\lambda}\times\frac{h}{2\pi}=\frac{h}{\lambda}$$

which is the de Broglie relation. The Schrödinger equation therefore has solutions consistent with the de Broglie relation.

7B.2 The Born interpretation

One piece of dynamical information contained in the wavefunction is the location of the particle. Max Born used an analogy with the wave theory of radiation, in which the square of the amplitude of an electromagnetic wave in a region is interpreted as its intensity and therefore (in quantum terms) as a measure of the probability of finding a photon present in the region. The **Born interpretation** of the wavefunction is:

> If the wavefunction of a particle has the value ψ at x, then the probability of finding the particle between x and $x + dx$ is proportional to $|\psi|^2 dx$ (Fig. 7B.1).

Born interpretation

The quantity $|\psi|^2 = \psi^* \psi$ allows for the possibility that ψ is complex (see *The chemist's toolkit* 14). If the wavefunction is real (such as $\cos kx$), then $|\psi|^2 = \psi^2$.

Because $|\psi|^2 dx$ is a (dimensionless) probability, $|\psi|^2$ is the **probability density**, with the dimensions of 1/length (for a one-dimensional system). The wavefunction ψ itself is called the **probability amplitude**. For a particle free to move in three dimensions (for example, an electron near a nucleus in an atom), the wavefunction depends on the coordinates x, y, and z and is denoted $\psi(\boldsymbol{r})$. In this case the Born interpretation is (Fig. 7B.2):

> If the wavefunction of a particle has the value ψ at $\boldsymbol{r}$, then the probability of finding the particle in an infinitesimal volume $d\tau = dxdydz$ at that position is proportional to $|\psi|^2 d\tau$.

The chemist's toolkit 14 Complex numbers

Complex numbers have the general form

$$z = x + iy$$

where $i = \sqrt{-1}$. The real number x is the 'real part of z', denoted $\mathrm{Re}(z)$; likewise, the real number y is 'the imaginary part of z', $\mathrm{Im}(z)$. The **complex conjugate** of z, denoted z^*, is formed by replacing i by −i:

$$z^* = x - iy$$

The product of z^* and z is denoted $|z|^2$ and is called the **square modulus** of z. From the definition of z and z^* and $i^2 = -1$ it follows that

$$|z|^2 = z^*z = (x + iy)(x - iy) = x^2 + y^2$$

The square modulus is a real, non-negative number. The **absolute value** or **modulus** is denoted $|z|$ and is given by:

$$|z| = (z^*z)^{1/2} = (x^2 + y^2)^{1/2}$$

For further information about complex numbers, see *The chemist's toolkit* 16 in Topic 7C.

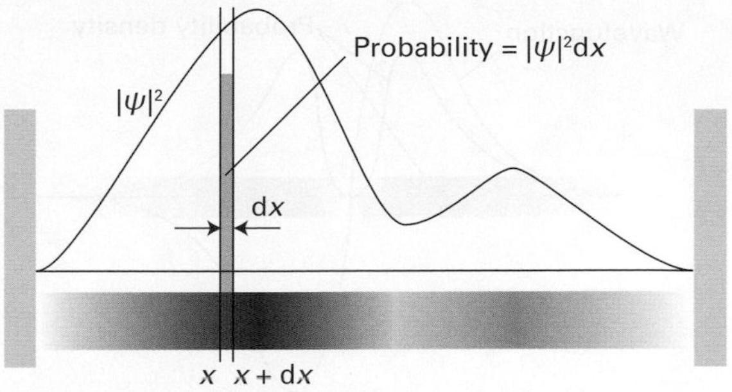

Figure 7B.1 The wavefunction ψ is a probability amplitude in the sense that its square modulus ($\psi^* \psi$ or $|\psi|^2$) is a probability density. The probability of finding a particle in the region between x and $x + dx$ is proportional to $|\psi|^2 dx$. Here, the probability density is represented by the density of shading in the superimposed band.

In this case, $|\psi|^2$ has the dimensions of 1/length^3 and the wavefunction itself has dimensions of 1/length$^{3/2}$ (and units such as $m^{-3/2}$).

The Born interpretation does away with any worry about the significance of a negative (and, in general, complex) value of ψ because $|\psi|^2$ is always real and nowhere negative. There is no *direct* significance in the negative (or complex) value of a wavefunction: only the square modulus is directly physically significant, and both negative and positive regions of a wavefunction may correspond to a high probability of finding a particle in a region (Fig. 7B.3). However, the presence of positive and negative regions of a wavefunction is of great *indirect* significance, because it gives rise to the possibility of constructive and destructive interference between different wavefunctions.

A wavefunction may be zero at one or more points, and at these locations the probability density is also zero. It is important to distinguish a point at which the wavefunction is zero (for instance, far from the nucleus of a hydrogen atom) from the point at which it passes *through* zero. The latter is called a **node**. A location where the wavefunction approaches zero without actually passing through zero is not a node. Thus, the

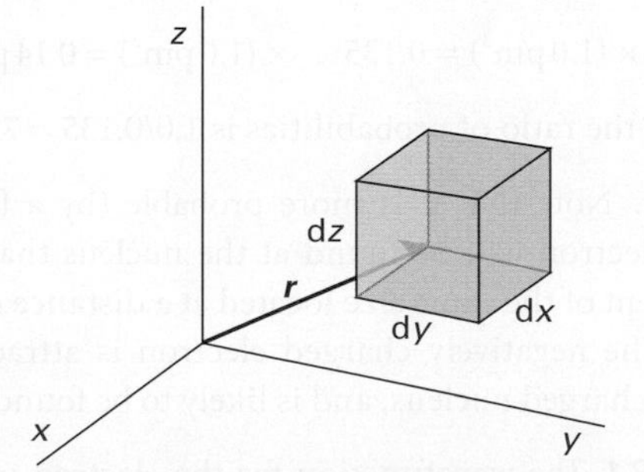

Figure 7B.2 The Born interpretation of the wavefunction in three-dimensional space implies that the probability of finding the particle in the volume element $d\tau = dxdydz$ at some position $\boldsymbol{r}$ is proportional to the product of $d\tau$ and the value of $|\psi|^2$ at that position.

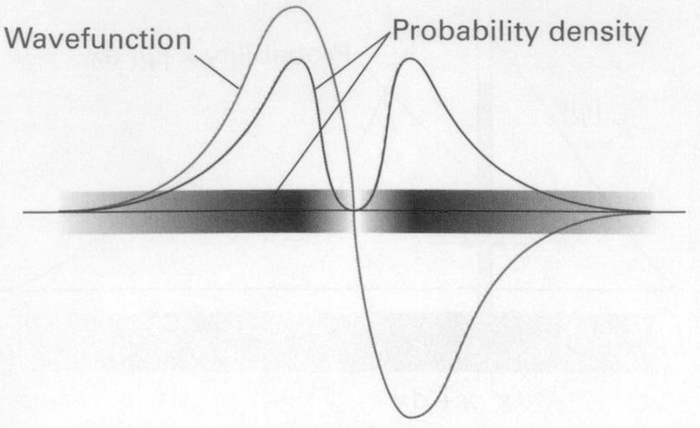

Figure 7B.3 The sign of a wavefunction has no direct physical significance: the positive and negative regions of this wavefunction both correspond to the same probability distribution (as given by the square modulus of ψ and depicted by the density of the shading).

wavefunction $\cos kx$ has nodes wherever kx is an odd integral multiple of $\frac{1}{2}\pi$ (where the wave passes through zero), but the wavefunction e^{-kx} has no nodes, despite becoming zero as $x \to \infty$.

Example 7B.1 Interpreting a wavefunction

The wavefunction of an electron in the lowest energy state of a hydrogen atom is proportional to e^{-r/a_0}, where a_0 is a constant and r the distance from the nucleus. Calculate the relative probabilities of finding the electron inside a region of volume $\delta V = 1.0\,\text{pm}^3$, which is small even on the scale of the atom, located at (a) the nucleus, (b) a distance a_0 from the nucleus.

Collect your thoughts The region of interest is so small on the scale of the atom that you can ignore the variation of ψ within it and write the probability, P, as proportional to the probability density (ψ^2; note that ψ is real) evaluated at the point of interest multiplied by the volume of interest, δV. That is, $P \propto \psi^2\delta V$, with $\psi^2 \propto e^{-2r/a_0}$.

The solution In each case $\delta V = 1.0\,\text{pm}^3$. (a) At the nucleus, $r = 0$, so

$$P \propto e^0 \times (1.0\,\text{pm}^3) = 1 \times (1.0\,\text{pm}^3) = 1.0\,\text{pm}^3$$

(b) At a distance $r = a_0$ in an arbitrary direction,

$$P \propto e^{-2} \times (1.0\,\text{pm}^3) = 0.135\ldots \times (1.0\,\text{pm}^3) = 0.14\,\text{pm}^3$$

Therefore, the ratio of probabilities is 1.0/0.135 = 7.4.

Comment. Note that it is more probable (by a factor of 7) that the electron will be found at the nucleus than in a volume element of the same size located at a distance a_0 from the nucleus. The negatively charged electron is attracted to the positively charged nucleus, and is likely to be found close to it.

Self-test 7B.1 The wavefunction for the electron in its lowest energy state in the ion He^+ is proportional to e^{-2r/a_0}. Repeat the calculation for this ion and comment on the result.

Answer: 55; the wavefunction is more compact

(a) Normalization

A mathematical feature of the Schrödinger equation is that if ψ is a solution, then so is $N\psi$, where N is any constant. This feature is confirmed by noting that because ψ occurs in every term in eqn 7B.1, it can be replaced by $N\psi$ and the constant factor N cancelled to recover the original equation. This freedom to multiply the wavefunction by a constant factor means that it is always possible to find a **normalization constant**, N, such that rather than the probability density being *proportional* to $|\psi|^2$ it becomes *equal* to $|\psi|^2$.

A normalization constant is found by noting that, for a normalized wavefunction $N\psi$, the probability that a particle is in the region dx is equal to $(N\psi^*)(N\psi)dx$ (N is taken to be real). Furthermore, the sum over all space of these individual probabilities must be 1 (the probability of the particle being somewhere is 1). Expressed mathematically, the latter requirement is

$$N^2\int_{-\infty}^{\infty}\psi^*\psi\,dx = 1 \qquad (7B.2)$$

and therefore

$$N = \frac{1}{\left(\int_{-\infty}^{\infty}\psi^*\psi\,dx\right)^{1/2}} \qquad (7B.3)$$

Provided this integral has a finite value (that is, the wavefunction is 'square integrable'), the normalization factor can be found and the wavefunction 'normalized' (and specifically 'normalized to 1'). From now on, unless stated otherwise, all wavefunctions are assumed to have been normalized to 1, in which case in one dimension

$$\int_{-\infty}^{\infty}\psi^*\psi\,dx = 1 \qquad (7B.4a)$$

and in three dimensions

$$\int_{-\infty}^{\infty}\int_{-\infty}^{\infty}\int_{-\infty}^{\infty}\psi^*\psi\,dx\,dy\,dz = 1 \qquad (7B.4b)$$

In quantum mechanics it is common to write all such integrals in a short-hand form as

$$\int\psi^*\psi\,d\tau = 1 \qquad (7B.4c)$$

where $d\tau$ is the appropriate volume element and the integration is understood as being over all space.

Example 7B.2 Normalizing a wavefunction

Carbon nanotubes are thin hollow cylinders of carbon with diameters between 1 nm and 2 nm, and lengths of several micrometres. According to one simple model, the lowest-energy electrons of the nanotube are described by the wavefunction $\sin(\pi x/L)$, where L is the length of the nanotube. Find the normalized wavefunction.

Collect your thoughts Because the wavefunction is one-dimensional, you need to find the factor N that guarantees that the integral in eqn 7B.4a is equal to 1. The wavefunction is real, so $\psi^* = \psi$. Relevant integrals are found in the *Resource section*.

The solution Write the wavefunction as $\psi = N\sin(\pi x/L)$, where N is the normalization factor. The limits of integration are $x = 0$ to $x = L$ because the wavefunction spans the length of the tube. It follows that

$$\int \psi^*\psi \,\mathrm{d}\tau = N^2 \overbrace{\int_0^L \sin^2\frac{\pi x}{L}\,\mathrm{d}x}^{\text{Integral T.2}} = \tfrac{1}{2}N^2 L$$

For the wavefunction to be normalized, this integral must be equal to 1; that is, $\tfrac{1}{2}N^2L = 1$, and hence

$$N = \left(\frac{2}{L}\right)^{1/2}$$

The normalized wavefunction is therefore

$$\psi = \left(\frac{2}{L}\right)^{1/2} \sin\frac{\pi x}{L}$$

Comment. Because L is a length, the dimensions of ψ are 1/length$^{1/2}$, and therefore those of ψ^2 are 1/length, as is appropriate for a probability density in one dimension.

Self-test 7B.2 The wavefunction for the next higher energy level for the electrons in the same tube is $\sin(2\pi x/L)$. Normalize this wavefunction.

Answer: $N = (2/L)^{1/2}$

To calculate the probability of finding the system in a finite region of space the probability density is summed (integrated) over the region of interest. Thus, for a one-dimensional system, the probability P of finding the particle between x_1 and x_2 is given by

$$P = \int_{x_1}^{x_2} |\psi(x)|^2 \,\mathrm{d}x \qquad \text{Probability [one-dimensional region]} \qquad (7B.5)$$

Example 7B.3 Determining a probability

As seen in *Example* 7B.2, the lowest-energy electrons of a carbon nanotube of length L can be described by the normalized wavefunction $(2/L)^{1/2}\sin(\pi x/L)$. What is the probability of finding the electron between $x = L/4$ and $x = L/2$?

Collect your thoughts Use eqn 7B.5 and the normalized wavefunction to write an expression for the probability of finding the electron in the region of interest. Relevant integrals are given in the *Resource section*.

The solution From eqn 7B.5 the probability is

$$P = \frac{2}{L}\overbrace{\int_{L/4}^{L/2} \sin^2(\pi x/L)\,\mathrm{d}x}^{\text{Integral T.2}}$$

It follows that

$$P = \frac{2}{L}\left(\frac{x}{2} - \frac{\sin(2\pi x/L)}{4\pi/L}\right)\Bigg|_{L/4}^{L/2} = \frac{2}{L}\left(\frac{L}{4} - \frac{L}{8} - 0 + \frac{L}{4\pi}\right) = 0.409$$

Comment. There is a chance of about 41 per cent that the electron will be found in the region.

Self-test 7B.3 As remarked in *Self-test* 7B.2, the normalized wavefunction of the next higher energy level of the electron in this model of the nanotube is $(2/L)^{1/2}\sin(2\pi x/L)$. What is the probability of finding the electron between $x = L/4$ and $x = L/2$?

Answer: 0.25

(b) Constraints on the wavefunction

The Born interpretation puts severe restrictions on the acceptability of wavefunctions. The first constraint is that ψ must not be infinite over a finite region, because if it were, the Born interpretation would fail. This requirement rules out many possible solutions of the Schrödinger equation, because many mathematically acceptable solutions rise to infinity and are therefore physically unacceptable. The Born interpretation also rules out solutions of the Schrödinger equation that give rise to more than one value of $|\psi|^2$ at a single point because it would be absurd to have more than one value of the probability density for the particle at a point. This restriction is expressed by saying that the wavefunction must be *single-valued*; that is, it must have only one value at each point of space.

The Schrödinger equation itself also implies some mathematical restrictions on the type of functions that can occur. Because it is a second-order differential equation (in the sense that it depends on the second derivative of the wavefunction), $\mathrm{d}^2\psi/\mathrm{d}x^2$ must be well-defined if the equation is to be applicable everywhere. The second derivative is defined only if the first derivative is continuous: this means that (except as specified below) there can be no kinks in the function. In turn, the first derivative is defined only if the function is continuous: no sharp steps are permitted.

Overall, therefore, the constraints on the wavefunction, which are summarized in Fig. 7B.4, are that it

- must not be infinite over a finite region;
- must be single-valued;
- must be continuous;
- must have a continuous first derivative (slope).

Constraints on the wavefunction

The last of these constraints does not apply if the potential energy has abrupt, infinitely high steps (as in the particle-in-a-box model treated in Topic 7D).

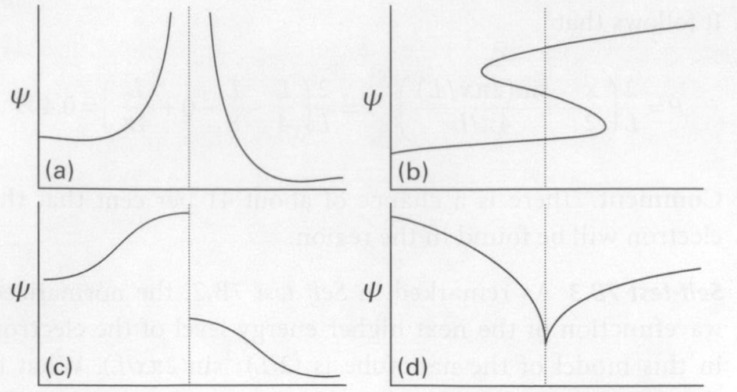

Figure 7B.4 The wavefunction must satisfy stringent conditions for it to be acceptable: (a) unacceptable because it is infinite over a finite region; (b) unacceptable because it is not single-valued; (c) unacceptable because it is not continuous; (d) unacceptable because its slope is discontinuous.

(c) Quantization

The constraints just noted are so severe that acceptable solutions of the Schrödinger equation do not in general exist for arbitrary values of the energy E. In other words, a particle may possess only certain energies, for otherwise its wavefunction would be physically unacceptable. That is,

> As a consequence of the restrictions on its wavefunction, the energy of a particle is quantized.

These acceptable energies are found by solving the Schrödinger equation for motion of various kinds, and selecting the solutions that conform to the restrictions listed above.

Checklist of concepts

☐ 1. A **wavefunction** is a mathematical function that contains all the dynamical information about a system.

☐ 2. The **Schrödinger equation** is a second-order differential equation used to calculate the wavefunction of a system.

☐ 3. According to the **Born interpretation**, the probability density at a point is proportional to the square of the wavefunction at that point.

☐ 4. A **node** is a point where a wavefunction passes through zero.

☐ 5. A wavefunction is **normalized** if the integral over all space of its square modulus is equal to 1.

☐ 6. A wavefunction must be single-valued, continuous, not infinite over a finite region of space, and (except in special cases) have a continuous slope.

☐ 7. The quantization of energy stems from the constraints that an acceptable wavefunction must satisfy.

Checklist of equations

Property	Equation	Comment	Equation number
The time-independent Schrödinger equation	$-(\hbar^2/2m)(\mathrm{d}^2\psi/\mathrm{d}x^2) + V(x)\psi = E\psi$	One-dimensional system*	7B.1
Normalization	$\int \psi^*\psi \,\mathrm{d}\tau = 1$	Integration over all space	7B.4c
Probability of a particle being between x_1 and x_2	$P = \int_{x_1}^{x_2} \lvert\psi(x)\rvert^2 \,\mathrm{d}x$	One-dimensional region	7B.5

* Higher dimensions are treated in Topics 7D, 7F, and 8A.

TOPIC 7C Operators and observables

> ➤ **Why do you need to know this material?**
>
> To interpret the wavefunction fully it is necessary to be able to extract dynamical information from it. The predictions of quantum mechanics are often very different from those of classical mechanics, and those differences are essential for understanding the structures and properties of atoms and molecules.
>
> ➤ **What is the key idea?**
>
> The dynamical information in the wavefunction is extracted by calculating the expectation values of hermitian operators.
>
> ➤ **What do you need to know already?**
>
> You need to know that the state of a system is fully described by a wavefunction (Topic 7B), and that the probability density is proportional to the square modulus of the wavefunction.

A wavefunction contains all the information it is possible to obtain about the dynamical properties of a particle (for example, its location and momentum). The Born interpretation (Topic 7B) provides information about location, but the wavefunction contains other information, which is extracted by using the methods described in this Topic.

7C.1 Operators

The Schrödinger equation can be written in the succinct form

$$\hat{H}\psi = E\psi \qquad \text{Operator form of Schrödinger equation} \qquad (7C.1a)$$

Comparison of this expression with the one-dimensional Schrödinger equation

$$-\frac{\hbar^2}{2m}\frac{d^2\psi}{dx^2} + V(x)\psi = E\psi$$

shows that in one dimension

$$\hat{H} = -\frac{\hbar^2}{2m}\frac{d^2}{dx^2} + V(x) \qquad \text{Hamiltonian operator} \qquad (7C.1b)$$

The quantity $\hat{H}$ (commonly read h-hat) is an **operator**, an expression that carries out a mathematical operation on a function. In this case, the operation is to take the second derivative of ψ, and (after multiplication by $-\hbar^2/2m$) to add the result to the outcome of multiplying ψ by $V(x)$.

The operator $\hat{H}$ plays a special role in quantum mechanics, and is called the **hamiltonian operator** after the nineteenth century mathematician William Hamilton, who developed a form of classical mechanics which, it subsequently turned out, is well suited to the formulation of quantum mechanics. The hamiltonian operator (and commonly simply 'the hamiltonian') is the operator corresponding to the total energy of the system, the sum of the kinetic and potential energies. In eqn 7C.1b the second term on the right is the potential energy, so the first term (the one involving the second derivative) must be the operator for kinetic energy.

In general, an operator acts on a function to produce a new function, as in

(operator)(function) = (new function)

In some cases the new function is the same as the original function, perhaps multiplied by a constant. Combinations of operators and functions that have this property are of great importance in quantum mechanics.

Brief illustration 7C.1

For example, when the operator d/dx, which means 'take the derivative of the following function with respect to x', acts on the function $\sin ax$, it generates the new function $a\cos ax$. However, when d/dx operates on e^{-ax} it generates $-ae^{-ax}$, which is the original function multiplied by the constant $-a$.

(a) Eigenvalue equations

The Schrödinger equation written as in eqn 7C.1a is an **eigenvalue equation**, an equation of the form

(operator)(function) = (constant factor) × (same function) (7C.2a)

In an eigenvalue equation, the action of the operator on the function generates the *same* function, multiplied by a constant. If a general operator is denoted $\hat{\Omega}$ (where Ω is uppercase omega) and the constant factor by ω (lowercase omega), then an eigenvalue equation has the form

$$\hat{\Omega}\psi = \omega\psi \qquad \text{Eigenvalue equation} \qquad (7C.2b)$$

If this relation holds, the function ψ is said to be an **eigenfunction** of the operator $\hat{\Omega}$, and ω is the **eigenvalue** associated with that eigenfunction. With this terminology, eqn 7C.2a can be written

$$\text{(operator)(eigenfunction)} = \text{(eigenvalue)} \times \text{(eigenfunction)} \quad (7C.2c)$$

Equation 7C.1a is therefore an eigenvalue equation in which ψ is an eigenfunction of the hamiltonian and E is the associated eigenvalue. It follows that 'solving the Schrödinger equation' can be expressed as 'finding the eigenfunctions and eigenvalues of the hamiltonian operator for the system'.

Just as the hamiltonian is the operator corresponding to the total energy, there are operators that represent other **observables**, the measurable properties of the system, such as linear momentum or electric dipole moment. For each such operator $\hat{\Omega}$ there is an eigenvalue equation of the form $\hat{\Omega}\psi=\omega\psi$, with the following significance:

> If the wavefunction is an eigenfunction of the operator $\hat{\Omega}$ corresponding to the observable Ω, then the outcome of a measurement of the property Ω will be the eigenvalue corresponding to that eigenfunction.

Quantum mechanics is formulated by constructing the operator corresponding to the observable of interest and then predicting the outcome of a measurement by examining the eigenvalues of the operator.

(b) The construction of operators

A basic postulate of quantum mechanics specifies how to set up the operator corresponding to a given observable.

Observables are represented by operators built from the following position and linear momentum operators:

$$\hat{x}=x\times \qquad \hat{p}_x=\frac{\hbar}{\mathrm{i}}\frac{\mathrm{d}}{\mathrm{d}x} \qquad \text{Specification of operators} \quad (7C.3)$$

That is, the operator for location along the x-axis is multiplication (of the wavefunction) by x, and the operator for linear momentum parallel to the x-axis is $\hbar/\mathrm{i}$ times the derivative (of the wavefunction) with respect to x.

The definitions in eqn 7C.3 are used to construct operators for other spatial observables. For example, suppose the potential energy has the form $V(x)=\frac{1}{2}k_\mathrm{f}x^2$, where k_f is a constant (this potential energy describes the vibrations of atoms in molecules). Because the operator for x is multiplication by x, by extension the operator for x^2 is multiplication by x and then by x again, or multiplication by x^2. The operator corresponding to $\frac{1}{2}k_\mathrm{f}x^2$ is therefore

$$\hat{V}(x)=\tfrac{1}{2}k_\mathrm{f}x^2\times \quad (7C.4)$$

In practice, the multiplication sign is omitted and multiplication is understood. To construct the operator for kinetic energy, the classical relation between kinetic energy and linear momentum, $E_\mathrm{k}=p_x^2/2m$ is used. Then, by using the operator for p_x from eqn 7C.3:

$$\hat{E}_\mathrm{k}=\frac{1}{2m}\overbrace{\left(\frac{\hbar}{\mathrm{i}}\frac{\mathrm{d}}{\mathrm{d}x}\right)}^{\hat{p}_x}\overbrace{\left(\frac{\hbar}{\mathrm{i}}\frac{\mathrm{d}}{\mathrm{d}x}\right)}^{\hat{p}_x}=-\frac{\hbar^2}{2m}\frac{\mathrm{d}^2}{\mathrm{d}x^2} \quad (7C.5)$$

It follows that the operator for the total energy, the hamiltonian operator, is

$$\hat{H}=\hat{E}_\mathrm{k}+\hat{V}=-\frac{\hbar^2}{2m}\frac{\mathrm{d}^2}{\mathrm{d}x^2}+\hat{V}(x) \qquad \text{Hamiltonian operator} \quad (7C.6)$$

where $\hat{V}(x)$ is the operator corresponding to whatever form the potential energy takes, exactly as in eqn 7C.1b.

Example 7C.1 Determining the value of an observable

What is the linear momentum of a free particle described by the wavefunctions (a) $\psi(x)=\mathrm{e}^{ikx}$ and (b) $\psi(x)=\mathrm{e}^{-ikx}$?

Collect your thoughts You need to operate on ψ with the operator corresponding to linear momentum (eqn 7C.3), and inspect the result. If the outcome is the original wavefunction multiplied by a constant (that is, if the application of the operator results in an eigenvalue equation), then you can identify the constant with the value of the observable.

The solution (a) For $\psi(x)=\mathrm{e}^{ikx}$,

$$\hat{p}_x\psi=\frac{\hbar}{\mathrm{i}}\frac{\mathrm{d}\psi}{\mathrm{d}x}=\frac{\hbar}{\mathrm{i}}\frac{\mathrm{d}\mathrm{e}^{ikx}}{\mathrm{d}x}=\frac{\hbar}{\mathrm{i}}\times ik\mathrm{e}^{ikx}=\overbrace{+k\hbar}^{\text{Eigenvalue}}\psi$$

This is an eigenvalue equation, with eigenvalue $+k\hbar$. It follows that a measurement of the momentum will give the value $p_x=+k\hbar$.

(b) For $\psi(x)=\mathrm{e}^{-ikx}$,

$$\hat{p}_x\psi=\frac{\hbar}{\mathrm{i}}\frac{\mathrm{d}\psi}{\mathrm{d}x}=\frac{\hbar}{\mathrm{i}}\frac{\mathrm{d}\mathrm{e}^{-ikx}}{\mathrm{d}x}=\frac{\hbar}{\mathrm{i}}\times(-ik)\mathrm{e}^{-ikx}=\overbrace{-k\hbar}^{\text{Eigenvalue}}\psi$$

Now the eigenvalue is $-k\hbar$, so $p_x=-k\hbar$. In case (a) the momentum is positive, meaning that the particle is travelling in the positive x-direction, whereas in (b) the particle is moving in the opposite direction.

Comment. A general feature of quantum mechanics is that taking the complex conjugate of a wavefunction reverses the direction of travel. An implication is that if the wavefunction is real (such as cos kx), then taking the complex conjugate leaves the wavefunction unchanged: there is no net direction of travel.

Self-test 7C.1 What is the kinetic energy of a particle described by the wavefunction cos kx?

Answer: $E_\mathrm{k}=\hbar^2k^2/2m$

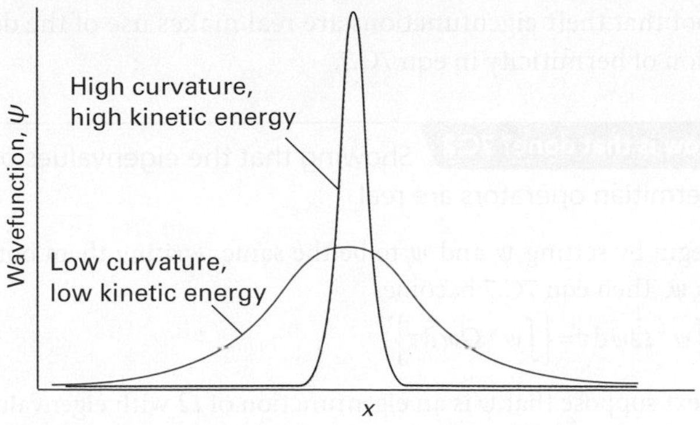

Figure 7C.1 The average kinetic energy of a particle can be inferred from the average curvature of the wavefunction. This figure shows two wavefunctions: the sharply curved function corresponds to a higher kinetic energy than the less sharply curved function.

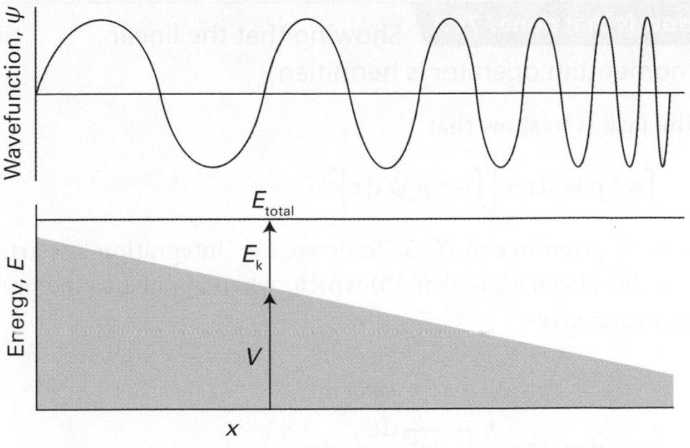

Figure 7C.3 The wavefunction of a particle with a potential energy V that decreases towards the right. As the total energy is constant, the kinetic energy E_k increases to the right, which results in a faster oscillation and hence greater curvature of the wavefunction.

The expression for the kinetic energy operator (eqn 7C.5) reveals an important point about the Schrödinger equation. In mathematics, the second derivative of a function is a measure of its curvature: a large second derivative indicates a sharply curved function (Fig. 7C.1). It follows that a sharply curved wavefunction is associated with a high kinetic energy, and one with a low curvature is associated with a low kinetic energy.

The curvature of a wavefunction in general varies from place to place (Fig. 7C.2): wherever a wavefunction is sharply curved, its contribution to the total kinetic energy is large; wherever the wavefunction is not sharply curved, its contribution to the overall kinetic energy is low. The observed kinetic energy of the particle is an average of all the contributions of the kinetic energy from each region. Hence, a particle can be expected to have a high kinetic energy if the average curvature of its wavefunction is high. Locally there can be both positive and negative contributions to the kinetic energy (because the curvature can be either positive, ∪, or negative, ∩) locally, but the average is always positive.

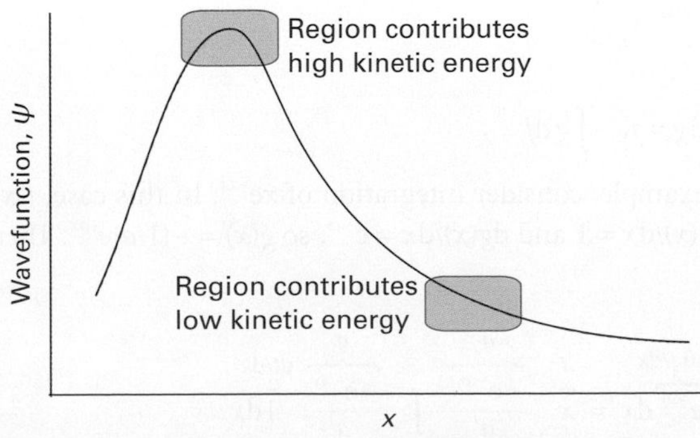

Figure 7C.2 The observed kinetic energy of a particle is an average of contributions from the entire space covered by the wavefunction. Sharply curved regions contribute a high kinetic energy to the average; less sharply curved regions contribute only a small kinetic energy.

The association of high curvature with high kinetic energy is a valuable guide to the interpretation of wavefunctions and the prediction of their shapes. For example, suppose the wavefunction of a particle with a given total energy and a potential energy that decreases with increasing x is required. Because the difference $E - V = E_k$ increases from left to right, the wavefunction must become more sharply curved by oscillating more rapidly as x increases (Fig. 7C.3). It is therefore likely that the wavefunction will look like the function sketched in the illustration, and more detailed calculation confirms this to be so.

(c) Hermitian operators

All the quantum mechanical operators that correspond to observables have a very special mathematical property: they are 'hermitian'. A **hermitian operator** is one for which the following relation is true:

$$\int \psi_i^* \hat{\Omega} \psi_j \, d\tau = \left\{ \int \psi_j^* \hat{\Omega} \psi_i \, d\tau \right\}^* \quad \text{Hermiticity [definition]} \quad (7C.7)$$

As stated in Topic 7B, in quantum mechanics $\int \ldots d\tau$ implies integration over the full range of all relevant spatial variables.

It is easy to confirm that the position operator ($x \times$) is hermitian because in this case the order of the factors in the integrand can be changed:

$$\int \psi_i^* x \psi_j \, d\tau = \int \psi_j \, x \psi_i^* d\tau = \left\{ \int \psi_j^* x \psi_i \, d\tau \right\}^*$$

The final step uses $(\psi^*)^* = \psi$. The demonstration that the linear momentum operator is hermitian is more involved because the order of functions being differentiated cannot be changed.

How is that done? 7C.1 Showing that the linear momentum operator is hermitian

The task is to show that

$$\int \psi_i^* \hat{p}_x \psi_j \mathrm{d}\tau = \left\{\int \psi_j^* \hat{p}_x \psi_i \mathrm{d}\tau\right\}^*$$

with $\hat{p}_x$ given in eqn 7C.3. To do so, use 'integration by parts' (see *The chemist's toolkit* 15) which, when applied to the present case, gives

$$\int \psi_i^* \hat{p}_x \psi_j \mathrm{d}\tau = \frac{\hbar}{\mathrm{i}}\int_{-\infty}^{\infty} \overbrace{\psi_i^*}^{f} \overbrace{\frac{\mathrm{d}\psi_j}{\mathrm{d}x}}^{\mathrm{d}g/\mathrm{d}x} \mathrm{d}x$$

$$= \overbrace{\frac{\hbar}{\mathrm{i}} \overbrace{\psi_i^* \psi_j}^{fg}\Big|_{-\infty}^{\infty}}^{0} - \frac{\hbar}{\mathrm{i}}\int_{-\infty}^{\infty} \overbrace{\psi_j}^{g} \overbrace{\frac{\mathrm{d}\psi_i^*}{\mathrm{d}x}}^{\mathrm{d}f/\mathrm{d}x} \mathrm{d}x$$

The blue term is zero because all wavefunctions are either zero at $x=\pm\infty$ (see Topic 7B) or the product $\psi_i^* \psi_j$ converges to the same value at $x=+\infty$ and $x=-\infty$. As a result

$$\int \psi_i^* \hat{p}_x \psi_j \mathrm{d}\tau = -\frac{\hbar}{\mathrm{i}}\int_{-\infty}^{\infty} \psi_j \frac{\mathrm{d}\psi_i^*}{\mathrm{d}x} \mathrm{d}x = \left\{\frac{\hbar}{\mathrm{i}}\int_{-\infty}^{\infty} \psi_j^* \frac{\mathrm{d}\psi_i}{\mathrm{d}x} \mathrm{d}x\right\}^*$$

$$= \left\{\int \psi_j^* \hat{p}_x \psi_i \mathrm{d}\tau\right\}^*$$

as was to be proved. The final line uses $(\psi^*)^* = \psi$ and $\mathrm{i}^* = -\mathrm{i}$.

Hermitian operators are enormously important in quantum mechanics because their eigenvalues are *real*: that is, $\omega^* = \omega$. Any measurement must yield a real value because a position, momentum, or an energy cannot be complex or imaginary. Because the outcome of a measurement of an observable is one of the eigenvalues of the corresponding operator, those eigenvalues must be real. It therefore follows that an operator that represents an observable must be hermitian. The proof that their eigenfunctions are real makes use of the definition of hermiticity in eqn 7C.7.

How is that done? 7C.2 Showing that the eigenvalues of hermitian operators are real

Begin by setting ψ_i and ψ_j to be the same, writing them both as ψ. Then eqn 7C.7 becomes

$$\int \psi^* \hat{\Omega} \psi \mathrm{d}\tau = \left\{\int \psi^* \hat{\Omega} \psi \mathrm{d}\tau\right\}^*$$

Next suppose that ψ is an eigenfunction of $\hat{\Omega}$ with eigenvalue ω. That is, $\hat{\Omega}\psi = \omega\psi$. Now use this relation in both integrals on the left- and right-hand sides:

$$\int \psi^* \omega \psi \mathrm{d}\tau = \left\{\int \psi^* \omega \psi \mathrm{d}\tau\right\}^*$$

The eigenvalue is a constant that can be taken outside the integrals:

$$\omega \int \psi^* \psi \mathrm{d}\tau = \left\{\omega \int \psi^* \psi \mathrm{d}\tau\right\}^* = \omega^* \int \psi \psi^* \mathrm{d}\tau$$

Finally, the (blue) integrals cancel, leaving $\omega = \omega^*$. It follows that ω is real.

(d) Orthogonality

To say that two different functions ψ_i and ψ_j are **orthogonal** means that the integral (over all space) of $\psi_i^* \psi_j$ is zero:

$$\int \psi_i^* \psi_j \mathrm{d}\tau = 0 \quad \text{for } i \neq j \qquad \text{Orthogonality [definition]} \qquad (7C.8)$$

Functions that are both normalized and mutually orthogonal are called **orthonormal**. Hermitian operators have the important property that

> Eigenfunctions that correspond to different eigenvalues of a hermitian operator are orthogonal.

The proof of this property also follows from the definition of hermiticity (eqn 7C.7).

The chemist's toolkit 15 Integration by parts

Many integrals in quantum mechanics have the form $\int f(x)h(x)\mathrm{d}x$ where $f(x)$ and $h(x)$ are two different functions. Such integrals can often be evaluated by regarding $h(x)$ as the derivative of another function, $g(x)$, such that $h(x) = \mathrm{d}g(x)/\mathrm{d}x$. For instance, if $h(x) = x$, then $g(x) = \frac{1}{2}x^2$. The integral is then found using **integration by parts**:

$$\int f \frac{\mathrm{d}g}{\mathrm{d}x} \mathrm{d}x = fg - \int g \frac{\mathrm{d}f}{\mathrm{d}x} \mathrm{d}x$$

The procedure is successful only if the integral on the right turns out to be one that can be evaluated more easily than the one on the left. The procedure is often summarized by expressing this relation as

$$\int f \mathrm{d}g = fg - \int g \mathrm{d}f$$

As an example, consider integration of $x\mathrm{e}^{-ax}$. In this case, $f(x) = x$, so $\mathrm{d}f(x)/\mathrm{d}x = 1$ and $\mathrm{d}g(x)/\mathrm{d}x = \mathrm{e}^{-ax}$, so $g(x) = -(1/a)\mathrm{e}^{-ax}$. Then

$$\int \overbrace{x}^{f} \overbrace{\mathrm{e}^{-ax}}^{\mathrm{d}g/\mathrm{d}x} \mathrm{d}x = \overbrace{x}^{f} \overbrace{\frac{-\mathrm{e}^{-ax}}{a}}^{g} - \int \overbrace{\frac{-\mathrm{e}^{-ax}}{a}}^{g} \overbrace{1}^{\mathrm{d}f/\mathrm{d}x} \mathrm{d}x$$

$$= -\frac{x\mathrm{e}^{-ax}}{a} + \frac{1}{a}\int \mathrm{e}^{-ax} \mathrm{d}x = -\frac{x\mathrm{e}^{-ax}}{a} - \frac{\mathrm{e}^{-ax}}{a^2} + \text{constant}$$

How is that done? 7C.3 Showing that the eigenfunctions of hermitian operators are orthogonal

Start by supposing that ψ_j is an eigenfunction of $\hat{\Omega}$ with eigenvalue ω_j (i.e. $\hat{\Omega}\psi_j = \omega_j\psi_j$) and that ψ_i is an eigenfunction with a different eigenvalue ω_i (i.e. $\hat{\Omega}\psi_i = \omega_i\psi_i$, with $\omega_i \neq \omega_j$). Then eqn 7C.7 becomes

$$\int \psi_i^* \omega_j \psi_j \,\mathrm{d}\tau = \left\{\int \psi_j^* \omega_i \psi_i \,\mathrm{d}\tau\right\}^*$$

The eigenvalues are constants and can be taken outside the integrals; moreover, they are real (being the eigenvalues of hermitian operators), so $\omega_i^* = \omega_i$. Then

$$\omega_j \int \psi_i^* \psi_j \,\mathrm{d}\tau = \omega_i \left\{\int \psi_j^* \psi_i \,\mathrm{d}\tau\right\}^*$$

Next, note that $\left\{\int \psi_j^* \psi_i \,\mathrm{d}\tau\right\}^* = \int \psi_j \psi_i^* \,\mathrm{d}\tau$, so

$$\omega_j \int \psi_i^* \psi_j \,\mathrm{d}\tau = \omega_i \int \psi_j \psi_i^* \,\mathrm{d}\tau, \quad \text{hence} \quad (\omega_j - \omega_i)\int \psi_i^* \psi_j \,\mathrm{d}\tau = 0$$

The two eigenvalues are different, so $\omega_j - \omega_i \neq 0$; therefore it must be the case that $\int \psi_i^* \psi_j \,\mathrm{d}\tau = 0$. That is, the two eigenfunctions are orthogonal, as was to be proved.

The hamiltonian operator is hermitian (it corresponds to an observable, the energy, but its hermiticity can be proved specifically). Therefore, if two of its eigenfunctions correspond to different energies, the two functions must be orthogonal. The property of orthogonality is of great importance in quantum mechanics because it eliminates a large number of integrals from calculations. Orthogonality plays a central role in the theory of chemical bonding (Focus 9) and spectroscopy (Focus 11).

Example 7C.2 Verifying orthogonality

Two possible wavefunctions for a particle constrained to move along the x axis between $x = 0$ and $x = L$ are $\psi_1 = \sin(\pi x/L)$ and $\psi_2 = \sin(2\pi x/L)$. Outside this region the wavefunctions are zero. The wavefunctions correspond to different energies. Verify that the two wavefunctions are mutually orthogonal.

Collect your thoughts To verify the orthogonality of two functions, you need to integrate $\psi_2^* \psi_1 = \sin(2\pi x/L)\sin(\pi x/L)$ over all space, and show that the result is zero. In principle the integral is taken from $x = -\infty$ to $x = +\infty$, but the wavefunctions are zero outside the range $x = 0$ to L so you need integrate only over this range. Relevant integrals are given in the *Resource section.*

The solution To evaluate the integral, use Integral T.5 from the *Resource section* with $a = 2\pi/L$ and $b = \pi/L$:

$$\int_0^L \sin(2\pi x/L)\sin(\pi x/L)\,\mathrm{d}x = \left.\frac{\sin(\pi x/L)}{2(\pi/L)}\right|_0^L - \left.\frac{\sin(3\pi x/L)}{2(3\pi/L)}\right|_0^L = 0$$

The sine functions have been evaluated by using $\sin n\pi = 0$ for $n = 0, \pm1, \pm2, \ldots$. The two functions are therefore mutually orthogonal.

Self-test 7C.2 The next higher energy level has $\psi_3 = \sin(3\pi x/L)$. Confirm that the functions $\psi_1 = \sin(\pi x/L)$ and $\psi_3 = \sin(3\pi x/L)$ are mutually orthogonal.

Answer: $\int_0^L \sin(3\pi x/L)\sin(\pi x/L)\,\mathrm{d}x = 0$

7C.2 Superpositions and expectation values

The hamiltonian for a free particle moving in one dimension is

$$\hat{H} = -\frac{\hbar^2}{2m}\frac{\mathrm{d}^2}{\mathrm{d}x^2}$$

The particle is 'free' in the sense that there is no potential to constrain it, hence $V(x) = 0$. It is easily confirmed that $\psi(x) = \cos kx$ is an eigenfunction of this operator

$$\hat{H}\psi(x) = -\frac{\hbar^2}{2m}\frac{\mathrm{d}^2}{\mathrm{d}x^2}\cos kx = \frac{k^2\hbar^2}{2m}\cos kx$$

The energy associated with this wavefunction, $k^2\hbar^2/2m$, is therefore well defined, as it is the eigenvalue of an eigenvalue equation. However, the same is not necessarily true of other observables. For instance, $\cos kx$ is not an eigenfunction of the linear momentum operator:

$$\hat{p}_x\psi(x) = \frac{\hbar}{\mathrm{i}}\frac{\mathrm{d}\psi}{\mathrm{d}x} = \frac{\hbar}{\mathrm{i}}\frac{\mathrm{d}\cos kx}{\mathrm{d}x} = -\frac{k\hbar}{\mathrm{i}}\sin kx \qquad (7C.9)$$

This expression is not an eigenvalue equation, because the function on the right ($\sin kx$) is different from that on the left ($\cos kx$).

When the wavefunction of a particle is not an eigenfunction of an operator, the corresponding observable does not have a definite value. However, in the current example the momentum is not completely indefinite because the cosine wavefunction can be written as a **linear combination**, or sum,[1] of $\mathrm{e}^{\mathrm{i}kx}$ and $\mathrm{e}^{-\mathrm{i}kx}$: $\cos kx = \frac{1}{2}(\mathrm{e}^{\mathrm{i}kx} + \mathrm{e}^{-\mathrm{i}kx})$ (see *The chemist's toolkit* 16). As shown in *Example* 7C.1, these two exponential functions are eigenfunctions of $\hat{p}_x$ with eigenvalues $+k\hbar$ and $-k\hbar$, respectively. They therefore each correspond to a state of definite but different momentum. The wavefunction $\cos kx$ is said to be a **superposition** of the two individual wavefunctions $\mathrm{e}^{\mathrm{i}kx}$ and $\mathrm{e}^{-\mathrm{i}kx}$, and is written

$$\psi = \underbrace{\mathrm{e}^{+\mathrm{i}kx}}_{\substack{\text{Particle with linear}\\ \text{momentum } +k\hbar}} + \underbrace{\mathrm{e}^{-\mathrm{i}kx}}_{\substack{\text{Particle with linear}\\ \text{momentum } -k\hbar}}$$

The interpretation of this superposition is that if many repeated measurements of the momentum are made, then half the measurements would give the value $p_x = +k\hbar$, and half would give the value $p_x = -k\hbar$. The two values $\pm k\hbar$ occur equally often since $\mathrm{e}^{\mathrm{i}kx}$ and $\mathrm{e}^{-\mathrm{i}kx}$ contribute equally to the superposition. All that can be inferred from the wavefunction $\cos kx$ about the linear momentum is that the particle it describes is equally

[1] A linear combination is more general than a sum, for it includes weighted sums of the form $ax + by + \cdots$ where $a, b, \ldots$ are constants. A sum is a linear combination with $a = b = \cdots = 1$.

The chemist's toolkit 16 Euler's formula

A complex number $z = x + iy$ can be represented as a point in a plane, the **complex plane**, with Re(z) along the x-axis and Im(z) along the y-axis (Sketch 1). The position of the point can also be specified in terms of a distance r and an angle ϕ (the polar coordinates). Then $x = r\cos\phi$ and $y = r\sin\phi$, so it follows that

$$z = r(\cos\phi + i\sin\phi)$$

The angle ϕ, called the **argument** of z, is the angle that r makes with the x-axis. Because $y/x = \tan\phi$, it follows that

$$r = (x^2 + y^2)^{1/2} = |z| \qquad \phi = \arctan\frac{y}{x}$$

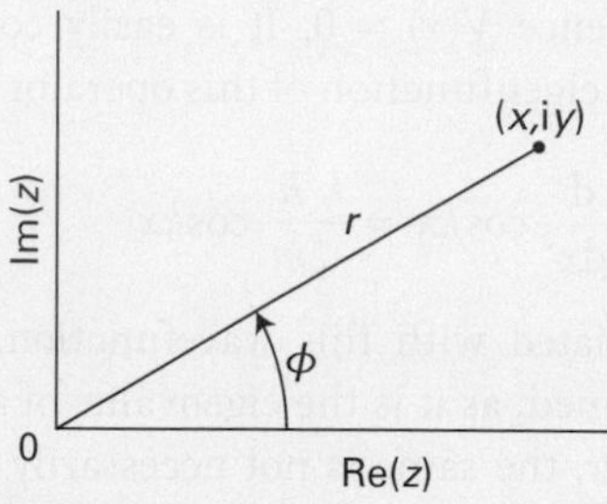

Sketch 1

One of the most useful relations involving complex numbers is **Euler's formula**:

$$e^{i\phi} = \cos\phi + i\sin\phi$$

from which it follows that $z = r(\cos\phi + i\sin\phi)$ can be written

$$z = re^{i\phi}$$

Two more useful relations arise by noting that $e^{-i\phi} = \cos(-\phi) + i\sin(-\phi) = \cos\phi - i\sin\phi$; it then follows that

$$\cos\phi = \tfrac{1}{2}(e^{i\phi} + e^{-i\phi}) \qquad \sin\phi = -\tfrac{1}{2}i(e^{i\phi} - e^{-i\phi})$$

The polar form of a complex number is commonly used to perform arithmetical operations. For instance, the product of two complex numbers in polar form is

$$z_1z_2 = (r_1e^{i\phi_1})(r_2e^{i\phi_2}) = r_1r_2e^{i(\phi_1+\phi_2)}$$

This construction is illustrated in Sketch 2.

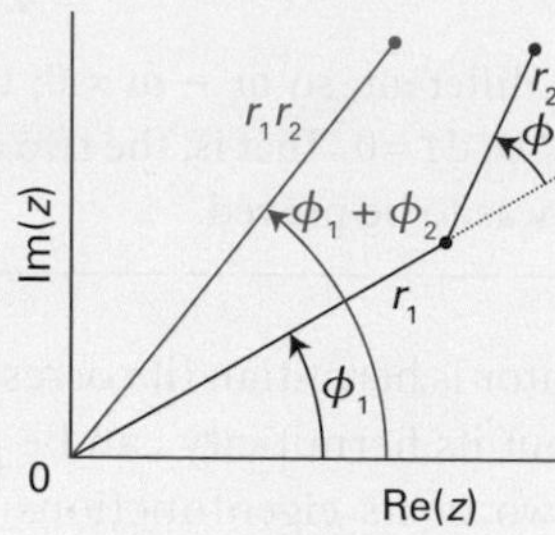

Sketch 2

likely to be found travelling in the positive and negative x directions, with the same magnitude, $k\hbar$, of the momentum.

A similar interpretation applies to any wavefunction written as a linear combination of eigenfunctions of an operator. In general, a wavefunction can be written as the following linear combination

$$\psi = c_1\psi_1 + c_2\psi_2 + \cdots = \sum_k c_k\psi_k \qquad \text{Linear combination of eigenfunctions} \qquad (7C.10)$$

where the c_k are numerical (possibly complex) coefficients and the ψ_k are different eigenfunctions of the operator $\hat{\Omega}$ corresponding to the observable of interest. The functions ψ_k are said to form a **complete set** in the sense that any arbitrary function can be expressed as a linear combination of them. Then, according to quantum mechanics:

Physical interpretation

- A single measurement of the observable corresponding to the operator $\hat{\Omega}$ will give one of the eigenvalues corresponding to the ψ_k that contribute to the superposition.
- The probability of measuring a specific eigenvalue in a series of measurements is proportional to the square modulus ($|c_k|^2$) of the corresponding coefficient in the linear combination.

The average value of a large number of measurements of an observable Ω is called the **expectation value** of the operator $\hat{\Omega}$, and is written $\langle\Omega\rangle$. For a normalized wavefunction ψ, the expectation value of $\hat{\Omega}$ is calculated by evaluating the integral

$$\langle\Omega\rangle = \int \psi^*\hat{\Omega}\psi\,d\tau \qquad \text{Expectation value [normalized wavefunction, definition]} \qquad (7C.11)$$

This definition can be justified by considering two cases, one where the wavefunction is an eigenfunction of the operator $\hat{\Omega}$ and another where the wavefunction is a superposition of that operator's eigenfunctions.

How is that done? 7C.4 Justifying the expression for the expectation value of an operator

If the wavefunction ψ is an eigenfunction of $\hat{\Omega}$ with eigenvalue ω (so $\hat{\Omega}\psi = \omega\psi$),

$$\langle\Omega\rangle = \int \psi^*\overbrace{\hat{\Omega}\psi}^{\omega\psi}\,d\tau = \int \psi^*\omega\psi\,d\tau \overset{\omega\text{ a constant}}{=} \omega\int\psi^*\psi\,d\tau \overset{\psi\text{ normalized}}{=} \omega$$

The interpretation of this expression is that, because the wavefunction is an eigenfunction of $\hat{\Omega}$, each observation of the property Ω results in the same value ω; the average value of all the observations is therefore ω.

Now suppose the (normalized) wavefunction is the linear combination of two eigenfunctions of the operator $\hat{\Omega}$, each of which is individually normalized to 1. Then

$$\begin{aligned}\langle\Omega\rangle &= \int(c_1\psi_1+c_2\psi_2)^*\hat{\Omega}(c_1\psi_1+c_2\psi_2)\mathrm{d}\tau \\ &= \int(c_1\psi_1+c_2\psi_2)^*\Big(c_1\overbrace{\hat{\Omega}\psi_1}^{\omega_1\psi_1}+c_2\overbrace{\hat{\Omega}\psi_2}^{\omega_2\psi_2}\Big)\mathrm{d}\tau \\ &= \int(c_1\psi_1+c_2\psi_2)^*(c_1\omega_1\psi_1+c_2\omega_2\psi_2)\mathrm{d}\tau \\ &= c_1^*c_1\omega_1\overbrace{\int\psi_1^*\psi_1\mathrm{d}\tau}^{1}+c_2^*c_2\omega_2\overbrace{\int\psi_2^*\psi_2\mathrm{d}\tau}^{1} \\ &\quad +c_1^*c_2\omega_2\overbrace{\int\psi_1^*\psi_2\mathrm{d}\tau}^{0}+c_2^*c_1\omega_1\overbrace{\int\psi_2^*\psi_1\mathrm{d}\tau}^{0}\end{aligned}$$

The first two integrals on the right are both equal to 1 because the wavefunctions ψ_1 and ψ_2 are individually normalized. Because ψ_1 and ψ_2 correspond to different eigenvalues of a hermitian operator, they are orthogonal, so the third and fourth integrals on the right are zero. Therefore

$$\langle\Omega\rangle = |c_1|^2\omega_1 + |c_2|^2\omega_2$$

The interpretation of this expression is that in a series of measurements each individual measurement yields either ω_1 or ω_2, but that the probability of ω_1 occurring is $|c_1|^2$, and likewise the probability of ω_2 occurring is $|c_2|^2$. The average is the sum of the two eigenvalues, but with each weighted according to the probability that it will occur in a measurement:

average = (probability of ω_1 occurring) × ω_1
+ (probability of ω_2 occurring) × ω_2

The expectation value therefore predicts the result of taking a series of measurements, each of which gives an eigenvalue, and then taking the weighted average of these values. This justifies the form of eqn 7C.11.

Example 7C.3 Calculating an expectation value

Calculate the average value of the position of an electron in the lowest energy state of a one-dimensional box of length L, with the (normalized) wavefunction $\psi=(2/L)^{1/2}\sin(\pi x/L)$ inside the box and zero outside it.

Collect your thoughts The average value of the position is the expectation value of the operator corresponding to position, which is multiplication by x. To evaluate $\langle x\rangle$, you need to evaluate the integral in eqn 7C.11 with $\hat{\Omega}=\hat{x}=x\times$

The solution The expectation value of position is

$$\langle x\rangle=\int_0^L\psi^*\hat{x}\psi\,\mathrm{d}x \qquad \text{with } \psi=\left(\frac{2}{L}\right)^{1/2}\sin\frac{\pi x}{L} \quad \text{and } \hat{x}=x\times$$

The integral is restricted to the region $x = 0$ to $x = L$ because outside this region the wavefunction is zero. Use Integral T.11 from the *Resources section* to obtain

$$\langle x\rangle=\frac{2}{L}\overbrace{\int_0^L x\sin^2\frac{\pi x}{L}\mathrm{d}x}^{\text{Integral T.11}}=\frac{2}{L}\frac{L^2}{4}=\tfrac{1}{2}L$$

Comment. This result means that if a very large number of measurements of the position of the electron are made, then the mean value will be at the centre of the box. However, each different observation will give a different and unpredictable individual result somewhere in the range $0 \le x \le L$ because the wavefunction is not an eigenfunction of the operator corresponding to x.

Self-test 7C.3 Evaluate the mean square position, $\langle x^2\rangle$, of the electron; you will need Integral T.12 from the *Resource section*.

Answer: $L^2\{\tfrac{1}{3}-\tfrac{1}{2}\pi^2\}=0.217L^2$

The mean kinetic energy of a particle in one dimension is the expectation value of the operator given in eqn 7C.5. Therefore,

$$\langle E_\mathrm{k}\rangle=\int_{-\infty}^{\infty}\psi^*\hat{E}_\mathrm{k}\psi\,\mathrm{d}x=-\frac{\hbar^2}{2m}\int_{-\infty}^{\infty}\psi^*\frac{\mathrm{d}^2\psi}{\mathrm{d}x^2}\mathrm{d}x \qquad (7\text{C}.12)$$

This conclusion confirms the previous assertion that the kinetic energy is a kind of average over the curvature of the wavefunction: a large contribution to the observed value comes from regions where the wavefunction is sharply curved (so $\mathrm{d}^2\psi/\mathrm{d}x^2$ is large) and the wavefunction itself is large (so that ψ^* is large there too).

7C.3 The uncertainty principle

The wavefunction $\psi=\mathrm{e}^{ikx}$ is an eigenfunction of $\hat{p}_x$ with eigenvalue $+k\hbar$: in this case the wavefunction describes a particle with a definite state of linear momentum. Where, though, is the particle? The probability density is proportional to $\psi^*\psi$, so if the particle is described by the wavefunction e^{ikx} the probability density is proportional to $(\mathrm{e}^{ikx})^*\mathrm{e}^{ikx}=\mathrm{e}^{-ikx}\mathrm{e}^{ikx}=\mathrm{e}^{-ikx+ikx}=\mathrm{e}^0=1$. In other words, the probability density is the same for all values of x: the location of the particle is completely unpredictable. In summary, if the momentum of the particle is known precisely, it is not possible to predict its location.

This conclusion is an example of the consequences of the **Heisenberg uncertainty principle**, one of the most celebrated results of quantum mechanics:

> It is impossible to specify simultaneously, with arbitrary precision, both the linear momentum and the position of a particle.

Heisenberg uncertainty principle

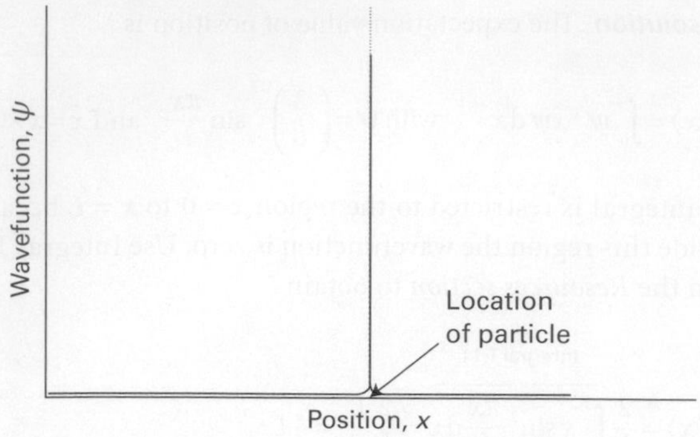

Figure 7C.4 The wavefunction of a particle at a well-defined location is a sharply spiked function that has zero amplitude everywhere except at the position of the particle.

Note that the uncertainty principle also implies that if the position is known precisely, then the momentum cannot be predicted. The argument runs as follows.

Suppose the particle is known to be at a definite location, then its wavefunction must be large there and zero everywhere else (Fig. 7C.4). Such a wavefunction can be created by superimposing a large number of harmonic (sine and cosine) functions, or, equivalently, a number of e^{ikx} functions (because $e^{ikx} = \cos kx + i\sin kx$). In other words, a sharply localized wavefunction, called a **wavepacket**, can be created by forming a linear combination of wavefunctions that correspond to many different linear momenta.

The superposition of a few harmonic functions gives a wavefunction that spreads over a range of locations (Fig. 7C.5). However, as the number of wavefunctions in the superposition increases, the wavepacket becomes sharper on account of the more complete interference between the positive and negative regions of the individual waves. When an infinite number of components are used, the wavepacket is a sharp, infinitely narrow spike, which corresponds to perfect localization of the particle. Now the particle is perfectly localized but all information about its momentum has been lost. A measurement of the momentum will give a result corresponding to any one of the infinite number of waves in the superposition, and which one it will give is unpredictable. Hence, if the location of the particle is known precisely (implying that its wavefunction is a superposition of an infinite number of momentum eigenfunctions), then its momentum is completely unpredictable.

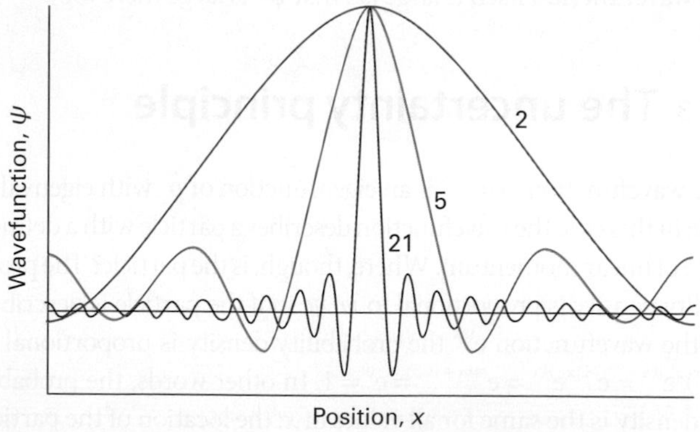

Figure 7C.5 The wavefunction of a particle with an ill-defined location can be regarded as a superposition of several wavefunctions of definite wavelength that interfere constructively in one place but destructively elsewhere. As more waves are used in the superposition (as given by the numbers attached to the curves), the location becomes more precise at the expense of uncertainty in the momentum of the particle. An infinite number of waves are needed in the superposition to construct the wavefunction of the perfectly localized particle.

The quantitative version of the uncertainty principle is

$$\Delta p_q \Delta q \geq \tfrac{1}{2}\hbar \qquad \text{Heisenberg uncertainty principle} \qquad (7C.13a)$$

In this expression Δp_q is the 'uncertainty' in the linear momentum parallel to the axis q, and Δq is the uncertainty in position along that axis. These 'uncertainties' are given by the root-mean-square deviations of the observables from their mean values:

$$\Delta p_q = \{\langle p_q^2\rangle - \langle p_q\rangle^2\}^{1/2} \qquad \Delta q = \{\langle q^2\rangle - \langle q\rangle^2\}^{1/2} \qquad (7C.13b)$$

If there is complete certainty about the position of the particle ($\Delta q = 0$), then the only way that eqn 7C.13a can be satisfied is for $\Delta p_q = \infty$, which implies complete uncertainty about the momentum. Conversely, if the momentum parallel to an axis is known exactly ($\Delta p_q = 0$), then the position along that axis must be completely uncertain ($\Delta q = \infty$).

The p and q that appear in eqn 7C.13a refer to the same direction in space. Therefore, whereas simultaneous specification of the position on the x-axis and momentum parallel to the x-axis are restricted by the uncertainty relation, simultaneous location of position on x and motion parallel to y or z are not restricted.

Example 7C.4 Using the uncertainty principle

Suppose the speed of a projectile of mass 1.0 g is known to within 1 μm s⁻¹. What is the minimum uncertainty in its position?

Collect your thoughts You can estimate Δp from $m\Delta v$, where Δv is the uncertainty in the speed; then use eqn 7C.13a to estimate the *minimum* uncertainty in position, Δq, by using it in the form $\Delta p\Delta q = \tfrac{1}{2}\hbar$ rearranged into $\Delta q = \tfrac{1}{2}\hbar/\Delta p$. You will need to use $1\,\text{J} = 1\,\text{kg}\,\text{m}^2\,\text{s}^{-2}$.

The solution The minimum uncertainty in position is

$$\Delta q = \frac{\hbar}{2m\Delta v}$$

$$= \frac{1.055\times10^{-34}\,\text{J s}}{2\times(1.0\times10^{-3}\,\text{kg})\times(1\times10^{-6}\,\text{m s}^{-1})} = 5\times10^{-26}\,\text{m}$$

Comment. This uncertainty is completely negligible for all practical purposes. However, if the mass is that of an electron, then the same uncertainty in speed implies an uncertainty in position far larger than the diameter of an atom (the analogous calculation gives $\Delta q = 60\,\text{m}$).

Self-test 7C.4 Estimate the minimum uncertainty in the speed of an electron in a one-dimensional region of length $2a_0$, the approximate diameter of a hydrogen atom, where a_0 is the Bohr radius, 52.9 pm.

Answer: $500\,\text{km s}^{-1}$

The Heisenberg uncertainty principle is more general than even eqn 7C.13a suggests. It applies to any pair of observables, called **complementary observables**, for which the corresponding operators $\hat{\Omega}_1$ and $\hat{\Omega}_2$ have the property

$$\hat{\Omega}_1\hat{\Omega}_2\psi \neq \hat{\Omega}_2\hat{\Omega}_1\psi \qquad \text{Complementarity of observables} \qquad (7\text{C}.14)$$

The term on the left implies that $\hat{\Omega}_2$ acts first, then $\hat{\Omega}_1$ acts on the result, and the term on the right implies that the operations are performed in the opposite order. When the effect of two operators applied in succession depends on their order (as this equation implies), they do not **commute**. The different outcomes of the effect of applying $\hat{\Omega}_1$ and $\hat{\Omega}_2$ in a different order are expressed by introducing the **commutator** of the two operators, which is defined as

$$[\hat{\Omega}_1,\hat{\Omega}_2] = \hat{\Omega}_1\hat{\Omega}_2 - \hat{\Omega}_2\hat{\Omega}_1 \qquad \text{Commutator [definition]} \qquad (7\text{C}.15)$$

By using the definitions of the operators for position and momentum, an explicit value of this commutator can be found.

How is that done? 7C.5 Evaluating the commutator of position and momentum

You need to consider the effect of $\hat{x}\hat{p}_x$ (i.e. the effect of $\hat{p}_x$ followed by the effect on the outcome of multiplication by x) on an arbitrary wavefunction ψ, which need not be an eigenfunction of either operator.

$$\hat{x}\hat{p}_x\psi = x\times\frac{\hbar}{\text{i}}\frac{\text{d}\psi}{\text{d}x}$$

Then you need to consider the effect of $\hat{p}_x\hat{x}$ on the same function (that is, the effect of multiplication by x followed by the effect of $\hat{p}_x$ on the outcome):

$$\hat{p}_x\hat{x}\psi = \frac{\hbar}{\text{i}}\frac{\text{d}(x\psi)}{\text{d}x} \overset{\text{d}(fg)/\text{d}x=(\text{d}f/\text{d}x)g+f(\text{d}g/\text{d}x)}{=} \frac{\hbar}{\text{i}}\left(\psi + x\frac{\text{d}\psi}{\text{d}x}\right)$$

The second expression is different from the first, so $\hat{p}_x\hat{x}\psi \neq \hat{x}\hat{p}_x\psi$ and therefore the two operators do not commute. You can infer the value of the commutator from the difference of the two expressions:

$$[\hat{x},\hat{p}_x]\psi = \hat{x}\hat{p}_x\psi - \hat{p}_x\hat{x}\psi = -\frac{\hbar}{\text{i}}\psi = \text{i}\hbar\psi, \quad \text{so } [\hat{x},\hat{p}_x]\psi = \text{i}\hbar\psi$$

This relation is true for any wavefunction ψ, so the commutator is

$$[\hat{x},\hat{p}_x] = \text{i}\hbar \qquad \text{Commutator of position and momentum operators} \qquad (7\text{C}.16)$$

The commutator in eqn 7C.16 is of such central significance in quantum mechanics that it is taken as a fundamental distinction between classical mechanics and quantum mechanics. In fact, this commutator may be taken as a postulate of quantum mechanics and used to justify the choice of the operators for position and linear momentum in eqn 7C.3.

Classical mechanics supposed, falsely as is now known, that the position and momentum of a particle could be specified simultaneously with arbitrary precision. However, quantum mechanics shows that position and momentum are complementary, and that a choice must be made: position can be specified, but at the expense of momentum, or momentum can be specified, but at the expense of position.

7C.4 The postulates of quantum mechanics

The principles of quantum theory can be summarized as a series of postulates, which will form the basis for chemical applications of quantum mechanics throughout the text.

The wavefunction: All dynamical information is contained in the wavefunction ψ for the system, which is a mathematical function found by solving the appropriate Schrödinger equation for the system.

The Born interpretation: If the wavefunction of a particle has the value ψ at some position $\boldsymbol{r}$, then the probability of finding the particle in an infinitesimal volume $\text{d}\tau = \text{d}x\text{d}y\text{d}z$ at that position is proportional to $|\psi|^2\text{d}\tau$.

Acceptable wavefunctions: An acceptable wavefunction must be single-valued, continuous, not infinite over a finite region of space, and (except in special cases) have a continuous slope.

Observables: Observables, Ω, are represented by hermitian operators, $\hat{\Omega}$, built from the position and momentum operators specified in eqn 7C.3.

Observations and expectation values: A single measurement of the observable represented by the operator $\hat{\Omega}$ gives one of the eigenvalues of $\hat{\Omega}$. If the wavefunction is not an eigenfunction of $\hat{\Omega}$, the average of many measurements is given by the expectation value, $\langle\Omega\rangle$, defined in eqn 7C.11.

The Heisenberg uncertainty principle: It is impossible to specify simultaneously, with arbitrary precision, both the linear momentum and the position of a particle and, more generally, any pair of observables represented by operators that do not commute.

Checklist of concepts

- ☐ 1. The Schrödinger equation is an **eigenvalue equation.**
- ☐ 2. An **operator** carries out a mathematical operation on a function.
- ☐ 3. The **hamiltonian operator** is the operator corresponding to the total energy of the system, the sum of the kinetic and potential energies.
- ☐ 4. The wavefunction corresponding to a specific energy is an **eigenfunction** of the hamiltonian operator.
- ☐ 5. Two different functions are **orthogonal** if the integral (over all space) of their product is zero.
- ☐ 6. **Hermitian operators** have real eigenvalues and orthogonal eigenfunctions.
- ☐ 7. **Observables** are represented by hermitian operators.
- ☐ 8. Sets of functions that are normalized and mutually orthogonal are called **orthonormal**.
- ☐ 9. When the system is not described by a single eigenfunction of an operator, it may be expressed as a **superposition** of such eigenfunctions.
- ☐ 10. The mean value of a series of observations is given by the **expectation value** of the corresponding operator.
- ☐ 11. The **uncertainty principle** restricts the precision with which complementary observables may be specified and measured simultaneously.
- ☐ 12. **Complementary observables** are observables for which the corresponding operators do not commute.

Checklist of equations

Property	Equation	Comment	Equation number
Eigenvalue equation	$\hat{\Omega}\psi=\omega\psi$	ψ eigenfunction; ω eigenvalue	7C.2b
Hermiticity	$\int\psi_i^*\hat{\Omega}\psi_j\mathrm{d}\tau=\left\{\int\psi_j^*\hat{\Omega}\psi_i\mathrm{d}\tau\right\}^*$	Hermitian operators have real eigenvalues and orthogonal eigenfunctions	7C.7
Orthogonality	$\int\psi_i^*\psi_j\mathrm{d}\tau=0$ for $i\neq j$	Integration over all space	7C.8
Expectation value	$\langle\Omega\rangle=\int\psi^*\hat{\Omega}\psi\mathrm{d}\tau$	Definition; assumes ψ normalized	7C.11
Heisenberg uncertainty principle	$\Delta p_q\Delta q\geq\frac{1}{2}\hbar$	For position and momentum	7C.13a
Commutator of two operators	$[\hat{\Omega}_1,\hat{\Omega}_2]=\hat{\Omega}_1\hat{\Omega}_2-\hat{\Omega}_2\hat{\Omega}_1$	The observables are complementary if $[\hat{\Omega}_1,\hat{\Omega}_2]\neq0$	7C.15
	Special case: $[\hat{x},\hat{p}_x]=\mathrm{i}\hbar$		7C.16

TOPIC 7D Translational motion

> ➤ **Why do you need to know this material?**
>
> The application of quantum theory to translational motion reveals the origin of quantization and non-classical features, such as tunnelling and zero-point energy. This material is important for the discussion of atoms and molecules that are free to move within a restricted volume, such as a gas in a container.
>
> ➤ **What is the key idea?**
>
> The translational energy levels of a particle confined to a finite region of space are quantized, and under certain conditions particles can pass into and through classically forbidden regions.
>
> ➤ **What do you need to know already?**
>
> You should know that the wavefunction is the solution of the Schrödinger equation (Topic 7B), and be familiar, in one instance, with the techniques of deriving dynamical properties from the wavefunction by using the operators corresponding to the observables (Topic 7C).

Translation, motion through space, is one of the basic types of motion. Quantum mechanics, however, shows that translation can have a number of non-classical features, such as its confinement to discrete energies and passage into and through classically forbidden regions.

7D.1 Free motion in one dimension

A free particle is unconstrained by any potential, which may be taken to be zero everywhere. In one dimension $V(x) = 0$ everywhere, so the Schrödinger equation becomes (Topic 7B)

$$-\frac{\hbar^2}{2m}\frac{d^2\psi(x)}{dx^2} = E\psi(x) \qquad \text{Free motion in one dimension} \qquad (7D.1)$$

The most straightforward way to solve this simple second-order differential equation is to take the known general form of solutions of equations of this kind, and then show that it does indeed satisfy eqn 7D.1.

How is that done? 7D.1 Finding the solutions to the Schrödinger equation for a free particle in one dimension

The general solution of a second-order differential equation of the kind shown in eqn 7D.1 is

$$\psi_k(x) = Ae^{ikx} + Be^{-ikx}$$

where k, A, and B are constants. You can verify that $\psi_k(x)$ is a solution of eqn 7D.1 by substituting it into the left-hand side of the equation, evaluating the derivatives, and then confirming that you have generated the right-hand side. Because $de^{\pm ax}/dx = \pm ae^{\pm ax}$, the left-hand side becomes

$$-\frac{\hbar^2}{2m}\frac{d^2}{dx^2}\overbrace{(Ae^{ikx}+Be^{-ikx})}^{\psi_k(x)} = -\frac{\hbar^2}{2m}\{A(ik)^2e^{ikx}+B(-ik)^2e^{-ikx}\}$$

$$= \overbrace{\frac{k^2\hbar^2}{2m}}^{E_k}\overbrace{(Ae^{ikx}+Be^{-ikx})}^{\psi_k(x)}$$

The left-hand side is therefore equal to a constant × $\psi_k(x)$, which is the same as the term on the right-hand side of eqn 7D.1 provided the constant, the term in blue, is identified with E. The value of the energy depends on the value of k, so henceforth it will be written E_k. The wavefunctions and energies of a free particle are therefore

$$\psi_k(x) = Ae^{ikx} + Be^{-ikx} \qquad E_k = \frac{k^2\hbar^2}{2m} \qquad (7D.2)$$

Wavefunctions and energies [one dimension]

The wavefunctions in eqn 7D.2 are continuous, have continuous slope everywhere, are single-valued, and do not go to infinity: they are therefore acceptable wavefunctions for all values of k. Because k can take any value, the energy can take any non-negative value, including zero. As a result, *the translational energy of a free particle is not quantized.*

In Topic 7C it is explained that in general a wavefunction can be written as a superposition (a linear combination) of the eigenfunctions of an operator. The wavefunctions of eqn 7D.2 can be recognized as superpositions of the two functions $e^{\pm ikx}$ which are eigenfunctions of the linear momentum operator with eigenvalues $\pm k\hbar$ (Topic 7C). These eigenfunctions correspond to states with definite linear momentum:

$$\psi_k(x) = \underbrace{A\,e^{+ikx}}_{\text{Particle with linear momentum } +k\hbar} + \underbrace{B\,e^{-ikx}}_{\text{Particle with linear momentum } -k\hbar}$$

According to the interpretation given in Topic 7C, if a system is described by the wavefunction $\psi_k(x)$, then repeated measurements of the momentum will give $+k\hbar$ (that is, the particle travelling in the positive x-direction) with a probability proportional to A^2, and $-k\hbar$ (that is, the particle travelling in the negative x-direction) with a probability proportional to B^2. Only if A or B is zero does the particle have a definite momentum of $-k\hbar$ or $+k\hbar$, respectively.

Brief illustration 7D.1

Suppose an electron emerges from an accelerator moving towards positive x with kinetic energy 1.0 eV (1 eV = 1.602×10^{-19} J). The wavefunction for such a particle is given by eqn 7D.2 with $B=0$ because the momentum is definitely in the positive x-direction. The value of k is found by rearranging the expression for the energy in eqn 7D.2 into

$$k=\left(\frac{2m_eE_k}{\hbar^2}\right)^{1/2}=\left(\frac{2\times(9.109\times10^{-31}\,\text{kg})\times(1.6\times10^{-19}\,\text{J})}{(1.055\times10^{-34}\,\text{J s})^2}\right)^{1/2}$$

$$=5.1\times10^{9}\,\text{m}^{-1}$$

or 5.1 nm^{-1} (with 1 nm = 10^{-9} m). Therefore, the wavefunction is $\psi(x)=Ae^{5.1ix/\text{nm}}$.

So far, the motion of the particle has been confined to the x-axis. In general, the linear momentum is a vector (see *The chemist's toolkit* 17) directed along the line of travel of the particle. Then $\boldsymbol{p}=\boldsymbol{k}\hbar$ and the magnitude of the vector is $p=k\hbar$ and its component on each axis is $p_q=k_q\hbar$, with the wavefunction for each component proportional to e^{ik_qq} with $q=x$, y, or z and overall equal to $e^{i(k_xx+k_yy+k_zz)}$ [1].

The chemist's toolkit 17 Vectors

A vector is a quantity with both magnitude and direction. The vector $\boldsymbol{v}$ shown in Sketch 1 has components on the x, y, and z axes with values v_x, v_y, and v_z, respectively, which may be positive or negative. For example, if $v_x=-1.0$, the x-component of the vector $\boldsymbol{v}$ has a magnitude of 1.0 and points in the $-x$ direction. The magnitude of a vector is denoted v or $|\boldsymbol{v}|$ and is given by

$$v=(v_x^2+v_y^2+v_z^2)^{1/2}$$

Thus, a vector with components $v_x=-1.0$, $v_y=+2.5$, and $v_z=+1.1$ has magnitude 2.9 and would be represented by an arrow of length 2.9 units and the appropriate orientation (as in the inset in the Sketch). Velocity and momentum are vectors; the magnitude of a velocity vector is called the speed. Force, too, is a vector. Electric and magnetic fields are two more examples of vectors.

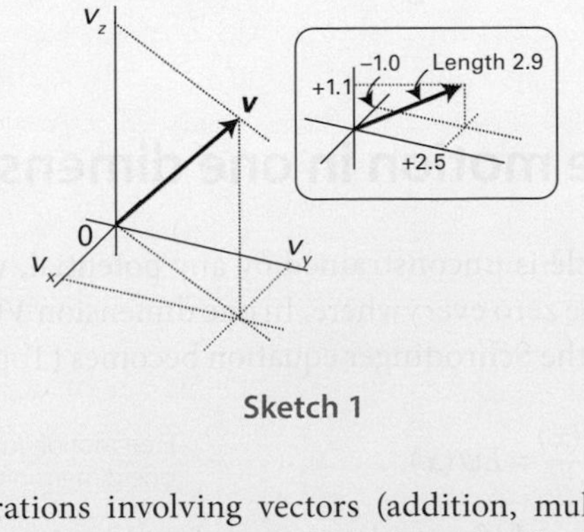

Sketch 1

The operations involving vectors (addition, multiplication, etc.) needed for this text are described in *The chemist's toolkit* 22 in Topic 8C.

[1] In terms of scalar products, this overall wavefunction would be written $e^{i\boldsymbol{k}\cdot\boldsymbol{r}}$.

7D.2 Confined motion in one dimension

Consider a **particle in a box** in which a particle of mass m is confined to a region of one-dimensional space between two impenetrable walls. The potential energy is zero inside the box but rises abruptly to infinity at the walls located at $x=0$ and $x=L$ (Fig. 7D.1). When the particle is between the walls, the Schrödinger equation is the same as for a free particle (eqn 7D.1), so the general solutions given in eqn 7D.2 are also the same. However, it will prove convenient to rewrite the wavefunction in terms of sines and cosines by using $e^{\pm ikx}=\cos kx\pm i\sin kx$ (*The chemist's toolkit* 16 in Topic 7C)

$$\begin{aligned}\psi_k(x)&=Ae^{ikx}+Be^{-ikx}\\&=A(\cos kx+i\sin kx)+B(\cos kx-i\sin kx)\\&=(A+B)\cos kx+i(A-B)\sin kx\end{aligned}$$

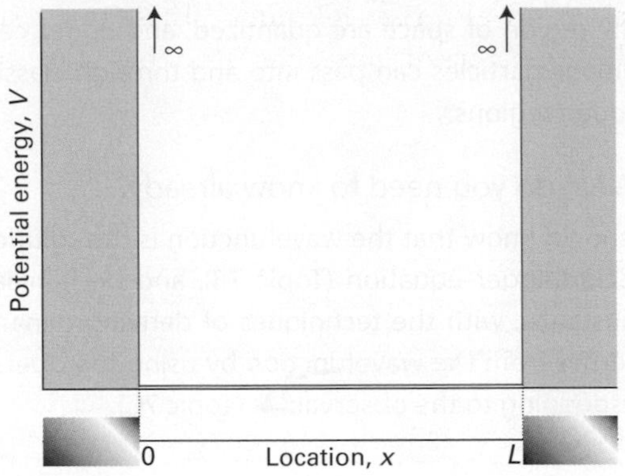

Figure 7D.1 The potential energy for a particle in a one-dimensional box. The potential is zero between $x=0$ and $x=L$, and then rises to infinity outside this region, resulting in impenetrable walls which confine the particle.

The constants $i(A - B)$ and $A + B$ can be denoted C and D, respectively, in which case

$$\psi_k(x) = C\sin kx + D\cos kx \tag{7D.3}$$

Outside the box the wavefunctions must be zero as the particle will not be found in a region where its potential energy would be infinite:

$$\text{For } x<0 \text{ and } x>L,\ \psi_k(x)=0 \tag{7D.4}$$

(a) The acceptable solutions

One of the requirements placed on a wavefunction is that it must be continuous. It follows that since the wavefunction is zero when $x < 0$ (where the potential energy is infinite) the wavefunction must be zero *at* $x = 0$ which is the point where the potential energy rises to infinity. Likewise, the wavefunction is zero where $x > L$ and so must be zero at $x = L$ where the potential energy also rises to infinity. These two restrictions are the **boundary conditions**, or constraints on the function:

$$\psi_k(0) = 0 \text{ and } \psi_k(L) = 0 \qquad \text{Boundary conditions} \tag{7D.5}$$

Now it is necessary to show that the requirement that the wavefunction must satisfy these boundary conditions implies that only certain wavefunctions are acceptable, and that as a consequence only certain energies are allowed.

How is that done? 7D.2 Showing that the boundary conditions lead to quantized levels

You need to start from the general solution and then explore the consequences of imposing the boundary conditions.

Step 1 *Apply the boundary conditions*

At $x = 0$, $\psi_k(0) = C\sin 0 + D\cos 0 = D$ (because $\sin 0 = 0$ and $\cos 0 = 1$). One boundary condition is $\psi_k(0) = 0$, so it follows that $D = 0$.

At $x = L$, $\psi_k(L) = C\sin kL$. The boundary condition $\psi_k(L) = 0$ therefore requires that $\sin kL = 0$, which in turn requires that $kL = n\pi$ with $n = 1, 2,\ldots$. Although $n = 0$ also satisfies the boundary condition it is ruled out because the wavefunction would be $C\sin 0 = 0$ for all values of x, and the particle would be found nowhere. Negative integral values of n also satisfy the boundary condition, but simply result in a change of sign of the wavefunction (because $\sin(-\theta) = -\sin\theta$). It therefore follows that the wavefunctions that satisfy the two boundary conditions are $\psi_k(x) = C\sin(n\pi x/L)$ with $n = 1, 2, \ldots$ and $k = n\pi/L$.

Step 2 *Normalize the wavefunctions*

To normalize the wavefunction, write it as $N\sin(n\pi x/L)$ and require that the integral of the square of the wavefunction over all space is equal to 1. The wavefunction is zero outside the range $0 \le x \le L$, so the integration needs to be carried out only inside this range:

$$\int_0^L \psi^2 dx = N^2 \overbrace{\int_0^L \sin^2\frac{n\pi x}{L}dx}^{\text{Integral T.2}} = N^2 \times \frac{L}{2} = 1, \text{ so } N = \left(\frac{2}{L}\right)^{1/2}$$

Step 3 *Identify the allowed energies*

According to eqn 7D.2, $E_k = k^2\hbar^2/2m$, but because k is limited to the values $k = n\pi/L$ with $n = 1, 2,\ldots$ the energies are restricted to the values

$$E_k = \frac{k^2\hbar^2}{2m} = \frac{(n\pi/L)^2(h/2\pi)^2}{2m} = \frac{n^2h^2}{8mL^2}$$

At this stage it is sensible to replace the label k by the label n, and to label the wavefunctions and energies as $\psi_n(x)$ and E_n. The allowed normalized wavefunctions and energies are therefore

$$\psi_n(x) = \left(\frac{2}{L}\right)^{1/2}\sin\left(\frac{n\pi x}{L}\right) \quad E_n = \frac{n^2h^2}{8mL^2} \quad n = 1, 2,\ldots \qquad \text{Particle in a one-dimensional box} \tag{7D.6}$$

The fact that n is restricted to positive integer values implies that the energy of the particle in a one-dimensional box is quantized. This quantization arises from the boundary conditions that ψ must satisfy. This is a general conclusion: *the need to satisfy boundary conditions implies that only certain wavefunctions are acceptable, and hence restricts the eigenvalues to discrete values.*

The integer n that has been used to label the wavefunctions and energies is an example of a 'quantum number'. In general, a **quantum number** is an integer (in some cases, Topic 8B, a half-integer) that labels the state of the system. For a particle in a one-dimensional box there are an infinite number of acceptable solutions, and the quantum number n specifies the one of interest (Fig. 7D.2).[2] As well as acting as a label, a

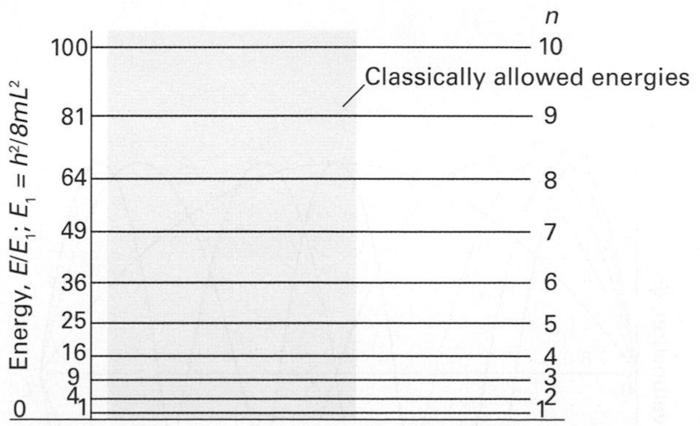

Figure 7D.2 The energy levels for a particle in a box. Note that the energy levels increase as n^2, and that their separation increases as the quantum number increases. Classically, the particle is allowed to have any value of the energy in the continuum shown as a tinted area.

[2] You might object that the wavefunctions have a discontinuous slope at the edges of the box, and so do not qualify as acceptable according to the criteria in Topic 7B. This is a rare instance where the requirement does not apply, because the potential energy suddenly jumps to an infinite value.

quantum number can often be used to calculate the value of a property, such as the energy corresponding to the state, as in eqn 7D.6b.

(b) The properties of the wavefunctions

Figure 7D.3 shows some of the wavefunctions of a particle in a one-dimensional box. The points to note are as follows.

- The wavefunctions are all sine functions with the same maximum amplitude but different wavelengths; the wavelength gets shorter as n increases.
- Shortening the wavelength results in a sharper average curvature of the wavefunction and therefore an increase in the kinetic energy of the particle (recall that, as $V = 0$ inside the box, the energy is entirely kinetic).
- The number of nodes (the points where the wavefunction passes through zero) also increases as n increases; the wavefunction ψ_n has $n - 1$ nodes.

The probability density for a particle in a one-dimensional box is

$$\psi_n^2(x) = \frac{2}{L}\sin^2\left(\frac{n\pi x}{L}\right) \qquad \text{(7D.7)}$$

and varies with position. The non-uniformity in the probability density is pronounced when n is small (Fig. 7D.4). The maxima in the probability density give the locations at which the particle has the greatest probability of being found.

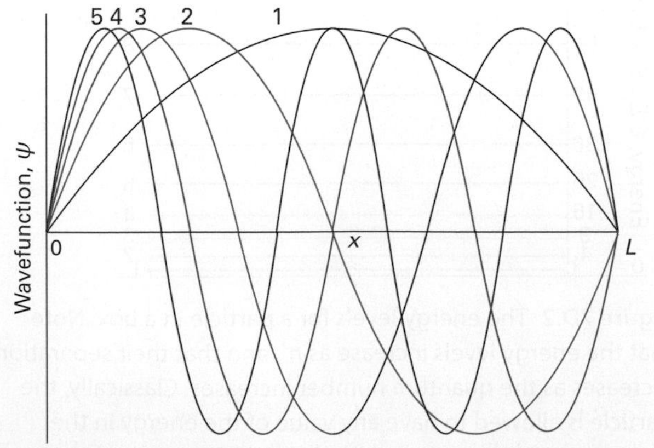

Figure 7D.3 The first five normalized wavefunctions of a particle in a box. As the energy increases the wavelength decreases, and successive functions possess one more half wave. The wavefunctions are zero outside the box.

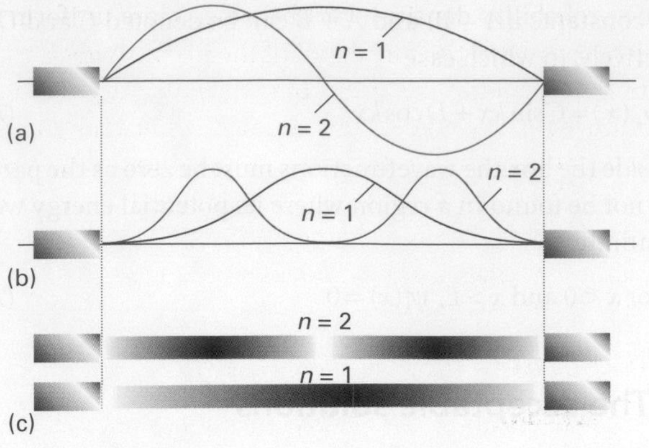

Figure 7D.4 (a) The first two wavefunctions for a particle in a box, (b) the corresponding probability densities, and (c) a representation of the probability density in terms of the darkness of shading.

Brief illustration 7D.2

As explained in Topic 7B, the total probability of finding the particle in a specified region is the integral of $\psi(x)^2\mathrm{d}x$ over that region. Therefore, the probability of finding the particle with $n = 1$ in a region between $x = 0$ and $x = L/2$ is

$$P = \int_0^{L/2} \psi_1^2 \mathrm{d}x = \overbrace{\frac{2}{L}\int_0^{L/2} \sin^2\left(\frac{\pi x}{L}\right)\mathrm{d}x}^{\text{Integral T.2}} = \frac{2}{L}\left[\frac{x}{2} - \frac{1}{4\pi/L}\sin\left(\frac{2\pi x}{L}\right)\right]_0^{L/2}$$

$$= \frac{2}{L}\left(\frac{L}{4} - \frac{1}{4\pi/L}\overbrace{\sin\pi}^{0}\right) = \tfrac{1}{2}$$

The result should not be a surprise, because the probability density is symmetrical around $x = L/2$. The probability of finding the particle between $x = 0$ and $x = L/2$ must therefore be half of the probability of finding the particle between $x = 0$ and $x = L$, which is 1.

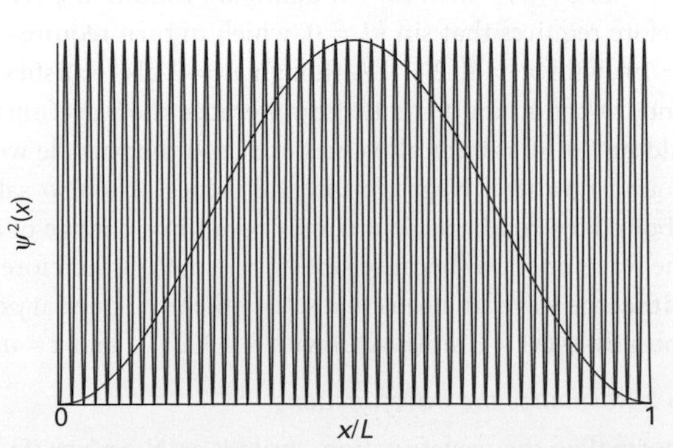

Figure 7D.5 The probability density $\psi^2(x)$ for large quantum number (here $n = 50$, blue, compared with $n = 1$, red). Notice that for high n the probability density is nearly uniform, provided the fine detail of the increasingly rapid oscillations is ignored.

The probability density $\psi_n^2(x)$ becomes more uniform as n increases provided the fine detail of the increasingly rapid oscillations is ignored (Fig. 7D.5). The probability density at high quantum numbers reflects the classical result that a particle bouncing between the walls spends equal times at all points. This conclusion is an example of the **correspondence principle**, which states that as high quantum numbers are reached, the classical result emerges from quantum mechanics.

(c) The properties of the energy

The linear momentum of a particle in a box is not well defined because the wavefunction sin kx is not an eigenfunction of the linear momentum operator. However, because $\sin kx = (e^{ikx} - e^{-ikx})/2i$,

$$\psi_n(x) = \left(\frac{2}{L}\right)^{1/2} \sin\left(\frac{n\pi x}{L}\right) = \frac{1}{2i}\left(\frac{2}{L}\right)^{1/2}(e^{in\pi x/L} - e^{-in\pi x/L}) \quad (7D.8)$$

It follows that, if repeated measurements are made of the linear momentum, half will give the value $+n\pi\hbar/L$ and half will give the value $-n\pi\hbar/L$. This conclusion is the quantum mechanical version of the classical picture in which the particle bounces back and forth in the box, spending equal times travelling to the left and to the right.

Because n cannot be zero, the lowest energy that the particle may possess is not zero (as allowed by classical mechanics, corresponding to a stationary particle) but

$$E_1 = \frac{h^2}{8mL^2} \quad \text{Zero-point energy} \quad (7D.9)$$

This lowest, irremovable energy is called the **zero-point energy**. The physical origin of the zero-point energy can be explained in two ways:

- The Heisenberg uncertainty principle states that $\Delta p_x \Delta x \geq \frac{1}{2}\hbar$. For a particle confined to a box, Δx has a finite value, therefore Δp_x cannot be zero, as that would violate the uncertainty principle. Therefore the kinetic energy cannot be zero.
- If the wavefunction is to be zero at the walls, but smooth, continuous, and not zero everywhere, then it must be curved, and curvature in a wavefunction implies the possession of kinetic energy.

Brief illustration 7D.3

The lowest energy of an electron in a region of length 100 nm is given by eqn 7D.6 with $n = 1$:

$$E_1 = \frac{(1)^2 \times (6.626\times10^{-34}\,\text{J s})^2}{8\times(9.109\times10^{-31}\,\text{kg})\times(100\times10^{-9}\,\text{m})^2} = 6.02\times10^{-24}\,\text{J}$$

where $1\,\text{J} = 1\,\text{kg m}^2\,\text{s}^{-2}$ has been used. The energy E_1 can be expressed as 6.02 yJ (1 yJ = 10^{-24} J).

The separation between adjacent energy levels with quantum numbers n and $n+1$ is

$$E_{n+1} - E_n = \frac{(n+1)^2h^2}{8mL^2} - \frac{n^2h^2}{8mL^2} = (2n+1)\frac{h^2}{8mL^2} \quad (7D.10)$$

This separation decreases as the length of the container increases, and is very small when the container has macroscopic dimensions. The separation of adjacent levels becomes zero when the walls are infinitely far apart. Atoms and molecules free to move in normal laboratory-sized vessels may therefore be treated as though their translational energy is not quantized.

Example 7D.1 Estimating an absorption wavelength

β-Carotene (**1**) is a linear polyene in which 10 single and 11 double bonds alternate along a chain of 22 carbon atoms. If each CC bond length is taken to be 140 pm, the length of the molecular box in β-carotene is $L = 2.94$ nm. Estimate the wavelength of the light absorbed by this molecule when it undergoes a transition from its ground state to the next higher excited state.

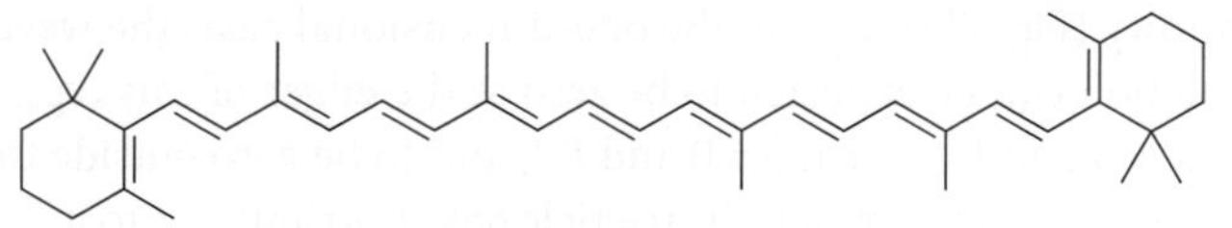

1 β-Carotene

Collect your thoughts For reasons that will be familiar from introductory chemistry, each π-bonded C atom contributes one p electron to the π-orbitals and two electrons occupy each state. Use eqn 7D.10 to calculate the energy separation between the highest occupied and the lowest unoccupied levels, and convert that energy to a wavelength by using the Bohr frequency condition (eqn 7A.9, $\Delta E = h\nu$).

The solution There are 22 C atoms in the conjugated chain; each contributes one p electron to the levels, so each level up to $n = 11$ is occupied by two electrons. The separation in energy between the ground state and the state in which one electron is promoted from $n = 11$ to $n = 12$ is

$$\begin{aligned}\Delta E &= E_{12} - E_{11} \\ &= (2\times11+1)\frac{(6.626\times10^{-34}\,\text{J s})^2}{8\times(9.109\times10^{-31}\,\text{kg})\times(2.94\times10^{-9}\,\text{m})^2} \\ &= 1.60\ldots\times10^{-19}\,\text{J}\end{aligned}$$

or 0.160 aJ. It follows from the Bohr frequency condition ($\Delta E = h\nu$) that the frequency of radiation required to cause this transition is

$$\nu = \frac{\Delta E}{h} = \frac{1.60\ldots\times10^{-19}\,\text{J}}{6.626\times10^{-34}\,\text{J s}} = 2.42\times10^{14}\,\text{s}^{-1}$$

or 242 THz (1 THz = 10^{12} Hz), corresponding to a wavelength $\lambda = 1240$ nm. The experimental value is 603 THz ($\lambda = 497$ nm), corresponding to radiation in the visible range of the electromagnetic spectrum.

Comment. The model is too crude to expect quantitative agreement, but the calculation at least predicts a wavelength in the right general range.

Self-test 7D.1 Estimate a typical nuclear excitation energy in electronvolts (1 eV = 1.602×10^{-19} J; 1 GeV = 10^9 eV) by calculating the first excitation energy of a proton confined to a one-dimensional box with a length equal to the diameter of a nucleus (approximately 1×10^{-15} m, or 1 fm).

Answer: 0.6 GeV

7D.3 Confined motion in two and more dimensions

Now consider a rectangular two-dimensional region, between 0 and L_1 along x, and between 0 and L_2 along y. Inside this region the potential energy is zero, but at the edges it rises to infinity (Fig. 7D.6). As in the one-dimensional case, the wavefunction can be expected to be zero at the edges of this region (at $x = 0$ and L_1, and at $y = 0$ and L_2), and to be zero outside the region. Inside the region the particle has contributions to its kinetic energy from its motion along both the x and y directions, and so the Schrödinger equation has two kinetic energy terms, one for each axis. For a particle of mass m the equation is

$$-\frac{\hbar^2}{2m}\left(\frac{\partial^2\psi}{\partial x^2}+\frac{\partial^2\psi}{\partial y^2}\right)=E\psi \qquad (7D.11)$$

Equation 7D.11 is a *partial* differential equation, and the resulting wavefunctions are functions of both x and y, denoted $\psi(x,y)$.

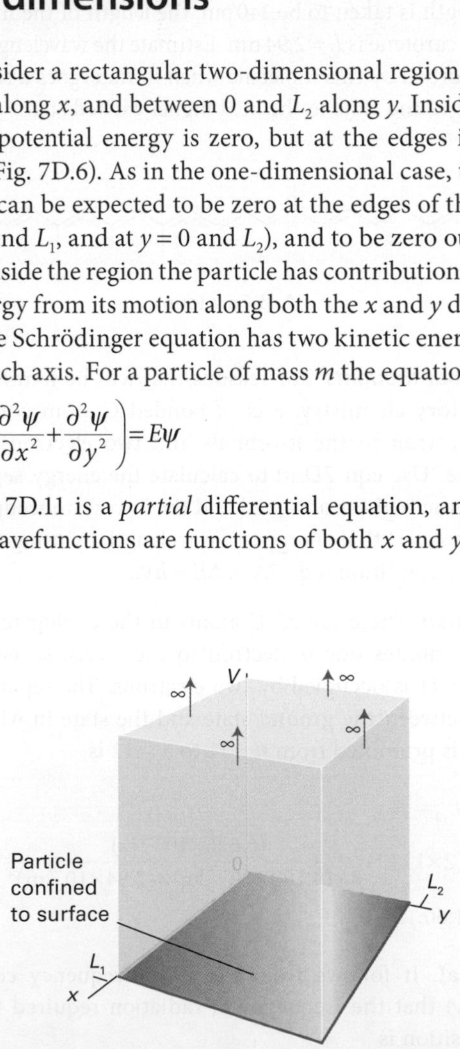

Figure 7D.6 A two-dimensional rectangular well. The potential goes to infinity at $x = 0$ and $x = L_1$, and $y = 0$ and $y = L_2$, but in between these values the potential is zero. The particle is confined to this rectangle by impenetrable walls.

(a) Energy levels and wavefunctions

The procedure for finding the allowed wavefunctions and energies involves starting with the two-dimensional Schrödinger equation, and then applying the 'separation of variables' technique to turn it into two separate one-dimensional equations.

How is that done? 7D.3 Constructing the wavefunctions for a particle in a two-dimensional box

The 'separation of variables' technique, which is explained and used here, is used in several cases in quantum mechanics.

Step 1 *Apply the separation of variables technique*
First, recognize the presence of two operators, each of which acts on functions of only x or y:

$$\hat{H}_x=-\frac{\hbar^2}{2m}\frac{\partial^2}{\partial x^2} \qquad \hat{H}_y=-\frac{\hbar^2}{2m}\frac{\partial^2}{\partial y^2}$$

Equation 7D.11, which is

$$\overbrace{-\frac{\hbar^2}{2m}\frac{\partial^2}{\partial x^2}}^{\hat{H}_x}\psi\overbrace{-\frac{\hbar^2}{2m}\frac{\partial^2}{\partial y^2}}^{\hat{H}_y}\psi=E\psi$$

then becomes

$$\hat{H}_x\psi+\hat{H}_y\psi=E\psi$$

Now suppose that the wavefunction ψ can be expressed as the product of two functions, $\psi(x,y)=X(x)Y(y)$, one depending only on x and the other depending only on y. This assumption is the central step of the procedure, and does not work for all partial differential equations: that it works here must be demonstrated. With this substitution the preceding equation becomes

$$\hat{H}_xX(x)Y(y)+\hat{H}_yX(x)Y(y)=EX(x)Y(y)$$

Then, because H_x operates on (takes the second derivatives with respect to x of) $X(x)$, and likewise for H_y and $Y(y)$, this equation is the same as

$$Y(y)\hat{H}_xX(x)+X(x)\hat{H}_yY(y)=EX(x)Y(y)$$

Division by both sides by $X(x)Y(y)$ then gives

$$\overbrace{\frac{1}{X(x)}\hat{H}_xX(x)}^{\text{Depends only on }x}+\overbrace{\frac{1}{Y(y)}\hat{H}_yY(y)}^{\text{Depends only on }y}=\overbrace{E}^{\text{A constant}}$$

If x is varied, only the first term can change; but the other two terms do not change, so the first term must be a constant for the equality to remain true. The same is true of the second term when y is varied. Therefore, denoting these constants as E_X and E_Y,

$$\frac{1}{X(x)}\hat{H}_xX(x)=E_X, \text{ so } \hat{H}_xX(x)=E_XX(x)$$

$$\frac{1}{Y(y)}\hat{H}_yY(y)=E_Y, \text{ so } \hat{H}_yY(y)=E_YY(y)$$

with $E_X + E_Y = E$. The procedure has successfully separated the partial differential equation into two ordinary differential equations, one in x and the other in y.

Step 2 *Recognize the two ordinary differential equations*

Each of the two equations is identical to the Schrödinger equation for a particle in a one-dimensional box, one for the coordinate x and the other for the coordinate y. The boundary conditions are also essentially the same (that the wavefunction must be zero at the walls). Consequently, the two solutions are

$$X_{n_1}(x)=\left(\frac{2}{L_1}\right)^{1/2}\sin\left(\frac{n_1\pi x}{L_1}\right)\quad E_{X,n_1}=\frac{n_1^2h^2}{8mL_1^2}$$

$$Y_{n_2}(y)=\left(\frac{2}{L_2}\right)^{1/2}\sin\left(\frac{n_2\pi y}{L_2}\right)\quad E_{Y,n_2}=\frac{n_2^2h^2}{8mL_2^2}$$

with each of n_1 and n_2 taking the values 1, 2, … independently.

Step 3 *Assemble the complete wavefunction*

Inside the box, which is when $0\le x\le L_1$ and $0\le x\le L_2$, the wavefunction is the product $X_{n_1}(x)Y_{n_2}(y)$, and is given by eqn 7D.12a below. Outside the box, the wavefunction is zero. The energies are the sum $E_{X,n_1}+E_{Y,n_2}$. The two quantum numbers take the values $n_1 = 1, 2, \ldots$ and $n_2 = 1, 2, \ldots$ independently. Overall, therefore,

$$\psi_{n_1,n_2}(x,y)=\frac{2}{(L_1L_2)^{1/2}}\sin\left(\frac{n_1\pi x}{L_1}\right)\sin\left(\frac{n_2\pi y}{L_2}\right)$$

Wavefunctions [two dimensions] (7D.12a)

$$E_{n_1,n_2}=\left(\frac{n_1^2}{L_1^2}+\frac{n_2^2}{L_2^2}\right)\frac{h^2}{8m}$$

Energy levels [two dimensions] (7D.12b)

Some of the wavefunctions are plotted as contours in Fig. 7D.7. They are the two-dimensional versions of the wavefunctions shown in Fig. 7D.3. Whereas in one dimension the wavefunctions resemble states of a vibrating string with ends fixed, in two dimensions the wavefunctions correspond to vibrations of a rectangular plate with fixed edges.

Brief illustration 7D.4

Consider an electron confined to a square cavity of side L (that is $L_1 = L_2 = L$), and in the state with quantum numbers $n_1 = 1$, $n_2 = 2$. Because the probability density is

$$\psi_{1,2}^2(x,y)=\frac{4}{L^2}\sin^2\left(\frac{\pi x}{L}\right)\sin^2\left(\frac{2\pi y}{L}\right)$$

the most probable locations correspond to $\sin^2(\pi x/L) = 1$ and $\sin^2(2\pi y/L) = 1$, or $(x,y) = (L/2,L/4)$ and $(L/2,3L/4)$. The least probable locations (the nodes, where the wavefunction passes through zero) correspond to zeroes in the probability density within the box, which occur along the line $y = L/2$.

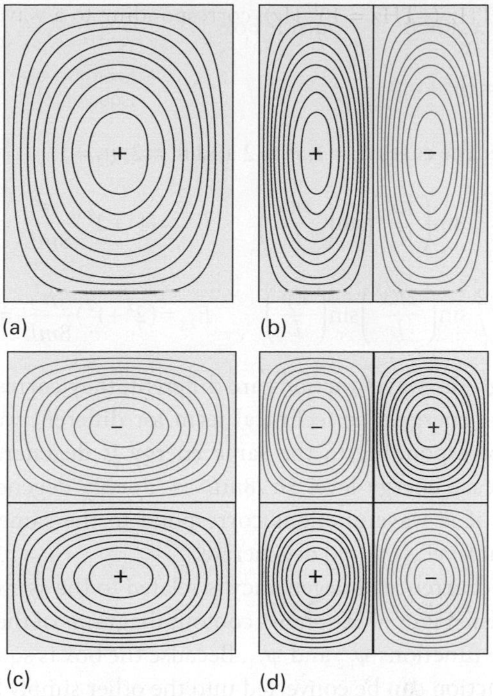

Figure 7D.7 The wavefunctions for a particle confined to a rectangular surface depicted as contours of equal amplitude. (a) $n_1 = 1$, $n_2 = 1$, the state of lowest energy; (b) $n_1 = 2$, $n_2 = 1$; (c) $n_1 = 1$, $n_2 = 2$; (d) $n_1 = 2$, $n_2 = 2$.

A three-dimensional box can be treated in the same way: the wavefunctions are products of three terms and the energy is a sum of three terms. As before, each term is analogous to that for the one-dimensional case. Overall, therefore,

$$\psi_{n_1,n_2,n_3}(x,y,z)=\left(\frac{8}{L_1L_2L_3}\right)^{1/2}\sin\left(\frac{n_1\pi x}{L_1}\right)\sin\left(\frac{n_2\pi y}{L_2}\right)\sin\left(\frac{n_3\pi z}{L_3}\right)$$

Wavefunctions [three dimensions] (7D.13a)

for $0\le x\le L_1, 0\le y\le L_2, 0\le z\le L_3$

$$E_{n_1,n_2,n_3}=\left(\frac{n_1^2}{L_1^2}+\frac{n_2^2}{L_2^2}+\frac{n_3^2}{L_3^2}\right)\frac{h^2}{8m}$$

Energy levels [three dimensions] (7D.13b)

The quantum numbers n_1, n_2, and n_3 are all positive integers 1, 2, … that can be chosen independently. The system has a zero-point energy, the value of $E_{1,1,1}$.

(b) Degeneracy

A special feature of the solutions arises when a two-dimensional box is not merely rectangular but square, with $L_1 = L_2 = L$. Then the wavefunctions and their energies are

$$\psi_{n_1,n_2}(x,y)=\frac{2}{L}\sin\left(\frac{n_1\pi x}{L}\right)\sin\left(\frac{n_2\pi y}{L}\right)$$

for $0\le x\le L, 0\le y\le L$

Wavefunctions [square] (7D.14a)

$$\psi_{n_1,n_2}(x,y)=0 \qquad \text{outside box}$$

$$E_{n_1,n_2}=(n_1^2+n_2^2)\frac{h^2}{8mL^2} \qquad \text{Energy levels [square]} \qquad (7D.14b)$$

Consider the cases $n_1 = 1$, $n_2 = 2$ and $n_1 = 2$, $n_2 = 1$:

$$\psi_{1,2}=\frac{2}{L}\sin\left(\frac{\pi x}{L}\right)\sin\left(\frac{2\pi y}{L}\right) \qquad E_{1,2}=(1^2+2^2)\frac{h^2}{8mL^2}=\frac{5h^2}{8mL^2}$$

$$\psi_{2,1}=\frac{2}{L}\sin\left(\frac{2\pi x}{L}\right)\sin\left(\frac{\pi y}{L}\right) \qquad E_{2,1}=(2^2+1^2)\frac{h^2}{8mL^2}=\frac{5h^2}{8mL^2}$$

Although the wavefunctions are different, they correspond to the same energy. The technical term for different wavefunctions corresponding to the same energy is **degeneracy**, and in this case energy level $5h^2/8mL^2$ is 'doubly degenerate'. In general, if N wavefunctions correspond to the same energy, then that level is 'N-fold degenerate'.

The occurrence of degeneracy is related to the symmetry of the system. Figure 7D.8 shows contour diagrams of the two degenerate functions $\psi_{1,2}$ and $\psi_{2,1}$. Because the box is square, one wavefunction can be converted into the other simply by rotating the plane by 90°. Interconversion by rotation through 90° is not possible when the plane is not square, and $\psi_{1,2}$ and $\psi_{2,1}$ are then not degenerate. Similar arguments account for the degeneracy of the energy levels of a particle in a cubic box. Other examples of degeneracy occur in quantum mechanical systems (for instance, in the hydrogen atom, Topic 8A), and all of them can be traced to the symmetry properties of the system.

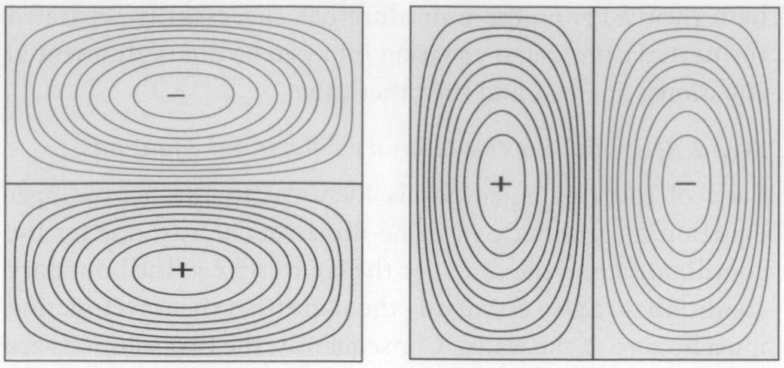

(a) (b)

Figure 7D.8 Two of the wavefunctions for a particle confined to a geometrically square well: (a) $n_1 = 1$, $n_2 = 2$; (b) $n_1 = 2$, $n_2 = 1$. The two functions correspond to the same energy and are said to be degenerate. Note that one wavefunction can be converted into the other by rotation of the box by 90°: degeneracy is always a consequence of symmetry.

Brief illustration 7D.5

The energy of a particle in a two-dimensional square box of side L in the energy level with $n_1 = 1$, $n_2 = 7$ is

$$E_{1,7}=(1^2+7^2)\frac{h^2}{8mL^2}=\frac{50h^2}{8mL^2}$$

The level with $n_1 = 7$ and $n_2 = 1$ has the same energy. Thus, at first sight the energy level $50h^2/8mL^2$ is doubly degenerate. However, in certain systems there may be levels that are not apparently related by symmetry but have the same energy and are said to be 'accidentally' degenerate. Such is the case here, for the level with $n_1 = 5$ and $n_2 = 5$ also has energy $50h^2/8mL^2$. The level is therefore actually three-fold degenerate. Accidental degeneracy is also encountered in the hydrogen atom (Topic 8A) and can always be traced to a 'hidden' symmetry, one that is not immediately obvious.

7D.4 Tunnelling

A new quantum-mechanical feature appears when the potential energy does not rise abruptly to infinity at the walls (Fig. 7D.9). Consider the case in which there are two regions where the potential energy is zero separated by a barrier where it rises to a finite value, V_0. Suppose the energy of the particle is less than V_0. A particle arriving from the left of the barrier has an oscillating wavefunction but inside the barrier the wavefunction decays rather than oscillates. Provided the barrier is not too wide the wavefunction emerges to the right, but with reduced amplitude; it then continues to oscillate once it is back in a region where it has zero potential energy. As a result of this behaviour the particle has a non-zero probability of passing through the barrier, which is forbidden classically because a particle cannot have a potential energy that exceeds its total energy. The ability of a particle to penetrate into, and possibly pass through, a classically forbidden region is called **tunnelling**.

The Schrödinger equation can be used to calculate the probability of tunnelling of a particle of mass m incident from the left on a rectangular potential energy barrier of width W. On the left of the barrier ($x < 0$) the wavefunctions are those of a particle with $V = 0$, so from eqn 7D.2,

$$\psi = Ae^{ikx} + Be^{-ikx} \qquad k\hbar = (2mE)^{1/2} \qquad \text{Wavefunction to left of barrier} \qquad (7D.15)$$

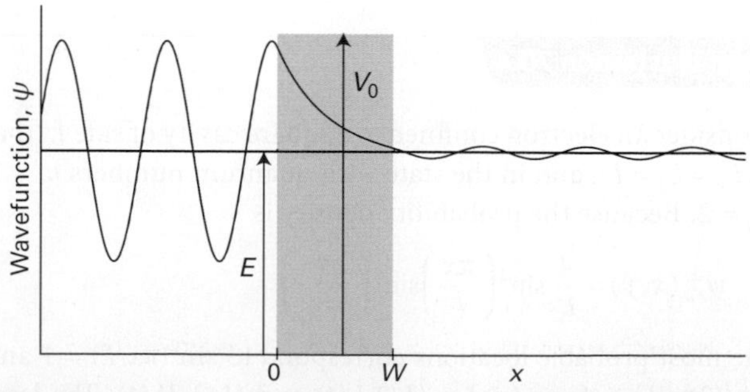

Figure 7D.9 The wavefunction for a particle encountering a potential barrier. Provided that the barrier is neither too wide nor too tall, the wavefunction will be non-zero as it exits to the right.

The Schrödinger equation for the region representing the barrier ($0 \le x \le W$), where the potential energy has the constant value V_0, is

$$-\frac{\hbar^2}{2m}\frac{d^2\psi(x)}{dx^2}+V_0\psi(x)=E\psi(x) \quad (7D.16)$$

Provided $E < V_0$ the general solutions of eqn 7D.16 are

$$\psi = Ce^{\kappa x}+De^{-\kappa x} \qquad \kappa\hbar=\{2m(V_0-E)\}^{1/2}$$

Wavefunction inside barrier (7D.17)

as can be verified by substituting this solution into the left-hand side of eqn 7D.16. The important feature to note is that the two exponentials in eqn 7D.17 are now real functions, as distinct from the complex, oscillating functions for the region where $V = 0$. To the right of the barrier ($x > W$), where $V = 0$ again, the wavefunctions are

$$\psi = A'e^{ikx} \qquad k\hbar=(2mE)^{1/2}$$

Wavefunction to right of barrier (7D.18)

Note that to the right of the barrier, the particle can be moving only to the right and therefore only the term e^{ikx} contributes as it corresponds to a particle with positive linear momentum (moving to the right).

The complete wavefunction for a particle incident from the left consists of (Fig. 7D.10):

- an incident wave (Ae^{ikx} corresponds to positive linear momentum);
- a wave reflected from the barrier (Be^{-ikx} corresponds to negative linear momentum, motion to the left);
- the exponentially changing amplitude inside the barrier (eqn 7D.17);
- an oscillating wave (eqn 7D.18) representing the propagation of the particle to the right after tunnelling through the barrier successfully.

Physical interpretation

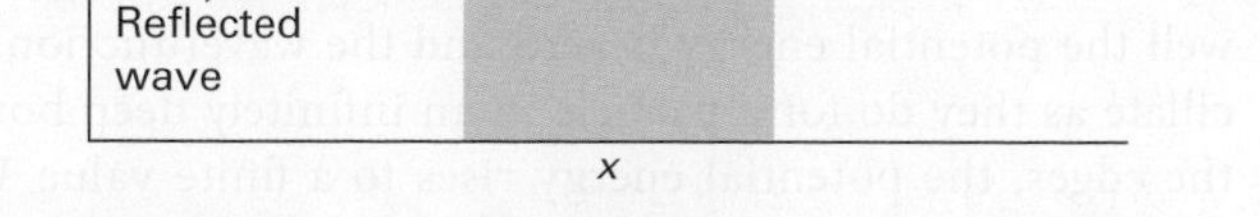

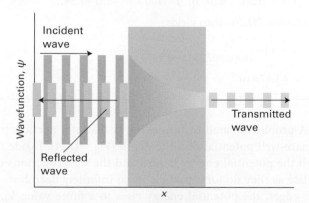

Figure 7D.10 When a particle is incident on a barrier from the left, the wavefunction consists of a wave representing linear momentum to the right, a reflected component representing momentum to the left, a varying but not oscillating component inside the barrier, and a (weak) wave representing motion to the right on the far side of the barrier.

The probability that a particle is travelling towards positive x (to the right) on the left of the barrier ($x < 0$) is proportional to $|A|^2$, and the probability that it is travelling to the right after passing through the barrier ($x > W$) is proportional to $|A'|^2$. The ratio of these two probabilities, $|A'|^2/|A|^2$, which expresses the probability of the particle tunnelling through the barrier, is called the **transmission probability**, T.

The values of the coefficients A, B, C, and D are found by applying the usual criteria of acceptability to the wavefunction. Because an acceptable wavefunction must be continuous at the edges of the barrier (at $x = 0$ and $x = W$)

$$\text{at } x=0\text{: } A+B=C+D \qquad \text{at } x=W\text{: } Ce^{\kappa W}+De^{-\kappa W}=A'e^{ikW} \quad (7D.19a)$$

Their slopes (their first derivatives) must also be continuous at these positions (Fig. 7D.11):

$$\begin{aligned}&\text{at } x=0\text{: } ikA-ikB=\kappa C-\kappa D\\&\text{at } x=W\text{: } \kappa Ce^{\kappa W}-\kappa De^{-\kappa W}=ikA'e^{ikW}\end{aligned} \quad (7D.19b)$$

After straightforward but lengthy algebraic manipulations of these four equations 7D.19 (see Problem P7D.12), the transmission probability turns out to be

$$T=\left\{1+\frac{(e^{\kappa W}-e^{-\kappa W})^2}{16\varepsilon(1-\varepsilon)}\right\}^{-1}$$

Transmission probability [rectangular barrier] (7D.20a)

where $\varepsilon = E/V_0$. This function is plotted in Fig. 7D.12. The transmission probability for $E > V_0$ is shown there

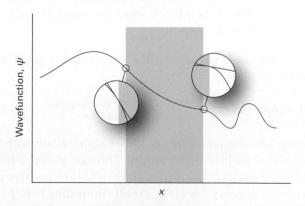

Figure 7D.11 The wavefunction and its slope must be continuous at the edges of the barrier. The conditions for continuity enable the wavefunctions at the junctions of the three zones to be connected and hence relations between the coefficients that appear in the solutions of the Schrödinger equation to be obtained.

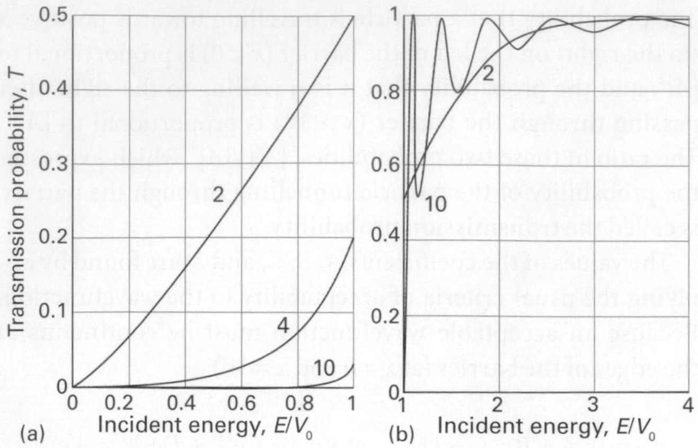

Figure 7D.12 The transmission probabilities T for passage through a rectangular potential barrier. The horizontal axis is the energy of the incident particle expressed as a multiple of the barrier height. The curves are labelled with the value of $W(2mV_0)^{1/2}/\hbar$. (a) $E < V_0$; (b) $E > V_0$.

too. The transmission probability has the following properties:

- $T \approx 0$ for $E \ll V_0$: there is negligible tunnelling when the energy of the particle is much lower than the height of the barrier;
- T increases as E approaches V_0: the probability of tunnelling increases as the energy of the particle rises to match the height of the barrier;
- T approaches 1 for $E > V_0$, but the fact that it does not immediately reach 1 means that there is a probability of the particle being reflected by the barrier even though according to classical mechanics it can pass over it;
- $T \approx 1$ for $E \gg V_0$, as expected classically: the barrier is invisible to the particle when its energy is much higher than the barrier.

Physical interpretation

For high, wide barriers (in the sense that $\kappa W \gg 1$), eqn 7D.20a simplifies to

$$T \approx 16\varepsilon(1-\varepsilon)e^{-2\kappa W} \quad \text{Rectangular potential barrier; } \kappa W \gg 1 \qquad (7D.20b)$$

The transmission probability decreases exponentially with the thickness of the barrier and with $m^{1/2}$ (because $\kappa \propto m^{1/2}$). It follows that particles of low mass are more able to tunnel through barriers than heavy ones (Fig. 7D.13). Tunnelling is very important for electrons and muons ($m_\mu \approx 207m_e$), and moderately important for protons ($m_p \approx 1840m_e$); for heavier particles it is less important.

A number of effects in chemistry depend on the ability of the proton to tunnel more readily than the deuteron. The very rapid equilibration of proton transfer reactions is also a manifestation of the ability of protons to tunnel through barriers and transfer quickly from an acid to a base. Tunnelling of protons between acidic and basic groups is also an important feature of the mechanism of some enzyme-catalysed reactions.

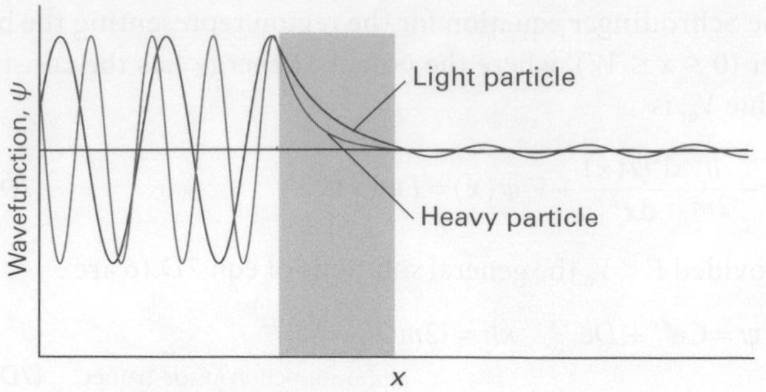

Figure 7D.13 The wavefunction of a heavy particle decays more rapidly inside a barrier than that of a light particle. Consequently, a light particle has a greater probability of tunnelling through the barrier.

Brief illustration 7D.6

Suppose that a proton of an acidic hydrogen atom is confined to an acid that can be represented by a barrier of height 2.000 eV and length 100 pm. The probability that a proton with energy 1.995 eV (corresponding to 0.3195 aJ) can escape from the acid is computed using eqn 7D.20a, with $\varepsilon = E/V_0 = 1.995\,\text{eV}/2.000\,\text{eV} = 0.9975$ and $V_0 - E = 0.005\,\text{eV}$ (corresponding to 8.0×10^{-22} J). The quantity κ is given by eqn 7D.17:

$$\kappa = \frac{\{2\times(1.67\times10^{-27}\,\text{kg})\times(8.0\times10^{-22}\,\text{J})\}^{1/2}}{1.055\times10^{-34}\,\text{J s}} = 1.54\cdots\times10^{10}\,\text{m}^{-1}$$

It follows that

$$\kappa W = (1.54\ldots\times10^{10}\,\text{m}^{-1})\times(100\times10^{-12}\,\text{m}) = 1.54\ldots$$

Equation 7D.20a then yields

$$T = \left\{1+\frac{(e^{1.54\ldots}-e^{-1.54\ldots})^2}{16\times0.9975\times(1-0.9975)}\right\}^{-1} = 1.97\times10^{-3}$$

A problem related to tunnelling is that of a particle in a square-well potential of finite depth (Fig. 7D.14). Inside the well the potential energy is zero and the wavefunctions oscillate as they do for a particle in an infinitely deep box. At the edges, the potential energy rises to a finite value V_0. If $E < V_0$ the wavefunction decays as it penetrates into the walls, just as it does when it enters a barrier. The wavefunctions are found by ensuring, as in the discussion of the potential barrier, that they and their slopes are continuous at the edges of the potential. The two lowest energy solutions are shown in Fig. 7D.15.

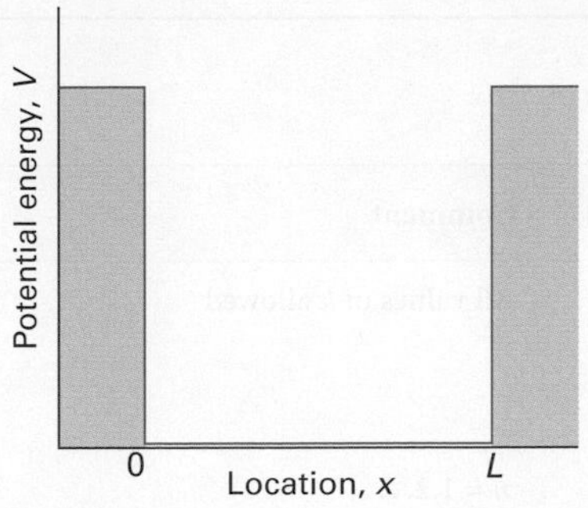

Figure 7D.14 A potential well with a finite depth.

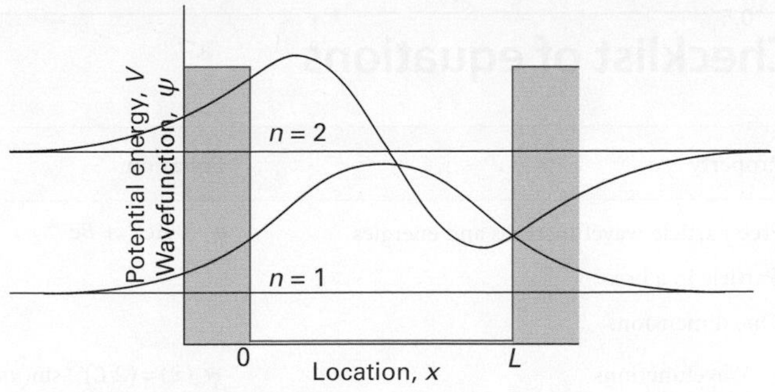

Figure 7D.15 Wavefunctions of the lowest two bound levels for a particle in the potential well shown in Fig. 7D.14.

For a potential well of finite depth, there are a finite number of wavefunctions with energy less than V_0: they are referred to as **bound states**, in the sense that the particle is mainly confined to the well. Detailed consideration of the Schrödinger equation for the problem shows that the number of bound states is equal to N, with

$$N-1<\frac{(8mV_0L)^{1/2}}{h}<N \qquad \text{(7D.21)}$$

where V_0 is the depth of the well and L is its width. This relation shows that the deeper and wider the well, the greater the number of bound states. As the depth becomes infinite, so the number of bound states also becomes infinite, as for the particle in a box treated earlier in this Topic.

Checklist of concepts

☐ **1.** The translational energy of a free particle is not quantized.

☐ **2.** The need to satisfy **boundary conditions** implies that only certain wavefunctions are acceptable and restricts observables, specifically the energy, to discrete values.

☐ **3.** A **quantum number** is an integer (in certain cases, a half-integer) that labels the state of the system.

☐ **4.** A particle in a box possesses a **zero-point energy**, an irremovable minimum energy.

☐ **5.** The **correspondence principle** states that the quantum mechanical result with high quantum numbers should agree with the predictions of classical mechanics.

☐ **6.** The wavefunction for a particle in a two- or three-dimensional box is the product of wavefunctions for the particle in a one-dimensional box.

☐ **7.** The energy of a particle in a two- or three-dimensional box is the sum of the energies for the particle in two or three one-dimensional boxes.

☐ **8.** Energy levels are ***N*-fold degenerate** if N wavefunctions correspond to the same energy.

☐ **9.** The occurrence of degeneracy is a consequence of the symmetry of the system.

☐ **10.** **Tunnelling** is penetration into or through a classically forbidden region.

☐ **11.** The probability of tunnelling decreases with an increase in the height and width of the potential barrier.

☐ **12.** Light particles are more able to tunnel through barriers than heavy ones.

Checklist of equations

Property	Equation	Comment	Equation number
Free-particle wavefunctions and energies	$\psi_k = Ae^{ikx} + Be^{-ikx}$ $\quad E_k = k^2\hbar^2/2m$	All values of k allowed	7D.2
Particle in a box			
One dimension:			
Wavefunctions	$\psi_n(x) = (2/L)^{1/2}\sin(n\pi x/L),\ 0 \le x \le L$ $\psi_n(x) = 0,\ x<0$ and $x>L$	$n = 1,2,\ldots$	7D.6
Energies	$E_n = n^2h^2/8mL^2$		
Two dimensions:			
Wavefunctions	$\psi_{n_1,n_2}(x,y) = \psi_{n_1}(x)\psi_{n_2}(y)$ $\psi_{n_1}(x) = (2/L_1)^{1/2}\sin(n_1\pi x/L_1),\ 0 \le x \le L_1$ $\psi_{n_2}(y) = (2/L_2)^{1/2}\sin(n_2\pi y/L_2),\ 0 \le y \le L_2$	$n_1, n_2 = 1, 2, \ldots$	7D.12a
Energies	$E_{n_1,n_2} = (n_1^2/L_1^2 + n_2^2/L_2^2)h^2/8m$		7D.12b
Three dimensions:			
Wavefunctions	$\psi_{n_1,n_2,n_3}(x,y,z) = \psi_{n_1}(x)\psi_{n_2}(y)\psi_{n_3}(z)$	$n_1, n_2, n_3 = 1, 2, \ldots$	7D.13a
Energies	$E_{n_1,n_2,n_3} = (n_1^2/L_1^2 + n_2^2/L_2^2 + n_3^2/L_3^2)h^2/8m$		7D.13b
Transmission probability	$T = \{1 + (e^{\kappa W} - e^{-\kappa W})^2/16\varepsilon(1-\varepsilon)\}^{-1}$	Rectangular potential barrier	7D.20a
	$T = 16\varepsilon(1-\varepsilon)e^{-2\kappa W}$	High, wide rectangular barrier	7D.20b

TOPIC 7E Vibrational motion

➤ **Why do you need to know this material?**

Molecular vibration plays a role in the interpretation of thermodynamic properties, such as heat capacities (Topics 2A and 13E), and of the rates of chemical reactions (Topic 18C). The measurement and interpretation of the vibrational frequencies of molecules is the basis of infrared spectroscopy (Topics 11C and 11D).

➤ **What is the key idea?**

The energy of vibrational motion is quantized.

➤ **What do you need to know already?**

You should know how to formulate the Schrödinger equation for a given potential energy. You should also be familiar with the concepts of tunnelling (Topic 7D) and the expectation value of an observable (Topic 7C).

Atoms in molecules and solids vibrate around their equilibrium positions as bonds stretch, compress, and bend. The simplest model for this kind of motion is the 'harmonic oscillator', which is considered in detail in this Topic.

7E.1 The harmonic oscillator

In classical mechanics a **harmonic oscillator** is a particle of mass m that experiences a restoring force proportional to its displacement, x, from the equilibrium position. As is shown in *The chemist's toolkit* 18, the particle oscillates about the equilibrium position at a characteristic frequency, ν. The potential energy of the particle is

$$V(x)=\tfrac{1}{2}k_f x^2 \qquad \text{Parabolic potential energy} \qquad (7E.1)$$

where k_f is the **force constant**, which characterizes the strength of the restoring force (Fig. 7E.1) and is expressed in newtons per metre ($N\,m^{-1}$). This form of potential energy is called a 'harmonic potential energy' or a 'parabolic potential energy'. The Schrödinger equation for the oscillator is therefore

$$-\frac{\hbar^2}{2m}\frac{d^2\psi(x)}{dx^2}+\tfrac{1}{2}k_f x^2\psi(x)=E\psi(x) \qquad \text{Schrödinger equation} \qquad (7E.2)$$

The potential energy becomes infinite at $x=\pm\infty$, and so the wavefunction is zero at these limits. However, as the potential energy rises smoothly rather than abruptly to infinity, as it does for a particle in a box, the wavefunction decreases

The chemist's toolkit 18 The classical harmonic oscillator

A harmonic oscillator consists of a particle of mass m that experiences a 'Hooke's law' restoring force, one that is proportional to the displacement of the particle from equilibrium. For a one-dimensional system,

$$F_x=-k_f x$$

From Newton's second law of motion ($F=ma=m(d^2x/dt^2)$; see *The chemist's toolkit* 3 in Topic 1B),

$$m\frac{d^2x}{dt^2}=-k_f x$$

If $x=0$ at $t=0$, a solution (as may be verified by substitution) is

$$x(t)=A\sin 2\pi\nu t \qquad \nu=\frac{1}{2\pi}\left(\frac{k_f}{m}\right)^{1/2}$$

This solution shows that the position of the particle oscillates *harmonically* (i.e. as a sine function) with frequency ν (units: Hz). The *angular frequency* of the oscillator is $\omega=2\pi\nu$ (units: radians per second). It follows that the angular frequency of a classical harmonic oscillator is $\omega=(k_f/m)^{1/2}$.

The potential energy V is related to force by $F=-dV/dx$ (*The chemist's toolkit* 6 in Topic 2A), so the potential energy corresponding to a Hooke's law restoring force is

$$V(x)=\tfrac{1}{2}k_f x^2$$

As the particle moves away from the equilibrium position its potential energy increases and so its kinetic energy, and hence its speed, decreases. At some point all the energy is potential and the particle comes to rest at a turning point. The particle then accelerates back towards and through the equilibrium position. The greatest probability of finding the particle is where it is moving most slowly, which is close to the turning points.

The turning point, x_{tp}, of a classical oscillator occurs when its potential energy $\tfrac{1}{2}k_f x^2$ is equal to its total energy, so

$$x_{tp}=\pm\left(\frac{2E}{k_f}\right)^{1/2}$$

The turning point increases with the total energy: in classical terms, the amplitude of the swing of a pendulum or the displacement of a mass on a spring increases.

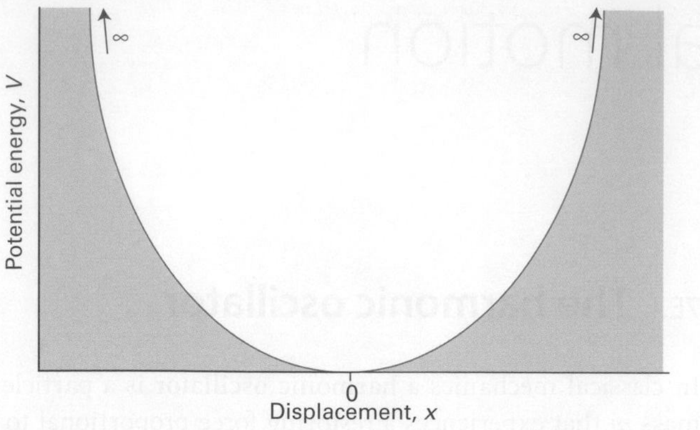

Figure 7E.1 The potential energy for a harmonic oscillator is the parabolic function $V_{HO}(x) = \frac{1}{2}k_f x^2$, where x is the displacement from equilibrium. The larger the force constant k_f the steeper the curve and narrower the curve becomes.

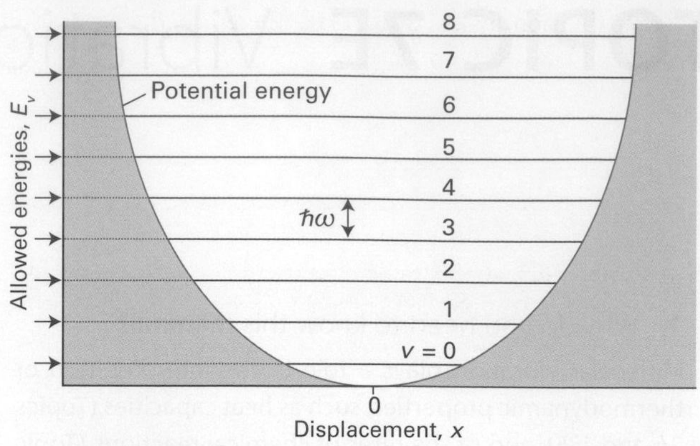

Figure 7E.2 The energy levels of a harmonic oscillator are evenly spaced with separation $\hbar\omega$, where $\omega = (k_f/m)^{1/2}$. Even in its lowest energy state, an oscillator has an energy greater than zero.

smoothly towards zero rather than becoming zero abruptly. The boundary conditions $\psi(\pm\infty)=0$ imply that only some solutions of the Schrödinger equation are acceptable, and therefore that the energy of the oscillator is quantized.

(a) The energy levels

Equation 7E.2 is a standard form of differential equation and its solutions are well known to mathematicians.[1] The energies permitted by the boundary conditions are

$$E_v = (v+\tfrac{1}{2})\hbar\omega \quad \omega = (k_f/m)^{1/2} \qquad \text{Energy levels} \qquad (7E.3)$$

$$v = 0, 1, 2, \ldots$$

where v is the **vibrational quantum number**. Note that the energies depend on ω, which has the same dependence on the mass and the force constant as the angular frequency of a classical oscillator (see *The chemist's toolkit* 18) and is high when the force constant is large and the mass small. The separation of adjacent levels is

$$E_{v+1} - E_v = \hbar\omega \qquad (7E.4)$$

for all v. The energy levels therefore form a uniform ladder with spacing $\hbar\omega$ (Fig. 7E.2). The energy separation $\hbar\omega$ is negligibly small for macroscopic objects (with large mass) but significant for objects with mass similar to that of an atom.

The energy of the lowest level, with $v = 0$, is not zero:

$$E_0 = \tfrac{1}{2}\hbar\omega \qquad \text{Zero-point energy} \qquad (7E.5)$$

The physical reason for the existence of this zero-point energy is the same as for the particle in a box (Topic 7D). The particle is confined, so its position is not completely uncertain. It follows that its momentum, and hence its kinetic energy, cannot be zero. A classical interpretation of the zero-point energy is that the quantum oscillator is never completely at rest and therefore has kinetic energy; moreover, because its motion samples the potential energy away from the equilibrium position, it also has non-zero potential energy.

The model of a particle oscillating in a parabolic potential is used to describe the vibrational motion of a diatomic molecule A–B (and, with elaboration, Topic 11D, polyatomic molecules). In this case both atoms move as the bond between them is stretched and compressed and the mass m is replaced by the **effective mass**, μ, given by

$$\mu = \frac{m_A m_B}{m_A + m_B} \qquad \text{Effective mass [diatomic molecule]} \qquad (7E.6)$$

When A is much heavier than B, m_B can be neglected in the denominator and the effective mass is $\mu \approx m_B$, the mass of the lighter atom. In this case, only the light atom moves and the heavy atom acts as a stationary anchor.

Brief illustration 7E.1

The effective mass of $^1H^{35}Cl$ is

$$\mu = \frac{m_H m_{Cl}}{m_H + m_{Cl}} = \frac{(1.0078m_u)\times(34.9688m_u)}{(1.0078m_u)+(34.9688m_u)} = 0.9796m_u$$

which is close to the mass of the hydrogen atom. The force constant of the bond is $k_f = 516.3\,\mathrm{N\,m^{-1}}$. It follows from eqn 7E.3 and $1\,\mathrm{N} = 1\,\mathrm{kg\,m\,s^{-2}}$, with μ in place of m, that

$$\omega = \left(\frac{k_f}{\mu}\right)^{1/2} = \left(\frac{516.3\,\mathrm{N\,m^{-1}}}{0.9796\times(1.660\,54\times10^{-27}\,\mathrm{kg})}\right)^{1/2} = 5.634\times10^{14}\,\mathrm{s^{-1}}$$

[1] For details, see our *Molecular quantum mechanics*, Oxford University Press, Oxford (2011).

or (after division by 2π) 89.67 THz. Therefore, the separation of adjacent levels is (eqn 7E.4)

$$E_{v+1} - E_v = (1.05457\times10^{-34}\,\text{J s})\times(5.634\times10^{14}\,\text{s}^{-1}) = 5.941\times10^{-20}\,\text{J}$$

or 59.41 zJ, about 0.37 eV. This energy separation corresponds to 36 kJ mol^{-1}, which is chemically significant. The zero-point energy (eqn 7E.5) of this molecular oscillator is 29.71 zJ, which corresponds to 0.19 eV, or 18 kJ mol^{-1}.

(b) The wavefunctions

The acceptable solutions of eqn 7E.2, all have the form

$\psi(x) = N \times$ (polynomial in x) $\times$ (bell-shaped Gaussian function)

where N is a normalization constant. A Gaussian function is a bell-shaped function of the form e^{-x^2} (Fig. 7E.3). The precise form of the wavefunctions is

$$\psi_v(x) = N_v H_v(y) e^{-y^2/2} \qquad \text{Wavefunctions} \qquad (7E.7)$$

$$y = \frac{x}{\alpha} \qquad \alpha = \left(\frac{\hbar^2}{mk_f}\right)^{1/4}$$

The factor $H_v(y)$ is a **Hermite polynomial**; the form of these polynomials and some of their properties are listed in Table 7E.1. Note that the first few Hermite polynomials are rather simple: for instance, $H_0(y) = 1$ and $H_1(y) = 2y$. Hermite polynomials, which are members of a class of functions called 'orthogonal polynomials', have a wide range of important properties which allow a number of quantum mechanical calculations to be done with relative ease.

The wavefunction for the ground state, which has $v = 0$, is

$$\psi_0(x) = N_0 e^{-y^2/2} = N_0 e^{-x^2/2\alpha^2} \qquad \text{Ground-state wavefunction} \qquad (7E.8a)$$

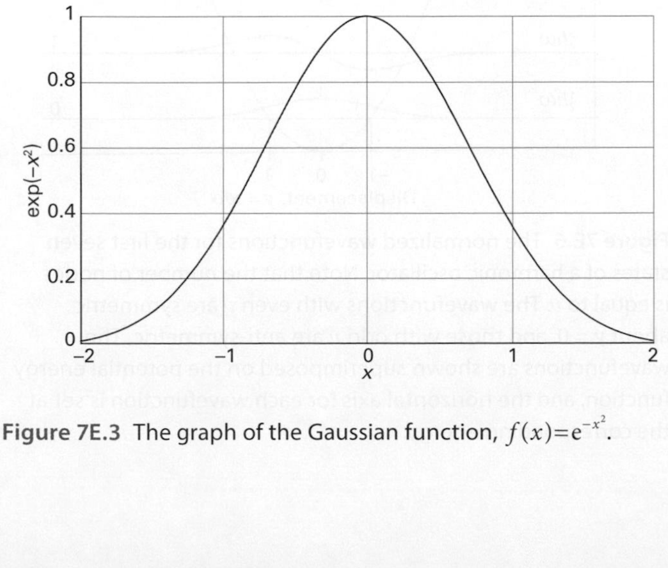

Figure 7E.3 The graph of the Gaussian function, $f(x) = e^{-x^2}$.

Table 7E.1 The Hermite polynomials

v	$H_v(y)$
0	1
1	$2y$
2	$4y^2 - 2$
3	$8y^3 - 12y$
4	$16y^4 - 48y^2 + 12$
5	$32y^5 - 160y^3 + 120y$
6	$64y^6 - 480y^4 + 720y^2 - 120$

The Hermite polynomials are solutions of the differential equation

$$H_v'' - 2yH_v' + 2vH_v = 0$$

where primes denote differentiation. They satisfy the recursion relation

$$H_{v+1} - 2yH_v + 2vH_{v-1} = 0$$

An important integral is

$$\int_{-\infty}^{\infty} H_{v'}H_v e^{-y^2}\,dy = \begin{cases} 0 & \text{if } v' \neq v \\ \pi^{1/2}2^v v! & \text{if } v' = v \end{cases}$$

and the corresponding probability density is

$$\psi_0^2(x) = N_0^2 e^{-y^2} = N_0^2 e^{-x^2/\alpha^2} \qquad \text{Ground-state probability density} \qquad (7E.8b)$$

The wavefunction and the probability density are shown in Fig. 7E.4. The probability density has its maximum value at $x = 0$, the equilibrium position, but is spread about this position. The curvature is consistent with the kinetic energy being non-zero and the spread is consistent with the potential energy also being non-zero, so resulting in a zero-point energy.

The wavefunction for the first excited state, $v = 1$, is

$$\psi_1(x) = N_1 2y e^{-y^2/2} = N_1\left(\frac{2}{\alpha}\right) x e^{-x^2/2\alpha^2} \qquad \text{First excited-state wavefunction} \qquad (7E.9)$$

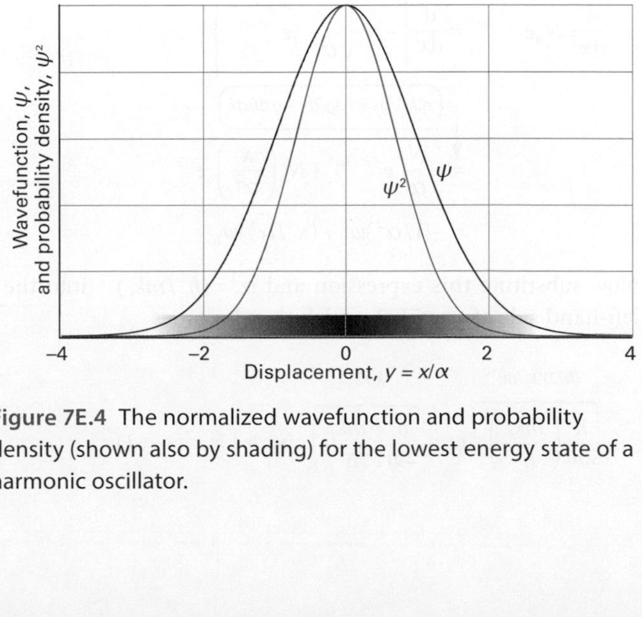

Figure 7E.4 The normalized wavefunction and probability density (shown also by shading) for the lowest energy state of a harmonic oscillator.

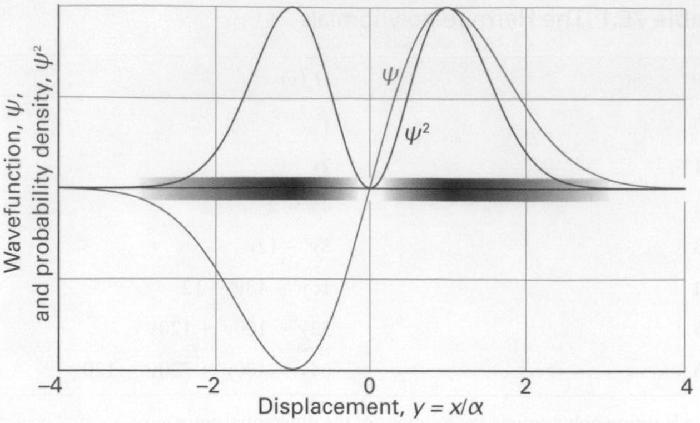

Figure 7E.5 The normalized wavefunction and probability density (shown also by shading) for the first excited state of a harmonic oscillator.

This function has a node at zero displacement ($x = 0$), and the probability density has maxima at $x = \pm\alpha$ (Fig. 7E.5).

Example 7E.1 Confirming that a wavefunction is a solution of the Schrödinger equation

Confirm that the ground-state wavefunction (eqn 7E.8a) is a solution of the Schrödinger equation (eqn 7E.2).

Collect your thoughts You need to substitute the wavefunction given in eqn 7E.8a into eqn 7E.2 and see that the left-hand side generates the right-hand side of the equation; use the definition of α in eqn 7E.7. Confirm that the factor that multiplies the wavefunction on the right-hand side agrees with eqn 7E.5.

The solution First, evaluate the second derivative of the ground-state wavefunction by differentiating it twice in succession:

$$\frac{d}{dx}N_0 e^{-x^2/2\alpha^2} = -N_0\left(\frac{x}{\alpha^2}\right)e^{-x^2/2\alpha^2}$$

$$\frac{d^2}{dx^2}N_0 e^{-x^2/2\alpha^2} = \frac{d}{dx}\left\{\overbrace{-N_0\left(\frac{x}{\alpha^2}\right)}^{f}\overbrace{e^{-x^2/2\alpha^2}}^{g}\right\}$$

d(fg)/dx = f dg/dx + g df/dx

$$= -\frac{N_0}{\alpha^2}e^{-x^2/2\alpha^2} + N_0\left(\frac{x}{\alpha^2}\right)^2 e^{-x^2/2\alpha^2}$$

$$= -(1/\alpha^2)\psi_0 + (x^2/\alpha^4)\psi_0$$

Now substitute this expression and $\alpha^2 = (\hbar^2/mk_f)^{1/2}$ into the left-hand side of eqn 7E.2, which then becomes

$$\overbrace{\frac{\hbar^2}{2m}\left(\frac{mk_f}{\hbar^2}\right)^{1/2}}^{(\hbar/2)(k_f/m)^{1/2}}\psi_0 - \overbrace{\frac{\hbar^2}{2m}\left(\frac{mk_f}{\hbar^2}\right)}^{k_f/2}x^2\psi_0 + \tfrac{1}{2}k_f x^2\psi_0 = E\psi_0$$

and therefore (keeping track of the blue terms)

$$\frac{\hbar}{2}\left(\frac{k_f}{m}\right)^{1/2}\psi_0 - \tfrac{1}{2}k_f x^2\psi_0 + \tfrac{1}{2}k_f x^2\psi_0 = E\psi_0$$

The blue terms cancel, leaving

$$\frac{\hbar}{2}\left(\frac{k_f}{m}\right)^{1/2}\psi_0 = E\psi_0$$

It follows that ψ_0 is a solution to the Schrödinger equation for the harmonic oscillator with energy $E = \frac{1}{2}\hbar(k_f/m)^{1/2}$, in accord with eqn 7E.5 for the zero-point energy.

Self-test 7E.1 Confirm that the wavefunction in eqn 7E.9 is a solution of eqn 7E.2 and evaluate its energy.

Answer: Yes, with $E_1 = \frac{3}{2}\hbar\omega$

The shapes of several of the wavefunctions are shown in Fig. 7E.6 and the corresponding probability densities are shown in Fig. 7E.7. These probability densities show that, as the quantum number increases, the positions of highest probability migrate towards the classical turning points (see *The chemist's toolkit* 18). This behaviour is another example of the correspondence principle (Topic 7D) in which at high quantum numbers the classical behaviour emerges from the quantum behaviour.

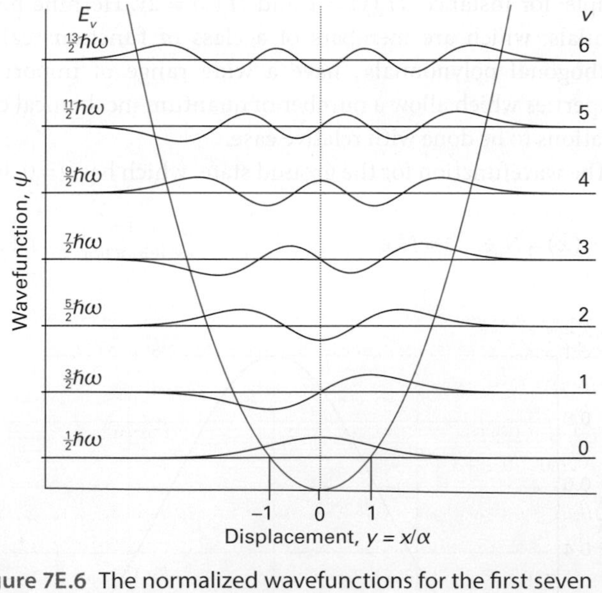

Figure 7E.6 The normalized wavefunctions for the first seven states of a harmonic oscillator. Note that the number of nodes is equal to v. The wavefunctions with even v are symmetric about $y = 0$, and those with odd v are anti-symmetric. The wavefunctions are shown superimposed on the potential energy function, and the horizontal axis for each wavefunction is set at the corresponding energy.

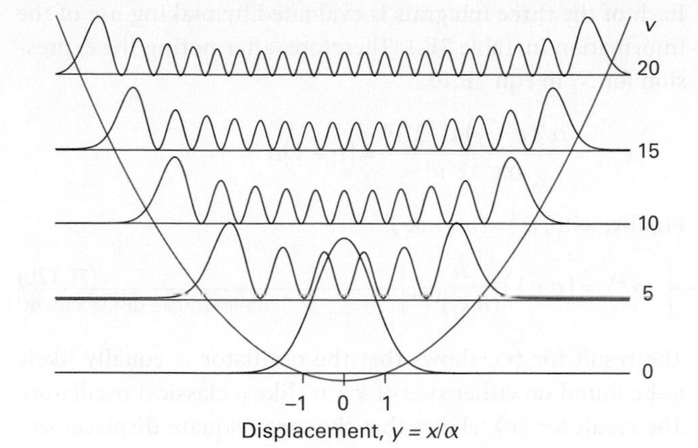

Figure 7E.7 The probability densities for the states of a harmonic oscillator with $v = 0$, 5, 10, 15, and 20. Note how the regions of highest probability density move towards the turning points of the classical motion as v increases.

The wavefunctions have the following features:

- The Gaussian function decays quickly to zero as the displacement in either direction increases, so all the wavefunctions approach zero at large displacements: the particle is unlikely to be found at large displacements.
- The wavefunction oscillates between the classical turning points but decays without oscillating outside them.
- The exponent y^2 is proportional to $x^2 \times (mk_f)^{1/2}$, so the wavefunctions decay more rapidly for large masses and strong restoring forces (stiff springs).
- As v increases, the Hermite polynomials become larger at large displacements (as x^v), so the wavefunctions grow large before the Gaussian function damps them down to zero: as a result, the wavefunctions spread over a wider range as v increases (Fig. 7E.6).

Physical interpretation

Example 7E.2 Normalizing a harmonic oscillator wavefunction

Find the normalization constant for the harmonic oscillator wavefunctions.

Collect your thoughts A wavefunction is normalized (to 1) by evaluating the integral of $|\psi|^2$ over all space and then finding the normalization factor from eqn 7B.3 ($N=1/(\int\psi^*\psi\mathrm{d}\tau)^{1/2}$). The normalized wavefunction is then equal to $N\psi$. In this one-dimensional problem, the volume element is $\mathrm{d}x$ and the integration is from $-\infty$ to $+\infty$. The wavefunctions are expressed in terms of the dimensionless variable $y = x/\alpha$, so begin by expressing the integral in terms of y by using $\mathrm{d}x = \alpha\mathrm{d}y$. The integrals required are given in Table 7E.1.

The solution The unnormalized wavefunction is

$$\psi_v(x)=H_v(y)\mathrm{e}^{-y^2/2}$$

It follows from the integrals given in Table 7E.1 that

$$\int_{-\infty}^{\infty}\psi_v^*\psi_v\mathrm{d}x=\alpha\int_{-\infty}^{\infty}\psi_v^*\psi_v\mathrm{d}y=\alpha\int_{-\infty}^{\infty}H_v^2(y)\mathrm{e}^{-y^2}\mathrm{d}y=\alpha\pi^{1/2}2^v v!$$

where $v! = v(v-1)(v-2)\ldots 1$ and $0! \equiv 1$. Therefore,

$$N_v=\left(\frac{1}{\alpha\pi^{1/2}2^v v!}\right)^{1/2} \quad \text{Normalization constant} \quad (7E.10)$$

Note that N_v is different for each value of v.

Self-test 7E.2 Confirm, by explicit evaluation of the integral, that ψ_0 and ψ_1 are orthogonal.

Answer: Show that $\int_{-\infty}^{\infty}\psi_0^*\psi_1\mathrm{d}x = 0$ by using the information in Table 7E.1

7E.2 Properties of the harmonic oscillator

The average value of a property is calculated by evaluating the expectation value of the corresponding operator (eqn 7C.11, $\langle\Omega\rangle=\int\psi^*\hat{\Omega}\psi\,\mathrm{d}\tau$ for a normalized wavefunction). For a harmonic oscillator,

$$\langle\Omega\rangle_v=\int_{-\infty}^{\infty}\psi_v^*\hat{\Omega}\psi_v\mathrm{d}x \quad (7E.11)$$

When the explicit wavefunctions are substituted, the integrals might look fearsome, but the Hermite polynomials have many features that simplify the calculation.

(a) Mean values

Equation 7E.11 can be used to calculate the mean displacement, $\langle x\rangle$, and the mean square displacement, $\langle x^2\rangle$, for a harmonic oscillator in a state with quantum number v.

How is that done? 7E.1 Finding the mean values of x and x^2 for the harmonic oscillator

The evaluation of the integrals needed to compute $\langle x\rangle$ and $\langle x^2\rangle$ is simplified by recognizing the symmetry of the problem and using the special properties of the Hermite polynomials.

Step 1 *Use a symmetry argument to find the mean displacement*

The mean displacement $\langle x\rangle$ is expected to be zero because the probability density of the oscillator is symmetrical about zero; that is, there is equal probability of positive and negative displacements.

Step 2 *Confirm the result by examining the necessary integral*

More formally, the mean value of x, which is the expectation value of x, is

$$\langle x\rangle_v=\int_{-\infty}^{\infty}\psi_v^{*}x\psi_v\,\mathrm{d}x=N_v^2\int_{-\infty}^{\infty}(H_v\mathrm{e}^{-y^2/2})x(H_v\mathrm{e}^{-y^2/2})\mathrm{d}x$$

$x=\alpha y \quad \mathrm{d}x=\alpha\mathrm{d}y$

An odd function

$$=\alpha^2N_v^2\int_{-\infty}^{\infty}y(H_v\mathrm{e}^{-y^2/2})^2\,\mathrm{d}y$$

The integrand is an odd function because when $y \rightarrow -y$ it changes sign (the squared term does not change sign, but the term y does). The integral of an odd function over a symmetrical range is necessarily zero, so

$$\langle x\rangle_v = 0 \text{ for all } v \qquad \text{Mean displacement} \qquad (7E.12a)$$

Step 3 *Find the mean square displacement*

The mean square displacement, the expectation value of x^2, is

$$\langle x^2\rangle_v=N_v^2\int_{-\infty}^{\infty}(H_v\mathrm{e}^{-y^2/2})x^2(H_v\mathrm{e}^{-y^2/2})\mathrm{d}x$$

$x=\alpha y \quad \mathrm{d}x=\alpha\mathrm{d}y$

$$=\alpha^3N_v^2\int_{-\infty}^{\infty}(H_v\mathrm{e}^{-y^2/2})y^2(H_v\mathrm{e}^{-y^2/2})\mathrm{d}y$$

You can develop the factor y^2H_v by using the recursion relation given in Table 7E.1 rearranged into $yH_v = vH_{v-1} + \frac{1}{2}H_{v+1}$. After multiplying this expression by y it becomes

$$y^2H_v = vyH_{v-1} + \tfrac{1}{2}yH_{v+1}$$

Now use the recursion relation (with v replaced by $v - 1$ or $v + 1$) again for both yH_{v-1} and yH_{v+1}:

$$yH_{v-1} = (v - 1)H_{v-2} + \tfrac{1}{2}H_v$$

$$yH_{v+1} = (v + 1)H_v + \tfrac{1}{2}H_{v+2}$$

It follows that

$$y^2H_v = vyH_{v-1} + \tfrac{1}{2}yH_{v+1} = v\{(v - 1)H_{v-2} + \tfrac{1}{2}H_v\} + \tfrac{1}{2}\{(v + 1)H_v + \tfrac{1}{2}H_{v+2}\}$$

$$= v(v - 1)H_{v-2} + \left(v+\tfrac{1}{2}\right)H_v + \tfrac{1}{4}H_{v+2}$$

Substitution of this result into the integral gives

$$\langle x^2\rangle_v=\alpha^3N_v^2\int_{-\infty}^{\infty}(H_v\mathrm{e}^{-y^2/2})\overbrace{\left\{v(v-1)H_{v-2}+\left(v+\tfrac{1}{2}\right)H_v+\tfrac{1}{4}H_{v+2}\right\}}^{y^2H_v}\mathrm{e}^{-y^2/2}\mathrm{d}y$$

$$=\alpha^3N_v^2v(v-1)\overbrace{\int_{-\infty}^{\infty}H_vH_{v-2}\mathrm{e}^{-y^2}\mathrm{d}y}^{0}$$

$$+\alpha^3N_v^2\left(v+\tfrac{1}{2}\right)\overbrace{\int_{-\infty}^{\infty}H_vH_v\mathrm{e}^{-y^2}\mathrm{d}y}^{\pi^{1/2}2^v v!}+\tfrac{1}{4}\alpha^3N_v^2\overbrace{\int_{-\infty}^{\infty}H_vH_{v+2}\mathrm{e}^{-y^2}\mathrm{d}y}^{0}$$

$$=\alpha^3N_v^2\left(v+\tfrac{1}{2}\right)\pi^{1/2}2^v v!$$

Each of the three integrals is evaluated by making use of the information in Table 7E.1. Therefore, after noting the expression for N_v in eqn 7E.10,

$$\langle x^2\rangle_v=\frac{\alpha^3\left(v+\frac{1}{2}\right)\pi^{1/2}2^v v!}{\alpha\pi^{1/2}2^v v!}=\left(v+\tfrac{1}{2}\right)\alpha^2$$

Finally, with, $\alpha^2=(\hbar^2/mk_\mathrm{f})^{1/2}$

$$\langle x^2\rangle_v=\left(v+\tfrac{1}{2}\right)\frac{\hbar}{(mk_\mathrm{f})^{1/2}} \qquad (7E.12b)$$

Mean square displacement

The result for $\langle x\rangle_v$ shows that the oscillator is equally likely to be found on either side of $x = 0$ (like a classical oscillator). The result for $\langle x^2\rangle_v$ shows that the mean square displacement increases with v. This increase is apparent from the probability densities in Fig. 7E.7, and corresponds to the amplitude of a classical harmonic oscillator becoming greater as its energy increases.

The mean potential energy of an oscillator, which is the expectation value of $V = \frac{1}{2}k_\mathrm{f}x^2$, can now be calculated:

$$\langle V\rangle_v=\tfrac{1}{2}\langle k_\mathrm{f}x^2\rangle_v=\tfrac{1}{2}k_\mathrm{f}\langle x^2\rangle_v=\tfrac{1}{2}\left(v+\tfrac{1}{2}\right)\hbar\left(\frac{k_\mathrm{f}}{m}\right)^{1/2}$$

or

$$\langle V\rangle_v=\tfrac{1}{2}\left(v+\tfrac{1}{2}\right)\hbar\omega \qquad \text{Mean potential energy} \qquad (7E.13a)$$

Because the total energy in the state with quantum number v is $(v + \frac{1}{2})\hbar\omega$, it follows that

$$\langle V\rangle_v=\tfrac{1}{2}E_v \qquad \text{Mean potential energy} \qquad (7E.13b)$$

The total energy is the sum of the potential and kinetic energies, $E_v=\langle V\rangle_v+\langle E_\mathrm{k}\rangle_v$, so it follows that the mean kinetic energy of the oscillator is

$$\langle E_\mathrm{k}\rangle_v=E_v-\langle V\rangle_v=E_v-\tfrac{1}{2}E_v=\tfrac{1}{2}E_v \qquad \text{Mean kinetic energy} \qquad (7E.13c)$$

The result that the mean potential and kinetic energies of a harmonic oscillator are equal (and therefore that both are equal to half the total energy) is a special case of the **virial theorem**:

> If the potential energy of a particle has the form $V = ax^b$, then its mean potential and kinetic energies are related by

$$2\langle E_\mathrm{k}\rangle = b\langle V\rangle \qquad \text{Virial theorem} \qquad (7E.14)$$

For a harmonic oscillator $b = 2$, so $\langle E_\mathrm{k}\rangle_v = \langle V\rangle_v$. The virial theorem is a short cut to the establishment of a number of useful results, and it is used elsewhere (e.g. in Topic 8A).

(b) Tunnelling

A quantum oscillator may be found at displacements with $V > E$, which are forbidden by classical physics because they correspond to negative kinetic energy. That is, a harmonic oscillator can tunnel into classically forbidden displacements. As shown in *Example* 7E.3, for the lowest energy state of the harmonic oscillator, there is about an 8 per cent chance of finding the oscillator at classically forbidden displacements in either direction. These tunnelling probabilities are independent of the force constant and mass of the oscillator.

Example 7E.3 Calculating the tunnelling probability for the harmonic oscillator

Calculate the probability that the ground-state harmonic oscillator will be found in a classically forbidden region.

Collect your thoughts Find the expression for the classical turning point, x_{tp}, where the kinetic energy goes to zero, by equating the potential energy to the total energy of the harmonic oscillator. You can then calculate the probability of finding the oscillator at a displacement beyond x_{tp} by integrating $\psi^2 dx$ between x_{tp} and infinity

$$P = \int_{x_{tp}}^{\infty} \psi_\upsilon^2 dx$$

By symmetry, the probability of the particle being found in the classically forbidden region from $-x_{tp}$ to $-\infty$ is the same.

The solution According to classical mechanics, the turning point, x_{tp}, of an oscillator occurs when its potential energy $\frac{1}{2}k_f x^2$ is equal to its total energy. When that energy is one of the allowed values E_υ, the turning point is at

$$E_\upsilon = \tfrac{1}{2}k_f x_{tp}^2 \quad \text{and therefore at} \quad x_{tp} = \pm\left(\frac{2E_\upsilon}{k_f}\right)^{1/2}$$

The variable of integration in the integral P is best expressed in terms of $y = x/\alpha$ with $\alpha = (\hbar^2/mk_f)^{1/4}$. With these substitutions, and also using $E_\upsilon = (\upsilon + \frac{1}{2})\hbar\omega$, the turning points are given by

$$y_{tp} = \frac{x_{tp}}{\alpha} = \left\{\frac{2(\upsilon+\frac{1}{2})\hbar\omega}{\alpha^2 k_f}\right\}^{1/2} \overset{\omega=(k_f/m)^{1/2}}{=} (2\upsilon+1)^{1/2}$$

For the state of lowest energy ($\upsilon = 0$), $y_{tp} = 1$ and the probability of being beyond that point is

$$P = \int_{x_{tp}}^{\infty} \psi_0^2 dx \overset{dx=\alpha dy}{=} \alpha\int_1^{\infty} \psi_0^2 dy = \alpha N_0^2 \int_1^{\infty} e^{-y^2} dy$$

with

$$N_0 = \left(\frac{1}{\alpha\pi^{1/2}2^0 0!}\right)^{1/2} \overset{2^0=1;\,0!\equiv 1}{=} \left(\frac{1}{\alpha\pi^{1/2}}\right)^{1/2}$$

Therefore

$$P = \frac{1}{\pi^{1/2}}\int_1^{\infty} e^{-y^2} dy$$

The integral must be evaluated numerically (by using mathematical software), and is equal to 0.139.... It follows that $P = 0.079$.

Comment. In 7.9 per cent of a large number of observations of an oscillator in the state with quantum number $\upsilon = 0$, the particle will be found beyond the (positive) classical turning point. It will be found with the same probability at negative forbidden displacements. The total probability of finding the oscillator in a classically forbidden region is about 16 per cent.

Self-test 7E.3 Calculate the probability that a harmonic oscillator in the state with quantum number $\upsilon = 1$ will be found at a classically forbidden extension. You will need to use mathematical software to evaluate the integral.

Answer: P = 0.056

The probability of finding the oscillator in classically forbidden regions decreases quickly with increasing υ, and vanishes entirely as υ approaches infinity, as is expected from the correspondence principle. Macroscopic oscillators (such as pendulums) are in states with very high quantum numbers, so the tunnelling probability is wholly negligible and classical mechanics is reliable. Molecules, however, are normally in their vibrational ground states, and for them the probability is very significant and classical mechanics is misleading.

Checklist of concepts

- ☐ 1. The energy levels of a quantum mechanical harmonic oscillator are evenly spaced.
- ☐ 2. The wavefunctions of a quantum mechanical harmonic oscillator are products of a **Hermite polynomial** and a Gaussian (bell-shaped) function.
- ☐ 3. A quantum mechanical harmonic oscillator has **zero-point energy**, an irremovable minimum energy.
- ☐ 4. The probability of finding a quantum mechanical harmonic oscillator at classically forbidden displacements is significant for the ground vibrational state ($\upsilon = 0$) but decreases quickly with increasing υ.

Checklist of equations

Property	Equation	Comment	Equation number
Energy levels	$E_v=(v+\frac{1}{2})\hbar\omega \quad \omega=(k_f/m)^{1/2}$	$v = 0, 1, 2, \ldots$	7E.3
Zero-point energy	$E_0=\frac{1}{2}\hbar\omega$		7E.5
Wavefunctions	$\psi_v(x)=N_v H_v(y)e^{-y^2/2}$	$v = 0, 1, 2, \ldots$	7E.7
	$y=x/\alpha \quad \alpha=(\hbar^2/mk_f)^{1/4}$		
Normalization constant	$N_v=(1/\alpha\pi^{1/2}2^v v!)^{1/2}$		7E.10
Mean displacement	$\langle x\rangle_v=0$		7E.12a
Mean square displacement	$\langle x^2\rangle_v=(v+\frac{1}{2})\hbar/(mk_f)^{1/2}$		7E.12b
Virial theorem	$2\langle E_k\rangle=b\langle V\rangle$	$V = ax^b$	7E.14

TOPIC 7F Rotational motion

➤ **Why do you need to know this material?**

Angular momentum is central to the description of the electronic structure of atoms and molecules and the interpretation of molecular spectra.

➤ **What is the main idea?**

The energy, angular momentum, and orientation of the angular momentum of a rotating body are quantized.

➤ **What do you need to know already?**

You should be aware of the postulates of quantum mechanics and the role of boundary conditions (Topics 7C and 7D). Background information on the description of rotation and the coordinate systems used to describe it are given in three *Toolkits*.

Rotational motion is encountered in many aspects of chemistry, including the electronic structures of atoms, because electrons orbit (in a quantum mechanical sense) around nuclei and spin on their axis. Molecules also rotate; transitions between their rotational states affect the appearance of spectra and their detection gives valuable information about the structures of molecules.

7F.1 Rotation in two dimensions

Consider a particle of mass m constrained to move in a circular path (a 'ring') of radius r in the xy-plane with constant potential energy, which may be taken to be zero (Fig. 7F.1); the energy is entirely kinetic. The Schrödinger equation is

$$-\frac{\hbar^2}{2m}\left(\frac{\partial^2}{\partial x^2}+\frac{\partial^2}{\partial y^2}\right)\psi(x,y)=E\psi(x,y) \qquad \text{(7F.1)}$$

with the particle confined to a path of constant radius r. The equation is best expressed in cylindrical coordinates r and ϕ with $z = 0$ (*The chemist's toolkit* 19) because they reflect the symmetry of the system. In cylindrical coordinates

$$\frac{\partial^2}{\partial x^2}+\frac{\partial^2}{\partial y^2}=\frac{\partial^2}{\partial r^2}+\frac{1}{r}\frac{\partial}{\partial r}+\frac{1}{r^2}\frac{\partial^2}{\partial \phi^2} \qquad \text{(7F.2)}$$

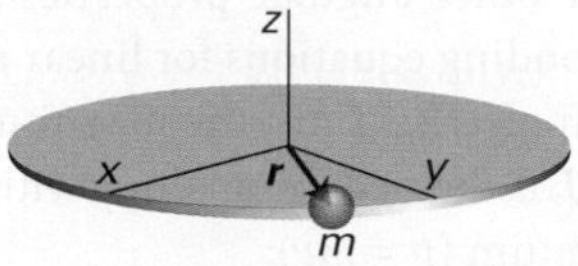

Figure 7F.1 A particle on a ring is free to move in the xy-plane around a circular path of radius r.

However, because the radius of the path is fixed, the (blue) derivatives with respect to r can be discarded. Only the last term in eqn 7F.2 then survives and the Schrödinger equation becomes

$$-\frac{\hbar^2}{2mr^2}\frac{d^2\psi(\phi)}{d\phi^2}=E\psi(\phi) \qquad \text{(7F.3a)}$$

The partial derivative has been replaced by a complete derivative because ϕ is now the only variable. The term mr^2 is the moment of inertia, $I = mr^2$ (*The chemist's toolkit* 20), and so the Schrödinger equation becomes

$$-\frac{\hbar^2}{2I}\frac{d^2\psi(\phi)}{d\phi^2}=E\psi(\phi) \qquad \text{Schrödinger equation [particle on a ring]} \qquad \text{(7F.3b)}$$

The chemist's toolkit 19 Cylindrical coordinates

For systems with cylindrical symmetry it is best to work in **cylindrical coordinates** r, ϕ, and z (Sketch 1), with

$$x = r\cos\phi \qquad y = r\sin\phi$$

and where

$$0 \le r \le \infty \qquad 0 \le \phi \le 2\pi \qquad -\infty \le z \le +\infty$$

The volume element is

$$d\tau = r\,dr\,d\phi\,dz$$

For motion in a plane, $z = 0$ and the volume element is

$$d\tau = r\,dr\,d\phi$$

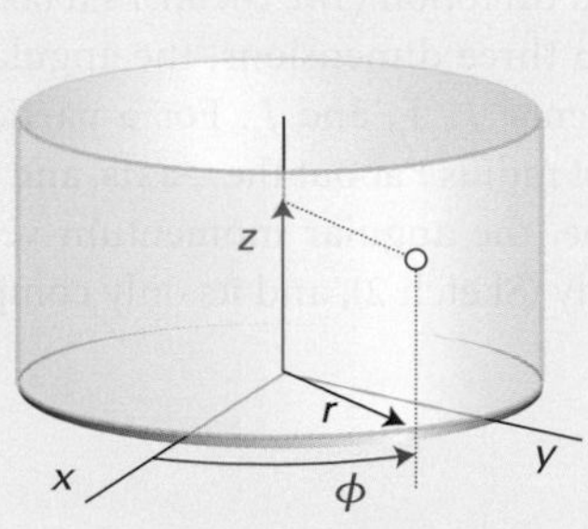

Sketch 1

The chemist's toolkit 20 Angular momentum

Angular velocity, ω (omega), is the rate of change of angular position; it is reported in radians per second (rad s^{-1}). There are 2π radians in a circle, so 1 cycle per second is the same as 2π radians per second. For convenience, the 'rad' is often dropped, and the units of angular velocity are denoted s^{-1}.

Expressions for other angular properties follow by analogy with the corresponding equations for linear motion (*The chemist's toolkit* 3 in Topic 1B). Thus, the magnitude, J, of the **angular momentum**, J, is defined, by analogy with the magnitude of the linear momentum ($p = mv$):

$$J = I\omega$$

The quantity I is the **moment of inertia** of the object. It represents the resistance of the object to a change in the state of rotation in the same way that mass represents the resistance of the object to a change in the state of translation. In the case of a rotating molecule the moment of inertia is defined as

$$I = \sum_i m_i r_i^2$$

where m_i is the mass of atom i and r_i is its perpendicular distance from the axis of rotation (Sketch 1). For a point particle of mass m moving in a circle of radius r, the moment of inertia about the axis of rotation is

$$I = mr^2$$

The SI units of moment of inertia are therefore kilogram metre^2 (kg m^2), and those of angular momentum are kilogram metre^2 per second (kg m^2 s^{-1}).

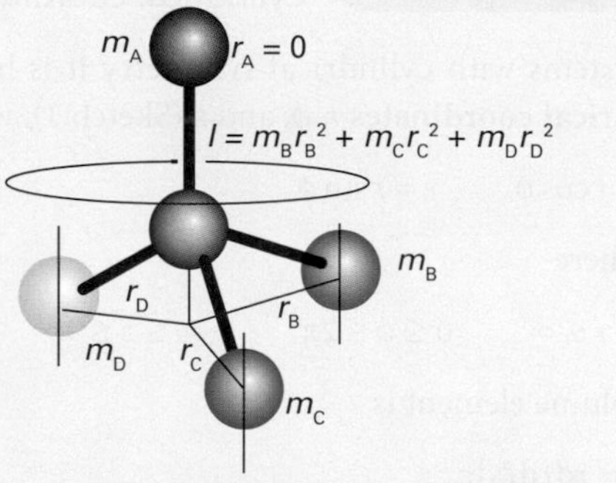

Sketch 1

The angular momentum is a vector, a quantity with both magnitude and direction (*The chemist's toolkit* 17 in Topic 7D). For rotation in three dimensions, the angular momentum has three components: J_x, J_y, and J_z. For a particle travelling on a circular path of radius r about the z-axis, and therefore confined to the xy-plane, the angular momentum vector points in the z-direction only (Sketch 2), and its only component is

$$J_z = \pm pr$$

where p is the magnitude of the linear momentum in the xy-plane at any instant. When $J_z > 0$, the particle travels in a clockwise direction as viewed from below; when $J_z < 0$, the motion is anticlockwise. A particle that is travelling at high speed in a circle has a higher angular momentum than a particle of the same mass travelling more slowly. An object with a high angular momentum (like a flywheel) requires a strong braking force (more precisely, a strong 'torque') to bring it to a standstill.

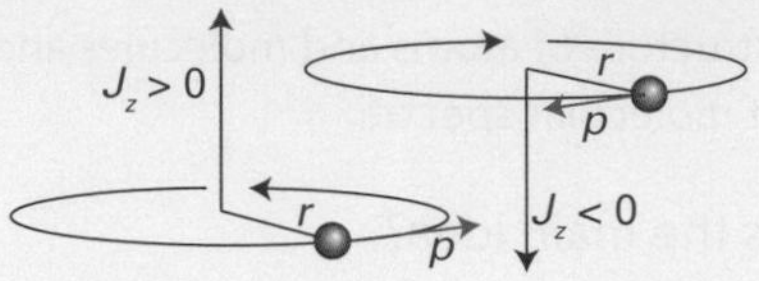

Sketch 2

The components of the angular momentum vector $\boldsymbol{J}$ when it lies in a general orientation are

$$J_x = yp_z - zp_y \qquad J_y = zp_x - xp_z \qquad J_z = xp_y - yp_x$$

where p_x is the component of the linear momentum in the x-direction at any instant, and likewise p_y and p_z in the other directions. The square of the magnitude of the angular momentum vector is given by

$$J^2 = J_x^2 + J_y^2 + J_z^2$$

By analogy with the expression for linear motion ($E_k = \frac{1}{2}mv^2 = p^2/2m$), the kinetic energy of a rotating object is

$$E_k = \tfrac{1}{2}I\omega^2 = \frac{J^2}{2I}$$

For a given moment of inertia, high angular momentum corresponds to high kinetic energy. As may be verified, the units of rotational energy are joules (J).

The analogous roles of m and I, of v and ω, and of p and J in the translational and rotational cases respectively provide a ready way of constructing and recalling equations. These analogies are summarized below:

Translation		Rotation	
Property	Significance	Property	Significance
Mass, m	Resistance to the effect of a force	Moment of inertia, I	Resistance to the effect of a twisting force (torque)
Speed, v	Rate of change of position	Angular velocity, ω	Rate of change of angle
Magnitude of linear momentum, p	$p = mv$	Magnitude of angular momentum, J	$J = I\omega$
Translational kinetic energy, E_k	$E_k = \frac{1}{2}mv^2 = p^2/2m$	Rotational kinetic energy, E_k	$E_k = \frac{1}{2}I\omega^2 = J^2/2I$

(a) The solutions of the Schrödinger equation

The most straightforward way of finding the solutions of eqn 7F.3b is to take the known general solution to a second-order differential equation of this kind and show that it does indeed satisfy the equation. Then find the allowed solutions and energies by imposing the relevant boundary conditions.

How is that done? 7F.1 Finding the solutions of the Schrödinger equation for a particle on a ring

A solution of eqn 7F.3b is

$$\psi(\phi)=e^{im_l\phi}$$

where, as yet, m_l is an arbitrary dimensionless number (the notation is explained later). This is not the most general solution, which would be $\psi(\phi)=Ae^{im_l\phi}+Be^{-im_l\phi}$, but is sufficiently general for the present purpose.

Step 1 *Verify that the function satisfies the equation*

To verify that $\psi(\phi)$ is a solution note that

$$\frac{d^2}{d\phi^2}e^{im_l\phi}=\frac{d}{d\phi}(im_l)e^{im_l\phi}=(im_l)^2e^{im_l\phi}=-m_l^2\overbrace{e^{im_l\phi}}^{\psi}=-m_l^2\psi$$

Then

$$-\frac{\hbar^2}{2I}\frac{d^2\psi}{d\phi^2}=-\frac{\hbar^2}{2I}(-m_l^2\psi)=\frac{m_l^2\hbar^2}{2I}\psi$$

which has the form constant × ψ, so the proposed wavefunction is indeed a solution and the corresponding energy is $m_l^2\hbar^2/2I$.

Step 2 *Impose the appropriate boundary conditions*

The requirement that a wavefunction must be single-valued implies the existence of a **cyclic boundary condition**, the requirement that the wavefunction must be the same after a complete revolution: $\psi(\phi+2\pi)=\psi(\phi)$ (Fig. 7F.2). In this case

$$\psi(\phi+2\pi)=e^{im_l(\phi+2\pi)}=e^{im_l\phi}e^{2\pi im_l}$$
$$=\psi(\phi)e^{2\pi im_l}=\psi(\phi)(e^{i\pi})^{2m_l}$$

As $e^{i\pi}=-1$, this relation is equivalent to

$$\psi(\phi+2\pi)=(-1)^{2m_l}\psi(\phi)$$

The cyclic boundary condition $\psi(\phi+2\pi)=\psi(\phi)$ requires $(-1)^{2m_l}=1$; this requirement is satisfied for any positive or negative integer value of m_l, including 0.

Step 3 *Normalize the wavefunction*

A one-dimensional wavefunction is normalized (to 1) by finding the normalization constant N given by eqn 7B.3 ($N=(\int_{-\infty}^{\infty}\psi^*\psi dx)^{-1/2}$). In this case, the wavefunction depends only on the angle ϕ and the range of integration is from $\phi=0$ to 2π, so the normalization constant is

$$N=\frac{1}{\left(\int_0^{2\pi}\psi^*\psi d\phi\right)^{1/2}}=\frac{1}{\left(\int_0^{2\pi}\underbrace{e^{-im_l\phi}e^{im_l\phi}}_{1}d\phi\right)^{1/2}}=\frac{1}{(2\pi)^{1/2}}$$

The normalized wavefunctions and corresponding energies are labelled with the integer m_l, which is playing the role of a quantum number, and are therefore

$$\psi_{m_l}(\phi)=\frac{e^{im_l\phi}}{(2\pi)^{1/2}}$$
$$E_{m_l}=\frac{m_l^2\hbar^2}{2I}\quad m_l=0,\pm1,\pm2,\ldots \tag{7F.4}$$

Wavefunctions and energy levels of a particle on a ring

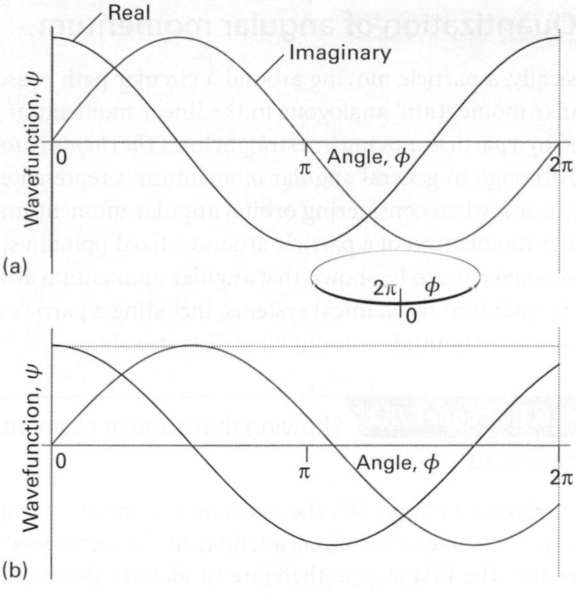

Figure 7F.2 Two possible solutions of the Schrödinger equation for a particle on a ring. The circumference has been opened out into a straight line; the points at $\phi=0$ and 2π are identical. The solution in (a), $e^{i\phi}=\cos\phi+i\sin\phi$, is acceptable because after a complete revolution the wavefunction has the same value. The solution in (b), $e^{0.9i\phi}=\cos(0.9\phi)+i\sin(0.9\phi)$ is unacceptable because its value, both for the real and imaginary parts, is not the same at $\phi=0$ and 2π.

Apart from the level with $m_l=0$, each of the energy levels is doubly degenerate because the dependence of the energy on m_l^2 means that two values of m_l (such as +1 and −1) correspond to the same energy.

A note on good practice Note that, when quoting the value of m_l, it is good practice always to give the sign, even if m_l is positive. Thus, write $m_l=+1$, not $m_l=1$.

(b) Quantization of angular momentum

Classically, a particle moving around a circular path possesses 'angular momentum' analogous to the linear momentum possessed by a particle moving in a straight line (*The chemist's toolkit* 20). Although in general angular momentum is represented by the vector $\boldsymbol{J}$, when considering **orbital angular momentum**, the angular momentum of a particle around a fixed point in space, it is denoted $\boldsymbol{l}$. It can be shown that angular momentum also occurs in quantum mechanical systems, including a particle on a ring, but its magnitude is confined to discrete values.

How is that done? 7F.2 Showing that angular momentum is quantized

As explained in Topic 7C, the outcome of a measurement of a property is one of the eigenfunctions of the corresponding operator. The first step is therefore to identify the operator corresponding to angular momentum, and then to identify its eigenvalues.

Step 1 *Construct the operator for angular momentum*

Because the particle is confined to the *xy*-plane, its angular momentum is directed along the *z*-axis, so only this component need be considered. According to *The chemist's toolkit* 20, the *z*-component of the orbital angular momentum is

$$l_z = xp_y - yp_x$$

where x and y specify the position and p_x and p_y are the components of the linear momentum of the particle. The corresponding operator is formed by replacing x, y, p_x, and p_y by their corresponding operators (Topic 7C; $\hat{q}=q\times$ and $\hat{p}_q=(\hbar/\mathrm{i})\partial/\partial q$, with $q = x$ and y), which gives

$$\hat{l}_z = \frac{\hbar}{\mathrm{i}}\left(x\frac{\partial}{\partial y} - y\frac{\partial}{\partial x}\right) \quad \text{Operator for the } z\text{-component of the angular momentum} \quad (7F.5a)$$

In cylindrical coordinates (see *The chemist's toolkit* 19) this operator becomes

$$\hat{l}_z = \frac{\hbar}{\mathrm{i}}\frac{\mathrm{d}}{\mathrm{d}\phi} \quad (7F.5b)$$

Step 2 *Verify that the wavefunctions are eigenfunctions of this operator*

To decide whether the wavefunctions in eqn 7F.4 are eigenfunctions of $\hat{l}_z$, allow it to act on the wavefunction:

$$\hat{l}_z\psi_{m_l} = \frac{\hbar}{\mathrm{i}}\frac{\mathrm{d}}{\mathrm{d}\phi}\mathrm{e}^{\mathrm{i}m_l\phi} = \frac{\hbar}{\mathrm{i}}\mathrm{i}m_l\overbrace{\mathrm{e}^{\mathrm{i}m_l\phi}}^{\psi_{m_l}} = m_l\hbar\psi_{m_l}$$

The wavefunction is an eigenfunction of the angular momentum, with the eigenvalue $m_l\hbar$. In summary,

$$\hat{l}_z\psi_{m_l}(\phi) = m_l\hbar\psi_{m_l}(\phi) \quad m_l = 0, \pm1, \pm2, \ldots \quad \text{Eigenfunctions of } \hat{l}_z \quad (7F.6)$$

Because m_l is confined to discrete values, the *z*-component of angular momentum is quantized. When m_l is positive, the *z*-component of angular momentum is positive (clockwise rotation when seen from below); when m_l is negative, the *z*-component of angular momentum is negative (anticlockwise when seen from below).

The important features of the results so far are:

Physical interpretation

- The energies are quantized because m_l is confined to integer values.
- The occurrence of m_l as its square means that the energy of rotation is independent of the sense of rotation (the sign of m_l), as expected physically.
- Apart from the state with $m_l = 0$, all the energy levels are doubly degenerate; rotation can be clockwise or anticlockwise with the same energy.
- There is no zero-point energy: the particle can be stationary.
- As m_l increases the wavefunctions oscillate with shorter wavelengths and so have greater curvature, corresponding to increasing kinetic energy (Fig. 7F.3).
- As pointed out in Topic 7D, a wavefunction that is complex represents a direction of motion, and taking its complex conjugate reverses the direction. The wavefunctions with $m_l > 0$ and $m_l < 0$ are each other's complex conjugate, and so they correspond to motion in opposite directions.

The probability density predicted by the wavefunctions of eqn 7F.4 is uniform around the ring:

$$\psi^*_{m_l}\psi_{m_l} = \left(\frac{\mathrm{e}^{\mathrm{i}m_l\phi}}{(2\pi)^{1/2}}\right)^*\left(\frac{\mathrm{e}^{\mathrm{i}m_l\phi}}{(2\pi)^{1/2}}\right)$$
$$= \left(\frac{\mathrm{e}^{-\mathrm{i}m_l\phi}}{(2\pi)^{1/2}}\right)\left(\frac{\mathrm{e}^{\mathrm{i}m_l\phi}}{(2\pi)^{1/2}}\right) = \frac{1}{2\pi}$$

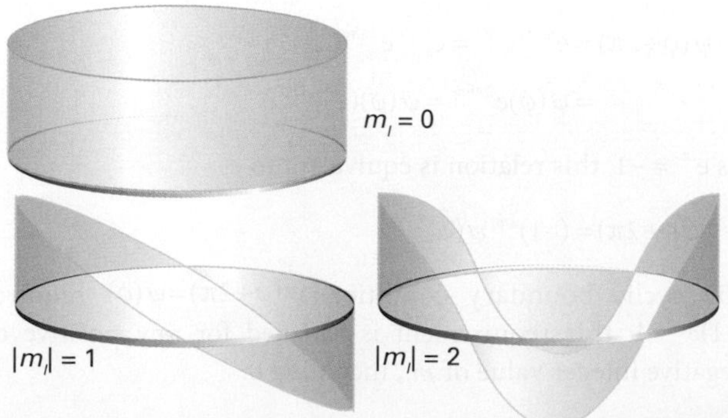

Figure 7F.3 The real parts of the wavefunctions of a particle on a ring. As the energy increases, so does the number of nodes and the curvature.

Angular momentum and angular position are a pair of complementary observables (in the sense defined in Topic 7C), and the inability to specify them simultaneously with arbitrary precision is another example of the uncertainty principle. In this case the z-component of angular momentum is known exactly (as $m_l\hbar$) but the location of the particle on the ring is completely unknown, which is reflected by the uniform probability density.

Example 7F.1 Using the particle on a ring model

The particle-on-a-ring is a crude but illustrative model of cyclic, conjugated molecular systems. Treat the π electrons in benzene as particles freely moving over a circular ring of six carbon atoms and calculate the minimum energy required for the excitation of a π electron. The carbon–carbon bond length in benzene is 140 pm.

Collect your thoughts Because each carbon atom contributes one π electron, there are six electrons to accommodate. Each state is occupied by two electrons, so only the $m_l = 0, +1$, and -1 states are occupied (with the last two being degenerate). The minimum energy required for excitation corresponds to a transition of an electron from the $m_l = +1$ (or -1) state to the $m_l = +2$ (or -2) state. Use eqn 7F.4, and the mass of the electron, to calculate the energies of the states. A hexagon can be inscribed inside a circle with a radius equal to the side of the hexagon, so take $r = 140$ pm.

The solution From eqn 7F.4, the energy separation between the states with $m_l = +1$ and $m_l = +2$ is

$$\Delta E = E_{+2} - E_{+1} = (4-1)\times\frac{(1.055\times10^{-34}\,\text{J s})^2}{2\times(9.109\times10^{-31}\,\text{kg})\times(1.40\times10^{-10}\,\text{m})^2}$$

$$= 9.35\times10^{-19}\,\text{J}$$

Therefore the minimum energy required to excite an electron is 0.935 aJ or 563 kJ mol^{-1}. This energy separation corresponds to an absorption frequency of 1410 THz (1 THz = 10^{12} Hz) and a wavelength of 213 nm; the experimental value for a transition of this kind is 260 nm. Such a crude model cannot be expected to give quantitative agreement, but the value is at least of the right order of magnitude.

Self-test 7F.1 Use the particle-on-a-ring model to calculate the minimum energy required for the excitation of a π electron in coronene, $C_{24}H_{12}$ (1). Assume that the radius of the ring is three times the carbon–carbon bond length in benzene and that the electrons are confined to the periphery of the molecule.

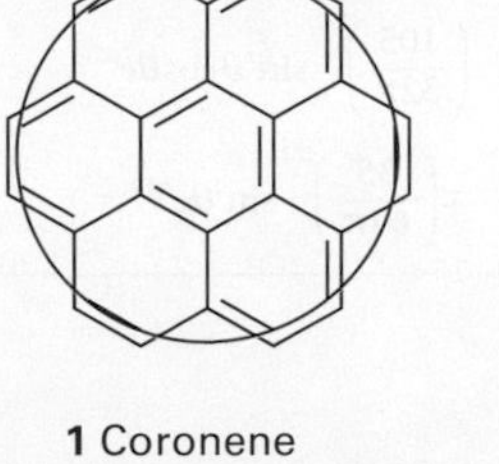

1 Coronene
(model ring in red)

Answer: For transition from $m_l = +3$ to $m_l = +4$: $\Delta E = 0.0147$ zJ or 8.83 J mol^{-1}

7F.2 Rotation in three dimensions

Now consider a particle of mass m that is free to move anywhere on the surface of a sphere of radius r.

(a) The wavefunctions and energy levels

The potential energy of a particle on the surface of a sphere is the same everywhere and may be taken to be zero. The Schrödinger equation is therefore

$$-\frac{\hbar^2}{2m}\nabla^2\psi = E\psi \tag{7F.7a}$$

where the sum of the three second derivatives, denoted ∇^2 and read 'del squared', is called the 'laplacian':

$$\nabla^2 = \frac{\partial^2}{\partial x^2}+\frac{\partial^2}{\partial y^2}+\frac{\partial^2}{\partial z^2} \qquad \text{Laplacian} \tag{7F.7b}$$

To take advantage of the symmetry of the problem it is appropriate to change to spherical polar coordinates (*The chemist's toolkit* 21) when the laplacian becomes

$$\nabla^2 = \frac{1}{r}\frac{\partial^2}{\partial r^2}r + \frac{1}{r^2}\Lambda^2$$

where the derivatives with respect to the colatitude θ and the azimuth ϕ are collected in Λ^2, which is called the 'legendrian' and is given by

$$\Lambda^2 = \frac{1}{\sin^2\theta}\frac{\partial^2}{\partial\phi^2} + \frac{1}{\sin\theta}\frac{\partial}{\partial\theta}\sin\theta\frac{\partial}{\partial\theta}$$

In the present case, r is fixed, so the derivatives with respect to r in the laplacian can be ignored and only the term Λ^2/r^2 survives. The Schrödinger equation then becomes

$$-\frac{\hbar^2}{2m}\frac{1}{r^2}\Lambda^2\psi(\theta,\phi) = E\psi(\theta,\phi)$$

The term mr^2 in the denominator can be recognized as the moment of inertia, I, of the particle, so the Schrödinger equation is

$$-\frac{\hbar^2}{2I}\Lambda^2\psi(\theta,\phi) = E\psi(\theta,\phi) \qquad \text{Schrödinger equation [particle on a sphere]} \tag{7F.8}$$

There are two cyclic boundary conditions to fulfil. The first is the same as for the two-dimensional case, where the wavefunction must join up on completing a circuit around the equator, as specified by the angle ϕ. The second is a similar requirement that the wavefunction must join up on encircling over the poles, as specified by the angle θ. These two conditions are illustrated in Fig. 7F.4. Once again, it can be shown that the need to satisfy them leads to the conclusion that the energy and the angular momentum are quantized.

The chemist's toolkit 21 Spherical polar coordinates

The mathematics of systems with spherical symmetry (such as atoms) is often greatly simplified by using **spherical polar coordinates** (Sketch 1): r, the distance from the origin (the radius), θ, the colatitude, and ϕ, the azimuth. The ranges of these coordinates are (with angles in radians, Sketch 2): $0 \le r \le +\infty$, $0 \le \theta \le \pi$, $0 \le \phi \le 2\pi$.

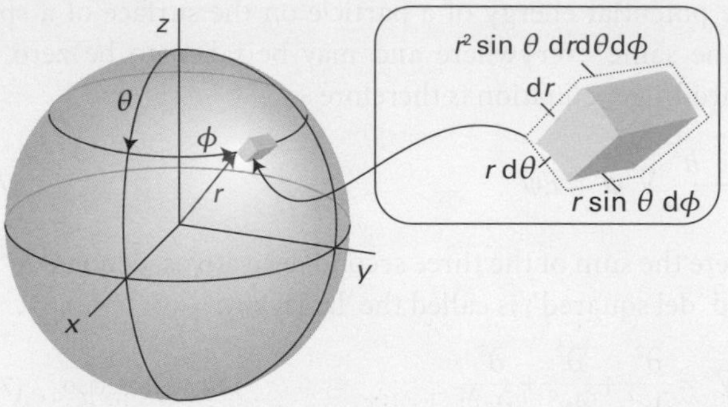

Sketch 1

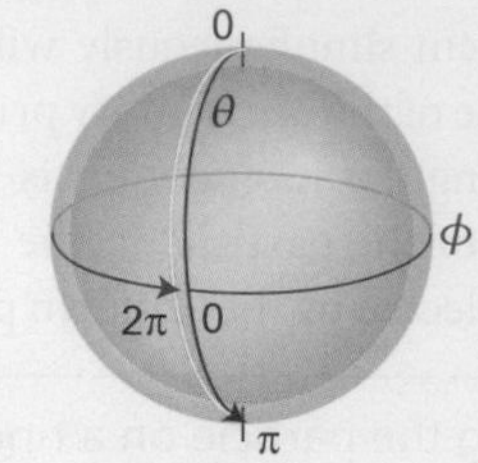

Sketch 2

Cartesian and polar coordinates are related by

$$x = r\sin\theta\cos\phi \quad y = r\sin\theta\sin\phi \quad z = r\cos\theta$$

The volume element in Cartesian coordinates is $d\tau = dxdydz$, and in spherical polar coordinates it becomes

$$d\tau = r^2 \sin\theta \, drd\theta d\phi$$

An integral of a function $f(r,\theta,\phi)$ over all space in polar coordinates therefore has the form

$$\int f d\tau = \int_{r=0}^{\infty}\int_{\theta=0}^{\pi}\int_{\phi=0}^{2\pi} f(r,\theta,\phi) r^2 \sin\theta \, drd\theta d\phi$$

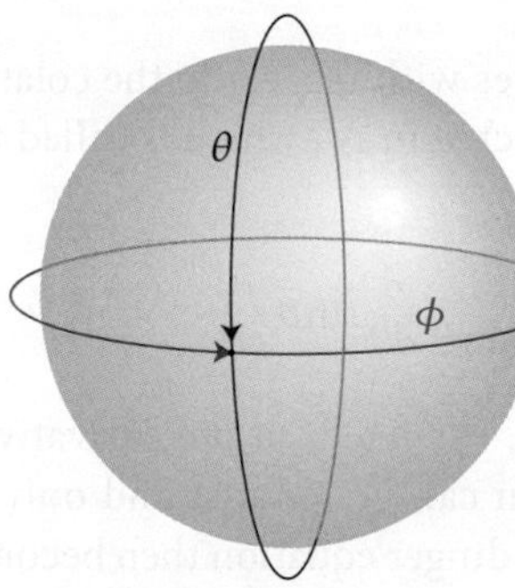

Figure 7F.4 The wavefunction of a particle on the surface of a sphere must satisfy two cyclic boundary conditions. This requirement leads to two quantum numbers for its state of angular momentum.

How is that done? 7F.3 Finding the solutions of the Schrödinger equation for a particle on a sphere

The functions known as **spherical harmonics**, $Y_{l,m_l}(\theta,\phi)$ (Table 7F.1), are well known to mathematicians and are the solutions of the equation[1]

$$\Lambda^2 Y_{l,m_l}(\theta,\phi) = -l(l+1)Y_{l,m_l}(\theta,\phi),$$
$$l = 0, 1, 2, \ldots \quad m_l = 0, \pm1, \ldots, \pm l \qquad (7F.9)$$

[1] See the first section of *A deeper look* 3 on the website for this text for details of how the separation of variables procedure is used to find the form of the spherical harmonics.

Table 7F.1 The spherical harmonics

l	m_l	$Y_{l,m_l}(\theta,\phi)$
0	0	$\left(\frac{1}{4\pi}\right)^{1/2}$
1	0	$\left(\frac{3}{4\pi}\right)^{1/2}\cos\theta$
	±1	$\mp\left(\frac{3}{8\pi}\right)^{1/2}\sin\theta e^{\pm i\phi}$
2	0	$\left(\frac{5}{16\pi}\right)^{1/2}(3\cos^2\theta-1)$
	±1	$\mp\left(\frac{15}{8\pi}\right)^{1/2}\cos\theta\sin\theta e^{\pm i\phi}$
	±2	$\left(\frac{15}{32\pi}\right)^{1/2}\sin^2\theta e^{\pm 2i\phi}$
3	0	$\left(\frac{7}{16\pi}\right)^{1/2}(5\cos^3\theta-3\cos\theta)$
	±1	$\mp\left(\frac{21}{64\pi}\right)^{1/2}(5\cos^2\theta-1)\sin\theta e^{\pm i\phi}$
	±2	$\left(\frac{105}{32\pi}\right)^{1/2}\sin^2\theta\cos\theta e^{\pm 2i\phi}$
	±3	$\mp\left(\frac{35}{64\pi}\right)^{1/2}\sin^3\theta e^{\pm 3i\phi}$

These functions satisfy the two cyclic boundary conditions and are normalized (to 1).

Step 1 *Show that the spherical harmonics solve the Schrödinger equation*

It follows from eqn 7F.8 that

$$-\frac{\hbar^2}{2I}\overbrace{\Lambda^2 Y_{l,m_l}(\theta,\phi)}^{-l(l+1)Y_{l,m_l}} = \overbrace{l(l+1)\frac{\hbar^2}{2I}}^{E} Y_{l,m_l}(\theta,\phi)$$

The spherical harmonics are therefore solutions of the Schrödinger equation with energies $E=l(l+1)\hbar^2/2I$. Note that the energies depend only on l and not on m_l.

Step 2 *Show that the wavefunctions are also eigenfunctions of the z-component of angular momentum*

The operator for the z-component of angular momentum is $\hat{l}_z=(\hbar/\mathrm{i})\partial/\partial\phi$. From Table 7F.1 note that each spherical harmonic is of the form $Y_{l,m_l}(\theta,\phi)=\mathrm{e}^{\mathrm{i}m_l\phi}f(\theta)$. It then follows that

$$\hat{l}_z Y_{l,m_l}(\theta,\phi)=\hat{l}_z \overbrace{\mathrm{e}^{\mathrm{i}m_l\phi}f(\theta)}^{Y_{l,m_l}(\theta,\phi)}=\frac{\hbar}{\mathrm{i}}\frac{\partial}{\partial\phi}\mathrm{e}^{\mathrm{i}m_l\phi}f(\theta)=m_l\hbar\times\mathrm{e}^{\mathrm{i}m_l\phi}f(\theta)$$
$$=m_l\hbar\times Y_{l,m_l}(\theta,\phi)$$

Therefore, the $Y_{l,m_l}(\theta,\phi)$ are eigenfunctions of $\hat{l}_z$ with eigenvalues $m_l\hbar$.

In summary, the $Y_{l,m_l}(\theta,\phi)$ are solutions to the Schrödinger equation for a particle on a sphere, with the corresponding energies given by

$$E_{l,m_l}=l(l+1)\frac{\hbar^2}{2I} \quad l=0,\,1,\,2\ldots \quad m_l=0,\pm1,\ldots\pm l \qquad (7F.10)$$

Energy levels [particle on a sphere]

The integers l and m_l are now identified as quantum numbers: l is the **orbital angular momentum quantum number** and m_l is the **magnetic quantum number**. The energy is specified by l alone, but for each value of l there are $2l+1$ values of m_l, so each energy level is $(2l+1)$-fold degenerate. Each wavefunction is also an eigenfunction of $\hat{l}_z$ and therefore corresponds to a definite value, $m_l\hbar$, of the z-component of the angular momentum.

Figure 7F.5 shows a representation of the spherical harmonics for $l=0$–4 and $m_l=0$. The use of different colours for different signs of the wavefunction emphasizes the location of the angular nodes (the positions at which the wavefunction passes through zero). Note that:

- There are no angular nodes around the z-axis for functions with $m_l=0$. The spherical harmonic with $l=0$, $m_l=0$ has no nodes: it has a constant value at all positions of the surface and corresponds to a stationary particle.
- The number of angular nodes for states with $m_l=0$ is equal to l. As the number of nodes increases, the wavefunctions become more buckled, and with this increasing curvature the kinetic energy of the particle increases.

Physical interpretation

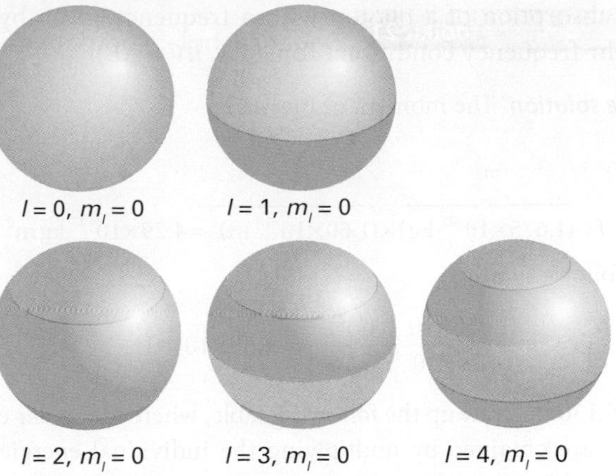

Figure 7F.5 A representation of the wavefunctions of a particle on the surface of a sphere that emphasizes the location of angular nodes: blue and grey shading correspond to different signs of the wavefunction. Note that the number of nodes increases as the value of l increases. All these wavefunctions correspond to $m_l=0$; a path round the vertical z-axis of the sphere does not cut through any nodes.

According to eqn 7F.10,

- Because l is confined to non-negative integral values, the energy is quantized.
- The energies are independent of the value of m_l, because the energy is independent of the direction of the rotational motion.
- There are $2l+1$ different wavefunctions (one for each value of m_l) that correspond to the same energy, so it follows that a level with quantum number l is $(2l+1)$-fold degenerate.
- There is no zero-point energy: $E_{0,0}=0$.

Physical interpretation

Example 7F.2 Using the rotational energy levels

The particle on a sphere is a good starting point for developing a model for the rotation of a diatomic molecule. Treat the rotation of $^1H^{127}I$ as a hydrogen atom rotating around a stationary I atom (this is a good first approximation as the I atom is so heavy it hardly moves). The bond length is 160 pm. Evaluate the energies and degeneracies of the lowest four rotational energy levels of $^1H^{127}I$. What is the frequency of the transition between the lowest two rotational levels?

Collect your thoughts The moment of inertia is $I=m_{^1H}R^2$, with $R=160$ pm; the rotational energies are given in eqn 7F.10. When describing the rotational energy levels of a molecule it is usual to denote the angular momentum quantum number by J rather than l; as a result the degeneracy is $2J+1$ (the analogue of $2l+1$). A transition between two rotational levels can be brought about by the emission or absorption of a photon with a frequency given by the Bohr frequency condition (Topic 7A, $h\nu = \Delta E$).

The solution The moment of inertia is

$$I = \overbrace{(1.675\times10^{-27}\,\text{kg})}^{m_{^1\text{H}}} \times \overbrace{(1.60\times10^{-10}\,\text{m})^2}^{R^2} = 4.29\times10^{-47}\,\text{kg}\,\text{m}^2$$

It follows that

$$\frac{\hbar^2}{2I} = \frac{(1.055\times10^{-34}\,\text{J s})^2}{2\times(4.29\times10^{-47}\,\text{kg}\,\text{m}^2)} = 1.30\times10^{-22}\,\text{J}$$

or 0.130 zJ. Draw up the following table, where the molar energies are obtained by multiplying the individual energies by Avogadro's constant:

J	E/zJ	E/(J mol^{-1})	Degeneracy
0	0	0	1
1	0.260	156	3
2	0.780	470	5
3	1.56	939	7

The energy separation between the two lowest rotational energy levels ($J = 0$ and 1) is 2.60×10^{-22} J, which corresponds to a photon of frequency

$$\nu = \frac{\Delta E}{h} = \frac{2.60\times10^{-22}\,\text{J}}{6.626\times10^{-34}\,\text{J s}} = 3.92\times10^{11}\,\overbrace{\text{s}^{-1}}^{\text{Hz}} = 392\,\text{GHz}$$

Comment. Radiation of this frequency belongs to the microwave region of the electromagnetic spectrum, so microwave spectroscopy is used to study molecular rotations (Topic 11B). Because the transition frequencies depend on the moment of inertia and frequencies can be measured with great precision, microwave spectroscopy is a very precise technique for the determination of bond lengths.

Self-test 7F.2 What is the frequency of the transition between the lowest two rotational levels in $^2\text{H}^{127}\text{I}$? (Assume that the bond length is the same as for $^1\text{H}^{127}\text{I}$ and that the iodine atom is stationary.)

Answer: 196 GHz

(b) Angular momentum

According to classical mechanics (*The chemist's toolkit* 20) the kinetic energy of a particle circulating on a ring is $E_k = J^2/2I$, where J is the magnitude of the angular momentum. By comparing this relation with eqn 7F.10, it follows that the square of the magnitude of the angular momentum is $l(l+1)\hbar^2$, so the magnitude of the angular momentum is

$$\text{Magnitude} = \{l(l+1)\}^{1/2}\hbar \qquad l = 0, 1, 2\ldots \qquad \text{Magnitude of angular momentum} \quad (7F.11)$$

The spherical harmonics are also eigenfunctions of $\hat{l}_z$ with eigenvalues

$$z\text{-Component} = m_l\hbar \qquad m_l = 0, \pm1, \ldots \pm l \qquad z\text{-Component of angular momentum} \quad (7F.12)$$

So, both the magnitude and the z-component of angular momentum are quantized.

Brief illustration 7F.1

The lowest four rotational energy levels of any object rotating in three dimensions correspond to $l = 0, 1, 2, 3$. The following table can be constructed by using eqns 7F.11 and 7F.12.

l	Magnitude of angular momentum/$\hbar$	Degeneracy	z-Component of angular momentum/$\hbar$
0	0	1	0
1	$2^{1/2}$	3	0, ±1
2	$6^{1/2}$	5	0, ±1, ±2
3	$12^{1/2}$	7	0, ±1, ±2, ±3

(c) The vector model

The result that m_l is confined to the values $0, \pm1, \ldots \pm l$ for a given value of l means that the component of angular momentum about the z-axis—the contribution to the total angular momentum of rotation around that axis—may take only $2l+1$ values. If the angular momentum is represented by a vector of length $\{l(l+1)\}^{1/2}$, it follows that this vector must be oriented so that its projection on the z-axis is m_l and that it can have only $2l+1$ orientations rather than the continuous range of orientations of a rotating classical body (Fig. 7F.6). The remarkable implication is that

The orientation of a rotating body is quantized.

The quantum mechanical result that a rotating body may not take up an arbitrary orientation with respect to some specified axis (e.g. an axis defined by the direction of an externally applied electric or magnetic field) is called **space quantization.**

The preceding discussion has referred to the z-component of angular momentum and there has been no reference to the x- and y-components. The reason for this omission is found by examining the operators for the three components, each one being given by a term like that in eqn 7F.5a:[2]

[2] Each one is in fact a component of the vector product of $\boldsymbol{r}$ and $\boldsymbol{p}$, $\boldsymbol{l} = \boldsymbol{r}\times\boldsymbol{p}$, and the replacement of $\boldsymbol{r}$ and $\boldsymbol{p}$ by their operator equivalents.

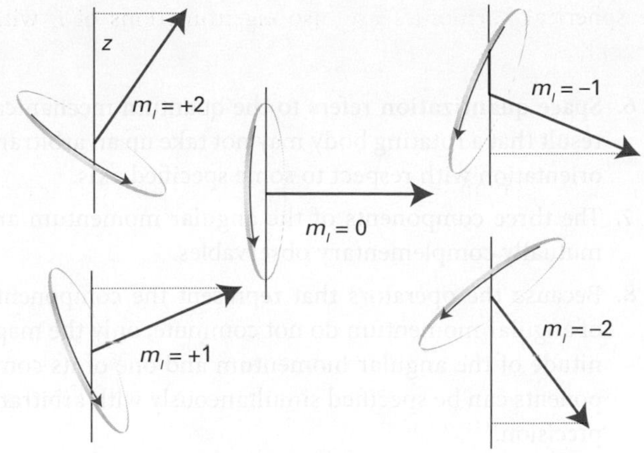

Figure 7F.6 The permitted orientations of angular momentum when $l = 2$. This representation is too specific because the azimuthal orientation of the vector (its angle around z) is indeterminate.

$$\hat{l}_x = \frac{\hbar}{i}\left(y\frac{\partial}{\partial z} - z\frac{\partial}{\partial y}\right)$$
$$\hat{l}_y = \frac{\hbar}{i}\left(z\frac{\partial}{\partial x} - x\frac{\partial}{\partial z}\right) \quad \text{Angular momentum operators} \quad (7F.13)$$
$$\hat{l}_z = \frac{\hbar}{i}\left(x\frac{\partial}{\partial y} - y\frac{\partial}{\partial x}\right)$$

Each of these expressions can be derived in the same way as eqn 7F.5a by converting the classical expressions for the components of the angular momentum into their quantum mechanical equivalents. The commutation relations among the three operators (Problem P7F.9), are

$$[\hat{l}_x, \hat{l}_y] = i\hbar\hat{l}_z \quad [\hat{l}_y, \hat{l}_z] = i\hbar\hat{l}_x \quad [\hat{l}_z, \hat{l}_x] = i\hbar\hat{l}_y \quad \text{Angular momentum commutation relations} \quad (7F.14)$$

Because the three operators do not commute, they represent complementary observables (Topic 7C). Therefore, the more precisely any one component is known, the greater the uncertainty in the other two. It is possible to have precise knowledge of only one of the components of the angular momentum, so if l_z is specified exactly (as in the preceding discussion), neither l_x nor l_y can be specified.

The operator for the square of the magnitude of the angular momentum is

$$\hat{l}^2 = \hat{l}_x^2 + \hat{l}_y^2 + \hat{l}_z^2 \quad \text{Operator for the square of the magnitude of angular momentum} \quad (7F.15)$$

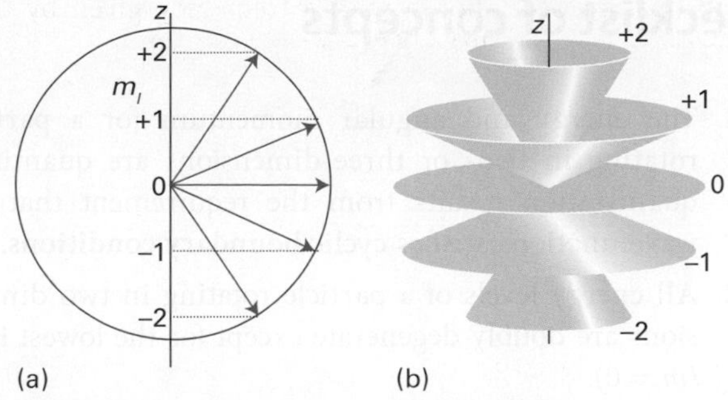

Figure 7F.7 (a) A summary of Fig. 7F.6. However, because the azimuthal angle of the vector around the z-axis is indeterminate, a better representation is as in (b), where each vector lies at an unspecified azimuthal angle on its cone.

This operator commutes with all three components (Problem P7F.11):

$$[\hat{l}^2, \hat{l}_q] = 0 \quad q = x, y, \text{and } z \quad \text{Commutators of angular momentum operators} \quad (7F.16)$$

It follows that both the square magnitude and one component, commonly the z-component, of the angular momentum can be specified precisely. The illustration in Fig. 7F.6, which is summarized in Fig. 7F.7(a), therefore gives a false impression of the state of the system, because it suggests definite values for the x- and y-components too. A better picture must reflect the impossibility of specifying l_x and l_y if l_z is known.

The **vector model** of angular momentum uses pictures like that in Fig. 7F.7(b). The cones are drawn with side $\{l(l+1)\}^{1/2}$ units, and represent the magnitude of the angular momentum. Each cone has a definite projection (of m_l units) on to the z-axis, representing the precisely known value of l_z. The projections of the vector on to the x- and y-axes, which give the values of l_x and l_y, are indefinite: the vector representing angular momentum can be thought of as lying with its tip on any point on the mouth of the cone. At this stage it should not be thought of as sweeping round the cone; that aspect of the model will be added when the picture is allowed to convey more information (Topics 8B and 8C).

Brief illustration 7F.2

If the wavefunction of a rotating molecule is given by the spherical harmonic $Y_{3,+2}$ then the angular momentum can be represented by a cone

- with a side of length $12^{1/2}$ (representing the magnitude of $12^{1/2}\hbar$); and
- with a projection of +2 on the z-axis (representing the z-component of $+2\hbar$).

Checklist of concepts

☐ 1. The energy and angular momentum for a particle rotating in two- or three-dimensions are quantized; quantization results from the requirement that the wavefunction satisfies **cyclic boundary conditions**.

☐ 2. All energy levels of a particle rotating in two dimensions are doubly-degenerate except for the lowest level ($m_l = 0$).

☐ 3. There is no zero-point energy for a rotating particle.

☐ 4. It is impossible to specify simultaneously the angular momentum and location of a particle with arbitrary precision.

☐ 5. For a particle rotating in three dimensions, the cyclic boundary conditions imply that the magnitude and z-component of the angular momentum are quantized.

☐ 6. **Space quantization** refers to the quantum mechanical result that a rotating body may not take up an arbitrary orientation with respect to some specified axis.

☐ 7. The three components of the angular momentum are mutually complementary observables.

☐ 8. Because the operators that represent the components of angular momentum do not commute, only the magnitude of the angular momentum and one of its components can be specified simultaneously with arbitrary precision.

☐ 9. In the **vector model** of angular momentum, the angular momentum is represented by a cone with a side of length $\{l(l+1)\}^{1/2}$ and a projection of m_l on the z-axis. The vector can be thought of as lying with its tip on an indeterminate point on the mouth of the cone.

Checklist of equations

Property	Equation	Comment	Equation number
Wavefunction of particle on a ring	$\psi_{m_l}(\phi)=e^{im_l\phi}/(2\pi)^{1/2}$	$m_l=0,\pm1,\pm2,\ldots$	7F.4
Energy of particle on a ring	$E_{m_l}=m_l^2\hbar^2/2I$	$m_l=0,\pm1,\pm2,\ldots$ $I=mr^2$	7F.4
z-Component of angular momentum of particle on a ring	$m_l\hbar$	$m_l=0,\pm1,\pm2,\ldots$	7F.6
Wavefunction of particle on a sphere	$\psi(\theta,\phi)=Y_{l,m_l}(\theta,\phi)$	Y is a spherical harmonic (Table 7F.1)	
Energy of particle on a sphere	$E_{l,m_l}=l(l+1)\hbar^2/2I$	$l=0, 1, 2,\ldots$	7F.10
Magnitude of angular momentum	$\{l(l+1)\}^{1/2}\hbar$	$l=0, 1, 2,\ldots$	7F.11
z-Component of angular momentum	$m_l\hbar$	$m_l=0,\pm1,\pm2,\ldots\pm l$	7F.12
Angular momentum commutation relations	$[\hat{l}_x,\hat{l}_y]=i\hbar\hat{l}_z$ $[\hat{l}_y,\hat{l}_z]=i\hbar\hat{l}_x$ $[\hat{l}_z,\hat{l}_x]=i\hbar\hat{l}_y$		7F.14
	$[\hat{l}^2,\hat{l}_q]=0,\ q=x, y,$ and z		7F.16

FOCUS 7 Quantum theory

TOPIC 7A The origins of quantum mechanics

Discussion questions

D7A.1 Summarize the evidence that led to the introduction of quantum mechanics.

D7A.2 Explain how Planck's introduction of quantization accounted for the properties of black-body radiation.

D7A.3 Explain how Einstein's introduction of quantization accounted for the properties of heat capacities at low temperatures.

D7A.4 Explain the meaning and summarize the consequences of wave–particle duality.

Exercises

E7A.1(a) Calculate the wavelength and frequency at which the intensity of the radiation is a maximum for a black body at 298 K.
E7A.1(b) Calculate the wavelength and frequency at which the intensity of the radiation is a maximum for a black body at 2.7 K.

E7A.2(a) The intensity of the radiation from an object is found to be a maximum at 2000 cm^{-1}. Assuming that the object is a black body, calculate its temperature.
E7A.2(b) The intensity of the radiation from an object is found to be a maximum at 282 GHz (1 GHz = 10^9 Hz). Assuming that the object is a black body, calculate its temperature.

E7A.3(a) Calculate the molar heat capacity of a monatomic non-metallic solid at 298 K which is characterized by an Einstein temperature of 2000 K. Express your result as a multiple of $3R$.
E7A.3(b) Calculate the molar heat capacity of a monatomic non-metallic solid at 500 K which is characterized by an Einstein temperature of 300 K. Express your result as a multiple of $3R$.

E7A.4(a) Calculate the energy of the quantum involved in the excitation of (i) an electronic oscillation of period 1.0 fs, (ii) a molecular vibration of period 10 fs, (iii) a pendulum of period 1.0 s. Express the results in joules and kilojoules per mole.
E7A.4(b) Calculate the energy of the quantum involved in the excitation of (i) an electronic oscillation of period 2.50 fs, (ii) a molecular vibration of period 2.21 fs, (iii) a balance wheel of period 1.0 ms. Express the results in joules and kilojoules per mole.

E7A.5(a) Calculate the energy of a photon and the energy per mole of photons for radiation of wavelength (i) 600 nm (red), (ii) 550 nm (yellow), (iii) 400 nm (blue).
E7A.5(b) Calculate the energy of a photon and the energy per mole of photons for radiation of wavelength (i) 200 nm (ultraviolet), (ii) 150 pm (X-ray), (iii) 1.00 cm (microwave).

E7A.6(a) Calculate the speed to which a stationary H atom would be accelerated if it absorbed each of the photons used in Exercise 7A.5(a).
E7A.6(b) Calculate the speed to which a stationary ^{4}He atom (mass 4.0026 m_u) would be accelerated if it absorbed each of the photons used in Exercise 7A.5(b).

E7A.7(a) A sodium lamp emits yellow light (550 nm). How many photons does it emit each second if its power is (i) 1.0 W, (ii) 100 W?
E7A.7(b) A laser used to read CDs emits red light of wavelength 700 nm. How many photons does it emit each second if its power is (i) 0.10 W, (ii) 1.0 W?

E7A.8(a) The work function of metallic caesium is 2.14 eV. Calculate the kinetic energy and the speed of the electrons ejected by light of wavelength (i) 700 nm, (ii) 300 nm.
E7A.8(b) The work function of metallic rubidium is 2.09 eV. Calculate the kinetic energy and the speed of the electrons ejected by light of wavelength (i) 650 nm, (ii) 195 nm.

E7A.9(a) A glow-worm of mass 5.0 g emits red light (650 nm) with a power of 0.10 W entirely in the backward direction. To what speed will it have accelerated after 10 y if released into free space and assumed to live?
E7A.9(b) A photon-powered spacecraft of mass 10.0 kg emits radiation of wavelength 225 nm with a power of 1.50 kW entirely in the backward direction. To what speed will it have accelerated after 10.0 y if released into free space?

E7A.10(a) To what speed must an electron be accelerated from rest for it to have a de Broglie wavelength of 100 pm? What accelerating potential difference is needed?
E7A.10(b) To what speed must a proton be accelerated from rest for it to have a de Broglie wavelength of 100 pm? What accelerating potential difference is needed?

E7A.11(a) To what speed must an electron be accelerated for it to have a de Broglie wavelength of 3.0 cm?
E7A.11(b) To what speed must a proton be accelerated for it to have a de Broglie wavelength of 3.0 cm?

E7A.12(a) The 'fine-structure constant', α, plays a special role in the structure of matter; its approximate value is 1/137. What is the de Broglie wavelength of an electron travelling at αc, where c is the speed of light?
E7A.12(b) Calculate the linear momentum of photons of wavelength 350 nm. At what speed does a hydrogen molecule need to travel for it to have the same linear momentum?

E7A.13(a) Calculate the de Broglie wavelength of (i) a mass of 1.0 g travelling at 1.0 cm s^{-1}; (ii) the same, travelling at 100 km s^{-1}; (iii) a He atom travelling at 1000 m s^{-1} (a typical speed at room temperature).
E7A.13(b) Calculate the de Broglie wavelength of an electron accelerated from rest through a potential difference of (i) 100 V; (ii) 1.0 kV; (iii) 100 kV.

Problems

P7A.1 Calculate the energy density in the range 650 nm to 655 nm inside a cavity at (a) 25 °C, (b) 3000 °C. For this relatively small range of wavelength it is acceptable to approximate the integral of the energy spectral density $\rho(\lambda,T)$ between λ_1 and λ_2 by $\rho(\lambda,T)\times(\lambda_2-\lambda_1)$.

P7A.2 Calculate the energy density in the range $1000\,\text{cm}^{-1}$ to $1010\,\text{cm}^{-1}$ inside a cavity at (a) 25 °C, (b) 4 K.

P7A.3 Demonstrate that the Planck distribution reduces to the Rayleigh–Jeans law at long wavelengths.

P7A.4 The wavelength $\lambda_{\max}$ at which the Planck distribution is a maximum can be found by solving $\mathrm{d}\rho(\lambda,T)/\mathrm{d}T = 0$. Differentiate $\rho(\lambda,T)$ with respect to T and show that the condition for the maximum can be expressed as $xe^x - 5(e^x - 1) = 0$, where $x = hc/\lambda kT$. There are no analytical solutions to this equation, but a numerical approach gives $x = 4.965$ as a solution. Use this result to confirm *Wien's law*, that $\lambda_{\max}T$ is a constant, deduce an expression for the constant, and compare it to the value quoted in the text.

P7A.5 For a black body, the temperature and the wavelength of the emission maximum, $\lambda_{\max}$, are related by Wien's law, $\lambda_{\max}T = hc/4.965k$; see Problem 7A.4. Values of $\lambda_{\max}$ from a small pinhole in an electrically heated container were determined at a series of temperatures, and the results are given below. Deduce the value of Planck's constant.

θ/°C	1000	1500	2000	2500	3000	3500
$\lambda_{\max}$/nm	2181	1600	1240	1035	878	763

P7A.6‡ Solar energy strikes the top of the Earth's atmosphere at $343\,\text{W m}^{-2}$. About 30 per cent of this energy is reflected directly back into space. The Earth–atmosphere system absorbs the remaining energy and re-radiates it into space as black-body radiation at $5.672\times10^{-8}(T/\text{K})^4\,\text{W m}^{-2}$, where T is the temperature. Assuming that the arrangement has come to equilibrium, what is the average black-body temperature of the Earth? Calculate the wavelength at which the black-body radiation from the Earth is at a maximum.

P7A.7 The total energy density of black-body radiation is found by integrating the energy spectral density over all wavelengths, eqn 7A.2. Evaluate this integral for the Planck distribution. This is most easily done by making the substitution $x = hc/\lambda kT$; you will need the integral $\int_0^\infty \{x^3/(e^x-1)\}\mathrm{d}x = \pi^4/15$. Hence deduce the *Stefan–Boltzmann law* that the total energy density of black-body radiation is proportional to T^4, and find the constant of proportionality.

P7A.8‡ Prior to Planck's derivation of the distribution law for black-body radiation, Wien found empirically a closely related distribution function which is very nearly but not exactly in agreement with the experimental results, namely $\rho(\lambda,T) = (a/\lambda^5)e^{-b/\lambda kT}$. This formula shows small deviations from Planck's at long wavelengths. (a) Find a form of the Planck distribution which is appropriate for short wavelengths (*Hint:* consider the behaviour of the term $e^{hc/\lambda kT} - 1$ in this limit). (b) Compare your expression from (a) with Wien's empirical formula and hence determine the constants a and b. (c) Integrate Wien's empirical expression for $\rho(\lambda,T)$ over all wavelengths and show that the result is consistent with the Stefan–Boltzmann law (*Hint:* to compute the integral use the substitution $x = hc/\lambda kT$ and then refer to the *Resource section*). (d) Show that Wien's empirical expression is consistent with Wien's law.

P7A.9‡ The temperature of the Sun's surface is approximately 5800 K. On the assumption that the human eye evolved to be most sensitive at the wavelength of light corresponding to the maximum in the Sun's radiant energy distribution, identify the colour of light to which the eye is the most sensitive.

P7A.10 The Einstein frequency is often expressed in terms of an equivalent temperature θ_E, where $\theta_E = h\nu/k$. Confirm that θ_E has the dimensions of temperature, and express the criterion for the validity of the high-temperature form of the Einstein equation in terms of θ_E. Evaluate θ_E for (a) diamond, for which $\nu = 46.5\,\text{THz}$, and (b) for copper, for which $\nu = 7.15\,\text{THz}$. Use these values to calculate the molar heat capacity of each substance at 25 °C, expressing your answers as multiples of $3R$.

TOPIC 7B Wavefunctions

Discussion questions

D7B.1 Describe how a wavefunction summarizes the dynamical properties of a system and how those properties may be predicted.

D7B.2 Explain the relation between probability amplitude, probability density, and probability.

D7B.3 Identify the constraints that the Born interpretation puts on acceptable wavefunctions.

Exercises

E7B.1(a) A possible wavefunction for an electron in a region of length L (i.e. from $x = 0$ to $x = L$) is $\sin(2\pi x/L)$. Normalize this wavefunction (to 1).

E7B.1(b) A possible wavefunction for an electron in a region of length L is $\sin(3\pi x/L)$. Normalize this wavefunction (to 1).

E7B.2(a) Normalize (to 1) the wavefunction e^{-ax^2} in the range $-\infty \le x \le \infty$, with $a > 0$. Refer to the *Resource section* for the necessary integral.

E7B.2(b) Normalize (to 1) the wavefunction e^{-ax} in the range $0 \le x \le \infty$, with $a > 0$.

E7B.3(a) Which of the following functions can be normalized (in all cases the range for x is from $x = -\infty$ to ∞, and a is a positive constant): (i) e^{-ax^2}; (ii) e^{-ax}. Which of these functions are acceptable as wavefunctions?

E7B.3(b) Which of the following functions can be normalized (in all cases the range for x is from $x = -\infty$ to ∞, and a is a positive constant): (i) $\sin(ax)$; (ii) $\cos(ax)\,e^{-x^2}$? Which of these functions are acceptable as wavefunctions?

E7B.4(a) For the system described in Exercise E7B.1(a), what is the probability of finding the electron in the range $\mathrm{d}x$ at $x = L/2$?

E7B.4(b) For the system described in Exercise E7B.1(b), what is the probability of finding the electron in the range $\mathrm{d}x$ at $x = L/6$?

E7B.5(a) For the system described in Exercise E7B.1(a), what is the probability of finding the electron between $x = L/4$ and $x = L/2$?

E7B.5(b) For the system described in Exercise E7B.1(b), what is the probability of finding the electron between $x = 0$ and $x = L/3$?

E7B.6(a) What are the dimensions of a wavefunction that describes a particle free to move in both the x and y directions?

E7B.6(b) The wavefunction for a particle free to move between $x = 0$ and $x = L$ is $(2/L)^{1/2}\sin(\pi x/L)$; confirm that this wavefunction has the expected dimensions.

E7B.7(a) Imagine a particle free to move in the x direction. Which of the following wavefunctions would be acceptable for such a particle? In each case, give your reasons for accepting or rejecting each function. (i) $\psi(x)=x^2$; (ii) $\psi(x)=1/x$; (iii) $\psi(x)=e^{-x^2}$.

E7B.7(b) Imagine a particle confined to move on the circumference of a circle ('a particle on a ring'), such that its position can be described by an angle ϕ in

‡ These problems were supplied by Charles Trapp and Carmen Giunta.

the range 0–2π. Which of the following wavefunctions would be acceptable for such a particle? In each case, give your reasons for accepting or rejecting each function. (i) $\cos\phi$; (ii) $\sin\phi$; (iii) $\cos(0.9\phi)$.

E7B.8(a) For the system described in Exercise E7B.1(a), at what value or values of x is the probability density a maximum? Locate the positions of any nodes in the wavefunction. You need consider only the range $x = 0$ to $x = L$.

E7B.8(b) For the system described in Exercise E7B.1(b), at what value or values of x is the probability density a maximum? Locate the position or positions of any nodes in the wavefunction. You need consider only the range $x = 0$ to $x = L$.

Problems

P7B.1 Imagine a particle confined to move on the circumference of a circle ('a particle on a ring'), such that its position can be described by an angle ϕ in the range 0 to 2π. Find the normalizing factor for the wavefunctions: (a) $e^{i\phi}$ and (b) $e^{im_l\phi}$, where m_l is an integer.

P7B.2 For the system described in Problem P7B.1 find the normalizing factor for the wavefunctions: (a) $\cos\phi$; (b) $\sin m_l\phi$, where m_l is an integer.

P7B.3 A particle is confined to a two-dimensional region with $0 \le x \le L_x$ and $0 \le y \le L_y$. Normalize (to 1) the functions (a) $\sin(\pi x/L_x)\sin(\pi y/L_y)$ and (b) $\sin(\pi x/L)\sin(\pi y/L)$ for the case $L_x = L_y = L$.

P7B.4 Normalize (to 1) the wavefunction $e^{-ax^2}e^{-by^2}$ for a system in two dimensions with $a > 0$ and $b > 0$, and with x and y both allowed to range from $-\infty$ to ∞. Refer to the *Resource section* for relevant integrals.

P7B.5 Suppose that in a certain system a particle free to move along one dimension (with $0 \le x \le \infty$) is described by the unnormalized wavefunction $\psi(x) = e^{-ax}$ with $a = 2\,\text{m}^{-1}$. What is the probability of finding the particle at a distance $x \ge 1$ m? (*Hint:* You will need to normalize the wavefunction before using it to calculate the probability.)

P7B.6 Suppose that in a certain system a particle free to move along x (without constraint) is described by the unnormalized wavefunction $\psi(x) = e^{-ax^2}$ with $a = 0.2\,\text{m}^{-2}$. Use mathematical software to calculate the probability of finding the particle at $x \ge 1$ m.

P7B.7 A normalized wavefunction for a particle confined between 0 and L in the x direction is $\psi = (2/L)^{1/2}\sin(\pi x/L)$. Suppose that $L = 10.0$ nm. Calculate the probability that the particle is (a) between $x = 4.95$ nm and 5.05 nm, (b) between $x = 1.95$ nm and 2.05 nm, (c) between $x = 9.90$ nm and 10.00 nm, (d) between $x = 5.00$ nm and 10.00 nm.

P7B.8 A normalized wavefunction for a particle confined between 0 and L in the x direction, and between 0 and L in the y direction (that is, to a square of side L) is $\psi = (2/L)\sin(\pi x/L)\sin(\pi y/L)$. The probability of finding the particle between x_1 and x_2 along x, and between y_1 and y_2 along y is

$$P = \int_{y=y_1}^{y=y_2}\int_{x=x_1}^{x=x_2} \psi^2\,dx\,dy$$

Calculate the probability that the particle is: (a) between $x = 0$ and $x = L/2$, $y = 0$ and $y = L/2$ (i.e. in the bottom left-hand quarter of the square); (b) between $x = L/4$ and $x = 3L/4$, $y = L/4$ and $y = 3L/4$ (i.e. a square of side $L/2$ centred on $x = y = L/2$).

P7B.9 The normalized ground-state wavefunction of a hydrogen atom is $\psi(r) = (1/\pi a_0^3)^{1/2}e^{-r/a_0}$ where $a_0 = 53$ pm (the Bohr radius) and r is the distance from the nucleus. (a) Calculate the probability that the electron will be found somewhere within a small sphere of radius 1.0 pm centred on the nucleus. (b) Now suppose that the same sphere is located at $r = a_0$. What is the probability that the electron is inside it? You may approximate the probability of being in a small volume δV at position r by $\psi(r)^2\delta V$.

P7B.10 Atoms in a chemical bond vibrate around the equilibrium bond length. An atom undergoing vibrational motion is described by the wavefunction $\psi(x) = Ne^{-x^2/2a^2}$, where a is a constant and $-\infty \le x \le \infty$. (a) Find the normalizing factor N. (b) Use mathematical software to calculate the probability of finding the particle in the range $a \le x \le a$ (the result will be expressed in terms of the 'error function', erf(x)).

P7B.11 Suppose that the vibrating atom in Problem P7B.10 is described by the wavefunction $\psi(x) = Nxe^{-x^2/2a^2}$. Where is the most probable location of the atom?

TOPIC 7C Operators and observables

Discussion questions

D7C.1 How may the curvature of a wavefunction be interpreted?

D7C.2 Describe the relation between operators and observables in quantum mechanics.

D7C.3 Use the properties of wavepackets to account for the uncertainty relation between position and linear momentum.

Exercises

E7C.1(a) Construct the potential energy operator of a particle with potential energy $V(x) = \frac{1}{2}k_f x^2$, where k_f is a constant.

E7C.1(b) Construct the potential energy operator of a particle with potential energy $V(x) = D_e(1 - e^{-ax})^2$, where D_e and a are constants.

E7C.2(a) Identify which of the following functions are eigenfunctions of the operator d/dx: (i) $\cos(kx)$; (ii) e^{ikx}, (iii) kx, (iv) e^{-ax^2}. Give the corresponding eigenvalue where appropriate.

E7C.2(b) Identify which of the following functions are eigenfunctions of the operator d^2/dx^2: (i) $\cos(kx)$; (ii) e^{ikx}, (iii) kx, (iv) e^{-ax^2}. Give the corresponding eigenvalue where appropriate.

E7C.3(a) Functions of the form $\sin(n\pi x/L)$, where $n = 1, 2, 3 \ldots$, are wavefunctions in a region of length L (between $x = 0$ and $x = L$). Show that the wavefunctions with $n = 1$ and 2 are orthogonal; you will find the necessary integrals in the *Resource section*. (*Hint:* Recall that $\sin(n\pi) = 0$ for integer n.)

E7C.3(b) For the same system as in Exercise E7C.3(a) show that the wavefunctions with $n = 2$ and 4 are orthogonal.

E7C.4(a) Functions of the form $\cos(n\pi x/L)$, where $n = 1, 3, 5 \ldots$, can be used to model the wavefunctions of particles confined to the region between $x = -L/2$ and $x = +L/2$. The integration is limited to the range $-L/2$ to $+L/2$ because the wavefunction is zero outside this range. Show that the wavefunctions are orthogonal for $n = 1$ and 3. You will find the necessary integral in the *Resource section*.

E7C.4(b) For the same system as in Exercise E7C.4(a) show that the wavefunctions with $n = 3$ and 5 are orthogonal.

E7C.5(a) Imagine a particle confined to move on the circumference of a circle ('a particle on a ring'), such that its position can be described by an angle ϕ in the range 0–2π. The wavefunctions for this system are of the form $\psi_m(\phi)=e^{im_l\phi}$ with m_l an integer. Show that the wavefunctions with $m_l = +1$ and $+2$ are orthogonal. (*Hint:* Note that $(e^{ix})^*=e^{-ix}$, and that $e^{ix}=\cos x+i\sin x$.)

E7C.5(b) For the same system as in Exercise E7C.5(a) show that the wavefunctions with $m_l = +1$ and -2 are orthogonal.

E7C.6(a) An electron in a region of length L is described by the normalized wavefunction $\psi(x) = (2/L)^{1/2}\sin(2\pi x/L)$ in the range $x = 0$ to $x = L$; outside this range the wavefunction is zero. Evaluate $\langle x\rangle$. The necessary integrals will be found in the *Resource section.*

E7C.6(b) For the same system as in Exercise E7C.6(a) find $\langle x\rangle$ when the wavefunction is $\psi(x) = (2/L)^{1/2}\sin(\pi x/L)$.

E7C.7(a) An electron in a one-dimensional region of length L is described by the normalized wavefunction $\psi(x) = (2/L)^{1/2}\sin(2\pi x/L)$ in the range $x = 0$ to $x = L$; outside this range the wavefunction is zero. The expectation value of the momentum of the electron is found from eqn 7C.11, which in this case is

$$\langle p_x\rangle=\frac{2}{L}\int_0^L \sin(2\pi x/L)\hat{p}_x \sin(2\pi x/L)\mathrm{d}x=\frac{2\hbar}{\mathrm{i}L}\int_0^L \sin(2\pi x/L)\frac{\mathrm{d}}{\mathrm{d}x}\sin(2\pi x/L)\mathrm{d}x$$

Evaluate the differential and then the integral, and hence find $\langle p_x\rangle$. The necessary integrals will be found in the *Resource section.*

E7C.7(b) For the same system as in Exercise E7C.7(a) find $\langle p_x\rangle$ for the case where the normalized wavefunction is $\psi(x) = (2/L)^{1/2}\sin(\pi x/L)$.

E7C.8(a) For the 'particle on a ring' system described in Exercise E7C.5(a) the expectation value of a quantity represented by the operator $\hat{\Omega}$ is given by

$$\Omega_{m_l}=\int_0^{2\pi}\psi_{m_l}^*(\phi)\hat{\Omega}\psi_{m_l}(\phi)\mathrm{d}\phi$$

where $\psi_{m_l}(\phi)$ are the *normalized* wavefunctions $\psi_{m_l}(\phi)=(1/2\pi)^{1/2}e^{im_l\phi}$, with m_l an integer. Compute the expectation value of the position, specified by the angle ϕ, for the case $m_l = +1$, and then for the general case of integer m_l.

E7C.8(b) For the system described in Exercise E7C.8(a), evaluate the expectation value of the angular momentum represented by the operator $(\hbar/\mathrm{i})\mathrm{d}/\mathrm{d}\phi$ for the case $m_l = +1$, and then for the general case of integer m_l.

E7C.9(a) Calculate the minimum uncertainty in the speed of a ball of mass 500 g that is known to be within 1.0 μm of a certain point on a bat. What is the minimum uncertainty in the position of a bullet of mass 5.0 g that is known to have a speed somewhere between 350.000 01 m s^{-1} and 350.000 00 m s^{-1}?

E7C.9(b) An electron is confined to a linear region with a length of the same order as the diameter of an atom (about 100 pm). Calculate the minimum uncertainties in its position and speed.

E7C.10(a) The speed of a certain proton is 0.45 Mm s^{-1}. If the uncertainty in its momentum is to be reduced to 0.0100 per cent, what uncertainty in its location must be tolerated?

E7C.10(b) The speed of a certain electron is 995 km s^{-1}. If the uncertainty in its momentum is to be reduced to 0.0010 per cent, what uncertainty in its location must be tolerated?

Problems

P7C.1 Identify which of the following functions are eigenfunctions of the inversion operator $\hat{i}$, which has the effect of making the replacement $x \rightarrow -x$: (a) $x^3 - kx$, (b) $\cos kx$, (c) $x^2 + 3x - 1$. Identify the eigenvalue of $\hat{i}$ when relevant.

P7C.2 An electron in a one-dimensional region of length L is described by the wavefunction $\psi_n(x) = \sin(n\pi x/L)$, where $n = 1, 2, \ldots$, in the range $x = 0$ to $x = L$; outside this range the wavefunction is zero. The orthogonality of these wavefunctions is confirmed by considering the integral

$$I=\int_0^L \sin(n\pi x/L)\sin(m\pi x/L)\mathrm{d}x$$

(a) Use the identity $\sin A\sin B=\frac{1}{2}\{\cos(A-B)-\cos(A+B)\}$ to rewrite the integrand as a sum of two terms. (b) Consider the case $n = 2$, $m = 1$, and make separate sketch graphs of the two terms identified in (a) in the range $x = 0$ to $x = L$. (c) Make use of the properties of the cosine function to argue that the area enclosed between the curves and the x axis is zero in both cases, and hence that the integral is zero. (d) Generalize the argument for the case of arbitrary n and m ($n \neq m$).

P7C.3 Confirm that the kinetic energy operator, $-(\hbar^2/2m)\mathrm{d}^2/\mathrm{d}x^2$, is hermitian. (*Hint:* Use the same approach as in the text, but because a second derivative is involved you will need to integrate by parts twice; you may assume that the derivatives of the wavefunctions go to zero as $x \rightarrow \pm\infty$.)

P7C.4 The operator corresponding to the angular momentum of a particle is $(\hbar/\mathrm{i})\mathrm{d}/\mathrm{d}\phi$, where ϕ is an angle. For such a system the criterion for an operator $\hat{\Omega}$ to be hermitian is

$$\int_0^{2\pi}\psi_i^*(\phi)\hat{\Omega}\psi_j(\phi)\mathrm{d}\phi=\left[\int_0^{2\pi}\psi_j^*(\phi)\hat{\Omega}\psi_i(\phi)\mathrm{d}\phi\right]^*$$

Show that $(\hbar/\mathrm{i})\mathrm{d}/\mathrm{d}\phi$ is a hermitian operator. (*Hint:* Use the same approach as in the text; recall that the wavefunction must be single-valued, so $\psi_i(\phi)=\psi_i(\phi+2\pi)$.)

P7C.5 (a) Show that the sum of two hermitian operators $\hat{A}$ and $\hat{B}$ is also a hermitian operator. (*Hint:* Start by separating the appropriate integral into two terms, and then apply the definition of hermiticity.) (b) Show that the product of a hermitian operator with itself is also a hermitian operator. Start by considering the integral

$$I=\int\psi_i^*\hat{\Omega}\hat{\Omega}\psi_j\,\mathrm{d}\tau$$

Recall that $\hat{\Omega}\psi_j$ is simply another function, so the integral can be thought of as

$$I=\int\psi_i^*\hat{\Omega}\overbrace{(\hat{\Omega}\psi_j)}^{\text{a function}}\mathrm{d}\tau$$

Now apply the definition of hermiticity and complete the proof.

P7C.6 Calculate the expectation value of the linear momentum p_x of a particle described by the following normalized wavefunctions (in each case N is the appropriate normalizing factor, which you do not need to find): (a) Ne^{ikx}, (b) $N\cos kx$, (c) Ne^{-ax^2}, where in each one x ranges from $-\infty$ to $+\infty$.

P7C.7 A particle freely moving in one dimension x with $0 \leq x \leq \infty$ is in a state described by the normalized wavefunction $\psi(x) = a^{1/2}e^{-ax/2}$, where a is a constant. Evaluate the expectation value of the position operator.

P7C.8 The normalized wavefunction of an electron in a linear accelerator is $\psi = (\cos\chi)e^{ikx} + (\sin\chi)e^{-ikx}$, where χ (chi) is a parameter. (a) What is the probability that the electron will be found with a linear momentum (a) $+k\hbar$, (b) $-k\hbar$? (c) What form would the wavefunction have if it were 90 per cent certain that the electron had linear momentum $+k\hbar$? (d) Evaluate the kinetic energy of the electron.

P7C.9 (a) Show that the expectation value of a hermitian operator is real. (*Hint:* Start from the definition of the expectation value and then apply the definition of hermiticity to it.) (b) Show that the expectation value of an operator that can be written as the square of a hermitian operator is positive. (*Hint:* Start from the definition of the expectation value for the operator $\hat{\Omega}\hat{\Omega}$; recognize that $\hat{\Omega}\psi$ is a function, and then apply the definition of hermiticity.)

P7C.10 Suppose the wavefunction of an electron in a one-dimensional region is a linear combination of cos nx functions. (a) Use mathematical software or a spreadsheet to construct superpositions of cosine functions as

$$\psi(x)=\frac{1}{N}\sum_{k=1}^{N}\cos(k\pi x)$$

where the constant $1/N$ (not a normalization constant) is introduced to keep the superpositions with the same overall magnitude. Set $x = 0$ at the centre of the screen and build the superposition there; consider the range $x = -1$ to $+1$. (b) Explore how the probability density $\psi^2(x)$ changes with the value of N. (c) Evaluate the root-mean-square location of the packet, $\langle x^2\rangle^{1/2}$. (d) Determine the probability that a given momentum will be observed.

P7C.11 A particle is in a state described by the normalized wavefunction $\psi(x)=(2a/\pi)^{1/4}e^{-ax^2}$, where a is a constant and $-\infty \le x \le \infty$. (a) Calculate the expectation values $\langle x\rangle$, $\langle x^2\rangle$, $\langle p_x\rangle$, and $\langle p_x^2\rangle$; the necessary integrals will be found in the *Resource section*. (b) Use these results to calculate $\Delta p_x = \{\langle p_x^2\rangle - \langle p_x\rangle^2\}^{1/2}$ and $\Delta x = \{\langle x^2\rangle - \langle x\rangle^2\}^{1/2}$. (c) Hence verify that the value of the product $\Delta p_x\Delta x$ is consistent with the predictions from the uncertainty principle.

P7C.12 A particle is in a state described by the normalized wavefunction $\psi(x) = a^{1/2}e^{-ax/2}$, where a is a constant and $0 \le x \le \infty$. Evaluate the expectation value of the commutator of the position and momentum operators.

P7C.13 Evaluate the commutators of the operators (a) d/dx and $1/x$, (b) d/dx and x^2. (*Hint:* Follow the procedure in the text by considering, for case (a), $(d/dx)(1/x)\psi$ and $(1/x)(d/dx)\psi$; recall that ψ is a function of x, so it will be necessary to use the product rule to evaluate some of the derivatives.)

P7C.14 Evaluate the commutators of the operators $\hat{a}$ and $\hat{a}^+$ where $\hat{a}=(\hat{x}+i\hat{p}_x)/2^{1/2}$ and $\hat{a}^+=(\hat{x}-i\hat{p}_x)/2^{1/2}$.

P7C.15 Evaluate the commutators (a) $[\hat{H},\hat{p}_x]$ and (b) $[\hat{H},\hat{x}]$ where $\hat{H}=\hat{p}_x^2/2m + \hat{V}(x)$. Choose (i) $V(x) = V_0$, a constant, (ii) $V(x) = \frac{1}{2}k_f x^2$. (*Hint:* See the hint for Problem P7C.13.)

TOPIC 7D Translational motion

Discussion questions

D7D.1 Explain the physical origin of quantization for a particle confined to the interior of a one-dimensional box.

D7D.2 Describe the features of the solution of the particle in a one-dimensional box that appear in the solutions of the particle in two- and three-dimensional boxes. What feature occurs in the two- and three-dimensional box that does not occur in the one-dimensional box?

D7D.3 Explain the physical origin of quantum mechanical tunnelling. Why is tunnelling more likely to contribute to the mechanisms of electron transfer and proton transfer processes than to mechanisms of group transfer reactions, such as AB + C → A + BC (where A, B, and C are large molecular groups)?

Exercises

E7D.1(a) Evaluate the linear momentum and kinetic energy of a free electron described by the wavefunction e^{ikx} with $k = 3\,\mathrm{nm}^{-1}$.

E7D.1(b) Evaluate the linear momentum and kinetic energy of a free proton described by the wavefunction e^{-ikx} with $k = 5\,\mathrm{nm}^{-1}$.

E7D.2(a) Write the wavefunction for a particle of mass 2.0 g travelling to the left with kinetic energy 20 J.

E7D.2(b) Write the wavefunction for a particle of mass 1.0 g travelling to the right at $10\,\mathrm{m\,s^{-1}}$.

E7D.3(a) Calculate the energy separations in joules, kilojoules per mole, electronvolts, and reciprocal centimetres between the levels (i) $n = 2$ and $n = 1$, (ii) $n = 6$ and $n = 5$ of an electron in a box of length 1.0 nm.

E7D.3(b) Calculate the energy separations in joules, kilojoules per mole, electronvolts, and reciprocal centimetres between the levels (i) $n = 3$ and $n = 2$, (ii) $n = 7$ and $n = 6$ of an electron in a box of length 1.50 nm.

E7D.4(a) For a particle in a one-dimensional box, show that the wavefunctions ψ_1 and ψ_2 are orthogonal. The necessary integrals will be found in the *Resource section*.

E7D.4(b) For a particle in a one-dimensional box, show that the wavefunctions ψ_1 and ψ_3 are orthogonal.

E7D.5(a) Calculate the probability that a particle will be found between $0.49L$ and $0.51L$ in a box of length L for (i) ψ_1, (ii) ψ_2. You may assume that the wavefunction is constant in this range, so the probability is $\psi^2\delta x$.

E7D.5(b) Calculate the probability that a particle will be found between $0.65L$ and $0.67L$ in a box of length L for the case where the wavefunction is (i) ψ_1, (ii) ψ_2. You may make the same approximation as in Exercise E7D.5(a).

E7D.6(a) For a particle in a box of length L sketch the wavefunction corresponding to the state with the lowest energy and on the same graph sketch the corresponding probability density. Without evaluating any integrals, explain why the expectation value of x is equal to $L/2$.

E7D.6(b) Without evaluating any integrals, state the value of the expectation value of x for a particle in a box of length L for the case where the wavefunction has $n = 2$. Explain how you arrived at your answer. (*Hint:* Consider the approach used in Exercise E7D.6(a).)

E7D.7(a) For a particle in a box of length L sketch the wavefunction corresponding to the state with $n = 1$ and on the same graph sketch the corresponding probability density. Without evaluating any integrals, explain why for this wavefunction the expectation value of x^2 is not equal to $(L/2)^2$.

E7D.7(b) For a particle in a box of length L sketch the wavefunction corresponding to the state with $n = 1$ and on the same graph sketch the corresponding probability density. For this wavefunction, explain whether you would expect the expectation value of x^2 to be greater than or less than the square of the expectation value of x.

E7D.8(a) An electron is confined to a square well of length L. What would be the length of the box such that the zero-point energy of the electron is equal to its rest mass energy, m_ec^2? Express your answer in terms of the parameter $\lambda_C = h/m_ec$, the 'Compton wavelength' of the electron.

E7D.8(b) Repeat Exercise E7D.8(a) for the case of a cubic box of side L.

E7D.9(a) For a particle in a box of length L and in the state with $n = 3$, at what positions is the probability density a maximum? At what positions is the probability density zero?

E7D.9(b) For a particle in a box of length L and in the state with $n = 5$, at what positions is the probability density a maximum? At what positions is the probability density a minimum?

E7D.10(a) For a particle in a box of length L, write the expression for the energy levels, E_n, and then write a similar expression E_n' for the energy levels when the length of the box has increased to $1.1L$ (that is, an increase by 10 per cent). Calculate $(E_n'-E_n)/E_n$, the fractional change in the energy that results from extending the box.

E7D.10(b) Repeat the calculation in Exercise E7D.10(a) but this time for a cubical box of side L and for a decrease to $0.9L$ (that is, a decrease by 10 per cent).

E7D.11(a) Find an expression for the value of n of a particle of mass m in a one-dimensional box of length L such that the separation between neighbouring levels is equal to the mean energy of thermal motion ($\frac{1}{2}kT$). Calculate the value of n for the case of a helium atom in a box of length 1 cm at 298 K.

E7D.11(b) Find an expression for the value of n of a particle of mass m in a one-dimensional box of length L such that the energy of the level is equal to the mean energy of thermal motion ($\frac{1}{2}kT$). Calculate the value of n for the case of an argon atom in a box of length 0.1 cm at 298 K.

E7D.12(a) For a particle in a square box of side L, at what position (or positions) is the probability density a maximum if the wavefunction has $n_1 = 2$, $n_2 = 2$? Also, describe the position of any node or nodes in the wavefunction.

E7D.12(b) For a particle in a square box of side L, at what position (or positions) is the probability density a maximum if the wavefunction has $n_1 = 1$, $n_2 = 3$? Also, describe the position of any node or nodes in the wavefunction.

E7D.13(a) For a particle in a rectangular box with sides of length $L_1 = L$ and $L_2 = 2L$, find a state that is degenerate with the state $n_1 = n_2 = 2$. (*Hint:* You will need to experiment with some possible values of n_1 and n_2.) Is this degeneracy associated with symmetry?

E7D.13(b) For a particle in a rectangular box with sides of length $L_1 = L$ and $L_2 = 2L$, find a state that is degenerate with the state $n_1 = 2$, $n_2 = 8$. Would you expect there to be any degenerate states for a rectangular box with $L_1 = L$ and $L_2 = \sqrt{2}L$? Explain your reasoning.

E7D.14(a) Consider a particle in a cubic box. What is the degeneracy of the level that has an energy three times that of the lowest level?

E7D.14(b) Consider a particle in a cubic box. What is the degeneracy of the level that has an energy $\frac{14}{3}$ times that of the lowest level?

E7D.15(a) Suppose that the junction between two semiconductors can be represented by a barrier of height 2.0 eV and length 100 pm. Calculate the transmission probability of an electron with energy 1.5 eV.

E7D.15(b) Suppose that a proton of an acidic hydrogen atom is confined to an acid that can be represented by a barrier of height 2.0 eV and length 100 pm. Calculate the probability that a proton with energy 1.5 eV can escape from the acid.

Problems

P7D.1 Calculate the separation between the two lowest levels for an O_2 molecule in a one-dimensional container of length 5.0 cm. At what value of n does the energy of the molecule reach $\frac{1}{2}kT$ at 300 K, and what is the separation of this level from the one immediately below?

P7D.2 A nitrogen molecule is confined in a cubic box of volume 1.00 m^3. (i) Assuming that the molecule has an energy equal to $\frac{3}{2}kT$ at $T = 300$ K, what is the value of $n = (n_x^2 + n_y^2 + n_z^2)^{1/2}$ for this molecule? (ii) What is the energy separation between the levels n and $n + 1$? (iii) What is the de Broglie wavelength of the molecule?

P7D.3 Calculate the expectation values of x and x^2 for a particle in the state with $n = 1$ in a one-dimensional square-well potential.

P7D.4 Calculate the expectation values of p_x and p_x^2 for a particle in the state with $n = 2$ in a one-dimensional square-well potential.

P7D.5 When β-carotene (**1**) is oxidized *in vivo*, it breaks in half and forms two molecules of retinal (vitamin A), which is a precursor to the pigment in the retina responsible for vision. The conjugated system of retinal consists of 11 C atoms and one O atom. In the ground state of retinal, each level up to $n = 6$ is occupied by two electrons. Assuming an average internuclear distance of 140 pm, calculate (a) the separation in energy between the ground state and the first excited state in which one electron occupies the state with $n = 7$, and (b) the frequency of the radiation required to produce a transition between these two states. (c) Using your results, choose among the words in parentheses to generate a rule for the prediction of frequency shifts in the absorption spectra of linear polyenes:

> The absorption spectrum of a linear polyene shifts to (higher/lower) frequency as the number of conjugated atoms (increases/decreases).

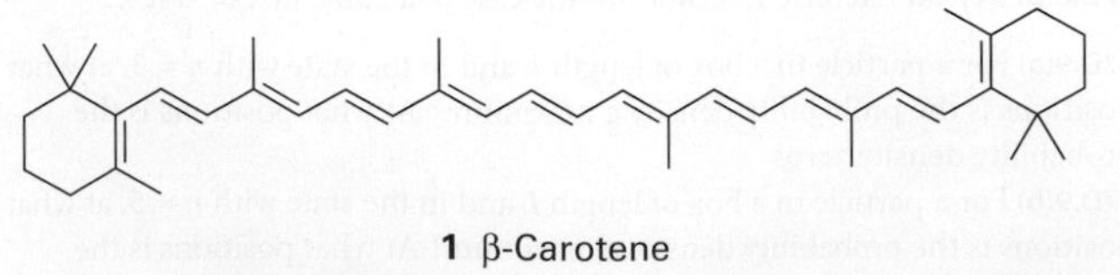

1 β-Carotene

P7D.6 Consider a particle of mass m confined to a one-dimensional box of length L and in a state with normalized wavefunction ψ_n. (a) Without evaluating any integrals, explain why $\langle x\rangle = L/2$. (b) Without evaluating any integrals, explain why $\langle p_x\rangle = 0$. (c) Derive an expression for $\langle x^2\rangle$ (the necessary integrals will be found in the *Resource section*). (d) For a particle in a box the energy is given by $E_n = n^2h^2/8mL^2$ and, because the potential energy is zero, all of this energy is kinetic. Use this observation and, without evaluating any integrals, explain why $\langle p_x^2\rangle = n^2h^2/4L^2$.

P7D.7 This problem requires the results for $\langle x\rangle$, $\langle x^2\rangle$, $\langle p_x\rangle$, and $\langle p_x^2\rangle$ obtained in Problem P7D.6. According to Topic 7C, the uncertainty in the position is $\Delta x = (\langle x^2\rangle - \langle x\rangle^2)^{1/2}$ and for the linear momentum $\Delta p_x = (\langle p_x^2\rangle - \langle p_x\rangle^2)^{1/2}$. (a) Use the results from Problem P7D.6 to find expressions for Δx and Δp_x. (b) Hence find an expression for the product $\Delta x\Delta p_x$. (c) Show that for $n = 1$ and $n = 2$ the result from (b) is in accord with the Heisenberg uncertainty principle, and infer that this is also true for $n \geq 1$.

P7D.8‡ A particle is confined to move in a one-dimensional box of length L. If the particle is behaving classically, then it simply bounces back and forth in the box, moving with a constant speed. (a) Explain why the probability density, $P(x)$, for the classical particle is $1/L$. (*Hint:* What is the total probability of finding the particle in the box?) (b) Explain why the average value of x^n is $\langle x^n\rangle = \int_0^L P(x)x^n\mathrm{d}x$. (c) By evaluating such an integral, find $\langle x\rangle$ and $\langle x^2\rangle$. (d) For a quantum particle $\langle x\rangle = L/2$ and $\langle x^2\rangle = L^2(\frac{1}{3} - 1/2n^2\pi^2)$. Compare these expressions with those you have obtained in (c), recalling that the correspondence principle states that, for very large values of the quantum numbers, the predictions of quantum mechanics approach those of classical mechanics.

P7D.9 (a) Set up the Schrödinger equation for a particle of mass m in a three-dimensional rectangular box with sides L_1, L_2, and L_3. Show that the Schrödinger equation is separable. (b) Show that the wavefunction and the energy are defined by three quantum numbers. (c) Specialize the result from part (b) to an electron moving in a cubic box of side $L = 5$ nm and draw an energy diagram resembling Fig. 7D.2 and showing the first 15 energy levels. Note that each energy level might be degenerate. (d) Compare the energy level diagram from part (c) with the energy level diagram for an electron in a one-dimensional box of length $L = 5$ nm. Are the energy levels become more or less sparsely distributed in the cubic box than in the one-dimensional box?

P7D.10 In the text the one-dimensional particle-in-a-box problem involves confining the particle to the range from $x = 0$ to $x = L$. This problem explores a similar situation in which the potential energy is zero between $x = -L/2$ and

$x = +L/2$, and infinite elsewhere. (a) Identify the boundary conditions that apply in this case. (b) Show that $\cos(kx)$ is a solution of the Schrödinger equation for the region with zero potential energy, find the values of k for which the boundary conditions are satisfied, and hence derive an expression for the corresponding energies. Sketch the three wavefunctions with the lowest energies. (c) Repeat the process, but this time with the trial wavefunction $\sin(k'x)$. (d) Compare the complete set of energies you have obtained in (b) and (c) with the energies for the case where the particle is confined between 0 and L: are they the same? (e) Normalize the wavefunctions (the necessary integrals are in the *Resource section*). (f) Without evaluating any integrals, explain why $\langle x\rangle=0$ for both sets of wavefunctions.

P7D.11 Many biological electron transfer reactions, such as those associated with biological energy conversion, may be visualized as arising from electron tunnelling between protein-bound co-factors, such as cytochromes, quinones, flavins, and chlorophylls. This tunnelling occurs over distances that are often greater than 1.0 nm, with sections of protein separating electron donor from acceptor. For a specific combination of donor and acceptor, the rate of electron tunnelling is proportional to the transmission probability, with $\kappa \approx 7\,\text{nm}^{-1}$ (eqn 7D.17). By what factor does the rate of electron tunnelling between two co-factors increase as the distance between them changes from 2.0 nm to 1.0 nm? You may assume that the barrier is such that eqn 7D.20b is appropriate.

P7D.12 Derive eqn 7D.20a, the expression for the transmission probability and show that when $\kappa W >> 1$ it reduces to eqn 7D.20b. The derivation proceeds by requiring that the wavefunction and its first derivative are continuous at the edges of the barrier, as expressed by eqns 7D.19a and 7D.19b.

P7D.13‡ A particle of mass m moves in one dimension in a region divided into three zones: zone 1 has $V = 0$ for $-\infty < x \leq 0$; zone 2 has $V = V_2$ for $0 \leq x \leq W$; zone 3 has $V = V_3$ for $W \leq x < \infty$. In addition, $V_3 < V_2$. In zone 1 the wavefunction is $A_1e^{ik_1x} + B_1e^{-ik_1x}$; the term e^{ik_1x} represents the wave incident on the barrier V_2, and the term e^{-ik_1x} represents the reflected wave. In zone 2 the wavefunction is $A_2e^{k_2x} + B_2e^{-k_2x}$. In zone 3 the wavefunction has only a forward component, $A_3e^{ik_3x}$, which represents a particle that has traversed the barrier. Consider a case in which the energy of the particle E is greater than V_3 but less than V_2, so that zone 2 represents a barrier. The transmission probability, T, is the ratio of the square modulus of the zone 3 amplitude to the square modulus of the incident amplitude, that is, $T = |A_3|^2/|A_1|^2$. (a) Derive an expression for T by imposing the requirement that the wavefunction and its slope must be continuous at the zone boundaries. You can simplify the calculation by assuming from the outset that $A_1 = 1$. (b) Show that this equation for T reduces to eqn 7D.20b in the high, wide barrier limit when $V_1 = V_3 = 0$. (c) Draw a graph of the probability of proton tunnelling when $V_3 = 0$, $W = 50\,\text{pm}$, and $E = 10\,\text{kJ mol}^{-1}$ in the barrier range $E < V_2 < 2E$.

P7D.14 A potential barrier of height V extends from $x = 0$ to positive x. Inside this barrier the normalized wavefunction is $\psi = Ne^{-\kappa x}$. Calculate (a) the probability that the particle is inside the barrier and (b) the average penetration depth of the particle into the barrier.

P7D.15 Use mathematical software or a spreadsheet for the following procedures:

(a) Plot the probability density for a particle in a box with $n = 1, 2, \ldots 5$, and $n = 50$. How do your plots illustrate the correspondence principle?

(b) Plot the transmission probability T against E/V for passage by (i) a hydrogen molecule, (ii) a proton, and (iii) an electron through a barrier of height V.

(c) Use mathematical software to generate three-dimensional plots of the wavefunctions for a particle confined to a rectangular surface with (i) $n_1 = 1, n_2 = 1$, the state of lowest energy, (ii) $n_1 = 1, n_2 = 2$, (iii) $n_1 = 2, n_2 = 1$, and (iv) $n_1 = 2, n_2 = 2$. Deduce a rule for the number of nodal lines in a wavefunction as a function of the values of n_1 and n_2.

TOPIC 7E Vibrational motion

Discussion questions

D7E.1 Describe the variation with the mass and force constant of the separation of the vibrational energy levels of a harmonic oscillator.

D7E.2 In what ways does the quantum mechanical description of a harmonic oscillator merge with its classical description at high quantum numbers?

D7E.3 To what quantum mechanical principle can you attribute the existence of a zero-point vibrational energy?

Exercises

E7E.1(a) Calculate the zero-point energy of a harmonic oscillator consisting of a particle of mass $2.33 \times 10^{-26}\,\text{kg}$ and force constant $155\,\text{N m}^{-1}$.
E7E.1(b) Calculate the zero-point energy of a harmonic oscillator consisting of a particle of mass $5.16 \times 10^{-26}\,\text{kg}$ and force constant $285\,\text{N m}^{-1}$.

E7E.2(a) For a certain harmonic oscillator of effective mass $1.33 \times 10^{-25}\,\text{kg}$, the difference in adjacent energy levels is 4.82 zJ. Calculate the force constant of the oscillator.
E7E.2(b) For a certain harmonic oscillator of effective mass $2.88 \times 10^{-25}\,\text{kg}$, the difference in adjacent energy levels is 3.17 zJ. Calculate the force constant of the oscillator.

E7E.3(a) Calculate the wavelength of the photon needed to excite a transition between neighbouring energy levels of a harmonic oscillator of effective mass equal to that of a proton ($1.0078 m_u$) and force constant $855\,\text{N m}^{-1}$.
E7E.3(b) Calculate the wavelength of the photon needed to excite a transition between neighbouring energy levels of a harmonic oscillator of effective mass equal to that of an oxygen atom ($15.9949 m_u$) and force constant $544\,\text{N m}^{-1}$.

E7E.4(a) Sketch the form of the wavefunctions for the harmonic oscillator with quantum numbers $\upsilon = 0$ and 1. Use a symmetry argument to explain why these two wavefunctions are orthogonal (do not evaluate any integrals).
E7E.4(b) Sketch the form of the wavefunctions for the harmonic oscillator with quantum numbers $\upsilon = 1$ and 2. Use a symmetry argument to explain why these two wavefunctions are orthogonal (do not evaluate any integrals).

E7E.5(a) Assuming that the vibrations of a $^{35}Cl_2$ molecule are equivalent to those of a harmonic oscillator with a force constant $k_f = 329\,\text{N m}^{-1}$, what is the zero-point energy of vibration of this molecule? Use $m(^{35}\text{Cl}) = 34.9688\,m_u$.
E7E.5(b) Assuming that the vibrations of a $^{14}N_2$ molecule are equivalent to those of a harmonic oscillator with a force constant $k_f = 2293.8\,\text{N m}^{-1}$, what is the zero-point energy of vibration of this molecule? Use $m(^{14}\text{N}) = 14.0031\,m_u$.

E7E.6(a) The classical turning points of a harmonic oscillator occur at the displacements at which all of the energy is potential energy; that is, when $E_\upsilon = \frac{1}{2}k_f x_{tp}^2$. For a particle of mass m_u undergoing harmonic motion with force constant $k_f = 1000\,\text{N m}^{-1}$, calculate the energy of the state with $\upsilon = 0$ and hence find the separation between the classical turning points. Repeat the calculation for an oscillator with $k_f = 100\,\mathrm{N\,m^{-1}}$.

E7E.6(b) Repeat the calculation in Exercise E7E.6(a) but for the first excited state, $v = 1$. Express your answers as a percentage of a typical bond length of 110 pm.

E7E.7(a) How many nodes are there in the wavefunction of a harmonic oscillator with (i) $v = 3$; (ii) $v = 4$?

E7E.7(b) How many nodes are there in the wavefunction of a harmonic oscillator with (i) $v = 5$; (ii) $v = 35$?

E7E.8(a) Locate the nodes of a harmonic oscillator wavefunction with $v = 2$. (Express your answers in terms of the coordinate y.)

E7E.8(b) Locate the nodes of the harmonic oscillator wavefunction with $v = 3$.

E7E.9(a) At what displacements is the probability density a maximum for a state of a harmonic oscillator with $v = 1$? (Express your answers in terms of the coordinate y.)

E7E.9(b) At what displacements is the probability density a maximum for a state of a harmonic oscillator with $v = 3$?

Problems

P7E.1 If the vibration of a diatomic A–B is modelled using a harmonic oscillator, the vibrational frequency is given by $\omega = (k_f/\mu)^{1/2}$, where μ is the effective mass, $\mu = m_A m_B/(m_A + m_B)$. If atom A is substituted by an isotope (for example ^{2}H substituted for ^{1}H), then to a good approximation the force constant remains the same. Why? (*Hint:* Is there any change in the number of charged species?) (a) Show that when an isotopic substitution is made for atom A, such that its mass changes from m_A to $m_{A'}$, the vibrational frequency of A′–B, $\omega_{A'B}$, can be expressed in terms of the vibrational frequency of A–B, ω_{AB} as $\omega_{A'B} = \omega_{AB}(\mu_{AB}/\mu_{A'B})^{1/2}$, where μ_{AB} and $\mu_{A'B}$ are the effective masses of A–B and A′–B, respectively. (b) The vibrational frequency of ^{1}H^{35}Cl is $5.63 \times 10^{14}\,\mathrm{s^{-1}}$. Calculate the vibrational frequency of (i) ^{2}H^{35}Cl and (ii) ^{1}H^{37}Cl. Use integer relative atomic masses.

P7E.2 Before attempting these calculations, see Problem P7E.1. Now consider the case where in the diatomic molecule A–B the mass of B is much greater than that of A. (a) Show that for an isotopic substitution of A, the ratio of vibrational frequencies is $\omega_{A'B} \approx \omega_{AB}(m_A/m_{A'})^{1/2}$. (b) Use this expression to calculate the vibrational frequency of ^{2}H^{35}Cl (the vibrational frequency of ^{1}H^{35}Cl is $5.63 \times 10^{14}\,\mathrm{s^{-1}}$). (c) Compare your answer with the value obtained in the previous Problem P7E.1. (d) In organic molecules it is commonly observed that the C–H stretching frequency is reduced by a factor of around 0.7 when ^{1}H is substituted by ^{2}H: rationalize this observation.

P7E.3 The vibrational frequency of ^{1}H$_2$ is 131.9 THz. What is the vibrational frequency of ^{2}H$_2$ and of ^{3}H$_2$? Use integer relative atomic masses for this estimate.

P7E.4 The force constant for the bond in CO is $1857\,\mathrm{N\,m^{-1}}$. Calculate the vibrational frequencies (in Hz) of ^{12}C^{16}O, ^{13}C^{16}O, ^{12}C^{18}O, and ^{13}C^{18}O. Use integer relative atomic masses for this estimate.

P7E.5 In infrared spectroscopy it is common to observe a transition from the $v = 0$ to $v = 1$ vibrational level. If this transition is modelled as a harmonic oscillator, the energy of the photon involved is $\hbar\omega$, where ω is the vibrational frequency. (a) Show that the wavenumber of the radiation corresponding to photons of this energy, $\tilde{\nu}$, is given by $\tilde{\nu} = \omega/2\pi c$, where c is the speed of light. (b) The vibrational frequency of ^{1}H^{35}Cl is $\omega = 5.63 \times 10^{14}\,\mathrm{s^{-1}}$; calculate $\tilde{\nu}$. (c) Derive an expression for the force constant k_f in terms of $\tilde{\nu}$. (d) For ^{12}C^{16}O the $v = 0 \rightarrow 1$ transition is observed at $2170\,\mathrm{cm^{-1}}$. Calculate the force constant and estimate the wavenumber at which the corresponding absorption occurs for ^{14}C^{16}O. Use integer relative atomic masses for this estimate.

P7E.6 Before attempting these calculations, see Problem P7E.5. The following data give the wavenumbers (wavenumbers in cm^{-1}) of the $v = 0 \rightarrow 1$ transition of a number of diatomic molecules. Calculate the force constants of the bonds and arrange them in order of increasing stiffness. Use integer relative atomic masses.

^{1}H^{35}Cl	^{1}H^{81}Br	^{1}H^{127}I	^{12}C^{16}O	^{14}N^{16}O
2990	2650	2310	2170	1904

P7E.7 Carbon monoxide binds strongly to the Fe^{2+} ion of the haem (heme) group of the protein myoglobin. Estimate the vibrational frequency of CO bound to myoglobin by using the data in Problem P7E.6 and by making the following assumptions: the atom that binds to the haem group is immobilized, the protein is infinitely more massive than either the C or O atom, the C atom binds to the Fe^{2+} ion, and binding of CO to the protein does not alter the force constant of the CO bond.

P7E.8 Of the four assumptions made in Problem P7E.7, the last two are questionable. Suppose that the first two assumptions are still reasonable and that you have at your disposal a supply of myoglobin, a suitable buffer in which to suspend the protein, ^{12}C^{16}O, ^{13}C^{16}O, ^{12}C^{18}O, ^{13}C^{18}O, and an infrared spectrometer. Assuming that isotopic substitution does not affect the force constant of the CO bond, describe a set of experiments that: (a) proves which atom, C or O, binds to the haem group of myoglobin, and (b) allows for the determination of the force constant of the CO bond for myoglobin-bound carbon monoxide.

P7E.9 A function of the form e^{-gx^2} is a solution of the Schrödinger equation for the harmonic oscillator (eqn 7E.2), provided that g is chosen correctly. In this problem you will find the correct form of g. (a) Start by substituting $\psi = e^{-gx^2}$ into the left-hand side of eqn 7E.2 and evaluating the second derivative. (b) You will find that in general the resulting expression is not of the form constant $\times\ \psi$, implying that ψ is not a solution to the equation. However, by choosing the value of g such that the terms in x^2 cancel one another, a solution is obtained. Find the required form of g and hence the corresponding energy. (c) Confirm that the function so obtained is indeed the ground state of the harmonic oscillator, as quoted in eqn 7E.7, and that it has the energy expected from eqn 7E.3.

P7E.10 Write the normalized form of the ground state wavefunction of the harmonic oscillator in terms of the variable y and the parameter α. (a) Write the integral you would need to evaluate to find the mean displacement $\langle y \rangle$, and then use a symmetry argument to explain why this integral is equal to 0. (b) Calculate $\langle y^2 \rangle$ (the necessary integral will be found in the *Resource section*). (c) Repeat the process for the first excited state.

P7E.11 The expectation value of the kinetic energy of a harmonic oscillator is most easily found by using the virial theorem, but in this Problem you will find it directly by evaluating the expectation value of the kinetic energy operator with the aid of the properties of the Hermite polynomials given in Table 7E.1. (a) Write the kinetic energy operator $\hat{T}$ in terms of x and show that it can be rewritten in terms of the variable y (introduced in eqn 7E.7) and the frequency ω as

$$\hat{T} = -\tfrac{1}{2}\hbar\omega\frac{d^2}{dy^2}$$

The expectation value of this operator for an harmonic oscillator wavefunction with quantum number v is

$$\langle T \rangle_v = -\tfrac{1}{2}\hbar\omega\alpha N_v^2 \int_{-\infty}^{\infty} H_v e^{-y^2/2} \frac{d^2}{dy^2} H_v e^{-y^2/2} dy$$

where N_v is the normalization constant (eqn 7E.10) and α is defined in eqn 7E.7 (the term α arises from $dx = \alpha dy$). (b) Evaluate the second derivative and then use the property $H_v'' - 2yH_v' + 2vH_v = 0$, where the prime indicates a derivative, to rewrite the derivatives in terms of the H_v (you should be able to eliminate all the derivatives). (c) Now proceed as in the text, in which terms of the form yH_v are rewritten by using the property $H_{v+1} - 2yH_v + 2vH_{v-1} = 0$; you will need to apply this twice. (d) Finally, evaluate the integral using the properties of the integrals of the Hermite polynomials given in Table 7E.1 and so obtain the result quoted in the text.

P7E.12 Calculate the values of $\langle x^3\rangle_v$ and $\langle x^4\rangle_v$ for a harmonic oscillator by using the properties of the Hermite polynomials given in Table 7E.1; follow the approach used in the text.

P7E.13 Use the same approach as in *Example* 7E.3 to calculate the probability that a harmonic oscillator in the first excited state will be found in the classically forbidden region. You will need to use mathematical software to evaluate the appropriate integral. Compare the result you obtain with that for the ground state and comment on the difference.

P7E.14 Use the same approach as in *Example* 7E.3 to calculate the probability that a harmonic oscillator in the states $v = 0, 1, \ldots 7$ will be found in the classically forbidden region. You will need to use mathematical software to evaluate the final integrals. Plot the probability as a function of v and interpret the result in terms of the correspondence principle.

P7E.15 The intensities of spectroscopic transitions between the vibrational states of a molecule are proportional to the square of the integral $\int\psi_{v'}x\psi_v dx$ over all space. Use the relations between Hermite polynomials given in Table 7E.1 to show that the only permitted transitions are those for which $v' = v \pm 1$ and evaluate the integral in these cases.

P7E.16 The potential energy of the rotation of one CH_3 group relative to its neighbour in ethane can be expressed as $V(\phi) = V_0 \cos 3\phi$. Show that for small displacements the motion of the group is harmonic and derive an expression for the energy of excitation from $v = 0$ to $v = 1$. (*Hint:* Use a series expansion for $\cos 3\phi$.) What do you expect to happen to the energy levels and wavefunctions as the excitation increases to high quantum numbers?

P7E.17 (a) Without evaluating any integrals, explain why you expect $\langle x\rangle_v = 0$ for all states of a harmonic oscillator. (b) Use a physical argument to explain why $\langle p_x\rangle_v = 0$. (c) Equation 7E.13c gives $\langle E_k\rangle_v = \frac{1}{2}E_v$. Recall that the kinetic energy is given by $p^2/2m$ and hence find an expression for $\langle p_x^2\rangle_v$. (d) Note from Topic 7C that the uncertainty in the position, Δx, is given by $\Delta x = (\langle x^2\rangle - \langle x\rangle^2)^{1/2}$ and likewise for the momentum $\Delta p_x = (\langle p_x^2\rangle - \langle p_x\rangle^2)^{1/2}$. Find expressions for Δx and Δp_x (the expression for $\langle x^2\rangle_v$ is given in the text). (e) Hence find an expression for the product $\Delta x\Delta p_x$ and show that the Heisenberg uncertainty principle is satisfied. (f) For which state is the product $\Delta x\Delta p_x$ a minimum?

P7E.18 Use mathematical software or a spreadsheet to gain some insight into the origins of the nodes in the harmonic oscillator wavefunctions by plotting the Hermite polynomials $H_v(y)$ for $v = 0$ through 5.

TOPIC 7F Rotational motion

Discussion questions

D7F.1 Discuss the physical origin of quantization of energy for a particle confined to motion on a ring.

D7F.2 Describe the features of the solution of the particle on a ring that appear in the solution of the particle on a sphere. What concept applies to the latter but not to the former?

D7F.3 Describe the vector model of angular momentum in quantum mechanics. What features does it capture?

Exercises

E7F.1(a) The rotation of a molecule can be represented by the motion of a particle moving over the surface of a sphere. Calculate the magnitude of its angular momentum when $l = 1$ and the possible components of the angular momentum along the z-axis. Express your results as multiples of $\hbar$.

E7F.1(a) The rotation of a molecule can be represented by the motion of a particle moving over the surface of a sphere with angular momentum quantum number $l = 2$. Calculate the magnitude of its angular momentum and the possible components of the angular momentum along the z-axis. Express your results as multiples of $\hbar$.

E7F.2(a) For a particle on a ring, how many nodes are there in the real part, and in the imaginary part, of the wavefunction for (i) $m_l = 0$ and (ii) $m_l = +3$? In both cases, find the values of ϕ at which any nodes occur.

E7F.2(b) For a particle on a ring, how many nodes are there in the real part, and in the imaginary part of the wavefunction for (i) $m_l = +1$ and (ii) $m_l = +2$? In both cases, find the values of ϕ at which any nodes occur.

E7F.3(a) The wavefunction for the motion of a particle on a ring is of the form $\psi = Ne^{im_l\phi}$. Evaluate the normalization constant, N.

E7F.3(b) The wavefunction for the motion of a particle on a ring can also be written $\psi = N\cos(m_l\phi)$, where m_l is integer. Evaluate the normalization constant, N.

E7F.4(a) By considering the integral $\int_0^{2\pi}\psi_{m_l}^*\psi_{m_l'}\,d\phi$, where $m_l \neq m_l'$, confirm that wavefunctions for a particle in a ring with different values of the quantum number m_l are mutually orthogonal.

E7F.4(b) By considering the integral $\int_0^{2\pi}\cos m_l\phi\cos m_l'\phi\,d\phi$, where $m_l \neq m_l'$, confirm that the wavefunctions $\cos m_l\phi$ and $\cos m_l'\phi$ for a particle on a ring are orthogonal. (*Hint:* To evaluate the integral, first apply the identity $\cos A\cos B = \frac{1}{2}\{\cos(A+B)+\cos(A-B)\}$.)

E7F.5(a) Calculate the minimum excitation energy (i.e. the difference in energy between the first excited state and the ground state) of a proton constrained to rotate in a circle of radius 100 pm around a fixed point.

E7F.5(b) Calculate the value of $|m_l|$ for the system described in the preceding Exercise corresponding to a rotational energy equal to the classical average energy at 25 °C (which is equal to $\frac{1}{2}kT$).

E7F.6(a) The moment of inertia of a CH_4 molecule is 5.27×10^{-47} kg m^2. What is the minimum energy needed to start it rotating?

E7F.6(b) The moment of inertia of an SF_6 molecule is 3.07×10^{-45} kg m^2. What is the minimum energy needed to start it rotating?

E7F.7(a) Use the data in Exercise E7F.6(a) to calculate the energy needed to excite a CH_4 molecule from a state with $l = 1$ to a state with $l = 2$.

E7F.7(b) Use the data in Exercise E7F.6(b) to calculate the energy needed to excite an SF_6 molecule from a state with $l = 2$ to a state with $l = 3$.

E7F.8(a) What is the magnitude of the angular momentum of a CH_4 molecule when it is rotating with its minimum energy?

E7F.8(b) What is the magnitude of the angular momentum of an SF_6 molecule when it is rotating with its minimum energy?

E7F.9(a) Draw scale vector diagrams to represent the states (i) $l = 1$, $m_l = +1$, (ii) $l = 2$, $m_l = 0$.

E7F.9(b) Draw the vector diagram for all the permitted states of a particle with $l = 6$.

E7F.10(a) How many angular nodes are there for the spherical harmonic $Y_{3,0}$ and at which values of θ do they occur?

E7F.10(b) Based on the pattern of nodes in Fig. 7F.5, how many angular nodes do you expect there to be for the spherical harmonic $Y_{4,0}$? Does it have a node at $\theta = 0$?

E7F.11(a) Consider the real part of the spherical harmonic $Y_{1,+1}$. At which values of ϕ do angular nodes occur? These angular nodes can also be described as planes: identify the positions of the corresponding planes (for example, the angular node with $\phi = 0$ is the *xz*-plane). Do the same for the imaginary part.

E7F.11(b) Consider the real part of the spherical harmonic $Y_{2,+2}$. At which values of ϕ do angular nodes occur? Identify the positions of the corresponding planes. Repeat the process for the imaginary part.

E7F.12(a) What is the degeneracy of a molecule rotating with $J = 3$?

E7F.12(b) What is the degeneracy of a molecule rotating with $J = 4$?

E7F.13(a) Draw diagrams to scale, and similar to Fig. 7F.7a, representing the states (i) $l = 1$, $m_l = -1, 0, +1$, (ii) $l = 2$ and all possible values of m_l.

E7F.13(b) Draw diagrams to scale, and similar to Fig. 7F.7a, representing the states (i) $l = 0$, (ii) $l = 3$ and all possible values of m_l.

E7F.14(a) Derive an expression for the angle between the vector representing angular momentum l with *z*-component $m_l = +l$ (that is, its maximum value) and the *z*-axis. What is this angle for $l = 1$ and for $l = 5$?

E7F.14(b) Derive an expression for the angle between the vector representing angular momentum l with *z*-component $m_l = +l$ and the *z*-axis. What value does this angle take in the limit that l becomes very large? Interpret your result in the light of the correspondence principle.

Problems

P7F.1 The particle on a ring is a useful model for the motion of electrons around the porphyrin ring (2), the conjugated macrocycle that forms the structural basis of the haem (heme) group and the chlorophylls. The group may be modelled as a circular ring of radius 440 pm, with 22 electrons in the conjugated system moving along its perimeter. In the ground state of the molecule each state is occupied by two electrons. (a) Calculate the energy and angular momentum of an electron in the highest occupied level. (b) Calculate the frequency of radiation that can induce a transition between the highest occupied and lowest unoccupied levels.

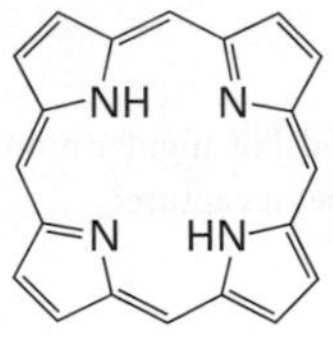

2 Porphyrin ring

P7F.2 Consider the following wavefunctions (i) $e^{i\phi}$, (ii) $e^{-2i\phi}$, (iii) $\cos\phi$, and (iv) $(\cos\chi)e^{i\phi} + (\sin\chi)e^{-i\phi}$ each of which describes a particle on a ring. (a) Decide whether or not each wavefunction is an eigenfunction of the operator $\hat{l}_z$ for the *z*-component of the angular momentum ($\hat{l}_z = (\hbar/i)(d/d\phi)$); where the function is an eigenfunction, give the eigenvalue. (b) For the functions that are not eigenfunctions, calculate the expectation value of l_z (you will first need to normalize the wavefunction). (c) Repeat the process but this time for the kinetic energy, for which the operator is $-(\hbar^2/2I)(d^2/d\phi^2)$. (d) Which of these wavefunctions describe states of definite angular momentum, and which describe states of definite kinetic energy?

P7F.3 Is the Schrödinger equation for a particle on an elliptical ring of semi-major axes a and b separable? (*Hint:* Although r varies with angle ϕ, the two are related by $r^2 = a^2\sin^2\phi + b^2\cos^2\phi$.)

P7F.4 Calculate the energies of the first four rotational levels of $^1H^{127}I$ free to rotate in three dimensions; use for its moment of inertia $I = \mu R^2$, with $\mu = m_H m_I/(m_H + m_I)$ and $R = 160$ pm. Use integer relative atomic masses for this estimate.

P7F.5 Consider the three spherical harmonics (a) $Y_{0,0}$, (b) $Y_{2,-1}$, and (c) $Y_{3,+3}$. (a) For each spherical harmonic, substitute the explicit form of the function taken from Table 7F.1 into the left-hand side of eqn 7F.8 (the Schrödinger equation for a particle on a sphere) and confirm that the function is a solution of the equation; give the corresponding eigenvalue (the energy) and show that it agrees with eqn 7F.10. (b) Likewise, show that each spherical harmonic is an eigenfunction of $\hat{l}_z = (\hbar/i)(d/d\phi)$ and give the eigenvalue in each case.

P7F.6 Confirm that $Y_{1,+1}$, taken from Table 7F.1, is normalized. You will need to integrate $Y_{1,+1}^{*}Y_{1,+1}$ over all space using the relevant volume element:

$$\int_{\theta=0}^{\pi}\int_{\phi=0}^{2\pi} Y_{1,+1}^{*}Y_{1,+1}\overbrace{\sin\theta\,d\theta\,d\phi}^{\text{volume element}}$$

P7F.7 Confirm that $Y_{1,0}$ and $Y_{1,+1}$, taken from Table 7F.1, are orthogonal. You will need to integrate $Y_{1,0}^{*}Y_{1,+1}$ over all space using the relevant volume element:

$$\int_{\theta=0}^{\pi}\int_{\phi=0}^{2\pi} Y_{1,0}^{*}Y_{1,+1}\overbrace{\sin\theta\,d\theta\,d\phi}^{\text{volume element}}$$

(*Hint:* A useful result for evaluating the integral is $(d/d\theta)\sin^3\theta = 3\sin^2\theta\cos\theta$.)

P7F.8 (a) Show that $\psi = c_1 Y_{l,m_l} + c_2 Y_{l,m_l'}$ is an eigenfunction of Λ^2 with eigenvalue $-l(l+1)$; c_1 and c_2 are arbitrary coefficients. (*Hint:* Apply Λ^2 to ψ and use the properties given in eqn 7F.9.) (b) The spherical harmonics $Y_{1,+1}$ and $Y_{1,-1}$ are complex functions (see Table 7F.1), but as they are degenerate eigenfunctions of Λ^2, any linear combination of them is also an eigenfunction, as was shown in (a). Show that the combinations $\psi_a = -Y_{1,+1} + Y_{1,-1}$ and $\psi_b = i(Y_{1,+1} + Y_{1,-1})$ are real. (c) Show that ψ_a and ψ_b are orthogonal (you will need to integrate using the relevant volume element, see Problem P7F.7). (d) Normalize ψ_a and ψ_b. (e) Identify the angular nodes in these two functions and the planes to which they correspond. (f) Is ψ_a an eigenfunction of $\hat{l}_z$? Discuss the significance of your answer.

P7F.9 In this problem you will establish the commutation relations, given in eqn 7E.14, between the operators for the *x*-, *y*-, and *z*-components of angular momentum, which are defined in eqn 7F.13. In order to manipulate the operators correctly it is helpful to imagine that they are acting on some arbitrary function f: it does not matter what f is, and at the end of the proof it is simply removed. Consider $[\hat{l}_x, \hat{l}_y] = \hat{l}_x\hat{l}_y - \hat{l}_y\hat{l}_x$. Consider the effect of the first term on some arbitrary function f and evaluate

$$\hat{l}_x\hat{l}_y f = -\hbar^2\left(\overbrace{y\frac{\partial}{\partial z}}^{A} - \overbrace{z\frac{\partial}{\partial y}}^{B}\right)\left(\overbrace{z\frac{\partial f}{\partial x}}^{C} - \overbrace{x\frac{\partial f}{\partial z}}^{D}\right)$$

The next step is to multiply out the parentheses, and in doing so care needs to be taken over the order of operations. (b) Repeat the procedure for the other term in the commutator, $\hat{l}_y\hat{l}_x f$. (c) Combine the results from (a) and (b) so as to evaluate $\hat{l}_x\hat{l}_y f - \hat{l}_y\hat{l}_x f$; you should find that many of the terms cancel. Confirm that the final expression you have is indeed $i\hbar\hat{l}_z f$, where $\hat{l}_z$ is given in eqn 7F.13. (d) The definitions in eqn 7F.13 are related to one another by cyclic permutation of the x, y, and z. That is, by making the permutation $x \rightarrow y$, $y \rightarrow z$, and $z \rightarrow x$, you can move from one definition to the next: confirm that this is so. (e) The same cyclic permutation can be applied to the commutators of these operators. Start with $[\hat{l}_x, \hat{l}_y] = i\hbar\hat{l}_z$ and show that cyclic permutation generates the other two commutators in eqn 7F.14.

P7F.10 Show that $\hat{l}_z$ and $\hat{l}^2$ both commute with the hamiltonian for a hydrogen atom. What is the significance of this result? Begin by noting that $\hat{l}^2 = \hat{l}_x^2 + \hat{l}_y^2 + \hat{l}_z^2$. Then show that $[\hat{l}_z, \hat{l}_q^2] = [\hat{l}_z, \hat{l}_q]\hat{l}_q + \hat{l}_q[\hat{l}_z, \hat{l}_q]$ and then use the angular momentum commutation relations in eqn 7F.14.

P7F.11 Starting from the definition of the operator $\hat{l}_z$ given in eqn 7F.13, show that in spherical polar coordinates it can be expressed as $\hat{l}_z = -i\hbar\partial/\partial\phi$. (*Hint:* You will need to express the Cartesian coordinates in terms of the spherical polar coordinates; refer to *The chemist's toolkit* 21.)

P7F.12 A particle confined within a spherical cavity is a starting point for the discussion of the electronic properties of spherical metal nanoparticles. Here, you are invited to show in a series of steps that the $l = 0$ energy levels of an electron in a spherical cavity of radius R are quantized and given by $E_n = n^2h^2/8m_eR^2$. (a) The hamiltonian for a particle free to move inside a spherical cavity of radius a is

$$\hat{H} = -\frac{\hbar^2}{2m}\nabla^2 \quad \text{with } \nabla^2 = \frac{1}{r}\frac{\partial^2}{\partial r^2}r + \frac{1}{r^2}\Lambda^2$$

Show that the Schrödinger equation is separable into radial and angular components. That is, begin by writing $\psi(r,\theta,\phi) = R(r)Y(\theta,\phi)$, where $R(r)$ depends only on the distance of the particle from the centre of the sphere, and $Y(\theta,\phi)$ is a spherical harmonic. Then show that the Schrödinger equation can be separated into two equations, one for $R(r)$, the radial equation, and the other for $Y(\theta,\phi)$, the angular equation. (b) Consider the case $l = 0$. Show by differentiation that the solution of the radial equation has the form

$$R(r) = (2\pi a)^{-1/2}\frac{\sin(n\pi r/a)}{r}$$

(c) Now go on to show (by acknowledging the appropriate boundary conditions) that the allowed energies are given by $E_n = n^2h^2/8ma^2$. With substitution of m_e for m and of R for a, this is the equation given above for the energy.

FOCUS 7 Quantum theory

Integrated activities

I7.1‡ A star too small and cold to shine has been found by S. Kulkarni et al. (*Science*, 1478 (1995)). The spectrum of the object shows the presence of methane which, according to the authors, would not exist at temperatures much above 1000 K. The mass of the star, as determined from its gravitational effect on a companion star, is roughly 20 times the mass of Jupiter. The star is considered to be a brown dwarf, the coolest ever found.

(a) Derive an expression for $\Delta_rG^\ominus$ for $CH_4(g) \rightarrow C(\text{graphite}) + 2\,H_2(g)$ at temperature T. Proceed by using data from the tables in the *Resource section* to find $\Delta_rH^\ominus$ and $\Delta_rS^\ominus$ at 298 K and then convert these values to an arbitrary temperature T by using heat capacity data, also from the tables (assume that the heat capacities do not vary with temperature). (b) Find the temperature above which $\Delta_rG^\ominus$ becomes positive. (The solution to the relevant equation cannot be found analytically, so use mathematical software to find a numerical solution or plot a graph). Does your result confirm the assertion that methane could not exist at temperatures much above 1000 K? (c) Assume the star to behave as a black body at 1000 K, and calculate the wavelength at which the radiation from it is maximum. (d) Estimate the fraction of the energy density of the star that it emitted in the visible region of the spectrum (between 420 nm and 700 nm). (You may assume that over this wavelength range $\Delta\lambda$ it is acceptable to approximate the integral of the Planck distribution by $\rho(\lambda,T)\Delta\lambda$.)

I7.2 Describe the features that stem from nanometre-scale dimensions that are not found in macroscopic objects.

I7.3 Explain why the particle in a box and the harmonic oscillator are useful models for quantum mechanical systems: what chemically significant systems can they be used to represent?

I7.4 Suppose that 1.0 mol of perfect gas molecules all occupy the lowest energy level of a cubic box. (a) How much work must be done to change the volume of the box by ΔV? (b) Would the work be different if the molecules all occupied a state $n \neq 1$? (c) What is the relevance of this discussion to the expression for the expansion work discussed in Topic 2A? (d) Can you identify a distinction between adiabatic and isothermal expansion?

I7.5 Evaluate $\Delta x = (\langle x^2\rangle - \langle x\rangle^2)^{1/2}$ and $\Delta p_x = (\langle p_x^2\rangle - \langle p_x\rangle^2)^{1/2}$ for the ground state of (a) a particle in a box of length L and (b) a harmonic oscillator. Discuss these quantities with reference to the uncertainty principle.

I7.6 Repeat Problem I7.5 for (a) a particle in a box and (b) a harmonic oscillator in a general quantum state (n and v, respectively).

FOCUS 8

Atomic structure and spectra

This Focus discusses the use of quantum mechanics to describe and investigate the 'electronic structure' of atoms, the arrangement of electrons around their nuclei. The concepts are of central importance for understanding the properties of atoms and molecules, and hence have extensive chemical applications.

8A Hydrogenic atoms

This Topic uses the principles of quantum mechanics introduced in Focus 7 to describe the electronic structure of a 'hydrogenic atom', a one-electron atom or ion of general atomic number Z. Hydrogenic atoms are important because their Schrödinger equations can be solved exactly and they provide a set of concepts that are used to describe the structures of many-electron atoms and molecules. Solving the Schrödinger equation for an electron in an atom involves the separation of the wavefunction into angular and radial parts and the resulting wavefunctions are the hugely important 'atomic orbitals' of hydrogenic atoms.

8A.1 **The structure of hydrogenic atoms;** 8A.2 **Atomic orbitals and their energies**

8B Many-electron atoms

A 'many-electron atom' is an atom or ion with more than one electron. Examples include all neutral atoms other than H; so even He, with only two electrons, is a many-electron atom. This Topic uses hydrogenic atomic orbitals to describe the structures of many-electron atoms. Then, in conjunction with the concept of 'spin' and the 'Pauli exclusion principle', it describes the origin of the periodicity of atomic properties and the structure of the periodic table.

8B.1 **The orbital approximation;** 8B.2 **The Pauli exclusion principle;** 8B.3 **The building-up principle;** 8B.4 **Self-consistent field orbitals**

8C Atomic spectra

The spectra of many-electron atoms are more complicated than that of hydrogen. Similar principles apply, but Coulombic and magnetic interactions between the electrons give rise to a variety of energy differences, which are summarized by constructing 'term symbols'. These symbols act as labels that display the total orbital and spin angular momentum of a many-electron atom and are used to express the selection rules that govern their spectroscopic transitions.

8C.1 **The spectra of hydrogenic atoms;** 8C.2 **The spectra of many-electron atoms**

Web resource What is an application of this material?

Impact 13 focuses on the use of atomic spectroscopy to examine stars. By analysing their spectra it is possible to determine the composition of their outer layers and the surrounding gases and to determine features of their physical state.

TOPIC 8A Hydrogenic atoms

> ➤ **Why do you need to know this material?**
>
> An understanding of the structure of hydrogenic atoms is central to the description of all other atoms, the periodic table, and bonding. All accounts of the structures of molecules are based on the language and concepts introduced here.
>
> ➤ **What is the key idea?**
>
> Atomic orbitals are one-electron wavefunctions for atoms and are labelled by three quantum numbers that specify the energy and angular momentum of the electron.
>
> ➤ **What do you need to know already?**
>
> You need to be aware of the concept of a wavefunction (Topic 7B) and its interpretation. You also need to know how to set up a Schrödinger equation and how boundary conditions result in only certain solutions being acceptable (Topic 7D).

When an electric discharge is passed through gaseous hydrogen, the H_2 molecules are dissociated and the energetically excited H atoms that are produced emit electromagnetic radiation at a number of discrete frequencies (and therefore discrete wavenumbers), producing a spectrum of a series of 'lines' (Fig. 8A.1).

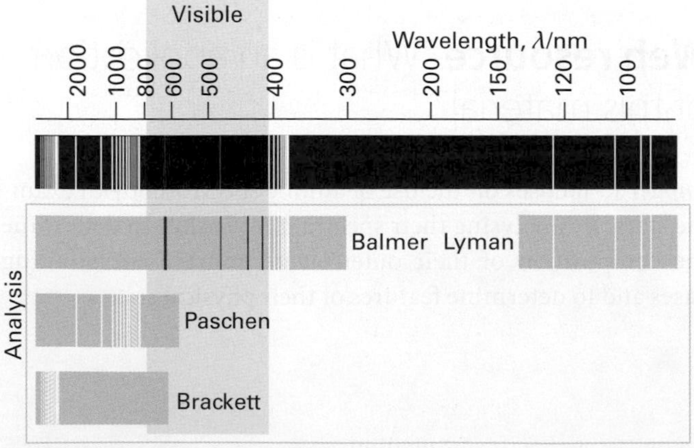

Figure 8A.1 The spectrum of atomic hydrogen. Both the observed spectrum and its resolution into overlapping series are shown. Note that the Balmer series lies in the visible region.

The Swedish spectroscopist Johannes Rydberg noted (in 1890) that the wavenumbers of all the lines are given by the expression

$$\tilde{\nu}=\tilde{R}_{\mathrm{H}}\left(\frac{1}{n_1^2}-\frac{1}{n_2^2}\right) \qquad \text{Spectral lines of a hydrogen atom} \qquad (8\text{A}.1)$$

with $n_1=1$ (the *Lyman series*), 2 (the *Balmer series*), and 3 (the *Paschen series*), and that in each case $n_2 = n_1 + 1, n_1 + 2, \ldots$. The constant $\tilde{R}_{\mathrm{H}}$ is now called the **Rydberg constant** for the hydrogen atom and is found empirically to have the value $109\,677\ \mathrm{cm}^{-1}$.

8A.1 The structure of hydrogenic atoms

Consider a **hydrogenic atom**, an atom or ion of arbitrary atomic number Z but having a single electron. Hydrogen itself is an example (with $Z=1$). The Coulomb potential energy of an electron in a hydrogenic atom of atomic number Z and therefore nuclear charge Ze is

$$V(r)=-\frac{Ze^2}{4\pi\varepsilon_0 r} \qquad (8\text{A}.2)$$

where r is the distance of the electron from the nucleus and ε_0 is the vacuum permittivity. The hamiltonian for the entire atom, which consists of an electron and a nucleus of mass m_{N}, is therefore

$$\hat{H}=\hat{E}_{\mathrm{k,electron}}+\hat{E}_{\mathrm{k,nucleus}}+\hat{V}(r) = -\frac{\hbar^2}{2m_{\mathrm{e}}}\nabla_{\mathrm{e}}^2-\frac{\hbar^2}{2m_{\mathrm{N}}}\nabla_{\mathrm{N}}^2-\frac{Ze^2}{4\pi\varepsilon_0 r} \qquad \text{Hamiltonian for a hydrogenic atom} \qquad (8\text{A}.3)$$

The subscripts e and N on ∇^2 indicate differentiation with respect to the electron or nuclear coordinates.

(a) The separation of variables

Physical intuition suggests that the full Schrödinger equation ought to separate into two equations, one for the motion of the atom as a whole through space and the other for the motion of the electron relative to the nucleus. The Schrödinger equation for the internal motion of the electron relative to the nucleus is[1]

$$-\frac{\hbar^2}{2\mu}\nabla^2\psi-\frac{Ze^2}{4\pi\varepsilon_0 r}\psi=E\psi \qquad \frac{1}{\mu}=\frac{1}{m_e}+\frac{1}{m_N} \qquad \text{Schrödinger equation for a hydrogenic atom} \quad (8A.4)$$

where differentiation is now with respect to the coordinates of the electron relative to the nucleus. The quantity μ is called the **reduced mass**. The reduced mass is very similar to the electron mass because m_N, the mass of the nucleus, is much larger than the mass of an electron, so $1/\mu \approx 1/m_e$ and therefore $\mu \approx m_e$. In all except the most precise work, the reduced mass can be replaced by m_e.

Because the potential energy is centrosymmetric (independent of angle), the equation for the wavefunction is expected to be separable into radial and angular components, as in

$$\psi(r,\theta,\phi)=R(r)Y(\theta,\phi) \quad (8A.5)$$

with $R(r)$ the **radial wavefunction** and $Y(\theta,\phi)$ the **angular wavefunction**. The equation does separate, and the two contributions to the wavefunction are solutions of two equations:

$$\Lambda^2 Y=-l(l+1)Y \quad (8A.6a)$$

$$-\frac{\hbar^2}{2\mu}\left(\frac{d^2R}{dr^2}+\frac{2}{r}\frac{dR}{dr}\right)+V_{eff}R=ER \quad (8A.6b)$$

where

$$V_{eff}(r)=-\frac{Ze^2}{4\pi\varepsilon_0 r}+\frac{l(l+1)\hbar^2}{2\mu r^2} \quad (8A.6c)$$

Equation 8A.6a is the same as the Schrödinger equation for a particle free to move at constant radius around a central point, and is considered in Topic 7F. The allowed solutions are the spherical harmonics (Table 7F.1), and are specified by the quantum numbers l and m_l. Equation 8A.6b is called the **radial wave equation**. The radial wave equation describes the motion of a particle of mass μ in a one-dimensional region $0 \le r < \infty$ where the potential energy is $V_{eff}(r)$.

(b) The radial solutions

Some features of the shapes of the radial wavefunctions can be anticipated by examining the form of $V_{eff}(r)$. The first term in eqn 8A.6c is the Coulomb potential energy of the electron in the field of the nucleus. The second term stems from what in classical physics would be called the centrifugal force arising from the angular momentum of the electron around the nucleus. When $l = 0$, the electron has no angular momentum, and the effective potential energy is purely Coulombic and the force exerted on the electron is attractive at all radii (Fig. 8A.2). When $l \neq 0$, the centrifugal term gives a positive contribution to the effective potential energy, corresponding to a repulsive force at all radii. When the electron is close to the nucleus ($r \approx 0$), the latter contribution to the potential energy, which is proportional to $1/r^2$, dominates the Coulombic contribution, which is proportional to $1/r$, and the net result is an effective repulsion of the electron from the nucleus. The two effective potential energies, the one for $l = 0$ and the one for $l \neq 0$, are therefore qualitatively very different close to the nucleus. However, they are similar at large distances because the centrifugal contribution tends to zero more rapidly (as $1/r^2$) than the Coulombic contribution (as $1/r$). Therefore, the solutions with $l = 0$ and $l \neq 0$ are expected to be quite different near the nucleus but similar far away from it.

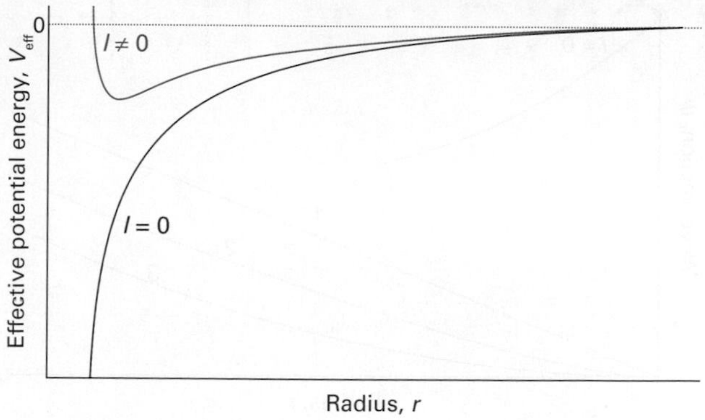

Figure 8A.2 The effective potential energy of an electron in the hydrogen atom. When the electron has zero orbital angular momentum, the effective potential energy is the Coulombic potential energy. When the electron has non-zero orbital angular momentum, the centrifugal effect gives rise to a positive contribution which is very large close to the nucleus. The $l = 0$ and $l \neq 0$ wavefunctions are therefore very different near the nucleus.

Two features of the radial wavefunction are important:

Physical interpretation

- Close to the nucleus the radial wavefunction is proportional to r^l, and the higher the orbital angular momentum, the less likely it is that the electron will be found there (Fig. 8A.3).
- Far from the nucleus all radial wavefunctions approach zero exponentially.

The detailed solution of the radial equation for the full range of radii shows how the form r^l close to the nucleus blends

[1] See the first section of *A deeper look* 3 on the website for this text for full details of this separation procedure and then the second section for the calculations that lead to eqn 8A.6.

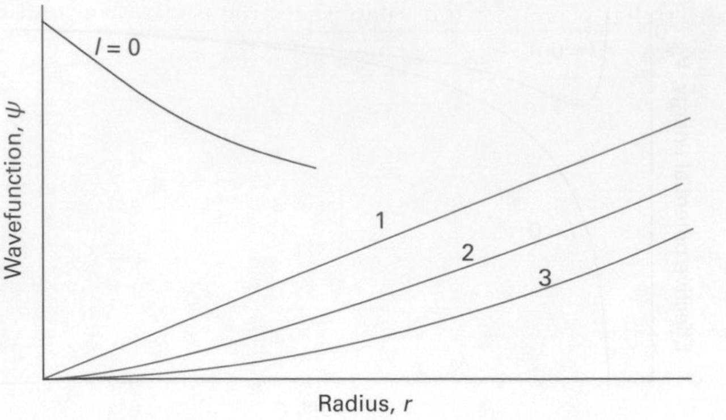

Figure 8A.3 Close to the nucleus, orbitals with $l = 1$ are proportional to r, orbitals with $l = 2$ are proportional to r^2, and orbitals with $l = 3$ are proportional to r^3. Electrons are progressively excluded from the neighbourhood of the nucleus as l increases. An orbital with $l = 0$ has a finite, non-zero value at the nucleus.

into the exponentially decaying form at great distances. It turns out that the two regions are bridged by a polynomial in r and that

$$R(r) = \underbrace{r^l}_{\substack{\text{Dominant}\\\text{close to the}\\\text{nucleus}}} \times \underbrace{(\text{polynomial in } r)}_{\substack{\text{Bridges the two}\\\text{ends of the function}}} \times \underbrace{(\text{decaying exponential in } r)}_{\substack{\text{Dominant far}\\\text{from the nucleus}}} \qquad (8A.7)$$

The radial wavefunction therefore has the form

$$R(r) = r^l L(r) e^{-r}$$

with various constants and where $L(r)$ is the bridging polynomial. Close to the nucleus ($r \approx 0$) the polynomial is a constant and $e^{-r} \approx 1$, so $R(r) \propto r^l$; far from the nucleus the dominant term in the polynomial is proportional to r^{n-l-1}, where n is an integer, so regardless of the value of l, all the wavefunctions of a given value of n are proportional to $r^{n-1}e^{-r}$ and decay exponentially to zero in the same way (exponential functions e^{-x} always dominate simple powers, x^n).

The detailed solution also shows that, for the wavefunction to be acceptable, the value of n that appears in the polynomial can take only positive integral values, and specifically $n = 1, 2, \ldots$. This number also determines the allowed energies through the expression:

$$E_n = -\frac{\mu e^4}{32\pi^2\varepsilon_0^2\hbar^2} \times \frac{Z^2}{n^2} \qquad \text{Bound-state energies} \qquad (8A.8)$$

So far, only the general form of the radial wavefunctions has been given. It is now time to show how they depend on various fundamental constants and the atomic number of the atom. They are most simply written in terms of the dimensionless quantity ρ (rho), where

$$\rho = \frac{2Zr}{na} \qquad a = \frac{m_e}{\mu} a_0 \qquad a_0 = \frac{4\pi\varepsilon_0\hbar^2}{m_e e^2} \qquad (8A.9)$$

The **Bohr radius**, a_0, has the value 52.9 pm; it is so called because the same quantity appeared in Bohr's early model of the hydrogen atom as the radius of the electron orbit of lowest energy. In practice, because $m_e \ll m_N$ (so $m_e/\mu \approx 1$) there is so little difference between a and a_0 that it is safe to use a_0 in the definition of ρ for all atoms (even for ^{1}H, $a = 1.0005a_0$). In terms of these quantities and with the various quantum numbers displayed, the radial wavefunctions for an electron with quantum numbers n and l are the (real) functions

$$R_{n,l}(r) = N_{n,l}\rho^l L_{n,l}(\rho) e^{-\rho/2} \qquad \text{Radial wavefunctions} \qquad (8A.10)$$

where $L_{n,l}(\rho)$ is an *associated Laguerre polynomial*. These polynomials have quite simple forms, such as 1, ρ, and $2 - \rho$ (they can be picked out in Table 8A.1). The factor $N_{n,l}$ ensures that the radial wavefunction is normalized to 1 in the sense that

$$\int_0^\infty R_{n,l}(r)^2 r^2 \mathrm{d}r = 1 \qquad (8A.11)$$

Table 8A.1 Hydrogenic radial wavefunctions

n	l	$R_{n,l}(r)$
1	0	$2\left(\frac{Z}{a}\right)^{3/2} e^{-\rho/2}$
2	0	$\frac{1}{8^{1/2}}\left(\frac{Z}{a}\right)^{3/2}(2-\rho)e^{-\rho/2}$
2	1	$\frac{1}{24^{1/2}}\left(\frac{Z}{a}\right)^{3/2}\rho e^{-\rho/2}$
3	0	$\frac{1}{243^{1/2}}\left(\frac{Z}{a}\right)^{3/2}(6-6\rho+\rho^2)e^{-\rho/2}$
3	1	$\frac{1}{486^{1/2}}\left(\frac{Z}{a}\right)^{3/2}(4-\rho)\rho e^{-\rho/2}$
3	2	$\frac{1}{2430^{1/2}}\left(\frac{Z}{a}\right)^{3/2}\rho^2 e^{-\rho/2}$

$\rho = (2Z/na)r$ with $a = 4\pi\varepsilon_0\hbar^2/\mu e^2$. For an infinitely heavy nucleus (or one that may be assumed to be), $\mu = m_e$ and $a = a_0$, the Bohr radius.

(The r^2 comes from the volume element in spherical coordinates; see *The chemist's toolkit* 21 in Topic 7F.) Specifically, the components of eqn 8A.10 can be interpreted as follows:

Physical interpretation

- The exponential factor ensures that the wavefunction approaches zero far from the nucleus.
- The factor ρ^l ensures that (provided $l > 0$) the wavefunction vanishes at the nucleus. The zero at $r = 0$ is not a radial node because the radial wavefunction does not pass through zero at that point (because r cannot be negative).
- The associated Laguerre polynomial is a function that in general oscillates from positive to negative values and accounts for the presence of radial nodes.

Expressions for some radial wavefunctions are given in Table 8A.1 and illustrated in Fig. 8A.4. Finally, with the form of the radial wavefunction established, the total wavefunction, eqn 8A.5, in full dress becomes

$$\psi_{n,l,m_l}(r,\theta,\phi) = R_{n,l}(r)Y_{l,m_l}(\theta,\phi) \qquad (8A.12)$$

Brief illustration 8A.1

To calculate the probability density at the nucleus for an electron with $n = 1$, $l = 0$, and $m_l = 0$, evaluate ψ at $r = 0$:

$$\psi_{1,0,0}(0,\theta,\phi) = R_{1,0}(0)Y_{0,0}(\theta,\phi) = 2\left(\frac{Z}{a_0}\right)^{3/2}\left(\frac{1}{4\pi}\right)^{1/2}$$

The probability density is therefore

$$\psi^2_{1,0,0}(0,\theta,\phi) = \frac{Z^3}{\pi a_0^3}$$

which evaluates to $2.15 \times 10^{-6}\,\text{pm}^{-3}$ when $Z = 1$.

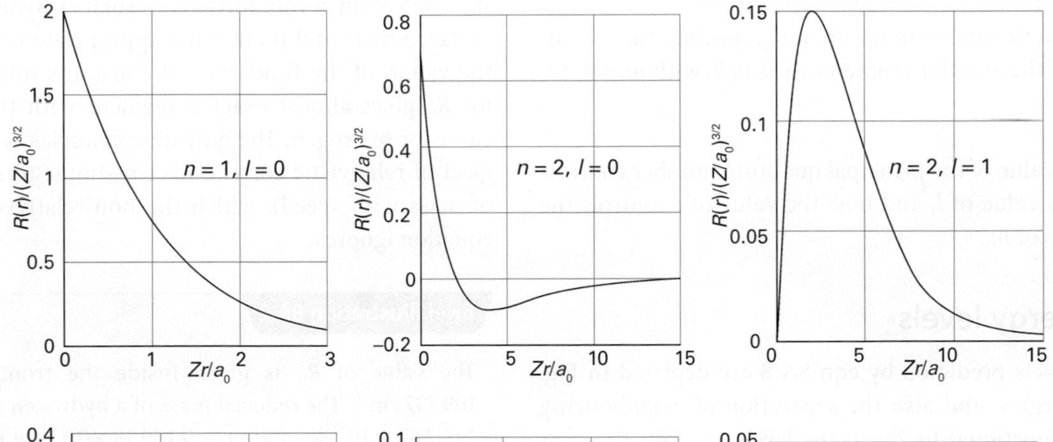

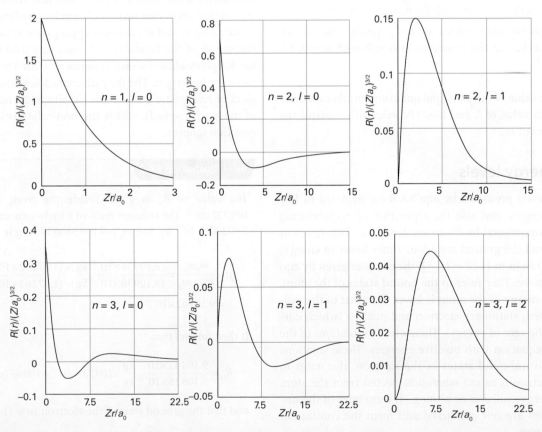

Figure 8A.4 The radial wavefunctions of the first few states of hydrogenic atoms of atomic number Z. Note that the orbitals with $l = 0$ have a non-zero and finite value at the nucleus. The horizontal scales are different in each case: as the principal quantum number increases, so too does the size of the orbital.

8A.2 Atomic orbitals and their energies

An **atomic orbital** is a one-electron wavefunction for an electron in an atom, and for hydrogenic atoms has the form specified in eqn 8A.12. Each hydrogenic atomic orbital is defined by three quantum numbers, designated n, l, and m_l. An electron described by one of the wavefunctions in eqn 8A.12 is said to 'occupy' that orbital. For example, an electron described by the wavefunction $\psi_{1,0,0}$ is said to 'occupy' the orbital with $n=1$, $l=0$, and $m_l=0$.

(a) The specification of orbitals

Each of the three quantum numbers specifies a different attribute of the orbital:

- The **principal quantum number**, n, specifies the energy of the orbital (through eqn 8A.8); it takes the values $n = 1, 2, 3, \ldots$.
- The **orbital angular momentum quantum number**, l, specifies the magnitude of the angular momentum of the electron as $\{l(l+1)\}^{1/2}\hbar$, with $l = 0, 1, 2, \ldots, n-1$.
- The **magnetic quantum number**, m_l, specifies the z-component of the angular momentum as $m_l\hbar$, with $m_l = 0, \pm1, \pm2, \ldots, \pm l$.

Note how the value of the principal quantum number controls the maximum value of l, and how the value of l controls the range of values of m_l.

(b) The energy levels

The energy levels predicted by eqn 8A.8 are depicted in Fig. 8A.5. The energies, and also the separation of neighbouring levels, are proportional to Z^2, so the levels are four times as wide apart (and the ground state four times lower in energy) in He^+ ($Z=2$) than in H ($Z=1$). All the energies given by eqn 8A.8 are negative. They refer to the **bound states** of the atom, in which the energy of the atom is lower than that of the infinitely separated, stationary electron and nucleus (which corresponds to the zero of energy). There are also solutions of the Schrödinger equation with positive energies. These solutions correspond to **unbound states** of the electron, the states to which an electron is raised when it is ejected from the atom by a high-energy collision or photon. The energies of the unbound electron are not quantized and form the continuum states of the atom.

Equation 8A.8, which can be written as

$$E_n = -\frac{hcZ^2\tilde{R}_N}{n^2} \qquad \tilde{R}_N = \frac{\mu e^4}{8\varepsilon_0^2 ch^3} \qquad \text{Bound-state energies} \qquad (8A.13)$$

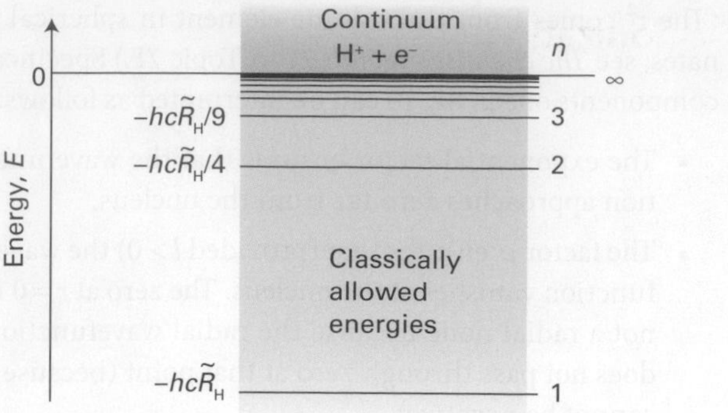

Figure 8A.5 The energy levels of a hydrogen atom. The values are relative to an infinitely separated, stationary electron and a proton.

is consistent with the spectroscopic result summarized by eqn 8A.1, with the Rydberg constant for the atom identified as

$$\tilde{R}_N = \frac{\mu}{m_e}\times\tilde{R}_\infty \qquad \tilde{R}_\infty = \frac{m_e e^4}{8\varepsilon_0^2 h^3 c} \qquad \text{Rydberg constant} \qquad (8A.14)$$

where μ is the reduced mass of the atom and $\tilde{R}_\infty$ is the **Rydberg constant**; the constant $\tilde{R}_N$ is the value that constant takes for a specified atom N (not nitrogen!), such as hydrogen, when N is replaced by H and μ takes the appropriate value. Insertion of the values of the fundamental constants into the expression for $\tilde{R}_H$ gives almost exact agreement with the experimental value for hydrogen. The only discrepancies arise from the neglect of relativistic corrections (in simple terms, the increase of mass with speed), which the non-relativistic Schrödinger equation ignores.

Brief illustration 8A.2

The value of $\tilde{R}_\infty$ is given inside the front cover and is 109 737 cm^{-1}. The reduced mass of a hydrogen atom with $m_p = 1.672\,62 \times 10^{-27}$ kg and $m_e = 9.109\,38 \times 10^{-31}$ kg is

$$\mu = \frac{m_e m_p}{m_e + m_p} = \frac{(9.10938\times10^{-31}\,\text{kg})\times(1.67262\times10^{-27}\,\text{kg})}{(9.10938\times10^{-31}\,\text{kg})+(1.67262\times10^{-27}\,\text{kg})}$$
$$= 9.10442\times10^{-31}\,\text{kg}$$

It then follows that

$$\tilde{R}_H = \frac{9.10442\times10^{-31}\,\text{kg}}{9.10938\times10^{-31}\,\text{kg}}\times109737\,\text{cm}^{-1} = 109677\,\text{cm}^{-1}$$

and that the ground state of the electron ($n=1$) lies at

$$E_1 = -hc\tilde{R}_H = -(6.62608\times10^{-34}\,\text{J s})\times(2.997945\times10^{10}\,\text{cm s}^{-1})$$
$$\times(109677\,\text{cm}^{-1}) = -2.17870\times10^{-18}\,\text{J}$$

or $-2.178\,70$ aJ. This energy corresponds to -13.598 eV.

(c) Ionization energies

The **ionization energy**, I, of an element is the minimum energy required to remove an electron from the **ground state**, the state of lowest energy, of one of its atoms in the gas phase. Because the ground state of hydrogen is the state with $n = 1$, with energy $E_1 = -hc\tilde{R}_H$ and the atom is ionized when the electron has been excited to the level corresponding to $n = \infty$ (see Fig. 8A.5), the energy that must be supplied is

$$I = hc\tilde{R}_H \quad (8A.15)$$

The value of I is 2.179 aJ (1 aJ = 10^{-18} J), which corresponds to 13.60 eV.

A note on good practice Ionization energies are sometimes referred to as *ionization potentials*. That is incorrect, but not uncommon. If the term is used at all, it should denote the electrical potential difference through which an electron must be moved for the change in its potential energy to be equal to the ionization energy, and reported in volts: the ionization energy of hydrogen is 13.60 eV; its ionization potential is 13.60 V.

Example 8A.1 Measuring an ionization energy spectroscopically

The emission spectrum of atomic hydrogen shows lines at 82 259, 97 492, 102 824, 105 292, 106 632, and 107 440 cm^{-1}, which correspond to transitions to the same lower state from successive upper states with $n = 2, 3, \ldots$. Determine the ionization energy of the lower state.

Collect your thoughts The spectroscopic determination of ionization energies depends on the identification of the 'series limit', the wavenumber at which the series terminates and becomes a continuum. If the upper state lies at an energy $-hc\tilde{R}_H/n^2$, then the wavenumber of the photon emitted when the atom makes a transition to the lower state, with energy E_{lower}, is

$$\tilde{\nu} = -\frac{\tilde{R}_H}{n^2} - \frac{E_{\text{lower}}}{hc} = -\frac{\tilde{R}_H}{n^2} + \frac{I}{hc} \qquad (I = -E_{\text{lower}})$$

A plot of the wavenumbers against $1/n^2$ should give a straight line of slope $-\tilde{R}_H$ and intercept I/hc. Use software to calculate a least-squares fit of the data in order to obtain a result that reflects the precision of the data.

The solution The wavenumbers are plotted against $1/n^2$ in Fig. 8A.6. From the (least-squares) intercept, it follows that $I/hc = 109\,679$ cm^{-1}, so the ionization energy is

$$I = hc \times (109\,679\,\text{cm}^{-1})$$
$$= (6.626\,08 \times 10^{-34}\,\text{J s}) \times (2.997\,945 \times 10^{10}\,\text{cm s}^{-1}) \times (109\,679\,\text{cm}^{-1})$$
$$= 2.1787 \times 10^{-18}\,\text{J}$$

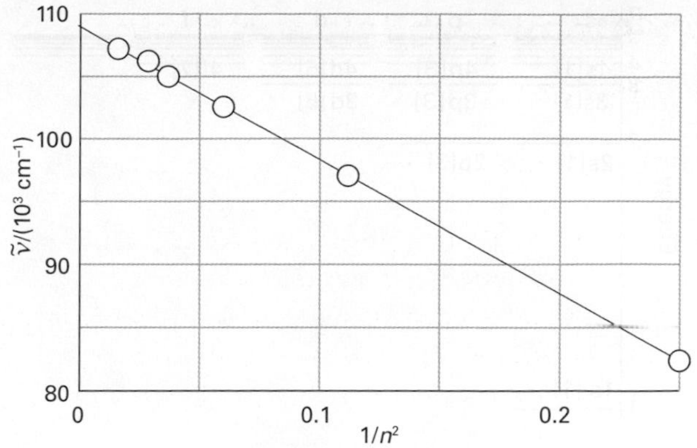

Figure 8A.6 The plot of the data in *Example* 8A.1 used to determine the ionization energy of an atom (in this case, of H).

or 2.1787 aJ, corresponding to 1312.1 kJ mol^{-1} (the negative of the value of E calculated in *Brief illustration* 8A.2).

Self-test 8A.1 The emission spectrum of atomic deuterium shows lines at 15 238, 20 571, 23 039, and 24 380 cm^{-1}, which correspond to transitions from successive upper states with $n = 3, 4, \ldots$ to the same lower state. Determine (a) the ionization energy of the lower state, (b) the ionization energy of the ground state, (c) the mass of the deuteron (by expressing the Rydberg constant in terms of the reduced mass of the electron and the deuteron, and solving for the mass of the deuteron).

Answer: (a) 328.1 kJ mol^{-1}, (b) 1312.4 kJ mol^{-1}, (c) 2.8×10^{-27} kg, a result very sensitive to $\tilde{R}_H$

(d) Shells and subshells

All the orbitals of a given value of n are said to form a single **shell** of the atom. In a hydrogenic atom (and only in a hydrogenic atom), all orbitals of given n, and therefore belonging to the same shell, have the same energy. It is common to refer to successive shells by letters:

$n =$	1	2	3	4…	Specification of shells
	K	L	M	N…	

Thus, all the orbitals of the shell with $n = 2$ form the L shell of the atom, and so on.

The orbitals with the same value of n but different values of l are said to form a **subshell** of a given shell. These subshells are also generally referred to by letters:

$l =$	0	1	2	3	4	5	6…	Specification of subshells
	s	p	d	f	g	h	i…	

All orbitals of the same subshell have the same energy in all kinds of atoms, not only hydrogenic atoms. After $l = 3$ the letters run alphabetically (j is not used because in some lan-

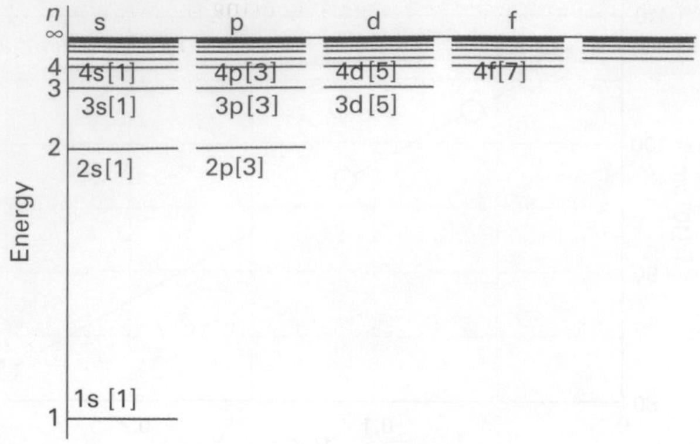

Figure 8A.7 The energy levels of a hydrogenic atom showing the subshells and (in square brackets) the numbers of orbitals in each subshell. All orbitals of a given shell have the same energy.

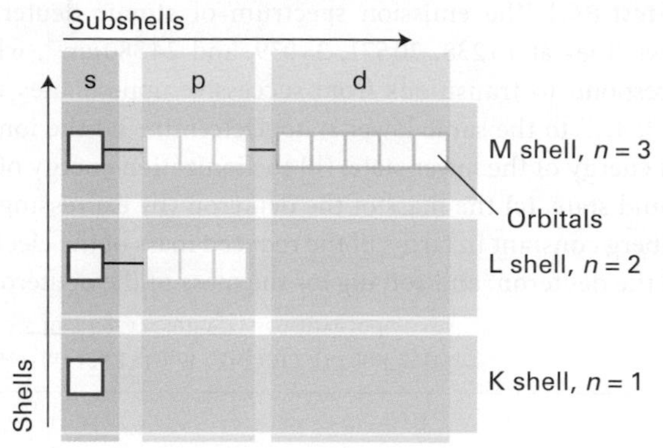

Figure 8A.8 The organization of orbitals (white squares) into subshells (characterized by l) and shells (characterized by n).

guages i and j are not distinguished). Figure 8A.7 is a version of Fig. 8A.5 which shows the subshells explicitly. Because l can range from 0 to $n-1$, giving n values in all, it follows that there are n subshells of a shell with principal quantum number n. The organization of orbitals in the shells is summarized in Fig. 8A.8. The number of orbitals in a shell of principal quantum number n is n^2, so in a hydrogenic atom each energy level is n^2-fold degenerate.

Brief illustration 8A.3

When $n=1$ there is only one subshell, that with $l=0$, and that subshell contains only one orbital, with $m_l=0$ (the only value of m_l permitted). When $n=2$, there are four orbitals, one in the s subshell with $l=0$ and $m_l=0$, and three in the $l=1$ subshell with $m_l=+1, 0, -1$. When $n=3$ there are nine orbitals (one with $l=0$, three with $l=1$, and five with $l=2$).

(e) s Orbitals

The orbital occupied in the ground state is the one with $n=1$ (and therefore with $l=0$ and $m_l=0$, the only possible values of these quantum numbers when $n=1$). From Table 8A.1 and with $Y_{0,0}=(1/4\pi)^{1/2}$ (Table 7F.1) it follows that (for $Z=1$):

$$\psi=\frac{1}{(\pi a_0^3)^{1/2}}\mathrm{e}^{-r/a_0} \qquad (8A.16)$$

This wavefunction is independent of angle and has the same value at all points of constant radius; that is, the 1s orbital (the s orbital with $n=1$, and in general ns) is 'spherically symmetrical'. The wavefunction decays exponentially from a maximum value of $1/(\pi a_0^3)^{1/2}$ at the nucleus (at $r=0$). It follows that the probability density of the electron is greatest at the nucleus itself.

The general form of the ground-state wavefunction can be understood by considering the contributions of the potential and kinetic energies to the total energy of the atom. The closer the electron is to the nucleus on average, the lower (more negative) its average potential energy. This dependence suggests that the lowest potential energy should be obtained with a sharply peaked wavefunction that has a large amplitude at the nucleus and is zero everywhere else (Fig. 8A.9). However, this shape implies a high kinetic energy, because such a wavefunction has a very high average curvature. The electron would have very low kinetic energy if its wavefunction had only a very low average curvature. However, such a wavefunction spreads to great distances from the nucleus and the average potential energy of the electron is correspondingly high. The actual ground-state wavefunction is a compromise between these two extremes: the wavefunction spreads away

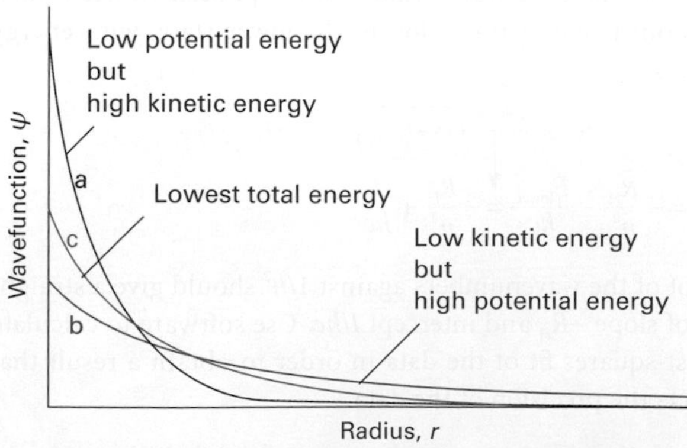

Figure 8A.9 The balance of kinetic and potential energies that accounts for the structure of the ground state of hydrogenic atoms. (a) The sharply curved but localized orbital has high mean kinetic energy, but low mean potential energy; (b) the mean kinetic energy is low, but the potential energy is not very favourable; (c) the compromise of moderate kinetic energy and moderately favourable potential energy.

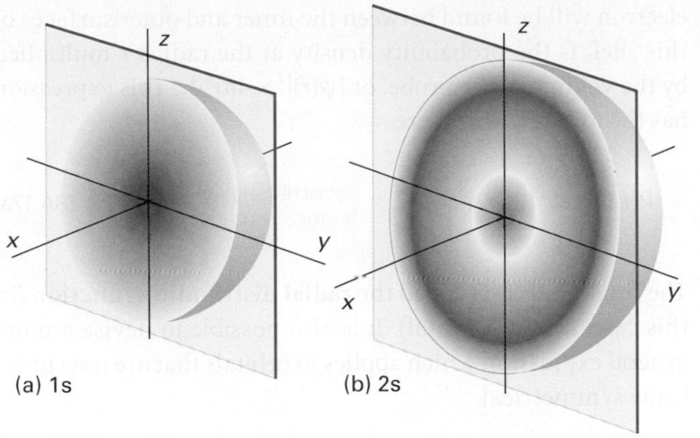

Figure 8A.10 Representations of cross-sections through the (a) 1s and (b) 2s hydrogenic atomic orbitals in terms of their electron probability densities (as represented by the density of shading).

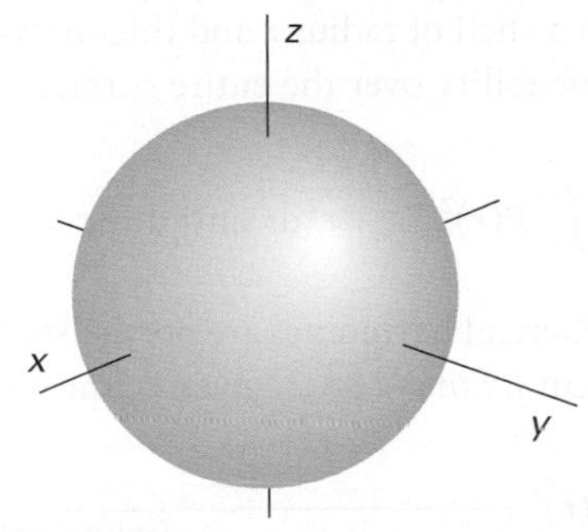

Figure 8A.11 The boundary surface of a 1s orbital, within which there is a 90 per cent probability of finding the electron. All s orbitals have spherical boundary surfaces.

from the nucleus (so the expectation value of the potential energy is not as low as in the first example, but nor is it very high) and has a reasonably low average curvature (so the expectation of the kinetic energy is not very low, but nor is it as high as in the first example).

One way of depicting the probability density of the electron is to represent $|\psi|^2$ by the density of shading (Fig. 8A.10). A simpler procedure is to show only the **boundary surface**, the surface that mirrors the shape of the orbital and captures a high proportion (typically about 90 per cent) of the electron probability. For the 1s orbital, the boundary surface is a sphere centred on the nucleus (Fig. 8A.11).

Example 8A.2 Calculating the mean radius of an orbital

Calculate the mean radius of a hydrogenic 1s orbital.

Collect your thoughts The mean radius is the expectation value

$$\langle r\rangle=\int\psi^* r\psi\,\mathrm{d}\tau=\int r|\psi|^2\,\mathrm{d}\tau$$

You need to evaluate the integral by using the wavefunctions given in Table 8A.1 and $\mathrm{d}\tau = r^2\mathrm{d}r\sin\theta\,\mathrm{d}\theta\,\mathrm{d}\phi$ (*The chemist's toolkit* 21 in Topic 7F). The angular parts of the wavefunction (Table 7F.1) are normalized in the sense that

$$\int_{\theta=0}^{\pi}\int_{\phi=0}^{2\pi}\left|Y_{l,m_l}\right|^2\sin\theta\,\mathrm{d}\theta\mathrm{d}\phi=1$$

The relevant integral over r is given in the *Resource section.*

The solution With the wavefunction written in the form $\psi = RY$, the integration (with the integral over the angular variables, which is equal to 1, in blue) is

$$\langle r\rangle=\int_0^{\infty}\int_0^{\pi}\int_0^{2\pi} rR_{n,l}^2\left|Y_{l,m_l}\right|^2 r^2\mathrm{d}r\sin\theta\,\mathrm{d}\theta\mathrm{d}\phi=\int_0^{\infty}r^3R_{n,l}^2\mathrm{d}r$$

For a 1s orbital

$$R_{1,0}=2\left(\frac{Z}{a_0}\right)^{3/2}\mathrm{e}^{-Zr/a_0}$$

Hence

$$\langle r\rangle=\frac{4Z^3}{a_0^3}\overbrace{\int_0^{\infty}r^3\mathrm{e}^{-2Zr/a_0}\mathrm{d}r}^{\text{Integral E.3}}=\frac{4Z^3}{a_0^3}\times\frac{3!}{(2Z/a_0)^4}=\frac{3a_0}{2Z}$$

Self-test 8A.2 Evaluate the mean radius of a 3s orbital by integration.

Answer: $27a_0/2Z$

All s orbitals are spherically symmetric, but differ in the number of radial nodes. For example, the 1s, 2s, and 3s orbitals have 0, 1, and 2 radial nodes, respectively. In general, an *ns* orbital has $n - 1$ radial nodes. As n increases, the radius of the spherical boundary surface that captures a given fraction of the probability also increases.

Brief illustration 8A.4

The radial nodes of a 2s orbital lie at the locations where the associated Laguerre polynomial factor (Table 8A.1) is equal to zero. In this case the factor is simply $2 - \rho$ so there is a node at $\rho = 2$. For a 2s orbital, $\rho = Zr/a_0$, so the radial node occurs at $r = 2a_0/Z$ (see Fig. 8A.4).

(f) Radial distribution functions

The wavefunction yields, through the value of $|\psi|^2$, the probability of finding an electron in any region. As explained in Topic 7B, $|\psi|^2$ is a probability *density* (dimensions: 1/volume) and can be interpreted as a (dimensionless) probability when multiplied by the (infinitesimal) volume of interest. Imagine a probe with a fixed volume $\mathrm{d}\tau$ and sensitive to electrons that

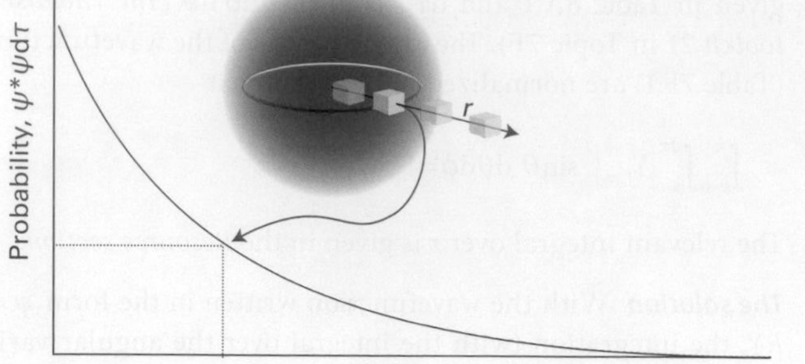

Figure 8A.12 A constant-volume electron-sensitive detector (the small cube) gives its greatest reading at the nucleus, and a smaller reading elsewhere. The same reading is obtained anywhere on a circle of given radius at any orientation: the s orbital is spherically symmetrical.

can move around near the nucleus of a hydrogenic atom. Because the probability density in the ground state of the atom is proportional to e^{-2Zr/a_0}, the reading from the detector decreases exponentially as the probe is moved out along any radius but is constant if the probe is moved on a circle of constant radius (Fig. 8A.12).

Now consider the total probability of finding the electron anywhere between the two walls of a spherical shell of thickness dr at a radius r. The sensitive volume of the probe is now the volume of the shell (Fig. 8A.13), which is $4\pi r^2 dr$ (the product of its surface area, $4\pi r^2$, and its thickness, dr). Note that the volume probed increases with distance from the nucleus and is zero at the nucleus itself, when $r = 0$. The probability that the

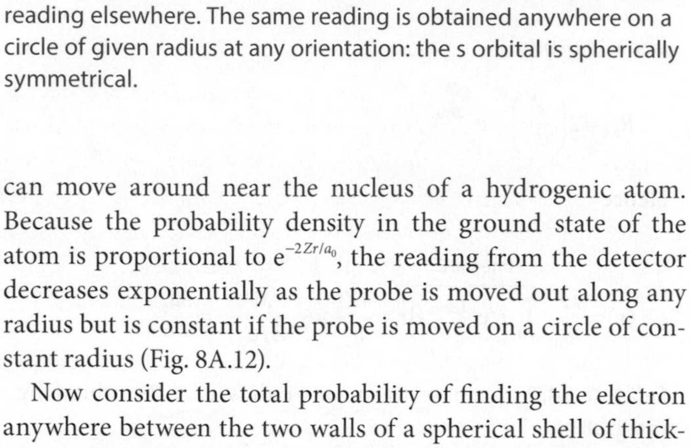

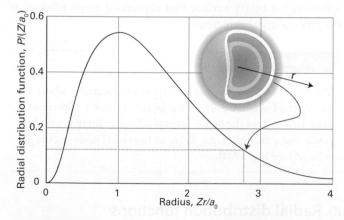

Figure 8A.13 The radial distribution function $P(r)$ is the probability density that the electron will be found anywhere in a shell of radius r; the probability itself is $P(r)$dr, where dr is the thickness of the shell. For a 1s electron in hydrogen, $P(r)$ is a maximum when r is equal to the Bohr radius a_0. The value of $P(r)$dr is equivalent to the reading that a detector shaped like a spherical shell of thickness dr would give as its radius is varied.

electron will be found between the inner and outer surfaces of this shell is the probability density at the radius r multiplied by the volume of the probe, or $|\psi(r)|^2 \times 4\pi r^2 dr$. This expression has the form $P(r)dr$, where

$$P(r) = 4\pi r^2 |\psi(r)|^2 \qquad \text{Radial distribution function [s orbitals only]} \qquad (8A.17a)$$

The function $P(r)$ is called the **radial distribution function** (in this case, for an s orbital). It is also possible to devise a more general expression which applies to orbitals that are not spherically symmetrical.

How is that done? 8A.1 Deriving the general form of the radial distribution function

The probability of finding an electron in a volume element dτ when its wavefunction is $\psi = RY$ is $|RY|^2 d\tau$ with $d\tau = r^2 dr \sin\theta d\theta d\phi$. The total probability of finding the electron at any angle in a shell of radius r and thickness dr is the integral of this probability over the entire surface, and is written $P(r)dr$; so

$$P(r)dr = \int_0^{\pi}\int_0^{2\pi} R(r)^2 |Y_{l,m_l}|^2 r^2 dr \sin\theta \, d\theta d\phi$$

Because the spherical harmonics are normalized to 1 (the blue integration, as in *Example* 8A.2, gives 1), the final result is

$$P(r) = r^2 R(r)^2 \qquad \text{Radial distribution function [general form]} \qquad (8A.17b)$$

The radial distribution function is a probability density in the sense that, when it is multiplied by dr, it gives the probability of finding the electron anywhere between the two walls of a spherical shell of thickness dr at the radius r. For a 1s orbital,

$$P(r) = \frac{4Z^3}{a_0^3} r^2 e^{-2Zr/a_0} \qquad (8A.18)$$

This expression can be interpreted as follows:

- Because $r^2 = 0$ at the nucleus, $P(0) = 0$. The volume of the shell is zero when $r = 0$ so the probability of finding the electron in the shell is zero.
- As $r \to \infty$, $P(r) \to 0$ on account of the exponential term. The wavefunction has fallen to zero at great distances from the nucleus and there is little probability of finding the electron even in a large shell.
- The increase in r^2 and the decrease in the exponential factor means that P passes through a maximum at an intermediate radius (see Fig. 8A.13); it marks the most probable radius at which the electron will be found regardless of direction.

Example 8A.3 Calculating the most probable radius

Calculate the most probable radius, r_{mp}, at which an electron will be found when it occupies a 1s orbital of a hydrogenic atom of atomic number Z, and tabulate the values for the one-electron species from H to Ne^{9+}.

Collect your thoughts You need to find the radius at which the radial distribution function of the hydrogenic 1s orbital has a maximum value by solving $dP/dr = 0$. If there are several maxima, you should choose the one corresponding to the greatest amplitude.

The solution The radial distribution function is given in eqn 8A.18. It follows that

$$\frac{dP}{dr} = \frac{4Z^3}{a_0^3}\left(2r - \frac{2Zr^2}{a_0}\right)e^{-2Zr/a_0} = \frac{8rZ^3}{a_0^3}\left(1 - \frac{Zr}{a_0}\right)e^{-2Zr/a_0}$$

This function is zero other than at $r = 0$ where the term in parentheses is zero, which is at

$$r_{mp} = \frac{a_0}{Z}$$

Then, with $a_0 = 52.9$ pm, the most probable radii are

	H	He^+	Li^{2+}	Be^{3+}	B^{4+}	C^{5+}	N^{6+}	O^{7+}	F^{8+}	Ne^{9+}
r_{mp}/pm	52.9	26.5	17.6	13.2	10.6	8.82	7.56	6.61	5.88	5.29

Comment. Notice how the 1s orbital is drawn towards the nucleus as the nuclear charge increases. At uranium the most probable radius is only 0.58 pm, almost 100 times closer than for hydrogen. (On a scale where $r_{mp} = 10$ cm for H, $r_{mp} = 1$ mm for U.) However, extending this result to very heavy atoms neglects important relativistic effects that complicate the calculation.

Self-test 8A.3 Find the most probable distance of a 2s electron from the nucleus in a hydrogenic atom.

Answer: $(3 + 5^{1/2})a_0/Z = 5.24a_0/Z$; this value reflects the expansion of the atom as its energy increases.

(g) p Orbitals

All three 2p orbitals have $l = 1$, and therefore the same magnitude of angular momentum; they are distinguished by different values of m_l, the quantum number that specifies the component of angular momentum around a chosen axis (conventionally taken to be the z-axis). The orbital with $m_l = 0$, for instance, has zero angular momentum around the z-axis. Its angular variation is given by the spherical harmonic $Y_{1,0}$, which is proportional to $\cos\theta$ (see Table 7F.1). Therefore, the probability density, which is proportional to $\cos^2\theta$, has its maximum value on either side of the nucleus along the z-axis (at $\theta = 0$ and 180°, where $\cos^2\theta = 1$). Specifically, the wavefunction of a 2p orbital with $m_l = 0$ is

$$\begin{aligned}\psi_{2,1,0} &= R_{2,1}(r)Y_{1,0}(\theta,\phi) = \frac{1}{4(2\pi)^{1/2}}\left(\frac{Z}{a_0}\right)^{5/2} r\cos\theta\, e^{-Zr/2a_0} \\ &= r\cos\theta f(r)\end{aligned} \quad (8A.19a)$$

where $f(r)$ is a function only of r. Because in spherical polar coordinates $z = r\cos\theta$ (*The chemist's toolkit* 21 in Topic 7F), this wavefunction may also be written

$$\psi_{2,1,0} = zf(r) \quad (8A.19b)$$

All p orbitals with $m_l = 0$ and any value of n have wavefunctions of this form, but $f(r)$ depends on the value of n. This way of writing the orbital is the origin of the name 'p_z orbital': its boundary surface is shown in Fig. 8A.14. The wavefunction is zero everywhere in the xy-plane, where $z = 0$, so the xy-plane is a **nodal plane** of the orbital: the wavefunction changes sign on going from one side of the plane to the other.

The wavefunctions of 2p orbitals with $m_l = \pm 1$ have the following form:

$$\begin{aligned}\psi_{2,1,\pm1} &= R_{2,1}(r)Y_{1,\pm1}(\theta,\phi) = \mp\frac{1}{8\pi^{1/2}}\left(\frac{Z}{a_0}\right)^{5/2} r\sin\theta\, e^{\pm i\phi}\, e^{-Zr/2a_0} \\ &= \mp\frac{1}{2^{1/2}} r\sin\theta\, e^{\pm i\phi} f(r)\end{aligned} \quad (8A.20)$$

In Topic 7D it is explained that a particle described by a complex wavefunction has net motion. In the present case, the

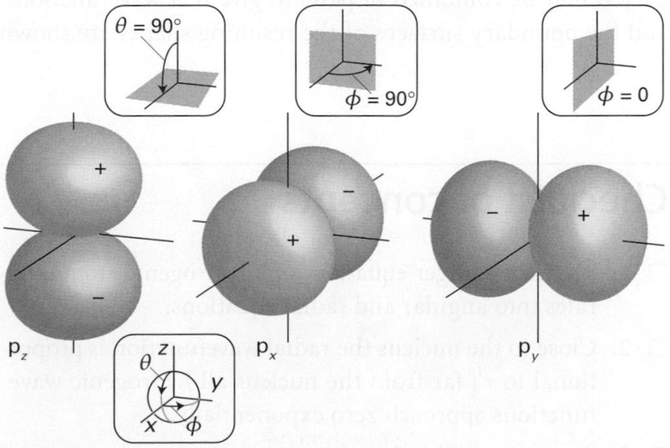

Figure 8A.14 The boundary surfaces of 2p orbitals. A nodal plane passes through the nucleus and separates the two lobes of each orbital. The dark and light lobes denote regions of opposite sign of the wavefunction. The angles of the spherical polar coordinate system are also shown. All p orbitals have boundary surfaces like those shown here.

functions correspond to non-zero angular momentum about the z-axis: $e^{+i\phi}$ corresponds to clockwise rotation when viewed from below, and $e^{-i\phi}$ corresponds to anticlockwise rotation (from the same viewpoint). They have zero amplitude where $\theta = 0$ and 180° (along the z-axis) and maximum amplitude at 90°, which is in the xy-plane. To draw the functions it is usual to represent them by forming the linear combinations

$$\psi_{2p_x} = \frac{1}{2^{1/2}}(-\psi_{2,1,+1} + \psi_{2,1,-1}) = r\sin\theta\cos\phi f(r) = x\,f(r) \qquad [e^{i\phi}+e^{-i\phi}=2\cos\phi]$$

$$\psi_{2p_y} = \frac{i}{2^{1/2}}(\psi_{2,1,+1} + \psi_{2,1,-1}) = r\sin\theta\sin\phi f(r) = y\,f(r) \qquad [e^{i\phi}-e^{-i\phi}=2i\sin\phi] \qquad (8A.21)$$

These linear combinations correspond to zero orbital angular momentum around the z-axis, as they are superpositions of states with equal and opposite values of m_l. The p_x orbital has the same shape as a p_z orbital, but it is directed along the x-axis (see Fig. 8A.14); the p_y orbital is similarly directed along the y-axis. The wavefunction of any p orbital of a given shell can be written as a product of x, y, or z and the same function f (which depends on the value of n).

(h) d Orbitals

When $n = 3$, l can be 0, 1, or 2. As a result, this shell consists of one 3s orbital, three 3p orbitals, and five 3d orbitals. Each value of the quantum number $m_l = 0, \pm1, \pm2$ corresponds to a different value of the component of angular momentum about the z-axis. As for the p orbitals, d orbitals with opposite values of m_l (and hence opposite senses of motion around the z-axis) may be combined in pairs to give real wavefunctions, and the boundary surfaces of the resulting shapes are shown in Fig. 8A.15. The real linear combinations have the following forms, with the function $f(r)$ depending on the value of n:

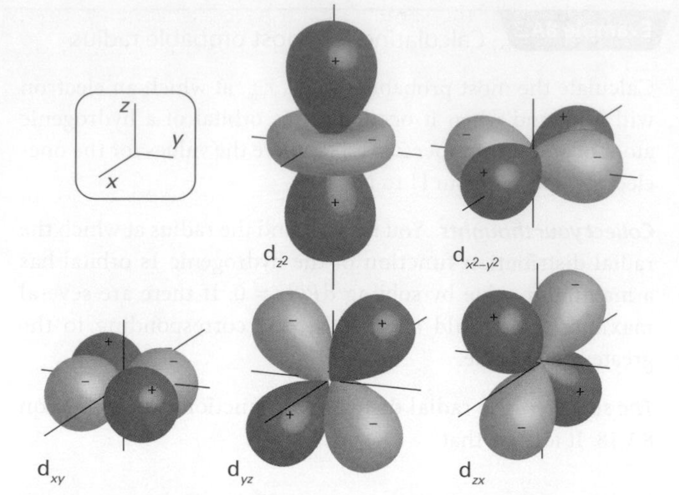

Figure 8A.15 The boundary surfaces of 3d orbitals. The purple and yellow areas denote regions of opposite sign of the wavefunction. All d orbitals have boundary surfaces like those shown here.

$$\psi_{d_{xy}} = xyf(r) \quad \psi_{d_{yz}} = yzf(r) \quad \psi_{d_{zx}} = zxf(r)$$
$$\psi_{d_{x^2-y^2}} = \tfrac{1}{2}(x^2-y^2)f(r) \quad \psi_{d_{z^2}} = \frac{12^{1/2}}{1}(3z^2-r^2)f(r) \qquad (8A.22)$$

These linear combinations give rise to the notation d_{xy}, d_{yz}, etc. for the d-orbitals. With the exception of the d_{z^2} orbital, each combination has two angular nodes which divide the orbital into four lobes. For the d_{z^2} orbital, the two angular nodes combine to give a conical surface that separates the main lobes from a smaller toroidal component encircling the nucleus.

Checklist of concepts

☐ 1. The Schrödinger equation for a hydrogenic atom separates into angular and radial equations.

☐ 2. Close to the nucleus the radial wavefunction is proportional to r^l; far from the nucleus all hydrogenic wavefunctions approach zero exponentially.

☐ 3. An **atomic orbital** is a one-electron wavefunction for an electron in an atom.

☐ 4. An atomic orbital is specified by the values of the **quantum numbers** n, l, and m_l.

☐ 5. The energies of the bound states of hydrogenic atoms are proportional to $-Z^2/n^2$.

☐ 6. The **ionization energy** of an element is the minimum energy required to remove an electron from the ground state of one of its atoms.

☐ 7. Orbitals of a given value of n form a **shell** of an atom, and within that shell orbitals of the same value of l form **subshells**.

☐ 8. Orbitals of the same shell all have the same energy in hydrogenic atoms; orbitals of the same subshell of a shell are degenerate in all types of atoms.

☐ 9. **s Orbitals** are spherically symmetrical and have non-zero probability density at the nucleus.

☐ 10. A **radial distribution function** is the probability density for the distribution of the electron as a function of distance from the nucleus.

☐ 11. There are three **p orbitals** in a given subshell; each one has one angular node.

☐ 12. There are five **d orbitals** in a given subshell; each one has two angular nodes.

Checklist of equations

Property	Equation	Comment	Equation number
Wavenumbers of the spectral lines of a hydrogen atom	$\tilde{\nu}=\tilde{R}_{\mathrm{H}}(1/n_1^2-1/n_2^2)$	$\tilde{R}_{\mathrm{H}}$ is the Rydberg constant for hydrogen (expressed as a wavenumber)	8A.1
Bohr radius	$a_0=4\pi\varepsilon_0\hbar^2/m_e e^2$	$a_0 = 52.9$ pm	8A.9
Wavefunctions of hydrogenic atoms	$\psi_{n,l,m_l}(r,\theta,\phi)=R_{n,l}(r)Y_{l,m_l}(\theta,\phi)$	Y_{l,m_l} are spherical harmonics	8A.12
Energies of hydrogenic atoms	$E_n=-hcZ^2\tilde{R}_{\mathrm{N}}/n^2$, $\tilde{R}_{\mathrm{N}}=\mu e^4/8\varepsilon_0^2ch^3$	$\tilde{R}_{\mathrm{N}}\approx\tilde{R}_\infty$, the Rydberg constant; $\mu=m_e m_{\mathrm{N}}/(m_e+m_{\mathrm{N}})$	8A.13
Radial distribution function	$P(r)=r^2R(r)^2$	$P(r)=4\pi r^2\psi^2$ for s orbitals	8A.17b

TOPIC 8B Many-electron atoms

➤ **Why do you need to know this material?**

Many-electron atoms are the building blocks of all compounds, and to understand their properties, including their ability to participate in chemical bonding, it is essential to understand their electronic structure. Moreover, a knowledge of that structure explains the structure of the periodic table and all that it summarizes.

➤ **What is the key idea?**

Electrons occupy the orbitals that result in the lowest energy of the atom, subject to the requirements of the Pauli exclusion principle.

➤ **What do you need to know already?**

This Topic builds on the account of the structure of hydrogenic atoms (Topic 8A), especially their shell structure.

A **many-electron atom** (or *polyelectron atom*) is an atom with more than one electron. The Schrödinger equation for a many-electron atom is complicated because all the electrons interact with one another. One very important consequence of these interactions is that orbitals of the same value of n but different values of l are no longer degenerate. Moreover, even for a helium atom, with just two electrons, it is not possible to find analytical expressions for the orbitals and energies, so it is necessary to use various approximations.

8B.1 The orbital approximation

The wavefunction of a many-electron atom is a very complicated function of the coordinates of all the electrons, written as $\Psi(\boldsymbol{r}_1,\boldsymbol{r}_2,\ldots)$, where $\boldsymbol{r}_i$ is the vector from the nucleus to electron i (uppercase psi, Ψ, is commonly used to denote a many-electron wavefunction). The **orbital approximation** states that a reasonable first approximation to this exact wavefunction is obtained by thinking of each electron as occupying its 'own' orbital, and writing

$$\Psi(\boldsymbol{r}_1,\boldsymbol{r}_2,\ldots)=\psi(\boldsymbol{r}_1)\psi(\boldsymbol{r}_2)\ldots \qquad \text{Orbital approximation} \quad (8B.1)$$

The individual orbitals can be assumed to resemble the hydrogenic orbitals based on nuclei with charges modified by the presence of all the other electrons in the atom. This assumption can be justified if, to a first approximation, electron–electron interactions are ignored.

How is that done? 8B.1 Justifying the orbital approximation

Consider a system in which the hamiltonian for the energy is the sum of two contributions, one for electron 1 and the other for electron 2: $\hat{H}=\hat{H}_1+\hat{H}_2$. In an actual two-electron atom (such as a helium atom), there is an additional term (proportional to $1/r_{12}$, where r_{12} is the distance between the two electrons) corresponding to their interaction:

$$\hat{H}=\overbrace{-\frac{\hbar^2}{2m_e}\nabla_1^2-\frac{2e^2}{4\pi\varepsilon_0 r_1}}^{\hat{H}_1}\overbrace{-\frac{\hbar^2}{2m_e}\nabla_2^2-\frac{2e^2}{4\pi\varepsilon_0 r_2}}^{\hat{H}_2}+\frac{e^2}{4\pi\varepsilon_0 r_{12}}$$

In the orbital approximation the final term is ignored. Then the task is to show that if $\psi(\boldsymbol{r}_1)$ is an eigenfunction of $\hat{H}_1$ with energy E_1, and $\psi(\boldsymbol{r}_2)$ is an eigenfunction of $\hat{H}_2$ with energy E_2, then the product $\Psi(\boldsymbol{r}_1,\boldsymbol{r}_2)=\psi(\boldsymbol{r}_1)\psi(\boldsymbol{r}_2)$ is an eigenfunction of the combined hamiltonian $\hat{H}$. To do so write

$$\begin{aligned}\hat{H}\Psi(\boldsymbol{r}_1,\boldsymbol{r}_2)&=(\hat{H}_1+\hat{H}_2)\psi(\boldsymbol{r}_1)\psi(\boldsymbol{r}_2)\\&=\overbrace{\hat{H}_1\psi(\boldsymbol{r}_1)\psi(\boldsymbol{r}_2)}^{\psi(\boldsymbol{r}_2)\hat{H}_1\psi(\boldsymbol{r}_1)}+\overbrace{\hat{H}_2\psi(\boldsymbol{r}_1)\psi(\boldsymbol{r}_2)}^{\psi(\boldsymbol{r}_1)\hat{H}_2\psi(\boldsymbol{r}_2)}\\&=\psi(\boldsymbol{r}_2)\overbrace{\hat{H}_1\psi(\boldsymbol{r}_1)}^{E_1\psi(\boldsymbol{r}_1)}+\psi(\boldsymbol{r}_1)\overbrace{\hat{H}_2\psi(\boldsymbol{r}_2)}^{E_2\psi(\boldsymbol{r}_2)}\\&=\psi(\boldsymbol{r}_2)E_1\psi(\boldsymbol{r}_1)+\psi(\boldsymbol{r}_1)E_2\psi(\boldsymbol{r}_2)=(E_1+E_2)\psi(\boldsymbol{r}_1)\psi(\boldsymbol{r}_2)\\&=E\Psi(\boldsymbol{r}_1,\boldsymbol{r}_2)\end{aligned}$$

where $E=E_1+E_2$, which is the desired result. Note how each hamiltonian operates on only its 'own' wavefunction. If the electrons interact (as they do in fact), then the term in $1/r_{12}$ must be included, and the proof fails. Therefore, this description is only approximate, but it is a useful model for discussing the chemical properties of atoms and is the starting point for more sophisticated descriptions of atomic structure.

The orbital approximation can be used to express the electronic structure of an atom by reporting its **configuration**, a statement of its occupied orbitals (usually, but not necessarily, in its ground state). Thus, as the ground state of a hydrogenic atom consists of the single electron in a 1s orbital, its configuration is reported as $1s^1$ (read 'one-ess-one').

A He atom has two electrons. The first electron occupies a 1s hydrogenic orbital, but because $Z=2$ that orbital is more compact than in H itself. The second electron joins the first in the 1s orbital, so the electron configuration of the ground state of He is $1s^2$.

Brief illustration 8B.1

According to the orbital approximation, each electron in He occupies a hydrogenic 1s orbital of the kind given in Topic 8A. Anticipating (see below) that the electrons experience an effective nuclear charge $Z_{\text{eff}}e$ rather than the actual charge on the nucleus with $Z = 2$ (specifically, as seen later, a charge $1.69e$ rather than $2e$), then the two-electron wavefunction of the atom is

$$\Psi(\boldsymbol{r}_1,\boldsymbol{r}_2)=\overbrace{\frac{Z_{\text{eff}}^{3/2}}{(\pi a_0^3)^{1/2}}\text{e}^{-Z_{\text{eff}}r_1/a_0}}^{\psi_{1s}(\boldsymbol{r}_1)}\times\overbrace{\frac{Z_{\text{eff}}^{3/2}}{(\pi a_0^3)^{1/2}}\text{e}^{-Z_{\text{eff}}r_2/a_0}}^{\psi_{1s}(\boldsymbol{r}_2)}$$

$$=\frac{Z_{\text{eff}}^3}{\pi a_0^3}\text{e}^{-Z_{\text{eff}}(r_1+r_2)/a_0}$$

There is nothing particularly mysterious about a two-electron wavefunction: in this case it is a simple exponential function of the distances of the two electrons from the nucleus.

8B.2 The Pauli exclusion principle

It is tempting to suppose that the electronic configurations of the atoms of successive elements with atomic numbers $Z =$ 3, 4,…, and therefore with Z electrons, are simply $1s^Z$. That, however, is not the case. The reason lies in two aspects of nature: that electrons possess 'spin' and that they must obey the very fundamental 'Pauli principle'.

(a) Spin

The quantum mechanical property of electron **spin**, the possession of an intrinsic angular momentum, was identified by an experiment performed by Otto Stern and Walther Gerlach in 1921, who shot a beam of silver atoms through an inhomogeneous magnetic field (Fig. 8B.1). The idea behind the experiment was that each atom possesses a certain electronic angular momentum and (because moving charges generate a magnetic field) as a result behaves like a small bar magnet aligned with the direction of the angular momentum vector. As the atoms pass through the inhomogeneous magnetic field they are deflected, with the deflection depending on the relative orientation of the applied magnetic field and the atomic magnet.

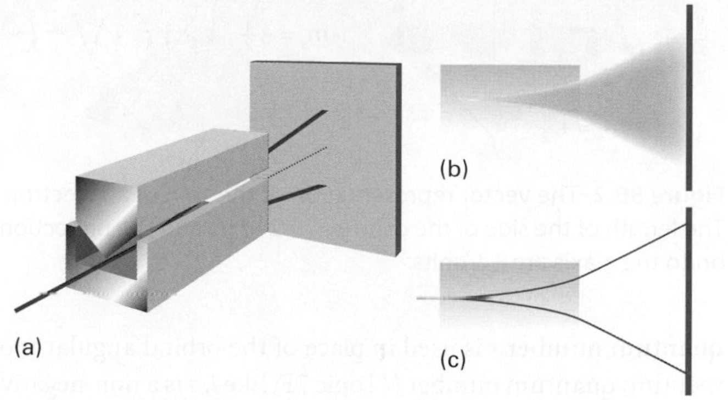

Figure 8B.1 (a) The experimental arrangement for the Stern–Gerlach experiment: the magnet provides an inhomogeneous field. (b) The classically expected result. (c) The observed outcome using silver atoms.

The classical expectation is that the electronic angular momentum, and hence the resulting magnet, can be oriented in any direction. Each atom would be deflected into a direction that depends on the orientation and the beam should spread out into a broad band as it emerges from the magnetic field. In contrast, the expectation from quantum mechanics is that the angular momentum, and hence the atomic magnet, has only discrete orientations (Topic 7F). Each of these orientations results in the atoms being deflected in a specific direction, so the beam should split into a number of sharp bands, each corresponding to a different orientation of the angular momentum of the electrons in the atom.

In their first experiment, Stern and Gerlach appeared to confirm the classical prediction. However, the experiment is difficult because collisions between the atoms in the beam blur the bands. When they repeated the experiment with a beam of very low intensity (so that collisions were less frequent), they observed discrete bands, and so confirmed the quantum prediction. However, Stern and Gerlach observed *two* bands of Ag atoms in their experiment. This observation seems to conflict with one of the predictions of quantum mechanics, because an angular momentum l gives rise to $2l + 1$ orientations, which is equal to 2 only if $l = \frac{1}{2}$, contrary to the requirement that l is an integer. The conflict was resolved by the suggestion that the angular momentum they were observing was not due to orbital angular momentum (the motion of an electron around the atomic nucleus) but arose instead from the rotation of the electron about its own axis, its 'spin'.

The spin of an electron does not have to satisfy the same boundary conditions as those for a particle circulating through space around a central point, so the quantum number for spin angular momentum is subject to different restrictions. The **spin**

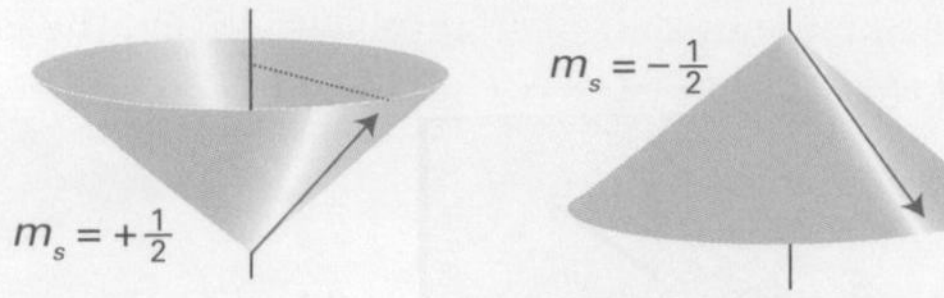

Figure 8B.2 The vector representation of the spin of an electron. The length of the side of the cone is $3^{1/2}/2$ units and the projections on to the z-axis are $\pm\frac{1}{2}$ units.

quantum number s is used in place of the orbital angular momentum quantum number l (Topic 7F; like l, s is a non-negative number) and m_s, the **spin magnetic quantum number**, is used in place of m_l for the projection on the z-axis. The magnitude of the spin angular momentum is $\{s(s+1)\}^{1/2}\hbar$ and the component $m_s\hbar$ is restricted to the $2s+1$ values $m_s = s, s-1, \ldots, -s$. To account for Stern and Gerlach's observation, $s = \frac{1}{2}$ and $m_s = \pm\frac{1}{2}$.

A note on good practice You will sometimes see the quantum number s used in place of m_s, and written $s = \pm\frac{1}{2}$. That is wrong: like l, s is never negative and denotes the magnitude of the spin angular momentum. For the z-component, use m_s.

The detailed analysis of the spin of a particle is sophisticated and shows that the property should not be taken to be an actual spinning motion. It is better to regard 'spin' as an intrinsic property like mass and charge: every electron has exactly the same value and the magnitude of the spin angular momentum of an electron cannot be changed. However, the picture of an actual spinning motion can be very useful when used with care. In the vector model of angular momentum (Topic 7F), the spin may lie in two different orientations (Fig. 8B.2). One orientation corresponds to $m_s = +\frac{1}{2}$ (this state is often denoted α or $\uparrow$); the other orientation corresponds to $m_s = -\frac{1}{2}$ (this state is denoted β or $\downarrow$).

Other elementary particles have characteristic spin. For example, protons and neutrons are spin-$\frac{1}{2}$ particles (i.e. $s = \frac{1}{2}$). Because the masses of a proton and a neutron are so much greater than the mass of an electron, yet they all have the same spin angular momentum, the classical picture would be of these two particles spinning much more slowly than an electron. Some mesons, another variety of fundamental particle, are spin-1 particles (i.e. $s = 1$), as are some atomic nuclei, but for our purposes the most important spin-1 particle is the photon. The importance of photon spin in spectroscopy is explained in Topic 11A; nuclear spin is the basis of nuclear magnetic resonance (Topic 12A).

Brief illustration 8B.2

The magnitude of the spin angular momentum, like any angular momentum, is $\{s(s+1)\}^{1/2}\hbar$. For any spin-$\frac{1}{2}$ particle, not only electrons, this angular momentum is $(\frac{3}{4})^{1/2}\hbar = 0.866\hbar$, or 9.13×10^{-35} J s. The component on the z-axis is $m_s\hbar$, which for a spin-$\frac{1}{2}$ particle is $\pm\frac{1}{2}\hbar$, or $\pm 5.27 \times 10^{-35}$ J s.

Particles with half-integral spin are called **fermions** and those with integral spin (including 0) are called **bosons**. Thus, electrons and protons are fermions; photons are bosons. It is a very deep feature of nature that all the elementary particles that constitute matter are fermions whereas the elementary particles that transmit the forces that bind fermions together are all bosons. Photons, for example, transmit the electromagnetic force that binds together electrically charged particles. Matter, therefore, is an assembly of fermions held together by forces conveyed by bosons.

(b) The Pauli principle

With the concept of spin established, it is possible to resume discussion of the electronic structures of atoms. Lithium, with $Z = 3$, has three electrons. The first two occupy a 1s orbital drawn even more closely than in He around the more highly charged nucleus. The third electron, however, does not join the first two in the 1s orbital because that configuration is forbidden by the **Pauli exclusion principle**:

> No more than two electrons may occupy any given orbital, and if two do occupy one orbital, then their spins must be paired.

Pauli exclusion principle

Electrons with paired spins, denoted $\uparrow\downarrow$, have zero net spin angular momentum because the spin of one electron is cancelled by the spin of the other. Specifically, one electron has $m_s = +\frac{1}{2}$ the other has $m_s = -\frac{1}{2}$ and in the vector model they are orientated on their respective cones so that the resultant spin is zero (Fig. 8B.3). The exclusion principle is the key to the structure of complex atoms, to chemical periodicity, and to molecular structure. It was proposed by Wolfgang Pauli in 1925 when he was trying to account for the absence of some lines in the spectrum of helium. Later he was able to derive a very general form of the principle from theoretical considerations.

The Pauli *exclusion* principle is a special case of a general statement called the **Pauli principle**:

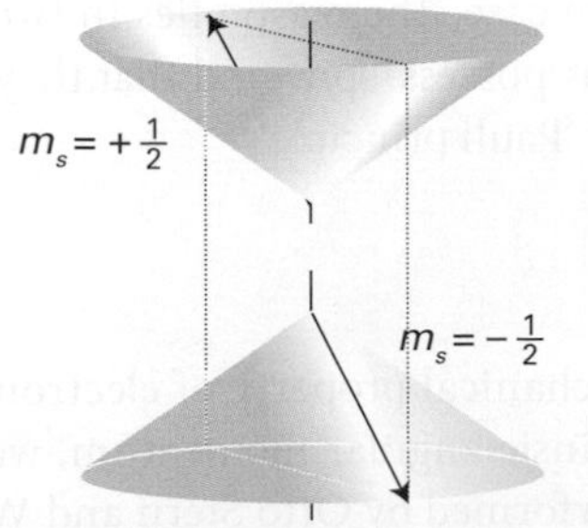

Figure 8B.3 Electrons with paired spins have zero resultant spin angular momentum. They can be represented by two vectors that lie at an indeterminate position on the cones shown here, but wherever one lies on its cone, the other points in the opposite direction; their resultant is zero.

> When the labels of any two identical fermions are exchanged, the total wavefunction changes sign; when the labels of any two identical bosons are exchanged, the sign of the total wavefunction remains the same.

Pauli principle

By 'total wavefunction' is meant the entire wavefunction, including the spin of the particles.

To see that the Pauli principle implies the Pauli exclusion principle, consider the wavefunction for two electrons, $\Psi(1,2)$. The Pauli principle implies that it is a fact of nature (which has its roots in the theory of relativity) that the wavefunction must change sign if the labels 1 and 2 are interchanged wherever they occur in the function:

$$\Psi(2,1) = -\Psi(1,2) \qquad (8B.2)$$

Suppose the two electrons in a two-electron atom occupy the same orbital ψ, then in the orbital approximation the overall *spatial* wavefunction is $\psi(r_1)\psi(r_2)$, which for simplicity will be denoted $\psi(1)\psi(2)$. To apply the Pauli principle, it is necessary to consider the *total* wavefunction, the wavefunction including spin. There are several possibilities for two electrons: both α, denoted $\alpha(1)\alpha(2)$, both β, denoted $\beta(1)\beta(2)$, and one α and the other β, denoted either $\alpha(1)\beta(2)$ or $\alpha(2)\beta(1)$. Because it is not possible to know which electron is α and which is β, in the last case it is appropriate to express the spin states as the (normalized) linear combinations[1]

$$\sigma_+(1,2) = \left(\frac{1}{2^{1/2}}\right)\{\alpha(1)\beta(2) + \beta(1)\alpha(2)\}$$
$$\sigma_-(1,2) = \left(\frac{1}{2^{1/2}}\right)\{\alpha(1)\beta(2) - \beta(1)\alpha(2)\} \qquad (8B.3)$$

These combinations allow one spin to be α and the other β with equal probability; the former corresponds to parallel spins (the individual spins do not cancel) and the latter to paired spins (the individual spins cancel). The total wavefunction of the system is therefore the product of the orbital part and one of the four spin states:

$$\psi(1)\psi(2)\alpha(1)\alpha(2) \qquad \psi(1)\psi(2)\beta(1)\beta(2)$$
$$\psi(1)\psi(2)\sigma_+(1,2) \qquad \psi(1)\psi(2)\sigma_-(1,2) \qquad (8B.4)$$

The Pauli principle says that for a wavefunction to be acceptable (for electrons), it must change sign when the electrons are exchanged. In each case, exchanging the labels 1 and 2 converts $\psi(1)\psi(2)$ into $\psi(2)\psi(1)$, which is the same, because the order of multiplying the functions does not change the value of the product. The same is true of $\alpha(1)\alpha(2)$ and $\beta(1)\beta(2)$. Therefore, $\psi(1)\psi(2)\alpha(1)\alpha(2)$ and $\psi(1)\psi(2)\beta(1)\beta(2)$ are not allowed, because they do not change sign. When the labels are exchanged the combination $\sigma_+(1,2)$ becomes

$$\sigma_+(2,1) = \left(\frac{1}{2^{1/2}}\right)\{\alpha(2)\beta(1) + \beta(2)\alpha(1)\} = \sigma_+(1,2)$$

because the central term is simply the original function written in a different order. The product $\psi(1)\psi(2)\sigma_+(1,2)$ is therefore also disallowed. Finally, consider $\sigma_-(1,2)$:

$$\sigma_-(2,1) = \left(\frac{1}{2^{1/2}}\right)\{\alpha(2)\beta(1) - \beta(2)\alpha(1)\}$$
$$= -\left(\frac{1}{2^{1/2}}\right)\{\alpha(1)\beta(2) - \beta(1)\alpha(2)\} = -\sigma_-(1,2)$$

The combination $\psi(1)\psi(2)\sigma_-(1,2)$ therefore does change sign (it is 'antisymmetric') and is acceptable.

In summary, only one of the four possible states is allowed by the Pauli principle: the one that survives has paired α and β spins. This is the content of the Pauli exclusion principle. The exclusion principle (but not the more general Pauli principle) is irrelevant when the orbitals occupied by the electrons are different, and both electrons may then have, but need not have, the same spin state. In each case the overall wavefunction must still be antisymmetric and must satisfy the Pauli principle itself.

Now returning to lithium, Li ($Z = 3$), the third electron cannot enter the 1s orbital because that orbital is already full: the K shell (the shell with $n = 1$, Topic 8A) is complete and the two electrons form a **closed shell**, a shell in which all the orbitals are fully occupied. Because a similar closed shell is characteristic of the He atom, it is commonly denoted [He]. The third electron cannot enter the K shell and must occupy the next available orbital, which is one with $n = 2$ and hence belonging to the L shell (which consists of the four orbitals with $n = 2$). It is now necessary to decide whether the next available orbital is the 2s orbital or a 2p orbital, and therefore whether the lowest energy configuration of the atom is [He]2s^1 or [He]2p^1.

8B.3 The building-up principle

Unlike in hydrogenic atoms, the 2s and 2p orbitals (and, in general, the subshells of a given shell) do not have the same energy in many-electron atoms.

(a) Penetration and shielding

An electron in a many-electron atom experiences a Coulombic repulsion from all the other electrons present. If the electron is at a distance r from the nucleus, it experiences an average repulsion that can be represented by a point negative charge located at the nucleus and equal in magnitude to the total charge

[1] A stronger justification for taking these linear combinations is that they correspond to eigenfunctions of the total spin operators S^2 and S_z, with $M_S = 0$ and, respectively, $S = 1$ and 0.

of all the other electrons within a sphere of radius r (Fig. 8B.4). This property is a conclusion of classical electrostatics, where the effect of a spherical distribution of charge can be represented by a point charge of the same magnitude located at its centre. The effect of this point negative charge is to reduce the full charge of the nucleus from Ze to $Z_{\text{eff}}e$, the **effective nuclear charge**. In everyday parlance, Z_{eff} itself is commonly referred to as the 'effective nuclear charge'. The electron is said to experience a **shielded** nuclear charge, and the difference between Z and Z_{eff} is called the **shielding constant**, σ:

$$Z_{\text{eff}} = Z - \sigma \quad \text{Nuclear shielding} \quad (8B.5)$$

The electrons do not actually 'block' the full Coulombic attraction of the nucleus: the shielding constant is simply a way of expressing the net outcome of the nuclear attraction and the electronic repulsions in terms of a single equivalent charge at the centre of the atom.

The shielding constant is different for s and p electrons because they have different radial distribution functions and therefore respond to the other electrons in the atom to different extents (Fig. 8B.5). An s electron has a greater **penetration** through inner shells than a p electron, in the sense that an s electron is more likely to be found close to the nucleus than a p electron of the same shell. Because only electrons inside the sphere defined by the location of the electron of interest contribute to shielding, an s electron experiences less shielding than a p electron. Consequently, as a result of the combined effects of penetration and shielding, an s electron is more tightly bound than a p electron of the same shell. Similarly, a d electron penetrates less than a p electron of the same shell (recall that a d orbital is proportional to r^2 close to the nucleus, whereas a p orbital is proportional to r, so the amplitude of a d orbital is smaller there than that of a p orbital), and therefore experiences more shielding.

Shielding constants for different types of electrons in atoms have been calculated from wavefunctions obtained by numerical solution of the Schrödinger equation (Table 8B.1). In general, valence-shell s electrons do experience higher effective nuclear charges than p electrons, although there are some discrepancies.

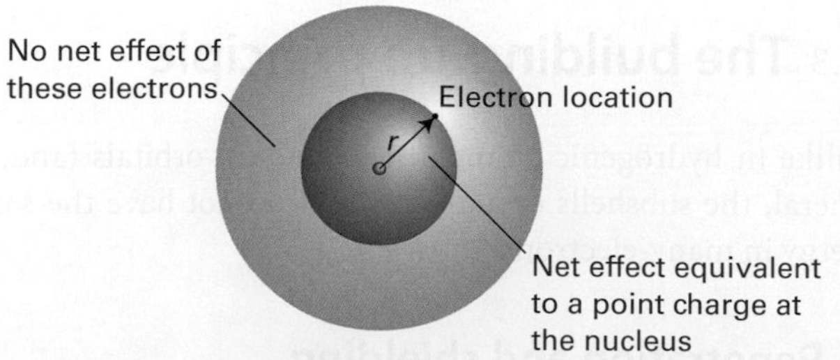

Figure 8B.4 An electron at a distance r from the nucleus experiences a Coulombic repulsion from all the electrons within a sphere of radius r. This repulsion is equivalent to that from a point negative charge located on the nucleus. The negative charge reduces the effective nuclear charge of the nucleus from Ze to $Z_{\text{eff}}e$.

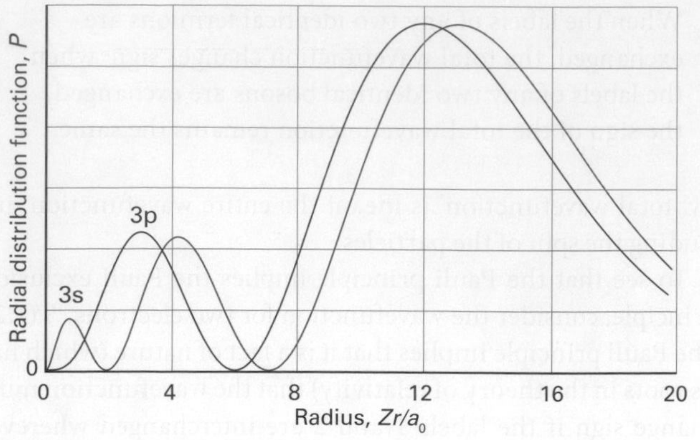

Figure 8B.5 An electron in an s orbital (here a 3s orbital) is more likely to be found close to the nucleus than an electron in a p orbital of the same shell (note the closeness of the innermost peak of the 3s orbital to the nucleus at $r = 0$). Hence an s electron experiences less shielding and is more tightly bound than a p electron of the same shell.

Table 8B.1 Effective nuclear charge*

Element	Z	Orbital	Z_{eff}
He	2	1s	1.6875
C	6	1s	5.6727
		2s	3.2166
		2p	3.1358

* More values are given in the *Resource section*.

Brief illustration 8B.3

The effective nuclear charge for 1s, 2s, and 2p electrons in a carbon atom are 5.6727, 3.2166, and 3.1358, respectively. The radial distribution functions for these orbitals (Topic 8A) are generated by forming $P(r) = r^2R(r)^2$, where $R(r)$ is the radial wavefunction, which are given in Table 8A.1. The three radial distribution functions are plotted in Fig. 8B.6. As can be seen (especially in the magnified view close to the nucleus), the s orbital has greater penetration than the p orbital. The average radii of the 2s and 2p orbitals are 99 pm and 84 pm, respectively, which shows that the average distance of a 2s electron from the nucleus is greater than that of a 2p orbital. To account for the lower energy of the 2s orbital, the extent of penetration is more important than the average distance from the nucleus.

The consequence of penetration and shielding is that the energies of subshells of a shell in a many-electron atom (those

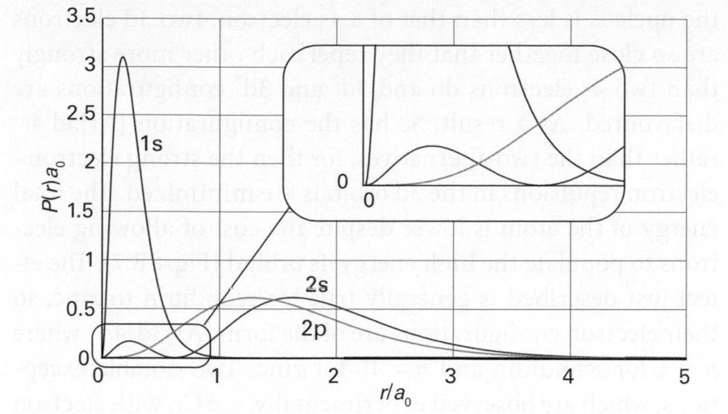

Figure 8B.6 The radial distribution functions for electrons in a carbon atom, as calculated in *Brief illustration* 8B.3.

with the same values of n but different values of l) in general lie in the order $s < p < d < f$. The individual orbitals of a given subshell (those with the same value of l but different values of m_l) remain degenerate because they all have the same radial characteristics and so experience the same effective nuclear charge.

To complete the Li story, consider that, because the shell with $n = 2$ consists of two subshells, with the 2s subshell lower in energy than the 2p subshell, the third electron occupies the 2s orbital (the only orbital in that subshell). This occupation results in the ground-state configuration $1s^22s^1$, with the central nucleus surrounded by a complete helium-like shell of two 1s electrons, and around that a more diffuse 2s electron. The electrons in the outermost shell of an atom in its ground state are called the **valence electrons** because they are largely responsible for the chemical bonds that the atom forms (and 'valence', as explained in Focus 9, refers to the ability of an atom to form bonds). Thus, the valence electron in Li is a 2s electron and its other two electrons belong to its core.

(b) Hund's rules

The extension of the argument used to account for the structures of H, He, and Li is called the **building-up principle**, or the *Aufbau principle*, from the German word for "building up", and should be familiar from introductory courses. In brief, imagine the bare nucleus of atomic number Z, and then feed into the orbitals Z electrons in succession. The order of occupation, following the shells and their subshells arranged in order of increasing energy, is

1s 2s 2p 3s 3p 4s 3d 4p 5s 4d 5p 6s

Each orbital may accommodate up to two electrons.

Brief illustration 8B.4

Consider the carbon atom, for which $Z = 6$ and there are six electrons to accommodate. Two electrons enter and fill the 1s orbital, two enter and fill the 2s orbital, leaving two electrons to occupy the orbitals of the 2p subshell. Hence the ground-state configuration of C is $1s^22s^22p^2$, or more succinctly $[He]2s^22p^2$, with [He] the helium-like $1s^2$ core.

It is possible to be more precise about the configuration of a carbon atom than in *Brief illustration* 8B.4. The last two electrons are expected to occupy different 2p orbitals because they are then farther apart on average and repel each other less than if they were in the same orbital. Thus, one electron can be thought of as occupying the $2p_x$ orbital and the other the $2p_y$ orbital (the x, y, z designation is arbitrary, and it would be equally valid to use the complex forms of these orbitals), and the lowest energy configuration of the atom is $[He]2s^22p_x^12p_y^1$. The same rule applies whenever degenerate orbitals of a subshell are available for occupation. Thus, another rule of the building-up principle is:

> Electrons occupy different orbitals of a given subshell before doubly occupying any one of them.

For instance, nitrogen ($Z = 7$) has the ground-state configuration $[He]2s^22p_x^12p_y^12p_z^1$, and only at oxygen ($Z = 8$) is a 2p orbital doubly occupied, giving $[He]2s^22p_x^22p_y^12p_z^1$.

When electrons occupy orbitals singly it is necessary to invoke **Hund's maximum multiplicity rule**:

> An atom in its ground state adopts a configuration with the greatest number of unpaired electrons.

Hund's maximum multiplicty rule

The explanation of Hund's rule is subtle, but it reflects the quantum mechanical property of **spin correlation**. In essence, the effect of spin correlation is to allow the atom to shrink slightly when the spins are parallel, so the electron–nucleus interaction is improved. As a consequence, in the ground state of the carbon atom, the two 2p electrons have parallel spins, all three 2p electrons in the N atoms have parallel spins, and the two 2p electrons in different orbitals in the O atom have parallel spins (the two in the $2p_x$ orbital are necessarily paired). The effect can be explained by considering the Pauli principle and showing that electrons with parallel spins behave as if they have a tendency to stay apart, and hence repel each other less.

How is that done? 8B.2 Exploring the origins of spin correlation

Suppose electron 1 is in orbital a and described by a wavefunction $\psi_a(r_1)$, and electron 2 is in orbital b with wavefunction $\psi_b(r_2)$. Then, in the orbital approximation, the joint spatial wavefunction of the electrons is the product $\Psi = \psi_a(r_1)\psi_b(r_2)$. However, this wavefunction is not acceptable, because it suggests that it is possible to know which electron is in which orbital. According to quantum mechanics, the correct description is either of the two following wavefunctions:

$$\Psi_{\pm} = \left(\frac{1}{2^{1/2}}\right)\{\psi_a(r_1)\psi_b(r_2) \pm \psi_b(r_1)\psi_a(r_2)\}$$

According to the Pauli principle, because Ψ_+ is symmetrical under particle interchange, it must be multiplied by an antisymmetric spin state (the one denoted σ_-). That combination corresponds to a spin-paired state. Conversely, Ψ_- is antisymmetric, so it must be multiplied by one of the three symmetric spin states. These three symmetric states correspond to electrons with parallel spins (see Topic 8C for an explanation of this point).

Now consider the behaviour of the two wavefunctions $\Psi_{\pm}$ when one electron approaches another, and $\boldsymbol{r}_1 = \boldsymbol{r}_2$. As a result, Ψ_- vanishes, which means that there is zero probability of finding the two electrons at the same point in space when they have parallel spins. In contrast, the wavefunction Ψ_+ does not vanish when the two electrons are at the same point in space. Because the two electrons have different relative spatial distributions depending on whether their spins are parallel or not, it follows that their Coulombic interaction is different, and hence that the two states described by these wavefunctions have different energies, with the spin-parallel state lower in energy than the spin-paired state.

Neon, with $Z = 10$, has the configuration [He]$2s^22p^6$, which completes the L shell. This closed-shell configuration is denoted [Ne], and acts as a core for subsequent elements. The next electron must enter the 3s orbital and begin a new shell, so an Na atom, with $Z = 11$, has the configuration [Ne]$3s^1$. Like lithium with the configuration [He]$2s^1$, sodium has a single s electron outside a complete core. This analysis hints at the origin of chemical periodicity. The L shell is completed by eight electrons, so the element with $Z = 3$ (Li) should have similar properties to the element with $Z = 11$ (Na). Likewise, Be ($Z =$ 4) should be similar to $Z = 12$ (Mg), and so on, up to the noble gases He ($Z = 2$), Ne ($Z = 10$), and Ar ($Z = 18$).

At potassium ($Z = 19$) the next orbital in line for occupation is 4s: this orbital is brought below 3d by the effects of penetration and shielding, and the ground state configuration is [Ar]$4s^1$. Calcium ($Z = 20$) is likewise [Ar]$4s^2$. At this stage the five 3d orbitals are in line for occupation, but there are complications arising from the energy changes arising from the interaction of the electrons in the valence shell, and penetration arguments alone are no longer reliable.

Calculations of the type discussed in Section 8B.4 show that for the atoms from scandium to zinc the energies of the 3d orbitals are always lower than the energy of the 4s orbital, in spite of the greater penetration of a 4s electron. However, spectroscopic results show that Sc has the configuration [Ar]$3d^14s^2$, not [Ar]$3d^3$ or [Ar]$3d^24s^1$. To understand this observation, consider the nature of electron–electron repulsions in 3d and 4s orbitals. Because the average distance of a 3d electron from the nucleus is less than that of a 4s electron, two 3d electrons are so close together that they repel each other more strongly than two 4s electrons do and $3d^2$ and $3d^3$ configurations are disfavoured. As a result, Sc has the configuration [Ar]$3d^14s^2$ rather than the two alternatives, for then the strong electron–electron repulsions in the 3d orbitals are minimized. The total energy of the atom is lower despite the cost of allowing electrons to populate the high energy 4s orbital (Fig. 8B.7). The effect just described is generally true for scandium to zinc, so their electron configurations are of the form [Ar]$3d^n4s^2$, where $n = 1$ for scandium and $n = 10$ for zinc. Two notable exceptions, which are observed experimentally, are Cr, with electron configuration [Ar]$3d^54s^1$, and Cu, with electron configuration [Ar]$3d^{10}4s^1$. At gallium, these complications disappear and the building-up principle is used in the same way as in preceding periods. Now the 4s and 4p subshells constitute the valence shell, and the period terminates with krypton. Because 18 electrons have intervened since argon, this row is the first 'long period' of the periodic table.

At this stage it becomes apparent that sequential occupation of the orbitals in successive shells results in periodic similarities in the electronic configurations. This periodicity of structure accounts for the formulation of the **periodic table** (see inside the back cover). The vertical columns of the periodic table are called **groups** and (in the modern convention) numbered from 1 to 18. Successive rows of the periodic table are called **periods**, the number of the period being equal to the principal quantum number of the valence shell.

The periodic table is divided into s, p, d, and f **blocks**, according to the subshell that is last to be occupied in the formulation of the electronic configuration of the atom. The members of the d block (specifically the members of Groups 3–11 in the d block) are also known as the **transition metals**; those of the f block (which is not divided into numbered groups) are sometimes called the **inner transition metals**. The upper row of the f block (Period 6) consists of the **lanthanoids**

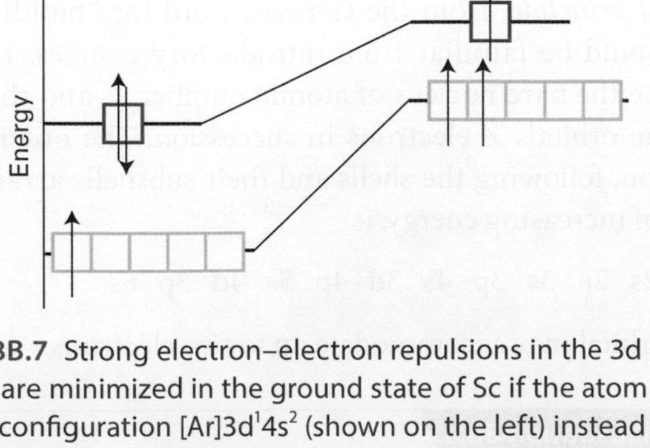

Figure 8B.7 Strong electron–electron repulsions in the 3d orbitals are minimized in the ground state of Sc if the atom has the configuration [Ar]$3d^14s^2$ (shown on the left) instead of [Ar]$3d^24s^1$ (shown on the right). The total energy of the atom is lower when it has the [Ar]$3d^14s^2$ configuration despite the cost of populating the high energy 4s orbital.

(still commonly the 'lanthanides') and the lower row (Period 7) consists of the **actinoids** (still commonly the 'actinides').

The configurations of cations of elements in the s, p, and d blocks of the periodic table are derived by removing electrons from the ground-state configuration of the neutral atom in a specific order. First, remove valence p electrons, then valence s electrons, and then as many d electrons as are necessary to achieve the specified charge. The configurations of anions of the p-block elements are derived by continuing the building-up procedure and adding electrons to the neutral atom until the configuration of the next noble gas has been reached.

Brief illustration 8B.5

Because the configuration of vanadium is $[Ar]3d^34s^2$, the V^{2+} cation has the configuration $[Ar]3d^3$. It is reasonable to remove the more energetic 4s electrons in order to form the cation, but it is not obvious why the $[Ar]3d^3$ configuration is preferred in V^{2+} over the $[Ar]3d^14s^2$ configuration, which is found in the isoelectronic Sc atom. Calculations show that the energy difference between $[Ar]3d^3$ and $[Ar]3d^14s^2$ depends on Z_{eff}. As Z_{eff} increases, transfer of a 4s electron to a 3d orbital becomes more favourable because the electron–electron repulsions are compensated by attractive interactions between the nucleus and the electrons in the spatially compact 3d orbital. Indeed, calculations reveal that, for a sufficiently large Z_{eff}, $[Ar]3d^3$ is lower in energy than $[Ar]3d^14s^2$. This conclusion explains why V^{2+} has a $[Ar]3d^3$ configuration and also accounts for the observed $[Ar]4s^03d^n$ configurations of the M^{2+} cations of Sc through Zn.

(c) Atomic and ionic radii

The **atomic radius** of an element is half the distance between the centres of neighbouring atoms in a solid (such as Cu) or, for non-metals, in a homonuclear molecule (such as H_2 or S_8). As seen in Table 8B.2 and Fig. 8B.8, atomic radii tend to decrease from left to right across a period of the periodic table, and increase down each group. The decrease across a period can be traced to the increase in nuclear charge, which draws the electrons in closer to the nucleus. The increase in nuclear charge is partly cancelled by the increase in the number of electrons, but because electrons are spread over a region of space, one electron does not fully shield one nuclear charge, so the increase in nuclear charge dominates. The increase in atomic radius down a group (despite the increase in nuclear charge) is explained by the fact that the valence shells of successive periods correspond to higher principal quantum numbers. That is, successive periods correspond to the start and then completion of successive (and more distant) shells of the atom that surround each other like the successive layers of an onion. The need to occupy a more distant shell leads to a larger atom despite the increased nuclear charge.

Table 8B.2 Atomic radii of main-group elements, r/pm

Li 157	Be 112	B 88	C 77	N 74	O 66	F 64
Na 191	Mg 160	Al 143	Si 118	P 110	S 104	Cl 99
K 235	Ca 197	Ga 153	Ge 122	As 121	Se 117	Br 114
Rb 250	Sr 215	In 167	Sn 158	Sb 141	Te 137	I 133
Cs 272	Ba 224	Tl 171	Pb 175	Bi 182	Po 167	

A modification of the increase down a group is encountered in Period 6, for the radii of the atoms in the d block and in the following atoms of the p block are not as large as would be expected by simple extrapolation down the group. The reason can be traced to the fact that in Period 6 the f orbitals are in the process of being occupied. An f electron is a very inefficient shielder of nuclear charge (for reasons connected with its radial extension), and as the atomic number increases from La to Lu, there is a considerable contraction in radius. By the time the d block resumes (at hafnium, Hf), the poorly shielded but considerably increased nuclear charge has drawn in the surrounding electrons, and the atoms are compact. They are so compact, that the metals in this region of the periodic table (iridium to lead) are very dense. The reduction in radius below that expected by extrapolation from preceding periods is called the **lanthanide contraction**.

The **ionic radius** of an element is its share of the distance between neighbouring ions in an ionic solid. That is, the distance between the centres of a neighbouring cation and anion is the sum of the two ionic radii. The size of the 'share' leads

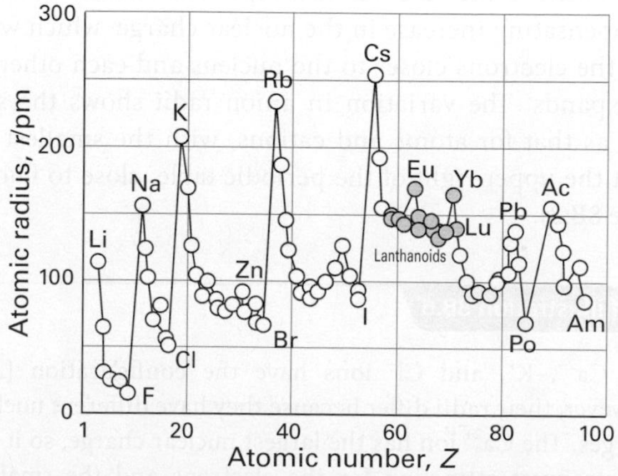

Figure 8B.8 The variation of atomic radius through the periodic table. Note the contraction of radius following the lanthanoids in Period 6 (following Lu, lutetium).

Table 8B.3 Ionic radii, *r*/pm*

Li^+(4)	Be^{2+}(4)	B^{3+}(4)	N^{3-}	O^{2-}(6)	F^-(6)
59	27	12	171	140	133
Na^+(6)	Mg^{2+}(6)	Al^{3+}(6)	P^{3-}	S^{2-}(6)	Cl^-(6)
102	72	53	212	184	181
K^+(6)	Ca^{2+}(6)	Ga^{3+}(6)	As^{3-}(6)	Se^{2-}(6)	Br^-(6)
138	100	62	222	198	196
Rb^+(6)	Sr^{2+}(6)	In^{3+}(6)		Te^{2-}(6)	I^-(6)
149	116	79		221	220
Cs^+(6)	Ba^{2+}(6)	Tl^{3+}(6)			
167	136	88			

*Numbers in parentheses are the *coordination numbers* of the ions, the numbers of species (for example, counterions, solvent molecules) around the ions. Values for ions without a coordination number stated are estimates. More values are given in the *Resource section.*

to some ambiguity in the definition. One common definition sets the ionic radius of O^{2-} equal to 140 pm, but there are other scales, and care must be taken not to mix them. Ionic radii also vary with the number of counterions (ions of opposite charge) around a given ion; unless otherwise stated, the values in this text have been corrected to correspond to an environment of six counterions.

When an atom loses one or more valence electrons to form a cation, the remaining atomic core is smaller than the parent atom. Therefore, a cation is invariably smaller than its parent atom. For example, the atomic radius of Na, with the configuration [Ne]3s^1, is 191 pm, but the ionic radius of Na^+, with the configuration [Ne], is only 102 pm (Table 8B.3). Like atomic radii, cation radii increase down each group because electrons are occupying shells with higher principal quantum numbers.

An anion is larger than its parent atom because the electrons added to the valence shell repel one another. Without a compensating increase in the nuclear charge, which would draw the electrons closer to the nucleus and each other, the ion expands. The variation in anion radii shows the same trend as that for atoms and cations, with the smallest anions at the upper right of the periodic table, close to fluorine (Table 8B.3).

Brief illustration 8B.6

The Ca^{2+}, K^+, and Cl^- ions have the configuration [Ar]. However, their radii differ because they have different nuclear charges. The Ca^{2+} ion has the largest nuclear charge, so it has the strongest attraction for the electrons and the smallest radius. The Cl^- ion has the lowest nuclear charge of the three ions and, as a result, the largest radius.

(d) Ionization energies and electron affinities

The minimum energy necessary to remove an electron from a many-electron atom in the gas phase is the **first ionization energy**, I_1, of the element. The **second ionization energy**, I_2, is the minimum energy needed to remove a second electron (from the singly charged cation). The variation of the first ionization energy through the periodic table is shown in Fig. 8B.9 and some numerical values are given in Table 8B.4.

The **electron affinity**, E_{ea}, is the energy released when an electron attaches to a gas-phase atom (Table 8B.5). In a common, logical (given its name), but not universal convention (which is adopted here), the electron affinity is positive if energy is released when the electron attaches to the atom. That is, $E_{ea} > 0$ implies that electron attachment is exothermic.

As will be familiar from introductory chemistry, ionization energies and electron affinities show periodicities. The former is more regular and concentrated on here. Lithium has a low first ionization energy because its outermost electron is well shielded from the nucleus by the core ($Z_{eff} = 1.3$, compared with $Z = 3$). The ionization energy of Be ($Z = 4$) is greater but that of B is lower because in the latter the outermost electron occupies a 2p orbital and is less strongly bound than if it had been a 2s electron. The ionization energy increases from B to N on account of the increasing nuclear charge. However, the ionization energy of O is less than would be expected by simple extrapolation. The explanation is that at oxygen a 2p orbital must become doubly occupied, and the electron–electron repulsions are increased above what would be expected by simple extrapolation along the row. In addition, the loss of a 2p electron results in a configuration with a half-filled subshell (like that of N), which is an arrangement of low energy, so the energy of $O^+ + e^-$ is lower than might be expected, and the ionization energy is correspondingly low too. (The kink is less pronounced in the next row, between phosphorus and sulfur

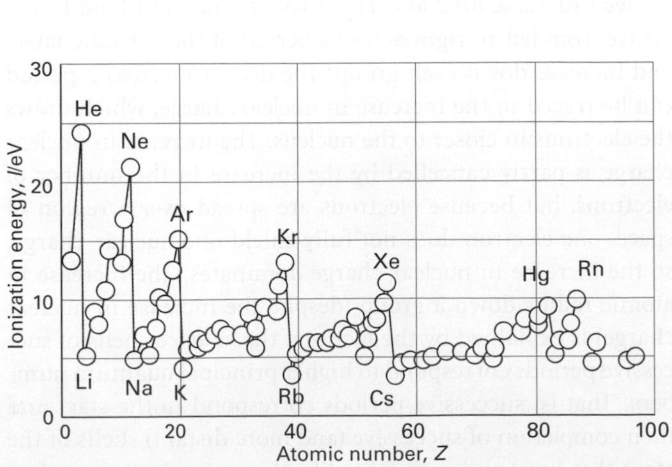

Figure 8B.9 The first ionization energies of the elements plotted against atomic number.

Table 8B.4 First and second ionization energies*

Element	I_1/(kJ mol^{-1})	I_2/(kJ mol^{-1})
H	1312	
He	2372	5251
Mg	738	1451
Na	496	4562

* More values are given in the *Resource section.*

Table 8B.5 Electron affinities, E_a/(kJ mol^{-1})*

Cl	349		
F	322		
H	73		
O	141	O^-	−844

* More values are given in the *Resource section.*

because their orbitals are more diffuse.) The values for O, F, and Ne fall roughly on the same line, the increase of their ionization energies reflecting the increasing attraction of the more highly charged nuclei for the outermost electrons.

The outermost electron in sodium (Z = 11) is 3s. It is far from the nucleus, and the latter's charge is shielded by the compact, complete neon-like core, with the result that $Z_{\text{eff}} \approx 2.5$. As a result, the ionization energy of Na is substantially lower than that of Ne (Z = 10, $Z_{\text{eff}} \approx 5.8$). The periodic cycle starts again along this row, and the variation of the ionization energy can be traced to similar reasons.

Electron affinities are greatest close to fluorine, for the incoming electron enters a vacancy in a compact valence shell and can interact strongly with the nucleus. The attachment of an electron to an anion (as in the formation of O^{2-} from O^-) is invariably endothermic, so E_{ea} is negative. The incoming electron is repelled by the charge already present. Electron affinities are also small, and may be negative, when an electron enters an orbital that is far from the nucleus (as in the heavier alkali metal atoms) or is forced by the Pauli principle to occupy a new shell (as in the noble gas atoms).

8B.4 Self-consistent field orbitals

The preceding treatment of the electronic configuration of many-electron species is only approximate because of the complications introduced by electron–electron interactions. However, computational techniques are available that give reliable approximate solutions for the wavefunctions and energies. The techniques were originally introduced by D.R. Hartree (before computers were available) and then modified by V. Fock to take into account the Pauli principle correctly. In broad outline, the **Hartree–Fock self-consistent field** (HF-SCF) procedure is as follows.

Start with an idea of the structure of the atom as suggested by the building-up principle. In the Ne atom, for instance, the principle suggests the configuration $1s^2 2s^2 2p^6$ with the orbitals approximated by hydrogenic atomic orbitals with the appropriate effective nuclear charges. Now consider one of the 2p electrons. A Schrödinger equation can be written for this electron by ascribing to it a potential energy due to the nuclear attraction and the average repulsion from the other electrons. Although the equation is for the 2p orbital, that repulsion, and therefore the equation, depends on the wavefunctions of all the other occupied orbitals in the atom. To solve the equation, guess an approximate form of the wavefunctions of all the other orbitals and then solve the Schrödinger equation for the 2p orbital. The procedure is then repeated for the 1s and 2s orbitals. This sequence of calculations gives the form of the 2p, 2s, and 1s orbitals, and in general they will differ from the set used to start the calculation. These improved orbitals can be used in another cycle of calculation, and a second improved set of orbitals and a better energy are obtained. The recycling continues until the orbitals and energies obtained are insignificantly different from those used at the start of the current cycle. The solutions are then self-consistent and accepted as solutions of the problem.

The outcomes of HF-SCF calculations are radial distribution functions that show the grouping of electron density into shells, as the building-up principle suggests. These calculations therefore support the qualitative discussions that are used to explain chemical periodicity. They also extend that discussion considerably by providing detailed wavefunctions and precise energies.

Checklist of concepts

☐ 1. In the **orbital approximation**, each electron is regarded as being described by its own wavefunction; the overall wavefunction of a many-electron atom is the product of the orbital wavefunctions.

☐ 2. The **configuration** of an atom is the statement of its occupied orbitals.

☐ 3. The **Pauli exclusion principle**, a special case of the Pauli principle, limits to two the number of electrons that can occupy a given orbital.

☐ 4. In many-electron atoms, s orbitals lie at a lower energy than p orbitals of the same shell due to the combined effects of **penetration** and **shielding**.

☐ 5. The **building-up principle** is a procedure for predicting the ground state electron configuration of an atom.

☐ 6. Electrons occupy different orbitals of a given subshell before doubly occupying any one of them.

☐ 7. An atom in its ground state adopts a configuration with the greatest number of unpaired electrons.

☐ 8. The **atomic radius** of an element is half the distance between the centres of neighbouring atoms in a solid or in a homonuclear molecule.

☐ 9. The **ionic radius** of an element is its share of the distance between neighbouring ions in an ionic solid.

☐ 10. The **first ionization energy** is the minimum energy necessary to remove an electron from a many-electron atom in the gas phase.

☐ 11. The **second ionization energy** is the minimum energy needed to remove an electron from a singly charged cation.

☐ 12. The **electron affinity** is the energy released when an electron attaches to a gas-phase atom.

☐ 13. The atomic radius, ionization energy, and electron affinity vary periodically through the periodic table.

☐ 14. The Schrödinger equation for many-electron atoms is solved numerically and iteratively until the solutions are self-consistent.

Checklist of equations

Property	Equation	Comment	Equation number
Orbital approximation	$\Psi(r_1,r_2,\ldots) = \psi(r_1)\psi(r_2)\ldots$		8B.1
Effective nuclear charge	$Z_{\text{eff}} = Z - \sigma$	The charge is this number times e	8B.5

TOPIC 8C Atomic spectra

> ➤ **Why do you need to know this material?**
>
> A knowledge of the energies of electrons in atoms is essential for understanding many chemical properties and chemical bonding.
>
> ➤ **What is the key idea?**
>
> The frequency and wavenumber of radiation emitted or absorbed when atoms undergo electronic transitions provide detailed information about their electronic energy states.
>
> ➤ **What do you need to know already?**
>
> This Topic draws on knowledge of the energy levels of hydrogenic atoms (Topic 8A) and the configurations of many-electron atoms (Topic 8B). In places, it uses the properties of angular momentum (Topic 7F).

The general idea behind atomic spectroscopy is straightforward: lines in the spectrum (in either emission or absorption) occur when the electron distribution in an atom undergoes a **transition**, a change of state, in which its energy changes by ΔE. This transition leads to the emission or is accompanied by absorption of a photon of frequency $\nu = |\Delta E|/h$ and wavenumber $\tilde{\nu} = |\Delta E|/hc$. In spectroscopy, transitions are said to take place between two **terms**. Broadly speaking, a term is simply another name for the energy level of an atom, but as this Topic progresses its full significance will become clear.

8C.1 The spectra of hydrogenic atoms

Not all transitions between the possible terms are observed. Spectroscopic transitions are **allowed**, if they can occur, or **forbidden**, if they cannot occur. A **selection rule** is a statement about which transitions are allowed.

The origin of selection rules can be identified by considering transitions in hydrogenic atoms. A photon has an intrinsic spin angular momentum corresponding to $s = 1$ (Topic 8B). Because total angular momentum is conserved in a transition, the angular momentum of the electron must change to compensate for the angular momentum carried away by the photon. Thus, an electron in a d orbital ($l = 2$) cannot make a transition into an s orbital ($l = 0$) because the photon cannot carry away enough angular momentum. Similarly, an s electron cannot make a transition to another s orbital, because there would then be no change in the angular momentum of the electron to make up for the angular momentum carried away by the photon. A more formal treatment of selection rules requires mathematical manipulation of the wavefunctions for the initial and final states of the atom.

How is that done? 8C.1 Identifying selection rules

The underlying classical idea behind a spectroscopic transition is that, for an atom or molecule to be able to interact with the electromagnetic field and absorb or create a photon of frequency ν, it must possess, at least transiently, a dipole oscillating at that frequency. The consequences of this idea are explored in the following steps.

Step 1 *Write an expression for the transition dipole moment*

The transient dipole is expressed quantum mechanically as the **transition dipole moment**, $\mu_{\rm fi}$, between the initial and final states i and f, where[1]

$$\mu_{\rm fi} = \int \psi_{\rm f}^* \hat{\mu} \psi_{\rm i}\, {\rm d}\tau \qquad (8C.1)$$

and $\hat{\mu}$ is the electric dipole moment operator. For a one-electron atom $\hat{\mu}$ is multiplication by $-e\boldsymbol{r}$. Because $\boldsymbol{r}$ is a vector with components x, y, and z, $\hat{\mu}$ is also a vector, with components $\mu_x = -ex$, $\mu_y = -ey$, and $\mu_z = -ez$. If the transition dipole moment is zero, then the transition is forbidden; the transition is allowed if the transition dipole moment is non-zero.

Step 2 *Formulate the integrand in terms of spherical harmonics*

To evaluate a transition dipole moment, consider each component in turn. For example, for the z-component,

$$\mu_{z,{\rm fi}} = -e\int \psi_{\rm f}^* z \psi_{\rm i}\, {\rm d}\tau$$

In spherical polar coordinates (see *The chemist's toolkit* 21 in Topic 7F) $z = r\cos\theta$. Then, according to Table 7F.1, $z = (4\pi/3)^{1/2} rY_{1,0}$. The wavefunctions for the initial and final states are atomic orbitals of the form $R_{n,l}(r)Y_{l,m_l}(\theta,\phi)$ (Topic 8A). With these substitutions the integral becomes

$$\int \psi_{\rm f}^* z \psi_{\rm i}\, {\rm d}\tau =$$

$$\int_0^\infty \int_0^\pi \int_0^{2\pi} \overbrace{R_{n_{\rm f},l_{\rm f}} Y_{l_{\rm f},m_{l,{\rm f}}}^*}^{\psi_{\rm f}^*} \overbrace{\left(\frac{4\pi}{3}\right)^{1/2} rY_{1,0}}^{z} \overbrace{R_{n_{\rm i},l_{\rm i}} Y_{l_{\rm i},m_{l,{\rm i}}}}^{\psi_{\rm i}} \overbrace{r^2 {\rm d}r \sin\theta\, {\rm d}\theta {\rm d}\phi}^{{\rm d}\tau}$$

[1] See our *Physical chemistry: Quanta, matter, and change* (2014) for a detailed development of the form of eqn 8C.1.

This multiple integral is the product of three factors, an integral over r and two integrals (in blue) over the angles, so the factors on the right can be grouped as follows:

$$\int \psi_f^* z \psi_i \mathrm{d}\tau =$$

$$\left(\frac{4\pi}{3}\right)^{1/2} \int_0^\infty R_{n_f,l_f} r^3 R_{n_i,l_i} \mathrm{d}r \int_0^\pi \int_0^{2\pi} Y_{l_f,m_{l,f}}^* Y_{1,0} Y_{l_i,m_{l,i}} \sin\theta \mathrm{d}\theta \mathrm{d}\phi$$

Step 3 *Evaluate the angular integral*

It follows from the properties of the spherical harmonics that the integral

$$I = \int_0^\pi \int_0^{2\pi} Y_{l_f,m_{l,f}}^* Y_{l,m} Y_{l_i,m_{l,i}} \sin\theta \, \mathrm{d}\theta \mathrm{d}\phi$$

is zero unless $l_f = l_i \pm l$ and $m_{l,f} = m_{l,i} + m$. Because in the present case $l = 1$ and $m = 0$, the angular integral, and hence the z-component of the transition dipole moment, is zero unless $\Delta l = \pm 1$ and $\Delta m_l = 0$, which is a part of the set of selection rules. The same procedure, but considering the x- and y-components, results in the complete set of rules:

$$\Delta l = \pm 1 \quad \Delta m_l = 0, \pm 1 \qquad \text{Selection rules for hydrogenic atoms} \quad (8C.2)$$

The principal quantum number n can change by any amount consistent with the value of Δl for the transition, because it does not relate directly to the angular momentum.

Brief illustration 8C.1

To identify the orbitals to which a 4d electron may make radiative transitions, first identify the value of l and then apply the selection rule for this quantum number. Because $l = 2$, the final orbital must have $l = 1$ or 3. Thus, an electron may make a transition from a 4d orbital to any np orbital (subject to $\Delta m_l = 0, \pm 1$) and to any nf orbital (subject to the same rule). However, it cannot undergo a transition to any other orbital, such as an ns or an nd orbital.

The selection rules and the atomic energy levels jointly account for the structure of a **Grotrian diagram** (Fig. 8C.1), which summarizes the energies of the states and the transitions between them. In some versions, the thicknesses of the transition lines in the diagram denote their relative intensities in the spectrum.

8C.2 The spectra of many-electron atoms

The spectra of atoms rapidly become very complicated as the number of electrons increases, in part because their energy levels, their terms, are not given solely by the energies of the orbitals but depend on the interactions between the electrons.

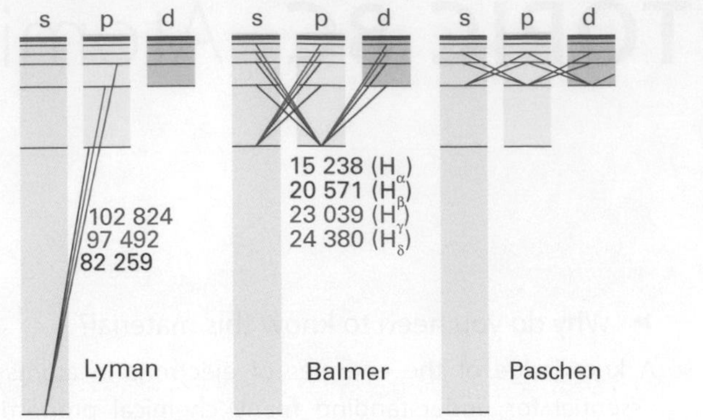

Figure 8C.1 A Grotrian diagram that summarizes the appearance and analysis of the spectrum of atomic hydrogen. The wavenumbers of some transitions (in cm^{-1}) are indicated. The colours of the lines are for reference only: they are not the colours of the transitions.

(a) Singlet and triplet terms

Consider the energy levels of a He atom, with its two electrons. The ground-state configuration is $1s^2$, and an excited configuration is one in which an electron has been promoted into a different orbital to give, for instance, the configuration $1s^1 2s^1$. The two electrons need not be paired because they occupy different orbitals. According to Hund's maximum multiplicity rule (Topic 8B), the state of the atom with the spins parallel lies lower in energy than the state in which they are paired. Both states are permissible, correspond to different terms, and can contribute to the spectrum of the atom.

Parallel and antiparallel (paired) spins differ in their total spin angular momentum. In the paired case, the two spin momenta cancel, and there is zero net spin (as depicted in Fig. 8C.2(a)). Its state is the one denoted σ_- in the discussion of the Pauli principle (Topic 8B):

$$\sigma_-(1,2) = \left(\frac{1}{2^{1/2}}\right)\{\alpha(1)\beta(2) - \beta(1)\alpha(2)\} \qquad (8C.3a)$$

The angular momenta of two parallel spins add to give a non-zero total spin. As illustrated in Fig. 8C.2(b), there are three ways of achieving non-zero total spin. The three spin states are the symmetric combinations introduced in Topic 8B:

$$\alpha(1)\alpha(2)$$

$$\sigma_+(1,2) = \left(\frac{1}{2^{1/2}}\right)\{\alpha(1)\beta(2) + \beta(1)\alpha(2)\} \qquad (8C.3b)$$

$$\beta(1)\beta(2)$$

The state of the He atom in which the two electrons are paired and their spins are described by eqn 8C.3a gives rise to a **singlet term**. The alternative arrangement, in which the spins are parallel and are described by any of the three expressions in eqn 8C.3b, gives rise to a **triplet term**. The fact that the parallel

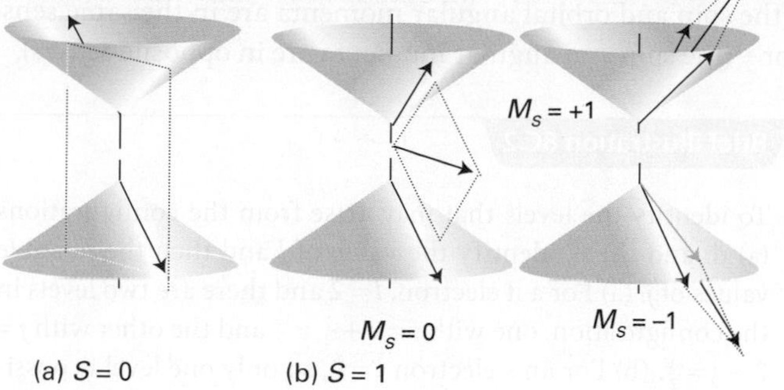

Figure 8C.2 (a) Electrons with paired spins have zero resultant spin angular momentum ($S = 0$). They can be represented by two vectors that lie at an indeterminate position on the cones shown here, but wherever one lies on its cone, the other points in the opposite direction; their resultant is zero. (b) When two electrons have parallel spins, they have a nonzero total spin angular momentum ($S = 1$). There are three ways of achieving this resultant, which are shown by these vector representations. The red vectors show the total spin angular momentum. Note that, whereas two paired spins are precisely antiparallel, two 'parallel' spins are not strictly parallel. The notation S, M_S is explained later.

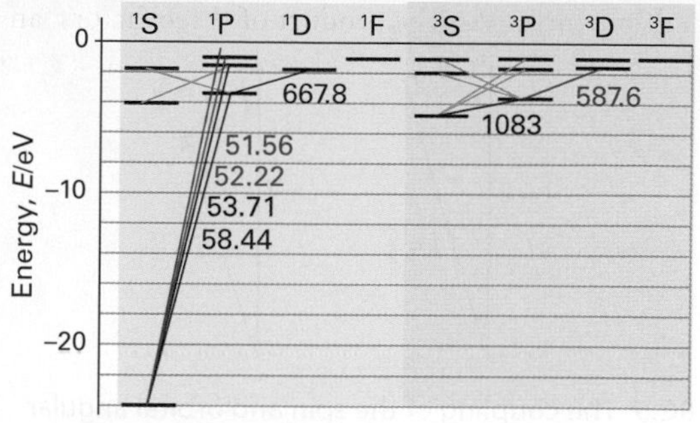

Figure 8C.3 Some of the transitions responsible for the spectrum of atomic helium. The labels give the wavelengths (in nanometres) of the transitions.

arrangement of spins in the triplet term of the $1s^12s^1$ configuration of the He atom lies lower in energy than the antiparallel arrangement, the singlet term, can now be expressed by saying that the triplet term of the $1s^12s^1$ configuration of He lies lower in energy than the singlet term. This is a general conclusion and applies to other atoms (and molecules):

> For states arising from the same configuration, the triplet term generally lies lower than the singlet term.

The origin of the energy difference lies in the effect of spin correlation on the Coulombic interactions between electrons, as in the case of Hund's maximum multiplicity rule for ground-state configurations (Topic 8B): electrons with parallel spins tend to avoid each other. Because the Coulombic interaction between electrons in an atom is strong, the difference in energies between singlet and triplet terms of the same configuration can be large. The singlet and triplet terms of the configuration $1s^12s^1$ of He, for instance, differ by $6421\ cm^{-1}$ (corresponding to 0.80 eV).

The spectrum of atomic helium is more complicated than that of atomic hydrogen, but there are two simplifying features. One is that the only excited configurations to consider are of the form $1s^1nl^1$; that is, only one electron is excited. Excitation of two electrons requires an energy greater than the ionization energy of the atom, so the He^+ ion is formed instead of the doubly excited atom. Second, and as seen later in this Topic, no radiative transitions take place between singlet and triplet terms because the relative orientation of the two electron spins cannot change during a transition. Thus, there is a spectrum arising from transitions between singlet terms (including the ground state) and between triplet terms, but not between the two. Spectroscopically, helium behaves like two distinct species. The Grotrian diagram for helium in Fig. 8C.3 shows the two sets of transitions.

(b) Spin–orbit coupling

An electron has a magnetic moment that arises from its spin. Similarly, an electron with orbital angular momentum (that is, an electron in an orbital with $l > 0$) is in effect a circulating current, and possesses a magnetic moment that arises from its orbital momentum. The interaction of the spin magnetic moment with the magnetic field arising from the orbital angular momentum is called **spin–orbit coupling**. The strength of the coupling, and its effect on the energy levels of the atom, depend on the relative orientations of the spin and orbital magnetic moments, and therefore on the relative orientations of the two angular momenta (Fig. 8C.4).

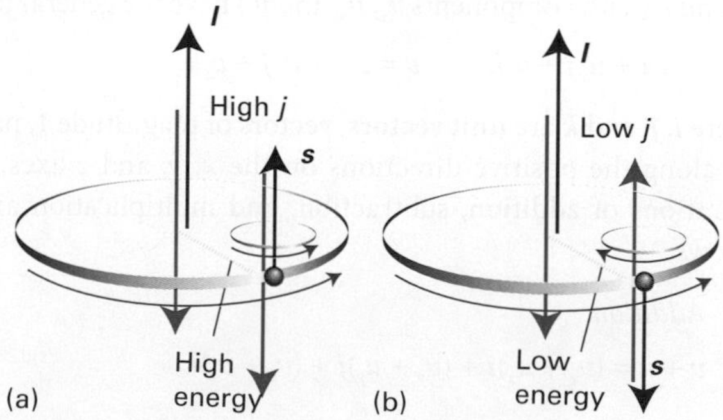

Figure 8C.4 Spin–orbit coupling is a magnetic interaction between spin and orbital magnetic moments; the black arrows show the direction of the angular momentum and the green arrows show the direction of the associated magnetic moments When the angular momenta are parallel, as in (a), the magnetic moments are aligned unfavourably; when they are opposed, as in (b), the interaction is favourable. This magnetic coupling is the cause of the splitting of a term into levels.

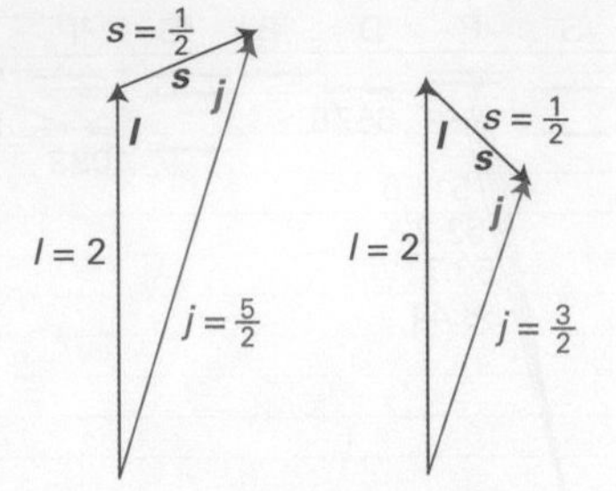

Figure 8C.5 The coupling of the spin and orbital angular momenta of a d electron ($l = 2$) gives two possible values of j depending on the relative orientations of the spin and orbital angular momenta of the electron.

One way of expressing the dependence of the spin–orbit interaction on the relative orientation of the spin and orbital momenta is to say that it depends on the total angular momentum of the electron, the vector sum of its spin and orbital momenta. Thus, when the spin and orbital angular momenta are nearly parallel, the total angular momentum is high; when the two angular momenta are opposed, the total angular momentum is low.

The total angular momentum of an electron is described by the quantum numbers j and m_j, with $j = l + \frac{1}{2}$ (when the orbital and spin angular momenta are in the same direction) or $j = l - \frac{1}{2}$ (when they are opposed; both cases are illustrated in Fig. 8C.5). The different values of j that can arise for a given value of l label the **levels** of a term. For $l = 0$, the only permitted value is $j = \frac{1}{2}$ (the total angular momentum is the same as the spin angular momentum because there is no other source of angular momentum in the atom). When $l = 1$, j may be either $\frac{3}{2}$ (the spin and orbital angular momenta are in the same sense) or $\frac{1}{2}$ (the spin and angular momenta are in opposite senses).

Brief illustration 8C.2

To identify the levels that may arise from the configurations (a) d^1 and (b) s^1, identify the value of l and then the possible values of j. (a) For a d electron, $l = 2$ and there are two levels in the configuration, one with $j = 2 + \frac{1}{2} = \frac{5}{2}$ and the other with $j = 2 - \frac{1}{2} = \frac{3}{2}$. (b) For an s electron $l = 0$, so only one level is possible, and $j = \frac{1}{2}$.

With a little work, it is possible to incorporate the effect of spin–orbit coupling on the energies of the levels.

How is that done? 8C.2 Deriving an expression for the energy of spin–orbit interaction

Classically, the energy of a magnetic moment $\boldsymbol{\mu}$ in a magnetic field $\boldsymbol{\mathcal{B}}$ is equal to their scalar product $-\boldsymbol{\mu}\cdot\boldsymbol{\mathcal{B}}$. Follow these steps to arrive at an expression for the spin–orbit interaction energy. The procedures for manipulating vectors are described in *The chemist's toolkit* 22.

Step 1 *Write an expression for the energy of interaction*

If the magnetic field arises from the orbital angular momentum of the electron, it is proportional to $\boldsymbol{l}$; if the magnetic moment $\boldsymbol{\mu}$ is that of the electron spin, then it is proportional to $\boldsymbol{s}$. It follows that the energy of interaction is proportional to the scalar product $\boldsymbol{s}\cdot\boldsymbol{l}$:

The chemist's toolkit 22 The manipulation of vectors

In three dimensions, the vectors $\boldsymbol{u}$ (with components u_x, u_y, and u_z) and $\boldsymbol{v}$ (with components v_x, v_y, and v_z) have the general form:

$$\boldsymbol{u} = u_x\boldsymbol{i} + u_y\boldsymbol{j} + u_z\boldsymbol{k} \qquad \boldsymbol{v} = v_x\boldsymbol{i} + v_y\boldsymbol{j} + v_z\boldsymbol{k}$$

where $\boldsymbol{i}$, $\boldsymbol{j}$, and $\boldsymbol{k}$ are **unit vectors**, vectors of magnitude 1, pointing along the positive directions on the x, y, and z axes. The operations of addition, subtraction, and multiplication are as follows:

1. *Addition*:

$$\boldsymbol{v} + \boldsymbol{u} = (v_x + u_x)\boldsymbol{i} + (v_y + u_y)\boldsymbol{j} + (v_z + u_z)\boldsymbol{k}$$

2. *Subtraction*:

$$\boldsymbol{v} - \boldsymbol{u} = (v_x - u_x)\boldsymbol{i} + (v_y - u_y)\boldsymbol{j} + (v_z - u_z)\boldsymbol{k}$$

3. *Multiplication*:

(a) The **scalar product**, or *dot product*, of the two vectors $\boldsymbol{u}$ and $\boldsymbol{v}$ is

$$\boldsymbol{u}\cdot\boldsymbol{v} = u_x v_x + u_y v_y + u_z v_z$$

The scalar product of a vector with itself gives the square magnitude of the vector.

$$\boldsymbol{u}\cdot\boldsymbol{u} = u_x^2 + u_y^2 + u_z^2 = u^2$$

(b) The **vector product**, or *cross product*, of two vectors is

$$\boldsymbol{u}\times\boldsymbol{v} = \begin{vmatrix} \boldsymbol{i} & \boldsymbol{j} & \boldsymbol{k} \\ u_x & u_y & u_z \\ v_x & v_y & v_z \end{vmatrix}$$

$$= (u_y v_z - u_z v_y)\boldsymbol{i} - (u_x v_z - u_z v_x)\boldsymbol{j} + (u_x v_y - u_y v_x)\boldsymbol{k}$$

(Determinants are discussed in *The chemist's toolkit* 23 in Topic 9D.) If the two vectors lie in the plane defined by the unit vectors $\boldsymbol{i}$ and $\boldsymbol{j}$, their vector product lies parallel to the unit vector $\boldsymbol{k}$.

Energy of interaction $= -\boldsymbol{\mu}\cdot\boldsymbol{\mathcal{B}} \propto \boldsymbol{s}\cdot\boldsymbol{l}$

Step 2 *Express the scalar product in terms of the magnitudes of the vectors*

Note that the total angular momentum is the vector sum of the spin and orbital momenta: $\boldsymbol{j} = \boldsymbol{l} + \boldsymbol{s}$. The magnitude of the vector $\boldsymbol{j}$ is calculated by evaluating

$$\overbrace{\boldsymbol{j}.\boldsymbol{j}}^{j^2}=(\boldsymbol{l}+\boldsymbol{s})\cdot(\boldsymbol{l}+\boldsymbol{s})=\overbrace{\boldsymbol{l}\cdot\boldsymbol{l}}^{l^2}+\overbrace{\boldsymbol{s}\cdot\boldsymbol{s}}^{s^2}+2\boldsymbol{s}\cdot\boldsymbol{l}$$

so

$$j^2 = l^2 + s^2 + 2\boldsymbol{s}\cdot\boldsymbol{l}$$

That is,

$$\boldsymbol{s}\cdot\boldsymbol{l}=\tfrac{1}{2}\{j^2-l^2-s^2\}$$

This equation is a classical result.

Step 3 *Replace the classical magnitudes by their quantum mechanical versions*

To derive the quantum mechanical version of this expression, replace all the quantities on the right with their quantum-mechanical values, which are of the form $j(j+1)\hbar^2$, etc (Topic 7F):

$$\boldsymbol{s}\cdot\boldsymbol{l}=\tfrac{1}{2}\{j(j+1)-l(l+1)-s(s+1)\}\hbar^2$$

Then, by inserting this expression into the formula for the energy of interaction ($E \propto \boldsymbol{s}\cdot\boldsymbol{l}$) and writing the constant of proportionality as $hc\tilde{A}/\hbar^2$, obtain an expression for the energy in terms of the quantum numbers and the **spin–orbit coupling constant**, $\tilde{A}$ (a wavenumber):

$$E_{l,s,j} = \tfrac{1}{2}hc\tilde{A}\{j(j+1) - l(l+1) - s(s+1)\} \quad (8C.4)$$

Spin–orbit interaction energy

Brief illustration 8C.3

The unpaired electron in the ground state of an alkali metal atom has $l = 0$, so $j = \tfrac{1}{2}$. Because the orbital angular momentum is zero in this state, the spin–orbit coupling energy is zero (as is confirmed by setting $j = s$ and $l = 0$ in eqn 8C.4). When the electron is excited to an orbital with $l = 1$, it has orbital angular momentum and can give rise to a magnetic field that interacts with its spin. In this configuration the electron can have $j = \tfrac{3}{2}$ or $j = \tfrac{1}{2}$, and the energies of these levels are

$$E_{1,1/2,3/2} = \tfrac{1}{2}hc\tilde{A}\{\tfrac{3}{2}\times\tfrac{5}{2} - 1\times 2 - \tfrac{1}{2}\times\tfrac{3}{2}\} = \tfrac{1}{2}hc\tilde{A}$$

$$E_{1,1/2,1/2} = \tfrac{1}{2}hc\tilde{A}\{\tfrac{1}{2}\times\tfrac{3}{2} - 1\times 2 - \tfrac{1}{2}\times\tfrac{3}{2}\} = -hc\tilde{A}$$

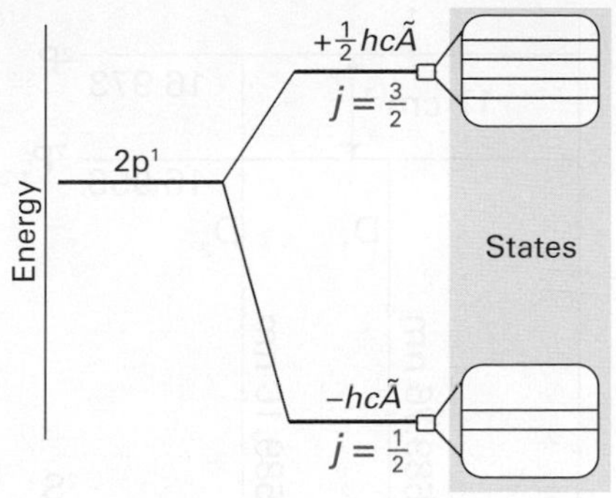

Figure 8C.6 The levels of a $2p^1$ configuration arising from spin–orbit coupling. Note that the low-j level lies below the high-j level in energy. The number of states in a level with quantum number j is $2j + 1$.

The corresponding energies are shown in Fig. 8C.6. Note that the barycentre (the 'centre of gravity') of the levels is unchanged, because there are four states of energy $\tfrac{1}{2}hc\tilde{A}$ and two of energy $-hc\tilde{A}$

The strength of the spin–orbit coupling depends on the nuclear charge. To understand why this is so, imagine riding on the orbiting electron and seeing a charged nucleus apparently orbiting around you (like the Sun rising and setting). As a result, you find yourself at the centre of a ring of current. The greater the nuclear charge, the greater is this current, and therefore the stronger is the magnetic field you detect. Because the spin magnetic moment of the electron interacts with this orbital magnetic field, it follows that the greater the nuclear charge, the stronger is the spin–orbit interaction. It turns out that the coupling increases sharply with atomic number (as Z^4) because not only is the current greater but the electron is drawn closer to the nucleus. Whereas the coupling is only weak in H (giving rise to shifts of energy levels of no more than about 0.4 cm^{-1}), in heavy atoms like Pb it is very strong (giving shifts of the order of thousands of reciprocal centimetres).

Two spectral lines are observed when the p electron of an electronically excited alkali metal atom undergoes a transition into a lower s orbital. One line is due to a transition starting in a $j = \tfrac{3}{2}$ level of the upper term and the other line is due to a transition starting in the $j = \tfrac{1}{2}$ level of the same term. The two lines are jointly an example of the **fine structure** of a spectrum, the structure due to spin–orbit coupling. Fine structure can be seen in the emission spectrum from sodium vapour excited by an electric discharge (for example, in one kind of street lighting). The yellow line at 589 nm (close to 17 000 cm^{-1}) is actually a doublet composed of one line at 589.76 nm (16 956.2 cm^{-1}) and another at 589.16 nm (16 973.4 cm^{-1}); the components of this doublet are the 'D lines' of the spectrum (Fig. 8C.7). Therefore, in Na, the spin–orbit coupling affects the energies by about 17 cm^{-1}.

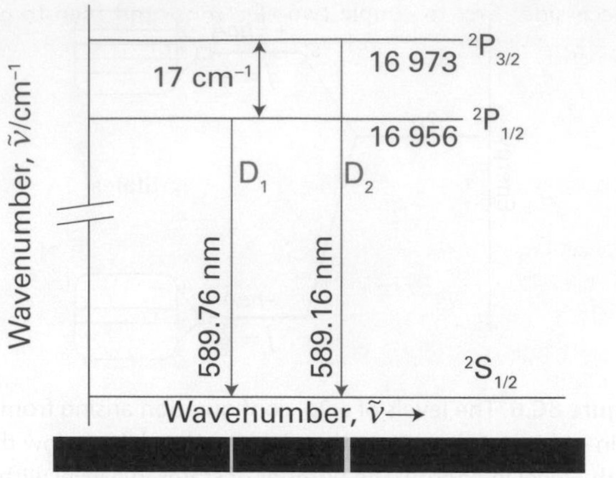

Figure 8C.7 The energy-level diagram for the formation of the sodium D lines. The splitting of the spectral lines (by 17 cm^{-1}) reflects the splitting of the levels of the ^2P term.

Example 8C.1 Analysing a spectrum for the spin–orbit coupling constant

The origin of the D lines in the spectrum of atomic sodium is shown in Fig. 8C.7. Calculate the spin–orbit coupling constant for the upper configuration of the Na atom.

Collect your thoughts It follows from Fig. 8C.7 that the splitting of the lines is equal to the energy separation of the $j = \frac{3}{2}$ and $\frac{1}{2}$ levels of the excited configuration. You need to express this separation in terms of $\tilde{A}$ by using eqn 8C.4.

The solution The two levels are split by

$$\Delta\tilde{\nu} = (E_{1,\frac{1}{2},\frac{3}{2}} - E_{1,\frac{1}{2},\frac{1}{2}})/hc = \tfrac{1}{2}\tilde{A}\{\tfrac{3}{2}(\tfrac{3}{2}+1) - \tfrac{1}{2}(\tfrac{1}{2}+1)\} = \tfrac{3}{2}\tilde{A}$$

The experimental value of $\Delta\tilde{\nu}$ is 17.2 cm^{-1}; therefore

$$\tilde{A} = \tfrac{2}{3} \times (17.2\,\text{cm}^{-1}) = 11.5\,\text{cm}^{-1}$$

Comment. The same calculation repeated for the atoms of other alkali metals gives Li: 0.23 cm^{-1}, K: 38.5 cm^{-1}, Rb: 158 cm^{-1}, Cs: 370 cm^{-1}. Note the increase of $\tilde{A}$ with atomic number (but more slowly than Z^4 for these many-electron atoms).

Self-test 8C.1 The configuration ...$4p^6 5d^1$ of rubidium has two levels at 25 700.56 cm^{-1} and 25 703.52 cm^{-1} above the ground state. What is the spin–orbit coupling constant in this excited state?

Answer: 1.18 cm^{-1}

(c) Term symbols

The discussion so far has used expressions such as 'the $j = \frac{3}{2}$ level of a doublet term with $l = 1$'. A **term symbol**, which is a symbol looking like $^2P_{3/2}$ or $^3D_2$, conveys this information, specifically the total spin, total orbital angular momentum, and total overall angular momentum, very succinctly.

A term symbol gives three pieces of information:

- The letter (P or D in the examples) indicates the total orbital angular momentum quantum number, L.
- The left superscript in the term symbol (the 2 in $^2P_{3/2}$) gives the multiplicity of the term.
- The right subscript on the term symbol (the $\frac{3}{2}$ in $^2P_{3/2}$) is the value of the total angular momentum quantum number, J, and labels the level of the term.

The meaning of these statements can be discussed in the light of the contributions to the energies summarized in Fig. 8C.8.

When several electrons are present, it is necessary to judge how their individual orbital angular momenta add together to augment or oppose each other. The **total orbital angular momentum quantum number**, L, gives the magnitude of the angular momentum through $\{L(L+1)\}^{1/2}\hbar$. It has $2L+1$ orientations distinguished by the quantum number M_L, which can take the values $0, \pm1, \ldots, \pm L$. Similar remarks apply to the **total spin quantum number**, S, and the quantum number M_S, and the **total angular momentum quantum number**, J, and the quantum number M_J.

The value of L (a non-negative integer) is obtained by coupling the individual orbital angular momenta by using the **Clebsch–Gordan series**:

$$L = l_1 + l_2, l_1 + l_2 - 1, \ldots, |l_1 - l_2| \quad \text{Clebsch–Gordan series} \quad (8C.5)$$

The modulus signs are attached to $l_1 - l_2$ to ensure that L is non-negative. The maximum value, $L = l_1 + l_2$, is obtained when the two orbital angular momenta are in the same direction; the lowest value, $|l_1 - l_2|$, is obtained when they are in opposite

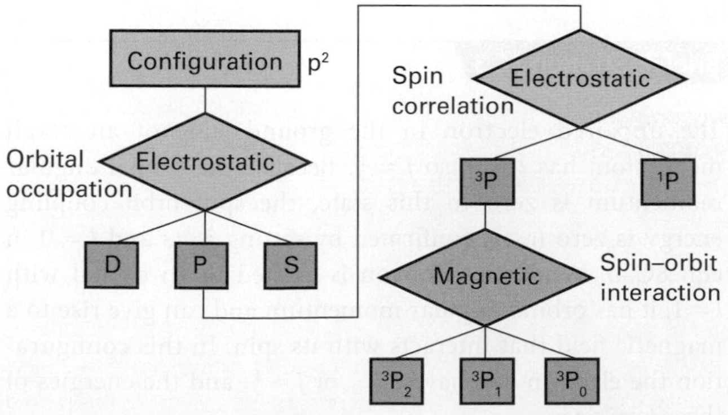

Figure 8C.8 A summary of the types of interaction that are responsible for the various kinds of splitting of energy levels in atoms. For light (low Z) atoms, magnetic interactions are small, but in heavy (high Z) atoms they may dominate the electrostatic (charge–charge) interactions.

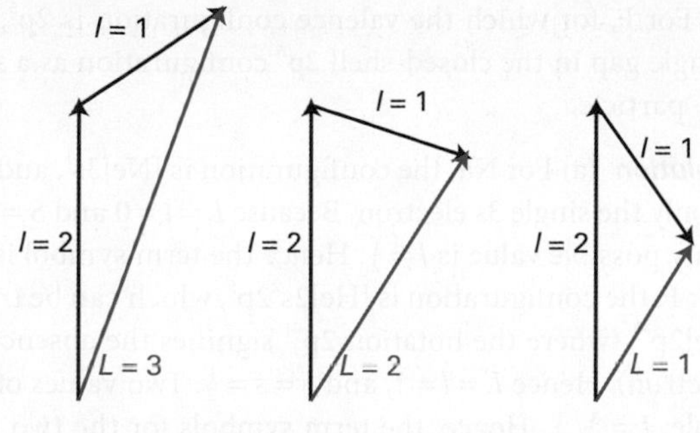

Figure 8C.9 The total orbital angular momenta of a p electron and a d electron correspond to $L = 3$, 2, and 1 and reflect the different relative orientations of the two momenta.

directions. The intermediate values represent possible intermediate relative orientations of the two momenta (Fig. 8C.9). For two p electrons (for which $l_1 = l_2 = 1$), $L = 2, 1, 0$. The code for converting the value of L into a letter is the same as for the s, p, d, f,... designation of orbitals, but uses uppercase Roman letters[2]:

L:	0	1	2	3	4	5	6...
	S	P	D	F	G	H	I...

Thus, a p^2 configuration has $L = 2, 1, 0$ and gives rise to D, P, and S terms. The terms differ in energy on account of the different spatial distribution of the electrons and the consequent differences in repulsion between them.

A note on good practice Throughout this discussion of atomic spectroscopy, distinguish italic S, the total spin quantum number, from Roman S, the term label.

A closed shell has zero orbital angular momentum because all the individual orbital angular momenta sum to zero. Therefore, when working out term symbols, only the electrons of the unfilled shell need to be considered. In the case of a single electron outside a closed shell, the value of L is the same as the value of l; so the configuration [Ne]$3s^1$ has only an S term.

Example 8C.2 Deriving the total orbital angular momentum of a configuration

Find the terms that can arise from the configurations (a) d^2, (b) p^3.

Collect your thoughts Use the Clebsch–Gordan series and begin by finding the minimum value of L (so that you know where the series terminates). When there are more than two electrons to couple together, you need to use two series in succession: first to couple two electrons, and then to couple the third to each combined state, and so on.

The solution (a) The minimum value is $|l_1 - l_2| = |2 - 2| = 0$. Therefore,

$$L = 2 + 2, 2 + 2 - 1, \ldots, 0 = 4, 3, 2, 1, 0$$

corresponding to G, F, D, P, and S terms, respectively. (b) Coupling two p electrons gives a minimum value of $|1 - 1| = 0$. Therefore,

$$L' = 1 + 1, 1 + 1 - 1, \ldots, 0 = 2, 1, 0$$

Now couple $l_3 = 1$ with $L' = 2$, to give $L = 3, 2, 1$; with $L' = 1$, to give $L = 2, 1, 0$; and with $L' = 0$, to give $L = 1$. The overall result is

$$L = 3, 2, 2, 1, 1, 1, 0$$

giving one F, two D, three P, and one S term.

Self-test 8C.2 Repeat the question for the configurations (a) f^1d^1 and (b) d^3.

Answer: (a) H, G, F, D, P; (b) I, 2H, 3G, 4F, 5D, 3P, S

When there are several electrons to be taken into account, their total spin angular momentum quantum number, S (a non-negative integer or half-integer), must be assessed. Once again the Clebsch–Gordan series is used, but now in the form

$$S = s_1 + s_2, s_1 + s_2 - 1, \ldots, |s_1 - s_2| \qquad (8C.6)$$

to decide on the value of S, noting that each electron has $s = \frac{1}{2}$. For two electrons the possible values of S are 1 and 0 (Fig. 8C.10). If there are three electrons, the total spin angular momentum is obtained by coupling the third spin to each of the values of S for the first two spins, which results in $S = \frac{3}{2}$ and $\frac{1}{2}$.

The **multiplicity** of a term is the value of $2S + 1$. When $S = 0$ (as for a closed shell, like $1s^2$) the electrons are all paired and there is no net spin: this arrangement gives a singlet term, ^{1}S.

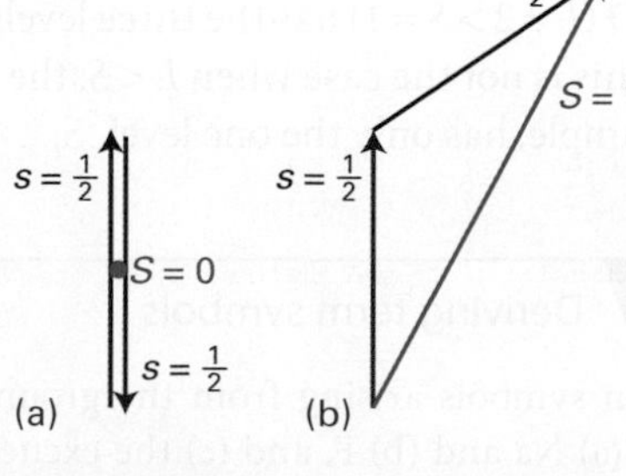

Figure 8C.10 For two electrons (each of which has $s = \frac{1}{2}$), only two total spin states are permitted ($S = 0, 1$). (a) The state with $S = 0$ can have only one value of M_S ($M_S = 0$) and gives rise to a singlet term; (b) the state with $S = 1$ can have any of three values of M_S (+1, 0, −1) and gives rise to a triplet term. The vector representations of the $S = 0$ and 1 states are shown in Fig. 8C.2.

[2] The convention of using lowercase letters to label orbitals and uppercase letters to label overall states applies throughout spectroscopy, not just to atoms.

A lone electron has $S = s = \frac{1}{2}$, so a configuration such as $[Ne]3s^1$ can give rise to a doublet term, ^2S. Likewise, the configuration $[Ne]3p^1$ is a doublet, ^2P. When there are two unpaired (parallel spin) electrons $S = 1$, so $2S + 1 = 3$, giving a triplet term, such as ^3D. The relative energies of singlets and triplets are discussed earlier in the Topic, where it is seen that their energies differ on account of spin correlation.

As already explained, the quantum number j gives the relative orientation of the spin and orbital angular momenta of a single electron. The **total angular momentum quantum number**, J (a non-negative integer or half-integer), does the same for several electrons. If there is a single electron outside a closed shell, $J = j$, with j either $l + \frac{1}{2}$ or $|l - \frac{1}{2}|$. The $[Ne]3s^1$ configuration has $j = \frac{1}{2}$ (because $l = 0$ and $s = \frac{1}{2}$), so the ^2S term has a single level, denoted $^2S_{1/2}$. The $[Ne]3p^1$ configuration has $l = 1$; therefore $j = \frac{3}{2}$ and $\frac{1}{2}$; the ^2P term therefore has two levels, $^2P_{3/2}$ and $^2P_{1/2}$. These levels lie at different energies on account of the spin–orbit interaction.

If there are several electrons outside a closed shell it is necessary to consider the coupling of all the spins and all the orbital angular momenta. This complicated problem can be simplified when the spin–orbit coupling is weak (for atoms of low atomic number), by using the **Russell–Saunders coupling** scheme. This scheme is based on the view that, if spin–orbit coupling is weak, then it is effective only when all the orbital momenta are operating cooperatively. That is, all the orbital angular momenta of the electrons couple to give a total L, and all the spins are similarly coupled to give a total S. Only at this stage do the two kinds of momenta couple through the spin–orbit interaction to give a total J. The permitted values of J are given by the Clebsch–Gordan series

$$J = L + S, L + S - 1, \ldots, |L - S| \qquad (8C.7)$$

For example, in the case of the ^3D term of the configuration $[Ne]2p^13p^1$, the permitted values of J are 3, 2, 1 (because ^3D has $L = 2$ and $S = 1$), so the term has three levels, $^3D_3$, $^3D_2$, and $^3D_1$.

When $L \geq S$, the multiplicity is equal to the number of levels. For example, a ^2P term ($L = 1 > S = \frac{1}{2}$) has the two levels $^2P_{3/2}$ and $^2P_{1/2}$, and ^3D ($L = 2 > S = 1$) has the three levels $^3D_3$, $^3D_2$, and $^3D_1$. However, this is not the case when $L < S$: the term ^2S ($L = 0 < S = \frac{1}{2}$), for example, has only the one level $^2S_{1/2}$.

Example 8C.3 Deriving term symbols

Write the term symbols arising from the ground-state configurations of (a) Na and (b) F, and (c) the excited configuration $1s^22s^22p^13p^1$ of C.

Collect your thoughts Begin by writing the configurations, but ignore inner closed shells. Then couple the orbital momenta to find L and the spins to find S. Next, couple L and S to find J. Finally, express the term as $^{2S+1}\{L\}_J$, where $\{L\}$ is the appropriate letter. For F, for which the valence configuration is $2p^5$, treat the single gap in the closed-shell $2p^6$ configuration as a single spin-$\frac{1}{2}$ particle.

The solution (a) For Na, the configuration is $[Ne]3s^1$, and consider only the single 3s electron. Because $L = l = 0$ and $S = s = \frac{1}{2}$, the only possible value is $J = \frac{1}{2}$. Hence the term symbol is $^2S_{1/2}$. (b) For F, the configuration is $[He]2s^22p^5$, which can be treated as $[Ne]2p^{-1}$ (where the notation $2p^{-1}$ signifies the absence of a 2p electron). Hence $L = l = 1$, and $S = s = \frac{1}{2}$. Two values of J are possible: $J = \frac{3}{2}, \frac{1}{2}$. Hence, the term symbols for the two levels are $^2P_{3/2}$ and $^2P_{1/2}$. (c) This is a two-electron problem, and $l_1 = l_2 = 1$, $s_1 = s_2 = \frac{1}{2}$. It follows that $L = 2, 1, 0$ and $S = 1, 0$. The terms are therefore ^3D and ^1D, ^3P and ^1P, and ^3S and ^1S. For ^3D, $L = 2$ and $S = 1$; hence $J = 3, 2, 1$ and the levels are $^3D_3$, $^3D_2$, and $^3D_1$. For ^1D, $L = 2$ and $S = 0$, so the single level is $^1D_2$. The triplet of levels of ^3P is $^3P_2$, $^3P_1$, and $^3P_0$, and the singlet is $^1P_1$. For the ^3S term there is only one level, $^3S_1$ (because $J = 1$ only), and the singlet term is $^1S_0$.

Comment. Fewer terms arise from a configuration like $\ldots 2p^2$ or $\ldots 3p^2$ than from a configuration like $\ldots 2p^13p^1$ because the Pauli exclusion principle forbids parallel arrangements of spins when two electrons occupy the same orbital. The analysis of the terms arising in such cases requires more detail than given here.

Self-test 8C.3 Identify the terms arising from the configurations (a) $2s^12p^1$, (b) $2p^13d^1$.

Answer: (a) $^3P_2$, $^3P_1$, $^3P_0$, $^1P_1$; (b) $^3F_4$, $^3F_3$, $^3F_2$, $^1F_3$, $^3D_3$, $^3D_2$, $^3D_1$, $^1D_2$, $^3P_2$, $^3P_1$, $^3P_0$, $^1P_1$

Russell–Saunders coupling fails when the spin–orbit coupling is large (in heavy atoms, those with high Z). In that case, the individual spin and orbital momenta of the electrons are coupled into individual j values; then these momenta are combined into a grand total, J, given by a Clebsch–Gordan series. This scheme is called ***jj*-coupling**. For example, in a p^2 configuration, the individual values of j are $\frac{3}{2}$ and $\frac{1}{2}$ for each electron. If the spin and the orbital angular momentum of each electron are coupled together strongly, it is best to consider each electron as a particle with angular momentum $j = \frac{3}{2}$ or $\frac{1}{2}$. These individual total momenta then couple as follows:

j_1	j_2	J
$\frac{3}{2}$	$\frac{3}{2}$	3, 2, 1, 0
$\frac{3}{2}$	$\frac{1}{2}$	2, 1
$\frac{1}{2}$	$\frac{3}{2}$	2, 1
$\frac{1}{2}$	$\frac{1}{2}$	1, 0

For heavy atoms, in which *jj*-coupling is appropriate, it is best to discuss their energies by using these quantum numbers.

Although *jj*-coupling should be used for assessing the energies of heavy atoms, the term symbols derived from Russell–Saunders coupling can still be used as labels. To see why this

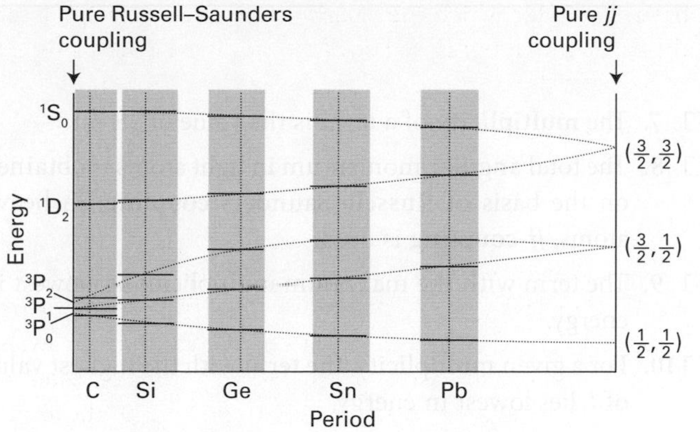

Figure 8C.11 The correlation diagram for some of the states of a two-electron system. All atoms lie between the two extremes, but the heavier the atom, the closer it lies to the pure *jj*-coupling case.

procedure is valid, it is useful to examine how the energies of the atomic states change as the spin–orbit coupling increases in strength. Such a **correlation diagram** is shown in Fig. 8C.11. It shows that there is a correspondence between the low spin–orbit coupling (Russell–Saunders coupling) and high spin–orbit coupling (*jj*-coupling) schemes, so the labels derived by using the Russell–Saunders scheme can be used to label the states of the *jj*-coupling scheme.

(d) Hund's rules

As already remarked, the terms arising from a given configuration differ in energy because they represent different relative orientations of the angular momenta of the electrons and therefore different spatial distributions. The terms arising from the ground-state configuration of an atom (and less reliably from other configurations) can be put into the order of increasing energy by using **Hund's rules**, which summarize the preceding discussion:

1. For a given configuration, the term of greatest multiplicity lies lowest in energy.

As discussed in Topic 8B, this rule is a consequence of spin correlation, the quantum-mechanical tendency of electrons with parallel spins to stay apart from one another.

2. For a given multiplicity, the term with the highest value of L lies lowest in energy.

This rule can be explained classically by noting that two electrons have a high orbital angular momentum if they circulate in the same direction, in which case they can stay apart. If they circulate in opposite directions, they meet. Thus, a D term is expected to lie lower in energy than an S term of the same multiplicity.

3. For atoms with less than half-filled shells, the level with the lowest value of J lies lowest in energy; for more than half-filled shells, the highest value of J lies lowest.

This rule arises from considerations of spin–orbit coupling. Thus, for a state of low J, the orbital and spin angular momenta lie in opposite directions, and so too do the corresponding magnetic moments. In classical terms the magnetic moments are then antiparallel, with the N pole of one close to the S pole of the other, which is a low-energy arrangement.

(e) Selection rules

Any state of the atom, and any spectral transition, can be specified by using term symbols. For example, the transitions giving rise to the yellow sodium doublet (which are shown in Fig. 8C.7) are

$$3p^1\ {}^2P_{3/2} \rightarrow 3s^1\ {}^2S_{1/2} \qquad 3p^1\ {}^2P_{1/2} \rightarrow 3s^1\ {}^2S_{1/2}$$

By convention, the upper term precedes the lower. The corresponding absorptions are therefore denoted ${}^2P_{3/2} \leftarrow {}^2S_{1/2}$ and ${}^2P_{1/2} \leftarrow {}^2S_{1/2}$. (The configurations have been omitted.)

As seen in Section 8C.1, selection rules arise from the conservation of angular momentum during a transition and from the fact that a photon has a spin of 1. They can therefore be expressed in terms of the term symbols, because the latter carry information about angular momentum. A detailed analysis leads to the following rules:

$$\Delta S = 0,$$
$$\Delta L = 0, \pm 1, \Delta l = \pm 1,$$
$$\Delta J = 0, \pm 1 \text{ but } J = 0 \not\leftrightarrow J = 0 \qquad \text{Selection rules for atoms} \qquad (8C.8)$$

where the symbol $\not\leftrightarrow$ denotes a forbidden transition. The rule about ΔS (no change of overall spin) stems from the fact that electromagnetic radiation does not affect the spin directly. The rules about ΔL and Δl express the fact that the orbital angular momentum of an individual electron must change (so $\Delta l = \pm 1$), but whether or not this results in an overall change of orbital momentum depends on the coupling.

The selection rules given above apply when Russell–Saunders coupling is valid (in light atoms, those of low Z). If labelling the terms of heavy atoms with symbols like ${}^3D$, then the selection rules progressively fail as the atomic number increases because the quantum numbers S and L become ill defined as *jj*-coupling becomes more appropriate. As explained above, Russell–Saunders term symbols are only a convenient way of labelling the terms of heavy atoms: they do not bear any direct relation to the actual angular momenta of the electrons in a heavy atom. For this reason, transitions between singlet and triplet states (for which $\Delta S = \pm 1$), while forbidden in light atoms, are allowed in heavy atoms.

Checklist of concepts

☐ 1. Two electrons with paired spins in a configuration give rise to a **singlet term**; if their spins are parallel, they give rise to a **triplet term**.

☐ 2. The orbital and spin angular momenta interact magnetically.

☐ 3. **Spin–orbit coupling** results in the levels of a term having different energies.

☐ 4. Fine structure in a spectrum is due to transitions to different levels of a term.

☐ 5. A **term symbol** specifies the angular momentum states of an atom.

☐ 6. Angular momenta are combined into a resultant by using the **Clebsch–Gordan series**.

☐ 7. The **multiplicity** of a term is the value of $2S + 1$.

☐ 8. The total angular momentum in light atoms is obtained on the basis of **Russell–Saunders coupling**; in heavy atoms, ***jj*-coupling** is used.

☐ 9. The term with the maximum multiplicity lies lowest in energy.

☐ 10. For a given multiplicity, the term with the highest value of L lies lowest in energy.

☐ 11. For atoms with less than half-filled shells, the level with the lowest value of J lies lowest in energy; for more than half-filled shells, the highest value of J lies lowest.

☐ 12. Selection rules for light atoms include the fact that changes of total spin do not occur.

Checklist of equations

Property	Equation	Comment	Equation number
Spin–orbit interaction energy	$E_{l,s,j} = \frac{1}{2}hc\tilde{A}\{j(j+1) - l(l+1) - s(s+1)\}$		8C.4
Clebsch–Gordan series	$J = j_1 + j_2, j_1 + j_2 - 1, \ldots, \lvert j_1 - j_2 \rvert$	J, j denote any kind of angular momentum	8C.5
Selection rules	$\Delta S = 0$, $\Delta L = 0, \pm1, \Delta l = \pm1$, $\Delta J = 0, \pm1$, but $J = 0 \nleftrightarrow J = 0$	Light atoms	8C.8

FOCUS 8 Atomic structure and spectra

TOPIC 8A Hydrogenic atoms

Discussion questions

D8A.1 Describe the separation of variables procedure as it is applied to simplify the description of a hydrogenic atom free to move through space.

D8A.2 List and describe the significance of the quantum numbers needed to specify the internal state of a hydrogenic atom.

D8A.3 Explain the significance of (a) a boundary surface and (b) the radial distribution function for hydrogenic orbitals.

Exercises

E8A.1(a) State the orbital degeneracy of the levels in a hydrogen atom that have energy (i) $-hc\tilde{R}_H$; (ii) $-\frac{1}{9}hc\tilde{R}_H$; (iii) $-\frac{1}{25}hc\tilde{R}_H$.
E8A.1(b) State the orbital degeneracy of the levels in a hydrogenic atom (Z in parentheses) that have energy (i) $-4hc\tilde{R}_N$, (2); (ii) $-\frac{1}{4}hc\tilde{R}_N$ (4), and (iii) $-hc\tilde{R}_N$ (5).

E8A.2(a) The wavefunction for the ground state of a hydrogen atom is Ne^{-r/a_0}. Evaluate the normalization constant N.
E8A.2(b) The wavefunction for the 2s orbital of a hydrogen atom is $N(2-r/a_0)e^{-r/2a_0}$. Evaluate the normalization constant N.

E8A.3(a) Evaluate the probability density at the nucleus of an electron with $n=2$, $l=0$, $m_l=0$.
E8A.3(b) Evaluate the probability density at the nucleus of an electron with $n=3$, $l=0$, $m_l=0$.

E8A.4(a) By differentiation of the 2s radial wavefunction, show that it has two extrema in its amplitude, and locate them.
E8A.4(b) By differentiation of the 3s radial wavefunction, show that it has three extrema in its amplitude, and locate them.

E8A.5(a) At what radius does the probability density of an electron in the H atom fall to 50 per cent of its maximum value?
E8A.5(b) At what radius in the H atom does the radial distribution function of the ground state have (i) 50 per cent, (ii) 75 per cent of its maximum value?

E8A.6(a) Locate the radial nodes in the 3s orbital of a hydrogenic atom.
E8A.6(b) Locate the radial nodes in the 4p orbital of a hydrogenic atom. You need to know that, in the notation of eqn 8A.10, $L_{4,1}(\rho) \propto 20 - 10\rho + \rho^2$, with $\rho = \frac{1}{2}Zr/a_0$.

E8A.7(a) The wavefunction of one of the d orbitals is proportional to $\cos\theta\sin\theta\cos\phi$. At what angles does it have nodal planes?
E8A.7(b) The wavefunction of one of the d orbitals is proportional to $\sin^2\theta\sin 2\phi$. At what angles does it have nodal planes?

E8A.8(a) Write down the expression for the radial distribution function of a 2s electron in a hydrogenic atom of atomic number Z and identify the radius at which it is a maximum. *Hint:* Use mathematical software.
E8A.8(b) Write down the expression for the radial distribution function of a 3s electron in a hydrogenic atom of atomic number Z and identify the radius at which the electron is most likely to be found. *Hint:* Use mathematical software.

E8A.9(a) Write down the expression for the radial distribution function of a 2p electron in a hydrogenic atom of atomic number Z and identify the radius at which the electron is most likely to be found.
E8A.9(b) Write down the expression for the radial distribution function of a 3p electron in a hydrogenic atom of atomic number Z and identify the radius at which the electron is most likely to be found. *Hint:* Use mathematical software.

E8A.10(a) What subshells and orbitals are available in the M shell?
E8A.10(b) What subshells and orbitals are available in the N shell?

E8A.11(a) What is the orbital angular momentum (as multiples of $\hbar$) of an electron in the orbitals (i) 1s, (ii) 3s, (iii) 3d? Give the numbers of angular and radial nodes in each case.
E8A.11(b) What is the orbital angular momentum (as multiples of $\hbar$) of an electron in the orbitals (i) 4d, (ii) 2p, (iii) 3p? Give the numbers of angular and radial nodes in each case.

E8A.12(a) Locate the radial nodes of each of the 2p orbitals of a hydrogenic atom of atomic number Z.
E8A.12(b) Locate the radial nodes of each of the 3d orbitals of a hydrogenic atom of atomic number Z.

Problems

P8A.1 At what point (not radius) is the probability density a maximum for the 2p electron?

P8A.2 Show by explicit integration that (a) hydrogenic 1s and 2s orbitals, (b) $2p_x$ and $2p_y$ orbitals are mutually orthogonal.

P8A.3 The value of $\tilde{R}_\infty$ is given inside the front cover and is 109 737 cm^{-1}. What is the energy of the ground state of a deuterium atom? Take $m_D = 2.013\,55m_u$.

P8A.4 Predict the ionization energy of Li^{2+} given that the ionization energy of He^+ is 54.36 eV.

P8A.5 Explicit expressions for hydrogenic orbitals are given in Tables 7F.1 (for the angular component) and 8A.1 (for the radial component). (a) Verify both that the $3p_x$ orbital is normalized (to 1) and that $3p_x$ and $3d_{xy}$ are mutually orthogonal. *Hint:* It is sufficient to show that the functions $e^{i\phi}$ and $e^{2i\phi}$ are mutually orthogonal. (b) Identify the positions of both the radial nodes and nodal planes of the 3s, $3p_x$, and $3d_{xy}$ orbitals. (c) Calculate the mean radius of the 3s orbital. *Hint:* Use mathematical software. (d) Draw a graph of the radial distribution function for the three orbitals (of part (b)) and discuss the significance of the graphs for interpreting the properties of many-electron atoms.

P8A.6 Determine whether the p_x and p_y orbitals are eigenfunctions of l_z. If not, does a linear combination exist that is an eigenfunction of l_z?

P8A.7 The 'size' of an atom is sometimes considered to be measured by the radius of a sphere within which there is a 90 per cent probability of finding the electron in the outermost occupied orbital. Calculate the 'size' of a hydrogen atom in its ground state according to this definition. Go on to explore how the 'size' varies as the definition is changed to other percentages, and plot your conclusion.

P8A.8 Some atomic properties depend on the average value of $1/r$ rather than the average value of r itself. Evaluate the expectation value of $1/r$ for (a) a hydrogenic 1s orbital, (b) a hydrogenic 2s orbital, (c) a hydrogenic 2p orbital. (d) Does $\langle 1/r\rangle = 1/\langle r\rangle$?

P8A.9 One of the most famous of the obsolete theories of the hydrogen atom was proposed by Niels Bohr. It has been replaced by quantum mechanics, but by a remarkable coincidence (not the only one where the Coulomb potential is concerned), the energies it predicts agree exactly with those obtained from the Schrödinger equation. In the Bohr atom, an electron travels in a circle around the nucleus. The Coulombic force of attraction ($Ze^2/4\pi\varepsilon_0 r^2$) is balanced by the centrifugal effect of the orbital motion. Bohr proposed that the angular momentum is limited to integral values of $\hbar$. When the two forces are balanced, the atom remains in a stationary state until it makes a spectral transition. Calculate the energies of a hydrogenic atom using the Bohr model.

P8A.10 The Bohr model of the atom is specified in Problem 8A.9. (a) What features of it are untenable according to quantum mechanics? (b) How does the ground state of the Bohr atom differ from the actual ground state? (c) Is there an experimental distinction between the Bohr and quantum mechanical models of the ground state?

P8A.11 Atomic units of length and energy may be based on the properties of a particular atom. The usual choice is that of a hydrogen atom, with the unit of length being the Bohr radius, a_0, and the unit of energy being the 'hartree', E_h, which is equal to twice the (negative of the) energy of the 1s orbital (specifically, and more precisely, $E_h = 2hc\tilde{R}_\infty$). Positronium consists of an electron and a positron (same mass, opposite charge) orbiting round their common centre of mass. If the positronium atom (e^+,e^-) were used instead, with analogous definitions of units of length and energy, what would be the relation between these two sets of atomic units?

TOPIC 8B Many-electron atoms

Discussion questions

D8B.1 Describe the orbital approximation for the wavefunction of a many-electron atom. What are the limitations of the approximation?

D8B.2 Outline the electron configurations of many-electron atoms in terms of their location in the periodic table.

D8B.3 Describe and account for the variation of first ionization energies along Period 2 of the periodic table. Would you expect the same variation in Period 3?

D8B.4 Describe the self-consistent field procedure for calculating the form of the orbitals and the energies of many-electron atoms.

Exercises

E8B.1(a) Construct the wavefunction for an excited state of the He atom with configuration $1s^12s^1$. Use $Z_{eff} = 2$ for the 1s electron and $Z_{eff} = 1$ for the 2s electron.

E8B.1(b) Construct the wavefunction for an excited state of the He atom with configuration $1s^13s^1$. Use $Z_{eff} = 2$ for the 1s electron and $Z_{eff} = 1$ for the 3s electron.

E8B.2(a) How many electrons can occupy subshells with $l = 3$?

E8B.2(b) How many electrons can occupy subshells with $l = 5$?

E8B.3(a) Write the ground-state electron configurations of the d-metals from scandium to zinc.

E8B.3(b) Write the ground-state electron configurations of the d-metals from yttrium to cadmium.

E8B.4(a) Write the electronic configuration of the Ni^{2+} ion.

E8B.4(b) Write the electronic configuration of the O^{2-} ion.

E8B.5(a) Consider the atoms of the Period 2 elements of the periodic table. Predict which element has the lowest first ionization energy.

E8B.5(b) Consider the atoms of the Period 2 elements of the periodic table. Predict which element has the lowest second ionization energy.

Problems

P8B.1 In 1976 it was mistakenly believed that the first of the 'superheavy' elements had been discovered in a sample of mica. Its atomic number was believed to be 126. What is the most probable distance of the innermost electrons from the nucleus of an atom of this element? (In such elements, relativistic effects are very important, but ignore them here.)

P8B.2 Why is the electronic configuration of the yttrium atom $[Kr]4d^15s^2$ and that of the silver atom $[Kr]4d^{10}5s^1$?

P8B.3 The d-metals iron, copper, and manganese form cations with different oxidation states. For this reason, they are found in many oxidoreductases and in several proteins of oxidative phosphorylation and photosynthesis. Explain why many d-metals form cations with different oxidation states.

P8B.4 One important function of atomic and ionic radius is in regulating the uptake of oxygen by haemoglobin, for the change in ionic radius that accompanies the conversion of Fe(II) to Fe(III) when O_2 attaches triggers a conformational change in the protein. Which do you expect to be larger: Fe^{2+} or Fe^{3+}? Why?

P8B.5 Thallium, a neurotoxin, is the heaviest member of Group 13 of the periodic table and is found most usually in the +1 oxidation state. Aluminium, which causes anaemia and dementia, is also a member of the group but its chemical properties are dominated by the +3 oxidation state. Examine this issue by plotting the first, second, and third ionization energies for the Group 13 elements against atomic number. Explain the trends you observe. *Hints*: The third ionization energy, I_3, is the minimum energy needed to remove an electron from the doubly charged cation: $E^{2+}(g) \rightarrow E^{3+}(g) + e^-(g)$, $I_3 = E(E^{3+}) - E(E^{2+})$. For data, see the links to databases of atomic properties provided in the text's website.

TOPIC 8C Atomic spectra

Discussion questions

D8C.1 Discuss the origin of the series of lines in the emission spectrum of hydrogen. What region of the electromagnetic spectrum is associated with each of the series shown in Fig. 8C.1?

D8C.2 Specify and account for the selection rules for transitions in (a) hydrogenic atoms, and (b) many-electron atoms.

D8C.3 Explain the origin of spin–orbit coupling and how it affects the appearance of a spectrum.

D8C.4 Why does the spin–orbit coupling constant depend so strongly on the atomic number?

Exercises

E8C.1(a) Identify the transition responsible for the shortest and longest wavelength lines in the Lyman series.
E8C.1(b) The Pfund series has $n_1 = 5$. Identify the transition responsible for the shortest and longest wavelength lines in the Pfund series.

E8C.2(a) Calculate the wavelength, frequency, and wavenumber of the $n = 2 \rightarrow n = 1$ transition in He^+.
E8C.2(b) Calculate the wavelength, frequency, and wavenumber of the $n = 5 \rightarrow n = 4$ transition in Li^{2+}.

E8C.3(a) Which of the following transitions are allowed in the electronic emission spectrum of a hydrogenic atom: (i) 2s → 1s, (ii) 2p → 1s, (iii) 3d → 2p?
E8C.3(b) Which of the following transitions are allowed in the electronic emission spectrum of a hydrogenic atom: (i) 5d → 2s, (ii) 5p → 3s, (iii) 6p → 4f?

E8C.4(a) Identify the levels of the configuration p^1.
E8C.4(b) Identify the levels of the configuration f^1.

E8C.5(a) What are the permitted values of j for (i) a d electron, (ii) an f electron?
E8C.5(b) What are the permitted values of j for (i) a p electron, (ii) an h electron?

E8C.6(a) An electron in two different states of an atom is known to have $j = \frac{3}{2}$ and $\frac{1}{2}$. What is its orbital angular momentum quantum number in each case?
E8C.6(b) What are the allowed total angular momentum quantum numbers of a composite system in which $j_1 = 5$ and $j_2 = 3$?

E8C.7(a) What information does the term symbol $^1D_2$ provide about the angular momentum of an atom?
E8C.7(b) What information does the term symbol $^3F_4$ provide about the angular momentum of an atom?

E8C.8(a) Suppose that an atom has (i) 2, (ii) 3 electrons in different orbitals. What are the possible values of the total spin quantum number S? What is the multiplicity in each case?
E8C.8(b) Suppose that an atom has (i) 4, (ii) 5, electrons in different orbitals. What are the possible values of the total spin quantum number S? What is the multiplicity in each case?

E8C.9(a) What are the possible values of the total spin quantum numbers S and M_S for the Ni^{2+} ion?
E8C.9(b) What are the possible values of the total spin quantum numbers S and M_S for the V^{2+} ion?

E8C.10(a) What atomic terms are possible for the electron configuration ns^1nd^1? Which term is likely to lie lowest in energy?
E8C.10(b) What atomic terms are possible for the electron configuration np^1nd^1? Which term is likely to lie lowest in energy?

E8C.11(a) What values of J may occur in the terms (i) ^1S, (ii) ^2P, (iii) ^3P? How many states (distinguished by the quantum number M_J) belong to each level?
E8C.11(b) What values of J may occur in the terms (i) ^3D, (ii) ^4D, (iii) ^2G? How many states (distinguished by the quantum number M_J) belong to each level?

E8C.12(a) Give the possible term symbols for (i) Li [He]$2s^1$, (ii) Na [Ne]$3p^1$.
E8C.12(b) Give the possible term symbols for (i) Zn [Ar]$3d^{10}4s^2$, (ii) Br [Ar]$3d^{10}4s^24p^5$.

E8C.13(a) Calculate the shifts in the energies of the two terms of a d^1 configuration that can arise from spin–orbit coupling.
E8C.13(b) Calculate the shifts in the energies of the two terms an f^1 configuration that can arise from spin–orbit coupling.

E8C.14(a) Which of the following transitions between terms are allowed in the electronic emission spectrum of a many-electron atom: (i) $^3D_2 \rightarrow {}^3P_1$, (ii) $^3P_2 \rightarrow {}^1S_0$, (iii) $^3F_4 \rightarrow {}^3D_3$?
E8C.14(b) Which of the following transitions between terms are allowed in the electronic emission spectrum of a many-electron atom: (i) $^2P_{3/2} \rightarrow {}^2S_{1/2}$, (ii) $^3P_0 \rightarrow {}^3S_1$, (iii) $^3D_3 \rightarrow {}^1P_1$?

Problems

P8C.1 The *Humphreys series* is a group of lines in the spectrum of atomic hydrogen. It begins at 12 368 nm and has been traced to 3281.4 nm. What are the transitions involved? What are the wavelengths of the intermediate transitions?

P8C.2 A series of lines involving a common level in the spectrum of atomic hydrogen lies at 656.46 nm, 486.27 nm, 434.17 nm, and 410.29 nm. What is the wavelength of the next line in the series? What is the ionization energy of the atom when it is in the lower state of the transitions?

P8C.3 The distribution of isotopes of an element may yield clues about the nuclear reactions that occur in the interior of a star. Show that it is possible to use spectroscopy to confirm the presence of both $^4He^+$ and $^3He^+$ in a star by calculating the wavenumbers of the $n = 3 \rightarrow n = 2$ and of the $n = 2 \rightarrow n = 1$ transitions for each ionic isotope.

P8C.4 The Li^{2+} ion is hydrogenic and has a Lyman series at 740 747 cm^{-1}, 877 924 cm^{-1}, 925 933 cm^{-1}, and beyond. Show that the energy levels are of the form $-hc\tilde{R}_{Li}/n^2$ and find the value of $\tilde{R}_{Li}$ for this ion. Go on to predict the wavenumbers of the two longest-wavelength transitions of the Balmer series of the ion and find its ionization energy.

P8C.5 A series of lines in the spectrum of neutral Li atoms rise from transitions between $1s^22p^1\ {}^2P$ and $1s^2nd^1\ {}^2D$ and occur at 610.36 nm, 460.29 nm, and 413.23 nm. The d orbitals are hydrogenic. It is known that the transition from the ^2P to the ^2S term (which arises from the ground-state configuration $1s^22s^1$) occurs at 670.78 nm. Calculate the ionization energy of the ground-state atom.

P8C.6‡ W.P. Wijesundera et al. (*Phys. Rev.* A **51**, 278 (1995)) attempted to determine the electron configuration of the ground state of lawrencium, element 103. The two contending configurations are [Rn]$5f^{14}7s^27p^1$ and [Rn]$5f^{14}6d7s^2$. Write down the term symbols for each of these configurations, and identify

‡These problems were supplied by Charles Trapp and Carmen Giunta.

the lowest level within each configuration. Which level would be lowest according to a simple estimate of spin–orbit coupling?

P8C.7 An emission line from K atoms is found to have two closely spaced components, one at 766.70 nm and the other at 770.11 nm. Account for this observation, and deduce what information you can.

P8C.8 Calculate the mass of the deuteron given that the first line in the Lyman series of ^{1}H lies at 82 259.098 cm^{-1} whereas that of ^{2}H lies at 82 281.476 cm^{-1}. Calculate the ratio of the ionization energies of ^{1}H and ^{2}H.

P8C.9 Positronium consists of an electron and a positron (same mass, opposite charge) orbiting round their common centre of mass. The broad features of the spectrum are therefore expected to be hydrogen-like, the differences arising largely from the mass differences. Predict the wavenumbers of the first three lines of the Balmer series of positronium. What is the binding energy of the ground state of positronium?

P8C.10 The *Zeeman effect* is the modification of an atomic spectrum by the application of a strong magnetic field. It arises from the interaction between applied magnetic fields and the magnetic moments due to orbital and spin angular momenta (recall the evidence provided for electron spin by the Stern–Gerlach experiment, Topic 8B). To gain some appreciation for the so-called *normal Zeeman effect*, which is observed in transitions involving singlet states, consider a p electron, with $l = 1$ and $m_l = 0, \pm1$. In the absence of a magnetic field, these three states are degenerate. When a field of magnitude $\mathcal{B}$ is present, the degeneracy is removed and it is observed that the state with $m_l = +1$ moves up in energy by $\mu_B\mathcal{B}$, the state with $m_l = 0$ is unchanged, and the state with $m_l = -1$ moves down in energy by $\mu_B\mathcal{B}$, where $\mu_B = e\hbar/2m_e = 9.274\times10^{-24}$ J T^{-1} is the 'Bohr magneton'. Therefore, a transition between a ^{1}S$_0$ term and a ^{1}P$_1$ term consists of three spectral lines in the presence of a magnetic field where, in the absence of the magnetic field, there is only one. (a) Calculate the splitting in reciprocal centimetres between the three spectral lines of a transition between a ^{1}S$_0$ term and a ^{1}P$_1$ term in the presence of a magnetic field of 2 T (where 1 T = 1 kg s^{-2} A^{-1}). (b) Compare the value you calculated in (a) with typical optical transition wavenumbers, such as those for the Balmer series of the H atom. Is the line splitting caused by the normal Zeeman effect relatively small or relatively large?

P8C.11 Some of the selection rules for hydrogenic atoms were derived in the text. Complete the derivation by considering the *x*- and *y*-components of the electric dipole moment operator.

P8C.12 Hydrogen is the most abundant element in all stars. However, neither absorption nor emission lines due to neutral hydrogen are found in the spectra of stars with effective temperatures higher than 25 000 K. Account for this observation.

FOCUS 8 Atomic structure and spectra

Integrated activities

I8.1 An electron in the ground-state He$^+$ ion undergoes a transition to a state specified by the quantum numbers $n = 4, l = 1, m_l = +1$. (a) Describe the transition using term symbols. (b) Calculate the wavelength, frequency, and wavenumber of the transition. (c) By how much does the mean radius of the electron change due to the transition? You need to know that the mean radius of a hydrogenic orbital is

$$r_{n,l,m_l} = \frac{n^2a_0}{Z}\left\{1+\tfrac{1}{2}\left[1-\frac{l(l+1)}{n^2}\right]\right\}$$

I8.2‡ Highly excited atoms have electrons with large principal quantum numbers. Such *Rydberg atoms* have unique properties and are of interest to astrophysicists. (a) For hydrogen atoms with large *n*, derive a relation for the separation of energy levels. (b) Calculate this separation for $n = 100$; also calculate the average radius (see the preceding activity), and the ionization energy. (c) Could a thermal collision with another hydrogen atom ionize this Rydberg atom? (d) What minimum velocity of the second atom is required? (e) Sketch the likely form of the radial wavefunction for a 100s orbital.

I8.3‡ Stern–Gerlach splittings of atomic beams are small and require either large magnetic field gradients or long magnets for their observation. For a beam of atoms with zero orbital angular momentum, such as H or Ag, the deflection is given by $x = \pm(\mu_B L^2/4E_k)\mathrm{d}\mathcal{B}/\mathrm{d}z$, where μ_B is the Bohr magneton (Problem P8C.10), *L* is the length of the magnet, E_k is the average kinetic energy of the atoms in the beam, and $\mathrm{d}\mathcal{B}/\mathrm{d}z$ is the magnetic field gradient across the beam. Calculate the magnetic field gradient required to produce a splitting of 1.00 mm in a beam of Ag atoms from an oven at 1000 K with a magnet of length 50 cm.

FOCUS 9

Molecular structure

The concepts developed in Focus 8, particularly those of orbitals, can be extended to a description of the electronic structures of molecules. There are two principal quantum mechanical theories of molecular electronic structure: 'valence-bond theory' is centred on the concept of the shared electron pair; 'molecular orbital theory' treats electrons as being distributed over all the nuclei in a molecule.

Prologue The Born–Oppenheimer approximation

The starting point for the theories discussed here and the interpretation of spectroscopic results (Focus 11) is the 'Born-Oppenheimer approximation', which separates the relative motions of nuclei and electrons in a molecule.

9A Valence-bond theory

The key concept of this Topic is the wavefunction for a shared electron pair, which is then used to account for the structures of a wide variety of molecules. The theory introduces the concepts of σ and π bonds, promotion, and hybridization, which are used widely in chemistry.

9A.1 **Diatomic molecules;** 9A.2 **Resonance;** 9A.3 **Polyatomic molecules**

9B Molecular orbital theory: the hydrogen molecule-ion

In molecular orbital theory the concept of an atomic orbital is extended to that of a 'molecular orbital', which is a wavefunction that spreads over all the atoms in a molecule. This Topic focuses on the hydrogen molecule-ion, setting the scene for the application of the theory to more complicated molecules.

9B.1 **Linear combinations of atomic orbitals;** 9B.2 **Orbital notation**

9C Molecular orbital theory: homonuclear diatomic molecules

The principles established for the hydrogen molecule-ion are extended to other homonuclear diatomic molecules and ions. The principal differences are that all the valence-shell atomic orbitals must be included and that they give rise to a more varied collection of molecular orbitals. The building-up principle for atoms is extended to the occupation of molecular orbitals and used to predict the electronic configurations of molecules and ions.

9C.1 **Electron configurations;** 9C.2 **Photoelectron spectroscopy**

9D Molecular orbital theory: heteronuclear diatomic molecules

The molecular orbital theory of heteronuclear diatomic molecules introduces the possibility that the atomic orbitals on the two atoms contribute unequally to the molecular orbital. As a result, the molecule is polar. The polarity can be expressed in terms of the concept of electronegativity. This Topic shows how quantum mechanics is used to calculate the form of a molecular orbital arising from the overlap of different atomic orbitals and its energy.

9D.1 **Polar bonds and electronegativity;** 9D.2 **The variation principle**

9E Molecular orbital theory: polyatomic molecules

Most molecules are polyatomic, so it is important to be able to account for their electronic structure. An early approach to the electronic structure of planar conjugated polyenes is the 'Hückel method', which uses severe approximations but sets the scene for more sophisticated procedures. The latter have given rise to the huge and vibrant field of computational theoretical chemistry in which elaborate computations are used to predict molecular properties. This Topic describes briefly how those calculations are formulated and displayed.

9E.1 **The Hückel approximation;** 9E.2 **Applications;** 9E.3 **Computational chemistry**

Web resources What is an application of this material?

The concepts introduced in this chapter pervade the whole of chemistry and are encountered throughout the text. Two biochemical aspects are discussed here. In *Impact* 14 simple concepts are used to account for the reactivity of small molecules that occur in organisms. *Impact* 15 provides a glimpse of the contribution of computational chemistry to the explanation of the thermodynamic and spectroscopic properties of several biologically significant molecules.

PROLOGUE The Born–Oppenheimer approximation

All theories of molecular structure make the same simplification at the outset. Whereas the Schrödinger equation for a hydrogen atom can be solved exactly, an exact solution is not possible for any molecule because even the simplest molecule consists of three particles (two nuclei and one electron). Therefore, it is common to adopt the **Born–Oppenheimer approximation** in which it is supposed that the nuclei, being so much heavier than an electron, move relatively slowly and may be treated as stationary while the electrons move in their field. That is, the nuclei are assumed to be fixed at arbitrary locations, and the Schrödinger equation is then solved for the wavefunction of the electrons alone.

To use the Born–Oppenheimer approximation for a diatomic molecule, the nuclear separation is set at a chosen value, the Schrödinger equation for the electrons is then solved and the energy calculated. Then a different separation is selected, the calculation repeated, and so on for other values of the separation. In this way the variation of the energy of the molecule with bond length is explored, and a **molecular potential energy curve** is obtained (see the illustration). It is called a potential energy curve because the kinetic energy of the stationary nuclei is zero. Once the curve has been calculated or determined experimentally (by using the spectroscopic techniques described in Focus 11), it is possible to identify the **equilibrium bond length**, R_e, the internuclear separation at the minimum of the curve, and the **bond dissociation energy**, $hc\tilde{D}_0$, which is closely related to the depth, $hc\tilde{D}_e$, of the minimum below the energy of the infinitely widely separated and stationary atoms. When more than one molecular parameter is changed in a polyatomic molecule, such as its various bond lengths and angles, a potential energy *surface* is obtained. The overall equilibrium shape of the molecule corresponds to the global minimum of the surface.

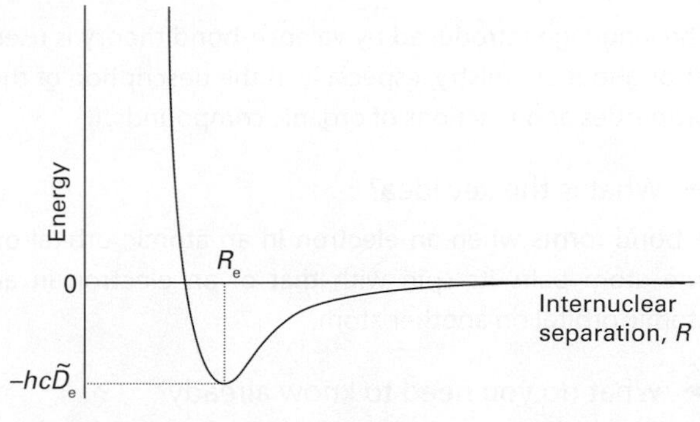

A molecular potential energy curve. The equilibrium bond length corresponds to the energy minimum.

TOPIC 9A Valence-bond theory

➤ **Why do you need to know this material?**

The language introduced by valence-bond theory is used throughout chemistry, especially in the description of the properties and reactions of organic compounds.

➤ **What is the key idea?**

A bond forms when an electron in an atomic orbital on one atom pairs its spin with that of an electron in an atomic orbital on another atom.

➤ **What do you need to know already?**

You need to know about atomic orbitals (Topic 8A) and the concepts of normalization and orthogonality (Topic 7C). This Topic also makes use of the Pauli principle (Topic 8B).

Valence-bond theory (VB theory) begins by considering the chemical bond in molecular hydrogen, H_2. The basic concepts are then applied to all diatomic and polyatomic molecules and ions.

9A.1 Diatomic molecules

The spatial wavefunction for an electron on each of two widely separated H atoms is

$$\Psi(1,2)=\psi_{\mathrm{H1s_A}}(\boldsymbol{r}_1)\psi_{\mathrm{H1s_B}}(\boldsymbol{r}_2) \qquad (9\mathrm{A}.1)$$

if electron 1 is in the H1s atomic orbital on atom A and electron 2 is in the H1s atomic orbital on atom B. For simplicity, this wavefunction will be written $\Psi(1,2)=\psi_A(1)\psi_B(2)$. When the atoms are close together, it is not possible to know whether it is electron 1 or electron 2 that is on A. An equally valid description is therefore $\Psi(1,2)=\psi_A(2)\psi_B(1)$, in which electron 2 is on A and electron 1 is on B. When two outcomes are equally probable in quantum mechanics, the true state of the system is described as a superposition of the wavefunctions for each possibility (Topic 7C), so a better description of the molecule than either wavefunction alone is one of the (unnormalized) linear combinations $\Psi(1,2)=\psi_A(1)\psi_B(2)\pm\psi_A(2)\psi_B(1)$. The combination with lower energy turns out to be the one with a + sign, so the valence-bond wavefunction of the electrons in an H_2 molecule is

$$\Psi(1,2)=\psi_A(1)\psi_B(2)+\psi_A(2)\psi_B(1) \qquad \text{A valence-bond wavefunction} \qquad (9\mathrm{A}.2)$$

The reason why this linear combination has a lower energy than either the separate atoms or the linear combination with a negative sign can be traced to the constructive interference between the wave patterns represented by the terms $\psi_A(1)\psi_B(2)$ and $\psi_A(2)\psi_B(1)$, and the resulting enhancement of the probability density of the electrons in the internuclear region (Fig. 9A.1).

Brief illustration 9A.1

The wavefunction in eqn 9A.2 might look abstract, but in fact it can be expressed in terms of simple exponential functions. Thus, if the wavefunction for an H1s orbital ($Z=1$) given in Topic 8A is used, then, with the distances r measured from their respective nuclei,

$$\begin{aligned}\Psi(1,2)&=\overbrace{\frac{1}{(\pi a_0^3)^{1/2}}\mathrm{e}^{-r_{A1}/a_0}}^{\psi_A(1)}\times\overbrace{\frac{1}{(\pi a_0^3)^{1/2}}\mathrm{e}^{-r_{B2}/a_0}}^{\psi_B(2)}\\&\quad+\overbrace{\frac{1}{(\pi a_0^3)^{1/2}}\mathrm{e}^{-r_{A2}/a_0}}^{\psi_A(2)}\times\overbrace{\frac{1}{(\pi a_0^3)^{1/2}}\mathrm{e}^{-r_{B1}/a_0}}^{\psi_B(1)}\\&=\frac{1}{\pi a_0^3}\{\mathrm{e}^{-(r_{A1}+r_{B2})/a_0}+\mathrm{e}^{-(r_{A2}+r_{B1})/a_0}\}\end{aligned}$$

where r_{A1} is the distance of electron 1 from nucleus A, etc.

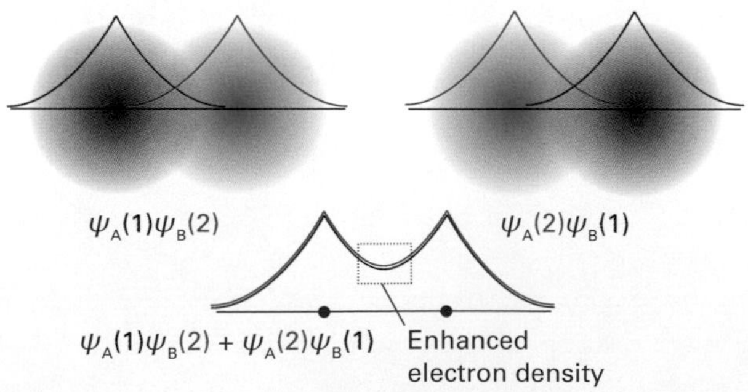

Figure 9A.1 It is very difficult to represent valence-bond wavefunctions because they refer to two electrons simultaneously. However, this illustration is an attempt. The atomic orbital for electron 1 is represented by the purple shading, and that of electron 2 is represented by the green shading. The left illustration represents $\psi_A(1)\psi_B(2)$ and the right illustration represents the contribution $\psi_A(2)\psi_B(1)$. When the two contributions are superimposed, there is interference between the purple contributions and between the green contributions, resulting in an enhanced (two-electron) density in the internuclear region.

The electron distribution described by the wavefunction in eqn 9A.2 is called a **σ bond**. A σ bond has cylindrical symmetry around the internuclear axis, and is so called because, when viewed along the internuclear axis, it resembles a pair of electrons in an s orbital (and σ is the Greek equivalent of s).

A chemist's picture of a covalent bond is one in which the spins of two electrons pair as the atomic orbitals overlap. It can be shown that the origin of the role of spin is that the wavefunction in eqn 9A.2 can be formed only by two spin-paired electrons.

How is that done? 9A.1 Establishing the origin of electron pairs in VB theory

The Pauli principle requires the overall wavefunction of two electrons, the wavefunction including spin, to change sign when the labels of the electrons are interchanged (Topic 8B). The overall VB wavefunction for two electrons is

$$\Psi(1,2) = \{\psi_A(1)\psi_B(2) + \psi_A(2)\psi_B(1)\}\sigma(1,2)$$

where σ represents the spin component of the wavefunction. When the labels 1 and 2 are interchanged, this wavefunction becomes

$$\begin{aligned}\Psi(2,1) &= \{\psi_A(2)\psi_B(1) + \psi_A(1)\psi_B(2)\}\sigma(2,1)\\ &= \{\psi_A(1)\psi_B(2) + \psi_A(2)\psi_B(1)\}\sigma(2,1)\end{aligned}$$

The Pauli principle requires that $\Psi(2,1) = -\Psi(1,2)$, which is satisfied only if $\sigma(2,1) = -\sigma(1,2)$. The combination of two spins that has this property is

$$\sigma_-(1,2) = \frac{1}{2^{1/2}}\{\alpha(1)\beta(2) - \beta(1)\alpha(2)\}$$

which corresponds to paired electron spins (Topic 8B). Therefore, the state of lower energy (and hence the formation of a chemical bond) is achieved if the electron spins are paired. Spin pairing is not an end in itself: it is a means of achieving a wavefunction, and the probability distribution it implies, that corresponds to a low energy.

The VB description of H_2 can be applied to other homonuclear diatomic molecules. The starting point for the discussion of N_2, for instance, is the valence electron configuration of each atom, which is $2s^2 2p_x^1 2p_y^1 2p_z^1$. It is conventional to take the z-axis to be the internuclear axis in a linear molecule, so each atom is imagined as having a $2p_z$ orbital pointing towards a $2p_z$ orbital on the other atom (Fig. 9A.2), with the $2p_x$ and $2p_y$ orbitals perpendicular to the axis. A σ bond is then formed by spin pairing between the two electrons in the two $2p_z$ orbitals. Its spatial wavefunction is given by eqn 9A.2, but now ψ_A and ψ_B stand for the two $2p_z$ orbitals.

The remaining N2p orbitals ($2p_x$ and $2p_y$) cannot merge to give σ bonds as they do not have cylindrical symmetry around the internuclear axis. Instead, they merge to form two 'π bonds'. A **π bond** arises from the spin pairing of electrons

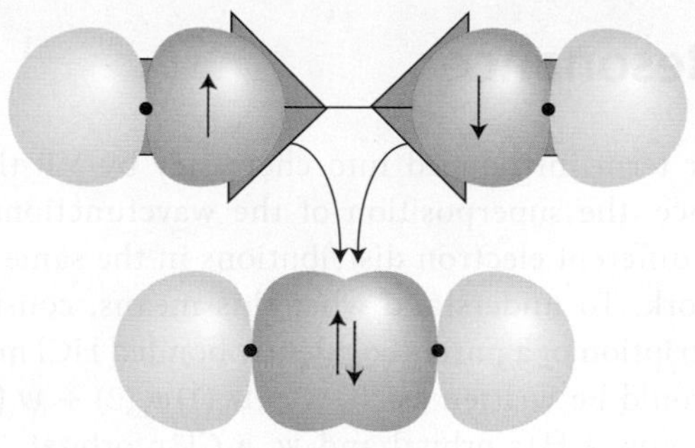

Figure 9A.2 The orbital overlap and spin pairing between electrons in two collinear p orbitals that results in the formation of a σ bond.

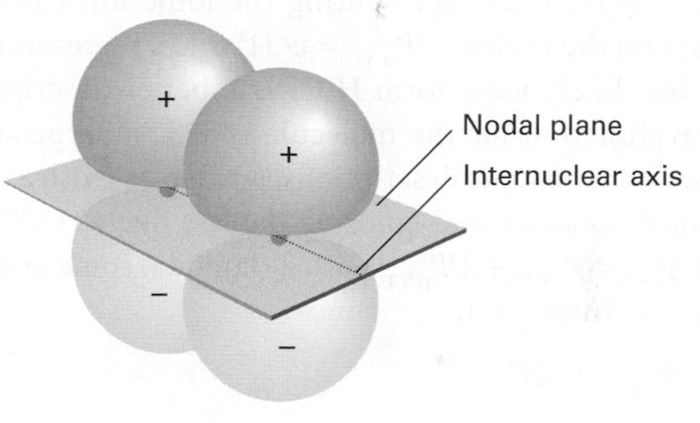

Figure 9A.3 A π bond results from orbital overlap and spin pairing between electrons in p orbitals with their axes perpendicular to the internuclear axis. The bond has two lobes of electron density separated by a nodal plane.

in two p orbitals that approach side-by-side (Fig. 9A.3). It is so called because, viewed along the internuclear axis, a π bond resembles a pair of electrons in a p orbital (and π is the Greek equivalent of p).[1]

There are two π bonds in N_2, one formed by spin pairing in two neighbouring $2p_x$ orbitals and the other by spin pairing in two neighbouring $2p_y$ orbitals. The overall bonding pattern in N_2 is therefore a σ bond plus two π bonds (Fig. 9A.4), which is consistent with the Lewis structure :N≡N: for dinitrogen.

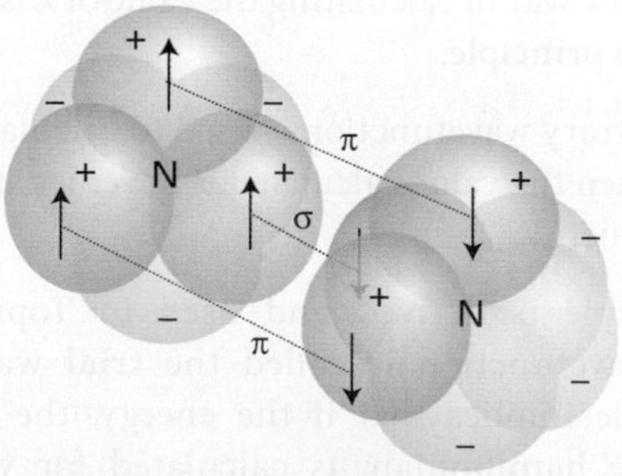

Figure 9A.4 The structure of bonds in a dinitrogen molecule, with one σ bond and two π bonds. The overall electron density has cylindrical symmetry around the internuclear axis.

[1] π bonds can also be formed from d orbitals in the appropriate orientation.

9A.2 Resonance

Another term introduced into chemistry by VB theory is **resonance**, the superposition of the wavefunctions representing different electron distributions in the same nuclear framework. To understand what this means, consider the VB description of a purely covalently bonded HCl molecule, which could be written as $\Psi_{\text{H–Cl}} = \psi_A(1)\psi_B(2) + \psi_A(2)\psi_B(1)$, with ψ_A now a H1s orbital and ψ_B a Cl3p orbital. This description allows electron 1 to be on the H atom when electron 2 is on the Cl atom, and vice versa, but it does not allow for the possibility that both electrons are on the Cl atom ($\Psi_{\text{H}^+\text{Cl}^-} = \psi_B(1)\psi_B(2)$, representing the ionic form H^+Cl^-) or both are on the H atom ($\Psi_{\text{H}^-\text{Cl}^+} = \psi_A(1)\psi_A(2)$, representing the much less likely ionic form H^-Cl^+). A better description of the wavefunction for the molecule is as a superposition of the covalent and ionic descriptions, written as (with a slightly simplified notation, and ignoring the less likely H^-Cl^+ possibility) $\Psi_{\text{HCl}} = \Psi_{\text{H–Cl}} + \lambda\Psi_{\text{H}^+\text{Cl}^-}$ with λ (lambda) some numerical coefficient. In general,

$$\Psi = \Psi_{\text{covalent}} + \lambda\Psi_{\text{ionic}} \tag{9A.3}$$

where Ψ_{covalent} is the two-electron wavefunction for the purely covalent form of the bond and Ψ_{ionic} is the two-electron wavefunction for the ionic form of the bond. In this case, where one structure is pure covalent and the other pure ionic, it is called **ionic–covalent resonance**. The interpretation of the (un-normalized) wavefunction, which is called a **resonance hybrid**, is that if the molecule is inspected, then the probability that it would be found with an ionic structure is proportional to λ^2. If $\lambda^2 \ll 1$, the covalent description is dominant. If $\lambda^2 \gg 1$, the ionic description is dominant. Resonance is not a flickering between the contributing states: it is a blending of their characteristics. It is only a mathematical device for achieving a closer approximation to the true wavefunction of the molecule than that represented by any single contributing electronic structure alone.

A systematic way of calculating the value of λ is provided by the **variation principle**:

> If an arbitrary wavefunction is used to calculate the energy, then the value calculated is never less than the true energy.

Variation principle

(This principle is derived and used in Topic 9C.) The arbitrary wavefunction is called the **trial wavefunction**. The principle implies that if the energy, the expectation value of the hamiltonian, is calculated for various trial wavefunctions with different values of the parameter λ, then the best value of λ is the one that results in the lowest energy. The ionic contribution to the resonance is then proportional to λ^2.

Brief illustration 9A.2

Consider a bond described by eqn 9A.3. If the lowest energy is reached when $\lambda = 0.1$, then the best description of the bond in the molecule is a resonance structure described by the wavefunction $\Psi = \Psi_{\text{covalent}} + 0.1\Psi_{\text{ionic}}$. This wavefunction implies that the probabilities of finding the molecule in its covalent and ionic forms are in the ratio 100:1 (because $0.1^2 = 0.01$).

9A.3 Polyatomic molecules

Each σ bond in a polyatomic molecule is formed by the spin pairing of electrons in atomic orbitals with cylindrical symmetry around the relevant internuclear axis. Likewise, π bonds are formed by pairing electrons that occupy atomic orbitals of the appropriate symmetry.

Brief illustration 9A.3

The VB description of H_2O is as follows. The valence-electron configuration of an O atom is $2s^2 2p_x^2 2p_y^1 2p_z^1$. The two unpaired electrons in the O2p orbitals can each pair with an electron in an H1s orbital, and each combination results in the formation of a σ bond (each bond has cylindrical symmetry about the respective O–H internuclear axis). Because the $2p_y$ and $2p_z$ orbitals lie at 90° to each other, the two σ bonds also lie at 90° to each other (Fig. 9A.5). Therefore, H_2O is predicted to be an angular molecule, which it is. However, the theory predicts a bond angle of 90°, whereas the actual bond angle is 104.5°.

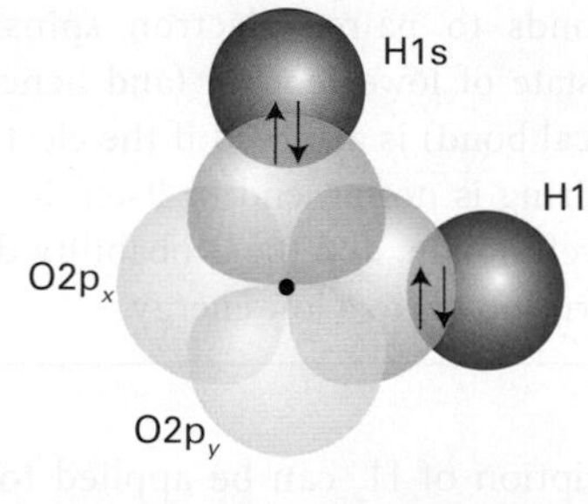

Figure 9A.5 In a primitive view of the structure of an H_2O molecule, each bond is formed by the overlap and spin pairing of an H1s electron and an O2p electron.

Resonance plays an important role in the VB description of polyatomic molecules. One of the most famous examples of resonance is in the VB description of benzene, where the wavefunction of the molecule is written as a superposition of the many-electron wavefunctions of the two covalent Kekulé structures:

$\Psi = \Psi($$) + \Psi($$)$ (9A.4)

The two contributing structures have identical energies, so they contribute equally to the superposition. The effect of resonance (which is represented by a double-headed arrow (1)), in this case, is to distribute double-bond character around the ring and to make the lengths and strengths of all the carbon–carbon bonds identical. The wavefunction is improved by allowing resonance because it allows the electrons to adjust into a distribution of lower energy. This lowering is called the **resonance stabilization** of the molecule and, in the context of VB theory, is largely responsible for the unusual stability of aromatic rings. Resonance always lowers the energy, and the lowering is greatest when the contributing structures have similar energies. The wavefunction of benzene is improved still further, and the calculated energy of the molecule is lowered still further, if ionic–covalent resonance is also considered, by allowing a small admixture of ionic structures, such as (2).

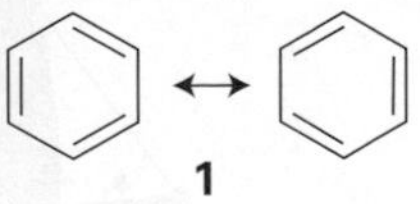

1

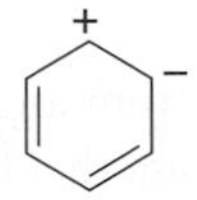

2

(a) Promotion

A deficiency of this initial formulation of VB theory is its inability to account for the common tetravalence of carbon (its ability to form four bonds). The ground-state configuration of carbon is $2s^2 2p_x^1 2p_y^1$, which suggests that a carbon atom should be capable of forming only two bonds, not four.

This deficiency is overcome by allowing for **promotion**, the excitation of an electron to an orbital of higher energy. In carbon, for example, the promotion of a 2s electron to a 2p orbital can be thought of as leading to the configuration $2s^1 2p_x^1 2p_y^1 2p_z^1$, with four unpaired electrons in separate orbitals. These electrons may pair with four electrons in orbitals provided by four other atoms (such as four H1s orbitals if the molecule is CH_4), and hence form four σ bonds. Although energy is required to promote the electron, it is more than recovered by the promoted atom's ability to form four bonds in place of the two bonds of the unpromoted atom.

Promotion, and the formation of four bonds, is a characteristic feature of carbon because the promotion energy is quite small: the promoted electron leaves a doubly occupied 2s orbital and enters a vacant 2p orbital, hence significantly relieving the electron–electron repulsion it experiences in the ground state. However, it is important to remember that promotion is not a 'real' process in which an atom somehow becomes excited and then forms bonds: it is a notional contribution to the overall energy change that occurs when bonds form.

Brief illustration 9A.4

Sulfur can form six bonds (an 'expanded octet'), as in the molecule SF_6. Because the ground-state electron configuration of sulfur is $[Ne]3s^2 3p^4$, this bonding pattern requires the promotion of a 3s electron and a 3p electron to two different 3d orbitals, which are nearby in energy, to produce the notional configuration $[Ne]3s^1 3p^3 3d^2$ with all six of the valence electrons in different orbitals and capable of bond formation with six electrons provided by six F atoms.

(b) Hybridization

The description of the bonding in CH_4 (and other alkanes) is still incomplete because it implies the presence of three σ bonds of one type (formed from H1s and C2p orbitals) and a fourth σ bond of a distinctly different character (formed from H1s and C2s). This problem is overcome by realizing that the electron density distribution in the promoted atom is equivalent to the electron density in which each electron occupies a **hybrid orbital** formed by interference between the C2s and C2p orbitals of the same atom. The origin of the hybridization can be appreciated by thinking of the four atomic orbitals centred on a nucleus as waves that interfere destructively and constructively in different regions, and give rise to four new shapes.

The specific linear combinations that give rise to four equivalent hybrid orbitals can be constructed by considering their tetrahedral arrangement.

How is that done? 9A.2 Constructing tetrahedral hybrid orbitals

Each tetrahedral bond can be regarded as directed to one corner of a unit cube (3). Suppose that each hybrid can be written in the form $h = as + b_x p_x + b_y p_y + b_z p_z$. The hybrid h_1 that points to the corner with coordinates (1,1,1) must have equal contributions from all three p orbitals, so the three b coefficients can be set equal to each other and $h_1 = as + b(p_x + p_y + p_z)$. The other three hybrids have the same composition (they are equivalent, apart from their direction in space), but are orthogonal to h_1. This orthogonality is achieved by choosing different signs for the p orbitals but the same overall composition. For instance, choosing $h_2 = as + b(-p_x - p_y + p_z)$, the orthogonality condition is

$$\int h_1 h_2 \, d\tau = \int \{as + b(p_x + p_y + p_z)\}\{as + b(-p_x - p_y + p_z)\} d\tau$$

$$= a^2 \overbrace{\int s^2 d\tau}^{1} - b^2 \overbrace{\int p_x^2 d\tau}^{1} - \cdots - ab \overbrace{\int s p_x d\tau}^{0} - \cdots - b^2 \overbrace{\int p_x p_y d\tau}^{0} + \cdots$$

$$= a^2 - b^2 - b^2 + b^2 = a^2 - b^2 = 0$$

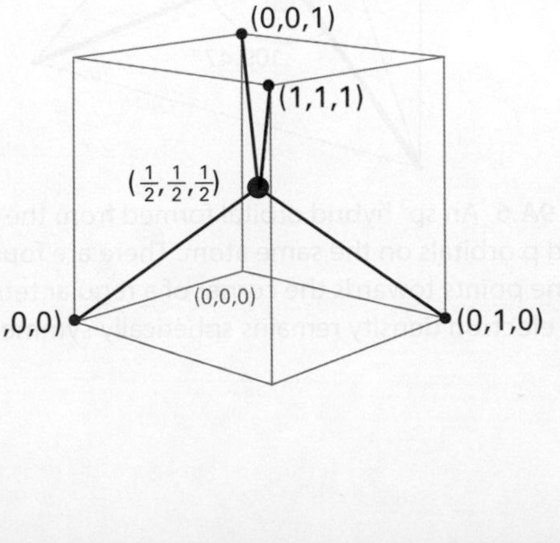

3

The values of the integrals come from the fact that the atomic orbitals are normalized and mutually orthogonal (Topic 7C). It follows that a solution is $a = b$ (the alternative solution, $a = -b$, simply corresponds to choosing different absolute phases for the p orbitals) and that the two hybrid orbitals are $h_1 = s + p_x + p_y + p_z$ and $h_2 = s - p_x - p_y + p_z$. A similar argument but with $h_3 = as + b(-p_x + p_y - p_z)$ or $h_4 = as + b(p_x - p_y - p_z)$ leads to two other hybrids. In sum,

$$h_1 = s + p_x + p_y + p_z \quad h_2 = s - p_x - p_y + p_z$$
$$h_3 = s - p_x + p_y - p_z \quad h_4 = s + p_x - p_y - p_z \qquad \text{(9A.5)}$$

sp^3 hybrid orbitals

As a result of the interference between the component orbitals, each hybrid orbital consists of a large lobe pointing in the direction of one corner of a regular tetrahedron (Fig. 9A.6). The angle between the axes of the hybrid orbitals is the tetrahedral angle, arccos$(-\frac{1}{3})$ = 109.47°. Because each hybrid is built from one s orbital and three p orbitals, it is called an **sp^3 hybrid orbital**.

It is now straightforward to see how the VB description of the CH_4 molecule leads to a tetrahedral molecule containing four equivalent C–H bonds. Each hybrid orbital of the promoted C atom contains a single unpaired electron; an H1s electron can pair with each one, giving rise to a σ bond pointing to a corner of a tetrahedron. For example, the (un-normalized) two-electron wavefunction for the bond formed by the hybrid orbital h_1 and the $1s_A$ orbital is

$$\Psi(1,2) = h_1(1)\psi_{H1s}(2) + h_1(2)\psi_{H1s}(1) \qquad \text{(9A.6)}$$

As for H_2, to achieve this wavefunction, the two electrons it describes must be paired. Because each sp^3 hybrid orbital has the same composition, all four σ bonds are identical apart from their orientation in space (Fig. 9A.7).

A hybrid orbital has enhanced amplitude in the internuclear region, which arises from the constructive interference between the s orbital and the positive lobes of the p orbitals. As a result, the bond strength is greater than for a bond formed from an s or p orbital alone. This increased bond strength is another factor that helps to repay the promotion energy.

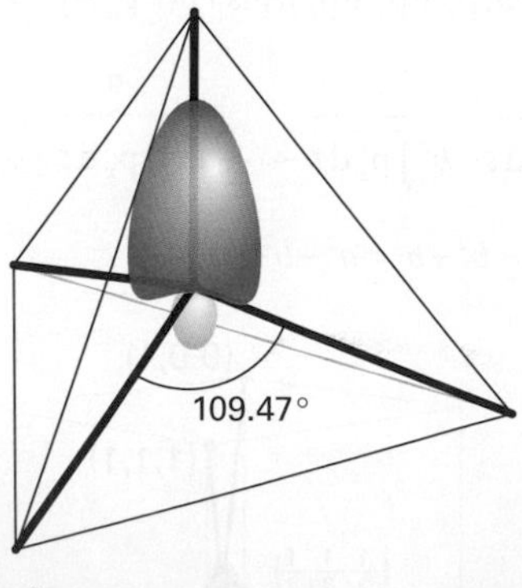

Figure 9A.6 An sp^3 hybrid orbital formed from the superposition of s and p orbitals on the same atom. There are four such hybrids: each one points towards the corner of a regular tetrahedron. The overall electron density remains spherically symmetrical.

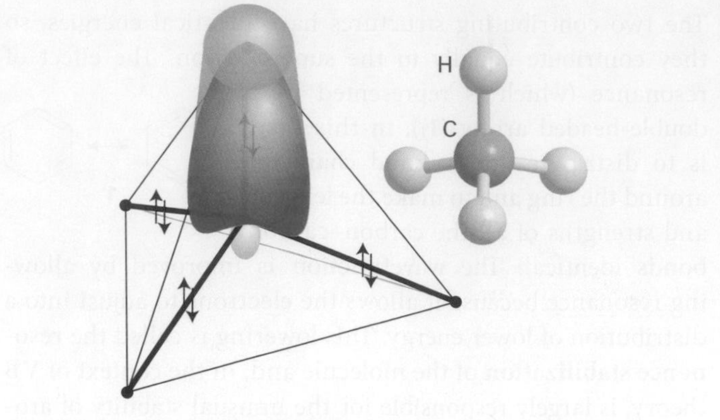

Figure 9A.7 Each sp^3 hybrid orbital forms a σ bond by overlap with an H1s orbital located at the corner of the tetrahedron. This model is consistent with the equivalence of the four bonds in CH_4.

The hybridization of N atomic orbitals always results in the formation of N hybrid orbitals, which may either form bonds or may contain **lone pairs** of electrons, pairs of electrons that do not participate directly in bond formation (but may influence the shape of the molecule).

Brief illustration 9A.5

To accommodate the observed bond angle of 104.5° in H_2O in VB theory it is necessary to suppose that the oxygen 2s and three 2p orbitals hybridize. As a first approximation, suppose they hybridize to form four equivalent sp^3 orbitals. Four electrons pair and occupy two of the hybrids, and so become lone pairs. The remaining two pair with the two electrons on the H atoms, and so form two O–H bonds at 109.5°. The actual hybridization will be slightly different to account for the observed bond angle not being exactly the tetrahedral angle.

Hybridization is also used to describe the structure of an ethene molecule, $H_2C{=}CH_2$, and the torsional rigidity of double bonds. An ethene molecule is planar, with HCH and HCC bond angles close to 120°. To reproduce the σ bonding structure, each C atom is regarded as being promoted to a $2s^12p^3$ configuration. However, instead of using all four orbitals to form hybrids, **sp^2 hybrid orbitals** are formed:

$$h_1 = s + 2^{1/2}p_y$$
$$h_2 = s + (\tfrac{3}{2})^{1/2}p_x - (\tfrac{1}{2})^{1/2}p_y \qquad \text{sp}^2\text{ hybrid orbitals} \quad \text{(9A.7)}$$
$$h_3 = s - (\tfrac{3}{2})^{1/2}p_x - (\tfrac{1}{2})^{1/2}p_y$$

These hybrids lie in a plane and point towards the corners of an equilateral triangle at 120° to each other (Fig. 9A.8). The

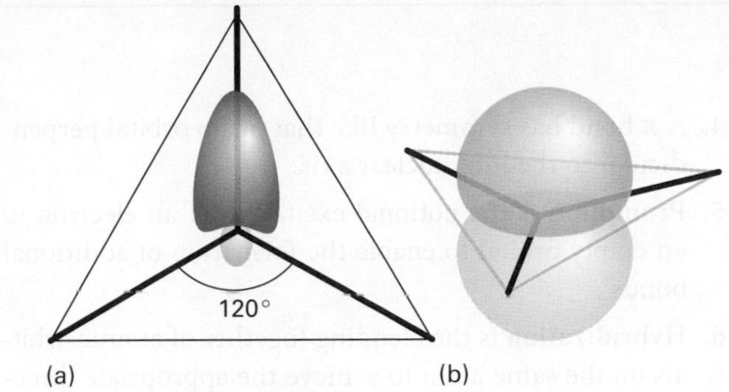

Figure 9A.8 (a) An s orbital and two p orbitals can be hybridized to form three equivalent orbitals that point towards the corners of an equilateral triangle. (b) The remaining, unhybridized p orbital is perpendicular to the plane.

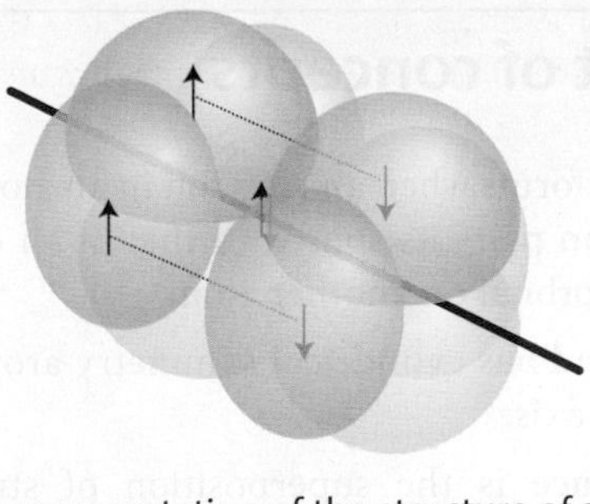

Figure 9A.10 A representation of the structure of a triple bond in ethyne; only the π bonds are shown explicitly. The overall electron density has cylindrical symmetry around the axis of the molecule.

third 2p orbital ($2p_z$) is not included in the hybridization; it lies along an axis perpendicular to the plane formed by the hybrids. The different signs of the coefficients, as well as ensuring that the hybrids are mutually orthogonal, also ensure that constructive interference takes place in different regions of space, so giving the patterns in the illustration. The sp^2-hybridized C atoms each form three σ bonds by spin pairing with either a hybrid orbital on the other C atom or with H1s orbitals. The σ framework therefore consists of C–H and C–C σ bonds at 120° to each other. When the two CH_2 groups lie in the same plane, each electron in the two unhybridized p orbitals can pair and form a π bond (Fig. 9A.9). The formation of this π bond locks the framework into the planar arrangement, because any rotation of one CH_2 group relative to the other leads to a weakening of the π bond (and consequently an increase in energy of the molecule).

A similar description applies to ethyne, HC≡CH, a linear molecule. Now the C atoms are **sp hybridized**, and the σ bonds are formed using hybrid atomic orbitals of the form

$$h_1 = s + p_z \qquad h_2 = s - p_z \qquad \text{sp hybrid orbitals} \qquad (9A.8)$$

These two hybrids lie along the internuclear axis (conventionally the z-axis in a linear molecule). The electrons in them pair either with an electron in the corresponding hybrid orbital on the other C atom or with an electron in one of the H1s orbitals. Electrons in the two remaining p orbitals on each atom, which are perpendicular to the molecular axis, pair to form two perpendicular π bonds (Fig. 9A.10).

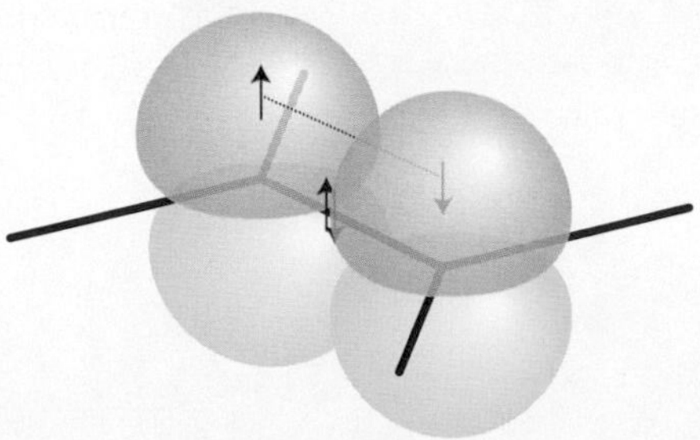

Figure 9A.9 A representation of the structure of a double bond in ethene; only the π bond is shown explicitly.

Other hybridization schemes, particularly those involving d orbitals, are often invoked in VB descriptions of molecular structure to be consistent with other molecular geometries (Table 9A.1).

Brief illustration 9A.6

Consider an octahedral molecule, such as SF_6. The promotion of sulfur's electrons as in *Brief illustration* 9A.4, followed by sp^3d^2 hybridization results in six equivalent hybrid orbitals pointing towards the corners of a regular octahedron.

Table 9A.1 Some hybridization schemes

Coordination number	Arrangement	Composition
2	Linear	sp, pd, sd
	Angular	sd
3	Trigonal planar	sp^2, p^2d
	Unsymmetrical planar	spd
	Trigonal pyramidal	pd^2
4	Tetrahedral	sp^3, sd^3
	Irregular tetrahedral	spd^2, p^3d, pd^3
	Square planar	p^2d^2, sp^2d
5	Trigonal bipyramidal	sp^3d, spd^3
	Tetragonal pyramidal	sp^2d^2, sd^4, pd^4, p^3d^2
	Pentagonal planar	p^2d^3
6	Octahedral	sp^3d^2
	Trigonal prismatic	spd^4, pd^5
	Trigonal antiprismatic	p^3d^3

Checklist of concepts

- ☐ 1. A bond forms when an electron in an atomic orbital on one atom pairs its spin with that of an electron in an atomic orbital on another atom.
- ☐ 2. A **σ bond** has cylindrical symmetry around the internuclear axis.
- ☐ 3. **Resonance** is the superposition of structures with different electron distributions but the same nuclear arrangement.
- ☐ 4. A **π bond** has symmetry like that of a p orbital perpendicular to the internuclear axis.
- ☐ 5. **Promotion** is the notional excitation of an electron to an empty orbital to enable the formation of additional bonds.
- ☐ 6. **Hybridization** is the blending together of atomic orbitals on the same atom to achieve the appropriate directional properties and enhanced overlap.

Checklist of equations

Property	Equation	Comment	Equation number
Valence-bond wavefunction	$\Psi = \psi_A(1)\psi_B(2) + \psi_A(2)\psi_B(1)$	Spins must be paired*	9A.2
Resonance	$\Psi = \Psi_{\text{covalent}} + \lambda\Psi_{\text{ionic}}$	Ionic–covalent resonance	9A.3
Hybridization	$h = as + bp + \cdots$	All atomic orbitals on the same atom; specific forms in the text	9A.5, 9A.7, and 9A.8

* The spin contribution is $\sigma\_(1,2) = \frac{1}{2^{1/2}}\{\alpha(1)\beta(2) - \beta(1)\alpha(2)\}$

TOPIC 9B Molecular orbital theory: the hydrogen molecule-ion

➤ **Why do you need to know this material?**

Molecular orbital theory is the basis of almost all descriptions of chemical bonding, in both individual molecules and solids.

➤ **What is the key idea?**

Molecular orbitals are wavefunctions that spread over all the atoms in a molecule and are commonly represented as linear combinations of atomic orbitals.

➤ **What do you need to know already?**

You need to be familiar with the shapes of atomic orbitals (Topic 8A) and how an energy is calculated from a wavefunction (Topic 7C). The entire discussion is within the framework of the Born–Oppenheimer approximation (see the *Prologue* for this Focus).

In **molecular orbital theory** (MO theory), electrons do not belong to particular bonds but spread throughout the entire molecule. This theory has been more fully developed than valence-bond theory (Topic 9A) and provides the language that is widely used in modern discussions of bonding. To introduce it, the strategy of Topic 8A is followed, where the one-electron H atom is taken as the fundamental species for discussing atomic structure and then developed into a description of many-electron atoms. This Topic uses the simplest molecular species of all, the hydrogen molecule-ion, H_2^+, to introduce the essential features of the theory, which are then used in subsequent Topics to describe the structures of more complex systems.

9B.1 Linear combinations of atomic orbitals

The hamiltonian for the single electron in H_2^+ is

$$\hat{H}=-\frac{\hbar^2}{2m_e}\nabla_1^2+V \qquad V=-\frac{e^2}{4\pi\varepsilon_0}\left(\frac{1}{r_{A1}}+\frac{1}{r_{B1}}-\frac{1}{R}\right) \qquad (9B.1)$$

where r_{A1} and r_{B1} are the distances of the electron from the two nuclei A and B (**1**) and R is the distance between the two nuclei. In the expression for V, the first two terms in parentheses are the attractive contribution from the interaction between the electron and the nuclei; the remaining term is the repulsive interaction between the nuclei. The collection of fundamental constants $e^2/4\pi\varepsilon_0$ occurs widely throughout this chapter, and is denoted j_0.

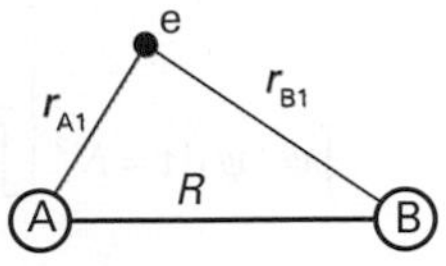

1

The one-electron wavefunctions obtained by solving the Schrödinger equation $\hat{H}\psi = E\psi$ are called **molecular orbitals**. A molecular orbital ψ gives, through the value of $|\psi|^2$, the distribution of the electron in the molecule. A molecular orbital is like an atomic orbital, but spreads throughout the molecule.

(a) The construction of linear combinations

The Schrödinger equation can be solved analytically for H_2^+ (within the Born–Oppenheimer approximation), but the wavefunctions are very complicated functions; moreover, the solution cannot be extended to polyatomic systems. The simpler procedure adopted here, while more approximate, can be extended readily to other molecules.

If an electron can be found in an atomic orbital ψ_A belonging to atom A and also in an atomic orbital ψ_B belonging to atom B, then the overall wavefunction is a superposition of the two atomic orbitals:

$$\psi_{\pm}=N_{\pm}(\psi_A \pm \psi_B) \qquad \text{Linear combination of atomic orbitals} \qquad (9B.2)$$

where, for H_2^+, ψ_A and ψ_B are 1s atomic orbitals on atom A and B, respectively, and $N_{\pm}$ is a normalization factor. The technical term for the type of superposition in eqn 9B.2 is a **linear combination of atomic orbitals** (LCAO). An approximate molecular orbital formed from a linear combination of atomic orbitals is called an **LCAO-MO**. A molecular orbital that has cylindrical symmetry around the internuclear axis, such as the one being discussed, is called a **σ orbital** because it resembles an s orbital when viewed along the axis and, more precisely, because it has zero orbital angular momentum around the internuclear axis.

Example 9B.1 Normalizing a molecular orbital

Normalize (to 1) the molecular orbital ψ_+ in eqn 9B.2.

Collect your thoughts You need to find the factor N_+ such that $\int\psi^*\psi\mathrm{d}\tau=1$, where the integration is over the whole of space. To proceed, you should substitute the LCAO into this integral and make use of the fact that the atomic orbitals are individually normalized.

The solution Substitution of the wavefunction gives

$$\int\psi^*\psi\mathrm{d}\tau=N_+^2\left\{\overbrace{\int\psi_A^2\mathrm{d}\tau}^{1}+\overbrace{\int\psi_B^2\mathrm{d}\tau}^{1}+2\overbrace{\int\psi_A\psi_B\mathrm{d}\tau}^{S}\right\}=2(1+S)N_+^2$$

where $S=\int\psi_A\psi_B\mathrm{d}\tau$ and has a value that depends on the nuclear separation (this 'overlap integral' will play a significant role later). For the integral to be equal to 1,

$$N_+=\frac{1}{\{2(1+S)\}^{1/2}}$$

For H_2^+ at its equilibrium bond length $S\approx 0.59$, so $N_+=0.56$.

Self-test 9B.1 Normalize the orbital ψ_- in eqn 9B.2 and evaluate N_- for $S=0.59$.

Answer: $N_-=1/\{2(1-S)\}^{1/2}$, so $N_-=1.10$

Figure 9B.1 shows the contours of constant amplitude for the molecular orbital ψ_+ in eqn 9B.2. Plots like these are readily obtained using commercially available software. The calculation is quite straightforward, because all that it is necessary to do is to feed in the mathematical forms of the two atomic orbitals and then let the software do the rest.

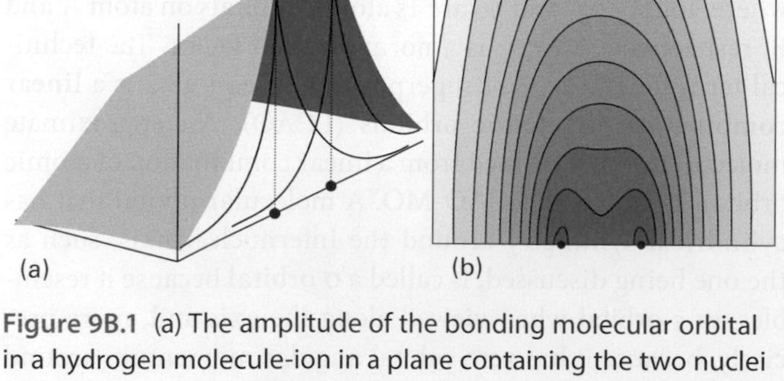

Figure 9B.1 (a) The amplitude of the bonding molecular orbital in a hydrogen molecule-ion in a plane containing the two nuclei and (b) a contour representation of the amplitude.

Brief illustration 9B.1

The surfaces of constant amplitude shown in Fig. 9B.2 have been calculated using the two H1s orbitals

$$\psi_A=\frac{1}{(\pi a_0^3)^{1/2}}\mathrm{e}^{-r_{A1}/a_0}\qquad \psi_B=\frac{1}{(\pi a_0^3)^{1/2}}\mathrm{e}^{-r_{B1}/a_0}$$

and noting that r_{A1} and r_{B1} are not independent (1). When expressed in Cartesian coordinates based on atom A (2), these radii are given by $r_{A1}=\{x^2+y^2+z^2\}^{1/2}$ and $r_{B1}=\{x^2+y^2+(z-R)^2\}^{1/2}$, where R is the bond length. A repeat of the analysis for ψ_- gives the results shown in Fig. 9B.3.

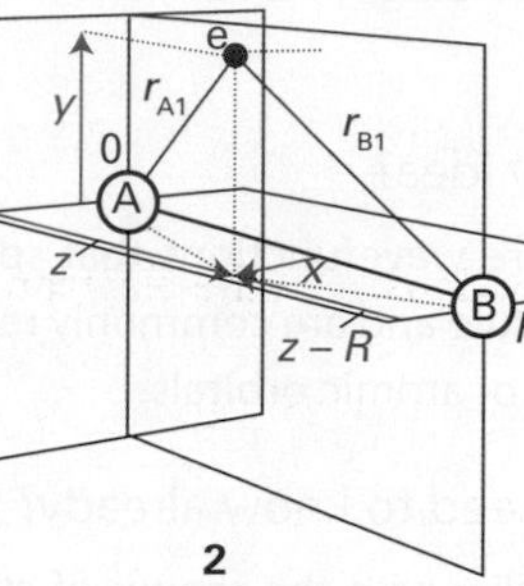

2

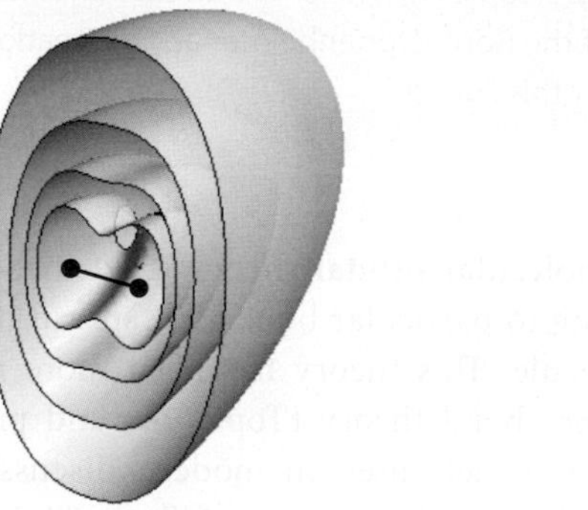

Figure 9B.2 Surfaces of constant amplitude of the wavefunction ψ_+ of the hydrogen molecule-ion.

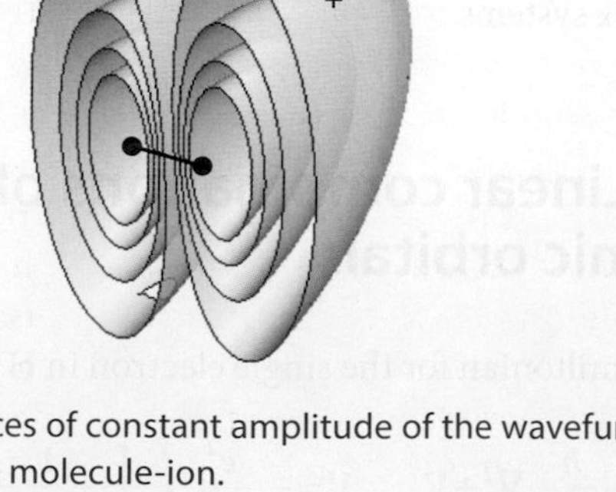

Figure 9B.3 Surfaces of constant amplitude of the wavefunction ψ_- of the hydrogen molecule-ion.

(b) Bonding orbitals

According to the Born interpretation, the probability density of the electron at each point in H_2^+ is proportional to the square modulus of its wavefunction at that point. The probability density corresponding to the (real) wavefunction ψ_+ in eqn 9B.2 is

$$\psi_+^2 \propto \psi_A^2 + \psi_B^2 + 2\psi_A\psi_B \qquad \text{Bonding probability density} \qquad (9B.3)$$

This probability density is plotted in Fig. 9B.4. An important feature becomes apparent in the internuclear region, where both atomic orbitals have similar amplitudes. According to eqn 9B.3, the total probability density is proportional to the sum of:

- ψ_A^2, the probability density if the electron were confined to atom A;
- ψ_B^2, the probability density if the electron were confined to atom B;
- $2\psi_A\psi_B$, an extra contribution to the density from both atomic orbitals.

Physical interpretation

The last contribution, the **overlap density**, is crucial, because it represents an enhancement of the probability of finding the electron in the internuclear region. The enhancement can be traced to the constructive interference of the two atomic orbitals: each has a positive amplitude in the internuclear region, so the total amplitude is greater there than if the electron were confined to a single atom. This observation is summarized as

> Bonds form as a result of the build-up of electron density where atomic orbitals overlap and interfere constructively.

The conventional explanation of this observation is based on the notion that accumulation of electron density between the nuclei puts the electron in a position where it interacts strongly with both nuclei. Hence, the energy of the molecule is lower than that of the separate atoms, where each electron can interact strongly with only one nucleus. This conventional explanation, however, has been called into question, because shifting an electron away from a nucleus into the internuclear region *raises* its potential energy. The modern (and still controversial) explanation does not emerge from the simple LCAO treatment given here. It seems that, at the same time as the electron shifts into the internuclear region, the atomic orbitals shrink. This orbital shrinkage improves the electron–nucleus attraction more than it is decreased by the migration to the internuclear region, so there is a net lowering of potential energy. The kinetic energy of the electron is also modified because the curvature of the wavefunction is changed, but the change

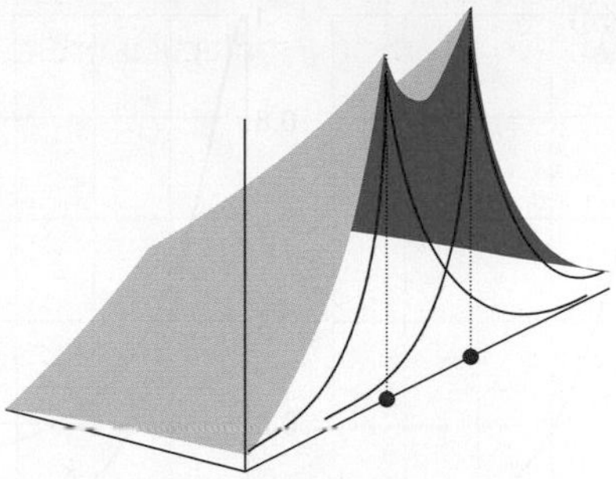

Figure 9B.4 The electron density calculated by forming the square of the wavefunction used to construct Fig. 9B.2. Note the accumulation of electron density in the internuclear region.

in kinetic energy is dominated by the change in potential energy. Throughout the following discussion the strength of chemical bonds is ascribed to the accumulation of electron density in the internuclear region. In molecules more complicated than H_2^+ the true source of energy lowering may be this accumulation of electron density or some indirect but related effect.

The σ orbital just described is an example of a **bonding orbital**, an orbital which, if occupied, helps to bind two atoms together. An electron that occupies a σ orbital is called a **σ electron**, and if that is the only electron present in the molecule (as in the ground state of H_2^+), then the configuration of the molecule is σ^1.

The energy E_σ of the σ orbital is:[1]

$$E_\sigma = E_{H1s} + \frac{j_0}{R} - \frac{j+k}{1+S} \qquad \text{Energy of bonding orbital} \qquad (9B.4)$$

where E_{H1s} is the energy of a H1s orbital, j_0/R is the potential energy of repulsion between the two nuclei (remember that j_0 is shorthand for $e^2/4\pi\varepsilon_0$), and

$$S = \int \psi_A\psi_B \, d\tau = \left\{1 + \frac{R}{a_0} + \frac{1}{3}\left(\frac{R}{a_0}\right)^2\right\} e^{-R/a_0} \qquad (9B.5a)$$

$$j = j_0 \int \frac{\psi_A^2}{r_B} \, d\tau = \frac{j_0}{R}\left\{1 - \left(1 + \frac{R}{a_0}\right) e^{-2R/a_0}\right\} \qquad (9B.5b)$$

$$k = j_0 \int \frac{\psi_A\psi_B}{r_B} \, d\tau = \frac{j_0}{a_0}\left(1 + \frac{R}{a_0}\right) e^{-R/a_0} \qquad (9B.5c)$$

Note that

$$\frac{j_0}{a_0} = \frac{e^2}{4\pi\varepsilon_0 a_0} = \frac{e^2}{4\pi\varepsilon_0} \times \frac{\pi m_e e^2}{\varepsilon_0 h^2} = \frac{m_e e^4}{4\varepsilon_0^2 h^2} = 2hc\tilde{R}_\infty \qquad (9B.5d)$$

[1] For a derivation of eqn 9B.4, see *A deeper look* 4 on the website for this text.

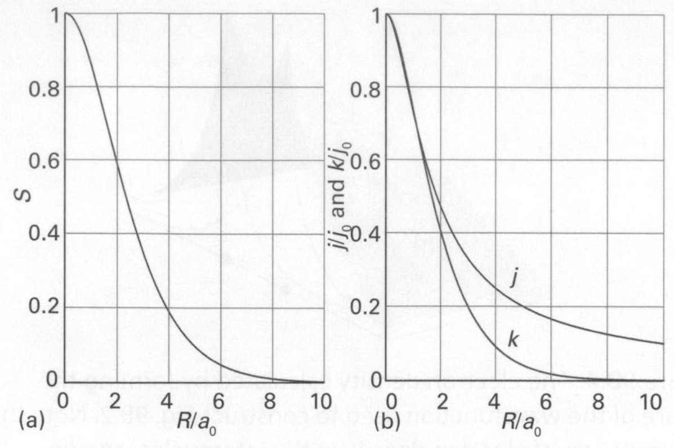

Figure 9B.5 The dependence of the integrals (a) *S*, (b) *j* and *k* on the internuclear distance, each calculated for H_2^+.

The numerical value of $2hc\tilde{R}_\infty$ (when expressed in electronvolts) is 27.21 eV. The integrals are plotted in Fig. 9B.5, and are interpreted as follows:

Physical interpretation

- All three integrals are positive and decline towards zero at large internuclear separations (*S* and *k* on account of the exponential term, *j* on account of the factor $1/R$). The integral *S* is discussed in more detail in Topic 9C.
- The integral *j* is a measure of the interaction between a nucleus and electron density centred on the other nucleus.
- The integral *k* is a measure of the interaction between a nucleus and the excess electron density in the internuclear region arising from overlap.

Brief illustration 9B.2

It turns out (see below) that the minimum value of E_σ occurs at $R = 2.49a_0$. At this separation

$$S=\left\{1+2.49+\frac{2.49^2}{3}\right\}e^{-2.49}=0.46$$

$$j=\frac{j_0/a_0}{2.49}\{1-3.49e^{-4.98}\}=0.39\,j_0/a_0$$

$$k=\frac{j_0}{a_0}(1+2.49)e^{-2.49}=0.29\,j_0/a_0$$

Therefore, with $j_0/a_0 = 27.21$ eV, $j = 10.7$ eV, and $k = 7.9$ eV. The energy separation between the bonding MO and the H1s atomic orbital (being cautious with rounding) is $E_\sigma - E_{H1s} = -1.76$ eV.

Figure 9B.6 shows a plot of E_σ against *R* relative to the energy of the separated atoms. The energy of the σ orbital decreases

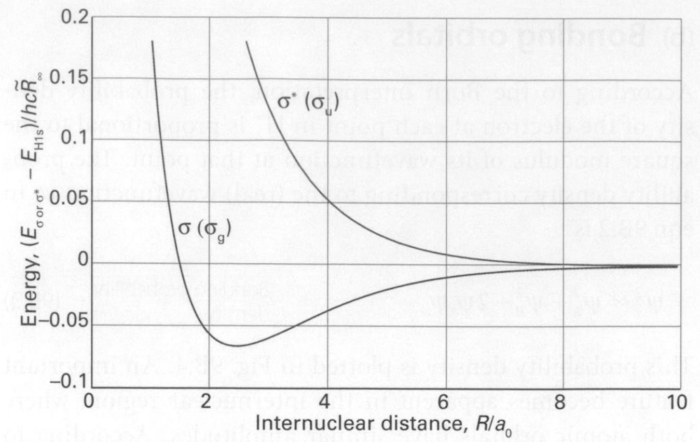

Figure 9B.6 The calculated molecular potential energy curves for a hydrogen molecule-ion showing the variation of the energies of the bonding and antibonding orbitals as the internuclear distance is changed. The energy E_σ is that of the σ orbital and E_{σ^*} is that of σ*.

as the internuclear separation is decreased from large values because electron density accumulates in the internuclear region as the constructive interference between the atomic orbitals increases (Fig. 9B.7). However, at small separations there is too little space between the nuclei for significant accumulation of electron density there. In addition, the nucleus–nucleus repulsion (which is proportional to $1/R$) becomes large. As a result, the energy of the molecular orbital rises at short distances, resulting in a minimum in the potential energy curve of depth $hc\tilde{D}_e$. Calculations on H_2^+ give $R_e = 2.49a_0 = 132$ pm and $hc\tilde{D}_e = 1.76$ eV (171 kJ mol^{-1}); the experimental values are 106 pm and 2.6 eV, so this simple LCAO-MO description of the molecule, while inaccurate, is not absurdly wrong.

(c) Antibonding orbitals

The linear combination ψ_- in eqn 9B.2 has higher energy than ψ_+, and for now it is labelled σ* because it is also a σ orbital. This orbital has a nodal plane perpendicular to the

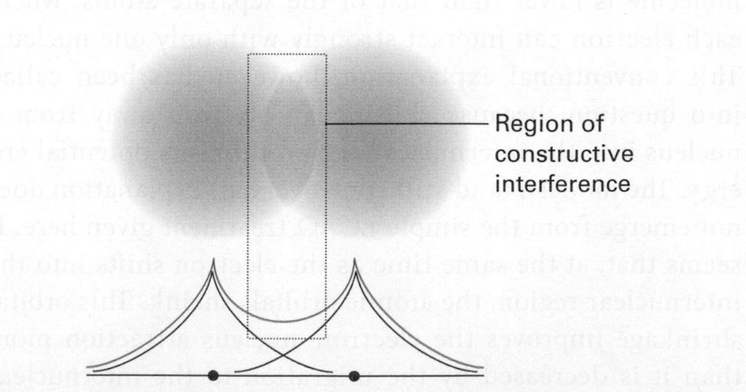

Figure 9B.7 A representation of the constructive interference that occurs when two H1s orbitals overlap and form a bonding σ orbital.

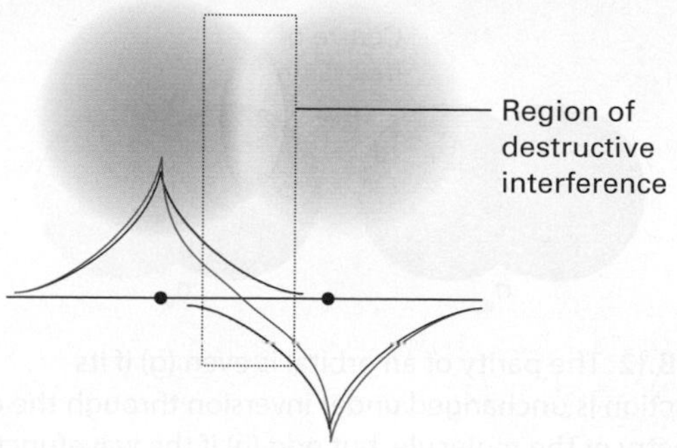

Figure 9B.8 A representation of the destructive interference that occurs when two H1s orbitals overlap and form an antibonding σ orbital.

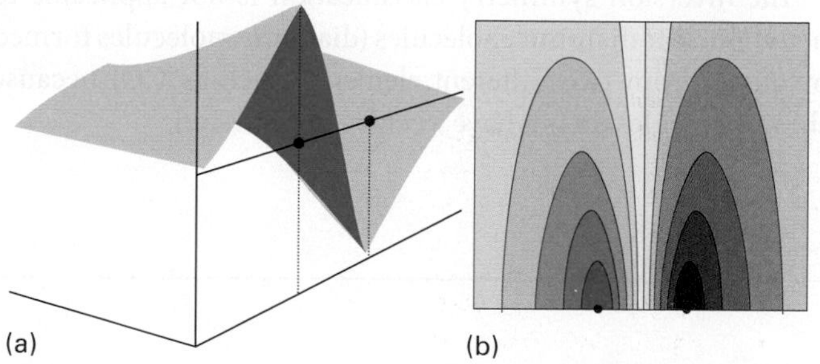

Figure 9B.9 (a) The amplitude of the antibonding molecular orbital in a hydrogen molecule-ion in a plane containing the two nuclei and (b) a contour representation of the amplitude. Note the internuclear nodal plane.

internuclear axis and passing through the mid-point of the bond where ψ_A and ψ_B cancel exactly (Figs. 9B.8 and 9B.9).

The probability density is

$$\psi_-^2 \propto \psi_A^2 + \psi_B^2 - 2\psi_A\psi_B \quad \text{Antibonding probability density} \quad (9B.6)$$

There is a reduction in probability density between the nuclei due to the term $-2\psi_A\psi_B$ (Fig. 9B.10); in physical terms, there is

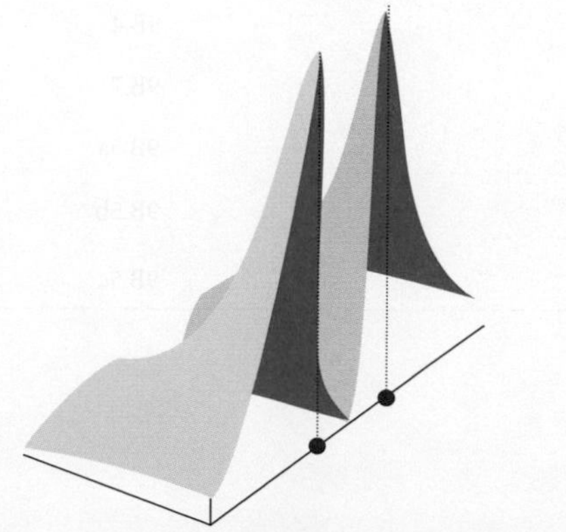

Figure 9B.10 The electron density calculated by forming the square of the wavefunction used to construct Fig. 9B.9. Note the reduction of electron density in the internuclear region.

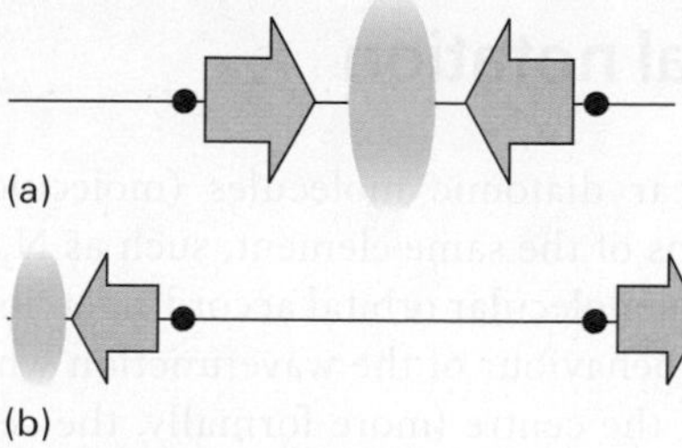

Figure 9B.11 A partial explanation of the origin of bonding and antibonding effects. (a) In a bonding orbital, the nuclei are attracted to the accumulation of electron density in the internuclear region. (b) In an antibonding orbital, the nuclei are attracted to an accumulation of electron density outside the internuclear region.

destructive interference where the two atomic orbitals overlap. The σ* orbital is an example of an **antibonding orbital**, an orbital that, if occupied, contributes to a reduction in the cohesion between two atoms and helps to raise the energy of the molecule relative to the separated atoms.

The energy E_{σ^*} of the σ* antibonding orbital is[2]

$$E_{\sigma^*} = E_{H1s} + \frac{j_0}{R} - \frac{j-k}{1-S} \quad (9B.7)$$

where the integrals S, j, and k are the same as in eqn 9B.5. The variation of E_{σ^*} with R is shown in Fig. 9B.6, which shows the destabilizing effect of an antibonding electron. The effect is partly due to the fact that an antibonding electron is excluded from the internuclear region and hence is distributed largely outside the bonding region. In effect, whereas a bonding electron pulls two nuclei together, an antibonding electron pulls the nuclei apart (Fig. 9B.11). The illustration also shows another feature drawn on later: $|E_{\sigma^*} - E_{H1s}| > |E_\sigma - E_{H1s}|$, which indicates that *the antibonding orbital is more antibonding than the bonding orbital is bonding*. This important conclusion stems in part from the presence of the nucleus–nucleus repulsion (j_0/R): this contribution raises the energy of both molecular orbitals.

Brief illustration 9B.3

At the minimum of the bonding orbital energy $R = 2.49a_0$, and, from *Brief illustration* 9B.2, $S = 0.46$, $j = 10.7$ eV, and $k = 7.9$ eV. It follows that at that separation, the energy of the antibonding orbital relative to that of a hydrogen atom 1s orbital is

$$(E_{\sigma^*} - E_{H1s})/\text{eV} = \frac{27.2}{2.49} - \frac{10.7-7.9}{1-0.46} = 5.7$$

That is, the antibonding orbital lies (5.7 + 1.76) eV = 7.5 eV above the bonding orbital at this internuclear separation.

[2] This result is obtained by applying the strategy in *A deeper look* 4 on the text's website.

9B.2 Orbital notation

For homonuclear diatomic molecules (molecules consisting of two atoms of the same element, such as N_2), it proves helpful to label a molecular orbital according to its **inversion symmetry**, the behaviour of the wavefunction when it is inverted through the centre (more formally, the centre of inversion, Topic 10A) of the molecule. Thus, any point on the bonding σ orbital that is projected through the centre of the molecule and out an equal distance on the other side leads to an identical value (and sign) of the wavefunction (Fig. 9B.12). This so-called **gerade symmetry** (from the German word for 'even') is denoted by a subscript g, as in σ_g. The same procedure applied to the antibonding σ* orbital results in the same amplitude but opposite sign of the wavefunction. This **ungerade symmetry** ('odd symmetry') is denoted by a subscript u, as in σ_u.

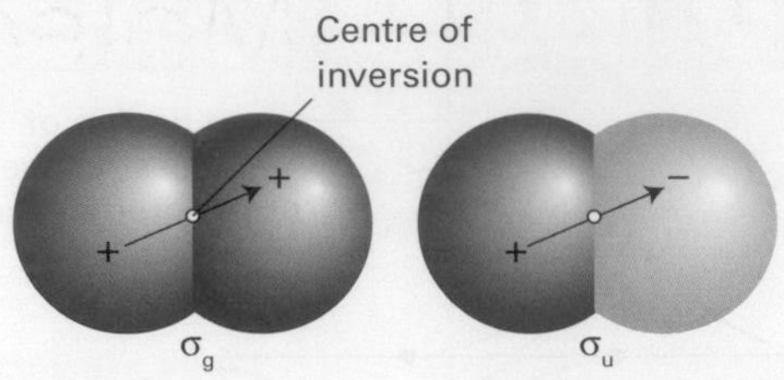

Figure 9B.12 The parity of an orbital is even (g) if its wavefunction is unchanged under inversion through the centre of symmetry of the molecule, but odd (u) if the wavefunction changes sign. Heteronuclear diatomic molecules do not have a centre of inversion, so for them the g, u classification is irrelevant.

The inversion symmetry classification is not applicable to heteronuclear diatomic molecules (diatomic molecules formed by atoms from two different elements, such as CO) because these molecules do not have a centre of inversion.

Checklist of concepts

☐ 1. A **molecular orbital** is constructed from a linear combination of atomic orbitals.

☐ 2. A **bonding orbital** arises from the constructive overlap of neighbouring atomic orbitals.

☐ 3. An **antibonding orbital** arises from the destructive overlap of neighbouring atomic orbitals.

☐ 4. **σ Orbitals** have cylindrical symmetry and zero orbital angular momentum around the internuclear axis.

☐ 5. A molecular orbital in a homonuclear diatomic molecule is labelled 'gerade' (g) or 'ungerade' (u) according to its behaviour under **inversion symmetry**.

Checklist of equations

Property	Equation	Comment	Equation number
Linear combination of atomic orbitals	$\psi_\pm = N_\pm(\psi_A \pm \psi_B)$	Homonuclear diatomic molecule	9B.2
Energies of σ orbitals formed from two 1s atomic orbitals	$E_\sigma = E_{H1s} + j_0/R - (j+k)/(1+S)$		9B.4
	$E_{\sigma^*} = E_{H1s} + j_0/R - (j-k)/(1-S)$		9B.7
Molecular integrals	$S = \int \psi_A \psi_B d\tau$,		9B.5a
	$j = j_0 \int (\psi_A^2 / r_B) d\tau$		9B.5b
	$k = j_0 \int (\psi_A \psi_B / r_B) d\tau$		9B.5c

TOPIC 9C Molecular orbital theory: homonuclear diatomic molecules

➤ Why do you need to know this material?

Almost all chemically significant molecules have more than one electron, so you need to see how to construct their electron configurations. This Topic shows how to use molecular orbital theory when more than one electron is present in a molecule.

➤ What is the key idea?

Each molecular orbital can accommodate up to two electrons, and the ground state of the molecule is the configuration of lowest energy.

➤ What do you need to know already?

You need to be familiar with the discussion of the bonding and antibonding linear combinations of atomic orbitals in Topic 9B and the building-up principle for atoms (Topic 8B).

Just as hydrogenic atomic orbitals and the building-up principle can be used as a basis for the discussion and prediction of the ground electronic configurations of many-electron atoms, the molecular orbitals for the one-electron hydrogen molecule-ion introduced in Topic 9B and a version of the building-up principle introduced in Topic 8B can be developed to account for the configurations of many-electron diatomic molecules and ions.

9C.1 Electron configurations

The starting point of the molecular orbital theory (MO theory) of bonding in diatomic molecules (and ions) is the construction of molecular orbitals as linear combinations of the available atomic orbitals. Once the molecular orbitals have been formed, a building-up principle, like that for atoms, can be used to establish their ground-state electron configurations (Topic 8B):

Building-up principle for molecules

- The electrons supplied by the atoms are accommodated in the molecular orbitals so as to achieve the lowest overall energy subject to the constraint of the Pauli exclusion principle that no more than two electrons may occupy a single orbital (and then their spins must be paired).
- If several degenerate molecular orbitals are available, electrons are added singly to each individual orbital before any one orbital is completed (because that minimizes electron–electron repulsions).
- According to Hund's maximum multiplicity rule (Topic 8B), if two electrons do occupy different degenerate orbitals, then a lower energy is obtained if their spins are parallel.

(a) σ Orbitals and π orbitals

Consider H_2, the simplest many-electron diatomic molecule. Each H atom contributes a 1s orbital (as in H_2^+), which combine to form bonding σ and antibonding σ* orbitals, as explained in Topic 9B. At the equilibrium nuclear separation these orbitals have the energies shown in Fig. 9C.1, which is called a

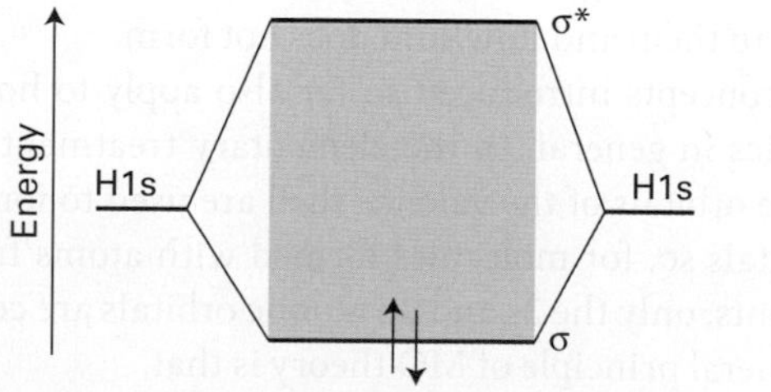

Figure 9C.1 A molecular orbital energy level diagram for orbitals constructed from the overlap of H1s orbitals. The energies of the atomic orbitals are indicated by the lines at the outer edges of the diagram, and the energies of the molecular orbitals are shown in the middle. The ground electronic configuration of H_2 is obtained by accommodating the two electrons in the lowest available orbital (the bonding orbital).

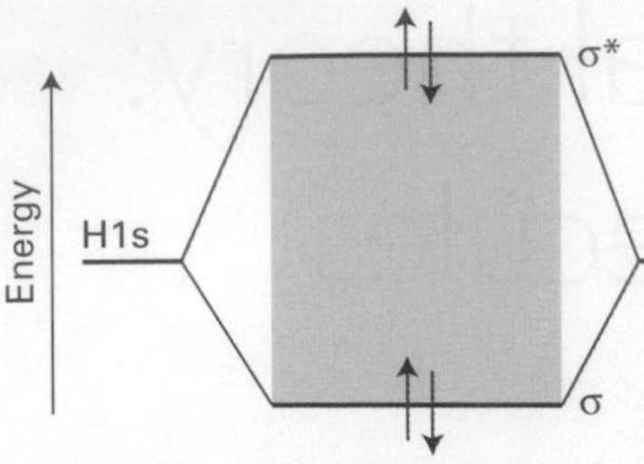

Figure 9C.2 The ground-state electronic configuration of the hypothetical four-electron molecule He_2 (at an arbitrary internuclear separation) has two bonding electrons and two antibonding electrons. It has a higher energy than the separated atoms, and so is unstable.

molecular orbital energy level diagram. Note that from two atomic orbitals two molecular orbitals are built. In general, from N atomic orbitals N molecular orbitals can be built.

There are two electrons to accommodate, and both can enter the σ orbital by pairing their spins, as required by the Pauli principle (just as for atoms, Topic 8B). The ground-state configuration is therefore σ^2 and the bond consists of an electron pair in a bonding σ orbital. This approach shows that an electron pair, which was the focus of Lewis's account of chemical bonding, represents the maximum number of electrons that can enter a bonding molecular orbital.

A straightforward extension of this argument explains why helium does not form diatomic molecules. Each He atom contributes a 1s orbital, so σ and σ* molecular orbitals can be constructed. Although these orbitals differ in detail from those in H_2, their general shapes are the same and the same qualitative energy level diagram can be used in the discussion. There are four electrons to accommodate. Two can enter the σ orbital, but then it is full, and the next two must enter the σ* orbital (Fig. 9C.2). The ground electronic configuration of He_2 is therefore $\sigma^2\sigma^{*2}$. Because σ* lies higher in energy above the separate atoms more than σ lies below them, an He_2 molecule has a higher energy than the separated atoms, so it is unstable relative to them and dihelium does not form.

The concepts introduced so far also apply to homonuclear diatomics in general. In the elementary treatment used here, only the orbitals of the valence shell are used to form molecular orbitals so, for molecules formed with atoms from Period 2 elements, only the 2s and 2p atomic orbitals are considered.

A general principle of MO theory is that

All orbitals of the appropriate symmetry contribute to a molecular orbital.

Thus, σ orbitals are built by forming linear combinations of all atomic orbitals that have cylindrical symmetry about the internuclear axis. These orbitals include the 2s orbitals on each atom and the $2p_z$ orbitals on the two atoms (Fig. 9C.3; the z-axis on each atom lies along the internuclear axis and points towards the neighbouring atom). The general form of the σ orbitals that may be formed is therefore

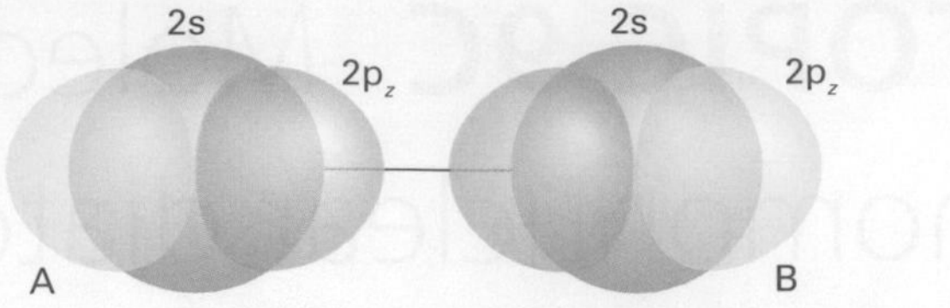

Figure 9C.3 According to molecular orbital theory, σ orbitals are built from all orbitals that have the appropriate symmetry. In homonuclear diatomic molecules of Period 2, that means that two 2s and two $2p_z$ orbitals should be used. From these four orbitals, four molecular orbitals can be built.

$$\psi = c_{A2s}\psi_{A2s} + c_{B2s}\psi_{B2s} + c_{A2p_z}\psi_{A2p_z} + c_{B2p_z}\psi_{B2p_z} \quad (9C.1)$$

From these four atomic orbitals four molecular orbitals of σ symmetry can be formed by an appropriate choice of the coefficients c.

Because the 2s and 2p orbitals on each atom have such different energies, they may be treated separately (this approximation is removed later). That is, the four σ molecular orbitals fall approximately into two sets, one consisting of two molecular orbitals formed from the 2s orbitals

$$\psi = c_{A2s}\psi_{A2s} + c_{B2s}\psi_{B2s} \quad (9C.2a)$$

and another consisting of two orbitals formed from the $2p_z$ orbitals

$$\psi = c_{A2p_z}\psi_{A2p_z} + c_{B2p_z}\psi_{B2p_z} \quad (9C.2b)$$

In a homonuclear diatomic molecule the energies of the 2s orbitals on atoms A and B are the same. Their coefficients are therefore equal (apart from a possible difference in sign). The same is true of the $2p_z$ orbitals on each atom. Therefore, the two sets of orbitals have the form $\psi_{A2s} \pm \psi_{B2s}$ and $\psi_{A2p_z} \pm \psi_{B2p_z}$, the + combination being bonding and the − combination antibonding in each case.

At this stage it is useful to adopt a more formal system for denoting molecular orbitals. First, the orbitals are labelled with g and u to indicate their inversion symmetry, as explained in Topic 9B. Then each set of orbitals of the same inversion symmetry is numbered separately. Therefore, the σ orbital formed from the 2s orbitals is labelled $1\sigma_g$ and the σ* orbital formed from the same atomic orbitals is denoted $1\sigma_u$.

The two $2p_z$ orbitals directed along the internuclear axis also overlap strongly. They may interfere either constructively or destructively, and give a bonding or antibonding σ orbital that lie higher in energy than the $1\sigma_g$ and $1\sigma_u$ orbitals because it has been supposed that the 2p atomic orbitals lie significantly higher in energy than the 2s orbitals (Fig. 9C.4). These two σ orbitals are labelled $2\sigma_g$ and $2\sigma_u$, respectively. Note how the numbering follows the order of increasing energy and orbitals of different symmetry are labelled separately.

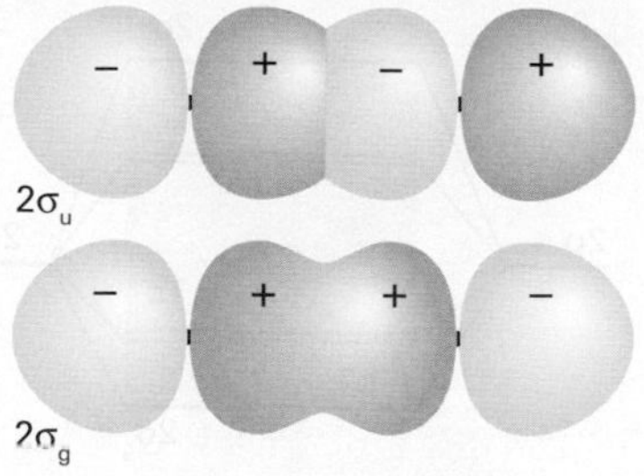

Figure 9C.4 A representation of the form of the bonding and antibonding σ orbitals built from the overlap of p orbitals. These illustrations are schematic.

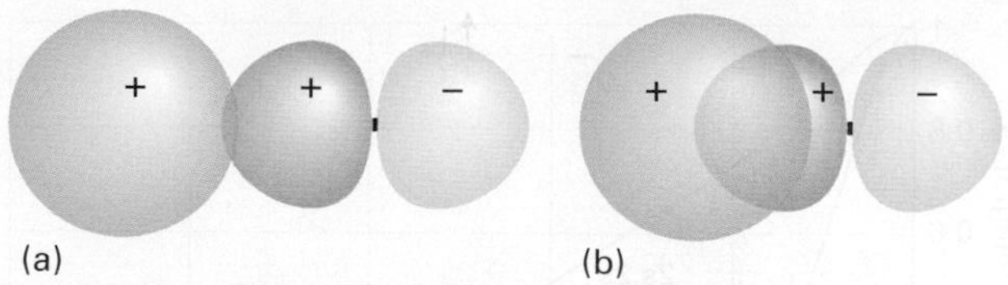

Figure 9C.6 (a) When two orbitals are on atoms that are far apart, the wavefunctions are small where they overlap, so *S* is small. (b) When the atoms are closer, both orbitals have significant amplitudes where they overlap, and *S* may approach 1. Note that *S* will decrease again as the two atoms approach more closely than shown here, because the region of negative amplitude of the p orbital starts to overlap the positive amplitude of the s orbital. When the centres of the atoms coincide, $S = 0$.

Now consider the $2p_x$ and $2p_y$ orbitals of each atom. These orbitals are perpendicular to the internuclear axis and overlap broadside-on when the atoms are close together. This overlap may be constructive or destructive and results in a bonding or an antibonding π orbital (Fig. 9C.5). The notation π is the analogue of p in atoms: when viewed along the axis of the molecule, a π orbital looks like a p orbital and has one unit of orbital angular momentum around the internuclear axis. The two neighbouring $2p_x$ orbitals overlap to give a bonding and antibonding π_x orbital, and the two $2p_y$ orbitals overlap to give two π_y orbitals. The π_x and π_y bonding orbitals are degenerate; so too are their antibonding partners. As seen in Fig. 9C.5, a bonding π orbital has odd parity (u) and the antibonding π orbital has even parity (g). The lower two doubly degenerate orbitals are therefore labelled $1\pi_u$ and their higher energy antibonding partners are labelled $1\pi_g$.

(b) The overlap integral

As in the discussion of the hydrogen molecule-ion, the lowering of energy that results from constructive interference between neighbouring atomic orbitals (and the raising of energy that results from destructive interference) correlates with the extent of overlap of the orbitals. As explained in Topic 9B, the extent to which two atomic orbitals overlap is measured by the **overlap integral**, *S*:

$$S = \int \psi_A^* \psi_B \, d\tau \qquad \text{Overlap integral [definition]} \qquad (9C.3)$$

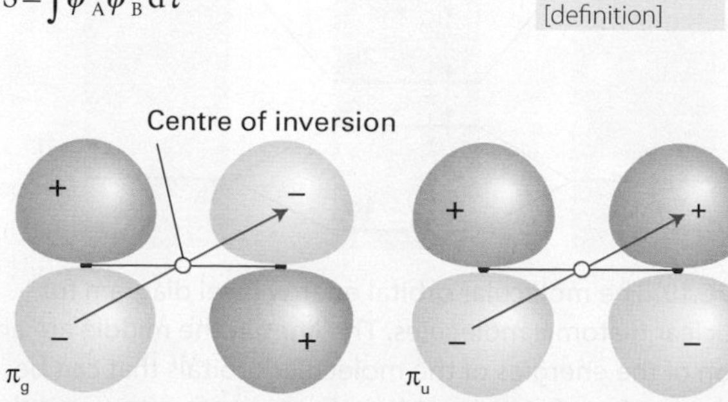

Figure 9C.5 The parity of π antibonding and bonding molecular orbitals.

If the atomic orbital ψ_A on A is small wherever the orbital ψ_B on B is large, or vice versa, then the product of their amplitudes is everywhere small and the integral—the sum of these products—is small (Fig. 9C.6). If ψ_A and ψ_B are both large in some region of space, then *S* may approach 1. If the two normalized atomic orbitals are identical (for instance, 1s orbitals on the same nucleus), then $S = 1$. In some cases, simple formulas can be given for overlap integrals (Table 9C.1) and illustrated in Fig. 9C.7.

Now consider the arrangement in which an s orbital spreads into the same region of space as a p_x orbital of a different atom (Fig. 9C.8). The integral over the region where the product of the wavefunctions is positive exactly cancels the integral over the region where the product is negative, so overall $S = 0$ exactly. Therefore, there is no net overlap between the s and p_x orbitals in this arrangement.

The extent of overlap as measured by the overlap integral is suggestive of the contribution that different kinds of orbital overlap makes to bond formation, but the value of the integral must be treated with caution. Thus, the overlap integral for broadside overlap of $2p_x$ or $2p_y$ orbitals is typically greater than that for the overlap of $2p_z$ orbitals, suggesting weaker σ than π bonding.

Table 9C.1 Overlap integrals between hydrogenic orbitals

Orbitals	Overlap integral
1s,1s	$S = \left\{1 + \frac{ZR}{a_0} + \frac{1}{3}\left(\frac{ZR}{a_0}\right)^2\right\} e^{-ZR/a_0}$
2s,2s	$S = \left\{1 + \frac{ZR}{2a_0} + \frac{1}{12}\left(\frac{ZR}{a_0}\right)^2 + \frac{1}{240}\left(\frac{ZR}{a_0}\right)^4\right\} e^{-ZR/2a_0}$
$2p_x$,$2p_x$ (π)	$S = \left\{1 + \frac{ZR}{2a_0} + \frac{1}{10}\left(\frac{ZR}{a_0}\right)^2 + \frac{1}{120}\left(\frac{ZR}{a_0}\right)^3\right\} e^{-ZR/2a_0}$
$2p_z$,$2p_z$ (σ)	$S = -\left\{1 + \frac{ZR}{2a_0} + \frac{1}{20}\left(\frac{ZR}{a_0}\right)^2 - \frac{1}{60}\left(\frac{ZR}{a_0}\right)^3 - \frac{1}{240}\left(\frac{ZR}{a_0}\right)^4\right\} e^{-ZR/2a_0}$

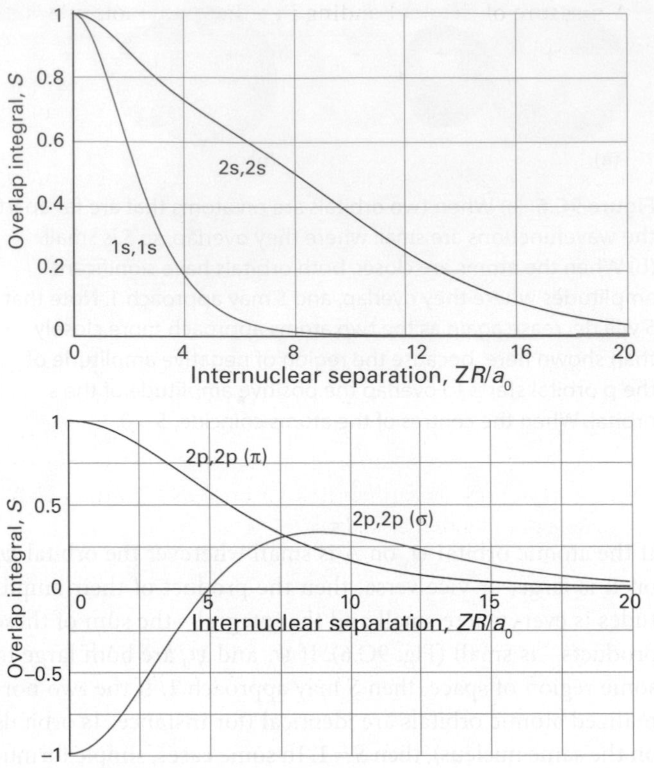

Figure 9C.7 The variation of the overlap integral, *S*, between two hydrogenic orbitals with the internuclear separation. A negative value of *S* corresponds to separations at which the contribution to the overlap of the positive region of one 2p orbital with the negative lobe of the other 2p orbital outweighs that from the regions where both have the same sign.

However, the constructive overlap in the region between the nuclei and on the axis is greater in σ interactions, and its effect on bonding is more important than the overall extent of overlap. As a result, the separation of $1\pi_u$ and $1\pi_g$ orbitals is likely to be smaller than the separation of $2\sigma_g$ and $2\sigma_u$ orbitals in the same molecule. The relative energies of these orbitals is therefore likely to be as shown in Fig. 9C.9, and electrons occupying π orbitals are likely to be less effective at bonding than those occupying the σ orbitals derived from the same p orbitals.

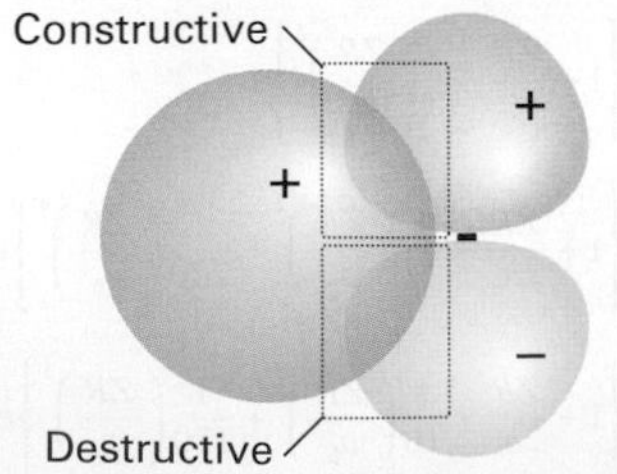

Figure 9C.8 A p orbital in the orientation shown here has zero net overlap (*S* = 0) with the s orbital at all internuclear separations.

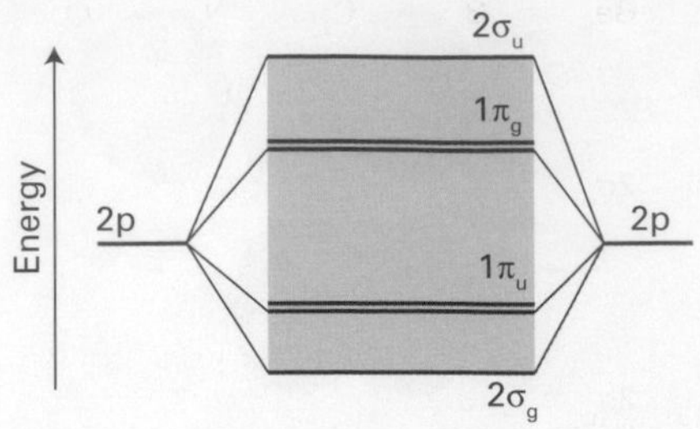

Figure 9C.9 As explained in the text, the separation of $1\pi_u$ and $1\pi_g$ orbitals is likely to be smaller than the separation of $2\sigma_g$ and $2\sigma_u$ orbitals in the same molecule, leading to the relative energies shown here.

(c) Period 2 diatomic molecules

To construct the molecular orbital energy level diagram for Period 2 homonuclear diatomic molecules, eight molecular orbitals are formed from the eight valence shell orbitals (four from each atom). The ordering suggested by the extent of overlap is shown in Fig. 9C.10. However, remember that this scheme assumes that the 2s and $2p_z$ orbitals contribute to different sets of molecular orbitals. In fact all four atomic orbitals have the same symmetry around the internuclear axis and contribute jointly to the four σ orbitals. Hence, there is no guarantee that this order of energies will be found, and detailed calculation shows that the order varies along Period 2 (Fig. 9C.11). The order shown in Fig. 9C.12 is appropriate as far as N_2, and Fig. 9C.10 is appropriate for O_2 and F_2. The relative order is controlled by the energy separation of the 2s and

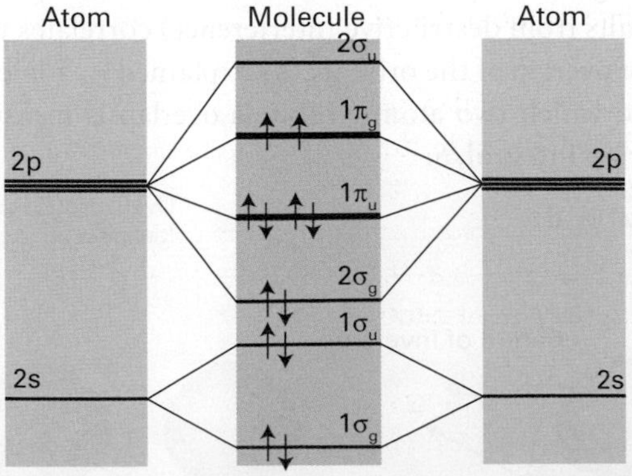

Figure 9C.10 The molecular orbital energy level diagram for homonuclear diatomic molecules. The lines in the middle are an indication of the energies of the molecular orbitals that can be formed by overlap of atomic orbitals. Energy increases upwards. As remarked in the text, this diagram is appropriate for O_2 (the configuration shown) and F_2.

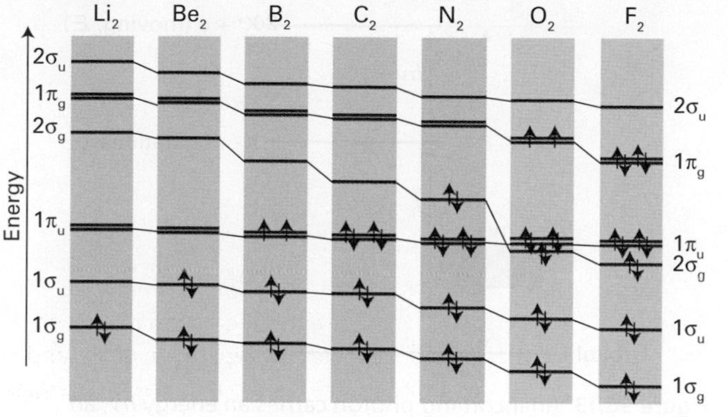

Figure 9C.11 The variation of the orbital energies of Period 2 homonuclear diatomics.

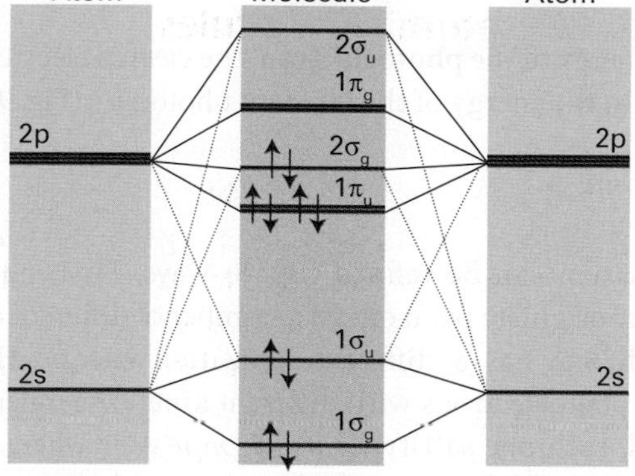

Figure 9C.12 An alternative molecular orbital energy level diagram for homonuclear diatomic molecules. Energy increases upwards. As remarked in the text, this diagram is appropriate for Period 2 homonuclear diatomics up to and including N_2 (the configuration shown).

2p orbitals in the atoms, which increases across the period. The change in the order of the $1\pi_u$ and $2\sigma_g$ orbitals occurs at about N_2.

With the molecular orbital energy level diagram established, the probable ground-state configurations of the molecules are deduced by adding the appropriate number of electrons to the orbitals and following the building-up rules. Anionic species (such as the peroxide ion, O_2^{2-}) need more electrons than the parent neutral molecules; cationic species (such as O_2^+) need fewer.

Consider N_2, which has 10 valence electrons. Two electrons pair, occupy, and fill the $1\sigma_g$ orbital; the next two occupy and fill the $1\sigma_u$ orbital. Six electrons remain. There are two $1\pi_u$ orbitals, so four electrons can be accommodated in them. The last two enter the $2\sigma_g$ orbital. Therefore, the ground-state configuration of N_2 is $1\sigma_g^2 1\sigma_u^2 1\pi_u^4 2\sigma_g^2$. It is sometimes helpful to include an asterisk to denote an antibonding orbital, in which case this configuration would be denoted $1\sigma_g^2 1\sigma_u^{*2} 1\pi_u^4 2\sigma_g^2$.

A measure of the net bonding in a diatomic molecule is its **bond order**, b:

$$b = \tfrac{1}{2}(N - N^*) \qquad \text{Bond order [definition]} \qquad (9C.4)$$

where N is the number of electrons in bonding orbitals and N^* is the number of electrons in antibonding orbitals.

Brief illustration 9C.1

Each electron pair in a bonding orbital increases the bond order by 1 and each pair in an antibonding orbital decreases b by 1. For H_2, $b = 1$, corresponding to a single bond, H–H, between the two atoms. In He_2, $b = 0$, and there is no bond. In N_2, $b = \tfrac{1}{2}(8 - 2) = 3$. This bond order accords with the Lewis structure of the molecule (:N≡N:).

The ground-state electron configuration of O_2, with 12 valence electrons, is based on Fig. 9C.10, and is $1\sigma_g^2 1\sigma_u^2 2\sigma_g^2 1\pi_u^4 1\pi_g^2$ (or $1\sigma_g^2 1\sigma_u^{*2} 2\sigma_g^2 1\pi_u^4 1\pi_g^{*2}$). The bond order is $b = \tfrac{1}{2}(8-4) = 2$. According to the building-up principle, however, the two $1\pi_g$ electrons occupy two different orbitals: one will enter $1\pi_{g,x}$ and the other will enter $1\pi_{g,y}$. Because the electrons are in different orbitals, they will have parallel spins. Therefore, an O_2 molecule is predicted to have a net spin angular momentum with $S = 1$ and, in the language introduced in Topic 8C, to be in a triplet state. As electron spin is the source of a magnetic moment, oxygen is also predicted to be paramagnetic, a substance that tends to be drawn into a magnetic field (see Topic 15C). This prediction, which VB theory does not make, is confirmed by experiment.

An F_2 molecule has two more electrons than an O_2 molecule. Its configuration is therefore $1\sigma_g^2 1\sigma_u^{*2} 2\sigma_g^2 1\pi_u^4 1\pi_g^{*4}$ and $b = 1$, so F_2 is a singly-bonded molecule, in agreement with its Lewis structure. The hypothetical molecule dineon, Ne_2, has two more electrons than F_2: its configuration is $1\sigma_g^2 1\sigma_u^{*2} 2\sigma_g^2 1\pi_u^4 1\pi_g^{*4} 2\sigma_u^{*2}$ and $b = 0$. The zero bond order is consistent with the fact that neon occurs as a monatomic gas.

The bond order is a useful parameter for discussing the characteristics of bonds, because it correlates with bond length and bond strength. For bonds between atoms of a given pair of elements:

- The greater the bond order, the shorter the bond.
- The greater the bond order, the greater the bond strength.

Physical interpretation

Table 9C.2 lists some typical bond lengths in diatomic and polyatomic molecules. The strength of a bond is measured by its bond dissociation energy, $hc\tilde{D}_0$, the energy required to separate the atoms to infinity or by the well depth, $hc\tilde{D}_e$, with $hc\tilde{D}_0 = hc\tilde{D}_e - \tfrac{1}{2}\hbar\omega$. Table 9C.3 lists some experimental values of $hc\tilde{D}_0$.

Table 9C.2 Bond lengths*

Bond	Order	R_e/pm
HH	1	74.14
NN	3	109.76
HCl	1	127.45
CH	1	*114*
CC	1	*154*
	2	*134*
	3	*120*

* More values will be found in the *Resource section*. Numbers in italics are mean values for polyatomic molecules.

Table 9C.3 Bond dissociation energies*

Bond	Order	$N_A hc\tilde{D}_0$/(kJ mol^{-1})
HH	1	432.1
NN	3	941.7
HCl	1	427.7
CH	1	*435*
CC	1	*368*
	2	*720*
	3	*962*

* More values will be found in the *Resource section*. Numbers in italics are mean values for polyatomic molecules.

Brief illustration 9C.2

From Fig. 9C.12, the electron configurations and bond orders of N_2 and N_2^+ are

N_2 $1\sigma_g^2 1\sigma_u^{*2} 1\pi_u^4 2\sigma_g^2$ $b=3$

N_2^+ $1\sigma_g^2 1\sigma_u^{*2} 1\pi_u^4 2\sigma_g^1$ $b=2\frac{1}{2}$

Because the cation has the smaller bond order, you should expect it to have the smaller dissociation energy. The experimental dissociation energies are 942 kJ mol^{-1} for N_2 and 842 kJ mol^{-1} for N_2^+.

9C.2 Photoelectron spectroscopy

So far, molecular orbitals have been regarded as purely theoretical constructs, but is there experimental evidence for their existence? **Photoelectron spectroscopy** (PES) measures the ionization energies of molecules when electrons are ejected from different orbitals by absorption of a photon of known energy, and uses the information to infer the energies of molecular orbitals.

Because energy is conserved when a photon ionizes a sample, the sum of the ionization energy, I, of the sample and the kinetic energy of the **photoelectron**, the ejected electron, must be equal to the energy of the incident photon $h\nu$ (Fig. 9C.13):

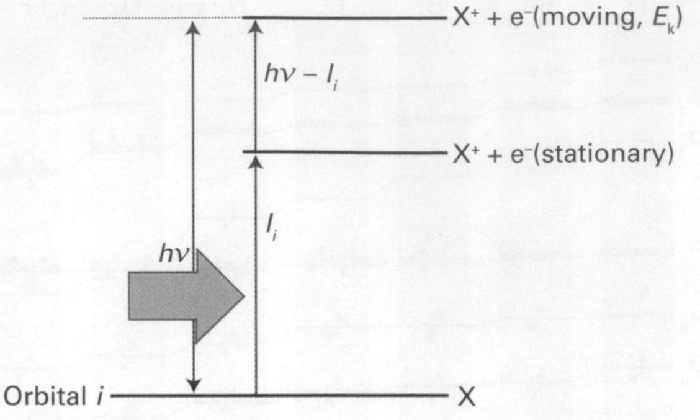

Figure 9C.13 An incoming photon carries an energy $h\nu$; an energy I_i is needed to remove an electron from an orbital i, and the difference appears as the kinetic energy of the electron.

$$h\nu = \tfrac{1}{2}m_e v^2 + I \qquad (9C.5)$$

This equation can be refined in two ways. First, photoelectrons may originate from one of a number of different orbitals, and each one has a different ionization energy. Hence, a series of photoelectrons with different kinetic energies will be obtained, each one satisfying $h\nu = \tfrac{1}{2}m_e v^2 + I_i$, where I_i is the ionization energy for ejection of an electron from an orbital i. Therefore, by measuring the kinetic energies of the photoelectrons, and knowing the frequency ν, these ionization energies can be determined. Photoelectron spectra are interpreted in terms of an approximation called **Koopmans' theorem**, which states that the ionization energy I_i is equal to the orbital energy of the ejected electron (formally: $I_i = -\varepsilon_i$). That is, the ionization energy can be identified with the energy of the orbital from which it is ejected. The theorem is only an approximation because it ignores the fact that the remaining electrons adjust their distributions when ionization occurs.

The ionization energies of molecules are several electronvolts even for valence electrons, so it is essential to work in at least the ultraviolet region of the spectrum and with wavelengths of less than about 200 nm. Much work has been done with radiation generated by a discharge through helium: the He(I) line ($1s^1 2p^1 \rightarrow 1s^2$) lies at 58.43 nm, corresponding to a photon energy of 21.22 eV. Its use gives rise to the technique of **ultraviolet photoelectron spectroscopy** (UPS). When core electrons are being studied, photons of even higher energy are needed to expel them: X-rays are used, and the technique is denoted XPS.

The kinetic energies of the photoelectrons are measured using an electrostatic deflector that produces different deflections in the paths of the photoelectrons as they pass between charged plates (Fig. 9C.14). As the field strength between the

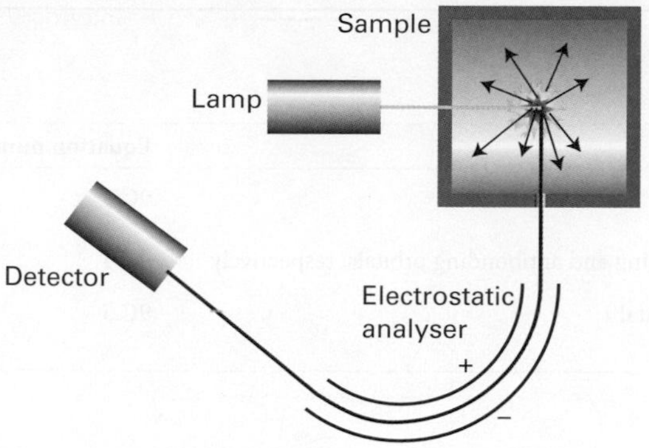

Figure 9C.14 A photoelectron spectrometer consists of a source of ionizing radiation (such as a helium discharge lamp for UPS and an X-ray source for XPS), an electrostatic analyser, and an electron detector. The deflection of the electron path caused by the analyser depends on the speed of the electrons.

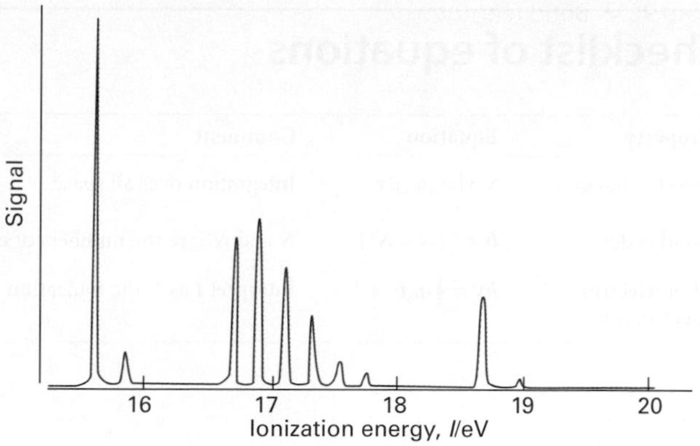

Figure 9C.15 The photoelectron spectrum of N_2 recorded using He(I) radiation.

plates is increased, electrons of different speeds, and therefore kinetic energies, reach the detector. The electron flux can be recorded and plotted against kinetic energy to obtain the photoelectron spectrum (Fig. 9C.15).

Brief illustration 9C.3

The photoelectrons of highest kinetic energy ejected from N_2 in a spectrometer using He(I) radiation have kinetic energies of 5.63 eV. Because photons of helium(I) radiation have energy 21.22 eV it follows that 21.22 eV = 5.63 eV + I_i, so I_i = 15.59 eV. This ionization energy is the energy needed to remove an electron from the occupied molecular orbital with the highest energy of the N_2 molecule, the $2\sigma_g$ bonding orbital. Photoelectrons are also detected at 4.53 eV, corresponding to an ionization energy of 16.7 eV. The likely origin of these electrons is the $1\pi_u$ orbital.

Photoejection commonly results in cations that are excited vibrationally. Because different energies are needed to excite different vibrational states of the ion, the photoelectrons appear with different kinetic energies. The result is **vibrational fine structure**, a progression of lines with a spacing in energy that corresponds to the vibrational frequency of the molecular ion. This fine structure occurs between 16.7 eV and 18 eV in the photoelectron spectrum of N_2 shown in Fig. 9C.15.

Checklist of concepts

☐ 1. Molecular orbitals are constructed as linear combinations of all valence orbitals of the appropriate symmetry.

☐ 2. As a first approximation, σ orbitals are constructed separately from valence s and p orbitals.

☐ 3. π **Orbitals** are constructed from the sideways overlap of p orbitals.

☐ 4. An **overlap integral** is a measure of the extent of orbital overlap.

☐ 5. According to the building-up principle, electrons occupy the available molecular orbitals so as to achieve the lowest total energy subject to the Pauli exclusion principle.

☐ 6. If electrons occupy different orbitals, the lowest energy is obtained if their spins are parallel.

☐ 7. The greater the **bond order** of a molecule or ion between the same two atoms, the shorter and stronger is the bond.

☐ 8. **Photoelectron spectroscopy** is a technique for determining the energies of electrons in molecular orbitals.

Checklist of equations

Property	Equation	Comment	Equation number
Overlap integral	$S=\int\psi_A^*\psi_B\mathrm{d}\tau$	Integration over all space	9C.3
Bond order	$b=\frac{1}{2}(N-N^*)$	N and N^* are the numbers of electrons in bonding and antibonding orbitals, respectively	9C.4
Photoelectron spectroscopy	$h\nu=\frac{1}{2}m_e v^2+I$	Interpret I as I_i, the ionization energy from orbital i	9C.5

TOPIC 9D Molecular orbital theory: heteronuclear diatomic molecules

➤ Why do you need to know this material?

Most diatomic molecules are heteronuclear, so you need to appreciate the differences in their electronic structure from homonuclear species, and how to treat those differences quantitatively.

➤ What is the key idea?

The bonding molecular orbital of a heteronuclear diatomic molecule is composed mostly of the atomic orbital of the more electronegative atom; the opposite is true of the antibonding orbital.

➤ What do you need to know already?

You need to know about the molecular orbitals of homonuclear diatomic molecules (Topic 9C) and the concepts of normalization and orthogonality (Topic 7C). This Topic makes light use of determinants (*The chemist's toolkit* 23) and the rules of differentiation (*The chemist's toolkit* 5 in Topic 1C).

The electrons in a covalent bond in a heteronuclear diatomic species are not distributed equally over the atoms because it is energetically favourable for the electron pair to be found closer to one atom than to the other. This imbalance results in a **polar bond**, a bond in which the bonding electron density is shared unequally between the bonded atoms. The bond in HF, for instance, is polar, with the bonding electron density greater near the F atom than the H atom. The accumulation of bonding electron density near the F atom results in that atom having a net negative charge, which is called a **partial negative charge** and denoted δ–. There is a matching **partial positive charge**, δ+, on the H atom (Fig. 9D.1).

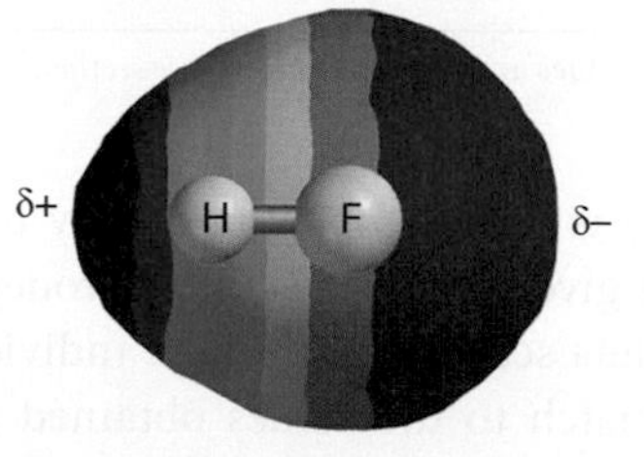

Figure 9D.1 The electron density of the molecule HF, computed using one of the methods described in Topic 9E. Different colours show the variation in the electrostatic potential and hence the net charge, with blue representing the region with largest partial positive charge, and red the region with largest partial negative charge.

9D.1 Polar bonds and electronegativity

The description of polar bonds is a straightforward extension of the molecular orbital theory of homonuclear diatomic molecules (Topic 9C). The principal difference is that the atomic orbitals on the two atoms have different energies and spatial extents.

A polar bond consists of two electrons in a bonding molecular orbital of the form

$$\psi = c_A\psi_A + c_B\psi_B \qquad \text{Wavefunction of a polar bond} \qquad (9D.1)$$

with unequal coefficients. It follows that the contribution of the atomic orbital ψ_A to the bond, in the sense that on inspection of the location of the electron the probability that it will be found on atom A, is $|c_A|^2$, and that of ψ_B is $|c_B|^2$. A nonpolar bond has $|c_A|^2 = |c_B|^2$, and a pure ionic bond has one coefficient equal to zero (so the species A^+B^- would have $c_A = 0$ and $c_B = 1$). The atomic orbital with the lower energy makes the larger contribution to the bonding molecular orbital. The opposite is true of the antibonding orbital, for which the dominant component comes from the atomic orbital with higher energy.

The distribution of partial charges in bonds is commonly discussed in terms of the **electronegativity**, χ (chi), of the elements involved. The electronegativity is a parameter introduced by Linus Pauling as a measure of the power of an atom in a bond to attract electrons to itself. Pauling used valence-bond arguments to suggest that an appropriate numerical scale of electronegativities could be defined in terms of bond dissociation energies, $hc\tilde{D}_0$, and proposed that the difference in electronegativities could be expressed as

$$|\chi_A - \chi_B| = \{hc\tilde{D}_0(AB) - \tfrac{1}{2}[hc\tilde{D}_0(AA) + hc\tilde{D}_0(BB)]\}^{1/2}/\text{eV}$$

Pauling electronegativity [definition] (9D.2)

Table 9D.1 Pauling electronegativities*

Element	χ_P
H	2.2
C	2.6
N	3.0
O	3.4
F	4.0
Cl	3.2
Cs	0.79

* More values are given in the *Resource section*.

where $hc\tilde{D}_0(XY)$ is the dissociation energy of an X–Y bond. This expression gives differences of electronegativities; to establish an absolute scale Pauling chose individual values that gave the best match to the values obtained from eqn 9D.2. Electronegativities based on this definition are called **Pauling electronegativities** (Table 9D.1). The most electronegative elements are those close to F (excluding the noble gases). It is found that the greater the difference in electronegativities, the greater the polar character of the bond. The difference for HF, for instance, is 1.8; a C–H bond, which is commonly regarded as almost nonpolar, has an electronegativity difference of 0.4.

Brief illustration 9D.1

The bond dissociation energies of H_2, Cl_2, and HCl are 4.52 eV, 2.51 eV, and 4.47 eV, respectively. From eqn 9D.2,

$$|\chi_{\text{Pauling}}(\text{H}) - \chi_{\text{Pauling}}(\text{Cl})| = \{4.47 - \tfrac{1}{2}(4.52 + 2.51)\}^{1/2} = 0.98 \approx 1.0$$

The spectroscopist Robert Mulliken proposed an alternative definition of electronegativity. He argued that an element is likely to be highly electronegative if it has a high ionization energy (so it will not release electrons readily) and a high electron affinity (so it is energetically favourable to acquire electrons). The **Mulliken electronegativity scale** is therefore based on the definition

$$\chi = \tfrac{1}{2}(I + E_{\text{ea}})/\text{eV} \quad \text{Mulliken electronegativity [definition]} \quad (9D.3)$$

where I is the ionization energy of the element and E_{ea} is its electron affinity. The greater the value of the Mulliken electronegativity the greater is the contribution of that atom to the electron distribution in the bond. There is one word of caution: the values of I and E_{ea} in eqn 9D.3 are strictly those for a special 'valence state' of the atom, not a true spectroscopic state, but that complication is ignored here. The Mulliken and Pauling scales are approximately in line with each other. A reasonably reliable conversion between the two is

$$\chi_{\text{Pauling}} = 1.35\chi_{\text{Mulliken}}^{1/2} - 1.37 \quad (9D.4)$$

9D.2 The variation principle

The systematic way of discussing bond polarity and finding the coefficients in the linear combinations used to build molecular orbitals is provided by the **variation principle**:

> If an arbitrary wavefunction is used to calculate the energy, the value calculated is never less than the true energy. *Variation principle*

It can be justified by setting up an arbitrary 'trial function' and showing that the corresponding energy is not less than the true energy (it might be the same).

How is that done? 9D.1 Justifying the variation principle

Any arbitrary function can be expressed as a linear combination of the eigenfunctions ψ_n of the exact hamiltonian for a molecule. In the present case, consider a normalized trial wavefunction written as a linear combination $\psi_{\text{trial}} = \sum_n c_n\psi_n$ and suppose that the ψ_n are themselves normalized and mutually orthogonal.

Step 1 *Write an expression for the difference between the calculated and true energy*

The energy associated with the normalized trial function is the expectation value

$$E = \int \psi_{\text{trial}}^* \hat{H} \psi_{\text{trial}} \,\text{d}\tau$$

The lowest energy of the system is E_0, the eigenvalue of $\hat{H}$ corresponding to ψ_0. Consider the following difference:

$$\begin{aligned} E - E_0 &= \int \psi_{\text{trial}}^* \hat{H} \psi_{\text{trial}} \,\text{d}\tau - E_0 \overbrace{\int \psi_{\text{trial}}^* \psi_{\text{trial}} \,\text{d}\tau}^{1} \\ &= \int \psi_{\text{trial}}^* \hat{H} \psi_{\text{trial}} \,\text{d}\tau - \int \psi_{\text{trial}}^* E_0 \psi_{\text{trial}} \,\text{d}\tau \\ &= \int \psi_{\text{trial}}^* (\hat{H} - E_0) \psi_{\text{trial}} \,\text{d}\tau \\ &= \int \Big(\sum_n c_n^* \psi_n^*\Big)(\hat{H} - E_0)\Big(\sum_{n'} c_{n'}\psi_{n'}\Big) \text{d}\tau \\ &= \sum_{n,n'} c_n^* c_{n'} \int \psi_n^* (\hat{H} - E_0)\psi_{n'} \,\text{d}\tau \end{aligned}$$

Step 2 *Simplify the expression*

Because $\int \psi_n^* \hat{H} \psi_{n'} \text{d}\tau = E_{n'} \int \psi_n^* \psi_{n'} \text{d}\tau$ and $\int \psi_n^* E_0 \psi_{n'} \text{d}\tau = E_0 \int \psi_n^* \psi_{n'} \text{d}\tau$, write

$$\int \psi_n^* (\hat{H} - E_0)\psi_{n'} \,\text{d}\tau = (E_{n'} - E_0)\int \psi_n^* \psi_{n'} \,\text{d}\tau$$

It follows that

$$E - E_0 = \sum_{n,n'} c_n^* c_{n'} (E_{n'} - E_0) \overbrace{\int \psi_n^* \psi_{n'} \,\text{d}\tau}^{1 \text{ if } n'=n;\ 0 \text{ otherwise}}$$

Step 3 *Analyse the final expression*

The eigenfunctions ψ_n are orthogonal, so only terms with $n'=n$ contribute to this sum. Because each eigenfunction is normalized, each surviving integral is 1. Consequently

$$E-E_0=\sum_n \overbrace{c_n^\star c_n}^{\geq 0}\overbrace{(E_n-E_0)}^{\geq 0}\geq 0$$

The quantity $c_n^\star c_n$ is necessarily real and greater than or equal to zero, and because E_0 is the lowest energy, E_n-E_0 is also greater than or equal to zero. It follows that the product of the two terms on the right is greater than or equal to zero. Therefore, $E\geq E_0$, as asserted.

The variation principle is the basis of all modern molecular structure calculations. The principle implies that, if the coefficients in the trial wavefunction are varied until the lowest energy is achieved (by evaluating the expectation value of the hamiltonian for the wavefunction in each case), then those coefficients will be the best for that particular form of trial function. A lower energy might be obtained with a more complicated wavefunction, for example, by taking a linear combination of several atomic orbitals on each atom. However, for a molecular orbital constructed from a given **basis set**, a given set of atomic orbits, the variation principle gives the optimum molecular orbital of that kind.

(a) The procedure

The practical application of the variation principle can be illustrated by applying it to the trial wavefunction in eqn 9D.1, where the coefficients define the trial function.

How is that done? 9D.2 Applying the variation principle to a heteronuclear diatomic molecule

The trial wavefunction in eqn 9D.1 is real but not normalized because at this stage the coefficients can take arbitrary values. Because it is real, write $\psi^\star=\psi$. To normalize it, multiply it by $N=1/(\int\psi^\star\psi\,\mathrm{d}\tau)^{1/2}$. So from now on use $\psi/(\int\psi^2\,\mathrm{d}\tau)^{1/2}$ as the trial function. Then follow these steps.

Step 1 *Write an expression for the energy*

The expectation value of the hamiltonian, the energy, using the normalized real trial function, is

$$E=\frac{\int\psi\hat{H}\psi\,\mathrm{d}\tau}{\int\psi^2\,\mathrm{d}\tau} \tag{9D.5}$$

The denominator is

$$\int\psi^2\,\mathrm{d}\tau=\int(c_A\psi_A+c_B\psi_B)^2\,\mathrm{d}\tau$$

$$=c_A^2\overbrace{\int\psi_A^2\,\mathrm{d}\tau}^{1}+c_B^2\overbrace{\int\psi_B^2\,\mathrm{d}\tau}^{1}+2c_Ac_B\overbrace{\int\psi_A\psi_B\,\mathrm{d}\tau}^{S}$$

$$=c_A^2+c_B^2+2c_Ac_BS$$

because the individual atomic orbitals are normalized to 1 and the third integral is the overlap integral S (eqn 9C.3, $S=\int\psi_A\psi_B\,\mathrm{d}\tau$). The numerator is

$$\int\psi\hat{H}\psi\,\mathrm{d}\tau=\int(c_A\psi_A+c_B\psi_B)\hat{H}(c_A\psi_A+c_B\psi_B)\,\mathrm{d}\tau$$

$$=c_A^2\overbrace{\int\psi_A\hat{H}\psi_A\,\mathrm{d}\tau}^{\alpha_A}+c_B^2\overbrace{\int\psi_B\hat{H}\psi_B\,\mathrm{d}\tau}^{\alpha_B}+c_Ac_B\overbrace{\int\psi_A\hat{H}\psi_B\,\mathrm{d}\tau}^{\beta}$$

$$+c_Ac_B\overbrace{\int\psi_B\hat{H}\psi_A\,\mathrm{d}\tau}^{\beta}$$

The significance of the quantities α_A, α_B, and β (which are all energies) is discussed shortly. Because the hamiltonian is hermitian (Topic 7C), the third and fourth integrals are equal. Therefore

$$\int\psi\hat{H}\psi\,\mathrm{d}\tau=c_A^2\alpha_A+c_B^2\alpha_B+2c_Ac_B\beta$$

At this point the complete expression for E is

$$E=\frac{c_A^2\alpha_A+c_B^2\alpha_B+2c_Ac_B\beta}{c_A^2+c_B^2+2c_Ac_BS}$$

Step 2 *Minimize the energy*

Now search for values of the coefficients in the trial function that minimize the value of E. This is a standard problem in calculus, and is solved by finding the coefficients for which

$$\frac{\partial E}{\partial c_A}=0 \qquad \frac{\partial E}{\partial c_B}=0$$

After some straightforward application of the rules of differentiation (*The chemist's toolkit* 5 in Topic 1C), the result is

$$\frac{\partial E}{\partial c_A}=\frac{2\{(\alpha_A-E)c_A+(\beta-SE)c_B\}}{c_A^2+c_B^2+2c_Ac_BS}$$

$$\frac{\partial E}{\partial c_B}=\frac{2\{(\alpha_B-E)c_B+(\beta-SE)c_A\}}{c_A^2+c_B^2+2c_Ac_BS}$$

For the derivatives to be equal to 0, the numerators, and specifically the terms in blue, of these expressions must vanish, leading to the **secular equations**:[1]

$$(\alpha_A-E)c_A+(\beta-SE)c_B=0 \tag{9D.6a}$$

$$(\alpha_B-E)c_B+(\beta-SE)c_A=0 \tag{9D.6b}$$

Secular equations

[1] The name 'secular' is derived from the Latin word for age or generation. The term comes from astronomy, where the same equations appear in connection with slowly accumulating modifications of planetary orbits.

The quantities α_A, α_B, β, and S in the secular equations are

$$\alpha_A = \int \psi_A \hat{H} \psi_A d\tau \quad \alpha_B = \int \psi_B \hat{H} \psi_B d\tau \quad \text{Coulomb integrals} \quad (9D.7a)$$

$$\beta = \int \psi_A \hat{H} \psi_B d\tau = \int \psi_B \hat{H} \psi_A d\tau \quad \text{Resonance integral} \quad (9D.7b)$$

$$S = \int \psi_A \psi_B d\tau \quad \text{Overlap integral} \quad (9D.7c)$$

The parameter α is called a **Coulomb integral**. It is negative and can be interpreted as the energy of the electron when it occupies ψ_A (for α_A) or ψ_B (for α_B). In a homonuclear diatomic molecule, $\alpha_A = \alpha_B$. The parameter β is called a **resonance integral** (for classical reasons). It vanishes when the orbitals do not overlap, and at equilibrium bond lengths it is normally negative. The overlap integral S is introduced and discussed in Topic 9C.

In order to solve the secular equations for the coefficients it is necessary to know the energy E and then use its value in eqn 9D.6. As for any set of simultaneous equations, the secular equations have a solution if the **secular determinant**, the determinant of the coefficients (*The chemist's toolkit* 23), is zero. That is, if

$$\begin{vmatrix} \alpha_A - E & \beta - SE \\ \beta - SE & \alpha_B - E \end{vmatrix} = (\alpha_A - E)(\alpha_B - E) - (\beta - SE)^2$$
$$= (1 - S^2)E^2 + \{2\beta S - (\alpha_A + \alpha_B)\}E + (\alpha_A\alpha_B - \beta^2) = 0 \quad (9D.8)$$

This is a quadratic equation for E. A quadratic equation of the form $ax^2 + bx + c = 0$ has the solutions

$$x = \frac{-b \pm (b^2 - 4ac)^{1/2}}{2a}$$

In the present case, $a = 1 - S^2$, $b = 2\beta S - (\alpha_A + \alpha_B)$, and $c = \alpha_A\alpha_B - \beta^2$, so the solutions (the energies) are

$$E_\pm = \frac{\alpha_A + \alpha_B - 2\beta S \pm \{(2\beta S - (\alpha_A + \alpha_B))^2 - 4(1 - S^2)(\alpha_A\alpha_B - \beta^2)\}^{1/2}}{2(1 - S^2)} \quad (9D.9a)$$

which, according to the variation principle, are the closest approximations to the true energy for a trial function of the form given in eqn 9D.1. They are the energies of the bonding and antibonding molecular orbitals formed from the two atomic orbitals.

Equation 9D.9a can be simplified. For a homonuclear diatomic, $\alpha_A = \alpha_B = \alpha$ and then

$$E_\pm = \frac{2\alpha - 2\beta S \pm \{\overbrace{(2\beta S - 2\alpha)^2 - 4(1 - S^2)(\alpha^2 - \beta^2)}^{(2\beta - 2\alpha S)^2}\}^{1/2}}{\underbrace{2(1 - S^2)}_{(1+S)(1-S)}}$$

$$= \frac{\alpha - \beta S \pm (\beta - \alpha S)}{(1+S)(1-S)} = \frac{(\alpha \pm \beta)(1 \mp S)}{(1+S)(1-S)}$$

That is,

$$E_+ = \frac{\alpha + \beta}{1 + S} \quad E_- = \frac{\alpha - \beta}{1 - S} \quad \text{Homonuclear diatomics} \quad (9D.9b)$$

For $\beta < 0$, E_+ is the lower energy solution.

For heteronuclear diatomic molecules, making the approximation that $S = 0$ (simply to obtain a more transparent expression) gives

$$E_\pm = \tfrac{1}{2}(\alpha_A + \alpha_B) \pm \tfrac{1}{2}(\alpha_A - \alpha_B)\left\{1 + \left(\frac{2\beta}{\alpha_A - \alpha_B}\right)^2\right\}^{1/2} \quad \text{Zero overlap approximation} \quad (9D.9c)$$

The values of the Coulomb integrals α_A and α_B may be estimated as follows. The extreme case of an atom X in a molecule is X^+ if it has lost control of the electron it supplied, X if it is sharing the electron pair equally with its bonded partner, and X^- if it has gained control of both electrons in the bond. If X^+ is taken as defining the energy 0, then X lies at $-I(X)$ and X^- lies at $-\{I(X) + E_{ea}(X)\}$, where I is the ionization energy and E_{ea} the electron affinity (Fig. 9D.2). The actual energy of the electron in the molecule lies at an intermediate value, and in the absence of further information, it is reasonable to estimate it as half-way down to the lowest of these values, namely at $-\tfrac{1}{2}\{I(X) + E_{ea}(X)\}$. This quantity should be recognized (apart from its sign) as the Mulliken definition of electronegativity.

The chemist's toolkit 23 Determinants

A 2×2 determinant is the entity

$$\begin{vmatrix} a & b \\ c & d \end{vmatrix} = ad - bc \quad \text{2×2 Determinant}$$

A 3×3 determinant is evaluated by expanding it as a sum of 2×2 determinants:

$$\begin{vmatrix} a & b & c \\ d & e & f \\ g & h & i \end{vmatrix} = a\begin{vmatrix} e & f \\ h & i \end{vmatrix} - b\begin{vmatrix} d & f \\ g & i \end{vmatrix} + c\begin{vmatrix} d & e \\ g & h \end{vmatrix}$$
$$= a(ei - fh) - b(di - fg) + c(dh - eg) \quad \text{3×3 Determinant}$$

Note the sign change in alternate columns (b occurs with a negative sign in the expansion). An important property of a determinant is that if any two rows or any two columns are interchanged, then the determinant changes sign:

Exchange columns: $\begin{vmatrix} b & a \\ d & c \end{vmatrix} = bc - ad = -(ad - bc) = -\begin{vmatrix} a & b \\ c & d \end{vmatrix}$

Exchange rows: $\begin{vmatrix} c & d \\ a & b \end{vmatrix} = cb - da = -(ad - bc) = -\begin{vmatrix} a & b \\ c & d \end{vmatrix}$

An implication is that if any two columns or rows are identical, then the determinant is zero.

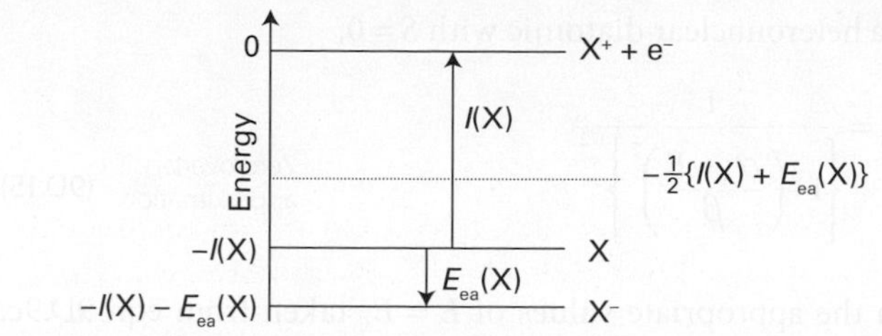

Figure 9D.2 The procedure for estimating the Coulomb integral in terms of the ionization energy and electron affinity.

Brief illustration 9D.2

Consider HF. The general form of the molecular orbital is $\psi = c_H\psi_H + c_F\psi_F$, where ψ_H is an H1s orbital and ψ_F is an F2p$_z$ orbital (with z along the internuclear axis, the convention for linear molecules). The relevant data are as follows:

	I/eV	E_{ea}/eV	$-\frac{1}{2}\{I + E_{ea}\}$/eV
H	13.6	0.75	−7.2
F	17.4	3.34	−10.4

Therefore set $\alpha_A = \alpha_H = -7.2$ eV and $\alpha_B = \alpha_F = -10.4$ eV. Taking $\beta = -1.0$ eV as a typical value and setting $S = 0$ for simplicity, substitution into eqn 9D.9c gives

$$E_\pm/\text{eV} = \tfrac{1}{2}(-7.2-10.4) \pm \tfrac{1}{2}(-7.2+10.4)\left\{1+\left(\frac{-2.0}{-7.2+10.4}\right)^2\right\}^{1/2}$$

$$= -8.8 \pm 1.9 = -10.7 \text{ and } -6.9$$

These values, representing a bonding orbital at −10.7 eV and an antibonding orbital at −6.9 eV, are shown in Fig. 9D.3.

(b) The features of the solutions

An important feature of eqn 9D.9c is that as the energy difference $|\alpha_A - \alpha_B|$ between the interacting atomic orbitals increases, the bonding and antibonding effects decrease (Fig. 9D.4). When $|\alpha_B - \alpha_A| >> 2|\beta|$ it is possible to use the approximation $(1 + x)^{1/2} \approx 1 + \frac{1}{2}x$ (see *The chemist's toolkit* 12 in Topic 5B) to obtain

$$E_+ \approx \alpha_A + \frac{\beta^2}{\alpha_A - \alpha_B} \qquad E_- \approx \alpha_B - \frac{\beta^2}{\alpha_A - \alpha_B} \qquad (9D.10)$$

As these expressions show, and as can be seen from the graph, when the energy difference $|\alpha_A - \alpha_B|$ is very large, the energies of the resulting molecular orbitals differ only slightly from those of the atomic orbitals, which implies in turn that the bonding and antibonding effects are small. That is:

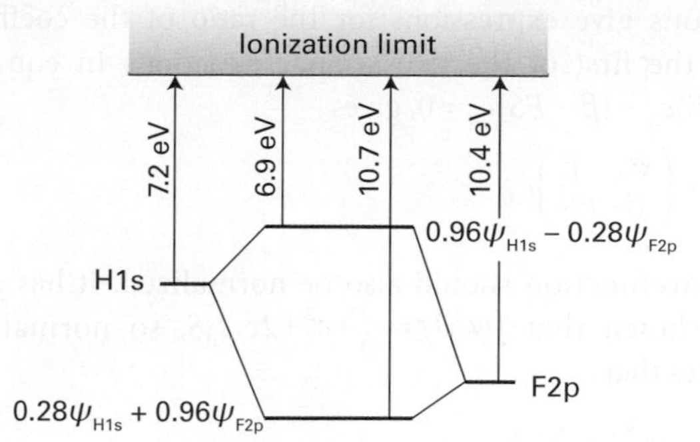

Figure 9D.3 The estimated energies of the Coulomb integrals α in HF and the molecular orbitals they form.

The strongest bonding and antibonding effects are obtained when the two contributing orbitals have similar energies.

Orbital contribution criterion

The large difference in energy between core and valence orbitals is the justification for neglecting the contribution of core orbitals to molecular orbitals constructed from valence atomic orbitals. Although the core orbitals of one atom have a similar energy to the core orbitals of the other atom, so might be expected to combine strongly, core–core interaction is largely negligible because the core orbitals are so contracted that the interaction between them, as measured by the value of $|\beta|$, is negligible. It is also a justification for treating the s and p$_z$ contributions to σ orbital formation separately, an approximation used in Topic 9C in the discussion of homonuclear diatomic molecules.

The values of the coefficients in the linear combination in eqn 9D.1 are obtained by solving the secular equations after substituting the two energies obtained from the secular determinant. The lower energy, E_+, gives the coefficients for the bonding molecular orbital, the upper energy, E_-, the coefficients for the antibonding molecular orbital. The secular

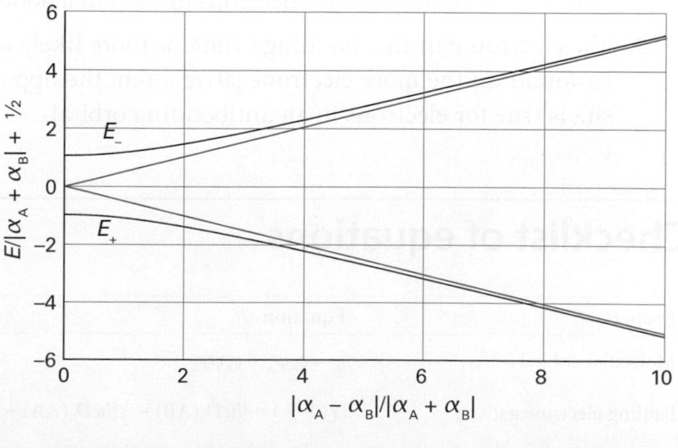

Figure 9D.4 The variation of the energies of the molecular orbitals as the energy difference of the contributing atomic orbitals is changed. The plots are for $\beta = -1$ eV; the blue lines are for the energies in the absence of mixing (i.e. $\beta = 0$).

equations give expressions for the ratio of the coefficients. Thus, the first of the two secular equations in eqn 9D.6a, $(\alpha_A - E)c_A + (\beta - ES)c_B = 0$, gives

$$c_B = -\left(\frac{\alpha_A - E}{\beta - ES}\right)c_A \qquad (9D.11)$$

The wavefunction should also be normalized. It has already been shown that $\int \psi^2 d\tau = c_A^2 + c_B^2 + 2c_A c_B S$, so normalization requires that

$$c_A^2 + c_B^2 + 2c_A c_B S = 1 \qquad (9D.12)$$

When eqn 9D.11 is substituted into this expression, the result is

$$c_A = \frac{1}{\left\{1 + \left(\frac{\alpha_A - E}{\beta - ES}\right)^2 - 2S\left(\frac{\alpha_A - E}{\beta - ES}\right)\right\}^{1/2}} \qquad (9D.13)$$

which, together with eqn 9D.11, gives explicit expressions for the coefficients once the appropriate values of $E = E_\pm$ given in eqn 9D.9a are substituted.

As before, this expression becomes more transparent in two cases. First, for a homonuclear diatomic, with $\alpha_A = \alpha_B = \alpha$ and $E_\pm$ given in eqn 9D.9b, the results are

$$E_+ = \frac{\alpha + \beta}{1 + S} \quad c_A = \frac{1}{\{2(1+S)\}^{1/2}} \quad c_B = c_A \qquad \text{Homonuclear diatomics} \quad (9D.14a)$$

$$E_- = \frac{\alpha - \beta}{1 - S} \quad c_A = \frac{1}{\{2(1-S)\}^{1/2}} \quad c_B = -c_A \qquad (9D.14b)$$

For a heteronuclear diatomic with $S = 0$,

$$c_A = \frac{1}{\left\{1 + \left(\frac{\alpha_A - E}{\beta}\right)^2\right\}^{1/2}} \qquad \text{Zero overlap approximation} \quad (9D.15)$$

with the appropriate values of $E = E_\pm$ taken from eqn 9D.9c. The coefficient c_B is then calculated from eqn 9D.11.

Brief illustration 9D.3

Consider HF again. In the previous *Brief illustration*, with $\alpha_H = -7.2$ eV, $\alpha_F = -10.4$ eV, $\beta = -1.0$ eV, and $S = 0$, the two orbital energies were found to be $E_+ = -10.7$ eV and $E_- = -6.9$ eV. When these values are substituted into eqn 9D.15 the following coefficients are found:

$$E_+ = -10.7 \text{ eV} \qquad \psi_+ = 0.28\psi_H + 0.96\psi_F$$

$$E_- = -6.9 \text{ eV} \qquad \psi_- = 0.96\psi_H - 0.28\psi_F$$

Notice how the lower energy orbital (the one with energy −10.7 eV) has a composition that is more F2p orbital than H1s, and that the opposite is true of the higher energy, antibonding orbital.

Checklist of concepts

☐ 1. A **polar bond** can be regarded as arising from a molecular orbital that is concentrated more on one atom than its partner.

☐ 2. The **electronegativity** of an element is a measure of the power of an atom to attract electrons to itself in a bond.

☐ 3. The electron pair in a bonding orbital is more likely to be found on the more electronegative atom; the opposite is true for electrons in an antibonding orbital.

☐ 3. The **variation principle** provides a criterion for optimizing a trial wavefunction.

☐ 4. A **basis set** is the set of atomic orbitals from which the molecular orbitals are constructed.

☐ 5. The bonding and antibonding effects are strongest when contributing atomic orbitals have similar energies.

Checklist of equations

Property	Equation	Comment	Equation number
Molecular orbital	$\psi = c_A\psi_A + c_B\psi_B$		9D.1
Pauling electronegativity	$\lvert\chi_A - \chi_B\rvert = \{hc\tilde{D}_0(AB) - \frac{1}{2}[hc\tilde{D}_0(AA) + hc\tilde{D}_0(BB)]\}^{1/2}/eV$		9D.2
Mulliken electronegativity	$\chi = \frac{1}{2}(I + E_{ea})/eV$		9D.3
Coulomb integral	$\alpha_A = \int \psi_A \hat{H} \psi_A d\tau$	Definition	9D.7a
Resonance integral	$\beta = \int \psi_A \hat{H} \psi_B d\tau$	Definition	9D.7b
Variation principle	$E = \int \psi^*_{trial} \hat{H} \psi_{trial} d\tau / \int \psi^*_{trial} \psi_{trial} d\tau;\ \partial E/\partial c = 0$		

TOPIC 9E Molecular orbital theory: polyatomic molecules

➤ Why do you need to know this material?

Most molecules of interest in chemistry are polyatomic, so it is important to be able to discuss their electronic structure. Although computational procedures are now widely available, to understand them it is helpful to see how they emerged from the more primitive approach described here.

➤ What is the key idea?

Molecular orbitals can be expressed as linear combinations of all the atomic orbitals of the appropriate symmetry.

➤ What do you need to know already?

This Topic extends the approach used for heteronuclear diatomic molecules in Topic 9D, particularly the concepts of secular equations and secular determinants. The principal mathematical technique used is matrix algebra (*The chemist's toolkits* 24 and 25). You should become familiar with the use of mathematical software to manipulate matrices numerically.

The molecular orbitals of polyatomic molecules are built in the same way as in diatomic molecules (Topic 9D), the only difference being that more atomic orbitals are used to construct them. As for diatomic molecules, polyatomic molecular orbitals spread over the entire molecule. A molecular orbital has the general form

$$\psi = \sum_i c_i \psi_i \quad \text{General form of LCAO-MO} \quad (9E.1)$$

where ψ_i is an atomic orbital and the sum extends over all the valence orbitals of all the atoms in the molecule. The coefficients are found by setting up the secular equations, just as for diatomic molecules, then solving them for the energies (Topic 9D). That step involves formulating the secular determinant and finding the values of the energy that ensure the determinant is equal to 0. Finally these energies are used in the secular equations to find the coefficients of the atomic orbitals for each molecular orbital.

The principal difference between diatomic and polyatomic molecules lies in the greater range of shapes that are possible: a diatomic molecule is necessarily linear, but a triatomic molecule, for instance, may be either linear or angular (bent) with a characteristic bond angle. The shape of a polyatomic molecule—the specification of its bond lengths and its bond angles—can be predicted by calculating the total energy of the molecule for a variety of nuclear positions, and then identifying the conformation that corresponds to the lowest energy. Such calculations are best done using software, which handles the minimization problem automatically and generates the molecular orbital coefficients. However, a more primitive approach gives useful insights into the electronic structure of polyatomic molecules and its interpretation.

Symmetry considerations play a central role in the construction of molecular orbitals of polyatomic molecules, for only atomic orbitals of matching symmetry have non-zero overlap and contribute to a molecular orbital. To discuss these symmetry requirements fully requires the machinery developed in Focus 10, especially Topic 10C. There is one type of symmetry, however, that is intuitive: the planarity of conjugated hydrocarbons. That symmetry provides a distinction between the σ and π orbitals of the molecule, and in elementary approaches such molecules are commonly discussed in terms of the characteristics of their π orbitals, with the σ bonds providing a rigid framework that determines the general shape of the molecule.

9E.1 The Hückel approximation

The π molecular orbital energy level diagrams of conjugated molecules can be constructed by using a set of approximations suggested by Erich Hückel in 1931. All the C atoms are treated identically, so all the Coulomb integrals α (Topic 9D) for the atomic orbitals that contribute to the π orbitals are set equal. For example, in ethene, which is used here to introduce the method, the σ bonds are regarded as fixed, and the calculation leads to the energies of π bonding and antibonding molecular orbitals.

(a) An introduction to the method

The π orbitals are expressed as linear combinations of the C2p orbitals that lie perpendicular to the molecular plane. In ethene, for instance,

$$\psi = c_A\psi_A + c_B\psi_B \quad (9E.2)$$

where ψ_A is a C2p orbital on atom A, and likewise for ψ_B. Next, the optimum coefficients and energies are found by the variation principle as explained in Topic 9D. That is, the appropriate secular determinant is set up, equated to 0, and the equation solved for the energies. For ethene, with $\alpha_A = \alpha_B = \alpha$, the secular determinant is

$$\begin{vmatrix} \alpha - E & \beta - ES \\ \beta - ES & \alpha - E \end{vmatrix} = 0 \qquad (9E.3)$$

In a modern computation all the resonance integrals and overlap integrals would be included, but an indication of the molecular orbital energy level diagram can be obtained more readily by making the following additional **Hückel approximations:**

- All overlap integrals are set equal to zero.
- All resonance integrals between non-neighbours are set equal to zero.
- All remaining resonance integrals are set equal (to β).

Hückel approximations

These approximations are obviously very severe, but they give at least a general picture of the molecular orbital energy levels. The approximations result in the following structure of the secular determinant:

- All diagonal elements: $\alpha - E$
- Off-diagonal elements between neighbouring atoms: β
- All other elements: 0

These approximations convert eqn 9E.3 into

$$\begin{vmatrix} \alpha - E & \beta \\ \beta & \alpha - E \end{vmatrix} = (\alpha - E)^2 - \beta^2 \overset{a^2 - b^2 = (a+b)(a-b)}{=} (\alpha - E + \beta)(\alpha - E - \beta) = 0 \qquad (9E.4)$$

where the determinant has been expanded as explained in *The chemist's toolkit* 23 in Topic 9D. The roots of the equation are $E = \alpha \pm \beta$. The + sign corresponds to the bonding combination (β is negative) and the − sign corresponds to the antibonding combination (Fig. 9E.1).

The building-up principle results in the configuration $1\pi^2$, because each carbon atom supplies one electron to the π system and both electrons can occupy the bonding orbital. The **highest occupied molecular orbital** in ethene, its HOMO, is the 1π orbital; the **lowest unoccupied molecular orbital**, its LUMO, is the 2π orbital (or, as it is sometimes denoted, the $2\pi^*$ orbital). These two orbitals jointly form the **frontier orbitals** of the molecule. The frontier orbitals are important because they are largely responsible for many of the chemical and spectroscopic properties of this and analogous molecules.

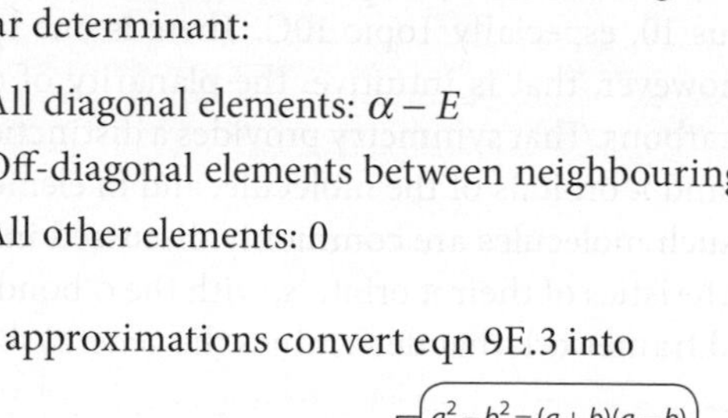

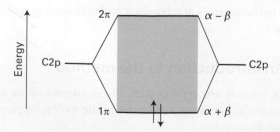

Figure 9E.1 The Hückel molecular orbital energy levels of ethene. Two electrons occupy the lower π orbital.

Brief illustration 9E.1

Within the Hückel framework the energy needed to excite a $\pi^* \leftarrow \pi$ transition is equal to the separation of the 1π and 2π orbitals, which is $2|\beta|$. This transition is known to occur at close to 40 000 cm^{-1}, corresponding to 5.0 eV. It follows that a plausible value of β is about −2.5 eV (−240 kJ mol^{-1}).

(b) The matrix formulation of the method

To make the Hückel theory readily applicable to bigger molecules, it helps to reformulate it in terms of matrices (see *The chemist's toolkit* 24). The starting point is the pair of secular equations developed for a heteronuclear diatomic molecule in Topic 9D:

$$(\alpha_A - E)c_A + (\beta - ES)c_B = 0$$
$$(\beta - ES)c_A + (\alpha_B - E)c_B = 0$$

To prepare to generalize this expression write $\alpha_J = H_{JJ}$ (with J = A or B), $\beta = H_{AB}$, and label the overlap integrals with their respective atoms, so S becomes S_{AB}. More symmetry can be introduced into the equations (which makes it simpler to generalize them) by replacing the E in $\alpha_J - E$ by ES_{JJ}, with $S_{JJ} = 1$. At this point, the two equations are

$$(H_{AA} - ES_{AA})c_A + (H_{AB} - ES_{AB})c_B = 0$$
$$(H_{BA} - ES_{BA})c_A + (H_{BB} - ES_{BB})c_B = 0$$

There is one further notational change. The coefficients c_J depend on the value of E, so it is necessary to distinguish the two sets corresponding to the two energies, denoted E_n with $n = 1$ and 2. The coefficients are written as $c_{n,J}$, with $n = 1$ (the coefficients $c_{1,A}$ and $c_{1,B}$ for energy E_1) or 2 (the coefficients $c_{2,A}$ and $c_{2,B}$ for energy E_2). With this notational change, the two equations become

$$(H_{AA} - E_n S_{AA})c_{n,A} + (H_{AB} - E_n S_{AB})c_{n,B} = 0 \qquad (9E.5a)$$
$$(H_{BA} - E_n S_{BA})c_{n,A} + (H_{BB} - E_n S_{BB})c_{n,B} = 0 \qquad (9E.5b)$$

with $n = 1$ and 2, giving four equations in all. Each pair of equations can be written in matrix form as

$$\begin{pmatrix} H_{AA} - E_n S_{AA} & H_{AB} - E_n S_{AB} \\ H_{BA} - E_n S_{BA} & H_{BB} - E_n S_{BB} \end{pmatrix} \begin{pmatrix} c_{n,A} \\ c_{n,B} \end{pmatrix} = 0 \qquad (9E.5c)$$

The chemist's toolkit 24 Matrices

A **matrix** is an array of numbers arranged in a certain number of rows and a certain number of columns; the numbers of rows and columns may be different. The rows and columns are numbered 1, 2, … so that the number at each position in the matrix, called the **matrix element**, has a unique row and column index. The element of a matrix M at row r and column c is denoted M_{rc}. For instance, a 3 × 3 matrix is

$$M=\begin{pmatrix} M_{11} & M_{12} & M_{13} \\ M_{21} & M_{22} & M_{23} \\ M_{31} & M_{32} & M_{33} \end{pmatrix}$$

The **trace** of a matrix, Tr M, is the sum of the diagonal elements. In this case

$$\mathrm{Tr}\,M = M_{11} + M_{22} + M_{33}$$

A **unit** matrix has diagonal elements equal to 1 and all other elements zero. A 3 × 3 unit matrix is therefore

$$\mathbf{1}=\begin{pmatrix} 1 & 0 & 0 \\ 0 & 1 & 0 \\ 0 & 0 & 1 \end{pmatrix}$$

Matrices are added by adding the corresponding matrix elements. Thus, to add the matrices A and B to give the sum $S = A + B$, each element of S is given by

$$S_{rc} = A_{rc} + B_{rc}$$

Only matrices of the same dimensions can be added together.

Matrices are multiplied to obtain the product $P = AB$; each element of P is given by

$$P_{rc} = \sum_n A_{rn}B_{nc}$$

Matrices can be multiplied only if the number of columns in A is equal to the number of rows in B. Square matrices (those with the same number of rows and columns) can therefore be multiplied only if both matrices have the same dimension (that is, both are $n \times n$). The products AB and BA are not necessarily the same, so matrix multiplication is in general 'non-commutative'.

An $n \times 1$ matrix (with n elements in one column) is called a **column vector**. It may be multiplied by a square $n \times n$ matrix to generate a new column vector, as in

$$\begin{pmatrix} P_1 \\ P_2 \\ P_3 \end{pmatrix} = \begin{pmatrix} A_{11} & A_{12} & A_{13} \\ A_{21} & A_{22} & A_{23} \\ A_{31} & A_{32} & A_{33} \end{pmatrix} \times \begin{pmatrix} B_1 \\ B_2 \\ B_3 \end{pmatrix}$$

The elements of the two column vectors need only one index to indicate their row. Each element of P is given by

$$P_r = \sum_n A_{rn}B_n$$

A $1 \times n$ matrix (a single row with n elements) is called a **row vector**. It may be multiplied by a square $n \times n$ matrix to generate a new row vector, as in

$$(P_1\ P_2\ P_3) = (B_1\ B_2\ B_3) \times \begin{pmatrix} A_{11} & A_{12} & A_{13} \\ A_{21} & A_{22} & A_{23} \\ A_{31} & A_{32} & A_{33} \end{pmatrix}$$

In general the elements of P are

$$P_c = \sum_n B_n A_{nc}$$

Note that a column vector is multiplied 'from the left' by the square matrix and a row vector is multiplied 'from the right'. The **inverse** of a matrix A, denoted A^{-1}, has the property that $AA^{-1} = A^{-1}A = \mathbf{1}$, where $\mathbf{1}$ is a unit matrix with the same dimensions as A.

as can be verified by multiplying out the matrices to give the two expressions in eqns 9E.5a and 9E.5b. Now introduce the **hamiltonian matrix** H and the **overlap matrix** S, and write the coefficients corresponding to the energy E_n as a column vector c_n:

$$H=\begin{pmatrix} H_{AA} & H_{AB} \\ H_{BA} & H_{BB} \end{pmatrix} \quad S=\begin{pmatrix} S_{AA} & S_{AB} \\ S_{BA} & S_{BB} \end{pmatrix} \quad c_n=\begin{pmatrix} c_{n,A} \\ c_{n,B} \end{pmatrix} \qquad (9E.6)$$

Then

$$H - E_nS = \begin{pmatrix} H_{AA} - E_nS_{AA} & H_{AB} - E_nS_{AB} \\ H_{BA} - E_nS_{BA} & H_{BB} - E_nS_{BB} \end{pmatrix}$$

and eqn 9E.5c may be written more succinctly as

$$(H - E_nS)c_n = 0 \quad \text{or} \quad Hc_n = Sc_nE_n \qquad (9E.7)$$

These two sets of equations (with $n = 1$ and 2) can be combined into a single matrix equation of the form

$$Hc = ScE \qquad (9E.8)$$

by introducing the matrices

$$c = (c_1\ \ c_2) = \begin{pmatrix} c_{1,A} & c_{2,A} \\ c_{1,B} & c_{2,B} \end{pmatrix} \qquad E = \begin{pmatrix} E_1 & 0 \\ 0 & E_2 \end{pmatrix} \qquad (9E.9)$$

How is that done? 9E.1 Justifying the matrix formulation

Substitution of the matrices defined in eqn 9E.9 into eqn 9E.8 gives

$$\underbrace{\begin{pmatrix} H_{AA} & H_{AB} \\ H_{BA} & H_{BB} \end{pmatrix}}_{H}\underbrace{\begin{pmatrix} c_{1,A} & c_{2,A} \\ c_{1,B} & c_{2,B} \end{pmatrix}}_{c} = \underbrace{\begin{pmatrix} S_{AA} & S_{AB} \\ S_{BA} & S_{BB} \end{pmatrix}}_{S}\underbrace{\begin{pmatrix} c_{1,A} & c_{2,A} \\ c_{1,B} & c_{2,B} \end{pmatrix}}_{c}\underbrace{\begin{pmatrix} E_1 & 0 \\ 0 & E_2 \end{pmatrix}}_{E}$$

The product on the left is

$$\begin{pmatrix} H_{AA} & H_{AB} \\ H_{BA} & H_{BB} \end{pmatrix}\begin{pmatrix} c_{1,A} & c_{2,A} \\ c_{1,B} & c_{2,B} \end{pmatrix} = \begin{pmatrix} H_{AA}c_{1,A}+H_{AB}c_{1,B} & H_{AA}c_{2,A}+H_{AB}c_{2,B} \\ H_{BA}c_{1,A}+H_{BB}c_{1,B} & H_{BA}c_{2,A}+H_{BB}c_{2,B} \end{pmatrix}$$

The product on the right is

$$\begin{pmatrix} S_{AA} & S_{AB} \\ S_{BA} & S_{BB} \end{pmatrix}\begin{pmatrix} c_{1,A} & c_{2,A} \\ c_{1,B} & c_{2,B} \end{pmatrix}\begin{pmatrix} E_1 & 0 \\ 0 & E_2 \end{pmatrix} = \begin{pmatrix} S_{AA} & S_{AB} \\ S_{BA} & S_{BB} \end{pmatrix}\begin{pmatrix} c_{1,A}E_1 & c_{2,A}E_2 \\ c_{1,B}E_1 & c_{2,B}E_2 \end{pmatrix}$$

$$= \begin{pmatrix} E_1S_{AA}c_{1,A}+E_1S_{AB}c_{1,B} & E_2S_{AA}c_{2,A}+E_2S_{AB}c_{2,B} \\ E_1S_{BA}c_{1,A}+E_1S_{BB}c_{1,B} & E_2S_{BA}c_{2,A}+E_2S_{BB}c_{2,B} \end{pmatrix}$$

Comparison of matching terms (like those in blue) recreates the four secular equations (two for each value of n) given in eqns 9E.5a and 9E.5b.

In the Hückel approximation, $H_{AA}=H_{BB}=\alpha$, $H_{AB}=H_{BA}=\beta$, and overlap is neglected by setting $\boldsymbol{S}=\boldsymbol{1}$, the unit matrix (with 1 on the diagonal and 0 elsewhere). Then the first two matrices in eqn 9E.6 become

$$\boldsymbol{H}=\begin{pmatrix} \alpha & \beta \\ \beta & \alpha \end{pmatrix} \qquad \boldsymbol{S}=\begin{pmatrix} 1 & 0 \\ 0 & 1 \end{pmatrix}$$

and because $\boldsymbol{S}$ is now a unit matrix, multiplication by which has no effect, eqn 9E.8 becomes

$$\boldsymbol{Hc}=\boldsymbol{cE}$$

At this point, multiplication from the left by the inverse matrix $\boldsymbol{c}^{-1}$ gives, after using $\boldsymbol{c}^{-1}\boldsymbol{c}=\boldsymbol{1}$,

$$\boldsymbol{c}^{-1}\boldsymbol{Hc}=\boldsymbol{E} \qquad (9E.10)$$

The matrix $\boldsymbol{E}$ is diagonal, with diagonal elements E_n, so an interpretation of this equation is that the energies are calculated by finding a transformation of $\boldsymbol{H}$, its conversion to $\boldsymbol{c}^{-1}\boldsymbol{Hc}$, that makes it diagonal. This procedure is called **matrix diagonalization**. The columns of the matrix $\boldsymbol{c}$ that brings about this diagonalization are the coefficients of the orbitals used as the basis set, and give the composition of the molecular orbitals.

Example 9E.1 Finding molecular orbitals by matrix diagonalization

Set up and solve the matrix equations within the Hückel approximation for the π orbitals of butadiene (1).

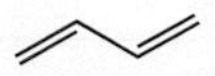

1 Butadiene

Collect your thoughts The matrices are four-dimensional for this four-atom system. You need to construct the matrix $\boldsymbol{H}$ by using the Hückel approximation and the parameters α and β. Once you have the hamiltonian matrix, you need to find the matrix $\boldsymbol{c}$ that diagonalizes it: for this step, use mathematical software. Full details are given in *The chemist's toolkit* 25, but note that if $\boldsymbol{H}=\alpha\boldsymbol{1}+\boldsymbol{M}$, where $\boldsymbol{M}$ is a non-diagonal matrix, then because $\alpha\boldsymbol{c}^{-1}\boldsymbol{1}\boldsymbol{c}=\alpha\boldsymbol{c}^{-1}\boldsymbol{c}=\alpha\boldsymbol{1}$, whatever matrix $\boldsymbol{c}$ diagonalizes $\boldsymbol{M}$ leaves $\alpha\boldsymbol{1}$ unchanged, so to achieve the overall diagonalization of $\boldsymbol{H}$ you need to diagonalize only $\boldsymbol{M}$.

The solution With C atoms labelled A, B, C, and D, the hamiltonian matrix $\boldsymbol{H}$ is

$$\boldsymbol{H}=\begin{pmatrix} \overbrace{H_{AA}}^{\alpha} & \overbrace{H_{AB}}^{\beta} & \overbrace{H_{AC}}^{0} & \overbrace{H_{AD}}^{0} \\ H_{BA} & H_{BB} & H_{BC} & H_{BD} \\ H_{CA} & H_{CB} & H_{CC} & H_{CD} \\ H_{DA} & H_{DB} & H_{DC} & H_{DD} \end{pmatrix} \xrightarrow{\text{Hückel approximation}} \begin{pmatrix} \alpha & \beta & 0 & 0 \\ \beta & \alpha & \beta & 0 \\ 0 & \beta & \alpha & \beta \\ 0 & 0 & \beta & \alpha \end{pmatrix}$$

which is written as

$$\boldsymbol{H}=\alpha\boldsymbol{1}+\beta\overbrace{\begin{pmatrix} 0 & 1 & 0 & 0 \\ 1 & 0 & 1 & 0 \\ 0 & 1 & 0 & 1 \\ 0 & 0 & 1 & 0 \end{pmatrix}}^{\boldsymbol{M}}$$

The diagonalized form of the matrix $\boldsymbol{M}$ (using software) is

$$\begin{pmatrix} +1.62 & 0 & 0 & 0 \\ 0 & +0.62 & 0 & 0 \\ 0 & 0 & -0.62 & 0 \\ 0 & 0 & 0 & -1.62 \end{pmatrix}$$

so the diagonalized hamiltonian matrix is

$$\boldsymbol{E}=\begin{pmatrix} \alpha+1.62\beta & 0 & 0 & 0 \\ 0 & \alpha+0.62\beta & 0 & 0 \\ 0 & 0 & \alpha-0.62\beta & 0 \\ 0 & 0 & 0 & \alpha-1.62\beta \end{pmatrix}$$

The matrix that achieves the diagonalization is

$$\boldsymbol{c}=\begin{pmatrix} 0.372 & 0.602 & 0.602 & 0.372 \\ 0.602 & 0.372 & -0.372 & -0.602 \\ 0.602 & -0.372 & -0.372 & 0.602 \\ 0.372 & -0.602 & 0.602 & -0.372 \end{pmatrix}$$

with each column giving the coefficients of the atomic orbitals for the corresponding molecular orbital. It follows that the energies and molecular orbitals are

$$\begin{aligned} E_1 &= \alpha+1.62\beta & \psi_1 &= 0.372\psi_A+0.602\psi_B+0.602\psi_C+0.372\psi_D \\ E_2 &= \alpha+0.62\beta & \psi_2 &= 0.602\psi_A+0.372\psi_B-0.372\psi_C-0.602\psi_D \\ E_3 &= \alpha-0.62\beta & \psi_3 &= 0.602\psi_A-0.372\psi_B-0.372\psi_C+0.602\psi_D \\ E_4 &= \alpha-1.62\beta & \psi_4 &= 0.372\psi_A-0.602\psi_B+0.602\psi_C-0.372\psi_D \end{aligned}$$

The chemist's toolkit 25 Matrix methods for solving eigenvalue equations

In matrix form, an **eigenvalue equation** is

$$\boldsymbol{Mx} = \lambda \boldsymbol{x} \qquad \text{Eigenvalue equation} \quad (1a)$$

where $\boldsymbol{M}$ is a square matrix with n rows and n columns, λ is a constant, the **eigenvalue**, and $\boldsymbol{x}$ is the **eigenvector**, an $n \times 1$ (column) matrix that satisfies the conditions of the eigenvalue equation and has the form:

$$\boldsymbol{x} = \begin{pmatrix} x_1 \\ x_2 \\ \vdots \\ x_n \end{pmatrix}$$

In general, there are n eigenvalues $\lambda^{(i)}$, $i = 1, 2, \ldots, n$, and n corresponding eigenvectors $\boldsymbol{x}^{(i)}$. Equation 1a can be rewritten as

$$(\boldsymbol{M} - \lambda \boldsymbol{1})\boldsymbol{x} = 0 \qquad (1b)$$

where $\boldsymbol{1}$ is an $n \times n$ unit matrix, and where the property $\boldsymbol{1x} = \boldsymbol{x}$ has been used. This equation has a solution only if the determinant $|\boldsymbol{M} - \lambda \boldsymbol{1}|$ of the matrix $\boldsymbol{M} - \lambda\boldsymbol{1}$ is zero. It follows that the n eigenvalues may be found from the solution of the secular equation:

$$|\boldsymbol{M} - \lambda \boldsymbol{1}| = 0 \qquad (2)$$

The n eigenvalues found by solving the secular equations are used to find the corresponding eigenvectors. To do so, begin by considering an $n \times n$ matrix $\boldsymbol{X}$ the columns of which are formed from the eigenvectors corresponding to all the eigenvalues. Thus, if the eigenvalues are $\lambda_1, \lambda_2, \ldots$, and the corresponding eigenvectors are

$$\boldsymbol{x}^{(1)} = \begin{pmatrix} x_1^{(1)} \\ x_2^{(1)} \\ \vdots \\ x_n^{(1)} \end{pmatrix} \quad \boldsymbol{x}^{(2)} = \begin{pmatrix} x_1^{(2)} \\ x_2^{(2)} \\ \vdots \\ x_n^{(2)} \end{pmatrix} \cdots \boldsymbol{x}^{(n)} = \begin{pmatrix} x_1^{(n)} \\ x_2^{(n)} \\ \vdots \\ x_n^{(n)} \end{pmatrix} \qquad (3a)$$

then the matrix $\boldsymbol{X}$ is

$$\boldsymbol{X} = (\boldsymbol{x}^{(1)}\ \boldsymbol{x}^{(2)} \cdots \boldsymbol{x}^{(n)}) = \begin{pmatrix} x_1^{(1)} & x_1^{(2)} & \cdots & x_1^{(n)} \\ x_2^{(1)} & x_2^{(2)} & \cdots & x_2^{(n)} \\ \vdots & \vdots & & \vdots \\ x_n^{(1)} & x_n^{(2)} & \cdots & x_n^{(n)} \end{pmatrix} \qquad (3b)$$

Similarly, form an $n \times n$ matrix $\boldsymbol{\Lambda}$ with the eigenvalues λ along the diagonal and zeroes elsewhere:

$$\boldsymbol{\Lambda} = \begin{pmatrix} \lambda_1 & 0 & \cdots & 0 \\ 0 & \lambda_2 & \cdots & 0 \\ \vdots & \vdots & & \vdots \\ 0 & 0 & \cdots & \lambda_n \end{pmatrix} \qquad (4)$$

Now all the eigenvalue equations $\boldsymbol{Mx}^{(i)} = \lambda_i \boldsymbol{x}^{(i)}$ may be combined into the single matrix equation

$$\boldsymbol{MX} = \boldsymbol{X\Lambda} \qquad (5)$$

Finally, form $\boldsymbol{X}^{-1}$ from $\boldsymbol{X}$ and multiply eqn 5 by it from the left:

$$\boldsymbol{X}^{-1}\boldsymbol{MX} = \boldsymbol{X}^{-1}\boldsymbol{X\Lambda} = \boldsymbol{\Lambda} \qquad (6)$$

A structure of the form $\boldsymbol{X}^{-1}\boldsymbol{MX}$ is called a **similarity transformation**. In this case the similarity transformation $\boldsymbol{X}^{-1}\boldsymbol{MX}$ makes $\boldsymbol{M}$ diagonal (because $\boldsymbol{\Lambda}$ is diagonal). It follows that if the matrix $\boldsymbol{X}$ that causes $\boldsymbol{X}^{-1}\boldsymbol{MX}$ to be diagonal is known, then the problem is solved: the diagonal matrix so produced has the eigenvalues as its only nonzero elements, and the matrix $\boldsymbol{X}$ used to bring about the transformation has the corresponding eigenvectors as its columns. In practice, the eigenvalues and eigenvectors are obtained by using mathematical software.

where the C2p atomic orbitals are denoted by $\psi_A, \ldots, \psi_D$. The molecular orbitals are mutually orthogonal and, with overlap neglected, normalized.

Comment. Note that $\psi_1, \ldots, \psi_4$ correspond to the 1π, …, 4π molecular orbitals of butadiene.

Self-test 9E.1 Repeat the exercise for the allyl radical, $\cdot CH_2{-}CH{=}CH_2$; assume that each carbon is sp^2 hybridized, and take as a basis one out-of-plane 2p orbital on each atom.

Answer: $E = \alpha + 1.41\beta, \alpha, \alpha - 1.41\beta$; $\psi_1 = 0.500\psi_A + 0.707\psi_B + 0.500\psi_C$, $\psi_2 = 0.707\psi_A - 0.707\psi_C$, $\psi_3 = 0.500\psi_A - 0.707\psi_B + 0.500\psi_C$

9E.2 Applications

Although the Hückel method is very primitive, it can be used to account for some of the properties of conjugated polyenes.

(a) π-Electron binding energy

As seen in *Example* 9E.1, the energies of the four LCAO-MOs for butadiene are

$$E = \alpha \pm 1.62\beta,\ \alpha \pm 0.62\beta \qquad (9E.11)$$

These orbitals and their energies are drawn in Fig. 9E.2. Note that:

- The greater the number of internuclear nodes, the higher the energy of the orbital.
- There are four electrons to accommodate, so the ground-state configuration is $1\pi^2 2\pi^2$.
- The frontier orbitals of butadiene are the 2π orbital (the HOMO, which is largely bonding) and the 3π orbital (the LUMO, which is largely antibonding).

'Largely bonding' means that an orbital has both bonding and antibonding interactions between various neighbours, but the

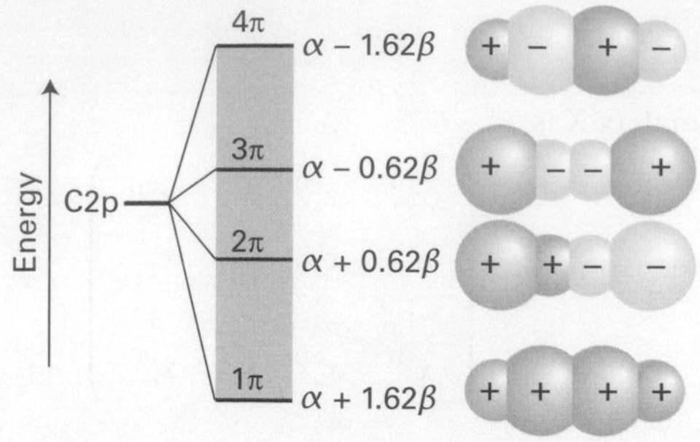

Figure 9E.2 The Hückel molecular orbital energy levels of butadiene and the top view of the corresponding π orbitals. The four p electrons (one supplied by each C) occupy the two lower π orbitals. Note that all the orbitals are delocalized.

bonding effects dominate. 'Largely antibonding' indicates that the antibonding effects dominate.

An important point emerges by calculating the total **π-electron binding energy**, E_π, the sum of the energies of each π electron, and comparing it with the value for ethene. In ethene the π-electron binding energy is

$$E_\pi = 2(\alpha+\beta) = 2\alpha + 2\beta$$

In butadiene it is

$$E_\pi = 2(\alpha+1.62\beta) + 2(\alpha+0.62\beta) = 4\alpha + 4.48\beta$$

Therefore, the energy of the butadiene molecule lies lower by 0.48β (about 115 kJ mol^{-1}) than the sum of two individual π bonds (recall that β is negative). This extra stabilization of a conjugated system compared with a set of localized π bonds is called the **delocalization energy** of the molecule.

A closely related quantity is the **π-bond formation energy**, $E_{\rm bf}$, the energy released when a π bond is formed. Because the contribution of α is the same in the molecule as in the atoms, the π-bond formation energy can be calculated from the π-electron binding energy by writing

$$E_{\rm bf} = E_\pi - N_{\rm C}\alpha \qquad \text{π-Bond formation energy [definition]} \qquad (9E.12)$$

where $N_{\rm C}$ is the number of carbon atoms in the molecule. The π-bond formation energy in butadiene, for instance, is 4.48β.

Example 9E.2 Estimating the delocalization energy

Use the Hückel approximation to find the energies of the π orbitals of cyclobutadiene, and estimate the delocalization energy.

Collect your thoughts Set up the hamiltonian matrix using the same basis as for butadiene, but note that atoms A and D are also now neighbours. Then diagonalize the matrix to find the energies. For the delocalization energy, subtract from the total π-bond energy the energy of two π-bonds.

The solution The hamiltonian matrix is

$$H = \begin{pmatrix} \alpha & \beta & 0 & \beta \\ \beta & \alpha & \beta & 0 \\ 0 & \beta & \alpha & \beta \\ \beta & 0 & \beta & \alpha \end{pmatrix}$$

$$= \alpha \mathbf{1} + \beta \begin{pmatrix} 0 & 1 & 0 & 1 \\ 1 & 0 & 1 & 0 \\ 0 & 1 & 0 & 1 \\ 1 & 0 & 1 & 0 \end{pmatrix} \xrightarrow{\text{Diagonalize}} \begin{pmatrix} 2 & 0 & 0 & 0 \\ 0 & 0 & 0 & 0 \\ 0 & 0 & 0 & 0 \\ 0 & 0 & 0 & -2 \end{pmatrix}$$

Diagonalization gives the energies of the orbitals as

$$E = \alpha + 2\beta,\ \alpha,\ \alpha,\ \alpha - 2\beta$$

Four electrons must be accommodated. Two occupy the lowest orbital (of energy $\alpha + 2\beta$), and two occupy the doubly degenerate orbitals (of energy α). The total energy is therefore $4\alpha + 4\beta$. Two isolated π bonds would have an energy $4\alpha + 4\beta$; therefore, in this case, the delocalization energy is zero.

Self-test 9E.2 Repeat the calculation for benzene (use software!).

Answer: See next subsection

(b) Aromatic stability

The most notable example of delocalization conferring extra stability is benzene and the aromatic molecules based on its structure. In elementary accounts, the structure of benzene, and other aromatic compounds, is often expressed in a mixture of valence-bond and molecular orbital terms, with typically valence-bond language (Topic 9A) used for its σ framework and molecular orbital language used to describe its π electrons.

First, the valence-bond component. The six C atoms are regarded as sp^2 hybridized, with a single unhydridized perpendicular 2p orbital. One H atom is bonded by (Csp^2,H1s) overlap to each C carbon, and the remaining hybrids overlap to give a regular hexagon of atoms (Fig. 9E.3). The internal angle of a

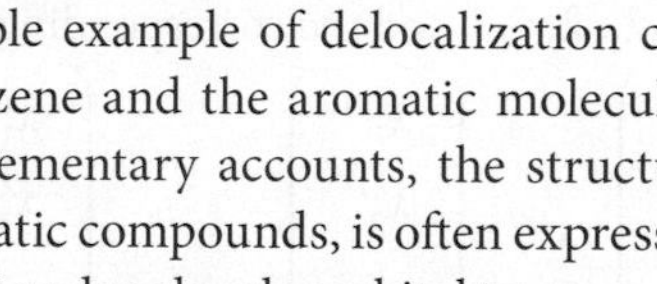

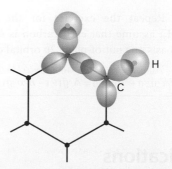

Figure 9E.3 The σ framework of benzene is formed by the overlap of Csp^2 hybrids, which fit without strain into a hexagonal arrangement.

regular hexagon is 120°, so sp^2 hybridization is ideally suited for forming σ bonds. The hexagonal shape of benzene permits strain-free σ bonding.

Now consider the molecular orbital component of the description. The six C2p orbitals overlap to give six π orbitals that spread all round the ring. Their energies are calculated within the Hückel approximation by diagonalizing the hamiltonian matrix

$$H=\begin{pmatrix} \alpha & \beta & 0 & 0 & 0 & \beta \\ \beta & \alpha & \beta & 0 & 0 & 0 \\ 0 & \beta & \alpha & \beta & 0 & 0 \\ 0 & 0 & \beta & \alpha & \beta & 0 \\ 0 & 0 & 0 & \beta & \alpha & \beta \\ \beta & 0 & 0 & 0 & \beta & \alpha \end{pmatrix}$$

$$=\alpha\mathbf{1}+\beta\begin{pmatrix} 0 & 1 & 0 & 0 & 0 & 1 \\ 1 & 0 & 1 & 0 & 0 & 0 \\ 0 & 1 & 0 & 1 & 0 & 0 \\ 0 & 0 & 1 & 0 & 1 & 0 \\ 0 & 0 & 0 & 1 & 0 & 1 \\ 1 & 0 & 0 & 0 & 1 & 0 \end{pmatrix} \xrightarrow{\text{Diagonalize}} \begin{pmatrix} 2 & 0 & 0 & 0 & 0 & 0 \\ 0 & 1 & 0 & 0 & 0 & 0 \\ 0 & 0 & 1 & 0 & 0 & 0 \\ 0 & 0 & 0 & -1 & 0 & 0 \\ 0 & 0 & 0 & 0 & -1 & 0 \\ 0 & 0 & 0 & 0 & 0 & -2 \end{pmatrix}$$

The MO energies, the diagonal elements of this matrix, are

$$E=\alpha\pm 2\beta,\ \alpha\pm\beta,\ \alpha\pm\beta, \qquad (9E.13)$$

as shown in Fig. 9E.4. The orbitals there have been given symmetry labels that are explained in Topic 10B. Note that the lowest energy orbital is bonding between all neighbouring atoms, the highest energy orbital is antibonding between each

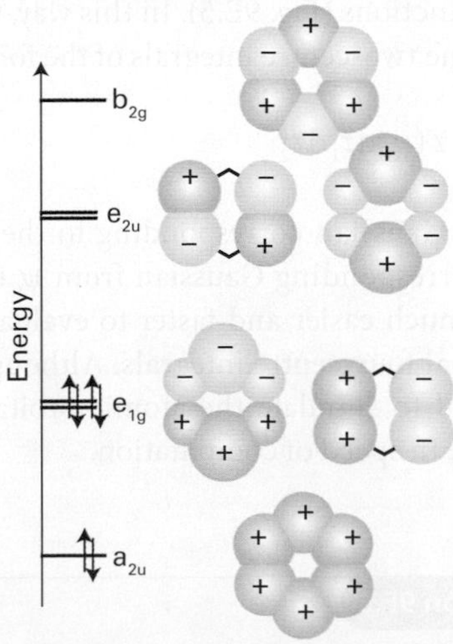

Figure 9E.4 The Hückel orbitals of benzene and the corresponding energy levels. The orbital labels are explained in Topic 10B. The bonding and antibonding character of the delocalized orbitals reflects the numbers of nodes between the atoms. In the ground state, only the bonding orbitals are occupied.

pair of neighbours, and the intermediate orbitals are a mixture of bonding, nonbonding, and antibonding character between adjacent atoms.

Now apply the building-up principle to the π system. There are six electrons to accommodate (one from each C atom), so the three lowest orbitals (a_{2u} and the doubly-degenerate pair e_{1g}) are fully occupied, giving the ground-state configuration $a_{2u}^2e_{1g}^4$. A significant point is that the only molecular orbitals occupied are those with net bonding character (the analogy with the strongly bonded N_2 molecule, Topic 9B, should be noted).

The π-electron binding energy of benzene is

$$E_\pi = 2(\alpha+2\beta)+4(\alpha+\beta)=6\alpha+8\beta$$

If delocalization is ignored and the molecule is thought of as having three isolated π bonds, it would be ascribed a π-electron energy of only $3(2\alpha+2\beta)=6\alpha+6\beta$. The delocalization energy is therefore $2\beta\approx -480\ \text{kJ mol}^{-1}$, which is considerably more than for butadiene. The π-bond formation energy in benzene is 8β.

This discussion suggests that aromatic stability can be traced to two main contributions. First, the shape of the regular hexagon is ideal for the formation of strong σ bonds: the σ framework is relaxed and without strain. Second, the π orbitals are such as to be able to accommodate all the electrons in bonding orbitals, and the delocalization energy is large.

Brief illustration 9E.2

The energies of the four molecular orbitals of cyclobutadiene are $E=\alpha\pm 2\beta, \alpha, \alpha$ (see *Example* 9E.2). There are four π electrons to accommodate in C_4H_4, so the total π-electron binding energy is $2(\alpha+2\beta)+2\alpha=4(\alpha+\beta)$. The energy of two localized π-bonds is $4(\alpha+\beta)$. Therefore, the delocalization energy is zero, so the molecule is not aromatic. There are only two π electrons to accommodate in $C_4H_4^{2+}$, so the total π-electron binding energy is $2(\alpha+2\beta)=2\alpha+4\beta$. The energy of a single localized π-bond is $2(\alpha+\beta)$, so the delocalization energy is 2β and the molecule-ion is aromatic.

9E.3 Computational chemistry

The severe assumptions of the Hückel method are now easy to avoid by using a variety of software packages that can be used not only to calculate the shapes and energies of molecular orbitals but also to predict with reasonable accuracy the structure and reactivity of molecules. The full treatment of molecular electronic structure has received an enormous amount of attention by chemists and has become a keystone of modern chemical research. However, the calculations are very complex, and all this section seeks to do is to provide a brief introduction.[1] In every case, the procedures focus on the calculation or estimation of integrals like H_{JJ} and H_{IJ} rather than setting them equal to the constants α or β, or ignoring them entirely.

In all cases the Schrödinger equation is solved iteratively and self-consistently, just as for the self-consistent field (SCF) approach to atoms (Topic 8B). First, the molecular orbitals for the electrons present in the molecule are formulated as LCAOs. One molecular orbital is then selected and all the others are used to set up an expression for the potential energy of an electron in the chosen orbital. The resulting Schrödinger equation is then solved numerically to obtain a better version of the chosen molecular orbital and its energy. The procedure is repeated for all the molecular orbitals and used to calculate the total energy of the molecule. The process is repeated until the computed orbitals and energy are constant to within some tolerance.

(a) Semi-empirical and *ab initio* methods

In a **semi-empirical method**, many of the integrals are estimated by appealing to spectroscopic data or physical properties such as ionization energies, and using a series of rules to set certain integrals equal to zero. A primitive form of this procedure is used in *Brief illustration* 9D.1 of Topic 9D where the integral α is identified with a combination of the ionization energy and electron affinity of an atom. In an ***ab initio* method** an attempt is made to calculate all the integrals, including overlap integrals. Both procedures employ a great deal of computational effort. The integrals that are required involve atomic orbitals that in general may be centred on different nuclei. It can be appreciated that, if there are several dozen atomic orbitals used to build the molecular orbitals, then there will be tens of thousands of integrals of this form to evaluate (the number of integrals increases as the fourth power of the number of atomic orbitals in the basis, so even for a 10-atom molecule there are 10^4 integrals to evaluate). Some kind of approximation scheme is necessary.

One severe semi-empirical approximation used in the early days of computational chemistry was called **complete neglect of differential overlap** (CNDO), in which all molecular integrals of the form

$$j_0\int\int \psi_A(r_1)\psi_B(r_1)\frac{1}{r_{12}}\psi_C(r_2)\psi_D(r_2)\,d\tau_1\,d\tau_2$$

are set to zero unless ψ_A and ψ_B are the same orbitals centred on the same nucleus, and likewise for ψ_C and ψ_D. The surviving integrals are then adjusted until the energy levels are in good agreement with experiment or the computed enthalpy of formation of the compound is in agreement with experiment.

[1] A more complete account with detailed examples will be found in our companion volume, *Physical chemistry: Quanta, matter, and change* (2014).

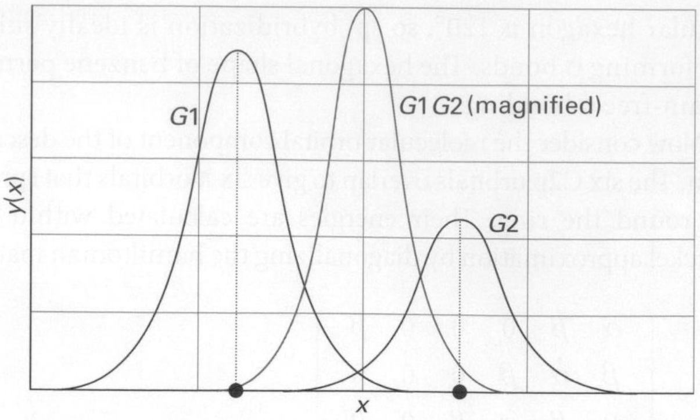

Figure 9E.5 The product of two Gaussian functions on different centres is itself a Gaussian function located at a point between the two contributing Gaussians. The scale of the product has been increased relative to that of its two components.

More recent semi-empirical methods make less severe decisions about which integrals are to be ignored, but they are all descendants of the early CNDO technique.

Commercial packages are also available for *ab initio* calculations. Here the problem is to evaluate as efficiently as possible thousands of integrals that arise from the Coulombic interaction between two electrons like that displayed above, with the possibility that each of the atomic orbitals is centred on a different atom, a so-called 'four-centre integral'. This task is greatly facilitated by expressing the atomic orbitals used in the LCAOs as linear combinations of Gaussian orbitals. A **Gaussian type orbital** (GTO) is a function of the form e^{-r^2}. The advantage of GTOs over the correct orbitals (which for hydrogenic systems are proportional to exponential functions of the form e^{-r}) is that the product of two Gaussian functions is itself a Gaussian function that lies between the centres of the two contributing functions (Fig. 9E.5). In this way, the four-centre integrals become two-centre integrals of the form

$$j_0\int\int X(r_1)\frac{1}{r_{12}}Y(r_2)\,d\tau_1\,d\tau_2$$

where X is the Gaussian corresponding to the product $\psi_A\psi_B$, and Y is the corresponding Gaussian from $\psi_C\psi_D$. Integrals of this form are much easier and faster to evaluate numerically than the original four-centre integrals. Although more GTOs have to be used to simulate the atomic orbitals, there is an overall increase in speed of computation.

Brief illustration 9E.3

Consider a one-dimensional 'homonuclear' system, with GTOs of the form e^{-ax^2} located at 0 and R. Then one of the integrals that would have to be evaluated would include the term

$$\psi_A(r_1)\psi_B(r_1) = e^{-ax^2}e^{-a(x-R)^2} = e^{-2ax^2+2axR-aR^2}$$

Next note that $-2a(x-\frac{1}{2}R)^2=-2ax^2+2axR-\frac{1}{2}aR^2$, so

$$\psi_A(r_1)\psi_B(r_1)=e^{-2a(x-R/2)^2-aR^2/2}=e^{-2a(x-R/2)^2}e^{-aR^2/2}$$

which is proportional to a single Gaussian (the term in blue) centred on the mid-point of the internuclear separation, at $x=\frac{1}{2}R$.

(b) Density functional theory

A technique that has gained considerable ground in recent years to become one of the most widely used techniques for the calculation of molecular structure is **density functional theory** (DFT). Its advantages include less demanding computational effort, less computer time, and—in some cases (particularly d-metal complexes)—better agreement with experimental values than is obtained from other procedures.

The central focus of DFT is the electron density, ρ, rather than the wavefunction, ψ. The 'functional' part of the name comes from the fact that the energy of the molecule is a function of the electron density, written $E[\rho]$, and the electron density is itself a function of position, $\rho(r)$: in mathematics a function of a function is called a 'functional'. The occupied orbitals are used to construct the electron density from

$$\rho(r)=\sum_{m,\text{occupied}}|\psi_m(r)|^2 \quad \text{Electron probability density} \quad (9E.14)$$

and are calculated from modified versions of the Schrödinger equation known as the **Kohn–Sham equations**.

The Kohn–Sham equations are solved iteratively and self-consistently. First, the electron density is guessed. For this step it is common to use a superposition of atomic electron densities. Next, the Kohn–Sham equations are solved to obtain an initial set of orbitals. This set of orbitals is used to obtain a better approximation to the electron density and the process is repeated until the density and the computed energy are constant to within some tolerance.

(c) Graphical representations

One of the most significant developments in computational chemistry has been the introduction of graphical representations of molecular orbitals and electron densities. The raw output of a molecular structure calculation is a list of the coefficients of the atomic orbitals in each molecular orbital and the energies of these orbitals. The graphical representation of a molecular orbital uses stylized shapes to represent the basis set, and then scales their size to indicate the coefficient in the linear combination. Different signs of the wavefunctions are represented by different colours.

Once the coefficients are known, it is possible to construct a representation of the electron density in the molecule by noting which orbitals are occupied and then forming the squares of those orbitals. The total electron density at any point is then the sum of the squares of the wavefunctions evaluated at that point. The outcome is commonly represented by an **isodensity surface**, a surface of constant total electron density (Fig. 9E.6). As shown in the illustration, there are several styles of representing an isodensity surface, as a solid form, as a transparent form with a ball-and-stick representation of the molecule within, or as a mesh. A related representation is a **solvent-accessible surface** in which the shape represents the shape of the molecule by imagining a sphere representing a solvent molecule rolling across the surface and plotting the locations of the centre of that sphere.

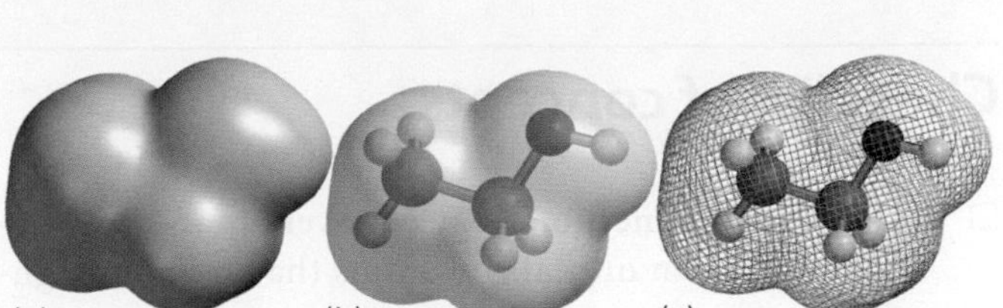
(a) (b) (c)

Figure 9E.6 Various representations of an isodensity surface of ethanol: (a) solid surface, (b) transparent surface, and (c) mesh surface.

One of the most important aspects of a molecule other than its geometrical shape is the distribution of charge over its surface, which is commonly depicted as an **electrostatic potential surface** (an 'elpot surface'). The potential energy, E_p, of an imaginary positive charge Q at a point is calculated by taking into account its interaction with the nuclei and the electron density throughout the molecule. Then, because $E_p = Q\phi$, where ϕ is the electric potential, the potential energy can be interpreted as a potential and depicted as an appropriate colour (Fig. 9E.7). Electron-rich regions usually have negative potentials and electron-poor regions usually have positive potentials.

Representations such as those illustrated here are of critical importance in a number of fields. For instance, they may be used to identify an electron-poor region of a molecule that is susceptible to association with or chemical attack by an electron-rich region of another molecule. Such considerations are important for assessing the pharmacological activity of potential drugs.

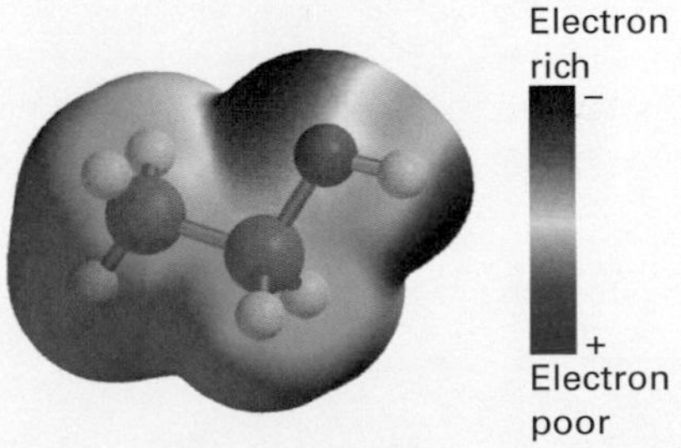

Figure 9E.7 An elpot diagram of ethanol; the molecule has the same orientation as in Fig. 9E.6. Red denotes regions of negative electrostatic potential and blue regions of positive potential (as in $^{\delta-}$O–H$^{\delta+}$).

Checklist of concepts

☐ 1. The **Hückel method** neglects overlap and interactions between orbitals on atoms that are not neighbours.

☐ 2. The highest occupied molecular orbital (HOMO) and the lowest unoccupied molecular orbital (LUMO) are the **frontier orbitals** of a molecule.

☐ 3. The Hückel method may be expressed in a compact manner by introducing matrices.

☐ 4. The **π-bond formation energy** is the energy released when a π bond is formed.

☐ 5. The **π-electron binding energy** is the sum of the energies of each π electron.

☐ 6. The **delocalization energy** is the difference between the π-electron binding energy and the energy of the same molecule with localized π bonds.

☐ 7. The stability of benzene arises from the geometry of the ring and the high delocalization energy.

☐ 8. **Semi-empirical calculations** approximate integrals by estimating them by using empirical data; ***ab initio* methods** evaluate all integrals numerically.

☐ 9. **Density functional theories** develop equations based on the electron density rather than the wavefunction itself.

☐ 10. Graphical techniques are used to plot a variety of surfaces based on electronic structure calculations.

Checklist of equations

Property	Equation	Comment	Equation number
LCAO-MO	$\psi = \sum_i c_i \psi_i$	ψ_i are atomic orbitals	9E.1
Hückel equations	$Hc = ScE$	Hückel approximations: $H_{AB} = 0$ except between neighbours; $S = 1$.	9E.8
Diagonalization	$c^{-1}Hc = E$		9E.10
π-Electron binding energy	E_π = sum of energies of π electrons	Definition	
π-Bond formation energy	$E_{bf} = E_\pi - N_C\alpha$	Definition; N_C is the number of carbon atoms	9E.12
π-Delocalization energy	$E_{deloc} = E_\pi - N_C(\alpha + \beta)$		

FOCUS 9 Molecular structure

TOPIC 9A Valence-bond theory

Discussion questions

D9A.1 Discuss the role of the Born–Oppenheimer approximation in the valence-bond calculation of a molecular potential energy curve or surface.

D9A.2 Why are promotion and hybridization invoked in valence-bond theory?

D9A.3 Describe the various types of hybrid orbitals and how they are used to describe the bonding in alkanes, alkenes, and alkynes. How does hybridization explain that in allene, $CH_2{=}C{=}CH_2$, the two CH_2 groups lie in perpendicular planes?

D9A.4 Why is spin-pairing so common a features of bond formation (in the context of valence-bond theory)?

D9A.5 What are the consequences of resonance?

Exercises

E9A.1(a) Write the valence-bond wavefunction for the single bond in HF.
E9A.1(b) Write the valence-bond wavefunction for the triple bond in N_2.

E9A.2(a) Write the valence-bond wavefunction for the resonance hybrid $HF \leftrightarrow H^+F^- \leftrightarrow H^-F^+$ (allow for different contributions of each structure).
E9A.2(b) Write the valence-bond wavefunction for the resonance hybrid $N_2 \leftrightarrow N^+N^- \leftrightarrow N^{2-}N^{2+} \leftrightarrow$ structures of similar energy.

E9A.3(a) Describe the structure of a P_2 molecule in valence-bond terms. Why is P_4 a more stable form of molecular phosphorus than P_2?
E9A.3(b) Describe the structures of SO_2 and SO_3 in terms of valence-bond theory.

E9A.4(a) Account for the ability of nitrogen to form four bonds, as in NH_4^+.
E9A.4(b) Account for the ability of phosphorus to form five bonds, as in PF_5.

E9A.5(a) Describe the bonding in 1,3-butadiene using hybrid orbitals.
E9A.5(b) Describe the bonding in 1,3-pentadiene using hybrid orbitals.

E9A.6(a) Describe the bonding in methylamine, CH_3NH_2, using hybrid orbitals.
E9A.6(b) Describe the bonding in pyridine, C_5H_5N, using hybrid orbitals.

E9A.7(a) Show that the linear combinations $h_1 = s + p_x + p_y + p_z$ and $h_2 = s - p_x - p_y + p_z$ are mutually orthogonal.
E9A.7(b) Show that the linear combinations $h_1 = (\sin\zeta)s + (\cos\zeta)p$ and $h_2 = (\cos\zeta)s - (\sin\zeta)p$ are mutually orthogonal for all values of the angle ζ (zeta).

E9A.8(a) Normalize to 1 the sp^2 hybrid orbital $h = s + 2^{1/2}p$ given that the s and p orbitals are each normalized to 1.
E9A.8(b) Normalize to 1 the linear combinations in Exercise E9A.7(b) given that the s and p orbitals are each normalized to 1.

Problems

P9A.1 Use the wavefunction for a H1s orbital to write a valence-bond wavefunction of the form $\Psi(1,2) = A(1)B(2) + A(2)B(1)$ in terms of the Cartesian coordinates of each electron, given that the internuclear separation (along the z-axis) is R.

P9A.2 An sp^2 hybrid orbital that lies in the xy-plane and makes an angle of 120° to the x-axis has the form

$$\psi = \frac{1}{3^{1/2}}\left(s - \frac{1}{2^{1/2}}p_x + \frac{3^{1/2}}{2^{1/2}}p_y\right)$$

Use a graphical argument to show that this function points in the specified direction. (*Hint:* Consider the p_x and p_y orbitals as being represented by unit vectors along x and y.)

P9A.3 Confirm that the hybrid orbitals in eqn 9A.7 make angles of 120° to each other. See the *Hint* to Problem P9A.2.

P9A.4 Show that if two equivalent hybrid orbitals of the form sp^λ make an angle θ to each other, then $\lambda = \pm(-1/\cos\theta)^{1/2}$. Plot a graph of λ against θ and confirm that $\theta = 180°$ when no s orbital is included and $\theta = 120°$ when $\lambda = 2$.

TOPIC 9B Molecular orbital theory: the hydrogen molecule-ion

Discussion questions

D9A.1 Discuss the role of the Born–Oppenheimer approximation in the molecular-orbital calculation of a molecular potential energy curve or surface.

D9B.2 What feature of molecular orbital theory is responsible for bond formation?

D9B.3 Why is spin-pairing so common a features of bond formation (in the context of molecular orbital theory)?

Exercises

E9B.1(a) Normalize to 1 the molecular orbital $\psi = \psi_A + \lambda\psi_B$ in terms of the parameter λ and the overlap integral S. Assume that ψ_A and ψ_B are normalized to 1.

E9B.1(b) A better description of the molecule in Exercise E9B.1(a) might be obtained by including more orbitals on each atom in the linear combination. Normalize to 1 the molecular orbital $\psi = \psi_A + \lambda\psi_B + \lambda'\psi_B'$ in terms of the parameters λ and λ' and the appropriate overlap integrals, where ψ_B and ψ_B' are mutually orthogonal and normalized orbitals on atom B.

E9B.2(a) Suppose that a molecular orbital has the (unnormalized) form $0.145A + 0.844B$. Find a linear combination of the orbitals A and B that is orthogonal to this combination and determine the normalization constants of both combinations using $S = 0.250$.

E9B.2(b) Suppose that a molecular orbital has the (unnormalized) form $0.727A + 0.144B$. Find a linear combination of the orbitals A and B that is orthogonal to this combination and determine the normalization constants of both combinations using $S = 0.117$.

E9B.3(a) The energy of H_2^+ with internuclear separation R is given by eqn 9B.4. The values of the contributions are given below. Plot the molecular potential energy curve and find the bond dissociation energy (in electronvolts) and the equilibrium bond length.

R/a_0	0	1	2	3	4
j/j_0	1.000	0.729	0.472	0.330	0.250
k/j_0	1.000	0.736	0.406	0.199	0.092
S	1.000	0.858	0.587	0.349	0.189

where $j_0 = 27.2$ eV, $a_0 = 52.9$ pm, and $E_{H1s} = -\frac{1}{2}j_0$.

E9B.3(b) The same data as in Exercise E9B.3(a) may be used to calculate the molecular potential energy curve for the antibonding orbital, which is given by eqn 9B.7. Plot the curve.

E9B.4(a) Identify the g or u character of bonding and antibonding π orbitals formed by side-by-side overlap of p atomic orbitals.

E9B.4(b) Identify the g or u character of bonding and antibonding δ orbitals formed by face-to-face overlap of d atomic orbitals.

Problems

P9B.1 Calculate the (molar) energy of electrostatic repulsion between two hydrogen nuclei at the separation in H_2 (74.1 pm). The result is the energy that must be overcome by the attraction from the electrons that form the bond. Does the gravitational attraction between the nuclei play any significant role? *Hint:* The gravitational potential energy of two masses is equal to $-Gm_1m_2/r$; the gravitational constant G is listed inside the front cover.

P9B.2 Imagine a small electron-sensitive probe of volume 1.00 pm^3 inserted into an H_2^+ molecule-ion in its ground state. Calculate the probability that it will register the presence of an electron at the following positions: (a) at nucleus A, (b) at nucleus B, (c) half way between A and B, (d) at a point 20 pm along the bond from A and 10 pm perpendicularly. Do the same for the molecule-ion the instant after the electron has been excited into the antibonding LCAO-MO. Take $R = 2.00a_0$.

P9B.3 Examine whether occupation of the bonding orbital in the H_2^+ molecule-ion by one electron has a greater or lesser bonding effect than occupation of the antibonding orbital by one electron. Is your conclusion true at all internuclear separations?

P9B.4 Use mathematical software or a spreadsheet to: (a) plot the amplitude of the σ wavefunction along the z-axis (eqn 9B.2, with the atomic orbitals given in *Brief illustration* 9B.1) for different values of the internuclear distance. Identify the features of the orbital that lead to bonding. (b) Plot the amplitude of the σ* wavefunction along the z-axis (eqn 9B.2, with the atomic orbitals given in *Brief illustration* 9B.1) for different values of the internuclear distance. Identify the features of the σ* orbital that lead to antibonding.

P9B.5 (a) Calculate the total amplitude of the normalized bonding and antibonding LCAO-MOs that may be formed from two H1s orbitals at a separation of $2a_0 = 106$ pm. Plot the two amplitudes for positions along the molecular axis both inside and outside the internuclear region. (b) Plot the probability densities of the two orbitals. Then form the *difference density*, the difference between ψ^2 and $\frac{1}{2}(\psi_A^2 + \psi_B^2)$.

TOPIC 9C Molecular orbital theory: homonuclear diatomic molecules

Discussion questions

D9C.1 Draw diagrams to show the various orientations in which a p orbital and a d orbital on adjacent atoms may form bonding and antibonding molecular orbitals.

D9C.2 Outline the rules of the building-up principle for homonuclear diatomic molecules.

D9C.3 What is the justification for treating s and p atomic orbital contributions to molecular orbitals separately?

D9C.4 To what extent can orbital overlap be related to bond strength? To what extent might that be a correlation rather than an explanation?

Exercises

E9C.1(a) Give the ground-state electron configurations and bond orders of (i) Li_2, (ii) Be_2, and (iii) C_2.

E9C.1(b) Give the ground-state electron configurations and bond orders of (i) F_2^-, (ii) N_2, and (iii) O_2^{2-}.

E9C.2(a) From the ground-state electron configurations of B_2 and C_2, predict which molecule should have the greater dissociation energy.

E9C.2(b) From the ground-state electron configurations of Li_2 and Be_2, predict which molecule should have the greater dissociation energy.

E9C.3(a) Which has the higher dissociation energy, F_2 or F_2^+?

E9C.3(b) Arrange the species O_2^+, O_2, O_2^-, O_2^{2-} in order of increasing bond length.

E9C.4(a) Evaluate the bond order of each Period 2 homonuclear diatomic molecule.

E9C.4(b) Evaluate the bond order of each Period 2 homonuclear diatomic cation, X_2^+, and anion, X_2^-.

E9C.5(a) For each of the species in Exercise E9C.4(b), specify which molecular orbital is the HOMO (the highest energy occupied orbital).

E9C.5(b) For each of the species in Exercise E9C.4(b), specify which molecular orbital is the LUMO (the lowest energy unoccupied orbital).

E9C.6(a) What is the speed of a photoelectron ejected from an orbital of ionization energy 12.0 eV by a photon of radiation of wavelength 100 nm?

E9C.6(b) What is the speed of a photoelectron ejected from a molecule with radiation of energy 21 eV and known to come from an orbital of ionization energy 12 eV?

Problems

P9C.1 Familiarity with the magnitudes of overlap integrals is useful when considering bonding abilities of atoms, and hydrogenic orbitals give an indication of their values. (a) The overlap integral between two hydrogenic 2s orbitals is

$$S(2s,2s)=\left\{1+\frac{ZR}{2a_0}+\frac{1}{12}\left(\frac{ZR}{a_0}\right)^2+\frac{1}{240}\left(\frac{ZR}{a_0}\right)^4\right\}e^{-ZR/2a_0}$$

Plot this expression. (b) For what internuclear distance is $S(2s,2s)=0.50$? (c) The side-by-side overlap of two 2p orbitals of atoms of atomic number Z is

$$S(2p,2p)=\left\{1+\frac{ZR}{2a_0}+\frac{1}{10}\left(\frac{ZR}{a_0}\right)^2+\frac{1}{120}\left(\frac{ZR}{a_0}\right)^3\right\}e^{-ZR/2a_0}$$

Plot this expression. (d) Evaluate $S(2s,2p)$ at the internuclear distance you calculated in part (b).

P9C.2 Before doing a calculation, sketch how the overlap between a 1s orbital and a 2p orbital directed towards it can be expected to depend on their separation. The overlap integral between an H1s orbital and an H2p orbital directed towards it on nuclei separated by a distance R is $S=(R/a_0)\{1+(R/a_0)+\frac{1}{3}(R/a_0)^2\}e^{-R/a_0}$. Plot this function, and find the separation for which the overlap is a maximum.

P9C.3‡ Use the $2p_x$ and $2p_z$ hydrogenic atomic orbitals to construct simple LCAO descriptions of $2p\sigma$ and $2p\pi$ molecular orbitals. (a) Make a probability density plot, and both surface and contour plots of the xz-plane amplitudes of the $2p_z\sigma$ and $2p_z\sigma$ molecular orbitals. (b) Plot the amplitude of the $2p_x\pi$ and $2p_x\pi$ molecular orbital wavefunctions in the xz-plane. Include plots for both an internuclear distance, R, of $10a_0$ and $3a_0$, where $a_0=52.9$ pm. Interpret the graphs, and explain why this graphical information is useful.

P9C.4 In a photoelectron spectrum using 21.21 eV photons, electrons were ejected with kinetic energies of 11.01 eV, 8.23 eV, and 15.22 eV. Sketch the molecular orbital energy level diagram for the species, showing the ionization energies of the three identifiable orbitals.

TOPIC 9D Molecular orbital theory: heteronuclear diatomic molecules

Discussion questions

D9D.1 Describe the Pauling and Mulliken electronegativity scales. Why should they be approximately in step?

D9D.2 Why do both ionization energy and electron affinity play a role in estimating the energy of an atomic orbital to use in a molecular orbital calculation?

D9D.3 Discuss the steps involved in the calculation of the energy of a system by using the variation principle. Are any assumptions involved?

D9D.4 What is the physical significance of the Coulomb and resonance integrals?

Exercises

E9D.1(a) Give the ground-state electron configurations of (i) CO, (ii) NO, and (iii) CN^-.

E9D.1(b) Give the ground-state electron configurations of (i) XeF, (ii) PN, and (iii) SO^-.

E9D.2(a) Sketch the molecular orbital energy level diagram for XeF and deduce its ground-state electron configuration. Is XeF likely to have a shorter bond length than XeF^+?

E9D.2(b) Sketch the molecular orbital energy level diagram for IF and deduce its ground-state electron configuration. Is IF likely to have a shorter bond length than IF^- or IF^+?

E9D.3(a) Use the electron configurations of NO^- and NO^+ to predict which is likely to have the shorter bond length.

E9D.3(b) Use the electron configurations of SO^- and SO^+ to predict which is likely to have the shorter bond length.

E9D.4(a) A reasonably reliable conversion between the Mulliken and Pauling electronegativity scales is given by eqn 9D.4. Use Table 9D.1 in the *Resource section* to assess how good the conversion formula is for Period 2 elements.

E9D.4(b) A reasonably reliable conversion between the Mulliken and Pauling electronegativity scales is given by eqn 9D.4. Use Table 9D.1 in the *Resource section* to assess how good the conversion formula is for Period 3 elements.

E9D.5(a) Estimate the orbital energies to use in a calculation of the molecular orbitals of HCl. For data, see Tables 8B.4 and 8B.5. Take $\beta=-1.00$ eV.

E9D.5(b) Estimate the orbital energies to use in a calculation of the molecular orbitals of HBr. For data, see Tables 8B.4 and 8B.5. Take $\beta=-1.00$ eV.

E9D.6(a) Use the values derived in Exercise 9D.5(a) to estimate the molecular orbital energies in HCl; use $S=0$.

E9D.6(b) Use the values derived in Exercise 9D.5(b) to estimate the molecular orbital energies in HBr; use $S=0$.

E9D.7(a) Now repeat Exercise 9D.6(a), but with $S=0.20$.

E9D.7(b) Now repeat Exercise 9D.6(b), but with $S=0.20$.

‡ These problems were supplied by Charles Trapp and Carmen Giunta.

Problems

P9D.1 Show, if overlap is ignored, (a) that if a molecular orbital is expressed as a linear combination of two atomic orbitals in the form $\psi = \psi_A \cos\theta + \psi_B \sin\theta$, where θ is a parameter that varies between 0 and π, with ψ_A and ψ_B are orthogonal and normalized to 1, then ψ is also normalized to 1. (b) To what values of θ do the bonding and antibonding orbitals in a homonuclear diatomic molecule correspond?

P9D.2 (a) Suppose that a molecular orbital of a heteronuclear diatomic molecule is built from the orbital basis A, B, and C, where B and C are both on one atom. Set up the secular equations for the values of the coefficients and the corresponding secular determinant. (b) Now let $\alpha_A = -7.2$ eV, $\alpha_B = -10.4$ eV, $\alpha_C = -8.4$ eV, $\beta_{AB} = -1.0$ eV, $\beta_{AC} = -0.8$ eV, and calculate the orbital energies and coefficients with both S_{AB} and S_{AC} equal to (i) 0, (ii) 0.2 (note that $S_{BC} = 0$ for orbitals on the same atom).

P9D.3 As a variation of the preceding problem explore the consequences of increasing the energy separation of the ψ_B and ψ_C orbitals (use S_{AB} and S_{AC} equal to 0 for this stage of the calculation). Are you justified in ignoring orbital ψ_C at any stage?

TOPIC 9E Molecular orbital theory: polyatomic molecules

Discussion questions

D9E.1 Discuss the scope, consequences, and limitations of the approximations on which the Hückel method is based.

D9E.2 Distinguish between delocalization energy, π-electron binding energy, and π-bond formation energy. Explain how each concept is employed.

D9E.3 Outline the computational steps used in the self-consistent field approach to electronic structure calculations.

D9E.4 Explain why the use of Gaussian-type orbitals is generally preferred over the use of hydrogenic orbitals in basis sets.

D9E.5 Identify the principal distinguishing features of semi-empirical, *ab initio*, and density functional theory methods of electronic structure determination.

Exercises

E9E.1(a) Set up the secular determinants for (i) linear H_3, (ii) cyclic H_3 within the Hückel approximation.

E9E.1(b) Set up the secular determinants for (i) linear H_4, (ii) cyclic H_4 within the Hückel approximation.

E9E.2(a) Predict the electron configurations of (i) the benzene anion, (ii) the benzene cation. Estimate the π-electron binding energy in each case.

E9E.2(b) Predict the electron configurations of (i) the allyl radical, $\cdot CH_2CHCH_2$, (ii) the cyclobutadiene cation $C_4H_4^+$. Estimate the π-electron binding energy in each case.

E9E.3(a) What is the delocalization energy and π-bond formation energy of (i) the benzene anion, (ii) the benzene cation?

E9E.3(b) What is the delocalization energy and π-bond formation energy of (i) the allyl radical, (ii) the cyclobutadiene cation?

E9E.4(a) Set up the secular determinants for (i) anthracene (1), (ii) phenanthrene (2) within the Hückel approximation and using the out-of-plane C2p orbitals as the basis set.

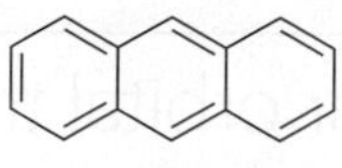

1 Anthracene

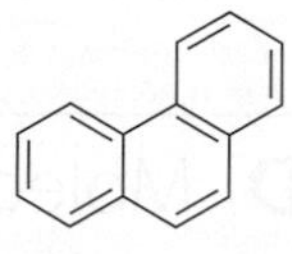

2 Phenanthrene

E9E.4(b) Set up the secular determinants for (i) azulene (3), (ii) acenaphthylene (4) within the Hückel approximation and using the out-of-plane C2p orbitals as the basis set.

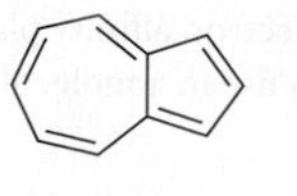

3 Azulene

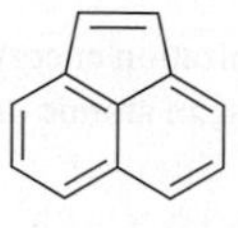

4 Acenaphthylene

E9E.5(a) Use mathematical software to estimate the π-electron binding energy of (i) anthracene (1), (ii) phenanthrene (2) within the Hückel approximation.

E9E.5(b) Use mathematical software to estimate the π-electron binding energy of (i) azulene (3), (ii) acenaphthylene (4) within the Hückel approximation.

E9E.6(a) Write the electronic hamiltonian for HeH^+.

E9E.6(b) Write the electronic hamiltonian for LiH^{2+}.

Problems

P9E.1 Set up and solve the Hückel secular equations for the π electrons of the triangular, planar CO_3^{2-} ion. Express the energies in terms of the Coulomb integrals α_O and α_C and the resonance integral β. Estimate the delocalization energy of the ion.

P9E.2 For monocyclic conjugated polyenes (such as cyclobutadiene and benzene) with each of N carbon atoms contributing an electron in a 2p orbital, simple Hückel theory gives the following expression for the energies E_k of the resulting π molecular orbitals (all are doubly degenerate except the lowest and highest values of k):

$$E_k = \alpha + 2\beta\cos\frac{2k\pi}{N} \quad k = 0, 1, \ldots, N/2 \text{ for } N \text{ even}$$

$$k = 0, 1, \ldots, (N-1)/2 \text{ for } N \text{ odd}$$

(a) Calculate the energies of the π molecular orbitals of benzene and cyclooctatetraene (5). Comment on the presence or absence of degenerate energy levels. (b) Calculate and compare the delocalization energies of benzene (using the expression above) and hexatriene (see Problem P9E.11). What do you conclude from your results? (c) Calculate and compare the delocalization energies of cyclooctatetraene and octatetraene. Are your conclusions for this pair of molecules the same as for the pair of molecules investigated in part (b)?

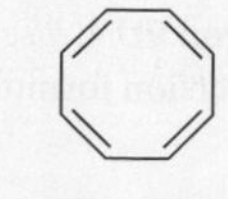

5 Cyclooctatetraene

P9E.3 Suppose that a molecular orbital of a heteronuclear diatomic molecule is built from the orbital basis ψ_A, ψ_B, and ψ_C, where ψ_B and ψ_C are both on one atom (they can be envisaged as F2s and F2p in HF, for instance). Set up the secular equations for the optimum values of the coefficients and set up the corresponding secular determinant.

P9E.4 Set up the secular determinants for the homologous series consisting of ethene, butadiene, hexatriene, and octatetraene and diagonalize them by using mathematical software. Use your results to show that the π molecular orbitals of linear polyenes obey the following rules:

- The π molecular orbital with lowest energy is delocalized over all carbon atoms in the chain.
- The number of nodal planes between C2p orbitals increases with the energy of the π molecular orbital.

P9E.5 Set up the secular determinants for cyclobutadiene, benzene, and cyclooctatetraene and diagonalize them by using mathematical software. Use your results to show that the π molecular orbitals of monocyclic polyenes with an even number of carbon atoms follow a pattern in which:

- The π molecular orbitals of lowest and highest energy are non-degenerate.
- The remaining π molecular orbitals exist as degenerate pairs.

P9E.6 Electronic excitation of a molecule may weaken or strengthen some bonds because bonding and antibonding characteristics differ between the HOMO and the LUMO. For example, a carbon–carbon bond in a linear polyene may have bonding character in the HOMO and antibonding character in the LUMO. Therefore, promotion of an electron from the HOMO to the LUMO weakens this carbon–carbon bond in the excited electronic state, relative to the ground electronic state. Consult Figs. 9E.2 and 9E.4 and discuss in detail any changes in bond order that accompany the $\pi^*\leftarrow\pi$ ultraviolet absorptions in butadiene and benzene.

P9E.7‡ In Exercise E9E.1(a) you are invited to set up the Hückel secular determinant for linear and cyclic H_3. The same secular determinant applies to the molecular ions H_3^+ and D_3^+. The molecular ion H_3^+ was discovered as long ago as 1912 by J.J. Thomson but the equilateral triangular structure was confirmed by M.J. Gaillard et al. (*Phys. Rev.* **A17**, 1797 (1978)) much more recently. The molecular ion H_3^+ is the simplest polyatomic species with a confirmed existence and plays an important role in chemical reactions occurring in interstellar clouds that may lead to the formation of water, carbon monoxide, and ethanol. The H_3^+ ion has also been found in the atmospheres of Jupiter, Saturn, and Uranus. (a) Solve the Hückel secular equations for the energies of the H_3 system in terms of the parameters α and β, draw an energy level diagram for the orbitals, and determine the binding energies of H_3^+, H_3, and H_3^-. (b) Accurate quantum mechanical calculations by G.D. Carney and R.N. Porter (*J. Chem. Phys.* **65**, 3547 (1976)) give the dissociation energy for the process $H_3^+ \rightarrow H + H + H^+$ as 849 kJ mol^{-1}. From this information and data in Table 9C.3, calculate the enthalpy of the reaction $H^+(g) + H_2(g) \rightarrow H_3^+(g)$. (c) From your equations and the information given, calculate a value for the resonance integral β in H_3^+. Then go on to calculate the binding energies of the other H_3 species in (a).

P9E.8‡ There is some indication that other hydrogen ring compounds and ions in addition to H_3 and D_3 species may play a role in interstellar chemistry. According to J.S. Wright and G.A. DiLabio (*J. Phys. Chem.* **96**, 10793 (1992)), H_5^-, H_6, and H_7^+ are particularly stable whereas H_4 and H_5^+ are not. Confirm these statements by Hückel calculations.

P9E.9 Use appropriate electronic structure software and basis sets of your or your instructor's choosing, perform self-consistent field calculations for the ground electronic states of H_2 and F_2. Determine ground-state energies and equilibrium geometries. Compare computed equilibrium bond lengths to experimental values.

P9E.10 Use an appropriate semi-empirical method to compute the equilibrium bond lengths and standard enthalpies of formation of (a) ethanol, (b) 1,4-dichlorobenzene. Compare to experimental values and suggest reasons for any discrepancies.

P9E.11 (a) For a linear conjugated polyene with each of N carbon atoms contributing an electron in a 2p orbital, the energies E_k of the resulting π molecular orbitals are given by:

$$E_k = \alpha + 2\beta\cos\frac{k\pi}{N+1} \quad k = 1, 2, \ldots, N$$

Use this expression to make a reasonable empirical estimate of the resonance integral β for the homologous series consisting of ethene, butadiene, hexatriene, and octatetraene given that $\pi\leftarrow\pi$ ultraviolet absorptions from the HOMO to the LUMO occur at 61 500, 46 080, 39 750, and 32 900 cm^{-1}, respectively. (b) Calculate the π-electron delocalization energy, $E_{\text{deloc}} = E_\pi - n(\alpha+\beta)$, of octatetraene, where E_π is the total π-electron binding energy and n is the total number of π-electrons. (c) In the context of this Hückel model, the π molecular orbitals are written as linear combinations of the carbon 2p orbitals. The coefficient of the jth atomic orbital in the kth molecular orbital is given by:

$$c_{kj} = \left(\frac{2}{N+1}\right)^{1/2} \sin\frac{jk\pi}{N+1} \quad j = 1, 2, \ldots, N$$

Evaluate the coefficients of each of the six 2p orbitals in each of the six π molecular orbitals of hexatriene. Match each set of coefficients (that is, each molecular orbital) with a value of the energy calculated with the expression given in part (a) of the molecular orbital. Comment on trends that relate the energy of a molecular orbital with its 'shape', which can be inferred from the magnitudes and signs of the coefficients in the linear combination that describes the molecular orbital.

FOCUS 9 Molecular structure

Integrated activities

I9.1 The languages of valence-bond theory and molecular orbital theory are commonly combined when discussing unsaturated organic compounds. Construct the molecular orbital energy level diagrams of ethene on the basis that the molecule is formed from the appropriately hybridized CH_2 or CH fragments.

I9.2 Here a molecular orbital theory treatment of the peptide group (6) is developed, a group that links amino acids in proteins, and establish the features that stabilize its planar conformation. (a) It will be familiar from introductory chemistry that valence-bond theory explains the planar conformation by invoking delocalization of the π bond over the oxygen, carbon, and nitrogen atoms by resonance:

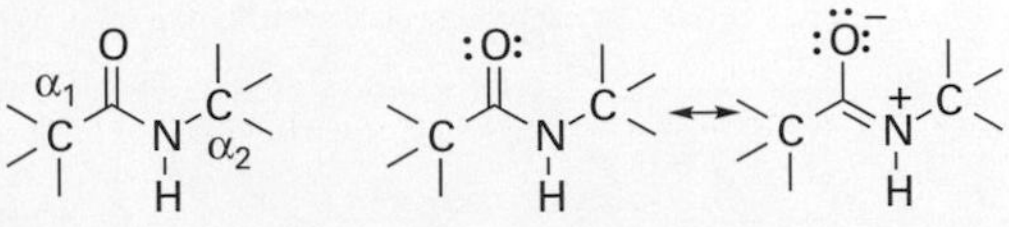

6 Peptide group

It follows that the peptide group can be modelled by using molecular orbital theory by constructing LCAO-MOs from 2p orbitals perpendicular to the plane defined by the O, C, and N atoms. The three combinations have the form:

$$\psi_1 = a\psi_O + b\psi_C + c\psi_N \quad \psi_2 = d\psi_O - e\psi_N \quad \psi_3 = f\psi_O - g\psi_C + h\psi_N$$

where the coefficients a to h are all positive. Sketch the orbitals ψ_1, ψ_2, and ψ_3 and characterize them as bonding, nonbonding, or antibonding. In a nonbonding molecular orbital, a pair of electrons resides in an orbital confined largely to one atom and not appreciably involved in bond formation. (b) Show that this treatment is consistent only with a planar conformation of the peptide link. (c) Draw a diagram showing the relative energies of these molecular orbitals and identify the occupancy of the orbitals. *Hint:* Convince yourself that there are four electrons to be distributed among the molecular orbitals. (d) Now consider a nonplanar conformation of the peptide link, in which the O2p and C2p orbitals are perpendicular to the plane defined by the O, C, and N atoms, but the N2p orbital lies on that plane. The LCAO-MOs are given by

$$\psi_4 = a\psi_O + b\psi_C \qquad \psi_5 = e\psi_N \qquad \psi_6 = f\psi_O - g\psi_C$$

Just as before, sketch these molecular orbitals and characterize them as bonding, nonbonding, or antibonding. Also, draw an energy level diagram and identify the occupancy of the orbitals. (e) Why is this arrangement of atomic orbitals consistent with a nonplanar conformation for the peptide link? (f) Does the bonding MO associated with the planar conformation have the same energy as the bonding MO associated with the nonplanar conformation? If not, which bonding MO is lower in energy? Repeat the analysis for the nonbonding and antibonding molecular orbitals. (g) Use your results from parts (a)–(f) to construct arguments that support the planar model for the peptide link.

I9.3 Molecular electronic structure methods may be used to estimate the standard enthalpy of formation of molecules in the gas phase. (a) Use a semi-empirical method of your or your instructor's choice to calculate the standard enthalpy of formation of ethene, butadiene, hexatriene, and octatetraene in the gas phase. (b) Consult a database of thermochemical data, and, for each molecule in part (a), calculate the difference between the calculated and experimental values of the standard enthalpy of formation. (c) A good thermochemical database will also report the uncertainty in the experimental value of the standard enthalpy of formation. Compare experimental uncertainties with the relative errors calculated in part (b) and discuss the reliability of your chosen semi-empirical method for the estimation of thermochemical properties of linear polyenes.

I9.4 The standard potential of a redox couple is a measure of the thermodynamic tendency of an atom, ion, or molecule to accept an electron (Topic 6D). Studies indicate that there is a correlation between the LUMO energy and the standard potential of aromatic hydrocarbons. Do you expect the standard potential to increase or decrease as the LUMO energy decreases? Explain your answer.

I9.5 Molecular orbital calculations may be used to predict trends in the standard potentials of conjugated molecules, such as the quinones and flavins, that are involved in biological electron transfer reactions. It is commonly assumed that decreasing the energy of the LUMO enhances the ability of a molecule to accept an electron into the LUMO, with an accompanying increase in the value of the molecule's standard potential. Furthermore, a number of studies indicate that there is a linear correlation between the LUMO energy and the reduction potential of aromatic hydrocarbons.

(a) The standard potentials at pH = 7 for the one-electron reduction of methyl-substituted 1,4-benzoquinones (**7**) to their respective semiquinone radical anions are:

R_2	R_3	R_5	R_6	$E^{\ominus}$/V
H	H	H	H	0.078
CH_3	H	H	H	0.023
CH_3	H	CH_3	H	−0.067
CH_3	CH_3	CH_3	H	−0.165
CH_3	CH_3	CH_3	CH_3	−0.260

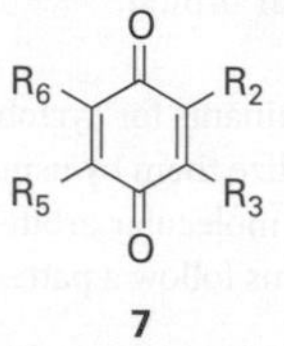

7

Use the computational method of your or your instructor's choice (semi-empirical, *ab initio*, or density functional theory methods) to calculate E_{LUMO}, the energy of the LUMO of each substituted 1,4-benzoquinone, and plot E_{LUMO} against $E^{\ominus}$. Do your calculations support a linear relation between E_{LUMO} and $E^{\ominus}$? (b) The 1,4-benzoquinone for which $R_2 = R_3 = CH_3$ and $R_5 = R_6 = OCH_3$ is a suitable model of ubiquinone, a component of the respiratory electron transport chain. Determine E_{LUMO} of this quinone and then use your results from part (a) to estimate its standard potential. (c) The 1,4-benzoquinone for which $R_2 = R_3 = R_5 = CH_3$ and $R_6 = H$ is a suitable model of plastoquinone, an electron carrier in photosynthesis. Determine E_{LUMO} of this quinone and then use your results from part (a) to estimate its standard potential. Is plastoquinone expected to be a better or worse oxidizing agent than ubiquinone?

I9.6 Molecular orbital calculations based on semi-empirical, *ab initio*, and DFT methods describe the spectroscopic properties of conjugated molecules better than simple Hückel theory. (a) Use the computational method of your or your instructor's choice (semi-empirical, *ab initio*, or density functional methods) to calculate the energy separation between the HOMO and LUMO of ethene, butadiene, hexatriene, and octatetraene. (b) Plot the HOMO–LUMO energy separations against the experimental frequencies for $\pi^*\leftarrow\pi$ ultraviolet absorptions for these molecules (61 500, 46 080, 39 750, and 32 900 cm^{-1}, respectively). Use mathematical software to find the polynomial equation that best fits the data. (b) Use your polynomial fit from part (b) to estimate the wavenumber and wavelength of the $\pi^*\leftarrow\pi$ ultraviolet absorption of decapentaene from the calculated HOMO–LUMO energy separation. (c) Discuss why the calibration procedure of part (b) is necessary.

I9.7 The variation principle can be used to formulate the wavefunctions of electrons in atoms as well as molecules. Suppose that the function $\psi_{trial} = N(\alpha)e^{-\alpha r^2}$ with $N(\alpha)$ the normalization constant and α an adjustable parameter, is used as a trial wavefunction for the 1s orbital of the hydrogen atom. Show that

$$E(\alpha) = \frac{3\hbar^2\alpha}{2\mu} - \frac{e^2\alpha^{1/2}}{2^{1/2}\pi^{3/2}\varepsilon_0}$$

where e is the fundamental charge, and μ is the reduced mass for the atom. What is the minimum energy associated with this trial wavefunction?

FOCUS 10

Molecular symmetry

In this Focus the concept of 'shape' is sharpened into a precise definition of 'symmetry'. As a result, symmetry and its consequences can be discussed systematically, thereby providing a very powerful tool for the prediction and analysis of molecular structure and properties.

10A Shape and symmetry

This Topic shows how to classify any molecule according to its symmetry. Two immediate applications of this classification are the identification of whether or not a molecule can have an electric dipole moment (and so be polar) and whether or not it can be chiral (and so be optically active).

10B Group theory

The systematic treatment of symmetry is an application of 'group theory'. This theory represents the outcome of symmetry operations (such as rotations and reflections) by matrices. This step is important, for once symmetry operations are expressed numerically they can be manipulated quantitatively. This Topic introduces 'character tables' which are exceptionally important in the application of group theory to chemical problems.

10C Applications of symmetry

Group theory provides simple criteria for deciding whether certain integrals necessarily vanish. One application is to decide whether the overlap integral between two atomic orbitals is necessarily zero and therefore to decide which atomic orbitals can contribute to molecular orbitals. Symmetry is also used to identify linear combinations of atomic orbitals that match the symmetry of the nuclear framework. By considering the symmetry properties of integrals, it is also possible to derive the selection rules that govern spectroscopic transitions.

TOPIC 10A Shape and symmetry

➤ Why do you need to know this material?

Symmetry arguments can be used to make immediate assessments of the properties of molecules; the initial step is to identify the symmetry a molecule possesses and then to classify it accordingly.

➤ What is the key idea?

Molecules can be classified into groups according to their symmetry elements.

➤ What do you need to know already?

This Topic does not draw on others directly, but it will be useful to be aware of the shapes of a variety of simple molecules and ions encountered in introductory chemistry courses.

Some objects are 'more symmetrical' than others. A sphere is more symmetrical than a cube because it looks the same after it has been rotated through any angle about any axis passing through the centre. A cube looks the same only if it is rotated through certain angles about specific axes, such as 90°, 180°, or 270° about an axis passing through the centres of any of its opposite faces (Fig. 10A.1), or by 120° or 240° about an axis passing through any of its opposite corners. Similarly, an NH_3 molecule is 'more symmetrical' than an H_2O molecule because NH_3 looks the same after rotations of 120° or 240° about the axis shown in Fig. 10A.2, whereas H_2O looks the same only after a rotation of 180°.

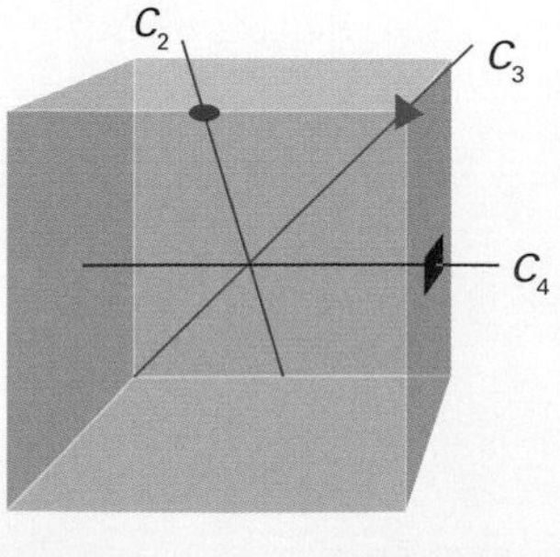

Figure 10A.1 Some of the symmetry elements of a cube. The twofold, threefold, and fourfold axes are labelled with the conventional symbols.

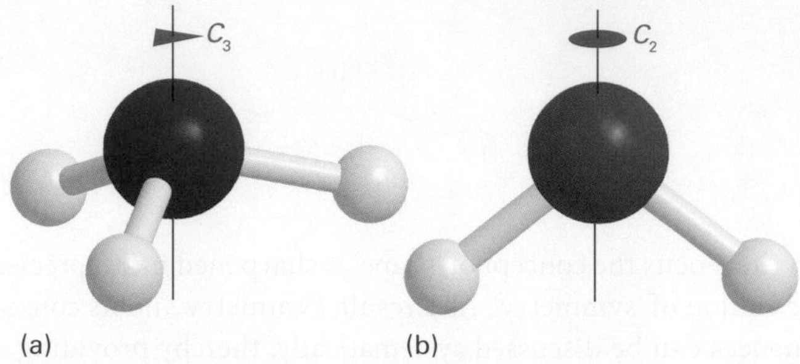

Figure 10A.2 (a) An NH_3 molecule has a threefold (C_3) axis and (b) an H_2O molecule has a twofold (C_2) axis. Both have other symmetry elements too.

This Topic puts these intuitive notions on a more formal foundation. It will be seen that molecules can be grouped together according to their symmetry, with the tetrahedral species CH_4 and SO_4^{2-} in one group and the pyramidal species NH_3 and SO_3^{2-} in another. It turns out that molecules in the same group share certain physical properties, so powerful predictions can be made about whole series of molecules once the group to which they belong has been identified.

10A.1 Symmetry operations and symmetry elements

An action that leaves an object looking the same after it has been carried out is called a **symmetry operation**. Typical symmetry operations include rotations, reflections, and inversions. There is a corresponding **symmetry element** for each symmetry operation, which is the point, line, or plane with respect to which the symmetry operation is performed. For instance, a rotation (a symmetry operation) is carried out around an axis (the corresponding symmetry element). Molecules can be classified by identifying all their symmetry elements, and then grouping together molecules that possess the same set of symmetry elements. This procedure, for example, puts the trigonal planar species BF_3 and CO_3^{2-} into one group and the species H_2O (bent) and ClF_3 (T-shaped) into another group.

An ***n*-fold rotation** (the operation) about an ***n*-fold axis of symmetry**, C_n (the corresponding element) is a rotation through 360°/n. An H_2O molecule has one twofold axis, C_2. An NH_3 molecule has one threefold axis, C_3, with which is associated two symmetry operations, one being a rotation by 120° in a clockwise sense, and the other a rotation by 120° in an anticlockwise sense. There is only one twofold rotation associated with a C_2 axis because clockwise and anticlockwise 180° rotations are identical. A pentagon has a C_5 axis, with two rotations (one clockwise, the other anticlockwise) through 72° associated with it. It also has an operation denoted C_5^2, corresponding to two successive C_5 rotations; there are two such operations, one through 144° in a clockwise sense and the other through 144° in an anticlockwise sense. A cube has three C_4 axes, four C_3 axes, and six C_2 axes. However, even this high symmetry is exceeded by that of a sphere, which possesses an infinite number of symmetry axes (along any axis passing through the centre) of all possible integral values of n.

If a molecule possesses several rotation axes, then the one with the highest value of n is called the **principal axis**. The principal axis of a benzene molecule is the sixfold axis perpendicular to the hexagonal ring (**1**). If a molecule has more than one rotation axis with this highest value of n, and it is wished to designate one of them as the principal axis, then it is common to choose the axis that passes through the greatest number of atoms or, in the case of a planar molecule (such as naphthalene, **2**, which has three C_2 axes competing for the title), to choose the axis perpendicular to the plane.

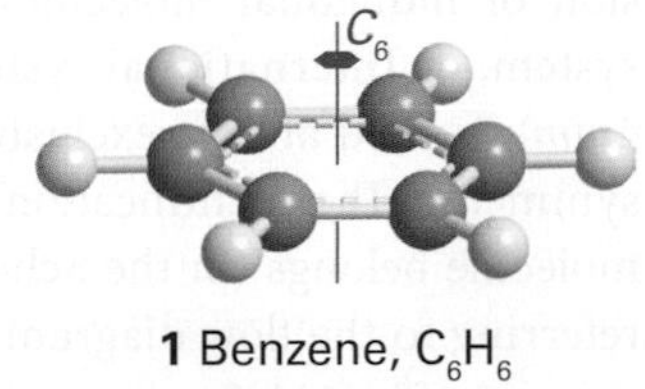

1 Benzene, C_6H_6

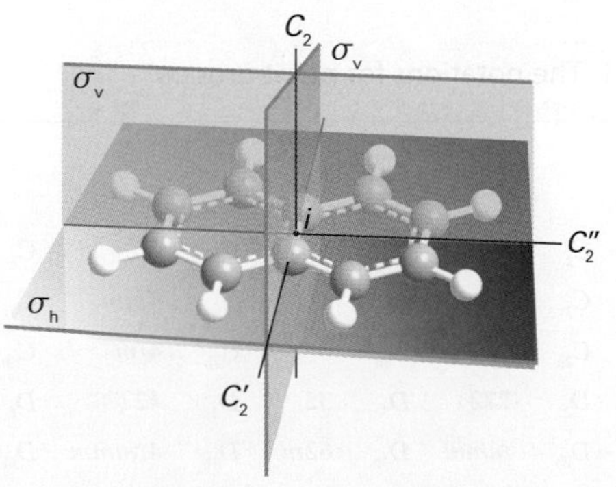

2 Naphthalene, $C_{10}H_8$

A **reflection** is the operation corresponding to a **mirror plane**, σ (the element). If the plane contains the principal axis, it is called 'vertical' and denoted σ_v. An H_2O molecule has two vertical mirror planes (Fig. 10A.3) and an NH_3 molecule has three. A vertical mirror plane that bisects the angle between two C_2 axes is called a 'dihedral plane' and is denoted

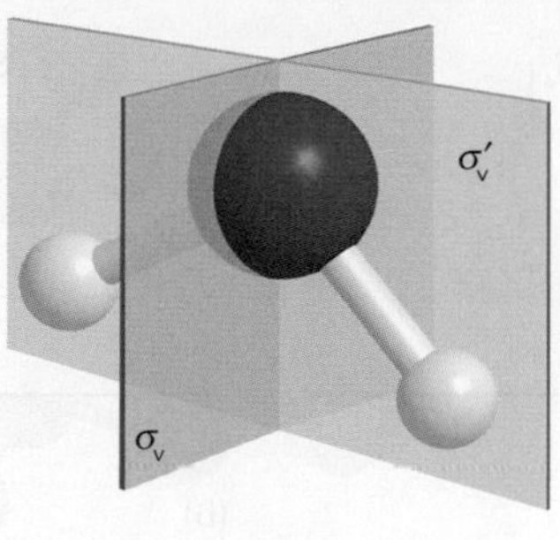

Figure 10A.3 An H_2O molecule has two mirror planes. They are both vertical (i.e. contain the principal axis), so are denoted σ_v and σ_v'.

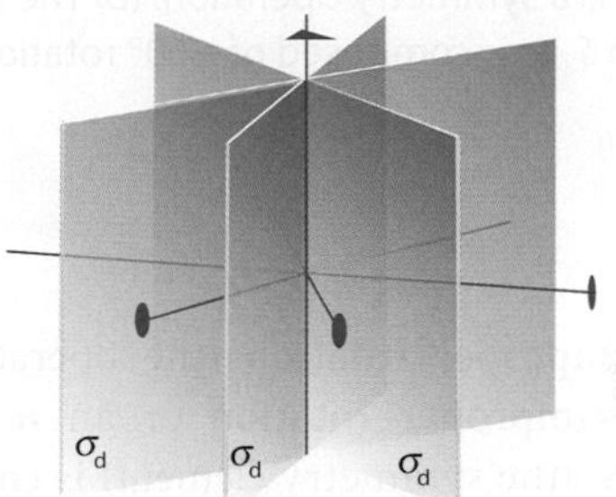

Figure 10A.4 Dihedral mirror planes (σ_d) bisect the C_2 axes perpendicular to the principal axis.

σ_d (Fig. 10A.4). When the mirror plane is perpendicular to the principal axis it is called 'horizontal' and denoted σ_h. The benzene molecule has such a horizontal mirror plane, perpendicular to the C_6 (principal) axis.

In an **inversion** (the operation) through a **centre of symmetry**, i (the element), each point in a molecule is imagined as being moved in a straight line to the centre of the molecule and then out the same distance on the other side; that is, the point (x, y, z) is taken into the point $(-x, -y, -z)$. Neither an H_2O molecule nor an NH_3 molecule has a centre of inversion, but a sphere and a cube do have one. A benzene molecule has a centre of inversion, as does a regular octahedron (Fig. 10A.5); a regular tetrahedron and a CH_4 molecule do not.

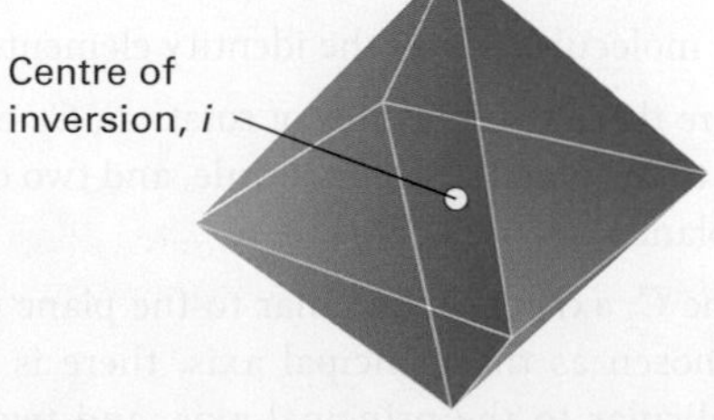

Figure 10A.5 A regular octahedron has a centre of inversion (i).

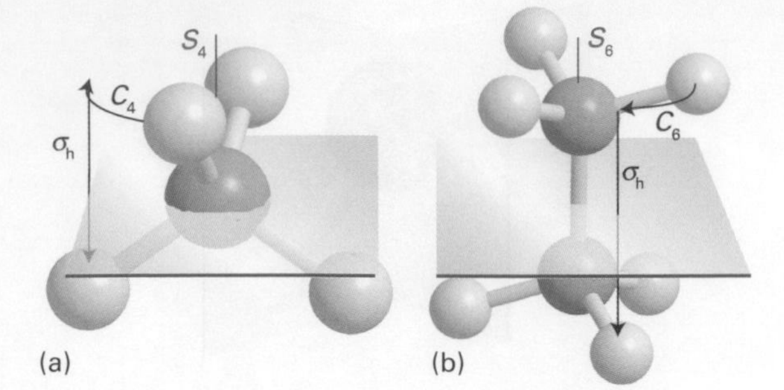

Figure 10A.6 (a) A CH_4 molecule has a fourfold improper rotation axis (S_4): the molecule is indistinguishable after a 90° rotation followed by a reflection across the horizontal plane, but neither operation alone is a symmetry operation. (b) The staggered form of ethane has an S_6 axis composed of a 60° rotation followed by a reflection.

An ***n*-fold improper rotation** (the operation) about an ***n*-fold axis of improper rotation** or an ***n*-fold improper rotation axis**, S_n (the symmetry element) is composed of two successive transformations. The first is a rotation through $360°/n$, and the second is a reflection through a plane perpendicular to the axis of that rotation. Neither transformation alone needs to be a symmetry operation. A CH_4 molecule has three S_4 axes, and the staggered conformation of ethane has an S_6 axis (Fig. 10A.6).

The **identity**, E, consists of doing nothing; the corresponding symmetry element is the entire object. Because every molecule is indistinguishable from itself if nothing is done to it, every object possesses at least the identity element. One reason for including the identity is that some molecules have only this symmetry element (**3**).

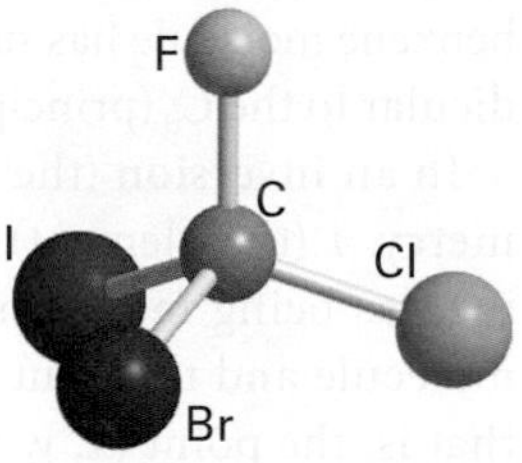

3 CBrClFI

Brief illustration 10A.1

To identify the symmetry elements of a naphthalene molecule (**2**), note that:

- Like all molecules, it has the identity element, E.
- There are three twofold axes of rotation, C_2: one perpendicular to the plane of the molecule, and two others lying in the plane.
- With the C_2 axis perpendicular to the plane of the molecule chosen as the principal axis, there is a σ_h plane perpendicular to the principal axis, and two σ_v planes which contain the principal axis.
- There is also a centre of inversion, i, through the midpoint of the molecule, which is mid-way along the C–C bond at the ring junction.

10A.2 The symmetry classification of molecules

Objects are classified into groups according to the symmetry elements they possess. **Point groups** arise when objects are classified according to symmetry elements that correspond to operations leaving at least one common point unchanged. The five kinds of symmetry element identified so far are of this kind. When crystals are considered (Topic 15A), symmetries arising from translation through space also need to be taken into account, and the classification according to these elements gives rise to the more extensive **space groups**.

All molecules with the same set of symmetry elements belong to the same point group, and the name of the group is determined by this set of symmetry elements. There are two systems of notation (Table 10A.1). The **Schoenflies system** (in which a name looks like C_{4v}) is more common for the discussion of individual molecules, and the **Hermann–Mauguin system**, or **International system** (in which a name looks like $4mm$), is used almost exclusively in the discussion of crystal symmetry. The identification of the point group to which a molecule belongs (in the Schoenflies system) is simplified by referring to the flow diagram in Fig. 10A.7 and to the shapes shown in Fig. 10A.8.

Table 10A.1 The notations for point groups*

C_i	$\bar{1}$								
C_s	m								
C_1	1	C_2	2	C_3	3	C_4	4	C_6	6
		C_{2v}	$2mm$	C_{3v}	$3m$	C_{4v}	$4mm$	C_{6v}	$6mm$
		C_{2h}	$2/m$	C_{3h}	$\bar{6}$	C_{4h}	$4/m$	C_{6h}	$6/m$
		D_2	222	D_3	32	D_4	422	D_6	622
		D_{2h}	mmm	D_{3h}	$\bar{6}2m$	D_{4h}	$4/mmm$	D_{6h}	$6/mmm$
		D_{2d}	$\bar{4}2m$	D_{3d}	$\bar{3}m$	S_4	$\bar{4}$	S_6	$\bar{3}$
T	23	T_d	$\bar{4}3m$	T_h	$m3$				
O	432	O_h	$m3m$						

* Schoenflies notation in black, Hermann–Mauguin (International system) in blue. In the Hermann–Mauguin system, a number n denotes the presence of an n-fold axis and m denotes a mirror plane. A slash (/) indicates that the mirror plane is perpendicular to the symmetry axis. It is important to distinguish symmetry elements of the same type but of different classes, as in $4/mmm$, in which there are three classes of mirror plane. A bar over a number indicates that the element is combined with an inversion. The only groups listed here are the so-called 'crystallographic point groups'.

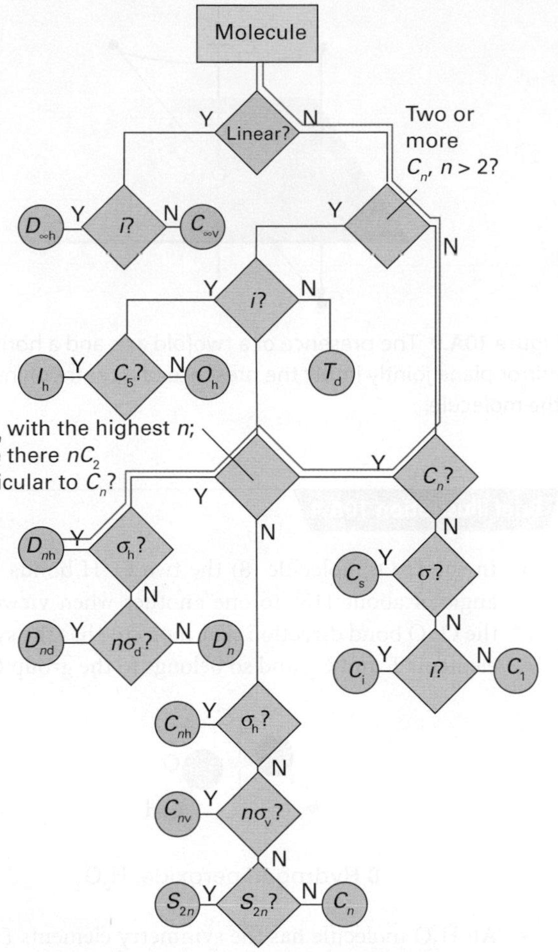

Figure 10A.7 A flow diagram for determining the point group of a molecule. Start at the top and answer the question posed in each diamond (Y = yes, N = no). The blue line refers to the path taken in *Brief illustration* 10A.2.

Brief illustration 10A.2

To identify the point group to which a ruthenocene molecule (4) belongs, first identify the symmetry elements present and the use the flow diagram in Fig. 10A.7. Note that:

- The molecule has a fivefold axis, and five twofold axes which pass through the Ru and are perpendicular to the C_5 axis.
- There is a mirror plane, σ_h, perpendicular to the C_5 axis and passing through the Ru.
- There are five σ_v planes containing the principal axis: each passes through one carbon in a ring and the mid-point of the C–C bond on the opposite side. Each one of these planes contains one of the twofold axes.

The path to trace in Fig. 10A.7 is shown by a blue line; it ends at D_{nh}, and because the molecule has a fivefold axis, it belongs to the point group D_{5h}.

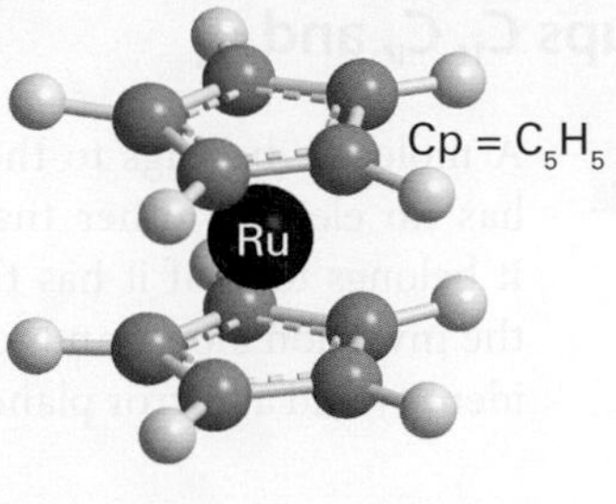

4 Ruthenocene, $Ru(Cp)_2$

If the rings are staggered, as they are in an excited state of ferrocene (5), the σ_h plane is absent. The other mirror planes are still present, but now they bisect the angles between twofold axes and so are described as σ_d. Tracing the appropriate path in Fig. 10A.7 gives the point group as D_{5d}.

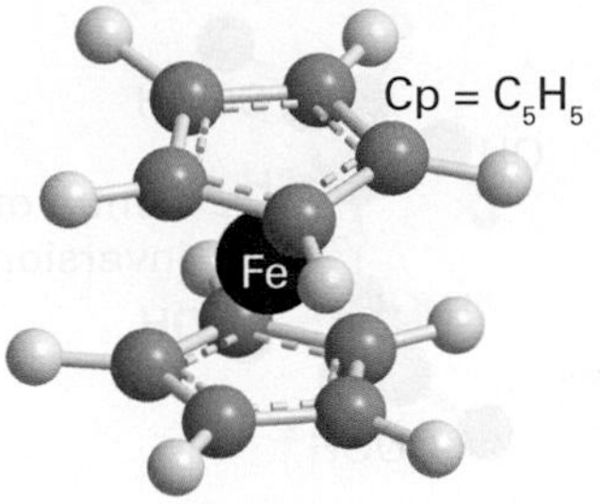

5 Ferrocene, $Fe(Cp)_2$
(excited state)

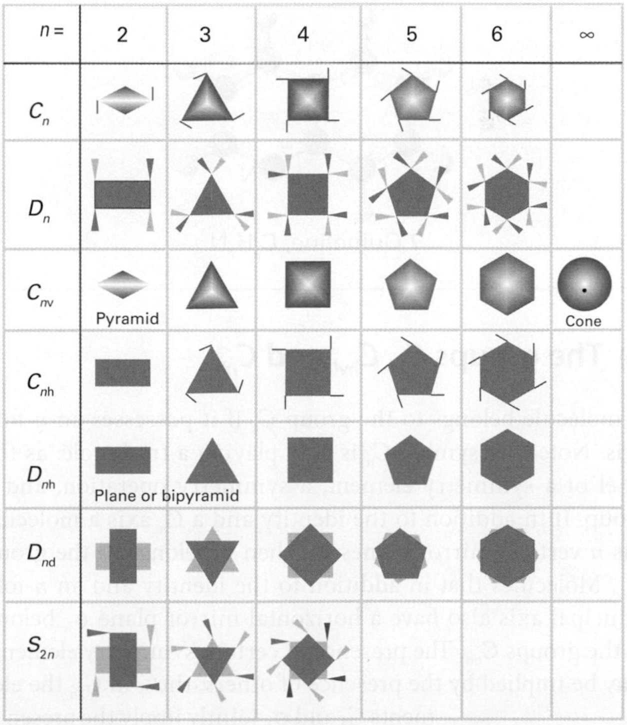

Figure 10A.8 A summary of the shapes corresponding to different point groups. The group to which a molecule belongs can often be identified from this diagram without going through the formal procedure in Fig. 10A.7.

(a) The groups C_1, C_i, and C_s

Name	Elements
C_1	E
C_i	E, i
C_s	E, σ

A molecule belongs to the group C_1 if it has no element other than the identity. It belongs to C_i if it has the identity and the inversion alone, and to C_s if it has the identity and a mirror plane alone.

Brief illustration 10A.3

- The CBrClFI molecule (**3**) has only the identity element, and so belongs to the group C_1.
- *Meso*-tartaric acid (**6**) has the identity and inversion elements, and so belongs to the group C_i.

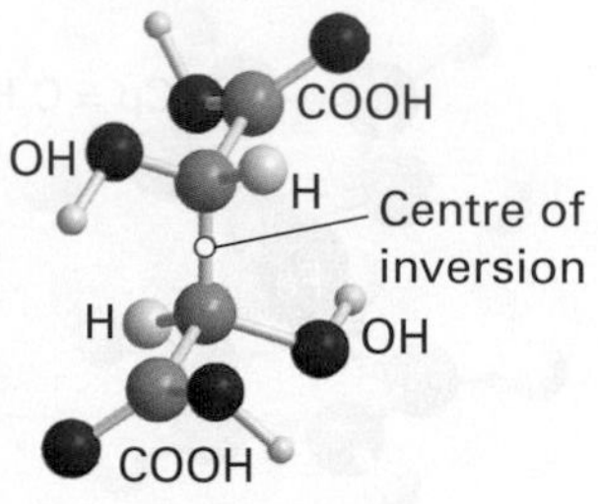

6 *Meso*-tartaric acid, HOOCCH(OH)CH(OH)COOH

- Quinoline (**7**) has the elements (E,σ), and so belongs to the group C_s.

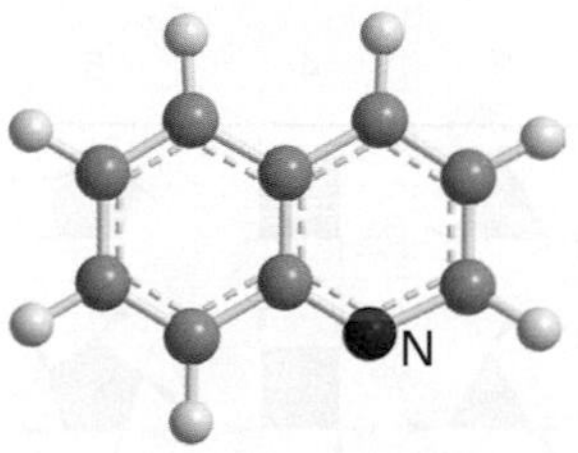

7 Quinoline, C_9H_7N

(b) The groups C_n, C_{nv}, and C_{nh}

A molecule belongs to the group C_n if it possesses an n-fold axis. Note that symbol C_n is now playing a triple role: as the label of a symmetry element, a symmetry operation, and a group. If in addition to the identity and a C_n axis a molecule has n vertical mirror planes σ_v, then it belongs to the group C_{nv}. Molecules that in addition to the identity and an n-fold principal axis also have a horizontal mirror plane σ_h belong to the groups C_{nh}. The presence of certain symmetry elements may be implied by the presence of others: thus, in C_{2h} the elements C_2 and σ_h jointly imply the presence of a centre of inversion (Fig. 10A.9). Note also that the tables specify the *elements*, not the *operations*: for instance, there are two operations associated with a single C_3 axis (rotations by +120° and −120°).

Name	Elements
C_n	E, C_n
C_{nv}	$E, C_n, n\sigma_v$
C_{nh}	E, C_n, σ_h

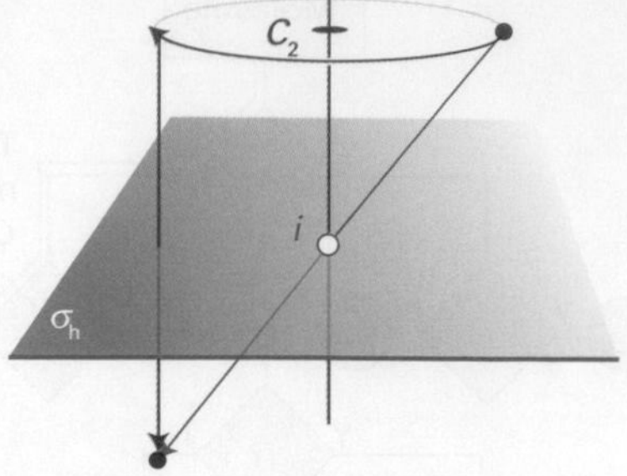

Figure 10A.9 The presence of a twofold axis and a horizontal mirror plane jointly imply the presence of a centre of inversion in the molecule.

Brief illustration 10A.4

- In the H_2O_2 molecule (**8**) the two O–H bonds make an angle of about 115° to one another when viewed down the O–O bond direction. The molecule has the symmetry elements E and C_2, and so belongs to the group C_2.

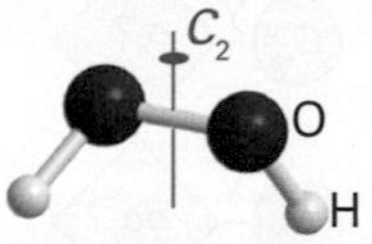

8 Hydrogen peroxide, H_2O_2

- An H_2O molecule has the symmetry elements E, C_2, and $2\sigma_v$, so it belongs to the group C_{2v}.
- An NH_3 molecule has the elements E, C_3, and $3\sigma_v$, so it belongs to the group C_{3v}.
- A heteronuclear diatomic molecule such as HCl belongs to the group $C_{\infty v}$ because rotations around the internuclear axis by any angle and reflections in any of the infinite number of planes that contain this axis are symmetry operations. Other members of the group $C_{\infty v}$ include the linear OCS molecule and a cone.
- The molecule *trans*-CHCl=CHCl (**9**) has the elements E, C_2, and σ_h, so belongs to the group C_{2h}.

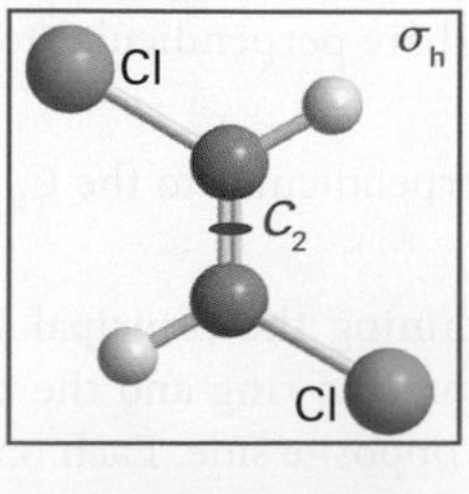

9 *trans*-CHCl=CHCl

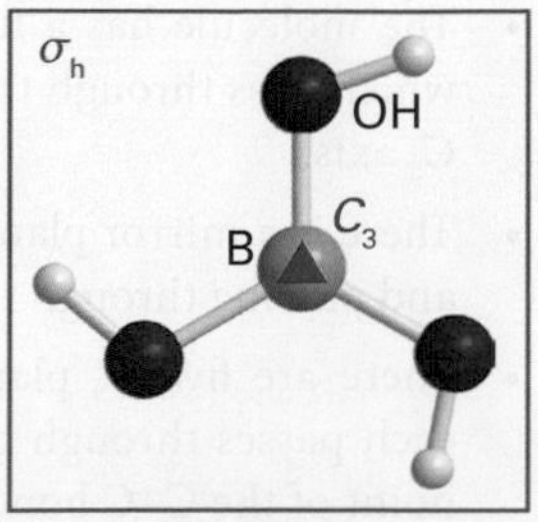

10 $B(OH)_3$

- The molecule $B(OH)_3$, in the planar conformation shown in (**10**), has a C_3 axis and a σ_h plane, and so belongs to the point group C_{3h}.

(c) The groups D_n, D_{nh}, and D_{nd}

Figure 10A.7 shows that a molecule that has an n-fold principal axis and n twofold axes perpendicular to C_n belongs to the group D_n. A molecule belongs to D_{nh} if it also possesses a horizontal mirror plane. The linear molecules OCO and HCCH, and a uniform cylinder, all belong to the group $D_{\infty h}$. A molecule belongs to the group D_{nd} if in addition to the elements of D_n it possesses n dihedral mirror planes σ_d.

Name	Elements
D_n	E, C_n, nC_2'
D_{nh}	E, C_n, nC_2', σ_h
D_{nd}	$E, C_n, nC_2', n\sigma_d$

Brief illustration 10A.5

- The trigonal planar BF_3 molecule (**11**) has the elements E, C_3, $3C_2$ (with one C_2 axis along each B–F bond), and σ_h, so belongs to D_{3h}.

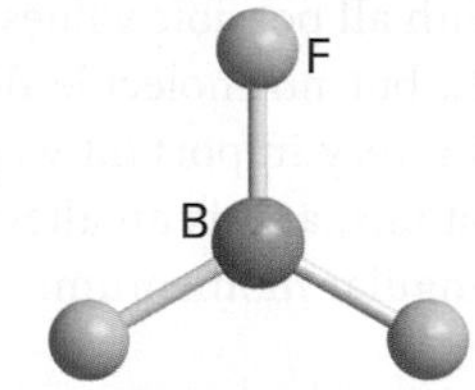

11 Boron trifluoride, BF_3

- The C_6H_6 molecule has the elements E, C_6, $3C_2$, $3C_2'$, and σ_h together with some others that these elements imply, so it belongs to D_{6h}. Three of the C_2 axes bisect C–C bonds on opposite sides of the hexagonal ring formed by the carbon atoms, and the other three pass through vertices on opposite sides of the ring. The prime on $3C_2'$ indicates that these axes are different from the other three C_2 axes.
- All homonuclear diatomic molecules, such as N_2, belong to the group $D_{\infty h}$ because all rotations around the internuclear axis are symmetry operations, as are end-over-end rotations by 180°.
- PCl_5 (**12**) is another example of a D_{3h} species.

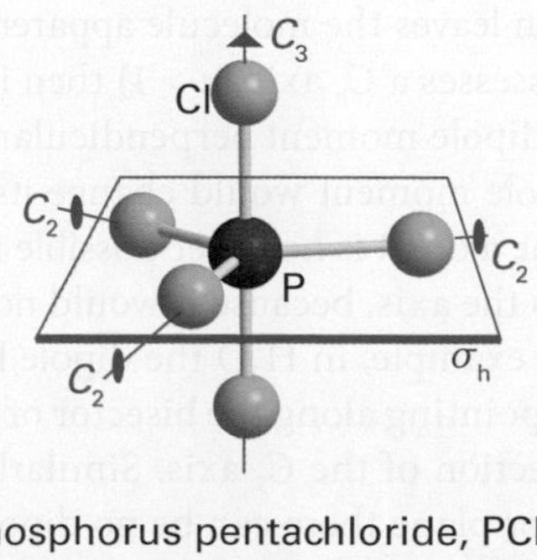

12 Phosphorus pentachloride, PCl_5 (D_{3h})

- Propadiene (an allene, **13**), in which the two CH_2 groups lie in perpendicular planes, belongs to the point group D_{2d}.

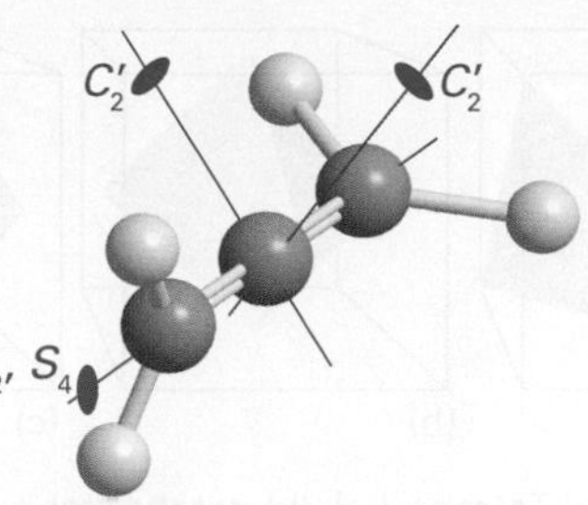

13 Propadiene, C_3H_4 (D_{2d})

(d) The groups S_n

Molecules that have not been classified into one of the groups mentioned so far, but which possess one S_n axis, belong to the groups S_n. Note that the group S_2 is the same as C_i, so such a molecule will already have been classified as C_i. Tetraphenylmethane (**14**) belongs to the point group S_4; molecules belonging to S_n with $n > 4$ are rare.

Name	Elements
S_n	E, S_n and not previously classified

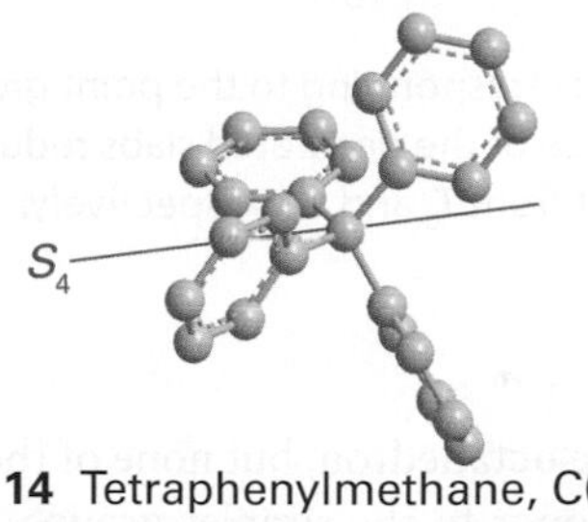

14 Tetraphenylmethane, $C(C_6H_5)_4$ (S_4)

(e) The cubic groups

A number of very important molecules possess more than one principal axis. Most belong to the **cubic groups**, and in particular to the **tetrahedral groups** T, T_d, and T_h (Fig. 10A.10a) or to the **octahedral groups** O and O_h (Fig. 10A.10b). A few icosahedral (20-faced) molecules belonging to the **icosahedral group**, I (Fig. 10A.10c), are also known. The groups T_d and O_h are the groups of the regular tetrahedron and the regular octahedron, respectively. If the object possesses the rotational symmetry of

Name	Elements
T	$E, 4C_3, 3C_2$
T_d	$E, 3C_2, 4C_3, 3S_4, 6\sigma_d$
T_h	$E, 3C_2, 4C_3, i, 4S_6, 3\sigma_h$
O	$E, 3C_4, 4C_3, 6C_2$
O_h	$E, 3S_4, 3C_4, 6C_2, 4S_6, 4C_3, 3\sigma_h, 6\sigma_d, i$
I	$E, 6C_5, 10C_3, 15C_2$
I_h	$E, 6S_{10}, 10S_6, 6C_5, 10C_3, 15C_2, 15\sigma, i$

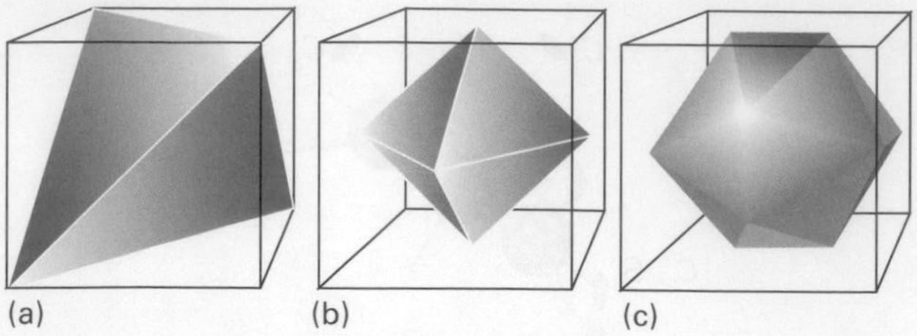

Figure 10A.10 (a) Tetrahedral, (b) octahedral, and (c) icosahedral shapes drawn to show their relation to a cube: they belong to the cubic groups T_d, O_h, and I_h, respectively.

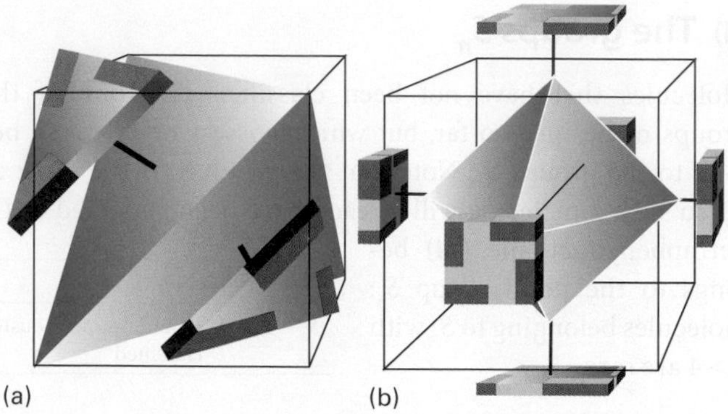

Figure 10A.11 Shapes corresponding to the point groups (a) T and (b) O. The presence of the decorated slabs reduces the symmetry of the object from T_d and O_h, respectively.

the tetrahedron or the octahedron, but none of their planes of reflection, then it belongs to the simpler groups T or O (Fig. 10A.11). The group T_h is based on T but also contains a centre of inversion (Fig. 10A.12).

Brief illustration 10A.6

- The molecules CH_4 and SF_6 belong, respectively, to the groups T_d and O_h.
- Molecules belonging to the icosahedral group I include some of the boranes and buckminsterfullerene, C_{60} (**15**).

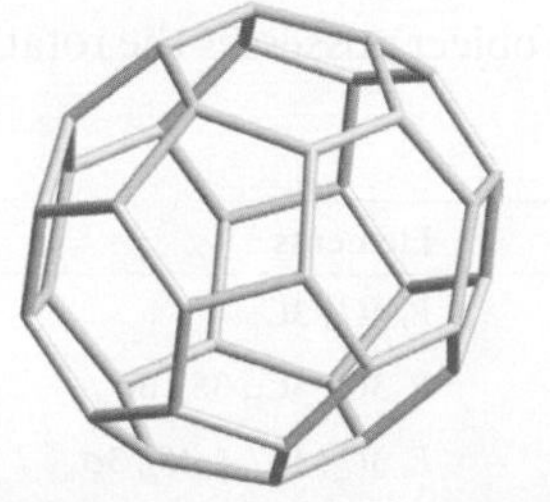

15 Buckminsterfullerene, C_{60} (I)

- The objects shown in Fig. 10A.11 belong to the groups T and O, respectively.

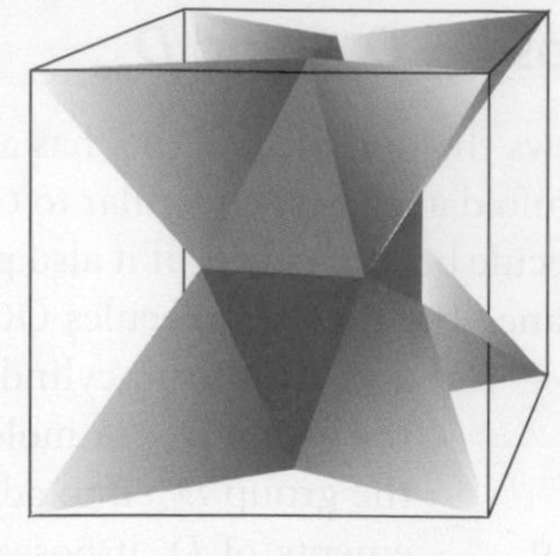

Figure 10A.12 The shape of an object belonging to the group T_h.

(f) The full rotation group

Name	Elements
R_3	$E, \infty C_2, \infty C_3, \ldots$

The **full rotation group**, R_3 (the 3 refers to rotation in three dimensions), consists of an infinite number of rotation axes with all possible values of n. A sphere and an atom belong to R_3, but no molecule does. Exploring the consequences of R_3 is a very important way of applying symmetry arguments to atoms, and is an alternative approach to the theory of orbital angular momentum.

10A.3 Some immediate consequences of symmetry

Some statements about the properties of a molecule can be made as soon as its point group has been identified.

(a) Polarity

A **polar molecule** is one with a permanent electric dipole moment (HCl, O_3, and NH_3 are examples). A dipole moment is a property of the molecule, so it follows that the dipole moment (which is represented by a vector) must be unaffected by any symmetry operation of the molecule because, by definition, such an operation leaves the molecule apparently unchanged. If a molecule possesses a C_n axis ($n > 1$) then it is not possible for there to be a dipole moment perpendicular to this axis because such a dipole moment would change its orientation on rotation about the axis. It is however possible for there to be a dipole parallel to the axis, because it would not be affected by the rotation. For example, in H_2O the dipole lies in the plane of the molecule, pointing along the bisector of the HOH bond, which is the direction of the C_2 axis. Similarly, if a molecule possesses a mirror plane there can be no dipole moment perpendicular to this plane, because reflection in the plane would reverse its direction. A molecule that possesses a centre of symmetry cannot have a dipole moment in any direction because the inversion operation would reverse it.

These considerations lead to the conclusion that

> Only molecules belonging to the groups C_n, C_{nv}, and C_s may have a permanent electric dipole moment.

For C_n and C_{nv}, the dipole moment must lie along the principal axis.

Brief illustration 10A.7

- Ozone, O_3, which has an angular structure, belongs to the group C_{2v} and is polar.
- Carbon dioxide, CO_2, which is linear and belongs to the group $D_{\infty h}$, is not polar.
- Tetraphenylmethane (**14**) belongs to the point group S_4 and so is not polar.

(b) Chirality

A **chiral molecule** (from the Greek word for 'hand') is a molecule that cannot be superimposed on its mirror image. An **achiral molecule** is a molecule that can be superimposed on its mirror image. Chiral molecules are **optically active** in the sense that they rotate the plane of polarized light. A chiral molecule and its mirror-image partner constitute an **enantiomeric pair** (from the Greek word for 'both') of isomers and rotate the plane of polarization by equal amounts but opposite directions.

> A molecule may be chiral, and therefore optically active, only if it does not possess an axis of improper rotation, S_n.

An S_n improper rotation axis may be present under a different name, and be implied by other symmetry elements that are present. For example, molecules belonging to the groups C_{nh} possess an S_n axis implicitly because they possess both C_n and σ_h, which are the two components of an improper rotation axis. A centre of inversion, i, is in fact the same as S_2 because the two corresponding operations achieve exactly the same result (Fig. 10A.13). Furthermore, a mirror plane is the same as S_1 (rotation through 360° followed by reflection). Thus molecules possessing a mirror plane or a centre of inversion effectively possess an axis of improper rotation and so, by the above rule, are achiral.

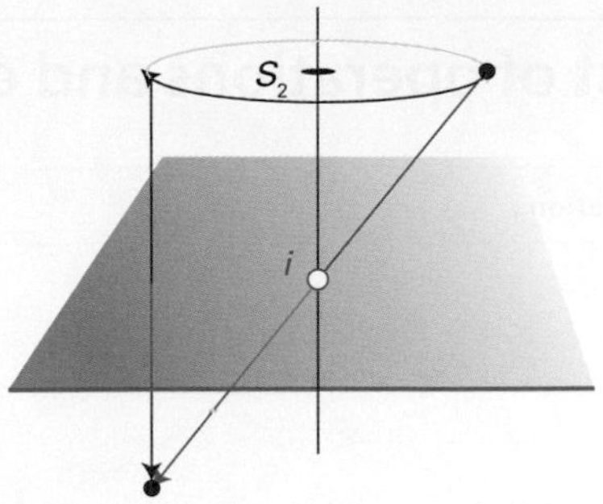

Figure 10A.13 The operations i and S_2 are equivalent in the sense that they achieve exactly the same outcome when applied to a point in the object.

Brief illustration 10A.8

- The amino acid alanine (**16**) does not possess a centre of inversion nor does it have any mirror planes: it is therefore chiral.

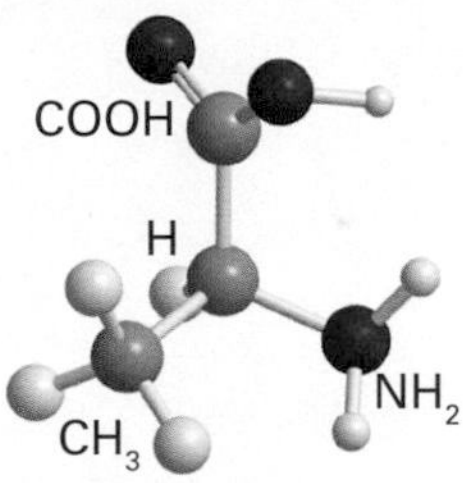

16 L-Alanine, $NH_2CH(CH_3)COOH$

- In contrast, glycine (**17**) has a mirror plane and so is achiral.

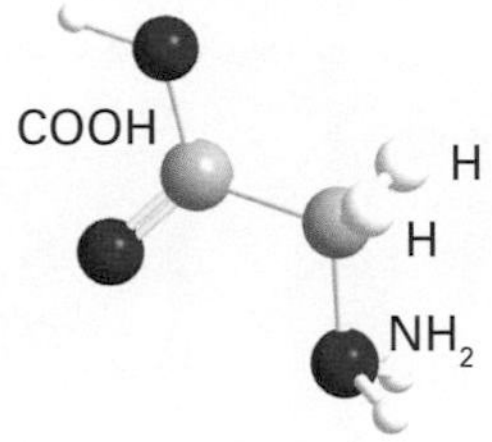

17 Glycine, NH_2CH_2COOH

- Tetraphenylmethane (**14**) belongs to the point group S_4; it does not possess a centre of inversion or any mirror planes, but it is still achiral since it possesses an axis of improper rotation (S_4).

Checklist of concepts

☐ 1. A **symmetry operation** is an action that leaves an object looking the same after it has been carried out.

☐ 2. A **symmetry element** is a point, line, or plane with respect to which a symmetry operation is performed.

☐ 3. The **notation** for point groups commonly used for molecules and solids is summarized in Table 10A.1.

☐ 4. To be **polar**, a molecule must belong to C_n, C_{nv}, or C_s (and have no higher symmetry).

☐ 5. A molecule will be **chiral** only if it does not possess an axis of improper rotation, S_n.

Checklist of operations and elements

Symmetry operation	Symbol	Symmetry element
n-Fold rotation	C_n	*n*-Fold axis of rotation
Reflection	σ	Mirror plane
Inversion	i	Centre of symmetry
n-Fold improper rotation	S_n	*n*-Fold improper axis of rotation
Identity	E	Entire object

TOPIC 10B Group theory

> ➤ Why do you need to know this material?
>
> Group theory expresses qualitative ideas about symmetry mathematically and can be applied systematically to a wide variety of problems. The theory is also the origin of the labelling of atomic and molecular orbitals that is used throughout chemistry.
>
> ➤ What is the key idea?
>
> Symmetry operations may be represented by the effect of matrices on a basis.
>
> ➤ What do you need to know already?
>
> You need to know about the types of symmetry operation and element introduced in Topic 10A. This discussion uses matrix algebra, and especially matrix multiplication, as set out in *The chemist's toolkit* 24 in Topic 9E.

The systematic discussion of symmetry is called **group theory**. Much of group theory is a summary of common sense about the symmetries of objects. However, because group theory is systematic, its rules can be applied in a straightforward, mechanical way. In most cases the theory gives a simple, direct method for arriving at useful conclusions with the minimum of calculation, and this is the aspect that is stressed here.

10B.1 The elements of group theory

A **group** in mathematics is a collection of transformations that satisfy four criteria. If the transformations are written as R, R', … (which might be the reflections, rotations, and so on introduced in Topic 10A), then they form a group if:

1. One of the transformations is the identity (i.e. 'do nothing').
2. For every transformation R, the inverse transformation R^{-1} is included in the collection so that the combination RR^{-1} (the transformation R^{-1} followed by R) is equivalent to the identity.
3. The combination RR' (the transformation R' followed by R) is equivalent to a single member of the collection of transformations.
4. The combination $R(R'R'')$, the transformation $(R'R'')$ followed by R, is equivalent to $(RR')R''$, the transformation R'' followed by (RR').

Example 10B.1 Showing that the symmetry operations of a molecule form a group

The point group C_{2v} consists of the elements $\{E, C_2, \sigma_v, \sigma_v'\}$ and correspond to the operations $\{E, C_2, \sigma_v, \sigma_v'\}$. Show that this set of operations is a group in the mathematical sense.

Collect your thoughts You need to show that combinations of the operations match the criteria set out above. The operations are specified in Topic 10A, and illustrated in Figs. 10A.2 and 10A.3 for H_2O, which belongs to this group.

The solution

- Criterion 1 is fulfilled because the collection of symmetry operations includes the identity E.
- Criterion 2 is fulfilled because in each case the inverse of an operation is the operation itself. Thus, two successive twofold rotations is equivalent to the identity: $C_2C_2 = E$ and likewise for the two reflections and the identity itself.
- Criterion 3 is fulfilled, because in each case one operation followed by another is the same as one of the four symmetry operations. For instance, a twofold rotation C_2 followed by the reflection σ_v is the same as the single reflection σ_v' (Fig. 10B.1); thus, $\sigma_v C_2 = \sigma_v'$. A 'group multiplication table' can be constructed in a similar way for all possible products of symmetry operations RR'; as required, each product is equivalent to another symmetry operation.

$R\downarrow\ R'\rightarrow$	E	C_2	σ_v	σ_v'
E	E	C_2	σ_v	σ_v'
C_2	C_2	E	σ_v'	σ_v
σ_v	σ_v	σ_v'	E	C_2
σ_v'	σ_v'	σ_v	C_2	E

- Criterion 4 is fulfilled, as it is immaterial how the operations are grouped together. Thus $(\sigma_v\sigma_v')C_2 = C_2C_2 = E$ and $\sigma_v(\sigma_v'C_2) = \sigma_v\sigma_v = E$, and likewise for all other combinations.

Self-test 10B.1 Confirm that C_{2h}, which has the elements $\{E, C_2, i, \sigma_h\}$ and hence the corresponding operations $\{E, C_2, i, \sigma_h\}$, is a group (construct the group multiplication table).

Answer: Criteria are fulfilled

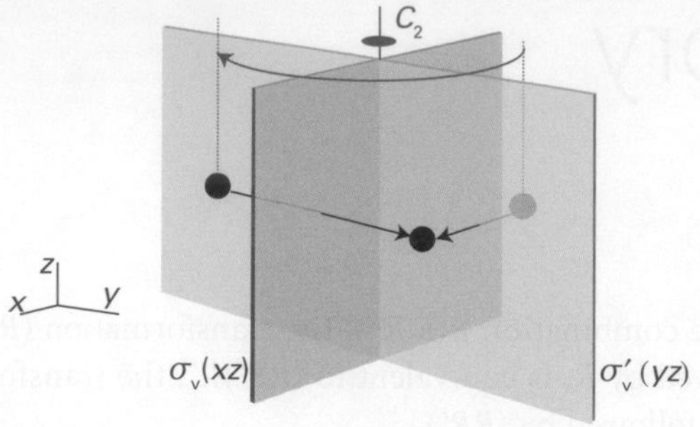

Figure 10B.1 A twofold rotation C_2 followed by the reflection σ_v gives the same result as the reflection σ'_v.

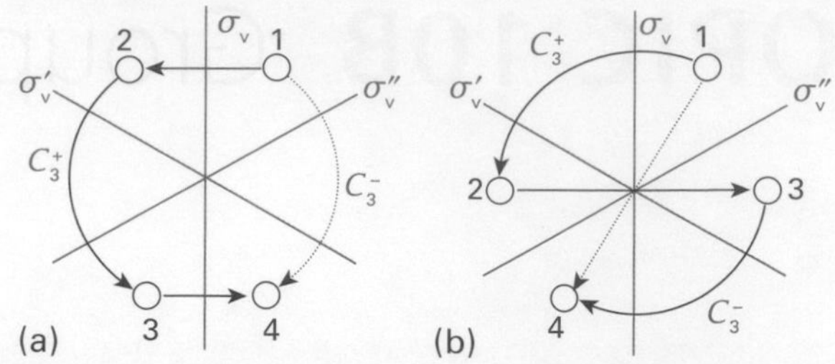

Figure 10B.3 (a) The sequence of operations $\sigma_v^{-1}C_3^+\sigma_v$ when applied to the point 1 takes it through the sequence $1 \rightarrow 2 \rightarrow 3 \rightarrow 4$ (σ_v^{-1} has the same effect as σ_v). The single operation C_3^- (dotted curve) takes point $1 \rightarrow 4$, so C_3^+ and C_3^- are in the same class. (b) The sequence of operations $(C_3^+)^{-1}\sigma_v C_3^+$ takes point $1 \rightarrow 4$, ($(C_3^+)^{-1}$ has the same effect as C_3^-) but the same transformation can be achieved with the single operation σ'_v (dotted line); so σ_v and σ'_v are in the same class.

One potentially confusing point needs to be clarified at the outset. The entities that make up a group are its 'elements'. For applications in chemistry, these elements are almost always symmetry operations. However, as explained in Topic 10A, 'symmetry operations' are distinct from 'symmetry elements', the latter being the points, axes, and planes with respect to which the operations are carried out. A third use of the word 'element' is to denote the number lying in a particular location in a matrix. Be very careful to distinguish *element* (of a group), *symmetry element*, and *matrix element*.

Symmetry operations fall into the same **class** if they are of the same type (for example, rotations) and can be transformed into one another by a symmetry operation of the group. The two threefold rotations in C_{3v} belong to the same class because one can be converted into the other by a reflection (Fig. 10B.2); the three reflections all belong to the same class because each can be rotated into another by a threefold rotation. The formal definition of a class is that two operations R and R' belong to the same class if there is a member S of the group such that

$$R' = S^{-1}RS \qquad \text{Membership of a class} \qquad (10B.1)$$

where S^{-1} is the inverse of S.

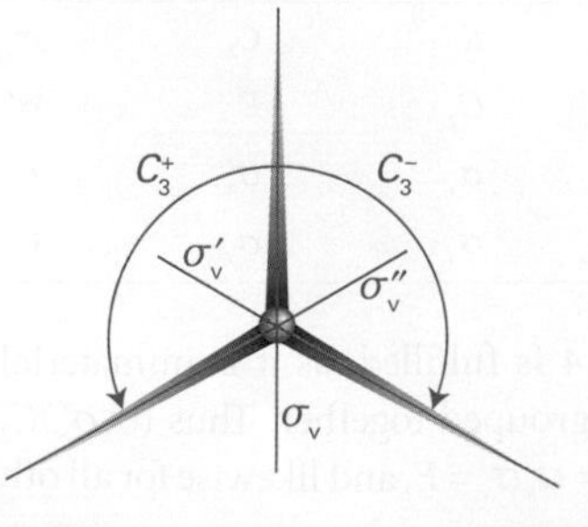

Figure 10B.2 Symmetry operations in the same class are related to one another by the symmetry operations of the group. Thus, the three mirror planes shown here are related by threefold rotations, and the two rotations shown here are related by reflection in σ_v.

Figure 10B.3(a) shows how eqn 10B.1 can be used to confirm that C_3^+ and C_3^- belong to the same class in the group C_{3v} by considering how an arbitrary point, 1, behaves under the various operations. The transformation of interest is $\sigma_v^{-1}C_3^+\sigma_v$. Start at 1: the operation σ_v moves the point to 2, and then C_3^+ moves the point to 3. The inverse of a reflection is itself, $\sigma_v^{-1} = \sigma_v$, so the effect of σ_v^{-1} is to move the point to 4. From the diagram it can be seen that point 4 can be reached by applying C_3^- to point 1, thus demonstrating that $\sigma_v^{-1}C_3^+\sigma_v = C_3^-$, and hence that C_3^+ and C_3^- do indeed belong in the same class.

Brief illustration 10B.1

To show that σ_v and σ'_v are in the same class in the group C_{3v}, consider the transformation $(C_3^+)^{-1}\sigma_v C_3^+$. Because C_3^- is the inverse of C_3^+, this transformation is the same as $C_3^-\sigma_v C_3^+$; the effect of this sequence of operations on an arbitrary point 1 is shown in Fig. 10B.3(b). The final position, 4, can also be reached from 1 by applying the operation σ'_v, thus showing that $C_3^-\sigma_v C_3^+ = \sigma'_v$ and hence that σ_v and σ'_v are in the same class.

10B.2 Matrix representations

Group theory takes on great power when the notional ideas presented so far are expressed in terms of collections of numbers in the form of matrices. For basic information about how to handle matrices, see *The chemist's toolkit* 24 in Topic 9E.

(a) Representatives of operations

Consider the set of five p orbitals shown on the C_{2v} SO_2 molecule in Fig. 10B.4 and how they are affected by the reflection operation σ_v. The corresponding symmetry element is the mirror plane perpendicular to the plane of the molecule and passing through the S atom. The effect of this reflection is to leave p_x and p_z unaffected, to change the sign of p_y, and to exchange p_A and p_B. Its effect can be written $(p_x\ {-p_y}\ p_z\ p_B\ p_A) \leftarrow (p_x\ p_y\ p_z\ p_A\ p_B)$. This transformation can be expressed by using matrix multiplication:

$$(p_x\ {-p_y}\ p_z\ p_B\ p_A) = (p_x\ p_y\ p_z\ p_A\ p_B)\overbrace{\begin{pmatrix} 1 & 0 & 0 & 0 & 0 \\ 0 & -1 & 0 & 0 & 0 \\ 0 & 0 & 1 & 0 & 0 \\ 0 & 0 & 0 & 0 & 1 \\ 0 & 0 & 0 & 1 & 0 \end{pmatrix}}^{D(\sigma_v)}$$

$$= (p_x\ p_y\ p_z\ p_A\ p_B)\,\boldsymbol{D}(\sigma_v) \qquad (10B.2a)$$

The matrix $\boldsymbol{D}(\sigma_v)$ is called a **representative** of the operation σ_v. Representatives take different forms according to the **basis**, the set of orbitals that has been adopted. In this case, the basis is the row vector $(p_x\ p_y\ p_z\ p_A\ p_B)$. Note that the matrix $\boldsymbol{D}$ appears to the right of the basis functions on which it acts.

The same technique can be used to find matrices that reproduce the other symmetry operations. For instance, C_2 has the effect $(-p_x\ {-p_y}\ p_z\ {-p_B}\ {-p_A}) \leftarrow (p_x\ p_y\ p_z\ p_A\ p_B)$, and its representative is

$$\boldsymbol{D}(C_2) = \begin{pmatrix} -1 & 0 & 0 & 0 & 0 \\ 0 & -1 & 0 & 0 & 0 \\ 0 & 0 & 1 & 0 & 0 \\ 0 & 0 & 0 & 0 & -1 \\ 0 & 0 & 0 & -1 & 0 \end{pmatrix} \qquad (10B.2b)$$

The effect of σ_v' (reflection in the plane of the molecule) is $(-p_x\ p_y\ p_z\ {-p_A}\ {-p_B}) \leftarrow (p_x\ p_y\ p_z\ p_A\ p_B)$; the oxygen orbitals remain in the same places, but change sign. The representative of this operation is

$$\boldsymbol{D}(\sigma_v') = \begin{pmatrix} -1 & 0 & 0 & 0 & 0 \\ 0 & 1 & 0 & 0 & 0 \\ 0 & 0 & 1 & 0 & 0 \\ 0 & 0 & 0 & -1 & 0 \\ 0 & 0 & 0 & 0 & -1 \end{pmatrix} \qquad (10B.2c)$$

The identity operation has no effect on the basis, so its representative is the 5×5 unit matrix:

$$\boldsymbol{D}(E) = \begin{pmatrix} 1 & 0 & 0 & 0 & 0 \\ 0 & 1 & 0 & 0 & 0 \\ 0 & 0 & 1 & 0 & 0 \\ 0 & 0 & 0 & 1 & 0 \\ 0 & 0 & 0 & 0 & 1 \end{pmatrix} \qquad (10B.2d)$$

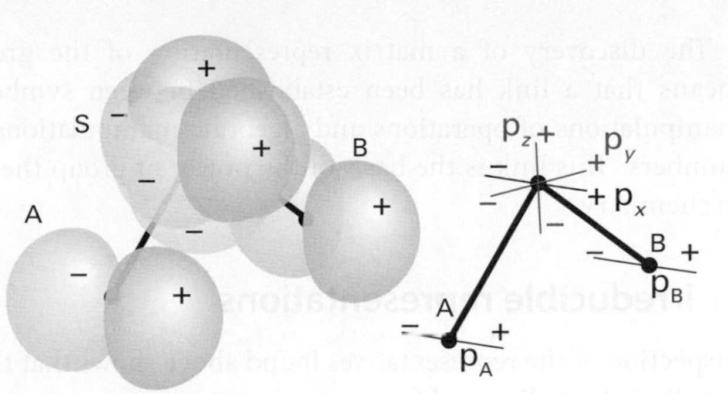

Figure 10B.4 The five p orbitals (three on the sulfur and one on each oxygen) that are used to illustrate the construction of a matrix representation in a C_{2v} molecule (SO_2).

(b) The representation of a group

The set of matrices that represents *all* the operations of the group is called a **matrix representation**, Γ (uppercase gamma), of the group in the basis that has been chosen. In the current example, there are five members of the basis and the representation is five-dimensional in the sense that the matrices are all 5×5 arrays. The matrices of a representation multiply together in the same way as the operations they represent. Thus, if for any two operations R and R', $RR' = R''$, then $\boldsymbol{D}(R)\boldsymbol{D}(R') = \boldsymbol{D}(R'')$ for a given basis.

Brief illustration 10B.2

In the group C_{2v}, a twofold rotation followed by a reflection in a mirror plane is equivalent to a reflection in the second mirror plane: specifically, $\sigma_v' C_2 = \sigma_v$. Multiplying out the representatives specified in eqn 10B.2 gives

$$\boldsymbol{D}(\sigma_v')\boldsymbol{D}(C_2) = \begin{pmatrix} -1 & 0 & 0 & 0 & 0 \\ 0 & 1 & 0 & 0 & 0 \\ 0 & 0 & 1 & 0 & 0 \\ 0 & 0 & 0 & -1 & 0 \\ 0 & 0 & 0 & 0 & -1 \end{pmatrix}\begin{pmatrix} -1 & 0 & 0 & 0 & 0 \\ 0 & -1 & 0 & 0 & 0 \\ 0 & 0 & 1 & 0 & 0 \\ 0 & 0 & 0 & 0 & -1 \\ 0 & 0 & 0 & -1 & 0 \end{pmatrix}$$

$$= \begin{pmatrix} 1 & 0 & 0 & 0 & 0 \\ 0 & -1 & 0 & 0 & 0 \\ 0 & 0 & 1 & 0 & 0 \\ 0 & 0 & 0 & 0 & 1 \\ 0 & 0 & 0 & 1 & 0 \end{pmatrix} = \boldsymbol{D}(\sigma_v)$$

As expected, this multiplication reproduces the same result as the group multiplication table. The same is true for multiplication of any two representatives, so the four matrices form a representation of the group.

The discovery of a matrix representation of the group means that a link has been established between symbolic manipulations of operations and algebraic manipulations of numbers. This link is the basis of the power of group theory in chemistry.

(c) Irreducible representations

Inspection of the representatives found above shows that they are all of **block-diagonal form**:

$$\boldsymbol{D}=\begin{pmatrix} \blacksquare & 0 & 0 & 0 & 0 \\ 0 & \blacksquare & 0 & 0 & 0 \\ 0 & 0 & \blacksquare & 0 & 0 \\ 0 & 0 & 0 & \blacksquare & \blacksquare \\ 0 & 0 & 0 & \blacksquare & \blacksquare \end{pmatrix} \quad \text{Block-diagonal form} \quad (10B.3)$$

The block-diagonal form of the representatives implies that the symmetry operations of C_{2v} never mix p_x, p_y, and p_z together, nor do they mix these three orbitals with p_A and p_B, but p_A and p_B are mixed together by the operations of the group. Consequently, the basis can be cut into four parts: three for the individual p orbitals on S and the fourth for the two oxygen orbitals (p_A p_B). The representations in these three one-dimensional bases are

For p_x: $\boldsymbol{D}(E)=1 \quad \boldsymbol{D}(C_2)=-1 \quad \boldsymbol{D}(\sigma_v)=1 \quad \boldsymbol{D}(\sigma'_v)=-1$

For p_y: $\boldsymbol{D}(E)=1 \quad \boldsymbol{D}(C_2)=-1 \quad \boldsymbol{D}(\sigma_v)=-1 \quad \boldsymbol{D}(\sigma'_v)=1$

For p_z: $\boldsymbol{D}(E)=1 \quad \boldsymbol{D}(C_2)=1 \quad \boldsymbol{D}(\sigma_v)=1 \quad \boldsymbol{D}(\sigma'_v)=1$

These representations will be called $\Gamma^{(1)}$, $\Gamma^{(2)}$, and $\Gamma^{(3)}$, respectively. The remaining two functions (p_A p_B) are a basis for a two-dimensional representation denoted Γ':

$$\boldsymbol{D}(E)=\begin{pmatrix}1&0\\0&1\end{pmatrix} \quad \boldsymbol{D}(C_2)=\begin{pmatrix}0&-1\\-1&0\end{pmatrix}$$

$$\boldsymbol{D}(\sigma_v)=\begin{pmatrix}0&1\\1&0\end{pmatrix} \quad \boldsymbol{D}(\sigma'_v)=\begin{pmatrix}-1&0\\0&-1\end{pmatrix}$$

The original five-dimensional representation has been **reduced** to the 'direct sum' of three one-dimensional representations 'spanned' by each of the p orbitals on S, and a two-dimensional representation spanned by (p_A p_B). The reduction is represented symbolically by writing[1]

$$\Gamma=\Gamma^{(1)}+\Gamma^{(2)}+\Gamma^{(3)}+\Gamma' \quad \text{Direct sum} \quad (10B.4)$$

[1] The symbol $\oplus$ is sometimes used to denote a direct sum to distinguish it from an ordinary sum, in which case eqn 10B.4 would be written $\Gamma=\Gamma^{(1)}\oplus\Gamma^{(2)}\oplus\Gamma^{(3)}\oplus\Gamma'$.

The representations $\Gamma^{(1)}$, $\Gamma^{(2)}$, and $\Gamma^{(3)}$ cannot be reduced any further, and each one is called an **irreducible representation** of the group (an 'irrep'). That the two-dimensional representation Γ' is reducible (for this basis, in this group) is demonstrated by switching attention to the linear combinations $p_1 = p_A + p_B$ and $p_2 = p_A - p_B$ in (Fig. 10B.5). The effect of the operation σ_v is to exchange p_A and p_B: $(p_B\ p_A) \leftarrow (p_A\ p_B)$. Therefore $(p_B + p_A) \leftarrow (p_A + p_B)$, corresponding to $(p_1) \leftarrow (p_1)$. Similarly, $(p_B - p_A) \leftarrow (p_A - p_B)$, corresponding to $(-p_2) \leftarrow (p_2)$. It follows from these results, and similar ones for the other operations, that the representation in the basis (p_1 p_2) is

$$\boldsymbol{D}(E)=\begin{pmatrix}1&0\\0&1\end{pmatrix} \quad \boldsymbol{D}(C_2)=\begin{pmatrix}-1&0\\0&1\end{pmatrix}$$

$$\boldsymbol{D}(\sigma_v)=\begin{pmatrix}1&0\\0&-1\end{pmatrix} \quad \boldsymbol{D}(\sigma'_v)=\begin{pmatrix}-1&0\\0&-1\end{pmatrix}$$

The new representatives are all in block-diagonal form, in this case in the form $\begin{pmatrix}\blacksquare&0\\0&\blacksquare\end{pmatrix}$, and the two combinations are not mixed with each other by any operation of the group. The representation Γ' has therefore been reduced to the sum of two one-dimensional representations. Thus, p_1 spans the one-dimensional representation

$$\boldsymbol{D}(E)=1 \quad \boldsymbol{D}(C_2)=-1 \quad \boldsymbol{D}(\sigma_v)=1 \quad \boldsymbol{D}(\sigma'_v)=-1$$

which is the same as the representation $\Gamma^{(1)}$ spanned by p_x. The combination p_2 spans

$$\boldsymbol{D}(E)=1 \quad \boldsymbol{D}(C_2)=1 \quad \boldsymbol{D}(\sigma_v)=-1 \quad \boldsymbol{D}(\sigma'_v)=-1$$

which is a new one-dimensional representation and denoted $\Gamma^{(4)}$. At this stage the original representation has been reduced into five one-dimensional representations as follows:

$$\Gamma=2\Gamma^{(1)}+\Gamma^{(2)}+\Gamma^{(3)}+\Gamma^{(4)}$$

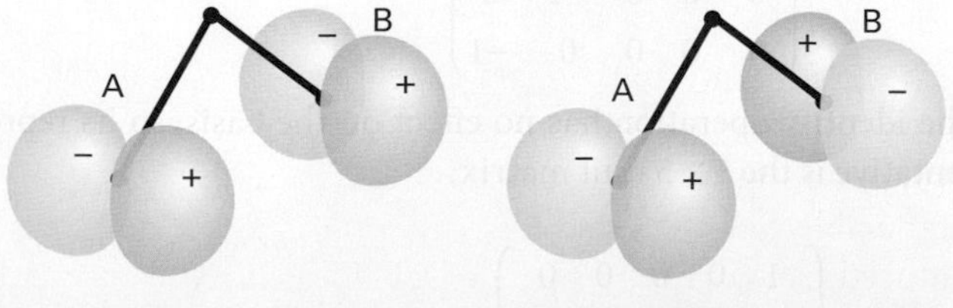

Figure 10B.5 Two symmetry-adapted linear combinations of the oxygen basis orbitals shown in Fig. 10B.4: on the left $p_1 = p_A + p_B$, on the right $p_2 = p_A - p_B$. The two combinations each span a one-dimensional irreducible representation, and their symmetry species are different.

(d) Characters

The **character**, χ (chi), of an operation in a particular matrix representation is the sum of the diagonal elements of the representative of that operation. Thus, in the original basis (p_x p_y p_z p_A p_B) the characters of the representatives are

R	E	C_2
$D(R)$	$\begin{pmatrix}1&0&0&0&0\\0&1&0&0&0\\0&0&1&0&0\\0&0&0&1&0\\0&0&0&0&1\end{pmatrix}$	$\begin{pmatrix}-1&0&0&0&0\\0&-1&0&0&0\\0&0&1&0&0\\0&0&0&0&-1\\0&0&0&-1&0\end{pmatrix}$
$\chi(R)$	5	−1

R	σ_v	σ'_v
$D(R)$	$\begin{pmatrix}1&0&0&0&0\\0&-1&0&0&0\\0&0&1&0&0\\0&0&0&0&1\\0&0&0&1&0\end{pmatrix}$	$\begin{pmatrix}-1&0&0&0&0\\0&1&0&0&0\\0&0&1&0&0\\0&0&0&-1&0\\0&0&0&0&-1\end{pmatrix}$
$\chi(R)$	1	−1

The characters of one-dimensional representatives are just the representatives themselves. For each operation, the sum of the characters of the reduced representations is the same as the character of the original representation (allowing for the appearance of $\Gamma^{(1)}$ twice in the reduction $\Gamma = 2\Gamma^{(1)} + \Gamma^{(2)} + \Gamma^{(3)} + \Gamma^{(4)}$):

R	E	C_2	σ_v	σ'_v
$\chi(R)$ for $\Gamma^{(1)}$	1	−1	1	−1
$\chi(R)$ for $\Gamma^{(1)}$	1	−1	1	−1
$\chi(R)$ for $\Gamma^{(2)}$	1	−1	−1	1
$\chi(R)$ for $\Gamma^{(3)}$	1	1	1	1
$\chi(R)$ for $\Gamma^{(4)}$	1	1	−1	−1
Sum for Γ:	5	−1	1	−1

At this point, four irreducible representations of the group C_{2v} have been found. Are these the only irreducible representations of the group C_{2v}? There are in fact no more irreducible representations in this group, a fact that can be deduced from a surprising theorem of group theory, which states that

$$\text{Number of irreducible representations} = \text{number of classes} \quad \text{Number of irreducible representations} \quad (10B.5)$$

In C_{2v} there are four classes of operations (the four columns in the table), so there must be four irreducible representations. The ones already found are the only ones for this group.

Another powerful result from group theory, which applies to all groups other than the pure rotation groups C_n with $n>2$, relates the sum of the squares of the dimensions, d_i, of all the irreducible representations $\Gamma^{(i)}$ to the **order** of the group, which is the total number of symmetry operations, h:

$$\sum_{\text{irreducible representations, } i} d_i^2 = h \quad \text{Dimensionality and order} \quad (10B.6)$$

The four irreducible representations of C_{2v} are all one-dimensional, so

$$\sum_{\text{irreducible representations, } i} d_i^2 = 1^2+1^2+1^2+1^2=4$$

and there are indeed four symmetry operations of the group.

Brief illustration 10B.3

The group C_{3v} has three classes of operations $\{E,2C_3,3\sigma_v\}$, so there are three irreducible representations. The order of the group is $1 + 2 + 3 = 6$, so if it is already known that two of the irreducible representations are one-dimensional, the remaining irreducible representation must be two-dimensional by using eqn 10B.6: $1^2 + 1^2 + d_3^2 = 6$, hence $d_3 = 2$.

10B.3 Character tables

Tables showing all the characters of the operations of a group are called **character tables** and from now on they move to the centre of the discussion. The columns of a character table are labelled with the symmetry operations of the group. Although the notation $\Gamma^{(i)}$ is used to label general irreducible representations, in chemical applications and for displaying character tables it is more common to distinguish different irreducible representations by the use of the labels A, B, E, and T to denote the **symmetry species** of each representation:

A: one-dimensional representation, character +1 under the principal rotation

B: one-dimensional representation, character −1 under the principal rotation

E: two-dimensional irreducible representation

T: three-dimensional irreducible representation

Subscripts are used to distinguish the irreducible representations if there is more than one of the same type: A_1 is reserved for the representation with character 1 for all operations

Table 10B.1 The C_{2v} character table*

C_{2v}, 2*mm*	*E*	C_2	$\sigma_v(xz)$	$\sigma_v'(yz)$	$h=4$	
A_1	1	1	1	1	z	z^2, y^2, x^2
A_2	1	1	−1	−1		xy
B_1	1	−1	1	−1	x	zx
B_2	1	−1	−1	1	y	yz

* More character tables are given in the *Resource section.*

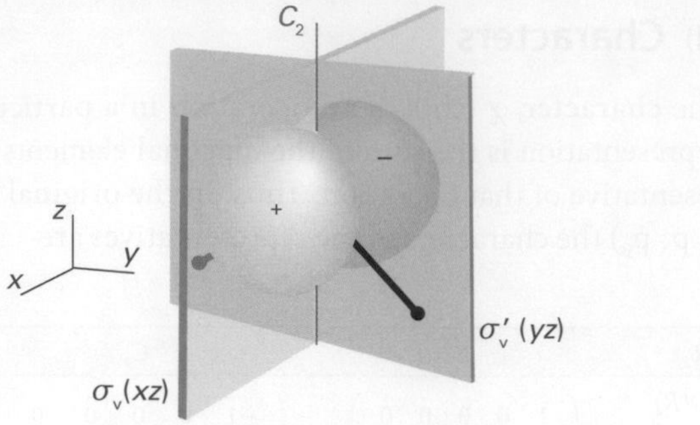

Figure 10B.6 A p_x orbital on the central atom of a C_{2v} molecule and the symmetry elements of the group.

(called the **totally symmetric irreducible representation**); A_2 has 1 for the principal rotation but −1 for reflections. There appears to be no systematic way of attaching subscripts to B symmetry species, so care must be used when referring to character tables from different sources.

Table 10B.1 shows the character table for the group C_{2v}, with its four symmetry species (irreducible representations) and its four columns of symmetry operations. Table 10B.2 shows the table for the group C_{3v}. The columns are headed E, $2C_3$, and $3\sigma_v$: the numbers multiplying each operation are the number of members of each class. As inferred in *Brief illustration* 10B.3, there are three symmetry species, with one of them two-dimensional (E).

Character tables, and some of the data contained in them, are constructed on the assumption that the axis system is arranged in a particular way and is specified in the character table when there is ambiguity. There is ambiguity in C_{2v} (and certain other groups), and so a more detailed specification of the symmetry operations is then necessary. The principal axis (a unique C_n axis with the greatest value of n), is taken to be the z-direction. If the molecule is planar the molecule is taken to lie in the yz-plane (referring to Fig. 10B.6). Then σ_v' is a reflection in the yz-plane and henceforth will be denoted $\sigma_v'(yz)$, and σ_v is a reflection in the xz-plane, and henceforth is denoted $\sigma_v(xz)$.

The irreducible representations are mutually orthogonal in the sense that if the set of characters is regarded as forming a row vector, the dot product (or scalar product) of the vectors corresponding to different irreducible representations is zero: the vectors are mutually perpendicular.[2] The vectors are also each normalized to 1, in the sense that the dot product of a vector with itself is equal to 1. Vectors that are both orthogonal and normalized (to 1) are said to be 'orthonormal'. Formally, this orthonormality is expressed as

$$\frac{1}{h}\sum_C N(C)\chi^{\Gamma^{(i)}}(C)\chi^{\Gamma^{(j)}}(C)=\begin{cases}0 \text{ for } i\neq j\\ 1 \text{ for } i=j\end{cases}$$

Orthonormality of irreducible representations (10B.7)

where the sum is over the classes of the group, $N(C)$ is the number of operations in class C, and h is the number of operations in the group (its order).

Brief illustration 10B.4

In the point group C_{3v} with elements $\{E,2C_3,3\sigma_v\}$ and $h=6$, for the two irreducible representations with labels A_2 (with characters $\{1,1,-1\}$) and E (with characters $\{2,-1,0\}$), eqn 10B.7 is

$$\tfrac{1}{6}\{1\times1\times2+2\times1\times(-1)+3\times(-1)\times0\}=0$$

If the two irreducible representations are both E, the sum in eqn 10B.7 is

$$\tfrac{1}{6}\{1\times2\times2+2\times(-1)\times(-1)+3\times0\times0\}=1$$

The sum is also 1 if both irreducible representations are A_2:

$$\tfrac{1}{6}\{1\times1\times1+2\times1\times1+3\times(-1)\times(-1)\}=1$$

Table 10B.2 The C_{3v} character table*

C_{3v}, 3*m*	*E*	$2C_3$	$3\sigma_v$	$h=6$	
A_1	1	1	1	z	z^2, x^2+y^2
A_2	1	1	−1		
E	2	−1	0	(x, y)	$(xy, x^2-y^2), (yz, zx)$

* More character tables are given in the *Resource section.*

[2] This result is a consequence of the 'great orthogonality theorem' of group theory; see our *Molecular quantum mechanics* (2011). In this Topic, the characters are taken to be real.

(a) The symmetry species of atomic orbitals

The characters in the rows of one-dimensional irreducible representations (the rows labelled A or B) and in the columns headed by symmetry operations indicate the behaviour of an orbital under the corresponding operations: a 1 indicates that an orbital is unchanged, and a −1 indicates that it changes sign. It follows that the symmetry label of the orbital can be identified by comparing the changes that occur to an orbital under each operation, and then comparing the resulting 1 or −1 with the entries in a row of the character table for the relevant point group. By convention, the orbitals are labelled with the lower case equivalent of the symmetry species label (so an orbital of symmetry species A_1 is called an a_1 orbital).

Brief illustration 10B.5

Consider an H_2O molecule, point group C_{2v}, shown in Fig. 10B.6. The effect of C_2 on the oxygen $2p_x$ orbital is to cause it to change sign, so the character is −1; $\sigma_v'(yz)$ has the same effect and so has character −1. In contrast, $\sigma_v(xz)$ leaves the orbital unaffected and so has character 1, and of course the same is true of the identity operation. The characters of the operations $\{E,C_2,\sigma_v,\sigma_v'\}$ are therefore {1,−1,1,−1}. Reference to the C_{2v} character table (Table 10B.1) shows that {1,−1,1,−1} are the characters for the symmetry species B_1; the orbital is therefore labelled b_1. A similar procedure gives the characters for oxygen $2p_y$ as {1,−1,−1,1}, which corresponds to B_2: the orbital is labelled b_2, therefore. Both the oxygen $2p_z$ and 2s are a_1.

The characters in a row of the table for irreducible representations of dimensionality greater than 1 (typically, but not only, the E and T symmetry species) are the sums of the characters for the behaviour of the individual orbitals in the basis. Thus, if one member of a pair remains unchanged under a symmetry operation but the other changes sign (Fig. 10B.7), then the entry is reported as $\chi = 1 - 1 = 0$.

The behaviour of s, p, and d orbitals on a central atom under the symmetry operations of the molecule is so important that the symmetry species of these orbitals are generally indicated in a character table. To make these assignments, identify the symmetry species of *x*, *y*, and *z*, which appear on the right hand side of the character table. Thus, the position of *z* in Table 10B.2 shows that p_z (which is proportional to $zf(r)$), has symmetry species A_1 in C_{3v}, whereas p_x and p_y (which are proportional to $xf(r)$ and $yf(r)$, respectively) are jointly of E symmetry. In technical terms, it is said that p_x and p_y jointly **span** an irreducible representation of symmetry species E. An s orbital on the central atom always spans the totally symmetric irreducible representation of a group as it is unchanged under all symmetry operations; in C_{3v} it has symmetry species A_1.

The five d orbitals of a shell are represented by *xy* for d_{xy} etc. and are also listed on the right of the character table. It can be seen at a glance that in C_{3v} d_{xy} and $d_{x^2-y^2}$ on a central atom jointly span E.

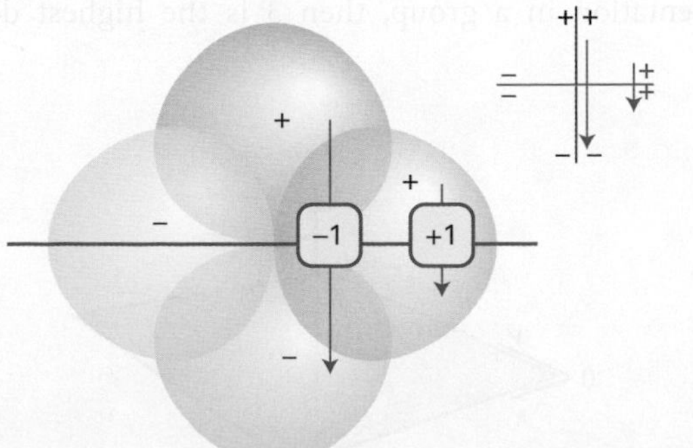

Figure 10B.7 The two orbitals shown here have different properties under reflection through the mirror plane: one changes sign (character −1), the other does not (character +1).

(b) The symmetry species of linear combinations of orbitals

The same technique may be applied to identify the symmetry species of linear combinations of orbitals, such as the combination $\psi_1 = s_A + s_B + s_C$ of the three H1s orbitals in the C_{3v} molecule NH_3 (Fig. 10B.8). This combination remains unchanged under a C_3 rotation and under any of the three vertical reflections of the group, so its characters are

$$\chi(E) = 1 \quad \chi(C_3) = 1 \quad \chi(\sigma_v) = 1$$

Comparison with the C_{3v} character table shows that ψ_1 is of symmetry species A_1, and therefore has the label a_1.

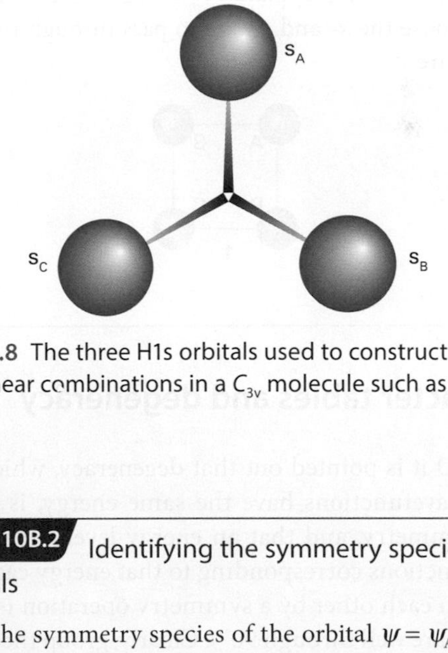

Figure 10B.8 The three H1s orbitals used to construct symmetry-adapted linear combinations in a C_{3v} molecule such as NH_3.

Example 10B.2 Identifying the symmetry species of orbitals

Identify the symmetry species of the orbital $\psi = \psi_A - \psi_B$ in a C_{2v} NO_2 molecule, where ψ_A is an $O2p_x$ orbital on one O atom (and perpendicular to the molecular plane) and ψ_B that on the other O atom.

Collect your thoughts The negative sign in ψ indicates that the sign of ψ_B is opposite to that of ψ_A. You need to consider how the combination changes under each operation of the group, and then write the character as 1, −1, or 0 as specified above.

Then compare the resulting characters with each row in the character table for the point group, and hence identify the symmetry species.

The solution The combination is shown in Fig. 10B.9. Under C_2, ψ changes into itself, implying a character of 1. Under the reflection $\sigma_v(xz)$ both atomic orbitals change sign, so $\psi \rightarrow -\psi$, implying a character of −1. Under $\sigma_v'(yz)$ $\psi \rightarrow -\psi$, so the character for this operation is also −1. The characters are therefore

$$\chi(E) = 1 \quad \chi(C_2) = 1 \quad \chi(\sigma_v(xz)) = -1 \quad \chi(\sigma_v'(yz)) = -1$$

These values match the characters of the A_2 symmetry species, so ψ is labelled a_2.

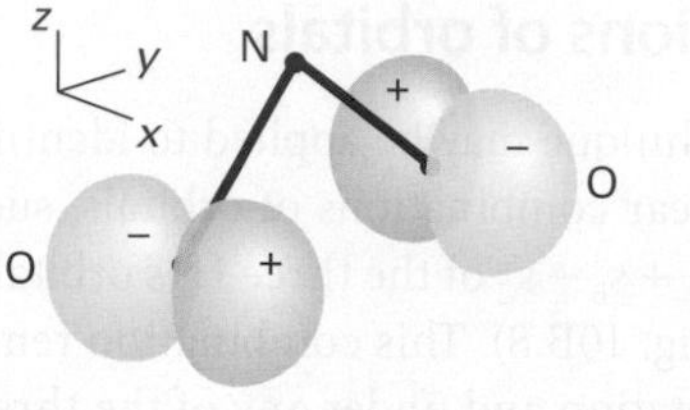

Figure 10B.9 One symmetry-adapted linear combination of $O2p_x$ orbitals in the C_{2v} NO_2 molecule.

Self-test 10B.2 Consider $PtCl_4^{2-}$, in which the Cl ligands form a square planar array and the ion belongs to the point group D_{4h} (1). Identify the symmetry species of the combination $\psi_A - \psi_B + \psi_C - \psi_D$. Note that in this group the C_2 axes coincide with the x and y axes, and σ_v planes coincide with the xz- and yz-planes; choose the x- and y-axes to pass through the corners of the square.

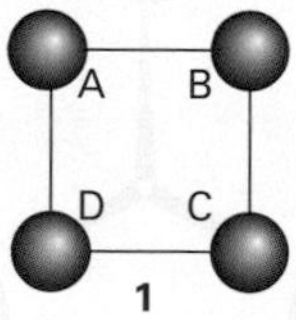

1

Answer: B_{1g}

(c) Character tables and degeneracy

In Topic 7D it is pointed out that degeneracy, which is when different wavefunctions have the same energy, is always related to symmetry, and that an energy level is degenerate if the wavefunctions corresponding to that energy can be transformed into each other by a symmetry operation (such as rotating a square well through 90°). Clearly, group theory should have a role in the identification of degeneracy.

A geometrically square well belongs to the group C_4 (Fig. 10B.10 and Table 10B.3), with the C_4 rotations (through 90°) converting x into y and vice versa.[3] As explained in Topic 7D, the two wavefunctions $\psi_{1,2} = (2/L)\sin(\pi x/L)\sin(2\pi y/L)$ and $\psi_{2,1} = (2/L)\sin(2\pi x/L)\sin(\pi y/L)$ both correspond to the energy $5h^2/8mL^2$, so that level is doubly degenerate. Under the operations of the group, these two functions transform as follows:

$$E: (\psi_{1,2}\ \psi_{2,1}) \rightarrow (\psi_{1,2}\ \psi_{2,1}) \qquad C_4^+: (\psi_{1,2}\ \psi_{2,1}) \rightarrow (\psi_{2,1}\ -\psi_{1,2})$$

$$C_4^-: (\psi_{1,2}\ \psi_{2,1}) \rightarrow (-\psi_{2,1}\ \psi_{1,2}) \qquad C_2: (\psi_{1,2}\ \psi_{2,1}) \rightarrow (-\psi_{1,2}\ -\psi_{2,1})$$

The corresponding matrix representatives are

$$D(E) = \begin{pmatrix} 1 & 0 \\ 0 & 1 \end{pmatrix} \qquad D(C_4^+) = \begin{pmatrix} 0 & 1 \\ -1 & 0 \end{pmatrix}$$

$$D(C_4^-) = \begin{pmatrix} 0 & -1 \\ 1 & 0 \end{pmatrix} \qquad D(C_2) = \begin{pmatrix} -1 & 0 \\ 0 & -1 \end{pmatrix}$$

and their characters are

$$\chi(E) = 2 \quad \chi(C_4^+) = 0 \quad \chi(C_4^-) = 0 \quad \chi(C_2) = -2$$

A glance at the character table in Table 10B.3 (noting that the rotations C_4^+ and C_4^- belong to the same class and appear in the column labelled $2C_4$) shows that the basis spans the irreducible representation of symmetry species E. The same is true of all the doubly-degenerate energy levels, and there are no triply-degenerate (or higher) energy levels in the system. Notice too that in the group C_4 there are no irreducible representations of dimension 3 or higher. These two observations illustrate the general principle that:

> The highest dimensionality of irreducible representation in a group is the maximum degree of degeneracy in the group.

Thus, if there is an E irreducible representation in a group, 2 is the highest degree of degeneracy; if there is a T irreducible representation in a group, then 3 is the highest degree of

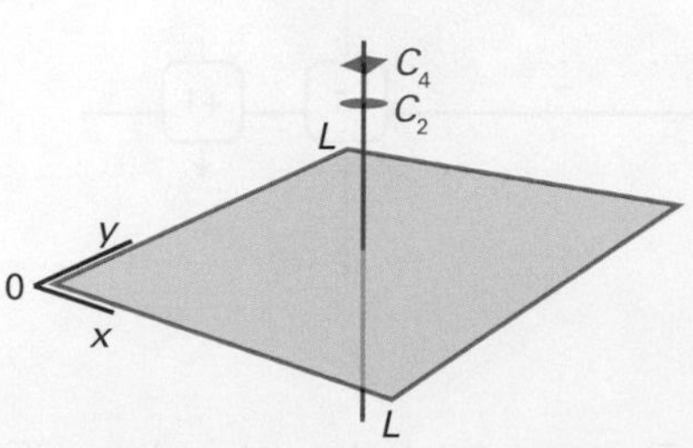

Figure 10B.10 A geometrically square well can be treated as belonging to the group C_4 (with elements $\{E, 2C_4, C_2\}$).

[3] More complicated groups could be used, such as C_{4v} or D_{4h}, but C_4 captures the symmetry sufficiently.

Table 10B.3 The C_4 character table*

C_4, 4	E	$2C_4$	C_2	$h = 4$	
A	1	1	1	z	$z^2, x^2 + y^2$
B	1	−1	1		$xy, x^2 - y^2$
E	2	0	−2	(x, y)	(yz, zx)

* More character tables are given in the *Resource section*.

degeneracy. Some groups have irreducible representations of higher dimension, and therefore allow higher degrees of degeneracy. Furthermore, because the character of the identity operation is always equal to the dimensionality of the representation, the maximum degeneracy can be identified by noting the maximum value of $\chi(E)$ in the relevant character table.

Brief illustration 10B.6

- A trigonal planar molecule such as BF_3 cannot have triply degenerate orbitals because its point group is D_{3h} and the character table for this group (in the *Resource section*) does not have a T symmetry species.
- A methane molecule belongs to the tetrahedral point group T_d and because that group has irreducible representations of T symmetry, it can have triply-degenerate orbitals. The same is true of tetrahedral P_4, which, with just four atoms, is the simplest kind of molecule with triply-degenerate orbitals.
- A buckminsterfullerene molecule, C_{60}, belongs to the icosahedral point group (I_h) and its character table (in the *Resource section*) shows that the maximum dimensionality of its irreducible representations is 5, so it can have five-fold degenerate orbitals.

Checklist of concepts

☐ 1. A **group** is a collection of transformations that satisfy the four criteria set out at the start of the Topic.

☐ 2. The **order** of a group is the number of its symmetry operations.

☐ 3. A **matrix representative** is a matrix that represents the effect of an operation on a basis.

☐ 4. The **character** is the sum of the diagonal elements of a matrix representative of an operation.

☐ 5. A **matrix representation** is the collection of matrix representatives for the operations in the group.

☐ 6. A **character table** consists of entries showing the characters of all the irreducible representations of a group.

☐ 7. A **symmetry species** is a label for an irreducible representation of a group.

☐ 8. The highest dimensionality of irreducible representation in a group is the maximum degree of degeneracy in the group.

Checklist of equations

Property	Equation	Comment	Equation number
Class membership	$R' = S^{-1}RS$	All elements members of the group	10B.1
Number of irreducible representations	Number of irreducible representations = number of classes		10B.5
Dimensionality and order	$\sum_{\text{irreps } i} d_i^2 = h$	For groups other than pure rotation groups with $n > 2$	10B.6
Orthonormality of irreducible representations	$\frac{1}{h}\sum_C N(C)\chi^{\Gamma^{(i)}}(C)\chi^{\Gamma^{(j)}}(C) = \begin{cases} 0 \text{ for } i \neq j \\ 1 \text{ for } i = j \end{cases}$	Sum over classes	10B.7

TOPIC 10C Applications of symmetry

> ➤ **Why do you need to know this material?**
> Group theory is a key tool for constructing molecular orbitals and formulating spectroscopic selection rules.
>
> ➤ **What is the key idea?**
> An integral can be non-zero only if the integrand is invariant under the symmetry transformations of a molecule.
>
> ➤ **What do you need to know already?**
> This Topic develops the material in Topic 10A, where the classification of molecules on the basis of their symmetry elements is introduced, and draws heavily on the properties of characters and character tables described in Topic 10B. You need to be aware that many quantum-mechanical properties, including transition dipole moments (Topic 8C), depend on integrals involving products of wavefunctions (Topic 7C).

Group theory shows its power when brought to bear on a variety of problems in chemistry, among them the construction of molecular orbitals and the formulation of spectroscopic selection rules.

10C.1 Vanishing integrals

Any integral, I, of a function $f(x)$ over a symmetric range around $x=0$ is zero if the function is antisymmetric in the sense that $f(-x)=-f(x)$. In two dimensions the integral (over a symmetrical range) of the integrand $f(x,y)$ has contributions from regions that are related by symmetry operations of the area of integration (Fig. 10C.1). If $f(x,y)$ changes sign under one of these operations, the contribution of the first region is cancelled by that from the symmetry-related region and the integral is zero. The integral may be non-zero only if the integrand is invariant (or at least can be expressed as a sum of terms at least one of which is invariant) under each symmetry operation of the group that reflects the shape of the area (and in general, the volume) of the range of integration. In group-theoretical terms:

> An integral over a region of space can be non-zero only if the integrand (or a contribution to it) spans the totally symmetric irreducible representation of the point group of the region.

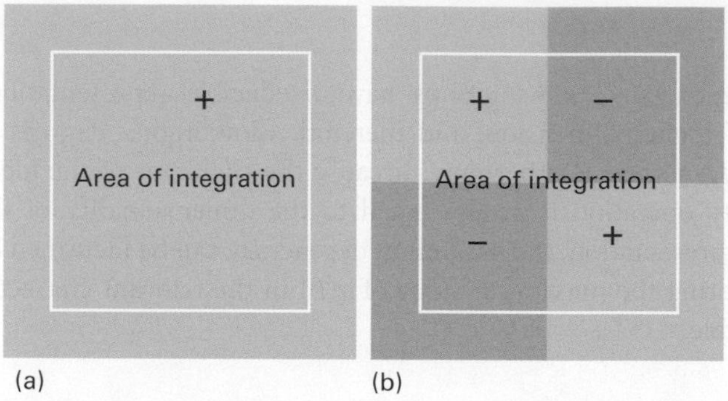

Figure 10C.1 (a) Only if the integrand is unchanged under each symmetry operation of the group (here C_4) can its integral over the region indicated be non-zero. (b) If the integrand changes sign under any operation, its integral is necessarily zero.

The totally symmetric irreducible representation has all characters equal to 1, and is typically the symmetry species denoted A_1.

Brief illustration 10C.1

To decide whether the integral of the function $f=xy$ may be non-zero when evaluated over a region the shape of an equilateral triangle centred on the origin (Fig. 10C.2), recognize that the triangle belongs to the group C_3. Reference to the character table of the group shows that xy is a member of a basis that spans the irreducible representation E. Therefore, its integral must be zero, because the integrand has no component that spans A_1.

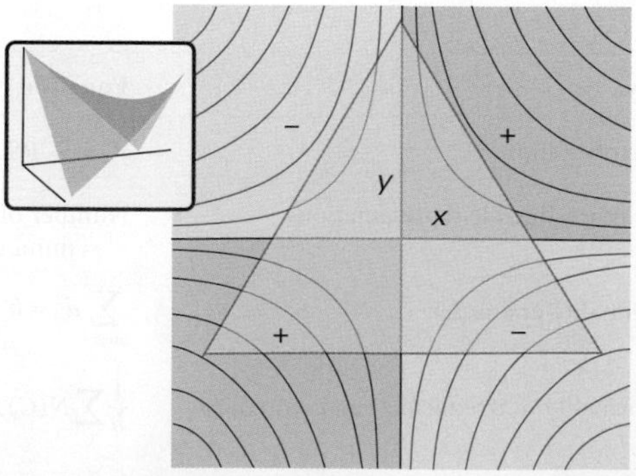

Figure 10C.2 The integral of the function $f=xy$ over the yellow region (of symmetry C_3) is zero. In this case, the result is obvious by inspection, but group theory can be used to establish similar results in less obvious cases. The insert shows the shape of the function in three dimensions.

(a) Integrals of the product of functions

Suppose the integral of interest is of a product of two functions, f_1 and f_2, taken over all space and over all relevant variables (represented, as is usual in quantum mechanics, by integration over $d\tau$):

$$I=\int f_1 f_2 \, d\tau \tag{10C.1}$$

For example, f_1 and f_2 might be atomic orbitals on different atoms, in which case I would be their overlap integral. The implication of such an integral being zero is that a molecular orbital does not result from the overlap of these two orbitals. It follows from the general point made above that the integral may be non-zero only if the integrand itself, the product $f_1 f_2$, is unchanged by any symmetry operation of the molecular point group and so spans the totally symmetric irreducible representation (typically the symmetry species with the label A_1). To decide whether the product $f_1 f_2$ does indeed span A_1, it is necessary to form the **direct product** of the symmetry species spanned by f_1 and f_2 separately. The procedure is as follows:

- Write down a table with columns headed by the symmetry operations, R, of the group.
- In the first row write down the characters of the symmetry species spanned by f_1; in the second row write down the characters of the symmetry species spanned by f_2.
- Multiply the numbers in the two rows together, column by column. The resulting set of numbers are the characters of the representation spanned by $f_1 f_2$.

Brief illustration 10C.2

Suppose that in the point group C_{2v} f_1 has the symmetry species A_2, and f_2 has the symmetry species B_1. From the character table the characters for these species are 1,1,–1,–1 and 1,–1,1,–1, respectively. The direct product of these two species is found by setting up the following table

	E	C_2	$\sigma_v(xz)$	$\sigma_v'(yz)$
A_2	1	1	−1	−1
B_1	1	−1	1	−1
product	1	−1	−1	1

Now recognize that the characters in the final row are those of the symmetry species B_2. It follows that the symmetry species of the product $f_1 f_2$ is B_2. Because the direct product does not contain A_1, the integral of $f_1 f_2$ over all space must be zero.

Direct products have some simplifying features.

- The direct product of the totally symmetric irreducible representation with any other representation is the latter irreducible representation itself: $A_1 \times \Gamma^{(i)} = \Gamma^{(i)}$.

All the characters of A_1 are 1, so multiplication by them leaves the characters of $\Gamma^{(i)}$ unchanged. It follows that if one of the functions in eqn 10C.1 transforms as A_1, then the integral will vanish if the other function is not A_1.

- The direct product of two irreducible representations is A_1 only if the two irreducible representations are identical: $\Gamma^{(i)} \times \Gamma^{(j)}$ contains A_1 only if $i = j$.

For one-dimensional irreducible representations the characters are either 1 or −1, and the character 1 is obtained only if the characters of $\Gamma^{(i)}$ and $\Gamma^{(j)}$ are the same (both 1 or both −1). For example, in C_{2v}, $A_1 \times A_1$, $A_2 \times A_2$, $B_1 \times B_1$, and $B_2 \times B_2$, but no other combination, all give A_1. That the requirement also holds for higher-dimensional representations requires more work and is demonstrated at the end of Section 10C.1b.

It follows that if f_1 and f_2 transform as different symmetry species, then the product cannot transform as the totally symmetric irreducible representation and so the integral of $f_1 f_2$ is necessarily zero. If, on the other hand, both functions transform as the same symmetry species, then the product transforms as the totally symmetric irreducible representation (and possibly has contributions from other symmetry species too) and the integral is not necessarily zero.

An important point is that group theory is specific about when an integral must be zero, but integrals that it allows to be non-zero may be zero for reasons unrelated to symmetry. For example, the N–H distance in ammonia may be so long that the $(s_N, s_1+s_2+s_3)$ overlap integral, in which f_1 is a 2s orbital on N and f_2 is a combination of 1s orbitals on the three H atoms with the same symmetry, is zero simply because the orbitals are so far apart.

Integrals of the form

$$I=\int f_1 f_2 f_3 \, d\tau \tag{10C.2}$$

are also common in quantum mechanics, and it is important to know when they are necessarily zero. For example, they appear in the calculation of transition dipole moments (Topic 8C). As for integrals over two functions, for I to be non-zero, *the product $f_1 f_2 f_3$ must span the totally symmetric irreducible representation or contain a component that spans that representation.* To test whether this is so, the characters of all three irreducible representations are multiplied together in the same way as in the rules set out above.

Example 10C.1 Deciding if an integral must be zero

Does the integral $\int (d_{z^2}) x (d_{xy}) d\tau$ vanish in a C_{2v} molecule?

Collect your thoughts Use the C_{2v} character table to find the characters of the irreducible representations spanned by $3z^2 - r^2$ (the form of the d_{z^2} orbital), x, and xy. Then set up a table to work out the triple direct product and identify whether the symmetry species it spans includes A_1.

The solution The C_{2v} character table shows that the function xy, and hence the orbital d_{xy}, transforms as A_2, that z^2 transforms as A_1, and that x transforms as B_1. The table is therefore

	E	C_2	$\sigma_v(xz)$	$\sigma_v'(yz)$	
A_2	1	1	−1	−1	$f_3 = d_{xy}$
B_1	1	−1	1	−1	$f_2 = x$
A_1	1	1	1	1	$f_1 = d_{z^2}$
	1	−1	−1	1	product

The characters in the bottom row are those of B_2, not of A_1. Therefore, the integral is necessarily zero.

Comment. A quicker solution involves noting that A_1 (for f_1) has no effect on the outcome of the triple direct product (by the first feature mentioned above), and therefore, by the second feature, the symmetry species of the two functions f_2 and f_3 must be the same for their direct product to be A_1; but in this example they are not the same.

Self-test 10C.1 Does the integral $\int (d_{xz})x(p_z)d\tau$ necessarily vanish in a C_{2v} molecule?

Answer: No

(b) Decomposition of a representation

In some cases, it turns out that the direct product is a sum of symmetry species, not just a single species. For instance, in C_{3v} the characters of the direct product E × E are {4,1,0}, which can be decomposed as A_1, A_2, and E:

	E	$2C_3$	$3\sigma_v$
A_1	1	1	1
A_2	1	1	−1
E	2	−1	0
sum	4	1	0

This decomposition is written symbolically E × E = A_1 + A_2 + E.[1]

In simple cases the decomposition can be done by inspection. Group theory, however, provides a systematic way of using the characters of the representation to find the symmetry species of the irreducible representations of which it is composed. The formal recipe for finding the number of times, $n(\Gamma)$, that irreducible representation Γ occurs is based on a general expression derived from a very deep result of group theory:[2]

$$n(\Gamma)=\frac{1}{h}\sum_{C} N(C)\chi^{(\Gamma)}(C)\chi(C) \qquad \text{Decomposition of a representation} \qquad (10C.3a)$$

Here Γ is the symmetry species of the irreducible representation of interest, h is the order of the group, $\chi^{(\Gamma)}(C)$ is the character of the members of class C of operations for that irreducible representation, and $\chi(C)$ is the corresponding character of the representation being decomposed. Note that the sum is over the classes of operations. In the character table the number of operations in each class, $N(C)$, is indicated in the header of the columns. All the characters of the totally symmetric irreducible representation of symmetry species A_1 are 1, so setting $\Gamma = A_1$ and $\chi^{(A_1)}(C) = 1$ for all C in eqn 10C.3a gives

$$n(A_1)=\frac{1}{h}\sum_{C} N(C)\chi(C) \qquad \text{Occurrence of } A_1 \qquad (10C.3b)$$

Brief illustration 10C.3

In the character table for C_{3v}, the columns are headed E, $2C_3$, and $3\sigma_v$ indicating that the numbers in each class are 1, 2, and 3, respectively and $h = 1 + 2 + 3 = 6$. To decide whether A_1 occurs in the representation with characters {4,1,0} in C_{3v} form

$$n(A_1) = \tfrac{1}{6}\{1\times\chi(E) + 2\times\chi(C_3) + 3\times\chi(\sigma_v)\}$$

$$= \tfrac{1}{6}\{1\times 4 + 2\times 1 + 3\times 0\} = 1$$

A_1 therefore occurs once in the decomposition.

It is asserted in Section 10C.1a that the direct product of two irreducible representations is A_1 only if the two irreducible representations are identical. That this is so can now be shown with the aid of eqn 10C.3b.

How is that done? 10C.1 Confirming the criterion for a direct product to contain the totally symmetric irreducible representation

Start by considering the characters of the direct product between irreducible representations $\Gamma^{(i)}$ and $\Gamma^{(j)}$. The character of a class of operations in a direct product is the product of the characters of the two contributing representations: $\chi(C) = \chi^{\Gamma^{(i)}}(C)\chi^{\Gamma^{(j)}}(C)$, where $\chi^{\Gamma^{(i)}}(C)$ is the character for the operation in class C of irreducible representation $\Gamma^{(i)}$ and likewise

[1] As mentioned in Topic 10B, a direct sum is sometimes denoted ⊕. The analogous symbol for a direct product is ⊗. The symbolic expression is then written E ⊗ E = A_1 ⊕ A_2 ⊕ E.

[2] This result arises from the 'great orthogonality theorem': see our *Molecular quantum mechanics* (2011). In this Topic, the characters are taken to be real.

for $\chi^{\Gamma^{(j)}}(C)$. The number of times that the totally symmetric irreducible representation (A_1) occurs in this direct-product representation is given by eqn 10C.3b as

$$n(A_1) = \frac{1}{h}\sum_C N(C)\chi^{\Gamma^{(i)}}(C)\chi^{\Gamma^{(j)}}(C)$$

Irreducible representations are orthonormal in the sense that (eqn 10B.7)

$$\frac{1}{h}\sum_C N(C)\chi^{\Gamma^{(i)}}(C)\chi^{\Gamma^{(j)}}(C) = \begin{cases} 0 \text{ if } i \neq j \\ 1 \text{ if } i = j \end{cases}$$

It follows that

$$n(A_1) = \begin{cases} 0 \text{ if } i \neq j \\ 1 \text{ if } i = j \end{cases}$$

In other words, the direct product of two irreducible representations has a component that spans A_1 only if the two irreducible representations belong to the same symmetry species. This result is independent of the dimensionality of the irreducible representations.

10C.2 Applications to molecular orbital theory

The rules outlined so far can be used to decide which atomic orbitals may have non-zero overlap in a molecule. Group theory also provides procedures for constructing linear combinations of atomic orbitals of a specified symmetry.

(a) Orbital overlap

The overlap integral, S, between orbitals ψ_1 and ψ_2 is

$$S = \int \psi_2^* \psi_1 \mathrm{d}\tau \quad \text{Overlap integral} \quad (10\text{C}.4)$$

It follows from the discussion of eqn 10C.1 that this integral can be non-zero only if the two orbitals span the same symmetry species. In other words,

> Only orbitals of the same symmetry species may have non-zero overlap ($S \neq 0$) and hence go on to form bonding and antibonding combinations.

The selection of atomic orbitals with non-zero overlap is the central and initial step in the construction of molecular orbitals as LCAOs.

Example 10C.2 Identifying which orbitals can contribute to bonding

The four H1s orbitals of methane span $A_1 + T_2$. With which of the C2s and C2p atomic orbitals can they overlap? What additional overlap would be possible if d orbitals on the C atom were also considered?

Collect your thoughts Refer to the T_d character table (in the *Resource section*) and look for s, p, and d orbitals spanning A_1 or T_2. Recall that the symmetry species can be identified by looking for the appropriate Cartesian functions listed on the right of the table.

The solution A C2s orbital spans A_1 in the group T_d, so it may have non-zero overlap with the A_1 combination of H1s orbitals. From the table (x,y,z) jointly span T_2, so the three C2p orbitals together transform in the same way; they may have non-zero overlap with the T_2 combination of H1s orbitals. The combinations (xy,yz,zx) span T_2, therefore the d_{xy}, d_{yz}, and d_{zx} orbitals do the same and so they may overlap with the T_2 combination of H1s orbitals. The other two d orbitals span E and so they cannot overlap with the A_1 or T_2 H1s orbitals and remain nonbonding. It follows that in methane there are a_1 orbitals arising from (C2s,H1s)-overlap and t_2 orbitals arising from (C2p,H1s)-overlap. The C3d orbitals might contribute to the latter. The lowest energy configuration is probably $a_1^2t_2^6$, with all bonding orbitals occupied.

Self-test 10C.2 Consider the octahedral SF_6 molecule, with the bonding arising from overlap of s orbitals and a 2p orbital on each fluorine directed towards the central sulfur atom. The latter span $A_{1g} + E_g + T_{1u}$. Which sulfur orbitals have non-zero overlap with these F orbitals? Suggest what the ground-state configuration is likely to be.

Answer: $3s(A_{1g})$, $3p(T_{1u})$, $(3d_{x^2-y^2}, 3d_{z^2})(E_g)$; $a_{1g}^2t_{1u}^6e_g^4$

(b) Symmetry-adapted linear combinations

Topic 10B introduces the idea of generating a combination of atomic orbitals designed to transform as a particular symmetry species. Such a combination is an example of a **symmetry-adapted linear combination** (SALC), which is a combination of orbitals constructed from equivalent atoms and having a specified symmetry. SALCs are very useful in constructing molecular orbitals because a given SALC has non-zero overlap only with other orbitals of the same symmetry.

The technique for building SALCs is derived by using the full power of group theory and involves the use of a **projection operator**, $P^{(\Gamma)}$, an operator that takes one of the basis orbitals and generates from it—projects from it—an SALC of the symmetry species Γ:

$$P^{(\Gamma)} = \frac{1}{h}\sum_R \chi^{(\Gamma)}(R)R \qquad \psi^{(\Gamma)} = P^{(\Gamma)}\psi_i \quad \text{Projection operator} \quad (10\text{C}.5)$$

Here ψ_i is one of the basis orbitals and $\psi^{(\Gamma)}$ is a SALC (there might be more than one) that transforms as the symmetry species Γ; the sum is over the operations (not the classes) of the group of order h. To implement this rule, do the following:

- Construct a table with the columns headed by each symmetry operation R of the group; include a column for each operation, not just for each class.
- Select a basis function and work out the effect that each operation has on it. Enter the resulting function beneath each operation.
- On the next row enter the characters of the symmetry species of interest, $\chi^{(\Gamma)}(R)$.
- Multiply the entries in the previous two rows, operation by operation.
- Sum the result, and divide it by the order of the group, h.

Brief illustration 10C.4

To construct the B_1 SALC from the two O2p_x orbitals in NO_2, point group C_{2v} (Fig. 10C.3), draw up the following table:

	E	C_2	$\sigma_v(xz)$	$\sigma'_v(yz)$
Effect on p_A	p_A	$-p_B$	p_B	$-p_A$
Characters for B_1	1	−1	1	−1
Product of rows 1 and 2	p_A	p_B	p_B	p_A

The sum of the final row, divided by the order of the group ($h = 4$), gives $\psi^{(B_1)} = \frac{1}{2}(p_A + p_B)$.

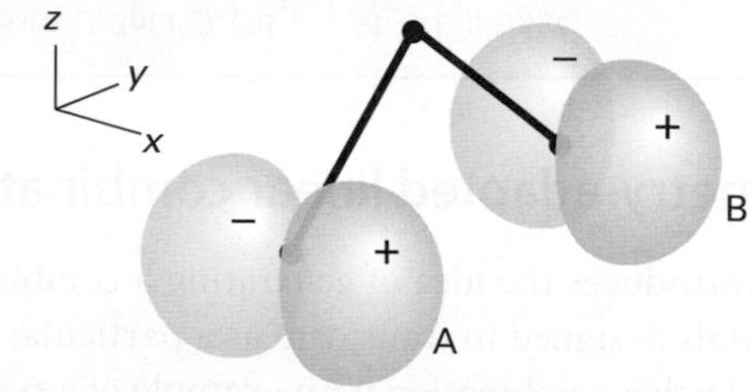

Figure 10C.3 The two O2p_x atomic orbitals in NO_2 (point group C_{2v}) can be used as a basis for forming SALCs.

If an attempt is made to generate a SALC with symmetry that is not spanned by the basis functions, the result is zero. For example, if in the *Brief illustration* an attempt is made to project an A_1 symmetry orbital, all the characters in the second row of the table will be 1, so when the product of rows 1 and 2 is formed the result is $p_A - p_B + p_B - p_A = 0$.

A difficulty is encountered when aiming to generate an SALC of symmetry species of dimension higher than 1, because then the rules generate sums of SALCs. Consider, for instance, the generation of SALCs from the three H1s atomic orbitals in NH_3, point group C_{3v}. The molecule and the orbitals are shown in Fig. 10C.4. The table below shows the effect of applying the projection operator to s_A, s_B, and s_C in turn to give the SALC of symmetry species E.

Row		E	C_3^+	C_3^-	σ_v	σ'_v	σ''_v
1	effect on s_A	s_A	s_B	s_C	s_A	s_C	s_B
2	characters for E	2	−1	−1	0	0	0
3	product of rows 1 and 2	$2s_A$	$-s_B$	$-s_C$			
4	effect on s_B	s_B	s_C	s_A	s_C	s_B	s_A
5	product of rows 4 and 2	$2s_B$	$-s_C$	$-s_A$			
6	effect on s_C	s_C	s_A	s_B	s_B	s_A	s_C
7	product of rows 6 and 2	$2s_C$	$-s_A$	$-s_B$			

Application of the projection operator to a different basis function gives a different SALC in each case (rows 3, 5, and 7).

$$\tfrac{1}{6}(2s_A - s_B - s_C) \qquad \tfrac{1}{6}(2s_B - s_C - s_A) \qquad \tfrac{1}{6}(2s_C - s_A - s_B)$$

However, any one of these SALCs can be expressed as a sum of the other two (the three are not 'linearly independent'). The difference of the first and second gives $\frac{1}{2}(s_A - s_B)$. This combination and the third, $\frac{1}{6}(2s_C - s_A - s_B)$, are the two (now linearly independent) SALCs that are used in the construction of the molecular orbitals.

According to the discussion in Topic 9E concerning the construction of molecular orbitals of polyatomic molecules, only orbitals with the same symmetry can overlap to give a molecular orbital. In the language introduced here, this means that only SALCs of the same symmetry species have non-zero overlap and contribute to a molecular orbital. In NH_3, for instance, the molecular orbitals will have the form

$$\psi(a_1) = c_{a1}s_N + c_{a2}(s_A + s_B + s_C)$$
$$\psi(e_x) = c_{e1}p_{Nx} + c_{e2}(2s_C - s_A - s_B)$$
$$\psi(e_y) = c_{e1}p_{Ny} + c_{e2}(s_A - s_B)$$

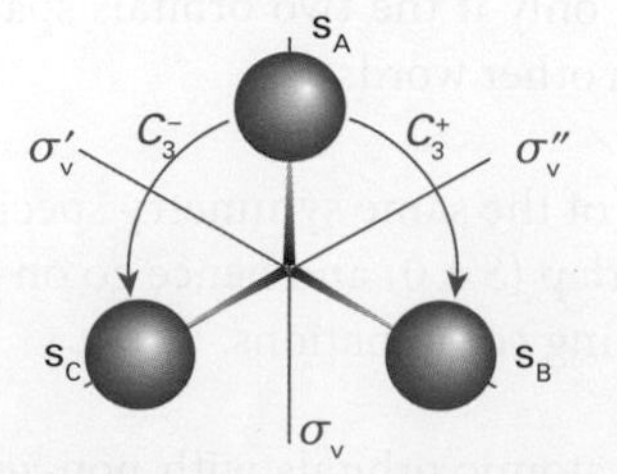

Figure 10C.4 The three H1s atomic orbitals in NH_3 (point group C_{3v}) can be used as a basis for forming SALCs.

Group theory is silent on the values of the coefficients: they have to be determined by one of the methods outlined in Topic 9E.

10C.3 Selection rules

The intensity of a spectral line arising from a molecular transition between some initial state with wavefunction ψ_i and a final state with wavefunction ψ_f depends on the (electric) transition dipole moment, μ_{fi} (Topic 8C). The q-component, where q is x, y, or z, of this vector, is defined through

$$\mu_{q,\mathrm{fi}} = -e\int \psi_\mathrm{f}^* q \psi_\mathrm{i}\,\mathrm{d}\tau \qquad \text{Transition dipole moment [definition]} \qquad (10C.6)$$

where $-e$ is the charge of the electron. The transition moment has the form of the integral over $f_1f_2f_3$ (eqn 10C.2). Therefore, once the symmetry species of the wavefunctions and the operator are known, group theory can be used to formulate the selection rules for the transitions.

Example 10C.3 Deducing a selection rule

Is $p_y \rightarrow p_z$ an allowed electric dipole transition in a molecule with C_{2v} symmetry?

Collect your thoughts You need to decide whether the product $p_z q p_y$, with $q = x$, y, or z, spans A_1. The symmetry species for p_y, p_z, and q can be read off from the right-hand side of the character table.

The solution The p_y orbital spans B_2 and p_z spans A_1, so the required direct product is $A_1 \times \Gamma^{(q)} \times B_2$, where $\Gamma^{(q)}$ is the symmetry species of x, y, or z. It does not matter in which order the direct products are calculated, so noting that $A_1 \times B_2 = B_2$ implies that $A_1 \times \Gamma^{(q)} \times B_2 = \Gamma^{(q)} \times B_2$. This direct product can be equal to A_1 only if $\Gamma^{(q)}$ is B_2, which is the symmetry species of y. Therefore, provided $q = y$ the integral may be non-zero and the transition allowed.

Comment. The analysis implies that the electromagnetic radiation involved in the transition has a component of its electric vector in the y-direction.

Self-test 10C.3 Are (a) $p_x \rightarrow p_y$ and (b) $p_x \rightarrow p_z$ allowed electric dipole transitions in a molecule with C_{2v} symmetry?

Answer: (a) No; (b) yes, with $q = x$

Checklist of concepts

- ☐ 1. Character tables can be used to decide whether an integral is necessarily zero.
- ☐ 2. For an integral to be non-zero, the integrand must include a component that is a basis for the totally symmetric irreducible representation (A_1).
- ☐ 3. Only orbitals of the same symmetry species may have non-zero overlap.
- ☐ 4. A **symmetry-adapted linear combination** (SALC) is a linear combination of atomic orbitals constructed from equivalent atoms and having a specified symmetry.

Checklist of equations

Property	Equation	Comment	Equation number
Decomposition of a representation	$n(\Gamma)=\frac{1}{h}\sum_C N(C)\chi^{(\Gamma)}(C)\chi(C)$	Real characters*	10C.3a
Presence of A_1 in a decomposition	$n(A_1)=\frac{1}{h}\sum_C N(C)\chi(C)$	Real characters*	10C.3b
Overlap integral	$S=\int \psi_2^*\psi_1\,\mathrm{d}\tau$	Definition	10C.4
Projection operator	$P^{(\Gamma)}=\frac{1}{h}\sum_R \chi^{(\Gamma)}(R)R$	To generate $\psi^{(\Gamma)}=P^{(\Gamma)}\psi_i$	10C.5
Transition dipole moment	$\mu_{q,\mathrm{fi}}=-e\int \psi_\mathrm{f}^* q\psi_\mathrm{i}\,\mathrm{d}\tau$	q-Component, $q = x$, y or z	10C.6

* In general, characters may have complex values; throughout this text only real values are encountered.

FOCUS 10 Molecular symmetry

TOPIC 10A Shape and symmetry

Discussion questions

D10A.1 Explain how a molecule is assigned to a point group.

D10A.2 List the symmetry operations and the corresponding symmetry elements that occur in point groups.

D10A.3 State and explain the symmetry criteria that allow a molecule to be polar.

D10A.4 State the symmetry criterion that allows a molecule to be optically active.

Exercises

E10A.1(a) The CH_3Cl molecule belongs to the point group C_{3v}. List the symmetry elements of the group and locate them in a drawing of the molecule.

E10A.1(b) The BF_3 molecule belongs to the point group D_{3h}. List the symmetry elements of the group and locate them in a drawing of the molecule.

E10A.2(a) Identify the group to which the naphthalene molecule belongs and locate the symmetry elements in a drawing of the molecule.

E10A.2(b) Identify the group to which the *trans*-difluoroethene molecule belongs and locate the symmetry elements in a drawing of the molecule.

E10A.3(a) Identify the point groups to which the following objects belong: (i) a sphere, (ii) an isosceles triangle, (iii) an equilateral triangle, (iv) an unsharpened cylindrical pencil.

E10A.3(b) Identify the point groups to which the following objects belong: (i) a sharpened cylindrical pencil, (ii) a box with a rectangular cross-section, (iii) a coffee mug with a handle, (iv) a three-bladed propeller (assume the sector-like blades are flat), (v) a three-bladed propeller (assume the blades are twisted out of the plane, all by the same amount).

E10A.4(a) List the symmetry elements of the following molecules and name the point groups to which they belong: (i) NO_2, (ii) PF_5, (iii) $CHCl_3$, (iv) 1,4-difluorobenzene.

E10A.4(b) List the symmetry elements of the following molecules and name the point groups to which they belong: (i) furan (**1**), (ii) γ-pyran (**2**), (iii) 1,2,5-trichlorobenzene.

1 Furan

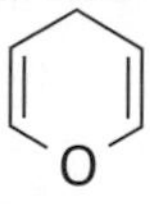

2 γ-Pyran

E10A.5(a) Assign (i) *cis*-dichloroethene and (ii) *trans*-dichloroethene to point groups.

E10A.5(b) Assign the following molecules to point groups: (i) HF, (ii) IF_7 (pentagonal bipyramid), (iii) ClF_3 (T-shaped), (iv) $Fe_2(CO)_9$ (**3**), (v) cubane, C_8H_8, (vi) tetrafluorocubane, $C_8H_4F_4$ (**4**).

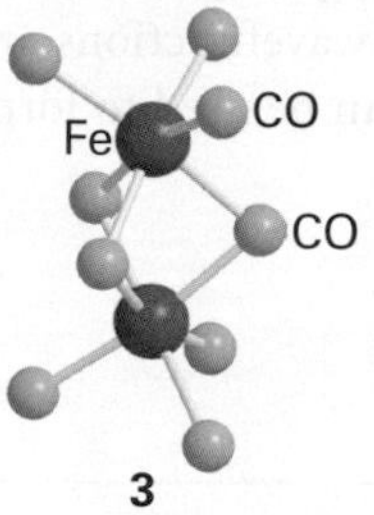

3

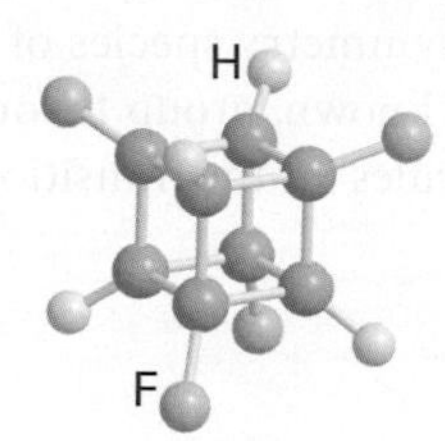

4

E10A.6(a) Which of the following molecules may be polar? (i) pyridine, (ii) nitroethane, (iii) gas-phase BeH_2 (linear), (iv) B_2H_6.

E10A.6(b) Which of the following molecules may be polar? (i) CF_3H, (ii) PCl_5, (iii) *trans*-difluoroethene, (iv) 1,2,4-trinitrobenzene.

E10A.7(a) Identify the point group to which each of the possible isomers of dichloronaphthalene belong.

E10A.7(b) Identify the point group to which each of the possible isomers of dichloroanthracene belong.

E10A.8(a) Can molecules belonging to the point groups D_{2h} or C_{3h} be chiral? Explain your answer.

E10A.8(b) Can molecules belonging to the point groups T_h or T_d be chiral? Explain your answer.

Problems

P10A.1 List the symmetry elements of the following molecules and name the point groups to which they belong: (a) staggered CH_3CH_3, (b) chair and boat cyclohexane, (c) B_2H_6, (d) $[Co(en)_3]^{3+}$, where en is 1,2-diaminoethane (ignore its detailed structure), (e) crown-shaped S_8. Which of these molecules can be (i) polar, (ii) chiral?

P10A.2 Consider the series of molecules SF_6, SF_5Cl, SF_4Cl_2, SF_3Cl_3. Assign each to the relevant point group and state whether or not the molecule is expected to be polar. If isomers are possible for any of these molecules, consider all possible structures.

P10A.3 (a) Identify the symmetry elements in ethene and in allene, and assign each molecule to a point group. (b) Consider the biphenyl molecule, Ph–Ph, in which different conformations are possible according to the value of the dihedral angle between the planes of the two benzene rings: if this angle is 0°, the molecule is planar, if it is 90°, the two rings are perpendicular to one another. For each of the following dihedral angles, identify the symmetry elements present and hence assign the point group: (i) 0°, (ii) 90°, (iii) 45°, (iv) 60°.

P10A.4 Find the point groups of all the possible geometrical isomers for the complex $MA_2B_2C_2$ in which there is 'octahedral' coordination around the central atom M and where the ligands A, B, and C are treated as structureless points. Which of the isomers are chiral?

P10A.5‡ In the square-planar complex anion [*trans*-$Ag(CF_3)_2(CN)_2]^-$, the Ag–CN groups are collinear. (a) Assume free rotation of the CF_3 groups (i.e. disregarding the AgCF and AgCN angles) and identify the point group

‡ These problems were provided by Charles Trapp and Carmen Giunta.

of this complex ion. (b) Now suppose the CF_3 groups cannot rotate freely (because the ion was in a solid, for example). Structure (5) shows a plane which bisects the NC–Ag–CN axis and is perpendicular to it. Identify the point group of the complex if each CF_3 group has a CF bond in that plane (so the CF_3 groups do not point to either CN group preferentially) and the CF_3 groups are (i) staggered, (ii) eclipsed.

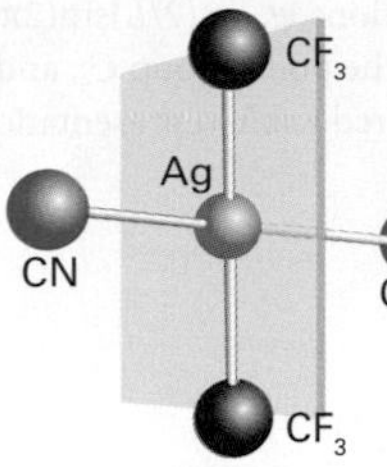

5

P10A.6[‡] B.A. Bovenzi and G.A. Pearse, Jr. (*J. Chem. Soc. Dalton Trans.*, 2763 (1997)) synthesized coordination compounds of the tridentate ligand pyridine-2,6-diamidoxime ($C_7H_9N_5O_2$, 6). Reaction with $NiSO_4$ produced a complex in which two of the essentially planar ligands are bonded at right angles to a single Ni atom. Identify the point group and the symmetry operations of the resulting $[Ni(C_7H_9N_5O_2)_2]^{2+}$ complex cation.

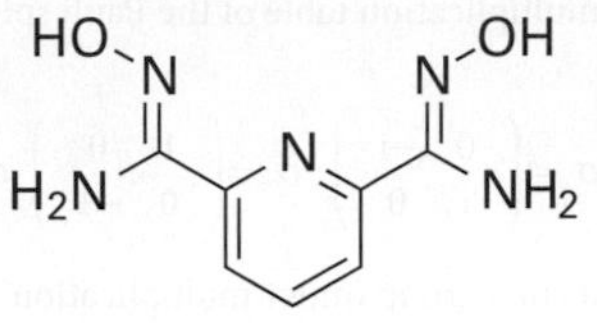

6

TOPIC 10B Group theory

Discussion questions

D10B.1 Explain what is meant by a 'group'.

D10B.2 Explain what is meant by (a) a representative and (b) a representation in the context of group theory.

D10B.3 Explain the construction and content of a character table.

D10B.4 Explain what is meant by the reduction of a representation to a direct sum of irreducible representations.

D10B.5 Discuss the significance of the letters and subscripts used to denote the symmetry species of an irreducible representation.

Exercises

E10B.1(a) Use as a basis the $2p_z$ orbitals on each atom in BF_3 to find the representative of the operation σ_h. Take z as perpendicular to the molecular plane.
E10B.1(b) Use as a basis the $2p_z$ orbitals on each atom in BF_3 to find the representative of the operation C_3. Take z as perpendicular to the molecular plane.

E10B.2(a) Use the matrix representatives of the operations σ_h and C_3 in a basis of $2p_z$ orbitals on each atom in BF_3 to find the operation and its representative resulting from $\sigma_h C_3$. Take z as perpendicular to the molecular plane.
E10B.2(b) Use the matrix representatives of the operations σ_h and C_3 in a basis of $2p_z$ orbitals on each atom in BF_3 to find the operation and its representative resulting from $C_3\sigma_h$. Take z as perpendicular to the molecular plane.

E10B.3(a) Show that all three C_2 operations in the group D_{3h} belong to the same class.
E10B.3(b) Show that all three σ_v operations in the group D_{3h} belong to the same class.

E10B.4(a) For the point group C_{2h}, confirm that all the irreducible representations are orthonormal according to the property defined in eqn 10B.7. The character table will be found in the online resources.
E10B.4(b) For the point group D_{3h}, confirm that the irreducible representation E′ is orthogonal (in the sense defined by eqn 10B.7) to the irreducible representations A_1', A_2', and E″.

E10B.5(a) By inspection of the character table for D_{3h}, state the symmetry species of the 3p and 3d orbitals located on the central Al atom in AlF_3.
E10B.5(b) By inspection of the character table for D_{4h}, state the symmetry species of the 4s, 4p, and 3d orbitals located on the central Ni atom in $Ni(CN)_4^{2-}$.

E10B.6(a) What is the maximum degeneracy of the wavefunctions of a particle confined to the interior of an octahedral hole in a crystal?
E10B.6(b) What is the maximum degeneracy of the wavefunctions of a particle confined to the interior of an icosahedral nanoparticle?

E10B.7(a) What is the maximum possible degree of degeneracy of the orbitals in benzene?
E10B.7(b) What is the maximum possible degree of degeneracy of the orbitals in 1,4-dichlorobenzene?

Problems

P10B.1 The group C_{2h} consists of the elements E, C_2, σ_h, i. Construct the group multiplication table. Give an example of a molecule that belongs to the group.

P10B.2 The group D_{2h} has a C_2 axis perpendicular to the principal axis and a horizontal mirror plane. Show that the group must therefore have a centre of inversion.

P10B.3 Consider the H_2O molecule, which belongs to the group C_{2v}. Take the molecule to lie in the yz-plane, with z directed along the C_2 axis; the mirror plane σ_v' is the yz-plane, and σ_v is the xz-plane. Take as a basis the two H1s orbitals and the four valence orbitals of the O atom and set up the 6×6 matrices that represent the group in this basis. (a) Confirm, by explicit matrix multiplication, that $C_2\sigma_v = \sigma_v'$ and $\sigma_v\sigma_v' = C_2$. (b) Show that the representation is reducible and spans $3A_1 + B_1 + 2B_2$.

P10B.4 Find the representatives of the operations of the group T_d in a basis of four H1s orbitals, one at each apex of a regular tetrahedron (as in CH_4). You need give the representative for only *one* member of each class.

P10B.5 Find the representatives of the operations of the group D_{2h} in a basis of the four H1s orbitals of ethene. Take the molecule as lying in the xy-plane, with x directed along the C–C bond.

P10B.6 Confirm that the representatives constructed in Problem P10B.5 reproduce the group multiplications $C_2^z C_2^y = C_2^x$, $\sigma^{xz} C_2^z = C_2^y$, and $iC_2^y = \sigma^{xz}$.

P10B.7 The (one-dimensional) matrices $D(C_3) = 1$ and $D(C_2) = 1$, and $D(C_3) = 1$ and $D(C_2) = -1$ both represent the group multiplication $C_3C_2 = C_6$ in the group C_{6v} with $D(C_6) = +1$ and -1, respectively. Use the character table to confirm these remarks. What are the representatives of σ_v and σ_d in each case?

P10B.8 Construct the multiplication table of the Pauli spin matrices, σ, and the 2×2 unit matrix:

$$\sigma_x = \begin{pmatrix} 0 & 1 \\ 1 & 0 \end{pmatrix} \quad \sigma_y = \begin{pmatrix} 0 & -i \\ i & 0 \end{pmatrix} \quad \sigma_z = \begin{pmatrix} 1 & 0 \\ 0 & -1 \end{pmatrix} \quad \sigma_0 = \begin{pmatrix} 1 & 0 \\ 0 & 1 \end{pmatrix}$$

Do the four matrices form a group under multiplication?

P10B.9 The algebraic forms of the f orbitals are a radial function multiplied by one of the factors (a) $z(5z^2 - 3r^2)$, (b) $y(5y^2 - 3r^2)$, (c) $x(5x^2 - 3r^2)$, (d) $z(x^2 - y^2)$, (e) $y(x^2 - z^2)$, (f) $x(z^2 - y^2)$, (g) xyz. Identify the irreducible representations spanned by these orbitals in the point group C_{2v}. (*Hint:* Because r is the radius, r^2 is invariant to any operation.)

P10B.10 Using the same approach as in Section 10B.3c find the representatives using as a basis two wavefunctions $\psi_{2,3} = (2/L)\sin(2\pi x/L)\sin(3\pi y/L)$ and $\psi_{3,2} = (2/L)\sin(3\pi x/L)\sin(2\pi y/L)$ in the point group C_4, and hence show that these functions span a degenerate irreducible representation.

TOPIC 10C Applications of symmetry

Discussion questions

D10C.1 Identify and list four applications of character tables.

D10C.2 Explain how symmetry arguments are used to construct molecular orbitals.

Exercises

E10C.1(a) Use symmetry properties to determine whether or not the integral $\int p_x z p_z \mathrm{d}\tau$ is necessarily zero in a molecule with symmetry C_{2v}.
E10C.1(b) Use symmetry properties to determine whether or not the integral $\int p_x z p_z \mathrm{d}\tau$ is necessarily zero in a molecule with symmetry D_{3h}.

E10C.2(a) Is the transition $A_1 \rightarrow A_2$ forbidden for electric dipole transitions in a C_{3v} molecule?
E10C.2(b) Is the transition $A_{1g} \rightarrow E_{2u}$ forbidden for electric dipole transitions in a D_{6h} molecule?

E10C.3(a) Show that the function xy has symmetry species B_{1g} in the group D_{2h}.
E10C.3(b) Show that the function xyz has symmetry species A_u in the group D_{2h}.

E10C.4(a) Consider the C_{2v} molecule OF_2; take the molecule to lie in the yz-plane, with z directed along the C_2 axis; the mirror plane σ_v' is the yz-plane, and σ_v is the xz-plane. The combination $p_z(A) + p_z(B)$ of the two F atoms spans A_1, and the combination $p_z(A) - p_z(B)$ of the two F atoms spans B_2. Are there any valence orbitals of the central O atom that can have a non-zero overlap with these combinations of F orbitals? How would the situation be different in SF_2, where 3d orbitals might be available?
E10C.4(b) Consider the same situation as in Exercise E10C.4(a). Find the irreducible representations spanned by the combinations $p_y(A) + p_y(B)$ and $p_y(A) - p_y(B)$. Are there any valence orbitals of the central O atom that can have a non-zero overlap with these combinations of F orbitals?

E10C.5(a) Consider the C_{2v} molecule NO_2. The combination $p_x(A) - p_x(B)$ of the two O atoms (with x perpendicular to the plane) spans A_2. Is there any valence orbital of the central N atom that can have a non-zero overlap with that combination of O orbitals? What would be the case in SO_2, where 3d orbitals might be available?
E10C.5(b) Consider BF_3 (point group D_{3h}). There are SALCs from the F valence orbitals which transform as A_2'' and E''. Are there any valence orbitals of the central B atom that can have a non-zero overlap with these SALCs? How would your conclusion differ for AlF_3, where 3d orbitals might be available?

E10C.6(a) The ground state of NO_2 is A_1 in the group C_{2v}. To what excited states may it be excited by electric dipole transitions, and what polarization of light is it necessary to use?
E10C.6(b) The ClO_2 molecule (which belongs to the group C_{2v}) was trapped in a solid. Its ground state is known to be B_1. Light polarized parallel to the y-axis (parallel to the OO separation) excited the molecule to an upper state. What is the symmetry species of that state?

E10C.7(a) A set of basis functions is found to span a reducible representation of the group C_{4v} with characters 5,1,1,3,1 (in the order of operations in the character table in the *Resource section*). What irreducible representations does it span?
E10C.7(b) A set of basis functions is found to span a reducible representation of the group D_2 with characters 6,–2,0,0 (in the order of operations in the character table in the *Resource section*). What irreducible representations does it span?

E10C.8(a) A set of basis functions is found to span a reducible representation of the group D_{4h} with characters 4,0,0,2,0,0,0,4,2,0 (in the order of operations in the character table in the *Resource section*). What irreducible representations does it span?
E10C.8(b) A set of basis functions is found to span a reducible representation of the group O_h with characters 6,0,0,2,2,0,0,0,4,2 (in the order of operations in the character table in the *Resource section*). What irreducible representations does it span?

E10C.9(a) What states of (i) benzene, (ii) naphthalene may be reached by electric dipole transitions from their (totally symmetrical) ground states?
E10C.9(a) What states of (i) anthracene, (ii) coronene (7) may be reached by electric dipole transitions from their (totally symmetrical) ground states?

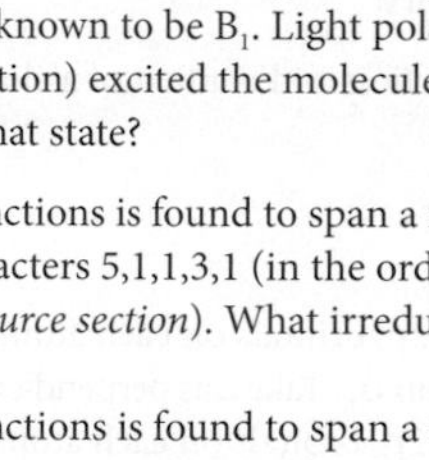

7 Coronene

Problems

P10C.1 What irreducible representations do the four H1s orbitals of CH_4 span? Are there s and p orbitals of the central C atom that may form molecular orbitals with them? In SiH_4, where 3d orbitals might be available, could these orbitals play a role in forming molecular orbitals by overlapping with the H1s orbitals?

P10C.2 Suppose that a methane molecule became distorted to (a) C_{3v} symmetry by the lengthening of one bond, (b) C_{2v} symmetry, by a kind of scissors action in which one bond angle opened and another closed slightly. Would more d orbitals on the carbon become available for bonding?

P10C.3 Does the integral of the function $3x^2 - 1$ necessarily vanish when integrated over a symmetrical range in (a) a cube, (b) a tetrahedron, (c) a hexagonal prism, each centred on the origin?

P10C.4‡ In a spectroscopic study of C_{60}, Negri et al. (*J. Phys. Chem.* **100**, 10849 (1996)) assigned peaks in the fluorescence spectrum. The molecule has icosahedral symmetry (I_h). The ground electronic state is A_{1g}, and the lowest-lying excited states are T_{1g} and G_g. (a) Are photon-induced transitions allowed from the ground state to either of these excited states? Explain your answer. (b) What if the molecule is distorted slightly so as to remove its centre of inversion?

P10C.5 In the square planar XeF_4 molecule, consider the symmetry-adapted linear combination $p_1 = p_A - p_B + p_C - p_D$, where p_A, p_B, p_C, and p_D are $2p_z$ atomic orbitals on the fluorine atoms (clockwise labelling of the F atoms). Decide which of the various s, p, and d atomic orbitals on the central Xe atom can form molecular orbitals with p_1.

P10C.6 The chlorophylls that participate in photosynthesis and the haem (heme) groups of cytochromes are derived from the porphine dianion group (**8**), which belongs to the D_{4h} point group. The ground electronic state is A_{1g} and the lowest-lying excited state is E_u. Is a photon-induced transition allowed from the ground state to the excited state? Explain your answer.

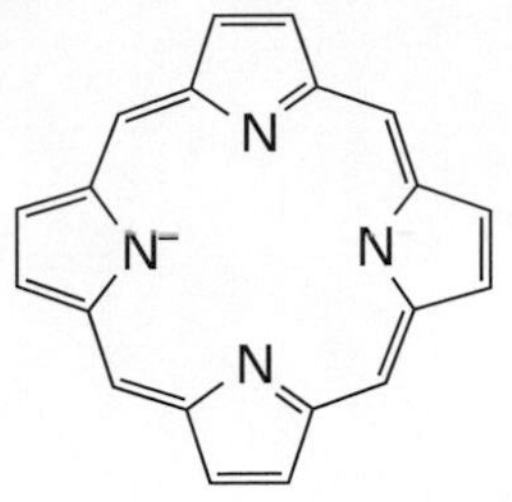

8

P10C.7 Consider the ethene molecule (point group D_{2h}), and take it as lying in the *xy*-plane, with *x* directed along the C–C bond. By applying the projection formula to one of the hydrogen 1s orbitals generate SALCs which have symmetry A_g, B_{2u}, B_{3u}, and B_{1g}. What happens when you try to project out a SALC with symmetry B_{1u}?

P10C.8 Consider the molecule $F_2C{=}CF_2$ (point group D_{2h}), and take it as lying in the *xy*-plane, with *x* directed along the C–C bond. (a) Consider a basis formed from the four $2p_z$ orbitals from the fluorine atoms: show that the basis spans B_{1u}, B_{2g}, B_{3g}, and A_u. (b) By applying the projection formula to one of the $2p_z$ orbitals, generate the SALCs with the indicated symmetries. (c) Repeat the process for a basis formed from four $2p_x$ orbitals (the symmetry species will be different from those for $2p_z$).

FOCUS 11

Molecular spectroscopy

The origin of spectral lines in molecular spectroscopy is the absorption, emission, or scattering of a photon accompanied by a change in the energy of a molecule. The difference from atomic spectroscopy (Topic 8C) is that the energy of a molecule can change not only as a result of electronic transitions but also because it can undergo changes of rotational and vibrational state. Molecular spectra are therefore more complex than atomic spectra. However, they contain information relating to more properties, and their analysis leads to values of bond strengths, lengths, and angles. They also provide a way of determining a variety of molecular properties, such as dissociation energies and dipole moments.

11A General features of molecular spectroscopy

This Topic begins with a discussion of the theory of absorption and emission of radiation, leading to the factors that determine the intensities and widths of spectral lines. The features of the instrumentation used to monitor the absorption, emission, and scattering of radiation spanning a wide range of frequencies are also described.

11B Rotational spectroscopy

This Topic shows how expressions for the values of the rotational energy levels of diatomic and polyatomic molecules are derived. The most direct procedure, which is used here, is to identify the expressions for the energy and angular momentum obtained in classical physics, and then to transform these expressions into their quantum mechanical counterparts. The Topic then focuses on the interpretation of pure rotational and rotational Raman spectra, in which only the rotational state of a molecule changes. The observation that not all molecules can occupy all rotational states is shown to arise from symmetry constraints resulting from the presence of nuclear spin.

11C Vibrational spectroscopy of diatomic molecules

The harmonic oscillator (Topic 7E) is a good starting point for modelling the vibrations of diatomic molecules, but it is shown that the description of real molecules requires deviations from harmonic behaviour to be taken into account. The vibrational spectra of gaseous samples show features due to the rotational transitions that accompany the excitation of vibrations.

11D Vibrational spectroscopy of polyatomic molecules

The vibrational spectra of polyatomic molecules may be discussed as though they consisted of a set of independent harmonic oscillators. Their spectra can then be understood in much the same way as those of diatomic molecules.

11E Symmetry analysis of vibrational spectra

The atomic displacements involved in the vibrations of polyatomic molecules can be classified according to the symmetry possessed by the molecule. This classification makes it possible to decide which vibrations can be studied spectroscopically.

11F Electronic spectra

This Topic introduces the key idea that electronic transitions occur within a stationary nuclear framework. The electronic spectra of diatomic molecules are considered first, and it is seen that in the gas phase it is possible to observe simultaneous vibrational and rotational transitions that accompany the electronic transition. The general features of the electronic spectra of polyatomic molecules are also described.

11G Decay of excited states

This Topic begins with an account of spontaneous emission by molecules, including the phenomena of 'fluorescence' and 'phosphorescence'. It is also seen how non-radiative decay of excited states can result in the transfer of energy as heat to the surroundings or result in molecular dissociation. The stimulated radiative decay of excited states is the key process responsible for the action of lasers.

Web resources What is an application of this material?

Molecular spectroscopy is also useful to astrophysicists and environmental scientists. *Impact* 16 discusses how the identities of molecules found in interstellar space can be inferred from their rotational and vibrational spectra. *Impact* 17 focuses back on Earth and shows how the vibrational properties of its atmospheric constituents can affect its climate.

TOPIC 11A General features of molecular spectroscopy

> ➤ Why do you need to know this material?
>
> To interpret data from the wide range of varieties of molecular spectroscopy you need to understand the experimental and theoretical features shared by them all.
>
> ➤ What is the key idea?
>
> A transition from a low energy state to one of higher energy can be stimulated by absorption of radiation; a transition from a higher to a lower state, resulting in emission of a photon, may be either spontaneous or stimulated by radiation.
>
> ➤ What do you need to know already?
>
> You need to be familiar with the fact that molecular energy is quantized (Topics 7E and 7F) and be aware of the concept of selection rules (Topic 8C).

In **emission spectroscopy** the electromagnetic radiation that arises from molecules undergoing a transition from a higher energy state to a lower energy state is detected and its frequency analysed. In **absorption spectroscopy**, the net absorption of radiation passing through a sample is monitored over a range of frequencies. It is necessary to specify the *net* absorption because not only can radiation be absorbed but it can also stimulate the emission of radiation, so the net absorption is detected. In **Raman spectroscopy**, the frequencies of radiation scattered by molecules is analysed to determine the changes in molecular states that accompany the scattering process. Throughout this discussion it is important to be able to express the characteristics of radiation variously as a frequency, ν, a wavenumber, $\tilde{\nu}=\nu/c$, or a wavelength, $\lambda = c/\nu$, as set out in *The chemist's toolkit* 13 in Topic 7A.

In each case, the emission, absorption, or scattering of radiation can be interpreted in terms of individual photons. When a molecule undergoes a transition between states of energy E_l, the energy of the lower state, and E_u, the energy of the upper state, the energy, $h\nu$, of the photon emitted or absorbed is given by the Bohr frequency condition (eqn 7A.9 of Topic 7A, $h\nu = E_u - E_l$), where ν is the frequency of the radiation emitted or absorbed. Emission and absorption spectroscopy give the same information about electronic, vibrational, or rotational energy level separations, but practical considerations generally determine which technique is employed.

In Raman spectroscopy the sample is exposed to monochromatic (single frequency) radiation and therefore photons of the same energy. When the photons encounter the molecules, most are scattered elastically (without change in their energy): this process is called **Rayleigh scattering.** About 1 in 10^7 of the photons are scattered inelastically (with different energy). In **Stokes scattering** the photons lose energy to the molecules and the emerging radiation has a lower frequency. In **anti-Stokes scattering**, a photon gains energy from a molecule and the emerging radiation has a higher frequency (Fig. 11A.1). By analysing the frequencies of the scattered radiation it is possible to gather information about the energy levels of the molecules. Raman spectroscopy is used to study molecular vibrations and rotations.

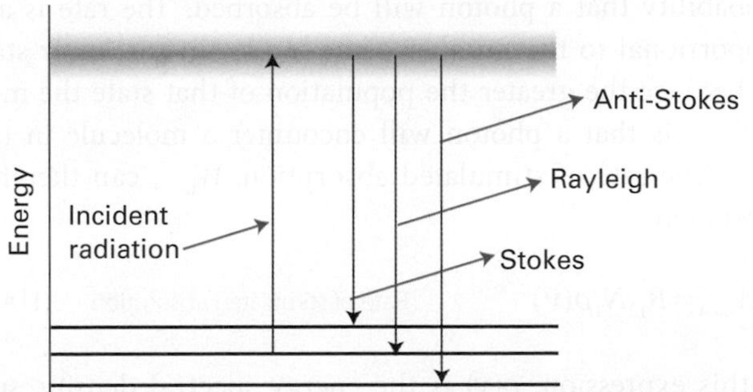

Figure 11A.1 In Raman spectroscopy, incident photons are scattered from a molecule. Most photons are scattered elastically and so have the same energy as the incident photons. Some photons lose energy to the molecule and so emerge as Stokes radiation; others gain energy and so emerge as anti-Stokes radiation. The scattering can be regarded as taking place by an excitation of the molecule from its initial state to a series of excited states (represented by the shaded band), and the subsequent return to a final state. Any net energy change is either supplied from, or carried away by, the photon.

11A.1 The absorption and emission of radiation

The separation of rotational energy levels (in small molecules, $\Delta E \approx 0.01$ zJ, corresponding to about $0.01\,\text{kJ mol}^{-1}$) is smaller than that of vibrational energy levels ($\Delta E \approx 10$ zJ, corresponding to $10\,\text{kJ mol}^{-1}$), which itself is smaller than that of electronic energy levels ($\Delta E \approx 0.1$–1 aJ, corresponding to about 10^2–$10^3\,\text{kJ mol}^{-1}$). From the Bohr frequency condition in the form $\nu = \Delta E/h$, the corresponding frequencies of the photons involved in these different kinds of transitions is about 10^{10} Hz for rotation, 10^{13} Hz for vibration, and in the range 10^{14}–10^{15} Hz for electronic transitions. It follows that rotational, vibrational, and electronic transitions result from the absorption or emission of microwave, infrared, and ultraviolet/visible radiation, respectively.

(a) Stimulated and spontaneous radiative processes

Albert Einstein identified three processes by which radiation could be either generated or absorbed by matter as a result of transitions between states. In **stimulated absorption** a transition from a lower energy state l to a higher energy state u is driven by an electromagnetic field oscillating at the frequency ν corresponding to the energy separation of the two states: $h\nu = E_u - E_l$. The rate of such transitions is proportional to the intensity of the incident radiation at the transition frequency: the more intense the radiation, the greater the number of photons impinging on the molecules and the greater the probability that a photon will be absorbed. The rate is also proportional to the number of molecules in the lower state, N_l, because the greater the population of that state the more likely it is that a photon will encounter a molecule in that state. The rate of stimulated absorption, $W_{u\leftarrow l}$, can therefore be written

$$W_{u\leftarrow l} = B_{u,l}N_l\rho(\nu) \quad \text{Rate of stimulated absorption} \quad (11A.1a)$$

In this expression, $\rho(\nu)$ is the energy spectral density, such that $\rho(\nu)\mathrm{d}\nu$ is the energy density of radiation in the frequency range from ν to $\nu + \mathrm{d}\nu$. The constant $B_{u,l}$ is the **Einstein coefficient of stimulated absorption.**

Einstein also supposed that the radiation could induce the molecule in an upper state to undergo a transition to a lower state and thereby generate a photon of frequency ν. This process is called **stimulated emission**, and its rate depends on the number of molecules in the upper level N_u and the intensity of the radiation at the transition frequency. As in eqn 11A.1a he wrote

$$W'_{u\rightarrow l} = B_{l,u}N_u\rho(\nu) \quad \text{Rate of stimulated emission} \quad (11A.1b)$$

In this expression $B_{l,u}$ is the **Einstein coefficient of stimulated emission.**

Einstein went on to suppose that a molecule could lose energy by **spontaneous emission** in which the molecule makes a transition to a lower state without it being driven by the presence of radiation. The rate of spontaneous emission is written

$$W''_{u\rightarrow l} = A_{l,u}N_u \quad \text{Rate of spontaneous emission} \quad (11A.1c)$$

where $A_{l,u}$ is the **Einstein coefficient of spontaneous emission.** When both stimulated and spontaneous emission are taken into account, the total rate of emission is

$$W_{u\rightarrow l} = B_{l,u}N_u\rho(\nu) + A_{l,u}N_u \quad \text{Total rate of emission} \quad (11A.1d)$$

When the molecules and radiation are in equilibrium, the rates given in eqns 11A.1a and 11A.1d must be equal, and the populations then have their equilibrium values N_l^{eq} and N_u^{eq}. Therefore

$$B_{u,l}N_l^{eq}\rho(\nu) = B_{l,u}N_u^{eq}\rho(\nu) + A_{l,u}N_u^{eq} \quad (11A.2a)$$

and so

$$\rho(\nu) = \frac{A_{l,u}/B_{u,l}}{N_l^{eq}/N_u^{eq} - B_{l,u}/B_{u,l}} \quad (11A.2b)$$

However, the ratio of the equilibrium populations must be in accord with the Boltzmann distribution (as specified in the *Prologue* of this text and Topic 13A):

$$\frac{N_u^{eq}}{N_l^{eq}} = e^{-(E_u - E_l)/kT} = e^{-h\nu/kT} \quad (11A.3)$$

and therefore, at equilibrium,

$$\rho(\nu) = \frac{A_{l,u}/B_{u,l}}{e^{h\nu/kT} - B_{l,u}/B_{u,l}} \quad (11A.4)$$

Moreover, at equilibrium the radiation density is given by the Planck distribution of radiation in equilibrium with a black body (eqn 7A.6b, Topic 7A):

$$\rho(\nu) = \frac{8\pi h\nu^3/c^3}{e^{h\nu/kT} - 1} \quad \text{Planck distribution} \quad (11A.5)$$

It then follows that

$$A_{l,u} = \left(\frac{8\pi h\nu^3}{c^3}\right)B_{l,u} \quad \text{and} \quad B_{l,u} = B_{u,l} \quad (11A.6a)$$

Although these relations have been derived on the assumption that the molecules and radiation are in equilibrium, they are properties of the molecules themselves and are independent of the spectral distribution of the radiation (that is, whether it is black-body or not) and can be used in eqn 11A.1 for any energy densities.

The ratio of the rate of spontaneous to stimulated emission can be found by combining eqns 11A.1b, 11A.1c, and 11A.6a to give

$$\frac{W''_{u\to l}}{W'_{u\to l}} = \frac{A_{l,u}}{B_{l,u}\rho(\nu)} = \frac{8\pi h\nu^3}{c^3\rho(\nu)} \qquad (11A.6b)$$

This relation shows that, for a given spectral density, the relative importance of spontaneous emission increases as the cube of the transition frequency and therefore that spontaneous emission is most likely to be of importance at high frequencies. Conversely, spontaneous emission can be ignored at low frequencies, in which case the intensities of such transitions can be discussed in terms of stimulated emission and absorption alone.

Brief illustration 11A.1

On going from infrared to visible radiation, the frequency increases by a factor of about 100, so the ratio of the rates of spontaneous to stimulated emission increases by a factor of 10^6 for the same spectral density. This strong increase accounts for the observation that whereas electronic transitions are often monitored by emission spectroscopy, vibrational spectroscopy is an absorption technique and spontaneous (but not stimulated) emission is negligible.

(b) Selection rules and transition moments

A 'selection rule' is a statement about whether a transition is forbidden or allowed (Topic 8C). The underlying idea is that, for the molecule to be able to interact with the electromagnetic field and absorb or create a photon of frequency ν, it must possess, at least transiently, an electric dipole oscillating at that frequency. This transient dipole is expressed quantum mechanically in terms of the **transition dipole moment**, μ_{fi}, between the initial and final states with wavefunctions ψ_i and ψ_f:

$$\mu_{fi} = \int \psi_f^* \hat{\mu} \psi_i \, d\tau \qquad \text{Transition dipole moment [definition]} \qquad (11A.7)$$

where $\hat{\mu}$ is the electric dipole moment operator. The magnitude of the transition dipole moment can be regarded as a measure of the charge redistribution that accompanies a transition and a transition is active (and generates or absorbs a photon) only if the accompanying charge redistribution is dipolar (Fig. 11A.2). It follows that, to identify the selection rules, the conditions for which $\mu_{fi} \neq 0$ must be established.

A **gross selection rule** specifies the general features that a molecule must have if it is to have a spectrum of a given kind. For instance, in Topic 11B it is shown that a molecule gives a rotational spectrum only if it has a permanent electric dipole

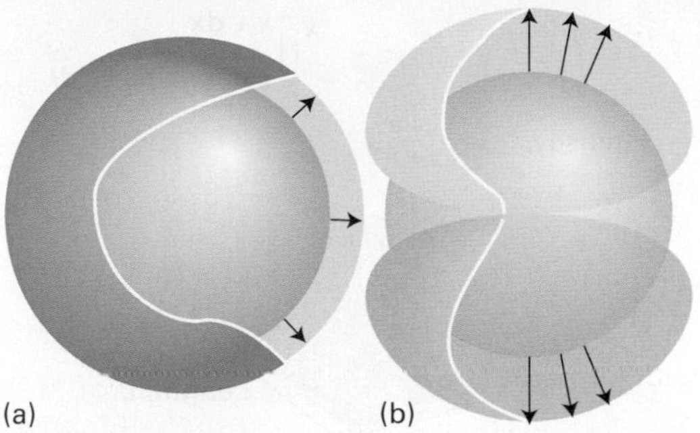

Figure 11A.2 (a) When a 1s electron becomes a 2s electron, there is a spherical migration of charge. There is no dipole moment associated with this migration of charge, so this transition is electric-dipole forbidden. (b) In contrast, when a 1s electron becomes a 2p electron, there is a dipole associated with the charge migration; this transition is allowed.

moment. This rule, and others like it for other types of transition, is explained in the relevant Topic. A detailed study of the transition moment leads to the **specific selection rules** which express the allowed transitions in terms of the changes in quantum numbers or various symmetry features of the molecule.

(c) The Beer–Lambert law

It is found empirically that when electromagnetic radiation passes through a sample of length L and molar concentration [J] of the absorbing species J, the incident and transmitted intensities, I_0 and I, are related by the **Beer–Lambert law**:

$$I = I_0 10^{-\varepsilon[\mathrm{J}]L} \qquad \text{Beer–Lambert law} \qquad (11A.8)$$

The quantity ε (epsilon) is called the **molar absorption coefficient** (formerly, and still widely, the 'extinction coefficient'); it depends on the frequency (or wavenumber and wavelength) of the incident radiation and is greatest where the absorption is most intense. The dimensions of ε are 1/(concentration × length), and it is normally convenient to express it in cubic decimetres per mole per centimetre ($dm^3\,mol^{-1}\,cm^{-1}$); in SI base units it is expressed in metre squared per mole ($m^2\,mol^{-1}$). The latter units imply that ε may be regarded as a (molar) cross-section for absorption, and that the greater the cross-sectional area of the molecule for absorption, the greater is its ability to block the passage of the incident radiation at a given frequency. The Beer–Lambert law is an empirical result. However, its form can be derived on the basis of a simple model.

How is that done? 11A.1 Justifying the Beer–Lambert law

You need to imagine the sample as consisting of a stack of infinitesimal slices, like sliced bread (Fig. 11A.3). The thickness of each layer is dx.

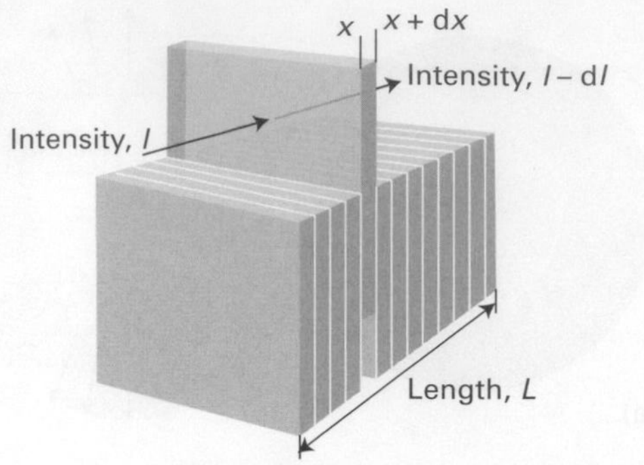

Figure 11A.3 To establish the Beer–Lambert law, the sample is supposed to be divided into a large number of thin slices. The reduction in intensity caused by one slice is proportional to the intensity incident on it (after passing through the preceding slices), the thickness of the slice, and the concentration of the absorbing species.

Step 1 *Calculate the change in intensity due to passage through one slice*

The change in intensity, dI, that occurs when electromagnetic radiation passes through one particular slice is proportional to the thickness of the slice, the molar concentration of the absorber J, and (because the absorption is stimulated) the intensity of the incident radiation at that slice of the sample, so $dI \propto [J]Idx$. The intensity is reduced by absorption, which means that dI is negative and can therefore be written

$$dI = -\kappa[J]Idx$$

where κ (kappa) is the proportionality coefficient. Division of both sides by I gives

$$\frac{dI}{I} = -\kappa[J]dx$$

This expression applies to each successive slice.

Step 2 *Evaluate the total change in intensity due to passage through successive slices*

To obtain the intensity that emerges from a sample of thickness L when the intensity incident on one face of the sample is I_0, you need the sum of all the successive changes. Assume that the molar concentration of the absorbing species is uniform and may be treated as a constant. Because a sum over infinitesimally small increments is an integral, it follows that:

$$\overbrace{\int_{I_0}^{I}\frac{dI}{I}}^{\text{Integral A.2}} = -\kappa\int_0^L [J]dx = -\kappa[J]\overbrace{\int_0^L dx}^{\text{Integral A.1}}$$

([J] a constant)

Therefore

$$\ln\frac{I}{I_0} = -\kappa[J]L$$

Now express the natural logarithm as a common logarithm (to base 10) by using $\ln x = (\ln 10)\log x$, and a new constant ε defined as $\varepsilon = \kappa/\ln 10$ to give

$$\log\frac{I}{I_0} = -\varepsilon[J]L$$

Raising each side as a power of 10 gives the Beer–Lambert law (eqn 11A.8).

The spectral characteristics of a sample are commonly reported as the **transmittance**, T, of the sample at a given frequency:

$$T = \frac{I}{I_0} \quad \text{Transmittance [definition]} \quad (11A.9a)$$

or its **absorbance**, A:

$$A = \log\frac{I_0}{I} \quad \text{Absorbance [definition]} \quad (11A.9b)$$

The two quantities are related by $A = -\log T$ (note the common logarithm) and the Beer–Lambert law becomes

$$A = \varepsilon[J]L \quad (11A.9c)$$

The product $\varepsilon[J]L$ was known formerly as the *optical density* of the sample.

Example 11A.1 Determining a molar absorption coefficient

Radiation of wavelength 280 nm passed through 1.0 mm of a solution that contained an aqueous solution of the amino acid tryptophan at a molar concentration of 0.50 mmol dm^{-3}. The intensity is reduced to 54 per cent of its initial value (so $T = 0.54$). Calculate the absorbance and the molar absorption coefficient of tryptophan at 280 nm. What would be the transmittance through a cell of thickness 2.0 mm?

Collect your thoughts From $A = -\log T = \varepsilon[J]L$, it follows that $\varepsilon = A/[J]L$. For the transmittance through the thicker cell, you need to calculate the absorbance by using $A = -\log T = \varepsilon[J]L$ and the computed value of ε; the transmittance is $T = 10^{-A}$.

The solution The absorbance is $A = -\log 0.54 = 0.27$, and so the molar absorption coefficient is

$$\varepsilon = \frac{-\log 0.54}{(5.0\times10^{-4}\,\text{mol dm}^{-3})\times(1.0\,\text{mm})} = 5.4\times10^{2}\,\text{dm}^3\,\text{mol}^{-1}\,\text{mm}^{-1}$$

These units are convenient for the rest of the calculation (but the outcome could be reported as $5.4\times10^3\,\text{dm}^3\,\text{mol}^{-1}\,\text{cm}^{-1}$ if desired or even as $5.4 \times 10^2\,m^2\,mol^{-1}$). The absorbance of a sample of length 2.0 mm is

$$A = (5.4 \times 10^2\,dm^3\,mol^{-1}\,mm^{-1}) \times (5.0 \times 10^{-4}\,mol\,dm^{-3}) \times (2.0\,mm) = 0.54$$

The transmittance is now $T = 10^{-A} = 10^{-0.54} = 0.29$.

Self-test 11A.1 The transmittance of an aqueous solution containing the amino acid tyrosine at a molar concentration of $0.10\,mmol\,dm^{-3}$ was measured as 0.14 at 240 nm in a cell of length 5.0 mm. Calculate the absorbance of the solution and the molar absorption coefficient of tyrosine at that wavelength. What would be the transmittance through a cell of length 1.0 mm?

Answer: $A = 0.85$, $1.7 \times 10^4\,dm^3\,mol^{-1}\,cm^{-1}$, $T = 0.67$

The maximum value of the molar absorption coefficient, ε_{max}, is an indication of the intensity of a transition. However, because absorption bands generally spread over a range of wavenumbers, quoting the absorption coefficient at a single wavenumber might not give a true indication of the intensity of a transition. The **integrated absorption coefficient**, $\mathcal{A}$, is the sum of the absorption coefficients over the entire band (Fig. 11A.4), and corresponds to the area under the plot of the molar absorption coefficient against wavenumber:

$$\mathcal{A} = \int_{band} \varepsilon(\tilde{\nu})d\tilde{\nu} \qquad \text{Integrated absorption coefficient [definition]} \qquad (11A.10)$$

For bands of similar widths, the integrated absorption coefficients are proportional to the heights of the bands. Equation 11A.10 also applies to the individual lines that contribute to a band: a *spectroscopic* line is not a geometrically thin line, but has a width.

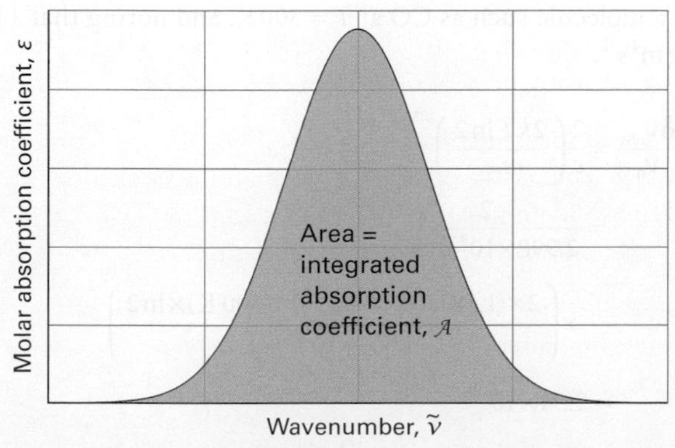

Figure 11A.4 The integrated absorption coefficient of a transition is the area under a plot of the molar absorption coefficient against the wavenumber of the incident radiation.

11A.2 Spectral linewidths

A number of effects contribute to the widths of spectroscopic lines. The design of the spectrometer itself affects the linewidth, and there are other contributions that arise from physical processes in the sample. Some of the latter can be minimized by altering the conditions, while others are intrinsic to the molecules and cannot be altered.

(a) Doppler broadening

One important broadening process in gaseous samples is the **Doppler effect**, in which radiation is shifted in frequency when its source is moving towards or away from the observer. When a molecule emitting electromagnetic radiation of frequency ν moves with a speed s relative to an observer, the observer detects radiation of frequency

$$\nu_{receding} = \left(\frac{1-s/c}{1+s/c}\right)^{1/2}\nu_0 \qquad \nu_{approaching} = \left(\frac{1+s/c}{1-s/c}\right)^{1/2}\nu_0 \qquad \text{Doppler shifts} \quad (11A.11a)$$

where c is the speed of light. For nonrelativistic speeds ($s \ll c$), these expressions simplify to

$$\nu_{receding} \approx (1-s/c)\nu_0 \qquad \nu_{approaching} \approx (1+s/c)\nu_0 \qquad (11A.11b)$$

Atoms and molecules reach high speeds in all directions in a gas, and a stationary observer detects the corresponding Doppler-shifted range of frequencies. Some molecules approach the observer, some move away; some move quickly, others slowly. The detected spectral 'line' is the absorption or emission profile arising from all the resulting Doppler shifts. The challenge is to relate the observed linewidth to the spread of speeds in the gas, and in turn to see how that spread depends on the temperature.

How is that done? 11A.2 Deriving an expression for Doppler broadening

You need to relate the spread of Doppler shifts to the distribution of molecular kinetic energy as expressed by the Boltzmann distribution.

Step 1 *Establish the relation between the observed frequency and the molecular speed*

It follows from the Boltzmann distribution (see the *Prologue* to the text) that the probability that an atom or molecule of mass m and speed s in a gas phase sample at a temperature T

has kinetic energy $E_k = \frac{1}{2}ms^2$ is proportional to $e^{-ms^2/2kT}$. When $s \ll c$, the Doppler shifts for receding and approaching molecules are given by the expressions in eqn 11A.11b. It follows that the shift between the observed frequency and the true frequency is $\nu_{obs} - \nu_0 \approx \pm\nu_0 s/c$. This expression can be rearranged to give

$$s = \pm c(\nu_{obs} - \nu_0)/\nu_0$$

Step 2 *Evaluate the distribution of frequencies arising from a distribution of speeds*

The intensity I of a transition at ν_{obs} is proportional to the probability of there being a molecule that emits or absorbs at ν_{obs}. Such a molecule would have a speed given by the above expression, so it follows from the Boltzmann distribution that

$$I(\nu_{obs}) \propto e^{-ms^2/2kT} = e^{-mc^2(\nu_{obs}-\nu_0)^2/2\nu_0^2 kT}$$

which has the form of a Gaussian function. The width at half-height, $\delta\nu_{obs}$, can be inferred directly from the general form of such a function (as specified in *The chemist's toolkit* 26):

$$\delta\nu_{obs} = \frac{2\nu_0}{c}\left(\frac{2kT\ln 2}{m}\right)^{1/2} \quad (11A.12a)$$

Doppler broadening

Doppler broadening increases with temperature (Fig. 11A.5) because the molecules then acquire a wider range of speeds. Conversely, reducing the temperature results in narrower lines. Note that the Doppler linewidth is proportional to the frequency, so Doppler broadening becomes more important as higher frequencies are observed.

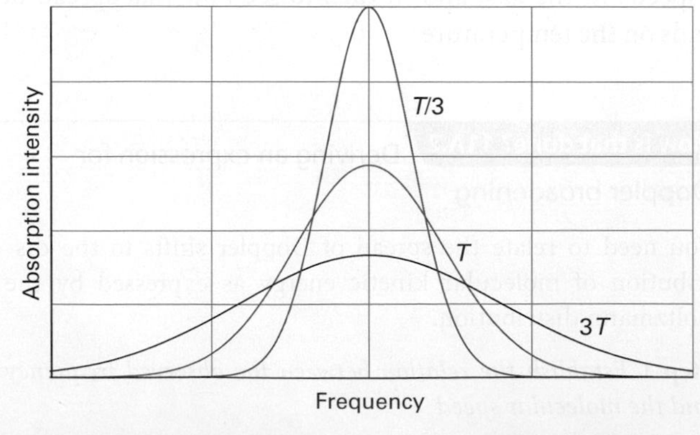

Figure 11A.5 The Gaussian shape of a Doppler-broadened spectral line reflects the Boltzmann distribution of translational kinetic energies in the sample at the temperature of the experiment. The line broadens as the temperature is increased.

The chemist's toolkit 26 Exponential and Gaussian functions

An **exponential function** is a function of the form

$$f(x) = ae^{-bx} \quad \text{Exponential function}$$

This function has the value a at $x = 0$ and decays toward zero as $x \to \infty$. This decay is faster when b is large than when it is small. The function rises rapidly to infinity as $x \to -\infty$. See Sketch 1.

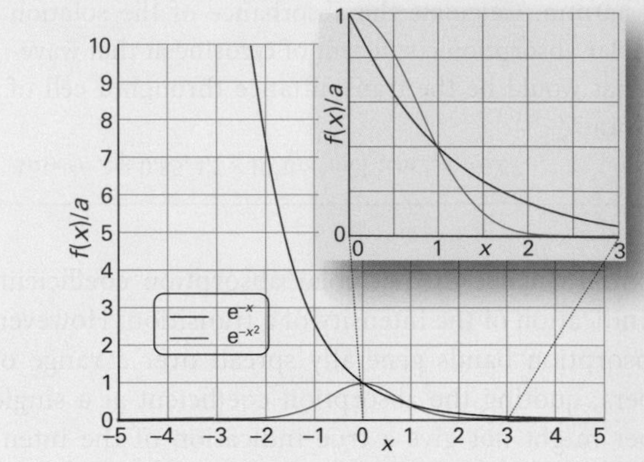

Sketch 1

The general form of a **Gaussian function** is

$$f(x) = ae^{-(x-b)^2/2\sigma^2} \quad \text{Gaussian function}$$

The graph of this function is a symmetrical bell-shaped curve centred on $x = b$; the function has is maximum values of a at its centre. The width of the function, measured at half its height, is $\delta x = 2\sigma(2\ln 2)^{1/2}$; the greater σ, the greater is the width at half-height. Sketch 1 also shows a Gaussian function with $b = 0$.

Brief illustration 11A.2

For a molecule such as CO at $T = 300$ K, and noting that 1 J = 1 kg m^2 s^{-2},

$$\frac{\delta\nu_{obs}}{\nu_0} = \frac{2}{c}\left(\frac{2kT\ln 2}{m_{CO}}\right)^{1/2}$$

$$= \frac{2}{2.998\times10^8 \text{ m s}^{-1}} \times\left(\frac{2\times(1.381\times10^{-23}\text{ J K}^{-1})\times(300\text{ K})\times\ln 2}{4.651\times10^{-26}\text{ kg}}\right)^{1/2}$$

$$= 2.34\times10^{-6}$$

For a transition wavenumber of 2150 cm^{-1} from the infrared spectrum of CO, corresponding to a frequency of 64.4 THz (1 THz = 10^{12} Hz), the linewidth is 151 MHz or 5.0×10^{-3} cm^{-1}.

The linewidth due to Doppler broadening can also be expressed in terms of wavelength as

$$\delta\lambda_{\text{obs}} = \frac{2\lambda_0}{c}\left(\frac{2kT\ln 2}{m}\right)^{1/2} \qquad \text{Doppler broadening} \qquad (11\text{A}.12\text{b})$$

(b) Lifetime broadening

At any instant a molecule exists in a specific state but it does not remain in that state indefinitely. For example, the molecule might collide with another and in the process change its state. Alternatively, the molecule may fall to a lower level and emit a photon by the process of spontaneous emission. How long the state persists depends on the rates of these processes and is characterized by a lifetime, τ. When the Schrödinger equation is analysed for a state that persists for a time τ, it turns out that the energy is uncertain to an extent $\delta E \approx \hbar/\tau$.[1] Therefore, spectroscopic transitions that involve this state have a linewidth of the order of $\delta E/h = 1/2\pi\tau$; the shorter the lifetime, the broader is the line. This process is called **lifetime broadening**. When the lifetime is limited by the process of spontaneous emission rather than external causes such as collisions, the resulting linewidth is called the **natural linewidth**.

Brief illustration 11A.3

Excited electronic states of molecules often have short lifetimes due to the high rate of spontaneous emission. A typical lifetime might be 10 ns, which would lead to a natural linewidth of $1/(2\pi \times 10 \times 10^{-9}\,\text{s}) = 16\,\text{MHz}$ or $5.3 \times 10^{-4}\,\text{cm}^{-1}$. As can be inferred from the previous *Brief illustration*, the Doppler linewidth is typically much greater than the natural linewidth.

Collisions between molecules are generally efficient at changing their rotational or vibrational energies, so a good estimate of the resulting **collisional lifetime**, τ_{col}, is to equate it to $1/z$, where z is the collision frequency (Topic 1B). If it is assumed that each collision results in a change of rotational or vibrational state, the lifetime of a state can be taken as τ_{col}, and hence the resulting broadening is $\delta E/h = 1/2\pi\tau_{\text{col}} = z/2\pi$; this contribution to the linewidth is often referred to as **collisional line broadening**. The collision frequency for two molecules with masses m_A and m_B is given by eqn 1B.12b as

$$z = \frac{\sigma v_{\text{rel}} p}{kT} \quad \text{with} \quad v_{\text{rel}} = \left(\frac{8kT}{\pi\mu}\right)^{1/2} \quad \text{and} \quad \mu = \frac{m_A m_B}{m_A + m_B}$$

Note that the collision frequency, and hence the linewidth, is proportional to the pressure, which is why the broadening due to collisions is sometimes referred to as **pressure broadening**. This contribution to the linewidth can be minimized by lowering the pressure as much as possible, although doing so decreases the intensity of the absorption because there are fewer molecules to absorb the radiation. In contrast to Doppler broadening, pressure broadening is independent of the transition frequency.

Brief illustration 11A.4

The linewidth due to pressure broadening in methane gas at 1 bar and 298 K can be estimated by using the expressions just quoted; the collision cross-section σ is $0.46\,\text{nm}^2$. Taking m_A and m_B both to be the mass of a methane molecule, v_{rel} is $888\,\text{m s}^{-1}$. Hence

$$\begin{aligned} z &= \frac{\sigma v_{\text{rel}} p}{kT} \\ &= \frac{(0.46\times10^{-18}\,\text{m}^2)\times(888\,\text{m s}^{-1})\times(1\times10^5\,\text{N m}^{-2})}{(1.381\times10^{-23}\,\text{J K}^{-1})\times(300\,\text{K})} \\ &= 9.9\times10^9\,\text{s}^{-1} \end{aligned}$$

The linewidth is therefore $z/2\pi = 1.6\,\text{GHz}$ or $0.053\,\text{cm}^{-1}$. In *Brief illustration* 11A.2 the Doppler linewidth for a transition in the infrared is estimated as 150 MHz, which is much less than the pressure broadening estimated for the present set of conditions. As the frequency is raised, the Doppler broadening increases in proportion, but the pressure broadening remains unchanged, so Doppler broadening might become dominant.

11A.3 Experimental techniques

Common to all spectroscopic techniques is a *spectrometer*, an instrument used to detect the characteristics of radiation scattered, emitted, or absorbed by atoms and molecules. Figure 11A.6 shows the general layout of an absorption spectrometer. Radiation from an appropriate source is directed

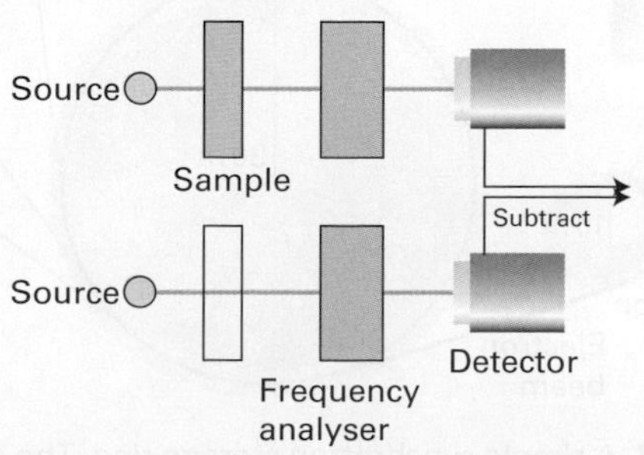

Figure 11A.6 The layout of a typical absorption spectrometer. Radiation from the source passes through the sample; the radiation is then dispersed according to frequency and the intensity at each frequency is measured with a detector.

[1] See our *Molecular quantum mechanics* (2011) for a discussion of the origin of this relation.

towards a sample and the radiation transmitted strikes a device that separates it into different frequencies or wavelengths. The intensity of radiation at each frequency is then analysed by a suitable detector. It is usual to record one spectrum with the sample in place, and one with the sample removed (the 'background spectrum'): the difference between these two spectra eliminates any absorption not due to the sample itself.

(a) Sources of radiation

Sources of radiation are either *monochromatic*, those spanning a very narrow range of frequencies around a central value, or *polychromatic*, those spanning a wide range of frequencies. In the microwave region *frequency synthesizers* and various solid state devices can be used to generate monochromatic radiation that can be tuned over a wide range of frequencies. Certain kinds of lasers and light-emitting diodes are often used to provide monochromatic radiation from the infrared to the ultraviolet region. Polychromatic black-body radiation from hot materials (Topic 7A) can be used over the same range. Examples include mercury arcs inside a quartz envelope (usable in the range 35–200 cm^{-1}), *Nernst filaments* and *globars* (200–4000 cm^{-1}), and *quartz-tungsten-halogen lamps* (320–2500 nm).

A *gas discharge lamp* is a common source of ultraviolet and visible radiation. In a *xenon discharge lamp*, an electrical discharge excites xenon atoms to excited states, which then emit ultraviolet radiation. In a *deuterium lamp*, excited D_2 molecules dissociate into electronically excited D atoms that emit intense radiation in the range 200–400 nm.

For certain applications, radiation is generated in a *synchrotron storage ring*, which consists of an electron beam travelling in a circular path with circumferences of up to several hundred metres. As electrons travelling in a circle are constantly accelerated by the forces that constrain them to their path, they generate radiation (Fig. 11A.7). This 'synchrotron radiation' spans a wide range of frequencies, including infrared radiation and X-rays. Except in the microwave region, synchrotron radiation is much more intense than can be obtained by most conventional sources.

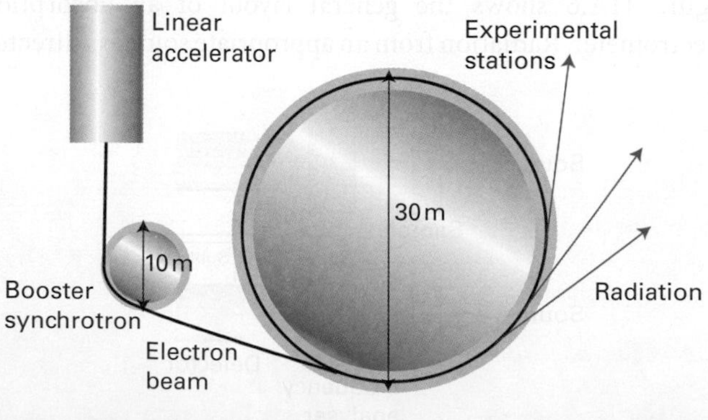

Figure 11A.7 A simple synchrotron storage ring. The electrons injected into the ring from the linear accelerator and booster synchrotron are accelerated to high speed in the main ring. An electron in a curved path is subject to constant acceleration, and an accelerated charge radiates electromagnetic energy.

(b) Spectral analysis

A common device for the analysis of the frequencies, wavenumbers, or wavelengths in a beam of radiation is a *diffraction grating*, which consists of a glass or ceramic plate into which fine grooves have been cut and covered with a reflective aluminium coating. For work in the visible region of the spectrum, the grooves are cut about 1000 nm apart (a spacing comparable to the wavelength of visible light). The grating causes interference between waves reflected from its surface, and constructive interference occurs at specific angles that depend on the wavelength of the radiation being used. Thus, each wavelength of light is diffracted into a specific direction (Fig. 11A.8). In a *monochromator*, a narrow exit slit allows only a narrow range of wavelengths to reach the detector. Turning the grating on an axis perpendicular to the incident and diffracted beams allows different wavelengths to be analysed; in this way, the absorption spectrum is built up one narrow wavelength range at a time. In a *polychromator*, there is no slit and a broad range of wavelengths can be analysed simultaneously by *array detectors*, such as those discussed below.

Currently, almost all spectrometers operating in the infrared and near-infrared use 'Fourier transform' techniques for spectral detection and analysis. (Fourier transforms are discussed, but in more detail than needed here, in *The chemist's toolkit* 28 in Topic 12C.) The heart of a Fourier transform spectrometer is a *Michelson interferometer*, a device for analysing the wavelengths present in a composite signal. The Michelson interferometer works by splitting the beam from the sample

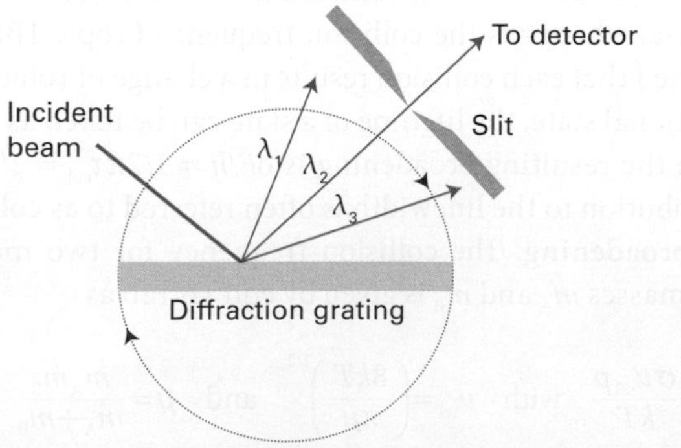

Figure 11A.8 A polychromatic beam is dispersed by a diffraction grating into three component wavelengths λ_1, λ_2, and λ_3. In the configuration shown, only radiation with λ_2 passes through a narrow slit and reaches the detector. Rotation of the diffraction grating (as shown by the arrows on the dotted circle) allows λ_1 or λ_3 to reach the detector.

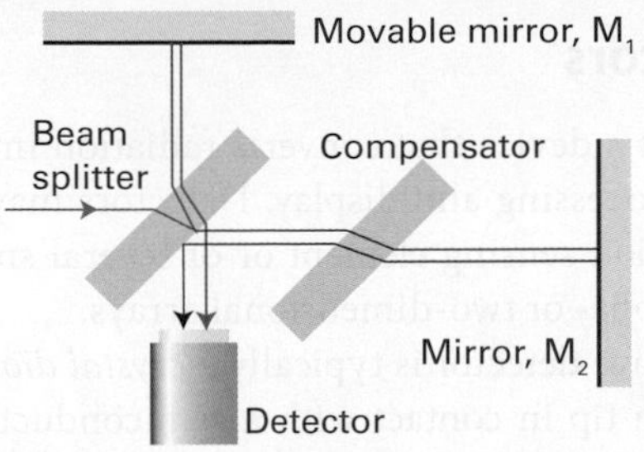

Figure 11A.9 A Michelson interferometer. The beam-splitting element divides the incident beam into two beams with a path difference that depends on the location of the movable mirror M_1. The compensator ensures that both beams pass through the same thickness of material. The beams have been kept separate and coloured to distinguish them: they do not indicate different wavelengths.

into two and arranging for them to take different routes through the instrument before eventually recombining at the detector (Fig. 11A.9). One beam is reflected from mirror M_1 and one from mirror M_2; by moving M_1 it is therefore possible to introduce a difference in the length of the path traversed by the two beams.

Consider first the simplest case in which a beam of monochromatic light of wavelength λ is passed into the interferometer. If the path length difference p is 0, the two beams interfere constructively; the same is true if p is an integer number of wavelengths: λ, 2λ, 3λ,…. If p is one half of a wavelength, the two beams interfere destructively and cancel; the same is true if p is an odd multiple of half-wavelengths: $\lambda/2$, $3\lambda/2$, $5\lambda/2$,…. Therefore, as the mirror M_1 is moved the detected signal goes through a series of peaks and troughs depending on whether the two beams interfere constructively or destructively, and the net signal varies as $1 + \cos(2\pi p/\lambda)$, or $1 + \cos(2\pi p\tilde{\nu})$ (Fig. 11A.10).

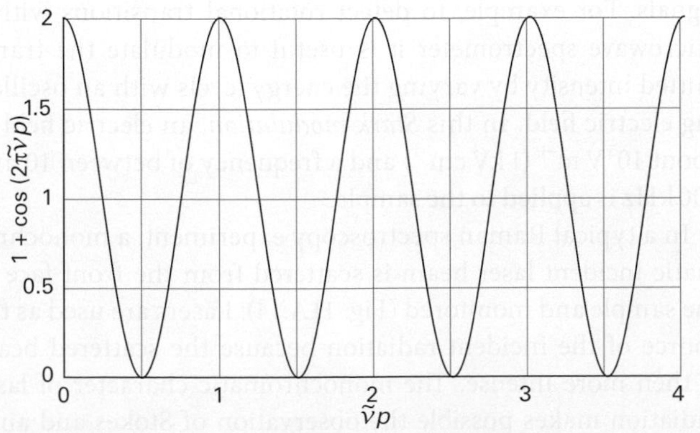

Figure 11A.10 An interferogram produced as the path length difference p is changed in the interferometer shown in Fig. 11A.9. Only a single wavelength component is present in the signal, so the graph is a plot of $1 + \cos(2\pi\tilde{\nu}p)$.

In a spectroscopic observation a mixture of radiation of different wavelengths and intensities is passed into the spectrometer. Each component gives rise an interference pattern proportional to $1 + \cos(2\pi p\tilde{\nu})$, and the signal recorded by the detector is their sum. Thus, if the intensity of the radiation entering the spectrometer consists of a mixture of wavenumber $\tilde{\nu}_i$ with intensities $I(\tilde{\nu}_i)$, the signal measured at the detector is given by the sum

$$\tilde{I}(p) = \sum_i I(\tilde{\nu}_i)\{1+\cos(2\pi\tilde{\nu}_i p)\} \qquad (11A.13)$$

A plot of $\tilde{I}(p)$ against p, which is detected by the system and recorded, is called an **interferogram**. The problem is to find $I(\tilde{\nu})$, the variation of the intensity with wavenumber, which is the spectrum, from $\tilde{I}(p)$. This conversion can be carried out by using a standard mathematical technique, called the *Fourier transform*, which involves evaluating the integral

$$I(\tilde{\nu}) = 4\int_0^\infty \left\{\tilde{I}(p) - \tfrac{1}{2}\tilde{I}(0)\right\}\cos(2\pi\tilde{\nu}p)\mathrm{d}p$$

Fourier transform (11A.14)

In practice, the measured values of $\tilde{I}(p)$ are digitized, stored in a computer attached to the spectrometer, and then the Fourier transform is computed numerically.

Example 11A.2 Relating a spectrum to an interferogram

Suppose the light entering the interferometer consists of three components with the following characteristics:

$\tilde{\nu}_i/\mathrm{cm}^{-1}$	150	250	450
$I(\tilde{\nu}_i)$	1	3	6

where the intensities are relative to the first value listed. Plot the interferogram associated with this signal. Then calculate and plot the Fourier transform of the interferogram.

Collect your thoughts For a signal consisting of just these three component beams, you can use eqn 11A.13 directly. Although in this case (where $\tilde{I}(p)$ is simply the sum of trigonometric functions) the Fourier transform $I(\tilde{\nu})$ can be carried out exactly, in general it is best done numerically by using mathematical software.

The solution From the data, the interferogram is

$$\begin{aligned}\tilde{I}(p) &= (1+\cos 2\pi\tilde{\nu}_1 p) + 3\times(1+\cos 2\pi\tilde{\nu}_2 p) + 6\times(1+\cos 2\pi\tilde{\nu}_3 p)\\ &= 10 + \cos 2\pi\tilde{\nu}_1 p + 3\cos 2\pi\tilde{\nu}_2 p + 6\cos 2\pi\tilde{\nu}_3 p\end{aligned}$$

This function is plotted in Fig. 11A.11. The result of evaluating the Fourier transform numerically is shown in Fig. 11A.12.

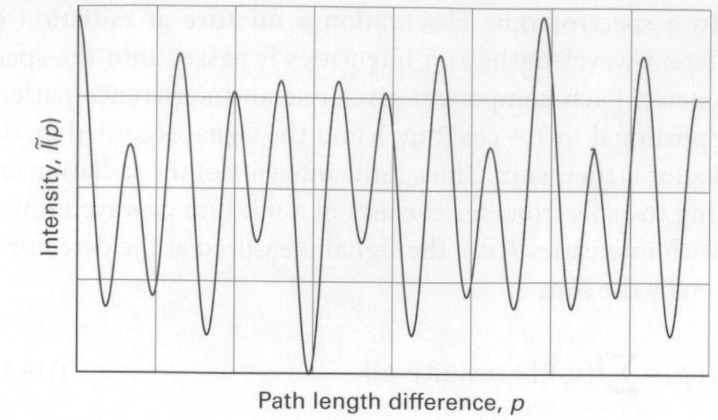

Figure 11A.11 The interferogram calculated from data in *Example* 11A.2.

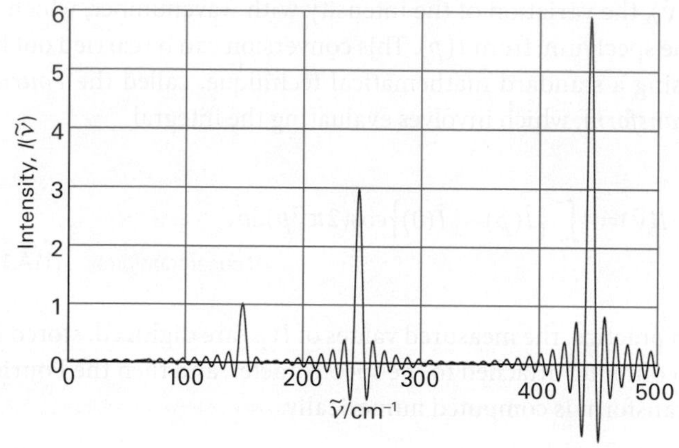

Figure 11A.12 The Fourier transform of the interferogram shown in Fig. 11A.11. The oscillations arise from the way that the signal in Fig. 11A.11 is sampled. As the sampling is extended to greater path-length differences, the oscillations disappear and the peaks become sharper.

Self-test 11A.2 Explore the effect of varying the wavenumbers of the three components of the radiation on the shape of the interferogram by changing the value of $\tilde{\nu}_3$ to 550 cm^{-1}.

Answer: Fig. 11A.13

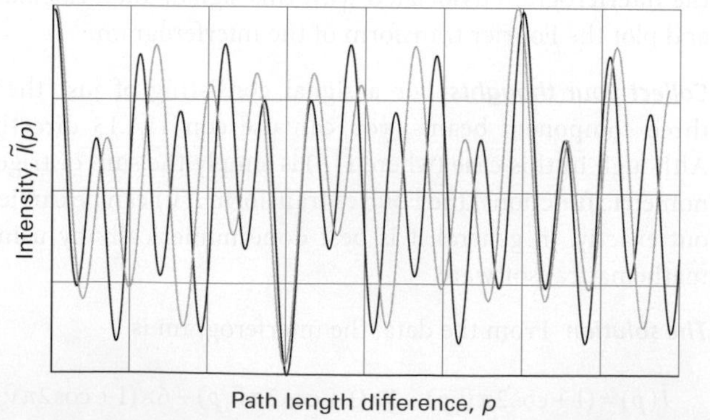

Figure 11A.13 The interferogram calculated from the data in *Self-test* 11A.2 superimposed on the interferogram obtained in the Example itself (in pale blue).

(c) Detectors

A **detector** is a device that converts radiation into an electric signal for processing and display. Detectors may consist of a single radiation sensing element or of several small elements arranged in one- or two-dimensional arrays.

A microwave detector is typically a *crystal diode* consisting of a tungsten tip in contact with a semiconductor. The most common detectors found in commercial infrared spectrometers are sensitive in the mid-infrared region. In a *photovoltaic device* the potential difference changes upon exposure to infrared radiation. In a *pyroelectric device* the capacitance is sensitive to temperature and hence to the presence of infrared radiation.

A common detector for work in the ultraviolet and visible ranges is a *photomultiplier tube* (PMT), in which the photoelectric effect (Topic 7A) is used to generate an electrical signal proportional to the intensity of light that strikes the detector. A common, but less sensitive, alternative to the PMT is a *photodiode*, a solid-state device that conducts electricity when struck by photons because light-induced electron transfer reactions in the detector material create mobile charge carriers (negatively charged electrons and positively charged 'holes').

A *charge-coupled device* (CCD) is a two-dimensional array of several million small photodiode detectors. With a CCD, a wide range of wavelengths that emerge from a polychromator are detected simultaneously, thus eliminating the need to measure the radiation intensity one narrow wavelength range at a time.

(d) Examples of spectrometers

With an appropriate choice of spectrometer, absorption spectroscopy can be used to probe electronic, vibrational, and rotational transitions in molecules. It is often necessary to modify the general design of Fig. 11A.6 in order to detect weak signals. For example, to detect rotational transitions with a microwave spectrometer it is useful to modulate the transmitted intensity by varying the energy levels with an oscillating electric field. In this *Stark modulation*, an electric field of about 10^5 V m^{-1} (1 kV cm^{-1}) and a frequency of between 10 and 100 kHz is applied to the sample.

In a typical Raman spectroscopy experiment, a monochromatic incident laser beam is scattered from the front face of the sample and monitored (Fig. 11A.14). Lasers are used as the source of the incident radiation because the scattered beam is then more intense. The monochromatic character of laser radiation makes possible the observation of Stokes and anti-Stokes lines with frequencies that differ only slightly from that of the incident radiation. Such high resolution is particularly useful for observing rotational transitions by Raman spectroscopy.

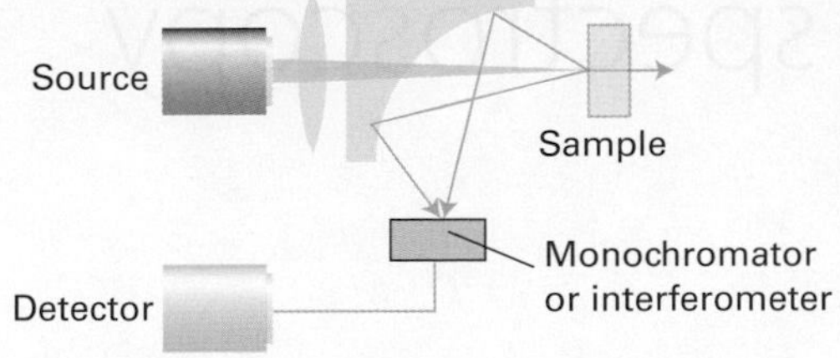

Figure 11A.14 A common arrangement adopted in Raman spectroscopy. A laser beam passes through a lens and then through a small hole in a mirror with a curved reflecting surface. The focused beam strikes the sample and scattered light is both deflected and focused by the mirror. The spectrum is analysed by a monochromator or an interferometer.

Checklist of concepts

☐ 1. In **emission spectroscopy** the electromagnetic radiation that arises from molecules undergoing a transition from a higher energy state to a lower energy state is detected.

☐ 2. In **absorption spectroscopy**, the net absorption of radiation passing through a sample is monitored.

☐ 3. In **Raman spectroscopy**, changes in molecular state are explored by examining the energies (frequencies) of the photons scattered by molecules.

☐ 4. Photons which are scattered elastically give rise to **Raleigh scattering**.

☐ 5. In **Stokes scattering** a photon gives up some of its energy to a molecule; in **anti-Stokes scattering** the photon gains energy from the molecule.

☐ 6. **Stimulated absorption** is a process in which a transition from a low energy state to one of higher energy is driven by an electromagnetic field oscillating at the transition frequency; its rate is determined in part by the **Einstein coefficient of stimulated absorption.**

☐ 7. **Stimulated emission** is a process in which a transition from a high energy state to one of lower energy is driven by an electromagnetic field oscillating at the transition frequency; its rate is determined in part by the **Einstein coefficient of stimulated emission.**

☐ 8. **Spontaneous emission** is the transition from a high energy state to a lower energy state at a rate independent of any radiation also present. The relative importance of spontaneous emission increases as the cube of the transition frequency.

☐ 9. A **gross selection rule** specifies the general features a molecule must have if it is to have a spectrum of a given kind; a **specific selection rule** expresses the allowed transitions in terms of the changes in quantum numbers.

☐ 10. **Collisional line broadening** arises from the shortened lifetime due to collisions. The broadening is proportional to the pressure, and is often termed **pressure broadening.**

Checklist of equations

Property	Equation	Comment	Equation number
Ratio of Einstein coefficients of spontaneous and stimulated emission	$A_{l,u}=(8\pi h\nu^3/c^3)B_{l,u}$	$B_{u,l}=B_{l,u}$	11A.6a
Transition dipole moment	$\mu_{fi}=\int\psi_f^*\hat{\mu}\psi_i\,d\tau$	Electric dipole transitions	11A.7
Beer–Lambert law	$I=I_0 10^{-\varepsilon[J]L}$	Uniform sample	11A.8
Absorbance and transmittance	$A=\log(I_0/I)=-\log T$	Definition	11A.9b
Integrated absorption coefficient	$\mathcal{A}=\int_{band}\varepsilon(\tilde{\nu})d\tilde{\nu}$	Definition	11A.10
Doppler broadening	$\delta\nu_{obs}=(2\nu_0/c)(2kT\ln 2/m)^{1/2}$		11A.12a
Lifetime broadening	$\delta E/h=1/2\pi\tau$	τ is the lifetime of the state	

TOPIC 11B Rotational spectroscopy

> ➤ Why do you need to know this material?
>
> Rotational spectroscopy provides very precise values of bond lengths, bond angles, and dipole moments of molecules in the gas phase.
>
> ➤ What is the key idea?
>
> The spacing of the lines in rotational spectra is used to determine the rotational constants of molecules and, through them, values of their bond lengths and angles.
>
> ➤ What do you need to know already?
>
> You need to be familiar with the classical description of rotational motion (*The chemist's toolkit* 20 in Topic 7F), the quantization of angular momentum (Topic 7F), the general principles of molecular spectroscopy (Topic 11A), and the Pauli principle (Topic 8B).

Pure rotational spectra, in which only the rotational state of a molecule changes, can be observed only in the gas phase. In spite of this limitation, rotational spectroscopy can provide a wealth of information about molecules, including precise bond lengths, angles, and dipole moments.

11B.1 Rotational energy levels

The classical expression for the energy of a body rotating about an axis q (*The chemist's toolkit* 20 in Topic 7F) is

$$E_q = \tfrac{1}{2} I_q \omega_q^2 \qquad (11B.1)$$

where ω_q is the angular velocity about the axis $q = x, y, z$ and I_q is the corresponding moment of inertia. The **moment of inertia**, I, of a molecule about an axis passing through the centre of mass is defined as (Fig. 11B.1)

$$I = \sum_i m_i x_i^2 \qquad \text{Moment of inertia [definition]} \qquad (11B.2)$$

where m_i is the mass of the atom i treated as a point and x_i is its perpendicular distance from the axis of rotation. In general, the rotational properties of any molecule can be expressed in terms of its three **principal moments of inertia** I_q about three mutually perpendicular axes, $q = x, y, z$. For linear molecules, the moment of inertia around the internuclear axis is zero (because $x_i = 0$ for all the atoms) and the two remaining moments of inertia, which are equal, are denoted simply I. Explicit expressions for the moments of inertia of some symmetrical molecules are given in Table 11B.1. The principal moments of inertia are also commonly recorded as I_a, I_b, and I_c, with $I_c \geq I_b \geq I_a$.

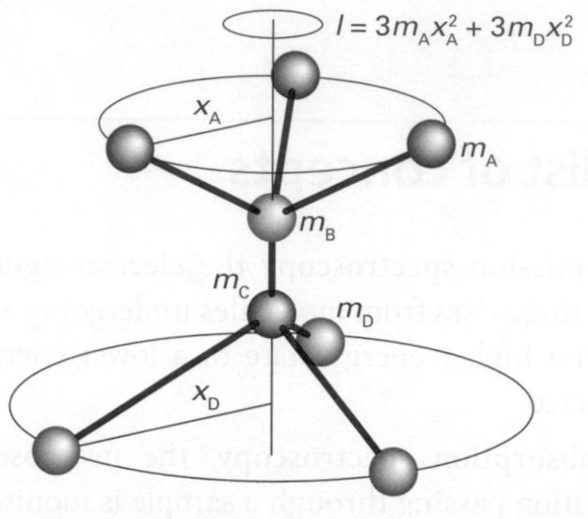

Figure 11B.1 The definition of moment of inertia. In this molecule there are three atoms with mass m_A attached to the B atom and three atoms with mass m_D attached to the C atom. The moment of inertia about the axis passing through the B and C atoms depends on the perpendicular distance x_A from this axis to the A atoms, and the perpendicular distance x_D to the D atoms.

A note on good practice The mass to use in the calculation of the moment of inertia is the actual atomic mass, not the element's molar mass; don't forget to convert from relative masses to actual masses by using the atomic mass constant m_u.

The energy of a body free to rotate about three axes is

$$E = \tfrac{1}{2} I_x \omega_x^2 + \tfrac{1}{2} I_y \omega_y^2 + \tfrac{1}{2} I_z \omega_z^2 \qquad (11B.3)$$

Because the classical angular momentum about the axis q is $J_q = I_q \omega_q$, it follows that

$$E = \frac{J_x^2}{2I_x} + \frac{J_y^2}{2I_y} + \frac{J_z^2}{2I_z} \qquad \text{Rotational energy: classical expression} \qquad (11B.4)$$

Table 11B.1 Moments of inertia*

Molecule	Moment of inertia
1. *Diatomic molecules*	
	$I=\mu R^2 \quad \mu=\frac{m_A m_B}{m}$
2. *Triatomic linear rotors*	
	$I=m_A R^2+m_C R'^2-\frac{(m_A R-m_C R')^2}{m}$
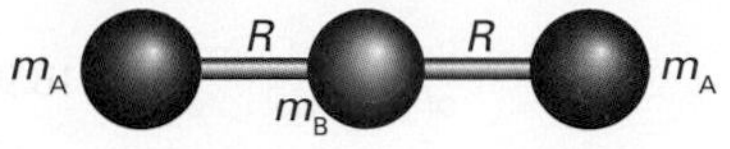	$I=2m_A R^2$
3. *Symmetric rotors*	
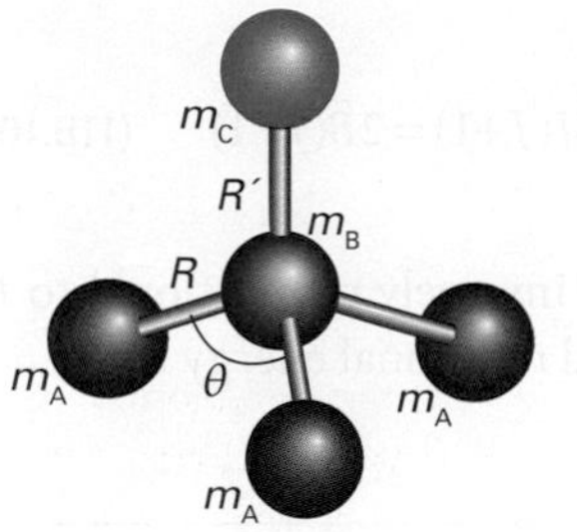	$I_\parallel=2m_A f_1(\theta)R^2$ $I_\perp=m_A f_1(\theta)R^2+\frac{m_A}{m}(m_B+m_C)f_2(\theta)R^2+\frac{m_C}{m}\{(3m_A+m_B)R'+6m_A R[\frac{1}{3}f_2(\theta)]^{1/2}\}R'$
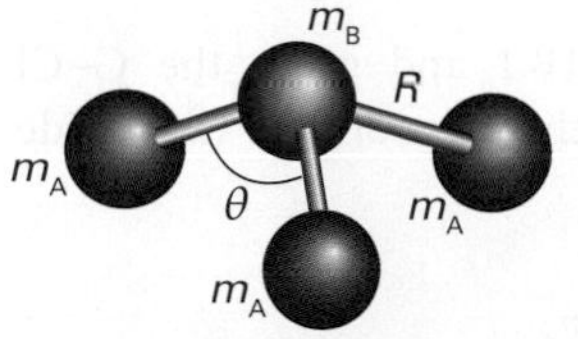	$I_\parallel=2m_A f_1(\theta)R^2$ $I_\perp=m_A f_1(\theta)R^2+\frac{m_A m_B}{m}f_2(\theta)R^2$
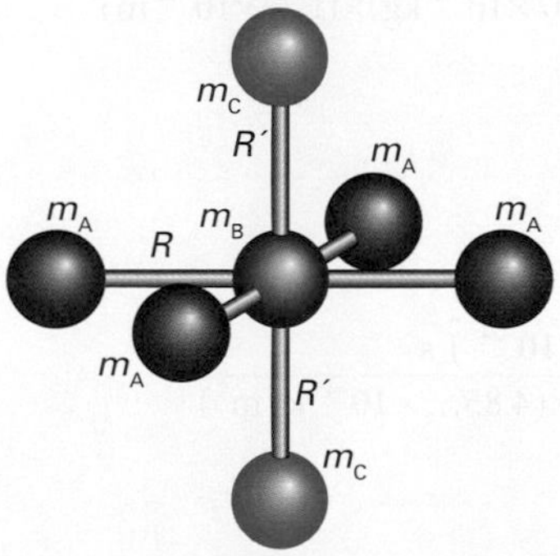	$I_\parallel=4m_A R^2$ $I_\perp=2m_A R^2+2m_C R'^2$
4. *Spherical rotors*	
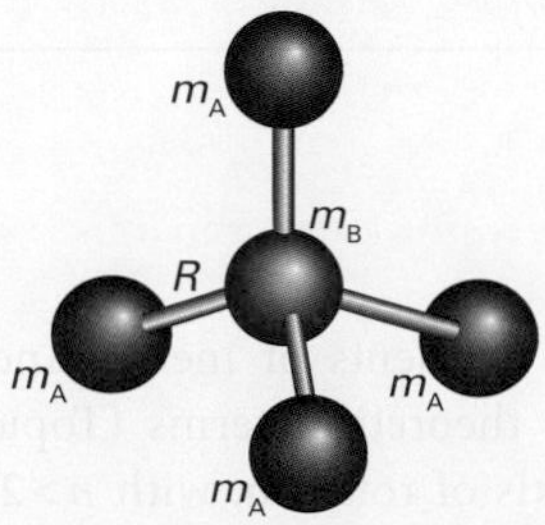	$I=\frac{8}{3}m_A R^2$
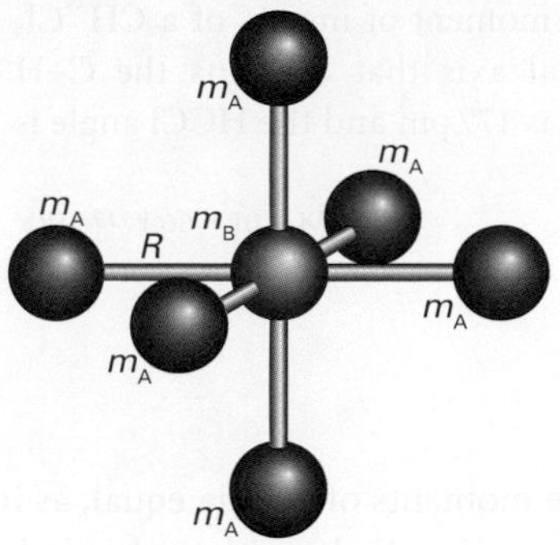	$I=4m_A R^2$

* $f_1(\theta)=1-\cos\theta$, $f_2(\theta)=1+2\cos\theta$; in each case, m is the total mass of the molecule.

Example 11B.1 Calculating the moment of inertia of a molecule

Calculate the moment of inertia of an H_2O molecule around the axis defined by the bisector of the HOH angle (**1**). The HOH bond angle is 104.5° and the OH bond length is 95.7 pm. Use $m(^1H) = 1.0078m_u$.

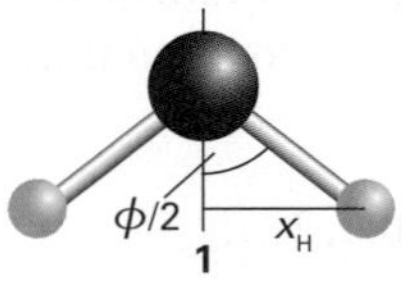

1

Collect your thoughts You can compute the moment of inertia using eqn 11B.2. In this equation the x_i are the perpendicular distances from each atom to the axis of rotation, and you will be able to calculate these distances by using trigonometry and the bond angle and bond length.

The solution From eqn 11B.2,

$$I=\sum_i m_i x_i^2 = m_H x_H^2 + 0 + m_H x_H^2 = 2m_H x_H^2$$

If the bond angle of the molecule is ϕ and the OH bond length is R, trigonometry gives $x_H = R\sin\frac{1}{2}\phi$. It follows that

$$I = 2m_H R^2 \sin^2\tfrac{1}{2}\phi$$

Substitution of the data gives

$$I = 2\times(1.0078\times1.6605\times10^{-27}\,\text{kg})\times(9.57\times10^{-11}\,\text{m})^2 \times\sin^2(\tfrac{1}{2}\times104.5°)$$
$$= 1.92\times10^{-47}\,\text{kg m}^2$$

Note that the mass of the O atom makes no contribution to the moment of inertia because the axis passes through this atom and so it does not move.

Self-test 11B.1 Calculate the moment of inertia of a $CH^{35}Cl_3$ molecule around a rotational axis that contains the C–H bond. The C–Cl bond length is 177 pm and the HCCl angle is 107°; $m(^{35}Cl) = 34.97m_u$.

Answer: $4.99 \times 10^{-45}\,\mathrm{kg\,m^2}$

(a) Spherical rotors

Spherical rotors have all three moments of inertia equal, as in CH_4 and SF_6. If these moments of inertia have the value I, the classical expression for the energy is

$$E = \frac{J_x^2 + J_y^2 + J_z^2}{2I} = \frac{\mathcal{J}^2}{2I} \qquad (11B.5)$$

where $\mathcal{J}^2$ is the square of the magnitude of the angular momentum. The corresponding quantum expression is generated by making the replacement

$$\mathcal{J}^2 \rightarrow J(J+1)\hbar^2 \quad J = 0, 1, 2, \ldots$$

where J is the angular momentum quantum number. Therefore, the energy of a spherical rotor is confined to the values

$$E_J = J(J+1)\frac{\hbar^2}{2I} \quad J = 0, 1, 2, \ldots \qquad \text{Energy levels of a spherical rotor} \qquad (11B.6)$$

The resulting ladder of energy levels is illustrated in Fig. 11B.2. The energy is normally expressed in terms of the **rotational constant**, $\tilde{B}$ (a wavenumber), of the molecule, where

$$hc\tilde{B} = \frac{\hbar^2}{2I} \quad \text{so} \quad \tilde{B} = \frac{\hbar}{4\pi cI} \qquad \text{Rotational constant [definition]} \qquad (11B.7)$$

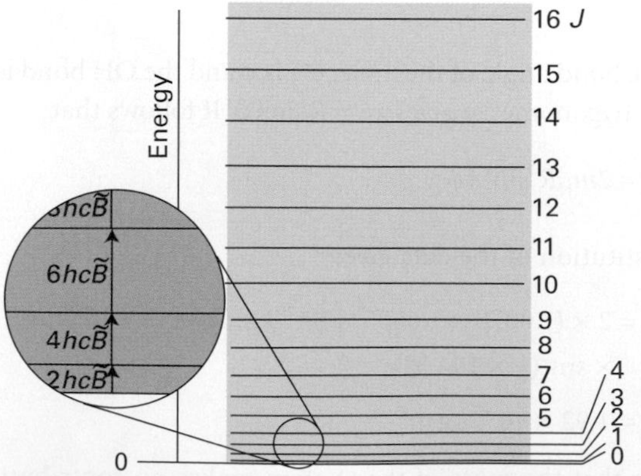

Figure 11B.2 The rotational energy levels of a linear or spherical rotor. Note that the energy separation between neighbouring levels increases as J increases.

The expression for the energy is then

$$E_J = hc\tilde{B}J(J+1) \quad J = 0, 1, 2, \ldots \qquad \text{Energy levels of a spherical rotor} \qquad (11B.8)$$

It is also common to express the rotational constant as a frequency and to denote it B. Then $B = \hbar/4\pi I$ and the energy is $E_J = hBJ(J+1)$. The two quantities are related by $B = c\tilde{B}$.

The energy of a rotational state is normally reported as the **rotational term**, $\tilde{F}(J)$, a wavenumber, by division of both sides of eqn 11B.8 by hc:

$$\tilde{F}(J) = \tilde{B}J(J+1) \qquad \text{Rotational terms of spherical rotor} \qquad (11B.9)$$

To express the rotational term as a frequency, use $F = c\tilde{F}$. The separation of adjacent terms is

$$\tilde{F}(J+1) - \tilde{F}(J) = \tilde{B}(J+1)(J+2) - \tilde{B}J(J+1) = 2\tilde{B}(J+1) \qquad (11B.10)$$

Because the rotational constant is inversely proportional to I, large molecules have closely spaced rotational energy levels.

Brief illustration 11B.1

Consider $^{12}C^{35}Cl_4$: from Table 11B.1 and given the C–Cl bond length ($R_{C-Cl} = 177$ pm) and the mass of the ^{35}Cl nuclide ($m(^{35}Cl) = 34.97m_u$), find

$$I = \tfrac{8}{3}m(^{35}\mathrm{Cl})R_{\mathrm{C-Cl}}^2 = \tfrac{8}{3} \times \overbrace{(5.807 \times 10^{-26}\,\mathrm{kg})}^{34.97m_u} \times (1.77 \times 10^{-10}\,\mathrm{m})^2$$
$$= 4.85\ldots \times 10^{-45}\,\mathrm{kg\,m^2}$$

and, from eqn 11B.7,

$$\tilde{B} = \frac{1.05457 \times 10^{-34}\,\overbrace{\mathrm{J}}^{\mathrm{kg\,m^2\,s^{-2}}}\,\mathrm{s}}{4\pi \times (2.998 \times 10^8\,\mathrm{m\,s^{-1}}) \times (4.85\ldots \times 10^{-45}\,\mathrm{kg\,m^2})}$$
$$= 5.77\,\mathrm{m^{-1}} = 0.0577\,\mathrm{cm^{-1}}$$

It follows from eqn 11B.10 that the separation between the $J = 0$ and $J = 1$ terms is $\tilde{F}(1) - \tilde{F}(0) = 2\tilde{B} = 0.1154\,\mathrm{cm^{-1}}$, corresponding to 3.46 GHz.

(b) Symmetric rotors

Symmetric rotors have two equal moments of inertia and a third that is non-zero. In group theoretical terms (Topic 10A), such rotors have an n-fold axis of rotation, with $n > 2$.

The unique axis of a symmetric rotor (such as CH_3Cl, NH_3, and C_6H_6) is its **principal axis** (or *figure axis*). If the moment of inertia about the principal axis is larger than the other two, the rotor is classified as **oblate** (like a pancake, and C_6H_6). If the moment of inertia around the principal axis is smaller than the other two, the rotor is classified as **prolate** (like a cigar, and CH_3Cl). The two equal moments of inertia (I_x and I_y) are denoted $I_\perp$ and I_z is denoted $I_\parallel$; then eqn 11B.4 becomes

$$E=\frac{J_x^2+J_y^2}{2I_\perp}+\frac{J_z^2}{2I_\parallel} \qquad (11B.11)$$

This expression can be written in terms of $\mathcal{J}^2=J_x^2+J_y^2+J_z^2$:

$$E=\frac{\mathcal{J}^2-J_z^2}{2I_\perp}+\frac{J_z^2}{2I_\parallel}=\frac{\mathcal{J}^2}{2I_\perp}+\left(\frac{1}{2I_\parallel}-\frac{1}{2I_\perp}\right)J_z^2 \qquad (11B.12)$$

The quantum expression is generated by replacing $\mathcal{J}^2$ by $J(J+1)\hbar^2$. The quantum theory of angular momentum (Topic 7F) also restricts the component of angular momentum about any axis to the values $K\hbar$, with $K=0, \pm1, \ldots, \pm J$. (The quantum number K is used to signify the component on the principal axis, as distinct from the quantum number M_J which is used to signify the component on an externally defined axis.) Then, after making the replacements $\mathcal{J}^2 \rightarrow J(J+1)\hbar^2$ and $J_z^2 \rightarrow K^2\hbar^2$ the rotational terms are

$$\tilde{F}(J,K)=\tilde{B}J(J+1)+(\tilde{A}-\tilde{B})K^2 \qquad \text{Rotational terms of a symmetric rotor} \qquad (11B.13a)$$
$$J=0, 1, 2, \ldots \quad K=0, \pm1, \ldots, \pm J$$

with

$$\tilde{A}=\frac{\hbar}{4\pi c I_\parallel} \qquad \tilde{B}=\frac{\hbar}{4\pi c I_\perp} \qquad (11B.13b)$$

Equation 11B.13a matches what is expected for the dependence of the energy levels on the two distinct moments of inertia of the molecule:

- When $K=0$, there is no component of angular momentum about the principal axis, and the energy levels depend only on $I_\perp$ (Fig. 11B.3a).
- When $K=\pm J$, almost all the angular momentum arises from rotation around the principal axis, and the energy levels are determined largely by $I_\parallel$ (Fig. 11B.3b).
- The sign of K does not affect the energy because opposite values of K correspond to opposite senses of rotation, and the energy does not depend on the sense of rotation.

Physical interpretation

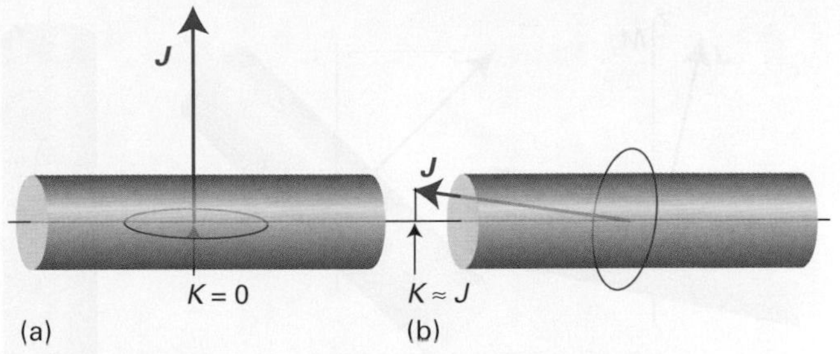

Figure 11B.3 The significance of the quantum number K. (a) When $K=0$ the molecule has no angular momentum about its principal axis: it is undergoing end-over-end rotation. (b) When $|K|$ is close to its maximum value, J, most of the molecular rotation is around the principal axis.

Example 11B.2 Calculating the rotational energy levels of a symmetric rotor

A $^{14}N^1H_3$ molecule is a symmetric rotor with bond length 101.2 pm and HNH bond angle 106.7°. Calculate its rotational terms.

Collect your thoughts Begin by calculating the moments of inertia by using the expressions given in Table 11B.1. Then use eqn 11B.13a to find the rotational terms. The rotational constants are found using eqn 11B.13b.

The solution Substitution of $m_A = 1.0078m_u$, $m_B = 14.0031m_u$, $R = 101.2$ pm, and $\theta = 106.7°$ into the second of the symmetric rotor expressions in Table 11B.1 gives $I_\parallel = 4.4128 \times 10^{-47}$ kg m^2 and $I_\perp = 2.8059 \times 10^{-47}$ kg m^2. The expressions in eqn. 11B.13b give $\tilde{A} = 6.344$ cm^{-1} and $\tilde{B} = 9.977$ cm^{-1}. It follows from eqn 11B.13a that

$$\tilde{F}(J,K)/\text{cm}^{-1}=9.977J(J+1)-3.633K^2$$

Multiplication by c converts $\tilde{F}(J,K)$ to a frequency, denoted $F(J,K)$:

$$F(J,K)/\text{GHz}=299.1J(J+1)-108.9K^2$$

For $J=1$, the energy needed for the molecule to rotate mainly about its principal axis ($K=\pm J$) is equivalent to 16.32 cm^{-1} (489.3 GHz), but end-over-end rotation ($K=0$) corresponds to 19.95 cm^{-1} (598.1 GHz).

Self-test 11B.2 A $^{12}C^1H_3{}^{35}Cl$ molecule has a C–Cl bond length of 178 pm, a C–H bond length of 111 pm, and an HCH angle of 110.5°. Identify whether the molecule is oblate or prolate, and calculate its rotational energy terms.

Answer: $I_\perp = 6.262 \times 10^{-46}$ kg m^2, $I_\parallel = 5.568 \times 10^{-47}$ kg m^2; prolate; $\tilde{A} = 5.0275$ cm^{-1} and $\tilde{B} = 0.4470$ cm^{-1}; $\tilde{F}(J,K)/\text{cm}^{-1} = 0.4470J(J+1) + 4.58K^2$

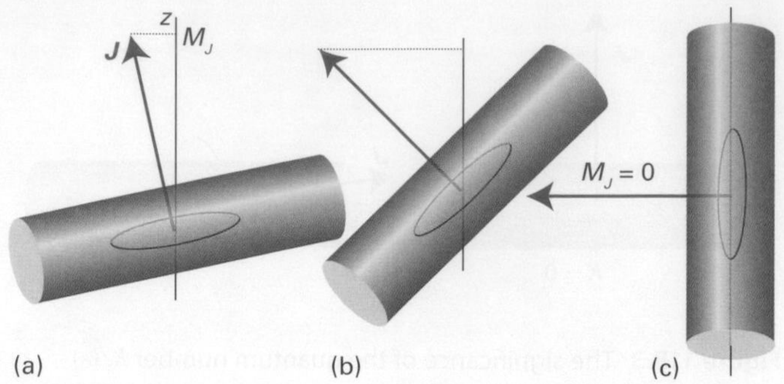

Figure 11B.4 The significance of the quantum number M_J. (a) When M_J is close to its maximum value, J, most of the molecular rotation is around the laboratory axis (taken as the z-axis). (b) An intermediate value of M_J. (c) When $M_J = 0$ the molecule has no angular momentum about the z-axis. All three diagrams correspond to a state with $K = 0$; there are corresponding diagrams for different values of K, in which the angular momentum makes a different angle to the principal axis of the molecule.

The energy of a symmetric rotor depends on J and K. Because the states with K and $-K$ have the same energy, each level, except those with $K = 0$, is doubly degenerate. In addition, the angular momentum of the molecule has a component on an external, laboratory-fixed axis. This component is quantized, and its permitted values are $M_J\hbar$, with $M_J = 0, \pm1, \ldots, \pm J$, giving $2J + 1$ values in all (Fig. 11B.4). The quantum number M_J does not appear in the expression for the energy, but it is necessary for a complete specification of the state of the rotor. Consequently, all $2J + 1$ orientations of the rotating molecule have the same energy. It follows that a symmetric rotor level is $2(2J + 1)$-fold degenerate for $K \neq 0$ and $(2J + 1)$-fold degenerate for $K = 0$.

A spherical rotor can be regarded as a version of a symmetric rotor in which $I_\perp = I_\parallel$ and therefore $\tilde{A} = \tilde{B}$. The quantum number K still takes any one of $2J + 1$ values, but the energy is independent of which value it takes. Therefore, as well as having a $(2J + 1)$-fold degeneracy arising from its orientation in space, the rotor also has a $(2J + 1)$-fold degeneracy arising from its orientation with respect to an arbitrary axis in the molecule. The overall degeneracy of a symmetric rotor energy level with quantum number J is therefore $(2J + 1)^2$. This degeneracy increases very rapidly: when $J = 10$, for instance, there are 441 states of the same energy.

(c) Linear rotors

For a linear rotor (such as CO_2, HCl, and C_2H_2), in which the atoms are regarded as mass points, the rotation occurs only about an axis perpendicular to the internuclear axis and there is no rotation around that axis. Therefore the component of angular momentum around the internuclear axis of a linear rotor is identically zero, and $K \equiv 0$ in eqn 11B.13a. The rotational terms of a linear molecule are therefore

$$\tilde{F}(J) = \tilde{B}J(J+1) \quad J = 0,1,2,\ldots \qquad \text{Rotational terms of linear rotor} \qquad (11B.14)$$

This expression is the same as eqn 11B.9 but arrived at it in a significantly different way: here $K \equiv 0$, but for a spherical rotor $\tilde{A} = \tilde{B}$ and K has a range of values. The angular momentum of a linear rotor has $2J + 1$ components on an external axis, so its degeneracy is just $2J + 1$ rather than the $(2J + 1)^2$-fold degeneracy of a spherical rotor.

Brief illustration 11B.2

Equation 11B.10 for the energy separation of adjacent levels of a spherical rotor also applies to linear rotors, so $\tilde{F}(3) - \tilde{F}(2) = 6\tilde{B}$. Spectroscopic measurements on $^1H^{35}Cl$ gives $\tilde{F}(3) - \tilde{F}(2) = 63.56\,\text{cm}^{-1}$, so it follows that $6\tilde{B} = 63.56\,\text{cm}^{-1}$, $\tilde{B} = 10.59\,\text{cm}^{-1}$, and therefore

$$I = \frac{\hbar}{4\pi c\tilde{B}} = \frac{1.05457\times10^{-34}\,\text{J s}}{4\pi\times(2.998\times10^{10}\,\text{cm s}^{-1})\times(10.59\,\text{cm}^{-1})}$$
$$= 2.643\times10^{-47}\,\text{kg m}^2$$

(d) Centrifugal distortion

In the discussion so far molecules have been treated as rigid rotors. However, the atoms of rotating molecules are subject to centrifugal forces which tend to distort the molecular geometry and change its moments of inertia (Fig. 11B.5). The effect of centrifugal distortion on a diatomic molecule is to stretch the bond and hence to increase the moment of inertia. As a result, the rotational constant is reduced and the energy levels are slightly closer together than the rigid-rotor expressions predict. The effect is usually taken into account by including in the energy expression a negative term that becomes more important as J increases:

$$\tilde{F}(J) = \tilde{B}J(J+1) - \tilde{D}_J J^2(J+1)^2 \qquad \text{Rotational terms affected by centrifugal distortion} \qquad (11B.15)$$

The parameter $\tilde{D}_J$ is the **centrifugal distortion constant.** The centrifugal distortion constant of a diatomic molecule is related to the vibrational wavenumber of the bond, $\tilde{\nu}$ (which, as seen in Topic 11C, is a measure of its stiffness), through the approximate relation (see Problem P11C.16)

$$\tilde{D}_J = \frac{4\tilde{B}^3}{\tilde{\nu}^2} \qquad \text{Centrifugal distortion constant} \qquad (11B.16)$$

As expected, a bond that is easily stretched, and therefore has a low vibrational wavenumber, has a high centrifugal distortion constant.

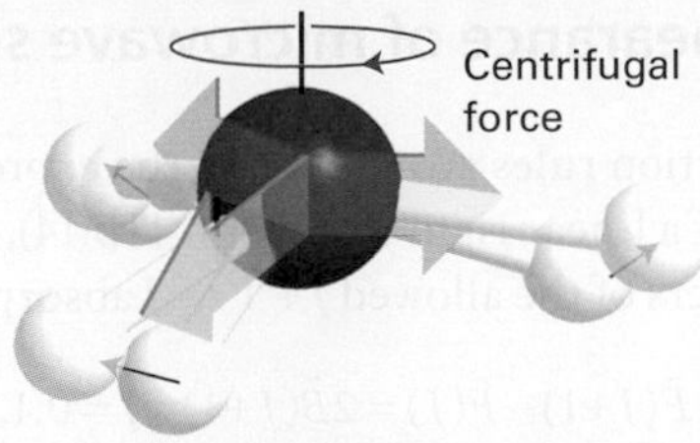

Figure 11B.5 The effect of rotation on a molecule. The centrifugal force arising from rotation distorts the molecule, opening out bond angles and stretching bonds slightly. The effect is to increase the moment of inertia of the molecule and hence to decrease its rotational constant.

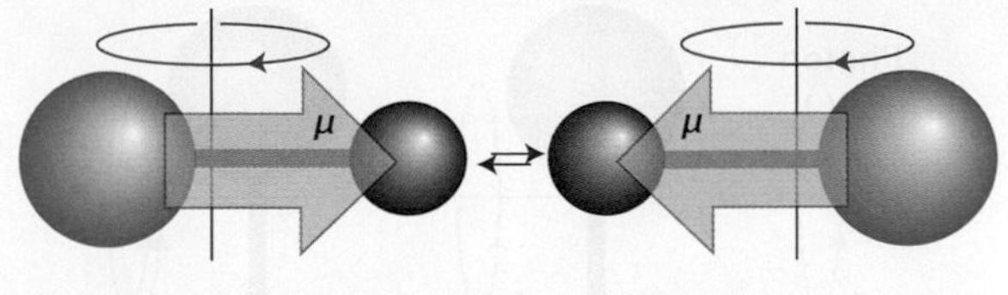

Figure 11B.6 To a stationary observer, a rotating polar molecule looks like an oscillating dipole which will generate an oscillating electromagnetic wave (or, in the case of absorption, interact with such a wave). This picture is the classical origin of the gross selection rule for rotational transitions.

Brief illustration 11B.3

For $^{12}C^{16}O$, $\tilde{B} = 1.931\,\text{cm}^{-1}$ and $\tilde{\nu} = 2170\,\text{cm}^{-1}$. It follows that

$$\tilde{D}_J = \frac{4\times(1.931\,\text{cm}^{-1})^3}{(2170\,\text{cm}^{-1})^2} = 6.116\times10^{-6}\,\text{cm}^{-1}$$

Because $\tilde{D}_J << \tilde{B}$, centrifugal distortion has a very small effect on the energy levels until J is large. For $J = 20$, $\tilde{D}_J J^2(J+1)^2 = 1.08\,\text{cm}^{-1}$ (corresponding to 32 GHz).

11B.2 Microwave spectroscopy

Typical values of the rotational constant $\tilde{B}$ for small molecules are in the region of 0.1–10 cm^{-1}; two examples are 0.356 cm^{-1} for NF_3 and 10.59 cm^{-1} for HCl. It follows that rotational transitions can be studied with **microwave spectroscopy**, a technique that monitors the absorption of radiation in the microwave region of the spectrum.

(a) Selection rules

As usual in spectroscopy, the selection rules can be established by considering the relevant transition dipole moment. The details of the calculation are shown in *A deeper look* 5 on the website of this text. The conclusion is that the gross selection rule for the observation of a pure rotational transition is that a molecule must have a permanent electric dipole moment. The classical basis of this rule is that a polar molecule appears to possess a fluctuating dipole when rotating, but a nonpolar molecule does not (Fig. 11B.6). The permanent dipole can be regarded as a handle with which the molecule stirs the electromagnetic field into oscillation (and vice versa for absorption).

Brief illustration 11B.4

Homonuclear diatomic molecules and nonpolar polyatomic molecules such as CO_2, $CH_2{=}CH_2$, and C_6H_6 do not give rise to microwave spectra. On the other hand, OCS and H_2O are polar and have microwave spectra. Spherical rotors cannot have electric dipole moments unless they become distorted by rotation, so they are rotationally inactive except in special cases. An example of a spherical rotor that does become sufficiently distorted for it to acquire a dipole moment is SiH_4, which has a dipole moment of about 8.3 μD by virtue of its rotation when $J \approx 10$ (for comparison, HCl has a permanent dipole moment of 1.1 D; molecular dipole moments and their units are discussed in Topic 14A).

The analysis also shows that, for a linear molecule, the transition moment vanishes unless the following conditions are fulfilled:

$$\Delta J = \pm 1 \quad \Delta M_J = 0, \pm 1 \qquad \text{Rotational selection rules: linear rotors} \quad (11B.17)$$

The transition $\Delta J = +1$ corresponds to absorption and the transition $\Delta J = -1$ corresponds to emission.

- The allowed change in J arises from the conservation of angular momentum when a photon, a spin-1 particle, is emitted or absorbed (Fig. 11B.7).
- The allowed change in M_J also arises from the conservation of angular momentum when a photon is emitted or absorbed in a specific direction.

Physical interpretation

When the transition moment is evaluated for all possible relative orientations of the molecule to the line of flight of the photon, it is found that the total $J + 1 \leftrightarrow J$ transition intensity is proportional to

$$|\mu_{J+1,J}|^2 = \left(\frac{J+1}{2J+1}\right)\mu_0^2 \qquad (11B.18)$$

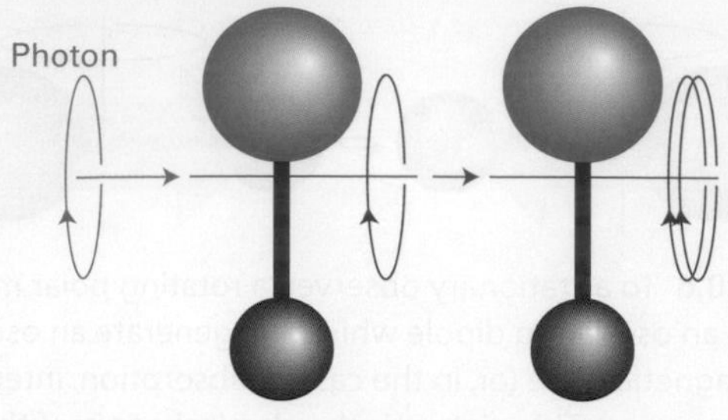

Figure 11B.7 When a photon is absorbed by a molecule, the angular momentum of the combined system is conserved. If the molecule is rotating in the same sense as the spin of the incoming photon, then J increases by 1.

where μ_0 is the permanent electric dipole moment of the molecule. The intensity is proportional to the square of μ_0, so strongly polar molecules give rise to much more intense rotational lines than less polar molecules.

Rotation of a symmetric rotor about its principal (figure) axis does not lead to any change in the orientation of the dipole; there is no fluctuating dipole to interact with the radiation, and therefore no change in K is possible. For symmetric rotors the selection rules are therefore:

$$\Delta J = \pm 1 \quad \Delta M_J = 0, \pm 1 \quad \Delta K = 0 \qquad \text{Rotational selection rules: symmetric rotors} \qquad (11B.19)$$

The degeneracy associated with the quantum number M_J (the orientation of the rotation in space) is partly removed when an electric field is applied to a polar molecule (Fig. 11B.8). The splitting of states by an electric field is called the **Stark effect**. The energy shift depends on the permanent electric dipole moment, μ_0, so the observation of the Stark effect in a rotational spectrum can be used to measure the magnitudes of electric dipole moments.

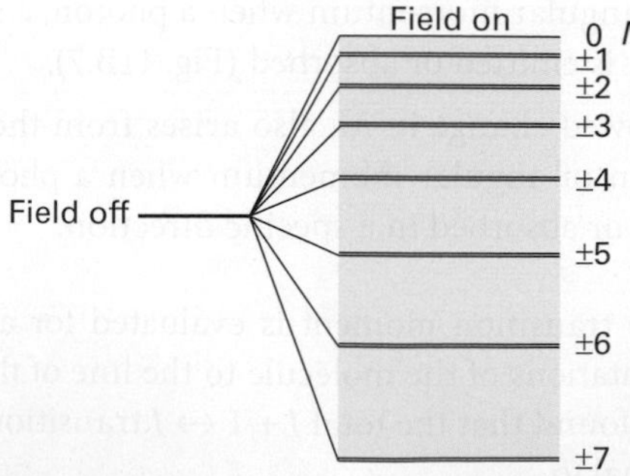

Figure 11B.8 The effect of an electric field on the energy level with $J = 7$ of a polar linear rotor. All levels are doubly degenerate except that with $M_J = 0$.

(b) The appearance of microwave spectra

When the selection rules are applied to the expressions for the energy levels of a linear rigid rotor (eqn 11B.14), it follows that the wavenumbers of the allowed $J+1 \leftarrow J$ absorptions are

$$\tilde{\nu}(J+1 \leftarrow J) = \tilde{F}(J+1) - \tilde{F}(J) = 2\tilde{B}(J+1) \quad J = 0,1,2,\ldots$$

$$\text{Wavenumbers of rotational transitions: linear rotor} \qquad (11B.20a)$$

When centrifugal distortion is taken into account, the corresponding expression obtained from eqn 11B.15 is

$$\tilde{\nu}(J+1 \leftarrow J) = 2\tilde{B}(J+1) - 4\tilde{D}_J(J+1)^3 \qquad (11B.20b)$$

However, because the second term is typically very small compared with the first, the appearance of the spectrum closely resembles that predicted from eqn 11B.20a.

Example 11B.3 Predicting the appearance of a rotational spectrum

Predict the form of the rotational spectrum of $^{14}NH_3$, which is an oblate symmetric rotor with $\tilde{B} = 9.977\,\text{cm}^{-1}$.

Collect your thoughts The rotational terms are given by eqn 11B.13a. Because $\Delta J = \pm 1$ and $\Delta K = 0$, the expression for the wavenumbers of the rotational transitions is identical to eqn 11B.20a and depends only on $\tilde{B}$.

The solution The following table can be drawn up for the $J + 1 \leftarrow J$ transitions.

J	0	1	2	3	…
$\tilde{\nu}/\text{cm}^{-1}$	19.95	39.91	59.86	79.82	…
ν/GHz	598.1	1196	1795	2393	…

The line spacing is $19.95\,\text{cm}^{-1}$ (598.1 GHz).

Self-test 11B.3 Repeat the problem for $CH_3{}^{35}Cl$, a prolate symmetric rotor for which $\tilde{B} = 0.444\,\text{cm}^{-1}$.

Answer: Lines of separation $0.888\,\text{cm}^{-1}$ (26.6 GHz)

The form of the spectrum predicted by eqn 11B.20a is shown in Fig. 11B.9. The most significant feature is that it consists of a series of lines with wavenumbers $2\tilde{B}$, $4\tilde{B}$, $6\tilde{B}$, … and of separation $2\tilde{B}$. The measurement of the line spacing therefore gives $\tilde{B}$, and hence the moment of inertia $I_\perp$ perpendicular to the principal axis of the molecule. Because the masses of the atoms are known, it is a simple matter to deduce the bond length of a diatomic molecule. However, in the case of a polyatomic molecule such as OCS or NH_3, a knowledge of one moment of inertia is insufficient data from which to infer, for example, the two bond lengths in OCS, or the bond length and bond angle in NH_3.

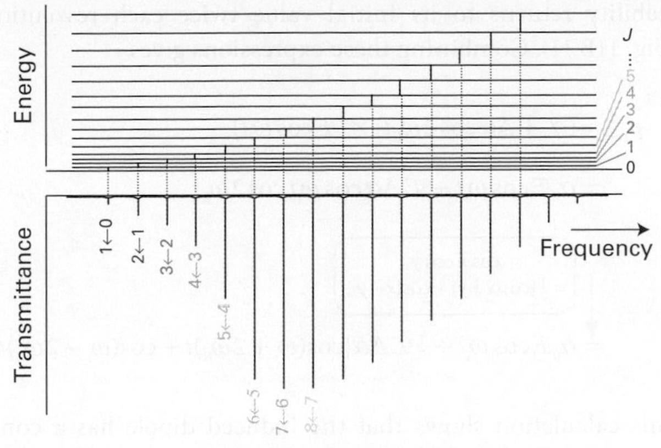

Figure 11B.9 The rotational energy levels of a linear rotor, the transitions allowed by the selection rule $\Delta J = +1$, and a typical pure rotational absorption spectrum (displayed here in terms of the radiation transmitted through the sample). The intensities reflect the populations of the initial level in each case and the strengths of the transition dipole moments.

This difficulty can be overcome by measuring the spectra of **isotopologues**, isotopically substituted molecules. The spectrum from each isotopologue gives a separate moment of inertia and, if it is assumed that the bond lengths and angles are unaffected by isotopic substitution, the extra data make it possible to extract values of the bond lengths and angles. A good example of this procedure is the study of OCS; the actual calculation is worked through in Problem P11B.7. The assumption that bond lengths are unchanged in isotopologues is only an approximation, but it is a good one in most cases. Nuclear spin (Topic 12A), which differs from one isotope to another, also affects the appearance of high-resolution rotational spectra because spin is a source of angular momentum and can couple with the rotation of the molecule itself and hence affect the rotational energy levels.

The intensities of spectral lines increase with increasing J and pass through a maximum before tailing off as J becomes large. The most important reason for this behaviour is the existence of a maximum in the population of rotational levels. The Boltzmann distribution (see the *Prologue* to this text and Topic 13A) implies that the population of a state decreases exponentially as its energy increases. However, the population of a *level* is also proportional to its degeneracy, and in the case of rotational levels this degeneracy increases with J. These two opposite trends result in the population of the energy levels (as distinct from the individual states) passing through a maximum. Specifically, the population N_J of a rotational energy level J is given by the Boltzmann expression

$$N_J \propto N g_J e^{-E_J/kT}$$

where N is the total number of molecules in the sample and g_J is the degeneracy of the level J. The value of J corresponding to a maximum of this expression is found by treating J as a continuous variable, differentiating with respect to J, and then setting the result equal to zero. The result for a linear rotor (see Problem P11B.11) is

$$J_{\max} \approx \left(\frac{kT}{2hc\tilde{B}}\right)^{1/2} - \tfrac{1}{2} \qquad \text{Rotational level with largest population: linear rotor} \qquad (11B.21)$$

For a typical molecule (e.g. OCS, with $\tilde{B} = 0.2\,\text{cm}^{-1}$) $kT/2hc\tilde{B} \approx 500$ at room temperature, so $J_{\max} \approx 22$. However, it must be recalled that the transition dipole moment depends on the value of J (eqn 11B.18) and, because the radiation can also cause stimulated emission (Topic 11A), the intensity also depends on the population *difference* between the two states involved in the transition. Hence the value of J corresponding to the most intense line is not quite the same as the value of J for the most highly populated level.

11B.3 Rotational Raman spectroscopy

Raman scattering (Topic 11A) can also arise as a result of rotational transitions. The gross selection rule for rotational Raman transitions is *that the molecule must be anisotropically polarizable.* To understand this criterion it is necessary to know that the distortion of a molecule in an electric field is determined by its polarizability, α (Topic 14A). More precisely, if the strength of the field is $\mathcal{E}$, then the molecule acquires an induced dipole moment of magnitude

$$\mu = \alpha\mathcal{E} \qquad (11B.22)$$

in addition to any permanent dipole moment it might have. An atom is isotropically polarizable: that is, the same distortion is induced whatever the direction of the applied field. The polarizability of a spherical rotor is also isotropic. However, non-spherical rotors have polarizabilities that do depend on the direction of the field relative to the molecule, so these molecules are anisotropically polarizable (Fig. 11B.10). The electron distribution in H_2, for example, is more distorted when the field is applied parallel to the bond than when it is applied perpendicular to it, and so $\alpha_{\parallel} > \alpha_{\perp}$.

All linear molecules, including both heteronuclear and homonuclear diatomics, have anisotropic polarizabilities and so are rotationally Raman active. This activity is one reason for the importance of rotational Raman spectroscopy, because the technique can be used to study many of the molecules that are inaccessible to microwave spectroscopy. Spherical rotors such as CH_4 and SF_6, however, are rotationally Raman inactive as well as microwave inactive. This inactivity does not mean that such molecules are never found in rotationally excited states. Molecular collisions do not have to obey such

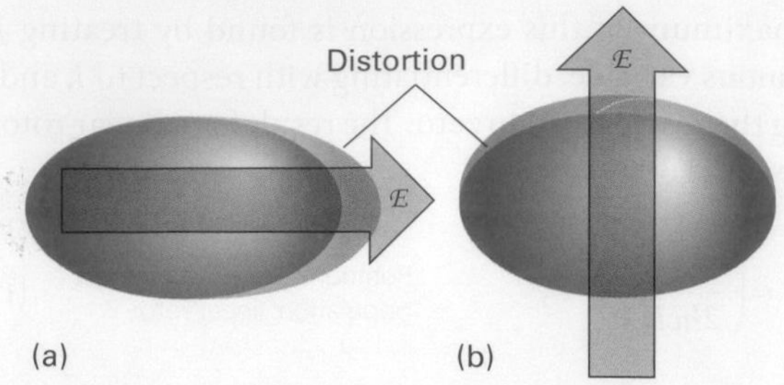

Figure 11B.10 An electric field $\mathcal{E}$ applied to a molecule results in its distortion, and the distorted molecule acquires a contribution to its dipole moment (even if it is nonpolar initially). The polarizability may be different when the field is applied (a) parallel or (b) perpendicular to the molecular axis (or, in general, in different directions relative to the molecule); if that is so, then the molecule has an anisotropic polarizability.

restrictive selection rules, and hence collisions between molecules can result in the population of any rotational state.

As usual, to establish the selection rules, it is necessary to consider the transition dipole moment. The full calculation can be found in *A deeper look* 5 on the website of this text and leads to the conclusion that the specific rotational Raman selection rules are

Linear rotors: $\Delta J = 0, \pm 2$

Symmetric rotors: $\Delta J = 0, \pm 1, \pm 2$ $\quad \Delta K = 0$ Rotational Raman selection rules (11B.23)

The $\Delta J = 0$ transitions do not lead to a shift in frequency of the scattered photon and therefore contribute to the unshifted radiation (the Rayleigh radiation, Topic 11A). A classical argument can be used to give physical insight into the quantum mechanical calculation.

How is that done? 11B.1 Justifying the rotational Raman selection rules

The incident electric field, $\mathcal{E}$, of a wave of electromagnetic radiation of frequency ω_i induces a molecular dipole moment given by

$$\mu_{\text{ind}} = \alpha \mathcal{E}(t) = \alpha \mathcal{E} \cos \omega_i t$$

If the molecule is rotating at an angular frequency ω_R, it appears to an external observer that the polarizability is also time dependent (if it is anisotropic). This dependence can be written

$$\alpha = \alpha_0 + \Delta\alpha \cos 2\omega_R t$$

where $\Delta\alpha = \alpha_\parallel - \alpha_\perp$ and α ranges from $\alpha_0 + \Delta\alpha$ to $\alpha_0 - \Delta\alpha$ as the molecule rotates. The $2\omega_R$ appears because the polarizability returns to its initial value twice each revolution (Fig. 11B.11). Combining these expressions gives

$$\mu_{\text{ind}} = (\alpha_0 + \Delta\alpha \cos 2\omega_R t) \times (\mathcal{E} \cos \omega_i t)$$
$$= \alpha_0 \mathcal{E} \cos \omega_i t + \mathcal{E}\, \Delta\alpha \cos \omega_i t \cos 2\omega_R t$$

$\cos x \cos y = \tfrac{1}{2}\{\cos(x+y) + \cos(x-y)\}$

$$= \alpha_0 \mathcal{E} \cos \omega_i t + \tfrac{1}{2} \mathcal{E}\, \Delta\alpha \{\cos(\omega_i + 2\omega_R)t + \cos(\omega_i - 2\omega_R)t\}$$

This calculation shows that the induced dipole has a component oscillating at the incident frequency (which results in Rayleigh scattering), and that it also has components at $\omega_i \pm 2\omega_R$, which give rise to the shifted Raman lines. These lines appear only if $\Delta\alpha \neq 0$; hence the polarizability must be anisotropic for there to be Raman lines. This is the gross selection rule for rotational Raman spectroscopy.

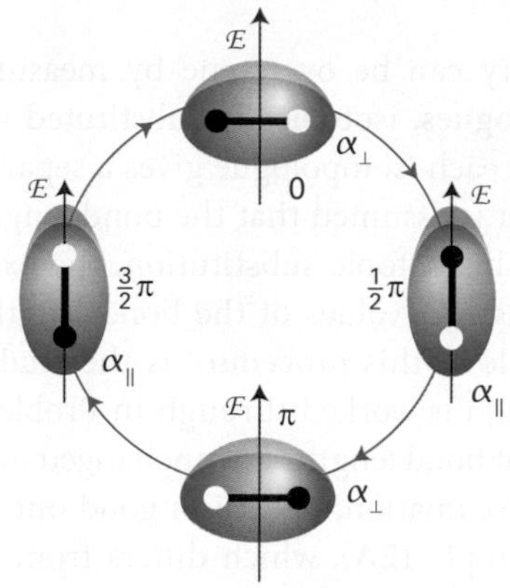

Figure 11B.11 The distortion induced in a molecule by an applied electric field returns the polarizability to its initial value after a rotation of only 180° (i.e. twice a revolution). This is the origin of the $\Delta J = \pm 2$ selection rule in rotational Raman spectroscopy.

The distortion induced in the molecule by the incident electric field returns to its initial value after a rotation of 180° (i.e. twice a revolution). This is the classical origin of the specific selection rule $\Delta J = \pm 2$.

To predict the form of the Raman spectrum of a linear rotor the selection rule $\Delta J = \pm 2$ is applied to the rotational energy levels (Fig. 11B.12). For Stokes lines, $\Delta J = +2$ and the scattered radiation is at a lower wavenumber than the incident radiation at $\tilde{\nu}_i$, the shift being the difference $\tilde{F}(J+2) - \tilde{F}(J)$

$$\tilde{\nu}(J+2 \leftarrow J) = \tilde{\nu}_i - \{\tilde{F}(J+2) - \tilde{F}(J)\}$$
$$= \tilde{\nu}_i - 2\tilde{B}(2J+3)$$

Wavenumbers of Stokes lines: linear rotor (11B.24a)

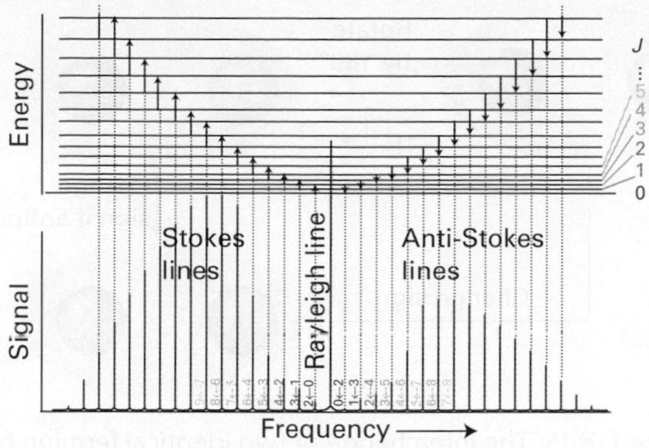

Figure 11B.12 The rotational energy levels of a linear rotor and the transitions allowed by the $\Delta J = \pm 2$ Raman selection rules. The form of a typical rotational Raman spectrum is also shown. In practice the Rayleigh line is much stronger than depicted in the figure.

For anti-Stokes lines $\Delta J = -2$ and the scattered radiation is at a higher wavenumber, the shift being the difference $\tilde{F}(J) - \tilde{F}(J-2)$:

$$\tilde{\nu}(J-2 \leftarrow J) = \tilde{\nu}_i + \{\tilde{F}(J) - \tilde{F}(J-2)\} = \tilde{\nu}_i + 2\tilde{B}(2J-1) \quad \text{Wavenumbers of anti-Stokes lines: linear rotor} \quad (11B.24b)$$

The Stokes lines appear to low frequency of the incident radiation and at displacements $6\tilde{B}$, $10\tilde{B}$, $14\tilde{B}$, ... from $\tilde{\nu}_i$ for $J = 0, 1, 2, \ldots$. The anti-Stokes lines occur at displacements of $6\tilde{B}$, $10\tilde{B}$, $14\tilde{B}$, ... (for $J = 2, 3, 4, \ldots$; $J = 2$ is the lowest state that can contribute under the selection rule $\Delta J = -2$) to high frequency of the incident radiation. The separation of adjacent lines in both the Stokes and the anti-Stokes regions is $4\tilde{B}$, so from the spacing $I_\perp$ can be determined and then used to find the bond length exactly as in the case of microwave spectroscopy.

Example 11B.4 Predicting the form of a Raman spectrum

Predict the form of the rotational Raman spectrum of $^{14}N_2$, for which $\tilde{B} = 1.99\,\text{cm}^{-1}$, when it is exposed to 336.732 nm laser radiation.

Collect your thoughts The molecule is rotationally Raman active because end-over-end rotation modulates its polarizability as viewed by a stationary observer. The wavenumbers of the Stokes and anti-Stokes lines are given by eqn 11B.24.

The solution The incident radiation with wavelength 336.732 nm corresponds to a wavenumber of $\tilde{\nu}_i = 29\,697.2\,\text{cm}^{-1}$; eqns 11B.24a and 11B.24b give the following line positions:

J	0	1	2	3
Stokes lines				
$\tilde{\nu}/\text{cm}^{-1}$	29 685.3	29 677.3	29 669.3	29 661.4
λ/nm	336.867	336.958	337.048	337.139
Anti-Stokes lines				
$\tilde{\nu}/\text{cm}^{-1}$			29 709.1	29 717.1
λ/nm			336.597	336.507

There will be a strong central line at 336.732 nm accompanied on either side by lines of increasing and then decreasing intensity (as a result of transition moment and population effects). The spread of the entire spectrum is very small, so the incident light must be highly monochromatic.

Self-test 11B.4 Repeat the calculation for the rotational Raman spectrum of $^{35}Cl_2$ ($\tilde{B} = 0.9752\,\text{cm}^{-1}$).

Answer: Stokes lines at 29 691.3, 29 687.4, 29 683.5, 29 679.6 cm^{-1}, anti-Stokes lines at 29 703.1, 29 707.0 cm^{-1}

11B.4 Nuclear statistics and rotational states

If eqn 11B.24 is used to analyse the rotational Raman spectrum of $C^{16}O_2$, the rotational constant derived from the spacing of the lines is inconsistent with other measurements of C–O bond lengths. The results are consistent if it is supposed that the molecule can exist in states with only even values of J, so the observed Stokes lines are $2 \leftarrow 0$, $4 \leftarrow 2, \ldots$; the lines $3 \leftarrow 1$, $5 \leftarrow 3, \ldots$ are missing.

The explanation of the missing lines lies in the Pauli principle (Topic 8B) and the fact that ^{16}O nuclei are spin-0 bosons: just as the Pauli principle excludes certain electronic states, so too does it exclude certain molecular rotational states. The Pauli principle states that, when two identical bosons are exchanged, the overall wavefunction must remain unchanged. When a $C^{16}O_2$ molecule rotates through 180°, two identical ^{16}O nuclei are interchanged, so the overall wavefunction of the molecule must remain unchanged. However, inspection of the form of the rotational wavefunctions (which have the same angular dependence as the s, p, etc. orbitals of atoms) shows that they change sign by $(-1)^J$ under such a rotation (Fig. 11B.13). Therefore, only even values of J are permissible for $C^{16}O_2$, and hence the Raman spectrum shows only alternate lines.

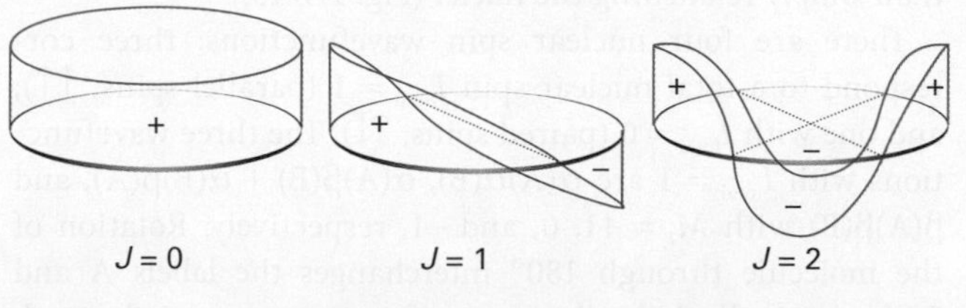

Figure 11B.13 The symmetries of rotational wavefunctions (shown here, for simplicity as a two-dimensional rotor) under a rotation through 180° depend on the value of J. Wavefunctions with J even do not change sign; those with J odd do change sign.

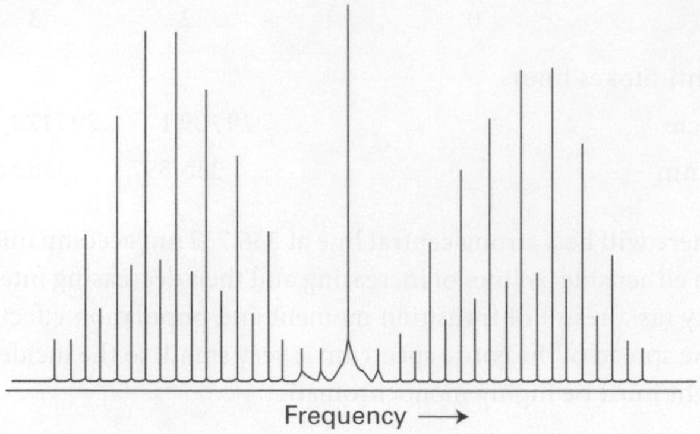

Figure 11B.14 The rotational Raman spectrum of a homonuclear diatomic molecule with two identical spin-$\frac{1}{2}$ nuclei shows an alternation in intensity as a result of nuclear statistics. In practice the Rayleigh line is much stronger than depicted in the figure.

The selective existence of rotational states that stems from the Pauli principle is termed **nuclear statistics**. Nuclear statistics must be taken into account whenever a rotation interchanges equivalent nuclei. However, the consequences are not always as simple as for $C^{16}O_2$ because there are complicating features when the nuclei have non-zero spin: it is found that there are several different relative nuclear spin orientations consistent with even values of J and a different number of spin orientations consistent with odd values of J. For $^1H_2$ and $^{19}F_2$, which have two identical spin-$\frac{1}{2}$ nuclei, by using the Pauli principle it can be shown that there are three times as many ways of achieving a state with odd J than with even J, and there is a corresponding 3:1 alternation in intensity in their rotational Raman spectra (Fig. 11B.14).

How is that done? 11B.2 Identifying the effect of nuclear statistics

Because ^{1}H nuclei have $I=\frac{1}{2}$, like electrons, they are fermions and the Pauli principle requires the overall wavefunction to change sign under particle interchange. However, the rotation of a $^1H_2$ molecule through 180° has a more complicated effect than simply relabelling the nuclei (Fig. 11B.15).

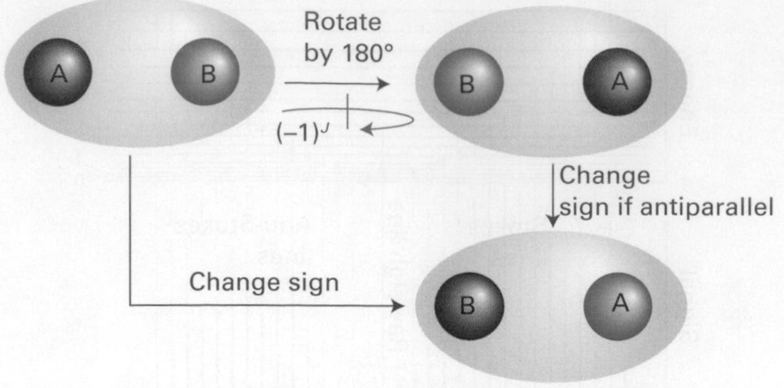

Figure 11B.15 The interchange of two identical fermion nuclei results in the change in sign of the overall wavefunction. The relabelling can be thought of as occurring in two steps: the first is a rotation of the molecule; the second is the interchange of unlike spins (represented by the different colours of the nuclei). The wavefunction changes sign in the second step if the nuclei have antiparallel spins.

There are four nuclear spin wavefunctions: three correspond to a total nuclear spin $I_{total}=1$ (parallel spins, ↑↑); and one with $I_{total}=0$ (paired spins, ↑↓). The three wavefunctions with $I_{total}=1$ are $\alpha(A)\alpha(B)$, $\alpha(A)\beta(B)+\alpha(B)\beta(A)$, and $\beta(A)\beta(B)$ with $M_I=+1$, 0, and -1, respectively. Rotation of the molecule through 180° interchanges the labels A and B, but overall these three wavefunctions are unchanged. Therefore, to achieve an overall change of sign, the rotational wavefunction must change sign, and so only odd values of J are allowed.

The fourth wavefunction, with $I_{total}=0$ and $M_I=0$, is $\alpha(A)\beta(B)-\alpha(B)\beta(A)$. When the labels A and B are interchanged the nuclear spin wavefunction changes sign: $\alpha(A)\beta(B)-\alpha(B)\beta(A)\rightarrow\alpha(B)\beta(A)-\alpha(A)\beta(B)\equiv-\{\alpha(A)\beta(B)-\alpha(B)\beta(A)\}$. Therefore, in this case for the overall wavefunction to change sign requires that the rotational wavefunction not change sign. Hence, only even values of J are allowed.

The analysis leads to the conclusion that there are three nuclear spin wavefunctions that can be combined with odd values of J, and one wavefunction that can be combined with even values of J. In accord with the prediction of eqn 11B.25, the ratio of the number of ways of achieving odd J to even J is 3:1. In general, for a homonuclear diatomic molecule with nuclei of spin I, the numbers of ways of achieving states of odd and even J are in the ratio

$$\frac{\text{Number of ways of achieving odd } J}{\text{Number of ways of achieving even } J}=\begin{cases}(I+1)/I & \text{for half-integral spin nuclei}\\ I/(I+1) & \text{for integral spin nuclei}\end{cases}$$

Nuclear statistics: homonuclear diatomics (11B.25)

For $^1H_2$, $I=\frac{1}{2}$ and the ratio is 3:1. For $^{14}N_2$, with $I=1$ the ratio is 1:2. Additional complications arise when the electronic state of the molecule is not totally symmetric (as for O_2, Topic 11F).

Nuclear statistics have consequences outside spectroscopy. Different relative nuclear spin orientations change into one another only very slowly, so a $^1H_2$ molecule with parallel nuclear spins remains distinct from one with paired nuclear spins for long periods. The form with parallel nuclear spins is called ***ortho*-hydrogen** and the form with paired nuclear spins is called ***para*-hydrogen**. Because *ortho*-hydrogen cannot exist in a state with $J=0$, it continues to rotate at very low temperatures and has an effective rotational zero-point energy.

Checklist of concepts

☐ 1. A **rigid rotor** is a body that does not distort under the stress of rotation.

☐ 2. Rigid rotors are classified as **spherical, symmetric, linear**, or **asymmetric** by noting the number of equal principal moments of inertia (or their symmetry).

☐ 3. Symmetric rotors are classified as **prolate** or **oblate**.

☐ 4. **Centrifugal distortion** arises from forces that change the geometry of a molecule.

☐ 5. The **gross selection rule** for a molecule to give a pure rotational spectrum is that it must be polar.

☐ 6. The **specific selection rules** for microwave spectroscopy are $\Delta J = \pm 1$, $\Delta M_J = 0, \pm 1$; for symmetric rotors the additional rule $\Delta K = 0$ also applies.

☐ 7. A molecule must be **anisotropically polarizable** for it to give rise to rotational Raman scattering.

☐ 8. The **specific selection rules** for rotational Raman spectroscopy are: (i) linear rotors, $\Delta J = 0, \pm 2$; (ii) symmetric rotors, $\Delta J = 0, \pm 1, \pm 2$; $\Delta K = 0$.

☐ 9. The appearance of rotational spectra is affected by **nuclear statistics**, the selective occupation of rotational states that stems from the Pauli principle.

Checklist of equations

Property	Equation	Comment	Equation number
Moment of inertia	$I = \sum_i m_i x_i^2$	x_i is the perpendicular distance of atom i from the axis of rotation	11B.2
Rotational terms of a spherical or linear rotor	$\tilde{F}(J) = \tilde{B}J(J+1)$	$J = 0, 1, 2, \ldots$ $\tilde{B} = \hbar/4\pi cI$	11B.9, 11B.14
Rotational terms of a symmetric rotor	$\tilde{F}(J,K) = \tilde{B}J(J+1) + (\tilde{A}-\tilde{B})K^2$	$J = 0, 1, 2, \ldots$ $K = 0, \pm 1, \ldots, \pm J$ $\tilde{A} = \hbar/4\pi cI_{\parallel}$ $\tilde{B} = \hbar/4\pi cI_{\perp}$	11B.13a 11B.13b
Centrifugal distortion	$\tilde{F}(J) = \tilde{B}J(J+1) - \tilde{D}_J J^2(J+1)^2$	Spherical or linear rotor	11B.15
Centrifugal distortion constant	$\tilde{D}_J = 4\tilde{B}^3/\tilde{\nu}^2$		11B.16
Wavenumbers of rotational transitions	$\tilde{\nu}(J+1 \leftarrow J) = 2\tilde{B}(J+1)$	$J = 0, 1, 2, \ldots$; linear rigid rotors	11B.20a
Rotational state with largest population	$J_{\max} \approx (kT/2hc\tilde{B})^{1/2} - \frac{1}{2}$	Linear rotors	11B.21
Wavenumbers of (i) Stokes and (ii) anti-Stokes lines in the rotational Raman spectrum of linear rotors	(i) $\tilde{\nu}(J+2 \leftarrow J) = \tilde{\nu}_i - 2\tilde{B}(2J+3)$ (ii) $\tilde{\nu}(J-2 \leftarrow J) = \tilde{\nu}_i + 2\tilde{B}(2J-1)$	(i) $J = 0, 1, 2, \ldots$ (ii) $J = 2, 3, 4, \ldots$	11B.24a 11B.24b

TOPIC 11C Vibrational spectroscopy of diatomic molecules

➤ **Why do you need to know this material?**

The observation of vibrational transition frequencies is used to determine the strengths and rigidities of bonds. Measurements in the gas phase can also be used to measure the bond lengths of diatomic molecules.

➤ **What is the key idea?**

The vibrational spectrum of a diatomic molecule can be interpreted by using the harmonic oscillator model, with modifications that account for bond dissociation and the coupling of rotational and vibrational motion.

➤ **What do you need to know already?**

You need to be familiar with the harmonic oscillator (Topic 7E) and rigid rotor (Topic 11B) models of molecular motion and the general principles of spectroscopy (Topic 11A).

One internal mode of motion of a diatomic molecule is its vibration, in which the internuclear separation increases and decreases periodically. This motion, and the transitions between the allowed quantum states, can be treated initially as an example of harmonic motion like that described in Topic 7E.

11C.1 Vibrational motion

Figure 11C.1 shows a typical potential energy curve of a diatomic molecule (it is essentially a reproduction of Fig. 7E.1 of Topic 7E). The potential energy $V(x)$, where $x = R - R_e$ (the displacement from equilibrium), can be expanded around its minimum by using a Taylor series (see *The chemist's toolkit* 12 in Topic 5B):

$$V(x)=V(0)+\left(\frac{\mathrm{d}V}{\mathrm{d}x}\right)_0 x+\tfrac{1}{2}\left(\frac{\mathrm{d}^2V}{\mathrm{d}x^2}\right)_0 x^2+\cdots \qquad (11\text{C}.1\text{a})$$

The notation $(\ldots)_0$ means that the derivative is evaluated at $x = 0$. The term $V(0)$ can be set arbitrarily to zero, and the first derivative of V is zero at the minimum. Therefore, the first surviving term is proportional to the square of the displacement. For small displacements all the higher terms can be ignored so the potential energy can be written

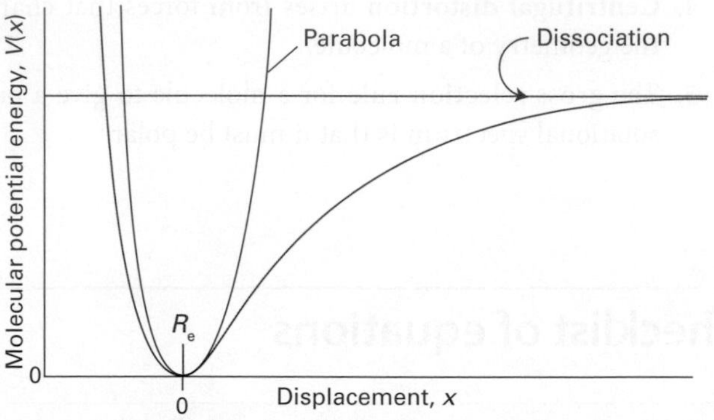

Figure 11C.1 A molecular potential energy curve can be approximated by a parabola near the bottom of the well. The parabolic potential energy results in harmonic oscillations. At high excitation energies the parabolic approximation is poor (the true potential energy is less confining), and is totally wrong near the dissociation limit.

$$V(x)\approx\tfrac{1}{2}\left(\frac{\mathrm{d}^2V}{\mathrm{d}x^2}\right)_0 x^2 \qquad (11\text{C}.1\text{b})$$

Therefore, the first approximation to a molecular potential energy curve is a parabolic potential of the form

$$V(x)=\tfrac{1}{2}k_f x^2 \qquad x=R-R_e \qquad \text{Parabolic potential energy} \qquad (11\text{C}.2\text{a})$$

where k_f is the **force constant** of the bond, a measure of its stiffness:

$$k_f=\left(\frac{\mathrm{d}^2V}{\mathrm{d}x^2}\right)_0 \qquad \text{Force constant [definition]} \qquad (11\text{C}.2\text{b})$$

If the potential energy curve is sharply curved close to its minimum, then k_f will be large and the bond stiff. Conversely, if the potential energy curve is wide and shallow, then k_f will be small and the bond easily stretched or compressed (Fig. 11C.2).

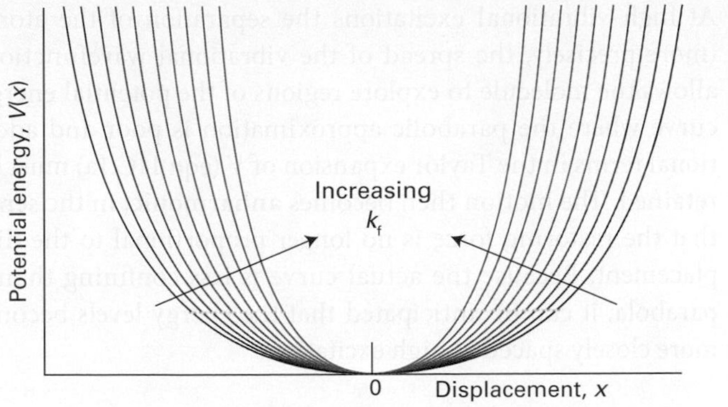

Figure 11C.2 The force constant is a measure of the curvature of the potential energy close to the equilibrium extension of the bond. A strongly confining well (one with steep sides, a stiff bond) corresponds to high values of k_f.

The Schrödinger equation for the relative motion of two atoms of masses m_1 and m_2 with a parabolic potential energy is

$$-\frac{\hbar^2}{2m_{\text{eff}}}\frac{d^2\psi}{dx^2}+\tfrac{1}{2}k_f x^2\psi=E\psi \quad (11C.3a)$$

where m_{eff} is the **effective mass**:

$$m_{\text{eff}}=\frac{m_1m_2}{m_1+m_2} \quad \text{Effective mass [definition]} \quad (11C.3b)$$

These equations are derived by using the separation of variables procedure (see *A deeper look* 3 on the website for this text) to separate the relative motion of the atoms from the motion of the molecule as a whole.

A note on good practice Distinguish *effective mass* from *reduced mass*. The former is a measure of the mass that is moved during a vibration. The latter is the quantity that emerges from the separation of relative internal and overall translational motion. For a diatomic molecule the two are the same, but that is not true in general for vibrations of polyatomic molecules. Many, however, do not make this distinction and refer to both quantities as the 'reduced mass'.

Apart from the appearance of the effective mass, the Schrödinger equation in eqn 11C.3a is the same as eqn 7E.2 for a particle of mass m undergoing harmonic motion. Therefore, the results from that Topic can be used to write the permitted vibrational energy levels:

$$E_v=(v+\tfrac{1}{2})\hbar\omega \qquad \omega=\left(\frac{k_f}{m_{\text{eff}}}\right)^{1/2} \qquad v=0,1,2,\ldots$$

Vibrational energy levels [diatomic molecule] (11C.4a)

The **vibrational terms** of a molecule, the energies of its vibrational states expressed as wavenumbers, are denoted $\tilde{G}(v)$, with $E_v = hc\tilde{G}(v)$. Therefore, with $\omega=2\pi\nu$ and $\tilde{\nu}=\nu/c=\omega/2\pi c$:

$$\tilde{G}(v)=(v+\tfrac{1}{2})\tilde{\nu} \qquad \tilde{\nu}=\frac{1}{2\pi c}\left(\frac{k_f}{m_{\text{eff}}}\right)^{1/2}$$

Vibrational terms [diatomic] (11C.4b)

The vibrational wavefunctions are the same as those discussed in Topic 7E for a harmonic oscillator.

The vibrational terms depend on the *effective* mass of the molecule, not directly on its total mass. This dependence is physically reasonable, because if atom 1 is very much heavier than atom 2, then the effective mass is close to m_2, the mass of the lighter atom. The vibration would then be that of the light atom relative to an essentially stationary heavy atom. For a homonuclear diatomic molecule $m_1 = m_2$, and the effective mass is half the total mass: $m_{\text{eff}}=\tfrac{1}{2}m$.

Brief illustration 11C.1

The force constant of the bond in HCl is 516 N m^{-1}, a reasonably typical value for a single bond. The effective mass of $^1H^{35}Cl$ is 1.63×10^{-27} kg (note that this mass is very close to the mass of the hydrogen atom, 1.67×10^{-27} kg, implying that the H atom is essentially vibrating against a stationary Cl atom). The vibrational frequency is therefore

$$\omega=\left(\frac{516\ \overbrace{\text{N}}^{\text{kg m s}^{-2}}\ \text{m}^{-1}}{1.63\times10^{-27}\ \text{kg}}\right)^{1/2}=5.63\times10^{14}\ \text{s}^{-1}$$

and the corresponding wavenumber is

$$\tilde{\nu}=\frac{\omega}{2\pi c}=\frac{5.63\times10^{14}\ \text{s}^{-1}}{2\pi\times(2.998\times10^{10}\ \text{cm s}^{-1})}=2.99\times10^{3}\ \text{cm}^{-1}$$

11C.2 Infrared spectroscopy

The gross and specific selection rules for vibrational transitions are established, as usual, by considering the properties of the electric transition dipole moment. The detailed calculation is shown in *A deeper look* 6 on the website of this text. The conclusion is that

> The gross selection rule for a change in vibrational state brought about by absorption or emission of radiation is that the electric dipole moment of the molecule must change when the atoms are displaced relative to one another.

Such vibrations are said to be **infrared active**. The classical basis of this rule is that an oscillating electric dipole generates an electromagnetic wave, and an oscillating electric field of such a wave generates an oscillating electric dipole.

Note that the molecule need not have a permanent dipole moment: the rule requires only a *change* in dipole moment. Some vibrations do not affect the dipole moment of the molecule (for instance, the stretching motion of a homonuclear diatomic molecule), so they neither absorb nor generate radiation: such vibrations are said to be **infrared inactive**. Weak infrared transitions can be observed from homonuclear diatomic molecules trapped within various nanomaterials. For instance, when incorporated into solid C_{60}, H_2 molecules interact through van der Waals forces with the surrounding C_{60} molecules and acquire dipole moments, with the result that they have observable infrared spectra.

The calculation also shows that the specific selection rule is

$$\Delta v = \pm 1 \quad \text{Specific selection rule [harmonic oscillator]} \quad (11C.5)$$

Transitions for which $\Delta v = +1$ correspond to absorption and those with $\Delta v = -1$ correspond to emission. It follows that the wavenumbers of allowed vibrational transitions, which are denoted $\Delta\tilde{G}_{v+\frac{1}{2}}$ for the transition $v+1 \leftarrow v$, are

$$\Delta\tilde{G}_{v+\frac{1}{2}} = \tilde{G}(v+1) - \tilde{G}(v) = \tilde{\nu} \quad \text{Harmonic oscillator} \quad (11C.6)$$

The wavenumbers of vibrational transitions correspond to radiation in the infrared region of the electromagnetic spectrum, so vibrational transitions absorb and generate infrared radiation.

At room temperature $kT/hc \approx 200\,\text{cm}^{-1}$, and because most vibrational wavenumbers are significantly greater than $200\,\text{cm}^{-1}$ it follows from the Boltzmann distribution that almost all the molecules are in their vibrational ground states. Hence, the dominant spectral transition will be the **fundamental transition**, $1 \leftarrow 0$. As a result, the spectrum is expected to consist of a single absorption line. If the molecules are formed in a vibrationally excited state, such as when vibrationally excited HF molecules are formed in the reaction $H_2 + F_2 \rightarrow 2\,HF^*$, where the star indicates a vibrationally 'hot' molecule, the transitions $5 \rightarrow 4$, $4 \rightarrow 3$, … may also appear (in emission). In the harmonic approximation, all these lines lie at the same frequency, and the spectrum is also a single line. However, the breakdown of the harmonic approximation causes the transitions to lie at slightly different frequencies, so several lines are observed.

11C.3 Anharmonicity

The vibrational terms in eqn 11C.4b are only approximate because they are based on a parabolic approximation to the actual potential energy curve. A parabola cannot be correct at all extensions because it does not allow a bond to dissociate. At high vibrational excitations the separation of the atoms (more precisely, the spread of the vibrational wavefunction) allows the molecule to explore regions of the potential energy curve where the parabolic approximation is poor and additional terms in the Taylor expansion of V (eqn 11C.1a) must be retained. The motion then becomes **anharmonic**, in the sense that the restoring force is no longer proportional to the displacement. Because the actual curve is less confining than a parabola, it can be anticipated that the energy levels become more closely spaced at high excitations.

(a) The convergence of energy levels

One approach to the calculation of the energy levels in the presence of anharmonicity is to use a function that resembles the true potential energy more closely. The **Morse potential energy** is

$$V(x) = hc\tilde{D}_e\{1 - e^{-ax}\}^2 \qquad a = \left(\frac{m_{\text{eff}}\omega^2}{2hc\tilde{D}_e}\right)^{1/2} \quad \text{Morse potential energy} \quad (11C.7)$$

At $x = 0$, $V(0) = 0$; at large displacements $V(x)$ approaches $hc\tilde{D}_e$ (Fig. 11C.3). Near the well minimum the variation of V with displacement resembles a parabola (as can be checked by expanding the exponential and retaining the first two terms). The Schrödinger equation can be solved for the Morse potential energy and the permitted levels are

$$\tilde{G}(v) = (v+\tfrac{1}{2})\tilde{\nu} - (v+\tfrac{1}{2})^2 x_e\tilde{\nu}$$

$$v = 0, 1, 2, \ldots, v_{\max} \quad \text{Vibrational terms [Morse potential energy]} \quad (11C.8)$$

$$x_e = \frac{a^2\hbar}{2m_{\text{eff}}\omega} = \frac{\tilde{\nu}}{4\tilde{D}_e}$$

The positive dimensionless parameter x_e is called the **anharmonicity constant**. The number of vibrational levels

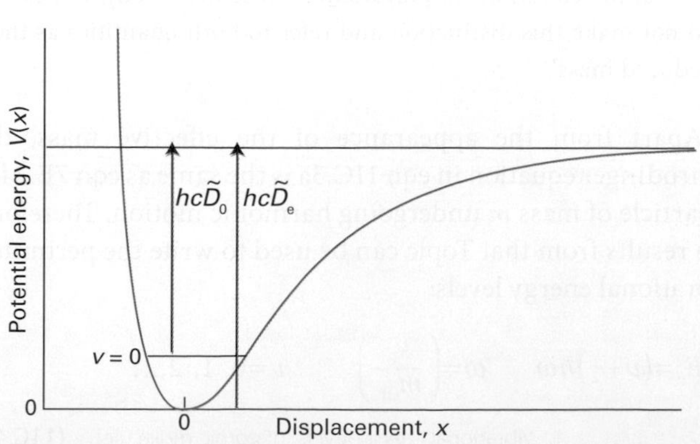

Figure 11C.3 The dissociation energy of a molecule, $hc\tilde{D}_0$, differs from the depth of the potential well, $hc\tilde{D}_e$, on account of the zero-point energy of the vibration of the bond.

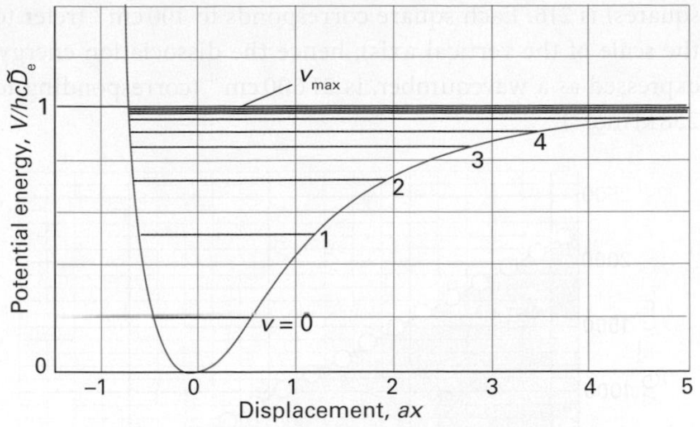

Figure 11C.4 The Morse potential energy curve reproduces the general shape of a molecular potential energy curve. The corresponding Schrödinger equation can be solved, and the values of the energies obtained. The number of bound levels is finite.

of a Morse oscillator is finite, as shown in Fig. 11C.4 (see also Problem P11C.8). The second term in the expression for $\tilde{G}$ subtracts from the first with increasing effect as v increases, and hence gives rise to the convergence of the levels at high quantum numbers. In addition to the depth of the well, $hc\tilde{D}_e$, the **dissociation energy** $hc\tilde{D}_0$ is the energy difference between the lowest vibrational state ($v = 0$) and the infinitely separated atoms. As can be seen in Fig. 11C.3, the two quantities are related by $\tilde{D}_e = \tilde{D}_0 + \tilde{G}(0)$.

Although the Morse oscillator is quite useful theoretically, in practice the more general expression

$$\tilde{G}(v) = (v+\tfrac{1}{2})\tilde{\nu} - (v+\tfrac{1}{2})^2 x_e\tilde{\nu} + (v+\tfrac{1}{2})^3 y_e\tilde{\nu} + \cdots \qquad (11C.9a)$$

where x_e, y_e, … are empirical dimensionless constants characteristic of the molecule, is used to fit the experimental data and to determine the dissociation energy of the molecule. In this case the wavenumbers of transitions with $\Delta v = +1$ are

$$\Delta\tilde{G}_{v+\frac{1}{2}} = \tilde{G}(v+1) - \tilde{G}(v) = \tilde{\nu} - 2(v+1)x_e\tilde{\nu} + \cdots \qquad (11C.9b)$$

Equation 11C.9b shows that, because $x_e > 0$, the transitions move to lower wavenumbers as v increases.

In addition to the strong fundamental transition 1←0, a set of weaker absorption lines are also seen and correspond to the transitions 2←0, 3←0, …. These transitions are forbidden for a harmonic oscillator, but become weakly allowed as a result of anharmonicity. The transition 2←0 is known as the **first overtone**, 3←0 is the **second overtone**, and so on. The wavenumber of the first overtone is given by

$$\tilde{G}(v+2) - \tilde{G}(v) = 2\tilde{\nu} - 2(2v+3)x_e\tilde{\nu} + \cdots \qquad (11C.10)$$

The reason for the appearance of overtones is that the selection rule is derived from the properties of harmonic oscillator wavefunctions, which are only approximately valid when anharmonicity is present. Therefore, the selection rule is also only an approximation. For an anharmonic oscillator, all values of Δv are allowed, but transitions with $\Delta v > 1$ are allowed only weakly if the anharmonicity is slight. Typically, the first overtone is only about one-tenth as intense as the fundamental.

Example 11C.1 Estimating an anharmonicity constant

Estimate the anharmonicity constant x_e for $^{35}Cl^{19}F$ given that the wavenumbers of the fundamental and first overtones are found to be 773.8 and 1535.3 cm^{-1}, respectively.

Collect your thoughts You can find an expression for the wavenumber of the fundamental transition 1←0, by using eqn 11C.9b with $v = 0$, and for the wavenumber of the first overtone 2←0 by using eqn 11C.10 with $v = 0$. You then need to solve the two equations to give values for $\tilde{\nu}$ and $x_e\tilde{\nu}$, and hence find x_e itself.

The solution From eqn 11C.9b the expression for the wavenumber of the fundamental is $\tilde{\nu} - 2x_e\tilde{\nu}$, and from eqn 11C.10 the expression for the first overtone is $2\tilde{\nu} - 6x_e\tilde{\nu}$. From the data it follows that $773.8\,\text{cm}^{-1} = \tilde{\nu} - 2x_e\tilde{\nu}$ and $1535.3\,\text{cm}^{-1} = 2\tilde{\nu} - 6x_e\tilde{\nu}$. The terms in $\tilde{\nu}$ are eliminated by noting that $(\tilde{\nu} - 2x_e\tilde{\nu}) - \frac{1}{2}(2\tilde{\nu} - 6x_e\tilde{\nu}) = x_e\tilde{\nu}$ to give $x_e\tilde{\nu} = 773.8\,\text{cm}^{-1} - \frac{1}{2}\times 1535.3\,\text{cm}^{-1} = 6.15\,\text{cm}^{-1}$. This value for $x_e\tilde{\nu}$ can then be substituted into $773.8\,\text{cm}^{-1} = \tilde{\nu} - 2x_e\tilde{\nu}$ to give $\tilde{\nu} = 773.8\,\text{cm}^{-1} + 2x_e\tilde{\nu} = 773.8\,\text{cm}^{-1} + 2\times 6.15\,\text{cm}^{-1} = 786.1\,\text{cm}^{-1}$. It follows that

$$x_e = \frac{x_e\tilde{\nu}}{\tilde{\nu}} = \frac{6.15\,\text{cm}^{-1}}{786.1\,\text{cm}^{-1}} = 7.82\times 10^{-3}$$

Self-test 11C.1 Predict the wavenumber of the second overtone for this molecule.

Answer: 2284.5 cm^{-1}

(b) The Birge–Sponer plot

When several vibrational transitions are detectable, a graphical technique called a **Birge–Sponer plot** can be used to determine the dissociation energy of the bond. The basis of the Birge–Sponer plot is that the sum of the successive intervals $\Delta\tilde{G}_{v+\frac{1}{2}}$ (eqn 11C.9b) from $v = 0$ to the dissociation limit is the dissociation wavenumber $\tilde{D}_0$:

$$\tilde{D}_0 = \Delta\tilde{G}_{1/2} + \Delta\tilde{G}_{3/2} + \cdots = \sum_v \Delta\tilde{G}_{v+\frac{1}{2}} \qquad (11C.11)$$

just as the height of a ladder is the sum of the separations of its rungs (Fig. 11C.5). The construction in Fig. 11C.6 shows that the area under the plot of $\Delta\tilde{G}_{v+\frac{1}{2}}$ against $v+\frac{1}{2}$ is equal to the

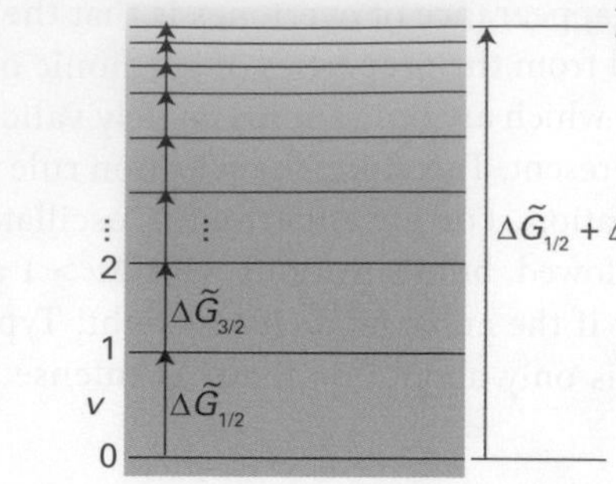

Figure 11C.5 The dissociation wavenumber is the sum of the separations $\Delta\tilde{G}_{v+\frac{1}{2}}$ of the vibrational terms up to the dissociation limit, just as the height of a ladder is the sum of the separations of its rungs.

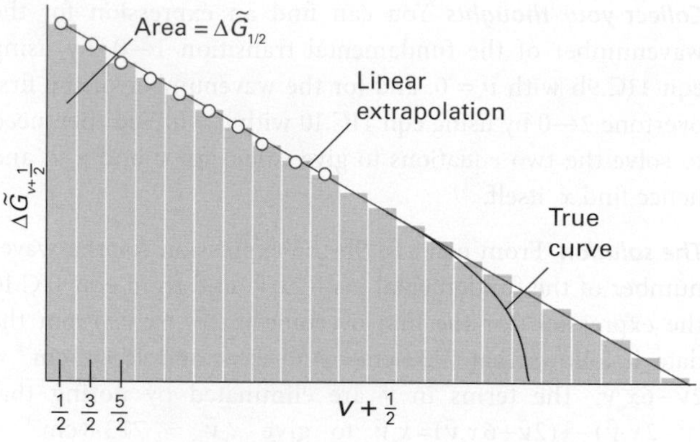

Figure 11C.6 The area under a plot of $\Delta\tilde{G}_{v+\frac{1}{2}}$ against vibrational quantum number is equal to the dissociation wavenumber of the molecule. The assumption that the differences approach zero linearly is the basis of the Birge–Sponer extrapolation.

sum, and therefore to $\tilde{D}_0$. The successive terms decrease linearly when only the x_e anharmonicity constant is taken into account and the inaccessible part of the spectrum can be estimated by linear extrapolation. Most actual plots differ from the linear plot as shown in Fig. 11C.6, so the value of $\tilde{D}_0$ obtained in this way is usually an overestimate of the true value.

Example 11C.2 Using a Birge–Sponer plot

The observed vibrational intervals of H_2^+ lie at the following values for 1←0, 2←1, …, respectively (in cm^{-1}): 2191, 2064, 1941, 1821, 1705, 1591, 1479, 1368, 1257, 1145, 1033, 918, 800, 677, 548, 411. Determine the dissociation energy of the molecule.

Collect your thoughts Plot the separations against $v+\frac{1}{2}$, extrapolate linearly to the point cutting the horizontal axis, and then measure the area under the curve.

The solution The points are plotted in Fig. 11C.7, and a linear extrapolation is shown as a blue line. The area under the curve (use the formula for the area of a triangle or count the squares) is 216. Each square corresponds to $100\,cm^{-1}$ (refer to the scale of the vertical axis); hence the dissociation energy, expressed as a wavenumber, is $21\,600\,cm^{-1}$ (corresponding to $258\,kJ\,mol^{-1}$).

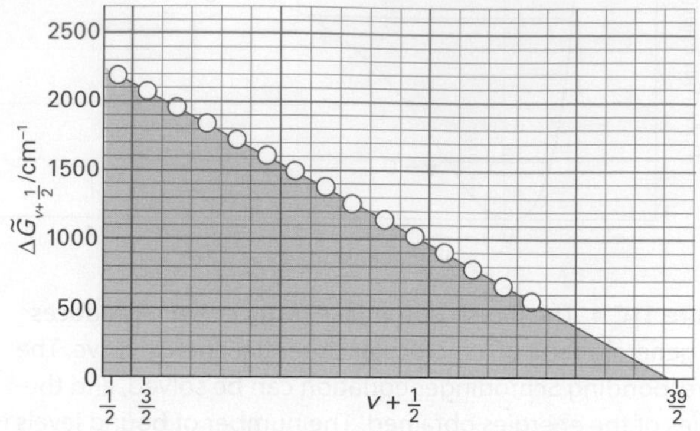

Figure 11C.7 The Birge–Sponer plot used in *Example* 11C.2. The area is obtained simply by counting the squares beneath the line or using the formula for the area of a triangle (area = $\frac{1}{2}$ × base × height).

Self-test 11C.2 The vibrational levels of HgH converge rapidly, and successive intervals are 1203.7 (which corresponds to the 1←0 transition), 965.6, 632.4, and $172\,cm^{-1}$. Estimate the molar dissociation energy.

Answer: $35.6\,kJ\,mol^{-1}$

11C.4 Vibration–rotation spectra

Each line of the high resolution vibrational spectrum of a gas-phase heteronuclear diatomic molecule is found to consist of a large number of closely spaced components (Fig. 11C.8). Hence, molecular spectra are often called **band spectra**. The separation between the components is less than $10\,cm^{-1}$, which

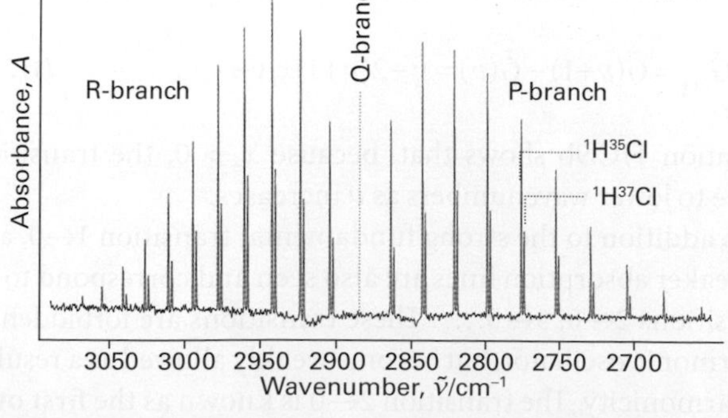

Figure 11C.8 A high-resolution vibration–rotation spectrum of HCl. The lines appear in pairs because $H^{35}Cl$ and $H^{37}Cl$ both contribute (their abundance ratio is 3:1). There is no Q branch (see below), because $\Delta J = 0$ is forbidden for this molecule.

suggests that the structure is due to rotational transitions accompanying the vibrational transition. A rotational change should be expected because classically a vibrational transition can be thought of as leading to a sudden increase or decrease in the instantaneous bond length. Just as ice-skaters rotate more rapidly when they bring their arms in, and more slowly when they throw them out, so the molecular rotation is either accelerated or retarded by a vibrational transition.

(a) Spectral branches

A detailed analysis of the quantum mechanics of simultaneous vibrational and rotational changes shows that the rotational quantum number J changes by ±1 during the vibrational transition of a diatomic molecule. If the molecule also possesses angular momentum about its axis, as in the case of the electronic orbital angular momentum of the molecule NO with its configuration $\ldots\pi^1$, then the selection rules also allow $\Delta J = 0$.

The appearance of the vibration–rotation spectrum of a diatomic molecule can be discussed by using the combined vibration–rotation terms, $\tilde{S}$:

$$\tilde{S}(v,J) = \tilde{G}(v) + \tilde{F}(J) \qquad (11C.12a)$$

If anharmonicity and centrifugal distortion are ignored, $\tilde{G}(v)$ can be replaced by the expression in eqn 11C.4b, and $\tilde{F}(J)$ can be replaced by the expression in eqn 11B.9 ($\tilde{F}(J) = \tilde{B}J(J+1)$) to give

$$\tilde{S}(v,J) = \left(v+\tfrac{1}{2}\right)\tilde{\nu} + \tilde{B}J(J+1) \qquad (11C.12b)$$

In a more detailed treatment, $\tilde{B}$ is allowed to depend on the vibrational state and written $\tilde{B}_v$.

In the vibrational transition $v+1 \leftarrow v$, J changes by ±1 and in some cases by 0 (when $\Delta J = 0$ is allowed). The absorptions then fall into three groups called **branches** of the spectrum. The **P branch** consists of all transitions with $\Delta J = -1$:

$$\tilde{\nu}_{\rm P}(J) = \tilde{S}(v+1,J-1) - \tilde{S}(v,J) = \tilde{\nu} - 2\tilde{B}J \qquad J = 1,2,3,\ldots \qquad \text{P branch transitions} \qquad (11C.13a)$$

This branch consists of lines extending to the low wavenumber side of $\tilde{\nu}$ at $\tilde{\nu} - 2\tilde{B}$, $\tilde{\nu} - 4\tilde{B}$, … with an intensity distribution reflecting both the populations of the rotational levels and the magnitude of the $J-1 \leftarrow J$ transition moment (Fig. 11C.9). The **Q branch** consists of all transitions with $\Delta J = 0$, and its wavenumbers are the same for all values of J:

$$\tilde{\nu}_{\rm Q}(J) = \tilde{S}(v+1,J) - \tilde{S}(v,J) = \tilde{\nu} \qquad \text{Q branch transitions} \qquad (11C.13b)$$

This branch, when it is allowed (as in NO), appears at the vibrational transition wavenumber $\tilde{\nu}$. In Fig. 11C.8 there is a gap at the expected location of the Q branch because it is forbidden in HCl because it has zero electronic angular momentum around its internuclear axis. The **R branch** consists of lines with $\Delta J = +1$:

$$\tilde{\nu}_{\rm R}(J) = \tilde{S}(v+1,J+1) - \tilde{S}(v,J) = \tilde{\nu} + 2\tilde{B}(J+1) \qquad J = 0,1,2,\ldots \qquad \text{R branch transitions} \qquad (11C.13c)$$

This branch consists of lines extending to the high-wavenumber side of $\tilde{\nu}$ at $\tilde{\nu}+2\tilde{B}$, $\tilde{\nu}+4\tilde{B}$, ….

The separation between the lines in the P and R branches of a vibrational transition gives the value of $\tilde{B}$. Therefore, the bond length can be deduced in the same way as from microwave spectra (Topic 11B). However, the latter technique gives more precise bond lengths because microwave frequencies can be measured with greater precision than infrared frequencies.

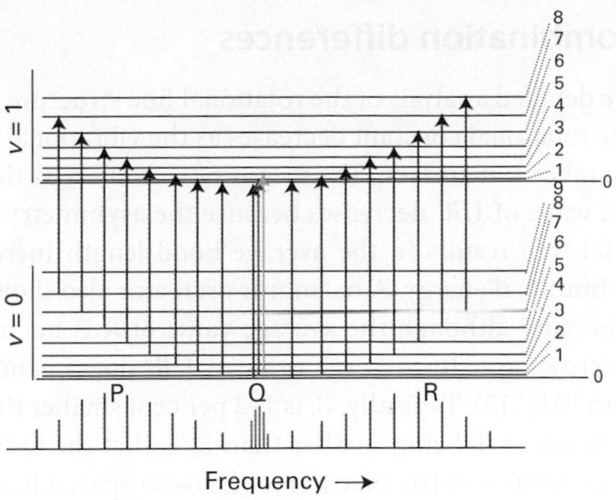

Figure 11C.9 The formation of P, Q, and R branches in a vibration–rotation spectrum. The intensities reflect the populations of the initial rotational levels and magnitudes of the transition moments.

Brief illustration 11C.2

The infrared absorption spectrum of $^1H^{81}Br$ contains a band arising from $v = 0$. It follows from eqn 11C.13c and the data in Table 11C.1 that the wavenumber of the line in the R branch originating from the rotational state with $J = 2$ is

$$\tilde{\nu}_{\rm R}(2) = \tilde{\nu} + 6\tilde{B} = 2648.98\,{\rm cm^{-1}} + 6\times(8.465\,{\rm cm^{-1}}) = 2699.77\,{\rm cm^{-1}}$$

Table 11C.1 Properties of diatomic molecules*

	$\tilde{\nu}$/cm^{-1}	R_e/pm	$\tilde{B}$/cm^{-1}	k_f/(N m^{-1})	$\tilde{D}_0$/(10^4 cm^{-1})
$^1H_2$	4400	74	60.86	575	3.61
$^1H^{35}Cl$	2991	127	10.59	516	3.58
$^1H^{127}I$	2308	161	6.51	314	2.46
$^{35}Cl_2$	560	199	0.244	323	2.00

* More values are given in the *Resource section*.

(b) Combination differences

A more detailed analysis of the rotational fine structure shows that the rotational constant decreases as the vibrational quantum number v increases. The origin of this effect is that the average value of $1/R^2$ decreases because the asymmetry of the potential well results in the average bond length increasing with vibrational energy. A harmonic oscillator also shows this effect because although the average value of R is unchanged with increasing v, the average value of $1/R^2$ does change (see Problem P11C.13). Typically, $\tilde{B}_1$ is 1–2 per cent smaller than $\tilde{B}_0$.

The result of $\tilde{B}_1$ being smaller than $\tilde{B}_0$ is that the Q branch (if it is present) consists of a series of closely spaced lines; the lines of the R branch converge slightly as J increases, and those of the P branch diverge. It follows from eqn 11C.12b with $\tilde{B}_v$ in place of $\tilde{B}$

$$\begin{aligned}\tilde{\nu}_P(J) &= \tilde{\nu} - (\tilde{B}_1 + \tilde{B}_0)J + (\tilde{B}_1 - \tilde{B}_0)J^2 \\ \tilde{\nu}_Q(J) &= \tilde{\nu} + (\tilde{B}_1 - \tilde{B}_0)J(J+1) \\ \tilde{\nu}_R(J) &= \tilde{\nu} + (\tilde{B}_1 + \tilde{B}_0)(J+1) + (\tilde{B}_1 - \tilde{B}_0)(J+1)^2\end{aligned} \quad (11C.14)$$

To determine the two rotational constants individually, the method of **combination differences** is used, which involves setting up expressions for the difference in the wavenumbers of transitions to a common state. The resulting expression then depends solely on properties of the other states.

As can be seen from Fig. 11C.10, the transitions $\tilde{\nu}_R(J-1)$ and $\tilde{\nu}_P(J+1)$ have a common upper state, and hence the difference between these transitions can be anticipated to depend on $\tilde{B}_0$. From the diagram it can be seen that $\tilde{\nu}_R(J-1) - \tilde{\nu}_P(J+1) = \tilde{S}(0,J+1) - \tilde{S}(0,J-1)$. The right-hand side is evaluated by using the expression for $\tilde{S}(v,J)$ in eqn 11C.12b (with $\tilde{B}_0$ in place of $\tilde{B}$) to give

$$\tilde{\nu}_R(J-1) - \tilde{\nu}_P(J+1) = 4\tilde{B}_0(J+\tfrac{1}{2}) \quad (11C.15a)$$

Therefore, a plot of the combination difference against $J+\tfrac{1}{2}$ should be a straight line of slope $4\tilde{B}_0$ and intercept (with the vertical axis) zero; the value of $\tilde{B}_0$ can therefore be determined from the slope. The presence of centrifugal distortion results in the intercept deviating from zero, but has little effect on the quality of the straight line.

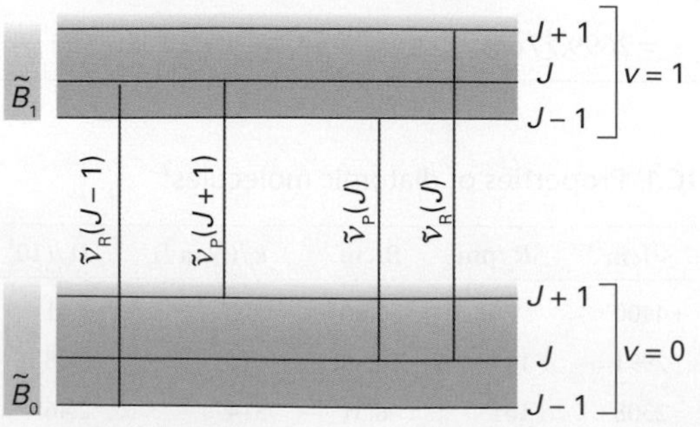

Figure 11C.10 The method of combination differences makes use of the fact that certain pairs of transitions share a common level.

The two lines $\tilde{\nu}_R(J)$ and $\tilde{\nu}_P(J)$ have a common lower state, and hence their combination difference depends on $\tilde{B}_1$. As before, from Fig. 11C.10 it can be seen that $\tilde{\nu}_R(J) - \tilde{\nu}_P(J) = \tilde{S}(1,J+1) - \tilde{S}(1,J-1)$ which is

$$\tilde{\nu}_R(J) - \tilde{\nu}_P(J) = 4\tilde{B}_1(J+\tfrac{1}{2}) \quad (11C.15b)$$

Brief illustration 11C.3

The rotational constants of $\tilde{B}_0$ and $\tilde{B}_1$ can be estimated from a calculation involving only a few transitions. For $^1H^{35}Cl$, $\tilde{\nu}_R(0) - \tilde{\nu}_P(2) = 62.6\,cm^{-1}$, and it follows from eqn 11C.15a, with J = 1, that $\tilde{B}_0 = 62.6/4(1+\tfrac{1}{2})\,cm^{-1} = 10.4\,cm^{-1}$. Similarly, $\tilde{\nu}_R(1) - \tilde{\nu}_P(1) = 60.8\,cm^{-1}$, and it follows from eqn 11C.15b, again with J = 1 that $\tilde{B}_1 = 60.8/4(1+\tfrac{1}{2})\,cm^{-1} = 10.1\,cm^{-1}$. If more lines are used to make combination difference plots, the values $\tilde{B}_0 = 10.440\,cm^{-1}$ and $\tilde{B}_1 = 10.136\,cm^{-1}$ are found. The two rotational constants differ by about 3 per cent of $\tilde{B}_0$.

11C.5 Vibrational Raman spectra

The gross and specific selection rules for vibrational Raman transitions are established, as usual, by considering the appropriate transition dipole moment. The details are set out in *A deeper look* 6 on the website of this text. The conclusion is that the gross selection rule for vibrational Raman transitions is that *the polarizability must change as the molecule vibrates*. The polarizability plays a role in vibrational Raman spectroscopy because the molecule must be squeezed and stretched by the incident radiation in order for vibrational excitation to occur during the inelastic photon–molecule collision. Both homonuclear and heteronuclear diatomic molecules swell and contract during a vibration, the control of the nuclei over the electrons varies, and hence the molecular polarizability changes. Both types of diatomic molecule are therefore vibrationally Raman active. The analysis also shows that the specific selection rule for vibrational Raman transitions in the harmonic approximation is $\Delta v = \pm 1$, just as for infrared transitions.

The lines to high frequency of the incident radiation, in the language introduced in Topic 11A, the 'anti-Stokes lines', are those for which $\Delta v = -1$. The lines to low frequency, the 'Stokes lines', correspond to $\Delta v = +1$. The intensities of the anti-Stokes and Stokes lines are governed largely by the Boltzmann populations of the vibrational states involved in the transition. It follows that anti-Stokes lines are usually weak because the populations of the excited vibrational states are very small.

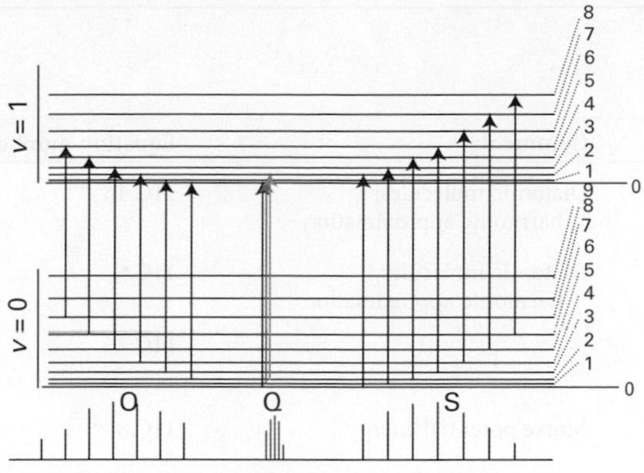

Figure 11C.11 The formation of O, Q, and S branches in a vibration–rotation Raman spectrum of a diatomic molecule (its Stokes lines). Note that the frequency scale runs in the opposite direction to that in Fig. 11C.9, because the higher energy transitions (on the right) extract more energy from the incident beam and leave it at lower frequency.

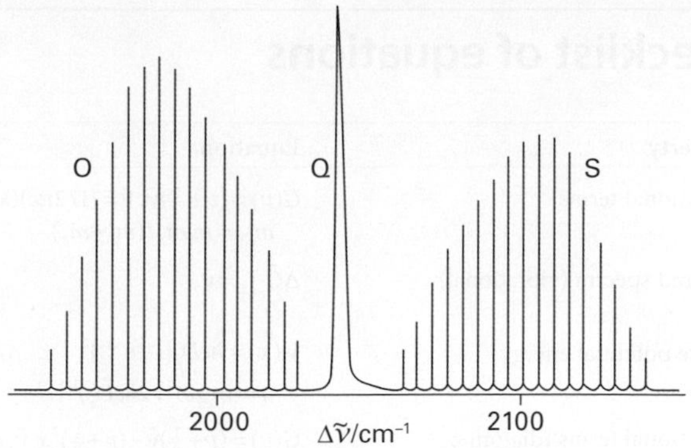

Figure 11C.12 The structure of a vibrational line in the vibrational Raman spectrum (the Stokes lines) of carbon monoxide, showing the O, Q, and S branches. The horizontal axis represents the wavenumber difference between the incident and scattered radiation. For these Stokes lines the wavenumber of the scattered radiation (as distinct from the difference, which represents the energy deposited in the molecule) increases to the left, as in Fig. 11C.11.

In gas-phase spectra, the Stokes and anti-Stokes lines have a branch structure arising from the simultaneous rotational transitions that accompany the vibrational excitation (Fig. 11C.11). The selection rules are $\Delta J = 0, \pm 2$ (as in pure rotational Raman spectroscopy), and give rise to the **O branch** ($\Delta J = -2$), the **Q branch** ($\Delta J = 0$), and the **S branch** ($\Delta J = +2$):

$$\tilde{\nu}_{\mathrm{O}}(J)=\tilde{\nu}_{\mathrm{i}}-\tilde{\nu}+4\tilde{B}(J-\tfrac{1}{2}) \quad J=2,3,4,\ldots \qquad \text{O branch transitions}$$

$$\tilde{\nu}_{\mathrm{Q}}(J)=\tilde{\nu}_{\mathrm{i}}-\tilde{\nu} \qquad \text{Q branch transitions} \qquad (11\text{C}.16)$$

$$\tilde{\nu}_{\mathrm{S}}(J)=\tilde{\nu}_{\mathrm{i}}-\tilde{\nu}-4\tilde{B}(J+\tfrac{3}{2}) \quad J=0,1,2,\ldots \qquad \text{S branch transitions}$$

where $\tilde{\nu}_{\mathrm{i}}$ is the wavenumber of the incident radiation. Note that, unlike in infrared spectroscopy, a Q branch is obtained for all linear molecules. The spectrum of CO, for instance, is shown in Fig. 11C.12: rather than being a single line, as implied by eqn 11C.16, the Q branch appears as a broad feature. Its breadth arises from the presence of several overlapping lines arising from the difference in rotational constants of the upper and lower vibrational states.

The information available from vibrational Raman spectra adds to that from infrared spectroscopy because homonuclear diatomics can also be studied. The spectra can be interpreted in terms of the force constants, dissociation energies, and bond lengths, and some of the information obtained is included in Table 11C.1.

Checklist of concepts

☐ 1. The vibrational energy levels of a diatomic molecule modelled as a harmonic oscillator depend on the **force constant** k_{f} (a measure of the stiffness of the bond) and the **effective mass** of the vibration.

☐ 2. The **gross selection rule** for infrared spectra is that the electric dipole moment of the molecule must depend on the bond length.

☐ 3. The **specific selection rule** for infrared spectra (within the harmonic approximation) is $\Delta v = \pm 1$.

☐ 4. The **Morse potential energy** can be used to model anharmonic vibration.

☐ 5. The strongest infrared transitions are the **fundamental transitions** ($v = 1 \leftarrow v = 0$).

☐ 6. Anharmonicity gives rise to weaker **overtone transitions** ($v = 2 \leftarrow v = 0$, $v = 3 \leftarrow v = 0$, …).

☐ 7. A **Birge–Sponer** plot may be used to determine the dissociation energy of a diatomic molecule.

☐ 8. In the gas phase, vibrational transitions have a **P, R branch structure** due to simultaneous rotational transitions; some molecules also have a **Q branch**.

☐ 9. For a vibration to be **Raman active**, the polarizability must change as the molecule vibrates.

☐ 10. The **specific selection rule** for vibrational Raman spectra (within the harmonic approximation) is $\Delta v = \pm 1$.

☐ 11. In gas-phase spectra, the Stokes and anti-Stokes lines in a Raman spectrum have an **O, Q, S branch structure.**

Checklist of equations

Property	Equation	Comment	Equation number
Vibrational terms	$\tilde{G}(v)=(v+\tfrac{1}{2})\tilde{\nu},\ \tilde{\nu}=(1/2\pi c)(k_f/m_{\text{eff}})^{1/2}$ $m_{\text{eff}}=m_1m_2/(m_1+m_2)$	Diatomic molecules; harmonic approximation	11C.4b
Infrared spectra (vibrational)	$\Delta\tilde{G}_{v+\frac{1}{2}}=\tilde{\nu}$	Diatomic molecules; harmonic approximation	11C.6
Morse potential energy	$V(x)=hc\tilde{D}_e\{1-e^{-ax}\}^2$ $a=(m_{\text{eff}}\omega^2/2hc\tilde{D}_e)^{1/2}$		11C.7
Vibrational terms (diatomic molecules)	$\tilde{G}(v)=(v+\tfrac{1}{2})\tilde{\nu}-(v+\tfrac{1}{2})^2x_e\tilde{\nu},\ x_e=\tilde{\nu}/4\tilde{D}_e$	Morse potential energy	11C.8
Infrared spectra (vibrational)	$\Delta\tilde{G}_{v+\frac{1}{2}}=\tilde{\nu}-2(v+1)x_e\tilde{\nu}+\cdots$	Anharmonic oscillator	11C.9b
	$\tilde{G}(v+2)-\tilde{G}(v)=2\tilde{\nu}-2(2v+3)x_e\tilde{\nu}+\cdots$	First overtone	11C.10
Dissociation wavenumber	$\tilde{D}_0=\Delta\tilde{G}_{1/2}+\Delta\tilde{G}_{3/2}+\cdots=\sum_v\Delta\tilde{G}_{v+\frac{1}{2}}$	Birge–Sponer plot	11C.11
Vibration–rotation terms (diatomic molecules)	$\tilde{S}(v,J)=(v+\tfrac{1}{2})\tilde{\nu}+BJ(J+1)$	Rotation coupled to vibration	11C.12b
Infrared spectra (vibration–rotation)	$\tilde{\nu}_P(J)=\tilde{S}(v+1,J-1)-\tilde{S}(v,J)=\tilde{\nu}-2\tilde{B}J$ $J=1,2,3,\ldots$	P branch ($\Delta J=-1$)	11C.13a
	$\tilde{\nu}_Q(J)=\tilde{S}(v+1,J)-\tilde{S}(v,J)=\tilde{\nu}$	Q branch ($\Delta J=0$)	11C.13b
	$\tilde{\nu}_R(J)=\tilde{S}(v+1,J+1)-\tilde{S}(v,J)=\tilde{\nu}+2\tilde{B}(J+1)$ $J=0,1,2,\ldots$	R branch ($\Delta J=+1$)	11C.13c
	$\tilde{\nu}_R(J-1)-\tilde{\nu}_P(J+1)=4\tilde{B}_0(J+\tfrac{1}{2})$ $\tilde{\nu}_R(J)-\tilde{\nu}_P(J)=4\tilde{B}_1(J+\tfrac{1}{2})$	Combination differences	11C.15
Raman spectra (vibration–rotation)	$\tilde{\nu}_O(J)=\tilde{\nu}_i-\tilde{\nu}+4\tilde{B}(J-\tfrac{1}{2})\quad J=2,3,4,\ldots$	O branch ($\Delta J=-2$)	11C.16
	$\tilde{\nu}_Q(J)=\tilde{\nu}_i-\tilde{\nu}$	Q branch ($\Delta J=0$)	
	$\tilde{\nu}_S(J)=\tilde{\nu}_i-\tilde{\nu}-4\tilde{B}(J+\tfrac{3}{2})\quad J=0,1,2,\ldots$	S branch ($\Delta J=+2$)	

TOPIC 11D Vibrational spectroscopy of polyatomic molecules

> ➤ **Why do you need to know this material?**
>
> The analysis of vibrational spectra is a widely used analytical technique that provides information about the identity and shapes of polyatomic molecules in the gas and condensed phases.
>
> ➤ **What is the key idea?**
>
> The vibrational spectrum of a polyatomic molecule can be interpreted in terms of its normal modes.
>
> ➤ **What do you need to know already?**
>
> You need to be familiar with the harmonic oscillator (Topic 7E), the general principles of spectroscopy (Topic 11A), and the selection rules for vibrational infrared and Raman spectroscopy (Topic 11C).

There is only one mode of vibration for a diatomic molecule: the periodic stretching and compression of the bond. In polyatomic molecules there are many bond lengths and angles that can change, and as a result the vibrational motion of the molecule is very complex. Some order can be brought to this complexity by introducing the concept of 'normal modes'.

11D.1 Normal modes

The first step in the analysis of the vibrations of a polyatomic molecule is to calculate the total number of vibrational modes.

How is that done? 11D.1 Counting the number of vibrational modes

The total number of coordinates needed to specify the locations of N atoms is $3N$. Each atom may change its location by varying each of its three coordinates (x, y, and z), so the total number of displacements available is $3N$. These displacements can be grouped together in a physically sensible way. For example, three coordinates are needed to specify the location of the centre of mass of the molecule, so three of the $3N$ displacements correspond to the translational motion of the molecule as a whole. The remaining $3N-3$ displacements are 'internal' modes of the molecule.

Two angles are needed to specify the orientation of a linear molecule in space: in effect, only the latitude and longitude of the direction in which the molecular axis is pointing need be specified (Fig. 11D.1a). However, three angles are needed for a nonlinear molecule because the orientation of the molecule around the direction defined by the latitude and longitude also needs to be specified (Fig. 11D.1b). Therefore, for linear molecules two of the $3N-3$ internal displacements are rotational, whereas for nonlinear molecules three of the displacements are rotational. That leaves $3N-5$ (linear) or $3N-6$ (nonlinear) non-rotational internal displacements of the atoms: these are the vibrational modes. It follows that the number of modes of vibration is:

$$\text{Linear molecule:} \quad N_{\text{vib}} = 3N-5$$
$$\text{Nonlinear molecule:} \quad N_{\text{vib}} = 3N-6 \qquad (11\text{D}.1)$$

Numbers of vibrational modes

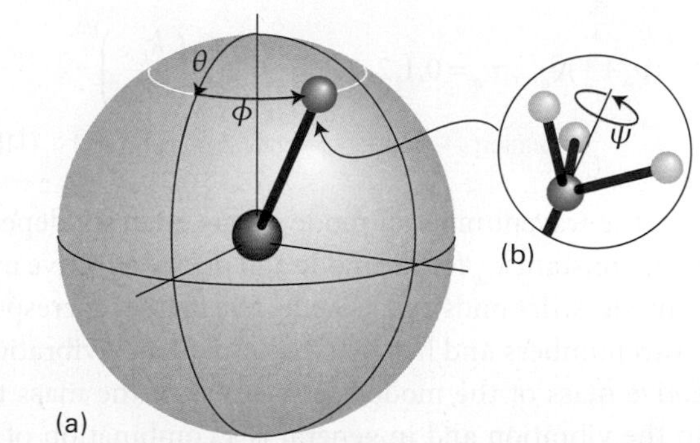

Figure 11D.1 (a) The orientation of a linear molecule requires the specification of two angles. (b) The orientation of a nonlinear molecule requires the specification of three angles.

Brief illustration 11D.1

Water, H_2O, is a nonlinear triatomic molecule with $N = 3$, and so has $3N-6 = 3$ modes of vibration; CO_2 is a linear triatomic molecule, and has $3N-5 = 4$ modes of vibration. Methylbenzene has 15 atoms and 39 modes of vibration.

The greatest simplification of the description of vibrational motion is obtained by analysing it in terms of 'normal modes'. A **normal mode** is a vibration of the molecule in which the centre of mass remains fixed, the orientation is unchanged, and the atoms move synchronously. When a normal mode is excited, the energy remains in that mode and does not migrate into other normal modes of the molecule.

A normal mode analysis is possible only if it is assumed that the potential energy is parabolic (as in a harmonic oscillator, Topic 11C). In reality, the potential energy is not parabolic, the vibrations are anharmonic (Topic 11C), and the normal modes are not completely independent. Nevertheless, a normal mode analysis remains a good starting point for the description of the vibrations of polyatomic molecules.

Figure 11D.2 shows the four normal modes of CO_2. Mode ν_1 is the **symmetric stretch** in which the two oxygen atoms move in and out synchronously but the carbon atom remains stationary. Mode ν_2, the **antisymmetric stretch**, in which the two oxygen atoms always move in the same direction and opposite to that of the carbon. Finally, there are two **bending modes** ν_3 in which the oxygen atoms move perpendicular to the internuclear axis in one direction and the carbon atom moves in the opposite direction: this bending motion can take place in either of two perpendicular planes. In all these modes, the position of the centre of mass and orientation of the molecule are unchanged by the vibration.

In the harmonic approximation, each normal mode, q, behaves like an independent harmonic oscillator and has an energy characterized by the quantum number v_q. Expressed as a wavenumber, these terms are

$$\tilde{G}_q(v)=(v_q+\tfrac{1}{2})\tilde{\nu}_q \quad v_q=0,1,2,\ldots \quad \tilde{\nu}_q=\frac{1}{2\pi c}\left(\frac{k_{f,q}}{m_q}\right)^{1/2}$$

Vibrational terms of normal modes [harmonic] (11D.2)

where $\tilde{\nu}_q$ is the wavenumber of mode q; this quantity depends on the force constant $k_{f,q}$ for the mode and on the effective mass m_q of the mode: stiff bonds and low effective masses correspond to high wavenumbers and hence to high frequency vibrations. The effective mass of the mode is a measure of the mass that moves in the vibration and in general is a combination of the masses of the atoms. For example, in the symmetric stretch of CO_2, the carbon atom is stationary, and the effective mass depends on the masses of only the oxygen atoms. In the antisymmetric stretch and in the bends all three atoms move, so the masses of all three atoms contribute (but to different extents) to the effective mass of each mode.

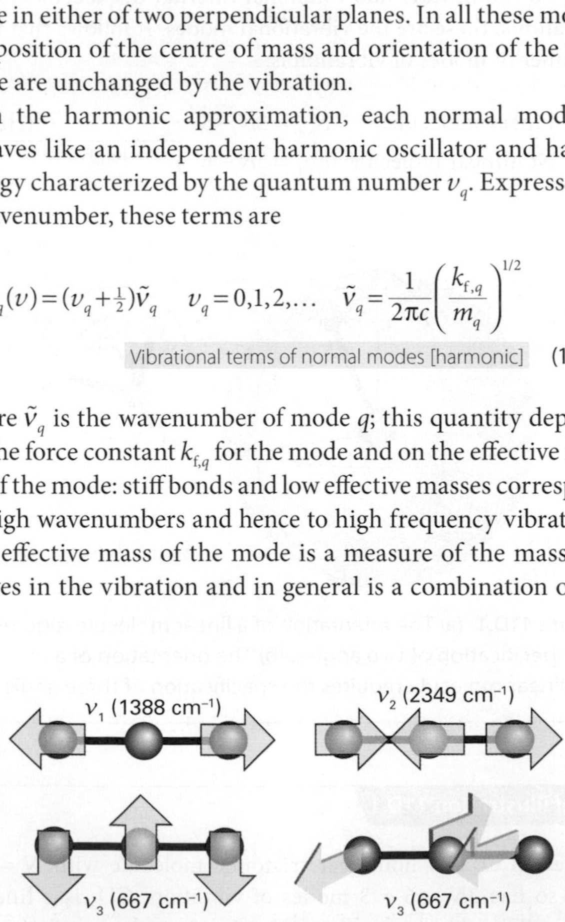

Figure 11D.2 The four normal modes of CO_2. The two bending motions (ν_3) have the same vibrational frequency.

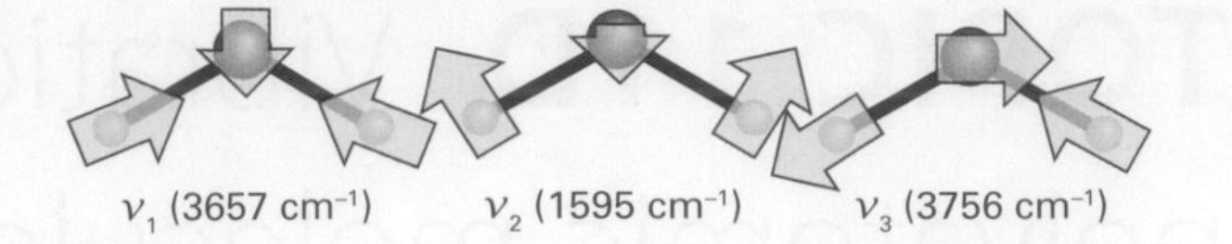

Figure 11D.3 The three normal modes of H_2O. The mode ν_2 is predominantly bending, and occurs at lower wavenumber than the other two.

The three normal modes of H_2O are shown in Fig. 11D.3: note that the predominantly bending mode (ν_2) has a lower frequency (and wavenumber) than the others, which are predominantly stretching modes. It is generally the case that the frequencies of bending motions are lower than those of stretching modes. Only in special cases (such as the CO_2 molecule) are the normal modes purely stretches or purely bends. In general, a normal mode is a composite motion involving simultaneous stretching of bonds and changes to bond angles. In a given normal mode, heavy atoms generally move less than light atoms.

The vibrational state of a polyatomic molecule is specified by the vibrational quantum number v_q for each of the normal modes. For example, for H_2O with three normal modes, the vibrational state is designated (v_1,v_2,v_3). The vibrational ground state of an H_2O molecule is therefore (0,0,0); the state (0,1,0) implies that modes 1 and 3 are in their ground states, and mode 2 is in the first excited state.

11D.2 Infrared absorption spectra

The gross selection rule for infrared activity is a straightforward generalization of the rule for diatomic molecules (Topic 11C):

> The motion corresponding to a normal mode must be accompanied by a change of electric dipole moment.

Simple inspection of atomic motions is sometimes all that is needed in order to assess whether a normal mode is infrared active. For example, the symmetric stretch of CO_2 leaves the dipole moment unchanged (at zero, see Fig. 11D.2), so this mode is infrared inactive. The antisymmetric stretch, however, changes the dipole moment because the molecule becomes unsymmetrical as it vibrates, so this mode is infrared active. Because the dipole moment change is parallel to the principal axis, the transitions arising from this mode are classified as **parallel bands** in the spectrum. Both bending modes are infrared active: they are accompanied by a changing dipole perpendicular to the principal axis, so transitions involving them lead to a **perpendicular band** in the spectrum.

Example 11D.1 Using the gross selection rule for infrared spectroscopy

State which of the following molecules are infrared active: N_2O, OCS, H_2O, $CH_2{=}CH_2$.

Collect your thoughts Molecules that are infrared active have a normal mode (or modes) in which there is a change in dipole moment during the course of the motion. Therefore, to work out if a molecule is infrared active you need to decide whether there is any distortion of the molecule that results in a change in its electric dipole moment (including changes from zero).

The solution The linear molecules N_2O and OCS both have permanent electric dipole moments that change as a result of stretching any of the bonds; in addition, bending perpendicular to the internuclear axis results in a dipole in that direction: both molecules are therefore infrared active. An H_2O molecule also has a permanent dipole moment which changes either by stretching the bonds or by altering the bond angle: the molecule is infrared active. A $CH_2{=}CH_2$ molecule does not have a permanent dipole moment (it possesses a centre of symmetry) but there are vibrations in which the symmetry is reduced and a dipole moment forms, for example, by stretching the two C–H bonds on one carbon atom and simultaneously compressing the two C–H bonds on the other carbon atom.

Comment. Topic 11E describes a systematic procedure based on group theory for deciding whether a vibrational mode is infrared active.

Self-test 11D.1 Identify an infrared inactive normal mode of $CH_2{=}CH_2$.

Answer: A 'breathing' mode in which all the C–H bonds contract and stretch synchronously

The specific selection rule in the harmonic approximation is $\Delta v_q = \pm 1$. In this approximation the quantum number of only one active mode can change in the interaction of a molecule with a photon. A **fundamental transition** is a transition from the ground state of the molecule to the next higher energy level of the specified mode. For example, in H_2O there are three such fundamentals corresponding to the excitation of each of the three normal modes: $(1,0,0) \leftarrow (0,0,0)$, $(0,1,0) \leftarrow (0,0,0)$, and $(0,0,1) \leftarrow (0,0,0)$.

Anharmonicity also allows transitions in which more than one quantum of excitation takes place: such transitions are referred to as **overtones**. A transition such as $(0,0,2) \leftarrow (0,0,0)$ in H_2O is described as a **first overtone**, and a transition such as $(0,0,3) \leftarrow (0,0,0)$ is a **second overtone** of the mode ν_3. **Combination bands** (or *combination lines*) corresponding to the simultaneous excitation of more than one normal mode in the transition, as in $(1,1,0) \leftarrow (0,0,0)$, are also possible in the presence of anharmonicity.

As for diatomic molecules (Topic 11C), transitions between vibrational levels can be accompanied by simultaneous changes in rotational state, so giving rise to band spectra rather than the single absorption line of a pure vibrational transition. The spectra of linear polyatomic molecules show branches similar to those of diatomic molecules. For nonlinear molecules, the rotational fine structure is considerably more complex and difficult to analyse: even in moderately complex molecules the presence of several normal modes gives rise to several fundamental transitions, many overtones, and many combination lines, each with associated rotational fine structure, and results in infrared spectra of considerable complexity.

These complications are eliminated (or at least concealed) by recording infrared spectra of samples in the condensed phase (liquids, solutions, or solids). Molecules in liquids do not rotate freely but collide with each other after only a small change of orientation. As a result, the lifetimes of rotational states in liquids are very short, which results in a broadening of the associated energies (Topic 11A). Collisions occur at a rate of about $10^{13}\,s^{-1}$ and, even allowing for only a 10 per cent success rate in changing the molecule into another rotational state, a lifetime broadening of more than $1\,cm^{-1}$ can easily result. The rotational structure of the vibrational spectrum is blurred by this effect, so the infrared spectra of molecules in condensed phases usually consist of bands without any resolved branch structure.

Infrared spectroscopy is commonly used in routine chemical analysis, most usually on samples in solution, prepared as a fine dispersion (a 'mull'), or compressed as solids into a very thin layer. The resulting spectra show many absorption bands, even for moderately complex molecules. There is no chance of analysing such complex spectra in terms of the normal modes. However, they are of great utility for it turns out that certain groups within a molecule (such as a carbonyl group or an $-NH_2$ group) give rise to absorption bands in a particular range of wavenumbers. The spectra of a very large number of molecules have been recorded and these data have been used to draw up charts and tables of the expected range of the wavenumbers of absorptions from different groups. Comparison of the features in the spectrum of an unknown molecule or the product of a chemical reaction with entries in these data tables is a common first step towards identifying the molecule.

11D.3 Vibrational Raman spectra

As for diatomic molecules, the normal modes of vibration of molecules are Raman active if they are accompanied by a changing polarizability. A closer analysis of infrared and

Raman activity of normal modes based on considerations of symmetry leads to the **exclusion rule**:

> If the molecule has a centre of symmetry, then no mode can be both infrared and Raman active. Exclusion rule

(A mode may be inactive in both.) Because it is often possible to judge intuitively if a mode changes the molecular dipole moment, this rule can be used to identify modes that are not Raman active.

Brief illustration 11D.2

The symmetric stretch of CO_2 alternately swells and contracts the molecule: this motion changes the size and hence the polarizability of the molecule, so the mode is Raman active. Because CO_2 has a centre of symmetry the exclusion rule applies, so the stretching mode cannot be infrared active. The antisymmetric stretch and the two bends are infrared active, and so cannot be Raman active.

The assignment of Raman lines to particular vibrational modes is aided by noting the state of polarization of the scattered light. The **depolarization ratio**, ρ, of a line is the ratio of the intensities of the scattered light with polarization perpendicular and parallel, $I_\perp$ and $I_\parallel$, to the plane of the incident radiation (Fig. 11D.4):

$$\rho=\frac{I_\perp}{I_\parallel} \qquad \text{Depolarization ratio [definition]} \quad (11D.3)$$

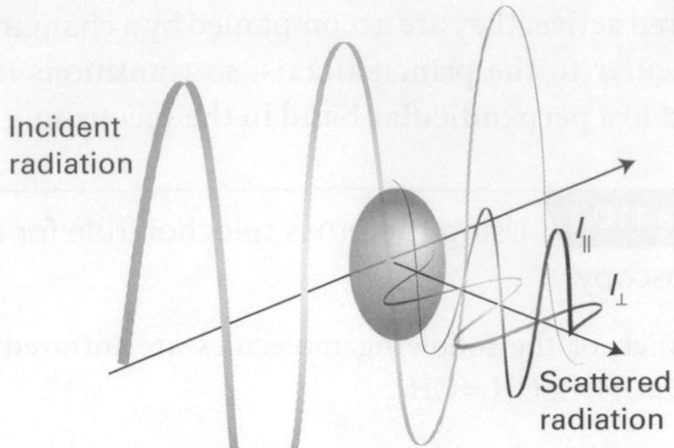

Figure 11D.4 The definition of the planes used for the specification of the depolarization ratio, ρ, in Raman scattering.

To measure ρ, the intensity of a Raman line is measured with a polarizing filter (a 'half-wave plate') first parallel and then perpendicular to the polarization of the incident beam. If the emergent light is not polarized, then both intensities are the same and ρ is close to 1; if the light retains its initial polarization, then $I_\perp=0$, so $\rho = 0$. A line is classified as **depolarized** if it has ρ close to or greater than 0.75 and as **polarized** if $\rho < 0.75$. Only totally symmetrical vibrations give rise to polarized lines in which the incident polarization is largely preserved. Vibrations that are not totally symmetrical give rise to depolarized lines because the incident radiation can give rise to radiation in the perpendicular direction too.

Checklist of concepts

- ☐ 1. A **normal mode** is a synchronous displacement of the atoms in which the centre of mass and orientation of the molecule remains fixed. In the harmonic approximation, normal modes are mutually independent.
- ☐ 2. A normal mode is **infrared active** if it is accompanied by a change of electric dipole moment; the specific selection rule is $\Delta v_q = \pm 1$.
- ☐ 3. A normal mode is **Raman active** if it is accompanied by a change in polarizability; the specific selection rule is $\Delta v_q = \pm 1$.
- ☐ 4. The **exclusion rule** states that, if the molecule has a centre of symmetry, then no mode can be both infrared and Raman active.
- ☐ 5. **Polarized lines** preserve the polarization of the incident radiation in the Raman spectrum and arise from totally symmetrical vibrations.

Checklist of equations

Property	Equation	Comment	Equation number
Number of normal modes	$N_{vib}=3N-5$ (linear) $N_{vib}=3N-6$ (nonlinear)	Independent if harmonic; N is the number of atoms	11D.1
Vibrational terms of normal modes	$\tilde{G}_q(v_q)=(v_q+\frac{1}{2})\tilde{\nu}_q$, $\tilde{\nu}_q=(1/2\pi c)(k_{f,q}/m_q)^{1/2}$	Harmonic approximation	11D.2
Depolarization ratio	$\rho=I_\perp/I_\parallel$	Depolarized lines: ρ close to or greater than 0.75 Polarized lines: $\rho < 0.75$	11D.3

TOPIC 11E Symmetry analysis of vibrational spectra

➤ **Why do you need to know this material?**

The analysis of vibrational spectra is aided by understanding the relationship between the symmetry of the molecule, its normal modes, and the selection rules that govern the transitions.

➤ **What is the key idea?**

The vibrational modes of a molecule can be classified according to the symmetry of the molecule.

➤ **What do you need to know already?**

You need to be familiar with the vibrational spectra of polyatomic molecules (Topic 11D) and the treatment of symmetry in Focus 10.

The classification of the normal modes of vibration of a polyatomic molecule according to their symmetry makes it possible to predict in a very straightforward way which are infrared or Raman active.

11E.1 Classification of normal modes according to symmetry

Each normal mode can be classified as belonging to one of the symmetry species of the irreducible representations of the molecular point group. The classification proceeds as follows:

1. The basis functions are the three displacement vectors (x, y, and z) on each atom: there are $3N$ such basis functions for a molecule with N atoms.
2. The character, $\chi(C)$, for each class, C, of operations in the group is found by considering the effect of one operation of the class and counting 1 for each basis function that is unchanged by the operation, −1 if the basis function changes sign, and 0 if it changes into some other displacement.
3. The resulting representation is decomposed into its component irreducible representations, denoted Γ with characters $\chi^{(\Gamma)}(C)$, by using the relevant character table in conjunction with eqn 10C.3a:

$$n(\Gamma)=\frac{1}{h}\sum_{C}N(C)\chi^{(\Gamma)}(C)\chi(C)$$

where h is the order of the group and $N(C)$ the number of operations in the class C.

4. The symmetry species corresponding to x, y, and z (corresponding to translations) and those corresponding to the rotations about x, y, and z (denoted R_x, R_y, and R_z) are removed. Their symmetry species are listed in the character tables.
5. The remaining symmetry species correspond to the normal modes.

Example 11E.1 Identifying the symmetry species of the normal modes of H_2O

Identify the symmetry species of the normal modes of H_2O, which belongs to the point group C_{2v}.

Collect your thoughts You need to identify the axes in the molecule and then refer to the character table (in the *Resource section*) for the symmetry operations and their characters. You need consider only one symmetry operation of each class, because all members of the same class have the same character. (In C_{2v}, there is only one member of each class anyway.) Then follow the five steps outlined in the text. Note from the character table the symmetry species of the translations and rotations, which are given in the right-hand column.

The solution The molecule lies in the yz-plane, with the z-axis bisecting the HOH bond angle. The three displacement vectors of each atom are shown in the Fig. 11E.1. From the character table for C_{2v}, the symmetry operations are E, C_2, $\sigma_v(xz)$, and $\sigma_v'(yz)$. None of the nine displacement vectors is affected by the operation E, so $\chi(E) = 9$. The C_2 operation moves all the displacement vectors on the H atoms to other positions, so these count 0; the x and y displacement vectors on the O atom change sign, giving a count of −1 each, whereas the z displacement vector is unaffected, giving a count of +1. Hence $\chi(C_2) = -1 - 1 + 1 = -1$. The operation $\sigma_v(xz)$ moves all the displacement vectors on the H atoms to other positions, changes the sign of the y displacement vector on the O atom, and leaves the x and z vectors unaffected: hence $\chi(\sigma_v) = -1 + 1 + 1 = 1$. The operation $\sigma_v'(yz)$ changes the sign of the x displacement vectors on both H atoms and leaves the sign of the y and z displacement vectors unaffected. For the O atom, the x vector changes sign and the y and z displacement vectors are unaffected: hence $\chi(\sigma_v') = -1 - 1 + 1 + 1 + 1 + 1 - 1 + 1 + 1 = 3$. The characters of the representation are therefore 9, −1, 1, 3. Decomposition by using eqn 10C.3a shows that the reducible representation spans the symmetry species $3A_1 + A_2 + 2B_1 + 3B_2$. The translations have symmetry species B_1, B_2, and A_1 and the rotations are B_1, B_2, and A_2; their removal leaves $2A_1 + B_2$ as the symmetry species of the normal modes. As expected, there are three such modes.

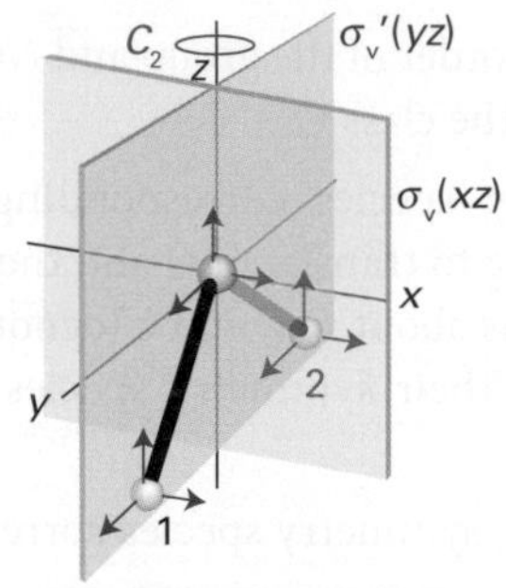

Fig. 11E.1 The atomic displacements of H_2O and the symmetry elements used to calculate the characters.

Comment. In Fig. 11D.3 (Topic 11D), ν_1 and ν_2 have symmetry A_1, ν_3 has symmetry B_2. This assignment is evident from the fact that the combination of displacements for both ν_1 and ν_2 are unchanged by any of the operations of the group, so the characters are all 1 as required for A_1. In contrast, for ν_3 the displacements shown change sign under C_2 and σ_v giving the characters 1, −1, −1, 1, which correspond to B_2.

Self-test 11E.1 Identify the symmetry species of the normal modes of methanal, $H_2C{=}O$, point group C_{2v} (orientate the molecule in the same way as H_2O, with the CH_2 group in the yz-plane).

Answer: $3A_1 + B_1 + 2B_2$

All the normal modes of H_2O are either A or B and therefore non-degenerate. There are no two- or higher-dimensional irreducible representations in C_{2v} molecules, so vibrational degeneracy never arises. Degeneracy can arise in molecules with higher symmetry, as illustrated in the following example.

Example 11E.2 Identifying the symmetry species of the normal modes of BF_3

Identify the symmetry species of the normal modes of vibration of BF_3, which is trigonal planar and belongs to the point group D_{3h}.

Collect your thoughts The overall procedure is the same as in *Example* 11E.1. However, because the molecule is D_{3h}, which has two-dimensional irreducible representations (E′ and E″), you need to be alert for the possibility that there are doubly degenerate pairs of normal modes. You can treat the displacement vectors on the B atom separately from those on the F atoms because no symmetry operation interconverts these two sets: this separation simplifies the calculations. Because the molecule is nonlinear with 4 atoms, there are 6 normal modes.

The solution The C_3 axis is the principal axis and defines the z-direction; the molecule lies in the xy-plane. The three C_2 axes pass along the B–F bonds, and the three σ_v planes contain the B–F bonds and are perpendicular to the plane of the molecule. The σ_h plane lies in the plane of the molecule, and the S_3 axis is coincident with the C_3 axis.

First, consider the displacement vectors on the B atom. Because this atom lies on the principal axis, the z displacement vector must transform as the function z, which from the character table has the symmetry species A_2''. Similarly, the x and y displacement vectors together transform as E′.

Next consider the nine displacement vectors on the F atoms. The identity operation has no effect, so $\chi(E) = 9$. The C_3 operation moves all these vectors, so $\chi(C_3) = 0$. A C_2 operation about a particular B–F bond has no effect on the displacement vector that points along the bond, but the other two vectors change sign; the displacement vectors on the other F atoms are moved, hence $\chi(C_2) = 1 - 1 - 1 = -1$. Under σ_h the z displacement vector on each F changes sign, but the x and y vectors do not. The character is therefore $\chi(\sigma_h) = 3 \times (-1 + 1 + 1) = 3$. The character for S_3 is the same as for C_3, $\chi(S_3) = 0$. A σ_v reflection in a plane containing a particular B–F bond has no effect on the displacement vector that points along the bond, nor on the z displacement vector; however, the other vector changes sign. The displacement vectors on the other atoms are moved, hence $\chi(\sigma_v) = 1 + 1 - 1 = 1$. The characters of the reducible representation are therefore 9, 0, −1, 3, 0, 1; this set can be decomposed into the symmetry species $A_1' + A_2' + 2E' + A_2'' + E''$ for the displacement vectors on the F atoms. The displacement vectors on the B atom transform as $A_2'' + E'$, so the complete set of symmetry species is $A_1' + A_2' + 3E' + 2A_2'' + E''$.

The character table shows that z transforms as A_2'', and x and y together span E′. The rotation about z, R_z, transforms as A_2' and rotations about x and y together (R_x, R_y) transform as E″. Removing these symmetry species from the complete set leaves $A_1' + 2E' + A_2''$ as the symmetry species of the vibrational modes. Figure 11E.2 shows these normal modes.

Comment. Because the E′ symmetry species is two-dimensional, the corresponding normal mode is doubly degenerate. The above analysis shows that there are two E′ symmetry species present, which correspond to two different doubly-degenerate normal modes. The total number of normal modes represented by $A_1' + 2E' + A_2''$ is therefore $1 + 2 \times 2 + 1 = 6$.

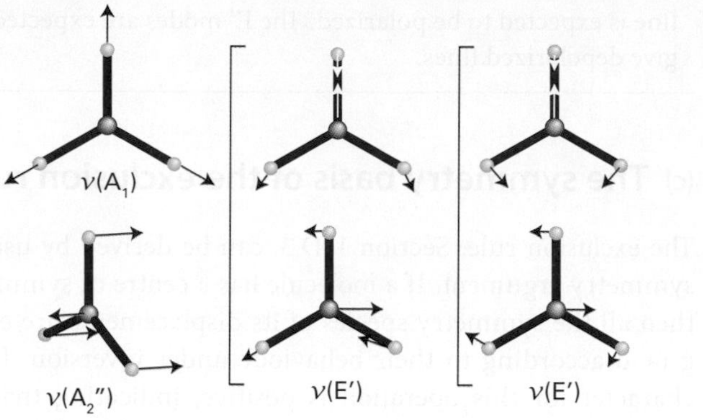

Fig. 11E.2 The normal modes of vibration of BF_3.

Self-test 11E.2 Identify the symmetry species of the normal modes of ammonia NH_3, point group C_{3v}.

Answer: $2A_1 + 2E$

11E.2 Symmetry of vibrational wavefunctions

For a one-dimensional harmonic oscillator the ground-state wavefunction (with $v = 0$) is proportional to $e^{-x^2/2\alpha^2}$, where x is the displacement from the equilibrium position and α is a constant (Topic 7E). For the first excited state, with $v = 1$, the wavefunction is proportional to $xe^{-x^2/2\alpha^2}$. The same wavefunctions apply to a normal mode q of a more complex molecule provided that x is replaced by the **normal coordinate**, Q_q, which is the combination of displacements that corresponds to the normal mode. For example, in the case of the symmetric stretch of CO_2 the normal coordinate is $z_{O,1} - z_{O,2}$, where $z_{O,i}$ is the z-displacement of oxygen atom i.

The effect of any symmetry operation on the normal coordinate of a non-degenerate normal mode is either to leave it unchanged or at most to change its sign. In other words, all the characters are either 1 or –1. The ground-state wavefunction is a function of the square of the normal coordinate, so regardless of whether $Q_q \to +Q_q$, or $Q_q \to -Q_q$ as a result of any symmetry operation, the effect on Q_q^2 is to leave it unaffected. Therefore, all the characters for Q_q^2 are 1, so the ground-state wavefunction transforms as the totally symmetric irreducible representation (typically A_1).

The first excited state wavefunction is a product of a part that depends on Q_q^2 (the exponential term) and a factor proportional to Q_q. As has already been seen, Q_q^2 transforms as the totally symmetric irreducible representation and Q_q has the same symmetry species as the normal mode. The direct product of the totally symmetric irreducible representation with any symmetry species leaves the latter unaffected, hence it follows that the symmetry of the first excited state wavefunction is the same as that of the normal mode.

(a) Infrared activity of normal modes

Once the symmetry of a particular normal mode is known it is a simple matter to determine from the appropriate character table whether or not the fundamental transition of that mode is allowed and therefore is infrared active.

How is that done? 11E.1 Determining the infrared activity of a normal mode

You need to note that the fundamental transition of a particular normal mode is the transition from the ground state, $v_q = 0$, to the first excited state, $v_q = 1$. You already know that the state with $v_q = 0$ transforms as the totally symmetric irreducible representation, and the state with $v_q = 1$ has the same symmetry as the corresponding normal mode.

Step 1 *Formulate the integral used to identify the selection rule*

Whether or not the transition between $v_q = 0$ and $v_q = 1$ is allowed is assessed by evaluating the transition dipole between ψ_0 and ψ_1, $\mu_{10} = \int \psi_1^* \hat{\mu} \psi_0 \mathrm{d}\tau$ (Topic 11A); the dipole moment operator transforms as x, y, or z (Topic 10C). As is shown in Topic 10C, this integral may be non-zero only if the product $\psi_1^* \hat{\mu} \psi_0$ spans the totally symmetric irreducible representation.

Step 2 *Identify the symmetry species spanned by the integrand*

You can find the symmetry species of $\psi_1^* \hat{\mu} \psi_0$ by taking the direct product of the symmetry species spanned by each of ψ_1, $\hat{\mu}$, and ψ_0 separately (Topic 10C). Because ψ_0 transforms as the totally symmetric irreducible representation it has no effect of the symmetry of $\psi_1^* \hat{\mu} \psi_0$, and so you need consider only the symmetry of the product $\psi_1^* \hat{\mu}$. As is shown in Topic 10C, the only way for this product to span the totally symmetric irreducible representation is for ψ_1^* and $\hat{\mu}$ to span the same symmetry species. In other words, the integral can be non-zero only if ψ_1^*, and hence the normal mode, has the same symmetry species as x, y, or z.

The result of the analysis can be summarized by the following rule:

A mode is infrared active only if its symmetry species is the same as the symmetry species of any of x, y, or z.

Symmetry test for IR activity

Brief illustration 11E.1

The normal modes of BF_3 (point group D_{3h}) have symmetry species $A_1' + 2E' + A_2''$ (*Example* 11E.2). From the character table it can be seen that z transforms as A_2'' and (x,y) jointly transform as E'. The A_2'' normal mode and the two doubly-degenerate E' normal modes are therefore infrared active. The A_1' mode is not.

(b) Raman activity of normal modes

Symmetry arguments also provide a systematic way of deciding whether or not the fundamental of a normal mode gives rise to Raman scattering; that is, whether or not the mode is Raman active. The argument is similar to that for assessing infrared activity except that it is based on the symmetry of the polarizability operator rather than the dipole moment operator. That operator transforms in the same way as the quadratic forms (x^2, xy, and so on, which are listed in the character tables), and leads to the following rule:

> A mode is Raman active only if its symmetry species is the same as the symmetry species of a quadratic form.

Symmetry test for Raman activity

Brief illustration 11E.2

The normal modes of BF_3 (point group D_{3h}) have symmetry species $A_1' + 2E' + A_2''$ (*Example* 11E.2). From the character table it can be seen that x^2, y^2, and z^2 all transform as A_1', and that (x^2-y^2, xy) jointly transform as E′. The A_1' normal mode and the two doubly-degenerate E′ normal modes are therefore Raman active. The A_2'' mode is not active because no quadratic form has this symmetry species. The E′ modes are both infrared and Raman active: the exclusion rule does not apply because BF_3 does not have a centre of symmetry. The A_1' normal mode is the highly-symmetrical breathing mode in which all the B–F bonds stretch together. The corresponding Raman line is expected to be polarized. The E′ modes are expected to give depolarized lines.

(c) The symmetry basis of the exclusion rule

The exclusion rule, Section 11D.3, can be derived by using a symmetry argument. If a molecule has a centre of symmetry, then all the symmetry species of its displacements are either g or u according to their behaviour under inversion. If the character for this operation is positive, indicating that the displacement or displacements are unchanged by the operation, the label is g, whereas if the character is negative, indicating that the sign of the displacement or displacements are changed, the label is u.

The functions x, y, and z (which occur in the transition dipole moment) all change sign under inversion, so they must correspond to symmetry species with a label u. In contrast, the quadratic forms (which govern the Raman activity) are all unchanged by inversion and so have the label g. For example, the effect of the inversion on xz is to transform it into $(-x)(-z) = xz$.

Any normal mode in a molecule with a centre of symmetry corresponds to a symmetry species that is either g or u. If the normal mode has the same symmetry species as x, y, or z, it is infrared active; such a mode must be u. If the normal mode has the same symmetry species as a quadratic form, it is Raman active; such a mode must be g. Because a normal mode cannot be both g and u, no mode can be both infrared and Raman active.

Checklist of concepts

☐ 1. A normal mode is infrared active if its symmetry species is the same as the symmetry species of x, y, or z.

☐ 2. A normal mode is Raman active if its symmetry species is the same as the symmetry species of a quadratic form.

TOPIC 11F Electronic spectra

> ➤ **Why do you need to know this material?**
>
> The study of spectroscopic transitions between different electronic states of molecules gives access to data on the electronic structure of molecules, and hence insight into bonding, as well as vibrational frequencies and bond lengths.
>
> ➤ **What is the key idea?**
>
> Electronic transitions occur within a stationary nuclear framework.
>
> ➤ **What do you need to know already?**
>
> You need to be familiar with the general features of spectroscopy (Topic 11A), the quantum mechanical origins of selection rules (Topics 8C, 11B, and 11C), and vibration–rotation spectra (Topic 11C). It would be helpful to be aware of atomic term symbols (Topic 8C).

Electronic spectra arise from transitions between the electronic energy levels of molecules. These transitions may also be accompanied by simultaneous changes in vibrational energy; for small molecules in the gas phase the resulting spectral features can be resolved (Fig. 11F.1a), but in a liquid or solid the individual lines usually merge together and result in a broad, almost featureless band (Fig. 11F.1b).

The energies needed to change the electron distributions of molecules are of the order of several electronvolts (1 eV is equivalent to about 8000 cm^{-1} or 100 kJ mol^{-1}). Consequently, the photons emitted or absorbed when such changes occur lie in the visible and ultraviolet regions of the spectrum (Table 11F.1).

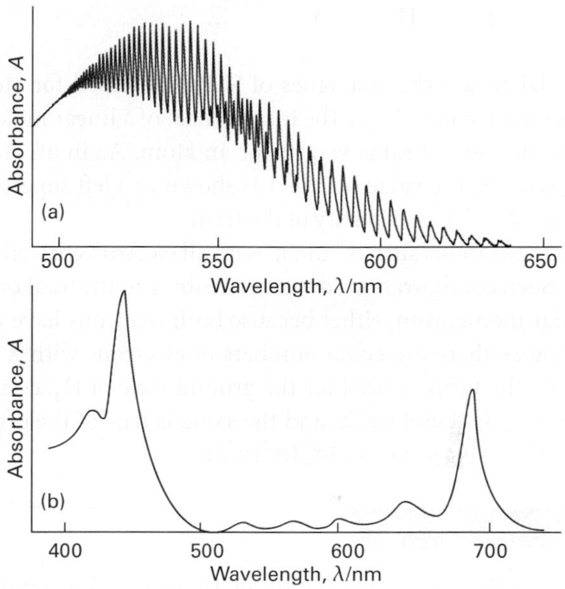

Figure 11F.1 Electronic absorption spectra recorded in the visible region. (a) The spectrum of I_2 in the gas phase shows resolved vibrational structure. (b) The spectrum of chlorophyll recorded in solution, shows only broad bands with no resolved structure. (Absorbance, *A*, is defined in Topic 11A.)

Table 11F.1 Colour, frequency, and energy of light*

Colour	λ/nm	$\nu/(10^{14}$ Hz)	E/(kJ mol^{-1})
Infrared	>1000	<3.0	<120
Red	700	4.3	170
Yellow	580	5.2	210
Blue	470	6.4	250
Ultraviolet	<400	>7.5	>300

* More values are given in the *Resource section*.

11F.1 Diatomic molecules

Topic 8C explains how the states of atoms are described by using term symbols. The electronic states of diatomic molecules are also specified by using term symbols, the key difference being that the full spherical symmetry of atoms is replaced by the cylindrical symmetry defined by the axis of the molecule.

(a) Term symbols

In a diatomic molecule only the component of the total orbital angular momentum around the internuclear axis can be specified: the quantum number for this component is Λ (uppercase lambda). Its value is found by adding together the component of the orbital angular momentum along the internuclear axis, λ_i, for each electron present:

$$\Lambda = \lambda_1 + \lambda_2 + \cdots \qquad (11F.1)$$

For an electron in a σ molecular orbital (which is cylindrically symmetric), $\lambda = 0$; for an electron in one of the degenerate pair of π orbitals $\lambda = \pm 1$. In the molecular term symbol the value of Λ is represented by an uppercase Greek letter in the following way

$\lvert\Lambda\rvert$	0	1	2	...
	Σ	Π	Δ	...

These labels are the analogues of S, P, D,... used for atomic states with $L = 0, 1, 2, \ldots$. The total spin, S, of a linear molecule is specified in the same way as for an atom. As in an atomic term symbol, the value of $2S + 1$ is shown as a left superscript and denotes the multiplicity of the term.

Configurations such σ^2 and π^4 with all electrons paired have $S = 0$. Such configurations do not contribute to the total orbital angular momentum, either because both electrons have $\lambda = 0$ or because there are equal numbers of electrons with $\lambda = +1$ and -1. The term symbol for the ground state of H_2, configuration $1\sigma_g^2$, is therefore $^1\Sigma$, and the same is true of the ground state of N_2, configuration $1\sigma_g^2 1\sigma_u^2 1\pi_u^4 2\sigma_g^2$.

Brief illustration 11F.1

The ground-state configuration of H_2^+ is $1\sigma_g^1$. The single σ electron has $\lambda = 0$, so $\Lambda = 0$; for a single electron $S = \frac{1}{2}$, so $2S + 1 = 2$. The term symbol is therefore $^2\Sigma$ (read as 'doublet sigma').

The ground-state configuration of NO is $\ldots 1\pi^1$, where... indicates completed orbitals that make no contribution to either S or Λ. The single electron can occupy either of the degenerate π orbitals, so $\Lambda = +1$ or -1; for a single electron, $S = \frac{1}{2}$. The term symbol is therefore $^2\Pi$ (read as 'doublet pi').

The ground-state configuration of O_2 is $\ldots 1\pi_g^2$. If the two electrons occupy the same π orbital, $\Lambda = (+1) + (+1) = +2$ (or $\Lambda = (-1) + (-1) = -2$); in this arrangement the electron spins must be paired, so $S = 0$. The resulting term is $^1\Delta$; a $^3\Delta$ term is not possible as it would require two electrons with parallel spins to occupy one of the π orbitals. If the electrons occupy different π orbitals, $\Lambda = (+1) + (-1) = 0$; in this arrangement the spins can be paired or parallel, so $S = 0$ or $S = 1$. Two further terms therefore arise: $^1\Sigma$ and $^3\Sigma$; the latter turns out to be the lowest in energy of all the three terms.

As explained in Topic 9B, homonuclear diatomic molecules (but not heteronuclear diatomic molecules) possess a centre of symmetry and their orbitals are labelled g or u according to their parity (the behaviour under inversion through a centre of symmetry). Orbitals that are unchanged upon inversion are g and orbitals that change sign are u. Parity labels also apply to centrosymmetric polyatomic linear molecules, such as CO_2 and HC≡CH. The overall parity of a configuration of a many-electron homonuclear diatomic molecule is found by noting the parity of each occupied orbital and using

$$g \times g = g \quad u \times u = g \quad u \times g = u \tag{11F.2}$$

for each electron. (These rules are generated by interpreting g as +1 and u as −1.) The resulting parity label g or u is added as a right-subscript to the term symbol. Any molecule in which the occupied orbitals are full (in the sense of being occupied by a pair of electrons) must have overall parity g because there is an even number of electrons. The term symbol for such a homonuclear diatomic molecule is therefore $^1\Sigma_g$.

Brief illustration 11F.2

Dinitrogen, N_2, has the configuration $1\sigma_g^2 1\sigma_u^2 1\pi_u^4 2\sigma_g^2$ in which all the occupied orbitals are full; the same is true of H_2 and F_2: all three therefore have the term symbol $^1\Sigma_g$.

The configuration of He_2^+ is $1\sigma_g^2 1\sigma_u^1$. There is one electron outside the doubly occupied bonding orbital, and the parity of that orbital is u. Because $S = \frac{1}{2}$ and $\Lambda = 0$, its term symbol is $^2\Sigma_u$.

The ground-state configuration of O_2 is $\ldots 1\pi_g^2$. Although the π_g orbitals may be both singly occupied, both electrons are in orbitals with g parity, so the overall parity is $g \times g = g$. The three terms arising from this configuration are therefore $^1\Sigma_g$, $^3\Sigma_g$, and $^1\Delta_g$ (see *Brief illustration* 11F.1).

Diatomic molecules (and all linear molecules) possess a mirror plane containing the internuclear axis. All σ orbitals (both bonding and antibonding) are symmetric with respect to reflection in this plane. The overall symmetry of a configuration is found by assigning +1 to an electron in a symmetric orbital and −1 to an electron in an orbital that changes sign under reflection and then multiplying the numbers for all the electrons. For example, for the ground state of H_2, in which both electrons are in σ orbitals, the overall symmetry is $(+1) \times (+1) = +1$. A + sign is added as a right-superscript to the term symbol: $^1\Sigma_g^+$. Any configuration consisting of electrons solely in σ orbitals necessarily has + overall reflection symmetry; for example, the ground state of He_2^+ (*Brief illustration* 11F.2) is $^2\Sigma_u^+$.

The behaviour of the degenerate pair of π molecular orbitals under reflection is more complex: as is shown in Fig. 11F.2 one of the orbitals changes sign but the other does not. The consequences of this observation can be explored by considering the mathematical form of the π orbitals and how they depend on the angle ϕ shown in Fig. 11F.3. The orbital π_x is proportional to $\cos\phi$ and therefore has a nodal plane at $\phi = \pi/2$ (the yz-plane) with positive and negative lobes on either side of this plane; it is unchanged by reflection in the xz-plane. The orbital π_y is proportional to $\sin\phi$, so the xz-plane at $\phi = 0$ is a nodal plane; the orbital changes sign on reflection in the xz-plane. These two wavefunctions are degenerate, so any linear combination of them is also an acceptable wavefunction. For the present discussion the combinations $\pi_+ = \cos\phi + i\sin\phi = e^{i\phi}$ and $\pi_- = \cos\phi - i\sin\phi = e^{-i\phi}$ are convenient for they correspond

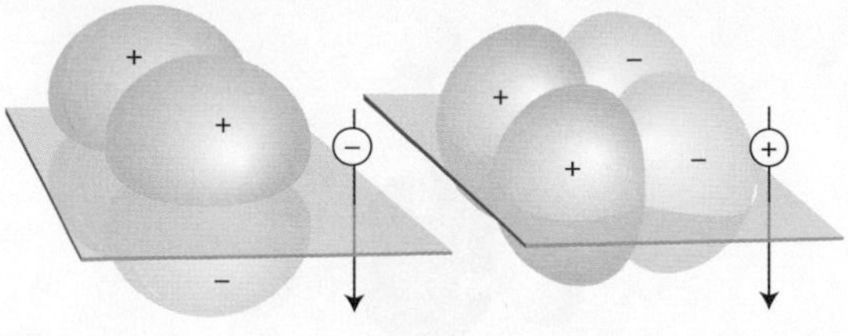

Figure 11F.2 A molecular orbital can be classified as symmetric (+) or antisymmetric (–) according to whether it changes sign under reflection in a plane containing the internuclear axis.

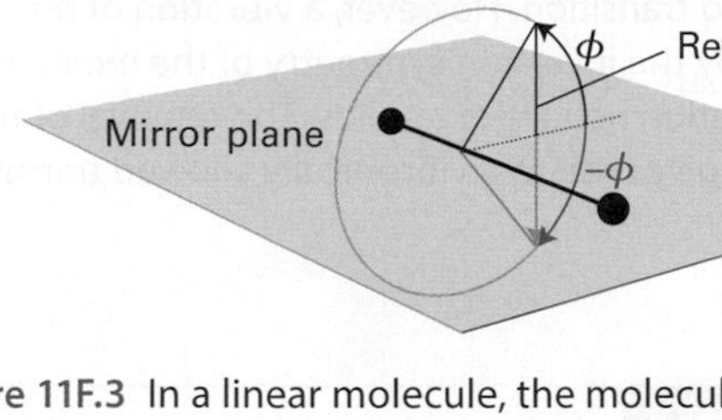

Figure 11F.3 In a linear molecule, the molecular orbital depends on the azimuthal angle ϕ. Reflection in the mirror plane is equivalent to reversing the sign of ϕ.

to $\lambda=+1$, and $\lambda=-1$, respectively. As shown in Fig. 11F.3, reflection in the xz-plane leaves $\cos\phi$ unchanged but changes the sign of $\sin\phi$. It follows that this reflection interconverts π_+ and π_-.

Now consider O_2, which has the configuration $\ldots 1\pi_g^2$. The triplet state ($S=1$), in which the two electrons have parallel spins and necessarily occupy different orbitals, is the state of lowest energy. The triplet spin wavefunction is symmetric with respect to the interchange of the two electrons (it consists of spin states $\alpha(1)\alpha(2)$, and so on), so it follows from the Pauli principle (Topic 8B) that the spatial part of the wavefunction must be antisymmetric with respect to interchange. Such a wavefunction, in which one electron occupies the π_+ orbital and the other occupies π_-, is $\Psi_-(1,2)=\pi_+(r_1)\pi_-(r_2)-\pi_+(r_2)\pi_-(r_1)$. Reflection in the mirror plane gives $\pi_-(r_1)\pi_+(r_2)-\pi_-(r_2)\pi_+(r_1)=-\Psi_+(1,2)$. That is, the spatial wavefunction of the triplet state is antisymmetric with respect to reflection in the mirror plane, and so a right-superscript – is attached to the term symbol, to give ${}^3\Sigma_g^-$.

Brief illustration 11F.3

An alternative, higher energy configuration of O_2 is with the outermost two electrons in separate π orbitals but with their spins paired (a ${}^1\Sigma_g$ term). The spin state, which is proportional to $\alpha(1)\beta(2)-\alpha(2)\beta(1)$, is antisymmetric with respect to the interchange of the electrons. The spatial function therefore must be symmetric. A suitable wavefunction is $\Psi_+(1,2)=\pi_+(r_1)\pi_-(r_2)+\pi_+(r_2)\pi_-(r_1)$. Reflection changes this function to $\pi_-(r_1)\pi_+(r_2)+\pi_-(r_2)\pi_+(r_1)$ which is $+\Psi_+(1,2)$. The state is symmetric with respect to this reflection, and a superscript + is added to the term symbol, to give ${}^1\Sigma_g^+$.

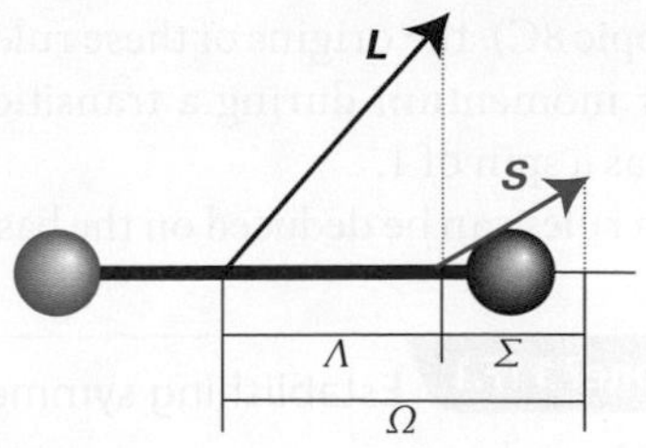

Figure 11F.4 The coupling of spin and orbital angular momenta in a linear molecule: only their components along the internuclear axis (Σ and Λ) are well defined.

As for atoms, it is sometimes necessary to specify the total electronic angular momentum, the sum of the orbital and spin contributions, and hence the different 'levels' of a term. In a linear molecule, only the total electronic angular momentum about the internuclear axis is well defined, and is specified by the quantum number Ω (uppercase omega). For light molecules, where the spin–orbit coupling is weak, Ω is obtained by adding together the components of orbital angular momentum around the axis (the value of Λ) and the component of the electron spin on that axis (Fig. 11F.4). The latter is denoted Σ, where $\Sigma=S, S-1, S-2, \ldots, -S$.[1] Then

$$\Omega=\Lambda+\Sigma \qquad (11F.3)$$

The value of $|\Omega|$ is then attached to the term symbol as a right subscript (just like J is used in atoms) to denote the different levels. These levels differ in energy, as in atoms, as a result of spin–orbit coupling.

Brief illustration 11F.4

The ground-state configuration of NO is $\ldots 1\pi^1$, so it is a ${}^2\Pi$ term with $\Lambda=\pm1$ and $S=\frac{1}{2}$; from the latter it follows that $\Sigma=\pm\frac{1}{2}$. There are two levels of the term, one with $\Omega=\pm\frac{1}{2}$ and the other with $\pm\frac{3}{2}$, denoted ${}^2\Pi_{1/2}$ and ${}^2\Pi_{3/2}$, respectively. Each level is doubly degenerate (corresponding to the opposite signs of Ω). It turns out that, in NO, ${}^2\Pi_{1/2}$ lies slightly lower in energy than ${}^2\Pi_{3/2}$.

(b) Selection rules

A number of selection rules govern which transitions can be observed in the electronic spectrum of a molecule. The selection rules concerned with changes in angular momentum in a linear molecule are

$$\Delta\Lambda=0,\pm1 \quad \Delta S=0 \quad \Delta\Sigma=0 \quad \Delta\Omega=0,\pm1$$

Selection rules for electronic spectra of linear molecules (11F.4)

[1] It is important to distinguish between the upright term symbol Σ and the sloping quantum number Σ.

As in atoms (Topic 8C), the origins of these rules are conservation of angular momentum during a transition and the fact that a photon has a spin of 1.

Two selection rules can be deduced on the basis of symmetry.

How is that done? 11F.1 Establishing symmetry-based selection rules

As usual when establishing selection rules, you need to consider properties of the electric-dipole transition moment introduced in Topic 8C, $\mu_{fi} = \int \psi_f^* \hat{\mu} \psi_i d\tau$, and to note that it vanishes unless the integrand is invariant under all symmetry operations of the molecule (Topic 10C).

The z-component (the component parallel to the axis of the molecules) of the electric dipole moment operator is responsible for $\Sigma \leftrightarrow \Sigma$ transitions (the other components of $\boldsymbol{\mu}$ perpendicular to the axis have Π symmetry and cannot make a contribution to this transition). The z-component of $\boldsymbol{\mu}$ has (+) symmetry with respect to reflection in a plane containing the internuclear axis. Therefore, for a $(+) \leftrightarrow (-)$ transition, the overall symmetry of the integrand is $(+) \times (+) \times (-) = (-)$, so the integral must be zero and hence $\Sigma^+ \leftrightarrow \Sigma^-$ transitions are not allowed. The integrands for $\Sigma^+ \leftrightarrow \Sigma^+$ and $\Sigma^- \leftrightarrow \Sigma^-$ transitions transform as $(+) \times (+) \times (+) = (+)$ and $(-) \times (+) \times (-) = (+)$, respectively. The integrals are therefore not necessarily zero and so both transitions are allowed.

The three components of the dipole moment operator transform like x, y, and z, and in a centrosymmetric molecule are all u. Therefore, for a $g \rightarrow g$ transition, the overall parity of the integrand is $g \times u \times g = u$, so the integral must be zero. Likewise, for a $u \rightarrow u$ transition, the overall parity is $u \times u \times u = u$, so the integral is again zero. Hence, transitions without a change of parity are forbidden. For a $g \leftrightarrow u$ transition the integrand transforms as $g \times u \times u = g$, so the transition is allowed.

The first part of this analysis can be summarized as follows:

For Σ terms, only $\Sigma^+ \leftrightarrow \Sigma^+$ and $\Sigma^- \leftrightarrow \Sigma^-$ are allowed.

The second part is in fact the **Laporte selection rule** for centrosymmetric molecules (those with a centre of inversion, not only linear molecules) which states that *the only transitions allowed are accompanied by a change of parity*. That is,

For centrosymmetric molecules, only $u \rightarrow g$ and $g \rightarrow u$ transitions are allowed.

Laporte selection rule

A forbidden $g \rightarrow g$ transition can become allowed if the centre of symmetry is eliminated by an asymmetrical vibration, such as the one shown in Fig. 11F.5. When the centre of symmetry is lost, $g \rightarrow g$ and $u \rightarrow u$ transitions are no longer parity-forbidden and become weakly allowed. A transition that derives its intensity from an asymmetrical vibration of a molecule is called a **vibronic transition**.

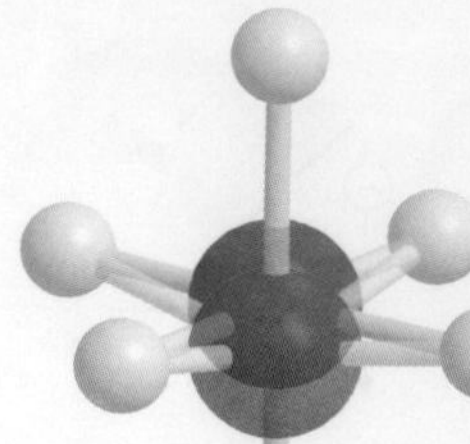

Figure 11F.5 A d–d transition is parity-forbidden because it corresponds to a g–g transition. However, a vibration of the molecule can destroy the inversion symmetry of the molecule and the g,u classification no longer applies. The removal of the centre of symmetry gives rise to a vibronically allowed transition.

Brief illustration 11F.5

Three possible transitions in the electronic spectrum of O_2, ${}^3\Sigma_g^- \leftarrow {}^3\Sigma_u^-$, ${}^3\Sigma_g^- \leftarrow {}^1\Delta_g$, ${}^3\Sigma_g^- \leftarrow {}^3\Sigma_u^+$, can be considered in the light of the selection rules in eqn 11F.4 to see which are allowed. A table can be drawn up, in which forbidden values are shown in red.

	ΔS	$\Delta\Lambda$	$\Sigma^\pm \leftarrow \Sigma^\pm$	Change of parity	
${}^3\Sigma_g^- \leftarrow {}^3\Sigma_u^-$	0	0	$\Sigma^- \leftarrow \Sigma^-$	$g \leftarrow u$	Allowed
${}^3\Sigma_g^- \leftarrow {}^1\Delta_g$	+1	−2	Not applicable	$g \leftarrow g$	Forbidden
${}^3\Sigma_g^- \leftarrow {}^3\Sigma_u^+$	0	0	$\Sigma^- \leftarrow \Sigma^+$	$g \leftarrow u$	Forbidden

(c) Vibrational fine structure

An electronic transition may be accompanied by a simultaneous change in the vibrational state of a molecule, giving rise to **vibrational fine structure** in the spectrum. In the case of absorption spectra, the transitions are from the ground electronic state, and typically it is only the ground vibrational level, $v''=0$, of this state that is occupied significantly. In some cases, the transition from $v''=0$ in the electronic ground state to $v'=0$ in the upper electronic state is found to be the strongest, with a sharp decline in intensity as v' increases. In other cases, transitions with significant intensity to a range of v' levels are seen (as in Fig. 11F.1a).

The **Franck–Condon principle** accounts for the vibrational fine structure in the electronic spectra of molecules:

Because nuclei are so much more massive than electrons, an electronic transition takes place very much faster than the nuclei can respond.

Franck–Condon principle

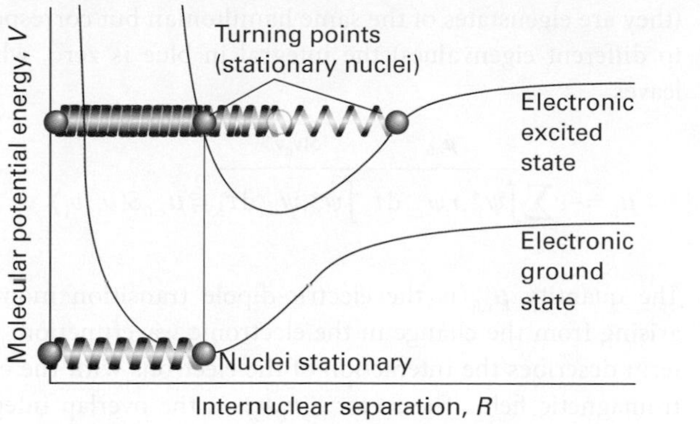

Figure 11F.6 According to the Franck–Condon principle, the most intense vibronic transition is from the ground vibrational state to the vibrational state lying vertically above it. As a result of the vertical transition, the nuclei suddenly experience a new force field, to which they respond through their vibrational motion. The equilibrium separation of the nuclei in the initial electronic state therefore becomes a turning point in the final electronic state. Transitions to other vibrational levels also occur, but with lower intensity.

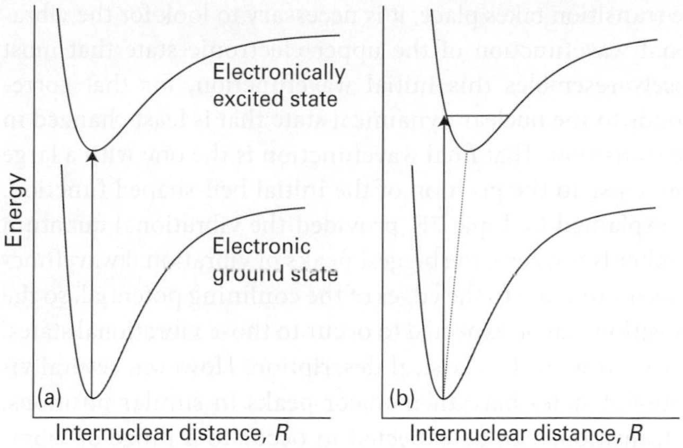

Figure 11F.7 (a) If the equilibrium bond lengths of the ground and excited electronic states are the same, a vertical transition leaves the vibrational state of the molecule unexcited. (b) If the equilibrium bond length is greater in the upper electronic state, the vertical transition ends at a compressed state of the bond and results in vibrational excitation.

The physical basis of this principle is as follows. As a result of the electronic transition, electron density is built up rapidly in new regions of the molecule and removed from others. In classical terms, the initially stationary nuclei suddenly experience a new force field, to which they respond by beginning to vibrate and (in classical terms) swing backwards and forwards from their original separation which was maintained during the rapid electronic excitation. The stationary equilibrium separation of the nuclei in the initial electronic state therefore becomes a stationary turning point in the final electronic state. The transition can be thought of as taking place up the vertical line in Fig. 11F.6. This interpretation is the origin of the expression **vertical transition**, which denotes an electronic transition that occurs without change of nuclear geometry and, in classical terms, with the nuclei remaining stationary.

Now consider the two potential energy curves shown in Fig. 11F.7a in which the equilibrium bond lengths are the same and initially the molecule is not vibrating. The vertical transition takes place from the minimum of the lower curve, the nuclei remain at the same separation, and ends at the minimum of the upper curve.

Next consider the case shown in Fig. 11F.7b, in which the equilibrium bond length in the upper state is greater than that in the ground electronic state, and the molecule is initially not vibrating. Preservation of the nuclear separation during the transition takes the molecule up the vertical line. The nuclei are not moving initially, and do not start to move during the transition, so the transition terminates at the turning point of the upper electronic state where the nuclei are still stationary.

The quantum mechanical version of the Franck–Condon principle refines this picture. Instead of saying that the nuclei stay at the same locations and are stationary during the transition, it replaces that statement by the assertion that *the nuclei retain their initial dynamical state*. In quantum mechanics, the dynamical state is expressed by the wavefunction, so an equivalent statement is that the vibrational wavefunction does not change during the electronic transition. Initially the molecule is in the lowest vibrational state of its ground electronic state with a bell-shaped wavefunction centred on the equilibrium bond length (Fig. 11F.8). To find the nuclear state to which

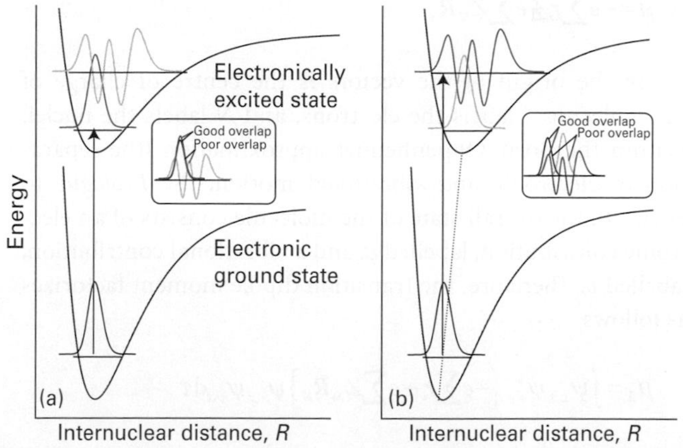

Figure 11F.8 (a) If the equilibrium bond lengths of the ground and excited electronic states are the same, the wavefunctions for $v''=0$ and $v'=0$ are similar and the most probable transition leaves the molecule vibrationally unexcited. (b) If the equilibrium bond length of the upper state is greater than that of the ground state, the wavefunction that most resembles the ground state vibrational wavefunction is that of an excited state. Other transitions also occur with lower intensity. In the inserts, the black curve is the initial vibrational wavefunction and the blue curves are those of the upper electronic state.

the transition takes place, it is necessary to look for the vibrational wavefunction of the upper electronic state that most closely resembles this initial wavefunction, for that corresponds to the nuclear dynamical state that is least changed in the transition. That final wavefunction is the one with a large peak close to the position of the initial bell-shaped function. As explained in Topic 7E, provided the vibrational quantum number is not zero, the biggest peaks of vibrational wavefunctions occur close to the edges of the confining potential, so the transition can be expected to occur to those vibrational states, in accord with the classical description. However, several vibrational states have their major peaks in similar positions, so transitions can be expected to occur to a range of vibrational states, to give rise to a **vibrational progression**, a series of transitions to different vibrational states of the upper electronic state. In a typical progression, the vertical transition is the most intense.

The quantitative version of the Franck–Condon principle involves considering how the transition dipole moment for a given electronic transition varies with the vibrational levels in the two electronic states.

How is that done? 11F.2 Expressing the Franck–Condon principle quantitatively

Once again, you need to consider the properties of the electric-dipole transition moment. First, note that the electric dipole moment operator is a sum over all nuclei and electrons in the molecule:

$$\hat{\mu}=-e\sum_i \boldsymbol{r}_i+e\sum_N Z_N\boldsymbol{R}_N$$

where the origin of the vectors is the centre of charge of the molecule, i labels the electrons, and N labels the nuclei. Within the Born–Oppenheimer approximation (the separation of electronic and vibrational motion, the *Prologue* to Focus 9), the overall state of the molecule consists of an electronic contribution, labelled ε, and a vibrational contribution, labelled v. Therefore, the transition dipole moment factorizes as follows:

$$\begin{aligned}\mu_{\rm fi}&=\int\psi^*_{\varepsilon,\rm f}\psi^*_{v,\rm f}\Big\{-e\sum_i\boldsymbol{r}_i+e\sum_N Z_N\boldsymbol{R}_N\Big\}\psi_{\varepsilon,\rm i}\psi_{v,\rm i}\,{\rm d}\tau\\&=-e\sum_i\int\psi^*_{\varepsilon,\rm f}\boldsymbol{r}_i\psi_{\varepsilon,\rm i}\,{\rm d}\tau_{\rm e}\int\psi^*_{v,\rm f}\psi_{v,\rm i}\,{\rm d}\tau_{\rm N}\\&\quad+e\sum_N Z_N\overbrace{\int\psi^*_{\varepsilon,\rm f}\psi_{\varepsilon,\rm i}\,{\rm d}\tau_{\rm e}}^{0}\int\psi^*_{v,\rm f}\boldsymbol{R}_N\psi_{v,\rm i}\,{\rm d}\tau_{\rm N}\end{aligned}$$

where ${\rm d}\tau_{\rm e}$ indicates integration over the electronic coordinates, and ${\rm d}\tau_{\rm N}$ integration over the nuclear coordinates. Because the two different electronic states are orthogonal (they are eigenstates of the same hamiltonian but correspond to different eigenvalues) the integral in blue is zero, which leaves

$$\mu_{\rm fi}=\overbrace{-e\sum_i\int\psi^*_{\varepsilon,\rm f}\boldsymbol{r}_i\psi_{\varepsilon,\rm i}\,{\rm d}\tau_{\rm e}}^{\mu_{\varepsilon,\rm fi}}\overbrace{\int\psi^*_{v,\rm f}\psi_{v,\rm i}\,{\rm d}\tau_{\rm N}}^{S(v_{\rm f},v_{\rm i})}=\mu_{\varepsilon,\rm fi}S(v_{\rm f},v_{\rm i})$$

The quantity $\mu_{\varepsilon,\rm fi}$ is the electric-dipole transition moment arising from the change in the electronic wavefunction: this term describes the interaction of the electrons with the electromagnetic field. The factor $S(v_{\rm f},v_{\rm i})$, is the overlap integral between the vibrational level with quantum number $v_{\rm i}$ in the initial electronic state of the molecule, and the vibrational level with quantum number $v_{\rm f}$ in the final electronic state of the molecule.

The transition intensity is proportional to the square of the magnitude of the transition dipole moment, so is proportional to the square of $S(v_{\rm f},v_{\rm i})$, and specifically

$$|S(v_{\rm f},v_{\rm i})|^2=\left(\int\psi^*_{v,\rm f}\psi_{v,\rm i}\,{\rm d}\tau_{\rm N}\right)^2 \qquad \text{Franck–Condon factor} \qquad (11F.5)$$

The integral on the right-hand side of eqn 11F.5 is the overlap between the two vibrational wavefunctions: the greater this overlap (physically, the greater the resemblance of the vibrational wavefunctions), the greater is the intensity of the transition.

Example 11F.1 Calculating a Franck–Condon factor

Consider the transition from one electronic state to another, their equilibrium bond lengths being $R_{\rm e}$ and $R'_{\rm e}$, and their force constants equal. Calculate the Franck–Condon factor for the $v''=0$ to $v'=0$ transition (the 0–0 transition) and show that the transition is most intense when the bond lengths are equal.

Collect your thoughts You need to calculate $S(0,0)$, the overlap integral of the two ground-state vibrational wavefunctions, and then take its square. The difference between harmonic and anharmonic vibrational wavefunctions is negligible for $v=0$, so it is safe for you to use the harmonic oscillator wavefunctions.

The solution The (real) wavefunctions are (Topic 7E)

$$\psi_0=\left(\frac{1}{\alpha\pi^{1/2}}\right)^{1/2}{\rm e}^{-x^2/2\alpha^2}\qquad \psi'_0=\left(\frac{1}{\alpha\pi^{1/2}}\right)^{1/2}{\rm e}^{-x'^2/2\alpha^2}$$

where $x=R-R_{\rm e}$ and $x'=R-R'_{\rm e}$, with $\alpha=(\hbar^2/\mu k_{\rm f})^{1/4}$. The overlap integral is

$$S(0,0)=\int_{-\infty}^{\infty}\psi'_0\psi_0\,{\rm d}R=\frac{1}{\alpha\pi^{1/2}}\int_{-\infty}^{\infty}{\rm e}^{-(x^2+x'^2)/2\alpha^2}\,{\rm d}x$$

Now recognize that

$$x^2+x'^2=(R-R_e)^2+(R-R_e')^2$$
$$=2R^2+R_e^2+R_e'^2-2R(R_e+R_e')$$
$$=2\{R-\tfrac{1}{2}(R_e+R_e')\}^2+\tfrac{1}{2}(R_e-R_e')^2$$

and write $\alpha z = R - \tfrac{1}{2}(R_e + R_e')$, so

$$\frac{x^2+x'^2}{2\alpha^2}=z^2+\frac{(R_e-R_e')^2}{4\alpha^2}$$

and $dR = \alpha dz$. Then

$$S(0,0)=\frac{1}{\pi^{1/2}}e^{-(R_e-R_e')^2/4\alpha^2}\overbrace{\int_{-\infty}^{\infty}e^{-z^2}dz}^{\text{Integral G.1, }\pi^{1/2}}=e^{-(R_e-R_e')^2/4\alpha^2}$$

and the Franck–Condon factor is

$$S(0,0)^2=e^{-(R_e-R_e')^2/2\alpha^2}$$

This factor is equal to 1 when $R_e' = R_e$ and decreases as the equilibrium bond lengths diverge from each other (Fig. 11F.9).

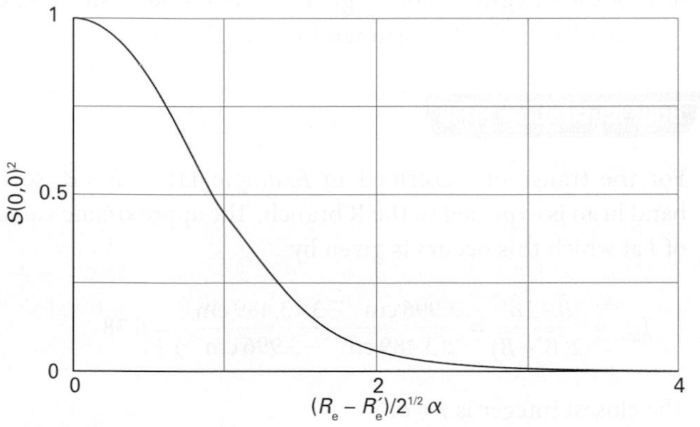

Figure 11F.9 The Franck–Condon factor for the arrangement discussed in *Example* 11F.1.

For $^{79}Br_2$, $R_e = 228$ pm and there is an upper state with $R_e' = 266$ pm. Taking the vibrational wavenumber as 250 cm^{-1} gives $\alpha^2 = 3.42 \times 10^{-23}$ m^2 and hence $S(0,0)^2 = 6.7 \times 10^{-10}$, so the intensity of the 0–0 transition is only 6.7×10^{-10} what it would have been if the potential curves had been directly above each other.

Self-test 11F.1 Suppose the normalized vibrational wavefunctions can be approximated by rectangular functions of width W and W', centred on the equilibrium bond lengths (Fig. 11F.10). Find the corresponding Franck–Condon factors when the centres are coincident and $W' < W$.

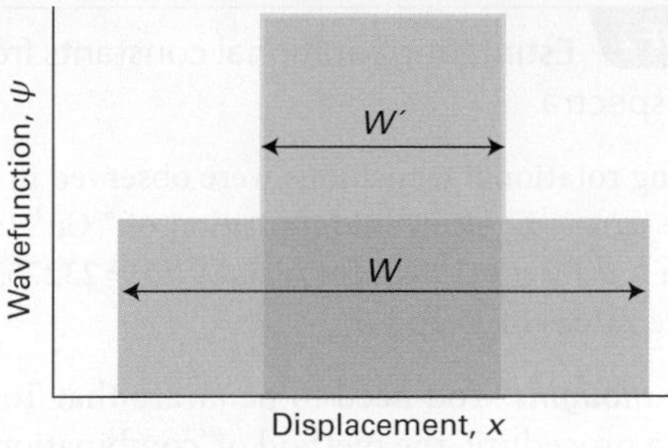

Figure 11F.10 The model wavefunctions used in *Self-test* 11F.1.

Answer: $S^2 = W'/W$

(d) Rotational fine structure

Electronic transitions may be accompanied by simultaneous changes in both vibrational and rotational energy. Therefore, when the lines dues to vibrational fine structure are inspected at higher resolution they are found to have **rotational fine structure** and to consist of P, Q, and R branches of the type discussed in Topic 11C. Because electronic excitation can result in much larger changes in bond length than vibrational excitation causes alone, the rotational branches have a more complex structure than in vibration–rotation spectra.

The rotational constants of the electronic ground and excited states are denoted $\tilde{B}$ and $\tilde{B}'$, respectively. The rotational terms of the initial and final states are

$$\tilde{F}(J)=\tilde{B}J(J+1) \qquad \tilde{F}(J')=\tilde{B}'J'(J'+1) \qquad (11F.6)$$

When a transition occurs with $\Delta J = -1$ the wavenumber of the vibrational contribution to the electronic transition is shifted from $\tilde{\nu}$ to

$$\tilde{\nu}+\tilde{B}'(J-1)J-\tilde{B}J(J+1)=\tilde{\nu}-(\tilde{B}'+\tilde{B})J+(\tilde{B}'-\tilde{B})J^2$$

This transition is a contribution to the P branch (just as in Topic 11C). There are corresponding transitions for the Q and R branches with wavenumbers that may be calculated in a similar way. All three branches are:

$$\text{P branch } (\Delta J=-1)\text{: } \tilde{\nu}_P(J)=\tilde{\nu}-(\tilde{B}'+\tilde{B})J+(\tilde{B}'-\tilde{B})J^2 \qquad (11F.7a)$$

$$\text{Q branch } (\Delta J=0)\text{: } \tilde{\nu}_Q(J)=\tilde{\nu}+(\tilde{B}'-\tilde{B})J(J+1) \qquad (11F.7b)$$

$$\text{R branch } (\Delta J=+1)\text{: } \tilde{\nu}_R(J)=\tilde{\nu}+(\tilde{B}'+\tilde{B})(J+1)+(\tilde{B}'-\tilde{B})(J+1)^2 \qquad (11F.7c)$$

Example 11F.2 Estimating rotational constants from electronic spectra

The following rotational transitions were observed in the 0–0 band of the ${}^1\Sigma^+ \leftarrow {}^1\Sigma^+$ electronic transition of ${}^{63}Cu^2H$: $\tilde{\nu}_R(3)$=23347.69 cm^{-1}, $\tilde{\nu}_P(3)$=23298.85 cm^{-1}, and $\tilde{\nu}_P(5)$=23275.77 cm^{-1}. Estimate the values of $\tilde{B}'$ and $\tilde{B}$.

Collect your thoughts You need to be aware that Topic 11C introduces a procedure, the method of 'combination differences', for analysing transitions that have a common state. According to that method, form the differences $\tilde{\nu}_R(J)-\tilde{\nu}_P(J)$ and $\tilde{\nu}_R(J-1)-\tilde{\nu}_P(J+1)$ from eqns 11F.7a and 11F.7c, then use the resulting expressions to calculate the rotational constants $\tilde{B}'$ and $\tilde{B}$ from the data provided.

The solution From eqns 11F.7a and 11F.7c it follows that

$$\begin{aligned}\tilde{\nu}_R(J)-\tilde{\nu}_P(J) &= (\tilde{B}'+\tilde{B})(J+1)+(\tilde{B}'-\tilde{B})(J+1)^2\\ &\quad -\{-(\tilde{B}'+\tilde{B})J+(\tilde{B}'-\tilde{B})J^2\} = 4\tilde{B}'(J+\tfrac{1}{2})\\ \tilde{\nu}_R(J-1)-\tilde{\nu}_P(J+1) &= (\tilde{B}'+\tilde{B})J+(\tilde{B}'-\tilde{B})J^2\\ &\quad -\{-(\tilde{B}'+\tilde{B})(J+1)+(\tilde{B}'-\tilde{B})(J+1)^2\}\\ &= 4\tilde{B}(J+\tfrac{1}{2})\end{aligned}$$

(These equations are analogous to eqn 11C.14.) The data provided can be used in the following way:

$$\text{For } J=3:\quad \tilde{\nu}_R(3)-\tilde{\nu}_P(3)=\overbrace{48.84}^{23\,347.69-23\,298.85}\text{ cm}^{-1}=14\tilde{B}'$$

$$\text{For } J=4:\quad \tilde{\nu}_R(3)-\tilde{\nu}_P(5)=\overbrace{71.92}^{23\,347.69-23\,275.77}\text{ cm}^{-1}=18\tilde{B}$$

Hence $\tilde{B}'$=3.489 cm^{-1} and $\tilde{B}$=3.996 cm^{-1}.

Self-test 11F.2 The following rotational transitions were observed in the ${}^1\Sigma^+ \leftarrow {}^1\Sigma^+$ electronic transition of RhN: $\tilde{\nu}_R(5)$=22387.06 cm^{-1}, $\tilde{\nu}_P(5)$=22376.87 cm^{-1}, and $\tilde{\nu}_P(7)$=22373.95 cm^{-1}. Estimate the values of $\tilde{B}'$ and $\tilde{B}$.

Answer: $\tilde{B}'$=0.4632 cm^{-1}, $\tilde{B}$=0.5042 cm^{-1}

Suppose that the bond length in the electronically excited state is greater than that in the ground state; it follows that $\tilde{B}'<\tilde{B}$ and hence $\tilde{B}'-\tilde{B}<0$. In this case the lines of the R branch converge with increasing J and at sufficiently high values of J the negative term in $(J+1)^2$ in eqn 11F.7c will dominate the positive term in $(J+1)$ and the lines will start to appear at successively decreasing wavenumbers. That is, the R branch has a **band head** (Fig. 11F.11a).

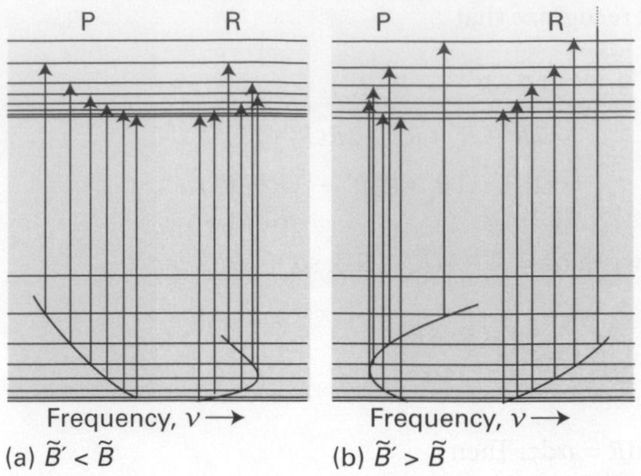

Figure 11F.11 (a) The formation of a head in the R branch when $\tilde{B}'<\tilde{B}$; (b) the formation of a head in the P branch when $\tilde{B}'>\tilde{B}$. The red curve shows the wavenumbers of the lines in the two branches as they spread away from the centre of the band.

The value of J at which the band head appears can be found by finding the maximum wavenumber of the lines in the R branch, in other words an integral value of J close to where $d\tilde{\nu}_R(J)/dJ=0$. This maximum occurs at $J_{max}\approx(\tilde{B}-3\tilde{B}')/2(\tilde{B}'-\tilde{B})$. When the bond is shorter in the excited state than in the ground state, $\tilde{B}'>\tilde{B}$ and $\tilde{B}'-\tilde{B}>0$. In this case, the lines of the P branch begin to converge and form a band head when $J_{max}\approx(\tilde{B}'+\tilde{B})/2(\tilde{B}'-\tilde{B})$, as shown in Fig. 11F.11b.

Brief illustration 11F.6

For the transition described in *Example* 11F.2, $\tilde{B}'<\tilde{B}$, so a band head is expected in the R branch. The approximate value of J at which this occurs is given by

$$J_{max}\approx\frac{\tilde{B}-3\tilde{B}'}{2(\tilde{B}'-\tilde{B})}=\frac{3.996\text{ cm}^{-1}-3\times3.489\text{ cm}^{-1}}{2(3.489\text{ cm}^{-1}-3.996\text{ cm}^{-1})}=6.38$$

The closest integer is $J=6$.

11F.2 Polyatomic molecules

The absorption of a photon can often be traced to the excitation of electrons that belong to a small group of atoms in a polyatomic molecule. For example, when a carbonyl group ($\gt$C=O) is present, an absorption at about 290 nm is normally observed, although its precise location depends on the nature of the rest of the molecule. Groups with characteristic optical absorption bands are called **chromophores** (from the Greek for 'colour bringer'), and their presence often accounts for the colours of substances (Table 11F.2).

Table 11F.2 Absorption characteristics of some groups and molecules*

Group	$\tilde{\nu}$/cm^{-1}	λ_{max}/nm	ε_{max}/(dm^3 mol^{-1} cm^{-1})
C=C ($\pi^* \leftarrow \pi$)	61 000	163	15 000
C=O ($\pi^* \leftarrow$ n)	35 000–37 000	270–290	10–20
H_2O	60 000	167	7000

* More values are given in the *Resource section*. ε_{max} is the molar absorption coefficient (see Topic 11A). The wavenumbers and wavelengths are the values for maximum absorption.

(a) d-Metal complexes

In a free atom, all five d orbitals of a given shell are degenerate. In a d-metal complex, where the immediate environment of the atom is no longer spherical, the d orbitals are not all degenerate, and electrons can absorb energy by making transitions between them.

To see the origin of this splitting in an octahedral complex such as $[Ti(OH_2)_6]^{3+}$ (**1**), the six ligands can be regarded as point negative charges that repel the d electrons of the central ion (Fig. 11F.12). As a result, the orbitals fall into two groups, with $d_{x^2-y^2}$ and d_{z^2} pointing directly towards the ligand positions, and d_{xy}, d_{yz}, and d_{zx} pointing between them. An electron occupying an orbital of the former group has a less favourable potential energy than when it occupies any of the three orbitals of the other group, and so the d orbitals split into the two sets shown in (**2**): a triply degenerate set comprising the d_{xy}, d_{yz}, and d_{zx} orbitals and labelled t_{2g}, and a doubly degenerate set comprising the $d_{x^2-y^2}$ and d_{z^2} orbitals and labelled e_g (these symmetry labels are discussed in Topic 10B). The t_{2g} orbitals lie below the e_g orbitals in energy; the difference in energy Δ_o is called the **ligand-field splitting parameter** (the o denotes octahedral symmetry). The ligand field splitting is typically about 10 per cent of the overall energy of interaction between the ligands and the central metal atom, which is largely responsible for the existence of the complex. The d orbitals also divide into two sets in a tetrahedral complex, but in this case the two e orbitals lie below the three t_2 orbitals (no g or u label is given because a tetrahedral complex has no centre of inversion); the separation of these groups of orbitals is written Δ_t.

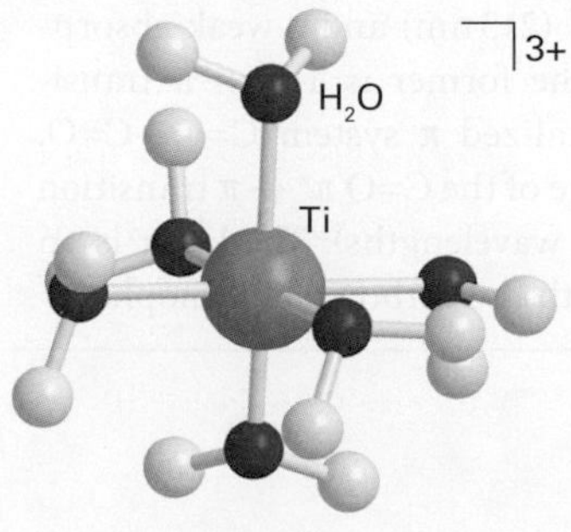

1 $[Ti(OH_2)_6]^{3+}$

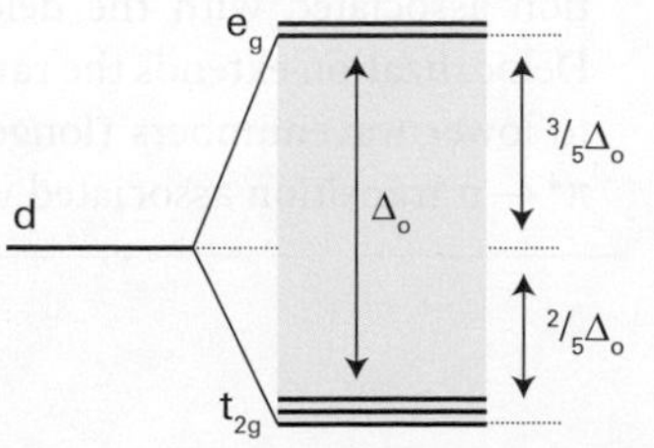

2

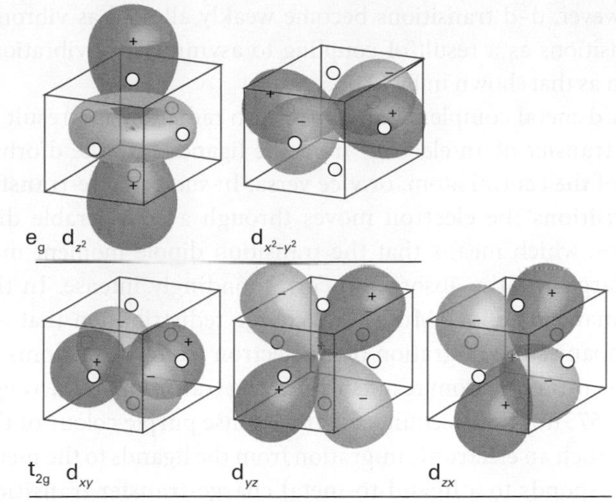

Figure 11F.12 The classification of d orbitals in an octahedral environment. The open circles represent the positions of the six (point-charge) ligands.

The values of Δ_o and Δ_t are such that transitions between the two sets of orbitals typically occur in the visible region of the spectrum. The transitions are responsible for many of the colours that are so characteristic of d-metal complexes.

Brief illustration 11F.7

The spectrum of $[Ti(OH_2)_6]^{3+}$ near 24 000 cm^{-1} (500 nm) is shown in Fig. 11F.13, and can be ascribed to the promotion of its single d electron from a t_{2g} orbital to an e_g orbital. The wavenumber of the absorption maximum suggests that $\Delta_o \approx 24\,000$ cm^{-1} for this complex, which corresponds to about 3.0 eV.

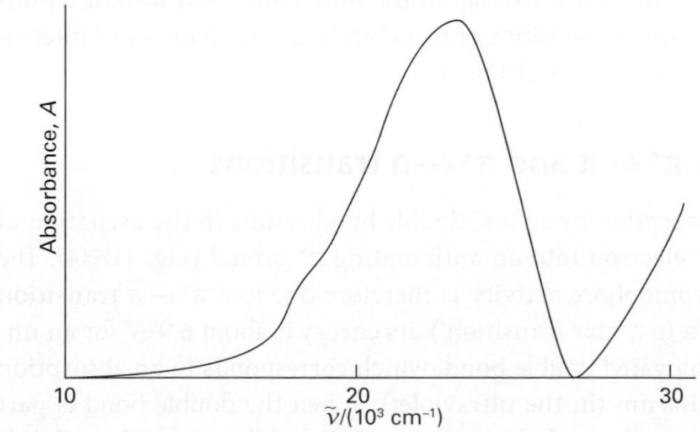

Figure 11F.13 The electronic absorption spectrum of $[Ti(OH_2)_6]^{3+}$ in aqueous solution.

According to the Laporte rule (Section 11F.1b), d–d transitions are parity-forbidden in octahedral complexes because they are g→g transitions (more specifically, $e_g \leftarrow t_{2g}$ transitions).

However, d–d transitions become weakly allowed as vibronic transitions as a result of coupling to asymmetrical vibrations such as that shown in Fig. 11F.5.

A d-metal complex may also absorb radiation as a result of the transfer of an electron from the ligands into the d orbitals of the central atom, or vice versa. In such **charge-transfer transitions** the electron moves through a considerable distance, which means that the transition dipole moment may be large and the absorption correspondingly intense. In the permanganate ion, MnO_4^-, the charge redistribution that accompanies the migration of an electron from the O atoms to the central Mn atom results in a strong transition in the range 475–575 nm that accounts for the intense purple colour of the ion. Such an electronic migration from the ligands to the metal corresponds to a **ligand-to-metal charge-transfer transition** (LMCT). The reverse migration, a **metal-to-ligand charge-transfer transition** (MLCT), can also occur. An example is the migration of a d electron onto the antibonding π orbitals of an aromatic ligand. The resulting excited state may have a very long lifetime if the electron is extensively delocalized over several aromatic rings.

In common with other transitions, the intensities of charge-transfer transitions are proportional to the square of the transition dipole moment. The transition moment can be thought of as a measure of the distance moved by the electron as it migrates from metal to ligand or vice versa, with a large distance of migration corresponding to a large transition dipole moment and therefore a high intensity of absorption. However, when calculating the transition dipole moment the integrand is proportional to the product of the initial and final wavefunctions; this product is zero unless the two wavefunctions have non-zero values in the same region of space. Therefore, although large distances of migration favour high intensities, the diminished overlap of the initial and final wavefunctions for large separations of metal and ligands favours low intensities (see Problem P11F.9).

(b) $\pi^* \leftarrow \pi$ and $\pi^* \leftarrow n$ transitions

Absorption by a C=C double bond results in the excitation of a π electron into an antibonding π* orbital (Fig. 11F.14). The chromophore activity is therefore due to a **$\pi^* \leftarrow \pi$ transition** (a 'π to π-star transition'). Its energy is about 6.9 eV for an unconjugated double bond, which corresponds to an absorption at 180 nm (in the ultraviolet). When the double bond is part of a conjugated chain, the energies of the molecular orbitals lie closer together and the $\pi^* \leftarrow \pi$ transition moves to a longer wavelength; it may even lie in the visible region if the conjugated system is long enough.

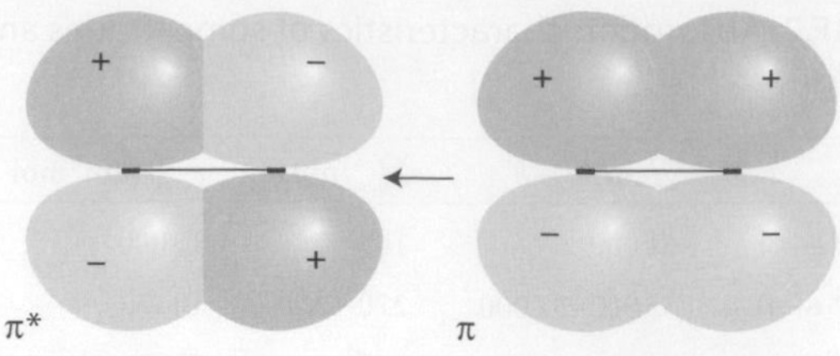

Figure 11F.14 A C=C double bond acts as a chromophore. In the $\pi^* \leftarrow \pi$ transition illustrated here, an electron is promoted from a π orbital to the corresponding antibonding orbital.

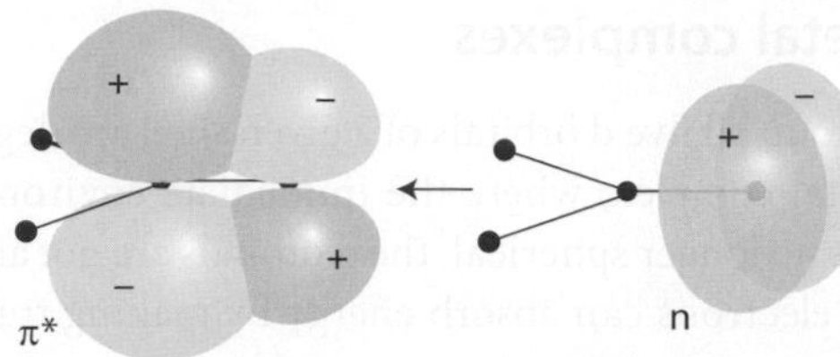

Figure 11F.15 A carbonyl group (C=O) acts as a chromophore partly on account of the excitation of a nonbonding O lone-pair electron to an antibonding CO π* orbital; this transition is denoted $\pi^* \leftarrow n$.

One of the transitions responsible for absorption in carbonyl compounds can be traced to the lone pairs of electrons on the O atom. The Lewis concept of a 'lone pair' of electrons is represented in molecular orbital theory by a pair of electrons in an orbital confined largely to one atom and not appreciably involved in bond formation. One of these electrons may be excited into an empty π* orbital of the carbonyl group (Fig. 11F.15), which gives rise to an **$\pi^* \leftarrow n$ transition** (an 'n to π-star transition'). Typical absorption energies are about 4.3 eV (290 nm). Because $\pi^* \leftarrow n$ transitions in carbonyls are symmetry forbidden, the absorptions are weak. By contrast, the $\pi^* \leftarrow \pi$ transition in a carbonyl, which corresponds to excitation of a π electron of the C=O double bond, is allowed by symmetry and results in relatively strong absorption.

Brief illustration 11F.8

The compound $CH_3CH=CHCHO$ has a strong absorption in the ultraviolet at 46 950 cm^{-1} (213 nm) and a weak absorption at 30 000 cm^{-1} (330 nm). The former is a $\pi^* \leftarrow \pi$ transition associated with the delocalized π system C=C—C=O. Delocalization extends the range of the C=O $\pi^* \leftarrow \pi$ transition to lower wavenumbers (longer wavelengths). The latter is an $\pi^* \leftarrow n$ transition associated with the carbonyl chromophore.

Checklist of concepts

☐ 1. The **term symbols** of linear molecules give the components of various kinds of angular momentum around the internuclear axis along with relevant symmetry labels.

☐ 2. The **Laporte selection rule** states that, for centrosymmetric molecules, only u→g and g→u transitions are allowed.

☐ 3. The **Franck–Condon principle** asserts that electronic transitions occur within an unchanging nuclear framework.

☐ 4. **Vibrational fine structure** is the structure in a spectrum that arises from changes in vibrational energy accompanying an electronic transition.

☐ 5. **Rotational fine structure** is the structure in a spectrum that arises from changes in rotational energy accompanying an electronic transition.

☐ 6. In gas phase samples, rotational fine structure can be resolved and under some circumstances **band heads** are formed.

☐ 7. In d-metal complexes, the presence of ligands removes the degeneracy of d orbitals and vibrationally allowed **d–d transitions** can occur between them.

☐ 8. **Charge-transfer transitions** typically involve the migration of electrons between the ligands and the central metal atom.

☐ 9. A **chromophore** is a group with a characteristic optical absorption band.

Checklist of equations

Property	Equation	Comment	Equation number
Selection rules (angular momentum)	$\Delta\Lambda = 0, \pm1; \Delta S = 0; \Delta\Sigma = 0; \Delta\Omega = 0, \pm1$	Linear molecules	11F.4
Franck–Condon factor	$\lvert S(\upsilon_f,\upsilon_i)\rvert^2 = \left(\int \psi^*_{\upsilon,f}\psi_{\upsilon,i}\,d\tau_N\right)^2$		11F.5
Rotational structure of electronic spectra (diatomic molecules)	$\tilde{\nu}_P(J) = \tilde{\nu} - (\tilde{B}' + \tilde{B})J + (\tilde{B}' - \tilde{B})J^2$	P branch ($\Delta J = -1$)	11F.7a
	$\tilde{\nu}_Q(J) = \tilde{\nu} + (\tilde{B}' - \tilde{B})J(J+1)$	Q branch ($\Delta J = 0$)	11F.7b
	$\tilde{\nu}_R(J) = \tilde{\nu} + (\tilde{B}' + \tilde{B})(J+1) + (\tilde{B}' - \tilde{B})(J+1)^2$	R branch ($\Delta J = +1$)	11F.7c

TOPIC 11G Decay of excited states

> ➤ Why do you need to know this material?
>
> Much information about the electronic structure of a molecule can be obtained by observing the radiative decay of excited electronic states back to the ground state. Such decay is also used in lasers, which are of exceptional technological importance.
>
> ➤ What is the key idea?
>
> Molecules in excited electronic states discard their excess energy by emission of electromagnetic radiation, transfer as heat to the surroundings, or fragmentation.
>
> ➤ What do you need to know already?
>
> You need to be familiar with electronic transitions in molecules (Topic 11F), the difference between spontaneous and stimulated emission of radiation (Topic 11A), and the general features of spectroscopy (Topic 11A). You need to be aware of the difference between singlet and triplet states (Topic 8C) and of the Franck–Condon principle (Topic 11F).

Radiative decay is a process in which a molecule discards its excitation energy as a photon (Topic 11A); depending on the nature of the excited state, this process is classified as either fluorescence or phosphorescence. A more common fate of an electronically excited molecule is **non-radiative decay**, in which the excess energy is transferred into the vibration, rotation, and translation of the surrounding molecules. This thermal degradation converts the excitation energy into thermal motion of the environment (i.e. to 'heat'). An excited molecule may also dissociate or take part in a chemical reaction (Topic 17G). Stimulated emission from excited states is the key process that can lead to laser action.

11G.1 Fluorescence and phosphorescence

In **fluorescence**, spontaneous emission of radiation occurs while the sample is being irradiated and ceases as soon as the exciting radiation is extinguished (Fig. 11G.1). In **phosphorescence**, the spontaneous emission may persist for long periods (even hours, but more commonly seconds or fractions of seconds). The difference suggests that fluorescence is a fast conversion of absorbed radiation into re-emitted energy, and that phosphorescence involves the storage of energy in a reservoir from which it slowly leaks.

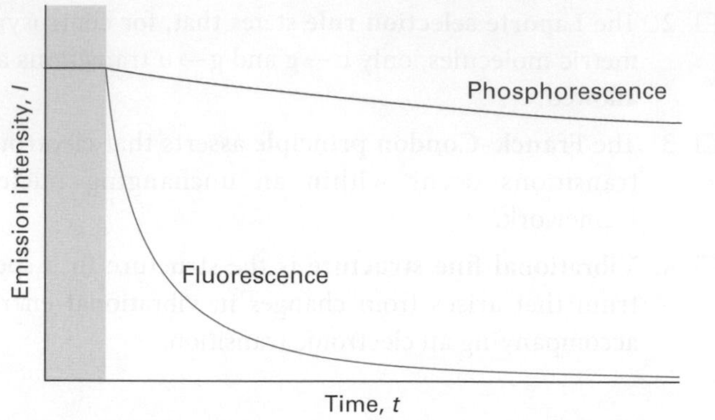

Figure 11G.1 The empirical (observation-based) distinction between fluorescence and phosphorescence is that the former is extinguished very quickly after the exciting radiation is removed, whereas the latter continues with relatively slowly diminishing intensity.

Figure 11G.2 shows the sequence of steps involved in fluorescence from molecules in solution. The initial stimulated absorption takes the molecule to an excited electronic state; if the absorption spectrum were monitored it would look like the

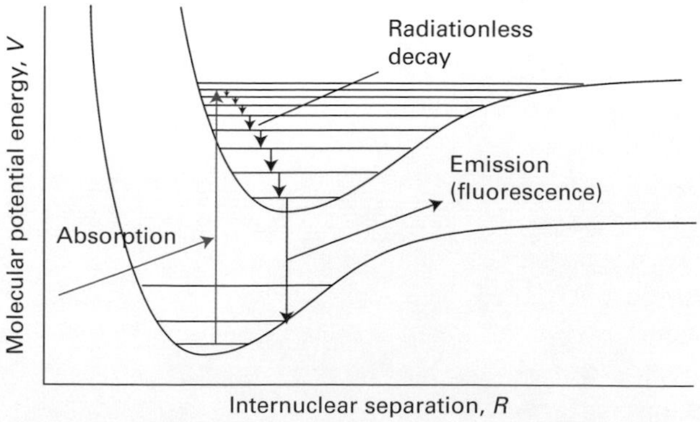

Figure 11G.2 The sequence of steps leading to fluorescence by molecules in solution. After the initial absorption, the upper vibrational states undergo radiationless decay by giving up energy to the surrounding molecules. A radiative transition then occurs from the vibrational ground state of the upper electronic state.

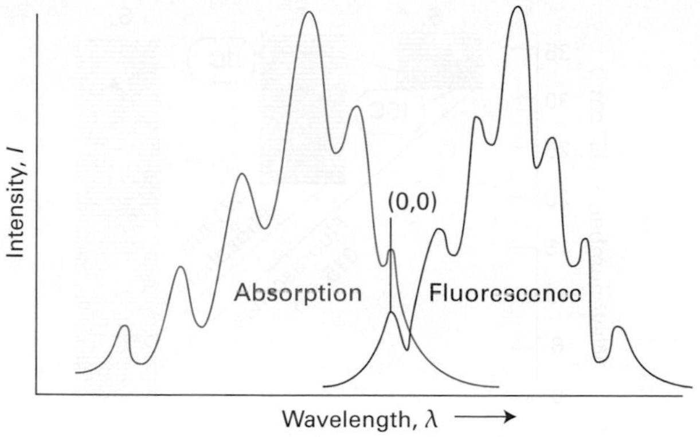

Figure 11G.3 An absorption spectrum (blue) shows vibrational structure characteristic of the upper electronic state. A fluorescence spectrum (purple) shows structure characteristic of the lower state; it is also displaced to lower frequencies (but the 0–0 transitions are coincident) and is often a mirror image of the absorption spectrum.

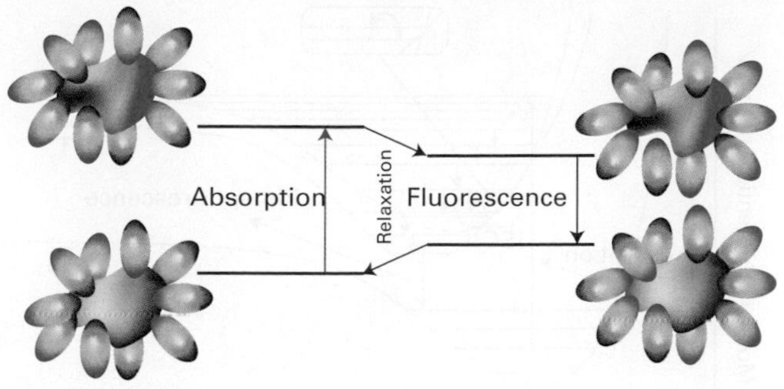

Figure 11G.4 The solvent can shift the fluorescence spectrum relative to the absorption spectrum. On the left absorption occurs with the solvent (depicted by the ellipses) in the arrangement characteristic of the ground electronic state of the molecule (the central blob). However, before fluorescence occurs, the solvent molecules relax into a new arrangement, and that arrangement is preserved during the subsequent radiative transition, the fluorescence.

one shown in Fig. 11G.3a. The excited molecule is subjected to collisions with the surrounding molecules, and as it gives up energy to them non-radiatively it steps down (typically within picoseconds) the ladder of vibrational levels to the lowest vibrational level of the excited electronic state. The surrounding molecules, however, might now be unable to accept the larger energy difference needed to lower the molecule to the ground electronic state. The excited electronic state might therefore survive long enough to undergo spontaneous emission and emit the remaining excess energy as radiation. The downward electronic transition is vertical, in accord with the Franck–Condon principle (Topic 11F), and the fluorescence spectrum has vibrational structure characteristic of the *lower* electronic state (Fig. 11G.3b).

Provided they can be seen, the 0–0 absorption and fluorescence transitions (where the numbers are the values of v_f and v_i, the vibrational quantum numbers for the final and initial states) can be expected to be coincident. The absorption spectrum arises from $0 \leftarrow 0$, $1 \leftarrow 0$, $2 \leftarrow 0, \ldots$ transitions which occur at progressively higher wavenumbers (shorter wavelengths) and with intensities governed by the Franck–Condon principle. The fluorescence spectrum arises from $0 \rightarrow 0$, $0 \rightarrow 1, \ldots$ downward transitions which occur with decreasing wavenumbers (longer wavelengths). The 0–0 absorption and fluorescence peaks are not always exactly coincident, however, because the solvent may interact differently with the solute in the ground and excited electronic states (for instance, the hydrogen bonding pattern might differ). Because the solvent molecules do not have time to rearrange during the transition, the absorption occurs in an environment characteristic of the solvated ground state; however, the fluorescence occurs in an environment characteristic of the solvated excited state (Fig. 11G.4).

Fluorescence occurs at lower frequencies than the incident radiation because the emissive transition occurs after some vibrational energy has been discarded into the surroundings. The vivid oranges and greens of fluorescent dyes are an everyday manifestation of this effect: they absorb in the ultraviolet and blue, and fluoresce in the visible. The mechanism also suggests that the intensity of the fluorescence ought to depend on the ability of the solvent molecules to accept the electronic and vibrational quanta. It is indeed found that a solvent composed of molecules with widely spaced vibrational levels (such as water) can in some cases accept the large quantum of electronic energy and so extinguish, or 'quench', the fluorescence. The rate at which fluorescence is quenched by other molecules also gives valuable kinetic information (Topic 17G).

Figure 11G.5 shows the sequence of events leading to phosphorescence for a molecule with a singlet ground state (denoted S_0). The first steps are the same as in fluorescence, but the presence of a triplet excited state (T_1) at an energy close to that of the singlet excited state (S_1) plays a decisive role. The singlet and triplet excited states share a common geometry at the point where their potential energy curves intersect. Hence, if there is a mechanism for unpairing two electron spins (and achieving the conversion of ↑↓ to ↑↑), the molecule may undergo **intersystem crossing**, a non-radiative transition between states of different multiplicity, and become a triplet state. As in the discussion of atomic spectra (Topic 8C), singlet–triplet transitions may occur in the presence of spin–orbit coupling. Intersystem crossing is expected to be important when a molecule contains a moderately heavy atom (such as sulfur), because then the spin–orbit coupling is large.

Once an excited molecule has crossed into a triplet state, it continues to discard energy into the surroundings. However, it is now stepping down the triplet's vibrational ladder and ends

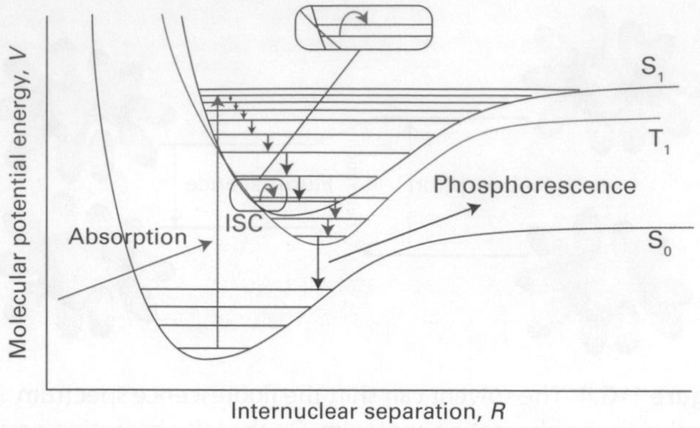

Figure 11G.5 The sequence of steps leading to phosphorescence. The important step is the intersystem crossing (ISC), the switch from a singlet state (S_1) to a triplet state (T_1) brought about by spin–orbit coupling. The triplet state acts as a slowly radiating reservoir because the return to the ground state is spin-forbidden.

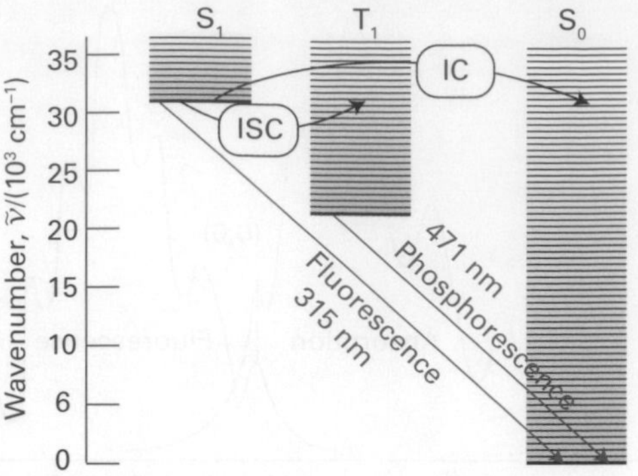

Figure 11G.6 A Jablonski diagram (here, for naphthalene) is a simplified portrayal of the relative positions of the electronic energy levels of a molecule. Vibrational levels of states of a given electronic state lie above each other, but the relative horizontal locations of the columns bear no relation to the nuclear separations in the states. The ground vibrational states of each electronic state are correctly located vertically but the other vibrational states are shown only schematically. (IC: internal conversion; ISC: intersystem crossing.)

in its lowest vibrational level. The triplet state is lower in energy than the corresponding singlet state (Hund's rule, Topic 8B). The solvent cannot absorb the final, large quantum of electronic excitation energy, and the molecule cannot radiate its energy because return to the ground state is spin-forbidden. The radiative transition, however, is not totally forbidden because the spin–orbit coupling that was responsible for the intersystem crossing also weakens the selection rule. The molecules are therefore able to emit weakly, and the emission may continue long after the original excited state was formed.

This mechanism accounts for the observation that the excitation energy seems to get trapped in a slowly leaking reservoir. It also suggests (as is confirmed experimentally) that phosphorescence should be most intense from solid samples: energy transfer is then less efficient and intersystem crossing has time to occur as the singlet excited state steps slowly past the intersection point. The mechanism also suggests that the phosphorescence efficiency should depend on the presence of a moderately heavy atom (with strong spin–orbit coupling), which is in fact the case.

The various types of non-radiative and radiative transitions that can occur in molecules are often represented on a schematic **Jablonski diagram** of the type shown in Fig. 11G.6.

Brief illustration 11G.1

Fluorescence efficiency decreases, and the phosphorescence efficiency increases, in the series of compounds: naphthalene, 1-chloronaphthalene, 1-bromonaphthalene, 1-iodonaphthalene. The replacement of an H atom by successively heavier atoms enhances intersystem crossing from S_1 into T_1, thereby decreasing the efficiency of fluorescence. The rate of the radiative transition from T_1 to S_0 is also enhanced by the presence of heavier atoms, thereby increasing the efficiency of phosphorescence.

11G.2 Dissociation and predissociation

A chemically important fate for an electronically excited molecule is **dissociation**, the breaking of bonds (Fig. 11G.7). The onset of dissociation can be detected in an absorption spectrum by noting that the vibrational fine structure of a band terminates at a certain frequency. Absorption occurs in a continuous band above this **dissociation limit** because the final state is an unquantized translational motion of the fragments. Locating the dissociation limit is a valuable way of determining the bond dissociation energy.

In some cases, the vibrational structure disappears but resumes at higher frequencies of the incident radiation. This

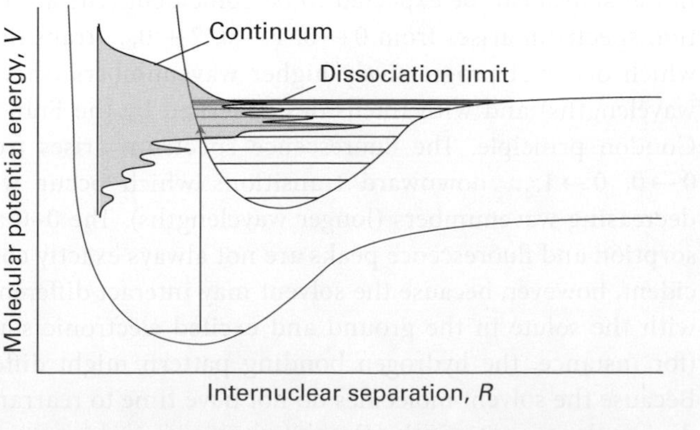

Figure 11G.7 When absorption occurs to unbound states of the upper electronic state, the molecule dissociates and the absorption spectrum is a continuum. Below the dissociation limit the electronic spectrum shows a normal vibrational structure.

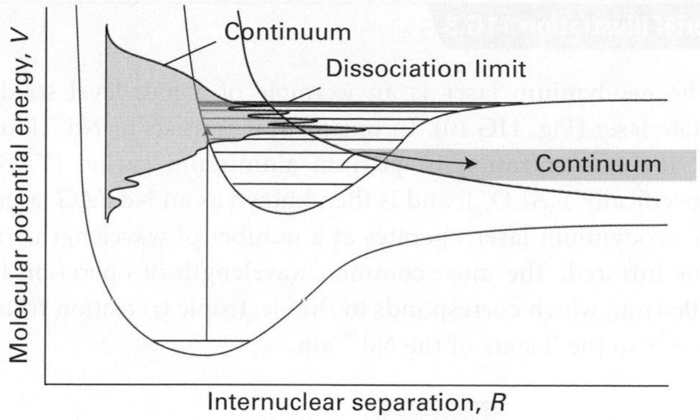

Figure 11G.8 When a dissociative state crosses a bound state, as in the upper part of the illustration, molecules excited to levels near the crossing may dissociate. This predissociation is detected in the spectrum as a loss of vibrational structure that resumes at higher frequencies.

effect provides evidence of **predissociation**, which can be interpreted in terms of the molecular potential energy curves shown in Fig. 11G.8. When a molecule is excited to a high vibrational level of the upper electronic state, its electrons may undergo a redistribution that results in it undergoing an **internal conversion**, a radiationless conversion to another electronic state of the same multiplicity. An internal conversion occurs most readily at the point of intersection of the two molecular potential energy curves, because there the nuclear geometries of the two electronic states are the same. The state into which the molecule converts may be dissociative, so the states near the intersection have a finite lifetime and hence their energies are imprecisely defined (as a result of lifetime broadening, Topic 11A). As a result, the absorption spectrum is blurred. When the incoming photon brings enough energy to excite the molecule to a vibrational level high above the intersection, the internal conversion does not occur (the nuclei are unlikely to have the same geometry). Consequently, the levels resume their well-defined, vibrational character with correspondingly well-defined energies, and the line structure resumes on the high-frequency side of the blurred region.

Brief illustration 11G.2

The O_2 molecule absorbs ultraviolet radiation in a transition from its ${}^3\Sigma_g^-$ ground electronic state to a ${}^3\Sigma_u^-$ excited state that is energetically close to a dissociative ${}^3\Pi_u$ state. In this case, the effect of predissociation is more subtle than the abrupt loss of vibrational–rotational structure in the spectrum; instead, the vibrational structure simply broadens rather than being lost completely. As before, the broadening is explained by the short lifetimes of the excited vibrational states near the intersection of the curves describing the bound and dissociative excited electronic states.

11G.3 Lasers

An excited state can be driven to discard its excess energy by using radiation to induce stimulated emission. The word laser is an acronym formed from **l**ight **a**mplification by **s**timulated **e**mission of **r**adiation. In stimulated emission (Topic 11A), an excited state is stimulated to emit a photon by radiation of the same frequency: the more photons that are present, the greater is the probability of emission.

Laser radiation has a number of striking characteristics (Table 11G.1). Each of them (sometimes in combination with the others) opens up interesting opportunities in physical chemistry. Raman spectroscopy (Topics 11B–D) has flourished on account of the high intensity of monochromatic radiation available from lasers, and the ultrashort pulses that lasers can generate make possible the study of light-initiated reactions on timescales of femtoseconds and even attoseconds.

One requirement of laser action is the existence of a **metastable excited state**, an excited state with a long enough lifetime for it to undergo stimulated emission. For stimulated emission to dominate absorption, it is necessary for there to be a **population inversion** in which the population of the excited state is greater than that of the lower state. Figure 11G.9 illustrates one way to achieve population inversion indirectly through an intermediate state I. Thus, the molecule or atom is excited to I, which then gives up some of its energy non-radiatively (for example, by passing energy on to vibrations of the surroundings) and changes into a lower state B. The laser

Table 11G.1 Characteristics of laser radiation and their chemical applications

Characteristic	Advantage	Application
High power	Multiphoton process	Spectroscopy
	Low detector noise	Improved sensitivity
	High scattering intensity	Raman spectroscopy (Topics 11B–11D)
Monochromatic	High resolution	Spectroscopy
	State selection	Photochemical studies (Topic 17G)
		Reaction dynamics (Topic 18D)
Collimated beam	Long path lengths	Improved sensitivity
	Forward-scattering observable	Raman spectroscopy (Topics 11B–11D)
Pulsed	Precise timing of excitation	Fast reactions (Topics 17G, 18C)
		Relaxation (Topic 17C)
		Energy transfer (Topic 17G)

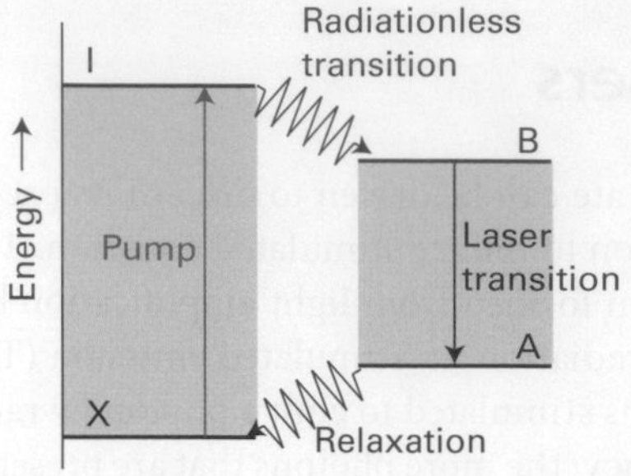

Figure 11G.9 The transitions involved in a four-level laser. Because the laser transition terminates in an excited state (A), the population inversion between A and B is much easier to achieve than when the lower state of the laser transition is the ground state, X.

transition is the return of B to a lower state A. Because four levels are involved overall, this arrangement leads to a **four-level laser**. One advantage of this arrangement is that the population inversion of the A and B levels is easier to achieve than one involving the heavily populated ground state. The transition from X to I is caused by irradiation with intense light (either continuously or as a flash) in the process called **pumping**. In some cases pumping is achieved with an electric discharge through xenon or with the radiation from another laser.

Brief illustration 11G.3

The neodymium laser is an example of a four-level solid-state laser (Fig. 11G.10). In one form it consists of Nd^{3+} ions at low concentration in yttrium aluminium garnet (YAG, specifically $Y_3Al_5O_{12}$), and is then known as an Nd:YAG laser. A neodymium laser operates at a number of wavelengths in the infrared. The most common wavelength of operation is 1064 nm, which corresponds to the electronic transition from the ^4F to the ^4I state of the Nd^{3+} ion.

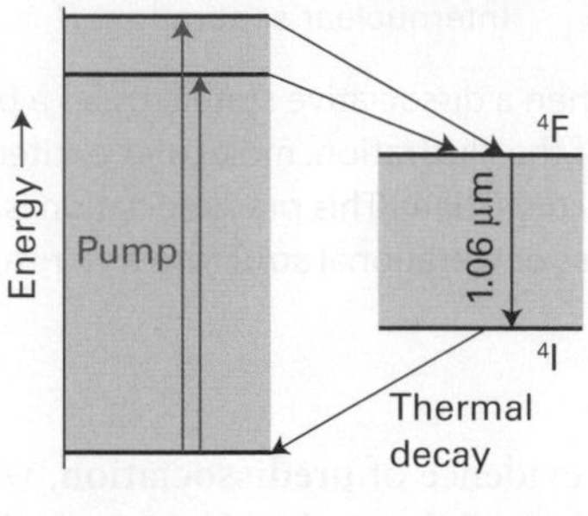

Figure 11G.10 The transitions involved in a neodymium laser.

Many of the most important laser systems are solid-state devices; they are discussed in Topic 15G.

Checklist of concepts

☐ 1. **Fluorescence** is radiative decay between states of the same multiplicity; it ceases soon after the exciting radiation is removed.

☐ 2. **Phosphorescence** is radiative decay between states of different multiplicity; it persists after the exciting radiation is removed.

☐ 3. **Intersystem crossing** is the non-radiative conversion to an electronic state of different multiplicity.

☐ 4. A **Jablonski diagram** is a schematic diagram showing the types of non-radiative and radiative transitions that can occur in molecules.

☐ 5. An additional fate of an electronically excited species is **dissociation**.

☐ 6. **Internal conversion** is a non-radiative conversion to an electronic state of the same multiplicity.

☐ 7. **Predissociation** is the observation of the effects of dissociation before the dissociation limit is reached.

☐ 8. **Laser action** is the stimulated emission of coherent radiation between states related by a population inversion.

☐ 9. A **metastable excited state** is an excited state with a long enough lifetime for it to undergo stimulated emission.

☐ 10. A **population inversion** is a condition in which the population of an upper state is greater than that of a relevant lower state.

☐ 11. **Pumping**, the stimulation of an absorption with an external source of intense radiation, is a process by which a population inversion is created.

FOCUS 11 Molecular spectroscopy

Note: The masses of nuclides are listed in Table 0.2 of the *Resource section*.

TOPIC 11A General features of molecular spectroscopy

Discussion questions

D11A.1 What is the physical origin of a selection rule?

D11A.2 Describe the physical origins of linewidths in absorption and emission spectra. Do you expect the same contributions for species in condensed and gas phases?

D11A.3 Describe the basic experimental arrangements commonly used for absorption, emission, and Raman spectroscopy.

Exercises

E11A.1(a) Calculate the ratio A/B for transitions with the following characteristics: (i) 70.8 pm X-rays, (ii) 500 nm visible light, (iii) 3000 cm^{-1} infrared radiation.

E11A.1(b) Calculate the ratio A/B for transitions with the following characteristics: (i) 500 MHz radiofrequency radiation, (ii) 3.0 cm microwave radiation.

E11A.2(a) The molar absorption coefficient of a substance dissolved in hexane is known to be 723 dm^3 mol^{-1} cm^{-1} at 260 nm. Calculate the percentage reduction in intensity when ultraviolet radiation of that wavelength passes through 2.50 mm of a solution of concentration 4.25 mmol dm^{-3}.

E11A.2(b) The molar absorption coefficient of a substance dissolved in hexane is known to be 227 dm^3 mol^{-1} cm^{-1} at 290 nm. Calculate the percentage reduction in intensity when ultraviolet radiation of that wavelength passes through 2.00 mm of a solution of concentration 2.52 mmol dm^{-3}.

E11A.3(a) A solution of a certain component of a biological sample when placed in an absorption cell of path length 1.00 cm transmits 18.1 per cent of ultraviolet radiation of wavelength 320 nm incident upon it. If the concentration of the component is 0.139 mmol dm^{-3}, what is the molar absorption coefficient?

E11A.3(b) When ultraviolet radiation of wavelength 400 nm passes through 2.50 mm of a solution of an absorbing substance at a concentration 0.717 mmol dm^{-3}, the transmission is 61.5 per cent. Calculate the molar absorption coefficient of the solute at this wavelength. Express your answer in square centimetres per mole (cm^2 mol^{-1}).

E11A.4(a) The molar absorption coefficient of a solute at 540 nm is 386 dm^3 mol^{-1} cm^{-1}. When light of that wavelength passes through a 5.00 mm cell containing a solution of the solute, 38.5 per cent of the light was absorbed. What is the molar concentration of the solute?

E11A.4(b) The molar absorption coefficient of a solute at 440 nm is 423 dm^3 mol^{-1} cm^{-1}. When light of that wavelength passes through a 6.50 mm cell containing a solution of the solute, 48.3 per cent of the light was absorbed. What is the molar concentration of the solute?

E11A.5(a) The following data were obtained for the absorption at 450 nm by a dye in carbon tetrachloride when using a 2.0 mm cell. Calculate the molar absorption coefficient of the dye at the wavelength employed:

[dye]/(mol dm^{-3})	0.0010	0.0050	0.0100	0.0500
T/(per cent)	81.4	35.6	12.7	3.0×10^{-3}

E11A.5(b) The following data were obtained for the absorption at 600 nm by a dye dissolved in methylbenzene using a 2.50 mm cell. Calculate the molar absorption coefficient of the dye at the wavelength employed:

[dye]/(mol dm^{-3})	0.0010	0.0050	0.0100	0.0500
T/(per cent)	72	20	4.0	1.00×10^{-5}

E11A.6(a) A 2.0 mm cell was filled with a solution of benzene in a non-absorbing solvent. The concentration of the benzene was 0.010 mol dm^{-3} and the wavelength of the radiation was 256 nm (where there is a maximum in the absorption). Calculate the molar absorption coefficient of benzene at this wavelength given that the transmission was 48 per cent. What will the transmittance be through a 4.0 mm cell at the same wavelength?

E11A.6(b) A 5.00 mm cell was filled with a solution of a dye. The concentration of the dye was 18.5 mmol dm^{-3}. Calculate the molar absorption coefficient of the dye at this wavelength given that the transmission was 29 per cent. What will the transmittance be through a 2.50 mm cell at the same wavelength?

E11A.7(a) A swimmer enters a gloomier world (in one sense) on diving to greater depths. Given that the mean molar absorption coefficient of sea water in the visible region is 6.2×10^{-5} dm^3 mol^{-1} cm^{-1}, calculate the depth at which a diver will experience (i) half the surface intensity of light, (ii) one tenth the surface intensity.

E11A.7(b) Given that the maximum molar absorption coefficient of a molecule containing a carbonyl group is 30 dm^3 mol^{-1} cm^{-1} near 280 nm, calculate the thickness of a sample that will result in (i) half the initial intensity of radiation, (ii) one tenth the initial intensity. Take the absorber concentration to be 10 mmol dm^{-3}.

E11A.8(a) The absorption associated with a particular transition begins at 220 nm, peaks sharply at 270 nm, and ends at 300 nm. The maximum value of the molar absorption coefficient is 2.21×10^4 dm^3 mol^{-1} cm^{-1}. Estimate the integrated absorption coefficient of the transition assuming a triangular lineshape.

E11A.8(b) The absorption associated with a certain transition begins at 167 nm, peaks sharply at 200 nm, and ends at 250 nm. The maximum value of the molar absorption coefficient is 3.35×10^4 dm^3 mol^{-1} cm^{-1}. Estimate the integrated absorption coefficient of the transition assuming an inverted parabolic lineshape (Fig. 11.1).

E11A.9(a) What is the Doppler-broadened linewidth of the electronic transition at 821 nm in atomic hydrogen at 300 K?

E11A.9(b) What is the Doppler-broadened linewidth of the vibrational transition at 2308 cm^{-1} in $^1H^{127}I$ at 400 K?

E11A.10(a) What is the Doppler-shifted wavelength of a red (680 nm) traffic light approached at 60 km h^{-1}?

E11A.10(b) At what speed of approach would a red (680 nm) traffic light appear green (530 nm)?

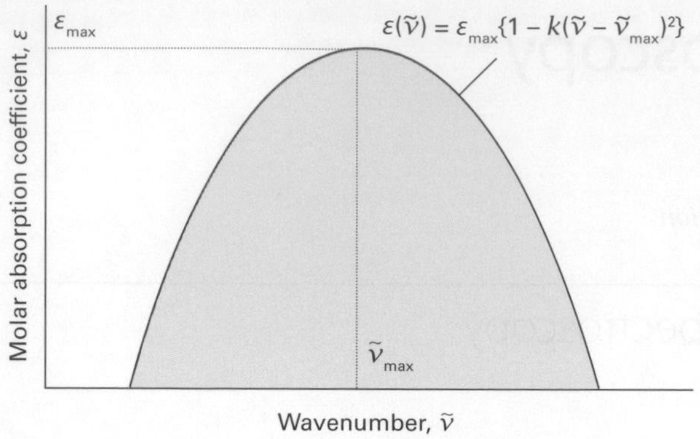

Figure 11.1 The parabolic lineshape considered in Exercise E11A.8(b).

E11A.11(a) Estimate the lifetime of a state that gives rise to a line of width (i) 0.20 cm^{-1}, (ii) 2.0 cm^{-1}.

E11A.11(b) Estimate the lifetime of a state that gives rise to a line of width (i) 200 MHz, (ii) 2.45 cm^{-1}.

E11A.12(a) A molecule in a liquid undergoes about 1.0×10^{13} collisions in each second. Suppose that (i) every collision is effective in deactivating the molecule vibrationally and (ii) that one collision in 100 is effective. Calculate the width (in cm^{-1}) of vibrational transitions in the molecule.

E11A.12(b) A molecule in a gas undergoes about 1.0×10^{9} collisions in each second. Suppose that (i) every collision is effective in deactivating the molecule rotationally and (ii) that one collision in 10 is effective. Calculate the width (in hertz) of rotational transitions in the molecule.

Problems

P11A.1 The flux of visible photons reaching Earth from the North Star is about 4×10^{3} mm^{-2} s^{-1}. Of these photons, 30 per cent are absorbed or scattered by the atmosphere and 25 per cent of the surviving photons are scattered by the surface of the cornea of the eye. A further 9 per cent are absorbed inside the cornea. The area of the pupil at night is about 40 mm^{2} and the response time of the eye is about 0.1 s. Of the photons passing through the pupil, about 43 per cent are absorbed in the ocular medium. How many photons from the North Star are focused onto the retina in 0.1 s? For a continuation of this story, see R.W. Rodieck, *The first steps in seeing*, Sinauer, Sunderland (1998).

P11A.2 The Beer–Lambert law is derived on the basis that the concentration of absorbing species is uniform. Suppose, instead, that the concentration falls exponentially as $[J]=[J]_0 e^{-x/x_0}$. Develop an expression for the variation of I with sample length; suppose that $L >> x_0$.

P11A.3 It is common to make measurements of absorbance at two wavelengths and use them to find the individual concentrations of two components A and B in a mixture. Show that the molar concentrations of A and B in a cell of length L are

$$[A]=\frac{\varepsilon_{B2}A_1-\varepsilon_{B1}A_2}{(\varepsilon_{A1}\varepsilon_{B2}-\varepsilon_{A2}\varepsilon_{B1})L} \qquad [B]=\frac{\varepsilon_{A1}A_2-\varepsilon_{A2}A_1}{(\varepsilon_{A1}\varepsilon_{B2}-\varepsilon_{A2}\varepsilon_{B1})L}$$

where A_1 and A_2 are absorbances of the mixture at wavelengths λ_1 and λ_2, and the molar extinction coefficients of A (and B) at these wavelengths are ε_{A1} and ε_{A2} (and ε_{B1} and ε_{B2}).

P11A.4 When pyridine is added to a solution of iodine in carbon tetrachloride the 520 nm band of absorption shifts toward 450 nm. However, the absorbance of the solution at 490 nm remains constant: this feature is called an *isosbestic point*. Show that an isosbestic point should occur when two absorbing species are in equilibrium. *Hint:* Assume that pyridine and iodine form a 1:1 complex, and that the absorption is due to pyridine and the complex only.

P11A.5‡ Ozone, uniquely among other abundant atmospheric constituents, absorbs ultraviolet radiation in a part of the electromagnetic spectrum energetic enough to disrupt DNA in biological organisms. This spectral range, which is denoted UV-B, spans from about 290 nm to 320 nm. The molar extinction coefficient of ozone over this range is given in the table below (DeMore et al., *Chemical kinetics and photochemical data for use in stratospheric modeling: Evaluation Number 11*, JPL Publication 94–26 (1994)).

λ/nm	292.0	296.3	300.8	305.4	310.1	315.0	320.0
ε/(dm^3 mol^{-1} cm^{-1})	1512	865	477	257	135.9	69.5	34.5

Evaluate the integrated absorption coefficient of ozone over the wavelength range 290–320 nm. *Hint:* $\varepsilon(\tilde{\nu})$ can be fitted to an exponential function quite well.

‡ These problems were supplied by Charles Trapp and Carmen Giunta.

P11A.6 In many cases it is possible to assume that an absorption band has a Gaussian lineshape (one proportional to e^{-x^2}) centred on the band maximum. Assume such a lineshape, and show that $\mathcal{A} = \int\varepsilon(\tilde{\nu})d\tilde{\nu} \approx 1.0645\varepsilon_{max}\Delta\tilde{\nu}_{1/2}$, where $\Delta\tilde{\nu}_{1/2}$ is the width at half-height. The absorption spectrum of azoethane ($CH_3CH_2N_2$) between 24 000 cm^{-1} and 34 000 cm^{-1} is shown in Fig. 11.2. First, estimate $\mathcal{A}$ for the band by assuming that it is Gaussian. Then use mathematical software to fit a polynomial (or a Gaussian) to the absorption band, and integrate the result analytically.

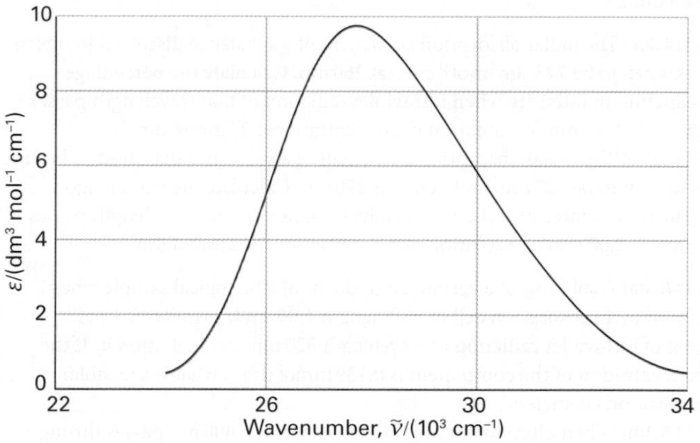

Figure 11.2 The electronic absorption spectrum of azomethane.

P11A.7‡ Wachewsky et al. (*J. Phys. Chem.* **100**, 11559 (1996)) examined the ultraviolet absorption spectrum of CH_3I, a species of interest in connection with stratospheric ozone chemistry. They found the integrated absorption coefficient to be dependent on temperature and pressure to an extent inconsistent with internal structural changes in isolated CH_3I molecules. They explained the changes as due to dimerization of a substantial fraction of the CH_3I, a process which would naturally be pressure and temperature dependent. (a) Compute the integrated absorption coefficient of CH_3I over a triangular lineshape in the range 31 250–34 483 cm^{-1} and a maximal molar absorption coefficient of 150 dm^3 mol^{-1} cm^{-1} at the mid-point of the range. (b) Suppose 1.0 per cent of the CH_3I units in a sample at 2.4 Torr and 373 K exists as dimers. Evaluate the absorbance expected at the mid-point of the absorption lineshape in a sample cell of length 12.0 cm. (c) Suppose 18 per cent of the CH_3I units in a sample at 100 Torr and 373 K exists as dimers. Calculate the absorbance expected at the wavenumber corresponding to the mid-point of the lineshape for a sample cell of length 12.0 cm; compute the molar absorption coefficient that would be inferred from this absorbance if dimerization is not considered.

P11A.8 When a star emitting electromagnetic radiation of frequency ν moves with a speed s relative to an observer, the observer detects radiation of frequency $\nu_{\text{receding}} = \nu f$ or $\nu_{\text{approaching}} = \nu/f$, where $f = \{(1 - s/c)/(1 + s/c)\}^{1/2}$ and c is the speed of light. (a) Three Fe I lines of the star HDE 271 182, which belongs to the Large Magellanic Cloud, occur at 438.882 nm, 441.000 nm, and 442.020 nm. The same lines occur at 438.392 nm, 440.510 nm, and 441.510 nm in the spectrum of an Earth-bound iron arc. Decide whether HDE 271 182 is receding from or approaching the Earth and estimate the star's radial speed with respect to the Earth. (b) What additional information would you need to calculate the radial velocity of HDE 271 182 with respect to the Sun?

P11A.9 In Problem 11A.8, it is remarked that Doppler shifts of atomic spectral lines are used to estimate the speed of recession or approach of a star. A spectral line of $^{48}Ti^{8+}$ (of mass $47.95m_u$) in a distant star was found to be shifted from 654.2 nm to 706.5 nm and to be broadened to 61.8 pm. What is the speed of recession and the surface temperature of the star?

P11A.10 The Gaussian shape of a Doppler-broadened spectral line reflects the Maxwell distribution of speeds (see Topic 1B) in the sample at the temperature of the experiment. In a spectrometer that makes use of *phase-sensitive detection* the output signal is proportional to the first derivative of the signal intensity, $dI/d\nu$. Plot the resulting lineshape for various temperatures. How is the separation of the peaks related to the temperature?

P11A.11 The collision frequency z of a molecule of mass m in a gas at a pressure p is $z = 4\sigma(kT/\pi m)^{1/2}p/kT$, where σ is the collision cross-section. Find an expression for the collision-limited lifetime of an excited state assuming that every collision is effective. Estimate the width of a rotational transition at 63.56 cm^{-1} in HCl ($\sigma = 0.30$ nm^2) at 25 °C and 1.0 atm. To what value must the pressure of the gas be reduced in order to ensure that collision broadening is less important than Doppler broadening?

P11A.12 Refer to Fig. 11A.9, which depicts a Michelson interferometer. The mirror M_1 moves in discrete distance increments, so the path difference p is also incremented in discrete steps. Explore the effect of increasing the step size on the shape of the interferogram for a monochromatic beam of wavenumber $\tilde{\nu}$ and intensity I_0. That is, draw plots of $I(p)/I_0$ against $\tilde{\nu}p$, each with a different number of data points spanning the same total distance path taken by the movable mirror M_1.

P11A.13 Use mathematical software to elaborate on the results of *Example* 11A.2 by: (a) exploring the effect of varying the wavenumbers and intensities of the three components of the radiation on the shape of the interferogram; and (b) calculating the Fourier transforms of the functions you generated in part (a).

TOPIC 11B Rotational spectroscopy

Discussion questions

D11B.1 Account for the rotational degeneracy of the various types of rigid rotor. Would the loss of rigidity affect your conclusions?

D11B.2 Does centrifugal distortion increase or decrease the separation between adjacent rotational energy levels?

D11B.3 Distinguish between an oblate and a prolate symmetric rotor and give several examples of each.

D11B.4 Describe the physical origins of the gross selection rule for microwave spectroscopy.

D11B.5 Describe the physical origins of the gross selection rule for rotational Raman spectroscopy.

D11B.6 Does $Be^{19}F_2$ exist in *ortho* and *para* forms? *Hints:* (a) Determine the geometry of BeF_2, then (b) decide whether fluorine nuclei are fermions or bosons.

D11B.7 Describe the role of nuclear statistics in the occupation of energy levels in $^1H^{12}C{\equiv}^{12}C^1H$, $^1H^{13}C{\equiv}^{13}C^1H$, and $^2H^{12}C{\equiv}^{12}C^2H$. For nuclear spin data, see Table 12A.2.

D11B.8 Account for the existence of a rotational zero-point energy in molecular hydrogen.

Exercises

E11B.1(a) Calculate the moment of inertia around the bisector of the OOO angle and the corresponding rotational constant of an $^{16}O_3$ molecule (bond angle 117°; OO bond length 128 pm).

E11B.1(b) Calculate the moment of inertia around the threefold symmetry axis and the corresponding rotational constant of a $^{31}P^1H_3$ molecule (bond angle 93.5°; PH bond length 142 pm).

E11B.2(a) Plot the expressions for the two moments of inertia of a pyramidal symmetric top version of an AB_4 molecule (Table 11B.1) with equal bond lengths but with the angle θ increasing from 90° to the tetrahedral angle.

E11B.2(b) Plot the expressions for the two moments of inertia of a pyramidal symmetric top version of an AB_4 molecule (Table 11B.1) with θ equal to the tetrahedral angle but with one A–B bond varying. *Hint:* Write $\rho = R'_{AB}/R_{AB}$, and allow ρ to vary from 2 to 1.

E11B.3(a) Classify the following rotors: (i) O_3, (ii) CH_3CH_3, (iii) XeO_4, (iv) $FeCp_2$ (Cp denotes the cyclopentadienyl group, C_5H_5).

E11B.3(b) Classify the following rotors: (i) $CH_2{=}CH_2$, (ii) SO_3, (iii) ClF_3, (iv) N_2O.

E11B.4.(a) Calculate the HC and CN bond lengths in HCN from the rotational constants $B(^1H^{12}C^{14}N) = 44.316$ GHz and $B(^2H^{12}C^{14}N) = 36.208$ GHz.

E11B.4(b) Calculate the CO and CS bond lengths in OCS from the rotational constants $B(^{16}O^{12}C^{32}S) = 6081.5$ MHz, $B(^{16}O^{12}C^{34}S) = 5932.8$ MHz.

E11B.5(a) Estimate the centrifugal distortion constant for $^1H^{127}I$, for which $\tilde{B} = 6.511$ cm^{-1} and $\tilde{\nu} = 2308$ cm^{-1}. By what factor would the constant change when ^2H is substituted for ^1H?

E11B.5(b) Estimate the centrifugal distortion constant for $^{79}Br^{81}Br$, for which $\tilde{B} = 0.0809$ cm^{-1} and $\tilde{\nu} = 323.2$ cm^{-1}. By what factor would the constant change when the ^{79}Br is replaced by ^{81}Br?

E11B.6(a) Which of the following molecules may show a pure rotational microwave absorption spectrum: (i) H_2, (ii) HCl, (iii) CH_4, (iv) CH_3Cl, (v) CH_2Cl_2?

E11B.6(b) Which of the following molecules may show a pure rotational microwave absorption spectrum: (i) H_2O, (ii) H_2O_2, (iii) NH_3, (iv) N_2O?

E11B.7(a) Calculate the frequency and wavenumber of the $J = 3 \leftarrow 2$ transition in the pure rotational spectrum of $^{14}N^{16}O$. The equilibrium bond length is 115 pm. Would the frequency increase or decrease if centrifugal distortion is considered?

E11B.7(b) Calculate the frequency and wavenumber of the $J = 2 \leftarrow 1$ transition in the pure rotational spectrum of $^{12}C^{16}O$. The equilibrium bond length is

112.81 pm. Would the frequency increase or decrease if centrifugal distortion is considered?

E11B.8(a) The wavenumber of the $J = 3 \leftarrow 2$ rotational transition of $^1H^{35}Cl$ considered as a rigid rotor is 63.56 cm^{-1}; what is the H–Cl bond length?
E11B.8(b) The wavenumber of the $J = 1 \leftarrow 0$ rotational transition of $^1H^{81}Br$ considered as a rigid rotor is 16.93 cm^{-1}; what is the H–Br bond length?

E11B.9(a) The spacing of lines in the microwave spectrum of $^{27}Al^1H$ is 12.604 cm^{-1}; calculate the moment of inertia and bond length of the molecule.
E11B.9(b) The spacing of lines in the microwave spectrum of $^{35}Cl^{19}F$ is 1.033 cm^{-1}; calculate the moment of inertia and bond length of the molecule.

E11B.10(a) What is the most highly populated rotational level of Cl_2 at (i) 25 °C, (ii) 100 °C? Take $\tilde{B} = 0.244$ cm^{-1}.
E11B.10(b) What is the most highly populated rotational level of Br_2 at (i) 25 °C, (ii) 100 °C? Take $\tilde{B} = 0.0809$ cm^{-1}.

E11B.11(a) Which of the following molecules may show a pure rotational Raman spectrum: (i) H_2, (ii) HCl, (iii) CH_4, (iv) CH_3Cl?
E11B.11(b) Which of the following molecules may show a pure rotational Raman spectrum: (i) CH_2Cl_2, (ii) CH_3CH_3, (iii) SF_6, (iv) N_2O?

E11B.12(a) The wavenumber of the incident radiation in a Raman spectrometer is 20 487 cm^{-1}. What is the wavenumber of the scattered Stokes radiation for the $J = 2 \leftarrow 0$ transition of $^{14}N_2$? Take $\tilde{B} = 1.9987$ cm^{-1}.
E11B.12(b) The wavenumber of the incident radiation in a Raman spectrometer is 20 623 cm^{-1}. What is the wavenumber of the scattered Stokes radiation for the $J = 4 \leftarrow 2$ transition of $^{16}O_2$? Take $\tilde{B} = 1.4457$ cm^{-1}.

E11B.13(a) The rotational Raman spectrum of $^{35}Cl_2$ shows a series of Stokes lines separated by 0.9752 cm^{-1} and a similar series of anti-Stokes lines. Calculate the bond length of the molecule.
E11B.13(b) The rotational Raman spectrum of $^{19}F_2$ shows a series of Stokes lines separated by 3.5312 cm^{-1} and a similar series of anti-Stokes lines. Calculate the bond length of the molecule.

E11B.14(a) What is the ratio of weights of populations due to the effects of nuclear statistics for $^{35}Cl_2$?
E11B.14(b) What is the ratio of weights of populations due to the effects of nuclear statistics for $^{12}C^{32}S_2$? What effect would be observed when ^{12}C is replaced by ^{13}C? For nuclear spin data, see Table 12A.2.

Problems

P11B.1 Show that the moment of inertia of a diatomic molecule composed of atoms of masses m_A and m_B and bond length R is equal to $m_{\text{eff}}R^2$, where $m_{\text{eff}} = m_Am_B/(m_A + m_B)$.

P11B.2 Confirm the expression given in Table 11B.1 for the moment of inertia of a linear ABC molecule. *Hint:* Begin by locating the centre of mass.

P11B.3 The rotational constant of NH_3 is 298 GHz. Calculate the separation of the pure rotational spectrum lines as a frequency (in GHz) and a wavenumber (in cm^{-1}), and show that the value of B is consistent with an N–H bond length of 101.4 pm and a bond angle of 106.78°.

P11B.4 Rotational absorption lines from $^1H^{35}Cl$ gas were found at the following wavenumbers (R.L. Hausler and R.A. Oetjen, *J. Chem. Phys.* **21**, 1340 (1953)): 83.32, 104.13, 124.73, 145.37, 165.89, 186.23, 206.60, 226.86 cm^{-1}. Calculate the moment of inertia and the bond length of the molecule. Predict the positions of the corresponding lines in $^2H^{35}Cl$.

P11B.5 Is the bond length in ^{1}HCl the same as that in ^{2}HCl? The wavenumbers of the $J = 1 \leftarrow 0$ rotational transitions for $^1H^{35}Cl$ and $^2H^{35}Cl$ are 20.8784 and 10.7840 cm^{-1}, respectively. Accurate atomic masses are 1.007 825m_u and 2.0140m_u for ^{1}H and ^{2}H, respectively. The mass of ^{35}Cl is 34.968 85m_u. Based on this information alone, can you conclude that the bond lengths are the same or different in the two molecules?

P11B.6 Thermodynamic considerations suggest that the copper monohalides CuX should exist mainly as polymers in the gas phase, and indeed it proved difficult to obtain the monomers in sufficient abundance to detect spectroscopically. This problem was overcome by flowing the halogen gas over copper heated to 1100 K (Manson et al. (*J. Chem. Phys.* **63**, 2724 (1975))). For $^{63}Cu^{79}Br$ the $J = 14 \leftarrow 13$, $15 \leftarrow 14$, and $16 \leftarrow 15$ transitions occurred at 84 421.34, 90 449.25, and 96 476.72 MHz, respectively. Calculate the rotational constant and bond length of $^{63}Cu^{79}Br$. The mass of ^{63}Cu is 62.9296m_u.

P11B.7 The microwave spectrum of $^{16}O^{12}CS$ gave absorption lines (in GHz) as follows:

J	1	2	3	4
^{32}S	24.325 92	36.488 82	48.651 64	60.814 08
^{34}S	23.732 33		47.462 40	

Use the expressions for moments of inertia in Table 11B.1, assuming that the bond lengths are unchanged by substitution, to calculate the CO and CS bond lengths in OCS.

P11B.8 Equation 11B.20b may be rearranged into

$$\tilde{\nu}(J+1 \leftarrow J)/\{2(J+1)\} = \tilde{B} - 2\tilde{D}_J(J+1)^2$$

which is the equation of a straight line when the left-hand side is plotted against $(J + 1)^2$. The following wavenumbers of transitions (in cm^{-1}) were observed for $^{12}C^{16}O$:

J	0	1	2	3	4
$\tilde{\nu}$/cm^{-1}	3.845 033	7.689 919	11.534 510	15.378 662	19.222 223

Evaluate $\tilde{B}$ and $\tilde{D}_J$ for CO.

P11B.9‡ In a study of the rotational spectrum of the linear FeCO radical, Tanaka et al. (*J. Chem. Phys.* **106**, 6820 (1997)) report the following $J + 1 \leftarrow J$ transitions:

J	24	25	26	27	28	29
ν/MHz	214 777.7	223 379.0	231 981.2	240 584.4	249 188.5	257 793.5

Evaluate the rotational constant of the molecule. Also, estimate the value of J for the most highly populated rotational energy level at 298 K and at 100 K.

P11B.10 The rotational terms of a symmetric top, allowing for centrifugal distortion, are commonly written

$$\tilde{F}(J,K) = \tilde{B}J(J+1) + (\tilde{A} - \tilde{B})K^2 - \tilde{D}_J J^2(J+1)^2 - \tilde{D}_{JK}J(J+1)K^2 - \tilde{D}_K K^4$$

(a) Develop an expression for the wavenumbers of the allowed rotational transitions. (b) The following transition frequencies (in gigahertz, GHz) were observed for CH_3F:

51.0718	102.1403	102.1421	153.1987	153.2068	153.2095

Evaluate as many constants in the expression for the rotational terms as these values permit.

P11B.11 Develop an expression for the value of J corresponding to the most highly populated rotational energy level of a diatomic rotor at a temperature T remembering that the degeneracy of each level is $2J + 1$. Evaluate the expression for ICl (for which $\tilde{B} = 0.1142$ cm^{-1}) at 25 °C. Repeat the problem for the most highly populated level of a spherical rotor, taking note of the fact that each level is $(2J+1)^2$-fold degenerate. Evaluate the expression for CH_4 (for which $\tilde{B} = 5.24\,\text{cm}^{-1}$) at 25 °C. *Hint:* To develop the expression, recall that the first derivative of a function is zero when the function reaches either a maximum or minimum value.

P11B.12 A. Dalgarno, in *Chemistry in the interstellar medium*, *Frontiers of Astrophysics*, ed. E.H. Avrett, Harvard University Press, Cambridge, MA (1976), notes that although both CH and CN spectra show up strongly in the interstellar medium in the constellation Ophiuchus, the CN spectrum has become the standard for the determination of the temperature of the cosmic microwave background radiation. Demonstrate through a calculation why CH would not be as useful for this purpose as CN. The rotational constants $\tilde{B}$ for CH and CN are $14.190\,\text{cm}^{-1}$ and $1.891\,\text{cm}^{-1}$, respectively.

P11B.13 The space immediately surrounding stars, the *circumstellar space*, is significantly warmer because stars are very intense black-body emitters with temperatures of several thousand kelvin. Discuss how such factors as cloud temperature, particle density, and particle velocity may affect the rotational spectrum of CO in an interstellar cloud. What new features in the spectrum of CO can be observed in gas ejected from and still near a star with temperatures of about 1000 K, relative to gas in a cloud with temperature of about 10 K? Explain how these features may be used to distinguish between circumstellar and interstellar material on the basis of the rotational spectrum of CO.

P11B.14 Pure rotational Raman spectra of gaseous C_6H_6 and C_6D_6 yield the following rotational constants: $\tilde{B}(C_6H_6) - 0.189\,60\,\text{cm}^{-1}$, $\tilde{B}(C_6D_6) = 0.156\,81\,\text{cm}^{-1}$. The moments of inertia of the molecules about any axis perpendicular to the C_6 axis were calculated from these data as $I(C_6H_6) = 1.4759\times10^{-45}\,\text{kg m}^2$, $I(C_6D_6) = 1.7845\times10^{-45}\,\text{kg m}^2$. Calculate the CC and CH bond lengths.

TOPIC 11C Vibrational spectroscopy of diatomic molecules

Discussion questions

D11C.1 Discuss the strengths and limitations of the parabolic and Morse functions as approximations to the true potential energy curve of a diatomic molecule.

D11C.2 Describe the effect of vibrational excitation on the rotational constant of a diatomic molecule.

D11C.3 How is the method of combination differences used in rotation–vibration spectroscopy to determine rotational constants?

D11C.4 In what ways may the rotational and vibrational spectra of molecules change as a result of isotopic substitution?

Exercises

E11C.1(a) An object of mass 100 g suspended from the end of a rubber band has a vibrational frequency of 2.0 Hz. Calculate the force constant of the rubber band.
E11C.1(b) An object of mass 1.0 g suspended from the end of a spring has a vibrational frequency of 10.0 Hz. Calculate the force constant of the spring.

E11C.2(a) Calculate the percentage difference in the fundamental vibrational wavenumbers of $^{23}Na^{35}Cl$ and $^{23}Na^{37}Cl$ on the assumption that their force constants are the same. The mass of ^{23}Na is $22.9898m_u$.
E11C.2(b) Calculate the percentage difference in the fundamental vibrational wavenumbers of $^1H^{35}Cl$ and $^2H^{37}Cl$ on the assumption that their force constants are the same.

E11C.3(a) The wavenumber of the fundamental vibrational transition of $^{35}Cl_2$ is $564.9\,\text{cm}^{-1}$. Calculate the force constant of the bond.
E11C.3(b) The wavenumber of the fundamental vibrational transition of $^{79}Br^{81}Br$ is $323.2\,\text{cm}^{-1}$. Calculate the force constant of the bond.

E11C.4(a) The hydrogen halides have the following fundamental vibrational wavenumbers: $4141.3\,\text{cm}^{-1}$ ($^1H^{19}F$); $2988.9\,\text{cm}^{-1}$ ($^1H^{35}Cl$); $2649.7\,\text{cm}^{-1}$ ($^1H^{81}Br$); $2309.5\,\text{cm}^{-1}$ ($H^{127}I$). Calculate the force constants of the hydrogen–halogen bonds.
E11C.4(b) From the data in Exercise E11C.4(a), predict the fundamental vibrational wavenumbers of the deuterium halides.

E11C.5(a) Calculate the relative numbers of Cl_2 molecules ($\tilde{\nu} = 559.7\,\text{cm}^{-1}$) in the ground and first excited vibrational states at (i) 298 K, (ii) 500 K.
E11C.5(b) Calculate the relative numbers of Br_2 molecules ($\tilde{\nu} = 321\,\text{cm}^{-1}$) in the second and first excited vibrational states at (i) 298 K, (ii) 800 K.

E11C.6(a) For $^{16}O_2$, $\Delta\tilde{G}$ values for the transitions $v = 1 \leftarrow 0$, $2 \leftarrow 0$, and $3 \leftarrow 0$ are, respectively, 1556.22, 3088.28, and $4596.21\,\text{cm}^{-1}$. Calculate $\tilde{\nu}$ and x_e. Assume y_e to be zero.
E11C.6(b) For $^{14}N_2$, $\Delta\tilde{G}$ values for the transitions $v = 1 \leftarrow 0$, $2 \leftarrow 0$, and $3 \leftarrow 0$ are, respectively, 2329.91, 4631.20, and $6903.69\,\text{cm}^{-1}$. Calculate $\tilde{\nu}$ and x_e. Assume y_e to be zero.

E11C.7(a) The first five vibrational energy levels of HCl are at 1481.86, 4367.50, 7149.04, 9826.48, and $12\,399.8\,\text{cm}^{-1}$. Calculate the dissociation energy of the molecule in reciprocal centimetres and electronvolts.
E11C.7(b) The first five vibrational energy levels of HI are at 1144.83, 3374.90, 5525.51, 7596.66, and $9588.35\,\text{cm}^{-1}$. Calculate the dissociation energy of the molecule in reciprocal centimetres and electronvolts.

E11C.8(a) Infrared absorption by $^1H^{127}I$ gives rise to an R branch from $v = 0$. What is the wavenumber of the line originating from the rotational state with $J = 2$? *Hint:* Use data from Table 11C.1.
E11C.8(b) Infrared absorption by $^1H^{81}Br$ gives rise to a P branch from $v = 0$. What is the wavenumber of the line originating from the rotational state with $J = 2$? *Hint:* Use data from Table 11C.1.

Problems

P11C.1 Use molecular modelling software and the computational method of your choice to construct molecular potential energy curves like the one shown in Fig. 11C.1. Consider the hydrogen halides (HF, HCl, HBr, and HI): (a) plot the calculated energy of each molecule against the bond length, and (b) identify the order of force constants of the H–Hal bonds.

P11C.2 Derive an expression for the force constant of an oscillator that can be modelled by a Morse potential energy (eqn 11C.7).

P11C.3 Suppose a particle confined to a cavity in a microporous material has a potential energy of the form $V(x) = V_0(e^{-a^2/x^2} - 1)$. Sketch $V(x)$. What is the value of the force constant corresponding to this potential energy? Would the particle undergo simple harmonic motion? Sketch the likely form of the first two vibrational wavefunctions.

P11C.4 The vibrational levels of $^{23}Na^{127}I$ lie at the wavenumbers 142.81, 427.31, 710.31, and 991.81 cm^{-1}. Show that they fit the expression $(v+\frac{1}{2})\tilde{\nu}-(v+\frac{1}{2})^2x_e\tilde{\nu}$, and deduce the force constant, zero-point energy, and dissociation energy of the molecule.

P11C.5 The $^1H^{35}Cl$ molecule is quite well described by the Morse potential energy with $hc\tilde{D}_e = 5.33$ eV, $\tilde{\nu} = 2989.7$ cm^{-1}, and $x_e\tilde{\nu} = 52.05$ cm^{-1}. Assuming that the potential is unchanged on deuteration, predict the dissociation energies ($hc\tilde{D}_0$, in electronvolts) of (a) $^1H^{35}Cl$, (b) $^2H^{35}Cl$.

P11C.6 The Morse potential energy (eqn 11C.7) is very useful as a simple representation of the actual molecular potential energy. When $^{85}Rb^1H$ was studied, it was found that $\tilde{\nu} = 936.8$ cm^{-1} and $x_e\tilde{\nu} = 14.15$ cm^{-1}. Plot the potential energy curve from 50 pm to 800 pm around $R_e = 236.7$ pm. Then go on to explore how the rotation of a molecule may weaken its bond by allowing for the kinetic energy of rotation of a molecule and plotting $V^* = V + hc\tilde{B}J(J+1)$ with $\tilde{B}=\hbar/4\pi c\mu R^2$. Plot these curves on the same diagram for $J = 40$, 80, and 100, and observe how the dissociation energy is affected by the rotation. *Hints:* Taking $\tilde{B} = 3.020$ cm^{-1} as the equilibrium bond length will greatly simplify the calculation. The mass of ^{85}Rb is $84.9118m_u$.

P11C.7‡ Luo et al. (*J. Chem. Phys.* **98**, 3564 (1993)) reported the observation of the He_2 complex, a species which had escaped detection for a long time. The fact that the observation required temperatures in the neighbourhood of 1 mK is consistent with computational studies which suggest that $hc\tilde{D}_e$ for He_2 is about 1.51×10^{-23} J, $hc\tilde{D}_0 \approx 2 \times 10^{-26}$ J, and R_e about 297 pm. (a) Estimate the fundamental vibrational wavenumber, force constant, moment of inertia, and rotational constant based on the harmonic oscillator and rigid-rotor approximations. (b) Such a weakly bound complex is hardly likely to be rigid. Estimate the vibrational wavenumber and anharmonicity constant based on the Morse potential energy.

P11C.8 Confirm that a Morse oscillator has a finite number of bound states, the states with $V < hc\tilde{D}_e$. Determine the value of v_{max} for the highest bound state.

P11C.9 Provided higher order terms are neglected, eqn 11C.9b for the vibrational wavenumbers of an anharmonic oscillator, $\Delta\tilde{G}_{v+1/2} = \tilde{\nu}-2(v+1)x_e\tilde{\nu}+\cdots$, is the equation of a straight line when the left-hand side is plotted against $v + 1$. Use the following data on CO to determine the values of $\tilde{\nu}$ and $x_e\tilde{\nu}$ for CO:

v	0	1	2	3	4
$\Delta\tilde{G}_{v+1/2}$/cm^{-1}	2143.1	2116.1	2088.9	2061.3	2033.5

P11C.10 The rotational constant for CO is 1.9225 cm^{-1} and 1.9050 cm^{-1} in the ground and first excited vibrational states, respectively. By how much does the internuclear distance change as a result of this transition?

P11C.11 The average spacing between the rotational lines of the P and R branches of $^{12}C_2{}^1H_2$ and $^{12}C_2{}^2H_2$ is 2.352 cm^{-1} and 1.696 cm^{-1}, respectively. Estimate the CC and CH bond lengths.

P11C.12 Absorptions in the $v = 1 \leftarrow 0$ vibration–rotation spectrum of $^1H^{35}Cl$ were observed at the following wavenumbers (in cm^{-1}):

2998.05	2981.05	2963.35	2944.99	2925.92
2906.25	2865.14	2843.63	2821.59	2799.00

Assign the rotational quantum numbers and use the method of combination differences to calculate the rotational constants of the two vibrational levels.

P11C.13 Suppose that the internuclear distance may be written $R = R_e + x$ where R_e is the equilibrium bond length. Also suppose that the potential well is symmetrical and confines the oscillator to small displacements. Deduce expressions for $1/\langle R\rangle^2$, $1/\langle R^2\rangle$, and $\langle 1/R^2\rangle$ to the lowest non-zero power of $\langle x^2\rangle/R_e^2$ and confirm that the values are not the same.

P11C.14 Continue the development of Problem P11C.13 by using the virial theorem (Topic 7E) to relate $\langle x^2\rangle$ to the vibrational quantum number. Does your result imply that the rotational constant increases or decreases as the oscillator becomes excited to higher quantum states? What would be the effect of anharmonicity?

P11C.15 The rotational constant for a diatomic molecule in the vibrational state with quantum number v typically fits the expression $\tilde{B}_v = \tilde{B}_e - a(v+\frac{1}{2})$, where $\tilde{B}_e$ is the rotational constant corresponding to the equilibrium bond length. For the interhalogen molecule IF it is found that $\tilde{B}_e = 0.27971$ cm^{-1} and $a = 0.187$ m^{-1} (note the change of units). Calculate $\tilde{B}_0$ and $\tilde{B}_1$ and use these values to calculate the wavenumbers of the transitions originating from $J = 3$ of the P and R branches. You will need the following additional information: $\tilde{\nu} = 610.258$ cm^{-1} and $x_e\tilde{\nu} = 3.141$ cm^{-1}. Estimate the dissociation energy of the IF molecule.

P11C.16 Develop eqn 11B.16 ($\tilde{D}_J = 4\tilde{B}^3/\tilde{\nu}^2$) for the centrifugal distortion constant $\tilde{D}_J$ of a diatomic molecule of effective mass m_{eff}. Treat the bond as an elastic spring with force constant k_f and equilibrium length R_e that is subjected to a centrifugal distortion to a new length R_c. Begin the derivation by letting the particles experience a restoring force of magnitude $k_f(R_c - R_e)$ that is countered perfectly by a centrifugal force $m_{eff}\omega^2 R_c$, where ω is the angular velocity of the rotating molecule. Then introduce quantum mechanical effects by writing the angular momentum as $\{J(J+1)\}^{1/2}\hbar$. Finally, write an expression for the energy of the rotating molecule, compare it with eqn 11B.15, and infer an expression for $\tilde{D}_J$.

P11C.17 At low resolution, the strongest absorption band in the infrared absorption spectrum of $^{12}C^{16}O$ is centred at 2150 cm^{-1}. Upon closer examination at higher resolution, this band is observed to be split into two sets of closely spaced peaks, one on each side of the centre of the spectrum at 2143.26 cm^{-1}. The separation between the peaks immediately to the right and left of the centre is 7.655 cm^{-1}. Make the harmonic oscillator and rigid rotor approximations and calculate from these data: (a) the vibrational wavenumber of a CO molecule, (b) its molar zero-point vibrational energy, (c) the force constant of the CO bond, (d) the rotational constant $\tilde{B}$, and (e) the bond length of CO.

P11C.18 For $^{12}C^{16}O$, $\tilde{\nu}_R(0) = 2147.084$ cm^{-1}, $\tilde{\nu}_R(1) = 2150.858$ cm^{-1}, $\tilde{\nu}_P(1) = 2139.427$ cm^{-1}, and $\tilde{\nu}_P(2) = 2135.548$ cm^{-1}. Estimate the values of $\tilde{B}_0$ and $\tilde{B}_1$.

P11C.19 The analysis of combination differences summarized in the text considered the R and P branches. Extend the analysis to the O and S branches of a Raman spectrum.

TOPIC 11D Vibrational spectroscopy of polyatomic molecules

Discussion questions

D11D.1 Describe the physical origin of the gross selection rule for infrared spectroscopy.

D11D.2 Describe the physical origin of the gross selection rule for vibrational Raman spectroscopy.

D11D.3 Can a linear, nonpolar molecule like CO_2 have a Raman spectrum?

Exercises

E11D.1(a) Which of the following molecules may show infrared absorption spectra: (i) H_2, (ii) HCl, (iii) CO_2, (iv) H_2O?

E11D.1(b) Which of the following molecules may show infrared absorption spectra: (i) CH_3CH_3, (ii) CH_4, (iii) CH_3Cl, (iv) N_2?

E11D.2(a) How many normal modes of vibration are there for the following molecules: (i) H_2O, (ii) H_2O_2, (iii) C_2H_4?

E11D.2(b) How many normal modes of vibration are there for the following molecules: (i) C_6H_6, (ii) $C_6H_5CH_3$, (iii) HC≡C–C≡C–H?

E11D.3(a) How many vibrational modes are there for the molecule NC–(C≡C–C≡C–)$_{10}$CN detected in an interstellar cloud?

E11D.3(b) How many vibrational modes are there for the molecule NC–(C≡C–C≡C–)$_8$CN detected in an interstellar cloud?

E11D.4(a) Write an expression for the vibrational term for the ground vibrational state of H_2O in terms of the wavenumbers of the normal modes. Neglect anharmonicities, as in eqn 11D.2.

E11D.4(b) Write an expression for the vibrational term for the ground vibrational state of SO_2 in terms of the wavenumbers of the normal modes. Neglect anharmonicities, as in eqn 11D.2.

E11D.5(a) Which of the three vibrations of an AB_2 molecule are infrared or Raman active when it is (i) angular, (ii) linear?

E11D.5(b) Is the out-of-plane mode of a planar AB_3 molecule infrared or Raman active?

E11D.6(a) Consider the vibrational mode that corresponds to the uniform expansion of the benzene ring. Is it (i) Raman, (ii) infrared active?

E11D.6(b) Consider the vibrational mode that corresponds to the boat-like bending of a benzene ring. Is it (i) Raman, (ii) infrared active?

E11D.7(a) Does the exclusion rule apply to H_2O?

E11D.7(b) Does the exclusion rule apply to C_2H_4?

Problems

P11D.1 Suppose that the out-of-plane distortion of an AB_3 planar molecule is described by a potential energy $V=V_0(1-e^{-bh^4})$, where h is the distance by which the central atom A is displaced. Sketch this potential energy as a function of h (allow h to be both negative and positive). What could be said about (a) the force constant, (b) the vibrations? Sketch the form of the ground-state wavefunction.

P11D.2 Predict the shape of the nitronium ion, NO_2^+, from its Lewis structure and the VSEPR model. It has one Raman active vibrational mode at 1400 cm^{-1}, two strong IR active modes at 2360 and 540 cm^{-1}, and one weak absorption in the IR at 3735 cm^{-1}. Are these data consistent with the predicted shape of the molecule? Assign the vibrational wavenumbers to the modes from which they arise.

P11D.3 The computational methods discussed in Topic 9E can be used to simulate the vibrational spectrum of a molecule, and it is then possible to determine the correspondence between a vibrational frequency and the atomic displacements that give rise to a normal mode. (a) Using molecular modelling software and the computational method of your choice, calculate the fundamental vibrational wavenumbers and depict the vibrational normal modes of SO_2 in the gas phase graphically. (b) The experimental values of the fundamental vibrational wavenumbers of SO_2 in the gas phase are 525 cm^{-1}, 1151 cm^{-1}, and 1336 cm^{-1}. Compare the calculated and experimental values. Even if agreement is poor, is it possible to establish a correlation between an experimental value of the vibrational wavenumber with a specific vibrational normal mode?

TOPIC 11E Symmetry analysis of vibrational spectra

Discussion question

D11E.1 Suppose that you wish to characterize the normal modes of benzene in the gas phase. Why is it important to obtain both infrared absorption and Raman spectra of the molecule?

Exercises

E11E.1(a) The molecule CH_2Cl_2 belongs to the point group C_{2v}. The displacements of the atoms span $5A_1 + 2A_2 + 4B_1 + 4B_2$. What are the symmetry species of the normal modes of vibration?

E11E.1(b) A carbon disulfide molecule belongs to the point group $D_{\infty h}$. The nine displacements of the three atoms span $A_{1g} + 2A_{1u} + 2E_{1u} + E_{1g}$. What are the symmetry species of the normal modes of vibration?

E11E.2(a) Which of the normal modes of CH_2Cl_2 (Exercise E12E.1a) are infrared active? Which are Raman active?

E11E.2(b) Which of the normal modes of carbon disulfide (Exercise E11E.1b) are infrared active? Which are Raman active?

E11E.3(a) Which of the normal modes of (i) H_2O, (ii) H_2CO are infrared active?

E11E.3(b) Which of the normal modes of (i) H_2O, (ii) H_2CO are Raman active?

Problems

P11E.1 Consider the molecule CH_3Cl. (a) To what point group does the molecule belong? (b) How many normal modes of vibration does the molecule have? (c) What are the symmetry species of the normal modes of vibration of this molecule? (d) Which of the vibrational modes of this molecule are infrared active? (e) Which of the vibrational modes of this molecule are Raman active?

P11E.2 Suppose that three conformations are proposed for the nonlinear molecule H_2O_2 (1, 2, and 3). The infrared absorption spectrum of gaseous H_2O_2 has bands at 870, 1370, 2869, and 3417 cm^{-1}. The Raman spectrum of the same sample has bands at 877, 1408, 1435, and 3407 cm^{-1}. All bands correspond to fundamental vibrational wavenumbers and you may assume that: (a) the 870 and 877 cm^{-1} bands arise from the same normal mode, and (b) the 3417 and 3407 cm^{-1} bands arise from the same normal mode. (i) If H_2O_2 were linear, how many normal modes of vibration would it have? (ii) Give the symmetry point group of each of the three proposed conformations of nonlinear H_2O_2. (iii) Determine which of the proposed conformations is inconsistent with the spectroscopic data. Explain your reasoning.

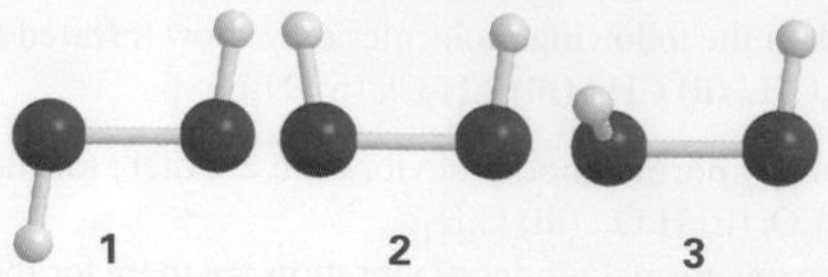

1 2 3

TOPIC 11F Electronic spectra

Discussion questions

D11F.1 Explain the origin of the term symbol $^3\Sigma_g^-$ for the ground state of a dioxygen molecule.

D11F.2 Explain the basis of the Franck–Condon principle and how it leads to the formation of a vibrational progression.

D11F.3 How do the band heads in P and R branches arise? Could the Q branch show a head?

D11F.4 Explain how colour can arise from molecules.

D11F.5 Suppose that you are a colour chemist and had been asked to intensify the colour of a dye without changing the type of compound, and that the dye in question was a conjugated polyene. (a) Would you choose to lengthen or to shorten the chain? (b) Would the modification to the length shift the apparent colour of the dye towards the red or the blue?

D11F.6 Can a complex of the Zn^{2+} ion have a d–d electronic transition? Explain your answer.

Exercises

E11F.1(a) What is the value of S and the term symbol for the ground state of H_2?
E11F.1(b) The term symbol for one of the lowest excited states of H_2 is $^3\Pi_u$. To which excited-state configuration does this term symbol correspond?

E11F.2(a) What is the full term symbol of the ground electronic state of Li_2^+?
E11F.2(b) What are the levels of the term for the ground electronic state of O_2^-?

E11F.3(a) One of the excited states of the C_2 molecule has the valence electron configuration $1\sigma_g^2 1\sigma_u^2 1\pi_u^3 1\pi_g^1$. Give the multiplicity and parity of the term.
E11F.3(b) Another of the excited states of the C_2 molecule has the valence electron configuration $1\sigma_g^2 1\sigma_u^2 1\pi_u^2 1\pi_g^2$. Give the multiplicity and parity of the term.

E11F.4(a) Which of the following transitions are electric-dipole allowed? (i) $^2\Pi \leftrightarrow {}^2\Pi$, (ii) $^1\Sigma \leftrightarrow {}^1\Sigma$, (iii) $\Sigma \leftrightarrow \Delta$, (iv) $\Sigma^+ \leftrightarrow \Sigma^-$, (v) $\Sigma^+ \leftrightarrow \Sigma^+$.
E11F.4(b) Which of the following transitions are electric-dipole allowed? (i) $^1\Sigma_g^+ \leftrightarrow {}^1\Sigma_u^+$, (ii) $^3\Sigma_g^+ \leftrightarrow {}^3\Sigma_u^+$, (iii) $\pi^* \leftrightarrow n$.

E11F.5(a) The ground-state wavefunction of a certain molecule is described by the vibrational wavefunction $\psi_0 = N_0 e^{-ax^2/2}$. Calculate the Franck–Condon factor for a transition to a vibrational state described by the wavefunction $\psi_0' = N_0 e^{-a(x-x_0)^2/2}$. The normalization constants are given by eqn 7E.10.
E11F.5(b) The ground-state wavefunction of a certain molecule is described by the vibrational wavefunction $\psi_0 = N_0 e^{-ax^2/2}$. Calculate the Franck–Condon factor for a transition to a vibrational state described by the wavefunction $\psi_1' = N_1(x-x_0)e^{-a(x-x_0)^2/2}$. The normalization constants are given by eqn 7E.10.

E11F.6(a) Suppose that the ground vibrational state of a molecule is modelled by using the particle-in-a-box wavefunction $\psi_0 = (2/L)^{1/2}\sin(\pi x/L)$ for $0 \le x \le L$ and 0 elsewhere. Calculate the Franck–Condon factor for a transition to a vibrational state described by the wavefunction $\psi' = (2/L)^{1/2}\sin\{\pi(x - L/4)/L\}$ for $L/4 \le x \le 5L/4$ and 0 elsewhere.
E11F.6(b) Suppose that the ground vibrational state of a molecule is modelled by using the particle-in-a-box wavefunction $\psi_0 = (2/L)^{1/2}\sin(\pi x/L)$ for $0 \le x \le L$ and 0 elsewhere. Calculate the Franck–Condon factor for a transition to a vibrational state described by the wavefunction $\psi' = (2/L)^{1/2}\sin\{\pi(x - L/2)/L\}$ for $L/2 \le x \le 3L/2$ and 0 elsewhere.

E11F.7(a) Use eqn 11F.7a to infer the value of J corresponding to the location of the band head of the P branch of a transition.
E11F.7(b) Use eqn 11F.7c to infer the value of J corresponding to the location of the band head of the R branch of a transition.

E11F.8(a) The following parameters describe the electronic ground state and an excited electronic state of SnO: $\tilde{B} = 0.3540\ cm^{-1}$, $\tilde{B}' = 0.3101\ cm^{-1}$. Which branch of the transition between them shows a head? At what value of J will it occur?
E11F.8(b) The following parameters describe the electronic ground state and an excited electronic state of BeH: $\tilde{B} = 10.308\ cm^{-1}$, $\tilde{B}' = 10.470\ cm^{-1}$. Which branch of the transition between them shows a head? At what value of J will it occur?

E11F.9(a) The R-branch of the $^1\Pi_u \leftarrow {}^1\Sigma_g^+$ transition of H_2 shows a band head at the very low value of $J = 1$. The rotational constant of the ground state is 60.80 cm^{-1}. What is the rotational constant of the upper state? Has the bond length increased or decreased in the transition?
E11F.9(b) The P-branch of the $^2\Pi \leftarrow {}^2\Sigma^+$ transition of CdH shows a band head at $J = 25$. The rotational constant of the ground state is 5.437 cm^{-1}. What is the rotational constant of the upper state? Has the bond length increased or decreased in the transition?

E11F.10(a) The complex ion $[Fe(OH_2)_6]^{3+}$ has an electronic absorption spectrum with a maximum at 700 nm. Estimate a value of Δ_o for the complex.
E11F.10(b) The complex ion $[Fe(CN)_6]^{3-}$ has an electronic absorption spectrum with a maximum at 305 nm. Estimate a value of Δ_o for the complex.

E11F.11(a) Suppose that a charge-transfer transition in a one-dimensional system can be modelled as a process in which a rectangular wavefunction that is non-zero in the range $0 \le x \le a$ makes a transition to another rectangular wavefunction that is non-zero in the range $\frac{1}{2}a \le x \le b$. Evaluate the transition moment $\int \psi_f x \psi_i dx$. (Assume $a < b$.) *Hint:* Don't forget to normalize each wavefunction to 1.
E11F.11(b) Suppose that a charge-transfer transition in a one-dimensional system can be modelled as a process in which an electron described by a rectangular wavefunction that is non-zero in the range $0 \le x \le a$ makes

a transition to another rectangular wavefunction that is non-zero in the range $ca \le x \le a$ where $0 \le c \le 1$. Evaluate the transition moment $\int \psi_f x \psi_i dx$ and explore its dependence on c. *Hint:* Don't forget to normalize each wavefunction to 1.

E11F.12(a) Suppose that a charge-transfer transition in a one-dimensional system can be modelled as a process in which a Gaussian wavefunction centred on $x = 0$ and width a makes a transition to another Gaussian wavefunction of the same width centred on $x = \frac{1}{2}a$. Evaluate the transition moment $\int \psi_f x \psi_i dx$. *Hint:* Don't forget to normalize each wavefunction to 1.
E11F.12(b) Suppose that a charge-transfer transition can be modelled in a one-dimensional system as a process in which an electron described by a Gaussian wavefunction centred on $x = 0$ and width a makes a transition to another Gaussian wavefunction of width $a/2$ and centred on $x = 0$. Evaluate the transition moment $\int \psi_f x \psi_i dx$. *Hint:* Don't forget to normalize each wavefunction to 1.

E11F.13(a) The two compounds 2,3-dimethyl-2-butene (4) and 2,5-dimethyl-2,4-hexadiene (5) are to be distinguished by their ultraviolet absorption spectra. The maximum absorption in one compound occurs at 192 nm and in the other at 243 nm. Match the maxima to the compounds and justify the assignment.

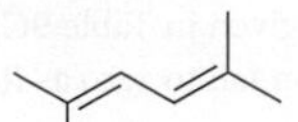

4 2,3-Dimethyl-2-butene **5** 2,5-Dimethyl-2,4-hexadiene

E11F.13(b) 3-Buten-2-one (6) has a strong absorption at 213 nm and a weaker absorption at 320 nm. Assign the ultraviolet absorption transitions, giving your reasons.

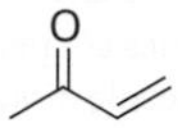

6 3-Buten-2-one

Problems

P11F.1 Which of the following electronic transitions are allowed in O_2: ${}^3\Sigma_g^- \leftrightarrow {}^1\Sigma_g^+$, and ${}^3\Sigma_g^- \leftrightarrow {}^3\Delta_u$?

P11F.2[‡] J.G. Dojahn et al. (*J. Phys. Chem.* **100**, 9649 (1996)) characterized the potential energy curves of the ground and electronic states of homonuclear diatomic halogen anions. These anions have a ${}^2\Sigma_u^+$ ground state and ${}^2\Pi_g$, ${}^2\Pi_u$, and ${}^2\Sigma_g^+$ excited states. To which of the excited states are electric-dipole transitions allowed from the ground state? Explain your conclusion.

P11F.3 The vibrational wavenumber of the oxygen molecule in its electronic ground state is 1580 cm^{-1}, whereas that in the excited state ($B^3\Sigma_u^-$), to which there is an allowed electronic transition, is 700 cm^{-1}. Given that the separation in energy between the minima in their respective potential energy curves of these two electronic states is 6.175 eV, what is the wavenumber of the lowest energy transition in the band of transitions originating from the $v = 0$ vibrational state of the electronic ground state to this excited state? Ignore any rotational structure or anharmonicity.

P11F.4 A transition of particular importance in O_2 gives rise to the Schumann–Runge band in the ultraviolet region. The wavenumbers (in cm^{-1}) of transitions from the ground state to the vibrational levels of the first excited state (${}^3\Sigma_u^-$) are 50 062.6, 50 725.4, 51 369.0, 51 988.6, 52 579.0, 53 143.4, 53 679.6, 54 177.0, 54 641.8, 55 078.2, 55 460.0, 55 803.1, 56 107.3, 56 360.3, 56 570.6. What is the dissociation energy of the upper electronic state? (Use a Birge–Sponer plot, Topic 11C.) The same excited state is known to dissociate into one ground-state O atom and one excited-state atom with an energy 190 kJ mol^{-1} above the ground state. (This excited atom is responsible for a great deal of photochemical mischief in the atmosphere.) Ground-state O_2 dissociates into two ground-state atoms. Use this information to calculate the dissociation energy of ground-state O_2 from the Schumann–Runge data.

P11F.5 You are now ready to understand more deeply the features of photoelectron spectra (Topic 9B). Figure 11.3 shows the photoelectron spectrum of HBr. Disregarding for now the fine structure, the HBr lines fall into two main groups. The least tightly bound electrons (with the lowest ionization energies and hence highest kinetic energies when ejected) are those in the lone pairs of the Br atom. The next ionization energy lies at 15.2 eV, and corresponds to the removal of an electron from the HBr σ bond. (a) The spectrum shows that ejection of a σ electron is accompanied by a lot of vibrational excitation. Use the Franck–Condon principle to account for this observation. (b) Go on to explain why the lack of much vibrational structure in the other band is consistent with the nonbonding role of the Br4p$_x$ and Br4p$_y$ lone-pair electrons.

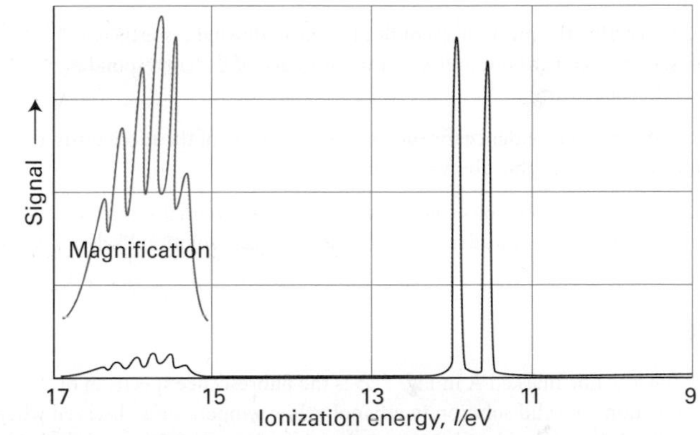

Figure 11.3 The photoelectron spectrum of HBr.

P11F.6 The highest kinetic energy electrons in the photoelectron spectrum of H_2O using 21.22 eV radiation are at about 9 eV and show a large vibrational spacing of 0.41 eV. The symmetric stretching mode of the neutral H_2O molecule lies at 3652 cm^{-1}. (a) What conclusions can be drawn from the nature of the orbital from which the electron is ejected? (b) In the same spectrum of H_2O, the band near 7.0 eV shows a long vibrational series with spacing 0.125 eV. The bending mode of H_2O lies at 1596 cm^{-1}. What conclusions can you draw about the characteristics of the orbital occupied by the photoelectron?

P11F.7 Assume that the states of the π electrons of a conjugated molecule can be approximated by the wavefunctions of a particle in a one-dimensional box, and that the magnitude of the dipole moment can be related to the displacement along this length by $\mu = -ex$. Show that the transition probability for the transition $n = 1 \rightarrow n = 2$ is non-zero, whereas that for $n = 1 \rightarrow n = 3$ is zero. *Hints:* The following relation will be useful: $\sin x \sin y = \frac{1}{2}\cos(x - y) - \frac{1}{2}\cos(x + y)$. Relevant integrals are given in the *Resource section*.

P11F.8 1,3,5-Hexatriene (a kind of 'linear' benzene) was converted into benzene itself. On the basis of a free-electron molecular orbital model (in which hexatriene is treated as a linear box and benzene as a ring), would you expect the lowest energy absorption to rise or fall in energy as a result of the conversion?

P11F.9 Estimate the magnitude of the transition dipole moment of a charge-transfer transition modelled as the migration of an electron from a H1s orbital on one atom to another H1s orbital on an atom a distance R away. Approximate the transition moment by $-eRS$ where S is the overlap integral of the two orbitals. Sketch the transition moment as a function of R using the expression for S given in Table 9C.1. Why does the intensity of a charge-transfer transition fall to zero as R approaches 0 and infinity?

P11F.10 Figure 11.4 shows the UV-visible absorption spectra of a selection of amino acids. Suggest reasons for their different appearances in terms of the structures of the molecules.

P11F.11 Propanone (acetone, $(CH_3)_2CO$) has a strong absorption at 189 nm and a weaker absorption at 280 nm. Identify the chromophore and assign the absorptions to $\pi^* \leftarrow n$ or $\pi^* \leftarrow \pi$ transitions.

P11F.12 Spin angular momentum is conserved when a molecule dissociates into atoms. What atom multiplicities are permitted when the ground state of (a) an O_2 molecule, (b) an N_2 molecule dissociates into atoms?

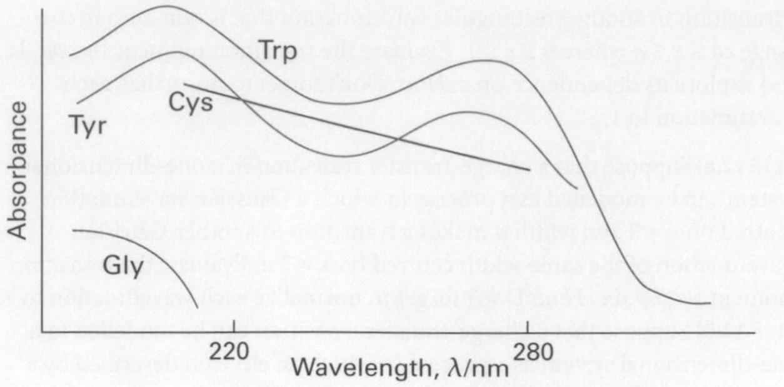

Figure 11.4 The UV-visible absorption spectra of several amino acids.

TOPIC 11G Decay of excited states

Discussion questions

D11G.1 Describe the mechanism of fluorescence. In what respects is a fluorescence spectrum not the exact mirror image of the corresponding absorption spectrum?

D11G.2 What is the evidence for the usual explanation of the mechanism of (a) fluorescence, (b) phosphorescence?

D11G.3 Consider an aqueous solution of a chromophore that fluoresces strongly. Is the addition of iodide ion to the solution likely to increase or decrease the efficiency of phosphorescence of the chromophore?

D11G.4 What can be estimated from the wavenumber of the onset of predissociation?

D11G.5 Describe the principles of a four-level laser.

Exercises

E11G.1(a) The line marked A in Fig. 11.5 is the fluorescence spectrum of benzophenone in solid solution in ethanol at low temperatures observed when the sample is illuminated with 360 nm ultraviolet radiation. What can be said about the vibrational energy levels of the carbonyl group in (i) its ground electronic state and (ii) its excited electronic state?

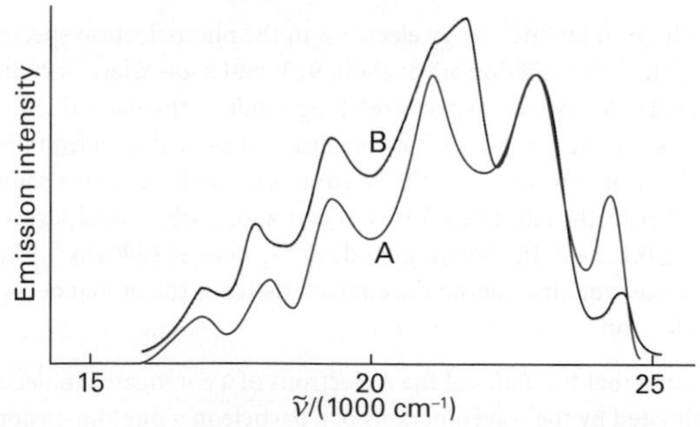

Figure 11.5 The fluorescence (A) spectrum of benzophenone and the phosphorescence (B) spectrum of of a mixture of naphthalene and benzophenone.

E11G.1(b) When naphthalene is illuminated with 360 nm ultraviolet radiation it does not absorb, but the line marked B in Fig. 11.5 is the phosphorescence spectrum of a frozen solution of a mixture of naphthalene and benzophenone in ethanol. Now a component of fluorescence from naphthalene can be detected. Account for this observation.

E11G.2(a) An oxygen molecule absorbs ultraviolet radiation in a transition from its ${}^3\Sigma_g^-$ ground electronic state to an excited state that is energetically close to a dissociative ${}^5\Pi_u$ state. The absorption band has a relatively large experimental linewidth. Account for this observation.

E11G.2(b) A hydrogen molecule absorbs ultraviolet radiation in a transition from its ${}^1\Sigma_g^+$ ground electronic state to an excited state that is energetically close to a dissociative ${}^1\Sigma_u^+$ state. The absorption band has a relatively large experimental linewidth. Account for this observation.

Problems

P11G.1 The fluorescence spectrum of anthracene vapour shows a series of peaks of increasing intensity with individual maxima at 440 nm, 410 nm, 390 nm, and 370 nm followed by a sharp cut-off at shorter wavelengths. The absorption spectrum rises sharply from zero to a maximum at 360 nm with a trail of peaks of lessening intensity at 345 nm, 330 nm, and 305 nm. Account for these observations.

P11G.2 The Beer–Lambert law states that the absorbance of a sample at a wavenumber $\tilde{\nu}$ is proportional to the molar concentration [J] of the absorbing species J and to the length L of the sample (eqn 11A.8). In this problem you are asked to show that the intensity of fluorescence emission from a sample of J is also proportional to [J] and L. Consider a sample of J that is illuminated with a beam of intensity $I_0(\tilde{\nu})$ at the wavenumber $\tilde{\nu}$. Before fluorescence can occur, a fraction of $I_0(\tilde{\nu})$ must be absorbed and an intensity $I(\tilde{\nu})$ will be transmitted. However, not all the absorbed intensity is re-emitted and the intensity of fluorescence depends on the *fluorescence quantum yield*, ϕ_F the efficiency of photon emission. The fluorescence quantum yield ranges from 0 to 1 and is proportional to the ratio of the integral of the fluorescence spectrum over the integrated absorption coefficient. Because of a shift of magnitude $\Delta\tilde{\nu}$, fluorescence occurs at a wavenumber $\tilde{\nu}_f$, with $\tilde{\nu}_f + \Delta\tilde{\nu} = \tilde{\nu}$. It follows that the fluorescence intensity at $\tilde{\nu}_f$, $I_f(\tilde{\nu}_f)$, is proportional to ϕ_f and to the intensity of exciting radiation that is absorbed by J, $I_{abs}(\tilde{\nu}) = I_0(\tilde{\nu}) - I(\tilde{\nu})$. (a) Use the Beer–Lambert law to express $I_{abs}(\tilde{\nu})$ in terms of $I_0(\tilde{\nu})$, [J], L, and $\varepsilon(\tilde{\nu})$, the molar absorption coefficient of J at $\tilde{\nu}$. (b) Use your result from part (a) to show that $I_f(\tilde{\nu}_f) \propto I_0(\tilde{\nu})\varepsilon(\tilde{\nu})\phi_f[J]L$.

P11G.3 A laser medium is confined to a cavity that ensures that only certain photons of a particular frequency, direction of travel, and state of polarization are generated abundantly. The cavity is essentially a region between two mirrors, which reflect the light back and forth. This arrangement can be regarded as a version of the particle in a box, with the particle now being a photon. As in the treatment of a particle in a box (Topic 7D), the only wavelengths that can be sustained satisfy $n \times \frac{1}{2}\lambda = L$, where n is an integer and L is the length of the cavity. That is, only an integral number of half-wavelengths fit into the cavity; all other waves undergo destructive interference with themselves. These wavelengths characterize the *resonant modes* of the laser. For a laser cavity of length 1.00 m, calculate (a) the allowed frequencies and (b) the frequency difference between successive resonant modes.

P11G.4 Laser radiation is spatially coherent in the sense that the electromagnetic waves are all in step across the cross-section of the beam emerging from the laser cavity (see Problem P11G.3). The *coherence length*, l_C, is the distance across the beam over which the waves remain coherent, and is related to the range of wavelengths, $\Delta\lambda$, present in the beam by $l_C = \lambda^2/2\Delta\lambda$. When many wavelengths are present, and $\Delta\lambda$ is large, the waves get out of step in a short distance and the coherence length is small. (a) How does the coherence length of a typical light bulb (l_C = 400 nm) compare with that of a He–Ne laser with λ = 633 nm and $\Delta\lambda$ = 2.0 pm? (b) What is the condition that would lead to an infinite coherence length?

P11G.5 A *continuous-wave laser* emits a continuous beam of radiation, whereas a *pulsed laser* emits pulses of radiation. The peak power, P_{peak}, of a pulse is defined as the energy delivered in a pulse divided by its duration. The average power, $P_{average}$, is the total energy delivered by a large number of pulses divided by the duration of the time interval over which that total energy is measured. Suppose that a certain laser can generate radiation in 3.0 ns pulses, each of which delivers an energy of 0.10 J, at a pulse repetition frequency of 10 Hz. Calculate the peak power and the average power of this laser.

P11G.6 Light-induced degradation of molecules, also called *photobleaching*, is a serious problem in applications that require very high intensities. A molecule of a fluorescent dye commonly used to label biopolymers can withstand about 10^6 excitations by photons before light-induced reactions destroy its π system and the molecule no longer fluoresces. For how long will a single dye molecule fluoresce while being excited by 1.0 mW of 488 nm radiation from a continuous-wave laser? You may assume that the dye has an absorption spectrum that peaks at 488 nm and that every photon delivered by the laser is absorbed by the molecule.

FOCUS 11 Molecular spectroscopy

Integrated activities

I11.1 In the group theoretical language developed in Focus 10, a spherical rotor is a molecule that belongs to a cubic or icosahedral point group, a symmetric rotor is a molecule with at least a threefold axis of symmetry, and an asymmetric rotor is a molecule without a threefold (or higher) axis. Linear molecules are linear rotors. Classify each of the following molecules as a spherical, symmetric, linear, or asymmetric rotor and justify your answers with group theoretical arguments: (a) CH_4, (b) CH_3CN, (c) CO_2, (d) CH_3OH, (e) benzene, (f) pyridine.

I11.2‡ The H_3^+ ion has been found in the interstellar medium and in the atmospheres of Jupiter, Saturn, and Uranus. The rotational energy levels of H_3^+, an oblate symmetric rotor, are given by eqn 11B.13a, with $\tilde{C}$ replacing $\tilde{A}$, when centrifugal distortion and other complications are ignored. Experimental values for vibrational–rotational constants are $\tilde{\nu}(E') = 2521.6\,\text{cm}^{-1}$, $\tilde{B} = 43.55\,\text{cm}^{-1}$, and $\tilde{C} = 20.71\,\text{cm}^{-1}$. (a) Show that for a planar molecule (such as H_3^+) $I_{\parallel} = 2I_{\perp}$. The rather large discrepancy with the experimental values is due to factors ignored in eqn 11B.13. (b) Calculate an approximate value of the H–H bond length in H_3^+. (c) The value of R_e obtained from the best quantum mechanical calculations by J.B. Anderson (*J. Chem. Phys.* **96**, 3702 (1991)) is 87.32 pm. Use this result to calculate the values of the rotational constants $\tilde{B}$ and $\tilde{C}$. (d) Assuming that the geometry and force constants are the same in D_3^+ and H_3^+, calculate the spectroscopic constants of D_3^+. The molecular ion D_3^+ was first produced by Shy et al. (*Phys. Rev. Lett* **45**, 535 (1980)) who observed the $\nu_2(E')$ band in the infrared.

I11.3 Use appropriate electronic structure software to perform calculations on H_2O and CO_2 with basis sets of your or your instructor's choosing. (a) Compute ground-state energies, equilibrium geometries and vibrational frequencies for each molecule. (b) Compute the magnitude of the dipole moment of H_2O; the experimental value is 1.854 D. (c) Compare computed values to experiment and suggest reasons for any discrepancies.

I11.4 The protein haemerythrin is responsible for binding and carrying O_2 in some invertebrates. Each protein molecule has two Fe^{2+} ions that are in very close proximity and work together to bind one molecule of O_2. The Fe_2O_2 group of oxygenated haemerythrin is coloured and has an electronic absorption band at 500 nm. The Raman spectrum of oxygenated haemerythrin obtained with laser excitation at 500 nm has a band at 844 cm^{-1} that has been attributed to the O–O stretching mode of bound $^{16}O_2$. (a) Proof that the 844 cm^{-1} band arises from a bound O_2 species may be obtained by conducting experiments on samples of haemerythrin that have been mixed with $^{18}O_2$, instead of $^{16}O_2$. Predict the fundamental vibrational wavenumber of the ^{18}O–^{18}O stretching mode in a sample of haemerythrin that has been treated with $^{18}O_2$. (b) The fundamental vibrational wavenumbers for the O–O stretching modes of O_2, O_2^- (superoxide anion), and O_2^{2-} (peroxide anion) are 1555, 1107, and 878 cm^{-1}, respectively. Explain this trend in terms of the electronic structures of O_2, O_2^-, and O_2^{2-}. *Hint:* Review Topic 9C. What are the bond orders of O_2, O_2^-, and O_2^{2-}? (c) Based on the data given above, which of the following species best describes the Fe_2O_2 group of haemerythrin: $Fe^{2+}{}_2O_2$, $Fe^{2+}Fe^{3+}O_2^-$, or $Fe^{3+}{}_2O_2^{2-}$? Explain your reasoning. (d) The Raman spectrum of haemerythrin mixed with $^{16}O^{18}O$ has two bands that can be attributed to the O–O stretching mode of bound oxygen. Discuss how this observation may be used to exclude one or more of the four proposed schemes (**7–10**) for binding of O_2 to the Fe_2 site of haemerythrin.

7

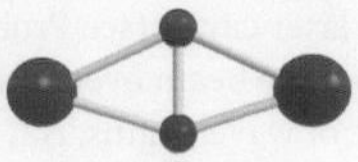

8

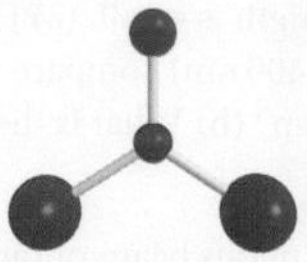

9

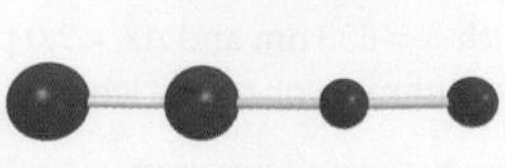

10

I11.5 The moments of inertia of the linear mercury(II) halides are very large, so the O and S branches of their vibrational Raman spectra show little rotational structure. Nevertheless, the position of greatest intensity in each branch can be identified and these data have been used to measure the rotational constants of the molecules (R.J.H. Clark and D.M. Rippon, *J. Chem. Soc. Faraday Soc. II* **69**, 1496 (1973)). Show, from a knowledge of the value of J corresponding to the intensity maximum, that the separation of the peaks of the O and S branches is given by the Placzek–Teller relation $\delta = (32\tilde{B}kT/hc)^{1/2}$. The following widths were obtained at the temperatures stated:

	$Hg^{35}Cl_2$	$Hg^{79}Br_2$	$Hg^{127}I_2$
θ/°C	282	292	292
δ/cm^{-1}	23.8	15.2	11.4

Calculate the bond lengths in the three molecules.

I11.6‡ A mixture of carbon dioxide (2.1 per cent) and helium, at 1.00 bar and 298 K in a gas cell of length 10 cm has an infrared absorption band centred at 2349 cm^{-1} with absorbances, $A(\tilde{\nu})$, described by:

$$A(\tilde{\nu}) = \frac{a_1}{1 + a_2(\tilde{\nu} - a_3)^2} + \frac{a_4}{1 + a_5(\tilde{\nu} - a_6)^2}$$

where the coefficients are $a_1 = 0.932$, $a_2 = 0.005050$ cm^2, $a_3 = 2333$ cm^{-1}, $a_4 = 1.504$, $a_5 = 0.01521$ cm^2, $a_6 = 2362$ cm^{-1}. (a) Draw graphs of $A(\tilde{\nu})$ and $\varepsilon(\tilde{\nu})$. What is the origin of both the band and the band width? What are the allowed and forbidden transitions of this band? (b) Calculate the transition wavenumbers and absorbances of the band with a simple harmonic oscillator–rigid rotor model and compare the result with the experimental spectra. The CO bond length is 116.2 pm. (c) Within what height, h, is basically all the infrared emission from the Earth in this band absorbed by atmospheric carbon dioxide? The mole fraction of CO_2 in the atmosphere is 3.3×10^{-4} and $T/K = 288 - 0.0065(h/m)$ below 10 km. Draw a surface plot of the atmospheric transmittance of the band as a function of both height and wavenumber.

I11.7‡ One of the principal methods for obtaining the electronic spectra of unstable radicals is to study the spectra of comets, which are almost entirely due to radicals. Many radical spectra have been detected in comets, including that due to CN. These radicals are produced in comets by the absorption of far-ultraviolet solar radiation by their parent compounds. Subsequently, their fluorescence is excited by sunlight of longer wavelength. The spectra of comet Hale–Bopp (C/1995 O1) have been the subject of many recent studies. One such study is that of the fluorescence spectrum of CN in the comet at large heliocentric distances by R.M. Wagner and D.G. Schleicher (*Science* **275**, 1918 (1997)), in which the authors determine the spatial distribution and rate of production of CN in the coma (the cloud constituting the major part of the head of the comet). The (0–0) vibrational band is centred on 387.6 nm and the weaker (1–1) band with relative intensity 0.1 is centred on 386.4 nm. The band heads for (0–0) and (0–1) are known to be 388.3 and 421.6 nm, respectively. From these data, calculate the energy of the excited S_1 state relative to the ground S_0 state, the vibrational wavenumbers and the difference in the vibrational wavenumbers of the two states, and the relative populations of the $v = 0$ and $v = 1$ vibrational levels of the S_1 state. Also estimate the effective temperature of the molecule in the excited S_1 state. Only eight rotational levels of the S_1 state are thought to be populated. Is that observation consistent with the effective temperature of the S_1 state?

I11.8 Use a group theoretical argument to decide which of the following transitions are electric-dipole allowed: (a) the $\pi^* \leftarrow \pi$ transition in ethene, (b) the $\pi^* \leftarrow n$ transition in a carbonyl group in a C_{2v} environment.

I11.9 Use molecule (**11**) as a model of the *trans* conformation of the chromophore found in rhodopsin. In this model, the methyl group bound to the nitrogen atom of the protonated Schiff's base replaces the protein. (a) Use molecular modelling software and the computational method of your instructor's choice, to calculate the energy separation between the HOMO and LUMO of (**11**). (b) Repeat the calculation for the 11-*cis* form of (**11**). (c) Based on your results from parts (a) and (b), do you expect the experimental frequency for the $\pi^* \leftarrow \pi$ visible absorption of the *trans* form of (**11**) to be higher or lower than that for the 11-*cis* form of (**11**)?

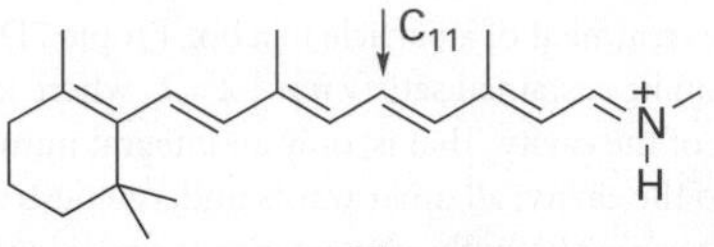

11

I11.10 Aromatic hydrocarbons and I_2 form complexes from which charge-transfer electronic transitions are observed. The hydrocarbon acts as an electron donor and I_2 as an electron acceptor. The energies $h\nu_{max}$ of the charge-transfer transitions for a number of hydrocarbon–I_2 complexes are given below:

Hydro-carbon	benzene	biphenyl	naphthalene	phenan-threne	pyrene	anthracene
$h\nu_{max}$/eV	4.184	3.654	3.452	3.288	2.989	2.890

Investigate the hypothesis that there is a correlation between the energy of the HOMO of the hydrocarbon (from which the electron comes in the charge-transfer transition) and $h\nu_{max}$. Use one of the computational methods discussed in Topic 9E to determine the energy of the HOMO of each hydrocarbon in the data set.

I11.11 A lot of information about the energy levels and wavefunctions of small inorganic molecules can be obtained from their ultraviolet spectra. An example of a spectrum with considerable vibrational structure, that of gaseous SO_2 at 25 °C, is shown in Fig. 11.6. Estimate the integrated absorption coefficient for the band centred at 280 nm. What electronic states are accessible from the A_1 ground state of this C_{2v} molecule by electric-dipole transitions?

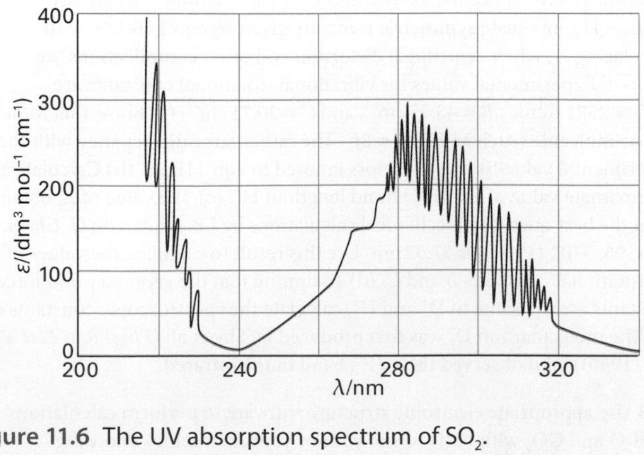

Figure 11.6 The UV absorption spectrum of SO_2.

FOCUS 12

Magnetic resonance

The techniques of 'magnetic resonance' observe transitions between spin states of nuclei and electrons in molecules. 'Nuclear magnetic resonance' (NMR) spectroscopy observes nuclear spin transitions and is one of the most widely used spectroscopic techniques for the exploration of the structures and dynamics of molecules ranging from simple organic species to biopolymers. 'Electron paramagnetic resonance' (EPR) spectroscopy is a similar technique that probes electron spin transitions in species with unpaired electrons.

12A General principles

This Topic gives an account of the principles that govern the energies and spectroscopic transitions between spin states of nuclei and electrons in molecules when a magnetic field is present. It describes simple experimental arrangements for the detection of these transitions.

12B Features of NMR spectra

This Topic contains a discussion of conventional NMR spectroscopy, showing how the properties of a magnetic nucleus are affected by its electronic environment and the presence of magnetic nuclei in its vicinity. These concepts explain how molecular structure governs the appearance of NMR spectra both in solution and in the solid state.

12C Pulse techniques in NMR

The modern implementation of NMR spectroscopy employs pulses of radiofrequency radiation followed by analysis of the resulting signal. This approach opens up many possibilities for the development of more sophisticated experiments. The Topic includes a discussion of spin relaxation in NMR and how it can be exploited, through the 'nuclear Overhauser effect', for structural studies.

12D Electron paramagnetic resonance

The detailed form of an EPR spectrum reflects the molecular environment of the unpaired electron and the nuclei with which it interacts. From an analysis of the spectrum it is possible to infer how the electron spin density is distributed.

Web resources What is an application of this material?

Magnetic resonance is ubiquitous in chemistry, as it is an enormously powerful analytical and structural technique, especially in organic chemistry and biochemistry. One of the most striking applications of nuclear magnetic resonance is in medicine. 'Magnetic resonance imaging' (MRI) is a portrayal of the distribution of protons in a solid object (*Impact* 18), and this technique has proved to be particularly useful for diagnosing disease. *Impact* 19 highlights an application of electron paramagnetic resonance in materials science and biochemistry: the use of a 'spin probe', a radical that interacts with biopolymers or nanostructures, and has an EPR spectrum that is sensitive to the local structure and dynamics of its environment.

TOPIC 12A General principles

> ➤ Why do you need to know this material?
>
> Nuclear magnetic resonance spectroscopy and electron paramagnetic resonance spectroscopy are used widely in chemistry to identify molecules and to determine their structures. To understand these techniques, you need to understand how a magnetic field affects the energies of the spin states of nuclei and electrons.
>
> ➤ What is the key idea?
>
> The spin states of nuclei and electrons have different energies in a magnetic field, and resonant transitions can take place between them when electromagnetic radiation of the appropriate frequency is applied.
>
> ➤ What do you need to know already?
>
> You need to be familiar with the quantum mechanical concept of spin (Topic 8B), the Boltzmann distribution (see the *Prologue* to this text and Topic 13A), and the general features of spectroscopy (Topic 11A).

Electrons and many nuclei have the property called 'spin', an intrinsic angular momentum. This spin gives rise to a magnetic moment and results in them behaving like small bar magnets. The energies of these magnetic moments depend on their orientation with respect to an applied magnetic field.

Spectroscopic techniques that measure transitions between nuclear and electron spin energy levels rely on the phenomenon of **resonance**, the strong coupling of oscillators of the same frequency. In fact, all spectroscopy is a form of resonant coupling between the electromagnetic field and the molecules, but in magnetic resonance, at least in its original form, the energy levels are adjusted to match the electromagnetic field rather than vice versa.

12A.1 Nuclear magnetic resonance

The **nuclear spin quantum number**, I, is a fixed characteristic property of a nucleus[1] and, depending on the nuclide, it is either an integer (including zero) or a half-integer (Table 12A.1). The angular momentum associated with nuclear spin has the same properties as other kinds of angular momentum (Topic 7F):

- The magnitude of the angular momentum is $\{I(I+1)\}^{1/2}\hbar$.
- The component of the angular momentum on a specified axis ('the z-axis') is $m_I\hbar$ where $m_I = I, I-1, \ldots, -I$.
- The orientation of the angular momentum, and hence of the magnetic moment, is determined by the value of m_I.

Physical interpretation

According to the second property, the angular momentum, and hence the magnetic moment, of the nucleus may lie in $2I+1$ different orientations relative to an axis. A ^{1}H nucleus has $I=\frac{1}{2}$ so its magnetic moment may adopt either of two orientations ($m_I=+\frac{1}{2}, -\frac{1}{2}$). The $m_I=+\frac{1}{2}$ state is commonly denoted α and the $m_I=-\frac{1}{2}$ state is commonly denoted β. A ^{14}N nucleus has $I=1$ so there are three orientations ($m_I=+1, 0, -1$). Examples of nuclei with $I=0$, and hence no magnetic moment, are ^{12}C and ^{16}O.

(a) The energies of nuclei in magnetic fields

The energy of a magnetic moment $\boldsymbol{\mu}$ in a magnetic field $\boldsymbol{\mathcal{B}}$ is equal to their scalar product (see *The chemist's toolkit* 22 in Topic 8C):

$$E = -\boldsymbol{\mu}\cdot\boldsymbol{\mathcal{B}} \tag{12A.1}$$

More formally, $\mathcal{B}$ is the 'magnetic induction' and is measured in tesla, T; $1\,\mathrm{T} = 1\,\mathrm{kg\,s^{-2}\,A^{-1}}$. The (non-SI) unit gauss, G, is also occasionally used: $1\,\mathrm{T} = 10^4\,\mathrm{G}$. The corresponding expression for the hamiltonian is

$$\hat{H} = -\hat{\boldsymbol{\mu}}\cdot\boldsymbol{\mathcal{B}} \tag{12A.2}$$

Table 12A.1 Nuclear constitution and the nuclear spin quantum number*

Number of protons	Number of neutrons	I
Even	even	0
Odd	odd	integer (1, 2, 3, ...)
Even	odd	half-integer ($\frac{1}{2}, \frac{3}{2}, \frac{5}{2}, \ldots$)
Odd	even	half-integer ($\frac{1}{2}, \frac{3}{2}, \frac{5}{2}, \ldots$)

* For nuclear ground states.

[1] Excited nuclear states, which are states in which the nucleons are arranged differently from the ground state, can have different spin from the ground state. Only ground states are considered here.

The magnetic moment operator of a nucleus is proportional to its spin angular momentum operator and is written

$$\hat{\mu}=\gamma_{\rm N}\hat{\boldsymbol{I}} \qquad (12A.3a)$$

The constant of proportionality, $\gamma_{\rm N}$, is the **nuclear magnetogyric ratio** (also called the 'gyromagnetic ratio'); its value depends on the identity of the nucleus and is determined empirically (Table 12A.2). If the magnetic field defines the z-direction and has magnitude $\mathcal{B}_0$, then eqn 12A.2 becomes

$$\hat{H}=-\hat{\mu}_z\mathcal{B}_0=-\gamma_{\rm N}\mathcal{B}_0\hat{I}_z \qquad (12A.3b)$$

The eigenvalues of the operator $\hat{I}_z$ for the z-component of the spin angular momentum are $m_I\hbar$. The eigenvalues of the hamiltonian in eqn 12A.3b, the allowed energy levels of the nucleus in a magnetic field, are therefore

$$E_{m_I}=-\gamma_{\rm N}\hbar\mathcal{B}_0 m_I \qquad \text{Energies of a nuclear spin in a magnetic field} \qquad (12A.4a)$$

It is common to rewrite this expression in terms of the **nuclear magneton**, $\mu_{\rm N}$,

$$\mu_{\rm N}=\frac{e\hbar}{2m_{\rm p}} \qquad \text{Nuclear magneton [definition]} \qquad (12A.4b)$$

(where $m_{\rm p}$ is the mass of the proton) and an experimentally determined dimensionless constant called the **nuclear *g*-factor**, g_I,

$$g_I=\frac{\gamma_{\rm N}\hbar}{\mu_{\rm N}} \qquad \text{Nuclear } g\text{-factor [definition]} \qquad (12A.4c)$$

Equation 12A.4a then becomes

$$E_{m_I}=-g_I\mu_{\rm N}\mathcal{B}_0 m_I \qquad \text{Energies of a nuclear spin in a magnetic field} \qquad (12A.4d)$$

The value of the nuclear magneton is $\mu_{\rm N}=5.051\times10^{-27}\,{\rm J\,T^{-1}}$. Typical values of nuclear g-factors range between −6 and +6 (Table 12A.2). Positive values of g_I and $\gamma_{\rm N}$ denote a magnetic moment that lies in the same direction as the spin angular momentum; negative values indicate that the magnetic moment and spin lie in opposite directions.

When $\gamma_{\rm N}>0$, as is the case for the most commonly observed nuclei ^{1}H and ^{13}C, in a magnetic field the energies of states with $m_I>0$ lie below states with $m_I<0$. For a **spin-$\frac{1}{2}$ nucleus**, a nucleus for which $I=\frac{1}{2}$, the α state lies lower in energy than the β state, and the separation between them is

$$\Delta E=E_{-1/2}-E_{+1/2}=\tfrac{1}{2}\gamma_{\rm N}\hbar\mathcal{B}_0-\left(-\tfrac{1}{2}\gamma_{\rm N}\hbar\mathcal{B}_0\right)=\gamma_{\rm N}\hbar\mathcal{B}_0 \qquad (12A.5)$$

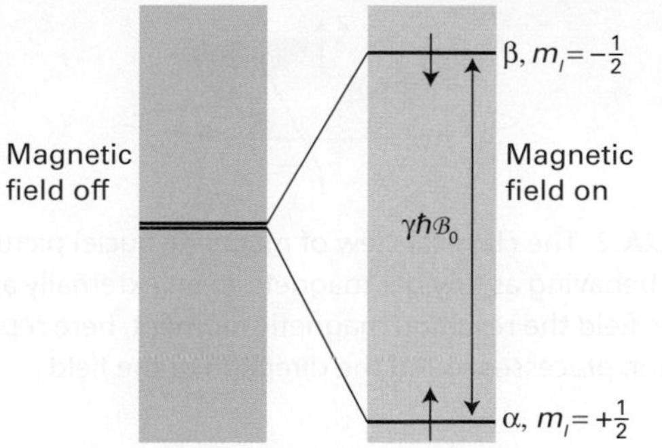

Figure 12A.1 The nuclear spin energy levels of a spin-$\frac{1}{2}$ nucleus with positive magnetogyric ratio (e.g. ^{1}H or ^{13}C) in a magnetic field. Resonant absorption of radiation occurs when the energy separation of the levels matches the energy of the photons.

The corresponding frequency of electromagnetic radiation for a transition between these states is given by the Bohr frequency condition $\Delta E=h\nu$ (Fig. 12A.1). Therefore,

$$h\nu=\gamma_{\rm N}\hbar\mathcal{B}_0 \quad \text{or} \quad \nu=\frac{\gamma_{\rm N}\mathcal{B}_0}{2\pi} \qquad \text{Resonance condition in NMR} \qquad (12A.6)$$

This relation is called the **resonance condition**, and ν is called the **NMR frequency** for that nucleus. Although eqn 12A.6 has been derived for a spin-$\frac{1}{2}$ nucleus, the same expression applies for any nucleus with non-zero spin.

It is sometimes useful to compare the quantum mechanical treatment with the classical picture in which magnetic nuclei are pictured as tiny bar magnets. A bar magnet in a magnetic field undergoes the motion called **precession** as it twists round the direction of the field and sweeps out the surface of a cone (Fig. 12A.2). The rate of precession $\nu_{\rm L}$ is called the **Larmor precession frequency**:

$$\nu_{\rm L}=\frac{\gamma_{\rm N}\mathcal{B}_0}{2\pi} \qquad \text{Larmor frequency of a nucleus [definition]} \qquad (12A.7)$$

Table 12A.2 Nuclear spin properties*

Nucleus	Natural abundance/%	Spin, I	g-factor, g_I	Magnetogyric ratio, $\gamma_{\rm N}/(10^7\,{\rm T^{-1}\,s^{-1}})$	NMR frequency at 1 T, ν/MHz
^{1}H	99.98	$\frac{1}{2}$	5.586	26.75	42.576
^{2}H	0.02	1	0.857	4.11	6.536
^{13}C	1.11	$\frac{1}{2}$	1.405	6.73	10.708
^{11}B	80.4	$\frac{3}{2}$	1.792	8.58	13.663
^{14}N	99.64	1	0.404	1.93	3.078

* More values are given in the *Resource section*.

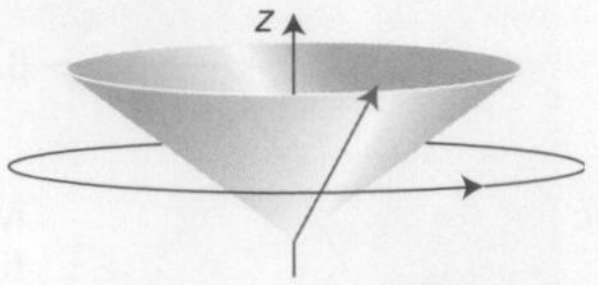

Figure 12A.2 The classical view of magnetic nuclei pictures them as behaving as tiny bar magnets. In an externally applied magnetic field the resulting magnetic moment, here represented as a vector, precesses round the direction of the field.

The Larmor precession frequency is the same as the resonance frequency given by eqn 12A.6. In other words, the frequency of radiation that causes resonant transitions between the α and β states is the same as the Larmor precession frequency. The achievement of resonance absorption can therefore be pictured as changing the applied magnetic field until the bar magnet representing the nuclear magnetic moment precesses at the same frequency as the magnetic component of the electromagnetic field to which it is exposed.

Brief illustration 12A.1

The NMR frequency for ^{1}H nuclei ($I=\frac{1}{2}$) in a 12.0 T magnetic field can be found using eqn 12A.6, with the relevant value of γ_N taken from Table 12A.1:

$$\nu = \frac{\overbrace{(2.6752\times10^8\,\mathrm{T^{-1}\,s^{-1}})}^{\gamma_N}\times\overbrace{(12.0\,\mathrm{T})}^{\mathcal{B}_0}}{2\pi} = 5.11\times10^8\,\mathrm{s^{-1}} = 511\,\mathrm{MHz}$$

This radiation lies in the radiofrequency region of the electromagnetic spectrum, close to frequencies used for radio communication.

(b) The NMR spectrometer

The key component of an NMR spectrometer (Fig. 12A.3) is the magnet into which the sample is placed. Most modern spectrometers use superconducting magnets capable of producing fields of 12 T or more. Such magnets have the advantages that the field they produce is stable over time and no electrical power is needed to maintain the field. With the currently available magnets, all NMR frequencies fall in the radiofrequency range (see the previous *Brief illustration*). Therefore, a radiofrequency transmitter and receiver are needed to excite and detect the transitions taking place between nuclear spin states. The details of how the transitions are excited and detected are discussed in Topic 12C.

The sample being studied is most commonly in the form of a solution contained in a glass tube placed within the magnet. It is also possible to study solid samples by using more specialized techniques. Although the superconducting magnet itself has to be held close to the temperature of liquid helium (4 K), the magnet is designed so as to have a room-temperature clear space into which the sample can be placed.

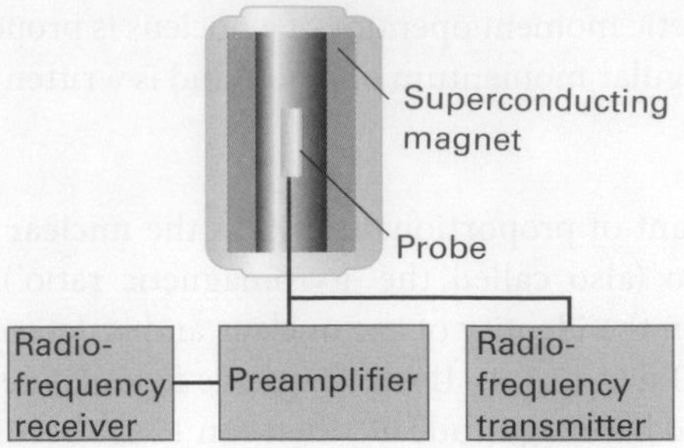

Figure 12A.3 The layout of a typical NMR spectrometer. The sample is held within the probe, which is placed at the centre of the magnetic field.

The intensity of an NMR transition depends on a number of factors which can be identified by considering the populations of the two spin states.

How is that done? 12A.1 Identifying the contributions to the absorption intensity

The rate of absorption of electromagnetic radiation is proportional to the population of the lower energy state (N_α in the case of a spin-$\frac{1}{2}$ nucleus) and the rate of stimulated emission is proportional to the population of the upper state (N_β). At the low frequencies typical of magnetic resonance, spontaneous emission can be neglected as it is very slow. Therefore, the net rate of absorption is proportional to the difference in populations:

$$\text{Rate of absorption} \propto N_\alpha - N_\beta$$

Step 1 *Write an expression for the intensity of absorption in terms of the population difference*

The intensity of absorption, the rate at which energy is absorbed, is proportional to the product of the rate of absorption (the rate at which photons are absorbed) and the energy of each photon. The latter is proportional to the frequency ν of the incident radiation (through $E=h\nu$). At resonance, this frequency is proportional to the applied magnetic field (through $\nu=\gamma_N\mathcal{B}_0/2\pi$), so it follows that

$$\begin{aligned}\text{Intensity of absorption} &\propto \text{rate of absorption}\\ &\quad\times\text{energy of photon}\\ &\propto (N_\alpha - N_\beta)\times h\gamma_N\mathcal{B}_0/2\pi\end{aligned}\qquad(12A.8a)$$

Step 2 *Write an expression for the ratio of populations*

Now use the Boltzmann distribution (see the *Prologue* to this text and Topic 13A) to write an expression for the ratio of populations:

$$\frac{N_\beta}{N_\alpha} = \mathrm{e}^{-\overbrace{\gamma_N\hbar\mathcal{B}_0}^{\Delta E}/kT} \approx 1 - \frac{\gamma_N\hbar\mathcal{B}_0}{kT} \qquad \boxed{\mathrm{e}^{-x}=1-x+\cdots}$$

The expansion of the exponential term (see *The chemist's toolkit* 12 in Topic 5B) is appropriate for $\Delta E=\gamma_N\hbar\mathcal{B}_0 << kT$, a condition usually met for nuclear spins.

Step 3 *Use that ratio to write an expression for the population difference*

Consider the ratio of the population difference to the total number of spins, N: $(N_\alpha-N_\beta)/N$ and use the expression for the ratio of populations to write this ratio as

$$\frac{N_\alpha-N_\beta}{\underbrace{N_\alpha+N_\beta}_{N}}=\frac{N_\alpha(1-N_\beta/N_\alpha)}{N_\alpha(1+N_\beta/N_\alpha)}=\frac{1-\overbrace{N_\beta/N_\alpha}^{1-\gamma_N\hbar\mathcal{B}_0/kT}}{1+\underbrace{N_\beta/N_\alpha}_{1-\gamma_N\hbar\mathcal{B}_0/kT}}$$

$$\approx\frac{1-(1-\gamma_N\hbar\mathcal{B}_0/kT)}{\underbrace{1+(1-\gamma_N\hbar\mathcal{B}_0/kT)}_{\approx 1}}=\frac{\gamma_N\hbar\mathcal{B}_0/kT}{2}$$

Therefore

$$N_\alpha-N_\beta\approx\frac{N\gamma_N\hbar\mathcal{B}_0}{2kT}$$ Population difference [spin-$\frac{1}{2}$ nuclei] (12A.8b)

The intensity is now obtained by substituting this expression into eqn 12A.8a which gives, after discarding constants that do not refer to the spins,

$$\text{Intensity}\propto\frac{N\gamma_N^2\mathcal{B}_0^2}{T}$$ (12A.8c) Absorption intensity

Because the intensity is proportional to $\mathcal{B}_0^2$ it follows that the signal can be enhanced significantly by increasing the strength of the applied magnetic field. The use of high magnetic fields also simplifies the appearance of spectra (a point explained in Topic 12B) and so allows them to be interpreted more readily. The intensity is also proportional to γ_N^2, so, all other things being equal, nuclei with large magnetogyric ratios (^{1}H, for instance) give more intense signals than those with small magnetogyric ratios (^{13}C, for instance).

Brief illustration 12A.2

For ^{1}H nuclei $\gamma_N = 2.675\times10^8\,\mathrm{T^{-1}\,s^{-1}}$. Therefore, for 1 000 000 protons in a field of 10 T at 20 °C,

$$N_\alpha-N_\beta\approx\frac{\overbrace{1000000}^{N}\times\overbrace{(2.675\times10^8\,\mathrm{T^{-1}\,s^{-1}})}^{\gamma_N}\times\overbrace{(1.055\times10^{-34}\,\mathrm{J\,s})}^{\hbar}\times\overbrace{(10\,\mathrm{T})}^{\mathcal{B}_0}}{2\times\underbrace{(1.381\times10^{-23}\,\mathrm{J\,K^{-1}})}_{k}\times\underbrace{(293\,\mathrm{K})}_{T}}$$
$$\approx 35$$

Even in such a strong field there is only a tiny imbalance of population of about 35 in a million. This small population difference means that special techniques had to be developed before NMR became a viable technique.

12A.2 Electron paramagnetic resonance

The observation of resonant transitions between the energy levels of an electron in a magnetic field is the basis of **electron paramagnetic resonance** (EPR; or *electron spin resonance*, ESR). This kind of spectroscopy has several features in common with NMR.

(a) The energies of electrons in magnetic fields

The magnetic moment of an electron is proportional to its spin angular momentum. Its magnetic moment operator and the hamiltonian for its interaction with a magnetic field are

$$\hat{\mu}=\gamma_e\hat{s}\quad\text{and}\quad\hat{H}=-\gamma_e\hat{s}\cdot\mathcal{B}$$ (12A.9a)

where $\hat{s}$ is the spin angular momentum operator and γ_e is the **magnetogyric ratio** of the electron:

$$\gamma_e=-\frac{g_e e}{2m_e}$$ Magnetogyric ratio of electron (12A.9b)

with $g_e = 2.002\,319\ldots$ as the **g-value** of the free electron. (Note that the current convention is to include the g-value in the definition of the magnetogyric ratio.) Dirac's relativistic theory, his modification of the Schrödinger equation to make it consistent with Einstein's special relativity, gives $g_e = 2$; the additional $0.002\,319\ldots$ arises from interactions of the electron with the electromagnetic fluctuations of the vacuum that surrounds it. The negative sign of γ_e (arising from the sign of the charge on the electron) shows that the magnetic moment is opposite in direction to the vector representing its angular momentum.

The hamiltonian when the magnetic field lies in the z-direction and has magnitude $\mathcal{B}_0$ is

$$\hat{H}=-\gamma_e\mathcal{B}_0\hat{s}_z$$ (12A.10)

where $\hat{s}_z$ is the operator for the z-component of the spin angular momentum. It follows that the energies of an electron spin in a magnetic field are

$$E_{m_s}=-\gamma_e\hbar\mathcal{B}_0 m_s$$ Energies of an electron spin in a magnetic field (12A.11a)

with $m_s=\pm\frac{1}{2}$. It is common to write this expression in terms of the **Bohr magneton**, μ_B, defined as

$$\mu_B=\frac{e\hbar}{2m_e}$$ Bohr magneton (12A.11b)

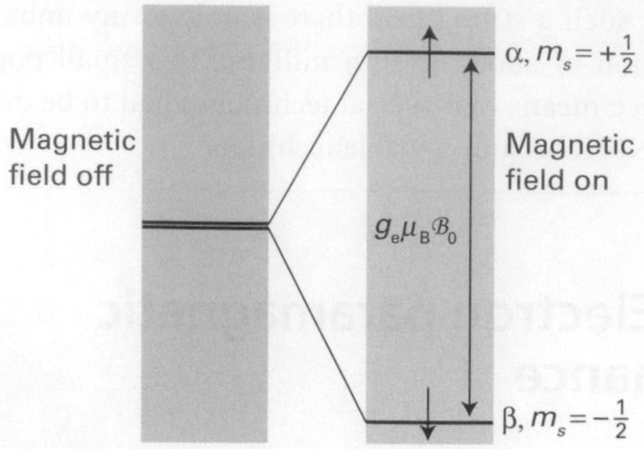

Figure 12A.4 Electron spin levels in a magnetic field. Note that the β state is lower in energy than the α state (because the magnetogyric ratio of an electron is negative). Resonant absorption occurs when the frequency of the incident radiation matches the frequency corresponding to the energy separation.

Its value is $9.274 \times 10^{-24}\,\mathrm{J\,T^{-1}}$. This positive quantity is often regarded as the fundamental quantum of magnetic moment. Note that the Bohr magneton is about 2000 times bigger than the nuclear magneton, so electron magnetic moments are that much bigger than nuclear magnetic moments. By using eqn 12A.9b the term $\gamma_e\hbar$ in eqn 12A.11a can be expressed as $-g_e e\hbar/2m_e$, which in turn can be written $-g_e\mu_B$ by introducing the definition of the Bohr magneton from eqn 12A.11b. It then follows that

$$E_{m_s} = g_e\mu_B\mathcal{B}_0 m_s \quad \text{Energies of an electron spin in a magnetic field} \quad (12A.11c)$$

and the energy separation between the $m_s = +\frac{1}{2}$ (α) and $m_s = -\frac{1}{2}$ (β) states is

$$\Delta E = E_{+1/2} - E_{-1/2} = \tfrac{1}{2}g_e\mu_B\mathcal{B}_0 - \left(-\tfrac{1}{2}g_e\mu_B\mathcal{B}_0\right) = g_e\mu_B\mathcal{B}_0 \quad (12A.12a)$$

with β the lower state. This energy separation comes into resonance with electromagnetic radiation of frequency ν when (Fig. 12A.4)

$$h\nu = g_e\mu_B\mathcal{B}_0 \quad \text{Resonance condition for EPR} \quad (12A.12b)$$

Brief illustration 12A.3

A typical commercial EPR spectrometer uses a magnetic field of about 0.33 T. The EPR resonance frequency is

$$\nu = \frac{\overbrace{(2.0023)}^{g_e}\times\overbrace{(9.274\times10^{-24}\,\mathrm{J\,T^{-1}})}^{\mu_B}\times\overbrace{(0.33\,\mathrm{T})}^{\mathcal{B}_0}}{\underbrace{6.626\times10^{-34}\,\mathrm{J\,s}}_{h}}$$

$$= 9.2\times10^{9}\,\mathrm{s^{-1}} = 9.2\,\mathrm{GHz}$$

This frequency corresponds to a wavelength of 3.2 cm, which is in the microwave region, and specifically in the 'X band' of frequencies.

(b) The EPR spectrometer

Most commercial EPR spectrometers use magnetic field strengths that result in EPR frequencies in the microwave region (see the preceding *Brief illustration*). The layout of a typical EPR spectrometer is shown in Fig. 12A.5. It consists of a fixed-frequency microwave source (typically a Gunn oscillator, based on a solid-state device), a cavity into which the sample (held in a glass or quartz tube) is inserted, a microwave detector, and an electromagnet with a variable magnetic field. The sample in an EPR observation must have unpaired electrons, so is either a radical or a d-metal complex. For technical reasons related to the detection procedure, the spectrum shows the first derivative of the absorption line (Fig. 12A.6).

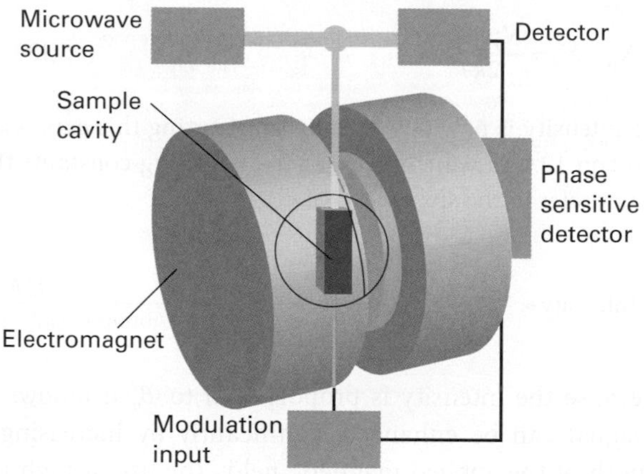

Figure 12A.5 The layout of a typical EPR spectrometer. A typical magnetic field is 0.3 T, which requires 9 GHz (3 cm) microwaves for resonance.

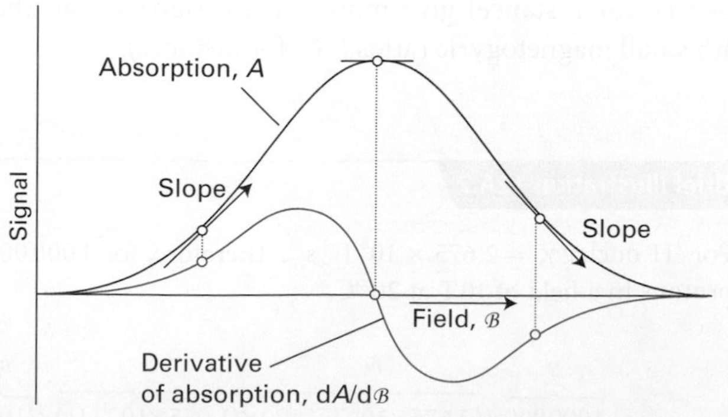

Figure 12A.6 When phase-sensitive detection is used, the signal is the first derivative of the absorption intensity. Note that the peak of the absorption corresponds to the point where the derivative passes through zero.

As in NMR, the intensities of spectral lines in EPR depend on the difference in populations between the ground and excited states. For an electron, the β state lies below the α state in energy and, by a similar argument that led to eqn 12A.8b for nuclei,

$$N_\beta - N_\alpha \approx \frac{Ng_e\mu_B\mathcal{B}_0}{2kT} \quad \text{Population difference [electrons]} \quad (12A.13)$$

where N is the total number of electron spins.

Brief illustration 12A.4

When 1000 electron spins experience a magnetic field of 1.0 T at 20 °C (293 K), the population difference is

$$N_\beta - N_\alpha \approx \frac{\overbrace{1000}^{N}\times\overbrace{2.0023}^{g_e}\times\overbrace{(9.274\times10^{-24}\,\mathrm{J\,T^{-1}})}^{\mu_B}\times\overbrace{(1.0\,\mathrm{T})}^{\mathcal{B}_0}}{2\times\underbrace{(1.381\times10^{-23}\,\mathrm{J\,K^{-1}})}_{k}\times\underbrace{(293\,\mathrm{K})}_{T}}$$

$$\approx 2.3$$

There is an imbalance of populations of only about two electrons in a thousand. However, the imbalance is much larger for electron spins than for nuclear spins (*Brief illustration* 12A.2) because the energy separation between the spin states of electrons is larger than that for nuclear spins even at the lower magnetic field strengths normally employed for EPR.

Checklist of concepts

- ☐ 1. The **nuclear spin quantum number**, I, of a nucleus is either a non-negative integer or half-integer; I can be zero.
- ☐ 2. In the presence of a magnetic field a nucleus has $2I + 1$ energy levels characterized by different values of m_I.
- ☐ 3. **Nuclear magnetic resonance** (NMR) is the observation of the absorption of radiofrequency electromagnetic radiation by nuclei in a magnetic field.
- ☐ 4. In NMR the absorption intensity increases with the strength of the applied magnetic field (as $\mathcal{B}_0^2$) and is also proportional to the square of the magnetogyric ratio of the nucleus.
- ☐ 5. In the presence of a magnetic field, an electron has two energy levels corresponding to the α and β spin states.
- ☐ 6. **Electron paramagnetic resonance** (EPR) is the observation of the resonant absorption of microwave electromagnetic radiation by unpaired electrons in a magnetic field.

Checklist of equations

Property	Equation	Comment	Equation number
Energies of a nuclear spin in a magnetic field	$E_{m_I} = -\gamma_N\hbar\mathcal{B}_0 m_I$		12A.4a
	$= -g_I\mu_N\mathcal{B}_0 m_I$		12A.4d
Nuclear magneton	$\mu_N = e\hbar/2m_p$	$\mu_N = 5.051\times10^{-27}\,\mathrm{J\,T^{-1}}$	12A.4b
Resonance condition (spin-$\frac{1}{2}$ nuclei)	$h\nu = \gamma_N\hbar\mathcal{B}_0$		12A.6
Larmor frequency	$\nu_L = \gamma_N\mathcal{B}_0/2\pi$		12A.7
Magnetogyric ratio (electron)	$\gamma_e = -g_e e/2m_e$	$g_e = 2.002\,319\ldots$	12A.9b
Energies of an electron spin in a magnetic field	$E_{m_s} = -\gamma_e\hbar\mathcal{B}_0 m_s$		12A.11a
	$= g_e\mu_B\mathcal{B}_0 m_s$		12A.11c
Bohr magneton	$\mu_B = e\hbar/2m_e$	$\mu_B = 9.274\times10^{-24}\,\mathrm{J\,T^{-1}}$	12A.11b
Resonance condition (electrons)	$h\nu = g_e\mu_B\mathcal{B}_0$		12A.12b

TOPIC 12B Features of NMR spectra

➤ **Why do you need to know this material?**

To analyse NMR spectra and extract the wealth of information they contain you need to understand how the appearance of a spectrum correlates with molecular structure.

➤ **What is the key idea?**

The resonance frequency of a magnetic nucleus is affected by its electronic environment and the presence of magnetic nuclei in the vicinity.

➤ **What do you need to know already?**

You need to be familiar with the general principles of magnetic resonance (Topic 12A).

Nuclear magnetic moments interact with the *local* magnetic field, the field at the location of the nucleus in question. The local field may differ from the applied field due to the effects of the electrons surrounding the nucleus and the presence of other magnetic nuclei in the molecule. The overall effect is that the NMR frequency of a given nucleus is sensitive to its molecular environment.

12B.1 The chemical shift

The applied magnetic field can be thought of as causing a circulation of electrons through the molecule. This circulation is analogous to an electric current and so gives rise to a magnetic field. The **local magnetic field,** $\mathcal{B}_{loc}$, the total field experienced by the nucleus, is the sum of the applied field $\mathcal{B}_0$ and the additional field $\delta\mathcal{B}$ due to the circulation of the electrons

$$\mathcal{B}_{loc} = \mathcal{B}_0 + \delta\mathcal{B} \qquad (12B.1)$$

The additional field is proportional to the applied field, and it is conventional to write

$$\delta\mathcal{B} = -\sigma\mathcal{B}_0 \qquad \text{Shielding constant [definition]} \qquad (12B.2)$$

where the dimensionless quantity σ is called the **shielding constant** of the nucleus. The ability of the applied field to induce an electronic current in the molecule, and hence affect the strength of the resulting local magnetic field, depends on the details of the electronic structure near the magnetic nucleus of interest, so nuclei in different chemical groups have different shielding constants. As a result, the Larmor frequency ν_L of the nucleus (and therefore its resonance frequency) changes from $\gamma_N\mathcal{B}_0/2\pi$ to

$$\nu_L = \frac{\gamma_N\mathcal{B}_{loc}}{2\pi} = \frac{\gamma_N(\mathcal{B}_0 + \delta\mathcal{B})}{2\pi} = \frac{\gamma_N\mathcal{B}_0}{2\pi}(1-\sigma) \qquad (12B.3)$$

The Larmor frequency is different for nuclei in different environments, even if those nuclei are of the same element.

The **chemical shift** of a nucleus is the difference between its resonance frequency and that of a reference standard. The standard for ^{1}H and ^{13}C is the resonance in tetramethylsilane, $Si(CH_3)_4$, commonly referred to as TMS. The frequency separation between the resonance from a particular nucleus and that from the standard increases with the strength of the applied magnetic field because the Larmor frequency (eqn 12B.3) is proportional to the applied field.

Chemical shifts are reported on the **δ scale**, which is defined as

$$\delta = \frac{\nu - \nu^\circ}{\nu^\circ} \times 10^6 \qquad \delta \text{ scale [definition]} \qquad (12B.4a)$$

where ν° is the resonance frequency (Larmor frequency) of the standard. Because ν° is very close to the operating frequency of the spectrometer ν_{spect}, which is typically chosen to be in the middle of the range of Larmor frequencies exhibited by the nucleus being studied, the ν° in the denominator of eqn 12B.4a can safely be replaced by ν_{spect} to give

$$\delta = \frac{\nu - \nu^\circ}{\nu_{spect}} \times 10^6 \qquad (12B.4b)$$

The advantage of the δ scale is that shifts reported on it are independent of the applied field (because both numerator and denominator are proportional to the applied field). The resonance frequencies themselves, however, do depend on the applied field through

$$\nu = \nu^\circ + \left(\frac{\nu_{spect}}{10^6}\right)\delta \qquad (12B.5)$$

Brief illustration 12B.1

In an NMR spectrometer operating at 500.130 00 MHz the resonance from TMS is found to be at a frequency of 500.127 50 MHz. The chemical shift of a resonance at a frequency of 500.128 25 MHz is

$$\delta = \frac{\nu - \nu^{\circ}}{\nu_{\text{spect}}} \times 10^6 = \frac{(500.12825\,\text{MHz}) - (500.12750\,\text{MHz})}{500.13000\,\text{MHz}} \times 10^6$$
$$= 1.5$$

On the same spectrometer the frequency separation of two resonances with chemical shifts $\delta_1 = 1.25$ and $\delta_2 = 5.75$ is found using eqn 12B.5

$$\nu_2 - \nu_1 = \left(\frac{\nu_{\text{spect}}}{10^6}\right)(\delta_2 - \delta_1) = \left(\frac{500.13000 \times 10^6\,\text{Hz}}{10^6}\right)(5.75 - 1.25)$$
$$= 2250\,\text{Hz}$$

A note on good practice In much of the literature, chemical shifts are reported in parts per million, ppm, in recognition of the factor of 10^6 in the definition; this is unnecessary. If you see '$\delta = 10$ ppm', interpret it, and use it in eqn 12B.5, as $\delta = 10$.

The relation between δ and σ is obtained by substituting eqn 12B.3 into eqn 12B.4a:

$$\delta = \frac{(1-\sigma)\mathcal{B}_0 - (1-\sigma^{\circ})\mathcal{B}_0}{(1-\sigma^{\circ})\mathcal{B}_0} \times 10^6$$
$$= \frac{\sigma^{\circ} - \sigma}{1 - \sigma^{\circ}} \times 10^6 \overset{|\sigma^{\circ}| \ll 1}{\approx} (\sigma^{\circ} - \sigma) \times 10^6 \quad \text{Relation between } \delta \text{ and } \sigma \quad (12\text{B}.6)$$

where σ° is the shielding constant of the reference standard. A decrease in σ (reduction in shielding) therefore leads to an increase in δ. Therefore, nuclei with large chemical shifts are said to be strongly **deshielded**. Some typical chemical shifts are given in Fig. 12B.1. As can be seen from the illustration, the nuclei of different elements have very different ranges of chemical shifts. The ranges exhibit the variety of electronic environments of the nuclei in molecules: the higher the atomic number of the element, the greater the number of electrons around the nucleus and hence the greater the range of the extent of shielding. By convention, NMR spectra are plotted with δ decreasing from left to right.

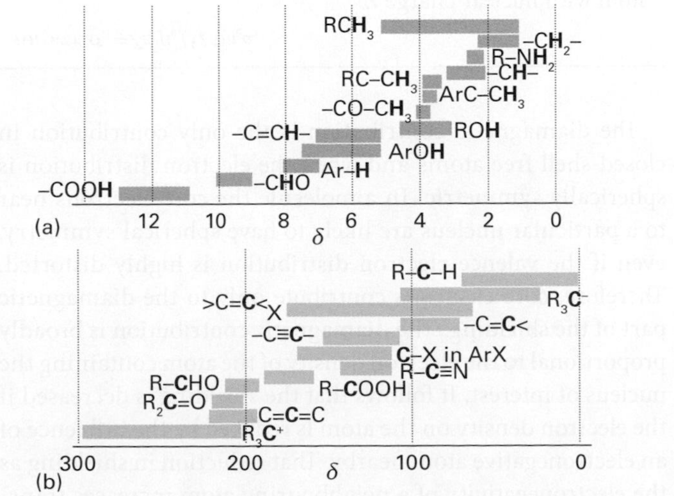

Figure 12B.1 The range of typical chemical shifts for (a) ^{1}H resonances and (b) ^{13}C resonances.

Example 12B.1 Interpreting a ^{1}H NMR spectrum

Figure 12B.2 shows the ^{1}H (proton) NMR spectrum of 1-methoxy-2-propanone, $CH_3OCH_2COCH_3$. Account for the observed chemical shifts.

Collect your thoughts You need to consider the effect of any electron-withdrawing atom: it deshields strongly the protons to which it is bound and has a diminishing effect on more distant protons. To identify which of the B and C resonances correspond to H atoms 2 and 3 you can take either of two approaches. One is to look at the large compilations of chemical shift data available. The second approach is to make use of the 'integral' of a line, the area under the resonance peak, which is proportional to the number of nuclei giving rise to the peak. These integrals are commonly shown by step-like curves superimposed on the spectrum, as is the case in Fig. 12B.2: the integral is proportional to the height of the step.

The solution The H atoms labelled 2 and 3 are all attached to a C atom that is attached to the strongly electron-withdrawing O atom, whereas the H atoms labelled 1 are further away from any O atoms. You can expect the deshielding of H atoms 2 and 3 to be greater than that of H atoms 1, and so the chemical shifts of 2 and 3 will be larger than that of 1. Resonance A at $\delta = 2.2$ can therefore confidently be assigned to the H atoms at position 1. It is evident from the spectrum that the integral of peak B is greater than that of C (in fact they are in the ratio 3:2), immediately identifying peak B as corresponding to the H atoms at position 3, and peak C to those at position 2.

Self-test 12B.1 The NMR spectrum of ethanal (acetaldehyde) has lines at $\delta = 2.20$ and $\delta = 9.80$. Which feature can be assigned to the CHO proton?

Answer: $\delta = 9.80$

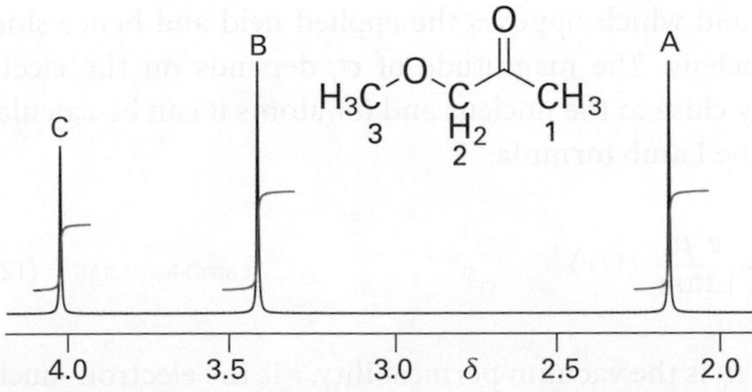

Figure 12B.2 The ^{1}H (proton) NMR spectrum of 1-methoxy-2-propanone. The step-like curve indicates the integral of the peak (the area under the peak) with the height of the step being proportional to the integral.

12B.2 The origin of shielding constants

The calculation of shielding constants (and hence chemical shifts) is difficult because it requires detailed knowledge of the distribution of electron density in the ground and excited states and the electronic excitation energies of the molecule. Nevertheless, it is helpful to understand the different contributions to chemical shifts so that patterns and trends can be identified.

A useful approach is to assume that the observed shielding constant is the sum of three contributions:

$$\sigma = \sigma(\text{local}) + \sigma(\text{neighbour}) + \sigma(\text{solvent}) \qquad (12B.7)$$

The **local contribution**, σ(local), is essentially the contribution of the electrons of the atom that contains the nucleus in question. The **neighbouring group contribution**, σ(neighbour), is the contribution from the groups of atoms that form the rest of the molecule. The **solvent contribution**, σ(solvent), is the contribution from the solvent molecules.

(a) The local contribution

It is convenient to regard the local contribution to the shielding constant as the sum of a **diamagnetic contribution**, σ_d, and a **paramagnetic contribution**, σ_p:

$$\sigma(\text{local}) = \sigma_d + \sigma_p \qquad \text{Local contribution to the shielding constant} \qquad (12B.8)$$

The diamagnetic contribution arises from additional fields that oppose the applied magnetic field and hence shield the nucleus; σ_d is therefore positive. The paramagnetic contribution arises from additional fields that reinforce the applied field and hence lead to deshielding; σ_p is therefore negative.

The diamagnetic contribution arises from the ability of the applied field to generate a circulation of charge in the ground-state electron distribution. The circulation generates a magnetic field which opposes the applied field and hence shields the nucleus. The magnitude of σ_d depends on the electron density close to the nucleus and for atoms it can be calculated from the **Lamb formula**:[1]

$$\sigma_d = \frac{e^2\mu_0}{12\pi m_e}\langle 1/r\rangle \qquad \text{Lamb formula} \qquad (12B.9)$$

where μ_0 is the vacuum permeability, r is the electron–nucleus distance, and the angle brackets $\langle\ldots\rangle$ indicate an expectation value.

[1] For a derivation, see our *Molecular quantum mechanics* (2011).

Example 12B.2 Using the Lamb formula

Calculate the shielding constant for the nucleus in a free H atom.

Collect your thoughts To calculate σ_d from the Lamb formula, you need to calculate the expectation value of $1/r$ for a hydrogen 1s orbital. The radial part of the wavefunction can be found from Table 8A.1 and the angular part from Table 7F.1.

The solution The normalized wavefunction for a hydrogen 1s orbital is, in spherical polar coordinates (see *The chemist's toolkit* 21 in Topic 7F),

$$\psi = \left(\frac{1}{\pi a_0^3}\right)^{1/2} e^{-r/a_0}$$

In this coordinate system the volume element is $d\tau = \sin\theta\, r^2 dr d\theta d\phi$, so the expectation value of $1/r$ is

$$\langle 1/r\rangle = \int\frac{\psi^*\psi}{r}d\tau = \frac{1}{\pi a_0^3}\overbrace{\int_0^{2\pi}d\phi}^{2\pi}\overbrace{\int_0^{\pi}\sin\theta\, d\theta}^{2}\overbrace{\int_0^{\infty}re^{-2r/a_0}dr}^{\text{Integral E.2}}$$

$$= \frac{1}{\pi a_0^3}\times 4\pi\times\frac{a_0^2}{4} = \frac{1}{a_0}$$

Therefore,

$$\sigma_d = \frac{e^2\mu_0}{12\pi m_e a_0} = \frac{(1.602\times10^{-19}\,\text{C})^2\times(4\pi\times10^{-7}\,\text{J}\,\text{s}^2\,\text{C}^{-2}\,\text{m}^{-1})}{12\pi\times(9.109\times10^{-31}\,\text{kg})\times(5.292\times10^{-11}\,\text{m})}$$

$$= \frac{(1.602\times10^{-19})^2\times(4\pi\times10^{-7})}{12\pi\times(9.109\times10^{-31})\times(5.292\times10^{-11})}\times\frac{\text{C}^2\,\text{J}\,\text{s}^2\,\text{C}^{-2}\,\text{m}^{-1}}{\text{kg}\,\text{m}}$$

$$= 1.775\times10^{-5}$$

where $1\,\text{J} = 1\,\text{kg}\,\text{m}^2\,\text{s}^{-2}$ has been used.

Self-test 12B.2 Derive an expression for σ_d for a hydrogenic atom with nuclear charge Z.

Answer: $\sigma_d = Ze^2\mu_0/12\pi m_e a_0$

The diamagnetic contribution is the only contribution in closed-shell free atoms and when the electron distribution is spherically symmetric. In a molecule the core electrons near to a particular nucleus are likely to have spherical symmetry, even if the valence electron distribution is highly distorted. Therefore, core electrons contribute only to the diamagnetic part of the shielding. The diamagnetic contribution is broadly proportional to the electron density of the atom containing the nucleus of interest. It follows that the shielding is decreased if the electron density on the atom is reduced by the influence of an electronegative atom nearby. That reduction in shielding as the electronegativity of a neighbouring atom increases translates into an increase in the chemical shift δ (Fig. 12B.3).

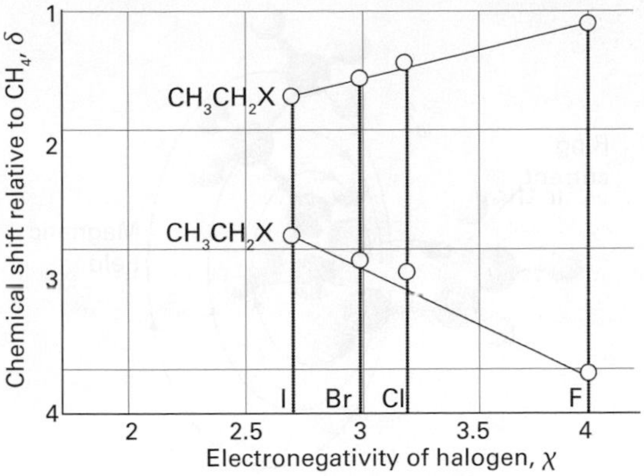

Figure 12B.3 The variation of chemical shielding with electronegativity. The shifts for the methyl protons follow the simple expectation that increasing the electronegativity of the halogen will increase the chemical shift. However, to emphasize that chemical shifts are subtle phenomena, notice that the trend for the methylene protons is opposite to that expected. For these protons another contribution (the magnetic anisotropy of C–H and C–X bonds) is dominant.

The local paramagnetic contribution, σ_p, arises from the ability of the applied field to force electrons to circulate through the molecule by making use of orbitals that are unoccupied in the ground state. It is absent in free atoms and also in linear molecules (such as ethyne, HC≡CH) when the applied field lies along the symmetry axis; in this arrangement the electrons can circulate freely and the applied field is unable to force them into other orbitals. Large paramagnetic contributions can be expected for light atoms (because the valence electrons, and hence the induced currents, are close to the nucleus) and in molecules with low-lying excited states (because an applied field can then induce significant currents). In fact, the paramagnetic contribution is the dominant local contribution for atoms other than hydrogen.

(b) Neighbouring group contributions

The neighbouring group contribution arises from the currents induced in nearby groups of atoms. Consider the influence of the neighbouring group X on the hydrogen atom in a molecule such as H–X. The applied field generates currents in the electron distribution of X and gives rise to an induced magnetic moment (an induced magnetic dipole) proportional to the applied field; the constant of proportionality is the magnetic susceptibility, χ (chi), of the group X: $\mu_{\text{induced}} = \chi\mathcal{B}_0$. The susceptibility is negative for a diamagnetic group because the induced moment is opposite to the direction of the applied field.

The induced moment gives rise to a magnetic field which is experienced by neighbouring nuclei. As is explained in *The chemist's toolkit* 27, a nucleus at distance R and angle θ (defined in **1**) from the induced moment experiences a local field that has the form

The chemist's toolkit 27 Dipolar magnetic fields

Standard electromagnetic theory gives the magnetic field at a point $\boldsymbol{r}$ from a point magnetic dipole $\boldsymbol{\mu}$ as

$$\mathcal{B} = -\frac{\mu_0}{4\pi r^3}\left(\boldsymbol{\mu} - \frac{3(\boldsymbol{\mu}\cdot\boldsymbol{r})\boldsymbol{r}}{r^2}\right)$$

where μ_0 is the vacuum permeability (a fundamental constant with the defined value $4\pi \times 10^{-7}\,\text{T}^2\,\text{J}^{-1}\,\text{m}^3$). The component of magnetic field in the z-direction is

$$\mathcal{B}_z = -\frac{\mu_0}{4\pi r^3}\left(\mu_z - \frac{3(\boldsymbol{\mu}\cdot\boldsymbol{r})z}{r^2}\right)$$

with $z = r\cos\theta$, the z-component of the distance vector $\boldsymbol{r}$. If the magnetic dipole is also parallel to the z-direction, it follows that

$$\mathcal{B}_z = -\frac{\mu_0}{4\pi r^3}\left(\overbrace{\mu}^{\mu_z} - \frac{3\overbrace{(\mu r\cos\theta)}^{\boldsymbol{\mu}\cdot\boldsymbol{r}}\overbrace{(r\cos\theta)}^{z}}{r^2}\right) = -\frac{\mu\mu_0}{4\pi r^3}(1-3\cos^2\theta)$$

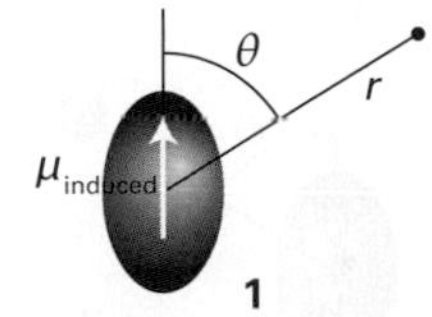

1

$$\mathcal{B}_{\text{local}} \propto \frac{\mu_{\text{induced}}}{r^3}(1-3\cos^2\theta) \qquad \text{Local dipolar field} \qquad (12\text{B}.10\text{a})$$

This local field is parallel to the applied field, and the angle θ is measured from the direction of the applied field. Note that the strength of the field is inversely proportional to the cube of the distance r between H and X. If the magnetic susceptibility is independent of the orientation of the molecule (that is, it is 'isotropic'), the local field averages to zero because, when averaged over a sphere, $1 - 3\cos^2\theta$ is zero (see Problem 12B.8). However, if the magnetic susceptibility varies with the orientation of the molecule with respect to the magnetic field, the local field may average to a non-zero value. For instance, suppose that the neighbouring group has axial symmetry (as might be the case for a triple bond): when the applied field is parallel to the symmetry axis the susceptibility is $\chi_\parallel$, and when it is perpendicular the susceptibility is $\chi_\perp$. After averaging over all orientations of the molecule the contribution to the shielding constant of a nucleus at a distance R has the following form

$$\sigma(\text{neighbour}) \propto (\chi_\parallel - \chi_\perp)\left(\frac{1-3\cos^2\Theta}{R^3}\right) \qquad \text{Neighbouring group contribution} \qquad (12\text{B}.10\text{b})$$

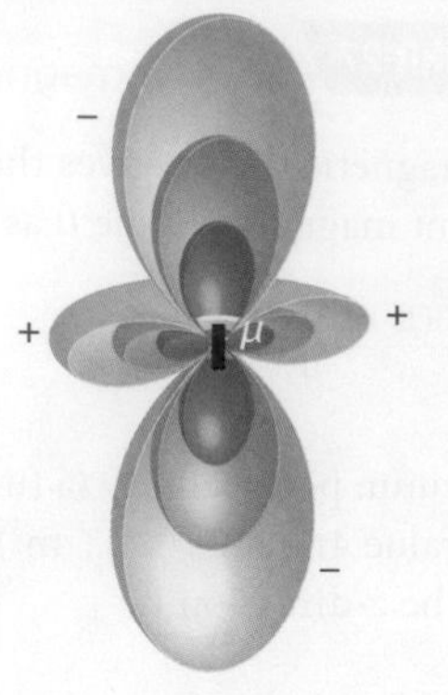

Figure 12B.4 A depiction of the field arising from a point magnetic dipole. The three shades of colour represent the strength of field declining with distance (as $1/R^3$), and each surface shows the angle dependence of the z-component of the field for each distance.

where Θ (uppercase theta) is the angle between the symmetry axis and the vector to the nucleus (2). Equation 12B.10b shows that the neighbouring group contribution may be positive or negative according to the relative magnitudes of the two magnetic susceptibilities and the direction given by Θ. If $54.7° < \Theta < 125.3°$, then $1 - 3\cos^2\Theta$ is positive, but it is negative otherwise (Figs. 12B.4 and 12B.5).

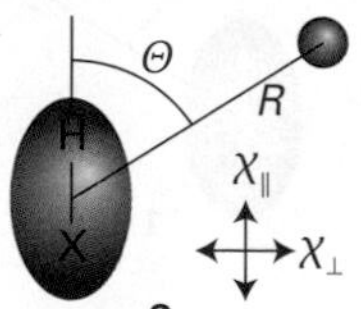

2

A special case of a neighbouring group effect is found in aromatic compounds. The strong anisotropy of the magnetic susceptibility of the benzene ring is ascribed to the ability of the field to induce a **ring current**, a circulation of electrons around the ring, when the field is applied perpendicular to the molecular plane. Protons in the plane are deshielded (Fig. 12B.6), but any that happen to lie above or below the plane (as members of substituents of the ring) are shielded.

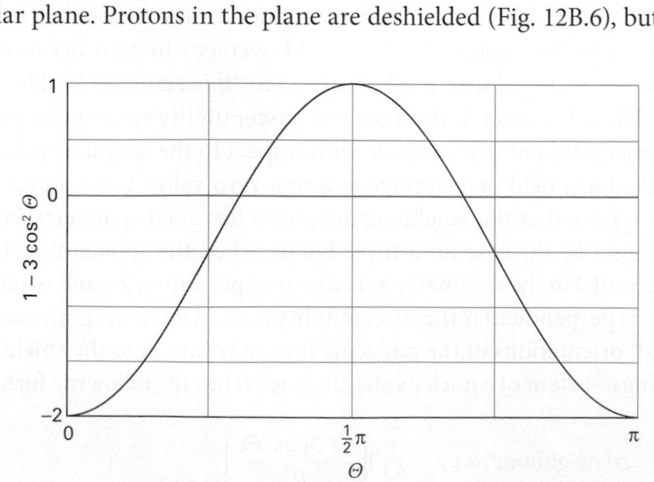

Figure 12B.5 The variation of the function $1 - 3\cos^2\Theta$ with the angle Θ.

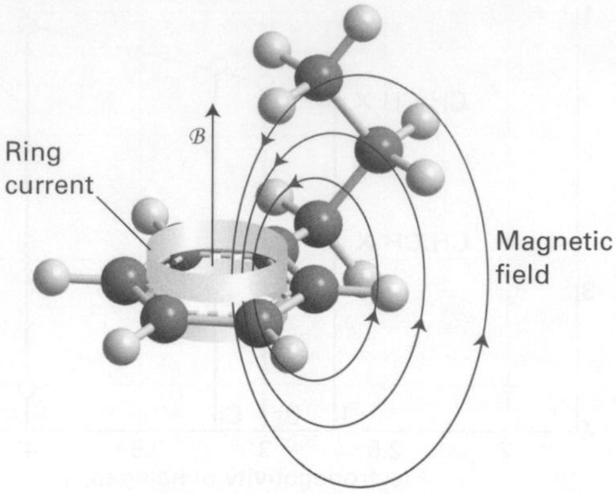

Figure 12B.6 The shielding and deshielding effects of the ring current induced in the benzene ring by the applied field. Protons attached to the ring are deshielded but a proton attached to a substituent that projects above the ring is shielded.

(c) The solvent contribution

A solvent can influence the local magnetic field experienced by a nucleus in a variety of ways. Some of these effects arise from specific interactions between the solute and the solvent (such as hydrogen bond formation and other forms of Lewis acid–base complex formation). The anisotropy of the magnetic susceptibility of the solvent molecules, especially if they are aromatic, can also be the source of a local magnetic field. Moreover, if there are steric interactions that result in a loose but specific interaction between a solute molecule and a solvent molecule, then protons in the solute molecule may experience shielding or deshielding effects according to their location relative to the solvent molecule. An aromatic solvent such as benzene can give rise to local currents that shield or deshield a proton in a solute molecule. The arrangement shown in Fig. 12B.7 leads to shielding of a proton on the solute molecule.

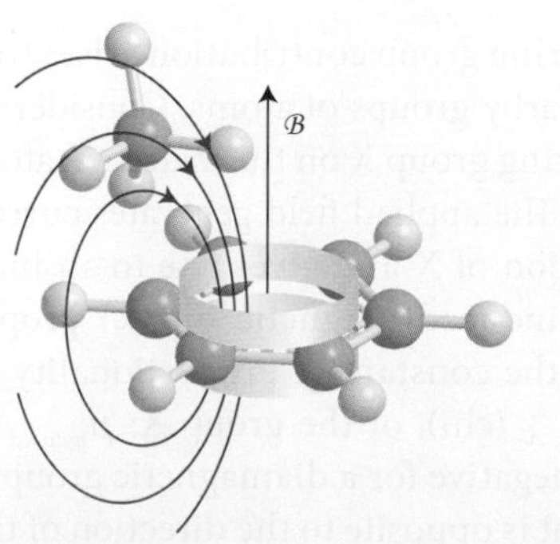

Figure 12B.7 An aromatic solvent (benzene here) can give rise to local currents that shield or deshield a proton in a solute molecule. In this relative orientation of the solvent and solute, the proton on the solute molecule is shielded.

12B.3 The fine structure

Figure 12B.8 shows the ^{1}H (proton) NMR spectrum of chloroethane. In this molecule there are two different types of ^{1}H, the methylene (CH_2) and the methyl (CH_3) protons, each with a characteristic chemical shift. In addition, the spectrum shows **fine structure**, the splitting of a resonance line into several components. The groups of lines are called **multiplets**.

This fine structure arises from **scalar coupling** in which the resonance frequency of one nucleus is affected by the spin state of another nucleus. Qualitatively, the effect of scalar coupling arises when the local magnetic field at one nucleus depends on the relative orientation of the other spin. If the spin is in one state (α, for instance) the local field is increased, whereas when the spin is in the other state (β in this case), the local field is decreased. Therefore, rather than there being one line in the spectrum, there are two because there are two possible values for the local field, corresponding to the second nucleus being either α or β.

The scalar coupling interaction is represented by a term $(hJ/\hbar^2)\hat{\boldsymbol{I}}_1\cdot\hat{\boldsymbol{I}}_2$ in the hamiltonian, where $\hat{\boldsymbol{I}}_N$, with $N = 1$ or 2, is the operator for the nuclear spin angular momentum of nucleus N. That the coupling term is a scalar product simply expresses the fact that the energy of interaction depends on the relative orientation of the spins of the two nuclei. The strength of the interaction is given by the value of the **scalar coupling constant**, J. The presence of $\hbar^2$ in $(hJ/\hbar^2)\hat{\boldsymbol{I}}_1\cdot\hat{\boldsymbol{I}}_2$ cancels the $\hbar^2$ arising from the eigenvalues of the two angular momenta, leaving the energy as hJ, so J is a frequency (measured in hertz, Hz). The coupling constant can be positive or negative, and it is independent of the field strength.

If the Larmor frequencies of the two coupled nuclei are significantly different, they precess at very different rates and the x- and y-components of their magnetic moments are never in step. Only the z-components remain in alignment whatever the precession rates, and so the only surviving term in the scalar product is $(hJ/\hbar^2)\hat{I}_{1z}\hat{I}_{2z}$. The eigenvalues of each $\hat{I}_{Nz}$ are $m_{I_N}\hbar$,

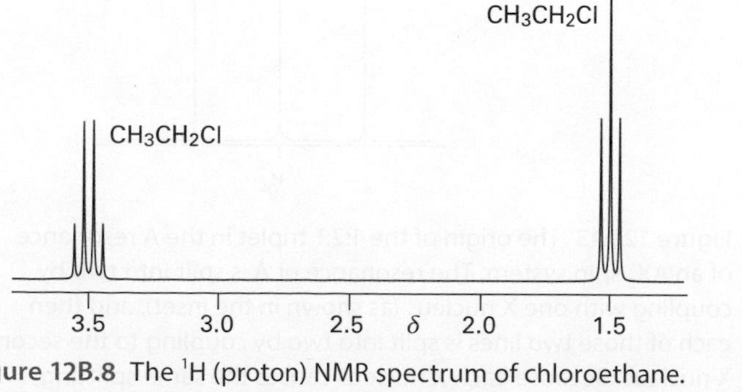

Figure 12B.8 The ^{1}H (proton) NMR spectrum of chloroethane. The coloured letters denote the protons giving rise to each multiplet; the multiplets arise due to scalar coupling between the protons.

so it follows that the eigenvalues (the energies) of the coupling term are

$$E_{m_{I_1}m_{I_2}} = hJm_{I_1}m_{I_2} \qquad \text{Spin–spin coupling energy} \qquad (12B.11)$$

(a) The appearance of the spectrum

In NMR, letters far apart in the alphabet (typically A and X) are used to indicate nuclei with very different chemical shifts in the sense that the difference in chemical shift corresponds to a frequency that is large compared to J; letters close together (such as A and B) are used for nuclei with similar chemical shifts.

Consider first an AX system, a molecule that contains two spin-$\frac{1}{2}$ nuclei A and X with very different chemical shifts, so eqn 12B.11 can be used for the spin–spin coupling energy. Nucleus A has two spin states with $m_A = \pm\frac{1}{2}$ corresponding to the states denoted α_A and β_A. The X nucleus also has two spin states with $m_X = \pm\frac{1}{2}$ (α_X and β_X). In the AX system there are therefore four spin states: $\alpha_A\alpha_X$, $\alpha_A\beta_X$, $\beta_A\alpha_X$, and $\beta_A\beta_X$. The energies of these states, neglecting any scalar coupling, are therefore

$$\begin{aligned} E_{m_A m_X} &= -\gamma_N\hbar(1-\sigma_A)\mathcal{B}_0 m_A - \gamma_N\hbar(1-\sigma_X)\mathcal{B}_0 m_X \\ &= -h\nu_A m_A - h\nu_X m_X \end{aligned} \qquad (12B.12a)$$

where ν_A and ν_X are the Larmor frequencies of A and X (eqn 12B.3). This expression gives the four levels illustrated on the left of Fig. 12B.9. When spin–spin coupling is included (by using eqn 12B.11) the energy levels are

$$E_{m_A m_X} = -h\nu_A m_A - h\nu_X m_X + hJm_A m_X \qquad (12B.12b)$$

The resulting energy level diagram (for $J > 0$) is shown on the right of Fig. 12B.9. The $\alpha_A\alpha_X$ and $\beta_A\beta_X$ states are both raised by $\frac{1}{4}hJ$ and the $\alpha_A\beta_X$ and $\beta_A\alpha_X$ states are both lowered by $\frac{1}{4}hJ$. For $J > 0$, the effect of the coupling term is to lower the energy of the $\alpha_A\beta_X$ and $\beta_A\alpha_X$ states, and raise the energy of the other two states. The opposite is the case for $J < 0$.

In a transition, only one nucleus changes its orientation, so the selection rule is that either m_A or m_X can change by ±1, but not both. There are two transitions in which the spin state of A changes while that of X remains fixed: $\beta_A\alpha_X \leftarrow \alpha_A\alpha_X$ and $\beta_A\beta_X \leftarrow \alpha_A\beta_X$. They are shown in Fig. 12B.9 and in a slightly different form in Fig. 12B.10. The energies of the transitions are

$$\Delta E = h\nu_A \pm \tfrac{1}{2}hJ \qquad (12B.13a)$$

The spectrum due to A transitions therefore consists of a doublet of separation J centred on the Larmor frequency of A (Fig. 12B.11). Similar remarks apply to the transitions in which the spin state of X changes while that of A remains fixed. These are also shown in Figs 12B.9 and 12B.10, and the transition energies are

$$\Delta E = h\nu_X \pm \tfrac{1}{2}hJ \qquad (12B.13b)$$

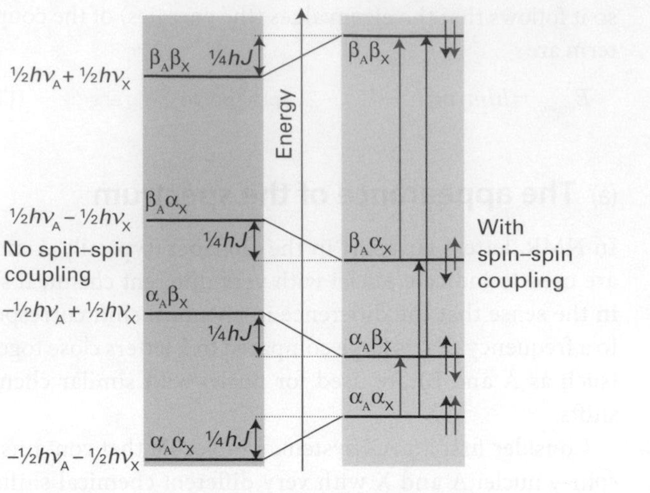

Figure 12B.9 The energy levels of an AX spin system. The four levels on the left are those in the case of no spin–spin coupling. The four levels on the right show how a positive spin–spin coupling constant affects the energies. The red arrows show the allowed transitions in which A goes from the α to the β spin state, while the spin state of X remains unchanged; the blue arrows show the corresponding transitions of the X nucleus. The effect of the coupling on the energy levels has been exaggerated greatly for clarity; in practice, the change in energy caused by spin–spin coupling is much smaller than that caused by the applied field.

It follows that there is a doublet with the same separation J, but now centred on the Larmor frequency of X (as shown in Fig. 12B.11). Overall, the spectrum of an AX spin system consists of two doublets.

If there is another X nucleus in the molecule with the same chemical shift as the first X (giving an AX_2 spin system), the X resonance is split into a doublet by A, just as for AX (Fig. 12B.12). The resonance of A is split into a doublet by one X, and each line of the doublet is split again by the same amount by the second X (Fig. 12B.13). This splitting results in three

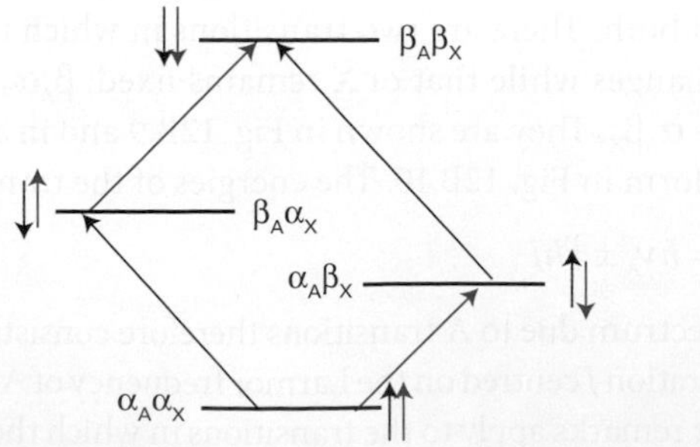

Figure 12B.10 An alternative depiction of the energy levels and transitions shown in Fig. 12B.9. Once again, the effect of spin–spin coupling has been exaggerated.

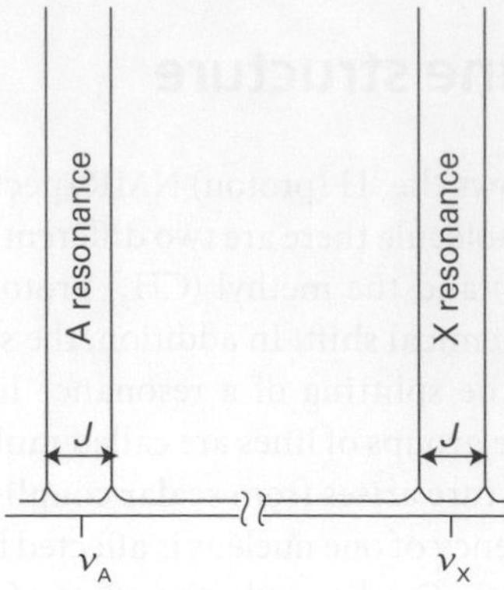

Figure 12B.11 The effect of spin–spin coupling on an AX spectrum. Each resonance is split into two lines, a doublet, separated by J. There is a doublet centred on the Larmor frequency (chemical shift) of A, and one centred on the Larmor frequency of B.

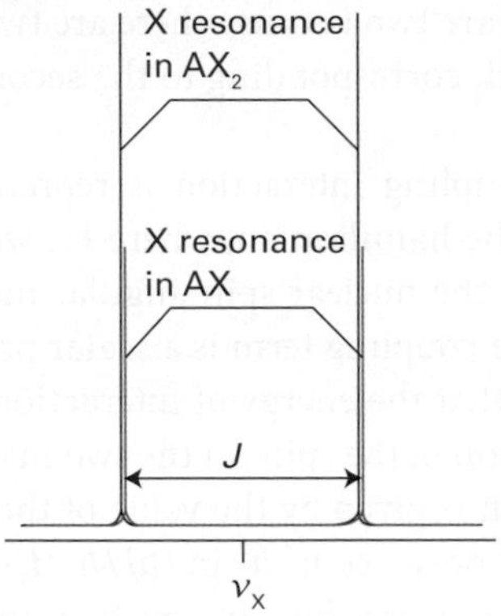

Figure 12B.12 The X resonance of an AX_2 spin system is also a doublet, because the two equivalent X nuclei behave like a single nucleus; however, the overall absorption is twice as intense as that of an AX spin system.

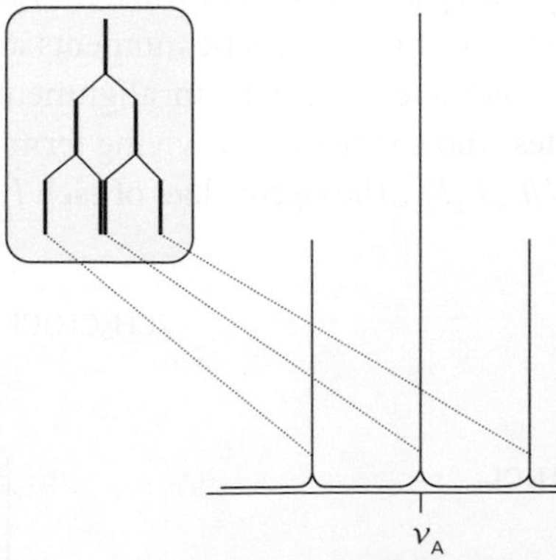

Figure 12B.13 The origin of the 1:2:1 triplet in the A resonance of an AX_2 spin system. The resonance of A is split into two by coupling with one X nucleus (as shown in the inset), and then each of those two lines is split into two by coupling to the second X nucleus. Because each X nucleus causes the same splitting, the two central transitions are coincident and give rise to an absorption line of double the intensity of the outer lines.

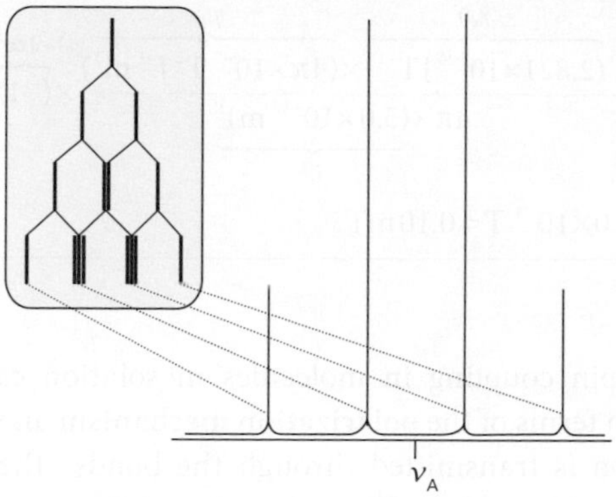

Figure 12B.14 The origin of the 1:3:3:1 quartet in the A resonance of an AX_3 species. The third X nucleus splits each of the lines shown in Fig. 12B.13 for an AX_2 species into a doublet, and the intensity distribution reflects the number of transitions that have the same energy.

lines in the intensity ratio 1:2:1 (because the central frequency can be obtained in two ways).

Three equivalent X nuclei (an AX_3 spin system) split the resonance of A into four lines of intensity ratio 1:3:3:1 (Fig. 12B.14). The X resonance remains a doublet as a result of the splitting caused by A. In general, N equivalent spin-$\frac{1}{2}$ nuclei split the resonance of a nearby spin or group of equivalent spins into $N+1$ lines with an intensity distribution given by Pascal's triangle (3). Successive rows of this triangle are formed by adding together the two adjacent numbers in the line above.

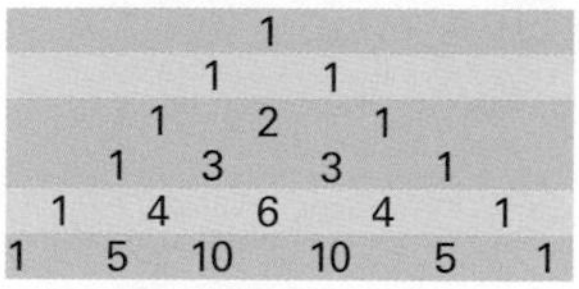

3

Example 12B.3 Accounting for the fine structure in a spectrum

Account for the fine structure in the ^1H (proton) NMR spectrum of chloroethane shown in Fig. 12B.8.

Collect your thoughts You need to consider how each group of equivalent protons (for instance, the three methyl protons) splits the resonances of the other groups of protons. There is no splitting within groups of equivalent protons. You can identify the pattern of intensities within a multiplet by referring to Pascal's triangle.

The solution The three protons of the CH_3 group split the resonance of the CH_2 protons into a 1:3:3:1 quartet with a splitting J. Likewise, the two protons of the CH_2 group split the resonance of the CH_3 protons into a 1:2:1 triplet with the same splitting J.

Self-test 12B.3 What fine-structure can be expected in the ^1H spectrum, and in the ^{15}N spectrum, of $^{15}NH_4^+$? Nitrogen-15 is a spin-$\frac{1}{2}$ nucleus.

Answer: The ^1H spectrum is a 1:1 doublet and the ^{15}N spectrum is a 1:4:6:4:1 quintet

(b) The magnitudes of coupling constants

The scalar coupling constant of two nuclei separated by N bonds is denoted ^NJ, with subscripts to indicate the types of nuclei involved. Thus, $^1J_{CH}$ is the coupling constant for a proton joined directly to a ^{13}C atom, and $^2J_{CH}$ is the coupling constant when the same two nuclei are separated by two bonds (as in ^{13}C–C–H). A typical value of $^1J_{CH}$ is in the range 120–250 Hz; $^2J_{CH}$ is between 10 and 20 Hz. Both ^3J and ^4J can give detectable effects in a spectrum, but couplings over larger numbers of bonds can generally be ignored.

As remarked in the discussion following eqn 12B.12b, the sign of J_{XY} determines whether a particular energy level is raised or lowered as a result of the coupling interaction. If $J > 0$, the levels with antiparallel spins are lowered in energy, whereas if $J < 0$ the levels with parallel spins are lowered. Experimentally, it is found that $^1J_{CH}$ is invariably positive, $^2J_{HH}$ is often negative, and $^3J_{HH}$ is often positive. An additional point is that J varies with the dihedral angle between the bonds (Fig. 12B.15). Thus, a $^3J_{HH}$ coupling constant is often found to depend on the dihedral angle ϕ (4) according to the **Karplus equation**:

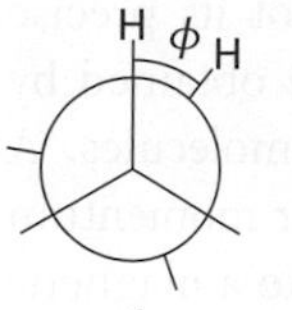

4

$$^3J_{HH} = A + B\cos\phi + C\cos 2\phi \quad \text{Karplus equation} \quad (12B.14)$$

with A, B, and C empirical constants with values close to +7 Hz, −1 Hz, and +5 Hz, respectively, for an HCCH fragment. It follows that the measurement of $^3J_{HH}$ in a series of related compounds can be used to determine their conformations. The coupling constant $^1J_{CH}$ also depends on the hybridization of the C atom, as the following values indicate:

	sp	sp^2	sp^3
$^1J_{CH}$/Hz	250	160	125

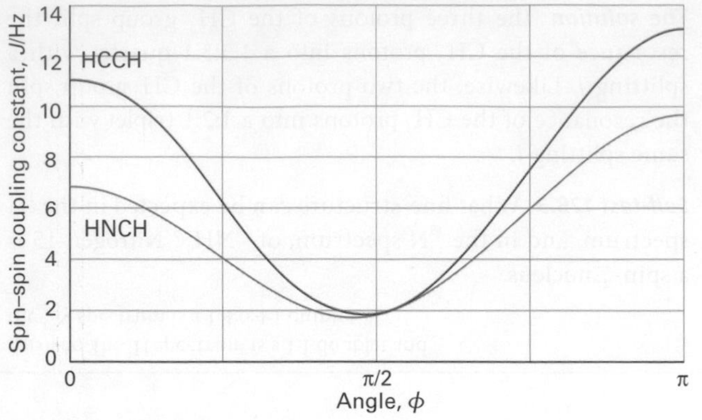

Figure 12B.15 The variation of the spin–spin coupling constant with dihedral angle predicted by the Karplus equation for an HCCH group and an HNCH group.

Brief illustration 12B.2

The investigation of H–N–C–H couplings in polypeptides can help reveal their conformation. For $^3J_{HH}$ coupling in such a group, $A = +5.1\,Hz$, $B = -1.4\,Hz$, and $C = +3.2\,Hz$. For a helical polymer, ϕ is close to 120°, which gives $^3J_{HH} \approx 4\,Hz$. For the sheet-like conformation, ϕ is close to 180°, which gives $^3J_{HH} \approx 10\,Hz$. Experimental measurements of the value of $^3J_{HH}$ should therefore make it possible to distinguish between the two possible structures.

(c) The origin of spin–spin coupling

Some insight into the origin of coupling, if not its precise magnitude—or always reliably its sign—can be obtained by considering the magnetic interactions within molecules. A nucleus with the z-component of its spin angular momentum specified by the quantum number m_I gives rise to a magnetic field with z-component $\mathcal{B}_{nuc}$ at a distance R, where, to a good approximation,

$$\mathcal{B}_{nuc} = -\frac{\gamma_N \hbar \mu_0}{4\pi R^3}(1-3\cos^2\theta)m_I \qquad (12B.15)$$

The angle θ is defined in (1); this expression is a version of eqn 12B.10a. However, in solution molecules tumble rapidly so it is necessary to average $\mathcal{B}_{nuc}$ over all values of θ. As has already been noted, the average of $1-3\cos^2\theta$ is zero, therefore the direct dipolar interaction between spins cannot account for the fine structure seen in the spectra of molecules in solution.

Brief illustration 12B.3

There can be a direct dipolar interaction between nuclei in solids, where the molecules do not rotate. The z-component of the magnetic field arising from a ^{1}H nucleus with $m_I = +\frac{1}{2}$, at $R = 0.30\,nm$, and at an angle $\theta = 0$ is

$$\mathcal{B}_{nuc} = -\frac{\overbrace{(2.821\times10^{-26}\,J\,T^{-1})}^{\gamma_N\hbar}\times\overbrace{(4\pi\times10^{-7}\,T^2\,J^{-1}\,m^3)}^{\mu_0}}{\underbrace{4\pi\times(3.0\times10^{-10}\,m)^3}_{R^3}}\overbrace{\times(-1)}^{(1-3\cos^2\theta)m_I}$$

$$=1.0\times10^{-4}\,T = 0.10\,mT$$

Spin–spin coupling in molecules in solution can be explained in terms of the **polarization mechanism**, in which the interaction is transmitted through the bonds. The simplest case to consider is that of $^1J_{XY}$, where X and Y are spin-$\frac{1}{2}$ nuclei joined by an electron-pair bond. The coupling mechanism depends on the fact that the energy depends on the relative orientation of the bonding electrons and the nuclear spins. This electron–nucleus coupling is magnetic in origin, and may be either a dipolar interaction or a **Fermi contact interaction**. A pictorial description of the latter is as follows. First, regard the magnetic moment of the nucleus as arising from the circulation of a current in a tiny loop with a radius similar to that of the nucleus (Fig. 12B.16). Far from the nucleus the field generated by this loop is indistinguishable from the field generated by a point magnetic dipole. Close to the loop, however, the field differs from that of a point dipole. The magnetic interaction between this non-dipolar field and the electron's magnetic moment is the contact interaction. The contact interaction—essentially the failure of the point-dipole approximation—depends on the very close approach of an electron to the nucleus and hence can occur only if the electron occupies an s orbital (which is the reason why $^1J_{CH}$ depends on the hybridization ratio).

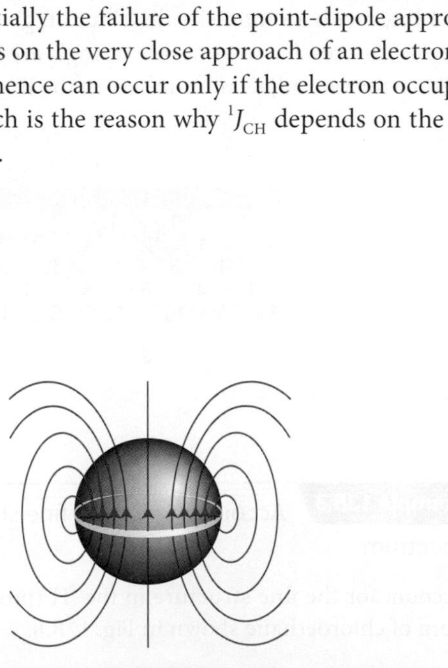

Figure 12B.16 The origin of the Fermi contact interaction. From far away, the magnetic field pattern arising from a ring of current (representing the rotating charge of the nucleus, the pale grey sphere) is that of a point dipole. However, if an electron can sample the field close to the region indicated by the sphere, the field distribution differs significantly from that of a point dipole. For example, if the electron can penetrate the sphere, then the spherical average of the field it experiences is not zero.

Suppose that it is energetically favourable for an electron spin and a nuclear spin to be antiparallel (as is the case for a proton and an electron in a hydrogen atom). If the X nucleus is α, a β electron of the bonding pair will tend to be found nearby, because that is an energetically favourable arrangement (Fig. 12B.17). The second electron in the bond, which must have α spin if the other is β (by the Pauli principle; Topic 8B), will be found mainly at the far end of the bond because electrons tend to stay apart to reduce their mutual repulsion. Because it is energetically favourable for the spin of Y to be antiparallel to an electron spin, a Y nucleus with β spin has a lower energy than when it has α spin. The opposite is true when X is β, for now the α spin of Y has the lower energy. In other words, the antiparallel arrangement of nuclear spins lies lower in energy than the parallel arrangement as a result of their magnetic coupling with the bond electrons. That is, ${}^1J_{CH}$ is positive.

To account for the value of ${}^2J_{XY}$, as for ${}^2J_{HH}$ in H–C–H, a mechanism is needed that can transmit the spin alignments through the central C atom (which may be ${}^{12}C$, with no nuclear spin of its own). In this case (Fig. 12B.18), an X nucleus with α spin polarizes the electrons in its bond, and the α electron is likely to be found closer to the C nucleus. The more favourable arrangement of two electrons on the same atom is with their spins parallel (Hund's rule, Topic 8B), so the more favourable arrangement is for the α electron of the neighbouring bond to be close to the C nucleus. Consequently, the β electron of that bond is more likely to be found close to the Y nucleus, and therefore that nucleus will have a lower energy if it is α. Hence, according to this mechanism, the lower energy will be obtained if the Y spin is parallel to that of X. That is, ${}^2J_{HH}$ is negative.

The coupling of nuclear spin to electron spin by the Fermi contact interaction is most important for proton spins, but it is not necessarily the most important mechanism for other nuclei. These nuclei may also interact by a dipolar mechanism with the electron magnetic moments and with their orbital motion, and there is no simple way of specifying whether J will be positive or negative. The dipolar interaction does not

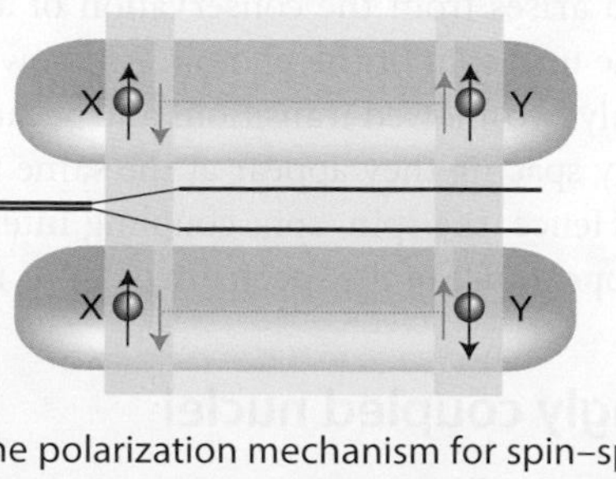

Figure 12B.17 The polarization mechanism for spin–spin coupling (${}^1J_{CH}$). The two arrangements have slightly different energies. In this case, J is positive, corresponding to a lower energy when the nuclear spins are antiparallel.

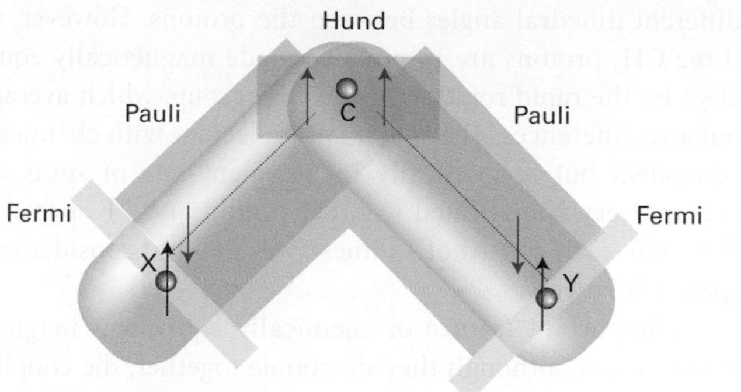

Figure 12B.18 The polarization mechanism for ${}^2J_{HH}$ spin–spin coupling. The spin information is transmitted from one bond to the next by a version of the mechanism that accounts for the lower energy of electrons with parallel spins in different atomic orbitals (Hund's rule of maximum multiplicity). In this case, $J < 0$, corresponding to a lower energy when the nuclear spins are parallel.

average to zero as the molecule tumbles if it accounts for the interaction of *both* nuclei with their surrounding electrons because then $1 - 3\cos^2\theta$ appears as its square and therefore with a non-negative value at all orientations, and its average value is no longer zero.

(d) Equivalent nuclei

A group of identical nuclei are **chemically equivalent** if they are related by a symmetry operation of the molecule and have the same chemical shifts. Chemically equivalent nuclei are from atoms that would be regarded as 'equivalent' according to ordinary chemical criteria. Nuclei are **magnetically equivalent** if, as well as being chemically equivalent, they also have identical spin–spin interactions with any other magnetic nuclei in the molecule.

Brief illustration 12B.4

The difference between chemical and magnetic equivalence is illustrated by CH_2F_2 and $H_2C{=}CF_2$ (recall that ${}^{19}F$ is a spin-$\frac{1}{2}$ nucleus). In each of these molecules the ${}^1H$ nuclei (protons) are chemically equivalent because they are related by symmetry. The protons in CH_2F_2 are magnetically equivalent, but those in $CH_2{=}CF_2$ are not. One proton in the latter has a *cis* spin-coupling interaction with a given F nucleus whereas the other proton has a *trans* interaction with the same nucleus. In contrast, in CH_2F_2 each proton has the same coupling to both fluorine nuclei since the bonding pathway between them is the same.

Strictly speaking, in a molecule such as CH_3CH_2Cl the three CH_3 protons are magnetically inequivalent because each may have a different coupling to the CH_2 protons on account of the different dihedral angles between the protons. However, the three CH_3 protons are in practice made magnetically equivalent by the rapid rotation of the CH_3 group, which averages out any differences. The spectra of molecules with chemically equivalent but magnetically inequivalent sets of spins can become very complicated (e.g. the proton and ^{19}F spectra of $H_2C{=}CF_2$ each consist of 12 lines); we shall not consider such spectra further.

An important feature of chemically equivalent magnetic nuclei is that, although they do couple together, the coupling has no effect on the appearance of the spectrum. How this comes about can be illustrated by considering the case of an A_2 spin system. The first step is to establish the energy levels.

How is that done? 12B.1 Deriving the energy levels of an A_2 system

Consider an A_2 system of two spin-$\frac{1}{2}$ nuclei, and first consider the energy levels in the absence of spin–spin coupling. When considering spin–spin coupling, be prepared to use the complete expression for the energy (the one proportional to $\boldsymbol{I}_1 \cdot \boldsymbol{I}_2$), because the Larmor frequencies are the same and the approximate form ($I_{1z}I_{2z}$) cannot be used as it is applicable only when the Larmor frequencies are very different.

Step 1 *Identify the states and their energies in the absence of spin–spin coupling*

There are four energy levels; they can be classified according to their *total* spin angular momentum I_{tot} (the analogue of S for several electrons) and its projection on to the z-axis, given by the quantum number M_I. There are three states with $I_{\text{tot}} = 1$, and one further state with $I_{\text{tot}} = 0$:

Spins parallel, $I_{\text{tot}} = 1$:	$M_I = +1$	$\alpha\alpha$
	$M_I = 0$	$(1/2^{1/2})\{\alpha\beta + \beta\alpha\}$
	$M_I = -1$	$\beta\beta$
Spins paired, $I_{\text{tot}} = 0$:	$M_I = 0$	$(1/2^{1/2})\{\alpha\beta - \beta\alpha\}$

The effect of a magnetic field on these four states is shown on the left-hand side of Fig. 12B.19: the two states with $M_I = 0$ are unaffected by the field as they are composed of equal proportions of α and β spins, and both spins have the same Larmor frequency.

Step 2 *Allow for* spin–spin interaction

The scalar product in the expression $E = (hJ/\hbar^2)\boldsymbol{I}_1 \cdot \boldsymbol{I}_2$ can be expressed in terms of the total nuclear spin $\boldsymbol{I}_{\text{tot}} = \boldsymbol{I}_1 + \boldsymbol{I}_2$ by noting that

$$I_{\text{tot}}^2 = (\boldsymbol{I}_1 + \boldsymbol{I}_2)\cdot(\boldsymbol{I}_1 + \boldsymbol{I}_2) = I_1^2 + I_2^2 + 2\boldsymbol{I}_1 \cdot \boldsymbol{I}_2$$

Rearranging this expression to

$$\boldsymbol{I}_1 \cdot \boldsymbol{I}_2 = \tfrac{1}{2}\{I_{\text{tot}}^2 - I_1^2 - I_2^2\}$$

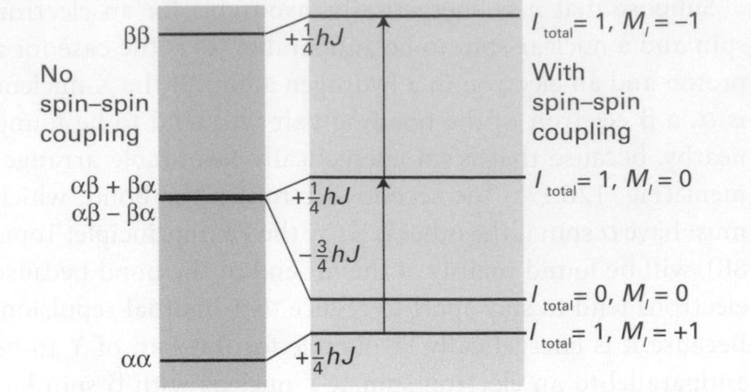

Figure 12B.19 The energy levels of an A_2 spin system in the absence of spin–spin coupling are shown on the left. When spin–spin coupling is taken into account, the energy levels on the right are obtained. Note that the effect of spin–spin coupling is to raise the three states with total nuclear spin $I_{\text{total}} = 1$ (the triplet) by the same amount (J is positive); in contrast, the one state with $I_{\text{total}} = 0$ (the singlet) is lowered in energy. The only allowed transitions, indicated by red arrows, are those for which $\Delta I_{\text{total}} = 0$ and $\Delta M_I = \pm 1$. These two transitions occur at the same resonance frequency as they would have in the absence of spin–spin coupling.

and replacing the square magnitudes by their quantum mechanical values gives:

$$\boldsymbol{I}_1 \cdot \boldsymbol{I}_2 = \tfrac{1}{2}\{I_{\text{tot}}(I_{\text{tot}} + 1) - I_1(I_1 + 1) - I_2(I_2 + 1)\}\hbar^2$$

Then, because $I_1 = I_2 = \frac{1}{2}$, it follows that

$$E = \tfrac{1}{2}hJ\{I_{\text{tot}}(I_{\text{tot}} + 1) - \tfrac{3}{2}\}$$

For parallel spins, $I_{\text{tot}} = 1$ and $E = +\frac{1}{4}hJ$; for antiparallel spins $I_{\text{tot}} = 0$ and $E = -\frac{3}{4}hJ$, as shown on the right-hand side of Fig. 12B.19.

The calculation shows that the three states with $I_{\text{total}} = 1$ all move in energy in the same direction and by the same amount. The single state with $I_{\text{total}} = 0$ moves three times as much in the opposite direction. In the resonance transition, the relative orientation of the nuclei cannot change, so there are no transitions between states of different I_{total}. The selection rule $\Delta M_I = \pm 1$ also applies, and arises from the conservation of angular momentum and the unit spin of the photon. As shown in Fig. 12B.19, there are only two allowed transitions and because they have the same energy spacing they appear at the same frequency in the spectrum. Hence, the spin–spin coupling interaction does not affect the appearance of the spectrum of an A_2 molecule.

(e) Strongly coupled nuclei

The multiplets seen in NMR spectra due to the presence of spin–spin coupling are relatively simple to analyse provided the difference in chemical shifts between any two coupled spins is much greater than the value of the spin–spin coupling constant between them. This limit is often described as **weak coupling**, and the resulting spectra are described as **first-order spectra**.

When the difference in chemical shifts is comparable to the value of the spin–spin coupling constant, the multiplets take on a more complex form. Such spin systems are said to be **strongly coupled**, and the spectra are described as **second-order**. In such spectra the lines shift from where they are expected in the weak coupling case, their intensities change, and in some cases additional lines appear. Strongly coupled spectra are more difficult to analyse in the sense that the relation between the frequencies of the lines in the spectrum and the values of chemical shifts and coupling constants is not as straightforward as in the weakly coupled case.

Figure 12B.20 shows NMR spectra for two coupled spins as a function of the difference in chemical shift between the two spins. In Fig. 12B.20a (an AX species) this difference is large enough for the weak coupling limit to apply and two doublets are detected, with all lines having the same intensity. As the shift difference decreases the inner two lines gain intensity at the expense of the outer lines, and in the limit that the shift difference is zero (an A_2 species), the outer lines disappear and the inner lines converge.

If the two nuclei belong to different elements (e.g. ^{1}H and ^{13}C), or different isotopes of the same element (e.g. ^{1}H and ^{2}H), the fact that they have widely different Larmor frequencies means that the spin system will always be weakly coupled, and hence described as AX. If the two nuclei are of the same element the spin system is described as **homonuclear**, whereas if they are of different elements the system is described as **heteronuclear**.

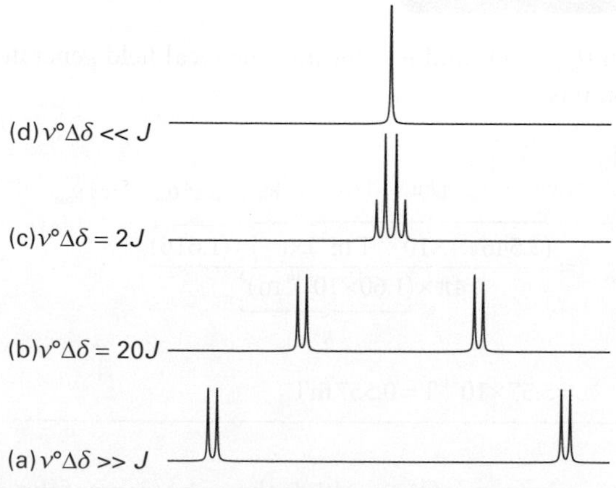

Figure 12B.20 The NMR spectra of (a) an AX system and (d) a 'nearly A_2' system are simple 'first-order' spectra (for an actual A_2 system, $\Delta\delta = 0$). At intermediate relative values of the chemical shift difference and the spin–spin coupling (b and c), more complex 'strongly coupled' spectra are obtained. Note how the inner two lines of the AX spectrum move together, grow in intensity, and form the single central line of the A_2 spectrum.

12B.4 Exchange processes

The appearance of an NMR spectrum is changed if magnetic nuclei can jump rapidly between different environments. For example, consider the molecule *N,N*-dimethylmethanamide, $HCON(CH_3)_2$, in which the O–C–N fragment is planar, and there is restricted rotation about the C–N bond. The lowest energy conformation is shown in Fig. 12B.21. In this conformation the two methyl groups are not equivalent because one is *cis* and the other is *trans* to the carbonyl group. The two groups therefore have different environments and hence different chemical shifts.

Rotation by 180° about the C–N bond gives the same conformation, but it exchanges the CH_3 groups between the two environments. When the jumping rate of this process is low, the spectrum shows a distinct line for each CH_3 environment. When the rate is fast, the spectrum shows a single line at the mean of the two chemical shifts. At intermediate rates, the lines start to broaden and eventually coalesce into a single broad line. Coalescence of the two lines occurs when

$$\tau = \frac{2^{1/2}}{\pi \delta \nu} \quad \text{Condition for coalescence of two NMR lines} \quad (12B.16)$$

where τ is the lifetime of an environment and $\delta\nu$ is the difference between the Larmor frequencies of the two environments.

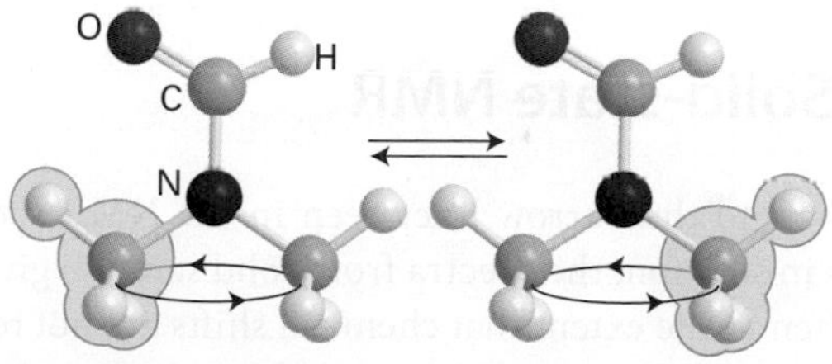

Figure 12B.21 In this molecule the two methyl groups are in different environments and so will have different chemical shifts. Rotation about the C–N bond interchanges the two groups, so that a particular methyl group is swapped between environments.

Brief illustration 12B.5

The NO group in *N,N*-dimethylnitrosamine, $(CH_3)_2N$–NO (**5**), rotates about the N–N bond and, as a result, the magnetic environments of the two CH_3 groups are interchanged. The two CH_3 resonances are separated by 390 Hz in a 600 MHz spectrometer. According to eqn 12B.16,

$$\tau = \frac{2^{1/2}}{\pi \times (390\,\mathrm{s^{-1}})} = 1.2\,\mathrm{ms}$$

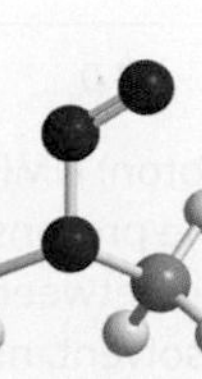

5 *N,N*-Dimethylnitrosamine

Such a lifetime corresponds to a (first-order) rate constant of $1/\tau = 870\,\text{s}^{-1}$. It follows that the signal will collapse to a single line when the rate constant for interconversion exceeds this value.

A similar explanation accounts for the loss of fine structure for protons that can exchange with the solvent. For example, the resonance from the OH group in the spectrum of ethanol appears as a single line (Fig. 12B.22). In this molecule the hydroxyl protons are able to exchange with the protons in water which, unless special precautions are taken, is inevitably present as an impurity in the (organic) solvent. When this **chemical exchange**, an exchange of atoms, occurs, a molecule ROH with an α-spin proton (written as ROH_α) rapidly converts to ROH_β and then perhaps to ROH_α again because the protons provided by the water molecules in successive exchanges have random spin orientations.

If the rate constant for the exchange process is fast compared to the value of the coupling constant J, in the sense $1/\tau >> J$, the two lines merge and no splitting is seen. Because the values of coupling constants are typically just a few hertz, even rather slow exchange leads to the loss of the splitting. In the case of OH groups, only by rigorously excluding water from the solvent can the exchange rate be made slow enough that splittings due to coupling to OH protons are observed.

12B.5 Solid-state NMR

In contrast to the narrow lines seen in the NMR spectra of samples in solution, the spectra from solid samples give broad lines, often to the extent that chemical shifts are not resolved. Nevertheless, there are good reasons for seeking to overcome these difficulties. They include the possibility that a compound is unstable in solution or that it is insoluble. Moreover, many species, such as polymers (both synthetic and naturally occurring), are intrinsically interesting as solids and might not be open to study by X-ray diffraction: in these cases, solid-state NMR provides a useful alternative way of probing both structure and dynamics.

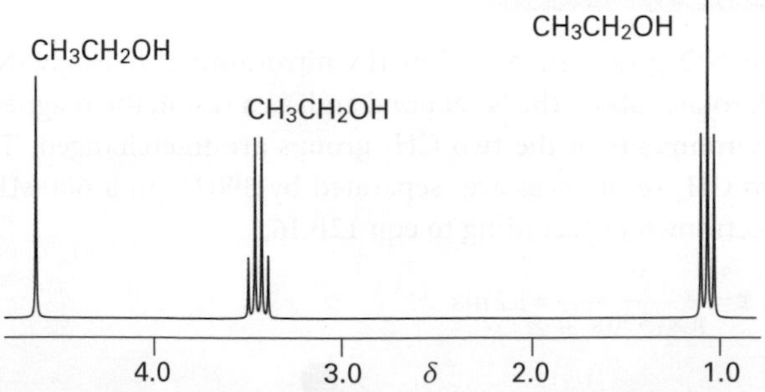

Figure 12B.22 The ^{1}H (proton) NMR spectrum of ethanol. The coloured letters denote the protons giving rise to each multiplet. Due to chemical exchange between the OH proton and water molecules present in the solvent, no splittings due to coupling to the OH proton are seen.

There are three principal contributions to the linewidths of solids. One is the direct magnetic dipolar interaction between nuclear spins. As pointed out in the discussion of spin–spin coupling, a nuclear magnetic moment gives rise to a local magnetic field which points in different directions at different locations around the nucleus. If the only component of interest is parallel to the direction of the applied magnetic field (because only this component has a significant effect), then provided certain subtle effects arising from transformation from the static to the rotating frame are neglected, the classical expression in *The chemist's toolkit* 27 can be used to write the magnitude of the local magnetic field as

$$\mathcal{B}_{\text{loc}} = -\frac{\gamma_N \hbar \mu_0 m_I}{4\pi R^3}(1 - 3\cos^2\theta) \qquad (12\text{B}.17\text{a})$$

Unlike in solution, in a solid this field is not averaged to zero by the molecular motion. Many nuclei may contribute to the total local field experienced by a nucleus of interest, and different nuclei in a sample may experience a wide range of fields. Typical dipole fields are of the order of 1 mT, which corresponds to splittings and linewidths of the order of 10 kHz for ^{1}H. When the angle θ can vary only between 0 and θ_{max}, the average value of $1–3\cos^2\theta$ can be shown to be $-(\cos^2\theta_{\text{max}} + \cos\theta_{\text{max}})$. This result, in conjunction with eqn 12B.17a, gives the average local field as

$$\mathcal{B}_{\text{loc,av}} = \frac{\gamma_N \hbar \mu_0 m_I}{4\pi R^3}(\cos^2\theta_{\text{max}} + \cos\theta_{\text{max}}) \qquad (12\text{B}.17\text{b})$$

Brief illustration 12B.6

When $\theta_{\text{max}} = 30°$ and $R = 160\,\text{pm}$, the local field generated by a proton is

$$\mathcal{B}_{\text{loc,av}} = \frac{\overbrace{(3.546\ldots\times10^{-32}\,\text{T m}^3)}^{\gamma_N\hbar\mu_0}\times\overbrace{(\tfrac{1}{2})}^{m_I}\times\overbrace{(1.616)}^{\cos^2\theta_{\text{max}}+\cos\theta_{\text{max}}}}{4\pi\times\underbrace{(1.60\times10^{-10}\,\text{m})^3}_{R^3}}$$

$$= 5.57\times10^{-4}\,\text{T} = 0.557\,\text{mT}$$

A second source of linewidth is the anisotropy of the chemical shift. Chemical shifts arise from the ability of the applied field to generate electron currents in molecules. In general, this ability depends on the orientation of the molecule relative to the applied field. In solution, when the molecule is tumbling rapidly, only the average value of the chemical shift is relevant. However, the anisotropy is not averaged to zero for stationary molecules in a solid, and molecules in different orientations have resonances at different frequencies. The chemical shift anisotropy also varies with the angle θ between the applied field and the principal axis of the molecule as $1-3\cos^2\theta$.

The third contribution is the electric quadrupole interaction. Nuclei with $I>\frac{1}{2}$ have an 'electric quadrupole moment', a measure of the extent to which the distribution of charge over the nucleus is not uniform (for instance, the positive charge may be concentrated around the equator or at the poles). An electric quadrupole interacts with an electric field gradient, such as may arise from a non-spherical distribution of charge around the nucleus. This interaction also varies as $1-3\cos^2\theta$.

Fortunately, there are techniques available for reducing the linewidths of solid samples. One technique, **magic-angle spinning** (MAS), takes note of the $1-3\cos^2\theta$ dependence of the dipole–dipole interaction, the chemical shift anisotropy, and the electric quadrupole interaction. The 'magic angle' is the angle at which $1-3\cos^2\theta=0$, and corresponds to 54.74°. In the technique, the sample is spun at high speed around an axis at the magic angle to the applied field (Fig. 12B.23). All the dipolar interactions and the anisotropies average to the value they would have at the magic angle, but at that angle they are zero. In principle, MAS therefore removes completely the line-broadening due to dipole–dipole interactions and chemical shift anisotropy. The difficulty with MAS is that the spinning frequency must not be less than the width of the spectrum, which is of the order of kilohertz. However, gas-driven sample spinners that can be rotated at up to 50 kHz are now routinely available.

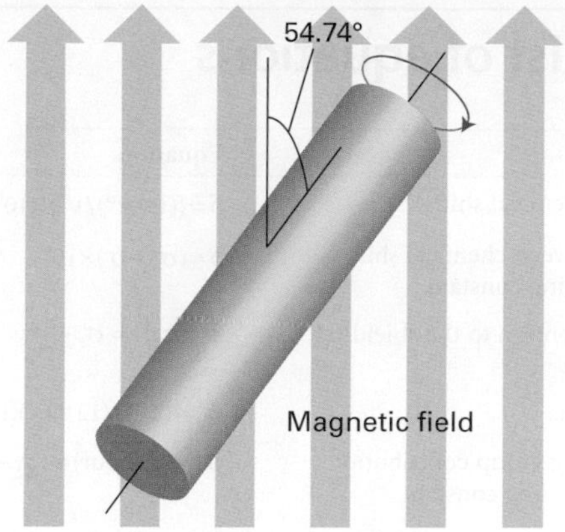

Figure 12B.23 In magic-angle spinning, the sample spins on an axis at 54.74° (i.e. arccos $1/3^{1/2}$) to the applied magnetic field. Rapid motion at this angle averages dipole-dipole interactions and chemical shift anisotropies to zero.

Checklist of concepts

☐ 1. The **chemical shift** of a nucleus is the difference between its resonance frequency and that of a reference standard.

☐ 2. The **shielding constant** is the sum of a local contribution, a neighbouring group contribution, and a solvent contribution.

☐ 3. The **local contribution** is the sum of a diamagnetic contribution and a paramagnetic contribution.

☐ 4. The **neighbouring group contribution** arises from the currents induced in nearby groups of atoms.

☐ 5. The **solvent contribution** can arise from specific molecular interactions between the solute and the solvent.

☐ 6. **Fine structure** is the splitting of resonances into individual lines by spin–spin coupling; these splittings give rise to **multiplets**.

☐ 7. **Spin–spin coupling** is expressed in terms of the **spin–spin coupling constant** J; coupling leads to the splitting of lines in the spectrum.

☐ 8. The coupling constant decreases as the number of bonds separating two nuclei increases.

☐ 9. Spin–spin coupling can be explained in terms of the **polarization mechanism** and the **Fermi contact interaction**.

☐ 10. If the shift difference between two nuclei is large compared to the coupling constant between the nuclei the spin system is said to be **weakly coupled**; if the shift difference is small compared to the coupling, the spin system is **strongly coupled**.

☐ 11. **Chemically equivalent** nuclei have the same chemical shifts; the same is true of **magnetically equivalent** nuclei, but in addition the coupling constant to any other nucleus is the same for each of the equivalent nuclei.

☐ 12. Coalescence of two NMR lines occurs when nuclei are exchanged rapidly between environments by either conformational or chemical process.

☐ 13. **Magic-angle spinning** (MAS) is a technique in which the NMR linewidths in a solid sample are reduced by spinning at an angle of 54.74° to the applied magnetic field.

Checklist of equations

Property	Equation	Comment	Equation number
δ-Scale of chemical shifts	$\delta=\{(\nu-\nu^\circ)/\nu^\circ\}\times10^6$	Definition	12B.4a
Relation between chemical shift and shielding constant	$\delta\approx(\sigma^\circ-\sigma)\times10^6$		12B.6
Local contribution to the shielding constant	$\sigma(\text{local}) = \sigma_d + \sigma_p$		12B.8
Lamb formula	$\sigma_d=(e^2\mu_0/12\pi m_e)\langle 1/r\rangle$	Applies to atoms	12B.9
Neighbouring group contribution to the shielding constant	$\sigma(\text{neighbour})\propto(\chi_\parallel-\chi_\perp)\{(1-3\cos^2\Theta)/R^3\}$		12B.10b
Karplus equation	${}^3J_{HH} = A + B\cos\phi + C\cos 2\phi$	A, B, and C are empirical constants	12B.14
Condition for coalescence of two NMR lines	$\tau=2^{1/2}/\pi\delta\nu$	τ is the lifetime of the exchange process	12B.16

TOPIC 12C Pulse techniques in NMR

➤ Why do you need to know this material?

To appreciate the power and scope of modern nuclear magnetic resonance techniques you need to understand how radiofrequency pulses can be used to obtain spectra.

➤ What is the key idea?

Sequences of pulses of radiofrequency radiation manipulate nuclear spins, leading to efficient acquisition of NMR spectra and the measurement of relaxation times.

➤ What do you need to know already?

You need to be familiar with the general principles of magnetic resonance (Topics 12A and 12B), and the vector model of angular momentum (Topic 7F). The development makes use of the concept of precession at the Larmor frequency (Topic 12A).

In modern forms of NMR spectroscopy the nuclear spins are first excited by a short, intense burst of radiofrequency radiation (a 'pulse'), applied at or close to the Larmor frequency. As a result of the excitation caused by the pulse, the spins emit radiation as they return to equilibrium. This time-dependent signal is recorded and its 'Fourier transform' computed (as will be described) to give the spectrum. The technique is known as **Fourier-transform NMR** (FT-NMR). An analogy for the difference between conventional spectroscopy and pulsed NMR is the detection of the frequencies at which a bell vibrates. The 'conventional' option is to connect an audio oscillator to a loudspeaker and direct the sound towards the bell. The frequency of the sound source is then scanned until the bell starts to ring in resonance. The 'pulse' analogy is to strike the bell with a hammer and then Fourier transform the signal to identify the resonance frequencies of the bell.

One advantage of FT-NMR over conventional NMR is that it improves the sensitivity. However, the real power of the technique comes from the possibility of manipulating the nuclear spins by applying a sequence of several pulses. In this way it is possible to record spectra in which particular features are emphasized, or from which other properties of the molecule can be determined.

12C.1 The magnetization vector

To understand the pulse procedure, consider a sample composed of many identical spin-$\frac{1}{2}$ nuclei. According to the vector model of angular momentum (Topic 7F), a nuclear spin can be represented by a vector of length $\{I(I+1)\}^{1/2}$ with a component of length m_I along the z-axis. As the three components of the angular momentum are complementary variables, the x- and y-components cannot be specified if the z-component is known, so the vector lies anywhere on a cone around the z-axis. For $I = \frac{1}{2}$, the length of the vector is $3^{1/2}/2$ and when $m_I = +\frac{1}{2}$ it makes an angle of $\arccos\{\frac{1}{2}/(3^{1/2}/2)\} = 54.7°$ to the z-axis (Fig. 12C.1); when $m_I = -\frac{1}{2}$ the cone makes the same angle to the $-z$-axis.

In the absence of a magnetic field, the sample consists of equal numbers of α and β nuclear spins with their vectors lying at random, stationary positions on their cones. The **magnetization**, $\boldsymbol{M}$, of the sample, its net nuclear magnetic moment, is zero (Fig. 12C.2a). There are two changes when a magnetic field of magnitude $\mathcal{B}_0$ is applied along the z-direction:

- The energies of the two spin states change, the α spins moving to a lower energy and the β spins to a higher energy (provided $\gamma_N > 0$).

In the vector model, the two vectors are pictured as precessing at the Larmor frequency (Topic 12A, $\nu_L = \gamma_N\mathcal{B}_0/2\pi$). At 10 T, the Larmor frequency for ^{1}H nuclei (commonly referred to as 'protons') is 427 MHz. As the strength of the field is increased, the Larmor frequency increases and the precession becomes faster.

- The populations of the two spin states (the numbers of α and β spins) at thermal equilibrium change, with slightly more α spins than β spins (Topic 12A).

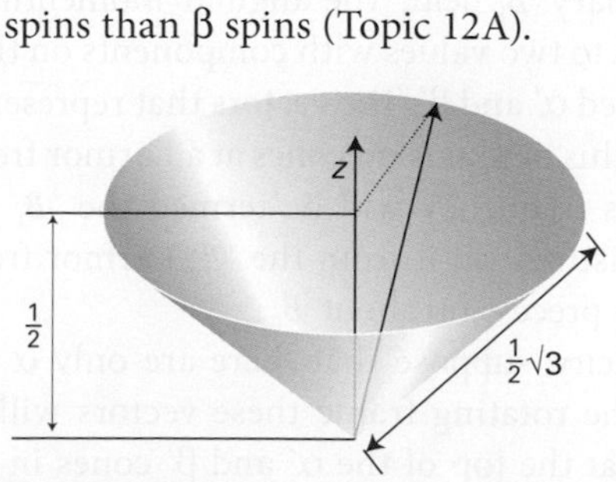

Figure 12C.1 The vector model of angular momentum for a single spin-$\frac{1}{2}$ nucleus with $m_I = +\frac{1}{2}$. The position of the vector on the cone is indeterminate.

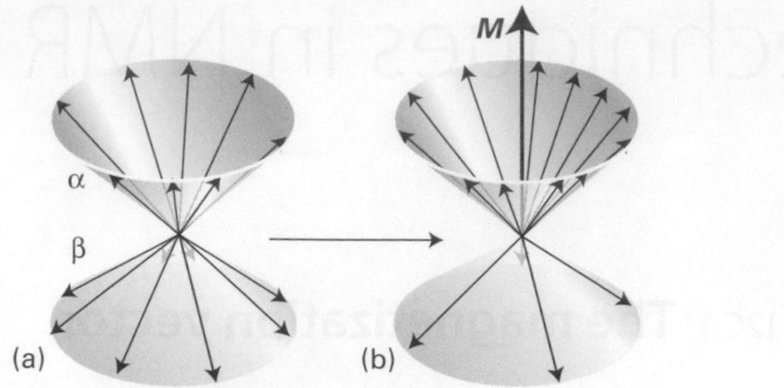

Figure 12C.2 The magnetization of a sample of spin-$\frac{1}{2}$ nuclei is the resultant of all their magnetic moments. (a) In the absence of an externally applied field, there are equal numbers of α and β spins lying at random angles around the cones: the magnetization is zero. (b) In the presence of a field there are slightly more α spins than β spins. As a result, there is a net magnetization, represented by the vector ***M***, along the *z*-axis. There is no magnetization in the transverse plane (the *xy*-plane) because the spins still lie at random angles around the cones.

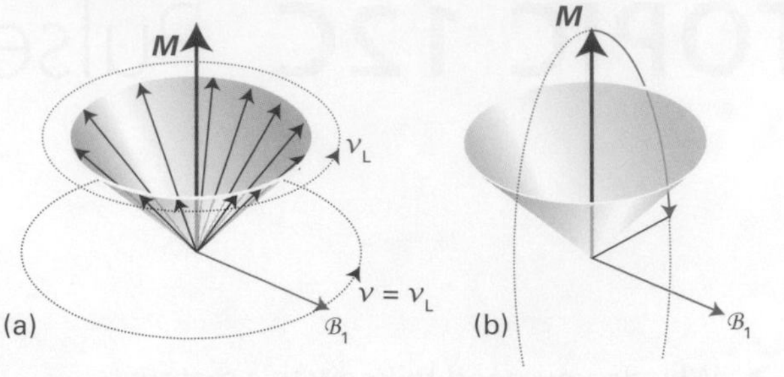

Figure 12C.3 (a) In a pulsed NMR experiment the net magnetization is rotated away from the *z*-axis by applying radiofrequency radiation with its magnetic component $\mathcal{B}_1$ rotating in the *xy*-plane at the Larmor frequency. (b) When viewed in a frame also rotating about *z* at the Larmor frequency, $\mathcal{B}_1$ appears to be stationary. The magnetization rotates around the $\mathcal{B}_1$ field, thus moving the magnetization away from the *z*-axis and generating transverse components.

This imbalance results in a net magnetization in the *z*-direction. It can be represented by a vector ***M*** lying along the *z*-axis with a length proportional to the population difference (Fig. 12C.2b). The manipulation of this net magnetization vector is the central feature of pulse techniques.

(a) The effect of the radiofrequency field

The magnetization vector can be rotated away from its equilibrium position by applying radiofrequency radiation that provides a magnetic field $\mathcal{B}_1$ lying in the *xy*-plane and rotating at the Larmor frequency (as determined by $\mathcal{B}_0$, Fig. 12C.3a). To understand this process, it is best to imagine stepping on to a **rotating frame**, a platform that rotates around the *z*-axis at the Larmor frequency: the $\mathcal{B}_1$ field is stationary in this frame (Fig. 12C3.b).

In the laboratory frame, the applied field defines the axis of quantization and the spins are either α or β with respect to that axis. In the rotating frame, the applied field has effectively disappeared and the new axis of quantization is the direction of the stationary $\mathcal{B}_1$ field. The angular momentum states are still confined to two values with components on that axis, and will be denoted α′ and β′. The vectors that represent them precess around this new axis on cones at a Larmor frequency $\nu_L' = \gamma_N \mathcal{B}_1/2\pi$. This frequency will be termed the '$\mathcal{B}_1$ Larmor frequency' to distinguish it from the '$\mathcal{B}_0$ Larmor frequency' associated with precession about $\mathcal{B}_0$.

For simplicity, suppose that there are only α spins in the sample. In the rotating frame these vectors will seem to be bunched up at the top of the α′ and β′ cones in the rotating frame (Fig. 12C.4). They precess around $\mathcal{B}_1$ and therefore migrate towards the *xy*-plane. Of course, there are β nuclei present too, which in the rotating frame are bunched together at the bottom of the α′ and β′ cones, but precess similarly. At thermal equilibrium there are fewer β spins than α spins, so the net effect is a magnetization vector initially along the *z*-axis that rotates around the $\mathcal{B}_1$ direction at the $\mathcal{B}_1$ Larmor frequency and into the *xy*-plane.

When the radiofrequency field is applied in a pulse of duration $\Delta\tau$ the magnetization rotates through an angle (in radians) of $\phi = \Delta\tau \times (\gamma_N \mathcal{B}_1/2\pi) \times 2\pi$; this angle is known as the **flip angle** of the pulse. Therefore, to achieve a flip angle ϕ, the duration of the pulse must be $\Delta\tau = \phi/\gamma_N \mathcal{B}_1$. A **90° pulse** (with 90°

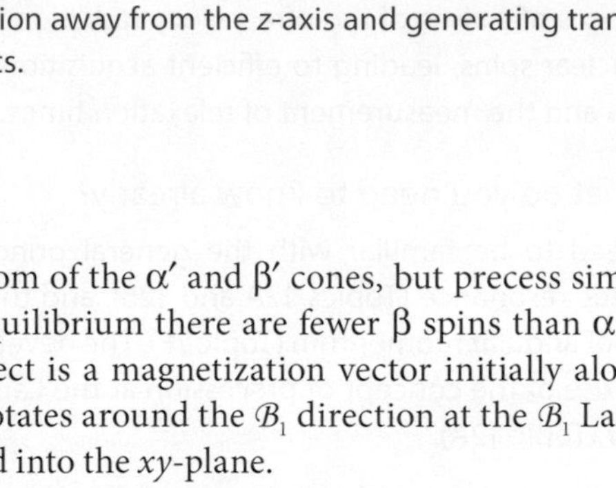

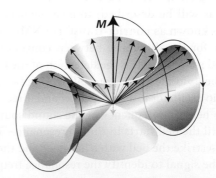

Figure 12C.4 When attention switches to the rotating frame, the vectors representing the spins are in states referring to the axis defined by $\mathcal{B}_1$, and precess on their cones. A uniform distribution in the $\mathcal{B}_0$ frame (blue) is actually a superposition of α' and β' states that seem to be bunched together on their cones (pink). As the latter precess on the pink cones, the magnetization vector rotates into the *xy*-plane. Vectors representing β spins in the original frame behave similarly.

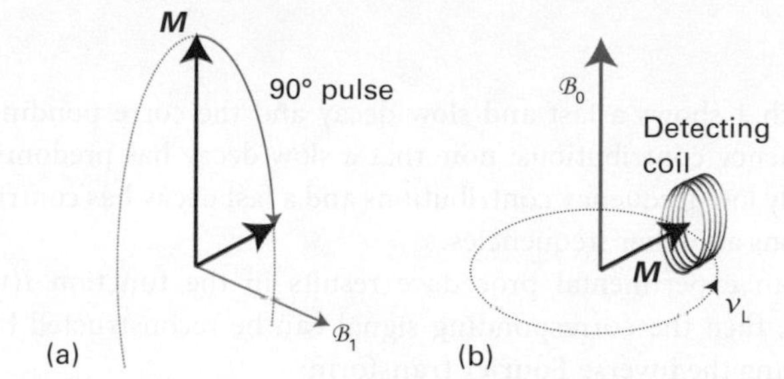

Figure 12C.5 (a) If the radiofrequency field is applied for the appropriate time, the magnetization vector is rotated into the *xy*-plane; this is termed a 90° pulse. (b) Once the magnetization is in the transverse plane it rotates about the $\mathcal{B}_0$ field at the Larmor frequency (when observed in a static frame). The magnetization vector periodically rotates past a small coil, inducing in it an oscillating current, which is the detected signal.

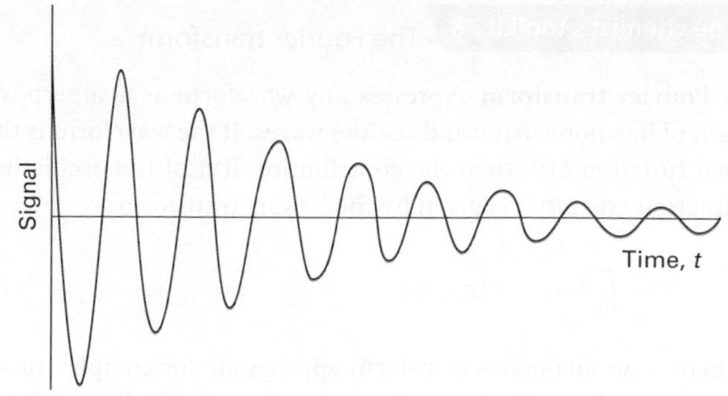

Figure 12C.6 A free-induction decay from of a sample of spins with a single resonance frequency.

corresponding to $\phi = \pi/2$ radians) of duration $\Delta\tau_{90} = \pi/2\gamma_N\mathcal{B}_1$) rotates the magnetization from the *z*-axis into the *xy*-plane (Fig. 12C.5a).

Brief illustration 12C.1

The duration of a radiofrequency pulse depends on the strength of the $\mathcal{B}_1$ field. If a 90° pulse requires 10 μs, then for protons

$$\mathcal{B}_1 = \frac{\pi}{2\times\underbrace{(2.675\times10^8\,\mathrm{T^{-1}\,s^{-1}})}_{\gamma_N}\times\underbrace{(1.0\times10^{-5}\,\mathrm{s})}_{\Delta\tau}} = 5.9\times10^{-4}\,\mathrm{T}$$

or 0.59 mT.

Immediately after a 90° pulse the magnetization lies in the *xy*-plane. Next, imagine stepping out of the rotating frame. The magnetization vector is now rotating in the *xy*-plane at the $\mathcal{B}_0$ Larmor frequency (Fig. 12C.5b). In an NMR spectrometer a small coil is wrapped around the sample and perpendicular to the $\mathcal{B}_0$ field in such a way that the precessing magnetization vector periodically 'cuts' the coil, thereby inducing in it a small current oscillating at the $\mathcal{B}_0$ Larmor frequency. This oscillating current is detected by a radiofrequency receiver.

As time passes the magnetization returns to an equilibrium state in which it has no transverse components; as it does so the oscillating signal induced in the coil decays to zero. This decay is exponential with a time constant denoted T_2. The overall form of the signal is therefore an decaying-oscillating **free-induction decay** (FID) like that shown in Fig. 12C.6 and of the form

$$S(t) = S_0\cos(2\pi\nu_L t)\mathrm{e}^{-t/T_2} \quad \text{Free-induction decay} \quad (12C.1)$$

So far it has been assumed that the radiofrequency radiation is exactly at the $\mathcal{B}_0$ Larmor frequency. However, virtually the same effect is obtained if the separation of the radiofrequency from the Larmor frequency is small compared to the inverse of the duration of the 90° pulse. In practice, a spectrum with several peaks, each with a slightly different Larmor frequency, can be excited by selecting a radiofrequency somewhere in the centre of the spectrum and then making sure that the 90° pulse is sufficiently short (which entails using an intense radiofrequency field so that $\mathcal{B}_1$ is still large enough to achieve rotation through 90°).

(b) Time- and frequency-domain signals

Each line in an NMR spectrum can be thought of as arising from its own magnetization vector. Once that vector has been rotated into the *xy*-plane it precesses at the frequency of the corresponding line. Each vector therefore contributes a decaying-oscillating term to the observed signal and the FID is the sum of many such contributions. If there is only one line it is possible to determine its frequency simply by inspecting the FID, but that is rarely possible when the signal is composite. In such cases, Fourier transformation (*The chemist's toolkit* 28), as mentioned in the introduction, is used to analyse the signal.

The input to the Fourier transform is the decaying-oscillating 'time-domain' function $S(t)$. The output is the absorption spectrum, the 'frequency-domain' function, $I(\nu)$, which is obtained by computing the integral

$$I(\nu) = \int_0^\infty S(t)\cos(2\pi\nu t)\mathrm{d}t \quad (12C.2)$$

where $I(\nu)$ is the intensity at the frequency ν. The complete frequency-domain function, which is the spectrum, is built up by evaluating this integral over a range of frequencies.

The chemist's toolkit 28 The Fourier transform

A **Fourier transform** expresses any waveform as a superposition of harmonic (sine and cosine) waves. If the waveform is the real function $S(t)$, then the contribution $I(\nu)$ of the oscillating function $\cos(2\pi\nu t)$ is given by the 'cosine transform'

$$I(\nu)=\int_0^\infty S(t)\cos(2\pi\nu t)\mathrm{d}t \quad (1)$$

There is an analogous transform appropriate for complex functions: see the additional information for this *Toolkit* available on the website. If the signal varies slowly, then the greatest contribution comes from low-frequency waves; rapidly changing features in the signal are reproduced by high-frequency contributions. If the signal is a simple exponential decay of the form $S(t) = S_0e^{-t/\tau}$, the contribution of the wave of frequency ν is

$$I(\nu)=S_0\int_0^\infty e^{-t/\tau}\cos(2\pi\nu t)\mathrm{d}t=\frac{S_0\tau}{1+(2\pi\nu\tau)^2} \quad (2)$$

Sketch 1 shows a fast and slow decay and the corresponding frequency contributions: note that a slow decay has predominantly low-frequency contributions and a fast decay has contributions at higher frequencies.

If an experimental procedure results in the function $I(\nu)$ itself, then the corresponding signal can be reconstructed by forming the **inverse Fourier transform**:

$$S(t)=\frac{2}{\pi}\int_0^\infty I(\nu)\cos(2\pi\nu t)\mathrm{d}\nu \quad (3)$$

Fourier transforms are applicable to spatial functions too. Their interpretation is similar but it is more appropriate to think in terms of the wavelengths of the contributing waves. Thus, if the function varies only slowly with distance, then its Fourier transform has mainly long-wavelength contributions. If the features vary quickly with distance (as in the electron density in a crystal), then short-wavelength contributions feature.

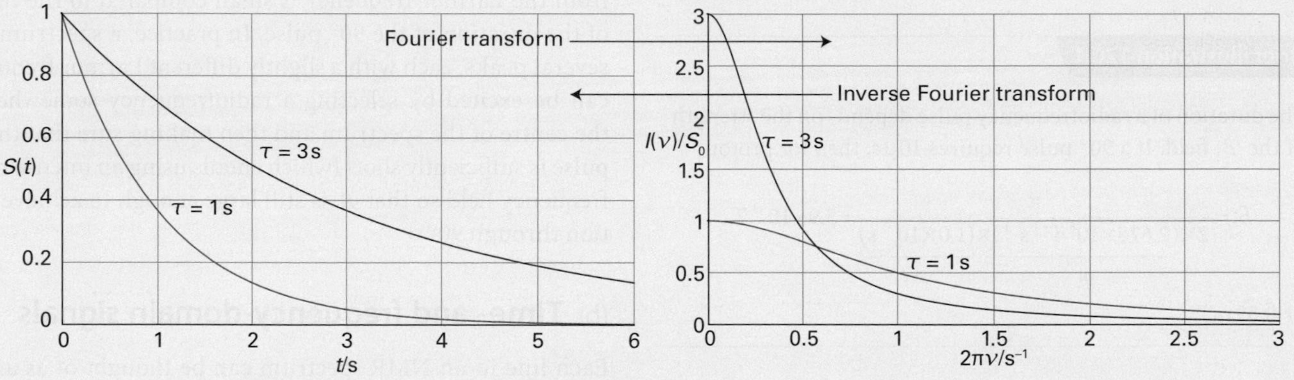

Sketch 1

If the time-domain contains a single decaying-oscillating term, as in eqn 12C.1, the Fourier transform (see the extended *Chemist's toolkit* 28 on the website) is

$$I(\nu)=\frac{S_0T_2}{1+(\nu_L-\nu)^2(2\pi T_2)^2} \quad (12C.3)$$

The graph of this expression has a so-called 'Lorentzian' shape, a symmetrical peak centred at $\nu = \nu_L$ and height S_0T_2; its width at half the peak height is $1/\pi T_2$ (Fig. 12C.7). If the FID consists of a sum of decaying-oscillating functions, Fourier transformation gives a spectrum consisting of a series of peaks at the various frequencies.

In practice, the FID is sampled digitally and the integral of eqn 12C.2 is evaluated numerically by a computer in the NMR spectrometer. Figure 12C.8 shows the time- and frequency-domain functions for three different FIDs of increasing complexity.

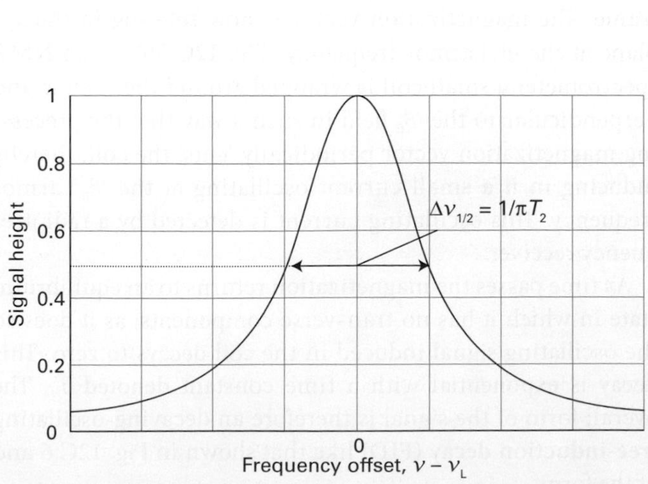

Figure 12C.7 A Lorentzian absorption line. The width at half-height depends of the time constant T_2 which characterizes the decay of the time-domain signal.

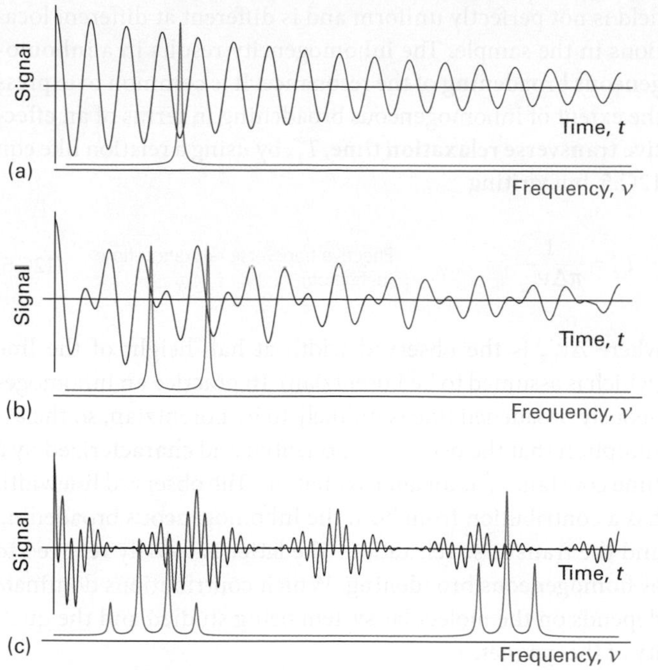

Figure 12C.8 Free-induction decays (the time domain) and the corresponding spectra (the frequency domain) obtained by Fourier transformation. (a) An uncoupled A resonance, (b) the A resonance of an AX system, (c) the A and X resonances of an A_2X_3 system.

12C.2 Spin relaxation

Relaxation is the process by which the magnetization returns to its equilibrium value, at which point it is entirely along the *z*-axis, with no *x*- and no *y*-component (no transverse components). In terms of the behaviour of individual spins, the approach to equilibrium involves transitions between the two spin states in order to establish the thermal equilibrium populations of α and β. The attainment of equilibrium also requires the magnetic moments of individual nuclei becoming distributed at random angles on their two cones.

As already explained, after a 90° pulse the magnetization vector lies in the *xy*-plane. This orientation implies that, from the viewpoint of the laboratory quantization axis, there are now equal numbers of α and β spins because otherwise there would be a component of the magnetization in the *z*-direction. At thermal equilibrium, however, there is a Boltzmann distribution of spins, with more α spins than β spins (provided $\gamma_N > 0$) and a non-zero *z*-component of magnetization. The return of the *z*-component of magnetization to its equilibrium value is termed **longitudinal relaxation**. It is usually assumed that this process follows an exponential recovery curve, with a time constant called the **longitudinal relaxation time**, T_1. Because longitudinal relaxation involves the transfer of energy between the spins and the surroundings (the 'lattice'), the time constant T_1 is also called the *spin–lattice relaxation time*. If the *z*-component of magnetization at time *t* is $M_z(t)$, then the recovery to the equilibrium magnetization M_0 takes the form

$$M_z(t) - M_0 \propto e^{-t/T_1}$$ Longitudinal relaxation time [definition] (12C.4)

Immediately after a 90° pulse the fact that the magnetization vector lies in the *xy*-plane implies that the α and β spins are aligned in a particular way so as to give a net (and rotating) component of magnetization in the *xy*-plane. At thermal equilibrium, however, the spins are at random angles on their cones and there is no transverse component of magnetization. The return of the transverse magnetization to its equilibrium value of zero is termed **transverse relaxation**. It is usually assumed that this process is an exponential decay with a time constant called the **transverse relaxation time**, T_2 (or *spin–spin relaxation time*). If the transverse magnetization at time *t* is $M_{xy}(t)$, then the decay takes the form

$$M_{xy}(t) \propto e^{-t/T_2}$$ Transverse relaxation time [definition] (12C.5)

This T_2 is the same as that describing the decay of the free induction signal which is proportional to $M_{xy}(t)$.

(a) The mechanism of relaxation

The return of the *z*-component of magnetization to its equilibrium value involves transitions between α and β spin states so as to achieve the populations required by the Boltzmann distribution. These transitions are caused by local magnetic fields that fluctuate at a frequency close to the resonance frequency of the $\beta \leftrightarrow \alpha$ transition. The local fields can have a variety of origins, but commonly arise from nearby magnetic nuclei or unpaired electrons. These fields fluctuate due to the tumbling motion of molecules in a fluid sample. If molecular tumbling is too slow or too fast compared with the resonance frequency, it gives rise to a fluctuating magnetic field with a frequency that is either too low or too high to stimulate a transition between β and α, so T_1 is long. Only if the molecule tumbles at about the resonance frequency is the fluctuating magnetic field able to induce spin flips effectively, and then T_1 is short.

The rate of molecular tumbling increases with temperature and with reducing viscosity of the solvent, so a dependence of the relaxation time like that shown in Fig. 12C.9 can be expected. The quantitative treatment of relaxation times depends on setting up models of molecular motion and using, for instance, the diffusion equation (Topic 16C) adapted for rotational motion.

Transverse relaxation is the result of the individual magnetic moments of the spins losing their relative alignment as they spread out on their cones. One contribution to this randomization is any process that involves a transition between the

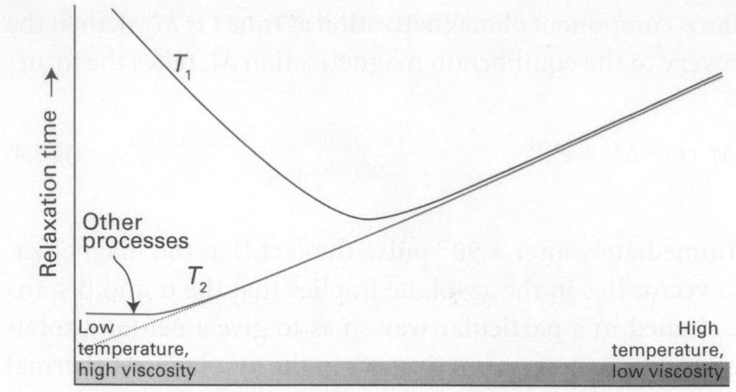

Figure 12C.9 The variation of the two relaxation times with the rate at which the molecules tumble in solution. The horizontal axis can be interpreted as representing temperature or viscosity. Note that the two relaxation times coincide when the motion is fast.

two spin states. That is, any process that causes longitudinal relaxation also contributes to transverse relaxation. Another contribution is the variation in the local magnetic fields experienced by the nuclei. When the fluctuations in these fields are slow, each molecule lingers in its local magnetic environment, and because the precessional rates depend on the strength of the field, the spin orientations randomize quickly around their cones. In other words, slow molecular motion corresponds to short T_2. If the molecules move rapidly from one magnetic environment to another, the effects of differences in local magnetic field average to zero: individual spins do not precess at very different rates, remain bunched for longer, and transverse relaxation does not take place as quickly. This fast motion corresponds to long T_2 (as shown in Fig. 12C.9). Calculations show that, when the motion is fast, transverse and longitudinal relaxation have similar time constants.

Brief illustration 12C.2

For a small molecule dissolved in a non-viscous solvent the value of T_2 for ^{1}H can be as long as several seconds. The width (measured at half the peak height) of the corresponding line in the spectrum is $1/\pi T_2$. For $T_2 = 3.0\,\text{s}$ the width is

$$\Delta\nu_{1/2} = \frac{1}{\pi T_2} = \frac{1}{\pi\times(3.0\,\text{s})} = 0.11\,\text{Hz}$$

In contrast, for a larger molecule such as a protein dissolved in water T_2 is much shorter, and a value of 30 ms is not unusual. The corresponding linewidth is 11 Hz.

So far, it has been assumed that the applied magnetic field is homogeneous (uniform) in the sense of having the same value across the sample so that the differences in Larmor frequencies arise solely from interactions within the sample. In practice, due to the limitations in the design of the magnet, the field is not perfectly uniform and is different at different locations in the sample. The inhomogeneity results in a **inhomogeneous broadening** of the resonance. It is common to express the extent of inhomogeneous broadening in terms of an **effective transverse relaxation time**, T_2^*, by using a relation like eqn 12C.5, but writing

$$T_2^* = \frac{1}{\pi\Delta\nu_{1/2}} \qquad \text{Effective transverse relaxation time [definition]} \qquad (12\text{C}.6)$$

where $\Delta\nu_{1/2}$ is the observed width at half-height of the line (which is assumed to be Lorentzian). In practice an inhomogeneously broadened line is unlikely to be Lorentzian, so the assumption that the decay is exponential and characterized by a time constant T_2^* is an approximation. The observed linewidth has a contribution from both the inhomogeneous broadening and the transverse relaxation. The latter is usually referred to as **homogeneous broadening**. Which contributions dominate depends on the molecular system being studied and the quality of the magnet.

(b) The measurement of T_1 and T_2

The longitudinal relaxation time T_1 can be measured by the **inversion recovery technique**. The first step is to apply a 180° pulse to the sample by applying the $\mathcal{B}_1$ field for twice as long as for a 90° pulse. As a result of the pulse, the magnetization vector is rotated into the $-z$-direction (Fig. 12C.10a). The effect of the pulse is to invert the population of the two levels and to result in more β spins than α spins.

Immediately after the 180° pulse no signal can be detected because the magnetization has no transverse component. The β spins immediately begin to relax back into α spins, and the magnetization vector first shrinks towards zero and then increases in the opposite direction until it reaches its thermal equilibrium value. Before that has happened, after an interval

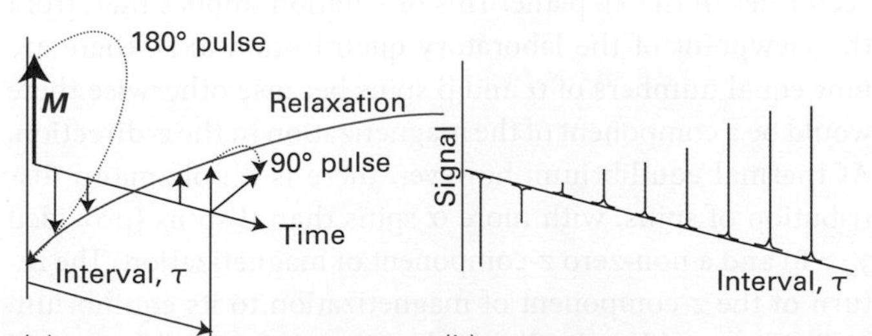

Figure 12C.10 (a) The result of applying a 180° pulse to the magnetization in the rotating frame, and the effect of a subsequent 90° pulse. (b) The amplitude of the frequency-domain spectrum varies with the interval between the two pulses because there has been time for longitudinal relaxation to occur.

τ, a 90° pulse is applied. That pulse rotates the remaining z-component of magnetization into the xy-plane, where it generates an FID signal. The frequency-domain spectrum is then obtained by Fourier transformation in the usual way.

The intensity of the resulting spectrum depends on the magnitude of the magnetization vector that has been rotated into the xy-plane. That magnitude changes exponentially with a time constant T_1 as the interval τ is increased, so the intensity of the spectrum also changes exponentially with increasing τ. The longitudinal relaxation time can therefore be measured by fitting an exponential curve to the series of spectra obtained with different values of τ.

The measurement of T_2 (as distinct from T_2^*) depends on being able to eliminate the effects of inhomogeneous broadening. The cunning required is at the root of some of the most important advances made in NMR since its introduction.

A **spin echo** is the magnetic analogue of an audible echo. The sequence of events is shown in Fig. 12C.11. The overall magnetization can be regarded as made up of a number of different magnetizations, each of which arises from a **spin packet** of nuclei with very similar precession frequencies. The spread in these frequencies arises from the inhomogeneity of $\mathcal{B}_0$ (which is responsible for inhomogeneous broadening), so different parts of the sample experience different fields. The precession frequencies also differ if there is more than one chemical shift present.

First, a 90° pulse is applied to the sample. The subsequent events are best followed in a rotating frame in which $\mathcal{B}_1$ is stationary along the x-axis and causes the magnetization to rotate on to the y-axis of the xy-plane. Immediately after the

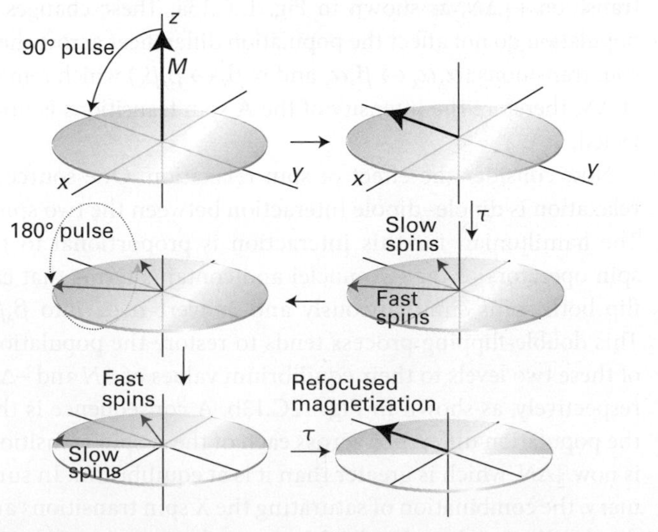

Figure 12C.11 The action of the spin echo pulse sequence 90°–τ–180°–τ, viewed in a rotating frame at the Larmor frequency. Note that the 90° pulse is applied about the x-axis, but the 180° pulse is applied about the y-axis. 'Slow' and 'Fast' refer to the speed of the spin packet relative to the rotating frame frequency.

pulse, the spin packets begin to fan out because they have different Larmor frequencies. In Fig. 12C.11 the magnetization vectors of two representative packets are shown and are described as 'fast' and 'slow', indicating their frequency relative to the rotating frame frequency (the nominal Larmor frequency). Because the rotating frame is at the Larmor frequency, the 'fast' and 'slow' vectors rotate in opposite senses when viewed in this frame.

First, suppose that there is no transverse relaxation but that the field is inhomogeneous. After an evolution period τ, a 180° pulse is applied along the y-axis of the rotating frame. The pulse rotates the magnetization vectors around that axis into mirror-image positions with respect to the yz-plane. Once there, the packets continue to move in the same direction as they did before, and so migrate back towards the y-axis. After an interval τ all the packets are again aligned along the axis. The resultant signal grows in magnitude, reaching a maximum, the 'spin echo', at the end of the second period τ. The fanning out caused by the field inhomogeneity is said to have been 'refocused'.

The important feature of the technique is that the size of the echo is independent of any local fields that remain constant during the two τ intervals. If a spin packet is 'fast' because it happens to be composed of spins in a region of the sample that experience a higher than average field, then it remains fast throughout both intervals, and so the angle through which it rotates is the same in the two intervals. Hence, the size of the echo is independent of inhomogeneities in the magnetic field, because these remain constant.

Now consider the consequences of transverse relaxation. This relaxation arises from fields that vary on a molecular scale, and there is no guarantee that an individual 'fast' spin will remain 'fast' in the refocusing phase: the spins within the packets therefore spread with a time constant T_2. Consequently, the effects of the relaxation are not refocused, and the size of the echo decays with the time constant T_2. The intensity of the signal after a spin echo is measured for a series of values of the delay τ, and the resulting data analysed to determine T_2.

12C.3 Spin decoupling

Carbon-13 has a natural abundance of only 1.1 per cent and is therefore described as a **dilute-spin species**. The probability that any one molecule contains more than one ^{13}C nucleus is rather low, and so the possibility of observing the effects of ^{13}C–^{13}C spin–spin coupling can be ignored. In contrast the abundance of the isotope ^1H is very close to 100 per cent and is therefore described as an **abundant-spin species**. All the hydrogen nuclei in a molecule can be assumed to be ^1H and the effects of coupling between them is observed.

Dilute-spin species are observed in NMR spectroscopy and show coupling to abundant-spin species present in the molecule. Generally speaking, ^{13}C-NMR spectra are very complex on account of the spin couplings of each ^{13}C nucleus with the numerous ^{1}H nuclei in the molecule. However, a dramatic simplification of the spectrum can be obtained by the use of **proton decoupling**. In this technique radiofrequency radiation is applied at (or close to) the ^{1}H Larmor frequency while the ^{13}C FID is being observed. This stimulation of the ^{1}H nuclei causes their spins state to change rapidly, so averaging the ^{1}H–^{13}C couplings to zero. As a result, each ^{13}C nucleus gives a single line rather than a complex multiplet. Not only is the spectrum simplified, but the sensitivity is also improved as all the intensity is concentrated into a single line.

12C.4 The nuclear Overhauser effect

A common source of the local magnetic fields that are responsible for relaxation is the **dipole–dipole interaction** between two magnetic nuclei (see *The chemist's toolkit* 27 in Topic 12B). In this interaction the magnetic dipole of the first spin generates a magnetic field that interacts with the magnetic dipole of the second spin. The strength of the interaction is proportional to $1/R^3$, where R is the distance between the two spins, and is also proportional to the product of the magnetogyric ratios of the spins. As a result, the interaction is characterized as being short-range and significant for nuclei with high magnetogyric ratios. In typical organic and biological molecules, which have many ^{1}H nuclei, the local fields due to the dipole–dipole interaction are likely to be the dominant source of relaxation.

The **nuclear Overhauser effect** (NOE) makes use of the relaxation caused by the dipole–dipole interaction of nuclear spins. In this effect, irradiation of one spin leads to a change in the intensity of the resonance from a second spin provided the two spins are involved in mutual dipole–dipole relaxation. Because the dipole–dipole interaction has only a short range, the observation of an NOE is indicative of the closeness of the two nuclei involved and can be interpreted in terms of the structure of the molecule.

To understand the effect, consider the populations of the four levels of a homonuclear AX spin system shown in Fig. 12C.12. At thermal equilibrium, the population of the $\alpha_A\alpha_X$ level is the greatest, and that of the $\beta_A\beta_X$ level is the least; the other two levels have the same energy and an intermediate population. For the purposes of this discussion it is sufficient to consider the deviations of the populations from the average value for all four levels: the $\alpha_A\alpha_X$ level has a greater population than the average, the $\beta_A\beta_X$ level has a smaller population, and the other two levels have a population equal to the average. The deviations from the average are ΔN for $\alpha_A\alpha_X$, $-\Delta N$ for $\beta_A\beta_X$, and 0 for the other two levels. These population differences are represented in Fig. 12C.12. The intensity of a transition reflects the difference in the population of the two energy levels involved, and for all four transitions this population difference (lower − upper) is ΔN, implying that all four transitions have the same intensity.

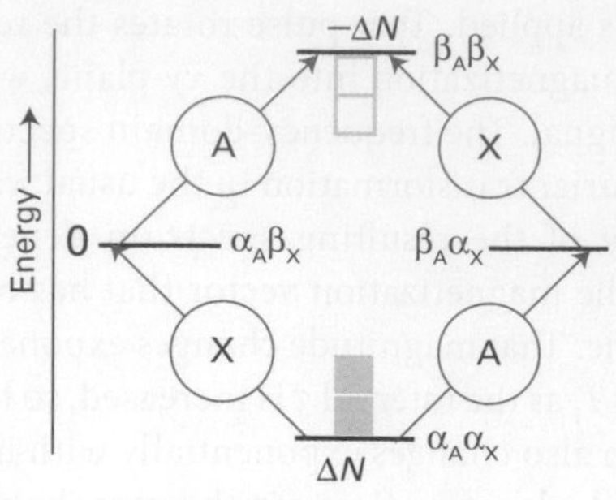

Figure 12C.12 The energy levels of an AX system and an indication of their relative populations. Each green square above the line represents an excess population above the average, and each white square below the line represents a lower population than the average. The symbols in red show the deviations in population from their average value.

The NOE experiment involves irradiating the two X spin transitions ($\alpha_A\alpha_X \leftrightarrow \alpha_A\beta_X$ and $\beta_A\alpha_X \leftrightarrow \beta_A\beta_X$) with a radiofrequency field, but making sure that the field is sufficiently weak that the two A spin transitions are not affected. When applied for a long time this field saturates the X spin transitions in the sense that the populations of the two energy levels are equalized. In the case of the $\alpha_A\alpha_X \leftrightarrow \alpha_A\beta_X$ transition the populations of the two levels become $\frac{1}{2}\Delta N$, and for the other X spin transition $-\frac{1}{2}\Delta N$, as shown in Fig. 12C.13a. These changes in population do not affect the population differences across the A spin transitions ($\alpha_A\alpha_X \leftrightarrow \beta_A\alpha_X$ and $\alpha_A\beta_X \leftrightarrow \beta_A\beta_X$) which remain at ΔN, therefore the intensity of the A spin transitions is unaffected.

Now consider the effect of spin relaxation. One source of relaxation is dipole–dipole interaction between the two spins. The hamiltonian for this interaction is proportional to the spin operators of the two nuclei and contains terms that can flip both spins simultaneously and convert $\alpha_A\alpha_X$ into $\beta_A\beta_X$. This double-flipping process tends to restore the populations of these two levels to their equilibrium values of ΔN and $-\Delta N$, respectively, as shown in Fig. 12C.13b. A consequence is that the population difference across each of the A spin transitions is now $\frac{3}{2}\Delta N$, which is greater than it is at equilibrium. In summary, the combination of saturating the X spin transitions and dipole–dipole relaxation leads to an enhancement of the intensity of the A spin transitions.

The hamiltonian for the dipole–dipole interaction also contains combinations of spin operators that flip the spins in op-

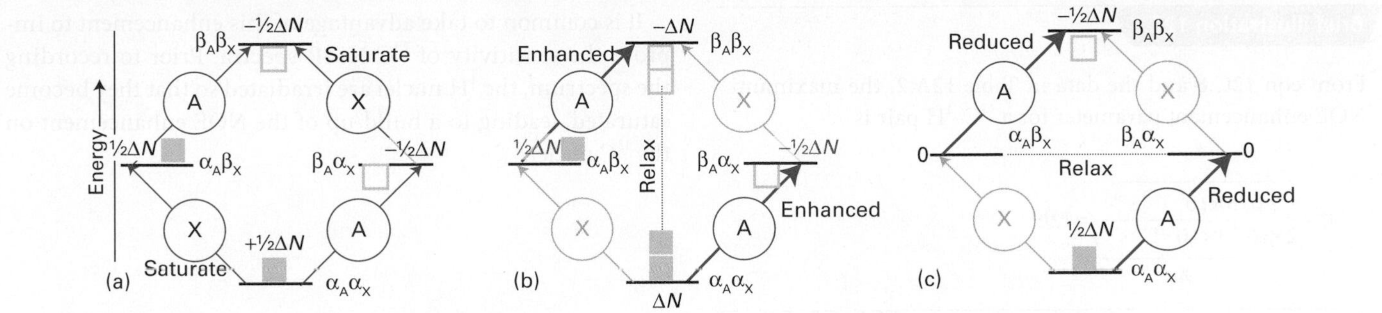

Figure 12C.13 (a) When an X transition is saturated, the populations of the two states are equalized, leading to the populations shown (using the same symbols as in Fig. 12C.12). (b) Dipole–dipole relaxation can cause transitions between the $\alpha\alpha$ and $\beta\beta$ states, such that they return to their original populations: the result is that the population difference across the A transitions is increased. (c) Dipole–dipole relaxation can also cause the populations of the $\alpha\beta$ and $\beta\alpha$ states to return to their equilibrium values: this decreases the population difference across the A transitions.

posite directions, so taking the system from $\alpha_A\beta_X$ to $\beta_A\alpha_X$. This process drives the populations of these states to their equilibrium values, as shown in Fig. 12C.13c. As before, the population differences across the A spins transitions are affected, but now they are reduced to $+\frac{1}{2}\Delta N$, meaning that the A spin transitions are less intense than they would have been in the absence of dipole–dipole relaxation.

It should be clear that there are two opposing effects: the relaxation induced transitions between $\alpha_A\alpha_X$ and $\beta_A\beta_X$ which enhance the A spin transitions, and the transitions between $\alpha_A\beta_X$ and $\beta_A\alpha_X$ which reduce the intensity of these transitions. Which effect dominates depends on the relative rates of these two relaxation pathways. As in the discussion of relaxation times in Section 12C.2, the efficiency of the $\beta_A\beta_X \leftrightarrow \alpha_A\alpha_X$ relaxation is high if the dipole field oscillates close to the transition frequency, which in this case is about $2\nu_L$; likewise, the efficiency of the $\alpha_A\beta_X \leftrightarrow \beta_A\alpha_X$ relaxation is high if the dipole field is stationary (in this case there is no frequency difference between the initial and final states). Small molecules tumble rapidly and have substantial motion at $2\nu_L$. As a result, the $\beta_A\beta_X \leftrightarrow \alpha_A\alpha_X$ pathway dominates and results in an increase in the intensity of the A spin transitions. This increase is called a 'positive NOE enhancement'. On the other hand, large molecules tumble more slowly so there is less motion at $2\nu_L$. In this case, the $\alpha_A\beta_X \leftrightarrow \beta_A\alpha_X$ pathway dominates and results in a decrease in the intensity of the A spin transitions. This decrease is called a 'negative NOE enhancement'.

The NOE enhancement is usually reported in terms of the parameter η (eta), where

$$\eta = \frac{I_A - I_A^\circ}{I_A^\circ} \quad \text{NOE enhancement parameter} \quad (12C.7)$$

Here I_A° is the intensity of the signals due to nucleus A before saturation, and I_A is the intensity after the X spins have been saturated for long enough for the NOE to build up (typically several multiples of T_1). For a homonuclear system, and if the only source of relaxation is due to the dipole–dipole interaction, η lies between −1 (a negative enhancement) for slow tumbling molecules and $+\frac{1}{2}$ (a positive enhancement) for fast tumbling molecules. In practice, other sources of relaxation are invariably present so these limiting values are rarely achieved.

The utility of the NOE in NMR spectroscopy comes about because only dipole–dipole relaxation can give rise to the effect: only this type of relaxation causes both spins to flip simultaneously. Thus, if nucleus X is saturated and a change in the intensity of the transitions from nucleus A is observed, then there must be a dipole–dipole interaction between the two nuclei. As that interaction is proportional to $1/R^3$, where R is the distance between the two spins, and the relaxation it causes is proportional to the square of the interaction, the NOE is proportional to $1/R^6$. Therefore, for there to be significant dipole–dipole relaxation between two spins they must be close (for ^1H nuclei, not more than 0.5 nm apart), so the observation of an NOE is used as qualitative indication of the proximity of nuclei. In principle it is possible to make a quantitative estimate of the distance from the size of the NOE, but to do so requires the effects of other kinds of relaxation to be taken into account.

The value of the NOE enhancement η also depends on the values of the magnetogyric ratios of A and X, because these properties affect both the populations of the levels and the relaxation rates. For a heteronuclear spin system the maximal enhancement is

$$\eta = \frac{\gamma_X}{2\gamma_A} \quad (12C.8)$$

where γ_A and γ_X are the magnetogyric ratios of nuclei A and X, respectively.

Brief illustration 12C.3

From eqn 12C.8 and the data in Table 12A.2, the maximum NOE enhancement parameter for a ^{13}C–^{1}H pair is

$$\eta = \frac{\overbrace{2.675\times10^{8}\,\mathrm{T^{-1}\,s^{-1}}}^{\gamma_{\mathrm{H}}}}{2\times\underbrace{(6.73\times10^{7}\,\mathrm{T^{-1}\,s^{-1}})}_{\gamma_{^{13}\mathrm{C}}}} = 1.99$$

It is common to take advantage of this enhancement to improve the sensitivity of ^{13}C NMR spectra. Prior to recording the spectrum, the ^{1}H nuclei are irradiated so that they become saturated, leading to a build-up of the NOE enhancement on the ^{13}C nuclei.

Checklist of concepts

☐ 1. The **free-induction decay** (FID) is the time-domain signal resulting from the precession of transverse magnetization.

☐ 2. Fourier transformation of the FID (the time domain) gives the NMR spectrum (the frequency domain).

☐ 3. **Longitudinal** (or *spin–lattice*) **relaxation** is the process by which the *z*-component of the magnetization returns to its equilibrium value.

☐ 4. **Transverse** (or *spin–spin*) **relaxation** is the process by which the *x*- and *y*-components of the magnetization return to their equilibrium values of zero.

☐ 5. The **longitudinal relaxation time** T_1 can be measured by the **inversion recovery technique**.

☐ 6. The **transverse relaxation time** T_2 can be measured by observing spin echoes.

☐ 7. In **proton decoupling** of ^{13}C-NMR spectra, the protons are continuously irradiated; the effect is to collapse the splittings due to the ^{13}C–^{1}H couplings.

☐ 8. The **nuclear Overhauser effect** is the modification of the intensity of one resonance by the saturation of another: it occurs only if the two spins are involved in mutual dipole–dipole relaxation.

Checklist of equations

Property	Equation	Comment	Equation number
Free-induction decay	$S(t)=S_0\cos(2\pi\nu_{\mathrm{L}}t)\mathrm{e}^{-t/T_2}$	T_2 is the transverse relaxation time	12C.1
Width at half-height of an NMR line	$\Delta\nu_{1/2}=1/\pi T_2$	Assumed Lorentzian	
Longitudinal relaxation	$M_z(t)-M_0\propto\mathrm{e}^{-t/T_1}$	T_1 is the longitudinal relaxation time	12C.4
Transverse relaxation	$M_{xy}(t)\propto\mathrm{e}^{-t/T_2}$		12C.5
NOE enhancement parameter	$\eta=(I_{\mathrm{A}}-I_{\mathrm{A}}^{\circ})/I_{\mathrm{A}}^{\circ}$	Definition	12C.7

TOPIC 12D Electron paramagnetic resonance

> **➤ Why do you need to know this material?**
>
> Many materials and biological systems contain species bearing unpaired electrons and some chemical reactions generate intermediates with unpaired electrons. Electron paramagnetic resonance is a key spectroscopic tool for studying them.
>
> **➤ What is the key idea?**
>
> The details of an EPR spectrum give information on the distribution of the density of the unpaired electron.
>
> **➤ What do you need to know already?**
>
> You need to be familiar with the concepts of electron spin (Topic 8B) and the general principles of magnetic resonance (Topic 12A). The discussion refers to spin–orbit coupling in atoms (Topic 8C) and the Fermi contact interaction in molecules (Topic 12B).

Electron paramagnetic resonance (EPR), which is also known as electron spin resonance (ESR), is used to study species that contain unpaired electrons. Both solids and liquids can be studied, but the study of gas-phase samples is complicated by the free rotation of the molecules.

12D.1 The *g*-value

According to the discussion in Topic 12A, the resonance frequency for a transition between the $m_s = -\frac{1}{2}$ and the $m_s = +\frac{1}{2}$ levels of a free electron is

$$h\nu = g_e \mu_B \mathcal{B}_0 \qquad \text{Resonance condition [free electron]} \qquad (12D.1)$$

where $g_e \approx 2.0023$. If the electron is in a radical the field it experiences differs from the applied field due to the presence of local magnetic fields arising from electronic currents induced in the molecular framework. This difference is taken into account by replacing g_e by g and expressing the resonance condition as

$$h\nu = g \mu_B \mathcal{B}_0 \qquad \text{EPR resonance condition} \qquad (12D.2)$$

where g is the ***g*-value** of the radical.

Electron paramagnetic resonance spectra are usually recorded by keeping the frequency of the microwave radiation fixed and then varying the magnetic field so as to bring the electron into resonance with the microwave frequency. The positions of the peaks, and the horizontal scale on spectra, are therefore specified in terms of the magnetic field.

Brief illustration 12D.1

The centre of the EPR spectrum of the methyl radical occurs at 329.40 mT in a spectrometer operating at 9.2330 GHz (radiation belonging to the X band of the microwave region). Its g-value is therefore

$$g = \frac{\overbrace{(6.62608\times10^{-34}\,\text{J s})}^{h}\times\overbrace{(9.2330\times10^{9}\,\text{s}^{-1})}^{\nu}}{\underbrace{(9.2740\times10^{-24}\,\text{J T}^{-1})}_{\mu_B}\times\underbrace{(0.32940\,\text{T})}_{\mathcal{B}_0}} = 2.0027$$

The g-value is related to the ease with which the applied field can generate currents through the molecular framework and the strength of the magnetic field these currents generate. Therefore, the g-value gives some information about electronic structure and plays a similar role in EPR to that played by shielding constants in NMR. A g-value smaller than g_e implies that in the molecule the electron experiences a magnetic field smaller than the applied field, whereas a value greater than g_e implies that the magnetic field is greater. Both outcomes are possible, depending on the details of the electronic excited states.

Two factors are responsible for the difference of the g-value from g_e. Electrons migrate through the molecular framework by making use of excited states (Fig. 12D.1). This circulation gives rise to a local magnetic field that can add to or subtract from the applied field. The extent to which these currents are induced is inversely proportional to the separation of energy levels, ΔE, in the radical or complex. Secondly, the strength of the field experienced by the electron spin as a result of these orbital currents is proportional to the spin–orbit coupling constant, ξ (Topic 8C). It follows that the difference of the g-value from g_e is proportional to $\xi/\Delta E$. This proportionality is widely

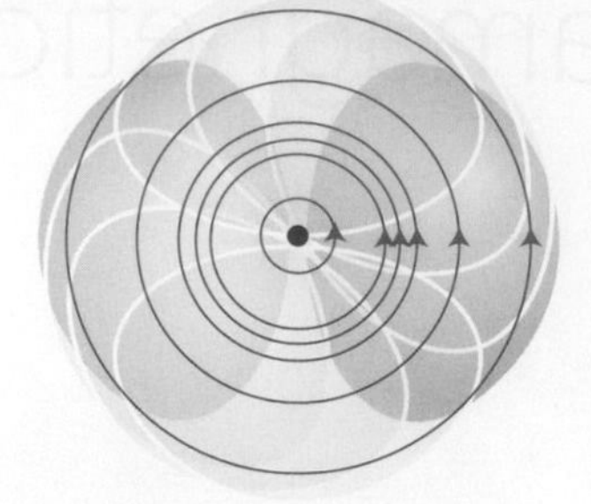

Figure 12D.1 An applied magnetic field can induce circulation of electrons that makes use of excited state orbitals (shown with a white line).

observed. Many organic radicals, for which ΔE is large and ξ (for carbon) is small, have g-values close to 2.0027, not far removed from g_e itself. Inorganic radicals, which commonly are built from heavier atoms and therefore have larger spin–orbit coupling constants, have g-values typically in the range 1.9–2.1. The g-values of paramagnetic d-metal complexes often differ considerably from g_e, varying from 0 to 6, because in them ΔE is small on account of the small splitting of d-orbitals brought about by interactions with ligands (Topic 11F).

The g-value is anisotropic: that is, its magnitude depends on the orientation of the radical with respect to the applied field. The anisotropy arises from the fact that the extent to which an applied field induces currents in the molecule, and therefore the magnitude of the local field, depends on the relative orientation of the molecules and the field. In solution, when the molecule is tumbling rapidly, only the average value of the g-value is observed. Therefore, the anisotropy of the g-value is observed only for radicals trapped in solids and crystalline d-metal complexes.

12D.2 Hyperfine structure

The most important feature of an EPR spectrum is its **hyperfine structure**, the splitting of individual resonance lines into components. In general in spectroscopy, the term 'hyperfine structure' means the structure of a spectrum that can be traced to interactions of the electrons with nuclei other than as a result of the point electric charge of the nucleus. The source of the hyperfine structure in EPR is the magnetic interaction between the electron spin and the magnetic dipole moments of the nuclei present in the radical that give rise to local magnetic fields.

(a) The effects of nuclear spin

Consider the effect on the EPR spectrum of a single ^{1}H nucleus located somewhere in a radical. The proton spin is a source of magnetic field and, depending on the orientation of the nuclear spin, the field it generates either adds to or subtracts from the applied field. The total local field is therefore

$$\mathcal{B}_{\text{loc}} = \mathcal{B}_0 + am_I \qquad m_I = \pm\tfrac{1}{2} \qquad (12\text{D}.3)$$

where a is the **hyperfine coupling constant** (or *hyperfine splitting constant*); from eqn 12D.3 it follows that a has the same units as the magnetic field, for example tesla. Half the radicals in a sample have $m_I = +\tfrac{1}{2}$, so half resonate when the applied field satisfies the condition

$$h\nu = g\mu_B(\mathcal{B}_0 + \tfrac{1}{2}a), \quad \text{or } \mathcal{B}_0 = \frac{h\nu}{g\mu_B} - \tfrac{1}{2}a \qquad (12\text{D}.4\text{a})$$

The other half (which have $m_I = -\tfrac{1}{2}$) resonate when

$$h\nu = g\mu_B(\mathcal{B}_0 - \tfrac{1}{2}a), \quad \text{or } \mathcal{B}_0 = \frac{h\nu}{g\mu_B} + \tfrac{1}{2}a \qquad (12\text{D}.4\text{b})$$

Therefore, instead of a single line, the spectrum shows two lines of half the original intensity separated by a and centred on the field determined by g (Fig. 12D.2).

If the radical contains an ^{14}N atom ($I = 1$), its EPR spectrum consists of three lines of equal intensity, because the ^{14}N nucleus has three possible spin orientations, and each spin orientation is possessed by one-third of all the radicals in the sample. In general, a spin-I nucleus splits the spectrum into $2I+1$ hyperfine lines of equal intensity.

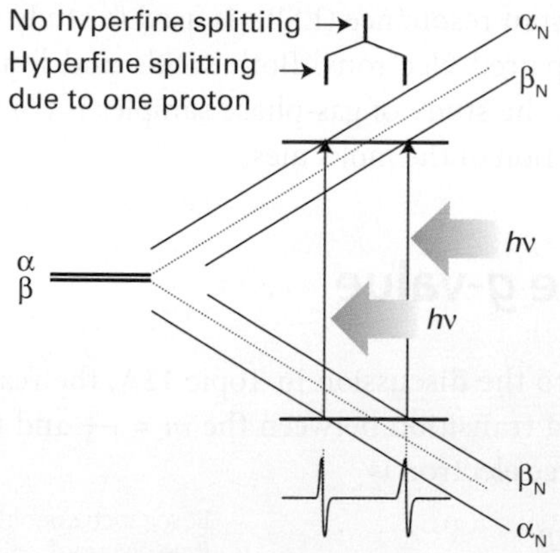

Figure 12D.2 The hyperfine interaction between an electron and a spin-$\tfrac{1}{2}$ nucleus results in four energy levels in place of the original two; α_N and β_N indicate the spin states of the nucleus. As a result, the spectrum consists of two lines (of equal intensity) instead of one. The intensity distribution can be summarized by a simple stick diagram. The diagonal lines show the energies of the states as the applied field is increased, and resonance occurs when the separation of states matches the fixed energy of the microwave photon.

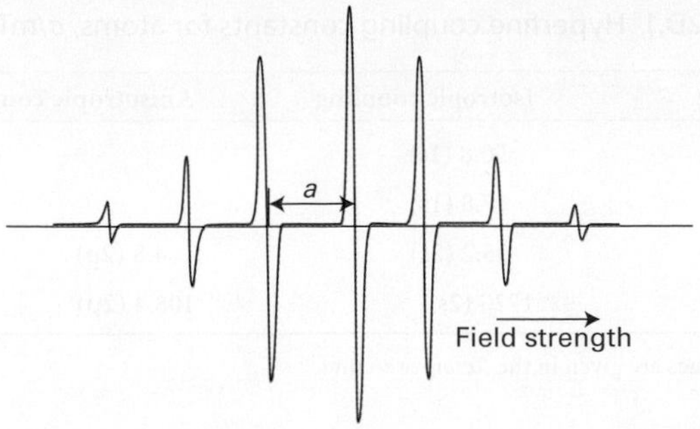

Figure 12D.3 The EPR spectrum of the benzene radical anion, $C_6H_6^-$, in solution; a is the hyperfine coupling constant. The centre of the spectrum is determined by the g-value of the radical.

When there are several magnetic nuclei present in the radical, each one contributes to the hyperfine structure. In the case of equivalent protons (for example, the two CH_2 protons in the radical CH_3CH_2) some of the hyperfine lines are coincident. If the radical contains N equivalent protons, then there are $N + 1$ hyperfine lines with an intensity distribution given by Pascal's triangle (**1**). The spectrum of the benzene radical anion in Fig. 12D.3, which has seven lines with intensity ratio 1:6:15:20:15:6:1, is consistent with a radical containing six equivalent protons. More generally, if the radical contains N equivalent nuclei with spin quantum number I, then there are $2NI + 1$ hyperfine lines.

1
1 1
1 2 1
1 3 3 1
1 4 6 4 1

1

Example 12D.1 Predicting the hyperfine structure of an EPR spectrum

A radical contains one ^{14}N nucleus ($I = 1$) with hyperfine constant 1.61 mT and two equivalent protons ($I = \frac{1}{2}$) with hyperfine constant 0.35 mT. Predict the form of the EPR spectrum.

Collect your thoughts You will need to consider the hyperfine structure that arises from each type of nucleus or group of equivalent nuclei in succession. First, split a line with one nucleus, then split each of the lines again by a second nucleus (or group of nuclei), and so on. It is best to start with the nucleus with the largest hyperfine splitting; however, any choice could be made, and the order in which nuclei are considered does not affect the conclusion.

The solution The ^{14}N nucleus gives three hyperfine lines of equal intensity separated by 1.61 mT. Each line is split into a doublet of spacing 0.35 mT by the first proton, and each line of these doublets is split into a doublet with the same 0.35 mT splitting by the second proton (Fig. 12D.4). Two of the lines in the centre coincide, so splitting by the two protons gives a 1:2:1 triplet of internal splitting 0.35 mT. Overall the spectrum consists of three identical 1:2:1 triplets.

Self-test 12D.1 Predict the form of the EPR spectrum of a radical containing three equivalent ^{14}N nuclei.

Answer: Fig. 12D.5

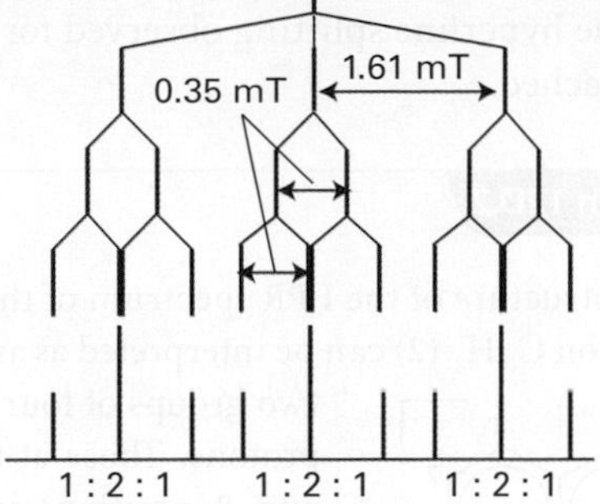

Figure 12D.4 The analysis of the hyperfine structure of a radical containing one ^{14}N nucleus ($I = 1$) and two equivalent protons.

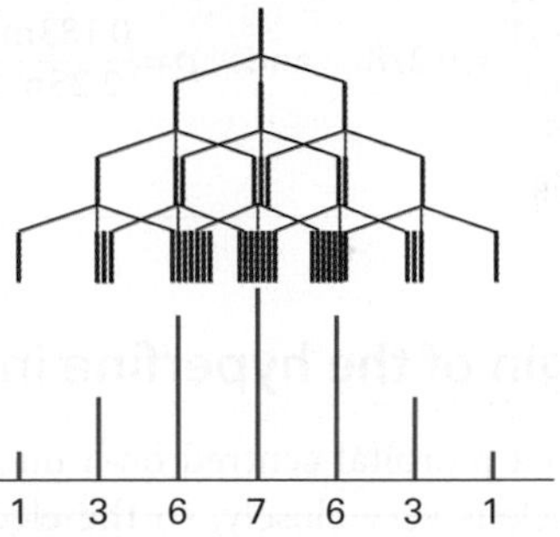

Figure 12D.5 The analysis of the hyperfine structure of a radical containing three equivalent ^{14}N nuclei.

(b) The McConnell equation

The hyperfine structure of an EPR spectrum is a kind of fingerprint that helps to identify the radicals present in a sample. Moreover, because the magnitude of the splitting depends on the distribution of the unpaired electron in the vicinity of the magnetic nuclei, the spectrum can be used to map the molecular orbital occupied by the unpaired electron.

The hyperfine splitting observed in $C_6H_6^-$ is 0.375 mT. If it is assumed that the unpaired electron is in an orbital with equal probability at each C atom, this hyperfine splitting can be attributed to the interaction between a proton and one-sixth of the unpaired electron spin density. If all the electron density were located on the neighbouring C atom, a hyperfine coupling of 6 × 0.375 mT = 2.25 mT would be expected. If in another aromatic radical the hyperfine coupling constant is found to be a, then the **spin density**, ρ, the probability that an unpaired electron is on the neighbouring C atom, can be calculated from the **McConnell equation**:

$$a = Q\rho \quad \text{McConnell equation} \quad (12D.5)$$

with $Q = 2.25\,\text{mT}$. In this equation, ρ is the spin density on a C atom and a is the hyperfine splitting observed for the H atom to which it is attached.

Brief illustration 12D.2

The hyperfine structure of the EPR spectrum of the naphthalene radical anion $C_{10}H_8^-$ (2) can be interpreted as arising from two groups of four equivalent protons. Those at the 1, 4, 5, and 8 positions in the ring have $a = 0.490\,\text{mT}$ and those in the 2, 3, 6, and 7 positions have $a = 0.183\,\text{mT}$. The spin densities obtained by using the McConnell equation are, respectively,

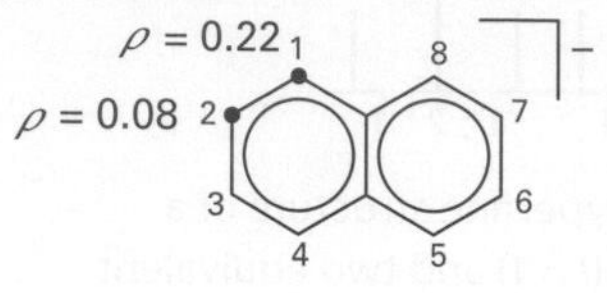

2

$$\rho = \underbrace{\frac{\overbrace{0.490\,\text{mT}}^{a}}{2.25\,\text{mT}}}_{Q} = 0.218 \quad \text{and} \quad \rho = \frac{0.183\,\text{mT}}{2.25\,\text{mT}} = 0.0813$$

(c) The origin of the hyperfine interaction

An electron in a p orbital centred on a nucleus does not approach the nucleus very closely, so the electron experiences a magnetic field that appears to arise from a point magnetic dipole. The resulting interaction is called the **dipole–dipole interaction**. The contribution of a magnetic nucleus to the local field experienced by the unpaired electron is given by an expression like that in eqn 12B.15 (a dependence proportional to $(1 - 3\cos^2\theta)/r^3$). A characteristic of this type of interaction is that it is anisotropic and averages to zero when the radical is free to tumble. Therefore, hyperfine structure due to the dipole–dipole interaction is observed only for radicals trapped in solids.

There is a second contribution to the hyperfine splitting. An s electron is spherically distributed around a nucleus and so has zero average dipole–dipole interaction with the nucleus even in a solid sample. However, because an s electron has a non-zero probability of being at the nucleus itself, it is incorrect to treat the interaction as one between two point dipoles. As explained in Topic 12B, an s electron has a 'Fermi contact interaction' with the nucleus, a magnetic interaction that occurs when the point dipole approximation fails. The contact interaction is isotropic (that is, independent of the orientation of the radical), and consequently is shown even by rapidly tumbling molecules in fluids (provided the spin density has some s character).

The dipole–dipole interactions of p electrons and the Fermi contact interaction of s electrons can be quite large. For example, a 2p electron in a nitrogen atom experiences an average field of about 3.4 mT from the ^{14}N nucleus. A 1s electron in a hydrogen atom experiences a field of about 50 mT as a result of its Fermi contact interaction with the central proton. More values are listed in Table 12D.1. The magnitudes of the contact interactions in radicals can be interpreted in terms of the s orbital character of the molecular orbital occupied by the unpaired electron, and the dipole–dipole interaction can be interpreted in terms of the p character. The analysis of hyperfine structure therefore gives information about the composition of the orbital, and especially the hybridization of the atomic orbitals.

Table 12D.1 Hyperfine coupling constants for atoms, a/mT*

Nuclide	Isotropic coupling	Anisotropic coupling
^{1}H	50.8 (1s)	
^{2}H	7.8 (1s)	
^{14}N	55.2 (2s)	4.8 (2p)
^{19}F	1720 (2s)	108.4 (2p)

*More values are given in the *Resource section*.

Brief illustration 12D.3

From Table 12D.1, the hyperfine interaction between a 2s electron and the nucleus of a nitrogen atom is 55.2 mT. The EPR spectrum of NO_2 shows an isotropic hyperfine interaction of 5.7 mT. The s character of the molecular orbital occupied by the unpaired electron is the ratio 5.7/55.2 = 0.10. For a continuation of this story, see Problem P12D.7.

Neither interaction appears to account for the hyperfine structure of the $C_6H_6^-$ anion and other aromatic radical anions. The sample is fluid, and as the radicals are tumbling the hyperfine structure cannot be due to the dipole–dipole interaction. Moreover, the protons lie in the nodal plane of the π orbital occupied by the unpaired electron, so the structure cannot be due to a Fermi contact interaction. The explanation lies in a **polarization mechanism** similar to the one responsible for spin–spin coupling in NMR. The magnetic interaction between a proton and the electrons favours one of the electrons

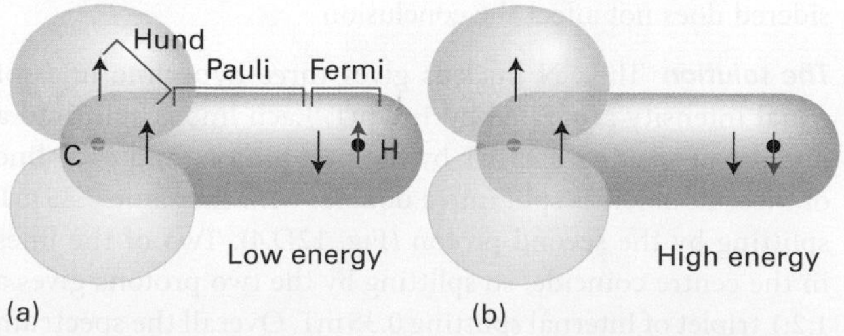

Figure 12D.6 The polarization mechanism for the hyperfine interaction in π-electron radicals. The arrangement in (a) is lower in energy than that in (b), so there is an effective coupling between the unpaired electron and the proton.

being found with a greater probability nearby (Fig. 12D.6). The electron with opposite spin is therefore more likely to be close to the C atom at the other end of the bond. The unpaired electron on the C atom has a lower energy if it is parallel to that electron (Hund's rule favours parallel electrons on atoms), so the unpaired electron can detect the spin of the proton indirectly. Calculation using this model leads to a hyperfine interaction in agreement with the observed value of 2.25 mT.

Checklist of concepts

- ☐ 1. The EPR resonance condition is expressed in terms of the **g-value** of the radical.
- ☐ 2. The value of g depends on the ability of the applied field to induce local electron currents in the radical and the magnetic field experienced by the electron as a result of these currents.
- ☐ 3. The **hyperfine structure** of an EPR spectrum is the splitting of individual resonance lines into components by the magnetic interaction between the electron and nuclei with spin.
- ☐ 4. If a radical contains N equivalent nuclei with spin quantum number I, then there are $2NI + 1$ hyperfine lines.
- ☐ 5. Hyperfine structure arises from **dipole–dipole interactions, Fermi contact interactions**, and the **polarization mechanism**.
- ☐ 6. The **spin density** on an atom is the probability that an unpaired electron is on that atom.

Checklist of equations

Property	Equation	Comment	Equation number
EPR resonance condition	$h\nu = g\mu_B \mathcal{B}_0$	No hyperfine interaction	12D.2
	$h\nu = g\mu_B(\mathcal{B}_0 \pm \tfrac{1}{2}a)$	Hyperfine interaction between an electron and a proton	12D.4
McConnell equation	$a = Q\rho$	$Q = 2.25$ mT	12D.5

FOCUS 12 Magnetic resonance

TOPIC 12A General principles

Discussion questions

D12A.1 Why do chemists and biochemists require spectrometers that operate at the highest available fields and frequencies to determine the structures of macromolecules by NMR spectroscopy?

D12A.2 Describe the effects of magnetic fields on the energies of nuclei and the energies of electrons. Explain the differences.

D12A.3 What is the Larmor frequency? What is its significance in magnetic resonance?

Exercises

E12A.1(a) Given that the nuclear g-factor, g_I, is a dimensionless number, what are the units of the nuclear magnetogyric ratio γ_N when it is expressed in tesla and hertz?
E12A.1(b) Given that the nuclear g-factor, g_I, is a dimensionless number, what are the units of the nuclear magnetogyric ratio γ_N when it is expressed in SI base units?

E12A.2(a) For a ^{1}H nucleus (a proton), what are the magnitude of the spin angular momentum and what are its allowed components along the z-axis? Express your answer in multiples of $\hbar$. What angles does the angular momentum make with the z-axis?
E12A.2(b) For a ^{14}N nucleus, what are the magnitude of the spin angular momentum and what are its allowed components along the z-axis? Express your answer in multiples of $\hbar$. What angles does the angular momentum make with the z-axis?

E12A.3(a) What is the NMR frequency of a ^{1}H nucleus (a proton) in a magnetic field of 13.5 T? Express your answer in megahertz.
E12A.3(b) What is the NMR frequency of a ^{19}F nucleus in a magnetic field of 17.1 T? Express your answer in megahertz.

E12A.4(a) The nuclear spin quantum number of ^{33}S is $\frac{3}{2}$ and its g-factor is 0.4289. Calculate (in joules) the energies of the nuclear spin states in a magnetic field of 6.800 T.
E12A.4(b) The nuclear spin quantum number of ^{14}N is 1 and its g-factor is 0.404. Calculate (in joules) the energies of the nuclear spin states in a magnetic field of 10.50 T.

E12A.5(a) Calculate the frequency separation (in megahertz) of the nuclear spin levels of a ^{13}C nucleus in a magnetic field of 15.4 T given that its magnetogyric ratio is $6.73 \times 10^{-7}\,\mathrm{T^{-1}\,s^{-1}}$.
E12A.5(b) Calculate the frequency separation (in megahertz) of the nuclear spin levels of a ^{14}N nucleus in a magnetic field of 14.4 T given that its magnetogyric ratio is $1.93 \times 10^{-7}\,\mathrm{T^{-1}\,s^{-1}}$.

E12A.6(a) In which of the following systems is the energy level separation larger for a given magnetic field? (i) A ^{15}N nucleus, (ii) a ^{31}P nucleus.
E12A.6(b) In which of the following systems is the energy level separation larger? (i) A ^{14}N nucleus in a magnetic field that corresponds to an NMR frequency for ^{1}H of 600 MHz, (ii) an electron in a field of 0.300 T.

E12A.7(a) Calculate the relative population differences $(N_\alpha - N_\beta)/N$ for ^{1}H nuclei in fields of (i) 0.30 T, (ii) 1.5 T, and (iii) 10 T at 25 °C.
E12A.7(b) Calculate the relative population differences $(N_\alpha - N_\beta)/N$ for ^{13}C nuclei in fields of (i) 0.50 T, (ii) 2.5 T, and (iii) 15.5 T at 25 °C.

E12A.8(a) By what factor must the applied magnetic field be increased for the relative population difference $(N_\alpha - N_\beta)/N$ to be increased by a factor of 5 for (i) ^{1}H nuclei, (ii) ^{13}C nuclei?
E12A.8(b) By what factor must the temperature be changed for the relative population difference $(N_\alpha - N_\beta)/N$ to be increased by a factor of 5 for ^{1}H nuclei relative to its value at room temperature? Is changing the temperature of the sample a practical way of increasing the sensitivity?

E12A.9(a) Some commercial EPR spectrometers use 8 mm microwave radiation (the 'Q band'). What magnetic field is needed to satisfy the resonance condition?
E12A.9(b) What is the EPR resonance frequency in a magnetic field for which the NMR frequency for ^{1}H nuclei (protons) is 500 MHz? Express your answer in gigahertz.

Problems

P12A.1 A scientist investigates the possibility of neutron spin resonance, and has available a commercial NMR spectrometer operating at 300 MHz for ^{1}H nuclei. What is the NMR frequency of the neutron in this spectrometer? What is the relative population difference at room temperature? Which is the lower energy spin state of the neutron?

P12A.2‡ The relative sensitivity of NMR spectroscopy, R, for equal numbers of different nuclei at constant temperature for a given magnetic field is $R \propto \{I(I+1)\}\gamma_N^3$. (a) From the data in Table 12A.2, calculate these sensitivities for ^{2}H, ^{13}C, ^{14}N, ^{15}N, and ^{11}B relative to that of ^{1}H. (b) For a given number of nuclei of a particular element, the fraction present as a particular isotope is affected by the natural abundance of that isotope. Recalculate these results taking this dependence into account.

P12A.3 The intensity of the NMR signal is given by eqn 12A.8c. The intensity can be increased further by 'isotopic labelling', which involves increasing the proportion of the atoms present that are of the desired NMR-active isotope. The degree of labelling is expressed by giving the fractional enrichment. For example, 'a 10 per cent enrichment in ^{15}N' would imply that 10 per cent of all the N atoms are ^{15}N. (a) What level of enrichment is needed for the ^{15}N signal to have the same intensity as that from ^{13}C with its natural abundance? (b) What is the intensity achievable, relative to natural abundance ^{13}C, by 100 per cent enrichment of ^{17}O?

‡These problems were supplied by Charles Trapp and Carmen Giunta.

P12A.4 With special techniques, known collectively as 'magnetic resonance imaging' (MRI), it is possible to obtain NMR spectra of entire organisms. A key to MRI is the application of a magnetic field that varies linearly across the specimen. If the field varies in the z-direction according to $\mathcal{B}_0 + \mathcal{G}_z z$, where $\mathcal{G}_z$ is the field gradient along the z-direction, the ^{1}H nuclei have NMR frequencies given by

$$\nu_L(z)=\frac{\gamma_N}{2\pi}(\mathcal{B}_0+\mathcal{G}_z z)$$

Similar equations may be written for gradients along the x- and y-directions. The NMR signal at frequency $\nu = \nu(z)$ is proportional to the numbers of protons at the position z. Suppose a uniform disk-shaped organ is placed in such a linear field gradient, and that in this case the NMR signal is proportional to the number of protons in a slice of width δz at each horizontal distance z from the centre of the disk. Sketch the shape of the absorption intensity for the MRI image of the disk.

TOPIC 12B Features of NMR spectra

Discussion questions

D12B.1 The earliest NMR spectrometers measured the spectrum by keeping the frequency fixed and then scanning the magnetic field to bring peaks successively into resonance. Peaks that came into resonance at higher magnetic fields were described as 'up field' and those at lower magnetic fields as 'down field'. Discuss what the terms 'up field' and 'down field' imply about chemical shifts and shielding.

D12B.2 Discuss in detail the origins of the local, neighbouring group, and solvent contributions to the shielding constant.

D12B.3 Explain why the resonance from two equivalent ^{1}H nuclei does not exhibit any splitting due to the spin–spin coupling that exists between the nuclei, but that the resonance is split by the coupling to a third (inequivalent) spin.

D12B.4 Explain the difference between magnetically equivalent and chemically equivalent nuclei, and give two examples of each.

D12B.5 Discuss how the Fermi contact interaction and the polarization mechanism contribute to spin–spin coupling in NMR.

Exercises

E12B.1(a) The ^{1}H resonance from TMS is found to occur at 500.130 000 MHz. What is the chemical shift (on the δ scale) of a peak at 500.132 500 MHz?
E12B.1(b) The ^{13}C resonance from TMS is found to occur at 125.130 000 MHz. What is the chemical shift (on the δ scale) of a peak at 125.148 750 MHz?

E12B.2(a) In a spectrometer operating at 500.130 000 MHz for ^{1}H, a resonance is found to occur 750 Hz higher in frequency than TMS. What is the chemical shift (on the δ scale) of this peak?
E12B.2(b) In a spectrometer operating at 125.130 000 MHz for ^{13}C, a resonance is found to occur 1875 Hz lower in frequency than TMS. What is the chemical shift (on the δ scale) of this peak?

E12B.3(a) What is the frequency separation, in hertz, between two peaks in a ^{1}H NMR spectrum with chemical shifts $\delta = 9.80$ and $\delta = 2.2$ in a spectrometer operating at 400.130 000 MHz for ^{1}H?
E12B.3(b) What is the frequency separation, in hertz, between two peaks in the ^{13}C spectrum with chemical shifts $\delta = 50.0$ and $\delta = 25.5$ in a spectrometer operating at 100.130 000 MHz for ^{13}C?

E12B.4(a) In a spectrometer operating at 400.130 000 MHz for ^{1}H, on the δ scale what separation of peaks in the ^{1}H spectrum corresponds to a frequency difference of 550 Hz?
E12B.4(b) In a spectrometer operating at 200.130 000 MHz for ^{13}C, on the δ scale what separation of peaks in the ^{13}C spectrum corresponds to a frequency difference of 25 000 Hz?

E12B.5(a) The chemical shift of the CH_3 protons in ethanal (acetaldehyde) is $\delta = 2.20$ and that of the CHO proton is 9.80. What is the difference in local magnetic field between the two regions of the molecule when the applied field is (i) 1.5 T, (ii) 15 T?
E12B.5(b) The chemical shift of the CH_3 protons in ethoxyethane (diethyl ether) is $\delta = 1.16$ and that of the CH_2 protons is 3.36. What is the difference in local magnetic field between the two regions of the molecule when the applied field is (i) 1.9 T, (ii) 16.5 T?

E12B.6(a) Make a sketch, roughly to scale, of the ^{1}H NMR spectrum expected for an AX spin system with $\delta_A = 1.00$, $\delta_X = 2.00$, and $J_{AX} = 10$ Hz recorded on a spectrometer operating at (i) 250 MHz, (ii) 800 MHz. The horizontal scale should be in hertz, taking the resonance from TMS as the origin.
E12B.6(b) Make a sketch, roughly to scale, of the ^{1}H NMR spectrum expected for an AX_2 spin system with $\delta_A = 1.50$, $\delta_X = 4.50$, and $J_{AX} = 5$ Hz recorded on a spectrometer operating at 500 MHz. The horizontal scale should be in hertz, taking the resonance from TMS as the origin.

E12B.7(a) Sketch the form of the ^{19}F NMR spectrum and the ^{10}B NMR spectrum of $^{10}BF_4^-$.
E12B.7(b) Sketch the form of the ^{19}F NMR spectrum and the ^{11}B NMR spectrum of $^{11}BF_4^-$.

E12B.8(a) Sketch the form of the ^{31}P NMR spectra of a sample of $^{31}PF_6^-$.
E12B.8(b) Sketch the form of the ^{1}H NMR spectra of $^{14}NH_4^+$ and of $^{15}NH_4^+$.

E12B.9(a) Use an approach similar to that shown in Figs. 12B.13 and 12B.14 to predict the multiplet you would expect for coupling to four equivalent spin-$\frac{1}{2}$ nuclei.
E12B.9(b) Predict the multiplet you would expect for coupling to two equivalent spin-1 nuclei.

E12B.10(a) Use an approach similar to that shown in Figs. 12B.13 and 12B.14, to predict the multiplet you would expect for coupling to two spin-$\frac{1}{2}$ nuclei when the coupling to the two nuclei is not the same.
E12B.10(b) Predict the multiplet you would expect for coupling of protons to two inequivalent spin-1 nuclei.

E12B.11(a) Use an approach similar to that shown in Figs. 12B.13 and 12B.14 to predict the multiplet you would expect for coupling to two equivalent spin-$\frac{5}{2}$ nuclei.
E12B.11(b) Predict the multiplet you would expect for coupling to three equivalent spin-$\frac{5}{2}$ nuclei.

E12B.12(a) Classify the ^{1}H nuclei in 1-chloro-4-bromobenzene into chemically or magnetically equivalent groups. Give your reasoning.
E12B.12(b) Classify the ^{1}H nuclei in 1,2,3-trichlorobenzene into chemically or magnetically equivalent groups. Give your reasoning.

E12B.13(a) Classify the ^{19}F nuclei in PF_5 into chemically or magnetically equivalent groups. Give your reasoning.
E12B.13(b) Classify the ^{19}F nuclei in SF_5^- (which is square-pyramidal) into chemically or magnetically equivalent groups. Give your reasoning.

E12B.14(a) A proton jumps between two sites with $\delta = 2.7$ and $\delta = 4.8$. What rate constant for the interconversion process is needed for the two signals to collapse to a single line in a spectrometer operating at 550 MHz?

E12B.14(b) A proton jumps between two sites with $\delta = 4.2$ and $\delta = 5.5$. What rate constant for the interconversion process is needed for the two signals to collapse to a single line in a spectrometer operating at 350 MHz?

Problems

P12B.1 Explain why the ^{129}Xe NMR spectrum of XeF^+ is a doublet with $J = 7600$ Hz but the ^{19}F NMR spectrum appears to be a triplet with $J = 3800$ Hz. *Hints:* ^{19}F has spin $\frac{1}{2}$ and 100 per cent natural abundance; ^{129}Xe has spin $\frac{1}{2}$ and 26 per cent natural abundance.

P12B.2 The ^{19}F NMR spectrum of IF_5 consists of two lines of equal intensity and a quintet (five lines with intensity ratio 1:4:6:4:1). Suggest a structure for IF_5 that is consistent with this spectrum, explaining how you arrive at your result. *Hint:* You do not need to consider possible interaction with the I nucleus.

P12B.3 The Lewis structure of SF_4 has four bonded pairs of electrons and one lone pair. Propose two structures for SF_4 based a trigonal bipyramidal coordination at S, and a further structure based on a square pyramid. For each structure, describe the expected form of the ^{19}F NMR spectrum, giving your reasons. *Hint:* You do not need to consider possible interaction with the S nucleus.

P12B.4 Refer to Fig. 12B.15 and use mathematical software or a spreadsheet to draw a family of curves showing the variation of $^3J_{HH}$ with ϕ using $A = +7.0$ Hz, $B = -1.0$ Hz, and allowing C to vary slightly from a typical value of +5.0 Hz. Explore the effect of changing the value of the parameter C on the shape of the curve. In a similar fashion, explore the effect of the values of A and B on the shape of the curve.

P12B.5‡ Various versions of the Karplus equation (eqn 12B.14) have been used to correlate data on three-bond proton coupling constants $^3J_{HH}$ in systems of the type $XYCHCHR_3R_4$. The original version (M. Karplus, *J. Am. Chem.* Soc. **85**, 2870 (1963)) is $^3J_{HH} = A\cos^2\phi_{HH} + B$. Experimentally it is found that when $R_3 = R_4 = H$, $^3J_{HH} = 7.3$ Hz; when $R_3 = CH_3$ and $R_4 = H$, $^3J_{HH} = 8.0$ Hz; when $R_3 = R_4 = CH_3$, $^3J_{HH} = 11.2$ Hz. Assuming that only staggered conformations are important, determine which version of the Karplus equation fits the data better. *Hint:* You will need to consider which conformations to include, and average the couplings predicted by the Karplus equation over them; assume that X and Y are 'bulky' groups.

P12B.6‡ It might be unexpected that the Karplus equation, which was first derived for $^3J_{HH}$ coupling constants, should also apply to three-bond coupling between the nuclei of metallic elements such as tin. T.N. Mitchell and B. Kowall (*Magn. Reson. Chem.* **33**, 325 (1995)) have studied the relation between $^3J_{HH}$ and $^3J_{SnSn}$ in compounds of the type $Me_3SnCH_2CHRSnMe_3$ and find that $(^3J_{SnSn}/\text{Hz}) = 78.86 \times (^3J_{HH}/\text{Hz}) + 27.84$. (a) Does this result support a Karplus type equation for tin nuclei? Explain your reasoning. (b) Obtain the Karplus equation for $^3J_{SnSn}$ and plot it as a function of the dihedral angle. (c) Draw the preferred conformation.

P12B.7 Show that the coupling constant as expressed by the Karplus equation (eqn 12B.14) passes through a minimum when $\cos\phi = B/4C$.

P12B.8 In a liquid, the dipolar magnetic field averages to zero: show this result by evaluating the average of the field given in eqn 12B.15. *Hint:* The relevant volume element in polar coordinates is $\sin\theta\, d\theta d\phi$.

P12B.9 Account for the following observations: (a) The ^{1}H NMR spectrum of cyclohexane shows a single peak at room temperature, but when the temperate is lowered significantly the peak starts to broaden and then separates into two. (b) At room temperature, the ^{19}F NMR spectrum of PF_5 shows two lines, and even at the lowest experimentally accessible temperatures the spectrum is substantially unchanged. (c) In the ^{1}H NMR spectrum of a casually prepared sample of ethanol a triplet and a quartet are seen. These multiplets show additional splittings if the sample is prepared with the careful exclusion of water.

TOPIC 12C Pulse techniques in NMR

Discussion questions

D12C.1 Discuss in detail the effects of a 90° pulse and of a 180° pulse on a system of spin-$\frac{1}{2}$ nuclei in a static magnetic field.

D12C.2 Suggest a reason why the relaxation times of ^{13}C nuclei are typically much longer than those of ^{1}H nuclei.

D12C.3 Suggest a reason why the spin–lattice relaxation time of a small molecule (like benzene) in a mobile, deuterated hydrocarbon solvent increases as the temperature increases, whereas that of a large molecule (like a polymer) decreases.

D12C.4 Discuss the origin of the nuclear Overhauser effect and how it can be used to identify nearby protons in a molecule.

D12C.5 Distinguish between homogeneous and inhomogeneous broadening.

Exercises

E12C.1(a) The duration of a 90° or 180° pulse depends on the strength of the $\mathcal{B}_1$ field. If a 180° pulse applied to ^{1}H requires 12.5 μs, what is the strength of the $\mathcal{B}_1$ field? How long would the corresponding 90° pulse require?

E12C.1(b) The duration of a 90° or 180° pulse depends on the strength of the $\mathcal{B}_1$ field. If a 90° pulse applied to ^{1}H requires 5 μs, what is the strength of the $\mathcal{B}_1$ field? How long would the corresponding 180° pulse require?

E12C.2(a) What is the effective transverse relaxation time when the width of a Lorentzian resonance line is 1.5 Hz?

E12C.2(b) What is the effective transverse relaxation time when the width of a Lorentzian resonance line is 12 Hz?

E12C.3(a) The envelope of a free induction decay is observed to decrease to half its initial amplitude in 1.0 s. What is the value of the transverse relaxation time, T_2?

E12C.3(b) If the transverse relaxation time, T_2, is 50 ms, after what time will the envelope of the free induction decay decrease to half its initial amplitude?

E12C.4(a) The ^{13}C NMR spectrum of ethanoic acid (acetic acid) shows a quartet centred at $\delta = 21$ with a splitting of 130 Hz. When the same spectrum is recorded using proton decoupling, the multiplet collapses to a single line. Another quartet, but with a much smaller spacing, is also seen centred at $\delta = 178$; this quartet collapses when decoupling is used. Explain these observations.

E12C.4(b) The ^{13}C NMR spectrum of fluoroethanoic acid shows a multiplet centred at $\delta = 79$. When the same spectrum is recorded using proton decoupling, the multiplet collapses to a doublet with a splitting of 160 Hz. Another multiplet, but with much smaller splittings, is also seen centred at $\delta = 179$; this multiplet collapses to a doublet when decoupling is used. Explain these observations.

E12C.5(a) Predict the maximum NOE enhancement (as the value of η) that could be obtained for ^{31}P as a result of dipole–dipole relaxation with ^{1}H.

E12C.5(b) Predict the maximum NOE enhancement (as the value of η) that could be obtained for ^{19}F as a result of dipole–dipole relaxation with ^{1}H.

Problems

P12C.1 An NMR spectroscopist performs a series of experiments in which a pulse of a certain duration is applied, the free-induction decay recorded and then Fourier transformed to give the spectrum. A pulse of duration 2.5 μs gave a satisfactory spectrum, but when the pulse duration was increased to 5.0 μs more intense peaks were seen. A further increase to 7.5 μs resulted in weaker signals, and increasing the duration to 10.0 μs gave no detectable spectrum. (a) By considering the effect of varying the flip angle of the pulse, rationalize these observations. Calculate (b) the duration of a 90° pulse and (c) the $\mathcal{B}_1$ Larmor frequency $\nu_L' = \gamma_N \mathcal{B}_1/2\pi$.

P12C.2 In a practical NMR spectrometer the free-induction decay is digitized at regular intervals before being stored in computer memory ready for subsequent processing. Technically, it is difficult to digitize a signal at the frequencies typical of NMR, so in practice a fixed reference frequency, close to the Larmor frequency, is subtracted from the NMR frequency. The resulting difference frequency, called the *offset frequency*, is of the order of several kilohertz, rather than the hundreds of megahertz typical of NMR resonance frequencies. This lower frequency can be digitized by currently available technology. For ^{1}H, if this reference frequency is set at the NMR frequency of TMS, then a peak with chemical shift δ will give rise to a contribution to the free-induction decay at $\delta \times (\nu_L/10^6)$. Use mathematical software to construct the FID curve for a set of three ^{1}H nuclei with resonances of equal intensity at $\delta = 3.2$, 4.1, and 5.0 in a spectrometer operating at 800 MHz. Assume that the reference frequency is set at the NMR frequency of TMS, that $T_2 = 0.5$ s, and plot the FID out to a maximum time of 1.5 s. Explore the effect of varying the relative amplitude of the three resonances.

P12C.3 First read the preamble to Problem P12C.2. The FID, $F(t)$, of a signal containing many frequencies, each corresponding to a different chemical shift, is given by

$$F(t) = \sum_j S_{0j} \cos(2\pi \nu_j t) e^{-t/T_{2j}}$$

where, for each resonance j, S_{0j} is the maximum intensity of the signal, ν_j is the offset frequency, and T_{2j} is the spin–spin relaxation time. (a) Use mathematical software to plot the FID (out to a maximum time of 3 s) for the case

$S_{01} = 1.0$	$\nu_1 = 50$ Hz	$T_{21} = 0.50$ s
$S_{02} = 3.0$	$\nu_2 = 10$ Hz	$T_{22} = 1.0$ s

(b) Explore how the form of the FID changes as ν_1 and T_{21} are changed. (c) Use mathematical software to calculate and plot the Fourier transforms of the FID curves you generated in parts (a) and (b). How do spectral linewidths vary with the value of T_2? *Hint:* Most software packages offer a 'fast Fourier transform' routine with which these calculations can be made: refer to the user manual for details. You should select the cosine Fourier transform.

P12C.4 (a) In many instances it is possible to approximate the NMR lineshape by using a Lorentzian function of the form

$$I_{\text{Lorentzian}}(\omega) = \frac{S_0 T_2}{1 + T_2^2(\omega - \omega_0)^2}$$

where $I(\omega)$ is the intensity as a function of the angular frequency $\omega = 2\pi\nu$, ω_0 is the resonance frequency, S_0 is a constant, and T_2 is the spin–spin relaxation time. Confirm that for this lineshape the width at half-height is $1/\pi T_2$. (b) Under certain circumstances, NMR lines are Gaussian functions of the frequency, given by

$$I_{\text{Gaussian}}(\omega) = S_0 T_2 e^{-T_2^2(\omega - \omega_0)^2}$$

Confirm that for the Gaussian lineshape the width at half-height is equal to $2(\ln 2)^{1/2}/T_2$. (c) Compare and contrast the shapes of Lorentzian and Gaussian lines by plotting two lines with the same values of S_0, T_2, and ω_0.

P12C.5 The shape of a spectral line, $I(\omega)$, is related to the free induction decay signal $G(t)$ by

$$I(\omega) = a \,\text{Re} \int_0^{\infty} G(t) e^{i\omega t}\, dt$$

where a is a constant and 'Re' means take the real part of what follows. Calculate the lineshape corresponding to an oscillating, decaying function $G(t) = \cos \omega t\, e^{-t/\tau}$. *Hint:* Write $\cos \omega t$ as $\frac{1}{2}(e^{-i\omega t} + e^{i\omega t})$.

P12C.6 In the language of Problem 12C.5, show that if $G(t) = (a \cos \omega_1 t + b \cos \omega_2 t) e^{-t/\tau}$, then the spectrum consists of two lines with intensities proportional to a and b and located at $\omega = \omega_1$ and ω_2, respectively.

P12C.7 The exponential relaxation of the z-component of the magnetization $M_z(t)$ back to its equilibrium value M_0 is described by the differential equation

$$\frac{dM_z(t)}{dt} = -\frac{M_z(t) - M_0}{T_1}$$

(a) In the inversion recovery experiment the initial condition (at time zero) is that $M_z(0) = -M_0$, which corresponds to inversion at the magnetization by the 180° pulse. Integrate the differential equation (it is separable), impose this initial condition, and hence show that $M_z(\tau) = M_0(1 - 2e^{-\tau/T_1})$, where τ is the delay between the 180° and 90° pulses. (b) Use mathematical software or a spreadsheet to plot $M_z(\tau)/M_0$ as a function of τ, taking $T_1 = 1.0$ s; explore the effect of increasing and decreasing T_1. (c) Show that a plot of $\ln\{(M_0 - M_z(\tau))/2M_0\}$ against τ is expected to be a straight line with slope $-1/T_1$. (d) In an experiment the following data were obtained; use them to determine a value for T_1.

τ/s	0.000	0.100	0.200	0.300	0.400	0.600	0.800	1.000	1.200
$M_z(\tau)/M_0$	−1.000	−0.637	−0.341	−0.098	0.101	0.398	0.596	0.729	0.819

P12C.8 Derive an expression for the time τ in an inversion recovery experiment at which the magnetization passes through zero. In an experiment it is found that this time for zero magnetization is 0.50 s; evaluate T_1. *Hint:* This Problem requires a result from Problem P12C.7.

P12C.9 The exponential relaxation of the transverse component of the magnetization $M_{xy}(t)$ back to its equilibrium value of zero is described by the differential equation

$$\frac{dM_{xy}(t)}{dt} = -\frac{M_{xy}(t)}{T_2}$$

(a) Integrate this differential equation (it is separable) between $t = 0$ and $t = \tau$ with the initial condition that the transverse magnetization is $M_{xy}(0)$ at $t = 0$ to obtain $M_{xy}(\tau) = M_{xy}(0)e^{-\tau/T_2}$. (b) Hence, show that a plot of $\ln\{M_{xy}(\tau)/M_{xy}(0)\}$ against τ is expected to be a straight line of slope $-1/T_2$. (c) The following data were obtained in a spin-echo experiment; use the data to evaluate T_2.

τ/ms	10.0	20.0	30.0	50.0	70.0	90.0	110	130
$M_{xy}(\tau)/M_{xy}(0)$	0.819	0.670	0.549	0.368	0.247	0.165	0.111	0.074

P12C.10 In the spin echo experiment analysed in Fig. 12C.11, the 180° pulse is applied about the *y*-axis, resulting in the magnetization vectors being reflected in the *yz*-plane. The experiment works just as well when the 180° pulse is applied about the *x*-axis, in which case the magnetization vectors are reflected in the *xz*-plane. Analyse the outcome of the spin echo experiment for the case where the 180° pulse is applied about the *x*-axis.

P12C.11 The *z*-component of the magnetic field at a distance *R* from a magnetic moment parallel to the *z*-axis is given by eqn 12B.17a. In a solid, a proton at a distance *R* from another can experience such a field and the measurement of the splitting it causes in the spectrum can be used to calculate *R*. In gypsum, for instance, the splitting in the H_2O resonance can be interpreted in terms of a magnetic field of 0.715 mT generated by one proton and experienced by the other. What is the separation of the hydrogen nuclei in the H_2O molecule?

P12C.12 In a liquid crystal a molecule might not rotate freely in all directions and the dipolar interaction might not average to zero. Suppose a molecule is trapped so that, although the vector separating two protons may rotate freely around the *z*-axis, the colatitude may vary only between 0 and θ'. Use mathematical software to average the dipolar field over this restricted range of orientation and confirm that the average vanishes when $\theta' = \pi$ (corresponding to free rotation over a sphere). What is the average value of the local dipolar field for the H_2O molecule in Problem P12C.11 if it is dissolved in a liquid crystal that enables it to rotate up to $\theta' = 30°$?

TOPIC 12D Electron paramagnetic resonance

Discussion questions

D12D.1 Describe how the Fermi contact interaction and the polarization mechanism contribute to hyperfine interactions in EPR.

D12D.2 Explain how the EPR spectrum of an organic radical can be used to identify and map the molecular orbital occupied by the unpaired electron.

Exercises

E12D.1(a) The centre of the EPR spectrum of atomic hydrogen lies at 329.12 mT in a spectrometer operating at 9.2231 GHz. What is the *g*-value of the electron in the atom?

E12D.1(b) The centre of the EPR spectrum of atomic deuterium lies at 330.02 mT in a spectrometer operating at 9.2482 GHz. What is the *g*-value of the electron in the atom?

E12D.2(a) A radical containing two equivalent ^{1}H nuclei shows a three-line spectrum with an intensity distribution 1:2:1. The lines occur at 330.2 mT, 332.5 mT, and 334.8 mT. What is the hyperfine coupling constant for each proton? What is the *g*-value of the radical given that the spectrometer is operating at 9.319 GHz?

E12D.2(b) A radical containing three equivalent protons shows a four-line spectrum with an intensity distribution 1:3:3:1. The lines occur at 331.4 mT, 333.6 mT, 335.8 mT, and 338.0 mT. What is the hyperfine coupling constant for each proton? What is the *g*-value of the radical given that the spectrometer is operating at 9.332 GHz?

E12D.3(a) A radical containing two inequivalent protons with hyperfine coupling constants 2.0 mT and 2.6 mT gives a spectrum centred on 332.5 mT. At what fields do the hyperfine lines occur and what are their relative intensities?

E12D.3(b) A radical containing three inequivalent protons with hyperfine coupling constants 2.11 mT, 2.87 mT, and 2.89 mT gives a spectrum centred on 332.8 mT. At what fields do the hyperfine lines occur and what are their relative intensities?

E12D.4(a) Predict the intensity distribution in the hyperfine lines of the EPR spectra of the radicals (i) $\cdot{}^1\mathrm{C}{}^1\mathrm{H}_3$, (ii) $\cdot{}^1\mathrm{C}{}^2\mathrm{H}_3$.

E12D.4(b) Predict the intensity distribution in the hyperfine lines of the EPR spectra of the radicals (i) $\cdot{}^1\mathrm{C}{}^1\mathrm{H}_2{}^1\mathrm{C}{}^1\mathrm{H}_3$, (ii) $\cdot{}^1\mathrm{C}{}^2\mathrm{H}_2{}^1\mathrm{C}{}^2\mathrm{H}_3$.

E12D.5(a) The benzene radical anion has $g = 2.0025$. At what field should you search for resonance in a spectrometer operating at (i) 9.313 GHz, (ii) 33.80 GHz?

E12D.5(b) The naphthalene radical anion has $g = 2.0024$. At what field should you search for resonance in a spectrometer operating at (i) 9.501 GHz, (ii) 34.77 GHz?

E12D.6(a) The EPR spectrum of a radical with a single magnetic nucleus is split into four lines of equal intensity. What is the nuclear spin of the nucleus?

E12D.6(b) The EPR spectrum of a radical with two equivalent nuclei of a particular kind is split into five lines of intensity ratio 1:2:3:2:1. What is the spin of the nuclei?

Problems

P12D.1 It is possible to produce very high magnetic fields over small volumes by special techniques. What would be the resonance frequency of an electron spin in an organic radical in a field of 1.0 kT? How does this frequency compare to typical molecular rotational, vibrational, and electronic energy-level separations?

P12D.2 The angular NO_2 molecule has a single unpaired electron and can be trapped in a solid matrix or prepared inside a nitrite crystal by radiation damage of NO_2^- ions. When the applied field is parallel to the OO direction the centre of the spectrum lies at 333.64 mT in a spectrometer operating at 9.302 GHz. When the field lies along the bisector of the ONO angle, the resonance lies at 331.94 mT. What are the *g*-values in the two orientations?

P12D.3 (a) The hyperfine coupling constant is proportional to the gyromagnetic ratio of the nucleus in question, γ_N. Rationalize this observation. (b) The hyperfine coupling constant in $\cdot\mathrm{C}{}^1\mathrm{H}_3$ is 2.3 mT. Use the information in Table 12D.1 to predict the splitting between the hyperfine lines of the spectrum of $\cdot\mathrm{C}{}^2\mathrm{H}_3$. What are the overall widths of the multiplet in each case?

P12D.4 The 1,4-dinitrobenzene radical anion can be prepared by reduction of 1,4-dinitrobenzene. The radical anion has two equivalent N nuclei ($I = 1$) and four equivalent protons. Predict the form of the EPR spectrum using $a(\mathrm{N}) = 0.148$ mT and $a(\mathrm{H}) = 0.112$ mT.

P12D.5 The hyperfine coupling constants for the anthracene radical anion are 0.274 mT (protons 1, 4, 5, 8), 0.151 mT (protons 2, 3, 6, 7), and 0.534 mT (protons 9, 10). Use the McConnell equation to estimate the spin density at carbons 1, 2, and 9 (use $Q = 2.25\,\text{mT}$).

P12D.6 The hyperfine coupling constants observed in the radical anions (1), (2), and (3) are shown (in millitesla, mT). Use the value for the benzene radical anion to map the probability of finding the unpaired electron in the π orbital on each C atom.

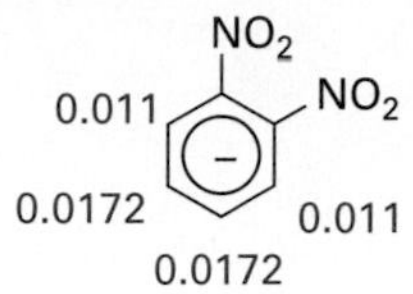

1

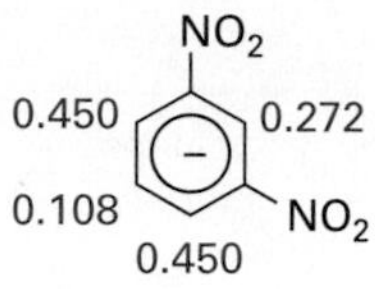

2

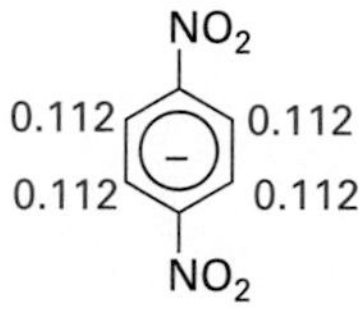

3

P12D.7 When an electron occupies a 2s orbital on an N atom it has a hyperfine interaction of 55.2 mT with the nucleus. The spectrum of NO_2 shows an isotropic hyperfine interaction of 5.7 mT. For what proportion of its time is the unpaired electron of NO_2 occupying a 2s orbital? The hyperfine coupling constant for an electron in a 2p orbital of an N atom is 3.4 mT. In NO_2 the anisotropic part of the hyperfine coupling is 1.3 mT. What proportion of its time does the unpaired electron spend in the 2p orbital of the N atom in NO_2? What is the total probability that the electron will be found on (a) the N atoms, (b) the O atoms? What is the hybridization ratio of the N atom? Does the hybridization support the view that NO_2 is angular?

P12D.8 Sketch the EPR spectra of the di-*tert*-butyl nitroxide radical (4) at 292 K in the limits of very low concentration (at which the averaging effect of electron exchange is negligible), moderate concentration (at which electron exchange effects begin to be observed), and high concentration (at which electron exchange effects predominate).

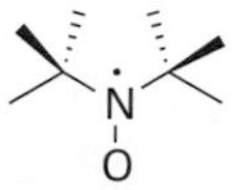

4 di-*tert*-butyl nitroxide

FOCUS 12 Magnetic resonance

Integrated activities

I12.1 Consider the following series of molecules: benzene, methylbenzene, trifluoromethylbenzene, benzonitrile, and nitrobenzene in which the substituents *para* to the C atom of interest are H, CH_3, CF_3, CN, and NO_2, respectively. (a) Use the computational method of your or your instructor's choice to calculate the net charge at the C atom *para* to these substituents in this series of organic molecules. (b) It is found empirically that the ^{13}C chemical shift of the *para* C atom increases in the order: methylbenzene, benzene, trifluoromethylbenzene, benzonitrile, nitrobenzene. Is there a correlation between the behaviour of the ^{13}C chemical shift and the computed net charge on the ^{13}C atom? (c) The ^{13}C chemical shifts of the *para* C atoms in each of the molecules that you examined computationally are as follows:

Substituent	CH_3	H	CF_3	CN	NO_2
δ	128.4	128.5	128.9	129.1	129.4

Is there a linear correlation between net charge and ^{13}C chemical shift of the *para* C atom in this series of molecules? (d) If you did find a correlation in part (c), explain the physical origins of the correlation.

I12.2 The computational techniques described in Topic 9E have shown that the amino acid tyrosine participates in a number of biological electron transfer reactions, including the processes of water oxidation to O_2 in plant photosynthesis and of O_2 reduction to water in oxidative phosphorylation. During the course of these electron transfer reactions, a tyrosine radical forms with spin density delocalized over the side chain of the amino acid. (a) The phenoxy radical shown in (5) is a suitable model of the tyrosine radical. Using molecular modelling software and the computational method of your or your instructor's choice, calculate the spin densities at the O atom and at all of the C atoms in (5). (b) Predict the form of the EPR spectrum of (5).

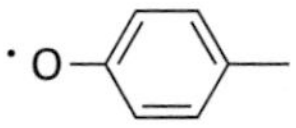

5 Phenoxy radical

I12.3 Two groups of protons have $\delta = 4.0$ and $\delta = 5.2$ and are interconverted by a conformational change of a fluxional molecule. In a 60 MHz spectrometer the spectrum collapsed into a single line at 280 K but at 300 MHz the collapse did not occur until the temperature had been raised to 300 K. Calculate the exchange rate constant at the two temperatures and hence find the activation energy of the interconversion (Topic 17D).

I12.4 NMR spectroscopy may be used to determine the equilibrium constant for dissociation of a complex between a small molecule, such as an enzyme inhibitor I, and a protein, such as an enzyme E:

$$\text{EI} \rightleftharpoons \text{E} + \text{I} \qquad K_I = [\text{E}][\text{I}]/[\text{EI}]$$

In the limit of slow chemical exchange, the NMR spectrum of a proton in I would consist of two resonances: one at ν_I for free I and another at ν_{EI} for bound I. When chemical exchange is fast, the NMR spectrum of the same proton in I consists of a single peak with a resonance frequency ν given by $\nu = f_I\nu_I + f_{EI}\nu_{EI}$, where $f_I = [\text{I}]/([\text{I}] + [\text{EI}])$ and $f_{EI} = [\text{EI}]/([\text{I}] + [\text{EI}])$ are, respectively, the fractions of free I and bound I. For the purposes of analysing the data, it is also useful to define the frequency differences $\delta\nu = \nu - \nu_I$ and $\Delta\nu = \nu_{EI} - \nu_I$. Show that when the initial concentration of I, $[\text{I}]_0$, is much greater than the initial concentration of E, $[\text{E}]_0$, a plot of $[\text{I}]_0$ against $(\delta\nu)^{-1}$ is a straight line with slope $[\text{E}]_0\Delta\nu$ and y-intercept $-K_I$.

FOCUS 13

Statistical thermodynamics

Statistical thermodynamics provides the link between the microscopic properties of matter and its bulk properties. It provides a means of calculating thermodynamic properties from structural and spectroscopic data and gives insight into the molecular origins of chemical properties.

13A The Boltzmann distribution

The 'Boltzmann distribution', which is used to predict the populations of states in systems at thermal equilibrium, is among the most important equations in chemistry for it summarizes the populations of states. It also provides insight into the nature of 'temperature'.

13A.1 **Configurations and weights;** 13A.2 **The relative populations of states**

13B Molecular partition functions

The Boltzmann distribution introduces the central mathematical concept of a 'partition function'. The Topic shows how to interpret the partition function and how to calculate it in a number of simple cases.

13B.1 **The significance of the partition function;** 13B.2 **Contributions to the partition function**

13C Molecular energies

A partition function is the thermodynamic version of a wavefunction, and contains all the thermodynamic information about a system. In this Topic partition functions are used to calculate the mean values of the energy of the basic modes of motion of a collection of independent molecules.

13C.1 **The basic equations;** 13B.2 **Contributions of the fundamental modes of motion**

13D The canonical ensemble

Molecules do interact with one another, and statistical thermodynamics would be incomplete without being able to take these interactions into account. This Topic shows how that is done in principle by introducing the 'canonical ensemble', and hints at how this concept can be used.

13D.1 **The concept of ensemble;** 13D.2 **The mean energy of a system;** 13D.3 **Independent molecules revisited;** 13D.4 **The variation of the energy with volume**

13E The internal energy and the entropy

This Topic shows how molecular partition functions are used to calculate (and give insight into) the two basic thermodynamic functions, the internal energy and the entropy. The latter is based on another central equation introduced by Boltzmann, his definition of 'statistical entropy'.

13E.1 **The internal energy;** 13E.2 **The entropy**

13F Derived functions

With expressions relating internal energy and entropy to partition functions, it is possible to develop expressions for the derived thermodynamic functions, such as the Helmholtz and Gibbs energies. Then, with the Gibbs energy available, the final step is taken into the calculations of chemically significant expressions by showing how equilibrium constants can be calculated from structural and spectroscopic data.

13F.1 **The derivations;** 13F.2 **Equilibrium constants**

Web resource What is an application of this material?

There are numerous applications of statistical arguments in biochemistry. One of the most directly related to partition functions is explored in *Impact* 20: the helix–coil equilibrium in a polypeptide and the role of cooperative behaviour.

TOPIC 13A The Boltzmann distribution

➤ **Why do you need to know this material?**

The Boltzmann distribution is the key to understanding a great deal of chemistry. All thermodynamic properties can be interpreted in its terms, as can the temperature dependence of equilibrium constants and the rates of chemical reactions. There is, perhaps, no more important unifying concept in chemistry.

➤ **What is the key idea?**

The most probable distribution of molecules over the available energy levels subject to certain restraints depends on a single parameter, the temperature.

➤ **What do you need to know already?**

You need to be aware that molecules can exist only in certain discrete energy levels (Topic 7A) and that in some cases more than one state has the same energy.

The problem addressed in this Topic is the calculation of the populations of states for any type of molecule in any mode of motion at any temperature. The only restriction is that the molecules should be independent, in the sense that the total energy of the system is a sum of their individual energies. In a real system a contribution to the total energy may arise from interactions between molecules, but that possibility is discounted at this stage. The development is based on the **principle of equal *a priori* probabilities**, the assumption that all possibilities for the distribution of energy are equally probable. '*A priori*' means loosely in this context 'as far as one knows'. There is no reason to presume otherwise than that for a collection of molecules at thermal equilibrium, a vibrational state of a certain energy, for instance, is as likely to be populated as a rotational state of the same energy.

One very important conclusion that will emerge from the following analysis is that the overwhelmingly most probable populations of the available states depend on a single parameter, the 'temperature'. That is, the work done here provides a molecular justification for the concept of temperature and some insight into this crucially important quantity.

13A.1 Configurations and weights

Any individual molecule may exist in states with energies ε_0, $\varepsilon_1, \ldots$. For reasons that will become clear, the lowest available state is always taken as the zero of energy (that is, $\varepsilon_0 \equiv 0$), and all other energies are measured relative to that state. To obtain the actual energy of the system it is necessary to add a constant to the energy calculated on this basis. For example, when considering the vibrational contribution to the energy, the total zero-point energy of any oscillators in the system must be added.

(a) Instantaneous configurations

At any instant there will be N_0 molecules in the state 0 with energy ε_0, N_1 in the state 1 with ε_1, and so on, with $N_0 + N_1 + \cdots = N$, the total number of molecules in the system. The specification of the set of populations $N_0, N_1, \ldots$ in the form $\{N_0, N_1, \ldots\}$ is a statement of the instantaneous **configuration** of the system. The instantaneous configuration fluctuates with time because the populations change, perhaps as a result of collisions.

Initially suppose that all the states have exactly the same energy. The energies of all the configurations are then identical, so there is no restriction on how many of the N molecules are in each state. Now picture a large number of different instantaneous configurations. One, for example, might be $\{N,0,0,\ldots\}$, corresponding to every molecule being in state 0. Another might be $\{N-2,2,0,0,\ldots\}$, in which two molecules are in state 1. The latter configuration is intrinsically more likely to be found than the former because it can be achieved in more ways: $\{N,0,0,\ldots\}$ can be achieved in only one way, but $\{N-2,2,0,\ldots\}$ can be achieved in $\frac{1}{2}N(N-1)$ different ways. Thus, one candidate for migration to state 1 can be selected in N ways. There are $N-1$ candidates for the second choice, so the total number of choices is $N(N-1)$. However, the choice (Jack, Jill) cannot be distinguished from the choice (Jill, Jack) because they lead to the same configuration. Therefore, only half the choices lead to distinguishable configurations, and the total number of distinguishable choices is $\frac{1}{2}N(N-1)$. If, as a result of collisions, the system were to fluctuate between the configurations $\{N,0,0,\ldots\}$ and $\{N-2,2,0,\ldots\}$, it would almost always be found in the second, more likely configuration, especially if N were large. In other words, a system free to switch between the two configurations would show properties characteristic almost exclusively of the second configuration.

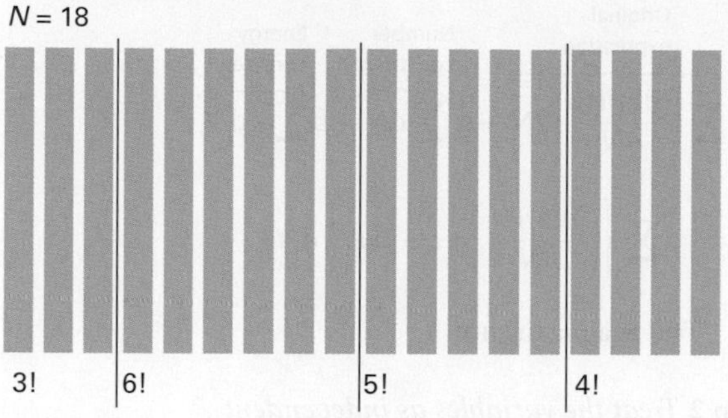

Figure 13A.1 Eighteen molecules (the vertical bars) shown here are distributed into four receptacles (distinguished by the three vertical lines) such that there are 3 molecules in the first, 6 in the second, and so on. There are 18! different ways in which this distribution can be achieved. However, there are 3! equivalent ways of putting three molecules in the first receptacle, and likewise 6! equivalent ways of putting six molecules into the second receptacle, and so on. Hence the number of *distinguishable* arrangements is 18!/3!6!5!4!, or about 515 million.

The next step is to develop an expression for the number of ways that a general configuration $\{N_0,N_1,\ldots\}$ can be achieved. This number is called the **weight** of the configuration and denoted $\mathcal{W}$.

How is that done? 13A.1 Evaluating the weight of a configuration

Consider the number of ways of distributing N balls into bins. The first ball can be selected in N different ways, the next ball in $N-1$ different ways from the balls remaining, and so on. Therefore, there are $N(N-1)\ldots 1 = N!$ ways of selecting the balls for distribution over the bins. However, if there are N_0 balls in the bin labelled ε_0, there would be $N_0!$ different ways in which the same balls could have been chosen (Fig. 13A.1). Similarly, there are $N_1!$ ways in which the N_1 balls in the bin labelled ε_1 can be chosen, and so on. Therefore, the total number of distinguishable ways of distributing the balls so that there are N_0 in bin ε_0, N_1 in bin ε_1, etc. regardless of the order in which the balls were chosen is

$$\mathcal{W}=\frac{N!}{N_0!N_1!N_2!\cdots} \qquad (13A.1)$$

Weight of a configuration

Brief illustration 13A.1

To calculate the number of ways of distributing 20 identical objects with the arrangement 1, 0, 3, 5, 10, 1, note that the configuration is $\{1,0,3,5,10,1\}$ with $N = 20$. Remember that $0! \equiv 1$, Therefore the weight is

$$\mathcal{W}=\frac{20!}{1!0!3!5!10!1!}=9.31\times10^{8}$$

It will turn out to be more convenient to deal with the natural logarithm of the weight, $\ln\mathcal{W}$, rather than with the weight itself:

$$\ln\mathcal{W}=\ln\frac{N!}{N_0!N_1!N_2!\cdots}\overset{\ln(x/y)=\ln x-\ln y}{=}\ln N!-\ln N_0!N_1!N_2!\cdots$$

$$\overset{\ln xy=\ln x+\ln y}{=}\ln N!-\ln N_0!-\ln N_1!-\ln N_2!-\cdots=\ln N!-\sum_i \ln N_i!$$

One reason for introducing $\ln\mathcal{W}$ is that it is easier to make approximations. In particular, the factorials can be simplified by using *Stirling's approximation*[1]

$$\ln x!\approx x\ln x-x \qquad (13A.2)$$

Stirling's approximation [$x >> 1$]

Then the approximate expression for the weight is

$$\begin{aligned}\ln\mathcal{W}&=\{N\ln N-N\}-\sum_i\{N_i\ln N_i-N_i\}\\&=N\ln N-N-\sum_i N_i\ln N_i+N\quad[\text{because}\textstyle\sum_i N_i=N]\\&=N\ln N-\sum_i N_i\ln N_i\end{aligned} \qquad (13A.3)$$

(b) The most probable distribution

The configuration $\{N-2,2,0,\ldots\}$ has much greater weight than $\{N,0,0,\ldots\}$, and it should be easy to believe that there may be other configurations that have a much greater weight than both. In fact, for large N there is a configuration with so great a weight that it overwhelms all the rest in importance to such an extent that the system will almost always be found in it. The properties of the system will therefore be characteristic of that particular dominating configuration. This dominating configuration can be found by looking for the values of N_i that lead to a maximum value of $\mathcal{W}$. Because $\mathcal{W}$ is a function of all the N_i, this search is done by varying the N_i and looking for the values that correspond to $\mathrm{d}\mathcal{W}=0$ (just as in the search for the maximum of any function), or equivalently a maximum value of $\ln\mathcal{W}$. Because $\ln\mathcal{W}$ depends on all the N_i, when a configuration changes and the N_i change to $N_i+\mathrm{d}N_i$, the function $\ln\mathcal{W}$ changes to $\ln\mathcal{W}+\mathrm{d}\ln\mathcal{W}$, where

$$\mathrm{d}\ln\mathcal{W}=\sum_i\left(\frac{\partial\ln\mathcal{W}}{\partial N_i}\right)\mathrm{d}N_i\overset{\text{when }\mathcal{W}\text{ is a maximum}}{=}0 \qquad (13A.4)$$

The derivative $\partial\ln\mathcal{W}/\partial N_i$ expresses how $\ln\mathcal{W}$ changes when N_i changes: if N_i changes by $\mathrm{d}N_i$, then $\ln\mathcal{W}$ changes by $(\partial\ln\mathcal{W}/\partial N_i)\times\mathrm{d}N_i$. The total change in $\ln\mathcal{W}$ is then the sum of

[1] A more precise form of this approximation is $\ln x!\approx\ln(2\pi)^{1/2}+(x+\tfrac{1}{2})\ln x-x$.

all these changes. However, there are two difficulties with this procedure.

Until now it has been assumed that all the states have the same energy. That restriction must now be removed and only configurations that correspond to the specified, constant, total energy of the system must be retained. This requirement rules out many configurations; $\{N,0,0,\ldots\}$ and $\{N-2,2,0,\ldots\}$, for instance, have different energies (unless ε_0 and ε_1 happen to have the same energy), so both cannot occur in the same isolated system. It follows that the configuration with the greatest weight must also satisfy the condition

$$\sum_i N_i\varepsilon_i = E \qquad \text{Energy constraint [constant total energy]} \qquad (13A.5a)$$

where E is the total energy of the system. Therefore, when the N_i change by dN_i, the total energy must not change, so

$$\sum_i \varepsilon_i dN_i = 0 \qquad \text{Energy constraint} \qquad (13A.5b)$$

The second constraint is that, because the total number of molecules present is also fixed (at N), not all the populations can be varied independently. Thus, increasing the population of one state by 1 demands that the population of another state must be reduced by 1. Therefore, the search for the maximum value of $\mathcal{W}$ is also subject to the condition

$$\sum_i N_i = N \qquad \text{Number constraint [constant total number of molecules]} \qquad (13A.6a)$$

It follows that when the N_i change by dN_i, this sum too must not change, so

$$\sum_i dN_i = 0 \qquad \text{Number constraint} \qquad (13A.6b)$$

The challenge now is to see how to solve eqn 13A.4 subject to these two constraints.

How is that done? 13A.2 Imposing constraints

The way to take constraints into account was devised by the mathematician Joseph-Louis Lagrange, and is called the 'method of undetermined multipliers':

- Multiply each constraint by a constant and then add it to the main variation equation.
- Now treat the variables as though they are all independent.
- Evaluate the constants at a later stage of the calculation.

Step 1 *Introduce the constants*

There are two constraints, so introduce two constants α and $-\beta$ (the negative sign will be helpful later), leading to

$$\underbrace{\sum_i\left(\frac{\partial \ln \mathcal{W}}{\partial N_i}\right)dN_i}_{\text{Original expression}} + \underbrace{\alpha\sum_i dN_i}_{\text{Number constraint}} - \underbrace{\beta\sum_i \varepsilon_i dN_i}_{\text{Energy constraint}}$$

$$= \sum_i\left\{\left(\frac{\partial \ln \mathcal{W}}{\partial N_i}\right) + \alpha - \beta\varepsilon_i\right\}dN_i$$

$$= 0 \text{ at a maximum}$$

Step 2 *Treat the variables as independent*

The dN_i are now treated as independent. Hence, the only way of satisfying $d\ln\mathcal{W} = 0$ is to require that for each i,

$$\left(\frac{\partial \ln \mathcal{W}}{\partial N_i}\right) + \alpha - \beta\varepsilon_i = 0 \qquad \text{Condition for maximum } \mathcal{W} \qquad (13A.7)$$

The next step in this lengthy derivation is to insert the expression for $\ln\mathcal{W}$ (eqn 13A.3) into this equation. That involves evaluating the differentiation of $\ln\mathcal{W}$ with respect to N_i.

How is that done? 13A.3 Evaluating the derivative of $\ln\mathcal{W}$

In preparation for this calculation it is useful to change eqn 13A.3 from

$$\ln\mathcal{W} = N\ln N - \sum_i N_i \ln N_i$$

to

$$\ln\mathcal{W} = N\ln N - \sum_j N_j \ln N_j$$

by using j instead of i as the 'name' of the states. In this way the i in the differentiation variable (N_i) will not be confused with the i in the summation. Differentiation of this expression gives

$$\frac{\partial \ln\mathcal{W}}{\partial N_i} = \underbrace{\frac{\partial(N\ln N)}{\partial N_i}}_{\text{First term}} - \underbrace{\sum_j \frac{\partial(N_j \ln N_j)}{\partial N_i}}_{\text{Second term}}$$

Step 1 *Evaluate the first term in the expression*

The first term on the right is obtained (by using the product rule) as follows:

$$\frac{\partial(N\ln N)}{\partial N_i} = \left(\frac{\partial N}{\partial N_i}\right)\ln N + N\left(\frac{\partial \ln N}{\partial N_i}\right) \qquad [df g/dx = f dg/dx + g df/dx]$$

Now note that

$$\frac{\partial N}{\partial N_i} = \frac{\partial}{\partial N_i}(N_1 + N_2 + \cdots) = 1$$

because N_i will match one and only one term in the sum whatever the value of i. Next, note that

d ln y/dx = (1/y)dy/dx

$$\frac{\partial \ln N}{\partial N_i}=\frac{1}{N}\overbrace{\frac{\partial N}{\partial N_i}}^{1}=\frac{1}{N}$$

Therefore

$$\frac{\partial (N \ln N)}{\partial N_i}=\ln N+1$$

Step 2 *Evaluate the second term*

For the derivative of the second term, first note that

dfg/dx = fdg/dx + gdf/dx

$$\sum_j \frac{\partial (N_j \ln N_j)}{\partial N_i}=\sum_j\left\{\left(\frac{\partial N_j}{\partial N_i}\right)\ln N_j+N_j\left(\frac{\partial \ln N_j}{\partial N_i}\right)\right\}$$

d ln y/dx = (1/y)dy/dx

$$=\sum_j\left\{\left(\frac{\partial N_j}{\partial N_i}\right)\ln N_j+\left(\frac{\partial N_j}{\partial N_i}\right)\right\}$$

$$=\sum_j\{\ln N_j+1\}\left(\frac{\partial N_j}{\partial N_i}\right)$$

All the N_j are independent, so the only term that survives in the differentiation $\partial N_j/\partial N_i$ is the one with $j = i$, and then $\partial N_i/\partial N_i = 1$. It follows that

$$\sum_j \frac{\partial (N_j \ln N_j)}{\partial N_i}=\ln N_i+1$$

Step 3 *Bring the two terms together*

Bringing the first and second terms together gives

$$\frac{\partial \ln \mathcal{W}}{\partial N_i}=\ln N+1-(\ln N_i+1)$$

That is,

$$\frac{\partial \ln \mathcal{W}}{\partial N_i}=-\ln\frac{N_i}{N}$$

It now follows from eqn 13A.7 that

$$-\ln\frac{N_i}{N}+\alpha-\beta\varepsilon_i=0$$

and therefore that

$$\frac{N_i}{N}=e^{\alpha-\beta\varepsilon_i} \qquad (13A.8)$$

which is very close to being the Boltzmann distribution.

(b) The values of the constants

At this stage note that

$$N=\sum_i N_i=\sum_i Ne^{\alpha-\beta\varepsilon_i}=Ne^{\alpha}\sum_i e^{-\beta\varepsilon_i}$$

Because the N cancels on each side of this equality, it follows that

$$e^{\alpha}=\frac{1}{\sum_i e^{-\beta\varepsilon_i}} \qquad (13A.9)$$

and therefore

$$\frac{N_i}{N}=e^{\alpha-\beta\varepsilon_i}=e^{\alpha}e^{-\beta\varepsilon_i}=\frac{e^{-\beta\varepsilon_i}}{\sum_i e^{-\beta\varepsilon_i}} \qquad \text{Boltzmann distribution} \qquad (13A.10a)$$

which is the **Boltzmann distribution**. This distribution is commonly written

$$\frac{N_i}{N}=\frac{e^{-\beta\varepsilon_i}}{q} \qquad \text{Boltzmann distribution} \qquad (13A.10b)$$

where q is called the **partition function**:

$$q=\sum_i e^{-\beta\varepsilon_i} \qquad \text{Partition function [definition]} \qquad (13A.11)$$

At this stage the partition function is no more than a convenient abbreviation for the sum; but Topic 13B shows that it is central to the statistical interpretation of thermodynamic properties.

Equation 13A.10 is the justification of the remark that a single parameter, here denoted β, governs the most probable populations of the states of the system, which strongly suggests that it is related to the temperature. The formal deduction of the value of β depends on using the Boltzmann distribution to deduce the perfect gas equation of state (that is done in Topic 13F, and specifically in *Example* 13F.1), which confirms this relation and shows that

$$\beta=\frac{1}{kT} \qquad (13A.12)$$

where T is the thermodynamic temperature and k is Boltzmann's constant. In other words:

> The temperature is the unique parameter that governs the most probable populations of states of a system at thermal equilibrium.

Brief illustration 13A.2

Suppose that two conformations of a molecule differ in energy by 5.0 kJ mol^{-1} (corresponding to 8.3 zJ for a single molecule; 1 zJ = 10^{-21} J), so conformation A lies at energy 0 and conformation B lies at ε = 8.3 zJ. At 20 °C (293 K) the denominator in eqn 13A.10a is

$$\sum_i e^{-\beta\varepsilon_i} = 1 + e^{-\varepsilon/kT} = 1 + e^{-(8.3\times10^{-21}\,\mathrm{J})/(1.381\times10^{-23}\,\mathrm{JK^{-1}})\times(293\,\mathrm{K})} = 1.12\ldots$$

The proportion of molecules in conformation B at this temperature is therefore

$$\frac{N_B}{N} = \frac{e^{-(8.3\times10^{-21}\,\mathrm{J})/(1.381\times10^{-23}\,\mathrm{JK^{-1}})\times(293\,\mathrm{K})}}{1.12\ldots} = 0.11$$

or 11 per cent of the molecules.

13A.2 The relative population of states

When considering only the relative populations of states by using eqn 13A.10b, the partition function need not be evaluated, because it cancels when the ratio is taken:

$$\frac{N_i}{N_j} = \frac{e^{-\beta\varepsilon_i}}{e^{-\beta\varepsilon_j}} = e^{-\beta(\varepsilon_i-\varepsilon_j)}$$ Boltzmann population ratio [thermal equilibrium] (13A.13a)

Note that for a given energy separation the ratio of populations N_1/N_0 decreases as β increases (and the temperature decreases). At $T = 0$ ($\beta = \infty$) all the population is in the ground state and the ratio is zero. Equation 13A.13a is enormously important for understanding a wide range of chemical phenomena and is the form in which the Boltzmann distribution is commonly employed (for instance, in the discussion of the intensities of spectral transitions, Topic 11A). It implies that the relative population of two states decreases exponentially with their difference in energy.

A very important point to note is that the Boltzmann distribution gives the relative populations of *states*, not energy *levels*. Several states might have the same energy, and each state has a population given by eqn 13A.13a. When calculating the relative populations of energy levels rather than states, it is necessary to take into account this degeneracy. Thus, if the level of energy ε_i is g_i-fold degenerate (in the sense that there are g_i states with that energy), and the level of energy ε_j is g_j-fold degenerate, then the relative total populations of the levels are given by

$$\frac{N_i}{N_j} = \frac{g_i e^{-\beta\varepsilon_i}}{g_j e^{-\beta\varepsilon_j}} = \frac{g_i}{g_j} e^{-\beta(\varepsilon_i-\varepsilon_j)}$$ Boltzmann population ratio [thermal equilibrium, degeneracies] (13A.13b)

Example 13A.1 Calculating the relative populations of rotational states

Calculate the relative populations of the $J = 1$ and $J = 0$ rotational levels of HCl at 25 °C; for HCl, $\tilde{B} = 10.591\,\mathrm{cm^{-1}}$.

Collect your thoughts Although the ground state is non-degenerate, you need to note that the level with $J = 1$ is triply degenerate ($M_J = 0, \pm1$); the energy of the state with quantum number J is $\varepsilon_J = hc\tilde{B}J(J+1)$ (Topic 11B). A useful relation is $kT/hc = 207.22\,\mathrm{cm^{-1}}$ at 298.15 K.

The solution The energy separation of states with $J = 1$ and $J = 0$ is $\varepsilon_1 - \varepsilon_0 = 2hc\tilde{B}$. The ratio of the population of a level with $J = 1$ and any *one* of its three states M_J to the population of the single state with $J = 0$ is therefore

$$\frac{N_{J,M_J}}{N_0} = e^{-2hc\tilde{B}\beta}$$

The relative populations of the *levels*, taking into account the three-fold degeneracy of the upper level, is

$$\frac{N_J}{N_0} = 3e^{-2hc\tilde{B}\beta}$$

Insertion of $hc\tilde{B}\beta = hc\tilde{B}/kT = (10.591\,\mathrm{cm^{-1}})/(207.22\,\mathrm{cm^{-1}}) = 0.0511\ldots$ then gives

$$\frac{N_J}{N_0} = 3e^{-2\times0.0511\ldots} = 2.708$$

Comment. Because the $J = 1$ level is triply degenerate, it has a higher population than the level with $J = 0$, despite being of higher energy. As the example illustrates, it is very important to take note of whether you are asked for the relative populations of individual states or of a (possibly degenerate) energy level.

Self-test 13A.1 What is the ratio of the populations of the levels with $J = 2$ and $J = 1$ of HCl at the same temperature?

Answer: 1.359

Checklist of concepts

- ☐ 1. The **principle of equal *a priori* probabilities** assumes that all possibilities for the distribution of energy are equally probable in the sense that the distribution is blind to the type of motion involved.
- ☐ 2. The **instantaneous configuration** of a system of N molecules is the specification of the set of populations $N_0, N_1, \ldots$ of the energy levels $\varepsilon_0, \varepsilon_1, \ldots$.
- ☐ 3. The **Boltzmann distribution** gives the numbers of molecules in each state of a system at any temperature.
- ☐ 4. The **relative populations** of energy levels, as opposed to states, must take into account the degeneracies of the energy levels.

Checklist of equations

Property	Equation	Comment	Equation number
Boltzmann distribution	$N_i/N = e^{-\beta\varepsilon_i}/q$	$\beta = 1/kT$	13A.10b
Partition function	$q = \sum_i e^{-\beta\varepsilon_i}$	See Topic 13B	13A.11
Boltzmann population ratio	$N_i/N_j = (g_i/g_j)e^{-\beta(\varepsilon_i - \varepsilon_j)}$	g_i, g_j are degeneracies	13A.13b

TOPIC 13B Molecular partition functions

➤ **Why do you need to know this material?**

Through the partition function, statistical thermodynamics provides the link between thermodynamic data and molecular properties that have been calculated or derived from spectroscopy. Therefore, this material is an essential foundation for understanding physical and chemical properties of bulk matter in terms of the properties of the constituent molecules.

➤ **What is the key idea?**

The partition function is calculated by drawing on calculated or spectroscopically derived structural information about molecules.

➤ **What do you need to know already?**

You need to know that the Boltzmann distribution expresses the most probable distribution of molecules over the available energy levels (Topic 13A). The concept of the partition function is introduced in that Topic, and is developed here. You need to be aware of the expressions for the rotational and vibrational levels of molecules (Topics 11B–11D) and the energy levels of a particle in a box (Topic 7D).

The partition function $q=\sum_i e^{-\beta\varepsilon_i}$ is introduced in Topic 13A simply as a symbol to denote the sum over states that occurs in the denominator of the Boltzmann distribution (eqn 13A.10b, $p_i=e^{-\beta\varepsilon_i}/q$, with $p_i = N_i/N$). But it is far more important than that might suggest. For instance, it contains all the information needed to calculate the bulk properties of a system of independent particles. In this respect q plays a role for bulk matter very similar to that played by the wavefunction in quantum mechanics for individual molecules: q is a kind of thermal wavefunction.

13B.1 The significance of the partition function

The **molecular partition function** is

$$q=\sum_{\text{states } i} e^{-\beta\varepsilon_i} \qquad \text{Molecular partition function [definition]} \qquad (13B.1a)$$

where $\beta = 1/kT$. As emphasized in Topic 13A, the sum is over *states*, not energy *levels*. If g_i states have the same energy ε_i (so the level is g_i-fold degenerate), then

$$q=\sum_{\text{levels } i} g_i e^{-\beta\varepsilon_i} \qquad \text{Molecular partition function [alternative definition]} \qquad (13B.1b)$$

where the sum is now over energy levels (sets of states with the same energy), not individual states. Also as emphasized in Topic 13A, the lowest available state is taken as the zero of energy, so $\varepsilon_0 \equiv 0$.

Brief illustration 13B.1

Suppose a molecule is confined to the following non-degenerate energy levels: 0, ε, 2ε, ... (Fig. 13B.1). Then the molecular partition function is

$$q=1+e^{-\beta\varepsilon}+e^{-2\beta\varepsilon}+\cdots=1+e^{-\beta\varepsilon}+(e^{-\beta\varepsilon})^2+\cdots$$

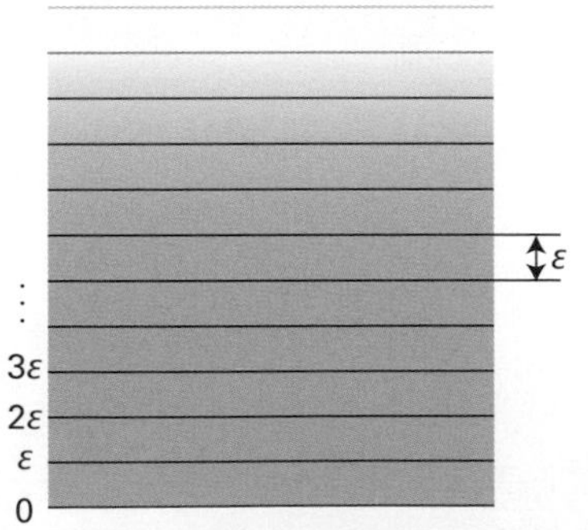

Figure 13B.1 The equally spaced infinite array of energy levels used in the calculation of the partition function. A harmonic oscillator has the same array of levels.

Provided $|x|<1$, the sum to infinity of the geometrical series $1+x+x^2+\cdots$ is $1/(1-x)$. In this case $x=e^{-\beta\varepsilon}<1$, so the series evaluates to

$$q=\frac{1}{1-e^{-\beta\varepsilon}}$$

This function is plotted in Fig. 13B.2 (with $\beta=1/kT$).

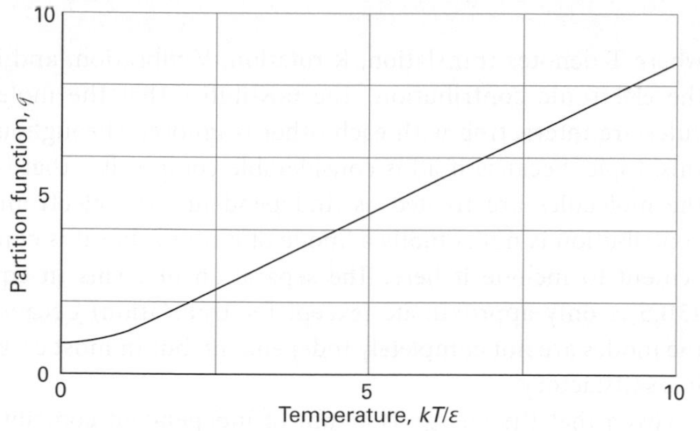

Figure 13B.2 The partition function for the system shown in Fig. 13B.1 (a harmonic oscillator) as a function of temperature.

The result in the *Brief illustration* is an important expression for the partition function for a uniform ladder of states of spacing ε:

$$q=\frac{1}{1-e^{-\beta\varepsilon}} \quad \text{Partition function [uniform ladder]} \quad (13B.2a)$$

This expression can be used to interpret the physical significance of a partition function. To do so, first note that the Boltzmann distribution for this arrangement of energy levels gives the fraction, $p_i=N_i/N$, of molecules in the state with energy ε_i as

$$p_i=\frac{e^{-\beta\varepsilon_i}}{q}=(1-e^{-\beta\varepsilon})e^{-\beta\varepsilon_i} \quad \text{Fractional population [uniform ladder]} \quad (13B.2b)$$

Figure 13B.3 shows how p_i varies with temperature. At very low temperatures (high β), where q is close to 1, only the lowest state is significantly populated. As the temperature is raised, the population breaks out of the lowest state, and the upper states become progressively more highly populated. At the same time, the partition function rises from 1, so its value gives an indication of the range of states populated at any given temperature. The name 'partition function' reflects the sense in which q measures how the total number of molecules is distributed—partitioned—over the available states.

The corresponding expressions for a system in which there are just two states, with energies $\varepsilon_0=0$ and $\varepsilon_1=\varepsilon$ (a 'two-level system'), are

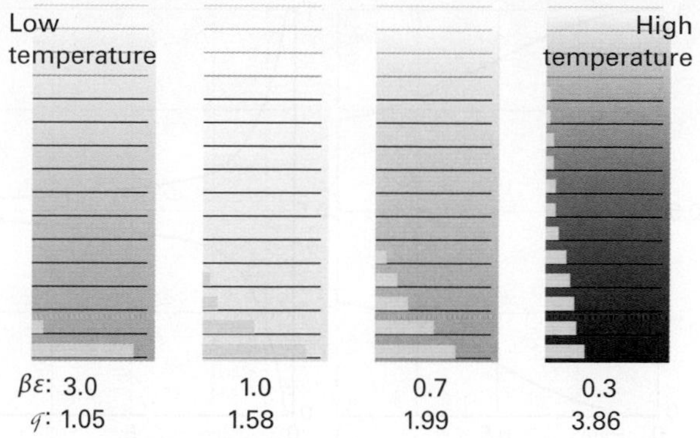

Figure 13B.3 The populations of the energy levels of the system shown in Fig. 13B.1 at different temperatures, and the corresponding values of the partition function calculated from eqn 13B.2a. Note that $\beta=1/kT$.

$$q=1+e^{-\beta\varepsilon} \quad \text{Partition function [two-level system]} \quad (13B.3a)$$

$$p_i=\frac{e^{-\beta\varepsilon_i}}{q}=\frac{e^{-\beta\varepsilon_i}}{1+e^{-\beta\varepsilon}} \quad \text{Fractional population [two-level system, } i=0,1] \quad (13B.3b)$$

The fractional populations of the two states are therefore

$$p_0=\frac{1}{1+e^{-\beta\varepsilon}} \qquad p_1=\frac{e^{-\beta\varepsilon}}{1+e^{-\beta\varepsilon}} \quad (13B.4)$$

Figure 13B.4 shows the variation of the partition function with temperature and Fig. 13B.5 shows how the fractional populations change. Notice how at $T=0$ the fractional populations are $p_0=1$ and $p_1=0$, and the partition function is $q=1$ (one state occupied). However, the fractional populations tend towards equality ($p_0=\frac{1}{2}, p_1=\frac{1}{2}$) and $q=2$ (two states occupied) as $T\to\infty$ ($\beta\to 0$).

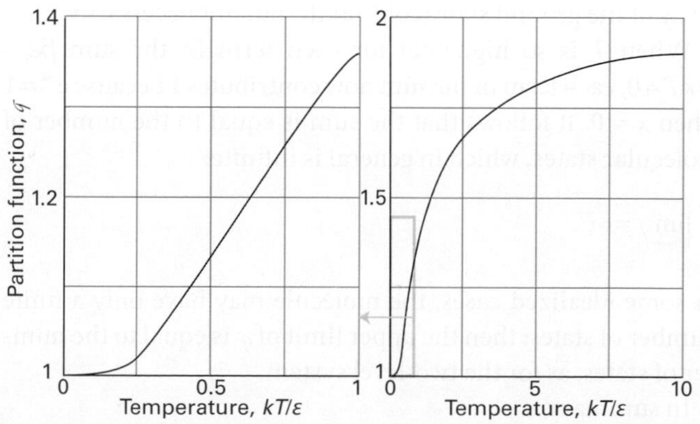

Figure 13B.4 The dependence of the partition function of a two-level system on temperature. The two graphs differ in the scale of the temperature axis to show the approach to 1 as $T\to 0$ and the slow approach to 2 as $T\to\infty$.

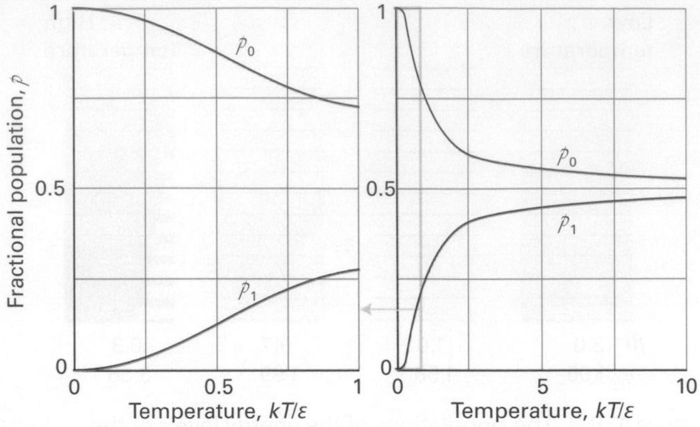

Figure 13B.5 The variation with temperature of the fractional populations of the two states of a two-level system (eqn 13B.4). Note that as the temperature approaches infinity, the populations of the two states become equal (and the fractional populations both approach 0.5).

A note on good practice A common error is to suppose that when $T = \infty$ all the molecules in the system will be found in the upper energy state. However, as seen from eqn 13B.4, as $T \to \infty$ the populations of states become equal. The same conclusion is also true of multi-level systems: as $T \to \infty$, all states become equally populated.

Now consider the general case of a system with an infinite number of energy levels as T approaches zero and therefore as the parameter $\beta = 1/kT$ approaches infinity. Then every term except one in the sum defining q in eqn 13B.1a is zero because each one has the form e^{-x} with $x \to \infty$. The exception is the term with $\varepsilon_0 \equiv 0$ (or the g_0 states at zero energy if this level is g_0-fold degenerate), because then $\varepsilon_0/kT = 0$ whatever the temperature, including zero. As there is only one surviving term when $T = 0$, and its value is g_0, it follows that

$$\lim_{T\to 0} q = g_0$$

That is, at $T = 0$, the partition function is equal to the degeneracy of the ground state (commonly, but not necessarily, 1).

When T is so high that for each term in the sum $\beta\varepsilon_i = \varepsilon_i/kT \approx 0$, each term in the sum now contributes 1 because $e^{-x} = 1$ when $x = 0$. It follows that the sum is equal to the number of molecular states, which in general is infinite:

$$\lim_{T\to\infty} q = \infty$$

In some idealized cases, the molecule may have only a finite number of states; then the upper limit of q is equal to the number of states, as for the two-level system.

In summary,

The molecular partition function gives an indication of the number of states that are thermally accessible to a molecule at the temperature of the system.

13B.2 Contributions to the partition function

The energy of an isolated molecule is the sum of contributions from its different modes of motion:

$$\varepsilon_i = \varepsilon_i^{\mathrm{T}} + \varepsilon_i^{\mathrm{R}} + \varepsilon_i^{\mathrm{V}} + \varepsilon_i^{\mathrm{E}} \qquad (13\mathrm{B}.5)$$

where T denotes translation, R rotation, V vibration, and E the electronic contribution. The possibility that the molecules are interacting with each other is ignored throughout this Topic, because it adds considerable complexity: that is, the molecules are treated as 'independent'. The electronic contribution is not actually a 'mode of motion', but it is convenient to include it here. The separation of terms in eqn 13B.5 is only approximate (except for translation) because the modes are not completely independent, but in most cases it is satisfactory.

Given that the energy is a sum of independent contributions, the partition function factorizes into a product of contributions:

$$\begin{aligned} q &= \sum_i e^{-\beta\varepsilon_i} = \sum_{i(\text{all states})} e^{-\beta\varepsilon_i^{\mathrm{T}}-\beta\varepsilon_i^{\mathrm{R}}-\beta\varepsilon_i^{\mathrm{V}}-\beta\varepsilon_i^{\mathrm{E}}} \\ &= \sum_{i(\text{translational})}\ \sum_{i(\text{rotational})}\ \sum_{i(\text{vibrational})}\ \sum_{i(\text{electronic})} e^{-\beta\varepsilon_i^{\mathrm{T}}-\beta\varepsilon_i^{\mathrm{R}}-\beta\varepsilon_i^{\mathrm{V}}-\beta\varepsilon_i^{\mathrm{E}}} \\ &= \left(\sum_{i(\text{translational})} e^{-\beta\varepsilon_i^{\mathrm{T}}}\right)\left(\sum_{i(\text{rotational})} e^{-\beta\varepsilon_i^{\mathrm{R}}}\right)\left(\sum_{i(\text{vibrational})} e^{-\beta\varepsilon_i^{\mathrm{V}}}\right) \\ &\quad \times\left(\sum_{i(\text{electronic})} e^{-\beta\varepsilon_i^{\mathrm{E}}}\right) \end{aligned}$$

That is,

$$q = q^{\mathrm{T}} q^{\mathrm{R}} q^{\mathrm{V}} q^{\mathrm{E}} \qquad \text{Factorization of the partition function} \qquad (13\mathrm{B}.6)$$

This factorization means that each contribution can be investigated separately. In general, exact analytical expressions for partition functions cannot be obtained. However, approximate expressions can often be found and prove to be very important for understanding chemical phenomena; they are derived in the following sections and collected at the end of this Topic.

(a) The translational contribution

The translational partition function for a particle of mass m free to move in a one-dimensional container of length X can be evaluated by making use of the fact that the separation of energy levels is very small and that large numbers of states are accessible at normal temperatures.

How is that done? 13B.1 Deriving an expression for the translational partition function

The starting point of the derivation is eqn 7D.6 ($E_n = n^2h^2/8mL^2$) for the energy levels of a particle in a one-dimensional box. For a molecule of mass m in a container of length X it follows that:

$$E_n = \frac{n^2h^2}{8mX^2}$$

Step 1 *Write an expression for the sum in eqn 13B.1a*

The lowest level ($n = 1$) has energy $h^2/8mX^2$, so the energies relative to that level are

$$\varepsilon_n = (n^2 - 1)\varepsilon \qquad \varepsilon = h^2/8mX^2$$

The sum to evaluate is therefore

$$q_X^{\mathrm{T}} = \sum_{n=1}^{\infty} \mathrm{e}^{-(n^2-1)\beta\varepsilon}$$

Step 2 *Convert the sum to an integral*

The translational energy levels are very close together in a container the size of a typical laboratory vessel. Therefore, the sum can be approximated by an integral:

$$q_X^{\mathrm{T}} = \int_1^{\infty} \mathrm{e}^{-(n^2-1)\beta\varepsilon}\mathrm{d}n \overset{n \gg 1}{\approx} \int_0^{\infty} \mathrm{e}^{-n^2\beta\varepsilon}\mathrm{d}n$$

The extension of the lower limit to $n = 0$ and the replacement of n^2-1 by n^2 introduces negligible error but turns the integral into a standard form.

Step 3 *Evaluate the integral*

Make the substitution $x^2 = n^2\beta\varepsilon$, implying $\mathrm{d}n = \mathrm{d}x/(\beta\varepsilon)^{1/2}$, and therefore that

$$q_X^{\mathrm{T}} = \left(\frac{1}{\beta\varepsilon}\right)^{1/2} \overbrace{\int_0^{\infty} \mathrm{e}^{-x^2}\mathrm{d}x}^{\text{Integral G.1: } \pi^{1/2}/2} = \left(\frac{1}{\beta\varepsilon}\right)^{1/2} \frac{\pi^{1/2}}{2} \overset{\varepsilon = h^2/8mX^2}{=} \left(\frac{2\pi m}{h^2\beta}\right)^{1/2} X$$

With $\beta = 1/kT$ this relation has the form

$$q_X^{\mathrm{T}} = \frac{X}{\Lambda} \qquad \Lambda = \frac{h}{(2\pi mkT)^{1/2}} \qquad \text{Translational partition function} \qquad (13\text{B}.7)$$

The quantity Λ (uppercase lambda) has the dimensions of length and is called the **thermal wavelength** (sometimes the 'thermal de Broglie wavelength') of the molecule. The thermal wavelength decreases with increasing mass and temperature. This expression shows that:

- The partition function for translational motion increases with the length of the box and the mass of the particle, because in each case the separation of the energy levels becomes smaller and more levels become thermally accessible.
- For a given mass and length of the box, the partition function also increases with increasing temperature (decreasing β), because more states become accessible.

Physical interpretation

The total energy of a molecule free to move in three dimensions is the sum of its translational energies in all three directions:

$$\varepsilon_{n_1n_2n_3} = \varepsilon_{n_1}^{(X)} + \varepsilon_{n_2}^{(Y)} + \varepsilon_{n_3}^{(Z)} \qquad (13\text{B}.8)$$

where n_1, n_2, and n_3 are the quantum numbers for motion in the x-, y-, and z-directions, respectively. Therefore, because $\mathrm{e}^{a+b+c} = \mathrm{e}^a\mathrm{e}^b\mathrm{e}^c$, the partition function factorizes as follows:

$$q^{\mathrm{T}} = \sum_{\text{all } n} \mathrm{e}^{-\beta\varepsilon_{n_1}^{(X)} - \beta\varepsilon_{n_2}^{(Y)} - \beta\varepsilon_{n_3}^{(Z)}} = \sum_{\text{all } n} \mathrm{e}^{-\beta\varepsilon_{n_1}^{(X)}} \mathrm{e}^{-\beta\varepsilon_{n_2}^{(Y)}} \mathrm{e}^{-\beta\varepsilon_{n_3}^{(Z)}}$$

$$= \left(\sum_{n_1} \mathrm{e}^{-\beta\varepsilon_{n_1}^{(X)}}\right)\left(\sum_{n_2} \mathrm{e}^{-\beta\varepsilon_{n_2}^{(Y)}}\right)\left(\sum_{n_3} \mathrm{e}^{-\beta\varepsilon_{n_3}^{(Z)}}\right)$$

That is,

$$q^{\mathrm{T}} = q_X^{\mathrm{T}} q_Y^{\mathrm{T}} q_Z^{\mathrm{T}} \qquad (13\text{B}.9)$$

Equation 13B.7 gives the partition function for translational motion in the x-direction. The only change for the other two directions is to replace the length X by the lengths Y or Z. Hence the partition function for motion in three dimensions is

$$q^{\mathrm{T}} = \left(\frac{2\pi m}{h^2\beta}\right)^{3/2} XYZ = \frac{(2\pi mkT)^{3/2}}{h^3} XYZ \qquad (13\text{B}.10\text{a})$$

The product of lengths XYZ is the volume, V, of the container, so

$$q^{\mathrm{T}} = \frac{V}{\Lambda^3} \qquad \text{Translational partition function [three-dimensional]} \qquad (13\text{B}.10\text{b})$$

with Λ defined in eqn 13B.7. As in the one-dimensional case, the partition function increases with the mass of the particle (as $m^{3/2}$) and the volume of the container (as V); for a given mass and volume, the partition function increases with temperature (as $T^{3/2}$). Moreover, $q^{\mathrm{T}} \to \infty$ as $T \to \infty$ because there is no limit to the number of states that become accessible as the temperature is raised. Even at room temperature, $q^{\mathrm{T}} \approx 2 \times 10^{28}$ for an O_2 molecule in a vessel of volume 100 cm^3.

Brief illustration 13B.2

To calculate the translational partition function of an H_2 molecule confined to a $100\,cm^3$ vessel at 25 °C, use $m = 2.016m_u$. Then, from $\Lambda = h/(2\pi mkT)^{1/2}$,

$$\Lambda = \frac{\overbrace{6.626\times10^{-34}\,\text{J s}}^{\text{kg m}^2\,\text{s}^{-2}}}{\left\{2\pi\times(2.016\times1.6605\times10^{-27}\,\text{kg})\times\left(1.381\times10^{-23}\,\underbrace{\text{J}}_{\text{kg m}^2\,\text{s}^{-2}}\,\text{K}^{-1}\right)\times(298\,\text{K})\right\}^{1/2}}$$

$$= 7.12\ldots\times10^{-11}\,\text{m}$$

Therefore,

$$q^{\text{T}} = \frac{1.00\times10^{-4}\,\text{m}^3}{(7.12\ldots\times10^{-11}\,\text{m})^3} = 2.77\times10^{26}$$

About 10^{26} quantum states are thermally accessible, even at room temperature and for this light molecule. Many states are occupied if the thermal wavelength (which in this case is 71.2 pm) is small compared with the linear dimensions of the container.

Equation 13B.10b can be interpreted in terms of the average separation, d, of the particles in the container. Because q is the total number of accessible states, the average number of translational states per molecule is q^{T}/N. For this quantity to be large, the condition $V/N\Lambda^3 \gg 1$ must be met. However, V/N is the volume occupied by a single particle, and therefore the average separation of the particles is $d = (V/N)^{1/3}$. The condition for there being many states available per molecule is therefore $d^3/\Lambda^3 \gg 1$, and therefore $d \gg \Lambda$. That is, for eqn 13B.10b to be valid, *the average separation of the particles must be much greater than their thermal wavelength*. For 1 mol H_2 molecules at 1 bar and 298 K, the average separation is 3 nm, which is significantly larger than their thermal wavelength (71.2 pm).

The validity of eqn 13B.10b can be expressed in a different way by noting that the approximations that led to it are valid if many states are occupied, which requires V/Λ^3 to be large. That will be so if Λ is small compared with the linear dimensions of the container. For H_2 at 298 K, $\Lambda = 71$ pm, which is far smaller than any conventional container is likely to be (but comparable to pores in zeolites or cavities in clathrates). For O_2, a heavier molecule, $\Lambda = 18$ pm.

(b) The rotational contribution

The energy levels of a linear rotor are $\varepsilon_J = hc\tilde{B}J(J+1)$, with $J = 0, 1, 2, \ldots$ (Topic 11B). The state of lowest energy has zero energy, so no adjustment need be made to the energies given by this expression. Each level consists of $2J + 1$ degenerate states. Therefore, the partition function of a non-symmetrical (AB) linear rotor is

$$q^{\text{R}} = \sum_J \overbrace{(2J+1)}^{g_J}\mathrm{e}^{-\beta\overbrace{hc\tilde{B}J(J+1)}^{\varepsilon_J}} \qquad (13\text{B}.11)$$

The direct method of calculating q^{R} is to substitute the experimental values of the rotational energy levels into this expression and to sum the series numerically. (The case of symmetrical A_2 molecules is dealt with later.)

Example 13B.1 Evaluating the rotational partition function explicitly

Evaluate the rotational partition function of $^1H^{35}Cl$ at 25 °C, given that $\tilde{B} = 10.591\,cm^{-1}$.

Collect your thoughts You need to evaluate eqn 13B.11 term by term, using $kT/hc = 207.224\,cm^{-1}$ at 298.15 K. The sum is readily evaluated by using mathematical software.

The solution To show how successive terms contribute, draw up the following table by using $hc\tilde{B}/kT = 0.051\,11$ (Fig. 13B.6):

J	0	1	2	3	4	…	10
$(2J+1)\mathrm{e}^{-0.05111J(J+1)}$	1	2.71	3.68	3.79	3.24	…	0.08

The sum required by eqn 13B.11 (the sum of the numbers in the second row of the table) is 19.9, hence $q^{\text{R}} = 19.9$ at this temperature. Taking J up to 50 gives $q^{\text{R}} = 19.903$.

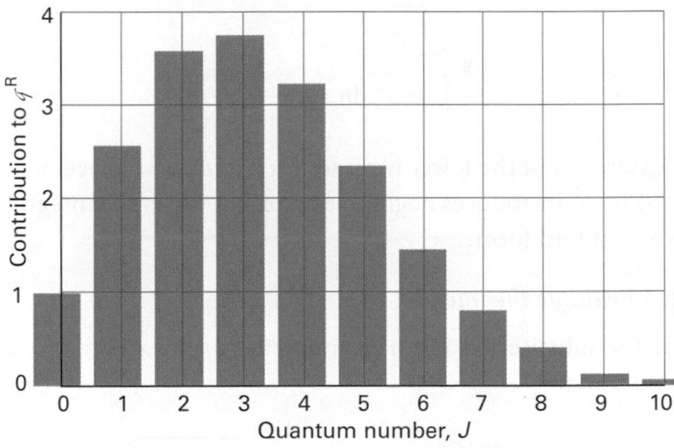

Figure 13B.6 The contributions to the rotational partition function of an HCl molecule at 25 °C. The vertical axis is the value of $(2J+1)\mathrm{e}^{-\beta hc\tilde{B}J(J+1)}$. Successive terms (which are proportional to the populations of the levels) pass through a maximum because the population of individual states decreases exponentially, but the degeneracy of the levels increases with J.

Comment. Notice that about ten J-levels are significantly populated but the number of populated *states* is larger on account of the $(2J + 1)$-fold degeneracy of each level.

Self-test 13B.1 Evaluate the rotational partition function for $^1H^{35}Cl$ at 0 °C.

Answer: 18.26

At room temperature, $kT/hc \approx 200\,\text{cm}^{-1}$. The rotational constants of many molecules are close to $1\,\text{cm}^{-1}$ (Table 11C.1) and often smaller (though the very light H_2 molecule, for which $\tilde{B} = 60.9\,\text{cm}^{-1}$, is one important exception). It follows that many rotational levels are populated at normal temperatures. When that is the case, explicit expressions for the rotational partition function can be derived.

Consider a linear rotor. If many rotational states are occupied and kT is much larger than the separation between neighbouring states, the sum that defines the partition function can be approximated by an integral:

$$q^{\text{R}} = \int_0^{\infty} (2J+1)\text{e}^{-\beta hc\tilde{B}J(J+1)}\text{d}J$$

This integral can be evaluated without much effort by making the substitution $x = \beta hc\tilde{B}J(J+1)$, so that $\text{d}x/\text{d}J = \beta hc\tilde{B}(2J+1)$ and therefore $(2J+1)\text{d}J = \text{d}x/\beta hc\tilde{B}$. Then

$$q^{\text{R}} = \frac{1}{\beta hc\tilde{B}} \overbrace{\int_0^{\infty} \text{e}^{-x}\text{d}x}^{\text{Integral E.1: } 1} = \frac{1}{\beta hc\tilde{B}}$$

which (because $\beta = 1/kT$) is

$$q^{\text{R}} = \frac{kT}{hc\tilde{B}} \qquad \text{Rotational partition function [unsymmetrical linear molecule]} \qquad (13\text{B}.12\text{a})$$

A similar but more elaborate approach can be used for nonlinear molecules.

How is that done? 13B.2 Deriving an expression for the rotational partition function of a nonlinear molecule

Consider a symmetric rotor (Topic 11B) for which the energy levels are

$$E_{J,K,M_J} = hc\tilde{B}J(J+1) + hc(\tilde{A} - \tilde{B})K^2$$

with $J = 0, 1, 2,\ldots$, $K = J, J-1,\ldots,-J$, and $M_J = J, J-1,\ldots, -J$. Instead of considering these ranges, the same values can be covered by allowing K to range from $-\infty$ to ∞, with J confined to $|K|, |K|+1,\ldots, \infty$ for each value of K (Fig. 13B.7).

Step 1 *Write an expression for the sum over energy states*

Because the energy is independent of M_J, and there are $2J+1$ values of M_J for each value of J, each value of J is $(2J+1)$-fold degenerate. It follows that the partition function

$$q = \sum_{J=0}^{\infty}\sum_{K=-J}^{J}\sum_{M_J=-J}^{J} \text{e}^{-\beta E_{J,K,M_J}}$$

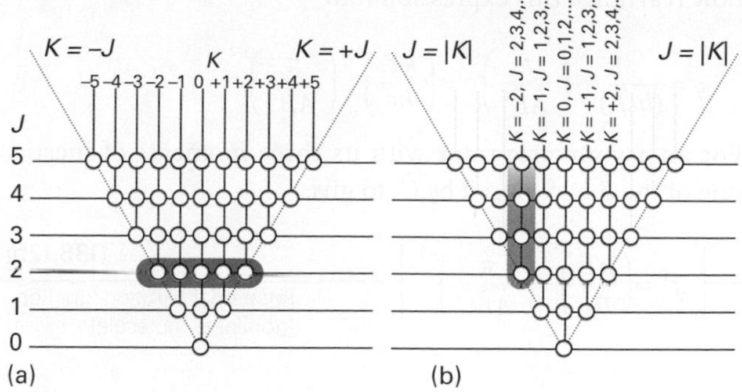

Figure 13B.7 The calculation of the rotational partition function includes a contribution, indicated by the circles, for all possible combinations of J and K. The sum is formed either (a) by allowing J to take the values 0, 1, 2,… and then for each J allowing K to range from J to $-J$, or (b) by allowing K to range from $-\infty$ to ∞, and for each value of K allowing J to take the values $|K|, |K|+1,\ldots, \infty$. The tinted areas show in (a) the sum from $K = -2$ to $+2$ for $J = 2$, and in (b) the sum with $J = 2, 3\ldots$ for $K = -2$.

can be written equivalently as

$$q = \sum_{J=0}^{\infty}\sum_{K=-J}^{J}(2J+1)\text{e}^{-\beta E_{J,K,M_J}} = \sum_{K=-\infty}^{\infty}\sum_{J=|K|}^{\infty}(2J+1)\text{e}^{-\beta E_{J,K,M_J}}$$

$$= \sum_{K=-\infty}^{\infty}\text{e}^{-hc\beta(\tilde{A}-\tilde{B})K^2}\sum_{J=|K|}^{\infty}(2J+1)\text{e}^{-hc\beta\tilde{B}J(J+1)}$$

Step 2 *Convert the sums to integrals*

As for linear molecules, assume that the temperature is so high that numerous states are occupied, in which case the sums may be approximated by integrals. Then

$$q = \int_{-\infty}^{\infty}\text{e}^{-hc\beta(\tilde{A}-\tilde{B})K^2}\int_{|K|}^{\infty}(2J+1)\text{e}^{-hc\beta\tilde{B}J(J+1)}\text{d}J\text{d}K$$

Step 3 *Evaluate the integrals*

You should recognize the integral over J as the integral of the derivative of a function, which is the function itself, so

$$\int_{|K|}^{\infty}(2J+1)\text{e}^{-hc\beta\tilde{B}J(J+1)}\text{d}J = \frac{1}{hc\beta\tilde{B}}\text{e}^{-hc\beta\tilde{B}|K|(|K|+1)} \overset{|K|\gg 1 \text{ for most allowed values}}{\approx} \frac{1}{hc\beta\tilde{B}}\text{e}^{-hc\beta\tilde{B}K^2}$$

Use this result in the integral over J in Step 2:

$$q = \frac{1}{hc\beta\tilde{B}}\int_{-\infty}^{\infty}\text{e}^{-hc\beta(\tilde{A}-\tilde{B})K^2}\text{e}^{-hc\beta\tilde{B}K^2}\text{d}K = \frac{1}{hc\beta\tilde{B}}\overbrace{\int_{-\infty}^{\infty}\text{e}^{-hc\beta\tilde{A}K^2}\text{d}K}^{\text{Integral G.1}}$$

$$= \frac{1}{hc\beta\tilde{B}}\left(\frac{\pi}{hc\beta\tilde{A}}\right)^{1/2}$$

now rearrange this expression into

$$q=\frac{1}{(hc\beta)^{3/2}}\left(\frac{\pi}{\tilde{A}\tilde{B}^2}\right)^{1/2}=\left(\frac{kT}{hc}\right)^{3/2}\left(\frac{\pi}{\tilde{A}\tilde{B}^2}\right)^{1/2}$$

For an asymmetric rotor with its three moments of inertia, one of the $\tilde{B}$ is replaced by $\tilde{C}$, to give

$$q^{\mathrm{R}}=\left(\frac{kT}{hc}\right)^{3/2}\left(\frac{\pi}{\tilde{A}\tilde{B}\tilde{C}}\right)^{1/2} \quad \text{Rotational partition function [nonlinear molecule]} \quad (13\text{B}.12\text{b})$$

Brief illustration 13B.3

For $^1\mathrm{H}^{35}\mathrm{Cl}$ at 298.15 K, use $kT/hc=207.224\,\mathrm{cm}^{-1}$ and $\tilde{B}=10.591\,\mathrm{cm}^{-1}$. Then

$$q^{\mathrm{R}}=\frac{kT}{hc\tilde{B}}=\frac{207.224\,\mathrm{cm}^{-1}}{10.591\,\mathrm{cm}^{-1}}=19.57$$

The value is in good agreement with the exact value (19.903) and obtained with much less effort.

A useful way of expressing the temperature above which eqns 13B.12a and 13B.12b are valid is to introduce the **characteristic rotational temperature**, $\theta^{\mathrm{R}}=hc\tilde{B}/k$. Then 'high temperature' means $T>>\theta^{\mathrm{R}}$ and under these conditions the rotational partition function of a linear molecule is simply T/θ^{R}. Some typical values of θ^{R} are given in Table 13B.1. The value for $^1\mathrm{H}_2$ (87.6 K) is abnormally high, so the approximation must be used carefully for this molecule. However, before using eqn 13B.12a for symmetrical molecules, such as H_2, read on (to eqn 13B.13a).

The general conclusion at this stage is that

> Molecules with large moments of inertia (and hence small rotational constants and low characteristic rotational temperatures) have large rotational partition functions.

A large value of q^{R} reflects the closeness in energy (compared with kT) of the rotational levels in large, heavy molecules, and the large number of rotational states that are accessible at normal temperatures.

It is important not to include too many rotational states in the sum that defines the partition function. For a homonuclear diatomic molecule or a symmetrical linear molecule

Table 13B.1 Rotational temperatures of diatomic molecules*

	θ^{R}/K
$^1\mathrm{H}_2$	87.6
$^1\mathrm{H}^{35}\mathrm{Cl}$	15.2
$^{14}\mathrm{N}_2$	2.88
$^{35}\mathrm{Cl}_2$	0.351

* More values are given in the *Resource section*, Table 11C.1.

(such as CO_2 or HC≡CH), a rotation through 180° results in an indistinguishable state of the molecule. Hence, the number of thermally accessible states is only half the number that can be occupied by a heteronuclear diatomic molecule, where rotation through 180° does result in a distinguishable state. Therefore, for a symmetrical linear molecule,

$$q^{\mathrm{R}}=\frac{kT}{2hc\tilde{B}}=\frac{T}{2\theta^{\mathrm{R}}} \quad \text{Rotational partition function [symmetrical linear rotor]} \quad (13\text{B}.13\text{a})$$

The equations for symmetrical and non-symmetrical molecules can be combined into a single expression by introducing the **symmetry number**, σ, which is the number of indistinguishable orientations of the molecule. Then

$$q^{\mathrm{R}}=\frac{T}{\sigma\theta^{\mathrm{R}}} \quad \text{Rotational partition function [linear rotor]} \quad (13\text{B}.13\text{b})$$

For a heteronuclear diatomic molecule $\sigma=1$; for a homonuclear diatomic molecule or a symmetrical linear molecule, $\sigma=2$. The formal justification of this rule depends on assessing the role of the Pauli principle.

How is that done? 13B.3 Identifying the origin of the symmetry number

The Pauli principle forbids the occupation of certain states. It is shown in Topic 11B, for example, that $^1\mathrm{H}_2$ may occupy rotational states with even J only if its nuclear spins are paired (*para*-hydrogen), and odd J states only if its nuclear spins are parallel (*ortho*-hydrogen). In *ortho*-H_2 there are three nuclear spin states for each value of J (because there are three 'parallel' spin states of the two nuclei); in *para*-H_2 there is just one nuclear spin state for each value of J.

To set up the rotational partition function and take into account the Pauli principle, note that 'ordinary' molecular hydrogen is a mixture of one part *para*-H_2 (with only its even-J rotational states occupied) and three parts *ortho*-H_2 (with only its odd-J rotational states occupied). Therefore, the average partition function for each molecule is

$$q^{\mathrm{R}}=\tfrac{1}{4}\sum_{\text{even }J}(2J+1)\mathrm{e}^{-\beta hc\tilde{B}J(J+1)}+\tfrac{3}{4}\sum_{\text{odd }J}(2J+1)\mathrm{e}^{-\beta hc\tilde{B}J(J+1)}$$

The odd-J states are three times more heavily weighted than the even-J states (Fig. 13B.8). The illustration shows that approximately the same answer would be obtained for the partition function (the sum of all the populations) if each J term contributed half its normal value to the sum. That is, the last equation can be approximated as

$$q^{\mathrm{R}}=\tfrac{1}{2}\sum_{J}(2J+1)\mathrm{e}^{-\beta hc\tilde{B}J(J+1)}$$

and this approximation is very good when many terms contribute (at high temperatures, $T>>\theta^{\mathrm{R}}$). At such high temperatures the sum can be approximated by the integral that led to

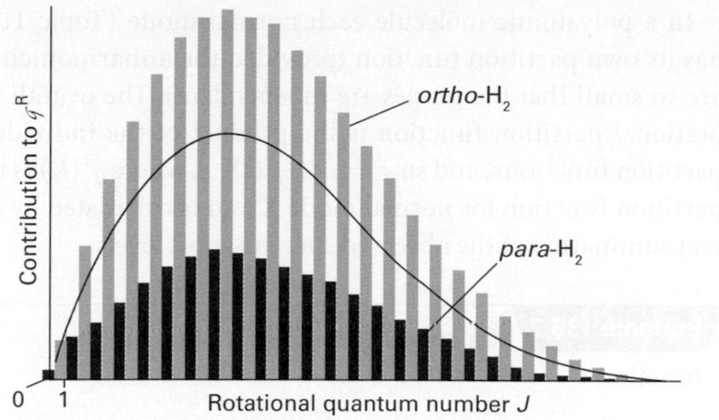

Figure 13B.8 The values of the individual terms $(2J+1)e^{-\beta hc\tilde{B}J(J+1)}$ contributing to the mean partition function of a 3:1 mixture of *ortho*- and *para*-H_2, but with a much smaller rotational constant than in reality (so as to illustrate a procedure that is in fact valid only for heavier homonuclear diatomic molecules). The partition function is the sum of all these terms. At high temperatures, the sum is approximately equal to the sum of the terms over all values of J, each with a weight of $\frac{1}{2}$. This sum is indicated by the curve.

eqn 13B.12a. Therefore, on account of the factor of $\frac{1}{2}$ in this expression, the rotational partition function for 'ordinary' molecular hydrogen at high temperatures is one-half this value, as in eqn 13B.13a.

The same type of argument may be used for linear symmetrical molecules in which identical bosons are interchanged by rotation (such as CO_2). As pointed out in Topic 11B, if the nuclear spin of the bosons is 0, then only even-J states are admissible. Because only half the rotational states are occupied, the rotational partition function is only half the value of the sum obtained by allowing all values of J to contribute (Fig. 13B.9).

The same care must be exercised for other types of symmetrical molecules, and for a nonlinear molecule eqn 13B.12b is replaced by

$$q^R = \frac{1}{\sigma}\left(\frac{kT}{hc}\right)^{3/2}\left(\frac{\pi}{\tilde{A}\tilde{B}\tilde{C}}\right)^{1/2} \quad \text{Rotational partition function [nonlinear rotor]} \quad (13B.14)$$

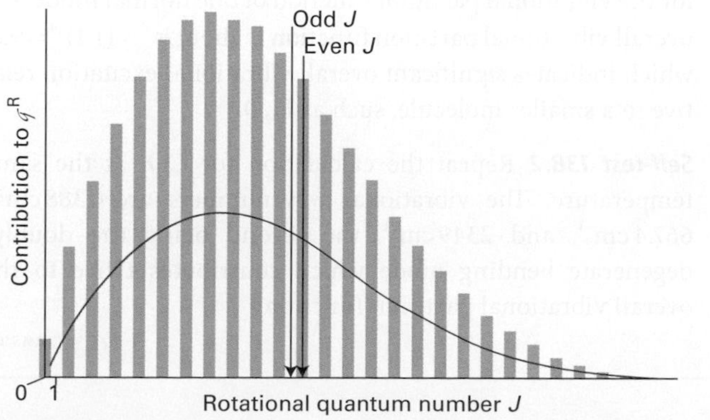

Figure 13B.9 The values of the individual terms contributing to the rotational partition function of CO_2. Only states with even J values are allowed. The full line shows the smoothed, averaged contributions of the levels.

Table 13B.2 Symmetry numbers of molecules*

	σ
$^1H_2$	2
$^1H^2H$	1
NH_3	3
C_6H_6	12

* More values are given in the *Resource section*, Table 11C.1.

Some typical values of the symmetry numbers are given in Table 13B.2. To see how group theory is used to identify the value of the symmetry number, see *Integrated activity* I13.1; the following *Brief illustration* outlines the approach.

Brief illustration 13B.4

The value $\sigma(H_2O) = 2$ reflects the fact that a 180° rotation about the bisector of the H–O–H angle interchanges two indistinguishable atoms. In NH_3, there are three indistinguishable orientations around the axis shown in (**1**). For CH_4, any of three 120° rotations about any of its four C–H bonds leaves the molecule in an indistinguishable state (**2**), so the symmetry number is $3 \times 4 = 12$.

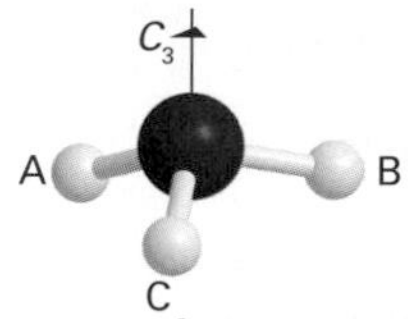

1

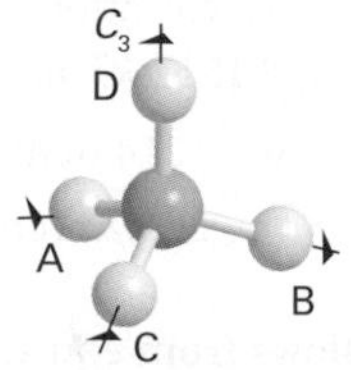

2

For benzene, any of six orientations around the axis perpendicular to the plane of the molecule leaves it apparently unchanged (Fig. 13B.10), as does a rotation of 180° around any of six axes in the plane of the molecule (three of which pass through C atoms diametrically opposite across the ring and the remaining three pass through the mid-points of C–C bonds on opposite sides of the ring).

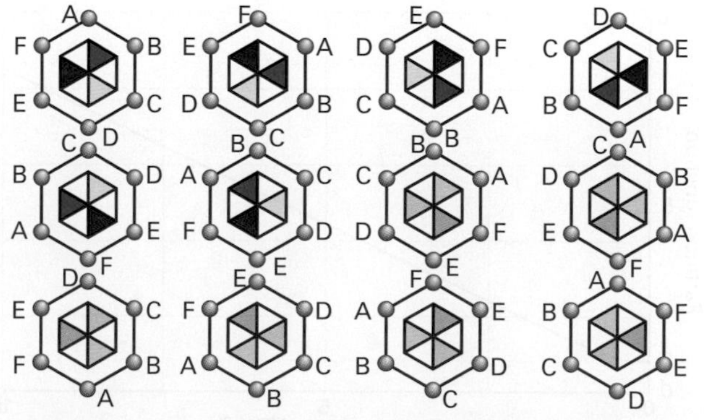

Figure 13B.10 The 12 equivalent orientations of a benzene molecule that can be reached by pure rotations and give rise to a symmetry number of 12. The six pale colours are the underside of the hexagon after that face has been rotated into view.

(c) The vibrational contribution

The vibrational partition function of a molecule is calculated by substituting the measured vibrational energy levels into the definition of q^V, and summing them numerically. However, provided it is permissible to assume that the vibrations are harmonic, there is a much simpler way. In that case, the vibrational energy levels form a uniform ladder of separation $hc\tilde{\nu}$ (Topics 7E and 11C), which is exactly the problem treated in *Brief illustration* 13B.1 and lead to eqn 13B.2a. Therefore that result can be used by setting $\varepsilon = hc\tilde{\nu}$, giving

$$q^V = \frac{1}{1-e^{-\beta hc\tilde{\nu}}} \quad \text{Vibrational partition function [harmonic approximation]} \quad (13B.15)$$

This function is plotted in Fig. 13B.11 (which is essentially the same as Fig. 13B.1). Similarly, the population of each state is given by eqn 13B.2b.

Brief illustration 13B.5

To calculate the partition function of I_2 molecules at 298.15 K note that their vibrational wavenumber is 214.6 cm^{-1}. Then, because at 298.15 K, $kT/hc = 207.224$ cm^{-1},

$$\beta hc\tilde{\nu} = \frac{hc\tilde{\nu}}{kT} = \frac{214.6\,\text{cm}^{-1}}{207.224\,\text{cm}^{-1}} = 1.035\ldots$$

Then it follows from eqn 13B.15 that

$$q^V = \frac{1}{1-e^{-1.035\ldots}} = 1.55$$

From this value it can be inferred that only the ground and first excited states are significantly populated.

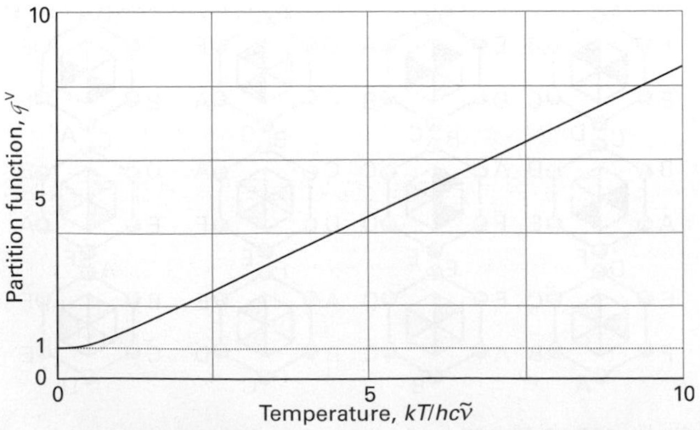

Figure 13B.11 The vibrational partition function of a molecule in the harmonic approximation. Note that the partition function is proportional to the temperature when the temperature is high ($T >> \theta^V$).

In a polyatomic molecule each normal mode (Topic 11D) has its own partition function (provided the anharmonicities are so small that the modes are independent). The overall vibrational partition function is the product of the individual partition functions, and so $q^V = q^V(1)q^V(2)\ldots$, where $q^V(K)$ is the partition function for normal mode K and is calculated by direct summation of the observed spectroscopic levels.

Example 13B.2 Calculating a vibrational partition function

The wavenumbers of the three normal modes of H_2O are 3656.7 cm^{-1}, 1594.8 cm^{-1}, and 3755.8 cm^{-1}. Evaluate the vibrational partition function at 1500 K.

Collect your thoughts You need to use eqn 13B.15 for each mode, and then form the product of the three contributions. At 1500 K, $kT/hc = 1042.6$ cm^{-1}.

The solution Draw up the following table displaying the contributions of each mode:

Mode	1	2	3
$\tilde{\nu}$/cm^{-1}	3656.7	1594.8	3755.8
$hc\tilde{\nu}/kT$	3.507	1.530	3.602
q^V	1.031	1.276	1.028

The overall vibrational partition function is therefore

$$q^V = 1.031 \times 1.276 \times 1.028 = 1.352$$

The three normal modes of H_2O are at such high wavenumbers that even at 1500 K most of the molecules are in their vibrational ground state.

Comment. There may be so many normal modes in a large molecule that their overall contribution may be significant even though each mode is not appreciably excited. For example, a nonlinear molecule containing 10 atoms has $3N-6=24$ normal modes (Topic 11D). If a value of about 1.1 is assumed for the vibrational partition function of one normal mode, the overall vibrational partition function is about $q^V \approx (1.1)^{24} = 9.8$, which indicates significant overall vibrational excitation relative to a smaller molecule, such as H_2O.

Self-test 13B.2 Repeat the calculation for CO_2 at the same temperature. The vibrational wavenumbers are 1388 cm^{-1}, 667.4 cm^{-1}, and 2349 cm^{-1}, the second being the doubly-degenerate bending mode which contributes twice to the overall vibrational partition function.

Answer: 6.79

In many molecules the vibrational wavenumbers are so great that $\beta hc\tilde{\nu} > 1$. For example, the lowest vibrational wavenumber of CH_4 is 1306 cm^{-1}, so $\beta hc\tilde{\nu} = 6.3$ at room temperature. Most C–H stretches normally lie in the range 2850–2960 cm^{-1},

Table 13B.3 Vibrational temperatures of diatomic molecules*

	θ^V/K
$^1H_2$	6332
$^1H^{35}Cl$	4304
$^{14}N_2$	3393
$^{35}Cl_2$	805

* More values are given in the *Resource section*, Table 11C.1

so for them $\beta hc\tilde{\nu} \approx 14$. In these cases, $e^{-\beta hc\tilde{\nu}}$ in the denominator of q^V is very close to zero (e.g. $e^{-6.3} = 0.002$), and the vibrational partition function for a single mode is very close to 1 ($q^V = 1.002$ when $\beta hc\tilde{\nu} = 6.3$), implying that only the lowest level is significantly occupied.

Now consider the case of modes with such low vibrational frequencies that $\beta hc\tilde{\nu} << 1$. When this condition is satisfied, the partition function may be approximated by expanding the exponential ($e^x = 1 + x + \cdots$):

$$q^V = \frac{1}{1-e^{-\beta hc\tilde{\nu}}} = \frac{1}{1-(1-\beta hc\tilde{\nu}+\cdots)}$$

That is, for low-frequency modes at high temperatures,

$$q^V \approx \frac{kT}{hc\tilde{\nu}} \qquad \text{Vibrational partition function [high-temperature approximation]} \qquad (13B.16)$$

The temperatures for which eqn 13B.16 is valid can be expressed in terms of the **characteristic vibrational temperature,** $\theta^V = hc\tilde{\nu}/k$ (Table 13B.3). The value for H_2 (6332 K) is abnormally high because the atoms are so light and the vibrational frequency is correspondingly high. In terms of the vibrational temperature, 'high temperature' means $T >> \theta^V$, and when this condition is satisfied eqn 13B.16 is valid and can be written $q^V = T/\theta^V$ (the analogue of the rotational expression).

(d) The electronic contribution

Electronic energy separations from the ground state are usually very large, so for most cases $q^E = 1$ because only the ground state is occupied. An important exception arises in the case of atoms and molecules having electronically degenerate ground states, in which case $q^E = g^E$, where g^E is the degeneracy of the electronic ground state. Alkali metal atoms, for example, have doubly degenerate ground states (corresponding to the two orientations of their electron spin), so $q^E = 2$.

Brief illustration 13B.6

Some atoms and molecules have degenerate ground states and low-lying electronically excited degenerate states. An example is NO, which has a configuration of the form $\ldots\pi^1$ (Topic 11F). The energy of the two degenerate states in which the orbital and spin momenta are parallel (giving the $^2\Pi_{3/2}$ term, Fig. 13B.12) is slightly greater than that of the two degenerate states in which they are antiparallel (giving the $^2\Pi_{1/2}$ term). The separation, which arises from spin–orbit coupling, is only 121 cm^{-1}.

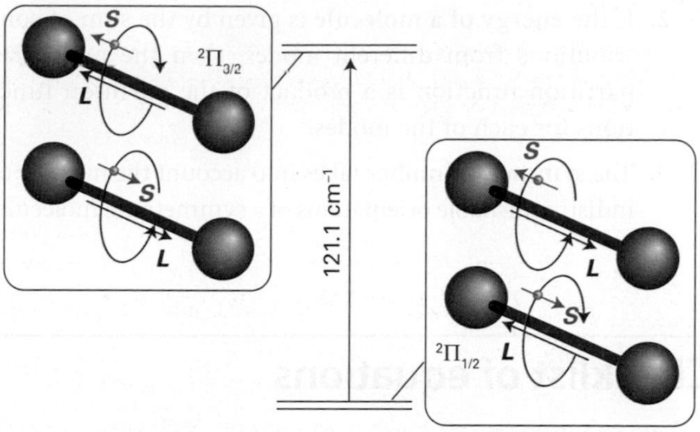

Figure 13B.12 The doubly-degenerate ground electronic level of NO (with the spin and orbital angular momentum around the axis in opposite directions) and the doubly-degenerate first excited level (with the spin and orbital momenta parallel). The upper level is thermally accessible at room temperature.

If the energies of the two levels are denoted as $E_{1/2} = 0$ and $E_{3/2} = \varepsilon$, then the partition function is

$$q^E = \sum_{\text{levels } i} g_i e^{-\beta\varepsilon_i} = 2 + 2e^{-\beta\varepsilon}$$

This function is plotted in Fig. 13B.13. At $T = 0$, $q^E = 2$, because only the doubly degenerate ground state is accessible. At high temperatures, q^E approaches 4 because all four states are accessible. At 25 °C, $q^E = 3.1$.

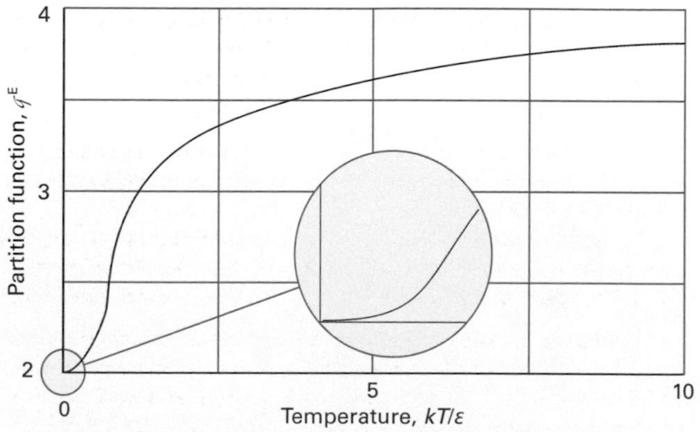

Figure 13B.13 The variation with temperature of the electronic partition function of an NO molecule. Note that the curve resembles that for a two-level system (Fig.13B.4), but rises from 2 (the degeneracy of the lower level) and approaches 4 (the total number of states) at high temperatures.

Checklist of concepts

☐ 1. The **molecular partition function** is an indication of the number of thermally accessible states at the temperature of interest.

☐ 2. If the **energy of a molecule** is given by the sum of contributions from different modes, then the molecular partition function is a product of the partition functions for each of the modes.

☐ 3. The **symmetry number** takes into account the number of indistinguishable orientations of a symmetrical molecule.

☐ 4. The **vibrational partition function** of a molecule is found by evaluating the contribution from each normal mode treated as a harmonic oscillator.

☐ 5. Because electronic energy separations from the ground state are usually very large, in most cases the **electronic partition function** is equal to the degeneracy of the electronic ground state.

Checklist of equations

Property	Equation	Comment	Equation number
Molecular partition function	$q=\sum_{\text{states } i} e^{-\beta\varepsilon_i}$	Definition, independent molecules	13B.1a
	$q=\sum_{\text{levels } i} g_i e^{-\beta\varepsilon_i}$	Definition, independent molecules	13B.1b
Uniform ladder	$q=1/(1-e^{-\beta\varepsilon})$		13B.2a
Two-level system	$q=1+e^{-\beta\varepsilon}$		13B.3a
Thermal wavelength	$\Lambda=h/(2\pi mkT)^{1/2}$		13B.7
Translation	$q^{\mathrm{T}}=V/\Lambda^3$		13B.10b
Rotation	$q^{\mathrm{R}}=kT/\sigma hc\tilde{B}$	$T \gg \theta^{\mathrm{R}}$, linear rotor	13B.13a
	$q^{\mathrm{R}}=(1/\sigma)(kT/hc)^{3/2}(\pi/\tilde{A}\tilde{B}\tilde{C})^{1/2}$	$T \gg \theta^{\mathrm{R}}$, nonlinear rotor, $\theta^{\mathrm{R}}=hc\tilde{X}/k$, $\tilde{X}=\tilde{A}$, $\tilde{B}$, or $\tilde{C}$	13B.14
Vibration	$q^{\mathrm{V}}=1/(1-e^{-\beta hc\tilde{\nu}})$	Harmonic approximation, $\theta^{\mathrm{V}}=hc\tilde{\nu}/k$	13B.15

TOPIC 13C Molecular energies

> ➤ **Why do you need to know this material?**
>
> For statistical thermodynamics to be useful, you need to know how to extract thermodynamic information from a partition function.
>
> ➤ **What is the key idea?**
>
> The average energy of a molecule in a collection of independent molecules can be calculated from the molecular partition function alone.
>
> ➤ **What do you need to know already?**
>
> You need to know how to calculate a molecular partition function from calculated or spectroscopic data (Topic 13B) and its significance as a measure of the number of accessible states. The Topic also draws on expressions for the rotational and vibrational energies of molecules (Topics 11B–11D).

A partition function in statistical thermodynamics is like a wavefunction in quantum mechanics. A wavefunction contains all the *dynamical* information about a system; a partition function contains all the *thermodynamic* information about a system. As in quantum mechanics, it is important to know how to extract that information. One of the simplest thermodynamic properties is the mean energy, and the equations are simplest for a system composed of non-interacting molecules.

13C.1 The basic equations

Consider a collection of N molecules that do not interact with one another. Any member of the collection can exist in a state i of energy ε_i measured from the lowest energy state of the molecule. The mean energy of a molecule, $\langle\varepsilon\rangle$, relative to its energy in its ground state, is the total energy of the collection, E, divided by the total number of molecules:

$$\langle\varepsilon\rangle=\frac{E}{N}=\frac{1}{N}\sum_i N_i\varepsilon_i \qquad (13\text{C}.1)$$

where N_i is the population of state i. In Topic 13A it is shown that the overwhelmingly most probable population of a state in a collection at a temperature T is given by the Boltzmann distribution, eqn 13A.10b ($N_i/N=(1/q)e^{-\beta\varepsilon_i}$), so

$$\langle\varepsilon\rangle=\frac{1}{q}\sum_i \varepsilon_i e^{-\beta\varepsilon_i} \qquad (13\text{C}.2)$$

with $\beta=1/kT$. This expression can be manipulated into a form involving only q. First note that

$$\varepsilon_i e^{-\beta\varepsilon_i}=-\frac{d}{d\beta}e^{-\beta\varepsilon_i}$$

It follows that

$$\langle\varepsilon\rangle=-\frac{1}{q}\sum_i\frac{d}{d\beta}e^{-\beta\varepsilon_i}=-\frac{1}{q}\frac{d}{d\beta}\overbrace{\sum_i e^{-\beta\varepsilon_i}}^{q}=-\frac{1}{q}\frac{dq}{d\beta} \qquad (13\text{C}.3)$$

Several points need to be made in relation to eqn 13C.3. Because $\varepsilon_0\equiv 0$, (all energies are measured from the lowest available level), $\langle\varepsilon\rangle$ is the value of the energy relative to the actual ground-state energy. If the lowest energy of the molecule is in fact ε_{gs} rather than 0, then the true mean energy is $\varepsilon_{gs}+\langle\varepsilon\rangle$. For instance, for a harmonic oscillator, ε_{gs} is the zero-point energy, $\frac{1}{2}hc\tilde{\nu}$. Secondly, because the partition function might depend on variables other than the temperature (e.g. the volume), the derivative with respect to β in eqn 13C.3 is actually a *partial* derivative with these other variables held constant. The complete expression relating the molecular partition function to the mean energy of a molecule is therefore

$$\langle\varepsilon\rangle=\varepsilon_{gs}-\frac{1}{q}\left(\frac{\partial q}{\partial\beta}\right)_V \qquad \text{Mean molecular energy} \quad (13\text{C}.4\text{a})$$

An equivalent form is obtained by noting that $dx/x=d\ln x$:

$$\langle\varepsilon\rangle=\varepsilon_{gs}-\left(\frac{\partial \ln q}{\partial\beta}\right)_V \qquad \text{Mean molecular energy} \quad (13\text{C}.4\text{b})$$

These two equations confirm that only the partition function (as a function of temperature) is needed in order to calculate the mean energy.

Brief illustration 13C.1

If a molecule has only two available states, one at 0 and the other at an energy ε, its partition function is

$$q = 1 + e^{-\beta\varepsilon}$$

Therefore, the mean energy of a collection of these molecules at a temperature T is

$$\langle\varepsilon\rangle = -\frac{1}{1+e^{-\beta\varepsilon}}\frac{d(1+e^{-\beta\varepsilon})}{d\beta} = \frac{\varepsilon e^{-\beta\varepsilon}}{1+e^{-\beta\varepsilon}} = \frac{\varepsilon}{e^{\beta\varepsilon}+1}$$

This function is plotted in Fig. 13C.1. Notice how the mean energy is zero at $T = 0$, when only the lower state (at the zero of energy) is occupied, and rises to $\frac{1}{2}\varepsilon$ as $T \to \infty$, when the two states become equally populated.

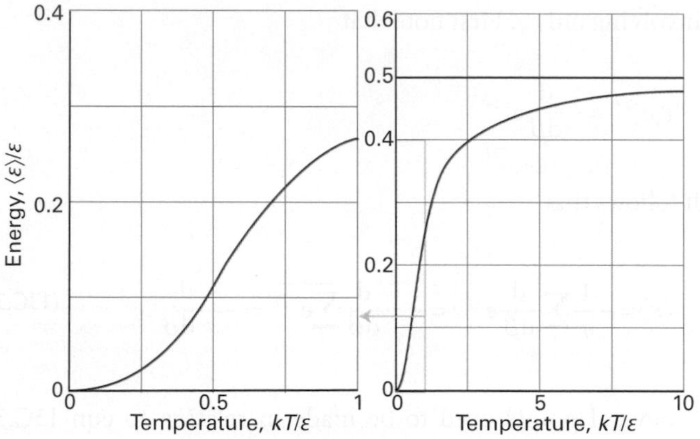

Figure 13C.1 The variation with temperature of the mean energy of a two-level system. The graph on the left shows the slow rise away from zero energy at low temperatures; the slope of the graph at $T = 0$ is 0. The graph on the right shows the slow approach to 0.5 as $T \to \infty$ as both states become equally populated.

13C.2 Contributions of the fundamental modes of motion

The remainder of this Topic explains how to write expressions for the contributions to the energy of three fundamental types of motion, namely translation (T), rotation (R), and vibration (V). It also shows how to incorporate the contribution of the electronic states of molecules (E) and electron spin (S).

(a) The translational contribution

For a one-dimensional container of length X, the partition function is $q_X^{T} = X/\Lambda$ with $\Lambda = h/(2\pi m/\beta)^{1/2}$ (Topic 13B, with 'constant volume V' replaced by 'constant length X'). The partition function can be written as constant $\times\ \beta^{-1/2}$, a form convenient for the calculation of the average energy using eqn 13C.4b:

Λ = constant × $\beta^{-1/2}$, and X a constant

$$\langle\varepsilon_X^{T}\rangle = -\left(\frac{\partial \ln q}{\partial\beta}\right)_X = -\left(\frac{\partial \ln(\text{constant}\times\beta^{-1/2})}{\partial\beta}\right)_X$$

$$= -\left(\frac{\partial(\ln(\text{constant}) + \ln\beta^{-1/2})}{\partial\beta}\right)_X$$

$$= -\left(\frac{\partial(-\frac{1}{2}\ln\beta)}{\partial\beta}\right)_X = \frac{1}{2\beta}$$

That is,

$$\langle\varepsilon_X^{T}\rangle = \tfrac{1}{2}kT \qquad \text{Mean translational energy [one dimension]} \qquad (13C.5a)$$

For a molecule free to move in three dimensions, the analogous calculation leads to

$$\langle\varepsilon^{T}\rangle = \tfrac{3}{2}kT \qquad \text{Mean translational energy [three dimensions]} \qquad (13C.5b)$$

(b) The rotational contribution

The mean rotational energy of a linear molecule is obtained from the rotational partition function (eqn 13B.11):

$$q^{R} = \sum_J (2J+1)e^{-\beta hc\tilde{B}J(J+1)}$$

When the temperature is not high (in the sense that it is not true that $T >> \theta^{R} = hc\tilde{B}/k$) the series must be summed term by term, which for a heteronuclear diatomic molecule or other non-symmetrical linear molecule gives

$$q^{R} = 1 + 3e^{-2\beta hc\tilde{B}} + 5e^{-6\beta hc\tilde{B}} + \cdots$$

Hence, because

$$\frac{dq^{R}}{d\beta} = -hc\tilde{B}(6e^{-2\beta hc\tilde{B}} + 30e^{-6\beta hc\tilde{B}} + \cdots)$$

(q^{R} is independent of V, so the partial derivative has been replaced by a complete derivative) it follows that

$$\langle\varepsilon^{R}\rangle = -\frac{1}{q^{R}}\frac{dq^{R}}{d\beta} = \frac{hc\tilde{B}(6e^{-2\beta hc\tilde{B}} + 30e^{-6\beta hc\tilde{B}} + \cdots)}{1 + 3e^{-2\beta hc\tilde{B}} + 5e^{-6\beta hc\tilde{B}} + \cdots}$$

Mean rotational energy [unsymmetrical linear molecule] (13C.6a)

This ungainly function is plotted in Fig. 13C.2. At high temperatures ($T >> \theta^{R}$), q^{R} is given by eqn 13B.13b ($q^{R} = T/\sigma\theta^{R}$) in the form $q^{R} = 1/\sigma\beta hc\tilde{B}$, where $\sigma = 1$ for a heteronuclear diatomic molecule. It then follows that

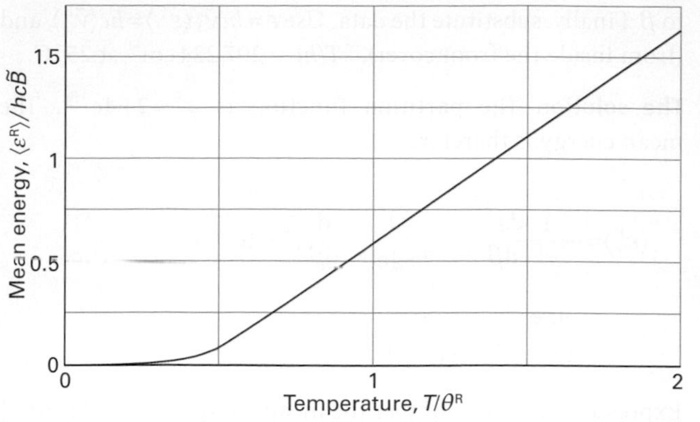

Figure 13C.2 The variation with temperature of the mean rotational energy of an unsymmetrical linear rotor. At high temperatures ($T >> \theta^R$), the energy is proportional to the temperature, in accord with the equipartition theorem.

$$\langle \varepsilon^R \rangle = -\frac{1}{q^R}\frac{dq^R}{d\beta} = -\sigma\beta hc\tilde{B}\frac{d}{d\beta}\left(\frac{1}{\sigma\beta hc\tilde{B}}\right) = -\beta\overbrace{\frac{d}{d\beta}\frac{1}{\beta}}^{-1/\beta^2}$$

and therefore that

$$\langle \varepsilon^R \rangle = \frac{1}{\beta} = kT \quad \text{Mean rotational energy [linear molecule, } T >> \theta^R] \quad (13C.6b)$$

The high-temperature result, which is valid when many rotational states are occupied, is also in agreement with the equipartition theorem (*The chemist's toolkit* 7 in Topic 2A), because the classical expression for the energy of a linear rotor is $E_k = \frac{1}{2}I_\perp\omega_a^2 + \frac{1}{2}I_\perp\omega_b^2$ and therefore has two quadratic contributions. (There is no rotation around the line of atoms.) It follows from the equipartition theorem that the mean rotational energy is $2 \times \frac{1}{2}kT = kT$.

Brief illustration 13C.2

To estimate the mean energy of a nonlinear molecule in the high-temperature limit recognize that its rotational kinetic energy (the only contribution to its rotational energy) is $E_k = \frac{1}{2}I_a\omega_a^2 + \frac{1}{2}I_b\omega_b^2 + \frac{1}{2}I_c\omega_c^2$. As there are three quadratic contributions, its mean rotational energy is $\frac{3}{2}kT$. The molar contribution is $\frac{3}{2}RT$. At 25 °C, this contribution is 3.7 kJ mol^{-1}, the same as the translational contribution, giving a total of 7.4 kJ mol^{-1}. A monatomic gas has no rotational contribution.

(c) The vibrational contribution

The vibrational partition function in the harmonic approximation is given in eqn 13B.15 ($q^V = 1/(1-e^{-\beta hc\tilde{\nu}})$). Because q^V is independent of the volume, it follows that

$$\frac{dq^V}{d\beta} = \frac{d}{d\beta}\left(\frac{1}{1-e^{-\beta hc\tilde{\nu}}}\right) \overset{d(1/f)/dx = -(1/f^2)df/dx}{=} -\frac{hc\tilde{\nu}e^{-\beta hc\tilde{\nu}}}{(1-e^{-\beta hc\tilde{\nu}})^2} \quad (13C.7)$$

and hence

$$\langle \varepsilon^V \rangle = -\frac{1}{q^V}\frac{dq^V}{d\beta} = (1-e^{-\beta hc\tilde{\nu}})\frac{hc\tilde{\nu}e^{-\beta hc\tilde{\nu}}}{(1-e^{-\beta hc\tilde{\nu}})^2} = \frac{hc\tilde{\nu}e^{-\beta hc\tilde{\nu}}}{1-e^{-\beta hc\tilde{\nu}}}$$

The final result, after multiplying the numerator and denominator by $e^{\beta hc\tilde{\nu}}$ is

$$\langle \varepsilon^V \rangle = \frac{hc\tilde{\nu}}{e^{\beta hc\tilde{\nu}}-1} \quad \text{Mean vibrational energy [harmonic approximation]} \quad (13C.8)$$

The zero-point energy, $\frac{1}{2}hc\tilde{\nu}$, can be added to the right-hand side if the mean energy is to be measured from 0 rather than the lowest attainable level (the zero-point level). The variation of the mean energy with temperature is illustrated in Fig. 13C.3. At high temperatures, when $T >> \theta^V = hc\tilde{\nu}/k$ or $\beta hc\tilde{\nu} << 1$, the exponential function can be expanded ($e^x = 1 + x + \cdots$) and all but the leading terms discarded. This approximation leads to

$$\langle \varepsilon^V \rangle = \frac{hc\tilde{\nu}}{(1+\beta hc\tilde{\nu}+\cdots)-1} \approx \frac{1}{\beta} = kT \quad \text{Mean vibrational energy } [T >> \theta^V] \quad (13C.9)$$

This result is in agreement with the value predicted by the classical equipartition theorem, because the energy of a one-dimensional oscillator is $E = \frac{1}{2}mv_x^2 + \frac{1}{2}k_f x^2$ and the mean energy of each quadratic term is $\frac{1}{2}kT$. Bear in mind, however, that the condition $T >> \theta^V$ is rarely satisfied.

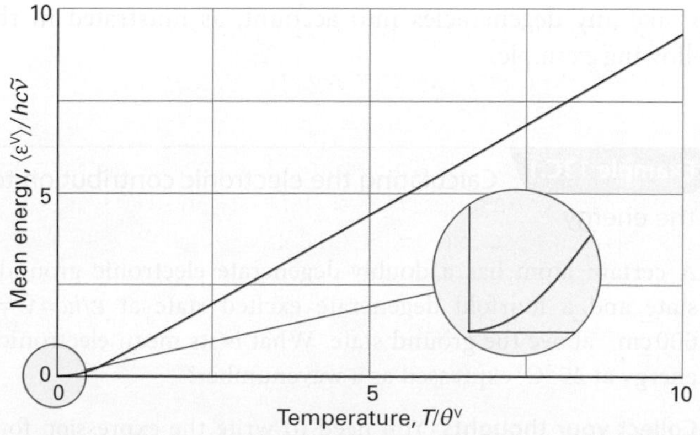

Figure 13C.3 The variation with temperature of the mean vibrational energy of a molecule in the harmonic approximation. At high temperatures ($T >> \theta^V$), the energy is proportional to the temperature, in accord with the equipartition theorem.

Brief illustration 13C.3

To calculate the mean vibrational energy of I_2 molecules at 298.15 K note that their vibrational wavenumber is 214.6 cm^{-1}. At 298.15 K, $kT/hc = 207.224\,\text{cm}^{-1}$,

$$\beta hc\tilde{\nu} = \frac{hc\tilde{\nu}}{kT} = \frac{214.6\,\text{cm}^{-1}}{207.224\,\text{cm}^{-1}} = 1.035\ldots$$

Because $\beta hc\tilde{\nu} = \theta^{\text{V}}/T$ is not large compared with 1 (implying that T is not high compared with θ^{V}), equipartition cannot be used. It then follows from eqn 13C.8 that

$$\langle \varepsilon^{\text{V}} \rangle / hc = \frac{214.6\,\text{cm}^{-1}}{\text{e}^{1.035\ldots} - 1} = 118.1\,\text{cm}^{-1}$$

The addition of the zero-point energy (corresponding to $\frac{1}{2}\times 214.6\,\text{cm}^{-1}$) increases this value to 225.4 cm^{-1}.

When there are several normal modes that can be treated as harmonic, the overall vibrational partition function is the product of each individual partition function, and the total mean vibrational energy is the sum of the mean energy of each mode.

(d) The electronic contribution

In most cases of interest, the electronic states of atoms and molecules are so widely separated that only the electronic ground state is occupied. Because all energies are measured from the ground state of each mode, it follows that

$$\langle \varepsilon^{\text{E}} \rangle = 0 \qquad \text{Mean electronic energy} \qquad (13\text{C}.10)$$

In certain cases, there are thermally accessible states at the temperature of interest. In that case, the partition function and hence the mean electronic energy are best calculated by direct summation over the available states. Care must be taken to take any degeneracies into account, as illustrated in the following example.

Example 13C.1 Calculating the electronic contribution to the energy

A certain atom has a doubly-degenerate electronic ground state and a fourfold degenerate excited state at $\varepsilon/hc = \tilde{\nu} = 600\,\text{cm}^{-1}$ above the ground state. What is its mean electronic energy at 25 °C, expressed as a wavenumber?

Collect your thoughts You need to write the expression for the partition function at a general temperature T (in terms of β) and then derive the mean energy by using eqn 13C.3. Doing so involves differentiating the partition function with respect to β. Finally, substitute the data. Use $\varepsilon = hc\tilde{\nu}$, $\langle \varepsilon^{\text{E}} \rangle = hc\langle \tilde{\nu}^{\text{E}} \rangle$, and (from inside the front cover), $kT/hc = 207.224\,\text{cm}^{-1}$ at 25 °C.

The solution The partition function is $q^{\text{E}} = 2 + 4\text{e}^{-\beta\varepsilon}$. The mean energy is therefore

$$\langle \varepsilon^{\text{E}} \rangle = -\frac{1}{q^{\text{E}}}\frac{\text{d}q^{\text{E}}}{\text{d}\beta} = -\frac{1}{2+4\text{e}^{-\beta\varepsilon}}\overbrace{\frac{\text{d}}{\text{d}\beta}(2+4\text{e}^{-\beta\varepsilon})}^{-4\varepsilon\text{e}^{-\beta\varepsilon}}$$

$$= \frac{4\varepsilon\text{e}^{-\beta\varepsilon}}{2+4\text{e}^{-\beta\varepsilon}} = \frac{\varepsilon}{\frac{1}{2}\text{e}^{\beta\varepsilon}+1}$$

Expressed as a wavenumber the mean energy is $\langle \varepsilon^{\text{E}} \rangle / hc = \langle \tilde{\nu}^{\text{E}} \rangle$

$$\varepsilon = hc\tilde{\nu}$$

$$\langle \tilde{\nu}^{\text{E}} \rangle = \frac{\tilde{\nu}}{\frac{1}{2}\text{e}^{hc\tilde{\nu}/kT}+1}$$

From the data,

$$\langle \tilde{\nu}^{\text{E}} \rangle = \frac{600\,\text{cm}^{-1}}{\frac{1}{2}\text{e}^{(600\,\text{cm}^{-1})/(207.224\,\text{cm}^{-1})}+1} = 59.7\,\text{cm}^{-1}$$

Self-test 13C.1 Repeat the problem for an atom that has a threefold degenerate ground state and a sevenfold degenerate excited state 400 cm^{-1} above.

Answer: 101 cm^{-1}

(e) The spin contribution

An electron spin in a magnetic field $\mathcal{B}$ has two possible energy states that depend on its orientation as expressed by the magnetic quantum number m_s, and which are given by

$$E_{m_s} = g_{\text{e}}\mu_{\text{B}}\mathcal{B}m_s \qquad \text{Electron spin energies} \qquad (13\text{C}.11)$$

where μ_{B} is the Bohr magneton (see inside front cover) and $g_{\text{e}} = 2.0023$. These energies are discussed in more detail in Topic 12A. The lower state has $m_s = -\frac{1}{2}$, so the two energy levels available to the electron lie (according to the convention that $\varepsilon_0 \equiv 0$) at $\varepsilon_{-1/2} = 0$ and at $\varepsilon_{+1/2} = g_{\text{e}}\mu_{\text{B}}\mathcal{B}$. The spin partition function is therefore

$$q^{\text{S}} = \sum_{m_s} \text{e}^{-\beta\varepsilon_{m_s}} = 1 + \text{e}^{-\beta g_{\text{e}}\mu_{\text{B}}\mathcal{B}} \qquad \text{Spin partition function} \qquad (13\text{C}.12)$$

The mean energy of the spin is therefore

$$\langle \varepsilon^{\text{S}} \rangle = -\frac{1}{q^{\text{S}}}\frac{\text{d}q^{\text{S}}}{\text{d}\beta} = -\frac{1}{1+\text{e}^{-\beta g_{\text{e}}\mu_{\text{B}}\mathcal{B}}}\overbrace{\frac{\text{d}}{\text{d}\beta}(1+\text{e}^{-\beta g_{\text{e}}\mu_{\text{B}}\mathcal{B}})}^{-g_{\text{e}}\mu_{\text{B}}\mathcal{B}\text{e}^{-\beta g_{\text{e}}\mu_{\text{B}}\mathcal{B}}}$$

$$= \frac{g_{\text{e}}\mu_{\text{B}}\mathcal{B}\text{e}^{-\beta g_{\text{e}}\mu_{\text{B}}\mathcal{B}}}{1+\text{e}^{-\beta g_{\text{e}}\mu_{\text{B}}\mathcal{B}}}$$

The final expression, after multiplying the numerator and denominator by $e^{\beta g_e \mu_B \mathcal{B}}$, is

$$\langle \varepsilon^S \rangle = \frac{g_e \mu_B \mathcal{B}}{e^{\beta g_e \mu_B \mathcal{B}}+1} \quad \text{Mean spin energy} \quad (13C.13)$$

This function is essentially the same as that plotted in Fig. 13C.1.

Brief illustration 13C.4

Suppose a collection of radicals is exposed to a magnetic field of 2.5 T (T denotes tesla) at 25 °C. With $\mu_B = 9.274 \times 10^{-24}\,\mathrm{J\,T^{-1}}$,

$$g_e \mu_B \mathcal{B} = 2.0023 \times (9.274 \times 10^{-24}\,\mathrm{J\,T^{-1}}) \times 2.5\,\mathrm{T} = 4.6\ldots \times 10^{-23}\,\mathrm{J}$$

$$\beta g_e \mu_B \mathcal{B} = \frac{2.0023 \times (9.274 \times 10^{-24}\,\mathrm{J\,T^{-1}}) \times (2.5\,\mathrm{T})}{(1.381 \times 10^{-23}\,\mathrm{J\,K^{-1}}) \times (298\,\mathrm{K})} = 0.011\ldots$$

The mean energy is therefore

$$\langle \varepsilon^S \rangle = \frac{4.6\ldots \times 10^{-23}\,\mathrm{J}}{e^{0.011\ldots}+1} = 2.3 \times 10^{-23}\,\mathrm{J}$$

This energy is equivalent to $14\,\mathrm{J\,mol^{-1}}$ (note joules, not kilojoules).

Checklist of concepts

☐ 1. The **mean molecular energy** can be calculated from the molecular partition function.

☐ 2. Individual contributions to the mean molecular energy from each mode of motion are calculated from the relevant partition functions,

☐ 3. In the high temperature limit, the results obtained for the mean molecular energy are in accord with the equipartition principle.

Checklist of equations

Property	Equation	Comment	Equation number
Mean energy	$\langle \varepsilon \rangle = \varepsilon_{gs} - (1/q)(\partial q/\partial \beta)_V$	$\beta = 1/kT$	13C.4a
	$\langle \varepsilon \rangle = \varepsilon_{gs} - (\partial \ln q/\partial \beta)_V$	Alternative version	13C.4b
Translation	$\langle \varepsilon^T \rangle = \frac{d}{2}kT$	In d dimensions, d = 1, 2, 3	13C.5
Rotation	$\langle \varepsilon^R \rangle = kT$	Linear molecule, $T >> \theta^R$	13C.6b
Vibration	$\langle \varepsilon^V \rangle = hc\tilde{\nu}/(e^{\beta hc\tilde{\nu}}-1)$	Harmonic approximation	13C.8
	$\langle \varepsilon^V \rangle = kT$	$T >> \theta^V$	13C.9
Spin	$\langle \varepsilon^S \rangle = g_e \mu_B \mathcal{B}/(e^{\beta g_e \mu_B \mathcal{B}}+1)$	Electron in a magnetic field	13C.13

TOPIC 13D The canonical ensemble

> ➤ **Why do you need to know this material?**
>
> Whereas Topics 13B and 13C deal with independent molecules, in practice molecules do interact. Therefore, this material is essential for constructing models of real gases, liquids, and solids and of any system in which intermolecular interactions cannot be neglected.
>
> ➤ **What is the key idea?**
>
> A system composed of interacting molecules is described in terms of a canonical partition function, from which its thermodynamic properties may be deduced.
>
> ➤ **What do you need to know already?**
>
> The calculations here, which are not carried through in detail, are essentially the same as in Topic 13A. Calculations of mean energies are also essentially the same as in Topic 13C and are not repeated in detail.

The crucial concept needed in the treatment of systems of interacting particles, as in real gases and liquids, is the 'ensemble'. Like so many scientific terms, the term has basically its normal meaning of 'collection', but in statistical thermodynamics it has been sharpened and refined into a precise significance.

13D.1 The concept of ensemble

To set up an ensemble, take a closed system of specified volume, composition, and temperature, and think of it as replicated $\tilde{N}$ times (Fig. 13D.1). All the identical closed systems are regarded as being in thermal contact with one another, so they can exchange energy. The total energy of the ensemble is $\tilde{E}$. Because the members of the ensemble are all in thermal equilibrium with each other, they have the same temperature, T. The volume of each member of the ensemble is the same, so the energy levels available to the molecules are the same in each system, and each member contains the same number of molecules, so there is a fixed number of molecules to distribute within each system. This imaginary collection of replications of the actual system with a common temperature is called the **canonical ensemble**.[1]

[1] The word 'canon' means 'according to a rule'.

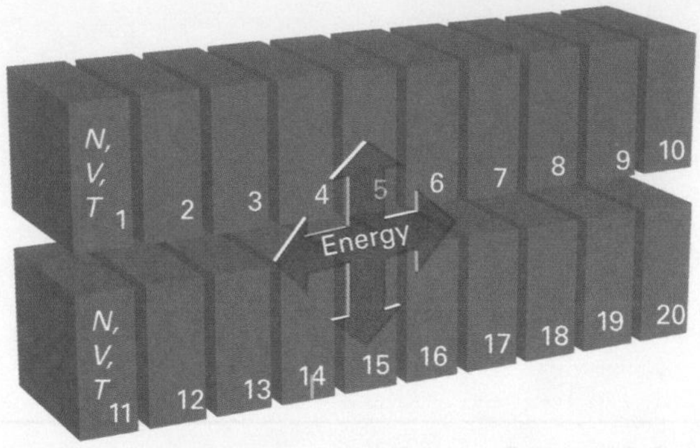

Figure 13D.1 A representation of the canonical ensemble, in this case for $\tilde{N} = 20$. The individual replications of the actual system all have the same composition and volume. They are all in mutual thermal contact, and so all have the same temperature. Energy may be transferred between them as heat, and so they do not all have the same energy. The total energy of all 20 replications is a constant because the ensemble is isolated overall.

There are two other important types of ensembles. In the **microcanonical ensemble** the condition of constant temperature is replaced by the requirement that all the systems should have exactly the same energy: each system is individually isolated. In the **grand canonical ensemble** the volume and temperature of each system is the same, but they are open, which means that matter can be imagined as able to pass between them; the composition of each one may fluctuate, but now the property known as the chemical potential (μ, Topic 5A) is the same in each system. In summary:

Ensemble	Common properties
Microcanonical	V, E, N
Canonical	V, T, N
Grand canonical	V, T, μ

The microcanonical ensemble is the basis of the discussion in Topic 13A; the grand canonical ensemble will not be considered explicitly.

The important point about an ensemble is that it is a collection of *imaginary* replications of the system, so the number of members can be as large as desired; when appropriate, $\tilde{N}$ can be taken as infinite. The number of members of the ensemble in a state with energy E_i is denoted $\tilde{N}_i$, and it is possible to speak of the configuration of the ensemble (by analogy with the configuration of the system used in Topic 13A) and its weight, $\tilde{W}$, the number of ways of achieving the configuration $\{\tilde{N}_0,\tilde{N}_1,\ldots\}$. Note that $\tilde{N}$ is unrelated to N, the number of molecules in the actual system; $\tilde{N}$ is the number of *imaginary* replications of that system. In summary:

N is the number of molecules in the system, the same in each member of the ensemble.

T is the common temperature of the members.

E_i is the total energy of one member of the ensemble (the one labelled i).

$\tilde{N}_i$ is the number of replicas that have the energy E_i.

$\tilde{E}$ is the total energy of the entire ensemble.

$\tilde{N}$ is the total number of replicas (the number of members of the ensemble).

$\tilde{W}$ is the weight of the configuration $\{\tilde{N}_0,\tilde{N}_1,\ldots\}$.

(a) Dominating configurations

Just as in Topic 13A, some of the configurations of the canonical ensemble are very much more probable than others. For instance, it is very unlikely that the whole of the total energy of the ensemble will accumulate in one system to give the configuration $\{\tilde{N},0,0,\ldots\}$. By analogy with the discussion in Topic 13A, there is a dominating configuration, and thermodynamic properties can be calculated by taking the average over the ensemble using that single, most probable, configuration. In the **thermodynamic limit** of $\tilde{N} \to \infty$, this dominating configuration is overwhelmingly the most probable, and dominates its properties.

The quantitative discussion follows the argument in Topic 13A (leading to the Boltzmann distribution) with the modification that N and N_i are replaced by $\tilde{N}$ and $\tilde{N}_i$. The weight $\tilde{W}$ of a configuration $\{\tilde{N}_0,\tilde{N}_1,\ldots\}$ is

$$\tilde{W} = \frac{\tilde{N}!}{\tilde{N}_0!\tilde{N}_1!\ldots} \quad \text{Weight} \quad (13D.1)$$

The configuration of greatest weight, subject to the constraints that the total energy of the ensemble is constant at $\tilde{E}$ and that the total number of members is fixed at $\tilde{N}$, is given by the **canonical distribution**:

$$\frac{\tilde{N}_i}{\tilde{N}} = \frac{e^{-\beta E_i}}{Q} \qquad Q = \sum_i e^{-\beta E_i} \quad \text{Canonical distribution} \quad (13D.2)$$

where the sum is over all members of the ensemble. The quantity Q, which is a function of the temperature, is called the **canonical partition function**, and $\beta = 1/kT$. Like the molecular partition function, the canonical partition function contains all the thermodynamic information about a system but allows for the possibility of interactions between the constituent molecules.

(b) Fluctuations from the most probable distribution

The canonical distribution in eqn 13D.2 is only apparently an exponentially decreasing function of the energy of the system. Just as the Boltzmann distribution gives the occupation of a single state of a molecule, the canonical distribution function gives the probability of occurrence of members in a single state i of energy E_i. There may in fact be numerous states with almost identical energies. For example, in a gas the identities of the molecules moving slowly or quickly can change without necessarily affecting the total energy. The **energy density of states**, the number of states in an energy range divided by the width of the range (Fig. 13D.2), is a very sharply increasing function of energy. It follows that the probability of a member of an ensemble having a specified energy (as distinct from being in a specified state) is given by eqn 13D.2, a sharply decreasing function, multiplied by a sharply increasing function (Fig. 13D.3). Therefore, the overall distribution is a sharply peaked function. That is, most members of the ensemble have an energy very close to the mean value.

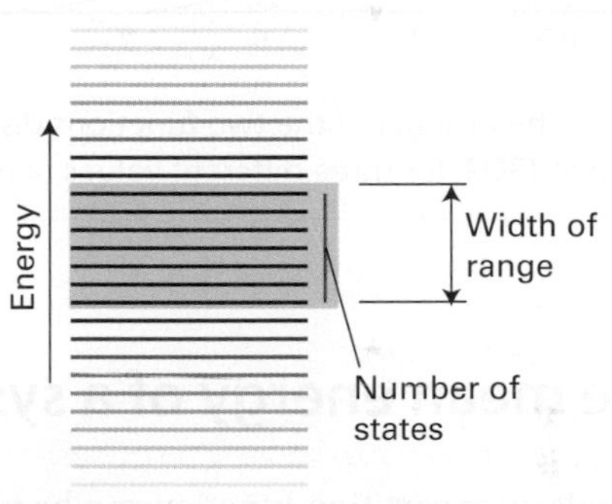

Figure 13D.2 The energy density of states is the number of states in an energy range divided by the width of the range.

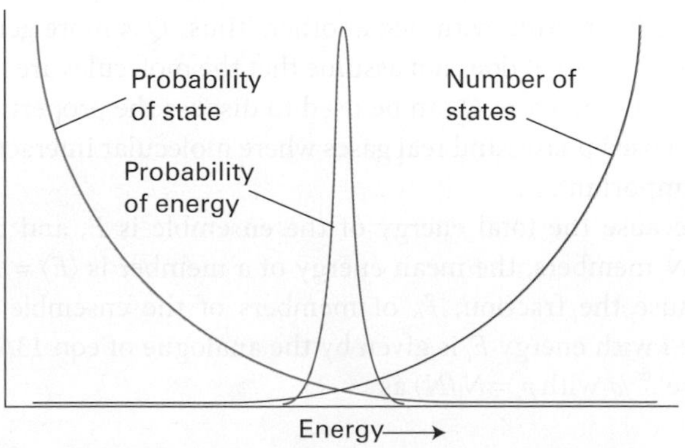

Figure 13D.3 To construct the form of the distribution of members of the canonical ensemble in terms of their energies, the probability that any one is in a state of given energy (eqn 13D.2) is multiplied by the number of states corresponding to that energy (a steeply rising function). The product is a sharply peaked function at the mean energy (here considerably magnified), which shows that almost all the members of the ensemble have that energy.

Brief illustration 13D.1

A function that increases rapidly is x^n, with n large. A function that decreases rapidly is e^{-nx}, once again, with n large. The product of these two functions, normalized so that the maxima for different values of n all coincide,

$$f(x)=e^n x^n e^{-nx}$$

is plotted for three values of n in Fig. 13D.4. Note that the width of the product does indeed decrease as n increases.

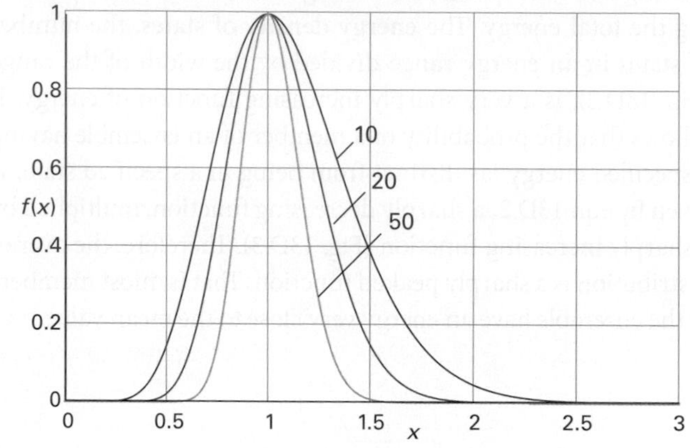

Figure 13D.4 The product of the two functions discussed in *Brief illustration* 13D.1, for three different values of n.

13D.2 The mean energy of a system

Just as the molecular partition function can be used to calculate the mean value of a molecular property, so the canonical partition function can be used to calculate the mean energy of an entire system composed of molecules which might or might not be interacting with one another. Thus, Q is more general than q because it does not assume that the molecules are independent. Therefore Q can be used to discuss the properties of condensed phases and real gases where molecular interactions are important.

Because the total energy of the ensemble is $\tilde{E}$, and there are $\tilde{N}$ members, the mean energy of a member is $\langle E\rangle = \tilde{E}/\tilde{N}$. Because the fraction, $\tilde{p}_i$, of members of the ensemble in a state i with energy E_i is given by the analogue of eqn 13A.10b ($p_i = e^{-\beta\varepsilon_i}/q$ with $p_i = N_i/N$) as

$$\tilde{p}_i = \frac{e^{-\beta E_i}}{Q} \quad (13D.3)$$

it follows that

$$\langle E\rangle = \sum_i \tilde{p}_i E_i = \frac{1}{Q}\sum_i E_i e^{-\beta E_i} \quad (13D.4)$$

By the same argument that led to eqn 13C.4a ($\langle\varepsilon\rangle = -(1/q)(\partial q/\partial\beta)_V$, when $\varepsilon_{gs} \equiv 0$),

$$\langle E\rangle = -\frac{1}{Q}\left(\frac{\partial Q}{\partial\beta}\right)_V = -\left(\frac{\partial \ln Q}{\partial\beta}\right)_V \quad \text{Mean energy of a system} \quad (13D.5)$$

As in the case of the mean molecular energy, the ground-state energy of the entire system must be added to this expression if it is not zero.

13D.3 Independent molecules revisited

When the molecules are in fact independent of each other, Q can be shown to be related to the molecular partition function q.

How is that done? 13D.1 Establishing the relation between Q and q

There are two cases you need to consider because it turns out to be important to consider initially a system in which the particles are distinguishable (such as when they are at fixed locations in a solid) and then one in which they are indistinguishable (such as when they are in a gas and able to exchange places).

Step 1 *Consider a system of independent, distinguishable molecules*

The total energy of a collection of N independent molecules is the sum of the energies of the molecules. Therefore, the total energy of a state i of the system is written as

$$E_i = \varepsilon_i(1) + \varepsilon_i(2) + \cdots + \varepsilon_i(N)$$

In this expression, $\varepsilon_i(1)$ is the energy of molecule 1 when the system is in the state i, $\varepsilon_i(2)$ the energy of molecule 2 when the system is in the same state i, and so on. The canonical partition function is then

$$Q = \sum_i e^{-\beta\varepsilon_i(1)-\beta\varepsilon_i(2)-\cdots-\beta\varepsilon_i(N)}$$

Provided the molecules are distinguishable (in a sense described below), the sum over the states of the system can be reproduced by letting each molecule enter all its own individual states. Therefore, instead of summing over the states i of the system, sum over all the individual states j of molecule 1, all the states j of molecule 2, and so on. This rewriting of the original expression leads to

$$Q = \underbrace{\left(\sum_j e^{-\beta\varepsilon_j}\right)}_{q}\underbrace{\left(\sum_j e^{-\beta\varepsilon_j}\right)}_{q}\cdots\underbrace{\left(\sum_j e^{-\beta\varepsilon_j}\right)}_{q}$$

and therefore to $Q = q^N$.

Step 2 *Consider a system of independent, indistinguishable molecules*

If all the molecules are identical and free to move through space, it is not possible to distinguish them and the relation $Q = q^N$ is not valid. Suppose that molecule 1 is in some state a, molecule 2 is in b, and molecule 3 is in c, then one member of the ensemble has an energy $E = \varepsilon_a + \varepsilon_b + \varepsilon_c$. This member, however, is indistinguishable from one formed by putting molecule 1 in state b, molecule 2 in state c, and molecule 3 in state a, or some other permutation. There are six such permutations in all, and $N!$ in general. In the case of indistinguishable molecules, it follows that too many states have been counted in going from the sum over system states to the sum over molecular states, so writing $Q = q^N$ overestimates the value of Q. The detailed argument is quite involved, but at all except very low temperatures it turns out that the correction factor is $1/N!$, so $Q = q^N/N!$.

Step 3 *Summarize the results*

It follows that:

For distinguishable independent molecules:
$$Q = q^N \tag{13D.6a}$$

For indistinguishable independent molecules:
$$Q = q^N/N! \tag{13D.6b}$$

For molecules to be indistinguishable, they must be of the same kind: an Ar atom is never indistinguishable from a Ne atom. Their identity, however, is not the only criterion. Each identical molecule in a crystal lattice, for instance, can be 'named' with a set of coordinates. Identical molecules in a lattice can therefore be treated as distinguishable because their sites are distinguishable, and eqn 13D.6a can be used. On the other hand, identical molecules in a gas are free to move to different locations, and there is no way of keeping track of the identity of a given molecule; therefore eqn 13D.6b must be used.

Brief illustration 13D.2

For a gas of N indistinguishable molecules, $Q = q^N/N!$. The energy of the system is therefore

$$\langle E\rangle = -\left(\frac{\partial \ln(q^N/N!)}{\partial\beta}\right)_V = -\left(\frac{\partial(\ln q^N - \ln N!)}{\partial\beta}\right)_V = -\left(\frac{\partial(N\ln q)}{\partial\beta}\right)_V$$
$$= -N\left(\frac{\partial \ln q}{\partial\beta}\right)_V = N\langle\varepsilon\rangle$$

That is, the mean energy of the gas is N times the mean energy of a single molecule.

13D.4 The variation of the energy with volume

When there are interactions between molecules, the energy of a collection depends on the average distance between them, and therefore on the volume that a fixed number occupy. This dependence on volume is particularly important for the discussion of real gases (Topic 1C).

To discuss the dependence of the energy on the volume at constant temperature it is necessary to evaluate $(\partial\langle E\rangle/\partial V)_T$. (In Topics 2D and 3E, this quantity is identified as the 'internal pressure' of a gas and denoted π_T.) To proceed, substitute eqn 13D.5 and obtain

$$\left(\frac{\partial\langle E\rangle}{\partial V}\right)_T = -\left(\frac{\partial}{\partial V}\left(\frac{\partial \ln Q}{\partial\beta}\right)_V\right)_T \tag{13D.7}$$

If the gas were perfect, $Q = q^N/N!$, with q the partition function for the translational and any internal (such as rotational) modes. If the gas is monatomic, only the translational mode is present and $q = V/\Lambda^3$ with $\Lambda = h/(2\pi mkT)^{1/2}$ and the canonical partition function would be $V^N/\Lambda^{3N}N!$. The presence of interactions is taken into account by replacing $V^N/N!$ by a factor called the **configuration integral**, Z, which depends on the intermolecular potentials (don't confuse this Z with the compression factor Z in Topic 1C), and writing

$$Q = \frac{Z}{\Lambda^{3N}} \tag{13D.8}$$

It then follows that

$$\left(\frac{\partial\langle E\rangle}{\partial V}\right)_T = -\left(\frac{\partial}{\partial V}\left(\frac{\partial \ln(Z/\Lambda^{3N})}{\partial\beta}\right)_V\right)_T$$
$$= -\left(\frac{\partial}{\partial V}\left(\frac{\partial \ln Z}{\partial\beta}\right)_V\right)_T - \left(\frac{\partial}{\partial V}\left(\frac{\partial \ln(1/\Lambda^{3N})}{\partial\beta}\right)_V\right)_T$$

$\partial^2 f/\partial x\partial y = \partial^2 f/\partial y\partial x$ for the second (blue) term

$$= -\left(\frac{\partial}{\partial V}\left(\frac{\partial \ln Z}{\partial\beta}\right)_V\right)_T - \left(\frac{\partial}{\partial \beta}\overbrace{\left(\frac{\partial \ln(1/\Lambda^{3N})}{\partial V}\right)_T}^{0}\right)_V$$

$\partial \ln y/\partial x = (1/y)\partial y/\partial x$

$$= -\left(\frac{\partial}{\partial V}\left(\frac{\partial \ln Z}{\partial\beta}\right)_V\right)_T = -\left(\frac{\partial}{\partial V}\frac{1}{Z}\left(\frac{\partial Z}{\partial\beta}\right)_V\right)_T \tag{13D.9}$$

In the third line, Λ is independent of volume, so its derivative with respect to volume is zero.

For a real gas of atoms (for which the intermolecular interactions are isotropic), Z is related to the total potential energy E_p of interaction of all the particles, which depends on all their relative locations, by

$$Z=\frac{1}{N!}\int e^{-\beta E_p}\,d\tau_1 d\tau_2 \dots d\tau_N \quad \text{Configuration integral} \quad (13D.10)$$

where $d\tau_i$ is the volume element for atom i and the integration is over all the variables. The physical origin of this term is that the probability of occurrence of each arrangement of molecules possible in the sample is given by a Boltzmann distribution in which the exponent is given by the potential energy corresponding to that arrangement.

Brief illustration 13D.3

Equation 13D.10 is very difficult to manipulate in practice, even for quite simple intermolecular potentials, except for a perfect gas for which $E_p = 0$. In that case, the exponential function becomes 1 and

$$Z=\frac{1}{N!}\int d\tau_1 d\tau_2 \dots d\tau_N=\frac{1}{N!}\left(\int d\tau\right)^N=\frac{V^N}{N!}$$

just as it should be for a perfect gas.

If the potential energy has the form of a central hard sphere surrounded by a shallow attractive well (Fig. 13D.5), then detailed calculation, which is too involved to reproduce here (see *A deeper look* 7 on the website of this text), leads to

$$\left(\frac{\partial\langle E\rangle}{\partial V}\right)_T=\frac{an^2}{V^2} \quad \text{Attractive potential} \quad (13D.11)$$

where n is the amount of molecules present in the volume V and a is a constant that is proportional to the area under the attractive part of the potential. In *Example* 3E.2 of Topic 3E exactly the same expression (in the form $\pi_T = an^2/V^2$) was derived from the van der Waals equation of state. The conclusion at this point is that if there are attractive interactions between molecules in a gas, then its energy increases as it expands isothermally (because $(\partial\langle E\rangle/\partial V)_T > 0$, and the slope of $\langle E\rangle$ with respect to V is positive). The energy rises because, at greater average separations, the molecules spend less time in regions where they interact favourably.

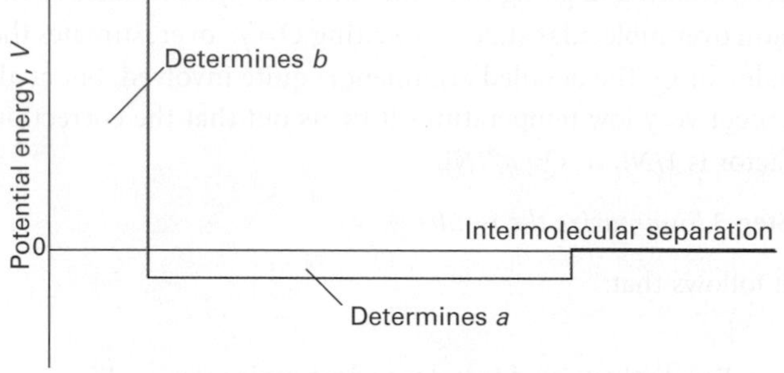

Figure 13D.5 The intermolecular potential energy of molecules in a real gas can be modelled by a central hard sphere that determines the van der Waals parameter b surrounded by a shallow attractive well that determines the parameter a. As mentioned in the text, calculations of the canonical partition function based on this are consistent with the van der Waals equation of state (Topic 1C).

Checklist of concepts

- ☐ 1. The **canonical ensemble** is a collection of imaginary replications of the actual system with a common temperature and number of particles.
- ☐ 2. The **canonical distribution** gives the most probable number of members of the ensemble with a specified total energy.
- ☐ 3. The mean energy of the members of the ensemble can be calculated from the **canonical partition function**.

Checklist of equations

Property	Equation	Comment	Equation number
Canonical partition function	$Q=\sum_i e^{-\beta E_i}$	Definition	13D.2
Canonical distribution	$\tilde{N}_i/\tilde{N}=e^{-\beta E_i}/Q$		13D.2
Mean energy	$\langle E\rangle=-(1/Q)(\partial Q/\partial\beta)_V=-(\partial\ln Q/\partial\beta)_V$		13D.5
Canonical partition function	$Q=Z/\Lambda^{3N}$		13D.8
Configuration integral	$Z=(1/N!)\int e^{-\beta E_p}d\tau_1 d\tau_2 \dots d\tau_N$	Isotropic interaction	13D.10
Variation of mean energy with volume	$(\partial\langle E\rangle/\partial V)_T=an^2/V^2$	Potential energy as specified in Fig. 13D.5	13D.11

TOPIC 13E The internal energy and the entropy

> **➤ Why do you need to know this material?**
>
> The importance of this discussion is the insight that a molecular interpretation provides into thermodynamic properties.
>
> **➤ What is the key idea?**
>
> The partition function contains all the thermodynamic information about a system and thus provides a bridge between spectroscopy and thermodynamics.
>
> **➤ What do you need to know already?**
>
> You need to know how to calculate a molecular partition function from structural data (Topic 13B); you should also be familiar with the concepts of internal energy (Topic 2A) and entropy (Topic 3A). This Topic makes use of the calculations of mean molecular energies in Topic 13C.

Any thermodynamic function can be obtained once the partition function is known. The two fundamental properties of thermodynamics are the internal energy, U, and the entropy, S. Once these two properties have been calculated, it is possible to turn to the derived functions, such as the Gibbs energy, G (Topic 13F) and all the chemically interesting properties that stem from them.

13E.1 The internal energy

The first example of the importance of the molecular partition function, q, is the derivation of an expression for the internal energy.

(a) The calculation of internal energy

It is established in Topic 13C that the mean energy of a molecule in a system composed of independent molecules is related to the molecular partition function by

$$\langle\varepsilon\rangle=-\frac{1}{q}\left(\frac{\partial q}{\partial\beta}\right)_V \qquad (13E.1)$$

with $\beta=1/kT$. The total energy of a system composed of N molecules is therefore $N\langle\varepsilon\rangle$. This total energy is the energy above the ground state (the calculation of q is based on the convention $\varepsilon_0\equiv 0$) and so the internal energy is $U(T)=U(0)+N\langle\varepsilon\rangle$, where $U(0)$ is the internal energy when only the ground state is occupied, which is the case at $T=0$. It follows that the internal energy is related to the molecular partition function by

$$U(T)=U(0)+N\langle\varepsilon\rangle=U(0)-\frac{N}{q}\left(\frac{\partial q}{\partial\beta}\right)_V$$

Internal energy [independent molecules] (13E.2a)

In many cases, the expression for $\langle\varepsilon\rangle$ established for each mode of motion in Topic 13C can be used and it is not necessary to go back to q itself except for some formal manipulations. For instance, it is established in Topic 13C (eqn 13C.8) that the mean energy of a harmonic oscillator is $\langle\varepsilon^{\rm V}\rangle=hc\tilde{\nu}/(e^{\beta hc\tilde{\nu}}-1)$. It follows that the molar internal energy of a system composed of oscillators is

$$U_{\rm m}^{\rm V}(T)=U_{\rm m}^{\rm V}(0)+\frac{N_{\rm A}hc\tilde{\nu}}{e^{\beta hc\tilde{\nu}}-1}$$

Brief illustration 13E.1

The vibrational wavenumber of an I_2 molecule is 214.6 cm^{-1}. At 298.15 K, kT/hc = 207.224 cm^{-1} and $hc\tilde{\nu}$ = 4.26… zJ. With these values

$$\beta hc\tilde{\nu}=\frac{hc\tilde{\nu}}{kT}=\frac{214.6\,{\rm cm}^{-1}}{207.224\,{\rm cm}^{-1}}=1.035\ldots$$

It follows that the vibrational contribution to the molar internal energy of I_2 is

$$U_{\rm m}^{\rm V}(T)=U_{\rm m}^{\rm V}(0)+\frac{(6.022\times10^{23}\,{\rm mol}^{-1})\times(4.26\ldots\times10^{-21}\,{\rm J})}{e^{1.035\ldots}-1}$$

$$=U_{\rm m}^{\rm V}(0)+1.41\,{\rm kJ\,mol}^{-1}$$

An alternative form of eqn 13E.2a is

$$U(T)=U(0)-N\left(\frac{\partial \ln q}{\partial\beta}\right)_V$$

Internal energy [independent molecules] (13E.2b)

A very similar expression is used for a system of interacting molecules. In that case the canonical partition function (Topic 13D), Q, is used to write

$$U(T)=U(0)-\left(\frac{\partial \ln Q}{\partial \beta}\right)_V \quad \text{Internal energy [interacting molecules]} \quad (13E.2c)$$

(b) Heat capacity

The constant-volume heat capacity (Topic 2A) is defined as $C_V = (\partial U/\partial T)_V$. Then, because the mean vibrational energy of a harmonic oscillator (eqn 13C.8, quoted above as $\langle\varepsilon^V\rangle = hc\tilde{\nu}/(e^{\beta hc\tilde{\nu}}-1)$) can be written in terms of the vibrational temperature $\theta^V = hc\tilde{\nu}/k$ as

$$\langle\varepsilon^V\rangle = \frac{k\theta^V}{e^{\theta^V/T}-1}$$

it follows that the vibrational contribution to the molar constant-volume heat capacity is

$$C_{V,m}^V \overset{U_m(T)=U_m(0)+N_A\langle\varepsilon^V\rangle}{=} \frac{dN_A\langle\varepsilon^V\rangle}{dT} = R\theta^V \frac{d}{dT}\frac{1}{e^{\theta^V/T}-1} \overset{d(1/f)/dx=-(1/f^2)df/dx \text{ used twice}}{=} R\left(\frac{\theta^V}{T}\right)^2 \frac{e^{\theta^V/T}}{(e^{\theta^V/T}-1)^2}$$

By noting that $e^{\theta^V/T} = (e^{\theta^V/2T})^2$, this expression can be rearranged into

$$C_{V,m}^V = Rf(T) \quad f(T) = \left(\frac{\theta^V}{T}\right)^2 \left(\frac{e^{-\theta^V/2T}}{1-e^{-\theta^V/T}}\right)^2 \quad \text{Vibrational contribution to } C_{V,m} \quad (13E.3)$$

The graph in Fig. 13E.1 shows how the vibrational heat capacity depends on temperature. Note that even when the temperature is only slightly above θ^V the heat capacity is close to its equipartition value. Equation 13E.3 is essentially the same as the Einstein formula for the heat capacity of a solid (eqn 7A.8a) with θ^V the Einstein temperature, θ_E. The only difference is that in a solid the vibrations take place in three dimensions.

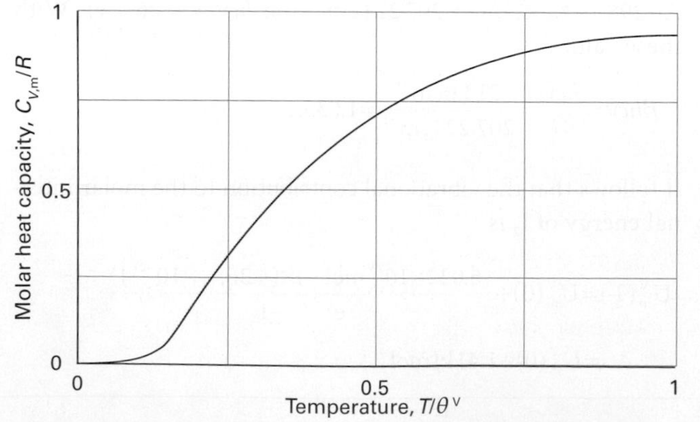

Figure 13E.1 The temperature dependence of the vibrational heat capacity of a molecule in the harmonic approximation calculated by using eqn 13E.3. Note that the heat capacity is within 10 per cent of its classical value for temperatures greater than θ^V.

It is sometimes more convenient to convert the derivative with respect to T into a derivative with respect to $\beta = 1/kT$ by using

$$\frac{d}{dT} = \frac{d\beta}{dT}\frac{d}{d\beta} = -\frac{1}{kT^2}\frac{d}{d\beta} = -k\beta^2\frac{d}{d\beta} \quad (13E.4)$$

It follows that

$$C_V = -k\beta^2\left(\frac{\partial U}{\partial \beta}\right)_V = -Nk\beta^2\left(\frac{\partial\langle\varepsilon\rangle}{\partial\beta}\right)_V \overset{\text{eqn 13E.1}}{=} Nk\beta^2\left(\frac{\partial^2 \ln q}{\partial\beta^2}\right)_V \quad \text{Heat capacity} \quad (13E.5)$$

There is a much simpler route to finding C_V when the equipartition principle can be applied, which is the case when $T >> \theta^M$, where θ^M is the characteristic temperature of the mode M ($\theta^V = hc\tilde{\nu}/k$ for vibration, $\theta^R = hc\tilde{B}/k$ for rotation). The heat capacity can then be estimated simply by counting the number of modes that are active, noting that the average molar energy of each mode is a multiple of $\frac{1}{2}RT$, differentiating that energy with respect to T, and getting that same multiple of $\frac{1}{2}R$. In gases, all three translational modes are always active and contribute $\frac{3}{2}R$ to the molar heat capacity. If the number of active rotational modes is denoted by ν^{R*} (so for most molecules at normal temperatures $\nu^{R*} = 2$ for linear molecules, and 3 for nonlinear molecules), then the rotational contribution is $\frac{1}{2}\nu^{R*}R$. If the temperature is high enough for ν^{V*} vibrational modes to be active, then the vibrational contribution to the molar heat capacity is $\nu^{V*}R$. In most cases, $\nu^{V*} \approx 0$. It follows that the total molar heat capacity of a gas is approximately

$$C_{V,m} = \tfrac{1}{2}(3+\nu^{R*}+2\nu^{V*})R \quad \text{Total heat capacity } [T >> \theta^M] \quad (13E.6)$$

Brief illustration 13E.2

The characteristic temperatures (in round numbers) of the vibrations of H_2O are 5300 K, 2300 K, and 5400 K; the vibrations are therefore not excited at 373 K. The three rotational modes of H_2O have characteristic temperatures 40 K, 21 K, and 13 K, so they are fully excited, like the three translational modes. The translational contribution is $\frac{3}{2}R = 12.5\,\text{J K}^{-1}\,\text{mol}^{-1}$. Fully excited rotations contribute a further $12.5\,\text{J K}^{-1}\,\text{mol}^{-1}$. Therefore, a value close to $25\,\text{J K}^{-1}\,\text{mol}^{-1}$ is predicted. The experimental value is $26.1\,\text{J K}^{-1}\,\text{mol}^{-1}$. The discrepancy is probably due to deviations from perfect gas behaviour.

13E.2 The entropy

One of the most celebrated equations in statistical thermodynamics, the 'Boltzmann formula' for the entropy, is obtained by establishing the relation between the entropy and the weight of the most probable configuration.

How is that done? 13E.1 Deriving the Boltzmann formula for the entropy

The starting point for this derivation is the expression for the internal energy, $U(T) = U(0) + N\langle\varepsilon\rangle$, which, with $\langle\varepsilon\rangle=(1/N)\sum_i N_i\varepsilon_i$ can be written

$$U(T)=U(0)+\sum_i N_i\varepsilon_i$$

The strategy is to use classical thermodynamics to establish a relation between dS and dU, and then to use this relation to express dU in terms of the weight of the most probable configuration.

Step 1 *Write an expression for* d$U(T)$

A change in $U(T)$ may arise from either a modification of the energy levels of a system (when ε_i changes to $\varepsilon_i + \mathrm{d}\varepsilon_i$) or from a modification of the populations (when N_i changes to $N_i + \mathrm{d}N_i$). The most general change is therefore

$$\mathrm{d}U(T)=\mathrm{d}U(0)+\sum_i N_i\mathrm{d}\varepsilon_i+\sum_i \varepsilon_i\mathrm{d}N_i$$

Step 2 *Write an expression for* dS

Because neither $U(0)$ nor the energy levels change when a system is heated at constant volume (Fig. 13E.2), in the absence of all changes other than heating, only the third (blue) term on the right survives. Moreover, from eqn 3E.1 ($\mathrm{d}U = T\mathrm{d}S - p\mathrm{d}V$), $\mathrm{d}U = T\mathrm{d}S$ under the same conditions. Therefore,

$$\mathrm{d}S=\frac{\mathrm{d}U}{T}=\frac{1}{T}\sum_i\varepsilon_i\mathrm{d}N_i=k\beta\sum_i\varepsilon_i\mathrm{d}N_i$$

Step 3 *Write an expression for* dS *in terms of* $\mathcal{W}$

Changes in the most probable configuration (the only one to consider) are given by eqn 13A.7 ($\partial(\ln\mathcal{W})/\partial N_i + \alpha - \beta\varepsilon_i = 0$). It follows that $\beta\varepsilon_i = \partial(\ln\mathcal{W})/\partial N_i + \alpha$ and, because the system contains a fixed number of molecules,

$$\mathrm{d}S=k\sum_i\overbrace{\beta\varepsilon_i}^{\partial\ln\mathcal{W}/\partial N_i+\alpha}\mathrm{d}N_i=k\overbrace{\sum_i\left(\frac{\partial\ln\mathcal{W}}{\partial N_i}\right)\mathrm{d}N_i}^{\mathrm{d}\ln\mathcal{W}}+k\alpha\overbrace{\sum_i\mathrm{d}N_i}^{0}=k(\mathrm{d}\ln\mathcal{W})$$

This relation strongly suggests that

$$S=k\ln\mathcal{W} \qquad (13\text{E}.7)$$

Boltzmann formula for the entropy

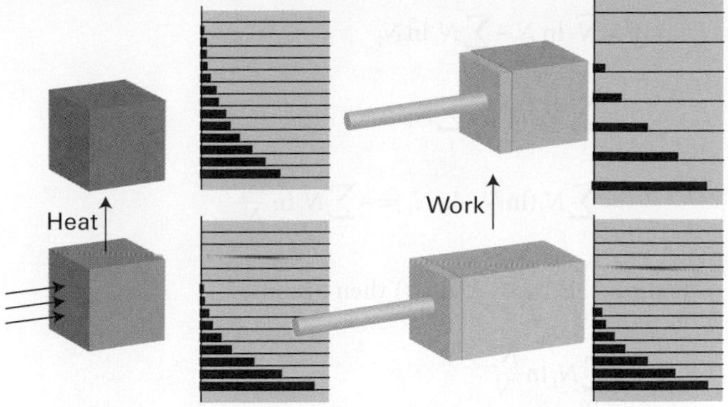

Figure 13E.2 (a) When a system is heated, the energy levels are unchanged but their populations are changed. (b) When work is done on a system, the energy levels themselves are changed. The levels in this case are the one-dimensional particle-in-a-box energy levels of Topic 7D: they depend on the size of the container and move apart as its length is decreased.

This very important expression is the **Boltzmann formula** for the entropy. In it, $\mathcal{W}$ is the weight of the most probable configuration of the system (as discussed in Topic 13A).

(a) Entropy and the partition function

The **statistical entropy**, the entropy calculated from the Boltzmann formula, behaves in exactly the same way as the thermodynamic entropy. Thus, as the temperature is lowered, the value of $\mathcal{W}$, and hence of S, decreases because fewer configurations are consistent with the total energy. In the limit $T \to 0$, $\mathcal{W} = 1$, so $\ln\mathcal{W} = 0$, because only one configuration (every molecule in the lowest level) is compatible with $U(T) = U(0)$. It follows that $S \to 0$ as $T \to 0$, which is compatible with the Third Law of thermodynamics, that the entropies of all perfect crystals approach the same value as $T \to 0$ (Topic 3C).

The challenge now is to establish a relation between the Boltzmann formula and the partition function.

How is that done? 13E.2 Relating the statistical entropy to the partition function

It is necessary to separate the calculation into two parts, one for distinguishable independent molecules and the other for indistinguishable independent molecules. Interactions can be taken into account, in principle at least, by a simple generalization.

Step 1 *Derive the relation for distinguishable independent molecules*

For a system composed of N distinguishable molecules, $\ln\mathcal{W}$ is given by eqn 13A.3 ($\ln\mathcal{W}= N\ln N - \sum_i N_i\ln N_i$); the substitution $N = \sum_i N_i$ then gives

$$\ln \mathcal{W} = \overbrace{N}^{\sum N_i} \ln N - \sum_i N_i \ln N_i$$

$$= \sum_i N_i \ln N - \sum_i N_i \ln N_i$$

$$= \sum_i N_i (\ln N - \ln N_i) = -\sum_i N_i \ln \frac{N_i}{N}$$

Equation 13E.7 ($S = k \ln \mathcal{W}$) then becomes

$$S = -k \sum_i N_i \ln \frac{N_i}{N}$$

The value of N_i/N for the most probable distribution is given by the Boltzmann distribution, $N_i/N = e^{-\beta\varepsilon_i}/q$, and so

$$\ln \frac{N_i}{N} = \ln e^{-\beta\varepsilon_i} - \ln q = -\beta\varepsilon_i - \ln q$$

Therefore,

$$S = k\beta \overbrace{\sum_i N_i \varepsilon_i}^{N\langle\varepsilon\rangle} + k \sum_i N_i \ln q = Nk\beta\langle\varepsilon\rangle + Nk \ln q$$

Finally, because $N\langle\varepsilon\rangle = U(T) - U(0)$ and $\beta = 1/kT$, it follows that

$$S = \frac{U(T) - U(0)}{T} + Nk \ln q \quad (13E.8a)$$

The entropy [independent, distinguishable molecules]

Step 2 *Derive the relation for indistinguishable molecules*

To treat a system composed of N indistinguishable molecules, the weight $\mathcal{W}$ is reduced by a factor of $N!$ because the $N!$ permutations of the molecules among the states result in the same state of the system. Therefore, $\ln \mathcal{W}$, and therefore the entropy, is reduced by $k \ln N!$ from the 'distinguishable' value. Because N is so large, Stirling's approximation ($\ln N! = N \ln N - N$) can be used to convert eqn 13E.8a into

$$S(T) = \frac{U(T) - U(0)}{T} + Nk \ln q - k\overbrace{(N \ln N - N)}^{\ln N!}$$

$$= \frac{U(T) - U(0)}{T} + Nk \ln \frac{q}{N} + kN$$

The kN can be combined with the logarithm by writing it as $kN \ln e$, to give

$$S = \frac{U(T) - U(0)}{T} + Nk \ln \frac{qe}{N} \quad (13E.8b)$$

The entropy [independent, indistinguishable molecules]

Step 3 *Generalize to interacting molecules*

For completeness, the corresponding expression for interacting molecules, based on the canonical partition function in place of the molecular partition function, is

$$S = \frac{U(T) - U(0)}{T} + k \ln Q \quad (13E.8c)$$

The entropy [interacting molecules]

Equation 13E.8a expresses the entropy of a collection of independent molecules in terms of the internal energy and the molecular partition function. However, because the energy of a molecule is a sum of contributions, such as translational (T), rotational (R), vibrational (V), and electronic (E), the partition function factorizes into a product of contributions. As a result, the entropy is also the sum of the individual contributions. In a gas, the molecules are free to change places, so they are indistinguishable; therefore, for the translational contribution to the entropy, use eqn 13E.8b. For the other modes (specifically R, V, and E), which do not involve the exchange of molecules and therefore do not require the weight to be reduced by $N!$, use eqn 13E.8a.

Brief illustration 13E.3

For a system with two states, with energies 0 and ε, it is shown in Topics 13B and 13C that the partition function and mean energy are $q = 1 + e^{-\beta\varepsilon}$ and $\langle\varepsilon\rangle = \varepsilon/(e^{\beta\varepsilon} + 1)$. The contribution to the molar entropy, with $1/T = k\beta$, is therefore

$$S_m = R\left\{\frac{\beta\varepsilon}{1 + e^{\beta\varepsilon}} + \ln(1 + e^{-\beta\varepsilon})\right\}$$

This awkward function is plotted in Fig. 13E.3. It should be noted that as $T \to \infty$ (corresponding to $\beta \to 0$), the molar entropy approaches $R \ln 2$.

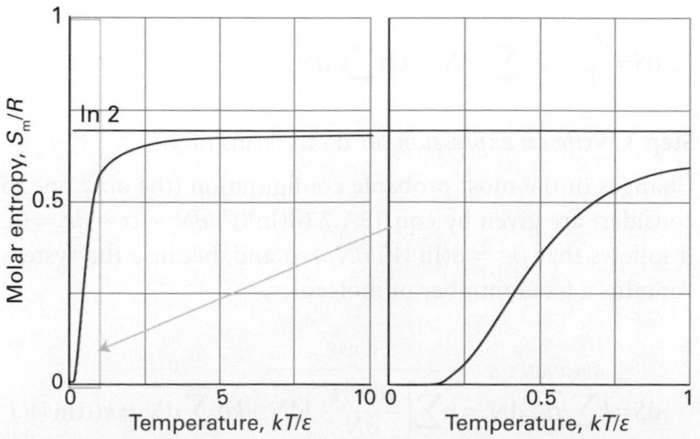

Figure 13E.3 The temperature variation of the molar entropy of a collection of two-level systems expressed as a multiple of $R = N_A k$. As $T \to \infty$ the two states become equally populated and S_m approaches $R \ln 2$.

(b) The translational contribution

The expressions derived for the entropy are in line with what is expected for entropy as a measure of the spread of the populations of molecules over the available states. This interpretation can be illustrated by deriving an expression for the molar entropy of a monatomic perfect gas.

How is that done? 13E.3 Deriving the expression for the entropy of a monatomic perfect gas

You need to start with eqn 13E.8b for a collection of independent, indistinguishable atoms and write $N = nN_A$, where N_A is Avogadro's constant and n is their amount (in moles). The only mode of motion for a gas of atoms is translation and $U(T) - U(0) = \frac{3}{2}nRT$. The partition function is $q = V/\Lambda^3$ (eqn 13B.10b), where Λ is the thermal wavelength. Therefore,

$$S = \overbrace{\frac{U(T)-U(0)}{T}}^{\frac{3}{2}nRT} + nN_A k \ln\frac{qe}{nN_A} = \tfrac{3}{2}nR + \overbrace{nN_A k}^{nR}\ln\frac{Ve}{nN_A\Lambda^3}$$

$$= nR\left\{\overbrace{\tfrac{3}{2}}^{\ln e^{3/2}} + \ln\frac{V_m e}{N_A\Lambda^3}\right\} = nR\ln\frac{V_m e^{5/2}}{N_A\Lambda^3}$$

where $V_m = V/n$ is the molar volume of the gas and $\frac{3}{2}$ has been replaced by $\ln e^{3/2}$. Division of both sides by n then results in the **Sackur–Tetrode equation:**

$$S_m = R\ln\left(\frac{V_m e^{5/2}}{N_A\Lambda^3}\right) \qquad (13E.9a)$$

Sackur–Tetrode equation [monatomic perfect gas]

where Λ is the thermal wavelength ($\Lambda = h/(2\pi mkT)^{1/2}$). To calculate the standard molar entropy, note that $V_m = RT/p$, and set $p = p^\ominus$:

$$S_m^\ominus = R\ln\left(\frac{RTe^{5/2}}{p^\ominus N_A\Lambda^3}\right) \overset{R/N_A=k}{=} R\ln\left(\frac{kTe^{5/2}}{p^\ominus\Lambda^3}\right) \qquad (13E.9b)$$

These expressions are based on the high-temperature approximation of the partition functions, which assumes that many levels are occupied; therefore, they do not apply when T is equal to or very close to zero.

Brief illustration 13E.4

The mass of an Ar atom is $m = 39.95m_u$. At 25 °C, the thermal wavelength of Ar is 16.0 pm and $kT = 4.12 \times 10^{-21}$ J. Therefore, the molar entropy of argon at this temperature is

$$S_m^\ominus = R\ln\left\{\frac{(4.12\times10^{-21}\,\mathrm{J})\times e^{5/2}}{(10^5\,\mathrm{N\,m^{-2}})\times(1.60\times10^{-11}\,\mathrm{m})^3}\right\}$$
$$= 18.6R = 155\,\mathrm{J\,K^{-1}\,mol^{-1}}$$

On the basis that there are fewer accessible translational states for a lighter atom than for a heavy atom under the same conditions (see below), it can be anticipated that the standard molar entropy of Ne is likely to be smaller than for Ar; its actual value is 17.60R at 298 K.

The physical interpretation of these equations is as follows:

- Because the molecular mass appears in the numerator (because it appears in the denominator of Λ), the molar entropy of a perfect gas of heavy molecules is greater than that of a perfect gas of light molecules under the same conditions. This feature can be understood in terms of the energy levels of a particle in a box being closer together for heavy particles than for light particles, so more states are thermally accessible.
- Because the molar volume appears in the numerator, the molar entropy increases with the molar volume of the gas. The reason is similar: large containers have more closely spaced energy levels than small containers, so once again more states are thermally accessible.
- Because the temperature appears in the numerator (because, like m, it appears in the denominator of Λ), the molar entropy increases with increasing temperature. The reason for this behaviour is that more energy levels become accessible as the temperature is raised.

Physical interpretation

The Sackur–Tetrode equation written in the form

$$S = nR\ln\frac{Ve^{5/2}}{nN_A\Lambda^3} = nR\ln aV, \quad a = \frac{e^{5/2}}{nN_A\Lambda^3}$$

implies that when a monatomic perfect gas expands isothermally from V_i to V_f, its entropy changes by

$$\Delta S = nR\ln aV_f - nR\ln aV_i$$
$$= nR\ln\frac{V_f}{V_i} \qquad (13E.10)$$

Change of entropy on expansion [perfect gas, isothermal]

This expression is the same as that obtained starting from the thermodynamic definition of entropy (Topic 3B).

(c) The rotational contribution

The rotational contribution to the molar entropy, S_m^R, can be calculated once the molecular partition function is known.

For a linear molecule, the high-temperature limit of $q^{\rm R}$ is $kT/\sigma hc\tilde{B}$ (eqn 13B.13b, $q^{\rm R}=T/\sigma\theta^{\rm R}$ with $\theta^{\rm R}=hc\tilde{B}/k$) and the equipartition theorem gives the rotational contribution to the molar internal energy as RT; therefore, from eqn 13E.8a:

$$S_{\rm m}^{\rm R}=\frac{\overbrace{U_{\rm m}(T)-U_{\rm m}(0)}^{RT}}{T}+R\ln \overbrace{q^{\rm R}}^{kT/\sigma hc\tilde{B}}$$

and the contribution at high temperatures is

$$S_{\rm m}^{\rm R}=R\left\{1+\ln\frac{kT}{\sigma hc\tilde{B}}\right\} \quad \text{Rotational contribution [linear molecule, high temperature } (T>>\theta^{\rm R})] \quad (13E.11a)$$

In terms of the rotational temperature,

$$S_{\rm m}^{\rm R}=R\left\{1+\ln\frac{T}{\sigma\theta^{\rm R}}\right\} \quad \text{Rotational contribution [Linear molecule, high temperature } (T>>\theta^{\rm R})] \quad (13E.11b)$$

This function is plotted in Fig. 13E.4. It is seen that:

Physical interpretation

- The rotational contribution to the entropy increases with temperature because more rotational states become accessible.
- The rotational contribution is large when $\tilde{B}$ is small, because then the rotational energy levels are close together.

It follows that large, heavy molecules have a large rotational contribution to their entropy. As shown in the following *Brief illustration*, the rotational contribution to the molar entropy of $^{35}Cl_2$ is 58.6 J K^{-1} mol^{-1} whereas that for H_2 is only 12.7 J K^{-1} mol^{-1}. That is, it is appropriate to regard Cl_2 as a more rotationally disordered gas than H_2, in the sense that at a given temperature Cl_2 occupies a greater number of rotational states than H_2 does.

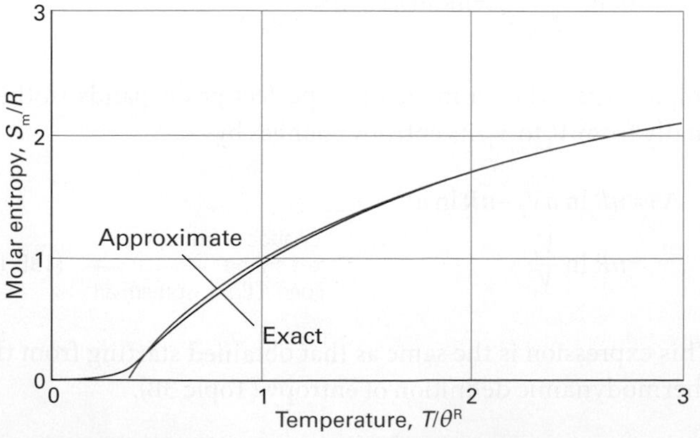

Figure 13E.4 The variation of the rotational contribution to the molar entropy of a linear molecule ($\sigma=1$) using the high-temperature approximation and the exact expression (the latter evaluated up to $J=20$).

Brief illustration 13E.5

The rotational contribution for $^{35}Cl_2$ at 25 °C, for instance, is calculated by noting that $\sigma=2$ for this homonuclear diatomic molecule and taking $\tilde{B}=0.2441\,{\rm cm}^{-1}$ (corresponding to 24.41 m^{-1}). The rotational temperature of the molecule is

$$\theta^{\rm R}=\frac{(6.626\times10^{-34}\,{\rm J\,s})\times(2.998\times10^{8}\,{\rm m\,s^{-1}})\times(24.41\,{\rm m^{-1}})}{1.381\times10^{-23}\,{\rm J\,K^{-1}}}=0.351\,{\rm K}$$

Therefore,

$$S_{\rm m}^{\rm R}=R\left\{1+\ln\frac{298\,{\rm K}}{2\times(0.351\,{\rm K})}\right\}=7.05R=58.6\,{\rm J\,K^{-1}\,mol^{-1}}$$

Equation 13E.11 is valid at high temperatures ($T>>\theta^{\rm R}$). To track the rotational contribution down to low temperatures it is necessary to use the full form of the rotational partition function (Topic 13B; see Problem P13E.12). The resulting graph has the form shown in Fig. 13E.4, and it is seen that the approximate curve matches the exact curve very well for $T/\theta^{\rm R}$ greater than about 1.

(d) The vibrational contribution

The vibrational contribution to the molar entropy, $S_{\rm m}^{\rm V}$, is obtained by combining the expression for the molecular partition function (eqn 13B.15, $q^{\rm V}=1/(1-{\rm e}^{-\beta hc\tilde{\nu}})=1/(1-{\rm e}^{-\beta\varepsilon})$ for $\varepsilon=hc\tilde{\nu}$) with the expression for the mean energy (eqn 13C.8, $\langle\varepsilon^{\rm V}\rangle=\varepsilon/({\rm e}^{\beta\varepsilon}-1)$), to obtain

$$S_{\rm m}^{\rm V}=\frac{\overbrace{U_{\rm m}(T)-U_{\rm m}(0)}^{N_{\rm A}\langle\varepsilon^{\rm V}\rangle}}{\underbrace{T}_{1/k\beta}}+R\ln q^{\rm V}=\frac{\overbrace{N_{\rm A}k}^{R}\beta\varepsilon}{{\rm e}^{\beta\varepsilon}-1}+R\ln\frac{1}{1-{\rm e}^{-\beta\varepsilon}}$$

$$=R\left\{\frac{\beta\varepsilon}{{\rm e}^{\beta\varepsilon}-1}-\ln(1-{\rm e}^{-\beta\varepsilon})\right\}$$

That is,

$$S_{\rm m}^{\rm V}=R\left\{\frac{\beta hc\tilde{\nu}}{{\rm e}^{\beta hc\tilde{\nu}}-1}-\ln(1-{\rm e}^{-\beta hc\tilde{\nu}})\right\} \quad \text{Vibrational contribution to the molar entropy} \quad (13E.12a)$$

Once again it is convenient to express this formula in terms of a characteristic temperature, in this case the vibrational temperature $\theta^{\rm V}=hc\tilde{\nu}/k$:

$$S_{\rm m}^{\rm V}=R\left\{\frac{\theta^{\rm V}/T}{{\rm e}^{\theta^{\rm V}/T}-1}-\ln(1-{\rm e}^{-\theta^{\rm V}/T})\right\} \quad \text{Vibrational contribution to the molar entropy} \quad (13E.12b)$$

This function is plotted in Fig. 13E.5. As usual, it is helpful to interpret it, with the graph in mind:

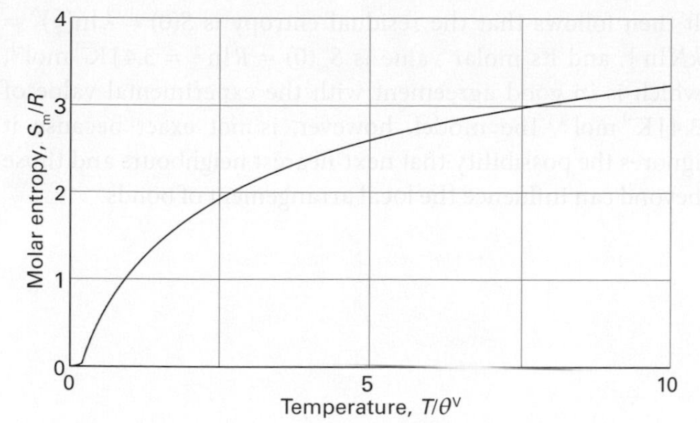

Figure 13E.5 The temperature variation of the molar entropy of a collection of harmonic oscillators expressed as a multiple of $R = N_A k$. The molar entropy approaches zero as $T \to 0$, and increases without limit as $T \to \infty$.

Physical interpretation

- Both terms multiplying R become zero as $T \to 0$, so the entropy is zero at $T = 0$.
- The molar entropy rises as the temperature is increased as more vibrational states become accessible.
- The molar entropy is higher at a given temperature for molecules with heavy atoms or low force constant than one with light atoms or high force constant. The vibrational energy levels are closer together in the former case than in the latter, so more are thermally accessible.

Brief illustration 13E.6

The vibrational wavenumber of I_2 is 214.6 cm^{-1}, corresponding to 2.146×10^4 m^{-1}, so its vibrational temperature is 309 K. At 25 °C

$$S_m^V = R\left\{\frac{(309\,K)/(298\,K)}{e^{(309\,K)/(298\,K)}-1} - \ln(1-e^{-(309\,K)/(298\,K)})\right\} = 1.01R$$
$$= 8.38\,J\,K^{-1}\,mol^{-1}$$

(e) Residual entropies

Entropies may be calculated from spectroscopic data; they may also be measured experimentally (Topic 3C). In many cases there is good agreement, but in some the experimental entropy is less than the calculated value. One possibility is that the experimental determination failed to take a phase transition into account and a contribution of the form $\Delta_{trs}H/T_{trs}$ was incorrectly omitted from the sum. Another possibility is that some disorder is present in the solid even at $T = 0$. The entropy at $T = 0$ is then greater than zero and is called the **residual entropy**.

The origin and magnitude of the residual entropy can be explained by considering a crystal composed of AB molecules, where A and B are similar atoms (such as CO, with its very small electric dipole moment). There may be so little energy difference between ... AB AB AB AB ..., ... AB BA BA AB ..., and other arrangements that the molecules adopt the orientations AB and BA at random in the solid. The entropy arising from residual disorder can be calculated readily by using the Boltzmann formula, $S = k \ln \mathcal{W}$. To do so, suppose that two orientations are equally probable, and that the sample consists of N molecules. Because the same energy can be achieved in 2^N different ways (because each molecule can take either of two orientations), the total number of ways of achieving the same energy is $\mathcal{W} = 2^N$. It follows that

$$S = k \ln 2^N = Nk \ln 2 = nR \ln 2 \qquad (13E.13a)$$

and that a residual molar entropy of $R \ln 2 = 5.8$ J K^{-1} mol^{-1} is expected for solids composed of molecules that can adopt either of two orientations at $T = 0$. If s orientations are possible, the residual molar entropy is

$$S_m(0) = R \ln s \qquad \text{Residual entropy} \qquad (13E.13b)$$

For CO, the measured residual entropy is 5 J K^{-1} mol^{-1}, which is close to $R \ln 2$, the value expected for a random structure of the form ...CO CO OC CO OC OC....

Similar arguments apply to more complicated cases. Consider a sample of ice with N H_2O molecules. Each O atom is surrounded tetrahedrally by four H atoms, two of which are attached by short σ bonds, the other two being attached by long hydrogen bonds (Fig. 13E.6). It follows that each of the $2N$ H atoms can be in one of two positions (either close to or far from an O atom as shown in Fig. 13E.7), resulting in 2^{2N} possible arrangements. However, not all these arrangements are acceptable. Indeed, of the $2^4 = 16$ ways of arranging four H atoms around one O atom, only 6 have two short and two long OH distances and hence are acceptable (Fig. 13E.7). Therefore, the number of permitted arrangements is $\mathcal{W} = 2^{2N}(\frac{6}{16})^N = (\frac{3}{2})^N$.

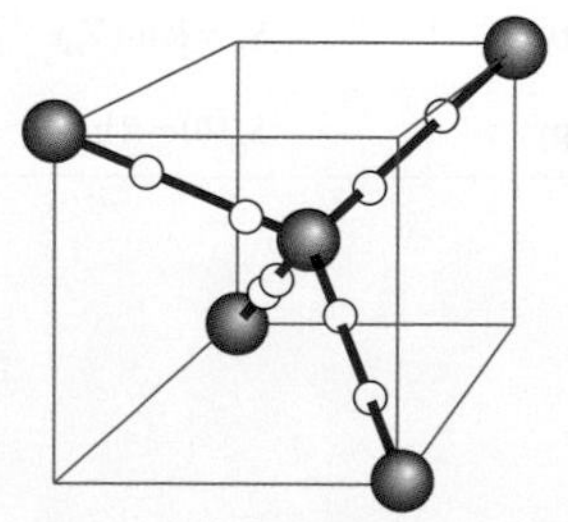

Figure 13E.6 The possible locations of H atoms around a central O atom in an ice crystal are shown by the white spheres. Only one of the locations on each bond may be occupied by an atom, and two H atoms must be close to the O atom and two H atoms must be distant from it.

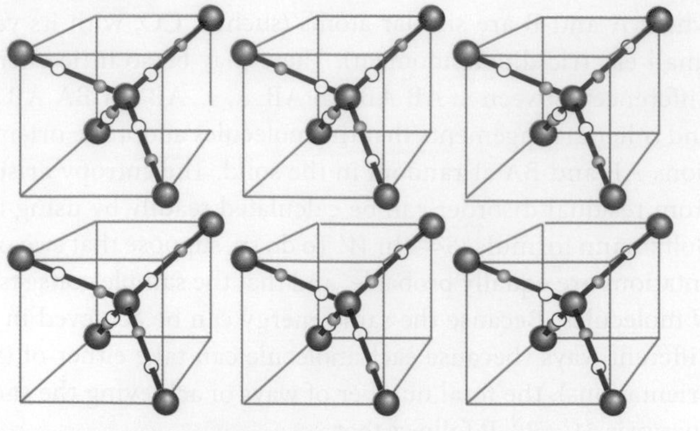

Figure 13E.7 The six possible arrangements of H atoms in the locations identified in Fig.13E.6. Occupied locations are denoted by grey spheres and unoccupied locations by white spheres.

It then follows that the residual entropy is $S(0) \approx k\ln(\frac{3}{2})^N = kN\ln\frac{3}{2}$, and its molar value is $S_m(0) \approx R\ln\frac{3}{2} = 3.4\,J\,K^{-1}\,mol^{-1}$, which is in good agreement with the experimental value of $3.4\,J\,K^{-1}\,mol^{-1}$. The model, however, is not exact because it ignores the possibility that next-nearest neighbours and those beyond can influence the local arrangement of bonds.

Checklist of concepts

☐ 1. The **internal energy** is proportional to the derivative of the logarithm of the partition function with respect to temperature.

☐ 2. The total **heat capacity** of a molecular substance is the sum of the contribution of each mode.

☐ 3. The **statistical entropy** is defined by the Boltzmann formula and expressed in terms of the molecular partition function.

☐ 4. The **residual entropy** is a non-zero entropy at $T = 0$ arising from molecular disorder.

Checklist of equations

Property	Equation	Comment	Equation number
Internal energy	$U(T)=U(0)-(N/q)(\partial q/\partial\beta)_V=U(0)-N(\partial\ln q/\partial\beta)_V$	Independent molecules	13E.2a
Heat capacity	$C_V=Nk\beta^2(\partial^2\ln q/\partial\beta^2)_V$	Independent molecules	13E.5
	$C_{V,m}=\frac{1}{2}(3+\nu^{R*}+2\nu^{V*})R$	$T \gg \theta^M$	13E.6
Boltzmann formula for the entropy	$S=k\ln\mathcal{W}$	Definition	13E.7
Entropy	$S=\{U(T)-U(0)\}/T+Nk\ln q$	Distinguishable molecules	13E.8a
	$S=\{U(T)-U(0)\}/T+Nk\ln(qe/N)$	Indistinguishable molecules	13E.8b
Sackur–Tetrode equation	$S_m=R\ln(V_m e^{5/2}/N_A\Lambda^3)$	Molar entropy of a monatomic perfect gas	13E.9a
Residual molar entropy	$S_m(0)=R\ln s$	s is the number of equivalent orientations	13E.13b

TOPIC 13F Derived functions

➤ **Why do you need to know this material?**

The power of chemical thermodynamics stems from its deployment of a variety of derived functions, particularly the enthalpy and Gibbs energy. It is therefore important to relate these functions to structural features through partition functions.

➤ **What is the key idea?**

The partition function provides a link between spectroscopic and structural data and the derived functions of thermodynamics, particularly the equilibrium constant.

➤ **What do you need to know already?**

This Topic develops the discussion of internal energy and entropy (Topic 13E). You need to know the relations between those properties and the enthalpy (Topic 2B) and the Helmholtz and Gibbs energies (Topic 3D). The final section makes use of the relation between the standard reaction Gibbs energy and the equilibrium constant (Topic 6A). Although the equations are introduced in terms of the canonical partition function (Topic 13D), all the applications are in terms of the molecular partition function (Topic 13B).

Classical thermodynamics makes extensive use of various derived functions. Thus, in thermochemistry the focus is on the enthalpy and, provided the pressure and temperature are constant, in discussions of spontaneity the focus is on the Gibbs energy. All these functions are derived from the internal energy and the entropy, which in terms of the canonical partition function, Q, are given by

$$U(T)=U(0)-\left(\frac{\partial \ln Q}{\partial \beta}\right)_V \quad \text{with } \beta=\frac{1}{kT} \qquad \text{Internal energy} \quad (13\text{F}.1\text{a})$$

$$S(T)=\frac{U(T)-U(0)}{T}+k\ln Q$$

$$=-k\beta\left(\frac{\partial \ln Q}{\partial \beta}\right)_V+k\ln Q \qquad \text{Entropy} \quad (13\text{F}.1\text{b})$$

There is no need to worry about this appearance of the canonical partition function (Topic 13D): the only applications in this Topic make use of the molecular partition function (q, Topic 13B). Equation 13F.1 can be regarded simply as a succinct way of expressing the relations for collections of independent molecules by writing $Q=q^N$ for distinguishable molecules and $Q=q^N/N!$ for indistinguishable molecules (as in a gas). The advantage of using Q is that the equations to be derived are simple and, if necessary, can be used to calculate thermodynamic properties when there are interactions between the molecules.

13F.1 The derivations

The Helmholtz energy, A, is defined as $A = U - TS$. This relation implies that at $T = 0$, $A(0) = U(0)$, so substitution of the expressions for $U(T)$ and $S(T)$ in eqn 13F.1 gives

$$A(T)=A(0)-\left(\frac{\partial \ln Q}{\partial \beta}\right)_V-T\left\{-k\beta\left(\frac{\partial \ln Q}{\partial \beta}\right)_V+k\ln Q\right\}$$

$$=A(0)-kT\ln Q$$

That is,

$$A(T)=A(0)-kT\ln Q \qquad \text{Helmholtz energy} \quad (13\text{F}.2)$$

An infinitesimal change in conditions changes the Helmholtz energy by $\mathrm{d}A = -p\mathrm{d}V - S\mathrm{d}T$ (this is the analogue of the expression for $\mathrm{d}G$ derived in Topic 3E (eqn 3E.7, $\mathrm{d}G = V\mathrm{d}p - S\mathrm{d}T$)). It follows that on imposing constant temperature ($\mathrm{d}T = 0$), the pressure and the Helmholtz energy are related by $p=-(\partial A/\partial V)_T$. It then follows from eqn 13F.2 that

$$p=kT\left(\frac{\partial \ln Q}{\partial V}\right)_T \qquad \text{Pressure} \quad (13\text{F}.3)$$

This relation is entirely general, and may be used for any type of substance, including perfect gases, real gases, and liquids. Because Q is in general a function of the volume, temperature, and amount of substance, eqn 13F.3 is an equation of state of the kind discussed in Topics 1A and 3E.

Example 13F.1 Deriving an equation of state

Derive an expression for the pressure of a gas of independent particles.

Collect your thoughts You should suspect that the pressure is that given by the perfect gas law, $p = nRT/V$. To proceed systematically, substitute the explicit formula for Q for a gas of independent, indistinguishable molecules. Only the translational partition function depends on the volume, so there is no need to include the partition functions for the internal modes of the molecules.

The solution For a gas of independent molecules, $Q = q^N/N!$ with $q = V/\Lambda^3$:

$$p = kT\left(\frac{\partial \ln Q}{\partial V}\right)_T = kT\left(\frac{\partial \ln(q^N/N!)}{\partial V}\right)_T$$

$$= kT\left(\frac{\partial(\ln q^N - \ln N!)}{\partial V}\right)_T = kT\left(\frac{\partial(N\ln q - \ln N!)}{\partial V}\right)_T$$

$$= NkT\left(\frac{\partial \ln q}{\partial V}\right)_T - kT\overbrace{\left(\frac{\partial \ln N!}{\partial V}\right)_T}^{0}$$

$d\ln f/dx = (1/f)\,df/dx$

$$= \frac{NkT}{q}\left(\frac{\partial q}{\partial V}\right)_T$$

$N_A k = R$

$$= \frac{NkT}{V/\Lambda^3}\left(\frac{\partial(V/\Lambda^3)}{\partial V}\right)_T = \frac{NkT}{V} = \frac{nN_A kT}{V} = \frac{nRT}{V}$$

The calculation shows that the equation of state of a gas of independent particles is indeed the perfect gas law, $pV = nRT$.

Comment. If β is left undefined in eqn 13F.1, then the same calculation carried through results in $p = N/\beta V$. For this result to be the perfect gas law, it follows that $\beta = 1/kT$, as anticipated and used in Topics 13A–13E. This is the formal proof of that relation. A similar approach can be used to derive an equation of state resembling the van der Waals equation: see *A deeper look* 7 on the website of this text.

Self-test 13F.1 Derive the equation of state of a gas for which $Q = q^N f/N!$, with $q = V/\Lambda^3$, where f depends on the volume.

Answer: $p = nRT/V + kT(\partial \ln f/\partial V)_T$

At this stage the expressions for U and p and the definition $H = U + pV$, with $H(0) = U(0)$, can be used to obtain an expression for the enthalpy, H, of any substance:

$$H(T) = H(0) - \left(\frac{\partial \ln Q}{\partial \beta}\right)_V + kTV\left(\frac{\partial \ln Q}{\partial V}\right)_T \quad \text{Enthalpy} \quad (13F.4)$$

The fact that eqn 13F.4 is rather cumbersome is a sign that the enthalpy is not a fundamental property: as shown in Topic 2B, it is more of an accounting convenience. For a gas of independent structureless particles $U(T) - U(0) = \tfrac{3}{2}nRT$ and $pV = nRT$. Therefore, for such a gas, it follows directly from $H = U + pV$ that

$$H(T) - H(0) = \tfrac{5}{2}nRT \quad (13F.5)$$

One of the most important thermodynamic functions for chemistry is the Gibbs energy, $G = H - TS$. This definition implies that $G = U + pV - TS$ and therefore that $G = A + pV$. Note that $G(0) = A(0)$, both being equal to $U(0)$. The Gibbs energy can now be written in terms of the partition function by combining the expressions for A and p:

$$G(T) = G(0) - kT\ln Q + kTV\left(\frac{\partial \ln Q}{\partial V}\right)_T \quad \text{Gibbs energy} \quad (13F.6)$$

This expression takes a simple form for a gas of independent molecules because pV in the expression $G = A + pV$ can be replaced by nRT:

$$G(T) = G(0) - kT\ln Q + nRT \quad (13F.7)$$

Furthermore, because $Q = q^N/N!$ for the indistinguishable particles in a gas and therefore $\ln Q = N\ln q - \ln N!$, it follows by using Stirling's approximation ($\ln N! = N\ln N - N$) that

$$G(T) = G(0) - \overbrace{NkT}^{nN_A k = nR}\ln q + kT\overbrace{\ln N!}^{N\ln N - N} + nRT$$

$$= G(0) - nRT\ln q + kT(N\ln N - N) + nRT$$

$$= G(0) - nRT\ln q + \overbrace{NkT}^{nRT}\ln N$$

$$= G(0) - nRT\ln\frac{q}{N} \quad (13F.8)$$

Note that a statistical interpretation of the Gibbs energy is now possible: because q is the number of thermally accessible states and N is the number of molecules, the difference $G(T) - G(0)$ is proportional to the logarithm of the average number of thermally accessible states available to a molecule. As this average number increases, the Gibbs energy falls further and further below $G(0)$. The thermodynamic tendency to lower Gibbs energy can now be seen to be a tendency to maximize the number of thermally accessible states.

It turns out to be convenient to define the **molar partition function**, $q_m = q/n$ (with units mol^{-1} and $n = N/N_A$), for then

$$G(T) = G(0) - nRT\ln\frac{q_m}{N_A} \quad \text{Gibbs energy [indistinguishable, independent molecules]} \quad (13F.9a)$$

To use this expression, $G(0) = U(0)$ is identified with the energy of the system when all the molecules are in their ground state, E_0. To calculate the standard Gibbs energy, the partition function has its standard value, $q_m^{\ominus}$, which is evaluated by setting the molar volume in the translational contribution equal to the standard molar volume, so $q_m^{\ominus} = (V_m^{\ominus}/\Lambda^3)q^R q^V$ with $V_m^{\ominus} = RT/p^{\ominus}$. The standard *molar* Gibbs energy is then obtained with these substitutions and after dividing through by n:

$$G_m^{\ominus}(T) = G_m^{\ominus}(0) - RT\ln\frac{q_m^{\ominus}}{N_A} \qquad (13F.9b)$$

Standard molar Gibbs energy [indistinguishable, independent molecules]

where $G_m^{\ominus}(0) = E_{0,m}$, the molar ground-state energy of the system.

Example 13F.2 Calculating a standard Gibbs energy of formation from partition functions

Calculate the standard Gibbs energy of formation of $H_2O(g)$ at 25 °C.

Collect your thoughts Write the chemical equation for the formation reaction, and then the expression for the standard Gibbs energy of formation in terms of the Gibbs energy of each molecule; then express those Gibbs energies in terms of the molecular partition functions. Ignore molecular vibration as it is unlikely to be excited at 25 °C. The rotational constant for H_2 is 60.864 cm^{-1}, that for O_2 is 1.4457 cm^{-1}, and those for H_2O are 27.877, 14.512, and 9.285 cm^{-1}. Before using the approximate form of the rotational partition functions, you need to judge whether the temperature is high enough: if it is not, use the full expression. For the values of $E_{0,m}$, use bond enthalpies from Table 9C.3; for precise calculations, bond energies (at $T = 0$) should be used instead. You will need the following expressions from Topic 13B:

$$\Lambda(\mathrm{J}) = \frac{h}{\{2\pi m(\mathrm{J})kT\}^{1/2}} \quad \overbrace{q^R = \frac{kT}{2hc\tilde{B}}}^{\text{Linear molecule, } \sigma=2} \quad \text{and} \quad \overbrace{q^R = \tfrac{1}{2}\left(\frac{kT}{hc}\right)^{3/2}\left(\frac{\pi}{\tilde{A}\tilde{B}\tilde{C}}\right)^{1/2}}^{\text{Nonlinear molecule, } \sigma=2}$$

The solution The chemical reaction is $H_2(g) + \tfrac{1}{2}O_2(g) \rightarrow H_2O(g)$. Therefore,

$$\Delta_f G^{\ominus} = G_m^{\ominus}(H_2O,g) - G_m^{\ominus}(H_2,g) - \tfrac{1}{2}G_m^{\ominus}(O_2,g)$$

Now write the standard molar Gibbs energies in terms of the standard molar partition functions of each species J:

$$G_m^{\ominus}(\mathrm{J}) = E_{0,m}(\mathrm{J}) - RT\ln\frac{q_m^{\ominus}(\mathrm{J})}{N_A},\; q_m^{\ominus}(\mathrm{J}) = q_m^{T\ominus}(\mathrm{J})q^R(\mathrm{J}) = \frac{V_m^{\ominus}}{\Lambda(\mathrm{J})^3}q^R(\mathrm{J})$$

Therefore

$$\Delta_f G^{\ominus} = \left\{E_{0,m}(H_2O) - RT\ln\frac{q_m^{\ominus}(H_2O)}{N_A}\right\} - \left\{E_{0,m}(H_2) - RT\ln\frac{q_m^{\ominus}(H_2)}{N_A}\right\} - \tfrac{1}{2}\left\{E_{0,m}(O_2) - RT\ln\frac{q_m^{\ominus}(O_2)}{N_A}\right\}$$

$$= \Delta E_{0,m} - RT\ln\frac{\overbrace{\{V_m^{\ominus}/N_A\Lambda(H_2O)^3\}q^R(H_2O)}^{q_m^{\ominus}(H_2O)/N_A}}{\underbrace{\left[\{V_m^{\ominus}/N_A\Lambda(H_2)^3\}q^R(H_2)\right]}_{q_m^{\ominus}(H_2)/N_A}\underbrace{\left[\{V_m^{\ominus}/N_A\Lambda(O_2)^3\}q^R(O_2)\right]^{1/2}}_{q_m^{\ominus}(O_2)/N_A}}$$

$$= \Delta E_{0,m} - RT\ln\frac{N_A^{1/2}\{\Lambda(H_2)\Lambda(O_2)^{1/2}/\Lambda(H_2O)\}^3}{V_m^{\ominus\,1/2}\{q^R(H_2)q^R(O_2)^{1/2}/q^R(H_2O)\}}$$

Now substitute the data. The molar energy difference (with bond dissociation energies in kilojoules per mole indicated) is

$$\Delta E_{0,m} = \overbrace{E_{0,m}(H_2O)}^{-492-428=-920} - \overbrace{E_{0,m}(H_2)}^{-436} - \tfrac{1}{2}\overbrace{E_{0,m}(O_2)}^{-497} = -236\,\mathrm{kJ\,mol^{-1}}$$

Next, establish whether the temperature is high enough for the approximate expressions for the rotational partition functions (Topic 13B) to be reliable:

	H_2O	H_2O	H_2O	H_2	O_2
$\tilde{X}$/cm^{-1}, $X = A, B,$ or C	27.877	14.512	9.285	60.864	1.4457
θ^R/K	40.1	20.9	13.4	87.6	2.1

Only H_2 is marginal at 298 K, so for it, use the full calculation of the rotational partition function; for the others, use the approximate forms quoted above:

$\Lambda(H_2) = 71.21$ pm $\quad \Lambda(O_2) = 17.87$ pm $\quad \Lambda(H_2O) = 23.82$ pm

$q^R(H_2) = 1.88 \quad q^R(O_2) = 71.60 \quad q^R(H_2O) = 42.13$

It then follows that

$$\Delta_f G^{\ominus} = -236\,\mathrm{kJ\,mol^{-1}} - \overbrace{RT}^{2.48\,\mathrm{kJ\,mol^{-1}}}\ln 0.0291 = -227\,\mathrm{kJ\,mol^{-1}}$$

The value quoted in Table 2C.1 of the *Resource section* is −228.57 kJ mol^{-1}.

Self-test 15E.2 Estimate the standard Gibbs energy of formation of $NH_3(g)$ at 25 °C. The rotational constants of NH_3 are $\tilde{B} = 10.001$ cm^{-1} and $\tilde{A} = 6.449$ cm^{-1}.

Answer: −16 kJ mol^{-1}

13F.2 Equilibrium constants

The following discussion is focused on gas-phase reactions in which the equilibrium constant is defined in terms of the partial pressures of the reactants and products.

(a) The relation between *K* and the partition function

The Gibbs energy of a gas of independent molecules is given by eqn 13F.9 in terms of the molar partition function, $q_m = q/n$. The equilibrium constant K of a reaction is related to the standard reaction Gibbs energy. The task is to combine these two relations and so obtain an expression for the equilibrium constant in terms of the molecular partition functions of the reactants and products.

How is that done? 13F.1 Relating the equilibrium constant to partition functions

You need to use the expressions for the standard molar Gibbs energies, $G^{\ominus}/n$, of each species to find an expression for the standard reaction Gibbs energy. Then find the equilibrium constant K by using eqn 6A.15 ($\Delta_r G^{\ominus} = -RT\ln K$).

Step 1 *Write an expression for* $\Delta_r G^{\ominus}$

From eqn 13F.9b, the standard molar reaction Gibbs energy for the reaction $aA + bB \rightarrow cC + dD$ is

$$\Delta_r G^{\ominus} = cG^{\ominus}_m(C) + dG^{\ominus}_m(D) - \{aG^{\ominus}_m(A) + bG^{\ominus}_m(B)\}$$

$$= cG^{\ominus}_m(C,0) + dG^{\ominus}_m(D,0) - \{aG^{\ominus}_m(A,0) + bG^{\ominus}_m(B,0)\}$$

$$-RT\left\{c\ln\frac{q^{\ominus}_{C,m}}{N_A} + d\ln\frac{q^{\ominus}_{D,m}}{N_A} - a\ln\frac{q^{\ominus}_{A,m}}{N_A} - b\ln\frac{q^{\ominus}_{B,m}}{N_A}\right\}$$

Because $G_m(J,0) = E_{0,m}(J)$, the molar ground-state energy of the species J, the first (blue) term on the right is

$$cE_{0,m}(C,0) + dE_{0,m}(D,0) - \{aE_{0,m}(A,0) + bE_{0,m}(B,0)\} = \Delta_r E_0$$

Then, by using $a\ln x = \ln x^a$ and $\ln x + \ln y = \ln xy$,

$$\Delta_r G^{\ominus} = \Delta_r E_0 - RT\ln\frac{(q^{\ominus}_{C,m}/N_A)^c(q^{\ominus}_{D,m}/N_A)^d}{(q^{\ominus}_{A,m}/N_A)^a(q^{\ominus}_{B,m}/N_A)^b}$$

$$= -RT\left\{-\frac{\Delta_r E_0}{RT} + \ln\frac{(q^{\ominus}_{C,m}/N_A)^c(q^{\ominus}_{D,m}/N_A)^d}{(q^{\ominus}_{A,m}/N_A)^a(q^{\ominus}_{B,m}/N_A)^b}\right\}$$

Step 2 *Write an expression for K*

At this stage pick out an expression for K by comparing this equation with $\Delta_r G^{\ominus} = -RT\ln K$, which gives

$$\ln K = -\frac{\Delta_r E_0}{RT} + \ln\frac{(q^{\ominus}_{C,m}/N_A)^c(q^{\ominus}_{D,m}/N_A)^d}{(q^{\ominus}_{A,m}/N_A)^a(q^{\ominus}_{B,m}/N_A)^b}$$

Finally, by forming the exponential of both sides, this equation becomes

$$K = \frac{(q^{\ominus}_{C,m}/N_A)^c(q^{\ominus}_{D,m}/N_A)^d}{(q^{\ominus}_{A,m}/N_A)^a(q^{\ominus}_{B,m}/N_A)^b}e^{-\Delta_r E_0/RT} \quad \text{The equilibrium constant} \quad (13F.10a)$$

where $\Delta_r E_0$ is the difference in molar energies of the ground states of the products and reactants and is calculated from the bond dissociation energies of the species (Fig. 13F.1). In terms of the (signed) stoichiometric numbers introduced in Topic 2C, eqn 13F.10a can be written

$$K = \left\{\prod_J \left(\frac{q^{\ominus}_{J,m}}{N_A}\right)^{\nu_J}\right\}e^{-\Delta_r E_0/RT} \quad (13F.10b)$$

The equilibrium constant

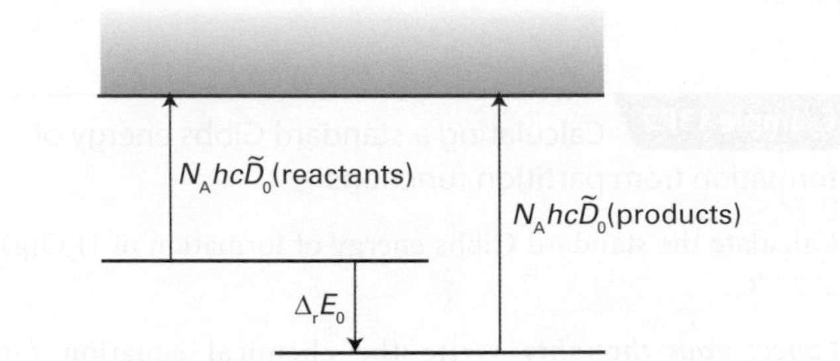

Figure 13F.1 The definition of $\Delta_r E_0$ for the calculation of equilibrium constants. The reactants are imagined as dissociating into atoms and then forming the products from the atoms.

(b) A dissociation equilibrium

Equation 13F.10a can be used to write an expression for the equilibrium constant for the dissociation of a diatomic molecule X_2:

$$X_2(g) \rightleftharpoons 2X(g) \quad K = \frac{p_X^2}{p_{X_2}p^{\ominus}}$$

According to eqn 13F.10a (with $a = 1$, $b = 0$, $c = 2$, and $d = 0$):

$$K = \frac{(q^{\ominus}_{X,m}/N_A)^2}{q^{\ominus}_{X_2,m}/N_A}e^{-\Delta_r E_0/RT} = \frac{(q^{\ominus}_{X,m})^2}{q^{\ominus}_{X_2,m}N_A}e^{-\Delta_r E_0/RT} \quad (13F.11a)$$

with

$$\Delta_r E_0 = 2E_{0,m}(X,0) - E_{0,m}(X_2,0) = N_A hc\tilde{D}_0(X\text{–}X) \quad (13F.11b)$$

where $N_A hc\tilde{D}_0$(X–X) is the (molar) dissociation energy of the X–X bond. The standard molar partition functions of the atoms X are

$$q^{\ominus}_{X,m} = \overbrace{g_X}^{q^E} \times \overbrace{\frac{V^{\ominus}_m}{\Lambda_X^3}}^{q^T} = \frac{g_X RT}{p^{\ominus}\Lambda_X^3} \quad (V^{\ominus}_m = RT/p^{\ominus})$$

where g_X is the degeneracy of the electronic ground state of X. The diatomic molecule X_2 also has rotational and vibrational degrees of freedom, so its standard molar partition function is

$$q^{\ominus}_{X_2,m} = g_{X_2}\frac{V^{\ominus}_m}{\Lambda^3_{X_2}}q^R_{X_2}q^V_{X_2} = \frac{g_{X_2}RTq^R_{X_2}q^V_{X_2}}{p^{\ominus}\Lambda^3_{X_2}}$$

where g_{X_2} is the degeneracy of the electronic ground state of X_2. It follows that

$$K = \frac{(g_X RT/p^{\ominus}\Lambda^3_X)^2}{g_{X_2}N_A RT q^R_{X_2}q^V_{X_2}/p^{\ominus}\Lambda^3_{X_2}}e^{-N_A hc\tilde{D}_0(X-X)/RT}$$

$R = N_A k$

$$= \frac{g^2_X kT\Lambda^3_{X_2}}{g_{X_2}p^{\ominus}q^R_{X_2}q^V_{X_2}\Lambda^6_X}e^{-hc\tilde{D}_0(X-X)/kT} \quad (13F.12)$$

All the quantities in this expression can be calculated from spectroscopic data.

Example 13F.3 Evaluating an equilibrium constant

Evaluate the equilibrium constant for the dissociation $Na_2(g) \rightarrow 2Na(g)$ at 1000 K. The data for Na_2 are: $\tilde{B} = 0.1547\,cm^{-1}$, $\tilde{\nu} = 159.2\,cm^{-1}$, and $N_A hc\tilde{D}_0 = 70.4\,kJ\,mol^{-1}$.

Collect your thoughts Recall that Na has a doublet ground state (term symbol $^2S_{1/2}$). You need to use eqn 13F.12 and the expressions for the partition functions assembled in Topic 13B. For such a heavy diatomic molecule it is safe to use the approximate expression for its rotational partition function (but check that assumption for consistency). Remember that for a homonuclear diatomic molecule, $\sigma = 2$.

The solution The partition functions and other quantities required are as follows:

$\Lambda(Na_2) = 8.14\,pm$	$\Lambda(Na) = 11.5\,pm$
$q^R(Na_2) = 2246$	$q^V(Na_2) = 4.885$
$g(Na) = 2$	$g(Na_2) = 1$
$hc\tilde{D}/kT = 8.47\ldots$	

There are many rotational states occupied, so the use of the approximate formula for $q^R(Na_2)$ is valid. Then, from eqn 13F.12,

$$K = \frac{2^2\times(1.381\times10^{-23}\,\overbrace{J}^{kg\,m^2\,s^{-2}}\,K^{-1})\times(1000\,K)\times(8.14\times10^{-12}\,m)^3}{(10^5\,\underbrace{Pa}_{kg\,m^{-1}\,s^{-2}})\times2246\times4.885\times(1.15\times10^{-11}\,m)^6}\times e^{-8.47\ldots}$$

$$= 2.45$$

Self-test 13F.3 Evaluate K at 1500 K. Is the answer consistent with the dissociation being endothermic?

Answer: 52; yes

(c) Contributions to the equilibrium constant

To appreciate the physical basis of equilibrium constants, consider a simple R ⇌ P gas-phase equilibrium (R for reactants, P for products).

Figure 13F.2 shows two sets of energy levels; one set of states belongs to R, and the other belongs to P. The populations of the states are given by the Boltzmann distribution, and are independent of whether any given state happens to belong to R or to P. A single Boltzmann distribution spreads, without distinction, over the two sets of states. If the spacings of R and P are similar (as in Fig. 13F.2), and the ground state of P lies above that of R, the diagram indicates that R will dominate in the equilibrium mixture. However, if P has a high density of states (a large number of states in a given energy range, as in Fig. 13F.3), then, even though its ground-state energy lies above that of R, the species P might still dominate at equilibrium.

It is quite easy to show that the ratio of numbers of R and P molecules at equilibrium is given by a Boltzmann-like expression.

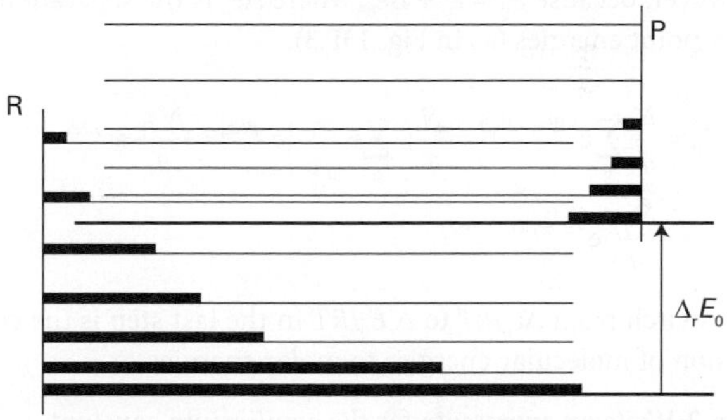

Figure 13F.2 The array of R(eactant) and P(roduct) energy levels. At equilibrium all are accessible (to differing extents, depending on the temperature), and the equilibrium composition of the system reflects the overall Boltzmann distribution of populations. As $\Delta_r E_0$ increases, R becomes dominant.

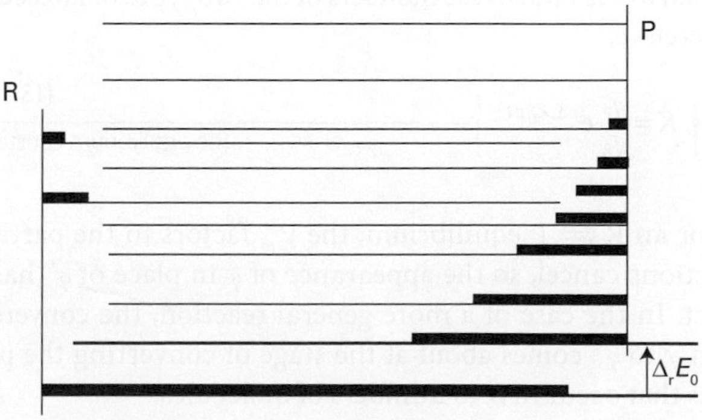

Figure 13F.3 It is important to take into account the densities of states of the molecules. Even though P might lie above R in energy (that is, $\Delta_r E_0$ is positive), P might have so many states that its total population dominates in the mixture. In classical thermodynamic terms, when considering equilibria, entropies must be taken into account as well as enthalpies.

How is that done? 13F.2 Relating the equilibrium constant to state populations

You need to start the derivation by noting that the population in a state i of the composite (R,P) system is $N_i = Ne^{-\beta\varepsilon_i}/q$, where N is the total number of molecules.

Step 1 *Write expressions for the numbers of R and P molecules*

The total number of R molecules is the sum of the populations of the (R,P) system taken over the states belonging to R; these states are labelled r with energies ε_r. The total number of P molecules is the sum over the states belonging to P; these states are labelled p with energies ε'_p (the prime is explained in a moment):

$$N_R = \sum_r N_r = \frac{N}{q}\sum_r e^{-\beta\varepsilon_r} \quad N_P = \sum_p N_p = \frac{N}{q}\sum_p e^{-\beta\varepsilon'_p}$$

The sum over the states of R is its partition function, q_R, so $N_R = Nq_R/q$. The sum over the states of P is also a partition function, but the energies are measured from the ground state of the combined system, which is the ground state of R. However, because $\varepsilon'_p = \varepsilon_p + \Delta\varepsilon_0$, where $\Delta\varepsilon_0$ is the separation of zero-point energies (as in Fig. 13F.3),

$$N_P = \frac{N}{q}\sum_p e^{-\beta(\varepsilon_p+\Delta\varepsilon_0)} = \frac{N}{q}\left(\sum_p e^{-\beta\varepsilon_p}\right)e^{-\beta\Delta\varepsilon_0} = \frac{Nq_P}{q}e^{-\beta\Delta\varepsilon_0}$$

$$= \frac{Nq_P}{q}e^{-\Delta_r E_0/RT}$$

The switch from $\Delta\varepsilon_0/kT$ to $\Delta_r E_0/RT$ in the last step is the conversion of molecular energies to molar energies.

Step 2 *Write an expression for the equilibrium constant*

The ratio of populations is

$$\frac{N_P}{N_R} = \frac{q_P}{q_R}e^{-\Delta_r E_0/RT}$$

The equilibrium constant of the R ⇌ P reaction is proportional to the ratio of the numbers of the two types of molecule. Therefore,

$$K = \frac{q_P}{q_R}e^{-\Delta_r E_0/RT} \qquad (13F.13)$$

The equilibrium constant

For an R ⇌ P equilibrium, the $V_m^\ominus$ factors in the partition functions cancel, so the appearance of q in place of $q^\ominus$ has no effect. In the case of a more general reaction, the conversion from q to $q^\ominus$ comes about at the stage of converting the pressures that occur in K to numbers of molecules.

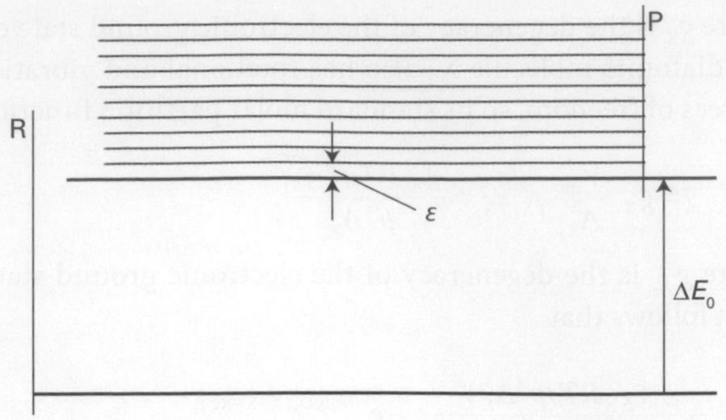

Figure 13F.4 The model used in the text for exploring the effects of energy separations and densities of states on equilibria. The products P can dominate provided $\Delta_r E_0$ is not too large and P has an appreciable density of states.

The implications of eqn 13F.13 can be seen most clearly by exaggerating the molecular features that contribute to it. Suppose that R has only a single accessible level, which implies that $q_R = 1$. Also suppose that P has a large number of evenly, closely spaced levels, with spacing ε (Fig. 13F.4). The partition function for such an array is calculated in *Brief illustration* 13B.1, and is $q = 1/(1-e^{-\beta\varepsilon})$. Provided the levels are close (in the sense $\varepsilon << kT$), $q \approx kT/\varepsilon$ in the high-temperature limit (Topic 13B). In this model system, the equilibrium constant is

$$K = \frac{kT}{\varepsilon}e^{-\Delta_r E_0/RT} \qquad (13F.14)$$

When $\Delta_r E_0$ is very large, the exponential term dominates and $K << 1$, which implies that very little P is present at equilibrium. When $\Delta_r E_0$ is small but still positive, K can exceed 1 because the factor kT/ε may be large enough to overcome the small size of the exponential term. The size of K then reflects the predominance of P at equilibrium on account of its high density of states. At low temperatures $K << 1$ and the system consists entirely of R. At high temperatures the exponential function approaches 1 and the factor kT/ε is large. Now P is dominant. In this endothermic reaction (endothermic because P lies above R) a rise in temperature favours P, because its states become accessible. This is the behaviour described, from a macroscopic perspective, in Topic 6B.

The model also shows why the Gibbs energy, G, and not just the enthalpy, determines the position of equilibrium. It shows that the density of states (and hence the entropy) of each species as well as their relative energies controls the distribution of populations and hence the value of the equilibrium constant.

Checklist of concepts

- ☐ 1. The **thermodynamic functions** A, p, H, and G can be calculated from the canonical partition function.
- ☐ 2. The **equilibrium constant** can be written in terms of the partition functions of the reactants and products.
- ☐ 3. The equilibrium constant for dissociation of a diatomic molecule in the gas phase may be calculated from spectroscopic data.
- ☐ 4. The **physical basis of chemical equilibrium** can be understood in terms of a competition between energy separations and densities of states.

Checklist of equations

Property	Equation	Comment	Equation number
Helmholtz energy	$A(T)=A(0)-kT\ln Q$		13F.2
Pressure	$p=kT(\partial \ln Q/\partial V)_T$		13F.3
Enthalpy	$H(T)=H(0)-(\partial \ln Q/\partial \beta)_V+kTV(\partial \ln Q/\partial V)_T$		13F.4
Gibbs energy	$G(T)=G(0)-kT\ln Q+kTV(\partial \ln Q/\partial V)_T$		13F.6
	$G(T)=G(0)-nRT\ln(q_m/N_A)$	Perfect gas	13F.9a
Equilibrium constant	$K=\left\{\prod_J (q^{\ominus}_{J,m}/N_A)^{\nu_J}\right\}e^{-\Delta_r E_0/RT}$	Perfect gas	13F.10b

FOCUS 13 Statistical thermodynamics

Assume that all gases are perfect and that data refer to 298 K unless otherwise stated.

TOPIC 13A The Boltzmann distribution

Discussion questions

D13A.1 Discuss the relations between 'population', 'configuration', and 'weight'. What is the significance of the most probable configuration?

D13A.2 What is the significance and importance of the principle of equal *a priori* probabilities?

D13A.3 What is temperature?

D13A.4 Summarize the role of the Boltzmann distribution in chemistry.

Exercises

E13A.1(a) Calculate the weight of the configuration in which 16 objects are distributed in the arrangement 0, 1, 2, 3, 8, 0, 0, 0, 0, 2.
E13A.1(b) Calculate the weight of the configuration in which 21 objects are distributed in the arrangement 6, 0, 5, 0, 4, 0, 3, 0, 2, 0, 0, 1.

E13A.2(a) Evaluate 8! by using (i) the exact definition of a factorial, (ii) Stirling's approximation (eqn 13A.2), and (iii) the more accurate version of Stirling's approximation, $x! \approx (2\pi)^{1/2}x^{x+1/2}e^{-x}$.
E13A.2(b) Evaluate 10! by using (i) the definition of a factorial; (ii) Stirling's approximation (eqn 13A.2), and (iii) the more accurate version of Stirling's approximation given in the preceding exercise.

E13A.3(a) What are the relative populations of the states of a two-level system when the temperature is infinite?
E13A.3(b) What are the relative populations of the states of a two-level system as the temperature approaches zero?

E13A.4(a) What is the temperature of a two-level system of energy separation equivalent to 400 cm^{-1} when the population of the upper state is one-third that of the lower state?
E13A.4(b) What is the temperature of a two-level system of energy separation equivalent to 300 cm^{-1} when the population of the upper state is one-half that of the lower state?

E13A.5(a) Calculate the relative populations of a linear rotor at 298 K in the levels with $J = 0$ and $J = 5$, given that $\tilde{B} = 2.71$ cm^{-1}.
E13A.5(b) Calculate the relative populations of a spherical rotor at 298 K in the levels with $J = 0$ and $J = 5$, given that $\tilde{B} = 2.71$ cm^{-1}.

E13A.6(a) A certain molecule has a non-degenerate excited state lying at 540 cm^{-1} above the non-degenerate ground state. At what temperature will 10 per cent of the molecules be in the upper state?
E13A.6(b) A certain molecule has a doubly degenerate excited state lying at 360 cm^{-1} above the non-degenerate ground state. At what temperature will 15 per cent of the molecules be in the upper state?

Problems

P13A.1 A sample consisting of five molecules has a total energy 5ε. Each molecule is able to occupy states of energy $j\varepsilon$, with $j = 0, 1, 2, \ldots$. (a) Calculate the weight of the configuration in which the molecules are distributed evenly over the available states with the stated total energy. (b) Draw up a table with columns headed by the energy of the states and write beneath them all configurations that are consistent with the total energy. Calculate the weights of each configuration and identify the most probable configurations.

P13A.2 A sample of nine molecules is numerically tractable but on the verge of being thermodynamically significant. Draw up a table of configurations for $N = 9$, total energy 9ε in a system with energy levels $j\varepsilon$ (as in Problem P13A.1). Before evaluating the weights of the configurations, guess (by looking for the most 'exponential' distribution of populations) which of the configurations will turn out to be the most probable. Go on to calculate the weights and identify the most probable configuration.

P13A.3 Use mathematical software to evaluate $\mathcal{W}$ for $N = 20$ for at least ten distributions over a uniform ladder of energy levels with separation ε, ensuring that the total energy is constant at 10ε. Identify the configuration of greatest weight, expressing the temperature as a multiple of ε, and compare it to the distribution predicted by the Boltzmann expression. Explore what happens as the value of the total energy is changed.

P13A.4 Suppose that two conformations A and B of a molecule differ in energy by 5.0 kJ mol^{-1}, and a third conformation C lies 0.50 kJ mol^{-1} above B. What proportion of molecules will be in conformation B at 273 K, with each conformation treated as a single energy level?

P13A.5 A certain atom has a doubly degenerate ground state and an upper level of four degenerate states at 450 cm^{-1} above the ground level. In an atomic beam study of the atoms it was observed that 30 per cent of the atoms were in the upper level, and the translational temperature of the beam was 300 K. Are the electronic states of the atoms in thermal equilibrium with the translational states? In other words, does the distribution of electronic states correspond to the same temperature as the distribution of translational states?

P13A.6 Explore the consequences of using the full version of Stirling's approximation, $x! \approx (2\pi)^{1/2}x^{x+1/2}e^{-x}$, in the development of the expression for the configuration of greatest weight. Does the more accurate approximation have a significant effect on the form of the Boltzmann distribution?

P13A.7‡ The variation of the atmospheric pressure p with altitude h is predicted by the *barometric formula* to be $p = p_0 e^{-h/H}$ where p_0 is the pressure at sea level and $H = RT/Mg$ with M the average molar mass of air and T the average temperature. Obtain the barometric formula from the Boltzmann distribution. Recall that the potential energy of a particle at height h above the surface of the Earth is mgh. Convert the barometric formula from pressure to number density, $\mathcal{N}$. Compare the relative number densities, $\mathcal{N}(h)/\mathcal{N}(0)$, for O_2 and H_2O at $h = 8.0$ km, a typical cruising altitude for commercial aircraft.

TOPIC 13B Molecular partition functions

Discussion questions

D13B.1 Describe the physical significance of the partition function.

D13B.2 What is the difference between a 'state' and an 'energy level'? Why is it important to make this distinction?

D13B.3 Why and when is it necessary to include a symmetry number in the calculation of a partition function?

Exercises

E13B.1(a) Calculate (i) the thermal wavelength, (ii) the translational partition function at 300 K and 3000 K of a molecule of molar mass 150 g mol^{-1} in a container of volume 1.00 cm^3.
E13B.1(b) Calculate (i) the thermal wavelength, (ii) the translational partition function of a Ne atom in a cubic box of side 1.00 cm at 300 K and 3000 K.

E13B.2(a) Calculate the ratio of the translational partition functions of H_2 and He at the same temperature and volume.
E13B.2(b) Calculate the ratio of the translational partition functions of Ar and Ne at the same temperature and volume.

E13B.3(a) The bond length of O_2 is 120.75 pm. Use the high-temperature approximation to calculate the rotational partition function of the molecule at 300 K.
E13B.3(b) The bond length of N_2 is 109.75 pm. Use the high-temperature approximation to calculate the rotational partition function of the molecule at 300 K.

E13B.4(a) The NOF molecule is an asymmetric rotor with rotational constants 3.1752 cm^{-1}, 0.3951 cm^{-1}, and 0.3505 cm^{-1}. Calculate the rotational partition function of the molecule at (i) 25 °C, (ii) 100 °C.
E13B.4(b) The H_2O molecule is an asymmetric rotor with rotational constants 27.877 cm^{-1}, 14.512 cm^{-1}, and 9.285 cm^{-1}. Calculate the rotational partition function of the molecule at (i) 25 °C, (ii) 100 °C.

E13B.5(a) The rotational constant of CO is 1.931 cm^{-1}. Evaluate the rotational partition function explicitly (without approximation) and plot its value as a function of temperature. At what temperature is the value within 5 per cent of the value calculated by using eqn 13B.12a, which gives the high-temperature limit? *Hint:* Use mathematical software or a spreadsheet.
E13B.5(b) The rotational constant of HI is 6.511 cm^{-1}. Evaluate the rotational partition function explicitly (without approximation) and plot its value as a function of temperature. At what temperature is the value within 5 per cent of the value calculated by using eqn 13B.12a, which gives the high-temperature limit? *Hint:* Use mathematical software or a spreadsheet.

E13B.6(a) The rotational constant of CH_4 is 5.241 cm^{-1}. Evaluate the rotational partition function explicitly (without approximation but ignoring the role of nuclear spin) and plot its value as a function of temperature. At what temperature is the value within 5 per cent of the value calculated by using eqn 13B.12b, which gives the high-temperature limit? *Hint:* Use mathematical software or a spreadsheet.
E13B.6(b) The rotational constant of CCl_4 is 0.0572 cm^{-1}. Evaluate the rotational partition function explicitly (without approximation but ignoring the role of nuclear spin) and plot its value as a function of temperature. At what temperature is the value within 5 per cent of the value calculated by using eqn 13B.12b, which gives the high-temperature limit? *Hint:* Use mathematical software or a spreadsheet.

E13B.7(a) Give the symmetry number for each of the following molecules: (i) CO, (ii) O_2, (iii) H_2S, (iv) SiH_4, and (v) $CHCl_3$.
E13B.7(b) Give the symmetry number for each of the following molecules: (i) CO_2, (ii) O_3, (iii) SO_3, (iv) SF_6, and (v) Al_2Cl_6.

E13B.8(a) Estimate the rotational partition function of ethene at 25 °C given that $\tilde{A} = 4.828$ cm^{-1}, $\tilde{B} = 1.0012$ cm^{-1}, and $\tilde{C} = 0.8282$ cm^{-1}. Take the symmetry number into account.
E13B.8(b) Evaluate the rotational partition function of pyridine, C_5H_5N, at 25 °C given that $\tilde{A} = 0.2014$ cm^{-1}, $\tilde{B} = 0.1936$ cm^{-1}, $\tilde{C} = 0.0987$ cm^{-1}. Take the symmetry number into account.

E13B.9(a) The vibrational wavenumber of Br_2 is 323.2 cm^{-1}. Evaluate the vibrational partition function explicitly (without approximation) and plot its value as a function of temperature. At what temperature is the value within 5 per cent of the value calculated from eqn 13B.16, which is valid at high temperatures?
E13B.9(b) The vibrational wavenumber of I_2 is 214.5 cm^{-1}. Evaluate the vibrational partition function explicitly (without approximation) and plot its value as a function of temperature. At what temperature is the value within 5 per cent of the value calculated from eqn 13B.16, which is valid at high temperatures?

E13B.10(a) Calculate the vibrational partition function of CS_2 at 500 K given the wavenumbers 658 cm^{-1} (symmetric stretch), 397 cm^{-1} (bend; doubly degenerate), 1535 cm^{-1} (asymmetric stretch).
E13B.10(b) Calculate the vibrational partition function of HCN at 900 K given the wavenumbers 3311 cm^{-1} (symmetric stretch), 712 cm^{-1} (bend; doubly degenerate), 2097 cm^{-1} (asymmetric stretch).

E13B.11(a) Calculate the vibrational partition function of CCl_4 at 500 K given the wavenumbers 459 cm^{-1} (symmetric stretch, non-degenerate), 217 cm^{-1} (deformation, doubly degenerate), 776 cm^{-1} (deformation, triply degenerate), 314 cm^{-1} (deformation, triply degenerate).
E13B.11(b) Calculate the vibrational partition function of CI_4 at 500 K given the wavenumbers 178 cm^{-1} (symmetric stretch, non-degenerate), 90 cm^{-1} (deformation, doubly degenerate), 555 cm^{-1} (deformation, triply degenerate), 125 cm^{-1} (deformation, triply degenerate).

E13B.12(a) A certain atom has a fourfold degenerate ground level, a non-degenerate electronically excited level at 2500 cm^{-1}, and a twofold degenerate level at 3500 cm^{-1}. Calculate the partition function of these electronic states at 1900 K. What is the relative population of each level at 1900 K?
E13B.12(b) A certain atom has a triply degenerate ground level, a non-degenerate electronically excited level at 850 cm^{-1}, and a fivefold degenerate level at 1100 cm^{-1}. Calculate the partition function of these electronic states at 2000 K. What is the relative population of each level at 2000 K?

‡ These problems were supplied by Charles Trapp and Carmen Giunta.

Problems

P13B.1 Consider a three-level system with levels 0, ε, and 2ε. Plot the partition function against kT/ε.

P13B.2 Plot the temperature dependence of the vibrational contribution to the molecular partition function for several values of the vibrational wavenumber. Estimate from your plots the temperature at which the partition function falls to within 10 per cent of the value expected at the high-temperature limit.

P13B.3 This problem is best done using mathematical software. Equation 13B.15 is the partition function for a harmonic oscillator. Consider a Morse oscillator (Topic 11C) in which the energy levels are given by

$$E_v = (v+\tfrac{1}{2})hc\tilde{\nu} - (v+\tfrac{1}{2})^2 hcx_e\tilde{\nu}$$

Evaluate the partition function for this oscillator, remembering to measure energies from the lowest level and to note that there is only a finite number of bound-state levels. Plot the partition function against $kT/hc\tilde{\nu}$ for values of x_e ranging from 0 to 0.1, and—on the same graph—compare your results with that for a harmonic oscillator.

P13B.4 Explore the conditions under which the 'integral' approximation for the translational partition function is not valid by considering the translational partition function of an H atom in a one-dimensional box of side comparable to that of a typical nanoparticle, 100 nm. Estimate the temperature at which, according to the integral approximation, $q = 10$ and evaluate the exact partition function at that temperature.

P13B.5 (a) Calculate the electronic partition function of a tellurium atom at (i) 298 K, (ii) 5000 K by direct summation using the following data:

Term	Degeneracy	Wavenumber/cm^{-1}
Ground	5	0
1	1	4707
2	3	4751
3	5	10 559

(b) What proportion of the Te atoms are in the ground term and in the term labelled 2 at the two temperatures?

P13B.6 The four lowest electronic levels of a Ti atom are $^3F_2$, $^3F_3$, $^3F_4$, and $^5F_1$, at 0, 170, 387, and 6557 cm^{-1}, respectively. There are many other electronic states at higher energies. The boiling point of titanium is 3287 °C. What are the relative populations of these levels at the boiling point? *Hint:* The degeneracies of the levels are $2J + 1$.

P13B.7‡ J. Sugar and A. Musgrove (*J. Phys. Chem. Ref. Data* **22**, 1213 (1993)) have published tables of energy levels for germanium atoms and cations from Ge^+ to Ge^{31+}. The lowest-lying energy levels in neutral Ge are as follows:

	$^3P_0$	$^3P_1$	$^3P_2$	$^1D_2$	$^1S_0$
(E/hc)/cm^{-1}	0	557.1	1410.0	7125.3	16 367.3

Calculate the electronic partition function at 298 K and 1000 K by direct summation. *Hint:* The degeneracy of a level J is $2J + 1$.

P13B.8 The pure rotational microwave spectrum (Topic 11B) of HCl has absorption lines at the following wavenumbers (in cm^{-1}): 21.19, 42.37, 63.56, 84.75, 105.93, 127.12 148.31 169.49, 190.68, 211.87, 233.06, 254.24, 275.43, 296.62, 317.80, 338.99, 360.18, 381.36, 402.55, 423.74, 444.92, 466.11, 487.30, 508.48. Calculate the rotational partition function at 25 °C by direct summation.

P13B.9 The rotational constants of CH_3Cl are $\tilde{A} = 5.097$ cm^{-1} and $\tilde{B} = 0.443$ cm^{-1}. Evaluate the rotational partition function explicitly (without approximation but ignoring the role of nuclear spin) and plot its value as a function of temperature. At what temperature is the value within 5 per cent of the value calculated by using eqn 13B.12a, which applies to the high-temperature limit? *Hint:* Use mathematical software or a spreadsheet.

P13B.10 Calculate, by explicit summation, the vibrational partition function of I_2 molecules at (a) 100 K, (b) 298 K given that its vibrational energy levels lie at the following wavenumbers above the zero-point energy level: 0, 215.30, 425.39, 636.27, 845.93 cm^{-1}. What proportion of I_2 molecules are in the ground and first two excited levels at the two temperatures?

TOPIC 13C Molecular energies

Discussion questions

D13C.1 Identify the conditions under which energies predicted from the equipartition theorem coincide with energies computed by using partition functions.

D13C.2 Describe how the mean energy of a system composed of two levels varies with temperature.

Exercises

E13C.1(a) Compute the mean energy at 298 K of a two-level system of energy separation equivalent to 500 cm^{-1}.

E13C.1(b) Compute the mean energy at 400 K of a two-level system of energy separation equivalent to 600 cm^{-1}.

E13C.2(a) Use mathematical software or a spreadsheet to evaluate, by explicit summation, the mean rotational energy of CO and plot its value as a function of temperature. At what temperature is the equipartition value within 5 per cent of the accurate value? $\tilde{B}(CO) = 1.931$ cm^{-1}.

E13C.2(b) Use mathematical software or a spreadsheet to evaluate, by explicit summation, the mean rotational energy of HI and plot its value as a function of temperature. At what temperature is the equipartition value within 5 per cent of the accurate value? $\tilde{B}(HI) = 6.511$ cm^{-1}.

E13C.3(a) Use mathematical software or a spreadsheet to evaluate, by explicit summation, the mean rotational energy of CH_4 and plot its value as a function of temperature. At what temperature is the equipartition value within 5 per cent of the accurate value? $\tilde{B}(CH_4) = 5.241$ cm^{-1}.

15C.3(b) Use mathematical software or a spreadsheet to evaluate, by explicit summation, the mean rotational energy of CCl_4 and plot its value as a function of temperature. At what temperature is the equipartition value within 5 per cent of the accurate value? $\tilde{B}(CCl_4) = 0.0572$ cm^{-1}.

E13C.4(a) Use mathematical software or a spreadsheet to evaluate the mean vibrational energy of Br_2 and plot its value as a function of temperature. At what temperature is the equipartition value within 5 per cent of the accurate value? Use $\tilde{\nu} = 323.2\,cm^{-1}$.

E13C.4(b) Use mathematical software or a spreadsheet to evaluate the mean vibrational energy of I_2 and plot its value as a function of temperature. At what temperature is the equipartition value within 5 per cent of the accurate value? Use $\tilde{\nu} = 214.5\,cm^{-1}$.

E13C.5(a) Use mathematical software or a spreadsheet to evaluate the mean vibrational energy of CS_2 and plot its value as a function of temperature. At what temperature is the equipartition value within 5 per cent of the accurate value? Use the wavenumbers 658 cm^{-1} (symmetric stretch), 397 cm^{-1} (bend; doubly degenerate), 1535 cm^{-1} (asymmetric stretch).

E13C.5(b) Use mathematical software or a spreadsheet to evaluate the mean vibrational energy of HCN and plot its value as a function of temperature. At what temperature is the equipartition value within 5 per cent of the accurate value? Use the wavenumbers 3311 cm^{-1} (symmetric stretch), 712 cm^{-1} (bend; doubly degenerate), 2097 cm^{-1} (asymmetric stretch).

E13C.6(a) Evaluate, by explicit summation, the mean vibrational energy of CCl_4 and plot its value as a function of temperature. At what temperature is the equipartition value within 5 per cent of the accurate value? Use the wavenumbers 459 cm^{-1} (symmetric stretch, non-degenerate), 217 cm^{-1} (deformation, doubly degenerate), 776 cm^{-1} (deformation, triply degenerate), 314 cm^{-1} (deformation, triply degenerate).

E13C.6(b) Evaluate, by explicit summation, the mean vibrational energy of CI_4 and plot its value as a function of temperature. At what temperature is the equipartition value within 5 per cent of the accurate value? Use the wavenumbers 178 cm^{-1} (symmetric stretch, non-degenerate), 90 cm^{-1} (deformation, doubly degenerate), 555 cm^{-1} (deformation, triply degenerate), 125 cm^{-1} (deformation, triply degenerate).

E13C.7(a) Calculate the mean contribution to the electronic energy at 1900 K for a sample composed of the atoms specified in Exercise E13B.12(a).

E13C.7(b) Calculate the mean contribution to the electronic energy at 2000 K for a sample composed of the atoms specified in Exercise E13B.12(b).

Problems

P13C.1 Evaluate, by explicit summation, the mean rotational energy of CH_3Cl and plot its value as a function of temperature. At what temperature is the equipartition value within 5 per cent of the accurate value? Use $\tilde{A} = 5.097\,cm^{-1}$ and $\tilde{B} = 0.443\,cm^{-1}$.

P13C.2 Deduce an expression for the mean energy when each molecule can exist in states with energies 0, ε, and 2ε.

P13C.3 How much energy does it take to raise the temperature of 1.0 mol $H_2O(g)$ from 100 °C to 200 °C at constant volume? Consider only translational and rotational contributions to the heat capacity.

P13C.4 What must the temperature be before the energy estimated from the equipartition theorem is within 2 per cent of the energy given by $\langle\varepsilon^V\rangle = hc\tilde{\nu}/(e^{\beta hc\tilde{\nu}}-1)$?

P13C.5 Suppose a collection of species with total spin $S=1$ is exposed to a magnetic field of 2.5 T. Calculate the mean energy of this system at 298 K. Use $g = 2.0$.

P13C.6 An electron trapped in an infinitely deep spherical well of radius R, such as may be encountered in the investigation of nanoparticles, has energies given by the expression $E_{nl} = \hbar^2 X_{nl}^2/2m_eR^2$, with X_{nl} the value obtained by searching for the zeroes of the spherical Bessel functions. The first six values (with a degeneracy of the corresponding energy level equal to $2l+1$) are as follows:

n	1	1	1	2	1	2
l	0	1	2	0	3	1
X_{nl}	3.142	4.493	5.763	6.283	6.988	7.725

Evaluate the partition function and mean energy of an electron as a function of temperature. *Hints:* Remember to measure energies from the lowest level. Note that $\hbar^2/2m_eR^2k$ has dimensions of temperature, so can be used as the characteristic temperature θ of the system. It follows that the partition function can be expressed in terms of $E_{nl} = X_{nl}^2 k\theta$ and the dimensionless parameter T/θ. Let T/θ range from 0 to 25.

P13C.7 The NO molecule has a doubly degenerate excited electronic level 121.1 cm^{-1} above the doubly degenerate electronic ground term. Calculate and plot the electronic partition function of NO from $T = 0$ to 1000 K. Evaluate (a) the populations of the levels and (b) the mean electronic energy at 300 K.

P13C.8 Consider a system of N molecules with energy levels $\varepsilon_j = j\varepsilon$ and $j = 0, 1, 2, \ldots$ (a) Show that if the mean energy per molecule is $a\varepsilon$, then the temperature is given by

$$\beta = \frac{1}{\varepsilon}\ln\left(1+\frac{1}{a}\right)$$

Evaluate the temperature for a system in which the mean energy is ε, taking ε equivalent to 50 cm^{-1}. (b) Calculate the molecular partition function q for the system when its mean energy is $a\varepsilon$.

P13C.9 Deduce an expression for the root mean square energy, $\langle\varepsilon^2\rangle^{1/2}$, in terms of the partition function and hence an expression for the root mean square deviation from the mean, $\Delta\varepsilon = (\langle\varepsilon^2\rangle - \langle\varepsilon\rangle^2)^{1/2}$. Evaluate the resulting expression for a harmonic oscillator. *Hint:* Use $\langle\varepsilon^2\rangle = (1/q)\sum_j e^{-\beta\varepsilon_j}\varepsilon_j^2$.

TOPIC 13D The canonical ensemble

Discussion questions

E13D.1 Why is the concept of a canonical ensemble required?

E13D.2 Explain what is meant by an ensemble and why it is useful in statistical thermodynamics.

E13D.3 Under what circumstances may identical particles be regarded as distinguishable?

E13D.4 What is meant by the 'thermodynamic limit'?

Exercises

E13D.1(a) Identify the systems for which it is essential to include a factor of $1/N!$ on going from Q to q: (i) a sample of helium gas, (ii) a sample of carbon monoxide gas, (iii) a solid sample of carbon monoxide, (iv) water vapour.

E13D.1(b) Identify the systems for which it is essential to include a factor of $1/N!$ on going from Q to q: (i) a sample of carbon dioxide gas, (ii) a sample of graphite, (iii) a sample of diamond, (iv) ice.

Problems

P13D.1‡ For a perfect gas, the canonical partition function, Q, is related to the molecular partition function q by $Q = q^N/N!$. In Topic 13F it is established that $p = kT(\partial \ln Q/\partial V)_T$. Use the expression for q to derive the perfect gas law $pV = nRT$.

P13D.2 Use statistical thermodynamic arguments to show that for a perfect gas, $(\partial E/\partial V)_T = 0$.

TOPIC 13E The internal energy and the entropy

Discussion questions

D13E.1 Describe the molecular features that affect the magnitudes of the constant-volume molar heat capacity of a molecular substance.

D13E.2 Discuss and illustrate the proposition that $1/T$ is a more natural measurement of temperature than T itself.

D13E.3 Discuss the relationship between the thermodynamic and statistical definitions of entropy.

D13E.4 Justify the differences between the partition-function expression for the entropy for distinguishable particles and the expression for indistinguishable particles.

D13E.5 Account for the temperature and volume dependence of the entropy of a perfect gas in terms of the Boltzmann distribution.

D13E.6 Explain the origin of residual entropy.

Exercises

E13E.1(a) Use the equipartition theorem to estimate the constant-volume molar heat capacity of (i) I_2, (ii) CH_4, (iii) C_6H_6 in the gas phase at 25 °C.

E13E.1(b) Use the equipartition theorem to estimate the constant-volume molar heat capacity of (i) O_3, (ii) C_2H_6, (iii) CO_2 in the gas phase at 25 °C.

E13E.2(a) Estimate the values of $\gamma = C_{p,m}/C_{V,m}$ for gaseous ammonia and methane. Do this calculation with and without the vibrational contribution to the energy. Which is closer to the experimental value at 25 °C? *Hint:* Note that $C_{p,m} - C_{V,m} = R$ for a perfect gas.

E13E.2(b) Estimate the value of $\gamma = C_{p,m}/C_{V,m}$ for carbon dioxide. Do this calculation with and without the vibrational contribution to the energy. Which is closer to the experimental value at 25 °C? *Hint:* Note that $C_{p,m} - C_{V,m} = R$ for a perfect gas.

E13E.3(a) The ground level of Cl is $^2P_{3/2}$ and a $^2P_{1/2}$ level lies 881 cm^{-1} above it. Calculate the electronic contribution to the heat capacity of Cl atoms at (i) 500 K and (ii) 900 K.

E13E.3(b) The first electronically excited state of O_2 is $^1\Delta_g$ (doubly degenerate) and lies 7918.1 cm^{-1} above the ground state, which is $^3\Sigma_g^-$ (triply degenerate). Calculate the electronic contribution to the heat capacity of O_2 at 400 K.

E13E.4(a) Plot the molar heat capacity of a collection of harmonic oscillators as a function of T/θ^V, and predict the vibrational heat capacity of ethyne at (i) 298 K, (ii) 500 K. The normal modes (and their degeneracies in parentheses) occur at wavenumbers 612(2), 729(2), 1974, 3287, and 3374 cm^{-1}.

E13E.4(b) Plot the molar entropy of a collection of harmonic oscillators as a function of T/θ^V, and predict the standard molar entropy of ethyne at (i) 298 K, (ii) 500 K. For data, see the preceding exercise.

E13E.5(a) Calculate the standard molar entropy at 298 K of (i) gaseous helium, (ii) gaseous xenon.

E13E.5(b) Calculate the translational contribution to the standard molar entropy at 298 K of (i) $H_2O(g)$, (ii) $CO_2(g)$.

E13E.6(a) At what temperature is the standard molar entropy of helium equal to that of xenon at 298 K?

E13E.6(b) At what temperature is the translational contribution to the standard molar entropy of $CO_2(g)$ equal to that of $H_2O(g)$ at 298 K?

E13E.7(a) Calculate the rotational partition function of H_2O at 298 K from its rotational constants 27.878 cm^{-1}, 14.509 cm^{-1}, and 9.287 cm^{-1} and use your result to calculate the rotational contribution to the molar entropy of gaseous water at 25 °C.

E13E.7(b) Calculate the rotational partition function of SO_2 at 298 K from its rotational constants 2.027 36 cm^{-1}, 0.344 17 cm^{-1}, and 0.293 535 cm^{-1} and use your result to calculate the rotational contribution to the molar entropy of sulfur dioxide at 25 °C.

E13E.8(a) The ground state of the Co^{2+} ion in $CoSO_4\cdot 7H_2O$ may be regarded as $^4T_{9/2}$. The entropy of the solid at temperatures below 1 K is derived almost entirely from the electron spin. Estimate the molar entropy of the solid at these temperatures.

E13E.8(b) Estimate the contribution of the spin to the molar entropy of a solid sample of a d-metal complex with $S = \frac{5}{2}$.

E13E.9(a) Predict the vibrational contribution to the standard molar entropy of methanoic acid (formic acid, HCOOH) vapour at (i) 298 K, (ii) 500 K. The normal modes occur at wavenumbers 3570, 2943, 1770, 1387, 1229, 1105, 625, 1033, 638 cm^{-1}.

E13E.9(b) Predict the vibrational contribution to the standard molar entropy of ethyne at (i) 298 K, (ii) 500 K. The normal modes (and their degeneracies in parentheses) occur at wavenumbers 612(2), 729(2), 1974, 3287, and 3374 cm^{-1}.

Problems

P13E.1 An NO molecule has a doubly degenerate electronic ground state and a doubly degenerate excited state at 121.1 cm^{-1}. Calculate and plot the electronic contribution to the molar heat capacity of the molecule up to 500 K.

P13E.2 Explore whether a magnetic field can influence the heat capacity of a paramagnetic molecule by calculating the electronic contribution to the heat capacity of an NO_2 molecule in a magnetic field. Estimate the total constant-volume heat capacity using equipartition, and calculate the percentage change in heat capacity brought about by a 5.0 T magnetic field at (a) 50 K, (b) 298 K. *Hints:* Recall that NO_2 has one unpaired electron. Assume that the sample is in the gas phase at 50 K and 298 K.

P13E.3 The energy levels of a CH_3 group attached to a larger fragment are given by the expression for a particle on a ring, provided the group is rotating freely. What is the high-temperature contribution to the heat capacity and entropy of such a freely rotating group at 25 °C? *Hint:* The moment of inertia of CH_3 about its threefold rotation axis (the axis that passes through the C atom and the centre of the equilateral triangle formed by the H atoms) is 5.341×10^{-47} kg m^2.

P13E.4 Calculate the temperature dependence of the heat capacity of p-H_2 (in which only rotational states with even values of J are populated) at low temperatures on the basis that its rotational levels $J = 0$ and $J = 2$ constitute a system that resembles a two-level system except for the degeneracy of the upper level. Use $\tilde{B} = 60.864$ cm^{-1} and sketch the heat capacity curve. The experimental heat capacity of p-H_2 does in fact show a peak at low temperatures.

P13E.5‡ In a spectroscopic study of buckminsterfullerene C_{60}, F. Negri et al. (*J. Phys. Chem.* **100**, 10849 (1996)) reviewed the wavenumbers of all the vibrational modes of the molecule:

Mode	Number	Degeneracy	Wavenumber/cm^{-1}
A_u	1	1	976
T_{1u}	4	3	525, 578, 1180, 1430
T_{2u}	5	3	354, 715, 1037, 1190, 1540
G_u	6	4	345, 757, 776, 963, 1315, 1410
H_u	7	5	403, 525, 667, 738, 1215, 1342, 1566

How many modes have a vibrational temperature θ^V below 1000 K? Estimate the molar constant-volume heat capacity of C_{60} at 1000 K, counting as active all modes with θ^V below this temperature.

P13E.6 Plot the function dS/dT for a two-level system, the temperature coefficient of its entropy, against kT/ε. Is there a temperature at which this coefficient passes through a maximum? If you find a maximum, explain its physical origins.

P13E.7 Derive an expression for the molar entropy of an equally spaced three-level system; taking the spacing as ε.

P13E.8 Although expressions like $\langle\varepsilon\rangle = -d\ln q/d\beta$ are useful for formal manipulations in statistical thermodynamics, and for expressing thermodynamic functions in neat formulas, they are sometimes more trouble than they are worth in practical applications. When presented with a table of energy levels, it is often much more convenient to evaluate the following sums directly (the dots simply identify the different functions):

$$q = \sum_j e^{-\beta\varepsilon_j} \qquad \dot{q} = \sum_j \beta\varepsilon_j e^{-\beta\varepsilon_j} \qquad \ddot{q} = \sum_j (\beta\varepsilon_j)^2 e^{-\beta\varepsilon_j}$$

(a) Derive expressions for the internal energy, heat capacity, and entropy in terms of these three functions. (b) Apply the technique to the calculation of the electronic contribution to the constant-volume molar heat capacity of magnesium vapour at 5000 K using the following data:

Term	^1S	$^3P_0$	$^3P_1$	$^3P_2$	$^1P_1$	$^3S_1$
Degeneracy	1	1	3	5	3	3
$\tilde{\nu}$/cm^{-1}	0	21 850	21 870	21 911	35 051	41 197

P13E.9 Use the accurate expression for the rotational partition function calculated in Problem 13B.8 for HCl(g) to calculate the rotational contribution to the molar entropy over a range of temperature and plot the contribution as a function of temperature.

P13E.10 Calculate the standard molar entropy of N_2(g) at 298 K from its rotational constant $\tilde{B} = 1.9987$ cm^{-1} and its vibrational wavenumber $\tilde{\nu} = 2358$ cm^{-1}. The thermochemical value is 192.1 J K^{-1} mol^{-1}. What does this suggest about the solid at $T = 0$?

P13E.11‡ J.G. Dojahn et al. (*J. Phys. Chem.* **100**, 9649 (1996)) characterized the potential energy curves of the ground and electronic states of homonuclear diatomic halogen anions. The ground state of F_2^- is $^2\Sigma_u^+$ with a fundamental vibrational wavenumber of 450.0 cm^{-1} and equilibrium internuclear distance of 190.0 pm. The first two excited states are at 1.609 and 1.702 eV above the ground state. Compute the standard molar entropy of F_2^- at 298 K.

P13E.12‡ Treat carbon monoxide as a perfect gas and apply equilibrium statistical thermodynamics to the study of its properties, as specified below, in the temperature range 100–1000 K at 1 bar. (a) Examine the probability distribution of molecules over available rotational and vibrational states. (b) Explore numerically the differences, if any, between the rotational molecular partition function as calculated with the discrete energy distribution with that predicted by the high-temperature limit. (c) Plot against temperature the individual contributions to $U_m(T) - U_m(100\,K)$, $C_{V,m}(T)$, and $S_m(T) - S_m(100\,K)$ made by the translational, rotational, and vibrational degrees of freedom. *Hints:* Let $\tilde{\nu} = 2169.8$ cm^{-1} and $\tilde{B} = 1.931$ cm^{-1}; neglect anharmonicity and centrifugal distortion.

P13E.13 The energy levels of a Morse oscillator are given in Problem P13B.3. Set up the expression for the molar entropy of a collection of Morse oscillators and plot it as a function of $kT/hc\tilde{\nu}$ for a series of anharmonicity constants ranging from 0 to 0.01. Take into account only the finite number of bound states. On the same graph plot the entropy of a harmonic oscillator and investigate how the two diverge.

P13E.14 Explore how the entropy of a collection of two-level systems behaves when the temperature is formally allowed to become negative. You should also construct a graph in which the temperature is replaced by the variable $\beta = 1/kT$. Account for the appearance of the graphs physically.

P13E.15 Derive the Sackur–Tetrode equation for a monatomic gas confined to a two-dimensional surface, and hence derive an expression for the standard molar entropy of condensation to form a mobile surface film.

P13E.16 The heat capacity ratio of a gas determines the speed of sound in it through the formula $c_s = (\gamma RT/M)^{1/2}$, where $\gamma = C_{p,m}/C_{V,m}$ and M is the molar mass of the gas. Deduce an expression for the speed of sound in a perfect gas of (a) diatomic, (b) linear triatomic, (c) nonlinear triatomic molecules at high temperatures (with translation and rotation active). Estimate the speed of sound in air at 25 °C. *Hint:* Note that $C_{p,m} - C_{V,m} = R$ for a perfect gas.

P13E.17 An average human DNA molecule has 5×10^8 binucleotides (rungs on the DNA ladder) of four different kinds. If each rung were a random choice of one of these four possibilities, what would be the residual entropy associated with this typical DNA molecule?

P13E.18 It is possible to write an approximate expression for the partition function of a protein molecule by including contributions from only two states: the native and denatured forms of the polymer. Proceeding with this crude model gives insight into the contribution of denaturation to the heat capacity of a protein. According to this model, the total energy of a system of N protein molecules is $E = N\varepsilon e^{-\varepsilon/kT}/(1+e^{-\varepsilon/kT})$, where ε is the energy separation between the denatured and native forms. (a) Show that the constant-volume molar heat capacity is

$$C_{V,m} = f(T)R \quad f(T) = \frac{(\varepsilon/kT)^2 e^{-\varepsilon/kT}}{(1+e^{-\varepsilon/kT})^2}$$

(b) Plot the variation of $C_{V,m}$ with temperature. (c) If the function $C_{V,m}(T)$ has a maximum or minimum, calculate the temperature at which it occurs.

TOPIC 13F Derived functions

Discussion questions

D13F.1 Suggest a physical interpretation of the relation between pressure and the partition function.

D13F.2 Suggest a physical interpretation of the relation between equilibrium constant and the partition functions of the reactants and products in a reaction.

D13F.3 How does a statistical analysis of the equilibrium constant account for the latter's temperature dependence?

Exercises

E13F.1(a) A CO_2 molecule is linear, and its vibrational wavenumbers are 1388.2 cm^{-1}, 2349.2 cm^{-1}, and 667.4 cm^{-1}, the last being doubly degenerate and the others non-degenerate. The rotational constant of the molecule is 0.3902 cm^{-1}. Calculate the rotational and vibrational contributions to the molar Gibbs energy at 298 K.

E13F.1(b) An O_3 molecule is angular, and its vibrational wavenumbers are 1110 cm^{-1}, 705 cm^{-1}, and 1042 cm^{-1}. The rotational constants of the molecule are 3.553 cm^{-1}, 0.4452 cm^{-1}, and 0.3948 cm^{-1}. Calculate the rotational and vibrational contributions to the molar Gibbs energy at 298 K.

E13F.2(a) Use the information in Exercise E13E.3(a) to calculate the electronic contribution to the molar Gibbs energy of Cl atoms at (i) 500 K and (ii) 900 K.

E13F.2(b) Use the information in Exercise E13E.3(b) to calculate the electronic contribution to the molar Gibbs energy of O_2 at 400 K.

E13F.3(a) Calculate the equilibrium constant of the reaction $I_2(g) \rightleftharpoons 2I(g)$ at 1000 K from the following data for I_2, $\tilde{\nu} = 214.36$ cm^{-1}, $\tilde{B} = 0.0373$ cm^{-1}, $hc\tilde{D}_e = 1.5422$ eV. The ground state of the I atoms is $^2P_{3/2}$, implying fourfold degeneracy.

E13F.3(b) Calculate the equilibrium constant at 298 K for the gas-phase isotopic exchange reaction $2\,{}^{79}Br^{81}Br \rightleftharpoons {}^{79}Br^{79}Br + {}^{81}Br^{81}Br$. The Br_2 molecule has a non-degenerate ground state, with no other electronic states nearby. Base the calculation on the wavenumber of the vibration of $^{79}Br^{81}Br$, which is 323.33 cm^{-1}.

Problems

P13F.1 Use mathematical software and work in the high-temperature limit to calculate and plot the equilibrium constant for the reaction $CD_4(g) + HCl(g) \rightleftharpoons CHD_3(g) + DCl(g)$ as a function of temperature, in the range 300 K to 1000 K. Use the following data (numbers in parentheses are degeneracies):

Molecule	$\tilde{\nu}$/cm^{-1}	$\tilde{B}$/cm^{-1}	$\tilde{A}$/cm^{-1}
CHD_3	2993(1), 2142(1), 1003(3), 1291(2), 1036(2);	3.28	2.63
CD_4	2109(1), 1092(2), 2259(3), 996(3)	2.63	
HCl	2991 (1)	10.59	
DCl	2145 (1)	5.445	

P13F.2 The exchange of deuterium between acid and water is an important type of equilibrium, and you can examine it using spectroscopic data on the molecules. Use mathematical software and work in the high-temperature limit, calculate the equilibrium constant for the exchange reaction $H_2O(g) + DCl(g) \rightleftharpoons HDO(g) + HCl(g)$ at (a) 298 K and (b) 800 K. Use the following data:

Molecule	$\tilde{\nu}$/cm^{-1}	$\tilde{A}$/cm^{-1}	$\tilde{B}$/cm^{-1}	$\tilde{C}$/cm^{-1}
H_2O	3656.7, 1594.8, 3755.8	27.88	14.51	9.29
HDO	2726.7, 1402.2, 3707.5	23.38	9.102	6.417
HCl	2991		10.59	
DCl	2145		5.449	

P13F.3 Here you are invited to decide whether a magnetic field can influence the value of an equilibrium constant. Consider the equilibrium $I_2(g) \rightleftharpoons 2I(g)$ at 1000 K, and calculate the ratio of equilibrium constants $K(\mathcal{B})/K$, where $K(\mathcal{B})$ is the equilibrium constant when a magnetic field $\mathcal{B}$ is present and removes the degeneracy of the four states of the $^2P_{3/2}$ level. Data on the species are given in Exercise 13F.3(a). The electronic g-value of the atoms is $\frac{4}{3}$. Calculate the field required to change the equilibrium constant by 1 per cent.

P13F.4‡ R. Viswanathan et al. (*J. Phys. Chem.* **100**, 10784 (1996)) studied thermodynamic properties of several boron–silicon gas-phase species experimentally and theoretically. These species can occur in the high-temperature chemical vapour deposition (CVD) of silicon-based semiconductors. Among the computations they reported was computation of the Gibbs energy of BSi(g) at several temperatures based on a $^4\Sigma^-$ ground state with equilibrium internuclear distance of 190.5 pm, a fundamental vibrational wavenumber of 772 cm^{-1}, and a $^2\Pi$ first excited level 8000 cm^{-1} above the ground level. Calculate the value of $G_m^\ominus(2000\,\text{K}) - G_m^\ominus(0)$.

P13F.5‡ The molecule Cl_2O_2, which is believed to participate in the seasonal depletion of ozone over Antarctica, has been studied by several means. M. Birk et al. (*J. Chem. Phys.* **91**, 6588 (1989)) report its rotational constants (B) as 13 109.4, 2409.8, and 2139.7 MHz. They also report that its rotational spectrum indicates a molecule with a symmetry number of 2. J. Jacobs et al. (*J. Amer. Chem. Soc.* **116**, 1106 (1994)) report its vibrational wavenumbers as 753, 542, 310, 127, 646, and 419 cm^{-1} (all non-degenerate). Calculate the value of $G_m^\ominus(200\,\text{K}) - G_m^\ominus(0)$ of Cl_2O_2.

P13F.6‡ J. Hutter et al. (*J. Amer. Chem. Soc.* **116**, 750 (1994)) examined the geometric and vibrational structure of several carbon molecules of formula C_n. Given that the ground state of C_3, a molecule found in interstellar space and in flames, is an angular singlet-state species with moments of inertia $39.340m_u\,\text{Å}^2$ $39.032m_u\,\text{Å}^2$, and $0.3082m_u\,\text{Å}^2$ (where 1 Å = 10^{-10} m) and with vibrational wavenumbers 63.4, 1224.5, and 2040 cm^{-1}, calculate the value of $G_m^\ominus(300.0\,\text{K}) - G_m^\ominus(0)$ for C_3.

FOCUS 13 Statistical thermodynamics

Integrated activities

I13.1 A formal way of arriving at the value of the symmetry number is to note that σ is the order (the number of elements) of the *rotational subgroup* of the molecule, the point group of the molecule with all but the identity and the rotations removed. The rotational subgroup of H_2O is $\{E,C_2\}$, so $\sigma = 2$. The rotational subgroup of NH_3 is $\{E,2C_3\}$, so $\sigma = 3$. This recipe makes it easy to find the symmetry numbers for more complicated molecules. The rotational subgroup of CH_4 is obtained from the T character table as $\{E,8C_3,3C_2\}$, so $\sigma = 12$. For benzene, the rotational subgroup of D_{6h} is $\{E,2C_6,2C_3,C_2,3C_2',3C_2''\}$ so $\sigma = 12$. (a) Estimate the rotational partition function of ethene at 25 °C given that $\tilde{A} = 4.828\,\text{cm}^{-1}$, $\tilde{B} = 1.0012\,\text{cm}^{-1}$, and $\tilde{C} = 0.8282\,\text{cm}^{-1}$. (b) Evaluate the rotational partition function of pyridine, C_5H_5N, at room temperature ($\tilde{A} = 0.2014\,\text{cm}^{-1}$, $\tilde{B} = 0.1936\,\text{cm}^{-1}$, $\tilde{C} = 0.0987\,\text{cm}^{-1}$).

I13.2 A feature of the rotational molar heat capacity of H_2 is that it rises to above the classical value of R before settling back to approach that value as the temperature is increased from 0. To understand this behaviour, the heat capacity can be treated as arising from the sum of transitions between the available levels. Show that the heat capacity of a linear rotor is related to the following sum:

$$\xi(\beta) = \frac{1}{q^2}\sum_{J,J'}\{\varepsilon(J)-\varepsilon(J')\}^2 g(J')g(J)\mathrm{e}^{-\beta\{\varepsilon(J)+\varepsilon(J')\}}$$

by

$$C = \tfrac{1}{2}Nk\beta^2\xi(\beta)$$

where the $\varepsilon(J)$ are the rotational energy levels and $g(J)$ their degeneracies. Then go on to show graphically that the total contribution to the heat capacity of a linear rotor can be regarded as a sum of contributions due to transitions $0 \rightarrow 1$, $0 \rightarrow 2$, $1 \rightarrow 2$, $1 \rightarrow 3$, etc. In this way, construct Fig. 13.1 for the rotational heat capacities of a linear molecule.

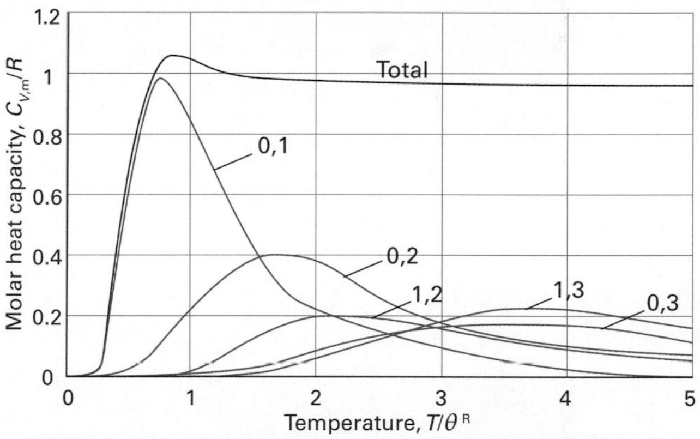

Figure 13.1 The variation of the constant-volume molar heat capacity with temperature, as calculated in Integrated activity I13.2.

I13.3 Set up a calculation like that in Integrated activity I13.2 to analyse the vibrational contribution to the heat capacity in terms of excitations between levels and illustrate your results graphically in terms of a diagram like that in Fig. 13.1.

FOCUS 14

Molecular interactions

The electric properties of molecules give rise to molecular interactions. In turn, those interactions govern the formation of condensed phases and the structures and functions of macromolecules and molecular assemblies.

Macromolecules are built from covalently linked components. They are everywhere, inside us and outside us. Some are natural: they include polysaccharides (such as cellulose), polypeptides (such as protein enzymes), and polynucleotides (such as deoxyribonucleic acid, DNA). Others are synthetic (such as nylon and polystyrene). Molecules both large and small may also gather together in a process called 'self-assembly' and give rise to aggregates that to some extent behave like macromolecules.

14A The electric properties of molecules

Important electric properties of molecules include 'electric dipole moments' and 'polarizabilities'. These properties reflect the degree to which the nuclei of atoms exert control over the electrons in a molecule, either by causing electrons to accumulate in particular regions, or by permitting them to respond more or less strongly to the effects of external electric fields.

14B Interactions between molecules

This Topic describes the basic theory of several important molecular interactions, with a special focus on 'van der Waals interactions' between closed-shell molecules and 'hydrogen bonding'. Many liquids and solids are bound together by one or more of the cohesive interactions explored in this Topic. These interactions are also important for the structural organization of macromolecules.

14C Liquids

This Topic begins with the basic theory of molecular interactions in liquids, and then turns to a description of the properties of liquid surfaces. It is seen that important effects, such as 'surface tension', 'capillary action', the formation of 'surface films', and condensation, can be explained by thermodynamics arguments.

14D Macromolecules

Macromolecules adopt shapes that are governed by molecular interactions. This Topic considers a range of shapes, but focuses on the structureless 'random coil' and partially structured coils. It also explores the connection between structure and the mechanical and thermal properties of macromolecules.

14E Self-assembly

Atoms, small molecules, and macromolecules can form large aggregates, sometimes by processes involving self-assembly. This Topic explores 'colloids', 'micelles', and biological membranes, which are assemblies with some of the typical properties of molecules but also with their own characteristic features. It also introduces an important type of molecular interaction, the 'hydrophobic interaction', which is driven by changes in entropy of the solvent.

Web resources What is an application of this material?

Molecular interactions play important roles in biochemistry and biomedicine. Natural macromolecules differ in certain respects from synthetic macromolecules, particularly in their composition and the resulting structure. The different levels of structure in proteins and nucleic acids are explored in *Impact* 21. In *Impact* 22 attention shifts to the binding of a drug, a small molecule or protein, to a specific receptor site of a target molecule, such as a larger protein or nucleic acid. The chemical result of the formation of this assembly is the inhibition of the progress of disease.

TOPIC 14A The electric properties of molecules

> ➤ **Why do you need to know this material?**
>
> Because the molecular interactions responsible for the formation of condensed phases and large molecular assemblies arise from the electric properties of molecules, you need to know how the electronic structures of molecules account for these interactions.
>
> ➤ **What is the key idea?**
>
> The nuclei of atoms exert control over the electrons in a molecule, and can cause electrons to accumulate in various regions, or permit them to respond to external fields.
>
> ➤ **What do you need to know already?**
>
> You need to be familiar with the Coulomb law (*The chemist's toolkit* 6 of Topic 2A), molecular geometry, and molecular orbital theory, especially the existence of the energy gap between a HOMO and LUMO (Topic 9E). One calculation draws on the Boltzmann distribution (Topic 13A).

The electric properties of molecules are responsible for many of the properties of bulk matter. The small imbalances of charge distributions in molecules and the ability of electron distributions to be distorted allow molecules to interact with one another and to respond to externally applied fields.

14A.1 Electric dipole moments

An **electric dipole** consists of two electric charges $+Q$ and $-Q$ with a vector separation $\boldsymbol{R}$. A **point electric dipole** is an electric dipole in which R is very small compared with its distance to the observer. The **electric dipole moment** is a vector $\boldsymbol{\mu}$ (1) that points from the negative charge to the positive charge and has a magnitude given by

$$\mu = QR \quad \text{Magnitude of the electric dipole moment [definition]} \qquad (14A.1)$$

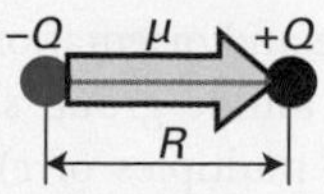

1 Electric dipole

Although the SI unit of dipole moment is coulomb metre (C m), it is still commonly reported in the non-SI unit debye, D, named after Peter Debye, a pioneer in the study of dipole moments of molecules:

$$1\,\text{D} = 3.335\,64 \times 10^{-30}\,\text{C m} \qquad (14A.2)$$

The magnitude of the dipole moment formed by a pair of charges $+e$ and $-e$ separated by 100 pm is 1.6×10^{-29} C m, which corresponds to 4.8 D. The magnitudes of the dipole moments of small molecules are typically about 1 D.

A **polar molecule** is a molecule with a permanent electric dipole moment. A **permanent dipole moment** stems from the partial charges on the atoms in the molecule, which arise from differences in electronegativity or, in more sophisticated treatments, variations in electron density through the molecule (Topic 9E). Nonpolar molecules acquire an **induced dipole moment** in an electric field on account of the distortion the field causes in their electronic distributions and nuclear positions. However, this induced moment is only temporary, and disappears as soon as the perturbing field is removed. Polar molecules also have their permanent dipole moments temporarily modified by an applied field.

All heteronuclear diatomic molecules are polar and typical values of μ are 1.08 D for HCl and 0.42 D for HI (Table 14A.1). Molecular symmetry is of the greatest importance in deciding whether a polyatomic molecule is polar or not. Indeed, molecular symmetry is more important than the question of whether or not the atoms in the molecule belong to the same element. For this reason, and as seen in the following *Brief illustration*, homonuclear polyatomic molecules may be polar if they have low symmetry and the atoms are in inequivalent positions.

Table 14A.1 Dipole moments and polarizability volumes*

	μ/D	$\alpha'/(10^{-30}\,\text{m}^3)$
CCl_4	0	10.5
H_2	0	0.819
H_2O	1.85	1.48
HCl	1.08	2.63
HI	0.42	5.45

* More values are given in the *Resource section*.

Brief illustration 14A.1

The angular molecule ozone (2) is homonuclear. However, it is polar because the central O atom is different from the outer two (it is bonded to two atoms, which are bonded only to one). Moreover, the dipole moments associated with each bond make an angle to each other and do not cancel. The heteronuclear linear triatomic molecule CO_2 is nonpolar because, although there are partial charges on all three atoms, the dipole moment associated with the OC bond points in the opposite direction to the dipole moment associated with the CO bond, and the two cancel (3).

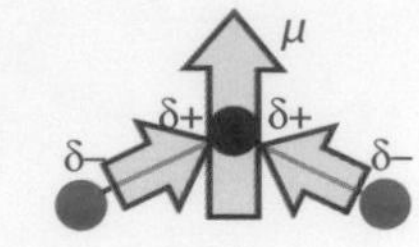

2 Ozone, O_3

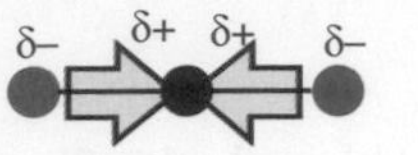

3 Carbon dioxide, CO_2

The dipole moment of a polyatomic molecule can be resolved into contributions from various groups of atoms in the molecule and their relative locations (Fig. 14A.1). Thus, 1,4-dichlorobenzene is nonpolar by symmetry on account of the cancellation of two equal but opposing C–Cl moments (exactly as in carbon dioxide). 1,2-Dichlorobenzene, however, has a dipole moment which is approximately the resultant of two chlorobenzene dipole moments arranged at 60° to each other. This technique of 'vector addition' can be applied with fair success to other series of related molecules. The magnitude of the resultant moment μ_{res} of μ_1 and μ_2 that make an angle Θ to each other (4) is approximately (see *The chemist's toolkit* 22 of Topic 8C)

$$\mu_{res} \approx (\mu_1^2 + \mu_2^2 + 2\mu_1\mu_2\cos\Theta)^{1/2} \qquad (14A.3a)$$

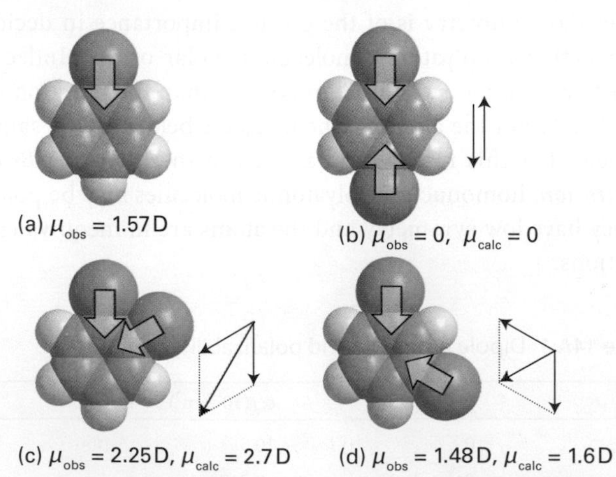

Figure 14A.1 The resultant dipole moments (red in (c) and (d)) of the dichlorobenzene isomers, (b) to (d), can be obtained approximately by vectorial addition of two chlorobenzene dipole moments (shown in (a), with $\mu_{obs} = 1.57$ D).

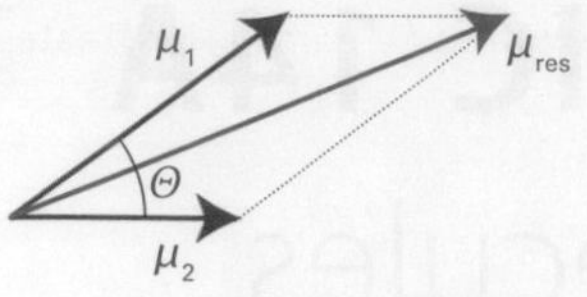

4 Addition of dipole moments

When the two contributing dipole moments have the same magnitude (as in the dichlorobenzenes), this equation simplifies to

$1 + \cos\Theta = 2\cos^2\tfrac{1}{2}\Theta$

$$\mu_{res} \approx \{2\mu_1^2(1+\cos\Theta)\}^{1/2} = 2\mu_1\cos\tfrac{1}{2}\Theta \qquad (14A.3b)$$

Brief illustration 14A.2

Consider *ortho* (1,2-) and *meta* (1,3-) disubstituted benzenes, for which $\Theta_{ortho} = 60°$ and $\Theta_{meta} = 120°$. It follows from eqn 14A.3b that the ratio of the magnitudes of the electric dipole moments is:

$$\frac{\mu_{res,\,ortho}}{\mu_{res,meta}} = \frac{\cos\frac{1}{2}\Theta_{ortho}}{\cos\frac{1}{2}\Theta_{meta}} = \frac{\cos\frac{1}{2}(60°)}{\cos\frac{1}{2}(120°)} = \frac{\frac{3^{1/2}}{2}}{\frac{1}{2}} = 3^{1/2} \approx 1.7$$

A more reliable approach to the calculation of dipole moments is to take into account the locations and magnitudes of the partial charges on all the atoms. These partial charges are included in the output of many molecular structure software packages. To calculate the *x*-component of the dipole moment, for instance, it is necessary to know the partial charge on each atom and the atom's *x*-coordinate relative to a point in the molecule and form the sum

$$\mu_x = \sum_J Q_J x_J \qquad (14A.4a)$$

Here Q_J is the partial charge of atom J, x_J is the *x*-coordinate of atom J, and the sum is over all the atoms in the molecule. Analogous expressions are used for the *y*- and *z*-components. For an electrically neutral molecule, the origin of the coordinates is arbitrary, so it is best chosen to simplify the calculations. In common with all vectors, the magnitude of μ is related to the three components μ_x, μ_y, and μ_z by

$$\mu = (\mu_x^2 + \mu_y^2 + \mu_z^2)^{1/2} \qquad (14A.4b)$$

Example 14A.1 **Calculating a molecular dipole moment**

Estimate the magnitude and orientation of the electric dipole moment of the planar amide group shown in (5) by using the partial charges (as multiples of *e*) and the locations of the atoms shown as (*x*, *y*, *z*) coordinates, with distances in picometres.

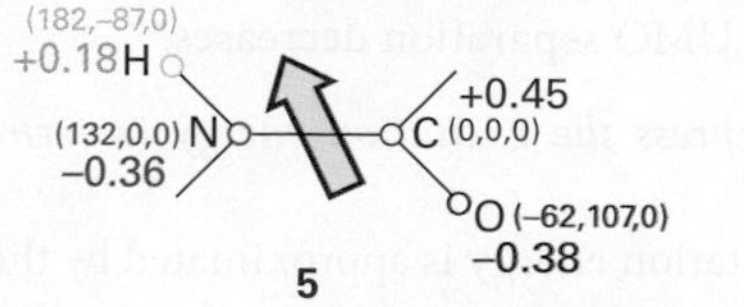

5

Collect your thoughts You need to use eqn 14A.4a to calculate each of the components of the dipole moment and then eqn 14A.4b to assemble the three components into the magnitude of the dipole moment. Note that the partial charges are multiples of the fundamental charge, $e = 1.602 \times 10^{-19}$ C. This group is a fragment of a neutral molecule, so the choice of origin of the coordinates is not arbitrary. Here the origin is taken to be coincident with the carbon atom.

The solution The expression for μ_x is

$$\begin{aligned}\mu_x &= (-0.36e)\times(132\,\text{pm}) + (0.45e)\times(0\,\text{pm}) + (0.18e)\times(182\,\text{pm})\\ &\quad + (-0.38e)\times(-62.0\,\text{pm})\\ &= 8.8e\,\text{pm}\\ &= 8.8\times(1.602\times10^{-19}\,\text{C})\times(10^{-12}\,\text{m}) = 1.4\times10^{-30}\,\text{C m}\end{aligned}$$

corresponding to $\mu_x = +0.42$ D. The expression for μ_y is:

$$\begin{aligned}\mu_y &= (-0.36e)\times(0\,\text{pm}) + (0.45e)\times(0\,\text{pm}) + (0.18e)\times(-87\,\text{pm})\\ &\quad + (-0.38e)\times(107\,\text{pm})\\ &= -56.3e\,\text{pm}\\ &= -9.02\times10^{-30}\,\text{C m}\end{aligned}$$

It follows that $\mu_y = -2.7$ D. The amide group is planar, so $\mu_z = 0$ and

$$\mu = \{(0.42\ \text{D})^2 + (-2.7\ \text{D})^2\}^{1/2} = 2.7\ \text{D}$$

The orientation of the dipole moment is found by arranging an arrow of length 2.7 units of length to have *x*, *y*, and *z* components of 0.42, −2.7, and 0 units; the orientation is superimposed on (5).

Self-test 14A.1 Estimate the magnitude of the electric dipole moment of methanal (formaldehyde) by using the information in (6).

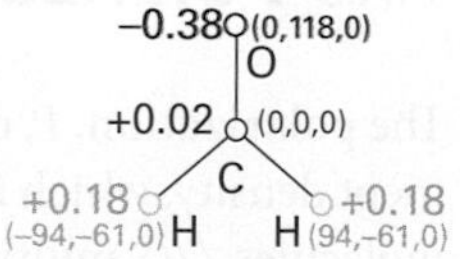

6

Answer: 3.2 D; experimental: 2.3 D

Molecules may have higher **multipoles**, or arrays of point charges (Fig. 14A.2). Specifically, an ***n*-pole** is an array of point charges with an *n*-pole moment but no lower moment. Thus, a **monopole** ($n = 1$) is a point charge, and the monopole moment is what is normally called the overall charge. A dipole ($n = 2$), as already seen, is an array of charges that has no monopole moment (no net charge). A **quadrupole** ($n = 3$) consists of

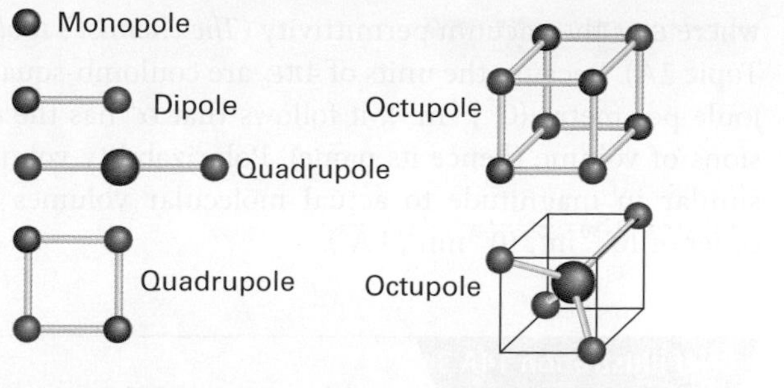

Figure 14A.2 Typical charge arrays corresponding to electric multipoles. The field arising from an arbitrary charge distribution can be expressed as the superposition of the fields arising from a superposition of multipoles.

an array of point charges that has neither net charge nor dipole moment (as for a CO_2 molecule, 3). An **octupole** ($n = 4$) consists of an array of point charges that sum to zero and which has neither a dipole moment nor a quadrupole moment (as for a CH_4 molecule, 7).

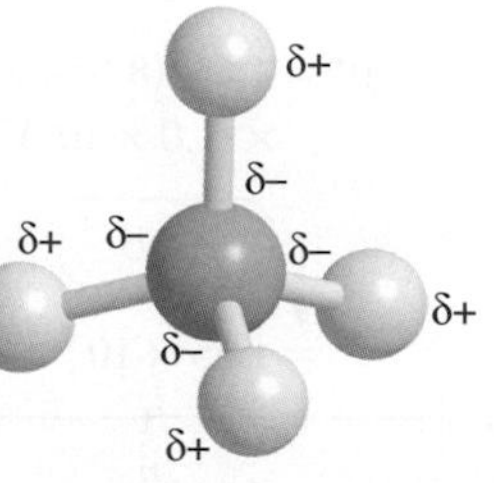

7 Methane, CH_4

14A.2 Polarizabilities

The failure of nuclear charges to control the surrounding electrons totally means that those electrons can respond to external fields. Therefore, an applied electric field can distort a molecule as well as align its permanent electric dipole moment. The magnitude μ^* of the **induced dipole moment,** μ^*, is proportional to the electric field strength, $\mathcal{E}$, so

$$\mu^* = \alpha\mathcal{E} \qquad \text{Polarizability [definition]} \qquad (14\text{A}.5\text{a})$$

The constant of proportionality α is the **polarizability** of the molecule. The greater the polarizability, the larger is the induced dipole moment for a given applied field. In a formal treatment, vector quantities are used to allow for the possibility that the induced dipole moment might not lie parallel to the applied field, in which case the scalar α is replaced by $\boldsymbol{\alpha}$, a 3×3 matrix.

When the applied field is very strong (as in tightly focused laser beams), the magnitude of the induced dipole moment is not strictly linear in the strength of the field, and

$$\mu^* = \alpha\mathcal{E} + \tfrac{1}{2}\beta\mathcal{E}^2 + \cdots \qquad \text{Hyperpolarizability [definition]} \qquad (14\text{A}.5\text{b})$$

The coefficient β is the (first) **hyperpolarizability** of the molecule.

Polarizability has the units (coulomb metre)^2 per joule ($C^2\,m^2\,J^{-1}$). That collection of units is awkward, so α is often expressed as a **polarizability volume,** α', by using the relation

$$\alpha' = \frac{\alpha}{4\pi\varepsilon_0} \qquad \text{Polarizability volume [definition]} \qquad (14\text{A}.6)$$

where ε_0 is the vacuum permittivity (*The chemist's toolkit* 6 of Topic 2A). Because the units of $4\pi\varepsilon_0$ are coulomb-squared per joule per metre ($C^2 J^{-1} m^{-1}$), it follows that α' has the dimensions of volume (hence its name). Polarizability volumes are similar in magnitude to actual molecular volumes (of the order of $10^{-30}\,m^3$, $10^{-3}\,nm^3$, 1 Å^3).

Brief illustration 14A.3

The polarizability volume of H_2O is $1.48 \times 10^{-30}\,m^3$. It follows from eqns 14A.5a and 14A.6 that $\mu^* = 4\pi\varepsilon_0\alpha'\mathcal{E}$ and the magnitude of the dipole moment of the molecule (in addition to the permanent dipole moment) induced by an applied electric field of strength $1.0 \times 10^5\,V\,m^{-1}$ is

$$\mu^* = 4\pi \times (8.854 \times 10^{-12}\,J^{-1}\,C^2\,m^{-1}) \times (1.48 \times 10^{-30}\,m^3) \times (1.0 \times 10^5\,V\,m^{-1})$$

$$\overset{1\,V = 1\,J\,C^{-1}}{=} 1.65 \times 10^{-35}\,C\,m = 4.9 \times 10^{-6}\,D = 4.9\,\mu D$$

The polarizability volumes of some molecules are given in Table 14A.1. It is possible to establish a correlation between these values and the electronic structure of atoms and molecules.

How is that done? 14A.1 Correlating polarizability and molecular structure

The argument starts from the quantum mechanical expression for the molecular polarizability in the *z*-direction:[1]

$$\alpha = 2\sum_{n\neq 0} \frac{|\mu_{z,0n}|^2}{E_n^{(0)} - E_0^{(0)}}$$

where $\mu_{z,0n} = \int \psi_n^* \hat{\mu}_z \psi_0 d\tau$ is the *z*-component of the *transition electric dipole moment*, a measure of the extent to which electric charge is shifted when an electron migrates from the ground state to create an excited state. The sum is over the excited states, with energies $E_n^{(0)}$.

Step 1 *Introduce approximations*

Now approximate the excitation energies by a mean value ΔE (an indication of the HOMO–LUMO separation). Also suppose that the most important transition dipole moment is approximately equal to the charge of an electron multiplied by the molecular radius *R*. Then

$$\alpha \approx \frac{2e^2R^2}{\Delta E}$$

This expression shows that α increases with the size of the molecule and with the ease with which it can be excited electronically. Therefore, the polarizability increases as the HOMO–LUMO separation decreases.

Step 2 *Express the excitation energy in terms of the atomic radius*

If the excitation energy is approximated by the energy needed to remove an electron to infinity from a distance *R* from a single positive charge, then $\Delta E \approx e^2/4\pi\varepsilon_0 R$. When you insert this value into the expression for α you find that

$$\alpha \approx 2e^2R^2 \times \frac{4\pi\varepsilon_0 R}{e^2} = 2 \times 4\pi\varepsilon_0 R^3$$

To obtain the polarizability volume, divide α by $4\pi\varepsilon_0$, and ignore the factor of 2 in this approximation. The result is $\alpha' \approx R^3$, which is of the same order of magnitude as the molecular volume.

As just shown, polarizability volumes correlate with the HOMO–LUMO separations in atoms and molecules. The electron distribution can be distorted readily if the LUMO lies close to the HOMO in energy, so the polarizability is then large. If the LUMO lies high above the HOMO, an applied field cannot perturb the electron distribution significantly, and the polarizability is low. Molecules with small HOMO–LUMO gaps are typically large, and have numerous electrons.

For most molecules, the polarizability is 'anisotropic', which means that its value depends on the orientation of the molecule relative to the applied field. The polarizability volume of benzene when the field is applied perpendicular to the ring is $0.0067\,nm^3$ and it is $0.0123\,nm^3$ when the field is applied in the plane of the ring. The anisotropy of the polarizability determines whether a molecule is rotationally Raman active (Topic 11B).

14A.3 Polarization

The **polarization**, *P*, of a bulk sample is the electric dipole moment density, which is the mean electric dipole moment of the molecules, $\langle\mu\rangle$, multiplied by the number density, $\mathcal{N} = N/V$:

$$P = \langle\mu\rangle\mathcal{N} \qquad \text{Polarization [definition]} \qquad (14A.7)$$

A **dielectric** is a polarizable, non-conducting medium.

(a) The frequency dependence of the polarization

The polarization of a fluid dielectric is zero in the absence of an applied field because the molecules adopt ceaselessly changing random orientations due to thermal motion, so $\langle\mu\rangle = 0$. In the presence of a weak electric field, the energy depends on the orientation of the dipole with respect to the field, with

[1] For a derivation of this equation see our *Physical chemistry: Quanta, matter, and change* (2014).

lower energy orientations being more populated; as a result, the mean dipole moment is no longer zero. The Boltzmann distribution can be used to find an expression for $\langle\mu\rangle$.

How is that done? 14A.2 Deriving an expression for the mean dipole moment

In spherical polar coordinates, the area of an infinitesimal patch of surface (at constant radius) is $\sin\theta\,\mathrm{d}\theta\mathrm{d}\phi$. The number of molecules with their electric dipole moments pointing into this patch is proportional to this area multiplied by the Boltzmann factor $e^{-E(\theta)/kT}$. If the electric field is in the z-direction, the energy is independent of the azimuthal angle ϕ. The probability $\mathrm{d}p$ that a dipole moment has an orientation in the range θ to $\theta+\mathrm{d}\theta$ and at any azimuthal angle ϕ around the direction of the applied field is therefore

$$\mathrm{d}p=\frac{e^{-E(\theta)/kT}\sin\theta\,\mathrm{d}\theta}{\int_0^\pi e^{-E(\theta)/kT}\sin\theta\,\mathrm{d}\theta}$$

where $0\le\theta\le\pi$. If the applied electric field $\mathcal{E}$ is in the z-direction, then the dipole moment is also aligned in the z-direction, and its mean value is

$$\langle\mu_z\rangle=\int\mu_z\,\mathrm{d}p$$

Step 1 *Write an expression for the energy of a dipole in an electric field*

The energy $E(\theta)$ of a dipole depends on the angle θ it makes with the electric field $\mathcal{E}$ as

$$E(\theta)=-\mu\mathcal{E}\cos\theta$$

Step 2 *Set up an expression for the average of the z-component of the dipole moment*

The average value of the component of the dipole moment parallel to the applied electric field is

$$\langle\mu_z\rangle=\int\overbrace{\mu\cos\theta}^{\mu_z}\mathrm{d}p=\mu\frac{\int_0^\pi\cos\theta\, e^{-E(\theta)/kT}\sin\theta\,\mathrm{d}\theta}{\int_0^\pi e^{-E(\theta)/kT}\sin\theta\,\mathrm{d}\theta}$$

$E(\theta)=-\mu\mathcal{E}\cos\theta$

$$=\mu\frac{\int_0^\pi e^{\mu\mathcal{E}\cos\theta/kT}\cos\theta\sin\theta\,\mathrm{d}\theta}{\int_0^\pi e^{\mu\mathcal{E}\cos\theta/kT}\sin\theta\,\mathrm{d}\theta}$$

Step 3 *Evaluate the integrals*

To simplify the appearance of this expression, write $x=\mu\mathcal{E}/kT$ and obtain

$$\langle\mu_z\rangle=\frac{\mu\int_0^\pi e^{x\cos\theta}\cos\theta\sin\theta\,\mathrm{d}\theta}{\int_0^\pi e^{x\cos\theta}\sin\theta\,\mathrm{d}\theta}$$

Then write $y=\cos\theta$ and $\mathrm{d}y=-\sin\theta\,\mathrm{d}\theta$, and change the limits of integration to $y=-1$ (at $\theta=\pi$) and $y=1$ (at $\theta=0$):

$$\langle\mu_z\rangle=\frac{\mu\overbrace{\int_{-1}^1 ye^{xy}\mathrm{d}y}^{\text{Integral E.2}}}{\underbrace{\int_{-1}^1 e^{xy}\mathrm{d}y}_{\text{Integral E.1}}}=\mu\frac{\left\{\frac{e^x+e^{-x}}{x}-\frac{e^x-e^{-x}}{x^2}\right\}}{\left\{\frac{e^x-e^{-x}}{x}\right\}}$$

$$=\mu\left\{\frac{e^x+e^{-x}}{e^x-e^{-x}}-\frac{1}{x}\right\}$$

That is, noting again that $x=\mu\mathcal{E}/kT$,

$$\langle\mu_z\rangle=\mu L(x)\quad L(x)=\frac{e^x+e^{-x}}{e^x-e^{-x}}-\frac{1}{x}\qquad(14A.8a)$$

Mean electric dipole moment

The function $L(x)$ is called the **Langevin function** (Fig. 14A.3). Under most circumstances, x is very small. For example, if $\mu=1\,\mathrm{D}$ and $T=300\,\mathrm{K}$, then x exceeds 0.01 only if the field strength exceeds $100\,\mathrm{kV\,cm^{-1}}$, and most measurements are done at much lower strengths. The exponentials in the Langevin function can be expanded when the field is so weak that $x\ll1$, and the largest term that survives is $L(x)=\frac{1}{3}x$. Therefore, the average molecular dipole moment is

$$\langle\mu_z\rangle=\frac{\mu^2\mathcal{E}}{3kT}\qquad(14A.8b)$$

Mean value of the dipole moment [weak electric field]

As the electric field strength is increased to very high values, the orientations of molecular dipole moments fluctuate less about the field direction and the mean dipole moment approaches its maximum value of $\langle\mu_z\rangle=\mu$.

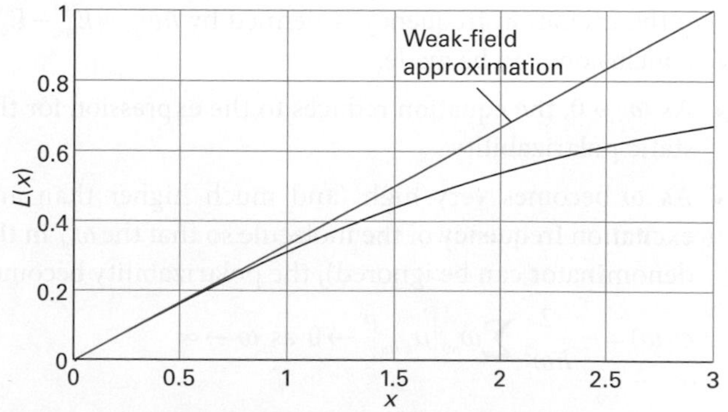

Figure 14A.3 The Langevin function (in purple) used in the calculation of the mean electric dipole moment. When x is small the weak-field approximation (in blue) is appropriate.

When the applied field changes direction slowly, the orientation of the permanent dipole moment has time to change—the whole molecule rotates into a new orientation—and follows the field. However, when the electric field changes direction rapidly, a molecule cannot change orientation fast enough to follow and the permanent dipole moment then makes no contribution to the polarization of the sample. The **orientation polarization**, the polarization arising from the permanent dipole moments, is lost at such high frequencies. Because a molecule takes about 1 ps to turn through about 1 radian in a fluid, the loss of the contribution of orientation polarization to the total polarization occurs when measurements are made at frequencies greater than about 10^{11} Hz (in the microwave region).

The next contribution to the polarization to be lost as the frequency increases is the **distortion polarization**, the polarization that arises from the distortion of the positions of the nuclei by the applied field. The molecule is bent and stretched by the applied field, and the molecular dipole moment changes accordingly. The time taken for a molecule to bend is approximately the inverse of the molecular vibrational frequency, so the distortion polarization disappears when the frequency of the radiation is increased through the infrared.

Polarization disappears in stages: each successive stage occurs as the frequency of oscillation of the electric field rises above the frequency of a particular mode of vibration. At even higher frequencies, in the visible region, only the electrons are mobile enough to respond to the rapidly changing direction of the applied field. The polarization that remains is now due entirely to the distortion of the electron distribution, and the surviving contribution to the molecular polarizability is called the **electronic polarizability**. This behaviour can be explored by noting that the quantum mechanical expression for the polarizability of a molecule in the presence of an electric field that is oscillating at a frequency ω in the z-direction is[2]

$$\alpha(\omega)=\frac{2}{\hbar}\sum_{n\neq 0}\frac{\omega_{n0}\left|\mu_{z,0n}\right|^2}{\omega_{n0}^2-\omega^2}$$

where the excitation frequency is defined by $\hbar\omega_{n0}=E_n^{(0)}-E_0^{(0)}$. Two conclusions can be made:

- As $\omega\rightarrow 0$, the equation reduces to the expression for the static polarizability
- As ω becomes very high (and much higher than any excitation frequency of the molecule so that the ω_{n0}^2 in the denominator can be ignored), the polarizability becomes

$$\alpha(\omega)=-\frac{2}{\hbar\omega^2}\sum_n\omega_{n0}\left|\mu_{z,0n}\right|^2\rightarrow 0 \text{ as } \omega\rightarrow\infty$$

The conclusion applies to each type of excitation, vibrational as well as electronic, and accounts for the successive decreases in polarizability as the frequency is increased.

(b) Molar polarization

When two charges Q_1 and Q_2 are separated by a distance r in a medium, the Coulomb potential energy of their interaction is

$$V=\frac{Q_1Q_2}{4\pi\varepsilon r} \tag{14A.9}$$

where ε is the permittivity of the medium, which is reported by introducing the relative permittivity and writing $\varepsilon=\varepsilon_r\varepsilon_0$ (*The chemist's toolkit* 6 in Topic 2A). The relative permittivity of a substance is measured by comparing the capacitance of a capacitor with and without the sample present (C and C_0, respectively) and using $\varepsilon_r = C/C_0$. The relative permittivity can have a very significant effect on the strength of the interactions between ions in solution. For instance, water has a relative permittivity of 78 at 25 °C, so the interionic Coulombic interaction energy is reduced by nearly two orders of magnitude from its vacuum value.

The relative permittivity of a substance is large if its molecules are polar or highly polarizable. The quantitative relation between the relative permittivity and the electric properties of the molecules is obtained by considering the polarization of a medium, and is expressed by the **Debye equation**:

$$\frac{\varepsilon_r-1}{\varepsilon_r+2}=\frac{\rho P_m}{M} \quad \text{Debye equation} \tag{14A.10}$$

where ρ is the mass density of the sample, M is the molar mass of the molecules, and P_m is the **molar polarization**, which is defined as

$$P_m=\frac{N_A}{3\varepsilon_0}\left(\alpha+\frac{\mu^2}{3kT}\right) \quad \text{Molar polarization [definition]} \tag{14A.11}$$

where α is the polarizability. The term $\mu^2/3kT$ stems from the thermal averaging of the electric dipole moment in the presence of the applied field (eqn 14A.8b). The corresponding expression without the contribution from the permanent dipole moment is called the **Clausius–Mossotti equation**:

$$\frac{\varepsilon_r-1}{\varepsilon_r+2}=\frac{\rho N_A\alpha}{3M\varepsilon_0} \quad \text{Clausius–Mossotti equation} \tag{14A.12}$$

The Clausius–Mossotti equation is used when there is no contribution from permanent electric dipole moments to the polarization, either because the molecules are nonpolar or because the frequency of the applied field is so high that the molecules cannot orientate quickly enough to follow the change in direction of the field.

[2] For a derivation of this equation, see our *Physical chemistry: Quanta, matter, and change* (2014).

Example 14A.2 Determining dipole moment and polarizability

The relative permittivity of camphor (**8**) was measured at a series of temperatures with the results given below. Determine the dipole moment and the polarizability volume of the molecule.

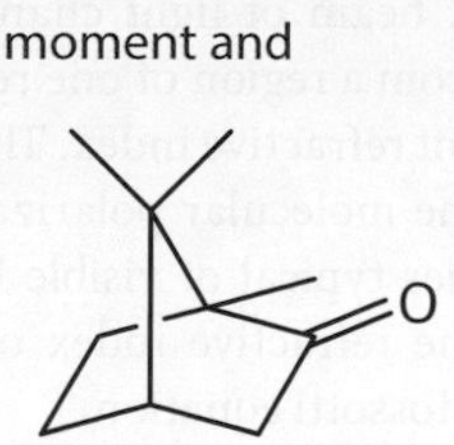

8 Camphor

θ/°C	ρ/(g cm^{-3})	ε_r
0	0.99	12.5
20	0.99	11.4
40	0.99	10.8
60	0.99	10.0
80	0.99	9.50
100	0.99	8.90
120	0.97	8.10
140	0.96	7.60
160	0.95	7.11
200	0.91	6.21

Collect your thoughts The relative permittivity depends on the molar polarization (eqn 14A.10), which in turn depends on the temperature, polarizability, and the magnitude of the permanent dipole moment (eqn 14A.11). These relations suggest that you should

- Calculate $(\varepsilon_r - 1)/(\varepsilon_r + 2)$ at each temperature, and then multiply by M/ρ to form P_m from eqn 14A.10.
- Plot P_m against $1/T$.

because eqn 14A.11 rearranges to

$$P_m = \overbrace{\frac{N_A\alpha}{3\varepsilon_0}}^{\text{intercept}} + \overbrace{\frac{N_A\mu^2}{9\varepsilon_0 k}}^{\text{slope}} \times \frac{1}{T}$$

The slope of the graph is $N_A\mu^2/9\varepsilon_0 k$ and its intercept at $1/T = 0$ is $N_A\alpha/3\varepsilon_0$.

The solution Use the data to draw up the following table, with $M = 152.23\,\text{g mol}^{-1}$ for camphor.

θ/°C	(10^3 K)/T	ε_r	$(\varepsilon_r - 1)/(\varepsilon_r + 2)$	P_m/(cm^3 mol^{-1})
0	3.66	12.5	0.793	122
20	3.41	11.4	0.776	119
40	3.19	10.8	0.766	118
60	3.00	10.0	0.750	115
80	2.83	9.50	0.739	114
100	2.68	8.90	0.725	111
120	2.54	8.10	0.703	110
140	2.42	7.60	0.688	109
160	2.31	7.11	0.671	108
200	2.11	6.21	0.635	106

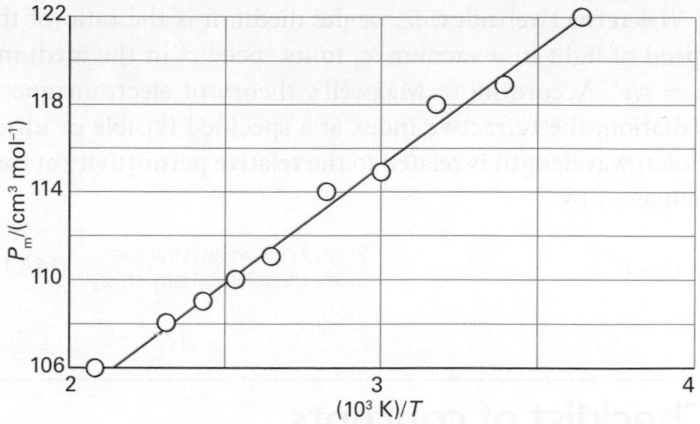

Figure 14A.4 The plot of P_m/(cm^3 mol^{-1}) against (10^3 K)/T used in *Example* 14A.2 for the determination of the polarizability and the magnitude of the dipole moment of camphor.

The points are plotted in Fig. 14A.4. The intercept on the vertical axis lies at P_m/(cm^3 mol^{-1}) = 83.5, so $N_A\alpha/3\varepsilon_0$ = 83.5 cm^3 mol^{-1} = 8.35 × 10^{-5} m^3 mol^{-1}. It then follows that

$$\alpha = \frac{3\times\overbrace{(8.854\times10^{-12}\,\text{J}^{-1}\text{C}^2\text{m}^{-1})}^{\varepsilon_0}}{\underbrace{6.02\times10^{23}\,\text{mol}^{-1}}_{N_A}}\times\overbrace{8.35\times10^{-5}\,\text{m}^3\,\text{mol}^{-1}}^{\text{intercept}}$$

$$=3.68\times10^{-39}\,\text{C}^2\,\text{m}^2\,\text{J}^{-1}$$

From eqn 14A.6, it follows that $\alpha' = 3.31 \times 10^{-29}$ m^3. The slope is 10.55, so $N_A\mu^2/9\varepsilon_0 k = 1.055 \times 10^4$ cm^3 mol^{-1} K = 1.055 × 10^{-2} m^3 mol^{-1} K, so from the expression for P_m it follows that

$$\mu = \left(\frac{9\times\overbrace{(8.854\times10^{-12}\,\text{J}^{-1}\,\text{C}^2\,\text{m}^{-1})}^{\varepsilon_0}\times\overbrace{(1.381\times10^{-23}\,\text{JK}^{-1})}^{k}}{\underbrace{6.022\times10^{23}\,\text{mol}^{-1}}_{N_A}}\right)^{1/2}$$

$$\times\,\overbrace{(1.055\times10^{-2}\,\text{m}^3\,\text{mol}^{-1}\,\text{K})}^{\text{slope}}{}^{1/2}$$

$$=4.39\times10^{-30}\,\text{C m} = 1.32\,\text{D}$$

Because the Debye equation describes molecules that are free to rotate, the data show that camphor, which does not melt until 175 °C, is rotating even in the solid. It is an approximately spherical molecule.

Self-test 14A.2 The relative permittivity of chlorobenzene is 5.71 at 20 °C and 5.62 at 25 °C. Assuming a constant density (1.11 g cm^{-3}), estimate its polarizability volume and the magnitude of its dipole moment.

Answer: 1.4×10^{-29} m^3, 1.2 D

The refractive index, n_r, of the medium is the ratio of the speed of light in a vacuum, c, to its speed c' in the medium: $n_r = c/c'$. According to Maxwell's theory of electromagnetic radiation, the refractive index at a specified (visible or ultraviolet) wavelength is related to the relative permittivity at that frequency by

$$n_r = \varepsilon_r^{1/2} \quad \text{Relation between refractive index and relative permittivity} \quad (14A.13)$$

A beam of light changes direction ('bends') when it passes from a region of one refractive index to a region with a different refractive index. Therefore, the molar polarization, P_m, and the molecular polarizability, α, can be measured at frequencies typical of visible light (about 10^{14}–10^{15} Hz) by measuring the refractive index of the sample and using the Clausius–Mossotti equation.

Checklist of concepts

- ☐ 1. An **electric dipole** consists of two electric charges $+Q$ and $-Q$ separated by a vector $\boldsymbol{R}$.
- ☐ 2. The **electric dipole moment** μ is a vector that points from the negative charge to the positive charge of a dipole; its magnitude is $\mu = QR$.
- ☐ 3. A **polar molecule** is a molecule with a permanent electric dipole moment.
- ☐ 4. Molecules may have higher electric multipoles: an ***n*-pole** is an array of point charges with an n-pole moment but no lower moment.
- ☐ 5. The **polarizability** is a measure of the ability of an electric field to induce a dipole moment in a molecule.
- ☐ 6. **Polarizabilities** (and **polarizability volumes**) correlate with the HOMO–LUMO separations in molecules.
- ☐ 7. The **polarization** of a medium is the electric dipole moment density.
- ☐ 8. **Orientation polarization** is the polarization arising from the permanent dipole moments.
- ☐ 9. **Distortion polarization** is the polarization arising from the distortion of the positions of the nuclei by the applied field.
- ☐ 10. **Electronic polarizability** is the polarizability due to the distortion of the electron distribution.

Checklist of equations

Property	Equation	Comment	Equation number
Magnitude of the electric dipole moment	$\mu = QR$	Definition	14A.1
Magnitude of the resultant of two dipole moments	$\mu_{res} \approx (\mu_1^2+\mu_2^2+2\mu_1\mu_2\cos\Theta)^{1/2}$		14A.3a
Magnitude of the induced dipole moment	$\mu^* = \alpha\mathcal{E}$	Linear approximation; α is the polarizability	14A.5a
	$\mu^* = \alpha\mathcal{E} + \frac{1}{2}\beta\mathcal{E}^2$	Quadratic approximation; β is the hyperpolarizability	14A.5b
Polarizability volume	$\alpha' = \alpha/4\pi\varepsilon_0$	Definition	14A.6
Polarization	$P = \langle\mu\rangle\mathcal{N}$	Definition	14A.7
Debye equation	$(\varepsilon_r-1)/(\varepsilon_r+2) = \rho P_m/M$		14A.10
Molar polarization	$P_m = (N_A/3\varepsilon_0)(\alpha+\mu^2/3kT)$		14A.11
Clausius–Mossotti equation	$(\varepsilon_r-1)/(\varepsilon_r+2) = \rho N_A\alpha/3M\varepsilon_0$		14A.12

TOPIC 14B Interactions between molecules

> ➤ **Why do you need to know this material?**
>
> Many types of molecular interactions are responsible for the formation of condensed phases and large molecular assemblies.
>
> ➤ **What is the key idea?**
>
> Attractive interactions result in cohesion but repulsive interactions prevent the complete collapse of matter to nuclear densities.
>
> ➤ **What do you need to know already?**
>
> You need to be familiar with elementary aspects of electrostatics, specifically the Coulomb interaction (*The chemist's toolkit* 6 of Topic 2A), and the relationships between the structure and electric properties of a molecule, specifically its dipole moment and polarizability (Topic 14A).

A **van der Waals interaction** is an attractive (energy lowering) interaction between closed-shell molecules that depends on the separation of the molecules as the inverse sixth power ($V \propto 1/r^6$). This precise criterion is often relaxed to include all nonbonding interactions. They occur in various guises and can all be traced to the interaction of partial charges. The underlying interaction throughout this discussion is the Coulomb potential energy, $V = Q_1Q_2/4\pi\varepsilon r$, discussed in *The chemist's toolkit* 6 of Topic 2A.

14B.1 The interactions of dipoles

A **point dipole** is a dipole in which the separation between the charges l is much smaller than the distance r from which the dipole is being observed ($l \ll r$).

(a) Charge–dipole interactions

The Coulomb potential energy of one charge near another can be adapted to find the potential energy of a point charge and a dipole, and extended to the interaction between two dipoles.

How is that done? 14B.1 Deriving the expression for the interaction between a point charge and a point dipole

You need to consider the interaction between the two charges $\pm Q_1$ of a point dipole, with a dipole moment of magnitude $\mu_1 = Q_1 l$, and the point charge Q_2 as shown in (**1**). Suppose that the arrangement is in a vacuum, so use $\varepsilon = \varepsilon_0$.

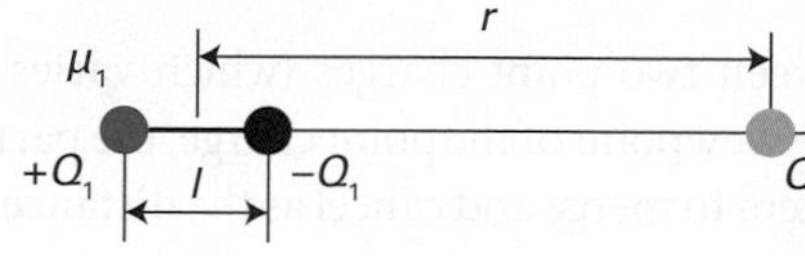

1

Step 1 *Write an expression for the potential energy due to interaction with both charges*

The sum of the potential energies due to repulsion between like charges and attraction between opposite charges is

$$V = \frac{1}{4\pi\varepsilon_0}\left(-\frac{Q_1Q_2}{r-\frac{1}{2}l} + \frac{Q_1Q_2}{r+\frac{1}{2}l}\right) \overset{x=l/2r}{=} \frac{Q_1Q_2}{4\pi\varepsilon_0 r}\left(-\frac{1}{1-x} + \frac{1}{1+x}\right)$$

Step 2 *Treat the dipole as a point dipole*

Because $l \ll r$ for a point dipole, this expression can be simplified by expanding the terms in x by using

$$\frac{1}{1+x} = 1 - x + x^2 - \cdots \qquad \frac{1}{1-x} = 1 + x + x^2 + \cdots$$

and retaining only the first two terms:

$$V = \frac{Q_1Q_2}{4\pi\varepsilon_0 r}\{-(1+x+\cdots) + (1-x+\cdots)\} \approx -\frac{2xQ_1Q_2}{4\pi\varepsilon_0 r} = -\frac{Q_1Q_2 l}{4\pi\varepsilon_0 r^2}$$

With $\mu_1 = Q_1 l$ this expression becomes

$$V = -\frac{\mu_1 Q_2}{4\pi\varepsilon_0 r^2} \qquad (14B.1)$$

Point dipole–point charge interaction [as in (**1**)]

With μ in coulomb metres, Q_2 in coulombs, and r in metres, V is obtained in joules. In the orientation shown in (**1**), V is negative, representing a net attraction. The expression should be multiplied by $\cos\Theta$ when the point charge lies at an angle Θ to the axis of the dipole and ε_0 replaced by ε if the medium is not a vacuum.

The potential energy approaches zero (the value at infinite separation of the charge and the dipole) more rapidly (as $1/r^2$)

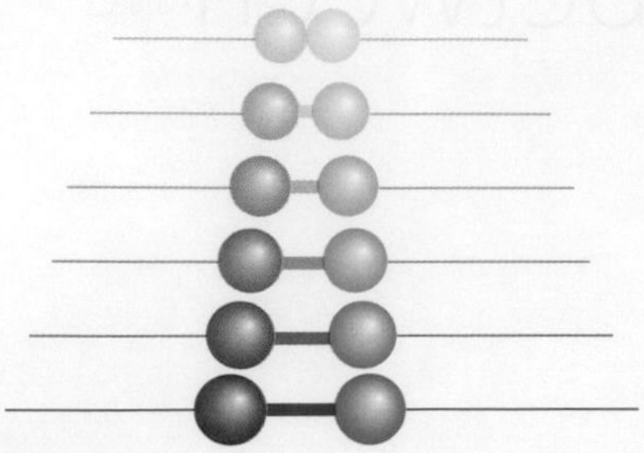

Figure 14B.1 There are two contributions to the diminishing field of an electric dipole with distance (here seen from the side). The potentials of the charges decrease (shown here by a fading intensity) and the two charges appear to merge, so their combined effect approaches zero more rapidly than by the distance effect alone.

than that between two point charges (which varies as $1/r$) because, from the viewpoint of the point charge, the partial charges of the dipole seem to merge and cancel as the distance r increases (Fig. 14B.1).

Brief illustration 14B.1

Consider a Li^+ ion and a water molecule ($\mu = 1.85$ D) separated by 1.0 nm in a vacuum, with the point charge on the ion and the dipole of the molecule arranged as in (**1**). The energy of interaction is given by eqn 14B.1 as

$$V = -\frac{\overbrace{(1.602\times10^{-19}\,\text{C})}^{Q_{\text{Li}^+}}\times\overbrace{(1.85\times3.336\times10^{-30}\,\text{C m})}^{\mu_{\text{H}_2\text{O}}}}{4\pi\times\underbrace{(8.854\times10^{-12}\,\text{J}^{-1}\,\text{C}^2\,\text{m}^{-1})}_{\varepsilon_0}\times\underbrace{(1.0\times10^{-9}\,\text{m})^2}_{r}}$$

$$= -8.9\times10^{-21}\,\text{J}$$

This energy corresponds to -5.4 kJ mol^{-1}.

(b) Dipole–dipole interactions

The preceding discussion can be extended to the interaction of two dipoles arranged as in (**2**).

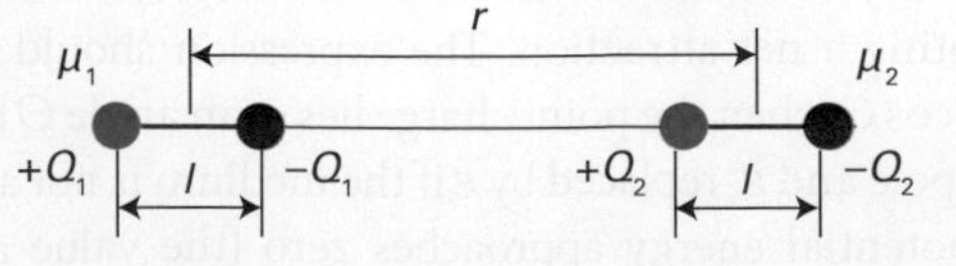

2

How is that done? 14B.2 Deriving the expression for the interaction energy of two point dipoles

To calculate the potential energy of interaction of two point dipoles separated by r in a vacuum in the arrangement shown in (**2**) proceed in exactly the same way as before. In this case the total interaction energy is the sum of four pairwise terms. Two are attractions between opposite charges, which contribute negative terms to the potential energy, and two are repulsions between like charges, which contribute positive terms.

Step 1 *Write an expression for the potential energy due to interaction of the charges on the two dipoles*

The sum of the four contributions is

$$V = \frac{1}{4\pi\varepsilon_0}\left(-\frac{Q_1Q_2}{r+l}+\frac{Q_1Q_2}{r}+\frac{Q_1Q_2}{r}-\frac{Q_1Q_2}{r-l}\right)$$

$$\overset{x=l/r}{=} -\frac{Q_1Q_2}{4\pi\varepsilon_0 r}\left(\frac{1}{1+x}-2+\frac{1}{1-x}\right)$$

Step 2 *Treat the dipoles as point dipoles*

As before, provided $l \ll r$ the two terms in x may be expanded, leading to

$$V = -\frac{Q_1Q_2}{4\pi\varepsilon_0 r}(\overbrace{1-x+x^2+\cdots}^{1/(1+x)}-2+\overbrace{1+x+x^2+\cdots}^{1/(1-x)})$$

The terms in blue sum to zero, so the only surviving term is $2x^2$. It follows that

$$V = -\frac{2x^2Q_1Q_2}{4\pi\varepsilon_0 r} \overset{x=l/r}{=} -\frac{2l^2Q_1Q_2}{4\pi\varepsilon_0 r^3}$$

Because $\mu_1 = Q_1 l$ and $\mu_2 = Q_2 l$, it follows that the potential energy of interaction in the alignment shown in (**2**) is given by

$$V = -\frac{\mu_1\mu_2}{2\pi\varepsilon_0 r^3} \quad (14\text{B}.2)$$

Point dipole–point dipole interaction [as in (**2**)]

This interaction energy approaches zero more rapidly (as $1/r^3$) than for the previous case: now both interacting entities appear neutral to each other at large separations.

Equation 14B.2 applies only to the arrangement in (**2**). More generally, as in the arrangement in (**3**), the potential energy of interaction between two polar molecules separated by a vector $\boldsymbol{r}$ is

$$V = \frac{1}{4\pi\varepsilon_0 r^3}\left\{\boldsymbol{\mu}_1\cdot\boldsymbol{\mu}_2 - \frac{3(\boldsymbol{\mu}_1\cdot\boldsymbol{r})(\boldsymbol{r}\cdot\boldsymbol{\mu}_2)}{r^2}\right\} \quad (14\text{B}.3\text{a})$$

Point dipole–point dipole interaction [as in (**3**)]

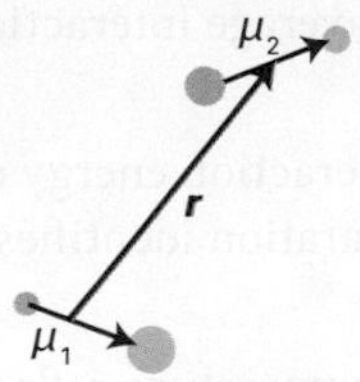

3

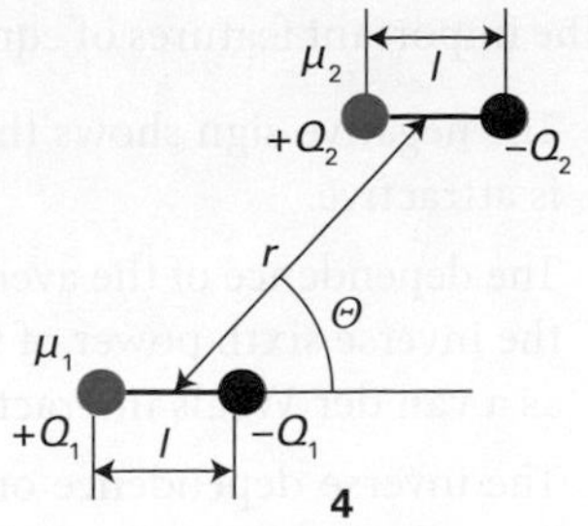

4

(For the origin of this expression, see *A deeper look* 8 on the website of this book.) When the two dipoles are parallel and arranged as in (4), the potential energy is simply

$$V = \frac{\mu_1\mu_2 f(\Theta)}{4\pi\varepsilon_0 r^3} \qquad f(\Theta) = 1 - 3\cos^2\Theta$$

Point dipole–point dipole interaction [as in (**4**)] (14B.3b)

Brief illustration 14B.2

Equation 14B.3b can be used to calculate the potential energy of the dipolar interaction between two amide groups. Supposing that the groups are separated by 3.0 nm with $\Theta = 180°$ (so $\cos\Theta = -1$ and $1 - 3\cos^2\Theta = -2$). Take $\mu_1 = \mu_2 = 2.7\,\text{D}$, corresponding to $9.0 \times 10^{-30}\,\text{C m}$, and find

$$V = \frac{\overbrace{(9.0\times10^{-30}\,\text{C m})^2}^{\mu_1\mu_2} \times \overbrace{(-2)}^{1-3\cos^2\Theta}}{4\pi\times\underbrace{(8.854\times10^{-12}\,\text{J}^{-1}\,\text{C}^2\,\text{m}^{-1})}_{\varepsilon_0}\times\underbrace{(3.0\times10^{-9}\,\text{m})^3}_{r^3}}$$

$$= \frac{(9.0\times10^{-30})^2\times(-2)}{4\pi\times(8.854\times10^{-12})\times(3.0\times10^{-9})^3}\,\frac{\text{C}^2\,\text{m}^2}{\text{J}^{-1}\text{C}^2\,\text{m}^{-1}\,\text{m}^3}$$

$$= -5.4\times10^{-23}\,\text{J}$$

This value corresponds to $-33\,\text{J mol}^{-1}$.

Equation 14B.3b applies to polar molecules in a fixed, parallel, orientation in a solid. In a fluid of freely rotating molecules, the interaction between dipoles averages to zero because like partial charges of two freely rotating molecules are close together as much as the two opposite partial charges, and the repulsion of the former is cancelled by the attraction of the latter. For instance, if the dipoles are arranged as in (5) with the second free to rotate,

$$V = \frac{1}{4\pi\varepsilon_0 r^3}\{\mu_1\mu_2\cos\theta - 3\mu_1\mu_2\cos\theta\} = -\frac{\mu_1\mu_2}{2\pi\varepsilon_0 r^3}\cos\theta$$

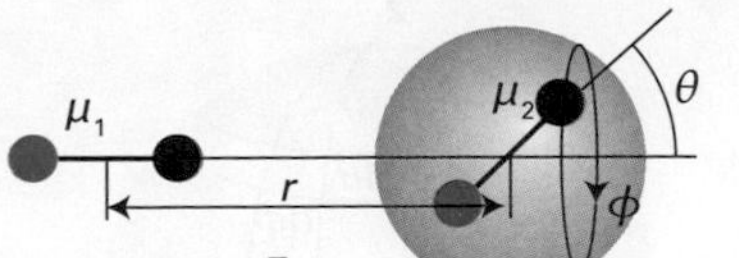

5

The average value of $\cos\theta$ is zero, because

$$\int_0^{2\pi}\int_0^{\pi}\cos\theta\,\overbrace{\sin\theta\,\mathrm{d}\theta\mathrm{d}\phi}^{\text{Surface area element in spherical polar coordinates}} = \overbrace{\int_0^{2\pi}\mathrm{d}\phi}^{2\pi}\,\overbrace{\int_{-1}^{1}x\mathrm{d}x}^{0} = 0$$

$\cos\theta = x$; $\mathrm{d}\cos\theta = -\sin\theta\,\mathrm{d}\theta$

Therefore, the average energy of interaction is also zero. This conclusion is applicable at any relative location of the two dipoles.

The average interaction energy of two *freely* rotating dipoles is zero. However, because their mutual potential energy depends on their relative orientation, the molecules do not in fact rotate completely freely, even in a gas. The lower energy orientations are marginally favoured, so there is a non-zero average interaction between polar molecules. The detailed calculation of the interaction energy between two polar molecules is quite complicated, but the form of the final answer can be constructed quite simply.

How is that done? 14B.3 Deriving the expression for the energy of interaction between rotating polar molecules

You can use the simplified model of the interaction for the arrangement shown in (5), with the second dipole free to rotate, but not sampling all orientations equally.

Step 1 *Write an expression for the average interaction energy*
The average interaction energy of two polar molecules rotating at a fixed separation r is given by

$$\langle V\rangle = -\frac{\mu_1\mu_2}{2\pi\varepsilon_0 r^3}\langle\cos\theta\rangle$$

where $\langle\cos\theta\rangle$ now includes a weighting factor in the averaging that recognizes that not all orientations are equally probable.

Step 2 *Write an expression for the probability of finding a particular orientation*
The probability that dipole 2 lies in a patch of orientation $\sin\theta\,\mathrm{d}\theta\mathrm{d}\phi$ of the surface of a sphere (Fig. 14B.2) is

$$\mathrm{d}p = \frac{\mathrm{e}^{-V(\theta)/kT}\sin\theta\,\mathrm{d}\theta\mathrm{d}\phi}{\int_0^{2\pi}\int_0^{\pi}\mathrm{e}^{-V(\theta)/kT}\sin\theta\,\mathrm{d}\theta\mathrm{d}\phi} \qquad V(\theta) = -\frac{\mu_1\mu_2}{2\pi\varepsilon_0 r^3}\cos\theta$$

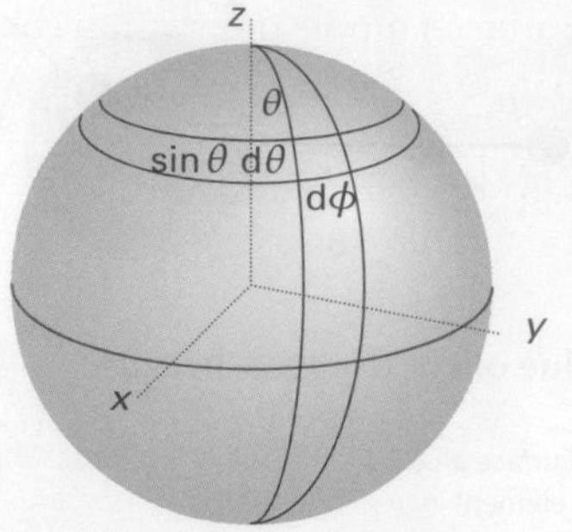

Figure 14B.2 The surface of a sphere showing the area element $\sin\theta\,d\theta\,d\phi$.

where $e^{-V(\theta)/kT}$ is the Boltzmann factor. The (weighted) average value of $\cos\theta$ is

$$\langle\cos\theta\rangle = \int \cos\theta\,dp = \frac{\int_0^{2\pi}\int_0^{\pi} e^{-V(\theta)/kT}\cos\theta\sin\theta\,d\theta\,d\phi}{\int_0^{2\pi}\int_0^{\pi} e^{-V(\theta)/kT}\sin\theta\,d\theta\,d\phi}$$

Step 3 *Evaluate the integrals*

With $a = \mu_1\mu_2/2\pi kT\varepsilon_0 r^3$ the denominator (after integration over ϕ, which gives a factor of 2π, and writing $x = \cos\theta$) gives

$$2\pi\int_0^{\pi} e^{a\cos\theta}\underbrace{\sin\theta\,d\theta}_{-d\cos\theta} = 2\pi\int_{-1}^{1} e^{ax}\,dx = \frac{2\pi}{a}(e^{a}-e^{-a})$$

Likewise, the numerator is

$$2\pi\int_0^{\pi} e^{a\cos\theta}\cos\theta\sin\theta\,d\theta = 2\pi\int_{-1}^{1} xe^{ax}\,dx$$
$$= -\frac{2\pi}{a^2}(e^{a}-e^{-a}) + \frac{2\pi}{a}(e^{a}+e^{-a})$$

It follows that

$$\langle\cos\theta\rangle = -\frac{1}{a} + \frac{e^{a}+e^{-a}}{e^{a}-e^{-a}} = L(a)$$

where $L(a)$ is the Langevin function introduced in Topic 14A. Therefore,

$$\langle V\rangle = -\frac{\mu_1\mu_2}{2\pi\varepsilon_0 r^3}L(a)$$

As in Topic 14A, for $a \ll 1$, $L(a) = a/3$, so

$$\langle V\rangle = -\frac{a\mu_1\mu_2}{6\pi\varepsilon_0 r^3} = -\frac{\mu_1^2\mu_2^2}{12\pi^2\varepsilon_0^2 kTr^6}$$

In a more realistic calculation, where the second dipole is allowed to roam around the first (at the same distance), a further factor of $\frac{1}{2}$ is introduced, and the final outcome is the **Keesom interaction**:

$$\langle V\rangle = -\frac{C}{r^6} \qquad C = \frac{2\mu_1^2\mu_2^2}{3(4\pi\varepsilon_0)^2 kT} \qquad (14B.4)$$

Keesom interaction

The important features of eqn 14B.4 are:

- The negative sign shows that the average interaction is attractive.
- The dependence of the average interaction energy on the inverse sixth power of the separation identifies it as a van der Waals interaction.
- The inverse dependence on the temperature reflects how the greater thermal motion overcomes the mutual orientating effects of the dipoles at higher temperatures.
- The inverse sixth power arises from the inverse third power of the interaction potential energy weighted by the energy in the Boltzmann term, which is also proportional to the inverse third power of the separation.

Physical interpretation

Brief illustration 14B.3

Suppose a water molecule ($\mu_1 = 1.85$ D) can rotate 1.0 nm from an amide group ($\mu_2 = 2.7$ D). The average energy of their interaction at 25 °C (298 K) is

$$V = -\frac{2\times\overbrace{(1.85\times3.336\times10^{-30}\,\mathrm{C\,m})^2}^{\mu_1}\times\overbrace{(2.7\times3.336\times10^{-30}\,\mathrm{C\,m})^2}^{\mu_2}}{3\times\underbrace{(1.710\times10^{-43}\,\mathrm{J^{-1}\,C^4\,m^{-2}\,K^{-1}})}_{(4\pi\varepsilon_0)^2 k}\times\underbrace{(298\,\mathrm{K})}_{T}\times\underbrace{(1.0\times10^{-9}\,\mathrm{m})^6}_{r}}$$

$$= -4.0\times10^{-23}\,\mathrm{J}$$

This interaction energy corresponds (after multiplication by Avogadro's constant) to $-24\,\mathrm{J\,mol^{-1}}$, and it is much smaller than the energies involved in the making and breaking of chemical bonds.

Table 14B.1 summarizes the various expressions for the interaction of charges and dipoles. It is quite easy to extend the formulas given there to obtain expressions for the energy of interaction of higher multipoles (electric multipoles are described in Topic 14A). The feature to remember is that the interaction energy approaches zero more rapidly the higher the order of the multipole. For the interaction of a stationary n-pole with a stationary m-pole, the potential energy varies with distance as

$$V \propto \frac{1}{r^{n+m-1}} \qquad (14B.5)$$

Energy of interaction between stationary multipoles

Table 14B.1 Interaction potential energies

Interaction type	Distance dependence of potential energy	Typical energy/ (kJ mol⁻¹)	Comment
Ion–ion	$1/r$	250	Only between ions
Hydrogen bond		20	Occurs in X–H···Y, where X, Y = N, O, or F
Ion–dipole	$1/r^2$	15	
Dipole–dipole	$1/r^3$	2	Between stationary polar molecules
	$1/r^6$	0.3	Between rotating polar molecules
London (dispersion)	$1/r^6$	2	Between all types of molecules and ions

The reason for the even steeper decrease with distance is the same as before: the array of charges appears to blend together into neutrality more rapidly with distance the higher the number of individual charges that contribute to the multipole. Note that a given molecule may have a charge distribution that corresponds to a combination of several different multipoles, and in such cases the energy of interaction is the sum of terms given by eqn 14B.5.

(c) Dipole–induced dipole interactions

A polar molecule can induce a dipole in a neighbouring polarizable molecule (Fig. 14B.3). The induced dipole interacts with the permanent dipole of the first molecule, and the two are attracted together. The average interaction energy when the separation of the centres of the molecules is r is

$$V=-\frac{C}{r^6} \qquad C=\frac{\mu_1^2\alpha_2'}{4\pi\varepsilon_0} \qquad \text{Potential energy of a polar molecule and a polarizable molecule} \qquad (14B.6)$$

where α_2' is the polarizability volume (Topic 14A) of molecule 2 and μ_1 is the magnitude of the permanent dipole moment of molecule 1. Note that the C in this expression is different from the C in eqn 14B.4 and other expressions below: the use of the same symbol in C/r^6 emphasizes the similarity of form of each expression.

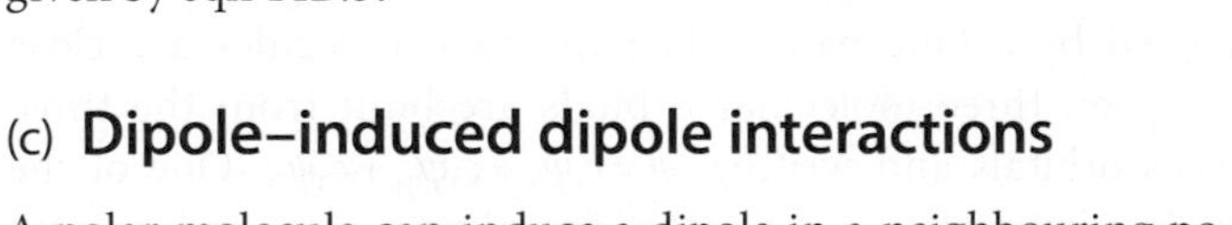

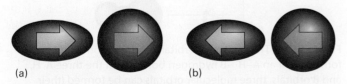

Figure 14B.3 (a) A polar molecule (yellow arrow) can induce a dipole (grey arrow) in a nonpolar molecule, and (b) the orientation of the latter follows that of the former, so the interaction does not average to zero.

The **dipole–induced dipole interaction** energy is independent of the temperature because thermal motion has no effect on the averaging process. Moreover, like the dipole–dipole interaction, the potential energy depends on $1/r^6$: this distance dependence stems from the $1/r^3$ dependence of the distorting electric field of molecule 1 (and hence the magnitude of the dipole induced in molecule 2) and the $1/r^3$ dependence of the potential energy of interaction between the permanent and induced dipoles.

Brief illustration 14B.4

For a molecule with μ = 1.0 D (3.3 × 10⁻³⁰ C m, such as HCl) separated by 0.30 nm from a molecule of polarizability volume α' = 10 × 10⁻³⁰ m³ (such as benzene, Table 14A.1), the average interaction energy is

$$V=-\frac{(3.3\times10^{-30}\,\text{C m})^2\times(10\times10^{-30}\,\text{m}^3)}{4\pi\times(8.854\times10^{-12}\,\text{J}^{-1}\,\text{C}^2\,\text{m}^{-1})\times(3.0\times10^{-10}\,\text{m})^6}$$
$$=-1.4\times10^{-21}\,\text{J}$$

which, upon multiplication by Avogadro's constant, corresponds to −0.83 kJ mol⁻¹.

(d) Induced dipole–induced dipole interactions

Nonpolar molecules (including closed-shell atoms, such as Ar) attract one another even though neither has a permanent dipole moment. The abundant evidence for the existence of interactions between nonpolar molecules is their ability to exist as condensed phases, such as liquid hydrogen or argon and the fact that benzene is a liquid at normal temperatures.

The interaction between nonpolar molecules arises from the transient dipoles which all molecules possess as a result of fluctuations in the instantaneous positions of electrons. To appreciate the origin of the interaction, suppose that the electrons in one molecule flicker into an arrangement that gives the molecule an instantaneous dipole moment μ_1^*. This dipole generates an electric field which polarizes the other molecule, and induces in that molecule an instantaneous dipole moment μ_2. The two dipoles attract each other and the potential energy of the pair is lowered. Although the first molecule will go on to change the size and direction of its instantaneous dipole, the electron distribution of the second molecule will follow; that is, the two dipoles are correlated in direction (Fig. 14B.4). Because of this correlation, the attraction between the two instantaneous dipoles does not average to zero, and gives rise to an **induced dipole–induced dipole interaction**. This interaction is called either the **dispersion**

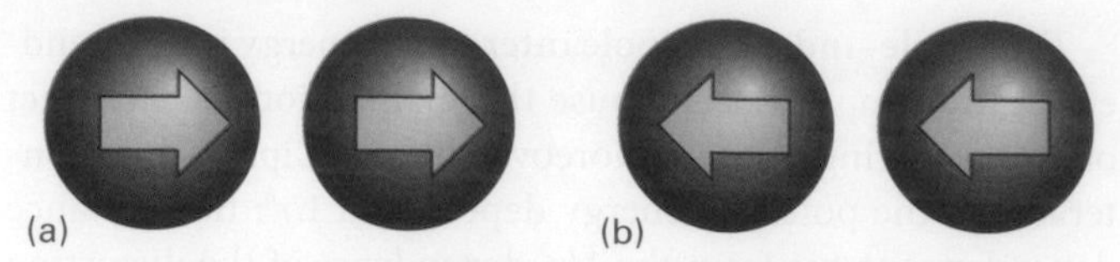

Figure 14B.4 (a) In the dispersion interaction, an instantaneous dipole on one molecule induces a dipole on another molecule, and the two dipoles then interact to lower the energy. (b) The two instantaneous dipoles are correlated, and although they occur in different orientations at different instants, the interaction does not average to zero.

interaction or the **London interaction** (for Fritz London, who first described it).

The strength of the dispersion interaction depends on the polarizability of the first molecule because the instantaneous dipole moment of magnitude μ_1^* depends on the looseness of the control that the nuclear charge exercises over the outer electrons. The strength of the interaction also depends on the polarizability of the second molecule, for that polarizability determines how readily a dipole can be induced by the electric field of another molecule. The actual calculation of the dispersion interaction is quite involved, but a reasonable approximation to the interaction energy is given by the **London formula**:

$$V = -\frac{C}{r^6} \quad C = \tfrac{3}{2}\alpha_1'\alpha_2'\frac{I_1 I_2}{I_1+I_2} \qquad \text{London formula} \qquad (14B.7)$$

where I_1 and I_2 are the ionization energies of the two molecules. This interaction energy is also proportional to the inverse sixth power of the separation of the molecules, which identifies it as a third contribution to the van der Waals interaction. The dispersion interaction generally dominates all the interactions between molecules other than hydrogen bonds.

Brief illustration 14B.5

For two CH_4 molecules separated by 0.30 nm, use eqn 14B.7 with $\alpha' = 2.6 \times 10^{-30}\,m^3$ and $I \approx 700\,kJ\,mol^{-1}$ and obtain

$$V = -\frac{\tfrac{3}{2}\times(2.6\times10^{-30}\,m^3)^2}{(0.30\times10^{-9}\,m)^6} \times \frac{(7.00\times10^5\,J\,mol^{-1})^2}{2\times(7.00\times10^5\,J\,mol^{-1})}$$
$$= -4.9\,kJ\,mol^{-1}$$

A very approximate check on this figure is the enthalpy of vaporization of methane, which is $8.2\,kJ\,mol^{-1}$. However, this comparison is questionable, partly because the total energy of interaction between molecules in a liquid is not due only to pairwise interactions and partly because the long-distance assumption breaks down.

14B.2 Hydrogen bonding

The interactions described so far are universal in the sense that they are possessed by all molecules independent of their specific identity. However, there is a type of interaction possessed by molecules that have a particular constitution. A **hydrogen bond** is an attractive interaction between two species that arises from a link of the form A–H···B, where A and B are highly electronegative elements and B possesses a lone pair of electrons. Hydrogen bonding is conventionally regarded as being limited to N, O, and F but if B is an anionic species (such as Cl^-) it may also participate in hydrogen bonding. There is no strict cut-off for an ability to participate in hydrogen bonding, but N, O, and F participate most effectively.

The formation of a hydrogen bond can be regarded either as the approach between a partial positive charge of H and a partial negative charge of B or as a particular example of delocalized molecular orbital formation in which A, H, and B each supply one atomic orbital from which three molecular orbitals are constructed (Fig. 14B.5). Experimental evidence and theoretical arguments have been presented in favour of both views and the matter has not yet been resolved.

In the molecular orbital model, the A–H bond is regarded as formed from the overlap of an orbital on A, ψ_A, and a hydrogen 1s orbital, ψ_H, and the orbital on B, ψ_B, which is occupied by a lone pair. When the two molecules are close together, three molecular orbitals are built from the three basis orbitals and writing: $\psi = c_1\psi_A + c_2\psi_H + c_3\psi_B$. One of the molecular orbitals is bonding, one almost nonbonding, and the third antibonding (Topic 9E). These three orbitals need to accommodate four electrons (two from the original A–H bond and two from the lone pair of B), so two enter the bonding orbital and two enter the nonbonding orbital. Because the

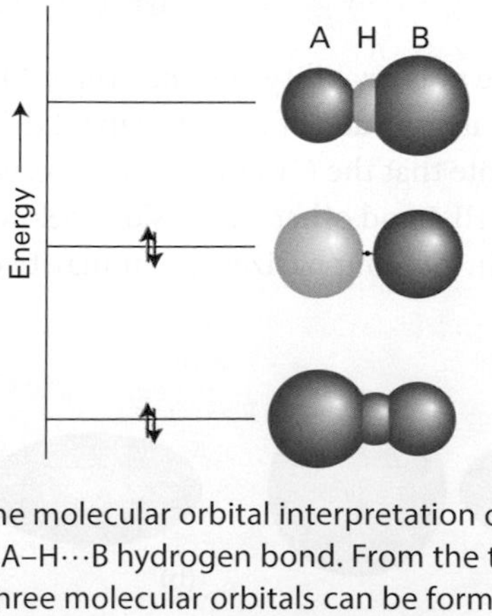

Figure 14B.5 The molecular orbital interpretation of the formation of an A–H···B hydrogen bond. From the three A, H, and B orbitals, three molecular orbitals can be formed (their relative contributions are represented by the sizes of the spheres). Only the two lower energy orbitals are occupied, and there may therefore be a net lowering of energy compared with the separate AH and B species.

antibonding orbital remains empty, the net effect—depending on the precise energy of the almost nonbonding orbital—may be a lowering of energy.

In practice, the strength of the bond is found to be about 20 kJ mol^{-1} (there are two hydrogen bonds per molecule in liquid water, and its standard enthalpy of vaporization, from Table 2C.1, is 44 kJ mol^{-1}). Because the bonding depends on orbital overlap, it is a contact-like interaction that is turned on when AH touches B and is zero as soon as the contact is broken. If hydrogen bonding is present, it dominates the other intermolecular interactions. The properties of liquid and solid water, for example, are dominated by the hydrogen bonding between H_2O molecules. The structural evidence for hydrogen bonding comes from noting that the internuclear distance between formally non-bonded atoms is less than expected on the basis of their **van der Waals radii**, the radii based on the closest approach of non-bonded atoms, which suggests that a dominating attractive interaction is present. For example, the O–O distance in O–H···O is expected to be 280 pm on the basis of van der Waals radii, but is found to be 270 pm in typical compounds. Moreover, the H···O distance is expected to be 260 pm but is found to be only 170 pm.

Hydrogen bonds may be either symmetric or non-symmetric. In a symmetric hydrogen bond, the H atom lies midway between the two other atoms. This arrangement is rare, but occurs in F–H···F$^-$, where both bond lengths are 120 pm. More common is the non-symmetric arrangement, where the A–H bond is shorter than the H···B bond. Simple electrostatic arguments, treating A–H···B as an array of point charges (partial negative charges on A and B, partial positive on H), suggest that the lowest energy is achieved when the bond is linear, because then the two partial negative charges are farthest apart. The experimental evidence from structural studies supports a linear or near-linear arrangement.

Brief illustration 14B.6

A common hydrogen bond is that formed between O–H groups and O atoms, as in liquid water and ice. In Problem P14B.8, you are invited to use the electrostatic model to calculate the dependence of the potential energy of interaction on the OOH angle, denoted Θ in (6), and the results are plotted in Fig. 14B.6. The strength of bonding is greatest at $\Theta = 0$ when the OHO atoms lie in a straight line; the molar potential energy is then -19 kJ mol^{-1}. Note that the interaction energy is negative (and the interaction is attractive) only between $-12°$ and $+12°$, so the atoms adopt a nearly linear arrangement.

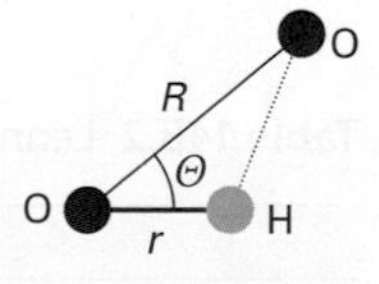

6

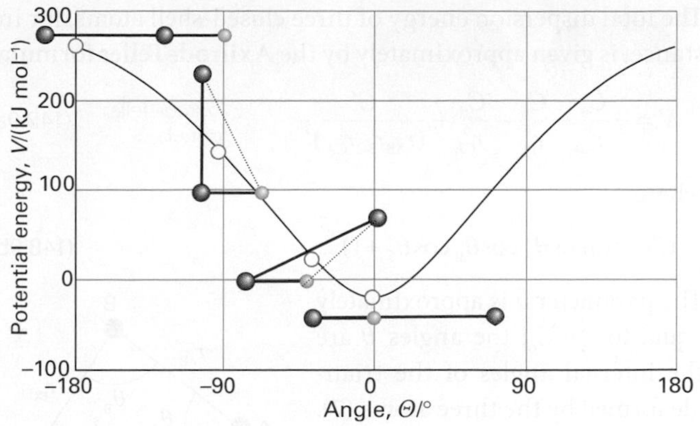

Figure 14B.6 The variation of the energy of interaction (according to the electrostatic model) of a hydrogen bond as the angle between the O–H and :O groups is changed.

14B.3 The total interaction

Consider molecules that are unable to participate in the formation of a hydrogen bond. The total favourable (energy lowering) interaction energy between rotating molecules is then the sum of the dipole–dipole, dipole–induced dipole, and dispersion interactions. Only the dispersion interaction contributes if both molecules are nonpolar. In a fluid phase, all three contributions to the potential energy vary as the inverse sixth power of the separation of the molecules, so for all of them and their sum

$$V = -\frac{C_6}{r^6} \qquad (14B.8)$$

where C_6 is a coefficient that depends on the identity of the molecules.

Although attractive interactions between molecules are often expressed as in eqn 14B.8, remember that this equation has only limited validity. First, only dipolar interactions of various kinds are taken into account, for they have the longest range and are dominant if the average separation of the molecules is large. However, in a complete treatment, quadrupolar and higher-order multipole interactions should also be considered, particularly if the molecules do not have permanent dipole moments. Secondly, the expressions have been derived by assuming that the molecules can rotate reasonably freely. That is not the case in most solids, and in rigid media the dipole–dipole interaction is proportional to $1/r^3$ (as in eqn 14B.3b) because the Boltzmann averaging procedure is irrelevant when the molecules are trapped into a fixed orientation.

A different kind of limitation is that eqn 14B.8 relates to the interactions of pairs of molecules. There is no reason to suppose that the energy of interaction of three (or more) molecules is the sum of the pairwise interaction energies alone.

The total dispersion energy of three closed-shell atoms, for instance, is given approximately by the **Axilrod–Teller formula**:

$$V = -\frac{C_6}{r_{AB}^6} - \frac{C_6}{r_{BC}^6} - \frac{C_6}{r_{CA}^6} + \frac{C'}{(r_{AB}r_{BC}r_{CA})^3} \quad \text{Axilrod–Teller formula} \quad (14B.9a)$$

where

$$C' = a(3\cos\theta_A \cos\theta_B \cos\theta_C + 1) \quad (14B.9b)$$

The parameter a is approximately equal to $\frac{3}{4}\alpha' C_6$; the angles θ are the internal angles of the triangle formed by the three atoms (7). The term in C' (which represents the non-additivity of the pairwise interactions) is negative for a linear arrangement of atoms (so that arrangement is stabilized) and positive for an equilateral triangular cluster (so that arrangement is destabilized). The three-body term contributes about 10 per cent of the total interaction energy in liquid argon.

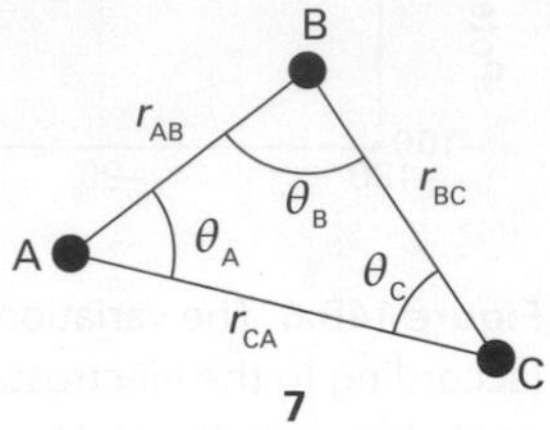

7

When molecules are squeezed together, the nuclear and electronic repulsions begin to dominate the attractive forces. The repulsions increase steeply with decreasing separation in a way that can be deduced only by very extensive, complicated molecular structure calculations of the kind described in Topic 9E (Fig. 14B.7).

In many cases, however, progress can be made by using a greatly simplified representation of the potential energy, where the details are ignored and the general features expressed by a few adjustable parameters. One such approximation is the **hard-sphere potential energy**, in which it is assumed that the potential energy rises abruptly to infinity as soon as the particles come within a separation d:

$$V = \begin{cases} \infty \text{ for } r \le d \\ 0 \text{ for } r > d \end{cases} \quad \text{Hard-sphere potential energy} \quad (14B.10)$$

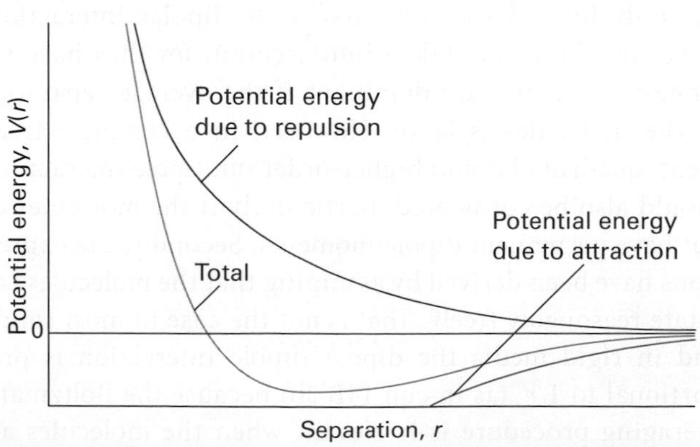

Figure 14B.7 The general form of an intermolecular potential energy curve (the graph of the potential energy of two closed shell species as the distance between them is changed). The attractive (negative) contribution has a long range, but the repulsive (positive) interaction increases more sharply once the molecules come into contact.

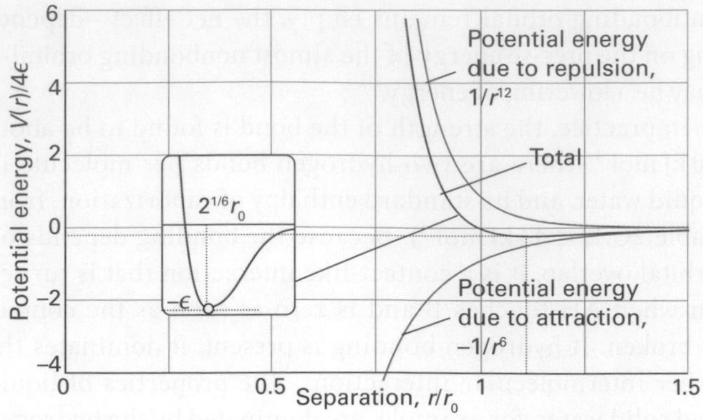

Fig 14B.8 The Lennard-Jones potential energy.

This very simple expression for the potential energy is surprisingly useful for assessing a number of properties. Another widely used approximation is the **Mie potential energy**:

$$V = \frac{C_n}{r^n} - \frac{C_m}{r^m} \quad \text{Mie potential energy} \quad (14B.11)$$

with $n > m$. The first term represents repulsions and the second term attractions. The **Lennard-Jones potential energy** is a special case of the Mie potential energy with $n = 12$ and $m = 6$ (Fig. 14B.8); it is often written in the form

$$V = 4\varepsilon\left\{\left(\frac{r_0}{r}\right)^{12} - \left(\frac{r_0}{r}\right)^{6}\right\} \quad \text{Lennard-Jones potential energy} \quad (14B.12)$$

The two parameters are ε, the depth of the well, and r_0, the separation other than infinity at which $V = 0$ (Table 14B.2).

Although the Lennard-Jones potential energy has been used in many calculations, there is plenty of evidence to show that $1/r^{12}$ is a very poor representation of the repulsive potential energy, and that an exponential form, e^{-r/r_0}, is greatly superior. An exponential function is more faithful to the exponential decay of atomic wavefunctions at large distances, and hence to the overlap that is responsible for repulsion. The potential energy with an exponential repulsive term and a $1/r^6$ attractive term is known as an **exp-6 potential energy**. These expressions for the

Table 14B.2 Lennard-Jones-(12,6) potential energy parameters*

	(ε/k)/K	r_0/pm
Ar	111.84	362.3
F_2	104.29	357.1
C_6H_6	377.46	617.4
Cl_2	296.27	448.5
N_2	91.85	391.9
O_2	113.27	365.4
Xe	213.96	426

* More values are given in the *Resource section.*

potential energy can be used to calculate the virial coefficients of gases, as explained in Topics 1C and 13D, and through them various properties of real gases. They are also used to model the structures of condensed fluids.

With the advent of **atomic force microscopy** (AFM), in which the force between a molecular-sized probe and a surface is monitored (Topic 19A), it has become possible to measure directly the forces acting between molecules. The force, F, is the negative slope of the potential energy, so for the Lennard-Jones potential energy between individual molecules

$$F = -\frac{dV}{dr} = \frac{24\varepsilon}{r_0}\left\{2\left(\frac{r_0}{r}\right)^{13} - \left(\frac{r_0}{r}\right)^{7}\right\} \qquad \text{(14B.13)}$$

Example 14B.1 Calculating an intermolecular force from the Lennard-Jones potential energy

Use the expression for the Lennard-Jones potential energy to estimate the greatest net attractive force between two N_2 molecules.

Collect your thoughts The force is greatest when $dF/dr = 0$. Therefore you need to differentiate eqn 14B.13 with respect to r, set the resulting expression to zero, and then solve for r. Finally, use the value of r in eqn 14B.13 to calculate the corresponding value of F.

The solution Because $dx^n/dx = nx^{n-1}$,

$$\frac{dF}{dr} = \frac{24\varepsilon}{r_0}\left\{2\left(-\frac{13r_0^{13}}{r^{14}}\right) - \left(-\frac{7r_0^{7}}{r^{8}}\right)\right\} = 24\varepsilon r_0^6\left\{\frac{7}{r^8} - \frac{26r_0^6}{r^{14}}\right\}$$

It follows that $dF/dr = 0$ when

$$\frac{7}{r^8} - \frac{26r_0^6}{r^{14}} = 0 \quad \text{or} \quad 7r^6 - 26r_0^6 = 0$$

That is, when

$$r = \left(\frac{26}{7}\right)^{1/6} r_0 = 1.244r_0$$

At this separation the force is

$$F = \frac{24\varepsilon}{r_0}\left\{2\left(\frac{r_0}{1.244r_0}\right)^{13} - \left(\frac{r_0}{1.244r_0}\right)^{7}\right\} = -\frac{2.396\varepsilon}{r_0}$$

From Table 14B.2, $\varepsilon = 1.268 \times 10^{-21}$ J and $r_0 = 3.919 \times 10^{-10}$ m. It follows that

$$F = -\frac{2.396 \times (1.268 \times 10^{-21}\,\text{J})}{3.919 \times 10^{-10}\,\text{m}} = -7.752 \times 10^{-12}\,\text{N} \qquad [1\,\text{N} = 1\,\text{J m}^{-1}]$$

That is, the magnitude of the force is about 8 pN.

Self-test 14B.1 At what separation r does a Lennard-Jones potential energy have its minimum value?

Answer: $r = 2^{1/6}r_0$

Checklist of concepts

☐ 1. A **van der Waals interaction** is an attractive interaction between closed-shell molecules; the corresponding potential energy is inversely proportional to the sixth power of their separation.

☐ 2. The following molecular interactions are important: **charge–dipole, dipole–dipole, dipole–induced dipole, dispersion (London)**, and **hydrogen bonding**.

☐ 3 The **van der Waals radius** of an atom is based on the closest approach of non-bonded atoms.

☐ 4. A **hydrogen bond** is an interaction of the form A–H···B, where A and B are typically N, O, or F.

☐ 5. The **Lennard-Jones potential energy** is a model of the total intermolecular potential energy, including repulsion.

Checklist of equations

Property	Equation	Comment	Equation number
Energy of interaction between a point dipole and a point charge	$V = -\mu_1 Q_2/4\pi\varepsilon_0 r^2$	Linear arrangement	14B.1
Energy of interaction between two fixed dipoles	$V = \mu_1\mu_2 f(\Theta)/4\pi\varepsilon_0 r^3$, $f(\Theta) = 1 - 3\cos^2\Theta$	Parallel dipoles	14B.3b
Energy of interaction between two rotating dipoles	$V = -2\mu_1^2\mu_2^2/3(4\pi\varepsilon_0)^2 kTr^6$		14B.4
Energy of interaction between a polar molecule and a polarizable molecule	$V = -\mu_1^2\alpha_2'/4\pi\varepsilon_0 r^6$		14B.6
London formula	$V = -\frac{3}{2}\alpha_1'\alpha_2'(I_1 I_2/(I_1 + I_2))/r^6$		14B.7
Lennard-Jones potential energy	$V = 4\varepsilon\{(r_0/r)^{12} - (r_0/r)^6\}$		14B.12

TOPIC 14C Liquids

> ➤ **Why do you need to know this material?**
>
> Many substances are liquids under normal conditions and many chemical reactions take place in liquids, so it is important to be able to describe and understand the structure of the liquid phase and its interface with its vapour.
>
> ➤ **What is the key idea?**
>
> The properties of liquids reflect the short-range order of their molecules in the bulk and the behaviour of their molecules at the mobile surface.
>
> ➤ **What do you need to know already?**
>
> You need to be aware of the ways in which molecules interact with each other (Topic 14B), and to be familiar with the Helmholtz and Gibbs energies (Topic 3D) and their significance. The Topic makes light use of the Boltzmann distribution (the *Prologue* and Topic 13A).

Molecules attract each other when they are less than a few diameters apart, but as soon as they come into contact they repel each other. The attraction is responsible for the formation of condensed phases, including liquids, and the repulsion is responsible for the fact that liquids (and solids) have a definite bulk. In a liquid the kinetic energies of the molecules are comparable to their potential energies and, as a result, although the molecules of a liquid are not free to escape completely from the bulk at low temperatures, the structure is very mobile. The cohesive forces responsible for the formation of liquids also result in the interface between the liquid and another phase having an effect on the thermodynamic and physical properties of the liquid.

14C.1 Molecular interactions in liquids

The starting point for the discussion of solids is the well-ordered structure of a perfect crystal (Topic 15A). The starting point for the discussion of gases is the completely disordered distribution of the molecules of a perfect gas (Topic 1A). Liquids lie between these two extremes. The structural and thermodynamic properties of liquids depend on the nature of intermolecular interactions and, just as for a real gas, a model based on these interactions can be used to develop an equation of state.

(a) The radial distribution function

The average relative locations of the particles of a liquid are expressed in terms of the **radial distribution function** (or *pair distribution function*), $g(r)$. This function is defined so that $4\pi\mathcal{N}g(r)r^2\mathrm{d}r$ is the number of molecules in a shell of thickness dr at radius r from a given molecule; $\mathcal{N}=N/V$ is the overall number density. For a uniform material, $g(r) = 1$. When defined in this way the radial distribution function is also the ratio of the number of molecules in a shell to the number expected in the same shell for a uniform material. If, at a particular distance, $g(r) > 1$ there are more molecules than in a uniform material, whereas if $g(r) < 1$, there are fewer.

In a perfect crystal, $g(r)$ is a periodic array of sharp spikes, representing the certainty (in the absence of defects and thermal motion) that molecules (or ions) lie at definite locations. This regularity continues out to the edges of the crystal, so crystals are said to have **long-range order**. When the crystal melts, the long-range order is lost and the probability of finding a second molecule at long distances from the first is independent of the distance. However, close to the first molecule the nearest neighbours might still adopt approximately their original relative positions and, even if thermal motion drives them away, incoming molecules adopt the vacated positions. It is therefore still possible to detect a shell of nearest neighbours at a distance r_1, and perhaps beyond them a shell of next-nearest neighbours at r_2. The existence of this **short-range order** means that the radial distribution function in a liquid can be expected to oscillate at short distances, with a peak at r_1, a smaller peak at r_2, and perhaps some more structure beyond that.

The experimentally determined radial distribution function of the oxygen atoms in liquid water is shown in Fig. 14C.1. Close analysis of a more elaborate form of the distribution function shows that any given H_2O molecule is surrounded by other molecules at the corners of a tetrahedron. The form of $g(r)$ at 100 °C shows that the intermolecular interactions (in this case, principally hydrogen bonds) are strong enough to affect the local structure right up to the boiling point. Raman spectra indicate that in liquid water most molecules participate in either three or four hydrogen bonds. Infrared spectra show that about 90 per cent of hydrogen bonds are intact at the

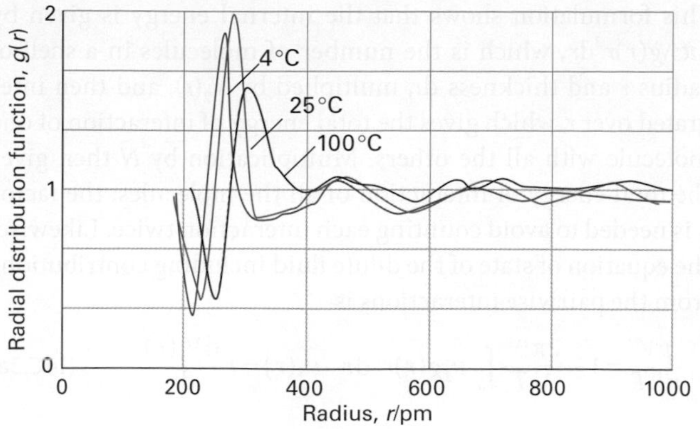

Figure 14C.1 The radial distribution function of the oxygen atoms in liquid water at three temperatures. Note the expansion as the temperature is raised. (Based on A.H. Narten, M.D. Danford, and H.A. Levy, *Discuss. Faraday Soc.* **43**, 97 (1967).)

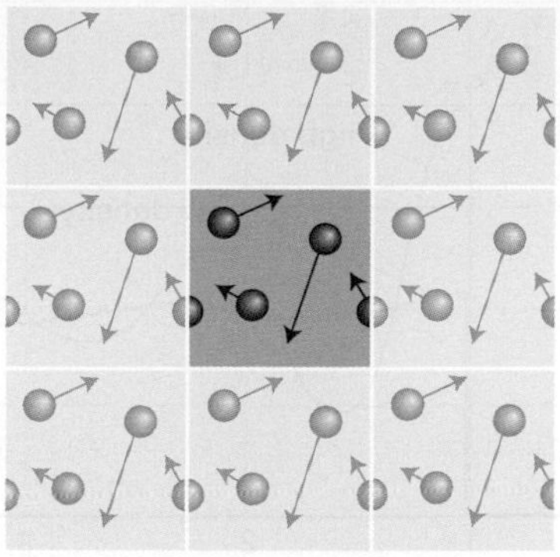

Figure 14C.2 In a two-dimensional simulation of a liquid that uses periodic boundary conditions, when one particle leaves the cell (on the left here) its mirror image enters through the opposite face (on the right).

melting point of ice, falling to about 20 per cent at the boiling point.

The formal expression for the radial distribution function for molecules 1 and 2 in a fluid consisting of N particles is

$$g(r_{12})=\frac{1}{(N-2)!\mathcal{N}^2 Z}\int e^{-\beta V_N}\,d\tau_3 d\tau_4 \ldots d\tau_N$$

Radial distribution function (14C.1a)

where $d\tau_i$ is the volume element for molecule i, $\beta = 1/kT$, V_N is the N-particle potential energy, and Z is the 'configuration integral' (this quantity is introduced in Topic 13D):

$$Z=\frac{1}{N!}\int e^{-\beta V_N}\,d\tau_1 d\tau_2 \ldots d\tau_N$$

Configuration integral [definition] (14C.1b)

Equation 14C.1a is nothing more than the Boltzmann distribution for the relative locations of two molecules in a field provided by all the molecules in the system. Thus, if there are no interactions between molecules (so $V_N = 0$), $Z = V^N/N!$ and

$$g(r_{12})=\frac{N!}{(N-2)!\mathcal{N}^2 V^N}\overbrace{\int d\tau_3 d\tau_4 \ldots d\tau_N}^{V^{N-2}}$$

$$=\frac{N(N-1)}{\mathcal{N}^2 V^2}=\frac{N(N-1)}{N^2}\overset{N\gg 1}{=}1$$

In the absence of interactions the fluid is expected to be uniform, which is consistent with this value of $g(r_{12})$.

(b) The calculation of $g(r)$

The integrals in eqn 14C.1 are very difficult to evaluate so various numerical procedures are used to calculate the radial distribution function. Such calculations involve specifying the form of the intermolecular potential energy, for example by specifying it as a pairwise Lennard-Jones interaction (Topic 14B).

Numerical methods typically approach the calculation of the distribution function by considering a box containing about 10^3 particles (the number increases as computers grow more powerful), and then mimicking the rest of the liquid by surrounding the box with replications of the original box (Fig. 14C.2). Whenever a particle leaves the box through one of its faces, its image arrives through the opposite face. When calculating the interactions of a molecule in a box, it interacts with all the molecules in the box and all the periodic replications of those molecules (and of itself) in the other boxes.

In the **Monte Carlo method**, the particles in the box are first distributed at random and then moved through small random distances. The change in total potential energy of the N particles in the box, ΔV_N, is calculated, and if the new arrangement has lower potential energy than the original one, it is accepted. If the potential energy increases, implying that ΔV_N is positive, the new arrangement is accepted only if the Boltzmann-type factor $e^{-\Delta V_N/kT}$ is greater than a random number (chosen to lie somewhere in the range 0 to 1). If this criterion is not met, a new arrangement is generated from the original one and tested in the same way.

The result of applying this selection process is that moving to an arrangement in which the energy is lower is always allowed but moving to one of higher energy, although possible, become increasing unlikely the higher its energy. As a result, the simulation explores a wide range of arrangements, but tends to include fewer with high energies. For each accepted arrangement the number of pairs of molecules with a separation r is counted, and the result is then averaged over the whole collection of accepted arrangements; by repeating this for different values of r, the form of $g(r)$ is built up.

In the **molecular dynamics** approach, the history of an initial arrangement is followed by calculating the trajectories of all the molecules under the influence of the intermolecular

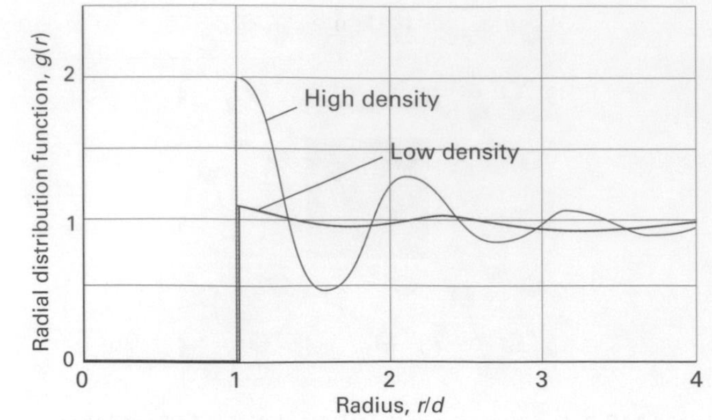

Figure 14C.3 The radial distribution function for a simulation of a liquid using impenetrable hard spheres (such as ball bearings).

potentials and the forces they exert. The calculation gives a series of snapshots of the liquid, and $g(r)$ can be calculated as before. The temperature of the system is inferred by computing the mean translational kinetic energy of the molecules and using the equipartition result (*The chemist's toolkit* 7, in Topic 2A) that $\langle \frac{1}{2}mv_q^2\rangle = \frac{1}{2}kT$, for each coordinate q.

Such numerical calculations for a fluid of hard spheres without attractive interactions (a collection of ball-bearings in a container) give a radial distribution function that oscillates for small separations of the molecules (Fig. 14C.3). It appears that one of the factors influencing, and sometimes dominating, the structure of a liquid is the geometrical problem of stacking together reasonably hard spheres. Indeed, the radial distribution function of a liquid composed of hard spheres shows more pronounced oscillations at a given temperature than that of any other model liquid. The attractive part of the potential modifies this basic structure, and one of the reasons behind the difficulty of describing actual liquids theoretically is the similar importance of both the attractive and repulsive (hard core) components of the potential energy.

(c) The thermodynamic properties of liquids

Once $g(r)$ is known, it can be used to calculate the thermodynamic properties of liquids, but for dense fluids the calculations are very complicated. The calculations are simpler for fluids that are so dilute that they are little more than real gases. In such a case, if it is assumed that the interaction of each pair of molecules is given by an isotropic pairwise potential energy function, $V_2(r)$, the result is a contribution to the internal energy given by

$$U_{\text{interaction}}(T) = \frac{2\pi N^2}{V}\int_0^\infty V_2(r)g(r)r^2\,\mathrm{d}r$$

Contribution of pairwise interactions to the internal energy (14C.2a)

which it is more revealing to write as

$$U_{\text{interaction}}(T) = \tfrac{1}{2}N\int_0^\infty V_2(r)\{4\pi\mathcal{N}g(r)\}r^2\,\mathrm{d}r \qquad (14\text{C}.2\text{b})$$

This formulation shows that the internal energy is given by $4\pi\mathcal{N}g(r)r^2\mathrm{d}r$, which is the number of molecules in a shell of radius r and thickness dr, multiplied by $V_2(r)$, and then integrated over r, which gives the total energy of interaction of one molecule with all the others. Multiplication by N then gives the total energy of interaction of all the molecules; the factor $\frac{1}{2}$ is needed to avoid counting each interaction twice. Likewise, the equation of state of the dilute fluid including contributions from the pairwise interactions is

$$\frac{pV}{nRT} = 1 - \frac{2\pi\mathcal{N}}{3kT}\int_0^\infty v_2 g(r)r^2\,\mathrm{d}r \quad v_2(r) = r\frac{\mathrm{d}V_2(r)}{\mathrm{d}r} \qquad (14\text{C}.3\text{a})$$

The quantity $v_2(r)$ is called the **virial** (hence the term 'virial equation of state'). To understand the physical significance of this expression, it can be rewritten as

$$p = \frac{nRT}{V} - \frac{2\pi}{3}\mathcal{N}^2\int_0^\infty v_2(r)g(r)r^2\,\mathrm{d}r$$

Pressure in terms of $g(r)$ (14C.3b)

and interpreted as follows:

Physical interpretation

- The first term on the right is the **kinetic pressure**, the contribution to the pressure from the impact of the molecules in free flight, as in a perfect gas.
- The second term, as explained below, is essentially the internal pressure, $\pi_T = (\partial U/\partial V)_T$ (Topic 2D), representing the contribution of the intermolecular forces to the pressure.

To see the connection to the internal pressure, the term $-\mathrm{d}V_2/\mathrm{d}r$ (in v_2) should be recognized as the force required to move two molecules apart, and therefore $-r(\mathrm{d}V_2/\mathrm{d}r)$ is the work required to separate the molecules through a distance r. The second term is therefore the average of this work over the range of pairwise separations in the fluid, with the contributions to the average weighted by the probability of finding two molecules at separations between r and $r + \mathrm{d}r$, which is $4\pi g(r)r^2\mathrm{d}r$. That is, the integral, when multiplied by the square of the number density, is the change in internal energy of the system as it expands, $(\partial U/\partial V)_T$, and is therefore equal to the internal pressure. *A deeper look* 9 on the website of this text explains how this interpretation can be used to infer the virial equation of state of a real gas and interpret the van der Waals parameters.

The pressure given by eqn 14C.3b has nothing to do with the **hydrostatic pressure**, the pressure experienced at the foot of a column of incompressible liquid. The mass of a column of liquid of mass density ρ, height h, and cross-sectional area A is ρhA. In a gravitational field the downward force is ρhAg_{acc}, where g_{acc} is the acceleration of free fall (it is normally denoted simply g, but in this Topic it is necessary to distinguish it from the radial distribution function). The hydrostatic pressure, the force divided by the area A on which it is exerted, is therefore

$$p = \rho g_{\text{acc}}h$$

Hydrostatic pressure [incompressible fluid] (14C.4)

The molecular origin of this pressure is as follows. Bear in mind that the fluid is incompressible, so if a molecule moves, space must be made available for it. If a molecule moves up through a molecular diameter from a certain point at a location a distance h down in the liquid, the entire column of height h above it must move up by a molecular diameter. The force required is $\rho g_{acc} ha$, where a is the cross-sectional area of that molecular column: this force corresponds to a pressure $\rho g_{acc} h$. If a molecule moves to one side through a molecular diameter, the incompressible column above the new position must move up to make room for it. Once again, the force required corresponds to a pressure $\rho g_{acc} h$. Even if a molecule moves down, a molecule must move to one side to make room for it, and therefore a whole column must move up to allow that movement, and the force once again corresponds to a pressure $\rho g_{acc} h$. This interpretation shows why the hydrostatic pressure is isotropic even though gravity operates downwards. It also shows why the interior of a solid column does not have an analogous hydrostatic pressure: the molecules cannot move past each other, so there are no consequent forces involved.

14C.2 The liquid–vapour interface

The distinctive feature of the interface between the liquid and its vapour is that it is mobile and molecules there experience attractive forces that no longer pull equally in all directions.

(a) Surface tension

Liquids tend to adopt shapes that minimize their surface area, because this maximizes the number of molecules that are in the bulk and hence are surrounded by and interact with neighbours. Droplets of liquids therefore tend to be spherical, because a sphere is the shape with the smallest surface-to-volume ratio. However, the presence of other forces distort this ideal shape: the combined effect of gravity and adhesion to a surface results in small droplets on a surface becoming flattened, and gravity results in larger volumes of liquid adopting the shape of the lower part of its container.

Surface effects can be analysed by using the properties of the Helmholtz and Gibbs energies, A and G (Topic 3D). As shown there, under appropriate conditions, dA and dG are equal to the work done on the system, including the work done when the surface area changes. The work needed to change the surface area, σ, of a sample by an infinitesimal amount $d\sigma$ is proportional to $d\sigma$, and is written

$$dw = \gamma d\sigma \quad \text{Surface tension [definition]} \quad (14C.5)$$

The constant of proportionality, γ, is called the **surface tension**; its dimensions are energy/area and so in SI its units are joules per metre squared ($J\,m^{-2}$). However, for reasons that will become clear, values of γ are usually reported in newtons per metre (because $1\,J = 1\,N\,m$, it follows that $1\,J\,m^{-2} = 1\,N\,m^{-1}$); Table 14C.1 gives some typical values. The maximum work of surface formation at constant volume and temperature can be identified with the change in the Helmholtz energy:

$$dA = \gamma d\sigma \quad (14C.6)$$

The Helmholtz energy decreases ($dA < 0$) if the surface area decreases ($d\sigma < 0$). A process in which A decreases is spontaneous, so it follows that surfaces have a natural tendency to contract, which is the thermodynamic explanation of the observations made at the start of this section.

Table 14C.1 Surface tensions of liquids at 293 K*

	$\gamma/(mN\,m^{-1})$
Benzene	28.88
Mercury	472
Methanol	22.6
Water	72.75

* More values are given in the *Resource section*. Note that $1\,mN\,m^{-1} = 1\,mJ\,m^{-2}$.

Example 14C.1 Using the surface tension

Consider the arrangement shown in Fig. 14C.4 in which a wire frame of width l is raised out of a liquid to a height h, thereby generating a rectangular film of liquid within the frame. Calculate the work needed to draw a frame of width 5.0 cm out of water at 20 °C (when $\gamma = 72.75\,mJ\,m^{-2}$) through 2.0 cm; disregard gravitational potential energy.

Collect your thoughts If you assume that the surface tension does not vary with the area, eqn 14C.5 becomes $w = \gamma\Delta\sigma$ for an increase in surface area $\Delta\sigma$. The increase in surface area is from zero to the area of the rectangle, but you need to recognize that *two* surfaces are created, one on each side of the frame. Once you have the appropriate expression, insert the data.

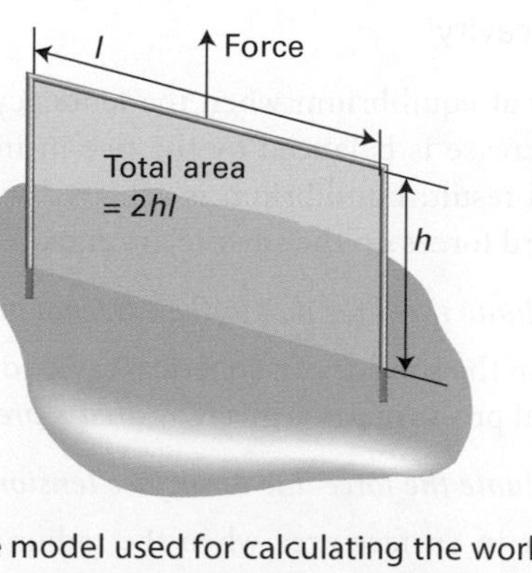

Figure 14C.4 The model used for calculating the work of forming a liquid film when a wire frame of width l is raised from a liquid to a height h.

The solution The area of the rectangle is lh, and hence the increase in surface area of the film is $2lh$; the work done is therefore $2\gamma lh$. A frame of width 5.0 cm pulled out of water at 20 °C through 2.0 cm requires

$$w = 2 \times (72.75\,\text{mJ m}^{-2}) \times (5 \times 10^{-2}\,\text{m}) \times (2 \times 10^{-2}\,\text{m}) = 0.15\,\text{mJ}$$

Comment. The expression $2\gamma lh$ can be thought of as $2\gamma l \times h$, which is force × distance. The term $2\gamma l$ can be identified as the opposing force on the top of the frame, which has length l. This interpretation is why γ is called a tension and why its units are often chosen to be newtons per metre (so γl is a force in newtons).

Self-test 14C.1 Derive an expression for the work of creating a spherical cavity of radius r in a liquid of surface tension γ; evaluate this work for a cavity of radius 1.0 cm in water at 20 °C.

Answer: $4\pi r^2\gamma$, 0.091 mJ

(b) Curved surfaces

The minimization of the surface area of a liquid commonly results in the formation of a curved surface. A **bubble** is a region in which vapour (and possibly air too) is trapped by a thin film; a **cavity** is a vapour-filled hole in a liquid. What are widely called 'bubbles' in liquids are therefore strictly cavities. True bubbles have two surfaces (one on each side of the film); cavities have only one. The treatments of both are similar, but a factor of 2 is required for bubbles to take into account the doubled surface area. A **droplet** is a small volume of liquid surrounded by its vapour (and possibly also air).

The tendency for a cavity to minimize its surface area results in the pressure inside the cavity (the concave side of the surface) being greater than that outside (the convex side of the surface). The challenge is to find the relation between these two pressures.

How is that done? 14C.1 Relating the pressures inside and outside a cavity

A cavity is at equilibrium when the tendency for its surface area to decrease is balanced by the rise in internal pressure that would result. Equilibrium is achieved when the inward and outward forces on the surface are equal.

Step 1 *Evaluate the force due to the external pressure*

The force on the surface of a spherical cavity of radius r due to the external pressure p_{out} is given by *area × pressure* $= 4\pi r^2 p_{\text{out}}$.

Step 2 *Evaluate the force due to surface tension*

The change in surface area when the radius changes from r to $r + \text{d}r$ is

$$\text{d}\sigma = 4\pi(r+\text{d}r)^2 - 4\pi r^2 = 8\pi r\text{d}r$$

(The second-order infinitesimal, $(\text{d}r)^2$, has been ignored.) The work done when the surface is stretched by this amount is therefore

$$\text{d}w = \gamma\text{d}\sigma = 8\pi\gamma r\text{d}r$$

As *force × distance* is work, the force opposing an increase in the radius by dr is

$$F = 8\pi\gamma r$$

The total inward force is therefore $4\pi r^2 p_{\text{out}} + 8\pi\gamma r$.

Step 3 *Balance the inward and outward forces*

If the pressure inside the cavity is p_{in}, the outward force on the surface is $4\pi r^2 p_{\text{in}}$. At equilibrium, the outward and inward forces are balanced:

$$4\pi r^2 p_{\text{in}} = 4\pi r^2 p_{\text{out}} + 8\pi\gamma r$$

Division of both sides by $4\pi r^2$ results in the **Laplace equation:**

$$p_{\text{in}} = p_{\text{out}} + \frac{2\gamma}{r} \qquad (14\text{C}.7)$$

Laplace equation

The Laplace equation shows that the difference in pressure decreases to zero as the radius of curvature becomes infinite (when the surface is flat, Fig. 14C.5). Small cavities have small radii of curvature, so the pressure difference across their surface is quite large.

Brief illustration 14C.1

The pressure difference across the surface of a spherical droplet of water of radius 200 nm at 20 °C can be calculated using the Laplace equation:

$$p_{\text{in}} - p_{\text{out}} = \frac{2\times\overbrace{(72.75\times10^{-3}\,\text{N m}^{-1})}^{\gamma_{\text{water}}\text{ at }20\,°\text{C}}}{\underbrace{2.00\times10^{-7}\,\text{m}}_{r}}$$

$$= 7.28\times10^{5}\,\text{N m}^{-2} = 728\,\text{kPa}$$

(c) Capillary action

When a narrow-bore tube (a 'capillary tube'; the name comes from the Latin word for 'hair') is dipped in a liquid there is a tendency for the liquid to rise up the tube. This tendency is called **capillary action** and can be understood as a consequence of surface tension.

Water has a tendency to adhere to the surface of glass, so when a glass capillary tube is first immersed in water a film of

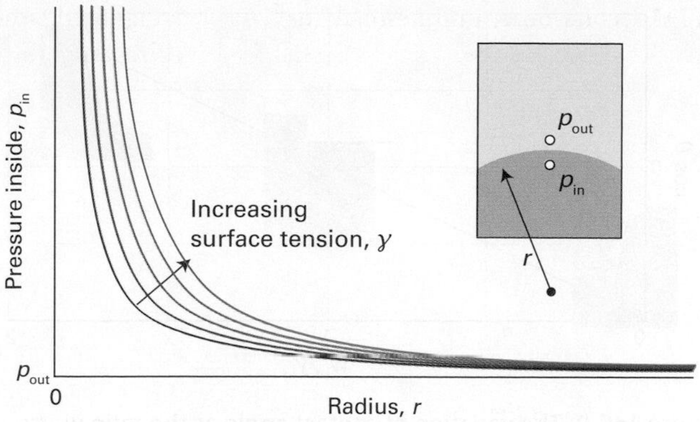

Figure 14C.5 The dependence of the pressure inside a curved surface on the radius of the surface, for increasing values of the surface tension.

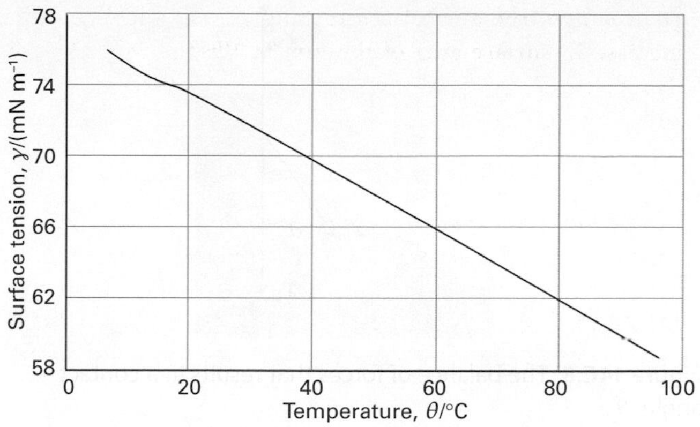

Figure 14C.7 The variation of the surface tension of water with temperature.

solvent spreads along the surface: the further the film spreads, the lower is the energy due to the interaction. As the film spreads up the inside walls of the capillary tube the surface of the liquid becomes curved and, as has just been discussed, this curvature results in a pressure difference across the surface. The pressure just above the meniscus, the concave side, is greater than that just below, the convex side.

If it is assumed that the surface is hemispherical and that the tube has radius r, the pressure difference is given by the Laplace equation as $2\gamma/r$. The pressure just above the surface is the atmospheric pressure, p, so the pressure just below the surface is $p - 2\gamma/r$. The atmospheric pressure pushing on the liquid outside the tube causes the liquid inside the tube to rise until hydrostatic equilibrium is achieved, which is when there are equal pressures at equal depths (Fig. 14C.6). When the liquid has risen to a height h the column exerts a hydrostatic pressure given by eqn 14C.4, $\rho g_{acc}h$, where ρ is the mass density of the liquid and g_{acc} is the acceleration of free fall. The total pressure at the base of the column is therefore $(p - 2\gamma/r) + \rho g_{acc}h$. At the same level outside the tube the pressure is p, but at equilibrium these two pressures must be the same: $p = p - 2\gamma/r + \rho g_{acc}h$. It therefore follows that $2\gamma/r = \rho g_{acc}h$ which rearranges to

$$h = \frac{2\gamma}{\rho g_{acc} r} \qquad (14C.8)$$

This simple expression provides a reasonably accurate way of measuring the surface tension of liquids. Surface tension decreases with increasing temperature (Fig. 14C.7), which can be understood as arising from the increase in thermal motion moving molecules more rapidly between the surface and the bulk.

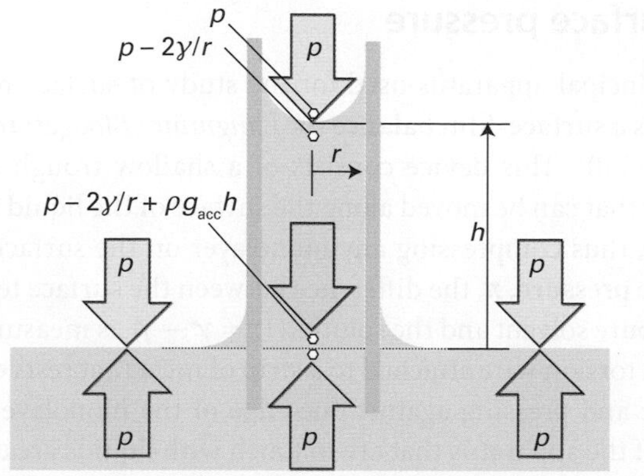

Figure 14C.6 When a capillary tube is first stood in a liquid, the liquid climbs up the walls, so curving the surface. The pressure just under the meniscus is less than that arising from the atmosphere by $2\gamma/r$. The pressure is equal at equal heights throughout the liquid provided the hydrostatic pressure (which is equal to $\rho g_{acc}h$) cancels the pressure difference arising from the curvature.

Brief illustration 14C.2

If water at 25 °C (with mass density 997.1 kg m^{-3}) rises through 7.36 cm in a capillary of radius 0.20 mm, its surface tension at that temperature is

$$\gamma = \tfrac{1}{2}\rho g_{acc}hr$$

$$= \tfrac{1}{2}\times(997.1\,\mathrm{kg\,m^{-3}})\times(9.81\,\mathrm{m\,s^{-2}})\times(7.36\times10^{-2}\,\mathrm{m})\times(2.0\times10^{-4}\,\mathrm{m})$$

($\mathrm{kg\,m\,s^{-2}} = \mathrm{N}$)

$$= 72\,\mathrm{mN\,m^{-1}}$$

When the adhesive forces between the liquid and the material of the capillary wall are weaker than the cohesive forces within the liquid (as for mercury in glass), it is energetically favourable for the liquid in the tube to retract from the walls. This retraction curves the surface with the concave, high pressure side downwards. To equalize the pressure at the same

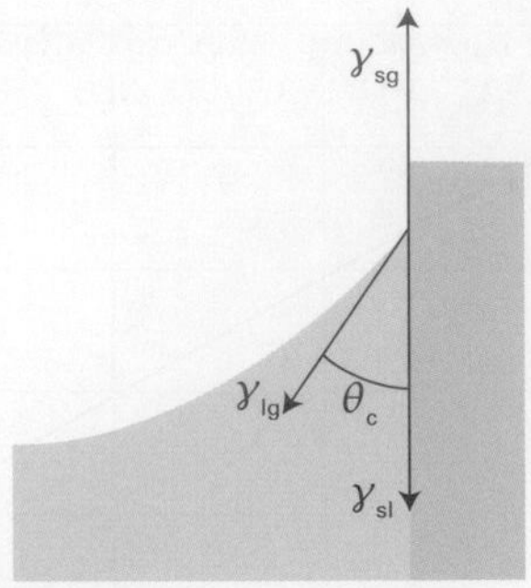

Figure 14C.8 The balance of forces that results in a contact angle, θ_c.

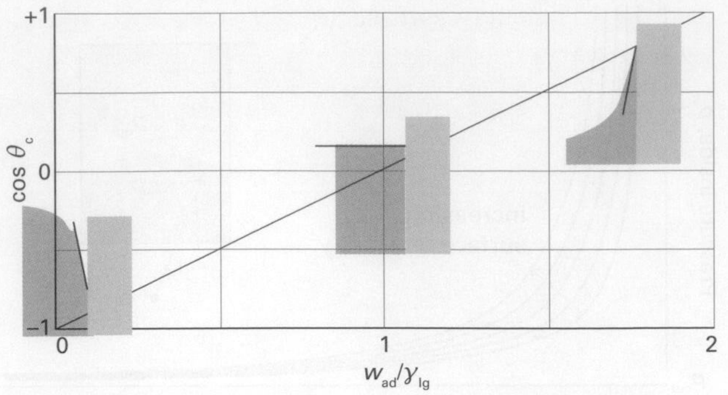

Figure 14C.9 The variation of contact angle as the ratio w_{ad}/γ_{lg} changes.

depth throughout the liquid the surface must fall to compensate for the increased pressure arising from its curvature. This compensation results in a capillary depression.

The angle between the surface and the wall, where the two meet, is called the **contact angle**, θ_c (Fig. 14C.8). In many cases this angle is found to be non-zero, and eqn 14C.8 must then be modified by multiplying the right-hand side by $\cos\theta_c$. The origin of the contact angle can be traced to the balance of forces at the line of contact between the liquid and the solid (Fig. 14C.8).

If the solid–gas, solid–liquid, and liquid–gas surface tensions are denoted γ_{sg}, γ_{sl}, and γ_{lg}, respectively, then the vertical forces are in balance if

$$\gamma_{sg} = \gamma_{sl} + \gamma_{lg}\cos\theta_c \qquad (14C.9a)$$

and therefore

$$\cos\theta_c = \frac{\gamma_{sg} - \gamma_{sl}}{\gamma_{lg}} \qquad (14C.9b)$$

The 'superficial work' of adhesion of the liquid to the solid, w_{ad}, is the work of adhesion divided by the area of contact. As the liquid–solid interface expands it does so at the expense of the solid–gas and liquid–gas interfaces, and the net work involved is

$$w_{ad} = \gamma_{sg} + \gamma_{lg} - \gamma_{sl} \qquad (14C.10a)$$

and therefore, from the previous equation,

$$\cos\theta_c = \frac{w_{ad}}{\gamma_{lg}} - 1 \qquad \text{Contact angle} \qquad (14C.10b)$$

It is now seen that:

- When the contact angle is between 0 and 90° (implying that $0 < \cos\theta_c < 1$) the liquid 'wets' the surface, meaning that it spreads over the surface. From eqn 14C.10b wetting occurs when $1 < w_{ad}/\gamma_{lg} < 2$ (Fig. 14C.9).
- When the contact angle is between 90° and 180° (implying that $-1 < \cos\theta_c < 0$) the liquid does not wet the surface; this condition corresponds to $0 < w_{ad}/\gamma_{lg} < 1$.

Physical interpretation

For mercury in contact with glass, $\theta_c = 140°$, which corresponds to $w_{ad}/\gamma_{lg} = 0.23$, indicating a relatively low work of adhesion of the mercury to glass on account of the strong cohesive forces within the liquid.

14C.3 Surface films

The compositions of surface layers have been investigated by the simple but technically elegant procedure of slicing thin layers off the surfaces of solutions and analysing their compositions. The physical properties of surface films have also been investigated. Surface films one molecule thick are called **monolayers**, and when such a monolayer has been transferred to a solid support, it is called a **Langmuir–Blodgett film**, after Irving Langmuir and Katharine Blodgett, who developed experimental techniques for studying them.

(a) Surface pressure

The principal apparatus used for the study of surface monolayers is a **surface-film balance** (or *Langmuir–Blodgett trough*, Fig. 14C.10). This device consists of a shallow trough and a barrier that can be moved along the surface of the liquid in the trough, thus compressing any monolayer on the surface. The **surface pressure**, π, the difference between the surface tension of the pure solvent and the solution ($\pi = \gamma^* - \gamma$), is measured by using a torsion wire attached to a strip of mica that rests on the surface and pressing against one edge of the monolayer. The parts of the apparatus that are in touch with liquids are coated in polytetrafluoroethene to eliminate effects arising from the liquid–solid interface. In an actual experiment, a small amount (about 0.01 mg) of the substance under investigation is dissolved in a volatile solvent and then poured on to the surface of the water; the compression barrier is then moved across the surface and the surface pressure exerted on the mica bar is monitored.

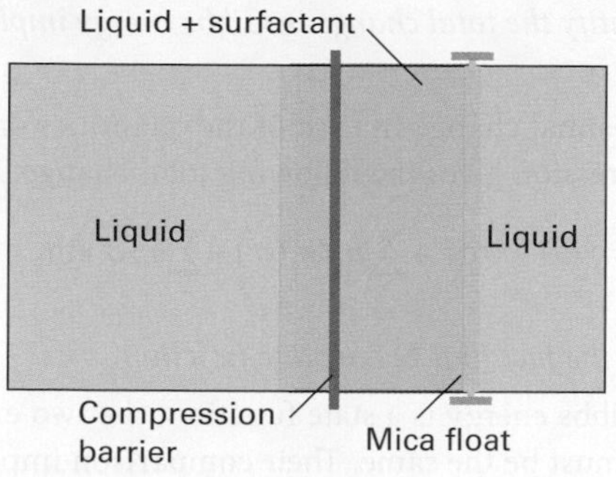

Figure 14C.10 A schematic diagram of the apparatus used to measure the surface pressure and other characteristics of a surface film. The surfactant is spread on the surface of the liquid in the trough, and then compressed horizontally by moving the compression barrier towards the mica float. The latter is connected to a torsion wire, so the net force on the float can be monitored.

When the surface coverage is low it is found that the surface pressure is inversely proportional to the total area of the surface. This behaviour is the analogue in two dimensions of a perfect gas (where $p \propto 1/V$), and can be interpreted as arising when the average separation between the molecules at the surface is so large that the interactions between them are not important. As the area is further decreased, the surface pressure eventually starts to increase more rapidly, as is illustrated in Fig. 14C.11. This behaviour can be thought of as arising from the formation of a monolayer in which the molecules are in relatively close contact; like a liquid, such a layer is almost incompressible. The area corresponding to a complete close-packed monolayer is found by extrapolating the steepest part of the isotherm.

As can be seen from Fig. 14C.11, even though stearic acid (**1**) and isostearic acid (**2**) are chemically very similar (they differ only in the location of a methyl group at the end of a long hydrocarbon chain), they occupy significantly different areas in the monolayer. Neither, though, occupies as much area as the tri-*p*-cresyl phosphate molecule (**3**), which is like a wide bush rather than a lanky tree.

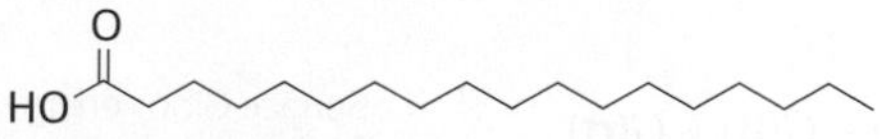

1 Stearic acid, $C_{17}H_{35}COOH$

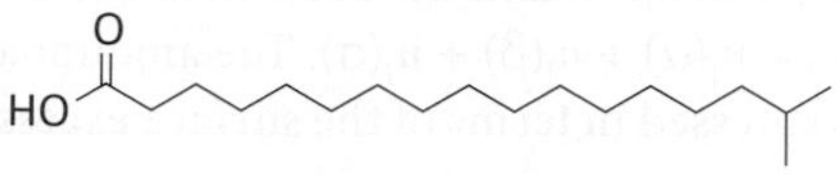

2 Isostearic acid, $C_{17}H_{35}COOH$

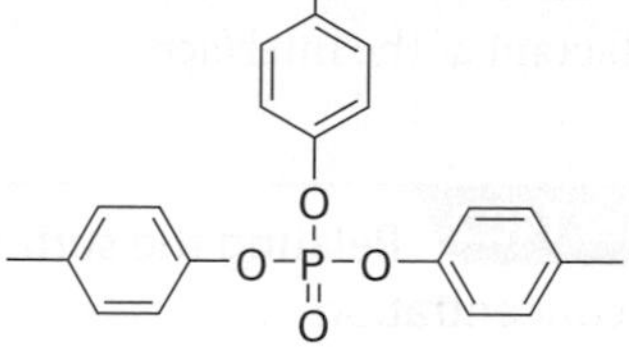

3 Tri-*p*-cresyl phosphate

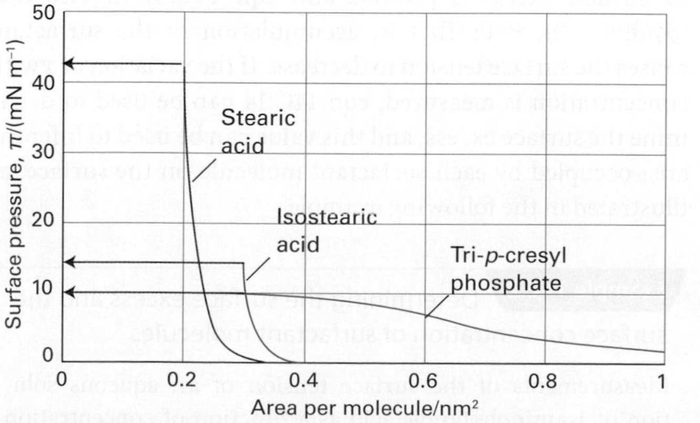

Figure 14C.11 The variation of surface pressure with the average area occupied by a surfactant molecule. The collapse pressures are indicated by the horizontal arrows.

The second feature to note from Fig. 14C.11 is that the tri-*p*-cresyl phosphate isotherm is much less steep than the stearic acid isotherms. This difference indicates that the tri-*p*-cresyl phosphate film is more compressible than the stearic acid films, which is consistent with their different molecular structures.

A third feature of the isotherms is the **collapse pressure**, the highest surface pressure at which a monolayer can be maintained. When the monolayer is compressed beyond the collapse pressure, the monolayer buckles and collapses into a film several molecules thick. As can be seen from the isotherms in Fig. 14C.11, stearic acid has a high collapse pressure, but that of tri-*p*-cresyl phosphate is significantly lower, indicating a much weaker film.

(b) The thermodynamics of surface layers

A **surfactant** is a species that accumulates at the interface between two phases, such as that between hydrophilic and hydrophobic phases, and modifies the surface tension. The relation between the concentration of surfactant at the surface and the change in surface tension it brings about can be established by considering a model in which two phases α and β come into contact, therefore creating an interface. Within each bulk phase the composition is constant, but the composition in the interfacial region may be different due to the accumulation of the surfactant.

The total Gibbs energy can be thought of as having a contribution from the two phases, $G(\alpha)$ and $G(\beta)$, together with a contribution $G(\sigma)$, the **surface Gibbs energy**, from the interfacial region

$$G = G(\alpha) + G(\beta) + G(\sigma) \quad \text{Surface Gibbs energy [definition]} \quad (14C.11)$$

In a similar way, the total amount of substance J, n_J, can be thought of as being divided between the amounts in phases α and β, $n_J(\alpha)$ and $n_J(\beta)$, and the amount at the interfacial region, $n_J(\sigma)$: $n_J = n_J(\alpha) + n_J(\beta) + n_J(\sigma)$. The amount at the interface can be expressed in terms of the **surface excess**, Γ_J:

$$\Gamma_J = \frac{n_J(\sigma)}{\sigma} \quad \text{Surface excess [definition]} \quad (14C.12)$$

where σ is the area of the surface. It is possible to relate the surface tension to the surface excess and therefore to the concentration of surfactant at the interface.

How is that done? 14C.2 Relating the surface tension to the surfactant concentration

The change in G brought about by changes in T, p, and n_J is given by eqn 5A.6:

$$dG = -SdT + Vdp + \sum_J \mu_J dn_J$$

where μ_J is the chemical potential of substance J. At constant pressure, $Vdp = 0$ and at constant temperature $SdT = 0$, so the first two terms on the right vanish.

Step 1 *Write an expression for the change in Gibbs energy of the interface*

To apply what remains of this relation to the interface, it is necessary to introduce an additional term $\gamma d\sigma$ (eqn 14C.5) arising from the work done expanding the interface. The expression for the change in Gibbs energy of the interface, $dG(\sigma)$, becomes

$$dG(\sigma) = \gamma d\sigma + \sum_J \mu_J dn_J(\sigma)$$

At equilibrium the chemical potential of each component is the same in each phase and is written μ_J.

Step 2 *Integrate the infinitesimal change*

Following the same argument as in the discussion of partial molar quantities (Topic 5A), this equation can be integrated at constant temperature, surface tension, and composition to give

$$G(\sigma) = \gamma\sigma + \sum_J \mu_J n_J(\sigma)$$

Step 3 *Identify the total change in Gibbs energy implied by this expression*

An infinitesimal change in each of the quantities on the right of this expression gives the following total change in $G(\sigma)$:

$$dG(\sigma) = \gamma d\sigma + \sigma d\gamma + \sum_J \mu_J dn_J(\sigma) + \sum_J n_J(\sigma) d\mu_J$$

Step 4 *Use the fact that G is a state function*

Because Gibbs energy is a state function, the two expressions for $dG(\sigma)$ must be the same. Their comparison implies that

$$\sigma d\gamma + \sum_J n_J(\sigma) d\mu_J = 0$$

Division by σ and introduction of the definition of the surface excess ($\Gamma_J = n_J(\sigma)/\sigma$) gives the **Gibbs isotherm**, which relates the change in surface tension to the changes in the chemical potentials of the substances present in the interface

$$d\gamma = -\sum_J \Gamma_J d\mu_J \quad \text{Gibbs isotherm} \quad (14C.13)$$

Step 5 *Relate the change in chemical potentials to the composition*

If just one species, the surfactant S, accumulates at the surface, the Gibbs isotherm becomes

$$d\gamma = -\Gamma_S d\mu_S$$

The chemical potential for species J in a dilute solution can be written as $\mu_J = \mu_J^{\ominus} + RT\ln(c_J/c^{\ominus})$, where c_J is the molar concentration and $c^{\ominus}$ is its standard value. It follows that at constant temperature $d\mu_S = RT d\ln(c/c^{\ominus})$. This expression for $d\mu_S$ is used in the equation above to give

$$d\gamma = -RT\Gamma_S \, d\ln(c/c^{\ominus})$$

which can be rearranged to

$$\left(\frac{\partial\gamma}{\partial\ln(c/c^{\ominus})}\right)_T = -RT\Gamma_S \quad \text{Dependence of the surface tension on surfactant concentration} \quad (14C.14)$$

When the surfactant accumulates at the interface, its surface excess is positive and eqn 14C.14 implies that $(\partial\gamma/\partial\ln(c/c^{\ominus}))_T < 0$. That is, accumulation of the surfactant causes the surface tension to decrease. If the variation of γ with concentration is measured, eqn 14C.14 can be used to determine the surface excess, and this value can be used to infer the area occupied by each surfactant molecule on the surface, as illustrated in the following example.

Example 14C.2 Determining the surface excess and the surface concentration of surfactant molecules

Measurements of the surface tension of an aqueous solution of 1-aminobutanoic acid as a function of concentration give $d\gamma/d\ln(c/c^{\ominus}) = -40\,\mu N\,m^{-1}$ at 20 °C. Calculate the surface excess of 1-aminobutanoic acid and the number of molecules per square metre.

Collect your thoughts Equation 14C.14 relates the measured value of $d\gamma/d\ln(c/c^{\ominus})$ directly to the surface excess. Multiplication of the surface excess by Avogadro's constant gives the number of molecules per square metre.

The solution From eqn 14C.14 it follows that

$$\Gamma_S = -\frac{1}{RT}\left(\frac{\partial\gamma}{\partial\ln(c/c^{\ominus})}\right)_T$$
$$= -\frac{1}{(8.3145\,\mathrm{J\,K^{-1}\,mol^{-1}})\times(293\,\mathrm{K})}\times(-4.0\times10^{-5}\,\mathrm{N\,m^{-1}})$$
$$= 1.6\times10^{-8}\,\mathrm{mol\,m^{-2}}$$

The number of molecules per square metre is $N_A\Gamma_S$:

$$N_A\Gamma_S = (6.022\times10^{23}\,\mathrm{mol^{-1}})\times(1.6\times10^{-8}\,\mathrm{mol\,m^{-2}}) = 9.6\times10^{15}\,\mathrm{m^{-2}}$$

Self-test 14C.2 Use the result obtained to calculate the area occupied by each molecule of 1-aminobutanoic acid at the surface.

Answer: $1.0\times10^2\,\mathrm{nm^2}$

14C.4 Condensation

The concepts from this Topic together with some from Topic 4B can be used to explain aspects of the condensation of a gas to a liquid. In Topic 4B it is shown that the vapour pressure of a liquid, p, is increased when additional pressure ΔP is applied to the liquid: according to eqn 4B.2, $p = p^*e^{V_m(l)\Delta P/RT}$, where p^* is the vapour pressure when no additional pressure is applied and $V_m(l)$ is the molar volume of the liquid. Because of its curved surface, a droplet experiences an additional pressure, given by the Laplace equation (eqn 14C.7) as $2\gamma/r$, where r is the radius of the surface. When this value is used for ΔP in eqn 4B.2 the result is the **Kelvin equation** for the vapour pressure of a liquid when it is dispersed as spherical droplets:

$$p = p^*e^{2\gamma V_m(l)/rRT} \qquad \text{Kelvin equation} \qquad (14C.15)$$

For a cavity, the pressure of the liquid outside is less than the pressure inside, so the sign of the exponent in eqn 14C.15 is changed to obtain an expression for the vapour pressure in a cavity. For droplets of water of radius 1 μm and 1 nm the ratios p/p^* at 25 °C are about 1.001 and 3, respectively. The second figure, although quite large, is unreliable because at that radius the droplet is less than about 10 molecules in diameter and the basis of the calculation is suspect. The first figure shows that the effect is usually small; nevertheless it may have important consequences.

For instance, consider the formation of a cloud. Warm, moist air rises into the cooler regions higher in the atmosphere. At some altitude the temperature drops to the point that the vapour becomes thermodynamically unstable with respect to the liquid and it is then expected that the vapour will condense into a cloud of liquid droplets. The initial step can be imagined as a swarm of water molecules congregating into a microscopic droplet. Because the initial droplet is so small it has an enhanced vapour pressure; therefore, instead of growing it evaporates. This effect stabilizes the vapour because an initial tendency to condense is overcome by a heightened tendency to evaporate. The vapour phase is then said to be **supersaturated**. It is thermodynamically unstable with respect to the liquid but not unstable with respect to the small droplets that need to form before the bulk liquid phase can appear, so the formation of the latter by a simple, direct mechanism is hindered.

Two processes are responsible for overcoming this tendency of droplets to evaporate and thus for allowing clouds to form. The first is that a sufficiently large number of molecules might congregate into a droplet so big that the enhanced evaporative effect is unimportant. The chance of one of these **spontaneous nucleation centres** forming is low, and in rain formation it is not a dominant mechanism. The more important process depends on the presence of minute dust particles or other kinds of foreign matter. These **nucleate** the condensation (that is, provide centres at which it can occur) by providing surfaces to which the water molecules can attach.

Liquids may be **superheated** above their boiling temperatures and **supercooled** below their freezing temperatures. In each case the thermodynamically stable phase is not achieved on account of the kinetic stabilization that occurs in the absence of nucleation centres. For example, superheating occurs because the vapour pressure inside a cavity is artificially low, so any cavity that does form tends to collapse. This instability is encountered when an unstirred beaker of water is heated, for its temperature may be raised above its boiling point. Violent bumping often ensues as spontaneous nucleation leads to bubbles big enough to survive. To ensure smooth boiling at the true boiling temperature, nucleation centres, such as small pieces of sharp-edged glass or bubbles (cavities) of air, should be introduced.

Checklist of concepts

☐ 1. The **radial distribution function**, $g(r)$, is defined such that $4\pi\mathcal{N}g(r)r^2\mathrm{d}r$ is the number of molecules in a shell of thickness dr at radius r from a given molecule; $\mathcal{N}$ is the overall number density.

☐ 2. The radial distribution function may be calculated numerically by using **Monte Carlo** and **molecular dynamics** techniques.

☐ 3. Liquids tend to adopt shapes that minimize their surface area.

☐ 4. **Capillary action** is the tendency of liquids to rise up (and in some cases, drop down) narrow tubes.

☐ 5. The **surface pressure** is the difference between the surface tension of a pure solvent and a solution.

☐ 6. The **collapse pressure** is the highest surface pressure that a surface film can sustain.

☐ 7. A **surfactant** is a species that accumulates at the interface between phases and modifies the surface tension and surface pressure.

☐ 8. **Nucleation** provides surfaces to which molecules can attach and thereby induce condensation.

Checklist of equations

Property	Equation	Comment	Equation number
Hydrostatic pressure	$p = \rho g_{\text{acc}} h$	Incompressible fluid	14C.4
Laplace equation	$p_{\text{in}} = p_{\text{out}} + 2\gamma/r$	γ is the surface tension	14C.7
Contact angle	$\cos\theta_{\text{c}} = w_{\text{ad}}/\gamma_{\text{lg}} - 1$		14C.10b
Surface Gibbs energy	$G = G(\alpha) + G(\beta) + G(\sigma)$	Definition	14C.11
Surface excess	$\Gamma_{\text{J}} = n_{\text{J}}(\sigma)/\sigma$	Definition	14C.12
Gibbs isotherm	$\mathrm{d}\gamma = -\sum_{\text{J}}\Gamma_{\text{J}}\mathrm{d}\mu_{\text{J}}$		14C.13
Dependence of the surface tension on surfactant concentration	$(\partial\gamma/\partial\ln(c/c^{\ominus}))_T = -RT\Gamma_{\text{S}}$		14C.14
Kelvin equation	$p = p^{*}\mathrm{e}^{2\gamma V_{\text{m}}(\text{l})/rRT}$		14C.15

TOPIC 14D Macromolecules

> **Why do you need to know this material?**
>
> Macromolecules give rise to special problems that include the investigation and description of their molar masses and shapes. You need to know how to describe the structural features of macromolecules in order to understand their physical and chemical properties.
>
> **What is the key idea?**
>
> The structure of a macromolecule takes on different meanings at the different levels at which the arrangement of the chain or network of its building blocks is considered.
>
> **What do you need to know already?**
>
> The discussion of the shapes of macromolecules depends on an understanding of the nonbonding interactions between molecules (Topic 14B). You also need to be familiar with the statistical interpretation of entropy (Topic 13E) and the concept of internal energy (Topic 2A). Some of the calculations draw on statistical arguments like those used in the discussion of the Boltzmann distribution (Topic 13A).

Macromolecules are very large molecules assembled from smaller molecules biosynthetically in organisms, by chemists in the laboratory, or in an industrial reactor. Naturally occurring macromolecules include polysaccharides such as cellulose, polypeptides such as protein enzymes, and polynucleotides such as deoxyribonucleic acid (DNA). This Topic deals principally with synthetic macromolecules. They include **polymers**, such as nylon and polystyrene, that are manufactured by stringing together, and in some cases crosslinking, smaller units known as **monomers** (Fig. 14D.1).

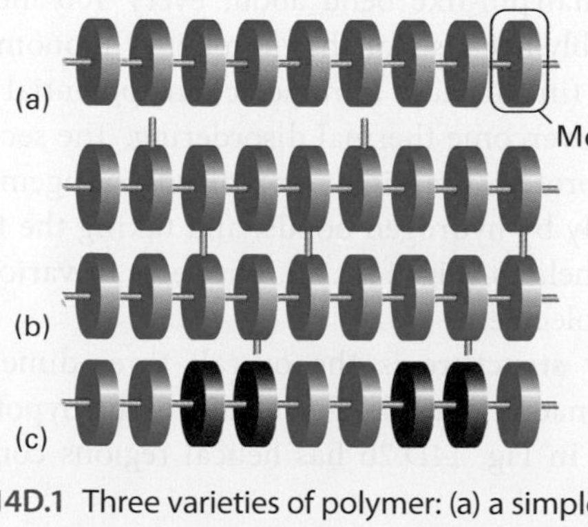

Figure 14D.1 Three varieties of polymer: (a) a simple linear polymer, (b) a cross-linked polymer, and (c) one variety of copolymer.

14D.1 Average molar masses

A **monodisperse** polymer has a single, definite molar mass. A synthetic polymer, however, is **polydisperse**, in the sense that a sample is a mixture of molecules with various chain lengths and molar masses. The various techniques that are used to measure molar masses result in different types of mean values of polydisperse systems.

The **number-average molar mass**, $\bar{M}_n$, is obtained by weighting each molar mass by the number of molecules of that mass present in the sample:

$$\bar{M}_n = \frac{1}{N_{total}}\sum_i N_i M_i \quad \text{Number-average molar mass [definition]} \quad (14D.1a)$$

where N_i is the number of molecules with molar mass M_i and N_{total} is the total number of molecules. This type of average is typically obtained by mass spectroscopic determinations of molar mass. The **weight-average molar mass** is the average calculated by weighting the molar masses of the molecules by the mass present in the sample:

$$\bar{M}_w = \frac{1}{m_{total}}\sum_i m_i M_i \quad \text{Weight-average molar mass [definition]} \quad (14D.1b)$$

In this expression, m_i is the total mass of molecules of molar mass M_i and m_{total} is the total mass of the sample. This type of average is typically obtained by measurements that make use of the ability of molecules to scatter light and by measurements that make use of the distribution of particles in solutions rotated at high speed in an ultracentrifuge.

Example 14D.1 Calculating number and weight averages

Evaluate the number-average and the weight-average molar masses of a sample of poly(vinyl chloride) from the following data:

Interval	1	2	3	4	5	6
M_i/(kg mol^{-1})	7.5	12.5	17.5	22.5	27.5	32.5
m_i/g	9.6	8.7	8.9	5.6	3.1	1.7

Collect your thoughts The relevant equations are eqns 14D.1a and 14D.1b. Note that because $N_i = n_i N_A$, you can express the number average in terms of amounts (in moles):

$$\bar{M}_n = \frac{1}{n_{total}N_A}\sum_i n_i N_A M_i = \frac{1}{n_{total}}\sum_i n_i M_i$$

where n_i is the amount (in moles) of molecules with molar mass M_i and n_{total} is the total amount of molecules. Calculate the amounts in each interval by dividing the mass of the sample in each interval by the average molar mass for that interval; $n_i = m_i/M_i$. Then calculate the two averages by weighting the molar mass M_i within each interval by the amount (n_i) and mass (m_i), respectively, of the molecules in each interval.

The solution The amounts in each interval are as follows:

Interval	1	2	3	4	5	6
M_i/(kg mol^{-1})	7.5	12.5	17.5	22.5	27.5	32.5
n_i/mmol	1.3	0.70	0.51	0.25	0.11	0.052

The total amount is $n_{\text{total}} = 2.92$ mmol and the number-average molar mass is

$$\bar{M}_{\text{n}}/(\text{kg mol}^{-1}) = \frac{1}{2.92}(1.3\times 7.5 + 0.70\times 12.5 + 0.51\times 17.5 + 0.25\times 22.5 + 0.11\times 27.5 + 0.052\times 32.5) = 13$$

The weight-average molar mass is calculated directly from the data after noting that the total mass of the sample is 37.6 g:

$$\bar{M}_{\text{w}}/(\text{kg mol}^{-1}) = \frac{1}{37.6}(9.6\times 7.5 + 8.7\times 12.5 + 8.9\times 17.5 + 5.6\times 22.5 + 3.1\times 27.5 + 1.7\times 32.5) = 16$$

Comment. Note the different values of the two averages. In this instance, $\bar{M}_{\text{w}}/\bar{M}_{\text{n}} = 1.2$.

Self-test 14D.1 The *Z*-average molar mass, which is obtained in certain sedimentation experiments, is defined as $\bar{M}_Z = \sum_i N_i M_i^3 / \sum_i N_i M_i^2$. Evaluate its value for the sample in this example.

Answer: 19 kg mol^{-1}

The ratio $\bar{M}_{\text{w}}/\bar{M}_{\text{n}}$ is called the (molar-mass) **dispersity, *Đ*** (previously the 'polydispersity index', PDI), read 'd-stroke' and defined as

$$Đ = \frac{\bar{M}_{\text{w}}}{\bar{M}_{\text{n}}} \quad \text{Dispersity [definition]} \quad (14\text{D}.2)$$

The term 'monodisperse' is conventionally applied to synthetic polymers for which the dispersity is less than 1.1; commercial polyethene samples might be much more heterogeneous, with a dispersity close to 30. One feature of a narrow molar-mass distribution for synthetic polymers is often a higher degree of long-range order in the solid and therefore higher density and melting point. The spread of values is controlled by the choice of catalyst and reaction conditions.

A note on good practice The masses of macromolecules are often reported in daltons (Da), where 1 Da = m_{u} (with $m_{\text{u}} = 1.661 \times 10^{-27}$ kg). Note that daltons are used to report *molecular* mass not *molar* mass. So the mass (not the molar mass) of a certain macromolecule may be reported as 100 kDa (i.e. its mass is $100 \times 10^3 \times m_{\text{u}}$), and its molar mass as 100 kg mol^{-1}. But it should not be said (even though it is common practice) that its molar mass is 100 kDa.

14D.2 The different levels of structure

The concept of the 'structure' of a macromolecule takes on different meanings at the different levels of the arrangement of the chain or network of monomers. The **primary structure** of a macromolecule is the sequence of small molecular residues making up the polymer. The residues may form either a chain, as in polyethene, or a more complex network in which cross-links connect different chains, as in cross-linked polyacrylamide. In a synthetic polymer, virtually all the residues are identical and it is sufficient to name the monomer used in the synthesis. Thus, the repeating unit of polyethene and its derivatives is $-CHXCH_2-$, and the primary structure of the chain is specified by denoting it as $-(CHXCH_2)_n-$. The concept of primary structure ceases to be trivial in the case of synthetic copolymers and biological macromolecules, because in general these substances are chains formed from different molecules. For example, proteins are **polypeptides** formed from different amino acids (about twenty occur naturally) strung together by the **peptide link**, –CONH– (1). The **degradation** of a biological macromolecule is a disruption of its primary structure, when the chain breaks into shorter components.

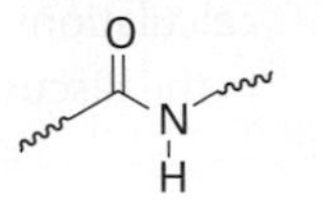

1 Peptide link

The term **conformation** refers to the spatial arrangement of the different parts of a chain, and one conformation can be changed into another by rotating one part of a chain around a bond. The conformation of a macromolecule is relevant at three levels of structure. The **secondary structure** of a macromolecule is the (often local) spatial arrangement of a chain. The secondary structure of a molecule of polyethene in some solvents may be a random coil (see below). In the absence of a solvent, polyethene forms crystals consisting of stacked sheets with a hairpin-like bend about every 100 monomer units, presumably because for that number of monomers the intermolecular (in this case *intra*molecular) potential energy is sufficient to overcome thermal disordering. The secondary structure of a protein is a highly organized arrangement determined largely by hydrogen bonds, and taking the form of random coils, helices (Fig. 14D.2a), or sheets in various segments of the molecule.

The **tertiary structure** is the overall three-dimensional structure of a macromolecule. For instance, the hypothetical protein shown in Fig. 14D.2b has helical regions connected

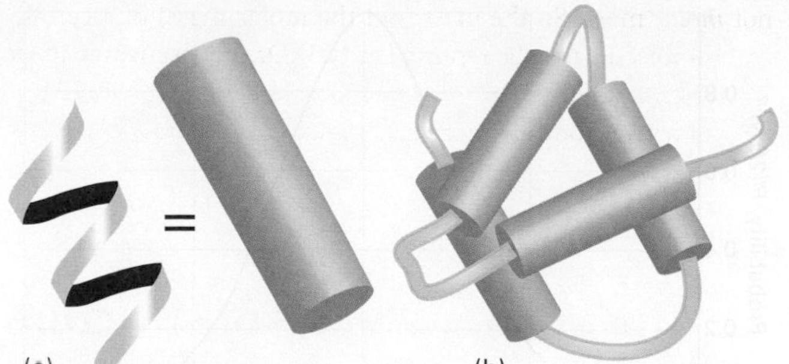

Figure 14D.2 (a) A polymer may adopt a highly organized helical conformation, an example of a secondary structure. The helix is represented as a cylinder. (b) Several helical segments connected by short random coils pack together, an example of tertiary structure.

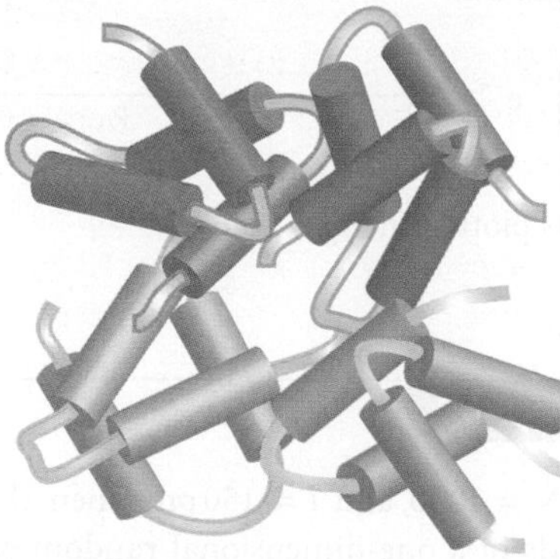

Figure 14D.3 Several subunits with specific tertiary structures pack together, an example of quaternary structure.

by short random-coil sections. The helices interact to form a compact tertiary structure.

The **quaternary structure** of a macromolecule is the manner in which large molecules are formed by the aggregation of others. Figure 14D.3 shows how four molecular subunits, each with a specific tertiary structure, aggregate. Quaternary structure can be very important in biology. For example, the oxygen-transport protein haemoglobin consists of four myoglobin-like subunits that work cooperatively to take up and release O_2.

14D.3 Random coils

The most likely conformation of a chain of identical units not capable of forming hydrogen bonds or any other type of specific bond is a **random coil**. Polyethene is a simple example. The simplest model of a random coil is a 'freely jointed chain', in which any bond is free to make any angle with respect to the preceding one (Fig. 14D.4). It is also assumed that the monomer units occupy zero volume, so different parts of the chain can occupy the same region of space. The model is obviously an oversimplification because a bond is actually constrained to a cone of angles around a direction defined by its neighbour (Fig. 14D.5) and real chains are self-avoiding in the sense that distant parts of the same chain cannot fold back and occupy the same space. In a hypothetical one-dimensional freely jointed chain all the monomer units lie in a straight line, and the angle between neighbours is either 0° or 180°. The units of a three-dimensional freely jointed chain are not restricted to lie in a line or a plane.

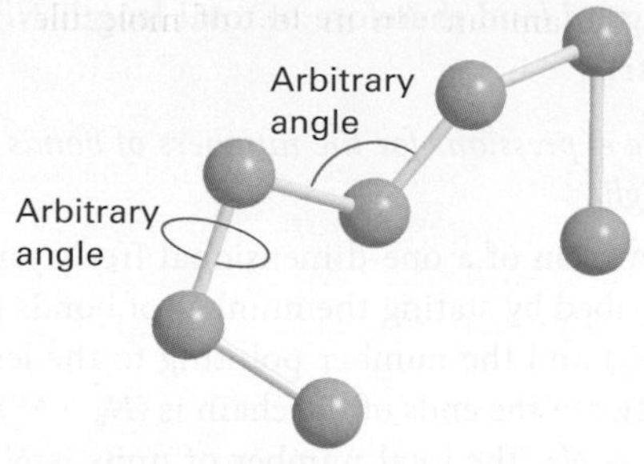

Figure 14D.4 A freely jointed chain is like a three-dimensional random walk, each step being in an arbitrary direction but of the same length.

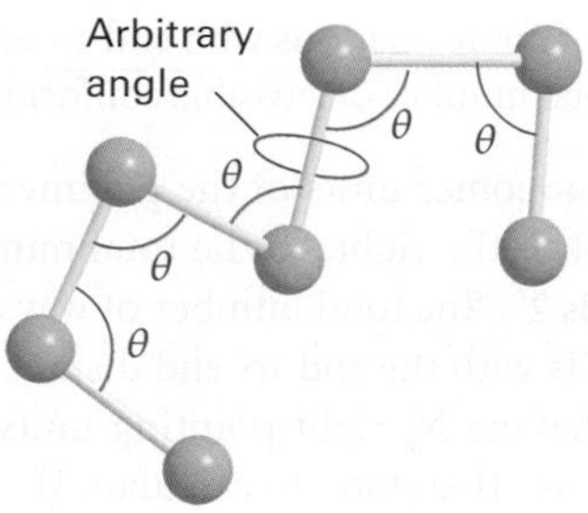

Figure 14D.5 A better description is obtained by fixing the bond angle (for example, at the tetrahedral angle) and allowing free rotation about a bond direction.

(a) Measures of size

The size of a freely jointed chain is related to the probability that its ends are a certain distance apart. That probability can be calculated by considering a one-dimensional random coil.

How is that done? 14D.1 Calculating the probability distribution in a one-dimensional random coil

Your goal is to calculate the probability, P, that the ends of a long one-dimensional freely jointed chain composed of N

units of length l (and therefore of total length Nl) are a distance nl apart.

Step 1 *Write expressions for the numbers of bonds pointing to the left or right*

The conformation of a one-dimensional freely jointed chain can be described by stating the number of bonds pointing to the right (N_R) and the number pointing to the left (N_L). The distance between the ends of the chain is $(N_R - N_L)l$; it follows that $n = N_R - N_L$. The total number of units is $N = N_R + N_L$, therefore, $N_R = \frac{1}{2}(N + n)$ and $N_L = \frac{1}{2}(N - n)$.

Step 2 *Write an expression for the probability that a polymer has a specified end-to-end separation*

The probability, P, that the end-to-end separation of a randomly selected polymer is nl is

$$P = \frac{\text{number of conformations with end-to-end distance } nl}{\text{total number of possible conformations}}$$

Each of the N monomer units of the polymer may in principle lie to the left or the right, so the total number of possible conformations is 2^N. The total number of ways, W, of forming a chain of N units with the end-to-end distance nl is the number of ways of having N_R right-pointing units, the rest being left-pointing units. Therefore, to calculate W, count the number of ways of achieving N_R right-pointing units given a total of N units. This is the same problem as selecting N_R objects from a collection of N objects (see Topic 13A), and is

$$W = \frac{N!}{N_R!(N-N_R)!} = \frac{N!}{N_R!N_L!} = \frac{N!}{\{\frac{1}{2}(N+n)\}!\{\frac{1}{2}(N-n)\}!}$$

It follows that

$$P = \frac{W}{2^N} = \frac{N!}{\{\frac{1}{2}(N+n)\}!\{\frac{1}{2}(N-n)\}!2^N}$$

Step 3 *Consider the case of compact chains*

When the chain is compact in the sense that $n \ll N$, it is more convenient to evaluate ln P: the factorials are then large and it is possible to use Stirling's approximation (Topic 13A). Although the approximation used there is $\ln x! = x\ln x - x$ (with $x = N$), here it is appropriate to use the more precise form

$$\ln x! \approx \ln(2\pi)^{1/2} + (x + \tfrac{1}{2})\ln x - x$$

The result, after quite a lot of algebra, is

$$\ln P \approx \ln\left(\frac{2}{\pi N}\right)^{1/2} - \tfrac{1}{2}(N+n+1)\ln(1+\nu) - \tfrac{1}{2}(N-n+1)\ln(1-\nu)$$

where $\nu = n/N$. For a compact coil ($\nu \ll 1$), use the approximation $\ln(1 \pm \nu) \approx \pm\nu - \frac{1}{2}\nu^2$ and obtain

$$\ln P \approx \ln\left(\frac{2}{\pi N}\right)^{1/2} - \tfrac{1}{2}N\nu^2$$

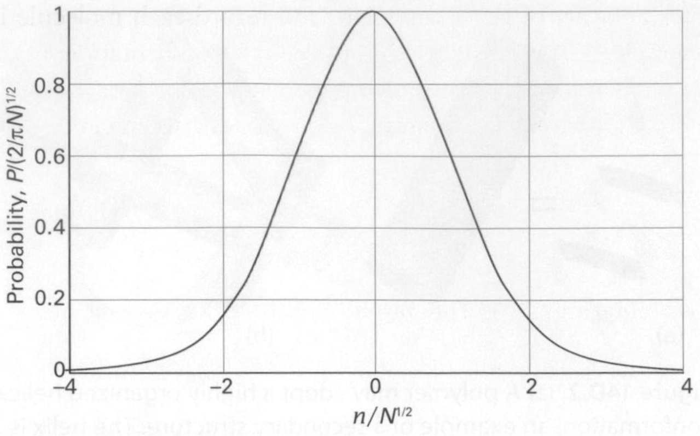

Figure 14D.6 The probability distribution for the separation of the ends of a one-dimensional random coil. The separation of the ends is nl, where l is the length of each monomer unit.

which rearranges into

$$P = \left(\frac{2}{\pi N}\right)^{1/2} e^{-n^2/2N} \quad (14D.3)$$

Probability distribution [1D random coil]

This function is plotted in Fig. 14D.6.

Brief illustration 14D.1

Suppose that $N = 1000$ and $l = 150$ pm, then the probability that the ends of a one-dimensional random coil are $nl = 3.00$ nm apart is given by eqn 14D.3 by setting $n = (3.00 \times 10^3\,\text{pm})/(150\,\text{pm}) = 20.0$:

$$P = \left(\frac{2}{\pi \times 1000}\right)^{1/2} e^{-20.0^2/(2\times 1000)} = 0.0207$$

meaning that there is a 1 in 48 chance of being found there.

Equation 14D.3 can be adapted to calculate the probability that the ends of a long three-dimensional freely jointed chain lie in the range r to $r + dr$. The probability is written as $f(r)dr$, where

$$f(r) = 4\pi\left(\frac{a}{\pi^{1/2}}\right)^3 r^2 e^{-a^2r^2} \qquad a = \left(\frac{3}{2Nl^2}\right)^{1/2} \quad (14D.4)$$

Probability distribution [3D random coil]

Here and elsewhere the fact that the chain cannot be longer than Nl is ignored. Although eqn 14D.4 gives a non-zero probability for $r > Nl$, the values are so small that the errors in pretending that r can range up to infinity are negligible. For a narrow range of distances δr, the probability density can be treated as a constant and the probability calculated from $f(r)\delta r$. An alternative interpretation of this expression is to regard each molecule in a sample as ceaselessly writhing from one conformation to another; then $f(r)\mathrm{d}r$ is the probability that at any instant the chain will be found with the separation of its ends between r and $r+\mathrm{d}r$.

Brief illustration 14D.2

Consider the chain described in *Brief illustration* 14D.1, with $N = 1000$ and $l = 150$ pm but now in three dimensions. Then

$$a=\left(\frac{3}{2\times1000\times(150\,\mathrm{pm})^2}\right)^{1/2}=2.58\ldots\times10^{-4}\,\mathrm{pm}^{-1}$$

Then the probability density at $r = 3.00$ nm is given by eqn 14D.4 as

$$f(3.00\,\mathrm{nm})=4\pi\times\left(\frac{2.58\ldots\times10^{-4}\,\mathrm{pm}^{-1}}{\pi^{1/2}}\right)^3$$
$$\times(3.00\times10^3\,\mathrm{pm})^2\times e^{-(2.58\ldots\times10^{-4}\,\mathrm{pm}^{-1})^2(3.00\times10^3\,\mathrm{pm})^2}$$
$$=1.92\times10^{-4}\,\mathrm{pm}^{-1}$$

The probability that the ends will be found in a narrow range of width $\delta r = 10.0$ pm at 3.00 nm (regardless of direction) is therefore

$$f(3.00\,\mathrm{nm})\delta r = (1.92\times10^{-4}\,\mathrm{pm}^{-1})\times(10.0\,\mathrm{pm}) = 1.92\times10^{-3}$$

or about 1 in 520.

There are several measures of the geometrical size of a random coil. The **contour length**, R_c, is the length of the polymer (not only a random coil) measured along its backbone from atom to atom. For a polymer of N monomer units each of length l, the contour length is

$$R_c = Nl \qquad \text{Contour length} \qquad (14D.5)$$

The **root-mean-square separation**, R_{rms}, is the square root of the mean value of the square of the separation of the ends of the coil. Thus, if the vector joining the ends of the coil is $\boldsymbol{R}$, and each monomer is represented by the vector $\boldsymbol{r}_i$, then $\boldsymbol{R}=\sum_{i=1}^{N}\boldsymbol{r}_i$ and

$$\langle R^2\rangle=\langle\boldsymbol{R}\cdot\boldsymbol{R}\rangle=\sum_{i,j=1}^{N}\langle\boldsymbol{r}_i\cdot\boldsymbol{r}_j\rangle=\sum_{i=1}^{N}\overbrace{\langle r_i^2\rangle}^{l^2}+\sum_{i\neq1}^{N}\langle\boldsymbol{r}_i\cdot\boldsymbol{r}_j\rangle$$

When N is large (as assumed throughout), the term in blue is zero because the individual vectors lie in random directions. The remaining term is Nl^2. It follows that for a random coil of any dimensionality

$$R_{\mathrm{rms}} = N^{1/2}l \qquad \text{Root-mean-square separation [random coil]} \qquad (14D.6)$$

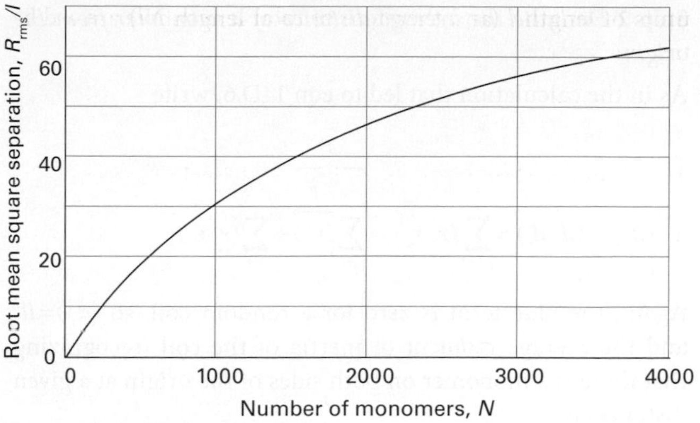

Figure 14D.7 The variation of the root-mean-square separation of the ends of a three-dimensional random coil, R_{rms}, with the number of monomers.

As the number of monomer units increases, the root-mean-square separation of the ends of the polymer increases as $N^{1/2}$ (Fig. 14D.7), and consequently the volume of a three-dimensional coil increases as $N^{3/2}$. The result must be multiplied by a factor when the chain is not freely jointed (see below).

Another convenient measure of size is the **radius of gyration**, R_g, which is the radius of a hollow sphere that has the same mass and moment of inertia (and therefore rotational characteristics) as the actual molecule. Once again, a one-dimensional random coil can be used to illustrate the procedure for the calculation of R_g.

How is that done? 14D.2 Deriving an expression for the radius of gyration

You need to set up an expression for the moment of inertia of the random one-dimensional coil of N monomer units each of mass m and then equate it to $m_{\mathrm{total}}R_g^2$, where m_{total} is the total mass of the polymer molecule, $m_{\mathrm{total}} = Nm$.

Step 1 *Set up the expression for the moment of inertia*

For a one-dimensional random coil with N identical monomers each of mass m, the moment of inertia around the centre of the chain (which is at the origin of the vector representing the first monomer, because the vectors point in equal numbers to left and right) is

$$I=\sum_{i=1}^{N}m_i d_i^2=m\sum_{i=1}^{N}d_i^2$$

where d_i is the distance of mass m_i from the origin. This distance is the length of the vector $\boldsymbol{d}_i$, the sum of i steps from the origin, $\boldsymbol{d}_i=\sum_{j=1}^{i}\boldsymbol{r}_j$.

Step 2 *Evaluate the average distance of a monomer from the origin*

As in the calculation that led to eqn 14D.6, write

$$\langle d_i^2\rangle = \langle \mathbf{d}_i\cdot\mathbf{d}_i\rangle = \sum_{j,k=1}^{i}\langle \mathbf{r}_j\cdot\mathbf{r}_k\rangle = \overbrace{\sum_{j=1}^{i}\overbrace{\langle r_j^2\rangle}^{l^2}}^{il^2} + \sum_{j\neq k}^{i}\langle \mathbf{r}_j\cdot\mathbf{r}_k\rangle$$

Again, the blue term is zero for a random coil, so $\langle d_i^2\rangle = il^2$ and the average moment of inertia of the coil (recognizing that there is a monomer on both sides of the origin at a given distance) is

$$\langle I\rangle = 2m\sum_{i=1}^{N}\langle d_i^2\rangle = 2ml^2\overbrace{\sum_{i=1}^{N} i}^{1+2+\cdots N=\frac{1}{2}N(N+1)\approx\frac{1}{2}N^2} = N^2ml^2 = Nm_{\text{total}}l^2$$

Step 3 *Identify the radius of gyration*

Finally, set this moment of inertia equal to $m_{\text{total}}R_g^2$, which implies that $R_g^2 = Nl^2$ and therefore that

$$R_g = N^{1/2}l \qquad \text{(14D.7a)}$$

Radius of gyration [1D random coil]

A similar calculation for a three-dimensional random coil gives

$$R_g = \left(\frac{N}{6}\right)^{1/2} l \qquad \text{(14D.7b)}$$

Radius of gyration [3D random coil]

The radius of gyration is smaller in this case because the extra dimensions enable the coil to be more compact.

The radius of gyration may also be calculated for other geometries. For example, a solid uniform sphere of radius R has $R_g = (\frac{3}{5})^{1/2}R$, and a long thin uniform rod of length L has $R_g = L/12^{1/2}$ for rotation about an axis perpendicular to the long axis. A solid sphere with the same radius and mass as a random coil has a greater radius of gyration as it is entirely dense throughout.

Brief illustration 14D.3

Consider a polymer that writhes as if it were a three-dimensional random coil. However, suppose that small segments of the macromolecule resist bending, so it is more appropriate to visualize it as a freely jointed chain with N and l as the number and length, respectively, of these rigid units. With the length $l = 45\,\text{nm}$ and $N = 200$ (and using $10^3\,\text{nm} = 1\,\mu\text{m}$),

From eqn 14D.5: $R_c = 200 \times 45\,\text{nm} = 9.0\,\mu\text{m}$

From eqn 14D.6: $R_{\text{rms}} = (200)^{1/2} \times 45\,\text{nm} = 0.64\,\mu\text{m}$

From eqn 14D.7b: $R_g = \left(\dfrac{200}{6}\right)^{1/2} \times 45\,\text{nm} = 0.26\,\mu\text{m}$

The random coil model ignores the role of the solvent: a poor solvent tends to cause the coil to tighten so that solute–solvent contacts are minimized; a good solvent does the opposite. Therefore, calculations based on this model are better regarded as lower bounds to the dimensions for a polymer in a good solvent and as an upper bound for a polymer in a poor solvent. The model is most reliable for a polymer in a bulk solid sample, where the coil is likely to have its natural dimensions.

(b) Constrained chains

The freely jointed chain model is improved by removing the freedom of bond angles to take any value. For long chains, it is convenient to take groups of neighbouring bonds and consider the direction of their resultant. Although each successive individual bond is constrained to a single cone of angle θ relative to its neighbour, the resultant of several bonds lies in a random direction. By concentrating on such groups rather than individuals, it turns out that for long chains the expressions for the root-mean-square separation and the radius of gyration given above should be multiplied by

$$F = \left(\frac{1-\cos\theta}{1+\cos\theta}\right)^{1/2} \qquad \text{(14D.8)}$$

For a tetrahedral arrangement of bonds, for which $\cos\theta = -\frac{1}{3}$ (i.e. $\theta = 109.5°$), $F = 2^{1/2}$. Therefore:

$$R_{\text{rms}} = (2N)^{1/2}l \qquad R_g = \left(\frac{N}{3}\right)^{1/2} l \qquad \text{(14D.9)}$$

Dimensions of a constrained tetrahedral chain

The model of a randomly coiled molecule is still an approximation, even after the bond angles have been restricted, because it does not take into account the impossibility of two or more atoms occupying the same place. Such self-avoidance tends to swell the coil, so (in the absence of solvent effects) it is better to regard R_{rms} and R_g as lower bounds to the actual values.

(c) Partly rigid coils

An important measure of the flexibility of a chain is the **persistence length**, l_p, a measure of the length over which the direction of the first monomer–monomer direction is sustained. If the chain is a rigid rod, then the persistence length is the same as the contour length. For a freely-jointed random coil, the persistence length is just the length of one monomer unit. Therefore, the persistence length can be regarded as a measure of the stiffness of the chain.

The mean square distance between the ends of a chain that has a persistence length greater than the monomer length can be expected to be greater than for a random coil because the partial rigidity of the coil does not let it roll up so tightly. A detailed calculation shows that

$$R_{\mathrm{rms}} = N^{1/2} l F \quad \text{where } F = \left(\frac{2l_{\mathrm{p}}}{l} - 1\right)^{1/2} \qquad (14\mathrm{D}.10)$$

For a random coil, $l_{\mathrm{p}} = l$, so $R_{\mathrm{rms}} = N^{1/2} l$, as already found. For $l_{\mathrm{p}} > l$, $F > 1$, so the coil has swollen, as anticipated.

Example 14D.2 Calculating the root-mean-square separation of a partly rigid coil

By what percentage does the root-mean-square separation of the ends of a polymer chain with $N = 1000$ increase or decrease when the persistence length changes from l (the length of one monomer unit) to 2.5 per cent of the contour length?

Collect your thoughts The contour length is $R_{\mathrm{c}} = Nl$. When $l_{\mathrm{p}} = l$, the chain is a random coil and $R_{\mathrm{rms,random\ coil}} = N^{1/2} l$, so eqn 14D.10 can be expressed as $R_{\mathrm{rms}} = F R_{\mathrm{rms,random\ coil}}$. The fractional change in root-mean-square separation is therefore

$$\frac{R_{\mathrm{rms}} - R_{\mathrm{rms,random\ coil}}}{R_{\mathrm{rms,random\ coil}}} = \frac{R_{\mathrm{rms}}}{R_{\mathrm{rms,random\ coil}}} - 1 = \left(\frac{2l_{\mathrm{p}}}{l} - 1\right)^{1/2} - 1$$

In the final step you should express this fractional change as a percentage.

The solution Because $l_{\mathrm{p}} = 0.025R_{\mathrm{c}} = 0.025Nl$, the fractional change is

$$\frac{R_{\mathrm{rms}} - R_{\mathrm{rms,random\ coil}}}{R_{\mathrm{rms,random\ coil}}} = \left(\frac{2 \times 0.025Nl}{l} - 1\right)^{1/2} - 1$$
$$= (0.050N - 1)^{1/2} - 1$$

With $N = 1000$, the fractional change is 6.00, so the root-mean-square separation increases by 600 per cent.

Self-test 14D.2 Calculate the fractional change in the volume of the same three-dimensional coil.

Answer: 340

14D.4 Mechanical properties

Insight into the consequences of stretching and contracting a polymer can be obtained on the basis of the freely jointed chain as a model.

(a) Conformational entropy

A random coil is the least structured conformation of a polymer chain and therefore corresponds to the state of greatest entropy. Any stretching of the coil reduces disorder and reduces the entropy. Conversely, the formation of a random coil from a more extended form is spontaneous (provided enthalpy contributions do not interfere). The same model can be used to deduce an expression for the change in **conformational entropy**, the statistical entropy arising from the arrangement of bonds, when a one-dimensional chain is stretched or compressed.

How is that done? 14D.3 Deriving an expression for the conformational entropy of a freely jointed chain

Consider a freely jointed one-dimensional chain containing N units of length l that is stretched or compressed through a distance x. You then need to use the Boltzmann formula (eqn 13E.7, $S = k \ln W$) to calculate the conformational entropy of the chain, which involves assessing the value of W, the number of ways of achieving a particular conformation.

Step 1 *Calculate W*

To achieve an extension, the number of steps to the right (N_{R}) must be greater than the number to the left (N_{L}), so with $N_{\mathrm{L}} + N_{\mathrm{R}} = N$ write $N_{\mathrm{R}} - N_{\mathrm{L}} = \lambda N$, with λ between -1 (all to the left) and 1 (all to the right). Then $N_{\mathrm{R}} = \frac{1}{2}(1 + \lambda)N$ and $N_{\mathrm{L}} = \frac{1}{2}(1 - \lambda)N$ and the distance stretched is $x = \lambda Nl$, or λR_{c}. The number of ways of taking these numbers of steps (as in the earlier discussion of the random coil) is

$$W = \frac{N!}{N_{\mathrm{R}}!N_{\mathrm{L}}!} = \frac{N!}{\{\frac{1}{2}(1+\lambda)N\}!\{\frac{1}{2}(1-\lambda)N\}!}$$

Step 2 *Write an expression for S*

It follows from the expression for W and the Boltzmann formula that

$$S/k = \ln N! - \ln\{\tfrac{1}{2}(1+\lambda)N\}! - \ln\{\tfrac{1}{2}(1-\lambda)N\}!$$

Because the factorials are large (except for large extensions), use Stirling's approximation in the form $\ln x! \approx (x+\frac{1}{2})\ln x - x + \frac{1}{2}\ln(2\pi)$ to obtain

$$S/k = -\ln(2\pi)^{1/2} + (N+1)\ln 2$$
$$+ (N+\tfrac{1}{2})\ln N - \tfrac{1}{2}\ln\{N^{2N+2}(1+\lambda)^{N(1+\lambda)+1}(1+\lambda)^{N(1-\lambda)+1}\}$$

Step 3 *Write an expression for the change in entropy*

When the coil is not extended, and adopts its most random conformation ($\lambda = 0$), the entropy is

$$S/k = -\ln(2\pi)^{1/2} + (N+1)\ln 2 - \tfrac{1}{2}\ln N$$

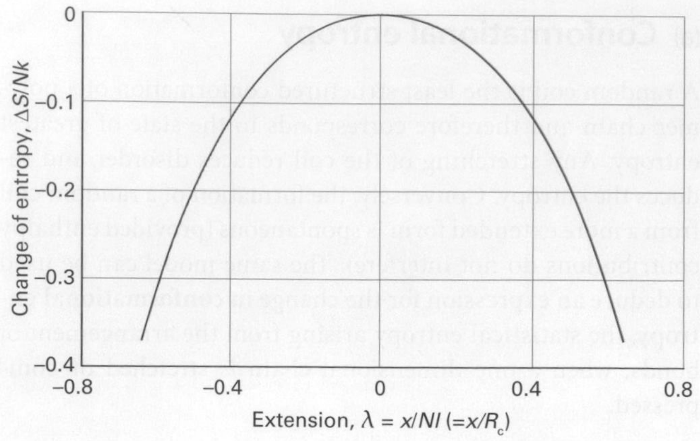

Figure 14D.8 The change in entropy of a perfect elastomer as its extension changes; $\lambda = \pm 1$ corresponds to complete extension in either direction; $\lambda = 0$, the conformation of highest entropy, corresponds to a random coil.

The change in entropy when the chain is stretched or compressed by the distance λR_c is therefore the difference between this quantity and that from Step 2. The resulting expression, after some algebraic manipulation and using $N >> 1$, is

$$\Delta S = -\tfrac{1}{2}kN \ln\{(1+\lambda)^{1+\lambda}(1-\lambda)^{1-\lambda}\} \text{ with } \lambda = x/R_c \qquad (14D.11)$$

Conformational entropy change [1D random coil]

This function is plotted in Fig. 14D.8, and it is seen that minimum extension corresponds to maximum entropy.

Brief illustration 14D.4

Suppose that $N = 1000$ and $l = 150\,\text{pm}$, so $R_c = 150\,\text{nm}$. The change in entropy when the (one-dimensional) random coil is stretched through 1.5 nm (corresponding to $\lambda = 1/100$) is

$$\Delta S = -\tfrac{1}{2}k\times(1000)\times\ln\left\{\left(1+\frac{1}{100}\right)^{1+(1/100)}\left(1-\frac{1}{100}\right)^{1-(1/100)}\right\}$$
$$= -0.050k$$

Because $R = N_A k$, the change in molar entropy is $\square S_m = -0.050R$, or $-0.42\,\text{J}\,\text{K}^{-1}\,\text{mol}^{-1}$.

(b) Elastomers

An **elastomer** is a flexible polymer that can expand or contract easily upon application of an external force. Elastomers are polymers with numerous crosslinks that pull them back into their original shape when a stress is removed. The weak directional constraints on silicon–oxygen bonds are responsible for the high elasticity of silicones. Even a freely jointed chain behaves as an elastomer for small extensions. It is a model of a 'perfect elastomer', a polymer in which the internal energy is independent of the extension, and can be used to deduce the restoring force associated with stretching or compression of the chain.

How is that done? 14D.4 Deriving an expression for the restoring force of a perfect elastomer

Your goal is to find an expression for the restoring force, F, of an elastomer, modelled as a one-dimensional random coil composed of N units each of length l, when the chain is stretched or compressed by a distance $x = vl$.

Step 1 *Use thermodynamics to relate the restoring force to the entropy*

The work done on an elastomer when it is extended reversibly through a distance dx is Fdx, The change in internal energy, from $dU = dw_{rev} + dq_{rev}$ with $dq_{rev} = TdS$ is therefore $dU = Fdx + TdS$. It follows that for an isothermal extension

$$\left(\frac{\partial U}{\partial x}\right)_T = F + T\left(\frac{\partial S}{\partial x}\right)_T$$

In a perfect elastomer, as in a perfect gas, the internal energy is independent of the dimensions (at constant temperature), so $(\partial U/\partial x)_T = 0$. The restoring force is therefore

$$F = -T\left(\frac{\partial S}{\partial x}\right)_T$$

Step 2 *Evaluate the force from the change in conformational entropy*

The conformational entropy (eqn 14D.11) is expressed in terms of the parameter λ used to express the extension x as $x = \lambda Nl$ (or $x = \lambda R_c$). Therefore, replace the derivative with respect to x by the derivative with respect to λ by noting that $dx = Nl d\lambda$. Then

$$F = -\frac{T}{Nl}\left(\frac{\partial S}{\partial \lambda}\right)_T = -\frac{T}{Nl}\left(\frac{\partial \Delta S}{\partial \lambda}\right)_T$$

The replacement of S by the change ΔS is valid because the initial value of the entropy is independent of the extension being applied. Now use eqn 14D.11 to obtain

$$F = \frac{T}{Nl}\times\frac{Nk}{2}\times\frac{d}{d\lambda}\ln\{(1+\lambda)^{1+\lambda}(1-\lambda)^{1-\lambda}\}$$
$$= \frac{kT}{2l}\frac{d}{d\lambda}\{(1+\lambda)\ln(1+\lambda)+(1-\lambda)\ln(1-\lambda)\}$$
$$= \frac{kT}{2l}\{\ln(1+\lambda)-\ln(1-\lambda)\}$$

That is,

$$F = \frac{kT}{2l}\ln\left(\frac{1+\lambda}{1-\lambda}\right) \qquad \lambda = \frac{x}{Nl} \qquad (14D.12a)$$

Restoring force [1D random coil]

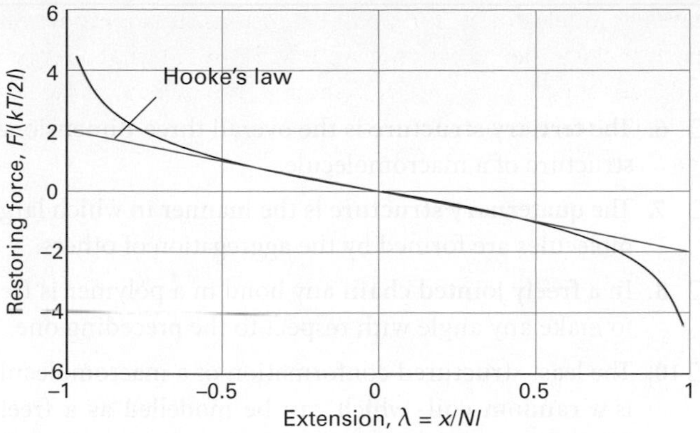

Figure 14D.9 The restoring force, F, of a one-dimensional perfect elastomer. For small extensions, F is proportional to the extension, corresponding to Hooke's law.

For small displacements ($\lambda << 1$, corresponding to $x << Nl$ and therefore $x << R_c$) the logarithms can be expanded by using $\ln(1+\lambda) \approx \lambda$ and $\ln(1-\lambda) \approx -\lambda$, to give

$$F \approx \frac{\lambda kT}{l} = \frac{kT}{Nl^2}x \quad \text{Restoring force [1D random coil]} \quad (14D.12b)$$

That is, for small displacements the sample obeys Hooke's law (Fig. 14D.9): the restoring force is proportional to the displacement and the force constant k_f (the constant of proportionality between the force and the displacement) is

$$k_f = \frac{kT}{Nl^2} \quad (14D.12c)$$

Brief illustration 14D.5

Consider a polymer chain with $N = 5000$ and $l = 0.15$ nm. If the ends of the chain are moved apart by $x = 1.5$ nm, then $\lambda = (1.5\,\text{nm})/(5000 \times 0.15\,\text{nm}) = 2.0 \times 10^{-3}$. Because $\lambda << 1$, the restoring force at 293 K is given by eqn 14D.12b as

$$F = \frac{(1.381\times10^{-23}\,\overbrace{\text{J}}^{\text{N m}}\,\text{K}^{-1})\times(293\,\text{K})}{5000\times(1.5\times10^{-10}\,\text{m})^2}\times1.5\times10^{-9}\,\text{m} = 5.4\times10^{-14}\,\text{N}$$

or 54 fN.

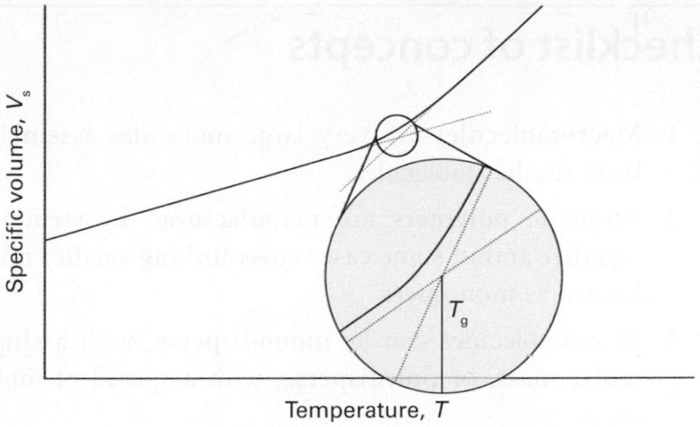

Figure 14D.10 The variation of specific volume with temperature of a synthetic polymer. The glass transition temperature, T_g, is at the point of intersection of extrapolations of the two linear parts of the curve.

14D.5 Thermal properties

The crystallinity of synthetic polymers can be destroyed by thermal motion at sufficiently high temperatures. This change in crystallinity may be thought of as a kind of intramolecular melting from a crystalline solid to a more fluid random coil. Polymer melting also occurs at a specific **melting temperature**, T_m, which increases with the strength and number of intermolecular interactions in the material. Thus, polyethene, which has chains that interact only weakly in the solid, has $T_m = 414$ K and nylon-66 fibres, in which there are strong hydrogen bonds between chains, have $T_m = 530$ K. High melting temperatures are desirable in most practical applications involving fibres and plastics.

All synthetic polymers undergo a transition from a state of high to low chain mobility at the **glass transition temperature**, T_g. To visualize the glass transition, consider what happens to an elastomer as its temperature is lowered. There is sufficient energy available at normal temperatures for limited bond rotation to occur and the flexible chains writhe. At lower temperatures, the amplitudes of the writhing motion decrease until a specific temperature, T_g, is reached at which motion is frozen completely and the sample forms a glass. Glass transition temperatures well below 300 K are desirable in elastomers that are to be used at normal temperatures. Both the glass transition temperature and the melting temperature of a polymer may be measured by calorimetric methods. Because the motion of the segments of a polymer chain increase at the glass transition temperature, T_g may also be determined from a plot of the specific volume of a polymer (the reciprocal of its mass density) against temperature (Fig. 14D.10).

Checklist of concepts

☐ 1. **Macromolecules** are very large molecules assembled from smaller molecules.

☐ 2. Synthetic **polymers** are manufactured by stringing together and in some cases cross-linking smaller units known as **monomers**.

☐ 3. Macromolecules can be **monodisperse**, with a single molar mass, or **polydisperse**, with a spread of molar mass.

☐ 4. The **conformation** of a macromolecule is the spatial arrangement of the different parts of a chain.

☐ 5. The **primary structure** of a macromolecule is the sequence of small molecular residues making up the polymer.

☐ 5. The **secondary structure** is the spatial arrangement of a chain of residues.

☐ 6. The **tertiary structure** is the overall three-dimensional structure of a macromolecule.

☐ 7. The **quaternary structure** is the manner in which large molecules are formed by the aggregation of others.

☐ 8. In a **freely jointed chain** any bond in a polymer is free to make any angle with respect to the preceding one.

☐ 10. The least structured conformation of a macromolecule is a **random coil**, which can be modelled as a freely jointed chain.

☐ 11. An **elastomer** is a flexible polymer that can expand or contract easily upon application of an external force.

☐ 12. The disruption of long-range order in a polymer occurs at a **melting temperature**.

☐ 13. Synthetic polymers undergo a transition from a state of high to low chain mobility at the **glass transition temperature**.

Checklist of equations

Property	Equation	Comment	Equation number
Number-average molar mass	$\bar{M}_n = \dfrac{1}{N_{\text{total}}}\sum_i N_i M_i$	Definition	14D.1a
Weight-average molar mass	$\bar{M}_w = \dfrac{1}{m_{\text{total}}}\sum_i m_i M_i$	Definition	14D.1b
Dispersity	$Đ = \bar{M}_w / \bar{M}_n$	Definition	14D.2
Probability distribution	$P = (2/\pi N)^{1/2} e^{-n^2/2N}$	One-dimensional random coil	14D.3
	$f(r) = 4\pi(a/\pi^{1/2})^3 r^2 e^{-a^2r^2}$ $a = (3/2Nl^2)^{1/2}$	Three-dimensional random coil	14D.4
Contour length of a random coil	$R_c = Nl$		14D.5
Root-mean-square separation of a random coil	$R_{\text{rms}} = N^{1/2}l$	Unconstrained chain	14D.6
Radius of gyration of a random coil	$R_g = N^{1/2}l$	Unconstrained one-dimensional chain	14D.7a
	$R_g = (N/6)^{1/2}l$	Unconstrained three-dimensional chain	14D.7b
Root-mean-square separation of a random coil	$R_{\text{rms}} = (2N)^{1/2}l$	Constrained tetrahedral chain	14D.9
Change in conformational entropy on extending a random coil	$\Delta S = -\tfrac{1}{2}kN \ln\{(1+\lambda)^{1+\lambda}(1-\lambda)^{1-\lambda}\}$		14D.11
Restoring force of a one-dimensional random coil	$F = (kT/2l)\ln\{(1+\lambda)/(1-\lambda)\}$		14D.12a
	$F \approx (kT/Nl^2)x$	$x \ll R_c$	14D.12b

TOPIC 14E Self-assembly

➤ **Why do you need to know this material?**

Aggregates of small and large molecules form the basis of many established and emerging technologies. To see why this is the case, you need to understand their structures and properties.

➤ **What is the key idea?**

Colloids and micelles form spontaneously by self-assembly of molecules or macromolecules and are held together by molecular interactions.

➤ **What do you need to know already?**

You need to be familiar with molecular interactions (Topic 14B) and interactions between ions (Topic 5E).

Self-assembly is the spontaneous formation of complex structures of molecules or macromolecules that are held together by molecular interactions, such as Coulombic, dispersion, hydrogen bonding, and hydrophobic interactions. Examples of self-assembly include the formation of liquid crystals, and of protein quaternary structures from two or more polypeptide chains (Topic 14C).

14E.1 Colloids

A **colloid**, or **disperse phase**, is a dispersion of small particles of one material in another that does not settle out under gravity. In this context, 'small' means that one dimension at least is smaller than about 500 nm (about the wavelength of visible light). Many colloids are suspensions of nanoparticles (particles of diameter up to about 100 nm). In general, colloidal particles are aggregates of numerous atoms or molecules, but are commonly but not universally too small to be seen with an ordinary optical microscope.

(a) Classification and preparation

The name given to the colloid depends on the two phases involved:

- A **sol** is a dispersion of a solid in a liquid (such as clusters of gold atoms in water) or of a solid in a solid (such as ruby glass, which is a gold-in-glass sol, and achieves its colour by light scattering).
- An **aerosol** is a dispersion of a liquid in a gas (like fog and many sprays) or a solid in a gas (such as smoke): the particles are often large enough to be seen with a microscope.
- An **emulsion** is a dispersion of a liquid in a liquid (such as milk).
- A **foam** is a dispersion of a gas in a liquid.

A further classification of colloids is as **lyophilic**, or solvent attracting, and **lyophobic**, solvent repelling. If the solvent is water, the terms **hydrophilic** and **hydrophobic**, respectively, are used instead. Lyophobic colloids include the metal sols. Lyophilic colloids generally have some chemical similarity to the solvent, such as –OH groups able to form hydrogen bonds. A **gel** is a semi-rigid mass of a lyophilic sol.

The preparation of aerosols can be as simple as sneezing (which produces an imperfect aerosol). Laboratory and commercial methods make use of several techniques. Material (e.g. quartz) may be ground in the presence of the dispersion medium. Passing a heavy electric current through a cell may lead to the sputtering (crumbling) of an electrode into colloidal particles. Arcing between electrodes immersed in the support medium also produces a colloid. Chemical precipitation sometimes results in a colloid. A precipitate (e.g. silver iodide) already formed may be dispersed by the addition of a 'peptizing agent' (e.g. potassium iodide). Clays may be peptized by alkalis, the OH^- ion being the active agent.

Emulsions are normally prepared by shaking the two components together vigorously, although some kind of emulsifying agent usually has to be added to stabilize the product. This emulsifying agent may be a soap (the salt of a long-chain carboxylic acid) or other **surfactant** (surface active) species, or a lyophilic sol that forms a protective film around the dispersed phase. In milk, which is an emulsion of fats in water, the emulsifying agent is casein, a protein containing phosphate groups. It is clear from the formation of cream on the surface of milk that casein is not completely successful in stabilizing milk: the dispersed fats coalesce into oily droplets which float to the surface. This coagulation may be prevented by ensuring that the emulsion is dispersed very finely initially: intense agitation with ultrasonics brings this dispersion about, the product being 'homogenized' milk.

One way to form an aerosol is to tear apart a spray of liquid with a jet of gas. The dispersal is aided if a charge is applied to the liquid, for then electrostatic repulsions help to blast it apart into droplets. This procedure may also be used to produce emulsions, for the charged liquid phase may be directed into another liquid.

Colloids are often purified by **dialysis**, the process of squeezing the solution though a membrane. The aim is to remove much (but not all, for reasons explained later) of the ionic material that may have accompanied their formation. A membrane (for example, cellulose) is selected that is permeable to solvent and ions, but not to the colloid particles. Dialysis is very slow, and is normally accelerated by applying an electric field and making use of the charges carried by many colloidal particles; the technique is then called **electrodialysis**.

(b) Structure and stability

Colloids are thermodynamically unstable with respect to the bulk. This instability can be expressed thermodynamically by noting that because the change in Helmholtz energy, $\mathrm{d}A$, when the surface area of the sample changes by $\mathrm{d}\sigma$ at constant temperature and pressure is $\mathrm{d}A = \gamma\mathrm{d}\sigma$, where γ is the interfacial surface tension (Topic 14C), it follows that $\mathrm{d}A < 0$ if $\mathrm{d}\sigma < 0$. That is, the contraction of the surface ($\mathrm{d}\sigma < 0$) is spontaneous ($\mathrm{d}A < 0$). The survival of colloids must therefore be a consequence of the kinetics of collapse: colloids are thermodynamically unstable but kinetically non-labile.

At first sight, even the kinetic argument seems to fail: colloidal particles attract each other over large distances, so there is a long-range force that tends to condense them into a single blob. The reasoning behind this remark is as follows. The energy of attraction between two individual atoms i and j separated by a distance R_{ij}, one in each colloidal particle, varies with their separation as $1/R_{ij}^6$ (Topic 14B). The sum of all these pairwise interactions, however, decreases only as approximately $1/R^2$ (the precise variation depending on the shape of the particles and their closeness), where R is the separation of the centres of the particles. The change in the power from 6 to 2 stems from the fact that at short distances only a few molecules interact but at large distances many individual molecules are at about the same distance from one another, and contribute equally to the sum (Fig. 14E.1), so the total interaction does not fall off as fast as the single molecule–molecule interaction.

Several factors oppose the long-range dispersion attraction. For example, there may be a protective film at the surface of the colloid particles that stabilizes the interface and cannot be penetrated when two particles touch. Thus, the surface atoms of a platinum sol in water react chemically and are turned into $-\mathrm{Pt(OH)_3H_3}$; this layer encases the particle like a shell. A fat can be emulsified by a soap because the long hydrocarbon tails penetrate the oil droplet but the carboxylate head groups (or other hydrophilic groups in synthetic detergents) surround the surface, form hydrogen bonds with water, and give rise to a shell of negative charge that repels a possible approach from another similarly charged particle.

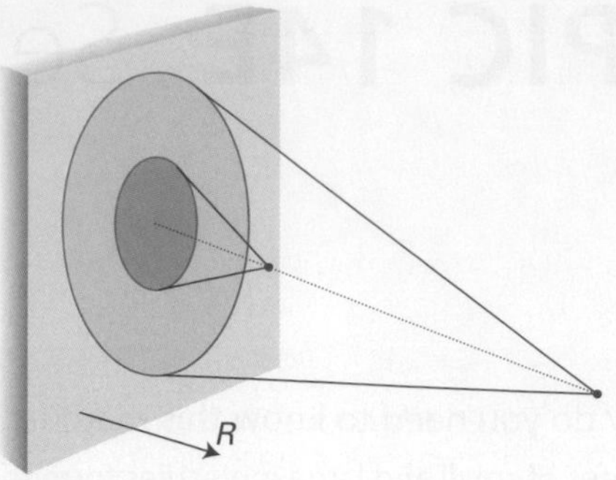

Figure 14E.1 Although the attraction between individual molecules is proportional to $1/R^6$, more molecules are within range at large separations (pale region) than at small separation (dark region), so the total interaction energy declines more slowly and is proportional to a lower power of $1/R$.

(c) The electrical double layer

A major source of kinetic non-lability of colloids is the existence of an electric charge on the surfaces of the particles. Ions of opposite charge tend to cluster near each other, and form an ionic atmosphere around the particles, just as for individual ions (Topic 5F).

There are two regions of charge. First, there is a fairly immobile layer of ions that adhere tightly to the surface of the colloidal particle, and which may include water molecules (if that is the support medium). The radius of the sphere that captures this rigid layer is called the **radius of shear** and is the major factor determining the mobility of the particles. The electric potential at the radius of shear relative to its value in the distant, bulk medium is called the **electrokinetic potential**, ζ (or the *zeta potential*). Second, the charged unit attracts an oppositely charged atmosphere of mobile ions. The inner shell of charge and the outer ionic atmosphere constitute the **electrical double layer**.

The theory of the stability of lyophobic dispersions was developed by B. Derjaguin and L. Landau and independently by E. Verwey and J.T.G. Overbeek, and is known as the **DLVO theory**. It assumes that there is a balance between the repulsive interaction between the charges of the electrical double layers on neighbouring particles and the attractive interactions arising from van der Waals interactions between the molecules in the colloidal particles. The potential energy arising from the repulsion of double layers on particles of radius a has the form

$$V_{\text{repulsion}} = +\frac{Aa^2\zeta^2}{R}\mathrm{e}^{-s/r_\mathrm{D}} \qquad (14\text{E}.1)$$

where A is a constant, ζ is the zeta potential, R is the separation of centres, s is the separation of the surfaces of the two particles ($s = R - 2a$ for spherical particles of radius a), and r_D is the thickness of the double layer. This expression is valid for small particles with a thick double layer ($r_D >> a$). When the double layer is thin ($r_D << a$), the expression is replaced by

$$V_{\text{repulsion}} = +\tfrac{1}{2}Ba^2\zeta^2 \ln(1+e^{-s/r_D}) \tag{14E.2}$$

where B is another constant. In each case, the thickness of the double layer can be estimated from an expression like that derived for the thickness of the ionic atmosphere in the Debye–Hückel theory (Topic 5F and *A deeper look* 1 on the website for this text) in which there is a competition between the assembling influences of the attraction between opposite charges and the disruptive effect of thermal motion:

$$r_D = \left(\frac{\varepsilon RT}{2\rho F^2 I b^{\ominus}}\right)^{1/2} \quad \text{Thickness of the electrical double layer} \tag{14E.3}$$

where I is the ionic strength of the solution (eqn 5F.28, $I = \tfrac{1}{2}\sum_i z_i^2 b_i / b^{\ominus}$ with $b^{\ominus} = 1\,\text{mol kg}^{-1}$) and ρ its mass density. As usual, F is Faraday's constant and ε is the permittivity, $\varepsilon = \varepsilon_r\varepsilon_0$. The potential energy arising from the attractive interaction has the form

$$V_{\text{attraction}} = -\frac{C}{s} \tag{14E.4}$$

where C is yet another constant. The variation of the total potential energy with separation is shown in Fig. 14E.2.

At high ionic strengths, the ionic atmosphere is dense and the potential shows a secondary minimum at large separations. Aggregation of the particles arising from the stabilizing effect of this secondary minimum is called **flocculation**. The flocculated material can often be redispersed by agitation because the well is so shallow. **Coagulation**, the irreversible aggregation of distinct particles into large particles, occurs when the separation of the particles is so small that they enter the primary minimum of the potential energy curve and van der Waals forces are dominant.

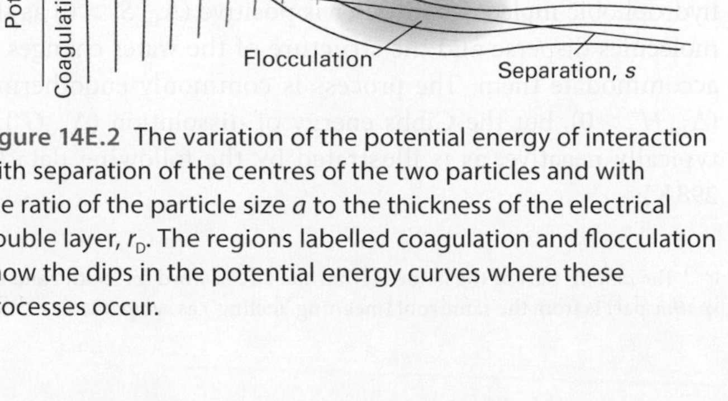

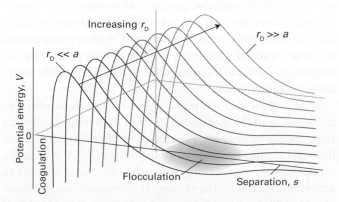

Figure 14E.2 The variation of the potential energy of interaction with separation of the centres of the two particles and with the ratio of the particle size a to the thickness of the electrical double layer, r_D. The regions labelled coagulation and flocculation show the dips in the potential energy curves where these processes occur.

The ionic strength is increased by the addition of ions, particularly those of high charge type, so such ions act as flocculating agents. This increase is the basis of the empirical **Schulze–Hardy rule**, that hydrophobic colloids are flocculated most efficiently by ions of opposite charge type and high charge number. The Al^{3+} ions in alum are very effective, and are used to induce the congealing of blood. When river water containing colloidal clay flows into the sea, the salt water induces flocculation and coagulation, and is a major cause of silting in estuaries.

Metal oxide sols tend to be positively charged whereas sulfur and the noble metals tend to be negatively charged. Naturally occurring macromolecules also acquire a charge when dispersed in water, and an important feature of proteins and other natural macromolecules is that their overall charge depends on the pH of the medium. For instance, in acidic environments protons attach to basic groups, and the net charge of the macromolecule is positive; in basic media the net charge is negative as a result of proton loss. At the **isoelectric point** the pH is such that there is no net charge on the macromolecule.

Example 14E.1 Determining the isoelectric point of a protein

The velocity with which the protein bovine serum albumin (BSA) moves through water under the influence of an electric field was monitored at several values of pH, and the data are listed below. What is the isoelectric point of the protein?

pH	4.20	4.56	5.20	5.65	6.30	7.00
Velocity/(μm s^{-1})	0.50	0.18	−0.25	−0.65	−0.90	−1.25

Collect your thoughts Plot velocity against pH, then use interpolation to find the pH at which the velocity is zero, which is the pH at which the molecule has zero net charge.

The solution The data are plotted in Fig.14E.3. The velocity passes through zero at pH = 4.8; hence pH = 4.8 is the isoelectric point.

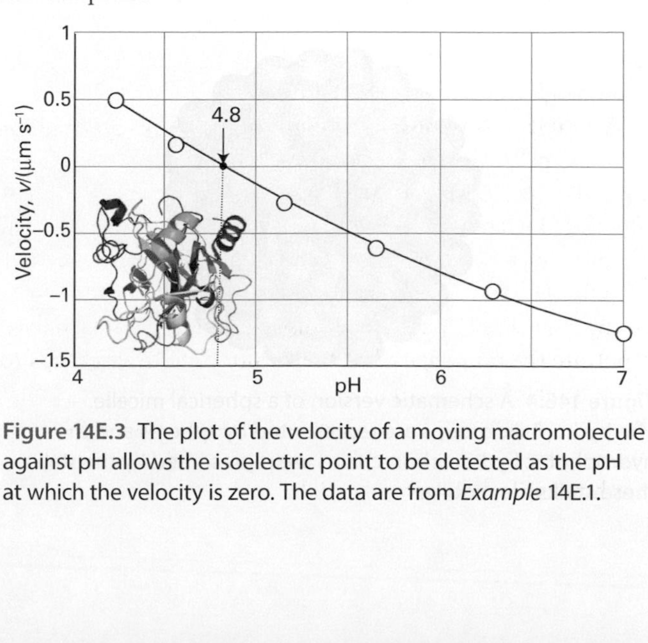

Figure 14E.3 The plot of the velocity of a moving macromolecule against pH allows the isoelectric point to be detected as the pH at which the velocity is zero. The data are from *Example* 14E.1.

Self-test 14E.1 The following data were obtained for another protein:

pH	3.5	4.5	5.0	5.5	6.0
Velocity/(μm s^{-1})	0.10	−0.10	−0.20	−0.30	−0.40

Estimate the pH of the isoelectric point.

Answer: 4.0

The primary role of the electrical double layer is to confer kinetic non-lability. Colliding colloidal particles break through the double layer and coalesce only if the collision is sufficiently energetic to disrupt the layers of ions and solvating molecules, or if thermal motion has stirred away the surface accumulation of charge. This disruption may occur at high temperatures, which is one reason why sols precipitate when they are heated.

14E.2 Micelles and biological membranes

In aqueous solutions surfactant molecules or ions can cluster together as **micelles**, which are colloid-sized clusters of molecules, for their hydrophobic tails tend to congregate, and their hydrophilic head groups provide protection (Fig. 14E.4).

(a) The hydrophobic interaction

Consider a long-chained alcohol, such as pentan-1-ol ($CH_3CH_2CH_2CH_2CH_2OH$). The hydrocarbon chain is hydrophobic and the –OH group is hydrophilic. A species with both

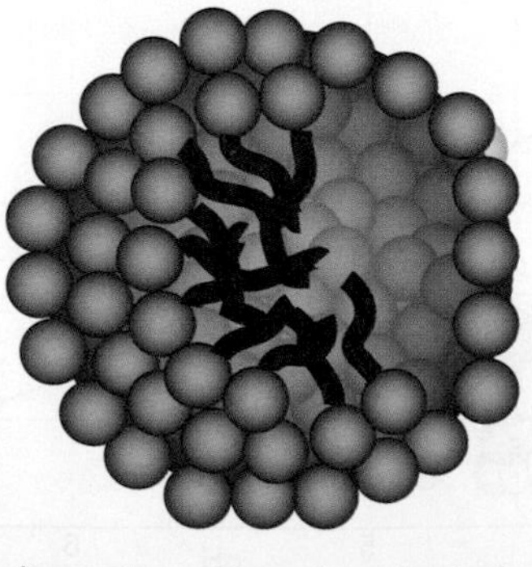

Figure 14E.4 A schematic version of a spherical micelle. The hydrophilic groups are represented by spheres and the hydrophobic hydrocarbon chains are represented by the stalks; these stalks are mobile.

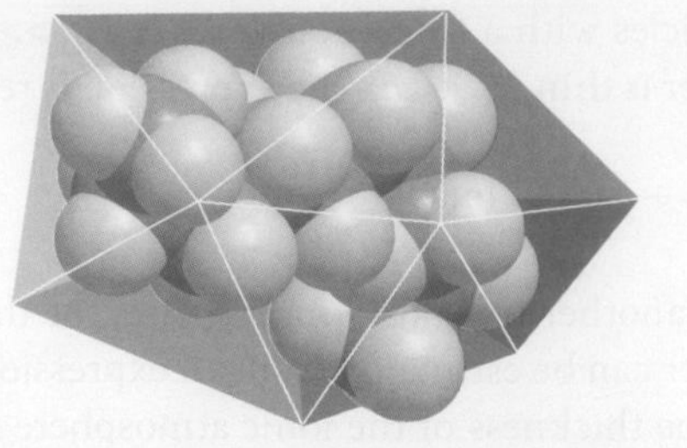

Figure 14E.5 When a hydrocarbon molecule is surrounded by water, the H_2O molecules form a cage. As a result of this acquisition of structure, the entropy of the water decreases, so the dispersal of the hydrocarbon into the water is accompanied by a local decrease in entropy. However, the aggregation of these individual caged hydrocarbon molecules into a micelle releases many of the caging water molecules back into the bulk and results in an increase in entropy.

hydrophobic and hydrophilic regions is called **amphipathic.**[1] Amphipathic substances do dissolve slightly in water, and an understanding of the process gives insight into the formation of micelles and biological structures in general.

To understand the dissolution process in more detail, imagine a hypothetical initial state in which the alcohol is present in water as individual molecules. Each hydrophobic chain is surrounded by a cage of water molecules (Fig. 14E.5). This order reduces the entropy of the water below its 'pure' value. Now consider the final state, in which the hydrophobic chains have clustered together. Although the clustering contributes to the lowering of the entropy of the system, fewer (but larger) cages are required, and more water molecules are free to move. The net effect of the formation of clusters of hydrophobic chains is therefore a decrease in the organization of water molecules and therefore a *net increase* in entropy of the system. This increase in entropy of the solvent (water) means that the association of hydrophobic groups in an aqueous environment is spontaneous (provided there are no overwhelming enthalpy effects). This spontaneous clustering of hydrophobic groups in the presence of water *gives the appearance* of it being the outcome of an actual intermolecular force and is called the **hydrophobic interaction.**

Some insight into the processes involved can be obtained from studies of the thermodynamics of dissolving (as distinct from micelle formation). The entropy of dissolution of largely hydrophobic molecules in water is positive ($\Delta_{diss}S^{\ominus} > 0$) as the molecules disperse and the structure of the water changes to accommodate them. The process is commonly endothermic ($\Delta_{diss}H^{\ominus} > 0$), but the Gibbs energy of dissolution ($\Delta_{diss}G^{\ominus}$) is typically negative, as is illustrated by the following data (at 298 K):

[1] The *amphi-* part of the name is from the Greek word for 'both', and the *-pathic* part is from the same root (meaning 'feeling') as *sympathetic*.

	$\Delta_{diss}G^{\ominus}$/ (kJ mol^{-1})	$\Delta_{diss}H^{\ominus}$/ (kJ mol^{-1})	$\Delta_{diss}S^{\ominus}$/ (J K^{-1} mol^{-1})
$CH_3CH_2CH_2CH_2OH$	−10	+8	+61
$CH_3CH_2CH_2CH_2CH_2OH$	−13	+8	+70

In other words, the tendency to dissolve (at least to a small extent) is entropy-driven, with contributions from the dispersion of the solute molecules and the restructuring of the water. Once dissolved, further reorganization of the water occurs to drive the formation of micelles. The experimental values are consistent with a general rule that each additional $-CH_2-$ group contributes a further $-3\,kJ\,mol^{-1}$ to the standard Gibbs energy of dissolution.

A further aspect of this discussion is that it is possible to establish a scale of hydrophobicities. The hydrophobicity of a small molecular group R is reported by defining the **hydrophobicity constant**, π, as

$$\pi = \log \frac{s(RX)}{s(HX)} \quad \text{Hydrophobicity constant [definition]} \quad (14E.5)$$

where s(RX) is the ratio of the molar solubility of the hydrophobic compound RX in the largely hydrocarbon solvent octan-1-ol to that in water, and s(HX) is the analogous ratio for the compound HX. A positive value of π indicates that RX is more hydrophobic than RH.

It is found that the π values of most compounds do not depend on the identity of X (which might be OH, NH_2, and so on). However, measurements suggest that π increases by the same amount each time a CH_2 group is added:

−R	$-CH_3$	$-CH_2CH_3$	$-(CH_2)_2CH_3$	$-(CH_2)_3CH_3$	$-(CH_2)_4CH_3$
π	0.5	1.0	1.5	2.0	2.5

It follows that acyclic saturated hydrocarbons become more hydrophobic as the carbon chain length increases. This trend can be rationalized by noting that $\Delta_{diss}G^{\ominus}$ becomes more negative as the number of carbon atoms in the chain increases, with the data on butan-1-ol and pentan-1-ol (see above) suggesting that the principal effect is due to the entropy.

(b) Micelle formation

Micelles form only above the **critical micelle concentration** (CMC) and above the **Krafft temperature**. The CMC is detected by noting a pronounced change in physical properties of the solution, particularly the molar conductivity (Fig. 14E.6). There is no abrupt change in some properties at the CMC; rather, there is a transition region corresponding to a range of concentrations around the CMC where physical properties vary smoothly but nonlinearly with the concentration. The hydrocarbon interior of a micelle is like a droplet of oil. Nuclear magnetic resonance shows that the hydrocarbon tails are mobile, but slightly more restricted than in the bulk. Micelles are important in industry and biology on account of their solubilizing function: matter can be transported by water after it has been dissolved in their hydrocarbon interiors. For this reason, micellar systems are used as detergents, for organic synthesis, froth flotation for the treatment of ores, and petroleum recovery.

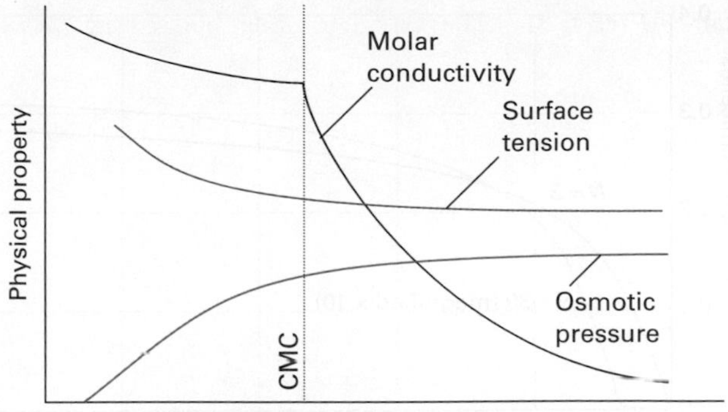

Figure 14E.6 The typical variation of some physical properties of an aqueous solution of sodium dodecyl sulfate close to the critical micelle concentration (CMC).

The self-assembly of a micelle has the characteristics of a cooperative process in which the addition of a surfactant molecule to a cluster that is forming becomes more probable the larger the size of the aggregate, so after a slow start there is a cascade of formation of micelles. If it is supposed that the dominant micelle M_N consists of N monomers M, then the dominant equilibrium to consider is

$$NM \rightleftharpoons M_N \qquad K = \frac{[M_N]/c^{\ominus}}{([M]/c^{\ominus})^N} \quad (14E.6a)$$

where it has been assumed, probably dangerously on account of the large sizes of monomers, that the solution is ideal and that activities can be replaced by molar concentrations. The total concentration of surfactant, $[M]_{total}$, is $[M] + N[M_N]$ because each micelle consists of N monomer molecules. Therefore (omitting the $c^{\ominus}$ for clarity),

$$K = \frac{[M_N]}{([M]_{total} - N[M_N])^N} \quad (14E.6b)$$

Brief illustration 14E.1

Equation 14E.6b can be solved numerically for the variation of the fraction of molecules present as micelles with the number of molecules present in a micelle and some results for $K = 1$ are shown in Fig. 14E.7. For large N, there is a reasonably sharp transition in the fractions of surfactant molecules that are present in micelles, which corresponds to the existence of a CMC.

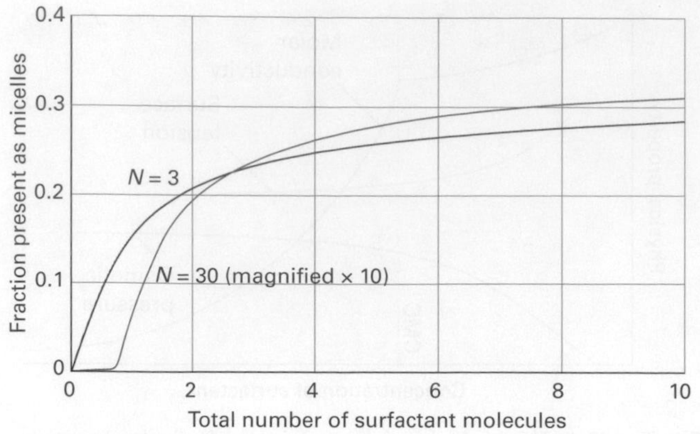

Figure 14E.7 The dependence of the fraction of surfactant molecules present as micelles on the number of molecules in the micelle for $K = 1$.

Non-ionic surfactant molecules may cluster together in clumps of 1000 or more, but ionic species tend to be disrupted by the electrostatic repulsions between head groups and are normally limited to groups of less than about 100. However, the disruptive effect depends more on the effective size of the head group than the charge. For example, ionic surfactants such as sodium dodecyl sulfate (SDS) and cetyltrimethylammonium bromide (CTAB) form rods at moderate concentrations whereas sugar surfactants form small, approximately spherical micelles. The micelle population commonly spans a wide range of particle sizes (i.e. it is polydisperse), and the shapes of the individual micelles vary with shape of the constituent surfactant molecules, surfactant concentration, and temperature. A useful predictor of the shape of the micelle is the **surfactant parameter**, N_s, defined as

$$N_s = \frac{V}{Al} \qquad \text{Surfactant parameter [definition]} \qquad (14E.7)$$

where V is the volume of the hydrophobic surfactant tail, A is the area of the hydrophilic surfactant head group, and l is the maximum length of the surfactant tail. Table 14E.1 summarizes the dependence of aggregate structure on the surfactant parameter.

In aqueous solutions spherical micelles form, as shown in Fig. 14E.4, with the polar head groups of the surfactant molecules on the micellar surface and interacting favourably with solvent and ions in solution. Hydrophobic interactions stabilize the aggregation of the hydrophobic surfactant tails in the micellar core. Under certain experimental conditions, a **liposome** may form, with an inward pointing inner surface of molecules surrounded by an outward pointing outer layer (Fig. 14E.8). Liposomes may be used to carry nonpolar drug molecules in blood.

Table 14E.1 Micelle shape and the surfactant parameter

N_s	Micelle shape
<0.33	Spherical
0.33–0.50	Cylindrical rods
0.50–1.00	Vesicles
1.00	Planar bilayers
>1.00	Reverse micelles and other shapes

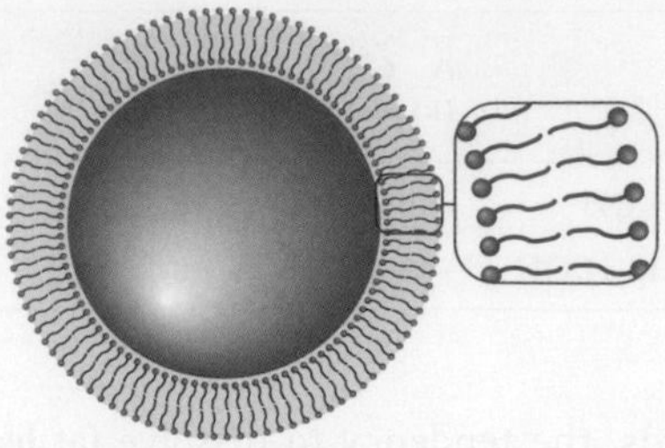

Figure 14E.8 The cross-sectional structure of a spherical liposome.

Increasing the ionic strength of the aqueous solution reduces repulsions between surface head groups, and cylindrical micelles can form. These cylinders may stack together in reasonably close-packed (hexagonal) arrays, forming **lyotropic mesomorphs** and, more colloquially, 'liquid crystalline phases'.

Reverse micelles form in nonpolar solvents, with small polar surfactant head groups in a micellar core and more voluminous hydrophobic surfactant tails extending into the organic bulk phase. These spherical aggregates can solubilize water in organic solvents by creating a pool of trapped water molecules in the micellar core. As aggregates arrange at high surfactant concentrations to yield long-range positional order, many other types of structures are possible including cubic and hexagonal shapes.

As already noted, micelle formation is driven by hydrophobic interactions. The enthalpy of formation reflects contributions of interactions between micelle chains within the micelles and between the polar head groups and the surrounding medium. Consequently, enthalpies of micelle formation display no readily discernible pattern and may be positive (endothermic) or negative (exothermic). Many non-ionic micelles form endothermically, with ΔH of the order of 10 kJ per mole of surfactant molecules. That such micelles do form above the CMC indicates that the entropy change accompanying their formation must then be positive, and measurements suggest a value of about $+140\,\mathrm{J\,K^{-1}\,mol^{-1}}$ at room temperature.

(c) Bilayers, vesicles, and membranes

Some micelles at concentrations well above the CMC form extended parallel sheets two molecules thick, called **planar bilayers**. The individual molecules lie perpendicular to the sheets, with hydrophilic groups on the outside in aqueous solution and on the inside in nonpolar media. When segments of planar bilayers fold back on themselves, **unilamellar vesicles** may form where the spherical hydrophobic bilayer shell separates an inner aqueous compartment from the external aqueous environment.

Bilayers show a close resemblance to biological membranes, and are often a useful model on which to base investigations of biological structures. However, actual membranes are highly sophisticated structures. The basic structural element of a membrane is a phospholipid, such as phosphatidyl choline (**1**), which contains long hydrocarbon chains (typically in the range C_{14}–C_{24}) and a variety of polar groups, such as $-CH_2CH_2N(CH_3)_3^+$. The hydrophobic chains stack together to form an extensive layer about 5 nm across. The lipid molecules form layers instead of micelles because the hydrocarbon chains are too bulky to allow packing into nearly spherical clusters.

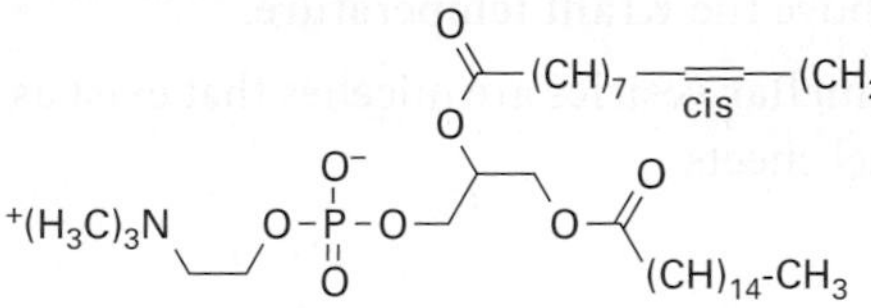

1 Phosphatidyl choline

The bilayer is a highly mobile structure. Not only are the hydrocarbon chains ceaselessly twisting and turning in the region between the polar groups, but the phospholipid molecules migrate over the surface. It is better to think of the membrane as a viscous fluid rather than a permanent structure, with a viscosity about 100 times that of water. Typically, a phospholipid molecule in a membrane migrates through about 1 μm in about 1 min.

All lipid bilayers undergo a transition from a state of high to low chain mobility at a temperature that depends on the structure of the lipid. To visualize the transition, consider what happens to a membrane as its temperature is lowered (Fig. 14E.9). There is sufficient energy available at normal temperatures for limited bond rotation to occur and the flexible chains writhe. However, the membrane is still highly organized in the sense that the bilayer structure does not come apart and the system is best described as a liquid crystal. At lower temperatures, the amplitudes of the writhing motion decrease until a specific temperature is reached at which motion is largely frozen. The membrane then exists as a gel. Biological membranes exist as liquid crystals at physiological temperatures.

Phase transitions in membranes are often observed as 'melting' from gel to liquid crystal by calorimetric methods. The data show relations between the structure of the lipid and the melting temperature. Interspersed among the phospholipids of biological membranes are sterols, such as cholesterol (**2**), which is largely hydrophobic but does contain a hydrophilic –OH group. Sterols, which are present in different proportions in different types of cells, prevent the hydrophobic chains of lipids from 'freezing' into a gel and, by disrupting the packing of the chains, spread the melting point of the membrane over a range of temperatures.

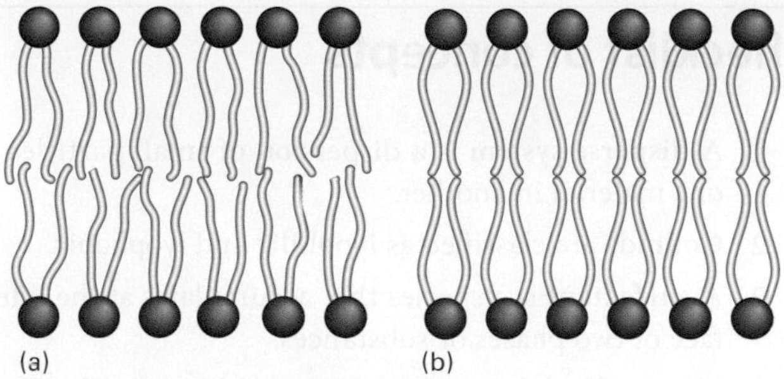

Figure 14E.9 A depiction of the variation with temperature of the flexibility of hydrocarbon chains in a lipid bilayer. (a) At physiological temperature, the bilayer exists as a liquid crystal, in which some order exists but the chains writhe. (b) At a specific temperature, the chains are largely frozen and the bilayer exists as a gel.

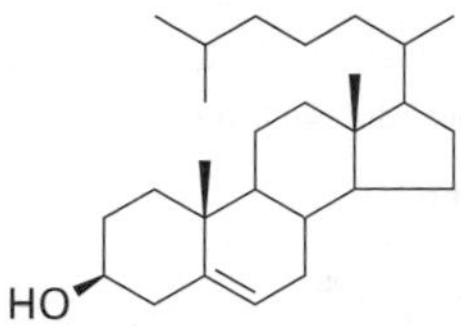

2 Cholesterol

Brief illustration 14E.2

To predict trends in melting temperatures you need to assess the strengths of the interactions between molecules. Longer chains can be expected to be held together more strongly by hydrophobic interactions than shorter chains, so you should expect the melting temperature to increase with the length of the hydrophobic chain of the lipid. On the other hand, any structural elements that prevent alignment of the hydrophobic chains in the gel phase lead to low melting temperatures. Indeed, lipids containing unsaturated chains, those containing some C=C bonds, form membranes with lower melting temperatures than those formed from lipids with fully saturated chains, those consisting of C–C bonds only.

Checklist of concepts

☐ 1. A **disperse system** is a dispersion of small particles of one material in another.

☐ 2. **Colloids** are classified as lyophilic and lyophobic.

☐ 3. A **surfactant** is a species that accumulates at the interface of two phases or substances.

☐ 4. Many colloidal particles are thermodynamically unstable but kinetically non-labile.

☐ 5. The **radius of shear** is the radius of the sphere that captures the rigid layer of charge attached to a colloid particle.

☐ 6. The **electrokinetic potential** is the electric potential at the radius of shear relative to its value in the distant, bulk medium.

☐ 7. The inner shell of charge and the outer atmosphere jointly constitute the **electrical double layer.**

☐ 8. **Flocculation** is the reversible aggregation of colloidal particles.

☐ 9. **Coagulation** is the irreversible aggregation of colloidal particles.

☐ 10. The **Schulze–Hardy rule** states that hydrophobic colloids are flocculated most efficiently by ions of opposite charge type and high charge number.

☐ 11. An **amphipathic** species has both hydrophobic and hydrophilic regions.

☐ 12. The **hydrophobic interaction** results in the clustering of nonpolar solutes in water.

☐ 13. A **micelle** is a colloid-sized cluster of molecules that forms at and above the **critical micelle concentration** and above the **Krafft temperature.**

☐ 14. **Unilamellar vesicles** are micelles that exist as extended parallel sheets.

Checklist of equations

Property	Equation	Comment	Equation number
Thickness of the electrical double layer	$r_D = (\varepsilon RT/2\rho F^2 I b^\ominus)^{1/2}$	Debye–Hückel theory	14E.3
Hydrophobicity constant	$\pi = \log\{s(\mathrm{RX})/s(\mathrm{HX})\}$	Definition	14E.5
Surfactant parameter	$N_s = V/Al$	Definition	14E.7

FOCUS 14 Molecular interactions

TOPIC 14A The electric properties of molecules

Discussion questions

D14A.1 Explain how the permanent dipole moment and the polarizability of a molecule arise.

D14A.2 Explain why the polarizability of a molecule decreases at high frequencies.

D14A.3 Describe the experimental procedures available for determining the electric dipole moment of a molecule.

Exercises

E14A.1(a) Which of the following molecules may be polar: ClF_3, O_3, H_2O_2?
E14A.1(b) Which of the following molecules may be polar: SO_3, XeF_4, SF_4?

E14A.2(a) Calculate the resultant of two dipole moments of magnitude 1.5 D and 0.80 D that make an angle of 109.5° to each other.
E14A.2(b) Calculate the resultant of two dipole moments of magnitude 2.5 D and 0.50 D that make an angle of 120° to each other.

E14A.3(a) Calculate the magnitude and direction of the dipole moment of the following arrangement of charges in the *xy*-plane: 3*e* at (0, 0), −*e* at (0.32 nm, 0), and −2*e* at an angle of 20° from the *x*-axis and a distance of 0.23 nm from the origin.
E14A.3(b) Calculate the magnitude and direction of the dipole moment of the following arrangement of charges in the *xy*-plane: 4*e* at (0, 0), −2*e* at (162 pm, 0), and −2*e* at an angle of 30° from the *x*-axis and a distance of 143 pm from the origin.

E14A.4(a) What strength of electric field is required to induce an electric dipole moment of magnitude 1.0 μD in a molecule of polarizability volume $2.6 \times 10^{-30}\,m^3$ (like CO_2)?
E14A.4(b) What strength of electric field is required to induce an electric dipole moment of magnitude 2.5 μD in a molecule of polarizability volume $1.05 \times 10^{-29}\,m^3$ (like CCl_4)?

E14A.5(a) The molar polarization of fluorobenzene vapour varies linearly with T^{-1}, and is 70.62 $cm^3\,mol^{-1}$ at 351.0 K and 62.47 $cm^3\,mol^{-1}$ at 423.2 K. Calculate the polarizability and dipole moment of the molecule.
E14A.5(b) The molar polarization of the vapour of a compound was found to vary linearly with T^{-1}, and is 75.74 $cm^3\,mol^{-1}$ at 320.0 K and 71.43 $cm^3\,mol^{-1}$ at 421.7 K. Calculate the polarizability and dipole moment of the molecule.

E14A.6(a) At 0 °C, the molar polarization of liquid chlorine trifluoride is 27.18 $cm^3\,mol^{-1}$ and its mass density is 1.89 $g\,cm^{-3}$. Calculate the relative permittivity of the liquid.
E14A.6(b) At 0 °C, the molar polarization of a liquid is 32.16 $cm^3\,mol^{-1}$ and its mass density is 1.92 $g\,cm^{-3}$. Calculate the relative permittivity of the liquid. Take $M = 85.0\,g\,mol^{-1}$.

E14A.7(a) The refractive index of CH_2I_2 is 1.732 for 656 nm light. Its mass density at 20 °C is 3.32 $g\,cm^{-3}$. Calculate the polarizability of the molecule at this wavelength.
E14A.7(b) The refractive index of a compound is 1.622 for 643 nm light. Its mass density at 20 °C is 2.99 $g\,cm^{-3}$. Calculate the polarizability of the molecule at this wavelength. Take $M = 65.5\,g\,mol^{-1}$.

E14A.8(a) The polarizability volume of H_2O at optical frequencies is $1.5 \times 10^{-24}\,cm^3$. Estimate the refractive index of water. The experimental value is 1.33.
E14A.8(b) The polarizability volume of a liquid of molar mass 72.3 $g\,mol^{-1}$ and mass density 865 $kg\,m^{-3}$ at optical frequencies is $2.2 \times 10^{-30}\,m^3$. Estimate the refractive index of the liquid.

E14A.9(a) The dipole moment of chlorobenzene is 1.57 D and its polarizability volume is $1.23 \times 10^{-23}\,cm^3$. Estimate its relative permittivity at 25 °C, when its mass density is 1.173 $g\,cm^{-3}$.
E14A.9(b) The dipole moment of bromobenzene is 5.17×10^{-30} C m and its polarizability volume is approximately $1.5 \times 10^{-29}\,m^3$. Estimate its relative permittivity at 25 °C, when its mass density is 1491 $kg\,m^{-3}$.

Problems

P14A.1 The electric dipole moment of methylbenzene (toluene) is 0.4 D. Estimate the dipole moments of the three isomers of dimethylbenzene (the xylenes). About which answer can you be sure?

P14A.2 Plot the magnitude of the electric dipole moment of hydrogen peroxide as the H–O–O–H (azimuthal) angle ϕ changes from 0 to 2π. Use the dimensions and partial charges shown in (1).

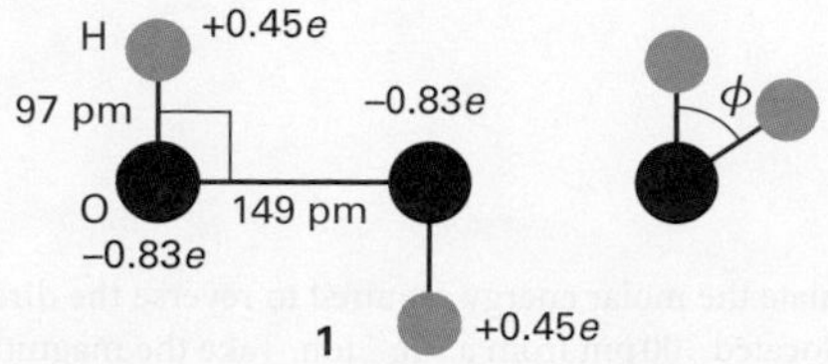

1

P14A.3 Ethanoic (acetic) acid vapour contains a proportion of planar, hydrogen-bonded dimers (2). The apparent dipole moment of molecules in pure gaseous ethanoic acid has a magnitude that increases with increasing temperature. Suggest an interpretation of this observation.

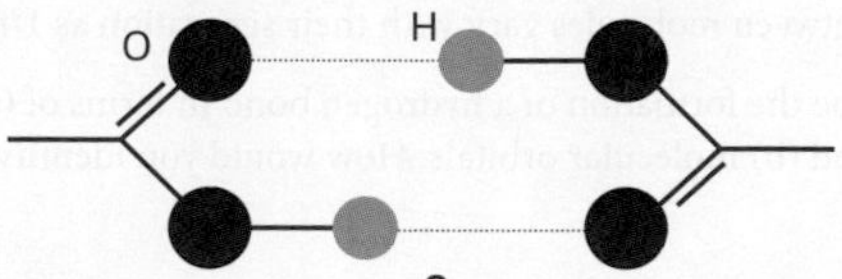

2

P14A.4‡ D.D. Nelson et al. (*Science* 238, 1670 (1987)) examined several weakly bound gas-phase complexes of ammonia in search of examples in which the

‡ These problems were supplied by Charles Trapp and Carmen Giunta.

H atoms in NH_3 formed hydrogen bonds, but found none. For example, they found that the complex of NH_3 and CO_2 has the carbon atom nearest the nitrogen (299 pm away): the CO_2 molecule is at right angles to the C–N 'bond', and the H atoms of NH_3 are pointing away from the CO_2. The magnitude of the permanent dipole moment of this complex is reported as 1.77 D. If the N and C atoms are the centres of the negative and positive charge distributions, respectively, what is the magnitude of those partial charges (as multiples of e)?

P14A.5 The polarizability volume of NH_3 is $2.22 \times 10^{-30}\,m^3$; calculate the contribution to the dipole moment of the molecule induced by an applied electric field of strength $15.0\,kV\,m^{-1}$.

P14A.6 The magnitude of the electric field at a distance r from a point charge Q is equal to $Q/4\pi\varepsilon_0 r^2$. How close to a water molecule (of polarizability volume $1.48 \times 10^{-30}\,m^3$) must a proton approach before the dipole moment it induces has a magnitude equal to that of the permanent dipole moment of the molecule (1.85 D)?

P14A.7 The relative permittivity of trichloromethane (chloroform) was measured over a range of temperatures with the following results:

θ/°C	−80	−70	−60	−40	−20	0	20
ε_r	3.1	3.1	7.0	6.5	6.0	5.5	5.0
ρ/(g cm^{-3})	1.65	1.64	1.64	1.61	1.57	1.53	1.50

The freezing point of trichloromethane is −64 °C. Account for these results and calculate the dipole moment and polarizability volume of the molecule.

P14A.8 The relative permittivities of methanol (with a melting point of −95 °C) corrected for density variation are given below. What molecular information can be deduced from these values? Take $\rho = 0.791\,g\,cm^{-3}$.

θ/°C	−185	−170	−150	−140	−110	−80	−50	−20	0	20
ε_r	3.2	3.6	4.0	5.1	67	57	49	43	38	34

P14A.9 In his classic book *Polar molecules*, Debye reports some early measurements of the polarizability of ammonia. From the selection below, determine the dipole moment and the polarizability volume of the molecule.

T/K	292.2	309.0	333.0	387.0	413.0	446.0
P_m/(cm^3 mol^{-1})	57.57	55.01	51.22	44.99	42.51	39.59

The refractive index of ammonia at 273 K and 100 kPa is 1.000 379 (for yellow sodium light). Calculate the molar polarization of the gas at this temperature. Combine the value calculated with the static molar polarization at 292.2 K and deduce from this information alone the molecular dipole moment.

P14A.10 Values of the molar polarization of gaseous water at 100 kPa as determined from capacitance measurements are given below as a function of temperature.

T/K	384.3	420.1	444.7	484.1	521.0
P_m/(cm^3 mol^{-1})	57.4	53.5	50.1	46.8	43.1

Calculate the dipole moment of H_2O and its polarizability volume.

P14A.11 From data in Table 14A.1 calculate the molar polarization, relative permittivity, and refractive index of methanol at 20 °C. Its mass density at that temperature is $0.7914\,g\,cm^{-3}$.

P14A.12 Show that, in a gas (for which the refractive index is close to 1), the refractive index depends on the pressure as $n_r = 1 + \text{constant} \times p$, and find the constant of proportionality. Go on to show how to deduce the polarizability volume of a molecule from measurements of the refractive index of a gaseous sample.

P14A.13 Ethanoic (acetic) acid vapour contains a proportion of planar, hydrogen-bonded dimers. The relative permittivity of pure liquid ethanoic acid is 7.14 at 290 K and increases with increasing temperature. Suggest an interpretation of the latter observation. What effect should isothermal dilution have on the relative permittivity of solutions of ethanoic acid in benzene?

TOPIC 14B Interactions between molecules

Discussion questions

D14B.1 Identify the terms in and of the following expressions and specify the conditions under which they are valid: (a) $V = -Q_2\mu_1/4\pi\varepsilon_0 r^2$, (b) $V = -Q_2\mu_1\cos\theta/4\pi\varepsilon_0 r^2$, and (c) $V = \mu_2\mu_1(1-3\cos^2\theta)/4\pi\varepsilon_0 r^3$.

D14B.2 Draw examples of the arrangements of electrical charges that correspond to monopoles, dipoles, quadrupoles, and octupoles. Suggest a reason for the different distance dependencies of their electric fields.

D14B.3 Account for the theoretical conclusion that many attractive interactions between molecules vary with their separation as $1/r^6$.

D14B.4 Describe the formation of a hydrogen bond in terms of (a) electrostatic interactions and (b) molecular orbitals. How would you identify the better model?

D14B.5 Some polymers have unusual properties. For example, Kevlar (3) is strong enough to be the material of choice for bulletproof vests and is stable at temperatures up to 600 K. What molecular interactions contribute to the formation and thermal stability of this polymer?

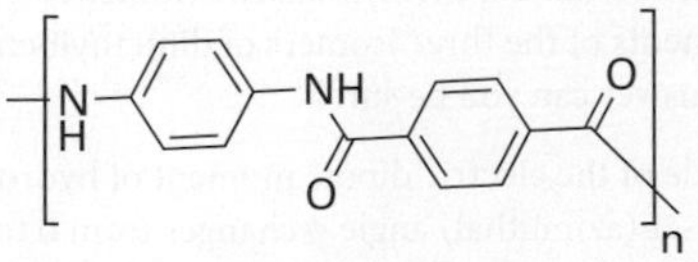

3 Kevlar

Exercises

E14B.1(a) Calculate the molar energy required to reverse the direction of an H_2O molecule located 100 pm from a Li^+ ion. Take the magnitude of the dipole moment of water as 1.85 D.

E14B.1(b) Calculate the molar energy required to reverse the direction of an HCl molecule located 300 pm from a Mg^{2+} ion. Take the magnitude of the dipole moment of HCl as 1.08 D.

E14B.2(a) Use eqn 14B.3b to calculate the molar potential energy of the dipolar interaction between two amide groups separated by 3.0 nm with $\theta = 45°$ in a vacuum. Take $\mu_1 = \mu_2 = 2.7\,\text{D}$.

E14B.2(b) Use eqn 14B.3b to calculate the molar potential energy of the dipolar interaction between an amide group ($\mu = 2.7\,\text{D}$) and a water molecule ($\mu = 1.85\,\text{D}$) separated by 3.0 nm with $\theta = 45°$ in a medium with relative permittivity of 3.5.

E14B.3(a) Calculate the potential energy of the interaction between two linear quadrupoles when they are collinear and their centres are separated by a distance r.

E14B.3(b) Calculate the potential energy of the interaction between two linear quadrupoles when they are parallel and separated by a distance r.

E14B.4(a) Calculate the average interaction energy for pairs of molecules in the gas phase with $\mu = 1\,\text{D}$ when the separation is 0.5 nm at 298 K. Compare this energy with the average molar kinetic energy of the molecules.

E14B.4(b) Calculate the average interaction energy for pairs of molecules in the gas phase with $\mu = 2.5\,\text{D}$ when the separation is 1.0 nm at 273 K. Compare this energy with the average molar kinetic energy of the molecules.

E14B.5(a) Calculate the average dipole–induced-dipole interaction energy, in joules per mole (J mol^{-1}), between a water molecule and a benzene molecule separated by 1.0 nm.

E14B.5(b) Calculate the average dipole–induced-dipole interaction energy, in joules per mole (J mol^{-1}), between a water molecule and a CCl_4 molecule separated by 1.0 nm.

E14B.6(a) Estimate the energy of the dispersion interaction (use the London formula) for two He atoms separated by 1.0 nm. Relevant data can be found in the *Resource section*.

E14B.6(b) Estimate the energy of the dispersion interaction (use the London formula) for two Ar atoms separated by 1.0 nm. Relevant data can be found in the *Resource section*.

Problems

P14B.1 In general, atoms in molecules have partial charges arising from the spatial variation in electron density in the ground state. If these charges were separated by a medium, they would attract or repel each other in accord with Coulomb's law: $V = Q_1Q_2/4\pi\varepsilon r$ where Q_1 and Q_2 are the partial charges, r is their separation, and ε is the permittivity of the medium lying between the charges. Different values of the permittivity of the medium take into account the possibility that other parts of the molecule, or other molecules, lie between the charges. (a) Calculate the energy of interaction between a partial charge of -0.36 (that is, $Q_1 = -0.36e$) on the N atom of an amide group and the partial charge of $+0.45$ ($Q_2 = +0.45e$) on the carbonyl C atom at a distance of 3.0 nm, on the assumption that the medium between them is a vacuum. (b) Repeat the calculation for bulk water as the medium.

P14B.2 Use the electrostatic model to calculate the energy (in kJ mol^{-1}) required to break an O···H hydrogen bond in a vacuum ($\varepsilon_r = 1$) and in water ($\varepsilon_r \approx 80.0$). Let the O···H distance be 170 pm, and partial charges (see Problem P14B.1) on the H and O atoms be $+0.42e$ and $-0.84e$, respectively.

P14B.3 An H_2O molecule ($\mu = 1.85\,\text{D}$) is aligned by an external electric field of strength $1.0\,\text{kV m}^{-1}$ (so that the dipole moment vector is parallel to the direction of the electric field) and an Ar atom ($\alpha' = 1.66 \times 10^{-30}\,\text{m}^3$) is brought up slowly from one side. At what separation is it energetically favourable for the H_2O molecule to rotate by 90° and the dipole moment point towards the approaching Ar atom?

P14B.4 Suppose an H_2O molecule ($\mu = 1.85\,\text{D}$) approaches an anion. What is the favourable orientation of the molecule? Calculate the magnitude of the electric field (in volts per metre) experienced by the anion when the water dipole is (a) 1.0 nm, (b) 0.3 nm, (c) 30 nm from the ion.

P14B.5 Phenylalanine (Phe, **4**) is a naturally occurring amino acid. What is the energy of interaction between its phenyl group and the electric dipole moment of a neighbouring peptide group? Take the distance between the groups as 0.4 nm and treat the phenyl group as a benzene molecule. The dipole moment of the peptide group is $\mu = 2.7\,\text{D}$ and the polarizability volume of benzene is $\alpha' = 1.04 \times 10^{-29}\,\text{m}^3$.

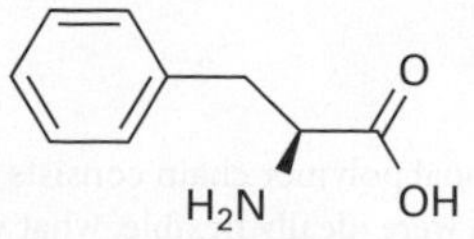

4 Phenylalanine

P14B.6 Now consider the London interaction between the phenyl groups of two Phe residues (see Problem P14B.5). (a) Estimate the potential energy of interaction between two such rings (treated as benzene molecules) separated by 0.4 nm. For the ionization energy, use $I = 5.0\,\text{eV}$. (b) Given that force is the negative slope of the potential, calculate the distance-dependence of the force acting between two non-bonded groups of atoms, such as the phenyl groups of Phe, in a polypeptide chain that can have a London dispersion interaction with each other. What is the separation at which the force between the phenyl groups (treated as benzene molecules) of two Phe residues is zero? *Hint:* Calculate the slope by considering the potential energy at r and $r + \delta r$, with $\delta r << r$, and evaluating $\{V(r + \delta r) - V(r)\}/\delta r$. At the end of the calculation, let δr become vanishingly small.

P14B.7 Given that $F = -\mathrm{d}V/\mathrm{d}r$, calculate the distance dependence of the force acting between two non-bonded groups of atoms in a polymer chain that have a London dispersion interaction with each other.

P14B.8 Consider the arrangement shown in **5** for a system consisting of an O–H group and an O atom, and then use the electrostatic model of the hydrogen bond to calculate the dependence of the molar potential energy of interaction on the angle θ.

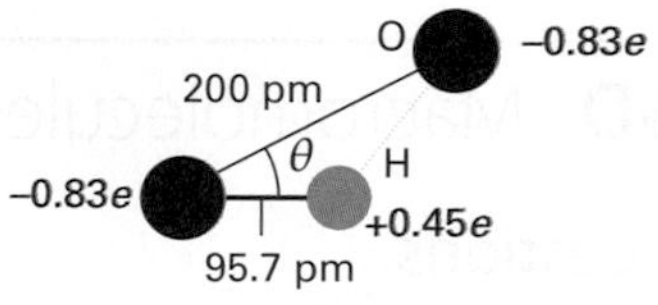

5

P14B.9 Suppose you distrusted the Lennard-Jones (12,6) potential for assessing a particular polypeptide conformation, and replaced the repulsive term by an exponential function of the form e^{-r/r_0}. (a) Sketch the form of the potential energy and locate the distance at which it is a minimum. *Hint:* Use mathematical software.

P14B.10 The *cohesive energy density*, $\mathcal{U}$, is defined as U/V, where U is the mean potential energy of attraction within the sample and V its volume. Show that $\mathcal{U} = \frac{1}{2}\mathcal{N}^2\int V(R)\mathrm{d}\tau$ where $\mathcal{N}$ is the number density of the molecules and $V(R)$ is their attractive potential energy and where the integration ranges from d to infinity and over all angles. Go on to show that the cohesive energy density of a uniform distribution of molecules that interact by a van der Waals attraction of the form $-C_6/R^6$ is equal to $-(2\pi/3)(N_A^2/d^3M^2)\rho^2C_6$, where ρ is the mass density of the solid sample and M is the molar mass of the molecules.

TOPIC 14C Liquids

Discussion questions

D14C.1 Explain the Monte Carlo and molecular dynamics methods for the calculation of the radial distribution function in liquids.

D14C.2 Describe the process of condensation.

Exercises

E14C.1(a) Calculate the vapour pressure of a spherical droplet of water of radius 10 nm at 20 °C. The vapour pressure of bulk water at that temperature is 2.3 kPa and its mass density is 0.9982 g cm^{-3}.

E14C.1(b) Calculate the vapour pressure of a spherical droplet of water of radius 20.0 nm at 35.0 °C. The vapour pressure of bulk water at that temperature is 5.623 kPa and its mass density is 994.0 kg m^{-3}.

E14C.2(a) The contact angle for water on clean glass is close to zero. Calculate the surface tension of water at 20 °C given that at that temperature water climbs to a height of 4.96 cm in a clean glass capillary tube of internal radius 0.300 mm. The mass density of water at 20 °C is 998.2 kg m^{-3}.

E14C.2(b) The contact angle for water on clean glass is close to zero. Calculate the surface tension of water at 30 °C given that at that temperature water climbs to a height of 9.11 cm in a clean glass capillary tube of internal diameter 0.320 mm. The mass density of water at 30 °C is 0.9956 g cm^{-3}.

E14C.3(a) Calculate the pressure differential of water across the surface of a spherical droplet of radius 200 nm at 20 °C.

E14C.3(b) Calculate the pressure differential of ethanol across the surface of a spherical droplet of radius 220 nm at 20 °C. The surface tension of ethanol at that temperature is 22.39 mN m^{-1}.

E14C.4(a) The contact angle for water on glass is close to zero. Calculate the surface tension of water at 25 °C given that at that temperature, water climbs to a height of 5.89 cm in a clean glass capillary tube of internal diameter 0.500 mm. The mass density of water at 25 °C is 9970 g cm^{-3}.

E14C.4(b) Calculate the surface tension of a liquid at 25 °C given that at that temperature, the liquid climbs to a height of 10.00 cm in a clean glass capillary tube of internal radius 0.300 mm. The mass density of the liquid at 25 °C is 0.9500 g cm^{-3}. Assume that the contact angle is zero.

Problems

P14C.1 A simple pair distribution function has the form

$$g(r)=1+\cos\left(\frac{4r}{r_0}-4\right)e^{-(r/r_0-1)}$$

for $r \geq r_0$ and $g(r)=0$ for $r<r_0$. Here the parameter r_0 is the separation at which the Lennard-Jones potential energy function (eqn 14B.12) $V=4\varepsilon\{(r_0/r)^{12}-(r_0/r)^6\}$ is equal to zero. (a) Plot the function $g(r)$. Does it resemble the form shown in Fig. 14C.1? (b) Plot the virial $v_2(r)=r(\mathrm{d}V/\mathrm{d}r)$.

P14C.2 The surface tensions of a series of aqueous solutions of a surfactant A were measured at 20 °C, with the following results:

[A]/(mol dm^{-3})	0	0.10	0.20	0.30	0.40	0.50
γ/(mN m^{-1})	72.8	70.2	67.7	65.1	62.8	59.8

Calculate the surface excess concentration.

TOPIC 14D Macromolecules

Discussion questions

D14D.1 Distinguish between number-average, weight-average, and *Z*-average molar masses. Which experiments give information about each one?

D14D.2 Distinguish between the four levels of structure of a macromolecule: primary, secondary, tertiary, and quaternary.

D14D.3 What are the consequences of there being partial rigidity in an otherwise random coil?

D14D.4 Define the terms in the following expressions and specify the conditions for their validity: (a) $R_c=Nl$, (b) $R_{rms}=N^{1/2}l$, (c) $R_{rms}=(2N)^{1/2}l$, (d) $R_{rms}=N^{1/2}lF$, (e) $R_g=N^{1/2}l$, (f) $R_g=(N/6)^{1/2}l$, (g) $R_g=(N/3)^{1/2}l$.

D14D.5 Distinguish between the melting temperature and the glass transition temperature of a polymer.

Exercises

E14D.1(a) Calculate the number-average molar mass and the weight-average molar mass of a mixture of equal amounts of two polymers, one having $M=$ 62 kg mol^{-1} and the other $M=78$ kg mol^{-1}.

E14D.1(b) Calculate the number-average molar mass and the weight-average molar mass of a mixture of two polymers, one having $M=62$ kg mol^{-1} and the other $M=78$ kg mol^{-1}, with their amounts in moles in the ratio 3:2.

E14D.2(a) A one-dimensional polymer chain consists of 700 segments, each 0.90 nm long. If the chain were ideally flexible, what would be the root-mean-square separation of the ends of the chain?

E14D.2(b) A one-dimensional polymer chain consists of 1200 segments, each 1.125 nm long. If the chain were ideally flexible, what would be the root-mean-square separation of the ends of the chain?

E14D.3(a) Calculate the contour length (the length of the extended chain) and the root-mean-square separation (the end-to-end distance) of polyethene with molar mass 280 kg mol^{-1}, modelled as a one-dimensional chain.
E14D.3(b) Calculate the contour length (the length of the extended chain) and the root-mean-square separation (the end-to-end distance) for polypropene of molar mass 174 kg mol^{-1}, modelled as a one-dimensional chain.

E14D.4(a) The radius of gyration of a long one-dimensional chain molecule is found to be 7.3 nm. The chain consists of C–C links. Assume that the chain is randomly coiled and estimate the number of links in the chain.
E14D.4(b) The radius of gyration of a long one-dimensional chain molecule is found to be 18.9 nm. The chain consists of links of length 450 pm. Assume that the chain is randomly coiled and estimate the number of links in the chain.

E14D.5(a) What is the probability that the ends of a polyethene chain of molar mass 65 kg mol^{-1} are 10 nm apart when the polymer is treated as a one-dimensional freely jointed chain?
E14D.5(b) What is the probability that the ends of a polyethene chain of molar mass 85 kg mol^{-1} are 15 nm apart when the polymer is treated as a one-dimensional freely jointed chain?

E14D.6(a) What is the probability that the ends of a polyethene chain of molar mass 65 kg mol^{-1} are between 10.0 nm and 10.1 nm apart when the polymer is treated as a three-dimensional freely jointed chain?
E14D.6(b) What is the probability that the ends of a polyethene chain of molar mass 75 kg mol^{-1} are between 14.0 nm and 14.1 nm apart when the polymer is treated as a three-dimensional freely jointed chain?

E14D.7(a) By what percentage does the radius of gyration of a one-dimensional polymer chain increase (+) or decrease (−) when the bond angle between units is limited to 109°? What is the percentage change in volume of the coil?
E14D.7(b) By what percentage does the root-mean-square separation of the ends of a one-dimensional polymer chain increase (+) or decrease (−) when the bond angle between units is limited to 120°? What is the percentage change in volume of the coil?

E14D.8(a) By what percentage does the root-mean-square separation of the ends of a one-dimensional polymer chain consisting of 1000 monomers increase (+) or decrease (−) when the persistence length changes from l (the bond length) to 5.0 per cent of the contour length? What is the percentage change in volume of the coil?
E14D.8(b) By what percentage does the root-mean-square separation of the ends of a one-dimensional polymer chain consisting of 1000 monomers increase (+) or decrease (−) when the persistence length changes from l (the bond length) to 2.5 per cent of the contour length? What is the percentage change in volume of the coil?

E14D.9(a) The radius of gyration of a three-dimensional partially rigid polymer of 1000 units each of length 150 pm was measured as 2.1 nm. What is the persistence length of the polymer?
E14D.9(b) The radius of gyration of a three-dimensional partially rigid polymer of 1500 units each of length 164 pm was measured as 3.0 nm. What is the persistence length of the polymer?

E14D.10(a) Calculate the restoring force when the ends of a one-dimensional polyethene chain of molar mass 65 kg mol^{-1} are moved apart by 1.0 nm at 20 °C.
E14D.10(b) Calculate the restoring force when the ends of a one-dimensional polyethene chain of molar mass 85 kg mol^{-1} are moved apart by 2.0 nm at 25 °C.

E14D.11(a) Calculate the change in molar entropy when the ends of a one-dimensional polyethene chain of molar mass 65 kg mol^{-1} are moved apart by 1.0 nm.
E14D.11(b) Calculate the change in molar entropy when the ends of a one-dimensional polyethene chain of molar mass 85 kg mol^{-1} are moved apart by 2.0 nm.

Problems

P14D.1 Evaluate the radius of gyration, R_g, of (a) a solid sphere of radius a, (b) a long straight rod of radius a and length l. Show that in the case of a solid sphere of specific volume v_s, $R_g/\text{nm} \approx 0.056902 \times \{(v_s/\text{cm}^3\,\text{g}^{-1})(M/\text{g mol}^{-1})\}^{1/3}$. Evaluate R_g for a species with $M = 100\,\text{kg mol}^{-1}$, $v_s = 0.750\,\text{cm}^3\,\text{g}^{-1}$, and, in the case of the rod, of radius 0.50 nm.

P14D.2 Use eqn 14D.4 to deduce expressions for (a) the root-mean-square separation of the ends of the chain, (b) the mean separation of the ends, and (c) their most probable separation. Evaluate these three quantities for a fully flexible chain with $N = 4000$ and $l = 154$ pm.

P14D.3 Deduce the relation $\langle r_i^2\rangle = Nl^2$ for the mean square distance of a monomer from the origin in a freely jointed chain of N units each of length l. *Hint*: Use the distribution in eqn 14D.4.

P14D.4 Derive expressions for the moments of inertia and hence the radii of gyration of (a) a uniform thin disk, (b) a long uniform rod, (c) a uniform sphere.

P14D.5 Construct a two-dimensional random walk by using a random number generating routine with mathematical software or spreadsheet. Construct a walk of 50 and 100 steps. If there are many people working on the problem, investigate the mean and most probable separations in the plots by direct measurement. Do they vary as $N^{1/2}$?

P14D.6 Show that it is possible to define the radius of gyration R_g as the average root-mean-square distance of the atoms or groups (all assumed to be of the same mass), that is, that $R_g^2 = (1/N)\Sigma_j R_j^2$, where R_j is the distance of atom j from the centre of mass.

P14D.7 Use the information below and the expression for R_g of a solid sphere quoted in the text (following eqn 14D.7b), to classify the species below as globular or rod-like.

A	$M/(\text{g mol}^{-1})$	$v_s/(\text{cm}^3\,\text{g}^{-1})$	R_g/nm
Serum albumin	66×10^3	0.752	2.98
Bushy stunt virus	10.6×10^6	0.741	12.0
DNA	4×10^6	0.556	117.0

P14D.8 Develop an expression for the fundamental vibrational frequency of a one-dimensional random coil that has been slightly stretched and then released. Evaluate this frequency for a sample of polyethene of molar mass 65 kg mol^{-1} at 20 °C. Account physically for the dependence of frequency on temperature and molar mass.

P14D.9 On the assumption that the tension, t, required to keep a sample at a constant length is proportional to the temperature ($t = aT$, the analogue of $p \propto T$), show that the tension can be ascribed to the dependence of the entropy on the length of the sample. Account for this result in terms of the molecular nature of the sample.

P14D.10 The following table lists the glass transition temperatures, T_g, of several polymers. Discuss the reasons why the structure of the monomer unit has an effect on the value of T_g.

Polymer	Poly-(oxymethylene)	Polyethene	Poly(vinyl chloride)	Polystyrene
Structure	$-(OCH_2)_n-$	$-(CH_2CH_2)_n-$	$-(CH_2{-}CHCl)_n-$	$-(CH_2{-}CH(C_6H_5))_n-$
T_g/K	198	253	354	381

TOPIC 14E Self-assembly

Discussion questions

D14E.1 Distinguish between a sol, an emulsion, and a foam. Provide examples of each.

D14E.2 Account for the hydrophobic interaction and discuss its manifestations.

D14E.3 It is observed that the critical micelle concentration of sodium dodecyl sulfate in aqueous solution decreases as the concentration of added sodium chloride increases. Explain this effect.

D14E.4 What effect is the inclusion of cholesterol likely to have on the transition temperatures of a lipid bilayer?

D14E.5 Why do bacterial and plant cells grown at low temperatures synthesize more phospholipids with unsaturated chains than do cells grown at higher temperatures?

Exercises

E14E.1(a) The velocity v with which a protein moves through water under the influence of an electric field varied with values of pH in the range 3.0 < pH < 7.0 according to the expression $v/(\mu m\ s^{-1}) = a + b(pH) + c(pH)^2 + d(pH)^3$ with $a = 0.50$, $b = -0.10$, $c = -3.0 \times 10^{-3}$, and $d = 5.0 \times 10^{-4}$. Identify the isoelectric point of the protein.

E14E.1(b) The velocity v with which a protein moves through water under the influence of an electric field varied with values of pH in the range 3.0 < pH < 5.0 according to the expression $v/(\mu m\ s^{-1}) = a + b(pH) + c(pH)^2$ with $a = 0.80$, $b = -4.0 \times 10^{-3}$, and $c = -5.0 \times 10^{-2}$. Identify the isoelectric point of the protein.

Problems

P14E.1 The binding of nonpolar groups of amino acid to hydrophobic sites in the interior of proteins is governed largely by hydrophobic interactions. (a) Consider a family of hydrocarbons R–H. The hydrophobicity constants, π, for R = CH_3, CH_2CH_3, $(CH_2)_2CH_3$, $(CH_2)_3CH_3$, and $(CH_2)_4CH_3$ are, respectively, 0.5, 1.0, 1.5, 2.0, and 2.5. Use these data to predict the π value for $(CH_2)_6CH_3$. (b) The equilibrium constants K_I for the dissociation of inhibitors (**6**) from the enzyme chymotrypsin were measured for different substituents R:

R	CH_3CO	CN	NO_2	CH_3	Cl
π	−0.20	−0.025	0.33	0.5	0.9
$\log K_I$	−1.73	−1.90	−2.43	−2.55	−3.40

Plot $\log K_I$ against π. Does the plot suggest a linear relationship? If so, what are the slope and intercept to the $\log K_I$ axis of the line that best fits the data? (c) Predict the value of K_I for the case R = H.

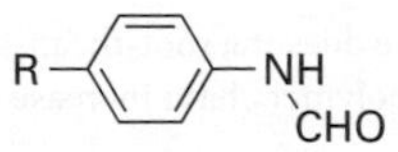

6

P14E.2 Use mathematical software to reproduce the features in Fig. 14E.7.

P14E.3 Equation 14E.6b is surprisingly tricky to solve, but it is possible to make good progress with simple cases. With $N = 2$ and $K = 1$, find an expression for $[M_2]$.

FOCUS 14 Molecular interactions

Integrated activities

I14.1 Show that the mean interaction energy of N atoms of diameter d interacting with a potential energy of the form C_6/R^6 is given by $U = -2N^2C_6/3Vd^3$, where V is the volume in which the molecules are confined and all effects of clustering are ignored. Hence, find a connection between the van der Waals parameter a and C_6, from $n^2a/V^2 = (\partial U/\partial V)_T$.

I14.2‡ F. Luo et al. (*J. Chem. Phys.* **98**, 3564 (1993)) reported experimental observation of the He_2 complex, a species that had escaped detection for a long time. The fact that the observation required temperatures in the neighbourhood of 1 mK is consistent with computational studies which suggest that $hc\tilde{D}_e$, for He_2 is about 1.51×10^{-23} J, $hc\tilde{D}_0$ about 2×10^{-26} J, and R about 297 pm. (a) Determine the Lennard-Jones parameters r_0, and ε and plot the Lennard-Jones potential for He–He interactions. (b) Plot the Morse potential given that $a = 5.79 \times 10^{10}\ m^{-1}$.

I14.3 Before attempting this problem, read *Impact* 21 on the website for this text. Molecular orbital calculations may be used to predict structures of intermolecular complexes. Hydrogen bonds between purine and pyrimidine bases are responsible for the double helix structure of DNA. Consider methyl-adenine (**7**, with R = CH_3) and methyl-thymine (**8**, with R = CH_3) as models of two bases that can form hydrogen bonds in DNA. (a) Use molecular modelling software and the computational method of your or your instructor's choice to calculate the atomic charges of all atoms in methyl-adenine and methyl-thymine. (b) Based on your tabulation of atomic charges, identify the atoms in methyl-adenine and methyl-thymine that are likely to participate in hydrogen bonds. (c) Draw all possible adenine–thymine pairs that can be linked by hydrogen bonds, keeping in mind that linear arrangements of the A–H···B fragments are preferred in DNA. For this step, you may want to use your molecular modelling software to align the molecules properly. (d) Consult *Impact* 21 and determine which of the pairs that you drew in part (c) occur naturally in DNA molecules. (e) Repeat parts (a)–(d) for cytosine and guanine, which also form base pairs in DNA.

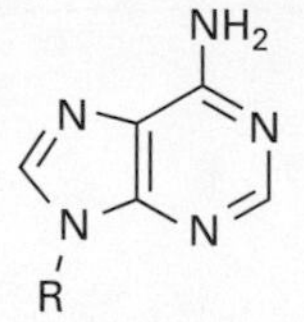

7 R = CH_3, Methyl adenine

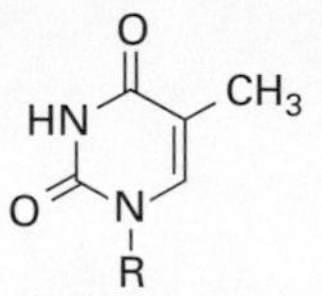

8 R = CH_3, Methyl thymine

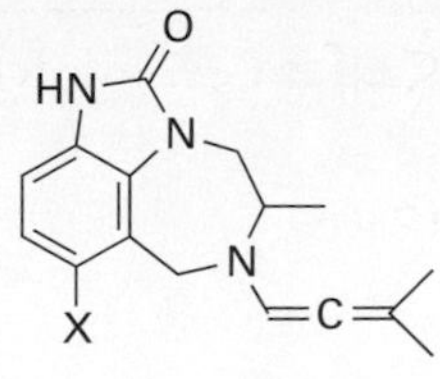

10 TIBO

I14.4 Molecular orbital calculations may be used to predict the dipole moments of molecules. (a) Using molecular modelling software and the computational method of your or your instructor's choice, calculate the dipole moment of the peptide link, modelled as a *trans-N*-methylacetamide (**9**). Plot the energy of interaction between these dipoles against the angle θ for r = 3.0 nm in the arrangement shown in structure **4** of Topic 14B. (b) Compare the maximum value of the dipole–dipole interaction energy from part (a) to 20 kJ mol^{-1}, a typical value for the energy of a hydrogen bonding interaction in biological systems.

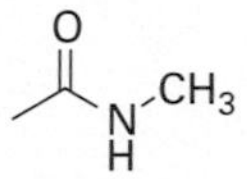

9 *trans-N*-methylacetamide

I14.5 Before attempting this problem, read *Impact* 22 on the website for this text. Derivatives of the compound TIBO (**10**) inhibit the enzyme reverse transcriptase, which catalyses the conversion of retroviral RNA to DNA. A quantitative structure–activity relationship (QSAR) analysis of the activity A of a number of TIBO derivatives suggests the following equation:

$$\log A = b_0 + b_1 S + b_2 W$$

where S is a parameter related to the drug's solubility in water and W is a parameter related to the width of the first atom in a substituent X shown in **10**. (a) Use the following data to determine the values of b_0, b_1, and b_2. *Hint:* The QSAR equation relates one dependent variable, $\log A$, to two independent variables, S and W. To fit the data, you must use the mathematical procedure of *multiple regression*, which can be performed with mathematical software or a spreadsheet.

X	H	Cl	SCH_3	OCH_3	CN	CHO	Br	CH_3	CCH
$\log A$	7.36	8.37	8.3	7.47	7.25	6.73	8.52	7.87	7.53
S	3.53	4.24	4.09	3.45	2.96	2.89	4.39	4.03	3.80
W	1.00	1.80	1.70	1.35	1.60	1.60	1.95	1.60	1.60

(b) What should be the value of W for a drug with S = 4.84 and $\log A$ = 7.60?

I14.6 Consider the thermodynamic description of stretching rubber. The observables are the tension, t, and length, l (the analogues of p and V for gases). Because $dw = tdl$, the basic equation is $dU = TdS + tdl$. If $G = U - TS - tl$, find expressions for dG and dA, and deduce the Maxwell relations

$$\left(\frac{\partial S}{\partial l}\right)_T = -\left(\frac{\partial t}{\partial T}\right)_l \qquad \left(\frac{\partial S}{\partial t}\right)_T = \left(\frac{\partial l}{\partial T}\right)_t$$

Go on to deduce the equation of state for rubber,

$$\left(\frac{\partial U}{\partial l}\right)_T = t - T\left(\frac{\partial t}{\partial T}\right)_l$$

I14.7 Before attempting this problem, read *Impact* 21 on the website for this text. Commercial software (more specifically 'molecular mechanics' or 'conformational search' software) automates the calculations that lead to Ramachandran plots. In this problem the model for the protein is the dipeptide (**11**) in which the terminal methyl groups replace the rest of the polypeptide chain. (a) Draw three initial conformers of the dipeptide with R = H: one with ϕ = +75°, ψ = −65°, a second with $\phi = \psi$ = +180°, and a third with ϕ = +65°, ψ = +35°. Use software of your or your instructor's choice to optimize the geometry of each conformer and find the final ϕ and ψ angles in each case. Did all the initial conformers converge to the same final conformation? If not, what do these final conformers represent? (b) Use the approach in part (a) to investigate the case R = CH_3, with the same three initial conformers as starting points for the calculations. Rationalize any similarities and differences between the final conformers of the dipeptides with R = H and R = CH_3.

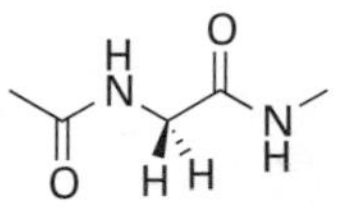

11

I14.8 The effective radius, a, of a random coil is related to its radius of gyration, R_g, by $a = \gamma R_g$, with γ = 0.85. Deduce an expression for the osmotic virial coefficient, B (Topic 5B), in terms of the number of chain units for (a) a freely jointed chain, (b) a chain with tetrahedral bond angles. Evaluate B for l = 154 pm and N = 4000. Estimate B for a randomly coiled polyethylene chain of arbitrary molar mass, M, and evaluate it for M = 56 kg mol^{-1}. *Hint:* Use $B = \frac{1}{2}N_A v_P$, where v_P is the excluded volume due to a single molecule.

FOCUS 15

Solids

This Focus explores the structures and physical properties of solids. The solid state includes most of the materials that make modern technology possible. It includes the wide varieties of steel that are used in architecture and engineering, the semiconductors and metallic conductors that are used in information technology and power distribution, the ceramics that increasingly are replacing metals, and the synthetic and natural polymers discussed in Focus 14 that are used in the textile industry and in the fabrication of many of the common objects of the modern world.

15A Crystal structure

The characteristic feature of a crystal is the regular arrangement of its constituents. This Topic describes how that regularity is described in terms of the symmetry of the arrangement and then explains how the arrangements are described quantitatively.

15B Diffraction techniques

Diffraction techniques enable the structures of solids to be determined in great detail. This Topic considers the basic principles of 'X-ray diffraction' and describes how the diffraction pattern can be interpreted in terms of the distribution of electron density. The diffraction of electrons and neutrons adds to the information that can be obtained about the structures of individual molecules and atoms in solids.

15C Bonding in solids

The constituents of solids are held together by a variety of interactions, which impart characteristic properties. This Topic explores the interactions and prepares the ground for a discussion of the resulting properties.

15D The mechanical properties of solids

The characteristic mechanical properties of a solid include various aspects of its rigidity. These properties are reported in terms of several parameters that can be related to the structure of the solid.

15E The electrical properties of solids

One very important property of a solid is its ability to transport an electric current. This Topic explores how solids are classified according to their electrical conductivity and how the different behaviours seen can be rationalized using the 'band theory' of electronic structure. It goes on to show how the introduction of low concentrations of impurities can have a profound effect on the properties of semiconductors, and how this effect is exploited in making the semiconductor devices which are ubiquitous in modern electronics.

15F The magnetic properties of solids

The magnetic properties of solids are reported in terms of their 'susceptibility'. If the magnetic centres are independent, this property can be traced to individual electron spins. If the centres interact, properties such as ferromagnetism emerge.

15G The optical properties of solids

Spectroscopy is a key tool for exploring the electronic structure of solids. As well as exploring the electronic band structure of a solid material, spectroscopic observations give insight into phenomena that arise from the interactions present in such materials. When subject to very intense radiation some solids respond nonlinearly, leading to useful phenomena such as frequency doubling.

Web resources What is an application of this material?

The deployment of X-ray diffraction techniques for the determination of the location of all the atoms in biological macromolecules has revolutionized the study of biochemistry and molecular biology. *Impact* 23 demonstrates the power of the techniques by exploring one of the most seminal X-ray images of all: the characteristic pattern obtained from strands of DNA and used in the construction of the double-helix model of DNA. Nanometre-sized assemblies that conduct electricity are currently of great technological interest, and *Impact* 24 describes their synthesis.

TOPIC 15A Crystal structure

➤ **Why do you need to know this material?**

Crystalline solids are important in many technologies, and to be able to account for their mechanical, electrical, optical, and magnetic properties you need to understand their microscopic structures.

➤ **What is the key idea?**

The regular arrangement of the atoms in periodic crystals can be described in terms of unit cells.

➤ **What do you need to know already?**

Light use is made of some of the language used to describe symmetry (Topic 10A).

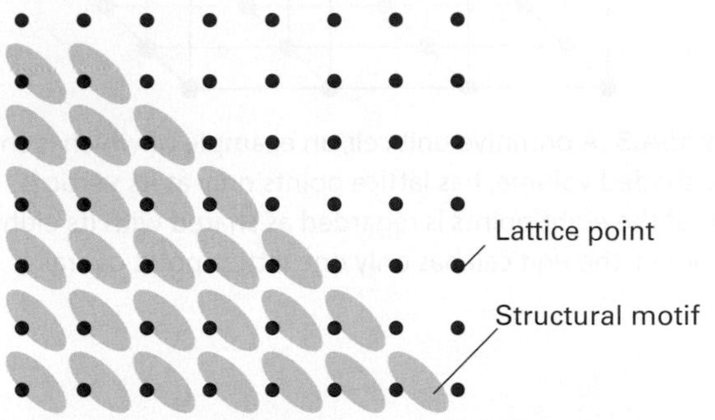

Figure 15A.1 Each lattice point specifies the location of a structural motif (e.g. a molecule or a group of molecules). The space lattice is the entire array of lattice points; the crystal structure is the collection of structural motifs arranged according to the lattice.

The internal structure of a crystal is a regular array of its atoms, ions, or molecules. The features of this regular array, such as the details of the stacking pattern and its characteristic dimensions, are a crucial aspect of the link between the structure and properties of the solid.

15A.1 Periodic crystal lattices

A **periodic crystal** is built up from regularly repeating 'structural motifs', which may be atoms, molecules, or groups of atoms, molecules, or ions. A **space lattice** is the pattern formed by points representing the locations of these motifs (Fig. 15A.1). A space lattice is, in effect, an abstract scaffolding for the crystal structure. More formally, a space lattice is a three-dimensional, infinite array of points, each of which is surrounded in an identical way by its neighbours, and which defines the basic structure of the crystal. In some cases there may be a structural motif centred on each lattice point, but that is not necessary. The crystal structure itself is obtained by associating with each lattice point an identical structural motif. The solids known as **quasicrystals** are 'aperiodic', in the sense that the space lattice, though still filling space, does not have translational symmetry. This Topic deals only with periodic crystals.

A **unit cell** is an imaginary parallelepiped (parallel-sided figure) from which the entire space lattice can be constructed by purely translational displacements (Fig. 15A.2). The cell is commonly formed by joining neighbouring lattice points by straight lines, and such unit cells are described as **primitive** (Fig. 15A.3). If each of the four points of a two-dimensional unit cell in Fig. 15A.2 is regarded as shared with its four neighbours, then the cell has only one lattice point overall. The same definition applies in three dimensions, where each of the eight points of a primitive unit cell is shared by eight neighbours, giving one lattice point overall. It is often more convenient to draw larger **non-primitive unit cells** that also have lattice

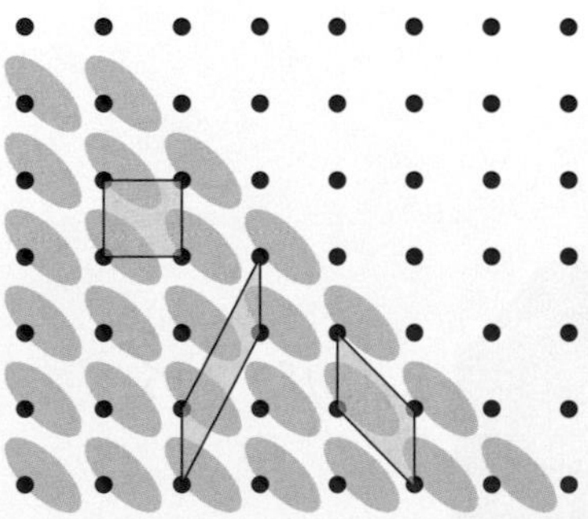

Figure 15A.2 A unit cell is a parallel-sided (but not necessarily rectangular) figure from which the entire space lattice can be constructed by using only translations (not reflections, rotations, or inversions). In the two-dimensional case shown here, each lattice point is shared by four neighbouring cells.

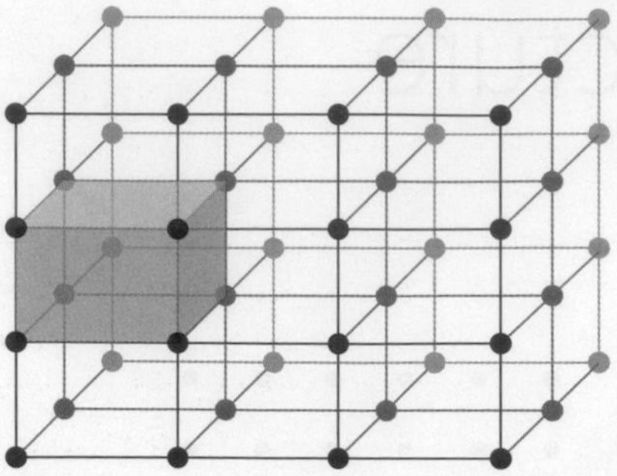

Figure 15A.3 A primitive unit cell, an example of which is shown by the shaded volume, has lattice points only at its vertices. If each of the eight points is regarded as shared with its eight neighbours, the unit cell has only one lattice point overall.

points at their centres or on pairs of opposite faces. An infinite number of different unit cells can describe the same lattice, but the one with sides that have the shortest lengths and that are most nearly perpendicular to one another is normally chosen. The lengths of the sides of a unit cell are denoted a, b, and c, and the angles between them are denoted α, β, and γ (Fig. 15A.4).

Unit cells are classified into seven **crystal systems** by noting the rotational symmetry elements they possess:

- A **cubic unit cell** has four threefold axes pointing to the corners of a tetrahedron, and passing through the centre of the cube (Fig. 15A.5).
- A **monoclinic unit cell** has one twofold axis (Fig. 15A.6).
- A **triclinic unit cell** has no rotational symmetry, and typically all three sides and angles are different (Fig. 15A.7).

Table 15A.1 lists the **essential symmetries**, the elements that must be present for the unit cell to belong to a particular crystal system.

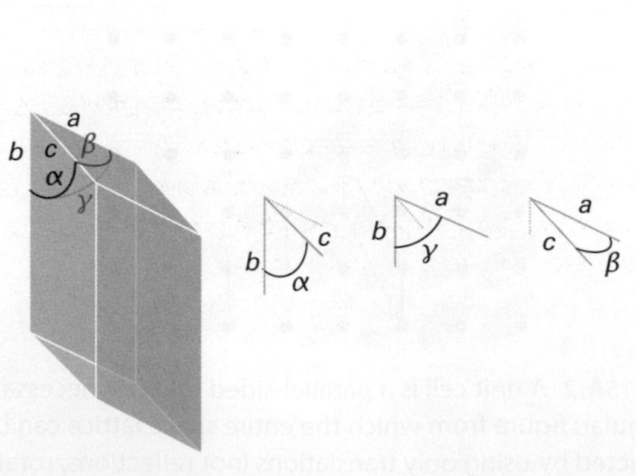

Figure 15A.4 The notation for the sides and angles of a unit cell. Note that the angle α lies in the plane (b,c).

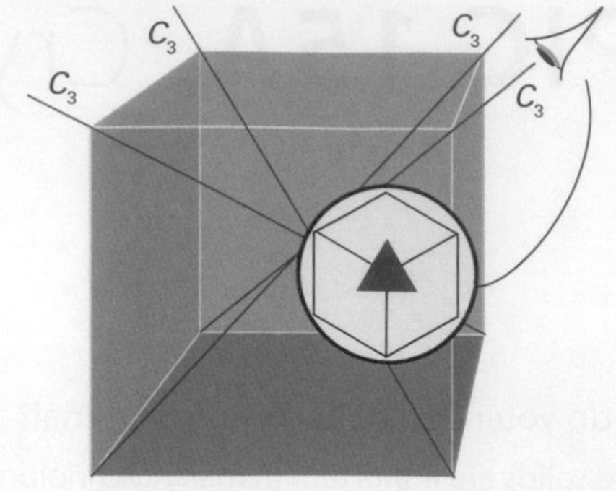

Figure 15A.5 A unit cell belonging to the cubic system has four threefold axes, denoted C_3, arranged tetrahedrally. The insert shows the threefold symmetry.

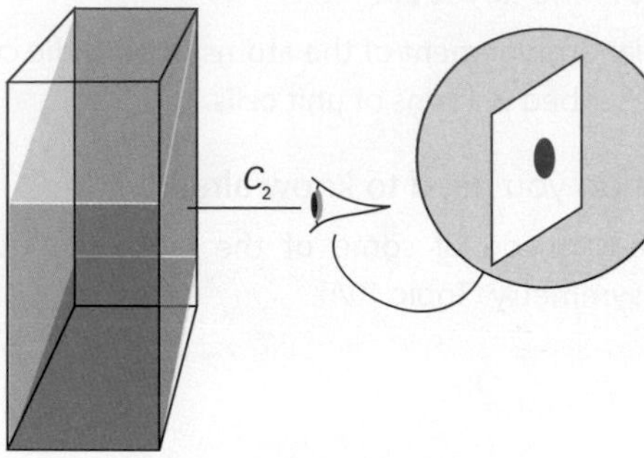

Figure 15A.6 A unit cell belonging to the monoclinic system has a twofold axis (denoted C_2 and shown in more detail in the insert).

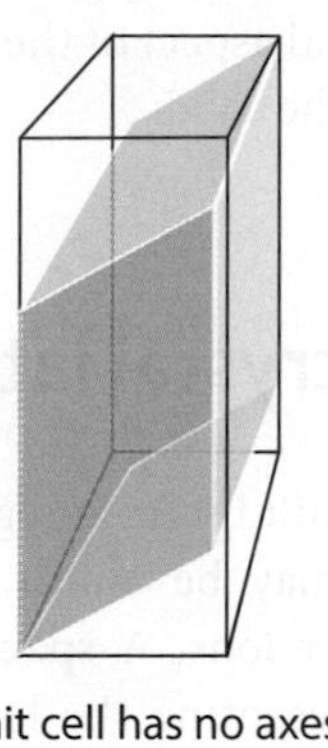

Figure 15A.7 A triclinic unit cell has no axes of rotational symmetry.

Table 15A.1 The seven crystal systems*

System	Essential symmetries
Triclinic	None
Monoclinic	One C_2 axis
Orthorhombic	Three perpendicular C_2 axes
Rhombohedral	One C_3 axis
Tetragonal	One C_4 axis
Hexagonal	One C_6 axis
Cubic	Four C_3 axes in a tetrahedral arrangement

*C_n denotes an n-fold rotation, in which identical structures are obtained after rotation by 360°/n.

In three dimensions there are only 14 distinct space lattices. The unit cells of these **Bravais lattices** are illustrated in Fig. 15A.8. It is conventional to portray the lattices by primitive unit cells in some cases and by non-primitive unit cells in others. The following notation is used:

- A **primitive unit cell** (P) has lattice points only at the corners.
- A **body-centred unit cell** (I) also has a lattice point at its centre.
- A **face-centred unit cell** (F) has lattice points at its corners and also at the centres of its six faces.
- A **side-centred unit cell** (A, B, or C) has lattice points at its corners and at the centres of two opposite faces.

For simple structures, it is often convenient to choose an atom belonging to the structural motif, or the centre of a molecule, as the location of a lattice point or the vertex of a unit cell, but that is not a necessary requirement. Symmetry-related lattice points within the unit cell of a Bravais lattice have identical surroundings.

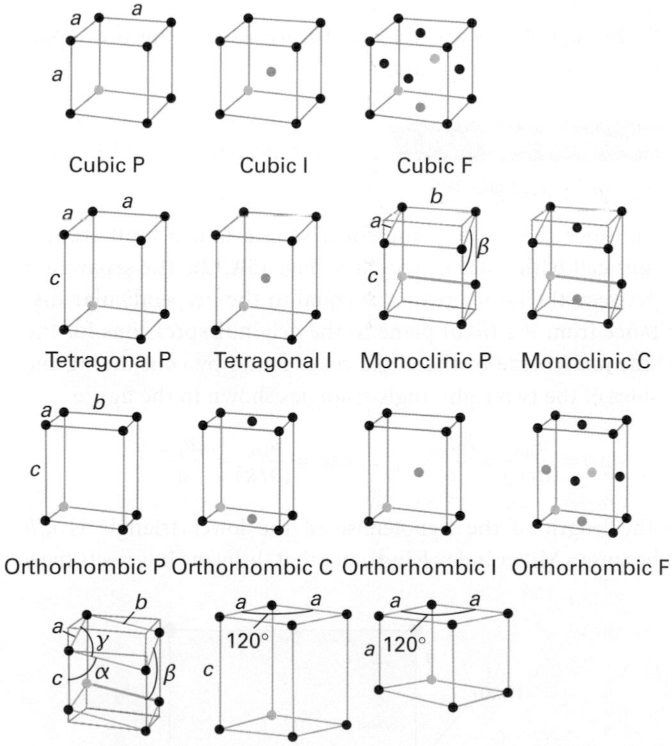

Figure 15A.8 The 14 Bravais lattices. The points are lattice points, and are not necessarily occupied by atoms. P denotes a primitive unit cell (R is used for a trigonal lattice), I a body-centred unit cell, F a face-centred unit cell, and C (or A or B) a cell with lattice points on two opposite faces. Trigonal lattices may belong to the rhombohedral or hexagonal systems (Table 15A.1).

Brief illustration 15A.1

The two-dimensional lattice shown in Fig. 15A.9 consists of a rectangular array of lattice points, with one additional point at the centre of each rectangle; a (non-primitive) unit cell is indicated. This cell has twofold axes of symmetry passing through the mid-points of opposite sides of the rectangle. Rotations about these axes interchange the lattice points at the corners of the rectangle, but the lattice point at the centre is not affected. It therefore follows that the lattice points at the corners are equivalent, but the lattice point at the centre is distinct.

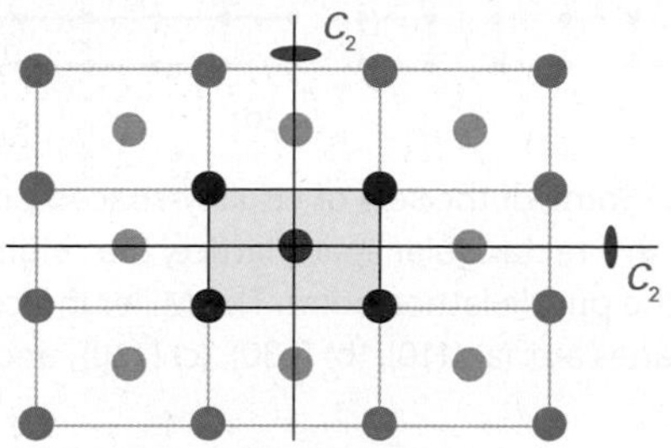

Figure 15A.9 The two-dimensional lattice used in *Brief illustration* 15A.1; a (non-primitive) unit cell is indicated by the shaded area. The lattice points at the corners of the unit cell are related by rotations about the twofold axes of symmetry shown; the lattice point at the centre is unaffected by these operations.

15A.2 The identification of lattice planes

The interpretation of the diffraction techniques that are used to measure the size of unit cells and the arrangement of molecules within them makes use of the orientation and separation of planes that pass through the crystal (Topic 15B). Two-dimensional lattices are easier to visualize than three-dimensional lattices, so in this discussion the concepts involved in identifying lattice planes are introduced for two dimensions initially, and then the results are extended by analogy to three dimensions. Note that lattice planes do not necessarily pass through lattice points.

(a) The Miller indices

Consider a two-dimensional rectangular lattice formed from a unit cell of sides a and b (Fig. 15A.10). Each panel in the illustration shows a set of evenly spaced planes that can be identified by considering, for each set, the plane lying closest to the origin (but not passing through it) and then quoting the distances at which this plane intersects the a and b axes. These distances are: (a) $(1a,1b)$, (b) $(\frac{1}{2}a,\frac{1}{3}b)$, (c) $(-1a,1b)$, and (d) $(\infty a,1b)$, with ∞ indicating that the plane is parallel to an axis and intersects it (notionally) at infinity. If it is agreed to quote distances along the axes as multiples of the

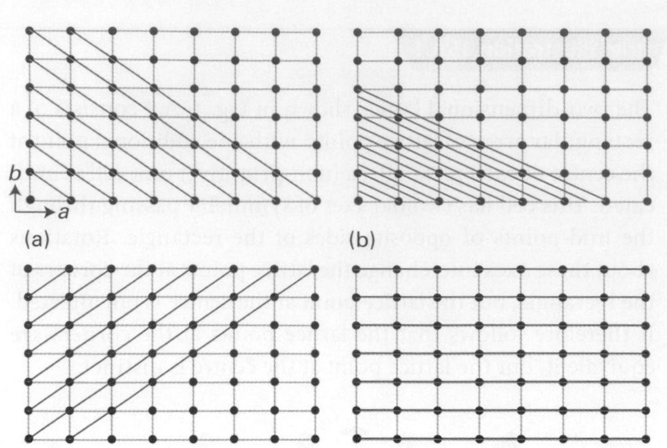

Figure 15A.10 Some of the sets of equally-spaced planes that can be drawn in a rectangular space lattice; the origin is indicated by the purple lattice point. The Miller indices {*hkl*} of each set of planes are: (a) {110}, (b) {230}, (c) {$\bar{1}$10}, and (d) {010}.

corresponding dimensions of the unit cell, then these intersections can be expressed more simply as (1,1), ($\frac{1}{2}$, $\frac{1}{3}$), (−1,1), and (∞,1), respectively. If the lattice in Fig. 15A.10 is the top view of a three-dimensional orthorhombic lattice, all four sets of planes intersect the *c* axis at infinity. Therefore, in the three-dimensional case the labels are (1,1,∞), ($\frac{1}{2}$, $\frac{1}{3}$,∞), (−1,1,∞), and (∞,1,∞).

The inconvenience of fractions and infinity in these labels can be avoided by specifying a plane by using its **Miller indices**, (*hkl*), where *h*, *k*, and *l* are the reciprocals of the intersection distances along the *a*, *b*, and *c* axes, respectively. For example, the plane ($\frac{1}{2}$,$\frac{1}{3}$,∞) has Miller indices (230). As will be seen, this notation brings with it additional advantages. The Miller notation has the following features:

- Negative indices are written with a bar over the number, as in ($\bar{1}$10).
- The notation (*hkl*) denotes an individual plane. A set of parallel planes with identical spacing, is denoted {*hkl*}.

For example,

Intersect axes at	(*a*,*b*,∞*c*)	($\frac{1}{2}a$,$\frac{1}{3}b$,∞*c*)	(−*a*,*b*,∞*c*)	(∞*a*,*b*,∞*c*)
Remove cell dimensions	(1,1,∞)	($\frac{1}{2}$,$\frac{1}{3}$,∞)	(−1,1,∞)	(∞,1,∞)
Take reciprocals	(1,1,0)	(2,3,0)	(−1,1,0)	(0,1,0)
Express as indices	(110)	(230)	($\bar{1}$10)	(010)
Sets of parallel planes	{110}	{230}	{$\bar{1}$10}	{010}

A helpful feature to remember is that the smaller the absolute value of *h* in {*hkl*}, the more nearly parallel the set of planes is to the *a* axis (the {*h*00} planes are an exception). The same is true of *k* and the *b* axis, and *l* and the *c* axis. When *h* = 0, the planes intersect the *a* axis at infinity, so the {0*kl*} planes are parallel to the *a* axis. Similarly, the {*h*0*l*} planes are parallel to the *b* axis and the {*hk*0} planes are parallel to the *c* axis.

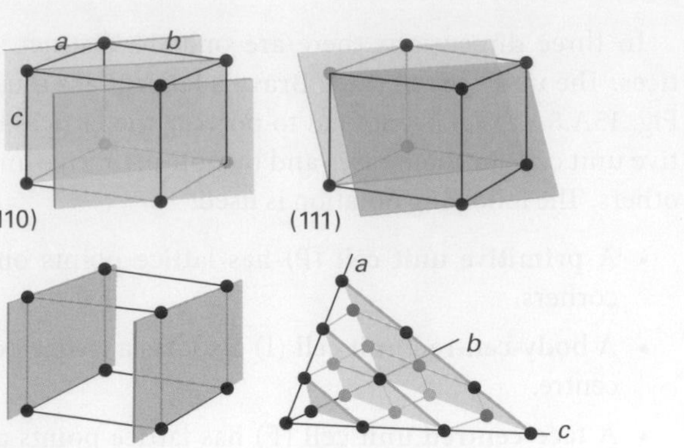

Figure 15A.11 Some representative planes in three dimensions and their Miller indices; the origin is indicated by the purple lattice point. Note that the index 0 indicates that a plane is parallel to the corresponding axis, and that the indexing may also be used for unit cells with non-orthogonal axes.

Figure 15A.11 shows a three-dimensional representation of a selection of planes, including one in a lattice with non-orthogonal axes.

(b) The separation of neighbouring planes

The Miller indices are very useful for expressing the separation of neighbouring planes.

How is that done? 15A.1 Deriving an expression for the separation of planes

Consider the {*hk*0} planes of a square lattice built from a unit cell with sides of length *a* (Fig. 15A.12). The separation between the lattice planes is equal to the perpendicular distance from the (*hk*0) plane to the origin. Expressions for the sine and cosine of the angle ϕ are found by considering the sides of the two right-angle triangles shown in the figure

$$\sin\phi = \frac{d_{hk0}}{(a/h)} = \frac{hd_{hk0}}{a} \qquad \cos\phi = \frac{d_{hk0}}{(a/k)} = \frac{kd_{hk0}}{a}$$

The length of the hypotenuse of the lower triangle is *a*/*h* because a Miller index *h* indicates that the plane intersects the *a*

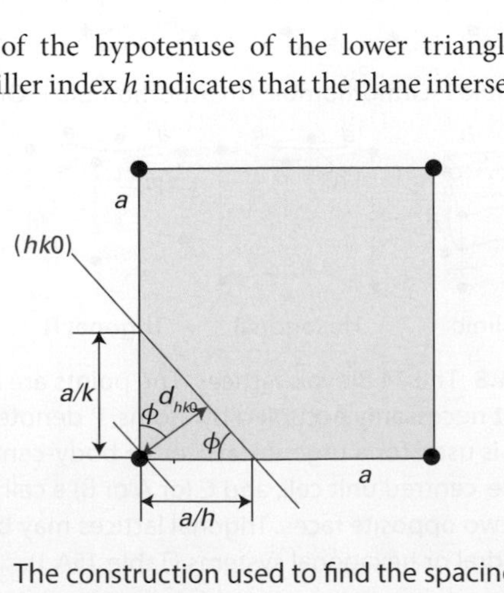

Figure 15A.12 The construction used to find the spacing of the (*hk*0) plane in a square unit cell.

axis at a distance a/h from the origin. Likewise, the hypotenuse of the upper triangle is a/k. Then, because $\sin^2\phi + \cos^2\phi = 1$, it follows that

$$\left(\frac{hd_{hk0}}{a}\right)^2 + \left(\frac{kd_{hk0}}{a}\right)^2 = 1$$

which can be rearranged by dividing both sides by d_{hk0}^2 into

$$\frac{1}{d_{hk0}^2} = \frac{h^2+k^2}{a^2} \quad \text{or} \quad d_{hk0} = \frac{a}{(h^2+k^2)^{1/2}}$$

By extension to three dimensions, the separation of the $\{hkl\}$ planes, d_{hkl}, of a cubic lattice is given by

$$\frac{1}{d_{hkl}^2} = \frac{h^2+k^2+l^2}{a^2} \qquad d_{hkl} = \frac{a}{(h^2+k^2+l^2)^{1/2}} \qquad \text{Separation of planes [cubic lattice]} \quad (15\text{A}.1\text{a})$$

The corresponding expression for a general orthorhombic lattice (one in which the axes are mutually perpendicular, but not equal in length) is the generalization of this expression:

$$\frac{1}{d_{hkl}^2} = \frac{h^2}{a^2} + \frac{k^2}{b^2} + \frac{l^2}{c^2} \qquad \text{Separation of planes [orthorhombic lattice]} \quad (15\text{A}.1\text{b})$$

Example 15A.1 Using the Miller indices

Calculate the separation of (a) the {123} planes and (b) the {246} planes of an orthorhombic unit cell with $a = 0.82$ nm, $b = 0.94$ nm, and $c = 0.75$ nm.

Collect your thoughts For the first part, all you need to do is substitute the given values into eqn 15A.1b. You could do the same for part (b), but note that the Miller indices for the second set of planes are just twice those of the first part. By referring to eqn 15A.1b you can see that multiplying the values of h, k, and l by n gives the following expression for the spacing of the $\{nh\ nk\ nl\}$ planes

$$\frac{1}{d_{nh,nk,nl}^2} = \frac{(nh)^2}{a^2} + \frac{(nk)^2}{b^2} + \frac{(nl)^2}{c^2} = n^2\overbrace{\left(\frac{h^2}{a^2} + \frac{k^2}{b^2} + \frac{l^2}{c^2}\right)}^{=1/d_{hkl}^2} = \frac{n^2}{d_{hkl}^2}$$

which implies that

$$d_{nh,nk,nl} = \frac{d_{hkl}}{n}$$

The solution Substituting the indices into eqn 15A.1b gives

$$\frac{1}{d_{123}^2} = \frac{1^2}{(0.82\,\text{nm})^2} + \frac{2^2}{(0.94\,\text{nm})^2} + \frac{3^2}{(0.75\,\text{nm})^2} = 22.0\ldots\,\text{nm}^{-2}$$

Hence, $d_{123} = 0.21$ nm. It then follows immediately that d_{246} is one-half this value, or 0.11 nm.

Self-test 15A.1 Calculate the separation of (a) the {133} planes and (b) the {399} planes in the same lattice.

Answer: 0.19 nm, 0.063 nm

Checklist of concepts

☐ 1. A **periodic crystal** is built up from regularly repeating structural motifs.

☐ 2. A **space lattice** is the pattern formed by points (lattice points) representing the locations of structural motifs (atoms, molecules, or groups of atoms, molecules, or ions).

☐ 3. A **unit cell** is an imaginary parallel-sided figure from which the entire space lattice can be constructed by purely translational displacements.

☐ 4. A **primitive unit cell** has lattice points only at its vertices and only one lattice point overall; **non-primitive unit cells** also have lattice points at their centres or on pairs of opposite faces.

☐ 5. Unit cells are classified into seven **crystal systems** according to their rotational symmetries: unit cells are classified as **cubic, monoclinic,** or **triclinic** according to the **essential symmetries** they possess.

☐ 6. The **Bravais lattices** are the 14 distinct space lattices in three dimensions (Fig. 15A.8).

☐ 7. The unit cells of the Bravais lattices are classed as **primitive** (P), **body-centred** (I), **face-centred** (F), and **side-centred** (A, B, or C).

☐ 8. A lattice plane is specified by a set of **Miller indices** (hkl); sets of planes are denoted $\{hkl\}$.

Checklist of equations

Property	Equation	Comment	Equation number
Separation of planes in a cubic lattice	$1/d_{hkl}^2 = (h^2 + k^2 + l^2)/a^2$	h, k, and l are Miller indices	15A.1a
Separation of planes in an orthorhombic lattice	$1/d_{hkl}^2 = h^2/a^2 + k^2/b^2 + l^2/c^2$		15A.1b

TOPIC 15B Diffraction techniques

➤ **Why do you need to know this material?**

To account for the properties of solids it is necessary to understand their detailed structures and how they are determined by a variety of diffraction techniques.

➤ **What is the key idea?**

The regular arrangement of the atoms in periodic crystals can be determined by techniques based on diffraction.

➤ **What do you need to know already?**

You need to be familiar with the description of crystal structures and the use of Miller indices to identify lattice planes (Topic 15A). You also need to be familiar with the wave description of electromagnetic radiation (*The chemist's toolkit* 13 in Topic 7A), and the basic properties of the Fourier transform (*The chemist's toolkit* 28 in Topic 12C). Light use is made of the de Broglie relation (Topic 7A) and the equipartition theorem (*The chemist's toolkit* 7 in Topic 2A).

Diffraction techniques can be used to determine the details of the arrangement of ions, atoms, and molecules in a crystalline solid to high precision. Such techniques are now so well developed that both the collection of the diffraction data and its interpretation in terms of a structure are automated to a high degree.

15B.1 X-ray crystallography

As explained in *The chemist's toolkit* 13 (in Topic 7A), a characteristic property of waves is that when they are present in the same region of space they interfere with one another. A greater displacement is obtained where peaks or troughs of the waves coincide and a smaller displacement where peaks coincide with troughs. Diffraction is the interference caused by an object in the path of waves; it occurs when the dimensions of the diffracting object are comparable to the wavelength of the radiation.

(a) X-ray diffraction

X-rays diffract when passed through a crystal because their wavelengths are comparable to the separation of lattice planes. Consequently, X-ray diffraction is a very powerful technique for structural studies of solid materials. The actual process of going from the observed diffraction pattern to a structure is rather involved, but such is the degree of integration of computers into the experimental apparatus that the technique is almost fully automated, even for large molecules and complex solids. The analysis is aided by molecular modelling techniques, which can guide the investigation towards a plausible structure.

X-rays are electromagnetic radiation with wavelengths of the order of 10^{-10} m. They are typically generated by bombarding a metal with high-energy electrons (Fig. 15B.1). The electrons decelerate as they plunge into the metal and generate radiation with a continuous range of wavelengths called **Bremsstrahlung** (*Bremse* is German for deceleration, *Strahlung* for ray). Superimposed on the continuum are a few high-intensity, sharp peaks (Fig. 15B.2). These peaks arise from collisions of the incoming electrons with the electrons in the inner shells of the atoms. The collision expels an electron from an inner shell, and an electron of higher energy drops into the vacancy, emitting the excess energy as an X-ray photon (Fig. 15B.3). If the electron falls into a K shell (a shell with $n = 1$), the X-rays are classified as 'K-radiation', and similarly for transitions into the L ($n = 2$) and M ($n = 3$) shells. Strong, distinct lines are labelled K_α, K_β, and so on. Synchrotrons (Topic 11A) generate high-intensity X-ray radiation which is increasingly used in diffraction experiments on account of the resulting greater intensity in the diffraction pattern, and hence higher sensitivity.

An early method of observing diffraction consisted of passing a beam containing X-rays with a range of wavelengths into

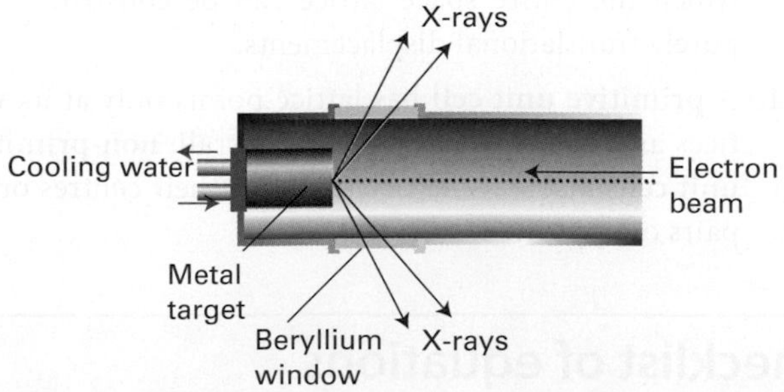

Figure 15B.1 X-rays are generated by directing an electron beam on to a cooled metal target. Beryllium is transparent to X-rays (on account of the small number of electrons in each atom) and is used for the windows.

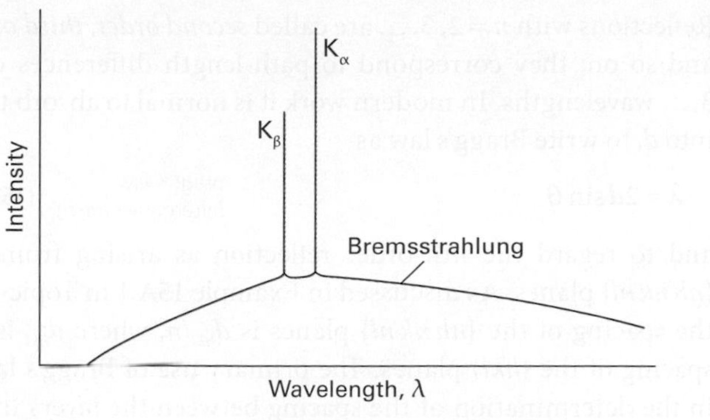

Figure 15B.2 The X-ray emission from a metal consists of a broad, featureless Bremsstrahlung background, with sharp peaks superimposed on it. The label K indicates that the radiation comes from a transition in which an electron falls into a vacancy in the K shell of the atom.

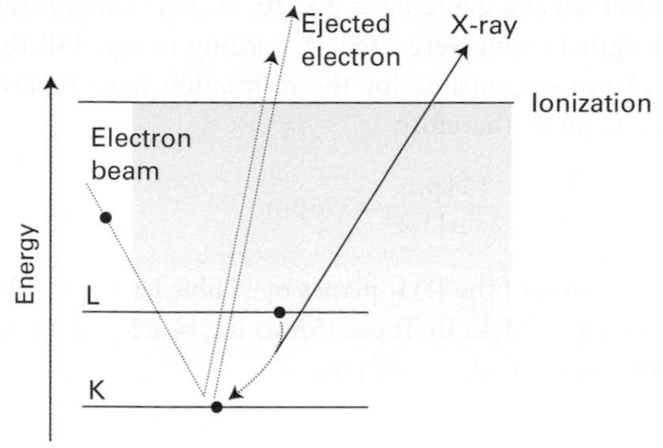

Figure 15B.3 The processes that contribute to the generation of X-rays. An incoming electron collides with an electron (in the K shell), and ejects it. Another electron (from the L shell in this illustration) falls into the vacancy and emits its excess energy as an X-ray photon.

a single crystal, and recording the diffraction pattern photographically. The idea behind this approach is that a crystal might not be suitably orientated to cause diffraction for a single wavelength but, whatever its orientation, diffraction would be achieved for at least one of the wavelengths present in the beam. There is currently a resurgence of interest in this approach because synchrotron radiation spans a range of X-ray wavelengths.

A more common technique uses monochromatic radiation and a powdered sample, which consists of many tiny crystallites, oriented at random. At least some of the crystallites will be appropriately orientated to give diffraction. In modern 'powder diffractometers' the intensities of the reflections are monitored electronically as the detector is rotated around the sample in a plane containing the incident ray. Powder diffraction techniques are used to identify the composition of a sample of a solid substance by comparison of the positions of the diffraction lines and their intensities with previously recorded

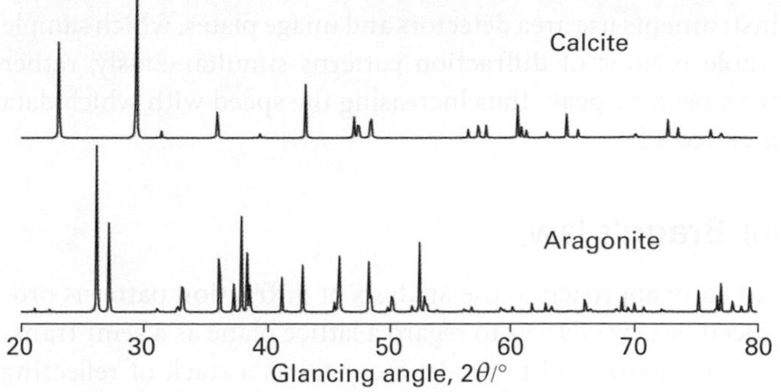

Figure 15B.4 X-ray powder diffraction patterns of two polymorphs of $CaCO_3$, calcite and aragonite. The patterns are distinctive and can be used to identify the polymorph present in an unknown sample.

patterns of known structures (Fig. 15B.4); large databases of such information are available. This approach can be used to determine the composition of mixed phases, and hence to construct a phase diagram. The technique is also used for the initial determination of the dimensions and symmetries of unit cells.

The method developed by the Braggs (William and his son Lawrence) is the foundation of almost all modern work in X-ray crystallography. They used a single crystal and a monochromatic beam of X-rays, and rotated the crystal until a reflection was detected. There are many different sets of planes in a crystal, so there are many angles at which a reflection occurs. The complete set of data consists of the list of angles at which reflections are observed and their intensities.

Single-crystal diffraction patterns are measured by using a 'four-circle diffractometer' (Fig. 15B.5). Once the dimensions and symmetry of the unit cell of the crystal in question have been identified, the angular setting of the detector on the four circles is adjusted so that the precise position and intensity of

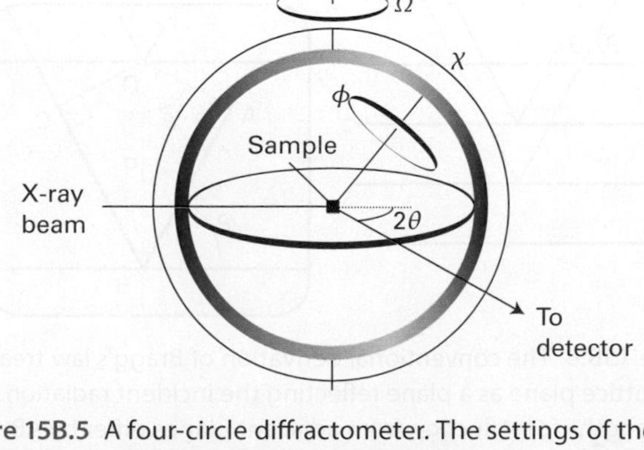

Figure 15B.5 A four-circle diffractometer. The settings of the orientations (ϕ, χ, θ, and Ω) of the components are controlled by computer; each (*hkl*) reflection is monitored in turn, and their intensities are recorded.

each peak in the diffraction pattern can be measured. Modern instruments use area detectors and image plates, which sample whole regions of diffraction patterns simultaneously, rather than peak by peak, thus increasing the speed with which data is collected.

(b) Bragg's law

An early approach to the analysis of diffraction patterns produced by crystals was to regard a lattice plane as a semi-transparent mirror and to model a crystal as a stack of reflecting lattice planes of separation d (Fig. 15B.6). The model makes it easy to calculate the angle the crystal must make to the incoming beam of X-rays for constructive interference to occur. It has also given rise to the name **reflection** to denote an intense beam arising from constructive interference.

Consider the reflection of two parallel rays of the same wavelength and phase by two adjacent planes of a lattice, as shown in Fig. 15B.6. One ray strikes point D on the upper plane but the other ray must travel an additional distance AB before striking the plane immediately below. The reflected rays also differ in path length by the distance BC. As is evident from the inset in Fig. 15B.6, both the lengths AB and BC are $d\sin\theta$; the total path length difference of the two rays is then

$$\mathrm{AB}+\mathrm{BC}=2d\sin\theta$$

where 2θ is the **glancing angle** (2θ rather than θ, because the beam is deflected through 2θ from its initial direction). For many glancing angles the path-length difference is not an integer number of wavelengths, and the waves interfere largely destructively. However, when the path-length difference is an integer number of wavelengths ($\mathrm{AB}+\mathrm{BC}=n\lambda$), the reflected waves are in phase and interfere constructively. It follows that a reflection should be observed when θ satisfies **Bragg's law**:

$$n\lambda=2d\sin\theta \qquad \text{Bragg's law} \qquad (15\text{B}.1\text{a})$$

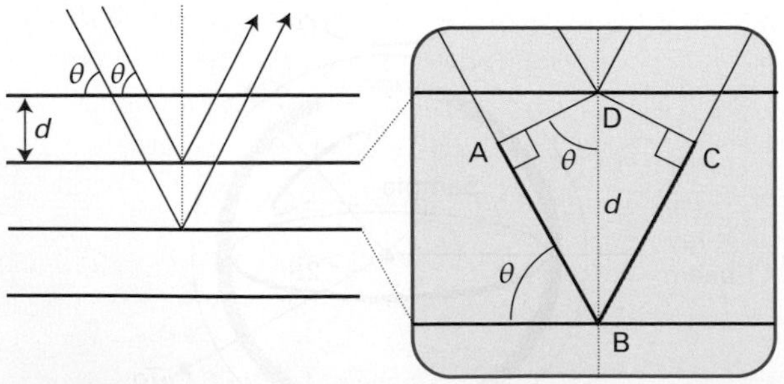

Figure 15B.6 The conventional derivation of Bragg's law treats each lattice plane as a plane reflecting the incident radiation. The path lengths for reflection from adjacent planes differ by AB + BC, which depends on the angle θ. Constructive interference (a 'reflection') occurs when AB + BC is equal to an integer number of wavelengths.

Reflections with $n = 2, 3, \ldots$ are called *second order, third order*, and so on; they correspond to path-length differences of 2, 3,... wavelengths. In modern work it is normal to absorb the n into d, to write Bragg's law as

$$\lambda=2d\sin\theta \qquad \text{Bragg's law [alternative form]} \qquad (15\text{B}.1\text{b})$$

and to regard the nth-order reflection as arising from the $\{nh\,nk\,nl\}$ planes. As discussed in Example 15A.1 in Topic 15A, the spacing of the $\{nh\,nk\,nl\}$ planes is d_{hkl}/n, where d_{hkl} is the spacing of the $\{hkl\}$ planes. The primary use of Bragg's law is in the determination of the spacing between the layers in the lattice because d may readily be calculated from a measured value of the angle θ.

Brief illustration 15B.1

A first-order reflection from the {111} planes of a cubic crystal was observed at a glancing angle, 2θ, of 22.4° when X-rays of wavelength 154 pm were used. According to eqn 15B.1b, the {111} planes responsible for the diffraction have separation $d_{111} = \lambda/(2\sin\theta)$, therefore

$$d_{111}=\frac{\lambda}{2\sin\theta}=\frac{154\,\text{pm}}{2\sin 11.2°}=396\,\text{pm}$$

The separation of the {111} planes of a cubic lattice of side a is given by eqn 15A.1a in Topic 15A as $d_{111} = a/3^{1/2}$. It therefore follows that $a = 3^{1/2}d_{111} = 687$ pm.

Some types of unit cell give characteristic patterns of lines. In a cubic lattice with dimension a, the spacing of the $\{hkl\}$ planes, d_{hkl}, is given by eqn 15A.1a in Topic 15A ($d_{hkl} = a/(h^2+k^2+l^2)^{1/2}$); the angles at which the $\{hkl\}$ planes give first-order reflections are given by

$$\sin\theta=\frac{\lambda}{2d_{hkl}}=(h^2+k^2+l^2)^{1/2}\frac{\lambda}{2a} \qquad [d_{hkl}=a/(h^2+k^2+l^2)^{1/2}]$$

Not all integral values of $h^2+k^2+l^2$ are obtained when integer values of the indices are substituted:

$\{hkl\}$	{100}	{110}	{111}	{200}	{210}
$h^2+k^2+l^2$	1	2	3	4	5
$\{hkl\}$	{211}	{220}	{300}	{221}	{310}
$h^2+k^2+l^2$	6	8	9	9	10

Notice that $h^2+k^2+l^2 = 7$ (and 15,...) does not appear. As a result, in the pattern of diffraction lines there is a larger gap between the {211} and {220} reflections than between nearby lines, and likewise between {321} (for which $h^2+k^2+l^2 = 14$) and {400} (for which $h^2+k^2+l^2 = 16$). The absence of these lines leads to a characteristic pattern which helps in identifying the type of unit cell.

(c) Scattering factors

X-ray scattering is caused by the oscillations an incoming electromagnetic wave generates in the electrons of atoms. Heavy, electron-rich atoms give rise to stronger scattering than light atoms. This dependence on the number of electrons is expressed in terms of the **scattering factor**, f, of the element. If the scattering factor is large, then the atoms scatter X-rays strongly. It turns out that, in a spherically symmetrical atom, the scattering factor for the atom is related to the electron density, $\rho(r)$, and the glancing angle, 2θ, by

$$f(\theta)=4\pi\int_0^{\infty}\rho(r)\frac{\sin kr}{kr}r^2\mathrm{d}r \quad k=\frac{4\pi}{\lambda}\sin\theta \quad \text{Scattering factor} \quad (15B.2)$$

The scattering factor is greatest in the forward direction ($\theta = 0$, Fig. 15B.7), and it can be shown that in that direction it is equal to the total number of electrons in the atom, N_e.

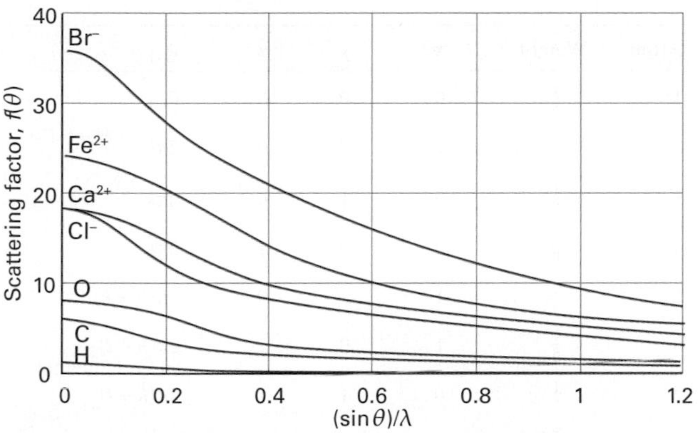

Figure 15B.7 The variation of the scattering factor of atoms and ions with atomic number and angle. The scattering factor in the forward direction (at $\theta = 0$, and hence at $(\sin\theta)/\lambda = 0$) is equal to the number of electrons present in the species.

How is that done? 15B.1 Evaluating the scattering factor in the forward direction

The first step is to note that because $\sin kr$ cannot exceed 1, the maximum value of $(\sin kr)/kr$ occurs as $k \to 0$, which corresponds to $\sin\theta \to 0$ and therefore $\theta \to 0$. Therefore, the scattering factor has its maximum value in the forward direction.

To evaluate the scattering factor in the forward direction you need to take the limit $k \to 0$, and therefore $kr \to 0$ in eqn 15B.2. Therefore, use $\sin x = x - \frac{1}{6}x^3 + \cdots$ and write

$$\lim_{kr\to 0}\frac{\sin kr}{kr}=\lim_{kr\to 0}\frac{kr-\frac{1}{6}(kr)^3+\cdots}{kr}=\lim_{kr\to 0}\{1-\tfrac{1}{6}(kr)^2+\cdots\}=1$$

In this limit eqn 15B.2 simplifies to

$$f(0)=\lim_{kr\to 0}4\pi\int_0^{\infty}\rho(r)\overbrace{\frac{\sin kr}{kr}}^{\to 1}r^2\mathrm{d}r$$

$$=4\pi\int_0^{\infty}\rho(r)r^2\mathrm{d}r=\int_0^{\infty}\rho(r)\overbrace{4\pi r^2\mathrm{d}r}^{\text{volume element}}$$

The factor $4\pi r^2\mathrm{d}r$ is the volume of a spherical shell of radius r and thickness $\mathrm{d}r$. The total number of electrons in this shell is the electron density at r multiplied by the volume of the shell, $4\pi r^2\rho(r)\mathrm{d}r$, and this number summed over shells of all radii is the total number of electrons in the atom. Hence, in the forward direction, $f = N_e$. For example, the scattering factors of Na^+, K^+, and Cl^- are 8, 18, and 18, respectively.

(d) The electron density

The **structure factor**, F_{hkl}, is the net amplitude of a given $\{hkl\}$ reflection that takes into account the positions and types of all the atoms in the unit cell. It can be expressed in terms of the locations and scattering factors of the atoms.

How is that done? 15B.2 Relating the structure factor to the location of the atoms and their scattering factors

Suppose that a unit cell contains several atoms with scattering factors f_j and coordinates (x_ja,y_jb,z_jc), where x_j is the coordinate of the atom j in the a direction, expressed as a fraction of the length a, and likewise for the other coordinates.

Step 1 *Consider the (h00) reflection with h = 1*

The reflection shown in Fig. 15B.8 corresponds to two waves from adjacent A planes; the phase difference of the waves is 2π. If there is a B atom at a fraction x of the distance between the two A planes, then it gives rise to a wave with a phase difference $2\pi x$ relative to an A reflection. To see this conclusion, note that, if $x = 0$, there is no phase difference; if $x = \frac{1}{2}$ the phase difference is π; if $x = 1$, the B atom lies where the lower A atom is and the phase difference is 2π.

Step 2 *Consider the (h00) reflection with h = 2*

In this case, there is a $2 \times 2\pi$ difference between the waves from the two A layers, and if B were to lie at $x = \frac{1}{2}$ it would give rise to a wave that differed in phase by 2π from the wave from the lower A layer. Thus, for a general fractional position x, the phase difference for a (200) reflection is $2 \times 2\pi x$.

Step 3 *Generalize these conclusions*

For a general ($h00$) reflection, the phase difference is $h \times 2\pi x$. For three dimensions, this result generalizes to $\phi_{hkl} = 2\pi(hx + ky + lz)$.

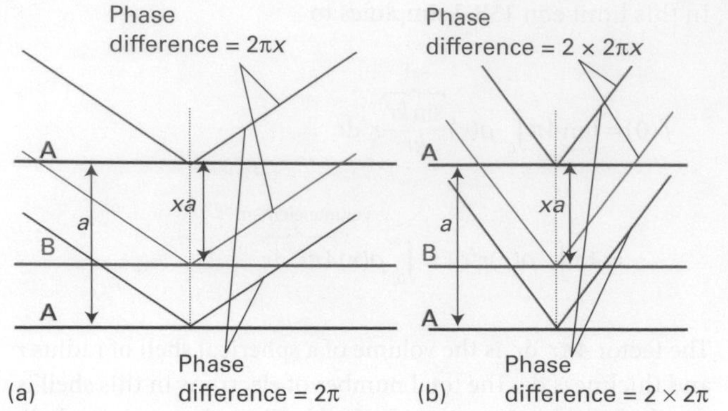

Figure 15B.8 Diffraction from a crystal containing two kinds of atoms. (a) For a (100) reflection from the A planes, there is a phase difference of 2π between waves reflected by neighbouring planes. (b) For a (200) reflection, the phase difference is 4π. The reflection from a B plane at a fractional distance *xa* from an A plane has a phase that is *x* times these phase differences.

Step 4 *Formulate the total amplitude of the scattered waves*

If the amplitude of the waves scattered from A is f_A at the detector, that of the waves scattered from B with phase difference ϕ_{hkl} is $f_B e^{i\phi_{hkl}}$. The total amplitude at the detector is therefore

$$F_{hkl} = f_A + f_B e^{i\phi_{hkl}}$$

When there are several atoms present, each with scattering factor f_j and phase $\phi_{hkl}(j) = 2\pi(hx_j + ky_j + lz_j)$, the total amplitude of the (*hkl*) reflection, the structure factor, is

$$F_{hkl} = \sum_j f_j e^{i\phi_{hkl}(j)} \qquad (15B.3)$$

Structure factor

Example 15B.1 Calculating a structure factor

Calculate the structure factor for the unit cell of NaCl depicted in Fig. 15B.9.

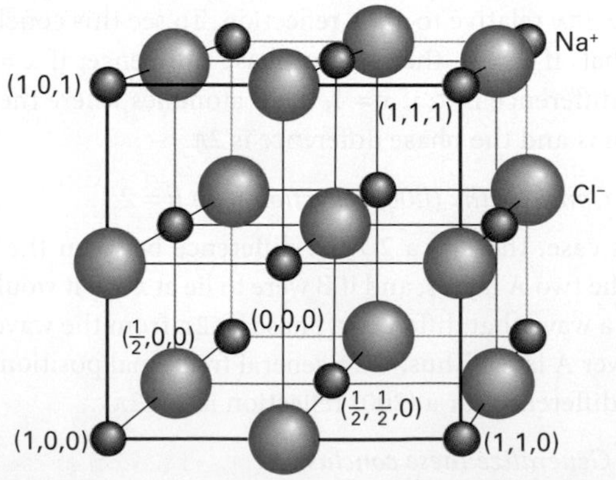

Figure 15B.9 The location of the atoms for the structure factor calculation in Example 15B.1. The red spheres are Na^+, the green spheres are Cl^-.

Collect your thoughts You need to evaluate the sum in eqn 15B.3. The sum is over all atoms in the unit cell, so you need to know the location of each atom, expressed as a fraction of the unit cell parameters. Several of the atomic coordinates are already marked in the figure. The atoms at the corners of the cube are shared between eight adjacent unit cells, so in the calculation of the structure factor these atoms are given a weight of $\frac{1}{8}$ and their scattering factors taken to be $\frac{1}{8}f$. The atoms on the faces are shared between two cells and so have a weight of $\frac{1}{2}$; those on the edges are shared between four cells, and so have a weight of $\frac{1}{4}$. Write f^+ for the Na^+ scattering factor and f^- for the Cl^- scattering factor; for simplicity, ignore the fact that scattering occurs in non-forward directions and suppose that all the Na^+ have the same scattering factor, and likewise for the Cl^- ions. The best way to proceed is to draw up a table showing the weights, positions, and phases. The phase factors $e^{i\phi_{hkl}}$ are evaluated by noting that *h*, *k*, and *l* are integers. A useful identity is that $e^{in\pi}$ is +1 for even *n*, and −1 for odd *n*, more succinctly expressed as $e^{in\pi} = (-1)^n$.

The solution The table for the Na^+ ions is

Atom	Weight	x	y	z	ϕ_{hkl}
1	$\frac{1}{8}$	0	0	0	0
2	$\frac{1}{8}$	1	0	0	$2\pi h$
3	$\frac{1}{8}$	0	1	0	$2\pi k$
4	$\frac{1}{8}$	0	0	1	$2\pi l$
5	$\frac{1}{8}$	1	1	0	$2\pi(h+k)$
6	$\frac{1}{8}$	1	0	1	$2\pi(h+l)$
7	$\frac{1}{8}$	0	1	1	$2\pi(k+l)$
8	$\frac{1}{8}$	1	1	1	$2\pi(h+k+l)$
9	$\frac{1}{2}$	$\frac{1}{2}$	$\frac{1}{2}$	0	$2\pi(\frac{1}{2}h+\frac{1}{2}k)$
10	$\frac{1}{2}$	$\frac{1}{2}$	0	$\frac{1}{2}$	$2\pi(\frac{1}{2}h+\frac{1}{2}l)$
11	$\frac{1}{2}$	0	$\frac{1}{2}$	$\frac{1}{2}$	$2\pi(\frac{1}{2}k+\frac{1}{2}l)$
12	$\frac{1}{2}$	1	$\frac{1}{2}$	$\frac{1}{2}$	$2\pi(h+\frac{1}{2}k+\frac{1}{2}l)$
13	$\frac{1}{2}$	$\frac{1}{2}$	1	$\frac{1}{2}$	$2\pi(\frac{1}{2}h+k+\frac{1}{2}l)$
14	$\frac{1}{2}$	$\frac{1}{2}$	$\frac{1}{2}$	1	$2\pi(\frac{1}{2}h+\frac{1}{2}k+l)$

The phase factors for the first eight Na atoms in the table are all +1, and as they each have a weight of $\frac{1}{8}$, the total contribution to the structure factor is f^+. The remaining six atoms all have weight $\frac{1}{2}$, and their contribution to the structure factor is

$$\tfrac{1}{2}[e^{i2\pi(h/2+k/2)} + e^{i2\pi(h/2+l/2)} + e^{i2\pi(k/2+l/2)} + e^{i2\pi(h+k/2+l/2)} + e^{i2\pi(h/2+k+l/2)} + e^{i2\pi(h/2+k/2+l)}]$$

The terms $e^{i2\pi h}$, $e^{i2\pi k}$, and $e^{i2\pi l}$ are all +1, so the last three terms can be simplified

$$\tfrac{1}{2}[e^{i2\pi(h/2+k/2)} + e^{i2\pi(h/2+l/2)} + e^{i2\pi(k/2+l/2)} + e^{i2\pi(k/2+l/2)} + e^{i2\pi(h/2+l/2)} + e^{i2\pi(h/2+k/2)}]$$

A further simplification is to use $e^{in\pi} = (-1)^n$:

$$\tfrac{1}{2}[(-1)^{h+k} + (-1)^{h+l} + (-1)^{k+l} + (-1)^{k+l} + (-1)^{h+l} + (-1)^{h+k}]$$
$$= (-1)^{h+k} + (-1)^{h+l} + (-1)^{k+l}$$

The overall contribution of the Na^+ ions to the structure factor is therefore

$$f^+[1 + (-1)^{h+k} + (-1)^{h+l} + (-1)^{k+l}]$$

The table for the Cl^- ions is

Atom	Weight	x	y	z	ϕ_{hkl}
1	$\tfrac{1}{4}$	$\tfrac{1}{2}$	0	0	πh
2	$\tfrac{1}{4}$	0	$\tfrac{1}{2}$	0	πk
3	$\tfrac{1}{4}$	1	$\tfrac{1}{2}$	0	$2\pi(h + \tfrac{1}{2}k)$
4	$\tfrac{1}{4}$	$\tfrac{1}{2}$	1	0	$2\pi(\tfrac{1}{2}h + k)$
5	$\tfrac{1}{4}$	0	0	$\tfrac{1}{2}$	πl
6	$\tfrac{1}{4}$	1	0	$\tfrac{1}{2}$	$2\pi(h + \tfrac{1}{2}l)$
7	$\tfrac{1}{4}$	0	1	$\tfrac{1}{2}$	$2\pi(k + \tfrac{1}{2}l)$
8	$\tfrac{1}{4}$	1	1	$\tfrac{1}{2}$	$2\pi(h + k + \tfrac{1}{2}l)$
9	1	$\tfrac{1}{2}$	$\tfrac{1}{2}$	$\tfrac{1}{2}$	$2\pi(\tfrac{1}{2}h + \tfrac{1}{2}k + \tfrac{1}{2}l)$
10	$\tfrac{1}{4}$	$\tfrac{1}{2}$	0	1	$2\pi(\tfrac{1}{2}h + l)$
11	$\tfrac{1}{4}$	0	$\tfrac{1}{2}$	1	$2\pi(\tfrac{1}{2}k + l)$
12	$\tfrac{1}{4}$	1	$\tfrac{1}{2}$	1	$2\pi(h + \tfrac{1}{2}k + l)$
13	$\tfrac{1}{4}$	$\tfrac{1}{2}$	1	1	$2\pi(\tfrac{1}{2}h + k + l)$

A similar procedure to that used for the Na^+ ions gives the following contribution of the Cl^- ions to the structure factor

$$f^-[(-1)^{h+k+l} + (-1)^h + (-1)^k + (-1)^l]$$

The structure factor is therefore

$$F_{hkl} = f^+[1 + (-1)^{h+k} + (-1)^{h+l} + (-1)^{k+l}] + f^-[(-1)^{h+k+l} + (-1)^h + (-1)^k + (-1)^l]$$

Now note that:

- if h, k, and l are all even, $F_{hkl} = f^+\{1 + 1 + 1 + 1\} + f^-\{1 + 1 + 1 + 1\} = 4(f^+ + f^-)$
- if h, k, and l are all odd, $F_{hkl} = 4(f^+ - f^-)$
- if one index is odd and two are even, or vice versa, $F_{hkl} = 0$

The hkl all-odd reflections are therefore less intense than the hkl all even, and some of the reflections are absent.

Comment. If $f^+ = f^-$, which is the case for identical atoms, the hkl all-odd reflections have zero intensity; such a structure would be a cubic P lattice with lattice parameter $a/2$.

Self-test 15B.1 Which reflections cannot be observed for a cubic I lattice?

Answer: for $h + k + l$ odd, $F_{hkl} = 0$

The intensity of a reflection is proportional to the square modulus of the amplitude of the wave, which is in turn proportional to the structure factor, F_{hkl}. If the structure factor is $f_A + f_B e^{i\phi_{hkl}}$, the intensity, I_{hkl}, is

$$I_{hkl} \propto F_{hkl}^\star F_{hkl} = (f_A + f_B e^{-i\phi_{hkl}})(f_A + f_B e^{i\phi_{hkl}})$$
$$= f_A^2 + f_B^2 + f_A f_B (e^{i\phi_{hkl}} + e^{-i\phi_{hkl}})$$

$e^{ix} + e^{-ix} = 2\cos x$

$$= f_A^2 + f_B^2 + 2 f_A f_B \cos\phi_{hkl}$$

The cosine term either adds to or subtracts from $f_A^2 + f_B^2$ depending on the value of ϕ_{hkl}, which in turn depends on h, k, and l and x, y, and z. Hence, there is a variation in the intensities of the reflections with different hkl. The A and B reflections interfere destructively when the phase difference is π, and in this case the total intensity is zero if the atoms have the same scattering power. For example, if the unit cell is cubic I with a B atom at $x = y = z = \tfrac{1}{2}$, then the A,B phase difference is $(h + k + l)\pi$. Therefore, all reflections for odd values of $h + k + l$ vanish if A and B are identical atoms because the waves from A and B are displaced in phase by π.

For a cubic P lattice diffraction is possible for all $\{hkl\}$, therefore the diffraction pattern for a cubic I lattice can be constructed from that for cubic P by striking out all reflections with odd values of $h + k + l$. Similarly, for a cubic F lattice the missing lines are ones with two out of h, k, and l odd, and the remaining one even, or two even and one odd. Recognition of these **systematic absences** in a powder spectrum can be used to assign the lattice type (Fig. 15B.10).

The intensity of the $\{hkl\}$ reflection is proportional to $|F_{hkl}|^2$, so in principle the structure factors can be determined experimentally by taking the square root of the corresponding intensities (but see below). Then, once all the structure factors F_{hkl} are known the electron density distribution, $\rho(r)$, in the unit cell can be calculated by using

$$\rho(\boldsymbol{r}) = \frac{1}{V}\sum_{hkl} F_{hkl} e^{-2\pi i(hx+ky+lz)} \quad \text{Fourier synthesis} \quad (15B.4)$$

where V is the volume of the unit cell. Equation 15B.4 is called a **Fourier synthesis** of the electron density. Fourier transforms occur throughout chemistry in a variety of guises, and are described in more detail in *The chemist's toolkit* 28 in Topic 12C.

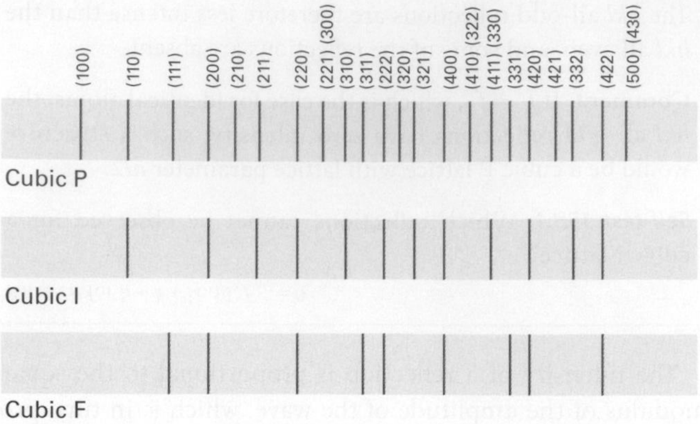

Figure 15B.10 The powder diffraction patterns and the systematic absences of three versions of a cubic cell as a function of angle: cubic F (fcc; *h*, *k*, *l* all even or all odd are present), cubic I (bcc; $h + k + l =$ odd are absent), cubic P. Comparison of the observed pattern with patterns like these enables the unit cell to be identified. The locations of the lines give the cell dimensions.

The essence of the procedure in this case is to express the varying electron density in a unit cell as a superposition of sine and cosine waves.

Example 15B.2 Calculating an electron density by Fourier synthesis

Consider the $\{h00\}$ planes of a crystal extending indefinitely in the x direction. In an X-ray analysis the structure factors were found as follows:

h	0	1	2	3	4	5	6	7	8	9
F_h	16	−10	2	−1	7	−10	8	−3	2	−3
h	10	11	12	13	14	15				
F_h	6	−5	3	−2	2	−3				

It was also found that $F_{-h} = F_h$. Construct a plot of the electron density projected on to the x-axis of the unit cell.

Collect your thoughts You need to substitute these values into eqn 15B.4, but because the problem is one-dimensional, the sum is over only the index h and only the terms $e^{-2\pi i h x}$ need be considered.

The solution Because $F_{-h} = F_h$, the sum, rather than running from $h = -\infty$ to $+\infty$, can be written as running from 1 to $+\infty$:

$$V\rho(x) = \sum_{h=-\infty}^{\infty} F_h e^{-2\pi i h x} = F_0 + \sum_{h=1}^{\infty}(F_h e^{-2\pi i h x} + \overbrace{F_{-h}}^{=F_h} e^{2\pi i h x})$$

$$= F_0 + \sum_{h=1}^{\infty} F_h (e^{-2\pi i h x} + e^{2\pi i h x})$$

$$e^{ix} + e^{-ix} = 2\cos x$$

$$= F_0 + 2\sum_{h=1}^{\infty} F_h \cos 2\pi h x$$

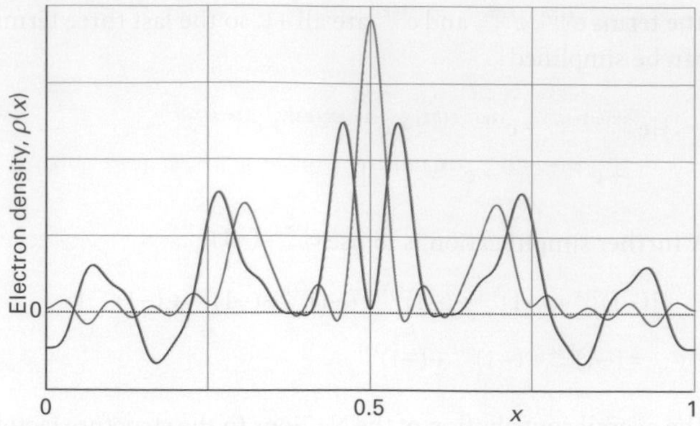

Figure 15B.11 The plot of the electron density calculated in *Example* 15B.2 (green) and *Self-test* 15B.2 (purple).

Only 15 values of F_h are given, so $\rho(x)$ will be approximate: the result (computed using mathematical software) is plotted in Fig. 15B.11 (green line). There are three clear maxima in this function, which can be identified as the positions of three atoms.

Comment. The more terms that are included (meaning the more reflections that are measured), the more accurate is the density plot. Terms corresponding to high values of h (which correspond to short-wavelength cosine terms in the sum) account for the finer details of the electron density; low values of h account for the broad features.

Self-test 15B.2 Use mathematical software to experiment with the result of altering the structure factors in the table: consider the effect of both changing the signs and amplitudes. For example, use the same values of F_h as above, but change all the signs for $h \geq 6$.

Answer: Fig. 15B.11 (purple line)

(e) The determination of structure

The structure factors used in eqn 15B.4 to compute the electron density are in general complex quantities that can be written $|F_{hkl}|e^{i\alpha}$, where $|F_{hkl}|$ is the amplitude and α the phase ('phase' in the sense used to express a complex number by a diagram in two dimensions; see *The chemist's toolkit* 16 in Topic 7C). However, the observed intensity I_{hkl} is proportional to the square modulus of the structure factor, $|F_{hkl}|^2$, so from the experiment no information is available about the phase, which may lie anywhere from 0 to 2π. This ambiguity is called the **phase problem**; its consequences are illustrated by comparing the two plots in Fig. 15B.11 in which the phases of the structure factors have been changed but the amplitudes kept the same. Some way must be found to assign phases to the structure factors, because unless these are known ρ cannot be evaluated using eqn 15B.4. The phase problem is less severe for centrosymmetric unit cells, for then the structure factors are real. It still remains a problem though, to decide whether F_{hkl} is positive or negative.

The phase problem can be overcome to some extent by a variety of methods. One procedure that is widely used for inorganic materials with a reasonably small number of atoms in a unit cell, and for organic molecules with a small number of heavy atoms, is the **Patterson synthesis**. Instead of the structure factors F_{hkl}, the values of $|F_{hkl}|^2$, which can be obtained without ambiguity from the intensities, are used in an expression that resembles eqn 15B.4:

$$P(\boldsymbol{r})=\frac{1}{V}\sum_{hkl}|F_{hkl}|^2\,\mathrm{e}^{-2\pi\mathrm{i}(hx+ky+lz)} \qquad \text{Patterson synthesis} \qquad (15\mathrm{B}.5)$$

where the r are the vector separations between the atoms in the unit cell; that is, the distances and directions between atoms. Whereas the electron density function $\rho(r)$ is the probability density of the positions of atoms, the function $P(r)$ is a map of the probability density of the separations between atoms; $P(r)$ is often called the **Patterson map**. In such a map, a peak at a position specified by a vector $\boldsymbol{r}$ from the origin arises from pairs of atoms that are separated by $\boldsymbol{r}$. Thus, if atom A is at the coordinates (x_A,y_A,z_A) and atom B is at (x_B,y_B,z_B), then there will be a peak at $(x_A-x_B, y_A-y_B, z_A-z_B)$ in the Patterson map. There will also be a peak at the negative of these coordinates, because there is a separation vector from B to A as well as a separation vector from A to B. The height of the peak in the map is proportional to the product of the atomic numbers of the two atoms, Z_AZ_B. The Patterson map also shows a strong feature at its origin arising from the separation between each atom and itself, which is necessarily zero.

Brief illustration 15B.2

For the electron density shown in Fig. 15B.12a, the corresponding Patterson map is shown in Fig. 15B.12b. The location of each peak, relative to the origin, corresponds to the separation and relative orientation of a pair of atoms in the cell. Note that the Patterson map is centrosymmetric and has a strong feature at the origin.

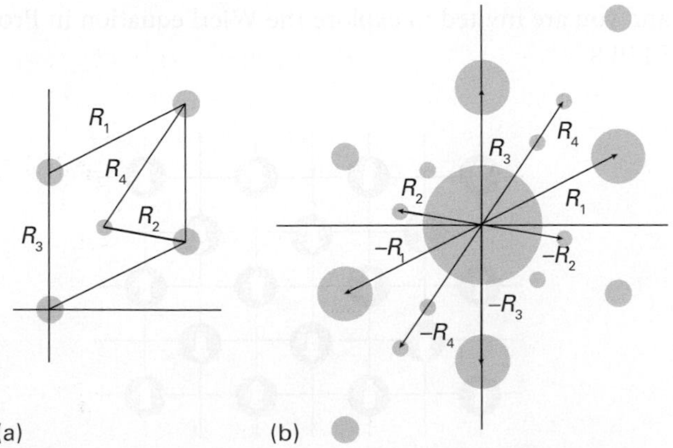

Figure 15B.12 The Patterson map corresponding to the electron density in (a) is shown in (b). The distance and orientation of each spot from the origin gives the orientation and separation of one atom–atom separation in (a); in addition, there is a large spot at the origin. Some of the typical distances and their contribution to (b) are shown as R_1, etc.

Heavy atoms dominate the scattering because their scattering factors are large, of the order of their atomic numbers, and their locations may be deduced quite readily. The sign of F_{hkl} can now be calculated from the known locations of the heavy atoms in the unit cell, and to a high probability the phase calculated for them will be the same as the phase for the entire unit cell. To see why this is so, consider a centrosymmetric cell for which each term in the structure factor is either positive or negative. The structure factor has the form

$$F=(\pm)f_{\text{heavy}}+(\pm)f_{\text{light}}+(\pm)f_{\text{light}}+\cdots \qquad (15\mathrm{B}.6)$$

where f_{heavy} is the scattering factor of the heavy atom and f_{light} the scattering factors of the light atoms. The f_{light} are all much smaller than f_{heavy}, and their phases are more or less random if the atoms are distributed throughout the unit cell. Therefore, the net effect of the f_{light} is to change F only slightly from f_{heavy}, and with reasonable confidence F will have the same sign as that calculated from the location of the heavy atoms. This phase can then be combined with the observed $|F|$ (from the reflection intensity) to perform a Fourier synthesis of the full electron density in the unit cell, and hence to locate the light atoms as well as the heavy atoms.

Modern structural analyses make extensive use of **direct methods**. Direct methods are based on the possibility of treating the atoms in a unit cell as being virtually randomly distributed (from the radiation's point of view), and then to use statistical techniques to compute the probabilities that the phases have a particular value. It is possible to deduce relations between some structure factors and sums (and sums of squares) of others, which have the effect of constraining the phases to particular values (with high probability, so long as the structure factors are large). For example, the **Sayre probability relation** has the form

$$\text{sign of } F_{h+h',k+k',l+l'} \text{ is probably equal to (sign of } F_{hkl}) \times (\text{sign of } F_{h'k'l'}) \qquad \text{Sayre probability relation} \qquad (15\mathrm{B}.7)$$

For example, if F_{122} and F_{232} are both large and negative, then it is highly likely that F_{354}, provided it is large, will be positive.

In the final stages of the determination of a crystal structure, the parameters describing the structure (atom positions, for instance) are adjusted systematically to give the best fit between the observed intensities and those calculated from the model of the structure deduced from the diffraction pattern. This process is called **structure refinement**. Not only does the procedure give accurate positions for all the atoms in the unit cell, but it also gives an estimate of the errors in those positions and in the bond lengths and angles derived from them. The procedure also provides information on the vibrational amplitudes of the atoms.

15B.2 Neutron and electron diffraction

Neutrons and electrons can also give rise to diffraction on account of their wave nature; their wavelength is given by the de Broglie relation (Topic 7A, $\lambda = h/p$). Neutrons generated in a nuclear reactor and then slowed to thermal velocities have wavelengths similar to those of X-rays and may also be used for diffraction studies. For instance, a neutron generated in a reactor and slowed to thermal velocities by repeated collisions with a moderator (such as graphite) until it is travelling at about 4 km s^{-1} has a wavelength of about 100 pm. In practice, a range of wavelengths occurs in a neutron beam, but a monochromatic beam can be selected by diffraction from a crystal, such as a single crystal of germanium.

Example 15B.3 Calculating the typical wavelength of thermal neutrons

Calculate the typical wavelength of neutrons after reaching thermal equilibrium with their surroundings at 373 K. For simplicity, assume that the particles are travelling in one dimension.

Collect your thoughts In order to use the de Broglie relation, you need to know the momentum of the neutrons, and therefore their velocity. The velocity can be computed from the kinetic energy, which you can assume has its equipartition value for translation in one dimension, $E_k = \frac{1}{2}kT$ (see *The chemist's toolkit* 7 in Topic 2A). The mass of a neutron is given inside the front cover.

The solution From the equipartition principle, the mean translational kinetic energy of a neutron at a temperature T travelling in the x-direction is $E_k = \frac{1}{2}kT$. The kinetic energy is also equal to $p^2/2m$, where p is the momentum of the neutron and m is its mass. Hence, $p = (mkT)^{1/2}$. It follows from the de Broglie relation $\lambda = h/p$ that the wavelength of the neutron is

$$\lambda = \frac{h}{(mkT)^{1/2}}$$

Therefore, at 373 K,

$$\lambda = \frac{6.626\times10^{-34}\,\mathrm{J\,s}}{\{(1.675\times10^{-27}\,\mathrm{kg})\times(1.381\times10^{-23}\,\mathrm{J\,K^{-1}})\times(373\,\mathrm{K})\}^{1/2}}$$

$1\,\mathrm{J} = 1\,\mathrm{kg\,m^2\,s^{-2}}$

$$= \frac{6.626\times10^{-34}}{(1.675\times10^{-27}\times1.381\times10^{-23}\times373)^{1/2}}\,\frac{\mathrm{kg\,m^2\,s^{-1}}}{(\mathrm{kg^2\,m^2\,s^{-2}})^{1/2}}$$

$$= 2.26\times10^{-10}\,\mathrm{m} = 226\,\mathrm{pm}$$

Self-test 15B.3 Calculate the temperature needed for the average wavelength of the neutrons to be 100 pm.

Answer: 1900 K

Neutron diffraction differs from X-ray diffraction in two main respects. First, the scattering of neutrons is a nuclear phenomenon. Neutrons pass through the extranuclear electrons of atoms and interact with the nuclei through the 'strong force' that is responsible for binding nucleons together. As a result, the intensity with which neutrons are scattered is independent of the number of electrons and neighbouring elements in the periodic table might scatter neutrons with markedly different intensities. Neutron diffraction can be used to distinguish atoms of elements such as Ni and Co that are present in the same compound and to study order–disorder phase transitions in FeCo. A second difference is that neutrons possess a magnetic moment due to their spin. This magnetic moment can couple to the magnetic fields of atoms or ions in a crystal (if the ions have unpaired electrons) and modify the diffraction pattern. One consequence is that neutron diffraction is well suited to the investigation of magnetically ordered lattices in which neighbouring atoms may be of the same element but have different orientations of their electronic spin (Fig. 15B.13).

Electrons accelerated through a potential difference of 40 kV have wavelengths of about 6 pm, and so are also suitable for diffraction studies of molecules. Consider the scattering of electrons from pairs of atoms with centres separated by a distance R_{ij} and orientated at a definite angle θ to an incident beam of electrons. When the molecule consists of a number of atoms, the scattering intensity can be calculated by summing over the contribution from all pairs. The total intensity $I(\theta)$ is given by the **Wierl equation**:

$$I(\theta) = \sum_{i,j} f_i f_j \frac{\sin sR_{ij}}{sR_{ij}} \qquad s = \frac{4\pi}{\lambda}\sin\tfrac{1}{2}\theta \qquad \text{Wierl equation} \qquad (15B.8)$$

where λ is the wavelength of the electrons in the beam, and f is the **electron scattering factor**, a measure of the electron scattering power of the atom. The main application of electron diffraction techniques is to the study of surfaces (Topic 19A), and you are invited to explore the Wierl equation in Problem P15B.8.

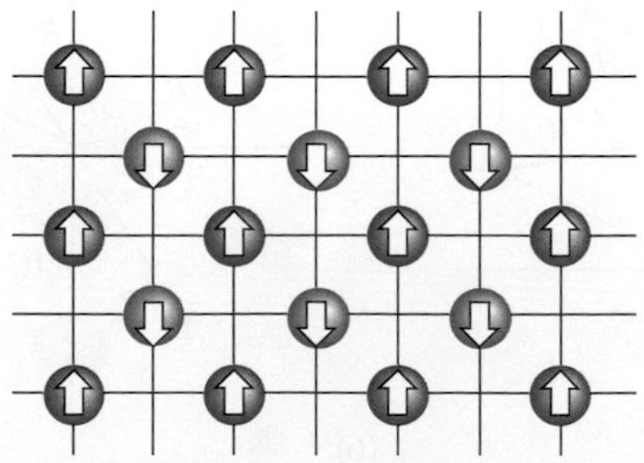

Figure 15B.13 If the spins of atoms at lattice points are orderly, as in this material, where the spins of one set of atoms are aligned antiparallel to those of the other set, neutron diffraction detects two interpenetrating simple cubic lattices on account of the magnetic interaction of the neutron with the atoms, but X-ray diffraction would see only a single bcc lattice.

Checklist of concepts

☐ 1. A **reflection** refers to an intense beam emerging in a particular direction and arising from constructive interference.

☐ 2. The **glancing angle**, 2θ, is the angle through which a beam is deflected.

☐ 3. **Bragg's law** relates the angle of a diffracted beam to the spacing of a given set of lattice planes.

☐ 4. The **scattering factor** is a measure of the ability of an atom to scatter electromagnetic radiation.

☐ 5. The **structure factor** is the overall amplitude of a wave diffracted by the $\{hkl\}$ planes and atoms distributed through the unit cell.

☐ 6. The electron density and the diffraction pattern are related by a Fourier transform.

☐ 7. **Fourier synthesis** is the construction of the electron density distribution from structure factors.

☐ 8. The **phase problem** arises because it is possible to measure only the intensity of the reflections and not their phases; as a result Fourier synthesis cannot be used in a straightforward way to determine the electron density.

☐ 9. A **Patterson map** is a map of the interatomic vectors.

☐ 10. **Direct methods** use statistical techniques to determine the likely phases of the reflections.

☐ 11. **Structure refinement** is the adjustment of structural parameters to give the best fit between the observed intensities and those calculated from the model of the structure deduced from the diffraction pattern.

☐ 12. The **Wierl equation** relates the intensity of electron scattering to the distances between pairs of atoms in the sample.

Checklist of equations

Property	Equation	Comment	Equation number
Bragg's law	$\lambda = 2d\sin\theta$	d is the lattice spacing, 2θ the glancing angle	15B.1b
Scattering factor	$f = 4\pi\int_0^\infty [\{\rho(r)\sin kr\}/kr]r^2\,dr,\ k = (4\pi/\lambda)\sin\theta$	Spherically symmetrical atom	15B.2
Structure factor	$F_{hkl} = \sum_j f_j e^{i\phi_{hkl}(j)},\quad \phi_{hkl}(j) = 2\pi(hx_j + ky_j + lz_j)$	Definition	15B.3
Fourier synthesis	$\rho(\boldsymbol{r}) = (1/V)\sum_{hkl} F_{hkl} e^{-2\pi i(hx+ky+lz)}$	V is the volume of the unit cell	15B.4
Patterson synthesis	$P(\boldsymbol{r}) = (1/V)\sum_{hkl} \lvert F_{hkl}\rvert^2 e^{-2\pi i(hx+ky+lz)}$		15B.5
Wierl equation	$I(\theta) = \sum_{i,j} f_i f_j (\sin sR_{ij}/sR_{ij}),\ s = (4\pi/\lambda)\sin\frac{1}{2}\theta$		15B.8

TOPIC 15C Bonding in solids

> ➤ **Why do you need to know this material?**
> To understand the properties and structures of solid materials you need to know about the type of bonding that holds together the atoms, ions, and molecules.
>
> ➤ **What is the key idea?**
> Four characteristic types of bonding result in metals, ionic solids, covalent solids, and molecular solids.
>
> ➤ **What do you need to know already?**
> You need to be familiar with molecular interactions (Topic 14B) and the general features of crystal structures (Topic 15A). For the discussion of metallic bonding you should be aware of the principles of Hückel molecular orbital theory (Topic 9E). The discussion of ionic bonding makes use of the concept of enthalpy (Topic 2B).

Solids may be classified into four broad types, namely metals, ionic solids, covalent (or network) solids, and molecular solids. Each is characterized by the nature of the bonding between the constituents.

15C.1 Metals

In a metal the electrons are delocalized over arrays of identical cations and bind the whole together into a rigid but ductile and malleable structure. The crystalline forms of metallic elements can be discussed in terms of a model in which their atoms are treated as identical hard spheres. Most metallic elements crystallize in one of three simple forms, two of which can be explained in terms of the hard spheres packing together in the closest possible arrangement.

(a) Close packing

Figure 15C.1 shows a **close-packed** layer of identical spheres, one with maximum utilization of space. A close-packed three-dimensional structure is obtained by stacking such layers on top of one another. However, this stacking can be done in different ways and results in close-packed **polytypes**, which are structures that are identical in two dimensions (the close-packed layers) but differ in the third dimension.

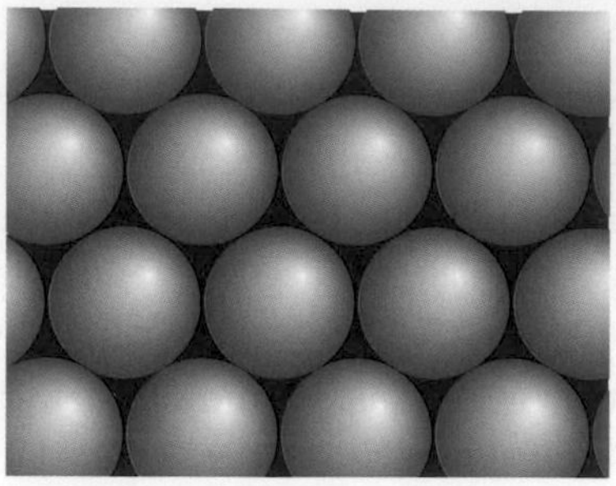

Figure 15C.1 A layer of close-packed spheres used to build a three-dimensional close-packed structure.

In all polytypes, the spheres of the second close-packed layer lie in the depressions of the first layer (Fig. 15C.2). The third layer may be added in either of two ways. In one, the spheres are placed directly above the first layer to give an ABA pattern of layers (Fig. 15C.3a). Alternatively, the spheres may be placed over the depressions in the first layer that are not occupied by the second layer (these depressions are visible in Fig. 15C.2), so giving an ABC pattern (Fig. 15C.3b). Two polytypes are formed if the two stacking patterns are repeated in the vertical direction:

- **Hexagonally close-packed** (hcp): the ABA pattern is repeated, to give the sequence of layers ABABAB....
- **Cubic close-packed** (ccp): the ABC pattern is repeated, to give the sequence ABCABC....

The origins of these names can be seen by referring to Fig. 15C.4. The ccp structure gives rise to a face-centred unit

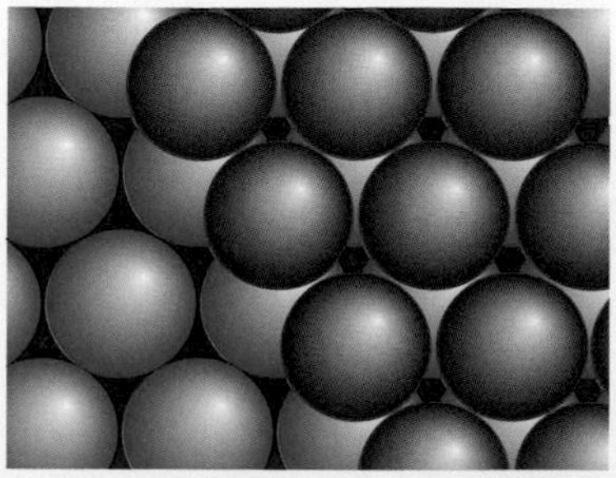

Figure 15C.2 To achieve the greatest packing density, the second layer of close-packed spheres must sit in the depressions of the first layer. The two layers are the AB component of the close-packed structure.

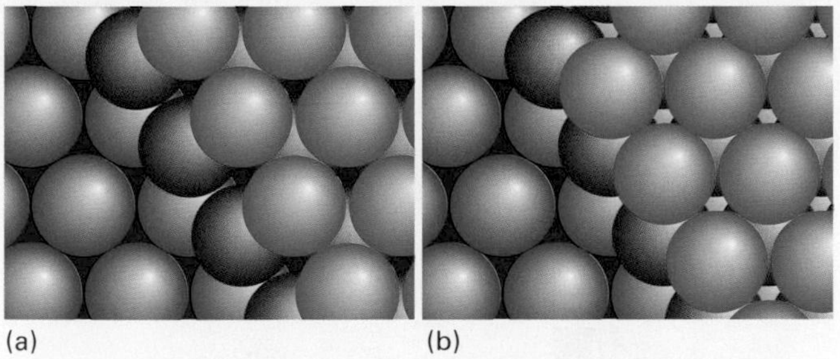

Figure 15C.3 (a) The third layer of close-packed spheres might occupy the depressions lying directly above the spheres in the first layer, resulting in an ABA structure, which corresponds to hexagonal close-packing. (b) Alternatively, the third layer might lie in the depressions that are not above the spheres in the first layer, resulting in an ABC structure, which corresponds to cubic close-packing.

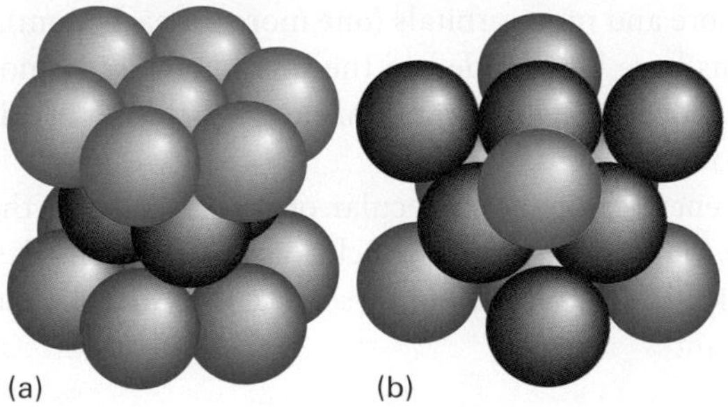

Figure 15C.4 Fragments of the structures shown in Fig. 15C.3 revealing the (a) hexagonal (b) cubic symmetry. The colours of the spheres are the same as for the layers in Fig. 15C.3.

cell, so may also be denoted cubic F (or fcc, for face-centred cubic). It is also possible to have random sequences of layers; however, the hcp and ccp polytypes are the most important. Table 15C.1 lists the structures adopted by a selection of elements.

The compactness of close-packed structures is indicated by their **coordination number**, the number of spheres immediately surrounding any selected sphere, which is 12 for both the ccp and hcp structures. Another measure of their compactness is the **packing fraction**, the fraction of space occupied by the spheres, which is 0.740 (see *Example* 15C.1). That is, in a close-packed solid of identical hard spheres, only 26.0 per cent of the volume is empty space. The fact that many metals are close-packed accounts for their high mass densities.

Table 15C.1 The crystal structures of some elements*

Structure	Element
hcp‡	Be, Cd, Co, He, Mg, Sc, Ti, Zn
fcc‡ (ccp, cubic F)	Ag, Al, Ar, Au, Ca, Cu, Kr, Ne, Ni, Pb, Pd, Pt, Rh, Rn, Sr, Xe
bcc (cubic I)	Ba, Cr, Cs, Fe, K, Li, Mo, Na, Rb, Ta, V, W
cubic P	Po

* The notation used to describe unit cells is introduced in Topic 15A. The structures are for the elements at 298 K and 1 bar.

‡ Close-packed structures.

Example 15C.1 Calculating a packing fraction

Calculate the packing fraction of a ccp structure formed from hard spheres.

Collect your thoughts You need to calculate the volume of the unit cell and the volume of the spheres that are wholly or partly contained within the cell, and then take the ratio of the two volumes. The key step is to establish a relation between the radius of the spheres, R, and the cell dimension, a. Figure 15C.5 shows that the sphere in the middle of a face just touches the two spheres at opposite corners of the face. The length of the face diagonal is therefore $4R$. To calculate the volume of the spheres in the cell, you need to account for the fraction of each sphere that is contained within the cell. Spheres at the corners contribute $\frac{1}{8}$ of their volume to the cell; those on the faces contribute $\frac{1}{2}$ to the cell.

The solution As can be seen from Fig. 15C.5, the length of the face diagonal is $4R$. From Pythagoras' theorem it follows that $a^2 + a^2 = (4R)^2$, so $2a^2 = 16R^2$ and therefore $a = 8^{1/2}R$. The volume of the unit cell is a^3, which is therefore $8^{3/2}R^3$. There are eight spheres at the corners, each contributing $\frac{1}{8}$ of their volume to the cell (for a net contribution of 1 sphere), and six spheres on the faces, each contributing $\frac{1}{2}$ (for a net contribution of 3 spheres). The total volume occupied by the spheres is equivalent to 4 complete spheres. Because the volume of each sphere is $\frac{4}{3}\pi R^3$, the total occupied volume is $\frac{16}{3}\pi R^3$. The fraction of space occupied is therefore

$$\frac{(\frac{16}{3})\pi R^3}{8^{3/2}R^3} \overset{8^{3/2}=16\sqrt{2}}{=} \frac{\pi}{3\sqrt{2}} = 0.740$$

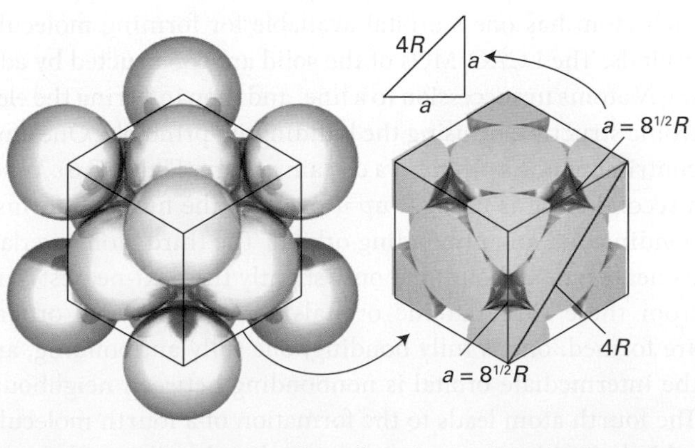

Figure 15C.5 In a ccp unit cell a sphere located on the face of the cube just touches the spheres at opposite corners of the face.

Because an hcp structure has the same coordination number, its packing fraction is the same.

Self-test 15C.1 Calculate the packing fraction of the cubic I (body-centred cubic, bcc) structure in which one sphere is at the centre of a cube formed by eight others. The spheres touch along the body diagonal of the cube.

Answer: $3^{1/2}\pi/8 = 0.680$

As shown in Table 15C.1, a number of common metals adopt structures that are less than close-packed. The departure from close packing suggests that factors, such as specific covalent bonding between neighbouring atoms, are beginning to influence the structure and impose a particular geometrical arrangement. One such arrangement results in a cubic I (bcc, for body-centred cubic) structure, with one sphere at the centre of a cube formed by eight others. The coordination number of a bcc structure is only 8, but there are six more atoms not much further away than the eight nearest neighbours. The packing fraction of 0.680 (*Self-test* 15C.1) is not much smaller than the value for a close-packed structure (0.740), and shows that about two-thirds of the available space is occupied by the atoms.

(b) Electronic structure of metals

The central aspect of solids that determines their electrical properties (Topic 15E) is the distribution of their electrons. There are two models of this distribution. In one, the **nearly free-electron approximation**, the valence electrons are assumed to be trapped in a box with a periodic potential, with low energy corresponding to the locations of cations. In the **tight-binding approximation**, the valence electrons are assumed to occupy molecular orbitals delocalized throughout the solid. The latter model is more in accord with the discussion of electrical properties of solids discussed in Topic 15E, and is described here.

As a starting point, consider a one-dimensional solid, which consists of a single, infinitely long line of atoms. Suppose that each atom has one s orbital available for forming molecular orbitals. The LCAO-MOs of the solid are constructed by adding N atoms in succession to a line, and then inferring the electronic structure by using the building-up principle. One atom contributes one s orbital at a certain energy (Fig. 15C.6). When a second atom is brought up it overlaps the first and forms a bonding and an antibonding orbital. The third atom overlaps its nearest neighbour (and only slightly the next-nearest), and from these three atomic orbitals, three molecular orbitals are formed: one is fully bonding, one fully antibonding, and the intermediate orbital is nonbonding between neighbours. The fourth atom leads to the formation of a fourth molecular orbital. At this stage, it can be seen that the effect of bringing up successive atoms is to spread the range of energies covered

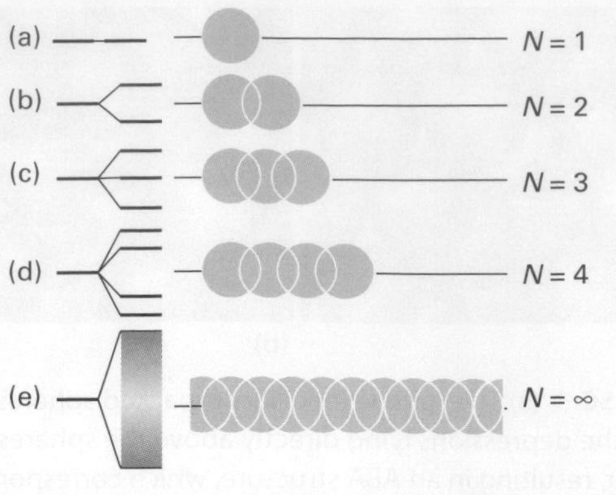

Figure 15C.6 The formation of a band of N molecular orbitals by successive addition of N atomic orbitals in a line. When N becomes infinite the band covers a finite range of energy and the orbitals within it are very closely spaced, but still discrete.

by the molecular orbitals and to fill in the range of energies with more and more orbitals (one more for each atom). When N atoms have been added to the line, there are N molecular orbitals covering a finite range of energies: this set of orbitals is said to form a **band**.

The energies of the molecular orbitals that form the band can be found by using the Hückel approximations described in Topic 9E. They are found by solving the Hückel secular determinant

$$\begin{vmatrix} \alpha-E & \beta & 0 & \dots & 0 \\ \beta & \alpha-E & \beta & \dots & 0 \\ 0 & \beta & \alpha-E & \dots & 0 \\ \vdots & \vdots & \vdots & \dots & \vdots \\ 0 & 0 & 0 & \dots & \alpha-E \end{vmatrix}=0$$

where α is the Coulomb integral and β is the (s,s) resonance integral. The general expression for the solutions of this 'tridiagonal determinant' gives the energies E_k of the molecular orbitals:

$$E_k=\alpha+2\beta\cos\frac{k\pi}{N+1} \quad k=1,2,\dots,N \qquad \text{Energy levels [linear array of s orbitals]} \qquad (15C.1)$$

It is not hard to show that this expression implies that when N is infinitely large, the separation between neighbouring levels, $E_{k+1}-E_k$, is infinitely small, but the band still has finite width overall, with $E_N-E_1\to-4\beta$.

How is that done? 15C.1 Evaluating the separation between neighbouring levels and width of a band

This calculation requires looking at a specific limiting case of eqn 15C.1.

Step 1 *Write an expression for the energy difference between two neighbouring levels*

From eqn 15C.1 it follows that the energy separation between neighbouring energy levels k and $k+1$ is

$$E_{k+1}-E_k=\left(\alpha+2\beta\cos\frac{(k+1)\pi}{N+1}\right)-\left(\alpha+2\beta\cos\frac{k\pi}{N+1}\right)$$

$$=2\beta\left(\cos\frac{(k+1)\pi}{N+1}-\cos\frac{k\pi}{N+1}\right)$$

Step 2 *Find the limit as $N\to\infty$*

By using the trigonometric identity $\cos(A+B)=\cos A\cos B-\sin A\sin B$, followed by $\cos 0=1$ and $\sin 0=0$, the first (blue) term in parentheses is

$$\cos\frac{(k+1)\pi}{N+1}=\cos\frac{k\pi}{N+1}\overbrace{\cos\frac{\pi}{N+1}}^{\to 1\text{ as }N\to\infty}-\sin\frac{k\pi}{N+1}\overbrace{\sin\frac{\pi}{N+1}}^{\to 0\text{ as }N\to\infty}$$

Therefore, as $N\to\infty$,

$$E_{k+1}-E_k\to 2\beta\left(\cos\frac{k\pi}{N+1}-\cos\frac{k\pi}{N+1}\right)=0$$

It follows that when N is infinitely large, the difference between neighbouring energy levels is infinitely small.

Step 3 *Write an expression for the width of the band for $N\to\infty$*

The width of the band is simply E_N-E_1. Each of the energies can be approximated as follows for the case that $N\to\infty$.

$$E_1=\alpha+2\beta\cos\frac{\pi}{N+1}$$

As $N\to\infty$, the term $\pi/(N+1)$ tends to zero so the cosine tends to 1; therefore in this limit

$$E_1=\alpha+2\beta$$

When k has its maximum value of N,

$$E_N=\alpha+2\beta\cos\frac{N\pi}{N+1}$$

As $N\to\infty$, the 1 in the denominator can be ignored, so the cosine term becomes $\cos\pi=-1$. Therefore, in this limit $E_N=\alpha-2\beta$, and so the band width is $E_N-E_1\to-4\beta$. Recall that β is negative, so that the band width, -4β, is positive.

The band can be thought of as consisting of N different molecular orbitals, the lowest-energy orbital ($k=1$) being fully bonding, and the highest-energy orbital ($k=N$) being fully antibonding between adjacent atoms (Fig. 15C.7). The molecular orbitals of intermediate energy have $k-1$ nodes distributed along the chain of atoms. Similar bands form in three-dimensional solids.

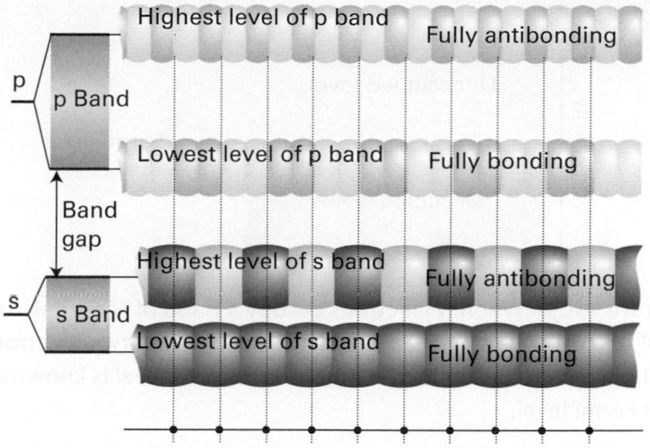

Figure 15C.7 The overlap of s orbitals gives rise to an s band and the overlap of p orbitals gives rise to a p band. In this case, the s and p orbitals of the atoms are so widely spaced in energy that there is a band gap; it is also possible that the separation will be less, resulting in the bands overlapping.

Brief illustration 15C.1

To illustrate the dependence of $E_{k+1}-E_k$ on N, note that

$$\text{for }N=3:\quad E_2-E_1=2\beta\left(\cos\frac{2\pi}{4}-\cos\frac{\pi}{4}\right)\approx-1.414\beta$$

$$\text{for }N=30:\quad E_2-E_1=2\beta\left(\cos\frac{2\pi}{31}-\cos\frac{\pi}{31}\right)\approx-0.0307\beta$$

$$\text{for }N=300:\quad E_2-E_1=2\beta\left(\cos\frac{2\pi}{301}-\cos\frac{\pi}{301}\right)\approx-0.000\,327\beta$$

The energy difference decreases with increasing N, as expected.

The band formed from overlap of s orbitals is called the **s band**. If the atoms have p orbitals available, the same procedure leads to a **p band** (as shown in the upper half of Fig. 15C.7). If the atomic p orbitals lie higher in energy than the s orbitals, then the p band lies higher than the s band, and there may be a **band gap**, a range of energies to which no orbital corresponds. However, it is also possible for the bands to touch, with the highest orbital of the s band coincident with the lowest level of the p band, or even overlap (as is the case for the 3s and 3p bands in magnesium).

Now consider the electronic structure of a solid formed from N atoms each able to contribute one electron (for example, the alkali metals). The N atomic orbitals give rise to a band consisting of N molecular orbitals. Each of these orbitals can accommodate two spin-paired electrons, so at $T=0$ only the lowest $\frac{1}{2}N$ molecular orbitals are occupied (Fig. 15C.8). The HOMO is called the **Fermi level**. Only the small number of electrons close to the Fermi level can undergo thermal excitation, so only these electrons contribute to the heat capacity of the metal. It is for this reason that Dulong and Petit's law

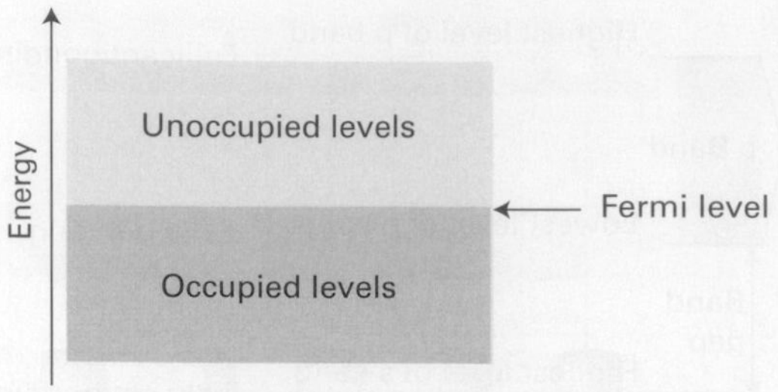

Figure 15C.8 When N electrons occupy a band of N orbitals, only half of the orbitals are occupied (at $T = 0$) because two electrons will occupy each orbital. The highest occupied level is known as the Fermi level.

for heat capacities (Topic 7A) gives reasonable agreement with experiment at normal temperatures by considering only the atoms in a sample, not the atoms plus the 'free' electrons. The presence of an incompletely filled band is responsible for electrical conductivity, as explained in Topic 15E.

15C.2 Ionic solids

An **ionic solid** consists of cations and anions held together by electrostatic interactions. Two key questions arise in considering such solids: the relative locations adopted by the ions and the energetics of the resulting structure.

(a) Structure

When crystals of compounds of monatomic ions (such as NaCl and MgO) are modelled by stacks of hard spheres it is necessary to allow for the possibility that the ions have different radii (typically with the cations smaller than the anions) and different electrical charges. The coordination number of an ion is the number of nearest neighbours of opposite charge; the structure itself is characterized as having **(N_+,N_-) coordination**, where N_+ is the coordination number of the cation and N_- that of the anion.

Even if, by chance, the ions have the same size, the requirement that the unit cells are electrically neutral make it impossible to achieve 12-coordinate close-packed ionic structures. As a result, ionic solids are generally less dense than metals. The best packing that can be achieved is the (8,8)-coordinate **caesium chloride structure** in which each cation is surrounded by eight anions and each anion is surrounded by eight cations (Fig. 15C.9). In this structure, an ion of one charge occupies the centre of a cubic unit cell with eight counter ions at its corners. The cell is electrically neutral because the ions at the corners of the cell are shared between eight cells and so contribute one eighth of their charge to each. The structure shown in Fig. 15C.9 is adopted by CsCl itself and also by CaS.

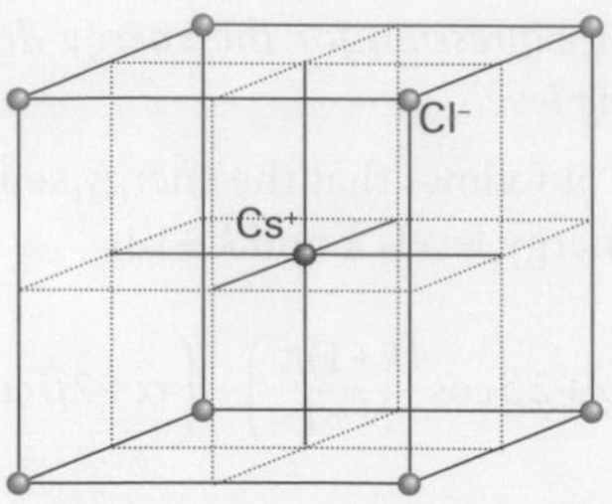

Figure 15C.9 The caesium chloride structure consists of two interpenetrating simple cubic arrays of ions, one of cations and the other of anions, so that each cube of ions of one kind has a counter-ion at its centre.

When the radii of the ions differ more than they do in CsCl, even eight-coordinate packing cannot be achieved. One common structure adopted is the (6,6)-coordinate **rock salt structure** typified by rock salt itself, NaCl (Fig. 15C.10). In this structure, each cation is surrounded by six anions and each anion is surrounded by six cations. The rock salt structure can be pictured as consisting of two interpenetrating slightly expanded cubic F (fcc) arrays, one composed of cations and the other of anions. This structure is adopted by NaCl itself and also by several other MX compounds, including KBr, AgCl, MgO, and ScN.

The switch from the caesium chloride structure to the rock salt structure is related to the value of the **radius ratio**, γ:

$$\gamma = \frac{r_{\text{smaller}}}{r_{\text{larger}}} \qquad \text{Radius ratio [definition]} \qquad (15C.2)$$

where r_{smaller} is the radius of the smaller ions in the crystal and r_{larger} that of the larger ions. The **radius-ratio rule**, which is derived by considering the geometrical problem of packing the maximum number of hard spheres of one radius around a hard sphere of a different radius, can be summarized as follows:

Radius ratio	Structural type
$\gamma < 2^{1/2} - 1 = 0.414$	sphalerite (Fig. 15C.11)
$0.414 < \gamma < 3^{1/2} - 1 = 0.732$	rock salt (Fig. 15C.10)
$\gamma > 0.732$	caesium chloride (Fig. 15C.9)

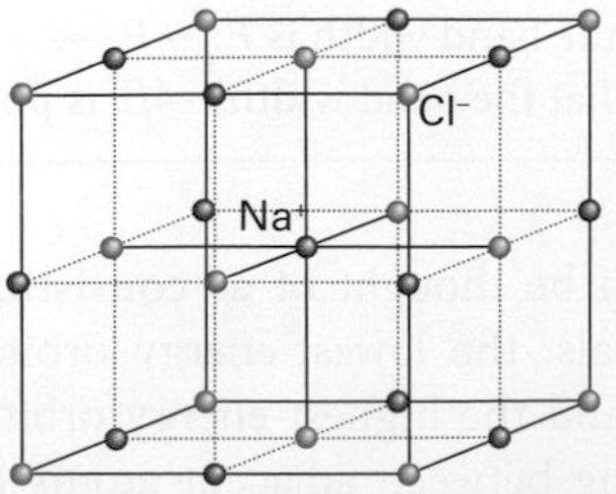

Figure 15C.10 The rock salt (NaCl) structure consists of two mutually interpenetrating slightly expanded face-centred cubic arrays of ions. The assembly shown here is a unit cell.

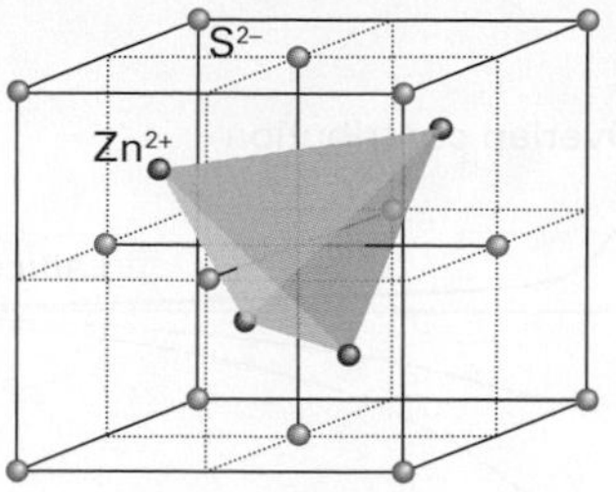

Figure 15C.11 The structure of the sphalerite form of ZnS showing the location of the Zn atoms in half the tetrahedral holes formed by the fcc array of S atoms.

The deviation of a structure from that expected on the basis of this rule is often taken to be an indication of a shift from ionic towards covalent bonding. A major source of unreliability, though, is the arbitrariness of ionic radii (as explained in a moment) and their variation with coordination number.

Experimental measurements give the distances between the centres of two ions, and a decision has to be made about how to apportion this difference between the two ions. One approach is simply to assign a value to the radius of one ion and then use this value to infer the radii of other ions. A scale based on the value 140 pm for the radius of the O^{2-} ion is widely used (Table 15C.2). Other scales are also available (such as one based on F^- for discussing halides), and it is essential not to mix values from different scales. Because ionic radii are so arbitrary, predictions based on them must be viewed cautiously.

Brief illustration 15C.2

From the values of ionic radii in the *Resource section*, the radius ratio for MgO is

$$\gamma = \frac{\overbrace{72\,\text{pm}}^{\text{radius of Mg}^{2+}}}{\underbrace{140\,\text{pm}}_{\text{radius of O}^{2-}}} = 0.51$$

which is consistent with the observed rock salt structure of MgO crystals.

Table 15C.2 Ionic radii, *r*/pm*

Na^+	102(6‡), 116(8)
K^+	138(6), 151(8)
F^-	128(2), 131(4)
Cl^-	181 (close packing)

* This scale is based on a value 140 pm for the radius of the O^{2-} ion. More values are given in the *Resource section*.
‡ Coordination number.

(b) Energetics

The **lattice energy** of a solid is the change in potential energy when the ions go from being packed together in a solid to being widely separated as a gas. All lattice energies are positive; a high lattice energy indicates that the ions interact strongly with one another to give a tightly bonded solid. The **lattice enthalpy**, ΔH_L, is the change in standard molar enthalpy for the process $MX(s) \rightarrow M^+(g) + X^-(g)$ and its equivalent for other charge types and stoichiometries. At $T = 0$ the lattice enthalpy is equal to the lattice energy; at normal temperatures they differ by only a few kilojoules per mole, an amount so small compared to typical lattice energies that the difference is normally neglected.

Each ion in a solid experiences favourable (energy lowering) electrostatic interactions from all the other oppositely charged ions and unfavourable (energy raising) interactions from all the other like-charged ions. The total Coulomb potential energy is the sum of all the electrostatic contributions. Each cation is surrounded by anions, and so there is a large negative contribution to the potential energy from the interaction of opposite charges. Beyond those nearest neighbours, there are cations that contribute a positive term to the total potential energy of the central cation. There is also a negative contribution from the anions beyond those cations, a positive contribution from the cations beyond them, and so on, to the edge of the solid. These favourable and unfavourable interactions become progressively weaker as the distance from the central ion increases, but the net outcome of all these contributions is dominated by the interaction between nearest neighbours and is therefore a lowering of the potential energy.

First, consider a simple one-dimensional model of an ionic solid that consists of a long line of uniformly spaced alternating cations and anions; the distance between neighbouring centres is d, the sum of the ionic radii (Fig. 15C.12). If the charge numbers of the ions have the same absolute value (+1 and −1, or +2 and −2, for instance), then $z_1 = +z$, $z_2 = -z$, and $z_1 z_2 = -z^2$. The potential energy of the central ion is calculated by summing all the terms, with negative terms representing favourable interactions with oppositely charged ions and positive terms representing unfavourable interactions with like-charged ions. For the interaction with ions extending in a line to the right of the central ion, the contribution of the Coulomb interaction to the lattice energy is

$$E_p = \frac{1}{4\pi\varepsilon_0} \times \left(-\frac{z^2 e^2}{d} + \frac{z^2 e^2}{2d} - \frac{z^2 e^2}{3d} + \frac{z^2 e^2}{4d} - \cdots \right)$$

$$= \frac{z^2 e^2}{4\pi\varepsilon_0 d} \times \overbrace{\left(-1 + \tfrac{1}{2} - \tfrac{1}{3} + \tfrac{1}{4} - \cdots\right)}^{-\ln 2}$$

$$= -\frac{z^2 e^2}{4\pi\varepsilon_0 d} \times \ln 2$$

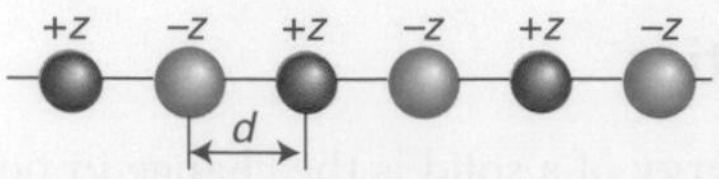

Figure 15C.12 A line of alternating cations and anions used in the calculation of the Madelung constant in one dimension.

To complete the calculation E_p is multiplied by 2 to obtain the total energy arising from interactions on both sides of the ion, and then by Avogadro's constant, N_A, to obtain an expression for the Coulomb contribution to the (molar) lattice energy. The outcome is

$$E_p = -2\ln 2 \times \frac{z^2 N_A e^2}{4\pi\varepsilon_0 d}$$

with $d = r_{cation} + r_{anion}$. This energy is negative, corresponding to a net favourable interaction. The calculation can be extended to three-dimensional arrays of ions with different charge numbers z_A and z_B:

$$E_p = -A \times \frac{|z_A z_B| N_A e^2}{4\pi\varepsilon_0 d} \qquad (15C.3)$$

The factor A is a positive numerical constant called the **Madelung constant**; its value depends on how the ions are arranged in the crystal. For a rock salt structure, $A = 1.748$; Table 15C.3 lists Madelung constants for other common structures.

The Coulomb interaction is not the only contribution to the lattice energy. When atomic orbitals overlap to form bonding and antibonding molecular orbitals and both kinds of orbitals are full, there is an increase in energy because the antibonding orbital is raised in energy more than the bonding orbital is lowered (Topic 9D). This positive contribution to the potential energy depends on the overlap of the atomic orbitals, and, because orbitals decay exponentially with distance, at large distances from the nucleus it is often modelled by writing

$$E_p^* = N_A C' e^{-d/d^*} \qquad (15C.4)$$

where d is the distance between the atoms, and C' and d^* are constants. It turns out that the value of C' is not needed (it cancels in expressions that make use of this formula; see below); d^* is commonly taken to be 34.5 pm.

The total potential energy is the sum of E_p and E_p^*, and passes through a minimum when $d(E_p + E_p^*)/dd = 0$ (Fig. 15C.13).

Table 15C.3 Madelung constants

Structural type	A
Caesium chloride	1.763
Fluorite	2.519
Rock salt	1.748
Rutile	2.408
Sphalerite (zinc blende)	1.638
Wurtzite	1.641

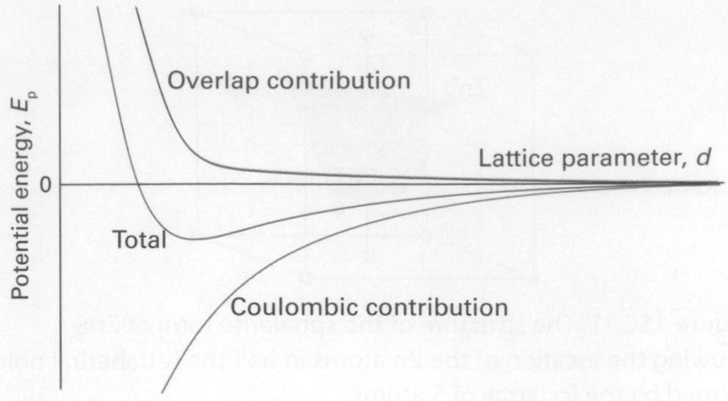

Figure 15C.13 The contributions to the total potential energy of an ionic crystal.

A short calculation leads to the **Born–Mayer equation** for the minimum total potential energy (see Problem P15C.9):

$$E_{p,min} = -\frac{N_A |z_A z_B| e^2}{4\pi\varepsilon_0 d}\left(1 - \frac{d^*}{d}\right)A \qquad \text{Born–Mayer equation} \qquad (15C.5)$$

Provided zero-point contributions to the energy are ignored, the negative of this potential energy can be identified with the lattice energy. The important features of this equation are:

Physical interpretation

- Because $E_{p,min} \propto |z_A z_B|$, the potential energy decreases (becomes more negative) with increasing charge number of the ions.
- Because the electrostatic (and dominant) contribution to $E_{p,min}$ is proportional to $1/d$, the potential energy decreases (becomes more negative) with decreasing ionic radius.

The second conclusion follows from the fact that the smaller the ionic radii, the smaller is the value of d. High lattice energies are expected when the ions are highly charged (so $|z_A z_B|$ is large) and small (so d is small).

Brief illustration 15C.3

To estimate $E_{p,min}$ for MgO, which has a rock salt structure ($A = 1.748$), use the following values: $d = r(Mg^{2+}) + r(O^{2-}) = 72 + 140\,pm = 212\,pm$. Note that

$$\frac{N_A e^2}{4\pi\varepsilon_0} = \frac{(6.022\,14\times10^{23}\,mol^{-1})\times(1.602\,176\times10^{-19}\,C)^2}{4\pi\times(8.854\,19\times10^{-12}\,J^{-1}\,C^2\,m^{-1})}$$

$$= 1.389\,35\times10^{-4}\,J\,m\,mol^{-1}$$

Then

$$E_{p,min} = -\frac{\overbrace{4}^{|z_{Mg^{2+}} z_{O^{2-}}|}}{\underbrace{2.12\times10^{-10}\,m}_{d}} \times \overbrace{(1.389\,35\times10^{-4}\,J\,m\,mol^{-1})}^{N_A e^2/4\pi\varepsilon_0}$$

$$\times \overbrace{\left(1 - \frac{34.5\,pm}{212\,pm}\right)}^{1-d^*/d} \times \overbrace{1.748}^{A}$$

$$= -3.84\times10^3\,kJ\,mol^{-1}$$

It is not possible to measure the lattice enthalpy directly, but values can be obtained by combining experimental values of other enthalpy changes by using a **Born–Haber cycle**. Such a cycle is a closed path of transformations starting and ending at the same point, one step of which is the formation of the solid compound from a gas of widely separated ions.

Example 15C.2 Using the Born–Haber cycle

Calculate the lattice enthalpy of KCl.

Collect your thoughts You need to construct a suitable Born–Haber cycle, such as the one shown in Fig. 15C.14. For the cycle to be useful, experimental values for the enthalpy changes for all the steps need to be available (for example, from tabulated data), apart, of course, from the step involving the formation of the lattice from the ions. For the cycle in Fig. 15C.14 the enthalpy changes are (for convenience, starting at the elements):

	$\Delta H/(\text{kJ mol}^{-1})$	
1. Sublimation of K(s)	+89	[dissociation enthalpy of K(s)]
2. Dissociation of $\frac{1}{2}Cl_2(g)$	+122	[$\frac{1}{2}$ × dissociation enthalpy of $Cl_2(g)$]
3. Ionization of K(g)	+418	[ionization enthalpy of K(g)]
4. Electron attachment to Cl(g)	−349	[electron-gain enthalpy of Cl(g)]
5. Formation of solid from gaseous ions	$-\Delta H_L/(\text{kJ mol}^{-1})$	[value to be determined]
6. Decomposition of compound to its elements in their reference states	+437	[negative of enthalpy of formation of KCl(s)]

Because this is a closed cycle, the sum of these enthalpy changes is equal to zero, and the lattice enthalpy can be inferred from the resulting equation.

The solution The sum of contributions around the cycle is

$$89 + 122 + 418 - 349 - \Delta H_L/(\text{kJ mol}^{-1}) + 437 = 0$$

It follows that $\Delta H_L = +717\,\text{kJ mol}^{-1}$.

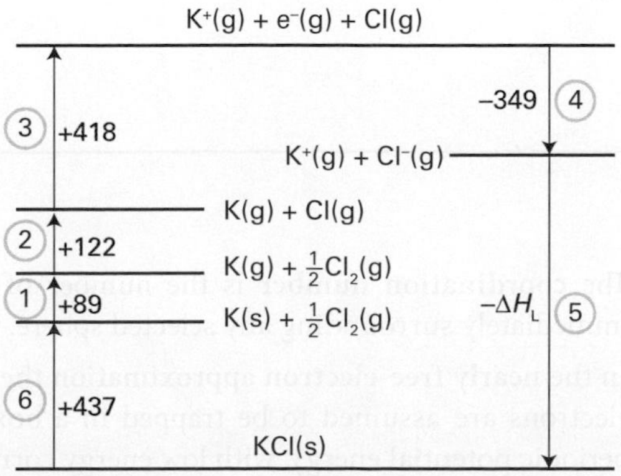

Figure 15C.14 The Born–Haber cycle for KCl at 298 K. Enthalpy changes are in kilojoules per mole.

Self-test 15C.2 Calculate the lattice enthalpy of CaO from the following data:

	$\Delta H/(\text{kJ mol}^{-1})$
Sublimation of Ca(s)	+178
Ionization of Ca(g) to $Ca^{2+}(g)$	+1735
Dissociation of $\frac{1}{2}O_2(g)$	+249
Electron attachment to O(g)	−141
Electron attachment to $O^-(g)$	+844
Formation of CaO(s) from Ca(s) and $\frac{1}{2}O_2(g)$	−635

Answer: +3500 kJ mol⁻¹

Some lattice enthalpies obtained by the Born–Haber cycle are listed in Table 15C.4. As can be seen from the data, the trends in values are in general accord with the predictions of the Born–Mayer equation. Agreement is typically taken to imply that the ionic model of bonding is valid for the substance; disagreement implies that there is a covalent contribution to the bonding. It is important, though, to be cautious, because numerical agreement might be coincidental and, as noted above, the values for ionic radii are subject to significant uncertainty.

15C.3 Covalent and molecular solids

X-ray diffraction studies of solids reveal a huge amount of information, including interatomic distances, bond angles, stereochemistry, and vibrational parameters. This section can do no more than hint at the diversity of types of solids found when molecules pack together or atoms link together in extended networks.

In **covalent solids** (or *covalent network solids*), covalent bonds in a definite spatial orientation link the atoms together into a network that extends through the crystal; effectively the crystal is a giant molecule. The demands of directional bonding, which have only a small effect on the structures of many metals, now override the geometrical problem of packing spheres together, resulting in a wide variety of often quite elaborate structures.

Table 15C.4 Lattice enthalpies at 298 K, $\Delta H_L/(\text{kJ mol}^{-1})$*

NaF	926
NaBr	752
MgO	3850
MgS	3406

* More values are given in the *Resource section*.

Brief illustration 15C.4

Diamond and graphite are two allotropes of carbon. In diamond each sp^3-hybridized carbon is bonded tetrahedrally to its four neighbours (Fig. 15C.15). The network of strong C–C bonds is repeated throughout the crystal and, as a result, diamond is very hard (in fact, the hardest known substance). In graphite, σ bonds between sp^2-hybridized carbon atoms form hexagonal rings which, when repeated throughout a plane, give rise to 'graphene' sheets (Fig. 15C.16). Because the sheets can slide against each other when impurities are present, impure graphite is used widely as a lubricant.

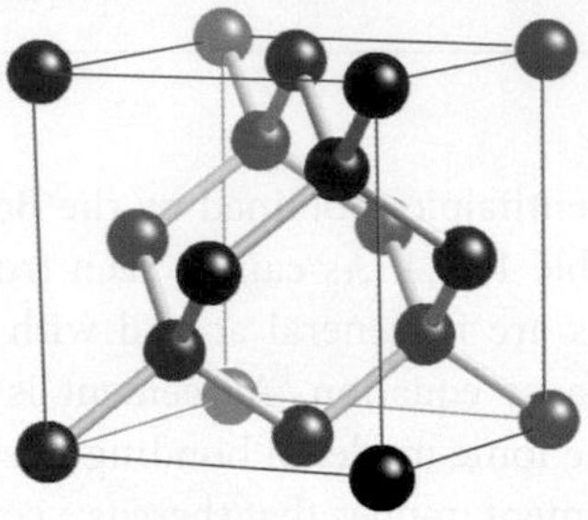

Figure 15C.15 A fragment of the structure of diamond. Each C atom is tetrahedrally bonded to four neighbours. This framework-like structure results in a rigid crystal.

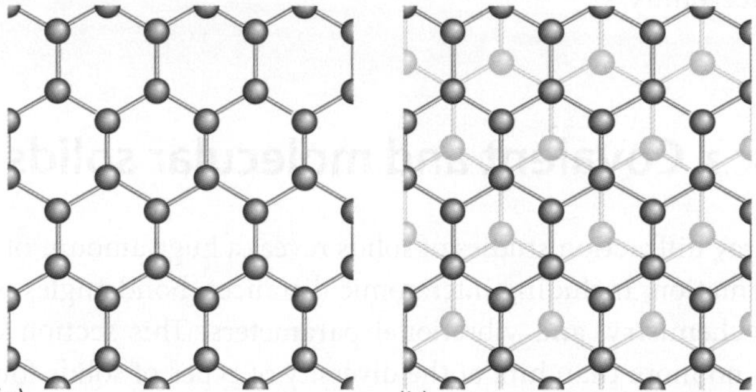

Figure 15C.16 Graphite consists of flat planes of hexagons of carbon atoms lying above one another. (a) The arrangement of carbon atoms in a 'graphene' sheet; (b) the relative arrangement of neighbouring sheets. The planes can slide over one another easily when impurities are present.

Molecular solids, which are the subject of the overwhelming majority of modern structural determinations, are held together by van der Waals interactions between the individual molecular components (Topic 14B). The observed crystal structure is nature's solution to the problem of condensing objects of various shapes into an aggregate of minimum energy (actually, for $T > 0$, of minimum Gibbs energy). The prediction of the structure is difficult, but software specifically designed to explore interaction energies can now make reasonably reliable predictions. The problem is made more complicated by the role of hydrogen bonds, which in some cases dominate the crystal structure, as in ice (Fig. 15C.17), but in others (for example, in solid phenol) distort a structure that is determined largely by the van der Waals interactions.

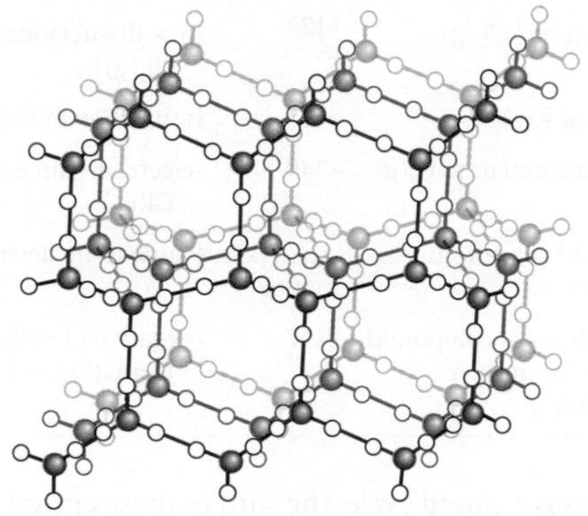

Figure 15C.17 A fragment of the crystal structure of ice (ice-I). Each O atom is at the centre of a tetrahedron of four O atoms at a distance of 276 pm. The central O atom is attached by two short O–H bonds to two H atoms and by two long hydrogen bonds to the H atoms of two of the neighbouring molecules. Both alternative H atoms locations are shown for each O–O separation. Overall, the structure consists of planes of hexagonal puckered rings of H_2O molecules (like the chair form of cyclohexane).

Checklist of concepts

☐ 1. A **close-packed** layer is a layer of spheres arranged so there is maximum utilization of space.

☐ 2. A **hexagonally close-packed** structure is one where the sequence of close-packed layers is ABABAB….

☐ 3. A **cubic close-packed** structure is one where the sequence of close-packed layers is ABCABC….

☐ 4. The **coordination number** is the number of spheres immediately surrounding any selected sphere.

☐ 5. In the **nearly free-electron approximation** the valence electrons are assumed to be trapped in a box with a periodic potential energy, with low energy corresponding to the locations of cations.

☐ 6. In the **tight-binding approximation** the valence electrons are assumed to occupy molecular orbitals delocalized throughout the solid.

☐ 7. In metals atomic orbitals overlap to form a **band**, which is a set of molecular orbitals that are closely spaced and cover a finite range of energy; electrons occupy the orbitals within the band.

☐ 8. A **band gap** is a range of energies to which no orbital corresponds.

☐ 9. The **Fermi level** is the highest occupied molecular orbital at $T = 0$.

☐ 10. The **coordination number** of an ionic lattice is denoted (N_+,N_-), where N_+ is the number of nearest neighbour anions around a cation and N_- the number of nearest neighbour cations around an anion.

☐ 11. The **lattice energy** of a solid is the change in potential energy when the ions go from being packed together in a solid to being widely separated as a gas.

☐ 12. A **Born–Haber cycle** is a closed path of transformations starting and ending at the same point, one step of which is the formation of the solid compound from a gas of widely separated ions.

☐ 13. A **molecular solid** is a solid consisting of discrete molecules held together by van der Waals interactions, and possibly hydrogen bonds.

Checklist of equations

Property	Equation	Comment	Equation number
Energy levels of a linear array of orbitals	$E_k = \alpha + 2\beta\cos(k\pi/(N+1)), k = 1, 2,\ldots, N$	Hückel approximation	15C.1
Band width	$E_N - E_1 \to -4\beta$ as $N \to \infty$	Hückel approximation	
Radius ratio	$\gamma = r_{\text{smaller}}/r_{\text{larger}}$	For criteria, see Section 15C.2	15C.2
Born–Mayer equation	$E_{\text{p,min}} = -\{N_A\lvert z_A z_B\rvert e^2/4\pi\varepsilon_0 d\}(1-d^*/d)A$		15C.5

TOPIC 15D The mechanical properties of solids

> ➤ Why do you need to know this material?
>
> An understanding of the mechanical properties of solid materials is crucial to the development of modern materials.
>
> ➤ What is the key idea?
>
> The mechanical properties of solids are expressed in terms of various 'moduli' that are related to the intermolecular potential energy of the constituents.
>
> ➤ What do you need to know already?
>
> You need to be familiar with the Lennard-Jones potential energy (Topic 14B).

The fundamental concepts needed for the discussion of the mechanical properties of solids are stress and strain. The **stress** on an object is the applied force divided by the area to which the force is applied. For instance, if a mass m hangs from a wire of radius r, and therefore cross-section πr^2, the mass exerts a gravitational force mg and the uniaxial stress (along the length of the wire) would be reported as $mg/\pi r^2$. The **strain** is the resulting fractional distortion of the object. The study of the relation between stress and strain is called **rheology** from the Greek word for 'flow'.

Stress may be applied in a number of different ways (Fig. 15D.1):

- **Uniaxial stress** is a simple compression or extension in one direction.
- **Hydrostatic stress** is a stress applied simultaneously in all directions, as in a body immersed in a fluid.
- **Pure shear** is a stress that tends to push opposite faces of the sample in opposite directions.

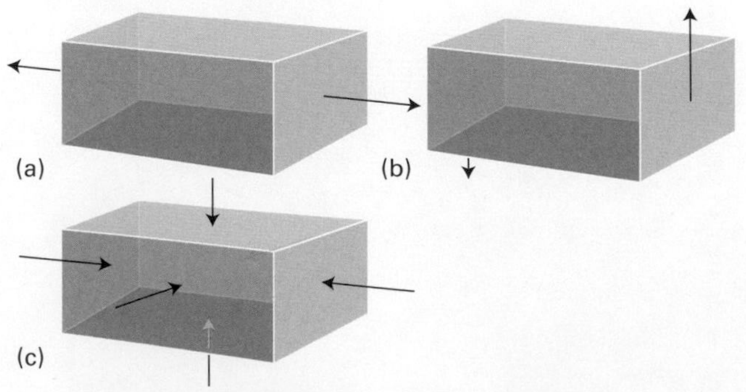

Figure 15D.1 Types of stress applied to a body. (a) Uniaxial stress, (b) shear stress, (c) hydrostatic pressure.

A sample subjected to a low stress typically undergoes **elastic deformation** in the sense that it recovers its original shape when the stress is removed. For low stresses, the strain is linearly proportional to the stress, and the stress–strain relation is a Hooke's law of force (Fig. 15D.2). The response becomes nonlinear at high stresses but may remain elastic. Above a certain threshold, the strain becomes **plastic** in the sense that recovery does not occur when the stress is removed. Plastic deformation occurs when bond breaking takes place and, in pure metals, typically takes place through the agency of dislocations. Brittle materials, such as ionic solids, exhibit sudden fracture as the stress focused by cracks causes them to spread catastrophically.

The response of a solid to an applied stress is commonly summarized by a number of coefficients of proportionality known as **moduli**:

$$\text{Young's modulus: } E = \frac{\text{uniaxial stress}}{\text{uniaxial strain}} \qquad (15\text{D}.1\text{a})$$

$$\text{Bulk modulus: } K = \frac{\text{pressure}}{\text{fractional change in volume}} \qquad (15\text{D}.1\text{b})$$

$$\text{Shear modulus: } G = \frac{\text{shear stress}}{\text{shear strain}} \qquad (15\text{D}.1\text{c})$$

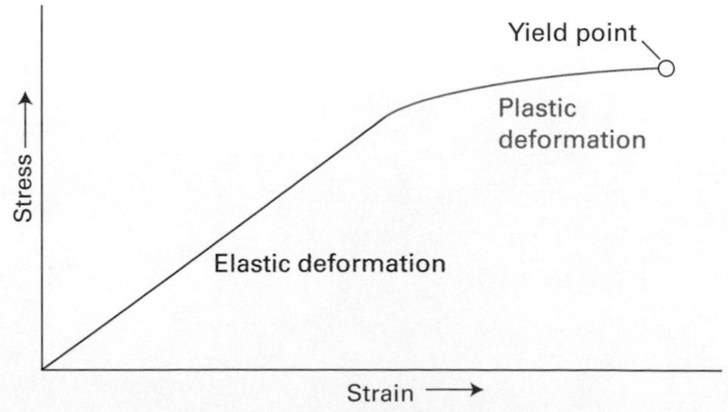

Figure 15D.2 At small strains, a body obeys Hooke's law (stress proportional to strain) and is elastic (recovers its shape when the stress is removed). At high strains, the body is no longer elastic, may become plastic, and finally yield.

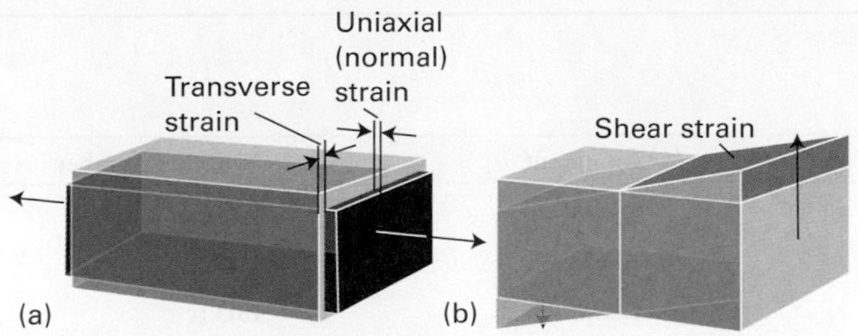

Figure 15D.3 (a) Uniaxial stress and the resulting uniaxial and transverse strain; Poisson's ratio indicates the extent to which a body changes shape when subjected to a uniaxial stress. (b) Shear stress and the resulting strain.

'Uniaxial strain' refers to stretching and compression of the material in one direction, as shown in Fig. 15D.3a, and 'shear strain' refers to the distortion arising from a shear stress, as depicted in Fig. 15D.3b. The fractional change in volume is $\delta V/V$, where δV is the change in volume of a sample of volume V; similarly, the uniaxial strain and the shear strain are (dimensionless) fractional changes in dimensions. The bulk modulus is the inverse of the isothermal compressibility, κ_T, discussed in Topic 2D (eqn 2D.7, $\kappa_T = -(\partial V/\partial p)_T/V$).

A third ratio, called **Poisson's ratio**, indicates how the sample changes its shape:

$$\nu_{\rm P} = \frac{\text{transverse strain}}{\text{normal strain}} \qquad \text{Poisson's ratio [definition]} \qquad (15\text{D}.2)$$

Transverse and normal strains are illustrated in Fig. 15D.3a: they are the mutually perpendicular uniaxial distortions arising from the 'normal' uniaxial stress. The three moduli introduced in eqn 15D.1 are interrelated in the following way (see Problem P15D.1):

$$G = \frac{E}{2(1+\nu_{\rm P})} \qquad K = \frac{E}{3(1-2\nu_{\rm P})} \qquad \text{Relations between moduli} \qquad (15\text{D}.3)$$

Brief illustration 15D.1

The uniaxial stress when a mass of m = 10.0 kg is suspended from an iron wire of radius r = 0.050 mm is

$$\text{uniaxial stress} = \frac{mg}{\pi r^2} = \frac{(10.0\,\text{kg})\times(9.81\,\text{m s}^{-2})}{\pi(5.0\times10^{-5}\,\text{m})^2}$$
$$= 1.24\ldots\times10^{10}\,\text{kg m}^{-1}\,\text{s}^{-2}$$

The Young's modulus of iron at room temperature is 215 GPa. Therefore

$$\text{uniaxial strain} = \frac{1.24\ldots\times10^{10}\,\text{kg m}^{-1}\,\text{s}^{-2}}{2.15\times10^{11}\,\underbrace{\text{kg m}^{-1}\,\text{s}^{-2}}_{\text{Pa}}} = 0.0581$$

which corresponds to elongation of the wire by 5.81 per cent.

If neighbouring molecules interact by a Lennard-Jones potential energy (Topic 14B), then the bulk modulus and the compressibility of the solid are related to the Lennard-Jones parameter ε (the depth of the potential well) by

$$K = \frac{8N_{\rm A}\varepsilon}{V_{\rm m}} \qquad \kappa_T = \frac{V_{\rm m}}{8N_{\rm A}\varepsilon} \qquad (15\text{D}.4)$$

For the derivation of these relations, see *A deeper look* 10 on the website of this text. The bulk modulus is large and the compressibility low (the solid stiff) if the potential well is deep and the solid is dense (its molar volume small).

The differing rheological characteristics of metals can be traced to the presence of **slip planes**, which are planes of atoms that, when under stress, may slip or slide relative to one another. The slip planes of a ccp structure are the close-packed planes, and careful inspection of a unit cell shows that there are eight sets of slip planes in different directions. As a result, metals with ccp structures, like copper, are malleable, meaning they can easily be bent, flattened, or hammered into shape. In contrast, a hexagonal close-packed structure has only one set of slip planes so that metals with hexagonal close packing, such as zinc or cadmium, tend to be more brittle.

Checklist of concepts

- ☐ 1. **Uniaxial stress** is a simple compression or extension applied to a sample in one direction.
- ☐ 2. **Hydrostatic stress** is a stress applied simultaneously in all directions, as in a body immersed in a fluid.
- ☐ 3. **Pure shear** is a stress that tends to push opposite faces of the sample in opposite directions.
- ☐ 4. A sample subjected to a small stress typically undergoes **elastic deformation**; as the stress increases the sample becomes **plastic**.
- ☐ 5. The response of a solid to an applied stress is summarized by the **Young's modulus**, the **bulk modulus**, the **shear modulus**, and **Poisson's ratio**.
- ☐ 6. The differing rheological characteristics of metals can be traced to the presence of **slip planes**.

Checklist of equations

Property	Equation	Comment	Equation number
Young's modulus	E = uniaxial stress / uniaxial strain	Definition	15D.1a
Bulk modulus	K = pressure / fractional change in volume	Definition	15D.1b
Shear modulus	G = shear stress / shear strain	Definition	15D.1c
Poisson's ratio	ν_P = transverse strain / normal strain	Definition	15D.2

TOPIC 15E The electrical properties of solids

> ➤ Why do you need to know this material?
>
> The electrical properties of solids underlie numerous technological applications on which the infrastructure of the modern world depends.
>
> ➤ What is the key idea?
>
> Electrons in solids occupy bands that determine the electrical conductivities of various types of solid.
>
> ➤ What do you need to know already?
>
> You need to be familiar with the formation of bands in solids (Topic 15C).

The electrical conductivity of common materials arises from the motion of electrons, but some ionic solids display ionic conductivity in which whole ions migrate through the lattice. Three types of solid are distinguished by the temperature dependence of their electrical conductivity (Fig. 15E.1):

- A **metallic conductor** is a substance with an electrical conductivity that decreases as the temperature is raised.

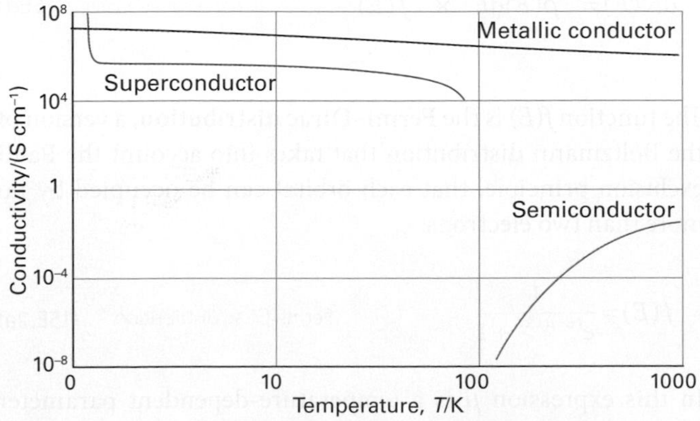

Figure 15E.1 The variation of the electrical conductivity of a substance with temperature is the basis of its classification as a metallic conductor, a semiconductor, or a superconductor. Conductivity is expressed in siemens per metre ($S\,m^{-1}$ or, as here, $S\,cm^{-1}$), where $1\,S = 1\,\Omega^{-1}$ (the resistance is expressed in ohms, Ω); note the log scale.

- A **semiconductor** is a substance with an electrical conductivity that increases as the temperature is raised.
- A **superconductor** is a solid that, below a critical temperature, conducts electricity without resistance.

A semiconductor generally has a lower conductivity than that typical of metallic conductors, but the magnitude of the conductivity is not the criterion for distinguishing between them. It is conventional to classify semiconductors with very low electrical conductivities, such as most synthetic polymers, as **insulators**. The term is one of convenience rather than one of fundamental significance.

15E.1 Metallic conductors

To understand the origins of the electric conductivity in conductors and semiconductors, it is necessary to explore the consequences of the formation of bands (Topic 15C). The starting point is Fig. 15C.8, which is repeated here for convenience as Fig. 15E.2. It shows the electronic structure of a solid formed from a line of N atoms, each of which contributes one electron (such as the alkali metals). At $T = 0$, only the lowest $\frac{1}{2}N$ molecular orbitals are occupied, up to the Fermi level. The levels are very closely spaced, so there are unoccupied molecular orbitals just above the Fermi level. A solid with a partially filled band is expected to be a metallic conductor, an observation which can be understood in the following way.

The key point is that each molecular orbital in a band can be regarded as the superposition of two waves travelling in

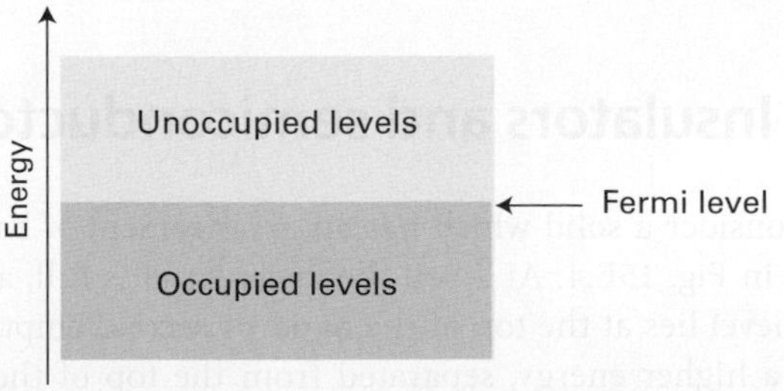

Figure 15E.2 (A reproduction of Fig. 15C.8.) When N electrons occupy a band of N orbitals at $T = 0$, it is only half full. The highest occupied level is the Fermi level.

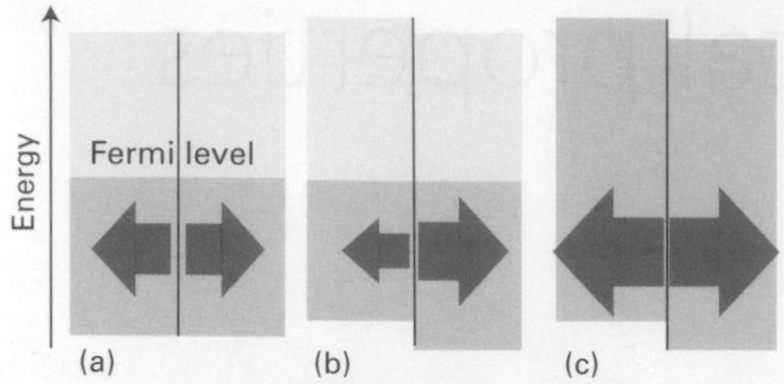

Figure 15E.3 Each part of this illustration separates out the waves travelling in opposite directions. (a) The two sets of waves have the same energy, are equally occupied, and there is no net motion. (b) When a potential difference is applied (positive on the right) the two sets no longer have the same energy. There are now more electrons moving to the right than to the left and therefore there is a net current. (c) If the band is full, the two sets remain equally populated and there is no net flow even when a potential difference is applied.

opposite directions (in the same sense that $\cos x \propto e^{ix} + e^{-ix}$). Figure 15E.3a is an adaptation of Fig. 15E.2 which separates the two contributions. In the absence of any applied field, electrons occupy both components equally and there is no net motion through the solid. This absence of net motion is true whether the band is full or incomplete. When a potential difference is applied the energies of the components differ, as it is energetically favourable for electrons to travel towards regions of positive potential. Now the two components are no longer equally occupied (Fig. 15E.3b) and provided the band is not full there are more electrons travelling in one direction than the other and electrical conduction occurs. If the band is full, however, the populations of the two components remain equal (Fig. 15E.3c) and there is no net motion in either direction. The material does not conduct: it is an insulator.

The conductivity of a metallic conductor decreases as the temperature is raised. This decrease is due to collisions between the moving electrons and the atoms. The greater the temperature, the more vigorous is the thermal motion of the atoms, so collisions between the moving electrons and an atom are more likely. That is, the electrons are scattered out of their paths through the solid, and are less efficient at transporting charge.

15E.2 Insulators and semiconductors

Now consider a solid which has an arrangement of bands as shown in Fig. 15E.4. At $T = 0$ the lower band is full, and the Fermi level lies at the top of the band. A second empty band lies at a higher energy, separated from the top of the lower band by an energy E_g, known as the **band gap**. At $T = 0$ this material is an insulator because there are no partially filled bands. If the temperature is high enough, though, electrons

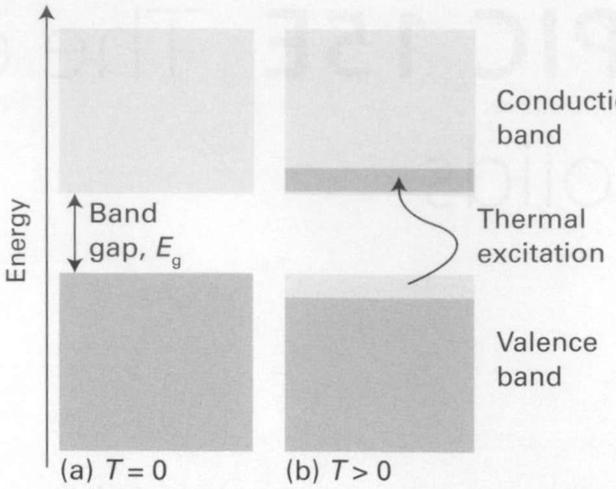

Figure 15E.4 (a) Typical band structure for a semiconductor: at $T = 0$ the valence band is full and the conduction band is empty. (b) At higher temperatures electrons populate the levels of the conduction band leading to electrical conductivity which increases with temperature.

are excited out of the lower band into the upper. There are now incomplete bands and conduction can occur.

The question that now arises is how the populations of the two bands, and therefore the conductivity of the semiconducting material, depend on the temperature. The discussion starts by introducing the **density of states**, $\rho(E)$, defined such that the number of states between E and $E + dE$ is $\rho(E)dE$. Note that the 'state' of an electron includes its spin, so each spatial orbital counts as two states. To obtain the number of electrons $dN(E)$ that occupy states between E and $E + dE$, $\rho(E)dE$ is multiplied by the probability $f(E)$ of occupation of the state with energy E. That is,

$$dN(E) = \overbrace{\rho(E)dE}^{\text{Number of states between } E \text{ and } E+dE} \times \overbrace{f(E)}^{\text{Probability of occupation of a state with energy } E} \quad (15E.1)$$

The function $f(E)$ is the **Fermi–Dirac distribution**, a version of the Boltzmann distribution that takes into account the Pauli exclusion principle, that each orbital can be occupied by no more than two electrons:

$$f(E) = \frac{1}{e^{(E-\mu)/kT} + 1} \qquad \text{Fermi–Dirac distribution} \quad (15E.2a)$$

In this expression μ is a temperature-dependent parameter known as the 'chemical potential' (it has a subtle relation to the familiar chemical potential of thermodynamics); provided $T > 0$, μ is the energy of the state for which $f = \frac{1}{2}$. At $T = 0$, only states up to a certain energy, known as the **Fermi energy**, E_F, are occupied (Fig. 15E.2). Provided the temperature is not so high that many electrons are excited to states above the Fermi energy, the chemical potential can be identified with E_F, in which case the Fermi–Dirac distribution becomes

$$f(E)=\frac{1}{e^{(E-E_F)/kT}+1} \quad \text{Fermi–Dirac distribution} \quad (15E.2b)$$

This expression implies that $f(E_F) = \frac{1}{2}$. For energies well above E_F, the exponential term is so large that the 1 in the denominator can be neglected, and then

$$f(E)\approx e^{-(E-E_F)/kT} \quad \text{Fermi–Dirac distribution [approximate form for } E > E_F] \quad (15E.2c)$$

The function now resembles a Boltzmann distribution, decaying exponentially with increasing energy; the higher the temperature, the longer is the exponential tail.

There is a distinction between the Fermi energy and the Fermi level:

- The Fermi level is the uppermost occupied level at $T = 0$.
- The Fermi energy is the energy level at which $f(E) = \frac{1}{2}$ at any temperature.

The Fermi energy coincides with the Fermi level as $T \to 0$.

Figure 15E.5 shows the form of $f(E)$ at different temperatures. At $T = 0$ the probability distribution is a step function, equal to 1 for $E < E_F$, and 0 at higher energies, as in Fig. 15E.2. At higher temperatures the probability of occupation of levels above E_F increases at the expense of those below E_F, with the greatest changes occurring in the energies close to E_F. As the temperature is raised, electrons are promoted from the lower band to the upper. This promotion is represented by the tail of the Fermi–Dirac distribution extending across the band gap and is significant only when kT is comparable to or greater than the band gap. The material, an insulator at $T = 0$, is now a conductor, because both bands are partially filled. As the temperature is increased, the conductivity increases as more electrons are promoted across the band gap, so the material is a semiconductor.

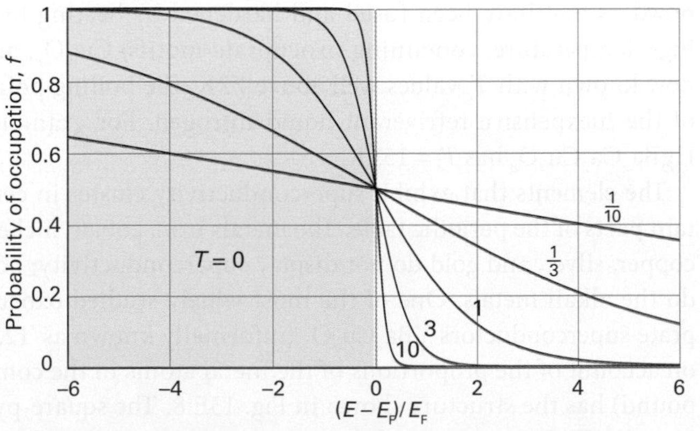

Figure 15E.5 The Fermi–Dirac distribution, which gives the probability of occupation of a state with energy E and at a temperature T. At higher energies the probability decays exponentially towards zero. The curves are labelled with the value of E_F/kT. The tinted region shows the occupation of levels at $T = 0$.

The lower band, which is full at $T = 0$, is called the **valence band** and the upper band, which is empty at $T = 0$ and to which electrons are thermally excited, is called the **conduction band**. When electrons leave the valence band they can be thought of as creating positively charged 'holes' in that band (i.e. the absence of an electron), and the electrical conductivity arises from the movement of these holes and the promoted electrons.

Figure 15E.4 depicts an **intrinsic semiconductor**, in which semiconduction is a property of the band structure of the pure material. Examples of intrinsic semiconductors include silicon and germanium. A **compound semiconductor** is an intrinsic semiconductor formed from a combination of different elements, such as GaN, CdS, and many d-metal oxides.

An **extrinsic semiconductor** is one in which charge carriers (electrons or holes) are present as a result of the replacement of some atoms (to the extent of about 1 in 10^9) by **dopant** atoms, the atoms of another element. If, for example, pure silicon (a Group 14 element) is doped with atoms of indium (a Group 13 element) an electron can be transferred from a Si atom to a neighbouring In atom, thereby creating a hole in the valence band and increasing the conductivity. The resulting semiconductor is described as **p-type**, the p indicating that the positive holes are responsible for conduction. Figure 15E.6a shows the band structure of such a semiconductor. The dopant atoms result in a set of empty levels, called **acceptor levels**, which lie just above the top of the valence band. Electrons from the valence band are transferred into these levels, so generating holes in the band.

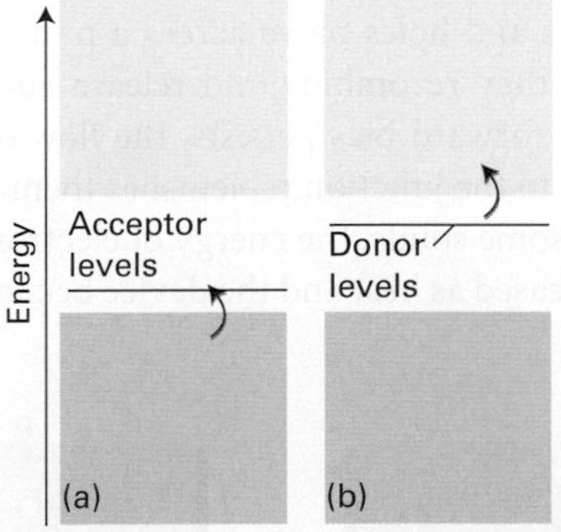

Figure 15E.6 (a) A dopant with fewer electrons than its host contributes levels that accept electrons from the valence band. The resulting holes in the band give rise to electrical conductivity; the doped semiconductor is classified as p-type. (b) A dopant with more electrons than its host contributes occupied levels that can supply electrons to the conduction band, thus giving rise to electrical conductivity; the substance is classed as an n-type semiconductor.

If the dopant atoms are from a Group 15 element (e.g. phosphorus), an electron can be transferred from a P atom into the otherwise empty conduction band, thereby increasing the conductivity. This type of doping results in an **n-type** semiconductor, where n refers to the negative charge of the carriers. The band structure is shown in Fig. 15E.6b. The dopant atoms create a set of occupied levels, called **donor levels**, just below the bottom of the conduction band, and electrons from these levels are transferred into the conduction band. In practical cases the level of doping is such that the charge carriers created by the dopant atoms are greatly in excess of those arising from thermal excitation across the band gap: electrical conductivity is therefore dominated by the type and extent of the doping.

Doped semiconductors are of great technological importance because they are the materials from which the active components of electronic circuits are made. The simplest example of an electronic device constructed from doped semiconductors is the 'p–n diode' which consists of a p-type semiconductor in contact with an n-type semiconductor, thereby creating a **p–n junction**. A p–n junction conducts electricity only in one direction. To understand this property consider first the arrangement shown in Fig. 15E.7a where the p-type semiconductor is attached to the negative electrode and the n-type is attached to the positive electrode; this arrangement is known as 'reverse bias'. The positively charged holes in the p-type semiconductor are attracted to the negative electrode, and the negatively charged electrons in the n-type semiconductor are attracted to the positive electrode. As a consequence, charge does not flow across the junction so the device does not conduct. Now consider what happens when the charges on the electrodes are reversed, as shown in Fig. 15E.7b, an arrangement known as 'forward bias'. Electrons in the n-type semiconductor move towards the positive electrode, and holes move in the opposite direction: as a result charge flows across the junction. The p–n junction therefore conducts only under forward bias.

As electrons and holes move across a p–n junction under forward bias, they recombine and release energy. However, as long as the forward bias persists, the flow of charge from the electrodes to the junction replenishes them with electrons and holes. In some solids, the energy of electron–hole recombination is released as heat and the device becomes warm. The reason lies in the fact that the return of the electron to a hole involves a change in the electron's linear momentum, which the atoms of the lattice must absorb, and therefore electron–hole recombination stimulates lattice vibrations. This is the case for silicon semiconductors, and is one reason why computers need efficient cooling systems.

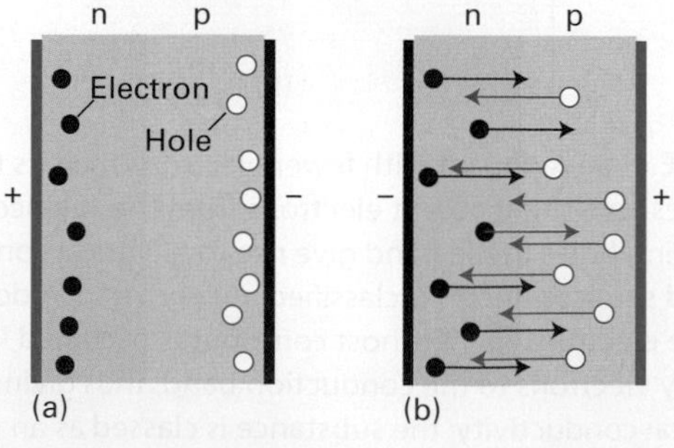

Figure 15E.7 A p–n junction under (a) reverse bias, (b) forward bias.

Another electronic device, a 'transistor', consists of a p-type semiconductor sandwiched between two n-type semiconductors, and as such has two p–n junctions. Under the correct conditions it is possible to control the current flowing between the two n-type semiconductors by varying the current flowing into the p-type semiconductor. Most significantly, the change in the current between the n-type semiconductors can be larger than the change in the current in the p-type semiconductor; in other words the device can act as an amplifier. It is the exploitation of this property that has led to the development of modern solid-state electronics.

15E.3 Superconductors

The resistance to flow of electrical current of a normal metallic conductor decreases smoothly with decreasing temperature but never vanishes. However, a superconductor conducts electricity without resistance once the temperature is below the critical temperature, T_c. Following the discovery in 1911 that mercury is a superconductor below 4.2 K, the normal boiling point of liquid helium, physicists and chemists made slow but steady progress in the discovery of superconductors with higher values of T_c. Metals, such as tungsten, mercury, and lead, have T_c values below about 10 K. Intermetallic compounds, such as Nb_3X (X = Sn, Al, or Ge), and alloys, such as Nb/Ti and Nb/Zr, have intermediate T_c values ranging between 10 K and 23 K. In 1986, **high-temperature superconductors** (HTSCs) were discovered. Several *ceramics*, inorganic powders that have been fused and hardened by heating to a high temperature, containing oxocuprate motifs, Cu_mO_n, are now known with T_c values well above 77 K, the boiling point of the inexpensive refrigerant liquid nitrogen. For example, $HgBa_2Ca_2Cu_3O_8$ has $T_c = 153$ K.

The elements that exhibit superconductivity cluster in certain parts of the periodic table. The metals iron, cobalt, nickel, copper, silver, and gold do not display superconductivity; nor do the alkali metals. One of the most widely studied oxocuprate superconductors $YBa_2Cu_3O_7$ (informally known as '123' on account of the proportions of the metal atoms in the compound) has the structure shown in Fig. 15E.8. The square-pyramidal CuO_5 units arranged as two-dimensional layers and the square planar CuO_4 units arranged in sheets are common structural features of oxocuprate HTSCs.

The mechanism of superconduction is well-understood for low-temperature materials, and is based on the properties of

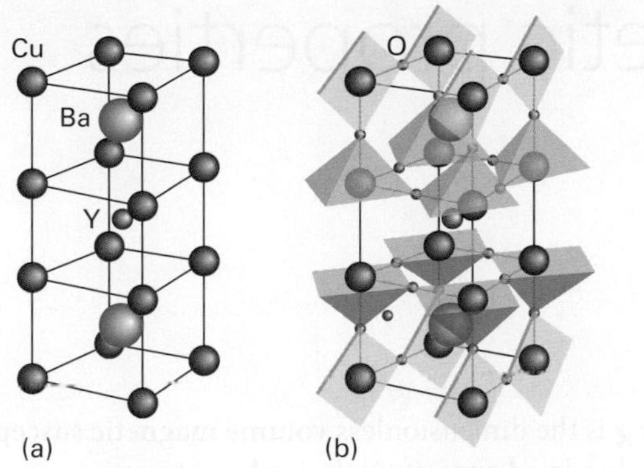

Figure 15E.8 Structure of the $YBa_2Cu_3O_7$ superconductor. (a) Metal atom positions. (b) The polyhedra show the positions of oxygen atoms and indicate that the Cu ions are either in square-planar or square-pyramidal coordination environments.

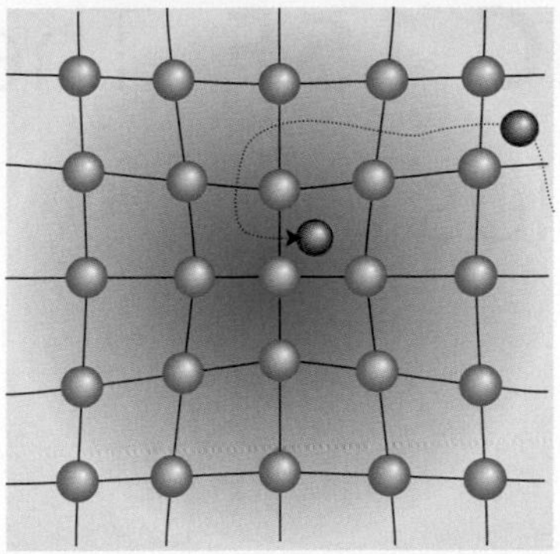

Figure 15E.9 The formation of a Cooper pair. One electron distorts the crystal lattice and the second electron has a lower energy if it goes to that region. These electron–lattice interactions effectively bind the two electrons into a pair.

a **Cooper pair**, a pair of electrons that exists on account of the indirect electron–electron interactions mediated by the nuclei of the atoms in the lattice. Thus, if one electron is in a particular region of a solid, the nuclei there move towards it to give a distorted local structure (Fig. 15E.9). Because that local distortion is rich in positive charge, it is favourable for a second electron to join the first. Hence, there is a virtual attraction between the two electrons and they move together as a pair. The local distortion is disrupted by thermal motion of the ions in the solid, so the virtual attraction occurs only at very low temperatures. A Cooper pair undergoes less scattering than an individual electron as it travels through the solid because the distortion caused by one electron can attract back the other electron should it be scattered out of its path in a collision. Because the Cooper pair is stable against scattering, it can carry charge freely through the solid, and hence give rise to superconduction.

The Cooper pairs responsible for low-temperature superconductivity are likely to be important in HTSCs, but the mechanism for pairing is hotly debated. There is evidence implicating the arrangement of CuO_5 layers and CuO_4 sheets in the mechanism. It is believed that movement of electrons along the linked CuO_4 units accounts for superconductivity, whereas the linked CuO_5 units act as 'charge reservoirs' that maintain an appropriate number of electrons in the superconducting layers.

Checklist of concepts

- ☐ 1. Electronic conductors are classified as **metallic conductors** or **semiconductors** according to the temperature dependence of their conductivities; an **insulator** is a semiconductor with very low conductivity.
- ☐ 2. **Superconductors** conduct electricity without resistance below a critical temperature T_c.
- ☐ 3. The **Fermi–Dirac distribution** gives the probability that a state with a particular energy is occupied by an electron.
- ☐ 4. The **Fermi energy** is the energy of the level for which the probability of occupation is $\frac{1}{2}$.
- ☐ 5. In a semiconductor at $T=0$ there is a full **valence band** and, at higher energy, an empty **conduction band**.
- ☐ 6. In an **intrinsic** semiconductor electrical conductivity is due to electrons thermally promoted from the valence band to the conduction band.
- ☐ 7. In an **extrinsic** semiconductor electrical conductivity is due to electrons or holes generated by the inclusion of **dopant** atoms.
- ☐ 8. Semiconductors are classified as **p-type** or **n-type** according to whether conduction is due to holes in the valence band or electrons in the conduction band.

Checklist of equations

Property	Equation	Comment	Equation number
Fermi–Dirac distribution	$f(E)=1/\{e^{(E-E_F)/kT}+1\}$	E_F is the Fermi energy	15E.2b

TOPIC 15F The magnetic properties of solids

➤ Why do you need to know this material?

The magnetic properties of solids give an indication of the electronic structures of individual molecules and many modern information storage devices make use of the additional properties that arise when the spins on different centres interact.

➤ What is the key idea?

The principal magnetic properties of solids arise from the spins of unpaired electrons and their interactions.

➤ What do you need to know already?

You need to be aware of the properties of electron angular momentum (Topic 8B) and the relation of magnetic moments to angular momenta (Topic 8C).

The magnetic properties of metallic solids and semiconductors depend strongly on the band structures of the material. In this section, attention is confined largely to the much simpler magnetic properties of collections of individual molecules or ions, such as d-metal complexes. Much of the discussion therefore applies to liquid and gas-phase samples, as well as to solids.

15F.1 Magnetic susceptibility

Some molecules possess permanent magnetic dipole moments. In the absence of an external magnetic field, the orientation of the dipole is random and the material has no net magnetic moment. That changes when a magnetic field is applied and certain orientations are favoured. The **magnetization**, $\mathcal{M}$, the net dipole-moment density, is the resulting average molecular magnetic dipole moment multiplied by the number density of molecules in the sample. The magnetization induced by a magnetic field of strength $\mathcal{H}$ is proportional to $\mathcal{H}$, and is written

$$\mathcal{M} = \chi\mathcal{H} \qquad \text{Volume magnetic susceptibility [definition]} \qquad (15F.1)$$

where χ is the dimensionless **volume magnetic susceptibility**. A closely related quantity is the **molar magnetic susceptibility**, χ_m:

$$\chi_m = \chi V_m \qquad \text{Molar magnetic susceptibility [definition]} \qquad (15F.2)$$

where V_m is the molar volume of the substance.

The magnetization can be thought of as contributing to the density of lines of force in the material (Fig. 15F.1). Materials for which $\chi > 0$ are called **paramagnetic**; they tend to move into a magnetic field and the density of lines of force within them is greater than in a vacuum. Those for which $\chi < 0$ are called **diamagnetic** and tend to move out of a magnetic field; the density of lines of force within them is lower than in a vacuum. A paramagnetic material consists of ions or molecules with unpaired electrons, such as radicals and many d-metal complexes; a diamagnetic substance (which is far more common) is one with no unpaired electrons.

The magnetic susceptibility is traditionally measured with a 'Gouy balance'. This instrument consists of a sensitive balance from which the sample, contained in a narrow tube, hangs between the poles of a magnet. If the sample is paramagnetic, it is drawn into the field and its apparent weight is greater when the field is turned on. A diamagnetic sample tends to be expelled from the field and appears to weigh less when the field is turned on. The balance is normally calibrated against a sample of known susceptibility. The modern version of the determination makes use of a 'superconducting quantum interference device' (a SQUID, Fig. 15F.2). A SQUID makes use

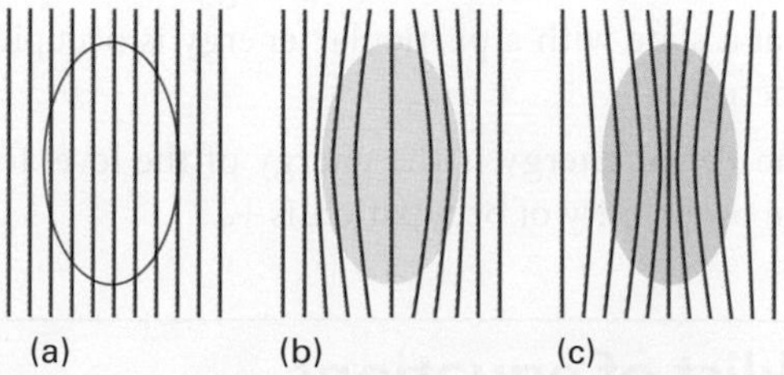

Figure 15F.1 (a) In a vacuum, the strength of a magnetic field can be represented by the density of lines of force; (b) in a diamagnetic material, the density is reduced; (c) in a paramagnetic material, the density is increased.

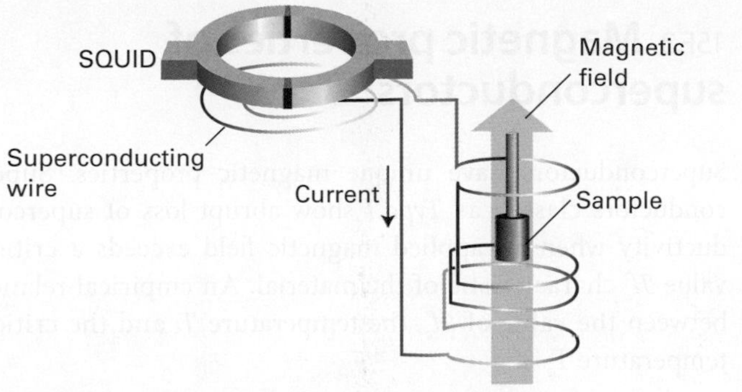

Figure 15F.2 The arrangement used to measure magnetic susceptibility with a SQUID. The sample is moved upwards in small increments and the potential difference across the SQUID is measured.

Table 15F.1 Magnetic susceptibilities at 298 K*

	$\chi/10^{-6}$	$\chi_m/(10^{-10}\ m^3\ mol^{-1})$
$H_2O(l)$	−9.02	−1.63
NaCl(s)	−16	−3.8
Cu(s)	−9.7	−0.69
$CuSO_4{\cdot}5H_2O(s)$	+167	+183

* More values are given in the *Resource section*.

of the quantization of magnetic flux and the property of current loops in superconductors that, as part of the circuit, include a weakly conducting link through which electrons must tunnel. The current that flows in the loop in a magnetic field depends on the value of the magnetic flux, and a SQUID can be exploited as a very sensitive magnetometer. Table 15F.1 lists some experimental values of the magnetic susceptibility.

15F.2 Permanent and induced magnetic moments

The permanent magnetic moment of a molecule arises from any unpaired electron spins in the molecule. The magnitude, m, of the magnetic moment of an electron is proportional to the magnitude of the spin angular momentum, $\{s(s+1)\}^{1/2}\hbar$:

$$m = g_e\{s(s+1)\}^{1/2}\mu_B \quad \mu_B = \frac{e\hbar}{2m_e} \qquad \text{Magnetic moment [magnitude]} \qquad (15F.3)$$

where $g_e = 2.0023$ and μ_B, the Bohr magneton, has the value $9.274 \times 10^{-24}\,J\,T^{-1}$. If there are several electron spins in each molecule, they combine to give a total spin S, and then $s(s+1)$ is replaced by $S(S+1)$.

The magnetization, and consequently the magnetic susceptibility, depends on the temperature because the orientations of the electron spins fluctuate. Some orientations have lower energy than others, and the magnetization depends on the randomizing influence of thermal motion. Thermal averaging of the permanent magnetic moments in the presence of an applied magnetic field results in a contribution to the magnetic susceptibility that is proportional to $m^2/3kT$.[1] It follows that the spin contribution to the molar magnetic susceptibility is

$$\chi_m = \frac{N_A g_e^2 \mu_0 \mu_B^2 S(S+1)}{3kT} \qquad \text{Molar magnetic susceptibility [spin contribution]} \qquad (15F.4a)$$

where μ_0 is the vacuum permeability. This susceptibility is positive, so the spin magnetic moments contribute to the paramagnetic susceptibilities of materials. Equation 15F.4a is commonly written as the **Curie law**:

$$\chi_m = \frac{C}{T} \quad C = \frac{N_A g_e^2 \mu_0 \mu_B^2 S(S+1)}{3k} \qquad \text{Curie law} \qquad (15F.4b)$$

The spin contribution to the susceptibility decreases with increasing temperature because the thermal motion randomizes the spin orientations. In practice, a contribution to the paramagnetism also arises from the orbital angular momenta of electrons: here only the spin contribution has been considered.

Brief illustration 15F.1

Consider a complex salt with three unpaired electrons per complex cation and molar volume $61.7\,cm^3\,mol^{-1}$; its molar magnetic susceptibility can be calculated using eqn 15F.4b. First, note that

$$\frac{N_A g_e^2 \mu_0 \mu_B^2}{3k} = 6.3001 \times 10^{-6}\,m^3\,K\,mol^{-1}$$

Then with $S = \frac{3}{2}$ eqn 15F.4b gives

$$\chi_m = (6.3001 \times 10^{-6}\,m^3\,K\,mol^{-1}) \times \frac{\frac{3}{2}(\frac{3}{2}+1)}{298\,K} = 7.92\ldots \times 10^{-8}\,m^3\,mol^{-1}$$

and from eqn 15F.2

$$\chi = \frac{\chi_m}{V_m} = \frac{7.92\ldots \times 10^{-8}\,m^3\,mol^{-1}}{6.17 \times 10^{-5}\,m^3\,mol^{-1}} = 1.28 \times 10^{-3}$$

At low temperatures, some paramagnetic solids make a phase transition to a state in which large domains of spins align with parallel orientations. This cooperative alignment gives rise to a very strong magnetization and is called **ferromagnetism** (Fig. 15F.3). In other cases, exchange interactions lead to alternating spin orientations: the spins are locked into a low-magnetization arrangement to give an **antiferromagnetic**

[1] See our *Physical chemistry: Quanta, matter, and change* (2014) for the derivation of this contribution.

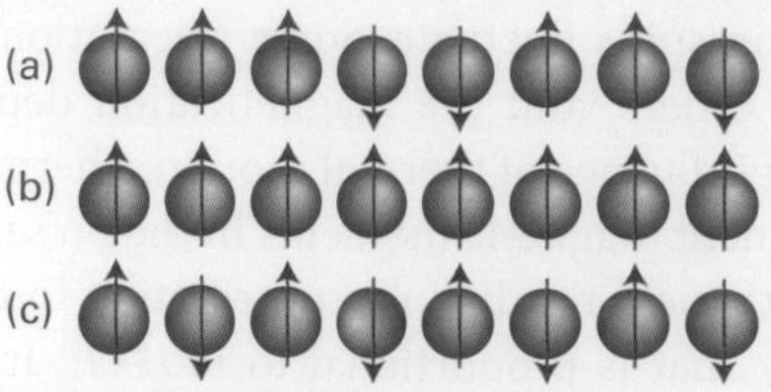

Figure15F.3 (a) In a paramagnetic material, the electron spins are aligned at random in the absence of an applied magnetic field. (b) In a ferromagnetic material, the electron spins are locked into a parallel alignment over large domains. (c) In an antiferromagnetic material, the electron spins are locked into an antiparallel arrangement. The latter two arrangements survive even in the absence of an applied field.

phase. The ferromagnetic phase has a non-zero magnetization in the absence of an applied field, but the antiferromagnetic phase has zero magnetization because the spin magnetic moments cancel. The ferromagnetic transition occurs at the **Curie temperature**, and the antiferromagnetic transition occurs at the **Néel temperature**. Which type of cooperative behaviour occurs depends on the details of the band structure of the solid.

Magnetic moments can also be induced in molecules. To see how this effect arises, it is necessary to note that the circulation of electronic currents induced by an applied field gives rise to a magnetic field which usually opposes the applied field, thus making the substance diamagnetic. In these cases, the induced electronic currents occur within the molecular orbitals that are occupied in its ground state. There are a few cases in which molecules are paramagnetic despite having no unpaired electrons. In these materials the induced electron currents flow in the opposite direction because they can make use of unoccupied orbitals that lie close to the HOMO in energy (a similar effect is the paramagnetic contribution to the chemical shift, Topic 12B). This orbital paramagnetism is distinguished from spin paramagnetism by the fact that it is temperature independent and is called **temperature-independent paramagnetism** (TIP).

These remarks can be summarized as follows. All molecules have a diamagnetic component to their susceptibility, but this contribution is dominated by spin paramagnetism if the molecules have unpaired electrons. In a few cases (where there are low-lying excited states) TIP is strong enough to make the molecules paramagnetic even though all their electrons are paired.

15F.3 Magnetic properties of superconductors

Superconductors have unique magnetic properties. Superconductors classed as *Type I* show abrupt loss of superconductivity when an applied magnetic field exceeds a critical value $\mathcal{H}_c$ characteristic of the material. An empirical relation between the value of $\mathcal{H}_c$, the temperature *T*, and the critical temperature T_c is

$$\mathcal{H}_c(T)=\mathcal{H}_c(0)\left(1-\frac{T^2}{T_c^2}\right) \quad \text{Dependence of } \mathcal{H}_c \text{ on } T \quad (15F.5)$$

provided $T \le T_c$. Note that the critical field falls as *T* rises from 0 towards T_c. Therefore, to maintain superconductivity in the presence of a magnetic field, it is best to keep *T* well below T_c and to select a material with a high $\mathcal{H}_c(0)$.

Brief illustration 15F.2

Lead has T_c = 7.19 K and $\mathcal{H}_c(0)$ = 63.9 kA m^{-1}. At *T* = 6.0 K the magnetic field that would quench its superconductivity would be

$$\mathcal{H}_c(6.0\,\text{K})=(63.9\,\text{kA}\,\text{m}^{-1})\overbrace{\left(1-\frac{(6.0\,\text{K})^2}{(7.19\,\text{K})^2}\right)}^{0.304}=19\,\text{kA}\,\text{m}^{-1}$$

The lead remains superconducting at 6.0 K for this and weaker applied field strengths. If the temperature is lowered to 5.0 K the corresponding calculation gives $\mathcal{H}_c(5.0\,\text{K})=33\,\text{kA}\,\text{m}^{-1}$, and superconductivity survives at higher field strengths. In each case, superconductivity would survive to higher field strengths for a material with a higher $\mathcal{H}_c(0)$.

Type I superconductors are also completely diamagnetic below $\mathcal{H}_c$, meaning that the magnetic field does not penetrate into the material. This complete exclusion of a magnetic field from a material is known as the **Meissner effect**, which can be demonstrated by the levitation of a superconductor above a magnet. *Type II* superconductors, which include the HTSCs, show a gradual loss of diamagnetism with increasing magnetic field.

Checklist of concepts

☐ 1. The **magnetization** of a material is the average molecular magnetic dipole moment multiplied by the number density of the molecules.

☐ 2. The **magnetic susceptibility** expresses the relation between the magnetization and the applied magnetic field strength.

☐ 3. **Diamagnetic materials** tend to move out of a magnetic field and have negative magnetic susceptibilities.

☐ 4. **Paramagnetic materials** tend to move into a magnetic field and have positive magnetic susceptibilities.

☐ 5. The **Curie law** describes the temperature dependence of the magnetic susceptibility.

☐ 6. **Ferromagnetism** is the cooperative alignment of electron spins in a material and gives rise to strong permanent magnetization.

☐ 7. **Antiferromagnetism** results from alternating spin orientations in a material and leads to weak magnetization.

☐ 8. **Temperature-independent paramagnetism** arises from induced electron currents.

☐ 9. The **Meissner effect** is the exclusion of a magnetic field from a Type I superconductor.

Checklist of equations

Property	Equation	Comment	Equation number
Magnetization	$\mathcal{M} = \chi\mathcal{H}$	Definition	15F.1
Molar magnetic susceptibility	$\chi_m = \chi V_m$	Definition	15F.2
Magnetic moment	$m = g_e\{s(s+1)\}^{1/2}\mu_B$	$\mu_B = e\hbar/2m_e$	15F.3
Curie law	$\chi_m = C/T$, $C = N_A g_e^2 \mu_0 \mu_B^2 S(S+1)/3k$	Paramagnetism	15F.4b
Dependence of $\mathcal{H}_c$ on T_c	$\mathcal{H}_c(T) = \mathcal{H}_c(0)(1 - T^2/T_c^2)$	Empirical	15F.5

TOPIC 15G The optical properties of solids

> ➤ **Why do you need to know this material?**
>
> The optical properties of solids are of ever increasing importance in modern technology, not only for the generation of light but for the propagation and manipulation of information.
>
> ➤ **What is the key idea?**
>
> The optical properties of molecules in solids differ from those of isolated molecules as a result of the interaction of their transition dipoles.
>
> ➤ **What do you need to know already?**
>
> You need to be familiar with the concept of transition dipole (Topics 8C and 11A) and of the band theory of solids (Topic 15C).

Topic 11A explains the factors that determine the energy and intensity of light absorbed by isolated atoms and molecules in the gas phase and in solution. However, significant differences arise when the molecules are neighbours in a solid.

15G.1 Excitons

Consider an electronic excitation of a molecule (or an ion) in a crystal. If the excitation corresponds to the removal of an electron from one orbital of a molecule and its elevation to an orbital of higher energy, then the excited state of the molecule can be envisaged as the coexistence of an electron and a hole. This electron–hole pair, which behaves as a particle-like **exciton**, migrates from molecule to molecule in the crystal (Fig. 15G.1). A migrating excitation of this kind is called a **Frenkel exciton**, and is commonly found in molecular solids. The electron and hole can also be on different molecules, but in each other's vicinity. A migrating excitation of this kind, which is now spread over several molecules (more usually ions), is called a **Wannier exciton**. Exciton formation causes spectral lines to shift, split, and change intensity.

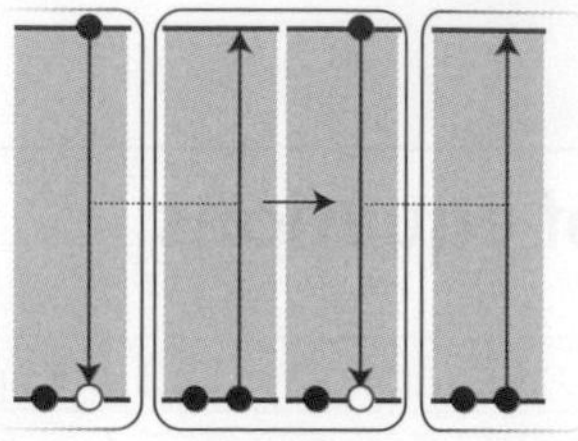

Figure 15G.1 The electron–hole pair shown on the left can migrate through a solid lattice as the excitation hops from molecule to molecule. The mobile excitation is called an exciton.

The migration of a Frenkel exciton (the only type considered here) implies that there is an interaction between the molecules that constitute the crystal: if this were not the case the excitation on one molecule could not move to another. This interaction affects the energy levels of the system. The strength of the interaction also governs the rate at which an exciton moves through the crystal: a strong interaction results in fast migration and a vanishingly small interaction leaves the exciton localized on its original molecule. The specific mechanism of interaction that leads to exciton migration is the interaction between the transition dipoles of the excitation (Topic 11A). Thus, an electric dipole transition in a molecule is accompanied by a shift of charge, and this transient dipole exerts a force on an adjacent molecule. The latter responds by shifting its charge. This process continues and the excitation migrates through the crystal.

The energy shift arising from the interaction between transition dipoles can be explained as follows. The potential energy of interaction between two parallel electric dipole moments μ_1 and μ_2 separated by a distance r is $V = \mu_1\mu_2(1 - 3\cos^2\theta)/4\pi\varepsilon_0 r^3$, where the angle θ is defined in (**1**). A head-to-tail alignment corresponds to $\theta = 0$, and a parallel alignment corresponds to $\theta = 90°$. From the expression for V it follows that $V < 0$ (a favourable, energy-lowering interaction) for $0 \le \theta < 54.7°$, $V = 0$ when $\theta = 54.7°$ (at this angle $1 - 3\cos^2\theta = 0$), and $V > 0$ (an unfavourable, energy-raising interaction) for $54.7° < \theta \le 90°$.

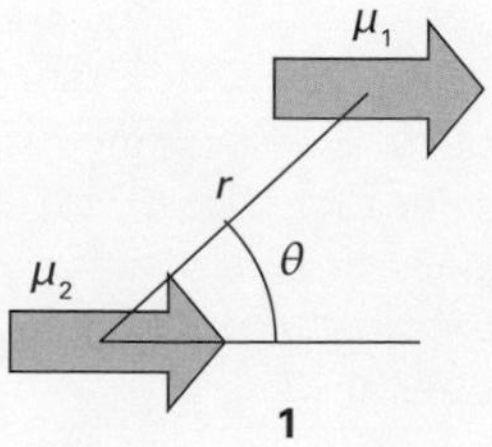

1

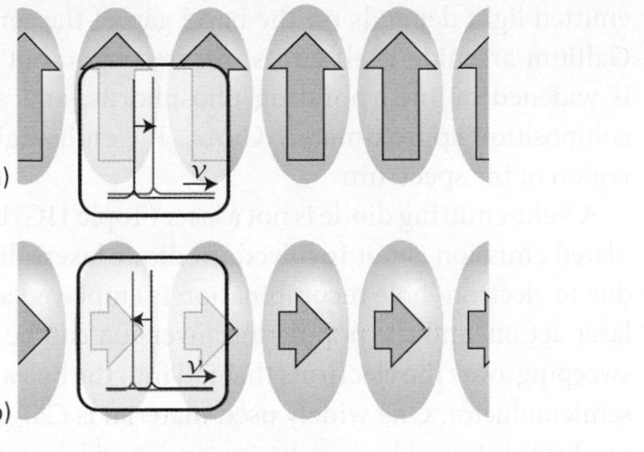

Figure 15G.2 (a) The alignment of transition dipoles (the yellow arrows) shown here is energetically unfavourable, and the exciton absorption is shifted to higher energy (higher frequency). (b) The alignment shown here is energetically favourable for a transition in this orientation, and the exciton band occurs at lower frequency than in the isolated molecules.

In a head-to-tail arrangement, there is a favourable interaction between the region of partial positive charge in one molecule and the region of partial negative charge in the other molecule. In contrast, in a parallel arrangement, the molecular interaction is unfavourable because of the close approach of regions of partial charge with the same sign. It follows that an all-parallel arrangement of the transition dipoles (Fig. 15G.2a) is energetically unfavourable, so the absorption occurs at a higher frequency than in the isolated molecule. Conversely, a head-to-tail alignment of transition dipoles (Fig. 15G.2b) is energetically favourable, and the transition occurs at a lower frequency than in the isolated molecules.

If there are N molecules per unit cell, there are N **exciton bands** in the spectrum (if all of them are allowed). The splitting between the bands is the **Davydov splitting**. To understand the origin of the splitting, consider the case $N = 2$ with the molecules arranged as in Fig. 15G.3 and suppose that the

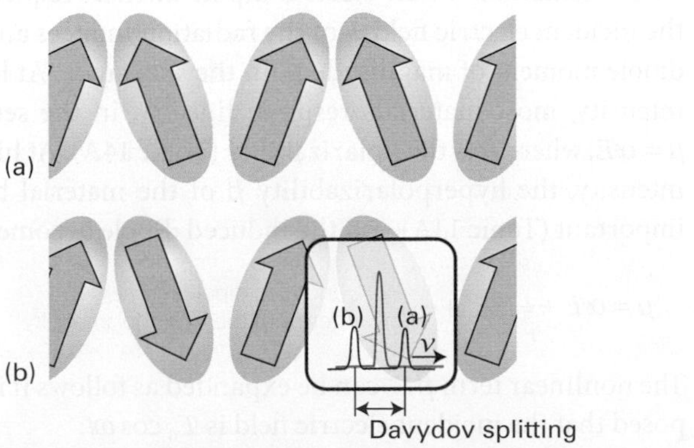

Figure 15G.3 When the transition moments within a unit cell lie in different relative directions, as depicted in (a) and (b), the energies of the transitions are shifted and give rise to the two bands labelled (a) and (b) in the spectrum. The separation of the bands is the Davydov splitting.

transition dipoles are along the length of the molecules. The radiation stimulates the collective excitation of the transition dipoles that are in-phase between neighbouring unit cells. Within each unit cell the transition dipoles may be arrayed in the two different ways shown in the illustration. The two orientations correspond to different interaction energies, with interaction being unfavourable in one and favourable in the other, so the two transitions appear in the spectrum as two bands of different frequencies. The magnitude of the Davydov splitting is determined by the energy of interaction between the transition dipoles within the unit cell.

15G.2 Metals and semiconductors

Figure 15C.8 shows the band structure in an idealized metallic conductor at $T = 0$. The absorption of a photon can excite electrons from the occupied levels to the unoccupied levels. There is a near continuum of unoccupied energy levels above the Fermi level, so absorption occurs over a wide range of frequencies. In metals, the bands are sufficiently wide that radiation is absorbed from the radiofrequency to the ultraviolet region of the electromagnetic spectrum but not to very high-frequency electromagnetic radiation, such as X-rays and γ-rays, so metals are transparent at these frequencies. Because this range of absorbed frequencies includes the entire visible spectrum, it might be expected that all metals should be black. However, metals are in fact lustrous (that is, they reflect light) and some are coloured (that is, they absorb light of certain wavelengths), so the model clearly needs some improvement.

(a) Light absorption

To explain the lustrous appearance of a smooth metal surface, it is important to realize that the absorbed energy can be re-emitted very efficiently as light, with only a small fraction of the energy being released into the bulk as heat. Because the atoms near the surface of the material absorb most of the radiation, emission also occurs primarily from the surface. In essence, if the sample is excited with visible light, then electrons near the surface are driven into oscillation at the same frequency, and visible light is emitted from the surface, so accounting for the lustre of the material.

The perceived colour of a metal depends on the frequency range of reflected light. That in turn depends on the frequency range of light that can be absorbed and, by extension, on the band structure. Silver reflects light with nearly equal efficiency across the visible spectrum because its band structure has many unoccupied energy levels that can be populated by absorption of, and depopulated by emission of, visible light. On the other hand, copper has its characteristic colour because it has relatively fewer unoccupied energy levels that can

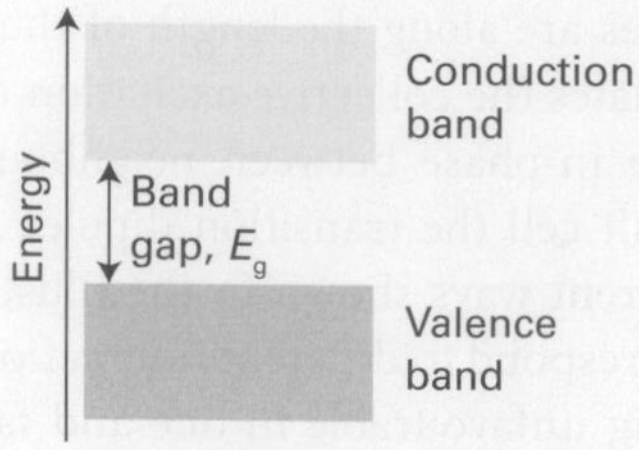

Figure 15G.4 In some materials, the band gap E_g is very large and electron promotion can occur only by excitation with electromagnetic radiation.

be excited with violet, blue, and green light. The material reflects at all wavelengths, but more light is emitted at lower frequencies (corresponding to yellow, orange, and red). Similar arguments account for the colours of other metals, such as the yellow of gold. It is interesting to note that the colour of gold can be accounted for only by including relativistic effects in the calculation of its band structure.

Now consider semiconductors. If the band gap E_g is comparable to kT the promotion of electrons from the conduction to the valence band of a semiconductor can be the result of thermal excitation. In some materials, the band gap is very large and electron promotion can occur only by excitation with electromagnetic radiation. However, as is seen from Fig. 15G.4, there is a frequency $\nu_{\text{min}} = E_g/h$ below which light absorption cannot occur. Above this frequency threshold, a wide range of frequencies can be absorbed by the material, as in a metal.

Brief illustration 15G.1

The energy of the band gap in the semiconductor cadmium sulfide (CdS) is 2.4 eV (equivalent to 3.8×10^{-19} J). It follows that the minimum electronic absorption frequency is

$$\nu_{\text{min}} = \frac{3.8 \times 10^{-19}\,\text{J}}{6.626 \times 10^{-34}\,\text{J s}} = 5.8 \times 10^{14}\,\text{s}^{-1}$$

This frequency, of 580 THz, corresponds to a wavelength of 520 nm (green light). Lower frequencies, corresponding to yellow, orange, and red, are not absorbed and consequently CdS appears yellow-orange.

(b) Light-emitting diodes and diode lasers

The unique electrical properties of p–n junctions between semiconductors (which are described in Topic 15E) can be put to good use in optical devices. In some materials, most notably gallium arsenide, GaAs, energy from electron–hole recombination is released not as heat but is carried away by photons as electrons move across the junction driven by the appropriate potential difference. Practical **light-emitting diodes** of this kind are widely used in electronic displays. The wavelength of emitted light depends on the band gap of the semiconductor. Gallium arsenide itself emits infrared light, but its band gap is widened by incorporating phosphorus, and a material of composition approximately $GaAs_{0.6}P_{0.4}$ emits light in the red region of the spectrum.

A light-emitting diode is not a laser (Topic 11G) because stimulated emission is not involved. In **diode lasers**, light emission due to electron–hole recombination is employed as the basis of laser action, and the population inversion can be sustained by sweeping away the electrons that fall into the holes of the p-type semiconductor. One widely used material is $Ga_{1-x}Al_xAs$, which produces infrared laser radiation and is widely used in CD and DVD players. High-power diode lasers are also used to pump other lasers. One example is the pumping of Nd:YAG lasers (Topic 11G) by $Ga_{0.91}Al_{0.09}As/Ga_{0.7}Al_{0.3}As$ diode lasers.

15G.3 Nonlinear optical phenomena

Nonlinear optical phenomena arise from changes in the optical properties of a material in the presence of intense electromagnetic radiation. In **frequency doubling** (or 'second harmonic generation'), an intense laser beam is converted to radiation with twice (and in general a multiple) of its initial frequency as it passes through a suitable material. It follows that frequency doubling and tripling of an Nd:YAG laser, which emits radiation at 1064 nm (Topic 11G), produce green light at 532 nm and ultraviolet radiation at 355 nm, respectively. Common materials that can be used for frequency doubling in laser systems include crystals of potassium dihydrogenphosphate (KH_2PO_4), lithium niobate ($LiNbO_3$), and β-barium borate (β-BaB_2O_4).

Frequency doubling can be explained by examining how a substance responds nonlinearly to incident radiation of frequency $\omega = 2\pi\nu$. Radiation of a particular frequency arises from oscillations of an electric dipole at that frequency and the incident electric field $\mathcal{E}$ of the radiation induces an electric dipole moment of magnitude μ, in the substance. At low light intensity, most materials respond linearly, in the sense that $\mu = \alpha\mathcal{E}$, where α is the polarizability (Topic 14A). At high light intensity, the hyperpolarizability β of the material becomes important (Topic 14A) and the induced dipole becomes

$$\mu = \alpha\mathcal{E} + \tfrac{1}{2}\beta\mathcal{E}^2 + \cdots \qquad \text{Induced dipole moment in terms of the hyperpolarizability} \qquad (15G.1)$$

The nonlinear term $\beta\mathcal{E}^2$ can be expanded as follows if it is supposed that the incident electric field is $\mathcal{E}_0 \cos \omega t$:

$$\beta\mathcal{E}^2 = \beta\mathcal{E}_0^2\cos^2\omega t = \tfrac{1}{2}\beta\mathcal{E}_0^2(1 + \cos 2\omega t) \qquad (15G.2)$$

Hence, the nonlinear term contributes an induced electric dipole that includes a component that oscillates at the frequency 2ω and that can act as a source of radiation of that frequency.

Checklist of concepts

☐ 1. An **exciton** is an electron-hole pair caused by optical excitation in a solid; **Frenkel excitons** are localized on a single molecule, whereas **Wannier excitons** are spread over several molecules.

☐ 2. If the unit cell contains N molecules, there are N **exciton bands** in the spectrum separated by the **Davydov splitting.**

☐ 3. **Nonlinear optical phenomena** arise from changes in the optical properties of a material in the presence of intense electromagnetic radiation; they can give rise to **frequency doubling.**

FOCUS 15 Solids

TOPIC 15A Crystal structure

Discussion questions

D15A.1 Describe the relationship between space lattice and unit cell.

D15A.2 Explain how planes in a lattice are labelled.

D15A.3 Draw unit cells representative of the three possible cubic lattices. State how many lattice points are in each of your cells and identify whether or not the cells you have drawn are primitive.

Exercises

E15A.1(a) The orthorhombic unit cell of $NiSO_4$ has the dimensions $a = 634$ pm, $b = 784$ pm, and $c = 516$ pm, and the mass density of the solid is estimated as 3.9 g cm^{-3}. Identify the number of formula units in a unit cell and calculate a more precise value of the mass density.

E15A.1(b) An orthorhombic unit cell of a compound of molar mass 135.01 g mol^{-1} has the dimensions $a = 589$ pm, $b = 822$ pm, and $c = 798$ pm. The mass density of the solid is estimated as 2.9 g cm^{-3}. Identify the number of formula units in a unit cell and calculate a more precise value of the mass density.

E15A.2(a) State the Miller indices of the planes that intersect the crystallographic axes at the distances ($2a$, $3b$, $2c$) and ($2a$, $2b$, ∞c).

E15A.2(b) State the Miller indices of the planes that intersect the crystallographic axes at the distances ($-a$, $2b$, $-c$) and (a, $4b$, $-4c$).

E15A.3(a) Calculate the separations of the planes {112}, {110}, and {224} in a crystal in which the cubic unit cell has side 562 pm.

E15A.3(b) Calculate the separations of the planes {123}, {222}, and {246} in a crystal in which the cubic unit cell has side 712 pm.

E15A.4(a) The unit cells of $SbCl_3$ are orthorhombic with dimensions $a = 812$ pm, $b = 947$ pm, and $c = 637$ pm. Calculate the spacing, d, of the {321} planes.

E15A.4(b) An orthorhombic unit cell has dimensions $a = 769$ pm, $b = 891$ pm, and $c = 690$ pm. Calculate the spacing, d, of the {312} planes.

Problems

P15A.1 Although the crystallization of large biological molecules may not be as readily accomplished as that of small molecules, their crystal lattices are no different. The protein tobacco seed globulin forms face-centred cubic crystals with unit cell dimension of 12.3 nm and a mass density of 1.287 g cm^{-3}. Determine its molar mass (assume there is one molecule associated with each lattice point).

P15A.2 Show that the volume of a monoclinic unit cell is $V = abc \sin \beta$.

P15A.3 Derive an expression for the volume of a hexagonal unit cell.

P15A.4 Show that the volume of a triclinic unit cell of sides a, b, and c and angles α, β, and γ is

$$V = abc(1 - \cos^2\alpha - \cos^2\beta - \cos^2\gamma + 2\cos\alpha\cos\beta\cos\gamma)^{1/2}$$

Use this expression to derive expressions for monoclinic and orthorhombic unit cells. For the derivation, it may be helpful to use the result from vector analysis that $V = \boldsymbol{a}\cdot\boldsymbol{b}\times\boldsymbol{c}$ and to calculate V^2 initially. The compound Rb_3TlF_6 has a tetragonal unit cell with dimensions $a = 651$ pm and $c = 934$ pm. Calculate the volume of the unit cell.

P15A.5 The volume of a monoclinic unit cell is $abc \sin \beta$ (see Problem P15A.2). Naphthalene has a monoclinic unit cell with two molecules in each cell and sides in the ratio 1.377:1:1.436. The angle β is 122.82° and the mass density of the solid is 1.152 g cm^{-3}. Calculate the dimensions of the cell.

P15A.6 Fully crystalline polyethene has its chains aligned in an orthorhombic unit cell of dimensions 740 pm × 493 pm × 253 pm. There are two repeating CH_2CH_2 units in each unit cell. Calculate the theoretical mass density of fully crystalline polyethene. The actual mass density ranges from 0.92 to 0.95 g cm^{-3}.

P15A.7‡ B.A. Bovenzi and G.A. Pearse, Jr. (*J. Chem. Soc. Dalton Trans.* 2793 (1997)) synthesized coordination compounds of the tridentate ligand pyridine-2,6-diamidoxime (**1**, $C_7H_9N_5O_2$). The compound they isolated from the reaction of the ligand with $CuSO_4$(aq) did not contain a $[Cu(C_7H_9N_5O_2)_2]^{2+}$ complex cation as expected. Instead, X-ray diffraction analysis revealed a linear polymer of formula $[\{Cu(C_7H_9N_5O_2)(SO_4)\}\cdot 2H_2O]_n$, which features bridging sulfate groups. The unit cell was primitive monoclinic with $a = 1.0427$ nm, $b = 0.8876$ nm, $c = 1.3777$ nm, and $\beta = 93.254°$. The mass density of the crystals is 2.024 g cm^{-3}. How many monomer units are there in the unit cell?

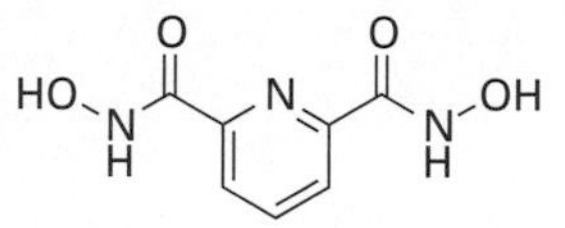

1 Pyridine-2,6-diamidoxime

P15A.8‡ D. Sellmann et al. (*Inorg. Chem.* **36**, 1397 (1997)) describe the synthesis and reactivity of the ruthenium nitrido compound $[N(C_4H_9)_4][Ru(N)(S_2C_6H_4)_2]$. The ruthenium complex anion has the two 1,2-benzenedithiolate ligands (**2**) at the base of a rectangular pyramid and the nitrido ligand at the apex. Compute the mass density of the compound given that it crystallizes with an orthorhombic unit cell with $a = 3.6881$ nm, $b = 0.9402$ nm, and $c = 1.7652$ nm and eight formula units in each cell. The replacement of the ruthenium with osmium results in a compound with the same crystal structure and a unit cell with a volume less than 1 per cent larger. Estimate the mass density of the osmium analogue.

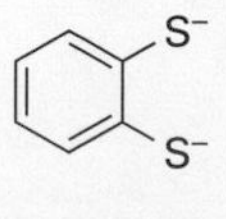

2 1,2-Benzenedithiolateion

P15A.9 Show that the separation of the {hkl} planes in an orthorhombic crystal with sides a, b, and c is given by eqn 15A.1b.

‡ These problems were supplied by Charles Trapp and Carmen Giunta.

TOPIC 15B Diffraction techniques

Discussion questions

D15B.1 What is meant by a systematic absence? How do they arise and how can they be helpful in identifying the type of unit cell?

D15B.2 Discuss what is meant by 'scattering factor'. How is it related to the number of electrons in the atoms scattering X-rays?

D15B.3 Describe the consequences of the phase problem in determining structure factors and how the problem is overcome.

Exercises

E15B.1(a) The angle of a Bragg reflection from a set of crystal planes separated by 99.3 pm is 20.85°. Calculate the wavelength of the X-rays.

E15B.1(b) The angle of a Bragg reflection from a set of crystal planes separated by 128.2 pm is 19.76°. Calculate the wavelength of the X-rays.

E15B.2(a) What are the values of the angle θ of the three diffraction lines with the smallest θ expected from a cubic I unit cell with lattice parameter 291 pm when the X-ray wavelength is 72 pm? *Hint:* Are all reflections possible for such a unit cell?

E15B.2(b) Repeat Exercise E15B.2(a) but for a cubic F unit cell with lattice parameter 407 pm and an X-ray wavelength of 129 pm.

E15B.3(a) Potassium nitrate crystals have orthorhombic unit cells of dimensions $a = 542$ pm, $b = 917$ pm, and $c = 645$ pm. Calculate the values of θ for the (100), (010), and (111) reflections using radiation of wavelength 154 pm.

E15B.3(b) Calcium carbonate crystals in the form of aragonite have orthorhombic unit cells of dimensions $a = 574.1$ pm, $b = 796.8$ pm, and $c = 495.9$ pm. Calculate the values of θ for the (100), (010), and (111) reflections using radiation of wavelength 83.42 pm.

E15B.4(a) Radiation from an X-ray source consists of two components of wavelengths 154.433 pm and 154.051 pm. Calculate the difference in glancing angles (2θ) of the diffraction lines arising from the two components in a diffraction pattern from planes of separation 77.8 pm.

E15B.4(b) Consider a source that emits X-radiation at a range of wavelengths, with two components of wavelengths 93.222 and 95.123 pm. Calculate the separation of the glancing angles (2θ) arising from the two components in a diffraction pattern from planes of separation 82.3 pm.

E15B.5(a) What is the value of the scattering factor in the forward direction for Br^-?

E15B.5(b) What is the value of the scattering factor in the forward direction for Mg^{2+}?

E15B.6(a) The coordinates, in units of a, of the atoms in a cubic P unit cell are (0,0,0), (0,1,0), (0,0,1), (0,1,1), (1,0,0), (1,1,0), (1,0,1), and (1,1,1). Calculate the structure factor F_{hkl} when all the atoms are identical. Where possible, simplify your expression by using $e^{in\pi} = (-1)^n$, as in *Example* 15B.1.

E15B.6(b) The coordinates, in units of a, of the atoms in a cubic I unit cell are (0,0,0), (0,1,0), (0,0,1), (0,1,1), (1,0,0), (1,1,0), (1,0,1), (1,1,1), and $(\frac{1}{2},\frac{1}{2},\frac{1}{2})$. Calculate the structure factor F_{hkl} when all the atoms are identical. Where possible, simplify your expression by using $e^{in\pi} = (-1)^n$, as in *Example* 15B.1.

E15B.7(a) Calculate the structure factors for an orthorhombic C structure in which the scattering factors of the two ions on the faces are twice that of the ions at the corners of the cube. Assume that $a = b = c$, that is the unit cell is a cube.

E15B.7(b) Calculate the structure factors for a body-centred cubic (cubic I) structure in which the scattering factor of the central ion is twice that of the ions at the corners of the cube.

E15B.8(a) In an X-ray investigation, the following structure factors were determined (with $F_{-h00} = F_{h00}$):

h	0	1	2	3	4	5	6	7	8	9
F_{h00}	10	−10	8	−8	6	−6	4	−4	2	−2

Construct the electron density along the corresponding direction.

E15B.8(b) In an X-ray investigation, the following structure factors were determined (with $F_{-h00} = F_{h00}$):

h	0	1	2	3	4	5	6	7	8	9
F_{h00}	10	10	4	4	6	6	8	8	10	10

Construct the electron density along the corresponding direction.

E15B.9(a) Construct the Patterson map from the information in Exercise E15B.8a.

E15B.9(b) Construct the Patterson map from the information in Exercise E15B.8b.

E15B.10(a) In a Patterson map, the spots correspond to the lengths and directions of the vectors joining the atoms in a unit cell. Sketch the pattern that would be obtained for a planar, triangular isolated BF_3 molecule.

E15B.10(b) In a Patterson map, the spots correspond to the lengths and directions of the vectors joining the atoms in a unit cell. Sketch the pattern that would be obtained from the carbon atoms in an isolated benzene molecule.

E15B.11(a) What speed should neutrons have if they are to have a wavelength of 65 pm?

E15B.11(b) What speed should electrons have if they are to have a wavelength of 105 pm?

E15B.12(a) Calculate the wavelength of neutrons that have reached thermal equilibrium by collision with a moderator at 350 K.

E15B.12(b) Calculate the wavelength of electrons that have reached thermal equilibrium by collision with a moderator at 380 K.

Problems

P15B.1 In the early days of X-ray crystallography there was an urgent need to know the wavelengths of X-rays. One technique was to measure the diffraction angle from a mechanically ruled grating. Another method was to estimate the separation of lattice planes from the measured density of a crystal. The mass density of NaCl is 2.17 g cm^{-3} and the (100) reflection using radiation of a certain wavelength occurred at 6.0°. Calculate the wavelength of the X-rays.

P15B.2 The element polonium crystallizes in a cubic system. Bragg reflections, with X-rays of wavelength 154 pm, occur at $\sin\theta = 0.225$, 0.316, and 0.388 from the {100}, {110}, and {111} sets of planes. The separation between the sixth and seventh lines observed in the diffraction pattern is larger than between the fifth and sixth lines. Is the unit cell cubic P, I, or F? Calculate the unit cell dimension.

P15B.3 Elemental silver reflects X-rays of wavelength 154.18 pm at angles of 19.076°, 22.171°, and 32.256°. However, there are no other reflections at angles of less than 33°. Assuming a cubic unit cell, determine its type and dimension. Calculate the mass density of silver. *Hint:* Calculate the expected reflections from different types of cubic unit cell and compare those with the data given.

P15B.4 In their book *X-rays and crystal structures* (which begins 'It is now two years since Dr. Laue conceived the idea …') the Braggs give a number of simple examples of X-ray analysis. For instance, they report that the reflection from {100} planes in KCl occurs at 5.38°, but for NaCl it occurs at 6.00° for X-rays of the same wavelength. If the side of the NaCl unit cell is 564 pm, what is the side of the KCl unit cell? The mass densities of KCl and NaCl are 1.99 g cm^{-3} and 2.17 g cm^{-3} respectively. Do these values support the X-ray analysis?

P15B.5 Use mathematical software to draw a graph of the scattering factor f against $(\sin\theta)/\lambda$ for an atom of atomic number Z for which $\rho(r) = 3Z/4\pi R^3$ for $0 \le r \le R$ and $\rho(r) = 0$ for $r > R$, with R a parameter that represents the radius of the atom. Explore how f varies with Z and R.

P15B.6 The coordinates of the four I atoms in the unit cell of KIO_4 are (0,0,0), $(0,\frac{1}{2},\frac{1}{2})$, $(\frac{1}{2},\frac{1}{2},\frac{1}{2})$, $(\frac{1}{2},0,\frac{3}{4})$. By calculating the phase of the I reflection in the structure factor, show that the I atoms contribute no net intensity to the (114) reflection.

P15B.7 The coordinates, as multiples of a, of the A atoms, with scattering factor f_A, in a cubic lattice are (0,0,0), (0,1,0), (0,0,1), (0,1,1), (1,0,0), (1,1,0), (1,0,1), and (1,1,1). There is also a B atom, with scattering factor f_B, at $(\frac{1}{2},\frac{1}{2},\frac{1}{2})$. Calculate the structure factors F_{hkl} and predict the form of the diffraction pattern when (a) $f_A = f$, $f_B = 0$, (b) $f_B = \frac{1}{2}f_A$, and (c) $f_A = f_B = f$.

P15B.8 Here we explore electron diffraction patterns. (a) Predict from the Wierl equation, eqn 15B.8, the positions of the first maximum and first minimum in the neutron and electron diffraction patterns of a Br_2 molecule obtained with neutrons of wavelength 78 pm wavelength and electrons of wavelength 4.0 pm. (b) Use the Wierl equation to predict the appearance of the electron diffraction pattern of CCl_4 with an (as yet) undetermined C–Cl bond length but of known tetrahedral symmetry; assume the electron energy to be 10 keV. Take $f_{Cl} = 17f$ and $f_C = 6f$ and note that $R(\text{Cl,Cl}) = (8/3)^{1/2}R(\text{C,Cl})$. Plot I/f^2 against positions of the maxima occurred at 3.17°, 5.37°, and 7.90° and minima occurred at 1.77°, 4.10°, 6.67°, and 9.17°. What is the C–Cl bond length in CCl_4?

TOPIC 15C Bonding in solids

Discussion questions

D15C.1 In what respects is the hard-sphere model of metallic solids deficient?

D15C.2 Describe the caesium-chloride and rock-salt structures in terms of the occupation of holes in expanded close-packed lattices.

Exercises

E15C.1(a) Calculate the packing fraction for close-packed cylinders; you need only consider one layer. *Hint:* Start by identifying a suitable unit cell. (For a generalization of this Exercise, see Problem P15C.2.)

E15C.1(b) Calculate the packing fraction for equilateral triangular rods stacked as shown in (3).

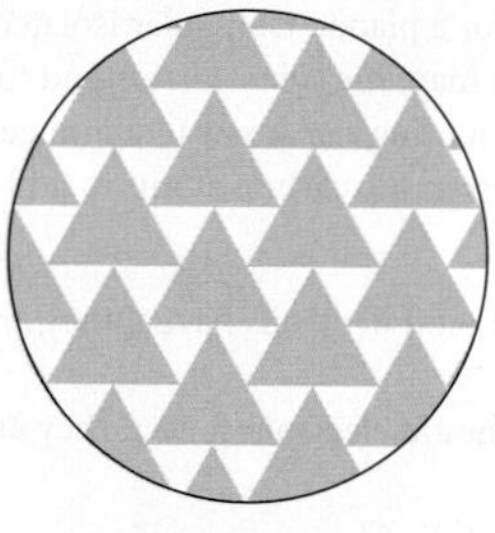

3

E15C.2(a) Calculate the packing fractions of (i) a primitive cubic unit cell, (ii) a bcc unit cell, (iii) an fcc unit cell, where each cell is composed of identical hard spheres. *Hint:* Start by identifying the unit cell and working out which atoms are in contact.

E15C.2(b) Calculate the packing fraction for an orthorhombic C cell in which all three sides are the same (assume that the spheres touch along one of the face diagonals which includes an atom on the face).

E15C.3(a) Determine the radius of the smallest cation that can have (i) sixfold and (ii) eightfold coordination with the Cl^- ion (radius 181 pm).

E15C.3(b) Determine the radius of the smallest anion that can have (i) sixfold and (ii) eightfold coordination with the Rb^+ ion (radius 149 pm).

E15C.4(a) Does titanium expand or contract as it transforms from hcp to bcc? The atomic radius of titanium is 145.8 pm in hcp but 142.5 pm in bcc. *Hint:* Consider the change in packing fraction.

E15C.4(b) Does iron expand or contract as it transforms from hcp to bcc? The atomic radius of iron is 126 pm in hcp but 122 pm in bcc.

E15C.5(a) Calculate the lattice enthalpy of CaO from the following data:

	$\Delta H/(\text{kJ mol}^{-1})$
Sublimation of Ca(s)	+178
Ionization of Ca(g) to Ca^{2+}(g)	+1735
Dissociation of O_2(g)	+249
Electron attachment to O(g)	−141
Electron attachment to O^-(g)	+844
Formation of CaO(s) from Ca(s) and $\frac{1}{2}O_2$(g) in their reference states.	−635

E15C.5(b) Calculate the lattice enthalpy of $MgBr_2$ from the following data:

	$\Delta H/(\text{kJ mol}^{-1})$
Sublimation of Mg(s)	+148
Ionization of Mg(g) to Mg^{2+}(g)	+2187
Vaporization of Br_2(l)	+31
Dissociation of Br_2(g)	+193
Electron attachment to Br(g)	−331
Formation of $MgBr_2$(s) from Mg(s) and Br_2(l) in their reference states.	−524

Problems

P15C.1 Calculate the atomic packing factor for diamond (refer to Fig. 15C.15); assume that the atoms touch along the body diagonal.

P15C.2 Rods of elliptical cross-section with semi-minor and -major axes a and b are close-packed as shown in (4). What is the packing fraction? Draw a graph of the packing fraction against the eccentricity ε of the ellipse. For an ellipse with semi-major axis a and semi-minor axis b, $\varepsilon = (1 - b^2/a^2)^{1/2}$.

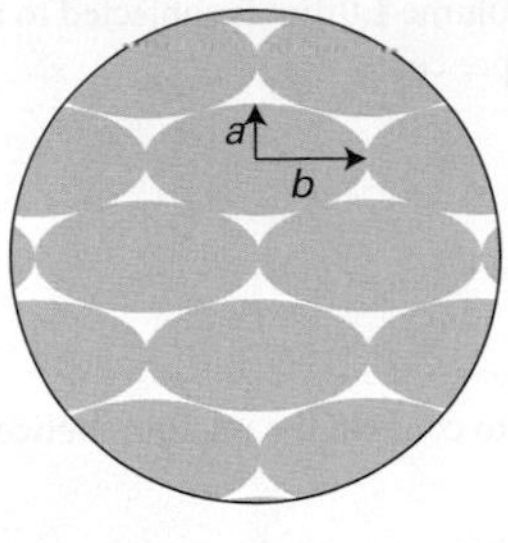

4

P15C.3 (a) Calculate the mass density of diamond assuming that it is a close-packed structure of hard spheres with radii equal to half the carbon–carbon bond length of 154.45 pm. (b) The diamond lattice is in fact based on a face-centred cubic lattice but with two atoms per lattice point, such that the structure consists of two interpenetrating fcc lattices, one with its origin at (0,0,0) and the other with its origin at (1/4,1/4,1/4). The experimentally determined mass density is 3.516 g cm^{-3}: can you explain the difference between this value and that in (a)?

P15C.4 When energy levels in a band form a continuum, the density of states $\rho(E)$, the number of levels in an energy range divided by the width of the range, may be written as $\rho(E) = dk/dE$, where dk is the change in the quantum number k and dE is the energy change. (a) Use eqn 15C.1 to show that

$$\rho(E) = -\frac{(N+1)/2\pi\beta}{\left\{1-\left(\dfrac{E-\alpha}{2\beta}\right)^2\right\}^{1/2}}$$

where k, N, α, and β have the meanings described in Topic 15C. (b) Use this expression to show that $\rho(E)$ becomes infinite as E approaches $\alpha \pm 2\beta$. That is, show that the density of states increases towards the edges of the bands in a one-dimensional metallic conductor.

P15C.5 The treatment in Problem P15C.4 applies only to one-dimensional solids. In three dimensions, the variation of density of states is more like that shown in Fig. 15.1. Account for the fact that in a three-dimensional solid the greatest density of states is near the centre of the band and the lowest density is at the edges.

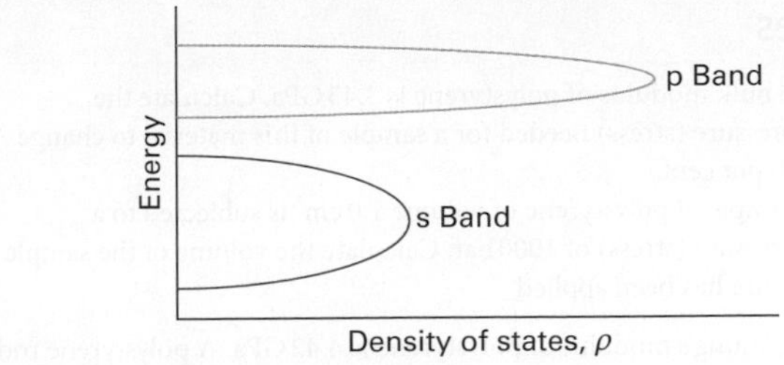

Figure 15.1 The variation of density of states in a three dimensional solid.

P15C.6 The energy levels of N atoms in the tight-binding Hückel approximation are the roots of a tridiagonal determinant (eqn 15C.1):

$$E_k = \alpha + 2\beta\cos\frac{k\pi}{N+1} \quad k = 1,2,\ldots,N$$

If the atoms are arranged in a ring, the solutions are the roots of a 'cyclic' determinant:

$$E_k = \alpha + 2\beta\cos\frac{2k\pi}{N} \quad k = 0,\pm1,\pm2,\ldots,\pm\tfrac{1}{2}N$$

(for N even). Discuss the consequences, if any, of joining the ends of an initially straight length of material.

P15C.7 Verify that the lowest value of the radius ratio for (a) sixfold coordination is 0.414, and (b) for eightfold coordination is 0.732.

P15C.8 (a) Use the Born–Mayer equation for the lattice enthalpy and a Born–Haber cycle to estimate the enthalpy of formation of CaCl(s). The sublimation enthalpy of Ca(s) is 176 kJ mol^{-1} and it can be assumed that the ionic radius of Ca^+ is close to that of K^+; other necessary data are to be found in *Example* 15C.2 or in the tables in the *Resource section*. (b) Show that an explanation for the nonexistence of CaCl(s) can be found in the reaction enthalpy for the disproportionation reaction $2\,CaCl(s) \rightarrow Ca(s) + CaCl_2(s)$.

P15C.9 Derive the Born–Mayer equation (eqn 15C.5) by calculating the energy at which $d(E_p + E_p^*)/dd = 0$, with E_p and E_p^* given by eqns 15C.3 and 15C.4, respectively.

P15C.10 Suppose that ions are arranged in a (somewhat artificial) two-dimensional lattice like the fragment shown in Fig. 15.2. Calculate the Madelung constant for the central ion in this array.

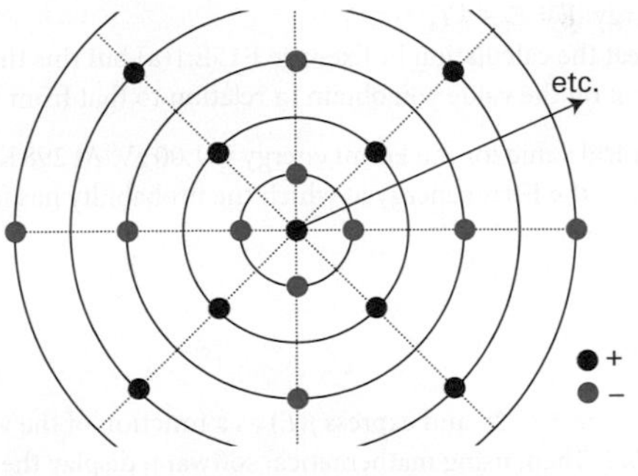

Figure 15.2 The two-dimensional lattice discussed in Problem P15C.10.

TOPIC 15D The mechanical properties of solids

Discussion question

D15D.1 Distinguish between the behaviour of a solid which undergoes elastic deformation when a stress is applied to one which undergoes plastic deformation.

Exercises

E15D.1(a) The bulk modulus of polystyrene is 3.43 GPa. Calculate the hydrostatic pressure (stress) needed for a sample of this material to change volume by 1.0 per cent.

E15D.1(b) A sample of polystyrene of volume 1.0 cm^3 is subjected to a hydrostatic pressure (stress) of 1000 bar. Calculate the volume of the sample after the pressure has been applied.

E15D.2(a) The Young's modulus of polystyrene is 4.42 GPa. A polystyrene rod of diameter 2.0 mm is subject to a force of 500 N along its length. Calculate the stress and hence the percentage increase in the length of the rod when the stress is applied.

E15D.2(b) Calculate the force which needs to be applied to a polystyrene rod of diameter 1.0 mm to increase its length from 10.00 cm to 10.05 cm.

E15D.3(a) Poisson's ratio for polyethene is 0.45. What change in volume takes place when a cube of polyethene of volume 1.0 cm^3 is subjected to a uniaxial stress that produces a strain of 1.0 per cent?

E15D.3(b) Poisson's ratio for lead is 0.41. What change in volume takes place when a cube of lead of volume 1.0 dm^3 is subjected to a uniaxial stress that produces a strain of 2.0 per cent?

Problems

P15D.1 For an isotropic substance, the moduli and Poisson's ratio may be expressed in terms of two parameters λ and μ called the *Lamé constants*:

$$E=\frac{\mu(3\lambda+2\mu)}{\lambda+\mu}\qquad K=\frac{3\lambda+2\mu}{3}\qquad G=\mu\qquad \nu_{\rm P}=\frac{\lambda}{2(\lambda+\mu)}$$

Use the Lamé constants to confirm the relations between G, K, and E given in eqn 15D.3.

P15D.2 The bulk modulus for liquid water at 298 K is 3.4 GPa. If it is assumed that the molecules are interacting with a Lennard-Jones potential energy, estimate the well depth ε (give your answer in kJ mol^{-1}).

TOPIC 15E The electrical properties of solids

Discussion question

D15E.1 Describe the characteristics of the Fermi–Dirac distribution; contrast it with the Boltzmann distribution.

Exercises

E15E.1(a) Calculate $f(E)$, the probability predicted by the Fermi–Dirac distribution, for the case where the energy E is the thermal energy, kT, above the Fermi energy: $E=E_{\rm F}+kT$.

E15E.1(b) Repeat the calculation in Exercise E15E.1(a) but this time for $E=E_{\rm F}-kT$. Comment on the value you obtain in relation to that from (a).

E15E.2(a) A typical value for the Fermi energy is 1.00 eV. At 298 K, calculate the energy above the Fermi energy at which the probability has fallen to 0.25; express your answer in eV.

E15E.2(b) Repeat the calculation in Exercise E15E.2(a) but this time for probability of 0.10 and Fermi energy of 2.00 eV.

E15E.3(a) Is arsenic-doped germanium a p-type or n-type semiconductor?

E15E.3(b) Is gallium-doped germanium a p-type or n-type semiconductor?

Problems

P15E.1 Refer to eqn 15E.2b and express $f(E)$ as a function of the variables $(E-E_{\rm F})/E_{\rm F}$ and $E_{\rm F}/kT$. Then, using mathematical software, display the set of curves shown in Fig. 15E.5 as a single surface.

P15E.2 In this and the following problem you are invited to explore further some of the properties of the Fermi–Dirac distribution, $f(E)$, eqn 15E.2a. (a) Show that at $T=0$, $f(E)=1$ for $E<\mu$, and $f(E)=0$ for $E>\mu$. (b) For a three-dimensional solid of volume V, it turns out that in eqn 15E.1 $\rho(E)=CE^{1/2}$, with $C=4\pi V(2m_{\rm e}/h^2)^{3/2}$. If the number of electrons in the solid of volume V is N, then this number must be equal to the result of integrating eqn 15E.1 over the full range of energy: $N=\int_0^\infty {\rm d}N(E)=\int_0^\infty \rho(E)f(E){\rm d}E$. Evaluate the integral at $T=0$. (*Hint:* It can be split into two integrals, one between $E=0$ and μ, and one between $E=\mu$ and ∞.) (c) Equate the expression obtained by evaluating the integral with N, and hence show that $\mu=(3\mathcal{N}/8\pi)^{2/3}(h^2/2m_{\rm e})$, where $\mathcal{N}=N/V$, the number density of electrons in the solid. (d) Evaluate μ for sodium, which has mass density 0.97 g cm^{-3}; assume that each atom contributes one electron.

P15E.3 By inspection of eqn 15E.2a and the expression for ${\rm d}N(E)$ in eqn 15E.1 (and without attempting to evaluate integrals explicitly), show that in order for N to remain constant as the temperature is raised, the chemical potential must decrease from its value at $T=0$.

P15E.4 In an intrinsic semiconductor, the band gap is so small that the Fermi–Dirac distribution results in some electrons populating the conduction band. It follows from the exponential form of the Fermi–Dirac distribution that the conductance G, the inverse of the resistance (with units of siemens, $1\,{\rm S}=1\,\Omega^{-1}$), of an intrinsic semiconductor should have an Arrhenius-like temperature dependence, shown in practice to have the form $G=G_0{\rm e}^{-E_{\rm g}/2kT}$, where $E_{\rm g}$ is the band gap. The conductance of a sample of germanium varied with temperature as indicated below. Estimate the value of $E_{\rm g}$.

T/K	312	354	420
G/S	0.0847	0.429	2.86

P15E.5 A sample of n-type semiconductor is found to be an insulator at very low temperatures. As the temperature is raised, there comes a point at which the conductivity increases markedly, but after this point the conductivity remains pretty much constant as the temperature is raised further. At much higher temperatures, the conductivity starts to increase steadily, with no sign of it reaching a plateau. Explain these observations.

P15E.6‡ P.G. Radaelli et al. (*Science* 265, 380 (1994)) reported the synthesis and structure of a material that becomes superconducting at temperatures below 45 K. The compound is based on a layered compound $Hg_2Ba_2YCu_2O_{8-\delta}$, which has a tetragonal unit cell with $a = 0.38606$ nm and $c = 2.8915$ nm; each unit cell contains two formula units. The compound is made superconducting by partially replacing Y by Ca, accompanied by a change in unit cell volume by less than 1 per cent. Estimate the Ca content x in superconducting $Hg_2Ba_2Y_{1-x}Ca_xCu_2O_{7.55}$ given that the mass density of the compound is 7.651 g cm^{-3}.

TOPIC 15F The magnetic properties of solids

Discussion question

D15F.1 Compare and contrast the polarization (Topic 14A) with the magnetization.

Exercises

E15F.1(a) The magnitude of the magnetic moment of $CrCl_3$ is $3.81\mu_B$. How many unpaired electrons does the Cr possess?
E15F.1(b) The magnitude of the magnetic moment of Mn^{2+} in its complexes is typically $5.3\mu_B$. How many unpaired electrons does the ion possess?

E15F.2(a) Calculate the molar susceptibility of benzene given that its volume susceptibility is -7.2×10^{-7} and its mass density is 0.879 g cm^{-3} at 25 °C.
E15F.2(b) Calculate the molar susceptibility of cyclohexane given that its volume susceptibility is -7.9×10^{-7} and its mass density is 811 kg m^{-3} at 25 °C.

E15F.3(a) Data on a single crystal of MnF_2 give $\chi_m = 0.1463$ cm^3 mol^{-1} at 294.53 K. Identify the effective number of unpaired electrons in this compound and compare your result with the theoretical value.
E15F.3(b) Data on a single crystal of $NiSO_4{\cdot}7H_2O$ give $\chi_m = 6.00 \times 10^{-8}$ m^3 mol^{-1} at 298 K. Identify the effective number of unpaired electrons in this compound and compare your result with the theoretical value.

E15F.4(a) Estimate the spin-only molar susceptibility of $CuSO_4{\cdot}5H_2O$ at 25 °C.
E15F.4(b) Estimate the spin-only molar susceptibility of $MnSO_4{\cdot}4H_2O$ at 298 K.

E15F.5(a) Nb has $T_c = 9.5$ K and $\mathcal{H}_c(0) = 158$ kA m^{-1}. Calculate the highest magnetic field at which superconductivity can be maintained at 6 K.
E15F.5(b) To what temperature must Nb be cooled for it to remain superconducting in a magnetic field of 150 kA m^{-1}? The necessary data are given in (a).

Problems

P15F.1‡ J.J. Dannenberg et al. (*J. Phys. Chem.* **100**, 9631 (1996)) carried out theoretical studies of organic molecules consisting of chains of unsaturated four-membered rings. The calculations suggest that such compounds have large numbers of unpaired spins, and that they should therefore have unusual magnetic properties. For example, the lowest-energy state of the compound shown as (5) is computed to have $S = 3$, but the energies of $S = 2$ and $S = 4$ structures are each predicted to be 50 kJ mol^{-1} higher in energy. Compute the molar magnetic susceptibility of these three low-lying levels at 298 K. Estimate the molar susceptibility at 298 K if each level is present in proportion to its Boltzmann factor (effectively assuming that the degeneracy is the same for all three of these levels).

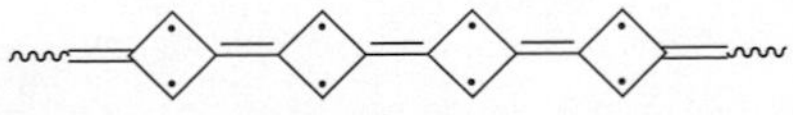

5

P15F.2 An NO molecule has thermally accessible electronically excited states. It also has an unpaired electron, and so may be expected to be paramagnetic. However, its ground state is not paramagnetic because the magnetic moment of the orbital motion of the unpaired electron almost exactly cancels the spin magnetic moment. The first excited state (at 121 cm^{-1}) is paramagnetic because the orbital magnetic moment adds to, rather than cancels, the spin magnetic moment. The upper state has a magnetic moment of magnitude $2\mu_B$. Because the upper state is thermally accessible, the paramagnetic susceptibility of NO shows a pronounced temperature dependence even near room temperature. Calculate the molar paramagnetic susceptibility of NO and plot it as a function of temperature.

TOPIC 15G The optical properties of solids

Discussion questions

D15G.1 Explain the origin of Davydov splitting in the exciton bands of a crystal.

D15G.2 Explain how the nonlinear response of a material to an electric field many give rise to frequency doubling. Why is frequency doubling typically seen only when using an intense beam from a laser as the light source?

Exercises

E15G.1(a) The promotion of an electron from the valence band into the conduction band in pure TiO_2 by light absorption requires a wavelength of less than 350 nm. Calculate the energy gap in electronvolts between the valence and conduction bands.

E15G.1(b) The band gap in silicon is 1.12 eV. Calculate the maximum wavelength of electromagnetic radiation that results in promotion of electrons from the valence to the conduction band.

Problems

P15G.1 This and the following problem explore quantitatively the spectra of molecular solids. First consider a dimer formed from two identical monomers. For the first monomer, the normalized ground state wavefunction is $\psi_a(1)$ and the normalized excited state wavefunction is $\psi_b(1)$; for the second monomer the wavefunctions are $\psi_a(2)$ and $\psi_b(2)$—the label in parenthesis identifies to which monomer the wavefunction refers, but otherwise ψ_a and ψ_b are the same for each monomer. In each monomer there is a transition between ψ_a and ψ_b with transition dipole moment μ_{mon} and wavenumber $\tilde{\nu}_{\text{mon}}$. For convenience the energy of the ground state is taken as zero, so the energy of the excited state, expressed as a wavenumber, is $\tilde{\nu}_{\text{mon}}$. It is assumed that dimerization does not affect the ground state wavefunctions, but the excited state wavefunctions become mixed so the excited state of the dimer has wavefunctions $\Psi_\pm = c_{1,\pm}\psi_b(1)+c_{2,\pm}\psi_b(2)$; the mixing of the two monomer wavefunctions gives two dimer wavefunctions, denoted $\Psi_\pm$, the coefficients $c_{i,\pm}$ are to be determined. In the basis $\psi_b(1),\psi_b(2)$ the hamiltonian matrix has the form

$$\hat{H}=\begin{pmatrix}\tilde{\nu}_{\text{mon}} & \tilde{\beta}\\ \tilde{\beta} & \tilde{\nu}_{\text{mon}}\end{pmatrix}$$

The diagonal elements are the energies (as a wavenumber) of the excited state of the monomer. The off-diagonal elements correspond to the energy of interaction between the transition dipoles. Using the arrangement illustrated in (1) of Topic 15G, this interaction energy (as a wavenumber) is:

$$\tilde{\beta}=\frac{\mu_{\text{mon}}^2}{4\pi\varepsilon_0 hcr^3}(1-3\cos^2\theta)$$

The eigenvectors of the hamiltonian matrix are the wavefunctions for the excited state of the dimer, and these can be written $\binom{c_{1,\pm}}{c_{2,\pm}}$. The eigenvalues are the energies corresponding to these wavefunctions, and because it has been assumed that the ground states of the dimer are the same as for the monomer, these energies will correspond to the transitions in the dimer.

(a) Show that $\binom{1}{1}$ and $\binom{1}{-1}$ are eigenvectors of the hamiltonian matrix and that the corresponding eigenvalues are $\tilde{\nu}_\pm=\tilde{\nu}_{\text{mon}}\pm\tilde{\beta}$.
(b) The first eigenvector corresponds to writing the wavefunction as $\Psi_+=c_{1,+}\psi_b(1)+c_{2,+}\psi_b(2)=\psi_b(1)+\psi_b(2)$. Normalize the wavefunction, assuming that $S=\int\psi_b^*(1)\psi_b(2)\mathrm{d}\tau$; do the same for the second eigenvector, which corresponds to $\Psi_-=\psi_b(1)-\psi_b(2)$. (c) The monomer transition dipole moment is $\mu_{\text{mon}}=\int\psi_b^*(1)\hat{\mu}\psi_a(1)\mathrm{d}\tau=\int\psi_b^*(2)\hat{\mu}\psi_a(2)\mathrm{d}\tau$. For the dimer the transition moment is $\mu_{\text{dim}}=\int\Psi_\pm^*\hat{\mu}\Psi_0\mathrm{d}\tau$, where Ψ_0 is the wavefunction of the dimer ground state. Because it is assumed that there is no interaction between the ground-state wavefunctions of the dimer, Ψ_0 can be written as $(1/2^{1/2})(\psi_a(1)+\psi_a(2))$. Find expressions for μ_{dim} for the two excited state wavefunctions, $\Psi_\pm$. In solving this problem it is helpful to realize that it is closely analogous to the overlap of two atomic orbitals to give molecular orbitals (Topic 9E).

P15G.2 Continues from the previous problem. (a) Consider a dimer formed of monomers which have $\mu_{\text{mon}}=4.00$ D, $\tilde{\nu}_{\text{mon}}=25\,000\ \text{cm}^{-1}$, and $r=0.50$ nm. Plot a graph to show how the energies (expressed as a wavenumber) of the excited states, $\tilde{\nu}_\pm$ vary with the angle θ. (b) Now expand the treatment given above to a chain of N monomers with $\mu_{\text{mon}}=4.00$ D, $\tilde{\nu}_{\text{mon}}=25\,000\ \text{cm}^{-1}$, and $r=0.50$ nm. For simplicity, assume that $\theta=0$ and that only nearest neighbours interact with interaction energy $\tilde{V}$ (expressed here as a wavenumber). For example the hamiltonian matrix for the case $N=4$ is

$$\hat{H}=\begin{pmatrix}\tilde{\nu}_{\text{mon}} & \tilde{V} & 0 & 0\\ \tilde{V} & \tilde{\nu}_{\text{mon}} & \tilde{V} & 0\\ 0 & \tilde{V} & \tilde{\nu}_{\text{mon}} & \tilde{V}\\ 0 & 0 & \tilde{V} & \tilde{\nu}_{\text{mon}}\end{pmatrix}$$

This matrix is analogous to the one that characterizes a band in a solid (Section 15C.1b) and so the eigenvalues (which in this case are the wavenumber of the transitions) can be written down by analogy with eqn 15C.1. Calculate the wavenumber of the lowest energy transition for $N=$ 5, 10, and 15, and then generalize your result for large N. (c) How does the transition dipole moment of the lowest energy transition vary with the size of the chain?

P15G.3 Show that if a substance responds nonlinearly to two sources of radiation, one of frequency ω_1 and the other of frequency ω_2, then it may give rise to radiation of the sum and difference of the two frequencies. This nonlinear optical phenomenon is known as *frequency mixing* and is used to expand the wavelength range of lasers in laboratory applications, such as spectroscopy and photochemistry.

FOCUS 15 Solids

Integrated activities

I15.1 Calculate the thermal expansion coefficient, $\alpha=(\partial V/\partial T)_p/V$, of diamond given that the (111) reflection shifts from 22.0403° to 21.9664° on heating a crystal from 100 K to 300 K and 154.0562 pm X-rays are used.

I15.2 Calculate the scattering factor for a hydrogenic atom of atomic number Z in which the single electron occupies (a) the 1s orbital, (b) the 2s orbital. Plot f as a function of $(\sin\theta)/\lambda$. *Hint:* Interpret $4\pi\rho(r)r^2$ as the radial distribution function $P(r)$; use mathematical software to evaluate the necessary integrals.

I15.3 Explore how the scattering factor of Integrated activity I15.2 changes when the actual 1s wavefunction of a hydrogenic atom is replaced by a Gaussian function. Use mathematical software to evaluate the necessary integrals.

FOCUS 16

Molecules in motion

This Focus is concerned with understanding how matter and other physical properties (such as energy and momentum) are transported from one place to another in both gases and liquids.

16A Transport properties of a perfect gas

The transport of matter and physical properties can be described by a set of closely related empirical equations. For a gas, it is possible to understand the form of these equations by building a model based on the kinetic theory of gases discussed in Topic 1B. With this approach, the rate of diffusion, the rate of thermal conduction, viscosity, and effusion can all be related to quantities arising from the kinetic theory.

16B Motion in liquids

Molecular motion in liquids is different from that in gases on account of the presence of significant intermolecular interactions and the much higher density typical of a liquid. One way to monitor motion in such systems is to explore the electrical resistance of electrolyte solutions and to analyse it in terms of the response of the ions to an applied electric field.

16C Diffusion

The diffusion of solutes and various physical properties is an important process in liquids. It can be discussed by introducing the concept of a general 'thermodynamic force' that can be regarded as being responsible for the motion of molecules. This apparent force can be used to construct the important 'diffusion equation', which describes how solutes spread out in space with increasing time. An alternative model of diffusion as a random walk gives further insight.

Web resource What is an application of this material?

A great deal of chemistry, chemical engineering, and biology depends on the ability of molecules and ions to migrate through media of various kinds. *Impact* 25 explains how conductivity measurements are used to analyse the motion of ions through biological membranes.

TOPIC 16A Transport properties of a perfect gas

> ➤ **Why do you need to know this material?**
>
> Many physical processes take place by the transfer of a property from one region to another, and gas-phase chemical reactions depend on the rate at which molecules collide. The material presented here also includes general aspects of transport in fluid systems of any kind, and which are applicable to reactions taking place in solution.
>
> ➤ **What is the key idea?**
>
> A molecule carries properties through space in steps of about the distance of its mean free path.
>
> ➤ **What do you need to know already?**
>
> This Topic builds on and extends the kinetic theory of gases (Topic 1B). You need to be familiar with the concepts of the mean speed of molecules and the mean free path and their dependence on pressure and temperature.

A **transport property** is a process by which matter or an attribute of matter, such as momentum, is carried through a medium from one location to another. The rate of transport is commonly expressed in terms of an equation that is an empirical summary of experimental observations. These equations apply to all kinds of properties and media, and can be adapted to the discussion of transport properties of gases. In such cases, the kinetic theory of gases provides simple expressions that show how the rates of transport of these properties depend on the pressure and the temperature. The most important concept from the kinetic theory developed in Topic 1B, and used throughout this Topic, is the *mean free path*, λ, the average distance a molecule travels between collisions. According to eqn 1B.14, at a temperature T and a pressure p

$$\lambda = \frac{kT}{\sigma p} \qquad \text{Mean free path [kinetic theory]} \qquad (16\text{A}.1\text{a})$$

The parameter σ is the collision cross-section of a molecule, a measure of the target area it presents in a collision. For the derivation and physical interpretation of this expression, see Topic 1B. Another important result from kinetic theory is the mean speed of molecules of molar mass M at a temperature T, which is given by eqn 1B.9:

$$v_{\text{mean}} = \left(\frac{8RT}{\pi M}\right)^{1/2} \qquad \text{Mean speed [kinetic theory]} \qquad (16\text{A}.1\text{b})$$

16A.1 The phenomenological equations

A 'phenomenological equation' is an equation that summarizes empirical observations on phenomena without, initially at least, being based on an understanding of the molecular processes responsible for the property. Such equations are encountered commonly in the study of fluids.

The rate of migration of a property is measured by its **flux**, J, the quantity of that property passing through a given area in a given time interval divided by the area and the duration of the interval. If matter is flowing (as in diffusion), the **matter flux** is reported as so many molecules per square metre per second (number or amount $\text{m}^{-2}\,\text{s}^{-1}$). If the property migrating is energy (as in thermal conduction), then the **energy flux** is expressed in joules per square metre per second ($\text{J}\,\text{m}^{-2}\,\text{s}^{-1}$), and so on. The total quantity of a property transferred through a given area A in a given time interval Δt is $|J|A\Delta t$. The flux J may be positive or negative: the significance of its sign is discussed below.

Experimental observations on transport properties show that the flux of a property is usually proportional to the first derivative of a related quantity. For example, the flux of matter diffusing parallel to the z-axis of a container is found to be proportional to the gradient of the concentration along the same direction:

$$J(\text{matter}) \propto \frac{\mathrm{d}\mathcal{N}}{\mathrm{d}z} \qquad \text{Fick's first law of diffusion} \qquad (16\text{A}.2)$$

where $\mathcal{N}$ is the number density of particles, with units number per metre cubed (m^{-3}). The proportionality of the flux of matter to the concentration gradient is sometimes called **Fick's first law of diffusion**: the law implies that diffusion is faster when the concentration varies steeply with position than when the concentration is nearly uniform. There is no net flux if the concentration is uniform ($d\mathcal{N}/dz = 0$). Similarly, the rate of thermal conduction (the flux of the energy associated with thermal motion) is found to be proportional to the temperature gradient:

$$J(\text{energy of thermal motion}) \propto \frac{dT}{dz} \qquad \text{Flux of energy} \qquad (16A.3)$$

A positive value of J signifies a flux towards positive z; a negative value of J signifies a flux towards negative z. Because matter flows down a concentration gradient, from high concentration to low concentration, J is positive if $d\mathcal{N}/dz$ is negative (Fig. 16A.1). Therefore, the coefficient of proportionality in eqn 16A.2 must be negative, and it is written as $-D$:

$$J(\text{matter}) = -D\frac{d\mathcal{N}}{dz} \qquad \text{Fick's first law in terms of the diffusion coefficient} \qquad (16A.4)$$

The constant D is the called the **diffusion coefficient**; its SI units are metre squared per second ($m^2\,s^{-1}$). Energy migrates down a temperature gradient, and the same reasoning leads to

$$J(\text{energy of thermal motion}) = -\kappa\frac{dT}{dz}$$

$$\text{Flux of energy in terms of the coefficient of thermal conductivity} \qquad (16A.5)$$

where κ (kappa) is the **coefficient of thermal conductivity**. The units of κ are joules per kelvin per metre per second ($J\,K^{-1}\,m^{-1}\,s^{-1}$) or, because $1\,J\,s^{-1} = 1\,W$, watts per kelvin per metre ($W\,K^{-1}\,m^{-1}$). Some experimental values are given in Table 16A.1.

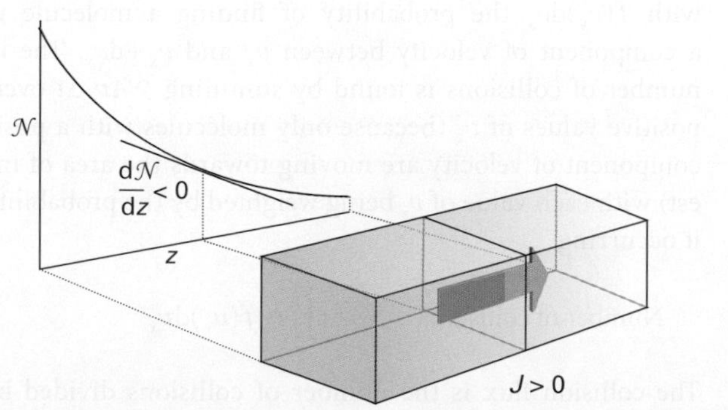

Figure 16A.1 The flux of particles down a concentration gradient. Fick's first law states that the flux of matter is proportional to the concentration gradient at that point.

Table 16A.1 Transport properties of gases at 1 atm*

	$\kappa/(mW\,K^{-1}\,m^{-1})$	$\eta/\mu P^{\ddagger}$	
	273 K	273 K	293 K
Ar	16.3	210	223
CO_2	14.5	136	147
He	144.2	187	196
N_2	24.0	166	176

* More values are given in the *Resource section*.
‡ $1\,\mu P = 10^{-7}\,kg\,m^{-1}\,s^{-1}$.

Brief illustration 16A.1

Suppose that two metal plates are placed perpendicular to the z-axis, with the first at $z = 0$ and the second at $z = +1.0$ cm, and that the temperature of the second plate is 10 K higher than that of the first plate. The temperature gradient is

$$\frac{dT}{dz} = \frac{10\,K}{1.0\times10^{-2}\,m} = +1.0\times10^{3}\,K\,m^{-1}$$

If the plates are separated by air, for which $\kappa = 24.1\,mW\,K^{-1}\,m^{-1}$, the energy flux is

$$J = -\kappa\frac{dT}{dz} = -(24.1\,mW\,K^{-1}\,m^{-1})\times(1.0\times10^{3}\,K\,m^{-1})$$
$$= -24\,W\,m^{-2}$$

The flux is negative because energy is transferred from the hotter plate at $z = +1.0$ cm to the cooler plate at $z = 0$, which is a flow in the $-z$ direction. The energy transferred through an area of $1.0\,cm^2$ between the two plates in 1 h (3600 s) is

$$\text{Transfer} = |J|A\Delta t = (24\,W\,m^{-2})\times(1.0\times10^{-4}\,m^2)\times(3600\,s) = 8.6\,J$$

Viscosity arises from the flux of linear momentum. To see the connection, consider a fluid in a state of **Newtonian flow**, in which a series of layers move past one another and, in this case, in the x-direction (Fig. 16A.2). The layer next to the wall of the vessel is stationary, and the velocity of successive layers varies linearly with distance, z, from the wall. Molecules ceaselessly move between the layers and bring with them the x-component of linear momentum they possessed in their original layer. A layer is retarded by molecules arriving from a more slowly moving layer because such molecules have a lower momentum in the x-direction. A layer is accelerated by molecules arriving from a more rapidly moving layer. The net retarding effect is interpreted as the viscosity of the fluid.

Because the retarding effect depends on the transfer of the x-component of linear momentum into the layer of interest,

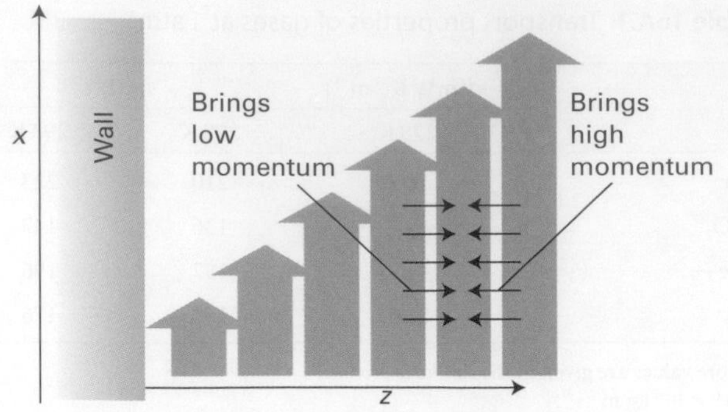

Figure 16A.2 The viscosity of a fluid arises from the transport of linear momentum. In this illustration the fluid is undergoing Newtonian (laminar) flow in the *x*-direction, and particles bring their initial momentum when they enter a new layer.

the viscosity depends on the flux of this *x*-component in the *z*-direction. The flux of the *x*-component of momentum is proportional to $\mathrm{d}v_x/\mathrm{d}z$, where v_x is the velocity in the *x*-direction; the flux can therefore be written

$$J(x\text{-component of momentum}) = -\eta \frac{\mathrm{d}v_x}{\mathrm{d}z}$$

Momentum flux in terms of the coefficient of viscosity (16A.6)

The constant of proportionality, η, is the **coefficient of viscosity** (or simply 'the viscosity'). Its units are kilograms per metre per second ($\mathrm{kg\,m^{-1}\,s^{-1}}$). Viscosities are often reported in the non-SI unit poise (P), with $1\,\mathrm{P} = 10^{-1}\,\mathrm{kg\,m^{-1}\,s^{-1}}$. Some experimental values are given in Table 16A.1.

Although it is not strictly a transport property, closely related to diffusion is **effusion**, the escape of matter through a small hole. The essential empirical observations on effusion are summarized by **Graham's law of effusion**, which states that the rate of effusion is inversely proportional to the square root of the molar mass, *M*.

16A.2 The transport parameters

The kinetic theory of gases (Topic 1B) can be used to derive expressions for the diffusion characteristics of a perfect gas. All the expressions depend on knowing the **collision flux**, Z_W, which is the rate at which molecules strike a region in the gas (this region may be an imaginary window, a part of a wall, or a hole in a wall). Specifically, the collision flux is the number of collisions divided by the area of the region and the duration of the time interval. Its dependence on pressure and temperature can be derived from the kinetic theory.

How is that done? 16A.1 Deriving an expression for the collision flux

Consider a wall of area *A* perpendicular to the *x*-axis (Fig. 16A.3). In the following calculation, note that for a perfect gas the equation of state $pV = nRT$ can be used to relate the number density, $\mathcal{N}$, to the pressure by $\mathcal{N} = N/V = nN_A/V = nN_Ap/nRT = p/kT$. In the final equality $R = N_Ak$ was used, where *k* is Boltzmann's constant.

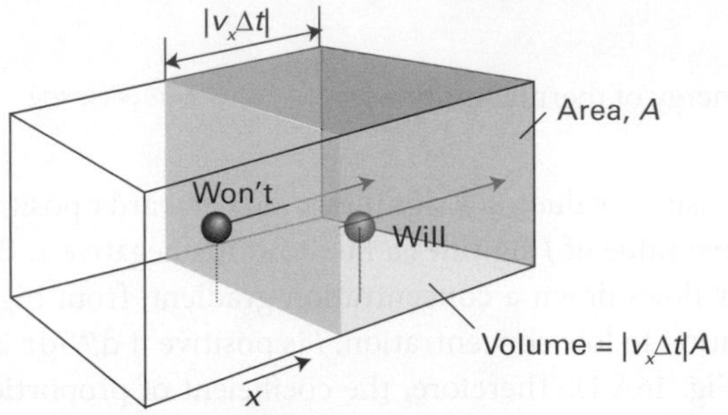

Figure 16A.3 A molecule will reach the wall on the right within an interval Δt if it is within a distance $v_x\Delta t$ of the wall and travelling to the right.

Step 1 *Identify the number of molecules that will strike an area*

If a molecule has $v_x > 0$ (that is, it is travelling in the direction of positive *x*), then it will strike the wall within an interval Δt if it lies within a distance $v_x\Delta t$ of the wall. Therefore, all molecules in the volume $Av_x\Delta t$, and with a positive *x*-component of velocity, will strike the wall in the interval Δt. The total number of collisions in this interval is therefore $\mathcal{N}Av_x\Delta t$, where $\mathcal{N}$ is the number density of molecules.

Step 2 *Take into account the range of velocities*

The velocity v_x has a range of values described by the probability distribution $f(v_x)$ given in eqn 1B.3:

$$f(v_x) = \left(\frac{m}{2\pi kT}\right)^{1/2} \mathrm{e}^{-mv_x^2/2kT}$$

with $f(v_x)\mathrm{d}v_x$ the probability of finding a molecule with a component of velocity between v_x and $v_x + \mathrm{d}v_x$. The total number of collisions is found by summing $\mathcal{N}Av_x\Delta t$ over all positive values of v_x (because only molecules with a positive component of velocity are moving towards the area of interest) with each value of v_x being weighted by the probability of it occurring:

$$\text{Number of collisions} = \mathcal{N}A\Delta t \int_0^\infty v_x f(v_x)\mathrm{d}v_x$$

The collision flux is the number of collisions divided by *A* and Δt, so

$$Z_W = \mathcal{N}\int_0^\infty v_x f(v_x)\mathrm{d}v_x$$

Step 3 *Evaluate the integral*

Because

$$\int_0^\infty v_x f(v_x)\mathrm{d}v_x = \left(\frac{m}{2\pi kT}\right)^{1/2} \overbrace{\int_0^\infty v_x \mathrm{e}^{-mv_x^2/2kT}\,\mathrm{d}v_x}^{\text{Integral G.2}} = \left(\frac{kT}{2\pi m}\right)^{1/2}$$

it follows that

$$Z_\mathrm{W} = \mathcal{N}\left(\frac{kT}{2\pi m}\right)^{1/2} \overset{\mathcal{N}=p/kT}{=} \frac{p}{kT}\left(\frac{kT}{2\pi m}\right)^{1/2}$$

and therefore

$$Z_\mathrm{W} = \frac{p}{(2\pi mkT)^{1/2}} \qquad (16\mathrm{A.7a})$$

Collision flux in terms of pressure [perfect gas]

Step 4 *Develop an alternative expression in terms of the mean speed*

The mean speed is given by eqn 16A.1b as

$$v_\mathrm{mean} = \left(\frac{8RT}{\pi M}\right)^{1/2} \overset{R=N_\mathrm{A}k\ \ M=N_\mathrm{A}m}{=} \left(\frac{8kT}{\pi m}\right)^{1/2}$$

It follows that

$$\left(\frac{kT}{2\pi m}\right)^{1/2} = \tfrac{1}{4}v_\mathrm{mean}$$

and therefore $Z_\mathrm{W} = \mathcal{N}(kT/2\pi m)^{1/2}$ can be expressed as

$$Z_\mathrm{W} = \tfrac{1}{4}\mathcal{N}v_\mathrm{mean} \qquad (16\mathrm{A.7b})$$

Collision flux [perfect gas]

According to eqn 16A.7a, the collision flux increases with pressure simply because increasing the pressure increases the number density and hence the number of collisions. The flux decreases with increasing mass of the molecules because heavy molecules move more slowly than light molecules. Caution, however, is needed with the interpretation of the role of temperature: it is wrong to conclude that because $T^{1/2}$ appears in the denominator that the collision flux decreases with increasing temperature. If the system has constant volume, the pressure increases with temperature ($p \propto T$), so the collision flux is in fact proportional to $T/T^{1/2} = T^{1/2}$, and increases with temperature (because the molecules are moving faster).

Brief illustration 16A.2

The collision flux of O_2 molecules, with $m = M/N_\mathrm{A}$ and $M = 32.00\,\mathrm{g\,mol^{-1}}$, at 25 °C and 1.00 bar is

$$Z_\mathrm{W} = \frac{\overbrace{1.00\times10^5\,\mathrm{Pa}}^{\mathrm{kg\,m^{-1}\,s^{-2}}}}{\left\{2\pi\times\dfrac{32.00\times10^{-3}\,\mathrm{kg\,mol^{-1}}}{6.022\times10^{23}\,\mathrm{mol^{-1}}}\times(1.381\times10^{-23}\,\mathrm{J\,K^{-1}})\times(298\,\mathrm{K})\right\}^{1/2}}$$

$$= 2.70\times10^{27}\,\mathrm{m^{-2}\,s^{-1}}$$

This flux corresponds to $0.45\,\mathrm{mol\,cm^{-2}\,s^{-1}}$.

(a) The diffusion coefficient

The first application of the result in eqn 16A.7b is to use it to find an expression for the net flux of molecules arising from a concentration gradient.

How is that done? 16A.2 Deriving an equation for the net flux of matter

Consider the arrangement depicted in Fig. 16A.4. The molecules passing through the area A at $z = 0$ have travelled an average of about one mean free path, λ, since their last collision.

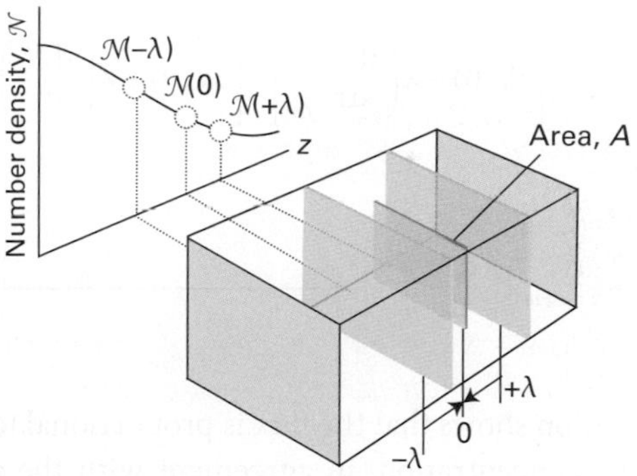

Figure 16A.4 The calculation of the rate of diffusion of a gas considers the net flux of molecules through a plane of area A as a result of molecules arriving from an average distance λ away in each direction.

Step 1 *Set up expressions for the flux in each direction*

If the number density at z is $\mathcal{N}(z)$, then the number density at $z = \lambda$ can be estimated by using a Maclaurin expansion of the form $f(x) = f(0) + (\mathrm{d}f/\mathrm{d}x)_0 x + \cdots$, truncated after the second term (see *The chemist's toolkit* 12 in Topic 5B):

$$\mathcal{N}(+\lambda) = \mathcal{N}(0) + \lambda\left(\frac{\mathrm{d}\mathcal{N}}{\mathrm{d}z}\right)_0$$

Similarly, the number density at $z = -\lambda$ is

$$\mathcal{N}(-\lambda) = \mathcal{N}(0) - \lambda\left(\frac{\mathrm{d}\mathcal{N}}{\mathrm{d}z}\right)_0$$

The average number of impacts on the imaginary region of area A during an interval Δt is $Z_W A\Delta t$, where Z_W is the collision flux. Therefore, the matter flux from left to right, $J(\text{L}\rightarrow\text{R})$, arising from the supply of molecules on the left, is this number of collisions divided by the time interval and the area:

$$J(\text{L}\rightarrow\text{R}) = \frac{Z_W A\Delta t}{A\Delta t} = Z_W \overset{Z_W = \frac{1}{4}\mathcal{N}v_{\text{mean}}}{=} \tfrac{1}{4}\mathcal{N}(-\lambda)v_{\text{mean}}$$

The number density is that at $z = -\lambda$, where the molecules originate before striking the area. There is also a flux of molecules from right to left. The molecules making this journey have originated from $z = +\lambda$ where the number density is $\mathcal{N}(\lambda)$. Therefore,

$$J(\text{L}\leftarrow\text{R}) = \tfrac{1}{4}\mathcal{N}(\lambda)v_{\text{mean}}$$

Step 2 *Evaluate the net flux*

The net flux from left to right is

$$\begin{aligned} J_z &= J(\text{L}\rightarrow\text{R}) - J(\text{L}\leftarrow\text{R}) \\ &= \tfrac{1}{4}v_{\text{mean}}\{\mathcal{N}(-\lambda) - \mathcal{N}(\lambda)\} \\ &= \tfrac{1}{4}v_{\text{mean}}\left\{\left[\mathcal{N}(0) - \lambda\left(\frac{d\mathcal{N}}{dz}\right)_0\right] - \left[\mathcal{N}(0) + \lambda\left(\frac{d\mathcal{N}}{dz}\right)_0\right]\right\} \end{aligned}$$

That is,

$$J_z = -\tfrac{1}{2}v_{\text{mean}}\lambda\left(\frac{d\mathcal{N}}{dz}\right)_0 \qquad \text{Net flux} \quad (16\text{A}.8)$$

This equation shows that the flux is proportional to the gradient of the concentration, in agreement with the empirical observation expressed by Fick's law, eqn 16A.2.

At this stage it looks as though a value of the diffusion coefficient can be picked out by comparing eqns 16A.8 and 16A.4, so obtaining $D = \frac{1}{2}\lambda v_{\text{mean}}$. It must be remembered, however, that the calculation is quite crude, and is little more than an assessment of the order of magnitude of D. One aspect that has not been taken into account is illustrated in Fig. 16A.5, which shows that although a molecule may have begun its journey very close to the window, it could have a long flight before it gets there. Because the path is long, the molecule is likely to collide before reaching the window, so it ought not to be counted as passing through the window. Taking this effect into account results in the appearance of a factor of $\frac{2}{3}$ representing the lower flux. The modification results in

$$D = \tfrac{1}{3}\lambda v_{\text{mean}} \qquad \text{Diffusion coefficient} \quad (16\text{A}.9)$$

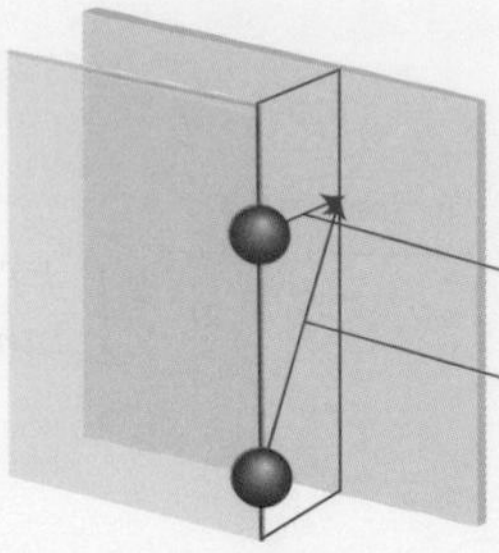

Figure 16A.5 One issue ignored in the simple treatment is that some molecules might make a long flight to the plane even though they are only a short perpendicular distance away from it. A molecule taking the longer flight has a higher chance of colliding during its journey.

Brief illustration 16A.3

In *Brief illustration* 1B.3 of Topic 1B it is established that the mean free path of N_2 molecules in a gas at 1.0 atm and 25 °C is 91 nm; in *Example* 1B.1 of the same Topic it is calculated that under the same conditions the mean speed of N_2 molecules is 475 m s^{-1}. Therefore, the diffusion coefficient for N_2 molecules under these conditions is

$$D = \tfrac{1}{3} \times (91 \times 10^{-9}\,\text{m}) \times (475\,\text{m s}^{-1}) = 1.4 \times 10^{-5}\,\text{m}^2\,\text{s}^{-1}$$

The experimental value is $1.5 \times 10^{-5}\,\text{m}^2\,\text{s}^{-1}$.

There are three points to note about eqn 16A.9:

Physical interpretation

- The mean free path, λ, decreases as the pressure is increased (eqn 16A.1a), so D decreases with increasing pressure and, as a result, the gas molecules diffuse more slowly.
- The mean speed, v_{mean}, increases with the temperature (eqn 16A.1b), so D also increases with temperature. As a result, molecules in a hot gas diffuse more quickly than those when the gas is cool (for a given concentration gradient).
- Because the mean free path increases when the collision cross-section σ of the molecules decreases, the diffusion coefficient is greater for small molecules than for large molecules.

(b) Thermal conductivity

According to the equipartition theorem (*The chemist's toolkit* 7 in Topic 2A), each molecule carries an average energy $\varepsilon = \nu kT$, where ν depends on the number of quadratic contributions to the energy of the molecule and is a number of the order of 1. For atoms, which have three translational degrees of freedom,

$\nu = \frac{3}{2}$. When one molecule passes through the imaginary window, it transports that average energy. An argument similar to that used for diffusion can be used to discuss the transport of energy through this window.

How is that done? 16A.3 Deriving an expression for the thermal conductivity

Assume that the number density is uniform but that the temperature, and hence the average energy of the molecules, is not. Molecules arrive from the left after travelling a mean free path from their last collision in a hotter region, and therefore arrive with a higher energy. Molecules also arrive from the right after travelling a mean free path from a cooler region, and hence arrive with lower energy.

Step 1 *Write expressions for the forward, reverse, and net energy flux*

If the average energy of the molecules at z is $\varepsilon(z)$, the two opposing energy fluxes are

$$J(\mathrm{L}\to\mathrm{R}) = \underbrace{\tfrac{1}{4}\mathcal{N}v_{\text{mean}}}_{z_\mathrm{W}}\varepsilon(-\lambda) \qquad J(\mathrm{L}\leftarrow\mathrm{R}) = \underbrace{\tfrac{1}{4}\mathcal{N}v_{\text{mean}}}_{z_\mathrm{W}}\varepsilon(\lambda)$$

and the net flux is

$$J_z = J(\mathrm{L}\to\mathrm{R}) - J(\mathrm{L}\leftarrow\mathrm{R})$$

$$= \tfrac{1}{4}v_{\text{mean}}\mathcal{N}\{\varepsilon(-\lambda) - \varepsilon(\lambda)\}$$

$\varepsilon(z) = \varepsilon(0) + z(\mathrm{d}\varepsilon/\mathrm{d}z)_0,\ z = \pm\lambda$

$$= \tfrac{1}{4}v_{\text{mean}}\mathcal{N}\left\{\left[\varepsilon(0) - \lambda\left(\frac{\mathrm{d}\varepsilon}{\mathrm{d}z}\right)_0\right] - \left[\varepsilon(0) + \lambda\left(\frac{\mathrm{d}\varepsilon}{\mathrm{d}z}\right)_0\right]\right\}$$

$$= -\tfrac{1}{2}v_{\text{mean}}\lambda\mathcal{N}\left(\frac{\mathrm{d}\varepsilon}{\mathrm{d}z}\right)_0$$

Step 2 *Express the energy gradient as a temperature gradient*

Because $\varepsilon = \nu kT$, a gradient of energy can be expressed as a gradient of temperature: $\mathrm{d}\varepsilon/\mathrm{d}z = \nu k(\mathrm{d}T/\mathrm{d}z)$; therefore

$$J_z = -\tfrac{1}{2}v_{\text{mean}}\lambda\mathcal{N}\left(\frac{\mathrm{d}\varepsilon}{\mathrm{d}z}\right)_0 = -\tfrac{1}{2}\nu k v_{\text{mean}}\lambda\mathcal{N}\left(\frac{\mathrm{d}T}{\mathrm{d}z}\right)_0$$

This equation shows that the energy flux is proportional to the temperature gradient, which is the desired result. As before, the constant is multiplied by $\frac{2}{3}$ to take long flight paths into account, and comparison of this equation with eqn 16A.5 shows that

$$\kappa = \tfrac{1}{3}\nu k v_{\text{mean}}\lambda\mathcal{N} \qquad \text{Thermal conductivity} \quad (16\text{A}.10\text{a})$$

The number density can be written in terms of the molar concentration [J] of the carrier particles J: $\mathcal{N} = nN_\mathrm{A}/V = [\mathrm{J}]N_\mathrm{A}$. Next, the quantity νkN_A can be identified as the molar constant-volume heat capacity of the gas: the molar energy is $N_\mathrm{A}\nu kT$, so it follows from $C_{V,\mathrm{m}} = (\partial U_\mathrm{m}/\partial T)_V$ that $C_{V,\mathrm{m}} = N_\mathrm{A}\nu k$. With these substitutions, eqn 16A.10a becomes

$$\kappa = \tfrac{1}{3}v_{\text{mean}}\lambda[\mathrm{J}]C_{V,\mathrm{m}} \qquad \text{Thermal conductivity} \quad (16\text{A}.10\text{b})$$

Yet another form is found by starting with eqn 16A.10a, recognizing that $\mathcal{N} = p/kT$ and then using the expression for D in eqn 16A.9:

$$\kappa = \tfrac{1}{3}\nu k v_{\text{mean}}\lambda\overbrace{\mathcal{N}}^{\mathcal{N}=p/kT} = \tfrac{1}{3}\nu k v_{\text{mean}}\lambda\frac{p}{kT} = \frac{\nu p \overbrace{D}^{D=\frac{1}{3}\lambda v_{\text{mean}}}}{T} \qquad \text{Thermal conductivity} \quad (16\text{A}.10\text{c})$$

Brief illustration 16A.4

In *Brief illustration* 16A.3 the value $D = 1.4 \times 10^{-5}\,\mathrm{m^2\,s^{-1}}$ was calculated for N_2 molecules at 25 °C and 1.0 atm. The thermal conductivity can be calculated by using eqn 16A.10c and noting that for N_2 molecules $\nu = \frac{5}{2}$ (three translational modes and two rotational modes; the vibrational mode is inactive for this 'stiff' molecule).

$$\kappa = \frac{\frac{5}{2}\times(1.01\times10^5\ \overbrace{\mathrm{Pa}}^{\mathrm{J\,m^{-3}}})\times(1.4\times10^{-5}\,\mathrm{m^2\,s^{-1}})}{298\,\mathrm{K}}$$
$$= 1.2\times10^{-2}\,\mathrm{J\,K^{-1}\,m^{-1}\,s^{-1}}$$

or $12\,\mathrm{mW\,K^{-1}\,m^{-1}}$. The experimental value is $26\,\mathrm{mW\,K^{-1}\,m^{-1}}$.

To interpret eqn 16A.10, note that:

- The mean free path is λ is inversely proportional to the pressure, but the number density $\mathcal{N}$ is proportional to the pressure ($\mathcal{N} = p/kT$). It follows that the product $\lambda\mathcal{N}$, which appears eqn 16A.10a, is independent of pressure, and so therefore is the thermal conductivity.
- The thermal conductivity is greater for gases with a high heat capacity (eqn 16A.10b) because a given temperature gradient then corresponds to a steeper energy gradient.

Physical interpretation

The physical reason for the pressure independence of the thermal conductivity is that the conductivity can be expected to be large when many molecules are available to transport the energy, but the presence of so many molecules limits their mean free path and they cannot carry the energy over a great distance.

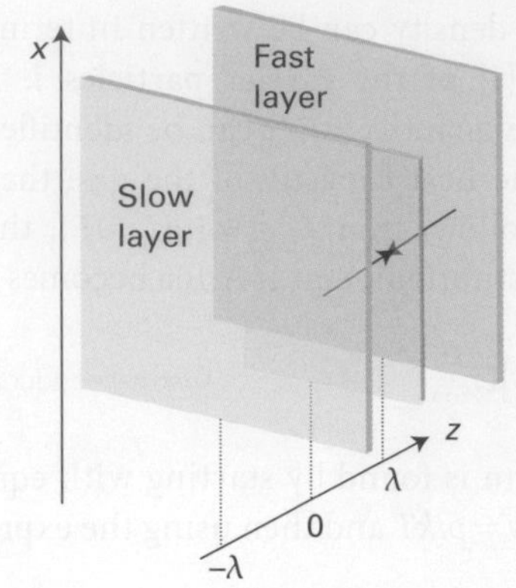

Figure 16A.6 The calculation of the viscosity of a gas examines the net x-component of momentum brought to a plane from faster and slower layers a mean free path away in each direction.

These two effects cancel. The thermal conductivity is indeed found experimentally to be independent of the pressure, except when the pressure is very low, and then $\kappa \propto p$. At very low pressures it is possible for λ to exceed the dimensions of the apparatus, and the distance over which the energy is transported is then determined by the size of the container and not by collisions with the other molecules present. The flux is still proportional to the number of carriers, but the length of the journey no longer depends on λ, so $\kappa \propto$ [J], which implies that $\kappa \propto p$.

(c) Viscosity

If the momentum in the x-direction depends on z as $mv_x(z)$, then molecules moving from the right in Fig. 16A.6 (from a fast layer to a slower one) transport a momentum $mv_x(\lambda)$ to their new layer at $z=0$; those travelling from the left transport $mv_x(-\lambda)$ to it. This picture can be used to build an expression for the coefficient of viscosity.

How is that done? 16A.4 Deriving an expression for the viscosity

The strategy is the same as in the previous derivations, but now the property being transported is the linear momentum of the layers.

Step 1 *Set up expressions for the flux of momentum in each direction and the net flux*

Molecules arriving from the right bring a momentum

$$mv_x(\lambda) = mv_x(0) + m\lambda\left(\frac{dv_x}{dz}\right)_0$$

Those arriving from the left bring a momentum

$$mv_x(-\lambda) = mv_x(0) - m\lambda\left(\frac{dv_x}{dz}\right)_0$$

The net flux of x-momentum in the z-direction is therefore

$$J_z = \tfrac{1}{4}v_{\text{mean}}\mathcal{N}\left\{\left[mv_x(0) - m\lambda\left(\frac{dv_x}{dz}\right)_0\right] - \left[mv_x(0) + m\lambda\left(\frac{dv_x}{dz}\right)_0\right]\right\}$$

$$= -\tfrac{1}{2}v_{\text{mean}}\lambda m\mathcal{N}\left(\frac{dv_x}{dz}\right)_0$$

Step 2 *Identify the coefficient of viscosity*

The flux is proportional to the velocity gradient, in line with the phenomenological equation. Comparison of this expression with eqn 16A.6, and multiplication by $\frac{2}{3}$ in the normal way, leads to

$$\eta = \tfrac{1}{3}v_{\text{mean}}\lambda m\mathcal{N} \quad (16A.11a)$$

Viscosity

Two alternative forms of this expression (after using $mN_A = M$ and eqn 16A.9, $D = \frac{1}{3}\lambda v_{\text{mean}}$) are

$$\eta = MD[\text{J}] \quad (16A.11b)$$

$$\eta = \frac{pMD}{RT} \quad (16A.11c)$$

where [J] is the molar concentration of the gas molecules J and M is their molar mass.

Brief illustration 16A.5

From *Brief illustration* 16A.3 the value of D for N_2 molecules at 25 °C and 1.0 atm is $1.4\times10^{-5}\,\text{m}^2\,\text{s}^{-1}$. Because $M = 28.02\,\text{g mol}^{-1}$, eqn 16A.11c gives

$$\eta = \frac{(1.01\times10^5\,\overbrace{\text{Pa}}^{\text{J m}^{-3}})\times(28.02\times10^{-3}\,\text{kg mol}^{-1})\times(1.4\times10^{-5}\,\text{m}^2\,\text{s}^{-1})}{(8.3145\,\text{J K}^{-1}\,\text{mol}^{-1})\times(298\,\text{K})}$$

$$= 1.6\times10^{-5}\,\text{kg m}^{-1}\,\text{s}^{-1}$$

or 160 μP. The experimental value is 177 μP.

The physical interpretation of eqn 16A.11a is as follows:

- As has already been noted for thermal conductivity, the product $\lambda\mathcal{N}$ is independent of p. Therefore, like the thermal conductivity, the viscosity is independent of the pressure.
- Because $v_{\text{mean}} \propto T^{1/2}$, $\eta \propto T^{1/2}$ at constant volume (and $\eta \propto T^{3/2}$ at constant pressure). That is, the viscosity of a gas *increases* with temperature.

Physical interpretation

The physical reason for the pressure-independence of the viscosity is the same as for the thermal conductivity: more molecules are available to transport the momentum, but they carry it less far on account of the decrease in mean free path.

The increase of viscosity with temperature is explained when it is recalled that at high temperatures the molecules travel more quickly, so the flux of momentum is greater. In contrast, as discussed in Topic 16B, the viscosity of a liquid *decreases* with increase in temperature because intermolecular interactions must be overcome.

(d) Effusion

Because the mean speed of molecules is inversely proportional to $M^{1/2}$, the rate at which they strike the area of the hole through which they are effusing is also inversely proportional to $M^{1/2}$, in accord with Graham's law. However, by using the expression for the collision flux, a more detailed expression for the rate of effusion can be obtained and used to interpret effusion data in a more sophisticated way.

When a gas at a pressure p and temperature T is separated from a vacuum by a small hole, the rate of escape of its molecules is equal to the rate at which they strike the area of the hole, which is the product of the collision flux and the area of the hole, A_0:

$$\text{Rate of effusion} = Z_W A_0 = \frac{pA_0}{(2\pi mkT)^{1/2}} \overset{m=M/N_A,\ k=R/N_A}{=} \frac{pA_0N_A}{(2\pi MRT)^{1/2}}$$

Rate of effusion (16A.12)

This rate is inversely proportional to $M^{1/2}$, in accord with Graham's law. Do not conclude, however, that because the expression includes a factor of $T^{-1/2}$ the rate of effusion decreases as the temperature increases. Because $p \propto T$, the rate is in fact proportional to $T^{1/2}$ and increases with temperature.

Equation 16A.12 is the basis of the **Knudsen method** for the determination of the vapour pressures of liquids and solids, particularly of substances with very low vapour pressures and which cannot be measured directly. Thus, if the vapour pressure of a sample is p, and it is enclosed in a cavity with a small hole, then the rate of loss of mass from the container is proportional to p.

Example 16A.1 Calculating the vapour pressure from a mass loss

Caesium (m.p. 29 °C, b.p. 686 °C) was introduced into a container and heated to 500 °C; the container is pierced by a hole of diameter 0.50 mm. It is found that in 100 s the mass of the container decreased by 385 mg. Calculate the vapour pressure of liquid caesium at 500 °C.

Collect your thoughts The pressure of vapour is constant inside the container despite the effusion of atoms because the hot liquid metal replenishes the vapour. The rate of effusion is therefore constant, and given by eqn 16A.12. To express the rate in terms of mass, multiply the number of atoms that escape by the mass of each atom.

The solution The mass loss Δm in an interval Δt is equal to the number of molecules that strike the area of the hole in this interval multiplied by the mass of each molecule:

$$\Delta m = \text{rate of effusion} \times \Delta t \times m = \overbrace{\frac{pA_0N_A}{(2\pi MRT)^{1/2}}}^{\text{eqn 16A.12}} \times \Delta t \times m$$

where A_0 is the area of the hole and M is the molar mass. Rearrangement of this equation gives an expression for p:

$$p = \left(\frac{2\pi RT}{M}\right)^{1/2} \frac{\Delta m}{A_0 \Delta t}$$

Substitution of the data and $M = 132.9\,\text{g mol}^{-1}$ gives

$$p = \left(\frac{2\pi \times (8.3145\,\text{J K}^{-1}\,\text{mol}^{-1}) \times (773\,\text{K})}{0.1329\,\text{kg mol}^{-1}}\right)^{1/2} \times \frac{385 \times 10^{-6}\,\text{kg}}{\underbrace{\pi \times (0.25 \times 10^{-3}\,\text{m})^2}_{A_0} \times (100\,\text{s})} = 1.1 \times 10^4\,\text{Pa or } 11\,\text{kPa}$$

Self-test 16A.1 How long would it take for 200 mg of Cs atoms to effuse out of the oven under the same conditions?

Answer: 52 s

Checklist of concepts

☐ 1. **Flux** is the quantity of a property passing through a given area in a given time interval divided by the area and the duration of the interval.

☐ 2. **Diffusion** is the migration of matter down a concentration gradient.

☐ 3. **Fick's first law of diffusion** states that the flux of matter is proportional to the concentration gradient.

☐ 4. **Thermal conduction** is the migration of energy down a temperature gradient; the flux of energy is proportional to the temperature gradient.

☐ 5. **Viscosity** is the migration of linear momentum down a velocity gradient; the flux of momentum is proportional to the velocity gradient.

☐ 6. **Effusion** is the emergence of a gas from a container through a small hole.

☐ 7. **Graham's law of effusion** states that the rate of effusion is inversely proportional to the square root of the molar mass.

Checklist of equations

Property	Equation	Comment	Equation number
Fick's first law of diffusion	$J=-D\mathrm{d}\mathcal{N}/\mathrm{d}z$		16A.4
Flux of energy of thermal motion	$J=-\kappa\mathrm{d}T/\mathrm{d}z$		16A.5
Flux of momentum along x	$J=-\eta\mathrm{d}v_x/\mathrm{d}z$		16A.6
Diffusion coefficient of a perfect gas	$D=\frac{1}{3}\lambda v_{\mathrm{mean}}$	KMT*	16A.9
Coefficient of thermal conductivity of a perfect gas	$\kappa=\frac{1}{3}v_{\mathrm{mean}}\lambda[\mathrm{J}]C_{V,\mathrm{m}}$	KMT and equipartition	16A.10b
Coefficient of viscosity of a perfect gas	$\eta=\frac{1}{3}v_{\mathrm{mean}}\lambda m\mathcal{N}$	KMT	16A.11a
Rate of effusion	Rate $\propto 1/M^{1/2}$	Graham's law	16A.12

* KMT indicates that the equation is based on the kinetic theory of gases.

TOPIC 16B Motion in liquids

> ➤ **Why do you need to know this material?**
>
> Many chemical reactions take place in liquids, so for a full understanding of them it is important to know how solute molecules and ions move through such environments.
>
> ➤ **What is the key idea?**
>
> Ions reach a terminal velocity when the electrical force on them is balanced by the drag due to the viscosity of the solvent.
>
> ➤ **What do you need to know already?**
>
> The discussion of viscosity starts with the definition of the coefficient of viscosity introduced in Topic 16A. Some calculations make use of the information about electrostatics set out in *The chemist's toolkit* 29.

The motion of ions and molecules is an important aspect of the properties of liquids and of the reactions taking place in them. It can be studied experimentally by a variety of methods. For example, bulk measurements of viscosity and its temperature dependence can be used to build models of the motion. At a more detailed level, relaxation time measurements in NMR (Topic 12C) and EPR can be used to show how molecules rotate. For instance, these observations show that large molecules in viscous fluids typically rotate in a series of small (about 5°) steps, whereas small molecules in non-viscous fluids typically jump through about 1 radian (57°) in each step. Another important technique is **inelastic neutron scattering**, in which the energy neutrons collect or discard as they pass through a sample is interpreted in terms of the motion of its molecules.

16B.1 Experimental results

There are two 'classical' methods of investigating molecular motion in liquids. One is the measurement of viscosity and its temperature dependence. Another involves inferring details about molecular motion by dragging ions through a solvent under the influence of an electric field.

(a) Liquid viscosity

The coefficient of viscosity, η (eta), is introduced in Topic 16A as the constant of proportionality in the relation between the flux of linear momentum and the velocity gradient in a fluid:

$$J_z(x\text{-component of momentum}) = -\eta\frac{\mathrm{d}v_x}{\mathrm{d}z} \qquad (16\text{B}.1)$$

The units of viscosity are kilograms per metre per second ($\mathrm{kg\,m^{-1}\,s^{-1}}$), but they may also be reported in the equivalent units of pascal seconds (Pa s). The non-SI units poise (P) and centipoise (cP) are still widely encountered: $1\,\mathrm{P} = 10^{-1}\,\mathrm{Pa\,s}$ and so $1\,\mathrm{cP} = 1\,\mathrm{mPa\,s}$. Table 16B.1 lists some values of η for liquids.

Unlike in a gas, for a molecule to move in a liquid it must acquire at least a minimum energy (an 'activation energy', E_a, in the language of Topic 17D) to escape from its neighbours. From the Boltzmann distribution it follows that the probability that a molecule has at least an energy E_a is proportional to $e^{-E_a/RT}$, so the mobility of the molecules in the liquid should follow this type of temperature dependence. As the temperature increases, the molecules become more mobile and so the viscosity is reduced; the expected temperature dependence of the viscosity, which decreases as the mobility of the molecules increases, is therefore of the form

$$\eta = \eta_0 e^{E_a/RT} \qquad \text{Temperature dependence of viscosity (liquid)} \qquad (16\text{B}.2)$$

(Note the positive sign of the exponent.) The activation energy typical of viscosity is comparable to the mean potential energy of intermolecular interactions. Equation 16B.2 implies that the viscosity should decrease sharply with increasing temperature. Such a variation is found experimentally, at

Table 16B.1 Viscosities of liquids at 298 K*

	$\eta/(10^{-3}\,\mathrm{kg\,m^{-1}\,s^{-1}})$
Benzene	0.601
Mercury	1.55
Pentane	0.224
Water‡	0.891

* More values are given in the *Resource section*.
‡ Note that $1\,\mathrm{cP} = 10^{-3}\,\mathrm{kg\,m^{-1}\,s^{-1}}$; the viscosity of water corresponds to 0.891 cP.

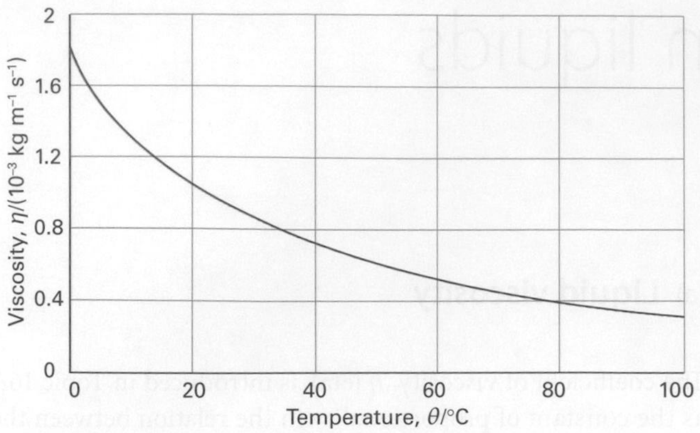

Figure 16B.1 The temperature dependence of the viscosity of water. As the temperature is increased, more molecules are able to escape from the potential wells provided by their neighbours, and so the liquid becomes less viscous.

least over reasonably small temperature ranges (Fig. 16B.1). Intermolecular interactions govern the magnitude of E_a, but the calculation of its value from an assumed form of the potential energy due to the interactions between the molecules is an immensely difficult and largely unsolved problem.

Brief illustration 16B.1

The viscosity of water at 25 °C and 50 °C is 0.891 mPa s and 0.547 mPa s, respectively. It follows from eqn 16B.2 that the activation energy for molecular migration is the solution of

$$\frac{\eta(T_2)}{\eta(T_1)} = e^{(E_a/R)(1/T_2-1/T_1)}$$

which, after taking logarithms of both sides, is

$$\begin{aligned} E_a &= \frac{R\ln\{\eta(T_2)/\eta(T_1)\}}{1/T_2-1/T_1} \\ &= \frac{(8.3145\,\mathrm{J\,K^{-1}\,mol^{-1}})\ln\{(0.547\,\mathrm{mPa\,s})/(0.891\,\mathrm{mPa\,s})\}}{1/(323\,\mathrm{K})-1/(298\,\mathrm{K})} \\ &= 1.56\times10^4\,\mathrm{J\,mol^{-1}} \end{aligned}$$

or 15.6 kJ mol^{-1}. This value is comparable to the strength of a hydrogen bond.

One problem with the interpretation of viscosity measurements is that the change in density of the liquid as it is heated makes a pronounced contribution to the temperature variation of the viscosity. Thus, the temperature dependence of viscosity at constant volume, when the density is constant, is much less than that at constant pressure. At low temperatures, the viscosity of water decreases as the pressure is increased. This behaviour is consistent with the need to rupture hydrogen bonds for migration to occur.

(b) Electrolyte solutions

When a potential difference is applied between two electrodes immersed in a solution containing ions there is a flow of current due to the migration of ions through the solution. The key electrical property of the solution is its resistance, R, which is expressed in ohms, Ω (1 Ω = 1 C^{-1} V s). It is often more convenient to work in terms of the **conductance**, G, which is the inverse of the resistance: $G = 1/R$, and therefore expressed in Ω^{-1}. The reciprocal ohm used to be called the mho, but its SI designation is now the siemens, S, and 1 S = 1 Ω^{-1} = 1 C V^{-1} s^{-1}. Electric current is expressed in amperes, A, with 1 A = 1 C s^{-1}, so a more physically revealing relation is 1 S = 1 A V^{-1}.

The conductance of a sample is proportional to its cross-sectional area, A, and inversely proportional to its length, l. Therefore

$$G = \kappa\frac{A}{l} \qquad \text{Conductivity [definition]} \quad (16B.3)$$

where the constant of proportionality κ (kappa) is the electrical **conductivity** of the sample. With the conductance in siemens and the dimensions in metres, it follows that the units of κ are siemens per metre (S m^{-1}). The conductivity is a property of the material, whereas the conductance depends both on the material and its dimensions. The conductivity depends on the concentration of charge carriers in the sample, which suggests that it is sensible to introduce the **molar conductivity**, Λ_m, defined as

$$\Lambda_m = \frac{\kappa}{c} \qquad \text{Molar conductivity [definition]} \quad (16B.4)$$

where c is the molar concentration of the added electrolyte. The units of molar conductivity are siemens metre-squared per mole (S m^2 mol^{-1}), and typical values are about 10 mS m^2 mol^{-1} (where 1 mS = 10^{-3} S).

Experimentally, it is found that the value of the molar conductivity varies with the concentration. One reason for this variation is that the number of ions in the solution might not be proportional to the nominal concentration of the electrolyte. For instance, the concentration of ions in a solution of a weak electrolyte depends on the degree of dissociation, which is a complicated function of the total amount of solute: doubling the total concentration of the solute does not double the number of ions. Secondly, even for fully dissociated strong electrolytes, because ions interact strongly with one another, the conductivity of a solution is not exactly proportional to the number of ions present.

In an extensive series of measurements during the nineteenth century, Friedrich Kohlrausch established the **Kohlrausch law**, that at low concentrations the molar conductivities of strong electrolytes depend on the square root of the concentration:

$$\Lambda_m = \Lambda_m^\circ - \mathcal{K}c^{1/2} \qquad \text{Kohlrausch law} \quad (16B.5)$$

where Λ_m° is the **limiting molar conductivity**, the molar conductivity in the limit of zero concentration when the ions are so far apart that they move independently of each other. Kohlrausch also established that this limiting molar conductivity is the sum of contributions from the individual ions present. If the limiting molar conductivity of the cations is denoted λ_+ and that of the anions λ_-, then his **law of the independent migration of ions** states that

$$\Lambda_m^\circ = \nu_+\lambda_+ + \nu_-\lambda_- \qquad \text{Law of the independent migration of ions} \qquad (16B.6)$$

where ν_+ and ν_- are the numbers of cations and anions provided by each formula unit of electrolyte. For example, for HCl, NaCl, and $CuSO_4$, $\nu_+ = \nu_- = 1$, but for $MgCl_2$ $\nu_+ = 1$, $\nu_- = 2$.

Example 16B.1 Determining the limiting molar conductivity

The conductivity of KCl(aq) at 25 °C is 14.688 mS m^{-1} when c = 1.0000 mmol dm^{-3}, and 71.740 mS m^{-1} when c = 5.0000 mmol dm^{-3}. Determine the values of the limiting molar conductivity Λ_m° and the Kohlrausch constant $\mathcal{K}$.

Collect your thoughts You need to use eqn 16B.4 to determine the molar conductivities at the two concentrations. Then, by using eqn 16B.5, you can express the difference between these two values as $\Lambda_m(c_2) - \Lambda_m(c_1) = \mathcal{K}(c_1^{1/2} - c_2^{1/2})$. From this relation you can determine $\mathcal{K}$ and then go on to find Λ_m° by using one of the values of the molar conductivity in eqn 16B.5, rearranged into $\Lambda_m^\circ = \Lambda_m + \mathcal{K}c^{1/2}$.

The solution The molar conductivity of KCl(aq) when c = 1.0000 mmol dm^{-3} (which is the same as 1.0000 mol m^{-3}) is

$$\Lambda_m = \frac{14.688\,\text{mS}\,\text{m}^{-1}}{1.0000\,\text{mol}\,\text{m}^{-3}} = 14.688\,\text{mS}\,\text{m}^2\,\text{mol}^{-1}$$

Similarly, when c = 5.0000 mol dm^{-3} its molar conductivity is 14.348 mS m^2 mol^{-1}. It then follows that

$$\mathcal{K} = \frac{\Lambda_m(c_2) - \Lambda_m(c_1)}{c_1^{1/2} - c_2^{1/2}} = \frac{(14.348 - 14.688)\,\text{mS}\,\text{m}^2\,\text{mol}^{-1}}{(0.001\,0000^{1/2} - 0.005\,0000^{1/2})\,(\text{mol}\,\text{dm}^{-3})^{1/2}}$$
$$= 8.698\,\text{mS}\,\text{m}^2\,\text{mol}^{-1}/(\text{mol}\,\text{dm}^{-3})^{1/2}$$

(For reasons that will become clear immediately below, it is best to keep this awkward but convenient array of units rather than converting them to the equivalent $10^{-3/2}$ S m$^{7/2}$ mol$^{-3/2}$.) The limiting molar conductivity is then found by using the data for c = 1.0000 mmol dm^{-3}:

$$\Lambda_m^\circ = 14.688\,\text{mS}\,\text{m}^2\,\text{mol}^{-1} + 8.698\,\frac{\text{mS}\,\text{m}^2\,\text{mol}^{-1}}{(\text{mol}\,\text{dm}^{-3})^{1/2}}$$
$$\times (1.0000\times10^{-3}\,\text{mol}\,\text{dm}^{-3})^{1/2} = 14.963\,\text{mS}\,\text{m}^2\,\text{mol}^{-1}$$

Comment. Although the value of $\mathcal{K}$ has been given to four significant figures in conformity with the data, that degree of precision is probably over-optimistic in practice.

Self-test 16B.1 The conductivity of $KClO_4$(aq) at 25 °C is 13.780 mS m^{-1} when c = 1.000 mmol dm^{-3} and 67.045 mS m^{-1} when c = 5.000 mmol dm^{-3}. Determine the values of the limiting molar conductivity Λ_m° and the Kohlrausch constant $\mathcal{K}$ for this system.

Answer: $\mathcal{K}$ = 9.491 mS m^2 mol^{-1}/(mol dm^{-3})$^{1/2}$, Λ_m° = 14.080 mS m^2 mol^{-1}

16B.2 The mobilities of ions

The reason why different ions have different molar conductivities in solution can be understood by analysing the motion of an ion subject to an electric field and at the same time surrounded by a viscous medium.

(a) The drift speed

An ion in a vacuum is accelerated by an electric field, but in a viscous liquid the motion of the ion is impeded by the need for it to push its way through the tightly packed solvent molecules. The latter effect is called **viscous drag**. As the ion accelerates under the influence of the field, the viscous drag increases and the ion quickly reaches a steady terminal speed, called the **drift speed**, which can be found by balancing the two forces.

How is that done? 16B.1 Deriving an expression for the drift speed

The starting point for this calculation is the result from electrostatics that when the potential difference between two planar electrodes a distance l apart is $\Delta\phi$, the ions in the solution between them experience a uniform electric field of magnitude $\mathcal{E} = \Delta\phi/l$. Here, and throughout this section, the sign of the charge number is disregarded so as to avoid notational complications.

Step 1 *Find the force on the ion due to the field*

In an electric field $\mathcal{E}$, an ion of charge ze experiences a force of magnitude $ze\mathcal{E}$ (see *The chemist's toolkit* 29). Therefore,

$$F_{\text{electric}} = \frac{ze\Delta\phi}{l}$$

Step 2 *Find the force on the ion due to viscous drag*

As the ion moves through the solvent it experiences a frictional retarding force proportional to its speed. For a spherical particle of radius a travelling at a speed s, this force is given by **Stokes' law**, which Stokes derived by considering the hydrodynamics of the passage of a sphere through a continuous fluid:

$$F_{\text{viscous}} = fs \qquad f = 6\pi\eta a \qquad \text{Stokes' law} \qquad (16B.7)$$

where η is the coefficient of viscosity. In this calculation it is assumed that Stokes' law applies on a molecular scale;

experimental evidence suggests that it often gives at least the right order of magnitude for the viscous force.

Step 3 *Find the drift speed by balancing the two forces*
The two forces act in opposite directions and the ions quickly reach a terminal speed, the drift speed, s, when they are in balance. This balance occurs when $fs = ze\mathcal{E}$, and therefore

$$s=\frac{ze\mathcal{E}}{f} \qquad \text{(16B.8a)}$$

Drift speed

Equation 16B.8a shows that the drift speed is proportional to the electric field strength. The constant of proportionality is called the **mobility** of the ion, u:

$$s = u\mathcal{E} \qquad \text{Mobility [definition]} \quad \text{(16B.8b)}$$

With the electric field strength in volts per metre ($\mathrm{V\,m^{-1}}$) and the drift speed in metres per second ($\mathrm{m\,s^{-1}}$) the slightly awkward units of u are metres-squared per volt per second ($\mathrm{m^2\,V^{-1}\,s^{-1}}$; note that $\mathrm{m^2\,V^{-1}\,s^{-1}} \times \mathrm{V\,m^{-1}} = \mathrm{m\,s^{-1}}$); a selection of values is given in Table 16B.2. Comparison of the last two equations shows that

$$u=\frac{ze}{f}=\frac{ze}{6\pi\eta a} \qquad (f=6\pi\eta a) \qquad \text{(16B.9)}$$

where the Stokes' law value for the frictional coefficient f has been used.

Brief illustration 16B.2

An order-of-magnitude estimate of the mobility can be found using eqn 16B.9 with $z = 1$ and $a = 130\,\mathrm{pm}$, which is typical of the radius of a hydrated ion; the viscosity of water at 25 °C is 0.9 cP, or 0.9 mPa s. Then

$$u=\frac{1.6\times10^{-19}\,\overset{\mathrm{J\,V^{-1}}}{\overbrace{\mathrm{C}}}}{6\pi\times(0.9\times10^{-3}\,\underbrace{\mathrm{Pa\,s}}_{\mathrm{J\,m^{-3}}})\times(130\times10^{-12}\,\mathrm{m})}=7.3\times10^{-8}\,\mathrm{m^2\,V^{-1}\,s^{-1}}$$

This value means that when there is a potential difference of 1.0 V across a solution of length 1.0 cm (so $\mathcal{E} = 100\,\mathrm{V\,m^{-1}}$), the drift speed is 7.3 μm s^{-1}. That speed might seem slow, but not when expressed on a molecular scale, because it corresponds to an ion passing about 10^4 solvent molecules per second.

Equation 16B.9 implies that the mobility of an ion decreases with increasing solution viscosity and ion size. Experiments confirm these predictions for bulky ions (such as R_4N^+ and RCO_2^-) but not for small ions. For example, the mobilities of the alkali metal ions in water increase from Li^+ to Rb^+ (Table 16B.2) even though the ionic radii increase. The paradox is resolved when it is realized that the radius a in the Stokes formula is the **hydrodynamic radius** (or 'Stokes radius') of the ion, its effective radius in the solution taking into account all the H_2O molecules it carries in its hydration shell. Small ions

The chemist's toolkit 29 Electrostatics

A charge Q_1 (units: coulomb, C) gives rise to a Coulomb **potential** ϕ (units: volt, V). The potential energy (units: joule, J, with 1 J = 1 V C) of a second charge Q in that potential is

$$E_p = -Q\phi$$

In one dimension, the **electric field strength** (units: volt per metre, $\mathrm{V\,m^{-1}}$), $\mathcal{E}$, is the negative of the gradient of the electric potential ϕ:

$$\mathcal{E}=-\frac{\mathrm{d}\phi}{\mathrm{d}x} \qquad \text{Electric field strength}$$

In three dimensions the electric field is a vector, and

$$\boldsymbol{\mathcal{E}}=-\nabla\phi$$

The electric field between two plane parallel plates separated by a distance l, and between which there is a potential difference $\Delta\phi$, is uniform and given by

$$\mathcal{E}=-\frac{\Delta\phi}{l}$$

A charge Q experiences a force proportional to the electric field strength at its location:

$$F_{\text{electric}}=Q\mathcal{E}$$

A potential gives rise to a force only if it varies with distance.

Table 16B.2 Ionic mobilities in water at 298 K*

	$u/(10^{-8}\,\mathrm{m^2\,V^{-1}\,s^{-1}})$		$u/(10^{-8}\,\mathrm{m^2\,V^{-1}\,s^{-1}})$
H^+	36.23	OH^-	20.64
Li^+	4.01	F^-	5.70
Na^+	5.19	Cl^-	7.91
K^+	7.62	Br^-	8.09
Rb^+	7.92	SO_4^{2-}	8.29

* More values are given in the *Resource section*.

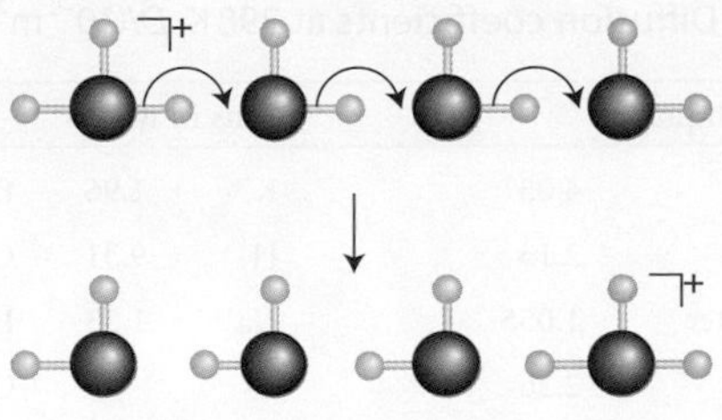

Figure 16B.2 A highly schematic diagram showing the effective motion of a proton in water.

give rise to stronger electric fields than large ones (the electric field at the surface of a sphere of radius r is proportional to z/r^2, so the smaller the radius the stronger the field). Consequently, small ions are more extensively solvated than big ions and a small ion may have a large hydrodynamic radius because it drags many solvent molecules through the solution as it migrates. The hydrating H_2O molecules are often very labile, however, and NMR and isotope studies have shown that the exchange between the coordination sphere of the ion and the bulk solvent is very rapid for ions of low charge but slow for ions of high charge.

The proton, although it is very small, has a very high mobility (Table 16B.2). Proton and ^{17}O-NMR show that the characteristic lifetime of protons hopping from one molecule to the next is about 1.5 ps, which is comparable to the time that inelastic neutron scattering shows it takes a water molecule to turn through about 1 radian (1 to 2 ps). According to the **Grotthuss mechanism**, there is an *effective* motion of a proton that involves the rearrangement of bonds in a group of water molecules (Fig. 16B.2). However, the actual mechanism is still highly contentious. The mobility of protons in liquid ammonia is also anomalous and presumably occurs by an analogous mechanism.

(b) Mobility and conductivity

The limiting molar conductivity of an ion is a measurable quantity, and the mobility of an ion can, in principle, be calculated on the basis of a model of its motion through the solvent. It should be possible to find a relation between these two quantities.

How is that done? 16B.2 Establishing the relation between ionic mobility and limiting molar conductivity

To keep things notationally simple, ignore the signs of quantities in what follows and focus on their magnitudes. Consider a solution of an electrolyte at a molar concentration c. Let each formula unit give rise to ν_+ cations of charge z_+e and ν_- anions of charge z_-e. The molar concentration of each type of ion is therefore νc (with $\nu = \nu_+$ or ν_-), and the number density of each type is $\mathcal{N} = \nu c N_A$.

Step 1 *Calculate the number of ions passing through an imaginary window*

By referring to Fig. 16B.3 you can see that the number of ions of one kind, moving at speed s, that pass through an imaginary window of area A during an interval Δt is equal to the number within the distance $s\Delta t$, and therefore to the number in the volume $s\Delta tA$. It follows that the number of ions passing through the window in that period is $\mathcal{N}s\Delta tA = s\Delta tA\nu cN_A$.

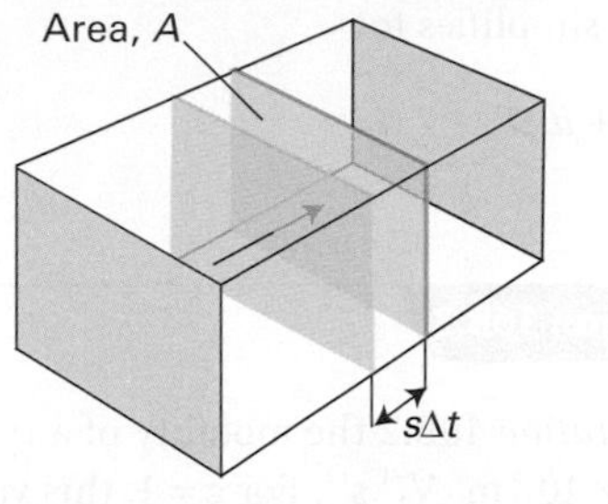

Figure 16B.3 In the calculation of the current, all the ions, moving at speed s, within a distance $s\Delta t$ (i.e. those in the volume $sA\Delta t$) will pass through the area A.

Step 2 *Calculate the charge passing through the window, and hence the current*

Each ion carries a charge ze, so the charge passing through the window is $zes\Delta tA\nu cN_A$ which, by introducing Faraday's constant $F = eN_A$, can be written $zs\Delta tA\nu cF$. The electrical current, I, is the rate of passage of charge, which in this case is the charge divided by the time interval Δt: so $I = zs\Delta tA\nu cF/\Delta t = zsA\nu cF$.

Step 3 *Set up expressions for the conductance, the conductivity, and the molar conductivity*

The conductance is given by $G = I/\Delta\phi$, where $\Delta\phi$ is the potential difference across the solution. It follows that

$$G = \frac{I}{\Delta\phi} = \frac{zsA\nu cF}{\Delta\phi}$$

The conductivity is

$$\kappa = \frac{Gl}{A} = \frac{zsA\nu cFl}{\Delta\phi A} = \frac{zs\nu cFl}{\Delta\phi}$$

The limiting molar ionic conductivity is

$$\lambda = \frac{\kappa}{\nu c} = \frac{zs\nu cFl}{\nu c\Delta\phi} = \frac{zsFl}{\Delta\phi}$$

Step 4 *Introduce the ionic mobility*

At this point you can identify $\Delta\phi/l$ as the electric field strength $\mathcal{E}$, and $s/\mathcal{E}$ as the mobility, u:

$$\lambda = \frac{zsFl}{\Delta\phi} \overset{\Delta\phi/l = \mathcal{E}}{=} \frac{zsF}{\mathcal{E}} \overset{s = u\mathcal{E}}{=} zuF \qquad \text{Ion molar conductivity in terms of mobility} \qquad (16B.10)$$

Equation 16B.10 applies to the cations and to the anions. For an electrolyte where there are ν_+ and ν_- cations and anions, respectively, from each formula unit, it follows from eqn 16B.6 that $\Lambda_m^\circ = \nu_+\lambda_+ + \nu_-\lambda_-$. Therefore

$$\Lambda_m^\circ = (z_+u_+\nu_+ + z_-u_-\nu_-)F \qquad (16B.11a)$$

Limiting molar conductivity in terms of mobilities

For a symmetrical $z{:}z$ electrolyte (e.g. $CuSO_4$ with $z_+ = z_- = 2$), this equation simplifies to

$$\Lambda_m^\circ = z(u_+ + u_-)F \qquad (16B.11b)$$

Brief illustration 16B.3

In *Brief illustration* 16B.2 the mobility of a typical ion is estimated as $7.3 \times 10^{-8}\,m^2\,V^{-1}\,s^{-1}$. For $z = 1$, this value can be used to estimate a typical limiting molar conductivity of the ion as

$$\lambda = 1 \times (7.3 \times 10^{-8}\,m^2\,V^{-1}\,s^{-1}) \times (9.649 \times 10^4\,C\,mol^{-1})$$
$$= 7.0 \times 10^{-3}\,m^2\,V^{-1}\,s^{-1}\,C\,mol^{-1}$$

Because $1\,V^{-1}\,C\,s^{-1} = 1\,S$, the value can be expressed as $7.0\,mS\,m^2\,mol^{-1}$. The experimental value for $K^+(aq)$ is $7.6\,mS\,m^2\,mol^{-1}$.

(c) The Einstein relations

The relation between drift speed and the electric field strength in eqn 16B.8a ($s = ze\mathcal{E}/f$) is a special case of a more general relation derived in Topic 16C (eqn 16C.5):

$$s = \frac{D\mathcal{F}}{RT} \qquad (16B.12)$$

Drift speed in terms of diffusion coefficient

where $\mathcal{F}$ is the force (per mole of ions) driving the ions through the viscous medium and D is the diffusion coefficient for the species (Table 16B.3). For an ion in solution the drift speed is $s = u\mathcal{E}$ (eqn 16B.8b), and the force on each ion in an electric field of strength $\mathcal{E}$ is $ez\mathcal{E}$. It follows that the force per mole of ions is $N_A ez\mathcal{E}$ which, by using $N_A e = F$, can be written $zF\mathcal{E}$. Substitution of these expressions for s and $\mathcal{F}$ into eqn 16B.12 gives, on cancelling the $\mathcal{E}$, the **Einstein relation**:

$$u = \frac{zDF}{RT} \qquad (16B.13)$$

Einstein relation

The mobility is high when the diffusion coefficient (which is inversely proportional to the viscosity) is high, indicating that the solute molecules are very mobile. Don't be misled by the presence of temperature in the denominator into thinking that the ion mobility decreases with increasing temperature: the diffusion coefficient increases more rapidly with temperature than T itself, so u increases with increasing temperature.

Table 16B.3 Diffusion coefficients at 298 K, $D/(10^{-9}\,m^2\,s^{-1})$*

Molecules in liquids		Ions in water			
I_2 in hexane	4.05	K^+	1.96	Br^-	2.08
in benzene	2.13	H^+	9.31	Cl^-	2.03
Glycine in water	1.055	Na^+	1.33	I^-	2.05
H_2O in water	2.26			OH^-	5.03
Sucrose in water	0.5216				

* More values are given in the *Resource section*.

Brief illustration 16B.4

From Table 16B.2, the mobility of SO_4^{2-} is $8.29 \times 10^{-8}\,m^2\,V^{-1}\,s^{-1}$. It follows from eqn 16B.13 in the form $D = uRT/zF$ that the diffusion coefficient for the ion in water at 25 °C is

$$D = \frac{(8.29\times10^{-8}\,m^2\,V^{-1}\,s^{-1})\times(8.3145\,J\,K^{-1}\,mol^{-1})\times(298\,K)}{2\times(9.649\times10^4\,\underbrace{C\,mol^{-1}}_{J\,V^{-1}})}$$
$$= 1.06\times10^{-9}\,m^2\,s^{-1}$$

The Einstein relation can be developed to provide a link between the limiting molar conductivity of an electrolyte and the diffusion coefficients of its ions. First, by using eqns 16B.10 and 16B.13 the limiting molar conductivity of an ion can be written

$$\lambda = zuF \overset{u=zDF/RT}{=} \frac{z^2DF^2}{RT} \qquad (16B.14)$$

Then, by using $\Lambda_m^\circ = \nu_+\lambda_+ + \nu_-\lambda_-$ (eqn 16B.6), the limiting molar conductivity is

$$\Lambda_m^\circ = (\nu_+z_+^2D_+ + \nu_-z_-^2D_-)\frac{F^2}{RT} \qquad (16B.15)$$

Nernst–Einstein equation

which is the **Nernst–Einstein equation**. One application of this equation is to the determination of ionic diffusion coefficients from conductivity measurements; another is to the prediction of conductivities based on models of ionic diffusion.

Checklist of concepts

☐ 1. The viscosity of a liquid decreases with increasing temperature.

☐ 2. **Kohlrausch's law** states that, at low concentrations, the molar conductivities of strong electrolytes vary as the square root of the concentration.

☐ 3. The **law of the independent migration of ions** states that the molar conductivity, in the limit of zero concentration, is the sum of contributions from its individual ions.

☐ 4. An ion reaches a **drift speed** when the acceleration due to the electrical force is balanced by the viscous drag.

☐ 5. The **hydrodynamic radius** of an ion may be greater than its ionic radius.

☐ 6. The high mobility of a proton in water is explained by the **Grotthuss mechanism.**

☐ 7. The **mobility** of an ion can be related to its limiting molar conductivity and, via the Einstein relation, to its diffusion coefficient.

Checklist of equations

Property	Equation	Comment	Equation number
Viscosity of a liquid	$\eta = \eta_0 e^{E_a/RT}$	Over a narrow temperature range	16B.2
Conductivity	$\kappa = Gl/A,\ G = 1/R$	Definition	16B.3
Molar conductivity	$\Lambda_m = \kappa/c$	Definition	16B.4
Kohlrausch's law	$\Lambda_m = \Lambda_m^\circ - \mathcal{K}c^{1/2}$	Empirical observation	16B.5
Law of independent migration of ions	$\Lambda_m^\circ = \nu_+\lambda_+ + \nu_-\lambda_-$	Limiting law	16B.6
Stokes' law	$F_{viscous} = fs,\ f = 6\pi\eta a$		16B.7
Drift speed	$s = u\mathcal{E}$	Defines u	16B.8b
Ion mobility	$u = ze/6\pi\eta a$	Assumes Stokes' law	16B.9
Conductivity and mobility	$\lambda = zuF$		16B.10
Molar conductivity and mobility	$\Lambda_m^\circ = (z_+u_+\nu_+ + z_-u_-\nu_-)F$		16B.11a
Drift speed	$s = D\mathcal{F}/RT$	$\mathcal{F}$ is a general (molar) force	16B.12
Einstein relation	$u = zDF/RT$		16B.13
Nernst–Einstein equation	$\Lambda_m^\circ = (\nu_+z_+^2D_+ + \nu_-z_-^2D_-)(F^2/RT)$		16B.15

FOCUS 16C Diffusion

➤ **Why do you need to know this material?**

The diffusion of chemical species through space determines the rates of many chemical reactions in chemical reactors, living cells, and the atmosphere.

➤ **What is the key idea?**

Molecules and ions tend to spread into a uniform distribution.

➤ **What do you need to know already?**

This Topic draws on arguments relating to the calculation of flux (Topic 16A) and the notion of drift speed introduced in Topic 16B. It also uses the concept of chemical potential to discuss the direction of spontaneous change (Topic 5A). The final section uses a statistical argument like that used to discuss the properties of a random coil in Topic 14D.

That solutes in gases, liquids, and solids have a tendency to spread can be discussed from three points of view. The thermodynamic viewpoint makes use of the Second Law of thermodynamics and the tendency for entropy to increase and, if the temperature and pressure are constant, for the Gibbs energy to decrease. The second approach is to set up a differential equation for the change in concentration in a region by considering the flux of material through its boundaries. The third approach is based on a model in which diffusion is imagined as taking place in a series of random small steps.

Several derivations in this Topic use Fick's first law of diffusion, which is discussed in Topic 16A and repeated here for convenience:

$$J(\text{number}) = -D\frac{\mathrm{d}\mathcal{N}}{\mathrm{d}x} \qquad \text{Fick's first law [number]} \qquad (16C.1a)$$

where $\mathcal{N}$ is the number density and D is the diffusion coefficient. In a number of cases it is more convenient to discuss the flux in terms of the amount of molecules and the molar concentration, c. Division by Avogadro's constant turns eqn 16C.1a into

$$J(\text{amount}) = -D\frac{\mathrm{d}c}{\mathrm{d}x} \qquad \text{Fick's first law [amount]} \qquad (16C.1b)$$

16C.1 The thermodynamic view

At constant temperature and pressure, the maximum non-expansion work that can be done by a spontaneous process is equal to the change in the Gibbs energy (Topic 3D). In this case the spontaneous process is the spreading of a solute, and the work it could achieve per mole of solute molecules can be identified with the change in the chemical potential of the solute: $\mathrm{d}w_\mathrm{m} = \mathrm{d}\mu$. The difference in chemical potential between the locations $x + \mathrm{d}x$ and x is

$$\mathrm{d}\mu = \mu(x+\mathrm{d}x) - \mu(x) = \left(\frac{\partial \mu}{\partial x}\right)_{T,p}\mathrm{d}x$$

so the molar work associated with migration through $\mathrm{d}x$ is

$$\mathrm{d}w_\mathrm{m} = \left(\frac{\partial \mu}{\partial x}\right)_{T,p}\mathrm{d}x$$

The work done in moving a distance $\mathrm{d}x$ against an opposing force $\mathcal{F}$ (in this context, a molar quantity) is $\mathrm{d}w_\mathrm{m} = -\mathcal{F}\mathrm{d}x$. By comparing the two expressions for $\mathrm{d}w_\mathrm{m}$ it is seen that the slope of the chemical potential with respect to position can be interpreted as an effective force per mole of molecules. This **thermodynamic force** is written as

$$\mathcal{F} = -\left(\frac{\partial \mu}{\partial x}\right)_{T,p} \qquad \text{Thermodynamic force [definition]} \qquad (16C.2)$$

There is not a real force pushing the molecules down the slope of the chemical potential: the apparent force represents the spontaneous tendency of the molecules to disperse as a consequence of the Second Law and the tendency towards greater entropy.

In a solution in which the activity of the solute is a, the chemical potential is $\mu = \mu^\ominus + RT\ln a$. The thermodynamic force can therefore be written in terms of the gradient of the logarithm of the activity:

$$\mathcal{F} = -\left(\frac{\partial \mu}{\partial x}\right)_{T,p} \overset{(\partial\mu/\partial x)_{T,p} = \partial(\mu^\ominus + RT\ln a)/\partial x}{=} -RT\left(\frac{\partial \ln a}{\partial x}\right)_{T,p} \qquad (16C.3a)$$

If the solution is ideal, a may be replaced by $c/c^\ominus$, where c is the molar concentration and $c^\ominus$ is its standard value ($1\,\mathrm{mol\,dm^{-3}}$):

$$\mathcal{F} = -RT\left(\frac{\partial \ln(c/c^\ominus)}{\partial x}\right)_{T,p} \overset{\mathrm{d}\ln y/\mathrm{d}x = (1/y)(\mathrm{d}y/\mathrm{d}x)}{=} -\frac{RT}{c}\left(\frac{\partial c}{\partial x}\right)_{T,p} \qquad (16C.3b)$$

Example 16C.1 Calculating the thermodynamic force

Suppose that the concentration of a solute varies linearly along x according to $c = c_0 + \alpha x$, where c_0 is the concentration at $x = 0$. Find an expression for the thermodynamic force at position x, and evaluate this force at $x = 0$ and $x = 1.0$ cm for the case where $c_0 = 1.0\,\text{mol dm}^{-3}$ and $\alpha = 10\,\text{mol dm}^{-3}\,\text{m}^{-1}$. Take $T = 298$ K.

Collect your thoughts You will need to use eqn 16C.3b to find the force, so begin by evaluating $\partial c/\partial x$.

The solution The gradient of the molar concentration is

$$\left(\frac{\partial c}{\partial x}\right)_{T,p} = \left(\frac{\partial(c_0+\alpha x)}{\partial x}\right)_{T,p} = \alpha$$

Then, from eqn 16C.3b, the thermodynamic force is

$$\mathcal{F} = -\frac{RT}{c}\left(\frac{\partial c}{\partial x}\right)_{T,p} \overset{(\partial c/\partial x)_{T,p}=\alpha;\ c=c_0+\alpha x}{=} -\frac{\alpha RT}{c_0+\alpha x}$$

At $x = 0$,

$$\mathcal{F} = -\frac{(10\,\text{mol dm}^{-3}\,\text{m}^{-1})\times(8.3145\,\text{J K}^{-1}\,\text{mol}^{-1})\times(298\,\text{K})}{1.0\,\text{mol dm}^{-3}}$$

$$= -2.5\times10^{4}\,\overbrace{\text{J m}^{-1}}^{\text{N}}\,\text{mol}^{-1}$$

or $-25\,\text{kN mol}^{-1}$. A similar calculation at $x = 1.0\,\text{cm} = 1.0 \times 10^{-2}\,\text{m}$ gives the force as $-23\,\text{kN mol}^{-1}$.

Comment. The negative sign indicates that the force is towards the left (towards negative x), because the concentration increases towards the right (as $c_0+\alpha x$) and there is therefore a tendency for the solute to migrate to the left under the influence of that apparent force. The magnitude of the thermodynamic force decreases as x increases because the gradient of $\ln(c/c_0)$, which is $\alpha/(c_0+\alpha x)$, becomes smaller on going to the right (Fig. 16C.1).

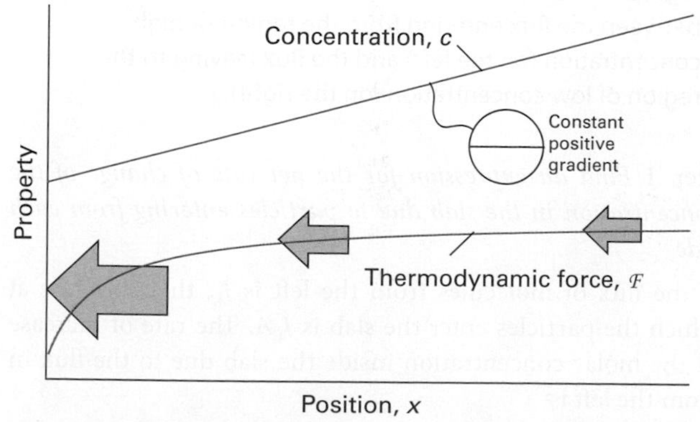

Figure 16C.1 The thermodynamic force is proportional to $-\partial \ln c/\partial x = -(1/c)\partial c/\partial x$. The force thus drives the molecules from a region with higher concentration to one with lower concentration, and becomes smaller in magnitude as the concentration increases.

Self-test 16C.1 Suppose that the concentration of a solute decreases exponentially to the right as $c(x)=c_0\mathrm{e}^{-x/l}$. Derive an expression for the thermodynamic force at any position.

Answer: $\mathcal{F}=RT/l$

The thermodynamic force acts in many respects like a real physical force. In particular it is responsible for accelerating solute molecules until the viscous drag they experience balances the apparent driving force and they settle down to a steady 'drift speed' through the medium. By considering the balance of apparent driving force and the retarding viscous force it is possible to derive Fick's first law of diffusion and relate the diffusion coefficient to the properties of the medium.

How is that done? 16C.1 Deriving Fick's first law of diffusion

The flux due to a concentration gradient is the amount (in moles) of molecules passing through an area A in an interval Δt divided by the area and the interval. This flux can be related to the drift speed, s, by using an approach like that used in Topic 16A for diffusion in a gas.

Step 1 *Find an expression for the flux due to molecules moving at the drift speed*

In a time interval Δt all the particles within a distance $s\Delta t$ can pass through the window, which means that all of the particles in a volume $s\Delta tA$ pass through the window (there is no reverse flux because, unlike in a gas, in this model all the solute molecules are moving down the concentration gradient). Hence, the amount (in moles) of solute molecules that can pass through the window is $s\Delta tAc$. The flux J is this number divided by the area A and by the time interval Δt:

$$J(\text{amount}) = sc$$

Step 2 *Find an expression for the drift speed*

The apparent driving force (per mole) acting on the solute molecules is $\mathcal{F} = -(RT/c)\mathrm{d}c/\mathrm{d}x$, eqn 16C.3b. The molecules also experience a viscous drag which is assumed to be proportional to the speed: expressed as a molar quantity this force is written $N_\mathrm{A}fs$, where f is a constant, the 'frictional constant', depending on the medium. When these two forces are in balance the molecules will be moving at the drift speed. It therefore follows that, with $R/N_\mathrm{A} = k$,

$$N_\mathrm{A}fs = -\frac{RT}{c}\frac{\mathrm{d}c}{\mathrm{d}x} \quad \text{hence} \quad s = -\frac{RT}{N_\mathrm{A}fc}\frac{\mathrm{d}c}{\mathrm{d}x} = -\frac{kT}{fc}\frac{\mathrm{d}c}{\mathrm{d}x}$$

The negative sign arises because the molecules are moving opposite to the direction of increasing concentration.

Step 3 *Combine the two expressions*

Now substitute the expression for the drift speed into that for the flux to give

$$J(\text{amount}) = sc = -\frac{kT}{fc}\frac{\mathrm{d}c}{\mathrm{d}x}c = -\frac{kT}{f}\frac{\mathrm{d}c}{\mathrm{d}x}$$

This expression has the same form as Fick's first law, eqn 16C.1b, that the flux is proportional to the concentration gradient. In addition, the diffusion constant D can be identified as kT/f, which is the **Stokes–Einstein relation**:

$$D = \frac{kT}{f} \qquad \text{Stokes–Einstein relation} \qquad (16C.4a)$$

The constant f that appears in the Stokes–Einstein relation can be inferred from the hydrodynamic result known as **Stokes' law** for the viscous drag (this law is used in Topic 16B). According to this law the magnitude of the viscous force is $6\pi\eta as$ for a spherical particle of radius a. It therefore follows that $f = 6\pi\eta a$, and substituting this into eqn 16C.4a gives the **Stokes–Einstein equation**

$$D = \frac{kT}{6\pi\eta a} \qquad \text{Stokes–Einstein equation} \qquad (16C.4b)$$

This equation is an explicit relation between the diffusion coefficient and the viscosity for a species of hydrodynamic radius a and confirms that D is inversely proportional to η.

The drift speed can be related to the diffusion constant and the thermodynamic force by equating two expressions for the flux, $J = sc$ and $J = -D\,\mathrm{d}c/\mathrm{d}x$, to obtain $s = -(D/c)\,\mathrm{d}c/\mathrm{d}x$. The concentration gradient can be expressed in terms of the thermodynamic force, that is, eqn 16C.3b rearranged in the form $\mathrm{d}c/\mathrm{d}x = -c\mathcal{F}/RT$ to give

$$s = \frac{D\mathcal{F}}{RT} \qquad (16C.5)$$

This relation provides a way to estimate the thermodynamic force from measurements of the drift speed and the diffusion coefficient.

Brief illustration 16C.1

Laser measurements show that a particular molecule has a drift speed of 1.0 μm s^{-1} in water at 25 °C, at which temperature the diffusion coefficient is 5.0×10^{-9} m^2 s^{-1}. The corresponding thermodynamic force calculated using eqn 16C.5, rearranged into the form $\mathcal{F} = sRT/D$, is

$$\mathcal{F} = \frac{(1.0\times10^{-6}\,\mathrm{m\,s^{-1}})\times(8.3145\,\mathrm{J\,K^{-1}\,mol^{-1}})\times(298\,\mathrm{K})}{5.0\times10^{-9}\,\mathrm{m^2\,s^{-1}}}$$

$$= 5.0\times10^{5}\,\overbrace{\mathrm{J\,m^{-1}}}^{\mathrm{N}}\,\mathrm{mol^{-1}}$$

or about 500 kN mol^{-1}. This thermodynamic force is many times that of gravity, which explains why solutions do not sediment.

16C.2 The diffusion equation

Diffusion results in the modification of the distribution of concentration of the solute (or of a physical property) as inhomogeneities disappear. The discussion is expressed in terms of the diffusion of molecules, but similar arguments apply to the diffusion of other entities, such as ions, and of various physical properties, such as temperature.

(a) Simple diffusion

The **diffusion equation**, one of the most important equations for discussing the properties of fluids, is an equation that expresses the rate of change of concentration of a species in terms of the inhomogeneity of its concentration. It is also called 'Fick's second law of diffusion', but that name is now rarely used. The diffusion equation can be derived on the basis of Fick's first law.

How is that done? 16C.2 Deriving the diffusion equation

The diffusion equation is developed by considering the net flux of particles entering a thin slab of cross-sectional area A that extends from x to $x + l$ (Fig. 16C.2) and therefore has volume Al.

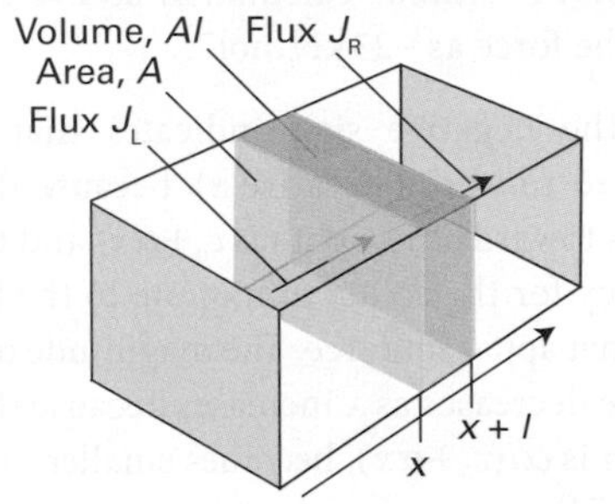

Figure 16C.2 The net flux in a thin slab is the difference between the flux entering from the region of high concentration (on the left) and the flux leaving to the region of low concentration (on the right).

Step 1 *Find an expression for the net rate of change of the concentration in the slab due to particles entering from each side*

If the flux of molecules from the left is J_L, then the rate at which the particles enter the slab is $J_L A$. The rate of increase of the molar concentration inside the slab due to the flux in from the left is

$$\left(\frac{\partial c}{\partial t}\right)_L = \frac{J_L A}{Al} = \frac{J_L}{l}$$

Molecules also flow out of the right face of the slab. If this flux is J_R, then by a similar argument

$$\left(\frac{\partial c}{\partial t}\right)_{\mathrm{R}} = -\frac{J_{\mathrm{R}}}{l}$$

Note the minus sign: this flux reduces the concentration. The net rate of change of concentration in the slab is

$$\frac{\partial c}{\partial t} = \left(\frac{\partial c}{\partial t}\right)_{\mathrm{L}} + \left(\frac{\partial c}{\partial t}\right)_{\mathrm{R}} = \frac{J_{\mathrm{L}} - J_{\mathrm{R}}}{l}$$

Step 2 *Relate the fluxes to the concentration gradients*

From Fick's first law (eqn 16C.1b), each flux can be expressed in terms of the diffusion coefficient and the concentration gradient at each face:

$J = -D(\partial c/\partial x)$

$$J_{\mathrm{L}} - J_{\mathrm{R}} = -D\left(\frac{\partial c}{\partial x}\right)_{\mathrm{L}} + D\left(\frac{\partial c}{\partial x}\right)_{\mathrm{R}} = D\left\{\left(\frac{\partial c}{\partial x}\right)_{\mathrm{R}} - \left(\frac{\partial c}{\partial x}\right)_{\mathrm{L}}\right\}$$

where $(\partial c/\partial x)_{\mathrm{L}}$ is the concentration gradient at the left face of the slab, and similarly $(\partial c/\partial x)_{\mathrm{R}}$ is that at the right face. The concentration gradients at the two faces can be expressed in terms of the gradient (the first derivative of the concentration) at the centre of the slab, $(\partial c/\partial x)_0$, and the first derivative of that gradient (which is the second derivative of the concentration), $\partial^2 c/\partial x^2$. The distances of the faces from the centre are $\frac{1}{2}l$ in each direction, so it follows that

$$\left(\frac{\partial c}{\partial x}\right)_{\mathrm{R}} - \left(\frac{\partial c}{\partial x}\right)_{\mathrm{L}} = \left\{\left(\frac{\partial c}{\partial x}\right)_0 + \frac{l}{2}\frac{\partial^2 c}{\partial x^2}\right\} - \left\{\left(\frac{\partial c}{\partial x}\right)_0 - \frac{l}{2}\frac{\partial^2 c}{\partial x^2}\right\} = l\frac{\partial^2 c}{\partial x^2}$$

This expression is then substituted into that for the difference of the fluxes to give

$(\partial c/\partial x)_{\mathrm{R}} - (\partial c/\partial x)_{\mathrm{L}} = l(\partial^2 c/\partial x^2)$

$$J_{\mathrm{L}} - J_{\mathrm{R}} = Dl\frac{\partial^2 c}{\partial x^2}$$

Step 3 *Combine the expression for the net flux with that for the time dependence of concentration*

Substitute the last expression into the equation for the rate of change of concentration, cancel the l, and obtain the **diffusion equation**:

$$\frac{\partial c}{\partial t} = D\frac{\partial^2 c}{\partial x^2} \qquad (16\mathrm{C}.6)$$

Diffusion equation

The diffusion equation shows that the rate of change of concentration in a region is proportional to the curvature (more precisely, to the second derivative) of the concentration with respect to distance in that region. If the concentration changes sharply from point to point (if the distribution is highly wrinkled), then the concentration changes rapidly with time. Specifically:

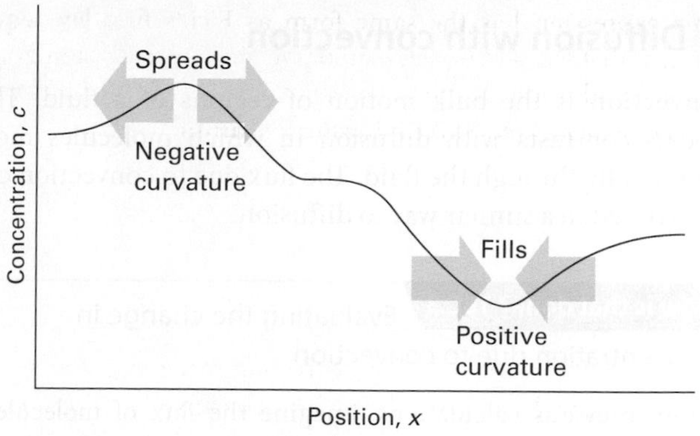

Figure 16C.3 The diffusion equation implies that, over time, peaks in a distribution (regions of negative curvature) spread and troughs (regions of positive curvature) fill in.

- Where the curvature is positive (a dip, Fig. 16C.3), the change in concentration with time is positive; the dip tends to fill.
- Where the curvature is negative (a heap), the change in concentration with time is negative; the heap tends to spread.
- If the curvature is zero, then the concentration is constant in time.

Physical interpretation

The diffusion equation can be regarded as a mathematical formulation of the intuitive notion that there is a natural tendency for the wrinkles in a distribution to disappear.

Brief illustration 16C.2

If a concentration across a small region of space varies linearly as $c = c_0 - \alpha x$ then it follows that $\partial^2 c/\partial x^2 = 0$ and so from eqn 16C.6 $\partial c/\partial t = 0$. The concentration does not vary with time because the flow into one face of a slab is exactly matched by the flow out from the opposite face (Fig. 16C.4a). If the concentration varies as $c = c_0 - \frac{1}{2}\beta x^2$ then $\partial^2 c/\partial x^2 = -\beta$ and consequently $\partial c/\partial t = -D\beta$. The concentration decreases with time because the flow out of the slab is greater than the flow into it (Fig. 16C.4b).

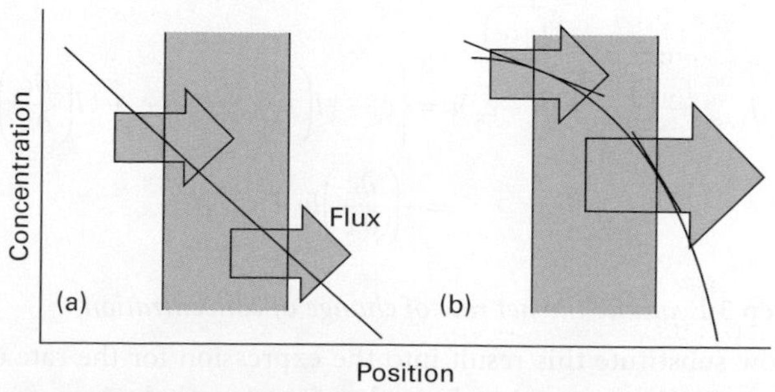

Figure 16C.4 The two instances treated in *Brief illustration* 16C.2: (a) linear concentration gradient, (b) parabolic concentration gradient.

(b) Diffusion with convection

Convection is the bulk motion of regions of a fluid. This process contrasts with diffusion in which molecules move individually through the fluid. The flux due to convection can be analysed in a similar way to diffusion.

How is that done? 16C.3 Evaluating the change in concentration due to convection

As in previous calculations, imagine the flux of molecules through an area A in an interval Δt but now due to convective flow in which the fluid moves at a speed v.

Step 1 *Express the rate of change of concentration in terms of the net flux*

As in the derivation of the diffusion equation,

$$\frac{\partial c}{\partial t}=\frac{J_{\mathrm{L,conv}}-J_{\mathrm{R,conv}}}{l}$$

where $J_{\mathrm{L,conv}}$ and $J_{\mathrm{R,conv}}$ are, respectively, the fluxes from the left into and on the right out of the slab, but here due to convection.

Step 2 *Evaluate the net flux*

In an interval Δt all the particles within a distance $v\Delta t$ and therefore in the volume $Av\Delta t$ pass through a face of the slab. If the molar concentration at the relevant face is c, then the amount passing through that face is $cAv\Delta t$. The 'convective flux' is this amount divided by the area of the face and the time interval:

$$J_{\mathrm{conv}}=\frac{cAv\Delta t}{A\Delta t}=cv \qquad \text{Convective flux} \quad (16\text{C}.7)$$

The concentrations on the left (c_{L}) and right (c_{R}) faces of the slab are related to the concentration at its centre, c_0, by

$$c_{\mathrm{R}}=c_0+\tfrac{1}{2}l\left(\frac{\partial c}{\partial x}\right) \qquad c_{\mathrm{L}}=c_0-\tfrac{1}{2}l\left(\frac{\partial c}{\partial x}\right)$$

where the first derivatives are evaluated at the centre of the slab. It follows from eqn 16C.7 that

$$J_{\mathrm{L,conv}}-J_{\mathrm{R,conv}}\overset{J=cv}{=}(c_{\mathrm{L}}-c_{\mathrm{R}})v=\left\{c_0-\tfrac{1}{2}l\left(\frac{\partial c}{\partial x}\right)\right\}v-\left\{c_0+\tfrac{1}{2}l\left(\frac{\partial c}{\partial x}\right)\right\}v$$
$$=-\left(\frac{\partial c}{\partial x}\right)lv$$

Step 3 *Evaluate the net rate of change of concentration*

Now substitute this result into the expression for the rate of change of concentration derived in Step 1, and obtain

$$\frac{\partial c}{\partial t}=-\left(\frac{\partial c}{\partial x}\right)v \qquad \text{Convection} \quad (16\text{C}.8)$$

Brief illustration 16C.3

If it is assumed that the concentration across a small region of space varies linearly as $c = c_0 - \alpha x$, then $\partial c/\partial x = -\alpha$. If there is convective flow at velocity v, it follows from eqn 16C.8 that $\partial c/\partial t = \alpha v$. The concentration in the slab increases because the convective flow into the left face outweighs the flow out from the right face; with this linear concentration dependence, there is no diffusion. If $\alpha = 0.010\,\mathrm{mol\,dm^{-3}\,m^{-1}}$ and $v = +1.0\,\mathrm{mm\,s^{-1}}$,

$$\frac{\partial c}{\partial t}=(0.010\,\mathrm{mol\,dm^{-3}\,m^{-1}})\times(1.0\times10^{-3}\,\mathrm{m\,s^{-1}})$$
$$=1.0\times10^{-5}\,\mathrm{mol\,dm^{-3}\,s^{-1}}$$

The concentration increases at the rate of $10\,\mu\mathrm{mol\,dm^{-3}\,s^{-1}}$.

When diffusion and convection occur together, the total rate of change of concentration in a region is the sum of the two effects, which is described by the **generalized diffusion equation:**

$$\frac{\partial c}{\partial t}=D\frac{\partial^2 c}{\partial x^2}-v\frac{\partial c}{\partial x} \qquad \text{Generalized diffusion equation} \quad (16\text{C}.9)$$

A further refinement, which is important in chemistry, is the possibility that the concentrations of molecules may change as a result of reaction. When reactions are included in eqn 16C.9 (in Topic 18B) a differential equation is obtained that can be used to discuss the properties of reacting, diffusing, convecting systems. This equation is the basis for modelling reactors in the chemical industry and the utilization of resources in living cells.

(c) Solutions of the diffusion equation

The diffusion equation (eqn 16C.6) is a second-order differential equation with respect to space and a first-order differential equation with respect to time. To find solutions it is necessary to know two boundary conditions for the spatial dependence and a single initial condition for the time dependence.

As an illustration, consider an arrangement in which there is a layer of a solute (such as sugar) at the bottom of a tall beaker of water (which for simplicity may be taken to be infinitely tall), with base area A; x is the distance measured up from the base. At $t = 0$ it is assumed that all N_0 particles are concentrated on the yz-plane at $x = 0$: this is the initial condition. The two boundary conditions are derived from the requirements that the concentration must everywhere be finite and the total amount of particles present is n_0 (with $n_0 = N_0/N_{\mathrm{A}}$) at all times. With these conditions,

$$c(x,t)=\frac{n_0}{A(\pi Dt)^{1/2}}\mathrm{e}^{-x^2/4Dt} \qquad \text{One-dimensional diffusion} \quad (16\text{C}.10)$$

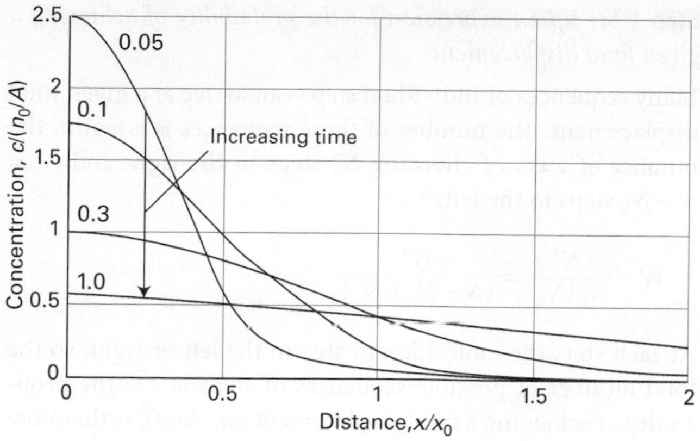

Figure 16C.5 The concentration profiles above a plane from which a solute is diffusing into pure solvent. The curves are labelled with the corresponding value of Dt, and $x_0 = \{4Dt\}^{1/2}$.

as may be verified by direct substitution. Figure 16C.5 shows the shape of the concentration distribution at various times and illustrates how the concentration spreads.

Another useful result is for the diffusion in three dimensions arising from an initially localized concentration of solute (a sugar lump suspended in an infinitely large flask of water). The concentration of diffused solute is spherically symmetrical, and at a radius r is

$$c(r,t) = \frac{n_0}{8(\pi Dt)^{3/2}} e^{-r^2/4Dt} \quad \text{Three-dimensional diffusion} \quad (16C.11)$$

Other chemically (and physically) interesting arrangements, such as transport of substances across biological membranes can be treated, but the mathematical forms of the solutions are more cumbersome.

The diffusion equation is useful for the experimental determination of diffusion coefficients. In the **capillary technique**, a capillary tube, open at one end and containing a solution, is immersed in a well-stirred larger quantity of solvent, and the change of concentration in the tube is monitored. The solute diffuses from the open end of the capillary at a rate that can be calculated by solving the diffusion equation with the appropriate boundary and initial conditions, so D may be determined. In the **diaphragm technique**, the diffusion occurs through the capillary pores of a sintered glass diaphragm separating the well-stirred solution and solvent. The concentrations are monitored and then related to the diffusion equation that has been solved for this arrangement. Diffusion coefficients may also be measured by a number of other techniques, including NMR spectroscopy (Topic 12C).

The diffusion equation can be used to predict the concentration of particles (or the value of some other physical quantity, such as the temperature in a non-uniform system) at any location. It can also be used to calculate the average displacement of the particles in a given time.

How is that done? 16C.4 Evaluating the average displacement in a one-dimensional system

To calculate the average value of the displacement x, denoted $\langle x\rangle$, you need to use eqn 16C.10 to find an expression for the probability density of the particles $P(x)$, defined such that $P(x)\mathrm{d}x$ is the probability of finding a particle between x and $x + \mathrm{d}x$. Then

$$\langle x\rangle = \int_0^\infty xP(x)\mathrm{d}x$$

Step 1 *Set up an expression for the probability density*

The number of particles in a slab of thickness $\mathrm{d}x$ at x and at time t is the volume of the slab, $A\mathrm{d}x$, multiplied by the molar concentration at that location (and time) and Avogadro's constant: $N(x,t) = c(x,t)N_A A\mathrm{d}x$. The molar concentration at the position and time of interest is given by eqn 16C.10. The total number of particles is $N_A n_0$, where n_0 is the total amount, so the probability of finding a molecule in the slab is

$$c(x,t) = \frac{n_0}{A(\pi Dt)^{1/2}} e^{-x^2/4Dt}$$

$$P(x)\mathrm{d}x = \frac{N_A c(x,t)A\mathrm{d}x}{N_A n_0} = \frac{A}{n_0}\frac{n_0}{A(\pi Dt)^{1/2}} e^{-x^2/4Dt}\mathrm{d}x = \frac{1}{(\pi Dt)^{1/2}} e^{-x^2/4Dt}\mathrm{d}x$$

Step 2 *Evaluate the integral*

The integration required is

$$\int_0^\infty x\frac{1}{(\pi Dt)^{1/2}} e^{-x^2/4Dt}\mathrm{d}x = \frac{1}{(\pi Dt)^{1/2}} \overbrace{\int_0^\infty xe^{-x^2/4Dt}\mathrm{d}x}^{\text{Integral G.2}} = 2\left(\frac{Dt}{\pi}\right)^{1/2}$$

That is, the average displacement of a diffusing particle in a time t in a one-dimensional system is

$$\langle x\rangle = 2\left(\frac{Dt}{\pi}\right)^{1/2} \quad \text{Mean displacement [one dimension]} \quad (16C.12)$$

A similar calculation shows that the root-mean-square displacement in the same time is

$$\langle x^2\rangle^{1/2} = (2Dt)^{1/2} \quad \text{Root-mean-square displacement [one dimension]} \quad (16C.13a)$$

This result is a useful measure of the spread of particles when they can diffuse in both directions from the origin, because in that case $\langle x\rangle = 0$ at all times. The time-dependence of the root-mean-square displacement for particles with a typical diffusion coefficient in a liquid ($D = 5 \times 10^{-10}\,\mathrm{m^2\,s^{-1}}$) is illustrated in Fig. 16C.6. The graph shows that diffusion is a very slow process (which is why solutions are stirred, to encourage mixing by convection).

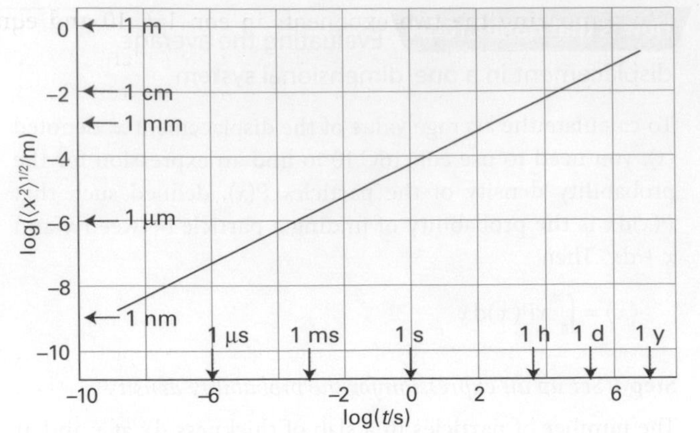

Figure 16C.6 The root-mean-square distance covered by particles with $D = 5\times10^{-10}\,\mathrm{m^2\,s^{-1}}$. Note the great slowness of diffusion.

In three dimensions the root-mean-square displacement is given by a similar expression:

$$\langle r^2\rangle^{1/2} = (6Dt)^{1/2} \qquad \text{Root-mean-square displacement [three dimensions]} \qquad (16C.13b)$$

16C.3 The statistical view

An intuitive picture of diffusion is of the particles moving in a series of small steps and gradually migrating from their original positions. This picture suggests a model in which each particle jumps through a distance d after a time τ. The total distance travelled by a particle in time t is therefore td/τ. However, it is most unlikely that a particle will end up at this distance from the origin because each jump might be in a different direction.

The discussion is simplified by allowing the particles to travel only along a straight line (the x-axis) with each step a jump through distance d to the left or to the right. This model is called the **one-dimensional random walk**. The probability that the walk will end up at a specific distance from the origin can be calculated by considering the statistics of the process.[1]

How is that done? 16C.5 Evaluating the probability distribution for a one-dimensional random walk

Imagine that a molecule has made N steps in total, N_R of which are to the right and N_L to the left. The displacement from the origin is therefore $(N_R - N_L)d$, which is written nd with $n = N_R - N_L$.

Step 1 *Set up an expression for the probability of achieving a given final displacement*

Many sequences of individual steps can arrive at a given final displacement. The number of these sequences is equal to the number of ways of choosing N_R steps to the right and $N_L = N - N_R$ steps to the left:

$$W = \frac{N!}{N_L!N_R!} = \frac{N!}{(N-N_R)!N_R!}$$

At each step, the molecule can step to the left or right, so the total number of possible sequences of steps is 2^N. The probability of achieving a final displacement nd, $P(nd)$, is therefore

$$P(nd) = \frac{W}{2^N} = \frac{N!}{(N-N_R)!N_R!2^N}$$

Step 2 *Simplify this expression by using Stirling's approximation*

This expression can be simplified by taking its logarithm to give

$$\ln P = \ln N! - \{\ln(N-N_R)! + \ln N_R! + \ln 2^N\}$$

and then using Stirling's approximation in the form

$$\ln x! \approx \ln(2\pi)^{1/2} + (x+\tfrac{1}{2})\ln x - x$$

to obtain

$$\ln P = -\ln(2\pi)^{1/2}2^N + (N+\tfrac{1}{2})\ln\frac{1}{1-N_R/N} + N_R\ln\frac{1-N_R/N}{N_R/N} - \tfrac{1}{2}\ln N_R$$

For reasons that will become clear shortly, it is convenient to introduce the new variable μ:

$$\mu = \frac{N_R}{N} - \frac{1}{2}$$

from which definition it follows that $1-N_R/N = \tfrac{1}{2}-\mu$ and $N_R/N = \tfrac{1}{2}+\mu$. With these substitutions the expression for $\ln P$ can be written in terms of N and μ alone:

$$\ln P = -\ln(2\pi)^{1/2}2^N - (N+\tfrac{1}{2})\ln(\tfrac{1}{2}-\mu) + N(\mu+\tfrac{1}{2})\ln(\tfrac{1}{2}-\mu) - N(\mu+\tfrac{1}{2})\ln(\mu+\tfrac{1}{2}) - \tfrac{1}{2}\ln N(\mu+\tfrac{1}{2})$$

Step 3 *Expand the logarithms*

Because the probability of taking a step to the right or to the left is the same, it is expected that after many steps the total number to the right will be very close to half the number of steps. That is $N_R/N \approx \tfrac{1}{2}$. It follows that $\mu << 1$, so you can use the series expansion

$$\ln(\tfrac{1}{2}\pm\mu) = -\ln 2 \pm 2\mu - 2\mu^2 + \cdots$$

and retain terms up to that in μ^2. After rather a lot of algebra (see *A deeper look* 11 on the website, where the details are given) you will obtain

[1] The calculation is essentially the same as in the discussion of the random coil structures of denatured polymers (Topic 14D).

$$\ln P = -\ln(2\pi N)^{1/2} 2^N + \ln 2^{N+1} - 2(N-1)\mu^2$$

At this point take antilogarithms of both sides and use $N >> 1$:

$$P = \frac{2^{N+1}\mathrm{e}^{-2(N-1)\mu^2}}{2^N(2\pi N)^{1/2}} = \frac{2\mathrm{e}^{-2(N-1)\mu^2}}{(2\pi N)^{1/2}} \approx \frac{2\mathrm{e}^{-2N\mu^2}}{(2\pi N)^{1/2}}$$

Step 4 *Recast the expression for the probability in terms of the time for each step*

The exponent $N\mu^2$ can be rewritten

$\mu = N_R/N - 1/2$ $\qquad$ $n = N_R - N_L$

$$N\mu^2 = \frac{(2N_R - N)^2}{4N} = \frac{(N_R - N_L)^2}{4N} = \frac{n^2}{4N}$$

The final distance from the origin, x, is equal to nd, and the number of steps taken in a time t is $N = t/\tau$. It follows that $N\mu^2 = n^2/4N = \tau x^2/4td^2$. Substitution of these expressions for N and $N\mu^2$ into the expression for P gives

$$P(x,t) = \left(\frac{2\tau}{\pi t}\right)^{1/2} \mathrm{e}^{-x^2\tau/2td^2} \qquad (16\text{C}.14)$$

One-dimensional random walk

The differences of detail between eqns 16C.10 (for one-dimensional diffusion) and 16C.14 arises from the fact that in the present calculation the particles can migrate in either direction from the origin. Moreover, they can be found only at discrete points separated by d instead of being anywhere on a continuous line. The fact that the two expressions are so similar suggests that diffusion can indeed be interpreted as the outcome of a large number of steps in random directions.

By comparing the two exponents in eqn 16C.10 and eqn 16C.14 it is possible to relate the diffusion coefficient D to the step length d and the time between jumps, τ. The result is the **Einstein–Smoluchowski equation**:

$$D = \frac{d^2}{2\tau} \qquad \text{Einstein–Smoluchowski equation} \qquad (16\text{C}.15)$$

Brief illustration 16C.4

Suppose that in aqueous solution an SO_4^{2-} ion jumps through its own diameter of 500 pm each time it makes a move, then because $D = 1.1 \times 10^{-9}\,\mathrm{m^2\,s^{-1}}$ (as deduced from mobility measurements, Topic 16B), it follows from eqn 16C.15 that

$$\tau = \frac{d^2}{2D} = \frac{(500\times10^{-12}\,\mathrm{m})^2}{2\times(1.1\times10^{-9}\,\mathrm{m^2\,s^{-1}})} = 1.1\times10^{-10}\,\mathrm{s}$$

or $\tau = 110$ ps. Because τ is the time for one jump, the ion makes about 1×10^{10} jumps per second.

The Einstein–Smoluchowski equation makes the connection between the microscopic details of particle motion and the macroscopic parameters relating to diffusion. It also brings the discussion back full circle to the properties of the perfect gas treated in Topic 16A. If d/τ as is interpreted as v_{mean}, the mean speed of the molecules, and d is interpreted as a mean free path λ, then the Einstein–Smoluchowski equation becomes $D = \frac{1}{2}d(d/\tau) = \frac{1}{2}\lambda v_{\text{mean}}$, which is essentially the same expression as obtained from the kinetic model of gases (eqn 16A.9 of Topic 16A, $D = \frac{1}{3}\lambda v_{\text{mean}}$). That is, the diffusion of a perfect gas is a random walk with an average step size equal to the mean free path.

Checklist of concepts

☐ 1. A **thermodynamic force** is an apparent force that mirrors the spontaneous tendency of the molecules to disperse as a consequence of the Second Law and the tendency towards greater entropy.

☐ 2. The **drift speed** is achieved when the viscous retarding force matches the thermodynamic force.

☐ 3. The **diffusion equation** (Fick's second law) can be regarded as a mathematical formulation of the notion that there is a natural tendency for concentration to become uniform.

☐ 4. **Convection** is the bulk motion of regions of a fluid.

☐ 5. A model of diffusion is of the particles moving in a series of small steps, a **random walk**, and gradually migrating from their original positions.

Checklist of equations

Property	Equation	Comment	Equation number
Fick's first law	$J(\text{amount}) = -D\mathrm{d}c/\mathrm{d}x$		16C.1b
Thermodynamic force	$\mathcal{F} = -(\partial\mu/\partial x)_{T,p}$	Definition	16C.2
Stokes–Einstein relation	$D = kT/f$	fs is the frictional drag	16C.4a
Drift speed	$s = D\mathcal{F}/RT$		16C.5
Diffusion equation	$\partial c/\partial t = D\partial^2 c/\partial x^2$	One dimension	16C.6
Generalized diffusion equation	$\partial c/\partial t = D\partial^2 c/\partial x^2 - v\partial c/\partial x$	One dimension	16C.9
Mean displacement	$\langle x\rangle = 2(Dt/\pi)^{1/2}$	One-dimensional diffusion	16C.12
Root-mean-square displacement	$\langle x^2\rangle^{1/2} = (2Dt)^{1/2}$	One-dimensional diffusion	16C.13a
	$\langle r^2\rangle^{1/2} = (6Dt)^{1/2}$	Three-dimensional diffusion	16C.13b
Probability of displacement	$P(x,t) = (2\tau/\pi t)^{1/2}\mathrm{e}^{-x^2\tau/2td^2}$	One-dimensional random walk	16C.14
Einstein–Smoluchowski equation	$D = d^2/2\tau$	One-dimensional random walk	16C.15

FOCUS 16 Molecules in motion

TOPIC 16A Transport properties of a perfect gas

Discussion questions

D16A.1 Explain how Fick's first law arises from considerations of the flux of molecules due to a concentration gradient in a perfect gas.

D16A.2 Provide molecular interpretations for the dependencies of the diffusion constant and the viscosity on the temperature, pressure, and size of gas molecules.

Exercises

E16A.1(a) A solid surface with dimensions 2.5 mm × 3.0 mm is exposed to argon gas at 90 Pa and 500 K. How many collisions do the Ar atoms make with this surface in 15 s?

E16A.1(b) A solid surface with dimensions 3.5 cm × 4.0 cm is exposed to helium gas at 111 Pa and 1500 K. How many collisions do the He atoms make with this surface in 10 s?

E16A.2(a) Calculate the diffusion constant of argon at 20 °C and (i) 1.00 Pa, (ii) 100 kPa, (iii) 10.0 MPa; take $\sigma = 0.36\,\text{nm}^2$. If a pressure gradient of $1.0\,\text{bar m}^{-1}$ is established in a pipe, what is the flow of gas due to diffusion at each pressure?

E16A.2(b) Calculate the diffusion constant of nitrogen at 20 °C and (i) 100.0 Pa, (ii) 100 kPa, (iii) 20.0 MPa; take $\sigma = 0.43\,\text{nm}^2$. If a pressure gradient of $1.20\,\text{bar m}^{-1}$ is established in a pipe, what is the flow of gas due to diffusion at each pressure?

E16A.3(a) Calculate the thermal conductivity of argon ($C_{V,\text{m}} = 12.5\,\text{J K}^{-1}\,\text{mol}^{-1}$, $\sigma = 0.36\,\text{nm}^2$) at 298 K.

E16A.3(b) Calculate the thermal conductivity of nitrogen ($C_{V,\text{m}} = 20.8\,\text{J K}^{-1}\,\text{mol}^{-1}$, $\sigma = 0.43\,\text{nm}^2$) at 298 K.

E16A.4(a) Use the experimental value of the thermal conductivity of neon (Table 16A.1) to estimate the collision cross-section of Ne atoms at 273 K.

E16A.4(b) Use the experimental value of the thermal conductivity of nitrogen (Table 16A.1) to estimate the collision cross-section of N_2 molecules at 298 K.

E16A.5(a) Calculate the flux of energy arising from a temperature gradient of $10.5\,\text{K m}^{-1}$ in a sample of argon in which the mean temperature is 280 K. The necessary data needed to calculate the thermal conductivity are in Exercise E16A.3(a).

E16A.5(b) Calculate the flux of energy arising from a temperature gradient of $8.5\,\text{K m}^{-1}$ in a sample of N_2 in which the mean temperature is 290 K. The necessary data needed to calculate the thermal conductivity are in Exercise E16A.3(b).

E16A.6(a) In a double-glazed window, the panes of glass are separated by 1.0 cm and the space is filled with a gas with thermal conductivity $24\,\text{mW K}^{-1}\,\text{m}^{-1}$. What is the rate of transfer of heat by conduction from the warm room (28 °C) to the cold exterior (−15 °C) through a window of area $1.0\,\text{m}^2$? You may assume that one pane of glass is at the same temperature as the inside and the other as the outside. What power of heater is required to make good the loss of heat?

E16A.6(b) Two sheets of copper of area $2.00\,\text{m}^2$ are separated by 5.00 cm in N_2(g). What is the rate of transfer of heat by conduction from the warm sheet (70 °C) to the cold sheet (0 °C)? Refer to the *Resource section* for any necessary data.

E16A.7(a) Calculate the viscosity of air at (i) 273 K, (ii) 298 K, (iii) 1000 K. Take σ as $0.40\,\text{nm}^2$ and M as $29.0\,\text{g mol}^{-1}$.

E16A.7(b) Calculate the viscosity of benzene vapour at (i) 273 K, (ii) 298 K, (iii) 1000 K. Take σ as $0.88\,\text{nm}^2$.

E16A.8(a) Use the experimental value of the coefficient of viscosity for neon (Table 16A.1) to estimate the collision cross-section of Ne atoms at 273 K.

E16A.8(b) Use the experimental value of the coefficient of viscosity for nitrogen (Table 16A.1) to estimate the collision cross-section of the molecules at 273 K.

E16A.9(a) An effusion cell has a circular hole of diameter 2.50 mm. If the molar mass of the solid in the cell is $260\,\text{g mol}^{-1}$ and its vapour pressure is 0.835 Pa at 400 K, by how much will the mass of the solid decrease in a period of 2.00 h?

E16A.9(b) An effusion cell has a circular hole of diameter 3.00 mm. If the molar mass of the solid in the cell is $300\,\text{g mol}^{-1}$ and its vapour pressure is 0.224 Pa at 450 K, by how much will the mass of the solid decrease in a period of 24.00 h?

E16A.10(a) A solid compound of molar mass $100\,\text{g mol}^{-1}$ was introduced into a container and heated to 400 °C. When a hole of diameter 0.50 mm was opened in the container for 400 s, a mass loss of 285 mg was measured. Calculate the vapour pressure of the compound at 400 °C.

E16A.10(b) A solid compound of molar mass $200\,\text{g mol}^{-1}$ was introduced into a container and heated to 300 °C. When a hole of diameter 0.50 mm was opened in the container for 500 s, a mass loss of 277 mg was measured. Calculate the vapour pressure of the compound at 300 °C.

E16A.11(a) A manometer was connected to a bulb containing an unknown gas under slight pressure. The gas was allowed to escape through a small pinhole, and the time for the manometer reading to drop from 75 cm to 50 cm was 52 s. When the experiment was repeated using nitrogen (for which $M = 28.02\,\text{g mol}^{-1}$) the same fall took place in 42 s. Calculate the molar mass of the unknown gas. *Hint:* The pressure changes and, as a consequence, so does the rate of effusion; note, however, that the change is the same in both cases.

E16A.11(b) A manometer was connected to a bulb containing nitrogen under slight pressure. The gas was allowed to escape through a small pinhole, and the time for the manometer reading to drop from 65.1 cm to 42.1 cm was 18.5 s. When the experiment was repeated using a fluorocarbon gas, the same fall took place in 82.3 s. Calculate the molar mass of the fluorocarbon.

E16A.12(a) A space vehicle of internal volume $3.0\,\text{m}^3$ is struck by a meteor and a hole of radius 0.10 mm is formed. If the oxygen pressure within the vehicle is initially 80 kPa and its temperature 298 K, how long will the pressure take to fall to 70 kPa assuming that the temperature is held constant?

E16A.12(b) A container of internal volume $22.0\,\text{m}^3$ was punctured, and a hole of radius 0.050 mm was formed. If the nitrogen pressure within the container is initially 122 kPa and its temperature 293 K, how long will the pressure take to fall to 105 kPa assuming that the temperature is held constant?

Problems

P16A.1‡ A. Fenghour et al. (*J. Phys. Chem. Ref. Data* **24**, 1649 (1995)) compiled an extensive table of viscosity coefficients for ammonia in the liquid and vapour phases. Deduce the effective molecular diameter of NH_3 based on each of the following vapour-phase viscosity coefficients: (a) $\eta = 9.08 \times 10^{-6}\,kg\,m^{-1}\,s^{-1}$ at 270 K and 1.00 bar; (b) $\eta = 1.749 \times 10^{-5}\,kg\,m^{-1}\,s^{-1}$ at 490 K and 10.0 bar.

P16A.2 Calculate the ratio of the thermal conductivities of gaseous hydrogen at 300 K to gaseous hydrogen at 10 K. *Hint:* Think about the modes of motion that are thermally active at the two temperatures.

P16A.3 Interstellar space is quite different from the gaseous environments we commonly encounter on Earth. For instance, a typical density of the medium is about 1 atom cm^{-3} and that atom is typically H; the effective temperature due to stellar background radiation is about 10 kK. Estimate the diffusion coefficient and thermal conductivity of H under these conditions. Compare your answers with the values for gases under typical terrestrial conditions. *Comment:* Energy is in fact transferred much more effectively by radiation.

P16A.4 A Knudsen cell was used to determine the vapour pressure of germanium at 1000 °C. During an interval of 7200 s the mass loss through a hole of radius 0.50 mm amounted to 43 μg. What is the vapour pressure of germanium at 1000 °C? Assume the gas to be monatomic.

P16A.5 An atomic beam is designed to function with (a) cadmium, (b) mercury. The source is an oven maintained at 380 K, and the vapour escapes though a small slit of dimensions 10 mm by 1.0×10^{-2} mm. The vapour pressure of cadmium is 0.13 Pa and that of mercury is 12 Pa at this temperature. What is the number of atoms per second in the beams?

P16A.6 Derive an expression that shows how the pressure of a gas inside an effusion oven (a heated chamber with a small hole in one wall) varies with time if the oven is not replenished as the gas escapes. Then show that $t_{1/2}$, the time required for the pressure to decrease to half its initial value, is independent of the initial pressure. *Hint:* Start from the expression for the rate of effusion and rewrite it as a differential equation relating dp/dt to p; recall that $pV = NkT$ can be used to relate the pressure to the number density.

TOPIC 16B Motion in liquids

Discussion questions

D16B.1 Discuss the difference between the hydrodynamic radius of an ion and its ionic radius. Explain how it is possible for the hydrodynamic radius of an ion to decrease down a group even though the ionic radius increases.

D16B.2 Discuss the mechanism of proton conduction in water. Might the same mechanism also occur in ice?

Exercises

E16B.1(a) The viscosity of water at 20 °C is 1.002 cP and 0.7975 cP at 30 °C. What is the energy of activation associated with viscosity?
E16B.1(b) The viscosity of mercury at 20 °C is 1.554 cP and 1.450 cP at 40 °C. What is the energy of activation associated with viscosity?

E16B.2(a) The limiting molar conductivities of NaI, $NaNO_3$, and $AgNO_3$ are $12.69\,mS\,m^2\,mol^{-1}$, $12.16\,mS\,m^2\,mol^{-1}$, and $13.34\,mS\,m^2\,mol^{-1}$, respectively (all at 25 °C). What is the limiting molar conductivity of AgI at this temperature? *Hint:* Each limiting molar conductivity can be expressed as a sum of two ionic conductivities.
E16B.2(b) The limiting molar conductivities of KF, KCH_3CO_2, and $Mg(CH_3CO_2)_2$ are $12.89\,mS\,m^2\,mol^{-1}$, $11.44\,mS\,m^2\,mol^{-1}$, and $18.78\,mS\,m^2\,mol^{-1}$, respectively (all at 25 °C). What is the limiting molar conductivity of MgF_2 at this temperature?

E16B.3(a) At 25 °C the molar ionic conductivities of Li^+, Na^+, and K^+ are $3.87\,mS\,m^2\,mol^{-1}$, $5.01\,mS\,m^2\,mol^{-1}$, and $7.35\,mS\,m^2\,mol^{-1}$, respectively. What are their mobilities?
E16B.3(b) At 25 °C the molar ionic conductivities of F^-, Cl^-, and Br^- are $5.54\,mS\,m^2\,mol^{-1}$, $7.635\,mS\,m^2\,mol^{-1}$, and $7.81\,mS\,m^2\,mol^{-1}$, respectively. What are their mobilities?

E16B.4(a) The mobility of a chloride ion in aqueous solution at 25 °C is $7.91 \times 10^{-8}\,m^2\,s^{-1}\,V^{-1}$. Calculate the molar ionic conductivity.
E16B.4(b) The mobility of an ethanoate (acetate) ion in aqueous solution at 25 °C is $4.24 \times 10^{-8}\,m^2\,s^{-1}\,V^{-1}$. Calculate the molar ionic conductivity.

E16B.5(a) The mobility of a Rb^+ ion in aqueous solution is $7.92 \times 10^{-8}\,m^2\,s^{-1}\,V^{-1}$ at 25 °C. The potential difference between two electrodes, separated by 7.00 mm and placed in the solution, is 25.0 V. What is the drift speed of the Rb^+ ion?
E16B.5(b) The mobility of a Li^+ ion in aqueous solution is $4.01 \times 10^{-8}\,m^2\,s^{-1}\,V^{-1}$ at 25 °C. The potential difference between two electrodes separated by 5.00 mm and placed in the solution, is 24.0 V. What is the drift speed of the ion?

E16B.6(a) The mobility of a NO_3^- ion in aqueous solution at 25 °C is $7.40 \times 10^{-8}\,m^2\,s^{-1}\,V^{-1}$. Calculate its diffusion coefficient in water at 25 °C.
E16B.6(b) The mobility of a $CH_3CO_2^-$ ion in aqueous solution at 25 °C is $4.24 \times 10^{-8}\,m^2\,s^{-1}\,V^{-1}$. Calculate its diffusion coefficient in water at 25 °C.

Problems

P16B.1 The viscosity of benzene varies with temperature as shown in the following table. Use the data to infer the activation energy associated with viscosity.

θ/°C	10	20	30	40	50	60	70
η/cP	0.758	0.652	0.564	0.503	0.442	0.392	0.358

‡ These problems were provided by Charles Trapp and Carmen Giunta.

P16B.2 An empirical expression that reproduces the viscosity of water in the range 20–100 °C is

$$\log\frac{\eta}{\eta_{20}} = \frac{1.3272(20-\theta/°C) - 0.001053(20-\theta/°C)^2}{\theta/°C + 105}$$

where η_{20} is the viscosity at 20 °C. Explore (by using mathematical software) the possibility of fitting these data to an expression of the form $\eta = \text{const} \times e^{E_a/RT}$ and hence identifying an activation energy for the viscosity.

P16B.3 The conductivity of aqueous ammonium chloride at a series of concentrations is listed in the following table. Calculate the molar conductivity at each concentration, and use the resulting data to determine the parameters that occur in the Kohlrausch law.

$c/(\text{mol dm}^{-3})$	1.334	1.432	1.529	1.672	1.725
$\kappa/(\text{mS cm}^{-1})$	131	139	147	156	164

P16B.4 Conductivities are often measured by comparing the resistance of a cell filled with the sample to its resistance when filled with some standard solution, such as aqueous potassium chloride. The conductivity of water is $76\,\text{mS m}^{-1}$ at 25 °C and the conductivity of $0.100\,\text{mol dm}^{-3}$ KCl(aq) is $1.1639\,\text{S m}^{-1}$. A cell had a resistance of $33.21\,\Omega$ when filled with $0.100\,\text{mol dm}^{-3}$ KCl(aq) and $300.0\,\Omega$ when filled with $0.100\,\text{mol dm}^{-3}$ CH_3COOH(aq). What is the molar conductivity of ethanoic (acetic) acid at that concentration and temperature?

P16B.5 A cell was used to measure the resistance R of a series of solutions. The cell has been calibrated against a standard solution and as a result the conductivity of the solution is given by $\kappa = C/R$, where $C = 0.2063\,\text{cm}^{-1}$. The following values of R were found:

$c/(\text{mol dm}^{-3})$	0.00050	0.0010	0.0050	0.010	0.020	0.050
R/Ω	3314	1669	342.1	174.1	89.08	37.14

(a) Verify that the molar conductivity follows the Kohlrausch law, find the limiting molar conductivity and the coefficient $\mathcal{K}$. (b) Consider a solution of $0.010\,\text{mol dm}^{-3}$ NaI(aq) at 25 °C placed in the cell. Assume that the same value of $\mathcal{K}$ applies to this solution as to that in (a), and that $\lambda(Na^+) = 5.01\,\text{mS m}^2\,\text{mol}^{-1}$ and $\lambda(I^-) = 7.68\,\text{mS m}^2\,\text{mol}^{-1}$. Predict (i) the molar conductivity, (ii) the conductivity, and (iii) the resistance of the solution in the cell.

P16B.6 (a) Calculate the drift speeds of Li^+, Na^+, and K^+ in water when a potential difference of 100 V is applied across a 5.00 cm conductivity cell. Refer to the *Resource section* for values of ion mobilities. (b) Calculate how long it takes each ion to move from one electrode to the other. (c) In conductivity measurements it is normal to use an alternating potential difference. Calculate the displacement of each of the ions in (i) centimetres, and (ii) solvent diameters (take as 300 pm), during a half cycle of an applied potential at a frequency of 2.0 kHz. *Hint:* The drift speed will vary over the half cycle, so you will need to integrate s over time to find the displacement.

P16B.7‡ G. Bakale et al. (*J. Phys. Chem.*, 12477 (1996)) measured the mobility of singly charged C_{60}^- ions in a variety of nonpolar solvents. In cyclohexane at 22 °C (viscosity is $0.93 \times 10^{-3}\,\text{kg m}^{-1}\,\text{s}^{-1}$), the mobility is $1.1 \times 10^{-4}\,\text{cm}^2\,\text{V}^{-1}\,\text{s}^{-1}$. Estimate the effective radius of the C_{60}^- ion. Suggest a reason why there is a substantial difference between this number and the van der Waals radius of neutral C_{60}.

P16B.8 Estimate the diffusion coefficients and the effective hydrodynamic radii of the alkali metal cations in water from their mobilities at 25 °C (refer to the *Resource section* for values of the mobilities). Estimate the approximate number of water molecules that are dragged along by the cations. Ionic radii are given Table 15C.2.

P16B.9‡ (a) A dilute solution of a weak electrolyte AB, which dissociates to $A^+ + B^-$, is prepared with an initial concentration c_{AB}. Suppose that a fraction α of AB dissociates. Assuming that activities can be approximated by concentrations, show that the equilibrium constant K for dissociation may be written

$$K = \frac{\alpha^2 c_{AB}}{(1-\alpha)c^{\ominus}}$$

(b) The conductivity of the solution described in (a) is measured as κ, and the molar conductivity is then calculated as $\Lambda_m = \kappa/c_{AB}$. However, because the degree of dissociation, and hence the concentration of the ions, varies strongly with the initial concentration c_{AB}, values of Λ_m calculated in this way also vary strongly with c_{AB}. Given that κ can be expected to be proportional to the concentration of the ions, explain why $\alpha = \Lambda_m/\Lambda_{m,1}$, where $\Lambda_{m,1}$ is the molar conductivity in the limit of complete dissociation of AB. (c) Substitute this expression for α into the above expression for K. You now have two expressions for K: one in terms of α and one in terms of $\Lambda_m/\Lambda_{m,1}$. Equate these two expressions and hence show (by rearranging your expression) that

$$\frac{1}{\Lambda_m} = \frac{1}{\Lambda_{m,1}} + \frac{\Lambda_m(1-\alpha)}{\Lambda_{m,1}^2\alpha^2}$$

TOPIC 16C Diffusion

Discussion questions

D16C.1 Describe the origin of the thermodynamic force. To what extent can it be regarded as an actual force?

D16C.2 Account physically for the form of the diffusion equation.

Exercises

E16C.1(a) The diffusion coefficient of glucose in water at 25 °C is $6.73 \times 10^{-10}\,\text{m}^2\,\text{s}^{-1}$. Estimate the time required for a glucose molecule to undergo a root-mean-square displacement of 5.0 mm.

E16C.1(b) The diffusion coefficient of H_2O in water at 25 °C is $2.26 \times 10^{-9}\,\text{m}^2\,\text{s}^{-1}$. Estimate the time required for an H_2O molecule to undergo a root-mean-square displacement of 1.0 cm.

E16C.2(a) A layer of 20.0 g of sucrose is spread uniformly over a surface of area $5.0\,\text{cm}^2$ and covered in water. What will be the molar concentration of sucrose molecules at 10 cm above the original layer after (i) 10 s, (ii) 24 h? Assume diffusion is the only transport process, take $D = 5.216 \times 10^{-9}\,\text{m}^2\,\text{s}^{-1}$ and assume that the layer of water in infinitely deep.

E16C.2(b) A layer of 10.0 g of iodine is spread uniformly over a surface of area $10.0\,\text{cm}^2$ and covered in hexane. What will be the molar concentration of iodine molecules at 5.0 cm above the original layer after (i) 10 s, (ii) 24 h? Assume diffusion is the only transport process, take $D = 4.05 \times 10^{-9}\,\text{m}^2\,\text{s}^{-1}$, and assume that the layer of hexane in infinitely deep.

E16C.3(a) Suppose the concentration of a solute decays linearly along the length of a container according to $c(x) = c_0 - \alpha c_0 x$, where c_0 is the concentration at $x = 0$. Calculate the thermodynamic force on the solute at 25 °C and at $x = 10$ cm and 20 cm given that the concentration falls to $\tfrac{1}{2}c_0$ when $x = 10$ cm. *Hint:* Start by finding the value of α.

E16C.3(b) Suppose the concentration of a solute varies along the length of a container according to $c(x) = c_0 - \beta c_0 x^2$, where c_0 is the concentration at $x = 0$. Calculate the thermodynamic force on the solute at 25 °C and $x = 8$ cm and 16 cm given that the concentration falls to $\tfrac{1}{2}c_0$ when $x = 15$ cm. *Hint:* Start by finding the value of β.

E16C.4(a) Suppose the concentration of a solute follows a Gaussian distribution, $c(x)=c_0e^{-\alpha x^2}$, where c_0 is the concentration at $x = 0$, along the length of a container. Calculate the thermodynamic force on the solute at 20 °C and at $x = 5.0$ cm given that the concentration falls to $\frac{1}{2}c_0$ when $x = 5.0$ cm.

E16C.4(b) For the same arrangement as in Exercise E16C.4(a) calculate the thermodynamic force on the solute at 18 °C and $x = 10.0$ cm given that the concentration falls to $\frac{1}{2}c_0$ when $x = 10.0$ cm.

E16C.5(a) The diffusion coefficient of CCl_4 in heptane at 25 °C is $3.17 \times 10^{-9}\,m^2\,s^{-1}$. Estimate the time required for a CCl_4 molecule to have a root mean square displacement of 5.0 mm.

E16C.5(b) The diffusion coefficient of I_2 in hexane at 25 °C is $4.05 \times 10^{-9}\,m^2\,s^{-1}$. Estimate the time required for an iodine molecule to have a root mean square displacement of 1.0 cm.

E16C.6(a) Estimate the effective radius of a sucrose molecule in water at 25 °C given that its diffusion coefficient is $5.2 \times 10^{-10}\,m^2\,s^{-1}$ and that the viscosity of water is 1.00 cP.

E16C.6(b) Estimate the effective radius of a glycine molecule in water at 25 °C given that its diffusion coefficient is $1.055 \times 10^{-9}\,m^2\,s^{-1}$ and that the viscosity of water is 1.00 cP.

E16C.7(a) The diffusion coefficient for molecular iodine in benzene is $2.13 \times 10^{-9}\,m^2\,s^{-1}$. How long does a molecule take to jump through about one molecular diameter (approximately the fundamental jump length for translational motion)?

E16C.7(b) The diffusion coefficient for CCl_4 in heptane is $3.17 \times 10^{-9}\,m^2\,s^{-1}$. How long does a molecule take to jump through about one molecular diameter (approximately the fundamental jump length for translational motion)? The viscosity of heptane is 0.386 cP.

E16C.8(a) What are the root-mean-square distances travelled (in one dimension) by an iodine molecule in benzene and by a sucrose molecule in water at 25 °C in 1.0 s? Refer to the *Resource section* for the necessary data.

E16C.8(b) About how long does it take for the molecules referred to in Exercise 16C.8(a) to drift to a point (i) 1.0 mm, (ii) 1.0 cm from their starting points?

Problems

P16C.1 A dilute solution of potassium permanganate in water at 25 °C was prepared. The solution was in a horizontal tube of length 10 cm, and at first there was a linear gradation of intensity of the purple solution from the left (where the concentration was $0.100\,mol\,dm^{-3}$) to the right (where the concentration was $0.050\,mol\,dm^{-3}$). What is the magnitude and sign of the thermodynamic force acting on the solute (a) close to the left face of the container, (b) in the middle, (c) close to the right face. Give the force per mole and force per molecule in each case.

P16C.2 A dilute solution of potassium permanganate in water at 25 °C was prepared. The solution was in a horizontal tube of length 10 cm, and at first there was a Gaussian distribution of concentration around the centre of the tube at $x = 0$, $c(x)=c_0e^{-\alpha x^2}$, with $c_0 = 0.100\,mol\,dm^{-3}$ and $a = 0.10\,cm^{-2}$. Evaluate the thermodynamic force acting on the solute as a function of location, x, and plot the result. Give the force per mole and force per molecule in each case. What do you expect to be the consequence of the thermodynamic force?

P16C.3 Instead of a Gaussian 'heap' of solute, as in Problem P16C.2, suppose that there is a Gaussian dip, a distribution of the form $c(x)=c_0(1-e^{-\alpha x^2})$. Repeat the calculation in Problem P16C.2 and describe its consequences. Comment on the behaviour at $x = 0$.

P16C.4 A lump of sucrose of mass 10.0 g is suspended in the middle of a spherical flask of water of radius 10 cm at 25 °C. What is the concentration of sucrose at the wall of the flask after (a) 1.0 h, (b) 1.0 week. Take $D = 5.22 \times 10^{-10}\,m^2\,s^{-1}$.

P16C.5 Confirm that eqn 16C.10 is a solution of the diffusion equation, eqn 16C.6, and that it has the correct initial value.

P16C.6 (a) Confirm that

$$c(x,t)=\frac{c_0}{(4\pi Dt)^{1/2}}e^{-(x-x_0-vt)^2/4Dt}$$

is a solution of the diffusion equation with convection (eqn 16C.19) with all the solute concentrated at $x = x_0$ at $t = 0$. (b) Using mathematical software or a spreadsheet, plot $c(x,t)/c_0$ as a function of t and separately as a function of x for some typical values of D and v. Recall that diffusion is a slow process so you will need to consider quite long times and short distances. Similarly, consider slow convection speeds. (c) A different way of plotting this function is first to define $x_c = x_0 + vt$; x_c is the position to which the solute would move if the only process was convection. Now define $z = (x - x_c)/(4D)^{1/2}$ and use this quantity to rewrite the expression for $c(x,t)$ as

$$\frac{(4\pi D)^{1/2}c(x,t)}{c_0}=\frac{1}{t^{1/2}}e^{-z^2/t}$$

Now plot the right-hand side as a function of z for some representative values of t.

P16C.7 Calculate the relation between $\langle x^2\rangle^{1/2}$ and $\langle x^4\rangle^{1/4}$ for diffusing particles at a time t if they have a diffusion constant D.

P16C.8 The diffusion equation is valid when many elementary steps are taken in the time interval of interest, but the random walk calculation makes it possible to discuss distributions for short times as well as for long. Use the expression $P(nd) = N!/(N - N_R)!N_R!2^N$ to calculate the probability of being six paces from the origin (i.e. at $x = 6d$) after (a) four, (b) six, (c) twelve steps. *Hint:* Recall that $n = N_R - N_L$ and $N = N_R + N_L$.

P16C.9 Use mathematical software to calculate $P(nd)$ in a one-dimensional random walk, and evaluate the probability of being at $x = 6d$ for $N = 6$, 10, 14,…, 60. Compare the numerical value with the analytical value in the limit of a large number of steps. At what value of N is the discrepancy no more than 0.1 per cent? *Hint:* Recall that $n = N_R - N_L$ and $N = N_R + N_L$.

P16C.10 The diffusion coefficient of a particular kind of t-RNA molecule is $D = 1.0 \times 10^{-11}\,m^2\,s^{-1}$ in the medium of a cell interior. How long does it take molecules produced in the cell nucleus to reach the walls of the cell at a distance 1.0 μm, corresponding to the radius of the cell?

P16C.11 Nuclear magnetic resonance can be used to determine the mobility of molecules in liquids. A set of measurements on methane in carbon tetrachloride showed that its diffusion coefficient is $2.05 \times 10^{-9}\,m^2\,s^{-1}$ at 0 °C and $2.89 \times 10^{-9}\,m^2\,s^{-1}$ at 25 °C. Deduce what information you can about the mobility of methane in carbon tetrachloride.

FOCUS 16 Molecules in motion

Integrated activities

I16.1 In Topic 17D it is shown that a general expression for the activation energy of a chemical reaction is $E_a = RT^2(\mathrm{d}\ln k_r/\mathrm{d}T)$. Confirm that the same expression may be used to extract the activation energy from eqn 16B.2 for the viscosity and then apply the expression to deduce the temperature-dependence of the activation energy when the viscosity of water is given by the empirical expression in Problem P16B.2. Plot this activation energy as a function of temperature. Suggest an explanation of the temperature dependence of E_a.

I16.2‡ In this problem, you are invited to use mathematical software to examine a model for the transport of oxygen from air in the lungs to blood. (a) Show that, for the initial and boundary conditions $c(x,t) = c(x,0) = c_0$, $(0 < x < \infty)$ and $c(0,t) = c_s$ $(0 \le t \le \infty)$ where c_0 and c_s are constants, the diffusion equation (eqn 16C.6) is solved by the following expression for $c(x,t)$

$$c(x,t) = c_0 + (c_s - c_0)\{1 - \mathrm{erf}(\xi)\}$$

where $\xi = x/(4Dt)^{1/2}$. In this expression $\mathrm{erf}(\xi)$ is the error function and the concentration $c(x,t)$ evolves by diffusion from the yz-plane of constant concentration, such as might occur if a condensed phase is absorbing a species from a gas phase. (b) Draw graphs of concentration profiles at several different times of your choice for the diffusion of oxygen into water at 298 K (when $D = 2.10 \times 10^{-9}\,\mathrm{m^2\,s^{-1}}$) on a spatial scale comparable to passage of oxygen from lungs through alveoli into the blood. Use $c_0 = 0$ and set c_s equal to the solubility of oxygen in water, $2.9 \times 10^{-4}\,\mathrm{mol\,dm^{-3}}$.

FOCUS 17

Chemical kinetics

This Focus introduces the principles of 'chemical kinetics', the study of reaction rates. The rate of a chemical reaction might depend on variables that can be controlled, such as the pressure, the temperature, and the presence of a catalyst, and it is possible to optimize the rate by the appropriate choice of conditions.

17A The rates of chemical reactions

This Topic discusses the definition of reaction rate and outlines the techniques for its measurement. The results of such measurements show that reaction rates depend on the concentration of reactants (and sometimes products) and 'rate constants' that are characteristic of the reaction. This dependence can be expressed in terms of differential equations known as 'rate laws'.

17A.1 **Monitoring the progress of a reaction;** 17A.2 **The rates of reactions**

17B Integrated rate laws

'Integrated rate laws' are the solutions of the differential equations that describe rate laws. They are used to predict the concentrations of species at any time after the start of the reaction and to provide procedures for measuring rate constants. This Topic explores some simple yet useful integrated rate laws that appear throughout the Focus.

17B.1 **Zeroth-order reactions;** 17B.2 **First-order reactions;** 17B.3 **Second-order reactions**

17C Reactions approaching equilibrium

In general, rate laws must take into account both the forward and reverse reactions and describe the approach to equilibrium, when the forward and reverse rates are equal. The results of the analysis are relations, which can be explored experimentally, between the equilibrium constant of the overall process and the rate constants of the forward and reverse reactions in the proposed mechanism.

17C.1 **First-order reactions approaching equilibrium;** 17C.2 **Relaxation methods**

17D The Arrhenius equation

The rate constants of most reactions increase with increasing temperature. This Topic introduces the 'Arrhenius equation', which captures this temperature dependence by using only two parameters that can be determined experimentally.

17D.1 **The temperature dependence of reaction rates;** 17D.2 **The interpretation of the Arrhenius parameters**

17E Reaction mechanisms

The study of reaction rates also leads to an understanding of the 'mechanisms' of reactions, their analysis into a sequence of elementary steps. This Topic shows how to construct rate laws from a proposed mechanism. The elementary steps themselves have simple rate laws which can be combined into an overall rate law by invoking the concept of the 'rate-determining step' of a reaction, by making the 'steady-state approximation', or by supposing that a 'pre-equilibrium' exists.

17E.1 **Elementary reactions;** 17E.2 **Consecutive elementary reactions;** 17E.3 **The steady-state approximation;** 17E.4 **The rate-determining step;** 17E.5 **Pre-equilibria;** 17E.6 **Kinetic and thermodynamic control of reactions**

17F Examples of reaction mechanisms

This Topic develops three examples of reaction mechanisms. The first describes a special class of reactions in the gas phase that depend on the collisions between reactants. The second gives insight into the formation of polymers and shows how the kinetics of their formation affects their properties. The third examines the general mechanism of action of 'enzymes', which are biological catalysts.

17F.1 **Unimolecular reactions;** 17A.2 **Polymerization kinetics;** 17F.3 **Enzyme-catalysed reactions**

17G Photochemistry

'Photochemistry' is the study of reactions that are initiated by light. This Topic explores the fate of the electronically excited molecules formed by the absorption of photons. One possible fate is energy transfer to another molecule; this process is particularly interesting as it is the basis for a method of estimating the distance between certain groups in large molecules.

17G.1 **Photochemical processes;** 17G.2 **The primary quantum yield;** 17G.3 **Mechanism of decay of excited singlet states;** 17G.4 **Quenching;** 17G.5 **Resonance energy transfer**

Web resource What is an application of this material?

Plants, algae, and some species of bacteria have evolved apparatus that performs 'photosynthesis', the capture of visible and near-infrared radiation for the purpose of synthesizing complex molecules in the cell. *Impact* 26 introduces the reaction steps involved.

TOPIC 17A The rates of chemical reactions

➤ Why do you need to know this material?

Studies of the rates of consumption of reactants and formation of products make it possible to predict how quickly a reaction mixture approaches equilibrium. They also lead to detailed descriptions of the molecular events that transform reactants into products.

➤ What is the key idea?

Reaction rates are expressed as rate laws, which are empirical summaries of the rates in terms of the concentrations of reactants and, in some cases, products.

➤ What do you need to know already?

This introductory Topic is the foundation of a sequence: all you need to be aware of initially is the significance of stoichiometric numbers (Topic 2C). For more background on the spectroscopic determination of concentration, refer to Topic 11A.

Chemical kinetics is the study of reaction rates. Experiments show that reaction rates depend on the concentration of reactants (and in some cases products) in characteristic ways that can be expressed in terms of differential equations known as 'rate laws'.

17A.1 Monitoring the progress of a reaction

The first step in the kinetic analysis of reactions is to establish the stoichiometry of the reaction and identify any side reactions. The basic data of chemical kinetics are then the concentrations of the reactants and products at different times after a reaction has been initiated.

(a) General considerations

The rates of most chemical reactions are sensitive to the temperature (as described in Topic 17D), so in conventional experiments the temperature of the reaction mixture must be held constant throughout the course of the reaction. This requirement puts severe demands on the design of an experiment. Gas-phase reactions, for instance, are often carried out in a vessel held in contact with a substantial block of metal. Liquid-phase reactions must be carried out in an efficient thermostat. Special efforts have to be made to study reactions at low temperatures, as in the study of the kinds of reactions that take place in interstellar clouds. Thus, supersonic expansion of the reacting gas can be used to attain temperatures as low as 10 K. For work in the liquid phase and the solid phase, very low temperatures are often reached by flowing cold liquid or cold gas around the reaction vessel. Alternatively, the entire reaction vessel is immersed in a thermally insulated container filled with a cryogenic liquid, such as liquid helium (for work at around 4 K) or liquid nitrogen (for work at around 77 K). Non-isothermal conditions are sometimes employed. For instance, the shelf life of an expensive pharmaceutical may be explored by slowly raising the temperature of a single sample.

Spectroscopy is widely applicable to the study of reaction kinetics, and is especially useful when one substance in the reaction mixture has a strong characteristic absorption in a conveniently accessible region of the electromagnetic spectrum. For example, the progress of the reaction $H_2(g) + Br_2(g) \rightarrow 2HBr(g)$ can be followed by measuring the absorption of visible radiation by bromine. A reaction that changes the number or type of ions present in a solution may be followed by monitoring the electrical conductivity of the solution. The replacement of neutral molecules by ionic products can result in dramatic changes in the conductivity, as in the reaction $(CH_3)_3CCl(aq) + H_2O(l) \rightarrow (CH_3)_3COH(aq) + H^+(aq) + Cl^-(aq)$. If hydrogen ions are produced or consumed, the reaction may be followed by monitoring the pH of the solution.

Other methods of determining composition include emission spectroscopy (Topic 11F), mass spectrometry, gas chromatography, nuclear magnetic resonance (Topics 12B and 12C), and electron paramagnetic resonance (for reactions involving radicals or paramagnetic d-metal ions; Topic 12D). A reaction in which at least one component is a gas might result in an overall change in pressure in a system of constant volume, so its progress may be followed by recording the variation of pressure with time.

Example 17A.1 Relating the variation in the total pressure to the partial pressures of the species present

The decomposition $N_2O_5(g) \rightarrow 2NO_2(g) + \frac{1}{2}O_2(g)$ is monitored by measuring the total pressure in a constant-volume reaction vessel held at constant temperature. If the initial pressure is p_0, and is due solely to N_2O_5, and the total pressure at any later time is p, derive expressions for the partial pressures of all three species in terms of p_0 and p.

Collect your thoughts You can assume perfect-gas behaviour, in which case the partial pressure of a gas is proportional to its amount. Then imagine that a certain amount of N_2O_5 decomposes such that its partial pressure falls from p_0 to $p_0 - \Delta p$. Because 1 mol N_2O_5 decomposes to give 2 mol NO_2, the partial pressure of NO_2 increases from 0 to $2\Delta p$. Likewise, the partial pressure of O_2 increases from 0 to $\frac{1}{2}\Delta p$. The total pressure p is the sum of the partial pressures of the three components.

The solution Draw up the following table:

	$p(N_2O_5)$	$p(NO_2)$	$p(O_2)$	p
Initially	p_0	0	0	p_0
Later	$p_0 - \Delta p$	$2\Delta p$	$\frac{1}{2}\Delta p$	$p_0 + \frac{3}{2}\Delta p$
Equivalent to	$\frac{5}{3}p_0 - \frac{2}{3}p$	$\frac{4}{3}(p - p_0)$	$\frac{1}{3}(p - p_0)$	

The bottom line follows from $p = p_0 + \frac{3}{2}\Delta p$, rearranged into $\Delta p = \frac{2}{3}(p - p_0)$, and then substituted into the line above.

Comment. A check of the calculation is to note that when all the N_2O_5 has been consumed its partial pressure is zero, which implies that $\frac{5}{3}p_0 - \frac{2}{3}p_{\text{final}} = 0$ and hence $p_{\text{final}} = \frac{5}{2}p_0$. This result is expected because 1 mol N_2O_5 is replaced by 2 mol NO_2 and $\frac{1}{2}$ mol O_2: the total amount of molecules therefore goes from 1 mol to $2\frac{1}{2}$ mol, resulting in the pressure increasing by a factor of $2\frac{1}{2} = \frac{5}{2}$.

Self-test 17A.1 Repeat the calculation of the partial pressures for the species in the reaction $2NOBr(g) \rightarrow 2NO(g) + Br_2(g)$, assuming that the initial pressure is p_0 and is due to NOBr alone.

Answer: $p_{NOBr} = \frac{3}{2}p_0 - \frac{1}{2}p$, $p_{NO} = p - p_0$, $p_{Br_2} = \frac{1}{2}(p - p_0)$

(b) Special techniques

The method used to monitor concentrations depends on the species involved and the rapidity with which their concentrations change. Many reactions reach equilibrium over periods of minutes or hours, and several techniques may then be used to follow the changing concentrations. In a **real-time analysis** the composition of the system is analysed while the reaction is in progress. Either a small sample is withdrawn or the bulk solution is monitored.

In the **flow method** the reactants are mixed as they flow together into a chamber (Fig. 17A.1). The reaction continues as the thoroughly mixed solutions flow through the outlet tube, and observation of the composition at different positions along the tube is equivalent to the observation of the reaction mixture at different times after mixing. The disadvantage of conventional flow techniques is that a large volume of reactant solution is necessary. This requirement makes the study of fast reactions particularly difficult because to spread the reaction over a length of tube the flow must be rapid. This disadvantage is avoided by the **stopped-flow technique**, in which the reagents are mixed very quickly in a small chamber fitted with a movable piston instead of an outlet tube (Fig. 17A.2). The flow pushes the piston back and ceases when it reaches a stop; the reaction continues in the mixed solutions. Observations, commonly using spectroscopic techniques such as ultraviolet–visible absorption and fluorescence emission, are made on the sample as a function of time. The technique permits the study of reactions that occur on the millisecond to second timescale. The suitability of the stopped-flow method for the study of small samples means that it is appropriate for many biochemical reactions; it has been widely used to study the kinetics of protein folding and enzyme action.

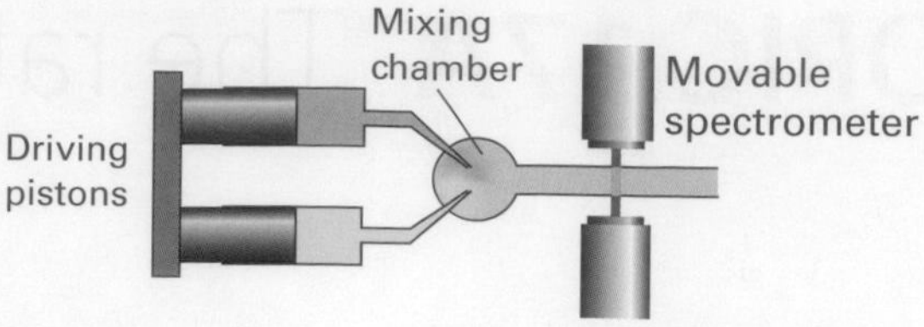

Figure 17A.1 The arrangement used in the flow technique for studying reaction rates. The reactants are injected into the mixing chamber at a steady rate. The location of the spectrometer corresponds to different times after initiation of the reaction.

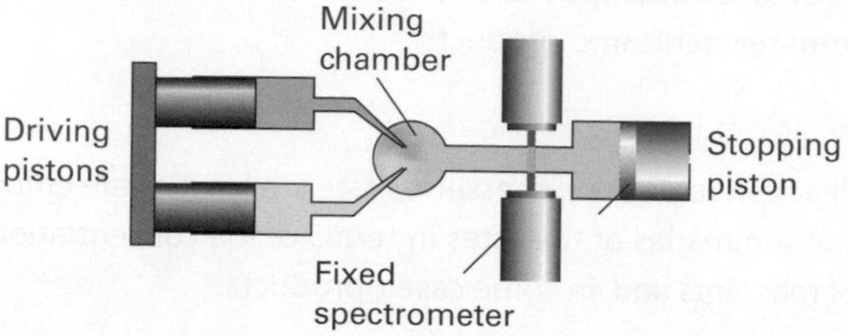

Figure 17A.2 In the stopped-flow technique the reagents are driven quickly into the mixing chamber by the driving pistons and then the time dependence of the concentrations is monitored.

Very fast reactions can be studied by **flash photolysis**, in which the sample is exposed to a brief flash of light, which initiates the reaction, and then the contents of the reaction chamber are monitored by electronic absorption or emission, infrared absorption, or Raman scattering. In the arrangement shown in Fig. 17A.3 a strong and short laser pulse, the *pump*, promotes a molecule A to an excited electronic state A* which can either emit a photon (as fluorescence or phosphorescence)

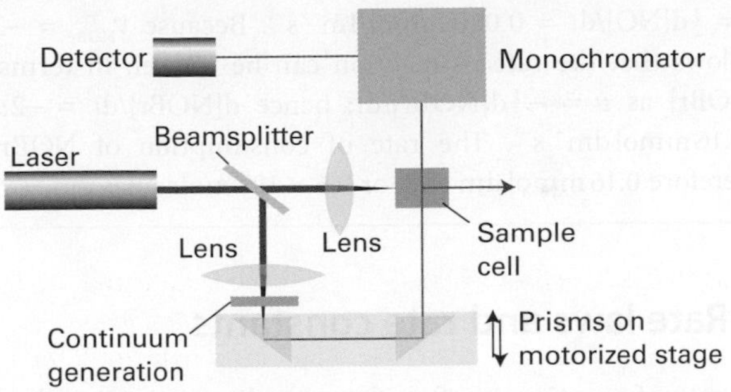

Figure 17A.3 A configuration used for flash photolysis, in which the same pulsed laser is used to generate a monochromatic pump pulse and, after continuum generation, a 'white' light probe pulse. The time delay between the pump and probe pulses may be varied.

or react with another species B to form first the intermediate AB and then the product C:

$A + h\nu \rightarrow A^*$ (absorption)

$A^* \rightarrow A$ (emission)

$A^* + B \rightarrow AB \rightarrow C$ (reaction)

The rates of appearance and disappearance of the various species are determined by observing time-dependent changes in the absorption spectrum of the sample during the course of the reaction. This monitoring is done by passing a weak pulse of white light, the *probe*, through the sample at different times after the laser pulse. Pulsed 'white' light can be generated directly from the laser pulse by the phenomenon of **continuum generation**, in which focusing a short laser pulse on sapphire or a vessel containing water or carbon tetrachloride results in an outgoing beam with a wide distribution of frequencies. A time delay between the strong laser pulse and the 'white' light pulse can be introduced by allowing one of the beams to travel a longer distance before reaching the sample. For example, a difference in travel distance of $\Delta d = 3\,\text{mm}$ corresponds to a time delay $\Delta t = \Delta d/c \approx 10\,\text{ps}$ between two beams, where c is the speed of light. The relative distances travelled by the two beams in Fig. 17A.3 are controlled by directing the 'white' light beam to a motorized stage carrying a pair of mirrors.

In contrast to real-time analysis, **quenching methods** are based on 'quenching', or stopping, the reaction after it has been allowed to proceed for a certain time. In this way the composition is analysed at leisure and reaction intermediates may be trapped. These methods are suitable only for reactions slow enough for there to be little reaction during the time it takes to quench the mixture. In the **chemical quench flow method**, the reactants are mixed in much the same way as in the flow method but the reaction is quenched by another reagent, such as a solution of acid or base, after the mixture has travelled along a fixed length of the outlet tube. Different reaction times can be selected by varying the flow rate along the outlet tube. An advantage of the chemical quench flow method over the stopped-flow method is that rapid spectroscopic measurements are not needed in order to measure the concentration of reactants and products. Once the reaction has been quenched, the solution may be examined by 'slow' techniques, such as mass spectrometry and chromatography. In the **freeze quench method**, the reaction is quenched by cooling the mixture within milliseconds and the concentrations of reactants, intermediates, and products are measured spectroscopically.

17A.2 The rates of reactions

Reaction rates depend on the composition and the temperature of the reaction mixture. The next few sections look at these observations in more detail.

(a) The definition of rate

Consider a reaction of the form $A + 2B \rightarrow 3C + D$, in which at some instant the molar concentration of a participant J is [J] and the volume of the system is constant. The **instantaneous rate** of consumption of a reactant or formation of a product is the slope of the tangent to the graph of concentration against time (expressed as a positive quantity). It follows that the instantaneous **rate of consumption** of one of the reactants at a given time is $-d[R]/dt$, where R is A or B. This rate is a positive quantity (Fig. 17A.4). The **rate of formation** of one of the products (C or D, denoted P) is $d[P]/dt$ (note the difference in sign). This rate is also positive.

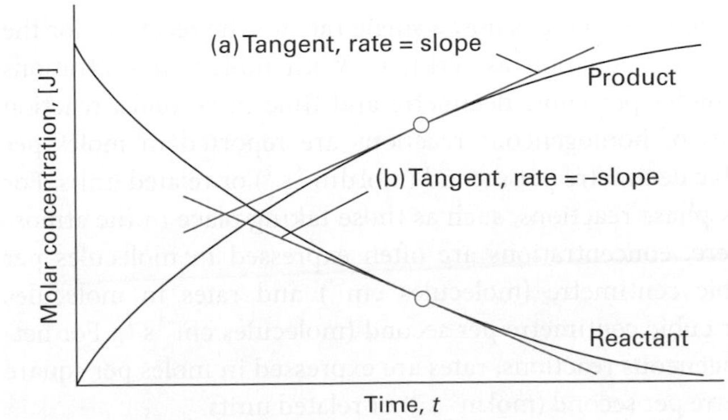

Figure 17A.4 The definition of (instantaneous) rate as the slope of the tangent drawn to the curve showing the variation of concentration of (a) products, (b) reactants with time. For negative slopes, the sign is changed when reporting the rate, so all reaction rates are positive.

It follows from the stoichiometry of the reaction A + 2 B → 3 C + D that

$$\frac{d[D]}{dt}=\frac{1}{3}\frac{d[C]}{dt}=-\frac{d[A]}{dt}=-\frac{1}{2}\frac{d[B]}{dt}$$

so there are several rates connected with the reaction. The undesirability of having different rates to describe the same reaction is avoided by introducing the **extent of reaction**, ξ (xi), which is defined so that for each species J in the reaction, the change in amount of J, dn_J, is

$$dn_J=\nu_J d\xi \qquad \text{Extent of reaction [definition]} \qquad (17A.1)$$

where ν_J is the stoichiometric number of the species (Topic 2C; remember that ν_J is negative for reactants and positive for products). The unique **rate of reaction**, v, is then defined as

$$v=\frac{1}{V}\frac{d\xi}{dt} \qquad \text{Rate of reaction [definition]} \qquad (17A.2)$$

where V is the volume of the system. For any species J, $d\xi = dn_J/\nu_J$, so

$$v=\frac{1}{\nu_J}\times\frac{1}{V}\frac{dn_J}{dt} \qquad (17A.3a)$$

For a homogeneous reaction in a constant-volume system the volume V can be taken inside the differential and n_J/V is written as the molar concentration [J] to give

$$v=\frac{1}{\nu_J}\frac{d[J]}{dt} \qquad (17A.3b)$$

For a heterogeneous reaction the (constant) surface area, A, occupied by the species is used in place of V. Then, because the surface concentration is $\sigma_J = n_J/A$ it follows that

$$v=\frac{1}{\nu_J}\frac{d\sigma_J}{dt} \qquad (17A.3c)$$

In each case there is now a single rate for the reaction (for the chemical equation as written). With molar concentrations in moles per cubic decimetre and time in seconds, reaction rates of homogeneous reactions are reported in moles per cubic decimetre per second ($mol\,dm^{-3}\,s^{-1}$) or related units. For gas-phase reactions, such as those taking place in the atmosphere, concentrations are often expressed in molecules per cubic centimetre (molecules cm^{-3}) and rates in molecules per cubic centimetre per second (molecules $cm^{-3}\,s^{-1}$). For heterogeneous reactions, rates are expressed in moles per square metre per second ($mol\,m^{-2}\,s^{-1}$) or related units.

Brief illustration 17A.1

The rate of formation of NO, d[NO]/dt, in the reaction $2\,NOBr(g) \rightarrow 2\,NO(g) + Br_2(g)$ is reported as $0.16\,mmol\,dm^{-3}\,s^{-1}$. Because $\nu_{NO} = +2$, the rate of the reaction is reported as $v = \frac{1}{2}d[NO]/dt = 0.080\,mmol\,dm^{-3}\,s^{-1}$. Because $\nu_{NOBr} = -2$ it follows that the rate of reaction can be written in terms of [NOBr] as $v = -\frac{1}{2}d[NOBr]/dt$; hence $d[NOBr]/dt = -2v = -0.16\,mmol\,dm^{-3}\,s^{-1}$. The rate of consumption of NOBr is therefore $0.16\,mmol\,dm^{-3}\,s^{-1}$, or 9.6×10^{16} molecules $cm^{-3}\,s^{-1}$.

(b) Rate laws and rate constants

The rate of reaction is often found to be proportional to the concentrations of the reactants raised to a power. For example, the rate of a reaction might be found to be proportional to the molar concentrations of two reactants A and B, so

$$v=k_r[A][B] \qquad (17A.4)$$

The constant of proportionality k_r is called the **rate constant** for the reaction; it is independent of the concentrations but depends on the temperature. An experimentally determined equation of this kind is called the **rate law** of the reaction. More formally, a rate law is an equation that expresses the rate of reaction in terms of the concentrations of all the species present in the overall chemical equation for the reaction at the time of interest:

$$v=f([A],[B],\ldots) \qquad \text{Rate law in terms of concentrations [general form]} \qquad (17A.5a)$$

For homogeneous gas-phase reactions, it is often more convenient to express the rate law in terms of partial pressures, which for perfect gases are related to molar concentrations by $p_J = RT[J]$. In this case,

$$v=f(p_A,p_B,\ldots) \qquad \text{Rate law in terms of partial pressures [general form]} \qquad (17A.5b)$$

The rate law of a reaction is determined experimentally, and cannot in general be inferred from the chemical equation for the reaction. The reaction of hydrogen and bromine, for example, has a very simple stoichiometry, $H_2(g) + Br_2(g) \rightarrow 2\,HBr(g)$, but its rate law is complicated:

$$v=\frac{k_a[H_2][Br_2]^{3/2}}{[Br_2]+k_b[HBr]} \qquad (17A.6)$$

In certain cases the rate law does reflect the stoichiometry of the reaction; but that is either a coincidence or reflects a feature of the underlying reaction mechanism (Topic 17E).

A note on good (or, at least, our) practice A general rate constant is denoted k_r to distinguish it from the Boltzmann constant k. In some texts k is used for the former and k_B for the latter. When expressing the rate constants in a more complicated rate law, such as that in eqn 17A.6, we use k_a, k_b, and so on.

The units of k_r are always such as to convert the product of concentrations, each raised to the appropriate power, into a rate expressed as a change in concentration divided by time. For example, if the rate law is the one shown in eqn 17A.4, with concentrations expressed in mol dm^{-3}, then the units of k_r will be dm^3 mol^{-1} s^{-1} because

$$\underbrace{\mathrm{dm^3\,mol^{-1}\,s^{-1}}}_{k_r}\times\underbrace{\mathrm{mol\,dm^{-3}}}_{[\mathrm{A}]}\times\underbrace{\mathrm{mol\,dm^{-3}}}_{[\mathrm{B}]}=\mathrm{mol\,dm^{-3}\,s^{-1}}$$

which are the units of v. If the concentrations are expressed in molecules cm^{-3}, and the rate in molecules cm^{-3} s^{-1}, then the rate constant is expressed in cm^3 molecule^{-1} s^{-1}. The approach just developed can be used to determine the units of the rate constant from rate laws of any form.

Brief illustration 17A.2

The rate constant for the reaction $O(g) + O_3(g) \rightarrow 2O_2(g)$ is 8.0×10^{-15} cm^3 molecule^{-1} s^{-1} at 298 K. To express this rate constant in dm^3 mol^{-1} s^{-1}, make use of the relation 1 cm = 10^{-1} dm to convert the volume:

$$\begin{aligned}k_r &= 8.0\times10^{-15}\,\overbrace{\mathrm{cm^3}}^{(10^{-1}\,\mathrm{dm})^3}\,\mathrm{molecule^{-1}\,s^{-1}}\\ &= 8.0\times10^{-18}\,\mathrm{dm^3\,molecule^{-1}\,s^{-1}}\end{aligned}$$

Now note that the number of molecules can be expressed as an amount in moles by division by Avogadro's constant expressed as molecules per mole:

$$\begin{aligned}k_r &= 8.0\times10^{-18}\,\mathrm{dm^3\,molecule^{-1}\,s^{-1}}\\ &= 8.0\times10^{-18}\,\mathrm{dm^3}\times\left(\frac{1\ \mathrm{molecule}}{6.022\times10^{23}\ \mathrm{molecules\ mol^{-1}}}\right)^{-1}\mathrm{s^{-1}}\\ &= 8.0\times10^{-18}\times6.022\times10^{23}\,\mathrm{dm^3\,mol^{-1}\,s^{-1}}\\ &= 4.8\times10^{6}\,\mathrm{dm^3\,mol^{-1}\,s^{-1}}\end{aligned}$$

A practical application of a rate law is that once the law and the value of the rate constant are known, it is possible to predict the rate of reaction from the composition of the mixture. Moreover, as demonstrated in Topic 17B, by knowing the rate law, it is also possible to predict the composition of the reaction mixture at a later stage of the reaction. A rate law also provides evidence used to assess the plausibility of a proposed mechanism of the reaction. This application is developed in Topic 17E.

(c) Reaction order

Many reactions are found to have rate laws of the form

$$v = k_r[\mathrm{A}]^a[\mathrm{B}]^b\ldots \qquad (17\mathrm{A}.7)$$

The power to which the concentration of a species (a product or a reactant) is raised in a rate law of this kind is the **order** of the reaction with respect to that species. A reaction with the rate law in eqn 17A.4 is first order in A and first order in B. The **overall order** of a reaction with a rate law like that in eqn 17A.7 is the sum of the individual orders, $a + b + \cdots$. The overall order of the rate law in eqn 17A.4 is $1 + 1 = 2$; the rate law is therefore said to be second-order overall.

A reaction need not have an integral order, and many gas-phase reactions do not. For example, a reaction with the rate law

$$v = k_r[\mathrm{A}]^{1/2}[\mathrm{B}] \qquad (17\mathrm{A}.8)$$

is half order in A, first order in B, and three-halves order overall.

Brief illustration 17A.3

The experimentally determined rate law for the gas-phase reaction $H_2(g) + Br_2(g) \rightarrow 2HBr(g)$ is given by eqn 17A.6. In the rate law the concentration of H_2 appears raised to the power +1, so the reaction is first order in H_2. However, the concentrations of Br_2 and HBr do not appear as a single term raised to a power, so the reaction has an indefinite order with respect to both Br_2 and HBr, and an indefinite order overall.

Some reactions obey a zeroth-order rate law, and therefore have a rate that is independent of the concentration of the reactant (so long as some is present). Thus, the catalytic decomposition of phosphine (PH_3) on hot tungsten at high pressures has the rate law

$$v = k_r \qquad (17\mathrm{A}.9)$$

This law means that PH_3 decomposes at a constant rate until it has entirely disappeared.

As seen in *Brief illustration* 17A.3, when a rate law is not of the form in eqn 17A.7, the reaction does not have an overall order and might not even have definite orders with respect to each participant.

These remarks point to three important tasks:

- To identify the rate law and obtain the rate constant from the experimental data. This aspect is discussed in this Topic.
- To account for the values of the rate constants and explain their temperature dependence. This task is undertaken in Topic 17D.
- To construct reaction mechanisms consistent with the rate law. The techniques for doing so are introduced in Topic 17E.

(d) The determination of the rate law

The determination of a rate law is simplified by the **isolation method**, in which all the reactants except one are present in large excess. The dependence of the rate on each of the reactants can be found by isolating each of them in turn—by having all the other substances present in large excess—and piecing together a picture of the overall rate law.

If a reactant B is in large excess its concentration is nearly constant throughout the reaction. Then, although the true rate law might be $v = k_r[A][B]^2$, the current value of [B] can be approximated by its initial value $[B]_0$ (from which it hardly changes in the course of the reaction) to give

$$v = k_{r,\text{eff}}[A], \text{ with } k_{r,\text{eff}} = k_r[B]_0^2 \quad \text{A pseudofirst-order reaction, B in excess} \quad (17A.10a)$$

Because the true rate law has been forced into first-order form by assuming a constant B concentration, the effective rate law is classified as a **pseudofirst-order rate law** and $k_{r,\text{eff}}$ is called the **effective rate constant** for a given, fixed concentration of B. If, instead, the concentration of A is in large excess, and hence effectively constant, then the original rate law simplifies to

$$v = k'_{r,\text{eff}}[B]^2, \text{ now with } k'_{r,\text{eff}} = k_r[A]_0 \quad \text{A pseudosecond-order reaction, A in excess} \quad (17A.10b)$$

This **pseudosecond-order rate law** is also much easier to analyse and identify than the complete law. Note that the order of the reaction and the form of the effective rate constant change according to whether A or B is in excess. In a similar manner, a reaction may even appear to be zeroth order. Many reactions in aqueous solution that are reported as first or second order are actually pseudofirst or pseudosecond order: the solvent water, for instance, might participate in a reaction but it is in such large excess that its concentration remains constant.

In the **method of initial rates**, which is often used in conjunction with the isolation method, the instantaneous rate is measured at the beginning of the reaction for several different initial concentrations of the isolated reactant. If the initial rate is doubled when the concentration of an isolated reactant A is doubled, then the reaction is first order in A; if the initial rate is quadrupled, then the reaction is second order in A. More formally, with an eye on developing a graphical method for determining the order, suppose the rate law for a reaction with A isolated is

$$v = k_r[A]^a$$

Then the initial rate of the reaction, v_0, is given by the initial concentration of A:

$$v_0 = k_{r,\text{eff}}[A]_0^a \quad \text{Initial rate of an } a\text{th-order reaction} \quad (17A.11a)$$

Taking (common) logarithms gives

$$\log v_0 = \log(k_{r,\text{eff}}[A]_0^a) = \log k_{r,\text{eff}} + \log[A]_0^a \quad (17A.11b)$$

(using $\log(xy) = \log x + \log y$)

$$= \log k_{r,\text{eff}} + a\log[A]_0$$

(using $\log x^a = a\log x$)

This equation has the form of the equation for a straight line:

$$\underbrace{\log v_0}_{y} = \underbrace{\log k_{r,\text{eff}}}_{\text{intercept}} + \underbrace{a\log[A]_0}_{\text{slope}\times x} \quad (17A.11c)$$

It follows that, for a series of initial concentrations, a plot of the logarithms of the initial rates against the logarithms of the initial concentrations of A should be a straight line, and that the slope of the graph is a, the order of the reaction with respect to A.

Example 17A.2 Using the method of initial rates

The recombination of I atoms in the gas phase in the presence of argon was investigated and the order of the reaction was determined by the method of initial rates. The initial rates of reaction of $2\,I(g) + Ar(g) \rightarrow I_2(g) + Ar(g)$ were as follows:

$[I]_0/(10^{-5}\,\text{mol dm}^{-3})$		1.0	2.0	4.0	6.0
$v_0/(\text{mol dm}^{-3}\,\text{s}^{-1})$	(a)	8.70×10^{-4}	3.48×10^{-3}	1.39×10^{-2}	3.13×10^{-2}
	(b)	4.35×10^{-3}	1.74×10^{-2}	6.96×10^{-2}	1.57×10^{-1}
	(c)	8.69×10^{-3}	3.47×10^{-2}	1.38×10^{-1}	3.13×10^{-1}

The Ar concentrations are (a) $1.0\times10^{-3}\,\text{mol dm}^{-3}$, (b) $5.0\times10^{-3}\,\text{mol dm}^{-3}$, and (c) $1.0\times10^{-2}\,\text{mol dm}^{-3}$. Find the orders of reaction with respect to I and Ar, and the rate constant.

Collect your thoughts You need to identify sets of data in which only one reactant is changing (such as each row of data for constant [Ar]). The identification of order from such data involves the application of eqn 17A.11c, plotting the logarithm of the rate against the logarithm of a concentration of one of the reactants (in this case, arbitrarily chosen to be I). So, first tabulate the logarithms of the concentrations of I and the rates for the values at constant $[Ar]_0$ (i.e. for each row of data). The slope gives the order with respect to [I] and the intercept at $\log[I]_0 = 0$ gives $\log k_{r,\text{eff}}$, with a different value for each $[Ar]_0$. The effective rate constant obtained in this way is $k_{r,\text{eff}} = k_r[Ar]_0^b$, so to extract k_r and b, take logarithms, as in the text, to obtain

$$\log k_{r,\text{eff}} = \log k_r + b\log[Ar]_0$$

Now realize that you need to plot the $\log k_{r,\text{eff}}$ found in the first part of the solution against $\log[Ar]_0$. Then the slope gives b and the intercept at $\log[Ar]_0 = 0$ gives $\log k_r$.

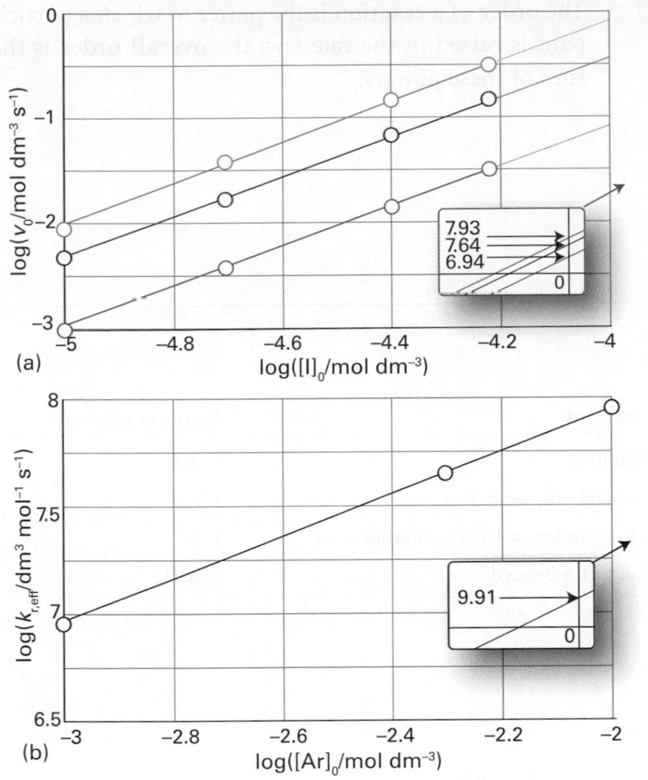

Figure 17A.5 Analysis of the data in *Example* 17A.2. (a) Plots for finding the order with respect to I. The intercepts at $\log [I]_0 = 0$ are far to the right, and are shown in the inset. (b) The plots for finding the order with respect to Ar and the rate constant k_r. The intercept at $\log [Ar]_0 = 0$ is far to the right, and is shown in the inset.

The solution The data give the following points for the graph:

$\log([I]_0/\text{mol dm}^{-3})$		−5.00	−4.70	−4.40	−4.22
$\log(v_0/\text{mol dm}^{-3}\,\text{s}^{-1})$	(a)	−3.060	−2.458	−1.857	−1.504
	(b)	−2.362	−1.759	−1.157	−0.804
	(c)	−2.061	−1.460	−0.860	−0.504

The graph of the data for varying [I] but constant [Ar] is shown in Fig. 17A.5a. The slopes of the lines are 2, so the reaction is second order with respect to I. The effective rate constants $k_{r,eff}$ are as follows:

$[Ar]_0/(\text{mol dm}^{-3})$	1.0×10^{-3}	5.0×10^{-3}	1.0×10^{-2}
$\log([Ar]_0/\text{mol dm}^{-3})$	−3.00	−2.30	−2.00
$\log(k_{r,eff}/\text{dm}^3\,\text{mol}^{-1}\,\text{s}^{-1})$	6.94	7.64	7.93

Figure 17A.5b shows the plot of $\log\{k_{r,eff}/(\text{dm}^3\,\text{mol}^{-1}\,\text{s}^{-1})\}$ against $\log\{[Ar]_0/(\text{mol dm}^{-3})\}$. The slope is 0.99, so $b = 0.99$ and the reaction is first order with respect to Ar. The intercept at $\log\{[Ar]_0/(\text{mol dm}^{-3})\} = 0$ is $\log\{k_{r,eff}/(\text{dm}^3\,\text{mol}^{-1}\,\text{s}^{-1})\} = 9.91$, so $k_r = 8.1 \times 10^9\,\text{dm}^6\,\text{mol}^{-2}\,\text{s}^{-1}$. The overall (initial) rate law is therefore $v = k_r[I]_0^2[Ar]_0$.

A note on good practice When taking the common logarithm of a number of the form $x.xx \times 10^n$ (with $n < 10$) there are *four* significant figures in the answer (for instance, $\log 1.23 \times 10^4 = 4.090$): the figure before the decimal point is simply the power of 10. Conversely, when taking the common antilogarithm of $y.yyy$, there are *three* significant figures in the answer (for instance, $10^{5.678} = 4.76 \times 10^5$).

Self-test 17A.2 The initial rate of a certain reaction depended on the concentration of a substance J as follows:

$[J]_0/(10^{-3}\,\text{mol dm}^{-3})$	5.0	10.2	17	30
$v_0/(10^{-7}\,\text{mol dm}^{-3}\,\text{s}^{-1})$	3.6	9.6	4	130

Find the order of the reaction with respect to J and the rate constant.

Answer: 2, $1.6 \times 10^{-2}\,\text{dm}^3\,\text{mol}^{-1}\,\text{s}^{-1}$

The method of initial rates might not reveal the full rate law because once the products have started to be generated they might participate in the reaction and affect its rate. For example, in the reaction between H_2 and Br_2, the rate law in eqn 17A.6 shows that the rate depends on the concentration of the product HBr. To avoid this difficulty, the rate law should be fitted to the data throughout the reaction. The fitting may be done, in simple cases at least, by using a proposed rate law to predict the concentration of any component at any time, and comparing it with the data; methods based on this procedure are described in Topic 17B. A rate law should also be tested by observing whether the addition of products or, for gas-phase reactions, a change in the surface-to-volume ratio in the reaction chamber affects the rate.

Checklist of concepts

☐ 1. The rates of chemical reactions are measured by using techniques that monitor the concentrations of species present in the reaction mixture. Examples include **real-time** and **quenching** procedures, **flow** and **stopped-flow** techniques, and **flash photolysis**.

☐ 2. The **instantaneous rate** of consumption of a reactant or formation of a product is the slope of the tangent to the graph of concentration against time (expressed as a positive quantity).

☐ 3. The **rate of reaction** is defined in terms of the extent of reaction in such a way that it is independent of the species being considered.

☐ 4. A **rate law** is an expression for the reaction rate in terms of the concentrations of the species that occur in the overall chemical reaction.

☐ 5. The **order of a reaction** is the power to which a participant is raised in the rate law; the **overall order** is the sum of these powers.

Checklist of equations

Property	Equation	Comment	Equation number
Rate of a reaction	$v = (1/V)(\mathrm{d}\xi/\mathrm{d}t)$	Definition	17A.2
	$v = (1/\nu_J)(\mathrm{d}[J]/\mathrm{d}t)$	Constant-volume system	17A.3b
Rate law (in some cases)	$v = k_r[A]^a[B]^b\cdots$	$a, b, \ldots$: orders; $a + b + \cdots$: overall order	17A.7
Method of initial rates	$\log v_0 = \log k_{r,\mathrm{eff}} + a \log [A]_0$	Reactant A isolated	17A.11c

TOPIC 17B Integrated rate laws

➤ **Why do you need to know this material?**

You need the integrated rate law if you want to predict the composition of a reaction mixture as it approaches equilibrium. The integrated rate law is also the basis of determining the order and rate constants of a reaction, which is a necessary step in the formulation of the mechanism of the reaction.

➤ **What is the key idea?**

A rate law is a differential equation that can be integrated to find how the concentrations of reactants and products change with time.

➤ **What do you need to know already?**

You need to be familiar with the concepts of rate law, reaction order, and rate constant (Topic 17A). The manipulation of simple rate laws requires only elementary techniques of integration (see the *Resource section* for standard integrals).

Rate laws (Topic 17A) are differential equations, which can be integrated to predict how the concentrations of the reactants and products change with time. Even the most complex rate laws may be integrated numerically. However, in a number of simple cases analytical solutions, known as **integrated rate laws**, are easily obtained and prove to be very useful.

17B.1 Zeroth-order reactions

The rate of a zeroth-order reaction of the type A → P is constant (so long as reactant remains), so

$$\frac{d[A]}{dt} = -k_r$$

It follows that the change in concentration of A is simply its rate of consumption (which is $-k_r$) multiplied by the time t for which the reaction has been in progress:

$$[A]-[A]_0 = -k_r t$$

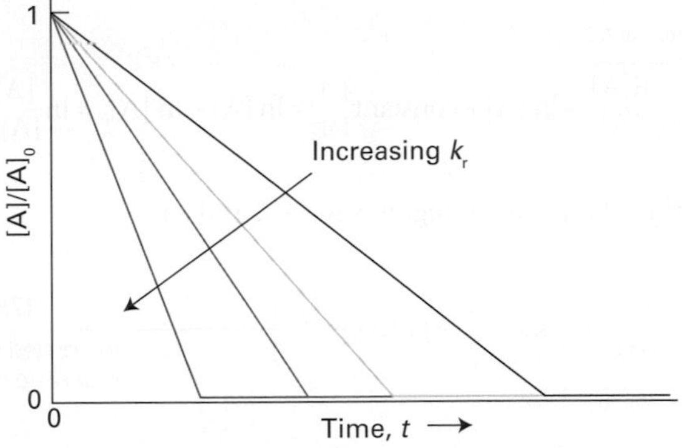

Figure 17B.1 The linear decay of the reactant in a zeroth-order reaction.

where [A] is the concentration of A at t and $[A]_0$ is the initial concentration of A. This expression rearranges to

$$[A]=[A]_0 - k_r t \qquad \text{Integrated zeroth-order rate law} \qquad (17B.1)$$

This expression applies until all the reactant has been used up at $t = [A]_0/k_r$, after which [A] remains zero (Fig. 17B.1).

17B.2 First-order reactions

Consider the first-order rate law

$$\frac{d[A]}{dt} = -k_r[A] \qquad (17B.2a)$$

This equation can be integrated to show how the concentration of A changes with time.

How is that done? 17B.1 Deriving the first-order integrated rate law

First, rearrange eqn 17B.2a into

$$\frac{d[A]}{[A]} = -k_r dt$$

and recognize that k_r is a constant independent of t. Initially (at $t = 0$) the concentration of A is $[A]_0$, and at a later time t it is

[A], so make these values the matching limits of the integrals and write

$$\int_{[A]_0}^{[A]} \frac{d[A]}{[A]} = -k_r \int_0^t dt$$

Because the integral of $1/x$ is $\ln x$ + constant, the left-hand side of this expression is

$$\overbrace{\int_{[A]_0}^{[A]} \frac{d[A]}{[A]} = \ln[A] + \text{constant}\Big|_{[A]_0}^{[A]}}^{\text{Integral A.2}} = \ln[A] - \ln[A]_0 = \ln\frac{[A]}{[A]_0}$$

The right-hand side integrates to $-k_r t$, and so

$$\ln\frac{[A]}{[A]_0} = -k_r t, \quad [A] = [A]_0 e^{-k_r t} \qquad (17B.2b)$$

Integrated first-order rate law

Equation 17B.2b shows that if $\ln([A]/[A]_0)$ is plotted against t, then a first-order reaction will give a straight line of slope $-k_r$. Some rate constants determined in this way are given in Table 17B.1. The second expression in eqn 17B.2b shows that in a first-order reaction the reactant concentration decreases exponentially with time with a rate determined by k_r (Fig. 17B.2).

Table 17B.1 Kinetic data for first-order reactions*

Reaction	Phase	θ/°C	k_r/s^{-1}	$t_{1/2}$
$2N_2O_5 \rightarrow 4NO_2 + O_2$	g	25	3.38×10^{-5}	5.70 h
	Br_2(l)	25	4.27×10^{-5}	4.51 h
$C_2H_6 \rightarrow 2CH_3$	g	700	5.36×10^{-4}	21.6 min

* More values are given in the *Resource section.*

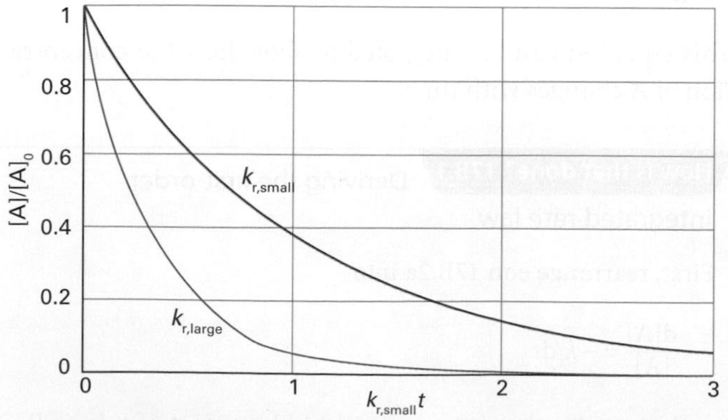

Figure 17B.2 The exponential decay of the reactant in a first-order reaction. The larger the rate constant, the more rapid is the decay: here $k_{r,large} = 3k_{r,small}$.

The integrated rate law in eqn 17B.2b can be expressed in terms of the concentration of the product P by noting that for a reaction A → P the increase in the concentration of P matches the decrease in concentration of A. If it is assumed that there is no P present at the start of the reaction, then $[P] = [A]_0 - [A]$, and hence $[A] = [A]_0 - [P]$. This expression for [A] can be substituted into eqn 17B.2b to give

$$\ln\frac{[A]_0 - [P]}{[A]_0} = -k_r t, \quad [P] = [A]_0(1 - e^{-k_r t}) \qquad (17B.2c)$$

A useful indication of the rate of a first-order chemical reaction is the **half-life**, $t_{1/2}$, of a substance, the time taken for the concentration of a reactant to fall to half its initial value. This quantity is readily obtained from the integrated rate law. Thus, the time for the concentration of A to decrease from $[A]_0$ to $\frac{1}{2}[A]_0$ in a first-order reaction is given by eqn 17B.2b as

$$k_r t_{1/2} = -\ln\frac{\frac{1}{2}[A]_0}{[A]_0} = -\ln\tfrac{1}{2} = \ln 2$$

Hence

$$t_{1/2} = \frac{\ln 2}{k_r} \qquad (17B.3)$$

Half-life [first-order reaction]

(Note that $\ln 2 = 0.693$.) The main point to note about this result is that for a first-order reaction, the half-life of a reactant is independent of its initial concentration. Therefore, if the concentration of A at some *arbitrary* stage of the reaction is [A], then it will have fallen to $\frac{1}{2}$[A] after a further interval of $(\ln 2)/k_r$. Some half-lives are given in Table 17B.1.

Example 17B.1 Analysing a first-order reaction

The variation in the partial pressure of azomethane with time was followed at 600 K, with the results given below. Confirm that the decomposition $CH_3N_2CH_3(g) \rightarrow CH_3CH_3(g) + N_2(g)$ is first-order in azomethane, and find the rate constant and half-life at 600 K.

t/s	0	1000	2000	3000	4000
p/Pa	10.9	7.63	5.32	3.71	2.59

Collect your thoughts To confirm that a reaction is first order, plot $\ln([A]/[A]_0)$ against time and expect a straight line. Because the partial pressure of a gas is proportional to its concentration, an equivalent procedure is to plot $\ln(p/p_0)$ against t. If a straight line is obtained, its slope can be identified with $-k_r$. The half-life is then calculated from k_r by using eqn 17B.3.

The solution Draw up the following table by using p_0 = 10.9 Pa:

t/s	0	1000	2000	3000	4000
p/p_0	1	0.700	0.488	0.340	0.238
$\ln(p/p_0)$	0	−0.357	−0.717	−1.078	−1.437

Figure 17B.3 shows the plot of $\ln(p/p_0)$ against $t/(10^3\ \mathrm{s})$. The plot is straight, confirming a first-order reaction, and its slope is −0.36. Therefore, the slope of the plot of $\ln(p/p_0)$ against t/s is -3.6×10^{-4} and $k_r = 3.6\times10^{-4}\ \mathrm{s}^{-1}$. It follows from eqn 17B.3 that the half-life is

$$t_{1/2} = \frac{\ln 2}{3.6\times10^{-4}\ \mathrm{s}^{-1}} = 1.9\times10^{3}\ \mathrm{s}$$

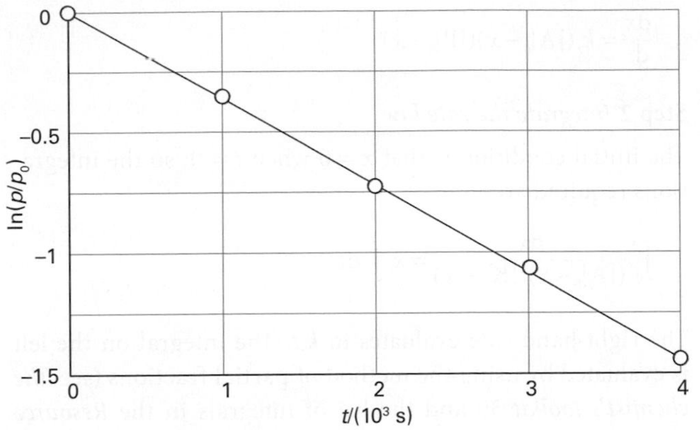

Figure 17B.3 The determination of the rate constant of a first-order reaction: a straight line is obtained when ln [A] (or, as here, ln p/p_0) is plotted against t; the slope is $-k_r$. The data plotted are from *Example* 17B.

Self-test 17B.1 In a particular experiment, it was found that the concentration of N_2O_5 in liquid bromine varied with time as follows:

t/s	0	200	400	600	1000
$[N_2O_5]/(\mathrm{mol\ dm^{-3}})$	0.110	0.073	0.048	0.032	0.014

Confirm that the reaction is first order in N_2O_5 and determine the rate constant.

Answer: $k_r = 2.1\times10^{-3}\ \mathrm{s}^{-1}$

17B.3 Second-order reactions

The integrated form of the second-order rate law,

$$\frac{\mathrm{d}[A]}{\mathrm{d}t} = -k_r[A]^2 \qquad (17B.4a)$$

can be found by much the same method as for first-order reactions.

How is that done? 17B.2 Deriving a second-order integrated rate law

To integrate eqn 17B.4a, first rearrange it into

$$\frac{\mathrm{d}[A]}{[A]^2} = -k_r \mathrm{d}t$$

The concentration is $[A]_0$ at $t = 0$ and at a later time t it is [A]. Therefore,

$$-\overbrace{\int_{[A]_0}^{[A]}\frac{\mathrm{d}[A]}{[A]^2}}^{\text{Integral A.1}} = k_r\int_0^t \mathrm{d}t$$

The integral on the left-hand side (including the minus sign) is

$$\left.\frac{1}{[A]}+\text{constant}\right|_{[A]_0}^{[A]} = \frac{1}{[A]}-\frac{1}{[A]_0}$$

and that of the right-hand side is $k_r t$. It follows that

$$\frac{1}{[A]}-\frac{1}{[A]_0} = k_r t, \quad [A] = \frac{[A]_0}{1+k_r t[A]_0} \qquad (17B.4b)$$

Integrated second-order rate law

Equation 17B.4b shows that for a second-order reaction a plot of 1/[A] against t is expected to be a straight line. The slope of the graph is k_r. Some rate constants determined in this way are given in Table 17B.2. The alternative form of the equation can be used to predict the concentration of A at any time after the start of the reaction. It shows that the concentration of A approaches zero more slowly than in a first-order reaction with the same initial rate (Fig. 17B.4).

Table 17B.2 Kinetic data for second-order reactions*

Reaction	Phase	$\theta/°\mathrm{C}$	$k_r/(\mathrm{dm^3\ mol^{-1}\ s^{-1}})$
$2\mathrm{NOBr} \rightarrow 2\mathrm{NO} + \mathrm{Br_2}$	g	10	0.80
$2\mathrm{I} \rightarrow \mathrm{I_2}$	g	23	7×10^9

* More values are given in the *Resource section*.

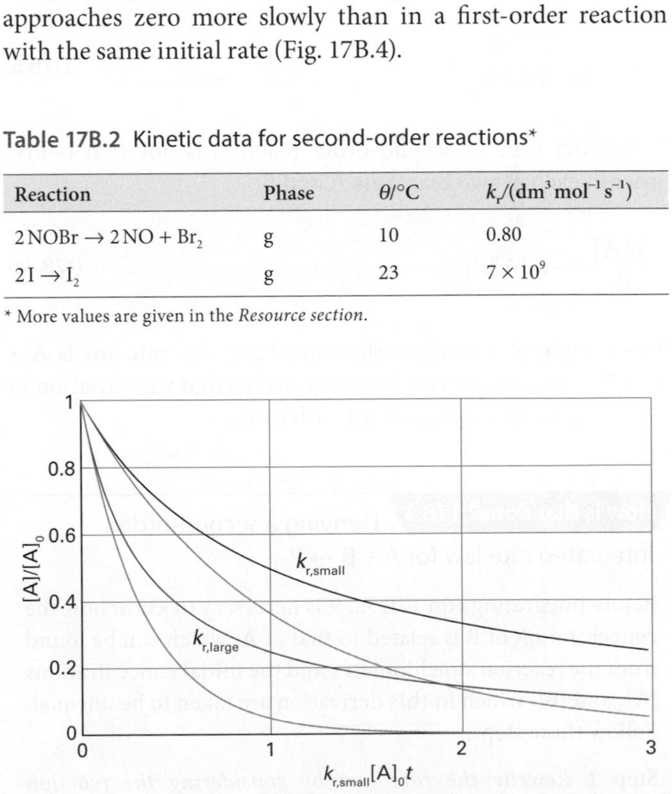

Figure 17B.4 The variation with time of the concentration of a reactant in a second-order reaction. The grey lines are the corresponding decays in a first-order reaction with the same initial rate. For this illustration, $k_{r,\text{large}} = 3k_{r,\text{small}}$.

As in the case of the first-order reactions, eqn 17B.4b can be rewritten in terms of the product P given the stoichiometry $A \rightarrow P$ by noting that $[A] = [A]_0 - [P]$. With this substitution, and after some rearrangement,

$$\frac{[P]}{([A]_0-[P])[A]_0}=k_r t, \quad [P]=\frac{k_r t[A]_0^2}{1+[A]_0 k_r t} \tag{17B.4c}$$

It follows from eqn 17B.4b by substituting $t = t_{1/2}$ and $[A] = \frac{1}{2}[A]_0$ that the half-life of a species A that is consumed in a second-order reaction is

$$t_{1/2}=\frac{1}{k_r[A]_0} \tag{17B.5}$$

Half-life [second-order reaction]

Therefore, unlike a first-order reaction, the half-life of a substance in a second-order reaction depends on the initial concentration. A practical consequence of this dependence is that species that decay by second-order reactions (which includes some environmentally harmful substances) may persist in low concentrations for long periods because their half-lives are long when their concentrations are low. In general, for an nth-order reaction (with $n > 1$) of the form $A \rightarrow P$, the half-life is related to the rate constant and the initial concentration of A by (see Problem P17B.15)

$$t_{1/2}=\frac{2^{n-1}-1}{(n-1)k_r[A]_0^{n-1}} \tag{17B.6}$$

Half-life [nth-order reaction, $n > 1$]

Another type of second-order reaction is one that is first order in each of two reactants A and B:

$$\frac{d[A]}{dt}=-k_r[A][B] \tag{17B.7a}$$

An example of a reaction that may have this rate law is $A + B \rightarrow P$. This rate law can be integrated to find the variation of the concentrations [A] and [B] with time.

How is that done? 17B.3 Deriving a second-order integrated rate law for $A + B \rightarrow P$

Before integrating eqn 17B.7a, it is necessary to know how the concentration of B is related to that of A, which can be found from the reaction stoichiometry and the initial concentrations $[A]_0$ and $[B]_0$ which in this derivation are taken to be unequal. Follow these steps.

Step 1 *Rewrite the rate law by considering the reaction stoichiometry*

It follows from the reaction stoichiometry that when the concentration of A has fallen to $[A]_0 - x$, the concentration of B will have fallen to $[B]_0 - x$ (because each A that disappears entails the disappearance of one B). Then eqn 17B.7a becomes

$$\frac{d[A]}{dt}=-k_r([A]_0-x)([B]_0-x)$$

Because $[A] = [A]_0 - x$, it follows that $d[A]/dt = -dx/dt$ and the rate law may be written

$$\frac{dx}{dt}=k_r([A]_0-x)([B]_0-x)$$

Step 2 *Integrate the rate law*

The initial condition is that $x = 0$ when $t = 0$; so the integrations required are

$$\int_0^x \frac{dx}{([A]_0-x)([B]_0-x)}=k_r\int_0^t dt$$

The right-hand side evaluates to $k_r t$. The integral on the left is evaluated by using the method of partial fractions (see *The chemist's toolkit* 30 and the list of integrals in the *Resource section*):

$$\overbrace{\int_0^x \frac{dx}{([A]_0-x)([B]_0-x)}}^{\text{Integral A.3}}=\frac{1}{[B]_0-[A]_0}\left\{\ln\frac{[A]_0}{[A]_0-x}-\ln\frac{[B]_0}{[B]_0-x}\right\}$$

The two logarithms can be combined as follows:

$$\ln\frac{[A]_0}{\underbrace{[A]_0-x}_{[A]}}-\ln\frac{[B]_0}{\underbrace{[B]_0-x}_{[B]}}=\ln\frac{[A]_0}{[A]}-\ln\frac{[B]_0}{[B]}$$

$$=\ln\frac{1}{[A]/[A]_0}-\ln\frac{1}{[B]/[B]_0}$$

$$=\ln\frac{[B]/[B]_0}{[A]/[A]_0}$$

Step 3 *Finalize the expression*

Combining all the results so far gives

$$\ln\frac{[B]/[B]_0}{[A]/[A]_0}=([B]_0-[A]_0)k_r t \tag{17B.7b}$$

Integrated rate law [second-order reaction of the type $A + B \rightarrow P$, with $[A]_0 \neq [B]_0$]

Therefore, a plot of the expression on the left against t should be a straight line from which k_r can be obtained. As shown in the following *Brief illustration*, the rate constant may be estimated quickly by using data from only two measurements.

Similar calculations may be carried out to find the integrated rate laws for other orders, and some are listed in Table 17B.3.

The chemist's toolkit 30 Integration by the method of partial fractions

To solve an integral of the form

$$I=\int\frac{1}{(a-x)(b-x)}\mathrm{d}x$$

where a and b are constants with $a\neq b$, use the **method of partial fractions** in which a fraction that is the product of terms (as in the denominator of this integrand) is written as a sum of fractions. To implement this procedure write the integrand as

$$\frac{1}{(a-x)(b-x)}=\frac{1}{b-a}\left(\frac{1}{a-x}-\frac{1}{b-x}\right)$$

Then integrate each term on the right. It follows that

$$I=\frac{1}{b-a}\left(\overbrace{\int\frac{\mathrm{d}x}{a-x}}^{\text{Integral A.2}}-\overbrace{\int\frac{\mathrm{d}x}{b-x}}^{\text{Integral A.2}}\right)$$
$$=\frac{1}{b-a}\left(\ln\frac{1}{a-x}-\ln\frac{1}{b-x}\right)+\text{constant}$$

Brief illustration 17B.1

Consider a second-order reaction of the type A + B → P carried out in a solution. Initially, the concentrations of reactants are $[\mathrm{A}]_0 = 0.075\,\mathrm{mol\,dm^{-3}}$ and $[\mathrm{B}]_0 = 0.050\,\mathrm{mol\,dm^{-3}}$. After 1.0 h the concentration of B has fallen to $[\mathrm{B}] = 0.020\,\mathrm{mol\,dm^{-3}}$. Because the change in the concentration of B is the same as that of A (and equal to x), it follows that during this time interval

$$x=(0.050-0.020)\,\mathrm{mol\,dm^{-3}}=0.030\,\mathrm{mol\,dm^{-3}}$$

Therefore, the concentration of A after 1.0 h is

$$[\mathrm{A}]=[\mathrm{A}]_0-x=(0.075-0.030)\,\mathrm{mol\,dm^{-3}}=0.045\,\mathrm{mol\,dm^{-3}}$$

and you are given that $[\mathrm{B}] = 0.020\,\mathrm{mol\,dm^{-3}}$. It follows from eqn 17B.7b that

$$k_\mathrm{r}=\frac{1}{((0.050-0.075)\,\mathrm{mol\,dm^{-3}})\times(3600\,\mathrm{s})}\ln\frac{0.020/0.050}{0.045/0.075}$$
$$=4.5\times10^{-3}\,\mathrm{dm^3\,mol^{-1}\,s^{-1}}$$

Table 17B.3 Integrated rate laws

Order	Reaction	Rate law and its integrated form*	$t_{1/2}$
0	A → P	$v=k_\mathrm{r}$	$[\mathrm{A}]_0/2k_\mathrm{r}$
		$k_\mathrm{r}t=[\mathrm{P}]$ for $0\le[\mathrm{P}]\le[\mathrm{A}]_0$,	
		$[\mathrm{A}]=[\mathrm{A}]_0-k_\mathrm{r}t$ for $0\le[\mathrm{A}]\le[\mathrm{A}]_0$	
1	A → P	$v=k_\mathrm{r}[\mathrm{A}]$	$(\ln 2)/k_\mathrm{r}$
		$k_\mathrm{r}t=\ln\frac{[\mathrm{A}]_0}{[\mathrm{A}]},\quad [\mathrm{A}]=[\mathrm{A}]_0\mathrm{e}^{-k_\mathrm{r}t},\quad [\mathrm{P}]=[\mathrm{A}]_0(1-\mathrm{e}^{-k_\mathrm{r}t})$	
2	A → P	$v=k_\mathrm{r}[\mathrm{A}]^2$	$1/k_\mathrm{r}[\mathrm{A}]_0$
		$k_\mathrm{r}t=\frac{[\mathrm{P}]}{[\mathrm{A}]_0([\mathrm{A}]_0-[\mathrm{P}])},\quad [\mathrm{A}]=\frac{[\mathrm{A}]_0}{1+k_\mathrm{r}t[\mathrm{A}]_0},\quad [\mathrm{P}]=\frac{k_\mathrm{r}t[\mathrm{A}]_0^2}{1+[\mathrm{A}]_0k_\mathrm{r}t}$	
	A + B → P	$v=k_\mathrm{r}[\mathrm{A}][\mathrm{B}]$	
		$k_\mathrm{r}t=\frac{1}{[\mathrm{B}]_0-[\mathrm{A}]_0}\ln\frac{[\mathrm{A}]_0([\mathrm{B}]_0-[\mathrm{P}])}{([\mathrm{A}]_0-[\mathrm{P}])[\mathrm{B}]_0}$,	
		$\ln\frac{[\mathrm{B}]/[\mathrm{B}]_0}{[\mathrm{A}]/[\mathrm{A}]_0}=([\mathrm{B}]_0-[\mathrm{A}]_0)k_\mathrm{r}t,\quad [\mathrm{P}]=\frac{[\mathrm{A}]_0[\mathrm{B}]_0(1-\mathrm{e}^{([\mathrm{B}]_0-[\mathrm{A}]_0)k_\mathrm{r}t})}{[\mathrm{A}]_0-[\mathrm{B}]_0\mathrm{e}^{([\mathrm{B}]_0-[\mathrm{A}]_0)k_\mathrm{r}t}}$	
	A + 2B → P	$v=k_\mathrm{r}[\mathrm{A}][\mathrm{B}]$	
		$k_\mathrm{r}t=\frac{1}{[\mathrm{B}]_0-2[\mathrm{A}]_0}\ln\frac{[\mathrm{A}]_0([\mathrm{B}]_0-2[\mathrm{P}])}{([\mathrm{A}]_0-[\mathrm{P}])[\mathrm{B}]_0},\quad [\mathrm{P}]=\frac{[\mathrm{A}]_0[\mathrm{B}]_0(1-\mathrm{e}^{([\mathrm{B}]_0-2[\mathrm{A}]_0)k_\mathrm{r}t})}{2[\mathrm{A}]_0-[\mathrm{B}]_0\mathrm{e}^{([\mathrm{B}]_0-2[\mathrm{A}]_0)k_\mathrm{r}t}}$	
3	A + 2B → P	$v=k_\mathrm{r}[\mathrm{A}][\mathrm{B}]^2$	
		$k_\mathrm{r}t=\frac{2[\mathrm{P}]}{(2[\mathrm{A}]_0-[\mathrm{B}]_0)([\mathrm{B}]_0-2[\mathrm{P}])[\mathrm{B}]_0}+\frac{1}{(2[\mathrm{A}]_0-[\mathrm{B}]_0)^2}\ln\frac{[\mathrm{A}]_0([\mathrm{B}]_0-2[\mathrm{P}])}{([\mathrm{A}]_0-[\mathrm{P}])[\mathrm{B}]_0}$	
		[P] must be determined graphically or numerically	
$n\ge2$	A → P	$v=k_\mathrm{r}[\mathrm{A}]^n$	$\frac{2^{n-1}-1}{(n-1)k_\mathrm{r}[\mathrm{A}]_0^{n-1}}$
		$k_\mathrm{r}t=\frac{1}{n-1}\left\{\frac{1}{([\mathrm{A}]_0-[\mathrm{P}])^{n-1}}-\frac{1}{[\mathrm{A}]_0^{n-1}}\right\}$	
		No simple general solution for [P] for $n>3$	

* $v=\mathrm{d}[\mathrm{P}]/\mathrm{d}t$

Checklist of concepts

- ☐ 1. An **integrated rate law** is an expression for the concentration of a reactant or product as a function of time (Table 17B.3).
- ☐ 2. The **half-life** of a reactant is the time it takes for its concentration to fall to half its initial value.
- ☐ 3. Analysis of experimental data using integrated rate laws allow for the prediction of the composition of a reaction system at any stage, the verification of the rate law, and the determination of the rate constant.

Checklist of equations

Property	Equation	Comment	Equation number
Integrated rate law	$[A] = [A]_0 - k_r t$	Zeroth order, A → P	17B.1
Integrated rate law	$\ln([A]/[A]_0) = -k_r t$ or $[A] = [A]_0 e^{-k_r t}$	First order, A → P	17B.2b
Half-life	$t_{1/2} = (\ln 2)/k_r$	First order, A → P	17B.3
Integrated rate law	$1/[A] - 1/[A]_0 = k_r t$ or $[A] = [A]_0/(1 + k_r t[A]_0)$	Second order, A → P	17B.4b
Half-life	$t_{1/2} = 1/k_r[A]_0$	Second order, A → P	17B.5
	$t_{1/2} = (2^{n-1} - 1)/(n-1)k_r[A]_0^{n-1}$	*n*th order, $n > 1$	17B.6
Integrated rate law	$\ln\{([B]/[B]_0)/([A]/[A]_0)\} = ([B]_0 - [A]_0)k_r t$	Second order, A + B → P	17B.7b

TOPIC 17C Reactions approaching equilibrium

➤ **Why do you need to know this material?**

All reactions tend towards equilibrium and the rate laws can be used to describe the changing concentrations as they approach that composition. The analysis of the time-dependence also reveals the connection between rate constants and equilibrium constants.

➤ **What is the key idea?**

Both forward and reverse reactions must be incorporated into a reaction scheme in order to account for the approach to equilibrium.

➤ **What do you need to know already?**

You need to be familiar with the concepts of rate law, reaction order, and rate constant (Topic 17A), integrated rate laws (Topic 17B), and equilibrium constants (Topic 6A). As in Topic 17B, the manipulation of simple rate laws requires only elementary techniques of integration.

In practice, most kinetic studies are made on reactions that are far from equilibrium and if products are in low concentration the reverse reactions are unimportant. Close to equilibrium, however, the products might be so abundant that the reverse reaction must be taken into account.

17C.1 First-order reactions approaching equilibrium

Considering a reaction in which A forms B and both forward and reverse reactions are first order (as in some isomerizations):

$$\mathrm{A}\rightarrow\mathrm{B} \qquad \frac{\mathrm{d}[\mathrm{A}]}{\mathrm{d}t}=-k_\mathrm{r}[\mathrm{A}]$$
$$\mathrm{B}\rightarrow\mathrm{A} \qquad \frac{\mathrm{d}[\mathrm{A}]}{\mathrm{d}t}=k_\mathrm{r}'[\mathrm{B}] \qquad (17\mathrm{C}.1)$$

The concentration of A is reduced by the forward reaction (at a rate $k_\mathrm{r}[\mathrm{A}]$) but it is increased by the reverse reaction (at a rate $k_\mathrm{r}'[\mathrm{B}]$). The net rate of change at any stage is therefore

$$\frac{\mathrm{d}[\mathrm{A}]}{\mathrm{d}t}=-k_\mathrm{r}[\mathrm{A}]+k_\mathrm{r}'[\mathrm{B}] \qquad (17\mathrm{C}.2)$$

If the initial concentration of A is $[\mathrm{A}]_0$, and no B is present initially, then at all times $[\mathrm{A}]+[\mathrm{B}]=[\mathrm{A}]_0$. Therefore,

$$\frac{\mathrm{d}[\mathrm{A}]}{\mathrm{d}t}=-k_\mathrm{r}[\mathrm{A}]+k_\mathrm{r}'([\mathrm{A}]_0-[\mathrm{A}]) = -(k_\mathrm{r}+k_\mathrm{r}')[\mathrm{A}]+k_\mathrm{r}'[\mathrm{A}]_0 \qquad (17\mathrm{C}.3)$$

The solution of this first-order differential equation (as may be checked by differentiation, Problem P17C.1) is

$$[\mathrm{A}]=\frac{k_\mathrm{r}'+k_\mathrm{r}\mathrm{e}^{-(k_\mathrm{r}+k_\mathrm{r}')t}}{k_\mathrm{r}+k_\mathrm{r}'}[\mathrm{A}]_0, \quad [\mathrm{B}]=[\mathrm{A}]_0-[\mathrm{A}] \qquad (17\mathrm{C}.4)$$

Figure 17C.1 shows the time dependence predicted by this equation.

As $t\rightarrow\infty$, the exponential term in eqn 17C.4 decreases to zero and the concentrations reach their equilibrium values, which are therefore

$$[\mathrm{A}]_\mathrm{eq}=\frac{k_\mathrm{r}'[\mathrm{A}]_0}{k_\mathrm{r}+k_\mathrm{r}'}, \quad [\mathrm{B}]_\mathrm{eq}=[\mathrm{A}]_0-[\mathrm{A}]_\mathrm{eq}=\frac{k_\mathrm{r}[\mathrm{A}]_0}{k_\mathrm{r}+k_\mathrm{r}'} \qquad (17\mathrm{C}.5)$$

It follows that the equilibrium constant of the reaction is

$$K=\frac{[\mathrm{B}]_\mathrm{eq}}{[\mathrm{A}]_\mathrm{eq}}=\frac{k_\mathrm{r}}{k_\mathrm{r}'} \qquad (17\mathrm{C}.6)$$

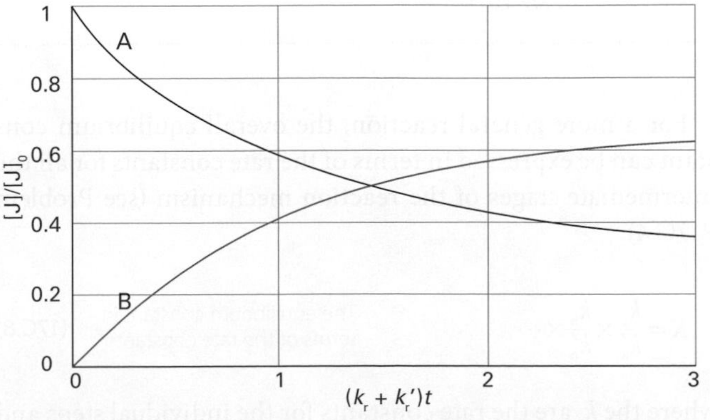

Figure 17C.1 The approach of concentrations to their equilibrium values as predicted by eqn 17C.4 for a reaction $\mathrm{A}\rightleftharpoons\mathrm{B}$ that is first order in each direction, and for which $k_\mathrm{r}=2k_\mathrm{r}'$.

(As explained in Topic 6A, the replacement of activities by the numerical values of molar concentrations is justified if the system is treated as ideal.) Exactly the same conclusion can be reached—more simply, in fact—by noting that, at equilibrium, the forward and reverse rates must be the same, so

$$k_r[A]_{eq} = k_r'[B]_{eq} \tag{17C.7}$$

This relation rearranges into eqn 17C.6. The theoretical importance of eqn 17C.6 is that it relates a thermodynamic quantity, the equilibrium constant, to quantities relating to rates. Its practical importance is that if one of the rate constants can be measured, then the other may be obtained if the equilibrium constant is known.

Equation 17C.6 is valid even if the forward and reverse reactions have different orders, but in that case more care needs to be taken with units. For instance, if the reaction A + B → C is second-order in the forward direction and first-order in the reverse direction, then the condition for equilibrium is $k_r[A]_{eq}[B]_{eq} = k_r'[C]_{eq}$ and the dimensionless equilibrium constant in full dress is

$$K = \frac{[C]_{eq}/c^{\ominus}}{([A]_{eq}/c^{\ominus})([B]_{eq}/c^{\ominus})} = \left(\frac{[C]}{[A][B]}\right)_{eq} c^{\ominus} = \frac{k_r}{k_r'} \times c^{\ominus}$$

The presence of $c^{\ominus} = 1\,\text{mol dm}^{-3}$ in the last term ensures that the ratio of second-order to first-order rate constants, with their different units, is turned into a dimensionless quantity.

Brief illustration 17C.1

The rate constants of the forward and reverse reactions for a dimerization reaction were found to be $8.0 \times 10^8\,\text{dm}^3\,\text{mol}^{-1}\,\text{s}^{-1}$ (second order) and $2.0 \times 10^6\,\text{s}^{-1}$ (first order). The equilibrium constant for the dimerization is therefore

$$K = \frac{8.0\times10^8\,\text{dm}^3\,\text{mol}^{-1}\,\text{s}^{-1}}{2.0\times10^6\,\text{s}^{-1}} \times 1\,\text{mol dm}^{-3} = 4.0\times10^2$$

For a more general reaction, the overall equilibrium constant can be expressed in terms of the rate constants for all the intermediate stages of the reaction mechanism (see Problem P17C.4):

$$K = \frac{k_a}{k_a'} \times \frac{k_b}{k_b'} \times \cdots \qquad \text{The equilibrium constant in terms of the rate constants} \tag{17C.8}$$

where the k_r are the rate constants for the individual steps and the k_r' are those for the corresponding reverse steps. The appropriate powers of $c^{\ominus}$ should be included in each factor if the orders of the forward and reverse reactions are different.

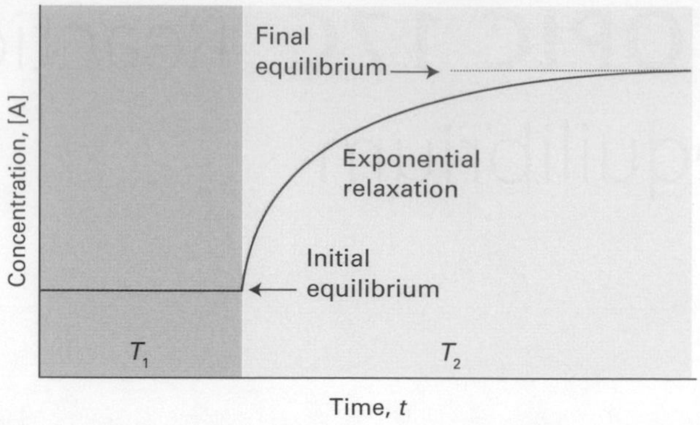

Figure 17C.2 The relaxation to the new equilibrium composition when a reaction initially at equilibrium at a temperature T_1 is subjected to a sudden change of temperature, which takes it to T_2.

17C.2 Relaxation methods

The term **relaxation** denotes the return of a system to equilibrium. It is used in chemical kinetics to indicate that an externally applied influence has shifted the equilibrium position of a reaction, often suddenly, and that the concentrations of the species involved then adjust towards the equilibrium values characteristic of the new conditions (Fig. 17C.2).

Consider the response of reaction rates to a **temperature jump**, a sudden change in temperature. As explained in Topic 6B, provided $\Delta_r H^{\ominus}$ is non-zero the equilibrium composition of a reaction depends on the temperature, so a sudden shift in temperature acts as a perturbation on the system. One way of achieving a temperature jump is to subject the sample that has been made conducting by the addition of ions to an electric discharge; bursts of microwave radiation or intense electromagnetic pulses from lasers can also be used. Electrical discharges can achieve temperature jumps of between 5 and 10 K in about 1 μs. The high energy output of a pulsed laser is sufficient to generate temperature jumps of between 10 and 30 K within nanoseconds in aqueous samples.

The response of a system to a sudden temperature increase can be analysed by considering the rate laws for the forward and reverse reactions and the temperature dependence of the rate constants.

How is that done? 17C.1 Exploring the response to a temperature jump

First consider a simple A ⇌ B equilibrium that is first order in each direction and then an equilibrium A ⇌ B + C that is first order forward and second order reverse. In each case, when the temperature is increased suddenly, the rate constants change from their original values to the new values k_r and k_r' characteristic of the new temperature, but the concentrations of A and B remain for an instant at their old equilibrium values.

(a) The equilibrium A ⇌ B (first order forward and reverse)

As the system immediately after the temperature jump is no longer at equilibrium, it readjusts to the new equilibrium concentrations, which are now given by $k_r[A]_{eq} = k_r'[B]_{eq}$, and it does so at a rate that depends on the new rate constants. Let the deviation of [A] from its new equilibrium value be x, so $[A] = [A]_{eq} + x$; then the reaction stoichiometry implies that $[B] = [B]_{eq} - x$. At the new temperature the concentration of A changes as follows:

$$\frac{d[A]}{dt} = -k_r[A] + k_r'[B]$$

$$= -k_r([A]_{eq} + x) + k_r'([B]_{eq} - x)$$

$$= \overbrace{-k_r[A]_{eq} + k_r'[B]_{eq}}^{\text{Cancel}} - (k_r + k_r')x$$

$$= -(k_r + k_r')x$$

Because $d[A]/dt = dx/dt$, this equation is a first-order differential equation with a solution that resembles eqn 17B.2b. If x_0 is the deviation from equilibrium immediately after the temperature jump, the time-dependence of x is

$$x = x_0 e^{-t/\tau} \qquad \tau = \frac{1}{k_r + k_r'} \qquad (17C.9a)$$

Relaxation after a temperature jump [first-order reactions]

(b) The equilibrium A ⇌ B + C (first order forward and second order reverse)

As in the previous derivation, the system immediately after the temperature jump is no longer at equilibrium, so it readjusts to the new equilibrium concentrations, which are now given by $k_r[A]_{eq} = k_r'[B]_{eq}[C]_{eq}$, and it does so at a rate that depends on the new rate constants. Let the deviation of [A] from its new equilibrium value be x, so $[A] = [A]_{eq} + x$; then the reaction stoichiometry implies that $[B] = [B]_{eq} - x$ and $[C] = [C]_{eq} - x$.

Step 1 *Set up and solve the rate equations*

At the new temperature the concentration of A changes as follows:

$$\frac{d[A]}{dt} = -k_r[A] + k_r'[B][C]$$

$$= -k_r([A]_{eq} + x) + k_r'([B]_{eq} - x)([C]_{eq} - x)$$

$$= -\{k_r + k_r'([B]_{eq} + [C]_{eq})\}x \overbrace{- k_r[A]_{eq} + k_r'[B]_{eq}[C]_{eq}}^{0} + \overbrace{k_r'x^2}^{\text{Neglect}}$$

$$= -\{k_r + k_r'([B]_{eq} + [C]_{eq})\}x$$

As before, $d[A]/dt = dx/dt$; the solution of this differential equation is an exponential decay proportional to $e^{-t/\tau}$ with τ given by

$$\frac{1}{\tau} = k_r + k_r'([B]_{eq} + [C]_{eq})$$

Step 2 *Relate the equilibrium concentrations by introducing an equilibrium constant*

The equilibrium constant for the reaction (assuming ideal solutions) is

$$K = \frac{([B]_{eq}/c^{\ominus})([C]_{eq}/c^{\ominus})}{([A]_{eq}/c^{\ominus})} = \frac{[B]_{eq}[C]_{eq}}{[A]_{eq}c^{\ominus}}$$

The reaction stoichiometry implies that the concentrations of B and C are the same, so

$$[B]_{eq} = [C]_{eq} = \overbrace{(K[A]_{eq}c^{\ominus})^{1/2}}^{a}$$

and the time constant becomes

$$\frac{1}{\tau} = k_r + 2ak_r' = k_r'\left(\frac{k_r}{k_r'} + 2a\right)$$

Step 3 *Identify the equilibrium constant of the reaction*

You should now recognize that the ratio of the rate constants is a form of the equilibrium constant at the new temperature. Specifically:

$$k_r[A]_{eq} = k_r'[B]_{eq}[C]_{eq} \text{ and } K = \frac{[B]_{eq}[C]_{eq}}{[A]_{eq}c^{\ominus}}$$

implying that

$$\frac{k_r}{k_r'} = \frac{[B]_{eq}[C]_{eq}}{[A]_{eq}} = Kc^{\ominus}$$

and therefore

$$\frac{1}{\tau} = k_r'(Kc^{\ominus} + 2a)$$

The time dependence of x is therefore

$$x = x_0 e^{-t/\tau} \qquad \tau = \frac{1}{k_r'(Kc^{\ominus} + 2a)} \qquad a = (K[A]_{eq}c^{\ominus})^{1/2} \qquad (17C.9b)$$

Relaxation after a temperature jump [mixed-order reaction]

Checklist of concepts

☐ 1. There is a relation between the equilibrium constant, a thermodynamic quantity, and the rate constants of the forward and reverse reactions (see below).

☐ 2. In **relaxation methods** of kinetic analysis, the equilibrium position of a reaction is shifted suddenly and then the time-dependence of the concentration of the species involved is followed.

Checklist of equations

Property	Equation	Comment	Equation number
Equilibrium constant in terms of rate constants	$K = k_a/k'_a \times k_b/k'_b \times \cdots$	Include $c^{\ominus}$ as appropriate	17C.8
Relaxation of an equilibrium $A \rightleftharpoons B$ after a temperature jump	$x = x_0 e^{-t/\tau}$ $\tau = 1/(k_r + k'_r)$	First order in each direction	17C.9a

TOPIC 17D The Arrhenius equation

> ➤ Why do you need to know this material?
>
> Exploration of the dependence of reaction rates on temperature leads to the formulation of theories that reveal the details of the processes that occur when reactant molecules meet and undergo reaction.
>
> ➤ What is the key idea?
>
> The temperature dependence of the rate of a reaction depends on the activation energy, the minimum energy needed for reaction to occur in an encounter between reactants.
>
> ➤ What do you need to know already?
>
> You need to know that the rate of a chemical reaction is expressed by a rate constant (Topic 17A).

Chemical reactions usually go faster as the temperature is raised. It is found experimentally for many reactions that a plot of $\ln k_r$ against $1/T$ gives a straight line with a negative slope, indicating that an increase in $\ln k_r$ (and therefore an increase in k_r) results from a decrease in $1/T$ (i.e. an increase in T).

17D.1 The temperature dependence of reaction rates

The temperature dependence characteristic of a reaction is normally expressed mathematically by introducing two parameters, one representing the intercept and the other the slope of the straight line of a so-called 'Arrhenius plot' of $\ln k_r$ against $1/T$ and writing the **Arrhenius equation**

$$\ln k_r = \ln A - \frac{E_a}{RT} \qquad \text{Arrhenius equation} \qquad (17D.1)$$

The parameter A, which is obtained from the intercept of the line at $1/T = 0$ (at infinite temperature, Fig. 17D.1), is called the **frequency factor** (and still commonly the *pre-exponential factor*). The parameter E_a, which is obtained from the slope of the line (which is equal to $-E_a/R$), is called the **activation energy**. Collectively the two quantities are called the **Arrhenius parameters** (Table 17D.1).

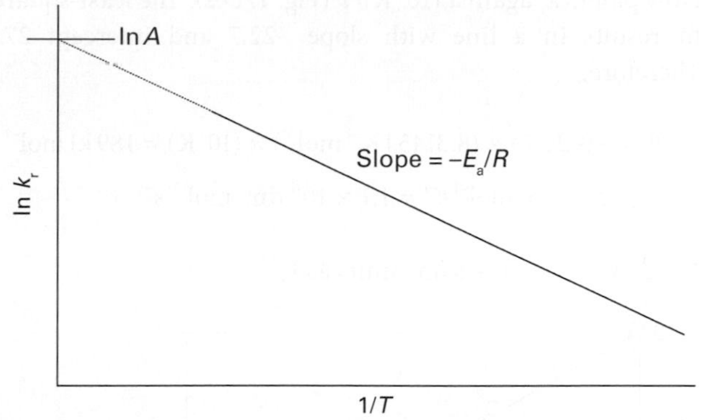

Figure 17D.1 An Arrhenius plot, a plot of $\ln k_r$ against $1/T$, is a straight line when the reaction follows the behaviour described by the Arrhenius equation (eqn 17D.1). The slope is $-E_a/R$ and the intercept at $1/T = 0$ is $\ln A$.

Table 17D.1 Arrhenius parameters*

(1) First-order reactions	Phase	$A/\mathrm{s^{-1}}$	$E_a/(\mathrm{kJ\,mol^{-1}})$
$CH_3NC \rightarrow CH_3CN$	gas	3.98×10^{13}	160
$2N_2O_5 \rightarrow 4NO_2 + O_2$	gas	4.94×10^{13}	103.4
(2) Second-order reactions	**Phase**	$A/(\mathrm{dm^3\,mol^{-1}\,s^{-1}})$	$E_a/(\mathrm{kJ\,mol^{-1}})$
$OH + H_2 \rightarrow H_2O + H$	gas	8.0×10^{10}	42
$NaC_2H_5O + CH_3I$	in ethanol	2.42×10^{11}	81.6

* More values are given in the *Resource section*.

Example 17D.1 Determining the Arrhenius parameters

The rate of the second-order decomposition of ethanal (acetaldehyde, CH_3CHO) was measured over the temperature range 700–1000 K, and the rate constants are reported below. Find E_a and A.

T/K	700	730	760	790	810	840	910	1000
$k_r/(\mathrm{dm^3\,mol^{-1}\,s^{-1}})$	0.011	0.035	0.105	0.343	0.789	2.17	20.0	145

Collect your thoughts According to eqn 17D.1, the data can be analysed by plotting $\ln(k_r/\mathrm{dm^3\,mol^{-1}\,s^{-1}})$ against $1/(T/\mathrm{K})$, or more conveniently $(10^3\,\mathrm{K})/T$, expecting to get a straight line. The activation energy is found from the dimensionless slope by writing $-E_a/R = slope$/units, where in this case 'units' = $1/(10^3\,\mathrm{K})$, so $E_a = -slope \times R \times 10^3\,\mathrm{K}$. The intercept at $1/T = 0$

is $\ln(A/\text{dm}^3\,\text{mol}^{-1}\,\text{s}^{-1})$. Use a least-squares procedure to determine the slope and the intercept.

The solution Draw up the following table:

$(10^3\,\text{K})/T$	1.43	1.37	1.32	1.27	1.23	1.19	1.10	1.00
$\ln(k_r/\text{dm}^3\,\text{mol}^{-1}\,\text{s}^{-1})$	−4.51	−3.35	−2.25	−1.07	−0.24	0.77	3.00	4.98

Now plot $\ln k_r$ against $(10^3\,\text{K})/T$ (Fig. 17D.2). The least-squares fit results in a line with slope −22.7 and intercept 27.7. Therefore,

$$E_a = -(-22.7) \times (8.3145\,\text{J K}^{-1}\,\text{mol}^{-1}) \times (10^3\,\text{K}) = 189\,\text{kJ mol}^{-1}$$
$$A = e^{27.7}\,\text{dm}^3\,\text{mol}^{-1}\,\text{s}^{-1} = 1.1 \times 10^{12}\,\text{dm}^3\,\text{mol}^{-1}\,\text{s}^{-1}$$

Note that A has the same units as k_r.

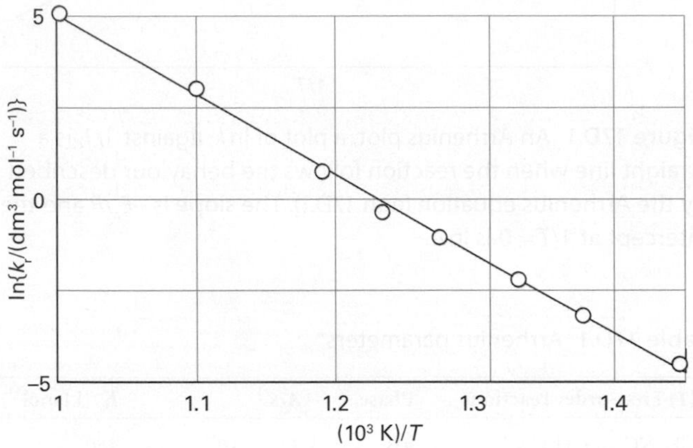

Figure 17D.2 The Arrhenius plot using the data in *Example* 17D.1.

Self-test 17D.1 Determine A and E_a from the following data:

T/K	300	350	400	450	500
$k_r/(\text{dm}^3\,\text{mol}^{-1}\,\text{s}^{-1})$	7.9×10^6	3.0×10^7	7.9×10^7	1.7×10^8	3.2×10^8

Answer: $8\times10^{10}\,\text{dm}^3\,\text{mol}^{-1}\,\text{s}^{-1}$, 23 kJ mol^{-1}

Once the activation energy of a reaction is known, the value of a rate constant $k_{r,2}$ at a temperature T_2 can be predicted from its value $k_{r,1}$ at another temperature T_1. To do so, write

$$\ln k_{r,1} = \ln A - \frac{E_a}{RT_1} \qquad \ln k_{r,2} = \ln A - \frac{E_a}{RT_2}$$

and then subtract the first from the second to obtain

$$\ln k_{r,2} - \ln k_{r,1} = -\frac{E_a}{RT_2} + \frac{E_a}{RT_1}$$

which can be rearranged into

$$\ln\frac{k_{r,2}}{k_{r,1}} = \frac{E_a}{R}\left(\frac{1}{T_1} - \frac{1}{T_2}\right) \qquad (17\text{D}.2)$$

Brief illustration 17D.1

For a reaction with an activation energy of 50 kJ mol^{-1}, an increase in the temperature from 25 °C to 37 °C (body temperature) corresponds to

$$\ln\frac{k_{r,2}}{k_{r,1}} = \frac{50\times10^3\,\text{J mol}^{-1}}{8.3145\,\text{J K}^{-1}\,\text{mol}^{-1}}\left(\frac{1}{298\,\text{K}} - \frac{1}{310\,\text{K}}\right)$$
$$= \frac{50\times10^3}{8.3145}\left(\frac{1}{298} - \frac{1}{310}\right) = 0.781\ldots$$

By taking natural antilogarithms (that is, by forming e^x), $k_{r,2} = 2.18k_{r,1}$. This result corresponds to slightly more than a doubling of the rate constant as the temperature is increased from 298 K to 310 K.

The fact that E_a is given by the slope of the plot of $\ln k_r$ against $1/T$ leads to the following conclusions:

- A high activation energy signifies that the rate constant depends strongly on temperature.
- If a reaction has zero activation energy, its rate is independent of temperature.
- A negative activation energy indicates that the rate decreases as the temperature is raised.

Physical interpretation

For some reactions it is found that a plot of $\ln k_r$ against $1/T$ does not give a straight line. It is still possible to define an activation energy for these 'non-Arrhenius reactions' at a particular temperature by writing

$$-\frac{E_a}{R} = \frac{d\ln k_r}{d(1/T)} \overset{d(1/T)/dT=-1/T^2}{=} -T^2\frac{d\ln k_r}{dT}$$

and therefore

$$E_a = RT^2\left(\frac{d\ln k_r}{dT}\right) \qquad \text{Activation energy [definition]} \qquad (17\text{D}.3)$$

This expression is the formal definition of activation energy. It reduces to the earlier one (as the slope of a straight line) for a temperature-independent activation energy (see Problem P17D.1). Non-Arrhenius behaviour is sometimes a sign that quantum mechanical tunnelling is playing a significant role in the reaction (Topic 7D). In biological reactions it might signal that an enzyme has undergone a structural change and has become less efficient.

17D.2 The interpretation of the Arrhenius parameters

For the present Topic the Arrhenius parameters are regarded as purely empirical quantities which summarize the variation of rate constants with temperature. Focus 18 provides a more elaborate interpretation.

(a) A first look at the energy requirements of reactions

To interpret E_a, consider how the molecular potential energy changes in the course of a chemical reaction that begins with a collision between A and B molecules (Fig. 17D.3). In the gas phase that step is an actual collision; in solution it is best regarded as a close encounter, possibly with excess energy, which might involve the solvent too. As the reaction event proceeds, A and B come into contact, distort, and begin to exchange or discard atoms. The **reaction coordinate** summarizes the collection of motions, such as changes in interatomic distances and bond angles, that are directly involved in the formation of products from reactants. The reaction coordinate is essentially a geometrical concept and quite distinct from the extent of reaction. The potential energy rises to a maximum and the cluster of atoms that corresponds to the region close to the maximum is called the **activated complex**.

After the maximum, the potential energy falls as the atoms in the cluster rearrange and eventually it reaches a value characteristic of the products. The climax of the reaction is at the peak of the potential energy curve, which corresponds to the activation energy E_a. Here two reactant molecules have come to such a degree of closeness and distortion that a small further distortion will send them in the direction of products. This crucial configuration is called the **transition state** of the reaction. Although some molecules entering the transition state might revert to reactants, if they pass through this configuration then it is inevitable that products will emerge from the encounter.

A note on good practice The terms 'activated complex' and 'transition state' are often used as synonyms; however, there is a distinction, which is best kept in mind. An activated complex is a cluster of atoms that corresponds to the region close to the maximum; a transition state is a conformation of the atoms in the activated complex that, after a small further distortion, leads inevitably to products.

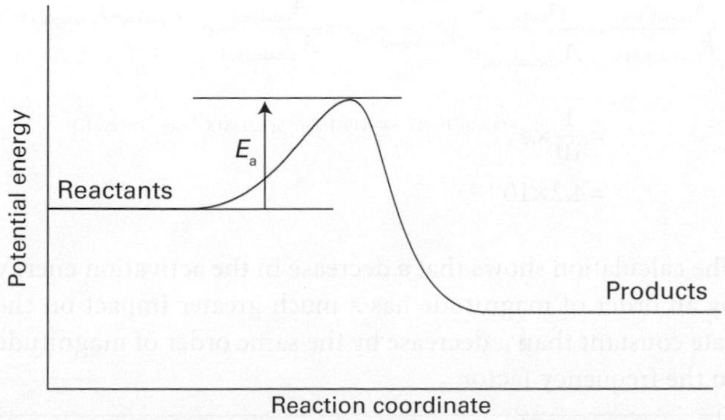

Figure 17D.3 A potential energy profile for an exothermic reaction. The height of the barrier between the reactants and products is the activation energy of the forward reaction.

The conclusion from the preceding discussion is that

The activation energy is the minimum energy reactants must have in order to form products.

For example, in a reaction mixture there are numerous molecular encounters each second, but only very few are sufficiently energetic to lead to reaction. The fraction of close encounters between reactants with energy in excess of E_a is given by the Boltzmann distribution (*Prologue* and Topic 13A) as $e^{-E_a/RT}$. This interpretation is confirmed by comparing this expression with the Arrhenius equation written in the form

$$k_r = Ae^{-E_a/RT} \quad \text{Arrhenius equation [alternative form]} \quad (17D.4)$$

which is obtained by taking antilogarithms of both sides of eqn 17D.1. Insight into this expression can be obtained by considering the role of the Boltzmann distribution for a simple model system.

How is that done? 17D.1 Interpreting the exponential factor in the Arrhenius equation

Suppose the energy levels available to the system form a uniform array of separation ε such that the energy levels are $i\varepsilon$, with $i = 0, 1, 2, \ldots$ (Fig. 17D.4).

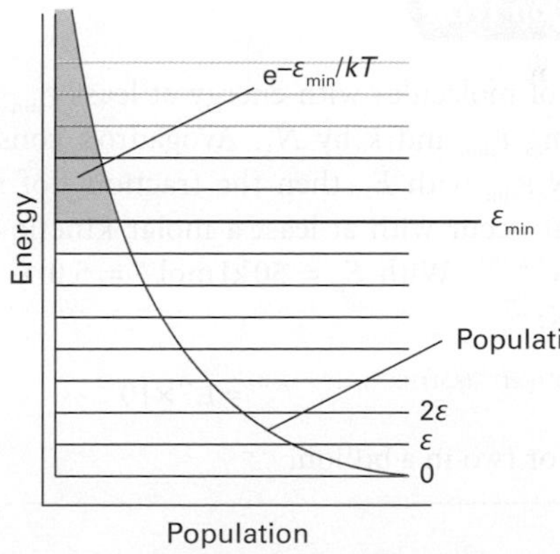

Figure 17D.4 Equally spaced energy levels of a model system. As shown in the text, the fraction of molecules with energy of at least ε_{min} is $e^{-\varepsilon_{min}/kT}$.

The Boltzmann distribution for this system is

$$\frac{N_i}{N} = \frac{e^{-i\varepsilon\beta}}{q} = (1-e^{-\varepsilon\beta})e^{-i\varepsilon\beta}$$

where N_i is the number of molecules in state i, N is the total number of molecules, $\beta = 1/kT$, and the partition function q comes from the result in eqn 13B.2a. The total number of molecules in states with energy greater than or equal to $i_{\min}\varepsilon$ is

$$\sum_{i=i_{\min}}^{\infty} N_i = \sum_{i=0}^{\infty} N_i - \sum_{i=0}^{i_{\min}-1} N_i = N - \frac{N}{q}\sum_{i=0}^{i_{\min}-1} e^{-i\varepsilon\beta}$$

$N_i = (N/q)e^{-i\varepsilon\beta}$

$$= N - \frac{N}{q}\sum_{i=0}^{i_{\min}-1} (e^{-\varepsilon\beta})^i$$

The sum in blue is a finite geometrical series of the form $1 + r + r^2 + \cdots$, with $r = e^{-\varepsilon\beta}$. The sum of $n-1$ terms of such a series is $(1-r^n)/(1-r)$. The term in blue can therefore be written

$$\sum_{i=0}^{i_{\min}-1} (e^{-\varepsilon\beta})^i = \frac{1-e^{-i_{\min}\varepsilon\beta}}{1-e^{-\varepsilon\beta}}$$

$1/(1-e^{-\varepsilon\beta}) = q$

$$= q(1-e^{-i_{\min}\varepsilon\beta})$$

Therefore, the fraction of molecules in states with energy of at least $\varepsilon_{\min} = i_{\min}\varepsilon$ is

$$\frac{1}{N}\sum_{i=i_{\min}}^{\infty} N_i = \frac{1}{N}\left\{N - \frac{N}{q}\times q(1-e^{-i_{\min}\varepsilon\beta})\right\}$$

$$= 1-(1-e^{-i_{\min}\varepsilon\beta}) = e^{-i_{\min}\varepsilon\beta} = e^{-\varepsilon_{\min}/kT}$$

which has the form of eqn 17D.4.

Brief illustration 17D.2

The fraction of molecules with energy at least $\varepsilon_{\min}$ is $e^{-\varepsilon_{\min}/kT}$. By multiplying $\varepsilon_{\min}$ and k by N_A, Avogadro's constant, and identifying $N_A\varepsilon_{\min}$ with E_a, then the fraction f of molecular collisions that occur with at least a molar kinetic energy E_a becomes $f = e^{-E_a/RT}$. With $E_a = 50\,\text{kJ mol}^{-1} = 5.0\times10^4\,\text{J mol}^{-1}$ and $T = 298\,\text{K}$:

$$f = e^{-(5.0\times10^4\,\text{J mol}^{-1})/(8.3145\,\text{J K}^{-1}\,\text{mol}^{-1}\times 298\,\text{K})} = 1.7\times10^{-9}$$

or about one or two in a billion.

If the activation energy is zero, each collision leads to reaction and, according to the Arrhenius equation, the rate constant is equal to the frequency factor A. This factor can therefore be identified as the rate constant in the limit that each collision is successful. The exponential factor, $e^{-E_a/RT}$, gives the fraction of collisions that are sufficiently energetic to be successful, so the rate constant is reduced from A to $Ae^{-E_a/RT}$.

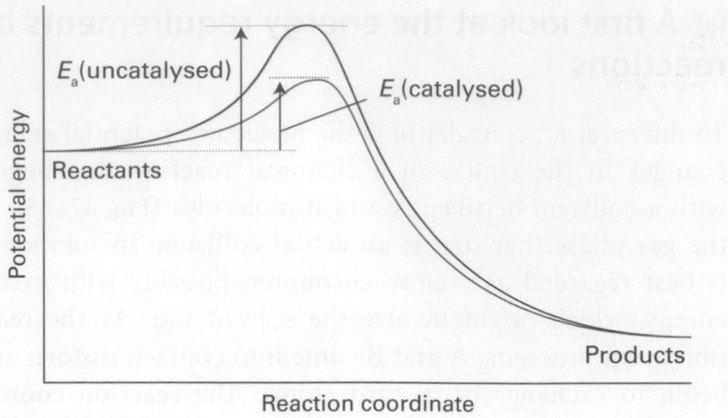

Figure 17D.5 A catalyst provides a different path with a lower activation energy. The result is an increase in the rate of the reaction (in both directions).

(b) The effect of a catalyst on the activation energy

The Arrhenius equation predicts that the rate constant of a reaction can be increased by increasing the temperature or by decreasing the activation energy. Changing the temperature of a reaction mixture is easy to do. Reducing the activation energy is more challenging, but is possible if a reaction takes place in the presence of a suitable **catalyst**, a substance that accelerates a reaction but undergoes no net chemical change. The catalyst lowers the activation energy of the reaction by providing an alternative path (Fig. 17D.5).

Brief illustration 17D.3

Enzymes are biological catalysts. Suppose that an enzyme reduces both the activation energy and the frequency factor of a reaction by a factor of ten. Letting the activation energy change from $80\,\text{kJ mol}^{-1}$ to $8\,\text{kJ mol}^{-1}$, and using eqn 17D.4, the ratio of rate constants at 298 K is

$$\frac{k_{r,\text{catalysed}}}{k_{r,\text{uncatalysed}}} = \frac{A_{\text{catalysed}}e^{-E_{a,\text{catalysed}}/RT}}{A_{\text{uncatalysed}}e^{-E_{a,\text{uncatalysed}}/RT}} = \frac{A_{\text{catalysed}}}{A_{\text{uncatalysed}}}e^{-(E_{a,\text{catalysed}}-E_{a,\text{uncatalysed}})/RT}$$

$$= \frac{1}{10}\times e^{-(8\times10^3\,\text{J mol}^{-1} - 80\times10^3\,\text{J mol}^{-1})/(8.3145\,\text{J K}^{-1}\,\text{mol}^{-1})\times(298\,\text{K})}$$

$$= 4.2\times10^{11}$$

The calculation shows that a decrease in the activation energy by an order of magnitude has a much greater impact on the rate constant than a decrease by the same order of magnitude in the frequency factor.

Checklist of concepts

☐ 1. The **activation energy**, the parameter E_a in the **Arrhenius equation**, is the minimum energy that a molecular encounter needs in order to result in reaction.

☐ 2. The higher the activation energy, the more sensitive the rate constant is to the temperature.

☐ 3. The **frequency factor** is the rate constant in the limit that all encounters, irrespective of their energy, lead to reaction.

☐ 4. A **catalyst** lowers the activation energy of a reaction.

Checklist of equations

Property or process	Equation	Comment	Equation number
Arrhenius equation	$\ln k_r = \ln A - E_a/RT$		17D.1
Activation energy	$E_a = RT^2(\mathrm{d}\ln k_r/\mathrm{d}T)$	General definition	17D.3
Arrhenius equation	$k_r = Ae^{-E_a/RT}$	Alternative form	17D.4

TOPIC 17E Reaction mechanisms

> ➤ Why do you need to know this material?
>
> The ability to construct the rate law for a reaction that takes place by a sequence of steps provides insight into chemical reactions at the molecular level and also suggests how the yield of desired products can be optimized.
>
> ➤ What is the key idea?
>
> Many chemical reactions occur as a sequence of simpler steps, with corresponding rate laws that can be combined into an overall rate law by applying a variety of approximations.
>
> ➤ What do you need to know already?
>
> You need to be familiar with the concept of rate laws (Topic 17A), and how to integrate them (Topics 17B and 17C). You also need to be familiar with the Arrhenius equation for the effect of temperature on the rate constant (Topic 17D).

The study of reaction rates leads to an understanding of the **mechanism** of a reaction, its analysis into a sequence of elementary steps. Simple elementary steps have simple rate laws, which can be combined into an overall rate law by invoking one or more approximations.

17E.1 Elementary reactions

Many reactions occur in a sequence of steps called **elementary reactions**, each of which involves only a small number of molecules or ions. A typical elementary reaction is

$$H + Br_2 \rightarrow HBr + Br$$

Note that the phase of the species is not specified in the chemical equation for an elementary reaction and the equation represents the specific process occurring to individual molecules. This equation, for instance, signifies that an H atom attacks a Br_2 molecule to produce an HBr molecule and a Br atom.

The **molecularity** of an elementary reaction is the number of molecules coming together to react in an elementary reaction. In a **unimolecular reaction**, a single molecule shakes itself apart or its atoms into a new arrangement, as in the isomerization of cyclopropane to propene. In a **bimolecular reaction**, a pair of molecules collide and exchange energy, atoms, or groups of atoms, or undergo some other kind of change. It is important to distinguish molecularity from order:

- *reaction order* is an empirical quantity, and obtained from the experimentally determined rate law;
- *molecularity* refers to an elementary reaction proposed as an individual step in a mechanism.

The rate law of a unimolecular elementary reaction is first-order in the reactant:

$$A \rightarrow P \qquad \frac{d[A]}{dt} = -k_r[A] \qquad \text{Unimolecular elementary reaction} \qquad (17E.1)$$

where P denotes products (several different species may be formed). A unimolecular reaction is first order because the number of A molecules that decay in a short interval is proportional to the number available to decay. For instance, ten times as many decay in the same interval when there are initially 1000 A molecules as when there are only 100 present. Therefore, the rate of decomposition of A is proportional to its concentration at any moment during the reaction.

An elementary bimolecular reaction has a second-order rate law:

$$A + B \rightarrow P \qquad \frac{d[A]}{dt} = -k_r[A][B] \qquad \text{Bimolecular elementary reaction} \qquad (17E.2)$$

A bimolecular reaction is second order because its rate is proportional to the rate at which the reactant species meet, which in turn is proportional to both their concentrations. Therefore, if there is evidence that a reaction is a single-step, bimolecular process, the rate law can simply be written down as in eqn 17E.2 (and then tested against experimental data).

Brief illustration 17E.1

Bimolecular elementary reactions are believed to account for many homogeneous reactions, such as the dimerizations of alkenes and dienes and reactions such as

$$CH_3I(alc) + CH_3CH_2O^-(alc) \rightarrow CH_3OCH_2CH_3(alc) + I^-(alc)$$

(where 'alc' signifies alcohol solution). There is evidence that the mechanism of this reaction is a single elementary step:

$$CH_3I + CH_3CH_2O^- \rightarrow CH_3OCH_2CH_3 + I^-$$

This mechanism is consistent with the observed rate law

$v = k_r[CH_3I][CH_3CH_2O^-]$

The following sections describe how a series of simple steps can be combined into a mechanism and how the corresponding overall rate law can be derived. For the present it is important to note that, *if the reaction is an elementary bimolecular process, then it has second-order kinetics, but if the kinetics is second order, then the reaction might be complex.* The postulated mechanism can be explored only by detailed detective work on the system and by investigating whether side products or intermediates appear during the course of the reaction. Detailed analysis of this kind was one of the ways, for example, in which the reaction $H_2(g) + I_2(g) \rightarrow 2\,HI(g)$ was shown to proceed by a complex mechanism. For many years the reaction had been accepted on good but insufficiently meticulous evidence as a fine example of a simple bimolecular reaction, $H_2 + I_2 \rightarrow HI + HI$, in which atoms exchanged partners during a collision.

17E.2 Consecutive elementary reactions

Some reactions proceed through the formation of an intermediate (denoted I), as in the consecutive unimolecular reactions

$$A \xrightarrow{k_a} I \xrightarrow{k_b} P$$

Note that the intermediate occurs in the reaction steps but does not appear in the overall reaction, which in this case is A → P. Any reverse reactions are ignored here, so the reaction proceeds from all A to all P, not to an equilibrium mixture of the two. An example of this type of mechanism is the decay of a radioactive family, such as

$$^{239}U \xrightarrow{23.5\text{ min}} {}^{239}Np \xrightarrow{2.35\text{ days}} {}^{239}Pu$$

(The times are half-lives.) The characteristics of this type of reaction are discovered by setting up the rate laws for the net rate of change of the concentration of each substance and then combining them in the appropriate manner.

The rate of unimolecular decomposition of A is

$$\frac{d[A]}{dt} = -k_a[A] \qquad (17E.3a)$$

The intermediate I is formed from A (at a rate $k_a[A]$) but decays to P (at a rate $k_b[I]$). The net rate of formation of I is therefore

$$\frac{d[I]}{dt} = k_a[A] - k_b[I] \qquad (17E.3b)$$

The product P is formed by the unimolecular decay of I:

$$\frac{d[P]}{dt} = k_b[I] \qquad (17E.3c)$$

If it is assumed that initially the molar concentration of A is $[A]_0$, the first-order rate law of eqn 17E.3a can be integrated (as in Topic 17B) to give

$$[A] = [A]_0 e^{-k_a t} \qquad (17E.4a)$$

When this equation is substituted into eqn 17E.3b, the result is, after rearrangement,

$$\frac{d[I]}{dt} + k_b[I] = k_a[A]_0 e^{-k_a t}$$

This differential equation has a standard form in the sense that it has been studied and its solution listed. With the initial condition $[I]_0 = 0$, because no intermediate is present initially, the solution of the differential equation (provided $k_a \neq k_b$) is

$$[I] = \frac{k_a}{k_b - k_a}(e^{-k_a t} - e^{-k_b t})[A]_0 \qquad (17E.4b)$$

At all times $[A] + [I] + [P] = [A]_0$, so it follows that

$$[P] = \left\{1 + \frac{k_a e^{-k_b t} - k_b e^{-k_a t}}{k_b - k_a}\right\}[A]_0 \qquad (17E.4c)$$

The concentration of the intermediate I rises to a maximum and then falls to zero (Fig. 17E.1). The concentration of the product P rises from zero towards $[A]_0$, when all A has been converted to P.

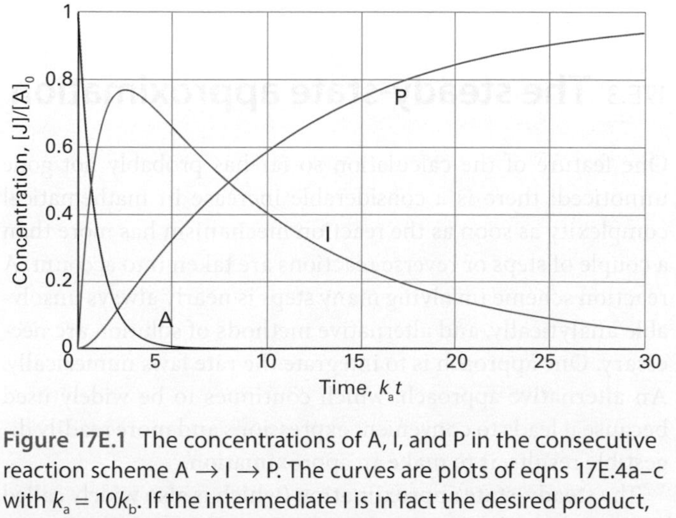

Figure 17E.1 The concentrations of A, I, and P in the consecutive reaction scheme A → I → P. The curves are plots of eqns 17E.4a–c with $k_a = 10k_b$. If the intermediate I is in fact the desired product, it is important to be able to predict when its concentration is greatest; see *Example* 17E.1.

Example 17E.1 Analysing consecutive reactions

Suppose that in an industrial batch process a substance A produces the desired compound I which goes on to decay to a worthless product P, each step of the reaction being first order. At what time will I be present in greatest concentration?

Collect your thoughts The time dependence of the concentration of I is given by eqn 17E.4b. Find the time at which [I] passes through a maximum, t_{max}, by calculating the derivative d[I]/dt and then setting it equal to zero.

The solution It follows from eqn 17E.4b that

$$\frac{d[I]}{dt}=-\frac{k_a(k_a e^{-k_a t}-k_b e^{-k_b t})[A]_0}{k_b-k_a}$$

This derivative is equal to zero when $t = t_{max}$ and $k_a e^{-k_a t_{max}} = k_b e^{-k_b t_{max}}$. Therefore, taking natural logarithms of both sides gives

$$\ln k_a - k_a t_{max} = \ln k_b - k_b t_{max}$$

which rearranges to

$$\overbrace{\ln k_a - \ln k_b}^{\ln(k_a/k_b)} = k_a t_{max} - k_b t_{max} = (k_a - k_b)t_{max}$$

It then follows that

$$t_{max}=\frac{1}{k_a-k_b}\ln\frac{k_a}{k_b}$$

Comment. For a given value of k_a, as k_b increases both the time at which [I] is a maximum and the yield of I decrease.

Self-test 17E.1 Calculate the maximum concentration of I and justify the last remark.

Answer: $[I]_{max}/[A]_0 = (k_a/k_b)^c$, $c = k_b/(k_b - k_a)$

17E.3 The steady-state approximation

One feature of the calculation so far has probably not gone unnoticed: there is a considerable increase in mathematical complexity as soon as the reaction mechanism has more than a couple of steps or reverse reactions are taken into account. A reaction scheme involving many steps is nearly always unsolvable analytically, and alternative methods of solution are necessary. One approach is to integrate the rate laws numerically. An alternative approach, which continues to be widely used because it leads to convenient expressions and more readily digestible results, is to make an approximation.

The **steady-state approximation** (which is also widely called the *quasi-steady-state approximation* to distinguish it from a true steady state) assumes that the intermediate, I, is in a low, constant concentration. More specifically, after an initial **induction period**, an interval during which the concentrations of intermediates rise from zero, the rates of change of the concentrations of all reaction intermediates are negligibly small during the major part of the reaction (Fig. 17E.2):

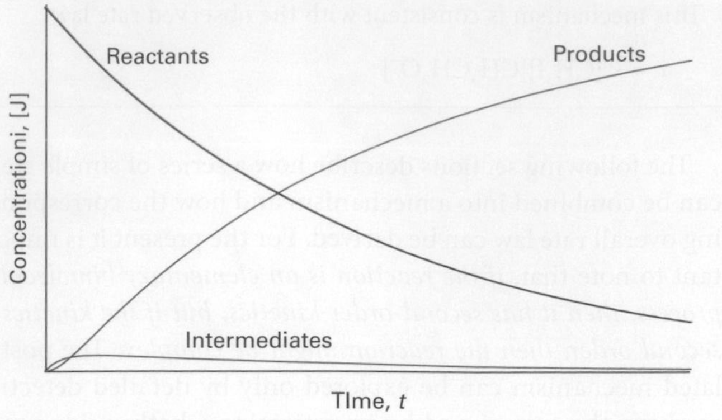

Figure 17E.2 The basis of the steady-state approximation. It is supposed that the concentrations of intermediates remain small and hardly change during most of the course of the reaction.

$$\frac{d[I]}{dt}\approx 0 \quad \text{Steady-state approximation} \quad (17E.5)$$

This approximation greatly simplifies the discussion of reaction schemes. For example, to apply the approximation to the consecutive first-order mechanism, d[I]/dt is set equal to 0 in eqn 17E.3b, which then becomes $k_a[A]-k_b[I]=0$. It then follows that

$$[I]=\frac{k_a}{k_b}[A] \quad (17E.6)$$

The steady-state approximation requires the concentration of the intermediate to be low relative to that of the reactants, which is the case when $k_a \ll k_b$. Equation 17E.6 implies that the concentration of the intermediate changes as the concentration of A changes, but if $k_a/k_b \ll 1$ it changes very little. Both requirements of the steady-state approximation, the low concentration of intermediate and its slow change, are therefore satisfied.

On substituting the value of [I] in eqn 17E.6 into eqn 17E.3c, that equation becomes

$$\frac{d[P]}{dt}=k_b\overbrace{[I]}^{[I]=(k_a/k_b)[A]}\approx k_a[A] \quad (17E.7)$$

It follows that the rate of formation of P is the same as the rate of loss of A (as given by eqn 17E.3a), and that both processes are governed by the rate constant k_a. In effect, the reactant A flows directly through to becoming P without any I accumulating on the way. The solution of eqn 17E.7 is found by

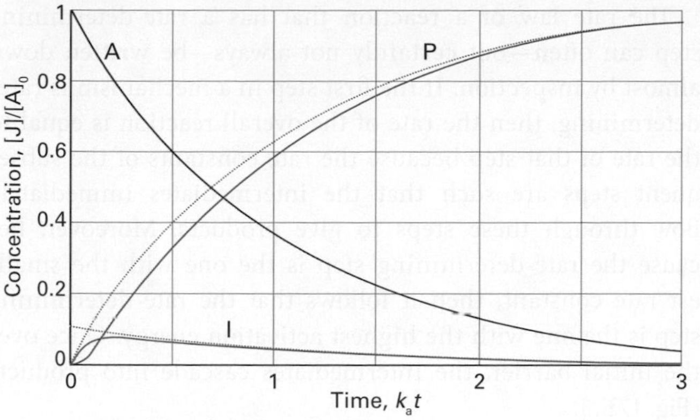

Figure 17E.3 A comparison of the exact result for the concentrations of a consecutive reaction and the concentrations obtained by using the steady-state approximation (dotted lines) for $k_b = 20k_a$. (The curve for [A] is unchanged.)

substituting the solution for [A], eqn 17E.4a, and integrating the resulting expression:

$$\int_0^{[P]} d[P] = \int_0^t k_a[A]dt = k_a[A]_0 \int_0^t e^{-k_a t} dt$$

([A] = [A]$_0$e$^{-k_a t}$; Integral E.1)

Hence

$$[P] = [A]_0(1-e^{-k_a t}) \qquad (17E.8)$$

This result is the same as in eqn 17E.4c when $k_a << k_b$; however, the use of the steady-state approximation is much simpler. Figure 17E.3 compares the approximate solutions found here with the exact solutions found earlier: k_a does not have to be very much smaller than k_b for the approach to be reasonably accurate.

Example 17E.2 Using the steady-state approximation

Devise the rate law for the decomposition of N_2O_5, $2N_2O_5(g) \rightarrow 4NO_2(g) + O_2(g)$ on the basis of the following mechanism:

$$N_2O_5 \rightarrow NO_2 + NO_3 \qquad k_a$$
$$NO_2 + NO_3 \rightarrow N_2O_5 \qquad k_a'$$
$$NO_2 + NO_3 \rightarrow NO_2 + O_2 + NO \qquad k_b$$
$$NO + N_2O_5 \rightarrow NO_2 + NO_2 + NO_2 \qquad k_c$$

A note on good practice Note that when writing the equation for an elementary reaction all the species are displayed individually; so write A → B + B, for instance, not A → 2B.

Collect your thoughts First identify the intermediates and for each of them write an expression for the net rate of formation. Then apply the steady-state approximation and set these net rates to zero. You can then solve the resulting equations algebraically to obtain expressions for the concentrations of the intermediates. Finally, use these solutions to obtain an expression for the overall rate of consumption of N_2O_5.

The solution The intermediates are NO and NO_3; the net rates of change of their concentrations are

$$\frac{d[NO]}{dt} = k_b[NO_2][NO_3] - k_c[NO][N_2O_5] \approx 0$$

$$\frac{d[NO_3]}{dt} = k_a[N_2O_5] - k_a'[NO_2][NO_3] - k_b[NO_2][NO_3] \approx 0$$

The solutions of these two simultaneous equations (in blue) are

$$[NO_3] = \frac{k_a[N_2O_5]}{(k_a'+k_b)[NO_2]} \qquad [NO] = \frac{k_b[NO_2][NO_3]}{k_c[N_2O_5]} = \frac{k_a k_b}{(k_a'+k_b)k_c}$$

(Use expression for [NO$_3$])

The net rate of change of concentration of N_2O_5 is then

$$\frac{d[N_2O_5]}{dt} = -k_a[N_2O_5] + k_a'[NO_2][NO_3] - k_c[NO][N_2O_5]$$

$$= -k_a[N_2O_5] + \frac{k_a k_a'[N_2O_5]}{k_a'+k_b} - \frac{k_a k_b}{k_a'+k_b}[N_2O_5]$$

$$= -\frac{2k_a k_b[N_2O_5]}{k_a'+k_b}$$

That is, N_2O_5 decays with a first-order rate law with a rate constant that depends on k_a, k_a', and k_b but not on k_c.

Self-test 17E.2 Derive the rate law for the decomposition of ozone in the reaction $2O_3(g) \rightarrow 3O_2(g)$ on the basis of the (incomplete) mechanism

$$O_3 \rightarrow O_2 + O \qquad k_a$$
$$O_2 + O \rightarrow O_3 \qquad k_a'$$
$$O + O_3 \rightarrow O_2 + O_2 \qquad k_b$$

Answer: $d[O_3]/dt = -2k_a k_b[O_3]^2/(k_a'[O_2] + k_b[O_3])$

17E.4 The rate-determining step

When the steady-state approximation is valid, which is when $k_a << k_b$ in the reaction A → I → P, the decrease in the concentration of A is matched by an increase in the concentration of P. It is important to realize that the rates of the steps A → I and I → P are the same: the concentration of I is so low compared to the concentration of A that even though $k_a << k_b$ (and therefore $k_b >> k_a$) the rate of the second step, k_b[I], matches that of the first, k_a[A].

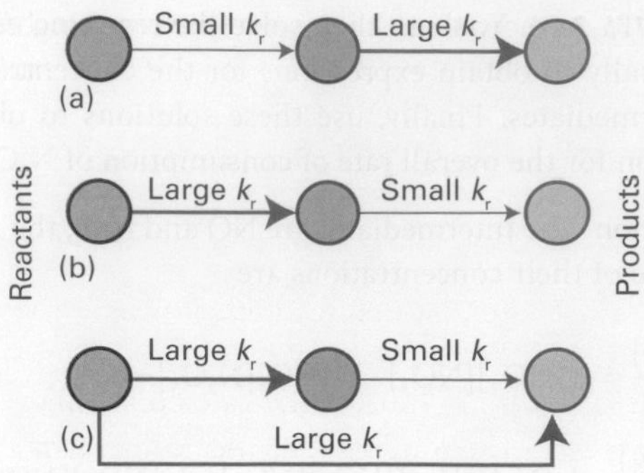

Figure 17E.4 In these diagrams of reaction schemes, heavy arrows represent steps with large rate constants and light arrows represent steps with small rate constants. (a) The first step is rate-determining; (b) the second step is rate-determining; (c) although one step has a small rate constant, it is not rate-determining because there is a route with a large rate constant that circumvents it.

A note on good practice It is commonly said that 'the first step is slow and the second is fast, so the first step is rate-determining'. Such a statement is incorrect: the two *rates* are equal; it is the *rate constants* that are different.

In general, the **rate-determining step** (RDS) is the step in a mechanism that controls the overall rate of the reaction (in the present example, the first step governed by k_a, with $k_a \ll k_b$). The rate-determining step must be a crucial gateway for the formation of products, and not just a reaction with a small rate constant. If another reaction with a larger rate constant can also lead to products, then the step with the small rate constant is irrelevant because it can be sidestepped (Fig. 17E.4). In some cases, when a first-order reaction is in competition with a second-order reaction, the criterion has to be expressed in terms of the relative sizes of the first-order (for one step) and a pseudofirst-order (for the second step) rate constants, for only then can their magnitudes be compared. This point is illustrated in the next *Brief illustration*.

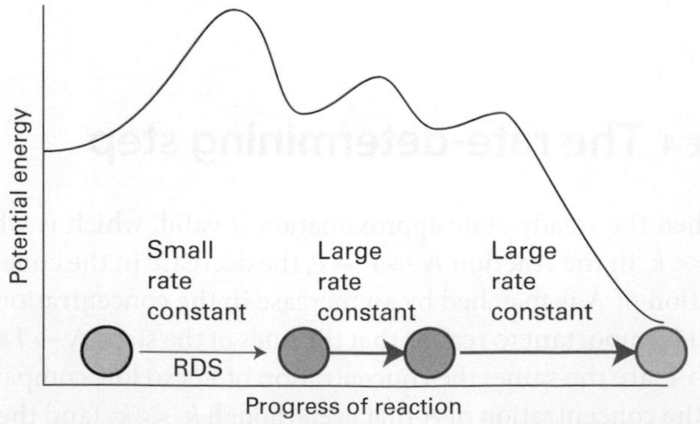

Figure 17E.5 The reaction profile for a mechanism in which the first step (RDS) is rate-determining.

The rate law of a reaction that has a rate-determining step can often—but certainly not always—be written down almost by inspection. If the first step in a mechanism is rate-determining, then the rate of the overall reaction is equal to the rate of that step because the rate constants of the subsequent steps are such that the intermediates immediately flow through these steps to give products. Moreover, because the rate-determining step is the one with the smallest rate constant, then it follows that the rate-determining step is the one with the highest activation energy. Once over the initial barrier, the intermediates cascade into products (Fig. 17E.5).

Brief illustration 17E.2

The oxidation of NO to NO_2, $2NO(g) + O_2(g) \rightarrow 2NO_2(g)$, proceeds by the following mechanism:

$$NO + NO \rightarrow N_2O_2 \qquad k_a$$

$$N_2O_2 \rightarrow NO + NO \qquad k_a'$$

$$N_2O_2 + O_2 \rightarrow NO_2 + NO_2 \qquad k_b$$

with rate law (see Problem P17E.6)

$$\frac{d[NO_2]}{dt} = \frac{2k_a k_b [NO]^2[O_2]}{k_a' + k_b[O_2]}$$

When the concentration of O_2 in the reaction mixture is so high that $k_a' \ll k_b[O_2]$, the rate law then simplifies to

$$\frac{d[NO_2]}{dt} = 2k_a[NO]^2$$

which shows that the formation of N_2O_2 in the first step is rate-determining. The rate law could also have been written by inspection of the mechanism, because the rate law for the overall reaction is simply the rate law of that rate-determining step.

17E.5 Pre-equilibria

Now consider a slightly more complicated mechanism in which an intermediate I reaches an equilibrium with the reactants A and B:

$$A+B \underset{k_a'}{\overset{k_a}{\rightleftharpoons}} I \xrightarrow{k_b} P \qquad \text{Pre-equilibrium} \qquad (17E.9)$$

This scheme involves a **pre-equilibrium**, in which an intermediate is in equilibrium with the reactants. A pre-equilibrium can arise when the rate of decay of the intermediate back into reactants is much faster than the rate at which it forms products; this condition is satisfied if $k_a' \gg k_b$. Because A, B, and I are assumed to be in equilibrium, it follows that (see Topic 17C)

$K = k_a c^{\ominus}/k'_a$

$$K = \frac{[\mathrm{I}]c^{\ominus}}{[\mathrm{A}][\mathrm{B}]} \quad \text{and} \quad [\mathrm{I}] = \frac{K}{c^{\ominus}}[\mathrm{A}][\mathrm{B}] = \frac{k_a}{k'_a}[\mathrm{A}][\mathrm{B}] \qquad (17\mathrm{E}.10)$$

In writing these equations, the rate of reaction of I to form P is presumed to be too slow to affect the maintenance of the pre-equilibrium (see the following *Example*). The rate of formation of P may now be written

$[\mathrm{I}] = (k_a/k'_a)[\mathrm{A}][\mathrm{B}]$

$$\frac{\mathrm{d}[\mathrm{P}]}{\mathrm{d}t} = k_b[\mathrm{I}] = k_b\frac{k_a}{k'_a}[\mathrm{A}][\mathrm{B}] \qquad (17\mathrm{E}.11)$$

This rate law has the form of a second-order rate law with a composite rate constant:

$$\frac{\mathrm{d}[\mathrm{P}]}{\mathrm{d}t} = k_r[\mathrm{A}][\mathrm{B}] \quad \text{with} \quad k_r = \frac{k_b k_a}{k'_a} \qquad (17\mathrm{E}.12)$$

In this pre-equilibrium mechanism the final step, I → P, is rate-determining. The preceding steps control the steady concentration of the intermediate.

Example 17E.3 Analysing a pre-equilibrium using the steady-state assumption

Analyse the scheme shown in eqn 17E.9 using the steady-state approximation.

Collect your thoughts Begin by writing the net rates of change of the concentrations of P and I, and then invoke the steady-state approximation for the intermediate I. Use the resulting expression to obtain the rate of change of the concentration of P.

The solution The net rates of change of P and I are

$$\frac{\mathrm{d}[\mathrm{P}]}{\mathrm{d}t} = k_b[\mathrm{I}]$$

Steady-state approximation

$$\frac{\mathrm{d}[\mathrm{I}]}{\mathrm{d}t} = k_a[\mathrm{A}][\mathrm{B}] - k'_a[\mathrm{I}] - k_b[\mathrm{I}] \approx 0$$

The second equation implies that

$$[\mathrm{I}] \approx \frac{k_a[\mathrm{A}][\mathrm{B}]}{k'_a + k_b}$$

Now substitute this result into the expression for the rate of formation of P:

$$\frac{\mathrm{d}[\mathrm{P}]}{\mathrm{d}t} = k_b[\mathrm{I}] \approx k_b\frac{k_a[\mathrm{A}][\mathrm{B}]}{k'_a + k_b} = k_r[\mathrm{A}][\mathrm{B}] \quad \text{with} \quad k_r = \frac{k_a k_b}{k'_a + k_b}$$

This expression reduces to that in eqn 17E.12 when the rate constant for the decay of I into products is much smaller than that for its decay into reactants, $k_b << k'_a$.

Self-test 17E.3 Show that the pre-equilibrium mechanism in which $2\mathrm{A} \rightleftharpoons \mathrm{I}$ (K) followed by I + B → P (k_b) results in an overall third-order reaction.

Answer: $\mathrm{d}[\mathrm{P}]/\mathrm{d}t = (k_b K/c^{\ominus})[\mathrm{A}]^2[\mathrm{B}]$

One feature to note is that although each of the rate constants in eqn 17E.12 increases with temperature, this might not be true of k_r itself. Thus, if the rate constant k'_a increases more rapidly than the product $k_a k_b$ increases, then $k_r = k_a k_b/k'_a$ decreases with increasing temperature and the reaction goes more slowly as the temperature is raised. Mathematically, the overall reaction is said to have a 'negative activation energy'. For example, suppose that each rate constant in eqn 17E.12 has an Arrhenius temperature dependence (Topic 17D). It follows from the Arrhenius equation (eqn 17D.4, $k_r = Ae^{-E_a/RT}$) that

$e^{x+y} = e^x e^y$, $e^{x-y} = e^x/e^y$

$$k_r = \frac{(A_a e^{-E_{a,a}/RT})(A_b e^{-E_{a,b}/RT})}{A_{a'} e^{-E_{a,a'}/RT}} = \frac{A_a A_b}{A_{a'}} e^{-(E_{a,a}+E_{a,b}-E_{a,a'})/RT}$$

The effective activation energy of the reaction is therefore

$$E_a = E_{a,a} + E_{a,b} - E_{a,a'} \qquad (17\mathrm{E}.13)$$

This activation energy is positive if $E_{a,a} + E_{a,b} > E_{a,a'}$ (Fig. 17E.6a) but negative if $E_{a,a'} > E_{a,a} + E_{a,b}$ (Fig. 17E.6b). An important consequence of this discussion is that it is necessary to be cautious when making predictions about the effect of temperature on reactions that are the outcome of several steps.

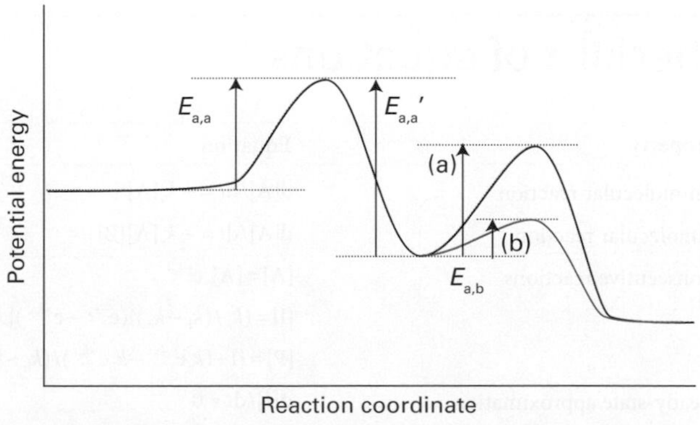

Figure 17E.6 For a reaction with a pre-equilibrium, there are three activation energies to take into account: two referring to the reversible steps of the pre-equilibrium and one for the final step. The relative magnitudes of the activation energies determine whether the overall activation energy is (a) positive or (b) negative.

17E.6 Kinetic and thermodynamic control of reactions

In some cases reactants can give rise to a variety of products, as in nitrations of mono-substituted benzene, when various proportions of the *ortho*-, *meta*-, and *para*-substituted products are obtained, depending on the directing power of the original substituent. Suppose two products, P_1 and P_2, are produced by the following competing reactions:

$$A + B \rightarrow P_1 \qquad v(P_1) = k_{r,1}[A][B]$$

$$A + B \rightarrow P_2 \qquad v(P_2) = k_{r,2}[A][B]$$

The relative proportion in which the two products have been produced at a given stage of the reaction (before it has reached equilibrium) is given by the ratio of the two rates, and therefore by the ratio of the two rate constants:

$$\frac{[P_2]}{[P_1]} = \frac{k_{r,2}}{k_{r,1}} \qquad \text{Kinetic control} \qquad (17E.14)$$

This ratio represents the **kinetic control** over the proportions of products, control that stems from relative rates rather than thermodynamic considerations about equilibrium. Kinetic control is a common feature of the reactions encountered in organic chemistry, where reactants are chosen that facilitate pathways favouring the formation of a desired product. If a reaction is allowed to reach equilibrium, then the proportion of products is determined by thermodynamic rather than kinetic considerations, and the ratio of concentrations is controlled by considerations of the standard reaction Gibbs energy.

Checklist of concepts

- ☐ 1. The **mechanism** of reaction is the sequence of elementary steps that leads from reactants to products.
- ☐ 2. The **molecularity** of an elementary reaction is the number of molecules coming together to react.
- ☐ 3. An elementary unimolecular reaction has first-order kinetics; an elementary bimolecular reaction has second-order kinetics.
- ☐ 4. The **rate-determining step** is the step in a reaction mechanism that controls the rate of the overall reaction.
- ☐ 5. In the **steady-state approximation**, it is assumed that the concentrations of all reaction intermediates remain constant and small throughout the reaction.
- ☐ 6. **Pre-equilibrium** is a state in which an intermediate is in equilibrium with the reactants and which arises when the rate of decay of the intermediate back to reactants is much faster than the rate at which products are formed from the intermediate.
- ☐ 7. **Kinetic control** over the proportions of products stems from relative rates rather than thermodynamic considerations about equilibrium.

Checklist of equations

Property	Equation	Comment	Equation number
Unimolecular reaction	$d[A]/dt = -k_r[A]$	$A \rightarrow P$	17E.1
Bimolecular reaction	$d[A]/dt = -k_r[A][B]$	$A + B \rightarrow P$	17E.2
Consecutive reactions	$[A] = [A]_0 e^{-k_a t}$	$A \xrightarrow{k_a} I \xrightarrow{k_b} P$	17E.4
	$[I] = (k_a/(k_b - k_a))(e^{-k_a t} - e^{-k_b t})[A]_0$		
	$[P] = \{1 + (k_a e^{-k_b t} - k_b e^{-k_a t})/(k_b - k_a)\}[A]_0$		
Steady-state approximation	$d[I]/dt \approx 0$	I is an intermediate	17E.5

TOPIC 17F Examples of reaction mechanisms

➤ What do you need to know this material?

Some important reactions have complex mechanisms and need special treatment, so you need to see how to make and implement assumptions about the relative rates of the steps in a mechanism.

➤ What is the key idea?

The steady-state approximation can often be used to derive rate laws for proposed mechanisms.

➤ What do you need to know already?

You need to be familiar with the concept of rate laws (Topic 17A) and the formulation of an overall rate law from a mechanism by using the steady-state approximation (Topic 17E).

Many reactions take place by mechanisms that involve several elementary steps. In each case it is possible to approach the setting up (and testing) of a rate law by proposing a mechanism and, when appropriate, applying the steady-state approximation.

17F.1 Unimolecular reactions

A number of gas-phase reactions follow first-order kinetics, as in the isomerization of cyclopropane to propene:

$$cyclo\text{-}C_3H_6(g) \rightarrow CH_3CH{=}CH_2(g) \qquad v = k_r[cyclo\text{-}C_3H_6]$$

The problem with the interpretation of first-order rate laws is that presumably a molecule acquires enough energy to react as a result of collisions with other molecules. However, collisions are simple bimolecular events, so how can they result in a first-order rate law? First-order gas-phase reactions are widely called 'unimolecular reactions' because they also involve an elementary unimolecular step in which the reactant molecule changes into the product. This term must be used with caution, however, because the overall mechanism has bimolecular as well as unimolecular steps.

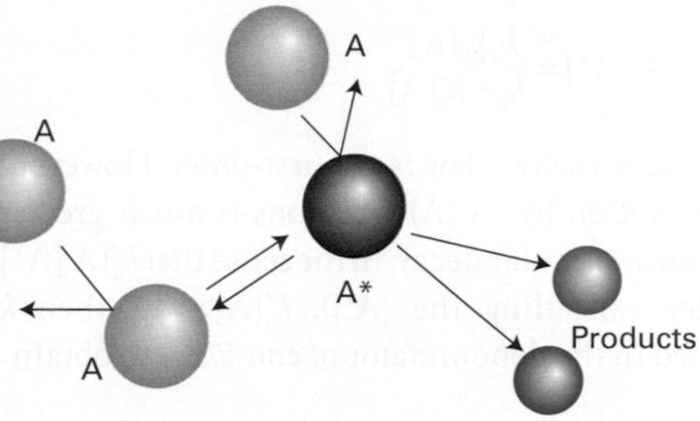

Figure 17F.1 A representation of the Lindemann–Hinshelwood mechanism of unimolecular reactions. The species A is excited by collision with A, and the energized A molecule (A*) may either be deactivated by a collision with A or go on to decay by a unimolecular process to form products.

The first successful explanation of unimolecular reactions was provided by Frederick Lindemann in 1921 and then elaborated by Cyril Hinshelwood. In the **Lindemann–Hinshelwood mechanism** it is supposed that a reactant molecule A becomes energetically excited by collision with another A molecule in a bimolecular step (Fig. 17F.1):

$$A + A \rightarrow A^* + A \qquad \frac{d[A^*]}{dt} = k_a[A]^2 \qquad (17F.1a)$$

The energized molecule (A*) might lose its excess energy by collision with another molecule:

$$A + A^* \rightarrow A + A \qquad \frac{d[A^*]}{dt} = -k_a'[A][A^*] \qquad (17F.1b)$$

Alternatively, the excited molecule might shake itself apart or into a different arrangement of its atoms and form products P. That is, it might undergo the unimolecular decay

$$A^* \rightarrow P \qquad \frac{d[A^*]}{dt} = -k_b[A^*] \qquad (17F.1c)$$

If the unimolecular step is the rate-determining step, the overall reaction will have first-order kinetics, as observed. This conclusion can be demonstrated explicitly by applying the steady-state approximation to the net rate of formation of A*:

$$\frac{d[A^*]}{dt}=k_a[A]^2-k_a'[A][A^*]-k_b[A^*]\approx 0 \qquad (17F.2)$$

Rearrangement of this equation gives

$$[A^*]=\frac{k_a[A]^2}{k_b+k_a'[A]} \qquad (17F.3)$$

so the rate law for the formation of P is

$$\frac{d[P]}{dt}=k_b[A^*]=\frac{k_ak_b[A]^2}{k_b+k_a'[A]} \qquad (17F.4)$$

At this stage the rate law is not first-order. However, if the rate of deactivation by (A*,A) collisions is much greater than the rate of unimolecular decay, in the sense that $k_a'[A][A^*]>>k_b[A^*]$, or (after cancelling the [A*]), $k_a'[A]>>k_b$, then k_b can be neglected in the denominator of eqn 17F.4 to obtain

$$\frac{d[P]}{dt}=k_r[A] \quad \text{with} \quad k_r=\frac{k_ak_b}{k_a'}$$

Lindemann–Hinshelwood rate law (17F.5)

Equation 17F.5 is a first-order rate law, as required.

The Lindemann–Hinshelwood mechanism can be tested because it predicts that, as the concentration (and therefore the partial pressure) of A is reduced, the reaction should switch to overall second-order kinetics. Thus, when $k_a'[A][A^*]<<k_b[A^*]$ or (after cancelling the [A*]) $k_a'[A]<<k_b$, the rate law in eqn 17F.4 becomes

$$\frac{d[P]}{dt}=k_a[A]^2 \qquad (17F.6)$$

The physical reason for the change of order is that as the pressure is reduced the rate of the bimolecular process in which A* loses its excess energy becomes negligible compared to the rate at which A* goes on to form products. The reaction mechanism is then a sequence of two steps, with the first step (which is bimolecular) being rate limiting. If the full rate law in eqn 17F.4 is written as

$$\frac{d[P]}{dt}=k_r[A] \quad \text{with} \quad k_r=\frac{k_ak_b[A]}{k_b+k_a'[A]} \qquad (17F.7)$$

then the expression for the effective rate constant, k_r, can be rearranged (by inverting each side) to

$$\frac{1}{k_r}=\frac{k_a'}{k_ak_b}+\frac{1}{k_a[A]}$$

Effective rate constant [Lindemann–Hinshelwood mechanism] (17F.8)

Hence, a test of the theory is to plot $1/k_r$ against 1/[A] and expect a straight line. This behaviour is observed often at low concentrations but deviations are common at high concentrations. Topic 18A develops the description of the mechanism further to take into account experimental results over a range of concentrations and pressures.

Example 17F.1 Analysing data in terms of the Lindemann–Hinshelwood mechanism

At 300 K the effective rate constant for a gaseous reaction A → P, which has a Lindemann–Hinshelwood mechanism, is $k_{r,1}$ = 2.50 × 10^{-4} s^{-1} at $[A]_1$ = 5.21 × 10^{-4} mol dm^{-3} and $k_{r,2}$ = 2.10 × 10^{-5} s^{-1} at $[A]_2$ = 4.81 × 10^{-6} mol dm^{-3}. Calculate the rate constant k_a for the activation step in the mechanism.

Collect your thoughts Use eqn 17F.8 to write an expression for the difference $1/k_{r,2} - 1/k_{r,1}$, rearrange the expression for k_a, and then insert the data.

The solution It follows from eqn 17F.8 that

$$\frac{1}{k_{r,2}}-\frac{1}{k_{r,1}}=\frac{1}{k_a}\left(\frac{1}{[A]_2}-\frac{1}{[A]_1}\right)$$

and so

$$k_a=\frac{1/[A]_2-1/[A]_1}{1/k_{r,2}-1/k_{r,1}}$$

$$=\frac{1/(4.81\times10^{-6}\,\text{mol dm}^{-3})-1/(5.21\times10^{-4}\,\text{mol dm}^{-3})}{1/(2.10\times10^{-5}\,\text{s}^{-1})-1/(2.50\times10^{-4}\,\text{s}^{-1})}$$

$$=4.72\,\text{dm}^3\,\text{mol}^{-1}\,\text{s}^{-1}$$

Self-test 17F.1 The effective rate constants for a gaseous reaction A→P, which has a Lindemann–Hinshelwood mechanism, are 1.70×10^{-3} s^{-1} and 2.20×10^{-4} s^{-1} at [A] = 4.37×10^{-4} mol dm^{-3} and 1.00 × 10^{-5} mol dm^{-3}, respectively. Calculate the rate constant for the activation step in the mechanism.

Answer: 24.7 dm^3 mol^{-1} s^{-1}

17F.2 Polymerization kinetics

There are two major classes of polymerization processes and in each one the average molar mass of the product varies with time in a distinctive way. In **stepwise polymerization** any two monomers present in the reaction mixture can link together at any time and growth of the polymer is not confined to chains that are already forming (Fig. 17F.2). As a result, monomers are consumed early in the reaction and, as will be seen, the average molar mass of the product grows linearly with time. In **chain polymerization** a monomer, M, attacks another monomer, links to it, then that unit attacks another monomer, and so on. The monomer is used up as it becomes linked to the growing chains (Fig. 17F.3). Polymers built from numerous monomers are formed rapidly and only the yield, not the average molar mass, of the polymer is increased by allowing long reaction times.

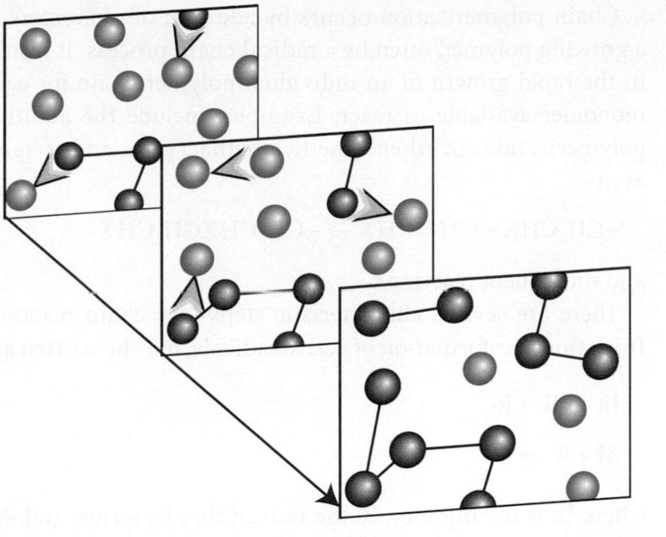

Figure 17F.2 In stepwise polymerization, growth can start at any pair of monomers (in green), and so new chains (in red) begin to form throughout the reaction.

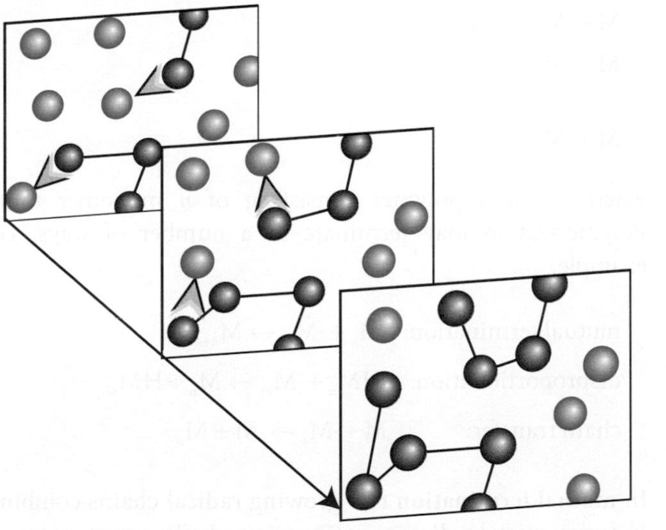

Figure 17F.3 The process of chain polymerization. Chains (in red) grow as each chain acquires additional monomers (in green).

(a) Stepwise polymerization

Stepwise polymerization commonly proceeds by a **condensation reaction**, in which a small molecule (typically H_2O) is eliminated in each step. Stepwise polymerization is the mechanism of production of polyamides, as in the formation of nylon-6,6:

$$H_2N(CH_2)_6NH_2 + HOOC(CH_2)_4COOH \rightarrow H_2N(CH_2)_6NHCO(CH_2)_4COOH + H_2O$$

continuing to

$$\rightarrow H\text{–}[HN(CH_2)_6NHCO(CH_2)_4CO]_n\text{–}OH$$

Polyesters and polyurethanes are formed similarly (the latter without elimination). A polyester, for example, can be regarded as the outcome of the stepwise condensation of a hydroxyacid HO–R–COOH. Consider the formation of a polyester from such a monomer. Its progress can be measured in terms of the concentration of the –COOH groups in the sample (denoted A), because these groups gradually disappear as the condensation proceeds. Because the condensation reaction can occur between molecules containing any number of monomer units, chains of many different lengths can grow in the reaction mixture.

In the absence of a catalyst, condensation is expected to be overall second-order in the concentration of the –OH and –COOH (or A) groups, so

$$\frac{d[A]}{dt} = -k_r[OH][A] \quad (17F.9a)$$

However, because there is one –OH group for each –COOH group, this equation is the same as

$$\frac{d[A]}{dt} = -k_r[A]^2 \quad (17F.9b)$$

If the rate constant for the condensation is independent of the chain length, k_r remains constant throughout the reaction. The solution of this rate law is then given by eqn 17B.4b, and is

$$[A] = \frac{[A]_0}{1+k_rt[A]_0} \quad (17F.10)$$

The fraction, p, of –COOH groups that have condensed at time t is therefore

$$p = \frac{[A]_0-[A]}{[A]_0} \overset{[A]=[A]_0/(1+k_rt[A]_0)}{=} \frac{k_rt[A]_0}{1+k_rt[A]_0} \quad \text{Fraction of condensed groups [stepwise polymerization]} \quad (17F.11)$$

The **degree of polymerization**, the average number of monomer residues per polymer molecule, can now be calculated. This quantity is the ratio of the initial concentration of A, $[A]_0$, to the concentration of end groups, [A], at the time of interest, because there is one A group per polymer molecule. For example, if there were initially 1000 A groups and there are now only 10, the average length of each polymer must be 100 units. Because [A] can be expressed in terms of p (the first part of eqn 17F.11), the average number of monomers per polymer molecule, $\langle N \rangle$, is

$$\langle N \rangle = \frac{[A]_0}{[A]} = \frac{1}{1-p} \quad \text{Degree of polymerization [stepwise polymerization]} \quad (17F.12a)$$

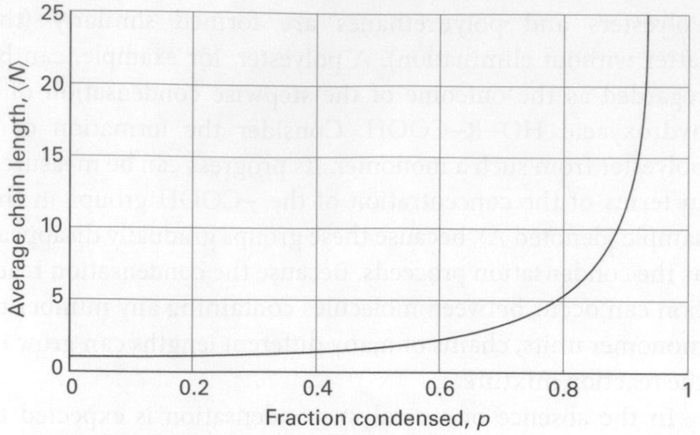

Figure 17F.4 The average chain length of a polymer as a function of the fraction of reacted monomers, p. Note that p must be very close to 1 for the chains to be long.

This result is illustrated in Fig. 17F.4. When p is expressed in terms of the rate constant k_r (the second part of eqn 17F.11), the result is

$$\langle N\rangle = 1 + k_r t[A]_0 \quad \text{Degree of polymerization in terms of the rate constant [stepwise polymerization]} \quad (17F.12b)$$

The average length grows linearly with time. Therefore, the longer a stepwise polymerization proceeds, the higher the average molar mass of the product.

Brief illustration 17F.1

Consider a polymer formed by a stepwise process with $k_r = 1.00\,dm^3\,mol^{-1}\,s^{-1}$ and an initial monomer concentration of $[A]_0 = 4.00 \times 10^{-3}\,mol\,dm^{-3}$. From eqn 17F.12b, the degree of polymerization at $t = 1.5 \times 10^4\,s$ is

$$\langle N\rangle = 1 + (1.00\,dm^3\,mol^{-1}\,s^{-1}) \times (1.5 \times 10^4\,s) \times (4.00 \times 10^{-3}\,mol\,dm^{-3}) = 61$$

From eqn 17F.12a, the fraction condensed, p, is

$$p = \frac{\langle N\rangle - 1}{\langle N\rangle} = \frac{61-1}{61} = 0.98$$

(b) Chain polymerization

Many gas-phase reactions and liquid-phase polymerization reactions are **chain reactions**. In a chain reaction, a reaction intermediate produced in one step generates an intermediate in a subsequent step, then that intermediate generates another intermediate, and so on. The intermediates in a chain reaction are called **chain carriers.** In a **radical chain reaction** the chain carriers are radicals (species with unpaired electrons).

Chain polymerization occurs by addition of monomers to a growing polymer, often by a radical chain process. It results in the rapid growth of an individual polymer chain for each monomer available to react. Examples include the addition polymerizations of ethene, methyl methacrylate, and styrene, as in

$$-CH_2\dot{C}HX + CH_2{=}CHX \rightarrow -CH_2CHXCH_2\dot{C}HX$$

and subsequent reactions.

There are several characteristic steps in a chain reaction. **Initiation**, the formation of active radicals, may be written as

$$In \rightarrow R\cdot + R\cdot$$
$$M + R\cdot \rightarrow \cdot M_1$$

where In is the initiator, R· the radical that In forms, and $\cdot M_1$ is a monomer radical. In this reaction a radical is produced, but in some polymerizations the initiation step leads to the formation of an ionic chain carrier. Initiation is followed by **propagation**, the continuation of the chain reaction:

$$M + \cdot M_1 \rightarrow \cdot M_2$$
$$M + \cdot M_2 \rightarrow \cdot M_3$$
$$\vdots$$
$$M + \cdot M_{n-1} \rightarrow \cdot M_n$$

where M_n is a polymer consisting of n monomer units. Polymerization may terminate in a number of ways. For example,

$$\text{mutual termination:} \quad \cdot M_n + \cdot M_m \rightarrow M_{n+m}$$
$$\text{disproportionation:} \quad \cdot HM_n + \cdot M_m \rightarrow M_n + HM_m$$
$$\text{chain transfer:} \quad M + \cdot M_n \rightarrow \cdot M + M_n$$

In **mutual termination** two growing radical chains combine. In termination by **disproportionation** a hydrogen atom transfers from one chain to another, corresponding to the oxidation of the donor and the reduction of the acceptor. In **chain transfer**, a new chain initiates at the expense of the one currently growing. As can be suspected, the mechanism is complicated, but can be explored by using the steady-state approximation.

How is that done? 17F.1 Deriving an expression for the rate of chain polymerization

The kinetic analysis of chain polymerization must take into account initiation, propagation, and termination.

Step 1 *Write an expression for the rate of initiation of the process*
If the initiation step is

$$In \rightarrow R\cdot + R\cdot \quad v_i = k_i[In]$$
$$M + R\cdot \rightarrow \cdot M_1$$

If the rate constants of the chain propagation steps are large enough, the first of these two steps is rate-determining for the overall polymerization process and the rate of initiation is equal to v_i.

Step 2 *Write an expression for propagation*

If the rate of propagation is independent of chain size for sufficiently large chains, then the rate of propagation, v_p, may be written

$$v_p = k_p[\text{M}][\cdot\text{M}]$$

where ·M stands for a polymer of any length. It follows from the remark in Step 1 that

$$\left(\frac{\text{d}[\cdot\text{M}]}{\text{d}t}\right)_{\text{production}} = 2fk_i[\text{In}]$$

where f is the fraction of radicals R· that successfully initiate a chain. The factor 2 recognizes that two radicals are formed in each initiation step.

Step 3 *Consider the termination of the process*

For the present analysis, suppose that only mutual termination occurs. If the rate of termination is assumed to be independent of the length of the chain, the rate law for termination is

$$v_t = k_t[\cdot\text{M}]^2$$

and the rate of change of radical concentration by this depletion process is

$$\left(\frac{\text{d}[\cdot\text{M}]}{\text{d}t}\right)_{\text{depletion}} = -2k_t[\cdot\text{M}]^2$$

In this case, the factor 2 recognizes that two radicals are removed in each depletion step.

Step 4 *Apply the steady-state approximation*

The net rate of formation of ·M is

$$\frac{\text{d}[\cdot\text{M}]}{\text{d}t} = \overbrace{2fk_i[\text{In}]}^{\text{production}} - \overbrace{2k_t[\cdot\text{M}]^2}^{\text{depletion}} \approx 0 \quad \text{(steady-state approximation)}$$

The steady-state concentration of radical chains is therefore

$$[\cdot\text{M}] = \left(\frac{fk_i}{k_t}\right)^{1/2}[\text{In}]^{1/2}$$

In Step 2 it is established that the overall rate of polymerization is equal to the rate of propagation, which is given by $v_p = k_p[\text{M}][\cdot\text{M}]$. The steady-state expression for [·M] can now be inserted into this expression to give

$$v_p = k_p[\cdot\text{M}][\text{M}] = k_p\left(\frac{fk_i}{k_t}\right)^{1/2}[\text{In}]^{1/2}[\text{M}]$$

The overall rate of polymerization is therefore proportional to the square root of the concentration of the initiator, In, and is given by

$$v = k_r[\text{In}]^{1/2}[\text{M}] \quad \text{with} \quad k_r = k_p\left(\frac{fk_i}{k_t}\right)^{1/2} \qquad (17\text{F}.13)$$

Rate of polymerization [chain polymerization]

The **kinetic chain length**, λ, is the ratio of the number of monomer units consumed to the number of radicals produced in the initiation step:

$$\lambda = \frac{\text{number of monomer units consumed}}{\text{number of radicals produced}} \qquad (17\text{F}.14\text{a})$$

Kinetic chain length [definition]

The kinetic chain length can be imagined as the average number of molecules in a chain produced by one initiating radical. The kinetic chain length can be expressed in terms of the rate expressions above. To do so, recognize that monomers are consumed at the rate that chains propagate. Then,

$$\lambda = \frac{\text{rate of propagation of chains}}{\text{rate of production of radicals}} \qquad (17\text{F}.14\text{b})$$

Kinetic chain length in terms of reaction rates

In applying the steady-state approximation, the rate of production of radicals is set equal to the termination rate (Step 4 in the discussion above). Therefore, the kinetic chain length may be written as

$$\lambda = \frac{k_p[\cdot\text{M}][\text{M}]}{2k_t[\cdot\text{M}]^2} = \frac{k_p[\text{M}]}{2k_t[\cdot\text{M}]}$$

The steady-state expression for [·M], $[\cdot\text{M}] = (fk_i/k_t)^{1/2}[\text{In}]^{1/2}$, is substituted for the radical concentration, to obtain

$$\lambda = k_r[\text{M}][\text{In}]^{-1/2} \quad \text{with} \quad k_r = k_p(4fk_ik_t)^{-1/2} \qquad (17\text{F}.14\text{c})$$

Kinetic chain length [chain polymerization]

In mutual termination, the average number of monomers in a polymer molecule, $\langle N\rangle$, produced by the reaction is the sum of the numbers of monomers in the two combining polymer chains. The average number of units in each chain is λ. Therefore,

$$\langle N\rangle = 2\lambda = 2k_r[\text{M}][\text{In}]^{-1/2} \qquad (17\text{F}.15)$$

Degree of polymerization [chain polymerization]

with k_r given in eqn 17F.14c. That is, the slower the initiation of the chain (the smaller the initiator concentration and the smaller the initiation rate constant), the greater is the kinetic chain length, and therefore the higher is the average molar mass of the polymer.

17F.3 Enzyme-catalysed reactions

A catalyst is a substance that accelerates a reaction but undergoes no net chemical change (Topic 17D): the catalyst lowers the activation energy of the reaction by providing an alternative path to that of the uncatalysed reaction (Fig. 17F.5). **Enzymes**, which are homogeneous biological catalysts, are very specific and can have a dramatic effect on the reactions they control. For example, the enzyme catalase accelerates the reaction it catalyses by a factor of 10^{12} at 298 K.

Enzymes contain an **active site**, which is responsible for binding the **substrates**, the reactants, and processing them into products. As is true of any catalyst, the active site returns to its original state after the products are released. Many enzymes consist primarily of proteins, some featuring organic or inorganic co-factors in their active sites. However, certain RNA molecules can also be biological catalysts, forming *ribozymes*.

The structure of the active site is specific to the reaction that it catalyses, with groups in the substrate attached to groups in the active site primarily by hydrogen bonding, electrostatic forces, and van der Waals interactions. Figure 17F.6 shows two models that explain the binding of a substrate to the active site of an enzyme. In the **lock-and-key model**, the active site and substrate have complementary three-dimensional structures and dock without the need for major structural change. Experimental evidence however favours the **induced fit model**, in which binding of the substrate induces a conformational change in the active site. Only after the change does the substrate fit snugly in the active site.

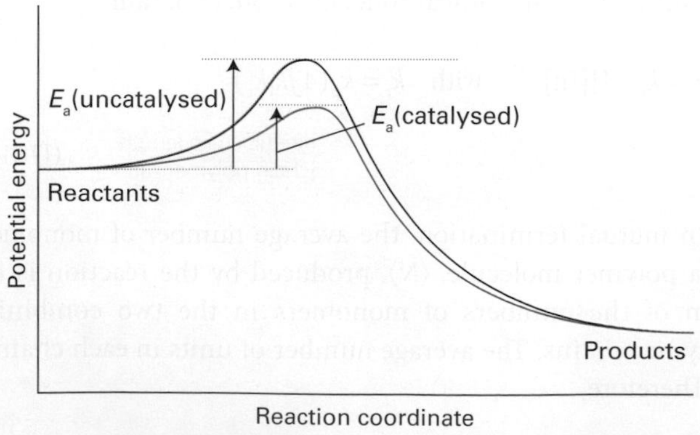

Figure 17F.5 A catalyst provides a different path with a lower activation energy. The result is an increase in the rates of the forward (and reverse) reaction.

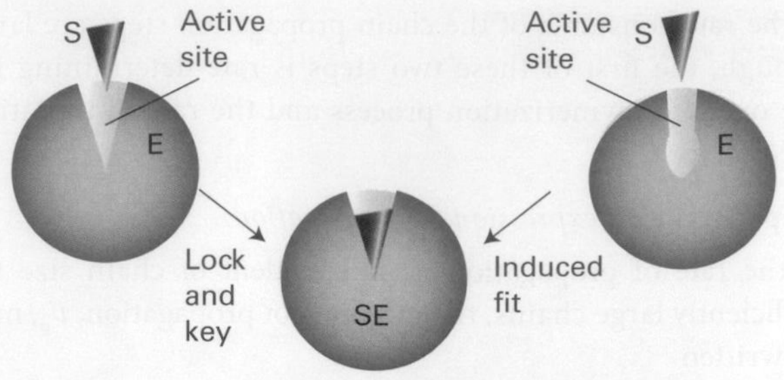

Figure 17F.6 Two models that explain the binding of a substrate to the active site of an enzyme. In the lock-and-key model, the active site and substrate have complementary three-dimensional structures and dock without the need for major atomic rearrangements. In the induced fit model, binding of the substrate induces a conformational change in the active site. The substrate fits well in the active site after the conformational change has taken place.

Experimental studies of enzyme kinetics are typically conducted by monitoring the initial rate of product formation in a solution in which the enzyme is present at very low concentration. Indeed, enzymes are such efficient catalysts that significant accelerations may be observed even when their concentration is more than three orders of magnitude smaller than that of the substrate.

The principal features of many enzyme-catalysed reactions are as follows:

- For a given initial concentration of substrate, $[S]_0$, the initial rate of product formation is proportional to the total concentration of enzyme, $[E]_0$.
- For a given $[E]_0$ and low values of $[S]_0$, the rate of product formation is proportional to $[S]_0$.
- For a given $[E]_0$ and high values of $[S]_0$, the rate of product formation becomes independent of $[S]_0$, reaching a maximum value known as the **maximum velocity**, v_{max}.

The **Michaelis–Menten mechanism** accounts for these features. According to this mechanism, an enzyme–substrate complex is formed in the first step and then either the substrate is released unchanged or, after modification, released as the product:

$$E + S \underset{k_a'}{\overset{k_a}{\rightleftharpoons}} ES$$

Michaelis–Menten mechanism

$$ES \xrightarrow{k_b} P + E$$

Again, the mechanism may be analysed by using the steady-state approximation.

How is that done? 17F.2 Deriving the Michaelis–Menten equation

The rate of product formation according to the Michaelis–Menten mechanism is

$$v = k_b[\mathrm{ES}]$$

so the strategy is centred on finding an expression for the concentration of the intermediate, ES.

Step 1 *Apply the steady-state approximation*

The concentration of the enzyme–substrate complex is found by invoking the steady-state approximation and writing

$$\frac{d[\mathrm{ES}]}{dt} = k_a[\mathrm{E}][\mathrm{S}] - k_a'[\mathrm{ES}] - k_b[\mathrm{ES}] \approx 0$$

It follows that

$$[\mathrm{ES}] = \frac{k_a[\mathrm{E}][\mathrm{S}]}{k_a' + k_b}$$

where [E] and [S] are the concentrations of *free* enzyme and substrate, respectively.

Step 2: *Simplify the expression for* [ES]

Now define the **Michaelis constant** as

$$K_M = \frac{k_a' + k_b}{k_a}$$

(Note that this constant has the dimensions of molar concentration.) To express the rate law in terms of the total concentrations of enzyme and the initial concentration of substrate first added, note that the total concentration of the enzyme is $[\mathrm{E}]_0 = [\mathrm{E}] + [\mathrm{ES}]$. It follows that $[\mathrm{E}] = [\mathrm{E}]_0 - [\mathrm{ES}]$. This expression for [E] is inserted into the steady-state expression for [ES] above to give

$$[\mathrm{ES}] = \frac{([\mathrm{E}]_0 - [\mathrm{ES}])[\mathrm{S}]}{K_M}$$

Because the substrate is typically in large excess relative to the enzyme, the free substrate concentration is approximately equal to the initial substrate concentration: $[\mathrm{S}] \approx [\mathrm{S}]_0$. The solution of the resulting expression for [ES] is

$$[\mathrm{ES}] = \frac{[\mathrm{S}]_0[\mathrm{E}]_0}{K_M + [\mathrm{S}]_0} = \frac{[\mathrm{E}]_0}{1 + K_M/[\mathrm{S}]_0}$$

Step 3 *Write an expression for the rate law*

The expression for [ES] can now be substituted into $v = k_b[\mathrm{ES}]$, to give the **Michaelis–Menten equation**

$$v = \frac{k_b[\mathrm{E}]_0}{1 + K_M/[\mathrm{S}]_0} \qquad (17F.16)$$

Michaelis–Menten equation

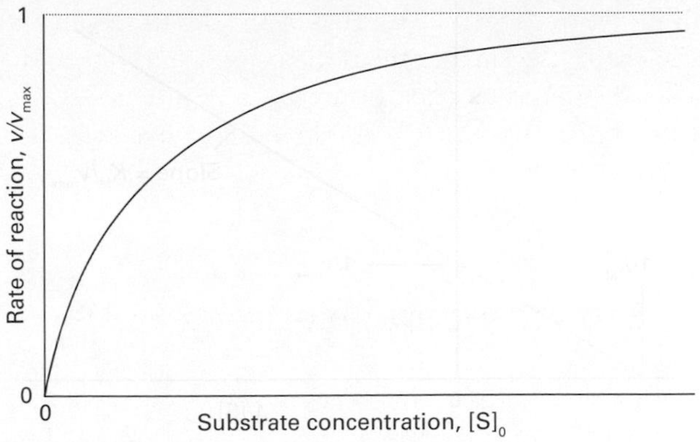

Figure 17F.7 The variation of the rate of an enzyme-catalysed reaction with substrate concentration. The approach to a maximum rate, v_{max}, for large $[\mathrm{S}]_0$ is explained by the Michaelis–Menten mechanism.

Equation 17F.16 predicts that, in accord with experimental observations (Fig. 17F.7):

- When $[\mathrm{S}]_0 \ll K_M$, the rate is proportional to $[\mathrm{S}]_0$:

$$v = \frac{k_b}{K_M}[\mathrm{S}]_0[\mathrm{E}]_0 \qquad (17F.17a)$$

- When $[\mathrm{S}]_0 \gg K_M$, the rate reaches its maximum value and is independent of $[\mathrm{S}]_0$:

$$v_{max} = k_b[\mathrm{E}]_0 \qquad (17F.17b)$$

Substitution of this definition of v_{max} into eqn 17F.16 gives

$$v = \frac{v_{max}}{1 + K_M/[\mathrm{S}]_0} \qquad (17F.18a)$$

which can be rearranged into a form suitable for data analysis by linear regression by taking reciprocals of both sides:

$$\frac{1}{v} = \frac{1}{v_{max}} + \left(\frac{K_M}{v_{max}}\right)\frac{1}{[\mathrm{S}]_0} \qquad \text{Lineweaver–Burk plot} \quad (17F.18b)$$

A **Lineweaver–Burk plot** is a plot of $1/v$ against $1/[\mathrm{S}]_0$. According to eqn 17F.18b, it should yield a straight line with slope of K_M/v_{max}, a y-intercept at $1/v_{max}$, and an x-intercept at $-1/K_M$ (Fig. 17F.8). The value of K_M can also be obtained from the ratio of the slope to the y-intercept. The value of k_b is then calculated from the y-intercept and eqn 17F.17b. However, the plot cannot give the individual rate constants k_a and k_a' that appear in the definition of K_M. The stopped-flow technique described in Topic 17A can give the additional data needed, because the rate of formation of the enzyme–substrate complex can be found by monitoring the concentration after mixing the enzyme and substrate. This procedure gives a value for k_a, and k_a' is then found by combining this result with the values of k_b and K_M.

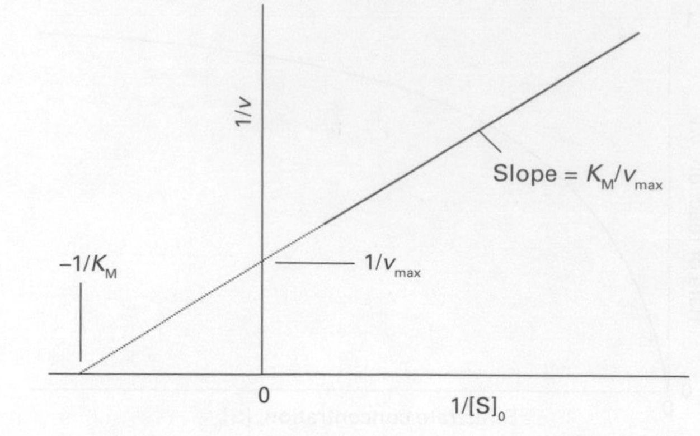

Figure 17F.8 The structure of a Lineweaver–Burk plot for the analysis of an enzyme-catalysed reaction that proceeds by a Michaelis–Menten mechanism and the significance of the intercepts and the slope.

Example 17F.2 Analysing data by using a Lineweaver–Burk plot

The enzyme carbonic anhydrase catalyses the hydration of CO_2 in red blood cells to give hydrogencarbonate (bicarbonate) ion: $CO_2(g) + H_2O(l) \rightarrow HCO_3^-(aq) + H^+(aq)$. The following data were obtained for the reaction at pH = 7.1, 273.5 K, and an enzyme concentration of 2.3 nmol dm^{-3}:

$[CO_2]/(\text{mmol dm}^{-3})$	1.25	2.5	5	20
$v/(\text{mmol dm}^{-3}\,\text{s}^{-1})$	2.78×10^{-2}	5.00×10^{-2}	8.33×10^{-2}	1.67×10^{-1}

Determine the maximum velocity and the Michaelis constant for the reaction.

Collect your thoughts Prepare a Lineweaver–Burk plot as explained in the text and determine the values of K_M and v_{max} by linear regression analysis.

The solution Draw up the following table:

$1/([CO_2]/(\text{mmol dm}^{-3}))$	0.800	0.400	0.200	0.0500
$1/(v/(\text{mmol dm}^{-3}\,\text{s}^{-1}))$	36.0	20.0	12.0	6.0

Figure 17F.9 shows the Lineweaver–Burk plot for the data. The slope is 40.0 and the y-intercept is 4.00. Hence,

$$v_{max}/(\text{mmol dm}^{-3}\,\text{s}^{-1}) = \frac{1}{\text{intercept}} = \frac{1}{4.00} = 0.250$$

so $v_{max} = 0.250\,\text{mmol dm}^{-3}\,\text{s}^{-1}$.

$$K_M/(\text{mmol dm}^{-3}) = \frac{\text{slope}}{\text{intercept}} = \frac{40.00}{4.00} = 10.0$$

and so $K_M = 10.0\,\text{mmol dm}^{-3}$.

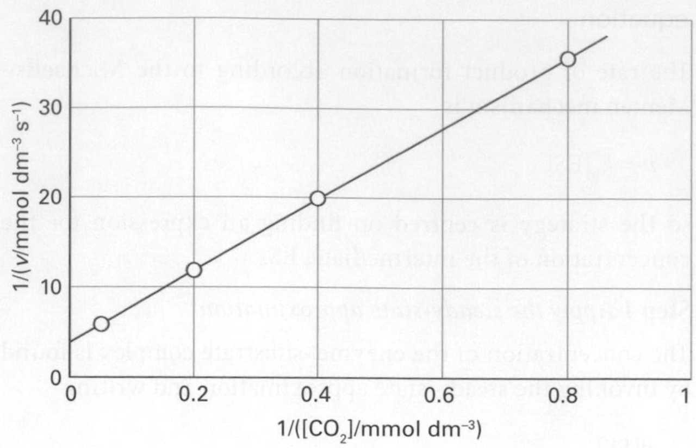

Figure 17F.9 The Lineweaver–Burk plot of the data for *Example* 17F.2.

A note on good practice The slope and the intercept are unit-less: all graphs should be plotted as pure numbers.

Self-test 17F.2 The enzyme α-chymotrypsin is secreted in the pancreas of mammals and cleaves peptide bonds between certain amino acids. Several solutions containing the small peptide *N*-glutaryl-L-phenylalanine-*p*-nitroanilide at different concentrations were prepared and the same small amount of α-chymotrypsin was added to each one. The following data were obtained on the initial rates of the formation of product:

$[S]/(\text{mmol dm}^{-3})$	0.334	0.450	0.667	1.00	1.33	1.67
$v/(\text{mmol dm}^{-3}\,\text{s}^{-1})$	0.150	0.199	0.285	0.406	0.516	0.619

Determine the maximum velocity and the Michaelis constant for the reaction.

Answer: $v_{max} = 2.80\,\text{mmol dm}^{-3}\,\text{s}^{-1}$, $K_M = 5.89\,\text{mmol dm}^{-3}$

The action of an enzyme may be partially suppressed by the presence of a foreign substance, which is called an **inhibitor**. An inhibitor may be a poison that has been administered to the organism, or it may be a substance that is naturally present in a cell and involved in its regulatory mechanism. An inhibitor typically works by blocking the active site or attaching elsewhere to the enzyme and forcing a change in geometry at the site so that it can no longer accommodate the substrate.[1]

[1] The use of kinetic criteria to distinguish different types of inhibition is described in our *Physical chemistry for the life sciences* (2012).

Checklist of concepts

☐ 1. The **Lindemann–Hinshelwood mechanism** of 'unimolecular' reactions accounts for the first-order kinetics of some gas-phase reactions.

☐ 2. In **stepwise polymerization** any two monomers in the reaction mixture can link together at any time.

☐ 3. The longer a stepwise polymerization proceeds, the higher the average molar mass of the product.

☐ 4. In **chain polymerization** an activated monomer attacks another monomer and links to it; the slower the initiation of the chain, the higher the average molar mass of the polymer.

☐ 5. The **kinetic chain length** is the ratio of the number of monomer units consumed to the number of radicals produced in the initiation step.

☐ 6. **Enzymes** are homogeneous biological catalysts.

☐ 7. The **Michaelis–Menten mechanism** of enzyme kinetics accounts for the dependence of rate on the concentrations of the substrate and the enzyme.

☐ 8. A **Lineweaver–Burk plot** is used to determine the parameters that occur in the Michaelis–Menten mechanism.

Checklist of equations

Property	Equation	Comment	Equation number
Lindemann–Hinshelwood rate law	$d[P]/dt = k_r[A]$ with $k_r = k_a k_b / k_a'$	$k_a'[A] >> k_b$	17F.5
Effective rate constant	$1/k_r = k_a'/k_a k_b + 1/k_a[A]$	Lindemann–Hinshelwood mechanism	17F.8
Fraction of condensed groups	$p = k_r t[A]_0/(1+k_r t[A]_0)$	Stepwise polymerization	17F.11
Degree of polymerization	$\langle N\rangle = 1/(1-p) = 1 + k_r t[A]_0$	Stepwise polymerization	17F.12
Rate of polymerization	$v = k_r[\mathrm{In}]^{1/2}[M]$	Chain polymerization	17F.13
Kinetic chain length	$\lambda = k_r[M][\mathrm{In}]^{-1/2}$, $k_r = k_p(4fk_ik_t)^{-1/2}$	Chain polymerization	17F.14c
Degree of polymerization	$\langle N\rangle = 2k_r[M][\mathrm{In}]^{-1/2}$	Chain polymerization	17F.15
Michaelis–Menten equation	$v = v_{max}/(1 + K_M/[S]_0)$		17F.18a
Lineweaver–Burk plot	$1/v = 1/v_{max} + (K_M/v_{max})(1/[S]_0)$		17F.18b

TOPIC 17G Photochemistry

> ➤ **Why do you need to know this material?**
>
> Many chemical and biological processes, including photosynthesis and vision, are initiated by the absorption of electromagnetic radiation, so you need to know how to include its effect in rate laws. The quantitative analysis of these processes provides insight into their mechanisms.
>
> ➤ **What is the key idea?**
>
> The mechanisms of many photochemical reactions lead to relatively simple rate laws that yield rate constants and quantitative measures of the efficiency with which radiant energy induces reactions.
>
> ➤ **What do you need to know already?**
>
> You need to be familiar with the concepts of singlet and triplet states (Topic 11F), modes of radiative decay (fluorescence and phosphorescence, Topic 11G), concepts of electronic spectroscopy (Topic 11F), and the formulation of a rate law from a proposed mechanism (Topic 17E).

Photochemical processes are initiated by the absorption of electromagnetic radiation. Among the most important of these processes are the ones that capture the radiant energy of the Sun. Some of these reactions lead to the heating of the atmosphere during the daytime by absorption of ultraviolet radiation. Others include the absorption of visible radiation during photosynthesis. Without photochemical processes, the Earth would probably be simply a warm, sterile, rock.

17G.1 Photochemical processes

Table 17G.1 summarizes common photochemical reactions. Photochemical processes are initiated by the absorption of radiation by at least one component of a reaction mixture. In a **primary process**, products are formed directly from the excited state of a reactant. Examples include fluorescence (Topic 11G) and *cis–trans* photoisomerizations. Products of a **secondary process** originate from intermediates that are formed directly from the excited state of a reactant, such as oxidative processes initiated by the oxygen atom formed by the photodissociation of ozone.

Table 17G.1 Examples of photochemical processes

Process	General form	Example
Ionization	$A^* \rightarrow A^+ + e^-$	$NO^* \rightarrow NO^+ + e^-$
Electron transfer	$A^* + B \rightarrow A^+ + B^-$ or $A^- + B^+$	$Ru(bpy)_3^{2+*} + Fe^{3+} \rightarrow Ru(bpy)_3^{3+} + Fe^{2+}$
Dissociation	$A^* \rightarrow B + C$	$O_3^* \rightarrow O_2 + O$
	$A^* + B{-}C \rightarrow A + B + C$	$Hg^* + CH_4 \rightarrow Hg + CH_3 + H$
Addition	$A^* + A^* \rightarrow B$	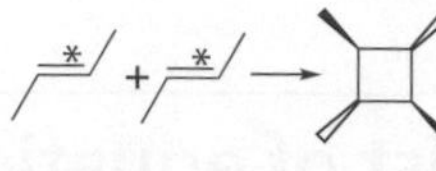 and isomers
	$A^* + B \rightarrow AB$	$Hg^* + H_2 \rightarrow HgH + H$
Abstraction	$A^* + B{-}C \rightarrow A{-}B + C$	$Hg^* + CH_3{-}H \rightarrow Hg{-}CH_3 + H$
Isomerization or rearrangement	$A^* \rightarrow A'$	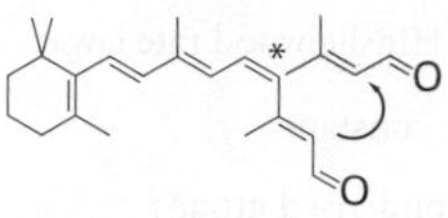

* Excited state.

Competing with the formation of photochemical products are numerous primary photophysical processes that can deactivate the excited state (Table 17G.2). Therefore, it is important to consider the timescales of the formation and decay of excited states before describing the mechanisms of photochemical reactions.

Electronic transitions caused by absorption of ultraviolet and visible radiation occur within 10^{-16}–10^{-15} s. The upper limit for the rate constant of a first-order photochemical reaction is then expected to be about $10^{16}\,s^{-1}$. Fluorescence is slower than absorption, with typical lifetimes of 10^{-12}–10^{-6} s. Therefore, the excited singlet state can initiate very fast photochemical reactions in the femtosecond (10^{-15} s) to picosecond (10^{-12} s) range. Examples of such ultrafast reactions are the initial events of vision and of photosynthesis. Internal conversion (IC) occurs on a timescale similar to that for the release of vibrational energy in molecules, so can occur in less than 10^{-12} s. Typical intersystem crossing (ISC, Topic 11G) and phosphorescence lifetimes for large organic molecules are 10^{-12}–10^{-4} s and 10^{-6}–10^{-1} s, respectively. As a consequence of their long lifetimes, excited triplet states are photochemically important. Indeed, because phosphorescence decay is several orders of magnitude slower than most typical reactions,

Table 17G.2 Common photophysical processes

Primary absorption	$S + h\nu \rightarrow S^*$
Excited-state absorption	$S^* + h\nu \rightarrow S^{**}$
	$T^* + h\nu \rightarrow T^{**}$
Fluorescence	$S^* \rightarrow S + h\nu$
Stimulated emission	$S^* + h\nu \rightarrow S + 2h\nu$
Intersystem crossing (ISC)	$S^* \rightarrow T^*$
Phosphorescence	$T^* \rightarrow S + h\nu$
Internal conversion (IC)	$S^* \rightarrow S$
Collision-induced emission	$S^* + M \rightarrow S + M + h\nu$
Collisional deactivation	$S^* + M \rightarrow S + M$
	$T^* + M \rightarrow S + M$
Electronic energy transfer:	
Singlet–singlet	$S^* + S \rightarrow S + S^*$
Triplet–triplet	$T^* + T \rightarrow T + T^*$
Excimer formation	$S^* + S \rightarrow (SS)^*$
Energy pooling	
Singlet–singlet	$S^* + S^* \rightarrow S^{**} + S$
Triplet–triplet	$T^* + T^* \rightarrow S^{**} + S$

* Denotes an excited state; S is a singlet state, T a triplet state, and M a 'third body'.

species in excited triplet states can undergo a very large number of collisions with other reactants before they lose their energy radiatively.

Brief illustration 17G.1

To judge whether the excited singlet or triplet state of the reactant is a suitable product precursor, the emission lifetimes are compared with the half-life of the relevant chemical reaction (Topic 17B). Consider a unimolecular photochemical reaction with rate constant $k_r = 1.7 \times 10^4\,s^{-1}$, and therefore a half-life of 41 μs. The observed fluorescence lifetime of the reactant is 1.0 ns and the observed phosphorescence lifetime is 1.0 ms. The excited singlet state is therefore too short-lived to be a major source of product in this reaction. On the other hand, the relatively long-lived excited triplet state is a good candidate for an intermediate.

17G.2 The primary quantum yield

The rates of deactivation of the excited state by radiative, non-radiative, and chemical processes determine the yield of product in a photochemical reaction. The **primary quantum yield**, ϕ, is defined as the number of photophysical or photochemical events that lead to primary products divided by the number of photons absorbed by the molecule in the same interval:

$$\phi = \frac{\text{number of events}}{\text{number of photons absorbed}} = \frac{N_{\text{events}}}{N_{\text{abs}}}$$

Primary quantum yield [definition] (17G.1a)

When both the numerator and denominator of this expression are divided by the time interval over which the events occur, the primary quantum yield is also seen to be the rate of radiation-induced primary events divided by the rate of photon absorption, I_{abs}:

$$\phi = \frac{\text{rate of process}}{\text{rate of photon absorption}} = \frac{v}{I_{\text{abs}}}$$

Primary quantum yield in terms of rates of processes (17G.1b)

Example 17G.1 Calculating a primary quantum yield

In an experiment to determine the quantum yield of a photochemical reaction, the absorbing substance was exposed to light of wavelength 490 nm from a 1.00 W laser source for 2700 s, with 60 per cent of the incident light being absorbed. As a result of irradiation, 3.44 mmol of the absorbing substance decomposed. What is the primary quantum yield?

Collect your thoughts You need to calculate the quantities in eqn 17G.1a. The number of photochemical events is simply the number of decomposed molecules, $N_{\text{events}} = N_{\text{decomposed}}$. To calculate the number of absorbed photons N_{abs}, note that:

- The energy absorbed by the substance is $E_{\text{abs}} = fPt$, where P is the incident power, t is the time of exposure, and the factor f (in this case $f = 0.60$) is the fraction of incident light that is absorbed.
- E_{abs} is also related to the number N_{abs} of absorbed photons through $E_{\text{abs}} = N_{\text{abs}}h\nu = N_{\text{abs}}hc/\lambda$, where hc/λ is the energy of a single photon of wavelength λ.

You can combine these two expressions for the absorbed energy to obtain N_{abs}. The primary quantum yield follows from $\phi = N_{\text{decomposed}}/N_{\text{abs}}$.

The solution From the two expressions for the absorbed energy, it follows that

$$fPt = N_{\text{abs}}\left(\frac{hc}{\lambda}\right)$$

and therefore that $N_{\text{abs}} = fPt\lambda/hc$. Now use eqn 17G.1a to write

$$\phi = \frac{N_{\text{decomposed}}}{N_{\text{abs}}} = \frac{N_{\text{decomposed}}hc}{fPt\lambda}$$

With $N_{\text{decomposed}} = (3.44 \times 10^{-3}\,\text{mol}) \times (6.022 \times 10^{23}\,\text{mol}^{-1}) = 2.07\ldots \times 10^{21}$, $P = 1.00\,\text{W} = 1.00\,\text{J s}^{-1}$, $t = 2700\,\text{s}$, $\lambda = 490\,\text{nm} = 4.90 \times 10^{-7}\,\text{m}$, and $f = 0.60$ it follows that

$$\phi = \frac{(2.07\ldots\times10^{21})\times(6.626\times10^{-34}\,\text{J s})\times(2.998\times10^{8}\,\text{m s}^{-1})}{0.60\times(1.00\,\text{J s}^{-1})\times(2700\,\text{s})\times(4.90\times10^{-7}\,\text{m})}$$
$$= 0.52$$

That is, about half the photons that are absorbed bring about photodissociation.

Self-test 17G.1 In an experiment to measure the quantum yield of a photochemical reaction, the absorbing substance was exposed to 320 nm radiation from an 87.5 mW laser source for 38 min. The intensity of the transmitted light was 0.35 that of the incident light. As a result of irradiation, 0.324 mmol of the absorbing substance decomposed. Determine the primary quantum yield.

Answer: $\phi = 0.93$

A molecule in an excited state must either decay to the ground state or form a photochemical product. Therefore, the total number of molecules deactivated by radiative processes, non-radiative processes, and photochemical reactions must be equal to the number of excited species produced by absorption of the incident radiation. It follows that the sum of primary quantum yields ϕ_i for *all* photophysical and photochemical events i must be equal to 1, regardless of the number of reactions involving the excited state:

$$\sum_i \phi_i = \sum_i \frac{v_i}{I_{\text{abs}}} = \frac{1}{I_{\text{abs}}}\sum_i v_i = 1 \qquad (17\text{G}.2\text{a})$$

Then, from eqn 17G.1b in the form $\phi_i = v_i / I_{\text{abs}}$ it follows that

$$\phi_i = \frac{v_i}{\sum_i v_i} \qquad (17\text{G}.2\text{b})$$

Therefore, the primary quantum yield of a particular process may be determined directly from the experimental rates of *all* photophysical and photochemical processes that deactivate the excited state.

If it is assumed that the only photophysical processes for the excited singlet state are fluorescence, internal conversion, and phosphorescence, then it follows that

$$\phi_{\text{F}} + \phi_{\text{IC}} + \phi_{\text{P}} = 1$$

where ϕ_{F}, ϕ_{IC}, and ϕ_{P} are the quantum yields of fluorescence, internal conversion, and phosphorescence, respectively (intersystem crossing from the singlet to the triplet state is taken into account by the presence of ϕ_{P}). The quantum yield of photon emission by fluorescence and phosphorescence is $\phi_{\text{emission}} = \phi_{\text{F}} + \phi_{\text{P}}$, which is less than 1. If the excited singlet state also participates in a primary photochemical reaction with quantum yield ϕ_{r}, then

$$\phi_{\text{F}} + \phi_{\text{IC}} + \phi_{\text{P}} + \phi_{\text{r}} = 1$$

17G.3 Mechanism of decay of excited singlet states

Consider the formation and decay of an excited singlet state in the absence of a chemical reaction:

Absorption:	$S + h\nu_i \rightarrow S^*$	$v_{\text{abs}} = I_{\text{abs}}$
Fluorescence:	$S^* \rightarrow S + h\nu_f$	$v_{\text{F}} = k_{\text{F}}[S^*]$
Internal conversion:	$S^* \rightarrow S$	$v_{\text{IC}} = k_{\text{IC}}[S^*]$
Intersystem crossing:	$S^* \rightarrow T^*$	$v_{\text{ISC}} = k_{\text{ISC}}[S^*]$

in which S is an absorbing singlet-state species, S* an excited singlet state, T* an excited triplet state, and $h\nu_i$ and $h\nu_f$ are the energies of the incident and fluorescent photons, respectively. From the methods presented in Topic 17E, the rate of formation of S* and its net rate of disappearance may be written as:

$$\text{Rate of formation of } S^* = I_{\text{abs}}$$

$$\text{Rate of disappearance of } S^* = k_{\text{F}}[S^*] + k_{\text{ISC}}[S^*] + k_{\text{IC}}[S^*]$$
$$= (k_{\text{F}} + k_{\text{ISC}} + k_{\text{IC}})[S^*]$$

It follows that the excited state decays by a first-order process so, when the light is turned off, the concentration of S* varies with time t as

$$[S^*](t) = [S^*]_0 e^{-t/\tau_0} \qquad (17\text{G}.3\text{a})$$

where the **observed lifetime**, τ_0, of the excited singlet state is defined as

$$\tau_0 = \frac{1}{k_{\text{F}} + k_{\text{ISC}} + k_{\text{IC}}} \qquad \text{Observed lifetime of the excited singlet state [definition]} \qquad (17\text{G}.3\text{b})$$

This expression can be used in a kinetic analysis of the decay of S* to find an expression for the quantum yield of fluorescence.

How is that done? 17G.1 Deriving an expression for the quantum yield of fluorescence

Most fluorescence measurements are conducted by illuminating a dilute sample with a continuous and intense beam of visible or ultraviolet radiation. It follows that [S*] is small and constant, so the steady-state approximation (Topic 17E) may be used for [S*]:

$$\frac{d[S^*]}{dt} = I_{\text{abs}} - k_{\text{F}}[S^*] - k_{\text{ISC}}[S^*] - k_{\text{IC}}[S^*]$$
$$= I_{\text{abs}} - (k_{\text{F}} + k_{\text{ISC}} + k_{\text{IC}})[S^*] \approx 0$$

Consequently,

$$I_{\text{abs}} = (k_{\text{F}} + k_{\text{ISC}} + k_{\text{IC}})[S^*]$$

The rate of fluorescence, v_F, is $k_F[S^*]$, so it follows from eqn 17G.1b that the quantum yield of fluorescence is

$$\phi_{F,0}=\frac{v_F}{I_{abs}}=\frac{k_F[S^*]}{(k_F+k_{ISC}+k_{IC})[S^*]}$$

which, by cancelling the [S*], simplifies to

$$\phi_{F,0}=\frac{k_F}{k_F+k_{ISC}+k_{IC}}$$

Then, by using the result for the lifetime in eqn 17G.3b,

$$\phi_{F,0}=k_F\tau_0 \quad \text{Quantum yield of fluorescence} \quad (17G.4)$$

The observed fluorescence lifetime can be measured by using a pulsed laser technique. First, the sample is excited with a short light pulse from a laser using a wavelength at which S absorbs strongly. Then, the exponential decay of the fluorescence intensity after the pulse is monitored.

Brief illustration 17G.2

At a certain wavelength, the fluorescence quantum yield and observed fluorescence lifetime of tryptophan in water are $\phi_{F,0}=0.20$ and $\tau_0=2.6\,\text{ns}$, respectively. It follows from eqn 17G.4 that the fluorescence rate constant k_F is

$$k_F=\frac{\phi_{F,0}}{\tau_0}=\frac{0.20}{2.6\times10^{-9}\,\text{s}}=7.7\times10^{7}\,\text{s}^{-1}$$

17G.4 Quenching

The shortening of the lifetime of the excited state by the presence of another species is called **quenching**. Quenching may be either a desired process, such as in energy or electron transfer, or an undesired side reaction that can decrease the quantum yield of a desired photochemical process. Quenching effects are studied by monitoring the emission from the excited state that is involved in the photochemical reaction.

The addition of a quencher, Q, opens an additional channel for deactivation of S*:

$$\text{Quenching: } S^*+Q\rightarrow S+Q \qquad v_Q=k_Q[Q][S^*]$$

The fluorescence quantum yields $\phi_{F,0}$ and ϕ_F measured in the absence and presence of Q, respectively, can be expressed in terms of the molar concentration of the quencher, [Q].

How is that done? 17G.2 Assessing the effect of a quencher on the fluorescence quantum yield

In the presence of quenching, the steady-state approximation for [S*] becomes

$$\frac{d[S^*]}{dt}=I_{abs}-(k_F+k_{ISC}+k_{IC}+k_Q[Q])[S^*]\approx0$$

and the fluorescence quantum yield is

$$\phi_F=\frac{k_F}{k_F+k_{ISC}+k_{IC}+k_Q[Q]}$$

The ratio of the quantum yields without and with a quencher present is

$$\frac{\phi_{F,0}}{\phi_F}=\frac{k_F}{k_F+k_{ISC}+k_{IC}}\times\frac{k_F+k_{ISC}+k_{IC}+k_Q[Q]}{k_F}$$
$$=\frac{k_F+k_{ISC}+k_{IC}+k_Q[Q]}{k_F+k_{ISC}+k_{IC}}$$
$$=1+\frac{k_Q}{k_F+k_{ISC}+k_{IC}}[Q]$$

By recognizing from eqn 17G.3b that $1/(k_F+k_{ISC}+k_{IC})=\tau_0$, this expression becomes the **Stern–Volmer equation**:

$$\frac{\phi_{F,0}}{\phi_F}=1+\tau_0k_Q[Q] \quad \text{Stern–Volmer equation} \quad (17G.5)$$

The Stern–Volmer equation implies that a plot of $\phi_{F,0}/\phi_F$ against [Q] should be a straight line with slope τ_0k_Q. Such a plot is called a **Stern–Volmer plot** (Fig. 17G.1). The method may also be applied to the quenching of phosphorescence.

Equation 17G.4 in the form $k_F=\phi_{F,0}/\tau_0$ shows that the rate constant for fluorescence, and hence the rate of fluorescence (which determines the intensity of fluorescence), is proportional to the quantum yield. The ratio $\phi_{F,0}/\phi_F$ is therefore equal to the ratio $I_{F,0}/I_F$, where $I_{F,0}$ is the intensity of fluorescence in the absence of quencher and I_F the intensity when quencher is present. Similarly, from the same equation in the form $\tau_0=\phi_{F,0}/k_F$, the fluorescence lifetime is also proportional to the quantum yield, so the ratio τ_0/τ (where τ is the lifetime in the presence of the quencher) is also equal to $\phi_{F,0}/\phi_F$.

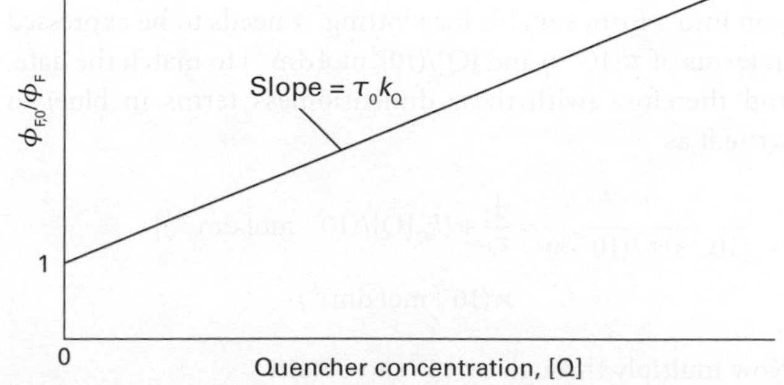

Figure 17G.1 The form of a Stern–Volmer plot and the interpretation of the slope in terms of the rate constant for quenching and the observed fluorescence lifetime in the absence of quenching.

Stern–Volmer plots can therefore be made by plotting either $I_{F,0}/I_F$ or τ_0/τ against the quencher concentration. The slope and intercept are the same as those shown for eqn 17G.5.

Example 17G.2 Determining the quenching rate constant

The molecule 2,2′-bipyridine (1, bpy) forms a complex with the Ru^{2+} ion. Tris-(2,2′-bipyridyl)ruthenium(II), $[Ru(bpy)_3]^{2+}$ (2), has a strong metal-to-ligand charge transfer (MLCT) transition (Topic 11F) at 450 nm.

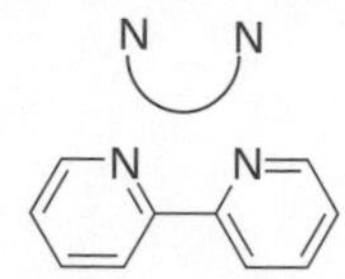

1 2,2′-Bipyridine (bpy)

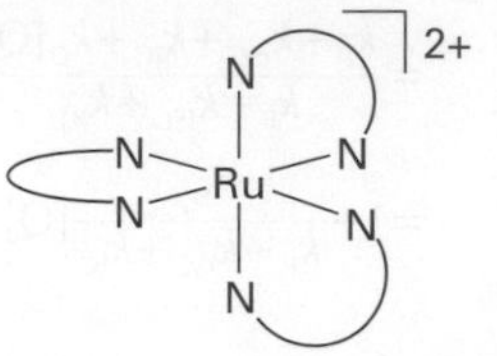

2 $[Ru(bpy)_3]^{2+}$

The quenching of the $^*[Ru(bpy)_3]^{2+}$ excited state by Fe^{3+} (present as the complex ion $[Fe(OH_2)_6]^{3+}$) in acidic solution was monitored by measuring emission lifetimes at 600 nm. Determine the quenching rate constant for this reaction from the following data:

$[[Fe(OH_2)_6]^{3+}]/(10^{-2}\,mol\,dm^{-3})$	0	1.6	4.7	7.0	9.4
$\tau/(10^{-7}\,s)$	6.00	4.05	3.37	2.96	2.17

Collect your thoughts Rewrite the Stern–Volmer equation (eqn 17G.5) for use with lifetime data; then fit the data to a straight line.

The solution Substitute τ_0/τ for $\phi_{F,0}/\phi_F$ in eqn 17G.5 and, after rearrangement, obtain

$$\frac{1}{\tau}=\frac{1}{\tau_0}+k_Q[Q]$$

Because the axes of plots should be labelled with pure numbers, it is necessary to introduce and handle units before using this equation for the analysis of the data. To bring the expression into a form suitable for plotting, it needs to be expressed in terms of $\tau/(10^{-7}\,s)$ and $[Q]/(10^{-2}\,mol\,dm^{-3})$ to match the data, and therefore (with these dimensionless terms in blue) to write it as

$$\frac{1}{(10^{-7}\,s)\tau/(10^{-7}\,s)}=\frac{1}{\tau_0}+\{k_Q[Q]/(10^{-2}\,mol\,dm^{-3})\}\times(10^{-2}\,mol\,dm^{-3})$$

Now multiply through by 10^{-7} s to obtain

$$\frac{1}{\tau/(10^{-7}\,s)}=\frac{10^{-7}\,s}{\tau_0}+k_Q\times(10^{-2}\,mol\,dm^{-3})\times(10^{-7}\,s)\times[Q]/(10^{-2}\,mol\,dm^{-3})$$

and collect terms:

$$\overbrace{\frac{1}{\tau/(10^{-7}\,s)}}^{y}=\overbrace{\frac{1}{\tau_0/(10^{-7}\,s)}}^{y\text{-intercept}}+\overbrace{(k_Q\times10^{-9}\,mol\,dm^{-3}\,s)}^{\text{slope}}\times\overbrace{[Q]/(10^{-2}\,mol\,dm^{-3})}^{x}$$

Note that because $slope = k_Q\times10^{-9}\,mol\,dm^{-3}\,s$, then $k_Q = slope\times10^{9}\,dm^3\,mol^{-1}\,s^{-1}$. Draw up the following table with Q = $[Fe(OH_2)_6]^{3+}$:

$[[Fe(OH_2)_6]^{3+}]/(10^{-2}\,mol\,dm^{-3})$	0	1.6	4.7	7.0	9.4
$1/(\tau/10^{-7}\,s)$	0.167	0.247	0.297	0.338	0.461

Figure 17G.2 shows a plot of $1/(\tau/10^{-7}\,s)$ against $[[Fe(OH_2)_6]^{3+}]/(10^{-2}\,mol\,dm^{-3})$ and the results of a fit to this expression. The slope of the line is 0.029, so $k_Q = 2.9\times10^{7}\,dm^3\,mol^{-1}\,s^{-1}$.

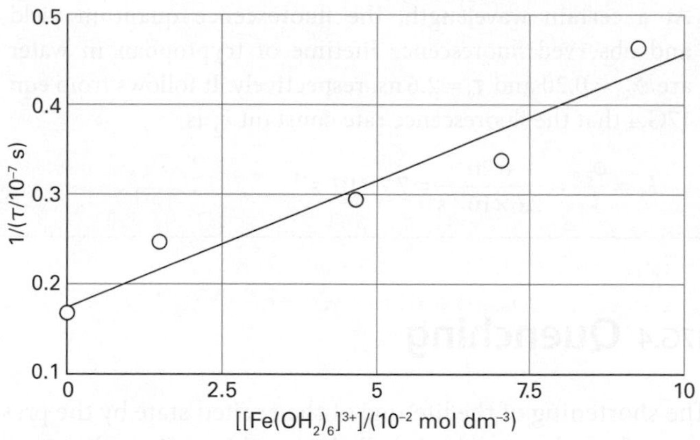

Figure 17G.2 The Stern–Volmer plot of the data for *Example* 17G.2.

Comment. Measurements of emission lifetimes are preferred because they yield the value of k_Q directly. To determine the value of k_Q from intensity or quantum yield measurements, it is necessary to make an independent measurement of τ_0.

Self-test 17G.2 The quenching of tryptophan fluorescence by dissolved O_2 gas was monitored by measuring emission lifetimes at 348 nm in aqueous solutions. Determine the quenching rate constant for this process from the following data:

$[O_2]/(10^{-2}\,mol\,dm^{-3})$	0	2.3	5.5	8	10.8
τ/ns	2.6	1.5	0.92	0.71	0.57

Answer: $1.3\times10^{10}\,dm^3\,mol^{-1}\,s^{-1}$

Three common mechanisms for bimolecular quenching of an excited singlet (or triplet) state are:

Collisional deactivation: $S^* + Q \rightarrow S + Q$

Resonance energy transfer: $S^* + Q \rightarrow S + Q^*$

Electron transfer: $S^* + Q \rightarrow S^{+/-} + Q^{-/+}$

The quenching rate constant itself does not give much insight into the mechanism of quenching. Collisional quenching is particularly efficient when Q is a species, such as iodide ion, which receives energy from S* and then decays to the ground state primarily by releasing energy as heat. For the system of *Example* 17G.2, it is known that the quenching of the excited state of $[Ru(bpy)_3]^{2+}$ is a result of electron transfer to Fe^{3+}, but the quenching data do not prove the mechanism.

17G.5 Resonance energy transfer

The energy transfer process $S^* + Q \rightarrow S + Q^*$ can be regarded as taking place as follows. The oscillating electric field of the incoming electromagnetic radiation induces an oscillating electric dipole moment (a transition dipole moment) in S. Energy is absorbed by S if the frequency of the incident radiation, ν, is such that $\nu = \Delta E_S/h$, where ΔE_S is the energy separation of the ground and excited electronic states of S and h is Planck's constant. This is the 'resonance condition' for absorption of radiation (essentially the Bohr frequency condition, eqn 7A.9). The oscillating dipole on S can now affect electrons bound to a nearby Q molecule by inducing an oscillating dipole moment (another transition dipole moment) in them. If the frequency of oscillation of the electric dipole moment in S is such that $\nu = \Delta E_Q/h$, where ΔE_Q is the energy separation of the ground and excited electronic states of Q, then Q will absorb energy from S. The coupling of the two transition moments can be regarded as an exchange of a photon, in which a photon generated by S is absorbed by Q.

The efficiency, η_T, of resonance energy transfer is defined as

$$\eta_T = 1 - \frac{\phi_F}{\phi_{F,0}} \quad \text{Efficiency of resonance energy transfer [definition]} \quad (17G.6)$$

According to the **Förster theory** of resonance energy transfer, energy transfer is efficient when:

- The energy donor and acceptor are separated by a short distance (of the order of nanometres).
- The photon is regarded as emitted by the excited state of the donor and then absorbed directly by the acceptor.

For donor–acceptor systems held rigidly either by covalent bonds or by a protein 'scaffold', η_T increases with decreasing distance, R, according to

$$\eta_T = \frac{R_0^6}{R_0^6 + R^6} \quad \text{Efficiency of energy transfer in terms of the donor–acceptor distance} \quad (17G.7)$$

where R_0 is a parameter (with dimensions of distance) that is characteristic of each donor–acceptor pair. It can be regarded as the distance at which energy transfer is 50 per cent efficient for a given donor–acceptor pair. (This assertion can be confirmed by using $R = R_0$ in eqn 17G.7.) Equation 17G.7 has been verified experimentally and values of R_0 are available for a number of donor–acceptor pairs (Table 17G.3).

The emission and absorption spectra of molecules span a range of wavelengths, so the second requirement of the Förster theory is met when the emission spectrum of the donor molecule overlaps significantly with the absorption spectrum of the acceptor. In the overlap region, a photon emitted by the donor has the appropriate energy to be absorbed by the acceptor (Fig. 17G.3).

Table 17G.3 Values of R_0 for some donor–acceptor pairs*

Donor‡	Acceptor	R_0/nm
Naphthalene	Dansyl	2.2
Dansyl	ODR	4.3
Pyrene	Coumarin	3.9
1,5-I AEDANS	FITC	4.9
Tryptophan	1,5-I AEDANS	2.2
Tryptophan	Haem (heme)	2.9

*Additional values may be found in J.R. Lacowicz, *Principles of fluorescence spectroscopy*, Kluwer Academic/Plenum, New York (1999).

‡Abbreviations:

Dansyl: 5-dimethylamino-1-naphthalenesulfonic acid

FITC: fluorescein 5-isothiocyanate

1,5-I AEDANS: 5-(((2-iodoacetyl)amino)ethyl)amino)naphthalene-1-sulfonic acid (3)

ODR: octadecyl-rhodamine

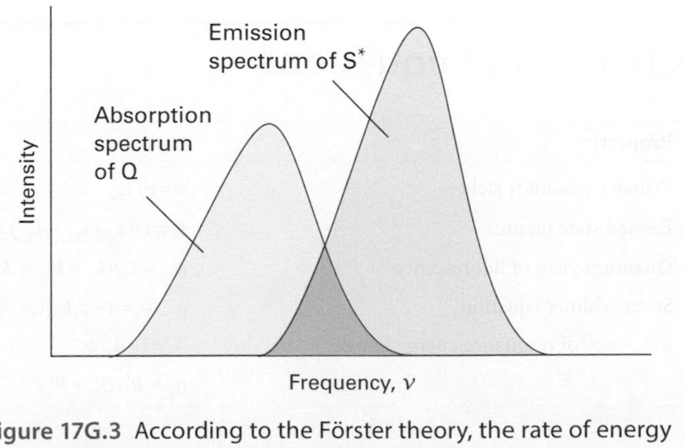

Figure 17G.3 According to the Förster theory, the rate of energy transfer from a molecule S* in an excited state to a quencher molecule Q is optimized at radiation frequencies for which the emission spectrum of S* overlaps the absorption spectrum of Q, as shown in the (dark green) shaded region.

Equation 17G.7 forms the basis of **fluorescence resonance energy transfer** (FRET), in which the dependence of the energy transfer efficiency, η_T, on the distance, R, between energy donor and acceptor is used to measure distances in biological systems. In a typical FRET experiment, a site on a biopolymer or membrane is labelled covalently with an energy donor and another site is labelled covalently with an energy acceptor. In certain cases, the donor or acceptor may be natural constituents of the system, such as amino acid groups, cofactors, or enzyme substrates. The distance between the labels is then calculated from the known value of R_0 and eqn 17G.7. Several tests have shown that the FRET technique is useful for measuring distances ranging from 1 to 9 nm.

Brief illustration 17G.3

As an illustration of the FRET technique, consider a study of the protein rhodopsin. When an amino acid on the surface of rhodopsin was labelled covalently with the energy donor 1,5-I AEDANS (3), the fluorescence quantum yield of the label decreased from 0.75 to 0.68 due to quenching by the visual pigment 11-*cis*-retinal (4), which is attached elsewhere in the protein. From eqn 17G.6 it follows that $\eta_T = 1 - 0.68/0.75 = 0.093$ and from eqn 17G.7 and the known value of $R_0 = 5.4\,\text{nm}$ for the 1.5-I AEDANS/11-*cis*-retinal, $R = 7.9\,\text{nm}$. Therefore, take 7.9 nm to be the distance between the surface of the protein and 11-*cis*-retinal.

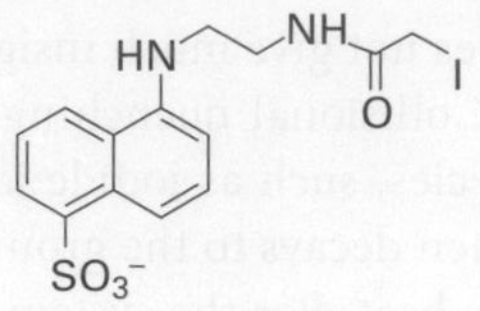

3 1,5-I AEDANS

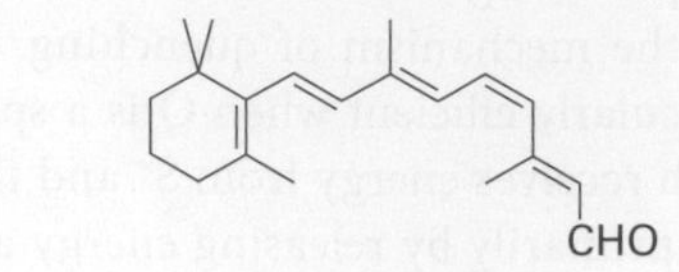

4 11-*cis*-Retinal

If donor and acceptor molecules diffuse in solution or in the gas phase, Förster theory predicts that the efficiency of quenching by energy transfer increases as the average distance travelled between collisions of donor and acceptor decreases. That is, the quenching efficiency increases with the concentration of quencher, as predicted by the Stern–Volmer equation.

Checklist of concepts

☐ 1. The **primary quantum yield** of a photochemical reaction is the number of reactant molecules producing specified primary products for each photon absorbed.

☐ 2. The **observed lifetime** of an excited state is related to the quantum yield and rate constant of emission.

☐ 3. A **Stern–Volmer plot** is used to analyse the kinetics of fluorescence quenching in solution.

☐ 4. Collisional deactivation, electron transfer, and resonance energy transfer are common fluorescence quenching processes.

☐ 5. The efficiency of resonance energy transfer decreases with increasing separation between donor and acceptor molecules.

Checklist of equations

Property	Equation	Comment	Equation number
Primary quantum yield	$\phi = v/I_{\text{abs}}$		17G.1b
Excited state lifetime	$\tau_0 = 1/(k_F + k_{ISC} + k_{IC})$	No quencher present	17G.3b
Quantum yield of fluorescence	$\phi_{F,0} = k_F/(k_F + k_{ISC} + k_{IC}) = k_F\tau_0$	Without quencher present	17G.4
Stern–Volmer equation	$\phi_{F,0}/\phi_F = 1 + \tau_0 k_Q[\text{Q}]$		17G.5
Efficiency of resonance energy transfer	$\eta_T = 1 - \phi_F/\phi_{F,0}$	Definition	17G.6
	$\eta_T = R_0^6/(R_0^6 + R^6)$	Förster theory	17G.7

FOCUS 17 Chemical kinetics

TOPIC 17A The rates of chemical reactions

Discussion questions

D17A.1 Summarize the characteristics of zeroth-order, first-order, second-order, and pseudofirst-order reactions.

D17A.2 When can a reaction order not be ascribed?

D17A.3 What are the advantages of ascribing an order to a reaction?

D17A.4 Summarize the experimental procedures that can be used to monitor the composition of a reaction system.

Exercises

E17A.1(a) Predict how the total pressure varies during the reaction $2\,ICl(g) + H_2(g) \rightarrow I_2(g) + 2\,HCl(g)$ in a constant-volume container. Assume that at the start of the reaction the partial pressures of the reactants are equal and that no products are present.

E17A.1(b) Predict how the total pressure varies during the reaction $N_2(g) + 3\,H_2(g) \rightarrow 2\,NH_3(g)$ in a constant-volume container. Assume that at the start of the reaction the partial pressures of H_2 and N_2 are in the ratio 3 to 1 and that no products are present.

E17A.2(a) The rate of formation of NO in the reaction $2\,NOBr(g) \rightarrow 2\,NO(g) + Br_2(g)$ was reported as $d[NO]/dt = 0.24\,mmol\,dm^{-3}\,s^{-1}$ under particular conditions. What is the rate of formation of Br_2?

E17A.2(b) The rate of change of molar concentration of CH_3 radicals in the reaction $2\,CH_3(g) \rightarrow CH_3CH_3(g)$ was reported as $d[CH_3]/dt = -1.2\,mol\,dm^{-3}\,s^{-1}$ under particular conditions. What is the rate of formation of CH_3CH_3?

E17A.3(a) The rate of the reaction $A + 2B \rightarrow 3C + D$ was reported as 2.7 mol $dm^{-3}\,s^{-1}$. State the rates of formation and consumption of the participants.

E17A.3(b) The rate of the reaction $A + 3B \rightarrow C + 2D$ was reported as 2.7 mol $dm^{-3}\,s^{-1}$. State the rates of formation and consumption of the participants.

E17A.4(a) The rate of formation of C in the reaction $2A + B \rightarrow 2C + 3D$ is $2.7\,mol\,dm^{-3}\,s^{-1}$. State the reaction rate, and the rates of formation or consumption of A, B, and D.

E17A.4(b) The rate of consumption of B in the reaction $A + 3B \rightarrow C + 2D$ is $2.7\,mol\,dm^{-3}\,s^{-1}$. State the reaction rate, and the rates of formation or consumption of A, C, and D.

E17A.5(a) The rate law for the reaction in Exercise E17A.3(a) was found to be $v = k_r[A][B]$. What are the units of k_r when the concentrations are in moles per cubic decimetre? Express the rate law in terms of (i) the rate of formation of C and (ii) the rate of consumption of A.

E17A.5(b) The rate law for the reaction in Exercise E17A.3(b) was found to be $v = k_r[A][B]^2$. What are the units of k_r when the concentrations are in moles per cubic decimetre? Express the rate law in terms of (i) the rate of formation of C and (ii) the rate of consumption of A.

E17A.6(a) The rate law for the reaction in Exercise E17A.4(a) was reported as $d[C]/dt = k_r[A][B][C]$. Express the rate law in terms of the reaction rate v. What are the units of k_r when the concentrations are in moles per cubic decimetre?

E17A.6(b) The rate law for the reaction in Exercise E17A.4(b) was reported as $d[C]/dt = k_r[A][B][C]^{-1}$. Express the rate law in terms of the reaction rate v. What are the units of k_r when the concentrations are in moles per cubic decimetre?

E17A.7(a) If the rate laws are expressed with (i) concentrations in moles per cubic decimetre, (ii) pressures in kilopascals, what are the units of a second-order and of a third-order rate constant?

E17A.7(b) If the rate laws are expressed with (i) concentrations in molecules per cubic metre, (ii) pressures in pascals, what are the units of a second-order and of a third-order rate constant?

E17A.8(a) The rate law $v=(k_{r1}[A][B])/(k_{r2}+k_{r3}[B]^{1/2})$ was established in a series of experiments. Identify the conditions under which (i) an order with respect to A, (ii) an order with respect to B, and (iii) an overall order, can be assigned.

E17A.8(b) Certain gas-phase reactions of the type $A \rightarrow P$ have rate laws of the form $v=k_a k_b[A]^2/(k_b+k_a'[A])$. What is the order with respect to A under a variety of conditions that you should specify?

E17A.9(a) At 400 K, the rate of decomposition of a gaseous compound was $9.71\,Pa\,s^{-1}$ when 10.0 per cent had reacted and $7.67\,Pa\,s^{-1}$ when 20.0 per cent had reacted. Identify the order of the reaction.

E17A.9(b) At 350 K, the rate of decomposition of a gaseous compound was $10.01\,Pa\,s^{-1}$ when 10.0 per cent had reacted and $8.90\,Pa\,s^{-1}$ when 20.0 per cent had reacted. Identify the order of the reaction.

Problems

P17A.1 The following initial-rate data were obtained on the rate of binding of glucose with the enzyme hexokinase present at $1.34\,mmol\,dm^{-3}$. What is (a) the order of reaction with respect to glucose, (b) the rate constant?

$[C_6H_{12}O_6]/(mmol\,dm^{-3})$	1.00	1.54	3.12	4.02
$v_0/(mol\,dm^{-3}\,s^{-1})$	5.0	7.6	15.5	20.0

P17A.2 The following data were obtained on the initial rates of a reaction of a d-metal complex with a reactant Y in aqueous solution. What is (a) the order of reaction with respect to the complex and Y, (b) the rate constant? For the experiments (i), $[Y] = 2.7\,mmol\,dm^{-3}$ and for experiments (ii) $[Y] = 6.1\,mmol\,dm^{-3}$.

$[complex]/(mmol\,dm^{-3})$		8.01	9.22	12.11
$v_0/(mol\,dm^{-3}\,s^{-1})$	(i)	125	144	190
	(ii)	640	730	960

P17A.3 The following kinetic data (v_0 is the initial rate) were obtained for the reaction $2\,ICl(g) + H_2(g) \rightarrow I_2(g) + 2\,HCl(g)$:

Experiment	$[ICl]_0/(mmol\,dm^{-3})$	$[H_2]_0/(mmol\,dm^{-3})$	$v_0/(mol\,dm^{-3}\,s^{-1})$
1	1.5	1.5	3.7×10^{-7}
2	3.0	1.5	7.4×10^{-7}
3	3.0	4.5	22×10^{-7}
4	4.7	2.7	?

(a) Write the rate law for the reaction. (b) From the data, determine the value of the rate constant. (c) Use the data to predict the reaction rate for experiment 4.

TOPIC 17B Integrated rate laws

Discussion questions

D17B.1 Describe the main features, including advantages and disadvantages, of the following experimental methods for determining the rate law of a reaction: the isolation method, the method of initial rates, and fitting data to integrated rate law expressions.

D17B.2 What is the origin of the classification of a reaction as having pseudofirst and pseudosecond order? Under what conditions can the apparent order of a reaction change?

D17B.3 Write the rate law that corresponds to each of the following expressions: (a) $[A] = [A]_0 - k_r t$, (b) $\ln([A]/[A]_0) = -k_r t$, and (c) $[A] = [A]_0/(1 + k_r t[A]_0)$.

Exercises

E17B.1(a) A number of reactions that take place on the surfaces of catalysts are zeroth order in the reactant. One example is the decomposition of ammonia on hot tungsten. In an experiment, the partial pressure of ammonia decreased from 21 kPa to 10 kPa in 770 s. (i) What is the rate constant for the zeroth-order reaction? (ii) How long will it take for all the ammonia to be consumed?

E17B.1(b) In a study of the enzyme-catalysed oxidation of ethanol, the molar concentration of ethanol decreased in a first-order reaction from 220 mmol dm^{-3} to 56.0 mmol dm^{-3} in 1.22×10^4 s. What is the rate constant of the reaction?

E17B.2(a) At 518 °C, the half-life for the decomposition of a sample of gaseous ethanal (acetaldehyde) initially at a partial pressure of 363 Torr was 410 s. When the partial pressure was 169 Torr, the half-life was 880 s. Identify the order of the reaction.

E17B.2(b) At 400 K, the half-life for the decomposition of a sample of a gaseous compound initially at a partial pressure of 55.5 kPa was 340 s. When the partial pressure was 28.9 kPa, the half-life was 178 s. Identify the order of the reaction.

E17B.3(a) The rate constant for the first-order decomposition of N_2O_5 in the reaction $2\,N_2O_5(g) \rightarrow 4\,NO_2(g) + O_2(g)$ is $k_r = 3.38 \times 10^{-5}\,s^{-1}$ at 25 °C. What is the half-life of N_2O_5? If the initial partial pressure of N_2O_5 is 500 Torr, what will its partial pressure be (i) 50 s, (ii) 20 min after initiation of the reaction?

E17B.3(b) The rate constant for the first-order decomposition of a compound A in the reaction $2\,A \rightarrow P$ is $k_r = 3.56 \times 10^{-7}\,s^{-1}$ at 25 °C. What is the half-life of A? If the initial partial pressure of A is 33.0 kPa, what will be its partial pressure (i) 50 s, (ii) 20 min after initiation of the reaction?

E17B.4(a) The second-order rate constant for the reaction $CH_3COOC_2H_5(aq) + OH^-(aq) \rightarrow CH_3CO_2^-(aq) + CH_3CH_2OH(aq)$ is 0.11 dm^3 mol^{-1} s^{-1}. What is the concentration of ester ($CH_3COOC_2H_5$) after (i) 20 s, (ii) 15 min when ethyl ethanoate is added to aqueous sodium hydroxide so that the initial concentrations are [NaOH] = 0.060 mol dm^{-3} and $[CH_3COOC_2H_5]$ = 0.110 mol dm^{-3}?

E17B.4(b) The second-order rate constant for the reaction $A + 2\,B \rightarrow C + D$ is 0.34 dm^3 mol^{-1} s^{-1}. What is the concentration of C after (i) 20 s, (ii) 15 min when the reactants are mixed with initial concentrations of [A] = 0.027 mol dm^{-1} and [B] = 0.130 mol dm^{-3}?

E17B.5(a) A reaction $2\,A \rightarrow P$ has a second-order rate law with $k_r = 4.30 \times 10^{-4}$ dm^3 mol^{-1} s^{-1}. Calculate the time required for the concentration of A to change from 0.210 mol dm^{-3} to 0.010 mol dm^{-3}.

E17B.5(b) A reaction $2\,A \rightarrow P$ has a third-order rate law with $k_r = 6.50 \times 10^{-4}$ dm^6 mol^{-2} s^{-1}. Calculate the time required for the concentration of A to change from 0.067 mol dm^{-3} to 0.015 mol dm^{-3}.

E17B.6(a) The reaction $A + B \rightarrow P$ is found to be first order in both A and B. The reaction was carried out in a solution that was initially 0.080 mol dm^{-3} in A and 0.060 mol dm^{-3} in B. After 1.0 h the concentration of B had fallen to 0.030 mol dm^{-3}. (i) Calculate the rate constant. (ii) What are the half-lives of the reactants?

E17B.6(b) A second-order reaction of the type $A + 2\,B \rightarrow P$ was carried out in a solution that was initially 0.050 mol dm^{-3} in A and 0.030 mol dm^{-3} in B. After 1.0 h the concentration of A had fallen to 0.040 mol dm^{-3}. (a) Calculate the rate constant. (b) What is the half-life of each reactant?

Problems

P17B.1 For a first-order reaction of the form $A \rightarrow nB$ (with n possibly fractional) and $[B]_0 = 0$, the concentration of the product varies with time as $[B]=n[A]_0(1-e^{-k_r t})$. Plot the time dependence of [A] and [B] for $n = \frac{1}{2}$, 1, and 2. *Hint:* To make your plots general, let the horizontal axis be $k_r t$ and plot $[A]/[A]_0$ or $[B]/[A]_0$ on the vertical axis.

P17B.2 For a second-order reaction of the form $A \rightarrow nB$ (with n possibly fractional) and $[B]_0 = 0$, the concentration of the product varies with time as $[B] = nk_r t[A]_0^2/(1 + k_r t[A]_0)$. Plot the time dependence of [A] and [B] for $n = \frac{1}{2}$, 1, and 2. *Hint:* See the hint to Problem P17B.1.

P17B.3 The data below apply to the formation of urea from ammonium cyanate, $NH_4CNO \rightarrow NH_2CONH_2$. Initially 22.9 g of ammonium cyanate was dissolved in enough water to prepare 1.00 dm^3 of solution. Identify the order of the reaction and calculate the rate constant and the mass of ammonium cyanate left after 300 min.

t/min	0	20.0	50.0	65.0	150
m(urea)/g	0	7.0	12.1	13.8	17.7

P17B.4 The data below apply to the reaction, $(CH_3)_3CBr(aq) + H_2O(l) \rightarrow (CH_3)_3COH(aq) + HBr(aq)$. Identify the order of the reaction and calculate the rate constant and the molar concentration of $(CH_3)_3CBr$ remaining after 43.8 h.

t/h	0	3.15	6.20	10.00	18.30	30.80
$[(CH_3)_3CBr]/(10^{-2}\,mol\,dm^{-3})$	10.39	8.96	7.76	6.39	3.53	2.07

P17B.5 The thermal decomposition of an organic nitrile produced the following data:

$t/(10^3\,s)$	0	2.00	4.00	6.00	8.00	10.00	12.00
[nitrile]/(mol dm^{-3})	1.50	1.26	1.07	0.92	0.81	0.72	0.65

Identify the order of the reaction and calculate the rate constant.

P17B.6[‡] The oxidation of HSO_3^- by O_2 in aqueous solution is a reaction of importance to the processes of acid rain formation and flue gas

‡ These problems were supplied by Charles Trapp and Carmen Giunta.

desulfurization. R.E. Connick et al. (*Inorg. Chem.* **34**, 4543 (1995)) report that the reaction $2\,HSO_3^-(aq) + O_2(g) \rightarrow 2\,SO_4^{2-}(aq) + 2\,H^+(aq)$ follows the rate law $v = k_r[HSO_3^-]^2[H^+]^2$. Given pH = 5.6 and an O_2 molar concentration of 0.24 mmol dm^{-3} (both presumed constant), an initial HSO_3^- molar concentration of 50 μmol dm^{-3}, and a rate constant of 3.6×10^6 dm^9 mol^{-3} s^{-1}, what is the initial rate of reaction? How long would it take for HSO_3^- to reach half its initial concentration?

P17B.7 Pharmacokinetics is the study of the rates of absorption and elimination of drugs by organisms. In most cases, elimination is slower than absorption and is a more important determinant of availability of a drug for binding to its target. A drug can be eliminated by many mechanisms, such as metabolism in the liver, intestine, or kidney followed by excretion of breakdown products through urine or faeces. As an example of pharmacokinetic analysis, consider the elimination of beta adrenergic blocking agents (beta blockers), which are used in the treatment of hypertension. After intravenous administration of a beta blocker, the blood plasma of a patient was analysed for remaining drug and the data are shown below, where c is the drug concentration measured at a time t after the injection.

t/min	30	60	120	150	240	360	480
c/(ng cm^{-3})	699	622	413	292	152	60	24

(a) Is the decay of the concentration of the drug first- or second-order in the drug? (b) Calculate the rate constant and half-life of the process. *Comment:* An essential aspect of drug development is the optimization of the half-life of elimination, which needs to be long enough to allow the drug to find and act on its target organ but not so long that harmful side effects become important.

P17B.8 The following data have been obtained for the decomposition of $N_2O_5(g)$ at 67 °C according to the reaction $2\,N_2O_5(g) \rightarrow 4\,NO_2(g) + O_2(g)$. Identify the order of the reaction with respect to N_2O_5 and calculate the rate constant and the half-life of N_2O_5. *Hint:* It is not necessary to obtain the result graphically; you may do a calculation by making estimates of the rates of change of concentration.

t/min	0	1	2	3	4	5
$[N_2O_5]$/(mol dm^{-3})	1.000	0.705	0.497	0.349	0.246	0.173

P17B.9 The gas phase decomposition of ethanoic acid at 1189 K proceeds by way of two parallel reactions:

(1) $CH_3COOH \rightarrow CH_4 + CO_2$ $\qquad k_1 = 3.74\,s^{-1}$

(2) $CH_3COOH \rightarrow CH_2CO + H_2O$ $\qquad k_2 = 4.65\,s^{-1}$

(a) What is the maximum theoretical yield of the ketene CH_2CO at this temperature? (b) Does the ratio of ketene to methane vary over time?

P17B.10 Sucrose is readily hydrolysed to glucose and fructose in acidic solution. The hydrolysis can be monitored by measuring the angle of rotation of plane-polarized light passing through the solution because the concentration of sucrose can be inferred from this angle. An experiment on the hydrolysis of sucrose in 0.50 M HCl(aq) produced the following data:

t/min	0	14	39	60	80	110	140	170	210
[sucrose]/(mol dm^{-3})	0.316	0.300	0.274	0.256	0.238	0.211	0.190	0.170	0.146

Assume that the reaction is first-order in sucrose, and determine the rate constant of the reaction and the half-life of sucrose.

P17B.11 The composition of a liquid phase reaction $2\,A \rightarrow B$ was monitored by a spectrophotometric method with the following results:

t/min	0	10	20	30	40	∞
[B]/(mol dm^{-3})	0	0.089	0.153	0.200	0.230	0.312

Identify the order of the reaction with respect to A and calculate its rate constant.

P17B.12 In the gas phase, the ClO radical decays rapidly by way of the reaction $2\,ClO(g) \rightarrow Cl_2(g) + O_2(g)$. The following data have been obtained:

t/ms	0.12	0.62	0.96	1.60	3.20	4.00	5.75
[ClO]/(μmol dm^{-3})	8.49	8.09	7.10	5.79	5.20	4.77	3.95

Calculate the rate constant of the reaction and the half-life of ClO.

P17B.13 Cyclopropane isomerizes into propene when heated to 500 °C in the gas phase. The extent of conversion for various initial pressures has been followed by gas chromatography by allowing the reaction to proceed for a time with various initial pressures:

p_0/Torr	200	200	400	400	600	600
t/s	100	200	100	200	100	200
p/Torr	186	173	373	347	559	520

where p_0 is the initial partial pressure and p is the final partial pressure of cyclopropane. What is the order and rate constant for the reaction under these conditions?

P17B.14 The addition of hydrogen halides to alkenes has played a fundamental role in the investigation of organic reaction mechanisms. In one study (M.J. Haugh and D.R. Dalton, *J. Amer. Chem. Soc.* **97**, 5674 (1975)), high pressures of hydrogen chloride (up to 25 atm) and propene (up to 5 atm) were examined over a range of temperatures and the amount of 2-chloropropane formed was determined by NMR. (a) Show that if the reaction $A + B \rightarrow P$ proceeds for a short time δt, the concentration of product follows $[P]/[A] = k_r[A]^{m-1}[B]^n\delta t$ if the reaction is mth-order in A and nth-order in B. (b) In a series of runs the ratio of [chloropropane] to [propene] was independent of [propene] but the ratio of [chloropropane] to [HCl] for constant amounts of propene depended on [HCl]. For $\delta t \approx 100$ h (which is short on the timescale of the reaction) the latter ratio rose from zero to 0.05, 0.03, 0.01 for p(HCl) = 10 atm, 7.5 atm, 5.0 atm, respectively. What are the orders of the reaction with respect to each reactant?

P17B.15 (a) Show that $t_{1/2}$ is given by eqn 17B.6 for a reaction that is nth order in A. (b) Derive an expression for the time it takes for the concentration of a substance to fall to one-third the initial value in an nth-order reaction.

P17B.16 Derive an integrated expression for a second-order rate law $v = k_r[A][B]$ for a reaction of stoichiometry $2\,A + 3\,B \rightarrow P$, with $[P]_0 = 0$. Express your rate law in terms of $[A]_0$, $[B]_0$, and x, where $[A] = [A]_0 - 2x$.

P17B.17 Derive the integrated form of a third-order rate law $v = k_r[A]^2[B]$ in which the stoichiometry is $2\,A + B \rightarrow P$ and the reactants are initially present in (a) their stoichiometric proportions ($[B]_0 = \frac{1}{2}[A]_0$); (b) with B present initially in twice that amount ($[B]_0 = [A]_0$). Express your rate law in terms of $[A]_0$, $[B]_0$, and x, where $[A] = [A]_0 - 2x$.

P17B.18 Show that the ratio $t_{1/2}/t_{3/4}$, where $t_{1/2}$ is the half-life and $t_{3/4}$ is the time for the concentration of A to decrease to $\frac{3}{4}$ of its initial value (implying that $t_{3/4} < t_{1/2}$), can be written as a function of n alone, and can therefore be used as a rapid assessment of the order of a reaction.

TOPIC 17C Reactions approaching equilibrium

Discussion questions

D17C.1 Describe the strategy of a temperature-jump experiment. What parameters of a reaction can be determined by this technique?

D17C.2 What feature of a reaction would ensure that its rate can respond to a pressure jump?

Exercises

E17C.1(a) The rates of the forward and reverse reactions for a reaction $A+B \rightleftarrows C$ were found to be $5.0\times10^6\,dm^3\,mol^{-1}\,s^{-1}$ (second order) and $2.0\times10^4\,s^{-1}$ (first order). What is the equilibrium constant of the reaction?

E17C.1(b) The equilibrium constant for the binding of a drug molecule to a protein was measured as 200. In a separate experiment, the rate constant for the binding process, which is second order overall, was found to be $1.5\times10^8\,dm^3\,mol^{-1}\,s^{-1}$. What is the rate constant for the first-order dissociation of the drug molecule from the protein–drug complex?

E17C.2(a) In a temperature-jump experiment to investigate the kinetics of an isomerization reaction that is first order in both directions, the relaxation time was measured as 27.6 μs. The rate constant for the forward reaction is known to be $12.4\,ms^{-1}$. Calculate the rate constant for the reverse reaction.

E17C.2(b) The half-lives for the forward and reverse reactions that are first order in both directions are 24 ms and 39 ms, respectively. Calculate the corresponding relaxation time for return to equilibrium after a temperature jump.

Problems

P17C.1 Show by differentiation that eqn 17C.4 is a solution of eqn 17C.3.

P17C.2 Set up the rate equations and plot the corresponding graphs for the approach to an equilibrium of a reaction of the form $A \rightleftarrows 2B$ (first-order forward, second-order reverse.)

P17C.3 The reaction $A \rightleftarrows B$ is first-order in both directions. (a) Derive an expression for the concentration of A as a function of time when the initial molar concentrations of A and B are $[A]_0$ and $[B]_0$. (b) What is the final composition of the system?

P17C.4 Show that eqn 17C.8 is an expression for the overall equilibrium constant in terms of the rate constants for the intermediate steps of a reaction mechanism. *Hint:* Begin with a mechanism containing three steps, and then argue that your expression may be generalized for any number of steps.

P17C.5 Consider the dimerization $2A \rightleftarrows A_2$, with forward rate constant k_a and reverse rate constant k_a'; the forward step is second-order in A, and the reverse step is first-order in A_2. (a) Derive the expression

$$\frac{1}{\tau^2} = k_a'^2 + 8k_a k_a'[A]_{tot}$$

for the relaxation time in terms of the total concentration of A, $[A]_{tot} = [A] + 2[A_2]$. (b) Describe a straight-line plot you could use to determine values of the rate constants k_a and k_a' from measurements of τ for different values of $[A]_{tot}$. (c) The following data refer to the dimerization of 2-pyridone, P. Analyse the data to obtain values of the rate constants k_a and k_a', and the equilibrium constant K for the dimerization reaction:

$[P]/(mol\,dm^{-3})$	0.500	0.352	0.251	0.151	0.101
τ/ns	2.3	2.7	3.3	4.0	5.3

P17C.6 The equilibrium $A \rightleftharpoons B + C$ at 25 °C is subjected to a temperature jump which slightly increases the concentrations of B and C. The measured relaxation time is 3.0 μs. The equilibrium constant for the system is 2.0×10^{-16} at the new temperature, and the equilibrium concentrations of B and C then are both $0.20\,mmol\,dm^{-3}$. Calculate the rate constants for the forward and reverse steps given that the forward step is first-order in A, and the reverse step is first-order in both B and C.

TOPIC 17D The Arrhenius equation

Discussion questions

D17D.1 Define the terms in $\ln k_r = \ln A - E_a/RT$ and discuss the conditions under which the expression is valid.

D17D.2 What might account for the failure of the Arrhenius equation to fit experimental data at low temperatures?

Exercises

E17D.1(a) Calculate the rate constant at 500 K for the second-order gas-phase reaction between Cl and H_2 given the frequency factor, $A=8.1\times10^{-10}\,dm^3\,mol^{-1}\,s^{-1}$ and activation energy $E_a=23\,kJ\,mol^{-1}$.

E17D.1(b) The Arrhenius parameters for the gas-phase decomposition of cyclobutane, $C_4H_8(g) \rightarrow 2C_2H_4(g)$, are $A = 4.00\times10^{15}\,s^{-1}$ and $E_a = 261\,kJ\,mol^{-1}$. What is the half-life of cyclobutane at (i) 20 °C, (ii) 500 °C?

E17D.2(a) The rate constant for the decomposition of a certain substance is $3.80\times10^{-3}\,dm^3\,mol^{-1}\,s^{-1}$ at 35 °C and $2.67\times10^{-2}\,dm^3\,mol^{-1}\,s^{-1}$ at 50 °C. Evaluate the Arrhenius parameters of the reaction.

E17D.2(b) The rate constant for the decomposition of a certain substance is $2.25\times10^{-2}\,dm^3\,mol^{-1}\,s^{-1}$ at 29 °C and $4.01\times10^{-2}\,dm^3\,mol^{-1}\,s^{-1}$ at 37 °C. Evaluate the Arrhenius parameters of the reaction.

E17D.3(a) The rate constant of a chemical reaction is found to triple when the temperature is raised from 24 °C to 49 °C. Evaluate the activation energy.

E17D.3(b) The rate constant of a chemical reaction is found to double when the temperature is raised from 25 °C to 35 °C. Evaluate the activation energy.

E17D.4(a) The activation energy of one of the reactions in a biochemical process is 87 kJ mol^{-1}. What is the change in rate constant when the temperature falls from 37 °C to 15 °C?

E17D.4(b) The activation energy for the decomposition of benzene diazonium chloride is 99.1 kJ mol^{-1}. At what temperature is the rate constant 10 per cent greater than at 25 °C?

E17D.5(a) At what temperature does the fraction of molecular collisions with enough energy to result in a bimolecular reaction reach 0.10 if E_a = 50 kJ mol^{-1}?

E17D.5(b) At 500 K, what is the fraction of molecular collisions with enough energy to result in a bimolecular reaction with E_a = 80 kJ mol^{-1}?

Problems

P17D.1 Show that the definition of E_a given in eqn 17D.3 reduces to eqn 17D.1 for a temperature-independent activation energy.

P17D.2 A first-order decomposition reaction is observed to have the following rate constants at the indicated temperatures. Estimate the activation energy.

$k_r/(10^{-3}\,\mathrm{s}^{-1})$	2.46	45.1	576
$\theta/°\mathrm{C}$	0	20.0	40.0

P17D.3 The rate constant for the gas-phase reaction of ethene and hydrogen, $C_2H_4(g)+H_2(g)\rightarrow C_2H_6(g)$, was measured at different temperatures. Use the following values to calculate the Arrhenius parameters.

T/K	1000	1200	1400	1600
$k_r/(\mathrm{dm^3\,mol^{-1}\,s^{-1}})$	8.35×10^{-10}	3.08×10^{-8}	4.06×10^{-7}	2.80×10^{-6}

P17D.4 The second-order rate constants for the reaction of oxygen atoms with aromatic hydrocarbons have been measured (R. Atkinson and J.N. Pitts, *J. Phys. Chem.* **79**, 295 (1975)). In the reaction with benzene the rate constants are 1.44×10^7 dm^3 mol^{-1} s^{-1} at 300.3 K, 3.03×10^7 dm^3 mol^{-1} s^{-1} at 341.2 K, and 6.9×10^7 dm^3 mol^{-1} s^{-1} at 392.2 K. Find the frequency factor and activation energy of the reaction.

P17D.5‡ Methane is a by-product of a number of natural processes (such as digestion of cellulose in ruminant animals, and anaerobic decomposition of organic waste matter), and industrial processes (such as food production and fossil fuel use). Reaction with the hydroxyl radical OH is the main path by which CH_4 is removed from the lower atmosphere. T. Gierczak et al. (*J. Phys. Chem. A* **101**, 3125 (1997)) measured the rate constants for the elementary bimolecular gas-phase reaction of methane with the hydroxyl radical over a range of temperatures of importance to atmospheric chemistry. Deduce the Arrhenius parameters A and E_a from the following measurements:

T/K	295	223	218	213	206	200	195
$k_r/(10^6\,\mathrm{dm^3\,mol^{-1}\,s^{-1}})$	3.55	0.494	0.452	0.379	0.295	0.241	0.217

P17D.6‡ As described in Problem P17D.5, reaction with the hydroxyl radical OH is the main path by which CH_4 is removed from the lower atmosphere. T. Gierczak et al. (*J. Phys. Chem. A* **101**, 3125 (1997)) measured the rate constants for the bimolecular gas-phase reaction $CH_4 + OH \rightarrow CH_3 + H_2O$ and found $A = 1.13\times10^9$ dm^3 mol^{-1} s^{-1} and E_a = 14.1 kJ mol^{-1} for the Arrhenius parameters. (a) Estimate the rate of consumption of CH_4 under the following conditions: take the average OH concentration to be 3.5×10^{-15} mol dm^{-3}, that of CH_4 to be 40 nmol dm^{-3}, and the temperature to be −10 °C. (b) Estimate the global annual mass of CH_4 consumed by this reaction (which is slightly less than the mass introduced to the atmosphere) given an effective volume for the Earth's lower atmosphere of 4×10^{21} dm^3.

TOPIC 17E Reaction mechanisms

Discussion questions

D17E.1 Distinguish between reaction order and molecularity.

D17E.2 Comment on the validity of the statement that the rate-determining step is the slowest step in a reaction mechanism.

D17E.3 Distinguish between the pre-equilibrium approximation and the steady-state approximation. Why might they lead to different conclusions?

D17E.4 Explain and illustrate how reaction orders may change under different circumstances.

D17E.5 Distinguish between kinetic and thermodynamic control of a reaction. Suggest criteria for expecting one rather than the other.

D17E.6 Explain how it is possible for the activation energy of a reaction to be negative.

Exercises

E17E.1(a) The reaction mechanism for the decomposition of A_2 is thought to be

$$A_2 \underset{k_a'}{\overset{k_a}{\rightleftharpoons}} A+A \qquad A+B \xrightarrow{k_b} P$$

where the dissociation of A_2 is first order in A_2, and the recombination of A is second order in A; the reaction of A with B is first order in both A and B. Deduce the rate law for the rate of formation of P in two ways: (i) by assuming a pre-equilibrium between A_2 and A, and (ii) by assuming that the steady-state approximation can be applied to A.

E17E.1(b) The reaction mechanism for renaturation of a double helix from its strands A and B is thought to be

$$A+B \underset{k_a'}{\overset{k_a}{\rightleftharpoons}} U \qquad U \xrightarrow{k_b} H$$

where U is an unstable helix, and H is the stable form of the helix. The reaction between A and B is first order in each species and the return of U to A + B is first order in U; the reaction of U to H is first order in U. Deduce the rate law for the rate of formation of H in two ways: (i) by assuming a pre-equilibrium and (ii) by assuming that the steady-state approximation can be applied to U.

E17E.2(a) The following mechanism has been proposed for the decomposition of ozone in the atmosphere:

$$O_3 \rightarrow O_2 + O \qquad k_a$$
$$O_2 + O \rightarrow O_3 \qquad k_a'$$
$$O + O_3 \rightarrow O_2 + O_2 \qquad k_b$$

Show that if the third step is rate limiting, then the rate law for the formation of O_2 is second-order in O_3 and of order −1 in O_2.

E17E.2(b) The mechanism for the reaction between 2-chloroethanol, CH_2ClCH_2OH, and hydroxide ions in aqueous solution to form ethylene oxide, $(CH_2CH_2)O$, is thought to consist of the steps

$$(1)\ CH_2ClCH_2OH + OH^- \rightleftarrows CH_2ClCH_2O^- + H_2O$$
$$(2)\ CH_2ClCH_2O^- \rightarrow (CH_2CH_2)O + Cl^-$$

Show that if it can be assumed that there is a pre-equilibrium involving step (1), the rate of formation of ethylene oxide is $v = k_2K[CH_2ClCH_2OH][OH^-]/c^{\ominus}$, where K is the equilibrium constant for the first step and k_2 is the rate constant for the second step.

E17E.3(a) The mechanism of a reaction consists of a pre-equilibrium step with forward and reverse activation energies of 25 kJ mol^{-1} and 38 kJ mol^{-1}, respectively, followed by a rate-limiting elementary step of activation energy 10 kJ mol^{-1}. What is the activation energy of the overall reaction?

E17E.3(b) The mechanism of a reaction consists of a pre-equilibrium step with forward and reverse activation energies of 27 kJ mol^{-1} and 35 kJ mol^{-1}, respectively, followed by a rate-limiting elementary step of activation energy 15 kJ mol^{-1}. What is the activation energy of the overall reaction?

Problems

P17E.1 Use mathematical software or a spreadsheet to examine the time dependence of [I] in the reaction mechanism $A \xrightarrow{k_a} I \xrightarrow{k_b} P$. In all the following calculations, use $[A]_0 = 1\ \text{mol dm}^{-3}$ and a time range of 0–5 s. (a) Plot [I] against t for $k_a = 10\ \text{s}^{-1}$ and $k_b = 1\ \text{s}^{-1}$. (b) Increase the ratio k_b/k_a steadily by decreasing the value of k_a and examine the plot of [I] against t at each turn. What approximation about d[I]/dt becomes increasingly valid?

P17E.2 Use mathematical software or a spreadsheet to investigate the effects on [A], [I], [P], and t_{max} of decreasing the ratio k_a/k_b from 10 (as in Fig. 17E.1) to 0.01.

P17E.3 Two radioactive nuclides decay by successive first-order processes: $X \xrightarrow{22.5\,\text{d}} Y \xrightarrow{33.0\,\text{d}} Z$ (the quantities over the arrows are the half-lives in days). Suppose that Y is an isotope that is required for medical applications. At what time after X is first formed will Y be most abundant?

P17E.4 Set up the rate equations for the reaction mechanism:

$$A \underset{k_a'}{\overset{k_a}{\rightleftarrows}} B \underset{k_b'}{\overset{k_b}{\rightleftarrows}} C$$

Show that, under specific circumstances which you should identify, the mechanism is equivalent to

$$A \underset{k_r'}{\overset{k_r}{\rightleftarrows}} C$$

P17E.5 Derive an equation for the steady-state rate of the sequence of reactions $A \rightleftarrows B \rightleftarrows C \rightleftarrows D$, with [A] maintained at a fixed value and the product D removed as soon as it is formed.

P17E.6 The oxidation of NO to NO_2, $2\,NO(g) + O_2(g) \rightarrow 2\,NO_2(g)$, proceeds by the following mechanism:

$$NO + NO \rightarrow N_2O_2 \qquad k_a$$
$$N_2O_2 \rightarrow NO + NO \qquad k_a'$$
$$N_2O_2 + O_2 \rightarrow NO_2 + NO_2 \qquad k_b$$

Verify that application of the steady-state approximation to the intermediate N_2O_2 results in the rate law

$$\frac{d[NO_2]}{dt} = \frac{2k_ak_b[NO]^2[O_2]}{k_a' + k_b[O_2]}$$

P17E.7 Show that the following mechanism can account for the rate law of the reaction in Problem P17B.14 (the final step is rate determining):

$$HCl + HCl \rightleftharpoons (HCl)_2 \qquad K_1$$
$$HCl + CH_3CH{=}CH_2 \rightleftharpoons \text{complex} \qquad K_2$$
$$(HCl)_2 + \text{complex} \rightarrow CH_3CHClCH_3 + HCl + HCl \qquad k_r$$

What further tests could you apply to verify this mechanism?

P17E.8 Polypeptides are polymers of amino acids. Suppose that a long polypeptide chain can undergo a transition from a helical conformation to a random coil. Consider a mechanism for a helix–coil transition that begins in the middle of the chain:

hhhh… ⇄ *hchh…*

hchh… ⇄ *cccc…*

in which *h* and *c* label, respectively, an amino acid in a helical or coil part of the chain. The first conversion from *h* to *c*, also called a nucleation step, is relatively slow, so neither step may be rate determining. (a) Set up the rate equations for this mechanism. (b) Apply the steady-state approximation and show that, under these circumstances, the mechanism is equivalent to *hhhh…* ⇄ *cccc…*.

P17F.9‡ J. Czarnowski and H.J. Schuhmacher (*Chem. Phys. Lett.* **17**, 235 (1972)) suggested the following mechanism for the thermal decomposition of F_2O in the reaction $2\,F_2O(g) \rightarrow 2\,F_2(g) + O_2(g)$:

$$(1)\ F_2O + F_2O \rightarrow F + OF + F_2O \qquad k_a$$
$$(2)\ F + F_2O \rightarrow F_2 + OF \qquad k_b$$
$$(3)\ OF + OF \rightarrow O_2 + F + F \qquad k_c$$
$$(4)\ F + F + F_2O \rightarrow F_2 + F_2O \qquad k_d$$

Use the steady-state approximation to show that this mechanism is consistent with the experimental rate law $-d[F_2O]/dt = k_r[F_2O]^2 + k_r'[F_2O]^{3/2}$.

P17E.10 Consider two products formed from reactant R in reactions for which: (a) product P_1 is thermodynamically more stable than product P_2; and (b) the activation energy E_a for the reaction leading to P_2 is greater than that leading to P_1. Derive an expression for the ratio $[P_2]/[P_1]$ when the reaction is under thermodynamic control. State your assumptions.

TOPIC 17F Examples of reaction mechanisms

Discussion questions

D17F.1 Discuss the conditions under which the expression $k_r = k_ak_b[A]/(k_b + k_a'[A])$ for the effective rate constant of a unimolecular reaction according to the Lindemann–Hinshelwood mechanism results in a (a) first-order, or (b) second-order rate law.

D17F.2 Bearing in mind distinctions between the mechanisms of stepwise and chain polymerization, describe how it is possible to control the molar mass of a polymer by manipulating the kinetic parameters of polymerization.

D17F.3 Discuss the features, advantages, and limitations of the Michaelis–Menten mechanism of enzyme action.

D17F.4 A plot of the rate of an enzyme-catalysed reaction against temperature has a maximum, in an apparent deviation from the behaviour predicted by the Arrhenius equation (Topic 17D). Suggest an interpretation.

Exercises

E17F.1(a) The effective rate constant for a gaseous reaction which proceeds by a Lindemann–Hinshelwood mechanism is $2.50 \times 10^{-4}\,s^{-1}$ at 1.30 kPa and $2.10 \times 10^{-5}\,s^{-1}$ at 12 Pa. Calculate the rate constant for the activation step in the mechanism.

E17F.1(b) The effective rate constant for a gaseous reaction which proceeds by a Lindemann–Hinshelwood mechanism is $1.7 \times 10^{-3}\,s^{-1}$ at 1.09 kPa and $2.2 \times 10^{-4}\,s^{-1}$ at 25 Pa. Calculate the rate constant for the activation step in the mechanism.

E17F.2(a) Calculate the fraction condensed and the degree of polymerization at $t = 5.00\,h$ of a polymer formed by a stepwise process with $k_r = 1.39\,dm^3\,mol^{-1}\,s^{-1}$ and an initial monomer concentration of 10.0 mmol dm^{-3}.

E17F.2(b) Calculate the fraction condensed and the degree of polymerization at $t = 10.00\,h$ of a polymer formed by a stepwise process with $k_r = 2.80 \times 10^{-2}\,dm^3\,mol^{-1}\,s^{-1}$ and an initial monomer concentration of 50.0 mmol dm^{-3}.

E17F.3(a) Consider a polymer formed by a chain process. By how much does the kinetic chain length change if the concentration of initiator is increased by a factor of 3.6 and the concentration of monomer is decreased by a factor of 4.2?

E17F.3(b) Consider a polymer formed by a chain process. By how much does the kinetic chain length change if the concentration of initiator is decreased by a factor of 10.0 and the concentration of monomer is increased by a factor of 5.0?

E17F.4(a) The enzyme-catalysed conversion of a substrate at 25 °C has a Michaelis constant of 0.046 mol dm^{-3}. The rate of the reaction is 1.04 mmol dm^{-3} s^{-1} when the substrate concentration is 0.105 mol dm^{-3}. What is the maximum velocity of this reaction?

E17F.4(b) The enzyme-catalysed conversion of a substrate at 25 °C has a Michaelis constant of 0.032 mol dm^{-3}. The rate of the reaction is 0.205 mmol dm^{-3} s^{-1} when the substrate concentration is 0.875 mol dm^{-3}. What is the maximum velocity of this reaction?

E17F.5(a) The ratio k_b/K_M is called the *catalytic efficiency* of an enzyme. Calculate the catalytic efficiency of carbonic anhydrase by using the data in *Example* 17F.2.

E17F.5(b) The enzyme-catalysed conversion of a substrate at 298 K has $K_M = 0.032$ mol dm^{-3} and $v_{max} = 4.25 \times 10^{-4}$ mol dm^{-3} s^{-1} when the enzyme concentration is 3.60×10^{-9} mol dm^{-3}. Calculate the catalytic efficiency of the enzyme, as defined in Exercise E17F.5(a).

Problems

P17F.1 The isomerization of cyclopropane over a limited pressure range was examined in Problem 17B.13. If the Lindemann–Hinshelwood mechanism of unimolecular reactions is to be tested data is also needed at low pressures. This information has been obtained (H.O. Pritchard et al., *Proc. R. Soc. A* **217**, 563 (1953)):

p/Torr	84.1	11.0	2.89	0.569	0.120	0.067
$10^4\,k_r/s^{-1}$	2.98	2.23	1.54	0.857	0.392	0.303

Test the Lindemann–Hinshelwood mechanism with these data.

P17F.2 Calculate the average polymer length in a polymer produced by a chain mechanism in which termination occurs by a disproportionation reaction of the form $\cdot HM_n + \cdot M_m \rightarrow M_n + HM_m$.

P17F.3 Derive an expression for the time dependence of the degree of polymerization for the stepwise polymerization of a hydroxyacid HO–R–COOH for which the rate law is $d[A]/dt = -k_r[A]^2[OH]$, where A denotes the carboxylic acid group.

P17F.4 Michaelis and Menten derived their rate law by assuming a rapid pre-equilibrium of E, S, and ES. Derive the rate law in this manner, and identify the conditions under which it becomes the same as that based on the steady-state approximation (eqn 17F.16).

P17F.5 Use the Michaelis–Menten equation (eqn 17F.16) to generate two families of curves showing the dependence of v on [S]: one in which K_M varies but v_{max} is constant, and another in which v_{max} varies but K_M is constant. *Hint:* Use mathematical software or a spreadsheet.

P17F.6 For many enzymes, the mechanism of action involves the formation of two intermediates:

$$E + S \rightarrow ES \qquad v = k_a[E][S]$$

$$ES \rightarrow E + S \qquad v = k_a'[ES]$$

$$ES \rightarrow ES' \qquad v = k_b[ES]$$

$$ES' \rightarrow E + P \qquad v = k_c[ES']$$

Show that the rate of formation of product has the same form as that shown in eqn 17F.16, but with v_{max} and K_M given by

$$v_{max} = \frac{k_b k_c [E]_0}{k_b + k_c} \qquad K_M = \frac{k_c(k_a' + k_b)}{k_a(k_b + k_c)}$$

P17F.7 The following results were obtained for the action of an ATPase on ATP at 20 °C, when the concentration of the ATPase was 20 nmol dm^{-3}:

[ATP]/(μmol dm^{-3})	0.60	0.80	1.4	2.0	3.0
v/(μmol dm^{-3} s^{-1})	0.81	0.97	1.30	1.47	1.69

Evaluate the Michaelis constant and the maximum velocity of the reaction.

P17F.8 There are different ways to represent and analyse data for enzyme-catalysed reactions. The text shows how to construct a linear Lineweaver–Burk plot of $1/v$ against $1/[S]_0$. (a) Show, by rearranging eqn 17F.16, that an *Eadie–Hofstee plot* of $v/[S]_0$ against v is also expected to be a straight line. Identify how the Michaelis constant and the maximum velocity of the reaction may be obtained from such a plot. (b) In the same way, show that a *Hanes plot of* $[S_0]/v$ against $[S]_0$ is also a straight line. Identify how the parameters may be obtained from such a plot. (c) The enzyme catalase, catalyses the decomposition of hydrogen peroxide, H_2O_2. By constructing Lineweaver–Burk, Eadie–Hofstee, and Hanes plots, use the following values for the rate of reaction for various initial concentrations of hydrogen peroxide to calculate the Michaelis constant and the maximum velocity of the reaction.

$[H_2O_2]$/(mol dm^{-3})	0.300	0.400	0.500	0.600	0.700
v/(mol dm^{-3} s^{-1})	4.431	4.518	4.571	4.608	4.634

TOPIC 17G Photochemistry

Discussion question

D17G.1 Consult literature sources and list the observed ranges of timescales during which the following processes occur: radiative decay of excited electronic states, molecular rotational motion, molecular vibrational motion, proton transfer reactions, energy transfer between fluorescent molecules used in FRET analysis, electron transfer events between complex ions in solution, and collisions in liquids.

Exercises

E17G.1(a) In a photochemical reaction $A \rightarrow 2B + C$, the quantum yield with 500 nm light is $210\,\mathrm{mmol\,einstein^{-1}}$ (1 einstein = 1 mol photons). After exposure of 300 mmol of A to the light, 2.28 mmol of B is formed. How many photons were absorbed by A?

E17G.1(b) In a photochemical reaction $A \rightarrow B + C$, the quantum yield with 500 nm light is $120\,\mathrm{mmol\,einstein^{-1}}$ (1 einstein = 1 mol photons). After exposure of 200 mmol A to the light, 1.77 mmol B is formed. How many photons were absorbed by A?

E17G.2(a) A substance has a fluorescence quantum yield of $\phi_{F,0} = 0.35$. In an experiment to measure the fluorescence lifetime of this substance, it was observed that the fluorescence emission decayed with a half-life of 5.6 ns. What is the fluorescence rate constant of this substance?

E17G.2(b) A substance has a fluorescence quantum yield of $\phi_{F,0} = 0.16$. In an experiment to measure the fluorescence lifetime of this substance, it was observed that the fluorescence emission decayed with a half-life of 1.5 ns. What is the fluorescence rate constant of this substance?

E17G.3(a) Consider the quenching of an organic fluorescent species with $\tau_0 = 6.0\,\mathrm{ns}$ by a d-metal ion with $k_Q = 3.0 \times 10^8\,\mathrm{dm^3\,mol^{-1}\,s^{-1}}$. Predict the concentration of quencher required to decrease the fluorescence intensity of the organic species to 50 per cent of the unquenched value.

E17G.3(b) Consider the quenching of an organic fluorescent species with $\tau_0 = 3.5\,\mathrm{ns}$ by a d-metal ion with $k_Q = 2.5 \times 10^9\,\mathrm{dm^3\,mol^{-1}\,s^{-1}}$. Predict the concentration of quencher required to decrease the fluorescence intensity of the organic species to 75 per cent of the unquenched value.

E17G.4(a) An amino acid on the surface of a protein was labelled covalently with 1.5-I AEDANS and another was labelled covalently with FITC. The fluorescence quantum yield of 1.5-I AEDANS decreased by 10 per cent due to quenching by FITC. What is the distance between the amino acids? (Refer to Table 17G.3 for the appropriate value of R_0.)

E17G.4(b) An amino acid on the surface of an enzyme was labelled covalently with 1.5-I AEDANS and it is known that the active site contains a tryptophan residue. The fluorescence quantum yield of tryptophan decreased by 15 per cent due to quenching by 1.5-I AEDANS. What is the distance between the active site and the surface of the enzyme?

Problems

P17G.1 In an experiment to measure the quantum yield of a photochemical reaction, the absorbing substance was exposed to 320 nm radiation from an 87.5 W source for 28.0 min. The intensity of the transmitted radiation was 0.257 that of the incident radiation. As a result of irradiation, 0.324 mol of the absorbing substance decomposed. Evaluate the quantum yield.

P17G.2‡ Ultraviolet radiation photolyses O_3 to O_2 and O. Determine the rate at which ozone is consumed by 305 nm radiation in a layer of the stratosphere of thickness 1.0 km. The quantum yield is 0.94 at 220 K, the concentration about $8\,\mathrm{nmol\,dm^{-3}}$, the molar absorption coefficient $260\,\mathrm{dm^3\,mol^{-1}\,cm^{-1}}$, and the flux of 305 nm radiation about $1 \times 10^{14}\,\mathrm{photons\,cm^{-2}\,s^{-1}}$. Data from W.B. DeMore et al. (*Chemical kinetics and photochemical data for use in stratospheric modeling: Evaluation Number 11*, JPL Publication 94–26 (1994)).

P17G.3 Dansyl chloride, which absorbs maximally at 330 nm and fluoresces maximally at 510 nm, can be used to label amino acids in fluorescence microscopy and FRET studies. Tabulated below is the variation of the fluorescence intensity of an aqueous solution of dansyl chloride with time after excitation by a short laser pulse (with I_0 the initial fluorescence intensity). The ratio of intensities is equal to the ratio of the rates of photon emission.

t/ns	5.0	10.0	15.0	20.0
I_F/I_0	0.45	0.21	0.11	0.05

(a) Calculate the observed fluorescence lifetime of dansyl chloride in water.
(b) The fluorescence quantum yield of dansyl chloride in water is 0.70. What is the fluorescence rate constant?

P17G.4 When benzophenone is exposed to ultraviolet radiation it is excited into a singlet state. This singlet changes rapidly into a triplet, which phosphoresces. Triethylamine acts as a quencher for the triplet. In an experiment in the solvent methanol, the phosphorescence intensity varied with amine concentration as shown below. A time-resolved laser spectroscopy experiment had also shown that the half-life of the phosphorescence in the absence of quencher is 29 μs. What is the value of k_Q?

$[Q]/(\mathrm{mmol\,dm^{-3}})$	1.0	5.0	10.0
I_P/(arbitrary units)	0.41	0.25	0.16

P17G.5 An electronically excited state of Hg can be quenched by N_2 according to $Hg^*(g) + N_2(g, v=0) \rightarrow Hg(g) + N_2(g, v=1)$ in which energy transfer from Hg* excites N_2 vibrationally. The data below give the measured time dependence of the intensity of fluorescence for samples of Hg with and without N_2 present (for $T = 300\,\mathrm{K}$):

$p_{N_2} = 0$

Relative fluorescence intensity	1.000	0.606	0.360	0.22	0.135
t/μs	0.0	5.0	10.0	15.0	20.0

$p_{N_2} = 9.74 \times 10^{-4}\,\mathrm{atm}$

Relative fluorescence intensity	1.000	0.585	0.342	0.200	0.117
t/μs	0.0	3.0	6.0	9.0	12.0

Evaluate the rate constant for the energy transfer process. You may assume that all gases are perfect.

P17G.6 The Förster theory of resonance energy transfer and the basis for the FRET technique can be tested by performing fluorescence measurements on a series of compounds in which an energy donor and an energy acceptor are covalently linked by a rigid molecular linker of variable and known length. L. Stryer and R.P. Haugland (*Proc. Natl. Acad. Sci. USA* **58**, 719 (1967)) collected the following data on energy transfer efficiencies, η_T, for a family of compounds with the general composition dansyl-(L-prolyl)$_n$-naphthyl, in which the distance R between the naphthyl donor and the dansyl acceptor was varied from 1.2 nm to 4.6 nm by increasing the number of prolyl units in the linker:

R/nm	1.2	1.5	1.8	2.8	3.1	3.4	3.7	4.0	4.3	4.6
η_T	0.99	0.94	0.97	0.82	0.74	0.65	0.40	0.28	0.24	0.16

Are the data described adequately by eqn 17G.7? If so, what is the value of R_0 for the naphthyl–dansyl pair?

P17G.7 The first step in plant photosynthesis is absorption of light by chlorophyll molecules bound to proteins known as 'light-harvesting complexes', where the fluorescence of a chlorophyll molecule is quenched by other nearby chlorophyll molecules. Given that for a pair of chlorophyll a molecules $R_0 = 5.6$ nm, by what distance should two chlorophyll a molecules be separated to shorten the fluorescence lifetime from 1 ns (a typical value for monomeric chlorophyll a in organic solvents) to 10 ps?

FOCUS 17 Chemical kinetics

Integrated activities

I17.1 Autocatalysis is the catalysis of a reaction by the products. For example, for a reaction A → P it may be found that the rate law is $v = k_r[A][P]$ and the reaction rate is proportional to the concentration of P. The reaction gets started because there are usually other reaction routes for the formation of some P initially, which then takes part in the autocatalytic reaction proper. (a) Integrate the rate equation for an autocatalytic reaction of the form A → P, with rate law $v = k_r[A][P]$, and show that

$$\frac{[P]}{[P]_0} = \frac{(1+b)e^{at}}{1+be^{at}}$$

where $[P]_0$ is the initial concentration of P, $a = ([A]_0 + [P]_0)k_r$ and $b = [P]_0/[A]_0$. *Hint:* Start from the expression $v = -d[A]/dt = k_r[A][P]$, write $[A] = [A]_0 - x$, $[P] = [P]_0 + x$, and then write the expression for the rate of change of either species in terms of x. To integrate the resulting expression, use integration by the method of partial fractions (see *The chemist's toolkit* 30 in Topic 17B). (b) Plot $[P]/[P]_0$ against at for several values of b. Discuss the effect of autocatalysis on the shape of a plot of $[P]/[P]_0$ against t by comparing your results with those for a first-order process, in which $[P]/[A]_0 = 1 - e^{-k_r t}$. (c) Show that for the autocatalytic process discussed in parts (a) and (b), the reaction rate reaches a maximum at $t_{max} = -(1/a)\ln b$. (d) An autocatalytic reaction A → P is observed to have the rate law $d[P]/dt = k_r[A]^2[P]$. Solve the rate law for initial concentrations $[A]_0$ and $[P]_0$. Calculate the time at which the rate reaches a maximum. (e) Another reaction with the stoichiometry A → P has the rate law $d[P]/dt = k_r[A][P]^2$; integrate the rate law for initial concentrations $[A]_0$ and $[P]_0$. Calculate the time at which the rate reaches a maximum.

I17.2 Many biological and biochemical processes involve autocatalytic steps (see Integrated activity I17.1). In the SIR model of the spread and decline of infectious diseases the population is divided into three classes; the 'susceptibles', S, who can catch the disease, the 'infectives', I, who have the disease and can transmit it, and the 'removed class', R, who have either had the disease and recovered, are dead, are immune or isolated. The model mechanism for this process, written as S → I → R, implies the following rate laws:

$$\frac{dS}{dt} = -rSI \qquad \frac{dI}{dt} = rSI - aI \qquad \frac{dR}{dt} = aI$$

Which are the autocatalytic steps of this mechanism? Find the conditions on the ratio a/r that decide whether the disease will spread (an epidemic) or die out. Show that a constant population is built into this system, namely that S + I + R = N, meaning that the timescales of births, deaths by other causes, and migration are assumed large compared to that of the spread of the disease.

I17.3 Acid- and base-catalysed reactions are common in organic transformations. (a) Deduce the rate law of the base-catalysed reaction in which AH goes to products according to the following scheme

$$AH + B \underset{k_a'}{\overset{k_a}{\rightleftharpoons}} BH^+ + A^-$$

$$A^- + AH \xrightarrow{k_b} \text{product (rate-determining)}$$

(b) Deduce the rate law of the acid-catalysed reaction in which HA goes to products according to the following scheme

$$HA + H^+ \underset{k_a'}{\overset{k_a}{\rightleftharpoons}} HAH^+$$

$$HAH^+ + B \xrightarrow{k_b} \text{product} + AH \text{ (rate-determining)}$$

I17.4 Express the root-mean-square deviation $\{\langle M^2\rangle - \langle M\rangle^2\}^{1/2}$ of the molar mass of a condensation polymer in terms of the fraction p, and deduce its time dependence.

I17.5 Calculate the ratio of the mean cube molar mass to the mean square molar mass in terms of (a) the fraction p, (b) the chain length.

I17.6 Conventional equilibrium considerations do not apply when a reaction is driven by light absorption and the steady-state concentration of products and reactants might differ significantly from equilibrium values. For instance, suppose the reaction A → B is driven by light absorption, and that its rate is I_a, but that the reverse reaction B → A is bimolecular and second order with a rate $k_r[B]^2$. What is the stationary state concentration of B? Why does this 'photostationary state' differ from the equilibrium state?

I17.7 The photochemical chlorination of trichloromethane (chloroform, $CHCl_3$) in the gas phase to give CCl_4 has been found to follow the rate law $d[CCl_4]/dt = k_r[Cl_2]^{1/2}I_a^{1/2}$. Devise a mechanism that leads to this rate law when the chlorine pressure is high.

FOCUS 18

Reaction dynamics

This Focus examines the details of what happens to molecules at the climax of reactions. Extensive changes of structure are taking place and energies the size of dissociation energies are being redistributed among bonds: old bonds are being ripped apart and new bonds are being formed. This is the heart of chemistry.

The calculation of the rates of such processes from first principles is very difficult. Nevertheless, like so many intricate problems, the broad features can be established quite simply. Only upon deeper inquiry do the complications emerge. Several approaches to the calculation of a rate constant for elementary bimolecular processes are explored here, ranging from electron transfer to chemical reactions involving bond breakage and formation. Although a great deal of information can be obtained from gas-phase reactions, many reactions of interest take place in condensed phases, and it is useful to attempt to predict their rates.

18A Collision theory

This Topic explores 'collision theory', the simplest quantitative account of reaction rates. The treatment can be used only for the discussion of reactions between simple species in the gas phase. Basic collision theory considers only the impact of one molecule on another. An elaboration considered in this Topic takes into account how the resulting excitation energy accumulates in the bond where it is needed.

18B Diffusion-controlled reactions

Reactions in solution are classified into two types: 'diffusion-controlled' where the rate is controlled by the frequency with which reactants meet, and 'activation-controlled', where the accumulation of sufficient energy in a pair that have met is rate-determining. The rate constants for the former can be expressed quantitatively in terms of the diffusional characteristics of species in liquids. A more detailed account of the space- and time-development of products is obtained by using the diffusion equation.

18C Transition-state theory

This Topic discusses 'transition-state theory', in which it is assumed that the reactant molecules form a complex that can be discussed in terms of the population of its energy levels. The theory inspires a thermodynamic approach to reaction rates, in which the rate constant is expressed in terms of thermodynamic parameters. This approach is useful for parametrizing the rates of reactions in solution.

18D The dynamics of molecular collisions

The highest level of sophistication in the theoretical study of chemical reactions is in terms of potential energy surfaces and the motion of molecules on these surfaces. As explained in this Topic, such an approach gives an intimate picture of the events that occur when molecules collide, and provides a basis for studying them by using molecular beams.

18E Electron transfer in homogeneous systems

In this Topic transition-state theory is used to examine the transfer of electrons in homogeneous systems, which include oxidation–reduction reactions in solution. One widely used theory, Marcus theory, establishes a relation between the activation parameters and the rate constant of electron transfer, and can be expressed in terms of structural parameters of the species involved.

TOPIC 18A Collision theory

> ➤ **Why do you need to know this material?**
>
> A major component of chemistry is the study of the detailed molecular mechanisms of chemical reactions. One of the earliest approaches, which continues to give insight into the details of mechanisms of gas-phase reactions, is collision theory.
>
> ➤ **What is the key idea?**
>
> According to collision theory a bimolecular gas-phase reaction takes place when reactants collide, provided their relative kinetic energy exceeds a threshold value and certain steric requirements are fulfilled.
>
> ➤ **What do you need to know already?**
>
> This Topic draws on the kinetic theory of gases, especially the expression for the mean speed of molecules (Topic 1B), and extends the account of the Lindemann–Hinshelwood mechanism of gas-phase reactions (Topic 17F). One argument draws on the Maxwell–Boltzmann distribution of molecular speeds (Topic 1B).

The rate constant of the bimolecular elementary reaction

$$\mathrm{A} + \mathrm{B} \rightarrow \mathrm{P} \qquad v = k_r[\mathrm{A}][\mathrm{B}] \qquad (18\mathrm{A}.1\mathrm{a})$$

depends on the temperature according to the Arrhenius equation (Topic 17D):

$$k_r = A\mathrm{e}^{-E_a/RT} \qquad \text{Arrhenius equation} \qquad (18\mathrm{A}.1\mathrm{b})$$

where A is the 'frequency factor' and E_a is the 'activation energy'. This form of the Arrhenius expression can be explained by a model in which molecules in the gas collide and in the process may acquire sufficient energy to undergo reaction. Like all models, this one can be improved, but it is a good starting point for the discussion of gas-phase reactions.

18A.1 Reactive encounters

The general form of the expression for k_r in eqn 18A.1a can be anticipated by considering the physical requirements for reaction. The rate v can be expected to be proportional to the frequency of collisions, and therefore to the mean speed of the molecules, $v_{\text{mean}} \propto (T/M)^{1/2}$ where M is some combination of the molar masses of A and B. The rate can also be expected to be proportional to the target area the molecules present, which is their collision cross-section, σ (Topic 1B), and to the number densities $\mathcal{N}_A$ and $\mathcal{N}_B$ of A and B:

$$v \propto \sigma(T/M)^{1/2}\mathcal{N}_A\mathcal{N}_B \overset{\mathcal{N}_J \propto [J]}{\propto} \sigma(T/M)^{1/2}[\mathrm{A}][\mathrm{B}]$$

However, a collision is likely to be successful only if the kinetic energy of the molecules exceeds a minimum value, denoted E'. This requirement suggests that the rate should also be proportional to a Boltzmann factor of the form $\mathrm{e}^{-E'/RT}$ representing the fraction of collisions with at least the minimum required energy (Topic 17D). Therefore,

$$v \propto \sigma(T/M)^{1/2}\mathrm{e}^{-E'/RT}[\mathrm{A}][\mathrm{B}]$$

and, by writing the reaction rate in the form given in eqn 18A.1a, it follows that

$$k_r \propto \sigma(T/M)^{1/2}\mathrm{e}^{-E'/RT}$$

At this point, the form of the Arrhenius equation, eqn 18A.1b, begins to emerge, with the minimum kinetic energy E' identified as the activation energy E_a of the reaction. This identification, however, should not be regarded as precise, because collision theory is only a rudimentary model of chemical reactivity.

Not every collision will lead to reaction even if the energy requirement is satisfied, because the reactants might need to collide in a certain relative orientation. This 'steric requirement' suggests that a further factor, P, should be introduced, and that

$$k_r \propto P\sigma(T/M)^{1/2}\mathrm{e}^{-E'/RT} \qquad (18\mathrm{A}.2)$$

As seen in detail below, this expression has the form predicted by collision theory. It reflects three aspects of a successful collision:

$$k_r \propto \overbrace{P}^{\text{Steric requirement}}\ \overbrace{\sigma(T/M)^{1/2}}^{\text{Encounter rate}}\ \overbrace{\mathrm{e}^{-E'/RT}}^{\text{Minimum energy requirement}}$$

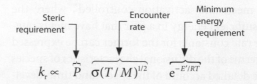

(a) Collision rates in gases

As remarked, the reaction rate, and hence k_r, is expected to depend on the frequency with which molecules collide. The **collision density**, Z_{AB}, is the number of (A,B) collisions in a region of the sample in an interval of time divided by the volume of the region and the duration of the interval. The frequency of collisions of a single molecule in a gas is calculated in Topic 1B (eqn 1B.12a, $z = \sigma v_{rel}\mathcal{N}$). That result can be adapted to derive an expression for Z_{AB}.

How is that done? 18A.1 Deriving an expression for the collision density

The parameter v_{rel} in the expression $z = \sigma v_{rel}\mathcal{N}$ is the mean relative speed of the colliding molecules and σ is the collision cross-section: $\sigma = \pi d^2$, with $d = \frac{1}{2}(d_A + d_B)$, as shown in Fig. 18A.1. For collisions between A molecules of mass m_A and B molecules of mass m_B, the mean relative speed is eqn 1B.11b ($v_{rel} = (8kT/\pi\mu)^{1/2}$, where $\mu = m_A m_B/(m_A + m_B)$). It follows that the collision rate of one A molecule with B molecules present at number density $\mathcal{N}_B$ is $\sigma v_{rel}\mathcal{N}_B$. The collision density is therefore this rate multiplied by the number density of A molecules, $\mathcal{N}_A$:

$$Z_{AB} = \sigma v_{rel}\mathcal{N}_A\mathcal{N}_B \quad (18A.3)$$

The number density of a species J is $\mathcal{N}_J = N_A[J]$, where [J] is its molar concentration and N_A is Avogadro's constant. It follows that

$$Z_{AB} = \sigma\left(\frac{8kT}{\pi\mu}\right)^{1/2} N_A^2[A][B] \quad (18A.4a)$$

Collision density [KMT]

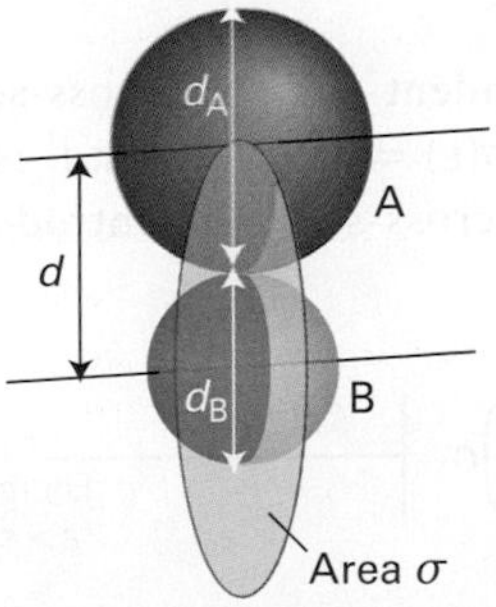

Figure 18A.1 The collision cross-section for two molecules can be regarded to be the area within which the projectile molecule (A) must enter around the target molecule (B) in order for a collision to occur. If the diameters of the two molecules are d_A and d_B, the radius of the target area is $d = \frac{1}{2}(d_A + d_B)$ and the cross-section is πd^2.

If the collision density is required in terms of the partial pressure of each gas J, then the molar concentrations in eqn 18A.4a are replaced by $[J] = n_J/V = p_J/RT$. For collisions between like molecules $\mu = \frac{1}{2}m_A$ and eqn 18A.4a becomes

$$Z_{AA} = \tfrac{1}{2}\sigma\left(\frac{16kT}{\pi m_A}\right)^{1/2} N_A^2[A]^2 = \sigma\left(\frac{4kT}{\pi m_A}\right)^{1/2} N_A^2[A]^2 \quad (18A.4b)$$

Collision density [identical molecules]

where the (blue) factor of $\frac{1}{2}$ has been introduced to avoid double counting of collisions (Smith with Jones and Jones with Smith, for instance).

Brief illustration 18A.1

In nitrogen at 25 °C and 1.0 bar, when $[N_2] \approx 40\,\text{mol m}^{-3}$, with $\sigma = 0.43\,\text{nm}^2$ and $m_{N_2} = 28.02\,m_u$ the collision density is

$$Z_{N_2N_2} = (4.3\times10^{-19}\,\text{m}^2)\times\left(\frac{4\times(1.381\times10^{-23}\,\text{J K}^{-1})\times(298\,\text{K})}{\pi\times28.02\times(1.661\times10^{-27}\,\text{kg})}\right)^{1/2}$$
$$\times(6.022\times10^{23}\,\text{mol}^{-1})^2\times(40\,\text{mol m}^{-3})^2$$
$$= 8.4\times10^{34}\,\text{m}^{-3}\,\text{s}^{-1}$$

This result shows that collision densities may be very large: even in 1 cm^3, there are over 8×10^{16} collisions in each picosecond.

(b) The energy requirement

According to collision theory, the rate of change of $\mathcal{N}_A$ due to reaction is the product of the collision density and the probability that a collision occurs with sufficient energy. The latter condition can be incorporated by writing the collision cross-section σ as a function of the kinetic energy ε of approach of the two colliding species, and setting the cross-section, $\sigma(\varepsilon)$, equal to zero if the kinetic energy of approach is below a certain threshold value, ε_a. Later, $N_A\varepsilon_a$ will be identified as E_a, the (molar) activation energy of the reaction. For a collision between A and B with a specific relative speed of approach v_{rel} (not, at this stage, a mean value) it follows from eqn 18A.3 that the rate of change of $\mathcal{N}_A$ is

$$\frac{d\mathcal{N}_A}{dt} = -\sigma(\varepsilon)v_{rel}\mathcal{N}_A\mathcal{N}_B \quad (18A.5a)$$

or, in terms of molar concentrations,

$$\frac{d[A]}{dt} = -\sigma(\varepsilon)v_{rel}N_A[A][B] \quad (18A.5b)$$

The kinetic energy associated with the relative motion of the two particles is $\varepsilon = \frac{1}{2}\mu v_{rel}^2$. Therefore the relative speed can also be expressed in terms of the relative kinetic energy as $v_{rel} = (2\varepsilon/\mu)^{1/2}$. Because there is a wide range of approach energies

ε in a sample, eqn 18A.5b must be averaged over a Boltzmann distribution of energies $f(\varepsilon)$ to give

$$\frac{d[A]}{dt} = -\left\{\int_0^\infty \sigma(\varepsilon)v_{rel}f(\varepsilon)d\varepsilon\right\}N_A[A][B] \quad (18A.6)$$

where $f(\varepsilon)d\varepsilon$ is the probability that the approach energy is between ε and $\varepsilon + d\varepsilon$. By comparison with eqn 18A.1a it follows that

$$k_r = N_A\int_0^\infty \sigma(\varepsilon)v_{rel}f(\varepsilon)d\varepsilon \quad \text{Rate constant} \quad (18A.7)$$

To evaluate this integral it is necessary to establish the energy dependence of the collision cross-section, $\sigma(\varepsilon)$.

How is that done? 18A.2 Deriving an expression for the energy dependence of the collision cross-section

The key aspect of this model is that in a collision only the kinetic energy associated with a head-on collision is effective at bringing about reaction.

Step 1 *Consider how the geometry of the collision affects the energy available for reaction*

Consider two molecules A and B colliding with relative speed v_{rel} and therefore relative kinetic energy $\varepsilon = \frac{1}{2}\mu v_{rel}^2$. Although a collision is counted when the centres of the molecules come within a distance d of each other, that might be more a glancing blow than a head-on collision. Intuitively a head-on collision between A and B will be most effective in bringing about a chemical reaction, so from now on suppose that only the kinetic energy associated with the head-on component of the collision leads to reaction. This contribution to the kinetic energy depends on $v_{rel,A-B}$, the magnitude of the relative velocity component parallel to an axis connecting the centres of A and B.

Step 2 *Find an expression for the head-on component of the velocity*

As shown in the arrangement in Fig. 18A.2, the distance a is the closest approach of the centres of the two molecules and d is the distance between the centres. From trigonometry and the definition of the angle θ given in the diagram, it follows that

$$v_{rel,A-B} = v_{rel}\cos\theta = v_{rel}\left(\frac{d^2-a^2}{d^2}\right)^{1/2} \quad [\cos\theta = (d^2-a^2)^{1/2}/d]$$

Step 3 *Relate the velocities to energies*

The kinetic energy associated with the head-on collision is $\varepsilon_{A-B} = \frac{1}{2}\mu v_{rel,A-B}^2$, and the total kinetic energy of the collision is $\varepsilon = \frac{1}{2}\mu v_{rel}^2$. You can relate these two quantities by using the result from Step 2:

$$\varepsilon_{A-B} = \tfrac{1}{2}\mu v_{rel,A-B}^2 = \tfrac{1}{2}\mu(v_{rel}\cos\theta)^2 = \overbrace{\tfrac{1}{2}\mu v_{rel}^2}^{\varepsilon}\cos^2\theta = \varepsilon\left(\frac{d^2-a^2}{d^2}\right)$$

[$v_{rel,A-B} = v_{rel}\cos\theta$; $\cos^2\theta = (d^2-a^2)/d^2$]

Step 4 *Introduce an energy threshold*

As a increases, the kinetic energy associated with the head-on collision decreases. The existence of an energy threshold, ε_a, for the formation of products implies that there is a maximum value of a, a_{max}, above which reaction does not occur. Therefore, set $a = a_{max}$ and $\varepsilon_{A-B} = \varepsilon_a$ and obtain

$$\varepsilon_a = \varepsilon\left(\frac{d^2-a_{max}^2}{d^2}\right) \text{ which rearranges to } a_{max}^2 = \left(1-\frac{\varepsilon_a}{\varepsilon}\right)d^2$$

Step 5 *Rewrite the expression in terms of the collision cross-section*

The energy-dependent collision cross-section is given in terms of a_{max} as $\sigma(\varepsilon) = \pi a_{max}^2$, and πd^2 is identified as the (simple) collision cross-section σ introduced in Fig. 18A.1. It follows that

$$\sigma(\varepsilon) = \left(1-\frac{\varepsilon_a}{\varepsilon}\right)\sigma \quad \text{Energy dependence of } \sigma(\varepsilon)\ [\varepsilon > \varepsilon_a] \quad (18A.8)$$

This form of the energy-dependence of $\sigma(\varepsilon)$ is broadly consistent with experimental determinations of the reaction between H and D_2 as determined by molecular beam measurements of the kind described in Topic 18D (Fig. 18A.3).

With the energy dependence of the collision cross-section established, the rate constant can now be calculated.

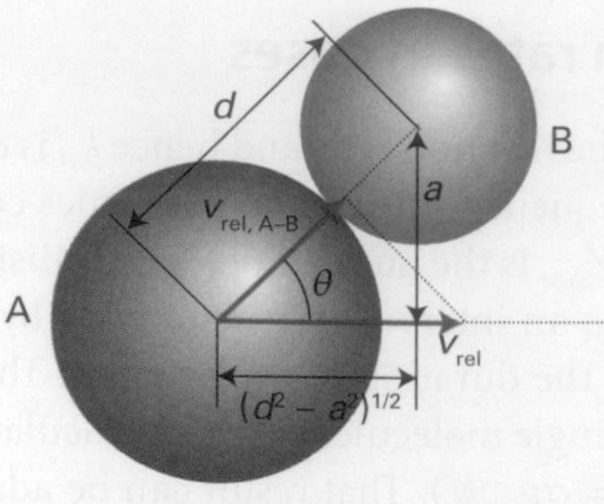

Figure 18A.2 The parameters used in the calculation of the relative kinetic energy associated with the head-on component of the collision of two molecules A and B.

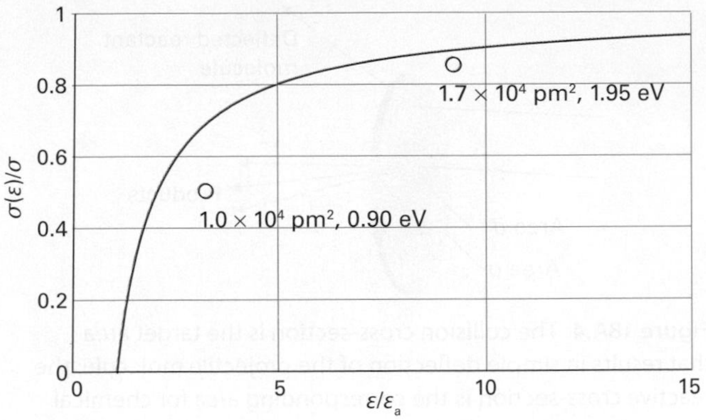

Figure 18A.3 The variation of the reactive cross-section with energy as expressed by eqn 18A.8. The data points are from experiments on the reaction $H + D_2 \rightarrow HD + D$ (K. Tsukiyama et al., *J. Chem. Phys.* **84**, 1934 (1986)).

How is that done? 18A.3 Deriving an expression for the rate constant

The calculation involves evaluating the integral in eqn 18A.7 with the energy-dependent collision cross section given in eqn 18A.8.

Step 1 *Use the Maxwell–Boltzmann distribution to write an expression for $f(\varepsilon)d\varepsilon$*

Adapt eqn 1B.4 in Topic 1B by replacing M/R with μ/k and so writing the distribution of *relative* molecular speeds as

$$f(v_{\rm rel})dv_{\rm rel} = 4\pi\left(\frac{\mu}{2\pi kT}\right)^{3/2} v_{\rm rel}^2 e^{-\mu v_{\rm rel}^2/2kT} dv_{\rm rel}$$

You can write this expression in terms of the relative kinetic energy, ε, by noting that $\varepsilon = \frac{1}{2}\mu v_{\rm rel}^2$. It follows that $v_{\rm rel} = (2\varepsilon/\mu)^{1/2}$ and so $dv_{\rm rel} = \frac{1}{2}(2/\mu)^{1/2}\varepsilon^{-1/2}d\varepsilon = d\varepsilon/(2\mu\varepsilon)^{1/2}$. With this substitution the distribution becomes

$$f(v_{\rm rel})dv_{\rm rel} = 4\pi\left(\frac{\mu}{2\pi kT}\right)^{3/2}\left(\frac{2\varepsilon}{\mu}\right)e^{-\varepsilon/kT}\frac{d\varepsilon}{(2\mu\varepsilon)^{1/2}}$$

$$= \underbrace{2\pi\left(\frac{1}{\pi kT}\right)^{3/2}\varepsilon^{1/2}e^{-\varepsilon/kT}d\varepsilon}_{f(\varepsilon)d\varepsilon}$$

Step 2 *Evaluate the integral*

Now evaluate the integral

$$\int_0^\infty \sigma(\varepsilon)\overbrace{v_{\rm rel}}^{(2\varepsilon/\mu)^{1/2}} f(\varepsilon)d\varepsilon = 2\pi\left(\frac{1}{\pi kT}\right)^{3/2}\int_0^\infty \sigma(\varepsilon)\left(\frac{2\varepsilon}{\mu}\right)^{1/2}\varepsilon^{1/2}e^{-\varepsilon/kT}d\varepsilon$$

$$= \left(\frac{8}{\pi\mu kT}\right)^{1/2}\left(\frac{1}{kT}\right)\int_0^\infty \varepsilon\sigma(\varepsilon)e^{-\varepsilon/kT}d\varepsilon$$

With $\sigma(\varepsilon)$ from eqn 18A.8, write

$$\int_0^\infty \varepsilon\sigma(\varepsilon)e^{-\varepsilon/kT}d\varepsilon = \sigma\int_{\varepsilon_a}^\infty \varepsilon\left(1-\frac{\varepsilon_a}{\varepsilon}\right)e^{-\varepsilon/kT}d\varepsilon$$

(annotation: $\sigma(\varepsilon) = (1-\varepsilon_a/\varepsilon)\sigma$; $\sigma = 0$ for $\varepsilon < \varepsilon_a$)

$$= \sigma\Big\{\underbrace{\int_{\varepsilon_a}^\infty \varepsilon e^{-\varepsilon/kT}d\varepsilon}_{\text{Integral E.2}} - \varepsilon_a\underbrace{\int_{\varepsilon_a}^\infty e^{-\varepsilon/kT}d\varepsilon}_{\text{Integral E.1}}\Big\} = (kT)^2\sigma e^{-\varepsilon_a/kT}$$

It follows that

$$\int_0^\infty \sigma(\varepsilon)v_{\rm rel}f(\varepsilon)d\varepsilon = \sigma\left(\frac{8kT}{\pi\mu}\right)^{1/2}e^{-\varepsilon_a/kT}$$

Step 3 *Finalize the expression for the rate constant*

Because $E_a = N_A\varepsilon_a$ it follows that $\varepsilon_a/kT = E_a/RT$. With these substitutions it follows from the integral just evaluated and eqn 18A.7 that

$$k_r = \sigma N_A\left(\frac{8kT}{\pi\mu}\right)^{1/2}e^{-E_a/RT} \qquad (18A.9)$$

Rate constant [collision theory]

This equation has the Arrhenius form $k_r = Ae^{-E_a/RT}$ provided the exponential temperature dependence dominates the weak square-root temperature dependence of the frequency factor. It follows that, within the constraints of collision theory, the activation energy, E_a, can be identified with the minimum kinetic energy along the line of approach that is needed for reaction, and that the frequency factor (after multiplication by [A][B]) determines the rate at which collisions occur.

The simplest procedure for calculating k_r is to use for σ the values obtained for non-reactive collisions (e.g. typically those obtained from viscosity measurements) or from tables of molecular radii. If the collision cross-sections of A and B are σ_A and σ_B, then an approximate value of the AB cross-section is estimated from $\sigma = \pi d^2$, with $d = \frac{1}{2}(d_A + d_B)$. That is,

$$\sigma \approx \tfrac{1}{4}(\sigma_A^{1/2} + \sigma_B^{1/2})^2$$

Brief illustration 18A.2

To estimate the rate constant for the reaction $H_2 + C_2H_4 \rightarrow C_2H_6$ at 628 K, first calculate μ by setting $m(H_2) = 2.016m_u$ and $m(C_2H_4) = 28.05m_u$. A straightforward calculation gives $\mu = 3.123 \times 10^{-27}$ kg. It then follows that

$$\left(\frac{8kT}{\pi\mu}\right)^{1/2} = \left(\frac{8\times(1.381\times10^{-23}\,{\rm J\,K^{-1}})\times(628\,{\rm K})}{\pi\times(3.123\times10^{-27}\,{\rm kg})}\right)^{1/2} = 2.65\ldots\,{\rm km\,s^{-1}}$$

From Table 1B.1, $\sigma(H_2) = 0.27\,nm^2$ and $\sigma(C_2H_4) = 0.64\,nm^2$, giving $\sigma(H_2,C_2H_4) \approx 0.44\,nm^2$. The activation energy for this reaction is $180\,kJ\,mol^{-1}$; therefore,

$$k_r = (4.4\times10^{-19}\,m^2)\times(6.022\times10^{23}\,mol^{-1})\times(2.65\ldots\times10^3\,m\,s^{-1})$$
$$\times e^{-(1.80\times10^5\,J\,mol^{-1})/(8.3145\,J\,K^{-1}\,mol^{-1})\times(628\,K)}$$
$$= \overbrace{7.04\ldots\times10^8\,m^3\,mol^{-1}\,s^{-1}}^{A}\times e^{-34.4\ldots} = 7.5\times10^{-7}\,m^3\,mol^{-1}\,s^{-1}$$

or $7.5\times10^{-4}\,dm^3\,mol^{-1}\,s^{-1}$.

(c) The steric requirement

Table 18A.1 compares some values of the frequency factor calculated from collision cross-sections determined in other measurements with values obtained from Arrhenius plots. One of the reactions shows fair agreement between theory and experiment, but for others there are major discrepancies. In some cases the experimental values are orders of magnitude smaller than those calculated, which suggests that the collision energy is not the only criterion for reaction and that some other feature, such as the relative orientation of the colliding species, is important. Moreover, one reaction in the table has a pre-exponential factor larger than theory, which seems to indicate that the reaction occurs more quickly than the particles collide!

The disagreement between experiment and theory can be eliminated by introducing a **steric factor**, P, and expressing the **reactive cross-section**, σ^*, the actual cross-section for reactive collisions, as a multiple of the collision cross-section, $\sigma^* = P\sigma$ (Fig. 18A.4). Then the rate constant becomes

$$k_r = P\sigma N_A\left(\frac{8kT}{\pi\mu}\right)^{1/2} e^{-E_a/RT} \qquad (18A.10)$$

This expression has the form anticipated in eqn 18A.2. The steric factor is normally found to be several orders of magnitude smaller than 1.

Table 18A.1 Arrhenius parameters for gas-phase reactions*

	$A/(dm^3\,mol^{-1}\,s^{-1})$		$E_a/(kJ\,mol^{-1})$	P
	Experiment	Theory		
$2NOCl \rightarrow 2NO + Cl_2$	9.4×10^9	5.9×10^{10}	102	0.16
$2ClO \rightarrow Cl_2 + O_2$	6.3×10^7	2.5×10^{10}	0	2.5×10^{-3}
$H_2 + C_2H_4 \rightarrow C_2H_6$	1.24×10^6	7.0×10^{11}	180	1.8×10^{-6}
$K + Br_2 \rightarrow KBr + Br$	1.0×10^{12}	2.1×10^{11}	0	4.8

* More values are given in the *Resource section*.

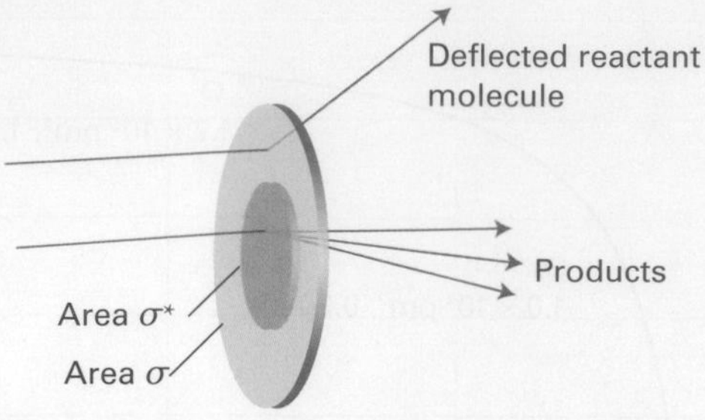

Figure 18A.4 The collision cross-section is the target area that results in simple deflection of the projectile molecule; the reactive cross-section is the corresponding area for chemical change to occur on collision.

Brief illustration 18A.3

It is found experimentally that the frequency factor for the reaction $H_2 + C_2H_4 \rightarrow C_2H_6$ at 628 K is $1.24\times10^6\,dm^3\,mol^{-1}\,s^{-1}$. The result in *Brief illustration* 18A.2 can be expressed as $A = 7.04\ldots\times10^{11}\,dm^3\,mol^{-1}\,s^{-1}$. It follows that the steric factor for this reaction is

$$P = \frac{A_{\text{experimental}}}{A_{\text{calculated}}} = \frac{1.24\times10^6\,dm^3\,mol^{-1}\,s^{-1}}{7.04\ldots\times10^{11}\,dm^3\,mol^{-1}\,s^{-1}} \approx 1.8\times10^{-6}$$

The very small value of P is one reason, the other being the high activation energy, why catalysts are needed to bring this reaction about at a reasonable rate. As a general guide, the more complex the reactant molecules, the smaller is the value of P.

An example of a reaction for which it is possible to estimate the steric factor is $K + Br_2 \rightarrow KBr + Br$, for which $P = 4.8$. In this reaction, the distance of approach at which reaction occurs appears to be considerably larger than the distance needed for deflection of the path of the approaching molecules in a non-reactive collision. It has been proposed that the reaction proceeds by a **harpoon mechanism**. This brilliant name is based on a model of the reaction in which the K atom is pictured as approaching a Br_2 molecule, and when the two are close enough an electron (the harpoon) flips across from K to Br_2. In place of two neutral particles there are now two ions, so there is a Coulombic attraction between them: this attraction is the line on the harpoon. Under its influence the ions move together (the line is wound in), the reaction takes place, and KBr + Br emerge. The harpoon extends the cross-section for the reactive encounter, and the reaction rate is significantly underestimated by taking for the collision cross-section the value for simple mechanical contact between K and Br_2.

Example 18A.1 Estimating a steric factor

Estimate the value of P for the harpoon mechanism by calculating the distance at which it becomes energetically favourable for the electron to leap from K to Br_2. Take the sum of the radii of the reactants (treating them as spherical) to be 400 pm.

Collect your thoughts Begin by identifying all the energy terms involved in the electron transfer process $K + Br_2 \rightarrow K^+ + Br_2^-$. There are three terms: the first is the ionization energy, I, of K; the second is the electron affinity, E_{ea}, of Br_2; and the third is the Coulombic interaction energy between the ions when they have been formed. When the separation of the ions is R, the Coulombic attraction energy is $-e^2/4\pi\varepsilon_0 R$. The electron flips across when the sum of these three contributions changes from positive to negative (that is, when the sum becomes zero) so making the process energetically favourable.

The solution The net change in energy when the transfer occurs at a separation R is

$$E = I - E_{ea} - \frac{e^2}{4\pi\varepsilon_0 R}$$

This energy is zero when R is equal to some critical value R^* (and is negative for smaller values of R)

$$0 = I - E_{ea} - \frac{e^2}{4\pi\varepsilon_0 R^*} \quad \text{rearranges to} \quad R^* = \frac{e^2}{4\pi\varepsilon_0 (I - E_{ea})}$$

When the particles are at this separation, the harpoon shoots across from K to Br_2. The reactive cross-section can therefore be identified as $\sigma^* = \pi R^{*2}$. The non-reactive collision cross-section is $\sigma = \pi d^2$, where $d = R(\mathrm{K}) + R(\mathrm{Br_2})$ is the sum of the radii of the (assumed) spherical reactants. These values of σ and σ^* imply that the steric factor is

$$P = \frac{\sigma^*}{\sigma} = \frac{\pi R^{*2}}{\pi d^2} = \left\{\frac{e^2}{4\pi\varepsilon_0 d(I - E_{ea})}\right\}^2$$

With $I = 420\,\mathrm{kJ\,mol^{-1}}$ (corresponding to 0.70 aJ), $E_{ea} \approx 250\,\mathrm{kJ\,mol^{-1}}$ (corresponding to 0.42 aJ), and $d = 400\,\mathrm{pm}$, the value of P is 4.2, in good agreement with the experimental value (4.8).

Self-test 18A.1 Estimate the value of P for the harpoon reaction between Na and Cl_2 for which $d \approx 350\,\mathrm{pm}$; take $E_{ea} \approx 230\,\mathrm{kJ\,mol^{-1}}$.

Answer: 2.2

Example 18A.1 illustrates two points about steric factors. First, the concept of a steric factor is not wholly useless because in some cases its numerical value can be estimated. Second, and more pessimistically, most reactions are much more complex than $K + Br_2$, and P cannot be obtained so easily.

18A.2 The RRK model

The rate constants of 'unimolecular' gas-phase reactions like those treated by the Lindemann–Hinshelwood mechanism (Topic 17F) can be estimated with a calculation based on the **Rice–Ramsperger–Kassel model** (RRK model). That model was proposed in 1926 by O.K. Rice and H.C. Ramsperger and almost simultaneously by L.S. Kassel. It has been elaborated, largely by R.A. Marcus, into the 'RRKM model'. The essential feature of the model is that although a molecule might have enough energy to react, that energy is distributed over all the modes of motion of the molecule, and reaction will occur only when enough of that energy has migrated into a particular location (such as a particular bond) in the molecule. The details are given in *A deeper look* 12 on the website of this book. Assuming that a molecule consists of s identical harmonic oscillators, the principal conclusion is that the **Kassel form** of the unimolecular rate constant for the decay of the energized molecule A* to products is

$$k_b(E) = \left(1 - \frac{E^*}{E}\right)^{s-1} k_b \quad \text{for } E \geq E^* \qquad \text{Unimolecular rate constant [Kassel form]} \qquad (18A.11)$$

where k_b is the rate constant used in the original Lindemann–Hinshelwood theory for the decomposition of the energized molecule (Topic 17F), and E^* is the minimum energy that must be accumulated in a bond in order for it to break.

The energy dependence of the rate constant given by eqn 18A.11 is shown in Fig. 18A.5 for various values of s. The equation can be interpreted as follows:

Physical interpretation

- The rate constant is smaller at a given excitation energy if s is large, as it takes longer for the excitation energy to migrate through all the oscillators of a large molecule and accumulate in the location needed for reaction.
- As E becomes very large, however, the term in parentheses approaches 1, and $k_b(E)$ becomes independent of the energy and the number of oscillators in the molecule, as there is now enough energy to accumulate immediately in the critical mode regardless of the size of the molecule.

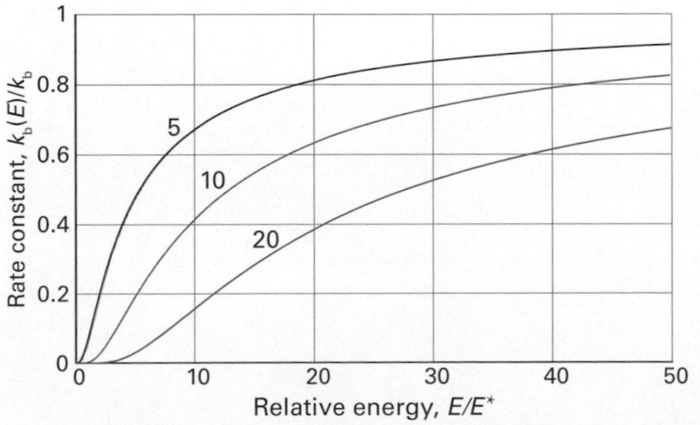

Figure 18A.5 The energy dependence of the rate constant given by eqn 18A.11 for three values of s.

Checklist of concepts

☐ 1. In **collision theory**, it is supposed that the rate is proportional to the collision frequency, a steric factor, and the fraction of collisions that occur with at least the kinetic energy E_a along their lines of centres.

☐ 2. The **collision density** is the number of collisions in a region of the sample in an interval of time divided by the volume of the region and the duration of the interval.

☐ 3. The **activation energy** is the minimum kinetic energy along the line of approach of reactant molecules that is required for reaction.

☐ 4. The **steric factor** is an adjustment that takes into account the orientational requirements for a successful collision.

☐ 5. The rate constant for decomposition of an energized molecule can be estimated by using the **RRK model.**

Checklist of equations

Property	Equation	Comment	Equation number
Collision density	$Z_{AB}=\sigma(8kT/\pi\mu)^{1/2}N_A^2[A][B]$	Unlike molecules, KMT (kinetic molecular theory)	18A.4a
Energy dependence of σ	$\sigma(\varepsilon)=(1-\varepsilon_a/\varepsilon)\sigma$	$\varepsilon\geq\varepsilon_a$, $\sigma=0$ otherwise	18A.8
Rate constant	$k_r=P\sigma N_A(8kT/\pi\mu)^{1/2}e^{-E_a/RT}$	KMT, collision theory	18A.10
Unimolecular rate constant	$k_b(E)=(1-E^*/E)^{s-1}k_b$	RRK theory ($E\geq E^*$)	18A.11

TOPIC 18B Diffusion-controlled reactions

➤ **Why do you need to know this material?**

Most chemical reactions take place in solution and for a thorough grasp of chemistry it is important to understand what controls their rates and how those rates can be modified.

➤ **What is the key idea?**

The rate of a chemical reaction in solution is controlled either by the rate of diffusion of the reactants or by the activation energy of the step that leads to products.

➤ **What do you need to know already?**

This Topic makes use of the steady-state approximation (Topic 17E) and draws on Fick's first law of diffusion (Topic 16C). At one point it uses the Stokes–Einstein relation (Topic 16C).

Reactions in solution are entirely different from those in gases. No longer are there collisions of molecules hurtling through space; now there is the jostling of one molecule through a dense but mobile collection of molecules making up the fluid environment.

18B.1 Reactions in solution

Encounters between reactants in solution occur in a very different manner from encounters in gases. The encounters of reactant molecules dissolved in a solvent are considerably less frequent than in a gas. However, because a molecule also migrates only slowly away from a location, two reactant molecules that encounter each other stay near each other for much longer than in a gas. This lingering of one molecule near another on account of the hindering presence of solvent molecules is called the **cage effect**. Such an **encounter pair** may accumulate enough energy to react even though it does not have enough energy to do so when it first forms. The activation energy of a reaction is a much more complicated quantity in solution than in a gas because the encounter pair is surrounded by solvent and the energy of the entire local assembly of reactant and solvent molecules must be considered.

(a) Classes of reaction

The complicated overall process can be divided into simpler parts by setting up a simple kinetic scheme. Suppose the rate of formation of an encounter pair AB is first order in each of the reactants A and B:

$$\mathrm{A+B \rightarrow AB} \quad v = k_\mathrm{d}[\mathrm{A}][\mathrm{B}]$$

As will be seen, k_d (where the d signifies diffusion) is determined by the diffusional characteristics of A and B. The encounter pair can break up without reaction or it can go on to form products P. If it is supposed that both processes are pseudofirst-order reactions (with the solvent perhaps playing a role), then the mechanism may be written

$$\mathrm{AB \rightarrow A+B} \quad v = k_\mathrm{d}'[\mathrm{AB}]$$

$$\mathrm{AB \rightarrow P} \quad v = k_\mathrm{a}[\mathrm{AB}]$$

The concentration of AB can now be found by applying the steady-state approximation (Topic 17E) to the equation for the net rate of change of concentration of AB:

$$\frac{\mathrm{d[AB]}}{\mathrm{d}t} = k_\mathrm{d}[\mathrm{A}][\mathrm{B}] - k_\mathrm{d}'[\mathrm{AB}] - k_\mathrm{a}[\mathrm{AB}] \overset{\text{Steady-state}}{=} 0$$

This expression solves to

$$[\mathrm{AB}] = \frac{k_\mathrm{d}[\mathrm{A}][\mathrm{B}]}{k_\mathrm{a}+k_\mathrm{d}'}$$

The rate of formation of products is therefore

$$\frac{\mathrm{d[P]}}{\mathrm{d}t} = k_\mathrm{a}[\mathrm{AB}] = k_\mathrm{r}[\mathrm{A}][\mathrm{B}] \qquad k_\mathrm{r} = \frac{k_\mathrm{a}k_\mathrm{d}}{k_\mathrm{a}+k_\mathrm{d}'} \qquad (18B.1)$$

Two limits can now be distinguished. If the rate of separation of the unreacted encounter pair is much slower than the rate at which it forms products, then $k_\mathrm{d}'[\mathrm{AB}] \ll k_\mathrm{a}[\mathrm{AB}]$ (or, after cancelling the [AB], $k_\mathrm{d}' \ll k_\mathrm{a}$), and the effective rate constant is

$$k_\mathrm{r} \approx \frac{k_\mathrm{a}k_\mathrm{d}}{k_\mathrm{a}} = k_\mathrm{d} \qquad \text{Diffusion-controlled limit} \qquad (18B.2a)$$

Table 18B.1 Arrhenius parameters for solvolysis reactions in solution

	Solvent	$A/(\text{dm}^3\,\text{mol}^{-1}\,\text{s}^{-1})$	$E_a/(\text{kJ mol}^{-1})$
$(CH_3)_3CCl$	Water	7.1×10^{16}	100
	Ethanol	3.0×10^{13}	112
	Chloroform	1.4×10^{4}	45
CH_3CH_2Br	Ethanol	4.3×10^{11}	90

In this **diffusion-controlled limit**, the rate of reaction is governed by the rate at which the reactant molecules diffuse through the solvent. Because the combination of radicals involves very little activation energy, radical and atom recombination reactions are often diffusion-controlled. An **activation-controlled reaction** arises when a substantial activation energy is involved in the reaction $AB \rightarrow P$. Then $k_a[AB] \ll k_d'[AB]$ (implying $k_a \ll k_d'$) and

$$k_r \approx \frac{k_a k_d}{k_d'} = k_a \frac{K}{c^{\ominus}} \qquad \text{Activation-controlled limit} \qquad (18B.2b)$$

where $k_d/k_d' = K/c^{\ominus}$ (see Topic 17C) and K is the equilibrium constant for $A + B \rightleftharpoons AB$. In this limit, the reaction proceeds at a rate that depends on the equilibrium concentration of encounter pairs and the rate at which energy accumulates in these pairs from the surrounding solvent. Some experimental data are given in Table 18B.1.

(b) Diffusion and reaction

The rate of a diffusion-controlled reaction is calculated by considering the rate at which the reactants diffuse together.

How is that done? 18B.1 Finding an expression for the rate constant of a diffusion-controlled reaction

Suppose that molecules of A and B in solution react immediately when they come within some critical distance R^* of one another and the rate of reaction is controlled by the rate of encounters between A and B molecules as they diffuse together. As a result of the reaction, the concentration of B molecules near A is decreased and a concentration gradient of B molecules is established. There is a diffusive flux of B towards A as a result of that gradient, and the flux is constant while the reaction is in progress.

Step 1 *Consider the rate at which B molecules cross the surface of a sphere centred on A*

If the (molar) flux of B molecules towards A is J_B, the rate (expressed as amount divided by time) at which B molecules pass through a shell of radius r and surface area $4\pi r^2$ centred on A is

$$v_B = 4\pi r^2 J_B \overset{\text{Fick's first law}}{=} 4\pi r^2 D_B \frac{d[B](r)}{dr}$$

An important point to recognize is that v_B is the same for a shell of any radius greater than or equal to R^*, because no B molecules are lost until they have reached R^*. Also keep in mind that v_B is a rate expressed as amount/time, not concentration/time.

Step 2 *Use the known values of [B] for the bulk to establish an expression for the variation of the concentration with distance*

For v_B to be independent of r, $r^2 d[B](r)/dr$ must be a constant, which implies that, provided $r > R^*$, $[B](r) = a + b/r$. Thus, $d[B](r)/dr = -b/r^2$, and $r^2 d[B](r)/dr = -b$, a constant, as required. You can find the values of the constants a and b by noting that as $r \rightarrow \infty$, $[B](r)$ tends to its bulk value, [B]. Therefore $a = [B]$ and hence $[B](r) = [B] + b/r$. When $r = R^*$, $[B](r) = 0$, which implies that $b = -[B]R^*$. It follows that

$$[B](r) = [B]\left(1 - \frac{R^*}{r}\right)$$

Figure 18B.1 illustrates the distance dependence of $[B](r)$ according to this equation. The first derivative of $[B](r)$ with respect to distance is $[B]R^*/r^2$, so

$$v_B = 4\pi r^2 D_B \frac{d[B](r)}{dr} \overset{d[B](r)/dr = [B]R^*/r^2}{=} 4\pi R^* D_B [B]$$

Step 3 *Write an expression for the overall rate of reaction*

To express the rate of reaction, v_B must be multiplied by the number of A molecules in the solution. If the bulk concentration of A is [A], then the number of A molecules in a solution of volume V is $N_A[A]V$. Therefore, the rate of reaction (still as amount/time) is

$$\text{rate} = v_B N_A [A] V = 4\pi R^* D_B N_A [A]\,[B] V$$

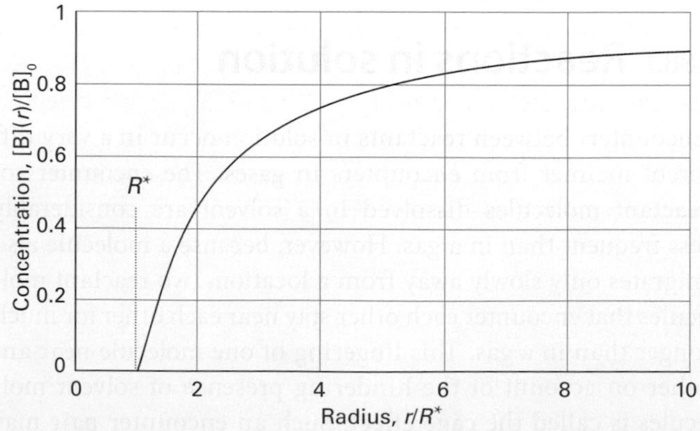

Figure 18B.1 The concentration profile for reaction in solution when a molecule B diffuses towards another reactant molecule and reacts if it reaches R^*.

It is unrealistic to suppose that all A molecules are stationary, so the diffusion coefficient D_B is now replaced by the sum of the diffusion coefficients of the two species, $D = D_A + D_B$. Because it is more convenient to express rates as concentration/time, both sides of the equation are divided by the volume V, in which case

$$v = \overbrace{4\pi R^* D N_A}^{k_d}[A][B]$$

from which it follows that the diffusion-controlled rate constant is

$$k_d = 4\pi R^* D N_A \quad \text{Rate constant of a diffusion-controlled reaction} \quad (18B.3)$$

Brief illustration 18B.1

The order of magnitude of R^* is 10^{-10} m (100 pm) and that of D for a species in water is $10^{-9}\,m^2\,s^{-1}$. It follows from eqn 18B.3 that

$$k_d \approx 4\pi \times (10^{-10}\,m) \times (10^{-9}\,m^2\,s^{-1}) \times (6.022 \times 10^{23}\,mol^{-1})$$
$$\approx 8 \times 10^5\,m^3\,mol^{-1}\,s^{-1}$$

which corresponds to about $10^9\,dm^3\,mol^{-1}\,s^{-1}$. An indication that a reaction is diffusion-controlled is therefore that its rate constant is of that order of magnitude.

Equation 18B.3 can be taken further by incorporating the Stokes–Einstein equation (eqn 16C.4b of Topic 16C, $D_J = kT/6\pi\eta R_J$) for the relation between the diffusion constant and the hydrodynamic radius R_A and R_B of each molecule in a medium of viscosity η. As this relation is approximate, little extra error is introduced by writing $R_A = R_B = \frac{1}{2}R^*$, which leads to

$$k_d = \frac{8RT}{3\eta} \quad \text{Diffusion-controlled rate constant} \quad (18B.4)$$

(The R in this equation is the gas constant.) The radii have cancelled because, although the diffusion constants are smaller when the radii are large, the reactive collision radius is larger and the particles need travel a shorter distance to meet. In this approximation, the rate constant is independent of the identities of the reactants, and depends only on the temperature and the viscosity of the solvent.

Brief illustration 18B.2

The rate constant for the recombination of I atoms in hexane at 298 K, when the viscosity of the solvent is 0.326 cP (with 1 P = $10^{-1}\,kg\,m^{-1}\,s^{-1}$) is

$$k_d = \frac{8 \times (8.3145\,J\,K^{-1}\,mol^{-1}) \times (298\,K)}{3 \times (3.26 \times 10^{-4}\,kg\,m^{-1}\,s^{-1})} = 2.0 \times 10^7\,m^3\,mol^{-1}\,s^{-1}$$

where $1\,J = 1\,kg\,m^2\,s^{-2}$. This result corresponds to $2.0 \times 10^{10}\,dm^3\,mol^{-1}\,s^{-1}$. The experimental value is $1.3 \times 10^{10}\,dm^3\,mol^{-1}\,s^{-1}$, so the agreement is very good considering the approximations involved.

18B.2 The material-balance equation

The diffusion of reactants plays an important role in many chemical processes, such as the diffusion of O_2 molecules into red blood cells and the diffusion of a gas towards a catalyst. To catch a glimpse of the kinds of calculations involved consider the diffusion equation (Topic 16C) generalized to take into account the possibility that the diffusing, convecting molecules are also reacting.

(a) The formulation of the equation

Consider a small volume element in a chemical reactor (or a biological cell). The net rate at which J molecules enter the region by diffusion and convection is given by eqn 16C.9 of Topic 16C:

$$\frac{\partial[J]}{\partial t} = D\frac{\partial^2[J]}{\partial x^2} - v\frac{\partial[J]}{\partial x} \quad \text{Diffusion equation} \quad (18B.5)$$

where v is the velocity of the convective flow of J and [J] in general depends on both position and time. If J disappears by a pseudofirst-order reaction, the net rate of change of molar concentration due to chemical reaction is

$$\frac{\partial[J]}{\partial t} = -k_r[J]$$

Therefore, the overall rate of change of the concentration of J is

$$\frac{\partial[J]}{\partial t} = \overbrace{D\frac{\partial^2[J]}{\partial x^2}}^{\text{Spread due to non-uniform distribution}} - \overbrace{v\frac{\partial[J]}{\partial x}}^{\text{Change due to convection}} - \overbrace{k_r[J]}^{\text{Loss due to reaction}} \quad \text{Material-balance equation} \quad (18B.6)$$

Equation 18B.6 is called the **material-balance equation**. If the rate constant is large, then [J] will decline rapidly. However, if the diffusion constant is large, then the decline can be replenished as J diffuses rapidly into the region. The convection term, which may represent the effects of stirring, can sweep material either into or out of the region according to the signs of v and the concentration gradient $\partial[J]/\partial x$.

(b) Solutions of the equation

The material-balance equation is a second-order partial differential equation and is far from easy to solve in general. Some idea of how it is solved can be obtained by considering the special case in which there is no convective motion (as in an unstirred reaction vessel):

$$\frac{\partial[\mathrm{J}]}{\partial t}=D\frac{\partial^2[\mathrm{J}]}{\partial x^2}-k_{\mathrm{r}}[\mathrm{J}] \quad (18\text{B}.7)$$

As may be verified by substitution (Problem 18B.1), if the solution of this equation in the absence of reaction (that is, for $k_r = 0$) is $[\mathrm{J}](x,t)$, then the solution $[\mathrm{J}]^*(x,t)$ in the presence of reaction ($k_r > 0$) is

$$[\mathrm{J}]^*(x,t) = [\mathrm{J}](x,t)\mathrm{e}^{-k_{\mathrm{r}}t} \quad \text{Diffusion with reaction} \quad (18\text{B}.8)$$

An example of a solution of the diffusion equation in the absence of reaction is that given in Topic 16C for a system in which initially a layer of n_0N_A molecules is spread over a plane of area A:

$$[\mathrm{J}](x,t)=\frac{n_0\mathrm{e}^{-x^2/4Dt}}{A(\pi Dt)^{1/2}} \quad (18\text{B}.9)$$

When this expression is substituted into eqn 18B.8, the result is an expression for the concentration of J as it diffuses away from its initial surface layer and undergoes reaction in the overlying solution (Fig. 18B.2).

Brief illustration 18B.3

Suppose 1.0 g of iodine (3.9 mmol I_2) is spread over a surface of area 5.0 cm^2 under a column of hexane ($D = 4.1 \times 10^{-9}\,\mathrm{m^2\,s^{-1}}$). As it diffuses upwards it reacts with a pseudofirst-order rate constant $k_r = 4.0 \times 10^{-5}\,\mathrm{s^{-1}}$. By substituting these values into

$$[\mathrm{J}]^*(x,t)=\frac{n_0\mathrm{e}^{-x^2/4Dt-k_{\mathrm{r}}t}}{A(\pi Dt)^{1/2}}$$

the following table of values can be constructed:

	$[\mathrm{J}]^*/(\mathrm{mol\,dm^{-3}})$ at x		
t	1 mm	5 mm	1 cm
100 s	3.72	0	0
1000 s	1.96	0.45	0.005
10 000 s	0.46	0.40	0.25

Even this relatively simple example has led to an equation that is difficult to solve, and only in some special cases can the full material-balance equation be solved analytically. Most modern work on reactor design and cell kinetics uses numerical methods to solve the equation, and detailed solutions for realistic environments, such as vessels of different shapes (which influence the boundary conditions on the solutions) and with a variety of inhomogeneously distributed reactants, can be obtained reasonably easily.

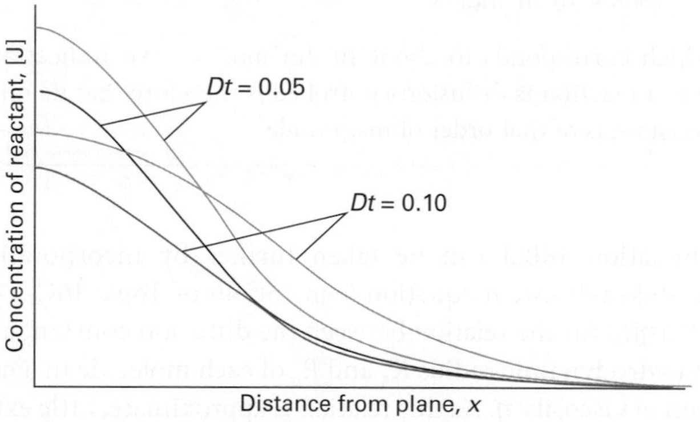

Figure 18B.2 The concentration profiles for a diffusing, reacting system (e.g. a column of solution) in which one reactant is initially in a layer at $x = 0$. In the absence of reaction (grey lines) the concentration profiles are similar to those in Fig. 16C.5.

Checklist of concepts

☐ 1. The **cage effect**, the lingering of one reactant molecule near another due to the hindering presence of solvent molecules, results in the formation of an **encounter pair** of reactant molecules.

☐ 2. A reaction in solution may be **diffusion controlled** if its rate is controlled by the rate at which reactant molecules encounter each other in solution.

☐ 3. The rate of an **activation-controlled reaction** is controlled by the rate at which the encounter pair accumulates sufficient energy.

☐ 4. The **material-balance equation** relates the overall rate of change of the concentration of a species to its rates of diffusion, convection, and reaction.

Checklist of equations

Property	Equation	Comment	Equation number
Diffusion-controlled limit	$k_r = k_d$	$v = k_d[A][B]$ for the encounter rate	18B.2a
Activation-controlled limit	$k_r = k_a(K/c^{\ominus})$	K for $A + B \rightleftharpoons AB$, k_a for the decomposition of AB	18B.2b
Diffusion-controlled rate constant	$k_d = 4\pi R^* D N_A$	$D = D_A + D_B$	18B.3
	$k_d = 8RT/3\eta$	Assumes Stokes–Einstein relation	18B.4
Material-balance equation	$\partial[J]/\partial t = D\partial^2[J]/\partial x^2 - v\partial[J]/\partial x - k_r[J]$	Diffusion and convection with first-order reaction	18B.6

TOPIC 18C Transition-state theory

➤ **Why do you need to know this material?**

Transition-state theory provides a way to relate the rate constant of reactions to models of the cluster of atoms supposed to form when reactants come together. It provides a link between information about the structures of reactants and the rate constant for their reaction.

➤ **What is the key idea?**

Reactants come together to form an activated complex, which decays into products.

➤ **What do you need to know already?**

This Topic makes use of two strands: one is the relation between equilibrium constants and partition functions (Topic 13F); the other is the relation between equilibrium constants and thermodynamic functions, such as the Gibbs energy, enthalpy, and entropy of reaction (Topic 6A). You need to be aware of the Arrhenius equation for the temperature dependence of the rate constant (Topic 17D).

In **transition-state theory** (which is also widely referred to as *activated complex theory*), the notion of the transition state is used in conjunction with concepts of statistical thermodynamics to provide a more detailed calculation of rate constants than collision theory provides (Topic 18A). Transition-state theory has the advantage that a quantity corresponding to the steric factor appears automatically and does not need to be grafted on to an equation as an afterthought; it is an attempt to identify the principal features governing the size of a rate constant in terms of a model of the events that take place during the reaction.

18C.1 The Eyring equation

In the course of a chemical reaction that begins with an encounter between molecules of A and molecules of B, the potential energy of the system typically changes in a manner shown in Fig. 18C.1. Although the illustration displays an exothermic reaction, a potential barrier is also common for endothermic reactions. As the reaction event proceeds, A and B come into contact, distort, and begin to exchange or discard atoms.

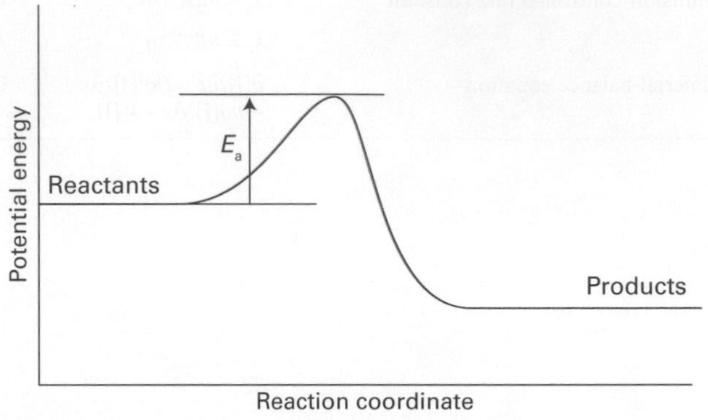

Figure 18C.1 A potential energy profile for an exothermic reaction. The height of the barrier between the reactants and products is the activation energy of the reaction.

(a) The formulation of the equation

The **reaction coordinate** is a representation of the atomic displacements, such as changes in interatomic distances and bond angles, that are directly involved in the formation of products from reactants. The potential energy rises to a maximum and the cluster of atoms that corresponds to the region close to the maximum is called the **activated complex**. After the maximum, the potential energy falls as the atoms rearrange in the cluster and reaches a value characteristic of the products. The climax of the reaction is at the peak of the potential energy, which can be identified with the activation energy E_a. However, as in collision theory, this identification should be regarded as approximate and is clarified later. At this peak, two reactant molecules have come to such a degree of closeness and distortion that a small further distortion will send them in the direction of products. This crucial configuration is called the **transition state** of the reaction. Although some molecules entering the transition state might revert to reactants, if they pass through this configuration then it is inevitable that products will emerge from the encounter.

A note on good practice The terms *activated complex* and *transition state* are often used as synonyms; however, it is best to preserve the distinction, with the former referring to the cluster of atoms in the vicinity of the peak of the potential energy curve, and the latter to their critical configuration.

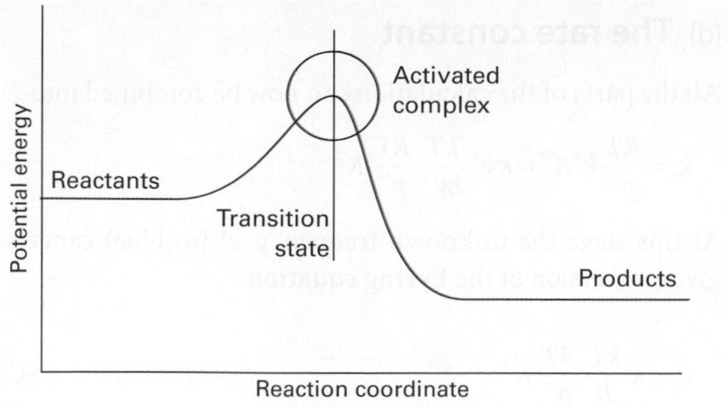

Figure 18C.2 A reaction profile (for an exothermic reaction). The horizontal axis is the reaction coordinate, and the vertical axis is potential energy. The activated complex is the region near the potential maximum, and the transition state corresponds to the maximum itself.

Transition-state theory pictures a reaction between A and B as proceeding through the formation of an activated complex, $C^‡$, in a rapid pre-equilibrium (Fig. 18C.2):

$$A+B \rightleftharpoons C^‡ \qquad K^‡ = \frac{p_{C^‡}p^\ominus}{p_A p_B} \qquad (18C.1)$$

where for this gas-phase reaction the activity of each species has been replaced by $p/p^\ominus$. The development of femtosecond (and even attosecond) pulsed lasers has made it possible to make observations on species that have such short lifetimes that in a number of respects they resemble an activated complex, which often survive for only a few picoseconds.

When the partial pressures, p_J, are expressed in terms of the molar concentrations, [J], by using $p_J = RT[J]$, the concentration of activated complex is related to the (dimensionless) equilibrium constant by

$$[C^‡] = \frac{RT}{p^\ominus}K^‡[A][B] \qquad (18C.2)$$

The activated complex falls apart by unimolecular decay into products, P, with a rate constant $k^‡$:

$$C^‡ \rightarrow P \qquad v = k^‡[C^‡] \qquad (18C.3)$$

It follows that

$$v = k_r[A][B] \qquad k_r = \frac{RT}{p^\ominus}k^‡K^‡ \qquad (18C.4)$$

The next task is the calculation of the unimolecular rate constant $k^‡$ and the equilibrium constant $K^‡$.

(b) The rate of decay of the activated complex

An activated complex forms products only if it passes through the transition state. As the reactant molecules approach the activated complex region, some bonds are forming and shortening while others are lengthening and breaking; therefore, along the reaction coordinate, there is a vibration-like motion of the atoms in the activated complex. If this motion occurs with a frequency $\nu^‡$, then the frequency with which the cluster of atoms forming the complex approaches the transition state is also $\nu^‡$. However, it is possible that not every oscillation along the reaction coordinate takes the complex through the transition state. For instance, the centrifugal effect of rotations might also be an important contribution to the break-up of the complex, and in some cases the complex might be rotating too slowly or rotating rapidly but about the wrong axis. Therefore, it is more appropriate to suppose that the rate of passage of the complex through the transition state is only proportional, not equal, to the vibrational frequency along the reaction coordinate, and to write

$$k^‡ = \kappa\nu^‡ \qquad (18C.5)$$

where κ (kappa) is the **transmission coefficient**. In the absence of information to the contrary, κ is assumed to be about 1.

Brief illustration 18C.1

Typical vibrations of small molecules occur at wavenumbers of the order of 10^3 cm^{-1} (C–H bends, for example, occur in the range 1340–1465 cm^{-1}) and therefore occur at frequencies of the order of 10^{13} Hz. Suppose that the loosely bound cluster vibrates at one or two orders of magnitude lower frequency, then $\nu^‡ \approx 10^{11}$–10^{12} Hz. These figures suggest that $k^‡ \approx 10^{11}$–10^{12} s^{-1}, with κ perhaps reducing that value further.

(c) The concentration of the activated complex

Topic 13F explains how to calculate equilibrium constants from structural data. Equation 13F.10b of that Topic (which expresses K in terms of the standard molar partition functions $q_J^\ominus$) can be used directly, which in this case gives

$$K^‡ = \frac{N_A q_{C^‡}^\ominus}{q_A^\ominus q_B^\ominus}e^{-\Delta E_0/RT} \qquad (18C.6)$$

with

$$\Delta E_0 = E_0(C^‡) - E_0(A) - E_0(B) \qquad (18C.7)$$

Note that the units of N_A and the $q_J^\ominus$ are mol^{-1}, so $K^‡$ is dimensionless (as is appropriate for an equilibrium constant).

The focus of the final step of this part of the calculation is the partition function of the activated complex. For the special vibration of the activated complex $C^‡$ that tips it through the transition state and has frequency $\nu^‡$ the partition function may be written from eqn 13B.15 of Topic 13B as

$$q = \frac{1}{1-e^{-h\nu^‡/kT}}$$

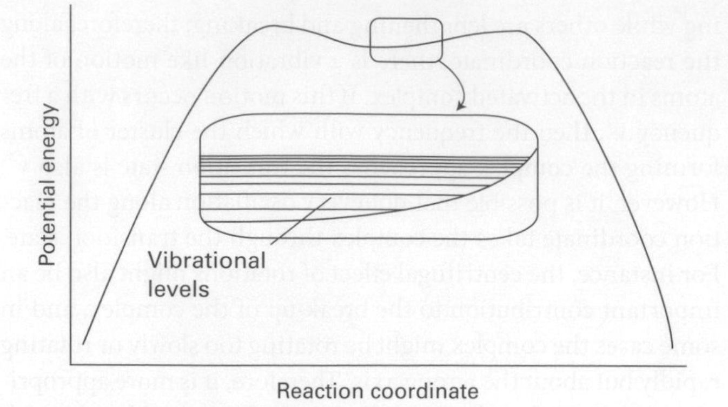

Figure 18C.3 In an elementary depiction of the activated complex close to the transition state, there is a broad, shallow dip in the potential energy surface along the reaction coordinate. The complex vibrates harmonically and almost classically in this well.

This frequency $\nu^\ddagger$ is much lower than for an ordinary molecular vibration because the oscillation corresponds to the complex falling apart (Fig. 18C.3), so the force constant is very low. Therefore, provided that $h\nu^\ddagger/kT \ll 1$, the exponential may be expanded and the partition function reduces to

$$q = \frac{1}{1-(1-h\nu^\ddagger/kT+\cdots)} \approx \frac{kT}{h\nu^\ddagger}$$

It follows that the partition function for the activated complex is

$$q_{C^\ddagger}^\ominus = \frac{kT}{h\nu^\ddagger}\bar{q}_{C^\ddagger}^\ominus \qquad (18C.8)$$

where the bar in $\bar{q}_{C^\ddagger}^\ominus$ denotes that the partition function is for all the other modes of the complex. The constant $K^\ddagger$ is therefore

$$K^\ddagger = \frac{kT}{h\nu^\ddagger}\bar{K}^\ddagger \quad \bar{K}^\ddagger = \frac{N_A \bar{q}_{C^\ddagger}^\ominus}{q_A^\ominus q_B^\ominus}e^{-\Delta E_0/RT} \qquad (18C.9)$$

with $\bar{K}^\ddagger$ a kind of equilibrium constant, but with one vibrational mode of $C^\ddagger$ discarded.

Brief illustration 18C.2

Consider the case of two structureless particles A and B colliding to give an activated complex that resembles a diatomic molecule. The activated complex is a diatomic cluster. It has one vibrational mode, but that mode corresponds to motion along the reaction coordinate and therefore does not appear in $\bar{q}_{C^\ddagger}^\ominus$. It follows that the standard molar partition function of the activated complex has only rotational and translational contributions.

(d) The rate constant

All the parts of the calculation can now be combined into

$$k_r = \frac{RT}{p^\ominus}k^\ddagger K^\ddagger = \kappa\nu^\ddagger\frac{kT}{h\nu^\ddagger}\frac{RT}{p^\ominus}\bar{K}^\ddagger$$

At this stage the unknown frequency $\nu^\ddagger$ (in blue) cancels to give one version of the **Eyring equation**:

$$k_r = \kappa\frac{kT}{h}\frac{RT}{p^\ominus}\bar{K}^\ddagger \qquad \text{Eyring equation} \qquad (18C.10)$$

The equilibrium constant $\bar{K}^\ddagger$, which here is expressed in terms of partial pressures, can be rewritten in terms of concentrations by using $[J] = p_J/RT$, and then with some rearrangement the equilibrium constant in terms of concentrations, $\bar{K}_c^\ddagger$ (the terms in blue), can be identified:

$$\bar{K}^\ddagger = \frac{p_{C^\ddagger}p^\ominus}{p_A p_B} = \frac{[C^\ddagger]}{[A][B]}\frac{p^\ominus}{RT} = \overbrace{\frac{[C^\ddagger]c^\ominus}{[A][B]}}^{\bar{K}_c^\ddagger}\times\frac{1}{c^\ominus}\times\frac{p^\ominus}{RT} = \bar{K}_c^\ddagger\frac{p^\ominus}{RTc^\ominus} \qquad (18C.11)$$

Substitution of this relation into eqn 18C.10, gives an alternative version of the Eyring equation:

$$k_r = \kappa\frac{kT}{hc^\ominus}\bar{K}_c^\ddagger \qquad \text{Eyring equation [alternative version]} \qquad (18C.12)$$

The equilibrium constant $\bar{K}^\ddagger$ can be computed from the partition functions of A, B, and $C^\ddagger$, so in principle the Eyring equation is an explicit expression for calculating the second-order rate constant for a bimolecular reaction in terms of the molecular parameters for the reactants and the activated complex, and the quantity κ.

The partition functions for the reactants can normally be calculated quite readily by using either spectroscopic information about their energy levels or the approximate expressions set out in the *Checklist* at the end of Topic 13B. The difficulty with the Eyring equation, however, lies in the calculation of the partition function of the activated complex: $C^\ddagger$ is difficult to investigate spectroscopically, and in general it is necessary to make assumptions about its size, shape, and structure.

Example 18C.1 Analysing the collision of structureless particles

Consider the case of two structureless (and different) particles A and B colliding to give an activated complex that resembles a diatomic molecule. Deduce an expression for the rate constant of the reaction $A + B \rightarrow P$.

Collect your thoughts Because the reactants are structureless 'atoms', the only contribution to their partition functions is from translation. The activated complex is a diatomic cluster of mass $m_{C^\ddagger} = m_A + m_B$ and moment of inertia I. It has one vibrational mode but, as explained in *Brief illustration* 18C.2, that mode corresponds to motion along the reaction coordinate. It follows that the standard molar partition function of the activated complex has only rotational and translational contributions. Expressions for the relevant partition functions are given at the end of Topic 13B.

The solution The translational partition functions are

$$q_J^\ominus = \frac{V_m^\ominus}{\Lambda_J^3} \qquad \Lambda_J = \frac{h}{(2\pi m_J kT)^{1/2}} \qquad V_m^\ominus = \frac{RT}{p^\ominus}$$

with J = A, B, and C‡, and with $m_{C^\ddagger} = m_A + m_B$. The expression for the partition function of the activated complex is

$$\bar{q}_{C^\ddagger}^\ominus = \overbrace{\frac{2IkT}{\hbar^2}}^{\text{rotation}} \times \overbrace{\frac{V_m^\ominus}{\Lambda_{C^\ddagger}^3}}^{\text{translation}}$$

where the high-temperature form of the rotational partition function has been used (Topic 13B). From eqn 18C.9, the constant $\bar{K}^\ddagger$ is expressed in terms of the partition functions as

$$\bar{K}^\ddagger = \frac{N_A \overbrace{(2IkT/\hbar^2)\, V_m^\ominus/\Lambda_{C^\ddagger}^3}^{\bar{q}_{C^\ddagger}^\ominus}}{\underbrace{(V_m^\ominus/\Lambda_A^3)}_{q_A^\ominus}\ \underbrace{(V_m^\ominus/\Lambda_B^3)}_{q_B^\ominus}} e^{-\Delta E_0/RT} = \left(\frac{N_A \Lambda_A^3 \Lambda_B^3}{\Lambda_{C^\ddagger}^3 V_m^\ominus}\right)\frac{2IkT}{\hbar^2} e^{-\Delta E_0/RT}$$

It follows from eqn 18C.10 that

$$k_r = \kappa \frac{kT}{h}\frac{RT}{p^\ominus} \overbrace{\left(\frac{N_A \Lambda_A^3 \Lambda_B^3}{\Lambda_{C^\ddagger}^3 V_m^\ominus}\right)\frac{2IkT}{\hbar^2} e^{-\Delta E_0/RT}}^{\bar{K}^\ddagger}$$

$$= \kappa \frac{kT}{h} N_A \left(\frac{\Lambda_A \Lambda_B}{\Lambda_{C^\ddagger}}\right)^3 \frac{2IkT}{\hbar^2} e^{-\Delta E_0/RT}$$

The moment of inertia of a diatomic molecule of bond length r is μr^2, where $\mu = m_A m_B/(m_A + m_B)$, so after introducing the expressions for the thermal wavelengths Λ and cancelling common terms, the result is

$$k_r = \kappa N_A \left(\frac{8kT}{\pi\mu}\right)^{1/2} \pi r^2 e^{-\Delta E_0/RT}$$

Finally, by identifying $\kappa\pi r^2$ as the reactive cross-section σ^*, the resulting expression is the same as that obtained from simple collision theory (eqn 18A.9):

$$k_r = N_A \left(\frac{8kT}{\pi\mu}\right)^{1/2} \sigma^* e^{-\Delta E_0/RT}$$

Self-test 18C.1 What additional contributions would there be to the partition functions of the reactants and of the activated complex if the reaction were AB + C → P, with a linear activated complex?

Answer: Rotation and vibration of AB; bends and symmetric stretch of the activated complex.

18C.2 Thermodynamic aspects

The statistical thermodynamic version of transition-state theory rapidly runs into difficulties because only in some cases is anything known about the structure of the activated complex. However, the concepts it introduces, principally that of an equilibrium between the reactants and the activated complex, have motivated a more general, empirical approach in which the activation process is expressed in terms of thermodynamic functions.

(a) Activation parameters

If $\bar{K}^\ddagger$ is taken as an equilibrium constant (despite one mode of C‡ having been discarded), then it can be expressed in terms of a **Gibbs energy of activation**, $\Delta^\ddagger G$, through the definition

$$\Delta^\ddagger G = -RT \ln \bar{K}^\ddagger \qquad \text{Gibbs energy of activation [definition]} \qquad (18C.13)$$

All the $\Delta^\ddagger X$ in this section are *standard* thermodynamic quantities, $\Delta^\ddagger X^\ominus$, but the standard state sign will be omitted to avoid overburdening the notation. Then from eqn 18C.10 the expression for the rate constant becomes

$$k_r = \kappa \frac{kT}{h}\frac{RT}{p^\ominus} e^{-\Delta^\ddagger G/RT} \qquad (18C.14)$$

Because $\Delta G = \Delta H - T\Delta S$, the Gibbs energy of activation can be divided into an **entropy of activation**, $\Delta^\ddagger S$, and an **enthalpy of activation**, $\Delta^\ddagger H$, by writing

$$\Delta^\ddagger G = \Delta^\ddagger H - T\Delta^\ddagger S \qquad \text{Entropy and enthalpy of activation [definition]} \qquad (18C.15)$$

When eqn 18C.15 is used in eqn 18C.14 and κ is absorbed into the entropy term, the result is

$$k_r = B e^{\Delta^\ddagger S/R} e^{-\Delta^\ddagger H/RT} \qquad B = \frac{kT}{h}\frac{RT}{p^\ominus} \qquad (18C.16)$$

To develop this expression further it is necessary to find a relation between the enthalpy of activation and the activation energy. The two are not the same, for two main reasons. One is that although it might be tempting to identify E_a with $\Delta^\ddagger U$, that is valid only at $T = 0$; at higher temperatures upper levels of all the species are occupied and contribute additional terms of the order of RT (a value suggested by the equipartition principle). Secondly, for gas-phase processes (but not for those in solution), $\Delta^{‡}H$ differs from $\Delta^{‡}U$ by another contribution RT. These additional contributions need to be identified.

How is that done? 18C.1 Relating the enthalpy of activation to the activation energy

The relation between the enthalpy of activation and the activation energy depends on two equations. One is the expression for the temperature dependence of the 'equilibrium constant' $K^{‡}$, which is eqn 6B.2 of Topic 6B in the form $\mathrm{d}\ln K^{‡}/\mathrm{d}T = \Delta^{‡}H/RT^2$, and the second is the definition of the activation energy, which is eqn 17D.3 of Topic 17D in the form $\mathrm{d}\ln k_r/\mathrm{d}T = E_a/RT^2$. The link between the two expressions is the alternative version of the Eyring equation, eqn 18C.12, $k_r = (\kappa kT/hc^{\ominus})\bar{K}_c^{‡}$. Differentiation of the corresponding expression for ln k_r with respect to T gives

$$\frac{\mathrm{d}\ln k_r}{\mathrm{d}T} = \frac{1}{T} + \frac{\mathrm{d}\ln \bar{K}_c^{‡}}{\mathrm{d}T}$$

Then, by using $\mathrm{d}\ln k_r/\mathrm{d}T = E_a/RT^2$, it follows that

$$E_a = RT + RT^2 \frac{\mathrm{d}\ln \bar{K}_c^{‡}}{\mathrm{d}T}$$

At this point it is necessary to distinguish between gas-phase and solution-phase reactions of the form $A + B \rightleftharpoons C^{‡}$. For the latter, the equilibrium constant is expressed in terms of concentrations and the second term in the preceding equation can be identified with $\Delta^{‡}H$ without further calculation. It follows that

For a solution-phase reaction: $E_a = RT + \Delta^{‡}H$

One further step is needed for a gas-phase reaction because $\bar{K}_c^{‡}$ is not the same as $\bar{K}^{‡}$(which is expressed in terms of partial pressures). According to eqn 18C.11 the two are related by $\bar{K}^{‡} = (p^{\ominus}/RTc^{\ominus})\bar{K}_c^{‡}$, which implies that

$$\frac{\mathrm{d}\ln \bar{K}_c^{‡}}{\mathrm{d}T} = \overbrace{\frac{\mathrm{d}}{\mathrm{d}T}\ln\frac{RTc^{\ominus}}{p^{\ominus}}}^{1/T} + \overbrace{\frac{\mathrm{d}\ln \bar{K}^{‡}}{\mathrm{d}T}}^{\Delta^{‡}H/RT^2} = \frac{1}{T} + \frac{\Delta^{‡}H}{RT^2}$$

Substitution of this expression into the previous expression for E_a in terms of $\mathrm{d}\ln \bar{K}_c^{‡}/\mathrm{d}T$ gives

$$E_a = RT + RT^2 \frac{\mathrm{d}\ln \bar{K}_c^{‡}}{\mathrm{d}T} = RT + RT^2\left(\frac{1}{T} + \frac{\Delta^{‡}H}{RT^2}\right) = RT + RT + \Delta^{‡}H$$

It therefore follows that

For a gas-phase reaction: $E_a = 2RT + \Delta^{‡}H$

In summary:

$$\text{(a) } \Delta^{‡}H = E_a - 2RT \quad \text{(bimolecular gas-phase reaction)}$$
$$\text{(b) } \Delta^{‡}H = E_a - RT \quad \text{(bimolecular reaction in solution)} \qquad (18C.17)$$

Relations between $\Delta^{‡}H$ and E_a

It now follows that

$$k_r = e^2 B e^{\Delta^{‡}S/R} e^{-E_a/RT} \qquad (18C.18a)$$

Rate constant [transition-state theory, bimolecular gas-phase reaction]

and

$$k_r = eBe^{\Delta^{‡}S/R} e^{-E_a/RT} \qquad (18C.18b)$$

Rate constant [transition-state theory, bimolecular reaction in solution]

where, from eqn 18C.16, $B = (kT/h)(RT/p^{\ominus})$. The Arrhenius frequency factors can be identified as

$$A = e^2 B e^{\Delta^{‡}S/R} \qquad (18C.19a)$$

Frequency factor [transition-state theory, bimolecular gas-phase reaction]

and

$$A = eBe^{\Delta^{‡}S/R} \qquad (18C.19b)$$

Frequency factor [transition-state theory, bimolecular reaction in solution]

The entropy of activation is negative because throughout the system reactant species are combining to form reactive pairs. However, if there is a reduction in entropy below what would be expected for the simple encounter of A and B, then the frequency factor A will be reduced further. Indeed, that *additional* reduction in entropy, $\Delta^{‡}S_{\text{steric}}$, can be identified as the origin of the steric factor P of collision theory (Topic 18A), so that

$$P = e^{\Delta^{‡}S_{\text{steric}}/R} \qquad (18C.20)$$

P-factor [transition-state theory]

Thus, the more complex the steric requirements of the encounter, the more negative the value of $\Delta^{‡}S_{\text{steric}}$, and the smaller the value of P.

Brief illustration 18C.3

The reaction of propylxanthate ion in ethanoic acid buffer solutions can be represented by the equation $A^- + H^+ \rightarrow P$. Near 30 °C, $A = 2.05 \times 10^{13}\,\mathrm{dm^3\,mol^{-1}\,s^{-1}}$. To evaluate the entropy of activation at 30 °C, use eqn 18C.19b, rearranged as

$$\Delta^{‡}S = R\ln\frac{A}{eB} \quad \text{with } B = \frac{kT}{h}\frac{RT}{p^{\ominus}} = 1.592\times10^{14}\,\mathrm{dm^3\,mol^{-1}\,s^{-1}}$$

Therefore,

$$\Delta^{‡}S = R\ln\frac{2.05\times10^{13}\,\mathrm{dm^3\,mol^{-1}\,s^{-1}}}{e\times1.592\times10^{14}\,\mathrm{dm^3\,mol^{-1}\,s^{-1}}} = R\ln 0.0473\ldots$$
$$= -25.4\,\mathrm{J\,K^{-1}\,mol^{-1}}$$

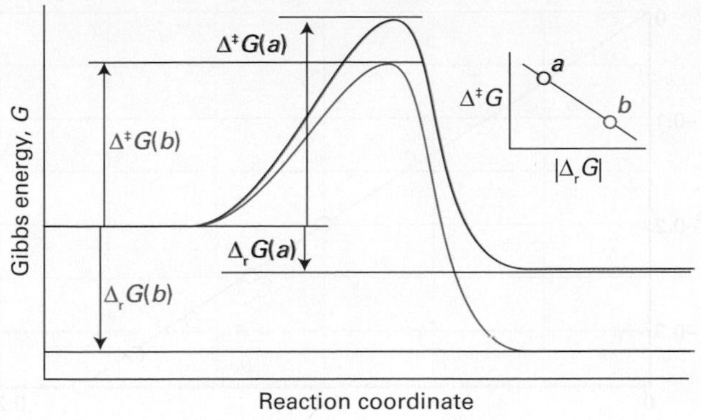

Figure 18C.4 For a related series of reactions here denoted *a* and *b*, as the standard reaction Gibbs energy becomes more negative on going from *a* to *b*, the activation Gibbs energy decreases and the rate constant increases. The approximate linear correlation between $\Delta^{\ddagger}G$ and $\Delta_r G^{\ominus}$ is the origin of 'linear free energy relations'.

Gibbs energies, enthalpies, and entropies of activation (and volumes and heat capacities of activation) are widely used to report experimental reaction rates, especially for organic reactions in solution. They are encountered when relationships between equilibrium constants and rates of reaction are explored by using **correlation analysis**, in which ln *K* (which is equal to $-\Delta_r G^{\ominus}/RT$) is plotted against ln k_r (which is proportional to $-\Delta^{\ddagger}G/RT$). In many cases the correlation is linear, signifying that as the reaction becomes thermodynamically more favourable, its rate constant increases (Fig. 18C.4). This linear correlation is the origin of the alternative name **linear free energy relation** (LFER).

(b) Reactions between ions

The full statistical thermodynamic theory is very complicated for reactions involving ions in solution because the solvent plays a role in the activated complex. The thermodynamic version of transition-state theory simplifies the discussion and is applicable to non-ideal systems. In the thermodynamic approach, the rate law

$$\frac{d[P]}{dt} = k^{\ddagger}[C^{\ddagger}]$$

is combined with the thermodynamic equilibrium constant (Topic 6A)

$$K = \frac{a_{C^{\ddagger}}}{a_A a_B} \overset{a_J = \gamma_J[J]/c^{\ominus}}{=} K_\gamma \frac{[C^{\ddagger}]c^{\ominus}}{[A][B]} \qquad K_\gamma = \frac{\gamma_{C^{\ddagger}}}{\gamma_A \gamma_B}$$

Then

$$\frac{d[P]}{dt} = k_r[A][B] \qquad k_r = \frac{k^{\ddagger}K}{K_\gamma c^{\ominus}} \qquad \text{(18C.21a)}$$

If k_r° is the rate constant when the activity coefficients are 1 (i.e. $k_r^{\circ} = k^{\ddagger}K/c^{\ominus}$), then

$$k_r = \frac{k_r^{\circ}}{K_\gamma} \qquad \log k_r = \log k_r^{\circ} - \log K_\gamma \qquad \text{(18C.21b)}$$

At low concentrations the activity coefficients can be expressed in terms of the ionic strength, *I*, of the solution by using the Debye–Hückel limiting law (Topic 5F, particularly eqn 5F.27, $\log \gamma_{\pm} = -\mathcal{A}|z_+z_-|I^{1/2}$). However, the expressions needed are those for the individual ions rather than the mean value, and so it is more appropriate to write $\log \gamma_J = -\mathcal{A}z_J^2 I^{1/2}$ and

$$\log \gamma_A = -\mathcal{A}z_A^2 I^{1/2} \qquad \log \gamma_B = -\mathcal{A}z_B^2 I^{1/2} \qquad \text{(18C.22a)}$$

with $\mathcal{A} = 0.509$ in aqueous solution at 298 K and z_A and z_B the (signed) charge numbers of A and B, respectively. Because the activated complex forms from reaction of one of the ions of A with one of the ions of B, the charge number of the activated complex is $z_A + z_B$ where z_J is positive for cations and negative for anions. Therefore

$$\log \gamma_{C^{\ddagger}} = -\mathcal{A}(z_A + z_B)^2 I^{1/2} \qquad \text{(18C.22b)}$$

When these expressions are inserted into eqn 18C.21b the result is

$$\begin{aligned} \log k_r &= \log k_r^{\circ} - \mathcal{A}\{z_A^2 + z_B^2 - (z_A + z_B)^2\}I^{1/2} \\ &= \log k_r^{\circ} + 2\mathcal{A}z_A z_B I^{1/2} \end{aligned} \qquad \text{(18C.23)}$$

Equation 18C.23 expresses the **kinetic salt effect**, the variation of the rate constant of a reaction between ions with the ionic strength of the solution (Fig. 18C.5). The equation is interpreted as follows:

Physical interpretation

- If the reactant ions have the same sign (as in a reaction between cations or between anions), then increasing the ionic strength by the addition of inert ions increases the rate constant.
- The formation of a single, highly charged ionic complex from two less highly charged ions is favoured by a high ionic strength because the new ion has a denser ionic atmosphere and interacts with that atmosphere more strongly.
- Conversely, ions of opposite charge react more slowly in solutions of high ionic strength. Now the charges cancel and the complex has a less favourable interaction with its atmosphere than the separated ions.

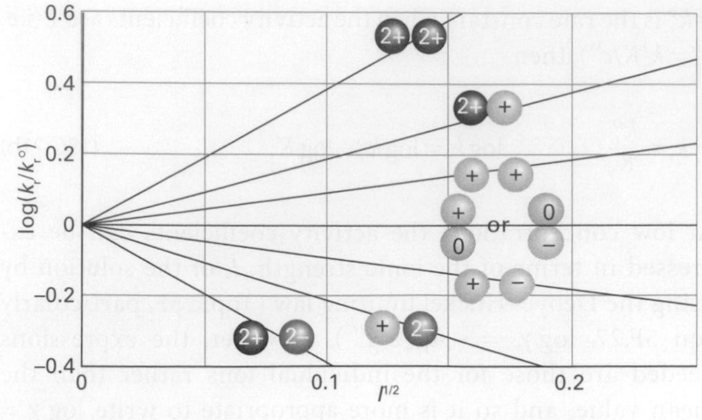

Figure 18C.5 Experimental tests of the kinetic salt effect for reactions in water at 298 K. The ion types are shown as spheres, and the slopes of the lines are those given by the Debye–Hückel limiting law and eqn 18C.23.

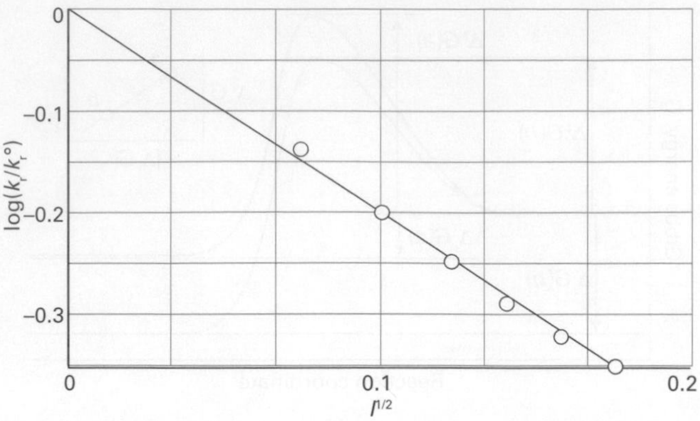

Figure 18C.6 The experimental ionic strength dependence of the rate constant of a hydrolysis reaction: the slope gives information about the charge types involved in the activated complex of the rate-determining step. The data plotted are from *Example* 18C.2.

Example 18C.2 Analysing data in terms of the kinetic salt effect

The rate constant (at 298 K) for the hydrolysis of $[CoBr(NH_3)_5]^{2+}$ under basic conditions in aqueous solution varies with ionic strength as in the following table. What can be deduced about the charge of the activated complex in the rate-determining step? What might this deduction imply about the mechanism?

I	0.0050	0.0100	0.0150	0.0200	0.0250	0.0300
k_r/k_r°	0.718	0.631	0.562	0.515	0.475	0.447

Collect your thoughts According to eqn 18C.23, a plot of $\log(k_r/k_r^{\circ})$ against $I^{1/2}$ should have a slope of $2\mathcal{A}z_Az_B$. Because $\mathcal{A} = 0.509$ for aqueous solutions at 298 K, the slope will be $1.02z_Az_B$. From the slope you can infer the charges of the ions involved in the formation of the activated complex.

The solution Form the following table:

I	0.0050	0.0100	0.0150	0.0200	0.0250	0.0300
$I^{1/2}$	0.071	0.100	0.122	0.141	0.158	0.173
$\log(k_r/k_r^{\circ})$	−0.14	−0.20	−0.25	−0.29	−0.32	−0.35

These values are plotted in Fig. 18C.6. The slope of the (least squares) straight line is −2.04, indicating that $z_Az_B = -2$. A possible explanation for this conclusion is that the two species involved in the formation of the activated complex are $[CoBr(NH_3)_5]^{2+}$, which has $z = +2$, and OH^-, which has $z = -1$. The product of the charges is therefore −2.

Comment. Although the point is not pursued here, you should be aware that the rate constant is also influenced by the relative permittivity of the medium.

Self-test 18C.2 An ion of charge number +1 is known to be involved in the activated complex of a reaction. Deduce the charge number of the other ion from the following data, recorded at 298 K in aqueous solution:

I	0.0050	0.0100	0.0150	0.0200	0.0250	0.0300
k_r/k_r°	0.847	0.791	0.750	0.717	0.690	0.666

Answer: −1

18C.3 The kinetic isotope effect

The postulation of a plausible reaction mechanism requires careful analysis of many experiments designed to determine the fate of atoms during the formation of products. Observation of the **kinetic isotope effect**, a decrease in the rate of a chemical reaction upon replacement of one atom in a reactant by a heavier isotope, facilitates the identification of bond-breaking events in the rate-determining step. A **primary kinetic isotope effect** is observed when the rate-determining step requires the scission of a bond involving the isotope. A **secondary kinetic isotope effect** is the reduction in reaction rate even though the bond involving the isotope is not broken to form product. In both cases, the effect arises from the change in activation energy that accompanies the replacement of an atom by a heavier isotope on account of changes in the zero-point vibrational energies. What follows is a description of the primary kinetic isotope effect.

Consider a reaction in which a C–H bond is cleaved. If scission of this bond is the rate-determining step (Topic 17E), then the reaction coordinate corresponds to the stretching of the C–H bond and the potential energy profile is shown in Fig. 18C.7. On deuteration, the dominant change is the reduction of the zero-point energy of the bond (because the deuterium atom is heavier). The whole reaction profile is not lowered, however, because the relevant vibration in the activated complex has a very low force constant, so there is little zero-point energy associated with the reaction coordinate in

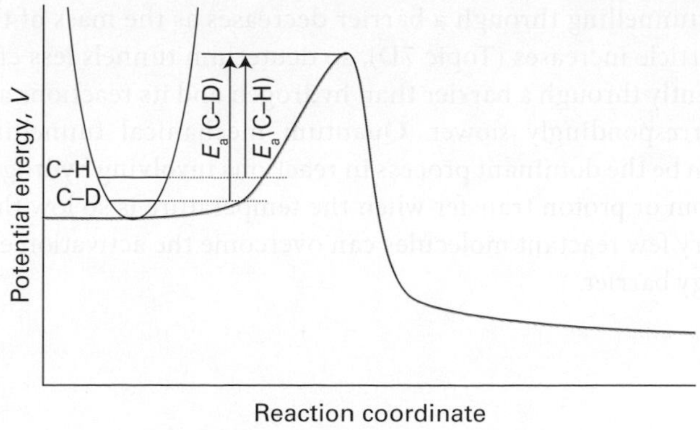

Figure 18C.7 Changes in the reaction profile when a C–H bond undergoing cleavage is deuterated. In this illustration the C–H and C–D bonds are modelled as harmonic oscillators. The only significant change is in the zero-point energy of the reactants, which is lower for C–D than for C–H. As a result, the activation energy is greater for C–D cleavage than for C–H cleavage.

either form of the activated complex. From these considerations it is possible to investigate the effect of deuteration on the activation energy.

How is that done? 18C.2 Exploring the primary kinetic isotope effect

Consider the cleavage of a C–H bond in a larger molecule. For such a reaction it is reasonable to assume that motion along the reaction coordinate is dominated by stretching and compression of the C–H fragment. Therefore, to a good approximation, a change in the activation energy arises only from the change in zero-point energy of the stretching vibration, $\frac{1}{2}\hbar\omega$. It follows from Fig. 18C.7 that

$$E_a(\text{C–D})-E_a(\text{C–H})=N_A\{\tfrac{1}{2}\hbar\omega(\text{C–H})-\tfrac{1}{2}\hbar\omega(\text{C–D})\}$$
$$=\tfrac{1}{2}N_A\hbar\{\omega(\text{C–H})-\omega(\text{C–D})\}$$

Then from Topic 11C $\omega(\text{C–D}) = (\mu_{CH}/\mu_{CD})^{1/2}\omega(\text{C–H})$, where μ is the relevant effective mass and therefore that

$$E_a(\text{C–D})-E_a(\text{C–H})=\tfrac{1}{2}N_A\hbar\omega(\text{C–H})\left\{1-\left(\frac{\mu_{CH}}{\mu_{CD}}\right)^{1/2}\right\}$$

Effect of deuteration on the activation energy (18C.24)

If the Arrhenius frequency factor does not change upon deuteration, the rate constants for the two species should be in the ratio

$$\frac{k_r(\text{C–D})}{k_r(\text{C–H})}=e^{-\{E_a(\text{C–D})-E_a(\text{C–H})\}/RT}=e^{-\{E_a(\text{C–D})-E_a(\text{C–H})\}/N_A kT}$$

where $R=N_Ak$. After using eqn 18C.24 for E_a(C–D) − E_a(C–H) in this expression, the result is

$$\frac{k_r(\text{C–D})}{k_r(\text{C–H})}=e^{-\zeta} \quad \text{with} \quad \zeta=\frac{\hbar\omega(\text{C–H})}{2kT}\left\{1-\left(\frac{\mu_{CH}}{\mu_{CD}}\right)^{1/2}\right\}$$

Effect of deuteration on the rate constant (18C.25)

Note that $\zeta > 0$ (ζ is zeta) because $\mu_{CD} > \mu_{CH}$ and so it follows that $k_r(\text{C–D})/k_r(\text{C–H}) < 1$. As expected from Fig. 18C.7, the rate constant decreases upon deuteration.

Brief illustration 18C.4

From infrared spectra, the fundamental vibrational wavenumber $\tilde{\nu}$ for stretching of a C–H bond is about 3000 cm^{-1}. To convert this wavenumber to an angular frequency, $\omega = 2\pi\nu$, use $\omega = 2\pi c\tilde{\nu}$, so that

$$\omega = 2\pi \times (2.998\times10^{10}\,\text{cm s}^{-1})\times(3000\,\text{cm}^{-1})$$
$$= 5.65\ldots\times10^{14}\,\text{s}^{-1}$$

The ratio of effective masses is

$$\frac{\mu_{CH}}{\mu_{CD}}=\left(\frac{m_C m_H}{m_C+m_H}\right)\times\left(\frac{m_C+m_D}{m_C m_D}\right)$$
$$=\left(\frac{12.01\times1.0078}{12.01+1.0078}\right)\times\left(\frac{12.01+2.0140}{12.01\times2.0140}\right)$$
$$=0.539\ldots$$

Now use eqn 18C.25 to calculate

$$\zeta=\frac{(1.055\times10^{-34}\,\text{J s})\times(5.65\ldots\times10^{14}\,\text{s}^{-1})}{2\times(1.381\times10^{-23}\,\text{J K}^{-1})\times(298\,\text{K})}\times(1-0.539\ldots^{1/2})$$
$$=1.92\ldots$$

and

$$\frac{k_r(\text{C–D})}{k_r(\text{C–H})}=e^{-1.92\ldots}=0.146$$

Therefore at room temperature cleavage of the C–H bond should be about seven times faster than cleavage of the C–D bond, other conditions being equal. Experimental values of $k_r(\text{C–D})/k_r(\text{C–H})$ can differ significantly from those predicted by eqn 18C.25 on account of the severity of the assumptions in the model.

In some cases, substitution of deuterium for hydrogen results in values of $k_r(\text{C–D})/k_r(\text{C–H})$ that are too low to be accounted for by eqn 18C.25, even when more complete models are used to predict ratios of rate constants. Such abnormal kinetic isotope effects are evidence for a path in which quantum mechanical tunnelling of hydrogen atoms takes place through the activation barrier (Fig. 18C.8). The probability

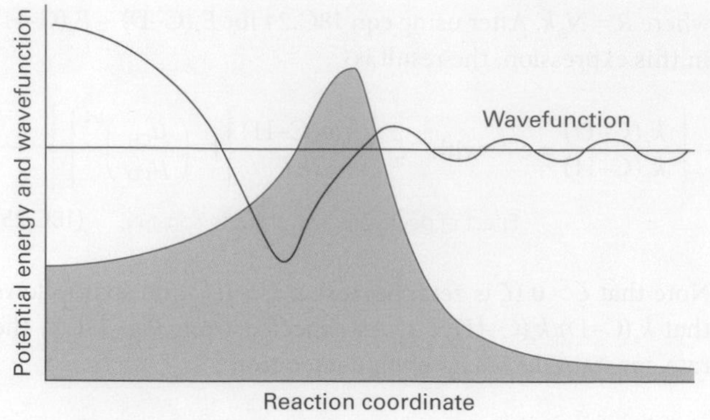

Figure 18C.8 A proton can tunnel through the activation energy barrier that separates reactants from products, so the effective height of the barrier is reduced and the rate of the proton transfer reaction increases. The effect is represented by drawing the wavefunction of the proton near the barrier. Proton tunnelling is important only at low temperatures, when most of the reactants are trapped on the left of the barrier.

of tunnelling through a barrier decreases as the mass of the particle increases (Topic 7D), so deuterium tunnels less efficiently through a barrier than hydrogen and its reactions are correspondingly slower. Quantum mechanical tunnelling can be the dominant process in reactions involving hydrogen atom or proton transfer when the temperature is so low that very few reactant molecules can overcome the activation energy barrier.

Checklist of concepts

☐ 1. In transition-state theory, it is supposed that an **activated complex** is in equilibrium with the reactants.

☐ 2. The rate at which the activated complex forms products depends on the rate at which it passes through a **transition state**.

☐ 3. The rate constant may be parametrized in terms of the **Gibbs energy, entropy,** and **enthalpy of activation**.

☐ 4. The **kinetic salt effect** is the effect of an added inert salt on the rate constant of a reaction between ions.

☐ 5. The **kinetic isotope effect** is the decrease in the rate constant of a chemical reaction upon replacement of one atom in a reactant by a heavier isotope.

Checklist of equations

Property	Equation	Comment	Equation number
'Equilibrium constant' for activated complex formation	$\bar{K}^{\ddagger} = (N_A \bar{q}^{\ominus}_{C^{\ddagger}} / q^{\ominus}_A q^{\ominus}_B) e^{-\Delta E_0/RT}$	Assume equilibrium; one vibrational mode of $C^{\ddagger}$ discarded	18C.9
Eyring equation	$k_r = \kappa(kT/h)(RT/p^{\ominus})\bar{K}^{\ddagger}$	Transition-state theory	18C.10
	$k_r = \kappa(kT/hc^{\ominus})\bar{K}^{\ddagger}_c$		18C.12
Gibbs energy of activation	$\Delta^{\ddagger}G = -RT \ln \bar{K}^{\ddagger}$	Definition	18C.13
Enthalpy and entropy of activation	$\Delta^{\ddagger}G = \Delta^{\ddagger}H - T\Delta^{\ddagger}S$	Definition	18C.15
Parametrization	$k_r = e^n B e^{\Delta^{\ddagger}S/R} e^{-E_a/RT}$	$n = 2$ for bimolecular gas-phase reactions; $n = 1$ for solution	18C.18
A-factor	$A = e^n B e^{\Delta^{\ddagger}S/R}$		18C.19
P-factor	$P = e^{\Delta^{\ddagger}S_{\text{steric}}/R}$		18C.20
Kinetic salt effect	$\log k_r = \log k_r^{\circ} + 2\mathcal{A} z_A z_B I^{1/2}$	Assumes Debye–Hückel limiting law valid	18C.23
Primary kinetic isotope effect	$k_r(\text{C–D})/k_r(\text{C–H}) = e^{-\zeta}$ $\zeta = (\hbar\omega(\text{C–H})/2kT) \times \{1-(\mu_{CH}/\mu_{CD})^{1/2}\}$	Cleavage of a C–H/D bond in the rate-determining step	18C.25

TOPIC 18D The dynamics of molecular collisions

> ➤ Why do you need to know this material?
>
> Chemists are interested in the details of chemical reactions, and there is no more detailed approach than that involved in the study of the dynamics of reactive encounters, when one molecule collides with another and atoms exchange partners.
>
> ➤ What is the key idea?
>
> The rates of reactions in the gas phase can be investigated by exploring the trajectories of molecules on potential energy surfaces.
>
> ➤ What do you need to know already?
>
> This Topic builds on the concept of rate constant (Topic 17A) and in one part of the discussion uses the concept of partition function (Topic 13B). The discussion of potential energy surfaces is qualitative, but the underlying calculations are those of self-consistent field theory (Topic 9E).

The investigation of the dynamics of the collisions between reactant molecules is the most detailed level of the examination of the factors that govern the rates of reactions. There are two approaches: an experimental one that uses molecular beams and a theoretical one that uses the results of computations.

18D.1 Molecular beams

Molecular beams, which consist of collimated, narrow streams of molecules travelling through an evacuated vessel, allow collisions between molecules in preselected states (e.g. specific rotational and vibrational states) to be studied, and can be used to identify the states of the products of a reactive collision. Information of this kind is essential if a full picture of the reaction is to be built, because the rate constant is an average over events in which reactants in different initial states evolve into products in their final states.

(a) Techniques

The basic arrangement of a molecular beam experiment is shown in Fig. 18D.1. Atoms or molecules emerge from a source chamber (which may be heated if the species is not already a gas) through a pinhole and out into a vacuum chamber. If the pressure of vapour in the source is increased so that the mean free path of the molecules in the emerging beam is much shorter than the diameter of the pinhole, many collisions take place even outside the source. The net effect of these collisions, which give rise to **hydrodynamic flow**, is to transfer momentum into the direction of the beam. The molecules in the beam then travel with very similar speeds, so further downstream few collisions take place between them. This condition is called **molecular flow**. If necessary, atoms or molecules moving with a given speed can be selected by using a velocity selector, as depicted in Fig. 18D.1.

In a molecular beam of this kind the spread of speeds is much smaller than that predicted by the Maxwell–Boltzmann distribution. The unexpected narrowness of the distribution is interpreted by assigning a low translational temperature to

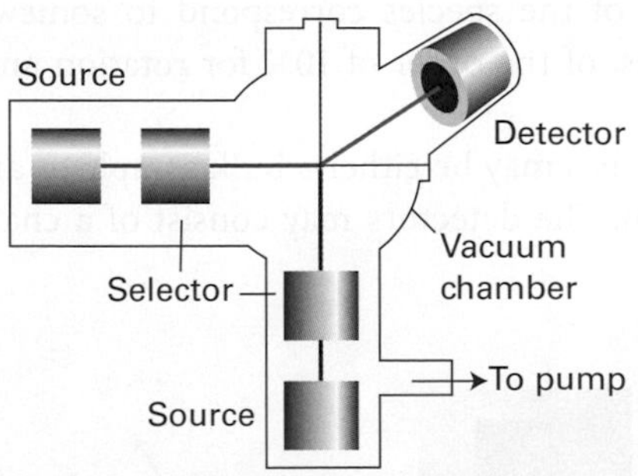

Figure 18D.1 The basic arrangement of a molecular beam apparatus. The atoms or molecules emerge from a source, and pass through a velocity selector, such as that discussed in Topic 1B. The scattering occurs from the target gas (which might take the form of another beam), and the flux of particles entering the detector set at some angle is recorded. In a crossed-beam experiment, state-selected molecules are generated in two separate sources, and are directed perpendicular to one another. The detector responds to molecules (which may be product molecules if a chemical reaction occurs) scattered into a chosen direction.

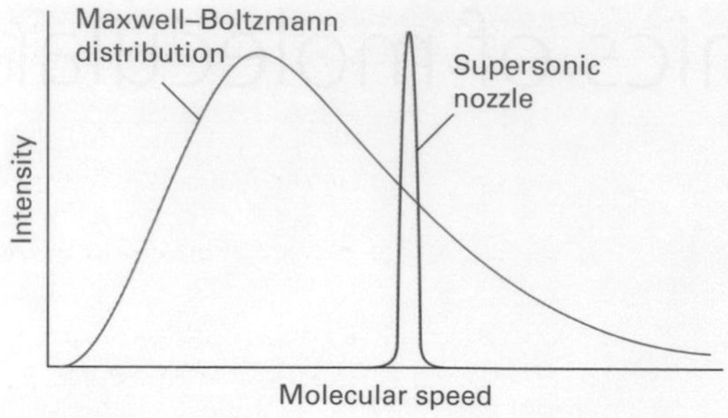

Figure 18D.2 The shift in the mean speed and the width of the distribution brought about by use of a supersonic nozzle.

the molecules in the beam (Fig. 18D.2), which may be as low as 1 K. Such jets are called **supersonic** because the mean speed of the molecules in the jet is much greater than the speed of sound in the jet.

A supersonic jet can be converted into a more parallel **supersonic beam** if it is 'skimmed' in the region of hydrodynamic flow and the excess gas pumped away. A skimmer consists of a conical nozzle shaped to avoid any supersonic shock waves spreading back into the gas and so increasing the translational temperature (Fig. 18D.3). A jet or beam may also be formed by using helium or neon as the principal gas, and injecting molecules of interest into it in the hydrodynamic region of flow.

As well as having a low translational temperature, the molecules in the beam also have low rotational and vibrational temperatures. In this context, a rotational or vibrational temperature means the temperature that should be used in the Boltzmann distribution to reproduce the observed populations of the states. However, as rotational states equilibrate more slowly than translational states, and vibrational states equilibrate even more slowly, the rotational and vibrational populations of the species correspond to somewhat higher temperatures, of the order of 10 K for rotation and 100 K for vibrations.

The target gas may be either a bulk sample or another molecular beam. The detectors may consist of a chamber fitted with a sensitive pressure gauge, a 'bolometer' (a detector that responds to the incident energy by making use of the temperature-dependence of resistance), or an ionization detector, in which the incoming molecule is first ionized and then detected electronically. The rotational and vibrational state of the scattered molecules may also be determined spectroscopically.

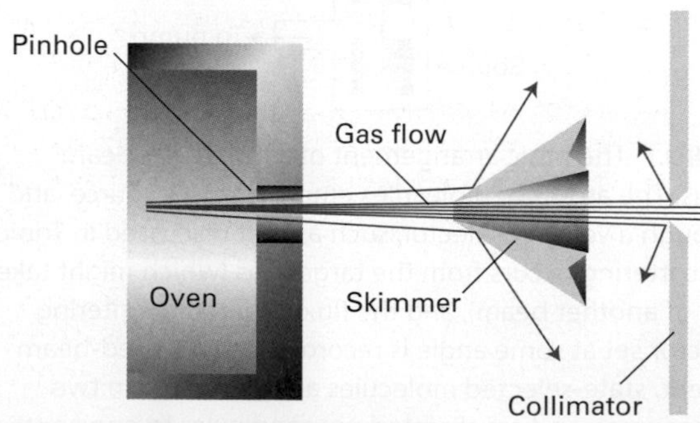

Figure 18D.3 A supersonic beam is generated by using a skimmer to remove some of the molecules from the beam, so leading to a greater degree of collimation.

(b) Experimental results

The primary experimental information from a molecular beam experiment is the fraction of the molecules in the incident beam that are scattered into a particular direction. The fraction is normally expressed in terms of dI, the number of molecules in a given time divided by the length of the interval, scattered into a cone (described by a solid angle $d\Omega$) that represents the area covered by the 'eye' of the detector (Fig. 18D.4). This rate is reported as the **differential scattering cross-section**, σ, the constant of proportionality between the value of dI and the intensity, I, of the incident beam, the number density of target molecules, $\mathcal{N}$, and the infinitesimal path length dx through the sample:

$$dI = \sigma I \mathcal{N} dx \quad \text{Differential scattering cross-section [definition]} \quad (18D.1)$$

The value of σ (which has the dimensions of area) depends on the **impact parameter**, b, the initial perpendicular separation of the paths of the colliding molecules (Fig. 18D.5), and the details of the intermolecular potential.

The role of the impact parameter is most easily seen by considering the impact of two hard spheres (Fig. 18D.6). If $b = 0$, the projectile is on a trajectory that leads to a head-on collision, so the only scattering intensity is detected when the detector is at $\theta = \pi$. When the impact parameter is so great that the spheres do not make contact ($b > R_A + R_B$), there is no

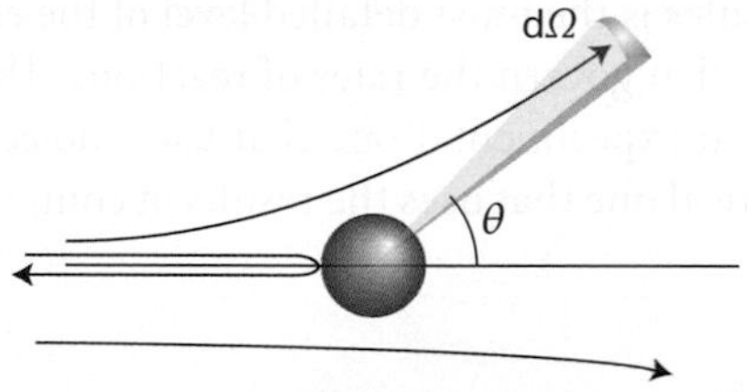

Figure 18D.4 The definition of the solid angle, $d\Omega$, for scattering.

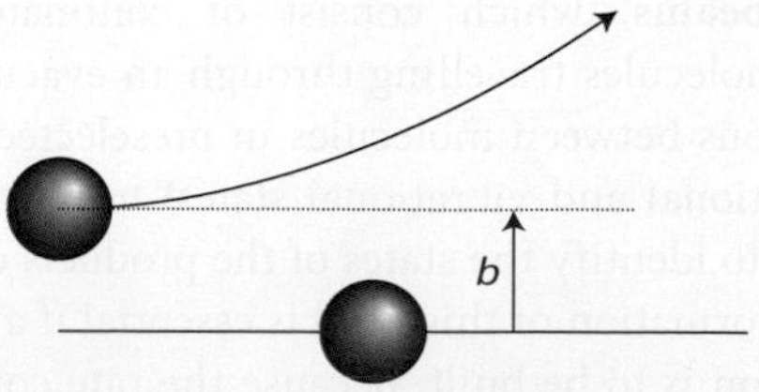

Figure 18D.5 The definition of the impact parameter, b, as the perpendicular separation of the initial paths of the particles.

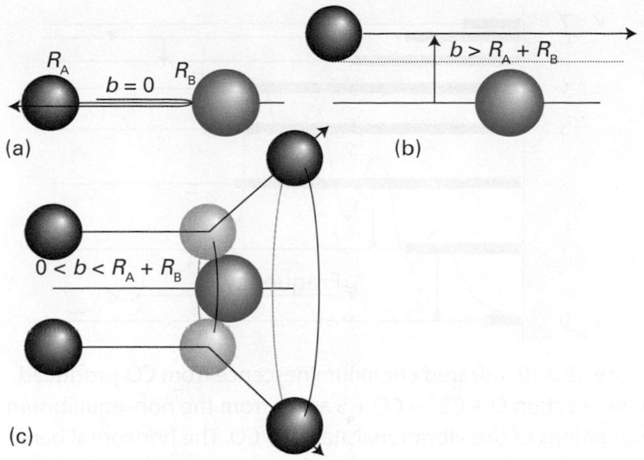

Figure 18D.6 Three typical cases for the collisions of two hard spheres: (a) $b = 0$, giving backward scattering; (b) $b > R_A + R_B$, giving forward scattering; (c) $0 < b < R_A + R_B$, leading to scattering into one direction on a ring of possibilities. (The target molecule is taken to be so heavy that it remains virtually stationary.)

scattering and the scattering cross-section is zero at all angles except $\theta = 0$. Glancing blows, with $0 < b \leq R_A + R_B$, lead to scattering intensity in cones around the forward direction.

The scattering pattern of real molecules, which are not hard spheres, depends on the details of the intermolecular potential energy and molecular shape. The scattering also depends on the relative speed of approach of the two molecules: a very fast molecule might pass through the interaction region without much deflection, whereas a slower one on the same path might be temporarily captured and undergo considerable deflection (Fig. 18D.7). The variation of the scattering cross-section with the relative speed of approach gives information about the strength and range of the intermolecular potential.

A further point is that the outcome of collisions is determined by quantum, not classical, mechanics. The wave nature of the molecules can be taken into account, at least to some extent, by drawing all classical trajectories that take the pro-

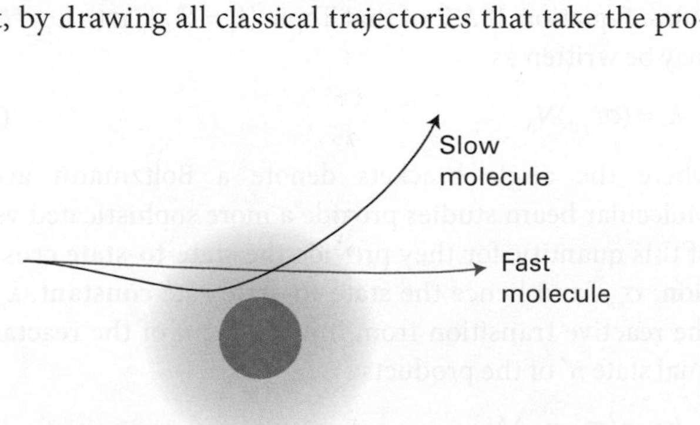

Figure 18D.7 The extent of scattering may depend on the relative speed of approach as well as the impact parameter. The dark central zone represents the repulsive core; the fuzzy outer zone represents the long-range attractive potential.

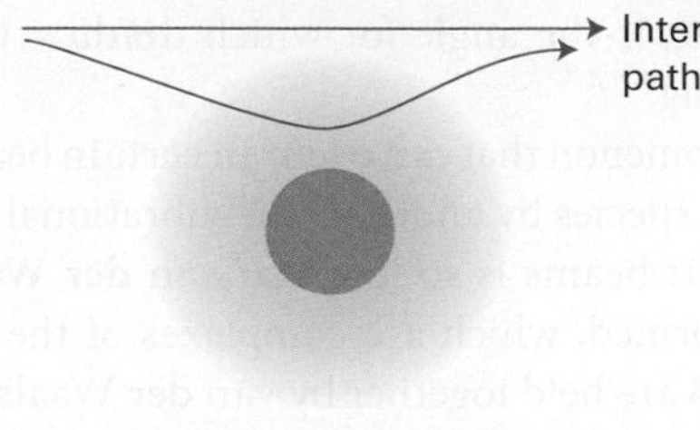

Figure 18D.8 Two paths leading to the same destination will interfere quantum mechanically; in this case they give rise to quantum oscillations in the forward direction.

jectile molecule from source to detector, and then considering the effects of interference between them.

Two quantum mechanical effects are of great importance. A molecule with a certain impact parameter might approach the attractive region of the potential in such a way that it is deflected towards the repulsive core (Fig. 18D.8), which then repels it out through the attractive region to continue its flight in the forward direction. Some molecules, however, also travel in the forward direction because they have impact parameters so large that they are undeflected. The wavefunctions of the molecules that take the two types of path interfere, and the intensity in the forward direction is modified. The effect is called **quantum oscillation**. The same phenomenon accounts for the optical 'glory effect', in which a bright halo can sometimes be seen surrounding an illuminated object. (The coloured rings around the shadow of an aircraft cast on clouds by the Sun, and often seen in flight, are an example of an optical glory.)

The second quantum effect is the observation of a strongly enhanced scattering in a non-forward direction. This effect is called **rainbow scattering** because the same mechanism accounts for the appearance of an optical rainbow. The origin of the phenomenon is illustrated in Fig. 18D.9. As the impact parameter decreases, there comes a stage at which the scattering angle passes through a maximum and the interference between the paths results in a strongly scattered beam. The

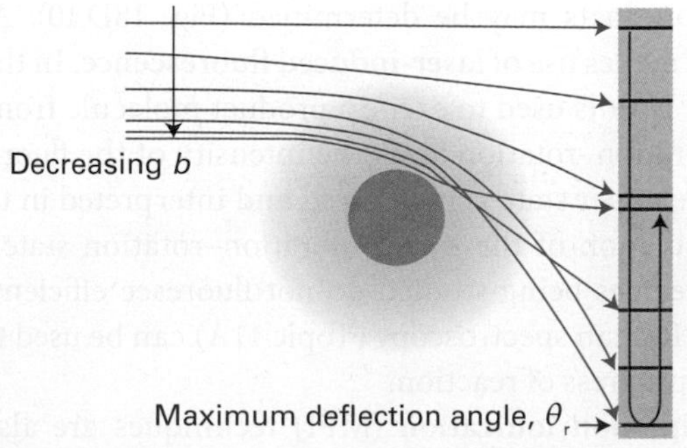

Figure 18D.9 The interference of paths leading to rainbow scattering. The rainbow angle, θ_r, is the maximum scattering angle reached as b is decreased. Interference between the numerous paths at that angle modifies the scattering intensity markedly.

rainbow angle, θ_r, is the angle for which $d\theta/db = 0$ and the scattering is strong.

Another phenomenon that can occur in certain beams is the capturing of one species by another. The vibrational temperature in supersonic beams is so low that **van der Waals molecules** may be formed, which are complexes of the form AB in which A and B are held together by van der Waals forces or hydrogen bonds. Large numbers of such molecules have been studied spectroscopically, including ArHCl, $(HCl)_2$, $ArCO_2$, and $(H_2O)_2$. The study of their spectroscopic properties gives detailed information about the intermolecular interactions involved.

18D.2 Reactive collisions

Detailed experimental information about the intimate processes that occur during reactive encounters comes from molecular beams, especially **crossed molecular beams** (see Fig. 18D.1). The detector for the products of the collision of molecules in the two beams can be moved to different angles to observe the angular distribution of the products. Because the molecules in the incoming beams can be prepared with different energies (e.g. with different translational energies by using rotating sectors and supersonic nozzles, with different vibrational energies by using selective excitation with lasers, and with different orientations by using electric fields), it is possible to study the dependence of the success of collisions on these variables and to study how they affect the properties of the product molecules.

(a) Probes of reactive collisions

One method for examining the energy distribution in the products is **infrared chemiluminescence**, in which vibrationally excited molecules emit infrared radiation as they return to their ground states. By studying the intensities of the infrared emission spectrum, the populations of the vibrational states of the products may be determined (Fig. 18D.10). Another method makes use of **laser-induced fluorescence**. In this technique, a laser is used to excite a product molecule from a specific vibration–rotation level; the intensity of the fluorescence from the upper state is monitored and interpreted in terms of the population of the initial vibration–rotation state. When the molecules being studied do not fluoresce efficiently, versions of Raman spectroscopy (Topic 11A) can be used to monitor the progress of reaction.

Multiphoton ionization (MPI) techniques are also good alternatives for the study of weakly fluorescing molecules. In MPI, the absorption by a molecule of several photons from one or more pulsed lasers results in ionization if the total photon energy is greater than the ionization energy of the molecule. An important variant of MPI is **resonant multiphoton ionization** (REMPI), in which one or more photons promote a molecule to an electronically excited state and then additional photons are used to generate ions from the excited state. The power of REMPI lies in the fact that the experimenter can choose which reactant or product to study by tuning the laser frequency to the electronic absorption band of a specific molecule.

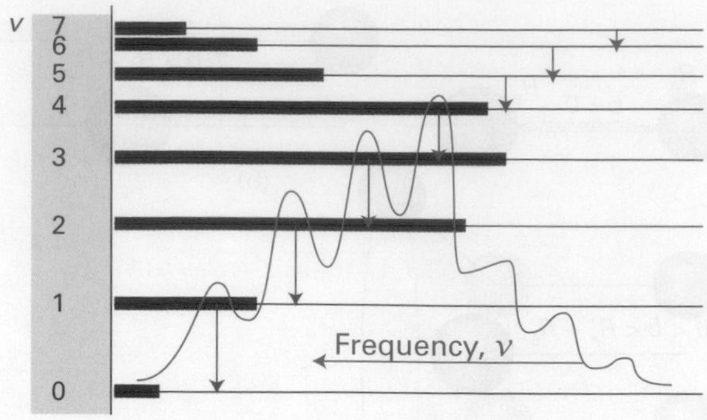

Figure 18D.10 Infrared chemiluminescence from CO produced in the reaction O + CS → CO + S arises from the non-equilibrium populations of the vibrational states of CO. The horizontal bars indicate the relative populations of the vibrational states and the blue arrows indicate the observed transitions.

The angular distribution of ions can be determined by **reaction product imaging**. In this technique, product ions are accelerated by an electric field towards a phosphorescent screen and the light emitted from specific spots where the ions struck the screen is imaged by a charge-coupled device (CCD).

(b) State-to-state reaction dynamics

The concept of collision cross-section is introduced in connection with collision theory in Topic 18A, where it is shown that the second-order rate constant, k_r, can be expressed as a Boltzmann-weighted average of the reactive cross-section and the relative speed of approach of the colliding reactant molecules. Equation 18A.7 of that Topic ($k_r = N_A \int_0^\infty \sigma(\varepsilon) v_{rel} f(\varepsilon) d\varepsilon$) may be written as

$$k_r = \langle \sigma v_{rel} \rangle N_A \qquad (18D.2)$$

where the angle brackets denote a Boltzmann average. Molecular beam studies provide a more sophisticated version of this quantity, for they provide the **state-to-state cross-section**, $\sigma_{nn'}$, and hence the **state-to-state rate constant**, $k_{nn'}$, for the reactive transition from initial state n of the reactants to final state n' of the products:

$$k_{nn'} = \langle \sigma_{nn'} v_{rel} \rangle N_A \qquad \text{State-to-state rate constant} \qquad (18D.3)$$

The rate constant k_r is the sum of the state-to-state rate constants over all final states (because a reaction is successful whatever the final state of the products) and over a Boltzmann-weighted sum of initial states (because the reactants are initially present with a characteristic distribution of populations at a temperature T):

$$k_{\mathrm{r}}=\sum_{n,n'}k_{nn'}(T)f_n(T) \qquad (18\mathrm{D}.4)$$

where $f_n(T)$ is the Boltzmann factor at a temperature T. It follows that if the state-to-state cross-sections can be determined or calculated for a wide range of approach speeds and initial and final states, then there is a route to the calculation of the rate constant for the reaction.

Brief illustration 18D.1

Suppose a harmonic oscillator collides with another oscillator of the same effective mass and force constant. If the state-to-state rate constant for the excitation of the latter's vibration is $k_{\upsilon\upsilon'}=k_{\mathrm{r}}^{\circ}\delta_{\upsilon\upsilon'}$ for all the states υ and υ', implying that an excitation can flow only from any level to the same level of the second oscillator, then at a temperature T, when $f_\upsilon(T)=\mathrm{e}^{-\upsilon h\nu/kT}/q$, where q is the molecular vibrational partition function (Topic 13B), the overall rate constant is

$$k_{\mathrm{r}}=\frac{k_{\mathrm{r}}^{\circ}}{q}\sum_{\upsilon,\upsilon'}\delta_{\upsilon\upsilon'}\mathrm{e}^{-\upsilon h\nu/kT}=\frac{k_{\mathrm{r}}^{\circ}}{q}\overbrace{\sum_{\upsilon'}\mathrm{e}^{-\upsilon' h\nu/kT}}^{q}=k_{\mathrm{r}}^{\circ}$$

18D.3 Potential energy surfaces

One of the most important concepts for discussing beam results and calculating the state-to-state collision cross-section is the **potential energy surface** of a reaction, the potential energy as a function of the relative positions of all the atoms taking part in the reaction. Potential energy surfaces may be constructed from experimental data and from results of quantum chemical calculations (Topic 9E). The theoretical method requires the systematic calculation of the energies of the system in a large number of geometrical arrangements. Special computational techniques, such as those described in Topic 9E, are used to take into account electron correlation, which arises from interactions between electrons as they move closer to and farther from each other in a molecule or molecular cluster. Techniques that incorporate electron correlation accurately are very time consuming and, consequently, the most reliable results are for reactions between relatively simple particles, such as the reactions $H + H_2 \rightarrow H_2 + H$ and $H + H_2O \rightarrow OH + H_2$. An alternative is to use semi-empirical methods, in which results of calculations and experimental parameters are used to construct the potential energy surface.

To illustrate the features of a potential energy surface, consider the collision between an H atom and an H_2 molecule. Detailed calculations show that the approach of an atom H_A

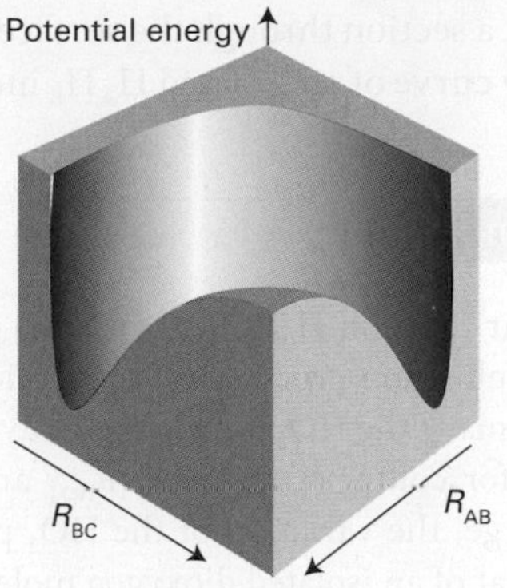

Figure 18D.11 The potential energy surface for the $H + H_2 \rightarrow H_2 + H$ reaction when the atoms are constrained to be collinear.

along the H_B–H_C axis requires less energy for reaction than any other approach, so initially it is convenient to confine attention to that collinear approach. Two parameters are required to define the nuclear separations: the H_A–H_B separation R_{AB} and the H_B–H_C separation R_{BC}.

At the start of the encounter R_{AB} is effectively infinite and R_{BC} is the H_2 equilibrium bond length. At the end of a successful reactive encounter R_{AB} is equal to the equilibrium bond length and R_{BC} is infinite. The total energy of the three-atom system depends on their relative separations, and can be found by doing an electronic structure calculation. The plot of the total energy of the system against R_{AB} and R_{BC} gives the potential energy surface of this collinear reaction (Fig. 18D.11). This surface is normally depicted as a contour diagram (Fig. 18D.12).

When R_{AB} is very large, the variation in potential energy represented by the surface as R_{BC} changes is that of an isolated H_2 molecule as its bond length is altered. A section through the surface at $R_{AB}=\infty$, for example, is the same as the H_2 bonding potential energy curve. At the edge of the diagram where

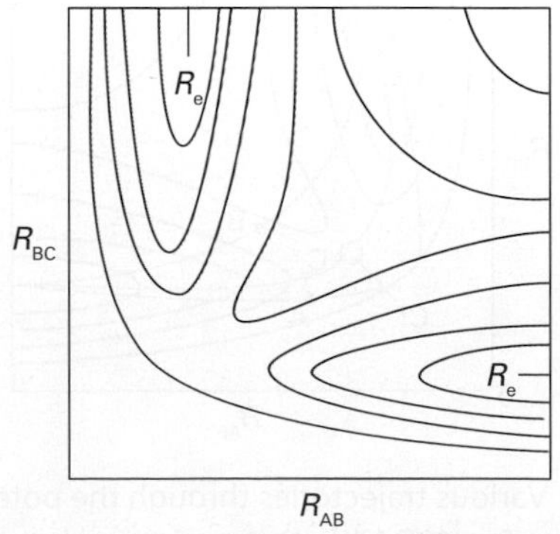

Figure 18D.12 The contour diagram (with contours of equal potential energy) corresponding to the surface in Fig. 18D.11. R_e marks the equilibrium bond length of an H_2 molecule (strictly, it relates to the arrangement when the third atom is at infinity).

R_{BC} is very large, a section through the surface is the molecular potential energy curve of an isolated H_AH_B molecule.

Brief illustration 18D.2

The bimolecular reaction $H + O_2 \rightarrow OH + O$ plays an important role in combustion processes. The reaction can be characterized in terms of the HO_2 potential energy surface and the two distances for collinear approach R_{HO_A} and $R_{O_AO_B}$. When R_{HO_A} is very large, the variation of the HO_2 potential energy with $R_{O_AO_B}$ is that of an isolated dioxygen molecule as its bond length is changed. Similarly, when $R_{O_AO_B}$ is very large, a section through the potential energy surface is the molecular potential energy curve of an isolated OH radical.

The actual path of the atoms in the course of the encounter depends on their total energy, the sum of their kinetic and potential energies. However, an initial idea of the paths available to the system can be obtained by identifying paths that correspond to least potential energy. For example, consider the changes in potential energy as H_A approaches H_BH_C. If the H_B–H_C bond length is constant during the initial approach of H_A, then the potential energy of the H_3 cluster rises along the path marked A in Fig. 18D.13. The potential energy reaches a high value as H_A is pushed into the molecule and then decreases sharply as H_C breaks off and separates to a great distance. An alternative reaction path can be imagined (B) in which the H_B–H_C bond length increases while H_A is still far away. Both paths, although feasible if the molecules have sufficient initial kinetic energy, take the three atoms to regions of high potential energy in the course of the encounter.

The path of least potential energy is the one marked C, corresponding to R_{BC} lengthening as H_A approaches and begins

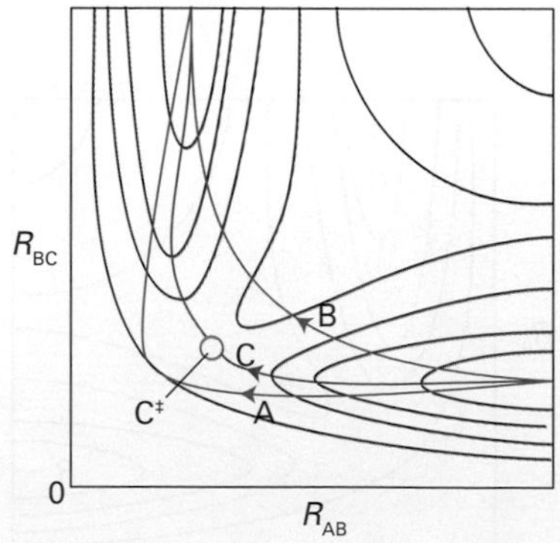

Figure 18D.13 Various trajectories through the potential energy surface shown in Fig. 18D.12. Path A corresponds to a route in which R_{BC} is held nearly constant as H_A approaches; path B corresponds to a route in which R_{BC} lengthens at an early stage during the approach of H_A; path C is the route along the floor of the potential valley.

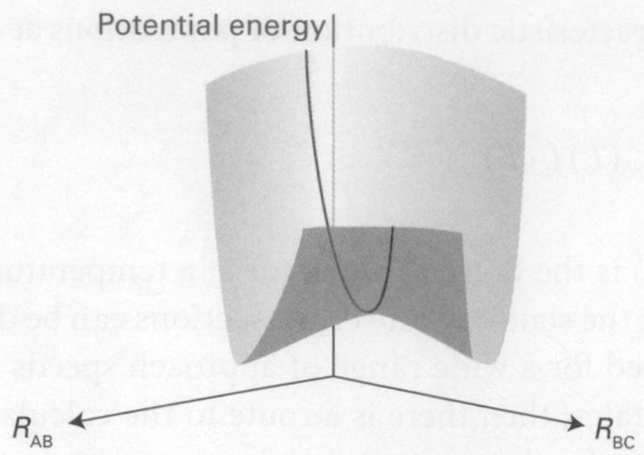

Figure 18D.14 The transition state is a set of configurations (here, marked by the red line across the saddle point) through which successful reactive trajectories must pass.

to form a bond with H_B. The H_B–H_C bond relaxes at the demand of the incoming atom, and the potential energy climbs only as far as the saddle-shaped region of the surface, to the **saddle point** marked $C^‡$. The encounter that requires the least increase in potential energy is the one in which the atoms take route C up the floor of the valley, through the saddle point, and down the floor of the other valley as H_C recedes and the new H_A–H_B bond achieves its equilibrium length. This path is the reaction coordinate.

It is now possible to make contact with the transition-state theory of reaction rates (Topic 18C). In terms of trajectories on potential surfaces with a total energy close to the saddle point energy, the transition state can be identified with a critical range of configurations such that every trajectory that goes through this configuration goes on to react (Fig. 18D.14). Most trajectories on potential energy surfaces do not go directly over the saddle point and therefore, to result in a reaction, they require a total energy significantly higher than the saddle-point energy. As a result, the experimentally determined activation energy is often significantly higher than the calculated saddle-point energy.

18D.4 Some results from experiments and calculations

Although quantum mechanical tunnelling can play an important role in reactivity, particularly in hydrogen atom and electron transfer reactions, this discussion begins with consideration of the classical trajectories of particles over surfaces. From this viewpoint, to travel successfully from reactants to products, the incoming molecules must possess enough kinetic energy to be able to climb to the saddle point of the potential surface. Therefore, the shape of the surface can be explored experimentally by changing the relative speed of approach (by selecting the beam velocity) and the degree of

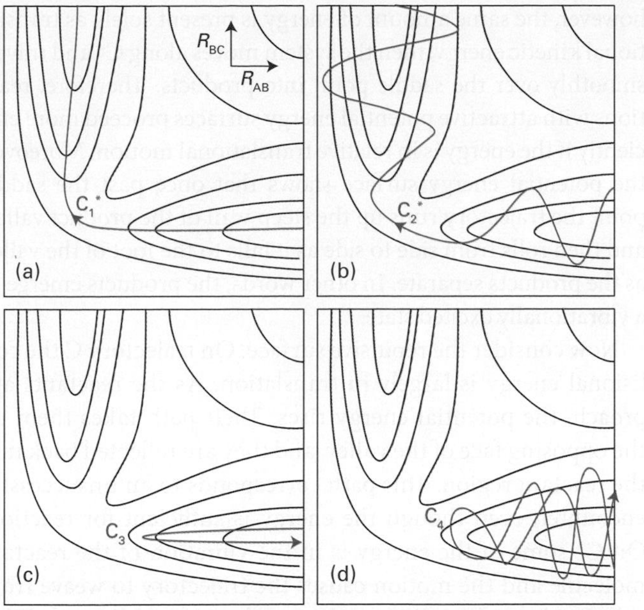

Figure 18D.15 Some successful (*) and unsuccessful encounters. (a) C_1^* corresponds to the path along the foot of the valley; (b) C_2^* corresponds to an approach of A to a vibrating BC molecule, and the formation of a vibrating AB molecule as C departs. (c) C_3 corresponds to A approaching a non-vibrating BC molecule, but with insufficient translational kinetic energy; (d) C_4 corresponds to A approaching a vibrating BC molecule, but still the energy, and the phase of the vibration, is insufficient for reaction.

vibrational excitation and observing whether reaction occurs and whether the products emerge in a vibrationally excited state (Fig. 18D.15). For example, one question that can be answered is whether it is better to smash the reactants together with a lot of translational kinetic energy or to ensure instead that they approach in highly excited vibrational states. Thus, is trajectory C_2^*, where the H_BH_C molecule is initially vibrationally excited, more efficient at leading to reaction than the trajectory C_1^*, in which the total energy is the same but reactants have a high translational kinetic energy?

(a) The direction of attack and separation

Figure 18D.16 shows the results of a calculation of the potential energy as an H atom approaches an H_2 molecule from different angles, the H_2 bond being allowed to relax to the optimum length in each case. The potential barrier is least for collinear attack, as assumed earlier. (But be aware that other lines of attack are feasible and contribute to the overall rate.) In contrast, Fig. 18D.17 shows the potential energy changes that occur as a Cl atom approaches an HI molecule. The lowest barrier occurs for approaches within a cone of half-angle 30° surrounding the H atom. The relevance of this result to the calculation of the steric factor of collision theory should be noted: not every collision is successful, because they do not all lie within the reactive cone.

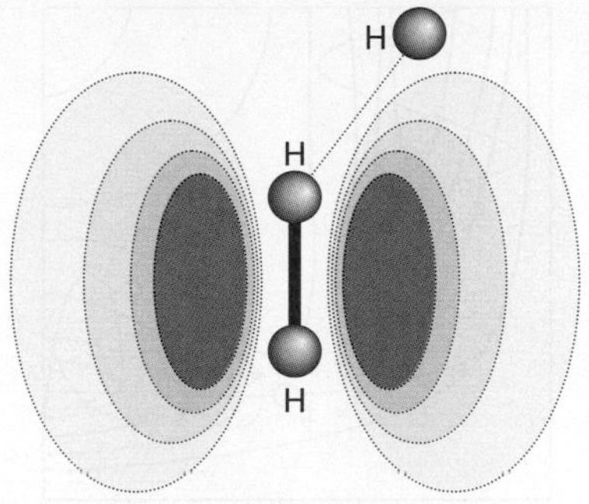

Figure 18D.16 An indication of how the anisotropy of the potential energy changes as H approaches H_2 with different angles of attack. The collinear attack has the lowest potential barrier to reaction. The surface indicates the potential energy profile along the reaction coordinate for each configuration.

If the collision is sticky, so that when the reactants collide they orbit around each other, the products can be expected to emerge in random directions because all memory of the approach direction has been lost. A rotation takes about 1 ps, so if the collision is over in less than that time the complex will not have had time to rotate and the products will be thrown off in a specific direction. In the collision of K and I_2, for example, most of the products are thrown off in the forward direction ('forward' and 'backward' refer to directions in a centre-of-mass coordinate system with the origin at the centre of mass of the colliding reactants and collision occurring when molecules are at the origin). This product distribution is consistent with the harpoon mechanism (Topic 18A) because the transition takes place at long range. In contrast, the collision of K with CH_3I leads to reaction only if the molecules approach each other very closely. In this mechanism, K effectively bumps into a brick wall, and the KI product bounces out in the backward direction. The detection of this anisotropy in the angular distribution of products gives an indication of the distance and orientation of approach needed for reaction, as well as showing that the event is complete in less than about 1 ps.

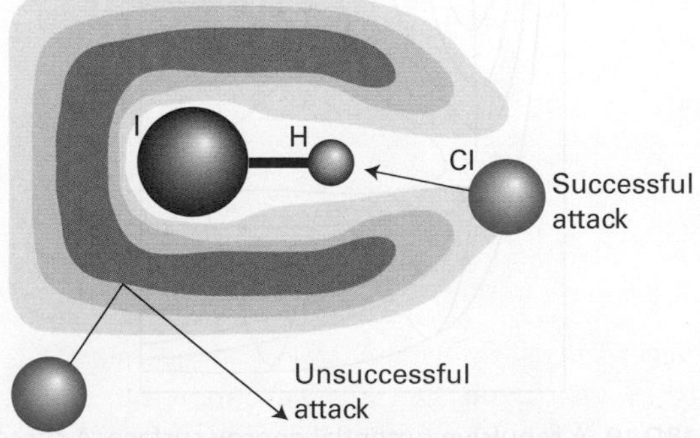

Figure 18D.17 The potential energy barrier for the approach of Cl to HI. In this case, successful encounters occur only when Cl approaches almost directly towards the H atom.

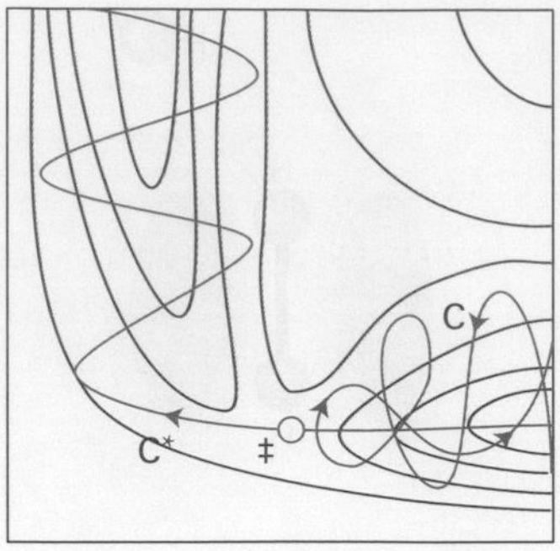

Figure 18D.18 An attractive potential energy surface. A successful encounter (C*) involves high translational kinetic energy and results in a vibrationally excited product.

(b) Attractive and repulsive surfaces

Some reactions are very sensitive to whether the energy has been deposited into a vibrational mode or left as the relative translational kinetic energy of the colliding molecules. For example, if two HI molecules are hurled together with more than twice the activation energy of the reaction, then no reaction occurs if all the energy is solely translational. For F + HCl → Cl + HF, for example, the reaction is about five times more efficient when the HCl is in its first vibrational excited state than when it is in its vibrational ground state, although HCl has the same total energy.

The origin of these requirements can be found by examining the potential energy surface. Figure 18D.18 shows an **attractive surface** in which the saddle point occurs early in the reaction coordinate. Figure 18D.19 shows a **repulsive surface** in which the saddle point occurs late. A surface that is attractive in one direction is repulsive in the reverse direction.

Consider first the attractive surface. If the original molecule is vibrationally excited, then a collision with an incoming molecule takes the system along C. This path is bottled up in the region of the reactants, and does not take the system to the saddle point. If, however, the same amount of energy is present solely as translational kinetic energy, then the system moves along C* and travels smoothly over the saddle point into products. Therefore, reactions with attractive potential energy surfaces proceed more efficiently if the energy is in relative translational motion. Moreover, the potential energy surface shows that once past the saddle point the trajectory runs up the steep wall of the product valley, and then rolls from side to side as it falls to the foot of the valley as the products separate. In other words, the products emerge in a vibrationally excited state.

Now consider the repulsive surface. On trajectory C the collisional energy is largely in translation. As the reactants approach, the potential energy rises. Their path takes them up the opposing face of the valley, and they are reflected back into the reactant region. This path corresponds to an unsuccessful encounter, even though the energy is sufficient for reaction. On C* some of the energy is in the vibration of the reactant molecule and the motion causes the trajectory to weave from side to side up the valley as it approaches the saddle point. This motion may be sufficient to tip the system round the corner to the saddle point and then on to products. In this case, the product molecule is expected to be in an unexcited vibrational state. Reactions with repulsive potential surfaces can therefore be expected to proceed more efficiently if the excess energy is present as vibrations. This is the case with the H + Cl_2 → HCl + Cl reaction, for instance.

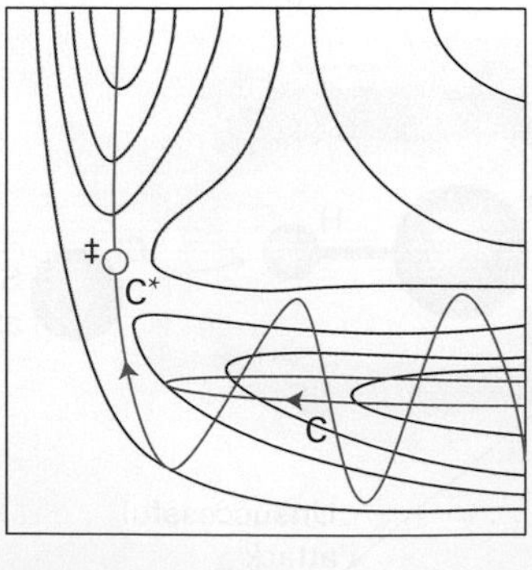

Figure 18D.19 A repulsive potential energy surface. A successful encounter (C*) involves initial vibrational excitation and the products have high translational kinetic energy. A reaction that is attractive in one direction is repulsive in the reverse direction.

Brief illustration 18D.3

The reaction H + Cl_2 → HCl + Cl has a repulsive potential surface. Of the following four reactive processes, H + $Cl_2(v)$ → $HCl(v')$ + Cl, denoted (v,v'), all at the same total energy, (a) (0,0), (b) (2,0), (c) (0,2), (d) (2,2), reaction (b) is most probable with reactants vibrationally excited and products vibrationally unexcited.

(c) Quantum mechanical scattering theory

A picture of the reaction event can be obtained by using classical mechanics to calculate the trajectories of the atoms taking place in a reaction from a set of initial conditions, such as velocities, relative orientations, and internal energies of the reacting particles. However, classical trajectory calculations do not recognize the fact that the motion of atoms, electrons, and nuclei is governed by quantum mechanics. The concept of trajectory then fades and is replaced by the unfolding of a wavefunction that represents initially the reactants and finally products.

Complete quantum mechanical calculations of trajectories and rate constants are very onerous because it is necessary to take into account all the allowed electronic, vibrational, and rotational states populated by each atom and molecule in the system at a given temperature. It is common to define a 'channel' as a group of molecules in well-defined quantum mechanically allowed states. Then, at a given temperature, there are many channels that represent the reactants and many channels that represent possible products, with some transitions between channels being allowed but others not allowed. Furthermore, not every transition leads to a chemical reaction. For example, the process $H_2^* + OH \rightarrow H_2 + OH^*$, where the asterisk denotes an excited state, amounts to energy transfer between H_2 and OH, whereas the process $H_2^* + OH \rightarrow H_2O + H$ represents a chemical reaction. What complicates a quantum mechanical calculation of rate constants even in this simple four-atom system is that many reacting channels present at a given temperature can lead to the desired products $H_2O + H$, which themselves may be formed as many distinct channels. The **cumulative reaction probability**, $\bar{P}(E)$, at a fixed total energy E is then written as

$$\bar{P}(E)=\sum_{i,j} P_{ij}(E) \qquad \text{Cumulative reaction probability} \qquad (18D.5)$$

where $P_{ij}(E)$ is the probability for a transition between a reacting channel i and a product channel j and the summation is over all possible transitions that lead to product. It is then possible to show that the rate constant is given by

$$k_r(T)=\frac{\int_0^\infty \bar{P}(E)e^{-E/kT}dE}{hQ_R(T)} \qquad \text{Rate constant} \qquad (18D.6)$$

where $Q_R(T)$ is the partition function density (the partition function divided by the volume) of the reactants at the temperature T. The significance of eqn 18D.6 is that it provides a direct connection between an experimental quantity, the rate constant, and a theoretical quantity, $\bar{P}(E)$.

Checklist of concepts

☐ 1. A **molecular beam** is a collimated, narrow stream of molecules travelling through an evacuated vessel.

☐ 2. In a molecular beam, the scattering pattern of molecules depends on quantum mechanical effects and the details of the intermolecular potential.

☐ 3. A **van der Waals molecule** is a complex of the form AB in which A and B are held together by van der Waals forces or hydrogen bonds.

☐ 4. Techniques for the study of reactive collisions include **infrared chemiluminescence**, **laser-induced fluorescence**, **multiphoton ionization** (MPI), **resonant multiphoton ionization** (REMPI), and **reaction product imaging**.

☐ 5. A **potential energy surface** maps the potential energy as a function of the relative positions of all the atoms taking part in a reaction.

☐ 6. In an **attractive surface**, the saddle point (the highest point on the valley between reactants and products) occurs early on the reaction coordinate.

☐ 7. In a **repulsive surface**, the saddle point occurs late on the reaction coordinate.

Checklist of equations

Property	Equation	Comment	Equation number
Rate of molecular scattering	$dI = \sigma I \mathcal{N} dx$	σ is the differential scattering cross-section	18D.1
Rate constant	$k_r = \langle \sigma v_{rel} \rangle N_A$		18D.2
State-to-state rate constant	$k_{nn'} = \langle \sigma_{nn'} v_{rel} \rangle N_A$		18D.3
Overall rate constant	$k_r = \sum_{n,n'} k_{nn'}(T) f_n(T)$		18D.4
Cumulative reaction probability	$\bar{P}(E) = \sum_{i,j} P_{ij}(E)$		18D.5
Rate constant	$k_r(T) = \int_0^\infty \bar{P}(E)e^{-E/kT}dE / hQ_R(T)$	$Q_R(T)$ is the partition function density	18D.6

TOPIC 18E Electron transfer in homogeneous systems

➤ **Why do you need to know this material?**

Electron transfer reactions between protein-bound cofactors or between proteins play an important role in a variety of biological processes. Electron transfer is also important in homogeneous, non-biological catalysis.

➤ **What is the key idea?**

The rate constant of electron transfer in a donor–acceptor complex depends on the distance between electron donor and acceptor, the standard reaction Gibbs energy, and the energy needed to reach a particular arrangement of atoms.

➤ **What do you need to know already?**

This Topic makes use of transition-state theory (Topic 18C). It also uses the concept of tunnelling (Topic 7D), the steady-state approximation (Topic 17E), and the Franck–Condon principle (Topic 11F).

Concepts of transition-state theory and quantum theory can be applied to the study of a deceptively simple process, electron transfer between molecules in homogeneous systems. The approach developed here can be used to predict the rates of electron transfer of this kind with reasonable accuracy.

18E.1 The rate law

Consider electron transfer from a donor species D to an acceptor species A in solution. The overall reaction is

$$D + A \rightarrow D^+ + A^- \quad v = k_r[D][A] \tag{18E.1}$$

In the first step of the mechanism, D and A must diffuse through the solution and on meeting form a complex DA:

$$D + A \underset{k_a'}{\overset{k_a}{\rightleftarrows}} DA \tag{18E.2a}$$

Suppose that in the complex D and A are separated by d, the distance between their outer surfaces. Then electron transfer occurs within the DA complex to yield D^+A^-:

$$DA \underset{k_{et}'}{\overset{k_{et}}{\rightleftarrows}} D^+A^- \tag{18E.2b}$$

The complex D^+A^- can also break apart and the ions diffuse through the solution:

$$D^+A^- \xrightarrow{k_d} D^+ + A^- \tag{18E.2c}$$

Now the techniques described in Topic 17E can be used to write an expression for the rate constant for the overall reaction $D + A \rightarrow D^+ + A^-$.

How is that done? 18E.1 Deriving an expression for the rate constant for electron transfer in solution

Identify the rate of the overall reaction (eqn 18E.1) with the rate of the step described by eqn 18E.2c because the products of the reaction are the separated ions:

$$v = k_d[D^+A^-]$$

Then follow these steps.

Step 1 *Apply the steady-state approximation to both intermediates*

There are two reaction intermediates, DA and D^+A^-, so apply the steady-state approximation (Topic 17E) to both. From

$$\frac{d[D^+A^-]}{dt} = k_{et}[DA] - k_{et}'[D^+A^-] - k_d[D^+A^-] = 0$$

it follows that

$$[DA] = \frac{k_{et}' + k_d}{k_{et}}[D^+A^-]$$

The steady-state expression for DA is

$$\frac{d[DA]}{dt} = k_a[D][A] - k_a'[DA] - k_{et}[DA] + k_{et}'[D^+A^-] = 0$$

Next, replace the terms in blue by the expression for [DA] from the preceding equation to give

$$k_a[D][A]-\frac{(k'_{et}+k_d)(k'_a+k_{et})}{k_{et}}[D^+A^-]+k'_{et}[D^+A^-]$$
$$=k_a[D][A]-[D^+A^-]\left\{\frac{(k'_{et}+k_d)(k'_a+k_{et})}{k_{et}}-k'_{et}\right\}$$
$$=k_a[D][A]-[D^+A^-]\frac{(k'_{et}+k_d)(k'_a+k_{et})-k'_{et}k_{et}}{k_{et}}=0$$

It follows that

$$[D^+A^-]=\frac{k_a k_{et}}{(k'_{et}+k_d)(k'_a+k_{et})-k'_{et}k_{et}}[D][A]$$
$$=\frac{k_a k_{et}}{k'_{et}k'_a+k_d k'_a+k_d k_{et}}[D][A]$$

Step 2 *Write an expression for the rate constant*

The overall rate of reaction is $v=k_d[D^+A^-]$, which now becomes

$$v=k_d\frac{k_a k_{et}}{k'_{et}k'_a+k_d k'_a+k_d k_{et}}[D][A]$$

This expression is of the form $v=k_r[D][A]$, with k_r given by

$$k_r=\frac{k_a k_{et}k_d}{k'_a k'_{et}+k_d k'_a+k_d k_{et}}=\frac{k_a k_{et}k_d}{k'_a(k'_{et}+k_d)+k_d k_{et}}$$

Step 3 *Rearrange the preceding expression*

You can obtain a more convenient form of the preceding expression by dividing the numerator and denominator on the right-hand side by $k_d k_{et}$ to obtain

$$k_r=\frac{k_a}{k'_a(k'_{et}+k_d)/k_{et}k_d+1}$$

The reciprocal of each side then gives

$$\frac{1}{k_r}=\frac{1}{k_a}+\frac{k'_a}{k_a k_{et}k_d}(k'_{et}+k_d)$$

and therefore

$$\frac{1}{k_r}=\frac{1}{k_a}+\frac{k'_a}{k_a k_{et}}\left(1+\frac{k'_{et}}{k_d}\right) \qquad (18E.3)$$

Electron transfer rate constant

To gain insight into this equation and the factors that determine the rate of electron transfer reactions in solution, assume that the main decay route for D^+A^- is dissociation of the complex into separated ions, and therefore that $k_d[D^+A^-] >> k'_{et}[D^+A^-]$, which implies that $k_d >> k'_{et}$. It follows that

$$\frac{1}{k_r}\approx\frac{1}{k_a}\left(1+\frac{k'_a}{k_{et}}\right)$$

Physical interpretation

- When $k_{et}[DA] >> k'_a[DA]$, which implies that $k_r \approx k_a$, the rate of product formation is controlled by diffusion of D and A in solution.
- When $k_{et}[DA] << k'_a[DA]$, it follows that $k_r \approx (k_a/k'_a)k_{et} = Kk_{et}$, where K is the equilibrium constant for the diffusive encounter. The process is controlled by k_{et} and therefore the activation energy of electron transfer in the DA complex.

18E.2 The role of electron tunnelling

This analysis can be taken further by introducing the implication from transition-state theory that, at a given temperature, $k_{et} \propto e^{-\Delta^{\ddagger}G/RT}$, where $\Delta^{\ddagger}G$ is the Gibbs energy of activation. The remaining task, therefore, is to write expressions for the proportionality constant and $\Delta^{\ddagger}G$. The discussion concentrates on the following two key aspects of the theory of electron transfer processes, which was developed independently by R.A. Marcus, N.S. Hush, V.G. Levich, and R.R. Dogonadze:

- Electrons are transferred by tunnelling through a potential energy barrier, the height of which is partly determined by the ionization energies of the DA and D^+A^- complexes. Electron tunnelling influences the magnitude of the proportionality constant in the expression for k_{et}.
- The complex DA and the solvent molecules surrounding it undergo structural rearrangements prior to electron transfer. The energy associated with these rearrangements and the standard reaction Gibbs energy determine $\Delta^{\ddagger}G$.

According to the Franck–Condon principle (Topic 11F), electronic transitions are so fast that they can be regarded as taking place in a stationary nuclear framework. This principle also applies to an electron transfer process in which an electron migrates from one energy surface, representing the dependence of the energy of DA on its geometry, to another representing the energy of D^+A^-. The potential energy (and the Gibbs energy) surfaces of the two complexes (the reactant complex, DA, and the product complex, D^+A^-) can be represented by the parabolas characteristic of harmonic oscillators, with the displacement coordinate corresponding to the changing geometries (Fig. 18E.1). This coordinate represents a collective mode of the donor, acceptor, and solvent.

According to the Franck–Condon principle, the nuclei do not have time to move when the system passes from the reactant to the product surface as a result of the transfer of an electron. Therefore, electron transfer can occur only after thermal fluctuations bring the geometry of DA to $q^{\ddagger}$ in Fig. 18E.1, the value of the nuclear coordinate at which the two parabolas intersect.

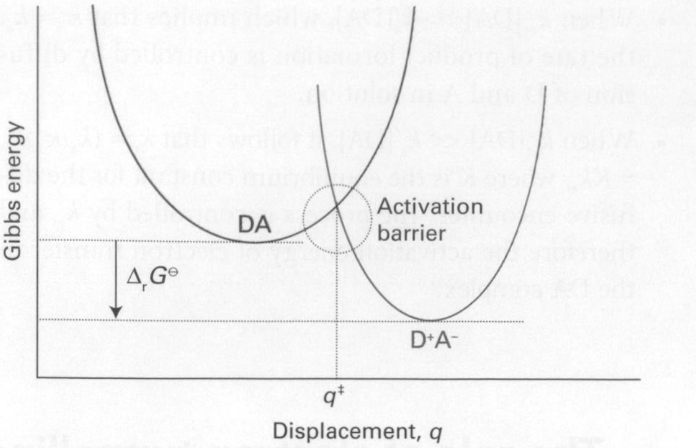

Figure 18E.1 The Gibbs energy surfaces of the complexes DA and D^+A^- involved in an electron transfer process are represented by parabolas characteristic of harmonic oscillators, with the displacement coordinate q corresponding to the changing geometries of the system.

The proportionality constant in the expression for k_{et} is a measure of the rate at which the system converts from reactants (DA) to products (D^+A^-) at $q^\ddagger$ by electron transfer within the thermally excited DA complex. To understand the process, consider the effect that the rearrangement of nuclear coordinates has on the electronic energy levels of DA and D^+A^- for a given distance d between D and A (Fig. 18E.2). Initially, the HOMO of DA is lower than the LUMO of D^+A^- (Fig. 18E.2a). As the nuclei rearrange into a configuration represented by $q^\ddagger$ in Fig. 18E.2b, the HOMO of DA and the LUMO of D^+A^- become similar in energy and electron transfer becomes feasible. Over reasonably short distances d, the main mechanism of electron transfer is tunnelling through the potential energy barrier depicted in Fig. 18E.2b. After an electron moves from the HOMO of DA to the LUMO of D^+A^-, the system relaxes to the configuration represented by q_0^P in Fig. 18E.2c. As shown in the illustration, now the energy of D^+A^- is lower than that of DA, reflecting the thermodynamic tendency for A to remain reduced (as A^-) and for D to remain oxidized (as D^+).

The tunnelling event responsible for electron transfer is similar to that described in Topic 7D, except that in this case the electron tunnels from an electronic level of DA, with wavefunction ψ_{DA}, to an electronic level of D^+A^-, with wavefunction $\psi_{D^+A^-}$. The rate of an electronic transition from a level described by the wavefunction ψ_{DA} to a level described by $\psi_{D^+A^-}$ is proportional to the square of the integral

$$H_{et} = \int \psi_{DA} \hat{h} \psi_{D^+A^-} d\tau$$

where $\hat{h}$ is a hamiltonian that describes the coupling of the electronic wavefunctions. The quantity H_{et} is often referred to as the 'electronic coupling matrix element'. The probability of tunnelling through a potential barrier typically has an exponential dependence on the width of the barrier (Topic 7D), suggesting that

$$H_{et}(d)^2 = H_{et}^{\circ 2} e^{-\beta d} \qquad (18E.4)$$

where d is the edge-to-edge distance between D and A, β is a parameter that measures the sensitivity of the electronic coupling matrix element to distance, and H_{et}° is the value of the electronic coupling matrix element at $d = 0$. The value of β depends on the medium through which the electron must travel from donor to acceptor. In a vacuum, $28\,nm^{-1} < \beta < 35\,nm^{-1}$, whereas $\beta \approx 9\,nm^{-1}$ when the intervening medium is a molecular link between donor and acceptor.

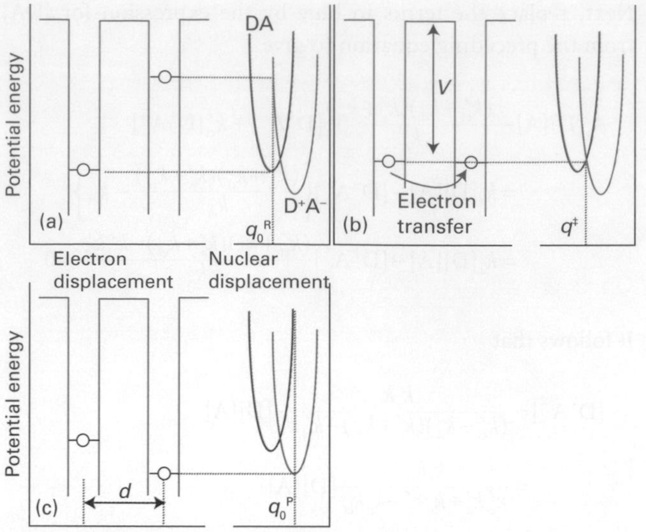

Figure 18E.2 (a) At the nuclear configuration denoted by q_0^R, the electron to be transferred in DA is in the HOMO; the LUMO of D^+A^- is too high in energy for efficient electron transfer. (b) As the nuclei rearrange to a configuration represented by $q^\ddagger$, the HOMO of DA and LUMO of D^+A^- become similar in energy and electron transfer occurs by tunnelling. (c) The system relaxes to the equilibrium nuclear configuration of D^+A^- denoted by q_0^P, in which the LUMO of DA is higher in energy than the HOMO of D^+A^-. Adapted from R.A. Marcus and N. Sutin, *Biochim. Biophys. Acta* **811**, 265 (1985).

18E.3 The rate constant

The proportionality constant in $k_{et} \propto e^{-\Delta^\ddagger G/RT}$ is proportional to $H_{et}(d)^2$, as expressed by eqn 18E.4. A detailed calculation (not reproduced here) shows that the full expression for k_{et} is

$$k_{et} = \frac{1}{h}\left(\frac{\pi^3}{RT\Delta E_R}\right)^{1/2} H_{et}(d)^2 e^{-\Delta^\ddagger G/RT} \qquad (18E.5)$$

where ΔE_R is the **reorganization energy**, the energy change associated with the molecular rearrangement that must take place so that DA can take on the equilibrium geometry of

D^+A^-. These molecular rearrangements include the relative reorientation of the D and A molecules in DA and the relative reorientation of the solvent molecules surrounding DA. To use eqn 18E.5 it is necessary to find an expression for the Gibbs energy of activation in terms of a simple model of the reaction.

How is that done? 18E.2 Establishing an expression for the Gibbs energy of activation

The simplest way to derive an expression for the Gibbs energy of activation of electron transfer is to construct a model in which the energy surfaces of DA (the 'reactant complex', denoted R) and D^+A^- (the 'product complex', denoted P) are plotted against the reaction coordinate q and assumed to be identical parabolic curves with displaced minima (Fig. 18E.3). For simplicity, q can be set equal to 0 at the minimum of the reactant parabola, and Gibbs energies measured from that minimum. Then

$$G_{m,R}(q) = k_e q^2 \text{ and } G_{m,P}(q) = k_e(q - q_0^P)^2 + \Delta_r G^\ominus$$

where q_0^P is the location of the minimum of the product parabola, k_e is a constant describing the curvature of the parabola, and $\Delta_r G^\ominus$ is the standard reaction Gibbs energy for the electron transfer process $DA \rightarrow D^+A^-$.

Write $\Delta E_R = G_{m,R}(q_0^P) = k_e q_0^{P2}$ as the difference in the Gibbs energy of R when the coordinate changes from $q = 0$ to the equilibrium value for P, q_0^P. The Gibbs energy of activation, $\Delta^\ddagger G = G_{m,R}(q^\ddagger) = k_e q^{\ddagger 2}$, is the change in Gibbs energy of R when the coordinate changes from $q = 0$ to $q^\ddagger$. Then follow these steps.

Step 1 *Identify the location of the activated complex*

The activated complex occurs at the point where the two parabolas intersect, which is where

$$k_e q^{\ddagger 2} = k_e(q^\ddagger - q_0^P)^2 + \Delta_r G^\ominus$$

$$= k_e q^{\ddagger 2} - 2k_e q_0^P q^\ddagger + \underbrace{k_e q_0^{P2}}_{\Delta E_R} + \Delta_r G^\ominus$$

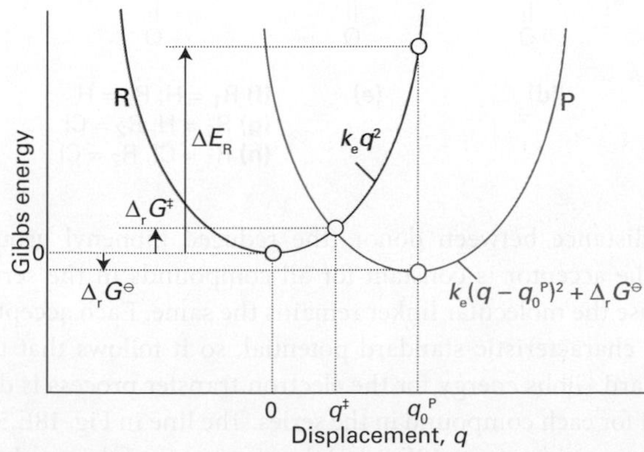

Figure 18E.3 The model system used in the calculation of the Gibbs energy of activation for an electron transfer process.

After rearrangement, the expression for $q^\ddagger$ is

$$q^\ddagger = \frac{\Delta E_R + \Delta_r G^\ominus}{2k_e q_0^P}$$

Step 2 *Use the expression for $q^\ddagger$ to find an expression for $\Delta^\ddagger G$*

Because Gibbs energies are measured from the minimum of the reactant parabola, where $q = 0$, it follows from the expression for $q^\ddagger$ that

$$\Delta^\ddagger G = k_e q^{\ddagger 2} = k_e\left(\frac{\Delta E_R + \Delta_r G^\ominus}{2k_e q_0^P}\right)^2 = \frac{(\Delta E_R + \Delta_r G^\ominus)^2}{4\underbrace{k_e q_0^{P2}}_{\Delta E_R}}$$

which simplifies to

$$\Delta^\ddagger G = \frac{(\Delta_r G^\ominus + \Delta E_R)^2}{4\Delta E_R} \qquad (18E.6)$$

Gibbs energy of activation

This equation shows that $\Delta^\ddagger G = 0$ when $\Delta_r G^\ominus = -\Delta E_R$, and the reorganization energy is cancelled by the standard reaction Gibbs energy. If this condition holds, the reaction is not slowed down by an activation barrier.

Equation 18E.6 has some limitations. For instance, it describes processes with weak electronic coupling between donor and acceptor. Weak coupling is observed when the electroactive species are sufficiently far apart ($d > 1$ nm) that the tunnelling is an exponential function of distance. The weak coupling limit applies to a large number of electron transfer reactions, including those between proteins during metabolism. Strong coupling is observed when the wavefunctions ψ_A and ψ_D overlap very extensively and, as well as other complications, the tunnelling rate is no longer a simple exponential function of distance. Examples of strongly coupled systems are mixed-valence, binuclear d-metal complexes with the general structure $L_m M^{n+}{-}B{-}M^{p+}L_m$, in which the electroactive metal ions are separated by a bridging ligand B. In these systems, $d < 1.0$ nm.

18E.4 Experimental tests of the theory

The most meaningful experimental tests of the dependence of k_{et} on d are those in which the same donor and acceptor are positioned at a variety of distances, perhaps by covalent attachment to molecular linkers (see **1** for an example, in which the biphenyl group is the donor and A are various acceptors). Under these conditions, the term $e^{-\Delta^\ddagger G/RT}$ becomes a constant and, after taking the natural logarithm of eqn 18E.5 and using eqn 18E.4, the result is

$$\ln k_{et} = -\beta d + \text{constant} \qquad (18E.7)$$

which implies that a plot of $\ln k_{\rm et}$ against d should be a straight line of slope $-\beta$.

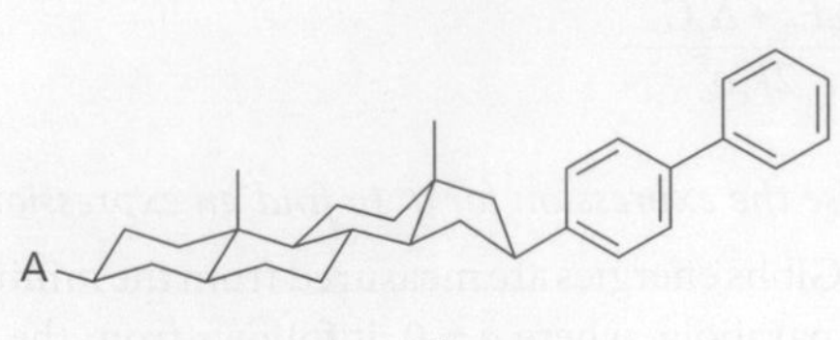

1

The dependence of $k_{\rm et}$ on the standard reaction Gibbs energy has been investigated in systems where the edge-to-edge distance and the reorganization energy are constant for a series of reactions. Then, because $k_{\rm et} \propto e^{-\Delta^{\ddagger}G/RT}$ implies that $\ln k_{\rm et} = -\Delta^{\ddagger}G/RT + \text{constant}$, it follows from eqn 18E.6 that

$$\ln k_{\rm et} = -\frac{(\Delta_{\rm r}G^{\ominus} + \Delta E_{\rm R})^2}{4RT\Delta E_{\rm R}} + \text{constant}$$

$$= -\frac{(\Delta_{\rm r}G^{\ominus})^2}{4RT\Delta E_{\rm R}} - \frac{\Delta_{\rm r}G^{\ominus}}{2RT} - \overbrace{\frac{\Delta E_{\rm R}}{4RT} + \text{constant}}^{\text{A constant}}$$

and therefore

$$\ln k_{\rm et} = -\frac{RT}{4\Delta E_{\rm R}}\left(\frac{\Delta_{\rm r}G^{\ominus}}{RT}\right)^2 - \tfrac{1}{2}\left(\frac{\Delta_{\rm r}G^{\ominus}}{RT}\right) + \text{constant} \qquad (18\text{E}.8)$$

A plot of $\ln k_{\rm et}$ (or $\log k_{\rm et} = \ln k_{\rm et}/\ln 10$) against $\Delta_{\rm r}G^{\ominus}$ (or $-\Delta_{\rm r}G^{\ominus}$) should therefore be a downward parabola (a curve of the form $y = ax^2 + bx + c$; Fig. 18E.4). Equation 18E.8 implies that the rate constant increases as $\Delta_{\rm r}G^{\ominus}$ decreases but only up to $-\Delta_{\rm r}G^{\ominus} = \Delta E_{\rm R}$. Beyond that, the reaction enters the **inverted region**, in which the rate constant decreases as the reaction becomes more exergonic ($\Delta_{\rm r}G^{\ominus}$ becomes more negative). The inverted region has been observed in a series of special compounds in which the electron donor and acceptor are linked covalently to a molecular spacer of known and fixed size (Fig. 18E.5).

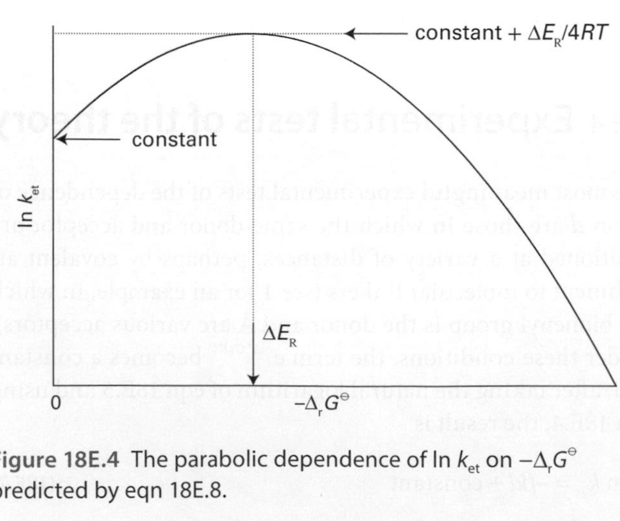

Figure 18E.4 The parabolic dependence of $\ln k_{\rm et}$ on $-\Delta_{\rm r}G^{\ominus}$ predicted by eqn 18E.8.

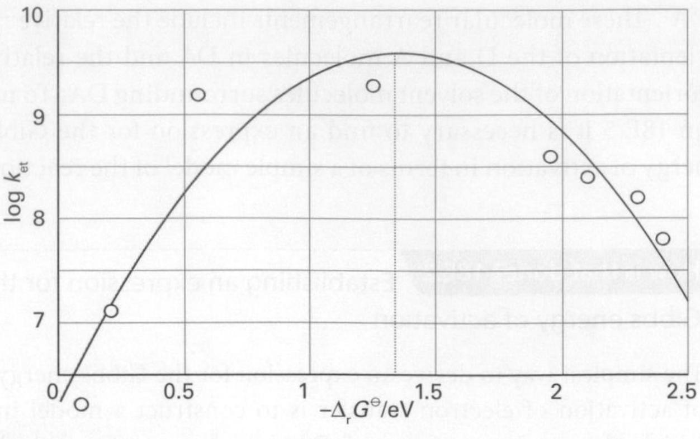

Figure 18E.5 Variation of $\log k_{\rm et}$ with $-\Delta_{\rm r}G^{\ominus}$ for a series of compounds with the structures given in (**1**) and as described in *Brief illustration* 18E.1. Based on J.R. Miller et al., *J. Am. Chem. Soc.* **106**, 3047 (1984).

Brief illustration 18E.1

Kinetic measurements were conducted in 2-methyltetrahydrofuran and at 296 K for a series of compounds with the structures given in (1) with A the following groups:

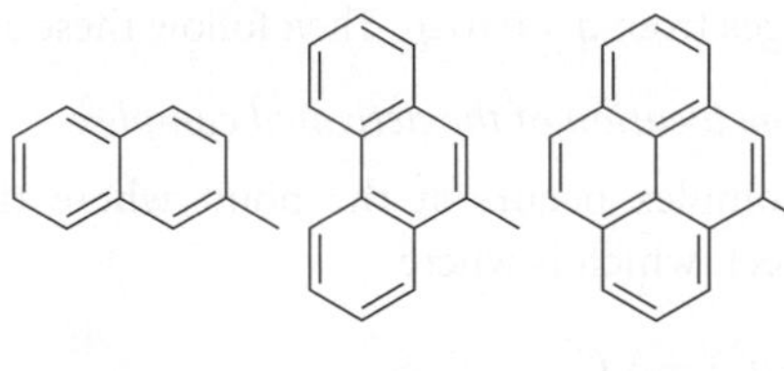

(a) **(b)** **(c)**

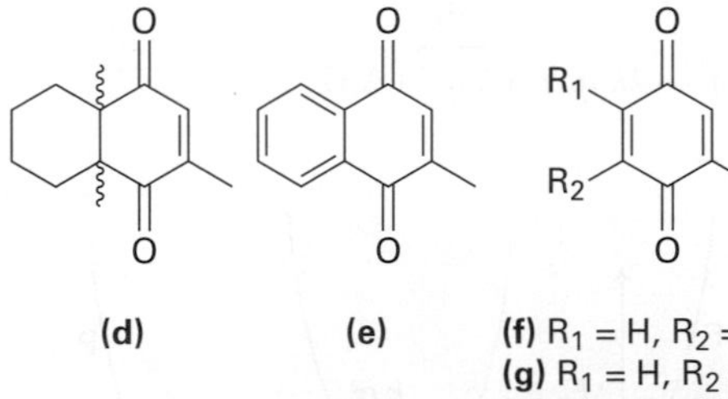

(d) **(e)** **(f)** R_1 = H, R_2 = H
(g) R_1 = H, R_2 = Cl
(h) R_1 = Cl, R_2 = Cl

The distance between donor (the reduced biphenyl group) and the acceptor is constant for all compounds in the series because the molecular linker remains the same. Each acceptor has a characteristic standard potential, so it follows that the standard Gibbs energy for the electron transfer process is different for each compound in the series. The line in Fig. 18E.5 is a fit to a version of eqn 18E.8 and the maximum of the parabola occurs at $-\Delta_{\rm r}G^{\ominus} = \Delta E_{\rm R} = 1.4\,\text{eV} = 1.4 \times 10^2\,\text{kJ mol}^{-1}$.

Checklist of concepts

☐ 1. Electron transfer can occur through tunnelling only after thermal fluctuations bring the nuclear coordinate to the point at which the donor and acceptor have the same configuration.

☐ 2. The **reorganization energy** is the energy change associated with molecular rearrangements that must take place so that DA can acquire the equilibrium geometry of D^+A^-.

Checklist of equations

Property	Equation	Comment	Equation number
Electron transfer rate constant	$1/k_r = 1/k_a + (k'_a/k_a k_{et})(1+k'_{et}/k_d)$	Steady-state assumption	18E.3
Tunnelling probability	$H_{et}(d)^2 = H_{et}^{o2}e^{-\beta d}$	Assumed	18E.4
Rate constant	$k_{et} = (1/h)(\pi^3/RT\Delta E_R)^{1/2} H_{et}(d)^2 e^{-\Delta^{\ddagger}G/RT}$	Transition-state theory	18E.5
Gibbs energy of activation	$\Delta^{\ddagger}G = (\Delta_r G^{\ominus} + \Delta E_R)^2/4\Delta E_R$	Assumes parabolic potential energy	18E.6

FOCUS 18 Reaction dynamics

TOPIC 18A Collision theory

Discussion questions

D18A.1 Discuss how the collision theory of gas-phase reactions builds on the kinetic molecular theory.

D18A.2 How might collision theory change for real gases?

D18A.3 Describe the essential features of the harpoon mechanism.

D18A.4 Explain how the complexity of the reacting molecules affects the value of the rate constant.

Exercises

E18A.1(a) Calculate the collision frequency, *z*, and the collision density, Z_{AA}, in ammonia, $d = 380$ pm, at 30 °C and 120 kPa. What is the percentage increase when the temperature is raised by 10 K at constant volume?
E18A.1(b) Calculate the collision frequency, *z*, and the collision density, Z_{AA}, in carbon monoxide, $d = 360$ pm, at 30 °C and 120 kPa. What is the percentage increase when the temperature is raised by 10 K at constant volume?

E18A.2(a) Collision theory depends on knowing the fraction of molecular collisions having at least the kinetic energy E_a along the line of flight. What is this fraction when (i) $E_a = 20\,\text{kJ mol}^{-1}$, (ii) $E_a = 100\,\text{kJ mol}^{-1}$ at (1) 350 K and (2) 900 K?
E18A.2(b) Collision theory depends on knowing the fraction of molecular collisions having at least the kinetic energy E_a along the line of flight. What is this fraction when (i) $E_a = 15\,\text{kJ mol}^{-1}$, (ii) $E_a = 150\,\text{kJ mol}^{-1}$ at (1) 300 K and (2) 800 K?

E18A.3(a) Calculate the percentage increase in the fractions in Exercise E18A.2(a) when the temperature is raised by 10 K in each case.
E18A.3(b) Calculate the percentage increase in the fractions in Exercise E18A.2(b) when the temperature is raised by 10 K in each case.

E18A.4(a) Use the collision theory of gas-phase reactions to calculate the theoretical value of the second-order rate constant for the elementary reaction $H_2 + I_2 \rightarrow HI + HI$ at 650 K. The collision cross-section is 0.36 nm^2, the reduced mass is 3.32×10^{-27} kg, and the activation energy is 171 kJ mol^{-1}. (Assume a steric factor of 1.)
E18A.4(b) Use the collision theory of gas-phase reactions to calculate the theoretical value of the second-order rate constant for the elementary reaction $D_2 + Br_2 \rightarrow DBr + DBr$ at 450 K. Take the collision cross-section as 0.30 nm^2, the reduced mass as $3.930m_u$, and the activation energy as 200 kJ mol^{-1}. (Assume a steric factor of 1.)

E18A.5(a) For the gaseous reaction $A + B \rightarrow P$, the reactive cross-section obtained from the experimental value of the frequency factor is 9.2×10^{-22} m^2. The collision cross-sections of A and B estimated from the transport properties are 0.95 nm^2 and 0.65 nm^2, respectively. Calculate the steric factor, *P*, for the reaction.
E18A.5(b) For the gaseous reaction $A + B \rightarrow P$, the reactive cross-section obtained from the experimental value of the frequency factor is 8.7×10^{-22} m^2. The collision cross-sections of A and B estimated from the transport properties are 0.88 nm^2 and 0.40 nm^2, respectively. Calculate the steric factor, *P*, for the reaction.

E18A.6(a) Consider the unimolecular decomposition of a nonlinear molecule containing five atoms according to RRK theory. If $k_b(E)/k_b = 3.0 \times 10^{-5}$, what is the value of E^*/E?
E18A.6(b) Consider the unimolecular decomposition of a linear molecule containing four atoms according to RRK theory. If $k_b(E)/k_b = 0.025$, what is the value of E^*/E?

E18A.7(a) Suppose that an energy of 250 kJ mol^{-1} is available in a collision but 200 kJ mol^{-1} is needed to break a particular bond in a molecule with $s = 10$. Use the RRK model to calculate the ratio $k_b(E)/k_b$.
E18A.7(b) Suppose that an energy of 500 kJ mol^{-1} is available in a collision but 300 kJ mol^{-1} is needed to break a particular bond in a molecule with $s = 12$. Use the RRK model to calculate the ratio $k_b(E)/k_b$.

Problems

P18A.1 In the dimerization of methyl radicals at 25 °C, the experimentally determined frequency factor is $2.4 \times 10^{10}\,\text{dm}^3\,\text{mol}^{-1}\,\text{s}^{-1}$. What are (a) the reactive cross-section, and (b) the steric factor for the reaction given that the C–H bond length is 154 pm?

P18A.2 Nitrogen dioxide reacts bimolecularly in the gas phase according to $NO_2 + NO_2 \rightarrow NO + NO + O_2$. The temperature dependence of the second-order rate constant for the rate law $d[P]/dt = k_r[NO_2]^2$ is given below. What are the steric factor and the reactive cross-section for the reaction?

T/K	600	700	800	1000
$k_r/(\text{cm}^3\,\text{mol}^{-1}\,\text{s}^{-1})$	4.6×10^2	9.7×10^3	1.3×10^5	3.1×10^6

Take $\sigma = 0.60$ nm^2.

P18A.3 The diameter of the methyl radical is about 308 pm. What is the maximum rate constant in the expression $d[C_2H_6]/dt = k_r[CH_3]^2$ for second-order recombination of radicals at 298 K? It is reported that 10 per cent of a sample of ethane of volume 1.0 dm^3 at 298 K and 100 kPa is dissociated into methyl radicals. What is the minimum time for 90 per cent recombination?

P18A.4 The reactive cross-sections for reactions between alkali metal atoms and halogen molecules are given in the table below (R.D. Levine and R.B. Bernstein, *Molecular reaction dynamics,* Clarendon Press, Oxford, 72 (1974)). Assess the data in terms of the harpoon mechanism.

σ^*/nm^2	Cl_2	Br_2	I_2
Na	1.24	1.16	0.97
K	1.54	1.51	1.27
Rb	1.90	1.97	1.67
Cs	1.96	2.04	1.95

Electron affinities are approximately 1.3 eV (Cl_2), 1.2 eV (Br_2), and 1.7 eV (I_2), and ionization energies are 5.1 eV (Na), 4.3 eV (K), 4.2 eV (Rb), and 3.9 eV (Cs).

P18A.5‡ R. Atkinson (*J. Phys. Chem. Ref. Data* **26**, 215 (1997)) has reviewed a large set of rate constants relevant to the atmospheric chemistry of volatile organic compounds. The recommended rate constant for the bimolecular reaction of O_2 with an alkyl radical R at 298 K is $4.7 \times 10^9\,dm^3\,mol^{-1}\,s^{-1}$ for $R = C_2H_5$ and $8.4 \times 10^9\,dm^3\,mol^{-1}\,s^{-1}$ for R = cyclohexyl. Assuming no energy barrier, compute the steric factor, *P*, for each reaction. *Hint:* Obtain collision diameters from collision cross-sections of similar molecules in the *Resource section*.

TOPIC 18B Diffusion-controlled reactions

Discussion questions

D18B.1 Distinguish between a diffusion-controlled reaction and an activation-controlled reaction. Do both have activation energies?

D18B.2 Describe the role of the encounter pair in the cage effect.

Exercises

E18B.1(a) A typical diffusion constant for small molecules in aqueous solution at 25 °C is $6 \times 10^{-9}\,m^2\,s^{-1}$. If the critical reaction distance is 0.5 nm, what value is expected for the second-order rate constant for a diffusion-controlled reaction?

E18B.1(b) Suppose that the typical diffusion coefficient for a reactant in aqueous solution at 25 °C is $5.2 \times 10^{-9}\,m^2\,s^{-1}$. If the critical reaction distance is 0.4 nm, what value is expected for the second-order rate constant for the diffusion-controlled reaction?

E18B.2(a) Calculate the magnitude of the diffusion-controlled rate constant at 298 K for a species in (i) water, (ii) pentane. The viscosities are $1.00 \times 10^{-3}\,kg\,m^{-1}\,s^{-1}$ and $2.2 \times 10^{-4}\,kg\,m^{-1}\,s^{-1}$, respectively.

E18B.2(b) Calculate the magnitude of the diffusion-controlled rate constant at 298 K for a species in (i) decylbenzene, (ii) concentrated sulfuric acid. The viscosities are 3.36 cP and 27 cP, respectively.

E18B.3(a) Calculate the magnitude of the diffusion-controlled rate constant at 320 K for the recombination of two atoms in water, for which $\eta = 0.89$ cP. Assuming the concentration of the reacting species is $1.5\,mmol\,dm^{-3}$ initially, how long does it take for the concentration of the atoms to fall to half that value? Assume the reaction is elementary.

E18B.3(b) Calculate the magnitude of the diffusion-controlled rate constant at 320 K for the recombination of two atoms in benzene, for which $\eta = 0.601$ cP. Assuming the concentration of the reacting species is $2.0\,mmol\,dm^{-3}$ initially, how long does it take for the concentration of the atoms to fall to half that value? Assume the reaction is elementary.

E18B.4(a) Two neutral species, A and B, with diameters 655 pm and 1820 pm, respectively, undergo the diffusion-controlled reaction $A + B \rightarrow P$ in a solvent of viscosity $2.93 \times 10^{-3}\,kg\,m^{-1}\,s^{-1}$ at 40 °C. Use eqn 18B.3 to calculate the initial rate d[P]/d*t*, given that the initial concentrations of A and B are 0.170 mol dm^{-3} and 0.350 mol dm^{-3}, respectively. Then repeat the calculation by using eqn 18B.4. Comment on the validity of the approximation that leads to eqn 18B.4.

E18A.4(b) Two neutral species, A and B, with diameters 421 pm and 945 pm, respectively, undergo the diffusion-controlled reaction $A + B \rightarrow P$ in a solvent of viscosity 1.35 cP at 20 °C. Use eqn 18B.3 to calculate the initial rate d[P]/d*t*, given that the initial concentrations of A and B are 0.155 mol dm^{-3} and 0.195 mol dm^{-3}, respectively. Then repeat the calculation by using eqn 18B.4. Comment on the validity of the approximation that leads to eqn 18B.4.

Problems

P18B.1 Confirm that eqn 18B.8 is a solution of eqn 18B.7, where [J] is a solution of the same equation but with $k_r = 0$ and the same initial conditions.

P18B.2 Use mathematical software or a spreadsheet to explore the effect of varying the value of the rate constant k_r on the spatial variation of [J]* (see eqn 18B.8 with [J] given in eqn 18B.9) for a constant value of the diffusion coefficient *D*.

P18B.3 Confirm that if the boundary condition is $[J] = [J]_0$ at $t > 0$ at all points on the *yz*-plane, and the initial condition is $[J] = 0$ at $t = 0$ everywhere else, then the solutions [J]* in the presence of a first-order reaction that removed J are related to those in the absence of reaction, [J], by

$$[J]^* = k_r \int_0^t [J] e^{-k_r t} dt + [J] e^{-k_r t}$$

Base you answer on eqn 18B.7.

P18B.4‡ The compound α-tocopherol, a form of vitamin E, is a powerful antioxidant that may help to maintain the integrity of biological membranes. R.H. Bisby and A.W. Parker (*J. Amer. Chem. Soc.* **117**, 5664 (1995)) studied the reaction of photochemically excited duroquinone with the antioxidant in ethanol. Once the duroquinone was photochemically excited, a bimolecular reaction took place at diffusion-limited rate. (a) Estimate the rate constant for a diffusion-limited reaction in ethanol. (b) The reported rate constant was $2.77 \times 10^9\,dm^3\,mol^{-1}\,s^{-1}$. Estimate the critical reaction distance if the sum of diffusion coefficients is $1 \times 10^{-9}\,m^2\,s^{-1}$.

TOPIC 18C Transition-state theory

Discussion questions

D18C.1 Which mode would be discarded for a reaction A + BC in which the activated complex is modelled as a linear triatomic cluster?

D18C.2 Describe in outline the formulation of the Eyring equation.

‡ These problems were provided by Charles Trapp and Carmen Giunta.

D18C.3 Explain the physical origin of the kinetic salt effect. How might the size of the effect be altered by a change in the relative permittivity of the medium?

D18C.4 How do kinetic isotope effects provide insight into the mechanism of a reaction?

Exercises

E18C.1(a) The reaction of propylxanthate anion, A^-, in ethanoic acid buffer solutions has the mechanism $A^- + H^+ \rightarrow P$. Near 30 °C the rate constant is given by the empirical expression $k_r = Ae^{-(8681\,K)/T}$ with $A = 2.05 \times 10^{13}\,dm^3\,mol^{-1}\,s^{-1}$. Evaluate the enthalpy and entropy of activation at 30 °C.

E18C.1(b) The reaction $A^- + H^+ \rightarrow P$ has a rate constant given by the empirical expression $k_r = Ae^{-(5925\,K)/T}$ with $A = 6.92 \times 10^{12}\,dm^3\,mol^{-1}\,s^{-1}$. Evaluate the enthalpy and entropy of activation at 25 °C.

E18C.2(a) When the reaction in Exercise E18C.1(a) occurs in a dioxane/water mixture which is 30 per cent dioxane by mass, the rate constant fits $k_r = Ae^{-(9134\,K)/T}$ with $A = 7.78 \times 10^{14}\,dm^3\,mol^{-1}\,s^{-1}$ near 30 °C. Calculate $\Delta^{\ddagger}G$ for the reaction at 30 °C; assume $\kappa = 1$.

E18C.2(b) A rate constant is found to fit the expression $k_r = Ae^{-(4972\,K)/T}$ with $A = 4.98 \times 10^{13}\,dm^3\,mol^{-1}\,s^{-1}$ near 25 °C. Calculate $\Delta^{\ddagger}G$ for the reaction at 25 °C; assume $\kappa = 1$.

E18C.3(a) The gas-phase reaction between F_2 and IF_5 is first order in each of the reactants. The energy of activation for the reaction is 58.6 kJ mol^{-1}. At 65 °C the rate constant is $7.84 \times 10^{-3}\,kPa^{-1}\,s^{-1}$. Calculate the entropy of activation at 65 °C.

E18C.3(b) A certain gas-phase reaction is first order in each of the reactants. The energy of activation for the reaction is 39.7 kJ mol^{-1}. At 65 °C the rate constant is $0.35\,m^3\,mol^{-1}\,s^{-1}$. Calculate the entropy of activation at 65 °C.

E18C.4(a) Calculate the entropy of activation for a collision between two structureless particles at 300 K, taking $M = 65\,g\,mol^{-1}$ and $\sigma = 0.35\,nm^2$. *Hint:* Refer to *Example* 18C.1.

E18C.4(b) Calculate the entropy of activation for a collision between two structureless particles at 450 K, taking $M = 92\,g\,mol^{-1}$ and $\sigma = 0.45\,nm^2$.

E18C.5(a) The frequency factor for the second-order gas-phase decomposition of ozone at low pressures is $4.6 \times 10^{12}\,dm^3\,mol^{-1}\,s^{-1}$ and its activation energy is 10.0 kJ mol^{-1}. What are (i) the entropy of activation, (ii) the enthalpy of activation, (iii) the Gibbs energy of activation at 298 K? Assume $\kappa = 1$.

E18C.5(b) The frequency factor for a second-order gas-phase decomposition of a species at low pressures is $2.3 \times 10^{13}\,dm^3\,mol^{-1}\,s^{-1}$ and its activation energy is 30.0 kJ mol^{-1}. What are (i) the entropy of activation, (ii) the enthalpy of activation, (iii) the Gibbs energy of activation at 298 K? Assume $\kappa = 1$.

E18C.6(a) The rate constant of the reaction $H_2O_2(aq) + I^-(aq) + H^+(aq) \rightarrow H_2O(l) + HIO(aq)$ is sensitive to the ionic strength of the aqueous solution in which the reaction occurs. At 25 °C, $k_r = 12.2\,dm^6\,mol^{-2}\,min^{-1}$ at an ionic strength of 0.0525. Use the Debye–Hückel limiting law to estimate the rate constant at zero ionic strength.

E18C.6(b) At 25 °C, $k_r = 1.55\,dm^6\,mol^{-2}\,min^{-1}$ at an ionic strength of 0.0241 for a reaction in which the rate-determining step involves the encounter of two singly charged cations. Use the Debye–Hückel limiting law to estimate the rate constant at zero ionic strength.

E18C.7(a) Estimate the magnitude of the primary kinetic isotope effect at 298 K on the relative rates of displacement of ^1H and ^3H in a C–H bond. Will raising the temperature enhance the difference? Take $k_f(C\text{–}H) = 450\,N\,m^{-1}$.

E18C.7(b) Estimate the magnitude of the primary isotope effect at 298 K on the relative rates of displacement of ^{16}O and ^{18}O in a C–O bond. Will raising the temperature enhance the difference? Take $k_f(C\text{–}O) = 1750\,N\,m^{-1}$.

Problems

P18C.1‡ For the gas-phase reaction $A + A \rightarrow A_2$, the experimental rate constant, k_r, has been fitted to the Arrhenius equation with a frequency factor $A = 4.07 \times 10^5\,dm^3\,mol^{-1}\,s^{-1}$ at 300 K and an activation energy of 65.4 kJ mol^{-1}. Calculate $\Delta^{\ddagger}S$, $\Delta^{\ddagger}H$, $\Delta^{\ddagger}U$, and $\Delta^{\ddagger}G$ for the reaction.

P18C.2 The rates of thermal decomposition of a variety of *cis*- and *trans*-azoalkanes have been measured over a range of temperatures in order to settle a controversy concerning the mechanism of the reaction. In ethanol an unstable *cis*-azoalkane decomposed at a rate that was followed by observing the N_2 evolution, and this led to the rate constants listed below (P.S. Engel and D.J. Bishop, *J. Amer. Chem. Soc.* **97**, 6754 (1975)). Calculate the enthalpy, entropy, energy, and Gibbs energy of activation at −20 °C.

θ/°C	−24.82	−20.73	−17.02	−13.00	−8.95
$10^4 \times k_r/s^{-1}$	1.22	2.31	4.39	8.50	14.3

P18C.3 Derive the expression for k_r given in *Example* 18C.1 starting from the point at which the thermal wavelengths are substituted.

P18C.4‡ Show that bimolecular reactions between nonlinear molecules are much slower than between atoms even when the activation energies of both reactions are equal. Use transition-state theory and make the following assumptions. (1) All vibrational partition functions are close to 1; (2) all rotational partition functions are approximately $1 \times 10^{1.5}$; (3) the translational partition function for each species is 1×10^{26}. *Hint:* Equation 18C.9 is a good starting point.

P18C.5 This exercise gives some familiarity with the difficulties involved in predicting the structure of activated complexes. It also demonstrates the importance of femtosecond spectroscopy for understanding chemical dynamics because direct experimental observation of a cluster resembling the activated complex removes much of the ambiguity of theoretical predictions. Consider the attack of H on D_2, which is one step in the $H_2 + D_2$ reaction. (a) Suppose that the H approaches D_2 from the side and forms a complex in the form of an isosceles triangle. Take the H–D distance as 30 per cent greater than in H_2 (74 pm) and the D–D distance as 20 per cent greater than in H_2. Let the critical coordinate be the antisymmetric stretching vibration in which one H–D bond stretches as the other shortens. Let all the vibrations be at about 1000 cm^{-1}. Estimate k_r for this reaction at 400 K using the experimental activation energy of about 35 kJ mol^{-1}. (b) Now change the model of the activated complex in part (a) and make it linear. Use the same estimated molecular bond lengths and vibrational frequencies to calculate k_r for this choice of model. (c) Clearly, there is much scope for modifying the parameters of the models of the activated complex. Use mathematical software to vary the structure of the complex and the parameters in a plausible way, and look for a model (or more than one model) that gives a value of k_r close to the experimental value, $4 \times 10^5\,dm^3\,mol^{-1}\,s^{-1}$.

P18C.6‡ M. Cyfert et al. (*Int. J. Chem. Kinet.* **28**, 103 (1996)) examined the oxidation of tris(1,10-phenanthroline)iron(II) by periodate in aqueous solution. To assess the kinetic salt effect, they measured rate constants at a variety of concentrations of Na_2SO_4 far in excess of reactant concentrations and reported the following data at 298 K:

$[Na_2SO_4]/(mol\,kg^{-1})$	0.2	0.15	0.1	0.05	0.025	0.0125	0.005
$k_r/(dm^{3/2}\,mol^{-1/2}\,s^{-1})$	0.462	0.430	0.390	0.321	0.283	0.252	0.224

What can be inferred about the charge of the activated complex of the rate-determining step? The ionic strength of a solution of Na_2SO_4 is $3[Na_2SO_4]/(mol\,kg^{-1})$.

P18C.7 The study of conditions that optimize the association of proteins in solution guides the design of protocols for formation of large crystals that are amenable to analysis by X-ray diffraction techniques. It is important to characterize protein dimerization because the process is considered to be the rate-determining step in the growth of crystals of many proteins. Consider the variation with ionic strength of the rate constant at 298 K of dimerization in aqueous solution of a cationic protein P:

I	0.0100	0.0150	0.0200	0.0250	0.0300	0.0350
k_r/k_r°	8.10	13.30	20.50	27.80	38.10	52.00

What can be deduced about the charge of P?

P18C.8 In an experimental study of a bimolecular reaction in aqueous solution, the second-order rate constant was measured at 25 °C and at a variety of ionic strengths; the results are tabulated below. It is known that a singly charged ion is involved in the rate-determining step. What is the charge on the other ion involved?

I	0.0025	0.0037	0.0045	0.0065	0.0085
$k_r/(\mathrm{dm^3\,mol^{-1}\,s^{-1}})$	1.05	1.12	1.16	1.18	1.26

P18C.9 The rate constant of the reaction $I^-(aq) + H_2O_2(aq) \rightarrow H_2O(l) + IO^-(aq)$ varies weakly with ionic strength, even though use of the Debye–Hückel limiting law predicts no effect. Assume that the rate-determining step involves a reaction between I^- and H_2O_2 and use the following data from 25 °C to find the dependence of log k_r on the ionic strength:

I	0.0207	0.0525	0.0925	0.1575
$k_r/(\mathrm{dm^3\,mol^{-1}\,min^{-1}})$	0.663	0.670	0.679	0.694

Evaluate the limiting value of k_r at zero ionic strength. What does the result suggest for the dependence of log γ on ionic strength for a neutral molecule in an electrolyte solution?

P18C.10 Use the Debye–Hückel limiting law to show that changes in ionic strength can affect the rate of reaction catalysed by H^+ from the deprotonation of a weak acid. Consider the mechanism: $H^+ + B \rightarrow P$, where H^+ is supplied by the weak acid, HA, which has a fixed concentration. First show that log $[H^+]$ depends on the activity coefficients of ions and thus depends on the ionic strength. Then find the relationship between log(rate) and log $[H^+]$ to show that the rate also depends on the ionic strength.

P18C.11 The bromination of a deuterated hydrocarbon at 298 K proceeds 6.4 times more slowly than the bromination of the undeuterated material. What value of the force constant for the cleaved bond can account for this difference?

TOPIC 18D The dynamics of molecular collisions

Discussion questions

D18D.1 Describe how the following techniques are used in the study of chemical dynamics: infrared chemiluminescence, laser-induced fluorescence, multiphoton ionization, resonant multiphoton ionization, and reaction product imaging.

D18D.2 Discuss the relationship between the saddle-point energy and the activation energy of a reaction.

D18D.3 Consider a reaction with an attractive potential energy surface. Discuss how the initial distribution of reactant energy affects how efficiently the reaction proceeds. Repeat the discussion for a repulsive potential energy surface.

D18D.4 Describe how molecular beams are used to investigate intermolecular forces.

Exercises

E18D.1(a) The interaction between an atom and a diatomic molecule is described by an 'attractive' potential energy surface. What distribution of vibrational and translational energies among the reactants is most likely to lead to a successful reaction? Describe the distribution of vibrational and translational energies among the products for these most successful reactions.

E18D.1(b) The interaction between an atom and a diatomic molecule is described by a 'repulsive' potential energy surface. What distribution of vibrational and translational energies among the reactants is most likely to lead to a successful reaction? Describe the distribution of vibrational and translational energies among the products for these most successful reactions.

E18D.2(a) If the cumulative reaction probability was independent of energy, what would be the temperature dependence of the rate constant predicted by the numerator of eqn 18D.6?

E18D.2(b) If the cumulative reaction probability equalled 1 for energies less than a barrier height V and vanished for higher energies, what would be the temperature dependence of the rate constant predicted by the numerator of eqn 18D.6?

Problems

P18D.1 Show that the intensities, I, of a molecular beam before and after passing through a chamber of length L containing inert scattering atoms are related by $I = I_0 e^{-\mathcal{N}\sigma L}$, where σ is the collision cross-section and $\mathcal{N}$ the number density of scattering atoms.

P18D.2 In a molecular beam experiment to measure collision cross-sections it was found that the intensity of a CsCl beam was reduced to 60 per cent of its intensity on passage through CH_2F_2 at 10 μTorr, but that when the target was Ar at the same pressure the intensity was reduced only by 10 per cent. What are the relative cross-sections of the two types of collision? Why is one much larger than the other?

P18D.3 Suppose a harmonic oscillator collides with another oscillator of the same effective mass and force constant. Evaluate k_r by assuming that the state-to-state rate constant for the excitation of the latter's vibration is $k_{\upsilon\upsilon'} = k_r^{\circ}\delta_{\upsilon\upsilon'}e^{-\lambda\upsilon}$, implying that the transfer becomes less efficient as the vibrational quantum number increases. *Hint:* Refer to *Brief illustration* 18D.1.

P18D.4 Use the approach in *Brief illustration* 18D.2 to analyse the reaction $H + OD \rightarrow OH + D$.

TOPIC 18E Electron transfer in homogeneous systems

Discussion questions

D18E.1 Discuss how the following factors affect the rate of electron transfer in homogeneous systems: the distance between electron donor and acceptor, the standard Gibbs energy of the process, and the reorganization energy of the redox active species and the surrounding medium.

D18E.2 What role does tunnelling play in electron transfer?

D18E.3 Explain why the rate constant for electron transfer decreases as the reaction becomes more exergonic in the inverted region.

Exercises

E18E.1(a) By how much does $H_{et}(d)^2$ change when d is increased from 1.0 nm to 2.0 nm, with $\beta \approx 9\,\text{nm}^{-1}$?

E18E.1(b) By how much does $H_{et}(d)^2$ change when d is increased from 1.0 nm to 2.0 nm, with $\beta \approx 30\,\text{nm}^{-1}$?

E18E.2(a) For an electron donor/acceptor pair at 298 K, $H_{et}(d) = 0.04\,\text{cm}^{-1}$, $\Delta_r G^\ominus = -0.185\,\text{eV}$, and $k_{et} = 37.5\,\text{s}^{-1}$. Use mathematical software to estimate the value of the reorganization energy.

E18E.2(b) For an electron donor/acceptor pair at 298 K, $k_{et} = 2.02 \times 10^5\,\text{s}^{-1}$ and $\Delta_r G^\ominus = -0.665\,\text{eV}$. The standard reaction Gibbs energy changes to $\Delta_r G^\ominus = -0.975\,\text{eV}$ when a substituent is added to the electron acceptor and the rate constant for electron transfer changes to $k_{et} = 3.33 \times 10^6\,\text{s}^{-1}$. Assume that the distance between donor and acceptor is the same in both experiments and estimate the values of $H_{et}(d)$ and ΔE_R.

E18E.3(a) For an electron donor/acceptor pair, $k_{et} = 2.02 \times 10^5\,\text{s}^{-1}$ when $d = 1.11\,\text{nm}$ and $k_{et} = 4.51 \times 10^4\,\text{s}^{-1}$ when $r = 1.23\,\text{nm}$. Assume that $\Delta_r G^\ominus$ and ΔE_R are the same in both experiments and estimate the value of β.

E18E.3(b) Refer to Exercise E18E.3(a). Estimate the value of k_{et} when $d = 1.59\,\text{nm}$.

Problems

P18E.1 Consider the reaction $D + A \rightarrow D^+ + A^-$. The rate constant k_r may be determined experimentally or may be predicted by the *Marcus cross-relation* $k_r = (k_{DD}k_{AA}K)^{1/2}f$, where k_{DD} and k_{AA} are the experimental rate constants for the electron self-exchange processes $^*D + D^+ \rightarrow {}^*D^+ + D$ and $^*A + A^+ \rightarrow {}^*A^+ + A$, respectively, and f is a function of $K = [D^+][A^-]/[D][A]$, k_{DD}, k_{AA}, and $\kappa\nu^\ddagger$. Derive the approximate form of the Marcus cross-relation by following these steps. (a) Use eqn 18E.6 to write expressions for $\Delta^\ddagger G$, $\Delta^\ddagger G_{DD}$, and $\Delta^\ddagger G_{AA}$, keeping in mind that $\Delta_r G^\ominus = 0$ for the electron self-exchange reactions. (b) Assume that the reorganization energy $\Delta E_{R,DA}$ for the reaction $D + A \rightarrow D^+ + A^-$ is the average of the reorganization energies $\Delta E_{R,DD}$ and $\Delta E_{R,AA}$ of the electron self-exchange reactions. Then show that in the limit of small magnitude of $\Delta_r G^\ominus$, or $|\Delta_r G^\ominus| << \Delta E_{R,DA}$, $\Delta^\ddagger G_{DA} = \frac{1}{2}(\Delta^\ddagger G_{DD} + \Delta^\ddagger G_{AA} + \Delta_r G^\ominus)$, where $\Delta_r G^\ominus$ is the standard Gibbs energy for the reaction $D + A \rightarrow D^+ + A^-$. (c) Use an equation of the form of eqn 18E.5 to write expressions for k_{DD} and k_{AA}. (d) Use eqn 18E.5 and the result you have derived to write an expression for k_r. (e) Complete the derivation by using the results from part (c), the relation $K = e^{-\Delta_r G^\ominus/RT}$), and assuming that all $\kappa\nu^\ddagger$ terms are identical.

P18E.2 Consider the reaction $D + A \rightarrow D^+ + A^-$. The rate constant k_r may be determined experimentally or may be predicted by the Marcus cross-relation (see Problem P18E.1). It is common to make the assumption that $f \approx 1$. Use the approximate form of the Marcus relation to estimate the rate constant for the reaction $[Ru(bpy)_3]^{3+} + [Fe(OH_2)_6]^{2+} \rightarrow [Ru(bpy)_3]^{2+} + [Fe(OH_2)_6]^{3+}$, where bpy stands for 4,4′-bipyridine. Use the following data:

$[Ru(bpy)_3]^{3+} + e^- \rightarrow [Ru(bpy)_3]^{2+}$	$E^\ominus = 1.26\,\text{V}$
$[Fe(OH_2)_6]^{3+} + e^- \rightarrow [Fe(OH_2)_6]^{2+}$	$E^\ominus = 0.77\,\text{V}$
$^*[Ru(bpy)_3]^{3+} + [Ru(bpy)_3]^{2+} \rightarrow {}^*[Ru(bpy)_3]^{2+} + [Ru(bpy)_3]^{3+}$	$k_{Ru} = 4.0 \times 10^8\,\text{dm}^3\,\text{mol}^{-1}\,\text{s}^{-1}$
$^*[Fe(OH_2)_6]^{3+} + [Fe(OH_2)_6]^{2+} \rightarrow {}^*[Fe(OH_2)_6]^{2+} + [Fe(OH_2)_6]^{3+}$	$k_{Fe} = 4.2\,\text{dm}^3\,\text{mol}^{-1}\,\text{s}^{-1}$

P18E.3 Some data in the inverted region on a series of donor–linker–acceptor complexes are as follows:

$-\Delta_r G^\ominus/\text{eV}$	0.20	0.60	1.0	1.3	1.6	2.0	2.4
$\log k_{et}$	8.2	9.7	10.2	10.1	9.4	7.7	5.1

Evaluate the reorganization energy.

P18E.4 A useful strategy for the study of electron transfer in proteins consists of attaching an electroactive species to the protein's surface and then measuring k_{et} between the attached species and an electroactive protein cofactor. J.W. Winkler and H.B. Gray (*Chem. Rev.* **92**, 369 (1992)) summarize data for cytochrome *c* modified by replacement of the haem (heme) iron by a zinc ion, resulting in a zinc–porphyrin (ZnP) group in the interior of the protein, and by attachment of a ruthenium ion complex to a surface histidine amino acid. The edge-to-edge distance between the electroactive species was thus fixed at 1.23 nm. A variety of ruthenium ion complexes with different standard potentials was used. For each ruthenium-modified protein, either $Ru^{2+} \rightarrow ZnP^+$ or $ZnP^* \rightarrow Ru^{3+}$, in which the electron donor is an electronically excited state of the zinc–porphyrin group formed by laser excitation, was monitored. This arrangement leads to different standard reaction Gibbs energies because the redox couples ZnP^+/ZnP and ZnP^+/ZnP^* have different standard potentials, with the electronically excited porphyrin being a more powerful reductant. Use the following data to estimate the reorganization energy for this system:

$-\Delta_r G^\ominus/\text{eV}$	0.665	0.705	0.745	0.975	1.015	1.055
$k_{et}/(10^6\,\text{s}^{-1})$	0.657	1.52	1.12	8.99	5.76	10.1

P18E.5 The photosynthetic reaction centre of the purple photosynthetic bacterium *Rhodopseudomonas viridis* contains a number of bound cofactors that participate in electron transfer reactions. The following table shows data compiled by Moser et al. (*Nature* 355, 796 (1992)) on the rate constants for electron transfer between different cofactors and their edge-to-edge distances:

Reaction	$BChl^- \rightarrow BPh$	$BPh^- \rightarrow BChl_2^+$	$BPh^- \rightarrow Q_A$	cyt $c_{559} \rightarrow BChl_2^+$
d/nm	0.48	0.95	0.96	1.23
k_{et}/s^{-1}	1.58×10^{12}	3.98×10^9	1.00×10^9	1.58×10^8

Reaction	$Q_A^- \rightarrow Q_B$	$Q_A^- \rightarrow BChl_2^+$
d/nm	1.35	2.24
k_{et}/s^{-1}	3.98×10^7	63.1

(BChl, bacteriochlorophyll; $BChl_2$, bacteriochlorophyll dimer, functionally distinct from BChl; BPh, bacteriophaeophytin; Q_A and Q_B, quinone molecules bound to two distinct sites; cyt c_{559}, a cytochrome bound to the reaction centre complex). Are these data in agreement with the behaviour predicted by eqn 18E.7? If so, evaluate the value of β.

P18E.6 The rate constant for electron transfer between a cytochrome *c* and the bacteriochlorophyll dimer of the reaction centre of the purple bacterium *Rhodobacter sphaeroides* (Problem P18E.5) decreases with decreasing temperature in the range 300 K to 130 K. Below 130 K, the rate constant becomes independent of temperature. Account for these results.

FOCUS 18 Reaction dynamics

Integrated activities

I18.1 According to the RRK model (see *A deeper look* 12 on the website of this book)

$$P=\frac{n!(n-n^*+s-1)}{(n-n^*)!(n+s-1)}$$

Use Stirling's approximation of the form $\ln x! \approx x\ln x - x$ to deduce that $P \approx \{(n-n^*)/n\}^{s-1}$ when $s-1 << n-n^*$. *Hint:* Replace terms of the form $n-n^*+s-1$ by $n-n^*$ inside logarithms but retain $n-n^*+s-1$ when it is a factor of a logarithm.

I18.2 Estimate the orders of magnitude of the partition functions involved in a rate expression. State the order of magnitude of q_m^T/N_A, q^R, q^V, q^E for typical molecules. Check that in the collision of two structureless molecules the order of magnitude of the pre-exponential factor is of the same order as that predicted by collision theory. Go on to estimate the steric factor for a reaction in which $A + B \rightarrow P$, and A and B are nonlinear triatomic molecules.

I18.3 Discuss the factors that govern the rates of electron transfer according to Marcus theory and that govern the rates of resonance energy transfer according to Förster theory (Topic 17G). Can you find similarities between the two theories?

FOCUS 19

Processes at solid surfaces

A great deal of chemistry occurs at solid surfaces. For example, the surfaces of solid catalysts provide sites where reactants can attach and then undergo chemical transformations. Even as simple an act as dissolving is intrinsically a surface phenomenon, with molecules or ions gradually escaping into the solvent from sites on its surface. Surface deposition, in which atoms are laid down on a surface to create layers, is crucial to the semiconductor industry as it is used in the fabrication of integrated circuits. Finally, in electrochemical processes, electron transfer takes place at the surface of the electrodes.

19A An introduction to solid surfaces

In many cases the surface of a solid has a different structure from a slice through a bulk solid. It can have various types of imperfections, which turn out to have a profound effect on how atoms and molecules interact with it. This Topic describes the structure of surfaces, the attachment of molecules to them, and some of the techniques used to study them.

19A.1 **Surface growth**; 19A.2 **Physisorption and chemisorption**; 19A.3 **Experimental techniques**

19B Adsorption and desorption

A knowledge of the extent to which molecules attach themselves to a surface is crucial to understanding the way in which a surface influences chemical processes. The extent of adsorption can be explored with the aid of some simple models that allow quantitative predictions to be made about how the extent of surface coverage varies with both pressure and temperature.

19B.1 **Adsorption isotherms**; 19B.2 **The rates of adsorption and desorption**

19C Heterogeneous catalysis

The chemical industry relies on the use of efficient catalysts to facilitate a wide variety of transformations, and the majority of these catalysts involve reactions at surfaces. This Topic describes how the concepts introduced in Topic 19B can be extended to provide a way of modelling surface reactions.

19C.1 **Mechanisms of heterogeneous catalysis**; 19C.2 **Catalytic activity at surfaces**

19D Processes at electrodes

The key process at an electrode is the transfer of electrons, which takes place at the interface between a solid surface and an electrolyte. The process can be modelled by using a version of transition-state theory and leads to an understanding of electron transfer as an activated process and how it is affected by factors that can be controlled.

19D.1 **The electrode–solution interface**; 19D.2 **The current density at an electrode**; 19D.3 **Voltammetry**; 19D.4 **Electrolysis**; 19D.5 **Working galvanic cells**

Web resources What is an application of this material?

Almost the whole of modern chemical industry depends on the development, selection, and application of catalysts, with heterogeneous catalysts being particularly important. *Impact* 27 gives a brief overview of the types of catalysts used and the way in which they are thought to act. The search for efficient, portable or small-scale methods of generating electrical power has led to the development of fuel cells which convert hydrogen or hydrocarbon fuels directly into electrical power. *Impact* 28 reviews some of the developments in this area.

TOPIC 19A An introduction to solid surfaces

➤ Why do you need to know this material?

To understand the processes occurring on a surface you need to know about its structure, how molecules are attached to it, and how it is studied experimentally.

➤ What is the key idea?

The attachment of molecules to a surface is influenced by structural features of the surface, including defects.

➤ What do you need to know already?

You need to be aware of the basic facts about the structures of solids (Topic 15A) and diffraction techniques (Topic 15B). This Topic draws on results from the kinetic theory of gases (Topic 1B).

Many events take place on surfaces. They include the growth of the surface itself as atoms or molecules condense on to it. They also include the attachment of other species, such as molecules from a gas. **Adsorption** is the attachment of particles to a solid surface; **desorption** is the reverse process. The substance that adsorbs is the **adsorbate** and the material to which it adsorbs is the **adsorbent** or **substrate**. Adsorption should be distinguished from absorption, the penetration of molecules into the interior of the solid; absorption is often preceded by adsorption.

19A.1 Surface growth

A simple picture of a perfect crystal surface is as a tray of oranges in a grocery store (Fig. 19A.1). A gas molecule that collides with the surface can be imagined as a ping-pong ball bouncing erratically over the oranges. The molecule loses energy as it bounces, but it is likely to escape from the surface before it has lost enough kinetic energy to be trapped. The same is true, to some extent, of an ionic crystal in contact with a solution. There is little energy advantage for an ion in solution to discard some of its solvating molecules and stick at an exposed position on the surface.

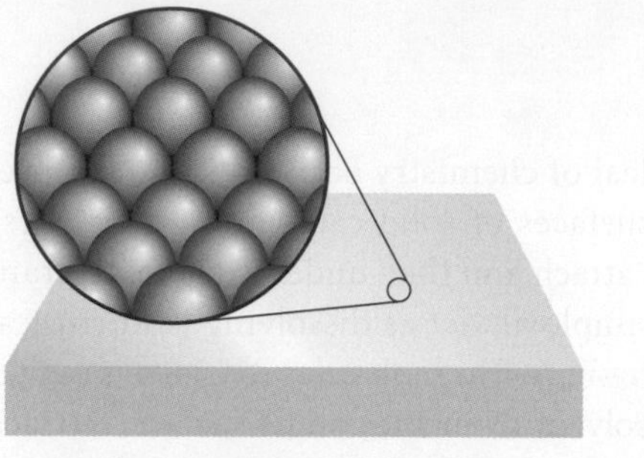

Figure 19A.1 A schematic diagram of the flat surface of a solid. This primitive model is largely supported by scanning tunnelling microscope images.

The surface of a solid is, however, rarely a flat plane: it is a more rugged landscape, featuring different kinds of defects arising from incomplete layers of atoms or ions. A common type of surface defect is a **step** between two otherwise flat layers of atoms called **terraces** (Fig. 19A.2). A step defect might itself have defects, such as kinks. The presence of defects can have a strong influence on the adsorption process. For example, when a molecule strikes a terrace it bounces across it under the influence of the intermolecular potential, and might then come to a step or a corner formed by a kink. Instead of interacting with a single terrace atom, the molecule now interacts with several, and the interaction may be strong enough to trap it. Likewise, when ions deposit from solution, the loss of the solvation interaction is offset by a strong Coulombic

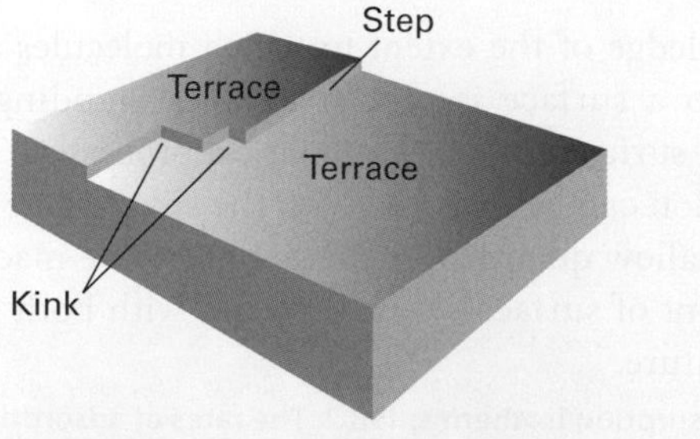

Figure 19A.2 Steps and kinks are two kinds of defects that may occur on otherwise perfect terraces. Defects play an important role in the adsorption of molecules and catalysis.

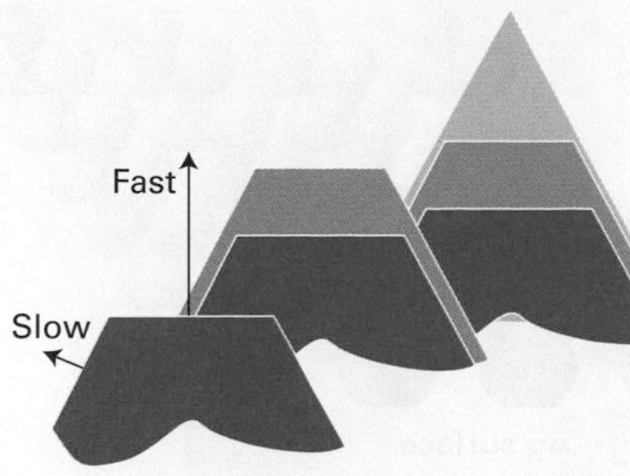

Figure 19A.3 The slower-growing faces of a crystal dominate its final external appearance. Three successive stages of the growth are shown.

interaction between the arriving ions and several ions at the surface defect.

A crystal grows as molecules or ions accumulate on its surface. Different crystal planes grow at different rates, and the slowest growing faces dominate the appearance of the crystal. This feature is explained in Fig. 19A.3, where it can be seen that, although the horizontal face grows forward most rapidly, it grows itself out of existence, and the slower-growing faces survive.

Under normal conditions, a surface exposed to a gas is constantly bombarded with molecules so a freshly prepared surface is covered very quickly. Just how quickly can be estimated by using the kinetic theory of gases to derive an expression for the collision flux Z_W (eqn 16A.7a), which is the number of collisions that occur on a region of surface in a given interval divided by the area of the region and the duration of the interval:

$$Z_W = \frac{p}{(2\pi mkT)^{1/2}} = \frac{p}{(2\pi MkT/N_A)^{1/2}} \quad \text{Collision flux} \quad (19A.1)$$

In this expression, m is the mass of the molecule and M is its molar mass.

Brief illustration 19A.1

For N_2 gas, for which $M = 28\,\text{g mol}^{-1}$, at 1.0 bar ($1.0 \times 10^5$ Pa) and 298 K the collision flux is

$$Z_W = \frac{\overbrace{1.0\times10^5\,\text{Pa}}^{\text{kg m}^{-1}\text{s}^{-2}}}{(2\pi\times(0.028\,\text{kg mol}^{-1}/6.022\times10^{23}\,\text{mol}^{-1})\times(1.381\times10^{-23}\,\text{J K}^{-1})\times(298\,\text{K}))^{1/2}}$$
$$= 2.9\times10^{27}\,\text{m}^{-2}\,\text{s}^{-1}$$

where $1\,\text{J} = 1\,\text{kg m}^2\,\text{s}^{-2}$ has been used. Because $1\,\text{m}^2$ of metal surface consists of about 10^{19} atoms, each atom is struck about 10^8 times each second. Even if only a few collisions result in successful adsorption, the time for which a freshly prepared surface remains clean is very short.

19A.2 Physisorption and chemisorption

Molecules and atoms can attach to surfaces in two ways, by 'physisorption' and by 'chemisorption'.

In **physisorption** (a contraction of 'physical adsorption'), there is a van der Waals interaction (e.g. a dispersion or a dipolar interaction, Topic 14B) between the adsorbate and the substrate. Such interactions have a long range but are weak, and the energy released when a particle is physisorbed is of the same order of magnitude as the enthalpy of condensation. For a molecule to remain bound to the surface, the energy released on binding needs to be dissipated in some way; if it is not lost, the molecule will leave the surface again. The energy involved in physisorption is sufficiently small that it can be dispersed into vibrations of the lattice and dissipated as thermal motion: this process is called **accommodation**.

The enthalpy of physisorption can be measured by monitoring the rise in temperature of a sample of known heat capacity; typical values are in the region of $-20\,\text{kJ mol}^{-1}$ (Table 19A.1). This small enthalpy change is insufficient to lead to bond breaking, so a physisorbed molecule retains its identity, although it might be distorted by the presence of the surface.

In **chemisorption** (a contraction of 'chemical adsorption'), the molecules (or atoms) attach to the surface by forming a chemical (usually covalent) bond, and tend to find sites that maximize their coordination number with the substrate. The enthalpy of chemisorption is very much greater than that for physisorption, and typical values are in the region of $-200\,\text{kJ mol}^{-1}$ (Table 19A.2). The distance between the surface

Table 19A.1 Maximum observed standard enthalpies of physisorption at 298 K*

Adsorbate	$\Delta_{ad}H^\ominus/(\text{kJ mol}^{-1})$
CH_4	−21
H_2	−84
H_2O	−59
N_2	−21

* More values are given in the *Resource section*.

Table 19A.2 Standard enthalpies of chemisorption, $\Delta_{ad}H^\ominus/(\text{kJ mol}^{-1})$, at 298 K*

Adsorbate	Adsorbent (substrate)		
	Cr	Fe	Ni
C_2H_4	−427	−285	−243
CO		−192	
H_2	−188	−134	
NH_3		−188	−155

* More values are given in the *Resource section*.

and the closest adsorbate atom is also typically shorter for chemisorption than for physisorption. Bonds within a chemisorbed molecule may be broken due to strong interactions with the surface atoms, resulting in molecular fragments bound to the surface. Such species often play a key role in the catalytic properties of solid surfaces (Topic 19C).

Except in special cases, chemisorption must be exothermic. A spontaneous process requires $\Delta G < 0$ at constant pressure and temperature. Because the translational freedom of the adsorbate is reduced when it is adsorbed, ΔS is negative. Therefore, in order for $\Delta G = \Delta H - T\Delta S$ to be negative, ΔH must be negative (i.e. the process must be exothermic). Exceptions may occur if the adsorbate dissociates and has high translational mobility on the surface. An example is the adsorption of H_2 on glass, which is an endothermic process and in which adsorption is believed to be accompanied by dissociation to mobile H atoms.

The enthalpy of adsorption depends on the extent of surface coverage, mainly because the adsorbed species interact with each other. If they repel each other (as for CO on palladium) the adsorption becomes less exothermic (the enthalpy of adsorption less negative) as coverage increases. Moreover, studies show that such species settle on the surface in a disordered way until packing requirements demand order. If the adsorbate species attract one another (as for O_2 on tungsten), then they tend to cluster together in islands, and growth occurs at the edges. These adsorbates also show order–disorder transitions when they are heated enough for thermal motion to overcome the interactions between the absorbed species, but not so much that they are desorbed.

The extent of surface coverage (by either physisorption or chemisorption) is normally expressed as the **fractional coverage**, θ:

$$\theta = \frac{\text{number of adsorption sites occupied}}{\text{number of adsorption sites available}} = \frac{N_{\text{occupied}}}{N_{\text{available}}}$$

Fractional coverage [definition] (19A.2)

The fractional coverage is often calculated from the relation $\theta = V/V_\infty$, where V is the volume of adsorbate adsorbed and V_∞ is the volume of adsorbate corresponding to complete monolayer coverage. The two volumes are of the free gas measured under the same conditions of temperature and pressure, not the volume that the adsorbed gas occupies when attached to the surface.

Brief illustration 19A.2

For the adsorption of CO on charcoal at 273 K, $V_\infty = 111\,\text{cm}^3$, a value corrected to 1 atm. When the charcoal is exposed to a mixture of gases in which the partial pressure of CO is 80.0 kPa, the value of V (also corrected to 1 atm) is $41.6\,\text{cm}^3$, so it follows that $\theta = (41.6\,\text{cm}^3)/(111\,\text{cm}^3) = 0.375$ under these conditions.

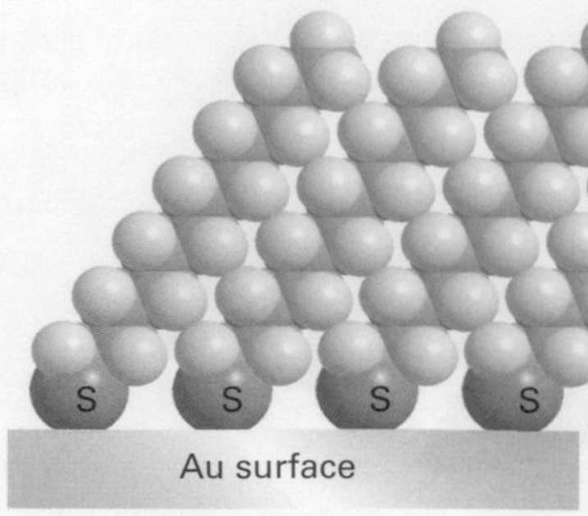

Figure 19A.4 Self-assembled monolayers of alkyl thiols formed onto a gold surface by reaction of the thiol groups with the surface and aggregation of the alkyl chains.

Chemisorption can be used as the basis for manipulation of surfaces on the nanometre scale. Of current interest are **self-assembled monolayers** (SAMs), ordered molecular aggregates that form a single layer of material on a surface. To understand the formation of SAMs, consider exposing molecules such as alkyl thiols, RSH, where R represents an alkyl chain, to an Au(0) surface. The thiols react with the surface, forming RS^- Au(I) adducts:

$$\text{RSH} + \text{Au(0)}_n \rightarrow \text{RS}^-\text{Au(I)}\cdot\text{Au(0)}_{n-1} + \tfrac{1}{2}\text{H}_2$$

If R is a sufficiently long chain, van der Waals interactions between the adsorbed RS units lead to the formation of a highly ordered monolayer on the surface (Fig. 19A.4). A self-assembled monolayer alters the properties of the surface. For example, a hydrophilic surface may be rendered hydrophobic once covered with a SAM. Furthermore, the attachment of functional groups to the exposed ends of the alkyl groups may impart specific chemical reactivity or ligand-binding properties to the surface, leading to applications in chemical (or biochemical) sensors and reactors.

19A.3 Experimental techniques

Many experimental techniques are available for the study of the structure of the solid surface as well as the properties and arrangement of any adsorbed molecules. Some of these techniques offer atomic level resolution and allow the direct visualization of changes to the surface as adsorption and chemical reactions take place.

Experimental procedures must begin with a clean surface. The obvious way to retain cleanliness of a surface is to reduce the pressure and thereby reduce the number of impacts on the surface. When the pressure is reduced to 0.1 mPa (as in a simple vacuum system) the collision flux falls to about $10^{18}\,\text{m}^{-2}\,\text{s}^{-1}$, corresponding to one hit on a surface atom in each 0.1 s. Even that is too frequent in most experiments, and in **ultrahigh vacuum** (UHV) techniques pressures as low as 0.1 μPa (when $Z_W \approx 10^{15}\,\mathrm{m^{-2}\,s^{-1}}$) are reached on a routine basis and as low as 1 nPa (when $Z_W \approx 10^{13}\,\mathrm{m^{-2}\,s^{-1}}$) are reached with special care. These collision fluxes correspond to each surface atom being hit once every 10^5–10^6 s, or about once a day.

(a) Microscopy

The basic approach of illuminating a small area of a sample and collecting light with a microscope has been used for many years to image small specimens. However, the resolution of a microscope, the minimum distance between two objects that leads to two distinct images, is on the order of the wavelength of the light being used. Therefore, conventional microscopes employing visible light have resolutions of the order of micrometres and are blind to features on a scale of nanometres.

One technique often used to image nanometre-sized objects is **electron microscopy**, in which a beam of electrons with a well-defined de Broglie wavelength (Topic 7A) replaces the light source found in traditional microscopes. Instead of glass or quartz lenses, magnetic fields are used to focus the beam. In **transmission electron microscopy** (TEM), the electron beam passes through the specimen and the image is collected on a screen. In **scanning electron microscopy** (SEM), electrons scattered back from a small area of the sample are detected and an image of the surface is then obtained by scanning the electron beam across the sample.

As in traditional light microscopy, the wavelength of the incident beam and the ability to focus it governs the resolution. It is now possible to achieve atomic resolution with TEM instruments, and SEM instruments can achieve resolution on the order of a few nanometres.

Scanning probe microscopy (SPM) is a collection of techniques that can be used to image and manipulate objects as small as atoms on surfaces. One version is **scanning tunnelling microscopy** (STM), in which a platinum–rhodium or tungsten needle is scanned across the surface of a conducting solid. When the tip of the needle is brought very close to the surface, electrons tunnel across the intervening space (Fig. 19A.5). In the 'constant-current mode' of operation, the tip moves up and down according to the form of the surface, and the topography of the surface, including any adsorbates, can be mapped on an atomic scale. The vertical motion of the tip is achieved by fixing it to a piezoelectric cylinder, which contracts or expands according to the potential difference it experiences. In the 'constant-*z* mode', the vertical position of the tip is held constant and the current is monitored. Because the tunnelling probability is very sensitive to the size of the gap, the microscope can detect tiny, atom-scale variations in the height of the surface.

Figure 19A.6 shows an example of the kind of image obtained from a surface, in this case that of copper atoms forming 'runways' on a surface. Each 'bump' on the surface corresponds to a single copper atom. In a further variation of the STM technique, the tip may be used to nudge single atoms around on the surface, making possible the fabrication of complex nanometre-sized materials and devices.

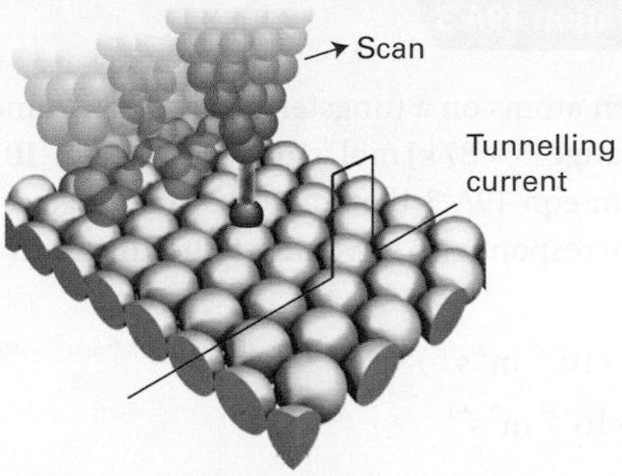

Figure 19A.5 A scanning tunnelling microscope makes use of the current due to electrons that tunnel between the surface and the tip. That current is very sensitive to the distance of the tip above the surface.

The diffusion characteristics of an adsorbate can be examined by using STM to follow the change in surface characteristics. Provided its path is not influenced by defects, an adsorbed atom makes a random walk across the surface, and the diffusion coefficient, D, is inferred from the mean distance, d, travelled in an interval τ by using the two-dimensional random walk expression $d = (D\tau)^{1/2}$ (which is derived like the one-dimensional random walk expression in Topic 16C). Values of D at different temperatures are interpreted by using an Arrhenius-like expression

$$D = D_0 e^{-E_{a,diff}/RT} \qquad \text{Temperature dependence of the diffusion coefficient} \qquad (19A.3)$$

where $E_{a,diff}$ is the activation energy for diffusion and D_0 is the diffusion coefficient in the limit of infinite temperature. The variation of D from one crystal plane to another can also be studied.

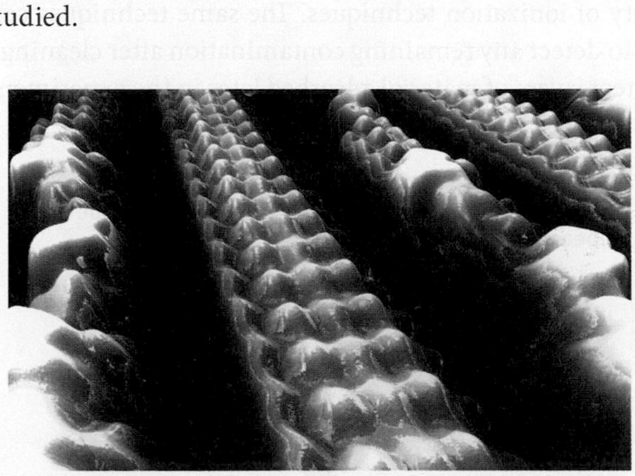

Figure 19A.6 An STM image of copper atoms forming 'runways' on the surface of metallic copper when a layer of CuO is formed on the surface. (Image provided by Stephen Driver and Stephen Jenkins, University of Cambridge.)

Brief illustration 19A.3

For tungsten atoms on a tungsten surface it is found that $E_{a,diff}$ is in the range 57–87 kJ mol^{-1} and $D_0 \approx 3.8 \times 10^{-11}$ m^2 s^{-1}. It follows from eqn 19A.3 that at 800 K this range of activation energies corresponds to a value of the diffusion coefficient between

$$D = (3.8\times10^{-11}\,\mathrm{m^2\,s^{-1}})\times e^{-5.7\times10^4\,\mathrm{J\,mol^{-1}}/((8.3145\,\mathrm{J\,K^{-1}\,mol^{-1}})\times(800\,\mathrm{K}))}$$
$$= 7.2\times10^{-15}\,\mathrm{m^2\,s^{-1}}$$

and

$$D = (3.8\times10^{-11}\,\mathrm{m^2\,s^{-1}})\times e^{-8.7\times10^4\,\mathrm{J\,mol^{-1}}/((8.3145\,\mathrm{J\,K^{-1}\,mol^{-1}})\times(800\,\mathrm{K}))}$$
$$= 7.9\times10^{-17}\,\mathrm{m^2\,s^{-1}}$$

In **atomic force microscopy** (AFM), a sharpened tip attached to a cantilever is scanned across the surface. The force exerted by the surface and any molecules attached to it pushes or pulls on the tip and deflects the cantilever (Fig. 19A.7). The deflection is monitored by using a laser beam. Because no current needs to pass between the sample and the probe, the technique can be applied to non-conducting surfaces and to the study of solid–liquid interfaces.

Two modes of operation of AFM are common. In 'contact mode', or 'constant-force mode', the force between the tip and surface is held constant and the tip makes contact with the surface. This mode of operation can damage fragile samples on the surface. In 'non-contact' mode, or 'tapping' mode, the tip bounces up and down with a specified frequency and never quite touches the surface. The amplitude of the oscillation of the tip changes when it passes over a species adsorbed on the surface.

(b) Ionization techniques

The chemical composition of a surface can be determined by a variety of ionization techniques. The same techniques can be used to detect any remaining contamination after cleaning and to detect layers of material adsorbed later in the experiment.

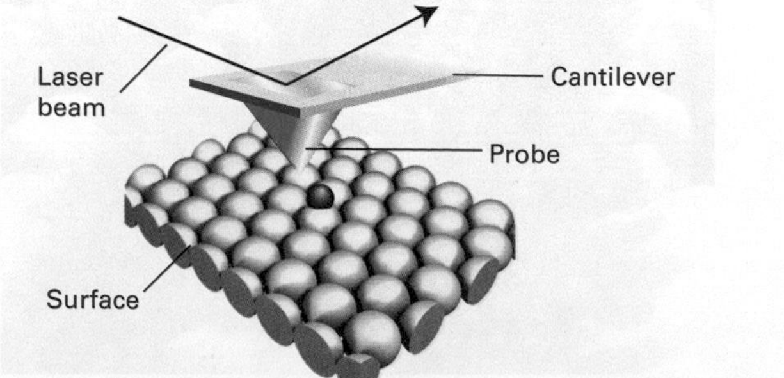

Figure 19A.7 In atomic force microscopy, a laser beam is used to monitor the tiny changes in position of a probe as it is attracted to or repelled by atoms on a surface.

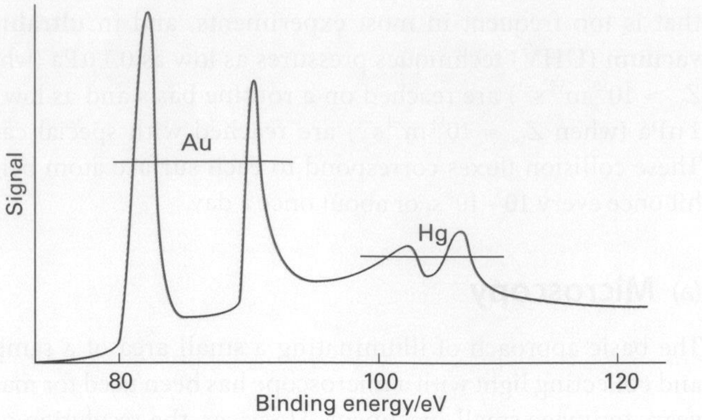

Figure 19A.8 The X-ray photoelectron emission spectrum of a sample of gold contaminated with a surface layer of mercury. (M.W. Roberts and C.S. McKee, *Chemistry of the metal–gas interface*, Oxford (1978).)

One technique is **photoemission spectroscopy**, the origins of which lie in the photoelectric effect (Topic 7A). In this technique the sample is irradiated with photons of sufficient energy to eject electrons from the absorbed species, and the energies of these electrons are measured. If the ionizing radiation is in the ultraviolet (when the technique is denoted UPS), the ejected electrons are from the valence shells and so the technique can be used to infer details of the bonding between the adsorbate and the substrate. If X-rays are used (the technique is then denoted XPS), core electrons are ionized and their energies are characteristic of the atoms present (Fig. 19A.8), thus providing a fingerprint for the material present.

Brief illustration 19A.4

The principal difference between the UPS of free benzene and benzene adsorbed on palladium is in the energies of the π electrons. This difference is interpreted as meaning that the benzene molecules interact with the surface through their π orbitals, implying that the molecules lie parallel to the surface.

A very important technique, which is widely used in the microelectronics industry, is **Auger electron spectroscopy** (AES). The **Auger effect** (pronounced oh-zhey) is the emission of a second electron after high energy radiation has expelled another. The first electron to depart leaves a hole in a low-lying orbital, and then an electron in a higher energy orbital falls into it. The energy this transition releases may result either in the generation of radiation, which is called **X-ray fluorescence** (Fig. 19A.9a) or in the ejection of another electron (Fig. 19A.9b). The latter is the 'secondary electron' of the Auger effect. The energies of the secondary electrons are characteristic of the material present, so AES effectively provides a fingerprint of the sample. In practice, the Auger spectrum is normally obtained by using an electron beam (with energy in the range

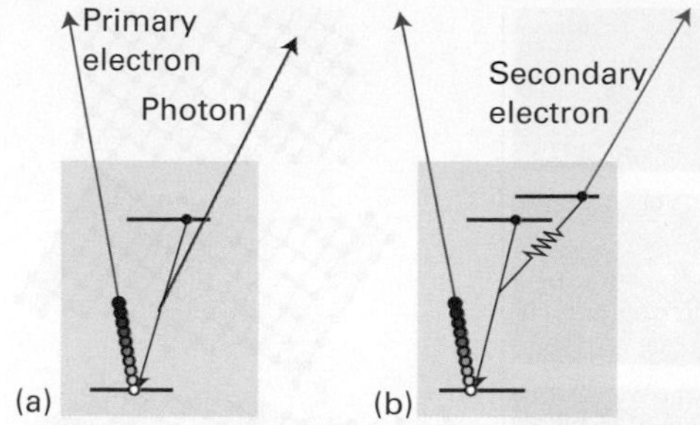

Figure 19A.9 When an electron is expelled from a solid (a) an electron of higher energy may fall into the vacated orbital and emit an X-ray photon (X-ray fluorescence). Alternatively (b) the electron falling into the orbital may give up its energy to another electron, which is ejected as the secondary electron in the Auger effect.

1–5 keV), rather than electromagnetic radiation, to eject the primary electron. In **scanning Auger electron microscopy** (SAM), the finely focused electron beam is scanned over the surface and a map of composition is compiled; the resolution can be less than about 50 nm.

(c) Diffraction techniques

A technique for determining the arrangement of the atoms close to the surface is **low energy electron diffraction** (LEED). This technique is similar to X-ray diffraction (Topic 15B), but makes use of the wavelike properties of electrons. The use of low energy electrons (with energies in the range 10–200 eV, corresponding to wavelengths in the range 400–100 pm) ensures that the diffraction is caused only by atoms at or close to the surface. The experimental arrangement is shown in Fig. 19A.10, and typical LEED patterns, obtained by photographing the fluorescent screen through the viewing port, are shown in Fig. 19A.11.

Observations using LEED show that the surface of a crystal rarely has exactly the same form as a slice through the bulk

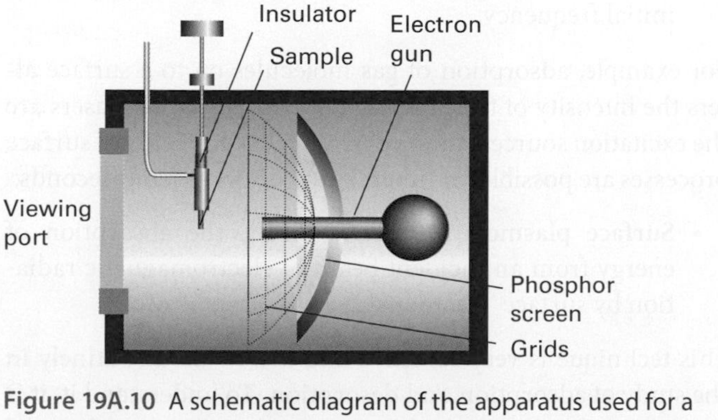

Figure 19A.10 A schematic diagram of the apparatus used for a LEED experiment. The electrons diffracted by the surface layers are detected by the fluorescence they cause on the phosphor screen.

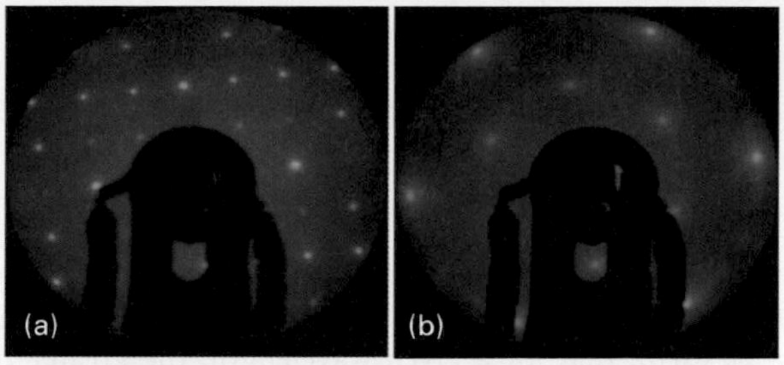

Figure 19A.11 LEED photographs of (a) a clean surface of FeS_2 and (b) after its exposure to Mo atoms; it is thought that MoS_2 forms on the surface. The black area is the shadow from the electron gun. (Photographs provided by Tao Liu, Israel Temprano, David King, Stephen Driver, and Stephen Jenkins, University of Cambridge.)

because surface and bulk atoms experience different forces. **Reconstruction** refers to processes by which atoms on the surface achieve their equilibrium structures. As a general rule, it is found that metal surfaces are simply truncations of the bulk lattice, but the distance between the top layer of atoms and the one below is contracted by around 5 per cent. Semiconductors generally have surfaces reconstructed to a depth of several layers. Reconstruction also occurs in ionic solids. For example, in lithium fluoride the Li^+ and F^- ions close to the surface apparently lie on slightly different planes. An actual example of the detail that can now be obtained from refined LEED techniques is shown in Fig. 19A.12 for CH_3C– adsorbed on a (110) plane of rhodium.

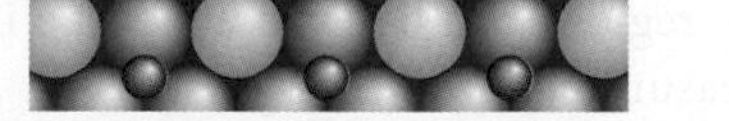

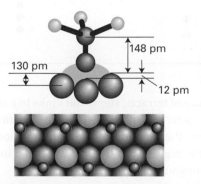

Figure 19A.12 The structure of a surface close to the point of attachment of CH_3C– to the (110) surface of rhodium at 300 K and the changes in positions of the metal atoms that accompany chemisorption.

Example 19A.1 Interpreting a LEED pattern

The LEED pattern from a clean (110) face of palladium is shown in (a) below. The reconstructed surface gives a LEED pattern shown as (b). What can be inferred about the structure of the reconstructed surface?

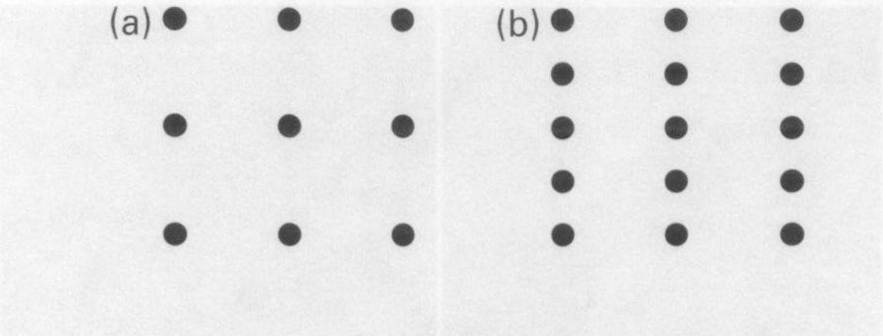

Collect your thoughts Recall from Bragg's law (Topic 15B, $\lambda = 2d\sin\theta$), that for a given wavelength, the greater the separation d of the layers, the smaller is the scattering angle (so that $2d\sin\theta$ remains constant). It follows that, in terms of the LEED pattern, the farther apart the atoms responsible for the pattern, the closer the spots appear in the pattern. Twice the separation between the atoms corresponds to half the separation between the spots, and vice versa. Therefore, by comparing the two patterns you can decide if there has been any change in the spacing between the atoms.

The solution The horizontal separation between spots is unchanged, which indicates that the atoms remain in the same position in that dimension when reconstruction occurs. However, the vertical spacing is halved, which suggests that the atoms are twice as far apart in that direction as they are in the unreconstructed surface.

Self-test 19A.1 Sketch the LEED pattern for a surface that differs from that shown in (a) above by tripling the vertical separation.

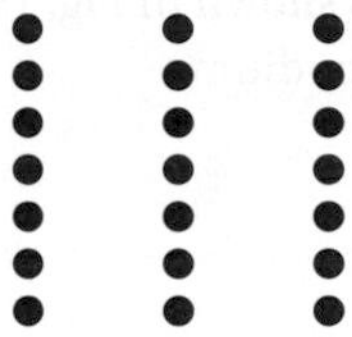

Answer:

The presence of terraces, steps, and kinks in a surface shows up in LEED patterns, and from such experiments the surface defect density (the number of defects in a region divided by the area of the region) can be estimated. The importance of this type of measurement will emerge later.

Brief illustration 19A.5

Three examples of how steps and kinks affect LEED patterns are shown in Fig. 19A.13. The samples were obtained by cleaving a crystal at different angles to a plane of atoms. Only terraces are produced when the cut is parallel to the plane, and the density of steps increases as the angle of the cut increases. The observation of additional structure in the LEED patterns, rather than blurring, shows that the steps are arrayed regularly.

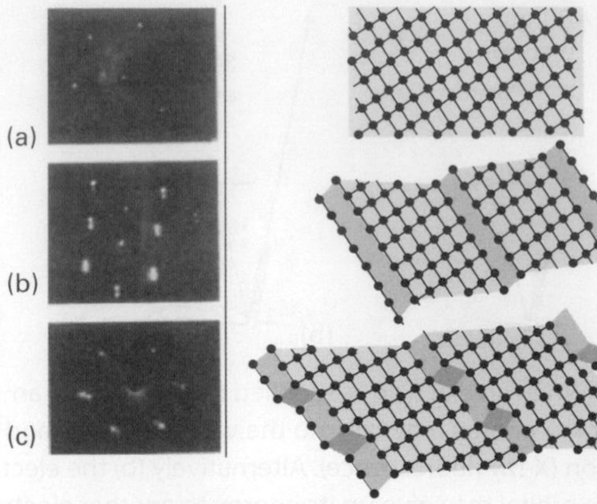

Figure 19A.13 LEED patterns may be used to assess the defect density of a surface. The photographs correspond to a platinum surface with (a) low defect density, (b) regular steps separated by about six atoms, and (c) regular steps with kinks. (Photographs provided by Professor G.A. Samorjai.)

(d) Determination of the extent and rates of adsorption and desorption

A common technique for measuring rates of processes on surfaces is to monitor the rates of flow of gas into and out of the system: the difference is the rate of gas uptake by the sample. Integration of this rate then gives the fractional coverage at any stage. Three other techniques for the determination of fractional coverage and the rate of adsorption are:

- **Gravimetry**, in which the sample is weighed on a microbalance during the experiment.

The technique commonly uses a **quartz crystal microbalance** (QCM), in which the mass of a sample adsorbed on the surface of a quartz crystal is related to changes in the characteristic vibrational frequency of the crystal. Masses as small as a few nanograms can be measured reliably in this way.

- **Second harmonic generation** (SHG), the conversion of an intense, pulsed laser beam to radiation with twice its initial frequency.

For example, adsorption of gas molecules on to a surface alters the intensity of the SHG signal. Because pulsed lasers are the excitation sources, time-resolved measurements of surface processes are possible on timescales as short as femtoseconds.

- **Surface plasmon resonance** (SPR), the absorption of energy from an incident beam of electromagnetic radiation by surface 'plasmons'.

This technique is very sensitive and is now used routinely in the study of adsorption and desorption. To understand it, it is necessary to understand what is meant by a 'surface plasmon' and what kind of 'resonance' is involved.

The mobile delocalized valence electrons of metals form a **plasma**, a dense gas-like collection of charged particles. Bombardment of this plasma by light or an electron beam can cause transient changes in the distribution of electrons, with some regions becoming slightly denser than others. Coulombic repulsion in the regions of high density causes electrons to move away from each other, so lowering their density. The resulting oscillations in electron density, the **plasmons**, can be excited both in the bulk and on the surface of a metal. A surface plasmon propagates away from the surface, but the amplitude of the wave, also called an **evanescent wave**, decreases sharply with distance from the surface. The term 'resonance' in this context refers to the absorption that can be observed with appropriate choice of the wavelength and angle of incidence of the excitation beam.

To detect surface plasmon resonance it is common practice to use a monochromatic beam and to vary the angle of incidence (the ϕ in Fig. 19A.14). The beam passes into a prism and reflects off a face which has been coated with a thin film of gold or silver. Because the evanescent wave interacts with material a short distance away from the surface, the angle at which resonant absorption occurs depends on the refractive index of the medium on the opposite side of the metallic film. Thus, changing the identity and quantity of material on the surface changes the resonance angle.

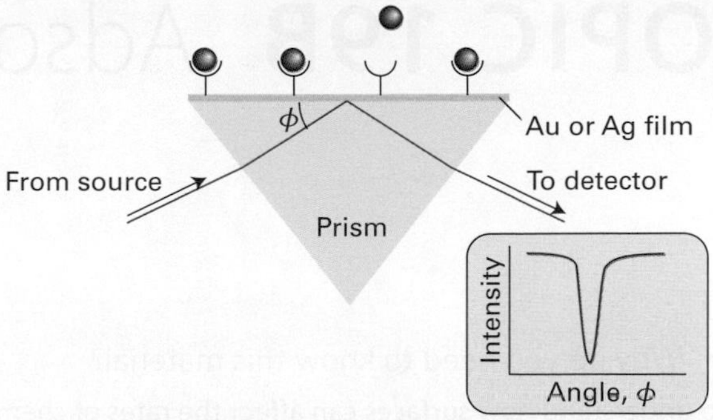

Figure 19A.14 The experimental arrangement for the observation of surface plasmon resonance (SPR), as explained in the text. Here receptors have been attached to the metallic film; the binding of ligands (red spheres) alters the angle at which SPR is detected.

The SPR technique can be used in the study of the binding of molecules to a surface or the binding of ligands to a biopolymer attached to the surface; this interaction mimics the biological recognition processes that occur in cells. Examples of complexes amenable to analysis include antibody–antigen and protein–DNA interactions. The most important advantage of SPR is its sensitivity: it is possible to measure the deposition of nanograms of material on to a surface. The main disadvantage of the technique is its requirement for immobilization of at least one of the components of the system under study.

Checklist of concepts

☐ 1. **Adsorption** is the attachment of molecules to a surface; the substance that adsorbs is the **adsorbate** and the underlying material is the **adsorbent** or **substrate**. The reverse of adsorption is **desorption**.

☐ 2. Surface defects play an important role in the process of adsorption.

☐ 3. In **physisorption** molecules are attached to the surface by relatively weak forces, such as van der Waals interactions; in **chemisorption** the interactions are much stronger, and chemical bonds are formed between adsorbate and substrate.

☐ 4. **Reconstruction** refers to processes by which atoms on the surface achieve their equilibrium structures.

☐ 5. Techniques for studying surfaces and their acronyms are:

AES	Auger electron spectroscopy
AFM	atomic force microscopy
LEED	low energy electron diffraction
QCM	quartz crystal microbalance
SEM	scanning electron microscopy
SHG	second harmonic generation
SPM	scanning probe microscopy
SPR	surface plasmon resonance
STM	scanning tunnelling microscopy
TEM	transmission electron microscopy
UPS	ultraviolet photoelectron spectroscopy
XPS	X-ray photoelectron spectroscopy

Checklist of equations

Property	Equation	Comment	Equation number
Collision flux	$Z_W = p/(2\pi mkT)^{1/2}$	Kinetic theory	19A.1
Fractional coverage	$\theta = N_{occupied}/N_{available}$	Definition	19A.2

TOPIC 19B Adsorption and desorption

> ➤ Why do you need to know this material?
>
> To understand how surfaces can affect the rates of chemical reactions, you need to know how to assess the extent of surface coverage and the factors that determine the rates at which molecules attach to and detach from solid surfaces.
>
> ➤ What is the key idea?
>
> The extent of surface coverage can be expressed in terms of isotherms derived on the basis of dynamic equilibria between adsorbed and free molecules.
>
> ➤ What do you need to know already?
>
> This Topic extends the discussion of adsorption in Topic 19A. You need to be familiar with the basic ideas of chemical kinetics (Topics 17A–17C) and the Arrhenius equation (Topic 17D). One argument makes use of the relation between an equilibrium constant and the standard Gibbs energy of reaction (Topic 6A) and also of the Gibbs–Helmholtz equation (Topic 3E).

When a gas is adsorbed on a surface (Topic 19A) there is a dynamic equilibrium between the free and the adsorbed molecules. The fractional coverage, θ, of the surface (eqn 19A.2) depends on the pressure of the overlying gas and the temperature; the expression describing its variation with pressure at a chosen temperature is called the **adsorption isotherm**.

19B.1 Adsorption isotherms

Many of the techniques discussed in Topic 19A can be used to measure θ. Another is **flash desorption**, in which the sample is suddenly heated electrically and the resulting rise of pressure is interpreted in terms of the amount of adsorbate originally on the sample.

(a) The Langmuir isotherm

An isotherm devised by Irving Langmuir is the simplest that is physically plausible. It is based on four assumptions:

- Adsorption cannot proceed beyond monolayer coverage.
- All sites on the surface are equivalent.
- A molecule can be adsorbed only at a vacant site.
- The probability of adsorption is independent of the occupation of neighbouring sites (that is, there are no interactions between adsorbed molecules).

From these assumptions it is possible to develop an expression for the dependence of the fractional coverage on the pressure.

How is that done? 19B.1 Deriving the Langmuir isotherm

You need to consider the dynamic equilibrium between the molecules (A) in the gas phase and those on the surface (denoted AM):

$$\mathrm{A(g)} + \mathrm{M(surface)} \underset{k_\mathrm{d}}{\overset{k_\mathrm{a}}{\rightleftarrows}} \mathrm{AM(surface)}$$

Step 1 *Write an expression for the rate of adsorption*

The rate of adsorption is proportional to the rate of collisions with the surface and therefore to the partial pressure p of A. The rate is also proportional to the number of vacant sites, because molecules can be adsorbed only at these sites. If the total number of sites is N and the fractional coverage is θ, the number of vacant sites is $N(1-\theta)$. The rate of change of surface coverage, $\mathrm{d}\theta/\mathrm{d}t$, due to adsorption is therefore

$$\frac{\mathrm{d}\theta}{\mathrm{d}t} = k_\mathrm{a} pN(1-\theta) \quad \text{Rate of adsorption} \quad (19B.1a)$$

Step 2 *Write an expression for the rate of desorption*

The rate of change of fractional coverage due to desorption is proportional to the number of adsorbed species already present, which is equal to the number of occupied sites, $N\theta$:

$$\frac{\mathrm{d}\theta}{\mathrm{d}t} = -k_\mathrm{d} N\theta \quad \text{Rate of desorption} \quad (19B.1b)$$

The term is negative because θ decreases as the molecules desorb.

Step 3 *Equate the rates and construct the isotherm*

At equilibrium there is no net change in θ, implying that the sum of these two rates must be zero: $k_\mathrm{a} pN(1-\theta) - k_\mathrm{d} N\theta = 0$. Rearranging this equation gives the following expression, the **Langmuir isotherm**, which relates the surface coverage to the pressure, and in which the parameter α has the dimensions of 1/pressure:

$$\theta=\frac{\alpha p}{1+\alpha p} \qquad \alpha=\frac{k_a}{k_d} \qquad (19B.2)$$

Langmuir isotherm

The Langmuir isotherm is tested by measuring the fractional coverage as a function of the pressure, and then plotting these data in a form expected to give a straight line, as illustrated in the following *Example*.

Example 19B.1 Using the Langmuir isotherm

The following data are for the adsorption of CO on charcoal at 273 K. Confirm that they conform to the Langmuir isotherm, find the value of the parameter α and the volume corresponding to complete coverage. In each case V has been corrected to 1 atm (101.325 kPa).

p/kPa	13.3	26.7	40.0	53.3	66.7	80.0	93.3
V/cm^3	10.2	18.6	25.5	31.5	36.9	41.6	46.1

Collect your thoughts The fractional coverage is given by $\theta = V/V_\infty$, where V_∞ is the volume corresponding to complete coverage (eqn 19A.2). You need to manipulate the Langmuir isotherm so that you can plot a straight-line graph and extract the required parameters from its slope and intercept.

The solution Multiply both sides of eqn 19B.2 by $(1 + \alpha p)$ to give $\theta(1+\alpha p)=\alpha p$, and then substitute $\theta = V/V_\infty$ to give

$$\frac{V}{V_\infty}+\frac{V\alpha p}{V_\infty}=\alpha p$$

Division of both sides by $V\alpha$ gives

$$\frac{1}{\alpha V_\infty}+\frac{p}{V_\infty}=\frac{p}{V} \quad \text{which rearranges to} \quad \frac{p}{V}=\overbrace{\frac{1}{\alpha V_\infty}}^{\text{Intercept}}+\overbrace{\frac{1}{V_\infty}}^{\text{Slope}}p$$

It follows that you should plot p/V against p and expect a straight line of slope $1/V_\infty$ and intercept $1/\alpha V_\infty$ at $p = 0$; note that *slope/intercept* $= (1/V_\infty)/(1/\alpha V_\infty) = \alpha$.

The data for the plot are as follows:

p/kPa	13.3	26.7	40.0	53.3	66.7	80.0	93.3
$(p$/kPa$)/(V$/cm$^3)$	1.30	1.44	1.57	1.69	1.81	1.92	2.02

The points are plotted in Fig. 19B.1. The (least squares) slope is 9.04×10^{-3}, so $V_\infty = 1/(9.04 \times 10^{-3}\ \text{cm}^{-3}) = 111\ \text{cm}^3$. The intercept at $p = 0$ is $(p/\text{kPa})/(V/\text{cm}^3) = 1.20$, or $p/V = 1.20\ \text{kPa cm}^{-3}$ hence $1/\alpha V_\infty = 1.20\ \text{kPa cm}^{-3}$. Therefore

$$\alpha=\frac{1/V_\infty}{1/\alpha V_\infty}=\frac{1/111\,\text{cm}^3}{1.20\,\text{kPa cm}^{-3}}=7.51\times10^{-3}\,\text{kPa}^{-1}$$

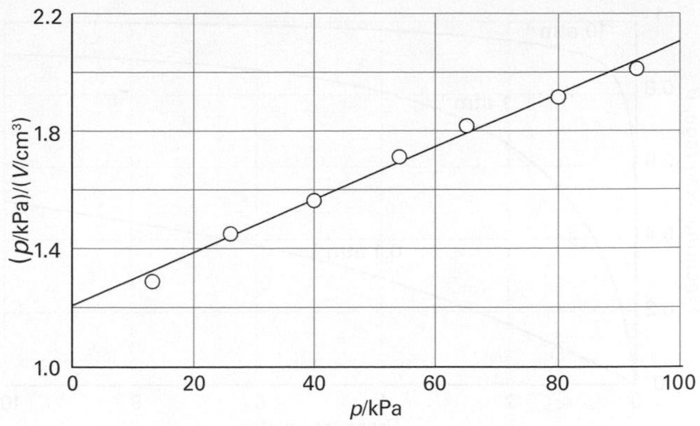

Figure 19B.1 The plot of the data in *Example* 19B.1. As illustrated here, the Langmuir isotherm predicts that a straight line should be obtained when p/V is plotted against p.

Self-test 19B.1 Repeat the calculation for the following data:

p/kPa	13.3	26.7	40.0	53.3	66.7	80.0	93.3
V/cm^3	10.3	19.3	27.3	34.1	40.0	45.5	48.0

Answer: 128 cm^3, 6.69×10^{-3} kPa^{-1}

If a molecule A_2 adsorbs with dissociation to give two A fragments on the surface, the rate of adsorption is proportional to the pressure and to the *square* of the number of vacant sites: two sites are needed to accommodate the two A. The rate of change of the fractional coverage due to adsorption is then

$$\frac{d\theta}{dt}=k_a p\{N(1-\theta)\}^2 \qquad (19B.3a)$$

Desorption requires that two of the species A encounter one another so they can leave as A_2. The rate of change of the fractional coverage is therefore second-order in the number of sites occupied:

$$\frac{d\theta}{dt}=-k_d(N\theta)^2 \qquad (19B.3b)$$

The condition for no net change, which means setting the rates in eqns 19B.3a and 19B.3b equal to each other, leads to the isotherm

$$\theta=\frac{(\alpha p)^{1/2}}{1+(\alpha p)^{1/2}} \qquad (19B.4)$$

Langmuir isotherm for adsorption with dissociation

The surface coverage depends more weakly on pressure than it does for non-dissociative adsorption.

The shapes of the Langmuir isotherms with and without dissociation are shown in Figs. 19B.2 and 19B.3. The fractional coverage increases with increasing pressure, and approaches 1 only at very high pressure when the gas is forced on to every available site of the surface.

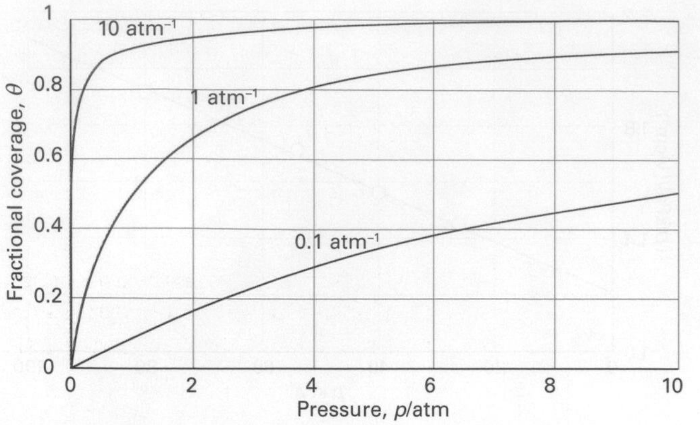

Figure 19B.2 The Langmuir isotherm for non-dissociative adsorption for different values of α.

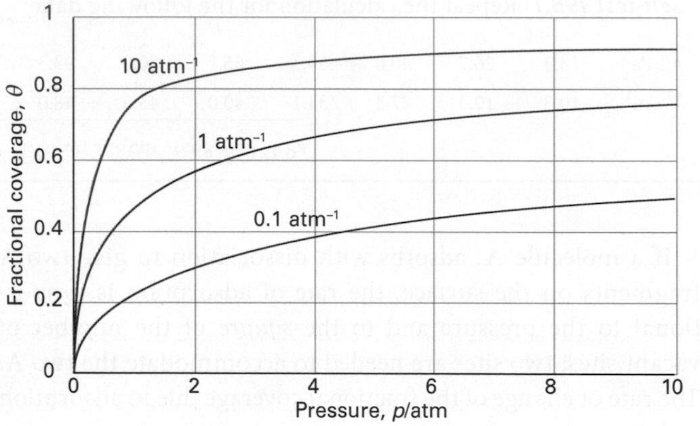

Figure 19B.3 The Langmuir isotherm for dissociative adsorption, $A_2(g) \rightarrow 2\,A(\text{surface})$, for different values of α.

(b) The isosteric enthalpy of adsorption

The Langmuir isotherm depends on the value of $\alpha = k_a/k_d$, which in turn depends on the temperature. It is possible to relate this temperature dependence to the **isosteric enthalpy of adsorption**, $\Delta_{ad}H^{\ominus}$, which is the standard enthalpy of adsorption at a fixed surface coverage.

How is that done? 19B.2 Relating the temperature dependence of α to the isosteric enthalpy of adsorption

The quantity $\alpha = k_a/k_d$ is the ratio of the rate constants for the forward and reverse reactions in the equilibrium $A(g) + M(\text{surface}) \rightleftharpoons AM(\text{surface})$. It follows from the discussion in Topic 17C that α is related to the equilibrium constant for this reaction, and so its temperature dependence can be developed in the same way as for any other equilibrium constant (Topic 6B).

Step 1 *Relate α to the equilibrium constant*

Because the dimensions of α are those of 1/pressure, its relation to the dimensionless equilibrium constant is $K = (k_a/k_d) \times p^{\ominus} = \alpha p^{\ominus}$.

Step 2 *Relate the equilibrium constant to the standard Gibbs energy of adsorption*

From eqn 6A.15 ($\Delta_r G^{\ominus} = -RT\ln K$) it follows that $\Delta_{ad}G^{\ominus} = -RT\ln(\alpha p^{\ominus})$, where $\Delta_{ad}G^{\ominus}$ is the standard Gibbs energy of adsorption. This expression can be rearranged to

$$-R\ln(\alpha p^{\ominus}) = \frac{\Delta_{ad}G^{\ominus}}{T}$$

Step 3 *Use the Gibbs–Helmholtz equation to relate the temperature dependence of $\Delta G^{\ominus}/T$ to the enthalpy of adsorption*

The derivative with respect to T of the last expression is

$$\frac{\mathrm{d}(-R\ln(\alpha p^{\ominus}))}{\mathrm{d}T} = \frac{\mathrm{d}}{\mathrm{d}T}\frac{\Delta_{ad}G^{\ominus}}{T}$$

Now use the Gibbs–Helmholtz equation (eqn 3E.11, $\mathrm{d}(\Delta G/T)/\mathrm{d}T = -\Delta H/T^2$) to write the right-hand side of this equation as $-\Delta_{ad}H^{\ominus}/T^2$, and therefore obtain

$$\frac{\mathrm{d}(-R\ln(\alpha p^{\ominus}))}{\mathrm{d}T} = \frac{-\Delta_{ad}H^{\ominus}}{T^2} \quad \text{hence} \quad \frac{\mathrm{d}\ln(\alpha p^{\ominus})}{\mathrm{d}T} = \frac{\Delta_{ad}H^{\ominus}}{RT^2}$$

There is a possibility that the standard enthalpy of adsorption depends on the fractional coverage, so this expression is restricted to constant θ. The derivative is therefore a partial derivative evaluated at constant θ and $\Delta_{ad}H^{\ominus}$ must be interpreted as the isosteric enthalpy of adsorption. The final result, therefore, is an expression for obtaining this quantity from the temperature dependence of α:

$$\left(\frac{\partial \ln(\alpha p^{\ominus})}{\partial T}\right)_{\theta} = \frac{\Delta_{ad}H^{\ominus}}{RT^2} \qquad (19B.5a)$$

This expression can be cast into a more useful form by using $\mathrm{d}(1/T)/\mathrm{d}T = -1/T^2$ to rewrite it as

$$\left(\frac{\partial \ln(\alpha p^{\ominus})}{\partial (1/T)}\right)_{\theta} = -\frac{\Delta_{ad}H^{\ominus}}{R} \qquad (19B.5b)$$

Isosteric enthalpy of adsorption

The following *Example* shows how eqn 19B.5b leads to a graphical method for determining a value of the isosteric enthalpy of adsorption.

Example 19B.2 Measuring the isosteric enthalpy of adsorption

The following data show the pressures of CO needed for the volume of adsorbed gas (corrected to 1 atm and 0 °C) to be 10.0 cm³ using the same sample as in *Example* 19B.1. In this case, there is no dissociation. Calculate the isosteric enthalpy of adsorption at this fractional coverage.

T/K	200	210	220	230	240	250
p/kPa	4.00	4.95	6.03	7.20	8.47	9.85

Collect your thoughts The same volume is adsorbed at each temperature, so the fractional coverage is the same at all temperatures; that is, the data are for isosteric conditions. You first need to relate the given pressures to a value of α by using the Langmuir isotherm (eqn 19B.2) arranged into $\alpha=\theta/p(1-\theta)$. However, because θ is constant, this expression reduces to $\alpha=C/p$, where C is a dimensionless constant. It follows that $\ln(\alpha p^{\ominus})=\ln(Cp^{\ominus}/p)=-\ln(p/p^{\ominus})+\ln C$ and therefore, from eqn 19B.5b, that a plot of $\ln(p/p^{\ominus})$ against $1/T$ should therefore be a straight line of slope $\Delta_{\rm ad}H^{\ominus}/R$.

The solution With $p^{\ominus} = 1\,\text{bar} = 10^2\,\text{kPa}$, draw up the following table:

T/K	200	210	220	230	240	250
$10^3/(T/\text{K})$	5.00	4.76	4.55	4.35	4.17	4.00
$(p/p^{\ominus}) \times 10^2$	4.00	4.95	6.03	7.20	8.47	9.85
$\ln(p/p^{\ominus})$	−3.22	−3.01	−2.81	−2.63	−2.47	−2.32

The points are plotted in Fig. 19B.4. The slope (of the least squares fitted line) is −0.901, so $(-\Delta_{\rm ad}H^{\ominus}/R)/10^3 = 0.901\,\text{K}$ and hence

$$\Delta_{\rm ad}H^{\ominus} = -(0.901\times10^3\,\text{K})\times(8.3145\,\text{J}\,\text{K}^{-1}\,\text{mol}^{-1}) = -7.5\,\text{kJ}\,\text{mol}^{-1}$$

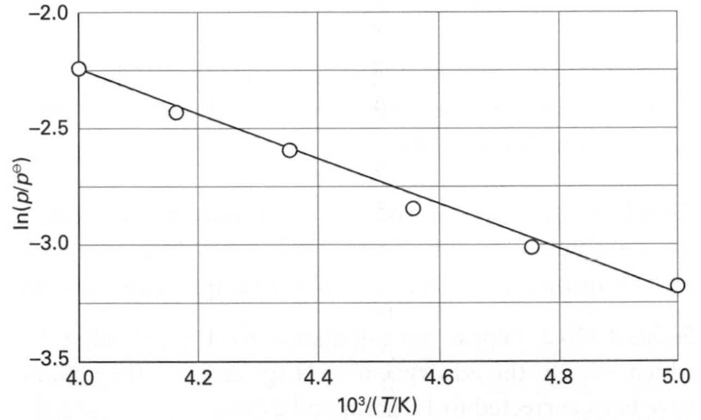

Figure 19B.4 The isosteric enthalpy of adsorption can be obtained from the slope of the plot of $\ln(p/p^{\ominus})$ against $1/T$, where p is the pressure needed to achieve the specified surface coverage. The data used are from *Example* 19B.2.

Self-test 19B.2 Repeat the calculation using the following data, which are isosteric:

T/K	200	210	220	230	240	250
p/kPa	4.32	5.59	7.07	8.80	10.67	12.80

Answer: $-9.0\,\text{kJ}\,\text{mol}^{-1}$

Two assumptions of the Langmuir isotherm are the independence and equivalence of the adsorption sites. Deviations from the isotherm can often be traced to the failure of these assumptions. For example, the enthalpy of adsorption often becomes less negative as θ increases, which suggests that the energetically most favourable sites are occupied first. Also interactions between the molecules already adsorbed on the surface can be important.

(c) The BET isotherm

A number of isotherms have been developed to deal with cases where deviations from the Langmuir isotherm are important. If the initial adsorbed layer can act as a substrate for further (e.g. physical) adsorption, then instead of the volume of gas adsorbed levelling off at high pressures to a value corresponding to a complete monolayer, it can be expected to rise indefinitely. The most widely used isotherm dealing with multilayer adsorption was derived by Stephen Brunauer, Paul Emmett, and Edward Teller and is called the **BET isotherm**:

$$\frac{V}{V_{\rm mon}} = \frac{cz}{(1-z)\{1-(1-c)z\}} \quad \text{with } z=\frac{p}{p^*} \qquad \text{BET isotherm} \qquad (19\text{B}.6)$$

where p^* is the vapour pressure of the pure liquid substrate, V is the volume of gas adsorbed, and $V_{\rm mon}$ is the volume of gas corresponding to a complete monolayer. The constant c is characteristic of the system: $c=\alpha_0/\alpha_1$ where $\alpha_0=k_{\rm a,0}/k_{\rm d,0}$ is the ratio of the rate constants for adsorption and desorption from the substrate, and $\alpha_1=k_{\rm a,1}/k_{\rm d,1}$ is the similar ratio for the subsequent layers. The rather fiddly derivation of this isotherm is given in *A deeper look* 13 on the website of this book.

Figure 19B.5 illustrates the shapes of BET isotherms. They rise indefinitely as the pressure is increased because there is no limit to the amount of material that may condense when multilayer coverage is possible. A BET isotherm is not accurate at all pressures, but it is widely used in industry to determine the surface areas of solids.

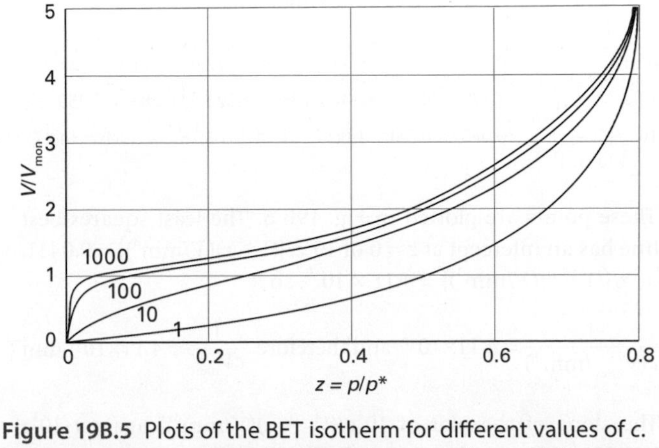

Figure 19B.5 Plots of the BET isotherm for different values of c. The value of $V/V_{\rm mon}$ rises indefinitely because the model permits the formation of multiple layers on the surface.

The form in which the BET isotherm is commonly used is obtained by inverting both sides of eqn 19B.6 to obtain

$$\frac{V_{\mathrm{mon}}}{V}=\frac{(1-z)\{1-(1-c)z\}}{cz}$$

and then multiplying both sides by $z/(1-z)V_{\mathrm{mon}}$ to obtain

$$\frac{z}{(1-z)V}=\frac{\{1-(1-c)z\}}{cV_{\mathrm{mon}}}$$

The right-hand side separates into two terms to give

$$\frac{z}{(1-z)V}=\overbrace{\frac{1}{cV_{\mathrm{mon}}}}^{\text{Intercept}}+\overbrace{\frac{(c-1)}{cV_{\mathrm{mon}}}}^{\text{Slope}}z \qquad (19B.7)$$

Therefore, a plot of $z/(1-z)V$ against z is expected to be a straight line with slope $(c-1)/cV_{\mathrm{mon}}$ and intercept $1/cV_{\mathrm{mon}}$ at $z=0$. Note that *slope/intercept* $=\{(c-1)/cV_{\mathrm{mon}}\}/(1/cV_{\mathrm{mon}})=c-1$.

Example 19B.3 Using the BET isotherm

The data below relate to the adsorption of N_2(g) on rutile (TiO_2) at 75 K.

p/kPa	0.160	1.87	6.11	11.67	17.02	21.92	27.29
V/mm^3	601	720	822	935	1046	1146	1254

The volumes have been corrected to 1.00 atm and 273 K and refer to 1.00 g of substrate. At 75 K, the vapour pressure of liquid nitrogen is $p^* = 76.0$ kPa. Confirm that these data fit a BET isotherm, and determine the values of V_{mon} and c.

Collect your thoughts Equation 19B.7 indicates that a plot of $z/(1-z)V$ against z, with $z=p/p^*$, gives a straight line of slope $(c-1)/cV_{\mathrm{mon}}$ and intercept $1/cV_{\mathrm{mon}}$ at $z=0$. As remarked in the text, the ratio of the slope to the intercept gives $c-1$. Make sure that the coordinates, slope, and intercept are all dimensionless and interpret them appropriately.

The solution Draw up the following table:

p/kPa	0.160	1.87	6.11	11.67	17.02	21.92	27.29
10^3z	2.11	24.6	80.4	154	224	288	359
$10^4z/\{(1-z)(V/\mathrm{mm}^3)\}$	0.035	0.350	1.06	1.94	2.76	3.54	4.47

These points are plotted in Fig. 19B.6. The least squares best line has an intercept at $z=0$ of $10^4z/\{(1-z)(V/\mathrm{mm}^3)\}=0.0411$, or $z/\{(1-z)(V/\mathrm{mm}^3)\}=4.11\times10^{-6}$, so

$$\frac{1}{c(V_{\mathrm{mon}}/\mathrm{mm}^3)}=4.11\times10^{-6}\ \text{and therefore}\ \frac{1}{cV_{\mathrm{mon}}}=4.11\times10^{-6}\,\mathrm{mm}^{-3}$$

The slope of the plot of $10^4z/\{(1-z)(V/\mathrm{mm}^3)\}$ against 10^3z is 1.22×10^{-2}, so the slope of $z/\{(1-z)(V/\mathrm{mm}^3)\}$ against z is $1.22\times10^{-2}\times10^{-4}\times10^3=1.22\times10^{-3}$. Therefore

$$\frac{c-1}{c(V_{\mathrm{mon}}/\mathrm{mm}^3)}=1.22\times10^{-3}\ \text{and so}\ \frac{c-1}{cV_{\mathrm{mon}}}=1.22\times10^{-3}\,\mathrm{mm}^{-3}$$

The ratio of $(c-1)/cV_{\mathrm{mon}}$ and $(1/cV_{\mathrm{mon}})$, from the previous two expressions, is

$$c-1=\frac{1.22\times10^{-3}\,\mathrm{mm}^{-3}}{4.11\times10^{-6}\,\mathrm{mm}^{-3}}=297$$

so $c=298$. Then

$$V_{\mathrm{mon}}=\frac{1}{298\times(4.11\times10^{-6}\,\mathrm{mm}^{-3})}=816\,\mathrm{mm}^3$$

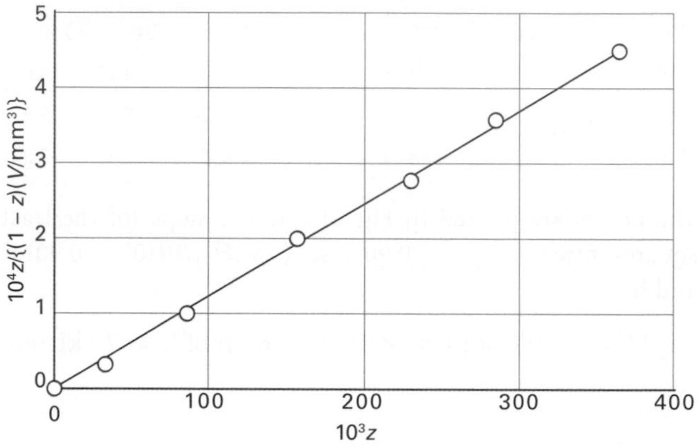

Figure 19B.6 The BET isotherm can be tested, and the parameters determined, by plotting $z/(1-z)V$ against z. The data are from *Example* 19B.3.

Comment. At 1.00 atm and 273 K, 816 mm^3 corresponds to 3.6×10^{-5} mol, or 2.2×10^{19} atoms. Because each atom occupies about 0.16 nm^2, the surface area of the sample is about 3.5 m^2.

Self-test 19B.3 Repeat the calculation for the following data which refer to the adsorption of N_2(g) at 75 K. The volumes have been corrected to 1.00 atm and 273 K.

p/kPa	0.160	1.87	6.11	11.67	17.02	21.92	27.29
V/cm^3	235	559	649	719	790	860	950

Answer: $c=370$, $V_{\mathrm{mon}}=615\,\mathrm{cm}^3$

The constant c depends on the temperature and can be related to the enthalpy changes associated with the formation of the first and subsequent monolayers.

How is that done? 19B.3 Relating the constant c in the BET isotherm to relevant enthalpy changes

Just as the Gibbs–Helmholtz equation can be used to express the temperature dependence of the 'equilibrium constant' α that appears in the Langmuir isotherm, it can also be used to express the temperature dependence of α_0 and α_1, and therefore of their ratio, c.

Step 1 *Write α_0 and α_1 in terms of the relevant Gibbs energy changes*

The parameter $\alpha_0 = k_{a,0}/k_{d,0}$ refers to the formation of the first monolayer (the one attached to the surface) which occurs in the Langmuir isotherm. It follows that α_0 is related to the standard Gibbs energy of adsorption, $\Delta_{ad}G^\ominus$: $\Delta_{ad}G^\ominus = -RT\ln(\alpha_0 p^\ominus)$. It will turn out to be convenient to replace the Gibbs energy of adsorption by the Gibbs energy of desorption, with $\Delta_{des}G^\ominus = -\Delta_{ad}G^\ominus$. It follows that $\Delta_{des}G^\ominus = RT\ln(\alpha_0 p^\ominus)$, or $\alpha_0 p^\ominus = e^{\Delta_{des}G^\ominus/RT}$.

The parameter $\alpha_1 = k_{a,1}/k_{d,1}$ refers to the formation of the second and subsequent monolayers, which is analogous to the condensation of a gas into a liquid. It follows that α_1 is related to the standard Gibbs energy of condensation, $\Delta_{con}G^\ominus = -RT\ln(\alpha_1 p^\ominus)$. In terms of the standard Gibbs energy of vaporization, $\Delta_{vap}G^\ominus = -\Delta_{con}G^\ominus$, it follows that $\Delta_{vap}G^\ominus = RT\ln(\alpha_1 p^\ominus)$ or $\alpha_1 p^\ominus = e^{\Delta_{vap}G^\ominus/RT}$.

Step 2 *Write c in terms of the relevant Gibbs energy changes*

You can now use the results from Step 1 to write $c = \alpha_0/\alpha_1$ in terms of the Gibbs energy changes

$$c = \frac{\alpha_0}{\alpha_1} = \frac{e^{\Delta_{des}G^\ominus/RT}}{e^{\Delta_{vap}G^\ominus/RT}} = \frac{e^{\Delta_{des}H^\ominus/RT}\, e^{-\Delta_{des}S^\ominus/R}}{e^{\Delta_{vap}H^\ominus/RT}\, e^{-\Delta_{vap}S^\ominus/R}}$$

where the Gibbs energies have been written in terms of the corresponding enthalpies and entropies.

Step 3 *Simplify the expression*

The entropies of desorption and vaporization can be assumed to be the same because they correspond to similar processes involving the escape of the condensed adsorbate to the gas phase. They cancel to give an expression for c in terms of the standard enthalpies of desorption and vaporization:

$$c = e^{(\Delta_{des}H^\ominus - \Delta_{vap}H^\ominus)/RT} \qquad \text{The } c \text{ constant from the BET isotherm in terms of enthalpy changes} \qquad (19B.8)$$

From eqn 19B.8 it follows that the constant c is large when the enthalpy of desorption for the first monolayer is large compared with the enthalpy of vaporization of the liquid adsorbate. In Fig. 19B.5 it is seen that full monolayer coverage (when $V/V_{mon} = 1$) is reached at lower pressures when c is large. This behaviour is consistent with the formation of the first layer becoming more favourable as $\Delta_{vap}H^\ominus$ becomes more negative (and $\Delta_{des}H^\ominus$ more positive).

Example 19B.3 indicates that c is of the order of 10^2. When $c \gg 1$, the BET isotherm takes the simpler form

$$\frac{V}{V_{mon}} = \frac{1}{1-z} \qquad \text{BET isotherm when } c \gg 1 \qquad (19B.9)$$

This expression is applicable to unreactive gases on polar surfaces, because $\Delta_{des}H^\ominus$ is then significantly greater than $\Delta_{vap}H^\ominus$. The BET isotherm fits experimental observations moderately well over restricted pressure ranges, but it errs by underestimating the extent of adsorption at low pressures and by overestimating it at high pressures.

(d) The Temkin and Freundlich isotherms

The Langmuir isotherm assumes that all sites are equivalent and independent, which implies that the enthalpy of adsorption is independent of the surface coverage. Experimentally, it is often found that the enthalpy of adsorption becomes less negative as θ increases, which suggests that the energetically most favourable sites are occupied first. Various attempts have been made to take these variations into account. The **Temkin isotherm**,

$$\theta = c_1 \ln(c_2 p) \qquad \text{Temkin isotherm} \qquad (19B.10)$$

where c_1 and c_2 are constants, corresponds to supposing that the adsorption enthalpy changes linearly with pressure. The **Freundlich isotherm**

$$\theta = c_1 p^{1/c_2} \qquad \text{Freundlich isotherm} \qquad (19B.11)$$

corresponds to a logarithmic change. This isotherm attempts to incorporate the role of interactions between the adsorbed molecules on the surface.

Different isotherms agree with experiment more or less well over restricted ranges of pressure, but they remain largely empirical. Empirical, however, does not mean useless for, if the parameters of a reliable isotherm are known, reasonably reliable results can be obtained for the extent of surface coverage under various conditions. This kind of information is essential for any discussion of heterogeneous catalysis (Topic 19C).

19B.2 The rates of adsorption and desorption

This section takes a more detailed look at adsorption and desorption at a molecular level, focusing in particular on the energetics of chemisorption.

(a) The precursor state

Figure 19B.7 shows how the potential energy of a molecule varies with its distance from the surface of the substrate. As the molecule approaches the surface its potential energy falls as it becomes physisorbed into the **precursor state** for chemisorption. Dissociation into fragments often takes place as a molecule moves into its chemisorbed state, and after an initial increase of energy as bonds in the molecule are distorted there

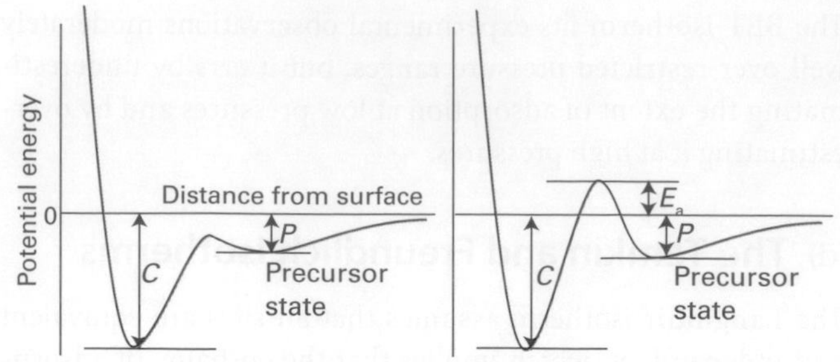

Figure 19B.7 The potential energy profiles for the dissociative chemisorption of an A_2 molecule. In each case, *P* is the enthalpy of (non-dissociative) physisorption and *C* that for chemisorption (at $T = 0$). The height of the intermediate peak determines whether the chemisorption is (a) not activated or (b) activated.

is a sharp decrease as the adsorbate–substrate bonds reach their full strength. Even if the molecule does not fragment, there is likely to be an initial increase of potential energy as the molecule approaches the surface and its bonds adjust.

In most cases, therefore, a potential energy barrier separating the precursor and chemisorbed states is expected. This barrier, though, might be low, and might not rise above the energy zero, which is the energy when the adsorbate is far away (Fig. 19B.7a). In this case, chemisorption is not an activated process and can be expected to be rapid, which is the case for many gas adsorptions on clean metals. In some cases, however, the barrier rises above zero (as in Fig. 19B.7b); such chemisorptions are activated and slower than the non-activated kind. An example is H_2 on copper, which has an activation energy in the region of 20–40 kJ mol^{-1}.

One point that emerges from this discussion is that rates are not good criteria for distinguishing between physisorption and chemisorption. Chemisorption can be fast if the activation energy is small or zero, but it may be slow if the activation energy is large. Physisorption is usually fast, but it can appear to be slow if adsorption is taking place on a porous medium.

Brief illustration 19B.1

Consider two adsorption experiments for hydrogen on different faces of a copper crystal. For adsorption on face 1 the activation energy is 28 kJ mol^{-1} and on face 2 the activation energy is 33 kJ mol^{-1}. If Arrhenius behaviour is assumed, and the frequency factors are the same, the ratio of the rates of adsorption on equal areas of the two faces at 250 K is

$$\frac{\text{Rate}(1)}{\text{Rate}(2)} = \frac{Ae^{-E_{a,ad}(1)/RT}}{Ae^{-E_{a,ad}(2)/RT}} = e^{-\{E_{a,ad}(1)-E_{a,ad}(2)\}/RT}$$
$$= e^{5\times10^3\,\text{J mol}^{-1}/(8.3145\,\text{J K}^{-1}\,\text{mol}^{-1})\times(250\,\text{K})} = 11$$

(b) Adsorption and desorption at the molecular level

The rate at which a surface is covered by adsorbate depends on the ability of the substrate to dissipate the energy of the incoming particle as it collides with the surface, the process of 'accommodation'. If the energy is not dissipated quickly, the particle migrates over the surface until it reaches an edge or until a vibration expels it into the overlying gas. The proportion of collisions with the surface that successfully lead to adsorption is called the **sticking probability**, *s*:

$$s = \frac{\text{rate of adsorption of particles by the surface}}{\text{rate of collision of particles with the surface}} \qquad \text{Sticking probability [definition]} \qquad (19B.12)$$

The denominator can be calculated from the kinetic model (from Z_W, Topic 19A), and the numerator can be measured by observing the rate of change of pressure.

Values of *s* vary widely. For example, at room temperature CO has *s* in the range 0.1–1.0 for several d-metal surfaces, but for N_2 on rhenium $s < 10^{-2}$, indicating that more than a hundred collisions are needed before one molecule becomes stuck to the surface. Studies on specific crystal planes show a pronounced specificity: for N_2 on tungsten at room temperature, *s* ranges from 0.74 on the (320) faces down to less than 0.01 on the (110) faces. The sticking probability decreases as the surface coverage increases (Fig. 19B.8). A simple assumption is that *s* is proportional to $1 - \theta$, the fraction uncovered, and it is common to write

$$s = (1-\theta)s_0 \qquad \text{Commonly used form of the sticking probability} \qquad (19B.13)$$

where s_0 is the sticking probability on a perfectly clean surface. The results in the illustration do not fit this expression because

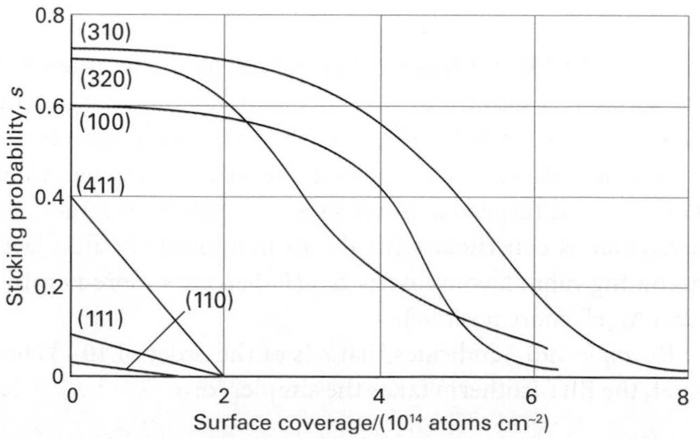

Figure 19B.8 The sticking probability of N_2 on various faces of a tungsten crystal and its dependence on surface coverage. Note the very low sticking probability for the (110) and (111) faces. (Data provided by Professor D.A. King.)

they show that s remains close to s_0 until the coverage has risen to about 6×10^{13} molecules cm^{-2}, and then falls steeply. The explanation is probably that the colliding molecule does not enter the chemisorbed state at once, but moves over the surface until it encounters an empty site.

Desorption is always activated because the particles have to be lifted from the bottom of a potential well. A physisorbed particle vibrates in its shallow potential well, and might shake itself off the surface after a short time. If the temperature dependence of the first-order rate constant for desorption follows Arrhenius behaviour, then $k_d = Ae^{-E_{a,des}/RT}$, where $E_{a,des}$ is the activation energy for desorption. Therefore, the temperature dependence of the **residence half-life**, the half-life for remaining on the surface, is

$$t_{1/2} = \frac{\ln 2}{k_d} = \tau_0 e^{E_{a,des}/RT} \qquad \tau_0 = \frac{\ln 2}{A} \qquad \text{Residence half-life} \qquad (19B.14)$$

The time τ_0 is the residence half-life in the limit of very high temperature, which is when the activation barrier has negligible effect; τ_0 is the lower limit of the residence half-life. Note the positive sign in the exponent: the greater the activation energy for desorption, the larger is the residence half-life.

Brief illustration 19B.2

If it is supposed that $1/\tau_0$ is approximately the same as the vibrational frequency of the weak adsorbate–surface bond (about 10^{12} Hz) and $E_{a,des} \approx 25$ kJ mol^{-1}, then residence half-lives of around 25 ns are predicted at room temperature. Lifetimes close to 1 s are obtained by lowering the temperature to about 100 K. For chemisorption, with $E_{a,des} = 100$ kJ mol^{-1} and guessing that $\tau_0 = 10^{-14}$ s (because the adsorbate–substrate bond is quite stiff), a residence half-life of about 3×10^3 s (about an hour) at room temperature is expected, decreasing to 1 s at about 370 K.

The desorption activation energy can be measured in several ways. However, such values must be interpreted with caution because they often depend on the fractional coverage, and so might change as desorption proceeds. Moreover, the transfer of concepts such as 'reaction order' and 'rate constant' from bulk studies to surfaces is hazardous, and there are few examples of strictly first-order or second-order desorption kinetics (just as there are few integral-order reactions in the gas phase too).

If the complications are disregarded, one way of measuring the desorption activation energy is to monitor the rate of increase in pressure when the sample is maintained at a series of temperatures, and then to attempt to make an Arrhenius plot. A more sophisticated technique is **temperature-programmed desorption** (TPD) or **thermal desorption spectroscopy** (TDS). In these experiments the temperature of the sample is raised

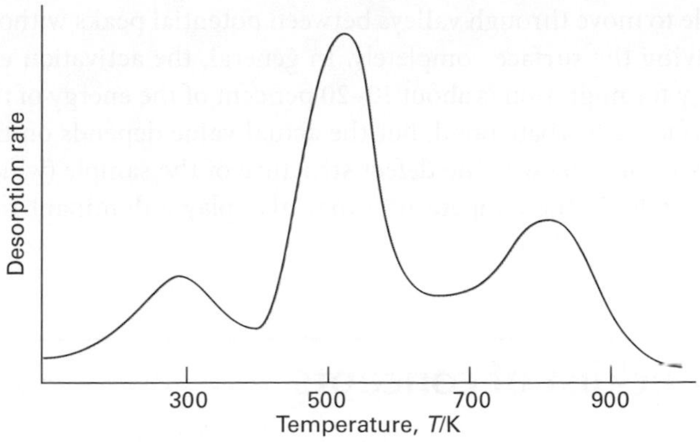

Figure 19B.9 The TPD spectrum of H_2 desorbing from a PtW_2 layer deposited on the surface of γ-alumina (γ-Al_2O_3). The profiles correspond to different fractional surface coverages of H_2. (Based on F. Lai, D-W. Kim, O.S. Alexeev, G.W. Graham, M. Shelef, and B.C. Gates, *Phys. Chem. Chem. Phys.* **2**, 1997 (2000).)

linearly and a surge in the desorption rate (as monitored by a mass spectrometer) is observed when the temperature reaches the point at which desorption occurs rapidly. However, once the desorption is complete (in the sense that there is no more adsorbate to escape from the surface), the desorption rate falls away as the temperature continues to rise. The TPD spectrum, the plot of desorption rate against temperature, therefore shows a peak, the location of which depends on the desorption activation energy (Fig. 19B.9).

In many cases only a single activation energy (and a single peak in the TPD spectrum) is observed. When several peaks are observed they might correspond to adsorption on different crystal planes or to multilayer adsorption. For instance, Cd atoms on tungsten show two activation energies, one of 18 kJ mol^{-1} and the other of 90 kJ mol^{-1}. The explanation is that the more tightly bound Cd atoms are attached directly to the substrate, and the less strongly bound are in a layer (or layers) above the first layer. Another example of a system showing two desorption activation energies is CO on tungsten, the values being 120 kJ mol^{-1} and 300 kJ mol^{-1}. The explanation is believed to be the existence of two types of metal–adsorbate binding site, one involving a simple M–CO bond, the other adsorption with dissociation into individually adsorbed C and O atoms.

(c) Mobility on surfaces

A further aspect of the strength of the interactions between adsorbate and substrate is the mobility of the adsorbate. Mobility is often a vital feature of a catalyst's activity, because a catalyst might be ineffective if the reactant molecules adsorb so strongly that they cannot migrate.

The activation energy for diffusion over a surface need not be the same as for desorption because the particles may be able to move through valleys between potential peaks without leaving the surface completely. In general, the activation energy for migration is about 10–20 per cent of the energy of the surface–adsorbate bond, but the actual value depends on the extent of coverage. The defect structure of the sample (which depends on the temperature) may also play a dominant role because the adsorbed molecules might find it easier to skip across a terrace than to roll along the foot of a step, and these molecules might become trapped in vacancies in an otherwise flat terrace. Diffusion may also be easier across one crystal face than another, and so the surface mobility depends on which lattice planes are exposed.

Checklist of concepts

- ☐ 1. An **adsorption isotherm** expresses the variation of the fractional coverage θ with pressure at constant temperature.
- ☐ 2. **Flash desorption** is a technique in which the sample is suddenly heated and the resulting rise of pressure is interpreted in terms of the amount of adsorbate originally on the substrate.
- ☐ 3. Examples of adsorption isotherms include the **Langmuir, BET, Temkin,** and **Freundlich** isotherms.
- ☐ 4. The **sticking probability** is the proportion of collisions with the surface that successfully lead to adsorption.
- ☐ 5. Desorption is an activated process; the desorption activation energy is measured by **temperature-programmed desorption** or **thermal desorption spectroscopy.**

Checklist of equations

Property	Equation	Comment	Equation number
Langmuir isotherm:			
(a) without dissociation	$\theta = \alpha p/(1+\alpha p)$	Independent and equivalent sites, monolayer coverage	19B.2
(b) with dissociation	$\theta = (\alpha p)^{1/2}/\{1+(\alpha p)^{1/2}\}$		19B.4
Isosteric enthalpy of adsorption	$(\partial \ln(\alpha p^{\ominus})/\partial(1/T))_{\theta} = -\Delta_{\mathrm{ad}}H^{\ominus}/R$		19B.5b
BET isotherm	$V/V_{\mathrm{mon}} = cz/(1-z)\{1-(1-c)z\}$, $z = p/p^{*}$, $c = \mathrm{e}^{(\Delta_{\mathrm{des}}H^{\ominus}-\Delta_{\mathrm{vap}}H^{\ominus})/RT}$	Multilayer adsorption	19B.6 and 19B.8
Temkin isotherm	$\theta = c_1 \ln(c_2 p)$	Enthalpy of adsorption varies with θ	19B.10
Freundlich isotherm	$\theta = c_1 p^{1/c_2}$	Adsorbate–adsorbate interactions	19B.11
Sticking probability	$s = (1-\theta)s_0$	Approximate form	19B.13

TOPIC 19C Heterogeneous catalysis

➤ **Why do you need to know this material?**

Because the chemical industry relies on heterogeneous catalysis for many of its most important large-scale processes, to see how they might be improved it is necessary to understand their mechanisms.

➤ **What is the key idea?**

Heterogeneous catalysis commonly involves chemisorption of one or more reactants and a consequent lowering of the activation energy.

➤ **What do you need to know already?**

Catalysis is introduced in Topic 17F. This Topic builds on the discussion of reaction mechanisms (Topic 17E), and uses the Arrhenius equation (Topic 17D) and adsorption isotherms (Topic 19B).

A **heterogeneous catalyst** is a catalyst in a different phase from that in which reactants and products are found. An example is the iron-based solid catalyst for the reaction of hydrogen and nitrogen to form ammonia. The metal provides a surface to which the reactants bind, so preparing them for reaction and facilitating their encounters.

19C.1 Mechanisms of heterogeneous catalysis

Many catalysts depend on **co-adsorption**, the adsorption of two or more species. One consequence of the presence of a second species may be the modification of the electronic structure at the surface of a metal. For instance, partial coverage of d-metal surfaces by alkali metals has a pronounced effect on the electron distribution at the surface and reduces the work function of the metal (the energy needed to remove an electron). Such modifiers can act as 'promoters' (to enhance the action of catalysts) or as 'poisons' (to inhibit catalytic action).

Figure 19C.1 shows the potential energy curve for a reaction in the presence of a heterogeneous catalyst. Differences between Fig. 19C.1 and Fig. 17F.5 arise from the fact that heterogeneous catalysis normally depends on at least one reactant being adsorbed (usually chemisorbed) and modified into a form in which it readily undergoes reaction, followed by desorption of products. Modification of the reactant often takes the form of a fragmentation of the reactant molecules. In practice, the catalyst is dispersed as very small particles of linear dimension less than 2 nm on a porous oxide support. **Shape-selective catalysts**, such as the zeolites, which have a pore size that can distinguish shapes and sizes at a molecular scale, have high internal specific surface areas, in the range of 100–500 $m^2\,g^{-1}$.

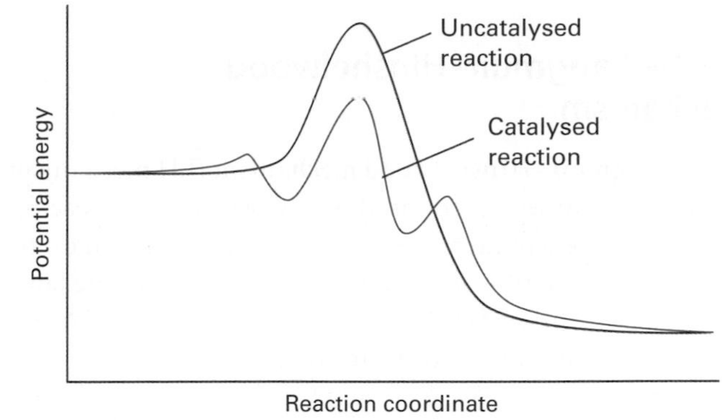

Figure 19C.1 The reaction profile for catalysed and uncatalysed reactions. The catalysed reaction path includes activation energies for adsorption and desorption as well as an overall lower activation energy for the process.

(a) Unimolecular reactions

A surface-catalysed unimolecular reaction is one in which an adsorbed molecule undergoes decomposition on a surface. Its rate law can be written in terms of an adsorption isotherm if it is assumed that the rate is proportional to the extent of surface coverage. For example, if the fractional coverage θ is given by the Langmuir isotherm (eqn 19B.2, $\theta = \alpha p/(1+\alpha p)$), the rate is

$$v = k_r\theta = \frac{k_r\alpha p}{1+\alpha p} \qquad (19C.1)$$

where p is the pressure of the adsorbing substance.

Brief illustration 19C.1

The decomposition of phosphine (PH_3) on tungsten is found to be first-order at low pressures. This order can be understood by using eqn 19C.1 and noting that when $\alpha p \ll 1$, $v = k_r\alpha p$, a first-order rate law. On the other hand, when $\alpha p >> 1$, the 1 in the denominator can be ignored and $v = k_r$, a zeroth-order law. A zeroth-order rate law is expected when the pressure is so high that the entire surface is covered; in this limit, the coverage, and hence the rate, is unaffected by a further rise in the pressure.

(b) The Langmuir–Hinshelwood mechanism

In the **Langmuir–Hinshelwood mechanism** (LH mechanism) of surface-catalysed reactions, it is proposed that the reaction takes place by encounters between molecules adsorbed on the surface. For a reaction between species A and B, the rate law is expected to be first order in the fractional coverage of A (θ_A), and of B (θ_B), and second-order overall:

$$A + B \rightarrow P \qquad v = k_r\theta_A\theta_B \qquad \text{Langmuir–Hinshelwood (LH) rate law} \qquad (19C.2a)$$

An example of a reaction thought to proceed by this mechanism is the catalytic oxidation of CO to CO_2. The LH rate law can be developed by using an isotherm to relate the fractional coverage of each species to its partial pressure.

How is that done? 19C.1 Developing the rate law of the LH mechanism

You can derive expressions for the fractional coverage of A and B by analysing the dynamic equilibrium between free and adsorbed molecules in much the same way as in the derivation of the Langmuir isotherm itself (Topic 19B), the difference being that two different species are now competing for the same adsorption sites.

Step 1 *Write expressions for the rates of adsorption and desorption of A and B*

The rate of adsorption of A is proportional to the partial pressure of A, p_A, and the number of vacant sites. If the number of surface sites is N, then the number of vacant sites is $N(1 - \theta_A - \theta_B)$. Therefore

$$\text{Rate of adsorption of A} = k_{a,A}p_AN(1 - \theta_A - \theta_B)$$

where $k_{a,A}$ is the rate constant for adsorption of A. The rate of desorption of A is proportional to the number of sites occupied by A molecules, $N\theta_A$:

$$\text{Rate of desorption of A} = k_{d,A}N\theta_A$$

where $k_{d,A}$ is the rate constant for desorption of A. The analogous expressions for B are

$$\text{Rate of adsorption of B} = k_{a,B}p_BN(1 - \theta_A - \theta_B)$$

$$\text{Rate of desorption of B} = k_{d,B}N\theta_B$$

Step 2 *Set the rates of adsorption and desorption to be equal*

At equilibrium, the rates of adsorption and desorption for each species are equal. For the species A

$$k_{a,A}p_AN(1 - \theta_A - \theta_B) = k_{d,A}N\theta_A$$

This expression is simplified by introducing $\alpha_A = k_{a,A}/k_{d,A}$ to give

$$\alpha_Ap_A(1 - \theta_A - \theta_B) = \theta_A$$

and therefore

$$(\alpha_Ap_A + 1)\theta_A + \alpha_Ap_A\theta_B = \alpha_Ap_A$$

Similarly for B, and with $\alpha_B = k_{a,B}/k_{d,B}$,

$$\alpha_Bp_B(1 - \theta_A - \theta_B) = \theta_B$$

and therefore

$$\alpha_Bp_B\theta_A + (\alpha_Bp_B + 1)\theta_B = \alpha_Bp_B$$

The solutions of these two simultaneous equations for θ_A and θ_B are

$$\theta_A = \frac{\alpha_Ap_A}{1+\alpha_Ap_A+\alpha_Bp_B} \qquad \theta_B = \frac{\alpha_Bp_B}{1+\alpha_Ap_A+\alpha_Bp_B}$$

Step 3 *Use the expressions for θ_A and θ_B in the rate law*

Now substitute the expressions for the fractional surface coverage into the rate law, eqn 19C.2a, to give

$$v = k_r\theta_A\theta_B = k_r\frac{\alpha_Ap_A}{1+\alpha_Ap_A+\alpha_Bp_B}\,\frac{\alpha_Bp_B}{1+\alpha_Ap_A+\alpha_Bp_B}$$

which, after minor rearrangement, gives the rate law in terms of the partial pressures and the parameters α_A and α_B:

$$v = \frac{k_r\alpha_A\alpha_Bp_Ap_B}{(1+\alpha_Ap_A+\alpha_Bp_B)^2} \qquad \text{Langmuir–Hinshelwood rate law} \qquad (19C.2b)$$

The parameters α and the rate constant k_r are all temperature-dependent, so the overall temperature dependence of the rate may be strongly non-Arrhenius (in the sense that the reaction rate constant is unlikely to be proportional to $e^{-E_a/RT}$). The LH mechanism has been analysed in terms of the adsorption and subsequent reaction between species A and B, but it may be that on adsorption either or both species dissociates to give fragments which then react.

(c) The Eley–Rideal mechanism

In the **Eley–Rideal mechanism** (ER mechanism) of a surface-catalysed reaction, it is proposed that a gas-phase molecule collides with another molecule already adsorbed on the surface. The rate of formation of product is proportional to the partial pressure, p_B, of the non-adsorbed gas B and the fractional surface coverage, θ_A, of the adsorbed gas A, to give the rate law

$$v = k_r p_B \theta_A \quad \text{Eley–Rideal rate law} \quad (19C.3)$$

The rate of the catalysed reaction might be much larger than for the uncatalysed gas-phase reaction because the reaction on the surface has a low activation energy and the adsorption itself is often not activated. If it is assumed that the Langmuir isotherm applies to species A, the fractional coverage is $\theta_A = \alpha p_A/(1+\alpha p_A)$ so the ER rate law becomes

$$v = \frac{k_r \alpha p_B p_A}{1+\alpha p_A} \quad (19C.4)$$

Brief illustration 19C.2

According to eqn 19C.4, when the partial pressure of A is high (in the sense $\alpha p_A >> 1$), the denominator is simply αp_A, and the rate is $k_r p_B$. At such high pressures the surface is completely covered with A, so increasing the pressure of A has no effect. The rate of reaction is then limited by the rate at which B reacts with adsorbed A. When the pressure of A is low ($\alpha p_A << 1$), the term αp_A in the denominator can be ignored and the rate of reaction becomes $k_r \alpha p_A p_B$. Now the coverage of the surface is low and so increasing the pressure of A increases the surface coverage, and hence the rate.

Almost all surface-catalysed reactions are thought to take place by the LH mechanism, but a number of reactions with an ER mechanism have also been identified from molecular beam investigations. For example, the reaction between gaseous H atoms and adsorbed D atoms to form gaseous HD is thought to proceed by the ER mechanism in which an H atom collides directly with an adsorbed D atom, picking it up to form HD. However, the two mechanisms should really be thought of as ideal limits with all reactions lying somewhere between the two and showing features of each one.

19C.2 Catalytic activity at surfaces

It has become possible to investigate how the catalytic activity of a surface depends on its structure as well as its composition. For instance, the cleavage of C–H and H–H bonds appears to depend on the presence of steps and kinks, and a terrace often has only minimal catalytic activity.

The reaction $H_2 + D_2 \rightarrow 2HD$ has been studied in detail. For this reaction, terrace sites are inactive but one molecule in ten reacts when it strikes a step. Although the step itself might be the important feature, it may be that the presence of the step merely exposes a more reactive crystal face (the step face itself). Likewise, the dehydrogenation of hexane to hexene depends strongly on the kink density, and it appears that kinks are needed to cleave C–C bonds. These observations suggest a reason why even small amounts of impurities may poison a catalyst: they are likely to attach to step and kink sites, and so impair the activity of the catalyst entirely. A constructive outcome is that the extent of dehydrogenation may be controlled relative to other types of reactions by seeking impurities that adsorb at kinks and act as specific poisons.

The activity of a catalyst depends on the strength of chemisorption as indicated by the 'volcano' curve in Fig. 19C.2 (which is so-called on account of its general shape). To be active, the catalyst should be extensively covered by adsorbate, which is the case if chemisorption is strong. On the other hand, if the strength of the substrate–adsorbate interaction becomes too great, the activity declines either because the other reactant molecules cannot react with the adsorbate or because the adsorbate molecules are immobilized on the surface. This pattern of behaviour suggests that the activity of a catalyst should initially increase with strength of adsorption (as measured, for instance, by the enthalpy of adsorption) and then decline, and that the most active catalysts should be those lying near the summit of the volcano. Most active metals are those that lie close to the middle of the d block. Many metals are suitable for adsorbing gases, and some trends are summarized in Table 19C.1.

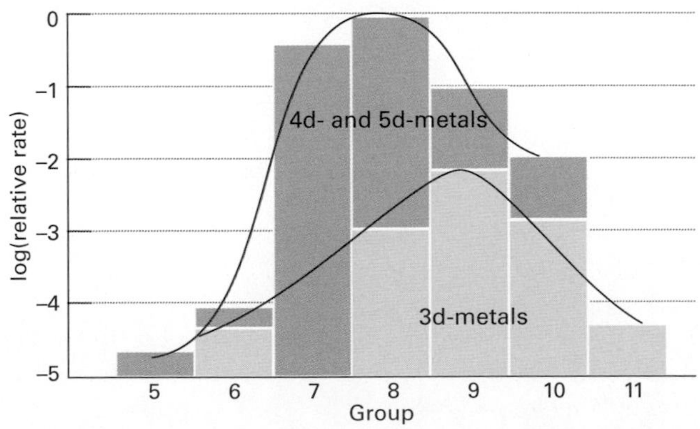

Figure 19C.2 A 'volcano curve' of catalytic activity arises because although the reactants must adsorb reasonably strongly, they must not adsorb so strongly that they are immobilized. The lower curve refers to the first series of d-block metals, the upper curve to the second and third series of d-block metals.

Table 19C.1 Chemisorption abilities*

	O_2	C_2H_2	C_2H_4	CO	H_2	CO_2	N_2
Ti, Cr, Mo, Fe	+	+	+	+	+	+	+
Ni, Co	+	+	+	+	+	+	−
Pd, Pt	+	+	+	+	+	−	−
Mn, Cu	+	+	+	+	±	−	−
Al, Au	+	+	+	−	−	−	−
Li, Na, K	+	+	−	−	−	−	−
Mg, Ag, Zn, Pb	+	−	−	−	−	−	−

* +, Strong chemisorption; ±, chemisorption; −, no chemisorption.

Brief illustration 19C.3

The data in Table 19C.1 show that for a number of metals the general order of chemisorption ability decreases along the series O_2, C_2H_2, C_2H_4, CO, H_2, CO_2, N_2. Some of these molecules adsorb dissociatively (e.g. H_2). Elements from the d block, such as iron, titanium, and chromium, adsorb all these gases, but manganese and copper are unable to adsorb N_2 and CO_2. Metals towards the left of the periodic table (e.g. magnesium) can adsorb (and, in fact, react with) only the most active gas (O_2).

Checklist of concepts

☐ 1. A **heterogeneous catalyst** is a catalyst in a different phase from the reaction mixture.

☐ 2. In the **Langmuir–Hinshelwood mechanism** of surface-catalysed reactions, the reaction takes place by encounters between molecules adsorbed on the surface.

☐ 3. In the **Eley–Rideal mechanism** of a surface-catalysed reaction, a gas-phase molecule collides with another molecule already adsorbed on the surface.

☐ 4. The activity of a catalyst depends on the strength of chemisorption.

Checklist of equations

Property	Equation	Comment	Equation number
Langmuir–Hinshelwood mechanism	$v = k_r\theta_A\theta_B$	A and B both adsorbed	19C.2a
Eley–Rideal mechanism	$v = k_r p_B\theta_A$	Only A adsorbed	19C.3

TOPIC 19D Processes at electrodes

> ➤ **Why do you need to know this material?**
>
> A knowledge of the factors that determine the rate of electron transfer at electrodes leads to a better understanding of the charging and discharging of batteries, the production of power using solar cells, and manufacturing using electrolysis, all of which are important technologies with wide impact.
>
> ➤ **What is the key idea?**
>
> The rate of oxidation and reduction at an electrode depends on the height of the activation barrier, which can be modified by applying a potential difference across the solution/electrode interface.
>
> ➤ **What do you need to know already?**
>
> You need to be familiar with electrochemical cells (Topic 6C), electrode potentials (Topic 6D), and the thermodynamic version of transition-state theory (Topic 18C), particularly the Gibbs energy of activation.

The surface of a solid electrode is in contact with the ions in an electrolyte solution. The rates of oxidation and reduction at this interface depend on how rapidly electrons can be transferred through it.

19D.1 The electrode–solution interface

An electrode in contact with a solution of an electrolyte acquires a charge as a result either of the escape of atoms into the solution as cations, leaving behind a negative charge, or as a result of ions becoming attached to the surface. As the electrode becomes charged, an electrical potential difference develops across the interface and makes that process more difficult. For example, if the charge arises from the escape of atoms as cations, the increasing negative charge on the electrode makes it more unfavourable for the cations to leave. Eventually equilibrium is reached with a characteristic potential difference between the electrode and the solution.

The charge on the electrode affects the composition of the surrounding electrolyte solution because it is energetically favourable for ions with the opposite charge to cluster nearby. This tendency, however, is disrupted by thermal motion and various models have been developed to describe the outcome of this competition, some simply by ignoring it. The modification of the local concentrations near an electrode implies that it might be misleading to use activity coefficients characteristic of the bulk to discuss the thermodynamic properties of ions near the interface. This is one of the reasons why measurements of the dynamics of electrode processes are almost always done by using a large excess of supporting electrolyte (e.g. a 1 mol dm^{-3} solution of a salt, an acid, or a base). Under such conditions, the activity coefficients are almost constant because the inert ions dominate the effects of local changes caused by any reactions taking place. The use of a concentrated solution also minimizes ion migration effects.

The most primitive model of the boundary between the electrode and the electrolyte solution is as an **electrical double layer**, in which it is supposed that there is a sheet of positive charge at the surface of the electrode and a sheet of negative charge next to it in the solution (or vice versa).

More sophisticated models for the interface introduce a more gradual change in the structure of the solution. In the **Helmholtz layer model** solvated ions lie along the surface of the electrode but are held away from it by their hydration spheres (Fig. 19D.1). The location of the sheet of ionic charge, which is called the **outer Helmholtz plane** (OHP), is identified

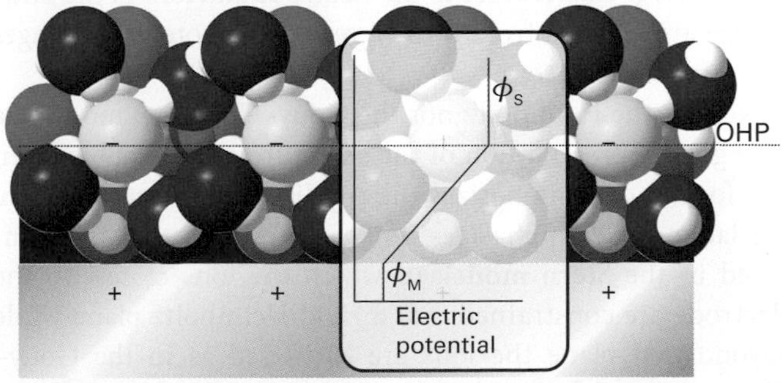

Figure 19D.1 A simple model of the electrode–solution interface treats it as two rigid planes of charge. One plane, the outer Helmholtz plane (OHP), is due to the ions with their solvating molecules and the other plane is that of the electrode itself. The plot shows the dependence of the electric potential with distance from the electrode surface according to this model. Between the electrode surface and the OHP, the potential varies linearly from ϕ_M, the value in the metal, to ϕ_S, the value in the bulk of the solution.

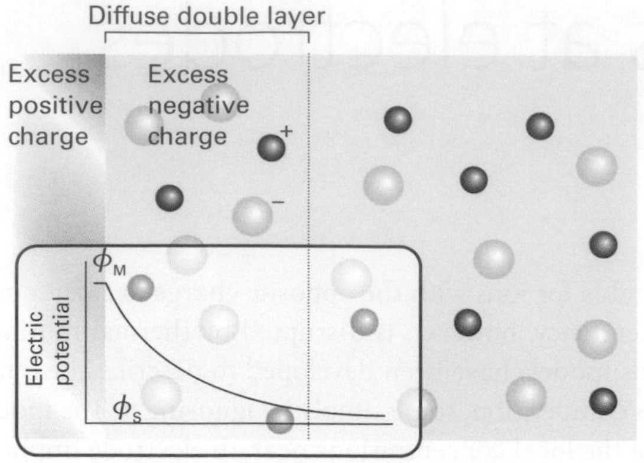

Figure 19D.2 The Gouy–Chapman model of the electrical double layer treats the outer region as an atmosphere of counter-charge, similar to the Debye–Hückel model of an ionic atmosphere.

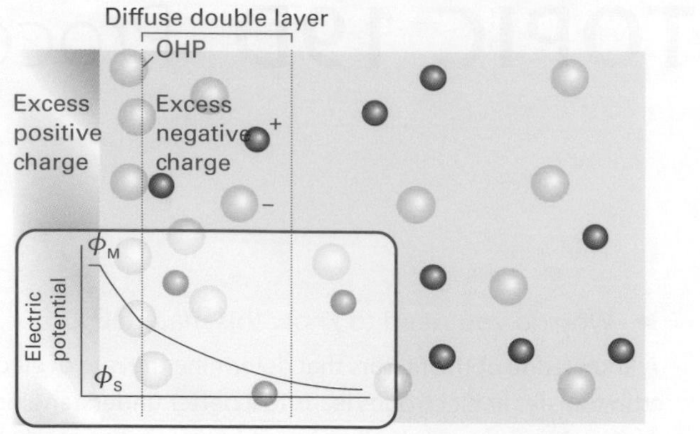

Figure 19D.3 A representation of the Stern model of the electrode–solution interface. The model incorporates the concepts of an outer Helmholtz plane near the electrode surface and of a diffuse double layer further away from the surface.

as the plane running through the solvated ions. In this simple model, the electrical potential changes linearly within the layer from ϕ_M at the metal to ϕ_S, the value characteristic of the solution, at the OHP. In a refinement of this model, ions that have discarded their solvating molecules and have become attached to the electrode surface by chemical bonds are regarded as forming the **inner Helmholtz plane** (IHP).

The Helmholtz layer model ignores the disrupting effect of thermal motion, which tends to break up and disperse the rigid outer plane of charge. In the **Gouy–Chapman model** of the **diffuse double layer**, the disordering effect of thermal motion is taken into account in much the same way as the Debye–Hückel model describes the ionic atmosphere of an ion (Topic 5F). The difference is that the central ion is replaced by an infinite plane electrode. Figure 19D.2 shows how, in the Gouy–Chapman model, the local concentrations of cations and anions differ from their bulk concentrations. Ions of opposite charge cluster close to the electrode and ions of the same charge are repelled from it. As a result, the potential changes smoothly from ϕ_M to ϕ_S.

Neither the Helmholtz nor the Gouy–Chapman model is a very good representation of the structure of the double layer. The former overemphasizes the rigidity of the local solution; the latter underemphasizes its structure. The two are combined in the **Stern model**, in which the ions closest to the electrode are constrained into a rigid Helmholtz plane while beyond that plane the ions are dispersed as in the Gouy–Chapman model (Fig. 19D.3). Yet another level of sophistication is found in the **Grahame model**, which adds an inner Helmholtz plane to the Stern model.

The potential difference between the bulk metal and the bulk solution is the **Galvani potential difference**, $\Delta\phi = \phi_M - \phi_S$. If the electrode is part of a cell from which no current is being drawn the Galvani potential difference can be identified with the electrode potential discussed in Topic 6D. However, the value of $\Delta\phi$ can be altered at will by the application of an external electrical potential difference to the cell, and when the cell is producing current the potential difference at the electrode/electrolyte interface also changes from its zero-current value.

19D.2 The current density at an electrode

The **current density**, j, is the electric current (in amperes, A; $1\,A = 1\,C\,s^{-1}$) flowing through a region of an electrode divided by the area of the region (in square metres or a submultiple, such as square centimetres). Current is the rate of flow of charge, so a current density of $1\,A\,cm^{-2}$ represents a flow of about $10\,\mu mol$ of electrons per second per square centimetre. The current density is a measure of the rate of the electron-transfer process occurring at the electrode.

(a) The Butler–Volmer equation

As explained in Topic 6C, a cathode is the site of reduction and an anode is the site of oxidation. This nomenclature is carried over into the classification of the current density. A flow of electrons from the electrode to bring about reduction of the electroactive species in the solution is called the **cathodic current density**, j_c. The opposite flow, from solution into the electrode due to oxidation of the electroactive species, is called the **anodic current density**, j_a. The **net current density**, j, is the difference of these two current densities, $j = j_a - j_c$. Reduction is dominant if $j_c > j_a$ and therefore $j < 0$: the current density is then said to be 'cathodic'. Oxidation is dominant if $j_c < j_a$, corresponding to $j > 0$: the current density is then 'anodic' (Fig. 19D.4).

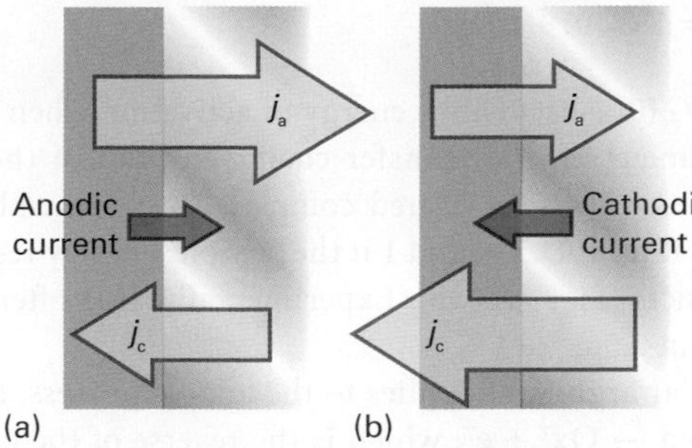

Figure 19D.4 The net current density is defined as the difference between the cathodic and anodic current densities. (a) When $j_a > j_c$, the net current is anodic, and there is a net oxidation of the species in solution. (b) When $j_c > j_a$, the net current is cathodic, and the net process is reduction.

When the electrode is at equilibrium there is no net current flow, and $\Delta\phi$ can be identified with the electrode potential E. This equilibrium is a dynamic one in which both the anodic and cathodic processes are taking place, but in such a way that their current densities are equal. The anodic (or cathodic) current density at equilibrium is called the **exchange-current density**, j_0.

If the potential difference at the interface, $\Delta\phi$, differs from E, net current flows through the electrode. The **overpotential** η is defined as

$$\eta = E' - E \qquad \text{Overpotential [definition]} \qquad (19D.1)$$

where E' is the potential difference applied to the cell or its potential difference under working conditions. It follows that $\Delta\phi = E + \eta$.

The energies of the charged species involved in the electron transfer process depend on the electrical potential on each side of the interface and are therefore affected by the potential difference across it. The rate of electron transfer is similarly affected, so the task is to relate the resulting current density to the overpotential.

How is that done? 19D.1 Deriving the relation between current density and the overpotential

The current density is determined by the rate of the electron-transfer process taking place at the electrode between an oxidized species Ox^+ and a reduced species Red. Both Red and Ox^+ are in solution; the electrons involved in the redox process are in the electrode. The rate constant for the reduction step (the cathodic process) is k_c, and for the reverse oxidation step (the anodic process) is k_a.

$$Ox^+(\text{solution}) + e^-(\text{electrode}) \underset{k_a}{\overset{k_c}{\rightleftharpoons}} \text{Red(solution)}$$

The model does not depend on this choice of the charges; they have been chosen for convenience.

Step 1 *Write expressions for the rate of oxidation and reduction*

An electrode reaction is heterogeneous, so its rate is specified by the flux of material. This flux is the amount of material produced over a region of the electrode surface in an interval of time, divided by the area of the region and the duration of the interval. A first-order heterogeneous rate law has the form

$$\text{Product flux} = k_r[\text{X}]$$

where [X] is the molar concentration of the relevant electroactive species in the solution. The rate constant has dimensions of length/time (with units, for example, of centimetres per second, cm s^{-1}). If the molar concentrations of the oxidized and reduced species are $[Ox^+]$ and [Red], respectively, then the rate of reduction of Ox^+ is $k_c[Ox^+]$ and the rate of oxidation of Red is k_a[Red].

Step 2 *Write expressions for the current density in terms of the rate*

The cathodic current density, j_c, is equal to the flux multiplied by Faraday's constant, $F = N_A e$, the magnitude of the charge per mole of electrons:

$$j_c = Fk_c[Ox^+] \text{ for } Ox^+ + e^- \rightarrow \text{Red}$$

Similarly, the anodic current density, j_a, is

$$j_a = Fk_a[\text{Red}] \text{ for Red} \rightarrow Ox^+ + e^-$$

The net current density at the electrode is the difference

$$j = j_a - j_c = Fk_a[\text{Red}] - Fk_c[Ox^+]$$

Step 3 *Write the rate constants in terms of the Gibbs energies of activation*

Now write the two rate constants in a form suggested by transition-state theory (Topic 18C) as

$$k_r = Be^{-\Delta^{\ddagger}G/RT}$$

where $\Delta^{\ddagger}G$ is the Gibbs energy of activation and B is a constant with the same dimensions as k_r. Then

$$j_a = FB_a[\text{Red}]e^{-\Delta^{\ddagger}G_a/RT} \qquad j_c = FB_c[Ox^+]e^{-\Delta^{\ddagger}G_c/RT}$$

Step 4 *Relate the Gibbs energies of activation to the electrical potential difference*

If a species of charge number z (for Ox^+, $z = +1$; for Red, $z = 0$; for e^-, $z = -1$) is present in a region of electrical potential ϕ its standard chemical potential is

$$\bar{\mu}^{\ominus} = \mu_0^{\ominus} + zF\phi$$

where $\mu_0^{\ominus}$ is the standard chemical potential in the absence of an electrical potential. The quantity $\bar{\mu}^{\ominus}$ is called the standard **electrochemical potential**. The species Ox^+ and Red are in the solution, and so experience the potential ϕ_S, whereas the electrons are in the metallic electrode and so experience the potential ϕ_M. The reduced species is electrically neutral (in this formalism). The standard electrochemical potentials of the three species are therefore

$$\bar{\mu}^{\ominus}(\mathrm{Ox}^+) = \mu_0^{\ominus}(\mathrm{Ox}^+) + F\phi_S \qquad \bar{\mu}^{\ominus}(\mathrm{Red}) = \mu_0^{\ominus}(\mathrm{Red})$$
$$\bar{\mu}^{\ominus}(\mathrm{e}^-) = \mu_0^{\ominus}(\mathrm{e}^-) - F\phi_M$$

In the reaction $\mathrm{Ox}^+ + \mathrm{e}^- \rightarrow \mathrm{Red}$ the standard Gibbs energy of the reactants is therefore

$$G_m^{\ominus}(\text{reactants}) = \bar{\mu}^{\ominus}(\mathrm{Ox}^+) + \bar{\mu}^{\ominus}(\mathrm{e}^-) = \mu_0^{\ominus}(\mathrm{Ox}^+) + \mu_0^{\ominus}(\mathrm{e}^-) + \overbrace{F\phi_S - F\phi_M}^{-F\Delta\phi}$$
$$= G_m^{\ominus}(0,\text{reactants}) - F\Delta\phi$$

where $G_m^{\ominus}(0, \text{reactants})$ is the standard molar Gibbs energy when no electrical potential is applied, and $\Delta\phi = \phi_M - \phi_S$.

If the activated complex appears early along the reaction pathway, meaning that it has a structure not too dissimilar from the reactants ($\mathrm{Ox}^+ + \mathrm{e}^-$), then its Gibbs energy is affected by the applied electrical potential in a similar way. Therefore, because the potential energy difference has the same effect on the Gibbs energies of both the reactants and activated complex, the Gibbs energy of activation is unaffected by the value of $\Delta\phi$ (Fig. 19D.5a). In contrast, if the activated complex appears late in the reaction pathway, and so resembles the electrically neutral product Red, then its Gibbs energy is unchanged by the potential difference. As $\Delta\phi$ increases, the standard molar Gibbs energy of the reactants ($\mathrm{Ox}^+ + \mathrm{e}^-$) is lowered by $F\Delta\phi$, so in this case the Gibbs energy of activation is increased by $F\Delta\phi$ (Fig. 19D.5b).

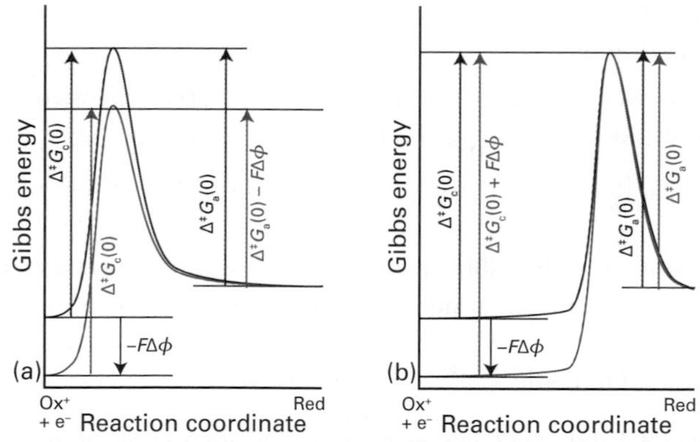

Figure 19D.5 Profiles of how the Gibbs energy varies between the oxidized species ($\mathrm{Ox}^+ + \mathrm{e}^-$) and the reduced species (Red) at an electrode. The purple line shows the profile, and the blue line shows how it is modified when a potential difference $\Delta\phi$ is applied across the electrode. In (a) the transition state resembles the oxidized species; in (b) the transition state resembles the reduced species.

These two special cases can be brought together if the Gibbs energy of activation for the cathodic (reduction) process is written

$$\Delta^{\ddagger}G_c = \Delta^{\ddagger}G_c(0) + \alpha F\Delta\phi$$

where $\Delta^{\ddagger}G_c(0)$ is the Gibbs energy of activation when $\Delta\phi = 0$. The parameter α, the **transfer coefficient**, lies in the range 0 to 1: it is 0 if the activated complex closely resembles the reactants (Fig. 19D.5a), and 1 if the complex closely resembles the products (Fig. 19D.5b). Experimentally, α is often found to be about 0.5.

A similar argument applies to the anodic process, the oxidation $\mathrm{Red} \rightarrow \mathrm{Ox}^+ + \mathrm{e}^-$, which is the reverse of the cathodic process. As is evident from Fig. 19D.5a, if the activated complex resembles $\mathrm{Ox}^+ + \mathrm{e}^-$ ($\alpha = 0$), the Gibbs energy of activation for the anodic step is decreased by $-F\Delta\phi$. On the other hand, if the activated complex resembles Red (Fig. 19D.5b, $\alpha = 1$), the Gibbs energy of activation for the anodic step is unaffected by a change in $\Delta\phi$. The overall effect on the Gibbs energy of activation for the anodic process can therefore be written

$$\Delta^{\ddagger}G_a = \Delta^{\ddagger}G_a(0) - (1-\alpha)F\Delta\phi$$

Step 5 *Write the rate constants using the expressions for the Gibbs energy of activation*

Now insert the Gibbs energies of activation into the expressions for j_a and j_c to give

$$j_a = FB_a[\mathrm{Red}]\mathrm{e}^{-\Delta^{\ddagger}G_a(0)/RT}\mathrm{e}^{(1-\alpha)f\Delta\phi} \qquad j_c = FB_c[\mathrm{Ox}^+]\mathrm{e}^{-\Delta^{\ddagger}G_c(0)/RT}\mathrm{e}^{-\alpha f\Delta\phi}$$

where the appearance of the expressions has been simplified by writing $F/RT = f$.

Step 6 *Consider the effect of the overpotential*

If a potential difference is applied such that the net current density is zero, $\Delta\phi$ can be identified as the electrode potential, E. The current densities are then both equal to the exchange-current density, j_0:

$$j_0 = FB_a[\mathrm{Red}]\mathrm{e}^{-\Delta^{\ddagger}G_a(0)/RT}\mathrm{e}^{(1-\alpha)fE} = FB_c[\mathrm{Ox}^+]\mathrm{e}^{-\Delta^{\ddagger}G_c(0)/RT}\mathrm{e}^{-\alpha fE}$$

The role of the overpotential can now be identified by substituting $\Delta\phi = E + \eta$:

$$j_c = FB_c[\mathrm{Ox}^+]\mathrm{e}^{-\Delta^{\ddagger}G_c(0)/RT}\mathrm{e}^{-\alpha f\Delta\phi} = FB_c[\mathrm{Ox}^+]\mathrm{e}^{-\Delta^{\ddagger}G_c(0)/RT}\mathrm{e}^{-\alpha f(E+\eta)}$$

$$= \overbrace{FB_c[\mathrm{Ox}^+]\mathrm{e}^{-\Delta^{\ddagger}G_c(0)/RT}\mathrm{e}^{-\alpha fE}}^{j_0}\mathrm{e}^{-\alpha f\eta} = j_0\mathrm{e}^{-\alpha f\eta}$$

A similar argument gives

$$j_a = j_0\mathrm{e}^{(1-\alpha)f\eta}$$

The net current density is $j = j_a - j_c$; therefore

$$j = j_0\{\mathrm{e}^{(1-\alpha)f\eta} - \mathrm{e}^{-\alpha f\eta}\} \qquad (19D.2)$$

Butler–Volmer equation

Equation 19D.2 is the **Butler–Volmer equation**. It can be interpreted as follows:

Physical interpretation

- When the overpotential η is zero, there is no net current density (there are equal and opposite flows).
- If $\alpha = 0$, the cathodic current density is equal to the exchange-current density and is independent of the overpotential.
- If $\alpha = 1$, the anodic current density is equal to the exchange-current density and is independent of the overpotential.
- Provided $0 < \alpha < 1$, as η becomes increasingly positive the anodic current density dominates the cathodic current density and the dominant process is the oxidation Red $\rightarrow$ $Ox^+ + e^-$. As η becomes increasingly negative, the cathodic current density dominates the anodic current density and the dominant process is the reduction $Ox^+ + e^- \rightarrow$ Red.

Figure 19D.6 shows how eqn 19D.2 predicts the dependence of the net current density on the overpotential for different values of the transfer coefficient. When the overpotential is so small that $f\eta \ll 1$ (in practice, η less than about 10 mV) the exponentials in eqn 19D.2 can be expanded by using $e^x = 1 + x + \cdots$ to give

$$j = j_0\{\overbrace{1+(1-\alpha)f\eta+\cdots}^{e^{(1-\alpha)f\eta}} - \overbrace{(1-\alpha f\eta+\cdots)}^{e^{-\alpha f\eta}}\} \approx j_0 f\eta \qquad (19D.3)$$

This equation shows that the net current density is proportional to the overpotential, so at low overpotentials the interface obeys Ohm's law. The relation can also be reversed to calculate the overpotential that must exist if a current density j has been established by some external circuit:

$$\eta = \frac{RTj}{Fj_0} \qquad (19D.4)$$

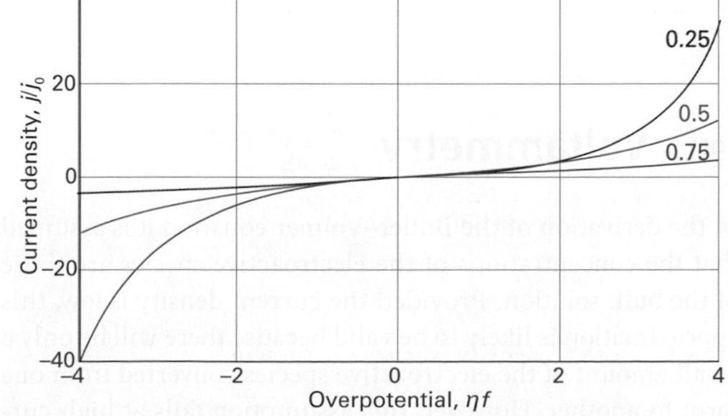

Figure 19D.6 The dependence of the current density on the overpotential for different values of the transfer coefficient.

Brief illustration 19D.1

The exchange-current density of a Pt(s)|H_2(g)|H^+(aq) electrode at 298 K is 0.79 mA cm^{-2}. The current density when the overpotential is +5.0 mV is obtained by using eqn 19D.3 and $f = F/RT = 1/(25.69\,\text{mV})$:

$$j = j_0 f\eta = \frac{(0.79\,\text{mA cm}^{-2})\times(5.0\,\text{mV})}{25.69\,\text{mV}} = 0.15\,\text{mA cm}^{-2}$$

The current through an electrode of total area 5.0 cm^2 is therefore 0.75 mA.

Some experimental values for the Butler–Volmer parameters are given in Table 19D.1. From them it is seen that exchange-current densities vary over a very wide range. Their values are generally large when the redox process involves no bond breaking (as in the $[Fe(CN)_6]^{3-}$,$[Fe(CN)_6]^{4-}$ couple) or if only weak bonds are broken (as in Cl_2,Cl^-). They are generally small when more than one electron needs to be transferred, or when multiple or strong bonds are broken, such as in the N_2,N_3^- couple and in redox reactions of organic compounds.

A further consequence of the Butler–Volmer equation is illustrated by the curves in Fig. 19D.7, in which the cathodic and anodic current densities are plotted separately against the overpotential. When the overpotential is zero the two currents are equal. As the overpotential increases, the cathodic current decreases and the anodic current increases. Note, however, that for modest values of the overpotential ($|\eta f| \le 3$) both currents are significant.

If the exchange-current density is decreased then, although the curves have the same general shape, a much greater overpotential is needed in order to achieve the same current density. This dependence leads to a distinction between different kinds of electrodes. If the exchange-current density is 'large', then modest values of the overpotential lead to significant net current flow. Such an electrode is described as **reversible** in the sense that both the cathodic and anodic processes are taking place to a significant extent. In contrast, if the exchange-current density is 'small' a much larger overpotential is needed to achieve the same current. With such a value for the overpotential either the anodic or the cathodic current dominates. Such an electrode is described as **irreversible**.

Table 19D.1 Exchange-current densities and transfer coefficients at 298 K*

Reaction	Electrode	j_0/(A cm^{-2})	α
$2H^+ + 2e^- \rightarrow H_2$	Pt	7.9×10^{-4}	
	Ni	6.3×10^{-6}	0.58
	Pb	5.0×10^{-12}	
$Fe^{3+} + e^- \rightarrow Fe^{2+}$	Pt	2.5×10^{-3}	0.58

* More values are given in the *Resource section*.

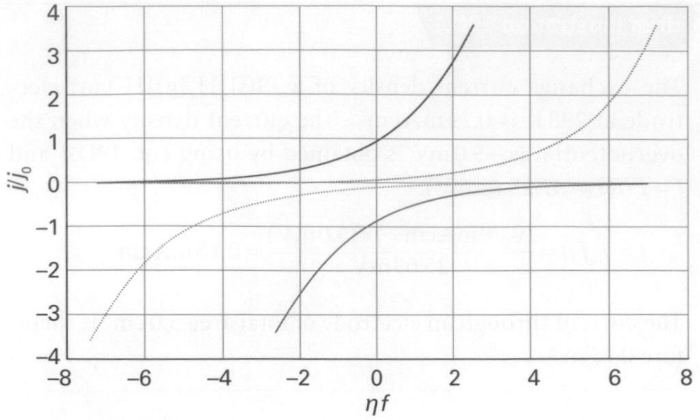

Figure 19D.7 The dependence of the anodic (purple) and cathodic (blue) current density on the overpotential (for $\alpha = 0.5$). The dotted lines are the corresponding current densities when the exchange-current density is one tenth of the value for the solid lines.

(b) Tafel plots

When the overpotential is large and positive (in practice, $\eta \geq 0.12\,\text{V}$), the anodic process is dominant. The current density is then given by the first term in eqn 19D.2:

$$j = j_0 e^{(1-\alpha)f\eta} \text{ hence } \ln j = \ln j_0 + (1-\alpha)f\eta \qquad \text{(19D.5a)}$$

A plot of the logarithm of the current density against the overpotential is called a **Tafel plot**. The slope, which is equal to $(1-\alpha)f$, gives the value of α and the intercept at $\eta = 0$ gives the exchange-current density. If the overpotential is large and negative (in practice, $\eta \leq -0.12\,\text{V}$), the cathodic process is dominant. The current density is then given by the second term in eqn 19D.2:

$$j = j_0 e^{-\alpha f\eta} \text{ hence } \ln j = \ln j_0 - \alpha f\eta \qquad \text{(19D.5b)}$$

In this case the slope of the Tafel plot is $-\alpha f$.

Example 19D.1 Analysing data using a Tafel plot

The following data refer to the anodic current through a platinum electrode of area $2.0\,\text{cm}^2$ in contact with an Fe^{3+},Fe^{2+} aqueous solution at 298 K. Determine the exchange-current density and the transfer coefficient for the electrode process.

η/mV	50	100	150	200	250
I/mA	8.8	25.0	58.0	131	298

Collect your thoughts Because the current is anodic, the appropriate plot is of $\ln j$ against η. The intercept at $\eta = 0$ is $\ln j_0$ and the slope is $(1-\alpha)f$. The current density is obtained by dividing the current by the area of the electrode.

The solution Draw up the following table:

η/mV	50	100	150	200	250
j/(mA cm^{-2})	4.4	12.5	29.0	65.5	149
$\ln(j/(\text{mA cm}^{-2}))$	1.48	2.53	3.37	4.18	5.00

The points are plotted in Fig. 19D.8. The data points for $\eta \geq 100\,\text{mV}$ give a straight line of extrapolated intercept 0.88 and slope 0.0165. From the intercept it follows that $\ln(j_0/(\text{mA cm}^{-2})) = 0.88$, so $j_0 = 2.4\,\text{mA cm}^{-2}$. From the slope it follows that $(1-\alpha)f = 0.0165\,\text{mV}^{-1}$, and because $f = F/RT = 38.9\,\text{V}^{-1}$, $\alpha = 0.58$.

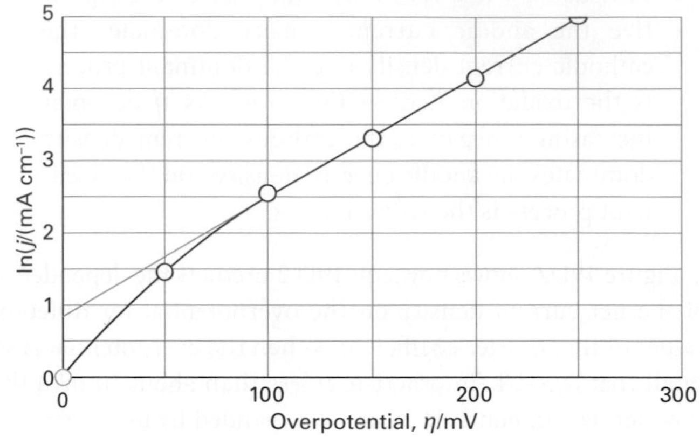

Figure 19D.8 A Tafel plot is used to measure the exchange-current density (given by the extrapolated intercept at $\eta = 0$) and the transfer coefficient (from the slope). The data are from *Example* 19D.1.

Comment. Note that the Tafel plot is nonlinear for $\eta < 100\,\text{mV}$; in this region $\alpha f\eta = 2.3$ and the condition $\alpha f\eta \gg 1$ is not satisfied.

Self-test 19D.1 Repeat the analysis using the following cathodic current data recorded at 298 K and for an electrode of area $2.0\,\text{cm}^2$:

η/mV	−50	−100	−150	−200	−250	−300
I/mA	0.3	1.5	6.4	27.6	118.6	510

Answer: $\alpha = 0.75$, $j_0 = 0.041\,\text{mA cm}^{-2}$

19D.3 Voltammetry

In the derivation of the Butler–Volmer equation it is assumed that the concentrations of the electroactive species are those of the bulk solution. Provided the current density is low, this approximation is likely to be valid because there will be only a small amount of the electroactive species converted from one form to another. However, this assumption fails at high current densities because the consumption of electroactive species close to the electrode results in a concentration gradient. The diffusion of the species towards the electrode from the bulk is slow and may become rate determining; if this is the case, increasing the overpotential leads to no further increase in the current. This effect is called **concentration polarization**. Concentration polarization is important in the interpretation of **voltammetry**, the measurement of the current through an electrode as the applied potential difference is changed.

In **linear-sweep voltammetry** the current is measured as the applied potential difference is increased linearly with time (Fig. 19D.9a); Fig. 19D.9b shows typical data obtained in this way. If the applied potential difference becomes more negative as the sweep proceeds, the cathodic current due to reduction increases and the anodic current decreases. As the applied potential difference becomes more negative than the electrode potential, the overpotential becomes negative and the Butler–Volmer equation predicts that the cathodic current will increase exponentially. This accounts for the rapid increase in the current seen in Fig. 19D.9b.

The Butler–Volmer equation predicts that the current will go on rising as the potential difference becomes more negative, but in practice the current reaches a maximum and then declines. The explanation for this decrease is that the concentration of the electroactive species (Ox^+ in this case) near the electrode is being depleted by the reduction process, thereby resulting in a decrease in the current. Diffusion of Ox^+ from the bulk solution towards the electrode replenishes its concentration there, and the balance between the rate of this process and the rate of the reduction determines the way in which the current declines at more negative potential differences.

If the bulk concentration of Ox^+ is increased, the peak current is increased, as shown in Fig. 19D.9b. The maximum current is proportional to the molar concentration of Ox^+, so its concentration can be determined from the peak height after subtraction of an extrapolated baseline. It is also found that increasing the sweep rate increases the peak current. This effect is a consequence of the balance between the rates of reduction and diffusion. The molecules close to the electrode become reduced, leading to a concentration gradient between the electrode surface and the bulk solution. This gradient drives the diffusion process: the greater the gradient, the faster is the diffusion and hence the greater is the current that can be sustained. A fast sweep leads to more rapid depletion of Ox^+ at the electrode, hence to a larger concentration gradient, faster diffusion, and therefore a larger current. According to the diffusion equation (Topic 16C), the net distance a molecule migrates is proportional to the square root of the time, and this dependence results in the peak current being proportional to the square root of the sweep rate.

In **cyclic voltammetry** the potential difference is applied with a triangular waveform (linearly up, then linearly down, Fig. 19D.10a) and the current is monitored. Cyclic voltammetry data are obtained at scan rates of about 50 mV s^{-1}, so a scan over a range of 2 V takes about 80 s. A typical cyclic voltammogram is shown in Fig. 19D.10b; note that only the oxidized species is present at the start of the experiment. For the first part of the sweep, up to the time t_3 when the direction of the sweep is reversed, the shape of the curve and its interpretation is just as for a linear-sweep experiment (Fig. 19D.9b). After t_3 the potential difference becomes less negative and consequently the rate of the cathodic process (reduction) decreases; however, the rate of the anodic process (oxidation) increases. The current therefore decreases, reflecting the reduced rate of the cathodic process and the growing rate of the anodic process in which the layer of the reduced species formed on the electrode during the first part of the sweep is progressively oxidized. Eventually, the anodic process dominates and the direction of the current reverses. There then follows a maximum in the current followed by a decrease, which are explained in the same way as for the first part of the sweep.

The voltammogram shown in Fig. 19D.10b is for a reversible process in which only a small overpotential is required to

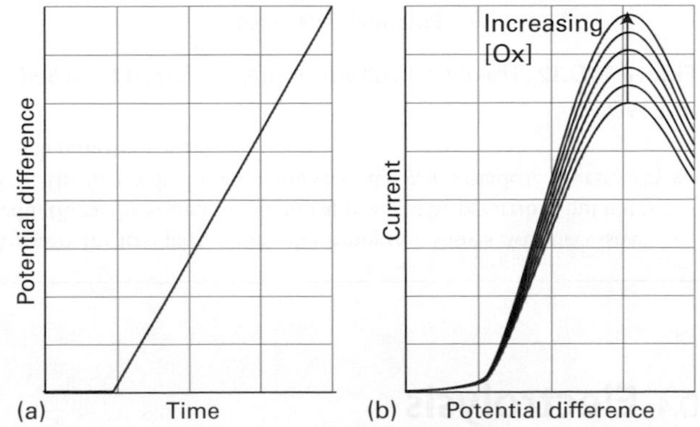

Figure 19D.9 (a) The change of potential difference with time and (b) the resulting current/potential curve in a voltammetry experiment. The peak value of the current density is proportional to the concentration of electroactive species (for instance, [Ox]) in solution.

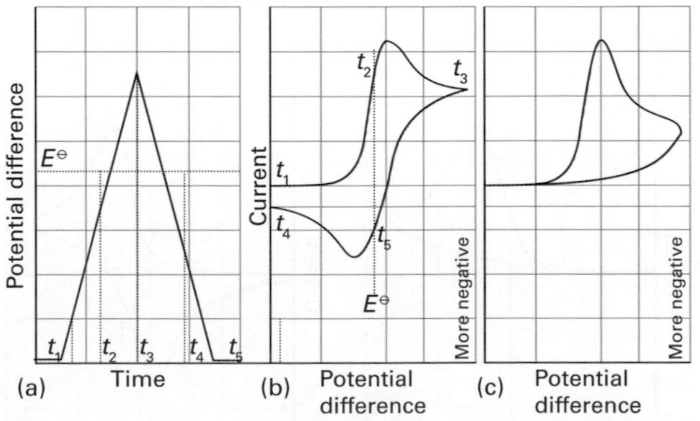

Figure 19D.10 (a) The change of potential difference with time used to record a cyclic voltammogram. The resulting current/potential curve for: (b) a reversible electrode process, and (c) an irreversible process.

give a significant current. Such a voltammogram is broadly symmetrical about the standard potential of the couple. The potential differences at which the peak current is found for forward and reverse sweeps are symmetrical about the standard potential of the couple, allowing it to be estimated. An example of such a system is the $[Fe(CN)_6]^{3-}$,$[Fe(CN)_6]^{4-}$ couple.

If the electrode process is irreversible (Fig. 19D.10c), a large overpotential is required to give a significant current, and the form of the cyclic voltammogram is affected. The first part of the voltammogram up to time t_3 is much the same as for the reversible case, except that it is shifted to a higher overpotential. When the sweep is reversed, the current decreases, reflecting the slower cathodic process. However, because a significant (positive) overpotential is needed for the anodic process to become important, the current falls back to zero as the cathodic process slows. The anodic process is not significant, and the current does not therefore change sign. The shape of the voltammogram therefore is very different from that for a reversible electrode.

The overall shape of a voltammogram gives details of the kinetics of the electrode process. Furthermore, the appearance of the curve may depend on the timescale of the sweep, because if the sweep is too fast some processes might not have time to occur.

Example 19D.2 Analysing a cyclic voltammetry experiment

The electroreduction of *p*-bromonitrobenzene in liquid ammonia is believed to occur by the following mechanism:

a $BrC_6H_4NO_2 + e^- \rightarrow BrC_6H_4NO_2^-$

b $BrC_6H_4NO_2^- \rightarrow \cdot C_6H_4NO_2 + Br^-$

c $\cdot C_6H_4NO_2 + e^- \rightarrow C_6H_4NO_2^-$

d $C_6H_4NO_2^- + H^+ \rightarrow C_6H_5NO_2$

Figure 19D.11 shows cyclic voltammograms recorded for this system at two different sweep rates. Interpret these in terms of the above mechanism.

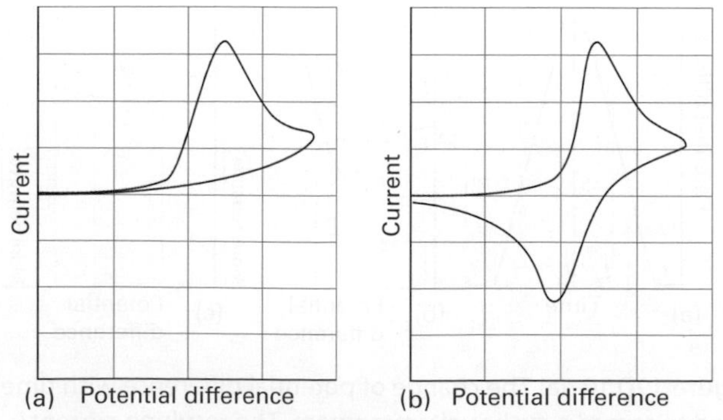

Figure 19D.11 Cyclic voltammograms referred to in *Example* 19D.2: (a) for a slow sweep rate, and (b) for a fast sweep rate.

Collect your thoughts As described in the text, the shape of the voltammogram is influenced by whether or not the process is reversible on the timescale of the sweep. That distinction depends on the relative rates of the steps involved, both the redox steps and the other reactions.

The solution The voltammogram for the slow sweep, Fig. 19D.11a, is reminiscent of the one already described for an irreversible electrode process. However, in this case the process is irreversible because the reduced species formed in step *a* goes on to react further, so $BrC_6H_4NO_2^-$ is not available to be oxidized in the second part of the sweep. The current therefore does not change sign. If step *b* is fast, then $\cdot C_6H_4NO_2$ may be reduced further in step *c*, but the process is made irreversible by the reduced species being removed in step *d*. At faster sweep rates, Fig. 19D.11b, the voltammogram is reminiscent of the one described for a reversible electrode process. In this case step *b* is not fast enough to remove $BrC_6H_4NO_2^-$, which remains available for oxidation during the second half of the sweep, leading to a change in the direction of the current.

Self-test 19D.2 Suggest an interpretation of the cyclic voltammogram shown in Fig. 19D.12 for the reduction of ClC_6H_4CN in acid solution in terms of the following reaction scheme:

a $ClC_6H_4CN + e^- \rightleftharpoons ClC_6H_4CN^-$

b $ClC_6H_4CN^- + H^+ + e^- \rightarrow C_6H_5CN + Cl^-$ (irreversible)

c $C_6H_5CN + e^- \rightleftharpoons C_6H_5CN^-$

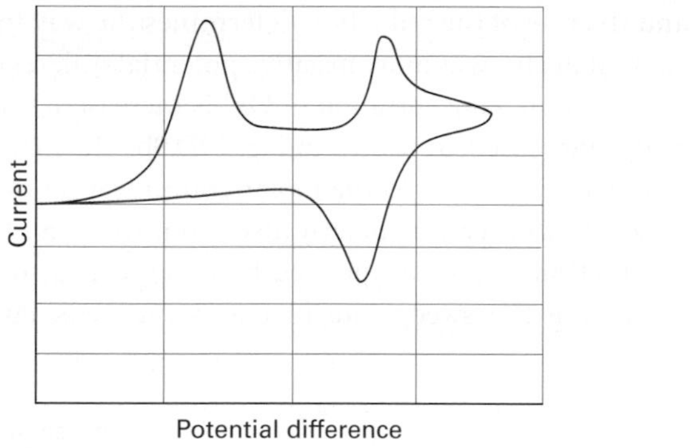

Figure 19D.12 The cyclic voltammogram referred to in *Self-test* 19D.2.

Answer: The first part of the voltammogram shows two successive reductions; the second reduction appears to be reversible, but not the first. The first reduction is *a*, the second is *c*; *a* is made irreversible by *b*, which removes $ClC_6H_4CN^-$.

19D.4 Electrolysis

To induce current to flow through an electrolytic cell and bring about a nonspontaneous cell reaction, the applied potential difference must exceed the zero-current potential by at least the **cell overpotential**, the sum of the overpotentials at the two electrodes and the ohmic drop (IR_s, where R_s is the internal resistance of the cell) due to the current through the electrolyte. The additional potential difference needed to achieve a detectable rate of reaction might need to be large when the exchange-current density at the electrodes is small. For similar reasons, a working galvanic cell generates a smaller potential difference than under zero-current conditions.

The relative rates of gas evolution or metal deposition during electrolysis can be estimated from the Butler–Volmer equation and tables of exchange-current densities. From eqn 19D.5b and assuming equal transfer coefficients, the ratio of the cathodic currents is

$$\frac{j'}{j}=\frac{j_0'}{j_0}\mathrm{e}^{(\eta-\eta')\alpha f} \qquad (19D.6)$$

where j' is the current density for electrodeposition, j is that for gas evolution, and j_0' and j_0 are the corresponding exchange-current densities. This equation shows that metal deposition is favoured by a large exchange-current density and relatively high gas evolution overpotential (so $\eta - \eta'$ is positive and large). Note that $\eta < 0$ for a cathodic process, so $-\eta' > 0$. The exchange-current density depends strongly on the nature of the electrode surface, and changes in the course of the electrodeposition of one metal on another. A very crude criterion is that significant evolution or deposition occurs only if the overpotential exceeds about 0.6 V.

A glance at Table 19D.1 shows the wide range of exchange-current densities for a metal/hydrogen electrode. The smallest exchange currents occur for lead and mercury: $1\,\mathrm{pA\,cm^{-2}}$ corresponds to a monolayer of atoms being replaced in about 5 years. For such systems, a high overpotential is needed to induce significant hydrogen evolution. In contrast, the value for platinum ($1\,\mathrm{mA\,cm^{-2}}$) corresponds to a monolayer being replaced in 0.1 s, so significant gas evolution occurs for a much lower overpotential.

The exchange-current density also depends on the crystal face exposed. For the deposition of copper on copper, the (100) face has $j_0 = 1\,\mathrm{mA\,cm^{-2}}$, so for the same overpotential the (100) face grows at 2.5 times the rate of the (111) face, for which $j_0 = 0.4\,\mathrm{mA\,cm^{-2}}$.

19D.5 Working galvanic cells

In working galvanic cells (those not balanced against an external potential), the overpotential leads to a smaller cell potential than under zero-current conditions. Furthermore, the cell potential is expected to decrease as current is generated because it is then no longer working reversibly and can therefore do less than maximum work.

Consider the cell $\mathrm{M|M^+(aq)||M'^+(aq)|M'}$, and ignore all the complications arising from liquid junctions. The potential difference generated by the cell is $E' = \Delta\phi_R - \Delta\phi_L$. Because the electrode potentials differ from their zero-current values by overpotentials, they can be written $\Delta\phi_X = E_X + \eta_X$ where X is L or R for the left or right electrode, respectively. The cell potential is therefore

$$E' = E + \eta_R - \eta_L \qquad (19D.7a)$$

To avoid confusion about signs (η_R is negative, η_L is positive), and to emphasize that a working cell generates a lower potential difference than a zero-current cell, this expression is written as

$$E' = E - |\eta_R| - |\eta_L| \qquad (19D.7b)$$

with E the cell potential. The ohmic potential difference IR_s, where R_s is the cell's internal resistance, should also be subtracted

$$E' = E - |\eta_R| - |\eta_L| - IR_s \qquad (19D.7c)$$

The ohmic term is a contribution to the cell's irreversibility—it is a thermal dissipation term—so the sign of IR_s is always such as to reduce the potential difference in the direction of zero.

The overpotentials in eqn 19D.7 can be calculated from the Butler–Volmer equation for a given current, I, being drawn. The equations are simplified by supposing that the areas, A, of the electrodes are the same, that only one electron is transferred in the rate-determining steps at the electrodes, that the transfer coefficients are both $\frac{1}{2}$, and that the high-overpotential limit of the Butler–Volmer equation may be used. Then from eqns 19D.5a, 19D.5b, and 19D.7c it follows that

$$E' = E - IR_s - \frac{4RT}{F}\ln\left(\frac{I}{A\bar{j}}\right) \qquad \bar{j} = (j_{0L}j_{0R})^{1/2} \qquad (19D.8)$$

where j_{0L} and j_{0R} are the exchange-current densities for the two electrodes.

Brief illustration 19D.2

Suppose that a cell consists of two electrodes each of area $10\,\mathrm{cm^2}$ with exchange-current densities $5\,\mu\mathrm{A\,cm^{-2}}$ and has internal resistance $10\,\Omega$. At 298 K, $RT/F = 25.7\,\mathrm{mV}$. The zero-current cell potential is 1.5 V. If the cell is producing a current of 10 mA, its working potential will be

$$E' = 1.5\,\mathrm{V} - \overbrace{(0.010\,\mathrm{A}\times 10\,\Omega)}^{0.10\,\mathrm{V}}$$

$$\overbrace{-4(0.0257\,\mathrm{V})\ln\left(\frac{1000\times 10\,\mu\mathrm{A}}{(10\,\mathrm{cm^2})\times(5\,\mu\mathrm{A\,cm^{-2}})}\right)}^{0.54\ldots\mathrm{V}} = 0.9\,\mathrm{V}$$

where 1 A Ω = 1 V has been used. Various other factors that reduce the cell potential, such as the inability of reactants to diffuse rapidly enough to the electrodes, have been ignored.

Electric storage cells operate as galvanic cells while they are producing electricity but as electrolytic cells while they are being charged by an external supply. The lead–acid battery is an old device, but one well suited to the job of starting cars (and the only one available). During charging the cathode reaction is the reduction of Pb^{2+} and its deposition as lead on the lead electrode. Deposition occurs instead of the reduction of the acid to hydrogen because the latter has a low exchange-current density on lead. The anode reaction during charging is the oxidation of Pb(II) to Pb(IV), which is deposited as the oxide PbO_2. On discharge, the two reactions run in reverse. Because they have such high exchange-current densities the discharge can occur rapidly, which is why the lead battery can produce large currents on demand.

Checklist of concepts

☐ 1. An **electrical double layer** consists of sheets of opposite charge at the surface of the electrode and next to it in the solution.

☐ 2. Descriptions of the double layer include the **Helmholtz layer model** and the **Gouy–Chapman model**.

☐ 3. The **Galvani potential difference** is the potential difference between the bulk of the metal electrode and the bulk of the solution.

☐ 4. The current density at an electrode is expressed by the **Butler–Volmer equation**.

☐ 5. A **Tafel plot** is the plot of the logarithm of the current density against the overpotential (see below).

☐ 6. **Voltammetry** is the study of the current through an electrode as a function of the applied potential difference.

☐ 7. To induce current to flow through an electrolytic cell and bring about a nonspontaneous cell reaction, the applied potential difference must exceed the cell potential by at least the **cell overpotential**.

☐ 8. In working galvanic cells the overpotential leads to a smaller potential difference.

Checklist of equations

Property	Equation	Comment	Equation number
Butler–Volmer equation	$j = j_0\{e^{(1-\alpha)f\eta} - e^{-\alpha f\eta}\}$		19D.2
Tafel plots	$\ln j = \ln j_0 + (1-\alpha)f\eta$	Anodic current density	19D.5a
	$\ln j = \ln j_0 - \alpha f\eta$	Cathodic current density	19D.5b
Potential of a working galvanic cell	$E' = E - IR_s - (4RT/F)\ln(I/A\bar{j})$	$\bar{j} = (j_{0L}j_{0R})^{1/2}$	19D.8

FOCUS 19 Processes at solid surfaces

TOPIC 19A An introduction to solid surfaces

Discussion questions

D19A.1 (a) What topographical features are found on clean surfaces? (b) Describe how steps and terraces might be formed.

D19A.2 What is the Auger effect and why is it useful for studying surfaces and the species deposited on them? Compare and contrast the techniques of scanning Auger microscopy (SAM) and scanning tunnelling microscopy (STM).

Exercises

E19A.1(a) Calculate the frequency of molecular collisions per square centimetre of surface in a vessel containing (i) hydrogen, (ii) propane at 25 °C when the pressure is 0.10 μTorr.
E19A.1(b) Calculate the frequency of molecular collisions per square centimetre of surface in a vessel containing (i) nitrogen, (ii) methane at 25 °C when the pressure is 10.0 Pa. Repeat the calculations for a pressure of 0.150 μTorr.

E19A.2(a) What pressure of argon gas is required to produce a collision rate of $4.5 \times 10^{20}\,\mathrm{s^{-1}}$ at 425 K on a circular patch of surface of diameter 1.5 mm?
E19A.2(b) What pressure of nitrogen gas is required to produce a collision rate of $5.00 \times 10^{19}\,\mathrm{s^{-1}}$ at 525 K on a circular patch of surface of diameter 2.0 mm?

E19A.3(a) At 0.10 bar it is found that a solid absorbs 10 cm^3 of a gas. At the same temperature and at 5.0 bar the volume absorbed is 22 cm^3, and this is thought to correspond to a complete monolayer. Convert the volumes to the same pressure and hence calculate the surface coverage θ at 0.1 bar.
E19A.3(b) At 0.30 bar it is found that a solid absorbs 11 cm^3 of a gas. At 5.0 bar the volume absorbed is 6.6 cm^3, and this is thought to correspond to a complete monolayer. Calculate the surface coverage θ at 0.30 bar.

E19A.4(a) Why is the adsorption of a gas onto a surface almost always an exothermic process?
E19A.4(b) If the adsorption of a gas onto a surface is found, somewhat unusually, to be endothermic, what can be said about the entropy change on adsorption?

Problems

P19A.1 This problem illustrates the way in which the binding energy of an adsorbed atom differs between a terrace and the corner of a step. Figure 19.1 shows a model consisting of a two-dimensional array of univalent cations and anions, spaced on a regular grid. In the arrangement shown in (a), a test ion sits above one edge of the lattice; in (b), the test ion sits in a corner (cations in white, anions in black). The Coulombic energy of interaction of the test ion with the lattice can be worked out by dividing the latter into sections, outlined by the boxes and denoted 'Type 1' and 'Type 2'. The energy of interaction of any two ions is simply $+C/r$ for like ions and $-C/r$ for unlike ions, where r is the distance between the ions and C is a constant; the distances r can all be expressed as multiples of the lattice spacing a_0. (a) Calculate the energy of interaction between the test ion and a 'Type 2' section of lattice, expressing your answer as multiple of C/a_0. Use a spreadsheet or mathematical software to compute the sum; it is sufficient to consider the interaction with the nearest 10 ions. (b) Similarly, calculate the energy of interaction between the test ion and a 'Type 1' section of lattice of dimension 10 atoms by 10 atoms. (c) Hence calculate the energy of interaction between the test atom and the lattice in arrangement (a) and then in (b). Which is the favoured arrangement?

P19A.2 In a study of the catalytic properties of a titanium surface it was necessary to maintain the surface free from contamination. Calculate the collision frequency per square centimetre of surface made by O_2 molecules at 300 K and (a) 100 kPa, (b) 1.00 Pa. Estimate the number of collisions made with a single surface atom in each second. The conclusions underline the importance of working at very low pressures (much lower than 1 Pa, in fact) in order to study the properties of uncontaminated surfaces. Take the nearest neighbour distance as 291 pm.

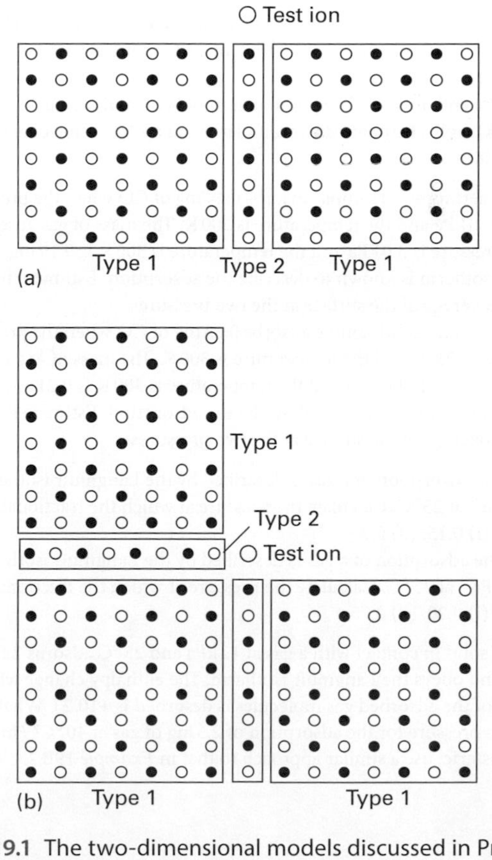

Figure 19.1 The two-dimensional models discussed in Problem P19A.1.

P19A.3 Nickel is face-centred cubic with a unit cell of side 352 pm. What is the number of atoms per square centimetre exposed on a surface formed by each of the following planes: (a) (100), (b) (110), (c) (111)? In each case, calculate the frequency of molecular collisions with a single atom in a vessel containing (a) hydrogen, (b) propane at 25 °C when the pressure is (i) 100 Pa, (ii) 0.10 μTorr.

P19A.4 The LEED pattern from a clean unreconstructed (110) face of a metal is shown below. Sketch the LEED pattern for a surface that was reconstructed by tripling the horizontal separation between the atoms.

TOPIC 19B Adsorption and desorption

Discussion questions

D19B.1 Distinguish between the following adsorption isotherms: Langmuir, BET, Temkin, and Freundlich. Indicate when and why each is likely to be appropriate.

D19B.2 What approximations underlie the formulation of the Langmuir isotherm and the BET isotherm?

Exercises

E19B.1(a) The volume of oxygen gas at 0 °C and 104 kPa adsorbed on the surface of 1.00 g of a sample of silica at 0 °C was 0.286 cm^3 at 145.4 Torr and 1.443 cm^3 at 760 Torr. Assume that the Langmuir isotherm applies estimate the value of V_∞.
E19B.1(b) The volume of gas at 20 °C and 1.00 bar adsorbed on the surface of 1.50 g of a sample of silica at 0 °C was 1.52 cm^3 at 56.4 kPa and 2.77 cm^3 at 108 kPa. Assume that the Langmuir isotherm applies and estimate the value of V_∞.

E19B.2(a) The enthalpy of adsorption of CO on a surface is found to be −120 kJ mol^{-1}. Estimate the mean lifetime of a CO molecule on the surface at 400 K; take $\tau_0 = 1.0 \times 10^{-14}$ s.
E19B.2(b) The enthalpy of adsorption of ammonia on a nickel surface is found to be −155 kJ mol^{-1}. Estimate the mean lifetime of an NH_3 molecule on the surface at 500 K; assume $\tau_0 = 1.0 \times 10^{-14}$ s.

E19B.3(a) A certain solid sample adsorbs 0.44 mg of CO when the pressure of the gas is 26.0 kPa and the temperature is 300 K. The mass of gas adsorbed when the pressure is 3.0 kPa and the temperature is 300 K is 0.19 mg. The Langmuir isotherm is known to describe the adsorption. Estimate the fractional coverage of the surface at the two pressures.
E19B.3(b) A certain solid sample adsorbs 0.63 mg of CO when the pressure of the gas is 36.0 kPa and the temperature is 300 K. The mass of gas adsorbed when the pressure is 4.0 kPa and the temperature is 300 K is 0.21 mg. The Langmuir isotherm is known to describe the adsorption. Estimate the fractional coverage of the surface at the two pressures.

E19B.4(a) The adsorption of a gas is described by the Langmuir isotherm with $\alpha = 0.75$ kPa^{-1} at 25 °C. Calculate the pressure at which the fractional surface coverage is (i) 0.15, (ii) 0.95.
E19B.4(b) The adsorption of a gas is described by the Langmuir isotherm with $\alpha = 0.548$ kPa^{-1} at 25 °C. Calculate the pressure at which the fractional surface coverage is (i) 0.20, (ii) 0.75.

E19B.5(a) A solid in contact with a gas at 12 kPa and 25 °C adsorbs 2.5 mg of the gas and obeys the Langmuir isotherm. The enthalpy change when 1.00 mmol of the adsorbed gas molecules is desorbed is +10.2 J. What is the equilibrium pressure for the adsorption of 2.5 mg of gas at 40 °C? *Hint:* The data are isosteric; use a similar approach to that in *Example* 19B.2.
E19B.5(b) A solid in contact with a gas at 8.86 kPa and 25 °C adsorbs 4.67 mg of the gas and obeys the Langmuir isotherm. The enthalpy change when 1.00 mmol of the adsorbed gas is desorbed is +12.2 J. What is the equilibrium pressure for the adsorption of the same mass of gas at 45 °C? *Hint:* See the hint for Exercise E19B.5(a).

E19B.6(a) Nitrogen gas adsorbed on charcoal to the extent of 0.921 cm^3 g^{-1} at 490 kPa and 190 K, but at 250 K the same amount of adsorption was achieved only when the pressure was increased to 3.2 MPa. What is the enthalpy of adsorption of nitrogen on charcoal? *Hint:* See the hint for Exercise E19B.5(a).
E19B.6(b) Nitrogen gas adsorbed on a surface to the extent of 1.242 cm^3 g^{-1} at 350 kPa and 180 K, but at 240 K the same amount of adsorption was achieved only when the pressure was increased to 1.02 MPa. What is the enthalpy of adsorption of nitrogen on the surface?

E19B.7(a) In an experiment on the adsorption of oxygen on tungsten it was found that the same volume of oxygen was desorbed in 27 min at 1856 K and 2.0 min at 1978 K. What is the activation energy of desorption? How long would it take for the same amount to desorb at (i) 298 K, (ii) 3000 K?
E19B.7(b) In an experiment on the adsorption of ethene on iron it was found that the same volume of the gas was desorbed in 1856 s at 873 K and 8.44 s at 1012 K. What is the activation energy of desorption? How long would it take for the same amount of ethene to desorb at (i) 298 K, (ii) 1500 K?

E19B.8(a) The average time for which an oxygen atom remains adsorbed to a tungsten surface is 0.36 s at 2548 K and 3.49 s at 2362 K. What is the activation energy for desorption?
E19B.8(b) The average time for which a hydrogen atom remains adsorbed on a manganese surface is 35 per cent shorter at 1000 K than at 600 K. What is the activation energy for desorption?

E19B.9(a) For how long on average would a hydrogen atom remain on a surface at 400 K if its desorption activation energy is (i) 15 kJ mol^{-1}, (ii) 150 kJ mol^{-1}? Take $\tau_0 = 0.10$ ps. Repeat both calculations at 1000 K.
E19B.9(b) For how long on average would an atom remain on a surface at 298 K if its desorption activation energy is (i) 20 kJ mol^{-1}, (ii) 200 kJ mol^{-1}? Take $\tau_0 = 0.12$ ps. Repeat both calculations at 800 K.

Problems

P19B.1 Use mathematical software or a spreadsheet to perform the following calculations. (a) Use eqn 19B.2 to generate a family of curves showing the dependence of $1/\theta$ on $1/p$ for several values of α. (b) Use eqn 19B.4 to generate a family of curves showing the dependence of $1/\theta$ on $1/p$ for several values of α. On the basis of your results from parts (a) and (b), discuss how plots of $1/\theta$ against $1/p$ can be used to distinguish between adsorption with and without dissociation. (c) Use eqn 19B.6 to generate a family of curves showing the dependence of $zV_{\text{mon}}/(1-z)V$ on z for different values of c.

P19B.2 The data below are for the chemisorption of hydrogen on copper powder at 25 °C. Confirm that they fit the Langmuir isotherm at low coverages (the volumes have been corrected so that they are all at the same pressure). Then find the value of α for the adsorption equilibrium and the adsorption volume corresponding to complete coverage.

p/Pa	25	129	253	540	1000	1593
V/cm^3	0.042	0.163	0.221	0.321	0.411	0.471

P19B.3 The data for the adsorption of ammonia on barium fluoride are reported below (the volumes have been corrected to be for the same pressure in each case). Confirm that they fit a BET isotherm and find values of c and V_{mon}.

(a) $\theta = 0\,°\text{C}$, $p^* = 429.6\,\text{kPa}$:

p/kPa	14.0	37.6	65.6	79.2	82.7	100.7	106.4
V/cm^3	11.1	13.5	14.9	16.0	15.5	17.3	16.5

(b) $\theta = 18.6\,°\text{C}$, $p^* = 819.7\,\text{kPa}$:

p/kPa	5.3	8.4	14.4	29.2	62.1	74.0	80.1	102.0
V/cm^3	9.2	9.8	10.3	11.3	12.9	13.1	13.4	14.1

P19B.4 The following data have been obtained for the adsorption of H_2 on the surface of 1.00 g of copper at 0 °C. The volume of hydrogen has been corrected to STP (0 °C and 1 atm).

p/atm	0.050	0.100	0.150	0.200	0.250
V/cm^3	1.22	1.33	1.31	1.36	1.40

Assume that the Langmuir isotherm applies, calculate the volume of H_2 necessary to form a monolayer, and then estimate the surface area of the copper sample. The mass density of liquid hydrogen is 0.708 g cm^{-3}. *Hint:* Use the mass density to estimate the area occupied by one H_2 molecule.

P19B.5‡ M.-G. Olivier and R. Jadot (*J. Chem. Eng. Data* **42**, 230 (1997)) studied the adsorption of butane on silica gel. They report the following amounts of absorption (in moles of C_4H_{10} per kilogram of silica gel) at 303 K:

p/kPa	31.00	38.22	53.03	76.38	101.97
n/(mol kg^{-1})	1.00	1.17	1.54	2.04	2.49
p/kPa	130.47	165.06	182.41	205.75	219.91
n/(mol kg^{-1})	2.90	3.22	3.30	3.35	3.36

Fit the data to a Langmuir isotherm, identify the value of n that corresponds to complete coverage, and evaluate the constant α.

P19B.6 The designers of a new industrial plant wanted to use a catalyst code-named CR-1 in a step involving the fluorination of butadiene. As a first step in the investigation they established the form of the adsorption isotherm. The volume of butadiene (corrected to the same pressure) adsorbed per gram of CR-1 at 15 °C varied with pressure as given below. (a) Investigate how well these data conform to the Langmuir isotherm.

p/kPa	13.3	26.7	40.0	53.3	66.7	80.0
V/cm^3	17.9	33.0	47.0	60.8	75.3	91.3

(b) Investigate whether the BET isotherm gives a better description of the adsorption of butadiene on CR-1; find V_{mon} and c. At 15 °C, p^*(butadiene) = 200 kPa.

P19B.7‡ C. Huang and W.P. Cheng (*J. Colloid Interface Sci.* **188**, 270 (1997)) examined the adsorption of the hexacyanoferrate(III) ion, $[Fe(CN)_6]^{3-}$, on γ-Al_2O_3 from aqueous solution. They modelled the adsorption with a Langmuir isotherm (modified to take into account a surface reaction between the $[Fe(CN)_6]^{3-}$ and the alumina), obtaining the following values of α at pH = 6.5:

T/K	283	298	308	318
$10^{-11}\,\alpha$/mol^{-1} dm^3	2.642	2.078	1.286	1.085

Evaluate the isosteric enthalpy of adsorption, $\Delta_{\text{ad}}H^{\ominus}$, at this pH. The researchers also reported $\Delta_{\text{ad}}S^{\ominus} = +146\,\text{J mol}^{-1}\,\text{K}^{-1}$ under these conditions. Determine $\Delta_{\text{ad}}G^{\ominus}$.

P19B.8‡ In a study relevant to automobile catalytic converters, C.E. Wartnaby et al. (*J. Phys. Chem.* **100**, 12 483 (1996)) measured the enthalpy of adsorption of CO, NO, and O_2 on initially clean platinum (110) surfaces. They report $\Delta_{\text{ad}}H^{\ominus}$ for NO to be $-160\,\text{kJ mol}^{-1}$. Calculate the ratio of the value of α at 500 °C to that at 400 °C.

P19B.9 The adsorption of solutes on solids from liquids often follows a Freundlich isotherm. Check the applicability of this isotherm to the following data for the adsorption of ethanoic acid on charcoal at 25 °C and find the values of the parameters c_1 and c_2.

[acid]/(mol dm^{-3})	0.05	0.10	0.50	1.0	1.5
w_a/g	0.04	0.06	0.12	0.16	0.19

where w_a is the mass adsorbed per gram of charcoal.

P19B.10‡ A. Akgerman and M. Zardkoohi (*J. Chem. Eng. Data* **41**, 185 (1996)) examined the adsorption of phenol from aqueous solution on to fly ash at 20 °C. They fitted their observations to a Freundlich isotherm of the form $c_{\text{ads}} = Kc_{\text{sol}}^{1/n}$, where c_{ads} is the concentration of adsorbed phenol and c_{sol} is the concentration of aqueous phenol. Among the data reported are the following:

c_{sol}/(mg g^{-1})	8.26	15.65	25.43	31.74	40.00
c_{ads}/(mg g^{-1})	4.41	9.2	35.2	52.0	67.2

Evaluate the constants K and n. What further information would be necessary in order to express the data in terms of fractional coverage, θ?

P19B.11‡ The following data were obtained for the extent of adsorption, s, of propanone (acetone) on charcoal from an aqueous solution of molar concentration, c, at 18 °C:

c/(mmol dm^{-3})	15.0	23.0	42.0	84.0	165	390	800
s/(mmol acetone/ g charcoal)	0.60	0.75	1.05	1.50	2.15	3.50	5.10

Which isotherm fits this data best, Langmuir, Freundlich, or Temkin?

P19B.12 Suppose it is known that ozone adsorbs on a certain surface in accord with a Langmuir isotherm. How could you use the pressure dependence of the fractional coverage to distinguish between adsorption (i) without dissociation, (ii) with dissociation into $O + O_2$, (c) with dissociation into $O + O + O$?

‡ These problems were supplied by Charles Trapp and Carmen Giunta.

TOPIC 19C Heterogeneous catalysis

Discussion questions

D19C.1 Describe the essential features of the Langmuir–Hinshelwood and Eley–Rideal mechanisms for surface-catalysed reactions.

D19C.2 Derive the expressions for θ_A and θ_B quoted in the derivation of the Langmuir–Hinshelwood rate equation.

Exercises

E19C.1(a) A monolayer of N_2 molecules is adsorbed on the surface of 1.00 g of an Fe/Al_2O_3 catalyst at 77 K, the boiling point of liquid nitrogen. Upon warming, the nitrogen occupies 3.86 cm^3 at 0 °C and 760 Torr. What is the surface area of the catalyst? Use the value of the collision cross-section for N_2 (from the *Resource section*) as an estimate for the area of a molecule.

E19C.1(b) A monolayer of CO molecules is adsorbed on the surface of 1.00 g of an Fe/Al_2O_3 catalyst at 77 K, the boiling point of liquid nitrogen. Upon warming, the carbon monoxide occupies 3.75 cm^3 at 0 °C and 1.00 bar. What is the surface area of the catalyst? Use the value of the collision cross-section for CO (from the *Resource section*) as an estimate for the area of a molecule.

Problems

P19C.1 (a) According to the Langmuir–Hinshelwood mechanism of surface-catalysed reactions, the rate of reaction between A and B depends on the rate at which the adsorbed species meet. Write the rate law for the reaction according to this mechanism. (b) Find the limiting form of this rate law for the case where the partial pressures of the reactants are low. What is the overall order in this case? (c) Could this mechanism ever account for zeroth-order kinetics?

P19C.2 Hydrogen iodide is very strongly adsorbed on gold but only slightly adsorbed on platinum. For all but the very lowest pressures, the rate of the decomposition of HI on gold is found to be independent of the pressure of HI. For the same process on platinum, the rate is found to be proportional to the partial pressure of HI. Explain these observations with the aid of the Langmuir isotherm.

P19C.3 In some catalytic reactions the products adsorb more strongly than the reacting gas. This is the case, for instance, in the catalytic decomposition of ammonia to N_2 and H_2 on platinum at 1000 °C, in which the H_2 is absorbed very strongly. The kinetics of such a process can be analysed with the aid of the Langmuir isotherm. (a) First show that when a gas J adsorbs very strongly, and its pressure is p_J, the fraction of *uncovered* sites is approximately $1/\alpha p_J$. (b) Assume that the ammonia is only weakly absorbed and use your result from (a) to argue that the rate of reaction is given by an expression of the form

$$\frac{dp_{NH_3}}{dt}=-k_c\frac{p_{NH_3}}{p_{H_2}}$$

(c) Integrate this rate equation by using the initial condition that at $t=0$ only ammonia is present at pressure p_0. *Hint:* You will need to write the pressure of hydrogen in terms of the current pressure of ammonia and p_0, taking into account the stoichiometric equation. (d) Hence determine the form of a suitable straight-line plot to determine k_c. Analyse the following data using such a plot and find a value for k_c, stating its units.

t/s	0	30	60	100	160	200	250
p/kPa	13.3	11.7	11.2	10.7	10.3	9.9	9.6

TOPIC 19D Processes at electrodes

Discussion questions

D19D.1 Describe the various models of the electrode–electrolyte interface.

D19D.2 Discuss the technique of cyclic voltammetry and account for the characteristic shape of a cyclic voltammogram, such as those shown in Figs. 19D.10b and 19B.10c.

Exercises

E19D.1(a) The transfer coefficient of a certain electrode in contact with M^{3+} and M^{4+} in aqueous solution at 25 °C is 0.39. The current density is found to be 55.0 mA cm^{-2} when the overpotential is 125 mV. What is the overpotential required for a current density of 75 mA cm^{-2}? *Hint:* With this overpotential the current is entirely anodic, and eqn 19D.5a applies.

E19D.1(b) The transfer coefficient of a certain electrode in contact with M^{2+} and M^{3+} in aqueous solution at 25 °C is 0.42. The current density is found to be 17.0 mA cm^{2} when the overpotential is 105 mV. What is the overpotential required for a current density of 72 mA cm^{-2}? *Hint:* See hint to Exercise E19D.1(a).

E19D.2(a) Calculate the exchange-current density from the information given in Exercise 19D.1(a).

E19D.2(b) Calculate the exchange-current density from the information given in Exercise 19D.1(b).

E19D.3(a) Significant evolution or deposition occurs in electrolysis only if the overpotential exceeds about 0.6 V. To illustrate this criterion calculate the effect that increasing the overpotential from 0.40 V to 0.60 V has on the current density in the electrolysis of 1.0 M NaOH(aq), which is 1.0 mA cm^{-2} at 0.4 V and 25 °C. Take $\alpha=0.5$. *Hint:* Assume that the current is entirely anodic.

E19D.3(b) Calculate the effect that increasing the overpotential from 0.50 V to 0.60 V has on the current density in the electrolysis of 1.0 M NaOH(aq), which is 1.22 mA cm^{-2} at 0.50 V and 25 °C. Take $\alpha=0.50$.

E19D.4(a) Use the data in Table 19D.1 for the exchange-current density and transfer coefficient for the reaction $2H^+ + 2e^- \rightarrow H_2$ on nickel at 25 °C to identify the current density needed to obtain an overpotential of 0.20 V as calculated (i) by using the Butler–Volmer equation, and (ii) by assuming that the current is entirely anodic. Comment on the validity of the latter assumption.

E19D.4(b) Use the data in Table 19D.1 for the exchange-current density and transfer coefficient for the reaction $Fe^{3+} + e^- \rightarrow Fe^{2+}$ on platinum at 25 °C to identify the current density needed to obtain an overpotential of 0.30 V as calculated (i) by using the Butler–Volmer equation, and (ii) by assuming that the current is entirely anodic. Comment on the validity of the latter assumption.

E19D.5(a) A typical exchange-current density, that for H^+ discharge at platinum, is 0.79 mA cm^{-2} at 25 °C. Use the Butler–Volmer equation to calculate the current density at an electrode when its overpotential is (a) 10 mV, (b) 100 mV, (c) −5.0 V. Take $\alpha = 0.5$.

E19D.5(b) The exchange-current density for a $Pt|Fe^{3+},Fe^{2+}$ electrode is 2.5 mA cm^{-2}. The standard potential of the electrode is +0.77 V. Derive an expression for the current flowing through an electrode of surface area 1.0 cm^2 as a function of the potential applied to the electrode; assume standard conditions.

E19D.6(a) How many electrons or protons are transported through the double layer per second for each of the electrodes $Pt,H_2|H^+$, $Pt|Fe^{3+},Fe^{2+}$, and $Pb,H_2|H^+$ when they are at equilibrium at 25 °C? Take the area as 1.0 cm^2 in each case. Estimate the number of times per second a single atom on the surface takes part in an electron transfer event, assuming an electrode atom occupies about (280 pm)^2 of the surface. Refer to Table 19D.1 for the necessary data. *Hint:* At equilibrium only the exchange current flows.

E19D.6(b) How many electrons or protons are transported through the double layer per second for each of the electrodes $Cu,H_2|H^+$ and $Pt|Ce^{4+},Ce^{3+}$ when they are at equilibrium at 25 °C? Take the area as 1.0 cm^2 in each case. Estimate the number of times each second a single atom on the surface takes part in an electron transfer event, assuming an electrode atom occupies about (260 pm)^2 of the surface.

E19D.7(a) When the overpotential is small, the current density is given by eqn 19D.4. Assume this to be the case, take the surface area of the electrode to be 1.0 cm^2, and calculate the effective resistance at 25 °C of the following electrodes: (i) $Pt,H_2|H^+$, and (ii) $Hg,H_2|H^+$. *Hint:* The resistance is given by the ratio of the potential to the current.

E19D.7(b) Repeat the calculation in Exercise 19D.7(a) for the electrodes: (i) $Pb,H_2|H^+$, and (ii) $Pt|Fe^{2+},Fe^{3+}$.

E19D.8(a) The exchange-current density for H^+ discharge at zinc is about 50 pA cm^{-2}. Can zinc be deposited from an aqueous solution of a zinc salt under standard conditions and at 25 °C? The standard potential of the $Zn^{2+}|Zn$ electrode is −0.76 V at 25 °C.

E19D.8(b) The exchange-current density for H^+ discharge at platinum is 0.79 mA cm^{-2}. Can zinc be plated on to platinum under standard conditions and at 25 °C? The standard potential of the $Zn^{2+}|Zn$ electrode is −0.76 V at 25 °C.

Problems

P19D.1 In an experiment on the $Pt|H_2|H^+$ electrode in dilute H_2SO_4 the following current densities were observed at 25 °C. (a) Evaluate α and j_0 for the electrode.

η/mV	50	100	150	200	250
j/(mA cm^{-2})	2.66	8.91	29.9	100	335

(b) For the same electrode, draw up a table of current densities for the case where the overpotentials have the same magnitude, but are of opposite sign, to those given above.

P19D.2 The standard potentials of $Pb^{2+}|Pb$ and $Sn^{2+}|Sn$ are −126 mV and −136 mV respectively at 25 °C, and the overpotential for their deposition are close to zero. What should the relative concentrations of Pb^{2+}(aq) and Sn^{2+}(aq) be in order to ensure simultaneous deposition from a mixture? You may assume that activities can be approximated by molar concentrations.

P19D.3‡ The rate of deposition of iron, v, on the surface of an iron electrode from an aqueous solution of Fe^{2+} has been studied as a function of potential, E', relative to the standard hydrogen electrode, by J. Kanya (*J. Electroanal. Chem.* **84**, 83 (1977)). The values in the table below are based on the data obtained with an electrode of surface area 9.1 cm^2 in contact with a solution of concentration 1.70 μmol dm^{-3} in Fe^{2+}. (a) Assume that activities can be approximated by molar concentrations, and calculate the zero-current potential of the Fe^{2+}/Fe cathode and hence the overpotential at each value of the potential given in the table. (b) Calculate the cathodic current density, j_c, from the rate of deposition of Fe^{2+} for each value of E'. (c) Analyse the data using a Tafel plot and hence determine the exchange-current density.

v/(pmol s^{-1})	1.47	2.18	3.11	7.26
$-E'$/mV	702	727	752	812

P19D.4‡ V.V. Losev and A.P. Pchel'nikov (*Soviet Electrochem.* **6**, 34 (1970)) obtained the following current–voltage data for an indium anode relative to a standard hydrogen electrode at 293 K:

$-E'$/V	0.388	0.365	0.350	0.335
j/(A m^{-2})	0	0.590	1.438	3.507

Use these data to calculate the transfer coefficient and the exchange-current density. What is the cathodic current density when the potential is −0.365 V? *Hint:* The value of E' with $j = 0$ is the equilibrium potential.

P19D.5‡ An early study of the hydrogen overpotential is that of H. Bowden and T. Rideal (*Proc. Roy. Soc.*, 59 (1928)), who measured the overpotential for H_2 evolution with a mercury electrode in dilute aqueous solutions of H_2SO_4 at 25 °C. Determine the exchange-current density and transfer coefficient, α, from their data:

j/(mA m^{-2})	2.9	6.3	28	100	250	630	1650	3300
η/V	0.60	0.65	0.73	0.79	0.84	0.89	0.93	0.96

P19D.6 If $\alpha = \frac{1}{2}$, an electrode interface is unable to rectify alternating current because the current density curve is symmetrical about $\eta = 0$. When $\alpha \neq \frac{1}{2}$, the magnitude of the current density depends on the sign of the overpotential, and so some degree of 'faradaic rectification' may be obtained. (a) Suppose that the overpotential varies as $\eta = \eta_0 \cos \omega t$. Derive an expression for the mean flow of current (averaged over a cycle) for general α, and confirm that the mean current is zero when $\alpha = \frac{1}{2}$. In your calculations work in the limit of small η_0 but to second order in $\eta_0 F/RT$ (that is, when expanding the exponentials in the Butler–Volmer equation, retain up to the second-order terms). (b) Calculate the mean direct current at 25 °C for a 1.0 cm^2 hydrogen–platinum electrode with $\alpha = 0.38$ when the overpotential varies between ±10 mV at 50 Hz.

P19D.7 (Continues from Problem P16D.6) Now suppose that the overpotential is in the high overpotential region at all times even though it is oscillating, and that it takes the form of a sawtooth between varying linearly between η_- and η_+ around an average of η_0, but in such a way that η is always positive. Derive an expression for the variation in the current density across the interface; take $\alpha = \frac{1}{2}$.

P19D.8 Figure 19.2 shows four different examples of voltammograms. Identify the processes occurring in each system. In each case the vertical axis is the current and the horizontal axis is the (negative) electrode potential.

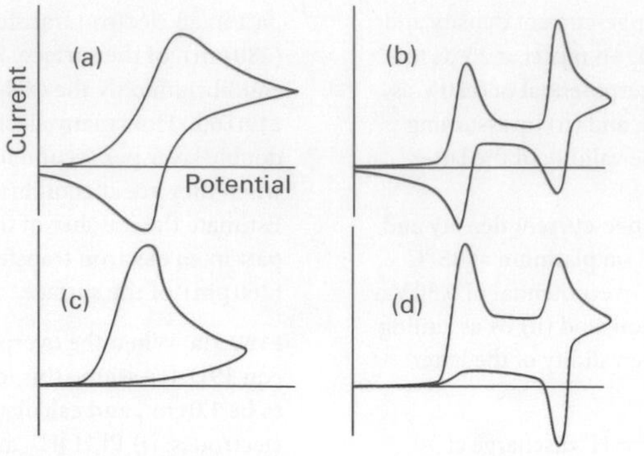

Figure 19.2 The voltammograms discussed in Problem P19D.8.

FOCUS 19 Processes at solid surfaces

Integrated activities

I19.1 Although the attractive van der Waals interaction between individual molecules varies as R^{-6} the interaction of a molecule with a nearby solid (a homogeneous collection of molecules) varies as R^{-3}, where R is its vertical distance above the surface. Confirm this assertion. Calculate the interaction energy between an Ar atom and the surface of solid argon on the basis of a Lennard-Jones-(6,12) potential. Estimate the equilibrium distance of an atom above the surface.

I19.2 Electron microscopes can obtain images with much higher resolution than optical microscopes because of the short wavelength obtainable from a beam of electrons. For electrons moving at speeds close to c, the speed of light, the expression for the de Broglie wavelength (eqn 7A.11, $\lambda = h/p$) needs to be corrected for relativistic effects:

$$\lambda = \frac{h}{\left\{2m_e e\Delta\phi\left(1+\dfrac{e\Delta\phi}{2m_e c^2}\right)\right\}^{1/2}}$$

where c is the speed of light in vacuum and $\Delta\phi$ is the potential difference through which the electrons are accelerated. (a) Use the expression above to calculate the de Broglie wavelength of electrons accelerated through 50 kV. (b) Is the relativistic correction important?

I19.3 The forces measured by AFM arise primarily from interactions between electrons of the tip and on the surface. To get an idea of the magnitudes of these forces, calculate the force acting between two electrons separated by 2.0 nm. To calculate the force between the electrons, use $F = -dV/dr$ where V is their mutual Coulombic potential energy and r is their separation.

I19.4 To appreciate the distance dependence of the tunnelling current in scanning tunnelling microscopy, suppose that the electron in the gap between sample and tip has an energy 2.0 eV less than the barrier height. By what factor would the current decrease if the needle is moved from $L_1 = 0.50$ nm to $L_2 = 0.60$ nm from the surface?

I19.5 Calculate the thermodynamic limit to the zero-current potential of fuel cells operating on (a) hydrogen and oxygen, (b) methane and air, and (c) propane and air. Refer to the *Resource section* to calculate relevant values of the reaction Gibbs energies, and take the species to be in their standard states at 25 °C.

RESOURCE SECTION

Contents

PART 1 Common integrals

	Indefinite integral*	Constraint	Definite integral
Algebraic functions			
A.1	$\int x^n \mathrm{d}x = \frac{1}{n+1}x^{n+1} + c$	$n \neq -1$	$\int_a^b x^n \mathrm{d}x = \frac{1}{n+1}(b^{n+1} - a^{n+1})$
A.2	$\int \frac{1}{x}\mathrm{d}x = \ln x + c$		$\int_a^b \frac{1}{x}\mathrm{d}x = \ln\frac{b}{a}$
A.3	$\int \frac{1}{(A-x)(B-x)}\mathrm{d}x = \frac{1}{B-A}\ln\frac{B-x}{A-x} + c$	$A \neq B$	$\int_a^b \frac{1}{(A-x)(B-x)}\mathrm{d}x = \frac{1}{B-A}\ln\frac{(B-b)(A-a)}{(A-b)(B-a)}$
Exponential functions			
E.1	$\int \mathrm{e}^{-kx}\mathrm{d}x = -\frac{1}{k}\mathrm{e}^{-kx} + c$		$\int_a^b \mathrm{e}^{-kx}\mathrm{d}x = \frac{1}{k}(\mathrm{e}^{-ka} - \mathrm{e}^{-kb})$
E.2	$\int x\mathrm{e}^{-kx}\mathrm{d}x = -\frac{1}{k^2}\mathrm{e}^{-kx} - \frac{x}{k}\mathrm{e}^{-kx} + c$		$\int_a^b x\mathrm{e}^{-kx}\mathrm{d}x = -\frac{1}{k^2}(\mathrm{e}^{-kb} - \mathrm{e}^{-ka}) - \frac{1}{k}(b\mathrm{e}^{-kb} - a\mathrm{e}^{-ka})$
E.3		$n \geq 0,\ k > 0$	$\int_0^\infty x^n \mathrm{e}^{-kx}\mathrm{d}x = \frac{n!}{k^{n+1}}$ $\quad n! = n(n-1)\ldots 1;\ 0! = 1$
E.4			$\int_0^\infty \frac{x^4\mathrm{e}^x}{(\mathrm{e}^x - 1)^2}\mathrm{d}x = \frac{4\pi^4}{15}$
Gaussian functions			
G.1			$\int_0^\infty \mathrm{e}^{-kx^2}\mathrm{d}x = \frac{1}{2}\left(\frac{\pi}{k}\right)^{1/2}$
G.2	$\int x\mathrm{e}^{-kx^2}\mathrm{d}x = -\frac{1}{2k}\mathrm{e}^{-kx^2} + c$		$\int_0^\infty x\mathrm{e}^{-kx^2}\mathrm{d}x = \frac{1}{2k}$
G.3			$\int_0^\infty x^2\mathrm{e}^{-kx^2}\mathrm{d}x = \frac{1}{4}\left(\frac{\pi}{k^3}\right)^{1/2}$
G.4	$\int x^3\mathrm{e}^{-kx^2}\mathrm{d}x = -\frac{1}{2k^2}\mathrm{e}^{-kx^2} - \frac{x^2}{2k}\mathrm{e}^{-kx^2} + c$	$k > 0$	$\int_0^\infty x^3\mathrm{e}^{-kx^2}\mathrm{d}x = \frac{1}{2k^2}$
G.5		$k > 0$	$\int_0^\infty x^4\mathrm{e}^{-kx^2}\mathrm{d}x = \frac{3}{8k^2}\left(\frac{\pi}{k}\right)^{1/2}$
G.6			$\mathrm{erf}\,z = \frac{2}{\pi^{1/2}}\int_0^z \mathrm{e}^{-x^2}\mathrm{d}x \quad \mathrm{erfc}\,z = 1 - \mathrm{erf}\,z$
G.7		$m \geq 0,\ k > 0$	$\int_0^\infty x^{2m+1}\mathrm{e}^{-kx^2}\mathrm{d}x = \frac{m!}{2k^{m+1}}$
G.8		$m \geq 1,\ k > 0$	$\int_0^\infty x^{2m}\mathrm{e}^{-kx^2}\mathrm{d}x = \frac{(2m-1)!!}{2^{m+1}k^m}\left(\frac{\pi}{k}\right)^{1/2}$
			$n!! = n \times (n-2) \times (n-4) \ldots 1$ or 2
Trigonometric functions			
T.1	$\int \sin kx\,\mathrm{d}x = -\frac{1}{k}\cos kx + c$		$\int_0^a \sin kx\,\mathrm{d}x = \frac{1}{k}(1 - \cos ka)$
T.2	$\int \sin^2 kx\,\mathrm{d}x = \frac{1}{2}x - \frac{1}{4k}\sin 2kx + c$		$\int_0^a \sin^2 kx\,\mathrm{d}x = \frac{1}{2}a - \frac{1}{4k}\sin 2ka$
T.3	$\int \sin^3 kx\,\mathrm{d}x = -\frac{1}{3k}(\sin^2 kx + 2)\cos kx + c$		$\int_0^a \sin^3 kx\,\mathrm{d}x = \frac{1}{3k}\{2 - (\sin^2 ka + 2)\cos ka\}$
T.4	$\int \sin^4 kx\,\mathrm{d}x = \frac{3x}{8} - \frac{3}{8k}\sin kx\cos kx - \frac{1}{4k}\sin^3 kx\cos kx + c$		$\int_0^a \sin^4 kx\,\mathrm{d}x = \frac{3a}{8} - \frac{3}{8k}\sin ka\cos ka - \frac{1}{4k}\sin^3 ka\cos ka$
T.5	$\int \sin Ax\sin Bx\,\mathrm{d}x = \frac{\sin(A-B)x}{2(A-B)} - \frac{\sin(A+B)x}{2(A+B)} + c$	$A^2 \neq B^2$	$\int_0^a \sin Ax\sin Bx\,\mathrm{d}x = \frac{\sin(A-B)a}{2(A-B)} - \frac{\sin(A+B)a}{2(A+B)}$
T.6	$\int \cos Ax\cos Bx\,\mathrm{d}x = \frac{\sin(A-B)x}{2(A-B)} + \frac{\sin(A+B)x}{2(A+B)} + c$	$A^2 \neq B^2$	$\int_0^a \cos Ax\cos Bx\,\mathrm{d}x = \frac{\sin(A-B)a}{2(A-B)} + \frac{\sin(A+B)a}{2(A+B)}$

	Indefinite integral*	Constraint	Definite integral
T.7	$\int \sin kx \cos kx \, \mathrm{d}x = \frac{1}{2k}\sin^2 kx + c$		$\int_0^a \sin kx \cos kx \, \mathrm{d}x = \frac{1}{2k}\sin^2 ka$
T.8	$\int \sin Ax \cos Bx \, \mathrm{d}x = \frac{\cos(B-A)x}{2(B-A)} - \frac{\cos(A+B)x}{2(A+B)} + c$	$A^2 \neq B^2$	$\int_0^a \sin Ax \cos Bx \, \mathrm{d}x = \frac{\cos(B-A)a-1}{2(B-A)} - \frac{\cos(A+B)a-1}{2(A+B)}$
T.9	$\int x \sin Ax \sin Bx \, \mathrm{d}x = -\frac{\mathrm{d}}{\mathrm{d}B}\underbrace{\int \sin Ax \cos Bx \, \mathrm{d}x}_{\text{T.7}}$	$A^2 \neq B^2$	
T.10	$\int \cos^2 kx \sin kx \, \mathrm{d}x = -\frac{1}{3k}\cos^3 kx + c$		$\int_0^a \cos^2 kx \sin kx \, \mathrm{d}x = \frac{1}{3k}\{1-\cos^3 ka\}$
T.11	$\int x \sin^2 kx \, \mathrm{d}x = \frac{x^2}{4} - \frac{1}{4k} x \sin 2kx - \frac{1}{8k^2}\cos 2kx + c$		$\int_0^a x \sin^2 kx \, \mathrm{d}x = \frac{a^2}{4} - \frac{1}{4k} a \sin 2ka - \frac{1}{8k^2}(\cos 2ka - 1)$
T.12	$\int x^2 \sin^2 kx \, \mathrm{d}x = \frac{x^3}{6} - \left(\frac{x^2}{4k} - \frac{1}{8k^3}\right)\sin 2kx - \frac{x}{4k^2}\cos 2kx + c$		$\int_0^a x^2 \sin^2 kx \, \mathrm{d}x = \frac{a^3}{6} - \left(\frac{a^2}{4k} - \frac{1}{8k^3}\right)\sin 2ka - \frac{a}{4k^2}\cos 2ka$
T.13	$\int x \cos kx \, \mathrm{d}x = \frac{1}{k^2}\cos kx + \frac{x}{k}\sin kx + c$		$\int_0^a x \cos kx \, \mathrm{d}x = \frac{1}{k^2}(\cos ka - 1) + \frac{a}{k}\sin ka$
T.14	$\int \cos^2 kx \, \mathrm{d}x = \frac{x}{2} + \frac{1}{4k}\sin 2kx + c$		$\int_0^a \cos^2 kx \, \mathrm{d}x = \frac{a}{2} + \frac{1}{4k}\sin 2ka$

In each case, c is a constant. Note that not all indefinite integrals have a simple closed form.

PART 2 Units

Table A.1 Some common units

Physical quantity	Name of unit	Symbol for unit	Value*
Time	minute	min	60 s
	hour	h	3600 s
	day	d	86 400 s
	year	a	31 556 952 s
Length	angstrom	Å	10^{-10} m
Volume	litre	L, l	1 dm^3
Mass	tonne	t	10^3 kg
Pressure	bar	bar	10^5 Pa
	atmosphere	atm	101.325 kPa
Energy	electronvolt	eV	$1.602\,177\,33 \times 10^{-19}$ J
			96.485 31 kJ mol^{-1}

*All values are exact, except for the definition of 1 eV, which depends on the measured value of *e*, and the year, which is not a constant and depends on a variety of astronomical assumptions.

Table A.2 Common SI prefixes

Prefix	y	z	a	f	p	n	μ	m	c	d
Name	yocto	zepto	atto	femto	pico	nano	micro	milli	centi	deci
Factor	10^{-24}	10^{-21}	10^{-18}	10^{-15}	10^{-12}	10^{-9}	10^{-6}	10^{-3}	10^{-2}	10^{-1}
Prefix	da	h	k	M	G	T	P	E	Z	Y
Name	deca	hecto	kilo	mega	giga	tera	peta	exa	zetta	yotta
Factor	10	10^2	10^3	10^6	10^9	10^{12}	10^{15}	10^{18}	10^{21}	10^{24}

Table A.3 The SI base units

Physical quantity	Symbol for quantity	Base unit
Length	l	metre, m
Mass	m	kilogram, kg
Time	t	second, s
Electric current	I	ampere, A
Thermodynamic temperature	T	kelvin, K
Amount of substance	n	mole, mol
Luminous intensity	I_v	candela, cd

Table A.4 A selection of derived units

Physical quantity	Derived unit*	Name and symbol of derived unit
Force	kg m s^{-2}	newton, N
Pressure	kg m^{-1} s^{-2}	pascal, Pa
	N m^{-2}	
Energy	kg m^2 s^{-2}	joule, J
	N m	
	Pa m^3	
Power	kg m^2 s^{-3}	watt, W
	J s^{-1}	

PART 3 Data

The following is a directory of all tables in the text; those included in this *Resource section* are marked with an asterisk. The remainder will be found on the pages indicated. These tables reproduce and expand the data given in the short tables in the text, and follow their numbering. Standard states refer to a pressure of $p^{\ominus} = 1$ bar. Data are for 298 K unless otherwise indicated. The general references are as follows:

AIP: D.E. Gray (ed.), *American Institute of Physics handbook*. McGraw Hill, New York (1972).

E: J. Emsley, *The elements*. Oxford University Press (1991).

HCP: D.R. Lide (ed.), *Handbook of chemistry and physics*. CRC Press, Boca Raton (2000).

JL: A.M. James and M.P. Lord, *Macmillan's chemical and physical data*. Macmillan, London (1992).

KL: G.W.C. Kaye and T.H. Laby (ed.), *Tables of physical and chemical constants*. Longman, London (1973).

LR: G.N. Lewis and M. Randall, revised by K.S. Pitzer and L. Brewer, *Thermodynamics*. McGraw Hill, New York (1961).

NBS: *NBS tables of chemical thermodynamic properties*, published as *J. Phys. Chem. Reference Data*, **11**, Supplement 2 (1982).

RS: R.A. Robinson and R.H. Stokes, *Electrolyte solutions*, Butterworth, London (1959).

TDOC: J.B. Pedley, J.D. Naylor, and S.P. Kirby, *Thermochemical data of organic compounds*. Chapman & Hall, London (1986).

Table 0.1 Physical properties of selected materials

	ρ/(g cm^{-3}) at 293 K†	T_f/K	T_b/K		ρ/(g cm^{-3}) at 293 K†	T_f/K	T_b/K
Elements				**Elements (Continued)**			
Aluminium(s)	2.698	933.5	2740	Gold(s)	19.320	1338	3080
Argon(g)	1.381	83.8	87.3	Helium(g)	0.125		4.22
Boron(s)	2.340	2573	3931	Hydrogen(g)	0.071	14.0	20.3
Bromine(l)	3.123	265.9	331.9	Iodine(s)	4.930	386.7	457.5
Carbon(s, gr)	2.260	3700^s		Iron(s)	7.874	1808	3023
Carbon(s, d)	3.513			Krypton(g)	2.413	116.6	120.8
Chlorine(g)	1.507	172.2	239.2	Lead(s)	11.350	600.6	2013
Copper(s)	8.960	1357	2840	Lithium(s)	0.534	453.7	1620
Fluorine(g)	1.108	53.5	85.0	Magnesium(s)	1.738	922.0	1363

Table 0.1 (Continued)

	ρ/(g cm^{-3}) at 293 K†	T_f/K	T_b/K		ρ/(g cm^{-3}) at 293 K†	T_f/K	T_b/K
Elements (Continued)				**Inorganic compounds (continued)**			
Mercury(l)	13.546	234.3	629.7	KCl(s)	1.984	1049	1773s
Neon(g)	1.207	24.5	27.1	NaCl(s)	2.165	1074	1686
Nitrogen(g)	0.880	63.3	77.4	H_2SO_4(l)	1.841	283.5	611.2
Oxygen(g)	1.140	54.8	90.2				
Phosphorus(s, wh)	1.820	317.3	553	**Organic compounds**			
Potassium(s)	0.862	336.8	1047	Aniline, $C_6H_5NH_2$(l)	1.026	267	457
Silver(s)	10.500	1235	2485	Anthracene, $C_{14}H_{10}$(s)	1.243	490	615
Sodium(s)	0.971	371.0	1156	Benzene, C_6H_6(l)	0.879	278.6	353.2
Sulfur(s, α)	2.070	386.0	717.8	Ethanal, CH_3CHO(l)	0.788	152	293
Uranium(s)	18.950	1406	4018	Ethanol, C_2H_5OH(l)	0.789	156	351.4
Xenon(g)	2.939	161.3	166.1	Ethanoic acid, CH_3COOH(l)	1.049	289.8	391
Zinc(s)	7.133	692.7	1180	Glucose, $C_6H_{12}O_6$(s)	1.544	415	
Inorganic compounds				Methanal, HCHO(g)		181	254.0
$CaCO_3$(s, calcite)	2.71	1612	1171[d]	Methane, CH_4(g)		90.6	111.6
$CuSO_4\cdot 5H_2O$(s)	2.284	383(−H_2O)	423(−5H_2O)	Methanol, CH_3OH(l)	0.791	179.2	337.6
HBr(g)	2.77	184.3	206.4	Naphthalene, $C_{10}H_8$(s)	1.145	353.4	491
HCl(g)	1.187	159.0	191.1	Octane, C_8H_{18}(l)	0.703	216.4	398.8
HI(g)	2.85	222.4	237.8	Phenol, C_6H_5OH(s)	1.073	314.1	455.0
H_2O(l)	0.997	273.2	373.2	Propanone, $(CH_3)_2CO$(l)	0.787	178	329
D_2O(l)	1.104	277.0	374.6	Sucrose, $C_{12}H_{22}O_{11}$(s)	1.588	457d	
NH_3(g)	0.817	195.4	238.8	Tetrachloromethane, CCl_4(l)	1.63	250	349.9
KBr(s)	2.750	1003	1708	Trichloromethane, $CHCl_3$(l)	1.499	209.6	334

d: decomposes; s: sublimes; Data: AIP, E, HCP, KL. † For gases, at their boiling points.

Table 0.2 Masses and natural abundances of selected nuclides

Nuclide		m/m_u	Abundance/%	Nuclide		m/m_u	Abundance/%
H	^{1}H	1.0078	99.985	O	^{16}O	15.9949	99.76
	^{2}H	2.0140	0.015		^{17}O	16.9991	0.037
He	^{3}He	3.0160	0.000 13		^{18}O	17.9992	0.204
	^{4}He	4.0026	100	F	^{19}F	18.9984	100
Li	^{6}Li	6.0151	7.42	P	^{31}P	30.9738	100
	^{7}Li	7.0160	92.58	S	^{32}S	31.9721	95.0
B	^{10}B	10.0129	19.78		^{33}S	32.9715	0.76
	^{11}B	11.0093	80.22		^{34}S	33.9679	4.22
C	^{12}C	12*	98.89	Cl	^{35}Cl	34.9688	75.53
	^{13}C	13.0034	1.11		^{37}Cl	36.9651	24.4
N	^{14}N	14.0031	99.63	Br	^{79}Br	78.9183	50.54
	^{15}N	15.0001	0.37		^{81}Br	80.9163	49.46
				I	^{127}I	126.9045	100

* Exact value.

Table 1B.1 Collision cross-sections, σ/nm^2

Ar	0.36
C_2H_4	0.64
C_6H_6	0.88
CH_4	0.46
Cl_2	0.93
CO_2	0.52
H_2	0.27
He	0.21
N_2	0.43
Ne	0.24
O_2	0.40
SO_2	0.58

Data: KL.

Table 1C.1 Second virial coefficients, $B/(cm^3\,mol^{-1})$

	100 K	273 K	373 K	600 K
Air	−167.3	−13.5	3.4	19.0
Ar	−187.0	−21.7	−4.2	11.9
CH_4		−53.6	−21.2	8.1
CO_2		−142	−72.2	−12.4
H_2	−2.0	13.7	15.6	
He	11.4	12.0	11.3	10.4
Kr		−62.9	−28.7	1.7
N_2	−160.0	−10.5	6.2	21.7
Ne	−6.0	10.4	12.3	13.8
O_2	−197.5	−22.0	−3.7	12.9
Xe		−153.7	−81.7	−19.6

Data: AIP, JL. The values relate to the expansion in eqn 1C.3b of Topic 1C; convert to eqn 1C.3a by using $B' = B/RT$.
For Ar at 273 K, $C = 1200\,cm^6\,mol^{-1}$.

Table 1C.2 Critical constants of gases

	p_c/atm	$V_c/(cm^3\,mol^{-1})$	T_c/K	Z_c	T_B/K
Ar	48.0	75.3	150.7	0.292	411.5
Br_2	102	135	584	0.287	
C_2H_4	50.50	124	283.1	0.270	
C_2H_6	48.20	148	305.4	0.285	
C_6H_6	48.6	260	562.7	0.274	
CH_4	45.6	98.7	190.6	0.288	510.0
Cl_2	76.1	124	417.2	0.276	
CO_2	72.9	94.0	304.2	0.274	714.8
F_2	55	144			
H_2	12.8	34.99	33.23	0.305	110.0
H_2O	218.3	55.3	647.4	0.227	
HBr	84.0	363.0			
HCl	81.5	81.0	324.7	0.248	
He	2.26	57.8	5.2	0.305	22.64
HI	80.8	423.2			
Kr	54.27	92.24	209.39	0.291	575.0
N_2	33.54	90.10	126.3	0.292	327.2
Ne	26.86	41.74	44.44	0.307	122.1
NH_3	111.3	72.5	405.5	0.242	
O_2	50.14	78.0	154.8	0.308	405.9
Xe	58.0	118.8	289.75	0.290	768.0

Data: AIP, KL.

Table 1C.3 van der Waals coefficients

	$a/(\text{atm dm}^6\,\text{mol}^{-2})$	$b/(10^{-2}\,\text{dm}^3\,\text{mol}^{-1})$		$a/(\text{atm dm}^6\,\text{mol}^{-2})$	$b/(10^{-2}\,\text{dm}^3\,\text{mol}^{-1})$
Ar	1.337	3.20	H_2S	4.484	4.34
C_2H_4	4.552	5.82	He	0.0341	2.38
C_2H_6	5.507	6.51	Kr	5.125	1.06
C_6H_6	18.57	11.93	N_2	1.352	3.87
CH_4	2.273	4.31	Ne	0.205	1.67
Cl_2	6.260	5.42	NH_3	4.169	3.71
CO	1.453	3.95	O_2	1.364	3.19
CO_2	3.610	4.29	SO_2	6.775	5.68
H_2	0.2420	2.65	Xe	4.137	5.16
H_2O	5.464	3.05			

Data: HCP.

Table 2B.1 Temperature variation of molar heat capacities, $C_{p,m}/(\text{J K}^{-1}\,\text{mol}^{-1}) = a + bT + c/T^2$

	a	$b/(10^{-3}\,\text{K}^{-1})$	$c/(10^5\,\text{K}^2)$
Monatomic gases			
	20.78	0	0
Other gases			
Br_2	37.32	0.50	−1.26
Cl_2	37.03	0.67	−2.85
CO_2	44.22	8.79	−8.62
F_2	34.56	2.51	−3.51
H_2	27.28	3.26	0.50
I_2	37.40	0.59	−0.71
N_2	28.58	3.77	−0.50
NH_3	29.75	25.1	−1.55
O_2	29.96	4.18	−1.67
Liquids (from melting to boiling)			
$C_{10}H_8$, naphthalene	79.5	0.4075	0
I_2	80.33	0	0
H_2O	75.29	0	0
Solids			
Al	20.67	12.38	0
C (graphite)	16.86	4.77	−8.54
$C_{10}H_8$, naphthalene	−100	0.936	0
Cu	22.64	6.28	0
I_2	40.12	49.79	0
NaCl	45.94	16.32	0
Pb	22.13	11.72	0.96

Source: Mostly LR.

Table 2C.1 Standard enthalpies of fusion and vaporization at the transition temperature, $\Delta_{trs}H^{\ominus}/(kJ\,mol^{-1})$

	T_f/K	Fusion	T_b/K	Vaporization		T_f/K	Fusion	T_b/K	Vaporization
Elements					**Inorganic compounds**				
Ag	1234	11.30	2436	250.6	CO_2	217.0	8.33	194.6	25.23^s
Ar	83.81	1.188	87.29	6.506	CS_2	161.2	4.39	319.4	26.74
Br_2	265.9	10.57	332.4	29.45	H_2O	273.15	6.008	373.15	40.656
Cl_2	172.1	6.41	239.1	20.41					44.016 at 298 K
F_2	53.6	0.26	85.0	3.16	H_2S	187.6	2.377	212.8	18.67
H_2	13.96	0.117	20.38	0.916	H_2SO_4	283.5	2.56		
He	3.5	0.021	4.22	0.084	NH_3	195.4	5.652	239.7	23.35
Hg	234.3	2.292	629.7	59.30	**Organic compounds**				
I_2	386.8	15.52	458.4	41.80	CH_4	90.68	0.941	111.7	8.18
N_2	63.15	0.719	77.35	5.586	CCl_4	250.3	2.47	349.9	30.00
Na	371.0	2.601	1156	98.01	C_2H_6	89.85	2.86	184.6	14.7
O_2	54.36	0.444	90.18	6.820	C_6H_6	278.61	10.59	353.2	30.8
Xe	161	2.30	165	12.6	C_6H_{14}	178	13.08	342.1	28.85
K	336.4	2.35	1031	80.23	$C_{10}H_{18}$	354	18.80	490.9	51.51
					CH_3OH	175.2	3.16	337.2	35.27
									37.99 at 298 K
					C_2H_5OH	158.7	4.60	352	43.5

Data: AIP; s denotes sublimation

Table 2C.3 Standard enthalpies of formation and combustion of organic compounds at 298 K. See Table 2C.6.

Table 2C.4 Standard enthalpies of formation of inorganic compounds at 298 K. See Table 2C.7.

Table 2C.5 Standard enthalpies of formation of organic compounds at 298 K. See Table 2C.6.

Table 2C.6 Thermodynamic data for organic compounds at 298 K

	$M/(g\,mol^{-1})$	$\Delta_f H^{\ominus}/(kJ\,mol^{-1})$	$\Delta_f G^{\ominus}/(kJ\,mol^{-1})$	$S_m^{\ominus}/(J\,K^{-1}\,mol^{-1})^{\dagger}$	$C_{p,m}^{\ominus}/(J\,K^{-1}\,mol^{-1})$	$\Delta_c H^{\ominus}/(kJ\,mol^{-1})$
C(s) (graphite)	12.011	0	0	5.740	8.527	−393.51
C(s) (diamond)	12.011	+1.895	+2.900	2.377	6.113	−395.40
CO_2(g)	44.040	−393.51	−394.36	213.74	37.11	
Hydrocarbons						
CH_4(g), methane	16.04	−74.81	−50.72	186.26	35.31	−890
CH_3(g), methyl	15.04	+145.69	+147.92	194.2	38.70	
C_2H_2(g), ethyne	26.04	+226.73	+209.20	200.94	43.93	−1300
C_2H_4(g), ethene	28.05	+52.26	+68.15	219.56	43.56	−1411
C_2H_6(g), ethane	30.07	−84.68	−32.82	229.60	52.63	−1560
C_3H_6(g), propene	42.08	+20.42	+62.78	267.05	63.89	−2058
C_3H_6(g), cyclopropane	42.08	+53.30	+104.45	237.55	55.94	−2091
C_3H_8(g), propane	44.10	−103.85	−23.49	269.91	73.5	−2220
C_4H_8(g), 1-butene	56.11	−0.13	+71.39	305.71	85.65	−2717
C_4H_8(g), *cis*-2-butene	56.11	−6.99	+65.95	300.94	78.91	−2710
C_4H_8(g), *trans*-2-butene	56.11	−11.17	+63.06	296.59	87.82	−2707
C_4H_{10}(g), butane	58.13	−126.15	−17.03	310.23	97.45	−2878

Table 1C.2 (Continued)

	$M/(\mathrm{g\,mol^{-1}})$	$\Delta_f H^{\ominus}/(\mathrm{kJ\,mol^{-1}})$	$\Delta_f G^{\ominus}/(\mathrm{kJ\,mol^{-1}})$	$S_m^{\ominus}/(\mathrm{J\,K^{-1}\,mol^{-1}})^{\dagger}$	$C_{p,m}^{\ominus}/(\mathrm{J\,K^{-1}\,mol^{-1}})$	$\Delta_c H^{\ominus}/(\mathrm{kJ\,mol^{-1}})$
C_5H_{12}(g), pentane	72.15	−146.44	−8.20	348.40	120.2	−3537
C_5H_{12}(l)	72.15	−173.1				
C_6H_6(l), benzene	78.12	+49.0	+124.3	173.3	136.1	−3268
C_6H_6(g)	78.12	+82.93	+129.72	269.31	81.67	−3302
C_6H_{12}(l), cyclohexane	84.16	−156	+26.8	204.4	156.5	−3920
C_6H_{14}(l), hexane	86.18	−198.7		204.3		−4163
$C_6H_5CH_3$(g), methylbenzene (toluene)	92.14	+50.0	+122.0	320.7	103.6	−3953
C_7H_{16}(l), heptane	100.21	−224.4	+1.0	328.6	224.3	
C_8H_{18}(l), octane	114.23	−249.9	+6.4	361.1		−5471
C_8H_{18}(l), iso–octane	114.23	−255.1				−5461
$C_{10}H_8$(s), naphthalene	128.18	+78.53				−5157
Alcohols and phenols						
CH_3OH(l), methanol	32.04	−238.66	−166.27	126.8	81.6	−726
CH_3OH(g)	32.04	−200.66	−161.96	239.81	43.89	−764
C_2H_5OH(l), ethanol	46.07	−277.69	−174.78	160.7	111.46	−1368
C_2H_5OH(g)	46.07	−235.10	−168.49	282.70	65.44	−1409
C_6H_5OH(s), phenol	94.12	−165.0	−50.9	146.0		−3054
Carboxylic acids, hydroxyacids, and esters						
HCOOH(l), methanoic	46.03	−424.72	−361.35	128.95	99.04	−255
CH_3COOH(l), ethanoic	60.05	−484.5	−389.9	159.8	124.3	−875
CH_3COOH(aq)	60.05	−485.76	−396.46	178.7		
$CH_3CO_2^-$(aq)	59.05	−486.01	−369.31	+86.6	−6.3	
$(COOH)_2$(s), oxalic	90.04	−827.2			117	−254
C_6H_5COOH(s), benzoic	122.13	−385.1	−245.3	167.6	146.8	−3227
$CH_3CH(OH)COOH$(s), lactic	90.08	−694.0				−1344
$CH_3COOC_2H_5$(l), ethyl ethanoate	88.11	−479.0	−332.7	259.4	170.1	−2231
Alkanals and alkanones						
HCHO(g), methanal	30.03	−108.57	−102.53	218.77	35.40	−571
CH_3CHO(l), ethanal	44.05	−192.30	−128.12	160.2		−1166
CH_3CHO(g)	44.05	−166.19	−128.86	250.3	57.3	−1192
CH_3COCH_3(l), propanone	58.08	−248.1	−155.4	200.4	124.7	−1790
Sugars						
$C_6H_{12}O_6$(s), α-D-glucose	180.16	−1274				−2808
$C_6H_{12}O_6$(s), β-D-glucose	180.16	−1268	−910	212		
$C_6H_{12}O_6$(s), β-D-fructose	180.16	−1266				−2810
$C_{12}H_{22}O_{11}$(s), sucrose	342.30	−2222	−1543	360.2		−5645
Nitrogen compounds						
$CO(NH_2)_2$(s), urea	60.06	−333.51	−197.33	104.60	93.14	−632
CH_3NH_2(g), methylamine	31.06	−22.97	+32.16	243.41	53.1	−1085
$C_6H_5NH_2$(l), aniline	93.13	+31.1				−3393
$CH_2(NH_2)COOH$(s), glycine	75.07	−532.9	−373.4	103.5	99.2	−969

Data: NBS, TDOC. † Standard entropies of ions may be either positive or negative because the values are relative to the entropy of the hydrogen ion.

Table 2C.7 Thermodynamic data for elements and inorganic compounds at 298 K

	$M/(\mathrm{g\,mol^{-1}})$	$\Delta_f H^\ominus/(\mathrm{kJ\,mol^{-1}})$	$\Delta_f G^\ominus/(\mathrm{kJ\,mol^{1}})$	$S_m^\ominus/(\mathrm{J\,K^{-1}\,mol^{1}})^\dagger$	$C_{p,m}^\ominus/(\mathrm{J\,K^{-1}\,mol^{-1}})$
Aluminium (aluminum)					
Al(s)	26.98	0	0	28.33	24.35
Al(l)	26.98	+10.56	+7.20	39.55	24.21
Al(g)	26.98	+326.4	+285.7	164.54	21.38
Al^{3+}(g)	26.98	+5483.17			
Al^{3+}(aq)	26.98	−531	−485	−321.7	
Al_2O_3(s, α)	101.96	−1675.7	−1582.3	50.92	79.04
$AlCl_3$(s)	133.24	−704.2	−628.8	110.67	91.84
Argon					
Ar(g)	39.95	0	0	154.84	20.786
Antimony					
Sb(s)	121.75	0	0	45.69	25.23
SbH_3(g)	124.77	+145.11	+147.75	232.78	41.05
Arsenic					
As(s, α)	74.92	0	0	35.1	24.64
As(g)	74.92	+302.5	+261.0	174.21	20.79
As_4(g)	299.69	+143.9	+92.4	314	
AsH_3(g)	77.95	+66.44	+68.93	222.78	38.07
Barium					
Ba(s)	137.34	0	0	62.8	28.07
Ba(g)	137.34	+180	+146	170.24	20.79
Ba^{2+}(aq)	137.34	−537.64	−560.77	+9.6	
BaO(s)	153.34	−553.5	−525.1	70.43	47.78
$BaCl_2$(s)	208.25	−858.6	−810.4	123.68	75.14
Beryllium					
Be(s)	9.01	0	0	9.50	16.44
Be(g)	9.01	+324.3	+286.6	136.27	20.79
Bismuth					
Bi(s)	208.98	0	0	56.74	25.52
Bi(g)	208.98	+207.1	+168.2	187.00	20.79
Bromine					
Br_2(l)	159.82	0	0	152.23	75.689
Br_2(g)	159.82	+30.907	+3.110	245.46	36.02
Br(g)	79.91	+111.88	+82.396	175.02	20.786
Br^-(g)	79.91	−219.07			
Br^-(aq)	79.91	−121.55	−103.96	+82.4	−141.8
HBr(g)	90.92	−36.40	−53.45	198.70	29.142
Cadmium					
Cd(s, γ)	112.40	0	0	51.76	25.98
Cd(g)	112.40	+112.01	+77.41	167.75	20.79
Cd^{2+}(aq)	112.40	−75.90	−77.612	−73.2	
CdO(s)	128.40	−258.2	−228.4	54.8	43.43
$CdCO_3$(s)	172.41	−750.6	−669.4	92.5	

Table 2C.7 (Continued)

	$M/(\mathrm{g\,mol^{-1}})$	$\Delta_f H^{\ominus}/(\mathrm{kJ\,mol^{-1}})$	$\Delta_f G^{\ominus}/(\mathrm{kJ\,mol^{1}})$	$S_m^{\ominus}/(\mathrm{J\,K^{-1}\,mol^{1}})^{\dagger}$	$C_{p,m}^{\ominus}/(\mathrm{J\,K^{-1}\,mol^{-1}})$
Caesium (cesium)					
Cs(s)	132.91	0	0	85.23	32.17
Cs(g)	132.91	+76.06	+49.12	175.60	20.79
Cs^{+}(aq)	132.91	−258.28	−292.02	+133.05	−10.5
Calcium					
Ca(s)	40.08	0	0	41.42	25.31
Ca(g)	40.08	+178.2	+144.3	154.88	20.786
Ca^{2+}(aq)	40.08	−542.83	−553.58	−53.1	
CaO(s)	56.08	−635.09	−604.03	39.75	42.80
$CaCO_3$(s) (calcite)	100.09	−1206.9	−1128.8	92.9	81.88
$CaCO_3$(s) (aragonite)	100.09	−1207.1	−1127.8	88.7	81.25
CaF_2(s)	78.08	−1219.6	−1167.3	68.87	67.03
$CaCl_2$(s)	110.99	−795.8	−748.1	104.6	72.59
$CaBr_2$(s)	199.90	−682.8	−663.6	130	
Carbon (for 'organic' compounds of carbon, see Table 2C.6)					
C(s) (graphite)	12.011	0	0	5.740	8.527
C(s) (diamond)	12.011	+1.895	+2.900	2.377	6.113
C(g)	12.011	+716.68	+671.26	158.10	20.838
C_2(g)	24.022	+831.90	+775.89	199.42	43.21
CO(g)	28.011	−110.53	−137.17	197.67	29.14
CO_2(g)	44.010	−393.51	−394.36	213.74	37.11
CO_2(aq)	44.010	−413.80	−385.98	117.6	
H_2CO_3(aq)	62.03	−699.65	−623.08	187.4	
HCO_3^{-}(aq)	61.02	−691.99	−586.77	+91.2	
CO_3^{2-}(aq)	60.01	−677.14	−527.81	−56.9	
CCl_4(l)	153.82	−135.44	−65.21	216.40	131.75
CS_2(l)	76.14	+89.70	+65.27	151.34	75.7
HCN(g)	27.03	+135.1	+124.7	201.78	35.86
HCN(l)	27.03	+108.87	+124.97	112.84	70.63
CN^{-}(aq)	26.02	+150.6	+172.4	+94.1	
Chlorine					
Cl_2(g)	70.91	0	0	223.07	33.91
Cl(g)	35.45	+121.68	+105.68	165.20	21.840
Cl^{-}(g)	34.45	−233.13			
Cl^{-}(aq)	35.45	−167.16	−131.23	+56.5	−136.4
HCl(g)	36.46	−92.31	−95.30	186.91	29.12
HCl(aq)	36.46	−167.16	−131.23	56.5	−136.4
Chromium					
Cr(s)	52.00	0	0	23.77	23.35
Cr(g)	52.00	+396.6	+351.8	174.50	20.79
CrO_4^{2-}(aq)	115.99	−881.15	−727.75	+50.21	
$Cr_2O_7^{2-}$(aq)	215.99	−1490.3	−1301.1	+261.9	

(Continued)

Table 2C.7 (Continued)

	$M/(\mathrm{g\,mol^{-1}})$	$\Delta_f H^\ominus/(\mathrm{kJ\,mol^{-1}})$	$\Delta_f G^\ominus/(\mathrm{kJ\,mol^{1}})$	$S_m^\ominus/(\mathrm{J\,K^{-1}\,mol^{1}})^\dagger$	$C_{p,m}^\ominus/(\mathrm{J\,K^{-1}\,mol^{-1}})$
Copper					
Cu(s)	63.54	0	0	33.150	24.44
Cu(g)	63.54	+338.32	+298.58	166.38	20.79
Cu^{+}(aq)	63.54	+71.67	+49.98	+40.6	
Cu^{2+}(aq)	63.54	+64.77	+65.49	−99.6	
Cu_2O(s)	143.08	−168.6	−146.0	93.14	63.64
CuO(s)	79.54	−157.3	−129.7	42.63	42.30
$CuSO_4$(s)	159.60	−771.36	−661.8	109	100.0
$CuSO_4\cdot H_2O$(s)	177.62	−1085.8	−918.11	146.0	134
$CuSO_4\cdot 5H_2O$(s)	249.68	−2279.7	−1879.7	300.4	280
Deuterium					
D_2(g)	4.028	0	0	144.96	29.20
HD(g)	3.022	+0.318	−1.464	143.80	29.196
D_2O(g)	20.028	−249.20	−234.54	198.34	34.27
D_2O(l)	20.028	−294.60	−243.44	75.94	84.35
HDO(g)	19.022	−245.30	−233.11	199.51	33.81
HDO(l)	19.022	−289.89	−241.86	79.29	
Fluorine					
F_2(g)	38.00	0	0	202.78	31.30
F(g)	19.00	+78.99	+61.91	158.75	22.74
F^{-}(aq)	19.00	−332.63	−278.79	−13.8	−106.7
HF(g)	20.01	−271.1	−273.2	173.78	29.13
Gold					
Au(s)	196.97	0	0	47.40	25.42
Au(g)	196.97	+366.1	+326.3	180.50	20.79
Helium					
He(g)	4.003	0	0	126.15	20.786
Hydrogen (see also deuterium)					
H_2(g)	2.016	0	0	130.684	28.824
H(g)	1.008	+217.97	+203.25	114.71	20.784
H^{+}(aq)	1.008	0	0	0	0
H^{+}(g)	1.008	+1536.20			
H_2O(s)	18.015			37.99	
H_2O(l)	18.015	−285.83	−237.13	69.91	75.291
H_2O(g)	18.015	−241.82	−228.57	188.83	33.58
H_2O_2(l)	34.015	−187.78	−120.35	109.6	89.1
Iodine					
I_2(s)	253.81	0	0	116.135	54.44
I_2(g)	253.81	+62.44	+19.33	260.69	36.90
I(g)	126.90	+106.84	+70.25	180.79	20.786
I^{-}(aq)	126.90	−55.19	−51.57	+111.3	−142.3
HI(g)	127.91	+26.48	+1.70	206.59	29.158

Table 2C.7 (Continued)

	$M/(\mathrm{g\,mol^{-1}})$	$\Delta_f H^{\ominus}/(\mathrm{kJ\,mol^{-1}})$	$\Delta_f G^{\ominus}/(\mathrm{kJ\,mol^{1}})$	$S_m^{\ominus}/(\mathrm{J\,K^{-1}\,mol^{1}})^{\dagger}$	$C_{p,m}^{\ominus}/(\mathrm{J\,K^{-1}\,mol^{-1}})$
Iron					
Fe(s)	55.85	0	0	27.28	25.10
Fe(g)	55.85	+416.3	+370.7	180.49	25.68
Fe^{2+}(aq)	55.85	−89.1	−78.90	−137.7	
Fe^{3+}(aq)	55.85	−48.5	−4.7	−315.9	
Fe_3O_4(s) (magnetite)	231.54	−1118.4	−1015.4	146.4	143.43
Fe_2O_3(s) (haematite)	159.69	−824.2	−742.2	87.40	103.85
FeS(s, α)	87.91	−100.0	−100.4	60.29	50.54
FeS_2(s)	119.98	−178.2	−166.9	52.93	62.17
Krypton					
Kr(g)	83.80	0	0	164.08	20.786
Lead					
Pb(s)	207.19	0	0	64.81	26.44
Pb(g)	207.19	+195.0	+161.9	175.37	20.79
Pb^{2+}(aq)	207.19	−1.7	−24.43	+10.5	
PbO(s, yellow)	223.19	−217.32	−187.89	68.70	45.77
PbO(s, red)	223.19	−218.99	−188.93	66.5	45.81
PbO_2(s)	239.19	−277.4	−217.33	68.6	64.64
Lithium					
Li(s)	6.94	0	0	29.12	24.77
Li(g)	6.94	+159.37	+126.66	138.77	20.79
Li^{+}(aq)	6.94	−278.49	−293.31	+13.4	68.6
Magnesium					
Mg(s)	24.31	0	0	32.68	24.89
Mg(g)	24.31	+147.70	+113.10	148.65	20.786
Mg^{2+}(aq)	24.31	−466.85	−454.8	−138.1	
MgO(s)	40.31	−601.70	−569.43	26.94	37.15
$MgCO_3$(s)	84.32	−1095.8	−1012.1	65.7	75.52
$MgCl_2$(s)	95.22	−641.32	−591.79	89.62	71.38
Mercury					
Hg(l)	200.59	0	0	76.02	27.983
Hg(g)	200.59	+61.32	+31.82	174.96	20.786
Hg^{2+}(aq)	200.59	+171.1	+164.40	−32.2	
Hg_2^{2+}(aq)	401.18	+172.4	+153.52	+84.5	
HgO(s)	216.59	−90.83	−58.54	70.29	44.06
Hg_2Cl_2(s)	472.09	−265.22	−210.75	192.5	102
$HgCl_2$(s)	271.50	−224.3	−178.6	146.0	
HgS(s, black)	232.65	−53.6	−47.7	88.3	
Neon					
Ne(g)	20.18	0	0	146.33	20.786
Nitrogen					
N_2(g)	28.013	0	0	191.61	29.125

(Continued)

Table 2C.7 (Continued)

	$M/(\text{g mol}^{-1})$	$\Delta_f H^\ominus/(\text{kJ mol}^{-1})$	$\Delta_f G^\ominus/(\text{kJ mol}^{1})$	$S^\ominus_m/(\text{J K}^{-1}\,\text{mol}^{1})^\dagger$	$C^\ominus_{p,m}/(\text{J K}^{-1}\,\text{mol}^{-1})$
N(g)	14.007	+472.70	+455.56	153.30	20.786
NO(g)	30.01	+90.25	+86.55	210.76	29.844
N_2O(g)	44.01	+82.05	+104.20	219.85	38.45
NO_2(g)	46.01	+33.18	+51.31	240.06	37.20
N_2O_4(g)	92.1	+9.16	+97.89	304.29	77.28
N_2O_5(s)	108.01	−43.1	+113.9	178.2	143.1
N_2O_5(g)	108.01	+11.3	+115.1	355.7	84.5
HNO_3(l)	63.01	−174.10	−80.71	155.60	109.87
HNO_3(aq)	63.01	−207.36	−111.25	146.4	−86.6
NO_3^-(aq)	62.01	−205.0	−108.74	+146.4	−86.6
NH_3(g)	17.03	−46.11	−16.45	192.45	35.06
NH_3(aq)	17.03	−80.29	−26.50	111.3	
NH_4^+(aq)	18.04	−132.51	−79.31	+113.4	79.9
NH_2OH(s)	33.03	−114.2			
HN_3(l)	43.03	+264.0	+327.3	140.6	
HN_3(g)	43.03	+294.1	+328.1	238.97	43.68
N_2H_4(l)	32.05	+50.63	+149.43	121.21	98.87
NH_4NO_3(s)	80.04	−365.56	−183.87	151.08	139.3
NH_4Cl(s)	53.49	−314.43	−202.87	94.6	84.1
Oxygen					
O_2(g)	31.999	0	0	205.138	29.355
O(g)	15.999	+249.17	+231.73	161.06	21.912
O_3(g)	47.998	+142.7	+163.2	238.93	39.20
OH^-(aq)	17.007	−229.99	−157.24	−10.75	−148.5
Phosphorus					
P(s, wh)	30.97	0	0	41.09	23.840
P(g)	30.97	+314.64	+278.25	163.19	20.786
P_2(g)	61.95	+144.3	+103.7	218.13	32.05
P_4(g)	123.90	+58.91	+24.44	279.98	67.15
PH_3(g)	34.00	+5.4	+13.4	210.23	37.11
PCl_3(g)	137.33	−287.0	−267.8	311.78	71.84
PCl_3(l)	137.33	−319.7	−272.3	217.1	
PCl_5(g)	208.24	−374.9	−305.0	364.6	112.8
PCl_5(s)	208.24	−443.5			
H_3PO_3(s)	82.00	−964.4			
H_3PO_3(aq)	82.00	−964.8			
H_3PO_4(s)	94.97	−1279.0	−1119.1	110.50	106.06
H_3PO_4(l)	94.97	−1266.9			
H_3PO_4(aq)	94.97	−1277.4	−1018.7	−222	
PO_4^{3-}(aq)	94.97	−1277.4	−1018.7	−221.8	
P_4O_{10}(s)	283.89	−2984.0	−2697.0	228.86	211.71
P_4O_6(s)	219.89	−1640.1			
Potassium					
K(s)	39.10	0	0	64.18	29.58

Table 2C.7 (Continued)

	$M/(\text{g mol}^{-1})$	$\Delta_f H^{\ominus}/(\text{kJ mol}^{-1})$	$\Delta_f G^{\ominus}/(\text{kJ mol}^{1})$	$S_m^{\ominus}/(\text{J K}^{-1}\text{ mol}^{1})^{\dagger}$	$C_{p,m}^{\ominus}/(\text{J K}^{-1}\text{ mol}^{-1})$
K(g)	39.10	+89.24	+60.59	160.336	20.786
K^+(g)	39.10	+514.26			
K^+(aq)	39.10	−252.38	−283.27	+102.5	21.8
KOH(s)	56.11	−424.76	−379.08	78.9	64.9
KF(s)	58.10	−576.27	−537.75	66.57	49.04
KCl(s)	74.56	−436.75	−409.14	82.59	51.30
KBr(s)	119.01	−393.80	−380.66	95.90	52.30
KI(s)	166.01	−327.90	−324.89	106.32	52.93
Silicon					
Si(s)	28.09	0	0	18.83	20.00
Si(g)	28.09	+455.6	+411.3	167.97	22.25
SiO_2(s, α)	60.09	−910.94	−856.64	41.84	44.43
Silver					
Ag(s)	107.87	0	0	42.55	25.351
Ag(g)	107.87	+284.55	+245.65	173.00	20.79
Ag^+(aq)	107.87	+105.58	+77.11	+72.68	21.8
AgBr(s)	187.78	−100.37	−96.90	107.1	52.38
AgCl(s)	143.32	−127.07	−109.79	96.2	50.79
Ag_2O(s)	231.74	−31.05	−11.20	121.3	65.86
$AgNO_3$(s)	169.88	−129.39	−33.41	140.92	93.05
Sodium					
Na(s)	22.99	0	0	51.21	28.24
Na(g)	22.99	+107.32	+76.76	153.71	20.79
Na^+(aq)	22.99	−240.12	−261.91	+59.0	46.4
NaOH(s)	40.00	−425.61	−379.49	64.46	59.54
NaCl(s)	58.44	−411.15	−384.14	72.13	50.50
NaBr(s)	102.90	−361.06	−348.98	86.82	51.38
NaI(s)	149.89	−287.78	−286.06	98.53	52.09
Sulfur					
S(s, α) (rhombic)	32.06	0	0	31.80	22.64
S(s, β) (monoclinic)	32.06	+0.33	+0.1	32.6	23.6
S(g)	32.06	+278.81	+238.25	167.82	23.673
S_2(g)	64.13	+128.37	+79.30	228.18	32.47
S^{2-}(aq)	32.06	+33.1	+85.8	−14.6	
SO_2(g)	64.06	−296.83	−300.19	248.22	39.87
SO_3(g)	80.06	−395.72	−371.06	256.76	50.67
H_2SO_4(l)	98.08	−813.99	−690.00	156.90	138.9
H_2SO_4(aq)	98.08	−909.27	−744.53	20.1	−293
SO_4^{2-}(aq)	96.06	−909.27	−744.53	+20.1	−293
HSO_4^-(aq)	97.07	−887.34	−755.91	+131.8	−84
H_2S(g)	34.08	−20.63	−33.56	205.79	34.23
H_2S(aq)	34.08	−39.7	−27.83	121	

(*Continued*)

Table 2C.7 (Continued)

	$M/(\mathrm{g\,mol^{-1}})$	$\Delta_f H^\ominus/(\mathrm{kJ\,mol^{-1}})$	$\Delta_f G^\ominus/(\mathrm{kJ\,mol^{1}})$	$S_m^\ominus/(\mathrm{J\,K^{-1}\,mol^{1}})^\dagger$	$C_{p,m}^\ominus/(\mathrm{J\,K^{-1}\,mol^{-1}})$
$HS^-(aq)$	33.072	−17.6	+12.08	+62.08	
$SF_6(g)$	146.05	−1209	−1105.3	291.82	97.28
Tin					
Sn(s, β)	118.69	0	0	51.55	26.99
Sn(g)	118.69	+302.1	+267.3	168.49	20.26
$Sn^{2+}(aq)$	118.69	−8.8	−27.2	−17	
SnO(s)	134.69	−285.8	−256.9	56.5	44.31
$SnO_2(s)$	150.69	−580.7	−519.6	52.3	52.59
Xenon					
Xe(g)	131.30	0	0	169.68	20.786
Zinc					
Zn(s)	65.37	0	0	41.63	25.40
Zn(g)	65.37	+130.73	+95.14	160.98	20.79
$Zn^{2+}(aq)$	65.37	−153.89	−147.06	−112.1	46
ZnO(s)	81.37	−348.28	−318.30	43.64	40.25

Source: NBS. † Standard entropies of ions may be either positive or negative because the values are relative to the entropy of the hydrogen ion.

Table 2D.1 Expansion coefficients (α) and isothermal compressibilities (κ_T) at 298 K

	$\alpha/(10^{-4}\,\mathrm{K^{-1}})$	$\kappa_T/(10^{-6}\,\mathrm{bar^{-1}})$
Liquids		
Benzene	12.4	92.1
Ethanol	11.2	76.8
Mercury	1.82	38.7
Tetrachloromethane	12.4	90.5
Water	2.1	49.6
Solids		
Copper	0.501	0.735
Diamond	0.030	0.187
Iron	0.354	0.589
Lead	0.861	2.21

The values refer to 20 °C.
Data: AIP (α), KL (κ_T).

Table 2D.2 Inversion temperatures (T_I), normal freezing (T_f) and boiling (T_b) points, and Joule–Thomson coefficients (μ) at 1 atm and 298 K

	T_I/K	T_f/K	T_b/K	$\mu/(\mathrm{K\,atm^{-1}})$
Air	603			0.189 at 50 °C
Argon	723	83.8	87.3	
Carbon dioxide	1500	194.7^s		1.11 at 300 K
Helium	40		4.22	−0.062
Hydrogen	202	14.0	20.3	−0.03
Krypton	1090	116.6	120.8	
Methane	968	90.6	111.6	
Neon	231	24.5	27.1	
Nitrogen	621	63.3	77.4	0.27
Oxygen	764	54.8	90.2	0.31

s: sublimes.
Data: AIP, JL, and M.W. Zemansky, *Heat and thermodynamics*. McGraw-Hill, New York (1957).

Table 3B.1* Standard entropies of phase transitions, $\Delta_{trs}S^{\ominus}/(J\,K^{-1}\,mol^{-1})$, at the corresponding normal transition temperatures

	Fusion (at T_f)	Vaporization (at T_b)
Ar	14.17 (at 83.8 K)	74.53 (at 87.3 K)
Br_2	39.76 (at 265.9 K)	88.61 (at 332.4 K)
C_6H_6	38.00 (at 278.6 K)	87.19 (at 353.2 K)
CH_3COOH	40.4 (at 289.8 K)	61.9 (at 391.4 K)
CH_3OH	18.03 (at 175.2 K)	104.6 (at 337.2 K)
Cl_2	37.22 (at 172.1 K)	85.38 (at 239.0 K)
H_2	8.38 (at 14.0 K)	44.96 (at 20.38 K)
H_2O	22.00 (at 273.2 K)	109.1 (at 373.2 K)
H_2S	12.67 (at 187.6 K)	87.75 (at 212.0 K)
He	4.8 (at 1.8 K and 30 bar)	19.9 (at 4.22 K)
N_2	11.39 (at 63.2 K)	75.22 (at 77.4 K)
NH_3	28.93 (at 195.4 K)	97.41 (at 239.73 K)
O_2	8.17 (at 54.4 K)	75.63 (at 90.2 K)

Data: AIP.

Table 3B.2 The standard enthalpies and entropies of vaporization of liquids at their boiling temperatures

	$\Delta_{vap}H^{\ominus}/(kJ\,mol^{-1})$	$\theta_b/°C$	$\Delta_{vap}S^{\ominus}/(J\,K^{-1}\,mol^{-1})$
Benzene	30.8	80.1	+87.2
Carbon disulfide	26.74	46.25	+83.7
Cyclohexane	30.1	80.7	+85.1
Decane	38.75	174	+86.7
Dimethyl ether	21.51	−23	+86
Ethanol	38.6	78.3	+110.0
Hydrogen sulfide	18.7	−60.4	+87.9
Mercury	59.3	356.6	+94.2
Methane	8.18	−161.5	+73.2
Methanol	35.21	65.0	+104.1
Tetrachloromethane	30.00	76.7	+85.8
Water	40.7	100.0	+109.1

Data: JL.

Table 3C.1 Standard Third-Law entropies at 298 K. See Tables 2C.6 and 2C.7.

Table 3D.1 Standard Gibbs energies of formation at 298 K. See Tables 2C.6 and 2C.7.

Table 5A.1 Henry's law constants for gases at 298 K, $K/(kPa\,kg\,mol^{-1})$

	Water	Benzene
CH_4	7.55×10^4	44.4×10^3
CO_2	3.01×10^3	8.90×10^2
H_2	1.28×10^5	2.79×10^4
N_2	1.56×10^5	1.87×10^4
O_2	7.92×10^4	

Data: converted from R.J. Silbey and R.A. Alberty, *Physical chemistry*. Wiley, New York (2001).

Table 5B.1 Freezing-point (K_f) and boiling-point (K_b) constants

	$K_f/(K\,kg\,mol^{-1})$	$K_b/(K\,kg\,mol^{-1})$
Benzene	5.12	2.53
Camphor	40	
Carbon disulfide	3.8	2.37
Ethanoic acid	3.90	3.07
Naphthalene	6.94	5.8
Phenol	7.27	3.04
Tetrachloromethane	30	4.95
Water	1.86	0.51

Data: KL.

Table 5F.2 Mean activity coefficients in water at 298 K

$b/b^{\ominus}$	HCl	KCl	$CaCl_2$	H_2SO_4	$LaCl_3$	$In_2(SO_4)_3$
0.001	0.966	0.966	0.888	0.830	0.790	
0.005	0.929	0.927	0.789	0.639	0.636	0.16
0.01	0.905	0.902	0.732	0.544	0.560	0.11
0.05	0.830	0.816	0.584	0.340	0.388	0.035
0.10	0.798	0.770	0.524	0.266	0.356	0.025
0.50	0.769	0.652	0.510	0.155	0.303	0.014
1.00	0.811	0.607	0.725	0.131	0.387	
2.00	1.011	0.577	1.554	0.125	0.954	

Data: RS, HCP, and S. Glasstone, *Introduction to electrochemistry*. Van Nostrand (1942).

Table 6D.1 Standard potentials at 298 K. (a) In electrochemical order

Reduction half-reaction	$E^\ominus$/V	Reduction half-reaction	$E^\ominus$/V
Strongly oxidizing		$ClO_4^- + H_2O + 2e^- \rightarrow ClO_3^- + 2OH^-$	+0.36
$H_4XeO_6 + 2H^+ + 2e^- \rightarrow XeO_3 + 3H_2O$	+3.0	$[Fe(CN)_6]^{3-} + e^- \rightarrow [Fe(CN)_6]^{4-}$	+0.36
$F_2 + 2e^- \rightarrow 2F^-$	+2.87	$Cu^{2+} + 2e^- \rightarrow Cu$	+0.34
$O_3 + 2H^+ + 2e^- \rightarrow O_2 + H_2O$	+2.07	$Hg_2Cl_2 + 2e^- \rightarrow 2Hg + 2Cl^-$	+0.27
$S_2O_8^{2-} + 2e^- \rightarrow 2SO_4^{2-}$	+2.05	$AgCl + e^- \rightarrow Ag + Cl^-$	+0.22
$Ag^{2+} + e^- \rightarrow Ag^+$	+1.98	$Bi^{3+} + 3e^- \rightarrow Bi$	+0.20
$Co^{3+} + e^- \rightarrow Co^{2+}$	+1.81	$Cu^{2+} + e^- \rightarrow Cu^+$	+0.16
$H_2O_2 + 2H^+ + 2e^- \rightarrow 2H_2O$	+1.78	$Sn^{4+} + 2e^- \rightarrow Sn^{2+}$	+0.15
$Au^+ + e^- \rightarrow Au$	+1.69	$NO_3^- + H_2O + 2e^- \rightarrow NO_2^- + 2OH^-$	+0.10
$Pb^{4+} + 2e^- \rightarrow Pb^{2+}$	+1.67	$AgBr + e^- \rightarrow Ag + Br^-$	+0.0713
$2HClO + 2H^+ + 2e^- \rightarrow Cl_2 + 2H_2O$	+1.63	$Ti^{4+} + e^- \rightarrow Ti^{3+}$	0.00
$Ce^{4+} + e^- \rightarrow Ce^{3+}$	+1.61	$2H^+ + 2e^- \rightarrow H_2$	0, by definition
$2HBrO + 2H^+ + 2e^- \rightarrow Br_2 + 2H_2O$	+1.60	$Fe^{3+} + 3e^- \rightarrow Fe$	−0.04
$MnO_4^- + 8H^+ + 5e^- \rightarrow Mn^{2+} + 4H_2O$	+1.51	$O_2 + H_2O + 2e^- \rightarrow HO_2^- + OH^-$	−0.08
$Mn^{3+} + e^- \rightarrow Mn^{2+}$	+1.51	$Pb^{2+} + 2e^- \rightarrow Pb$	−0.13
$Au^{3+} + 3e^- \rightarrow Au$	+1.40	$In^+ + e^- \rightarrow In$	−0.14
$Cl_2 + 2e^- \rightarrow 2Cl^-$	+1.36	$Sn^{2+} + 2e^- \rightarrow Sn$	−0.14
$Cr_2O_7^{2-} + 14H^+ + 6e^- \rightarrow 2Cr^{3+} + 7H_2O$	+1.33	$AgI + e^- \rightarrow Ag + I^-$	−0.15
$O_3 + H_2O + 2e^- \rightarrow O_2 + 2OH^-$	+1.24	$Ni^{2+} + 2e^- \rightarrow Ni$	−0.23
$O_2 + 4H^+ + 4e^- \rightarrow 2H_2O$	+1.23	$V^{3+} + e^- \rightarrow V^{2+}$	−0.26
$ClO_4^- + 2H^+ + 2e^- \rightarrow ClO_3^- + H_2O$	+1.23	$Co^{2+} + 2e^- \rightarrow Co$	−0.28
$MnO_2 + 4H^+ + 2e^- \rightarrow Mn^{2+} + 2H_2O$	+1.23	$In^{3+} + 3e^- \rightarrow In$	−0.34
$Pt^{2+} + 2e^- \rightarrow Pt$	+1.20	$Tl^+ + e^- \rightarrow Tl$	−0.34
$Br_2 + 2e^- \rightarrow 2Br^-$	+1.09	$PbSO_4 + 2e^- \rightarrow Pb + SO_4^{2-}$	−0.36
$Pu^{4+} + e^- \rightarrow Pu^{3+}$	+0.97	$Ti^{3+} + e^- \rightarrow Ti^{2+}$	−0.37
$NO_3^- + 4H^+ + 3e^- \rightarrow NO + 2H_2O$	+0.96	$Cd^{2+} + 2e^- \rightarrow Cd$	−0.40
$2Hg^{2+} + 2e^- \rightarrow Hg_2^{2+}$	+0.92	$In^{2+} + e^- \rightarrow In^+$	−0.40
$ClO^- + H_2O + 2e^- \rightarrow Cl^- + 2OH^-$	+0.89	$Cr^{3+} + e^- \rightarrow Cr^{2+}$	−0.41
$Hg^{2+} + 2e^- \rightarrow Hg$	+0.86	$Fe^{2+} + 2e^- \rightarrow Fe$	−0.44
$NO_3^- + 2H^+ + e^- \rightarrow NO_2 + H_2O$	+0.80	$In^{3+} + 2e^- \rightarrow In^+$	−0.44
$Ag^+ + e^- \rightarrow Ag$	+0.80	$S + 2e^- \rightarrow S^{2-}$	−0.48
$Hg_2^{2+} + 2e^- \rightarrow 2Hg$	+0.79	$In^{3+} + e^- \rightarrow In^{2+}$	−0.49
$AgF + e^- \rightarrow Ag + F^-$	+0.78	$O_2 + e^- \rightarrow O_2^-$	−0.56
$Fe^{3+} + e^- \rightarrow Fe^{2+}$	+0.77	$U^{4+} + e^- \rightarrow U^{3+}$	−0.61
$BrO^- + H_2O + 2e^- \rightarrow Br^- + 2OH^-$	+0.76	$Cr^{3+} + 3e^- \rightarrow Cr$	−0.74
$Hg_2SO_4 + 2e^- \rightarrow 2Hg + SO_4^{2-}$	+0.62	$Zn^{2+} + 2e^- \rightarrow Zn$	−0.76
$MnO_4^{2-} + 2H_2O + 2e^- \rightarrow MnO_2 + 4OH^-$	+0.60	$Cd(OH)_2 + 2e^- \rightarrow Cd + 2OH^-$	−0.81
$MnO_4^- + e^- \rightarrow MnO_4^{2-}$	+0.56	$2H_2O + 2e^- \rightarrow H_2 + 2OH^-$	−0.83
$I_2 + 2e^- \rightarrow 2I^-$	+0.54	$Cr^{2+} + 2e^- \rightarrow Cr$	−0.91
$I_3^- + 2e^- \rightarrow 3I^-$	+0.53	$Mn^{2+} + 2e^- \rightarrow Mn$	−1.18
$Cu^+ + e^- \rightarrow Cu$	+0.52	$V^{2+} + 2e^- \rightarrow V$	−1.19
$NiOOH + H_2O + e^- \rightarrow Ni(OH)_2 + OH^-$	+0.49	$Ti^{2+} + 2e^- \rightarrow Ti$	−1.63
$Ag_2CrO_4 + 2e^- \rightarrow 2Ag + CrO_4^{2-}$	+0.45	$Al^{3+} + 3e^- \rightarrow Al$	−1.66
$O_2 + 2H_2O + 4e^- \rightarrow 4OH^-$	+0.40	$U^{3+} + 3e^- \rightarrow U$	−1.79

Table 6D.1 (Continued)

Reduction half-reaction	$E^\ominus$/V	Reduction half-reaction	$E^\ominus$/V
$Be^{2+} + 2e^- \rightarrow Be$	−1.85	$Sr^{2+} + 2e^- \rightarrow Sr$	−2.89
$Sc^{3+} + 3e^- \rightarrow Sc$	−2.09	$Ba^{2+} + 2e^- \rightarrow Ba$	−2.91
$Mg^{2+} + 2e^- \rightarrow Mg$	−2.36	$Ra^{2+} + 2e^- \rightarrow Ra$	−2.92
$Ce^{3+} + 3e^- \rightarrow Ce$	−2.48	$Cs^+ + e^- \rightarrow Cs$	−2.92
$La^{3+} + 3e^- \rightarrow La$	−2.52	$Rb^+ + e^- \rightarrow Rb$	−2.93
$Na^+ + e^- \rightarrow Na$	−2.71	$K^+ + e^- \rightarrow K$	−2.93
$Ca^{2+} + 2e^- \rightarrow Ca$	−2.87	$Li^+ + e^- \rightarrow Li$	−3.05
		Strongly reducing	

Table 6D.1 Standard potentials at 298 K. (b) In alphabetical order

Reduction half-reaction	$E^\ominus$/V	Reduction half-reaction	$E^\ominus$/V
$Ag^+ + e^- \rightarrow Ag$	+0.80	$Cu^+ + e^- \rightarrow Cu$	+0.52
$Ag^{2+} + e^- \rightarrow Ag^+$	+1.98	$Cu^{2+} + 2e^- \rightarrow Cu$	+0.34
$AgBr + e^- \rightarrow Ag + Br^-$	+0.0713	$Cu^{2+} + e^- \rightarrow Cu^+$	+0.16
$AgCl + e^- \rightarrow Ag + Cl^-$	+0.22	$F_2 + 2e^- \rightarrow 2F^-$	+2.87
$Ag_2CrO_4 + 2e^- \rightarrow 2Ag + CrO_4^{2-}$	+0.45	$Fe^{2+} + 2e^- \rightarrow Fe$	−0.44
$AgF + e^- \rightarrow Ag + F^-$	+0.78	$Fe^{3+} + 3e^- \rightarrow Fe$	−0.04
$AgI + e^- \rightarrow Ag + I^-$	−0.15	$Fe^{3+} + e^- \rightarrow Fe^{2+}$	+0.77
$Al^{3+} + 3e^- \rightarrow Al$	−1.66	$[Fe(CN)_6]^{3-} + e^- \rightarrow [Fe(CN)_6]^{4-}$	+0.36
$Au^+ + e^- \rightarrow Au$	+1.69	$2H^+ + 2e^- \rightarrow H_2$	0, by definition
$Au^{3+} + 3e^- \rightarrow Au$	+1.40	$2H_2O + 2e^- \rightarrow H_2 + 2OH^-$	−0.83
$Ba^{2+} + 2e^- \rightarrow Ba$	−2.91	$2HBrO + 2H^+ + 2e^- \rightarrow Br_2 + 2H_2O$	+1.60
$Be^{2+} + 2e^- \rightarrow Be$	−1.85	$2HClO + 2H^+ + 2e^- \rightarrow Cl_2 + 2H_2O$	+1.63
$Bi^{3+} + 3e^- \rightarrow Bi$	+0.20	$H_2O_2 + 2H^+ + 2e^- \rightarrow 2H_2O$	+1.78
$Br_2 + 2e^- \rightarrow 2Br^-$	+1.09	$H_4XeO_6 + 2H^+ + 2e^- \rightarrow XeO_3 + 3H_2O$	+3.0
$BrO^- + H_2O + 2e^- \rightarrow Br^- + 2OH^-$	+0.76	$Hg_2^{2+} + 2e^- \rightarrow 2Hg$	+0.79
$Ca^{2+} + 2e^- \rightarrow Ca$	−2.87	$Hg_2Cl_2 + 2e^- \rightarrow 2Hg + 2Cl^-$	+0.27
$Cd(OH)_2 + 2e^- \rightarrow Cd + 2OH^-$	−0.81	$Hg^{2+} + 2e^- \rightarrow Hg$	+0.86
$Cd^{2+} + 2e^- \rightarrow Cd$	−0.40	$2Hg^{2+} + 2e^- \rightarrow Hg_2^{2+}$	+0.92
$Ce^{3+} + 3e^- \rightarrow Ce$	−2.48	$Hg_2SO_4 + 2e^- \rightarrow 2Hg + SO_4^{2-}$	+0.62
$Ce^{4+} + e^- \rightarrow Ce^{3+}$	+1.61	$I_2 + 2e^- \rightarrow 2I^-$	+0.54
$Cl_2 + 2e^- \rightarrow 2Cl^-$	+1.36	$I_3^- + 2e^- \rightarrow 3I^-$	+0.53
$ClO^- + H_2O + 2e^- \rightarrow Cl^- + 2OH^-$	+0.89	$In^+ + e^- \rightarrow In$	−0.14
$ClO_4^- + 2H^+ + 2e^- \rightarrow ClO_3^- + H_2O$	+1.23	$In^{2+} + e^- \rightarrow In^+$	−0.40
$ClO_4^- + H_2O + 2e^- \rightarrow ClO_3^- + 2OH^-$	+0.36	$In^{3+} + 2e^- \rightarrow In^+$	−0.44
$Co^{2+} + 2e^- \rightarrow Co$	−0.28	$In^{3+} + 3e^- \rightarrow In$	−0.34
$Co^{3+} + e^- \rightarrow Co^{2+}$	+1.81	$In^{3+} + e^- \rightarrow In^{2+}$	−0.49
$Cr^{2+} + 2e^- \rightarrow Cr$	−0.91	$K^+ + e^- \rightarrow K$	−2.93
$Cr_2O_7^{2-} + 14H^+ + 6e^- \rightarrow 2Cr^{3+} + 7H_2O$	+1.33	$La^{3+} + 3e^- \rightarrow La$	−2.52
$Cr^{3+} + 3e^- \rightarrow Cr$	−0.74	$Li^+ + e^- \rightarrow Li$	−3.05
$Cr^{3+} + e^- \rightarrow Cr^{2+}$	−0.41	$Mg^{2+} + 2e^- \rightarrow Mg$	−2.36
$Cs^+ + e^- \rightarrow Cs$	−2.92	$Mn^{2+} + 2e^- \rightarrow Mn$	−1.18

(*Continued*)

Table 6D.1 (Continued)

Reduction half-reaction	$E^{\ominus}$/V	Reduction half-reaction	$E^{\ominus}$/V
$Mn^{3+} + e^- \rightarrow Mn^{2+}$	+1.51	$PbSO_4 + 2e^- \rightarrow Pb + SO_4^{2-}$	−0.36
$MnO_2 + 4H^+ + 2e^- \rightarrow Mn^{2+} + 2H_2O$	+1.23	$Pt^{2+} + 2e^- \rightarrow Pt$	+1.20
$MnO_4^{2-} + 8H^+ + 5e^- \rightarrow Mn^{2+} + 4H_2O$	+1.51	$Pu^{4+} + e^- \rightarrow Pu^{3+}$	+0.97
$MnO_4^- + e^- \rightarrow MnO_4^{2-}$	+0.56	$Ra^{2+} + 2e^- \rightarrow Ra$	−2.92
$MnO_4^{2-} + 2H_2O + 2e^- \rightarrow MnO_2 + 4OH^-$	+0.60	$Rb^+ + e^- \rightarrow Rb$	−2.93
$Na^+ + e^- \rightarrow Na$	−2.71	$S + 2e^- \rightarrow S^{2-}$	−0.48
$Ni^{2+} + 2e^- \rightarrow Ni$	−0.23	$S_2O_8^{2-} + 2e^- \rightarrow 2SO_4^{2-}$	+2.05
$NiOOH + H_2O + e^- \rightarrow Ni(OH)_2 + OH^-$	+0.49	$Sc^{3+} + 3e^- \rightarrow Sc$	−2.09
$NO_3^- + 2H^+ + e^- \rightarrow NO_2 + H_2O$	+0.80	$Sn^{2+} + 2e^- \rightarrow Sn$	−0.14
$NO_3^- + 4H^+ + 3e^- \rightarrow NO + 2H_2O$	+0.96	$Sn^{4+} + 2e^- \rightarrow Sn^{2+}$	+0.15
$NO_3^- + H_2O + 2e^- \rightarrow NO_2^- + 2OH^-$	+0.10	$Sr^{2+} + 2e^- \rightarrow Sr$	−2.89
$O_2 + 2H_2O + 4e^- \rightarrow 4OH^-$	+0.40	$Ti^{2+} + 2e^- \rightarrow Ti$	−1.63
$O_2 + 4H^+ + 4e^- \rightarrow 2H_2O$	+1.23	$Ti^{3+} + e^- \rightarrow Ti^{2+}$	−0.37
$O_2 + e^- \rightarrow O_2^-$	−0.56	$Ti^{4+} + e^- \rightarrow Ti^{3+}$	0.00
$O_2 + H_2O + 2e^- \rightarrow HO_2^- + OH^-$	−0.08	$Tl^+ + e^- \rightarrow Tl$	−0.34
$O_3 + 2H^+ + 2e^- \rightarrow O_2 + H_2O$	+2.07	$U^{3+} + 3e^- \rightarrow U$	−1.79
$O_3 + H_2O + 2e^- \rightarrow O_2 + 2OH^-$	+1.24	$U^{4+} + e^- \rightarrow U^{3+}$	−0.61
$Pb^{2+} + 2e^- \rightarrow Pb$	−0.13	$V^{2+} + 2e^- \rightarrow V$	−1.19
$Pb^{4+} + 2e^- \rightarrow Pb^{2+}$	+1.67	$V^{3+} + e^- \rightarrow V^{2+}$	−0.26
		$Zn^{2+} + 2e^- \rightarrow Zn$	−0.76

Table 8B.1 Effective nuclear charge*

	H							He
1s	1							1.6875
	Li	Be	B	C	N	O	F	Ne
1s	2.6906	3.6848	4.6795	5.6727	6.6651	7.6579	8.6501	9.6421
2s	1.2792	1.9120	2.5762	3.2166	3.8474	4.4916	5.1276	5.7584
2p			2.4214	3.1358	3.8340	4.4532	5.1000	5.7584
	Na	Mg	Al	Si	P	S	Cl	Ar
1s	10.6259	11.6089	12.5910	13.5745	14.5578	15.5409	16.5239	17.5075
2s	6.5714	7.3920	8.3736	9.0200	9.8250	10.6288	11.4304	12.2304
2p	6.8018	7.8258	8.9634	9.9450	10.9612	11.9770	12.9932	14.0082
3s	2.5074	3.3075	4.1172	4.9032	5.6418	6.3669	7.0683	7.7568
3p			4.0656	4.2852	4.8864	5.4819	6.1161	6.7641

* The actual charge is $Z_{\text{eff}}e$.

Data: E. Clementi and D.L. Raimondi, *Atomic screening constants from SCF functions.* IBM Res. Note NJ-27 (1963). *J. Chem. Phys.* **38**, 2686 (1963).

Table 8B.3 Ionic radii, r/pm*

Li^+(4)	Be^{2+}(4)	B^{3+}(4)	N^{3-}	O^{2-}(6)	F^-(6)
59	27	12	171	140	133
Na^+(6)	Mg^{2+}(6)	Al^{3+}(6)	P^{3-}	S^{2-}(6)	Cl^-(6)
102	72	53	212	184	181
K^+(6)	Ca^{2+}(6)	Ga^{3+}(6)	As^{3-}(6)	Se^{2-}(6)	Br^-(6)
138	100	62	222	198	196
Rb^+(6)	Sr^{2+}(6)	In^{3+}(6)		Te^{2-}(6)	I^-(6)
149	116	79		221	220
Cs^+(6)	Ba^{2+}(6)	Tl^{3+}(6)			
167	136	88			

d-block elements (high-spin ions)

Sc^{3+}(6)	Ti^{4+}(6)	Cr^{3+}(6)	Mn^{3+}(6)	Fe^{2+}(6)	Co^{3+}(6)	Cu^{2+}(6)	Zn^{2+}(6)
73	60	61	65	63	61	73	75

*Numbers in parentheses are the coordination numbers of the ions. Values for ions without a coordination number stated are estimates.
Data: R.D. Shannon and C.T. Prewitt, *Acta Cryst.* **B25**, 925 (1969).

Table 8B.4 Ionization energies, I/(kJ mol^{-1})

H							He
1312.0							2372.3
							5250.4
Li	Be	B	C	N	O	F	Ne
513.3	899.4	800.6	1086.2	1402.3	1313.9	1681	2080.6
7298.0	1757.1	2427	2352	2856.1	3388.2	3374	3952.2
Na	Mg	Al	Si	P	S	Cl	Ar
495.8	737.7	577.4	786.5	1011.7	999.6	1251.1	1520.4
4562.4	1450.7	1816.6	1577.1	1903.2	2251	2297	2665.2
		2744.6		2912			
K	Ca	Ga	Ge	As	Se	Br	Kr
418.8	589.7	578.8	762.1	947.0	940.9	1139.9	1350.7
3051.4	1145	1979	1537	1798	2044	2104	2350
		2963	2735				
Rb	Sr	In	Sn	Sb	Te	I	Xe
403.0	549.5	558.3	708.6	833.7	869.2	1008.4	1170.4
2632	1064.2	1820.6	1411.8	1794	1795	1845.9	2046
		2704	2943.0	2443			
Cs	Ba	Tl	Pb	Bi	Po	At	Rn
375.5	502.8	589.3	715.5	703.2	812	930	1037
2420	965.1	1971.0	1450.4	1610			
		2878	3081.5	2466			

Data: E.

Table 8B.5 Electron affinities, $E_{ea}/(kJ\,mol^{-1})$

H							He
72.8							−21
Li	Be	B	C	N	O	F	Ne
59.8	≤0	23	122.5	−7	141	322	−29
					−844		
Na	Mg	Al	Si	P	S	Cl	Ar
52.9	≤0	44	133.6	71.7	200.4	348.7	−35
					−532		
K	Ca	Ga	Ge	As	Se	Br	Kr
48.3	2.37	36	116	77	195.0	324.5	−39
Rb	Sr	In	Sn	Sb	Te	I	Xe
46.9	5.03	34	121	101	190.2	295.3	−41
Cs	Ba	Tl	Pb	Bi	Po	At	Rn
45.5	13.95	30	35.2	101	186	270	−41

Data: E.

Table 9C.2 Bond lengths, R/pm

(a) Bond lengths in specific molecules

Br_2	228.3
Cl_2	198.75
CO	112.81
F_2	141.78
H_2^+	106
H_2	74.138
HBr	141.44
HCl	127.45
HF	91.680
HI	160.92
N_2	109.76
O_2	120.75

(b) Mean bond lengths from covalent radii*

H	37						
C	77(1)	N	74(1)	O	66(1)	F	64
	67(2)		65(2)		57(2)		
	60(3)						
Si	118	P	110	S	104(1)	Cl	99
					95(2)		
Ge	122	As	121	Se	104	Br	114
		Sb	141	Te	137	I	133

* Values are for single bonds except where indicated otherwise (values in parentheses). The length of an A–B covalent bond (of given order) is the sum of the corresponding covalent radii.

Table 9C.3a Bond dissociation enthalpies, $\Delta H^{\ominus}$(A–B)/(kJ mol^{-1}) at 298 K*

Diatomic molecules

H–H	436	F–F	155	Cl–Cl	242	Br–Br	193	I–I	151
O=O	497	C=O	1076	N≡N	945				
H–O	428	H–F	565	H–Cl	431	H–Br	366	H–I	299

Polyatomic molecules

H–CH_3	435	H–NH_2	460	H–OH	492	H–C_6H_5	469
H_3C–CH_3	368	H_2C=CH_2	720	HC≡CH	962		
HO–CH_3	377	Cl–CH_3	352	Br–CH_3	293	I–CH_3	237
O=CO	531	HO–OH	213	O_2N–NO_2	54		

* To a good approximation bond dissociation enthalpies and dissociation energies are related by $\Delta H^{\ominus} = N_A hc\tilde{D}_e + \frac{3}{2}RT$ with $hc\tilde{D}_e = hc\tilde{D}_0 + \frac{1}{2}\hbar\omega$. For precise values of $N_A hc\tilde{D}_0$ for diatomic molecules, see Table 11C.1.
Data: HCP, KL.

Table 9C.3b Mean bond enthalpies, $\Delta H^{\ominus}$(A–B)/(kJ mol^{-1})*

	H	C	N	O	F	Cl	Br	I	S	P	Si
H	436										
C	412	348(i)									
		612(ii)									
		838(iii)									
		518(a)									
N	388	305(i)	163(i)								
		613(ii)	409(ii)								
		890(iii)	946(iii)								
O	463	360(i)	157	146(i)							
		743(ii)		497(ii)							
F	565	484	270	185	155						
Cl	431	338	200	203	254	242					
Br	366	276				219	193				
I	299	238				210	178	151			
S	338	259			496	250	212		264		
P	322									201	
Si	318		374	466							226

* Mean bond enthalpies are such a crude measure of bond strength that they need not be distinguished from dissociation energies.
(i) Single bond, (ii) double bond, (iii) triple bond, (a) aromatic.
Data: HCP and L. Pauling, *The nature of the chemical bond*. Cornell University Press (1960).

Table 9D.1 Pauling (*italics*) and Mulliken electronegativities

H							He
2.20							
3.06							
Li	Be	B	C	N	O	F	Ne
0.98	*1.57*	*2.04*	*2.55*	*3.04*	*3.44*	*3.98*	
1.28	1.99	1.83	2.67	3.08	3.22	4.43	4.60
Na	Mg	Al	Si	P	S	Cl	Ar
0.93	*1.31*	*1.61*	*1.90*	*2.19*	*2.58*	*3.16*	
1.21	1.63	1.37	2.03	2.39	2.65	3.54	3.36
K	Ca	Ga	Ge	As	Se	Br	Kr
0.82	*1.00*	*1.81*	*2.01*	*2.18*	*2.55*	*2.96*	*3.0*
1.03	1.30	1.34	1.95	2.26	2.51	3.24	2.98
Rb	Sr	In	Sn	Sb	Te	I	Xe
0.82	*0.95*	*1.78*	*1.96*	*2.05*	*2.10*	*2.66*	*2.6*
0.99	1.21	1.30	1.83	2.06	2.34	2.88	2.59
Cs	Ba	Tl	Pb	Bi			
0.79	*0.89*	*2.04*	*2.33*	*2.02*			

Data: Pauling values: A.L. Allred, *J. Inorg. Nucl. Chem.* 17, 215 (1961); L.C. Allen and J.E. Huheey, *ibid.*, **42**, 1523 (1980). Mulliken values: L.C. Allen, *J. Am. Chem. Soc.* **111**, 9003 (1989). The Mulliken values have been scaled to the range of the Pauling values.

Table 10B.1 See Part 4

Table 10B.2 See Part 4

Table 11C.1 Properties of diatomic molecules

	$\tilde{\nu}/\text{cm}^{-1}$	θ^V/K	$\tilde{B}/\text{cm}^{-1}$	θ^R/K	R_e/pm	$k_f/(\text{N m}^{-1})$	$N_A hc\tilde{D}_0/(\text{kJ mol}^{-1})$	σ
$^{1}H_2^+$	2321.8	3341	29.8	42.9	106	160	255.8	2
$^{1}H_2$	4400.39	6332	60.864	87.6	74.138	574.9	432.1	2
$^{2}H_2$	3118.46	4487	30.442	43.8	74.154	577.0	439.6	2
$^{1}H^{19}F$	4138.32	5955	20.956	30.2	91.680	965.7	564.4	1
$^{1}H^{35}Cl$	2990.95	4304	10.593	15.2	127.45	516.3	427.7	1
$^{1}H^{81}Br$	2648.98	3812	8.465	12.2	141.44	411.5	362.7	1
$^{1}H^{127}I$	2308.09	3321	6.511	9.37	160.92	313.8	294.9	1
$^{14}N_2$	2358.07	3393	1.9987	2.88	109.76	2293.8	941.7	2
$^{16}O_2$	1580.36	2274	1.4457	2.08	120.75	1176.8	493.5	2
$^{19}F_2$	891.8	1283	0.8828	1.27	141.78	445.1	154.4	2
$^{35}Cl_2$	559.71	805	0.2441	0.351	198.75	322.7	239.3	2
$^{12}C^{16}O$	2170.21	3122	1.9313	2.78	112.81	1903.17	1071.8	1
$^{79}Br^{81}Br$	323.2	465	0.0809	10.116	283.3	245.9	190.2	1

Data: AIP.

Table 11F.1 Colour, frequency, and energy of light

Colour	λ/nm	$\nu/(10^{14}$ Hz)	$\tilde{\nu}/(10^4$ cm$^{-1})$	E/eV	E/(kJ mol^{-1})
Infrared	>1000	<3.00	<1.00	<1.24	<120
Red	700	4.28	1.43	1.77	171
Orange	620	4.84	1.61	2.00	193
Yellow	580	5.17	1.72	2.14	206
Green	530	5.66	1.89	2.34	226
Blue	470	6.38	2.13	2.64	254
Violet	420	7.14	2.38	2.95	285
Ultraviolet	<400	>7.5	>2.5	>3.10	>300

Data: J.G. Calvert and J.N. Pitts, *Photochemistry*. Wiley, New York (1966).

Table 11F.2 Absorption characteristics of some groups and molecules

Group	$\tilde{\nu}_{max}/(10^4$ cm$^{-1})$	λ_{max}/nm	$\varepsilon_{max}/(\text{dm}^3\,\text{mol}^{-1}\,\text{cm}^{-1})$
C=C ($\pi^* \leftarrow \pi$)	6.10	163	1.5×10^4
	5.73	174	5.5×10^3
C=O ($\pi^* \leftarrow$ n)	3.7–3.5	270–290	10–20
–N=N–	2.9	350	15
	>3.9	<260	Strong
$-NO_2$	3.6	280	10
	4.8	210	1.0×10^4
C_6H_5-	3.9	255	200
	5.0	200	6.3×10^3
	5.5	180	1.0×10^5
$[Cu(OH_2)_6]^{2+}$(aq)	1.2	810	10
$[Cu(NH_3)_4]^{2+}$(aq)	1.7	600	50
H_2O	6.0	167	7.0×10^3

Table 12A.2 Nuclear spin properties

Nuclide	Natural abundance/%	Spin I	Magnetic moment, μ/μ_N	g-value	$\gamma/(10^7\,T^{-1}\,s^{-1})$	NMR frequency at 1 T, ν/MHz
^{1}n*		$\frac{1}{2}$	−1.9130	−3.8260	−18.324	29.164
^{1}H	99.9844	$\frac{1}{2}$	2.792 85	5.5857	26.752	42.576
^{2}H	0.0156	1	0.857 44	0.857 44	4.1067	6.536
^{3}H*		$\frac{1}{2}$	2.978 96	−4.2553	−20.380	45.414
^{10}B	19.6	3	1.8006	0.6002	2.875	4.575
^{11}B	80.4	$\frac{3}{2}$	2.6886	1.7923	8.5841	13.663
^{13}C	1.108	$\frac{1}{2}$	0.7024	1.4046	6.7272	10.708
^{14}N	99.635	1	0.403 56	0.403 56	1.9328	3.078
^{17}O	0.037	$\frac{5}{2}$	−1.893 79	−0.7572	−3.627	5.774
^{19}F	100	$\frac{1}{2}$	2.628 87	5.2567	25.177	40.077
^{31}P	100	$\frac{1}{2}$	1.1316	2.2634	10.840	17.251
^{33}S	0.74	$\frac{3}{2}$	0.6438	0.4289	2.054	3.272
^{35}Cl	75.4	$\frac{3}{2}$	0.8219	0.5479	2.624	4.176
^{37}Cl	24.6	$\frac{3}{2}$	0.6841	0.4561	2.184	3.476

* Radioactive.
μ is the magnetic moment of the spin state with the largest value of m_I: $\mu = g_I\mu_N I$ and μ_N is the nuclear magneton (see inside front cover).
Data: KL and HCP.

Table 12D.1 Hyperfine coupling constants for atoms, a/mT

Nuclide	Spin	Isotropic coupling	Anisotropic coupling
^{1}H	$\frac{1}{2}$	50.8(1s)	
^{2}H	1	7.8(1s)	
^{13}C	$\frac{1}{2}$	113.0(2s)	6.6(2p)
^{14}N	1	55.2(2s)	4.8(2p)
^{19}F	$\frac{1}{2}$	1720(2s)	108.4(2p)
^{31}P	$\frac{1}{2}$	364(3s)	20.6(3p)
^{35}Cl	$\frac{3}{2}$	168(3s)	10.0(3p)
^{37}Cl	$\frac{3}{2}$	140(3s)	8.4(3p)

Data: P.W. Atkins and M.C.R. Symons, *The structure of inorganic radicals*. Elsevier, Amsterdam (1967).

Table 14A.1 Magnitudes of dipole moments (μ), polarizabilities (α), and polarizability volumes (α')

	$\mu/(10^{-30}$ C m)	μ/D	$\alpha'/(10^{-30}\,m^3)$	$\alpha/(10^{-40}\,J^{-1}\,C^2\,m^2)$
Ar	0	0	1.66	1.85
C_2H_5OH	5.64	1.69		
$C_6H_5CH_3$	1.20	0.36		
C_6H_6	0	0	10.4	11.6
CCl_4	0	0	10.3	11.7
CH_2Cl_2	5.24	1.57	6.80	7.57
CH_3Cl	6.24	1.87	4.53	5.04
CH_3OH	5.70	1.71	3.23	3.59
CH_4	0	0	2.60	2.89
$CHCl_3$	3.37	1.01	8.50	9.46
CO	0.390	0.117	1.98	2.20
CO_2	0	0	2.63	2.93
H_2	0	0	0.819	0.911
H_2O	6.17	1.85	1.48	1.65
HBr	2.67	0.80	3.61	4.01
HCl	3.60	1.08	2.63	2.93
He	0	0	0.20	0.22
HF	6.37	1.91	0.51	0.57
HI	1.40	0.42	5.45	6.06
N_2	0	0	1.77	1.97
NH_3	4.90	1.47	2.22	2.47
1,2-$C_6H_4(CH_3)_2$	2.07	0.62		

Data: HCP and C.J.F. Böttcher and P. Bordewijk, *Theory of electric polarization*. Elsevier, Amsterdam (1978).

Table 14B.2 Lennard-Jones-(12,6) potential energy parameters

	(ε/k)/K	r_0/pm
Ar	111.84	362.3
C_2H_2	209.11	463.5
C_2H_4	200.78	458.9
C_2H_6	216.12	478.2
C_6H_6	377.46	617.4
CCl_4	378.86	624.1
Cl_2	296.27	448.5
CO_2	201.71	444.4
F_2	104.29	357.1
Kr	154.87	389.5
N_2	91.85	391.9
O_2	113.27	365.4
Xe	213.96	426.0

Source: F. Cuadros, I. Cachadiña, and W. Ahamuda, *Molec. Eng.* **6**, 319 (1996).

Table 14C.1 Surface tensions of liquids at 293 K

	γ/(mN m^{-1})
Benzene	28.88
Ethanol	22.8
Hexane	18.4
Mercury	472
Methanol	22.6
Tetrachloromethane	27.0
Water	72.75
	72.0 at 25 °C
	58.0 at 100 °C

Data: KL.

Table 15C.2 Ionic radii, *r*/pm*

Li^+(4)	Be^{2+}(4)	B^{3+}(4)	N^{3-}	O^{2-}(6)	F^-(6)		
59	27	12	171	140	133		
Na^+(6)	Mg^{2+}(6)	Al^{3+}(6)	P^{3-}	S^{2-}(6)	Cl^-(6)		
102	72	53	212	184	181		
K^+(6)	Ca^{2+}(6)	Ga^{3+}(6)	As^{3-}(6)	Se^{2-}(6)	Br^-(6)		
138	100	62	222	198	196		
Rb^+(6)	Sr^{2+}(6)	In^{3+}(6)		Te^{2-}(6)	I^-(6)		
149	116	79		221	220		
Cs^+(6)	Ba^{2+}(6)	Tl^{3+}(6)					
167	136	88					
d-block elements (high-spin ions)							
Sc^{3+}(6)	Ti^{4+}(6)	Cr^{3+}(6)	Mn^{3+}(6)	Fe^{2+}(6)	Co^{3+}(6)	Cu^{2+}(6)	Zn^{2+}(6)
73	60	61	65	63	61	73	75

*Numbers in parentheses are the coordination numbers of the ions. Values for ions without a coordination number stated are estimates.
Data: R.D. Shannon and C.T. Prewitt, *Acta Cryst.* **B25**, 925 (1969).

Table 15C.4 Lattice enthalpies, $\Delta H_L^\ominus/(\text{kJ mol}^{-1})$ at 298 K

	F	Cl	Br	I				
Halides								
Li	1037	852	815	761				
Na	926	787	752	705				
K	821	717	689	649				
Rb	789	695	668	632				
Cs	750	676	654	620				
Ag	969	912	900	886				
Be		3017						
Mg		2524						
Ca		2255						
Sr		2153						
Oxides								
MgO	3850	CaO	3461	SrO	3283	BaO	3114	
Sulfides								
MgS	3406	CaS	3119	SrS	2974	BaS	2832	

Entries refer to $MX(s) \rightarrow M^+(g) + X^-(g)$.
Data: Principally D. Cubicciotti, *J. Chem. Phys.* **31**, 1646 (1959).

Table 15F.1 Magnetic susceptibilities at 298 K

	$\chi/10^{-6}$	$\chi_m/(10^{-10}\,\text{m}^3\,\text{mol}^{-1})$
$H_2O(l)$	−9.02	−1.63
$C_6H_6(l)$	−8.8	−7.8
$C_6H_{12}(l)$	−10.2	−11.1
$CCl_4(l)$	−5.4	−5.2
NaCl(s)	−16	−3.8
Cu(s)	−9.7	−0.69
S(rhombic)	−12.6	−1.95
Hg(l)	−28.4	−4.21
Al(s)	+20.7	+2.07
Pt(s)	+267.3	+24.25
Na(s)	+8.48	+2.01
K(s)	+5.94	+2.61
$CuSO_4{\cdot}5H_2O(s)$	+167	+183
$MnSO_4{\cdot}4H_2O(s)$	+1859	+1835
$NiSO_4{\cdot}7H_2O(s)$	+355	+503
$FeSO_4(s)$	+3743	+1558

Source: Principally HCP, with $\chi_m = \chi V_m = \chi\rho/M$.

Table 16A.1 Transport properties of gases at 1 atm

	$\kappa/(\text{mW K}^{-1}\,\text{m}^{-1})$	$\eta/\mu\text{P}$	
	273 K	273 K	293 K
Air	24.1	173	182
Ar	16.3	210	223
C_2H_4	16.4	97	103
CH_4	30.2	103	110
Cl_2	7.9	123	132
CO_2	14.5	136	147
H_2	168.2	84	88
He	144.2	187	196
Kr	8.7	234	250
N_2	24.0	166	176
Ne	46.5	298	313
O_2	24.5	195	204
Xe	5.2	212	228

Data: KL.

Table 16B.1 Viscosities of liquids at 298 K, $\eta/(10^{-3}\,\mathrm{kg\,m^{-1}\,s^{-1}})$

Benzene	0.601
Ethanol	1.06
Mercury	1.55
Methanol	0.553
Pentane	0.224
Sulfuric acid	27
Tetrachloromethane	0.880
Water†	0.891

† The viscosity of water over its entire liquid range is represented with less than 1 per cent error by the expression
$\log(\eta_{20}/\eta) = A/B$,
$A = 1.370\,23(t-20) + 8.36\times10^{-4}(t-20)^2$
$B = 109 + t \quad t = \theta/°\mathrm{C}$
Convert $\mathrm{kg\,m^{-1}\,s^{-1}}$ to centipoise (cP) by multiplying by 10^3 (so $\eta \approx 1$ cP for water).
Data: AIP, KL.

Table 16B.2 Ionic mobilities in water at 298 K, $u/(10^{-8}\,\mathrm{m^2\,s^{-1}\,V^{-1}})$

Cations		Anions	
Ag^+	6.24	Br^-	8.09
Ca^{2+}	6.17	$CH_3CO_2^-$	4.24
Cu^{2+}	5.56	Cl^-	7.91
H^+	36.23	CO_3^{2-}	7.46
K^+	7.62	F^-	5.70
Li^+	4.01	$[Fe(CN)_6]^{3-}$	10.5
Na^+	5.19	$[Fe(CN)_6]^{4-}$	11.4
NH_4^+	7.63	I^-	7.96
$N(CH_3)_4^+$	4.65	NO_3^-	7.40
Rb^+	7.92	OH^-	20.64
Zn^{2+}	5.47	SO_4^{2-}	8.29

Data: KL, RS.

Table 16B.3 Diffusion coefficients in liquids at 298 K, $D/(10^{-9}\,\mathrm{m^2\,s^{-1}})$

Molecules in liquids				Ions in water			
I_2 in hexane	4.05	H_2 in CCl_4(l)	9.75	K^+	1.96	Br^-	2.08
in benzene	2.13	N_2 in CCl_4(l)	3.42	H^+	9.31	Cl^-	2.03
CCl_4 in heptane	3.17	O_2 in CCl_4(l)	3.82	Li^+	1.03	F^-	1.46
Glycine in water	1.055	Ar in CCl_4(l)	3.63	Na^+	1.33	I^-	2.05
Dextrose in water	0.673	CH_4 in CCl_4(l)	2.89			OH^-	5.03
Sucrose in water	0.5216	H_2O in water	2.26				
		CH_3OH in water	1.58				
		C_2H_5OH in water	1.24				

Data: AIP.

Table 17B.1 Kinetic data for first-order reactions

	Phase	θ/°C	$k_r/\mathrm{s^{-1}}$	$t_{1/2}$
$2N_2O_5 \rightarrow 4NO_2 + O_2$	g	25	3.38×10^{-5}	5.70 h
	HNO_3(l)	25	1.47×10^{-6}	131 h
	Br_2(l)	25	4.27×10^{-5}	4.51 h
$C_2H_6 \rightarrow 2CH_3$	g	700	5.36×10^{-4}	21.6 min
Cyclopropane $\rightarrow$ propene	g	500	6.71×10^{-4}	17.2 min
$CH_3N_2CH_3 \rightarrow C_2H_6 + N_2$	g	327	3.4×10^{-4}	34 min
Sucrose $\rightarrow$ glucose + fructose	aq(H^+)	25	6.0×10^{-5}	3.2 h

g: High pressure gas-phase limit.
Data: Principally K.J. Laidler, *Chemical kinetics*. Harper & Row, New York (1987); M.J. Pilling and P.W. Seakins, *Reaction kinetics*. Oxford University Press (1995); J. Nicholas, *Chemical kinetics*. Harper & Row, New York (1976). See also JL.

Table 17B.2 Kinetic data for second-order reactions

	Phase	$\theta/°C$	$k_r/(dm^3 mol^{-1} s^{-1})$
$2NOBr \rightarrow 2NO + Br_2$	g	10	0.80
$2NO_2 \rightarrow 2NO + O_2$	g	300	0.54
$H_2 + I_2 \rightarrow 2HI$	g	400	2.42×10^{-2}
$D_2 + HCl \rightarrow DH + DCl$	g	600	0.141
$2I \rightarrow I_2$	g	23	7×10^9
	hexane	50	1.8×10^{10}
$CH_3Cl + CH_3O^-$	methanol	20	2.29×10^{-6}
$CH_3Br + CH_3O^-$	methanol	20	9.23×10^{-6}
$H^+ + OH^- \rightarrow H_2O$	water	25	1.35×10^{11}
	ice	−10	8.6×10^{12}

Data: Principally K.J. Laidler, *Chemical kinetics*. Harper & Row, New York (1987); M.J. Pilling and P.W. Seakins, *Reaction kinetics*. Oxford University Press (1995); J. Nicholas, *Chemical kinetics*. Harper & Row, New York (1976).

Table 17D.1 Arrhenius parameters

First-order reactions	A/s^{-1}	$E_a/(kJ mol^{-1})$
Cyclopropane → propene	1.58×10^{15}	272
$CH_3NC \rightarrow CH_3CN$	3.98×10^{13}	160
cis-CHD=CHD → *trans*-CHD=CHD	3.16×10^{12}	256
Cyclobutane → $2C_2H_4$	3.98×10^{13}	261
$C_2H_5I \rightarrow C_2H_4 + HI$	2.51×10^{17}	209
$C_2H_6 \rightarrow 2CH_3$	2.51×10^7	384
$2N_2O_5 \rightarrow 4NO_2 + O_2$	4.94×10^{13}	103.4
$N_2O \rightarrow N_2 + O$	7.94×10^{11}	250
$C_2H_5 \rightarrow C_2H_4 + H$	1.0×10^{13}	167
Second-order reactions, gas-phase	$A/(dm^3 mol^{-1} s^{-1})$	$E_a/(kJ mol^{-1})$
$O + N_2 \rightarrow NO + N$	1×10^{11}	315
$OH + H_2 \rightarrow H_2O + H$	8×10^{10}	42
$Cl + H_2 \rightarrow HCl + H$	8×10^{10}	23
$2CH_3 \rightarrow C_2H_6$	2×10^{10}	ca. 0
$NO + Cl_2 \rightarrow NOCl + Cl$	4.0×10^9	85
$SO + O_2 \rightarrow SO_2 + O$	3×10^8	27
$CH_3 + C_2H_6 \rightarrow CH_4 + C_2H_5$	2×10^8	44
$C_6H_5 + H_2 \rightarrow C_6H_6 + H$	1×10^8	ca. 25

Table 17D.1 (Continued)

Second-order reactions, solution	$A/(dm^3\,mol^{-1}\,s^{-1})$	$E_a/(kJ\,mol^{-1})$
$C_2H_5ONa + CH_3I$ in ethanol	2.42×10^{11}	81.6
$C_2H_5Br + OH^-$ in water	4.30×10^{11}	89.5
$C_2H_5I + C_2H_5O^-$ in ethanol	1.49×10^{11}	86.6
$C_2H_5Br + OH^-$ in ethanol	4.30×10^{11}	89.5
$CO_2 + OH^-$ in water	1.5×10^{10}	38
$CH_3I + S_2O_3^{2-}$ in water	2.19×10^{12}	78.7
Sucrose + H_2O in acidic water	1.50×10^{15}	107.9
$(CH_3)_3CCl$ solvolysis		
in water	7.1×10^{16}	100
in methanol	2.3×10^{13}	107
in ethanol	3.0×10^{13}	112
in ethanoic acid	4.3×10^{13}	111
in trichloromethane	1.4×10^{4}	45
$C_6H_5NH_2 + C_6H_5COCH_2Br$		
in benzene	91	34

Data: Principally J. Nicholas, *Chemical kinetics*. Harper & Row, New York (1976) and A.A. Frost and R.G. Pearson, *Kinetics and mechanism*. Wiley, New York (1961).

Table 18A.1 Arrhenius parameters for gas-phase reactions

	$A/(dm^3\,mol^{-1}\,s^{-1})$			
	Experiment	Theory	$E_a/(kJ\,mol^{-1})$	P
$2\,NOCl \rightarrow 2\,NO + Cl_2$	9.4×10^{9}	5.9×10^{10}	102.0	0.16
$2\,NO_2 \rightarrow 2\,NO + O_2$	2.0×10^{9}	4.0×10^{10}	111.0	5.0×10^{-2}
$2\,ClO \rightarrow Cl_2 + O_2$	6.3×10^{7}	2.5×10^{10}	0.0	2.5×10^{-3}
$H_2 + C_2H_4 \rightarrow C_2H_6$	1.24×10^{6}	7.4×10^{11}	180	1.7×10^{-6}
$K + Br_2 \rightarrow KBr + Br$	1.0×10^{12}	2.1×10^{11}	0.0	4.8

Data: Principally M.J. Pilling and P.W. Seakins, *Reaction kinetics*. Oxford University Press (1995).

Table 19A.1 Maximum observed standard enthalpies of physisorption, $\Delta_{ad}H^{\ominus}/(kJ\,mol^{-1})$ at 298 K

C_2H_2	−38	H_2	−84
C_2H_4	−34	H_2O	−59
CH_4	−21	N_2	−21
Cl_2	−36	NH_3	−38
CO	−25	O_2	−21
CO_2	−25		

Data: D.O. Haywood and B.M.W. Trapnell, *Chemisorption*. Butterworth (1964).

Table 19A.2 Standard enthalpies of chemisorption, $\Delta_{ad}H^{\ominus}/(\mathrm{kJ\,mol^{-1}})$ at 298 K

Adsorbate	Adsorbent (substrate)											
	Ti	Ta	Nb	W	Cr	Mo	Mn	Fe	Co	Ni	Rh	Pt
H_2		−188			−188	−167	−71	−134			−117	
N_2		−586						−293				
O_2						−720					−494	−293
CO	−640							−192	−176			
CO_2	−682	−703	−552	−456	−339	−372	−222	−225	−146	−184		
NH_3				−301				−188		−155		
C_2H_4		−577		−427	−427			−285		−243	−209	

Data: D.O. Haywood and B.M.W. Trapnell, *Chemisorption*. Butterworth (1964).

Table 19D.1 Exchange-current densities and transfer coefficients at 298 K

Reaction	Electrode	$j_0/(\mathrm{A\,cm^{-2}})$	α
$2\,H^+ + 2\,e^- \rightarrow H_2$	Pt	7.9×10^{-4}	
	Cu	1×10^{-6}	
	Ni	6.3×10^{-6}	0.58
	Hg	7.9×10^{-13}	0.50
	Pb	5.0×10^{-12}	
$Fe^{3+} + e^- \rightarrow Fe^{2+}$	Pt	2.5×10^{-3}	0.58
$Ce^{4+} + e^- \rightarrow Ce^{3+}$	Pt	4.0×10^{-5}	0.75

Data: Principally J.O'M. Bockris and A.K.N. Reddy, *Modern electrochemistry*. Plenum, New York (1970).

PART 4 Character tables

The groups C_1, C_s, C_i

C_1 (1)	E	$h = 1$	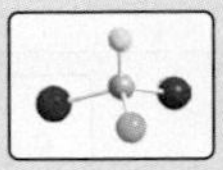
A	1		

$C_s = C_h$ (m)	E	σ_h	$h = 2$	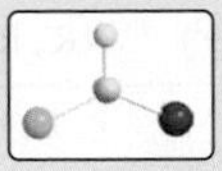
A′	1	1	x, y, R_z	x^2, y^2, z^2, xy
A″	1	−1	z, R_x, R_y	yz, zx

$C_i = S_2\,(\bar{1})$	E	i	$h = 2$	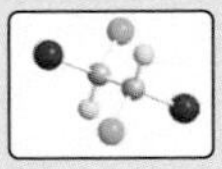
A_g	1	1	R_x, R_y, R_z	$x^2, y^2, z^2, xy, yz, zx,$
A_u	1	−1	x, y, z	

The groups C_{nv}

C_{2v}, $2mm$	E	C_2	σ_v	σ_v'	$h = 4$	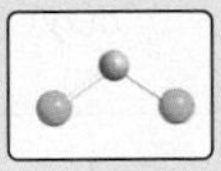
A_1	1	1	1	1	z, z^2, x^2, y^2	
A_2	1	1	−1	−1	xy	R_z
B_1	1	−1	1	−1	x, zx	R_y
B_2	1	−1	−1	1	y, yz	R_x

C_{3v}, $3m$	E	$2C_3$	$3\sigma_v$	$h = 6$	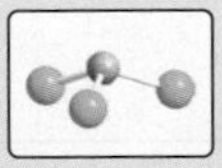
A_1	1	1	1	$z, z^2, x^2 + y^2$	
A_2	1	1	−1		R_z
E	2	−1	0	$(x, y), (xy, x^2 - y^2)\ (yz, zx)$	(R_x, R_y)

C_{4v}, $4mm$	E	C_2	$2C_4$	$2\sigma_v$	$2\sigma_d$	$h = 8$	
A_1	1	1	1	1	1	$z, z^2, x^2 + y^2$	
A_2	1	1	1	−1	−1		R_z
B_1	1	1	−1	1	−1	$x^2 - y^2$	
B_2	1	1	−1	−1	1	xy	
E	2	−2	0	0	0	$(x, y), (yz, zx)$	(R_x, R_y)

The σ_v planes coincide with the xz- and yz-planes.

C_{5v}	E	$2C_5$	$2C_5^2$	$5\sigma_v$	$h = 10$, $\alpha = 72°$	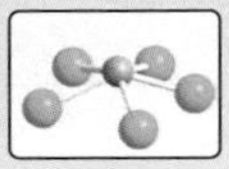
A_1	1	1	1	1	$z, z^2, x^2 + y^2$	
A_2	1	1	1	−1		R_z
E_1	2	$2\cos\alpha$	$2\cos 2\alpha$	0	$(x, y), (yz, zx)$	(R_x, R_y)
E_2	2	$2\cos 2\alpha$	$2\cos\alpha$	0	$(xy, x^2 - y^2)$	

C_{6v}, $6mm$	E	C_2	$2C_3$	$2C_6$	$3\sigma_d$	$3\sigma_v$	$h = 12$	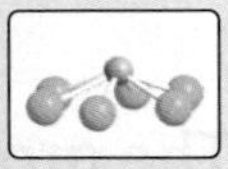
A_1	1	1	1	1	1	1	$z, z^2, x^2 + y^2$	
A_2	1	1	1	1	−1	−1		R_z
B_1	1	−1	1	−1	−1	1		
B_2	1	−1	1	−1	1	−1		
E_1	2	−2	−1	1	0	0	$(x, y), (yz, zx)$	(R_x, R_y)
E_2	2	2	−1	−1	0	0	$(xy, x^2 - y^2)$	

$C_{\infty v}$	E	$2C_\phi^\dagger$	…	$\infty\sigma_v$	$h = \infty$	
$A_1(\Sigma^+)$	1	1	…	1	$z, z^2, x^2 + y^2$	
$A_2(\Sigma^-)$	1	1	…	−1		R_z
$E_1(\Pi)$	2	$2\cos\phi$	…	0	$(x, y), (yz, zx)$	(R_x, R_y)
$E_2(\Delta)$	2	$2\cos 2\phi$	…	0	$(xy, x^2 - y^2)$	
…	…	…	…	…		

† There is only one member of this class if $\phi = \pi$.

The groups D_n

D_2, 222	E	C_2^z	C_2^y	C_2^x	$h = 4$	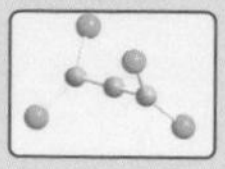
A	1	1	1	1	x^2, y^2, z^2	
B_1	1	1	−1	−1	z, xy	R_z
B_2	1	−1	1	−1	y, zx	R_y
B_3	1	−1	−1	1	x, yz	R_x

D_3, 32	E	$2C_3$	$3C_2'$	$h = 6$	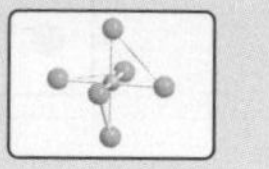
A_1	1	1	1	$z^2, x^2 + y^2$	
A_2	1	1	−1	z	R_z
E	2	−1	0	$(x, y), (yz, zx), (xy, x^2 - y^2)$	(R_x, R_y)

D_4, 422	E	C_2	$2C_4$	$2C_2'$	$2C_2''$	$h = 8$	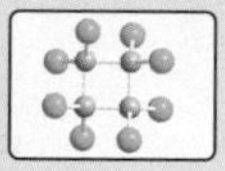
A_1	1	1	1	1	1	$z^2, x^2 + y^2$	
A_2	1	1	1	−1	−1	z	R_z
B_1	1	1	−1	1	−1	$x^2 - y^2$	
B_2	1	1	−1	−1	1	xy	
E	2	−2	0	0	0	$(x, y), (yz, zx)$	(R_x, R_y)

The groups D_{nh}

D_{2h} (mmm)	E	$C_2(z)$	$C_2(y)$	$C_2(x)$	i	$\sigma(xy)$	$\sigma(yz)$	$\sigma(zx)$	$h = 8$	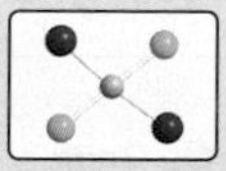
A_g	1	1	1	1	1	1	1	1	x^2, y^2, z^2	
B_{1g}	1	1	−1	−1	1	1	−1	−1	xy	R_z
B_{2g}	1	−1	1	−1	1	−1	−1	1	xz	R_y
B_{3g}	1	−1	−1	1	1	−1	1	−1	yz	R_x
A_u	1	1	1	1	−1	−1	−1	−1		
B_{1u}	1	1	−1	−1	−1	−1	1	1	z	
B_{2u}	1	−1	1	−1	−1	1	1	−1	y	
B_{3u}	1	−1	−1	1	−1	1	−1	1	x	

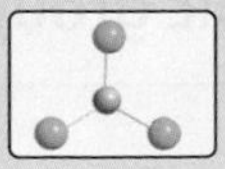

D_{3h}, $\bar{6}2m$	E	σ_h	$2C_3$	$2S_3$	$3C_2'$	$3\sigma_v$	$h = 12$	
A_1'	1	1	1	1	1	1	$z^2, x^2 + y^2$	
A_2'	1	1	1	1	−1	−1		R_z
A_1''	1	−1	1	−1	1	−1		
A_2''	1	−1	1	−1	−1	1	z	
E'	2	2	−1	−1	0	0	$(x, y), (xy, x^2 - y^2)$	
E''	2	−2	−1	1	0	0	(yz, zx)	(R_x, R_y)

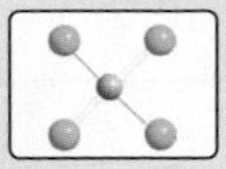

D_{4h}, $4/mmm$	E	$2C_4$	C_2	$2C_2'$	$2C_2''$	i	$2S_4$	σ_h	$2\sigma_v$	$2\sigma_d$	$h = 16$	
A_{1g}	1	1	1	1	1	1	1	1	1	1	$x^2 + y^2, z^2$	
A_{2g}	1	1	1	−1	−1	1	1	1	−1	−1		R_z
B_{1g}	1	−1	1	1	−1	1	−1	1	1	−1	$x^2 - y^2$	
B_{2g}	1	−1	1	−1	1	1	−1	1	−1	1	xy	
E_g	2	0	−2	0	0	2	0	−2	0	0	(yz, zx)	(R_x, R_y)
A_{1u}	1	1	1	1	1	−1	−1	−1	−1	−1		
A_{2u}	1	1	1	−1	−1	−1	−1	−1	1	1	z	
B_{1u}	1	−1	1	1	−1	−1	1	−1	−1	1		
B_{2u}	1	−1	1	−1	1	−1	1	−1	1	−1		
E_u	2	0	−2	0	0	−2	0	2	0	0	(x, y)	

The C_2' axes coincide with the x- and y-axes; the σ_v planes coincide with the xz- and yz-planes.

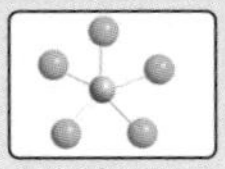

D_{5h}	E	$2C_5$	$2C_5^2$	$5C_2$	σ_h	$2S_5$	$2S_5^3$	$5\sigma_v$	$h = 20$, $\alpha = 72°$	
A_1'	1	1	1	1	1	1	1	1	$x^2 + y^2, z^2$	
A_2'	1	1	1	−1	1	1	1	−1		R_z
E_1'	2	$2\cos\alpha$	$2\cos 2\alpha$	0	2	$2\cos\alpha$	$2\cos 2\alpha$	0	(x, y)	
E_2'	2	$2\cos 2\alpha$	$2\cos\alpha$	0	2	$2\cos 2\alpha$	$2\cos\alpha$	0	$(x^2 - y^2, xy)$	
A_1''	1	1	1	1	−1	−1	−1	−1		
A_2''	1	1	1	−1	−1	−1	−1	1	z	
E_1''	2	$2\cos\alpha$	$2\cos 2\alpha$	0	−2	$-2\cos\alpha$	$-2\cos 2\alpha$	0	(yz, zx)	(R_x, R_y)
E_2''	2	$2\cos 2\alpha$	$2\cos\alpha$	0	−2	$-2\cos 2\alpha$	$-2\cos\alpha$	0		

$D_{\infty h}$	E	$2C_\phi$	…	$\infty\sigma_v$	i	$2S_\infty$	…	$\infty C_2'$	$h = \infty$	
$A_{1g}(\Sigma_g^+)$	1	1	…	1	1	1	…	1	$z^2, x^2 + y^2$	
$A_{1u}(\Sigma_u^+)$	1	1	…	1	−1	−1	…	−1	z	
$A_{2g}(\Sigma_g^-)$	1	1	…	−1	1	1	…	−1		R_z
$A_{2u}(\Sigma_u^-)$	1	1	…	−1	−1	−1	…	1		
$E_{1g}(\Pi_g)$	2	$2\cos\phi$	…	0	2	$-2\cos\phi$	…	0	(yz, zx)	(R_x, R_y)
$E_{1u}(\Pi_u)$	2	$2\cos\phi$	…	0	−2	$2\cos\phi$	…	0	(x, y)	
$E_{2g}(\Delta_g)$	2	$2\cos 2\phi$	…	0	2	$2\cos 2\phi$	…	0	$(xy, x^2 - y^2)$	
$E_{2u}(\Delta_u)$	2	$2\cos 2\phi$	…	0	−2	$-2\cos 2\phi$	…	0		
…	…	…	…	…	…	…	…	…		

The cubic groups

T_d, $\bar{4}3m$	E	$8C_3$	$3C_2$	$6\sigma_d$	$6S_4$	$h = 24$	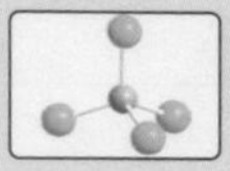
A_1	1	1	1	1	1	$x^2 + y^2 + z^2$	
A_2	1	1	1	−1	−1		
E	2	−1	2	0	0	$(3z^2 - r^2, x^2 - y^2)$	
T_1	3	0	−1	−1	1		(R_x, R_y, R_z)
T_2	3	0	−1	1	−1	(x, y, z), (xy, yz, zx)	

O_h, $m3m$	E	$8C_3$	$6C_2$	$6C_4$	$3C_2$ $(=C_4^2)$	i	$6S_4$	$8S_6$	$3\sigma_h$	$6\sigma_d$	$h = 48$	
A_{1g}	1	1	1	1	1	1	1	1	1	1	$x^2 + y^2 + z^2$	
A_{2g}	1	1	−1	−1	1	1	−1	1	1	−1		
E_g	2	−1	0	0	2	2	0	−1	2	0	$(2z^2 - x^2 - y^2, x^2 - y^2)$	
T_{1g}	3	0	−1	1	−1	3	1	0	−1	−1		(R_x, R_y, R_z)
T_{2g}	3	0	1	−1	−1	3	−1	0	−1	1	(xy, yz, zx)	
A_{1u}	1	1	1	1	1	−1	−1	−1	−1	−1		
A_{2u}	1	1	−1	−1	1	−1	1	−1	−1	1		
E_u	2	−1	0	0	2	−2	0	1	−2	0		
T_{1u}	3	0	−1	1	−1	−3	−1	0	1	1	(x, y, z)	
T_{2u}	3	0	1	−1	−1	−3	1	0	1	−1		

The icosahedral group

I	E	$12C_5$	$12C_5^2$	$20C_3$	$15C_2$	$h = 60$	
A	1	1	1	1	1	$x^2 + y^2 + z^2$	
T_1	3	$\frac{1}{2}(1 + 5^{1/2})$	$\frac{1}{2}(1 - 5^{1/2})$	0	−1	(x, y, z)	(R_x, R_y, R_z)
T_2	3	$\frac{1}{2}(1 - 5^{1/2})$	$\frac{1}{2}(1 + 5^{1/2})$	0	−1		
G	4	−1	−1	1	0		
H	5	0	0	−1	1	$(2z^2 - x^2 - y^2, x^2 - y^2, xy, yz, zx)$	

*The image illustrating the group I is in fact a representation of the group I_h, which is isomorphous with a dodecahedron. The group I_h includes reflections; its order is 120.
Further information: P.W. Atkins, M.S. Child, and C.S.G. Phillips, *Tables for group theory*. Oxford University Press (1970). In this source, which is available on the website for this text, other character tables such as D_{2d}, D_{3d}, D_{6h}, and I_h can be found.

INDEX

D

T